Ex Libris

Patterns of Reproductive Behaviour

By the same author

THE BIOLOGY OF ART
THE MAMMALS: A GUIDE TO THE LIVING SPECIES
PRIMATE ETHOLOGY (Editor)
THE NAKED APE
THE HUMAN ZOO

with Ramona Morris

MEN AND SNAKES
MEN AND APES
MEN AND PANDAS

for children

THE STORY OF CONGO
MONKEYS AND APES
THE BIG CATS
ZOOTIME

Patterns of Reproductive Behaviour

Collected Papers by

DESMOND MORRIS

McGraw-Hill Book Company
New York St. Louis San Francisco

This collection first published 1970 by Jonathan Cape Ltd., London

Library of Congress Catalog Card Number: 74-101165
43177

Printed and bound in Great Britain

Contents

Introduction 11

1. Homosexuality in the ten-spined stickleback (1952) 13
2. The reproductive behaviour of the zebra finch (1954) 42
3. The reproductive behaviour of the river bullhead (1954) 89
4. The causation of pseudofemale and pseudomale behaviour (1955) 118
5. The function and causation of courtship ceremonies (1956) 128
6. The feather postures of birds and the problem of the origin of social signals (1956) 153
7. 'Typical intensity' and its relationship to the problem of ritualization (1957) 187
8. The reproductive behaviour of the bronze mannikin (1957) 198
9. The courtship of pheasants (1957) 240
10. The reproductive behaviour of the ten-spined stickleback (1958) 245
11. The comparative ethology of grassfinches and mannikins (1958) 399
12. The behaviour of the green acouchi (1962) 454
13. The response of animals to a restricted environment (1964) 490
14. The rigidification of behaviour (1966) 512

Bibliography of scientific publications by the author 517

Index 520

Acknowledgments

I am extremely grateful to the following colleagues for their valuable help during the course of the research on which these papers are based: Dr M. Bastock, Dr D. Blest, Professor A. J. Cain, Dr F. J. Chatterly, Dr E. Cullen, Dr M. Cullen, Miss E. Eisner, Mr R. Fiennes, Mr O. Graham-Jones, Dr P. E. Guiton, Professor J. B. S. Haldane, Dr F. Hall, Sir Alister Hardy, Dr L. Harrison-Mathews, Dr O. W. Hill, Professor R. A. Hinde, Dr K. Hoffmann, Professor J. van Iersel, Dr J. Lecomte, Professor D. S. Lehrman, Professor K. Lorenz, Dr G. Manley, Dr A. Manning, Mrs R. Morris, Dr M. Moynihan, Dr H. Poulsen, Dr W. M. S. Russell, Dr P. Sevenster, Professor P. Shepherd, Professor N. Tinbergen, and Dr U. Weidmann.

I am also indebted to the Nature Conservancy and the Agricultural Research Council, who sponsored much of the research which led to the publications reprinted here; and to the editors and publishers of the following journals for permission to reprint the papers which first appeared in their pages: *Behaviour* (1, 2, 3, 4, 6, 7, 8 and 10); *Zoo Life* (9); *Proceedings of the Zoological Society of London* (11 and 12); *Symposia of the Zoological Society of London* (13); *Philosophical Transactions of the Royal Society of London* (14); also to the Fondation Singer-Polignac and Masson et Cie, Paris (5).

All drawings and photographs illustrating these papers are by the author, unless otherwise stated.

TO NIKO TINBERGEN

Introduction

The reproductive behaviour of vertebrate species has been a subject of interest to man for thousands of years. The human hunter, the myth-maker, the farmer and the naturalist have all viewed the matter, fancifully or factually, in their own ways, for centuries. Professional zoologists, by contrast, have only entered this field seriously in the last few decades. An explanation for their late arrival is not hard to find. The startlingly complex nature of vertebrate courtship and display systems and the incredibly elaborate patterns of nest construction and parental care offer a daunting challenge to the analytical scientist who wishes to unravel the component units of the behaviour involved. Darwin pointed the way, but the path was left virtually untrod until, in the 1930s, a new group of zoologically orientated behaviour students began to emerge in Europe. Led by Konrad Lorenz and Niko Tinbergen, they began to tackle more and more complex forms of behaviour, applying detailed observational techniques and simple experimental procedures in areas of animal activity which had previously been kept at a wary arm's length by academic zoology. They took the laboratory into the field and brought the field into the laboratory. Whole animals, indeed whole groups of animals became the basic raw material of these investigations, instead of bones, fossils and tissues. At first, this new development was met with some scepticism. How could anyone hope to carry out a respectable analysis of something so hopelessly complicated? But somehow they succeeded, and the new discipline of comparative ethology began to expand. The arrival of the Second World War almost smashed the impetus of the new movement, but it survived and re-asserted itself vigorously in the post-war period. Since then, it has gone from strength to strength.

Six years after the war, in 1951, I was fortunate enough to be able to join Niko Tinbergen's ethological group at Oxford University, where I began research into the reproductive behaviour of a number of species of fish and birds. My records show that, during my five years there, living specimens of no less than eighty-four vertebrate species were accommodated, at one time or another, in my laboratories. Under the enlightened professorship of Sir Alister Hardy this invasion was not only tolerated, but actively encouraged. The roof of the Department of Zoology became festooned with dozens of aviaries, the basement jammed with a hundred tanks of fish. While I and some of my colleagues were busily bringing the field into the laboratory, others of the group were reversing the process by taking their experiments out into the woodlands and sea-coasts. Under Niko Tinbergen's guidance we were all asking the same initial question: not what can we *make* animals do, but what *do* they do? Instead of applying a simple stimulus to produce a simple response and then varying the relationship between the two, in the manner of animal psychologists, we set out to observe how many different kinds of response a particular animal made to as wide a variety of natural stimuli as possible. From this we constructed an ethogram of the behaviour repertoire of a species. This was our starting point, and the fixed motor pattern was our basic unit of study. Then followed controlled observation periods, quantitatively analysed, of selected motor patterns, and a further probing into the relationships between the various responses. We wanted to know the causal influences and the survival values of the activities we had recorded and ultimately to understand their evolutionary significance.

Because earlier animal psychology studies had concentrated so much on simple stimulus-response relationships, the whole fascinating field of motivational conflict was open to us. Because of the complexity of the stimulus situations we encouraged, our animals were frequently faced with the problem of trying to respond in more than one way at one time. Confronted with a rival of their own species, they were stimulated both to flee and to attack. The threat behaviour resulting from this conflict became a major area of study. The courtship situation produced an even more complex conflict, involving the urge to mate as well as to flee and attack. Again, this behaviour received special attention. The particular interest of these patterns lay in the fact that the responses involved gave vital clues concerning the whole topic of the growth and evolution of animal communication. It is at these moments of conflict that animals have most to communicate to one another, and by making comparative studies of these activities it was possible to build up a picture of the way in which animal language has developed.

The fourteen papers collected in this volume cover a research period from 1951 to 1966, but the majority of them (the first eleven) stem from my five-year period in Tinbergen's group at Oxford. I have included only those papers that are exclusively concerned with, or have some general bearing on, problems of reproductive behaviour. As will emerge, my principal focus of attention has been on the development of visual signalling systems, a subject which, thanks to the work of many contemporary ethologists, is increasingly better understood, but is nevertheless still as tantalizing and fascinating as ever it was.

My debt to my teacher, Niko Tinbergen, and to my ethological colleagues is enormous. Without the stimulating atmosphere they created this research could never have been carried through.

DESMOND MORRIS
1969

Homosexuality in the ten-spined stickleback

(1952)

Introduction

Comparative ethologists have spent a number of years studying the behaviour of the three-spined stickleback, *Gasterosteus aculeatus* L., and its reproductive behaviour patterns are known in some detail (Leiner, 1929, 1930, 1931a, 1934; Wunder, 1928, 1930; ter Pelkwijk and Tinbergen, 1937; Tinbergen and van Iersel, 1947; van Iersel, unpublished; and also in a number of papers by Tinbergen, especially 1940, 1942, 1950, 1951). Other species of stickleback, of which there are several, have been little studied until very recently. Comparative studies are now being carried out, and this paper is part of a general analysis of the behaviour of the ten-spined species. It is being published separately from the main body of the work* because it deals with a piece of abnormal behaviour that has been observed, whilst the major investigation is concerned with the normal patterns of behaviour. However, as there is no modern description of the normal reproductive behaviour of this species available in English (there are two earlier papers in German by Leiner, 1931b and 1934, and a good short account in Dutch by Sevenster, 1949), it has been thought necessary to outline it briefly here. It will be reported in detail in later papers.

The ten-spined stickleback inhabits small rivers and streams in many parts of the British Isles. It differs from the more common three-spined stickleback in, amongst other things, the location of the nest site. Whereas the three-spined species constructs a covered depression in the sand or mud of the river bed, the ten-spined stickleback selects water-weeds in which to build its nest. This nest is roughly spherical and is pierced by a bent tunnel, the tunnel exit being higher than the entrance (see Figs 1 and 9).

The area around the nest site is patrolled and defended by the male owner, other fish, particularly males of the same species, being driven away. Normally there are a number of such territories together in a stretch of weed in the stream. At the beginning of the season the boundaries are not clearly defined, so that males are constantly trespassing on to one another's nesting areas and a considerable amount of fighting ensues. Gradually the boundaries

* See pp. 245–398.

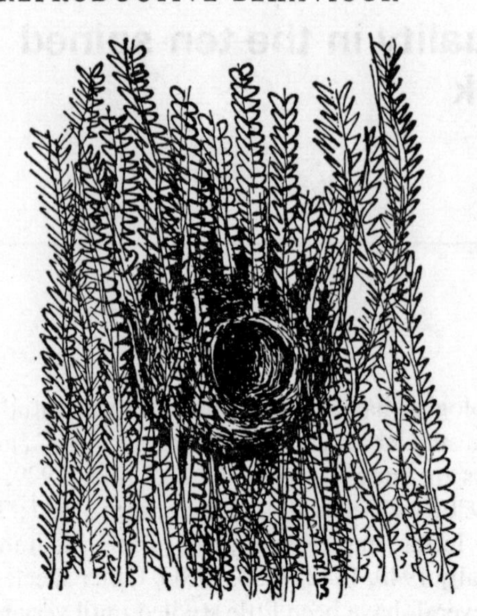

Fig. 1. The nest of the ten-spined stickleback.

become more clearly defined, and correlated with this there is less actual fighting and more threatening.

When the females are ripe they swim through the territories and are courted by those males which are ready to fertilize. If the females do not respond to the courting males in a certain sexual manner, they, like other males, are attacked also. So the area round the nest of a male is forbidden ground to all fish save a female which is ready to lay eggs.

As it is extremely difficult to make observations on the reproductive behaviour of these fish in their natural state, the following observations were carried out in the laboratory using glass aquarium tanks. A male was introduced into a tank that had been well planted with weeds, and when it had constructed its nest a female was then introduced. Once she had laid her eggs she was removed as she had then completed her sexual role, and the male was left to rear the young. If fighting and territory demarcation were to be observed, two males were introduced together into a tank. They were introduced together, for if one was already present it might have assumed the whole tank as its territory, in which case the second male would be so badly attacked that it would rarely settle and build anywhere in the tank. The size of the tanks used was normally in the region of 2 × 1 × 1 feet. Practically all observations were made under such conditions. In one tank, however, the population density was increased to study aggressive behaviour which, it was thought, would be more frequently displayed if the males were crowded. A tank of the dimensions 42 × 18 × 18 inches was employed for the purpose and five males were introduced together (Fig. 2). As was expected, a great deal of fighting and threatening was seen as a result of this crowding, but another particularly interesting behaviour pattern was also accidentally brought to light, apparently as a result of this particular population density: the phenomenon of *homosexuality* on the part of certain males. This homosexual behaviour took the form both of lack of discrimination where a sexual partner was concerned (on the part of dominant males), and of the inversion of the complete sexual pattern (on the part of dominated males). Before this

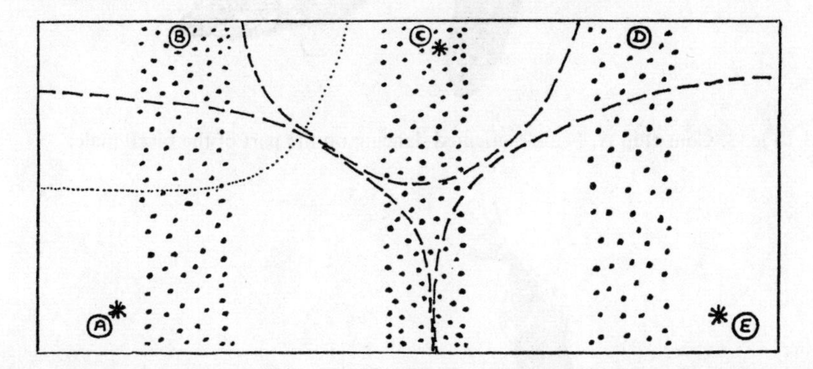

Fig. 2. Tank Plan showing territory areas at a particular stage of the experiment. The dotted areas are the three hedgerows of willowmoss. The five points A B C D and E are the positions of the five clumps of myriophyllum. The broken lines represent the territory boundaries, and the asterisks show the positions of the three nests present at the time. There were five male fish in the tank, and one of the nestless males spent much of its time at B. It was unable to build, as it was repeatedly molested by the male at A, but when the latter's nest was damaged and abandoned, the fish at B quickly built and formed the territory shown by the fine dotted line. Later a similar change-over occurred between fish at D and E.

is described in detail, the normal courtship pattern of the species will be reported.

It should perhaps be mentioned here that the term 'homosexual' is being used throughout this paper in its widest sense, to mean any sexual performance involving members of the same sex. For a discussion of the terminology, see last section of the paper.

Normal Courtship Behaviour

When a female is ripe it has a swollen belly which is silvery in colour. Its dorsal surface and sides are cryptically marked with a broken pattern. If it comes on the territory of a male which is ready to fertilize, it is courted by the male, which reacts to the swollen belly. There are four stages in the courtship of the male. They are as follows:

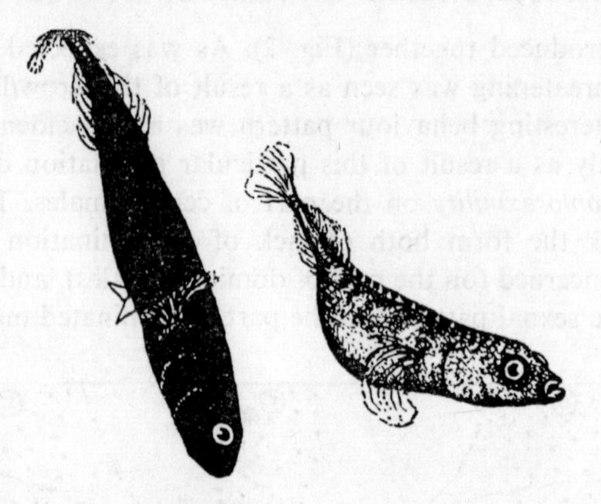

Fig. 3. Courtship A. Female-oriented dancing on the part of the black male.

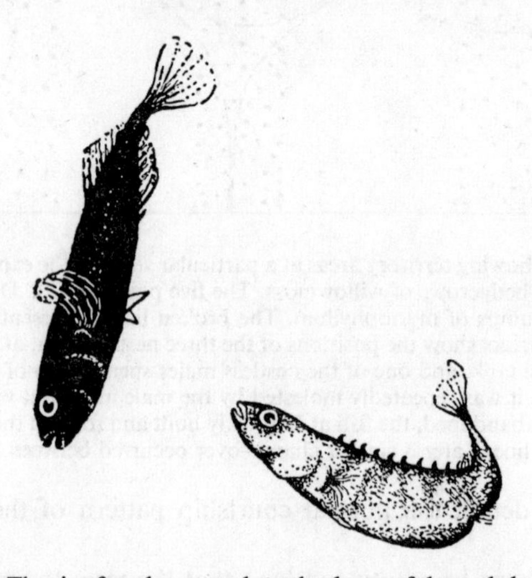

Fig. 4. Courtship B. The ripe female responds to the dance of the male by turning towards it.

1. Dancing oriented to the female (Fig. 3).
2. Dancing oriented to the nest (Fig. 6).
3. Showing the nest entrance (Fig. 7).
4. Shivering along the tail of the female (Fig. 8).

The dancing of the male consists of a number of short jumps with the body held head downwards, the body axis being at an angle of approximately 60 degrees from the horizontal. (This angle is very variable and is dependent on several factors that need not be discussed here.) If the female flees at this stage or any other, it is chased and bitten. If it remains still in the water, the male dances around it until it makes a movement to follow (Fig. 4). The male reacts to such a movement by dancing off in the direction of the nest. The female then follows close behind with its body tilted slightly upwards at the anterior end, so that its nose is near to the two white ventral spines of the male which show up clearly against the jet-black ventral surface (Fig. 5). These spines are erected throughout the courtship. Once the male has reached its nest it stands still, at the same angle, with its nose at the entrance hole. It then fans[1] in this position and the female responds to this by lowering its nose from the region of the ventral spines down to the nest entrance. The female then reacts to the entrance hole by pushing into it a number of times. The courtship usually breaks down at this stage and the female swims off. The male usually chases after it and bites it on the tail. If the female is highly motivated sexually, it will respond to this bite by turning back towards the male, whereupon the male dances back to the nest again. The sequence of events, up to the stage of the female pushing into the nest entrance, may be repeated in this way a number of times. After a while the female develops its pushing movement into entry of the nest, in which it lies with its head protruding from the exit, and its tail from the entrance. The male now shivers its nose along the side of the protruding tail of the female. After a few seconds the eggs are laid (Fig. 9) and the female leaves the nest by the exit. The male then enters the nest (Fig. 10) and passes straight through it without pausing. The eggs are now fertilized and the male stands guard over the nest and ventilates it by fanning movements of the pectoral fins. Later it protects the young fish during their first days away from the nest. The female plays no part in parental behaviour, and is driven away by the male after the eggs have been laid.

There are many minor variations in this courtship pattern, but these will be discussed more fully in subsequent papers. Sufficient detail of the normal behaviour has been given above to make possible a comparison here between it and the abnormal homosexual behaviour that has been observed.

[1] *Fanning:* This has already been described in detail, in the literature, for the Three-Spined Stickleback, and basically it is performed in the same way, and is functionally similar, in this species. It is a combination of fin movements that results in a current of water being passed through the nest, and is normally performed by a nest-owner during the parental phase. The pectoral fins throw a current of water forwards, and the tail beats hard at the same time, sending a current of water backwards, thus enabling the fish to hover at the nest. Fanning has the function of ventilating the nest and thereby aerating the developing eggs. It is also seen to occur during the sexual phase before eggs are present in the nest. At this stage it has been shown (van Iersel, unpublished) to act as a displacement activity, but I have recently observed that, for the three-spined species at least, it also has the function, during this phase, of preventing silting up of the nest. In courtship, when the male is showing the nest entrance to the female, fanning is certainly functioning as a displacement activity, but here it is combined with the tilted posture characteristic of the male dance, instead of being performed more or less horizontally, as it is at other times.

Fig. 5. The ventral surface of the dancing male as seen by the female, showing the conspicuous white spines against the black body.

Fig. 6. Courtship C. Nest-oriented dancing. The male swims to the nest in short inclined jumps, followed closely by the female.

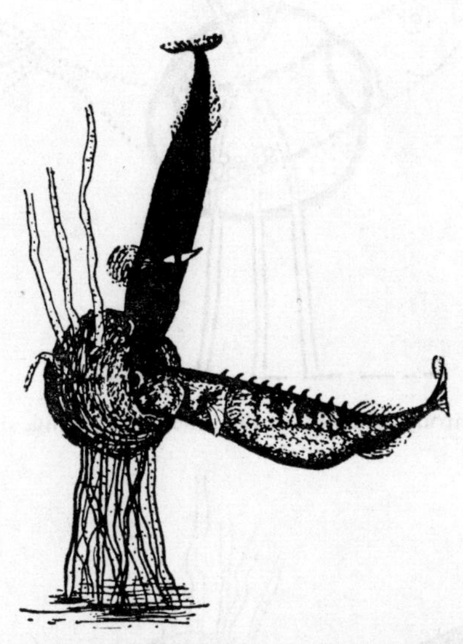

Fig. 7. Courtship D. The male shows the nest entrance to the female by fanning in the tilted position. The female pushes into the entrance with her nose.

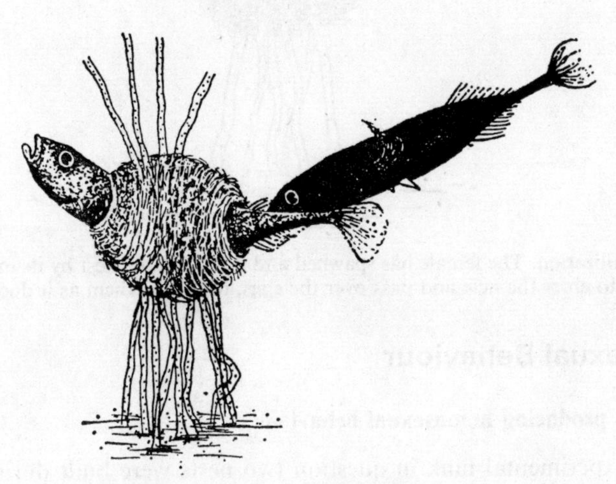

Fig. 8. Courtship E. Shivering. The female having entered the nest, the male vibrates its nose along her tail.

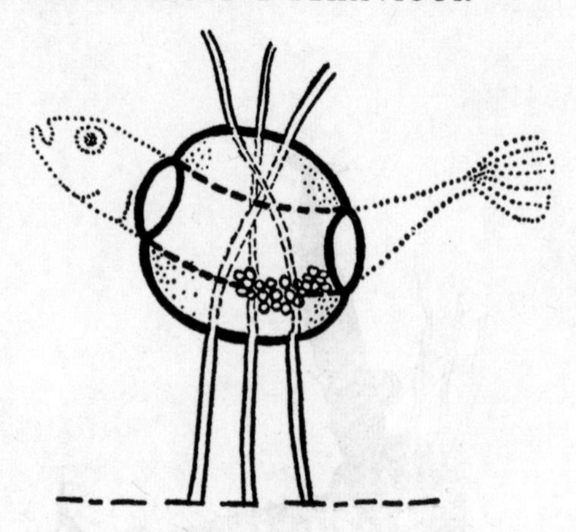

Fig. 9. A section through the nest at the time of spawning, showing the eggs.

Fig. 10. Fertilization. The female has spawned and is leaving the nest by its exit. The male is about to enter the nest and pass over the eggs, fertilizing them as it does so.

Homosexual Behaviour

Conditions producing homosexual behaviour

In the experimental tank in question two nests were built during the first few days, the owners of these nests taking on the full reproductive colour, a jet black which covers almost the entire body surface. The other three males were also seen to be aggressive, but to a lesser degree than the nest-owners.

These fish without nests were usually partially black but assumed cryptic coloration when repeatedly attacked and beaten. They did not, however, lose the white colour of the ventral spines and this could be used as a method of sex identification. The cryptic coloration is identical for both sexes, a fact which will be seen later to be of importance in the homosexual behaviour. It could be seen that these fish without nests were restricted in respect of the areas of the tank which they occupied, but their partially formed territories were only vaguely defined. During the following days two things became obvious concerning the nestless males. Firstly, they had little chance to settle and build nests because of the repeated attacks of the nest-owners, and secondly, they were highly motivated sexually. Several observations led to this second conclusion. On occasions when females were introduced for a short while, these nestless males were several times seen to court, and lead the females to imaginary nests. Also glueing,[1] an activity concerned exclusively with nest-building, was observed, although there was no nest to be glued. Boring[2] was seen, and this activity also is normally only performed into the nest entrance, but was in this case performed into a small lump of weed that looked like a nest, but which had, in fact, only just been put into the tank as nest material for the fish. Also, as mentioned, the fish were occasionally aggressive and were attempting to form territories.

It was concluded from these observations, and will be confirmed below, that the tank contained fish all of which were sexually motivated to a high degree, but some of which were sexually frustrated, being prevented from fully expressing themselves sexually by the presence of the dominant males.

Throughout the weeks during which the tank was under observation the number of nests, and the owners of the nests, varied from time to time, but at no stage were all five males nest-owners (Fig. 2). The volume of water in the tank was not great enough to permit the presence of five full territories. The result was that the tank always contained some sexually frustrated males. It was the interaction between these frustrated males and the nest-owners that provided the demonstration of homosexual behaviour. Had the tank contained fewer males there would have been no frustrated males, and the homosexual behaviour would not have occurred. Had there been a large number of

[1] *Glueing:* 'Glue' is a kidney secretion produced during the breeding season, which is used in nest-building. The nature of the glueing movement differs from species to species. The ten-spined species glues in two different ways. It can either deposit the glue directly on to the nest surface as it moves its body over the nest, or it can exude a blob of glue some distance from the nest, taking this blob up in its mouth with a circling movement of its body, and then swim with it to the nest. It was this second type of glueing that was observed in the instances mentioned above. When produced by these nestless males, and taken up into the mouth, it was then swallowed, there being no nest into which it could be pushed.

[2] *Boring:* This is a strenuous pushing movement of the snout into the nest entrance, and is nearly always performed in combination with fanning, which follows it. It has the function of keeping the tunnel clear.

males in the tank, it is probable that no territories would have been set up at all, as no single male would under those circumstances be able to keep part of the tank clear of fish long enough to settle and build. The reason why this homosexual behaviour has not been described before is probably because its occurrence depends on the presence of a particular population density, which is halfway between the two densities normally employed in the laboratory, for workers in this field normally use a very high density if they wish to prevent sexual behaviour, and a very low density if they wish to encourage it.

Concerning the conditions which provoke homosexual behaviour, it remains only to add two points. Firstly, the amount of water-weed present in the tank is a factor that must be taken into consideration, as this will control the amount of building space and cover. Wunder (1930) found that, with the three-spined species, artificial shelters put into a tank, in the form of screens, increased the number of territories. The tank employed here was planted with willowmoss (*Fontinalis antipyretica*) in three hedgerows across the tank, covering about one half of its floor space. Willowmoss provides dense vegetation to a height of about six inches. Amongst these hedgerows were planted a number of taller strands of *Myriophyllum spicatum*, which rose the eighteen inches to the surface of the water. The fish build in both types of vegetation, but appear to prefer the willowmoss. Secondly, the nest-owners were probably frustrated themselves, as females were only presented for short observation periods of an hour or so, during several weeks.

Indiscriminate courting

Homosexuality on the part of the nest-owners took the form of lack of discrimination between males and females. The nest-owning males danced not only to females when they were introduced, but also to other males. This occurred most frequently when they had been dancing to females which were then removed. If at that point a cryptically coloured male came into the territory, it was also danced to. This courtship of cryptic males by black nest-owners was also seen occasionally at times when no female had been introduced, although more often in such cases the intruding male was attacked in the normal way and driven off. Only black nest-owning males courted homosexually in this way, and they nearly always confined their attentions to other males that were not black, although on rare occasions one black male was seen to court another black male.

If these males were presented with male and female models there were times when they courted one as frequently as the other.

In these cases the cryptically coloured males never responded by following, so that the courting nest-owners were never able to reach the nest-entrance-showing stage of courtship. When the cryptic males did not follow they were invariably bitten, whereupon they fled.

Frustrated males behaving as females

So far there was no indication that there was any reciprocation of homosexual behaviour on the part of the courted males. It will become clear that this was because in cases of indiscriminacy the courting males did not reach stages three or four (showing or shivering) of the courtship. Only these later stages of courtship initiated homosexual responses on the part of the frustrated males. The first indication of this was when one of the cryptic males

Fig. 11. A pseudofemale fish (lowest one in figure) watches as a ripe female pushes into the nest entrance in response to a male's courtship movements.

was seen creeping stealthily through the weeds towards a nest at which the owner was showing the entrance to a female (Fig. 11). The cryptic male lay still in the weeds for some seconds, quite near to the nest, but was then seen to approach the entrance quickly. It then pushed the female out of the way, entered the nest, and passed through it.

It was at first thought that the intruding male was attempting to 'steal' a fertilization, and that it was trying to take the place of the male owner, but further observations and tests revealed that it was, in fact, taking on the role of a female. I shall call such a male showing the female pattern a *pseudofemale*.

This homosexual pattern inversion was repeated a number of times on different occasions and gradually a more detailed picture of its nature was constructed. This involved many observations, for the abnormality of the activity rendered it extremely variable, there being a minimum of three individuals involved – the male owner, the female, and one or more pseudofemales – and a slight change in the time sequences of the performances of their various actions produced an apparently different phenomenon. For example, if the female managed to get into the nest and lay its eggs before the pseudofemale advanced to the entrance, it appeared as if the latter was waiting for the eggs to be laid, and was then passing over them and fertilizing them before the owner male could interfere. If the pseudofemale pushed past the female into the nest, it appeared as if it was competing with the female for her role. If the pseudofemale pushed through the nest when both the male and the female had left the site, then the activity appeared to be something akin to masturbation.

From a number of such variations the basic pattern was deciphered. The first important clue came when, on an occasion when three pseudofemales

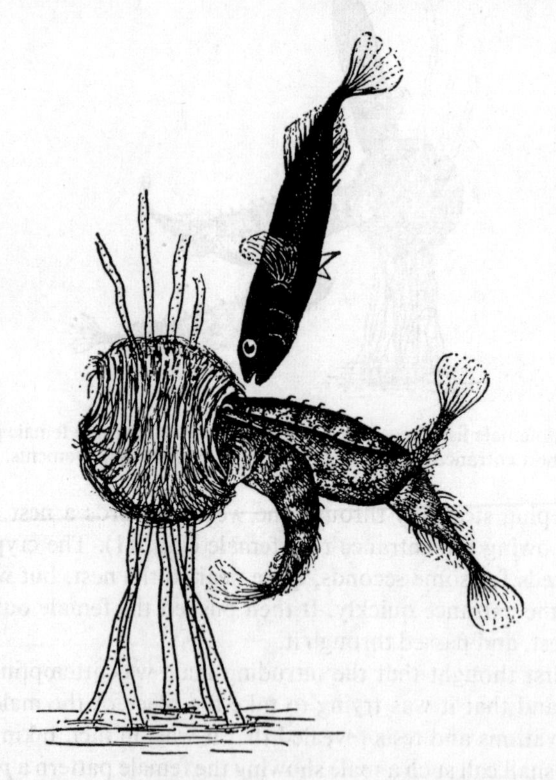

Fig. 12. Three pseudofemale individuals attempt to enter the nest, having driven away the ripe female as a result of competing with her for her position.

were advancing on a male and a female courting at the nest entrance, the female was frightened away by them. It swam off to the other end of the tank and left the owner male still showing the nest entrance, but showing it now, not to a female, but to a trio of pseudofemales (Fig. 12). These fish clustered around the entrance taking up the position vacated by the female. From above they appeared almost identical with the departed female, and the owner male treated them as if they were females. The movements of the pseudofemales were still stealthy, which is not surprising as they were in the most defended area of a territory belonging to an all-black male. They were seen to respond to the showing movements of the owner by pushing into the nest entrance with their noses, in exactly the same way as the female. On several occasions when they moved away from the nest a short distance, the owner ceased showing, and danced to them, whereupon they followed in the female manner with their noses between the white ventral spines. (It will be remembered that earlier it was stated that dancing alone was insufficient stimulus to initiate homosexual pattern-inversion, and yet here, following in response to dancing is reported. This is not, however, the contradiction it appears to be, for in this case the homosexual behaviour had already been initiated by the entrance-showing, and once initiated, it will then occur in response even to the dancing movements.) Also, when the pseudofemales reached the entrance again, they lowered their noses from the ventral spine region to the region of the nest entrance – again a female reaction. Furthermore, when one of them entered the nest, it lay in it while the owner shivered on its tail. This pausing in the nest is characteristic of the female passage through the nest when it lays its eggs, the fertilization movement of the male being performed without a pause inside the nest. Pauses of up to 40 seconds have been recorded for these pseudofemales, which means that the length of these pauses is similar to the length of the female egg-laying pauses.

This instance and subsequent similar instances revealed that the pseudofemales would perform the complete female pattern under these circumstances, with the single obvious exception of egg-laying. Furthermore, they were not taking on a passive role, but an active performance of six female movements, namely: following, nose-lowering to entrance, pushing into entrance, entering nest, pausing in nest, passing through nest.

The pseudofemale pattern was seen to take place both with and without the female being present. When the female was present, its own actions often confused the issue, and it was only when it accidentally left the scene that the real nature of the pseudofemale activities became clear. Unfortunately it was rarely that the female left the nest site at the required moment. Therefore, in order to produce this effect artificially, a female model was used.

Model experiments

A female model with a swollen belly was introduced near a nest, and was courted by the owner. When the model was made to follow the male to the nest the male showed it the nest entrance. At this point the cryptic males gathered round the nest site, lying still on the floor, and oriented to the nest entrance. When the female model was pushed into the nest entrance, the pseudofemales advanced and competed with it for position. The model was then removed, and the pseudofemales took its place, the courtship proceeding as before. This technique enabled me to control the time of departure of the 'female' from the scene.

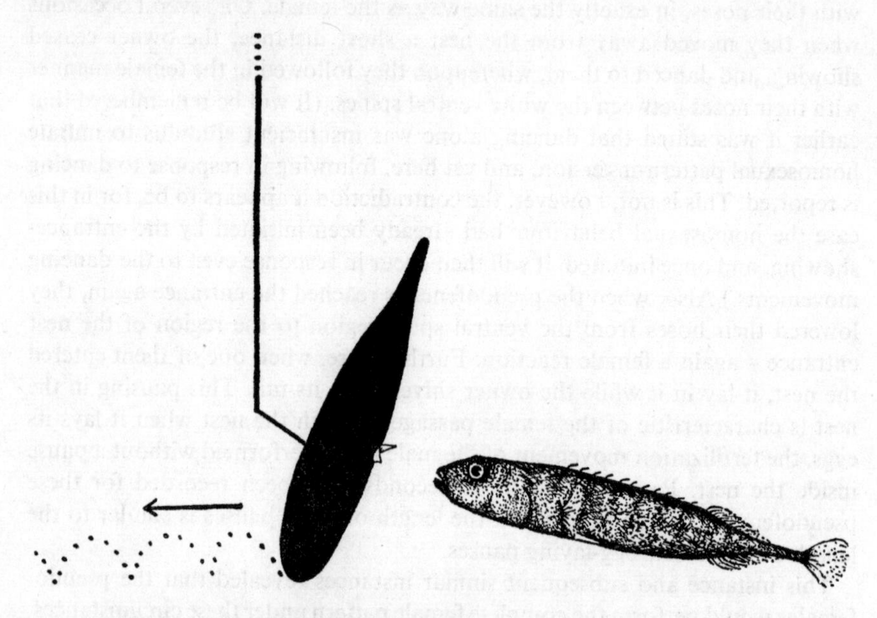

Fig. 13. A pseudofemale fish following a model of a male that is made to dance.

Model tests were also carried out using a male model with black coloration on the ventral surface and with white erected ventral spines. This was introduced and danced up to a nest where it was made to show the entrance. After this had been done several times, the pseudofemales followed the male model (Fig. 13) to the nest and pushed into the entrance a number of times, and in one or two cases a pseudofemale pushed into and through the nest. This was all carried out with no female in the tank. It shows that the presence of the female is not necessary to release the homosexual behaviour of the pseudofemales. The fact that, under non-experimental conditions, the female is invariably present when this behaviour occurs is purely incidental, as it is only when she is actively following that the real male owner reaches the later stages of courtship, at the nest, which produce the initial homosexual responses. Without these model tests it would have been impossible to say whether the presence of the female was essential or not.

Her presence, although proved to be inessential, does nevertheless appear to help produce the response, for if a female model is introduced and brought up to, and pushed into, an artificial nest, where there is no male owner, then

Fig. 14. A model of a ripe female is pushed into the entrance of an artificial nest. Pseudofemales competed with this model for position, despite the fact that the normal male position – as shown by the fish figured with dotted lines – was vacant in this case.

its presence does attract the pseudofemales. An artificial nest has to be used in this case, in order to ensure the absence of a male owner. Artificial nests used in these and other experiments were constructed from thin wire painted green, and green cotton. At the end of a length of wire a wire cage is made and green cotton is densely wound around this leaving a central tunnel.

In this case, where there is no male showing the entrance to the female model, which is being pushed into the nest entrance, it is possible to observe whether the presence of the showing male in the other cases was inhibiting the cryptic males from taking up the male position. But even here they compete with the female model for her position at the nest, despite the fact that the space just above the model, which a male would normally occupy, is vacant (Fig. 14). This helps to confirm the interpretation that the behaviour pattern of the cryptic males is essentially female and not male. It also appears to complicate the issue concerning what mechanism is involved (see later section on mechanism).

Frustrated females behaving as males

Ripe females kept in a tank without males do not show any homosexual behaviour, they simply shed their eggs after a while and then immediately eat them. Only one male activity has been seen to be performed by a female, and that is fanning. (This has been observed several times by the author. It was first seen by Tinbergen, who has a cinematic record of it.) Fanning is normally performed exclusively by the male, where, when autochthonously expressed, it is part of the parental pattern, and, as mentioned before, is concerned with egg ventilation.

The female is seen to fan only on rare occasions and under specifically frustrating conditions. When males are without nests they may court females if they are very highly motivated sexually. On these occasions they may even show the female the entrance to an imaginary nest, which they have yet to build, in the weeds. The female can be seen to respond to the imaginary-entrance-showing by pushing into the weeds. It is then that the female may be seen to fan sometimes. Such fanning may also occur if there is a nest present which is blocked in some way so that the female cannot enter. The female fanning movement is, in any case, a response to being unable to enter the nest at the appropriate moment.

The Function of Homosexual Pattern-Inversion

It seems almost certain that the function of this behaviour complex is that of an outlet which relieves sexual frustration, but two other possibilities must be considered:

1. Egg-stealing. Very often, as a result of the pushing through the nest by the pseudofemales, the eggs in the nest, if there were any present, fell into view. Often they fell completely out of the nest on to the floor of the tank. They were then eaten by the pseudofemales that remained at the nest site, despite the attacks of the owner. It might be assumed from this that the cryptic males behaved as females when a real courtship was in progress, because this was the only way in which they could approach close to the nest without being molested by its owner, and that they only came close to the nest at this point in search of eggs. To put it in another way, it might be assumed that they are egg-thieves that find themselves in a situation where submissive behaviour is called for, and this takes the form of female behaviour. Even if this is so, it does not alter the basic fact that they can and do behave as females.

It appears more likely, however, that the egg-stealing is brought about accidentally by the female behaviour. It is certainly true that whenever the female managed to lay her eggs before the pseudofemales invaded the nest, the eggs were always knocked from the nest and eaten. But it was more frequently the case that the pseudofemales competed with the female for her position below the male at the entrance, and pushed her out of the way and entered the nest themselves. If the function of pseudofemale behaviour was to steal eggs, the pseudofemales would certainly have waited until the female had been through the nest before pushing into it. Also, if they behaved as females until they reached the nest which contained eggs, or was to contain eggs, it would then be a simple matter to pull the eggs, when laid, from the nest without passing through it. Further, if egg-stealing is the function, there is no point in a pseudofemale reacting to a courting male owner in the absence of a female.

It should also be mentioned at this point that these fish were fed regularly with *Tubifex*, *Daphnia*, and chopped earthworm throughout the whole period during which the observations reported in this paper were made.

2. Female 'stealing'. This possibility is virtually ruled out by the evidence now assembled, although, as above, it may occur as the accidental outcome of the behaviour pattern. It is unfortunate that it is practically impossible to tell if such pseudofemale passages through the nest involve the ejaculation of sperm. If they do, and the owner manages to save the eggs from being eaten, which it may possibly do sometimes, then the pseudofemale will, in fact, have stolen a fertilization. But it is unlikely that this is the function of the pattern because of the decidedly female nature of the behaviour of the cryptic males, even, as model tests showed, in the absence of an owner male. Furthermore, if the same argument is applied here as before, that the males can only reach the nest where they can steal the fertilization by behaving as females, this possibility is still unlikely because of the pause in the nest made by the pseudofemale. It is true that males may pause in their own nests when passing

through them during nest-building, but they never pause in the nest when they are actually fertilizing eggs. So, if they do ejaculate sperm when they are 'pretending to lay eggs', then although the eggs may get fertilized in this way, it is practically certain to be an accidental outcome of the homosexual pattern.

It would appear to be the case, then, that pseudofemale behaviour on the part of sexually frustrated males is an outlet that serves to relieve frustration. In all essentials it is similar, as an outlet, to a displacement activity, except that it is more complex than most other recorded displacement activities. It is more than a displacement activity, it is a *displacement pattern*; but then it is not an act, but a whole pattern that is being thwarted. It should also be noticed that whereas other displacement activities 'spark-over' from Drive A-male to Drive B-male (or from Drive A-female to Drive B-female), in this case there is sparking-over from Drive A-male to Drive A-female.

The Possible Mechanism Involved

It might be expected that males frustrated in the manner described would attempt to express themselves sexually by stealing a nest and territory from a dominant male and driving it away, rather than employ a dormant female pattern. But the importance of relative dominance appears to be so great to such a fish that it is easier for it to call upon a contra-sexual pattern that it would never normally perform. It is well known for a number of species that an injection of female sex hormones into a male will result in the performance of the female sexual pattern, so that the nervous equipment for female sexual behaviour is present in the male, but is not used because the hormone balance favours the male pattern. In the case of the pseudofemale sticklebacks, the power of the dominant male sex hormone is overridden by the external inhibition of the activities normally resulting from the action of the male hormone.

The speed of the change-over from male to female behaviour on the part of the thwarted males is so great that it seems unlikely that a hormone change has taken place. For example, a cryptic, thwarted, nestless male may court a female for a few seconds sometimes, before it is interrupted by a black nest-owner. The latter, dancing more intensely, leads the female off to its nest, and straight away, in a matter of a few seconds, the previously courting cryptic male will follow the now courting couple and behave as a pseudofemale. This very quick change-over in a few seconds is almost certainly under nervous control, rather than as the result of some change in sex hormone balance.

It has been suggested to me by Tinbergen, however, that if this is so, it is strange that, in the case of the model experiment utilizing a female model and an artificial nest, the pseudofemales do not – in the absence of a male owner – at once switch over to male behaviour. At other times, as stated above, they will, if unmolested for a moment in the presence of a ripe female, snatch that moment and express themselves as males, even though they have no nest. Why then, when a female model is being pushed into an artificial nest, with no male owner in sight, should they compete with the model, instead of snatching this moment also for male expression? The answer, I feel, is not that the latter case reveals some hormonal change, but rather that it reveals that the 'nest which is not one's own' is sufficiently intimidating, even without the presence of its owner. And, in a way, the fact that a ripe female is pushing into the entrance indicates that there is an owner, even though no owner is to be seen.

Ford and Beach (1952), reporting instances of pseudofemale behaviour in rats (see section in this paper on other species), state that 'The physiological basis ... is not completely understood, but it does not appear to involve hormonal abnormalities.' One pseudofemale rat was castrated and the result was that both male and pseudofemale responses disappeared in a few days. Male hormone was then injected daily and the outcome of this was that *both* male and pseudofemale behaviour was shown again. One can conclude from this that pseudofemale behaviour *can* be under nervous rather than hormonal control. Further evidence that this is so in the case of the sticklebacks is that the individuals used were successful at mating as males and rearing young, both before and after their spell in the crowded tank where they behaved as pseudofemales. The rare cases of individuals of various species, where there is some hormonal abnormality that is causing pseudofemale (or pseudomale) behaviour, are usually incapable of functioning as their real sex, either before, or after, or both.

Ecological Significance

It may appear at first that the crowded conditions produced artificially in the laboratory would never exist in natural surroundings, the argument being that sticklebacks finding themselves this crowded would tend to spread out. But it must be remembered that not only do sticklebacks often become isolated in small stretches of water, but also that the areas of a river in which the ten-spined species can breed are limited. The conditions required are profuse fresh vegetation of a type in which nests can be built, water which is not very fast moving (as this would cause too much movement of the weeds and endanger the nest), water which is well oxygenated and of a particular pH, and water which does not dry up in the possible high temperatures of late spring and early summer.

The rivers from which the fish used were collected, for example the River Kennett in Wiltshire, are in most places too swift-moving for breeding purposes, and the fish are mostly found in slow-moving backwaters. It is quite possible that the population density in these backwaters does rise some years to the level employed in the laboratory in this case. Certainly, judging from the large numbers caught in one small backwater in the spring of 1952, the population density, it appears, is quite likely to rise to that level. It is true that the fish in such a backwater could migrate to another suitable area, but it will be seen that if the whole river population reaches a certain level, there will be a season or two when the homosexual pattern-inversion behaviour does take place in the natural state, and with it the accompanying phenomena of egg-stealing and eating. In one or two seasons this might bring the population density down considerably, to a level where homosexuality and egg-stealing would no longer occur. Although this species is difficult to watch in its natural state, it is hoped that the ecological significance of the phenomenon will be tested by field observations in the future.

Although there is little observational information concerning the behaviour of this species in the wild, there is a certain amount known about its ecology, and that of the three-spined stickleback. As will be mentioned in the next section, the latter species has been observed to exhibit pseudofemale behaviour also, but it has so far been seen in less detail. All indications, however, point to the phenomenon being much the same in both species.

It has been pointed out by Heuts (personal communication) that with the three-spined species there is often crowding on the breeding grounds in different parts of Europe. He reports having obtained, in certain seasons, several hundred sticklebacks with one haul of a net on the breeding grounds, although he admits that these were exceptional years. But the very fact that this happens, and what is more, does not happen each year, adds weight to the suggestion that pseudofemale behaviour could occur significantly in the wild state. Heuts also reports that examinations of stomach contents that he has carried out, often reveal the presence of eggs. Furthermore he has been able to identify these as being stickleback eggs. So that it can be said that both crowding and egg-eating do occur in the wild state.

A certain amount of quantitative evidence is available concerning this. Blegvad (1917) examined the stomach contents of 427 specimens of *Gasterosteus aculeatus*, and 112 specimens of *Pygosteus pungitius*, taken from Danish waters. In the case of *G. aculeatus*, 87 per cent of the food eaten was arthropodal, 8 per cent was stickleback eggs and larvae, and 5 per cent was various other food. For *P. pungitius* the figures were 72 per cent for arthropods, and 28 per cent for stickleback eggs and larvae. So it appears that not only do these fish eat their own eggs in nature, but that these eggs can make up an important part of their diet. More recently Hynes (1950) has investigated the problem, and his report provides two relevant points that are not available from Blegvad: namely, that although both eggs and larvae are eaten, the latter are relatively rarely found in stomach contents. This means that Blegvad's figure of 28 per cent for eggs and larvae for *P. pungitius* probably consisted mostly of eggs. Also Hynes found that the majority of fish that had eaten eggs were males. It is important to have this information because, as already mentioned, females may shed their eggs (if no mate is available at the required moment) and eat them, and it is necessary to be certain that all egg-eating in nature is not of this type. Another criticism might be that the eggs eaten in the wild by male fish could be devoured as a result of the habit of such fish of removing and swallowing an egg, from their own nest, that is not developing satisfactorily. But the evidence Hynes presents does not point to this, for many eggs were often found in one stomach, and this leads Hynes to say that it seems as if non-breeding males raid nests. Hynes also states that in the streams with which he was working, there was a very rich prey fauna, and apparently no scarcity of food.

This ecological evidence may be summarized as follows:

1. Crowding does occur in nature.
2. Egg-eating does occur in nature.
3. The egg-eating is probably egg-stealing.
4. There is plenty of food available other than eggs.

This, coupled with the laboratory observations described in this paper, points to the fact that pseudofemale behaviour probably occurs in nature. What the ecological evidence does not do is solve the problem of whether the pseudofemale behaviour is an elaborate egg-stealing device, or whether it is a displacement-like outlet relieving sexual frustration, with egg-stealing as an accidental outcome. It will be seen that it is a question of whether the level of crowding necessary to produce pseudofemale behaviour and egg-stealing is high enough to reduce the normally abundant food supply to a low enough level relatively, to make it an advantage to have a mechanism to obtain eggs as food. It is true that the laboratory observations have led me to the opinion that the phenomenon is a displacement-like outlet, but it would nevertheless be valuable to have some field observations on this point.

Homosexuality in Other Species

Similar homosexuality (both indiscriminacy and pattern-inversion types) has been observed for the three-spined stickleback, in similar circumstances. However, as mentioned above, in this species it has been observed in less detail. It was in this species that I first observed stickleback pseudofemale behaviour, but I did not at the time understand what I was seeing. I thought then that it was fertilization-stealing as a result of frustration. It is perhaps worth recording here what pseudofemale behaviour has been seen for the three-spined species. The following is a condensed version of notes made at the time:

Two males placed in a tank of dimensions 24 × 12 × 15 inches. Both take up territories straight away. A is smaller than B, which is constantly attacking it. B's territory is much larger than A's. (See Fig. 15 for tank plan.) B builds a nest in a corner of the tank. A builds no nest. A ripe female is introduced and is courted by A despite the fact that it has no nest. B takes over the female and courts her repeatedly. A never has another chance of courting the female, but watches the courtship of the other two. When the courtship between B and the female is at the nest-entrance stage, A is seen to creep stealthily along the floor of the tank in the direction of the nest, often lying quite still for some seconds. On such occasions B always notices A, and chases it away.

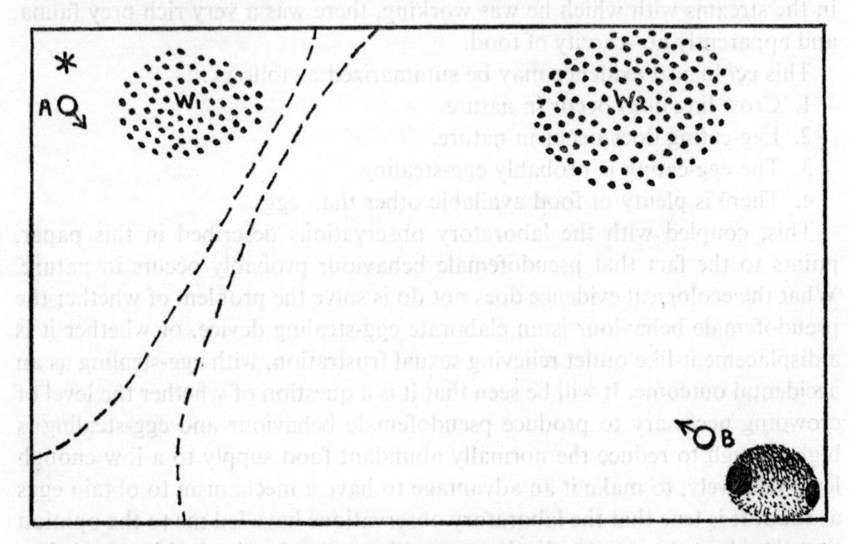

Fig. 15. Tank plan showing territories of two males A and B. The boundaries are shown by broken lines. W1 and W2 are two clumps of water-weed. Male B was dominant, and built a nest as shown. Male A attempted unsuccessfully to build at the point marked with an asterisk.

After a number of courtships the female enters the nest and lays her eggs. She is followed through the nest by the owner, B, but while B is passing through the nest, A rushes across the floor of the tank and pushes into the nest. B hovers outside the nest and watches A, who is now lying in it like a female. After some seconds, A leaves the nest and is chased and savagely bitten by B. The eggs are still safe in the nest. (They subsequently developed and were reared.) A, although nestless, returns to its corner of the tank and immediately performs the following movements: (a) sand-digging, (b) 'nest'-mending, (c) glueing. B, at the same time, was at its nest performing: (a) fanning, (b) boring, (c) nest-mending, (d) glueing.

It seems fairly clear from this that pseudofemale behaviour will prove to be much the same for both species. Two interesting new points arise from the above notes. Firstly, the eggs were not destroyed in this case, whereas the ten-spined pseudofemales always destroyed them. There are three factors that probably contribute to this difference. There was only one three-spined pseudofemale, whilst there were several ten-spined pseudofemales. Three-spined male B was very aggressive towards male A, probably because in this species the pseudofemale coloration is less like the real female coloration than it is in the ten-spined species. The nest of the three-spined species is a covered depression in the sand, and it is therefore less likely that eggs will be knocked out.

The second interesting point from these notes is the observation that immediately after behaving as a pseudofemale, male A performed nest-building movements. It had not built a nest, and had never been seen to perform any nest-building movements before, but as soon as it had behaved as a pseudofemale, there was an outburst of nest-building activity. This was never seen in the ten-spined species.

Homosexual behaviour has been observed for a number of other species, but it is fair to say that in most cases it takes the form of a sexually active male courting another male instead of a female. It would appear that where one sex performs the pattern belonging to the other sex, it is more frequently the female that performs the male pattern, rather than vice versa. Observations of males performing female sexual actions are not common, and reports of males performing the whole, or almost the whole, sexual pattern of the female are rare. (It must be remembered that the sexual behaviour of the male is nearly always more marked and easier to detect than the female sexual behaviour, so that pseudofemale activities may, in fact, occur as frequently as pseudomale activities, but be less often noticed as such.)

There is little information available concerning pseudofemale behaviour in other species of fish, although a few points can be extracted from the cichlid work of Baerends and Baerends (1950). It appears that similar territorial conditions can exist for a tank of cichlids as have been described here for sticklebacks, with several dominant territory owners and several territoryless males in one tank. Under these conditions homosexual pattern-inversion was seen on one occasion by Baerends and Baerends. On this occasion an intruding male behaved very much like a female, performing three active female courtship movements (following to pit, circling, and snapping up: see Baerends). Concerning the two male fish involved in this case, they mention that 'The resident male, as well as the intruding one, had been kept isolated during a rather long time.' There are two other points from this paper which, although not actually dealing with homosexual behaviour, are nevertheless relevant. Firstly, on one occasion when eggs had just been laid after a courtship, a second male rushed across the tank and took part in the fertilizing of them, and although attacked by the owner, could not be driven away. Secondly, it was observed that two frontally fighting male cichlids were often assisted by 'territoryless congeners'. When the fight was over, the territory owners were seen to drive their 'helpers' away. Concerning this the authors state that 'It looks as though the helpers take the opportunity to give expression to their fighting urge which is otherwise not possible to them, having no territories.' It is not intended to discuss these observations here except to say that it is quite obvious from the work of Baerends and Baerends that it would be rewarding to study this aspect of behaviour in more detail, not only in the sticklebacks, but also in the cichlids.

In certain species of bird the reversal of pairing position has been seen to occur during courtship. This has been reported for the great crested grebe, *Podiceps cristatus* (Selous, 1901; Huxley, 1914), the moorhen, *Galinula chloropus* (Selous, 1901), and the pigeon (Whitman, 1919). Homosexual 'pairs' are well known amongst the anatids.

Stone (1924) and Beach (1938) have independently reported rare cases of male rats exhibiting the female pattern, and in these reports there appears to be a situation akin to the one related here. In the first place, male rats with a high level of sexual motivation may attempt copulation with other males. In a few isolated cases the males that are being courted respond with the female pattern, although in most instances they fight off the other male. These pseudofemale rats behave as males when placed with a receptive female (as do the sticklebacks), and it has been observed that the switch-over from one pattern to the other may occur in a few seconds. This has led Beach to call the phenomenon 'reversible inversion' of the usual sexual role.

Pseudofemale behaviour has been reported in some detail for certain monkeys and man. Here the effects of experience complicate the issue. Williams (1944), in a study of homosexuality in man, employs the two designations FMH and MMH for Feminine Male Homosexual and Masculine Male Homosexual respectively (FMH = pseudofemale). It has been shown that for both types, homosexual behaviour may be *preferred* to heterosexual behaviour, even though the latter is available to the individuals concerned. In these cases the inversion of the usual sexual role has become irreversible as a result of conditioning. (It should be remembered that such irreversibility may also be caused, in certain individuals, by some hormonal abnormality.)

It would appear, then, that complex homosexual pseudofemale behaviour has been observed only in sticklebacks, cichlids, anatids, rats, monkeys, and man. The fact that the species in which the phenomenon has been observed are just those species which the comparative ethologists and the comparative psychologists have studied particularly elaborately, is perhaps an indication that such behaviour is of a much wider occurrence in the animal kingdom than was previously believed, and that it is only revealed after a detailed study of the animal concerned has been carried out.

Discussion of Terminology

Throughout this paper the term 'homosexual' has been used in its widest sense to refer to any sexual relationship between two individuals of the same sex, regardless of the details of the case. As will have been seen by now, there are several distinct ways in which homosexual behaviour can occur:

I. Only sensory field involved.
  A. *Lack of discrimination:* The male may fail to identify the sex of the intended partner.
  B. *Sufficiency of sub-optimal stimulus:* The male may identify the sex of its intended partner as being male, but still react sexually to it.
  C. *Preference for own sex:* The male may prefer a sexual partner of its own sex, as a pseudofemale partner.

II. Reversal of motor pattern involved.
  D. *Inversion of sexual pattern:* The male may exhibit the female sexual behaviour pattern in response to the advances of males of types A, B, or C.
  E. *Preference for inversion:* The male may prefer to behave as a pseudo-female (i.e. type D) despite the availability of heterosexual outlets.

(The above list applies equally well, of course, to the female sex.)

Type A homosexuality may be the result of the behaviour or appearance of the intended partner being, accidentally or otherwise, similar to that of the opposite sex. (For example, an over-fed male fish with a swollen belly may be mistaken for a female with her belly swollen with eggs.) Both type A and B homosexuality may be the result of an individual being very highly motivated sexually, so that the combination of this strong internal stimulus, with the small external stimulus provided by another member of the same sex, will be sufficient to release sexual behaviour. Type C homosexuality may be the result of the conditioning of a type A, or more probably, type B homosexual. The latter may modify its behaviour to type C, as the result of experiencing satisfaction from homosexual activity. Type D homosexuality may be caused, as shown in this paper, by the frustration of the reproductive behaviour in the presence of dominant males. Type E homosexuality may develop by conditioning from type D, in the same way in which type C may develop from type B. The occurrence of type E may also be the result of some physiological abnormality in an individual.

Types A, B and D homosexuality have all been shown to occur in male sticklebacks. It has not always been possible to distinguish cases as being definitely of type A or B, but on the rare occasions when black males courted other black males, type B was certainly occurring, and in instances where at the nest a pseudofemale had taken over from a female, then it is almost certain to have been type A.

Types C and E homosexuality are, as far as I know, restricted to primates. As already mentioned, the inversion of the sexual behaviour pattern may occur in heterosexual as well as homosexual situations, as it does for example, in the case of the female stickleback that fans.

Certain authors (Kinsey, Pomeroy and Martin, 1948, and Ford and Beach, 1952) have discussed in detail the difficulties and confusions involved in the terminology of reports of homosexual behaviour. I feel that matters would be simplified if the term 'homosexual' were retained and used in future for referring to all sexual relationships between two individuals of the same sex, and if all the synonyms for 'homosexual' which have been used in the past were now abandoned. Use of the word 'homosexual' as a general term, along with the classification of the five basic types presented here, would perhaps prevent further ambiguity. It is realized that in studies of human homosexual behaviour there will, naturally, be a need for the retention of a number of specialized terms such as, for example, transvestitism (Hirschfield, 1940), which refers to individuals that obtain sexual pleasure from dressing as members of the opposite sex.

References

BAERENDS, G. P. and J. M. BAERENDS (1950), 'An introduction to the study of the ethology of Cichlid Fishes'. *Behaviour*, Supplement I, pp. 1–242.

BEACH, F. A. (1938), 'Sex reversals in the mating pattern of the Rat'. *Genet. Psychol.* 53, pp. 329–34.

—— (1948). *Hormones and Behavior* (Hoeber, New York).

BLEGVAD, H. (1917), 'On the food of fish in the Danish waters within the Skaw'. *Rep. Danish Biol. Sta.* 24, pp. 19–72.

FORD, C. S. and F. A. BEACH (1952), *Patterns of Sexual Behavior* (Eyre & Spottiswoode, London).

HAMILTON, G. V. (1914), 'A study of the sexual tendencies in monkeys and baboons'. *J. Anim. Behav.* 4, pp. 295–319.

HIRSCHFIELD, M. (1940), *Sexual Pathology* (Emerson, New York).

HUXLEY, J. S. (1914), 'The courtship habits of the Great Crested Grebe (*Podiceps cristatus*)'. *Proc. Zool. Soc. London*, pp. 491–562.

HYNES, H. B. N. (1950), 'The food of freshwater sticklebacks'. *J. Anim. Ecol.* 19, pp. 36–58.

KINSEY, A. C., W. B. POMEROY and C. E. MARTIN (1948), *Sexual Behavior in the Human Male* (Saunders, Philadelphia and London).

LEINER, M. (1929), 'Ökologische Studien an *Gasterosteus aculeatus* L.'. *Zs. Morphol. Ökol. Tiere* 14, pp. 360–400.
—— (1930), 'Fortsetzung der ökologischen Studien an *Gasterosteus aculeatus*'. *Zs. Morphol. Ökol. Tiere* 16, pp. 499–540.
—— (1931a), 'Ökologisches von *Gasterosteus aculeatus* L.'. *Zool. Anz.* 93, pp. 317–33.
—— (1931b), 'Der Laich- und Brutpflegeinstinkt des Zwerchstichlings, *Pygosteus pungitius* L.'. *Zs. Morphol. Ökol. Tiere* 21, pp. 765–88.
—— (1934), 'Die drei europäischen Stichlinge und ihre Kreuzungsprodukte'. *Zs. Morphol. Ökol. Tiere* 28, pp. 107–54.
MARSHALL, F. H. A. (1922), *The Physiology of Reproduction* (Longmans, New York).
PELKWIJK, J. J. TER and N. TINBERGEN (1937), 'Eine reizbiologische Analyse einiger Verhaltensweisen von *Gasterosteus aculeatus* L.'. *Zs. Tierpsychol.* 1, pp. 201–18.
SELOUS, E. (1901), 'An observational diary of the habits of the Great Crested Grebe'. *Zoologist* 5, pp. 180–2.
SEVENSTER, P. (1949), 'Modderbaarsjes'. *De Levende Natuur* 52, pp. 160–8 and 184–9.
STONE, C. P. (1924), 'A note on feminine behaviour in adult male Rats'. *Am. J. Physiol* 68, pp. 39–41.
TINBERGEN, N. (1940), 'Die Übersprungbewegung'. *Zs. Tierpsychol.* 4, pp. 1–40.
—— (1942), 'An objectivistic study of the innate behaviour of animals'. *Biblioth. biotheor.* 1, pp. 39–98.
—— (1950), 'The hierarchical organisation of nervous mechanisms underlying instinctive behaviour'. *Sympos. Soc. exp. Biol.* 4, pp. 305–12.
—— (1951), *The Study of Instinct* (Oxford University Press).
—— and J. J. A. VAN IERSEL (1947), 'Displacement reactions in the Three-Spined Stickleback'. *Behaviour* 1, pp. 56–63.
WHITMAN, CH. O. (1919), *The Behavior of Pigeons* (Carnegie Inst. Wash. Publ. 257), pp. 1–161.
WILLIAMS, E. G. (1944), 'Homosexuality: A biological anomaly'. *J. Nerv. and Ment. Dis.* 99, pp. 65–73.
WUNDER, W. (1928), 'Experimentelle Untersuchungen an Stichlingen'. *Zool. Anz.* (Suppl.) 3 (Verh. Deutsch. Zool. Ges. 32), pp. 115–27.
—— (1930), 'Experimentelle Untersuchungen am dreistachligen Stichling (*Gasterosteus aculeatus* L.) während der Laichzeit'. *Zs. Morphol. Ökol. Tiere* 16, pp. 453–98.
ZUCKERMAN, S. (1932), *The Social Life of Monkeys and Apes* (Kegan Paul, London).

Author's Note, 1969

In this paper I drew attention to the strange pseudofemale behaviour of certain male sticklebacks that were sexually thwarted. In later observations I was puzzled to note that territorial nest-owners occasionally also performed pseudofemale patterns, but I was unable to pursue the investigation any further. A more detailed study of the phenomenon has recently been made, however, by Van den Assem (1967),* whose interpretations differ from mine in certain respects. (Although he studied the three-spined species, we both agree that there is little difference between the two species in this particular respect.) He believes that the main function of the pseudofemale pattern, which he calls 'sneaking', is fertilization-stealing, an explanation which I considered in my original paper and then rejected because it did not fit the facts. His view is that the behaviour of a 'sneaker' should not be thought of as pseudofemale. He states that 'A sneaker never follows another male in a female fashion'. This is not true. As I pointed out, a pseudofemale does not respond to male dancing by following unless the male has already performed the more advanced courtship action of showing-the-nest-entrance. Once it has done this, however, and the pseudofemale has become sufficiently aroused, the latter will follow further dancing just as if it were a courting female. Van den Assem also states that 'sneakers do not attempt to enter the nest as a female does when the owner shows the entrance'. Again this is not always true, as I observed on a number of occasions. Van den Assem further claims that once a true female is in the nest 'a sneaker may in fact take up the nest-owner's role'. I never saw this, although my pseudofemales had ample opportunity to behave in this way. I can only reiterate that I observed the performance of all the female patterns by my pseudofemale fish, with the single obvious exception of egg-laying.

Certain observations of Van den Assem's do, however, throw new light on the causation of stickleback pseudofemale behaviour. He found that, as I suspected from my later observations, this pattern can be shown by territory-owning males who are not inferior in breeding status and are not suffering from long-term sexual frustration. This I accept, but it should be remembered that any male stickleback, whatever his status, is in an immediate condition of sexual frustration when he is in the presence of another courting pair. He can witness the female sexual responses and he can also see that she is not giving them to him, but to another male. What Van den Assem has underlined, therefore, is that a male experiencing short-term sexual frustration will also respond with pseudofemale behaviour, as well as one suffering from long-term frustration. I still feel, however, that the latter cases produce a more marked response and this may account for the difference in our observations mentioned above.

As regards the nature of this strange pattern of behaviour I can see no reason to change the views expressed in my original paper. The 'sneakers' do perform patterns similar to courting females and can therefore justifiably be called pseudofemale. The courtship does proceed as an interaction between two males and can therefore justifiably be called homosexual. The

* Van den Assem, J. (1967): 'Territory in the Three-spined Stickleback.' *Behaviour*, Supplement 16, pp. 1–164.

function of the pattern is another matter. Here, Van den Assem has made two significant discoveries. One is that, if the pseudofemale manages to synchronize its actions in such a way that it passes into the foreign nest immediately after the female has laid her eggs in it, he does in fact fertilize them. The other is that in such cases, and where the pseudofemale is a nest-owner himself, he will try to steal the eggs, carry them to his own nest, insert them and rear them in the normal way. It is this that leads Van den Assem to conclude that the pattern is functionally primarily one of fertilization-stealing. Having drawn this conclusion, however, he then lists five extremely convincing reasons as to why this should *not* have survival value! It is clearly more efficient for each nest-owner to deal with his own eggs. The only explanation he feels is acceptable is that 'robbing has been selected as a means to synchronize the parental cycles in a population of settled males'. Considered as the primary selection pressure I must admit this fails to satisfy me. (Even if it were true, it still does not alter the fact that the 'sneaking' behaviour employed involves both pseudofemale and homosexual patterns.)

To sum up, it seems more likely that, in view of the great risk of damage to the eggs involved, the pattern is initiated by long- or short-term sexual thwarting. This leads the frustrated male on to the courting male's territory, where the invader's status sinks rapidly. He shifts suddenly from a dominant to a submissive condition, similar to that of the courted female. The combination of sexual state plus submissiveness is the blend that produces true female courtship behaviour and so, in him, it provokes similar behaviour, and he starts to perform as a pseudofemale. If, in the course of events that follow, he finds himself passing through a nest of unfertilized eggs, this automatically triggers off his ejaculation response. Having now fertilized a batch of eggs which are not in his own nest, he will be forced to try to put them where they 'belong', hence the egg-stealing. If he has no nest of his own, he will do what a normal male will do if his nest is destroyed, namely eat the eggs.

As to the survival value of the whole pattern, I must confess that neither Van den Assem's nor my own explanation seems very convincing. Perhaps future research will clarify the situation, or perhaps what we have witnessed is a weakness in a breeding system which, taken as a whole, is nevertheless highly successful.

The reproductive behaviour of the zebra finch (1954)

Introduction

The zebra finch is a small Australian ploceid which is ideally suitable for laboratory observations. It will nest and rear young in small indoor aviaries. New birds, transported to the laboratory in small boxes, will begin to nest-build and court within minutes of their release into an aviary. There are no seasonal difficulties, as it breeds all through the year. The species is exclusively a seed-eater and the nestlings require no special diet in captivity. The birds are not disturbed by the presence of an unconcealed human observer.

The subfamily Estrildinae, of the Ploceidae, to which the zebra finch belongs, has recently been revised taxonomically by Delacour (1943).[1] He divides it into three natural groups, or tribes, namely the waxbills, the grass-finches and the mannikins. The zebra finch is placed in the second of these, which is almost entirely confined to Australia. Several species of both the other tribes have also been kept under observation in the laboratory for comparative purposes, but these will only be mentioned in passing in the present paper.

In the zebra finch there is a strong sexual dimorphism and the fledglings are distinguishable from both adult males and females. The male possesses many distinct markings, the evolution of which will be discussed later in the paper. Here is a brief summary of the markings of the species, which will serve to supplement the accompanying illustrations.

Adult male: Bright red beak. Cheek white with black vertical margins. Ear-patch chestnut brown. Throat and breast finely barred with black and white, with a wide black band across the lower breast. Abdomen pure white. Flanks brown with white spots. Dorsal surface and wings grey. Upper tail coverts banded black and white. Rump white. Legs and feet orange.

Adult female: Grey over the whole body surface except for: Beak red, but less intense than male. Cheek white with black vertical margins. Upper tail coverts banded black and white. Rump white. Legs and feet orange.

Fledgling: As for female except: Beak black.

It will be seen from this that there are five distinct markings which are

[1] The names of the Estrildinae mentioned in this paper are based on Delacour's revision.

exclusively male and that these are situated on the ventral surface, from the throat to the abdomen, and the flanks and the ears. As will be shown later, these are the areas displayed in the male courtship. The cheek, tail and rump markings are the 'species markings' common to all individuals, while the colour of the beak denotes maturity in both sexes.

Twenty-three individuals were used for the following observations, fourteen of which were purchased from various aviculturalists and nine of which were bred in the laboratory.

The birds were housed in wire aviaries with dimensions of approximately 4 × 3 × 2 feet. Dense clumps of twigs were fixed in the corners of these to provide nesting sites. Nesting material, in the form of straw, grass and string, was scattered over the floor. Each aviary was screened from all the others. Wire passages which could be opened and closed by a panel, were inserted between certain of the aviaries (see Fig. 11). These were used in the study of territorial behaviour.

Agonistic Behaviour

Before describing the agonistic behaviour of the zebra finch, it is necessary to discuss the meaning of the term itself. A number of different names have been given, in the past, to the group of activities associated with intraspecific fighting. These activities include attack, threat, submissive and fleeing behaviour and any general name must embrace all of these. The terms 'hostile', 'aggressive', 'fighting', 'combat' and 'conflict', which have been used in the past, fail to do so. The words 'hostile' and 'aggressive' emphasize attack and threat, but ignore submission and flight. The words 'fighting' and 'combat' stress the actual bouts of physical contact involved, but ignore other aspects. The word 'conflict' is unfortunate because it is used in two senses; conflict behaviour may refer to actions resulting from a clash between two individuals, or from a clash between two tendencies in one individual.

Scott and Fredericson (1951), discussing the causes of fighting in mice and rats, have used the term 'agonistic behaviour' to describe the 'general group of behavioural adjustments' associated with fighting, and this would seem to be most suitable as it avoids the special associations of the more popular terms mentioned above.

Ethologists have shown that in agonistic situations a conflict in the individual, between the incompatible drives[1] to attack and to flee, results in the occurrence of threat postures or movements (see Tinbergen, 1953a and Moynihan, 1955). Furthermore, slight differences in the relative strengths of these incompatible drives give rise to a series of threat postures or movements,

[1] The term 'drive' is being used in its widest sense (see Hinde, 1953).

each of which represents a particular balance between the two drives. The result of this is that each species possesses a threat code.

It is obviously an advantage for individuals to settle disputes, if possible, without actual physical violence. Physical damage, which is the inevitable outcome of actual fighting, is avoided to a greater or lesser extent in most species by the learned acquisition of responses, or the evolution of innate responses, to the various signals of the threat code.

Similarly, where fleeing is for some reason impossible, or a disadvantage, responses exist to submissive postures or movements. Such postures or movements have an appeasement function. The threat and submissive codes together form an *agonistic code of signals*.

In general, the more dangerous the weapons possessed by a particular species, the more that species relies upon its agonistic code in the settlement of disputes. Lorenz (1952, chapter 12) has elaborated this point and it need only be mentioned here. The zebra finch, being a small seed-eater, possesses only mildly dangerous weapons. Therefore one neither expects, nor finds, the elaborate use of a complex agonistic code. Yet the beak can inflict some damage and such a code is not entirely absent, although it is comparatively unimportant. Nearly all disputes are settled by actual chasing and pecking.

Supplanting attacks (Hinde, 1953) are common in situations where one bird is dominant to another. The subordinate bird flees from the spot it occupied as the dominant one approaches. This may occur repeatedly in rapid succession resulting in a prolonged chase from branch to branch. If a subordinate bird is hit it is either pecked with closed beak, or it is snapped at. If, in the latter case, the dominant bird successfully catches hold of the other's plumage, a plucking may ensue. Owing to the speed of the movements it is difficult to be absolutely certain whether the dominant bird makes active plucking movements, or whether the subordinate bird plucks itself in its struggle to flee from the firm grip of its rival. It seems most probable that it is the fleeing of the caught bird which is the more important factor, especially in the light of the following observation. One male had snapped at and seized another by its wing. The next moment, the captor was hanging in mid-air from the wing of the attacked bird. The latter still clung to a branch. The captor made no attempt to fly off and thus tug out the feathers it was holding in its beak. Instead, it simply hung by its beak until the weight of its body and the captive's attempts at fleeing tore loose the feathers it was holding and, amid a cloud of feathers, it crashed down on to the top of a nest below. Not all pluckings are as spectacular as this, but, after one incident on the ground, the victor was seen to hop about collecting up in its beak a number of the feathers which it had just extracted from its rival. Then, sitting on a low branch, it began nibbling them, turning them back and forth with its tongue as it did so. After a while it let them drop, wiped its beak, and went about its business. The significance of this pattern is not understood.

When one bird is clearly dominant over another, it nearly always assumes a horizontal posture when attacking (see Fig. 1). The feathers are sleeked and the body is pointed directly at the enemy. This is the posture of the charge and is retained right up to the actual pecking or biting, with the result that a subordinate bird is attacked about the body rather than the head. Sometimes the horizontal forward posture is given without being followed through with an attack. In such cases it acts as a threat posture and may have an intimidating effect. This is an example of the way in which it is used: the dominant bird sits on a branch and the subordinate bird flies up to land near by. As it approaches, the former faces it with a horizontal forward posture and this results in the immediate flight of the subordinate. Actual attack, however, is far more common than threatening of this kind.

A different kind of attack is seen where two birds are approximately equally

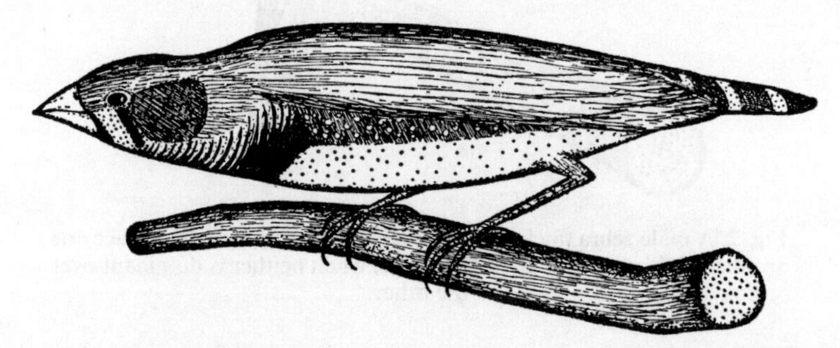

Fig. 1. A male zebra finch in the sleeked horizontal posture. A dominant bird takes up this posture when facing a rival.

balanced as regards dominance. Here, both assume a vertical posture (Fig. 2), face one another and beak-fence. All attacking is done with closed beaks in such cases and is confined to the head region. Usually many quickly repeated jabs are given by each bird at the beak of the other. When the match is a very even one, both birds lean slightly forward out of the vertical, but if one is slightly more on the defensive it tends to assume a completely vertical posture and may even lean backwards a little. Such a bird usually shows signs of fleeing, and judging by subsequent behaviour is invariably the loser of the fencing bout.

It is clear, therefore, that when one bird *faces* another in an agonistic situation the angle at which its body is inclined is an important indicator of the state of balance between the attacking and fleeing drives. The more horizontal the posture, then the more this balance is in favour of attack. The appreciation of this fact by the birds results in the existence of a simple threat code.

As in many other birds, fighting is carried on in complete silence, broken only by the sound of beating wings (supplanting) or the clicking of beaks

Fig. 2. A male zebra finch in the sleeked vertical posture. Birds face one another in this posture, and beak-fence, when neither is dominant over the other.

(fencing). Not only is there no extra noise when fighting occurs but there is an actual decrease in sound. An aviary containing several birds usually becomes quite silent when a fight is in progress. There is one exception to this, which occurs only occasionally, and that is a soft rasping call which may be given by a supplanting bird. Exactly what determines whether this call is given or not, is not yet clear. It was, however, noticed to be more frequent in one particular situation, namely when a female was incubating eggs. She often emerged from the nest and attacked any birds that happened to be sitting near by. As she attacked them, she nearly always gave the call, with the beak held open. Several times this was seen to act as a threat and the other birds fled simply on hearing her give the call as she emerged. On these occasions the call was given, not just as an addition to chasing, but as an alternative to it. It is not clear whether the response to it as a threat was innate or learnt.

When fleeing is thwarted by something other than the attack drive, special postures and movements can be observed. In aviaries, such thwarting may occur simply as the result of the physical impossibility of escape. In one aviary in which one pair had been kept for many months the male went through a phase, which lasted for a number of days, during which it was excessively aggressive towards its mate. Apart from repeated fleeing the female showed three special responses during this time: fluffing out all the feathers, squeaking, and begging. The fluffing response is interesting in this connection because (as has been pointed out by Hinde, 1953, for the chaffinch) it may here be comparable with head-flagging in the black-headed gull (Tinbergen and Moynihan, 1952) as an appeasement response.[1] In head-flagging, the gull turns away its black-masked face, thus performing the *opposite* of a threat action. When fluffed out, a zebra finch squats with all its feathers raised as shown in Fig. 3. This gives the opposite effect of an aggressive bird, which always has its feathers sleeked and its body tensed. But there is a difficulty here in that an alarmed bird, or one which is fleeing, is also sleeked and tense.

Fig. 3. A female zebra finch squatting in the fluffed-out submissive posture.

The answer to this lies in the fact that sleeked tension is characteristic of an active bird, whilst fluffed squatting is associated with inactivity, as when birds are resting or sleeping. A beaten bird, that cannot escape, signals its submission by assuming an 'inactive' posture even when it is actively feeding or moving about the aviary. This posture is therefore the opposite of the threat postures by virtue of being the opposite of the postures characteristic of activity.

The fluffed posture was adopted by the female nearly the whole time during the phase of extreme aggression by the male, whether he was actively persecuting her or not. After a prolonged bout of chasing and pecking, the

[1] The fluffed posture also occurs in subordinate great tits, where it appears to play the same role (Hinde, 1952, p. 41), and in various Fringillids, especially the bullfinch and greenfinch (Hinde, personal communication).

female, who had been silent previously, sometimes gave plaintive squeaks on being pecked, and squeaking therefore is characteristic of an intensely beaten bird. Begging, which was seen rarely, appeared to be the result of an even more intense subordination. Here is an extract made from notes at the time: 'The male chases the female viciously until she appears to be exhausted. Finally she stays still on the floor with her feathers sleeked. She crouches low and at the same time turns her head upwards and opens her beak, orienting it towards the male, who pecks her hard about the head for a while and then leaves her.'

From this it is clear that the female, who is posturing in a manner similar to young fledglings when food-begging, is submitting to the male in a way which has appeasement value. For, although he pecked her when she begged to him, he soon stopped and left her alone, giving her the first respite from his attacks for some time. Nor was this the result of the exhaustion of the male's attack drive, for he returned to the assault soon afterwards, when the female's behaviour changed.

This female, when removed to a small cage for safety, soon became sleeked again. When returned once more to her mate, she quickly resumed the fluffed posture without even being attacked. This was repeated several times with the same result. It is interesting to note that the begging female had her feathers sleeked. It seems that, at higher intensities of submission, begging is a substitute for, rather than an addition to, fluffing.

As stated above, these submissive actions can all be explained by the physical prevention of escape. Obviously such a situation is not going to occur in nature. Where submissive behaviour is found under natural circumstances, it must be the result of the fleeing drive being thwarted by some other incompatible drive. Sexual attraction is probably the most important of such conflicting drives. Even in the case described above, the sexual attraction resulting from the pair-bond, which existed between the two birds, probably played a part. This point of view seems more likely when the following observations are considered:

A single male bird was introduced into an aviary which contained a mated pair and their four fledgling offspring. It began to court the female, but was violently chased and attacked by her. After some minutes it was attacked also by the male parent and it then took refuge in the top far corner from the nest-site. This corner was then always employed as an escape headquarters by the intruder, despite the fact that its own, now empty, adjoining aviary was accessible. In the days that followed it began to defend the area around this corner more and more and it slept there at night. During this time the young birds became independent and the parental cycle of the pair was completed. Both males courted the female from time to time now, but the single male was usually chased off. During the next week there was an almost perpetual state of tension between the three birds, which was ultimately resolved when the intruder male succeeded in breaking the old pair-bond, won the female over, built her a new nest, and began a fresh cycle with her.

Just before the 'divorce' took place, the state of tension reached a particularly interesting phase during which the intruder male was dominant to the other male despite the fact that the old pair-bond was only partially broken. The female was still behaving as if paired to her original male, but was already beginning to show signs of the new pair formation. Her old mate was frequently attacked and chased by the intruder male.

It was under these special circumstances that the now subordinated male showed submissive displays. They took the form of a modified version of the fledgling food-begging response. The male crouched low with its head tilted upwards and its beak open. In this respect it was similar to the posture given by the submissive female described earlier. But it differed in that it was accompanied by a call similar to that of the young birds and also by a waving of the head from side to side. This waving was far more vigorous and exaggerated than anything seen in the young birds when begging.

The submissive display was always oriented towards the rival male when the latter approached, or, which is more interesting, when it showed signs of courting the female. For example, on one occasion all three birds were close together and the intruder male began to sing to the female and perform the initial courtship movements. Immediately the subordinate male gave the begging display and the courting ceased. This was all repeated several times. On other occasions, the intruder male had only to advance towards the female, without even courting her, for the begging displays to be given. It is clear, therefore, that this display had two functions. It was seen frequently to stem the attack of the dominant bird and was then functioning as an appeasement display. It also prevented the dominant bird from courting the female and in these instances served to prevent interference with the pair-bond. However, as was indicated above, it was only successful as a short-term measure and the intruder male finally succeeded in stealing the female. In this connection it should be mentioned that the intruder was much the finer specimen of the two males, the other being in very poor condition the whole time. This was almost certainly an important contributary factor and it is interesting that this gives such a 'divorce' a certain selective value.

Concerning the motivation of the begging display described above, it is almost certainly the result of a conflict between fleeing and sexual tendencies. The sexual tendencies involved were, however, of a special nature, namely, the maintenance of the pair-bond. Where fleeing comes into conflict with other – copulatory – sexual tendencies, the resulting behaviour is of a different kind and forms the courtship pattern of the species. In both these cases the drive to attack is not completely absent and probably plays a part in the motivational conflict. It should also be mentioned that, in the case of the begging display just described, the male, despite the aviary conditions, did not have his fleeing behaviour physically obstructed, as did the female described earlier. It was quite clear that it was his tendency to keep near his mate that thwarted his fleeing behaviour.

Nesting Behaviour

As is typical of the ploceids, the zebra finch constructs a domed nest (see Fig. 4). In the nest of this species the entrance is a simple side-hole. In the laboratory nearly all the nests were built in the clumps of twigs provided, although one was constructed on the floor in a corner.

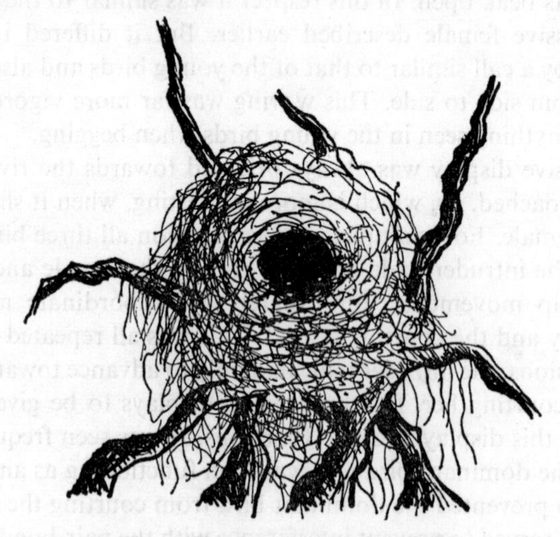

Fig. 4. The nest of the zebra finch.

The task of collecting nest material is almost exclusively performed by the males. Males kept in isolation will construct nests, but the presence of a female acts as a strong stimulus to nest-building, as is shown by Fig. 5. Both sexes have been observed to pick up and nibble fragments of material, but females have rarely been seen to carry a piece to the nest and incorporate it. The female is active, however, in helping to form the nest cavity. One specialized movement has developed in connection with this, namely, *neck-stretching*. The bird sits inside the nest and enlarges the cavity by repeatedly extending and withdrawing its neck. As its neck is extended, so the head is tilted back. This may be done upwards, which strengthens and raises the roof, or sideways, which scoops up the sides.

A simple lining may be included on the floor of the nest cavity, consisting of finer and softer fragments. Jones (1932) states that 'I find the nesting birds only like to line their nests just before the young hatch.' My observations disagree with this. If suitable materials are available, the nests are often lined even before the eggs are laid.

Although the nest-building drive waxes and wanes from day to day, it is nevertheless present all the year round, both in the laboratory and, apparently, in nature. Friedmann (1949), in a detailed review of the nesting habits of the Ploceidae, reports that weaver birds show nest-building behaviour 'equally avidly whether they are in breeding or non-breeding plumage; in other words,

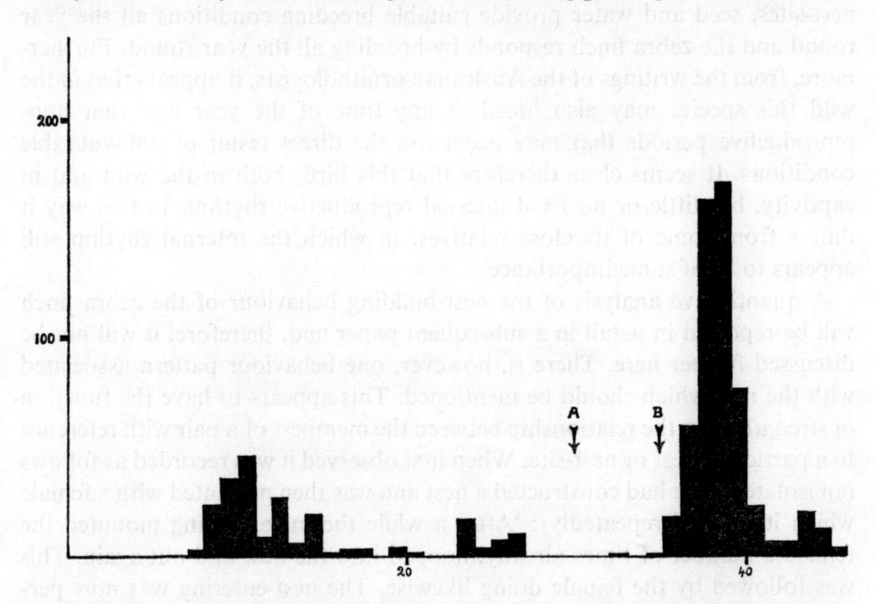

Fig. 5. A daily record, over 46 days, of the frequency of nest-material collection by one male. At the start of the test, the old nest and material were removed from the aviary. On each of the following days the male was provided with 200 pieces of material of standard size and texture, and the number of these which were incorporated into the nest was scored each day. When the collecting had fallen off to zero, a female was brought into the laboratory (Arrow A) and kept in a near-by cage out of sight. The two birds called to one another, but the auditory presence alone of the female did not have any effect on the building frequency of the male. The female was then introduced into the aviary of the male (Arrow B) and immediately the building increased tremendously. Ordinate: frequency of nest-building. Abscissa: time in days.

the urge to build, which in most birds is seasonal and is part of the cyclical sequence of behaviour patterns, is here extended far beyond its normal limits'. In the zebra finch, however, it seems that all reproductive activities are extended in this way. Individuals of this species kept in the laboratory for the past two years have seldom or never ceased to be sexually active, whilst other closely related species under the same conditions were reproductively inactive for long periods. Hediger (1950) has discussed the question of the effects of captivity on breeding seasons in some detail. This author distinguishes between two types of animals in this respect. The fixed type shows no modification of the time of appearance of its reproductive cycle in captivity. Thus, certain Australian birds, kept in captivity in Europe, laid their eggs in winter, with the result that the latter were nearly all frozen. The variable type, on the other hand, adapts to the changed conditions and may either change its rhythm, or, if circumstances are suitable, extend its season. The zebra finch is clearly of the latter type. Indoor aviaries well supplied with nesting material, nest-sites, seed and water provide suitable breeding conditions all the year round and the zebra finch responds by breeding all the year round. Furthermore, from the writings of the Australian ornithologists, it appears that in the wild this species may also breed at any time of the year and that non-reproductive periods that may occur are the direct result of unfavourable conditions. It seems clear therefore that this bird, both in the wild and in captivity, has little or no fixed internal reproductive rhythm. In this way it differs from some of its close relatives, in which the internal rhythm still appears to be of some importance.

A quantitative analysis of the nest-building behaviour of the zebra finch will be reported in detail in a subsequent paper and, therefore, it will not be discussed further here. There is, however, one behaviour pattern associated with the nest which should be mentioned. This appears to have the function of strengthening the relationship between the members of a pair with reference to a particular nest or nest-site. When first observed it was recorded as follows (an isolated male had constructed a nest and was then presented with a female which it courted repeatedly): 'After a while the male, having mounted the female a number of times already, hopped into the nest and out again. This was followed by the female doing likewise. The nest-entering was now performed again and again by the male and the female more or less alternately. Such behaviour can best be described in functional terms as the male showing the nest to the female and the female accepting the male's nest. During the first half-hour of this behaviour the sequence of entry was as follows (M = male entry; F = female entry; b = building movements were performed inside the nest): M F M M F F M F M F F F M Fb M Fb F F F F F Fb Mb M F Fb Mb Fb Mb Fb Mb F Mb ...

From this sequence it is noticeable that internal nest-building movements accompanied the entrances into the nest more and more as time went on. All this took place within an hour of the female's original introduction into the aviary.'

It was also recorded at the time that, in the above performances, when the female was in the nest there was on her part 'much turning round accompanied by a special call which can best be described as a hissing-chirping'.

The male showed similar behaviour on the next day, when the female was much more reluctant to enter the nest, although he had not done so previously. Similar behaviour has been noted by Boosey (1952), who records that 'During the construction of the nest the cock frequently enters the (nesting) box and, in order to encourage the female inside, makes a curious cosy little murmuring noise which I have always found quite impossible to describe.'

The same call has since been heard on many occasions, especially when one bird inside the nest is visited by its mate. Also, if a nestless male is given a female, he gives the call in the following manner. After the initial period of vigorous courtship, the male flies to a clump of twigs and moves round and round over it in repeated turns giving the nest-call and maintaining a special posture. This posture consists of fanning the tail and holding the body in a horizontal position. The legs are bent so that the body is kept low over the twigs. The beak is opened and shut extremely rapidly. This latter action has been called 'mandibulating' by certain authors and is commonly seen in many birds in various situations. In the zebra finch it is observed in connection with the testing of nest material and the nibbling of seeds. In the above case, however, it appears to be used purely as a signal.

This display by a nestless male inevitably precedes the construction of a nest on the site where it was performed, and its function appears to be to indicate to the female his intention to build there.

Sexual Behaviour

The courtship of the male zebra finch consists of auditory, static-visual, and dynamic-visual elements, whilst that of the female is primarily static-visual.

The courtship of the male

(a) *The auditory component*

During the courtship dance the male sings continuously, repeating over and over again his short song phrase. The dance is never performed without the song. The song, on the other hand, is often given when the male is not courting a female. As the courtship proceeds, the song may break down into loud, rapid, single-note calling which may be answered similarly by the female. The song phrase is unmusical to the human ear, and, when given repeatedly in quick succession, as it nearly always is, sounds not unlike the turning of a squeaky handle. Any such comparison as this, however, is bound to be inaccurate for the following, most interesting, reason: *each male possesses a unique song phrase*. Despite the fact that the basic pattern of the phrase can always be recognized as typical of the species, nevertheless, in each male, the song is not subtly,[1] but clearly, different from that of the other males. I hasten to add that this is based on observations of only a small number of males. It would be most interesting to compare song recordings of large numbers of males of this species. Analysis of such recordings would reveal what aspects of the song are common to all members of the species and also what is the basis of the individual variations. A possible function of these variations will be discussed later.

Fig. 6. The high intensity courtship posture of the male.

(b) *The static-visual component*

The singing male, as he dances towards the female, assumes a very characteristic courtship posture. The high intensity version of this posture is shown in Fig. 6. The legs are bent and the pure white belly feathers are fully fluffed out. The spotted flank feathers are fluffed outside the wings. The body is held in an upright position. The neck feathers are raised with the singing of the bird. The crown feathers are strongly depressed, making the top of the head appear completely flat, whilst the feathers at the back and sides of the head are raised. *It will be seen that this posturing serves to display to the female all those markings which are specifically male*. It is interesting to note that certain

[1] Slight individual variations in song patterns have been found in other species, but the zebra finch appears to be an extreme case.

closely related species, which do not possess special male markings, show similar courtship posturing. This suggests that the posturing is probably primary and that special markings have evolved on those parts of the body where they are most easily seen by the courted female.

(c) *The dynamic-visual component*

The singing, posturing male advances towards the female in a rhythmic, pivoting dance. The exact form of this dance varies and is often obscured by the particular arrangement of the branches on which the birds are situated. It is best understood when it occurs along one long, straight branch (see Fig. 7). As the male advances towards the female down the branch, it swings its body from side to side, turning first to the left and then to the right,

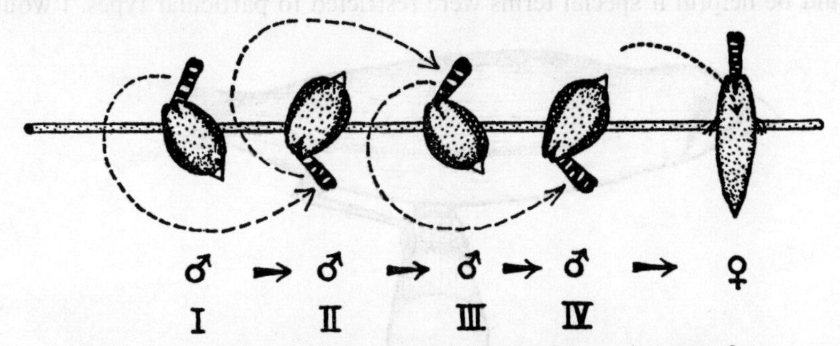

Fig. 7. The movements of the courtship dance of the male as seen from above. For explanation see text.

changing the position of its feet as it does so. Although its body turns through quite a wide arc, its tail swings even farther, as shown in the illustration. At any point in the dance, therefore, the male has its tail twisted sideways out of its normal position. This serves to display the black and white bands of the upper tail coverts.

The closely related cutthroat finch (*Amadina fasciata*), on the other hand, displays to the female frontally, bobbing quickly up and down without pivoting and, correlated with this, the male markings in this species appear more on the front of the body than on the sides.

During its dancing the male zebra finch often hurriedly wipes its beak. This action is most frequently seen in courtship when the female is very unresponsive and it appears to be a displacement activity. It differs in several ways from the form the movement takes when it is really cleaning its beak, but these differences will be dealt with in detail later in the paper.

The courtship of the female

A responsive female does not move about but stays still on the branch. Whereas the male faces down the branch towards her, she does not face him, but sits across the branch. Her display (see Fig. 8) consists of lowering her body into a horizontal position, with feathers rather sleeked and legs bent. She then vibrates her tail extremely rapidly in the vertical plane. Seen from behind, it is clear that, although the main impression is of vertical vibrations, the actual movement of the tail is not just a simple up-and-down. Without slow-motion cinematography, however, it is difficult to analyse further.

It is necessary to make a point here concerning the ethological terminology used in descriptions of vibratory movements. Many words have been employed to describe such movements, for example: shiver, tremble, quiver, shake, flutter, waggle, etc. As it is possible to distinguish to some extent between different types of vibratory movement by simple observation, it would be helpful if special terms were restricted to particular types. I would

Fig. 8. The tail-quivering display of a responsive female.

suggest tentatively that shiver, quiver and flutter could be retained and restricted to the following types of vibration:

Shiver: High-speed, small-amplitude vibrations.

Quiver: High-speed, small-amplitude vibrations, in which the proximal end of the vibrated organ does not move and the distal end moves most.

Flutter: Lower-speed vibrations, usually with greater amplitude.

On this basis, the vibrations which occur in the courtship of the female zebra finch will be called *tail-quivering*.

Copulation

The climax of courtship comes when the male, who has advanced along the branch to the female, mounts her. Sitting on her back, with his ventral feathers still raised, he spreads his tail and lowers it on one side of hers. In this way the cloacae are brought into contact and insemination takes place. Whilst he is mounted the male beats with his wings and in this way maintains his balance. After dismounting, both birds often preen themselves.

Variations in courtship and copulation

The behaviour just described may vary in a number of ways. Certain elements may be omitted, or may appear in different intensities, frequencies, or sequences. Also, the roles of the sexes may be reversed, but this will be dealt with later. All these variations can be thought of as the result of varying strengths, both relative and absolute, of the three drives to attack, flee and mate, which are present in both birds. A detailed analysis, from this point of view, of the variations in the courtship and copulation of the chaffinch has recently been made by Hinde (1953) and I can confirm that the general conclusions he reaches for that species are equally applicable to the zebra finch.

The variations in the courtship of the male can be conveniently discussed by distinguishing between low and high intensity displays, although in reality there is a whole scale of intensities. In the low intensity form the male tends to lean more forward, his ventral feathers are less raised and his pivoting turns are less extreme. He sings just as hard and his head is just as flat on top, but the advancing dance is more irregular and less rhythmic. This type of courtship sometimes develops into an attack on the female, if the latter does not respond.

In the high intensity form the male's body is pulled back more into the vertical plane and his pivoting turns become wide and strong. Although the dancing movements are more vigorous, he seems actually to advance at a slower pace. His ventral feathers are raised to the extreme. The whole display gives the effect of being more exaggerated, coherent and vigorous.

The display of the female varies little, except by virtue of complete omission. It was more often than not the case that the male performed his display and mounted the female without the latter having given the tail-quivering response. There are therefore four ways in which the female can respond to being courted. She can flee, attack the male, stay still, or stay still and display. In the third and fourth instances she may be mounted and in both these cases her drives to attack and flee must be more or less in balance. In the fourth instance her sex drive must be greater and the mounting is likely to be successful.

The sequence of courtship events may vary according to which sex is the initiator. In the vast majority of cases the sequence consisted of the male initiating proceedings and dancing towards the female. As he approaches, the female responds with the tail-quivering and this stimulates the male to mount and copulate. In a pair of Bengalese finches, which are closely related to the zebra finch, and which were well balanced sexually, the significance of the tail-quivering was clearly seen. In this species the nature of the sexual displays is almost identical to that of the zebra finch. The female nearly always gave her display in response to the male's dance and he then mounted her, but on the few occasions when she did not, the male stopped courting and did not mount. It was obvious from this that the tail-quivering was an important invitation signal to the male. However, although it was also stimulating to the male zebra finches, it appeared to be far less essential, and their sexual arousal was such that they frequently mounted without it.

Returning to the zebra finch, the usual sequence of events was broken sometimes by the female initiating courtship, in which case the male often gave his display as a response to that of the female, instead of vice versa.

The derivation of the courtship displays

Pre-copulatory postures and movements, which act as sexual signals, often take the form of ritualized intention movements (Daanje, 1950; Tinbergen, 1952). In the case of the zebra finch, the three drives to attack, flee and mate conflict with one another and mutually inhibit one another, but only partially. Thus, when the sex drive is strong enough to override the other two, so that sexual behaviour is the result, the drives to attack and flee may nevertheless show themselves as intention movements. In this species, the dance of the male can be thought of as principally the outcome of fleeing and mating movements. In the pivoting advance, each swing turns the male away from the female, as if he is going to flee from her, then brings him back towards her again. As he swings back towards her, he hops along the branch in her direction, but then swings away from her again in the other direction. The display of the female can be thought of as the result of a similar conflict, for she adopts the sleek, horizontal posture of a bird about to fly off, but does not actually do so. Instead, she makes special movements of her tail which, judging by the work of Daanje (1950), may well be ritualized intention movements of taking off.

The nature of the differences between the high and low intensity courtship displays of the male is revealing as regards the derivation of the courtship. In the high intensity form, the elements which are exaggerated are those associated with a tendency to flee. The body leans farther back, the feathers are raised more, the turns away from the female are wider, and the advance is slower. As regards the body leaning farther back, it will be remembered that in the section on agonistic behaviour it was pointed out that the more vertical the body of a bird *when facing* its opponent, the more the balance of attacking and fleeing drives was in favour of the latter. Speaking of the courtship of the chaffinch, Hinde (1953) says, 'Males which are often attacked by their mates also seemed to hold their bodies in a more erect posture, and this also may be connected with a strong fleeing drive.' I have also observed that in the cutthroat finch, in agonistic situations, the same relationship exists between horizontal and vertical frontal postures. (However, there are indications that this relationship is not a general one.) With regard to the increased raising of the feathers, it has already been pointed out that this is associated with submissive posturing when fleeing is thwarted, especially by sexual tendencies, and this provides a possible explanation of the fluffing out of a courting male. It is interesting to note that the feather-raising in courting males is differential in its extent, the white belly feathers being fluffed out more than the others and the crown and back being sleeked. As already mentioned, these modifications serve to exaggerate those parts of the body which are specifically male-marked. In connection with this it is significant that the Bengalese finch, which lacks elaborate male markings, also lacks to a large extent such modifications in feather-raising, under the same conditions.

In the low intensity version of the male courtship, the body points forwards at the female more, the feathers are less fluffed and the advance is quicker, all of which points to a switch in the conflict-balance in favour of the attack drive. Furthermore, a male is far more likely to develop this type of courtship (as opposed to the high intensity form) into actual attack, if the female is unresponsive.

Summing up, it is clear that, in courtship, when the approximate balance between the attacking and fleeing drives is in favour of the former, the result is the 'low intensity' display, and when it is in favour of the latter, the outcome is the 'high intensity' display. This is interesting, because it means that, although the intention movements involved have become ritualized in evolution – they are now stylized, regular and rhythmic – they have not yet become entirely 'emancipated' (see Tinbergen, 1952). They are, in fact, still susceptible to changes in the strengths of the attacking and fleeing drives. In other words they are *semi-ritualized*, whilst the tail-quivering of the female appears to be fully ritualized. It is intriguing that a similar state of affairs exists in the chaffinch despite the fact that the male and female displays of this species are different from those of the zebra finch. Thus, Hinde (1953) states that 'The full soliciting posture of the female has probably gone farther (towards emancipation) than the postures of the male.'

It is also interesting that the courtship of the zebra finch is made more intense when there are greater signs of fleeing rather than attacking. This suggests that the original conflict movements selected for ritualization were those more concerned causally with fleeing rather than attacking. This again is in agreement with Hinde's findings in the chaffinch: 'It seems, then, that in both sexes the courtship displays occur in conflict situations. Three drives are involved in this conflict – sex, attacking and fleeing. During normal courtship, however, the balance between attacking and fleeing shifts over in favour of fleeing, and we are concerned primarily with a conflict between sex and fleeing drives.'

Tinbergen (1952, p. 15) has suggested that there are three possible causal explanations of pre-coitional displays.

1. Sexual thwarting ('... the arousing of the sexual instinct while indispensable external stimuli are absent.')
2. Sex-attack conflict ('I even think that the tendency to withdraw may not be the only nor the main drive in conflict with the sexual drive, but that there may be a tendency to attack the female as well.')
3. Sex-flee conflict.

Hinde's work on the chaffinch and the present work on the zebra finch indicate that the third of these possibilities is the most important in these two species, although probably all three play some part. It is, of course, not suggested that this applies as a general rule. In different species, the relative importance of the three possibilities will vary. To give just two examples: firstly, an investigation of the reproductive behaviour of the river bullhead, *Cottus gobio* (Morris, 1954), has shown that the 'courtship' of the male is almost pure attack, which repels other males, but which attracts a ripe female and induces her to enter his nest and lay her eggs. The more readily she responds sexually to his attack, the more likely she is to inhibit further aggression. The pre-coitional conflict in the male in this species is therefore primarily one of sex and attack. The fact that the male may also threaten the female reveals that the fleeing drive is also aroused, but it is a less important contributory factor than attack. Thus, in the river bullhead, the second of the above three possibilities is the most important one in the shaping of the pre-coitional display. On the other hand, in the Mexican swordtail, *Xiphophorus helleri* (Morris, unpublished), the courtship dance of the male can be understood as movements resulting mainly from sexual thwarting. He darts around the female at high speed, shivering his sword and at the end of each movement has come to lie alongside the female, head to head. This is the position for fertilization in this fish, and if the female is responsive she remains still and fertilization takes place. If she is not responsive she keeps trying to move away from the male, and each time she moves the male has to make more darts and dashes to regain his fertilization position. These dashes are made around the front of the female, thus not only bringing the male into position but, at the same time, helping to block her line of escape. Little or no attempt is ever made to attack or flee from the female during courtship in this species and all the movements of the male courtship can be thought of as attempts to manœuvre into position for fertilization. Thus the males of the zebra finch, bullhead and swordtail illustrate clearly the three fundamental types, causally speaking, of pre-coitional display.

Parental Behaviour

All the female zebra finches kept in the laboratory laid eggs on a number of occasions, but, owing to various interference factors, only three parental cycles were both successfully completed and observed. In one case, which was studied in detail, four young birds were hatched and reared. The parental behaviour described briefly below is based mainly on observations of this brood and must be treated therefore with appropriate reservations.

Whereas the male performed most of the nest-building, the female played the more important role in the incubating of the eggs and the feeding of the young. But, just as the female helped in constructing the nest, so the male gave some assistance in rearing the young. At night both parents sat in the nest, and when the young had left the nest the male sometimes fed them. The majority of the day-time incubation and feeding of the young was nevertheless performed by the female. (From various avicultural reports it appears that considerable individual variation exists in this species concerning the extent to which males assist in the rearing of the young.)

Concerning the lengths of the various phases of the parental cycle, the pair in question was formed on May 14th and the young were present by June 7th, a period of twenty-four days. Of these twenty-four days, the first ten were spent completing the nest and, approximately, the last fourteen covered the incubation period. This last point agrees with Bourke (1941) for zebra finches in the wild state. The young birds fledged between June 22nd and 24th, giving a nestling period of just over two weeks, which again agrees with Bourke, but not with certain other authors who put this period as high as four weeks. This is due, I feel sure, to the fact that the young birds continue to use the old nest as a roost long after they have fledged and even after they have become independent of their parents who, by this time, are probably in the middle of another cycle in another nest.

Approximately two weeks after first emerging from the nest, the young were independent. There was, of course, an overlap period when they already fed themselves but still begged for and obtained food from their parents. The approximate date of their independence was July 7th, giving a total period of dependence of about thirty days. The whole reproductive cycle, from the formation of the pair to the independence of the young, lasted approximately eight weeks – a period which is apparently shortened sometimes (Bourke, 1941) by the construction of a second nest by the male a few feet away from the old one. This is done when the nestlings are still only half-grown, and the male is said to feed them whilst the female incubates the next clutch. By such means the reproductive cycle could be shortened considerably. The zebra finch is therefore potentially very prolific and this is one of the reasons for its great popularity with aviculturalists. Weston (1930) reports that he obtained fifty-five young from one pair in one year. In nature, the population densities are balanced, not so much by food supply, or nesting sites, both of which are usually plentiful, but, according to Australian reports, mainly by nest predators and drought. Macgillivray (1932) reports that in the wet seasons the populations increase enormously, but that in the dry seasons they die off in their thousands. Mathews (1925) reports that rain is a great stimulus to reproductive activities, and it appears that lack of it is one of the few things which prevents the birds from breeding all the year round in nature.

Returning to the parental behaviour observed in the laboratory, it has already been recorded that the markings of the fledglings are the same as those of the female except that the beaks of the former are black. These first show signs of changing into a dull orange colour when the fledglings have been out of the nest for nearly three weeks. The base of the beak changes colour first, then gradually the whole beak. At the same time the males begin to show the

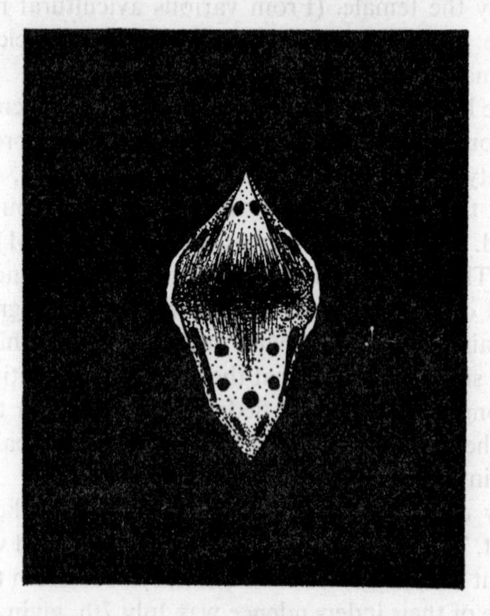

Fig. 9. The markings inside the mouth of the nestling. For explanation see text.

first signs of their special male markings. The bars of the throat seem to appear first, followed by the ear-patch and the spotted flanks.

Of especial interest is the mouth patterning associated with the feeding of the young. The nestlings possess dark-grey feathers and inside their dark domed nest are difficult to see. Despite this the parents have little difficulty in finding the throats into which food has to be pushed, for the young respond to the arrival of a parent bird at the nest by opening their beaks wide and exhibiting brightly marked mouths. The markings consist of a white background on which appear seven black spots arranged in a circle around the throat, thus making it an easy target (see Fig. 9). Five of the black spots form a semi-circle on the white roof of the mouth and two more appear side by side on the tongue. (In the illustration, which was drawn from a dead nestling, the mouth has been opened more than it would be in a begging bird, in order to show the markings more clearly, and this has distorted slightly the circle of spots around the throat.) The corners of the mouth are white and these white strips are visible even when the mouth is closed. In some of the close relatives of the zebra finch, such as the Gouldian finch (*Poephila gouldiae*), the region round the corners of the mouth shows incredibly elaborate and brightly coloured outgrowths. The markings inside the mouth of the young zebra finch gradually change as the bird grows older. For example, the twin spots on the tongue fuse together to form a black band across it at the fledgling stage.

The fledglings' begging movements of crouching, tilting up the head[1] and opening the beak, have already been discussed elsewhere. There is no wing- or tail-quivering. The parent responds by regurgitating seeds – a process accompanied by characteristic undulations of the neck. The adult bird then plunges its closed beak into the throat of the fledgling, pauses, and then withdraws again. The young survive well without being given any special foodstuffs.

Pseudofemale and Pseudomale Behaviour

In another paper (Morris, 1952*) I described the way in which male ten-spined sticklebacks (*Pygosteus pungitius*) behaved as females when sexually frustrated in a particular way. These males were seen to perform the female courtship pattern and I called this 'pseudofemale' behaviour. Since then I have observed sexually frustrated females of the same species performing part of the male courtship pattern. Both the pseudofemale and the pesudomale behaviour occurred in homosexual circumstances. The pseudofemales were courted by other males and the pseudomales courted other females. In a number of bird species (see review in Morris, 1952) reversal of sexual roles has been reported to occur in heterosexual situations. In these cases, reciprocal reversals between members of a pair lead to reversed mountings. Such mountings, which are comparatively rare, have hitherto simply been recorded rather than discussed and no causal explanations have been forthcoming. In the zebra finch it was possible to observe them, and the special circumstances associated with them, in sufficient detail to be able to form some idea of their probable causation.

The tail-quivering courtship display of the female, as stated earlier, was seen far less frequently than the courtship dance display of the male. The first

[1] The nature of this movement is rather unusual in this species, in that the neck is not stretched, but twisted round. This is undoubtedly adaptive in connection with the domed structure of the nest.

* See pp. 13–41.

time it was seen at all, it was performed by a male and its significance was not understood. Here is an excerpt from notes made at the time: 'Male A has been isolated for several months. He is now given a female and courts and mounts her repeatedly in a short space of time. She gives no special response of any kind. *Fast vertical tail-quivering* is then given by the male who assumes a horizontal body posture. It is not oriented to the female. It occurs only a few times and in rather quick succession. Could this be some kind of displacement activity as the result of the unresponsiveness of the female?'

Later, when it was observed that this action was, in fact, identical with the female courtship display, not only of the zebra finch, but also of the closely related Bengalese and cutthroat finches, it seemed probable that the male had been performing a pseudofemale display. In the same pair the male A was subsequently seen to perform this display on a number of occasions. The circumstances were always the same. This particular female was seldom responsive to the courtship of the male. Frequently, after a long bout of displaying, he would mount her regardless of the absence of the invitation display. Invariably these mountings were unsuccessful, owing to the posture or movements of the female. It was usually then, just after the male had dismounted from a thwarted copulation attempt, that he gave the female-like quivering display. The female never responded to this. Sometimes, instead of occurring as a postscript to unsuccessful mounting, it appeared as an alternative to mounting at the end of a long bout of courting to which the female had not responded.

From all this it seemed clear that a fundamental causal factor was the strong thwarting of the male sexual behaviour. In this respect the display was very suggestive of a displacement activity. Everything indicated that, when the male sexual pattern was strongly aroused and then thwarted, a displacement-like 'spark-over' to the dormant female sex pattern occurred in the male, with the result that pseudofemale behaviour was performed. It remained to establish with certainty that the tail-quivering display of the male was, in fact, a pseudofemale display. There were three difficulties that had to be overcome in connection with this:

1. There was a possibility that the display, when given by the male, was, in some way unnoticed by the observer, subtly different from the true female display. This difficulty was overcome by the fact that, in pairs B, C and D,[1] the displaying male was several times seen to be mounted by its mate. In pair B, the female only partially mounted the male, reluctantly placing one foot on its back for a second. In pairs C and D, however, the females not only mounted their males when the latter gave the female display, but also performed the male copulatory movements of wing-beating and spreading, twisting, and lowering the tail. Several of these reversed mountings were as

[1] And in several other pairs, since this paper was first written.

lengthy, as intense and *apparently* as complete as normal mountings. Whether they were functional and resulted in fertilization or not, it is not possible to say. In any case, these observations make it clear that the male's tail-quivering display is the same as the female's invitation display.

2. In some birds, the female invitation-to-copulation display is identical with the juvenile food-begging display. As this was known to be the case in the house sparrow (*Passer domesticus*), which is also a ploceid, it was thought possible that the pseudofemale display of the male zebra finch might actually be displacement food-begging, or, rather, that it would then be impossible to say whether it was derived primarily from the begging or the inviting display. This difficulty does not arise in the zebra finch, where the begging response is quite distinct from the female courtship display. Although the begging response was observed in adult birds under special circumstances, when they were subordinate, it was never seen to occur in pre-copulatory behaviour.

3. In some animals, the female courtship behaviour is similar to the behaviour of beaten, subordinate males. In the ten-spined stickleback, the tail-down posture of the sexually responsive female is the same as the tail-down submissive display of a beaten male. Yet the submissive displays of the zebra finch, which have already been described in this paper, in no way resemble the courtship display of the female. Furthermore, the female courtship display has not been observed to resemble *any* of the other displays of the zebra finch.

There can be no doubt, therefore, that this is a case of a *pseudofemale display resulting from thwarted male sexual behaviour*.

The pseudofemale display nearly always resulted from thwarting of the kind described above, in which the appropriate releasing stimuli from the female were absent, but in a few instances the circumstances were different. Sometimes, during the early phase of a cycle, when the nest was in the process of construction, the male of a pair was seen to hop a short distance from the nest and perform the pseudofemale display without a preceding courtship or copulation attempt. Also, on one occasion, the female was seen to give her courtship display, to which the male responded with his, but without mounting. The female soon stopped displaying and the male then gave the pseudo-female display. In these two situations it is not clear whether the pseudo-female display is the result of a similar causal state to that described already, or not. Where the display is given without any marked preceding activity, nothing can be told about the causal state of the animal concerned. Where the female displays and then stops, the male may be, just at that point, ready to mount and may thus be thwarted as before. It is also possible that in these cases thwarting did occur – since it so obviously occurred in the vast majority of cases – but that it was the result of a conflict between sex and some other incompatible drive, without involving the absence of releasing stimuli. It seems probable, from the earlier discussion, that fleeing was the incompatible drive concerned. This conflict between sex and fleeing drives appears to exist also in the typical examples of pseudofemale activity, but in a special way that will be discussed later. But first the mounting behaviour of the females must be considered.

When a female was not responding to the courtship of a male, her lack of responsiveness was usually accompanied by signs of fear of the advancing male. As he approached in his pivoting dance, the female often moved away from him down the branch, or showed the intention movements of taking off, or actually flew a short distance away. This frequently happened repeatedly during a phase of pre-coitional activity, resulting in a sort of pursuit-courtship. If, when the male stopped courting, he then did not attack her, the female ceased to show signs of fleeing. In the cases where reversed mounting took place, the male, after ceasing to court, gave the pseudofemale display and this, too, seemed to remove the signs of fleeing from the female's behaviour and she hopped towards the male and mounted him. A plausible explanation of the female's mounting seems to be that, when it occurred, the unresponsiveness of the female was due not to low sexual motivation but to a conflict between sexual and fleeing drives. The advance of the male towards her stimulated her to flee to such an extent that it inhibited the sexual response, and it appears that the cessation of the advance removed this inhibition, thus permitting the expression of her sexual tendencies. By this time, however, the male was performing the female display, which stimulated her to mount, rather than invite mounting herself.

At this point it is worth describing very briefly the circumstances under which the pseudofemale and pseudomale behaviour of the ten-spined stickleback takes place. The pseudofemale behaviour is shown by subordinate males which have been so badly persecuted by the dominant males that they have been unable to form territories or build nests. The fleeing drives of these males are very strong, but when one of the dominant males is given a female and begins to court her, the subordinate males gather round and join in the courtship, competing with the *female* for her role. Occasionally this frightens off the female and the courtship may then continue between the dominant male and one or more of the subordinate pseudofemales. Such a courtship may continue to the point of a pseudofemale lying in the nest as if it were laying eggs.

Pseudomale behaviour was seen in the sticklebacks in a tank containing only females. Several of the females were ripe and were swollen with eggs. The swollen belly of a ripe female is known to be a sign stimulus for the courtship of the male (Sevenster, 1949). Twice I observed that one ripe female gave the male courtship dance to another ripe female and that the latter responded and followed it.

Returning to the pseudofemales, the causation of the performance of the female pattern by the male sticklebacks appears to be similar to that of the zebra finches. The pseudofemale sticklebacks were sexually frustrated and their fleeing drives were strongly aroused by the presence of the dominant males. Yet their sexual motivation was so great that they were attracted to the scene of a courtship. Once there, however, they were intensely stimulated to flee. Thus there was a strong arousal of both the sexual and fleeing drives and the result was the pseudofemale behaviour. In the case of the pseudofemale zebra finches, in the typical instances, their sex drives were frustrated too, and their fleeing drives were aroused by the proximity of the female (as is shown by the nature of their male courting pattern). It seems, therefore, that in both cases the pseudofemale display is the result not simply of the frustration of the male sexual pattern, nor simply of a conflict between the sexual and fleeing drives, but of *the frustration of the male sexual pattern, in combination with the arousal of the drive to flee.*

Sexual frustration is also associated with the occurrence of pseudomale behaviour. The female sticklebacks which performed the male courtship dance did so when strongly frustrated sexually by the absence of males. The female zebra finches showed mounting behaviour when, apparently, they had been sexually stimulated by the courtship of their males, but were also stimulated to flee to such an extent that they were unable to respond in the usual way.

The fact that the reversal of the sexual role of the male takes place, in the sticklebacks in homosexual, and in the zebra finches in heterosexual situations, is probably due principally to differences in the nature of their courtship patterns. As already indicated, the courtship of the zebra finches appears to be due basically to a conflict between fleeing and mating. The courtship of the stickleback, on the other hand, is due basically to a conflict between *attacking* and mating. In a normal courtship between a male and a female stickleback, therefore, motivational states are not going to occur which are likely to give rise to pseudofemale behaviour.

It remains to point out that there are three distinct reactions which are the outcome of the frustration and conflict involved in the pre-copulatory behaviour of the male zebra finch:

1. The conflict between fleeing and mating results in the now semi-ritualized *pivot-dance*. During this the male is advancing towards the female, so that the balance of the conflict is slightly in favour of the mating drive.

2. Sexual frustration, as a result of the absence of releasing (tail-quivering) stimuli from the female, produces *displacement beak-wiping*. As it is not known whether the display of the female releases sexual behaviour in the male by direct stimulation, or by the removal of the inhibiting influence of the fleeing drive, it cannot be said for certain whether the sexual frustration

resulting in beak-wiping is due to simple thwarting of sexual behaviour, or whether it is due to the conflict between mating and fleeing, or whether it comes from a combination of sexual thwarting and fleeing drive together.

3. Finally, as a result of the last-named state, the pseudofemale display may occur.

It will be seen from this that, if the displacement beak-wiping is the result of the maintaining of the conflict between fleeing and mating, this conflict must differ in some way from the similar one which produces the pivot-dance. This difference may consist of some slight alteration in the balance between the mating and fleeing drives, or the balance may be the same, whilst the absolute levels of the two drives may change. Judging by the comparative evidence (see Tinbergen, 1952), an increase in the absolute levels seems more likely. Tinbergen (p. 12) states that '... the alternation of mere intention movements seems to be a sufficient outlet so long as the drives are not too strong ... Essentially the same situation, only in a more intense form, leads to displacement activities.'

If, on the other hand, the displacement beak-wiping results from the combination of sexual thwarting and fleeing, then this conflict must differ from the similar one which produces pseudofemale behaviour. Again, it is probable that the difference is that the latter occurs at a higher intensity, since it is seen at the end of a prolonged bout of courting, whilst the beak-wiping takes place actually during the courtship.

In conclusion it should be mentioned that this discussion of the causal factors underlying the performance of pseudofemale and pseudomale behaviour has been given in the awareness that, in many respects, it is highly speculative. It has, nevertheless, been presented here at this stage, despite its shortcomings, in the hope that it will stimulate further investigation of the problem.

Displacement Activities

Displacement activities have recently been defined and discussed at length (Tinbergen, 1951 and 1952; Bastock, Morris and Moynihan, 1953). A number have been seen to occur in the behaviour of the zebra finch. One of these, namely displacement beak-wiping, has been seen with sufficient frequency to make a quantitative analysis possible. Before this is discussed, however, here are brief notes on the other displacement activities which have been observed:

1. *Displacement preening*

A rapid, jerky preening of the feathers is sometimes seen in agonistic situations. It was often provoked in the following manner. The partition in the passage which had been built between two aviaries, each of which contained a nesting pair, was removed from time to time in order to make observations on territorial behaviour. Sooner or later it always happened that all four birds came to be in one aviary. There would then be outbursts of fighting and chasing. In between these outbursts, the birds sometimes sat a short distance from one another, eyeing each other and making quick swipes at their (own) plumage with their beaks. It is, perhaps, worth giving a short extract from notes made on one such occasion. Pair B were the intruders into aviary A: 'Now pair A sit together at one end of the uppermost branch which is about two feet long. Male B flies up to it and alights at the other end. He is followed by female B, who sits next to him on the far side from pair A. The two males are now about fifteen inches apart with their respective females sitting beyond them at the extreme ends of the branch. Pair A watch pair B intently and vice versa. Silence reigns. The four birds are much quieter now than when the pairs are separated in their two aviaries. Both the males begin to preen quickly and nervously, still watching each other closely. They preen almost in time with one another.'

Displacement preening was also seen to occur actually during fighting, interspersed with pecking and withdrawing.

Preening occurs as a displacement activity in agonistic situations in many other species of bird. Tinbergen (1951, p. 113) states that 'Fighting European starlings may vigorously preen their feathers.' Lorenz (1935) reports that various species of cranes do likewise, to mention only two examples from the literature. Many other species of bird, notably ducks and pigeons, show this displacement activity during pre-copulatory behaviour. When more data of this kind has accumulated in the literature, it will be most interesting to compare the list of species which show agonistic displacement preening with those which show courtship displacement preening.

2. *Displacement stretching?*

Like displacement preening, this was seen in the zebra finch during fighting. It took the form of the typical one-wing-one-leg stretch, but was performed rather quicker than usual. In one particular bout of fighting, a male was seen to stretch in this way immediately before chasing a rival, and also immediately after the cessation of chasing. In such cases it is, of course, difficult to disprove that the actions are autochthonous. Apparently autochthonous stretching often occurs before or after a bout of activity of any kind, but when it was seen during fighting, it seemed to be too intimately linked with it and to alternate too quickly with it to be anything but a displacement activity. Nevertheless, it must remain a doubtful example for the time being.

3. *Displacement shaking*

A strong ruffling and shaking of the feathers was sometimes seen in agonistic situations. The introduction of new birds into a territory occasionally resulted in the owners shaking their feathers before any actual fighting took place. this was more clearly seen in the cutthroat finch than in the zebra finch, but was also observed from time to time in the latter. It is perhaps significant that the fighting behaviour which followed the shaking in the cutthroat finches was much milder than almost all instances of fighting in the zebra finch.

4. *Displacement scratching*

Scratching movements also appeared during fights, but with less frequency than other comfort movements. It is interesting that during a long bout of cleaning proper – for example, after bathing – the scratching movement is also rarer in occurrence than the other cleaning movements.

5. *Displacement yawning*

This is yet another comfort movement which occurs during fighting bouts. Often, when intruding birds were present in a territory and the atmosphere was tense, one or more of the birds would relax for a second, yawn, and then become tense again.

It is worth mentioning that Heinroth (1930) does not consider that birds ever perform a true yawn. Various authors have disagreed with this from time to time. As the causation and function of yawning are still not entirely clear, it is difficult to take sides in this argument, but it can be stated that, in the Ploceids which have been studied, the movement of stretching wide the jaws, which I am calling yawning here, was seen to occur in just those situations in which true yawning occurs in mammals.

6. *Displacement sleep*

In one fight, one particular male appeared to be in a very intense conflict between attacking and fleeing, judging by his agonistic behaviour, and during the fight showed intense bouts of comfort movements, exhibiting almost his complete repertoire, with the interesting exception of beak-wiping which seems to be reserved for courtship periods. He performed the various actions rapidly and briefly and they were interspersed with short offensive and defensive agonistic movements. Shaking, preening, scratching, stretching and yawning were all seen, *and also sleep*. Displacement sleeping, which is a common occurrence in waders in agonistic situations, has rarely been seen in Passerines, although Tinbergen (1939) observed it in the snow bunting during fighting. In the zebra finch it was only seen once, but was then seen very clearly. The male had just been losing a beak-fencing match and its rival was only a few inches away. The male then made as if to fight again, but instead turned his head away through about 180 degrees. As he turned his head round on to his back, he tucked it in and crouched a little in the typical sleeping posture. He held this posture for one or two seconds and then continued the fighting. It is worth quoting a short passage from Makkink (1936) on the avocet, which shows the great similarity between displacement sleeping in the two species: '... it always happens at the most unexpected moment, namely just before or even during the very attack. Moreover, the bird may at the same time squat on its tarso-metatarsi ... After the assuming of this attitude the expected attack often does not take place, but sometimes it really does, in which case the bill is immediately produced.'

It must be pointed out that in the vast majority of fights, these particular comfort and sleep displacement activities did not take place. It is not clear what was peculiar about the fighting in which they were seen, but, as already mentioned, the conflict between attacking and fleeing seemed to be particularly intense. It is hoped that further investigation will reveal more about the relationship between particular conflict states and the particular displacement activities that may result from them. As Bastock, Morris and Moynihan (1953) have pointed out, little is yet known about the subject of alternative displacement activities.

7. *Displacement feeding*

Feeding is yet another displacement activity that was observed during fighting and was perhaps the most frequent. Displacement feeding was always performed rapidly and jerkily. Here is an extract from notes:

> The partition is removed from between aviaries A and B again at 6 p.m. At 6.10 p.m. male B goes through into aviary A. Once there, he hops around on the ground and male A now flies down very fast to the ground also. Both stand and hop around within a few inches of one another. The atmosphere is tense but neither gives any display. Both are very wary of getting *too* close to one another. Now, as the owner, male A, hops towards the intruder, male B, the latter advances instead of fleeing. The owner responds to this by immediately hopping to one side and quickly feeding. This feeding is very abbreviated, consisting of a few pecks at the ground, and then it stops and male A is hopping around again watching, and being watched by, male B. Now male B chases male A a little way and then it returns to the floor of its own aviary. The two males now feed nervously on the floors of their respective aviaries. Now male B flies into aviary A again and courts female A vigorously, but does not mount her. This attracts male A who rushes to the scene and chases male B hard. Male A then goes into aviary B and now the two males feed nervously in one another's aviaries!

The above extract is typical as regards the way in which displacement feeding occurs. Sometimes the two females joined in as well and then all four birds were seen together on the ground all feeding rapidly. Sometimes when the tension was extreme the feeding resulting from it was so careless that it is doubtful if much food was swallowed, but usually it seemed to be performed in a quite functional manner.

Displacement feeding is well known during fighting in a number of birds. To mention just one example, Hinde (1952, p. 74) found that the great tit (*Parus major*) pecked at twigs during fights above ground level and that, when fighting on the ground, 'In two skirmishes males were seen to turn leaves (an activity used in looking for food on the ground) at a stage in the skirmish when displacement pecking might have been expected to occur.' Thus, the great tit's type of displacement feeding seemed to be dependent to some extent on the external situation. This particular point, in respect of the zebra finch, will be discussed below.

8. *Displacement mounting*

This activity was seen quite clearly on a number of occasions during fighting. It consisted of an abbreviated mounting with no preceding courtship. A male simply jumped on to the back of a female which happened to be near the scene of a fight (although not taking part in it herself). A mounted male, on such occasion, usually only paused on the female's back for a second and then hopped off again and continued the fight. The first time displacement mounting was seen, it occurred at the point at which displacement feeding was seen on other occasions:

> Partition removed from passage between aviaries A and B at 3.35 p.m. Male B ventures into aviary A, apparently attracted by female A who was visible on the floor of A. The intrusion of male B brings the owner, male A, rushing down to investigate. It is the first time that male A has seen another male (or, for that matter, any other bird except his mate) for over a year. The two males come face to face on the floor of aviary A. Male A stands in a sleeked upright posture, pauses for a second looking at the intruder, and then hops quickly over to female A, his mate. She is turned away from them and is feeding. Male A hurriedly mounts her for a second and then turns back to male B and begins chasing him.

During the fighting which followed the above encounter, both females seemed to be attracted to the scene, although neither was seen to fight. When all four birds were in aviary A, with the females sitting around near the fighting males, displacement mounting was seen to take place a number of times. Both males hopped on to the backs of their own mates and of one another's mates! There was no courtship.

Armstrong (1947, p. 108) cites a number of examples of 'inappropriate' coition, in different species, stimulated apparently by various kinds of alarm or 'excitement'. In the zebra finch a male was more likely to court and mount a female after some kind of alarm, and this fact was made use of for observations of sexual behaviour. The introduction of a strange bird, followed quickly by its removal, or the waving of a stick near the aviary, or the spraying of water over birds, all seemed to increase the likelihood of the occurrence of sexual behaviour. The causal relationship between the behaviour seen under these circumstances and the displacement mounting described above is problematical and requires further investigation.

Before discussing any further displacement activities, it is worth looking back at those already described, all of which occur when a conflict between attacking and fleeing is the major causal factor, in order to ascertain why one is shown in one instance and another in another. It has already been mentioned that Hinde (1952) found that, in the great tit, the kind of displacement feeding that was shown during fighting was dependent on whether the fight took place on, or above, the ground. Armstrong (1950) has discussed this aspect of displacement activities and says, concerning the 'linkage' which exists beween certain specific causal circumstances and the resultant displacement activity, 'Innate linkages are apparently not so rigid that they cannot be modified by the intrusion of new patterns mediated or determined to some extent perceptually.' Bastock, Morris and Moynihan (1953) distinguish between true and apparent alternative displacement activities. The latter are those 'in which the external circumstances do determine which of the alternatives will appear at a given time' and the former are those in which they do not.

In the zebra finch, a conflict between attacking and fleeing may result in the following *Apparent alternative displacement activities*:

1. Feeding – when food is near at hand.
2. Mounting – when a female is near at hand.
3. Comfort actions or sleep – when neither of the above is so easily available.

The various comfort activities and sleep appear to be *true alternative displacement activities*, since they may all occur together under the same external conditions. It should be borne in mind that these distinctions are tentative; for example, there is a possibility that, in particular, the alternative between a mounting and a comfort activity may be determined by the special internal motivational condition of the displacing individual, as well as by the external circumstances.

9. *Displacement mandibulating?*

As already pointed out, the action of rapidly opening and shutting the beak is seen in connection with the nibbling of food particles and nest material. It is impossible to say definitely whether it is a nesting or a feeding action, and for the time being must be considered to be both. It is often seen to occur in association with various display movements when there is nothing at all held in the beak. It is then presumably a displacement activity. Unlike the other displacement activities, it appears to be almost an 'all-purpose' action. It is seen in connection with the nest-site display described earlier; it is seen during fighting when there is a conflict between attacking and fleeing; it is seen during courtship, when it may be performed by either the male or the female and both before and after copulation; it may be seen as part of a 'greeting ceremony' between members of a pair which have been separated and then put together again. It is often accompanied by a special soft call, but not always, and when it is seen in fighting it is usually quite silent. When it occurs with an empty beak, it is always performed 'to' another individual.

This must be considered a doubtful case of a displacement activity until more is known of the exact nature of its causation in all the circumstances in which it takes place.

10. *Displacement food-begging*

The food-begging actions of the fledgling zebra finch are sometimes shown by a submissive adult bird. As already described, the causation may be either physically thwarted fleeing, or fleeing thwarted by certain sexual tendencies. It occurs only in the most extreme situations. It may be done with the feathers fluffed out or sleeked and may, or may not, be accompanied by a waving of the head from side to side. It has appeasement function.

Tinbergen (1953b) has described the displacement food-begging of the herring gull at length. There it occurs as part of the pre-coitional behaviour.

11. *Displacement nest-building*

Although this is a common displacement action in many birds, it was not seen often in the zebra finch. When it was seen, it took the form of a male or a female fiddling with a piece of straw or grass during pre-copulatory behaviour. It was often impossible to be certain that such nesting actions were not autochthonous, but sometimes they were so intimately interspersed with courtship actions that there was little doubt as to their displacement nature. In the following extract from notes it will be seen that the male responded to the female's courtship display by first tugging at nest material and then courting:

> The female flies up to the nest, which is half-finished, and pulls at a pro-protruding piece of straw. The male, who was feeding, immediately flies up and lands near by. She fiddles with a piece of straw and begins to solicit. The male hops over and he too tugs at a piece of nest material. Then he begins to court. She stops displaying and appears to give all her attention to the straw in her beak, but then drops it and solicits again. The male courts again ...

It is interesting that, when the displacement nest-building occurs, it is performed by both sexes, whilst autochthonous nest-building is performed mostly by the male. In connection with this, Nice (1943, pp. 177–8) found that in the song sparrow, where the building is done mostly by the female, here, too: 'Both male and female indulge in this symbolic nest-building ... ' It appears, therefore, that whichever sex is the real builder, both may actively exhibit this displacement action.

As stated above, displacement nest-building was not common in the zebra finch. Delacour (1943) mentions that it is, on the other hand, a regular part of the pre-copulatory behaviour of the waxbills. I can confirm this for the red avadavat (*Estrilda amandava*). This waxbill was seen to perform deep bows to its mate whilst holding a stem of grass in its beak. This was done by either sex.

However, Delacour goes on to say that only waxbills and the members of the genus *Zonaeginthus* of the grassfinch group manipulate nest-material during courtship. My observations disagree with this. Also, Moynihan and Hall (1954) have found that the spice finch (*Lonchura punctulata*) shows this behaviour very commonly. It is, therefore, perhaps a little dangerous to use the presence or absence of this behaviour pattern as a taxonomic character in the way Delacour suggests.

12. *Displacement beak-wiping*

Apart from the nesting movements mentioned above, this is the only other definite displacement activity which occurs during the pre-copulatory displays. It may be performed by both sexes, but is done so with a much higher frequency by the dancing male. He may either pause in his dance to wipe his beak on the branch, or he may make a quick wipe without interrupting his pivoting advance towards the female. This activity was seen with such great frequency in the zebra finch that some quantitative statements, concerning the way in which it differs from its autochthonous 'example', can be made.

Tinbergen (1940) has pointed out that, in various birds of paradise, beak-wiping is a regular part of the male's courtship. The same can be said for the zebra finch, where it is more likely to occur than not. It is unusual for a male to show prolonged courtship without wiping his beak a number of times.

Hinde (1953), speaking of the displacement beak-wiping of the chaffinch male during courtship, says: 'It is usually not connected with feeding or any other possible contamination of the bill, and the movement is often incomplete, the beak not actually touching the branch. Further, although no precise data were obtained, it often seemed to be especially frequent when the female was unresponsive.' In all these respects the chaffinch and the zebra finch are identical. But Hinde goes on to say that it also occurs in purely aggressive – not sexual – situations, during fights in winter flocks. As I have already indicated, it was not seen in any but sexual situations in the zebra finch. However, my birds were never in 'winter flock' condition.

Tinbergen (1952) has discussed the way in which displacement activities differ from their autochthonous 'examples'. The displacement beak-wiping of the zebra finch differs from the true cleaning movement in the following ways:

(*a*) *Orientation.* In the typical cleaning movement, the bird turns from its perched position – across the branch – so that its body faces along the branch. It then lowers its neck, rotating its head slightly as it does so, thus bringing the beak down level with the branch and on one side of it. The beak is then scraped on the branch by a slight rotation of the head in the opposite direction as the neck is raised again. This process may then be repeated on the other side of the branch.

In the displacement form, the wiping movements may be made without the initial turning component, so that the bird still sits across the branch and 'wipes' its beak in mid-air. This particular incompleteness of the movement is not common.

(*b*) *Amplitude.* In the typical cleaning movement, the neck is always lowered sufficiently to enable the beak to make contact with the branch or twig. In the displacement form, it frequently happens that, although the bird makes the appropriate orientation to the branch, the amplitude of the neck-lowering component is inadequate. The result is that, again, the bird wipes its beak in mid-air, but in this case it does so just above the branch.

(*c*) *Velocity.* In both types of wiping the movements are made so quickly that it is difficult to be certain of differences in speed without cinematic analysis. The impression is, however, that the displacement action is the faster of the two and this is often the case in other displacement activities. Whether this is due solely to the two types of incompleteness mentioned above is also uncertain. It is nevertheless true that the whole wiping movement takes much less time to perform when it is a displacement activity, but this is mainly due to the following difference:

(*d*) *Frequency of wipes per wiping.* Each wiping, or wiping bout, consists of a number of separate wipes. In each bout the wipes are made on alternate sides of the branch. Although it is usually easy to distinguish between one bout of wiping and the next, it is necessary, when two bouts follow one another quickly, to have criteria for determining where one bout ends and the next begins. One such criterion is that the pause between one wipe and the next, within a wiping bout, is shorter than the pause between the last wipe of one bout and the first wipe of the next bout. Under usual conditions, the pause between bouts is immensely longer than that between wipes, but under conditions of very strong stimulation the wiping bouts may follow one another in quick succession. The other criterion which may be made use of is that, between wipes in one bout, the neck is not raised fully into its original position, whilst this position is nearly always resumed at the end of a bout.

Each wiping bout lasts only a second or two and consists of only a few wipes. The number of wipes per wiping is on the average less when the movement is performed as a displacement activity. This difference has been expressed numerically in Fig. 10. Figures were obtained for 200 displacement wipings and 200 autochthonous wipings. The results show that the typical displacement activity consists of one wipe only, whilst the typical autochthonous movement consists of two.

These results were collected from four different males. It was noticed that one of these males seemed generally to give more wipes per wiping bout than the other three. (It should be mentioned that approximately equal numbers of the two types of wiping were taken from each male, so that individual differences did not distort the results given in Fig. 10.) When the relationships between the average number of wipes per wiping bout were examined for each

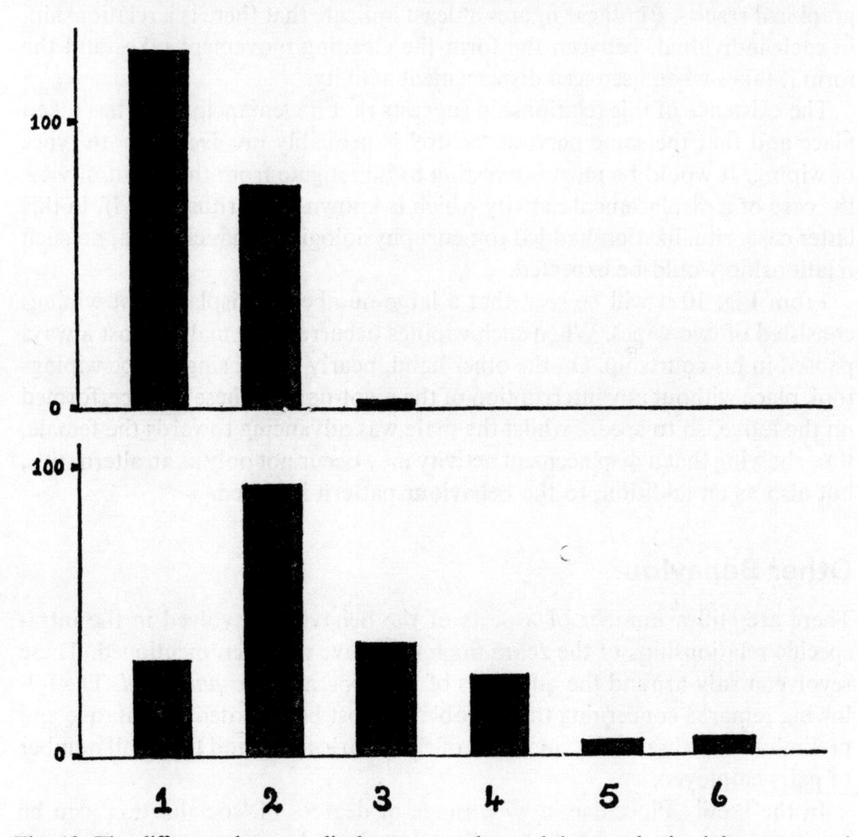

Fig. 10. The difference between displacement and autochthonous beak-wiping, expressed numerically. The number of wipes per wiping bout is on the average greater in autochthonous beak-wiping. Ordinate: Number of occurrences in 200 observations. Abscissa: Number of wipes per wiping bout.

individual male, it was found that the higher the autochthonous average, the higher the displacement average. This is expressed numerically in Table I, below. Unfortunately, males C and D, which were in the same aviary, were not always scored separately and the figures for these two have had to be lumped together:

TABLE I

	Autochthonous average	Displacement average
Male A:	2·9	1·6
Males C and D:	2·1	1·2
Males B:	1·8	1·0

It is hoped that analysis of further males will provide more interesting graphical results. But these figures at least indicate that there is a relationship, in each individual, between the form the cleaning movement takes and the form it takes when used as a displacement activity.

The existence of this relationship suggests that no 'emancipation' has taken place and that the same nervous 'centre' is probably involved in both types of wiping. It would be most interesting to investigate from this point of view the case of a displacement activity which is known to be ritualized. If, in this latter case, ritualization had led to neurophysiological emancipation, no such relationship would be expected.

From Fig. 10 it will be seen that a large number of displacement wipings consisted of two wipes. When such wipings occurred, the male almost always paused in his courtship. On the other hand, nearly all the single-wipe wipings took place without any interruption of the pivot-dance. These were performed on the move, so to speak, whilst the male was advancing towards the female, thus showing that a displacement activity may occur not only as an alternative, but also as an addition, to the behaviour pattern involved.

Other Behaviour

There are still a number of aspects of the behaviour involved in the intra-specific relationships of the zebra finch that have not been mentioned. These revolve mostly around the questions of *territory* and the *pair-bond*. The following remarks concerning these problems must be regarded as tentative and preliminary, owing to the small size of the aviaries used and the small number of pairs employed.

In the family Ploceidae, a wide range of degrees of 'socialization' can be found. This range extends from solitary, territorial nesting, on the one hand, to communal nesting on the other. In the latter, as many as eighty pairs may share one huge nest, the dimensions of which have been known to reach

25 × 15 × 10 feet (Friedmann, 1949). In an intermediate condition, one tree or bush may be loaded with a huge number of small separate nests. Even in the communal-nesting species, there still appears to be a territorial system of a sort. The nest has many openings, each leading to one small cavity and, since only one pair uses any one tunnel and cavity, each of these nests within a nest can be thought of as a territory.[1] The main process involved in the socialization of the family is probably, therefore, the greater and greater reduction in the minimum size of the territories. When this minimum size is equal to the size of the cavity and tunnel, then communal nesting occurs. The zebra finch has not gone this far, but in the wild state, a number of nests may occur in one small bush. Spencer and Gillen (1912, p. 41) report that, 'As you approach any water hole they rise in great flocks and even a small bush will contain eight or nine of their little grass nests.' In the laboratory they nested close to one another, but the region around the nest was defended by the pair.

Further observations on territorial behaviour in the laboratory are, of course, probably subject to considerable distortion, but such distortion is bound to be of a superficial nature. The first thing that was noticed was that there could be considerable fluctuations in the degree to which a territory was defended by the pair which owned it. Adjoining aviaries, in each of which a pair was nesting, when opened up to one another by a linking passage (see Fig. 11), were often defended by their owners with great vigour. Such defence, however, was sometimes dropped for periods varying from a few minutes to a few days. On one occasion, violent territorial fighting took place during the first day of two such territories being linked. The next day, the pairs were found to have exchanged nests and each pair was sitting in and reorganizing the other's nest. After a few more days, both pairs were back in their old territories again and from then on territorial stability seemed to be more or less maintained. This changing over of nests first manifests itself when an intruding bird, during a pause in the fighting, shows a strong attraction to the strange nest. An intruding pair will often go and sit together in the owners' nest as soon as they get the chance. If there are eggs in the nest which the female owner is incubating, or if nestlings are present, a ferocious attack from the female owner takes place. Once when a strange male intruded into a nest containing a female and her nestlings, the female and the intruder immediately emerged from the nest, their bodies locked together. They fell fluttering to the ground where the male only escaped at the expense of a large bunch of feathers. On another occasion a strange female was introduced into an aviary containing a nesting pair. The latter, however, did not possess any eggs or young. The following are extracts from notes made at the time:

[1] Although no data are available, it is almost certain that each nest cavity is defended by the owning pair.

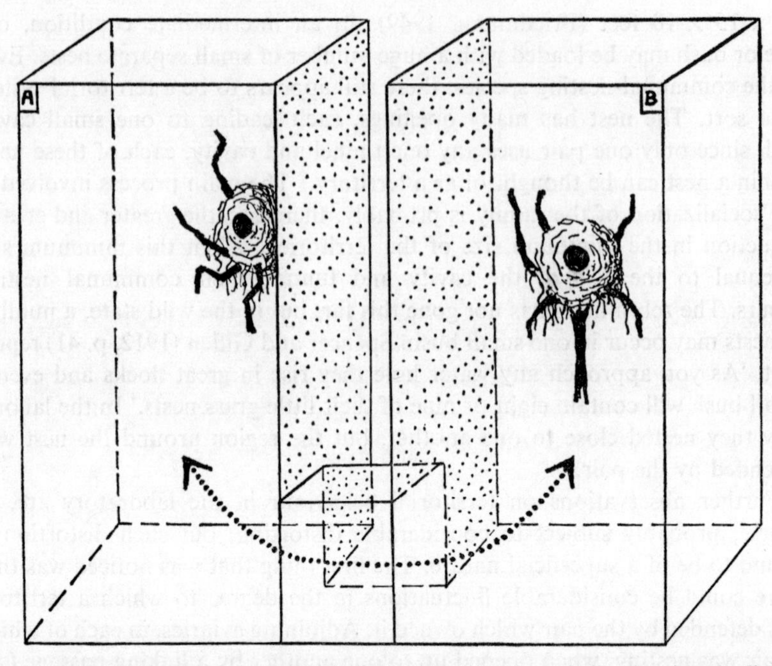

Fig. 11. Aviaries A and B, in each of which a nest has been built, showing the linking passage inserted between them. This passage could be opened or closed by a panel and was used in studying territorial behaviour. The dotted surfaces of the aviaries were opaque, as was the panel, so that, when the latter was in place, the birds in A and B were unable to see one another. The dotted line shows the flight path between the two aviaries when the passage was opened.

May 9th. Female C is brought into the laboratory and introduced into aviary B in which pair B have a centrally situated nest. The passage between this aviary and A is open and pair A are nesting in the latter. The result of the introduction of the new female is as follows: Both males lose interest in their mates. Female C is 'raped' by male A, then by male B, then by male A again, and once more by male B. Then male B manages to get her alone and courts her. He courts her fifteen times in quick succession. Each courtship song and dance ends in a mounting and attempted copulation. Then the next song and dance begin again immediately. The female does not flee but neither does she give the invitation display. The male B does not show displacement beak-wiping until the ninth courting, which is intriguing, and then he shows it in most of the courtings which follow. After a short pause, he courts and mounts for the sixteenth time. This has all taken place within the space of about five or ten minutes from the introduction of the new female. Now things calm down and female C goes into nest B and male B watches her. She fiddles around with the nest material of nest B, from the inside, whilst pair B look on. Every now and then male B tries to get into the nest, but he seems to be afraid of female C, and either stops just before entering, or else enters and then re-emerges again quickly.

May 10th. Today the new female, C, still monopolizes nest B, and the pair B daren't go anywhere near it now! At one point I saw female C sitting on top of nest B, holding a huge bunch of feathers in her beak, and just before I looked there had been a loud squeak of pain!

May 12th. Today pair B are trying to build a new nest in aviary B. They have chosen an unsuitable spot high up in one corner, so I have put a clump of twigs there for them. Female C still uses their old nest.

May 13th. This morning pair B have half built their new nest high up in aviary B. Female B is sitting in it all the time and male B is guarding the area against female C. The latter is now attacked by male B very viciously whenever she goes into the top half of the aviary. This includes the region of the old nest as well as the new one and female C can no longer use the former. She is hardly allowed to settle anywhere. Once a chase has begun, she is not left alone until she flies out of aviary B into A, where male A attacks her in the same way and chases her back into B and so on. This is kept up for some time, but sooner or later one of the males tires of chasing her and she is allowed to settle somewhere. But the initial spell she had cast over pair B is broken and she and they are now behaving in the manner expected of intruders and owners respectively.

As regards the territorial behaviour revealed by this account, it is clear that the arrival of strangers or a stranger has an initially intimidating effect, which, after a period of readjustment, is corrected. It is interesting to compare this situation with an equivalent one in sticklebacks, where the outcome of an introduction of an intruder is a foregone conclusion. The intruder stickleback never stands a chance and from the start is treated as female C was treated on May 13th, after the period of readjustment, when pair B had built a new nest for themselves.

But this account also reveals another interesting phenomenon. When the strange female was introduced, both the mated females were present and available to their two males. Courtship and copulation between the members of both pairs had been occurring from time to time. But when the new female arrived on the scene, she was 'raped' and courted repeatedly, and the other two females were ignored by their mates. For some reason the strange female provided a much stronger sexual stimulus than the familiar ones. Similar tests were repeated with the same results. This state of affairs lasted only for a short while after the introduction of a new female. After that, the males always returned to their mates and usually behaved aggressively towards the new females.

The first reaction of a male towards a new female was to 'rape' her. There was no display or courting in these cases and the males simply mounted and attempted to copulate with the females regardless of the posture or position of the latter. These incidents were always characterized by a sense of urgency and this led one male to mount a female the wrong way round, so that his tail was over the head of the female. Few, if any, of these mountings appeared to meet with success.

From these laboratory observations it is impossible to say anything of the functional significance of this behaviour. Tubman (1923) reports that, in the wild, nests containing as many as twenty-five eggs have been found and that 'I saw a considerable number of other nests with ten to fifteen eggs in them.' As the normal clutch size is four to six eggs, it appears that several females may lay in one nest. Despite this, monogamy appears to be the general rule for this species, and in the laboratory these relationships between a mated male and a second female were always temporary. Causally, these relationships must mean that a strange female is a greater sexual stimulus than a familiar mate. When she is no longer unfamiliar, the strength of the original pair-bond reasserts itself.

Leading on from this, a number of preliminary tests were made to investigate the nature of the pair-bond. Pairs A and B, which were of long standing, were separated. The females were removed to a separate cage. The males, which were in the linked aviaries, kept to their old territories. The next day the males were singing much more frequently and the females were calling to them. The males were only spasmodically territorial. Later, the partition between the two aviaries was inserted and two weeks after the removal of the females, male B was given a new female, female D. He treated her as a strange female in the usual way, raping her first and then courting her and attempting to copulate with her repeatedly. The two birds then settled down peacefully, after a little initial beak-fencing. Several weeks later, the original mate, female B, was returned and was immediately courted and mounted by her mate. During the following days, the old pair re-formed again and the male ceased to pay any more attention to female D. He now simply ignored her, but his reinstated mate, on the other hand, repeatedly attacked her and would not allow her anywhere near. The aggressiveness of female B was so violent that, after a few days, during which it showed no sign of decreasing, the female D had to be removed.

Male A had been isolated all this time and female D was now transferred to his aviary. Here, after the initial sexual responses, male A turned on her and prolonged chases ensued. She was removed and another female introduced with the same result. The original mate, female A, was then reintroduced. She was courted and mounted and this pair, too, re-formed and the male A showed no aggression towards her. Nevertheless he still attacked and chased the strange female which was still present. The latter was then removed.

These two tests revealed the durability of the pair-bond. In both cases it was maintained actively by one sex and passively by the other, although the active sex was male in one and female in the other.

Once again, I must stress the artificiality of the circumstances under which these tests were carried out. As far as I can ascertain from the literature, no behaviour study has yet been made of a large population of these or related species. The present results indicate that such a study would probably be most rewarding. It would be interesting to analyse the interpair relations and the way in which each pair is maintained, in a dense breeding population. The songs of the males are intriguing from this point of view. Each of the males used in the present study possessed a song pattern which was unique to that individual. Because of this and the fact that they all sang so frequently, it was unnecessary to ring the birds for identification. As was mentioned earlier, the basic song pattern was similar in all males, but owing to distinct individual variations, each male could be recognized auditorially. It seems quite possible that this has adaptive significance and that it serves to keep a female posted of the whereabouts of her mate.

In many species, the song has been shown to have a threat function, but in the zebra finch this does not appear to be the case. The zebra finch always sings during the courtship dance and it also sings at other times, when stationary. The stationary song is usually accompanied by a pivoting of the head, which looks like a low intensity version of the courtship pivot. Stationary singing is not oriented to the female or to any other bird. It decreases in frequency in the presence of another male, and increases in the absence of the mate. This increase also occurred in the related Bengalese finch. Some Java sparrows (*Padda oryzivora*) sang *only* when left alone by their companions.[1] The solitary, or stationary, song never had any repellent effect whatever and, if anything, sometimes the opposite. A recording was made of the song of one male zebra finch and when this was played back to it, it produced no response. Summing up, it seems that the only clear causal factor involved is the absence of the mate, and that pair maintenance is a likely function. The present evidence suggests that, in the zebra finch, the song is both motivationally and functionally a sexual, rather than an agonistic, activity.

A few observations were made on the social status of the fledglings when they had become independent. They were still distinguishable from the females by virtue of their black beaks and no attempt was made to mount them by the males in their aviary, although they were occasionally mildly attacked when

[1] Since this paper was first written, however, typical threat-song has been observed in the orange-cheeked waxbill (*Estrilda melpoda*).

they interfered with some adult activity. When one of these fledglings was introduced into an aviary containing a solitary male that had been without a female for several weeks, it was treated as a female and was courted and mounted.

A *redirection activity* (see Bastock, Morris and Moynihan, 1953) was observed once as a result of the presence of the fledglings. This was *redirected mounting*. The fledglings had already become independent and were present in one aviary with pair C. Male C began a high intensity courtship of female C. She moved away from him and he hopped after her, still courting, and followed her in and out of the group of fledglings, carefully avoiding the latter. Finally she flew to the other side of the aviary. Despite the fact that she was still clearly visible and available to the male, he then mounted one of the fledglings, which happened to be right in front of him at the moment when the female flew off. He had never previously courted or mounted a fledgling, nor did he do so subsequently. The entire courtship in this case had been oriented towards the female and only the mounting was redirected to the fledgling.

The Complex Markings of the Zebra Finch

The zebra finch possesses five distinct male markings, as well as several species markings. It has been shown that the courtship posturing of the male serves to display both those regions of its body which possess the male markings and the specific markings. In fact, the only parts of its body that are concealed during courtship are its back and wings, and it is significant that these are the only regions that are unmarked, being a neutral grey. This correlation between markings and display, coupled with the fact that they are conspicuous rather than cryptic, suggests that these markings act as *social releasers*. Lorenz and Tinbergen have, in various papers (see especially Tinbergen, 1948), discussed what characterizes a releaser. The general conclusion is that it is characterized by (*a*) conspicuousness; (*b*) specificity; and (*c*) simplicity. The markings of the zebra finch are both conspicuous and specific, and yet *they do not obey the rule of simplicity*. As simplicity has been shown to be the general rule for releasers, it seems worthwhile to discuss a possible reason for their complexity in this species.

Tinbergen (1951, p. 184) says that 'It is very interesting to notice how these requirements of simplicity, conspicuousness and specificity have been compromised by selection ... ' Thinking along these lines, it seemed probable that the markings of the zebra finch had to be as complicated as they are in order to satisfy the other two basic requirements. Furthermore, whilst conspicuousness can easily be achieved without complexity, specificity may be more difficult. There are about twenty-two good species of the sub-family Estrildinae in Australia. The zebra finch is distributed over the whole continent and is likely to co-exist with most of them. A study of skins and coloured plates of these close relatives of the zebra finch revealed that many of the component markings are shared between several species, but also that each species possesses a *unique combination of markings.* Only by such a combina-

Fig. 12. A diagram illustrating the occurrence of the same or similar markings to those of the zebra finch in the other Australian grass-finches. For explanation see text.

tion of different markings can the zebra finch meet the requirements of specificity, as is clearly shown by Fig. 12. Each segment of this diagram repersents one of the twenty-two related species. The eleven concentric rings represent the various regions of the body surface, divided up according to the markings of the zebra finch. A section of a segment is black where that region of the body surface bears the same marking as that of the zebra finch, grey where it is similar, and white where it is different. The all-black segment, therefore, represents the zebra finch itself. The following table acts as a key to the various markings or regions of the body surface. The numbers 1 to 11 read from the inside outwards.

TABLE II

BLACK	GREY	WHITE
1. Grey crown, back and wings.	Dull crown, back and wings.	Coloured or marked crown, back and/or wings.
2. White rump-patch	Bright rump-patch.	No rump-patch.
3. Chestnut ear-patch.	Bright ear-patch.	No ear-patch.
4. White belly.	Bright belly.	No distinct belly region.
5. Red beak.	Coloured beak.	Dull beak.
6. Orange-red legs.	Brightly coloured legs	Dull legs.
7. White-spotted chestnut flank feathers.	White-spotted flank feathers.	Unspotted flank feathers.
8. Black breast bar.	Coloured breast bar.	No breast bar.
9. Black-and-white finely barred throat.	Finely barred throat.	Throat unbarred.
10. Black-and-white banded tail.	Banded tail.	Unbanded tail.
11. Black-rimmed white cheek.	Rimmed cheek.	Cheek not rimmed.

A brief scrutiny of this diagram will show that there are thirty instances of individual markings of the zebra finch appearing in the same form elsewhere in the subfamily in Australia, and many more cases of markings which are similar. It should be borne in mind that this diagram is based on the markings of the zebra finch and that in certain other species there are special markings which do not appear in it. For example, certain species have a small bright patch across the forehead, but this fact is not revealed by the diagram. Nevertheless it does show well the complicated state of the markings of the Australian grassfinches which are shared with the zebra finch. Taking this into consideration, it seems probable that this species has evolved as simple a system of markings as is compatible with the demands of specificity. It may be argued that the demands of simplicity could better have been answered by the dropping out of the older, more commonly shared markings. The fact that this has not happened in many cases is possibly due to their being part of an indispensable gene-complex. This last point, however, requires further investigation.

References

Armstrong, E. A. (1947), *Bird Display and Behaviour* (Lindsay Drummond Ltd, London).
—— (1950), 'The nature and function of displacement activities'. *Sympos. Soc. exp. Biol.* 4, pp. 361–87.
Bastock, M., D. Morris and M. Moynihan (1953), 'Some comments on conflict and thwarting in animals'. *Behaviour* 6 (1), pp. 66–84.
Boosey, E. (1952), 'The Zebra Finch and its colour varieties'. *Avicult. Mag.* 58, pp. 29–31.
Bourke, P. A. (1941), 'Notes on two finches'. *Emu*, Melbourne 41, pp. 156–9.
Daanje, A. (1950), 'On locomotory movements in birds and the intention movements derived from them'. *Behaviour* 3, pp. 48–99.
Delacour, J. (1943), 'A revision of the subfamily Estrildinae of the family Ploceidae'. *Zoologica* 28, pp. 69–86.
Friedmann, H. (1949), 'The breeding habits of the weaver birds. A study in the biology of behaviour patterns'. *Ann. rep. Smithsonian Inst. for 1949*, pp. 293–316.
Hediger, H. (1950), *Wild Animals in Captivity* (Butterworths, London).
Heinroth, O. (1930), 'Über bestimmte Bewegungsweisen der Wirbeltiere'. *Sitzungsber. der Ges. naturforschender Freunde*, Berlin, Feb., pp. 333–42.
Hinde, R. A. (1952), 'The behaviour of the Great Tit (*Parus major*) and other related species'. *Behaviour*, Supplement 2.
—— (1953), 'The conflict between drives in the courtship and copulation of the Chaffinch'. *Behaviour* 5, pp. 1–31.
Jones, H. (1932), 'Notes on the breeding of the Zebra Finch'. *Avicult. Mag.* 10, pp. 169–72.
Lorenz, K. (1935), 'Der Kumpan in der Umwelt des Vogels'. *J. f. Ornithol.* 83, pp. 137–312, 289–413.
—— (1952), *King Solomon's Ring* (Methuen & Co., London).
MacGillivray, W. D. K. (1932), 'The Flock Pigeon'. *Emu*, Melbourne 31, pp. 169–74.
Makkink, G. F. (1936), 'An attempt at an ethogram of the European Avocet (*Recurvirostra avosetta* L.) with ethological and psychological remarks'. *Ardea* 25, pp. 1–60.
Mathews, G. M., *The Birds of Australia* (Witherby, London), vol. 12, pp. 173–80.
Morris, D. (1952), 'Homosexuality in the Ten-spined Stickleback (*Pygosteus pungitius* L.)'. *Behaviour* 4, pp. 233–61.
—— (1954), 'The reproductive behaviour of the River Bullhead (*Cottus gobio* L.), with special reference to the fanning activity'. *Behaviour* 7, pp. 1–30.
Moynihan, M. (1955), 'Some aspects of reproductive behaviour in the Black-Headed Gull (*Larus ridibundus* L.), and related species'. *Behaviour*, Supplement 4.
—— and F. Hall (1954), 'Hostile, sexual, and other social behaviour patterns of the Spice Finch (*Lonchura punctulata*), in captivity'. *Behaviour* 7, pp. 33–76.
Nice, M. M. (1943), 'Studies in the Life History of the Song Sparrow II'. *Trans. Linn. Soc. N.Y.*, vol. 6.
Scott, J. P. and E. Fredericson (1951), 'The causes of fighting in mice and rats'. *Physiol. Zool.* 24, pp. 273–309.

SEVENSTER, P. (1949), 'Modderbaares'. *De Lev. Nat.* 52, pp. 160–69 and 184–9.
SPENCER, B. and F. J. GILLEN (1912), *Across Australia* (Macmillan & Co., London).
TINBERGEN, N. (1939), 'The behaviour of the Snow Bunting in spring'. *Trans. Linn. Soc. N.Y.* 5, pp. 1–92.
—— (1940), 'Die Übersprungbewegung'. *Z. Tierpsychol.* 4, pp. 1–40.
—— (1948), 'Social releasers and the experimental method required for their study'. *Wilson ornith. Bull.* 60, pp. 6–51.
—— (1951), *The Study of Instinct* (Oxford University Press, Oxford).
—— (1952), 'Derived activities; their causation, biological significance, origin, and emancipation during evolution'. *Quart. Rev. Biol.* 27, pp. 1–32.
—— (1953a), 'Fighting and threat in animals'. *New Biology* 14, pp. 9–24.
—— (1953b), *The Herring Gull's World* (Collins, London).
TUBMAN, E. A. W. (1923), 'The house shortage again'. *Emu*, Melbourne 23, p. 50.
WESTON, D. (1930), 'Foreign finches that are easy to breed'. *Avicult. Mag.* 8, pp. 272–3.

The reproductive behaviour of the river bullhead
(1954)

Introduction

There are three common species of fresh-water fish in Britain which construct nests and rear their young. They are the three- and ten-spined stickleback (*Gasterosteus aculeatus* and *Pygosteus pungitius*) and the river bullhead (*Cottus gobio*). In the spring, these three species take up their breeding territories in the three basic ecological niches of most streams, namely the sandy, weedy and stony stretches, respectively. The reproductive behaviour of the sticklebacks has been the subject of intensive investigations by ethologists recently, but little or nothing has hitherto been recorded of the behaviour of the bullhead).[1] Brief notes on the habits of this latter species have appeared from time to time in the general literature, usually to the effect that it forms a cavity beneath stones in which the eggs are laid and guarded. Day (1880) reports that there has been much discussion as to whether the male, or the female, or both, perform the parental behaviour. He quotes Marsigli and various other authors as saying that it is the female which protects the eggs and young. However, this is not so and later writers, such as Meek (1916) and MacMahon (1946), report, correctly, that it is the male which assumes the role of parent, as in the stickleback species. It will be shown that there are other striking similarities between the behaviour of the sticklebacks and the bullhead, which make it a valuable species for comparative study.

All the observations reported here were carried out in the laboratory using aquarium tanks of 24 × 15 × 15 inches, with a substratum of sand and gravel. These tanks were unplanted, but were aerated profusely. Several pebbles and stones were placed in each tank and where possible the bullheads dug pits and hid underneath these. Unlike the sticklebacks, the bullhead is a solitary fish out of the breeding season, and hides under stones, swimming

[1] *Cottus gobio* must not be confused with the bullhead catfish (*Ameiurus nebulosus*), which American ichthyologists sometimes also refer to simply as 'the bullhead'. For example, Breder's (1932) paper on 'The breeding of Bullheads in the aquarium' refers to the catfish and not to *Cottus*. *Cottus gobio* is known by many popular names (e.g. Miller's thumb, tom-cull, nogglehead, bull-jub, bull-knob, horbeau, cod-pole, Tommy-logge, cob, stargazer, Tom-thumb, etc.), but despite this, it is preferable to retain the name bullhead, since this is the most widely recognized.

little except for sudden short darts after prey. Whilst it will accept almost any crack or crevice as a lair, it appears to be more particular when constructing a breeding cavity in the spring. Observations of the breeding behaviour were made possible by utilizing curved fragments of large broken flower-pots. These were pressed into the substratum to different degrees, with the potential entrances a few inches from and facing the front of the aquaria. It was found

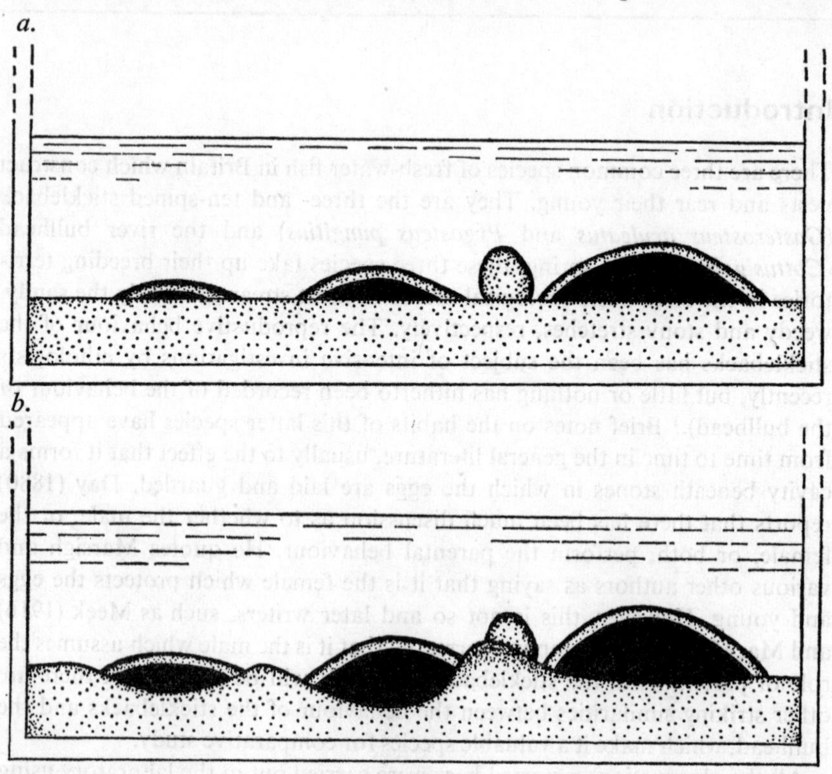

Fig. 1. Typical aquarium tank set-up for observing the reproductive behaviour of bullheads. Three pieces of broken flower-pot are pressed into the sand, leaving gaps facing the front of the tank. The one on the left, with the smallest gap, was preferred and eggs were laid in this. In a. the tank is shown before the fish were introduced and in b. the effects of the nest-digging can be seen.

that, if the curvature of a particular fragment left a gap of only about half an inch to an inch between it and the substratum, the fish were strongly stimulated to push into this and dig out a nesting cavity. If the height of the 'prefabricated' cavity was too great, it was less stimulating as a nest site (see Fig. 1). Once a cavity had been enlarged by a fish, the latter took up its position with its head at the entrance, thus facing the observer and facilitating observations.

Best results were obtained using fish which were caught at the beginning of the breeding season, in the second half of March. The males were then already in reproductive mood and the females ripe and swollen with eggs. Several males were placed in each tank, always one per potential nest-site. When the males had dug in, and set up territories, the ripe females were introduced. This simple method was found sufficient for the preliminary observations reported in the present paper.

Nesting Behaviour

The bullhead possesses a less complex nest-building pattern than the sticklebacks. It does not incorporate plant fragments as do the latter. It scoops out a cavity beneath a stone or rock and then maintains and modifies this cavity throughout the breeding cycle. This involves several forms of digging:

1. *Mouth-digging*

This form is comparable with the sand-digging of the three-spined stickleback (see Tinbergen, 1951, p. 140). The fish pushes into the cavity with its mouth held open and thus fills the latter with sand. It then shuts its mouth and turns round, emerges from the nest and spits out the sand a short distance away from it. It then returns quickly to the nest. The filling of the mouth with sand is accomplished by rapid beats of the tail which drive the open mouth of the animal into the sand at the back of the cavity. This method of digging is not the most common and is only used to remove sand from the very inside of the nest. Sand in this position cannot be removed in any other way. This type of digging therefore serves to enlarge the nest cavity in a horizontal plane.

2. *Carrying away obstructions*

Small pebbles and similar objects are picked up in the mouth, carried outside the nest, and spat away in a similar manner to sand being removed by mouth. In these cases, however, the objects are taken into the mouth simply by biting them, without the tail-beating and shovelling which is involved in the mouth-digging. Pests, such as small inedible crustacea, which are potential egg-eaters, are removed in the same way as small pebbles.

3. *Pectoral-fin-digging*

A bullhead, either inside the nest cavity, or out in the open, may lower itself into the substratum by special alternate beats of its strong pectoral fins. These, in combination with slight undulations of the body, serve to scoop away the sand beneath the fish sufficiently for it to sink partially into it. When performed in the open, this action, in association with colour change, provides an efficient means of concealment. Inside the nest, it may help to deepen the cavity.

4. *Tail-digging*

The bulk of nest-digging is performed by an intense and rapid lateral beating of the tail. The fish turns round so that its head points into the cavity and then proceeds to beat with its tail, shooting sand backwards and sideways. As it does this, the force of the beating drives the fish forwards into the inner wall of the nest, as in mouth-digging. But when tail-digging, the mouth is held shut and the beats of the tail are much more violent. So violent are they, that sometimes the nest is obscured by the great cloud of sand that is thrown up by the tail. When this subsides, the cavity can be seen to be much deeper and larger than before. Piles of sand accumulate outside the entrance as a result, especially on either side of it (see Fig. 1). Owing to the fact that, when tail-digging, the fish is pointing head-first into the nest, it is often difficult to ascertain what movements the pectoral fins are making. On certain occasions

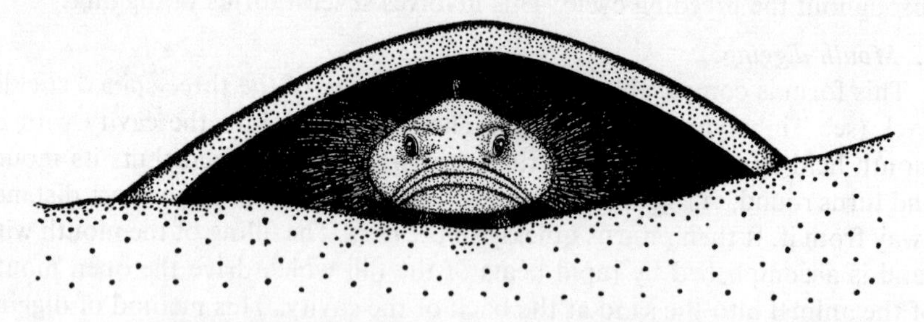

Fig. 2. The male bullhead in its nest-guarding position.

it appeared that they were beating sand back towards the tail where it was lashed out of the nest. In other instances they were seen to make furrows in the sand as the fish was forced forwards by the tail lashing and it appeared then that they were being used to oppose the forward force of the tail-beating.

Males were sometimes seen to turn upside down inside their nests and move about over the roofs, as if examining the surfaces. It is on these surfaces that the females will later lay the eggs.

Except for very brief sorties when chasing off a rival, removing sand, pebbles or pests, or pursuing prey, the male bullhead spends all its time inside its nest with its head emerging slightly from the entrance, or just inside it (Fig. 2). In this position it is ready and able to catch its food, court a female, intimidate a rival, rear its young and hide from its predators. It has one fundamental response to food, pests, females and rivals: *it bites them all.* If an object moves near the entrance of the nest of a male bullhead in spring, it is bitten. If the bitten object is food, it is then swallowed; if it is a pest, it is spat away from the entrance; if it is a female, it is spat into the entrance; if it is a rival, it is carried away from the entrance (although a rival nearly always flees before this can happen). The same kind of bite which stimulates a rival male to flee, stimulates a ripe female sexually. The bite is a multi-functional activity for the bullhead, and lying with the head at the entrance of the nest is a multi-functional posture. In this way the reproductive behaviour of the male is both efficient and inconspicuous.

Agonistic Behaviour

The usefulness of the term agonistic has been discussed elsewhere (Morris, 1954). Fighting and all behaviour associated with fighting can conveniently be described together under the general heading of agonistic behaviour.

The area in and around the nest-site is defended by a male owner against other males. A male, lying in its nest entrance, may leap out to attack an intruder, or threaten it, or both. An intruder, when attacked, usually flees so quickly that it is not caught and the owner then darts back to the shelter of its nest. At an advanced stage of the parental cycle, however, the male parent becomes excessively aggressive and may lie with its head protruding some way from the entrance of the nest. In this position it keeps a sharp look-out for the movement of rivals and may make comparatively lengthy attacking charges. Darts across the tank covering a distance of fifteen inches were recorded at this stage, whereas earlier in the cycle such darts were never more than a few inches. If the charged fish did not flee in time on these occasions, it was either bitten, or bitten and held and even carried round the tank. Once, when one male approached the nest of a male parent in the intensely aggressive phase, the latter emerged, grabbed the intruder by its pectoral fin and swam round the tank several times with it, finally spitting it out in the far corner. The parent then returned swiftly to its nest. This whole action lasted fifteen seconds and, had it been performed in a stream, would no doubt have removed the intruder some considerable distance from the nest. It is interesting to note that, whilst being carried, the intruder offered no resistance.

Apart from biting, chasing and carrying, another form of fighting has been observed on rare occasions, and that is mouth-fighting. This takes place when two males are more equally matched. Each bites the other's mouth so that one grips the upper jaw of its rival and the other the lower jaw. Firmly locked together, the two fish both attempt to swim forward with strong beats of both the pectoral fins and the tail, and thus carry the rival away from the respective nests. In this way they spectacularly lunge back and forth until one tears loose and flees. In one bout of mouth-fighting, which was observed particularly clearly, it was noticed that the male which appeared to be winning held the upper jaw of its rival. The winning fish held its dorsal fin fully erect the whole time whilst the other lowered its dorsal fin a number of times during the fight. When they separated, it was the fish that had appeared to be losing the fight that left the scene first. This is in agreement with the observations of

Baerends and Baerends (1950, p. 35) on mouth-fighting in Cichlid fish. They report that 'it seemed to us that the fish that gets hold of the upper jaw of its opponent has the best chance of winning the fight'.

Either as an alternative to, or in addition to, the actual bouts of fighting described above, the bullhead may perform a number of agonistic displays. All but one of these act as threat displays and function as intimidation signals. One is a submissive display and has appeasement function.

There are a number of threat actions, which are not mutually exclusive, and which may be performed separately or together. Each threat display consists of one or more of these threat actions. Each threat display appears to correspond to a particular causal state which is determined primarily by the relative and absolute levels of the attacking and fleeing drives (see Tinbergen, 1953 and Moynihan, 1955). In this way a code of threat signals exists. The various threat actions, taken separately, are as follows:

1. *Raising the gill-covers*

The gill-covers may be raised to the extreme, as shown in Figs 3 and 4. This has the effect of widening the head considerably. It is comparable with

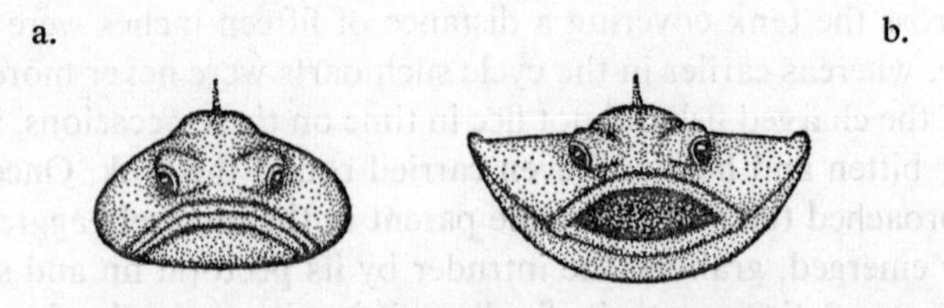

Fig. 3. Frontal view of a threat display consisting of two of the threat components, namely gill-cover raising and mouth opening. a: normal head-on attitude. b: threat attitude

the gill-cover displays of certain Anabantids and Cichlids. Tinbergen (1948) has pointed out that in different species of Cichlid the gill-covers are raised differently according to the location of releaser spots on them. The gill-cover raising of the bullhead does not appear to be so highly specialized. Its derivation is not clear.

2. *Lowering the head*

The head may be lowered independently of the body. When this is done in combination with the gill-cover raising, the visual effect, frontally, is of a greatly increased head size. In the other species of gill-cover displaying fish, the raising of the covers alone has this effect, and in these species there is no accompanying downward tilt of the head. However, a head-down posture does occur in the threat displays of a number of other species, notably sticklebacks (Tinbergen and van Iersel, 1947) and swordtails (*Xiphophorus helleri*) (Morris, unpublished). But in these species, the whole body is tilted down when the head is lowered. The derivation of these postures and their relationship to the head lowering of the bullhead is still not clear.

3. *Opening the mouth* (see Figs 3 and 4)

The mouth may be opened to a greater or lesser extent. When performed as part of a threat display, no attempt is necessarily made to carry this action on into a bite. Its function is visual rather than mechanical. The opening of the mouth in threat displays is of common occurrence throughout the vertebrates and in most cases it is probably derived from the intention movement of biting. Only two comparative examples, from other fish species, will be cited here. The sponge blenny (*Paraclinus marmoratus*), a fish with rather similar reproductive behaviour to that of the bullhead, possesses a similar mouth-opening display (see Breder, 1941b, Plate III, fig. 6). *Tilapia natalensis*, a Cichlid fish, opens its mouth wide in frontal display, revealing a white mouth against a black head (Baerends and Baerends, 1950, p. 41 and Fig. 14a).

4. *Darkening the head*

During an encounter of actual combat, or displays, the head of a male in reproductive condition may turn black. This occurs both in purely agonistic situations and also in courtship. The extent of this blackness is shown in Fig. 5. It is noticeable that only the head and the dorsal surface just behind the head blacken, and this is correlated with the position of the male at the nest entrance. Only the area exposed by the normal entrance-guarding posture changes colour. In the ten-spined stickleback the male becomes black all over in combat, but then its body is visible to the enemy over its whole surface. When this latter species is courting, however, only its ventral surface is visible to the female as the male leads her in the courtship dance, owing to his sexual posturing, and correlated with this the sexual colouring of the male is a black ventral surface only, with a light dorsal surface.

The darkening of the head in the bullhead results in a temporary form of sexual dimorphism. It is worth while mentioning at this point that there are two other sexual dimorphisms in this species. One is the width of the head and mouth. The head of the female is relatively less broad and the snout more pointed than that of the male. Also, the mouth of the male is larger. This last difference is best illustrated in Fig. 11. Further, the anterior dorsal fin of the male possesses markings (see below) which are absent in the female.

5. *Nodding*

The head may be nodded in short sharp rapid downward jerks. With increasing intensity, the jerks become more rapid and their amplitude increases slightly. Only the head is jerked in this way, but at higher intensities the whole body vibrates. The origin of this movement is obscure and possibly only a comparative study of closely related species will reveal its derivation.

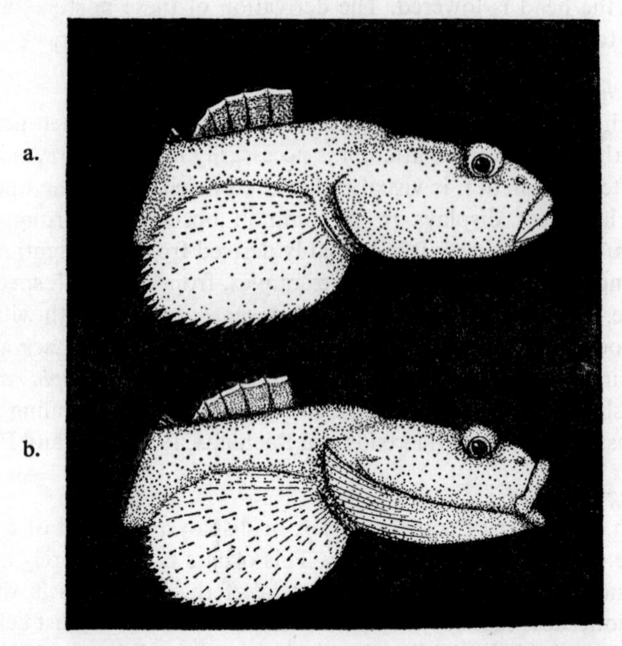

Fig. 4. Lateral view of the threat display shown in Fig. 3. Note also the erect anterior dorsal fin with a white band around its edge. a: normal lateral attitude. b: threat attitude.

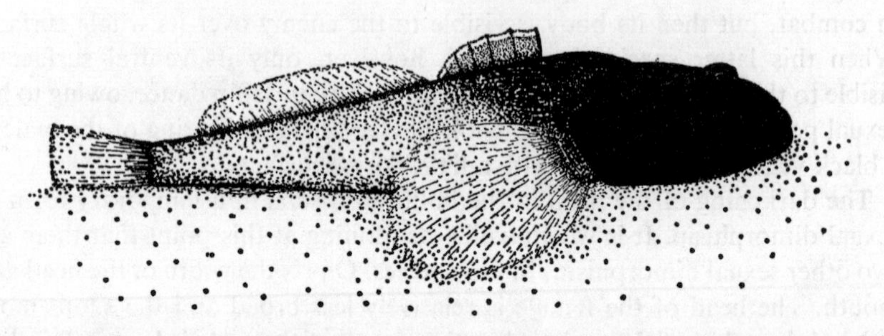

Fig. 5. Lateral view of a male bullhead in reproductive condition, showing the extent of the colour change which occurs during agonistic and sexual encounters.

6. *Raising the dorsal* (*and other*) *fins*

All the fins of a threatening fish are spread to the maximum. Seen from the side, this displays the yellowish-white band around the edge of the otherwise dark anterior dorsal fin of the male (see Fig. 4).

7. *Undulating the body laterally*

In a lateral display, the whole body may be undulated vigorously. This type of display movement is well known in many species of fish.

The actions described above form the basic threat components. Combinations of these components occur in every bout of threat displaying. Correlated with the nesting habits of the fish, frontal displays are by far the most common and the most intense version consists of a blackened, lowered, nodding head, with gill-covers raised and mouth open. This *Cephalic display* may vary in the following ways:

(*a*) It may be performed without any colour change. This happens when it is performed by a female, or by a male outside the reproductive season.

(*b*) It may be performed without a lowering of the head, or with the head only lowered slightly.

(*c*) It may be performed without the nodding movement, in which case the display is essentially a posture.

(*d*) It may be performed without the gill-covers being raised, but when this happens the display usually consists only of the nodding action.

(*e*) It may be performed with the mouth shut, but again this usually happens when only the nodding component is exhibited.

The raising of the fins occurs even in the frontal display and, conversely, various forms of the frontal display may occur when the fish is displaying laterally with body undulations.

It is not possible on the present evidence to construct a causal system, relating the various threat displays to various conflict states of the drives concerned. Certain fragments of evidence so far collected indicate that it should be possible to construct such a system when more data are available. For example, the display given by a male, in which the attack drive is particularly strong, consists of extreme mouth opening, accompanied by gill-cover raising, blackening and nodding, but with no lowering of the head at all. In less aggressive males, the same display is given, but with less extreme mouth opening and with the addition of a strong head-lowering component. This suggests that the head lowering is possibly associated with a greater tendency to flee. In this respect, it is interesting that the submissive posture consists of flattening all the fins and lowering the head, pressing it and the whole body flat on the floor.

Another distinct correlation concerns the nodding movement, which is often given by itself, as well as with the other components. In this way it differs from the other components, for although their relative strength varies and one or the other can be absent from a threat display, they are hardly ever given singly. When the nodding is given by itself as a display, it is nearly always performed by a male to a female, in a sexual situation. This suggests that perhaps the arousal of fleeing, attacking *and* sexual drives is the essential causal state which results in nodding, and that when the sex drive is relatively strong, only nodding occurs. But, on a few occasions, spent females and males out of the reproductive season, in both of which the sex drive must have been very low, were seen to perform threat displays in which nodding was the sole component. This leaves the possibility that the nodding display is the result of a particular state of the attacking and fleeing drives, and that this particular state is most likely to occur under the influence of a strong sex drive, although it may also occur at other times.

Further discussion of the causal basis of the threat code of this species would be too speculative at the present stage, but it can be said that any causal system must make allowances for the fact that there is a conflict state which permits the performance of all the threat components together, and also that the strength, or presence and absence, of each component seems to be able to vary independently of all the other components. Any causal system to explain this must therefore be more complex than any so far constructed for other species. Morris (1954) has pointed out that the angle (in the vertical plane) of the body of a zebra finch when facing its opponent is an indication of the relative balance between the attacking and fleeing drives. But then the zebra finch cannot be in a head-up and a head-forward posture at the same time! The bullhead, however, can perform any one of its threat actions at the same time as any one of the others.

Sexual Behaviour

The sexual behaviour of the bullhead can conveniently be divided into 'external' and 'internal' courtship, according to whether it is performed outside or inside the nest.

External courtship

As already mentioned, the reaction of a male to a ripe female is to bite her (Fig. 6) in the same way as any other moving object near the nest entrance. If she does not attempt to flee, she may be released, in which case she will probably swim into the nest, or the male may go back into the cavity 'walking' on his pectoral fins and release her directly into the nest. Apart from this, the only other courtship which occurs outside the nest consists of the male nodding at the female and blackening his head. The female responds to this with the submissive posture already described. (In these respects the bullhead clearly conforms to the courtship theory discussed by Tinbergen, 1953.)

It should be mentioned that the blackening of the head, which occurs as soon as the female appears and courtship begins, persists either until the female disappears, or until she enters the nest. Once she is inside with the male, his head colour lightens again, although he may continue to display to the female in other ways.

The only clear-cut displacement activity seen during the courtship phase was *displacement yawning*. When first presented with a female, one male interspersed his courtship actions with repeated yawning (i.e. wide stretching of the jaws, which may, or may not, be comparable with mammalian yawning). This was quite distinct from the mouth-opening component of the threat display. It did not appear to have any signal function.

It will be seen from all this that the external 'courtship' consists almost entirely of agonistic actions. There appear to be no special courtship displays at this stage, as there are in many other species.

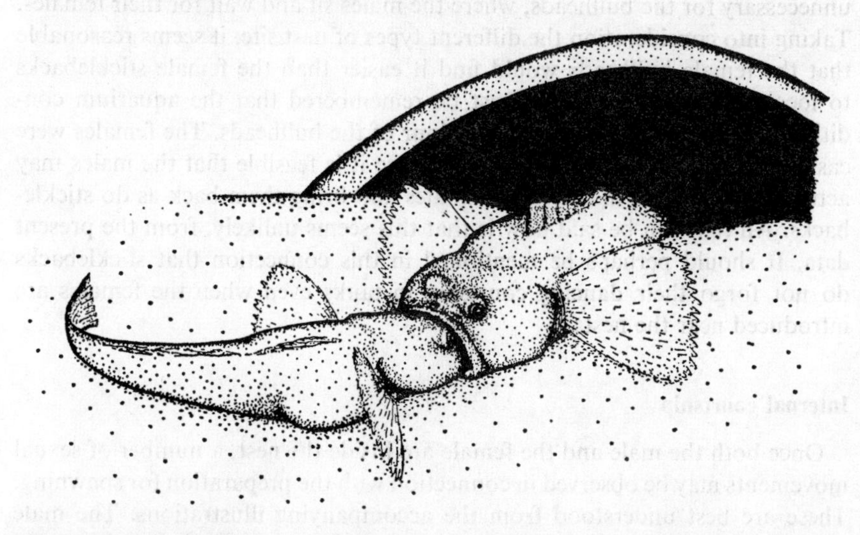

Fig. 6. The courtship bite. The male has emerged slightly from its nest and has bitten the head of a ripe female. The latter does not attempt to flee. It should be mentioned that, in this and the other illustrations accompanying this paper, the variable, broken, cryptic markings of the bullhead have been omitted, since (naturally) they only serve to obscure the shape of the fish.

The fact that a bite from a male fails to release fleeing only in a ripe female, means that sexual 'ripeness' must inhibit the female's fleeing drive. But this is not all, for although ripe females often respond to such biting simply by remaining still and passive, they may also respond actively by moving into the nest. This means that sexual 'ripeness' not only inhibits fleeing, but also transforms the stimuli received from being bitten, into sexual stimuli. The male therefore does not need to provide any special sexual stimuli and this permits the highly efficient system of the multi-functional bite.

The female's response to the bite, of moving into the nest, stimulates the male sexually and inhibits further aggression. In this way, a mate can be procured without special sexual displays. It is interesting to compare this situation with the equivalent one in sticklebacks, in which there is an elaborate male courtship dance. The nature of the stickleback courtship is only superficially different from that of the bullhead. The ripe female stickleback responds sexually to the bite of the male in just the same way, and this response of the female also stimulates the male stickleback sexually and inhibits his aggression. The reason why the courtship of the two species appears to be so different is because the male sticklebacks go out to look for their females and *have to escort them to the nest-sites*. The elaborate dancing of the male sticklebacks by which they lead their females to their nests is unnecessary for the bullheads, where the males sit and wait for their females. Taking into consideration the different types of nest site, it seems reasonable that the female bullheads would find it easier than the female sticklebacks to locate their mates. But it must be remembered that the aquarium conditions may have distorted the behaviour of the bullheads. The females were easily available to the males, and in nature it is feasible that the males may actually venture out in search of females and bring them back as do sticklebacks. All that can be said now is that this seems unlikely, from the present data. It should perhaps be mentioned in this connection that sticklebacks do not forgo their dancing courtship in tanks even when the females are introduced near the nests.

Internal courtship

Once both the male and the female are inside the nest, a number of sexual movements may be observed in connection with the preparation for spawning. These are best understood from the accompanying illustrations. The male may sidle over to one corner of the nest, making room for the female to enter (Fig. 7). Once she is inside, he may press up against her, holding her in a corner (Fig. 8). Or he may block her exit from the nest by lying across the entrance (Fig. 12). The female spends much time upside down (Fig. 9) apparently examining the roof of the nest, on which she is to lay her eggs. The male may sometimes lie with his head underneath her and with his body twisted so that his tail lies alongside hers (Fig. 10).

The preparations for spawning may take many hours and the male frefrequently displays laterally to the female inside the nest during this time. In doing this he erects all his fins, rearing up on his pelvic fins, and nods and undulates vigorously. The whitish band around the edge of the anterior dorsal fin can be seen even inside the dark nest, and must be clearly visible to the female during these displays. If more than one female is available and ripe, the male may accept them both into his nest (Figs 11 and 12). In one case two females both laid in the same nest, and in this respect too the bullhead is similar to the sticklebacks.

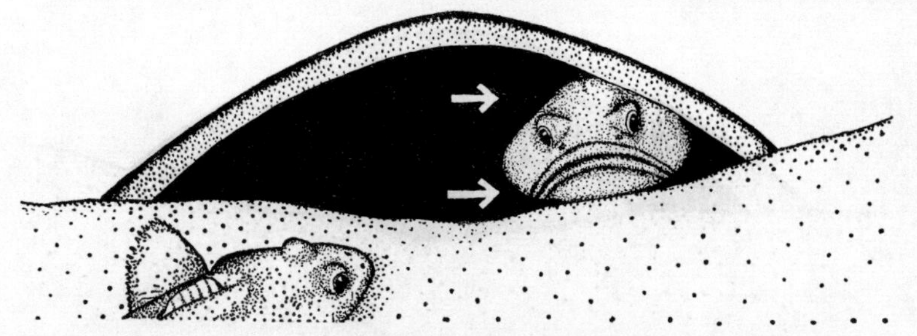

Fig. 7. The male moves over to one side of the nest cavity, making room for the female.

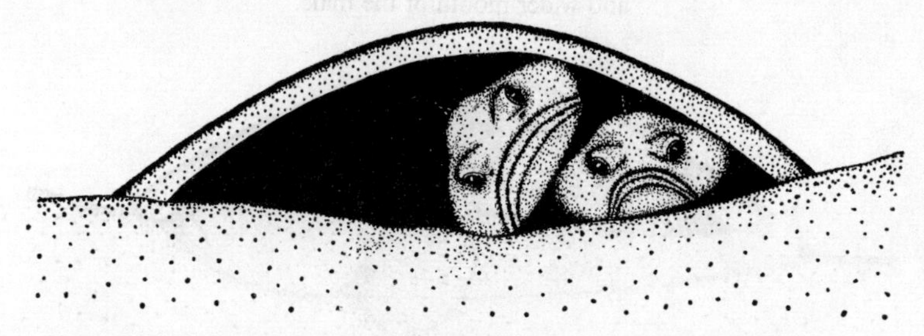

Fig. 8. The male (left) presses up to the female as she lies inside the nest.

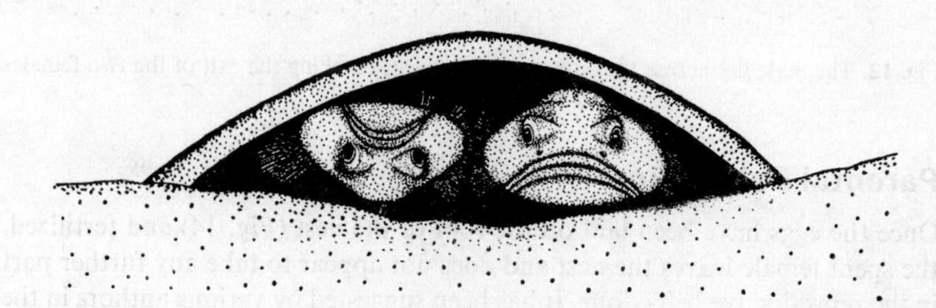

Fig. 9. The female turns upside down and examines the roof of the nest, on which she will lay her eggs.

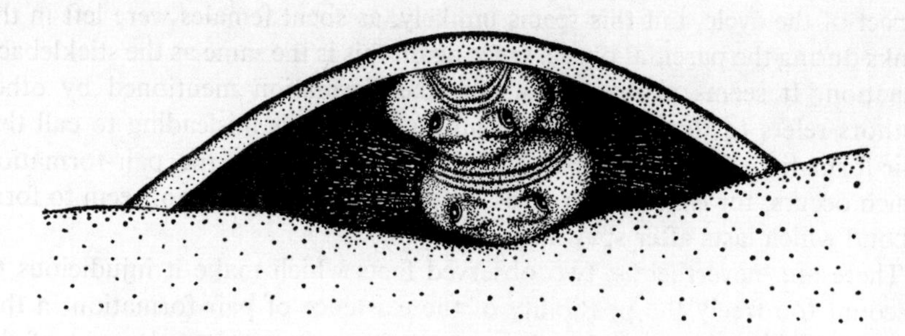

Fig. 10. The male lies upside down underneath the inverted female.

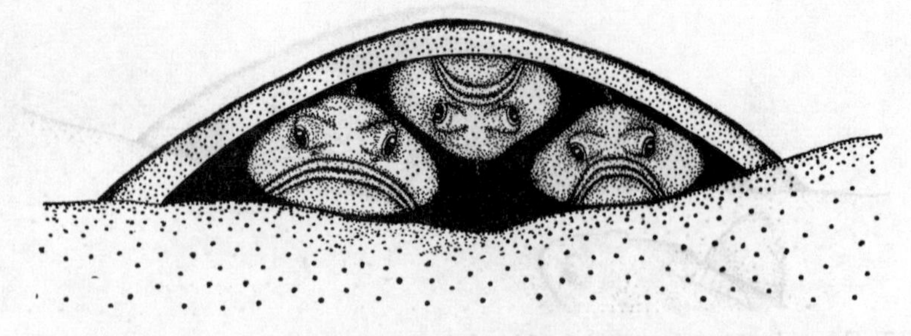

Fig. 11. A male (left) with two females inside its nest, one inverted. Note the broader head and wider mouth of the male.

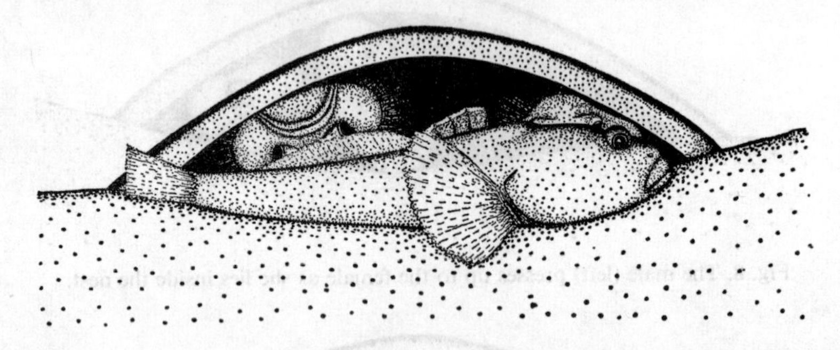

Fig. 12. The male lies across the entrance of the nest, blocking the exit of the two females inside.

Parental Behaviour

Once the eggs have been laid on the roof of the nest (Fig. 14) and fertilized, the spent female leaves the nest and does not appear to take any further part in the reproductive behaviour. It has been suggested by various authors in the past that this species forms pairs, but the laboratory evidence is mostly against this. It is just possible that the laboratory conditions distorted this aspect of the cycle, but this seems unlikely, as spent females were left in the tanks during the parental phase. Once again this is the same as the stickleback situation. It seems probable that the pair-formation mentioned by other authors refers to the long pre-spawning period. It is misleading to call this pair-formation, however, as it is not comparable with the pair-formation which occurs, for example, in birds, since the bullheads do not seem to form a bond which lasts after spawning.

There are, nevertheless, two observed facts which make it injudicious to discount too freely the possibility of the existence of pair-formation in this species. Firstly, the spent females several times returned to the nest of the egg-owning male. These 'returns' need not be thought of as anything more than the chance wandering of the females over the floor of the tank, and they usually resulted in the females fleeing after being bitten by the male. But on several occasions, the females responded to the biting by moving into the nest again, despite the fact that they had completed their spawning. The male ignored them when they did this, and went on with his parental duties of fanning and digging (see below). The female stayed inside the nest almost out of sight for some time, did not attempt to eat the eggs, and later left again without apparently having performed any parental actions. This is a state of affairs that could never happen in the behaviour of sticklebacks. A spent female stickleback is violently beaten off and flees just as violently. The male stickleback differentiates between a ripe and a spent female by virtue of how swollen her belly is, but the female bullhead, whether ripe or not, has her belly hidden behind her large pectoral fins when seen frontally – as she is, by the male, in the courtship phase. Therefore, the male bullhead is probably not so aware of the sexual state of a female, except as a result of her responses to his biting. But this still does not explain why the female should enter the nest after laying and the possibility must remain that the female does take part in parental duties in this species.

Such a conjecture is supported by the fact that spent females were seen to fan their pectorals whilst lying about on the floor of the tank. This fanning activity is a parental action performed by the male throughout the phase of the development of the eggs (see below). It was only performed at very low intensities and in very short bursts by the spent females, but unless they do in fact share the parental duties with the males, it is strange that they should exhibit this action at all. It is certainly the general rule amongst parental fish species that the male guards the eggs and young and assists in their development; but it is by no means always so. Breder (1935) reports that both sexes in the common catfish (*Ameiurus nebulosus*) take part in parental activities. But the same author reports that in *Opsanus beta*, *Paroclinus marmoratus* and *Gobiosoma robustum* – three species of fish which have similar reproductive habits to the bullhead – it is the male only that takes on the role of parent (Breder, 1941a, 1941b, 1942). Speaking of *Paraclinus*, the sponge blenny, Breder says, 'Wandering females are received by the male, adding their eggs to those already present ... ' Bearing all the evidence in mind, it seems likely that this will also prove to be a natural occurrence for the bullhead, despite the two difficulties discussed above. It is hoped that further investigations will clarify this point.

There is a long parental cycle in this species, lasting for approximately four weeks, during which time the male ventilates the nest and aerates the eggs by an almost constant fanning of its large pectoral fins. Typically, these beat alternately, but one may be fanned by itself if the movement of the other is obstructed in some way (e.g. by the nest-wall, or if the fish comes to lie on one side). With intense fanning, the whole body undulates slightly. The ventral fin rays of the pectoral fins hardly move at all in fanning, and the more dorsally situated a pectoral fin ray, the greater the amplitude of its movement. The current produced by the fanning action is often so strong that small particles can be observed being carried along by it. In this way it is possible to plot the fanning current. The male lies in its usual entrance-guarding position during fanning and the main force of the fin beats is backwards into the nest. During the early and very late phases of the parental cycle, the nest is usually open only at the front and the current then moves into the nest in the middle of the entrance, round the back of the cavity and out again from the corners of the entrance (see Fig. 13a). In the middle of the parental cycle, the male may open up the back of the nest, so that there is not only an

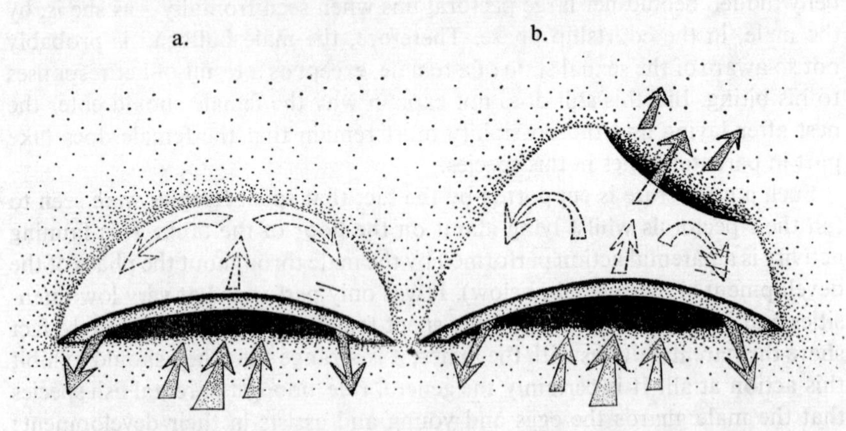

Fig. 13. The nest seen from the front and slightly above, showing the fanning currents. The male, which is not shown, would be laying with its head just protruding from the centre of the entrance, as shown in Fig. 2. In a. the current is shown circulating round a closed nest. In b. the back of the nest has been opened up and the main current passes through the nest.

entrance, but an exit as well. This permits a greater current, and the main stream passes straight through the nest (see Fig. 13b). (It must be borne in mind that under natural conditions such modifications may be either unnecessary or impossible.)

During the first part of the parental cycle, the male fans in his usual position, lying in the centre of the cavity facing outwards (Fig. 14), but later it may dig away the sand from directly beneath the eggs until there is room to fan lying immediately below them (Fig. 15). Towards the end of the parental cycle, the male becomes excessively aggressive: it not only attacks intruders which appear in front of the entrance, but even keeps a look-out for more distant enemies, and charges forth and fights them off some distance from the nest. During this phase, the male may lie across the entrance of the nest with its head in a position that enables it to keep such a look-out. Whilst in this position (see Fig. 16) it may continue to fan, passing a current across the entrance of the nest. It may even continue to fan a little when right away from the nest, so high is the fanning motivation.

A few days before the young hatch, the male can be seen to tilt on to one side and 'eye' the eggs (Fig. 17). No accurate records of the frequency of occurrence of this activity were made, but it was seen on a number of days during the late egg phase, and reached a peak one day, when it occurred

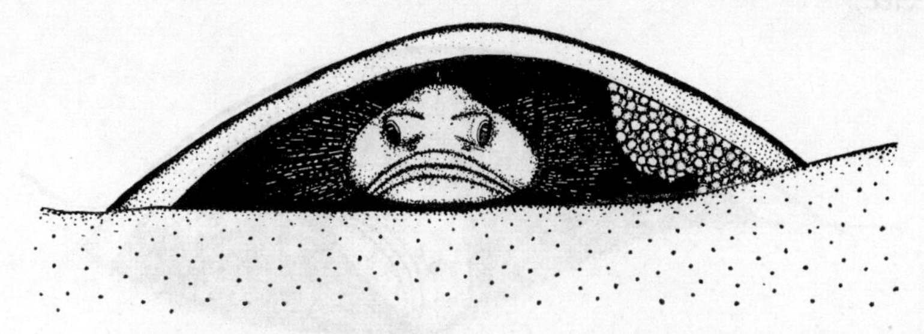

Fig. 14. A male with eggs. The male is in Fanning Position I.

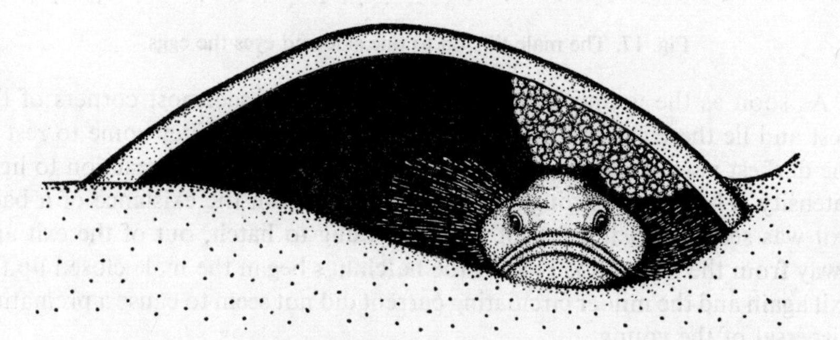

Fig. 15. Fanning Position II. The male has dug the sand away directly below the eggs.

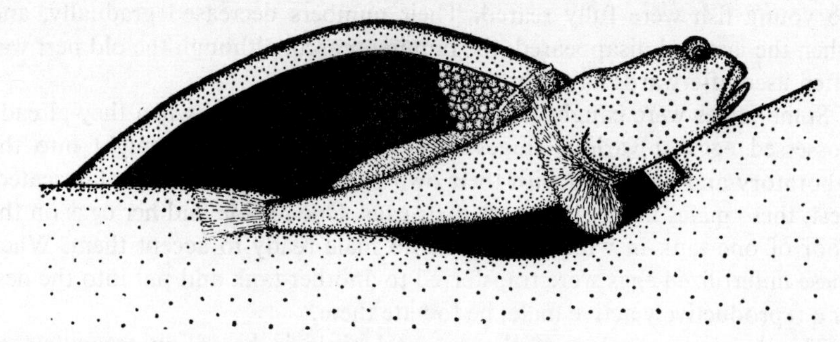

Fig. 16. Fanning Position III. The male is in the very aggressive phase, and fans and keeps a look-out at the same time.

every few minutes. The male appeared to be examining the eggs visually and the eye which, as a result of the tilting, was brought to bear on the egg mass, could be seen to be moved as if the male was scanning the surface of the clump of eggs. However, despite this action, and despite the fact that a number of other parental fish species clean their eggs or remove bad eggs, the bullhead was not seen to attempt to remove a small cluster of eggs which developed fungus in one nest. These bad eggs soon hung loose and finally were swept away by the fanning current, so that, even so, the remaining eggs were not infected.

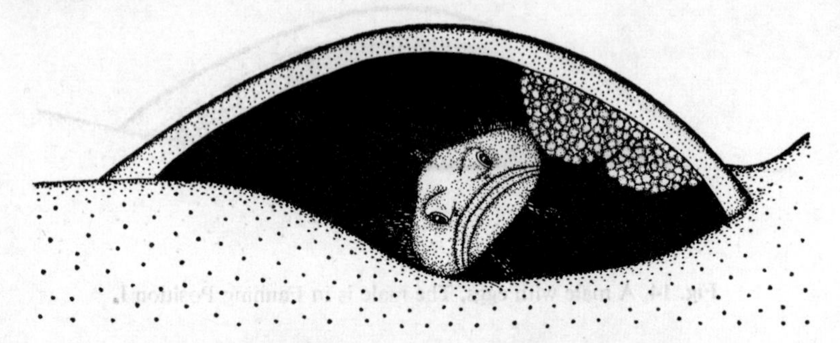

Fig. 17. The male tilts on to one side and eyes the eggs.

As soon as the young hatch, they make for the innermost corners of the nest and lie there around the inside edge of the cavity. They come to rest in the darkest part of the nest, and this is probably therefore a reaction to light intensity. The strong through-current resulting from the existence of a back exit was seen to wash a few of the first young to hatch, out of the exit and away from the nest. Shortly after the hatchings began the male closed up the exit again and the milder circulating current did not seem to cause a premature dispersal of the young.

Possibly because of an inadequate diet, the parents ate their offspring and no young fish were fully reared. Their numbers decreased gradually, and when the last had disappeared, the fanning ceased, although the old nest was often used afterwards as a lair by the parent male.

Some males were caught from under stones in streams when they already possessed eggs at various stages of development. When brought into the laboratory and presented with their own eggs again in a new 'prefabricated' nest, these males immediately ate their eggs. One female laid her eggs on the floor of one tank in which there was no male ready to accept them. When these unfertilized eggs were transferred to another tank and put into the nest of a reproductively active male, he too ate them.

This brief description of the parental cycle is based on fragments of information from a number of tanks and males, but in particular from one male, in which the entire reproductive cycle was clearly followed from beginning to end without interruption. In the case of this male it was possible to study the fanning activity quantitatively throughout the four weeks of the parental cycle.

The fanning activity

Van Iersel (1953) has analysed quantitatively in great detail the fanning activity of the three-spined stickleback. In that species, the length of time spent fanning per given period was used as a frequency measure. During the parental phase, a male stickleback swims round its territory, repeatedly returning to its nest to fan the eggs. When sticklebacks increase the frequency of their fanning, the increase is in the number of fanning bouts and in the lengths of the bouts, but there is no regular increase or decrease in the speed at which the pectoral fins beat as the fish fans. Van Iersel says 'it is slight and seems to be irregular, also it is difficult to measure. The total duration of fanning per time unit, however, varies in a striking and quite regular way.' In the bullhead, the situation is completely reversed. If this species varies the frequency of its fanning activity, it does so almost entirely by changing the speed of the beating of its pectoral fins. As regards the duration of fanning per time unit, the bullhead can be said to *fan all the time* during most of the parental cycle, except when interrupted. These interruptions are quite irregular – sudden arrival of an enemy, pest, food, etc. – so that the duration of fanning per time unit is almost useless as a measure. It should be pointed out, however, that at the very beginning and the very end of the parental cycle, there is an appreciable decrease in the duration of fanning. The male then fans only in short bursts, like a stickleback. Fig. 22 shows the fanning cycle of a three-spined stickleback. A peak is reached in a few days and falls off immediately in a few more days. The young hatch at or near the time of the fanning peak. It is quite clear that the heavily protected eggs of the bullhead, which take much longer to develop, require such a 'fanning-duration peak' to be higher and longer than those of the stickleback. Whilst the stickleback spends, very approximately, about half of its time fanning the eggs at the peak, the bullhead spends practically its whole time fanning – and, further, whereas the peak for the stickleback lasts only one day, it lasts about three weeks for the bullhead.

The speed of the beating of the pectoral fins was therefore used as the measure of the frequency of fanning for the bullhead. The number of fin-beats per minute was counted for ten separate minutes over a period of thirty minutes. The average of these ten counts was then scored as the figure for each test. There was little variation within a test. As mentioned before, the pectoral fins beat alternately during fanning and a beating of both fins was scored as one beat. A minute count was scrapped if during it the fish moved or was interrupted in any way. One or more tests were made on each day of the whole cycle, starting from the day after spawning. The results are plotted in Fig. 18.

The temperature in the laboratory increases considerably during the day, mainly owing to the overhead tank-lighting. The temperature of the water was taken at the time of each test and it was noticed that changes in temperature had a marked effect on the speed of fanning. Two rapid cooling tests were carried out and the results of these are shown in Fig. 19. Cold tap water was run into the tank and at the same time the warmer tank water was run off. This reduced the temperature considerably and the fanning rate also fell with it, revealing that there is no special daily fanning rhythm, but that the fanning rate adjusts quickly to any change in temperature.

In Fig. 20 the fanning speed is plotted against the temperature. It is clear from this that the speed of fanning varies directly with the temperature and that a rise of one degree Fahrenheit results in a rise of approximately six beats per minute. Fig. 21 shows the fanning cycle after this temperature correction has been made.

From the last figure, the following facts emerge:

1. There is a sharp rise in fanning speed during the first three days after spawning.
2. During the next ten days, while the eggs are developing, a more or less constant, very high level of the fanning activity is maintained.
3. There is a rapid fall-off of the level of fanning during the three days following the first hatching, during which all but two of the eggs have hatched.
4. A moderately high level of fanning is maintained during the next eight days, while the young are in the nest.
5. There is a sharp fall-off of fanning during the last three days at the end of which all the young have disappeared.

Before concluding this section, it should be mentioned that fanning was also seen to be performed by males which did not possess eggs. In the pre-parental phase, male bullheads often lay in nests and fanned weakly from time to time. The rate of beating of the pectoral fins was slow on such occasions, and the number of beats given in one bout of beating was small. Also the pauses between bouts of fanning were immensely longer than at any stage of parental fanning. Pre-parental fanning also occurs in sticklebacks. It has been suggested that pre-parental fanning is the result of the fanning motivation building up in readiness for the time when eggs will be obtained, so that, combined with the extra external stimulation from the eggs themselves, the fanning can be performed at high intensity as soon as the male becomes a parent. However, male ten-spined sticklebacks which were not able to fan at

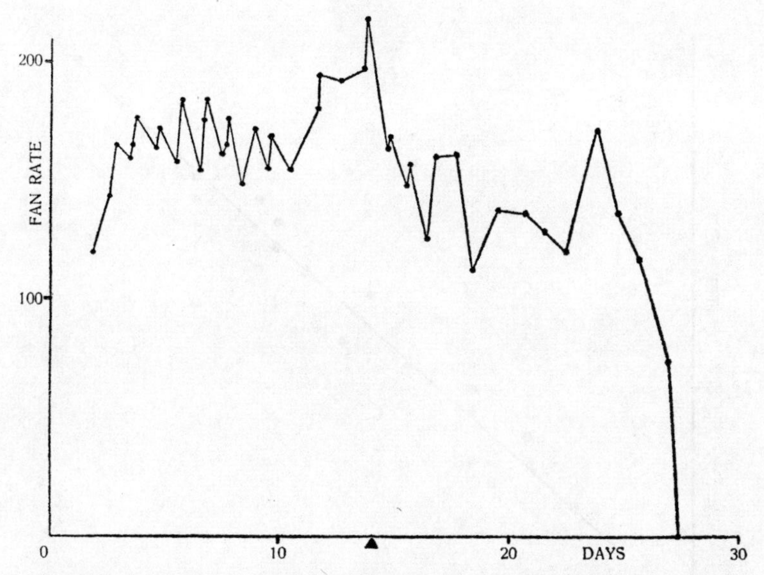

Fig. 18. The fanning cycle of a male bullhead. The fanning frequency is measured as the number of fin-beats per minute (ordinate), on one or more occasions on each day (abscissa) of the cycle. The black triangle marks the point at which the eggs hatched.

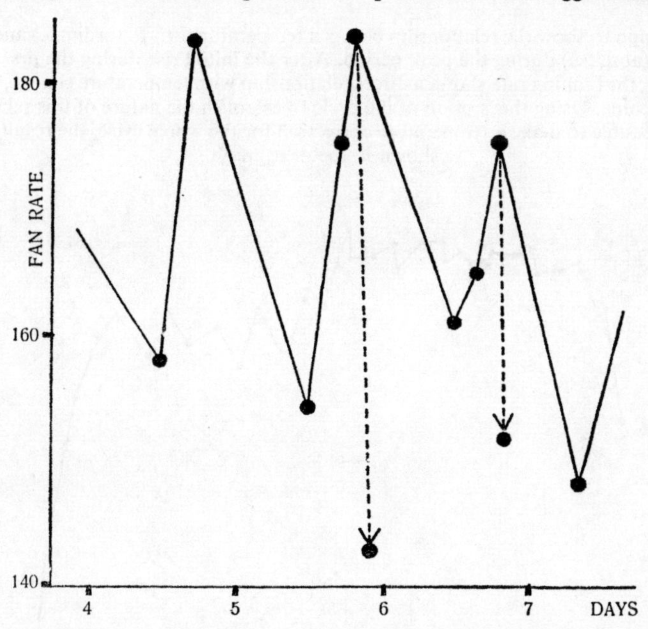

Fig. 19. Three days of the fanning cycle shown in Fig. 18, with the results of two cooling tests inserted. The rise and fall of the unbroken line is correlated with the day-night rise and fall in the temperature of the water. The broken lines show the rapid fall-off of the rate of fanning when cold water was run into the tank. Ordinate: rate of fanning. Abscissa: time in days.

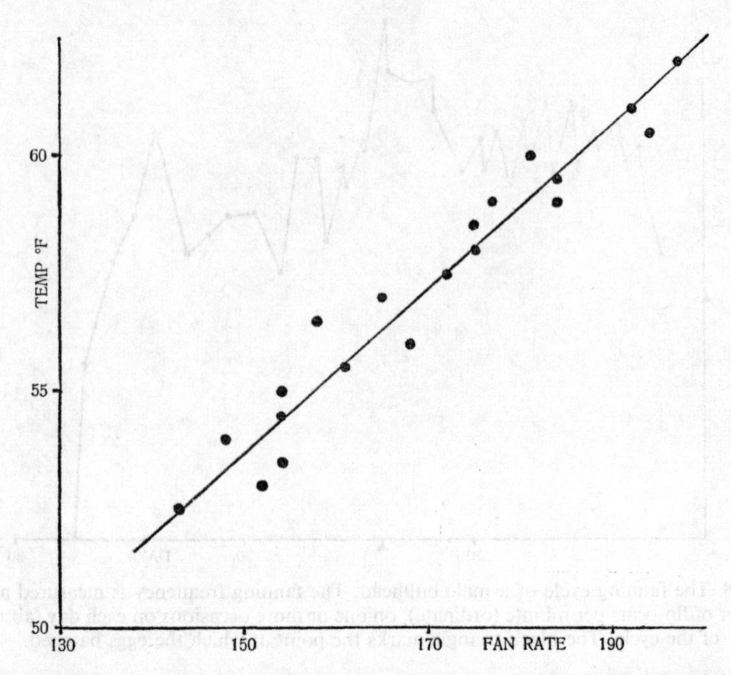

Fig. 20. Graph to show the relationship between temperature in ° F. (ordinate) and the rate of fanning (abscissa) during the peak period. After the initial rise during the first few days of the cycle, the fanning rate shows a direct relationship with temperature change, up to the date of hatching. Using this section of the cycle to establish the nature of this relationship, it is then possible to make a temperature correction for the whole cycle, the result of which is shown in the next graph.

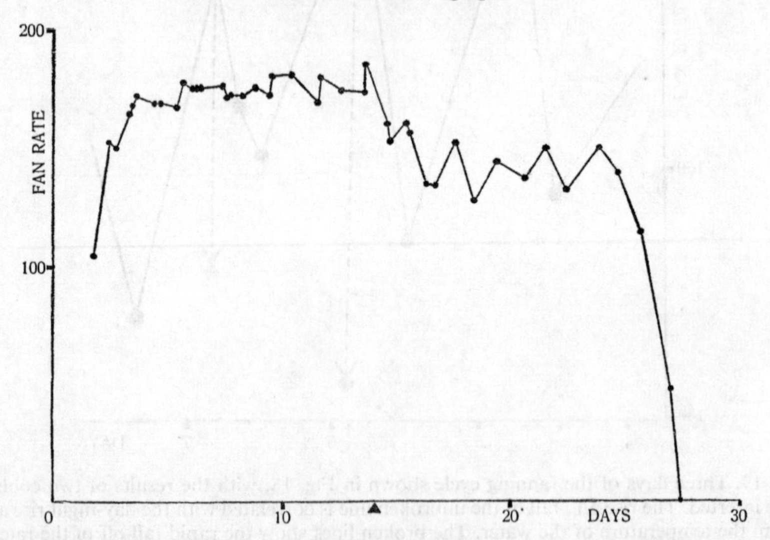

Fig. 21. The fanning cycle after a temperature correction has been made. The black triangle marks the time of hatching. Ordinate: rate of fanning at 58° F. Abscissa: time in days.

all in the pre-parental phase, were able to rear their eggs successfully (Morris, unpublished). It has been shown by Tinbergen and van Iersel (1947) that pre-parental fanning in the three-spined stickleback is largely displacement fanning due to a thwarted sex drive. This is possible in the bullhead also,

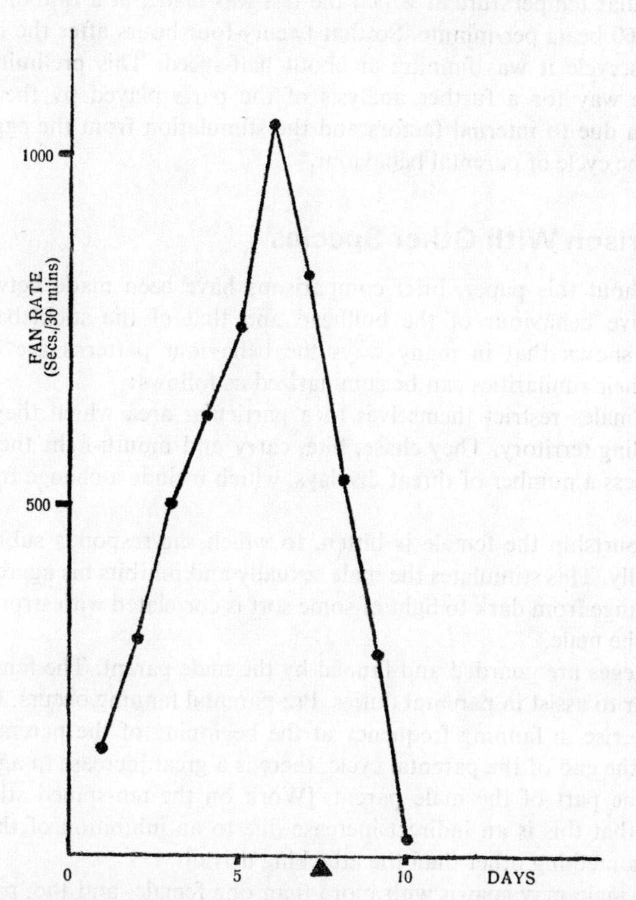

Fig. 22. The fanning cycle of the three-spined stickleback, for comparison with that of the bullhead. Note the immediate fall-off when the peak is reached, as compared with the prolonged bullhead peak. As before, the black triangle marks the time of hatching (after van Iersel).

since during the courtship of the latter species, both during the external and the internal phases, the males fans repeatedly and at a higher rate than at other (preparental) times.

One male bullhead was brought in from the river with its eggs and, twenty-four hours after being caught, was given its eggs again, and ate them all.

It was fanning in its new nest, and immediately after it ate its eggs, a fanning-rate test was made at a temperature of 55° F. It was fanning at an average rate of 81 beats per minute. Judging by the stage of development of its eggs, had it been uninterrupted in its cycle, it should have been fanning, at the particular temperature at which the test was made, at a rate of between 150 and 160 beats per minute. So that twenty-four hours after the interruption of its cycle it was fanning at about half-speed. This preliminary test points the way for a further analysis of the parts played by the fanning motivation due to internal factors and the stimulation from the eggs themselves in the cycle of parental behaviour.

Comparison With Other Species

Throughout this paper, brief comparisons have been made between the reproductive behaviour of the bullhead and that of the sticklebacks.* It has been shown that in many ways the behaviour patterns are basically similar. Their similarities can be summarized as follows:

1. The males restrict themselves to a particular area which they defend as a breeding territory. They chase, bite, carry and mouth-fight their rivals. They possess a number of threat displays, which include a change from light to dark.

2. In courtship the female is bitten, to which she responds submissively and sexually. This stimulates the male sexually and inhibits his aggression. A colour change from dark to light of some sort is correlated with strong sexual mood in the male.

3. The eggs are guarded and fanned by the male parent. The female does not appear to assist in parental duties. Pre-parental fanning occurs, but there is a sharp rise in fanning frequency at the beginning of the parental cycle. Towards the end of the parental cycle, there is a great increase in aggressiveness on the part of the male parent. [Work on the ten-spined stickleback indicates that this is an indirect increase due to an inhibition of the fleeing drive by something other than the attacking drive.]

4. One male may spawn with more than one female, and the 'pairing' in each case appears to last only as long as is necessary for the courtship.

The major differences between the sticklebacks and the bullhead seem to be that the latter does not encase its eggs in a nest of plant fragments and that its parental cycle is four or five times as long as that of the sticklebacks. I would suggest that there is a strong connection between these two differences. The eggs of the bullhead, stuck to the underside of stones and rocks, are far more likely to suffer mechanical damage than those of the stickleback.

* I refer here only to *Gasterosteus* and *Pygosteus*. For descriptions of their reproductive behaviour, see Ter Pelkwijk and Tinbergen (1936) and Morris (1952) respectively.

Correlated with this they are more heavily coated and thus better protected. Now although this probably protects them well from possible mechanical damage, it nevertheless must increase their aeration requirements, and this would account for the long fanning cycle with its extended peak. Breder (1935) has thought along these lines for the catfish, which appears to have gone even farther than the bullhead. Breder says, 'Possibly the heavy, gelatinous coating of these eggs serves to protect them from mechanical injury, on the one hand, and on the other causes a demand for an unusual amount of aeration.'

The fact that the reproductive behaviour of the bullhead is carried on more under cover than the sticklebacks', is almost certainly correlated with the fact that the bullhead is a bottom-living fish which can only swim in quick darts over short distances. Also, whereas the sticklebacks seem capable of a number of reproductive cycles in one season (they are certainly capable of this in the laboratory, and it is difficult to see why they should not be capable of it also in nature), it seems quite possible that the bullhead performs only one cycle per season. Laboratory bullheads could never be induced to start a second cycle, as the sticklebacks could. Also, incidental observations, made whilst collecting fish from streams at different times during the spring, indicated that the bullhead season is extremely short. It is hoped to make further observations of this last point, in the future, employing an underwater periscope.

A number of unrelated bottom-living fish species possess reproductive behaviour patterns which appear to be similar to that of the bullhead. Breder, in the papers already mentioned, has studied the behaviour of a number of these, and for a further comparative study the reader is referred to these.*

In conclusion, it is perhaps worth while to summarize briefly the various methods employed by fish in protecting their eggs. Since nothing is known of the way in which the bullhead protects its young after the eggs have hatched, apart from guarding its territory, this aspect of parental behaviour will not be stressed. The two major works from which much of the following information has been gathered are Gill (1907) and Innes (1951). For the sake of brevity, only one example of each type is given.

1. *Egg-shedding*

A large number of small eggs may simply be released into the water. These survive by virtue of their huge numbers (e.g. *Gadus morrhua*).

2. *Egg-sticking*

The adhesive eggs are stuck on to some surface or other, for example:

* Since the present paper was first written, a report on the reproductive behaviour of *Gobius microps* has been published (Nyman, 1953, in *Acta Soc. Faun. Flor. Fenn.* 69, pp. 1–11), which reveals striking behavioural convergencies between this goby and the bullhead.

(*a*) *Plants.* Many species simply stick their eggs on to the surface of aquatic vegetation and leave them. There is a preference for healthy growing plants which aid the development of the eggs. Such eggs can be laid singly (e.g. *Poecilobrycon auratus*), or together (e.g. *Pterophyllum eimekei*); or they may hang (e.g. *Cubanichthys cubensis*), or protrude (e.g. *Oryzias latipes*) from the vent of the female parent until she brushes against vegetation, when they adhere there. The eggs of the dogfish, on the other hand, cling to vegetation in a quite different way. They are surrounded by a capsule possessing tendrils which curl round seaweeds.

(*b*) *Rocks.* Many species stick their eggs to rocks, stones or other hard surfaces (e.g. *Loricaria pava*). Often, a concealed surface is selected (e.g. *Badis badis*).

(*c*) *Land.* The eggs are laid on some surface out of water. The parent fish then splashes water over the eggs at regular intervals until they hatch and fall back into the water (e.g. *Copeina arnoldi*).

(*d*) *Fish.* A number of species stick their eggs on to their own bodies. The Aspredinids press on to the newly fertilized eggs and stick them on to their ventral surfaces. Where each egg sticks, a small stalk develops from the parent and persists until the young hatch.

3. *Egg-nesting*

The eggs are encased in some way and, using the word 'nest' in its widest sense, these forms of encasement are as follows:

(*a*) *Bubbles.* A floating nest of bubbles is blown by the parents and the eggs are placed inside the bubble mass (e.g. Anabantids).

(*b*) *Weeds.* Fragments of vegetation are collected together and may be bound tight with glue threads by the parents. Such nests usually possess tunnels in which the eggs are laid and fertilized. Weed nests may float, be attached to plants (e.g. *Pygosteus*), or be dug in the floor (e.g. *Gasterosteus*).

(*c*) *Cavities.* A hole is dug out underneath some solid object resting on the floor and the eggs are usually stuck to the inside wall of the cavity (e.g. *Cottus*). Prefabricated cavities may also be used (e.g. *Paraclinus marmoratus*).

(*d*) *Depressions.* A simple depression is fanned or dug in the substratum, in which the eggs are placed (e.g. *Copeina guttata*).

(*e*) *Stones.* In one species of Australian catfish, the eggs are laid on the bottom and then covered over with stones collected from the surrounding area.

(*f*) *Animals.* The eggs are laid in some other species of animal, where they develop and hatch. The bitterling (*Rhodeus sericeus*) female possesses a long ovipositor with which she inserts her eggs in fresh-water mussels.

4. *Egg-carrying*

A number of species carry their eggs inside them in one way or another, as follows:

(*a*) *Mouth.* The eggs, and later the young, are carried inside the parents' mouths. They are constantly 'gargled' and thus aerated (e.g. *Haplochromis multicolor*).

(*b*) *Pouch.* The eggs develop in a brood pouch on the body from which the young finally emerge, or are expelled (see Breder, 1940) (e.g. Hippocampidae).

(*c*) *Body.* A number of species are viviparous and the young do not leave the bodies of the parents until able to fend for themselves (e.g. Poecilids).

5. *Egg-burying*

The eggs are buried in the mud or sand. The female may force each egg under the surface by use of a chute made by cupping her ventral fins together (e.g. *Aphyosemion coerulum*).

6. *Egg-incubating*

The parent fish lies on or curls round its eggs (e.g. *Ameiurus nebulosus*).

7. *Egg-guarding*

Many of the above types of fish guard their eggs against enemies of their own species and even of other species. This guarding involves one or both parents – usually only the male – remaining with the eggs until they develop, and attacking any other fish that ventures near.

8. *Egg-cleaning*

Some species clean their eggs. This may be done in a number of ways. The surface on which the eggs are to be laid may be cleaned prior to spawning (e.g. most Cichlids) and afterwards, from time to time, bad eggs may be removed and inedible pests may be caught and spat away from the egg-site (e.g. sticklebacks).

9. *Egg-aerating*

Eggs are often aerated by their parents, who either fan water over them with fin movements, or suck or blow water over them with modified respiratory movements. The pectoral fins are most commonly used in fanning and these may fan forwards (Gasterosteids) or backwards (*Cottus*). Fanning has been observed to occur in at least ten different families of fish.

The above list is almost certainly incomplete, but it serves to bring together the major categories of parental activities of fish, in respect of their eggs. It is difficult to describe any general evolutionary trends, and this will not be attempted here. Smaller trends have been discussed by various authors. The reader is referred, in particular, to Breder (1935), Leiner (1934), and Wunder (1931).

References

BAERENDS, G. P. and J. M. BAERENDS (1950), 'An introduction to the study of the ethology of Cichlid fishes'. *Behaviour*, Supplement 1.

BREDER, C. M. (1932), 'The breeding of Bullheads in the aquarium'. *Bull. N.Y.Z.S.* 35, pp. 129–31.

—— (1935), 'The reproductive habits of the Common Catfish, *Ameiurus nebulosus* (Le Sueur), with a discussion of their significance in ontogeny and phylogeny'. *Zoologica* 19, pp. 143–85.

—— (1940), 'The expulsion of young by the male of *Hippocampus zosterae*'. *Copeia* no. 2, pp. 137–8.

—— (1941a), 'On the reproduction of *Opsanus beta* Goode & Bean', *Zoologica* 26, pp. 229–32.

—— (1941b), 'On the reproductive behaviour of the Sponge Blenny, *Paraclinus marmoratus* (Steindachner)'. *Zoologica* 26, pp. 233–6.

—— (1942), 'On the reproduction of *Gobiosoma robustum* Ginsburg'. *Zoologica* 27, pp. 61–4.

DAY, F. (1880), *The Fishes of Great Britain and Ireland* (Williams & Norgate, London).

GILL, T. (1907), 'Parental care among fresh-water fishes'. *Rept. Smithsonian Instit. for 1905* (1688), pp. 403–531.

IERSEL, J. J. A. VAN (1953), 'An analysis of the parental behaviour of the male Three-spined Stickleback (*Gasterosteus aculeatus* L.)'. *Behaviour*, Supplement 3.

INNES, W. T. (1951), *Exotic Aquarium Fishes* (Innes, Philadelphia).

LEINER, M. (1934), 'Die drei europäischen Stichlinge und ihre Kreuzungsprodukte'. *Zs. Morphol. Ökol. Tiere* 28, pp. 107–54.

MACMAHON, A. F. M. (1946), *Fishlore* (Penguin Books, A 161).

MEEK, A. (1916), *The Migrations of Fishes* (Edward Arnold, London).

MORRIS, D. (1952), 'Homosexuality in the Ten-spined Stickleback (*Pygosteus pungitius* L.)'. *Behaviour* 4, pp. 233–61.

—— (1954), 'The reproductive behaviour of the Zebra Fish (*Poephila guttata*), with special reference to pseudofemale behaviour and displacement activities'. *Behaviour* 6, pp. 271–322.

MOYNIHAN, M. (1955), 'Some aspects of the reproductive behaviour of the Black-headed Gull (*Larus ridibundus* L.) and related species'. *Behaviour*, Supplement 4.

PELKWIJK, J. J. TER and N. TINBERGEN (1936), 'Roodkaakjes'. *De Lev. Nat.* 41, pp. 129–37.

TINBERGEN, N. (1948), 'Social releasers and the experimental method required for their study'. *Wilson Bull.* 60, pp. 6–52.

—— (1951), *The Study of Instinct* (Clarendon Press, Oxford).

—— (1953), 'Fighting and threat in animals'. *New Biology* 14, pp. 9–24.

—— and J. J. A. VAN IERSEL (1947), 'Displacement reactions in the Three-spined Stickleback'. *Behaviour* 1, pp. 56–63.

WUNDER, W. (1931), 'Brutpflege und Nestbau bei Fischen'. *Erg. der. Biol.* 7, pp. 118–95.

Author's Note, 1969

A film based on this paper was made in 1961 by the Granada Film Unit at the Zoological Society of London. Nearly all the patterns described in the paper were successfully recorded and a new discovery was made concerning the nature of one of the displays, namely nodding. This action had intrigued me for some time and in the paper I commented that 'the origin of this movement is obscure'. During the course of filming it emerged that this display is primarily concerned with sound production. By using a sensitive underwater microphone, the Granada unit was able to record the noises made by nodding bullheads. Each nod produces a grunting sound. At high intensities there are short bursts of multiple nodding, producing multiple, or 'segmented', grunts. The exact mechanism of the sound production is not known, but it probably takes the form of the grinding of certain hard parts against one another, in the region of the mouth or the gill-covers. It seems likely that the nodding action represents a ritualized, inhibited intention movement of biting that has become increasingly modified as a sound-producing signal, with its visual qualities taking second place.

The causation of pseudofemale and pseudomale behaviour

(1955)

Introduction

Pseudofemale behaviour has recently been reported for the ten-spined stickleback (*Pygosteus pungitius*) (Morris 1952)* and the zebra finch (*Poephila guttata*) (Morris 1954).† A male is said to be performing pseudo-female behaviour when it exhibits motor patterns belonging to the special sexual repertoire of the female. I refer here to the 'special' sexual repertoire of the female because in many species mutual sexual displays (Huxley 1914) occur in which both sex partners perform together the same movements (e.g. pre-copulatory head-tossing in gulls: Tinbergen 1953; and many examples in Armstrong 1947, Ch. XI). In such cases, even if the male were behaving as a pseudofemale, it would impossible to detect. The recognition of pseudofemale behaviour depends, therefore, on the existence of a distinct sexual 'di-ethism' (cf. sexual dimorphism) in a particular species. Furthermore, it must be a positive di-ethism. By this I mean that there must be sexual behaviour differences between the sexes which do not consist solely of the male doing something and the female doing nothing. For example, the females of many species flee if they are sexually unresponsive, and simply remain still if they are responsive. 'Staying still' acts as a sexual signal to the males concerned, but it cannot be used by ethologists in investigating pseudofemale behaviour. Those species in which both male and female perform special sexual actions are most valuable for such investigations.

Owing to the fact that males are, generally speaking, more elaborate in their sexual behaviour than females, it is much easier to study pseudomale than pseudofemale behaviour. Beach (1948) was able to assemble more evidence of pseudomale behaviour than of pseudofemale behaviour, and he concluded that 'the execution of feminine sexual patterns by genetic males occurs less frequently than does the display of male behaviour by the female'. Although this is probably true, it is safer, at this stage, to substitute for 'occurs less frequently', the words 'has been observed less frequently', if a general statement is to be made. Both the ten-spined stickleback and the zebra finch

* See pp. 13–41.
† See pp. 42–88.

exhibit positive di-ethism, and in both species I have now observed pseudofemale actions by males and pseudomale actions by females.

Additional Data

I have already reported the way in which male ten-spined sticklebacks have been observed to perform the entire mating pattern of the female (with the single exception of egg-laying!) under special circumstances. Since then I have observed females of this species performing part of the male sexual pattern, under the following circumstances.[1]

During the course of experiments in which ripe females are presented to isolated male sticklebacks, it is common practice to keep a number of females together in a tank in which they are fed liberally. Excessive feeding hastens the development of the eggs, and females are then taken from such a tank singly for experiments elsewhere when needed. In these 'female tanks' the females become sexually motivated to a very high degree and, since no males are present, they may become intensely sexually thwarted. This is revealed by the fact that they may even take up the initial female courting posture in the absence of the appropriate releasing stimuli. Under these circumstances, I have seen, on several occasions, one ripe female perform the male courtship dance (a series of head-down jumps) to another ripe female. On each occasion, the latter followed behind the dancing female and was led across the tank by it. Together the two females performed homosexual behaviour which was a replica of the first stages of the usual heterosexual courtship sequence of the species. As might be expected, these female homosexual courtships were comparatively brief, and were broken off, not by the following female, which was performing its usual motor patterns, but by the dancing pseudomale fish.

In the case of the zebra finch, male birds were seen to perform the female invitation-to-copulation display on many occasions. (This display consists of a rapid quivering of the tail while the body is held in a special crouched posture.) More rarely, the females responded to this by approaching the soliciting male and gingerly placing one foot on its back, as if to mount it. In a few instances, females mounted their soliciting males and performed the full male copulatory pattern (beating wings, lowered and twisted tail, etc.). Females have not yet been seen to perform the male courtship dance in this species.

In a previous report of this behaviour of the zebra finch (1954), I pointed out that the strong arousal of the sexual tendency of the male and its subsequent thwarting by the absence of appropriate releasing stimuli from an unresponsive female, were important causal factors in the pseudofemale display of this species. Since that time, I have recorded the exact behaviour

[1] This has been briefly mentioned elsewhere (Morris, 1954). (See pp. 66 and 67.)

sequences involved in a large number of courtships of this species, paying particular attention to the stage of the courtship sequence at which the pseudofemale display was given by the male. In twenty of the courtships, which were scored in this way, pseudofemale displays were seen, and most of these occurred at the same stage in the sequence and in the same type of courtship. Simplifying slightly, it can be said that there are four courtship situations in which these pseudofemale displays occurred:

1. In 17 out of the 20 cases, the following sequence was observed:
   (*a*) Male performs full song-and-dance display.
   (*b*) Female does not stay still, but hops about.
   (*c*) Male abandons song-and-dance, and performs pseudofemale display.
2. In 1 of the 20 cases, the following sequence was observed:
   (*a*) Male performs full song-and-dance display.
   (*b*) Female stays still, but does not solicit.
   (*c*) Male abandons song-and-dance, and performs pseudofemale display.
3. In 1 of the 20 cases, the following sequence was observed:
   (*a*) Male performs full song-and-dance display.
   (*b*) Female stays still, but does not solicit.
   (*c*) Male ignores absence of invitation display and mounts female.
   (*d*) Female flees from under male, thus preventing copulation.
   (*e*) Male performs pseudofemale display.
4. In 1 of the 20 cases, the following sequence was observed:
   (*a*) Male performs the full song-and-dance display.
   (*b*) Female stays still and solicits.
   (*c*) Male mounts.
   (*d*) Female flees from under male, thus preventing copulation.
   (*e*) Male performs pseudofemale display.

From these observations it can be concluded that the male performs male sexual behaviour at any stage, when the female is responsive at that stage in the sequence. Also the male may, to some extent, 'overlook' the unresponsiveness of the female and carry on to the next male stage in the sequence despite it. But the extent to which the male will do this is limited and there comes a point when the male abandons its male behaviour and switches to pseudofemale behaviour. The correlation between the sexual thwarting of the male and the appearance of the pseudofemale display seems very convincing.

Two points must be added here. Firstly, it must be stressed that after an apparently successful copulation, the male simply dismounts and then both birds usually shake and preen themselves, and there is no pseudofemale display. Secondly, the twenty cases described above were not selected in any special way. Many more instances of pseudofemale behaviour have been observed in this species, but these were the first twenty to be recorded in this particular series of observations. In conclusion, it should perhaps be mentioned that the data given here were obtained from four pairs of birds, each pair being isolated from the other three.

The Causation of Pseudofemale and Pseudomale Behaviour

In the light of the further data given here and certain other facts, it is interesting to re-examine the causal aspect of the problem of pseudofemale and pseudomale behaviour. In the past, I have stressed this or that causal factor, according to the particular case I have been studying. I now wish to postulate, tentatively, the four causal factors which I consider form the fundamental causal basis of such behaviour, and which I think may well be found to apply at a very general level. They are as follows:

A. Some hormonal and/or structural abnormality of the sexual system.

B. Intraspecific submissiveness or subordination in males; intraspecific aggressiveness or dominance in females.

C. The arousal and subsequent thwarting of the sex drive.

D. The presence of the releasing stimuli for the sexual behaviour of the opposite sex.

I suggest that one, or some combination, of the above factors, will probably be found to account for the occurrence of any pseudofemale behaviour. Some examples of the type of evidence on which this suggestion is based must now be given.

Causal factor A must be included if a general statement is to be made, but it does not concern us here. Beach (1948) has reviewed the evidence concerning this factor. The effects on sexual behaviour of alterations in the gonadal and hormonal systems of animals has been the subject of considerable experimentation in recent years. However, comparatively little experimental evidence exists concerning the purely neural causal factors B, C and D. This is undoubtedly the direction which future research into the subject of the reversal of sexual roles should take.

Causal factor B requires some explanation. I have already discussed in detail elsewhere the way in which courtship can be thought of as a three-point conflict between the tendencies to flee, attack and mate (Morris 1954 and 1956). The exact balance between these three tendencies varies from species to species. Speaking loosely, the male of some species are rather aggressive towards their mates when courting, whereas others are rather 'scared' of their mates. Clearly the mating tendency must ultimately become the strongest of the three, or sexual behaviour will not occur, but the relative strengths of attacking and fleeing can vary and the conflict between the stronger of these two and the mating tendency will then form the basis of the courtship pattern.

Not only is there a difference of this kind between species, but also between the male and female of each species. In many species it has been shown that the male takes the dominant role whilst the female is submissive. I shall deal with this type first. Zuckerman (1932), writing of baboons, has shown that in these animals causal factor B is all-important in producing pseudofemale and pseudomale behaviour. As he puts it (p. 289), 'In a particular situation, the animal assumes the dominant or male sexual role, while a fellow assumes the reciprocal and submissive role of female. Such behaviour might be either homosexual or heterosexual. Mounting behaviour therefore depends fundamentally upon degrees of dominance, and at this level of analysis it seems purely accidental whether a particular response is homosexual or heterosexual.' In other words, there is a particular balance of the fleeing, attacking and mating tendencies involved in the male sexual behaviour (which, for the sake of convenience can be described as an fAM-type[1]), and another balance in the female (FaM-type). When a male is unusually 'socially inferior', subordinate, or submissive, its usual flee-attack balance (fA) will shift in the direction of that typical of the female (Fa), its sexual behaviour will automatically and inevitably suffer a similar shift (fAM → FaM), and the male will show pseudofemale behaviour. If a female is unusually 'socially superior', dominant or aggressive, then the reverse process to the above will result in her performing pseudofemale behaviour. When, as often happens, these two shifts occur reciprocally, reversed mounting takes place.

The pseudofemale behaviour of the male ten-spined stickleback can be thought of in the same way. The male courtship is of the aggressive type (fAM) and intense pseudofemale responses were only observed in males which had been beaten into a submissive, subordinate state.

Before going on to discuss the next causal factor, it must be stressed that *active* subordination is not *always* an inevitable prelude to pseudofemale behaviour. To illustrate this point, it is necessary to compare the ten-spined stickleback with the zebra finch. In the former species, the male is so aggressive in its usual courtship that it has to be severely beaten up and subordinated before it can be in a state which permits it to behave as a pseudofemale. But the male zebra finch, far from being aggressive to its female during courtship, is rather 'scared' of her, and in this species the male behaves as a pseudofemale without previously being unusually subordinated. In other words, the very nature of the usual male sex pattern makes it easy for pseudofemale behaviour to occur, without the typical social reproductive organization of this species being first thrown out of balance.

[1] F = flee; A = attack; M = mate; capital letter = strong activation; small letter = weak activation. For full discussion see Morris, 1956.

Causal factor C must now be considered. I shall continue to use the zebra finch as an example here. If the male zebra finch does not have to be intensely subordinated to behave as a pseudofemale, then what determines whether it shall react to a particular sexual situation with typical male, or with atypical pseudofemale, responses? The answer to this has really already been given earlier in this paper, where it was clearly shown that the arousal and subsequent *thwarting* of the male sexual tendency leads to pseudofemale display. The same applies also to the male and female sticklebacks. In both cases, pseudofemale display occurred only when the fish were intensely sexually thwarted. In an earlier paper I have likened pseudofemale behaviour, in which this causal factor is important, to (an admittedly very aberrant form of) displacement activities, but I shall not pursue this comparison until further experimental data have become available.

Finally, *causal factor* D completes the list and is illustrated by the female zebra finch. The bird may respond to a pseudofemale display of a male, by mounting it and making the typical male copulatory actions. Such a female is giving a pseudomale response and yet is neither sexually thwarted nor unduly aggressive or dominant. The main causal factor here appears to be the presence of the releasing stimuli for the sexual responses of the opposite sex. The male is giving the female invitation display to the female and the latter, despite its sex, responds 'appropriately' to this, and accepts the invitation like a male.

Summarizing the foregoing, it is now possible to examine just which causal factors are at work in the various examples of pseudofemale and pseudomale behaviour mentioned already, and in any other cases in which some causal basis has been proved or suggested. In doing this, factor A will be ignored completely, and only the purely neural cases will be dealt with.

1. Male ten-spined stickleback (Morris 1952). Factors B, C and D are all important, especially C. I suspect that in very rare cases, pseudofemale behaviour occurs when B and D are only mildly involved.

2. Female ten-spined stickleback (in present paper). Factors C and D are important. Since there was no agonistic behaviour in the 'female tanks', factor B could not be involved.

3. Male zebra finch (Morris 1954 and in present paper). Factor B passively involved: factor C most important; factor D never operating.

4. Female zebra finch (Morris 1954). Factor D apparently the only one involved.

5. Male baboon (Zuckerman 1932). Factor B most important; factors C and D not essential.

6. Female baboon (Zuckerman 1932). Factor B most important; factors C and D not essential.

7. Woodlark (Niethammer 1937). It is reported that reversed mounting

takes place after the male has made unsuccessful attempts to copulate with the female. This suggests that this species shows pseudofemale and pseudomale behaviour which is causally similar to that of the zebra finch.

8. Pigeon (Whitman 1919). An example of pseudomale behaviour which involves causal factor C is illustrated by this passage from Whitman: 'Often the sexual impulses do not develop synchronously in a pair, and thus we may have a female dominated by passion mated with a male who is in a state of relative unreadiness. In such a situation, the female may make the advances and possibly assume the masculine role in courting and copulation.' (p. 98).

9. Dog and 10. Rat (Ford and Beach 1952, p. 143). Pseudomale behaviour involving factor C, in the same way as the pigeon cited above. Ford and Beach state: 'It is not at all uncommon to observe the temporary display of masculine behaviour on the part of receptive females confronted with sexually sluggish males. Female dogs, rats, and other animals in heat may mount the male repeatedly if he is slow to assume the initiative in the sexual relationship.'

The above list is not intended to be a complete one. But it is sufficient already to show that, although the causation of sex role reversal differs from sex to sex and species to species, there are, nevertheless, always the same few basic factors involved. Sometimes a single factor alone appears to account for a particular instance. Elsewhere, this or that combination of factors appears to be involved.

Conclusion

Reversed mounting has been observed in a number of bird species other than those mentioned here. In some species, reversed mounting is said to be the usual sequel to ordinary mounting, and it is possible that in such species the reversed act has become a functional ceremony. But until more evidence is available, it must be borne in mind that it is often extremely difficult, with birds, to observe whether a copulation has been successful or not. Reversal sequels may, in fact, be following *attempted* copulations as in the case of the zebra finch.

In connection with reversal sequels, which are known to have followed unsuccessful copulation attempts, it may seem strange that the female, which was not prepared to accept the male in the usual way, was immediately prepared to respond to him with pseudomale behaviour when he then behaved as a pseudofemale. I suggest that the answer is that the approaching male, giving the male display, aroused the fleeing tendency of the female too much to permit the sexual tendency of the female to dominate her mood. When the male then stopped advancing towards the female and gave the invitation display, the fleeing tendency of the female was less stimulated, and sexual behaviour was permitted expression. But now the presence of the releasing stimuli for the sexual behaviour of the opposite sex acted on the female to switch her response from female to pseudomale. This suggestion is based on observations of the fleeing intention movements made by female zebra finches when performing such behaviour (Morris 1954).* The strength of the fleeing tendency appears to decline when the male zebra finch switches from male to pseudofemale behaviour, but even then, the female may show such strong tendencies to flee that she will only gingerly place one foot on the back of the soliciting male. (It should also be mentioned that the occurrence of pseudomale behaviour is much rarer than that of pseudofemale behaviour in the zebra finch.)

Finally, a few points should be made concerning pseudofemale and pseudomale behaviour in our own species. Much has been written recently about homosexuality in human beings. Of particular interest are the contributions of Kinsey and his co-workers in America. Owing to the nature of their investigations, they were limited to making observations by interviewing techniques, but have done this on such a large scale as to produce important results (see Kinsey, Pomeroy and Martin 1948). In an earlier discussion of homosexuality, Kinsey (1941) had already dismissed the hormonal evidence that had been put forward to explain the general causation of homosexual behaviour in human beings, stating that 'the circumstances of the first sexual experience, psychic conditioning, and social pressures are obvious factors in determining the pattern of behaviour. It would appear that no similar correlation has as yet been shown between hormones and homosexual activity.' More recently, he and his colleagues (Kinsey, Pomeroy, Martin and Gebhard 1949) have emphasized the importance of the effects of experience in determining patterns of sexual behaviour: 'In brief, the psychosexual pattern in the human animal originates in indiscriminate sexual responses which, as a product of conditioning and social pressures, become increasingly restricted in the direction of traditional interpretations of what is normal or abnormal in sexual behaviour.' This, I feel, is satisfactory as far as choice of a male or a female as a sexual partner is concerned, but once such a partnership exists, be it homosexual or heterosexual, it still remains for something to determine who shall play the masculine role, and who the feminine role. I would tentatively suggest that factor C (submissiveness in the male; aggressiveness in the female), discussed earlier, is probably the most important causal factor in the production of pseudofemale or pseudomale behaviour in human beings (regardless of the *sex* of the partner).

It is necessary to point out here the relationship between pseudofemale and pseudomale behaviour (inversion), on the one hand, and homosexuality, on the other. Kinsey, Pomeroy and Martin (1948) have discussed this at length (pp. 612–17) and complain that 'It is unfortunate that students of

* See pp. 66 and 67.

animal behaviour have applied the term homosexual to a totally different sort of phenomenon ... ' They go on to say that pseudofemale and pseudomale behaviour is 'what the students of animal behaviour have referred to as homosexuality'. I have attempted to clarify this situation (Morris 1952) by introducing a classification of types of homosexuality, based on the idea that any sexual behaviour between two members of the same sex is homosexual behaviour (which is the way in which Kinsey and his colleagues use the term). Since the present paper is dealing with inversion as its central problem, it is useful to conclude with a tabulation of the relationship between inversion and homosexuality:

TABLE I

Motor Pattern of male	Male type	Partner	Relationship
Male	Masculine ♂	+ Feminine ♀	Heterosexual
Pseudofemale	Feminine ♂	+ Masculine ♀	Heterosexual
Pseudofemale	Feminine ♂	+ Masculine ♂	Homosexual
Male	Masculine ♂	+ Feminine ♂	Homosexual

The above table shows the four possible types of sexual behaviour available to the male. A similar table could be made for the female. It will be noticed that inversion is simply a matter of motor patterns, whereas homosexuality is simply a matter of partnership, and that although the two phenomena overlap, they can also be quite independent of one another.

References

ARMSTRONG, E. A. (1947), *Bird Display and Behaviour* (Lindsay Drummond Ltd, London).

BEACH, F. A. (1948), *Hormones and Behaviour* (Hoeber, New York).

FORD, C. S. and F. A. BEACH (1952), *Patterns of Sexual Behaviour* (Eyre & Spottiswoode, London).

HUXLEY, J. S. (1914), 'The courtship habits of the Great Crested Grebe (*Podiceps cristatus*)'. *Proc. Zool. Soc.*, London, pp. 491–562.

KINSEY, A. C. (1941), 'Homosexuality. Criteria for a hormonal explanation of the homosexual'. *J. Clin. Endocrin.* 1, pp. 424–8.

——, W. B. POMEROY and C. E. MARTIN (1948), *Sexual Behaviour in the Human Male* (Saunders, Philadelphia and London).

——, ——, —— and P. H. GEBHARD (1949), 'Concepts of normality and abnormality in sexual behaviour', pp. 11–32, in *Psychosexual Development in Health and Disease* (Grune and Stratton, New York).

MORRIS, D. (1952), 'Homosexuality in the Ten-spined Stickleback'. *Behaviour* 4, pp. 233–61.

—— (1954), 'The reproductive behaviour of the Zebra Finch (*Poephila guttata*), with special reference to pseudofemale behaviour and displacement activities'. *Behaviour* 6, pp. 271–322.

MORRIS, D. (1956), 'The function and causation of courtship ceremonies'. *Fondation Singer-Polignac: Colloque Internat. sur l'Instinct*, June 1954.
NIETHAMMER, G. (1938), *Handbuch der deutschen Vogelkunde* (Leipzig).
TINBERGEN, N. (1953), *The Herring Gull's World* (Collins, London).
WHITMAN, C. O. (1919), 'The Behaviour of Pigeons'. *Carnegie Inst. Wash. Publ.* 257, vol. 3, pp. 1–161.
ZUCKERMAN, S. (1932), *The Social Life of Monkeys and Apes* (Kegan Paul, London).

The function and causation of courtship ceremonies
(1956)

Introduction

Fifty years ago the 'love-play' of animals was usually described simply as 'comical'. Since that time, with the rapid growth of comparative ethology in Europe, the subject of courtship display has been more seriously analysed and many of the problems which it poses have been solved. Much still remains to be done, it is true, but already a number of general principles can be formulated, and these can be supported by an ever-widening range of facts.

Before going any farther, it is essential that I should provide a definition of the word 'courtship'. Courtship is *the heterosexual reproductive communication system leading up to the consummatory sexual act*. I am therefore using the term to include both pair-formation behaviour and pre-copulatory behaviour. The reasons for this will become clear later.

The first great step made by the ethologists, in the study of this and other problems, was the compiling of accurate ethograms. They began to tabulate minutely the elements of the behaviour patterns of animals and, amazingly enough, thus provided the first behaviour descriptions precise enough to stand comparison with the structural descriptions of comparative anatomists. (For years a spine had been something more than just a spine, but now, for the first time, a jump was something more than just a jump.)

Since display actions had been the worst described of all activities previously, this new attention to descriptive detail naturally was most effective in this realm of behaviour. The first really important formulation concerning signal movements, postures and structures was Lorenz's (1935, 1937) releaser concept, which is now too well known to warrant discussion here. Later, Tinbergen (1942, 1948; with Kuenen, 1939; with Perdeck, 1950) developed the releaser concept experimentally, and this same author's work on the behaviour of animals when thwarted, or in conflict situations (1940, 1952a; with van Iersel, 1947), led him to an analysis of the causation of courtship displays (1953a).

But firstly, the functional aspects of the problem must be considered.

The Function of Courtship

There are several very good reasons why courtship displays should not exist at all. Firstly, they render the animal performing them conspicuous, and therefore may attract predators. Secondly, they are usually performed with such intensity that they not only attract predators, but also claim the attention of the displaying animal to such a degree as to make it particularly vulnerable to attack from the predators. Thirdly, they may involve the expenditure of a considerable amount of time and energy without being 'mechanically effective' in any way.

These displays must therefore be of considerable advantage, if they are to outweigh such disadvantages.

Initial observations of courtship make it quite clear that the displays involved serve the function of releasing sexual responses in the mate; but it may be argued that this effect could, theoretically, be obtained by the simpler expedient of responding purely to the presence of the opposite sex, under special conditions of light, temperature, humidity and so forth. Yet, despite this, courtship displays are frequently very complex and elaborate.

However, there are other functions of display, which could not be served so easily in a more simple manner. Firstly, there are certain types of display which are primarily concerned with *finding* a mate. Secondly, the animal must make absolutely sure that it has found a mate of the *correct species*. Thirdly, it has to *arouse* the mate sexually, and the arousal of the male and female must be perfectly *synchronized*. For many species these problems cannot be solved without resort to the complicated communication system which we call courtship. But these various functions must be discussed one by one.

Attracting a mate is often a problem for a species, even though it possesses special responses to change of day-length, temperature and so forth. For these responses may lead it to a suitable breeding area, but within this area the problem of finding a mate may still be difficult. Many different auditory and chemical, as well as visual, displays have been evolved which help with this problem; bird song and mammalian scent, for example. Fish (with which I am mainly concerned in the present paper) commonly show two types of attraction display. Firstly, many species change their mode of swimming when in possession of a reproductive territory. Sticklebacks perform jerk-swimming, which consists of a forward progression of quick darts, interspersed with sudden pauses. They patrol their territories in this way, and there can be little doubt that it adds to their conspicuousness. Secondly, certain species of fish, in common with many other types of animal, adopt a breeding dress which contrasts strongly with their background. Sticklebacks, which are usually cryptically countershaded, reverse this countershading when in sexual condition.

These *attraction displays*, which operate from a distance, serve to bring the sexes together, and it might be argued that now, after responding to one another's specific attraction signals, nothing stands in the way of mating. Yet there is still more ceremony to come. For, having made contact with a potential mate, it is essential to make absolutely sure that it is the correct mate, and not a mate of a closely related species, or a species which is accidentally similar for some reason. This problem is solved in a number of species by the possession of a complex courtship reaction chain. Fig. 1 shows two such chains, for a goby (studied by Nyman, 1953) and the now well-known case of the three-spined stickleback (studied by Tinbergen, 1942). Each reaction is

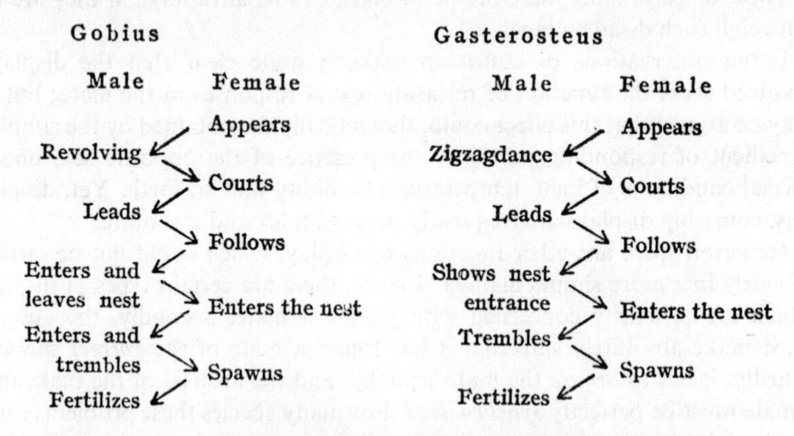

Fig. 1. The courtship reaction-chains of a goby (*Gobius microps*) and a stickleback (after Nyman). For explanation, see text.

the releaser for the next action of the opposite sex. In this way a series of responses and stimuli which, taken singly, are comparatively simple and might be mistaken, together form an intricate system which is the height of specificity and acts as a very efficient isolating mechanism preventing hybridization.

The case of the courtship of the ten-spined (Fig. 2) and three-spined sticklebacks may be given as an example. These two species often occur together in the wild and they have been hybridized artificially by Leiner (1940). The sexual stimuli provided by the females of both species are – in this instance – very similar (a head-up posture and a silvery swollen belly). Strongly motivated males will court ripe females of either species, but females will only respond to courting males of their own species. Not only do the males of the two species possess different breeding colours (three-spined: red; ten-spined: black), but the methods of dancing, leading to the nest, and showing the nest entrance are also different. If a female was strongly enough motivated to accept the wrong colour, she would then have to accept the

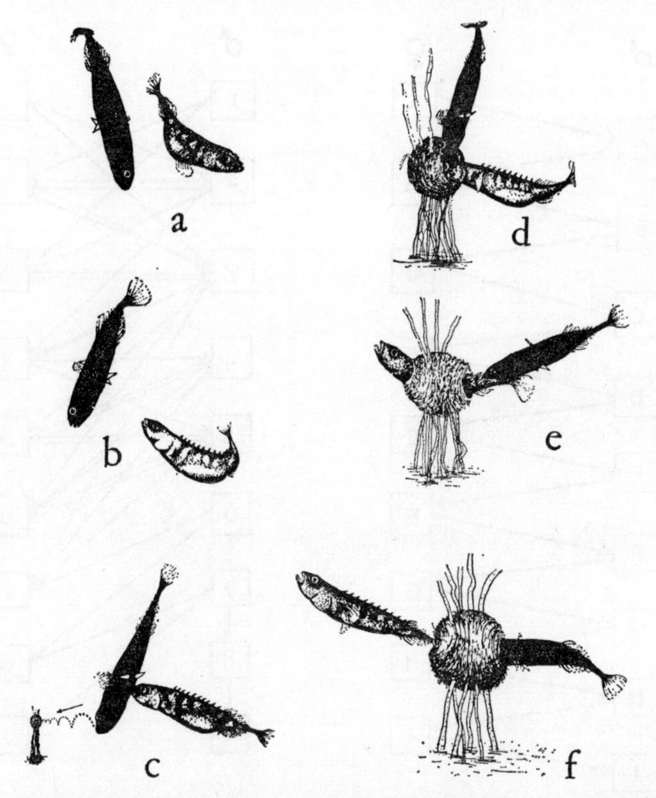

Fig. 2. The courtship of the ten-spined stickleback (*Pygosteus pungitius*). *a*, The male dances to the female. *b*, The female turns towards the male and the latter dances away from her. *c*, The male leads the female to the nest. *d*, The male shows the nest entrance to the female and the latter noses into it. *e*, The female enters the nest and the male shivers his snout on her tail. *f*, The female, having laid her eggs in the nest-tunnel, leaves by the exit and the male is about to pass through the nest and fertilize the eggs.

wrong dance, then the wrong type of leading, and then the wrong type of entrance-showing, before she could get into the wrong nest to lay her eggs. Each additional barrier makes it more unlikely that hybridization will occur.

A detailed study of a courtship reaction chain, which I have made in the case of the ten-spined stickleback, helps to illustrate the isolating function of such a chain even more clearly. The classical picture of a reaction chain, such as that shown in Figure 1 for the three-spined stickleback, is an idealized version of what actually happens. That picture gives the idea that each action releases only *one* particular reaction in the opposite sex. This is not the case. Fig. 3 shows two pictures of the courtship reaction chain of the ten-spined stickleback. On the left is the 'classical' version of this chain, and on the right is what really occurs. The right-hand picture is so confused that a new type of

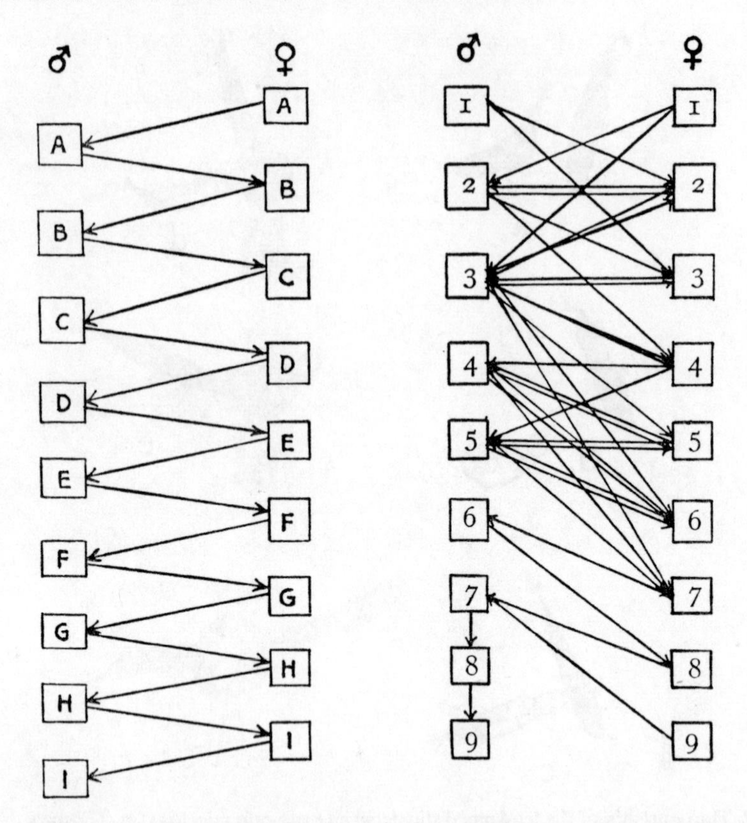

Fig. 3. On the left, is a hypothetical 'classical' version of the nine-stage courtship reaction chain of the ten-spined stickleback, in which each action releases only one reaction in the opposite sex. On the right, is the much more complex picture of what actually happens. It will be seen that a number of the actions in the courtship sequence can both release, or be released by, several actions of the opposite sex. This is more clearly demonstrated in Fig. 4.

diagram has to be employed (see Fig. 4). This reveals the difference more clearly. It shows that each reaction of the male can be released by several actions of the female and vice versa. One of these actions is always the typical one and from these typical releasing actions, the ideal type of diagram is constructed. It will be noticed that the atypical releasing actions are always grouped around the typical one, in the courtship sequence. In other words, there is an overlap at each stage of the courtship sequence. This is best shown by Fig. 5. Such an overlap means that the courtship releasers in the chain are not *absolutely* indispensable for their particular responses. But it will be noticed that an atypical releasing action is never more than two stages away from the typical one in question, and since the whole courtship sequence consists of nine stages, the overlap is never likely to have any detrimental effect. If the courtship sequence was somewhat shorter, with fewer stages, the chances of hybridization would be slightly increased.

Courtship sequences get out of phase when one partner is much more strongly motivated sexually than the other. If, in the stickleback case, a female was excessively ripe, she might be prepared to by-pass certain stages of the courtship, but the nine-stage ceremony will never permit her to provide

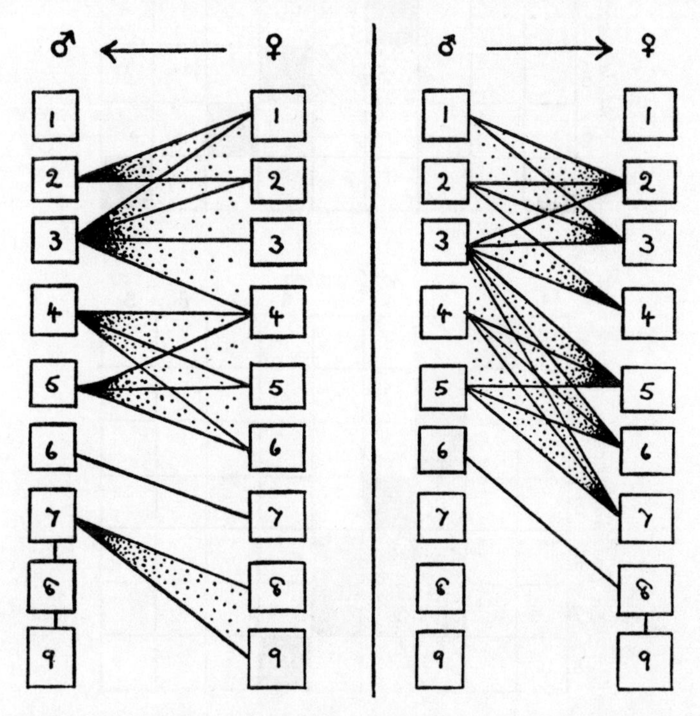

Fig. 4. Revised reaction-sequence diagrams. On the left, the actions of the female which are observed to release each of the reactions of the male are shown. On the right, the actions of the male which release each of the reactions of the female are shown. It will be noticed that most reactions can be released by several actions of the opposite sex and that the latter are grouped together in a courtship sequence.

a male of the wrong species with eggs. If the classical picture of a reaction-chain was really true, and each stage could only be released by the one preceding, then far fewer stages would be quite efficient.

This brings me to the third, and possibly the most important, function of courtship, namely the arousal of the sex drive of the mate, and the synchronization of the arousal of the male and female in time. A rough synchronization will already have been carried out by factors such as light and temperature, but split-second timing is often necessary for successful copulation.

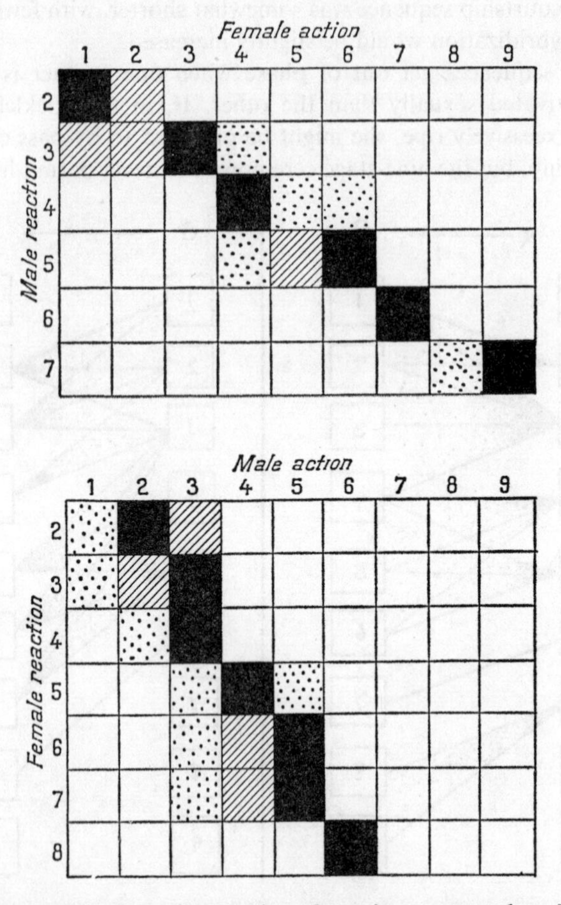

Fig. 5. An alternative presentation of reaction-sequence data. Here, where an action *typically* releases a reaction, a black square is shown. Where it *occasionally* does so, a shaded square is shown. Where it *rarely* does so, a stippled square is shown. From these diagrams, the stimulus-response 'overlap', which exists at most stages of the courtship, can be clearly seen.

As I mentioned above it often happens that the male and female are not at the same level of sexual motivation at the beginning of a courtship. Often a male has to keep on repeating one stage of his courtship before the female will respond and he can go on to the next stage. Sometimes the female is the more strongly aroused, and in the ten-spined stickleback, if the male is not ready to lead her to the nest, he proceeds to lead her round and round the water-weeds, all the time carefully avoiding the nest-site. Finally, when he is sufficiently aroused, he takes her to the nest.

I have attempted to analyse the arousal effect of the courtship of the ten-spined stickleback quantitatively. It is impossible to discuss this in detail here, but I will briefly mention some of the methods used. In the first place, there are two basic types of arousal: absolute or direct arousal on the one hand, and relative or indirect arousal on the other. The first is the more difficult to demonstrate, since it is necessary to show that there has been an increase in sexual responsiveness during the courtship, independent of a

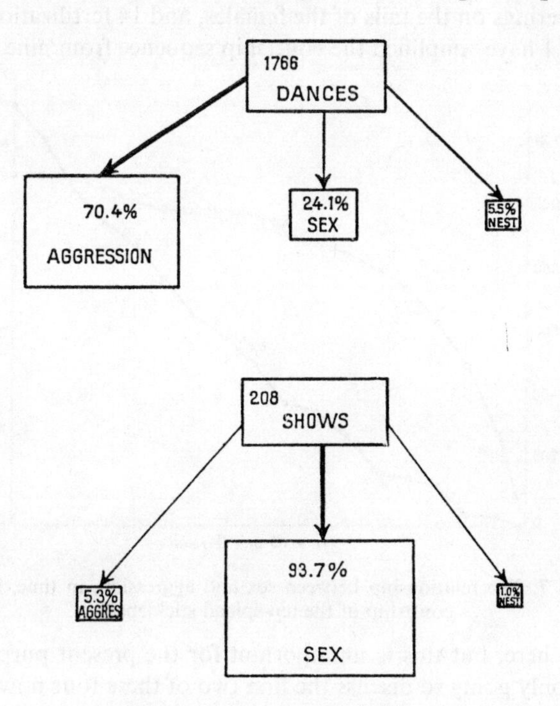

Fig. 6. For explanation see text.

change in any other tendencies. Indirect arousal is much easier to investigate and I have been able to demonstrate it in a number of ways.

It is becoming increasingly clear that courtship does not only involve a single, namely *sexual*, tendency, but also the tendencies to *attack* and to *flee* from the sexual partner. More will be said about this later, when the causal aspects of the problem are discussed, and it will suffice here to say simply that one of the principle functions of courtship is the suppression of the attack and fleeing tendencies in the sexual partners. In the case of the stickleback, the male repeatedly bites the female during courtship, and the female repeatedly flees from the male. It is possible to show in a number of ways that, as the courtship proceeds, so the tendency of the male to attack the female decreases and the tendency to behave sexually increases. Here are some of the methods which I have used:

1. A male in sexual condition and in possession of a nest is presented with a female whose belly is swollen with eggs, for a set observation period (usually thirty minutes). Using a certain section of my records I can say that, for a total observation time of 35 hours and 40 minutes, the males in question performed altogether 1,766 dances, 208 showing-the-nest-entrance ceremonies, 61 shiverings on the tails of the females, and 14 fertilizations. (It will be noticed that I have simplified the courtship sequence from nine down to four

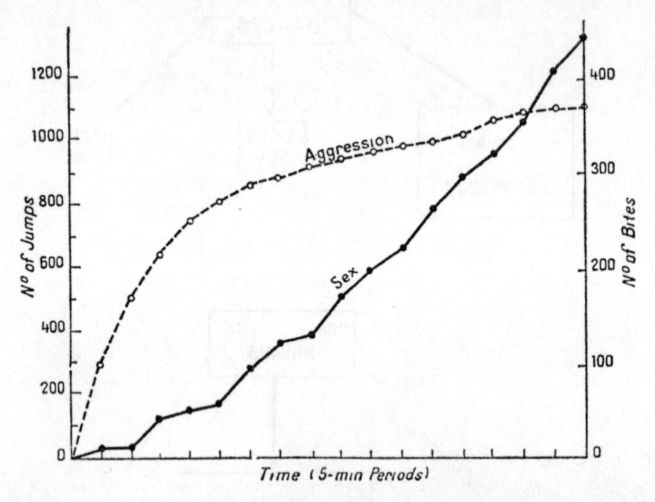

Fig. 7. The relationship between sex and aggression, in time, in the courtship of the ten-spined stickleback.

basic stages here, but this is unimportant for the present purpose. Furthermore I am only going to discuss the first two of these four now.) In Figure 6 I have shown the 1,766 different actions which the males performed immediately after the 1,766 dances, and also the 208 different actions which followed the 208 showing ceremonies. These are expressed as percentages and comprise the following: 70·4 per cent of the dances were followed by an aggressive action; 24·1 per cent were followed by some sexual action; and 5·5 by nesting actions. This compares with the showing ceremonies in the following way: 5·3 per cent only of the showings were followed by aggression; 93·7 per cent, however, were followed by some sexual action; and 1·0 per cent by nesting actions. (In the present analysis we can ignore the nesting actions.) It is quite clear from this that when a courtship has progressed beyond the dancing stage to the stage of showing-the-nest entrance, it has undergone a shift in the balance between sex and aggression in favour of the former. This shows itself in the form of a great decrease of the chance of a courtship breaking down into an attack, and a correspondingly great increase of it continuing with some further sexual activity.

2. Using the same basic method, of presenting a ripe female to a male, it is possible to show how sexual and aggressive actions by the male vary, in relation to one another, with time. As a particularly good example of this, I have selected a case in which a male was extremely aggressive to the female at the beginning of the courtship (see Fig. 7) because, in this instance, the shift, during courtship, from a predominantly aggressive mood to a predominantly sexual mood is demonstrated in an exaggerated form. The sexual tendency here is represented by response accumulation values for the number

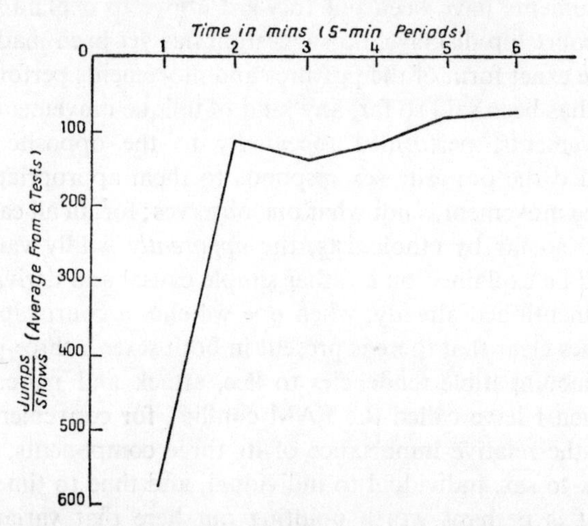

Fig. 8. Graph to show the increase in success, with time, of the male stickleback's courtship. For further explanation see text.

of dance-jumps performed by the male, and aggression by similar values for the number of times it bit the female.

3. A third method is as follows. If one uses the same kind of data as above, and divides the number of dance-jumps by the number of showing ceremonies for a given time unit, it is possible to show how much dancing the male had to do to get to the next stage of the courtship (Fig. 8). It is clear from this diagram that, as time went on, the courting male was able to progress to the second basic stage of the courtship more and more easily.

Before passing on to the question of the causation of courtship, I should perhaps summarize what I have said about the problem of function. It can be stated that there are indeed some good reasons for a courtship being prolonged and complicated; for complex courtships not only have releaser functions, but also arousal, synchronization and species isolation functions. These advantages must, in every species, be weighed against the disadvantages mentioned earlier. Where special display structures are involved, the compromise often takes the form of *transformation displays*, as they may be called, such as rapid colour changes, erection displays, or inflation displays, which ensure that a male, which is conspicuous while courting, may nevertheless be cryptic at other times. Another very general form of compromise is the restriction of bright courtship markings to those areas of the body which are not usually seen by the predators in question.

The Causation of Courtship Display

Although arguments have been put forward above to explain the existence of elaborate courtship displays, no suggestion has yet been made as to what determines the exact form of the postures and movements performed. On the basis of what has been said so far, any kind of unique movement, or series of different movements, performed repeatedly to the opposite sex, should suffice, provided the opposite sex responds to them appropriately. But any form of unique movement is not what one observes; for, in all cases that have been analysed so far by ethologists, the *apparently* wildly varied types of display can all be explained on a rather simple causal and derivational basis.

As I have mentioned already, when one watches a courtship in progress, it soon becomes clear that there is present in both sexes a three-point conflict between the incompatible tendencies to flee, attack and mate. This causal situation, which I have called the FAM conflict, for convenience, will vary in respect of the relative importance of its three components, from species to species, sex to sex, individual to individual, and time to time in the same individual. (It is perhaps worth pointing out here that variation between individuals of one sex of the same species seems to be the least marked of these.) Three basic courtship types can be categorized crudely as follows:

fAM TYPE (in which the animal tends to attack its mate during courtship).

FaM TYPE (in which the animal tends to flee from its mate during courtship).

faM TYPE (in which the animal neither flees from, nor attacks, its mate during courtship).

There are, of course, many gradations of courtship type, but nevertheless I feel that even the vastly over-simplified system given here is of some help. For example, in the past, various authors, depending almost entirely on which species they selected for study, interpreted courtship as fundamentally a conflict between, say, attack and sex, or fleeing and sex, or as something simply involving the repeated thwarting of, or perhaps the gradual building up of, the sexual tendency alone. I hope to be able to show that, with the basic concept of the FAM conflict, it is possible to fit all the different causal attitudes to courtship easily into one framework.

Because a conflict involving three variable tendencies is so complex to analyse, it is naturally a help to begin with a study of only part of it, and this is made possible by the fact that fighting behaviour involves only two of the three tendencies we are concerned with, namely fleeing and attacking.

Tinbergen (1952a, 1953b) has shown that, in fights, the enemy stimulates the two incompatible tendencies to flee and to attack, and that when these two tendencies are more or less in balance, one of two basic types of conflict reaction occurs. At low intensities, ambivalent movements or postures are seen. At higher intensities, displacement activities occur. Either of these may act as threat postures and intimidate the rival. I have attempted to represent Tinbergen's threat theory graphically in Figure 9. Moynihan (1953) has de-

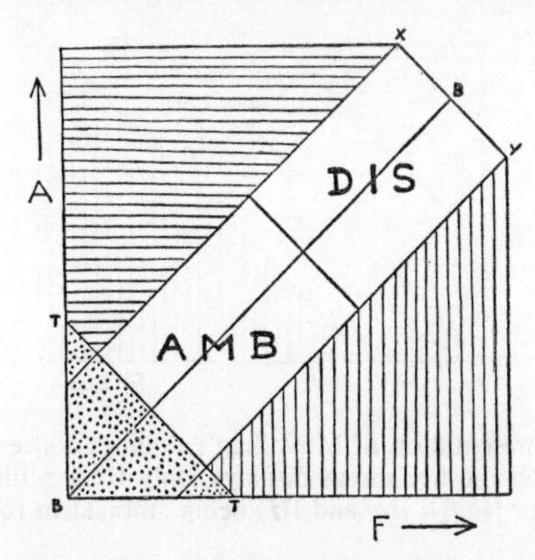

Fig. 9. A graphical representation of Tinbergen's threat theory. A and F are the attacking and fleeing tendencies; T-T is the threshold line below which no response is given; B-B is the line of precise balance between the two tendencies; X-Y is the width of the threat band, within which a conflict response is given; AMB represents the region of low intensity threat, where ambivalent responses are expected; DIS represents the region of high intensity, where displacement activities are expected; horizontal shading – actual attack; vertical shading – actual flight.

veloped this theory in his work on the black-headed gull, and in Figure 10 I have presented his results in a similar graphical form. It should be noted that, of the five displays of this gull, the so-called 'choking' display is the only one which is a displacement activity, and this occurs only at the highest intensities; also, the other four displays can all be thought of as ambivalent postures, thus confirming Tinbergen's theory.

An example of ambivalent movements is the back-and-forth pendulum-fighting of many fish species on their territory boundaries. Here, instead of part of the fish performing one action, whilst another part of it does something else at the same moment, as in ambivalent posturing, the whole fish performs first an intention movement of one action and then quickly performs the intention movement of another. In this way, it rapidly alternates, in the case of threat, between intention movements of attacking and fleeing.

A good example of ambivalent postures are the many types of 'lateral-undulation displays' which fish perform, and in which they beat backwards with their pectoral fins and at the same time beat forwards with their tails. Another example is, of course, the simple lateral display which is of such widespread occurrence in fish. Here, the fish takes up a posture which is a compromise between facing towards the enemy and facing away from it.

An example of a displacement activity (see Tinbergen 1940, 1952a;

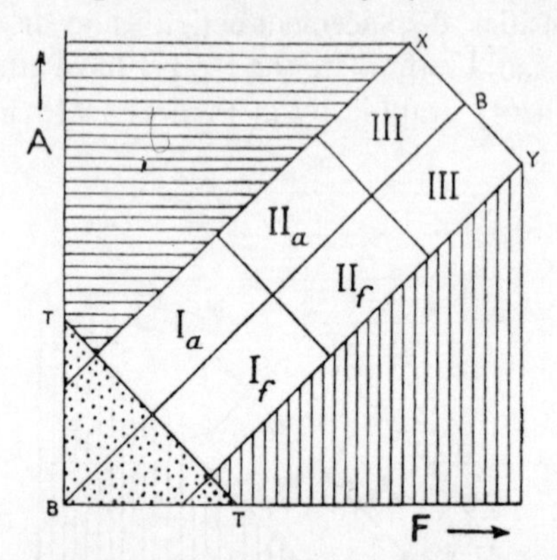

Fig. 10. A graphical representation of Moynihan's findings concerning the threat-code of the black-headed gull. Of the five threat displays, only III is a displacement activity, the other four (I*a*, I*f*, II*a*, and II*f*) being ambivalent responses.

Armstrong 1950; Bastock, Morris and Moynihan, 1953) (an activity performed 'out of context', as the result of the activation of two incompatible tendencies or the activation and subsequent thwarting of a single tendency), which occurs in threat behaviour, is the displacement sand-digging of the three-spined stickleback described by Tinbergen and van Iersel (1947). When the stickleback is strongly activated to attack and to escape simultaneously, it does neither, but instead performs the quite irrelevant action of sand-digging, which, in its mechanically effective form, is a nest-building activity.

In conflict situations during fighting, therefore, the observer is likely to encounter a number of basic kinds of action, namely, intention movements, ambivalent movements made up of alternating intention movements, ambivalent postures made up of simultaneous intention movements, and displacement activities. It is noticeable that in some instances these are little exaggerated, whereas elsewhere they may be amplified and elaborated, and are playing important roles as ritualized signals.

In instances where a beaten fish cannot escape from a dominant rival for some reason, and its fleeing pattern is thwarted, a submissive response may be given, which may be successful in having an appeasement function. Most submissive actions are the exact opposites of the special threat actions of the species concerned. Fishes which spread their fins to the maximum when threatening, flatten them hard when they are submissive. Fish which present themselves to their rivals broadside-on when threatening, roll over on to their sides when submissive. The ten-spined stickleback raises its ventral spines when threatening and its dorsal spines when submissive. Fish which lower their heads when threatening, lower their tails when submissive. Canadian rock-bass flatten their pelvic fins against their bodies when threatening, and these fins are then intensely black. When submissive, they raise their pelvic fins fully and these fins are then intensely white.

I have given a number of examples of submissive displays here, because there is some slight disagreement between Lorenz and Tinbergen concerning them. Lorenz (e.g. 1935 and 1952, Ch. 12) has suggested that submissive

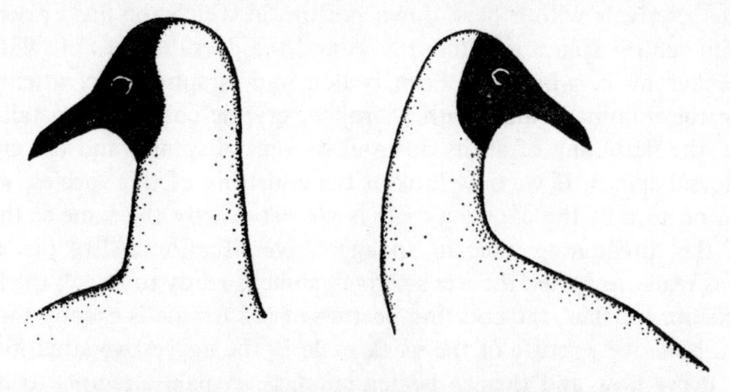

Fig. 11. Head-flagging in the black-headed gull. The birds turn their black face-masks away from one another.

postures consist of the demonstration of a vulnerable part of the body, as indeed they often do, particularly in birds and mammals, where the appeasement ceremony frequently consists of offering the neck to the rival. But Tinbergen and Moynihan (1952) have put forward the idea that offering the vulnerable spot is only, so to speak, an accidental outcome of the fact that the submission consists of turning away and concealing from the rival the threat or attack structures. When a wolf conceals its teeth from a rival, it automatically offers its neck. When a black-headed gull turns away its black threat-mask it automatically offers its neck, and so on. But, whilst I agree entirely with Tinbergen and Moynihan that the submissive gestures are the opposite of the threat postures, I cannot agree with them that as a general rule, 'In such cases conspicuous structures are concealed rather than displayed.' This may be true in the case of the black-headed gull which simply conceals its black face and red bill when making an appeasement gesture, and shows off only an unmarked neck (see Fig. 11). But in the various examples of submissive displays given by fish, which I mentioned above, only a few are simple concealments (e.g. the flattening of the fins). In the case of the rock-bass, for instance, there is a very positive kind of submission. Also, putting your tail down instead of your head can hardly be thought of as concealment. I suggest therefore that there is only one basic rule which applies to submission in general and that is that submissive gestures are fundamentally the *opposite* of some aspect of the threat repertoire of the particular species.

I have gone into the question of the behaviour associated with fighting in some detail because, as I said earlier, it is intricately involved in, and throws considerable light on, the nature of courtship displays.

Returning to the ten-spined stickleback again, it is noticeable that, in the fighting of this species, a dominant, aggressive male is black in colour and threatens its rivals with a head-down posture in which the fins are erected, the white ventral spines are fully raised and the dorsal spines are flattened.

A weaker male, which has been beaten and is submissive, attempts to appease the dominant male with a broken, cryptic coloration, a tail-down posture, the flattening of all its fins and its ventral spines, and the erection of its dorsal spines. If we now look at the courtship of this species, we find that the posture of the courting male is almost exactly the same as the posture of the threatening male in an aggressive situation. Also, just as the male was ready to attack the weaker rival, so he is ready to attack the female in courtship. Further, the courting posture of the female is exactly the same as the submissive posture of the weak male in the aggressive situation, and just as, every now and then, a beaten submissive male attempts to flee, so does a female attempt to flee, every now and then, from a courting male.

So far I have only explained the courtship *postures* of this species, and have said nothing about the courtship movements of dancing, showing and shivering, which do not occur in aggressive situations. But, before doing so, it is necessary to make one final point concerning the derivation of the courtship postures. It must be stressed that there are two derivational stages involved. The first is their primary derivation from displacement, ambivalent, or intention movements, in aggressive situations. Next there is their secondary derivation from fighting to courtship situations. For, in combination with different orientations and movements, these postures have become modified and changed both causally and functionally, so that they are now not only motivated sexually, but can also act as sexual stimuli. However, this secondary change is only partial, in a way which will be described later.

We must consider now the courtship movements of the ten-spined stickleback, and their causation. From what has been said already, it should not be difficult to understand these movements. The dances of both the ten- and three-spined sticklebacks consist of a number of jumps, as previously described. It would involve too much detail to analyse the exact nature of these jumps carefully here. Suffice it to say that basically the dances consist of a series of rhythmically alternating jumps, which have been interpreted by Tinbergen as ambivalent behaviour, involving first an aggressive intention movement towards the female and then a sexual intention movement towards the nest, this being repeated over and over again, often more than a thousand times in half an hour.

When the female follows the male to the nest, the latter begins to fan at the nest entrance (fanning is the parental action which ventilates the eggs) until the female enters. This is, of course, the 'showing' ceremony mentioned earlier, and can be thought of as displacement fanning resulting from the male's sexual tendency being aroused by the female's presence at the nest, and then thwarted by her refusal to enter. Once she has entered the nest, the male ceases to fan and begins to shiver his nose on the protruding tail of the female. This shivering also appears to be a modified form of displacement fanning, adapted to produce a tactile effect. A similar causal explanation applies here as before. The female has entered the nest and intensely stimulated the male sexually, but until she lays her eggs and leaves the nest, the male is sexually thwarted, and the result is a displacement activity.

But to return to the FAM concept, it is apparent from what I have said that the courtship of the male stickleback is predominantly a conflict between aggression and sex. The male repeatedly attacks the female but never flees from her. Why then should I insist on calling it an fAM type of courtship and not simply an AM courtship?

The reason for introducing the concept of a three-point courtship conflict is something more than just an attempt to lump together observations of quite different types of courtship under one rather artificial heading. The explanation is that although the male never flees from the female, the tendency to flee must play a part because otherwise the male would not adopt, during courtship, a posture derived from a *threat* posture. Threat postures have already been shown to be the result of a conflict between attacking and fleeing tendencies. A male stickleback that is actually attacking a rival, and is not hesitating about it, keeps in a horizontal posture and does not raise its spines. Only when a male is hesitant and shows some reluctance to attack, does it assume the head-down posture.

Many examples of fAM-type courtships can be found in the literature. The male bullhead, *Cottus gobio* (see pp. 89–117), for example, is extremely aggressive to the female, and the courtship often begins by the male swallowing the latter's head in one huge bite (Fig. 12). Even in this very aggressive courtship, however, the male does not only attack the female, but also threatens her, thus revealing that his fleeing drive is also activated, if only slightly.

Thus far I have concentrated on the first of the three types of FAM courtship, namely the fAM type. The best examples of the FaM type are furnished by certain passerine birds. Tinbergen (1939), in his analysis of the behaviour of the snow bunting, has described how the male repeatedly runs

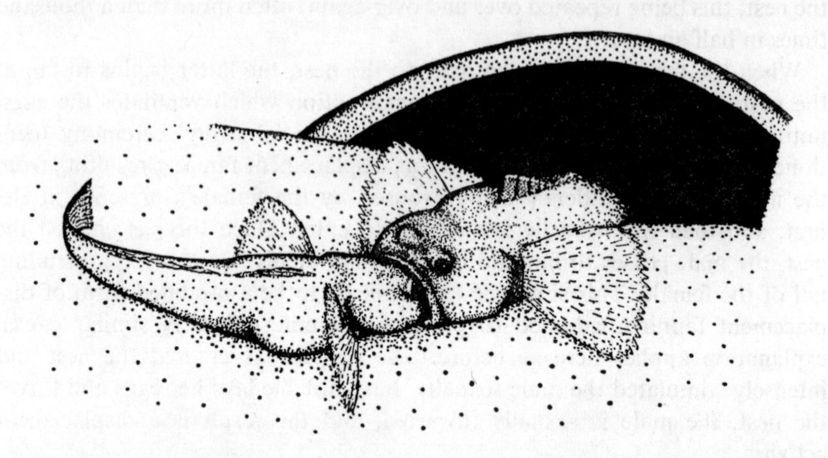

Fig. 12. fAM-type courtship. The courtship bite of the male bullhead (*Cottus gobio*). The female, if ripe with eggs, responds sexually to the assault of the male, and enters his nest, or allows herself to be carried in by him.

away from the female, displaying as he does so (Fig. 13). Many people must have noticed a cock blackbird displaying to a female in a similar way – running away from her with his feathers raised, wings drooped, and tail lowered and fanned. Hinde (1953) has made a detailed analysis of the courtship of the male chaffinch, which is of the FaM type. In this species, there is a

Fig. 13. FaM-type courtship. The display of the male snow bunting, which repeatedly flees from the female. (After Tinbergen.)

complete change of dominance between the male and female, the former becoming subordinate at the time when copulations occur. In an analysis of the courtship of the zebra finch (pp. 53–6), I have been able to show how the pivoting dance of the male can be thought of as the result of alternating fleeing and mating intention movements. Each time the male pivots, it turns away as if to flee from the female, then swings back again towards her and advances a little, then swings away from her again in the opposite lateral direction (see Fig. 14).

A third courtship type, namely the faM type, must now be mentioned. The best example of this is the viviparous Mexican swordtail (Morris, 1955). The males of this species do not hold territories, and their social organization

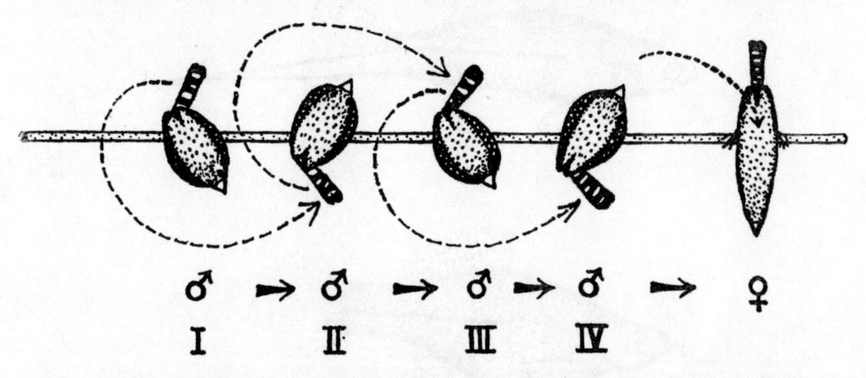

Fig. 14. FaM-type courtship. The pivot dance of the male zebra finch to the female, seen from above. As the male approaches the female, it repeatedly swings away, as if to flee from her.

consists of a straight-line hierarchy. Fighting to establish a position in this hierarchy is intense and long threat-battles take place. But despite this the courtship of the male swordtail is almost entirely free from any tendency to attack the female or to flee from her. Admittedly, the species is markedly dimorphic (Fig. 15), but then so are the sticklebacks. It seems most likely that it is the lack of territorial behaviour that is the cause of this (rather rare) type of courtship. Tinbergen (1952b) has suggested that the reason why there is so much aggression in courtship is because the males have to fight one another, and that the females, even when dimorphic, cannot help stimulating the males' aggression slightly. I would like to modify this a little and say that *territorial* males cannot help treating *intruding* females as objects to be attacked as well as courted, because such males respond aggressively primarily to intrusion on to their territories and only secondarily to the nature of the intruding object. This would then explain both the stickleback and the swordtail cases.

If the male swordtail never attacks the female, or flees from her during courtship, then there are two questions that must be answered: firstly, why must the courtship be thought of as an faM type instead of simply an M type; and, secondly, what is the causal significance of the elaborate male courtship dance?

Before answering these questions, a brief description of the courtship of this species must be given. The courtship behaviour of the female is simple to describe: if she is unresponsive, she flees from the male and if she is responsive she stays still. But even here there is a problem, for why should the female flee from the approaching male, if the latter never attacks her? Hediger

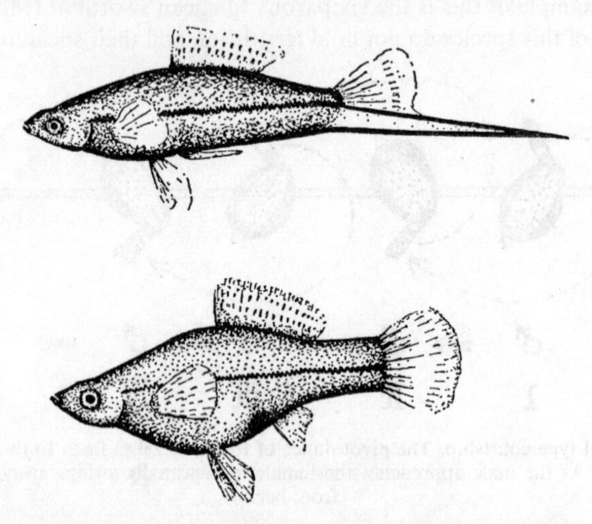

Fig. 15. Male (above) and female (below) Mexican swordtail (*Xiphophorus helleri*), showing the marked sexual dimorphism.

(1950) has discussed the phenomenon of 'individual distance' in animals in some detail, and has shown that in many species, bodily contact is strongly avoided. As far as I know, that applies to almost all species of fish, and certainly does so in the case of the swordtail. Therefore it is natural that an unresponsive female will flee when approached too closely by a male. It might be expected therefore that the courtship of the male would consist of attempts to prevent the flight of the female.

The male's courtship display begins by the male orienting towards a hovering female, some distance from her. The male then lowers its head-end and, in this dipped position, backs slightly away from the female. This slight retreat is performed slowly and then is followed by a sudden swoop straight at the female. The male does not halt alongside the female, but a little in front of her, and then immediately begins to swim slowly backwards, this time in a horizontal posture, shivering his long caudal appendage past the eye of the female. If, as usually happens, the female turns away from the male at this point, the latter, before she can manage to flee, darts around her in a very rapid, curved, forward path, which brings him back to his original position, that is, slightly in front of the female. He then begins the slow backward shiver again. If the female twists this way and that, in a frantic attempt to escape, the male darts round and round her, always blocking her way, until she pauses long enough for him to perform the backward shiver again. If the female is stimulated sufficiently by this display to respond sexually, the male draws close alongside her from one of his backwards shiverings and then, head to head and tail to tail, they both rotate longitudinally away from one another dorsally. This brings their ventral surfaces near enough together for the male's gonopodium (a modified anal fin) to make contact with the vent of the female and for the consummatory act to occur. This courtship sequence is illustrated diagrammatically in Figure 16. Without adding further to the above description, it is clear that the male's behaviour, as expected, consists of attempts to prevent the flight of the female. This is done by the physical obstruction of her flight paths.

It is impossible here to go into the elaborate fighting system of this species but in order to justify calling this courtship an faM type, rather than simply an M type, I must mention two points. Firstly, a moderately subordinate male often watches a dominant male from a distance and, in so doing, assumes a posture very similar to the initial courting posture. There are some minor fin differences, but they need not concern us here. It should be mentioned that this is not a case of mistaken identity, for such a male can not only distinguish between the sexes, but can also distinguish individual males with necessary ease. Without going into details, it can be said that this display indicates the weak activation of both the fleeing and the attacking tendency, but with a preponderance of the former. Secondly, males do shiver their caudal appendages – the so-called swords – when threatening one another, but they swim slowly forwards or hover when doing so and never swim backwards or make curved forward darts. Also, none of the six or more special threat displays employed in bouts of mutual intimidation occur in the courtship dance. So that, although there are traces of fighting behaviour in the courtship of this species, they do not in any way contribute to the form of the main body of the display. Nevertheless the, admittedly weak, tendencies to attack and to flee from the female do seem to be present, and I have therefore called this the faM type of courtship.

In conclusion, it is interesting to compare Lorenz's (1935) three courtship types with those put forward here using the FAM system. Lorenz proposed three types, called the 'Lizard' type, the 'Labyrinth' type and the 'Cichlid' type. Very briefly, the fundamental differences between these three were that, in the first, the female does not display at all in response to the male; in the

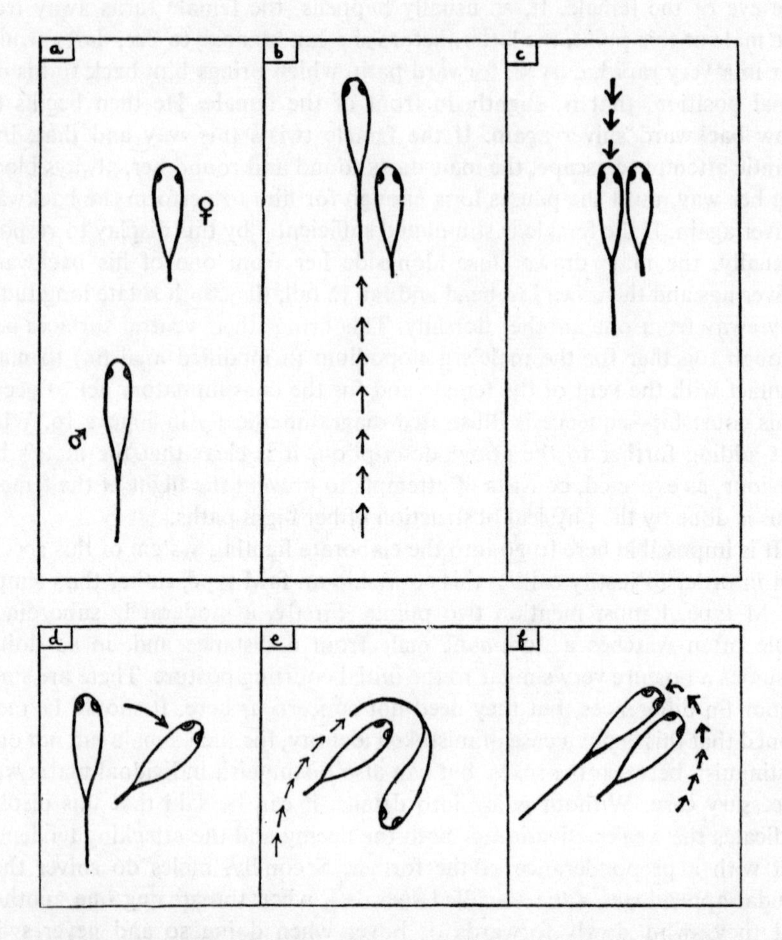

Fig. 16. faM-type courtship. The dance of the swordtail, showing the six basic steps: *a*, The male (longtail) hovers some distance from the female. *b*, The male swoops on the female and halts just in front of her. *c*, The male backs slowly alongside the female. *d*, The female turns away. *e*, The male makes a curved dart around the front of the female. *f*, The male backs slowly into his old position alongside the female.

second, the male is dominant and the female subordinate; and in the third both partners display aggressively. The important difference between Lorenz's system and the FAM system, which must be emphasized, is that Lorenz was primarily concerned with the *relationships* between the partners, whereas the FAM system which I have put forward is concerned with the conflict state that exists in *each* courting individual. (It is clear, however, that the two systems, although different, are quite compatible.)

The Problem of Pair Formation

I have already stated that by courtship I mean all heterosexual communication leading to fertilization. In some species this behaviour consists of one continuous complex – for example, the stickleback courtship. In other species, however, there are two quite distinct behaviour complexes – pair formation and precopulatory display – which are separated in time. I want to suggest that the stickleback type of courtship is the more original of the two and that the other type, which is so widespread in the higher vertebrates, has evolved from such a courtship type in the following way.

Wherever there is an advantage in the male and female staying together after fertilization has taken place, selection will favour any mechanism which facilitates a continuation of their relationship after their sexual tendencies have waned. As I have shown, sexual arousal to the peak which permits fertilization involves the suppression of attacking and fleeing tendencies. When the consummatory sexual act has been performed, nothing remains, in the case of the stickleback, to prevent the quick return of attacking and fleeing tendencies, and the male always attacks and drives away the female immediately after spawning. It is therefore a great advantage to many species to have a special learning process built into the courtship system, which renders the suppression of attacking and fleeing tendencies *irreversible in respect of one individual of the opposite sex.* Once this process is built into the system, the suppression of attacking and fleeing tendencies will be semi-permanent for the particular male and female concerned. There is then no longer any need to rush right through from the initial encounter to the consummatory act in one courtship complex. It is possible to perform the initial (pair-forming) courtship displays and then, having formed a pair-bond, to wait for some time before attempting to achieve fertilization. The time-lag will be greater or less according to the ecological circumstances of the species concerned. I suggest therefore that the basis of the pair-bond is the removal of the attacking and fleeing tendencies, which permits the now unsuppressed sexual tendency to keep the male and female together. (There certainly seems to be more to pairing than this in some species, but I am only putting this suggestion forward as the basis of the *origin* of pair formation.)

It should be mentioned here that Baerends and Baerends (1950), in their study of Cichlid fishes, have shown that in species where both partners take part in parental duties, there is a greater distinction between pair formation and pre-copulatory ceremonies, than in those species in which only one sex performs parental duties.

A criticism of the FAM concept is necessary at this point. I shall take the zebra finch as an example. This species has an FaM courtship which I have already described. Or so I have said. However, what I really meant was that the zebra finch has a *pre-copulatory* ceremony which is of the FaM type. But long before this occurs, the male and female of a pair have passed through a pairing stage in which both are aggressive to one another. The aggression having been removed, the pre-copulatory behaviour is left as a conflict basically between fleeing and mating as already described. (It is interesting to note that where the pair formation has removed aggression between the pair, it could not remove the tendency to flee completely, but here again I would remind you of Hediger's concept of individual distance.)

So it is really necessary to modify the FAM concept and to refer to a particular *display* as being this or that FAM-type of display, rather than to refer to a whole *courtship* in this way. In summing up, it should be pointed out that although the above criticism of the FAM concept is fundamentally correct, it need not be taken too far, because even if one compares the *whole* of one species' courtship with the whole of another's, it is still possible to distinguish general FAM differences between them.

The Ritualization of Courtship Display

Finally, some mention must be made of the ritualization of courtship movements and postures.

Many of the ambivalent, intention and displacement activities that arise in courtship situations take on a signal function during evolution. The perfection of their roles as signals nearly always involves some sort of modification, or ritualization, of the original movement or posture. When such a modification takes place it is often so drastic as to make it impossible to decide what was the original movement from which the signal has been derived. Lorenz (1950) has pointed out that, in such cases, only a comparative study of many closely related species can sometimes give the answer. It is not always so difficult, however, since many of the signals are only *semi-ritualized.* By this I mean that although they have become modified in their service as signals and have acquired a new motivational basis, they nevertheless still can be influenced to some degree by the tendencies that originally controlled their expression. An example will make this clearer. The courtship dance of the three-spined stickleback has clearly become ritualized and serves as a sexual signal to the female. Also it is now sexually motivated, although it originated as an ambivalent movement of alternating aggressive and sexual intention movements. Although, at the present state of evolution, it is sexually motivated and has a fixed rhythmic form, typical of ritualized activities, it can nevertheless be deformed slightly. If a courting male is unduly aggressive, Tinbergen's analysis has shown that the component of the dance leading the male *towards* the female is more pronounced than usual. If the male is unduly non-aggressive, then the component leading the male *away* from the female towards the nest is more pronounced than usual. But in both cases, the basic rhythmic form of the zigzag dance generally remains. This state of affairs is the case in a number of courtship dances. In the zebra finch, for example, if the male is unduly frightened of the female (as can be shown by independent factors), the pivoting swings of his approach dance turn him farther away from the female than usual. In other words, the component of the dance which has originated from the intention movement of fleeing is exaggerated, but this in no way affects the basic rhythmic character of the, therefore semi-ritualized, dance.

It would involve too much detail to discuss here the many different kinds of response modification associated with ritualization, and I can only refer you to the fascinating studies of this problem by Lorenz (e.g. 1941) and Daanje (1950).

References

ARMSTRONG, E. A. (1950), 'The nature and function of displacement activities'. *Sympos. Soc. Exp. Biol.* 4, pp. 361–87.

BAERENDS, G. P. and J. M. BAERENDS-VAN-ROON (1950), 'An introduction to the study of the ethology of Cichlid fishes'. *Behaviour*, Supplement 1.

BASTOCK, M., D. MORRIS and M. MOYNIHAN (1953), 'Some comments on conflict and thwarting in animals'. *Behaviour* 6, pp. 66–84.

DAANJE, A. (1950), 'On locomotory movements in birds and the intention movments derived from them'. *Behaviour* 3, pp. 48–99.

HEDIGER, H. (1950), *Wild Animals in Captivity* (Butterworths, London).

HINDE, R. A. (1953), 'The conflict between drives in the courtship and copulation of the Chaffinch'. *Behaviour* 5, pp. 1–31.

LEINER, M. (1940), 'Kurze Mitteilung über den Brutpflegeinstinkt von Stichlingbastarden'. *Zeitsch. f. Tierpsychol* 4, pp. 167–9.

LORENZ, K. (1935), 'Der Kumpan in der Umwelt des Vogels'. *Journ. f. Ornithol.* 83, pp. 137–212 and 289–413.

—— (1937), 'The companion in the birds' world'. *Auk* 54, pp. 245–73.

—— (1941), 'Vergleichende Bewegungsstudien an Anatinen'. *Journ. f. Ornithol.* 89, pp. 194–294.

—— (1950), 'The comparative method in studying innate behaviour patterns'. *Sympos. Soc. Exp. Biol.* 4, pp. 221–69.

—— (1952), *King Solomon's Ring* (Methuen & Co., London).

MORRIS, D. (1952), 'Homosexuality in the Ten-spined Stickleback (*Pygosteus pungitius* L.)'. *Behaviour* 4, pp. 233–61.

—— (1954a), 'The reproductive behaviour of the Zebra Finch (*Poephila guttata*) with special reference to pseudofemale behaviour and displacement activities'. *Behaviour* 6, pp. 271–322.

—— (1954b), 'The reproductive behaviour of the River Bullhead (*Cottus gobio* L.) with special reference to the fanning activity'. *Behaviour* 7, pp. 1–31.

MORRIS, D. (1955), 'The courtship dance of the Mexican Swordtail'. *The Aquarist* 19, pp. 247–9.
MOYNIHAN, M. (1953), 'Some aspects of reproductive behaviour in the Black-headed Gull (*Larus ridibundus*) and related species'. *Doctor Thesis, Oxford.*
NYMAN, K. (1953), 'Observations on the behaviour of *Gobius microps*'. *Acta Soc. Fawn. Flor. Fenn.* 69, pp. 1–11.
TINBERGEN, N. (1939), 'The behaviour of the Snow Bunting in spring'. *Trans. Linn. Soc. N.Y.* 5, pp. 1–92.
—— (1940), 'Die Übersprungbewegung'. *Zeitsch. f. Tierpsychol.* 4, pp. 1–40.
—— (1942), 'An objectivistic study of the innate behaviour of animals'. *Biblioth. biotheor.* 1, pp. 39–98.
—— (1948), 'Social releasers and the experimental method required for their study'. *Wilson Bull.* 60, pp. 6–51.
—— (1952a), 'Derived activities; their causation, biological significance, origin, and emancipation during evolution'. *Quart. Rev. Biol.* 27, pp. 1–32.
—— (1952b), 'A note on the origin and evolution of threat display'. *Ibis* 94, pp. 160–2.
—— (1953a), *Social Behaviour in Animals* (Methuen & Co., London).
—— (1953b), 'Fighting and threat in animals'. *New Biology* 14, pp. 9–24.
—— and J. VAN IERSEL (1947), 'Displacement reactions in the Three-spined Stickleback'. *Behaviour* 1, pp. 56–63.
—— and D. J. KUENEN (1939), 'Über die auslösenden und die richtunggebenden Reizsituationen der Sperrbewegung von jungen Drosseln'. *Zeitsch. f. Tierpsychol.* 3, pp. 37–60.
—— and M. MOYNIHAN (1952), 'Head-flagging in the Black-headed Gull; its function and origin'. *Brit. Birds* 45, pp. 19–22.
—— and A. C. PERDECK (1950), 'On the stimulus situation releasing the begging response in the newly hatched Herring Gull chick'. *Behaviour* 3, pp. 1–38.

The feather postures of birds and the problem of the origin of social signals (1956)

Introduction

From a functional point of view there are basically two kinds of feathers: those with which a bird flies and those which keep it warm. In the course of evolution either type may acquire additional functions in connection with concealment or display and, in the latter case, the new function may be such that it even obliterates the older one.

The origin of the secondary display-function of feathers is an important problem for the ethologist in interpreting the social behaviour of birds, and much progress has been made in certain directions. The work of Daanje (1950), in particular, has clarified the situation in connection with locomotory intention movements. This author has shown how the actions involved in incipient locomotion have become modified in different ways, in various bird species, as social signals. The modifications which accompany the development of signal functions involve a number of kinds of change such as component exaggeration, the shifting of component thresholds and component re-co-ordination. In addition, bright markings often occur on the most conspicuously displayed regions of the bird.

In the case of the locomotory intention movements, the feathers especially involved are only those of the wings and tail. The body feathers, whose primary function is temperature control, are not concerned. Nevertheless, these body feathers are often seen to be strangely depressed or erected as components of display patterns.

However, despite the fact that body-feather posturing, like wing and tail posturing, is quite accepted as a common and widespread bird display component, and despite the existence of the derivational analyses of the latter type of posturing, the question of the origin and evolution of the former appears to have attracted little or no attention. The present paper is an attempt to correct this omission and to focus attention on to this special category of bird signal.

The Primary Function of Body Feathers

First, it is essential to discuss in some detail the primary function of body feathers, namely that of insulation against heat loss and the control of the degree of this insulation.

The general problem of temperature regulation in birds has recently been reviewed by Sturkie (1954, Ch. 8). He points out that heat loss can be brought about by increased respiration and panting, and heat gain by shivering. In addition, he states that 'Heat loss is also regulated by the pilomotor and vasomotor nervous mechanisms. Erection of hair or feathers (pilomotor system) tends to conserve heat; in the non-erected state, heat loss is facilitated.' Concerning the vasomotor system, he points out that 'When the air temperature is high, the blood vessels in the skin dilate, thus increasing heat loss, and when the temperature is low, the vessels constrict, which tends to conserve heat.'

Wetmore, however, in an earlier paper (1921) on the subject of bird temperatures, after discussing the excellent insulatory properties of the plumage, proceeds to consider it as a *non-regulatory* heat-conserving system. He regards it as a system which retains heat for the bird, but which does not control how much heat it retains at any given moment, and goes on to say that 'This lack of heat regulation by means of the skin (including feathers) would throw the vital work of temperature control directly upon the respiratory system.' He then devotes considerable space to the question of the heat-reducing properties of the pulmonary air-sacs which, he postulates, are of fundamental importance in the temperature regulation of birds, and which compensate for the absence of sweat glands. He concludes by adding that, during extremely cold weather, special adjustment of the pulmonary system may result in the air-sacs being used as heat reservoirs, thus enabling certain species to withstand the bitter cold of winter weather.

Although Wetmore failed to recognize the regulatory significance of the pilomotor system, which can bring short-term influences to bear on the rate of heat loss, he nevertheless noted the existence of the long-term effect of plumage change: e.g. 'Birds that remain in regions where they are exposed to cold become more heavily feathered before the winter season, so that there is less radiation of heat externally. Correspondingly, in summer the feathered covering is thinner, and the feathers themselves often become worn so they are less burdensome.'

However, Wetmore, is not alone in ignoring the short-term possibilities of the pilomotor system. Certain other authors have also investigated differences in the insulatory efficiency of the plumage without referring to changes in feather postures. Hutt and Ball (1938), for example, using Wetmore's (1936) data on the body weights and numbers of feathers per individual in 97 species of Passerine birds, were able to show that although smaller birds have fewer feathers than larger ones (e.g. grebe, fifteen thousand; humming bird, fifteen hundred), they nevertheless have more relative to their body weight.

These authors conclude that 'While the increased metabolism of smaller birds is instrumental in their maintenance of high temperatures, the rapid increase in the number of feathers per unit of body weight with decreasing size of bird is an adaptation for retention of the heat produced.'

Neither Wetmore nor Hutt and Ball appear to have appreciated the importance of changes in body-feather postures in connection with rapid adjustments to fluctuations in the organism-environment temperature ratio. Even Sturkie's comment concerning the increase in insulation obtained from feather erection is only partially correct. The work of Moore (1945) gives a more complete picture. This author discusses the physics of heat loss from a bird's body in a cold environment. Concerning the loss of heat by conduction, he states that 'The coat of a bird, consisting of air entrapped by overlapping feathers, is an excellent insulator; that is, a poor conductor. Nevertheless it does conduct heat. When the feathers are 'normal' (neither pressed down, nor fluffed out) the air is almost perfectly entrapped. The heat is then conducted from warm skin to colder outer plumage both by means of the entrapped air and the feather material itself. Fluffing of the feathers is the very efficient means whereby the bird can secure a large increase in thickness of insulating coat. With the feathers erected, the coat becomes several times as thick as in the normal state.' Thus far Moore and Sturkie agree, but the important point which the latter missed is contained in Moore's next statement, to the effect that 'When the feathers are completely fluffed, they cannot do a perfect job of entrapping air among them, and movement of air through the feathers would reduce the insulating effect.'

The significance of this last point, in which Moore distinguishes between 'fluffed' and 'completely fluffed', will become clear below. It is surprising that, in his paper on the subject, Moore gives no supporting evidence for the above statements and indeed one gains the impression that his discussion is mainly hypothetical. If this is in fact the case, then his predictions are most accurate. However, before giving data to support them it is necessary to establish a simple but rigid terminology for feather posturings.

A survey of the literature reveals that many names have been used in the past to describe the state of the plumage; for example: depressed, flattened, sleeked, erected, raised, fluffed, ruffled, fluffled, etc. Since one name has been used for more than one feather posture and, conversely, more than one name has been given to one feather posture, in different cases, it is desirable to attempt to standardize the situation. To do this it is necessary to consider the way in which the feathers move.

Langley (1904), in his analysis of the physiology of feather movements, showed that there are distinct muscles for erecting and for depressing the plumage, the penna-erector and penna-depressor muscles, respectively, and that 'Both erector and depressor muscles are supplied with sympathetic nerve fibres and can be caused to contract separately.' There are, therefore, two distinct processes at work in feather posturing, and at first sight it might seem sufficient on this basis to consider only three basic feather postures, namely depressed, normal and erected. However, although erection and depression are terms which describe adequately the feather *movements*, the feather *postures* resulting from these movements require more specific terms. This is because a gradual increase in the *degree of erection* results in two quite distinct functional stages of 'erectedness'. As the feathers are raised from the normal, relaxed position, so the entrapped-air spaces between them increase in size, and greater insulation is achieved. This is made possible by the slight curvature of the feathers (see diagram in Fig. 1). But as they are raised even

Feather posture:	Sleeked	Relaxed	Fluffed	Ruffled
State of bird:	Active	Inactive	Cold	Hot
Feather movement:	Depression	—	Partial erection	Full erection
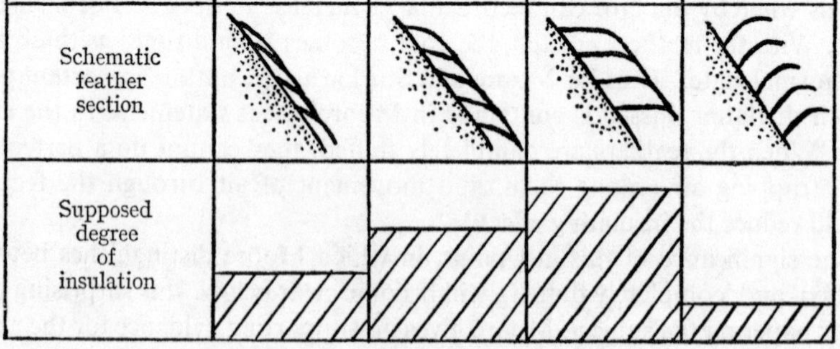 Schematic feather section				
Supposed degree of insulation				

Fig. 1. The four basic feather postures.

further, there comes a point when they cease to touch one another and they then stand out singly from the bird's body, giving it a ragged appearance. Immediately the feather-to-feather contact is broken, the entrapped-air spaces are lost and insulation with them.

Feather erection can, therefore, either help to keep a bird warm, or cool, according to its extent.

Feather depression, on the other hand, is simpler and can only help to keep a bird cool. The flattening of the feathers against the bird's body decreases the thickness of the insulating layer but, as it does not eliminate it completely, the cooling effect will be less drastic than that of complete erection. The primary function of feather depression, however, is that of streamlining. With its feathers depressed, the surface area of the bird is considerably reduced and it offers less resistance to the air when in flight. A bird with its feathers in the normal relaxed position is one which is inactive, or only moderately active; the same applies to birds with erect plumage, although the demands of temperature control may be so great that an extremely cold or hot bird may be observed to fly with its feathers erect.

There are, therefore, four basic feather postures that must be distinguished and for which I propose the following names. (It has been unnecessary to introduce new names, but only to standardize existing ones.)

(*a*) *Sleeked.* The feathers are fully depressed against the body, giving it a slim appearance. Characteristic of very active birds, or birds holding themselves in readiness for immediate action. The extra heat resulting from the high metabolic rate of the active bird is compensated for, to some extent, by the reduction of insulation produced by sleeking.

(*b*) *Relaxed.* The feathers are neither depressed nor erected. The body shape is intermediate between slim and rounded. This is the typical relaxed state of a bird, and is observed in individuals which are only moderately active or are inactive, and which are neither very hot nor very cold.

(*c*) *Fluffed.* The feathers are erected, but only partially, giving the body a very rounded appearance, with a smooth, unbroken outline. Characteristic of birds which are cold. Typically, fluffed birds are inactive; the increased heat production resulting from vigorous activity is usually sufficient to permit sleeking in flight, but in extremes of cold birds may be observed to be fluffed actually during flight.

(*d*) *Ruffled.* The feathers are fully erected and the body, although rounded, has a ragged appearance, with a broken outline. Characteristic of birds which are very hot and which are inactive.

Fig.1 summarizes the foregoing diagrammatically.

The correlation between these four feather postures and the four sets of conditions can be confirmed not only by simple observations at different times of the year, but also experimentally by controlling the temperature and the degree of activity artificially. This was done accidentally by the author in early attempts to photograph captive birds. The powerful electric lamps employed to illuminate the birds in question produced such a high temperature that the birds began to ruffle and even sunbathe. If they were particularly active in the photographic cage, the birds only responded to the heat by very intense sleeking, but as soon as they settled down and became inactive, the ruffling began and persisted. Naturally, when photographing courtship behaviour, these special temperature responses interfered with results, and in order to avoid this, the hot illumination was abandoned and all photographs (including the ones illustrating this paper) were taken with electronic flash apparatus.

In addition to the above observations, experiments have been carried out which show the loss in temperature control which occurs when the feathers are clipped short (Kendeigh and Baldwin, 1928), and, also, Davson (1951, p. 144) quotes investigations by Baldwin and Kendeigh, and Giaja, which reveal that the experimental restriction of feather movements interferes with the temperature regulatory system of birds.

Before passing on to consider the question of the secondary signal-function of the feather postures, it is necessary to mention briefly under what other non-signal conditions feather movements may occur.

A special secondary non-signal function of feather posturing is, according to Madsen (1941), that of the waterproofing of aquatic birds. This author believes that 'the secretion of the oil gland can hardly be the principle cause of the feathers repelling water' and considers that the main factor involved is the air entrapped in the plumage. He states that 'Notably swimming birds have a vigorous musculature under the skin, particularly on the neck, belly and sides. This skin musculature has the very important task of keeping the feathers at such a distance from one another that water cannot penetrate.' From this it appears that fluffing the submerged plumage is essential to prevent waterlogging, and it is interesting to note that in such a situation the feather posture has the double function of insulating against the cold of the water as well as the water itself.

A general fluffing of the feathers is observed in sleeping birds and birds which are ill. Both are conditions in which the increased insulation thus achieved is obviously an advantage.

Several kinds of feather movement can be seen during cleaning activities. The feathers are fully ruffled during bathing. Just as ruffling, when very hot, opens up the air spaces between the feathers and lets the outside air in, so does ruffling, when bathing, let the water in. In this way the skin and the plumage become efficiently soaked. Then, when the bird is drying and cleaning itself following the bath, the feathers are again ruffled during shaking and during preening. The ruffling that accompanies head and body shaking enables the bird to fling off the water from the places the latter reached as a result of the ruffling during bathing. The ruffling that occurs during preening facilitates the grasping of the individual feathers by the beak.

Ruffling of the feathers also occurs during dust-bathing, sand-bathing and smoke-bathing, apparently for similar reasons to those given above for water-bathing. The ruffling which occurs as a component of so-called sun-bathing is undoubtedly primarily nothing more than a simple temperature response.

A localized erection of the feathers in the cloacal region occurs during defecation. If this response is performed incompletely, the vent feathers become fouled with excreta. A similar localized feather erection occurs immediately before and during copulation; this appears to facilitate cloacal contact.

Finally, brooding behaviour also involves a rather specialized type of erection of the ventral feathers. This is most clearly seen as a bird settles down on its eggs at the beginning of a period of incubation.

The above conditions under which various forms of feather movement occur in non-signal situations (other than those concerned with temperature regulation) have been mentioned only briefly here. This is because they are not of main importance to the present study. It would be a mistake to ignore them completely, however, since in any discussion of derivational problems, it is essential to begin by surveying all possible derivational sources.

The Secondary Signal-Function of Body Feathers

Having surveyed in some detail the non-signal aspects of feather posturing, it is now possible to consider the way in which such responses have come to act as social signals.

It is valuable to make a broad distinction between specialized and unspecialized signals, according to whether or not the original motor patterns concerned have been modified in any way in connection with their new secondary function. This process of modification was termed 'ritualization' by Huxley (1923), and much has been written on the subject in recent years by ethologists. (See, in particular, Lorenz, 1935, 1937, 1941, 1954; Tinbergen, 1940, 1948a, 1951, 1952, 1953; Baerends, 1950; Daanje, 1950.)

Briefly, this process involves modifications which render the original response more conspicuous. Thus, movements and postures may become exaggerated in some ways and simplified in others. Not only is the response itself 'schematized' (Tinbergen, 1952), but also the organs involved in the response may be rendered more obvious, e.g. by the addition of bright markings. Also, the response may become more frequent, as a result of the development of a 'typical intensity' (Morris, 1957). In this latter process, the causal range (see Morris, 1954c) of a particular form of a response spreads as the signal function develops, so that although initially the reaction only occurred at a special stimulus-intensity level, it now occurs as a reaction to a whole range of stimulus intensities. It therefore becomes the typical form of the response, or, to put it in another way, the reaction develops a 'typical-response-intensity'. I consider that this is the fundamental change that occurs during ritualization and that it is the basis of the so-called 'fixation' of a behaviour pattern during evolution (and, for that matter, during ontogeny). However, this whole subject will be dealt with more fully in a later paper. In the present study, we are not primarily concerned with the *nature of the changes* that take place as a response acquires signal function, but rather with the causal problem of the *nature of the responses* that become used in this way. These two aspects are so closely linked, however, that the above brief comments on ritualization are relevent here.

The general question of the origin of all types of social signals will be discussed below (see pp. 178–84). In the present section, the discussion is limited to the problem of the use of feather posturing as social signals.

A. Feather signals in non-thwarting situations

Social signals can be divided up for discussion into two groups, according to the causal conditions involved. Most of the more elaborate signals arise in situations of considerable stress, resulting from the strong activation and subsequent thwarting of a response. The thwarting may be due to the absence of indispensable stimuli, or to the presence of another simultaneously activated, but incompatible, tendency (see Tinbergen, 1952). (Thus, I am including both thwarting proper and conflict under the single heading of thwarting.) I shall deal with this, the major category, second; first it is necessary to describe (generally simpler) signals which arise in situations where there is no frustration involved.

In many species, there is a marked tendency to 'do likewise'. As soon as one member of a group starts to perform an activity, the others quickly follow suit and group synchronization ensues. This is particularly noticeable with the very sociable Estrildine Finches which I am at present studying. They not only move as a group, but also flee, eat, drink, bathe, clean, sleep and rest as a group. Most of the visual signals involved are of the unspecialized type. For example, the bird that starts eating does not do so in an exaggerated manner; but nevertheless the other birds respond to his normal feeding movements by joining him. This inadvertent signalling, which occurs under normal, non-thwarting conditions, requires little further mention as far as the actor is concerned, but the behaviour of the reactors has been the cause of much controversy in the past and must be discussed briefly here.

This controversy has centred around the question of whether there is a 'social drive' or not. Tinbergen (1951) has stressed that, contrary to previous contentions, there is no evidence for one and that the 'social' responses are always connected with and are a part of some special activity or other. Moynihan and Hall (1954) disagree with this and state that their spice finches (*Lonchura punctulata*) 'certainly look as if they just want to be together'. They conclude from their observations that 'This seems to imply the existence of some (partially) independent general or social gregarious motivation', the consummatory act of which is 'being near other birds of the same species'.

Extending their argument, it would follow that birds eat as a group primarily because they have a strong tendency to stay together and secondarily because this puts them all at the same time into a situation where there are strong stimuli for feeding. In other words, the signal from the actor, which has the ultimate result of group feeding, would, according to them, be 'I am

Plate 1 (*above*): The Bicheno finch; the pair immediately before a bout of courtship activity.

Plate 2 (*below*): The courtship of the Bicheno finch. The courting male (*on the left*) has assumed the posture, characteristic of this species, in which there is a marked fluffing of all the body feathers.

Plate 3 (*above*): The cutthroat finch; the pair immediately before a bout of courtship activity.

Plate 4 (*below*): The courtship of the cutthroat finch. The courting male (*on the right*) has assumed the striking ruffled posture which is characteristic of this species. Although the ruffling is fairly generalized there is nevertheless slight specialization in the differentiated belly region.

moving to the feeding area' and not 'I am feeding'. When one thinks of the birds in the wild state, this does seem to be much more reasonable.

It is not, however, the whole story. I have observed that, in the laboratory, when birds in one aviary have started bathing, birds in other aviaries some distance away (which have been intently watching the bathers), will rush down to their own water trays and 'join in' the bathing. Clearly, there is no question of such synchronization being the result of a primary tendency to move about together. It seems as if there is simply a tendency on the part of individuals of communal groups to 'follow suit', regardless of whether this involves generalized locomotion or specialized activities. It would probably be more valuable in future to investigate the way in which this tendency develops in ontogeny, rather than to argue as to whether there is a 'social drive' or not. For the purpose of the present paper the matter must rest there, except to point out that the mechanism underlying the 'following suit' responsiveness of sociable species is most unlikely to be similar to the mechanisms normally referred to as drives or instincts, which underlie other groups of activities. As far as the present study is concerned, it is sufficient simply to accept the existence of these responses to 'non-thwarting', inadvertent signals.

The feather posturing which is most frequently observed to play such a role is the fluffing which occurs when a bird is at rest. As already pointed out, the resting bird compensates for the lowering of heat production, which results from its inactivity, by fluffing its plumage. This gives it a characteristic rounded appearance, which is further enhanced by the fact that the bird squats on its perch and retracts its neck. In this hunched position, it often approaches a spheroid shape contrasting strongly with its active posture, in which not only are the feathers sleeked, but the legs and neck are stretched.

Amongst the Estrildine finches, it is typical for a semi-reproductive or non-reproductive group to rest in a dense row or clump (see Moynihan and Hall, 1954). Also, if no roosting platforms, holes, or nests are available, a group will roost and sleep in such a clump. The signal which stimulates clumping is the 'spheroid posture', of which fluffing is the principle component. The actor, however, does no more than it would do if it were in complete isolation, and this signal is quite unspecialized. The reactors respond to the sight of a spheroid bird by moving up to it and pressing against its side, settling into a spheroid posture themselves as they do so. If only two birds are involved, they can be clearly seen to lean up against one another. As more birds join the clump, so the warming effect of the dense grouping increases. This undoubtedly helps these very small finches to withstand the colder nights and, in captivity, it is frequently also observed during daytime resting periods, unless the weather is particularly warm or unless the birds are in full reproductive condition.

At first sight, it seems remarkable that the normal individual-distance (see Hediger, 1950), which these birds usually maintain, should be eliminated so easily by this spheroid posture. The ease with which this is achieved is probably due to the fact that it is such a complete opposite of the thin 'activity posture'. Adoption of the spheroid posture, therefore, results in the resting bird giving out none of the usual activity-signals which might interfere with clumping by warding off the reactors: for to approach too closely to an active bird is to risk being attacked.

The power of these signals is demonstrated by the following observations. When individuals of a number of species of Estrildine finches have just been purchased, all the birds are sometimes placed together in one aviary as a preliminary to being introduced into experimental units. When the birds are in a mixed group of this kind, it is possible to observe interspecific reactions which are often most illuminating. Occasionally certain individuals of newly imported batches of finches are sick and, as mentioned earlier, sick birds insulate themselves by fluffing their plumage. They sit around hunched up in this manner even at times when all the other birds are very active. Now healthy birds of other species under these conditions are presented with, on the one hand, active members of their own kind, and on the other hand, members of a different species in the full spheroid posture. If the healthy individual is going to 'clump', it has to do so either with a bird which has the right markings but the wrong posture, or with one which has the wrong markings but the right posture. *In a number of cases I have seen the latter choice made.*

One example of this may be cited here. A mixed group of finches contained, amongst other things, one sick long-tailed grassfinch (*Poephila acuticauda*) and two healthy star finches (*Poephila ruficauda*).[1] The markings are very different in the two species. The former has a large black throat spot, whereas the latter has a bright red face mask and white chest and flank spots, to mention only a few of the characters. Despite the presence of a second (sleeked) member of its own species, one of the star finches clumped with the fluffed long-tailed grassfinch and repeatedly preened its head. This illustrates very clearly the power of this particular type of feather signal.

I mentioned above that the star finch preened the head of the long-tailed grassfinch, and this requires some explanation. The preening of the head of one bird by another is a typical subsidiary resting response. The birds, having come very close together during clumping, are often stimulated, by the close proximity of the plumage of the other individual, to preen its feathers. The feathers that offer the greatest stimulus for preening are those of the head and neck, because they cannot be kept preened by their owner, and it is here that

[1] The scientific names of the Estrildine finches mentioned in this paper are based on Delacour's (1943) revision of the group.

the preener concentrates its efforts. It was pointed out earlier that, during preening, the feathers are ruffled and, considering the biological advantage of mutual head preening, it is not surprising that a specialized signal has arisen in this connection. This consists of ruffling the plumage of the head and neck, but not the rest of the body. This helps to increase the stimulating effect of the head feathers and to concentrate the mutual preening in this region. The response has apparently become very distinct during evolution and now may be given as a special initiating signal itself. The full response consists of a more or less spheroid posture with the addition of a ruffling of the head and neck feathers, a rolling of the head to one side, away from the preener, and a half or full closing of the eyes. This response to being preened may now be seen to be performed by an individual as it approaches a resting bird, before the latter has made any attempt to preen it.

Whereas body fluffing was acting as an unspecialized signal in a non-thwarting situation, head ruffling has become a specialized signal, in a similar situation. It is worth noting that although the former occurs in isolated birds, the latter never does so. (The only exception to this last point is that parrots and certain other pet birds, which are kept in isolation, may attempt to stimulate their human companions to perform mutual preening by adopting the head ruffle and roll posture towards them.)

Moynihan and Hall (1954), in their study of the spice finch (*Lonchura punctulata*), have put forward the suggestion that this head-ruffling signal itself may have evolved into a more elaborate form of ruffling in this species in agonistic situations. They suggest that the ruffling of the head, neck, scapular and upper back in hostile encounters, in this species may have evolved from the simpler social preening invitation of head ruffling because 'it may thus help to release or stimulate a relatively friendly social reaction instead of an extremely unfriendly one.' As far as the evolutionary trend is concerned, I cannot agree with them. Ruffling in agonistic situations is of extremely widespread occurrence and many variations exist, both with regard to the motivational conditions of the performers, and to the areas of the plumage erected. Some much more general explanation is required and this is given in the following sections of the present paper. However, once evolved, agonistic ruffling might well function in the manner suggested by Moynihan and Hall in special cases such as the spice finch. Briefly, then, they conceive of the similarity in the spice finch between social-preening head-ruffling, on the one hand, and agonistic ruffling, on the other, as a divergence, whereas I contend that it is more probably a case of convergence. I will only give one example to support my view here because the later sections of this paper make it unnecessary to consider the question in any more detail. The male crimson finch (*Poephila phaeton*) possesses a specialized threat display, in which the head feathers are ruffled in a way almost identical with that seen in the social-preening invitation. Yet this display is given when the bird is extremely aggressive and may be observed to occur even whilst the performer is pecking or plucking other birds – hardly a situation in which to be 'stimulating friendly reactions'.

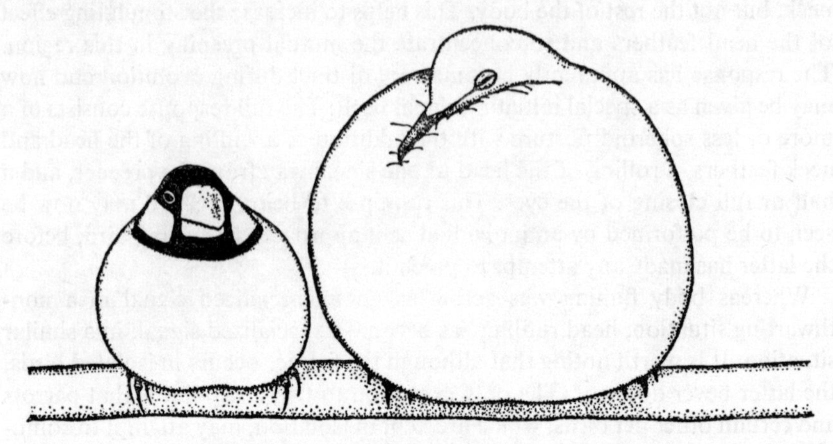

Fig. 2. The Java sparrow–dove relationship. The finches were strongly attracted to the doves and not only sat with them, as shown here, but also performed various other social activities with them. For explanation, see text.

Fig. 3. A Java sparrow clumping alongside a dove.

One further group of observations must now be mentioned to illustrate again the power of these clumping feather-signals. These observations concern a bizarre relationship that was observed to develop between some Java sparrows (*Padda oryzivora*) and some necklace doves (*Streptopelia chinensis*), which were kept together in the same aviary during the winter of 1954–5.

Goodwin (1952) had reported observing that Java sparrows kept in captivity with turtle doves (*Streptopelia turtur*) repeatedly roosted under incubating individuals of the latter species. This author found that 'On shining a torch on the nest one would see the two white eggs pushed to one side and, lifting up the turtle dove, would find the two Java sparrows side by side beneath her.'

The necklace doves, in my own observations, were not incubating and had no nest, but the Java sparrows were nevertheless seen to be attracted to them, and the following interspecific patterns were observed to occur:

1. A Java sparrow was frequently seen to sit next to a dove in preference to a member of its own species (Fig. 2).

2. On a number of occasions a Java sparrow was seen to push up against, or to lean against, a dove (Fig. 3).

3. A Java sparrow was seen to push up between the legs of a dove and plunge itself into the ventral plumage of the latter (Fig. 4). Once, at dusk, two doves were seen sitting near one another, with no Java sparrows in sight. The slight disturbance caused by the presence of the observer resulted in the appearance from beneath each dove of the head of a Java sparrow.

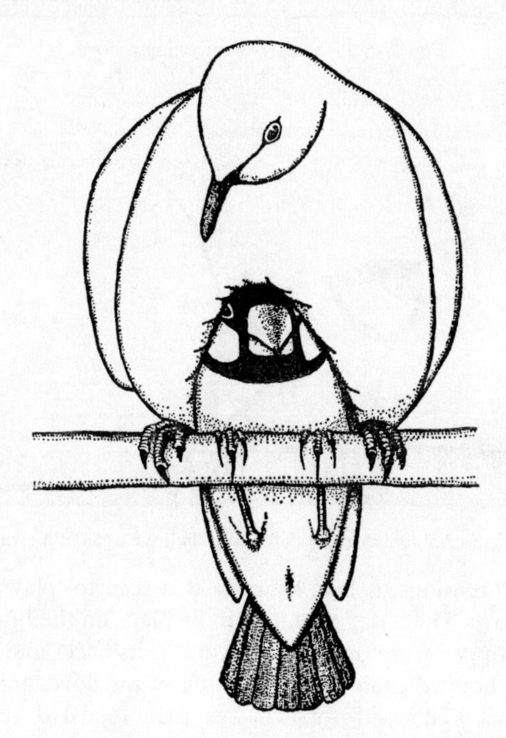

Fig. 4. A Java sparrow attempting to clump underneath a dove.

4. The Java sparrows were occasionally seen to preen the doves in preference to preening one another (Fig. 5).

5. When the number of Java sparrows was increased above that of the doves, the dominant Java sparrows defended against rivals the doves they had selected to clump with (Fig. 6).

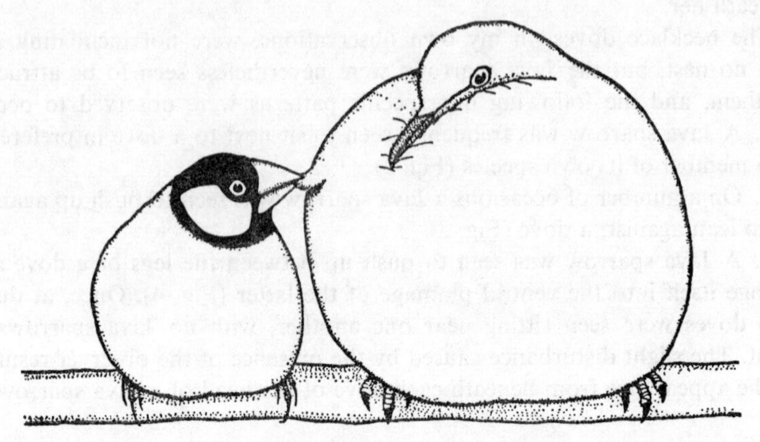

Fig. 5. A Java sparrow preening a dove.

Fig. 6. A Java sparrow defending its dove against a rival.

6. On a few occasions, a Java sparrow was seen to 'play leap-frog' over the back of a dove. The latter remained in its place on the branch, whilst the Java sparrow hopped from one side of it, on to its back, and then off again. Then it turned, hopped again on to the back of the dove, and again hopped off the other side. The performance was then repeated several times. It should be stressed that this was not a copulation attempt since, when the

Java sparrow was mounted on the dove, their long axes were at right angles to one another (Fig. 7).

These, then, are the observations which require explanation. At first sight the behaviour recorded appears to be extremely aberrant, to say the least. However, if the signal effect of the fluffed resting posture referred to earlier is called to mind, the situation becomes considerably simplified.

Fig. 7. A Java sparrow 'playing leap-frog' over the back of a dove.

It has already been pointed out that one species of finch will clump with and preen an individual of another species if the latter fluffs out into the spheroid posture. It is only one step further to the dove case. Doves, even when active, present an extremely rounded appearance. When doves assume the resting posture, it is a much larger and much rounder 'spheroid' that results, when compared with that of finches. It provides, in fact, a *supernormal*[1] *clumping stimulus* which is more attractive than even that given by other Java sparrows. It will be remembered that in the previous case the star finches had to decide between a fluffed long-tailed grassfinch and a sleeked star finch. Here, however, the doves provided such an intense stimulus that a Java sparrow would prefer them to one of its own kind *even though the latter was also in the fluffed spheroid posture*.

It was noticed that the colour of the plumage of the two species was rather similar in certain respects and in order to test whether this was a relevant stimulus, the doves were removed and a group of domestic pigeons was put

[1] For discussion of supernormal stimuli, see Tinbergen, 1951.

in their place. These were, of course, much larger than the doves and completely dwarfed the Java sparrows,[1] but even so the latter were attracted to them. Each pigeon was of a different colour and it was hoped that it would be possible to observe if there was a colour preference shown by the Java sparrows in selecting their companions. Unfortunately, the pigeons were so active in establishing a peck order amongst themselves that every time a Java sparrow began to creep up on one at dusk, it was frightened away, before it could make contact, by the sudden aggressive jerks and jumps of the large birds.

Although this made it impossible to obtain any detailed results, it is nevertheless interesting in itself that such comparatively huge birds should still stimulate the small finches to clump. (One cannot help thinking paradoxically of Tinbergen's (1948b) oyster-catcher trying to sit on a huge egg.)

Although the above explanation clarifies observations 1, 2 and 4 listed above, the others require further elucidation. The roosting underneath the perched doves which I observed, and the roosting under incubating birds recorded by Goodwin, can be taken together and explained as follows. Estrildines not only roost together in a tightly packed row along branches at night, but they may also roost on platforms or inside nests. In the latter instances they may roost tightly together, not only side to side, but also even in layer upon layer. Field observations on the roosting behaviour of African Estrildines to be found in Bannerman (1949, vol. 7, especially p. 227) reveal that as many as thirty of these small finches may pack together into a small roosting cavity. Getting underneath the doves, then, is a perfectly normal roosting pattern, once one has accepted the doves as supernormal Java sparrows.

A similar explanation applies to the 'leap-frog' phenomenon. When a group of Estrildines is clumped along a branch, the end birds of the clump occasionally attempt to get into a warmer position in the middle. To do this they hop along the backs of the row and push down in between two of the more centrally placed birds. In the case of the Java sparrows, the clump which consisted of one dove alongside one finch provided the stimuli for just such a manoeuvre. The Java sparrow, finding itself at the end of a 'row', tried to improve its position by moving to the middle of the 'row'. Its failure to find this mythical middle led to repeated trips back and forth over the 'backs of the row'. It should be noted that when two finches clump together they appear to know that there is no middle position and no attempts are then made to leap-frog to it. The Java sparrow, however, sitting on one side of the fluffed dove, could see nothing beyond it and appears to have been under the illusion that there must be a whole row of birds there.

This now explains all the listed observations on this interspecific relation-

[1] A pigeon weighs at least twenty times as much as a Java sparrow.

ship, with the exception of number 5, and this I cannot explain. Estrildine finches normally defend only the following objects: their nests, nest sites, roosting sites, mates and young. It is impossible to say into which, if any, of these categories the doves should be placed when they were being defended by the Java sparrows against one another.

These few examples of signal-functions of feather postures in non-thwarting situations must suffice for the present. Although they do not cover a wide range of either signals or species, they nevertheless serve to illustrate clearly the significance of this category of feather signals.

B. Feather signals in thwarting situations

When a bird is aroused in some way and the aroused tendency is then prevented from expressing itself fully, one of several things may happen (see Bastock, Morris and Moynihan, 1953). The types of response occurring in such a situation, which have received most attention from ethologists in the past, are thwarted intention movements, ambivalent responses, displacement activities and redirection activities. These will be discussed further in the last section of the present paper.

Although ethologists have recognized that, under such circumstances, other types of response may occur, they appear to have almost completely overlooked the behavioural significance of one whole group of effects, namely the *dramatic autonomic changes which inevitably accompany intense thwarting*.[1]

Physiologists have long been aware of the complexity and importance of these autonomic changes which accompany what they have termed 'emotional disturbances', and the physiological literature on the subject runs into many volumes. It will be convenient, at this point, to summarize briefly the physiological data, as far as they impinge on the ethological problem.

It is a little difficult to give an objective definition of the causal factor known as 'emotional disturbance'. In physiological writings many different words are used to describe such disturbances, e.g.: anxiety, tenseness, excitement, stress, fear, rage, resentment, shame and so on. However, whatever confusions may exist about the scientific meanings of these terms, one thing is quite clear: namely that, whatever else may or may not cause emotional disturbances, one of the best established causal factors, in a general sense, is intense thwarting. And this is sufficient for the present purpose. (I am still, of course, thinking of conflict as a double, or mutual, thwarting.)

The thwarting situation can be thought of as having two major *primary*

[1] This statement already requires modification. Since the present paper was completed, it has come to my notice that Dr R. Andrew (at Cambridge) has also been investigating the autonomic aspects of feather posturing.

effects: somatic[1] *and autonomic*. The somatic changes result in the performance of the motor patterns of the stimulated activity, or activities, and are concerned with the adjustment of the organism to its external environment. The autonomic changes, on the other hand, result in the adjustment of the internal environment of the animal to the requirements of these somatic responses which it has to perform.

Briefly, this adjustment of the internal environment, which mobilizes the animal for action, involves the following changes: The sympathetico-adrenal system is activated and this results in: (1) The cessation of processes of storing and digesting food. Thus, salivation is restrained; movements of the stomach, the secretion of gastric juices and the peristaltic movements of the intestine are all inhibited. Also, the rectum and bladder do not empty as easily as normally. (2) There are profound changes in the circulation of the blood. The heart beats faster and blood is transferred from the skin and viscera to the muscles and brain. This gives rise to an increase in blood pressure. In addition the number of red blood cells increases as a result of the excitation of the splenic nerves. (3) There are alterations in the rate of respiration; breathing is both quicker and deeper. (4) The temperature regulation mechanisms are violently activated and pilomotor and sudomotor responses are observed.

There are certain other physiological changes which also operate under the influence of the sympathetico-adrenal system, but all those which seem likely to have any bearing on ethological problems have been included in the above list. For further details, the reader is referred to the reviews in Cannon (1929, 1932), Morgan and Stellar (1950), and Gellhorn (1943, 1953).

This is not the whole of the autonomic story, however. There are still the parasympathetic responses to be considered. As is well known, the parasympathetic system works in opposition to the sympathetic system. Wherever a sympathetic influence causes a change in one direction, the parasympathetic causes one in the opposite direction. Under usual conditions, a balance is maintained between the two systems, but, under emergency conditions, when there is a sudden emotional disturbance, there is an immediate and powerful activation of the sympathetic system, which upsets this balance. The job of the parasympathetic system being to preserve and restore bodily reserves, it is quickly set in action to counteract the 'extravagant' effects of the sympathetic system mentioned above.

Knowing this, it might be expected that, when observing an emotional disturbance, all the above sympathetic symptoms would be observed first,

[1] The term somatic is employed with differing emphasis by geneticists, psychologists and physiologists. As is clear from the following discussion, it is used here in the sense employed by the physiologists.

then quickly all the opposite parasympathetic symptoms, and finally a 'calming down', or restoration of the autonomic balance. But this is not the case. Exactly what happens depends on the type of emotional disturbance, the particular individual concerned, and also the species involved. In some cases, the sympathetic dominates the scene, and few or no parasympathetic symptoms are observed. In other cases, the reverse is the case. Gellhorn (1953) mentions that certain studies of the problem 'suggest that fear causes reactions predominantly sympathetic, and feelings of hostility and anxiety predominantly parasympathetic discharges'. I must confess that I find it strange to see fear and anxiety contrasted in this way, but it is nevertheless worthwhile to note that fear and aggression may have opposite effects in this connection and we shall be returning to this point later. It must be added that individual variations may occur in a quite striking way and interfere with any completely clear-cut association of one type of autonomic balance with any one type of emotional disturbance.

These variations which occur under different circumstances, or with different individuals, are not necessarily all-or-none phenomena. Of the various changes that were listed above, some may go in a sympathetic direction, others in a parasympathetic direction, with the result that one may observe certain symptoms of both systems simultaneously. For example, when a bird is suddenly startled, its heart beats harder (sympathetic) and at the same time the bird may defecate (parasympathetic). Many such examples could be given, but from an ethological point of view they are not yet of primary importance. What really is important is to know the total range of easily observed symptoms which may occur as the result of the activation of the autonomic system during an emotional disturbance such as intense thwarting. (It was nevertheless relevant to point out that various combinations of these symptoms may be seen in different situations.)

It is valuable to conclude this brief survey of the physiological data, therefore, with a list of these symptoms, both sympathetic and parasympathetic, as follows (where pairs of alternatives are given the sympathetic effect is given first and then the parasympathetic): dryness in the mouth, or excessive salivation; defecation; urination; extreme pallor, or blushing and flushing; vasodilation of sex organs, including erection of penis; rapid deep breathing, or slow shallow breathing; gasping; sighing; panting; weeping; fainting; sweating; and, finally, the most important from the present point of view, namely, *pilomotor activity*. (It will be appreciated that not all the above responses occur in birds, but, in view of the general discussion to follow, it was thought advisable to include them here.)

It is my intention to show below how a number of these different symptoms have, in the course of evolution, become secondarily modified as social signals. The intimate association that exists between the various thwarting situations and these autonomic symptoms lends itself admirably to the selection of the latter as indicators of the mood of the performer.

However, the pilomotor responses must be considered separately first. When a bird prepares itself for action, it naturally streamlines itself in readiness for flight. Thus, the mechanically effective pilomotor response in an emergency situation is the sleeking of the plumage. This will also have the effect of reducing insulation and will thus help to reduce the overheating that will tend to occur with vigorous activity. This sleeking, rapid breathing and, in mammals, sweating, are particularly important as cooling devices here because vasodilation, which is normally a powerful cooling device, cannot occur in this context, the blood from the skin being needed elsewhere.

It is not surprising therefore that Langley (1904) found that 'The ordinary effect of stimulating any part of the sympathetic system or of stimulating a cutaneous nerve is depression of the feathers.' But it must also be pointed out that he goes on to say that (with sympathetic stimulation): 'At times a strong erection of feathers occurs instead of depression'; also, 'Frequently the feathers in one part are erected and in another depressed.' As far as I know, the exact causal differences which result in these different symptoms are still not understood, but it is nevertheless clear that in birds, the typical symptom of sympathetic stimulation is sleeking.

Despite this, in any general review of the subject, pilo-*erection* is said to occur as the result of sympathetic stimulation. At first sight this apparent contradiction is difficult to understand. It seems to be due to the fact that most physiologists are 'mammal-oriented'. (Even the term pilomotor itself bears evidence of this, for it was originally coined in connection with the erection and depression of hair, but is now used to include the movements of both hair and feathers.) It would appear that the basic organization of the erection and depression responses under sympathetic influences is different in mammals and birds. A full erection can occur in both groups, but only in birds, it seems, is the ordinary effect one of depression. In their general reviews, physiologists do not make any allowance for this difference. Also, it is unfortunate that they do not make any clear statements concerning the exact form of the pilo-erection under such conditions. Most physiologists refer to pilo-erection as a warming device, and then make no comment as to why an animal should set into action both cooling and warming systems at once as a result of sympathetic stimulation.

Of course, if the response was one of *full* pilo-erection it would give rise to ruffling in birds, which would have a cooling rather than a heating effect and would therefore be more in line with the general syndrome. Perhaps full erection of hair, in mammals, has the same effect. As was stressed above (p. 156), few authors have appreciated the functional differences between fluffing and ruffling in birds, but it is significant that Langley refers at various points to feathers being 'more or less erected', which contrasts with the 'strong erection' he mentions in connection with sympathetic stimulation. It seems likely that the latter form was indeed ruffling.

Further research is badly needed here. On the present evidence, I would predict that sleeking or ruffling would be expected as the usual or special sympathetic responses respectively, and that fluffing would be correlated with parasympathetic responses. It is important that the exact reasons should be found for the occurrence of sympathetic ruffling in birds, as opposed to sympathetic sleeking, since, as will be shown below, it is this form of the response that is of the greatest ethological interest.

Setting these details on one side for the moment, the general conclusion can be stated as follows: if a bird is intensely aroused and is then thwarted in some way, the autonomic changes which will accompany the somatic reactions to this situation will involve marked pilomotoric activity, which sometimes takes the form of pilo-erection. From this, it can be further stated that there is every reason to suppose that striking feather postures will become intimately associated with thwarting situations. As the latter circumstances are just those where social signals are most important, the use of feather postures as such signals is inevitable. Furthermore, the physiological data suggest that, in all probability, different types of thwarting will give rise to different basic types of feather posturing, although more physiological evidence is needed on this point.

If we now examine the feather-posture behaviour of birds under thwarting conditions, it soon becomes clear that, not only have the autonomic pilomotor responses been used in this way, but also they have, in many instances, undergone varying degrees of secondary modification during evolution. Many highly complex and specialized forms of feather posturing can be observed in displaying birds of many species. This specialization has taken the form of: (1) The restriction of the regions of the plumage where pilo-erection occurs to special areas of the body. There is typically a correlation between the regions thus differentially erected and those which are visible to the reactor. (2) The addition of bright markings and colours to the erected regions. (3) The enlargement of the body feathers on the displayed regions of the plumage. This enlargement may render the feathers useless as regards their original functions. (4) The loss of pilomotor control in special displayed regions. Thus, tufts or crests of feathers may become fixed in a permanently erected position. Here, as in number 3 above, the feathers concerned may lose their original functions.

The examples of the above phenomenon, of *the ritualization of autonomic pilomotor responses*, are so abundant that it is not possible to give more than a selection here. In selecting, I shall attempt to give one or a few examples of each type of display feather-signal, beginning with the less specialized and ending with the most highly specialized cases.

Firstly, the following examples of unspecialized cases may be mentioned: Morris (1954a) has noted that, in the zebra finch (*Poephila guttata*), thwarted escape may lead to a fluffing of the plumage and that this may tend to inhibit the attacks of the dominant individuals. This appeasement function of fluffing appears to operate successfully because of the 'inactivity' signals given out by the fluffed spheroid posture of the bird. Hinde (1953) records that subordinate chaffinches (*Fringilla coelebs*), which are very low in the peck order, sit around most of the time in the fluffed posture and he too has suggested that 'This latter posture may have the effect of reducing aggressive behaviour in other individuals.'

A rather generalized fluffing of the feathers occurs during the courtship of males of various species of grassfinch (Erythrurae) (see Pls. 1 and 2). This appears to be quite unspecialized in the case of the Bicheno finch (*Poephila bichenovi*), except that it is accompanied by an expansion of the black and white markings of the species. In the case of the zebra finch, there are further minor indications of specialization. The unmarked crown region is sleeked, rather than fluffed like the rest of the body feathers. The fluffing of the flank regions exposes the brightly spotted flank feathers.

Amongst another group of the Estrildine finches, namely the mannikins (Amadinae), the courting males often show a more or less generalized ruffling of all the body feathers, e.g. the bronze mannikin (*Lonchura cucullata*). The male cutthroat finch (*Amadina fasciata*) (see Morris 1954b) ruffles all its feathers during courtship, but some specialization does occur in the belly region. Here there is a large brown patch flanked with white, and there is a sudden sharp increase in the extent of pilo-erection just at the edges of this patch (see Pls. 3 and 4). Certain other mannikins, such as the spice finch (*Lonchura punctulata*), show an intermediate condition between fluffing and ruffling, part of the plumage being fluffed and part of it ruffled.

These rather unspecialized responses are not restricted to the finches. The carrion crow (*Corvus corone*) threatens with all its body feathers ruffled. In this case, the autonomic response is performed simultaneously with somatic posturings of the wings and tail.

Hingston (1933) gives many more examples of generalized feather erections occurring during agonistic encounters, e.g. 'Everyone knows how the fowl fluffs herself when rushing on a cat in defence of her chicks. Owls put up a tremendous show of threat, spreading their feathers so profusely, that they look twice the normal size.' Hingston, whose almost encyclopaedic knowledge of the whole range of animal displays is as admirable as his obsessional interpretation of them is deplorable, makes two generalizations (p. 114), as follows: Firstly, 'All birds ruffle the body feathers in anger'; and secondly,

'Were I to give all my examples of feather-fluffing under anger it would mean a list of every bird whose fighting behaviour I have seen recorded.' It is clear from these two statements that he is not distinguishing between fluffing and ruffling, and in scanning his many excellent examples the reader must be warned of this fact.

For the sake of brevity, we will now pass on to the more specialized displays and these will be categorized according to the region of the body where the particular exaggeration has occurred:

1. *Crests.* In many species, too numerous to mention, differential ruffling may take place, in which only the crown feathers are erected. The latter are not modified in many cases, but in others they are brightly marked or coloured and the erection displays these markings or colours. In still others, the crown feathers are not only brightly marked, but also enlarged into an erectile crest. In the latter cases, the normal pilo-erection movements involved in ruffling may be unaltered, but the elongated crown feathers, when raised in the normal way, nevertheless give a violently exaggerated display. In some species, such as the sulphur-crested cockatoo (*Kakatoe galerita*) (see Fig. 8), the enormous crest fans out so far forward that it seems as if the pilomotor movements themselves must also be exaggerated. In some species the crest has become rigidly fixed, leaving the bird in a permanently crown-ruffled state.

It would be pointless to include here a list of crested species, since there are so many and they are so well known. Suffice it to say that there is hardly any group of birds which does not possess certain members which are crested.

2. *Ruffs.* In a number of species, such as the great crested grebe (*Podiceps cristatus*), or the ruff (*Philomachus pugnax*) itself (see Fig. 8), there is a much enlarged erectile frill or circle of feathers around the neck, which is expanded in frontal displays.

3. *Ear-tufts.* A number of owls and various other species possess elongated erectile outgrowths of feathers on either side of the head, which form discrete tufts. Even when not erect, these structures are usually clearly visible. Hingston reports that 'Eagle Owls, when enraged, erect them so conspicuously that they look like black pointed horns.'

4. *Beards, or Chin-growths.* In some species there is a tuft of elongated feathers placed immediately beneath the beak and, in the display of the Capercaillie (*Tetrao urogallus*), for instance, the head is thrown back and these are fully erected (see Fig. 8). Hingston states that the purple fruit-crow also has such a 'beard'.

5. *Breast or Throat Plumes.* Many species possess brightly marked breast-patches which are differentially erected in display. One only need mention the well-known threat display of the robin (*Erithaceus rubecula*) (Lack, 1943), in which it ruffles its red breast. In certain birds, however, notably the herons,

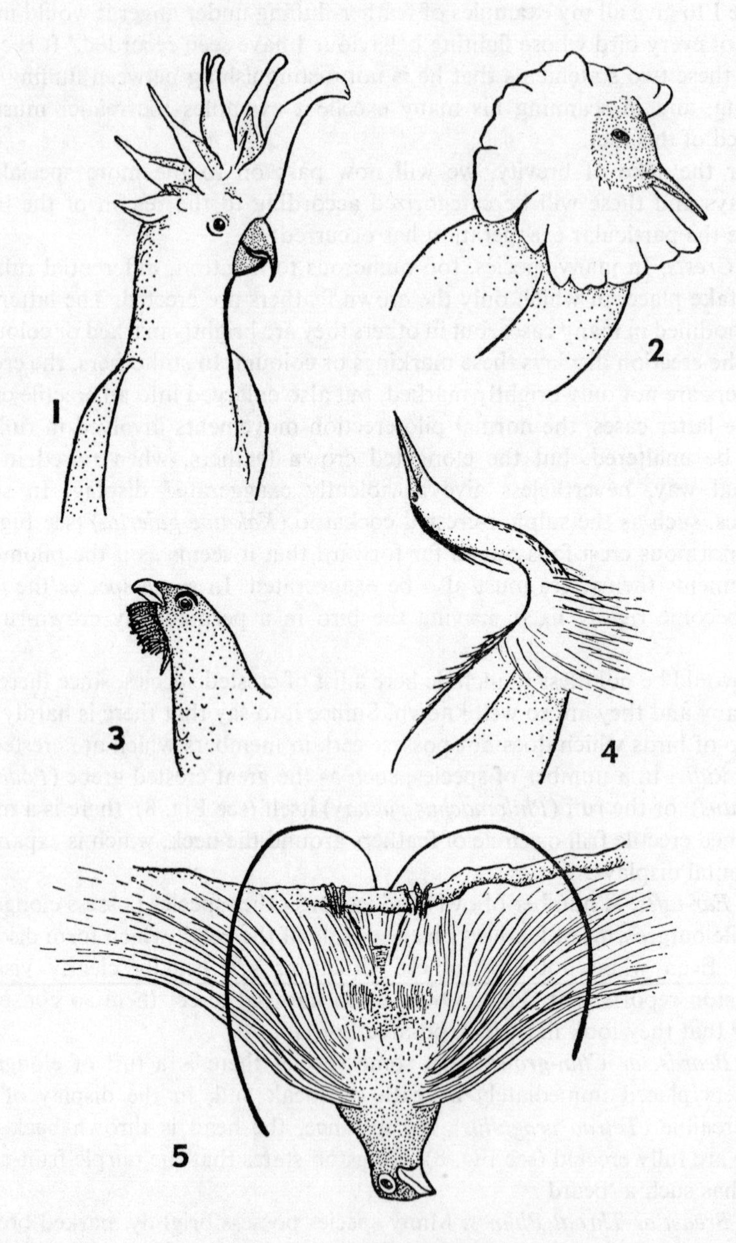

Fig. 8. Various specialized forms of feather signals. Nos. 1–5 respectively: crest, ruff, chin-growth, throat-plumes, flank-plumes.

the feathers of the throat region are drawn out into very long erectile plumes (see Fig. 8).

6. *Flank Plumes.* A fantastic development of patches of flank-feathers can be observed to have taken place in a number of species of birds of paradise. These 'paradise plumes', which spring from the sides of the body, just beneath the wings, are dramatically erected and even vibrated as the display climax of these species (see Fig. 8).

7. *Rump-Patches.* A number of the more elaborate displays involve the erection of a bright rump-patch, as, for example, in the courtship display of the golden pheasant (*Chrysolophus pictus*), where there is a fluffing out of a large and well-differentiated rump region.

8. '*Eye*' *Tufts.* Scattered throughout the bird world there are a number of unique forms of highly specialized feather signals, which will be represented here by the following example: The superb bird of paradise (*Lophorina superba*) possesses an extremely elaborate display which, amongst other things, involves the erection of 'two tiny patches of shining green feathers between the eyes'. These 'it raises in a manner so subtle that they catch the light and are transformed into a pair of scintillating green eyes, which, by their very brightness, mask the true eyes; they are so realistic that it is hard to believe they are no more than tiny spots of feathers' (Stonor, 1940).

Fragmentary as the above compilation may be, it nevertheless gives a clear picture of the immense variations and extreme developments which specialized feather signals may exhibit. For a further study, the reader is referred to the works of Hingston (1933), Stonor (1940) and Armstrong (1947).

It will be noticed that all the displays mentioned above take place in either agonistic or sexual contexts, during threatening, appeasing, pair-forming, greeting, or courting ceremonies. It will also be recalled that they have been grouped together here under the general heading of 'feather signals occurring under thwarting circumstances'. It is essential therefore to point out here that recent ethological research (see Tinbergen, 1952; Hinde, 1953; Moynihan, 1955; Morris, 1954c and 1956) has shown that it is a general principle of agonistic and sexual ceremonies that they involve causally an intense arousal coupled with subsequent thwarting, the latter usually being the result of the simultaneous activation of incompatible tendencies. In the next section, a more comprehensive sketch of the total effect of thwarting situations will be given.

Before leaving the present section, there is still one aspect of feather signals that must be dealt with. As stated earlier, there is some evidence that fear may be linked predominantly with sympathetic and aggression with parasympathetic discharges. Now, the temperature regulatory mechanisms set in motion by the sympathetic system are those concerned with cooling and those set in motion by the parasympathetic with warming the animal. It would follow from this that in a bird in which fear predominated, the appropriate pilo-erection response would be ruffling, and where aggression predominated, fluffing should occur. What little evidence there is so far suggests that this is not so. Ruffling occurs in many cases where there is a marked tendency to attack, but only a weak tendency to flee. Also, fluffing is generally more characteristic of submissive birds than hostile ones. Among the Estrildine finches, it is the males which are more aggressive towards their mates that ruffle while courting (e.g. bronze Mannikin), whereas the 'fluffed courters' are much less likely to attack their females (e.g. Bicheno finch). But I do not feel that this necessarily points to a simple connection in birds between fear and the parasympathetic and hostility and the sympathetic. The whole problem is much more complicated than this. Undoubtedly the suddenness of the reaction is important and a difference would be expected between the autonomic response to a sudden fright, for example, on the one hand, and a state of subordinate submission, on the other, although both cases would involve intense fear and thwarted escape tendencies. Also, of course, sympathetic sleeking is associated with fear, and, in any case, all kinds of complicated combinations of sleeking, ruffling and fluffing can occur simultaneously in the same situation. In conclusion, it can only be stated that this whole question of *the physio-ethology of thwarting* urgently requires further investigation. The current, largely non-physiological, ethological approach to the problem of thwarting has been immensely fruitful, but the stage has now been reached where any further advance will almost certainly be dependent on more strictly physiological analyses.

Conclusion: The General Problem of Thwarting and the Evolution of Signals

Since the foregoing discussion of feather signals has incorporated a somewhat new approach to the ethology of the responses that occur in thwarting conditions, it will perhaps be valuable here to review the more general aspects of the problem. For this purpose I propose to divide the reaction to thwarting into the *primary and secondary response*. The primary response is subdivisable into the *somatic response* and the *autonomic response* in the manner already put forward. The secondary response consists of the *alternatives to the primary somatic response* which may occur as a result of the failure of the latter to satisfy the requirements of the situation. They may arise in one of two ways. Either the thwarting situation is such that the animal is incapable of performing a primary somatic response at all, or the latter is performed but fails in its task.

It is convenient at this stage to tabulate briefly the various types of thwarting stimulus-situations and the primary and secondary responses that are given to them. They are as follows:

Thwarting stimulus-situations

I. Absence of indispensable stimuli, following intense arousal.
II. Simple physical obstruction of aroused activity.
III. Simultaneous arousal of two or more incompatible tendencies.

Responses to above situations

A. Primary response

a. Somatic response

1. 'Perseverance' (applies to I and II above, when animal attempts to continue with aroused response despite circumstances).
2. 'Snap decision' (applies to III above, when animal quickly responds to only one aspect of relevant stimulus-situation).
3. Thwarted intention movements (applies to I, II, III).
4. Ambivalent posturing (applies to III).
5. Alternating ambivalent movements (applies to III).

b. Autonomic response

1. Alimentary: Increase or decrease in salivation. Urination. Defecation.
2. Circulatory: Pallor. Flushing. Vasodilation of sex organs. Fainting.
3. Respiratory: Increase or decrease in respiratory rate or amplitude. Gasping. Sighing. Panting.
4. Thermoregulatory: Sweating. Pilomotoric activity.
5. Lacrimatory: Weeping.

B. Secondary response

a. Displacement activities.
b. Redirection activities.
c. Regressive activities.
d. 'Neurotic' inactivity.

The three thwarting stimulus-situations are self-explanatory and require no further explanation here. Also, little need be mentioned concerning the evolution as signals of primary somatic responses 3, 4 and 5. The ritualization of somatic posturing and locomotory intention movements has been dealt with fully elsewhere, especially in the works of Lorenz, Tinbergen and Daanje already cited.

A brief survey of the ritualization of the various primary autonomic responses, which appears to have occurred in many cases, is, however, highly relevant here. (Once again, only a concise selection can be attempted.)

The primary autonomic responses

Firstly, the alimentary effects: The most important of these are undoubtedly the many instances of the ritualized use of defecation and urination in mammals, as territorial marking systems. It is sufficient to cite as an example the elaborate territorial urinating of male dogs. Considering all that is known about the type of behaviour involved in territorial defence, it is highly probable that these methods of marking have evolved from the alimentary effects of the parasympathetic stimulation that occurs with thwarting.

Secondly, the circulatory effects: Parasympathetic vasodilation of various kinds of bare skin-patches in both birds and mammals appears to be of a quite widespread ritual significance. For example, Hingston states that 'Birds with naked head or neck skin or fleshy head appendages, such as the turkey, jungle-fowl, bateleur eagle, either flush up in these areas when angered, or cause the parts to become turgid' (p. 337). He also points out that man is not the only mammal to show a flushing of the face in emotional situations; it also occurs in the Bengal monkey and the mandrill, to give only two examples, and in both these cases their 'hind ends flush in unison with their reddening faces'.

Thirdly, the respiratory effects: Various forms of gaping and hissing may have evolved in this way. With gaping, however, there is a dangerous ambiguity, in that the opening wide of the mouth may also occur as a signal as a result of the ritualization of intention movements such as biting. More important is the question of vocalization. Spurway and Haldane (1953) have suggested that vocalization may be 'considered as a ritualization of breathing'. They refer to this ritualization as being a modification of *displacement* breathing. Since it is the autonomic changes in respiration, which accompany the primary somatic response, which seem most likely to have given rise to this ritualization, I cannot agree with them that it is necessary to invoke the concept of displacement activation here. This does not, however, really alter the significance of their suggestion. (The problem of displacement activities will be dealt with below.)

Another possible evolutionary trend originating in the respiratory disturbances that accompany thwarting is that leading to inflation displays. The latter are widespread in their occurrence, but we will consider only avian inflations here. Hingston states that: 'Male frigate-birds, when at courtship, blow their naked throats into scarlet distended bladders.' Stonor records that the sooty grouse (*Dendragapus richardsoni*) has brightly coloured throat pouches which it inflates with air from the lungs when displaying. In the sage hen (*Centrocerus urophasianus*), the throat pouches are enormous and can be swelled out to almost half the size of the owner's body. Certain species of bustard, especially the Australian bustard (*Eupodotis australis*), possess huge throat pouches which, in the latter species, are so large when distended with air that they reach down to the ground. In a number of cases such pouches are not only brightly coloured, and act as visual signals, but also resound in various ways to produce auditory signals. Inflation displays are, of course, also widespread amongst the lower vertebrates, but these will not be discussed in the present paper.

Fourthly, the thermoregulatory effects: Sweating in mammals in thwarting situations may well have been the starting point for the evolution of much 'scent signalling'. It is not unlikely that scent glands have evolved from sweat glands and scent from sweat. Thus, the territorial behaviour of many mammals involving scent signals may have arisen in a fundamentally similar way to that involving defecation and urination. The difference between them would, of course, be that whereas the latter would be parasympathetic in origin, the former would be sympathetic. Pilomotor responses in birds need not be referred to again here, but similar reactions in mammals may be noted. Hingston has compiled a formidable list of pilo-erection responses in threatening mammals of many species. They include the following: A rather generalized hair erection is seen in dogs, wolves, jackals, mongooses, weasels, skunks, badgers, bats and wild pigs, to mention only a few. More specialized hair erection is observed in the following: The lion spreads its mane, expanding the region of dark hairs; the squirrel whisks its tail while spreading the tail-fur; the giant ant-eater erects a crest along the middle of its back, and also its immense brush has its hairs spread out; Wart-hogs elevate their bristles into a crest. Amongst the most specialized forms of hair erection are those that occur in connection with the rump displays of certain ungulates, e.g. the prongbuck 'has a special circular muscle by which it can spread out the hairs on its rump into a pair of white flaming chrysanthemum-like discs' (Hingston, p. 59).

The widespread significance of ritualized autonomic signals in animal displays becomes immediately apparent when the above phenomena are grouped and considered in this way. In conclusion it need hardly be mentioned that this whole field of study offers considerable possibilities for future ethological research.

The secondary responses

Turning now to the problem of the secondary responses, observations of thwarted animals reveal that they may frequently intersperse their primary somatic responses with alternative, apparently irrelevant, somatic activities. These are of several kinds:

Firstly, there may appear motor patterns which are typically seen as integral parts of instinctive patterns other than those being thwarted. These have been called displacement activities (see Tinbergen, 1939, 1940, 1952; Kortlandt, 1940; Tinbergen and van Iersel, 1947; Armstrong, 1947, 1950; Moynihan, 1953; Bastock, Morris and Moynihan, 1953; Morris, 1954a). As regards their causation, it has been suggested that the 'nervous energy' which is being denied an outlet by the thwarting circumstances is displaced into new channels and that, as a result of this, they 'serve a function as outlets, through a safety valve, of dangerous surplus impulses' (Tinbergen, 1952).

Secondly, redirection activities (see Bastock, Morris and Moynihan, 1953) may occur. In these cases, the primary somatic response is continued as far as the fixed motor pattern is concerned, but the orientation is changed. (Since this secondary orientation alters the whole significance of the response, I propose to include redirection activities here under the heading of secondary responses, despite the fact that the motor pattern itself is the same as the primary one.) Typically, an animal, which is prevented from attacking one individual or object, turns and vents its aggression on some different, inoffensive individual or object, despite the continued presence of the former.

Thirdly, regressive activities may occur. This applies to special learning situations. When thwarted in psychological training apparatus, laboratory animals tend to show secondary responses which consist of a falling back to earlier behaviour patterns: that is, to patterns of responses given in the apparatus at an earlier stage of the training. This category of response has obvious implications for human behaviour.

Fourthly, with continued intense thwarting, a state of 'neurotic' inactivity may develop, which may continue until the animal ceases to be responsive in any way to almost any stimuli; it is then, of course, lethal. It has been suggested by Tinbergen (1952) that one of the possible functions of displacement activities is to prevent this from happening.

No more will be said about these last three types of secondary response, since they do not appear to be important from the point of view of ritualization. Ritualized displacement activities, however, do play a very important role in social signalling, as Tinbergen has emphasized, and it is worthwhile therefore to conclude with a re-examination of the causal aspects of this form of secondary response.

In the past, a number of reasons have been put forward to explain why one particular displacement activity, rather than another, should occur in a given situation (see Tinbergen, 1952, pp. 17–20). The three most important factors hitherto discussed are as follows: 1. The influence of the internal state of the thwarted animal. 2. The influence of the external stimuli present at the time of thwarting. 3. The influence of the initial posture of the thwarted animal. To give a simple, hypothetical example: If an animal has its aggression thwarted, it may show displacement feeding rather than any other displacement activity because it was rather hungry at the time its aggression was thwarted, or because there was food near by at the time, or because the thwarted posture of biting the enemy was very similar to that of biting food.

Little or no experimental evidence has been presented for the first of these three. Räber (1948) has provided conclusive evidence for the second. He was able to show that whether turkeys displacement fed or displacement drank was dependent of the relative availability of food and water. Tinbergen (1952) has discussed the third factor in some detail and states that 'the posture in which an animal finds itself when a drive is thwarted might decide which activity will be used as an outlet.' Although he points out that this has not yet been proved experimentally, he nevertheless cites a number of examples which can most plausibly be explained in this way.

I wish to add now a fourth causal factor which I consider to be of importance in determining which displacement activity will occur in a given situation, namely, *the influence of the primary autonomic responses.* It seems highly likely that many displacement activities are selected because the appropriate stimuli for them are provided by the sympathetic or parasympathetic discharges which accompany thwarting. For example, violent pilomotor, sudomotor, or vasomotor activity may provide skin stimuli which provoke the animal to perform displacement scratching, wiping, shaking or preening. Sleep is known to be intimately connected with parasympathetic activity and displacement sleeping (see for example, Makkink, 1936) may perhaps have been caused in this way.

It might be argued from the above that, if there is an autonomic causal link in a particular instance, then there can be no true displacement activity, because the autonomic response is primary and relevant to the thwarting situation. The stimuli that it provides, which determine the displacement activity, are relevant to the situation and therefore the so-called displacement activity is also relevant.

This might be contrasted with the other causal factors mentioned above, where the presence, at the moment of thwarting, of food, hunger, or the posture of feeding, is purely coincidental. It is not an *essential effect* of the thwarted aggression that the animal shall find itself hungry, near food, or in a feeding posture. (It can of course be a very common effect and thus lead to ritualization, without actually being an essential one.) The autonomic responses, on the other hand, are an integral, functional part of the response to the thwarting situation. Secondary somatic responses occurring as a result of stimuli produced by this autonomic discharge are *considerably less irrelevant* to the situation than displacement activities arising in other ways. It might be suggested, on this basis, that autonomically stimulated secondary somatic activities should not be called displacement activities, and that they should be separated from them conceptually. I think, however, that it would be unwise to do this at the present stage, when we know so little about the detailed causal mechanisms that operate in the production of any secondary response. For the present it is far better to leave the question open and to concentrate on the more detailed problems such as the more elaborate experimental analysis of the causal factors acting in special cases. In conclusion, it should however be pointed out that to call the primary autonomic responses *themselves* displacement activities, as certain authors have done in various cases, is quite unjustified.

Summing up the foregoing, it can be said that there are a number of aspects of the response to a thwarting situation that have been specialized as social signals. Ritualization has taken place, not only in connection with thwarted intention movements, ambivalent reactions, and displacement activities, but also with autonomic responses. It might be said, therefore, that there are three basic categories of signals arising from thwarting; namely, *primary somatic signals, secondary somatic signals* and *autonomic signals.* It is hoped that the present paper will serve to focus more attention on the latter category, the significance of which has been underestimated by ethologists in the past.

References

ARMSTRONG, E. A. (1947), *Bird Display and Behaviour* (Lindsay Drummond Ltd, London).

—— (1950), 'The nature and function of displacement activities'. *Sympos. Soc. exp. Biol.* 4, pp. 361–87.

BAERENDS, G. P. (1950), 'Specializations in organs and movements with a releasing function'. *Sympos. Soc. exp. Biol.* 4, pp. 337–60.

BANNERMAN, D. A. (1949), *The Birds of Tropical West Africa*, vol. 7 (London).

BASTOCK, M., D. MORRIS and M. MOYNIHAN (1953), 'Some comments on conflict and thwarting in animals'. *Behaviour* 6, pp. 66–84.

CANNON, W. B. (1929), *Bodily Changes in Pain, Hunger, Fear and Rage* (New York).

—— (1932), *The Wisdom of the Body* (New York).

DAANJE, A. (1950), 'On locomotory movements in birds and the intention movements derived from them'. *Behaviour* 3, pp. 48–99.

DAVSON, H. (1951). *A Textbook of General Physiology* (London).

DELACOUR, J. (1943), 'A revision of the subfamily Estrildinae of the family Ploceidae'. *Zoologica* 28, pp. 69–86.

GELLHORN, E. (1943), *Autonomic Regulations, their Significance for Physiology, Psychology and Neuropsychiatry* (New York).

—— (1953), *Physiological Foundations of Neurology and Psychiatry* (Minneapolis).

GOODWIN, D. (1952), 'Recollections of some small birds'. *Avic. Mag.* 58, pp. 24–9.

HEDIGER, H. (1950), *Wild Animals in Captivity* (Butterworths, London).
HINDE, R. (1953), 'The conflict between drives in the courtship and copulation of the Chaffinch'. *Behaviour* 5, pp. 1–31.
HINGSTON, R. W. G. (1933), *Animal Colour and Adornment* (London).
HUTT, F. B. and L. BALL (1938), 'Number of feathers and body size in passerines'. *Auk* 55, pp. 651–7.
HUXLEY, J. S. (1923), 'Courtship activities in the Red-throated Diver (*Colymbus stellatus* Pontopp); together with a discussion of the evolution of courtship in birds'. *J. Linn. Soc. Lond.* 25, pp. 253–92.
KENDEIGH, S. C. and S. P. BALDWIN (1928), 'Development of temperature control in nestling house wrens'. *Amer. Nat.* 42, pp. 249–78.
KORTLANDT, A. (1940), 'Wechselwirkung zwischen Instinkten'. *Arch. néerl. Zoöl.* 4, pp. 442–520.
LACK, D. (1943), *The Life of the Robin* (Witherby, London).
LANGLEY, J. N. (1904), 'On the sympathetic system of birds, and on the muscles which move the feathers'. *J. Physiol.* 30, pp. 221–52.
LORENZ, K. (1935), 'Der Kumpan in der Umwelt des Vogels'. *J. Ornithol.* 83, pp. 137–213 and 289–413.
—— (1937), 'The companion in the bird's world'. *Auk* 54, pp. 245–73.
—— (1941), 'Vergleichende Bewegungsstudien an Anatinen'. *J. Ornithol.* 89, pp. 194–294.
—— (1954), *Comparative Studies on the Behaviour of the Anatinae* (The Avicultural Society, London).
MADSEN, H. (1941), 'Hvad gør Fuglenes Fjer-Dragt vandskyende?' *Dansk Ornithol. Foren.* 35, pp. 49–59.
MAKKINK, G. F. (1936), 'An attempt at an ethogram of the European Avocet (*Recurvirostra avosetta* L.), with ethological and psychological remarks'. *Ardea* 25, pp. 1–63.
MOORE, A. D. (1945), 'Winter night habits of birds'. *Wilson Bull.* 57, pp. 253–60.
MORGAN, C. T. and E. STELLAR (1950), *Physiological Psychology* (New York).
MORRIS, D. (1954a), 'The reproductive behaviour of the Zebra Finch (*Poephila guttata*), with special reference to pseudofemale behaviour and displacement activities'. *Behaviour* 6, pp. 271–322.
—— (1954b), 'The courtship behaviour of the Cutthroat Finch (*Amadina fasciata*)'. *Avic. Mag.* 60, pp. 169–77.
—— (1954c), 'An analysis of the reproductive behaviour of the Ten-spined Stickleback (*Pygosteus pungitius* L.)'. Doctor Thesis, Oxford; also *Behaviour*, Supplement 6.
—— (1956), 'The function and causation of courtship ceremonies'. *Fondation Singer-Polignac: Colloque Internat. sur L'Instinct*, June 1954.
—— (1957), ' "Typical Intensity" and its relation to the problem of ritualization'. *Behaviour* 11, pp. 1–12.
MOYNIHAN, M. (1953), 'Some displacement activities of the Black-headed Gull'. *Behaviour* 5, pp. 58–80.
—— (1955), 'Some aspects of the reproductive behaviour of the Black-headed Gull (*Larus ridibundus* L.), and related species'. *Behaviour*, Supplement 4.
—— and F. HALL (1954), 'Hostile, sexual, and other social behaviour patterns of the Spice Finch (*Lonchura punctulata*), in captivity'. *Behaviour* 7, pp. 33–76.
RÄBER, H. (1948), 'Analyse des Balzverhaltens eines domestizierten Truthahns (*Meleagris*)'. *Behaviour* 1, pp. 237–66.

SPURWAY, H. and J. B. S. HALDANE (1953), 'The comparative ethology of vertebrate breathing. I. Breathing in Newts, with a general survey'. *Behaviour* 6, pp. 8–23.
STONOR, C. R. (1940), *Courtship and Display among Birds* (London).
STURKIE, P. D. (1954), *Avian Physiology* (New York).
TINBERGEN, N. (1939), 'On the analysis of social organisation among vertebrates, with special reference to birds'. *Amer. Midl. Nat.* 21, pp. 210–34.
—— (1940), 'Die Übersprungbewegung'. *Zeitsch. f. Tierpsychol.* 4, pp. 1–40.
—— (1948a), 'Social releasers and the experimental method required for their study'. *Wilson Bull.* 60, pp. 6–52.
—— (1948b), 'Dierkundeles in het meeuwenduin'. *De Levende Natuur* 51, pp. 49–56.
—— (1951), *The Study of Instinct* (Oxford University Press, Oxford).
—— (1952), 'Derived activities; their causation, biological significance, origin, and emancipation during evolution'. *Quart. Rev. Biol.* 27, pp. 1–32.
—— (1953), *Social Behaviour in Animals* (Methuen, London).
—— and J. J. A. VAN IERSEL (1947), 'Displacement reactions in the Three-spined Stickleback'. *Behaviour* 1, pp. 56–63.
WETMORE, A. (1921), 'A study of the body temperature of birds'. *Smithsonian Misc. Coll.* 72, pp. 1–52.
—— (1936), 'The number of contour feathers in Passeriformes and related birds'. *Auk* 53, pp. 159–69.

'Typical intensity' and its relation to the problem of ritualization
(1957)

Introduction

It is a basic property of simple signals, when these are contrasted with other types of response, that they remain constant in form regardless of any changes in the circumstances which cause them. For example, a car moves faster the harder the accelerator is pressed, but a telephone bell rings at the same speed regardless of how urgent the call may be.

Superficially, it might appear detrimental that such a feature as signal rigidity should interfere with the simple direct relationship that usually exists between the strength of a stimulus and the strength of the response it provokes. However, the reduction in the amount of information provided by the signal about the exact state of the signaller, that is involved with this stereotyping, is apparently more than compensated for by the elimination of signal ambiguity. A signal that is constant in form cannot be mistaken.

Now, it is a fundamental characteristic of the communicatory behaviour of animals that the signal patterns used are derived from non-signal sources (see, in particular, Daanje 1950, Tinbergen 1952, and Morris 1956a). This signalization may occur either in the phylogeny of an animal (when it is called ritualization), or in its ontogeny (when it may be termed stylization). In either case it results in the need for a fundamentally variable reaction becoming constant in form. These conflicting demands of variability and stability lead in many cases to a compromise: *typical intensity*.

The Relationship of Typical Intensity to Frequency

Frequency and intensity may be defined as follows: Frequency measurements refer to the number of times a specified response is shown in a given period; intensity measurements refer to the form which the response takes each time it occurs.

Fig. 1 shows three hypothetical graphs which illustrate: I, Variable intensity; II, Typical intensity; and III, Fixed intensity. In the first case, where there is an increase in stimulation, the response increases both in frequency (F) and intensity (I). In the second case, where a flattening of the graph has

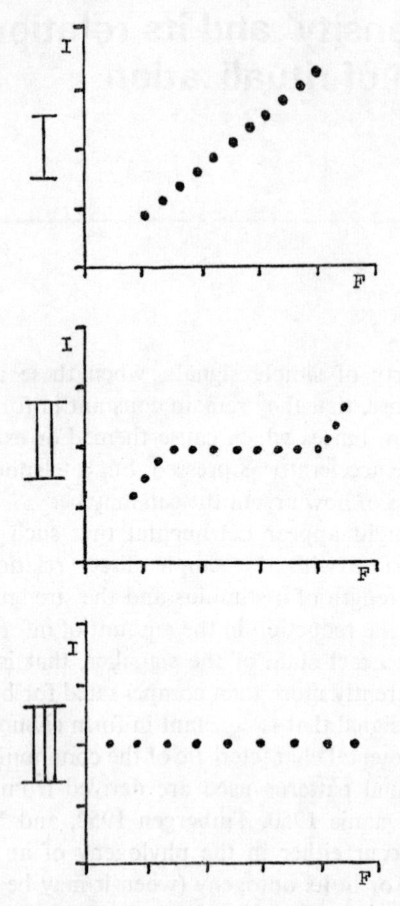

Fig. 1. The stereotyping of a response. I shows the response with a variable intensity. An increase in frequency (F) is accompanied by an increase in intensity (I). II shows the response with a typical intensity. At most frequencies, the intensity is the same; only at very high or very low frequencies do changes in intensity appear. III shows the response with a fixed intensity. Here, no matter how high or low the frequency, the intensity never varies.

occurred, the intensity of the response now varies only when the latter occurs with very low or very high frequencies. In the third case, the intensity has become constant at all frequencies; no matter how weak or how strong the stimuli now, the response, if it appears at all, will occur in a rigidly fixed form.

It should, however, be pointed out that a stereotyped response of the last-mentioned kind can still vary with varying stimulation, because its frequency of occurrence will still be varying. To return to the telephone analogy: if the call is urgent, the bell will ring no faster, but it will go on ringing for longer before the caller gives up. It is the existence of this other response variable, and its potential independence from the intensity variable, that make it possible for a signal response to become fixed and unambiguous, without completely ceasing to reflect quantitative fluctuations in the mood of the signaller.

For example, an animal may reveal that it is very hungry by (a) eating fast and (b) eating for a long time; or that it is not very hungry by (a) eating slowly and (b) eating only for a short while. On the other hand, it may reveal that it is very highly motivated sexually by (a) taking up a characteristic courtship posture and (b) displaying thus to the female for a long time; or that it is only mildly sexual by (a) taking up an *identical or nearly identical posture* and (b) displaying thus to the female for only a short period. The difference between the responses in the feeding and sexual contexts given here is obvious. In the sexual case, if the courtship display postures are identical at the two motivational levels, then a *fixed* intensity (Fig. 1, III) has developed. More commonly, however, slight differences can be detected which reveal that instead only a *typical* intensity has developed. Usually, if the motivational state is extremely low or extremely high, minute changes in the intensity of even the most fixed response can be detected. The photographs in Pls. 5 and 6 illustrate this point. They show the courtship posture of the cutthroat finch (*Amadina fasciata*) (see Morris 1954c, 1955). In Pl. 5 the low intensity form is seen and in Pl. 6 the high intensity version is shown. Note how minute are the differences between them (especially when either is compared with the non-displaying male on the left in Pl. 6). Once a male of this species has begun to court a female, he assumes this posture and no matter how long or short his period of courting may be, or how many times he may display in a given time period, he never shows much more variation in display intensity than is seen here. In other words, for enormous changes in response frequency, he will show only slight changes in response intensity. These latter changes are so slight that casual observation would lead to the conclusion that his display had a fixed intensity. However, as with most cases, a closer scrutiny reveals that it has, in fact, a typical intensity, albeit of an extreme kind.

Continuing with the cutthroat finch as an example, it is instructive to look for a moment at the nature of its courtship dance. Each dance consists of a number of 'inverted curtsies' (see Morris 1954c, 1957). With each of these curtsies, the bird makes a stilted upward jump. Each jump has a highly characteristic form with a very typical intensity, so that if a jump is given at all it nearly always involves exactly the same amplitude, speed, etc. This fact helps to give the dancing the distinctive rhythmic form so typical of most display movements. An important point emerges here, however, namely that although each inverted curtsy has a typical intensity, each dance, which is composed of a number of these curtsies, in this case does not. The bird may perform low intensity dances made up of only one curtsy, or high intensity dances made up of many. (It is nevertheless theoretically quite possible for the dance also to develop a typical intensity, so that it nearly always comprises the same number of curtsy movements. In many cases developments of this kind have taken place.) What is important here is the relationship, as regards frequency and intensity, between the courtship responses at the two levels of dancing and curtsying. For it will be seen that the frequency of curtsying is also referred to here as the intensity of dancing.

I am grateful to Dr W. M. S. Russell for bringing the significance of such relationships to my notice. He pointed out that the intensity of a response is also the frequency of the response at the level of integration immediately below it, and vice versa. (This point is briefly discussed in Russell, Mead and Hayes, 1954, pp. 197–8). As a further and more complex example of this, we may take the beak-cleaning pattern of birds:

A *beak-wipe* consists of (a) twist body, (b) lower and rotate head, (c) scrape, (d) raise head, and (e) twist body back.

A *beak-wiping* consists of a number of such wipes given in quick succession.

A *beak-wiping bout* consists of a number of such wipings (in a series uninterrupted by other activities).

A *full cleaning pattern* comprises not only a number of (a) wiping bouts, but also (b) shaking bouts, (c) preening bouts and (d) scratching bouts.

At each level of integration in a hierarchy of this kind, the frequency of the activity below is the intensity of a given response and the intensity of the activity above is the frequency. Thus, the intensity of a beak-wiping is given by the number of wipes it comprises, whilst the number of beak-wipings gives the intensity of a wiping bout. Now supposing, for the moment, that beak-wipings, as responses, developed a typical intensity; this would mean that beak-wipes would have developed a typical frequency. If four wipes nearly always occurred per wiping, giving the latter a typical intensity of four, then beak-wipes, which would now occur at this frequency regardless of fluctuations in their own intensities, would have a typical frequency, also of four.

It is, of course, a quite arbitrary choice as to whether one approaches each case as a typical intensity, or a typical frequency, since where there is one there must also be the other. For various reasons, however, I prefer the former. Frequency measures can sometimes be confusing owing to the fact that they may involve an artificial limiting time period, set by the experimenter. Intensity measures, on the other hand, provide a more reliable approach. An intensity measure is available every time an act occurs, but a frequency measure may be either a score of bout length, or the total number of actions per (arbitrary) time unit.

(See discussion on differences between intensity and tendency in Russell,

Mead and Hayes, 1954. It should be noted that 'act tendency' in their sense has four measures: frequency, latency, number and duration. These are all being lumped together as 'frequency' in the present paper. This does not, however, reflect any basic difference in approach, since, on page 166 of their paper, the above authors say of their four categories that 'Evidently all these measures can be reduced in principle to frequencies.')

Typical Intensity in Ambivalently Motivated Responses

Thus far, only those situations in which one type of response is stimulated have been considered. Common in signal-producing situations, however, is the state of motivational conflict. With such a state one finds various important forms of compromise; for instance, alternating intention movements, ambivalent posturing, etc. In many of these cases typical intensities have developed.

The fighting behaviour of the black-headed gulls studied recently by Moynihan (1955) must be mentioned here. Moynihan found that, with increasing or decreasing frequencies of particular conflict states, he did not observe a whole range of as many intensities. Instead, only a few quite distinct conflict responses occurred, each having a typical intensity covering part of the total motivational range.

He also found that conflict responses provided rather special motivational problems, as follows. Each response was the result of a combination of so much 'attack' and so much 'escape'. It could vary motivationally, therefore, in two distinct ways. Firstly, the ratio of attack to escape could remain constant, whilst their absolute levels changed; this type of variation gives rise to no new problems. Secondly, the ratio of attack to escape could change; this does provide a new problem. In the first case, one can relate simple frequency changes to the change in the form of the response, and deal with the situation as if it were an instance of simple single motivation. All the previous comments concerning typical intensities apply here as before. In the second case, however, this is not possible, and a new concept of *Typical Compromise* has to be introduced. This complication arose in the case of Moynihan's gulls, as it does in most cases, but a more suitable example for presentation here is offered by the lateral display common to so many species of fish during combat. The fighting males, which are both hostile towards and afraid of one another, either attack, flee, or show the lateral display. This means that for almost the entire time, when engaged in combat, the individuals are either facing directly towards or directly away from one another, or are precisely parallel. In other words, out of all the possible angles they could take up, only three basic ones are used. (There are a few species where this is not the case, but it is nevertheless the general rule.)

This may mean that the combatants are always either (1) fearlessly aggressive, or (2) completely afraid, or (3) in an exact state of balance between the two. Alternatively, the lateral display could have developed as a Typical Compromise such that if the fish are both afraid *and* aggressive, they take up this position, even though they may be, at one time, slightly more aggressive than afraid and, at another, slightly more afraid than aggressive. In order to determine which of these two possibilities is relevant in a given case, it is necessary to check the variations in the ratios of the attack-developments and flight-developments from the lateral postures. If all fish showing lateral display are just as likely to attack as to flee, then, if there are no intermediate postures, the first type of modification must have occurred. If, on the other hand, laterally displaying fish may in some instances show a greater tendency to flee than attack and in other instances a greater tendency to attack than flee, then a typical compromise must have developed. Judging by the analyses carried out so far, the latter appears to be the more usual occurrence.

The important difference between typical intensities and typical compromises is that in the former case the intensities could be compared with *frequency* differences, whereas in the latter case the compromises must be compared with *frequency-ratio* differences. Functionally, however, typical compromises are identical in significance with typical intensities. They fix characteristic conflict signals and render them unambiguous.

With the development of typical intensity spreading its causal range (see Morris 1954d) in one way, and typical compromise spreading it in another, a conflict response can easily become so readily evoked that no fight can occur without its appearing. In some species actual contact fighting itself may even become extremely rare and disputes are settled simply by prolonged bouts of performing the special conflict responses of the animals concerned. So far, only a single conflict response has been considered, but in most species there is a distinct set of these reactions, which form specific 'threat codes' (see Morris 1954d, 1956b). The Mexican swordtail (*Xiphophorus helleri*) provides a suitable example here. Whereas most species show only the three angles (towards; lateral; away) mentioned so far, when fighting, the swordtail shows five: towards; positive-sigmoid; lateral; negative-sigmoid; away. In the positive-sigmoid posture the fish holds its body parallel to the rival, as in the typical lateral display, but turns its head towards the rival and its tail away, as in the frontal posture. This gives its body a characteristic sigmoid shape, as does the negative-sigmoid display, in which the head points away from the rival, as if to flee from it. Fish which show the positive-sigmoid display are more likely to attack than flee, as might be expected, and the converse is true for the negative-sigmoid. In this species, then, there are three distinct responses where, in others, there is only one – the simple lateral display. This means that the swordtail responses have less

Plate 5. The low intensity ruffled courtship posture of the cutthroat finch male (*left*).

Plate 6. The high intensity ruffled courtship posture of the cutthroat finch male (*centre*). This posture has developed a typical intensity, great variations in response frequency being accompanied by no more than these very small intensity differences. The small extent of these intensity differences can best be judged by comparing the two display postures with the posture of the non-courting male, seen on the left.

chance of developing large causal ranges than the others. One way in which this difficulty is overcome in certain species (but not in the present case) is by the spreading of causal ranges to such an extent that they *overlap* with one another, giving rise to multiple conflict responses. This has been discussed fully elsewhere, however, and will not be dwelt on here (see Morris, 1954d, Chapter 4).

Typical Intensity and Ritualization

As will be clear by now, the development of typical intensity is a fundamental aspect of the process of ritualization. During the latter a number of types of change can be observed. These have been discussed to some extent by Daanje (1950). They will therefore be mentioned only briefly here and will only be discussed with regard to their relationship to typical intensity.

Firstly, there are several ways in which patterns can be modified as signals which do not touch on the problem of typical intensity. Included here are (1) change of stimulus-valance, (2) qualitative motivational changes ('neurophysiological emancipation' – see Tinbergen, 1952), (3) loss or modification of original orientation, and (4) supplementing of response by addition of display structures, colour, etc.

Then there are the following changes which may or may not be connected with the development of typical intensities: (5) threshold lowering, (6) development of rhythmic repetition, (7) differential exaggeration of components, (8) omission of components, (9) 'freezing' of movements, (10) change in sequence of components, (11) changes in component co-ordination, (12) increase or decrease in speed of performance, (13) change in vigour of movements.

Finally, of course, there is (14) development of typical intensity itself and, in special cases, (15) development of typical compromise.

No attempt has been made here to produce a comprehensive list of types of modification. Nor do the fifteen given above represent fifteen mutually exclusive categories. Rather, they present fifteen types of modification of the kind which is met with during the course of comparative ethological studies. The first group will not be mentioned further here, but some comments must be made on the second.

Firstly, lowering of response threshold, which so often goes hand in hand with other ritualizing changes, can apparently be the result of one, or both, of two kinds of basic development. On the one hand, a particular form of a response may become more frequent when it develops as the typical intensity of that response. It will now occur as the reaction to a wider variety of motivational states and will therefore be seen more often. Its frequency is increased, so to speak, at the expense of the other forms of the response. On the other hand, it may appear more often because *all* forms of the response appear more. In the latter case the response need not have developed a typical intensity.

Which of these two possibilities is operating in a given case is easy to determine. If the particular form of the response concerned shows an increase in frequency independent of an increase in the frequency of the response as a whole, then typical intensity is responsible. If its increase is accompanied by an increase in all other forms of the response as well, then typical intensity is not involved.

Next, rhythmic repetition, a particularly common development, can be partially, if not wholly, explained by the development of typical intensity. The existence of the latter has two effects here. First, it makes each separate action in a rhythmic repetition the same strength and, second, it thereby increases the likelihood of the time intervals being similar between each act.

Differential exaggeration of components applies to *multiform* responses as opposed to *uniform* ones. In the former, the components are qualitatively different (A-B-C-D); in the latter each component is the same (A-A-A-A). The beak-cleaning examples illustrate this: the beak-wipe response has multiform components (twist-lower-scrape-raise), whereas the beak-wiping response has uniform ones (wipe-wipe-wipe-wipe). In a ritualized multiform response, the development of a typical intensity at a certain stage through its sequence of components will give rise to differential exaggeration favouring that stage. For example, ritualized beak-wiping itself in certain species consists of little more than the initial bowing components. These may now be given with low or high frequency, but hardly ever develop into a full wiping. (It should be noted, however, that here the differential exaggeration occurs as a negative process; if a component is positively exaggerated beyond anything ever seen in the original response, then this of course is a special ritualizing development quite distinct from typical intensity.)

Leading on from this, the complete omission of components will take place as a further extension of the last process when differential exaggeration of the negative type becomes more extreme. An even further development of this is the 'freezing' of movements. A pattern begins, gets to a given stage and then fixes there. It seems likely that typical intensity may be involved here as before. Also, the other modifications mentioned in the second major group (i.e. changes in component sequence, co-ordination, speed, or vigour) may be associated, to a greater or lesser extent, with the development of typical intensities, but they will not be further discussed here. (They will be more fully dealt with by Blest, in his forthcoming review.)

Typical Intensity in Non-Signal Situations

Although the development of typical intensity is so intimately connected with the process of signalization, it is not altogether restricted to this context. Many non-signal responses have evolved typical intensities purely in relation to increased mechanical efficiency.

For example, in the fanning activity of the reproductive male stickleback (see van Iersel, 1953), the amplitude and speed of the fin beats is fairly constant, regardless of motivational fluctuations, giving fanning a typical intensity. The important and striking variable in this activity comes at the next highest level, in connection with the intensity of fanning bouts, or, to express it another way, with fanning frequencies.

In contrast with this, the fanning response of the river bullhead (*Cottus gobio*) (see Morris, 1954b*) shows no typical intensity at any level. The fin-beats vary considerably in amplitude and in speed, giving no typical intensity to fanning here. In this species, in fact, both the fanning intensity and the fanning frequency are extremely variable.

A possible reason for this difference between the fanning of the two species is that the bullhead fans while resting on the bottom, whereas the sticklebacks have to maintain themselves hovering in the water just in front of their nest entrances. In support of this view, there is the observation of van Iersel (1953, p. 24) that, in the case of the three-spined stickleback (*Gasterosteus aculeatus*), the very slight decreases in fanning intensity which can sometimes be seen are accompanied by maladjustments in the orientation of the male to its nest. Van Iersel concludes 'that the taxis component of the movement becomes incomplete when the intensity of the fixed motor pattern decreases'. The taxis component of the bullhead fanning would not suffer in this way, however, this species being supported by the substratum.

A second comparative example of this kind concerns the snail-eating behaviour of song thrushes (*Turdus ericetorum*) and blackbirds (*T. merula*) (see Morris, 1954a). The former species has evolved a special shell hammering motor pattern, thus enabling it to extract and devour the contents. This movement appears to have been derived from the more basic actions of shake, peck and wipe. All or some of these have been modified in such a way that a song thrush can now pick up a snail in its beak and, with repeated rhythmic swings of its neck, batter the shell on a hard stone surface until it cracks open. The strength and speed of each swing have become stabilized so that hammering has developed a typical intensity.

This appears to have arisen because stronger or faster blows would make the shell difficult to hold and weaker ones would not crack it open so efficiently. The blackbird, on the other hand, does not seem to have evolved

* See pp. 89–117.

this excellent technique. It tries hard sometimes to smash shells by using the old peck-and-shake actions in a vigorous and erratic fashion, but seldom with success. Although it is easy to see why the song thrush has evolved a typical intensity here, it is difficult to understand why the blackbird has not.

One final example concerns a purely ontogenetic case. When parrots – which are very active birds – are kept in captivity, they are often chained to a single perch or placed in a ridiculously small cage. At first, the birds make wild and erratic movements in attempting to escape from their strict confinement. After a while, however, they develop stereotyped versions of their escape intention movements (see Lorenz, 1950, p. 55), which then have extremely typical intensities – almost fixed intensities. Regardless of the frequency of the performance, the actions are repeated rhythmically over and over again ad nauseam with almost exactly the same amplitudes and at almost identical speeds. It is of interest to note that the exact form of the stereotyped movements differs considerably between individuals. Of the four types which I know personally one consists of a figure-of-8-on-its-side movement of the head and neck. Another involves a stilted form of bowing. A third is primarily a lateral pivoting and the fourth a tipping of the head first to one side then the other. It is most interesting to find individual variations occurring in a strictly ontogenetic development, contrasting, as one would expect, with the phylogenetic cases referred to previously, where such variations are infinitesimal.

In conclusion, the functional aspect of the typical intensities in the parrot case presents somewhat of a problem. It would seem that typical intensity is not only valuable in connection with certain signal requirements, and as an aid to non-signal mechanical efficiency, but is also in any case a basically convenient mode of response performance for the nervous system.

References

DAANJE, A. (1950), 'On locomotory movements in birds and the intention movements derived from them'. *Behaviour* 3, pp. 48–99.

IERSEL, J. J. A. VAN (1953), 'An analysis of the parental behaviour of the male Three-spined Stickleback (*Gasterosteus aculeatus* L.)'. *Behaviour*, Supplement 3.

LORENZ, K. Z. (1950), *King Solomon's Ring* (Methuen, London).

MORRIS, D. (1954a), 'The snail-eating behaviour of thrushes and blackbirds'. *Brit. Birds* 47, pp. 33–49.

—— (1954b), 'The reproductive behaviour of the River Bullhead (*Cottus gobio* L.), with special reference to the fanning activity'. *Behaviour* 7, pp. 1–31.

—— (1954c), 'The courtship behaviour of the Cutthroat Finch (*Amadina fasciata*)'. *Avic. Mag.* 60, pp. 169–77.

—— (1954d), 'An analysis of the reproductive behaviour of the Ten-spined Stickleback (*Pygosteus pungitius* L.)'. Doctor Thesis, Oxford; also *Behaviour*, Supplement 6.

MORRIS, D. (1955), 'The markings of the Cutthroat Finch'. *Birds Illustrated* 1, pp. 182–3.

—— (1956a), 'The feather postures of birds and the problem of the origin of social signals'. *Behaviour* 9, pp. 75–114.

—— (1956b), 'The function and causation of courtship ceremonies'. *Fondation Singer-Polignac: Colloque Internat. sur L'Instinct*, June 1954.

—— (1957), 'The reproductive behaviour of the Bronze Mannikin (*Lochura cucullata*)'. *Behaviour* 11, pp. 156–201.

MOYNIHAN, M. (1955), 'Some aspects of the reproductive behaviour of the Black-headed Gull'. *Behaviour*, Supplement 4.

RUSSELL, W. M. S., A. P. MEAD and J. S. HAYES (1954), 'A basis for the quantitative study of the structure of behaviour'. *Behaviour* 6, pp. 153–205.

TINBERGEN, N. (1952), 'Derived activities; their causation, biological significance, origin, and emancipation during evolution'. *Quart. Rev. Biol.* 27, pp. 1–32.

The reproductive behaviour of the bronze mannikin
(1957)

Introduction

There are approximately one hundred species of Estrildine finches (sub-family of Estrildinae of the Ploceidae). All the members of this exclusively Old World group are extremely small and many are comparatively easy to maintain in captivity. These and other considerations led to the selection of this sub-family as material for a detailed comparative behaviour study. Twenty-eight species have been acquired to date and their behaviour investigated in the laboratory. Of these, however, many have been studied only briefly, whereas a few have been selected for more detailed scrutiny. When a sufficient number have been added to this latter category it will be possible to compile a report on the comparative ethology of the group, but, in the interim, reports on particular species will be presented from time to time. It is hoped that this will assist in preventing the final analysis from becoming too cumbersome. Papers have already appeared on a small number of species (Morris 1954a, 1954b, 1955a, 1955b; Moynihan and Hall 1954) and to this list the present species is now added.

The bronze mannikin is an extremely common African species, with a wide distribution. Like most Estrildines, it is a bird of the grasslands, where it is encountered in large flocks, feeding on the ground or direct from the grasses. (For an analysis of its seed preferences, see Morris 1955c). Observers in Africa (e.g. Holman in Bannerman 1949) report that it is seldom or never to be seen in a solitary state, and the way in which this strong communal tendency interacts with certain incompatible reproductive tendencies will be discussed below. The species breeds throughout the rainy season in the wild state and, in captivity, will show reproductive behaviour at any time of the year.

The adult plumage colour is predominantly dark brown and white, arranged in the following way: back and wings dark brown; head, neck, throat and tail brownish-black; flanks, rump and tail coverts barred brown and white; lower breast and belly pure white. In addition, there are dark patches in the scapular region and on the anterior section of the flank region. When they catch the light, these patches glisten with a metallic green sheen. The head region also glistens in this way. The eyes, legs and feet are dark brown. The beak is distinctively two-coloured, the upper mandible being brownish-black whilst the lower is off-white. There is no sexual dimorphism. The general size of the bird is small, even for an Estrildine finch; of eight birds which were weighed, all were under ten grams, the average weight being 9.2 grams.

The present study is based on observations of twenty-nine individuals. Twenty-five of these were wild-caught imported birds, which were purchased from bird-dealers, and four were bred in the laboratory.

In order to avoid confusion, it should be pointed out that this species is better known by its earlier generic name of *Spermestes*, especially in the avicultural world. This was changed to *Lonchura* by Delacour (1943) in his taxonomic revision of the Estrildinae. There are indications that this particular change was perhaps not entirely justified, but, until much more comparative data are available, Delacour's names for this and other Estrildines will be employed.

Methods

Most of the work was purely observational and the birds were manipulated as little as possible. A number of simple experiments were carried out, however, but typically these only involved the separation and then the re-introduction of individuals to one another under various conditions.

When not under observation, birds were kept in small aviaries or cages in groups. From these they were transferred to the larger observational aviaries, or to experimental introduction cages. Two types of larger aviaries were used. One type was arranged as a set of nine units. Each unit measured approximately nine feet in length and height and three feet in width. These long thin enclosures had all four walls made of wood. Half the roof of each unit was also made of wood, but the other half was covered with movable wooden frames over which was stretched transparent 'Windowlite'. Under these frames was fixed wire netting. The frames were arranged in such a way that they could easily be slid off on warm summer days, leaving only a wire-netting roof. Down one side of the battery of units was a completely wood-covered corridor, with a door at one end. In the corridor-end of each of the units a small observation slit was cut in the wooden wall and fitted with a movable glass plate. The observer was then able to sit in the darkened corridor and, unseen by the birds, watch their behaviour through one of the slits, as if in a 'hide'.

The other type of aviary did not conceal the observer to the same extent. The observation wall was all wire netting, giving a better range of vision, but the conspicuousness of the observer was reduced by the use of stage lighting. A battery of bulbs, screened from the observer and pointing down into the aviary was placed just above his head, making it difficult for the birds to see any details beyond them. The perches used in the last type of unit were natural twigs and branches, but in this type they were thin wooden rods which were arranged symmetrically in grids. There were two such grids, one at the nest-box level, five feet from the ground, and one at the food-tray level, two-and-a-half feet from the ground. The nine nest-boxes and three food-trays were also arranged symmetrically. This made it possible to score accurately the position and movements of individuals and to analyse the spatial organization of a small colony of birds. For a photograph of part of the interior of this aviary, see Pl. 7.

The extremely rapid movements of the very small birds rendered it desirable to obtain detailed photographs of the various motor patterns. Initial attempts to photograph the birds in their various aviaries did not produce satisfactory results. A special photographic cage was therefore constructed. This was built in such a way that the birds were always seen against a plain white background, were always in camera range, and, if perched, were always in focus. On one side of the cage was attached a small 'introduction box'. With one bird in the cage and one in the box, it was then only necessary to raise a sliding panel in order to observe and photograph a social encounter and its outcome. An electronic flash apparatus was employed in order to prevent blurring from rapid movement. Almost all the photographs illustrating this paper were taken in this way.

Agonistic Behaviour

The spatial organization of a reproductive colony of bronze mannikins involves a considerable amount of fighting. This is because the very strong, non-reproductive tendency to keep together and act together as a group comes into immediate conflict with certain reproductive requirements. A group of non-reproductive individuals not only rests together and moves about together, but also feeds, drinks, bathes, cleans, flees, roosts and sleeps together. In several of the above activities the tendency to keep together is so strong that all or some of the individuals in a group come to be in close bodily contact with one another. This occurs when cleaning one another (Pl. 8), resting on a branch (Pls. 9 and 10), and when sleeping. In all these cases each bird acts as a strong attraction stimulus for its fellows. If the reproductive nesting, pairing, courting and parental patterns are to be successful, this general attraction of one individual for the others has to be modified drastically in certain respects. This is achieved in the present species by violent fighting of a rather special kind.

Many species possess elaborate 'threat codes' (see pp. 139–40) which consist of sets of ritualized signals evolved from conflict responses. The latter occur during fighting as a result of the simultaneous activation of the incompatible tendencies to attack and to flee. By the use of these specialized signals, the combatants can settle their disputes without recourse to actual physical contact. Such 'slanging matches', however, are not typical of the fighting behaviour of Estrildine finches. As has already been stressed in the case of the zebra finch (*Poephila guttata*) (see pp. 43–50), Estrildines nearly always come to blows when resolving an agonistic situation.

Typically, these finches possess three basic agonistic postures: one associated with fearless onslaught, one with defensive attack, and one with submission. In most of the species so far studied, these three are as follows:

An all-out assault, by an individual which shows no signs of being afraid of its enemy, is performed in a sleeked horizontal posture facing the opponent. If two birds are matched against one another and one is completely dominant over the other, this posture will appear repeatedly. If the weaker bird rebels, or if two equally matched birds meet in a fight, then defensive-attack postures will appear. In most Estrildine species the latter consists of a sleeked vertical posture. As before, the bird faces its opponent (see Morris 1955b). There are intermediates between the frontal-horizontal and the frontal-vertical postures, but they are less common than the two extremes. When in the vertical posture the birds are not only aggressive, but also afraid, and an increase in fear reveals itself by the bird leaning over slightly backwards. An individual behaving in this way is invariably the first to abandon the fight and flee.

The fighting itself consists of stabbing or plucking. The attacker either lunges out with closed beak in a strong stabbing movement, or else does so with its beak open slightly, grabs at the plumage of the rival, shuts its beak and retracts its neck. In the latter case a plucking results and the weaker bird, if attacked frequently, can be seriously denuded. In the case of the horizontal attack, the attacker is usually the only bird actually to stab or pluck, the rival typically being too concerned with fleeing. During vertical attacks, however, both birds are usually stabbing at one another alternately in very quick succession, giving the fight the appearance of a fencing match.

If an individual has been beaten repeatedly and has assumed a strongly subordinate role, it squats in a fully fluffed posture which appears to have at least some appeasement value. This posture is similar to other submissive gestures in that it presents an appearance which contrasts strongly with that of an aggressive bird. The rounded, fluffed posture of a submissive bird is just the opposite of the thin, sleeked posture of an aggressive individual.

The brief description given above of the three basic agonistic postures of typical Estrildines does not, in all respects, apply to the bronze mannikin. The frontal-horizontal posture (see Pl. 11) occurs in exactly the same form and under exactly the same conditions as with other species; so also does the fluffed submissive posture, although it does not appear to be used a great deal. But there is a striking specific difference concerning the frontal-vertical posture of defensive attack. Although this posture occurs in so many other Estrildines and, for that matter, in many other non-Estrildine species as well, *it never appears in the fighting behaviour of the bronze mannikin.*

This does not mean that this species does not indulge in bouts of beak-fencing. Such a mode of fighting is extremely common, more so even than with other species, but it is performed in a very characteristic latero-horizontal posture (see Pls. 12 and 13). The rivals crouch low over the branch in a sleeked horizontal position with their bodies parallel to one another. Their heads are twisted round towards one another and are thus held at right angles to their bodies. From this position they stab out at one another, striking in a snake-like manner with an extensive, vigorous and extremely rapid stretching and retraction of the neck. The force employed when striking out in this way appears to exceed anything achieved in the vertical beak-fencing bouts typical of other species.

Whilst raining a rapid succession of blows on one another, the contestants cling frantically on to the branch. Defeat is often achieved when one bird strikes so hard at the other that the latter loses its balance and falls from the perch (see Pl. 14). Alternatively, a weaker bird may flee before this actually happens (see Pl. 16).

Since the main objective of this type of fighting is to thrust the rival violently backwards, it is not surprising that an additional form of balancing has been called upon in an attempt to compensate for this. The birds employ their feet to steady themselves by clinging on to the branch with all their strength but, at high intensities, they supplement this by a special use of their wings (see Pls. 14 and 15). Each bird raises vertically, like a sail, the wing farthest from its enemy. The nearside wing remains closed and lowered. The raised wing is held in this vertical position whilte its owner stabs out at and is stabbed by the enemy (Pl. 14). Then, whenever the bird feels itself being thrust backwards particularly violently, it quickly swings its open wing downwards (Pl. 15). If this succeeds, and the bird maintains its balance, the wing is then rapidly returned to its vertical position.

This one-wing-up fighting posture of the bronze mannikin appears to have been ritualized (modified as a signal) to some extent. Firstly, it is often performed rather more than appears to be necessary from a strictly mechanical point of view. One gets the distinct impression that, if the bird merely wanted to balance itself with its far wing, it could do it with a much less spectacular action. Also, the one-wing-up pose is sometimes held after the rival has been defeated and has fled (see Pl. 17), or is assumed before a fight has actually commenced. Nevertheless, it is far less ritualized than it could be and a bird seldom relies on its use without the backing of a series of lunges.

It is perhaps worth noting that, of the many species of Estrildine finches, there are only a very few which possess special markings on the back or wing region. The bronze mannikin is one of the exceptions, having, as mentioned earlier, glossy green scapular patches. The one-wing-up posture displays both scapular patches simultaneously to the rival and this, coupled with the twisting of the head, shows off the maximum amount of glossy green area possible. Since so many species possess neither the one-wing-up posture, nor the dorsally situated markings, and the bronze mannikin possesses both, the correlation is probably a significant one. The amplification of a movement by the addition of conspicuous markings is probably a second way in which the action has become ritualized. This view is strengthened by the fact that the very closely related black-breasted mannikin (*Lonchura bicolor*) also has both the one-wing-up fighting posture and bright markings on its wings (see Pl. 18). In this species there are numerous small white rectangular wing markings which stand out clearly against the dark background of the general wing colour. (They are only just visible in the photograph in Pl. 18 because the leading edge of the wing concerned is facing the camera, but to the rival mannikin they must be conspicuously displayed.)

It is not at all clear at this stage why the bronze and black-breasted mannikins should have evolved this specialized fighting technique when so many other species have used the simple vertical body posture for beak-fencing. As far as is known at present, there are no important differences in the way in which these two species fly, hop, feed, or cling on to branches, when compared with the others. The only possible explanation seems to be that these two, for some reason unknown, having become more communal in their behaviour generally, needed a more violent and impressive method of counteracting this tendency. The very strong urge to keep close together can often be seen to interfere with reproductive activities, as will be described below, and perhaps a correspondingly strong method of repulsion was therefore evolved to balance the situation. These two species are certainly amongst the most sociable of all the Estrildines and, also, their method of striking when fencing is certainly more violent in its impact than that of the other species. (Butler, 1899, for example, reports that a bronze mannikin can even put the largest of the Estrildines, the Java sparrow, to flight. The respective body weights of these two species are 9.2 grams and 25.4 grams.)

Another apparent difference between the agonistic behaviour of the bronze mannikin and that of the other Estrildines may be mentioned at this point. This concerns what can only be described as the 'fluidity' of the aggressive encounters of the former when compared with the latter. If an agonistic situation arises, then it is, where the bronze mannikin is concerned, an immediate flare-up into actual attacking and actual fleeing. The competition between the two incompatible tendencies of attacking and fleeing is extremely unstable and the conflict states are quickly resolved. This does not mean, however, that bouts of aggression are necessarily shortened. They often are, it is true, because of one bird being quickly beaten, but they are just as likely to be prolonged and intense. The point is that during a long bout the birds are actually fighting the whole time, or chasing and fleeing. With other species, it can be observed that the birds concerned in an agonistic situation may hold themselves in a stiff, tense and rather static manner for some time. During such a period, when the birds are eyeing one another suspiciously, so to speak, and perhaps occasionally lunging out, typical conflict responses can often be observed (such as displacement activities, redirection activities, intention movements, or ambivalent behaviour). These periods of indecision, when the balance between attacking and fleeing has reached a state of temporary partial stability, are largely absent in the agonistic behaviour of the present species. So also are the typical conflict responses.

It should be stressed, however, that, although this difference exists within the Estrildines, it appears small when the Estrildines as a whole are compared with many other kinds of birds. Other finches, such as the Pope Cardinal (*Paroaria larvata*), for example, show a much higher ratio of periods of 'agonistic indecision' to periods of actual fighting, than do any of the Estrildines. (An increase in this ratio is usually, but not necessarily, coupled with an increase in complexity of the threat code of a species. However, this is discussed in more detail elsewhere [Morris, 1957].)*

In concluding this section, brief mention must be made of certain interspecific agonistic patterns. In the present study these all arose in connection with man-bird relationships, except for the simple observation that the bronze mannikin would attack almost any other finch with which it was housed. The only signs of actual aggression towards man occurred when certain individuals were held in the hand. In several cases, the sudden and unexpected pain received from the strong biting mandibles resulted in the involuntary release of the bird, and it is possible that a similar effect is achieved with certain natural predators. (It is strange that only certain species of Estrildines show this response.)

* See pp. 187–97.

No special display was observed on the part of brooding birds on or at the nest, when approached by man. They simply stayed on the eggs until the last moment and then fled. This is not the case with at least one other Estrildine, namely, the cutthroat finch (*Amadina fasciata*). Miss E. Eisner (personal communication) has observed the brooding female of the latter species to perform a bizarre 'snake-dance' on the nest when disturbed.

When alarmed by man, bronze mannikins show a number of characteristic patterns: 'freezing', vertical stretching, horizontal stretching, tail-flicking, wing-flicking, and alarm-calling. (These responses have been observed in most of the Estrildine species so far studied, but there are some interesting specific differences.)

'Freezing' absolutely still and remaining completely silent occurs as a response to the sudden cessation of a strong alarm stimulus. It is sometimes (most effectively) coupled with hiding. Hiding, by diving into a dark corner and 'freezing' there, is far from typical of other Estrildine species, but it is extremely common as a manoeuvre of escaping bronze mannikins.

As an individual relaxes after a fright, it becomes less static. After some minutes, the 'frozen' pose is dropped and usually the bird then shakes its plumage several times. Then, while it is still mildly alarmed, it begins twisting its body this way and that, calling with the cheep-cheep alarm note of the species, and rapidly flicking its tail and wings. The tail-flicking is primarily a lateral movement, but as the tail swings rapidly from side to side it is just possible to distinguish a slight vertical component. In this respect, it is very similar to the spice finch (*Lonchura punctulata*) and a number of other mannikins. Like them, it also flicks the wings slightly at the same time. (For a more detailed discussion of these movements, see Moynihan and Hall, 1954.) It is perhaps worth noting that certain of the Australian Estrildines (grassfinches) hardly ever show these flicking movements and, when they do, they appear in a much less conspicuous form. Also, certain other grassfinch species, such as the crimson finch (*Poephila phaeton*) show a conspicuous *vertical* tail-flick. The reasons for these specific differences are not yet clear.

This flicking and calling does not only occur during the period of partial recovery from a severe fright, but also at any time when the bird is mildly alarmed. Vertical and horizontal stretched postures also occur then with a high frequency. These both appear to be intention movements of flight and, coupled with the typical sleeking of the feathers which accompanies fear, give the bird a very long and thin appearance. The vertical stretched posture (see central bird in Pl. 10) appears to be an intention movement of jumping upwards, whereas the horizontal stretched posture is an intention movement of the usual horizontal flight take-off. It is perhaps worth noting here that the bronze mannikin is perfectly physically capable of performing a vertical stretched posture, when alarmed, which is as upright as that performed by any other Estrildine; also, it performs it just as readily. There is no structural reason, therefore, why this species should not beak-fence in the vertical posture. The explanation must, as already suggested, be sought elsewhere.

Nesting Behaviour

In the wild state the bronze mannikin is far from being fastidious when selecting a nest site, except that the latter is always well above the ground. A wide variety of host plants or objects is used, Chapin (1954, p. 453), for example, listing the following: '... between the leaf bases of oil palms, in forks of acacia, lemon, orange, mango, and many other trees, beneath the eaves of grass-thatched houses, or even in bunches of green plaintains hanging in gardens. Rather often, too, an old nest of some ploceine weaver will be taken over and relined.'

However, there is one point of especial interest in connection with the nest-site selection of this species: namely, its habit, as reported by Moreau (1942) and others, of building in the vicinity of wasps' nests. It is suggested that this may afford it some protection from predation, and Bannerman (1949, p. 228) suggests that a similar function may be served by its tendency to build in and around places of human habitation. Judging by various reports from the wild, this species is particularly heavily predated by a variety of animals, and it may well be that the superficially risky measure of building near wasps or humans improves a brood's chances of survival. It would certainly be most interesting to analyse the wasp association experimentally.

Structurally, the nest of the present species differes little from that of the zebra finch (see Morris 1954a),* or, for that matter, from that of any of the Estrildines. It is an untidy roughly spherical construction with a small entrance hole on one side. It is not woven in the manner of the true weaver birds, the motor patterns used when building being comparatively simple. The nearest approach to weaving which was observed was the synchronized use of the beak and one or both feet, when manoeuvring a piece of nest material. Having picked up a length of material in its beak, a bird would then bend forwards and downwards quickly until its beak came close to one of its feet. The latter would then grasp the strand and hold it firmly on the branch or substratum. The beak was thus free to examine the material at different places along its length. A great deal of this apparent testing was carried out at different times and it took the form of mandibulating (i.e. a rapid nibbling; the beak opens and shuts extremely quickly, whilst the tongue is moved about). Mandibulating appears to test the texture and perhaps other properties of a piece of material, such as elasticity, for example. After a few moments, the position of the material is usually shifted and a new section is then tested.

* See p. 50.

The shifting is performed by a rapid synchronized movement of the neck, beak and foot. The beak is brought down to near the foot which grasps the straw or grass. There it takes the material between its mandibles again, raises its head and then re-grasps the grass with its foot. Alternatively, it may quickly transfer the grass from one foot to the other via the beak in a similar way.

This manner of actually handling the lengths of material is similar to that of the true weavers, but the bronze mannikin does not seem to make use of it to weave the different pieces in amongst one another on the nest. The most of it appears to be able to do is to loop the pieces into and around the entrance of the nest, or on to and around the loose initial structure. The difference between the building methods of this species (and all other Estrildines) and the true weavers can be best understood by attempting to destroy their respective nests with one's hands. An Estrildine nest comes to pieces easily when it is pulled apart by hand, but a true weaver's nest can only be torn apart with difficulty. The woven knots of the latter have the effect of tightening the structure as it is pulled.

Apart from this major difference at the sub-family level of the Ploceidae, there are also differences within the Estrildinae concerning construction techniques. The bronze mannikin is apparently identical, in its building methods, with the spice finch (see Moynihan and Hall, 1954). It differs, however, from the zebra finch (Morris 1954) which never uses its feet when manipulating nesting material. This difference also promises to be taxonomically important, since there are a number of species which belong to each type; but this will be discussed in later, more primarily comparative, papers.

The nature of the material used seems to vary a great deal, according to reports from the wild, but basically it consists of fine long grasses. Lining may, or may not, be present. In captivity, the birds were given grass, straw and moss. There was a strong preference for the finer grasses, especially when freshly brought in from the fields. When they were older and drier they appeared to be too brittle for the birds, which repeatedly tested and then dropped them. Nevertheless, when given no choice, they would use almost anything (coloured strings, for example) to build with.

As with other Estrildines, the bronze mannikin pairs do not share the task of nest-building equally. The male performs all, or nearly all, of the actual collecting, while the female spends most of her time in or near the nest. A typical sequence involves the male flying up with a long piece of grass, entering the nest hole, depositing it inside and perhaps fiddling with it a little. Then he leaves to collect another piece while the female tidies up the last piece he brought by repeatedly grasping it with her beak and pushing it this way and that, up against the inside walls of the domed structure. The cavity is increased in size and form by a special movement of *neck-stretching*, as in the case of the zebra finch (Morris, 1954a) (= 'pushing', in the spice finch; Moynihan and Hall, 1954, p. 57). This scoops up the sides of the cavity or raises the roof, the material being pressed up or back by the beak and/or head. Both sexes perform this movement, but it is more often shown by the female, since she spends more time actually inside the incomplete nest than the male. It also occurs as part of a special nest ceremony (which will be described in full in the next section, on sexual behaviour) when it may be performed by both members of the pair at the same time in a strangely stilted and modified form.

The completed nest is hotly defended against all intruders. The latter may be permitted to sit within inches of the entrance without being molested, but, if they should stand in front of the entrance, peer into the nest hole, or attempt to enter, then a savage onslaught will develop and the offender may be pursued and attacked for some time and some distance away from the nest. It is remarkable how an aggressor in this and various other situations can keep track of its quarry as the latter moves about amongst a dense group of other individuals. If the escapee attempts to make itself inconspicuous by huddling with a group of other birds, it is nevertheless soon spotted and charged at and the chase begins again.

Territorial defence in this species, then, does not consist of the protecting of a large well-defined area, but of a small region in and immediately in front of the nest. Anyone except the mate is attacked there, but, away from this region, there may still, as described above, be a considerable show of aggression. The latter, however, is always a 'personal' and not a spatial dispute. In this way the colony can nest in dense groups, a number of nests occurring together in the same bush. Buttner (1949) even reports that 'Frequently these Mannikins nest in colonies comprising five or six nests built together as one large affair ... it is composed mostly of straw, with only small nesting spaces.' Communal nest-building of this sort is not, as far as is known, common amongst the Estrildines as a whole, although most of them will nest extremely close to one another. Nevertheless, it is interesting that the bronze mannikin shows such a tendency which, in a very mild way, is reminiscent of the nesting behaviour of the sociable weaver of South Africa (*Philetairus socius*). The latter constructs communal nests which may measure as much as 25 × 15 × 5 feet, and which may contain as many as ninety-five nest cavities (Friedmann, 1949).

Under the conditions of captivity, nest-building is often unsuccessful, for one of several reasons. Sometimes the birds attempt to build with material which is too brittle, or unsuitable in some other way, and this results in the dome of the nest being too fragile. Birds often sit on top of the nest and the result is that in certain cases the roof is crushed flat and the eggs squashed or abandoned because they are inaccessible. A low intensity bout of nest-building may also produce a flimsy nest roof, with the same result. If crowded,

Plate 7. Experimental aviary for bronze mannikins, showing grids of perches and symmetrically arranged nesting boxes.

Plate 8. Social preening. In picture on left, right-hand bird invites preening by clumping and tilting head. In picture on right, left-hand bird responds by preening the offered neck region.

Plate 9. A social clump. The birds rest in dense groups in this way; the central bird has assumed the fluffed 'spheroid' posture of sleep.

Plate 10. Clump disturbed by bird at end of row jumping along the backs of the others and attempting to squeeze down into more central position. Note central bird has now assumed vertical stretched posture of alarm.

Plate 11. Left-hand bird shows frontal-horizontal posture of all-out assault. Right-hand bird about to flee.

Plate 12. Left-hand bird and central bird show latero-horizontal postures with head-twisting, typical of more equally matched fighting.

Plate 13. Birds fighting; note crouched postures typical of this species, but different from that of other Estrildines. The latter typically beak-fence in vertical stretched postures.

Plate 14. High-intensity fighting. Note one-wing-up posture of the right-hand bird. The left-hand bird has been thrust backwards and is falling.

Plate 15. High-intensity fight, showing the use of the wings for balancing. Beating down with the wing farthest from the rival counteracts the force of the latter's blows.

Plate 16. Left-hand bird flees from attacks of the other. Note the stretched-leg posture of the fleeing bird.

Plate 17. The one-wing-up posture. This balancing posture has become modified slightly as a display. Here it is still being performed, although rival (*see top*) has fled.

Plate 18. The one-wing-up display, performed here by the very closely related black-breasted mannikin. Note white spotting on displayed wing.

Plate 19. Courtship posture of singing male (*right*) showing open beak with protruding tongue.

Plate 20. Courtship posture of singing male (*left*) at beginning of dance. Note erect plumage and twisting of head and tail towards female.

Plate 21. As above, but showing the legs-stretched phase of the inverted curtsy dance.

Plate 22. As for Plate 20, but showing the male (*left*) after having performed a lateral pivoting dance movement, through 180 degrees.

Plate 23. Leapfrog. On the left, one bird hops over another in a non-sexual situation. On the right, the courting male, in fully ruffled posture, performs repeated leapfrog ceremony over the back of the soliciting female. (Note: the nape gap in the plumage of the male is due to plucking during fights and not to some special feather-erection pattern.)

Plate 24. The courting male (*left*) showing displacement preening.

Plate 25. Copulting pair on left about to be attacked by bird on right. (Photograph taken in colony.)

and lacking in sufficient nesting sites, birds often invade a well-built nest despite the attacks of the owners and, since the invaders do not know the position of the entrance hole too well, this too can result in the flattening of a nest. Finally, nest destruction may occur in a very complete form by the collection *from* the already built nests of pieces of material which are then dropped on to the ground. Sometimes a whole colony of nests will gradually find its way to the floor of an aviary over a period of a few days. The reason for this last phenomenon is not the absence of nest material on the ground, but seems to be connected with a state of low intensity nesting coupled possibly with suboptimally stimulating nest material. The birds spend a great deal of time fiddling with the nearest pieces of grass available (those in the nests) and then, having pulled them out of the nests during the process, they lose interest and drop them to the floor. It is still possible, however, that nest destruction is a special social pattern.

Nest sanitation is apparently absent in the bronze mannikin, as it is with most other members of the subfamily. Correlated with this is the fact that, in healthy birds, the droppings are completely dry and the birds do not therefore foul their plumage. Typically, the birds always sit in a nest with their heads facing the entrance hole. When they defecate, they raise their bodies slightly and aim backwards, with the ultimate result that the back wall of the nest becomes a solid mass of excreta. Aldersparre (1931) reports the rather unusual finding that the black-breasted mannikin (*Lonchura bicolor*, in his case, race *nigriceps*, called by him *Spermestes nigriceps*) which, as already mentioned, is an extremely closely related species to the present one, does however possess a special form of nest sanitation. Other species, he states, attempt to stop up holes in their nests, but the black-breasted mannikin on the contrary makes special clefts in the nest walls opposite the entrance hole. He claims that these are used as latrines by the birds which therefore manage to keep the insides of their nests clean. One is tempted to suggest that the fissures in his nests (studied in captivity) were artifacts and that, because they occurred just where the birds would defecate in any case, they appeared to have been contrived especially in connection with this particular function. However, he does report that he repeatedly stopped up the holes and that the birds repeatedly opened them again, so the possibility that this species has evolved a special sanitation technique remains. (Unfortunately, the nest boxes used in the present study made it impossible to analyse this point, which was only discovered after the present observations had been terminated.) It is feasible that the droppings of the young nestlings require special treatment if the nest is not to be badly fouled during the parental phase.

Before concluding this section some comment must be made concerning the differences that exist between reproductive and non-reproductive nest-building in this species. During the non-reproductive phases of the cycles of most birds there is no nesting activity at all. However, during the non-reproductive phases of many of the Ploceidae, including most of the Estrildinae, this tendency persists at a level of low intensity. The fact that it is performed at low intensity does not, as might be expected, mean that the results are always disorganized and functionless. The bronze mannikin, when living as a non-reproductive group, constructs each evening a communal roosting platform or nest in which the whole group spends the night huddled together, often several layers deep. The task of collecting the material for this construction appears to be shared out between the different members of the group. Sometimes an old nest of its own or some other species is used as a roost, or some natural crevice may be adopted, but usually, whatever the nature of the site, there is a certain amount of nesting activity on the part of the roosting group. Holman (quoted in Bannerman, 1949, p. 227) gives the following account of the roosting behaviour of the bronze mannikin in the wild:

> The roost is similar to the breeding nest except that the structure is less elaborate and substantial. All the members of the family join in construction, and the grass is collected in the immediate vicinity of the intended roost, so the job is easily and quickly done, and as easily and quickly dismantled and reconstructed on another site. I watched the construction of one such roost between insulators on a cross-bearer at the top of a telegraph post. One evening I happened to see the party come home to roost. After a little friendly fussiness all entered. A minute later a second party came along, and then a third. All entered the roost, making a total of 34 birds in a space four inches in diameter. Immediately afterwards a fourth and then a fifth party came along, but when they settled on the wires some of the birds already inside the roost came out. Much entering and emerging followed so that any further noting of numbers was impossible. Some ill-temper might be expected in such circumstances, especially on the part of the rightful owners, but there was not the least sign of it.

Observations were made on the roosting behaviour of this species in captivity and the above statements were confirmed. The pattern was studied in the symmetrical aviary described earlier, which contained nine nest boxes and a flock of eighteen birds. The boxes were all of identical size and shape (see Pl. 7) and were equally spaced one from another around the back and side walls of the enclosure. The group of birds were, at the time in question, in an almost completely non-reproductive condition. Nest material was scattered on the floor, sufficient being provided for the construction of a number of domed nests. The colony had recently been a reproductive one, which was now at the end of the parental phase, the young having become fully fledged. The reproductive, domed nests had been destroyed gradually and the nest boxes emptied of all material. On each night, however, a platform of material was collected together by the birds and a simple domeless roosting nest was formed in one of the boxes. Into this the entire flock would settle at dusk. As far as could be ascertained in the dim light, each individual settled itself down in such a way that it came to lie facing the nest entrance. If a light was suddenly switched on, they could be observed to emerge from the nest, layer by layer, in rapid succession.

In Holman's description of roosting behaviour in the wild, the one aspect which seemed rather pointless was the dismantling and reconstructing of the roost on another site, which was said to take place. This was particularly watched for with the captive colony, and was indeed observed to occur repeatedly. Each day the position of the roost was checked and it was found to move about from place to place, a new site being selected almost every day. Even when an old site was used, the nest material was usually not the same on both occasions. During the non-reproductive phase, the nest boxes were always all empty except for one, in which the roosting platform existed. By noting the number of the box in which it was seen each day, the roosting movements were scored over a period of time, as follows (as mentioned earlier, the boxes were numbered one to nine from left to right): 5-4-4-6-6-6-?-8-?-2-?-4-6-3-3.

New sites were therefore used on seven of the twelve days recorded. The significance of these moves is not clear, either functionally or causally. From the latter point of view, it might be argued that the low intensity of the nest-building involved resulted in a great deal of low level 'fiddling' with material, which accidentally brought about a dismantling of the old structure. A new one would then have to be built and, without special site-attachment, it could be built anywhere. However, although this might explain the results in captivity, where the old roost material was the nearest and most easily available material for 'fiddling' with, it can hardly explain the occurrence of similar moves in the wild state.

It is interesting now to compare these roosting activities with those involved in reproductive nest-building. This can best be done by seeing the way in which a non-reproductive group gradually comes into reproductive mood. There are a number of gradual changes which co-vary, of course, there being an increase in sexual and aggressive behaviour in addition to an increase in nesting activity. The sexual changes will be discussed later, but the increased aggression must be mentioned here. As already noted, there is little or no aggression involved in the roosting behaviour, even under conditions of extraordinary crowding. However, coupled with a sudden increase in nesting intensity, there is the appearance of fighting at the nest, for the paired male and female not only have to construct a more substantial nest, but also have to discourage other individuals from using it. Unmated birds are attracted by the vigorous nesting activities, attempt to join in, and have to be driven away. If the mated pair are not intensely aggressive, they will find that by nightfall their carefully built reproductive nest has become a communal roosting site. The establishment of a private breeding nest would, in fact, be an extremely difficult task, were it not for the fact that the reproductive mood increases more or less synchronously in all the individuals in the group. The slight imperfections which do, however, exist in this synchronization give rise to quite a number of such disputes and nesting failures.

Once a colony has become fully reproductive, each pair having its own densely built domed nest, the communal roosts are no longer constructed. Each pair not only breeds in its private nest, but also sleeps in it at night-time. Unlike the typically territorial species, the bronze mannikin seems to prefer to build its private nests as near to one another as possible. This was easy to test in the symmetrical aviary which was being used. For example, at one stage in the colony's existence, when it was becoming more and more reproductive, the first fully successful private nest was observed to be built in nest box 1. At night, the remainder of the colony slept in two roosts, one in box 4 and one in box 7. A few days later the second and third successful private nests were completed. A typically territorial species would undoubtedly have nested, if at all, as far away as possible from the already existing nest in box 1. Yet the second and third nests were placed in boxes 2 and 3, as near as possible to the first nest and to one another. Five days later a fourth nest was finished, in box 4. During the next week the nest in box 1 was abandoned and crushed, but the fifth nest to be built was constructed in this box, on top of the old one. Later still, the nest in box 4 deteriorated into a communal roost. A few days afterwards, the sixth private nest was completed and this was placed in box 5.

It must be emphasized that, during the above period, all nine boxes were available and were potentially equally suitable as nest sites. (It will be recalled that, when building the single communal roosts, the colony selected a wide range of different boxes at different times.) Nevertheless, private site selection was always made as near to the others as possible. It is easy to see from this how, under special conditions, the multiple nests of the kind observed by Buttner (loc. cit.) could come into being.

Sexual Behaviour

Pair formation

Pair formation in the bronze mannikin is, to a large extent, a negative process. It is as much a matter of restricting already existing social responses as it is of performing new ones. A well-mated pair can be distinguished by the fact that they do not fight one another, that they nest-build together, move about together, rest together and sleep together, and that they preen one another's head plumage. But all these responses are typically performed by a non-reproductive individual to *all* the other members of the flock. The negative aspect of pair formation, therefore, consists of the restricting of these activities in such a way that they are performed only in connection with one other individual. Any attempt on the part of individuals other than the mate to respond in this way, is actively repulsed. The strong tendency to keep together and to behave together has to be repeatedly counteracted by fighting on the part of paired birds. In this way the pairs can segregate themselves from the flock and build their private reproductive nests.

Rather specialized aggressive responses can sometimes be observed in this respect. For example, if a pair is forming in such a way that the female is more strongly mated than the male, the latter may still be courting other females fairly frequently. It was observed that a strange female had only to be courted in this way for it to be violently attacked by the mated female. The new female had itself performed no distinct pattern of behaviour and had shown no change in behaviour when the male began to court it, and yet, because of the male's sexual approach towards it, it suddenly became a very strong attack stimulus for the female. Needless to say, if a strange male approached a mated female sexually, or a strange female followed a mated male around, the stranger was attacked and driven off by the mate of the approached bird, if not by the latter itself.

Turning now to the more positive side of pair-formation, there are two types of sexual ceremony which undoubtedly assist in cementing the pair-bond. These are two types of pre-copulatory display. One is the typical pre-copulatory performance which is comparable with those described already for certain other species (zebra finch: Morris, 1954a; cutthroat finch: Morris, 1954b; spice finch: Moynihan and Hall, 1954). It takes place in the open on a branch or twig. The other occurs inside the nest and is comparable with the pair-forming nest ceremony of the zebra finch (see Morris, 1954a). The latter, however, never leads up to a copulation, whereas the nest ceremony of the bronze mannikin often does so. This complication makes it difficult to divide up the courtship behaviour of the present species neatly into pair-formation and pre-copulatory patterns. Nor can it be divided up on the basis of a time sequence, since the two types of ceremony do not follow one another in a set order. However, as the nest ceremony of a number of other species does not seem to lead up to copulation, and as, in certain cases, its relationship with pair-bond maintenance has been well established (Morris, unpublished), it is best to consider it as a separate aspect of sexual behaviour and to discuss it before going on to deal with the typical pre-copulatory patterns.

The nest ceremony

The nest ceremony can be observed to occur between the members of a well-formed pair, or between two birds which are only just beginning to pair. It occurs most typically when the nest is under construction, but it may also be seen to take place before private building has commenced. In the latter case, the birds either use the remnants of a communal roost, someone else's temporarily unprotected nest, or even an empty nest-box. In the latter case, the ceremonial nesting movements are performed, so to speak, in vacuo.

The simplest form of the nest ceremony is adequately described by the following passage taken from protocols: 'The male leads in, then calls, and the female follows. Then both give a squeaky rasping nest-call together in the nest and there is a great deal of turning and twisting.' This basic pattern, including a soft nest-call, has been observed to occur in a number of other species of Estrildinae, but the more complicated high intensity versions which must now be described have so far been seen only in the present species.

In contrast to the above, here is a passage from protocols which describes a complicated high intensity version of the nest ceremony. It concerns the very early stages of the pairing of a female, Blueback, with a male, Blue, at a point where the former was taking more of the initiative:

> Blueblack is now seen clearly to (1) enter box 4 and to (2) start scooping up the sides of the old nest with neck-stretching movements. Then she (3) adds in the squeaky nest-call, which attracts male Blue, who looks into the nest from the perch outside. Blueblack then (4) stops calling and begins mandibulating, while still making scooping movements. The latter (5) become less intense as the mandibulating takes over and (6) finally the scooping stops and the bird then sits in the nest cup still maintaining her neck-stretched scooping posture, but not moving her head and neck any more, but holding her beak wide open, with her tongue protruding. This stimulates male Blue to (7) jump into the nest and (8) to fully raise his feathers and (9) to hop on to the female's back and (10) to copulate.

The bracketed figures above have been added here to emphasize the complexity of the behaviour sequence involved. It must further be stressed that the above sequence can be interfered with in various ways, if the male is more easily stimulated to enter the nest, or if he is himself the initiator. The above example was quoted here because, in this case, the female had to pass right through all the various stages of the ceremony to the highest intensity performance before the male would respond and enter. Many nest ceremonies were observed and in the different instances it was noticed that the male often responded to a displaying female at a much earlier stage. Sometimes, for example, the female would actually be accompanied or preceded by the male when she entered the nest, and both would then scoop up and call and mandibulate together in the nest cavity. If the female then gave her open-mouth display (or, sometimes, even if she did not), the male would mount and copulate. Often these last stages did not develop and the ceremony would end at the mandibulating phase by the departure of one of the birds.

It was often observed that, by the addition or substitution of a later stage in the above described ceremony, a displaying bird could coax a hitherto unwilling bird into the nest in question. For example, if a female started scooping inside a nest, this might be sufficient to attract a male in. Where this failed, the addition of the nest-call was seen to result in success in certain cases. If this still did not work, the calling might be changed to mandibulating (which, in this ceremonial form, is performed so hard that it can be heard as well as seen), and this change was also observed to bring success in certain instances. In the case previously quoted, Blueblack, failed even at this level of intensity, and only succeeded when she passed on to the highest peak, with the open-mouth display. When males took the initiative they never went beyond the mandibulating phase without some encouragement. If this was forthcoming, they then either mounted without further ceremony or, in a few cases, they sang and, in the latter instances, they too had their beaks wide open and their tongues protruding. (A special male pre-copulatory pattern called 'leapfrog' was also observed on rare occasions inside the nest, but this will not be discussed until the next section.)

The pairing significance of these ceremonies was emphasized by the fact that certain individuals which were following a particular member of the opposite sex around very closely, would also repeatedly attempt to attract the potential mate into a nest-box by giving the display in its proximity.

Functionally speaking, the nest ceremony seems to be both a way of achieving fertilization and also a method of strengthening pair-bonds, especially in connection with a particular nest site. The latter function it shares with the comparable ceremony of the zebra finch. (The nest ceremonies of the two species also have certain motor patterns, such as turning, mandibulating and the soft nest-call, in common, and also occur under very similar conditions.)

One way in which they are *not* functionally important any longer is nest-building itself. The scooping-up movements of neck-stretching are performed with an orientation component that frequently shows little adjustment to the specific surroundings. This is most clearly illustrated when one bird is displaying to another on a potential nest site (in the present study, this meant an empty nest-box). Here, in the absence of anything to be scooped, the bird nevertheless went through all the motions of scooping up imaginary nest material, with the full neck-stretching pattern. Obviously, this movement has become ritualized to the extent that it can occur purely as a signal, without the presence of the most important of its original stimuli. Similarly, the mandibulating action of the beak has become independent of any tactile stimuli. In non-signal situations, mandibulation occurs as a response to the presence of some solid material (in particular, of course, seed or nesting fragments) in the mouth, the object in question being crushed or certain of its physical properties being tested by the movement. Now, however, it can also occur in signal situations, with no object held in the beak at all.

When nesting, the birds are repeatedly seen mandibulating lengths of grass and it is not surprising that mandibulation also appears as part of the nest ceremony. The tongue is moved about a great deal during mandibulation of the mechanically effective kind, and the protrusion of the tongue during the open-mouth display makes it seem likely that this is yet a further, more modified signal version of mandibulating, in which the rapidly opening and shutting mandibles have 'fixed' in the open position. This will be discussed further in the next section, because these particular patterns are also shown in the typical pre-copulatory ceremonies.

The pre-copulatory ceremonies can be observed more clearly than the often obscured nest ceremonies, and it is possible that much remains to be learnt about even the most basic aspects of the latter, especially in other species. A further understanding of these nest ceremonies is, indeed, likely to provide valuable derivational clues concerning certain important aspects of the typical pre-copulatory ceremony, which must now be described.

Pre-copulatory behaviour

If a stranger is introduced into a colony, or if two individuals are placed together in an aviary for the first time, there is invariably an immediate outburst of aggressive or sexual behaviour. The newcomer to a colony may establish its sex without delay by courting the first bird it sees, or alternatively by giving the female invitation signal. The latter, however, is seldom performed unless the stranger is approached by one or more members of the group; even then it is typically only given when the approach consists of the male sex display.

If the newcomer does not reveal its sex to the group, it is likely to be courted, or attacked, according to the predominant mood of the colony members. If the new bird is a male and it is courted by the other males, it will attack them, with the result that violent beak-fencing and eventually prolonged pursuits develop. If it is a female, it may simply respond passively to being courted, by moving away, or staying still without performing any special motor patterns. As a result of this it may be attacked, but less violently than the new males, or it may be ignored after a while, or one of the old birds may clump with it.

If the sexually active newcomers succeed in finding a member of the opposite sex with which they can perform pre-copulatory patterns without too much interference, then the beginnings of pair formation may be observed. The male and female concerned will clump together, preen one another, prospect for nest sites together, and so on.

In an active colony, however, there are many interruptions in the development of the pair-bonds. Birds often lose or leave their mates and become involved with other individuals. Occasionally sexual triangles and other, more complicated relationships develop, but all these become smoothed out when the serious nest-building and then incubating stages are reached.

The pre-copulatory patterns, therefore, are extremely important and not only lead to fertilization, but also play a part in sex recognition, the setting up of social relationships between the members of reproductive colony, and the formation of the pairs. The male is typically the initiator of courtship sequences and, as is the case with all the other Estrildine species studied, his pattern is far more complex than that of the female. Both of these must now be described and analysed into their component parts.

The courtship of the male

(a) *The auditory component*

As in the case of the zebra finch (Morris, 1954a), the bronze mannikin male never courts without singing, but frequently sings without courting. The song is extremely faint; each bout of singing consists of a series of short identical phrases. To the human ear, the extent of individual variation of the song phrases is extremely small. It is difficult to tell which male is singing if the performer is unseen. In this respect the bronze mannikin contrasts strikingly with certain other species such as the zebra finch, where individual males can be recognized with ease by their song phrases.

There are certain distinct differences between the songs of non-courting males and those of courting ones. A non-courting singer, which is either isolated, or which does not orient itself towards any other bird when performing, typically sings in short bursts. These bursts consist of bouts of only a few quickly repeated song phrases, and they rarely develop into longer continuous responses. The singing bouts of a courting male, however, are often as long as the period of courting itself. There may be slight breaks in the singing of a courting male, correlated with short pauses in the dancing and posturing, but, even so, the continuous stretches of repeated song phrases are typically much longer than those of the non-courting male. Also, the singing posture of the male is more intense and elaborate when the bird is courting. It is fundamentally the same general posture, but it is much more extreme. This is true of most of the Estrildines studied, although the difference is greater in some than in others.

The motivation of Estrildine song appears to be predominantly sexual. The typical dual function of song, of repelling other males and attracting females, does not apply to these birds. It seems as if, with increasingly communal tendencies, not only has the volume and the carrying power of the songs diminished, but also the repelling function and the aggressive motivational component have disappeared. None of the grassfinches or mannikins studied has employed song in the classical repellent manner, and none has shown a co-variance between singing and aggressive tendencies. (One of the waxbills, *Estrilda melpoda*, provides a possible Estrildine exception to this rule; see Morris, 1954a.) There is a striking correlation, however, between the frequency of singing and that of pre-copulatory patterns.

The present evidence suggests that singing is the sexual pattern with the lowest threshold of all the pre-copulatory components.

The above view differs somewhat from that of Moynihan and Hall (1954), who consider that the auditory performance of courting spice finches should not be called song, and also that it has a 'more or less independent motivation'. They propose the term 'jingle' for what I am here calling the song. Their main reason for proposing this new term is the already mentioned absence of the dual function. However, since I feel sure that both the reduction in volume of the noise, and the loss of its repellent function, are specializations of the more usual song-type, in connection with increased communal tendencies, I also feel justified in calling the Estrildine performance a song. The fact that certain other kinds of Passerine birds, which have become more or less sociable during evolution, appear to have also lost to a greater or lesser extent the aggressive function of song, supports this view.

The reasons given by Moynihan and Hall for supposing that the motivation of the spice finch's jingle (song) is not sexual, are as follows. They point out that despite the connection between pre-copulatory patterns and the jingle, the non-courting version of the latter cannot be sexually motivated because it does not follow after, or lead to, other sexual patterns and because it may be performed by males with or without overt sexual tendencies. However, if non-courting singing is considered (as in the present paper) as the lowest intensity sexual response, these difficulties are reduced, if not eliminated.

In conclusion it must be pointed out that it is not intended here to give the impression that the problem of the song motivation of these species is a simple one. Much more experimental work is needed on this subject (e.g. sex hormone injections, or an investigation of the difference between song suppression and reduction). The present, rather general, evidence points to a strong connection between sex and singing, but future research may show that it is more complicated than this and that Moynihan and Hall were correct in assuming a partially independent motivation for song. Nevertheless, at the present stage, I do not feel that we are justified in concluding that the evidence collected so far definitely supports the latter view.

(b) *The static-visual component*

A courting male not only makes a characteristic noise, but also assumes a characteristic posture, as follows:

The bird crouches over the branch with its legs bent and its body at an angle of approximately 45 degrees from the vertical. It orients itself broadside on to the female, but twists both its head and its tail towards the latter. The posture of the head is complex; it involves three factors: (a) a lowering of the head so that the beak points downwards, (b) a slight turning of the head towards the female, and (c) a marked rotation of the head in the direction of the female. However, the fact that the head posture can be analysed into these three parts does not mean that they appear separately in any way. The head posture is, in fact, always assumed with a single movement, which is the resultant of the three factors described here.

In addition to its special position relative to the female, the beak of the male is held wide open and the tongue is protruded (see Pl. 19).

Finally, the posture includes dramatic plumage erection (see Morris, 1956b). The white ventral feathers are ruffled, as are the brown feathers of the nape, back, rump and flanks. The crown feathers may be slightly erected, but not conspicuously so, while the dark throat region remains comparatively sleeked. The general effect is to increase the apparent size of the male and to give it a ragged rounded appearance. All the various aspects of the static-visual display are illustrated in the accompanying photographs (see Pls. 19, 20, 21, 22 and 23). Their possible origins will be discussed below.

(c) *The dynamic-visual component*

Whilst rigidly maintaining the above described complex posture, the singing male adds further to its pre-copulatory display by dancing to the female. The dance consists of two basic movements, one vertical and the other horizontal.

The vertical movement consists simply of a stretching of the legs from the crouched position (cf. Pls. 20 and 21). The legs are stretched and bent again very rapidly, giving the effect of an inverted curtsy. The up component of this up-down movement is characteristically the more vigorous of the two.

The inverted curtsies may be performed singly, or in short bursts of several at a time. They may be made on the same spot, or interspersed with a hopping advance towards, or with a hopping retreat from the female. One, two, or three of these curtsies may occur per song-phrase.

The horizontal dance movement consists of a rhythmic lateral pivoting through a wide angle (cf. Pls. 20 and 22), as the male advances towards the female. When the courtship occurs along a single straight branch or perch, each pivoting hop involves a twisting of the body through 180 degrees, so that the left and right side of the male are presented alternately to the female, with the feet changing their position each time. In addition, the head and tail are re-twisted with each pivot, so that they maintain their orientation towards the female at each pause between pivoting. On the other hand, if a male faces the courted female across a gap between two perches, the pivoting may be performed without a change in the foot position, the body then being twisted from side to side through a rather smaller angle.

The exact nature of this pivoting dance movement is important when considering specific differences. Certain other Estrildines typically perform the 'foot-change' pivot through a much smaller angle than the 180 degrees of the bronze mannikin. Others will show both the 'foot-change' and the 'foot-still' pivoting along a straight branch towards a female, whereas the bronze mannikin usually performs the former only under such conditions.

Another apparently important comparative aspect of the dance movements concerns the relationship between the inverted curtsy and the pivot. At first, it was thought that certain species performed only one of these, whilst others displayed the other. However, after more detailed scrutiny, it was found that most species performed both, but usually with a greater emphasis on one of them. In the case of the present species, it is the vertical movement that has the lowest threshold. If a male begins to court, by singing to a female in the special pre-copulatory posture, it may do so without the addition of any dance movements at all. However, if there is no interference and the female does not flee or attack the performer, and the courtship continues, the male then typically begins to show the inverted curtsy movements. This is the most frequently observed form of the courtship complex. But, at higher intensities, the male may add on the pivoting movements as well. These are either combined with inverted curtsies simultaneously, or alternated with them. The former tends to occur when the pivoting is of the 'foot-still' variety, the vertical and the horizontal movements then combining in a one-to-one ratio to produce a body-swinging movement with a complex curved path. On the other hand, if a male is performing the display along a straight branch, it performs the 'foot-change' pivots not with, but in between, the inverted curtsies. Typically, one might observe: curtsy-curtsy pivot-to-left curtsy-curtsy pivot-to-right curtsy-curtsy pivot-to-left ... and so on; or, curtsy swing-to-left curtsy swing-to-right curtsy swing-to-left ... etc. Occasionally there are more than two curtsies in one position. Also, in one bout of dancing, the number of curtsies between each pivot is not always constant, although it tends to be so. Finally, one other irregularity that was sometimes seen involved the interspersing of a 'foot-still' pivot combined with an inverted curtsy when the latter was being performed in between two 'foot-change' pivots during a courtship along a straight branch.

This last point reveals that in the present species the position of the female is not the only controlling factor determining which type of pivot occurs. Whilst orientation undoubtedly plays a dominant role here, motivational differences also appear to be operating (as they most certainly are in other species), if only to a minor extent.

When comparing the relationship between the vertical and horizontal dance movements of the bronze mannikins with their relationships in other species, some interesting differences emerge. In the present species, the vertical has the low threshold and can occur by itself. The horizontal has a higher threshold and occurs only when the vertical is already present. The bronze mannikin, in fact, shows four basic stages (not necessarily sequential) of increasing complexity in its courtship pattern: a male may sing, unoriented to a female, in a posture which appears to be a very low intensity form of its full courting posture; next, a male may sing towards a female, the full courting posture being added in along with the orientation component; next, it may add the vertical dance to its song and posture; finally, it may add its horizontal dance to its vertical dance while it still sings and keeps its posture. Addition of elements is the rule here, rather than substitution. This is not the case, however, with certain other species.

With the bronze mannikin, pivoting does not occur in a bout of courting where there is no inverted curtsying (and the latter does not occur where there is no posturing, and the posturing does not occur where there is no singing). In certain other species, however, pivoting may occur by itself without the vertical movements, as the more frequent response, with the inverted curtsies appearing only as additions to it. In other species again, the vertical movements may be almost or completely absent. In still other species, the song and posture may occur either with the vertical movements alone, or with the horizontal movements alone, or with both together.

Taxonomically, these differences are interesting in themselves, but ethologically they pose complicated motivational problems, which will be discussed below.

In addition to the two dance movements described so far, the courting male bronze mannikin performs several other types of movement. One is a very rapid quivering of the protruded tongue and another is a fast and repeated opening and closing of the beak (mandibulation).

The exact conditions for the occurrence of courtship tongue-quivering are not yet certain, as it was only observable under ideal conditions. However, it is likely that it occurs at all times when the beak is held wide open and the tongue protruded; which means, in fact, during most of the courtship periods.

The circumstances for courtship mandibilating were rather more clearly seen. It was observed during most, but not all, of the courtship sequences and was always alternated with the song phrases. The fact that it has become exaggerated and more vigorous in its signal form than it is in its mechanically effective form has resulted in the production of a *brrrrrrr*-noise. The interspersing of this in between the song phrases has virtually turned mandibulation into part of the song itself. (The same auditory version of mandibulating is used inside the nests during the nest ceremonies already described, and birds could often be *heard* to be performing the nest ceremony, even when they could not be seen, as a result of this.)

In a bout of courting where mandibulation is appearing, the sequence is as follows. At the very beginning of the bout, it is the mandibulating that appears first. A fraction of a second after it has started, the male's feathers begin to ruffle and the characteristic courtship posture is assumed. This is achieved in about one second, and then the mandibulating ceases. The posture is, of course, retained from now on, during the courting, but the rapid opening and shutting of the beak stops, with the mandibles in the open position and with the tongue protruding and quivering. Simultaneously with the holding of the mouth open, the short song phrase begins. This lasts for only one or two seconds; as soon as it stops there is a very brief mandibulation and then it starts again. In this way, the posturing bird goes on alternating the *brrrrr* noise plus the mandibulating with the song phrase plus the open beak and protruding tongue.

Once again, this is a character which is likely to be very valuable for comparative purposes. In one species (*Lonchura ferruginosa*), for example, it has become even more ritualized and appears to be an important auditory signal. In this latter case, the opening and shutting of the beak occurs in exactly the same sequential position as in the bronze mannikin, but it has become much louder and slower. Instead of a vibrating *brrrrr* noise, there is now a regular and deliberate clapping together of the mandibles at about the speed of average human applause-clapping, producing a tick-tick-tick-tick noise. Other species, however, do not seem to have increased the intensity of the movement beyond that employed by the bronze mannikin, but have, on the other hand, increased the frequency with which they use it, and now employ it in several kinds of social encounters (e.g. certain *Poephila* species).

Under special conditions, yet another type of movement may be added to the courtship of the male bronze mannikin. The best name to describe this is 'leapfrog'. The male hops on to the female's back and pauses there for a split second, then hops off again. At this point he usually turns round, hops on again and off again, thus coming back to his starting point. Each time he sits on the back of the female, the long axis of his body is at right angles to hers (see Pl. 23). It is never parallel with it, as in a copulation attempt. The male, in fact, makes no attempt to turn into a copulatory position as he jumps on to the female, or while he pauses on her, and the whole pattern is quite distinct from typical mounting behaviour.

During a leapfrog session a male may jump back and forth over a female once or a number of times. While he is doing this the singing seems to stop, but the ruffled feather-posture persists, as can be seen from Pl. 23. Leapfrogging can occur before, after, or during a 'normal' bout of courting and is caused by the soliciting of a female to a male who is not strongly enough motivated sexually to mount and copulate. Instead of responding to her invitation signal in the usual way, he hops on and off her back in a rhythmic series of jumps. Whether his reluctance to mount correctly is the result of a too weak sexual tendency, or a stronger sexual tendency which has been suppressed by certain incompatible tendencies, is not yet known. However, here are six examples taken from protocols:

> ... male starts high intensity courtship, but without pivots ... female solicits ... he hops on and leapfrogs to her other side ... a few more inverted curtsies and a pivot and he mounts properly ...
>
> ... male dances to female ... nine inverted curtsies ... she solicits ... he leapfrogs over her once and back again ... then they begin to fight one another ...
>
> ... male starts courting and as soon as female solicits he mounts and shows full copulation attempt, which is apparently successful ... off her back ... mounts again ... off again ... mounts again ... off again and then quick preening movement ... leapfrog over her and then quick preen again ... then she attacks him and he fights back ...
>
> ... male performs song and dance with curtsy and pivot ... female quickly responds and male copulates ... then hops off ... three curtsies ... copulates again ... three more curtsies ... leapfrog over female ... eight more curtsies ... leapfrog over again ... quick preen movement ... leapfrog ... quick preen ... and then courtship ends and male feeds ...
>
> ... male dances ... nine curtsies ... then pivoting begins and interspersed with next twenty-three curtsies ... then mount and copulate, a full attempt, but female not responsive ... male off ... quick preen ... mount ... off and quick preen ... mount ... quick preen ... mount ... quick preen, preen, preen, preen, preen ... now the female solicits at last and the male responds by leapfrogging and more preening ...
>
> ... male approaches in ruffled posture but not singing or dancing and leapfrogs over the new female ... leapfrogs again and while sideways on her back, she solicits, but he hops off despite this ... after short pause he begins to sing and dance to her, but no copulation attempt ... soon he attacks her ...

A scrutiny of these cases reveals that the leapfrog ceremony is intimately connected with displacement preening (see Pl. 24). Since both these occur when the males are not being sexually thwarted by the absence of the appropriate stimuli from the female, it seems likely that they are both being caused by a conflict state in the male. (Of the two, the preening appears to have the lower threshold.)

The courtship of the female

The female possesses a comparatively simple pattern. If she is not responsive, she either flies away or attacks the courting male. To stay where she is is to subject herself to continual courtship, and remaining on the spot without displaying is a mild sexual response when performed by a courted female. It is also a sexual stimulus to the male. A strongly motivated female, however, not only stays still, but also performs the soliciting display referred to above. This consists of crouching low over the perch, erecting the feathers, stretching the neck forwards and a little upwards, opening the beak, protruding the tongue, and (most important of all) quivering the tail. A front view of the posture can be seen in Pl. 23. Slight variations in the above display have been observed from time to time; for example, the whole pattern was sometimes seen to take place but without the feather erection.

The tail-quivering is the most important sexual stimulus to the male, and is found in all the Estrildine species so far studied. It appears to be a most important character at the sub-family level, no other Ploceid groups possessing it, as far as is known at present. (Wing-drooping appears to be the most important female soliciting device amongst other finches in general.) The other aspects of the female bronze mannikin's invitation display are, on the other hand, rather more unique. The tongue and beak pattern, which she shares with the male as a sexual response, the oblique slope of the neck, and the ruffling of the feathers are components which have not been seen to appear in the soliciting displays of the females of any but a few of the other species observed.

The tail-quivering of the female bronze mannikin was seen to occur both as a response to being courted and also, on rare occasions, as a stimulus initiating courtship. In one case a newly introduced female, who was attacked by a male but not courted, responded to the pecks he delivered with a full soliciting display, which quickly changed his mood into a more predominantly sexual one and led to mounting. (It is interesting to note that, on this particular occasion, the soliciting display lacked the feather erection component and was also more horizontal.)

Frequently the tail-quivering only appeared after a female had been courted repeatedly for some time by a male, thus revealing the arousal effect of the latter's performance.

Displacement cleaning movements are performed by courted females as well as courting males. Displacement preening and, more rarely, displacement beak-wiping were observed in birds which, for some reason, did not solicit despite a strong interest in the courting male. This interest took the form of a stealthy approach towards the displaying bird, which often responded by continuing his display while slowly retreating from the on-coming preener. The preening bird advanced towards the male in a sleeked horizontal posture, while the latter hopped backwards still singing and dancing. The courted bird, on such occasions, usually became overtly aggressive after a while.

Copulation

Copulation was observed to occur in three situations: in the nest, on a branch, and on the ground. Perched copulation appeared to be the more usual procedure, although, as already stressed, it was difficult to know all that went on inside the nest cavities. A perched female, when mounted, shows readiness to copulate by leaning forwards and downwards (see Pl. 25) and thus counteracts the weight of a male as he manoeuvres into position for cloacal contact. Failure to lean forward in this way results in both birds tipping backwards off the perch. The male assists in achieving balance by beating with his wings as he leans backwards into position.

Copulation on the ground appeared in the present species only as a result of the presence in the colony of a bird with a damaged wing. This bird spent a great deal of its time on the ground when first introduced into the colony, and during this period, before its wing healed, it was repeatedly mounted by different males. The pattern was startlingly different when it occurred under these conditions. The mounted males held their wings open and down so that they touched the ground. Also they grabbed the mounted bird by the scruff of its neck with the beak and, whilst holding on in this way and balancing themselves with their open (but not beating) wings, they performed what appeared to be a highly efficient rape. What is so striking about this adaptation to the ground situation is that it results in a male copulation pattern which is virtually identical with that performed by quail and certain other Gallinaceous birds. It is also interesting to note that whereas successful rape is almost impossible for a male Estrildine when the female is perched, it is apparently comparatively easy on the ground.

Pl. 25 illustrates one final aspect of copulatory behaviour which must be mentioned. This concerns the interference by certain members of a colony with the mounted pair when performing the consummatory act. In the photograph, a mounted pair are seen on the left and, on the right, a third bird is seen approaching in the aggressive frontal-horizontal posture. A split second after the picture was taken the aggressor lunged forwards and struck the copulating birds, knocking them off balance. Such behaviour was seen not infrequently in the colony when it was in its more crowded state, but also was observed under conditions when there was no crowding. It is just possible that the nest-cavity copulations have arisen as an answer to this problem in these very colonial species. The nest cavity is the only really private place owned by a mated pair, where they can be undisturbed. A more intensely territorial species, on the other hand, does not have such problems.

The reason why individuals should attack a copulating pair is not at all clear. The attack usually comes just at the point at which the mounted male begins his wing-beating. Several birds sometimes attack simultaneously from all directions at this moment. Goodwin (personal communication) has noticed a similar interference with copulation in pigeons and states that 'The birds that interfere ... are usually sexually active males; sexually inactive pigeons, both cocks and hens, often, and probably always, ignore copulating pairs, even if these are within a few feet of them.' Since, with the bronze mannikins, the whole colony was sexually active, it is difficult to comment on this point for the present species. However, the impression gained was certainly that the most active individuals sexually were also most likely to be the ones which interfered, as in the case of the pigeons.

The derivation and motivation of the sexual signals

Thus far, I have concentrated on the forms of the various sexual display-components, on the circumstances under which they occur, and on the different ways in which they can be combined. It now remains to discuss briefly their possible origins and the motivational states underlying them.

As already mentioned, the nest ceremony is extremely valuable as a clue to the derivation of certain aspects of the pre-copulatory behaviour. If, on the one hand, only the pure nest-building behaviour were known and, on the other, only the usual pre-copulatory displays, the derivational link between the two would be almost certainly obscured. The existence of the nest ceremony, as a kind of intermediate, clarifies the situation considerably.

This latter ceremony, it will be remembered, involves a number of responses such as scooping, for example, which are quite obviously nest-building movements that have become used as sexual signals. Both the scooping and the lateral twisting and turning are nesting actions connected with the formation of the nest cavity, and both have become ritualized as sexual signals to the extent of being performed in the absence of a true cavity. Their form has, however, not been altered and they are therefore easy to trace to their origins. The same applies exactly to the signal mandibulation which has evolved from the nest-material-testing action.

In addition to this, however, the form of both the scooping and the mandibulating movements has been altered to give further display components. In the description of the nest ceremony it was mentioned that the scooping and mandibulating were replaced at higher intensities by a fixed neck-stretched beak-open position. Watching this change-over, one gained the impression that both the scooping and the mandibulating 'froze', and that it seems highly likely that the rigid 'neck-stretch and gape' posture has also originated from these ordinary nesting actions. The tongue-out-and-quiver element which accompanies the beak-open display is also probably derived from mandibulation, as the tongue is used a great deal in the latter action when an object is held in the beak.

In other words, almost all the signals used in the nest ceremony can be traced back to ordinary nesting patterns. Turning now to the typical pre-copulatory behaviour, we find a number of these signals appearing again. The courting male begins by mandibulating, quickly passes on to singing with the beak wide open and the tongue protruding and quivering, and then proceeds to alternate the two. It might be argued that the gape-and-tongue action, which is always accompanied by the song phrases, is a mechanical essential of song production. But this cannot be so, for unoriented singing is performed without it.

The full soliciting display of the female also involves the gape-and-tongue action and here, too, both this and her neck-stretched posture appear to be derived from the nesting patterns of mandibulating and scooping respectively.

All these nesting patterns seem to have become incorporated into the sexual repertoire as a result of the intimate connection in time between high intensity nest-building, on the one hand, and copulatory behaviour on the other. It is possible that some more complicated derivational explanation, such as displacement activation, or the 're-routing' of motivation (see Moynihan and Hall, 1954), is necessary. At the present stage it is perhaps best to leave this question open.

A second major derivational source for certain of the sexual signals is that of locomotory intention movements (see Daanje, 1950). Both the pivoting and the inverted curtsying of the courting male can be explained in this way. This has been discussed in detail elsewhere (Morris, 1954a, 1954b, and 1956a) for other species, and only deserves brief mention here.

It has been shown that in general the courtship dances of animals are frequently the outcome of a three-point conflict in the courting individual of the tendencies to flee, attack and mate (FAM conflict, see Morris, 1956a). Each tendency prevents the others from expressing themselves fully. This leads to a number of possible types of conflict response (see Bastock, Morris and Moynihan, 1953 and Morris, 1956b). One of these is thwarted intention movements, where one of the basic tendencies repeatedly begins to express itself, but only gets as far as its initial stages. A careful examination of the form of the inverted curtsy movements makes it clear that these actions are thwarted intention movements, probably of mounting the female. Pivoting, on the other hand, appears to be an ambivalent alternation of the intention movements of fleeing (turning away) and mating (turning towards the female).

There is a slight ambiguity concerning the nature of the intention movements involved in the inverted curtsying, which should be mentioned. A fleeing bird, in agonistic encounters, can be seen both to turn away and at the same time to stretch its legs for the take-off. In Pl. 16 the bird on the left is seen doing just this, and it will be noticed that the leg posture is very similar to that shown in the 'up' phase of the inverted curtsy in Pl. 21. It is possible therefore that this vertical dance component has a similar derivational source to that of the horizontal pivoting. Unfortunately the curtsy movements have become so stereotyped during evolution that it is difficult to come to any definite conclusion one way or the other. Sometimes it is posssible, despite the stereotyping, to observe motivational clues which make one source more likely than another. In the case of the present species, it was noticed that bouts of courtship which involved inverted curtsying alone were more likely to break down into aggressive encounters than those which also involved the pivoting movements. Thus a male was less likely to attack if he was in a pivoting mood, and this would agree with a derivational scheme in which pivoting was derived from fleeing and curtsying was not. Unfortunately, however, the situation was complicated by the performance of pivotless curtsying bouts *simultaneously* with a mild flight from on-coming females. (As already stated, certain males continued to court while retreating from advancing females.) This can only mean that the pivotless courter is simply less predominantly sexual than the pivoting one in the present species. A male which pivots as well as curtsying is less likely to show signs of attacking or of fleeing and must therefore be relatively more sexual.

It is important to stress that this does not necessarily apply to other Estrildine species, and that even in the bronze mannikin it must be considered as a tentative conclusion. In certain other species, ritualization has proceeded along slightly different lines, from a motivational point of view. The forms of the ritualized dance movements are fairly constant, but the exact motivational states which they represent differ from case to case. In the striated finch (*Lonchura striata*), for example, a mildly sexual male pivots, while a very sexual male both pivots and curtsies, so that here the situation is reversed. But this should not necessarily be interpreted as meaning that the inverted curtsy movement has been derived from a different source in the two species. It is much more likely that its origin was the same in both cases, but that the course taken by its emancipation as a sexual signal was different in the two instances.

Turning now to certain other elements of the male's pre-copulatory behaviour, the parallel orientation of the male to the female is also probably connected with the intention movements of fleeing. The twisting of the tail towards the female may be yet a further sign of this, for the tail is flicked from side to side, as already mentioned, as part of non-sexual 'scared' signalling. Alternatively, it may represent an anticipation of the copulatory tail posture, which includes a similar twisting to one side.

The leapfrog pattern may start out as an intention movement of mounting, in which the bird is sexually motivated enough to jump up in the direction of the female's back, but not sufficiently so for him to turn into the copulatory position as he lands. Balancing sideways for a second, he jumps off and then turns and repeats the process. This pattern is strikingly similar to a non-reproductive roosting activity (see Pls. 23 and 10). In the latter, one bird takes up a new position in a clump by leapfrogging over the backs of its neighbours and it seems likely that the predisposition to perform this type of response, which exists in the species, has facilitated the development of a sexual leap-frog ceremony. A male, about to land on the back of a female with whom he is not ready to copulate, does the only other thing that would come naturally to him under such circumstances, and vaults over her. It is a moot point as to whether displacement activation should be invoked in such a case.

The head posture of the male provides formidable derivational difficulties. A simple explanation would be that the twisting of the head towards the female is only a visual orientation to her, and that the lowering of the beak represents a 'frozen' stage of the scooping nest action, at the beginning of one of the upward stretches. This would be in line with the mandibulating and gape-and-tongue displays. However, comparative analysis provides difficulties for this possibility. In many species, displacement beak-wiping is a common activity during the male courtship. It has become modified in a number of different ways in different species. In some species it is only represented by a swinging down and round of the head, without the customary lowering of the body. Electronic flash photographs reveal that the initial stages of these wiping movements may involve just such a lowered and twisted head posture as is found in the courtship of a number of mannikins, including the present species. However, the bronze mannikin performs very little displacement wiping of the unritualized kind during its courtship, concentrating mainly on displacement preening. The other species, which show the highly modified versions of the wiping action, also show quite frequently the unaltered type, so that this explanation is not entirely satisfactory. A solution to this particular derivational problem must await further analysis.

Finally, the ruffling of the feathers, which occurs during all the male's pre-copulatory displays and much of the female's, must be mentioned. This appears to be the result of disturbances in the thermo-regulatory system caused by the dramatic autonomic discharge which characteristically accompanies a violent emotional disturbance, such as the intense state of conflict that is typical of courting animals. (For a detailed discussion of this problem, see Morris, 1956b.)

Displacement activities

Apart from the problematical cases mentioned in the last section there are three examples of displacement activities which occur in the courtship of the bronze mannikin. These are three cleaning patterns, which are often closely intermingled with intense courtship moments, but which are nevertheless totally irrelevant from the point of view of the basic tendencies involved in bouts of pre-copulatory behaviour. (I do not propose to discuss here the validity of the concept of 'irrelevance', as used in connection with displacement activation. I am aware of the logical pitfalls involved, but consider the concept useful, pending further investigation of the causation of such activities.)

In all Estrildines so far studied, a vigorous shaking of the body feathers accompanies, and appears to provide some relief to, moments of extreme tension. It is not surprising therefore that body shakes should accompany many courtship bouts. More interesting, perhaps, are the frequent beak-wiping and preening movements that appear during agonistic and/or sexual encounters in many of these species. In the zebra finch (see p. 75), displacement beak-wiping is restricted to sexual situations and is very common, and preening is restricted to agonistic situations but is less common. This seems to be fairly typical of most of the grassfinches and mannikins, but the bronze mannikin is strikingly different. As already reported, it never pauses long enough during fighting to perform any displacement activities. Also, during bouts of courtship it only beak-wipes on very rare occasions. However, it preens repeatedly. This it does in a rather rapid, stilted manner. The preens are of a simple form compared with those which occur during a standard bout of real cleaning. During the latter there is a certain percentage of preening movements which are aimed at parts of the body that are awkward to reach. The ratio of these more ungainly preens to the easier ones is reduced in the bouts of displacement preening. Also, the movement of the beak through the plumage is reduced or eliminated in the latter. These simplifications in the preening motor patterns do not necessarily reflect the existence of some degree of ritualization. They seem to be caused purely by the fact that the bird concerned is too busy with its sexual behaviour to give its full attention to the motor pattern it has begun. Nevertheless, as soon as distinct specific differences can be found in such cases, the incompleteness can be safely assumed to have begun to be ritualized.

The reason why the bronze mannikin exhibits displacement preening at those times when most other Estrildines show displacement beak-wiping is difficult to understand. A scrutiny of the normal post-bathing cleaning pattern of Estrildines makes this even more of a problem. After the bath is complete, an Estrildine flies up to a perch where it cleans and dries itself. This may take as long as half an hour and is a very complex but also very characteristic sequence of movements. It would take too much detail to describe all the aspects of this pattern here, but briefly the response complex is as follows.

As soon as it settles on the branch or perch, the bird shakes itself a few times, with the typical body-shake movement already mentioned. It then begins to wipe its beak a number of times on the perch. After a few minutes of wiping, it begins to preen. During this long bout of preening it uses its oil-gland several times. It then breaks the long preen-bout with a vigorous wing-and-tail-shake. This is followed by a few more preens and is then repeated. For some minutes to come it now alternates the wing-and-tail-shakes with short preening bouts. The latter get longer and longer as time goes on, until eventually it stops the shaking altogether and ends the whole complex with a final long bout of preening.

This is the typical course of the normal cleaning response (which has been somewhat simplified here). At its commencement, the bird is wet all over and all the stimuli for all the different cleaning motor patterns are present. Therefore, they must either all be given chaotically at once (which, of course, never happens), or they must occur in a set sequence with the lowest threshold responses occurring first and the highest threshold responses last. This means that body-shakes have the lowest threshold of any cleaning response, beak-wipes next and preens next. (The rest of the cleaning complex need not concern us here, since only these first three actions are involved as displacement activities.)

From this one would expect, perhaps, to see cleaning displacement activities occurring in that order, if at all. This is exactly what happens in other Estrildine species. As an example, for comparison with the present species, I shall select the Bicheno finch (*Poephila bichenovi*), which shows the highest frequency of these activities during courtship of any of the species so far studied. If a male and female of this species are isolated for some days and are then brought together for a short observation period, separated again, and then repeatedly brought together again in the same way over a period of weeks, it is possible to score accurately the sequence and frequency of these cleaning actions on each occasion. Briefly, the initial responses (ignoring all other types of behaviour for the moment) are typically body-shakes. After a few of these there is a long period of beak-wiping, which often exceeds that which would be expected in a normal cleaning pattern. Both the frequency and rate of wiping may be greater here, as many as ninety wipes being performed in a few minutes. Then, in those instances where the wipes have been extremely profuse, the birds switch, after five or ten minutes, to preening. Even in the most intense cases, however, the birds have never gone beyond this stage to the wing-and-tail-shaking stage.

In most other species, the body-shake stage has been observed and also the wiping stage, but no other species has ever been seen either to displacement-clean so intensely, or to pass over into the preening stage. The intriguing point about the bronze mannikin is that it does not pass over from wiping into preening, but instead passes *directly from body-shaking into prolonged preening*. The wiping is dramatically absent for some reason which is at present quite unknown. In order to test this accurately, a few simple experiments were carried out.

In these, a small observation enclosure was used from which the birds could not see the observer. A male was introduced and its behaviour was scored for a period of five minutes. A female was then introduced and the behaviour of the male was immediately scored again for the next five minutes. The birds were then removed and another pair used. The males and females employed were kept in groups in cages, but with the sexes segregated. The introductions were arranged in such a way that no male met a particular female repeatedly and in this way it was possible to prevent pair-formation. Both Bicheno finches and bronze mannikins were tested in this way. The former, which are rather shy birds, did not respond very intensely (for them) to the experimental situation, as it involved moving them from their usual quarters into the standard enclosure. This meant that in the present series, they never reached a sufficiently high intensity to show the preening stage of their displacement cleaning behaviour. However, their responses were sufficiently frequent to make a valid comparison, showing the way in which both species showed displacement shaking, but the one then passed on to wiping, whilst the other preened. Control experiments were carried out by introducing male bronze mannikins to female Bicheno finches and vice versa, and by going through the motions of introducing a female to a male and then not doing so, and scoring his behaviour for the second five minutes. These tests showed conclusively that the cleaning was the result of the meeting with the female and not a basic response to being handled. (Even without such controls, the specific difference the tests were concerned with could not have been explained by handling, since all Estrildines, including the bronze mannikin, clean in the same way when really dirty.)

Nine tests of the standard type, in which a male and female of the same species were introduced to one another, were carried out for both species, with the following results. In neither species was there a great deal of activity before the introduction of the female, but in the Bicheno (summating figures for the nine tests) there were 23 shakes, 127 wipes and only 3 preens, after the female was present. This contrasts strongly with the bronze mannikin figures, which show that there were 77 shakes, 280 preens, and only 20 wipes, after the female was introduced.

The exact relationship between the minute-by-minute frequency of wiping and preening, before and after the female's introduction, is shown graphically in Fig. 1 for the two species. More detailed figures for the bronze mannikin are given in Table 1. Here it will be seen that the peak in displacement preening follows immediately after the peak in sexual responses (measure in this instance by frequency of inverted curtsy movements).

TABLE 1

	Five mins. before female in					Five mins. after female in				
	1	2	3	4	5	1	2	3	4	5
Shake	0	3	2	2	1	20	22	13	11	11
Wipe	0	0	2	4	0	4	5	4	3	4
Preen	0	0	1	2	0	50	82	58	41	49
Curtsy	3	0	0	0	0	130	0	16	36	0

Only one small point needs some explanation here and that is the occurrence of three inverted curtsies in the first minute of introduction of the male, before the female was present. This happened during one of the nine tests, when the male concerned had become conditioned to the apparatus to the extent of 'expecting' a female. He performed a short burst of song and dance as soon as he was placed in the observation cage, despite the fact that he was in complete isolation.

An analysis of the individual tests seems to indicate that there is a tendency to perform more preening movements in five-minute periods where more courtship movements are shown. The preening actions may be seen either before the male begins to court, as soon as he has seen the female, or actually during his courtship of her, or immediately after he has stopped courting, or during the minutes following cessation of courtship. Preening occurs whether a male is being thwarted by an unresponsive female, or whether he is being invited to mount but for some reason cannot do so. Together, all these facts lead to the conclusion that the displacement preening in the present species must either be emancipated to the extent that it is now sexually motivated, or alternatively that it is caused by a particular balance of the flee-attack-mate conflict which exists during courtship. The former seems unlikely (though far from impossible) since the preening actions have not become modified in form to any great extent, as is typical of fully emancipated patterns. The latter is not incompatible with the evidence, but it is difficult to understand the exact nature of the relevant causal conflict balance, at the present time. One point, however, is very clear, and that is that this is a case where simple thwarting (as opposed to mutual thwarting of the conflict type) is definitely not essential as a causal factor. A male may respond to the most violent form of sexual stimulation (tail-quivering by the female) by displacement preening, so that only a strong conflict between the mating tendency and some other incompatible tendency, or tendencies, can explain this.

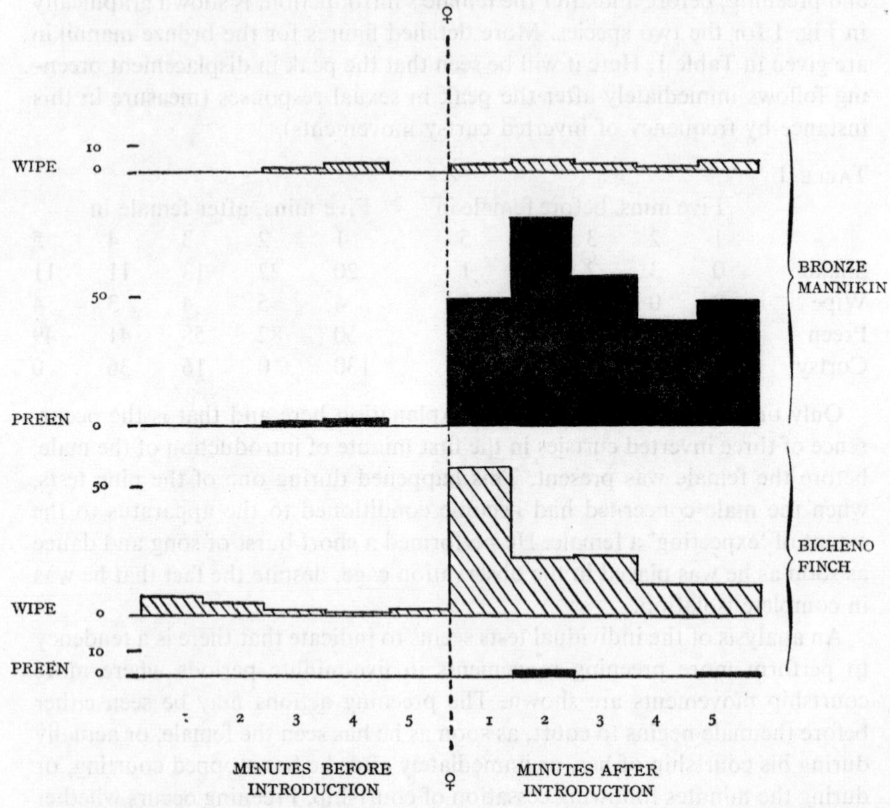

Fig. 1. Displacement beak-wiping and preening in the courtship of the bronze mannikin and the Bicheno finch. The response frequencies of the two actions differ markedly in the two species. For full explanation see text.

Colony arousal

One of the methods employed to provoke large numbers of sexual and aggressive responses for observational purposes was the setting up of a small artificial colony, individual by individual. As each new bird was added to the existing group, a violent outburst of reproductive patterns took place, thus facilitating repeated analysis of the nature of such responses. In addition to this, however, it was possible to study the effect of the newcomer on the state of the colony as a whole.

A large part of the increase in frequency of reproductive activities on these occasions was, naturally, due to interactions between the new bird and the old ones. Whether the new bird was a male or a female, it was attacked repeatedly and usually retaliated to some extent. If it was a male it also courted members of the colony a great deal or, if it was a female, it was courted repeatedly.

However, it was noticed that a certain amount of the increase was due to a greater number of interactions *between old colony members.* In other words the sudden presence of the new bird not only stimulated the established members of the group reproductively, but also aroused them, so that they were now prepared to perform more actions towards birds which had been present and available to them before. This effect, in its most extreme form, was over in the first hour or two after the introduction, but it usually took several days for a newcomer to become fully accepted into a colony. During the interim, he suffered repeated attacks by members of the group, who clearly knew him and all the other birds individually and sought him out over and over again. It is safe to assume that, during all this time, the fact that he stimulated them strongly also continued to have an arousal effect on their internal reproductive condition.

The importance of these observations lies in the fact that they probably reflect conditions in the wild state, where a population is bound to experience a certain amount of interference of this kind. The intensity would undoubtedly be much milder, since a stranger would never be forced to such a degree into so fixed a group as is produced by an aviary. But, if the comings and goings would be milder, they would also be more constant. This would probably lead to a mild but perpetual 'shaking up' of a group, with the advantageous effect of continuous arousal.

Under the artificial conditions of captivity, the forced strength of the interference, with the newcomer unable to leave the established group, is such that it often leads to reproductive inefficiences. For example, if the group contains some extremely aggressive individuals, they may now spend too much of their time chasing the new bird and neglect their other reproductive duties. If a new bird is a powerful male or a particularly attractive female, some of the old pair-bonds may break up and new ones form. The result is that eggs and even young nestlings may be abandoned. The new pairs forming may be so aggressive that they manage to drive out an established pair. Under these conditions, eggs may be seen rolling in all directions, and even young birds being built over and squashed as a new pair constructs a new nest on top of an old one (despite the presence of surplus nesting sites).

In the wild state, such extreme developments would no doubt be avoided by virtue of the greater freedom of movements of the birds. Even in captivity, it is not long, as already mentioned, before the newcomer is absorbed into the community. There is one interesting exception to this last point, however, which deserves brief mention here. If a newcomer is a weak bird and is badly attacked during its 'initiation' phase, it may lose a large number of feathers. These are plucked from it by the members of the group, especially from around the neck region (see male in Pl. 23). In a very badly plucked bird, feathers are removed from all parts of its body and it may soon assume a constantly dishevelled and ragged appearance. Once it has sunk to this condition, it is in serious difficulties. For, ironically, the ragged state is a strong stimulus releasing attack, so that the more ragged it gets beyond a certain stage, the more it is attacked, and the more it is attacked the more ragged it becomes. The strange thing about this type of aggression is that it develops into a widespread persecution, so that even the weakest and most peaceful members of a colony will join in the chase with vigour. If such behaviour occurs in the wild, it can only have the function of ruthlessly eliminating those individuals which cannot defend themselves sufficiently. (It differs markedly from the private beatings which birds may receive from their rivals, where other uninvolved group members more or less ignore the combatants, including the loser. In the latter type of bout, a bird may lose a certain number of feathers, but not enough to assume the persecution-producing raggedness.)

Returning now to the problem of colony arousal, some brief quantitative data may be given. On a number of the introductory occasions, a quantitative score of certain sexual and aggressive actions was made. The number of inverted curtsies and the number of supplanting attacks was noted in five-minute periods for half an hour before and also half an hour after the introduction of the new bird. The graphs in Fig. 2 show the great increase in both sexual and aggressive movements immediately following the arrival of the new bird, in a typical instance.

In this particular case, the new bird was a male and, in the half-hour period after it was present, the responses were scored separately according to whether they were performed by an old bird to the new bird, by the new bird to an old bird, or by an old bird to another old bird. In this way it was possible to separate increase (in the colony's reproductive condition) by stimulation from increase by arousal.

There were 127 supplanting attacks during the half-hour before the new male was present, as opposed to 221 afterwards. Of the latter, 100 were attacks by old birds on other old birds, 107 were attacks by old birds on the new bird, and 14 were attacks by the new bird on old birds. Turning to sex, there were 56 inverted curtsies performed in the first period, as opposed to 241 after the introduction. Of the latter, 106 were performed between old birds, but none were performed by the old birds to the new one (which established its maleness immediately upon introduction), the remaining 135 all being performed by the new male to old members of the group.

These figures reveal that the sexual behaviour of the old males towards the old females was doubled in frequency as an outcome of the intrusion of the new male into the group. Also, the frequency of their aggressive behaviour towards one another remained at the same (rather high) level, despite the fact that, in addition, they were now also performing as much again towards the newcomer.

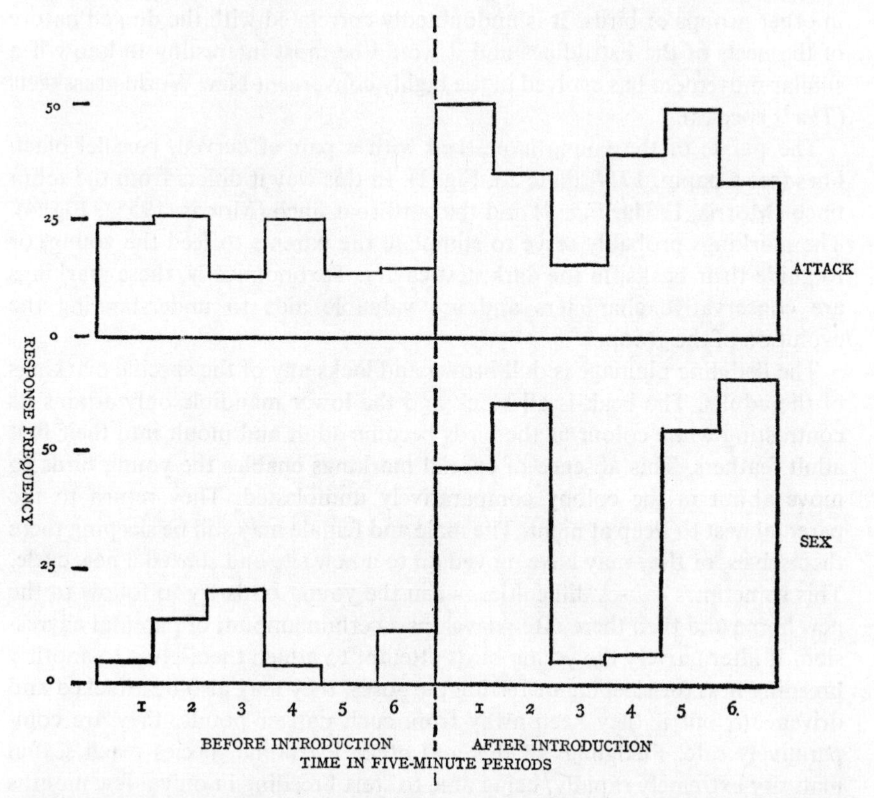

Fig. 2. Typical effect of the introduction of a new bird into an established colony. The frequencies of both sexual and aggressive patterns show an immediate increase. The latter (see text) is due partly to increased stimulation and partly to colony arousal.

Slightly different results were obtained on different days, depending on the initial state of the colony and the sex and vigour of the newcomer, but the test reported here is fairly typical, and reveals the basic characteristic of all such occasions: namely, colony arousal.

Parental Behaviour

The parental behaviour of the Estrildine finches does not appear to show a great deal of specific variation. In the present species, like most others, the clutch of (usually five) small white eggs was incubated for about two weeks. The incubation period was followed by a two-week nestling period and this by a four-week fledgling period of dependence on the parents for food.

The parents share the tasks of incubating and feeding the young. The latter beg with the characteristic Estrildine twisting of the neck. This contrasts with the more usual vertical stretching of the neck found in most gaping nestlings in other groups of birds. It is undoubtedly correlated with the domed nature of the nests of the Estrildines and it would be most interesting to know if a similar movement has evolved in the highly convergent New World grasskeets (*Tiaris* species).

The palate of the young is marked with a pair of curved, parallel black lines (see Chapin, 1954, Plate 23, Fig. 1). In this way it differs from the zebra finch (Morris, 1954a, Fig. 9) and the cutthroat finch (Morris, 1955a, Fig. 4). The markings probably serve to stimulate the parents to feed the young, or to guide their beaks, in the dark nest cavity. Taxonomically, these markings are conservative characters and are valuable aids to understanding the evolution of the group.

The fledgling plumage is dull brown and lacks any of the specific markings of the adults. The beak is all black and the lower mandible only attains its contrasting white colour as the birds become adult and moult into their first adult feathers. This absence of special markings enables the young birds to move about in the colony comparatively unmolested. They return to the parental nest to sleep at night. The male and female may still be sleeping there themselves, or they may have moved on to a new site and started a new cycle. This sometimes causes difficulties, when the young birds try to follow to the new home and then there often develops a certain amount of parental aggression. If alternatively the young birds attempt to attach themselves to another breeding nest for sleeping or resting purposes, they may also be attacked and driven off, but if they keep away from such danger points, they are comparatively safe. Fledglings of this and other Estrildine species reach sexual maturity extremely rapidly, being able to start breeding in only a few months from hatching.

References

ADLERSPARRE, A. (1931), 'Einiges über *Spermetes nigriceps* Cass. (Braunrücken-Elsterchen)'. *J. f. Ornithol.* 79, pp. 370–4.

BANNERMAN, D. A. (1949), *The Birds of Tropical West Africa* (London), vol. 7.

BASTOCK, M., D. MORRIS and M. MOYNIHAN (1953), 'Some comments on conflict and thwarting in animals'. *Behaviour* 6, pp. 66–84.

BUTLER, A. G. (1899), *Foreign Finches in Captivity* (London).

BUTTNER, E. E. (1949), 'Bronze Mannikins at home'. *Cage Birds* 96, pp. 319 and 328.

CHAPIN, J. P. (1954), 'The Birds of the Belgian Congo, Part 4'. *Bull. Amer. Mus. Nat. Hist.* 75b.

DAANJE, A. (1950), 'On locomotory movements in birds and the intention movements derived from them'. *Behaviour* 3, pp. 48–99.

DELACOUR, J. (1943), 'A revision of the subfamily Estrildinae of the family Ploceidae'. *Zoologica* 28, pp. 69–81.

FRIEDMANN, H. (1949), 'The breeding habits of the weaver birds. A study in the biology of behaviour patterns'. *Ann. rep. Smithsonian Inst. for 1949*, pp. 293–316.

MOREAU, R. E. (1942), 'The nesting of African birds in association with other living things'. *Ibis* 6, pp. 240–63.

MORRIS, D. (1954a), 'The reproductive behaviour of the Zebra Finch (*Poephila guttata*), with special reference to pseudofemale behaviour and displacement activities'. *Behaviour* 6, pp. 271–322.

—— (1954b), 'The courtship behaviour of the Cutthroat Finch'. *Avic. Mag.* 60, pp. 169–77.

—— (1955a), 'The markings of the Cutthroat Finch'. *Birds Illustrated* 1, pp. 182–3.

—— (1955b), 'The fighting postures of finches'. *Birds Illustrated* 1, pp. 232–4.

—— (1955c), 'The seed preferences of certain finches under controlled conditions'. *Avic. Mag.* 61, pp. 271–87.

—— (1956a), 'The function and causation of courtship ceremonies'. *Fondation Singer-Polignac: Colloque Internat. sur l'Instinct*, June 1954.

—— (1956b), 'The feather postures of birds and the problem of the origin of social signals'. *Behaviour* 9, pp. 75–117.

—— (1957), ' "Typical Intensity" and its relation to the problem of ritualization'. *Behaviour* 11, pp. 1–13.

MOYNIHAN, M. and F. HALL (1954), 'Hostile, sexual, and other social behaviour patterns of the Spice Finch (*Lonchura punctulata*) in captivity'. *Behaviour* 7, pp. 33–76.

The courtship of pheasants

(1957)

Nowhere in the animal world are there colours, patterns or displays to rival those found amongst the members of the pheasant family. When we first encounter the sexual ceremonies of these birds, we are aware only of their extraordinary complexity of form and intricacy of detail. Even the most academic mind is so numbed by the visual experience involved that any attempt to analyse the displays, or to trace their origins, appears to be an insult to their beauty, if not simply impossible. After the initial shock has died down, we are still left with a set of problems that are at the very least extremely puzzling.

However, in recent years great strides have been made in analysing courtship displays in general and, now that we know so much more about the evolution of many of the simpler rituals of other birds, we can start to tackle, albeit rather humbly, the incredible performances of the pheasants and their close relatives.

It is essential to consider how these ceremonies are carried out in the wild state. All pheasants which have been studied in the wild have been found to display only in special arenas. The birds are typically forest-dwellers and the males have to clear a special area as a display ground if their ceremonial performances are going to be seen to effect by the females.

The Argus pheasant may be taken as a typical example. A male selects a piece of level ground and clears a square with approximately 20-ft. sides. Every fragment of leaf, grass and twig is scratched or carried away until the floor is bare earth. He is then ready for a female and begins to call loudly. His call is audible as far as a mile away and after a while it attracts a female who approaches the courting site.

When she appears in the arena the male starts his display, and the courtship bouts and matings which follow may continue for several days. Pheasants never form true pair-bonds, but during the visit of one female to the arena of a male Argus pheasant, the latter can be said to be mildly faithful. If a second female appears when he is courting the first, he may drive the newcomer away, but as soon as the original female leaves him, he appears to forget her.

During the few days when the female is with the male, she is not only courted by day, but also roosts with him by night. While she is present the male ceases to call, but as soon as she leaves him to lay and incubate her eggs, he resumes his powerful shrieking until another female is attracted. Each arena is forbidden ground to all other males and intruders are savagely attacked.

Young males have no fixed display sites and can be heard calling as they wander about. Their calls, not being backed up by a display area for showing off their paces, seldom lead to sexual successes. Once a male has matured, however, and established an arena of his own, he often returns to it year after year, so that the initial task of selecting, defending and preparing a display ground is lessened after the first season.

This, then, is the basic story of the pheasant cycle in the wild. In most species, the males are solitary when breeding, as already described for the Argus pheasant, but in some cases groups of males will share a display ground. The African stone pheasant or partridge, for example, has social display areas where a group of males can be seen strutting and dancing to the females, who watch from near-by cover. In the case of the peafowl, too, several peacocks have been seen to share an arena which is visited from time to time by the peahens, who select a particular male from the group for mating.

The first feature thing in the rituals themselves is that there are two basic elements to be considered; namely, the dance movements and the display postures. One of the most fundamental aspects of pheasant ceremonies is the preliminary run. As soon as the female has made an appearance in the arena, the male has to approach her and manœuvre into a good position to adopt his eye-shattering display posture. But he must do this without scaring the female away, for, at first, she will probably not be sufficiently aroused sexually for a mating to take place and he will have to display to her repeatedly to build her up into a strong sexual mood. Therefore he must possess a dance step which will enable him continually to force his attentions upon her.

The way in which this is done can be seen clearly in the case of the Amherst or golden pheasants. The male runs up alongside the female and as she moves off he circles round in front of her and blocks her path. If she is to flee now, she has to turn away first, but, before she can do so, he starts to assume his display posture. If she is receptive and stays still when he displays he has only to complete his circle around her to mount and mate with her.

But this rarely happens at the beginning of a display bout. Invariably the female turns away and moves off. The male now makes his master move. He could, of course, simply follow and catch up with her, circle round to the front and display again. But this he does not do. Instead he makes a sharp turn *away* from her as if he is leaving her and then, in one extremely rapid movement, runs round in a tight circle which brings him quickly back into his vital position in front of her. The great advantage of this manœuvre is that he does not give the female a sense of being pursued and thus he does not stimulate her to flee at speed. If he chased straight after her each time, she would start to panic and perhaps flee from the arena altogether, to be lost in the undergrowth.

Each time the female moves away after one of his displays, he swings away from her and then runs quickly back round in a circle to confront her again until, after a long period of this repeated displaying, the female begins to move away less and less. Eventually the male changes his tactics and instead of turning away after a display, turns *towards* the slowly moving or stationary female and places one foot on her back. If she then moves off, he starts all over again, but, if she stays still, he mounts her and mating takes place.

Sometimes there is an intermediate phase of dancing, when the male has swung towards her and then the female has moved off more quickly than expected. The male then changes tactics in mid-run and quickly reverts to his earlier pattern, swinging away from her and then circling back again. When this happens, he performs a complicated double-curved dance movement.

These dance movements, as already pointed out, are the preliminaries to adopting special display postures. In the case of the Amherst pheasant, the male always displays broadside on to the female. The tail is spread and twisted round so that the upper surfaces of the feathers are shown off to the best advantage, and the extremely elaborate neck-ruff is fanned out and then swung round to the side on which the female is standing. At the moment of the display the male hisses violently at the female.

In this species and the golden pheasant, this appears to be the only display position employed, but in other species there is an additional frontal display. In the case of the Argus pheasant the male begins with dance movements around the female and then, when she stays still, he lowers the wing nearer to her. But in this species the broadside-on display is not so important and the male quickly changes it into a frontal display. He opens both wings to make a huge fan, faces the female and erects the fan until it is vertical. In the centre of the fan appear two huge feathers which he has also erected and which are two enormously exaggerated tail-coverts (the feathers which in the peacock make up the whole of the fan, the wings not being used at all there). This fan of ocellated feathers comes to rest, at the height of the display, in front of the head and the male has to peep at the female through the small gap that exists in the front of the screen, where the wings join the body.

In the case of the Tragopan pheasants, the broadside and frontal displays both occur, but here they are said to be quite distinct from one another and are not apparently performed in sequence. In their frontal displays these and certain other species show off erectile face wattles. These fill with blood when the birds become sexually excited and swell up into bizarre flaps or horns.

The peacock pheasant has converted the initial circling-dance phase of the typical pheasant courtship into a strange 'pile-up' ceremony. The male approaches the female after having scratched around for a tit-bit on the ground. He holds this in his beak and starts calling to her. The female is attracted by his offer and in this way he appears to be able to forgo the vigorous business of manoeuvring for position with all the violent dancing actions described earlier.

When she gets close, he quickly gets into his display posture, throwing the tit-bit at her. Both his wings and his tail are beautifully marked and both are fanned out for the female to see. The tail is spread obliquely, the near wing is lowered and the far wing raised, the bird crouching on the ground and throwing his head to one side, producing a powerful spectacle. If the female turns round, the male moves with her. This species is also said to possess a purely frontal display which is used on other occasions.

Finally the peafowl is a species in which both the preliminary run and the lateral display have been abandoned altogether and the male concentrates exclusively on its frontal display. This is too well known to need any description here.

Very briefly, then, these are some of the displays of pheasants, but there is still the problem of how they have arisen.

First, there is the arena itself to be considered. Many birds are known to tear at grass, peck at the ground, pick up twigs, peck up food, or scratch the earth, as signs of agitation when under the stresses and strains of reproductive encounters. In most instances these actions are rather desultory and the effect they produce is insignificant. But in various cases, of which the pheasants are one, these actions appear to have evolved into an organized system of activities, with the result that a whole patch of ground is stripped. This now functions as a display area and its existence has undoubtedly helped to shape the form of the dance movements of pheasants in the way described. Having evolved an arena, pheasants also evolved special dance sequences which kept the females in the arena.

The movements themselves are rather widespread in occurrence throughout the family and are probably ancient in origin. They involve rapid runs, turnings and brakings. When a bird turns or brakes suddenly, it uses wing and tail actions to balance itself. Spreading the wings, fanning and lowering or raising the tail can be seen in any bird that is trying to steady itself. Therefore the dance actions required to keep the female in the arena lead quite naturally to the use of much wing or tail spreading, or both, on the part of the dancing males. It is easy to see how these actions must gradually have become an integral part of the ritual and how, as they did so, the feathers concerned must have become more and more ornate. This still leaves the strange erections of certain body feathers and face wattles to be explained. These actions have a quite different origin.

When a bird is undergoing the stress of a bout of courtship, a violent autonomic discharge occurs in its system. This is characteristic of any strong emotional disturbance in animals. In man, the typical symptoms include sweating, flushing of the skin, gasping for breath, etc. Birds do not perspire, however. Their finer adjustments of body temperature are carried out by raising or lowering their body feathers. This action changes the thickness of their insulating coat and is highly effective.

Just as we may go 'hot and cold all over' when in an intensely emotional state, so, in a sense, do birds. This reveals itself in the form of exaggerated feather erections and depressions, some of which become highly developed in a secondary way as sexual display signals because they are always associated with intense sexual circumstances. The wattles, too, are erectile, blood being pumped into them when the males become sexually excited.

The reproductive behaviour of the ten-spined stickleback
(1958)

I

Introduction

Scope of the study

The investigation which is reported here is primarily of an observational rather than an experimental nature. It is fundamentally descriptive and attempts to give an all-round general picture of the reproductive behaviour of the ten-spined stickleback (*Pygosteus pungitius*). Certain experimental results are presented, however, but these are not prolonged unduly in any special direction. As a result, a comprehensive study of the reproductive cycle *as a whole system* has been possible and all its phases are described and discussed. It is hoped that this study will form the basis for future, more specialized, experimental studies.

Terminology

Contemporary taxonomy recognizes five genera of sticklebacks (family Gasterosteidae), with one species to each genus, as follows:

1. *Gasterosteus aculeatus* L. (three-spined stickleback).
2. *Apeltes quadracus* (Mitchill) (four-spined stickleback).
3. *Eucalia inconstans* (Kirtland) (five-spined or brook stickleback).
4. *Pygosteus pungitius* L. (ten-spined stickleback).
5. *Spinachia vulgaris* (Flem) (fifteen-spined or sea stickleback).

Since the number of spines in each species is variable, and since the popular names are more cumbersome than the generic names, I have used the latter in the present work. It seems that there is only one good species (perhaps one should talk of super-species) to each genus, and as no intrageneric behaviour differences appear to exist (except possibly in connection with migration), specific names have been omitted throughout.

The ethological terminology employed follows Tinbergen (1951a) and van Iersel (1953), with one or two minor modifications. The latter, however, will be discussed in their special contexts.

Review of the literature

The stickleback family (*Gasterosteidae*) has been the subject of many varied investigations in the past. During the last hundred years, the number of papers published, dealing with some aspect of the biology of this group, is somewhere in the region of one hundred and fifty. Since most of these are not concerned with the special subject of the present study, they will be reviewed only very briefly.

Just over half of the above papers deal with the anatomy and taxonomy of stickleback species. The most important of these is Bertin's (1925) monograph of the family. Certain of Bertin's conclusions have, however, been criticized by Heuts, who has recently developed the anatomical study of sticklebacks considerably (1944a, 1947a, 1947b, 1947c, 1949a, 1949b).

A number of studies dealing with various aspects of stickleback physiology have also been reported. The emphasis here has been on the osmoregulation associated with the adaptation to salt and fresh water which is involved in stickleback migrations (see Forbes, 1897; Giard, 1900; Siedlecki, 1903; Borcea, 1904; Heuts, 1942, 1943, 1944b; Koch and Heuts, 1943). The resistance of sticklebacks to the presence of toxic substances in the water, as the result of the stagnation or pollution of steams and ditches, has been studied by Erichsen Jones (1947a, 1947b, 1948, 1952). Endocrinological investigations have been made by Kerr (1943) on the pituitary, Koch and Heuts (1942) on the thyroid, and Stanworth (1953) on the reproductive endocrinology in general. Castration experiments have been carried out by Ikeda (1933) and these have been successfully developed by van Iersel. The effect of light and temperature on the reproductive cycle has been the subject of reports by Craig-Bennett (1931), Merriman and Schedl (1941) and Eeckhoudt (1947). Stickleback embryology has been investigated by Kunz and Radcliffe (1918).

The species studied by the above authors may be ascertained from the bibliography at the end of the present report. *Gasterosteus* was employed in the vast majority of cases. This species is undoubtedly the easiest to obtain and to handle and is the favourite experimental species for almost all types of stickleback investigation.

Stickleback behaviour studies have been largely confined to the reproductive activities. The feeding habits of the group appear to be almost the only non-reproductive behaviour that has been investigated. (See Blegvad, 1917; Markley, 1940; Hartley, 1948; and Hynes, 1950.) The defensive behaviour of sticklebacks against predators has, however, been dealt with by Hoogland (1951), Tinbergen (motion picture), Morris (1954e) and Hoogland, Morris and Tinbergen (1957).

From an early date, the interest of biologists has been aroused by the reproductive behaviour patterns of sticklebacks. Many brief notes appeared on this subject in the scientific literature of the nineteenth century. The more general of these will not be dealt with here, however, because all the phenomena recorded by these early authors have since been re-investigated in more detail and much more accurately by later workers. The focus of interest during this natural history phase was undoubtedly the nesting behaviour of the group. (See Coste, 1846, 1848 on *Gasterostues* and *Pygosteus*; Hancock, 1852, on *Gasterosteus* and *Spinachia*; Möbius, 1885a, 1885b, 1886, and Prince, 1885, 1886, on *Spinachia*; Ransom, 1865, and Landois, 1871, on *Pygosteus*; Ryder, 1881, 1887, on *Apeltes*; and Hess, 1918, on *Eucalia*.)

In the early 1920s, the secondary sexual characters of *Gasterosteus* and *Pygosteus* aroused some serious interest (see Titshack, 1922; van Oordt, 1924a, 1924b; Courrier, 1925), but it was not until the late 1920s and early 1930s that the first detailed studies of the reproductive behaviour as a whole appeared. These were carried out in Germany by Wunder (1928, 1930) and Leiner (1929, 1930, 1931a, 1931b, 1934) and in England by Craig-Bennett (1931). Nesting behaviour still claimed more attention than other aspects of the cycle, but already the territorial, sexual and parental behaviour patterns were being carefully scrutinized.

During the next few years, under the leadership of Konrad Lorenz (1935), comparative ethology began to develop rapidly and in 1936 and 1937 there appeared the first ethological studies of sticklebacks (*Gasterosteus*) by ter Pelkwijk and Tinbergen. This work was continued after the war by Tinbergen and van Iersel (1947), and the results have appeared since as sections of a number of general papers and books by Tinbergen (1939b, 1940, 1942, 1948, 1950, 1951a, 1952a, 1952b, 1953a, 1953b, and 1954), and as a thesis by van Iersel (1953).

Both van Iersel in Holland and Tinbergen in England, with co-workers, have organized research programmes to investigate further special points of interest in the behaviour of *Gasterosteus* and, especially, to make comparative studies of related species. Wunder, Leiner, Craig-Bennett, ter Pelkwijk, Tinbergen and van Iersel had all concentrated on *Gasterosteus* (although Leiner had also investigated *Pygosteus* and *Spinachia* – 1931a, 1943), and the results obtained with this species were so encouraging that it was decided to make the whole family the subject of an elaborate ethological study.

Further studies on *Gasterosteus* are being carried out by van Iersel and his co-workers, notably Sevenster, at Leiden University. Certain of the comparative studies have already appeared, short papers having been published by Sevenster, on *Pygosteus* (1949), and *Spinachia* (1951), and by Morris (1952) on *Pygosteus*. This present work forms the first detailed report of one of these comparative studies.

Materials

All the observations of the behaviour of *Pygosteus* that are reported in the following pages have been made in the laboratory, using aquarium tanks varying in size from 30 × 22 × 22 cms to 210 × 30 × 36 cms. It is regrettable that almost all the evidence obtained from field work is of an indirect nature. This limitation is imposed by the timidity of the species and the fact that in breeds in amongst the aquatic plants. I have, however, been able to make direct observations on *Gasterosteus* behaviour in its natural habitat, in streams near Oxford and Swindon (Wilts.). This species is neither so shy nor so retiring and appears to be quite undisturbed by an observer, so that I have been able to compile field-notes on its territorial, nesting and sexual behaviour. The reasons for this specific difference in timidity are discussed later.

It has been possible to obtain indirect evidence of the movements of *Pygosteus* in the river by systematic collections. Very rough estimates of territory formation and territory size can be made by studying the numbers and condition of fish caught in single sweeps of the net, when covering one strip of a river with a series of such sweeps.

The supply of fish used for the laboratory studies has always been plentiful. All the individuals employed for the present work were collected from rivers, streams and ditches in Oxfordshire and Wiltshire. All the waters collected from are ultimately fused with the River Thames, so that the stocks used are probably fairly similar. The vast majority of all fish collected were taken from the River Kennett in Wiltshire, west of Marlborough. This source of supply was utilized most because of the large numbers found there and because these fish were much more free from disease than in many places elsewhere.

Once collected, the fish were stored in large unplanted metal tanks from which they were selected and transferred to the smaller, planted, observational and experimental glass tanks. In the latter, a substratum of sand or gravel was provided. Many species of aquatic plant were used, of which the following were found to be most suitable:

Fontinalis antipyretica (willowmoss), *Myriophyllum spicatum, Vallisneria spiralis* (tape grass), *Hygrophila* species, *Nitella gracilis.*

All five of the above were used as host plants for the nests of *Pygosteus. Nitella*, however, was more often used as nesting material. Also supplied as nesting material were green algae, especially *Spirogyra.* In tanks planted with approximately equal areas of a number of different species, the sticklebacks most frequently selected the willowmoss in which to build their nests.

Methods

The special methods employed in the various experiments will be mentioned with those experiments in the text, but a few general remarks may be made at this point.

Collecting. The fish were caught in conical linen nets of quarter-inch mesh, stitched on to iron rings of 36 cm. diameter, attached to wooden poles five feet in length. These were dragged through the water against the current in a wide sweep through the aquatic plants. Plants that broke loose and stuck in the net were landed and sorted for fish on the bank. The fish were transferred from small field-cans to large flat travelling cans and were then transported by car to the laboratory.

Aquarium observations. There have been occasional objections to aquarium observations on the basis that they are 'unnatural' and do not give a true picture of the behaviour of the fish. Such criticisms are seldom more precise than this and the critics seem to overlook the fact that many valuable and coherent results have been obtained from aquarium observations in the past. Nevertheless, it is important to ask exactly how the behaviour of a fish in an aquarium tank is likely to be abnormal. There appear to be three main ways in which it is likely to be altered by the artificial conditions: (a) structurally, (b) spatially, (c) sequentially. For simply observations of the reproductive cycle, the laboratory methods must aim at reducing these three types of alteration. Firstly, the structure of the natural environment must be imitated as exactly as possible. This means, principally, the provision of well-aerated water of a suitable temperature, plants which provide cover, nest-sites and nesting-material and plenty of live food. (The fish were fed daily with *Tubifex* worms, also occasionally with chopped earthworms, *Daphnia*, and aquatic insect larvae.) Secondly, sufficient space must be allowed, per fish, for territory formation. A seven-foot-long tank was used to ascertain minimum and maximum territory sizes and the information gained from this tank was then borne in mind when setting up fish in smaller tanks. Thirdly, the sequence of appearance of the different phases of the reproductive cycle under natural circumstances was checked against the sequences obtained in the tanks, by means of a series of collections from the same area of the River Kennett at different times through the spring season. Also, a number of fish were set up in a large tank in the late winter and left undisturbed for several months to observe the gradual change-over from non-reproductive to reproductive condition. Bearing in mind the often violent fluctuations of the English spring climate, however, it is probable that, as regards the sequential problem, it is the behaviour in the river that is the irregular and distorted version of what is seen in the laboratory, rather than vice versa!

As far as experimental work is concerned, the criticisms that aquarium observations are 'unnatural' and invalid, hold true even less. For here the main interest lies in the effect on behaviour of controlled structural, spatial and sequential alterations of the environment. It is only by special 'unnatural distortions' of the usual patterns that they can be fully analysed. An important example of this type of approach is the dummy technique developed by Tinbergen, in which certain factors of the social environment of an animal are presented to it with particular elements missing. Lorenz, too, has always stressed the importance of the incomplete responses obtained from captive animals, as regards understanding the organization underlying their behaviour patterns. But if one is to eliminate a particular element from an environment, it is essential to know something of its significance in the natural habitat, and, furthermore, speaking of an incomplete response automatically implies knowledge of the complete response. To sum up, the observations recorded here have the dual aim of revealing as exactly as possible the course of the behaviour which actually does occur, on the one hand, and the organization underlying this behaviour by means of controlled alterations of various factors, on the other.

II

Non-Reproductive Behaviour

Although the primary concern of the present work is an investigation of the reproductive cycle of *Pygosteus,* it is nevertheless essential to include an ethogram in which *all* the actions that are performed by this species are described. A brief description of each non-reproductive activity is therefore given here.

Locomotion

Forward: Sticklebacks propel themselves with beats of their pectoral fins and their tails. They differ from many other fish in that their pectorals are more important as swimming organs than their tails. During slow and medium speed propulsion, the tail is not employed at all, but is held still and straight. Only in very high-speed dashes and darts, such as when fleeing from a predator, is the tail called upon to provide a propulsive force. Then it beats hard with rapid lateral undulations, as in other fish. In *Pygosteus* the tail-stem is much thinner than in *Gasterosteus* and this is correlated with the difference in migratory behaviour of the two species. *Gasterosteus* migrates annually from salt to fresh water and on these long journeys a stronger tail is necessary, for, in migratory swimming, tail-beating must be employed as the major propulsive force. In some species of the related pipe-fishes, the caudal fin has been lost altogether and progress is exclusively controlled by the pectoral fins. A pipe-fish moving through the water has the appearance of a long twig with twin paddles and this forms an extreme contrast with other types of elongate fishes, such as eels, where the marginal fins have become important, and the fish progress by tail-beats which have become exaggerated so that the whole fish wriggles through the water.

Under natural conditions, when the shapes of different river species cannot be distinguished, it is nevertheless possible to pick out the sticklebacks from the minnows and other small fry by virtue of the difference in the methods of locomotion.

Hovering: Pygosteus spends much of its time standing still in the water, which it does by alternate undulatory beating of its pectorals. In specialized forms of hovering, such as fanning (see later), the alternate beating of the pectorals is modified, producing a forward current of water, and their vigorous beating is counteracted by tail-beating which maintains the fish hovering on the spot. The speed of the beating of the pectorals in normal hovering varies a great deal. One curious correlation of this is with the degree of ripeness of reproductive females. If one scores the individuals in a group of females which are being 'ripened' (by over-feeding), for the degree of ripeness of each, it soon becomes clear that the riper a female is, the faster her pectorals beat in hovering. The degree of ripeness can be scored by an arbitrary measure of how swollen the belly is and when this has been done, then the rate of beating per minute can easily be obtained by counting and the use of a stop-watch. I have as yet no explanation for this correlation.

Flickering: A special form of hovering is a very high-speed alternate beating of the pectorals, at a rate too high to measure without special apparatus. I get the impression that flickering occurs in situations of what a psychologist would term 'emotional stress', but as yet I have no evidence that it forms an agonistic or sexual signal.

Sinking: This form of locomotion is seen in special agonistic situation and will be discussed fully there. The fish sinks from its hovering position to the floor of the tank, as a result of the complete cessation of all fin movement, including the beating of the pectoral fins.

Optomotor response: Sticklebacks, like most other fish, respond to a moving environment by keeping abreast of it. In this way a fish may maintain its position in a river in a strong current. It swims hard in fact, but, terrestrially speaking, it 'hovers'.

Schooling: In winter, *Pygosteus* shows social locomotory responses which result in the formation and maintenance of schools of fish. In the reproductive season these responses are not only lost but reversed, and a fish that was before followed closely and peacefully, is now attacked and driven away.

Defence against predators

The prey-predator relationships of sticklebacks with pike and perch are discussed in detail in Morris (1954e) and Hoogland, Morris and Tinbergen (1957). Also, in the chapter on agonistic behaviour, I have pointed out that normal escape from predators seems to involve the same actions as a particularly violent attempt to escape from rival sticklebacks. Sinking to the floor and freezing there, floating at the surface, darting for cover and hiding, are all described in the section on intraspecific agonstic behaviour and will not therefore be discussed here. The only difference between the reactions when they are given inter- and intraspecifically is that, in the former case, only a mild prey-predator situation will provoke the most intense escape response, but a very violent intraspecific situation is necessary to obtain the same effect.

Spine-raising is of great importance as a response to predators, and the attempts of various predators to swallow sticklebacks whose spines were raised is known, in a number of cases, to have resulted in the death of the predator. Associated with the raising of the spines is a special predator response which is performed by cornered sticklebacks at close range, where the predator is a larger fish. As the predator moves round to take the stickleback head-first, so the latter turns with it, thus always keeping its head turned away from the former's mouth.* The outcome of this is that, if the predator attempts to swallow the stickleback, it has to take it tail-first, and then the harder it tries to gulp it down, the more the spines are pressed out, and the more difficult the action becomes.

Feeding behaviour

Pygosteus and other sticklebacks live almost exclusively on live animal food. The main diet consists of aquatic insect larvae, small crustacea and worms. Hartley (1948), in an investigation of the stomach contents of a number of river fish, found that, in the River Cam near Cambridge, Chironomid larvae were by far the most important food of the sticklebacks. They were found in the stomachs of 95 (42 per cent of the 227 individuals examined. Also in evidence were Copepoda, which were found in 54 fish (24 per cent). Amongst the less important stomach contents were algal fragments. I have, however, never observed sticklebacks in tanks eating plants of any kind, and it seems possible that these fragments were taken in accidentally with animal foods.

The movements involved in feeding are as follows:

Searching: Sticklebacks which have not been fed for some time can be seen systematically searching the floor of the tank. They swim a short distance and then tilt downwards anteriorly so that their heads are about a centimetre above the substratum. In this head-down posture they may pause for several seconds before raising themselves into the horizontal position again and proceeding. This action is then repeated again and again all over the floor of the tank. The angle of the tilt may be so extreme that the fish are standing vertically in the water, but more usually they take up an angle of about 70 degrees out of the horizontal.

Biting: If, in such searching, they sight a prey protruding from the substratum, they bite into the substratum, taking up the prey and a certain amount of sand. The sand is then rejected and the prey swallowed. The bite is characteristically a twisting snap. As the snout moves down into the substratum the mouth is opened and shut and at the same time the body is rotated slightly around its long axis. The whole action takes a fraction of a

* This usually only applies where the predator is a perch (*Perca fluviatilis*), which has already snapped at the stickleback.

second. It should be noted that the twisting component of this feeding action is important when this action is being contrasted with the sand-digging movement employed in nest-building in *Gasterosteus*, in which there is no such twist.

Fig. 1 shows the relationship between the frequency of biting and the tilted angle of the biting fish. The ordinate shows the length of time taken to perform five bites, and the abscissa shows the mean angle at which the fish was inclined when it made the bites. It will be seen that the faster the fish was

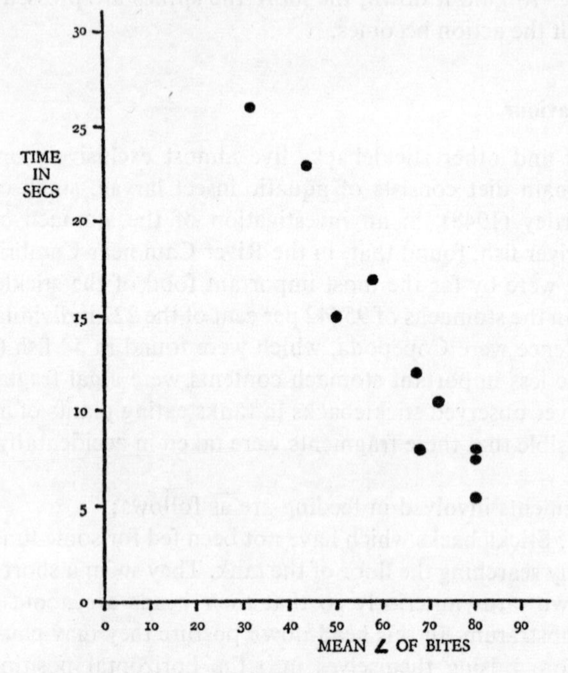

Fig. 1. The relationship between the time taken to perform five bites, when feeding, and the angle of the body when biting.

biting, the less it was tilted out of the horizontal. (These results were obtained by presenting a single live tubifex worm in a thin glass tube, so that the worm was seen at the level of the substratum. Attached to the tube was a scale of angles.)

Chasing: Bottom prey is dealt with as described above, but free-swimming prey is actively pursued by sticklebacks and snapped up in the open water without tilting.

Food-testing: Once a prey has been captured, it may be swallowed immediately, or it may be repeatedly taken into the mouth and spat out again. The movement has exactly the same appearance as the testing of nest material (see later). Often, as a result of this action, the food is rejected and falls to the bottom, instead of being eaten.

Shaking: Relatively large food objects, such as young earthworms, are sometimes shaken vigorously whilst held in the mouth. This shaking action, so common throughout the vertebrates, often serves the function of helping to kill the prey, or to break it up into smaller pieces, but in many species it does not seem to serve any obvious function at all, and such is the case in *Pygosteus.*

Gulping: After a large meal, a stickleback can often be seen to gulp repeatedly and vigorously and sometimes this results in a swallowed prey being regurgitated. The regurgitated prey is then usually tested as described above.

Comfort behaviour

The comfort repertoire of fish is small by comparison with terrestrial vertebrates and consists of only one special cleaning activity and stretching and yawning.

Chafing: Sticklebacks, and other species of fish which I have observed, possess a rubbing movement in which the body is thrown against some hard surface, so that the region of one, or other, of the gill-covers is scraped against it. Baerends and Baerends (1950, p. 28) have called this movement 'chafing'. When the hard surface is horizontal, this involves rotating the body around its long axis through a right angle. This is done by special movements of the pectoral fins and then, as fish comes to lie with its side facing the substratum, a strong tail-beat is given which knocks the nearside gill cover against the hard surface. The action usually takes a fraction of a second and may be repeated a number of times in quick succession. It probably serves to relieve some irritation of the gill region. (In the char, *Salmo alpinus*, the action of chafing appears to have been modified and used as a nest-digging action – see Fabricius, 1953, Fig. 3.)

Stretching: Sticklebacks stretch in two quite distinct ways. I am calling these S-bending and straight-stretching. The former is not very common in *Pygosteus*, but is seen frequently in *Gasterosteus*. Both species show straight-stretching regularly.

In S-bending, the fish bends its head to the left and at the same time bends its tail to the right (or vice versa), and then straightens out again. The spines are usually raised as the fish bends. Sometimes it may yawn as well. A low intensity version of the action seems to be the bending of the tail only.

'Yawning': Straight-stretching and yawning almost always occur at the same time and can be considered together. The fish opens its mouth wide, stretching its jaws to the full, and at the same time raises all its spines and

marginal fins. Although I have seen this action used as a displacement activity in the perch (*Perca fluviatalis*) and the bullhead (*Cottus gobio*), I have no such observations for *Pygosteus*. Yawning is most frequent when the fish is otherwise unoccupied. The function of mammalian yawning has long been controversial and discussions have centred around the intake of a deep breath of air. However, from my observations on *Pygosteus* and other animals I have been led to the conclusion that it is primarily a stretching movement and should be thought of as *jaw-stretching*. Not only do yawning and stretching occur together so frequently, but also I get the impression that species with strong, well-developed jaws tend to yawn more intensely and frequently than other species. (It may be that fish 'yawning' and mammalian yawning are not comparable activities, although it seems likely that they are.)

Pathological behaviour

Certain movements occur which are associated with the diseased condition of the individuals which perform them. A fish may sometimes be seen floating under the surface, or lying on the floor of the tank breathing rapidly, or rotating as it swims in an erratic course. It is not intended to go into the pathological causes of these actions here and it is unlikely that any of these movements will be confused with normal behaviour. Usually the fish performing them can be seen to be diseased in some way, or the erratic nature of the movements reveals their abnormality. One special type of pathological movement is worthy of mention and this is the 'shimmies' (Innes, 1951, p. 48). The fish appears to be able to swim only very slowly and does so by a quick series of little jerks, its whole body jerking back and forth as it does so. A number of causes are suspected, especially *Ichthyophthirius*.

Sticklebacks suffer from innumerable diseases, which are particularly prevalent at the end of the breeding season. It is perhaps significant that salt water is employed by aquarists for curing fresh-water fish of many of these ailments, when one thinks of the fact that it must be at about this time that *Gasterosteus* begins to migrate back to the sea.

III

The Phases of the Reproductive Cycle

Before dealing with the different aspects of the reproductive cycle of *Pygosteus* in detail, a brief summary of the phases of this cycle will be given. This will enable the reader to approach the more detailed sections with some idea of the cycle as a whole system. The question of the cycle as a whole will be returned to again and further discussed in the last chapter.

The pre-territorial phase

The first signs that the reproductive season has begun come with the cessation of schooling responses. Individual fish show spasmodic, low intensity, aggressive behaviour and the members of a group begin to spread out slightly. Certain fish become darker in colour. After a few days, high intensity attacks can be observed, but these are still spasmodic and may occur anywhere. No fish yet shows any tendency to restrict its movements to one area. During this phase already signs of sexual activity may be seen. Some males straight away, without territory or nest, begin to dance to the females which are just beginning to become swollen. These early dances usually only last for a few jumps, as compared with the later dances during the sexual phase which may consist of as many as several hundred jumps.

The territorial phase

After some days, certain individuals tend to restrict themselves to particular regions. Gradually more and more do so until a territorial society has been formed. In *Pygosteus* there is a heterosexual system of territories with the males and females interspersed. When the fish first take up territories they are still in winter dress and the sexes are indistinguishable from one another. As the spring advances, so certain territory owners become black all over (males) and others become black on the dorsal surfaces only and develop swollen silver bellies (females). There is no pair-formation in this species and no co-operation between individuals until the actual courtship and spawning. The territory of each individual has a headquarters, which is simply a position on it to which the owner flees, or returns after fighting a rival. Each territory also has a boundary line around it, which is the site of border fighting. As the territories become more firmly established, so the fighting becomes more and more restricted to the boundary regions. This is the outcome of the special learning process that spatially modifies the early, randomly scattered, agonistic behaviour, and turns it into territorial fighting. The nature of this modification, which makes each fish dominant in its own territory and subordinate in other territories, will be discussed later.

The nest-building phase

The males, having selected territories where there is vegetation, soon begin to construct their nests. During this phase, the females do not change their behaviour at all, but remain aggressive and territorial. The males may take anything from a few hours to several days to complete their nests, and during the actual collection and construction work they become slightly less aggressive. It should be stressed that agonistic and territorial behaviour persist throughout all the following phases of the cycle.

The sexual phase

Having constructed their nests, the males now become light coloured dorsally and intensely black ventrally. They begin to court the females on the adjoining territories, dancing to them now instead of attacking them. If, in a particular case, the female in question responds to the dance sexually, then she is led on to the male's territory and to his nest where spawning then takes place. After fertilization the male chases her back to her own territory, if she has not already returned there herself. Within a few minutes she is defending it again and the male and female who have just been courting are now fighting once more. The female soon begins eating voraciously and in a day or so is ready to lay another clutch of eggs. The male collects several clutches from the adjoining females and sometimes one male takes two clutches from the same female.

If the female on the adjoining territory is not sexually inclined, she is courted for a short while and then attacked by the male. The attempted courtship then breaks down into territorial fighting. If she is moderately sexually inclined, so that she follows the male's dancing, but gives up before laying her eggs, she is furiously attacked by the male and driven back to her own territory. This is because she has ceased to respond sexually to the male and, being on his territory, is therefore an intruder, and one who has intruded farther than any fish would be able to do under other circumstances.

During the time when the male with a territory and a nest is waiting for a ripe female, he is constantly maintaining the nest structure and in this way the nest becomes larger and stronger.

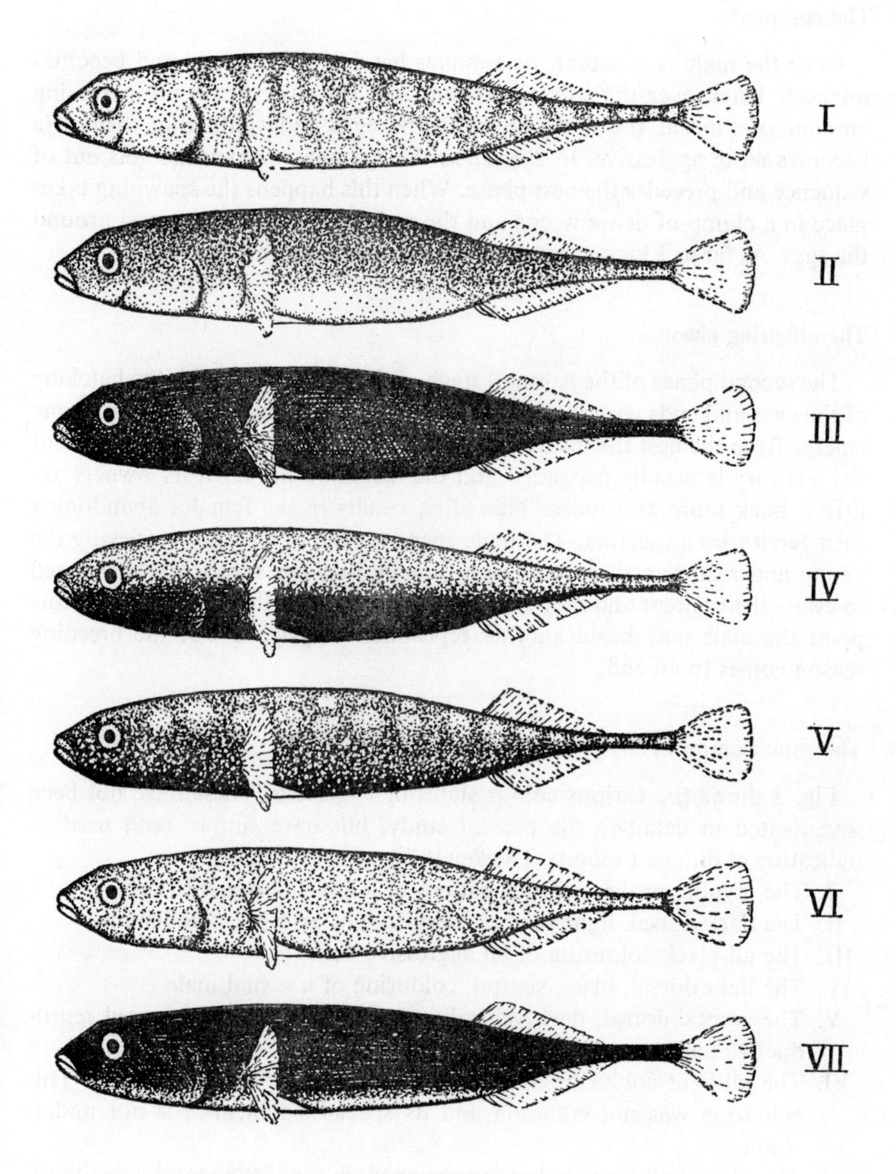

Fig. 2. The colour states of *Pygosteus*. For explanation see text.

The egg phase

Once the male is a parent, he resumes his dorsal blackness and becomes intensely black over the whole body surface. He now spends an increasing amount of time at the nest, adjusting the eggs and ventilating them. He becomes more aggressive. In *Pygosteus* the egg phase sometimes gets out of sequence and precedes the nest phase. When this happens the spawning takes place in a clump of dense weeds and the male then constructs a nest around the eggs. As far as I know, this does not occur in *Gasterosteus*.

The offspring phase

The second phase of the parental stage of the cycle begins with the hatching of the eggs and ends with the completion of the whole cycle. When the young emerge from the nest the male reaches a peak of aggressiveness. At this point the territory is usually expanded and the neighbouring territory owners are driven back more and more. This often results in the females abandoning their territories altogether. The male spends much of the time retrieving the young and retarding their scattering. Finally they are sufficiently developed to evade their parent and they then spread throughout the vegetation. At this point the male may begin another reproductive cycle, or else the breeding season comes to an end.

The colour states of 'Pygosteus'

Fig. 2 shows the various colour states of *Pygosteus*. These have not been investigated in detail in the present study, but have simply been used as indicators of different moods. The key to Fig. 2 is as follows:

I. The cryptic, broken, colour pattern of a non-reproductive fish.
II. The dark dorsal, light ventral, colouring of an aggressive female.
III. The all-black colouring of an aggressive male.
IV. The light dorsal, black ventral, colouring of a sexual male.
V. The cryptic dorsal, dark ventral, colouring of a recently scared reproductive male.
VI. The all-light colouring assumed by certain reproductive females. This colouring was not common and its special significance is not understood.
VII. The almost all-black colouring assumed, in rare instances, by territorial individuals. Such individuals did not seem to be able to get beyond the territorial phase of the cycle.

IV

Agonistic Behaviour

The organization underlying the complex code of agonistic signals which *Pygosteus* possesses will be discussed later. First, it is necessary to take each component action in turn, for descriptive purposes, without reference to its relationship with the other components. Some general reference to the state of the fish performing the action will, however, be given.

Agonistic ethogram

Charge: The fish darts through the water at the enemy at very high speed. Although the movement is too quick to analyse without cinematography, it appears that the high speed is obtained by a combination of caudal and pectoral beating. The charge is often braked suddenly, usually because the fish is getting too far away from its territorial headquarters.

Bite: The charge frequently leads to biting, in which any part of the enemy's body may be hit. Biting also may occur without the preceding charge, in which case it appears to be more accurately directed, usually at the pectoral or caudal fins, or the tail-stem. A badly beaten fish may lose the greater part of these fins.

Drag: Under intensely stimulating circumstances, such as when an intruder, for some reason, will not leave the territory on which it is trespassing despite the fact that it has been bitten repeatedly, the territory owner may bite the intruder on the fin and hold on. One of three things then happens. Either the intruder tries to swim away, in which case it drags the owner after it, tearing the held fin as it does so; or, the owner immediately makes a dart away from the intruder, still holding the latter's fin in its mouth, thus either dragging the intruder after it (see Fig. 3), or else tearing away part of its fin with the strength of the dart; or, both may happen at once.

Mouth-fighting: Many species of fish indulge in 'kissing' or mouth-fighting. In sticklebacks it is rare, but does occur. Two males, both intensely agressive, charge one another and bite simultaneously so that one is holding the other's upper jaw whilst the latter holds the former by its lower jaw. A trial of strength then follows in which the two fish, locked together, advance and retreat a little, until one can shake off the other.

Chase: If the enemy flees, it may be pursued around the tank for many minutes. The chasing fish follows the other as closely as it can and bites whenever the fleeing fish slows down or stops. In this way bite and chase may alternate over and over again. Under certain circumstances chasing may take the place of actual contact. Two fish with adjoining territories may be seen to repeatedly chase one another off when trespassing occurs, without ever biting one another. In this way territories can be maintained without actual contact. However, contact has always been necessary at some stage before such an 'agreement' can be arrived at, but this will be discussed more fully later.

Fig. 3. A dominant male dragging away a rival.

Roundabout fighting: The two fish concerned circle round and round one another, like the wooden horses on a fairground roundabout, never quite catching up with each other (see Fig. 4). The spines are raised during this circling and van Iersel (1953) has called it 'spine fighting'. This is a bad name, however, for two reasons. In the first place, this is only one of many circumstances in which the raising of the spines is used as intimidation-fighting. Secondly, it suggests rather that the fish may pierce one another with their spines, which is, of course, not the case. I shall therefore substitute the name 'roundabout fighting' for this activity. The activity ceases when either one or both fish swim off suddenly, or when one fish does catch up with the tail of the other and bites it.

Stationary charge: Like mouth-fighting, this is rarely seen in *Pygosteus*, but is quite common in many other fish. The tail and posterior part of the body beat rapidly in intense lateral movements, but the fish does not move forwards. This means, presumably, that the pectoral fins must be beating with an opposite and equal force, as in fanning (see later). The fish is pointed directly at its enemy and the stationary tail-beating usually develops into a charge and bite. This activity does, in fact, give the strange impression, in *Pygosteus*, of a 'mark-time charge'.

Spine-raising: The erection of the ten dorsal and/or two ventral spines is an extremely frequent activity of complex significance, which will be discussed

Fig. 4. Roundabout fighting.

fully later. It will suffice here to discuss the three basic forms which it may take:

(a) Only dorsal spines raised
(b) Only ventral spines raised
(c) All spines raised.

In intraspecific agonistic situations, the raising of dorsal spines alone is characteristic of subordinate fish, whilst the raising of ventral spines alone is associated with dominance. The raising of all spines may signify an intermediate status. According to van Iersel (1953, p. 7), in *Gasterosteus* the situation is exactly the reverse of this, with inferiority associated with ventral spine-raising and superiority with dorsal spine-raising.

The actual raising itself is most commonly with the spines either completely down, or completely up, but under special circumstances, the spines may only be raised through a part of the usual right-angle. All the ten dorsal spines are usually raised as one, but frequently only one of the two ventral spines will be raised. When this occurs, it is always the spine nearer the other fish that is raised. This last point applies equally to *Gasterosteus* (see Tinbergen, 1951a, frontispiece). The popular idea that sticklebacks pierce one another with their raised spines is, from all my observations, completely false. In intraspecific situations the spines play a purely signal role. If intraspecific piercing has ever occurred I feel sure that it must have been the completely accidental outcome of one fish impaling itself when charging another. One of the many possible examples of the 'piercing' fallacy is found in Bateman (1890, p. 145): 'The fish charge furiously at one another, and then, if no great harm has been done, continue to swim round and round, trying to bite with their strong mouths or to pierce each other with their sharp lateral or dorsal spines.' Lorenz (1952, p. 28) thinks that, in any case, piercing would be impossible because of the stickleback 'armour'. However, *Pygosteus* is an unarmoured form which it is quite easy to pierce with a needle, so that this point is obviously not important. (In fact, the most efficient method of removing a *Pygosteus* from a tank when it had just died is by impaling it on a needle attached to the end of a thin stick.)

Vertical posturing: There are two quite distinct vertical postures which are performed by *Pygosteus* in agonistic situations:

(a) Head-down posture (Fig. 5)

(b) Tail-down posture (Fig. 6).

In the head-down posture, the body is lowered quickly anteriorly until it is 60 to 70 degrees out of the horizontal. It is often lowered less than this, rarely more. In the tail-down posture the body may be lowered posteriorly until it is standing almost vertically in the water, but here too any angles may be assumed. The former posture is characteristic of fights involving approximately equally matched fish, whilst the latter is shown by subordinate individuals. (It should be noted that *Gasterosteus* workers have referred to the 'head-up' posture of that species, rather than the 'tail-down' posture. The latter name has been given here, because, in *Pygosteus*, it is the tail which is lowered and not the head which is raised when this posture is being assumed.)

The head-down posture occurs also in *Gasterosteus* where it was originally interpreted by Tinbergen (1940) as displacement feeding, and later as displacement sand-digging (Tinbergen and van Iersel, 1947). This led Sevenster (1949), on seeing the head-down posture in *Pygosteus*, to say that, as this species has no sand-digging pattern, it is probable that it evolved from an ancestor which built a nest in the sand. For a detailed criticism of these views, see section VIII. For the present I will call the head-down posture simply that, and not employ an interpretive name.

Rotation posturing: As with the vertical posturing, there are two distinct forms here, which I will call:

(a) The ventral roll

(b) The dorsal roll.

In the former, the ventral surface is rolled towards the other fish, whilst, in

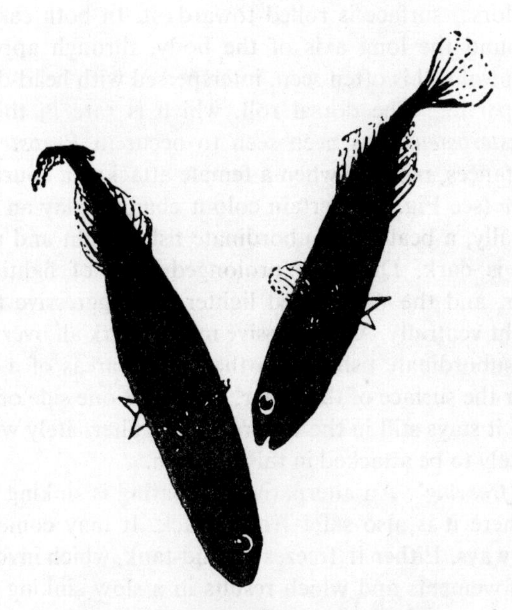

Fig. 5. The head-down threat posture.

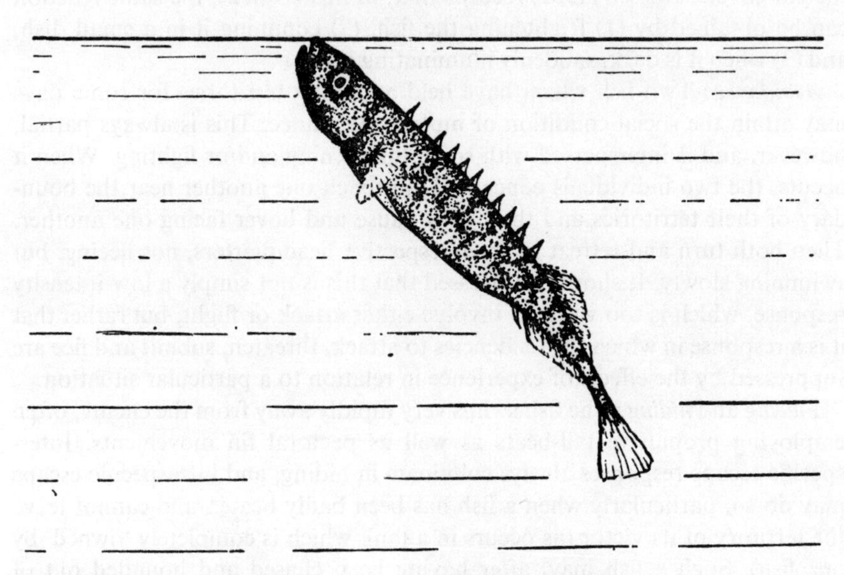

Fig. 6. The tail-down submissive posture. (This must not be confused with the emergency-respiration posture of a fish in badly oxygenated water.)

the latter, the dorsal surface is rolled towards it. In both cases, the rolling fish rotates around the long axis of the body, through approximately 90 degrees. The ventral roll is often seen, interspersed with head-down postures, in prolonged sparring. The dorsal roll, which is rare in this species, but common in *Gasterosteus*, has been seen to occur in *Pygosteus* only under special circumstances, namely, when a female attacked a courting male.

Colour change (see Fig. 2): Certain colour changes play an agonistic role. Speaking generally, a beaten or subordinate fish is light and a dominant or aggressive fish is dark. During a prolonged bout of fighting, the victor becomes darker, and the vanquished lighter. An aggressive female is dark dorsally and light ventrally. An aggressive male is dark all over.

Floating: A subordinate fish avoids the middle areas of a tank. Often it floats just under the surface of the water, usually at one side or in a corner of the tank, where it stays still in the water, beating alternately with its pectoral fins. It is less likely to be attacked in this position.

Sinking and *'freezing':* An alternative to floating is sinking to the bottom of the tank, where it is also safer from attack. It may come to lie on the bottom in two ways. Either it 'freezes' in mid-tank, which involves the cessation of all fin movements and which results in a slow sinking to the bottom where it then rests quite still, or else it swims to the bottom and then 'freezes' when it arrives there. With all the fins held still, including even the pectorals, the only remaining perceptible movements are the rise and fall of the gill-covers. Russell (1931) records that, in *Gasterosteus*, the same reaction can be obtained by (1) frightening the fish, (2) confining it in a small dish, and (3) when it is dark, suddenly illuminating it.

Avoidance: Two fish which have held adjoining territories for some days may attain the social condition of mutual avoidance. This is always partial, however, and is interspersed with some threatening and/or fighting. When it occurs, the two individuals concerned approach one another near the boundary of their territories and then both pause and hover facing one another. Then both turn and retreat to their respective headquarters, not fleeing, but swimming slowly. It should be stressed that this is not simply a low-intensity response, which is too weak to involve either attack or flight, but rather that it is a response in which the tendencies to attack, threaten, submit and flee are suppressed by the effects of experience in relation to a particular situation.

Fleeing and *hiding:* The fish swims very rapidly away from the enemy, often employing propulsive tail-beats as well as pectoral fin movements. Interspecific escape responses always culminate in hiding, and intraspecific escape may do so, particularly when a fish has been badly beaten and cannot leave the territory of its victor (as occurs in a tank which is completely 'owned' by one fish). Such a fish may, after having been chased and hounded out of hiding a number of times, simply hide again without preceding that action with the typical frantic dashing around the tank that characterizes the flight response. When it has been 'discovered' hiding in a clump of weeds after a beating, it may swim straight over to another clump and hide there. Such hiding often consists of burying only the head into the weeds and the contact of the weeds with the surface of the head appears to satisfy the fish despite the fact that its tail and half of its body are in full view. In this respect it is more ostrich-like than the ostrich itself.

Where two rival territory-owners are involved, the fleeing that occurs does not culminate in hiding. The flight paths are always directed towards the respective territorial headquarters. Once a fish nears or reaches its territorial headquarters, the fleeing ceases and either its behaviour changes to some quite different non-agonistic pattern, or it turns and advances once more on its rival. Where there are several adjoining territories, it is possible to map the positions of headquarters by tracing the different flight paths. Where these lines meet reveals the ethological centre of the territory, which may or may not coincide with its geographical centre. (This is illustrated, and discussed more fully, later.)

It is interesting to note that these intraspecific flight paths do not coincide with interspecific flight paths. The response given to such stimuli as vibration, strange moving object, predator-fish, etc., is different from the response given to a rival of same species. If a strange object is moved through the water towards a territory-owner from a direction in which it is used to being approached by a rival, then instead of fleeing towards its headquarters, it darts about in all directions, completely losing its sense of territory. This loss of territory sense applies to both its own and its rivals' territories. It now flees through regions which before it would not dare to enter, and also through regions which before it would have defended. The question as to whether the difference between the intra- and interspecific escape patterns is the result of differing intensities of stimulation of the same nervous 'centre', or of the stimulation of different 'centres', is answered by the fact that a completely subordinate fish flees from another stickleback in the same way that any stickleback flees from a predator or 'strange moving object'. This had led me to postulate – tentatively – a single escape centre for *Pygosteus*.

Territorial behaviour

The males of *Pygosteus* defend reproductive territories. By this is meant that the agonistic behaviour, the elements of which have just been described, is spatially restricted.

In order to understand clearly their relationship, it is necessary here to examine the two concepts of aggression and territory separately. Intraspecific reproductive aggression involves, functionally speaking, simply the removal of rivals. For the moment, the nature of this removal is unimportant. By physical damage, or threat of physical damage to the rivals, the aggressive individual in one way or another effectively removes them from its sphere of influence. In evolution, there will be an increasing tendency to respond to mere aggressive intention movements, since, if disputes can be settled without recourse to actual physical violence, the unscathed contestants will stand a better chance of survival. Thus, the most important factors in the aggressive removal of rivals are often the signal communication systems, or 'threat-codes' as I have called them (Morris, 1954b, 1954c), which may largely eclipse, in some species, the actual contact-fighting itself.

Thus far the aggressive system consists solely of a set of stimulus-response mechanisms without reference to any effects of experience. If the animal is in a particular internal chemical state (see Collias, 1944, for a review of the physiological mechanisms of aggression), it will, if this is the whole story, always respond to a standard sign stimulus in the same way. But in intra-specific reproductive aggression this is not the case, and it would not be very adaptive if it were, when one considers the time scale involved, because the activity of removing rivals has to be maintained often over a considerable period of time in the reproductive season. If aggressive behaviour were not capable of special modification by experience, then disputes would have to be settled over and over again, in a most inefficient manner, whenever and wherever two males met. Consequently, it is not surprising that one finds that the fundamental aggressiveness of a species is always inherently susceptible to one or other form of special modification by experience. Spatial modification is undoubtedly the most common and the most widespread amongst vertebrates, and it is that spatially modified aggression that has been called territorial behaviour (Howard, 1920).

As I shall describe below, *Pygosteus* males begin to show mild aggression to one another early in the season, and this aggression shows no modification either in respect of the position in space of the encounter, or the individual identity of the enemy. Gradually, however, each male becomes spatially restricted in his aggressive activities, and, in time, his aggressive space, or territory, may become an extremely accurately defined area, inside which he will be dominant, and outside which, he will not. But before describing the territorial behaviour of *Pygosteus* in detail, it is necessary to compare typical with atypical territorial behaviour and with other aggression-modification systems.

Firstly, it must be pointed out that the space that is defended, in territorial species, although usually more or less static, need not always be so. In the case of the bitterling fish (*Rhodeus amarus*) (Boeseman *et al.*, 1938) the male guards a mussel in which, ultimately, the female will lay her eggs. As the mussel moves, so the territory of the male moves. In the case of the anura, the male guards his female by kicking away other males. As the female moves, so the territory of the male moves. I have mentioned these two examples of atypical territorial behaviour alongside one another because, although no one questions the fact that the bitterling is territorial, the anura are not usually considered as being either aggressive or territorial. If the back of the mounted female is considered as the territory (which it should be, since it is a defended space), then the male frog or toad is displaying true territorial behaviour which is not fundamentally unlike that of the bitterling, or other territorial species, except that it is an extremely uncomplicated example.

An alternative aggression-modification system is the social hierarchy. Individual recognition is the basis of this system, which requires no particular spatial restriction of aggression on the part of the dominant animals. The latter may spend most of their time in a preferred region, but they will be just as able to intimidate weaker rivals, which they have learnt to recognize personally (and vice versa), wherever they happen to encounter them. The result of each threat or actual combat which takes place between two individuals establishes more firmly their dominant-submissive relationship, the weaker animal learning, usually after only a few encounters, to avoid the stronger. The efficiency of this system is obvious. It enables the dominant animal to maintain its status without persistently having to indulge in the time, energy and attention-consuming activity of combat.

There are a number of species which possess complicated social agonistic organizations involving combinations of territorial and hierarchical systems, but they cannot be discussed here. It is important, however, to consider what, in general, determines whether a species will modify its reproductive aggression territorially or hierarchically, in the simpler cases. As an example, it is convenient to compare the Mexican swordtail (*Xiphophorus helleri*) with the stickleback. Both have elaborate threat-codes, but the former is hierarchical, whilst the latter is territorial.

The swordtail is viviparous, and there is no parental behaviour for either male or female. The stickleback is oviparous, and there are elaborate parental duties for the male (see van Iersel, 1953). Both the male swordtail and the male stickleback are physically *capable* of fertilizing the eggs of a large number of females, but this is impracticable for the latter, since it can only *rear* the eggs from a limited number of females. This, however, is no problem for the swordtail, and natural selection will favour a less 'democratic' way of life for it than for the stickleback. Owing to the reproductive limitations of each individual male stickleback, natural selection cannot afford to favour individuals differentially to such an extreme as in the case of the swordtail. The social hierarchy system 'wastes' a large number of males that are reproductively capable, but the dominant males which benefit at their expense are almost certainly capable of fertilizing all the available females in a population.

(The system is probably not as rigid as this in nature,but this is certainly the fundamental tendency involved.) The territorial system, on the other hand, although it provides ample opportunity for the same kind of natural selection (stronger males own bigger and better territories), decreases the effect considerably. To sum up, if a 'super-male' swordtail was given twice as many fertilizations as it would normally get, then its 'super-male' qualities would be passed on twice as much. But if a 'super-male' stickleback was given twice as many fertilizations, then all its offspring would probably perish owing to its incapacity to rear such a huge brood. The existence of a reproductive hierarchical or a reproductive territorial system in any species is therefore primarily determined, I suggest, by the number of females that a male can fertilize efficiently, in relation to the number that are available. Limitations of this capacity, which will therefore favour the territorial system, are principally strenuous parental duties for the male, which will make extra fertilizations inefficient, but the incapacity to actually fertilize extra females may also be important, as in the case of the anura.

It is beyond the scope of the present discussion to enter into the interesting variants of the above scheme which exist (seals and primates, to mention only two examples), but it was felt necessary to give a brief general background to territorial behaviour, before describing the form which it takes in the case of *Pygosteus*. For further discussions of territorial theory, the reader is referred to Howard (1920), Collias (1944), Tinbergen (1936, 1951b, 1953a, 1953b) and Lack, D. and L. (1933).

As an introduction to a study of the territorial behaviour of *Pygosteus*, it is essential to look in some detail at the social organization of the species during the very early stages of the reproductive season. In order to do this, fish were collected from the river in February, before they had moved to their breeding grounds, and before they showed any signs of reproductive coloration.

In the laboratory, such fish were treated in one of two ways. Either they were kept together in a shoal in large, unplanted tanks, or they were placed into planted tanks, suitable for breeding, in much lower population densities. These latter population densities were aimed at being those most likely to have optimum breeding value.

Thus, in the former case the fish experienced light and temperature conditions suitable to bring them into reproductive mood, but social and other environmental conditions which were unsuitable in this way. The second type of population was reproductively favoured in all respects, as far as was possible.

The general difference between the two types of populations, as regards the course of development of reproductive behaviour, was that the crowded groups of fish were slower to begin. This is in agreement with van Iersel's (1953) findings for *Gasterosteus*, and furthermore, both in *Pygosteus* and

Gasterosteus, crowding, although it can retard the reproductive cycle, cannot completely prevent it from beginning. But before describing the special ways in which the behaviour of the two types of populations differed, it is necessary to report on the pre-territorial behaviour of the shoal.

Pre-territorial movement

In order to investigate this, a long (210 cms), thin (36 × 36 cms) tank was set up with an even substratum of fine gravel, but without any cover, plants or obstructions. There was no current. Into this blank tank of water were placed fourteen freshly caught, pre-reproductive *Pygosteus* of unknown sex. The tank was divided, on the outside, into fourteen regions by means of lines drawn on the frame and glass wall. From left to right there were seven sections, each 30 cms wide. Each of these was divided into an upper and lower half, so that each region was 30 cms wide, 15 cms high, and 36 cms deep. These regions were numbered on the metal frame. The observer sat centrally in front of the tank and glanced at each region in a set order, and with a standard short gap of time between each glance (five seconds was found to be most suitable). After a glance at a region of the tank, the number of fish present in that region was recorded in a table. In each such table, fifteen records were obtained for each region of the tank. Three such tables were compiled, one after the other, in every test. Nine such tests were made, under differing circumstances, as follows:

(a) Blank tank.
(b) Tank with centrally placed shelter (flower-pot).
(c) ditto, a day later.
(d) ditto, after another day.
(e) Tank with shelter replaced by clump of waterweeds.
(f) ditto, a day later.
(g) ditto, after two more days.
(h) Tank with a second clump of weeds added (in region 6).
(i) ditto, after five days, two males having taken up territories (in regions 1 and 6).

1A	2A	3A	4A	5A	6A	7A
1B	2B	3B	4B	5B	6B	7B

The above shows the numbering of the regions of the tank. The vertical movements of the fish were unimportant for most problems, and will be mentioned separately. For the following results, however, the A-B divisions will be ignored and the 1 to 7 divisions only will be taken into account. Here now is a table of the results obtained in the nine tests, for the seven regions:

	1	2	3	4	5	6	7		
(a)	142	75	73	72	75	77	147	—	blank tank
(b)	155	69	74	173	82	56	121	—	new pot
(c)	140	94	84	114	98	95	124	—	familiar pot
(d)	91	90	116	113	94	90	109	—	very familiar pot
(e)	124	72	70	185	74	42	95	—	new plant
(f)	79	71	123	175	101	81	92	—	familiar plant
(g)	116	94	104	160	58	65	111	—	very familiar plant
(h)	115	46	74	112	72	145	97	—	two clumps
(i)	123	249	117	117	45	9	12	—	two territories

These results are expressed graphically in Figs 7(a)–(i). They may be interpreted as follows.

In (a) there is a remarkably symmetrical pattern, showing that the central regions of the tank contained equal numbers of fish, but that the two end regions contained almost twice as many. This reflected the fact that the fish were swimming very actively 'up and down stream'. As mentioned earlier, there was, however, no current in the tank, other than a mild vertical one set up by the aerators, so that the fish were, so to speak, responding to the shape of the swimming space. Once they reached the glass at the end of the length of the tank, they often went on attempting to swim farther for some time, fluttering against the glass wall vigorously.

When, in (b), a shelter was placed in the central region (4) of the tank, in the shape of a flower-pot, the fish responded to this by increasing their occupation of that region considerably. They swam in and out of and around the pot with a high frequency and, although they never actually hid inside it, or used it as a serious shelter, they paid considerable attention to it. I can find no other words to describe such behaviour than 'exploratory' or 'investigatory'. More attention was paid to the pot than could be accounted for by thinking of it simply as an obstacle.

In (c) and (d) the same tank set-up was used, the pot being left untouched, but (c) was a test taken one day after the introduction of the pot and (d) two days after its introduction. It will be seen that the peak in the central (pot) region disappears. The fish paid less and less attention to the pot as time passed, as if a process of familiarization was taking place. Now, although this problem has been studied in the higher vertebrates, I am unaware of any report of its occurrence in fish. Since it was beyond the scope of the present study, the question was not pursued further, but it presents an interesting possibility for future research.

The fish having become familiarized with the pot to the extent of ignoring it, it was replaced by a single clump of *Myriophyllum*, in exactly the same spot. This plant was not dense enough to permit the fish to conceal themselves in

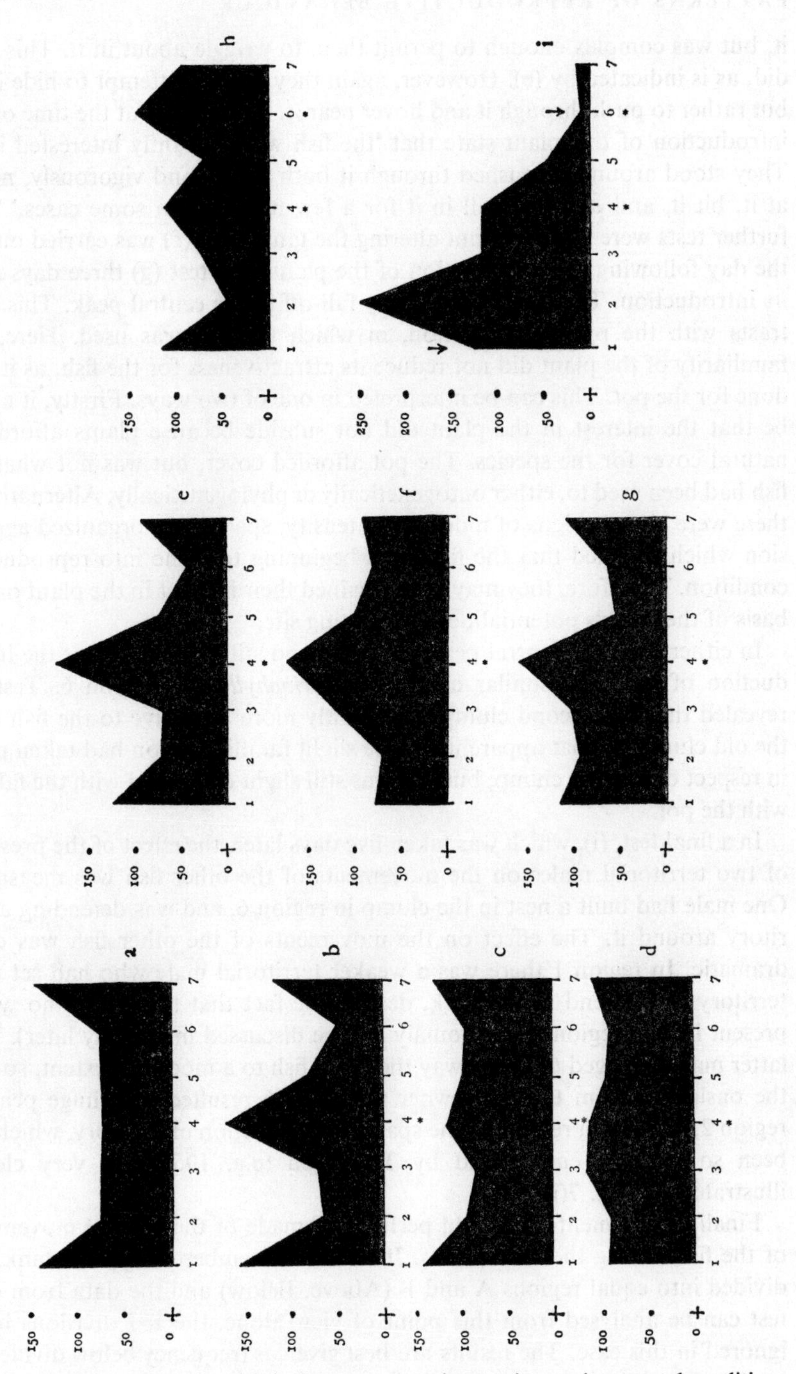

Fig. 7. The movements of a shoal of *Pygosteus* under varying environmental conditions, in a long aquarium tank. For explanation see text.

it, but was complex enough to permit them to wriggle about in it. This they did, as is indicated by (e). However, again they did not attempt to hide in it, but rather to push through it and hover near it. Notes made at the time of the introduction of this plant state that 'the fish were instantly interested in it. They stood around it, pushed through it both slowly and vigorously, nosed at it, bit it, and even lay still in it for a few moments in some cases.' Two further tests were done without altering the tank. Test (f) was carried out on the day following the introduction of the plant, and test (g) three days after its introduction. There was little if any fall-off in the central peak. This contrasts with the previous situation, in which the pot was used. Here, the familiarity of the plant did not reduce its attractiveness for the fish, as it had done for the pot. This can be interpreted in one of two ways. Firstly, it might be that the interest in the plant did not subside because plants afford the natural cover for the species. The pot afforded cover, but was not what the fish had been used to, either ontogenetically or phylogenetically. Alternatively, there were already signs of mild, low-intensity, spatially unorganized aggression which signified that the fish were beginning to come into reproductive condition. Therefore, they may have retained their interest in the plant on the basis of the latter's potentialities as a nesting site.

In either case, the central peak remained and still persisted after the introduction of a second similar clump of *Myriophyllum* in region 6. Test (h) revealed that this second clump was slightly more attractive to the fish than the old clump, so that apparently some slight familiarization had taken place in respect of the first clump, but this was still slight compared with the fall-off with the pot.

In a final test, (i), which was taken five days later, the effect of the presence of two territorial males on the movements of the other fish was measured. One male had built a nest in the clump in region 6, and was defending a territory around it. The effect on the movements of the other fish was quite dramatic. In region 1 there was a weaker territorial male who had set up a territory at that end of the tank, despite the fact that there were no weeds present in that region (this anomaly will be discussed more fully later). This latter male managed to keep away the other fish to a moderate extent, so that the onslaught from the nest-owner in region 6 resulted in a huge peak in region 2, and not in region 1. The spacing-out function of territory, which has been so strongly emphasized by Tinbergen (e.g. 1953a), is very clearly illustrated by Fig. 7(i).

Finally, brief mention should perhaps be made of the vertical movements of the fish during the above tests. It will be remembered that the tank was divided into equal regions A and B (Above, Below) and the data from each test can be analysed from this point of view alone, the 1–7 divisions being ignored in this case. The results are best given as frequency below divided by frequency above. From the following figures, it is clear that the fish strongly preferred the lower half of the tank to the upper half, and that for every 'fish presence' above there were several below. Only data from the first eight tests are used here, since in the ninth test (i), there were the territorial males present, which distorted the picture in a special way.

TEST	B/A	TEST	B/A
(a)	2·5	(e)	4·5
(b)	3·9	(f)	2·0
(c)	2·5	(g)	2·8
(d)	2·9	(h)	2·7

Lumping the above results together, it can be said that, for every occurrence in the top half of the tank, there were three in the bottom half. Therefore under the conditions of the tests the fish, which were not hiding, but swimming actively about, showed a distinct preference to keep low down in the water. The reason for this is not yet clear. None of the obvious factors which influence the vertical movements of sticklebacks were operating. For example, if the water had been foul the fish would have swum just under the surface, but the water was well aerated. If the fish had been deprived of food, they would have roamed the substratum searching for small prey, but they were always well fed before each test. If they had been in full reproductive condition, special activities might have detained them at a particular level of the water, but they were still in a pre-reproductive state. At the present stage, therefore, it can only be assumed that the vertical tendency described here is a basic characteristic of the locomotory behaviour of this species.

The first signs of aggression

It has been shown from the last series of tests that the first effect of territorial aggression is that of *spacing out*. The as yet non-territorial fish are scattered away from the selected site of the territorial individuals as a result of the aggressive activities of the latter. But before discussing the spatial organization of aggression in *Pygosteus*, it is necessary to describe the pre-territorial fighting that occurs in the very early stages of the reproductive cycle.

A male does not settle on a particular site and then commence to defend it but rather begins by being aggressive to any other individual anywhere in the tank. This aggression is at very low intensity and it is this mild fighting which gradually becomes spatially restricted. At the same time, it becomes more intense and the early sparring bouts turn into savage battles. It is the nature of the sparring bouts that must be reported first.

Adult sticklebacks in non-reproductive condition show no intraspecific aggressive responses. There are, in fact, very few intraspecific social responses of any kind shown outside the breeding season, in the case of *Pygosteus*. There appears to be no food-fighting in *Pygosteus*, but if one fish has found a prey, the agitation resulting from the special motor patterns involved in devouring the prey, are often responded to by any other fish in the neighbourhood. The latter may then swim quickly to the site and compete for the food object. Special non-reproductive social responses are also given which partially control the orientation of actively swimming fish. There is a tendency to keep near and follow the movements of an actively moving fish of the same species. It is counteracted by a tendency to keep at a certain minimum 'individual distance' from other fish of the same species (see Hediger, 1950). The two tendencies compete and result in the typical shoal formation which is so widespread amongst fish, and which appears to have evolved as a defence against predators. This is because a predator finds it extremely difficult to concentrate on one prey at a time when it finds a dense group swimming together. It dashes first at one and then at another, but they scatter and reform. If one or a few fish appeared in isolation, the predator would be able to concentrate on one individual at a time, without the constantly competing stimuli from other prey interfering with its attack manœuvres (see also Welty, 1934). It is noticeable that, when a predator is present in an unplanted tank with a shoal of *Pygosteus* and a shoal of *Gasterosteus* simultaneously, then the two prey species tend to keep distinct from one another. Also, the shoals become tighter when a predator is present, up to an optimum, when, under intense stimulation from a predator at very close quarters, the group scatters frantically, and later reforms. A group of *Pygosteus* in an unplanted tank, without a predator present, soon becomes very 'tame', and correlated with its lack of panic (when being fed, for example) is a greater and greater loss of any shoaling behaviour. Occasionally a few fish will be seen swimming along together, but many more will be seen moving about with complete independence. Under such conditions, at the beginning of the season, such 'independent' fish may still not show any signs of aggressiveness when they come close to another individual. This is the state of affairs that gradually develops, however.

It will be remembered that earlier I said that at the beginning of the season some tanks were set up with many fish and no vegetation, and others with few fish and much vegetation. So far I have discussed the former, but the latter must be mentioned at this point. The presence of the vegetation and the absence of a large number of other fish stimulates these fish reproductively much more quickly, whilst they are still in a 'scared' condition. In such cases it is possible to see a conflict arising between a tendency to shoal and, at the same time, to react aggressively to the other fish. The tendency to follow other fish, as in shoaling, can be seen to grade into aggressive following, or chasing. Also, passive shoaling can be seen to alternate with active repulsion. Gradually, the shoaling tendency fades away, and a stage is reached at which any meeting between two fish results in agonistic responses of some kind on the part of both individuals.

The earliest agonistic responses are characteristically weak and feeble. The muscular vigour involved in the motor patterns increases steadily as the cycle develops. Spine-raising appears as the first sign of aggressiveness in these encounters. The dorsal and/or ventral spines are raised when another individual comes into close proximity. Usually the two fish pause with their spines raised, and then slowly swim apart. The very earliest encounters seldom involve biting. By the time that the frequency of spine-raising has increased so that it accompanies nearly all encounters, mild biting can occasionally be seen. One fish approaches another slowly and if the latter does not move away it is weakly bitten, usually on the tail, and it then moves off. Also, rarely, a very slow roundabout fight is observed, but the two fish invariably part slowly and peacefully after circling around one another a few times.

At a slightly more advanced stage, it becomes apparent that a very loose hierarchy is formed. This occurs only in the tanks in which very few individuals are present. One male, which is coming into reproductive condition slightly in advance of the others, can be seen to intimidate the others with little effort. Another will be seen to intimidate all but the first male, and so on. But these hierarchies are so flexible and unstable, that territorial behaviour can develop out of them with ease. The positions of the different individuals in the hierarchy can change very quickly, and the gradual decrease in likelihood of occurrence of challenge by an inferior, which is the essential factor in the formation of a hierarchy, is far less marked than in truly hierarchical species. In the swordtail (*Xiphophorus helleri*), the very early stages of formation of a social hierarchy are similar to the hierarchical state which can exist in *Pygosteus*. In the swordtail, however, the system develops into a rigid straight-line hierarchy, whereas in *Pygosteus* it breaks down into a territorial system. This comparison is very suggestive, from an evolutionary point of view.

By the time that a loose hierarchy exists in a tank of a few *Pygosteus*, vertical displays of varying intensities can be seen. When a weaker male is approached by a stronger one, the former often lowers its tail-end in submission. The appeasement function of this action is illustrated by the following typical extract from observation notes: 'A comes right up to B and its head almost touches B's tailstalk and then B just slightly lowers its tail, and there is no bite and A stays still; then B flees and A chases it and the whole encounter repeats itself.' The more recently, frequently, or intensely a fish has been beaten, the more it lowers its tail in the submissive display.

The head-down threat response was seen to be particularly common, when a stronger male was challenged by a weaker male. Occasionally two males circled one another in the head-down posture, but generally speaking only one fish gave this display in any particular encounter. This changes later on, when it is much more likely to be a mutual display. It is noticeable that when the head-down display is given at very low intensities, it often lacks the special orientation which accompanies the high-intensity form. At high intensities the fish typically throws itself into a position broadside on to its enemy as it lowers its head end, when taking up the posture. At lower intensities the fish may simply lower its head end without any attempt at turning broadside on, so that it may be confronting its enemy dorsally, laterally or ventrally. Also, the low-intensity form of the head-down is frequently much shallower than the later, high-intensity versions.

In the sparsely populated tanks, the fish do not develop intense aggression before they have begun to form territories. As mentioned earlier, the intensity of the aggressive responses increases simultaneously with the formation of territories in such tanks. But in the densely populated tanks, the fish may develop high-intensity aggressiveness before they have spatially restricted it, or themselves, in any way. It is worth giving an example from the notes on this point: A male in a tank completely bare of vegetation which contained a large group of fish (most of which were in non-reproductive condition) was observed to exhibit spasmodic outbursts of aggression in which it attacked any fish with which it came into contact. The following is a record of a short (six-minute) period from one of the aggressive phases of this male: (C = chase; T = head-down threat; B = bite)

(16.24 hrs) CTTTTCTTTTCTTTBBBBTCTBBBBTBT (16.27)
CCBCTCBCCTBCCCBTBBBBTBTT (16.30)

It will be appreciated that an aggression score of 13 chases, 18 bites and 21 threat displays, in six minutes, clearly indicates the intense level of aggression that is possible in a pre-territorial fish.

The nature of the Pygosteus territory

The spatial restriction of aggression can be measured in a number of ways. Firstly, it is possible to map the positions of individual fish every few seconds over a set period of time. This involves a technique similar to that used in studying pre-territorial movements, with special modifications. Here, the exact position of the fish is marked on a map of the *plan* view of the tank. The presence of vegetation landmarks makes this possible, as does the fact that there are only a few fish present. Instead of recording how many fish are present in each region of the tank, in a given order, each fish is looked at in turn and its exact position marked down. The order of 'looking' is not, therefore, a set sequence of regions, but a set sequence of fish. For example, in a tank containing three territorial fish, A, B and C, fish A is looked at and its exact position marked on the map, then fish B is looked at and its position marked, then fish C, then fish A again, and so on. Each 'look' is made a set number of seconds after the last 'look'. Separate symbols are used for each of the three fish on the map. If the same time interval between 'looks' is always used, then the resultant data not only give an indication of the shapes and sizes of the territories, but also of the extent to which the fish are moving about within their territories.

Secondly, it is possible to map the exact position of the combat encounters which occur during an observation period. Again, if a set observation period is used, it is possible to gain information both about the position of fights and their frequency. Such encounters may, of course, be signal-combats or contact-combats. In the compilation of a territorial map, the marking of the positions of the fish, and of their fights, together make it possible to draw accurate boundary lines for the various territories. In Figs 8 I, II and III, the territories are shown marked in the way described above. Two fish were present in the tank in question. The vegetation is schematically figured in simplified shapes. The positions of fish A are shown by the spots, those of fish B by the triangles, and those of the combats by asterisks. Only twenty position marks were taken for each fish in each of these three maps, but it will be clear that, if a larger number were recorded, from a longer observation period, it would be possible to draw very accurate boundary lines for the two territories. This is exactly what has been done in the case of the maps in Figs 9 I to V, 10 I to III, and 11 I to III. Here, 90 observations of each fish were made for each map, always during an observation period of thirty minutes. This gave sufficient data to present these maps in the form of distinct boundaries. This method of presentation enables one to study clearly the nature of the fluctuations in the shape and size of territories at different times in the reproductive cycle.

A third characteristic of territories that can be recorded on these maps is their ethological centre, as distinct from their geographical centre. Each territorial fish has a particular place in its territory to which it returns after a boundary encounter, and also at which it spends most of its time when it is not actively swimming about. When a male has built a nest, the area immediately in front of the nest is invariably the ethological centre of the territory, but such a centre also exists even before a nest is present. The centre may change its position when nest-building begins, or, at other times, as the result of some other change in the environment. As a rule, however, the centre is more stable than the boundaries. Minor changes in boundaries may not affect the centre position, although major changes automatically do so. Tinbergen (1953b) has stressed that 'The nearer the intruder comes to the centre, the more vigorously it is attacked', and it is important to realize this fact and to know the exact

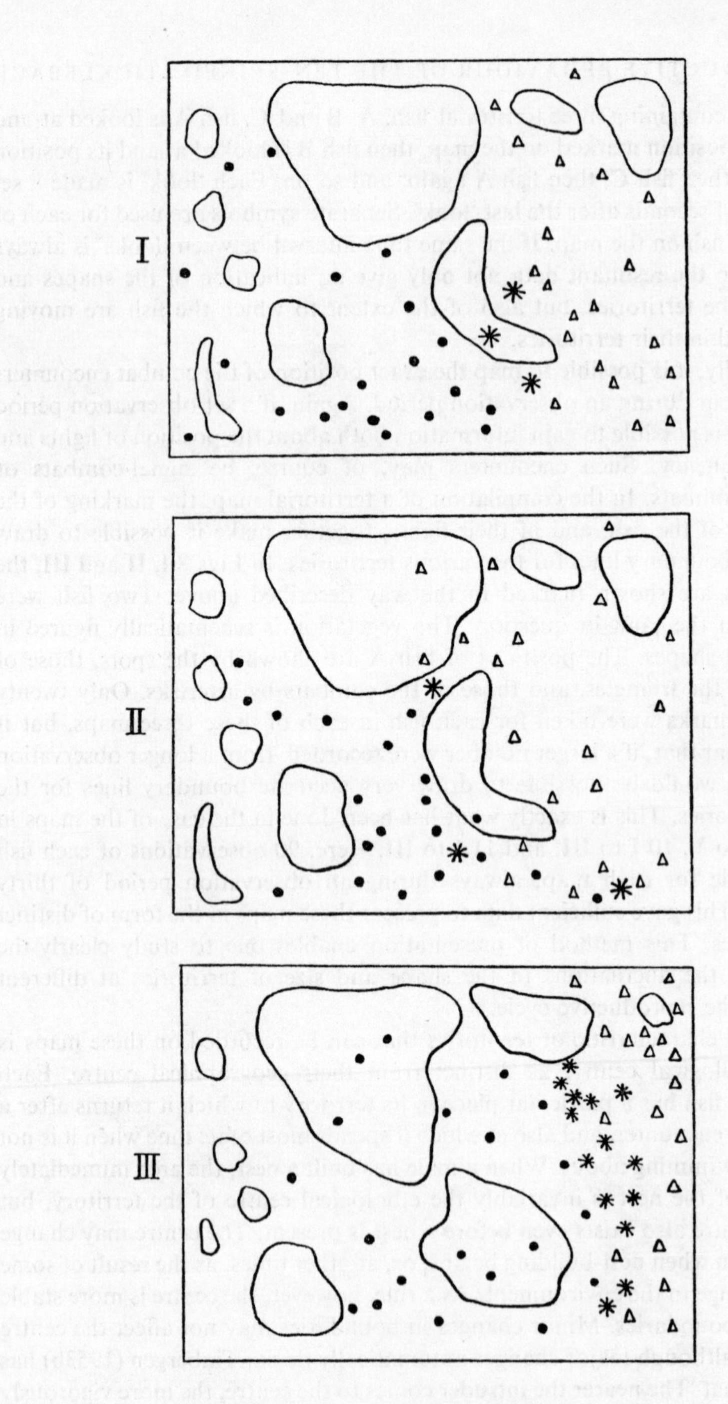

Fig. 8. Territorial maps of a tank containing two fish.

centre of a territory, before attempting to obtain quantitative evidence about the level of aggressiveness of a particular male. Since the ethological centre of a territory is only rarely identical with the geographical centre, and is often extremely different, it is dangerous to assume an approximate centre on common-sense grounds, without the necessary preparatory analysis.

Male and female territories

Having described what characterizes a *Pygosteus* territory, it is now important to discuss the differences and similarities between the sexes, with regard to their territorial behaviour. In the past, authors have either ignored (Leiner, 1913a) or denied (Sevenster, 1949) the existence of female territories in *Pygosteus.* As I have already indicated (Bastock, Morris and Moynihan, 1953, p. 77), such territories do in fact exist, and must be considered here in some detail.

The territory of the male stickleback has been well known for some time, and its significance clearly understood in terms of defence of the nest-site and defence of the eggs and young. Because the female stickleback neither nests nor performs parental behaviour of any kind, it would hardly be expected, therefore, that she would hold a territory. Yet this is what occurs.

Not only was this discovery extremely unexpected, but it was also completely out of line with the available comparative data on territorial systems. Lack (1943) found that in the robin both sexes held territories outside the reproductive season, but combined them at the beginning of the year. Once a pair formed, they defended a common territory until the end of the breeding period, in the early autumn, when they reverted to individual territories again. A similar state of affairs appears to exist in a number of other non-migratory birds. But this is very different from the *heterosexual reproductive territorial system* which exists in *Pygosteus.* I have been unable to find a report of a similar system in any other species. But before continuing to discuss why this species should exhibit such a system, it is necessary to describe the system itself in some detail.

If males and females are placed together in well-planted tanks *before* they have come into reproductive condition, then both sexes will develop territories. The populations must not be dense, only a few fish being present in each tank (examples of exact population densities will be given later). It is impossible to tell males and females apart before they are in reproductive condition, so that such tanks are set up without knowing the sex of the fish involved. As they become more and more aggressive, certain individuals develop a very dark dorsal surface and a silvery ventral surface (Fig. 2 II), whilst others develop both the dark dorsal surface and also 'dusky throats'. The dusky throats develop into an intense darkness all over the ventral

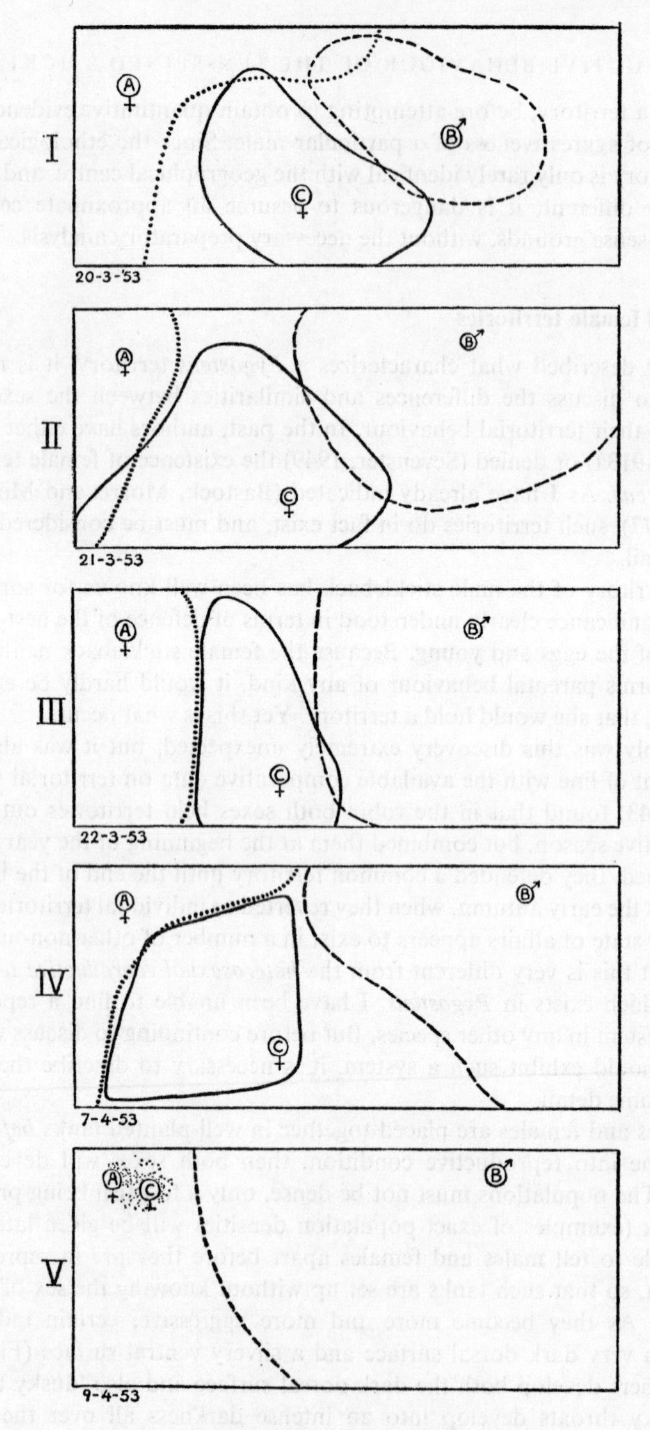

Fig. 9. Territorial maps showing the increase in size of a male territory during the parental phase.

surface, and ultimately these individuals become jet-black all over. These are the males. The other fish, with the silvery undersides, are the females. An aggressive female is always dark above and silvery below.

Both the dark females and the black males develop territorial behaviour. The aggressive actions and postures all occur in both sexes, in exactly the same way. Only the colouring is different. The result is that in this species the breeding grounds must be dotted with male and female territories alternating, very approximately, with one another. Any one male is likely to be in boundary contact with one or more females, and vice versa. (Additional field evidence is badly needed here.)

It seems that the existence of this system has been overlooked by previous authors because of the artificial nature of their laboratory procedure. Fish are usually caught when they have already taken on partial reproductive colouring, so that the sexes can be told apart. The males are then placed in separate tanks, and the females are placed all together in one large tank. When females are needed for tests, they are put into the territory of a male and the ensuing behaviour is then recorded. Such a procedure obviously prevents any female territorial behaviour and is based on the incorrect assumption that none exists. By bringing unsexed fish into the laboratory very early in the year and studying their gradual development into reproductive condition, the heterosexual territorial system was discovered. A brief summary of the first case recorded is as follows:

A large number of fish were collected from the River Kennett on February 21st, 1953. They were brought to the laboratory and placed together into an unplanted tank. They were not yet in reproductive condition. Three of these were taken and placed in a large planted tank (of dimensions 105 × 45 × 45 cms), on March 11th. After a few days, some mild aggression was observed and this grew stronger. By March 20th, high-intensity aggression was observed and there were indications of territorial behaviour. A territorial map was made (see Fig. 9 I), which revealed that each of the three fish was holding a distinct territory. It was then thought that all three fish must be males. Further territorial maps were made on the following two days (Figs 9 II and 9 III). It was noticed that fish B was enlarging its territory gradually and forcing A and C away. By March 22nd, fish B occupied half of the available space, whilst fishes A and C occupied approximately a quarter each. It had already been noted on March 20th that fish C had 'a very swollen belly today, but it must surely be food, for this one fights vigorously, although it certainly looks very ripe'. However, on March 24th, fish B was seen to be courting fish C, and the latter was observed to respond fully. Both before and after the courtship (which was unsuccessful because B did not possess a nest) the two participants defended their usual territories, the boundaries of which were by now very stable. Later, on March 28th, fish C was courted again by fish B, and this time C laid eggs and B fertilized them (and subsequently reared them successfully). A few minutes after spawning, female C was again defending her territory and was observed to fight both fish A (which also turned out to be a female) and her mate of a few moments ago, male B. The territories re-formed themselves as if nothing had happened. Some days later, on April 6th, young were seen to be present, and on April 7th the male was very actively concerned with keeping the young in the nest. At this point his behaviour changed and he became excessively aggressive. The effect of this on the female territories is shown in Fig. 9 IV. Two days later the male had practically ceased to ventilate the nest, and the young were clearly seen outside and around it. The last map of this series (9 V) was made then on April 9th, and showed that the male's aggression had now become so violent that the females were unable to maintain their territories any longer and were now hiding in the weeds in the far corner of the tank. They now always fled when attacked and their colour markings were cryptic once more. Since the behaviour of the females was no longer any measure of the male's aggression, this was tested by the use of models (discussed more fully in section VII). It persisted until April 18th, although it waned considerably after the male ceased to pay any attention to the young. The females did not re-form their territories again.

The above record was not unique. After it had been realized that the way in which tanks had been set up in the past was probably distorting the natural social organization of the species, a number of tanks were set up with small heterosexual populations, and the same general situation was found to develop in these. Figs 10 I to III show another example. In this particular case, there were two males and a female present in a smaller tank (60 × 36 × 36 cms). There was only one clump of weeds present and this was owned, and nested in, by male B. Female A and male C had no vegetation whatever on their territories, but landmarks were provided in the form of pieces of broken flower-pot, stuck vertically in the sandy substratum. This series of maps again shows clearly the increase in size of territory which occurs when a male's eggs hatch. In Fig. 10 I, the male B possessed a nest. The following day, this male's aggression had increased and the centre of male C's territory had been shifted away from the nest site. Also the territory of female A was more restricted. This is shown in Fig. 10 II. In a few days, the male B obtained eggs from female A and Fig. 10 III shows the territorial states of the three fish on the day when all these eggs had hatched. As before, the parental male increased its territory and the areas owned by the other two fish were considerably compressed, the headquarters of male C being shifted even farther from the nest site.

Before suggesting a possible explanation of the existence of a heterosexual territorial system in *Pygosteus*, it is necessary to look briefly at the state of

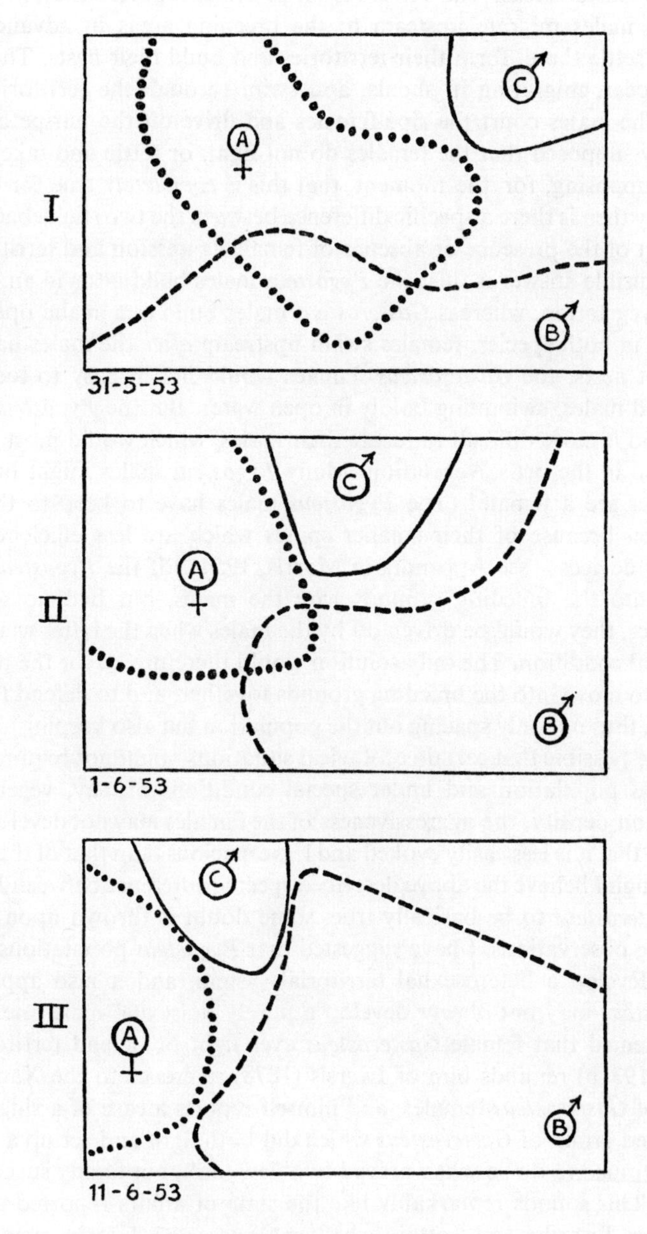

Fig. 10. Territorial maps showing the increase in size of a male territory during the parental phase.

affairs in *Gasterosteus*. Van Iersel (1953, p. 3) has reported that, in *Gasterosteus*, the males migrate upsteam to the breeding areas in advance of the females, settle there, form their territories and build their nests. The females soon appear, migrating in shoals, and swim through the territories of the males. The males court the ripe females and drive off the unripe ones. It is generally supposed that the females do not fight, or settle and take up territories. Supposing, for the moment, that this is *completely* true for *Gasterosteus*. Why then is there a specific difference between the two stickleback species in respect of the presence or absence of female aggression and territory? The most plausible answer is this: the *Pygosteus* males build nests in amongst the aquatic vegetation, whereas *Gasterosteus* males build out in the open, in the sand. If, in both species, females swam upstream *after* the males had settled and built nests, the *Gasterosteus* females would find it easy to locate their bright red males, swimming boldly in open water. But the *Pygosteus* females might find it more difficult to locate their males, which would most probably be hidden in the dense vegetation. Many *Pygosteus* males might build nests and never see a female! (The *Pygosteus* males have to keep to the denser vegetation because of their smaller spines which are less efficient as anti-predator devices – see Appendix in Morris, 1954e.) If the *Pygosteus* females moved into the breeding grounds *with* the males, but had no aggressive tendencies, they would be driven off by the males when the latter were in their pre-sexual condition. The only solution would therefore be for the males and females to move into the breeding grounds together, and to defend territories together, thus not only spacing out the population but also keeping it together. It is quite possible that certain ecological situations might not require this of a *Pygosteus* population and under special conditions of, say, vegetation, or population density, the aggressiveness of the females may not develop. (It will be noted that it is less easily evoked and less tenacious than that of the males.)

Although I believe the above described specific difference between *Pygosteus* and *Gasterosteus* to be basically true, some doubt is thrown upon it by the following observations. I have suggested that *Pygosteus* populations may not *always* develop a heterosexual territorial system, and it also appears that *Gasterosteus* may not *always* develop a purely unisexual male one. Wunder (1930) denied that female *Gasterosteus* ever fight or defend territories, but Leiner (1931b) reminds him of Evers's (1878) reference to the Xanthippean nature of *Gasterosteus* females, and himself reports a case of a single female in a mixed group of *Gasterosteus* which did both fight and set up a territory. She continued to do so when moved to a new tank, but finally succumbed to a male. This sounds remarkably like the state of affairs reported above for *Pygosteus*. In order to ascertain whether *Gasterosteus* females would in fact defend themselves as well as *Pygosteus* females, similar populations to those already described for the latter species were set up.

In one tank (of dimensions 105 × 45 × 45 cms) two males and one female *Gasterosteus* were placed. Both males built nests as shown in Fig. 11 I, and the female defended a territory in the middle of the tank. Although she drove off the males when they intruded, she seldom if ever trespassed herself. Male B obtained eggs and when these hatched, he showed the immediate increase in aggression that was known to be typical of *Pygosteus* (see Fig. 11 II). He compressed the other male's territory considerably, and utterly defeated the female A, which hardly fought back at all any more. The female

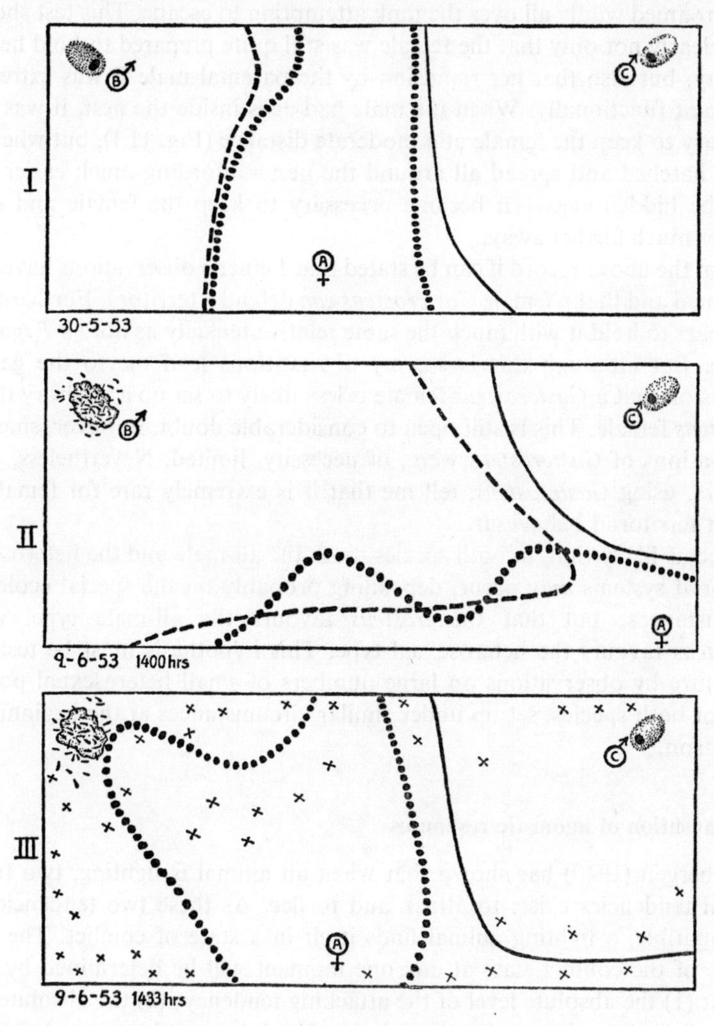

Fig. 11. Territorial maps for *Gasterosteus* showing similar increase in size of male territory during the parental phase (II) and also the effect of exchanging the parent male for a strange male (III).

had abandoned her old territory and had a 'flight headquarters' in the least coveted corner of the tank. In order to show conclusively that this change was the result of increased aggression on the part of the male B, and not simply decreased aggression on the part of female A, the male B was removed from the tank and replaced by a strange male from another tank. Fig. 11 III shows the result of this change. The female immediately took up her old territory again and also ventured over to the ownerless nest and *chased and ate* some of the young fish there. The new male was attacked by both A and C and it roamed wildly all over the tank attempting to escape. This test showed quite clearly not only that the female was still quite prepared to hold her old territory, but also that her repulsion by the parental male B was extremely important functionally. When the male had eggs inside the nest, it was only necessary to keep the female at a moderate distance (Fig. 11 I), but when the young hatched and spread all around the nest – affording much easier prey than the hidden eggs – it became necessary to keep the female and other enemies much farther away.

From the above record it can be stated that Leiner's observations have been confirmed and that a female *Gasterosteus can* defend a territory. Furthermore, it appears to hold it with much the same relative tenacity as does a *Pygosteus* female. But although this is so, my observations lead me to the general conclusion that a *Gasterosteus* female is less likely to set up a territory than a *Pygosteus* female. This is still open to considerable doubt, however, since my observations of *Gasterosteus* were, of necessity, limited. Nevertheless, other workers, using *Gasterosteus*, tell me that it is extremely rare for females to exhibit territorial behaviour.

It seems likely that, in both species, both the all-male and the heterosexual territorial systems *may* occur, depending probably on the special ecological circumstances; but that *Gasterosteus* favours the all-male type, whilst *Pygosteus* favours the heterosexual type. This hypothesis must be tested in the future by observations on large numbers of small heterosexual populations of both species, set up under similar circumstances at the beginning of the season.

The causation of agonistic responses

Tinbergen (1940) has shown that when an animal is fighting, two fundamental tendencies exist: to attack and to flee. As these two tendencies are incompatible, a fighting animal finds itself in a state of conflict. The exact nature of the conflict state at any one moment will be determined by three things: (1) the absolute level of the attacking tendency; (2) the absolute level of the fleeing tendency; (3) the relationship between these two levels (see Moynihan, 1953 a and b).

Tinbergen (1952a) has pointed out that when these tendencies to attack and to flee are more or less in balance, each prevents the other from expressing itself fully, so that the animal can neither attack its enemy, nor flee from it. In such a situation an animal performs one of two basic types of activity: *ambivalent actions*, or *displacement activities*. Tinbergen postulates that when the attacking and fleeing tendencies are both weakly activated, ambivalent autochthonous actions occur, and when they are more strongly activated, displacement activities occur. I have attempted to represent this graphically in Fig. 12. In this diagram, the tendency to attack (A) is shown vertically (as

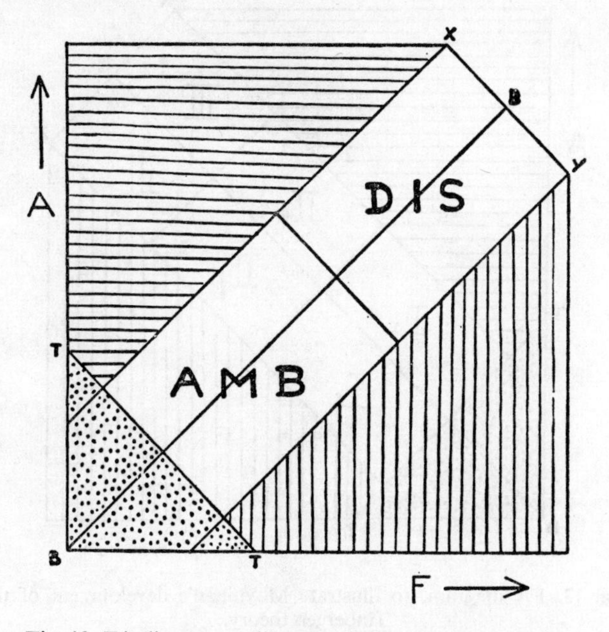

Fig. 12. FA diagram, to illustrate Tinbergen's threat theory.

the ordinate), and the tendency to flee (F) is shown horizontally (as the abscissa). The line T-T represents the threshold line, below which, it is supposed, the animal will be insufficiently activated to exhibit overt agonistic responses (dotted area). Above this threshold, one of three basic states may exist. The attacking tendency may be so much stronger than the fleeing tendency that the animal actually attacks (horizontally-lined area), or the reverse of this may occur and the animal will actually flee (vertically-lined area). If the two tendencies are more balanced, then conflict behaviour will be observed. X-Y represents the width of the 'conflict-band', which will vary from species to species. B-B represents the line of precise balance. Then, according to Tinbergen's theory, the region AMB is that in which ambivalent behaviour will occur, and DIS that in which displacement activities will occur. Moynihan (1953a), working with the black-headed gull, developed this theory, and I have represented his findings in a similar graphical form in Fig. 13. Although it is impossible here to go into the details of this work, it is nevertheless interesting to note that response III in Fig. 12 was a displacement activity, and responses Ia, If, IIa and IIf were all ambivalent actions, thus confirming Tinbergen's theory.

Before proceeding to apply these principles to the agonistic behaviour of

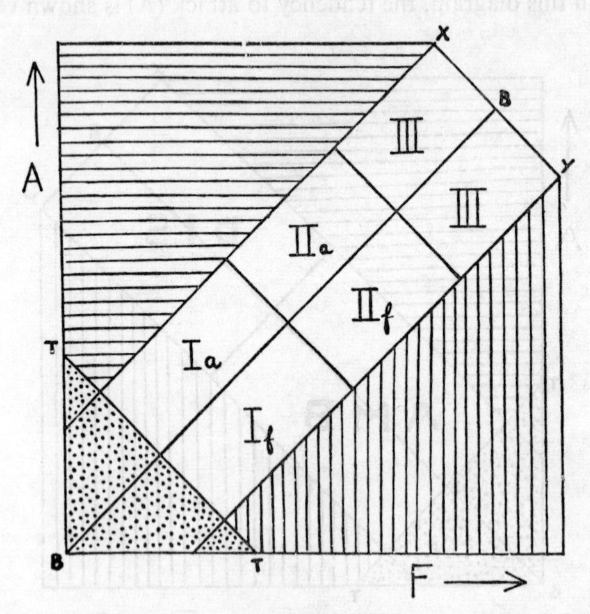

Fig. 13. FA diagram, to illustrate Moynihan's development of the Tinbergen theory.

Pygosteus, it is first necessary to discuss the nature of agonistic conflict responses in general terms, and also the way in which a special causal state can be inferred from such a response.

Ambivalent behaviour: When the two incompatible tendencies of attacking and fleeing are weakly activated and more or less of equal strength, the animal cannot fully obey either, but it can partially obey both. This ambivalence can show itself in one of three ways: (1) thwarted intention movements of one tendency; (2) rapidly alternating intention movements of both tendencies; (3) ambivalent posturing.

In the case of thwarted intention movements, one sees the beginning of an attack movement, or a flight movement, but this movement does not develop into a full response. The animal looks as if it is about to make a full response and then quickly checks itself. It is important here to distinguish between thwarted intention movements and simple intention movements. Heinroth (1911) introduced the term *Intentionsbewegungen* because he noted that when an animal performed an action at a very low intensity, it nevertheless revealed its intentions. But such movements occur not only as the result of very weak activation of the one tendency expressed, but also as a result of a slightly stronger activation which is thwarted by some incompatible tendency. It may be argued that it will be impossible to distinguish between the two in practice, and it must be admitted that sometimes this is so, but nevertheless it is often possible to recognize that a particular intention movement is ambivalently activated, by the way in which it ends. An ambivalently activated, and therefore thwarted, intention movement is checked more suddenly than an intention movement that is simply the result of the very weak activation of a single tendency. It stops with a jerk, so to speak, whereas the latter tends rather to 'tail off'.

In some cases of ambivalent behaviour, the checking of the intention movement may go farther than just 'stopping it with a jerk'. In such cases, the rapid alternation of the intention movements of both the attacking and fleeing behaviour patterns may be observed. The animal begins to attack, checks itself and begins to flee, checks itself and begins to attack again, and so on. These ambivalent movements which will typically be of a 'back-and-forth' nature are the result of a more evenly balanced conflict state than the thwarted intention movements described above, for obvious reasons.

It is also possible for an ambivalent response to be in the nature of a single posture. This occurs when the intention movements of fleeing and attacking both occur simultaneously. In ambivalent movements, the *whole* animal first performed one intention movement, and then the *whole* animal performed the other. In an ambivalent posture, *part* of the animal performs one intention movement, whilst another *part* performs the other (e.g. the tail-beating response mentioned earlier, in which the tail attempts to swim forwards whilst the pectoral fins attempt to swim backwards).

Displacement activities: So much has been written about displacement activities recently (Tinbergen, 1939, 1940, 1952a; Kortlandt, 1940; Tinbergen and van Iersel, 1947; Armstrong, 1950; Bastock, Morris and Moynihan, 1953; Moynihan, 1953b; Morris, 1954b) that they require little explanation here. Briefly, it can be stated that when a behaviour tendency is strongly activated and then thwarted, the result may be that the animal performs some apparently irrelevant action. This latter action is called a displacement activity. A strongly activated tendency can be thwarted either by the absence of the appropriate releasing stimuli, or by coming into conflict with a simultaneously activated, but incompatible, tendency. The latter case is the more common cause of displacement activities, and it is also the relevant one here.

In the flee-attack conflicts of a number of species it has been observed that, at high intensities, ambivalent behaviour gives way to displacement activities. It seems that each tendency is so strong that it prevents the other from expressing itself even as an intention movement, with the result that the intensely activated animal can only find an 'outlet' for its pent-up 'energy' by the performance of some out-of-context action. The zebra finch (*Poephila guttata*), for example (Morris, 1954b), when in an intense agonistic conflict may perform motor patterns belonging to feeding, sleeping, cleaning, juvenile or copulatory behaviour.

There is a theoretical objection to the concept of displacement activities, which I feel must be mentioned here, because it is important to the discussion of stickleback displacement activities in section VIII. This is that it is not always possible to make a hard-and-fast distinction as to when a particular motor pattern is relevant and when it is irrelevant. In some cases, such as when a bird repeatedly wipes its (quite clean) beak whilst it is actually courting a female, the irrelevance of the displacement activity is obvious enough; but it is not always as easy as this. The way in which the concept is formulated at present limits the investigator to the more obvious cases and, as it seems likely that the existence of displacement action is of a much wider significance, it is to be hoped that the current formulation will be improved in the near future. However, this will be discussed in more detail later.

These conflict-responses (ambivalent and displacement actions) frequently do not develop into actual attack or flight, but they often may do so, and it soon becomes clear that a particular action is more likely to develop into, say, attack, whereas another is more likely to develop into fleeing. The animals concerned, as well as the observer, are aware of this fact and a particular conflict-response may act as a signal to the enemy, telling it what the odds are that it will be attacked.

Since any system which reduces the danger of bodily damage, without denying the possibility of 'settlement by dispute', will be of advantage to a species, it is not surprising that many species have evolved elaborate threat-codes in which the conflict-responses have become greatly exaggerated. This process of modification in the service of signal function has been called 'ritualization' by Huxley (1923) and has been discussed fully by Lorenz (1941), Daanje (1950), Tinbergen (1952a) and Morris (1956). However, the subject will not be discussed further here except to say that one special aspect of ritualization is the evolution of conspicuous markings and structures which accentuate the signal effect of the movement or posture.

In the cases where two different conflict-responses are both more likely to lead to attack, say, rather than fleeing, it may be observed that one is more likely to lead to a *more violent attack* than the other. This provides a potential refinement of the agonistic signalling system, which permits the existence of quite a complex threat-code, as was found by Moynihan (1953a) in the black-headed gull (Fig. 13). The obvious assumption is that this is the result of conflict states in which the relative levels of the attacking and fleeing tendencies stay the same, but in which their absolute levels change.

It should be clear by now that there are three ways in which one can infer the particular causal state of an animal from observations of a particular conflict-response:

1. *Form of the response:* In some cases, the origins of the response can be detected and it can be seen that the postures or movements involved contain more elements of attacking than fleeing, or vice versa. This is not possible, of course, in the case of displacement activities, but it is often very useful when dealing with ambivalent responses. Sometimes it is also possible to see that, whilst giving the response, the animal is 'edging towards' or 'edging away from' the enemy, without actually attacking or fleeing. This can given an additional clue.

2. *Likelihood of particular development:* The frequency with which a particular action develops into actual attack or fleeing can be measured.

3. *Intensity of particular development:* The strength of the attack or flight response, into which a particular action develops, can be scored.

Together, this information enables the observer to construct a diagram such as those shown in Figs 12 and 13, for a given species. Each action can be given an approximate position on such a diagram and gradually the threat-code of a species can be pieced together.

My own observations on the threat-codes of a number of fish species, notably *Pygosteus*, *Gasterosteus*, *Xiphophorus helleri* (unpublished) and *Cottus gobio* (Morris, 1954c), whilst in general confirming the Tinbergen-Moynihan threat theory, also require some modifications of it. These must now be discussed.

Causal overlapping

When an animal performs a threat display, it frequently happens that the action is of a complex nature and is comprised of a number of quite distinct elements. If these elements *always* occur together in exactly the same way, then they can be considered together as one unit. If, however, such elements occur separately, or in different combinations, on different occasions, then each particular element warrants consideration as a response in its own right. In order to see how this will affect the construction of an FA diagram (Flee-Attack diagram, as in Figs 12 and 13), it is necessary to return for a moment to the question of what any space on such a diagram signifies.

Each exact point on an FA diagram represents an exact causal state, with reference to precisely 'how much' F and 'how much' A there is at the time. But since there are innumerable such points and yet only comparatively few distinct conflict-responses in any given species, each conflict-response must be the possible result of quite a wide range of exact causal states. Each conflict-response can be said to have a particular *causal range*, and this is amply borne out by observation (see later). Each point in the area covered by a particular causal range, although causally different from each other point, will have the same response-effect.

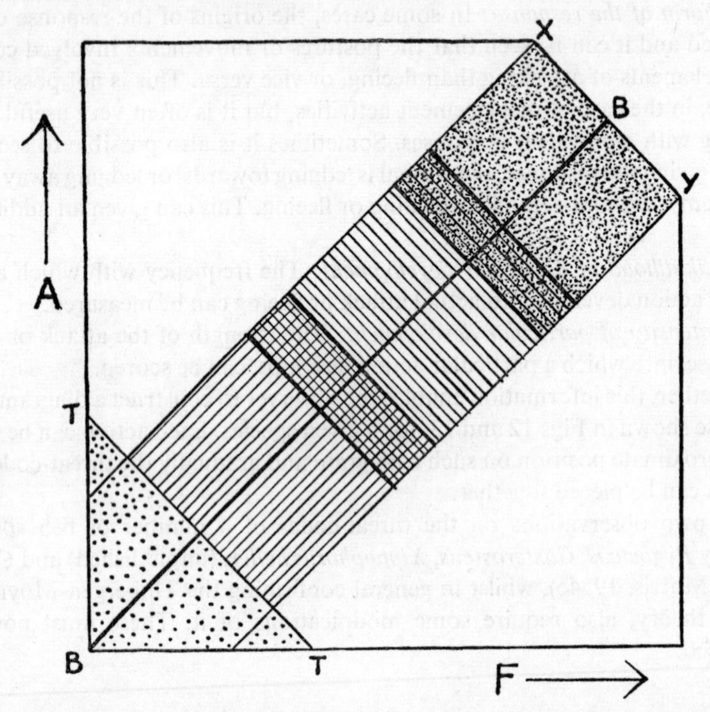

Fig. 14. FA diagram, to illustrate the concept of 'causal overlapping'.

In the FA diagrams presented here, the causal ranges of the various responses have been presented in a formal manner as geometrical rectangles. At this stage, such formality is simply a sign of ignorance, but it is hoped that on this basis it will be possible to construct more precise and realistic diagrams in the future, when more quantitative work has been done on this subject.

Returning now to the problem of independent elements of threat displays, it must be clear that, if a certain element can occur both simultaneously with and separately from another element, the causal ranges of these two elements must be *overlapping*. This does not seem to be an unreasonable suggestion, and it fits most of the facts which I have collected so far, perfectly well. The odd exceptions are not yet fully analysed and, in any case, do not concern *Pygosteus*, and will therefore not be discussed here.

Basically, this causal overlapping can be of three kinds:

1. *Absolute-level overlap:* Suppose there are three conflict-responses in a species and that these three are the result of (a) weak, (b) medium, and (c) strong activation of both F and A. The causal ranges of these three responses, instead of just touching, may just overlap, as shown in Fig. 14. Thus, as the

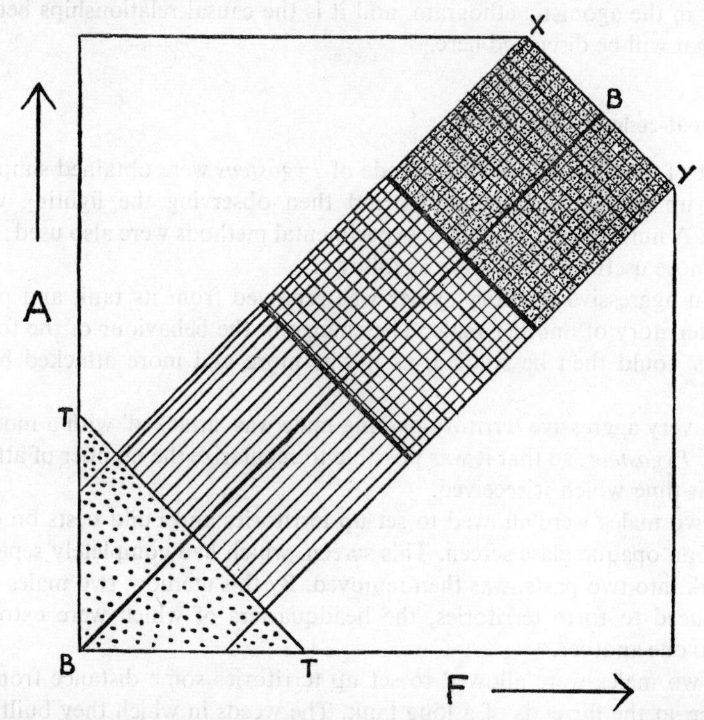

Fig. 15. FA diagram, to illustrate an alternative form of causal overlapping.

FA state of the animal becomes more and more intense, so the following responses would be observed: a, ab, b, bc, c.

2. *Relative-level overlap:* This is not illustrated, but is similar to the above, only at right angles to it. For example, if one action is very likely to be followed by attack, and another by fleeing, then the causal ranges of the two may overlap so that when the animal is equally likely to attack or flee it will perform the two actions simultaneously.

3. *Additive overlap:* Causal ranges may overlap in a special way, as illustrated by Fig. 15. Here, one response has a range which covers the whole

length of the threat-band. As the absolute levels of F and A rise, so another response is *added* to the display, and then another, so that the very high-intensity display has three elements. This addition of elements may also occur as the relative levels change, in which case one response may have a range which stretches the whole width of the threat-band (X-Y) or, perhaps, half of it (BX or BY), while other responses will be added as the balance becomes more and more unstable.

Starting with the above framework as a basis, it is now possible to discuss the threat-code of *Pygosteus*. The separate elements have already been described in the agonistic ethogram, and it is the causal relationships between them that will be discussed here.

The threat-code of 'Pygosteus'

Most of the data on the threat-code of *Pygosteus* were obtained simply by setting up territorial populations and then observing the fighting which ensued. A number of very simple experimental methods were also used; some of the more useful of these were as follows:

1. An aggressive territory-owner was removed from its tank and placed in the territory of another male. The change in the behaviour of the forced-intruder could then be studied, as it was more and more attacked by the owner.

2. A very aggressive territory-owning male was 'attacked' with a model of a black *Pygosteus*, so that it was possible to regularize the number of attacks-per-unit-time which it received.

3. Two males were allowed to set up territories and build nests on either side of an opaque glass screen. This screen, which had completely separated the tank into two parts, was then removed. By this method, two males could be induced to form territories, the headquarters of which were extremely close to one another.

4. Two males were allowed to set up territories some distance from one another, at the far ends of a long tank. The weeds in which they built their nests were planted in pots. These 'potted territories' were then moved close together, so that both males were on their own territories and their rival's simultaneously.

5. Two territorial males which had been rivals of long standing, were moved together into a new tank. The outcomes of such moves as this could be compared with the outcomes of placing two fish, which had been strangers before, together in a new tank. Differences between the two would reflect the existence and extent of individual recognition by rival territory-holders.

Spine-raising

By far the most frequent, and the most complicated, conflict-response of *Pygosteus* is the raising of its spines. The reason why it is a complicated response is as follows. In the evolution of *Pygosteus*, it appears that the spines were first of all ordinary fins and dealt with the problem of *balancing* the fish in the water. Next they became hardened into spiny fins, and ultimately spines, and they then acted as *anti-predator weapons*. Finally they came to act also as *intraspecific agonistic signalling devices*. The difficulty is that even today the spines may act in any of these capacities, so before it can be certain that, in a particular observation, a spine-raising has occurred as a special intraspecific signal, the other possibilities have first to be eliminated.

For example, when a male *Pygosteus* fans its nest, it may be seen to raise all its spines. This, however, is just the kind of situation in which a fish would call strongly upon its balancing devices, and I suggest that here the spines are being raised in their oldest capacity (for which, incidentally, they are no longer much use). The way in which they are actually raised in such a case supports this suggestion. They appear to be erected with less 'determination', so to speak.

When a *Pygosteus* is confronted with a predator, it raises all its spines and maintains them all stiffly erect whilst the danger persists. Also, when two *Pygosteus* are engaged in very high-intensity intraspecific fighting, all spines are raised. However, in ordinary fighting, several different kinds of spine-raising can be observed, and it is these that appear to act as special signals between rivals. Using data collected from a number of different kinds of observations, it is possible to make the following general statement about the relationship between a particular type of spine-raising and a particular *relative balance* of F and A (this statement excludes cases which involve a very high absolute level for F and A):

When a fish simply attacks ... it does so with all its spines unraised.

When it is very likely to attack, but with some hesitation ... ventral spines only are raised.

When it is slightly more likely to attack than flee ... dorsal spines are raised spasmodically and momentarily during the periods when the ventral spines are raised.

When it is equally likely to attack or to flee ... all spines are raised.

When it is slightly more likely to flee than attack ... ventral spines are raised spasmodically and momentarily during the periods when the dorsal spines are raised.

When it is very likely to flee, but with some hesitation ... dorsal spines only are raised.

When it simply flees ... all spines are unraised.

An alternative way of expressing this is as follows, where V = strong tendency to raise ventral spines; D = strong tendency to raise dorsal spines; v = weak tendency to raise ventral spines; and d = weak tendency to raise dorsal spines: ATTACK ... / - - / V- / Vd / VD / vD / -D / - - /... FLEE.

It is clear from this that the ventral spines are raised when the relative state of the FA balance is in favour of A, while the dorsal spines are raised when it favours F. It is also clear that there is an overlapping of the causal ranges

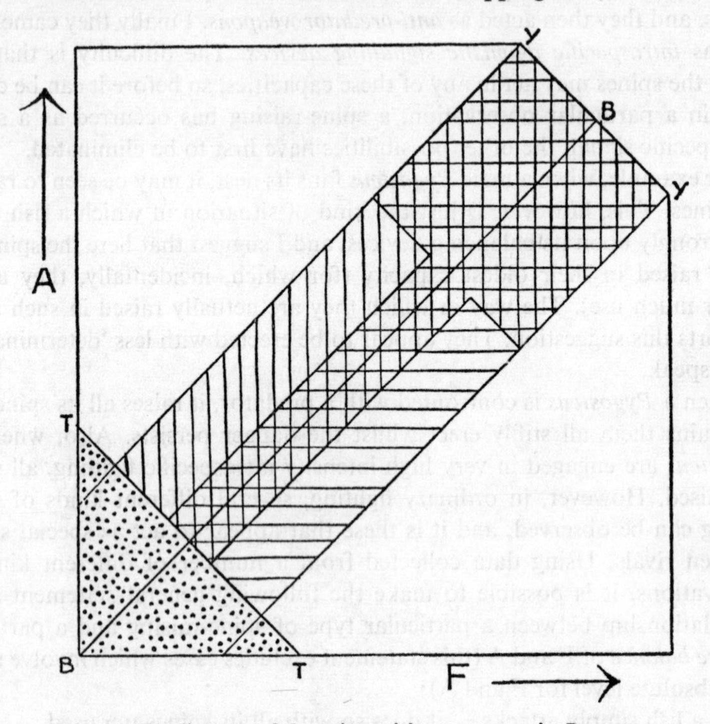

Fig. 16. FA diagram of the spine-raising responses in *Pygosteus*.

of these two responses. It has already been mentioned that, in very intense fighting (when the levels of both F and A are high), when spine-raising occurs, *all* spines are raised. These facts together make it possible to construct a tentative FA diagram for spine-raising in *Pygosteus* (see Fig. 16). In this diagram, the causal range of the response of dorsal spine-raising is shaded with horizontal lines, and that of ventral spine-raising with vertical lines. Where the two ranges overlap there is, of course, criss-cross shading. It must be emphasized that this is a very provisional diagram, but I feel that it is nevertheless justified.

It is difficult at present to understand why the ventral spines should have become especially associated with an FA balance favouring A, and the dorsal spines with one favouring F, rather than vice versa. One possible suggestion is as follows. It has not been mentioned above that there is a further complication in studying the spine-raising responses, due to the fact that there exists a tendency to raise the spine or spines *nearest* to the enemy. In the case of the paired ventral spines (which, it must be remembered, are modified pelvic fins), it can often be observed that only the one on the side nearest the enemy is raised. Fortunately this particular refinement of the signalling system does not interfere with the dorsal/ventral distinctions made above. However, a similar, although less marked, 'proximity factor' does occasionally play a role in such distinctions. For example, on rare occasions, a very beaten fish, which should only be raising its dorsal spines, or all its spines together, can be observed to raise its ventral spines despite the fact that it obviously is not in the 'correct' state to do so! Careful examination of a number of such cases revealed that the fish in question were always in a particular position relative to the aggressor; namely, above it. The ventral spines were only raised, in such a case, when they were the spines nearest to the enemy. If the latter was below and slightly to one side, then only the ventral spine on that side was raised. This discovery at once explained a number of previous interpretive difficulties, and also demanded special precautions for the future. It meant that, for an observation of the raising of the dorsal or ventral spines separately, where this was to be used in inferring the FA balance, the proximity influence had to be eliminated. In other words, such responses could only be used if the spines raised were *not* those nearest to the enemy.

Although this proximity factor proved to be a nuisance technically, it nevertheless provides a possible explanation of the association of ventral spine-raising with an FA balance favouring attack, and dorsal spine-raising with one favouring fleeing. For, supposing a 'superior' fish tends to keep at a higher water level, generally speaking, than an 'inferior' one, then the proximity factor will automatically associate ventral spine-raising with agonistic superiority, and dorsal spine-raising with agonistic inferiority, and it only remains for this association to become fixed and ultimately independent of the influence of proximity, during evolution. It is known from aquarium observations that a beaten fish is likely to go either to the bottom or to the surface, and always to avoid mid-tank. However, it seems more likely that, in nature, it is more adaptive for a beaten fish to lie on the bottom, rather than under the surface, when one takes the presence of a current into account. Also, the response of 'freezing' and sinking, which occurs in beaten animals, supports this. Further research is required here, however.

The spine-raising behaviour of *Gasterosteus* differs from that of *Pygosteus* in some respects. From my observations of *Gasterosteus* in agonistic situations, I have been able to ascertain that, whilst it differs from *Pygosteus*, it is nevertheless not the complete opposite of it in respect of spine-raising. The situation is not as simple as this. Wunder (1928, 1930), Tinbergen (1953a, Fig. 38), and van Iersel (1953, p. 7) have all stressed that, in *Gasterosteus*, a fish that raises its dorsal spines is more likely to attack than flee, and one that raises its ventral spines is more likely to flee than attack. If this were so, then the two species would have reversed, so to speak, the significance of these responses. I cannot believe that this is entirely true because, for example, there are a number of postures and movements which both species have in common and which, in both species, are accompanied by the same type of spine-raising (e.g. head-down, tail-down, dorsal roll, ventral roll). The vertical posturings of the two species can be used to illustrate this. The head-down posture is certainly more aggressive than the tail-down posture in both species. Also, in both species, the head-down posture is accompanied by the erection of the ventral spines alone, whilst the tail-down posture is accompanied by the erection of dorsal spines alone. (For illustrations of *Gasterosteus* showing head-down with erect ventral spines, see ter Pelkwijk and Tinbergen (1936, Fig. 1), and Tinbergen (1951a, frontispiece); for the same species showing tail-down with erect dorsal spines, see Fig. 7 in the 1936 paper.) Therefore in this respect the two species raise their spines in a similar manner.

The question of spine-raising has been considered carefully here because, apart from its intrinsic interest, once the causal significance of this very frequent response is well known in all its details, it can be used as a very helpful guide in interpreting other responses.

Other elements of the threat-code

It has already been mentioned that the *head-down threat posture* occurs when the ventral spines alone are raised, so that this already confines this response to a certain area of the FA diagram. It is even more restricted than this, however, because a fish which is *very* likely to attack maintains a horizontal posture.

The *ventral roll* is also accompanied by the erection of the ventral spines. This response occurs in vigorous fights and is interspersed with head-down postures. Exactly what determines whether a fish shall lower its head or roll its ventral surface towards the enemy is not clear at present. The ventral roll is less common than the head-down posture, generally speaking.

The *stationary-charge* response is given with all spines raised and is an activity in which the FA balance favours attacking, but in which both F and A are at very high absolute levels. This particular posture was only discovered when ‘potted territories’ were moved close together. The fish then had both a strong tendency to flee because of the proximity of the centre of the other territory, and also an even stronger tendency to attack because of the proximity of the centre of their own territory! The outcome was violent fighting which was interspersed with vigorous stationary-charge responses.

The *colour change* of *Pygosteus* was often rather slower than its change of posture, with the result that sometimes a colour state was out of tune with its accompanying posture or movement. Generally speaking, however, when the FA balance favoured the tendency to attack, the fish was all-black, if it was a male (see Fig. 2 III), or very dark dorsally, if it was a female (see Fig. 2 II). When the FA balance favoured the tendency to flee, the fish was cryptically marked in both sexes (see Fig. 2 I and V).

Submissive behaviour

Pygosteus has two submissive signals: the tail-down posture and the dorsal roll. The former is extremely common and the latter extremely rare.

The *tail-down posture* is accompanied either by the raising of the dorsal spines alone, or by the raising of all the spines. The latter occurs when a fish is being very badly beaten. Tail-down fish are always cryptically marked and their fins are often flattened. They may 'freeze' and sink when in this posture. A fish in this posture never attacks and is very likely to flee. It is interesting to note that this posture is the *opposite* of the more aggressive head-down threat posture. This is a characteristic of submissive postures in general, and I have discussed it at length elsewhere (Morris, 1956) (see also Lorenz, 1952, ch. 12, and Tinbergen and Moynihan, 1952).

The tail-down posture has an appeasement value in that it may prevent, or partially prevent, attack in some situations. I have called it a submissive posture partly because it never develops into attack. The FA balance which produces it is so much in favour of F that the animal can only stay still, or flee. The appreciation of this fact by other fish appears to give it its appeasement value. The raising of the dorsal spines, on the other hand, although it can occur with the tail-down posture, cannot be called a submissive action, because its causal range extends into a region where attack may develop.

The *dorsal roll* response is extremely rare in *Pygosteus*. It is much more common in *Gasterosteus* and in this latter species is a regular courtship ceremony. I have only ever seen the *full* response once in *Pygosteus*, but that one occasion was so striking that it is worth recording. A male was courting a group of seven females, two of which had spawned for him within the last three hours. The second one to spawn had done so forty minutes earlier, and this particular fish was rather unusual in that it was silvery all over (see Fig. 2 VI). This latter female, upon being courted again by the male, swam swiftly forward and bit him particularly viciously. The male's response was to sink slightly and perform the dorsal roll through a full right angle so that he was completely on his side, alongside the female, and with his dorsal spines erected and his ventral spines flattened. The female stopped her attack and the male then righted himself and swam off. The female concerned had no territorial rights in the tank, and was actually on the male's territory at the time when she bit him. A territorial female would never have dared to attack a male on his territory in this way, and non-territorial females are hardly ever aggressive to males under any circumstances – hence the extreme rarity of this response. This dorsal roll in *Pygosteus* had all the character of a submissive reaction, and it becomes important to consider it in relation to the very common dorsal rolls of *Gasterosteus*, when the courtships of the two species are compared (see section VI). It appears that whereas *Pygosteus* had to be *bitten* by a female when in a sexual situation, *Gasterosteus* has only to be *approached* by a female when in a sexual situation, to give the dorsal roll response. This suggests that the FA balance which exists in the males of the two species, when they are courting, is more strongly in favour of A in *Pygosteus* than in *Gasterosteus*, since in *Pygosteus* a stronger stimulus was needed to produce the submissive response. This will be discussed later. (For illustrations of the dorsal roll in *Gasterosteus*, see ter Pelkwijk and Tinbergen, 1936, Figs 8, 9 and 10.)

V

Nesting Behaviour

Introduction

Nest-building is performed exclusively by the males. I have never seen in the females even the slightest sign of a tendency to nest. A survey of the literature reveals that the complete absence of female nesting has been noted by other observers, not only in the case of *Pygosteus*, but *Gasterosteus* and *Spinachia* also. Nest-building usually begins a few days after a male has set up a territory, but this is very variable. If a male begins to construct its nest in the morning, it will have completed it on the same day. The bulk of the building takes only a few hours. Although the nest will be functionally complete at the end of the first day, the male does not then stop building altogether. On each of the following days a little more is added to the nest and nest-maintenance actions are performed which keep the nest structure firm and compact.

All five species of stickleback construct nests, in which spawning takes place and in which the eggs are reared. *Pygosteus, Spinachia, Apeltes* and *Eucalia* all build their nests in aquatic vegetation. *Gasterosteus*, however, as is well known, builds its nest in a sand-pit on the bottom.

The nest-building of *Pygosteus* has been described briefly by Ransom (1865), Landois (1871), Leiner (1931b, 1934), Sevenster (1949), and Morris (1952).

The nest structure

The nest of *Pygosteus* is composed of many fine plant fragments, which are collected together into a tubular to spheroid mass of approximately 4 cms in length. This mass is pierced by an inclined or bent tunnel, the entrance of which is always lower than the exit (see Figs 17, 18 and 19). Two basic types of nest can be distinguished, which I shall call *Tube-nests* and *Ball-nests* (see Figs 18a, b and c). The length of the nest, from tunnel entrance to tunnel exit, and the size of the tunnel is approximately the same in both types; but the ball-nest (Fig. 18c) is taller, and its walls are thicker than those of the tube-nest (Fig. 18a). Further, the tunnel of a tube-nest is straighter than that of a ball-nest.

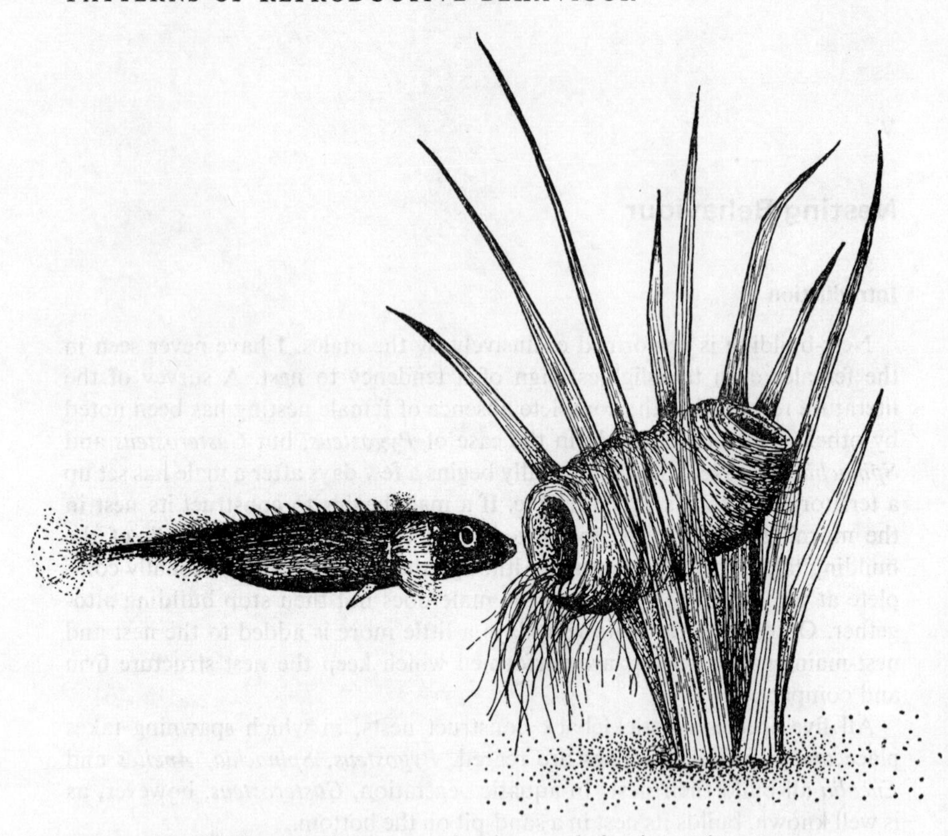

Fig. 17. The nest of *Pygosteus*, showing a male fanning at the entrance.

The tunnel of the latter is bent, so that although the exit is higher than the entrance, the entrance and the front half of the tunnel are more in the horizontal plane. The existence of these two types of nest explains the difference between the nest illustrations of Leiner (1931b) and Morris (1952). The former drew tube-nests, whereas the latter drew ball-nests. Intermediate forms, between the two extremes described above, are often observed (Fig. 18b). The factors determining which type of nest a male will build will be discussed later, after the nesting ethogram has been given.

The nest-site

It has already been stated that this species builds in aquatic vegetation, but it remains to be pointed out that the position of the nest in the vegetation and the type of host plant selected may vary. These two variations are not entirely dependent on one another. Naturally, if the only available plants which will hold a nest are very short, then nests will be built low down in the vegetation, but if suitable plants are provided at all heights in the water, then nests may appear at any level from the ground up to just under the surface of the water. What determines the selection of a particular height, in the choice of a nest site, is not clear. I found that from 10 to 15 cms off the ground was the average height. Sevenster gives 10 cms as the average, and Leiner (1931b) records between 2 and 20 cms.

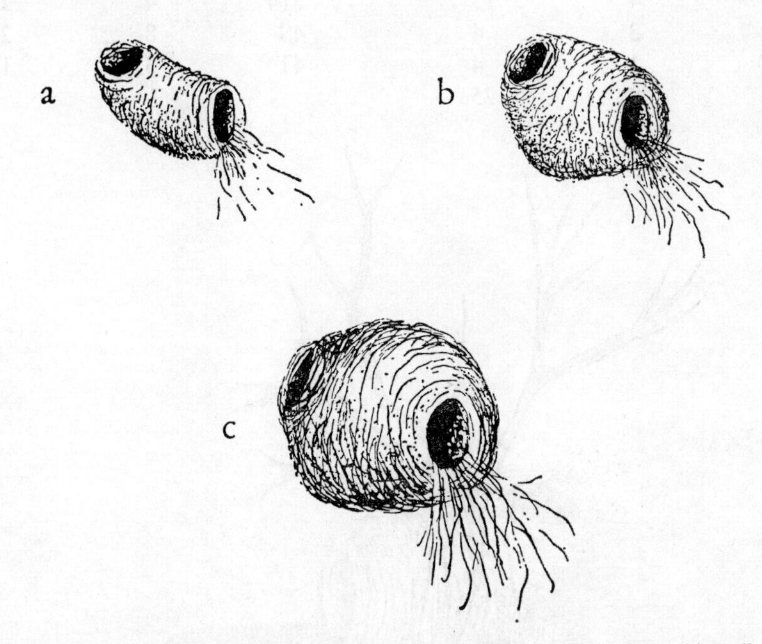

Fig. 18. Nests of *Pygosteus*, showing (a) Tube-nest, (b) Intermediate form, (c) Ball-nest. (N.B. The nests have been twisted slightly in order to show both entrance and exit.)

Quite a number of nests were actually built lying on the ground, and a few were even sunk slightly *into* the ground. These *Bottom-nests* (see Fig. 19) were first recorded by Leiner (1931b), who pointed out that they could be thought of as intermediate between the typical *Pygosteus* nest and that of *Gasterosteus*.

Host plant preferences appear to depend chiefly on the density of the foliage of any particular species. Of the species employed in the laboratory, willow-moss provided the densest foliage and was preferred to other types.

If *Pygosteus* is allowed to build a nest and the latter is then removed, it will usually build another. By repeated nest removal, it is possible to study the degree of site-attachment that exists in any individual. If there is no special site-attachment, then there will be as many nest sites as there are nests built. If there is complete site-attachment, then there will only be one site however many nests are built. Four *Pygosteus* males were tested in this way and the results are given in the table below. Three *Gasterosteus* were also tested and these results are also given:

PYGOSTEUS			*GASTEROSTEUS*		
Fish	No. of Sites	No. of Nests	Fish	No. of Sites	No. of Nests
La	4	13	314	4	23
Lb	3	4	4S	8	21
4D	1	4	4T	3	14
4Z	9	25			

Fig. 19. Bottom-nest of *Pygosteus*, showing sand-pit in front of entrance.

It is clear from the above that, in both species, there is a very strong tendency to build a new nest on an old site. This could, theoretically, be the result of the fact that very few suitable nest sites were provided in the tanks in question. In fact, however, there were innumerable sites available to both species. *Gasterosteus*, which nests in the sand, in particular had an almost infinite variety of sites from which to choose. Admittedly this species has a marked tendency to build at the base of plants, but a number of plants were available, and the males could have built 'near' each plant in many different ways. For example, a nest could be built north-north-east of a particular plant, or south-west of it, and in both cases it would be 'near' the plant, but two different sites would be involved. Furthermore, two nests could be built, both south-west of the plant in question, but if the nest-entrance-exit tunnel axis of the two nests were in different directions, the nests would still be considered to be on different sites. The same applies to *Pygosteus* also, but here, despite the added variable in the vertical plane, due to the fact that the nests can be built at any height in the foliage, the nature of the nest-building limits the number of suitable nest sites. Given a dense clump of waterweeds, *Pygosteus* could select a vast number of sites, all of which would be functionally successful, and the fact that it does not do so may be the result of certain sites being intrinsically more stimulating than others. However, in the cases in question, I feel certain that stimuli resulting from past experience of a particular site are far more important than stimuli resulting from the intrinsic qualities of that site.

It is interesting to note that if *Pygosteus* destroyed its nest itself, it never built its next nest on the old site, as it did so often when I removed the nest. I suggest that this difference was for the following reason: When a fish abandoned or destroyed its own nest, this was usually because either the host plant or nest material was rotting, or because the nest was coming adrift from the host plant. In any of the above cases, attempts to maintain the nest would gradually be less and less rewarding as the situation degenerated, and the positive site-attachment would gradually fade by a process of negative conditioning. By the time the nest was abandoned or destroyed the fish would have lost any attachment for the site, and might even be repelled by it. (In the case of a rotting host plant the fish would not be expected to use the old site again anyway, but where only the nest material, or the anchorage of the nest, was defective, the old site would still be intrinsically as attractive.) In the cases where I removed the nest myself, the fish had not suffered this gradual negative conditioning, and was left presumably with no nest, but a strong site attachment. The result was the repeated use of the same site in many instances. It might be expected, however, that the repeated removal of nests from a particular nest-site might itself negatively condition the fish to that site. This has ecological implications, because if a fish selects nest-sites in places where there is a strong current, the nests will be swept away, and the fish should be able to adapt to this by moving to a new site. An experiment was made to test this, which will be reported more fully in a later section, but it is worthwhile mentioning here one point that emerged from it. The male in question was given two clumps of weed at the far ends of a long (105 × 45 × 45 cms) tank, and ample nest material. The latter was placed in the centre of the tank. The tank was completely bare of vegetation except for the two (potted) clumps. Each day the male built a nest and each day I removed it. The first eleven nests were all built in one or other of the two clumps, but the twelfth 'nest' was 'built' in a corner of the tank, *where there were no weeds*! The male collected a huge pile of nest material in that corner, carrying each piece there and pushing it against the smooth metal upright of the tank frame. Naturally, each piece fell to the floor of the tank and a huge shapeless pile of material was formed in this way. Not only was 'nest' number 12 made in this way, but also numbers 15, 17, 18, 19, 20, 23 and 25! (See ringed points in Fig. 23.) After the twenty-fifth 'nest' the male did not build again and four days later it lost its reproductive coloration.

These results show the power of the negative conditioning that can occur in respect of nest-sites. The repeated lack of success when building in the clumps of weed (which consisted of willowmoss, the favourite host plant species), resulted in the weeds becoming less attractive than a bare and useless corner of the tank. Surprising as this is, it must be remembered that the building of the first eleven nests meant a great deal of work for the fish (their construction involved altogether the collection of 2,604 pieces of material), and it might be expected that the expenditure of so much energy without success would result in some drastic change in behaviour. Also, in nature, the fish would always be able to move on to a more suitable region.

Nesting ethogram

Collecting: The gathering together of the plant fragments which are used as nesting material involves six component actions: (1) Searching for material. (2) Biting material. (3) Breaking off material. (4) Picking up material. (5) Testing material. (6) Carrying material. These six actions do not, however, form a set sequence. The set sequence that exists consists of: (1) Searching for material. (2) Obtaining material. (3) Carrying material. Whilst searching and carrying will always involve the same actions, obtaining will vary according to the nature of the material and the motivational condition of the fish.

When searching for material, the male swims around actively, stopping here and there near plants, and often tilting into a head-down posture in which it surveys the fragments lying on the sand. This tilting is often indistinguishable from the tilting that occurs in the feeding pattern, and if both food and potential nest material are lying near the fish and it tilts its head down towards them and away again, without actually feeding or collecting material, then it is impossible to decide whether the tilt in question was a feeding action or a nesting action.

The areas searched for material are not always the most convenient. In a large tank two host plants were set up some distance apart, and close to each was placed a clump of green algae for nest material. Instead of building their nests exclusively out of the near-by algae, the males repeatedly made sorties to 'steal' material from the distant clump. Also, in a tank, the floor of which was completely covered with nest material, a male, which was building, nearly always swam round the tank on a searching trip before picking up a fragment.

The hunt for nest material frequently led to trespassing and fighting. Neighbouring nests were often plundered for nest material. The selection, out of all the available material, of one's rival's nest for the building of one's own nest, often seemed to me to be rather more than just coincidence. On the present evidence, however, I am not prepared to elaborate on this point.

Often, whilst searching, the male stopped and bit, or picked up and then immediately dropped, a plant fragment. Sometimes it carried a fragment a short distance and then dropped it. These are simple intensity differences (for similar observations in birds see Howard, 1920, and Tinbergen, 1951a, p. 139), but breaking off fragments with a special twisting dart forwards, and testing the material by regurgitation, are two responses which cannot be fitted into this scale. They may occur when collecting frequency is high or low.

A priori, it seems reasonable that a male will only break off a firmly attached fragment when more easily procured pieces are not available, but this is not the case. In a tank in which there are many suitable fragments lying loose, semi-attached and firmly attached, a male will collect from all three categories. Although a greater number of the first two types will always be taken, nevertheless, every now and then a firmly attached piece will be collected. This involves taking hold of it in the mouth and then immediately swimming with one or more vigorous darts. The strength of attachment of the piece in question often results in the fish being hurled and twisted round on the spot before success is achieved and the fragment breaks loose. It is difficult to see why this energetic action should be indulged in when more easily obtainable fragments are still available. The explanation must be that the fish selects a potential piece of nest material and having done so then collects it regardless of whether it has to break it off or not.

A similar problem arises with the testing of material. Occasionally, a piece of material that has been obtained is spat out of the mouth and taken in again repeatedly. This may result in the fragment either being abandoned or else finally taken in the mouth again and carried on to the nest. In *Gasterosteus*, van Iersel (1953, p. 9) believes that, as a result of this testing, pieces which rise or fall rapidly in the water when they have been spat out are abandoned and that only those pieces which float in the water are carried on to the nest. Although in general this seemed to be the case in *Pygosteus*, it was by no means always so. In any case, it is difficult to see why this testing action should occur at some times and not at others. It is not particularly frequent, and when it does occur I am usually unable to see any difference in the appearance of the fragment concerned. I can only suggest that a fragment is collected as a result of visual stimulation and then, as it is being carried, it may provide unfavourable chemical or tactile stimuli via the mouth, which cause it to be rejected. Once rejected it then floats in the water providing the attractive visual stimuli again and is taken into the mouth again, and so on, until either the visual or chemical-tactile stimuli become sufficiently dominant for it to be either carried to the nest, or abandoned.

Boring: Once a piece of material has been brought to the nest-site, it is pushed in amongst the vegetation, usually at a point where the leaves and stems grow densely together. The fish noses into the plants, still holding the material, then releases the latter and backs out again. This is repeated over and over again until there exists an indented bundle of fragments. The pushing into the centre of this mass is extended deeper, forming out of the indentation a tubular cavity. This cavity is still closed at the far end. Looping and other construction actions appear now (see later). When the nest is almost complete, high intensity boring can be seen in which the male pushes right into the cavity so that only its tail is visible. In this position it may often come to lie on its side before wriggling out backwards by means of tail-beats.

This movement persists throughout the rest of the reproductive cycle, taking on different functions at different phases. Its function during the nesting phase is essentially the formation and maintenance of the nest tunnel.

Pushing through: When the nest is nearing completion, the male can be seen to push into and through it, bursting out through the back wall and completing the nest tunnel formed by the boring. The nest is now pierced through by a slightly bent tunnel, the exit of which is higher than the entrance. From this point onwards, until the eggs are laid, the male repeats the pushing-through movement from time to time. The performance of this action, however, is always far less frequent than boring.

Looping: As soon as the nest bundle begins to take shape, the male not only pushes into it with a fragment, but then also loops that fragment into the cavity once or more times depending on the length of the piece in question. As it backs out of the cavity or, at a later stage, the tunnel, the male stops before swimming off in search of more material and takes hold of the middle of the 'tail' of the fragment which hangs down out of the entrance. It then repeats the action of pushing into the cavity, forming as it does so a loop of the fragment around the cavity wall.

Fanning: During the nesting phase, the action of boring is often followed by fanning. The male backs out of the nest until its nose is a few millimetres away from it and then, still facing the tunnel, fans a current of water through it. This activity has been described in detail elsewhere in the literature (e.g. van Iersel, 1953). Briefly, the pectoral fins, beating alternately, send a current of water forwards through the tunnel, whilst the male is held in position by strong counteracting beating of the tail, which sends a current of water backwards (see Fig. 113 in Tinbergen, 1951a).

Later, during the parental phase, the fanning action serves the very important function of ventilating the eggs. Its function during the nesting and sexual phases is still controversial and will be discussed later, in section VIII.

Tail-digging: Pygosteus always constructs its nests in waterweeds, but they are often built so low down in the plants that they are resting on the substratum. Where a male is the owner of one of these 'bottom-nests', two extra nesting activities can be observed. The first of these I have called tail-digging. The male pushes into the entrance in a typical boring action and then beats its tail rapidly and vigorously from side to side before backing out again. This has the effect of forming a large depression in the sand immediately in front of the nest entrance (see Fig. 19). In nature this would be useful in preventing a bottom-nest from becoming inundated with drifting sand or mud. Whilst tail-digging is obviously related to the second part of deep boring, it nevertheless exhibits modifications of this, in that the lateral tail-beats are more vigorous and more frequent.

Sand-digging: The second activity which appears only when a male owns a bottom-nest is the sand-digging pattern so familiar in the nesting male *Gasterosteus*, but which is very rare in *Pygosteus*. The reason for its rarity in *Pygosteus* is not only that males nesting above the bottom never show it, but also that the bottom-nesters employ this pattern exclusively for digging sand out of the bottom of the nest tunnel. The pits that are seen in front of *Pygosteus* bottom-nests are entirely due to tail-digging and only where the tunnel becomes clogged with sand is the sand-digging pattern used. The actual movements themselves are the same as in *Gasterosteus*, except that the sand-digging *Pygosteus* has to take the sand into its mouth whilst its head is inside the tunnel. By forward swimming movements the open mouth of the male is pushed into the sand in the bottom of the tunnel. The mouth is then closed and the fish backs out and swims a short distance away, its gill covers distended. It then spits out the sand in a cloud and returns to the nest. The carrying and ejection of the sand is performed in exactly the same way as in *Gasterosteus* (see Tinbergen, 1951a, Fig. 104).

Tearing off obstructions: The action employed in obtaining firmly attached pieces of nest material is also employed, in a modified form, just in front of the nest. It is the same as the collecting action, in that the male takes hold of a piece of weed and then swims suddenly and strongly whilst still grasping in it its mouth, but it differs here in that whereas before the sudden dart could be in any direction, now it is always away from the nest entrance. Further, whereas before the tearing at one piece of potential nest material was soon abandoned if unsuccessful, now it is repeated sometimes over and over again.

After a piece of weed, which has been obstructing the entrance, has been successfully broken off in this way, it is invariably only carried a short way and then dropped and is not treated as nest material.

The effect of this action is to remove obstructions in the region of the front of the nest so that there is formed there a clear space, which, during the courtship phase, will be of great importance. In the case of bottom-nests, the tail-digging action adds to this clearance. The tearing off of obstructions can be so vigorous that, if the nest is not securely anchored, it may be completely or partially dislodged by it. This is particularly likely if it is part of the host plant that is being torn off. However, a loosely lodged nest with an obstructed entrance is of so little use that, drastic as this measure may seem, it is nevertheless highly functional.

Pest removal: On a number of occasions a male has been seen to pick up a water snail in its mouth and carry it some way from the nest before dropping it. The basic pattern employed is similar to that used when removing sand from the nest tunnel. Snails are only dealt with in this way when they are on, or very near to, the nest. In the aquarium tanks, the removal of pests from nests is an activity little encountered, but in the streams it must be of very frequent occurrence and of great importance.

Nest mending: Once the nest is complete and until it is abandoned, the male maintains it, not only by consolidating it with more fragments and more glue, but also by repeatedly pushing the outside of it with its nose. This action I have called 'mending' to distinguish it from pushing *into* the nest which is called 'boring'.

Mending has the function of helping to keep the nest compact and dense. If a nest has many straggling fragments all over it, then the mending frequency is likely to be high. If a nest has come partially adrift, then the mending action may become particularly frequent and intense in the process of re-lodging it. In its high-intensity form, mending involves opening the mouth, grabbing, and then pushing the nest. In the usual, low-intensity forms, it is performed with the mouth shut, simply pushing at the nest with the snout.

Nest destruction: Frequently, pieces of material, that have been collected and incorporated in the nest and which appear to be satisfactory to the human observer, are taken by the fish, pulled off the nest, carried some distance away and dropped. Under special circumstances, this action may become so frequent as to lead to the destruction of the nest. Closer observation often revealed that the fragments taken from the nest were, in fact, rotting.

Glueing: During the nesting, sexual and early parental phases, the male exhibits patterns of behaviour associated with the secretion of a substance which is produced by a certain region of the kidney. This part of the kidney becomes specially modified during the reproductive season. This substance, or 'glue', which hardens on contact with water, is used by males of all species of stickleback in strengthening the nest structure. *Pygosteus* possesses two quite distinct methods of glueing. One of these, in slightly varying forms, it has in common with all the other species and I am calling this type *superficial glueing*. The other form has only been reported so far in the *Pygosteus*, and this I am calling *insertion glueing*.

Superficial glueing: The male swims over the surface of the nest in a characteristic slow, jerky movement, secreting a very fine thread of glue as it goes. This activity has often been likened to the spinning of a web by a spider. The length of one spinning journey and the thickness of the thread varies from species to species. In *Pygosteus* the male usually circles round and round the nest several times in one bout of superficial glueing. Its body is curled round the nest as it goes, and its ventral surface, which is in contact with the surface of the nest, is concave. The whole movement takes less than half a minute and the thread produced is so fine as to be practically invisible. Superficial glueing occurs during the phase of actual nest construction and is never seen at other times. It is far less frequent in its occurrence than insertion glueing. Its rarity may be due to the absence of any water current in the aquaria, and I suspect that in the river it may be more frequently and intensely expressed.

In *Gasterosteus*, the glueing movement is much shorter than in other species, and the ventral body surface is convexly curved instead of concave. This is due to the different location and shape of the nest. This was shown to be the case when bottom-nesting *Pygosteus* males were observed to perform superficial glueing. They had great difficulty in performing the circling of the nest and modified the usual movement, either by circling only the upper end of the nest around the exit hole, or by making short circling movements over the top and sides of the nest. In the latter modification the fish were performing movements very similar to those of *Gasterosteus*.

In *Spinachia*, the nest is bound round many times with a thicker, stronger, thread of glue. The whole glueing action is usually performed in one bout which may last many minutes. This greater development of the activity in *Spinachia* is correlated with its tidal nesting habitat. *Apeltes*, which is an estuarine form, but which has a much smaller nest than *Spinachia*, produces a glue 'cobweb' around its nest which appears intermediate in its strength between that of *Spinachia* on one hand and the river forms on the other.

Insertion glueing: The male suddenly stops swimming and begins to rotate in a horizontal plane, either clockwise or anti-clockwise. The head and tail are both bent to the same side as the fish rotates. A small opaque white blob of glue appears at the vent and this increases in size, stretching away from the vent as it does so, until it is about an inch or so long. The circling male then catches the end of the blob in its mouth and, carrying it inside its mouth, swims, with its gill covers distended, to the nest entrance. It bores into the nest tunnel and deposits the glue there, pushing it well in. Then it typically backs out of the tunnel in a series of short gulping jerks before swimming off. This is the usual pattern of insertion glueing, but it has a number of variations.

If the usual process is thought of as a five-stage sequence (Circle – Catch–Carry – Insert – Gulp), then the following table can be drawn up to show the frequency of the different omissions that occur from this pattern:

Circle – Catch – Carry – Insert – Gulp:	88
Circle – Catch – Carry – Insert:	28
Circle – Catch – Carry – Insert – Fan:	23
Circle – Catch – Carry – Insert – Fan – Gulp:	3
Catch – Carry – Insert:	1
Circle:	10

As can be seen from this table, the most frequent variation involves the omission of the gulping after insertion. On a number of occasions this, like other omissions, was due to interference from other fish at that point in the sequence. In a number of cases the male fanned after inserting the glue. This was not typical, however, and it seemed, usually, as if the gulping action inhibited the fanning which otherwise nearly always follows deep boring into the nest. In a few rare cases, the male both fanned and then gulped after inserting glue. Circling to obtain the blob of glue was rarely omitted, but if the male was fighting, or performing some other 'absorbing' activity at the time when the blob appeared, it was possible that it was secreted apparently unnoticed. If the male then caught sight of it floating in the water, it was seen, on rare occasions, to pick it up and complete the glueing pattern. In a similar way, if a male was circling and was just about to catch the glue in its mouth when it was stimulated by some other object, the whole glueing sequence was broken off at that point and sometimes the blob of glue would stream off behind it as it chased away an intruder. Once, it returned from such an encounter and picked up the fallen glue and carried on with and completed the sequence. A number of times it occurred that a male would circle and circle, but without any glue appearing, and this then caused the pattern to be broken off. On many occasions, however, it was seen that the glue blob appeared just *before* the circling, as if its appearance caused the circling to be performed. In this connection it should be mentioned that glueing did not appear to follow exclusively any particular action, nor to be followed by any particular action. However, if all activities are divided up into nesting activities and non-nesting activities, then it becomes clear that a male is more likely to glue after performing some nest action, and is less likely to perform a nesting, rather than some other, action after glueing. This is also true of *Gasterosteus*. Here are some figures for *Pygosteus*:

	Before Glueing	After Glueing
Nesting	42	13
Non-nesting	26	49

Whereas superficial glueing occurs during the construction of the nest, insertion glueing appears towards the end of nesting and persists until approximately the first day or two after fertilization. Under special circumstances (see Morris, 1952, p. 241 [p. 21 in this volume]) glueing was seen to be performed by a nestless male. When this happened, the glue was carried around the tank in the mouth for a while and then finally eaten with much gulping. This modification in the absence of a nest can be used tentatively as an indicator of what stage has been reached when a nest is in the process of being abandoned. Several times I have observed a male, with an old nest in its tank, eat its glue, but always, when I have seen this, the male in question very shortly afterwards built a new nest.

Whereas superficial glueing serves to hold the nest together from the outside, insertion glueing appears to strengthen the inner surfaces of the nest tunnel. The performance of this 'cementing' process just after the eggs have been laid indicates that a second function of insertion glueing is probably that of holding the eggs in place in the tunnel. It is interesting to note that, in the bottom-nesting *Gasterosteus*, where there is no danger of the eggs falling out of the nest, insertion glueing does not appear at all.

Glue-stealing was seen on a few occasions. A male was seen, for example, to push through the (eggless) nest of another fish and emerge from the exit with its gill covers raised. As it swam away it was observed to be gulping in and out a blob of glue, which it finally ate.

Comparison with 'Gasterosteus'

It is not intended to give here a detailed comparative account of the modes of nest-construction of *Pygosteus* and *Gasterosteus*, but there are certain aspects of such a comparison that must be mentioned at this point.

Since quantitative statements will be made concerning various nesting actions, in the following sections of this study, the limitations and dangers of using such data for certain comparative purposes must be pointed out here. The two actions to which this applies in particular are boring and mending.

The action of boring into the nest tunnel in the two species can usefully be compared at a qualitative level, but if fluctuations in the frequences of this action in *Pygosteus* are thought of in relation to *Gasterosteus*, then the comparison becomes meaningless. This is so for the following reasons. In *Pygosteus*, the boring action is far more important than in *Gasterosteus*, and occurs in many more situations. Nest material which is brought to the nest is bored into it. When looping long pieces into the nest, the male bores repeatedly. When glue is inserted into the nest, the male bores into the tunnel in the act of insertion. Also, when glue has come adrift in the tunnel, repairs are carried out by boring into the tunnel and re-packing the glue. The tunnel is created, enlarged, and its form is maintained by boring. Only this last sentence applies to *Gasterosteus*. In this latter species, there is no insertion glueing, or looping, and boring does not occur as an integral part of the addition of new material to the nest.

When *Pygosteus* adds new material to the nest it usually does so with a boring action, whereas *Gasterosteus* places it, or pushes it, on to the top of its nest. Pushing at the *surface* of the nest in this way, has been termed simply 'pushing' by van Iersel (1953, p. 10). When the nest is completed, *Gasterosteus* still exhibits pushing when repairing the nest, and it is then comparable with the action I have named nest mending in *Pygosteus*. Thus, whereas boring is more important to *Pygosteus*, pushing is more widely used by *Gasterosteus*. The comparative problem arising as a result of this difference is not completely solved simply by specifying the special causal factors involved in any particular occurrence of boring or pushing in the two species. Such a precaution is helpful, but encounters the following difficulties. Firstly, the causal factors are not always clearly distinguishable from one another. For example, if a *Pygosteus* male bores into its nest tunnel, holding neither nest material nor glue in its mouth, the causal factors involved may still be different from those occurring in *Gasterosteus* in apparently the same situation. The male *Pygosteus* may be re-packing glue, it may be re-fixing a piece of material recently brought to the nest or it may, of course, simply be improving the form of the tunnel – as in *Gasterosteus* – without respect to any such special stimuli. Secondly, if a male *Pygosteus* bores into its nest tunnel with a piece of material or glue in its mouth, despite the fact that the causal factors are clear in such cases, nevertheless, such borings are going to improve the form of the tunnel and thus automatically affect other causal factors.

Such difficulties as this do not arise in all cases. For instance, the collecting of material can be compared quantitatively in the two species. But amongst the nesting actions, this is really the only one which can, in fact, be compared in a straightforward manner. Even (pre-parental) fanning comparisons give rise to complications, since fanning is linked with boring and is thus involved in the difficulties described above. Furthermore, the sequential linkage between boring and pre-parental fanning differs in the two species (see p. 360).

Summing up, it must be stressed that it is in no way implied that one collection by *Pygosteus* is 'equal' to one collection by *Gasterosteus*, but rather that, in the case of this activity, it is possible to compare quantitatively the similarities and differences in its fluctuations during the reproductive cycle, in the two species. This also applies to many of the other activities such as biting and zig-zagging, but does not apply to boring and pushing, for the reasons given above.

The nest material

The preferred qualities of nest material have been analysed to some extent for *Gasterosteus* by Wunder (1928, 1930) and Leiner (1931b), but nothing has previously been recorded for *Pygosteus*. It is clear from simple observational data alone, that the latter species selects material of a particular *thickness* and *elasticity*. Only fine fragments are used, and stiff pieces can not be shaped into the nest. But other qualities such as colour, texture and length, can only be investigated experimentally. In order to do this a special tank arrangement was employed.

In an experimental tank, instead of the usual vegetation, there were placed a number of pieces of broken flower-pot. These were pushed vertically into the sand substratum here and there over the floor of the tank. This was done because the character 'open space' seems to inhibit territory formation in *Pygosteus*. In one corner of such a tank, one clump of *Hygrophila* was planted, as the potential host plant for the nest. This species of host plant was employed because of its comparatively large leaves, which prevent it from being used itself as nest material. Scattered over the floor of the tank were a known number of fragments of potential (artificial) nest material, which, in any one experiment, varied in one quality only. By collecting, with a hooked wire, those fragments remaining on the floor of the tank each day, it was possible to obtain a quantitative record of various preferences in nest material selection, without interfering with the nest itself. Alternatively, the nest itself could be hooked out with the wire. By the former method, it was possible to study the way in which collecting material changed quantitatively with the completion of the nest, in addition to studying the preferences for different kinds of material. Although this was an advantage, the presence of a more-or-less completed nest reduced the number of daily collection trips tremendously, so that when data on material preferences were more urgently required, the nests were removed each day.

Since the experimental tanks did not provide a very stimulating environment for building, the experimental males were usually given a neutral fish for company (the effect of the presence of a second fish is discussed later). The artificial nest material used was commercial machine cotton ('Sylko', Dewhurst's Three Shells Machine Twist). This was obtainable in various hues, shades and thicknesses, and could easily be cut to any length. It was readily accepted as building material by the fish.

Colour

Observational data alone give no clue concerning colour preference, since almost all natural nest material is green. It seemed likely that either the fish would have a preference for green, or would have no colour preference at all. Small-scale preliminary tests seemed to indicate that the second possibility was the case, but further large-scale tests disproved this.

(It must be pointed out that no elaborate attempt has been made in the following tests to analyse the exact sensory capacities of *Pygosteus* in respect of colour preferences. As Walls (1942) has stressed, it is exceedingly difficult to prove in any animal species that a particular colour preference has involved a response to differences in hue, rather than intensity, or saturation. Although some attention has been paid to this question here, it must be emphasized that the focus of interest in the present tests was on *differences* in colour preferences under different environmental circumstances. In the study of such differences, the more esoteric aspects of the visual colour perception of the species are comparatively irrelevant and can largely be ignored.)

A large number (100 in preliminary tests, 200 in all main tests) of 10 cm lengths of each of four types of cotton were placed in the tanks each day. Two series were used throughout the tests, a 'coloured' series and a 'grey' series, as follows:

'Coloured' series	'Grey' series
YELLOW (Lemon. D. 247)	WHITE (White, un-numbered)
BLUE (Cornflower. D. 65)	LIGHT GREY (Light Grey. D. 71)
RED (Oriental Poppy. D. 366)	DARK GREY (Elephant. D. 188)
GREEN (Frog Green. D. 340)	BLACK (Black. D. 14)

All lengths were of the same thickness (size 40). Before presenting the 800 pieces in any test, they were placed in a beaker of water and stirred up until the different coloured fragments were thoroughly mixed together.

The two series were never presented together, although some fish were given a spell of several days with one series, which was then followed by a spell with the other series. With the use of the Villalobos (1947) Colour Atlas, it was possible to ascertain that the 'greyness' of the 'coloured' series ran from light to dark in the order shown above (yellow being lightest and green darkest). On this basis, it was possible to predict which of the twenty-four possible orders of preference would be expected of a male, when given a test with the 'coloured' series, if the latter was in fact being seen as a series of 'greys'.

The following tables give the results obtained for a number of males tested with one, or other, or both series. All the daily tests have been added together for each male, to give the total number used, by each male, of each of the four qualities concerned. A number of *Gasterosteus* males were also tested in the same way, for comparative purposes, and the results obtained for this species are also given below:

PYGOSTEUS — 'Grey' Series

Fish No.	White	Light Grey	Dark Grey	Black
4Z	2616	2142	666	286
4D	266	238	81	45
La	777	764	188	69
TOTALS	3659	3144	935	400

GASTEROSTEUS — 'Grey' Series

Fish No.	White	Light Grey	Dark Grey	Black
3T4	315	316	366	267
4S	141	146	103	22
4T1	250	268	262	131
TOTALS	706	730	731	420

PYGOSTEUS — 'Coloured' Series

Fish No.	Yellow	Blue	Red	Green
F1	459	221	135	289
G2	121	88	58	4
C	108	10	6	21
La	691	266	96	79
Lb	531	12	64	38
TOTALS	1910	597	359	431

GASTEROSTEUS — 'Coloured' Series

Fish No.	Yellow	Blue	Red	Green
3a	90	14	28	5
3c	50	47	51	43
2T4	268	125	259	605
4S	510	188	271	253
4T1	168	132	55	101
4T2	75	50	65	38
TOTALS	1161	556	729	1045

The way in which the data are lumped together in the above tables conceals a number of important correlations, which will be taken separately below. Bearing this in mind, and making allowances for them, the general tables still tell us several interesting things, which can be summarized as follows:

(1) The whiter a piece of nest material, in a 'grey' series, the more it is preferred by *Pygosteus*. This applies to all three males tested.

(2) There is very little 'grey'-series preference in *Gasterosteus*, except that black is markedly disliked.

(3) All five *Pygosteus* males tested with the 'coloured' series showed a very strong preference for yellow. It cannot be said whether or not this was entirely due simply to yellow being the lightest of this series.

(4) In all *Pygosteus* males tested with the 'coloured' series, with the certain exception only of male F1, the order of preference is not inconsistent with the possibility of the 'coloured' series being seen as a 'grey' series.

(5) In *Pygosteus* male F1, the preference for green over red signifies that this individual must have been seeing green as a green hue. (That a male should prefer green to red, despite the fact that the green is darker than the red, is not, in any case, surprising; see discussion below.)

(6) In the six *Gasterosteus* males tested with the coloured series, the special variations discussed below were operating in such a way as to make these general figures valueless except to show that (a) they cannot be explained on

the basis of being seen as a scale of 'greys' and (b) they differ considerably from those for *Pygosteus.*

Summing up, it may be said that it seems likely that the sticklebacks may select nest material both by its lightness of shade and its special hue. Differences in the extent to which one or other of these qualities is involved in various males will give rise to individual differences. The causation of such differences is not clear.

The question of the function of the colour preferences of *Pygosteus* is an interesting one. It was expected, as mentioned earlier, that if a preference existed it would be for green. Clearly it is not as simple as this. Even in male F1, which preferred green more than it should have done in a 'grey' scale, yellow was still easily the favourite. In other words, the predominant interest was in the lightest material available. (Throughout this section, 'light' refers to shade and not, of course, to weight.) In nature this would probably lead to the construction of nests of very light yellow-green material. There are several possible adaptive advantages in this:

(a) It may make the nest conspicuous.
(b) It may make the nest cryptic.
(c) It may result in the selection only of the freshest material.
(d) It may result in the selection of material which is desirable for some other quality associated with lightness, such as particular texture or elasticity.

These possibilities must be considered in turn.

(a) and (b) *Conspicuousness:* If the males selected the most, or the least, conspicuous cottons, this may have been done by comparing them with either the substratum on which they lay, or with the host plant in which they were placed. The first of these possibilities was eliminated by providing males with substrata which were either dark, medium or light. The 'grey' series was used and in the light tank the light-grey cottons were extremely cryptic, whilst in the dark one, the dark-grey cottons were practically invisible. The results were as follows (for simplicity, the white and light grey have been taken together and the dark grey and black also, as light and dark respectively):

Background	Light Material Used	Dark Material Used
LIGHT	83	17
MEDIUM	86	14
DARK	80	20

The figures in the above table are expressed as percentages. It is clear from them that background has no effect on the colour preference of *Pygosteus.*

It was also found that the construction of the nest in dark-green or light-green host plants did not affect the lightness of the material selected. Also, nests built without the use of a host plant showed the typical colour preference.

(c) *Freshness:* If males selected the freshest young growing plant fragments, these would, in nature, undoubtedly be a very light green; this may well be the principal reason for the selection by *Pygosteus* of the lightest material available in the experimental tanks. As aquatic vegetation degenerates and rots, it generally tends to become darker and so would be avoided. This is of importance in supplying the eggs with a healthy environment. Considering the trouble to which *Pygosteus* goes to keep its eggs fresh by fanning them, or by transferring them to a new nest when the old one degenerates, it is not surprising that it should possess mechanisms for ensuring that the material used in the nest is the least likely to rot away and thereby interfere with the growth of its eggs.

(d) *Other associated qualities:* The white rootlets of aquatic vegetation are sometimes seen to be incorporated into the natural nests of non-experimental *Pygosteus* males. These rootlets are often much more tough and springy than other plant fragments and the incorporation in natural nests of material with such qualities would probably be encouraged by the preference of *Pygosteus* for the lightest material. It is doubtful if this is as important a factor as (c), but at present it cannot be excluded as a possibility.

The above discussion of the function of nest-material preferences in *Pygosteus* does not, of course, apply to *Gasterosteus*, as can be seen from the figures in the previous tables. *Gasterosteus* shows only a slight tendency to avoid the darker colours and, at the present stage of analysis, its preferences in the 'coloured' series are extremely variable from male to male. However, before discussing the possible functional significance of these differences between the two species, it is necessary to describe in some detail the very special way in which the colour preferences of *Gasterosteus* operate *during* the building of its nest.

Wunder (1930) reported a number of rather vague tests which he had made, using *Gasterosteus*, with different coloured nesting materials. No numerical data are given and the results are of little interest in respect of preferences themselves; however, Wunder mentions the extremely interesting fact that *Gasterosteus* males may *mark their nest entrances* with bright coloured materials. This was tested and confirmed in the preliminary *Gasterosteus* tests in the present investigation. Later, however, coloured nests of this species were removed whole from tanks and dissected carefully in shallow dishes. By this method it was found that the marking of the nest entrances was only a symptom of an extraordinary colour preference mechanism in *Gasterosteus*. In brief, a *Gasterosteus* male *drastically changes* its colour preferences *during* the building of its nest.

When several nests, which were built using the 'coloured' series of cottons, were carefully dissected out into layers, it was found that different colours predominated in different layers. *Gasterosteus* builds its nest by digging a shallow pit in the sand, and then laying down each piece of material in this

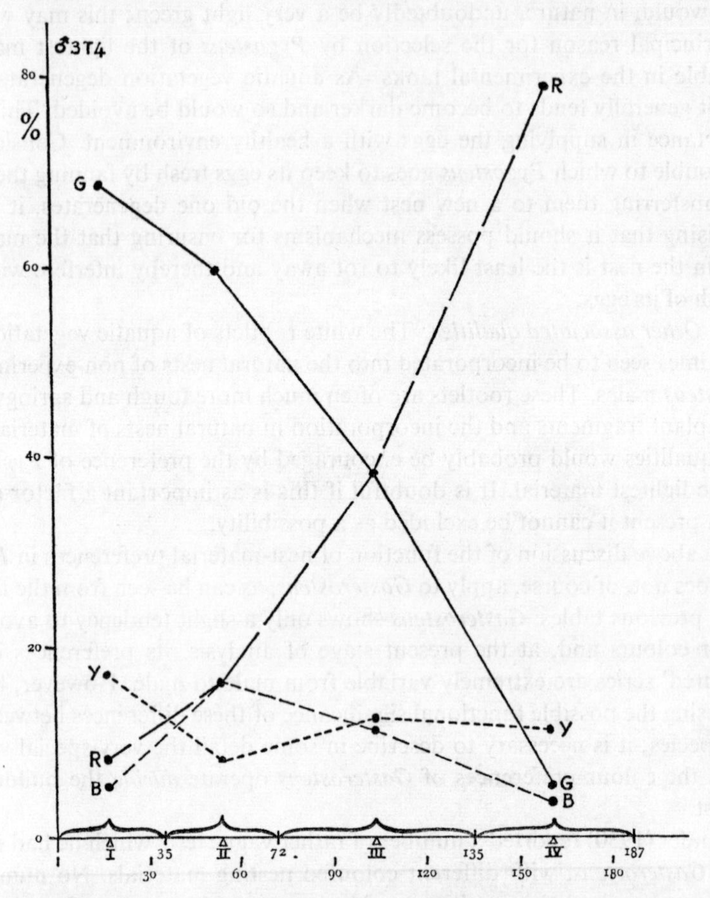

Fig. 20. Change in colour preference during the building of a *Gasterosteus* nest. Y, B, R, and G, represent yellow, blue, red and green respectively. I, II, III and IV represent the four stages of building. Ordinate values are percentages of a colour used at a particular stage of building.

pit until a floor of material is formed. This is added to, more and more, each piece being laid down on top of the pile. Each is dropped or pushed into the pile and then glued in place, until the nest is completed. About halfway through the building of the nest, the male begins to bore into one end of the pile and this boring develops the nest tunnel. (This description of the nest-building is an over-simplification, but is nevertheless valid as a generalization

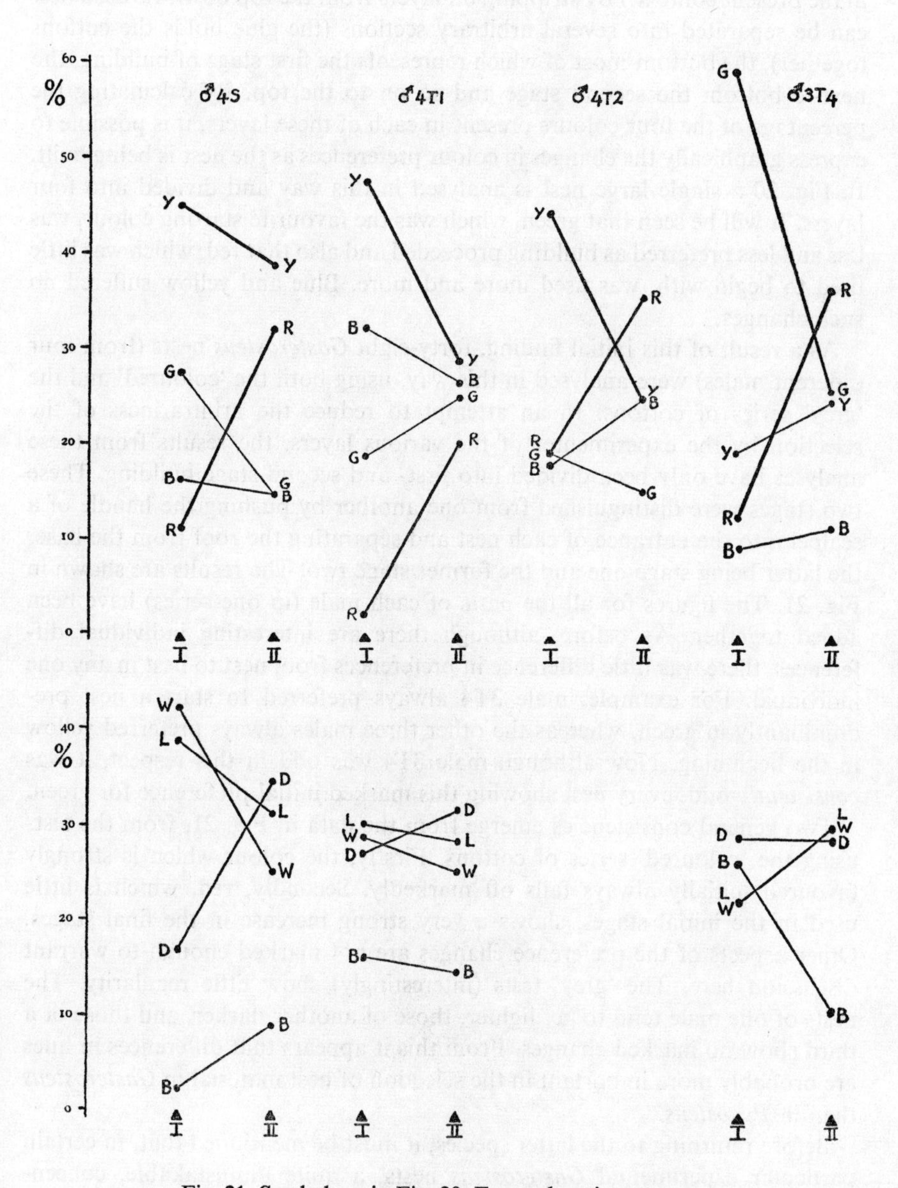

Fig. 21. Symbols as in Fig. 20. For explanation see text.

in the present context.) By stripping off layers from the top downwards, a nest can be separated into several arbitrary sections (the glue holds the cottons together), the bottom-most of which represents the first stage of building, the next-to-bottom the second stage and so on to the top. By calculating the percentage of the four colours present in each of these layers, it is possible to express graphically the changes in colour preferences as the nest is being built. In Fig. 20 a single large nest is analysed in this way and divided into four layers. It will be seen that green, which was the favourite starting colour, was less and less preferred as building proceeded and also that red, which was little used to begin with, was used more and more. Blue and yellow suffered no such changes.

As a result of this initial finding, forty-eight *Gasterosteus* nests (from four different males) were analysed in this way, using both the 'coloured' and the 'grey' series of cottons. In an attempt to reduce the arbitrariness of the selection by the experimenter of the various layers, the results from these analyses have only been divided into first- and second-stage building. These two stages were distinguished from one another by pushing the handle of a scalpel into the entrance of each nest and separating the roof from the base, the latter being stage one and the former stage two. The results are shown in Fig. 21. The figures for all the nests of each male (in one series) have been added together. As before, although there are interesting individual differences, there was little difference in preferences from nest to nest in any one individual. For example, male 3T4 always preferred to start a nest predominantly in green, whereas the other three males always preferred yellow in the beginning. Now although male 3T4 was odd in this respect, it was *consistently* odd, every nest showing this marked initial preference for green.

Two general consistencies emerge from the data in Fig. 21, from the tests using the 'coloured' series of cottons. Firstly, the colour which is strongly favoured initially always falls off markedly. Secondly, red, which is little used in the initial stages, shows a very strong increase in the final stages. Other aspects of the preference changes are not marked enough to warrant discussion here. The 'grey' tests (interestingly) show little regularity. The nests of one male tend to get lighter, those of another darker, and those of a third show no marked changes. From this it appears that differences in hues are probably more important in the selection of nest material in *Gasterosteus* than in *Pygosteus*.

Before returning to the latter species, it must be mentioned that, in certain particular experimental *Gasterosteus* nests, a quite unmistakable, concentrated *bright ring* of colour surrounded the entrance in such a way as to make it necessary, after all, to consider the possibility of the existence of a very special 'entrance-marking mechanism'.

I can only interpret these findings as follows: It appears to be an advantage for *Gasterosteus* to render its nest-entrance conspicuous. In evolution, this seems to have been achieved by changing the nest-material colour preference as the nest becomes more and more complete. Probably this was originally correlated with changes in nest-building motivation. It may still be, but my impression is that it is not. It seems now rather that stimuli from the less or more completed nest affect the changes in preference. However, further experiments are needed here. The final evolutionary stage would be – and partially already appears to be – a sudden sharp increase in preference for conspicuous, contrasting colours, at just the point when the nest entrance is being completed.

Returning now to *Pygosteus*, attempts were made to ascertain if similar changes occurred in this species during the building of the nest. It was known, from previous observations, that, if any such changes did occur, they were far less marked. The results were inconclusive, although the general impression was that some slight changes could be detected, when careful analyses were made. Here, the nesting method being different, the manner of separating the early from the late collected cottons also had to differ from that employed in the case of *Gasterosteus*. Briefly, the outer shell of a *Pygosteus* nest had to be separated from the inner shell, the former being the first stage of building and the latter the second stage. Using the 'coloured' series, results were not at all encouraging, but one male showed a distinct tendency, when tested with the 'grey' series, to build *whiter* in the second stage than in the first. The figures for the six nests of this male which were analysed in this way are given in the following table:[1]

No. of	STAGE ONE				STAGE TWO			
Nest	White	Light	Dark	Black	White	Light	Dark	Black
1	37	35	17	9	87	75	2	1
2	64	72	8	10	83	60	3	2
3	32	50	1	0	51	39	1	2
4	30	36	27	5	51	57	5	4
5	43	51	17	12	80	60	18	2
6	71	80	36	16	52	42	25	1
TOTALS	277	322	106	52	404	333	54	12

It is most interesting that *Pygosteus* should possess a very mild form of what is for *Gasterosteus* a dramatic change. Functionally speaking, I can only suggest that this is the result of the fact that *Gasterosteus* nests in more shallow, open spaces than *Pygosteus*, where landmarks are more likely to be scarce. A greater need might therefore exist to mark the position of the nest; and what better method of doing this than to make a nest conspicuous by

[1] This table is given in full because the difference discussed is much less obvious than in any of the previous instances. Nevertheless, it is statistically a highly significant result.

contrasting it *with itself* in just that region which is the focus of interest – the entrance? It is worthy of note that the more extreme preference in *Pygosteus* for lighter material, which probably results in this species building a 'healthier' nest for its eggs, is correlated with less intense ventilation activities in this species. (See section VII for a comparison of the fanning cycles of the two species.)

Before finally leaving this question of changes in nest-material collection preferences, a very special case of it in *Pygosteus* must be reported. It has been mentioned already that, in one experiment, a male had its nest removed repeatedly every day over a period of approximately a month. The ways in which this male attempted to deal with its lack of success in maintaining a nest were briefly as follows. Firstly the male tried re-building its nest in the original site. Then it tried building it on other sites and ultimately, as already described, built it in a bare corner of the tank. When this did not work, the male finally gave up nest-building, and a few days later it went out of reproductive condition. In addition to the above, however, this male also changed its behaviour, at one stage of the experiment, in respect of its colour preferences. It was being provided all the time with the 'grey' series and was, of course, preferring white to light grey, light to dark grey, and dark grey to black. But at one point, it began to take a greater and greater proportion of darker cottons, and the general appearance of the nest changed completely. Then, after reaching a peak on the tenth day of the experiment, this change gradually subsided and the male returned to its old preferences again. This is very clearly shown in Fig. 22. In this figure, the values for the collections of the various types of cottons are given as percentages of the total number collected in any one day (ordinate values). It seems as if the male, as a result of repeated lack of success with nests built on the 'lightest' preference system, tried out the opposite, found that this too was equally hopeless, and then returned to its original, natural preferences again. This is an extraordinarily 'plastic' piece of behaviour for a fish, and shows the dangers of generalizing about the 'stereotyped and fixed' nature of the behaviour of lower vertebrates, as certain authors have tended to do.

Finally, there is one small point which emerges from Fig. 22, which is perhaps worth noting; namely that, forgetting the drastic switch in preferences around the tenth day, there can also be discerned a gradual decrease in the preference for the white and light cottons, and a gradual increase in the preference for the dark and black ones, as the experiment proceeds. In other words, its colour preference in general is becoming less intense day by day through the experiment. This is not correlated with motivational changes, as is revealed by comparing Fig. 22 with Fig. 23 in which the total number of cottons collected each day over the whole experimental period is shown. A gradual decrease in degree of preference would be expected to be correlated

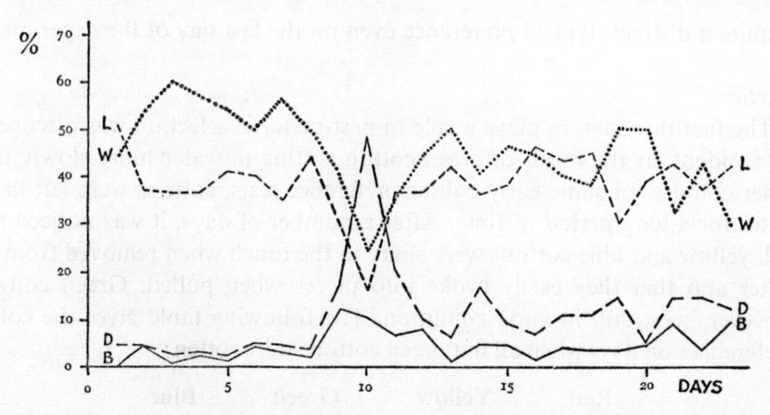

Fig. 22. Nest-material preferences with daily removal of nest. W, L, D and B represent white, light grey, dark grey and black respectively.

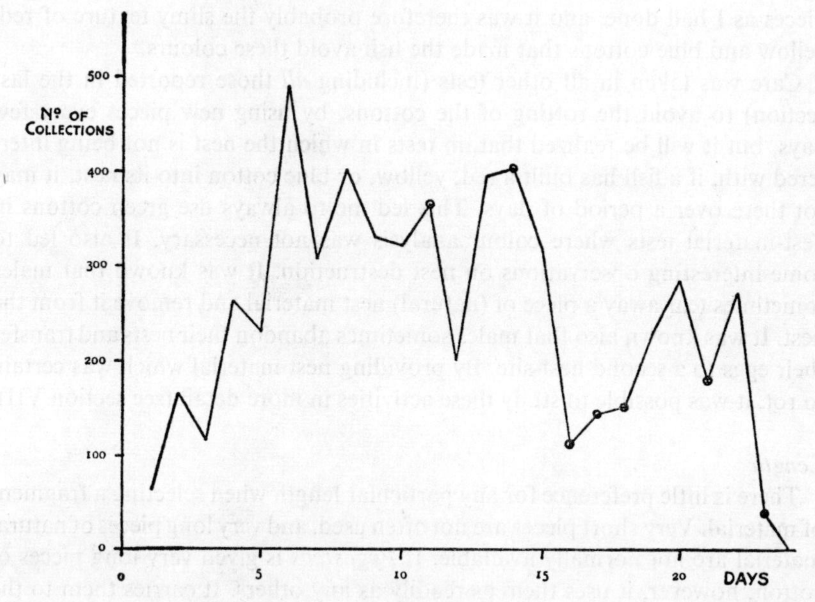

Fig. 23. The total number of collections made on each day of the above test. The ringed points represent those days when the nest was built in the corner of the tank away from the weeds. Explanation in text.

with a gradual increase in motivation, but this does not occur. It seems, therefore, that the gradual decrease in degree of colour preference by this male must be due to some learning process. It is as if the male, having failed to maintain its light-coloured nest over and over again, gets less and less inclined to 'bother' about the colour of a piece of material (but even so, there is quite a distinct, typical preference even on the last day of the experiment).

Texture

The fact that texture plays a role in nest-material selection was discovered by accident, as the result of green cotton rotting in water more slowly than other colours. In some early colour-preference tests, cottons were left in the water for a long period of time. After a number of days, it was noticed that red, yellow and blue cottons were slimy to the touch when removed from the water and that they easily broke into pieces when pulled. Green cottons, however, were still in good condition. The following table gives the colour preferences on days when all but green cottons were rotten:

Red	Yellow	Green	Blue
25	23	132	13

It should be pointed out that the fish could not pull the rotten cottons to pieces as I had done, and it was therefore probably the slimy texture of red, yellow and blue cottons that made the fish avoid these colours.

Care was taken in all other tests (including *all* those reported in the last section) to avoid the rotting of the cottons, by using new pieces every few days, but it will be realized that, in tests in which the nest is not being interfered with, if a fish has built a red, yellow, or blue cotton into its nest, it may rot there over a period of days. This led me to always use green cottons in nest-material tests where colour analysis was not necessary. It also led to some interesting observations on nest destruction. It was known that males sometimes tear away a piece of (natural) nest material and remove it from the nest. It was known also that males sometimes abandon their nests and transfer their eggs to a second nest-site. By providing nest material which was certain to rot, it was possible to study these activities in more detail (see section VII).

Length

There is little preference for any particular length when selecting a fragment of material. Very short pieces are not often used, and very long pieces of natural material are not normally available. If *Pygosteus* is given very long pieces of cotton, however, it uses them as readily as any others. It carries them to the nest entrance, pushes them in, and then loops them repeatedly into the tunnel. The only difference in the way in which they are treated is that the number of loops per fragment is greater in the case of the abnormally long pieces.

Density

Van Iersel (1953, p. 9) describes the way in which *Gasterosteus* males test material by repeatedly spitting it out and sucking it in again, and says that 'I have the impression that pieces that fall down rapidly, or rise quickly, are refused. Pieces that float in the water are usually brought to the nest.' This suggests that *Gasterosteus* selects nest material which has a density similar to that of water. *Pygosteus*, as pointed out earlier, is less prone to test material in this way, and density seems to play a less important role. This is strange, since it seems more reasonable for *Gasterosteus* to select heavier pieces for its bottom-nest, and for *Pygosteus* to select pieces with a density approximating to that of water for its suspended nest.

The nesting phase

As already described in section III, the nesting phase follows the setting up of a territory, and precedes the sexual phase, in the reproductive cycle. The nesting phase can best be considered in two stages: (1) nest-building, and (2) nest maintenance. By calculating the average frequencies of the various reproductive activities, from a number of half-hour observations, it is possible to make a quantitative statement concerning these two stages (Fig. 24). It will be seen that the (short) nest-building stage is characterized by a very high frequency of collecting and boring movements. Also, there is a considerable amount of fanning. Superficial glueing occurs, but there is little insertion

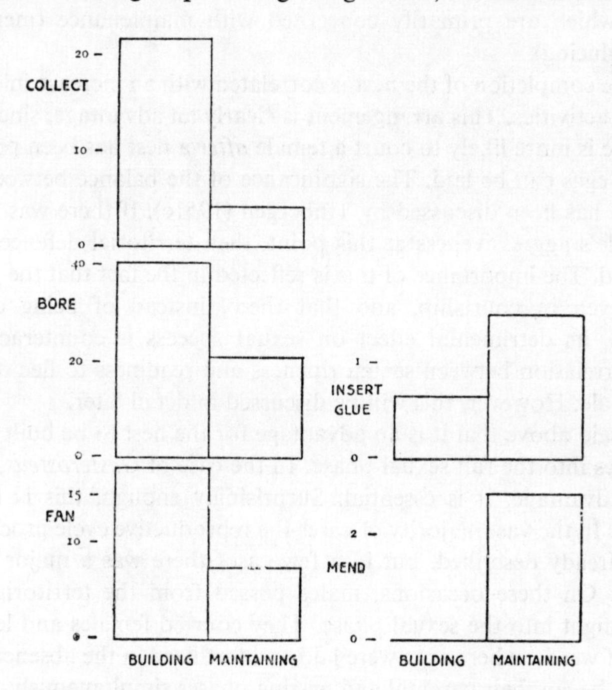

Fig. 24. Change in average frequency of nest-acts from nest-building to nest-maintaining phases. (Averages from 6 and 13 half-hour tests respectively.)

glueing. After the nest-building stage is ended, and the nest is complete, the male may maintain it in the absence of females for as long as a month, or, as is more usual, he abandons it and builds a second one, after a much shorter period. During this nest-maintenance stage, it is clear that collecting, boring and fanning are less frequent than they were in the building stage. Further, of these three movements, collecting had decreased in frequency more than boring and fanning. Insertion glueing shows an increase and superficial glueing disappears. Nest-mending movements appear for the first time. Correlated with these changes in nesting behaviour, there is a slight increase in aggression and sex. It should be mentioned that other fish – but no ripe females – were present during these tests, and that aggression refers to the number of bites, and sex the number of jumps, that were given. (That dance-jumps were given at all, to other males, reveals the presence of strong sexual motivation.)

From this comparison of the two stages of the nesting phase, it can be said that when a male passes from the one stage to the next, there is a decrease in those activities which are concerned primarily with construction (e.g. collecting), a slighter decrease in those activities which are concerned with both construction and maintenance (boring and fanning), and an increase in those activities which are primarily concerned with maintenance (mending and insertion glueing).

Also, the completion of the nest is correlated with an increase in sexual and aggressive activities. This arrangement is clearly an advantage, since it means that a male is more likely to court a female *after* a nest has been prepared, in which the eggs can be laid. The significance of the balance between sex and aggression has been discussed by Tinbergen (1951c). If there was a decrease in the male's aggressiveness at this point, then territorial defence would be endangered. The importance of this is reflected in the fact that the aggression persists, even in courtship, and that there, instead of being completely eliminated, its detrimental effect on sexual success is counteracted by an inverse correlation between sexual ripeness and readiness to flee on the part of the female. However, this will be discussed in detail later.

I have said above that it is an advantage for the nest to be built before the male passes into the full sexual phase. In the case of *Gasterosteus*, it is more than an advantage, it is essential. Surprisingly enough, this is not so for *Pygosteus*. In the vast majority of cases the reproductive cycle proceeds in the manner already described, but in a few cases there was a major sequential difference. On these occasions, males passed from the territorial, nestless phase, straight into the sexual phase. They courted females and led them to a clump of weeds, where eggs were laid and fertilized in the absence of a nest. They then began their parental and nesting phases simultaneously, building a nest around the eggs during the parental cycle. (*Gasterosteus* could never achieve this owing to the nature of its nest site.) A quantitative analysis of one such case will be given later, in the parental section, but it is worth noting here that, even under these conditions, young fish were successfully reared.

Whilst these unusual sequences are understandable from an evolutionary point of view (see section VIII), it is very difficult to explain them causally. To take the most accurately recorded case, for example: Three fish were placed in a large (105 × 45 × 45 cms) tank containing ample vegetation. Ideal nest-sites and nest material were present. The three fish slowly came into reproductive condition and set up territories. Their territorial fighting reached the usual intense level, and by this time it was apparent that two of the three fish were females and one was male. The females became swollen and ripe, but territorial fighting continued. Next, the male began to intersperse fighting with courting movements. To these the females responded by following. Finally one was led to a clump of weeds, where she laid her eggs and the male fertilized them. The female then returned to, and once again defended, her territory. The male then proceeded to build a nest around his eggs, and rear them. From this it will be seen that the male had a full opportunity to build a nest. All the external stimuli were apparently present and the cycle had not been artificially speeded up in any way. Furthermore, other males in the same tank, later on, performed their reproductive activities in the usual sequence. It is difficult to arrive at a causal explanation here. The only possibility seems to be that the presence of other territorial males is a strong stimulus in initiating nest-building. (Males kept in isolation will build nests in the usual way presumably because, although they do not have the extra stimulus of other males, they also avoid the early sexual stimulus of rival females.) Males kept with rival females only, would lack the nesting stimulus of the presence of other males, but would be subjected to sexual stimulation from the females, which might throw the usual sequence out of step. The presence of naked eggs then provides a strong stimulus for nest-building, so that the male builds a nest around them.

There is some evidence to support this hypothesis, although further large-scale experiments are still necessary. The effect of isolation on nesting activities during the *nest-maintenance* phase is shown in columns A and B of Fig. 25. It is clear from this that males which are kept alone (A) during this stage perform far fewer nesting activities than those kept with other males (B). This applies to tearing, collecting, boring, fanning, glueing and mending. It seems reasonable to assume that, if other males have this effect on nesting, during the nest-maintenance stage, they may also stimulate males to start building in the initial stages of nesting.

Before leaving this question, it is worth looking again at Fig. 25, from the point of view of the difference between tube-nests and ball-nests. It will be remembered that the latter are larger, rounder, and have thicker walls. It was found that tube-nests were built almost entirely by males kept in isolation, and ball-nests by those kept with other males. The obvious conclusion is that it is the extra building, which is performed in the maintenance stage, and which is due to the presence of the other males, that is resulting in the construction of a ball-nest.

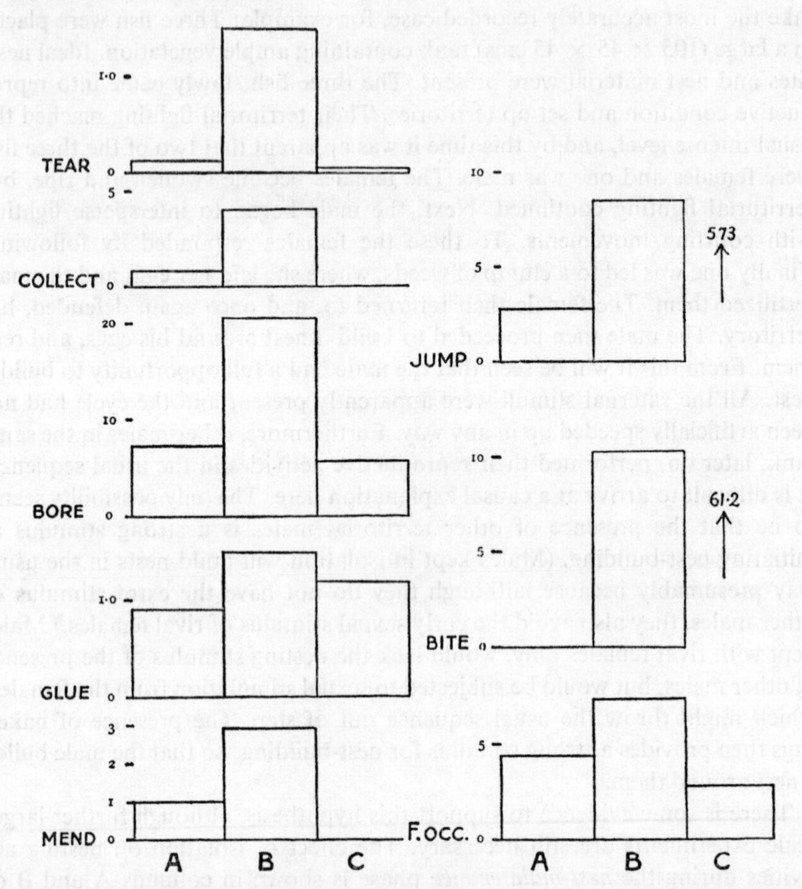

Fig. 25. Average frequencies of various actions. A: in isolated males with completed nests; B: in non-isolated males with completed nests; and C: in males presented with ripe females. (Averages from 10, 13 and 17 half-hour tests respectively.)

It should be borne in mind, however, that a ball-nest is not just a *bigger* tube-nest. This is because the boring movement has changed. In the nest-building stage, boring is intense and the male pushes deep into the nest tunnel, but in the nest-maintenance stage, despite the fact that there are more bores per collection (see Fig. 24), these bores are less penetrating. The fish only bores in to about the middle of the tunnel in most cases. This explains another difference between tube-nests and ball-nests. As stated earlier, the tunnel of a tube-nest is straighter and more inclined to the vertical than that of a ball-nest. The tunnel of the latter is bent, the front half being horizontal. This is explained perfectly by the change in the mode of boring. As stated above, there is relatively more boring in the maintenance phase. (Whereas there are twice as many bores as collections in the building stage, there are five times as many in the maintenance stage.) This accounts for the third difference between tube- and ball-nests, namely the fact that the latter are more tightly packed and compact, especially at the entrance.

VI

Sexual Behaviour

Introduction

Once a territorial male has completed its nest, its mood changes and it becomes intensely sexual. This change is characterized by the appearance of sexual (zig-zag) movements in response to sub-optimal stimulation from other males, in the absence of females. Even in the absence of all other fish, males in this condition may show sexual movements, 'courting' the slightest movement of a leaf, or shadow.

The special sexual coloration (Fig. 2 IV) appears at this time, the male losing its dorsal blackness and becoming, if possible, even more intensely black over the ventral surface. This change often does not occur until a ripe female has actually appeared and is being courted. Even then it does not always occur, and a male may remain black over its whole surface throughout the sexual phase.

As is the case in many other animals, there is in *Pygosteus* an elaborate ritual which always precedes and leads up to the consummatory sexual act. This ritual, or courtship, occurs in a rather set sequence, in which the reaction of one partner stimulates the next action of the other partner, and so on, to the final consummation. Such a reaction sequence has been analysed experimentally in the case of *Gasterosteus* (see Tinbergen, 1942 and 1951a, pp. 47–9). Tinbergen thinks that 'No doubt much the same state of affairs exists in most mating behaviour', and Nyman (1953), for example, has shown that this is so for the fish *Gobius microps*. But Hinde (1953), speaking of the courtship behaviour of the chaffinch (*Fringilla coelebs*), states that 'The diversity of the sequences of behaviour which occur show that the behaviour of the chaffinch is not a rigidly fixed chain of responses, each of which must be released by an appropriate stimulus from the partner.' Bastock and Manning (1955) have also shown that there is no set courtship sequence in the behaviour of the fruit-fly, *Drosophila*. In the present section, the reaction sequence which occurs in the courtship of *Pygosteus*, and which, although not rigid, nevertheless has a characteristic form, will be discussed at some length. The sexual ethogram has therefore been kept brief.

The courtship actions of the male

The Dance (Fig. 26a, b and c): A sexually motivated male reacts to a ripe female with a series of quick dancing movements. He assumes a head-down posture, similar to that described in the section dealing with agonistic behaviour (see Fig. 5). The white ventral spines are raised and the male, retaining this posture, makes a series of quick rhythmic jumps. Each jump consists of a forward, upward and sideways component. The strength of these three components may vary considerably, as may the angle of the body of the male in the head-down posture. The nature of these variations and the relationships between them will be discussed later, as will their orientation.

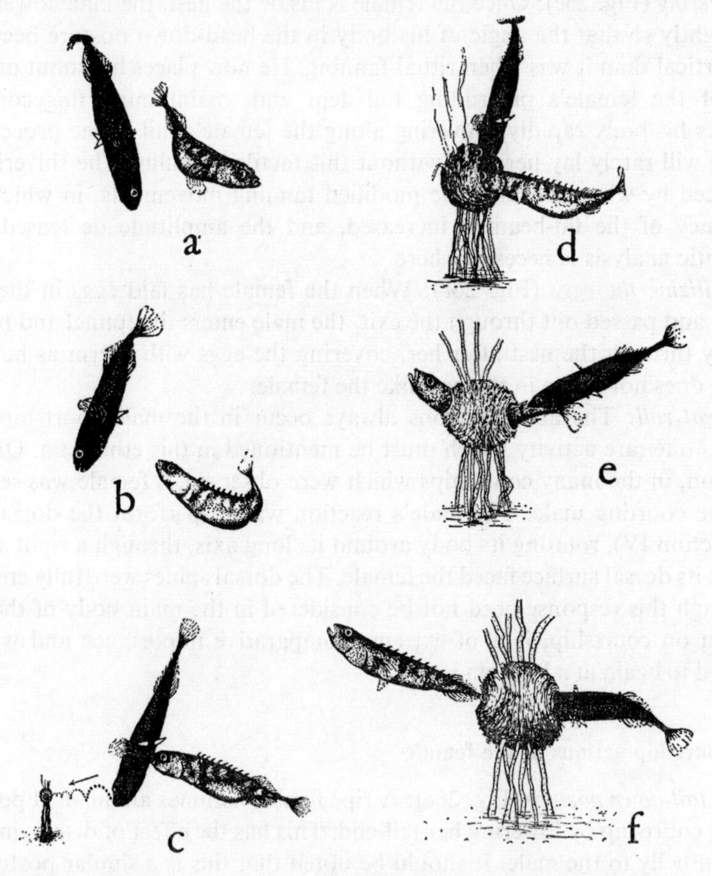

Fig. 26. The courtship pattern of *Pygosteus*. a. The male dances to the female. b. The female turns to the male. c. The male dances to the nest with the female following. d. The male shows the nest entrance to the female. e. The male shivers on the female's tail. f. The female leaves the nest and the male enters to fertilize the eggs.

Ritual-fanning (Fig. 26d): When they have arrived at the nest entrance, the male displays it to the female by ceasing to dance and commencing to fan. He does not, however, fan in the usual way. Instead of fanning in the usual horizontal posture (see Fig. 17), he retains his head-down posture. Also, his snout is not situated in front of the tunnel entrance, but just above it, thus leaving the entrance clear for the entry of the female.

This activity of the male must undoubtedly be thought of as ritualized displacement fanning, due originally to the thwarted sex drive of the male, when the female would not enter the nest. Now, however, it seems to have become a purely sexual movement and has a releasing function, stimulating the female to push into the nest tunnel.

Shivering (Fig. 26e): Once the female is inside the nest, the male lowers his tail slightly so that the angle of his body in the head-down posture becomes less vertical than it was when ritual-fanning. He now places his snout on one side of the female's protruding tail-stem and, maintaining this contact, vibrates his body rapidly, shivering along the female's tail in the process. A female will rarely lay her eggs without this tactile stimulus. The shivering is produced by what appear to be modified fanning movements, in which the frequency of the fin-beats is increased, and the amplitude decreased, but cinematic analysis is necessary here.

Fertilizing the eggs (Fig. 26f): When the female has laid eggs in the nest tunnel and passed out through the exit, the male enters the tunnel and passes quickly through the nest after her, covering the eggs with sperm as he does so. He does not pause in the nest like the female.

Dorsal roll: The above actions always occur in the male courtship, but there is one rare activity which must be mentioned in this ethogram. On one occasion, in the many courtships which were observed, a female was seen to bite the courting male. The male's reaction was to perform the dorsal roll (see section IV), rotating its body around its long axis, through a right angle, so that its dorsal surface faced the female. The dorsal spines were fully erected. Although this response need not be considered in the main body of the discussion on courtship, it is of extreme comparative importance and will be referred to again at a later stage.

The courtship actions of the female

The tail-down posture (Fig. 26c): A ripe female assumes an inclined posture during courtship by lowering her tail-end. This has the effect of displaying her swollen belly to the male. It should be noted that this is a similar posture to that assumed by a beaten, subordinate male in agonistic situations.

Following the male (Fig. 26c) A responsive female follows closely behind a dancing male. In this way they proceed together towards the nest. The female aims her snout approximately at a point between the erected white ventral spines of the male as she follows him.

Laying (Fig. 26e): The female invariably enters the nest hesitantly, but when she is finally inside the tunnel she lies with her head protruding from the exit and her tail from the entrance. The eggs are all laid together in one bunch. The actual extrusion of the eggs takes only about a second, but the female may lie in the nest for many seconds before extruding them. Once they are laid the female immediately leaves the nest and departs. Her belly is then typically hollow and crinkled. The crinkling soon disappears.

Courtship sequences

With that brief ethogram as a background, it is now possible to discuss the sequential aspects of the courtship. The reaction chain for *Gasterosteus* given by Tinbergen (1942), is given below, and it is clear that basically it is the same as that for *Pygosteus*:

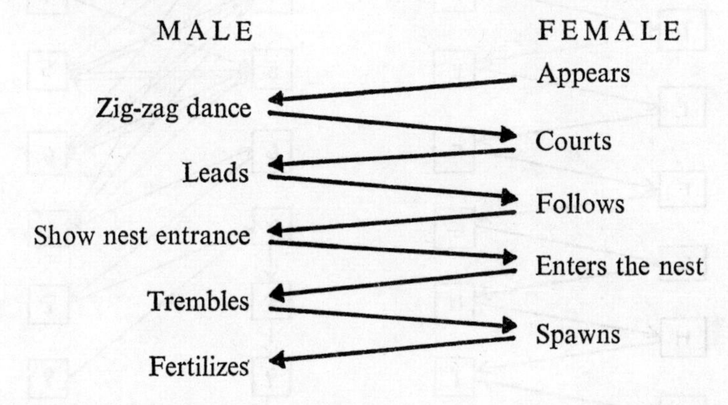

(Although it is not intended to enter into a comparative discussion here, it should be pointed out that 'zig-zag dance' is an unfortunate term to use when both species are considered together. This is because the *Pygosteus* male makes zig-zag movements both when dancing to the female, and also when leading her to the nest, whereas *Gasterosteus* only zig-zags when dancing to the female, and swims straight to the nest when leading. 'Female-oriented courtship' and 'Nest-oriented courtship' is the best way of expressing these two phases of the courtship sequence when *both* species have to be considered. As it is now so well known, however, the term 'zig-zag' must be retained in the case of *Gasterosteus*.)

It is important to emphasize that the *Gasterosteus* reaction-chain given above is an idealized version of what can actually be observed to occur. It represents a typical sequence which tends to occur, rather than a completely fixed sequence that has to occur in exactly that form. Bearing this in mind, the following *typical* sequence can be described for *Pygosteus*:

The male appears (M1). The female appears (F1) and the male dances to her (M2). The female responds by lowering her tail-end (F2) and usually also by turning towards the male (F3), to which the male responds by dancing off

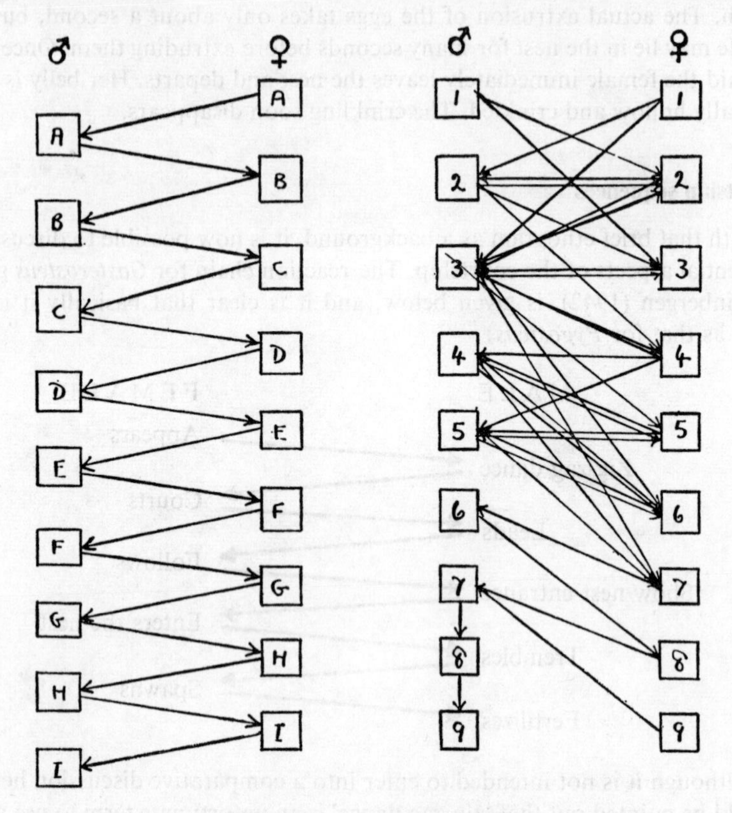

Fig. 27. The ideal (left) and the real (right) courtship sequence of *Pygosteus*. (After Morris, 1954f.)

towards the nest (M3). (If the female did not respond with F3 to M2, then she does so to M3.) The female follows the male to the nest (F4), where the male halts and points at the entrance (M4). The female responds by lowering her nose from between the male's ventral spines, to the entrance hole (F5), and she usually then bores into the entrance tunnel (F6). The male responds to this by ritual-fanning (M5). (If the female did not respond to M4 with F6, then she does so to M5.) The ritual-fanning of the male stimulates the female to enter the nest (F7) and the male responds to this by shivering on her tail (M6). This stimulates the female to lay (F8) and then leave the nest (F9). The male responds by entering the nest (M7), fertilizing the eggs (M8), and then passing out of the exit himself (M9). This completes the sequence.

The above is a typical complete sequence. Most sequences break down before they have gone a few stages, and then start again. Its typicality is based on composite evidence from a large number of sequences, and not on a 'general impression'.

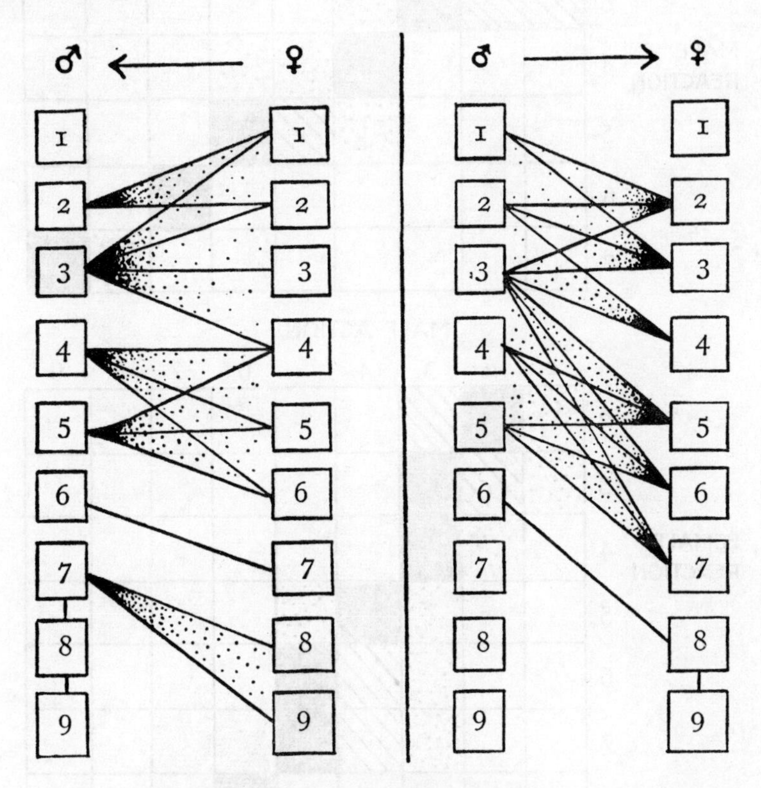

Fig. 28. Revised courtship sequence diagrams. For explanation see text. (After Morris, 1954f.)

In Fig. 27, the diagram on the left shows an imaginary 'classical' picture of the above nine-stage sequence. The diagram on the right shows what can really be observed. This picture, which is too confused to be easily studied, obviously has to be presented in another way. This has been done in Fig. 28, which shows a revised reaction-sequence presentation. On the left, here, are shown the female actions which produce male reactions, and on the right are shown the male actions which produce female reactions. It will be noticed that most reactions can be released by several different actions of the opposite sex. It will also be seen that the several actions which produce any particular

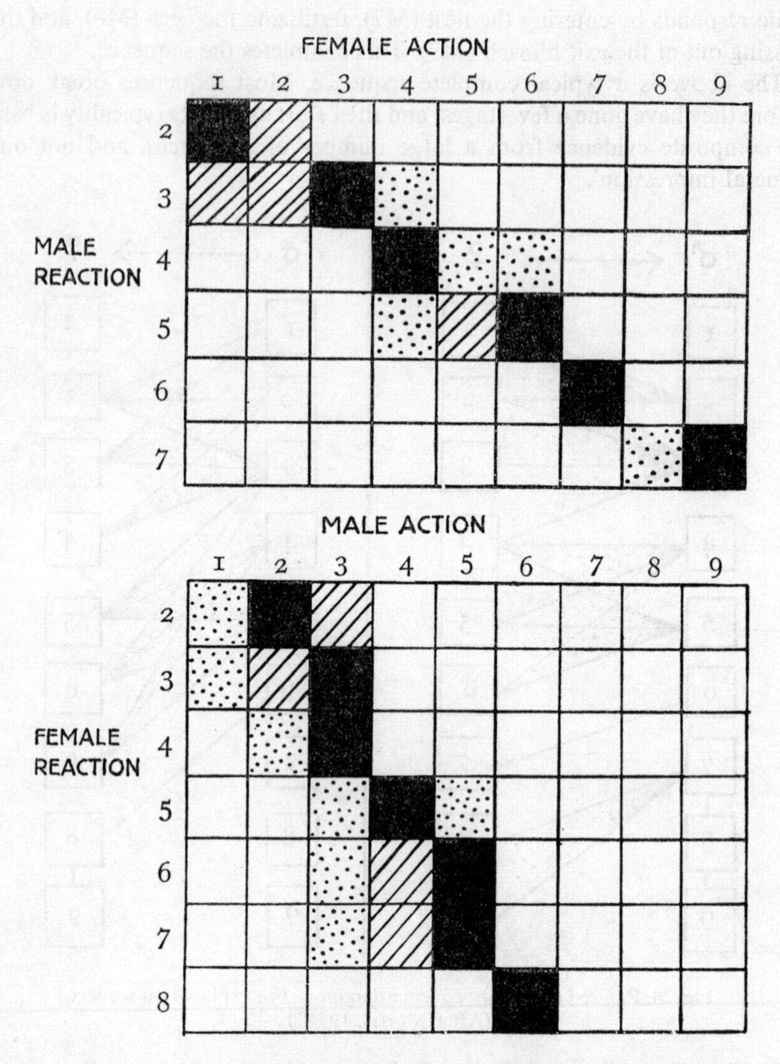

Fig. 29. Further-revised courtship sequence diagrams. For explanation see text. (After Morris, 1954f.)

reaction are always grouped together in one section of the total action sequence. There is therefore a stimulus-response overlap at most stages of the sequence. Causally, this is the result of the fact that the male and female are not usually perfectly synchronized sexually, one invariably being more highly motivated than the other. Functionally this overlap is interesting in two ways. Firstly, it shows that, even in a reaction-sequence as supposedly rigid as that of a stickleback, there is considerable plasticity. Secondly, this same lack of rigidity tends to work *against* reproductive isolation. If each stimulus-response unit of the sequence was completely fixed, the sequence would be a more efficient isolating mechanism. As it is, certain stages may be 'overlooked' by a very highly motivated individual, but the effect of this is not serious for the following reasons: None of the 'overlaps' are more than a few stages 'wide', and the total sequence consists of nine stages, so that the by-passing of stages is never likely to be detrimental. It is not necessary to infer from this, however, that the complexity of the courtship sequence has actually *evolved* in the service of maintaining reproductive isolation, but rather that once it evolved (for other reasons discussed below) it was far from being a cumbersome disadvantage, as it might otherwise have been, owing to its *assistance* in the problem of isolation.

Of the several actions which may produce a courtship reaction, one is always seen to do so more frequently than others. This is not shown by Fig. 28, and a still further revision of the method of presentation is necessary. This is given in Fig. 29, in which the same data as in the last figure (28) are shown, but here as a system of graph squares. Fig. 29 is self-explanatory except to point out that the black squares refer to the actions which *typically* produce a particular response, the shaded squares, actions which *occasionally* do so, and the stippled squares, those which do so more *rarely*. The addition of this extra information clarifies the situation considerably, and this method of presentation seems to be the most suitable for illustrating this type of study. It was from this figure that the 'typical complete sequence' described earlier was constructed.

The organization of the courtship sequence

The simplest possible sequential organization would be for each action rigidly to release only one response, and for this response then to operate as the next action in the chain. In this way, no special control system would be necessary. As is already evident, it is not as simple as this. A number of straightforward sequential observations can, however, be made, on the basis of which a reasonably simple hypothesis may be formed concerning the relationships between the successive stages of a courtship sequence. I shall attempt to put forward this hypothesis here, but it must be borne in mind that the problem is an exceedingly complex one and that great over-simplications are involved. However, even if the present hypothesis has ultimately to be modified drastically, or abandoned, as a result of future research, it is felt that, nevertheless, at the present stage, it is stimulating and helpful in clarifying some of the points concerning sequential organization.

The intersecting-gradients hypothesis

It is one of the fundamental observations of ethology that animals respond differently to the same stimulus at different times (see, in particular, Hinde, 1954); also, that they may respond in the same way to different stimuli at different times. It is therefore postulated that the nature of a response is dependent on a combination of variable internal and external factors. These combine in such a way that, if one is weak, the other has to be strong in order to produce a response. Where a stimulus-response sequence is involved, the combination of a particular stimulus and the internal state of the potential reactor has been seen, from the foregoing, to produce qualitatively different responses, at different times; also that the responses to a particular stimulus are grouped together in one section of the total response sequence. Similarly, different stimuli, which can produce the same response, are also grouped together in the total stimulus sequence.

From independent criteria concerning the initial (sexual) internal state of the fish (for example, the extent to which the belly of the female is swollen with eggs) it is possible to state that the stronger the internal factors, then the *later in the response sequence* will be the response to a particular stimulus. To give an example, a very ripe female will eagerly follow a dancing male, but a less ripe female may only turn towards such a male. Also the *stronger* the internal factors, the *earlier in the stimulus sequence* will be the stimulus which produces a particular response. For example, a very ripe female will respond sexually simply to the presence of a male, and turn towards him, but a less ripe female will not respond in this way unless the male dances to her, or until he starts to lead to the nest.

It would appear, therefore, that the later in the courtship sequence a response, the higher is its threshold value, and the later a stimulus, the more powerful is its releasing effect. Thus there appears to be both a response-threshold gradient and a stimulus-power gradient involved in the courtship sequence.

Supposing, for the moment, that the internal state does not vary *during* the performance of a courtship sequence. If the gradient of the increase in stimulus power and that of the increase in response-threshold level, through the sequence, were the same, then when the internal state was above a particular level, the whole sequence could go off from start to finish without a breakdown at any stage. If the internal state was below this level, then none of the stages of the courtship sequence would occur. This is not, however, what is always observed to take place. Usually a courtship sequence breaks down at some stage or other before the consummatory act of fertilization is reached. It would seem, therefore, that the response-threshold gradient is steeper than the stimulus-power gradient. This is shown diagrammatically in Fig. 30.

Four imaginary stages of the sequence are shown. S-S represents the stimulus-power gradient and T-T the response-threshold gradient. I-I shows the level of the internal state, which, for the time being, is still being taken as a constant during any one courtship. If the diagram is for a female, say, then her internal state will summate with stimulus 1 or 2 from the male to pass the response-threshold line and response 1 or 2 may occur. It does not summate with stimulus 3 or 4, however, to produce responses 3 or 4. How then, in such a situation, could the courtship develop further? Repeated presentation of

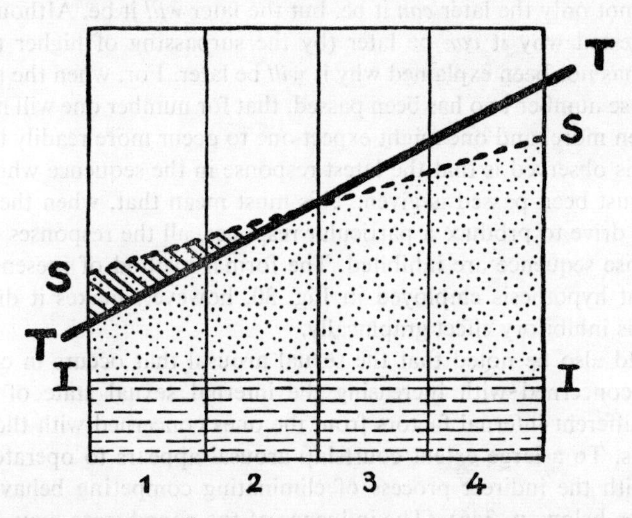

Fig. 30. The Intersecting-Gradients Hypothesis. Full explanation in text.

stimuli 1 and 2 by the male would meet with repeated success, but 3 and 4 would never do so. Clearly, the courtship of the male must *arouse* the female sexually, that is, it must increase the level of her internal sexual state, if it is to proceed any further. (That this arousal does in fact occur, will be demonstrated quantitatively in a later section.) In this way, as the courtship proceeds, the internal state of the female will be raised little by little and will summate successfully with the (sequentially) later and later stimuli in the sequence until the consummatory act is reached.

The over-simplification of Fig. 30 must be stressed here. Breakdown in the courtship sequence does not occur with equal frequency at each stage. The courtship of the male is most likely to break down at the leading stage. A male, who is leading an eagerly following female, frequently will not then show her the nest entrance. Once he has done so, however, he is almost always prepared to shiver on her tail if she enters. A female, who begins to follow a dancing male, frequently ceases to do so before she gets to the nest. Also, a female who has bored into the nest entrance often then swims off without entering. Once a female has entered the nest, however, she is much more likely to proceed to the next stage (and lay her eggs), although even at this point, she may occasionally pass through the nest without doing so. From all this it can be concluded that the gradients should probably not be represented by the simple straight lines shown in Fig. 30. Such a simplification is nevertheless justified at the present stage.

Another over-simplification in Fig. 30 is as follows. It has been said that the stronger the sex drive, the later in the sequence will be the response to a stimulus; not only the later *can* it be, but the later *will* it be. Although it has been suggested why it *can* be later (by the surpassing of higher threshold levels), it has not been explained why it *will* be later. For, when the threshold for response number two has been passed, that for number one will have been passed even more, and one might expect one to occur more readily than two. But what is observed is that the latest response in the sequence whose threshold has just been passed is given. This must mean that, when there is sufficient sex drive to produce a particular response, all the responses earlier in the response sequence are inhibited. The formal method of presentation of the present hypothesis employed in Fig. 30, however, makes it difficult to include this inhibitory effect graphically.

It should also be noted that the sexual arousal that occurs in courtship, although concerned with increasing the internal sexual state of the fish, involves different internal factors from the ones concerned with the state of the gonads. To a large extent courtship arousal appears to operate in connection with the indirect process of eliminating competing behaviour tendencies (see below, p. 346). (The influence of the gonad-state may, however, be so intense that no arousal is necessary. Indeed, females kept in isolation will even spawn in vacuo.)

It is essential to emphasize here that the scheme given above for the courtship sequence of *Pygosteus* cannot necessarily be applied as a general rule. Recent theories dealing with behaviour sequences (Tinbergen, 1950; Deutsch, 1953; Hayes, Russell, Hayes and Kohsen, 1953) have attempted to explain all behaviour sequences on the basis of one general system. The ethological evidence which is now available indicates that, in all probability, no such general scheme is acceptable. Some sequences appear to be quite rigid sets of stimulus-response reactions, which do not permit overlaps or the taking of short cuts; others are very plastic. The functional differences between different sequences may be very important in determining the way in which they are organized, and Hayes *et al.* (1953) have pointed out that the details of their hypothesis will undoubtedly 'be found to vary considerably not only between species but between systems in one individual'. Deutsch (1953) makes no such concession, although his theory would not have predicted the observations reported in the present section.

Finally, the phenomenon of 'short cuts', which were mentioned above, must be briefly discussed. A short cut is, so to speak, the opposite of a breakdown. In the latter, stage A occurs but stage B does not then occur. In the short cut, stage B occurs without stage A having occurred before it. How can this be explained on the intersecting-gradients hypothesis? One type of short cut is easy to explain. Supposing a male is given unfertilized eggs in its nest and it then fertilizes them, thus short-cutting the whole courtship sequence. This may simply be the result of the fact that the male's internal sexual state was so strong that it would have been prepared to go right through the whole sequence if necessary. This is straightforward enough, but supposing a male was *not* prepared to court a female, but was nevertheless prepared to fertilize eggs if the latter were placed in the nest. This is more difficult to explain on the present hypothesis. This type of observation (where stage B was performed when stage A, although available, was not performed) was *very rare* in the case of *Pygosteus* courtship, but the following specific example may be given:

A male with a nest had been repeatedly leading a female to his nest and showing her the nest entrance. Then, on several occasions, he led her towards the nest, but stopped leading before getting there, whereupon she swam on ahead and began boring into the entrance. Seeing her do this, the male then rushed to the nest and began showing her the entrance. Without special qualification, the present hypothesis would not have predicted this, because, if the male was not prepared to lead to the nest when the female followed, he should have been even less prepared to show her the entrance. There are, however, several possible explanations of this observation, which do not require any basic alteration in the hypothesis.

Firstly, there are often short pauses in the courtship behaviour over a long period of time, which indicate that very short-term minor fluctuations in sex drive occur. Such fluctuations could easily explain the above and similar such cases. The male had already shown the entrance to the female a number of times and may have just been pausing as a result of one of these momentary fluctuations in sex drive, the latter having passed by the time the female bored into the nest. Secondly, each response in the sequence may be specifically exhaustible. The male may have led the female over and over again until the leading response was partially extinguished, although the sex drive was still high. Thirdly, the boring into the nest by the female may not only have a releasing but also a strong arousal effect on the male, thus increasing his internal state sufficiently to pass the threshold level for showing the nest entrance. However, it would involve too much detail to discuss these aspects any further here, considering the scanty observational basis which exists for them at present. Much stronger observational data has been collected for stimulus-response overlap and courtship breakdown, and the present hypothesis has been formed especially to deal with these latter problems.

The arousal of the sex drive during courtship

I have already mentioned that courtship has an arousal effect, and the evidence for this must now be presented. A quantitative scoring of courtship periods is necessary as a basis for discussing this, and the method used for such scoring is as follows:

Apart from a large number of courtships which were observed and recorded qualitatively, many were scored quantitatively, using a simple shorthand code. In a standard test of the latter kind, a male with a completed nest was presented with a ripe female. The latter was placed in the tank of the male at the beginning of the test and was removed at the end of the test, which typically lasted thirty minutes. During the thirty minutes, all actions performed by the male, whether sexual or not, were noted. (The male was used here in preference to the female simply because his courtship actions lend themselves to quantitative analyses better than hers.) The record was divided up into five-minute periods for the purpose of temporal analysis. A number of variations of the above standard technique were employed from time to time, some of which will be discussed below. For example, the behaviour of the male, in the period immediately before the female was introduced, was often recorded quantitatively, in order to demonstrate the way in which the male's behaviour changed as a result of the introduction of the female. Also, in a number of tests, the thirty-minute period was extended over several hours, in order to study the effects of prolonged courting.

Two basic types of information are obtainable from the data procured by the above method: response sequences and response frequencies. These data can be analysed in many different ways, some of which are illustrated below, in order to demonstrate the arousal effect of courtship. Before presenting such analyses, however, it is necessary to point out that, theoretically, there are two forms of arousal. Arousal may be the result of an increase in sexual motivation independent of any change in any other motivation. Such arousal may be termed direct, or absolute, arousal. It may also be the result of an increase in sexual motivation which is dependent on a decrease in some competing motivation, or motivations. Arousal of this kind may be termed indirect, or relative, arousal. A number of ethological studies recently have shown that courtship does not usually involve only a single, namely sexual, tendency, but also involves the tendencies to flee from and to attack the mate. (See Tinbergen, 1952a, 1953a and b; Hinde, 1953; Morris, 1954b, c and d, 1955b, 1956.) This is certainly the case in *Pygosteus*. The male not only courts the female, but also repeatedly attacks her. His courtship primarily involves both sexual and aggressive tendencies. (For a fuller discussion of this, see next section.) Owing to this fact, it is difficult to demonstrate absolute arousal in this instance, and the following illustrations are of relative arousal.

1. *Arousal as shown by response sequences*

Using the data from all the typical tests of the kind described above (a total observation time of 35 hours 40 minutes), Fig. 31 has been constructed. It shows the number of times that each of the sex actions of the male was followed by any other action, either sexual or non-sexual. It will be noted that the sex actions of the male have been simplified down to four basic stages: dancing, showing, shivering and fertilizing. A dance consists of an uninterrupted series of jumps by the male. Since dancing to the female and

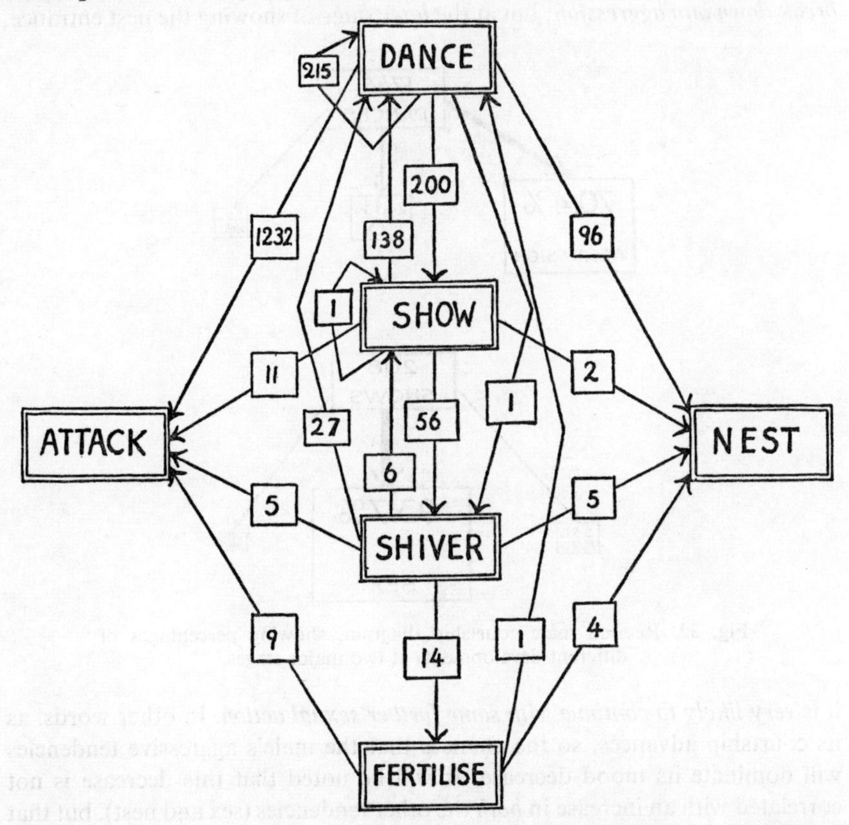

Fig. 31. Male courtship diagram, showing the number of times each action was followed by another action. (Data from 35 hours 40 mins total observation time, on ten males.)

dancing to the nest often merge smoothly into one another in a single series of jumps, they have been taken together for the present purpose. Also, all non-sexual activities at the nest have here been lumped under the one heading of nest actions. These include collecting, boring, mending, glueing and fanning. Aggression is always measured by the number of bites made by the male. Although there are other methods of scoring aggression, such as number of chases, etc., it has been found, particularly in the study of courtship behaviour, that the frequency of biting is by itself a highly efficient measure.

The figures for dancing and showing are sufficiently large to justify the construction of Fig. 32, which gives the percentage of dances and shows which were followed by each of the three basic kinds of action observed during the courtship periods. If dances were followed by any kind of sexual action, they are taken together. From this figure it is possible to postulate that, at the *early courtship stage* of dancing, the courtship of the male is *very likely to break down into aggression*; but at the *later stage* of showing the nest entrance,

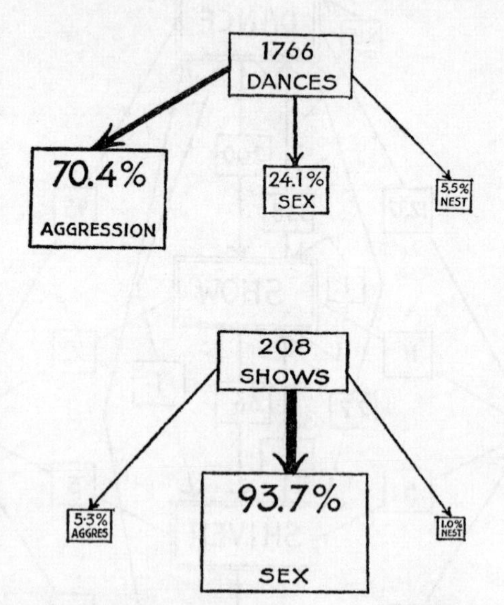

Fig. 32. Revised male courtship diagram, showing percentages of different developments at two major stages.

it is *very likely to continue with some further sexual action.* In other words, as its courtship advances, so the chances that the male's aggressive tendencies will dominate its mood decrease. It will be noted that this decrease is not correlated with an increase in *both* the other tendencies (sex and nest), but that only sexual tendencies show such a change, thus indicating that there is a distinct competition specifically between sex and aggression, the balance between the two undergoing a shift in favour of the former as courtship progresses.

The data for shivering given in Fig. 31, when analysed in the manner of Fig. 32, reveal that there is no further shift in the sex-aggression balance in favour of sex at this stage. This is unexpected, and it may only be the result of the small figures available for shivering. However, it will be remembered that earlier I stated that a male which dances to a very willing female may not be prepared to show her the nest entrance, whereas a male that shows the entrance is almost always prepared to shiver on her tail if the female enters. It may be that the greater chance of breakdown between the dancing and showing stages of courtship is significantly correlated with the greater arousal that may occur at that point. Both these changes might be expected if the response-threshold gradient was steeper than the stimulus-power gradient to a greater degree there.

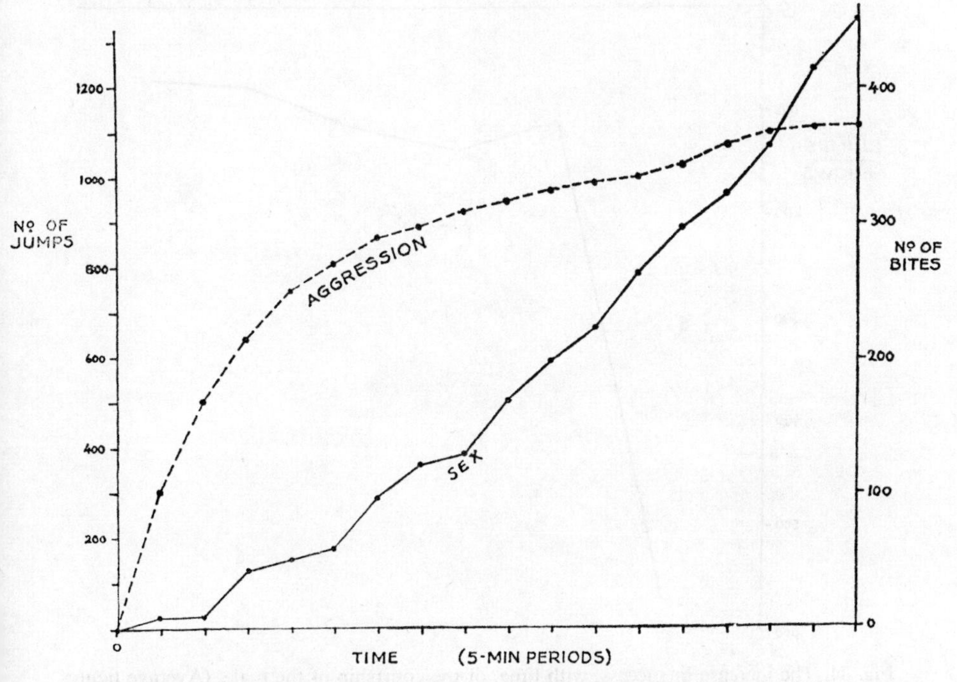

Fig. 33. Inverse relationship between sex and aggression in the male courtship.

2. *Arousal as shown by response frequencies*

By scoring the number of times a male makes a sexual action and the number of times it makes an aggressive action, in each five-minute period of a courtship test, it is possible to study the relationship between changes in the sexual and aggressive tendencies in time. Fig. 33 shows a particularly clear example of this relationship, the male in question being extremely aggressive at the beginning of the test. (Ordinate values show response-accumulation figures.) As time passed, the aggressiveness fell off and the sexual tendency increased, thus showing, in an entirely different manner, the same inverse correlation as was found above.

Another method is to relate the number of dance-jumps performed per five-minute period to the number of showing ceremonies for that period. The number of jumps per show gives an indication of the ease with which a unit of early courtship leads to a unit of later courtship. Fig. 34 illustrates this and shows that, as time passes, so less and less dancing is required to get to the stage of showing. In the first five-minute period, for example, nearly 600

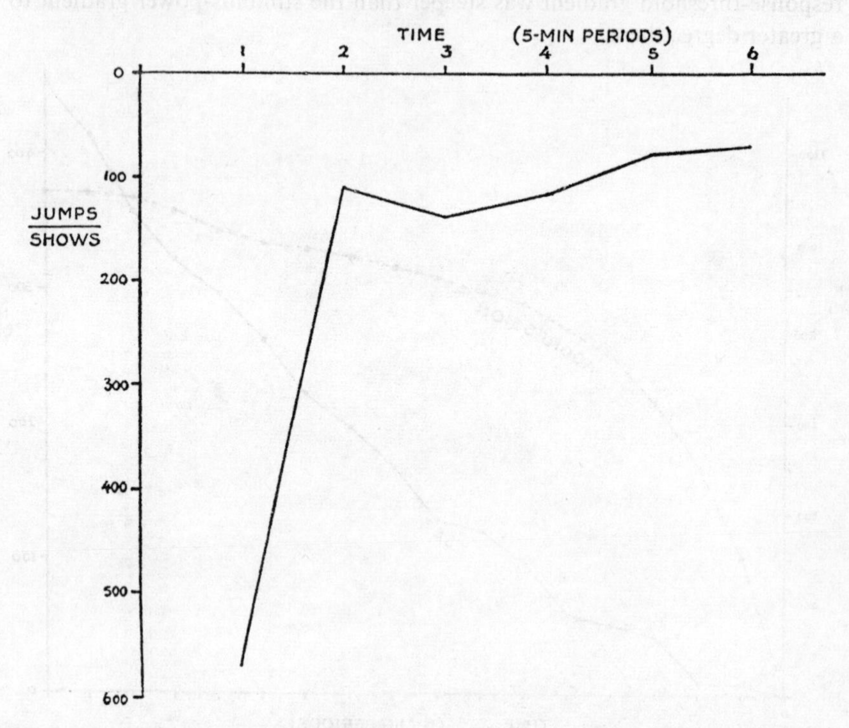

Fig. 34. The increase in success, with time, of the courtship of the male. (Average figures from 8 half-hour tests, using 4 males.)

jumps are necessary, on the average, to get to one showing ceremony, but in the sixth five-minute period, less than 100 are required. The data given represent averages from 8 half-hour tests. The value for the second five-minute period, it will be noticed, is lower than might be expected, and further data was obtained to investigate this. The averages from 40 half-hour tests, however, showed almost exactly the same graph, and, for the present, the high second value must remain unexplained. (This graph is presented in an inverted form owing to the fact that a decrease in the ordinate value represents an increase in arousal.)

The consummatory act

Thus far, the courtship behaviour leading up to the consummatory act has been discussed and it now remains to show what happens when fertilization is achieved. In order to investigate this quantitatively, a male with a completed nest was presented with several ripe females simultaneously. (Seven females were used in most tests of this kind.) This meant that, after the male had courted and fertilized one female, sexual stimuli were still present, in the form of the other ripe females. It was then possible to observe how long it took the

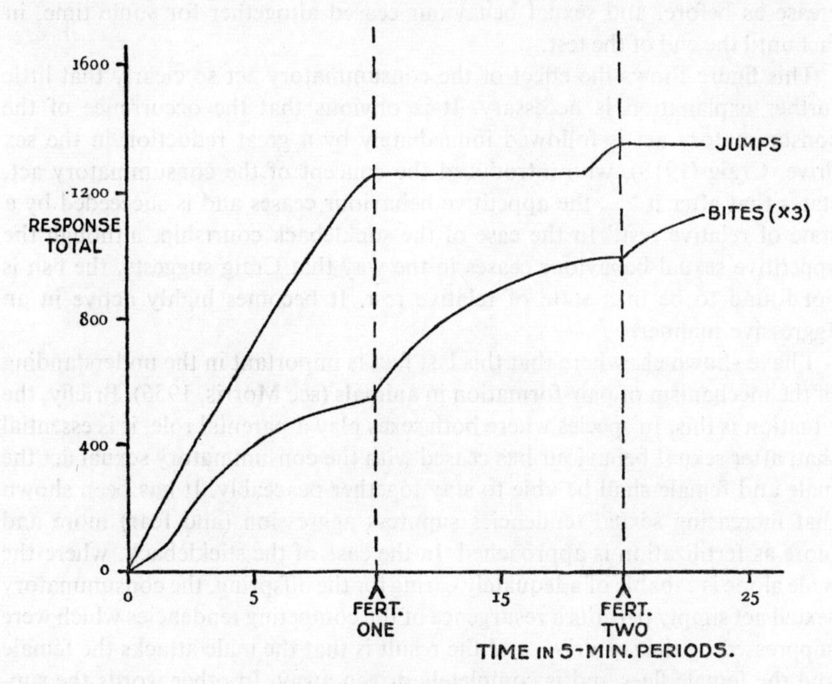

Fig. 35. A schematic representation of the effect of the consummatory sexual act on sexual and aggressive behaviour.

male, after the consummatory act, to become sexually responsive again. Fig. 35 shows a schematic version of the response-accumulation values for sex (number of jumps) and aggression (number of bites) that were obtained in one such test. Fert. one and two are the points at which there was a fertilization. The first fertilization occurred after 50 minutes of courtship, during which time aggressiveness fell off more and more. (The slight fall-off in sex immediately before the fertilization is merely the result of this method of presentation, and is due to the fact that at this point the later sexual responses of showing and shivering were taking up so much of the male's time.) After the fertilization, there is a sudden increase in aggression again, but no sexual behaviour shows itself for some time. Even before courtship begins again, however, aggression starts to fall off once more, thus suggesting that sexual motivation is building up already, before it passes the threshold that permits overt expression.

When the male began courting again, a female, which had apparently been strongly aroused by witnessing the previous courtship, responded immediately and eagerly to the new burst of dancing, and a second fertilization took place fifty minutes after the first one. After this, aggression showed a sudden increase as before, and sexual behaviour ceased altogether for some time, in fact until the end of the test.

This figure shows the effect of the consummatory act so clearly that little further explanation is necessary. It is obvious that the occurrence of the consummatory act is followed immediately by a great reduction in the sex drive. Craig (1918), who introduced the concept of the consummatory act, states that after it ' ... the appetitive behaviour ceases and is succeeded by a state of relative rest.' In the case of the stickleback courtship, although the appetitive sexual behaviour ceases in the way that Craig suggests, the fish is not found to be in a state of relative rest. It becomes highly active in an aggressive manner.

I have shown elsewhere that this last fact is important in the understanding of the mechanism of pair-formation in animals (see Morris, 1956). Briefly, the situation is this. In species where both sexes play a parental role, it is essential that, after sexual behaviour has ceased with the consummatory sexual act the male and female shall be able to stay together peaceably. It has been shown that increasing sexual tendencies suppress aggression (and fear) more and more as fertilization is approached. In the case of the stickleback, where the male alone is capable of adequately caring for the offspring, the consummatory sexual act simply permits a resurgence of the competing tendencies which were suppressed by the sex drive, and the result is that the male attacks the female and the female flees and is completely driven away. In other words the suppression in such a case is *reversible*. If the male and the female are to stay together after the consummatory act, then it is obvious that this suppression must be made *irreversible*, and, further, that this irreversibility must apply only to a particular member of the opposite sex. The basis of the pair-formation mechanism must therefore consist of a combination of the irreversibility of the suppression, on the one hand, and individual recognition, on the other. Once this has happened, the courtship can become split up in time if necessary, and in pair-forming species it usually occurs that there are two distinct sexual behaviour complexes, consisting of pair-forming ceremonies and, later on, pre-copulatory ceremonies. The early (dancing) stages of the stickleback courtship can be thought of as comparable with the pair-formation ceremonies of species which form pairs, and later stages of stickleback courtship (at the nest) can be thought of as comparable with precopulatory ceremonies. Owing to the reversibility of the suppression that occurs in the stickleback, the two stages of its courtship must be continuous, or any headway that has been made will be lost. (For a further discussion of this and other general courtship problems, the reader is referred to Morris, 1956.)

Before leaving the question of the consummatory sexual act, a few points must be made concerning the change-over from the sexual to the parental condition. The nature of this change-over has been beautifully analysed for *Gasterosteus* by van Iersel (1953). Both *Gasterosteus* and *Pygosteus* can rear more than one, but not more than a few, clutches of eggs. Clearly, if a fertilization resulted in a complete and permanent elimination of the sex drive, then only one clutch could be obtained by each male. On the other hand, if the sex drive were repeatedly able to build up again after each fertilization, a male could procure an unlimited number of clutches. If the latter occurred, the safety of all the clutches would be endangered. Not only would the nest be inadequate to hold more than a few clutches, but also the male would be incapable of aerating a larger number efficiently. In *Pygosteus*, especially, the acquisition of too many clutches may result in the disintegration of the nest and the subsequent devouring of the exposed eggs. The vigorous attempts of males to ram extra clutches (given to them by myself) out of sight into an already packed nest tunnel was several times seen to result in the breaking up of the nest, the tearing loose of it from its host plant, or the pushing of other clutches out of the nest-exit.

The way in which *Pygosteus* compromises is as follows: More than one clutch is obtained as a result of the temporary nature of the exhaustion of the sex drive after fertilization. Although, however, there is a resurgence of sexual behaviour, this is less and less marked with each successive fertilization, until, after several clutches have been obtained, the sex drive is so low – even when it has built up again after fertilization – that no further females can be induced to spawn. After a few days of parental behaviour the sex drive disappears altogether (see next section).

The derivation of the courtship behaviour

Thus far, the postures and movements of courtship themselves have been taken for granted, as behaviour units. The frequency, intensity and sequential variations of these units have been discussed, but the origin of the form of each of the units still remains to be explained. This has been dealt with more fully elsewhere (Morris, 1956), and a causo-derivational theory about types of courtship in general has been formulated. It is suggested that there are three fundamental relevant tendencies, namely *Fleeing*, *Attacking* and *Mating* (FAM, for convenience), involved in animal courtship. In different species, the relative strengths of these three tendencies will vary and such variations give rise to the different courtship types.

The male *Pygosteus*, as we have seen, repeatedly attacks the female during courtship, but never flees from her. This might, at first sight, be thought to prove that its courtship involves only the two tendencies, to attack and to mate (-AM). But if the nature of the courtship postures of the male *Pygosteus* are studied, it soon becomes clear that they are similar to certain agonistic postures which have been shown to be the result of a conflict between fleeing and attacking. The courting male stickleback, in other words, is not simply attacking its female, but also *threatening* her. This means that all three tendencies of fleeing, attacking and mating must be present in the male, but that the first of these is so weak that it does not find full expression (this can be shown as follows: fAM).

When the male dances to the female, it assumes a head-down posture and raises its ventral spines. This is a purely agonistic threat posture and has been described in section IV, but the male then proceeds to make forward jumps in this posture. Such jumps are not seen in agonistic situations. Their orientation is particularly interesting. They are at first directed towards or around the female, but when she is responding and following, they are directed away from her and towards the nest. It can be seen that the male is more likely to attack the female after making jumps towards or around her, than after making jumps away from her, and it seems clear that the direction of the jumping is determined by the nature of the balance between sex and aggression at any particular time.

Apart from the general direction of the jumping, there is also a variation in the particular direction of each separate jump. Thus, a male that is dancing away from a female may make jump one away and slightly to the right, jump two away and slightly to the left, jump three away and slightly to the right again, and so on, alternating from left to right in a zig-zag course. Although such a course may zig-zag from jump to jump, the jumping as a whole forms a straight line away from the female towards the nest. When a female follows a male very close, with her nose practically touching his ventral surface, the lateral component of each jump is reduced, or even eliminated, so that the male progresses in straight jumps forward. The straighter jumping here appears to be the result, as before, of a shift in the balance between sex and aggression in favour of the former. The close proximity of the female appears to sexually stimulate the male more, and with the subsequent suppression of his aggression the lateral components of the leading jumps drop out. This is borne out by the fact that a straight-jumping male is less likely to attack than a crooked-jumping one. From a derivational point of view, it seems that the lateral component of the jump represents an aggressive element of turning towards the female to attack her.

Another variation in the dancing postures is the angle of the head-down. In general, it seems that the more shallow the head-down in a male that is dancing towards a female, then the more likely the male is to attack the female. Comparing this with *Gasterosteus*, it seems that the latter therefore has a more aggressive courtship dance, since it is always horizontal when dancing.

When the male arrives with the female at the nest, it shows her the entrance by fanning whilst maintaining its head-down posture. The fanning may be thought of as displacement fanning resulting from sexual thwarting due to the fact that the female will not yet go into the nest. In its combination with the head-down posture it appears to have become ritualized as a sexual signal. In this connection, it is particularly interesting to note that the angle of the head-down posture in the showing ceremony is determined, not by any FAM balance, but now simply by the position and angle of the nest entrance and tunnel. Fig. 36 shows the way in which it is possible to alter the angle of the head-down posture in the showing ceremony, simply by altering the angle of the nest tunnel. When the nest is tilted back, the male stands more vertically in the water.

The shivering of the male on the tail of the female may also be thought of as displacement fanning, which has been modified here as a tactile signal. This has been done by increasing the frequency and decreasing the amplitude of the fin beats. Here, as with showing, the female has strongly stimulated the male sexually (by going into the nest), but thwarted him (by not yet laying).

The actions of the courting female need little derivational explanation, except for the tail-down posture. This, as has already been pointed out, is identical with the submissive agonistic posture of this species, and the raising of the dorsal spines, which occurs in submissive fish, also occurs in responsive females, thus indicating that a strong sex drive suppresses the attacking tendency more than the fleeing tendency in the female (while the reverse is true for the male). An aggressive, territorial, ripe female, who has recently been fighting a neighbouring male, will suddenly switch into the tail-down following posture when the same male begins to court her. She follows him on to his territory, where, if then attacked, she will not retaliate, but either stand her ground, or flee. The sudden assumption of the tail-down posture by such a female, without the latter previously being beaten into submission, suggests that the activation of the sex drive automatically has a differential effect on the two tendencies to attack and to flee. But it could also be postulated that the sexual stimuli attract the female on to the territory of the male where, even agonistically, she would be submissive. My observations indicate that the former possibility is more likely to be correct, since ripe females kept in the absence of males may, when very swollen with eggs, adopt the tail-down posture without ever 'trespassing'. But, very probably, the second possibility mentioned above gives the clue to the basic *origin* of the tail-down posture, and explains why the female of this species has a 'timid' courtship pattern.

When the forms of the various courtship displays of *Pygosteus* are compared with those of *Gasterosteus*, some puzzling contradictions arise. Firstly, as has already been mentioned, the dance of *Gasterosteus* may be thought of as a rather aggressive version of that of *Pygosteus*. Indeed, one extraordinarily aggressive *Pygosteus* individual was seen to dance in the typical *Gasterosteus* manner. But whereas *Pygosteus* still maintains its head-down posture during the ritual-fanning display when showing the nest entrance, *Gasterosteus* performs the *dorsal roll*. Now, we have seen that the dorsal roll only very rarely occurred in *Pygosteus* and that it was then the result of a female attacking a courting male. It appeared to be a submissive display there, being the opposite of the ventral roll and also concealing the sexually marked ventral surface. Both species have the ventral roll and both species are sexually marked ventrally. Further, the dorsal roll is accompanied by the raising of the dorsal spines in both species. All this suggests that, although *Gasterosteus* appears to have a more aggressive dance, it has a less aggressive showing ceremony, derivationally speaking. (It must be emphasized that the discussion here is concerned with derivations and origins, and that, in evolution, ritualization may have resulted in the partial or total emancipation of the activities concerned: see Tinbergen, 1952a.) Further research is needed here on the courtship of *Gasterosteus*, in order to make a more detailed comparison possible. At the present stage, the relationship between the two courtships appears to be complex; for example, why should a male *Pygosteus* that is not ready to show the nest entrance to a very responsive female, lead her around and around the vegetation (always avoiding the nest-site), when a male *Gasterosteus*, under similar conditions, 'pricks' the female? This pricking ceremony, which appears to be little more than a dorsal roll, has not been emphasized by *Gasterosteus* workers. It appears that, when a male *Gasterosteus* is not ready to go to the nest and a female comes up close to him and is waiting to be taken, he responds with a *submissive* display, but, as stated above, this subject needs further investigation.

Other activities during the sexual phase

A male which has completed its nest and is ready to court a female, still performs occasional nesting actions. From time to time, it may add another fragment to the nest, insert an additional blob of glue into the tunnel, or make nest-mending movements. It may also bore into the tunnel and fan at the entrance occasionally.

Once a ripe female is present, most of these actions cease to occur, or at least show a considerable drop in frequency of occurrence. *None of them shows any increase in frequency.* This last fact is important when comparing *Pygosteus* with *Gasterosteus*. Tinbergen and van Iersel (1947) have shown that there is an increase in pushing into the roof of the nest, superficial glueing, and fanning when a ripe female is introduced, in the case of *Gasterosteus*. Here, then, is an interesting specific difference.

This difference has already been mentioned by Leiner (1931a, 1934), who pointed out that *Pygosteus*, unlike *Gasterosteus*, does not rush to the nest to mend, fan or glue during the courtship. My own observations confirm this. The courting male *Pygosteus* almost completely ignores its nest while dancing to the female. The courting male *Gasterosteus*, on the other hand, dances to the female for a short while, swims quickly to the nest, performs a series of rapid nest activities there, and then rushes back to the female and continues courting. Fig. 25 in the present report shows (right-hand column) the decrease in nest actions with the introduction of a female. (Average values from twelve half-hour tests.) Glueing is the only activity that does not show a marked decrease. This particular activity appears to be independent, to a large extent, of rapid changes in other behaviour tendencies. Often a male was seen to be actively courting when a blob of glue was extruded. Such males frequently ignored this glue and continued with their courting. Sometimes they paused in their courting, took up the blob and inserted it into the nest. On other occasions, the glue simply fell away from them and was either never used or, rarely, was taken up after courting ceased, and inserted then. It appears that the glue is being produced in the kidney at a rather steady rate and when a certain amount is present, a blob is extruded almost automatically. This would explain the lack of short-term changes in glueing frequency. (It should be noted that superficial glueing in both *Pygosteus* and *Gasterosteus* can, however, show extremely short-term changes in frequency.)

Fig. 4 in Tinbergen and van Iersel (1947) contrasts remarkably with Fig. 25 here. There is a huge increase in glueing, pushing and fanning. These authors have postulated that these increases reflect the fact that 'pushing, fanning and glueing are outlets for the obstructed mating drive'. They suggest that they occur as displacement activities as a result of the unresponsiveness of the females in question. The female will not respond to the dancing of the male, and this thwarts his sex drive and the latter then finds an outlet by the performance of these nest actions. This certainly seems to be a reasonable explanation, but it is puzzling why *Pygosteus*, which can be just as thwarted as *Gasterosteus*, does not show these or any other displacement activities under the same conditions. It has already been mentioned that the ritualized fanning which is involved in the showing-the-nest-entrance ceremonies of both species appears to be a displacement activity with a signal function in both cases. Ritual fanning also seems to be the result of the thwarting of the sex drive of the male. Since both species show this type of displacement activity, it is even more difficult to understand why *Pygosteus* should show no displacement activity at the dancing stage of courtship.

One possible explanation is that, from a functional point of view, the performance of nesting actions by a courting male may be more disadvantageous for *Pygosteus* than for *Gasterosteus*. A courting male *Pygosteus* would be more likely to lose its female in the more densely vegetated *Pygosteus* environment, if it rushed off to the nest every now and then. The male *Gasterosteus*, dancing

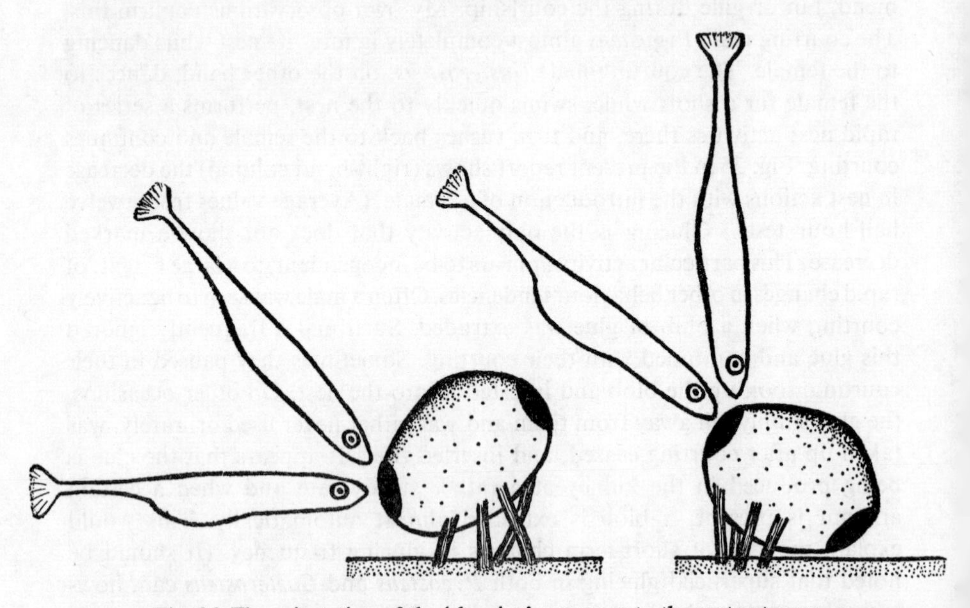

Fig. 36. The orientation of the 'showing' ceremony to the nest entrance.

in comparatively open water, would be much less likely to do so. Selection would therefore tend to eliminate *Pygosteus* males which were inclined to show nesting displacement actions during the stages of the courtship before the female had safely arrived at the nest. Once the female were there, however, there would be no such selection and this would explain why both species perform ritual fanning when showing the entrance. (This same argument explains why the leading to the nest by the male *Pygosteus* is slow and jerky, as compared with the rapid leading of *Gasterosteus*.)

It might be argued that, in *Pygosteus*, the rare nest actions which do occur during the courting phase, are in fact displacement activities, despite the fact that there is no general increase in these activities during the courtship. Even this does not seem to be the case, however. This conclusion is based on an analysis of a number of courtship tests made using both *Gasterosteus* and

Pygosteus. Fig. 37 shows that it was found that, in *Gasterosteus*, the average number of times that a male fanned during any five minutes of a courtship phase was *positively correlated* with the number of sexual actions (dance-jumps), during that period. In *Pygosteus*, the opposite was true, there being a *negative correlation* between the number of fans and the number of jumps.

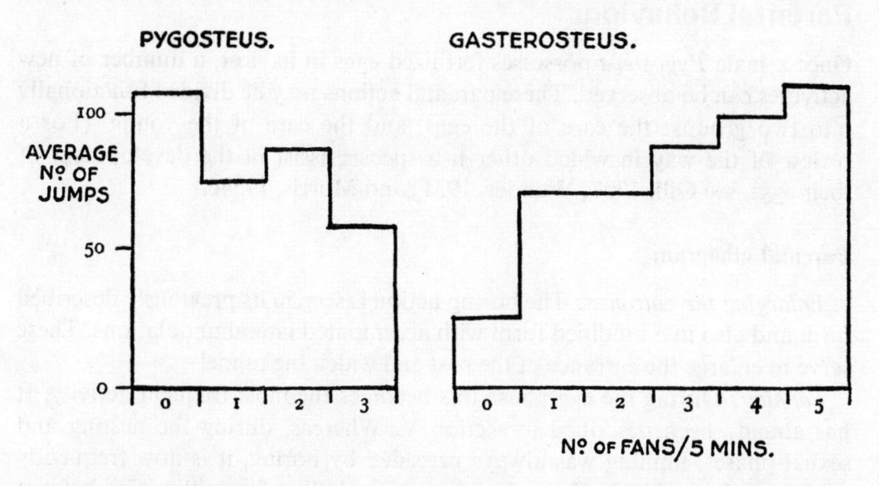

Fig. 37. Specific difference between *Pygosteus* and *Gasterosteus*, in respect of displacement fanning. (Averages based on 126 five-minute periods for *Pygosteus* and 42 for *Gasterosteus*.)

The *Gasterosteus* result is what would be expected if the fanning was, in fact, displacement fanning, but the *Pygosteus* result is what one would predict for autochthonous fanning.

Summing up, it can be said that the evidence for displacement fanning in *Gasterosteus* seems to be very sound, but that the evidence against this in *Pygosteus* is equally sound. Pre-parental fanning in *Pygosteus* in general appears to have little relationship with sexual states, but does, however, seem to vary with changes in nesting states. This last point will be discussed in section VIII.

VII

Parental Behaviour

Once a male *Pygosteus* possesses fertilized eggs in its nest, a number of new activities can be observed. These parental actions may be divided functionally into two groups: the care of the eggs, and the care of the young. (For a review of the way in which other fish species assist in the development of their eggs, see Gill, 1907; Wunder, 1931; and Morris, 1954c.)

Parental ethogram

Enlarging the entrance: The boring action is seen in its previously described form and also in a modified form with accentuated lateral undulations. These serve to enlarge the entrance of the nest and widen the tunnel.

Fanning: During the egg-phase this becomes the most frequent activity. It has already been described in section V. Whereas, during the nesting and sexual phases, fanning was always preceded by boring, it is now frequently performed by itself. The fish swims to the nest, halts a few millimetres in front of the aperture and, with the long axis of its body approximately in line with the axis of the nest tunnel, fans a current of water over the eggs (see Fig. 17). This ventilates the nest and aerates the eggs, thus assisting in their development. Often the fanning action is preceded by a slight movement towards the entrance. This forward movement, which is clearly an intention movement of boring, may be followed by prolonged and intense fanning, lasting ten to twenty seconds. In the pre-parental phases, the strength of the boring preceding fanning is directly related to the strength of the fanning which follows; but now the strength of the fanning has become independent of it. In the pre-parental phase an intention movement of boring would be followed by only an extremely brief fanning bout, and the pre-parental fanning bouts which were preceded by even the most intense borings, would never be of the intensity and duration of the independent parental fannings.

Examining the eggs: The male pushes gently into the tunnel of the nest until its nose is in contact with the eggs. It may stay still in this position for a second or two before withdrawing, or it may withdraw again immediately. This movement, which is obviously related to boring, is more gentle than boring and the lateral undulations are eliminated. There is no impression given of struggling into the tunnel, but rather of gliding into it and gliding back out again. Presumably the fish obtains information about the position and state of the eggs in this way, probably from both chemical and tactile stimuli. That such information is obtained in this way is borne out by the fact that this action is often followed by some readjustment to the eggs, of one sort or another.

Lodging the eggs: The individual eggs cling to one another in a moderately firm bunch, and unless some accidentally break loose, they are dealt with as one by the male. The bunch may become dislodged to a greater or lesser extent from time to time, and this always stimulates vigorous attempts to re-lodge them. This may be done in two ways. If the eggs have been only slightly shifted out of their correct position (in the bottom of the tunnel towards the exit end), then they are nosed into place by nudgings of the male, keeping its mouth shut. This is usually performed with the male facing in through the entrance of the nest, but I have also seen a male perform the action by pushing in through the exit (an otherwise rare occurrence), or, where a nest has become ragged and loose, by pushing in through the sides of the nest. If the male is repeatedly unsuccessful in its attempts to replace the eggs in their correct position, it may grab them in its mouth and then make little swimming movements, thus driving them into position.

This is a particularly important activity to *Pygosteus*, where the eggs may fall from the nest and be lost. Where several clutches are laid in the same nest, each new clutch endangers the old ones by virtue of the fact that the newly spawning females sometimes drag the old clutches towards the exit hole as they pass through the nest. Several times I have seen old clutches pushed completely out of the exit in this way. It is interesting that, in this connection, the male *Pygosteus*, with eggs in its nest, only pushes right through the nest tunnel when fertilizing. The pushing-through-the-nest activity seen in the earlier phases drops out completely as soon as eggs are present in this species, but this is not the case (van Iersel, 1953, pp. 25–6) in *Gasterosteus*. However, in the latter species there is little danger of eggs being lost in this way, owing to the position of the nest.

One point in the parental cycle, when the action of lodging the eggs is almost certain to be seen, is just after fertilization. The size of a nest in relation to the size of a female is nearly always such that the eggs are laid too near the entrance, and the male usually follows the act of chasing off the female with a vigorous ramming home of the eggs.

Retrieving eggs: Eggs that have fallen from the nest, either because it is faulty, or because the parents have dislodged them when spawning, or eggs that have been stolen from the nest by other fish, are often retrieved by the male parent. This it does by grabbing the bunch in its mouth and swimming steadily back to the nest where it inserts them and rams them home again. In cases where the eggs have been stolen, the male may chase after the thief and attempt to grab the eggs from its mouth. A number of eggs are broken off in the struggle that follows and these always get devoured (some even by the parent itself), but if the parent manages to secure a moderate-sized section, or more, of the original bunch, he swims back to the nest with them and reinstates them. A male parent will attack an egg-thief furiously as it tries to escape with its loot, alternating bites at the thief's body with grabs at the eggs themselves. If the eggs are dropped by the thief, even if some distance from the nest, the parent may pick them up and carry them back to the nest. Only if the eggs are broken up into single individuals (or groups of only a few) will the parent join in with the thief or thieves and devour them.

Transferring eggs: A similar pattern to the above is the actual transfer of eggs from one place to another by the male parent. The male picks up his own eggs in his mouth and takes them out of the nest and carries them over to a new spot in the weeds, where he pushes them in and lodges them. After a while he may transfer them back again to the nest. This activity is usually seen when the nest is deteriorating in some way. Often it coincides with the construction of a new nest. When this is the case the eggs may be taken out of the old nest and transferred to the new one. Alternatively, the new nest may be built around the eggs as they lie naked in the new position in the weeds. This activity is further discussed below.

Cleaning the eggs: The male pushes into the nest and, as a result of examining the eggs, may begin biting at them. Certain eggs can be seen by the human eye to change in appearance from the rest. These usually become opaquely white and it is these that the male bites at, removes, and eats. Such eggs are invariably fungus-infected and if they were not thus removed would infect the whole clutch. The importance of this activity is borne out by the fact that a clutch of eggs placed in a beaker under a constant stream of water, but without a male parent, will quickly become infected throughout if the fungus once successfully attacks a few eggs. In the nest, these few eggs would be removed by the male before this could happen.

Removing egg-cases: At the actual point of hatching, when the young fish emerge from their egg-cases, the male can be seen repeatedly nosing into the nest and then backing out again carrying the empty 'shells' which he removes from the nest tunnel.

Retrieving young: The young fish crawl up in the nest (which is by this time only loosely constructed) and sit on top of it. Occasionally one will make a sudden dart away from the nest. If the male happens to see this, it chases the young fish, catches it in its mouth, swims back to the entrance of the nest, and spits it back inside again. As the young grow older, the male's attempts to retrieve them become less and less successful and the young tend to scatter more and more from the nest-site, until finally the male ceases to pay any attention to them, except perhaps to treat them as food. If the male has constructed a nursery near the nest, the young are returned to this when they are retrieved. (For a full description of the nursery, see below, p. 379.)

The parental cycle

The course of the parental cycle of *Gasterosteus* has been studied in detail by van Iersel (1953, pp. 22–151). It is impossible to summarize such an elaborate analysis here, and I shall only refer to *Gasterosteus* where it appears to differ from *Pygosteus* in its parental activities. Fundamentally the two species appear to be extremely similar with regard to the manner in which their parental behaviour is organized. Generally speaking, the cycles show the same sort of changes from day to day, but the whole of the *Pygosteus* cycle is less intense. This appears to be correlated with the less ventilated environment with which *Gasterosteus* eggs have to contend.

There are a number of parental actions associated with the eggs, or the young, or the maintenance of the nest, which can be scored quantitatively during the days of the parental cycle. However, one of these actions, namely fanning, is far superior, as a measure of parental behaviour, to all the others. Van Iersel virtually uses no other measure in his study of *Gasterosteus*. As he points out, this action has no varying intensities and its frequency variations are extremely regular. It occurs in short bursts, so that there are three basic frequency measures available.

1. The number of bursts, or bouts, of fanning in a given time period, i.e. the frequency of *occurrence* of fanning (F. occ. in all graphs).

2. The total amount of time in seconds spent fanning in a given time period, i.e. the *duration* of fanning (F. secs in all graphs).

3. The average length of the fanning bouts (F. bout in all graphs). This third measure is obtained by dividing the second by the first.

Fig. 38 shows a typical *Pygosteus* parental fanning cycle. All three measures are shown. (The broken-line graph is the F. bout measure.) The black triangles mark the stage in the cycle at which the eggs hatched (this also applies to other graphs). Several interesting points emerge from this figure:

First, it can be seen that there is a day-by-day increase in all three measures of fanning, up to a peak day, and then a similar day-by-day decrease in all three measures, until fanning ceases altogether. Almost as soon as the parental cycle begins, there is an increase in fanning which takes it above its average pre-parental level, both in respect of the number of times the males go to the nest and fan, and the average length of the fanning bouts.

Secondly, a comparison with van Iersel's data for *Gasterosteus* shows that the average fanning peak for that species is much higher than in *Pygosteus*. *Pygosteus* males which had fertilized two clutches of eggs (2F2C, to use van Iersel's designation), spent on the average about 400 seconds per half-hour fanning on the peak days of their cycles. 2F2C *Gasterosteus* males spent on the average about 1,100 seconds per half-hour on their peak days. (All parental measures given here are based on half-hour observation periods,

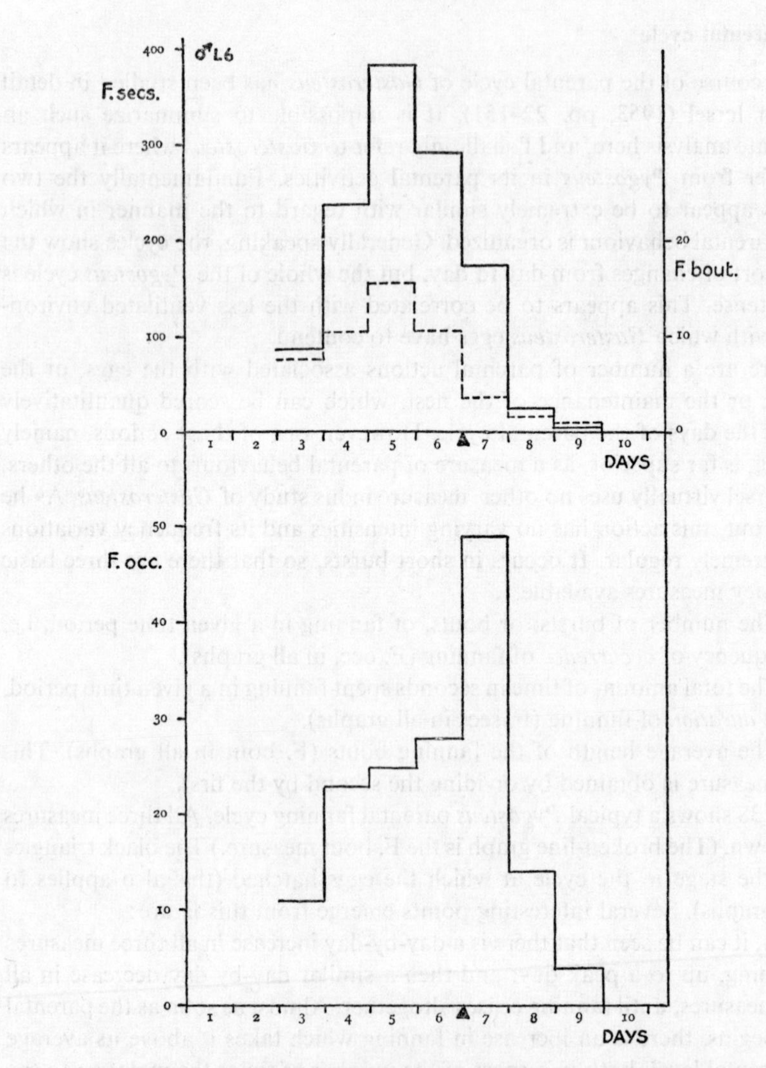

Fig. 38. The fanning cycle of *Pygosteus* measured in three different ways. See text.

unless otherwise stated.) This large difference between the species was not the result of any difference in experimental technique or conditions; I have recorded parental cycles of *Gasterosteus* under my own experimental conditions and find that the results obtained from these tests are the same as those obtained by van Iersel. This is a real specific difference, the significance and effects of which will become clear later. Despite this huge difference in general level, the graphs for the two species present the same general shapes and are otherwise similar.

Thirdly, Fig. 38 shows us that the peak in fanning duration precedes that for fanning occurrence. The fall-off in time spent fanning begins typically just *before* the eggs hatch, whereas the number of occurrences of fanning is likely to go on increasing *beyond* the hatching day. (An analysis of Table 1 in van Iersel reveals that this is also true of *Gasterosteus*.) The reason for this is clear. On the one hand, van Iersel has shown that 'In the course of the cycle the [fanning] drive becomes more autonomous and less responsive to the increasing stimulation from the growing eggs'; on the other hand, we know that when the eggs begin hatching, the time spent fanning begins to fall off, although the number of fanning bouts still increases. Now, the number of times a male returns to the nest to fan is obviously less dependent on stimuli coming from the eggs, than is the amount of time it spends fanning when it is at the nest. It appears therefore that the male begins to fan for shorter periods as an immediate response to the hatching eggs, but that when it is away from the nest, the autonomy of the fanning drive still persists in increasing the number of returns-to-the-nest-to-fan. Only after a day, or two days, does the inhibitory effect of the hatched eggs manage to reduce the internal fanning drive of the male. The time lag between the two peaks can, in fact, be taken as a measure of the 'degree of autonomy' of the fanning drive in any particular case. It is interesting to note that both in *Pygosteus* and, apparently, *Gasterosteus*, the time-lag between the two peaks is longer if the male has more clutches in its nest.

Length of the fanning cycle

The fanning cycle of *Pygosteus* is on the average 6–7 days long at a temperature of 15–16 degrees Centigrade and 4–5 days long at a temperature of 18–19 degrees Centigrade. Although the *Gasterosteus* cycles are longer at the same temperatures, they show a similar correlation with temperature changes (see van Iersel, 1953, Fig. 24). In connection with this it should be noted that the eggs of *Gasterosteus* are much larger than those of *Pygosteus*. (Furthermore, the river bullhead, *Cottus gobio*, which has a fanning cycle lasting for four weeks, has even larger eggs than *Gasterosteus*. Morris, 1954c.)

Undeveloped cycles

Of the parental cycles which were studied quantitatively, it was noticed that a large proportion did not develop successfully. The males in question began to show the typical increase in fanning rate, after normal fertilizations, but soon this rate fell off to the pre-parental level again. When the nests of

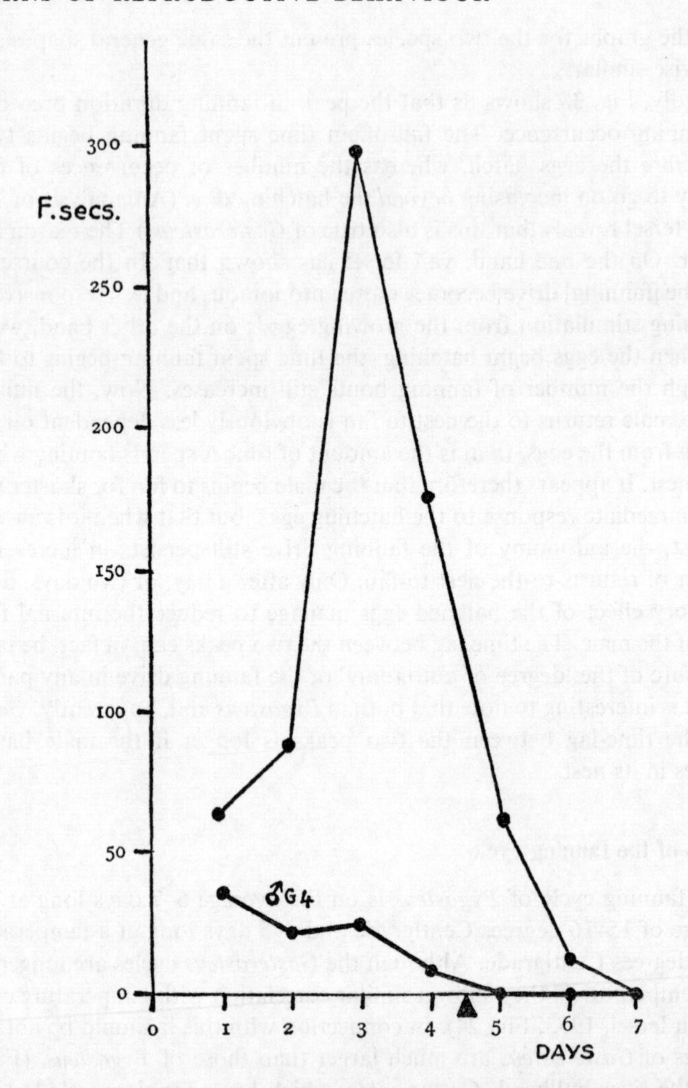

Fig. 39. Two successful fanning cycles of *Pygosteus*, one of which does not, however, develop properly.

such males were examined, they were found to contain no eggs. It appeared that the males, for some reason, had eaten the eggs. Since cycle failures of this kind were far more common when males had only had one fertilization and possessed only one clutch (1F1C males), than when they had had more than one fertilization and possessed more than one clutch, it seemed that the breakdown was due to a too weakly developed cycle. Had the failures been due to some kind of environmental 'accident', or inadequacy, it might have been expected that the cycles would have been equally likely to break down with either 1F1C males or 2F2C males. Having only one clutch appeared to result in an insufficient arousal of the fanning drive, in respect of the latter's suppressive effect on other competing tendencies (see below, p. 368).

In only one of the cases of undeveloped cycles did the eggs hatch. In Fig. 39 the fanning cycle of this male (G4) is compared with a typical fanning cycle of a male kept under similar conditions. It will be noticed that, despite the fact that there is a vast difference in the amount of fanning between the two, the eggs hatched at the same time. In other words, from the eggs' point of view, the environment of the fresh, growing, plant-filaments from which the nest was made, was almost adequate *by itself* to ensure successful development, and the great peak of fanning was superfluous. This does not, of course, mean that the existence of a parental fanning peak is stupid luxury. I have already mentioned that in all cases but the one above, in which the cycle did not develop properly, the eggs were destroyed by their parent-males. It is quite clear that the safety of the eggs is ensured by the strong development of a parental mood in the form of an intense fanning drive, which suppresses (as does any intensely activated drive) all competing tendencies, some of which would be harmful to the eggs. The mutual suppression of drives at high intensities is one of the basic findings of ethology. But it may be argued that, in the case in question, although the fanning drive is certainly important in this context, so would any other parental tendency be potentially equally as important and successful. However, a study of the breeding grounds of sticklebacks in nature soon reveals that the environment is often far inferior to that provided in the laboratory. In flowing streams, the sudden changes in rainfall in spring may bring, from upstream, large amounts of silt, dirt and detritus, which may change drastically and quickly the amount of environmental aeration from which the eggs can benefit. In stagnant stretches of water, equally dangerous changes may occur in spring, with its rapidly fluctuating temperature and rainfall. Van Iersel (1953, p. 149) has recognized that, in *Gasterosteus*, both the above-mentioned ('suppressive' and 'ecological') factors are important, but he has put much more emphasis on the first than I consider justified. He says: 'Even if it would appear from ecological studies that the high amount of fanning, while obviously a luxury under laboratory conditions, is necessary under certain rare emergency conditions, the function of the internal inhibitive effect remains a real fact.' I do not think that the emergency conditions he mentions are as rare as he appears to imagine. Unfortunately, my own field studies are limited mainly to the streams and ditches of the Thames basin, so that I can make no general statement on the matter. However, I am convinced that the ecological requirements have been of primary importance in producing the fanning peak, and that the suppressive effect of this peak on other activities is only a secondary consideration, as any 'parental peak' would have done as well from this point of view.

In conclusion, it should be mentioned that, in connection with the above argument, the *larger* fanning peak of *Gasterosteus* is not correlated with any *stronger* competing tendency, but is correlated with the existence of bigger eggs and much worse environmental conditions (on the river bed) for the eggs. This last statement must, however, be treated with caution, for reasons discussed earlier. It is only possible to make a rough general statement when comparing the strengths of whole tendencies between species. Sexual and nesting tendencies are the most relevant ones here. In both, *Pygosteus* both gives the appearance of being able to reach greater heights of intensity than *Gasterosteus* and also shows higher frequencies in any of the measures of these two tendencies, which I have used in quantitative records. Although this leads one to the common-sense conclusion that *Pygosteus* has just as much to suppress with its fanning-peak as *Gasterosteus*, as I said above, this conclusion must be treated with reserve. It is best to conclude, at the present stage, with the general statement that the fanning peak is important both for its effect in coping with unfortunate environmental conditions or changes, and for its suppressive effect on competing tendencies, especially nesting and sexual tendencies.

Sexual behaviour during the parental cycle

It has been mentioned above that sexual behaviour is suppressed by the development of the fanning cycle. Fig. 40 shows this clearly. The broken line represents the fanning cycle. The solid line shows the level of the sexual motivation, as measured by frequency of dance-jumps (J). Also included in this figure is the nest-building drive, as measured by the frequency of collection (col) trips (dotted line). It can be seen that, as the fanning cycle develops, so the sex drive falls away, until, around the hatching time, it is virtually non-existent. No material is collected during this period, but once the eggs have hatched, both sex and collecting nest material reappear. (Sex was tested here by presenting the male with a ripe female each day of its parental cycle, for half an hour, just after the half-hour parental observation period.) There is one unusual aspect of the behaviour of male C4 that is shown in Fig. 40, that must be mentioned. The resurgence of sexual activity on the seventh day which occurs there is a phenomenon which is atypical of fully developed parental cycles. When a male is slackening off its fanning after the hatching of its eggs, the parental drive as a whole is not finished. Whereas the parental state of a male expresses itself as fanning, in response to the presence of eggs, it expresses itself as *retrieving* in response to young fish.

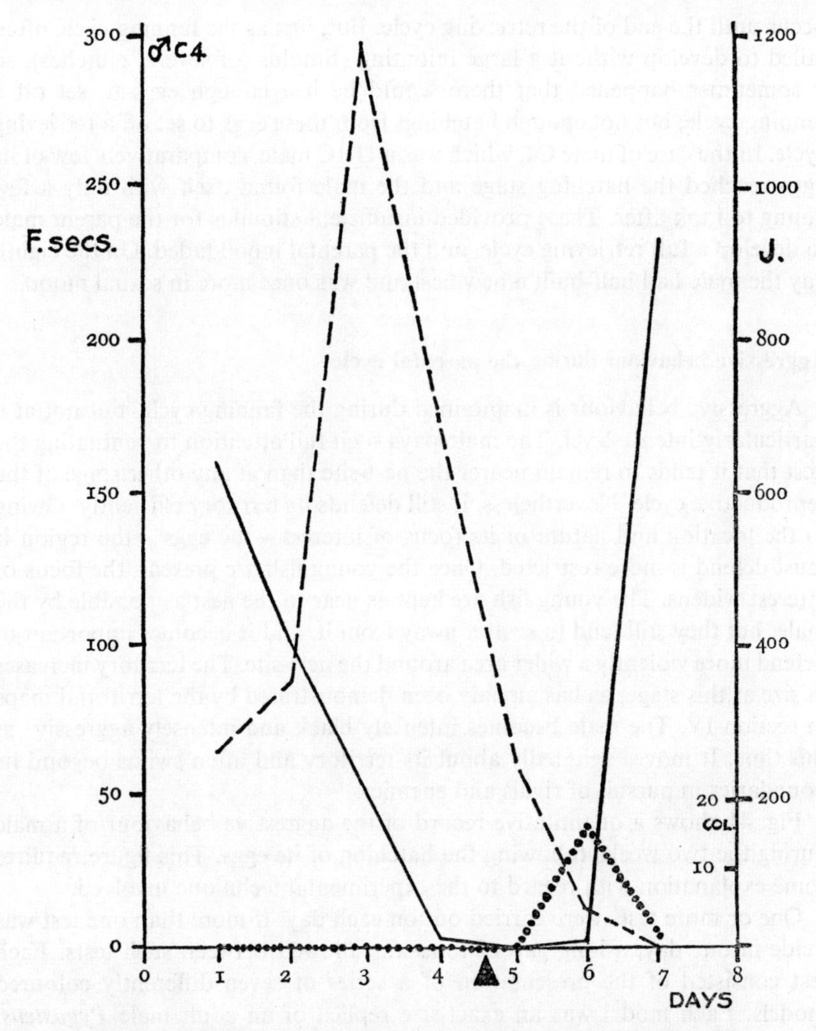

Fig. 40. The inverse relationship between the fanning cycle and sexual motivation.

The retrieving action has already been described. Its frequency increases as the frequency of fanning decreases. This is largely as a result of the fact that the young begin more and more to move about and scatter from the nest-site. The male tries more and more to catch them and return them to it. After some time – up to two weeks – the young develop so far that they cannot be caught and gradually the retrieving frequency falls off and the whole parental (and reproductive) cycle ends. Typically the male does not show any sudden resurgence of sexual behaviour at the end of the fanning cycle. This does not occur until the end of the retrieving cycle. But, just as the fanning cycle often failed to develop without a large initiating stimulus (of several clutches), so it sometimes happened that there would be just enough eggs to set off a fanning cycle, but not enough hatchings from these eggs to set off a retrieving cycle. In the case of male C4, which was a 1F1C male, comparatively few of its eggs reached the hatching stage and the male found itself with only a few young to look after. These provided insufficient stimulus for the parent male to develop a full retrieving cycle, and the parental mood faded. On the eighth day the male had half-built a new nest and was once more in sexual mood.

Aggressive behaviour during the parental cycle

Aggressive behaviour is maintained during the fanning cycle, but not at a particularly intense level. The male pays such full attention to ventilating the nest that it tends to remain nearer the nest-site than at any other stage of the reproductive cycle. Nevertheless, it still defends its territory efficiently. Owing to the location and nature of its focus of interest – the eggs – the region it must defend is more restricted. Once the young fish are present, the focus of interest widens. The young fish are kept as near to the nest as possible by the male, but they still tend to scatter away from it, and it becomes important to defend more violently a wider area around the nest-site. The territory increases in size at this stage, as has already been demonstrated by the territorial maps in section IV. The male becomes intensely black and intensely aggressive at this time. It moves agitatedly about its territory and often swims beyond its boundaries in pursuit of rivals and enemies.

Fig. 41 shows a quantitative record of the aggressive behaviour of a male during the two weeks following the hatching of its eggs. This figure requires some explanation with regard to the experimental technique involved.

One or more tests were carried out on each day. If more than one test was made in one day, a long gap of time was allowed between such tests. Each test consisted of the presentation of a series of seven differently coloured models. Each model was an exact-size replica of an adult male *Pygosteus*. These were made by placing a freshly dead male in plaster of Paris. A plaster cast having been made, moulds were taken from this, using commercial cement pastes. A thin wire was pressed into the top of each model while it was still soft. The models were then hung up by these wires and given several coats of oil paint when hard. The models were held by these wires when presenting them to the fish in the tanks. (For an illustration of such a model, see Fig. 14 in Morris, 1952.) These models were not painted in detail and no special markings, or features, such as eyes, were included. No attempt at reproducing fins was made. The males reacted violently to such models, responding to their general shape, size and colour.

In the particular series in question, a series consisting of black, blue, white, orange, silver, yellow and grey was given. Each model was presented in the same part of the territory, the same distance from the nest, for one minute. A pause of one minute was made between each presentation. The presentation sequences were randomized.

Each test made in this way gave two basic results. One concerned the degrees of preference for (biting) the different colours, and the other concerned the total number of bites per test; this second result is shown graphically as the broken line in Fig. 41. (The larger ordinate values apply to this second result.) The nature of colour preferences here is interesting. Briefly, black and blue were preferred approximately twice as much as any of the others. There were no marked preferences between these others. Throughout the two weeks of the experiment, which began three days after the young hatched, there was a gradual fall-off in the preference for black, blue taking over as the most preferred colour. This change is expressed graphically in Fig. 41 by the solid line. The points from which this graph is made are obtained by multiplying the number of times which the male bit the black model in one test by six, and dividing this figure by the total number of bites given to the other six models together, in the same test. This gives a figure which represents the preference for black over the mean of all the other colours. (Smaller ordinate values apply here.)

It will be seen that black was less and less preferred as time went on. It is interesting that Sevenster (1949) found also, in his tests on the aggression of *Pygosteus*, that at certain times in the cycle black was preferred, whilst at others blue was preferred. His tests were different in detail, but similar in principle, to the ones described here. He found that pre-parental males, in the very aggressive phase, preferred black, whereas parental, egg-owning males preferred blue. He concluded that this was due to a fall-off in the level of aggression and an increase in fear of the models by the males, in the second instance. He noted that when males were preferring blue they nevertheless paid a considerable amount of attention to black, but showed signs of hesitating when about to attack it, so that the 'biting' figures for black were not high. It seemed that the males were becoming more afraid of black than of the other colours, which is reasonable when one considers that black is the colour which is likely to be associated both with something to attack *and* something from which to flee. My observations, at different stages of the reproductive cycle, agree with those of Sevenster perfectly. As the parental cycle neared its end, the male became more and more afraid of the black model. The amount of threatening and hesitating before attacking increased considerably. It seems therefore that the male has two very aggressive peaks: one before the parental phase and one after the eggs hatch. The appearance of the young fish seems to set off a huge outburst of aggression, which gradually dies away again.

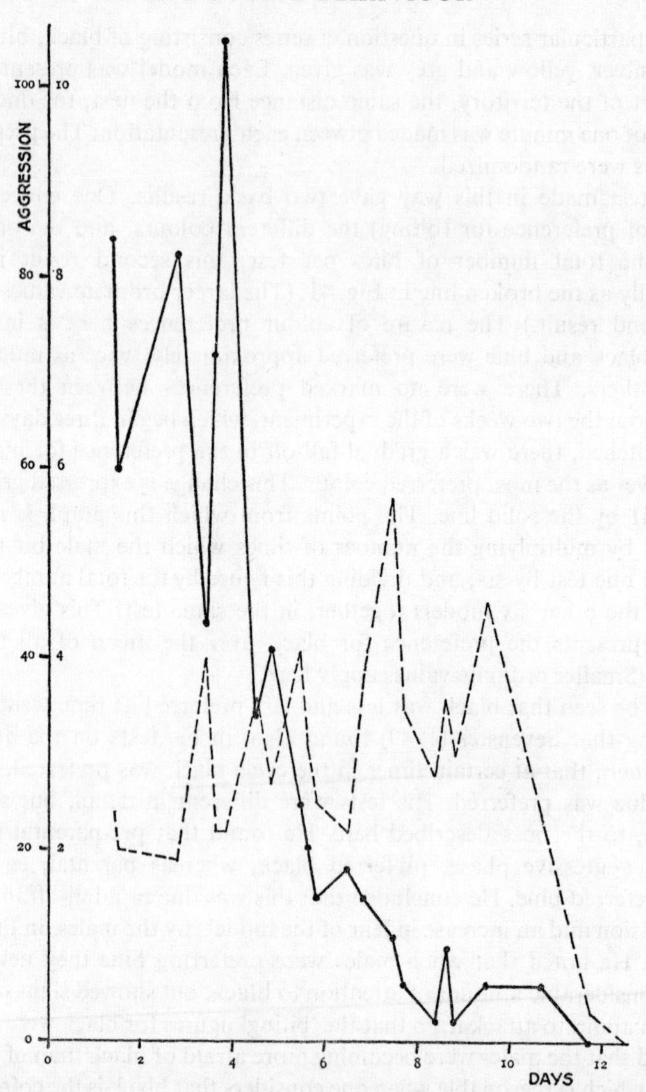

Fig. 41. The aggression of the male parent during the offspring phase. Explanation in text.

One puzzling aspect of Fig. 41 is the relationship between the changes in the two different measures of aggression. It will be noted that the fall-off in aggression, as measured by preference for black, occurs gradually throughout the two weeks, whereas the fall-off, as measured by the total number of bites per test, is sudden and does not come until right at the end of the experiment (when other parental activities had ceased). Compare the state of affairs on, say, the second and the eighth days. On both days the male was prepared to give a large total number of bites, but far more of these were delivered to the black model on the second day than on the eighth day. Since the total number of bites given in a test must be a measure of the degree of dominance of the tendency to attack over the tendency to flee, it is difficult to see how an apparent increase in fear of the black model can operate independently of this. I can only suggest that this experiment has revealed hitherto unsuspected complexities in the nature of aggressiveness as a response to different stimuli. Further research is needed here.

Nest-building behaviour during the parental cycle

Nest-building is suppressed during the parental phase. Collecting of nest material, and high intensities of boring into the nest entrance disappear. If this suppression did not occur, the eggs would be rammed out through the nest exit. Low-intensity boring takes the form of shallow nosing into the entrance, which, as described earlier, has the function of carefully examining, or adjusting the position of, the eggs.

Nest-maintenance activities persist in the parental phase, however. Nest-mending is not suppressed and parent males pay a good deal of attention to the state of repair of their nests. Even so, the nests sometimes become dilapidated: either they become too loosely woven, or they come partially adrift from their host plants, or the material of which they are built begins to rot. In any of these instances, a parent male is likely to take drastic action. This action takes the form of *egg transference*.

Egg transference was first noted briefly by Leiner (1931a). He observed that a male *Pungitius*, owing to the inadequacy of its old nest, built a second one and transferred its eggs successfully to this. I have made a further study of a number of cases of this phenomenon, which appears to be more fully developed in *Pygosteus* than *Gasterosteus*. (The reasons for this will be discussed below.)

Fig. 42 illustrates the transference of eggs to a new site in one particular case, where the whole process was recorded in detail. The following is a condensed version of notes made at the time (although this particular transference was unsuccessful, it nevertheless characterizes the activity very clearly):

10.30. Tank 4 (36 × 36 × 60 cms) contains one subordinate male, one female, and a nest-owning parental male. The latter obtained eggs from the female 2 days ago. In the first 45 minutes of observation today, the male was seen to push into the *exit* of its nest on six occasions. Each time it emerged with its clutch of eggs in its mouth! Each time it swam round to the entrance with them and replaced them in the nest tunnel, ramming them vigorously and with some difficulty back into the nest *entrance*. The journey from the exit to the entrance varied in length, the route becoming more and more indirect (as shown in Fig. 42). Hardly any fanning occurred, except for one or two brief bouts when the eggs were returned to the nest. The nest is loose and is made of *Nitella* fragments which seem to be rotting.

11.15. The male is fluttering against the glass now. Every now and then, it pauses, glues, and inserts the glue into the nest. Also, it occasionally pushes through another clump of weeds, which are about 45 cms from the host clump. It carries a piece of material to the old nest and bores it in, then goes to the exit, takes out the eggs and rams them into the entrance for the seventh time.

11.50. It carries the eggs 45 cms away from the old nest, after taking them from the entrance for the first time. On this, the eighth trip, it pushes them into the other clump of weeds. They fall to the ground. The other male approaches; a fight ensues. The egg-owner leaves its eggs to fight. It returns, finds them again and takes them back to the old nest and pushes them into the entrance. Now, for the ninth time, it takes the eggs out of the old nest, this time from the exit again, and in so doing tears away part of the dilapidated old nest. It leaves them hanging there from the broken nest, in full view, and goes off fighting for five minutes. Now it glues, but EATS the glue, as if it had no nest in which to insert it. It now returns to its eggs again and tries to put them back into the nest, but the latter is so damaged and disarranged that it is too difficult. Abandoning this attempt, the male carries the eggs to the 45-cm-distant clump again and pushes them into it several times. Now it brings them back to the nest again, and again tries to ram them into their old home, but without success. For the tenth time the eggs are carried away from the nest and once more they are pushed into the other clump. This time they are successfully lodged and left there by the male. (No fanning occurred during all this.) The male now leaves the eggs alone and fights and flutters again.

12.45. The male takes the eggs from their new position and replaces them in their old nest. It examines the new clump of weeds thoroughly. Now it takes the eggs out of the entrance again and puts them back in the new clump. Now they are replaced in the old nest again. This old nest is now just a disorganized mass of *Nitella* fragments, and the eggs are simply pushed by force into the centre of this ball, in the spot where there was once a good entrance. The male glues again but again eats the glue.

13.05. The male examines a third clump of weeds, pushing through it as if it were a nest. (This clump was not eventually used, however, the new nest being built in the one already mentioned.)

14.45. The male eats half its clutch of eggs just in front of the old nest, which is now completely dilapidated. No attempt has yet been made to build a new nest.

15.00. The male carries the second half of its clutch over to the 45-cm-distant clump of weeds and dumps it there.

On the morning of the following day, a second nest was found to have been built in the new clump of weeds, at just the place where the eggs were repeatedly placed during the transfer mood. The few remaining eggs, which were placed in this new nest, were, however, insufficient stimulus for a successful parental cycle and were eaten two days later. Fig. 42, it will be realized, is a schematized version of the above transfer. It shows the basic

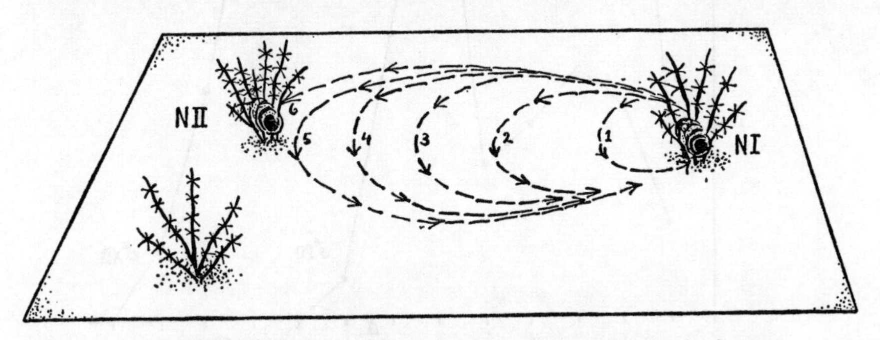

Fig. 42. Schematic diagram showing the transference of a clutch of eggs from an old (N I) to a new (N II) site.

character of the transference, namely the gradual shifting of the association of the eggs with the old site (N I) to the new site (N II).

The fact that the male ate some of its eggs during the above transfer reveals the danger involved in such an activity. The eggs are not removed from an old nest or nest-site until the latter has become so unsuitable that it has reduced the parental drive considerably. Fig. 43 shows the fanning cycles of two males which transferred their eggs successfully. The troughs, which occur where there should be peaks, in these fanning cycles, associated with the egg transferences, can be clearly seen. Once the eggs are removed from the unsuitable old site, the low parental state of the parent male immediately endangers their safety. If the male does not get them installed in their new home fairly quickly, they are likely to be eaten.

Apart from the above danger, there are also further difficulties involved in egg transference. Firstly, the new nest either has to be built before the eggs are placed in it, or it has to be constructed around the clutch. Both occur, but both are awkward for a parent male. If the nest is built first, then there is no difficulty in constructing it, but the eggs have to be taken care of in the

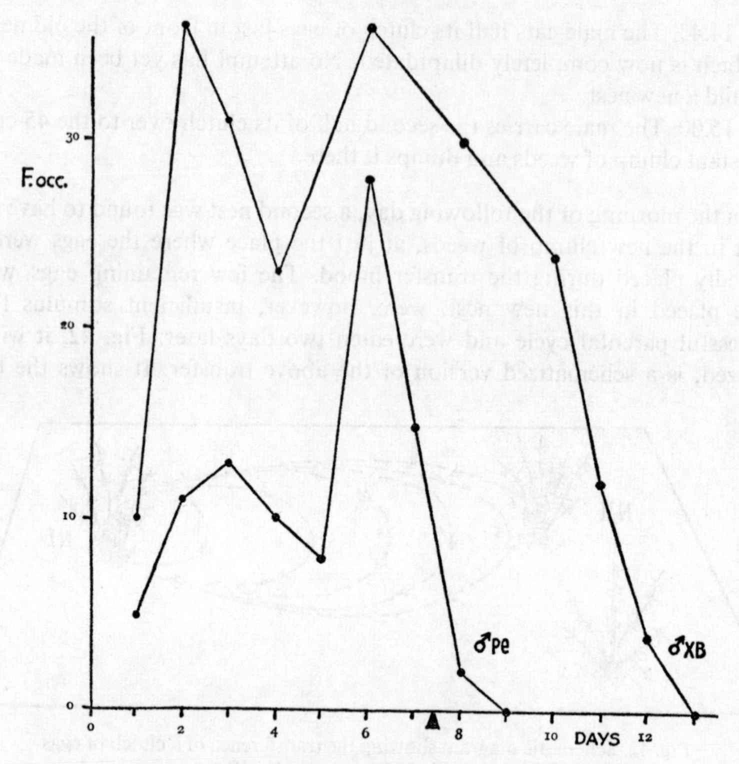

Fig. 43. Two interrupted fanning cycles resulting from clutch transference.

meantime. They may be dumped at the foot of the new host plant, or back into the old nest, but, in either case, they may suffer from the attacks of other fish while their owner's attention is riveted on the new nest-site. Under these conditions, even the owner himself may be stimulated to attack and eat the eggs. On the other hand, if the eggs are placed into their new nest-site position, and the nest is gradually built round them, then they have the status of 'eggs-in-nest' from the start, and are less in danger. However, it is not easy for a male to build a nest *round* anything, and such nests are invariably weak and loose, and often little better than the originals. The eggs only appear to survive in such cases because by the time the parent has become 'exasperated' enough to abandon the second nest, they have already hatched successfully. The difficulties in building such nests are as follows:

It will be remembered that the method of construction of the nest of *Pygosteus* involves carrying a piece of material to the nest-site, boring it into the centre of this site, ramming it home, and then looping-in any loose ends. It does not wind fragments around the site when building, or place them on the site, but rather *inserts* them. If eggs, which require careful handling (to prevent individual eggs from breaking loose from the clutch) are present in the centre of the nest-site, then the usual methods of building are obviously considerably hampered.

The nestless egg-owners overcome these difficulties partially, by repeatedly nosing at the nest and packing together as best they can the collected fragments which they have only been able to half-bore into the nest entrance. Looping occurs frequently and this does not interfere with the eggs, but tends to push them farther and farther to the back of the nest. Any intense boring that occurs has the same effect. The males were occasionally seen to link two actions in a way typical only of *Gasterosteus*. These were collecting and mending. The usual building of a nest never involves the addition of material

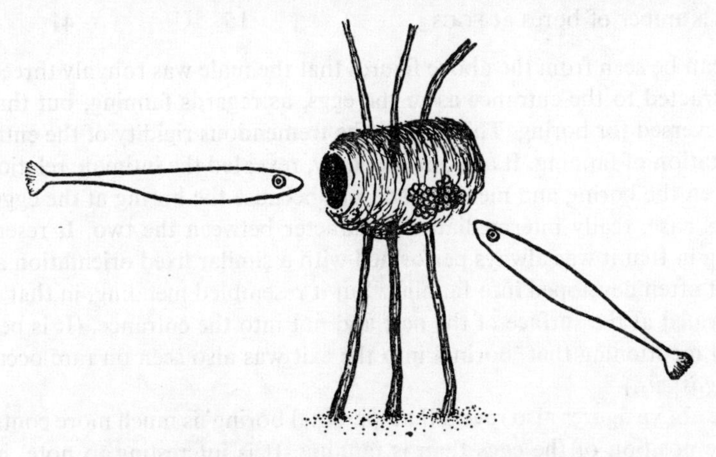

Fig. 44. Nest with eggs visible from below, showing the two fanning positions adopted by the parent male.

to its surface by the use of the nosing action which van Iersel has called 'pushing' and which I have, in *Pygosteus*, referred to as 'mending'. It is the typical nest-building method of *Gasterosteus*, however, and it is most interesting to see that *Pygosteus* under special circumstances is, on rare occasions, capable of collecting a piece of material and then pushing it on to the surface of the nest.

Several of the males which had to build nests round eggs managed to construct quite reasonable structures, although usually not without the loss of some of the eggs from the clutch. It was the males which did not manage so well, however, which provided some further interesting information. These males often ended up with their eggs hanging from the bottom of the back end of the nest, semi-exposed to view. The result was that, in several instances, these males were in a conflict as regards the fanning position they were to take up. One male, which had made a very distinct entrance to its second nest, developed two quite distinct fanning positions, as shown in Fig. 44. It either fanned the entrance, when it was out of touch with the eggs, or it fanned the eggs, when it had to take up an atypical fanning posture under the nest. The following are the figures for fanning and boring in the two postures, on the two days during which the conflict existed (results are from one half-hour period on each day):

	3rd day of cycle	4th day of cycle
Number of fan-bouts at ENTRANCE	13	29
Number of fan-bouts at EGGS	4	11
Number of bores at ENTRANCE	8	9
Number of bores at EGGS	15	41

It can be seen from the above figures that the male was roughly three times as attracted to the entrance as to the eggs, as regards fanning, but that this was reversed for boring. This reveals the tremendous rigidity of the entrance-orientation of fanning. It also, incidentally, revealed the intimate relationship between the boring and mending actions, because the boring at the eggs was, in this case, really intermediate in character between the two. It resembled boring in that it was always performed with a similar fixed orientation and in that it often developed into fanning. But it resembled mending, in that it was performed at the surface of the nest and not into the entrance. (It is perhaps worth mentioning that 'boring' into the exit was also seen on rare occasions in *Pygosteus*.)

The above figures also reveal that 'parental boring' is much more controlled by the position of the eggs than is fanning. It is interesting to note, in this connection, that one male which fertilized a clutch in a clump of weeds before it had ever built a nest, took some time to develop a fixed fanning position when it ultimately built a nest for them. It built a loose mass of material round the eggs and fanned at them through any likely looking crack. After a day or two it narrowed down its fanning positions to one side of the nest, and eventually to one special position. This reveals the extent to which fanning develops a rigid orientation during the pre-parental phase, when a male has a nest during that phase.

The above described phenomenon of egg-transference does not occur in *Gasterosteus*, as far as I know. If it does do so, then it is certainly far rarer than in *Pygosteus*. I suggest that the reasons for this are twofold. Firstly, as we have seen, the fanning cycle of *Gasterosteus* is far more intense than that of *Pygosteus*, and a major resurgence of nest-building is therefore far less likely to occur during the parental phase. Secondly, the nature of the construction and the position of the nest of *Gasterosteus* renders it less susceptible to the kind of deterioration which *Pygosteus* nests suffer and which calls for the transference of the eggs.

The nursery

I have shown above the way in which the fanning cycle shows a trough when nest-building occurs during the parental phase. Nest-building also may appear in the parental phase at a later stae, as the frequency of fanning falls off with the hatching of the eggs. This has already been illustrated in Fig. 40, which shows a small peak in collecting-of-material just as the fanning cycle tails off. This post-fanning nest-building did not always occur, but usually did so, and

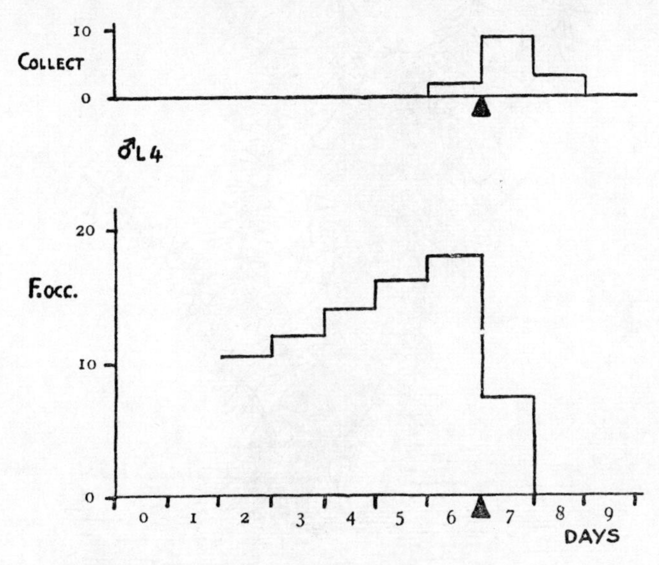

Fig. 45. Outburst of collecting which occurs at the end of the fanning cycle, resulting in the formation of the nursery.

when it was fully expressed revealed itself as a rather special activity: namely, the construction of a nursery for the young fish.

At about the hatching date, males were seen to begin collecting material again (see Fig. 45). This was carried, not to the nest, but to a site just above the nest. Here the material was placed, or gently pushed, into the host plant, rather than rammed in with the typical boring action. The result was that, after a considerable number of collections, there was formed a loose, homogeneous mass of material, without an entrance or tunnel, and without any distinct form (see Fig. 36). It must be emphasized that this was not simply a badly built nest, but a quite distinct construction. This was the result of the elimination of the boring and looping elements of nest-building.

Into this construction, the male eventually spat its young, when retrieving them. When the young have been present in the old nest for several days, they mostly come to lie on top of the nest. From here, one by one, they scatter. The male dashes after them, grabs them in its mouth and returns them. When there is a nursery present the male may return young fish either to it, or to the old nest. Young fish which are caught by the male at a position *below* the nest are returned to the nest. Those caught at a position *above* the nest are

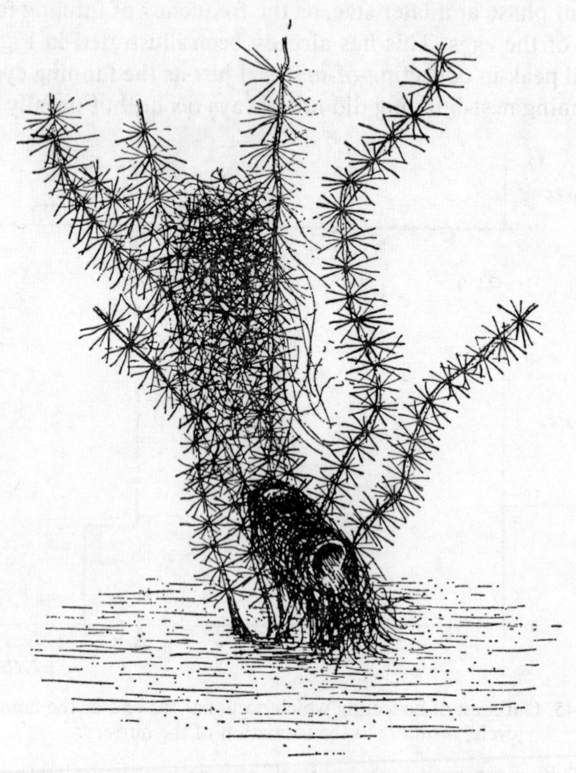

Fig. 46. The nursery of *Pygosteus*, in the host plant, above the old nest.

returned to the nursery. In the latter, the young fish can hide very efficiently in the loosely packed, but quite dense, plant fragments.

The nursery of *Pygosteus* appears to exist as a kind of alternative for the shoaling of *Gasterosteus*. In the latter species, the young keep together in a group which is controlled by the parent male. All attempts to produce shoaling behaviour in *Pygosteus* failed. As the young grew older they simply scattered more and more. Even in a 7-foot-long, sparsely planted tank, no shoals developed. The young fish from different nests often intermingled as they scattered, and males were seen to retrieve one another's young.

This is not so surprising when one considers the ecological situation. The young of *Gasterosteus* typically grow up in rather open stretches of water, where it is important that they should be guided and guarded by their parent males. The young of *Pygosteus* grow up in more densely vegetated regions, where, as they scatter, not only do they become lost to the parent males, but they also are more concealed from potential predators. Collections of young fish from streams and ditches revealed that there was a distinct correlation between the sparseness of the vegetation and the size of the fish. The smaller young fish tended to keep to the denser vegetation. In this way, they can pass that period of their lives when their spines are too small to be effective defence weapons, in comparative safety. This applies to both *Pygosteus* and to *Gasterosteus*. The latter, however, do not start out life in the weediest regions, and the shoaling behaviour of this species gets them over the difficult period before they take to the weeds. (It must be added here that a further ecological field study of this problem is badly needed. The above conclusions are based too much on indirect evidence. In this connection, an under-water 'periscope' is at present being designed with which, it is hoped, detailed field observations will be possible.)

VIII

The Reproductive Cycle as a Whole

Introduction

In the foregoing sections, the reproductive behaviour of *Pygosteus* has been dealt with piece by piece. The cycle must now be considered as a whole system. (The author has benefited from many valuable discussions with Dr P. Guiton on the subject of this chapter.)

One of the difficulties in describing and discussing the separate sections of such a cycle will have been seen to be the result of the fact that the phases and the functional units are not perfectly related. For example, in the 'nesting phase', there can be observed not only nesting behaviour but also agonistic behaviour and sexual behaviour; in the 'parental phase', similarly, there may be agonistic or nesting behaviour, and so on. This means that the cycle either has to be described phase by phase, or functional unit by functional unit. A compromise is made possible, however, by the fact that there is an approximate phase-unit relationship, and although the sections of the present report are based on units rather than phases, the latter are nevertheless characterized at each stage. Even so, it is helpful here to look at the cycle phase by phase. Each phase can be recorded quantitatively with respect to the frequency of the various reproductive activities. By placing together a number of such records, for all the important stages of the cycle, it is possible to gain a general impression of the way in which the different responses fluctuate as the cycle progresses.

The phases of the cycle

Figs 47 and 48 show the phase-by-phase change in the reproductive behaviour of *Pygosteus* throughout its cycle. The data given for each phase are based on averages from a number of tests and together involve a large number of fish. The figures given represent the average number of times each action occurred per half-hour.

It is, unfortunately, impossible to include values for all the different actions on one diagram; the resulting picture would be far too confused. To overcome this, the results are presented in two separate figures; also sexual and agonistic actions are omitted. (The reasons for these particular omissions will become clear from the following.)

In the Figs 47 and 48, an ideal cycle in time is shown. The stages A to J represent the following phases:

A: Pre-territorial aggressive phase. (At the beginning of the season, this may last several weeks, but is here shown as one day.)

B: Pre-nesting territorial phase. (May last a number of days, but here shown as one.)

C: Nest-building phase. (Active period of building lasts about half a day, but shown here as one day.)

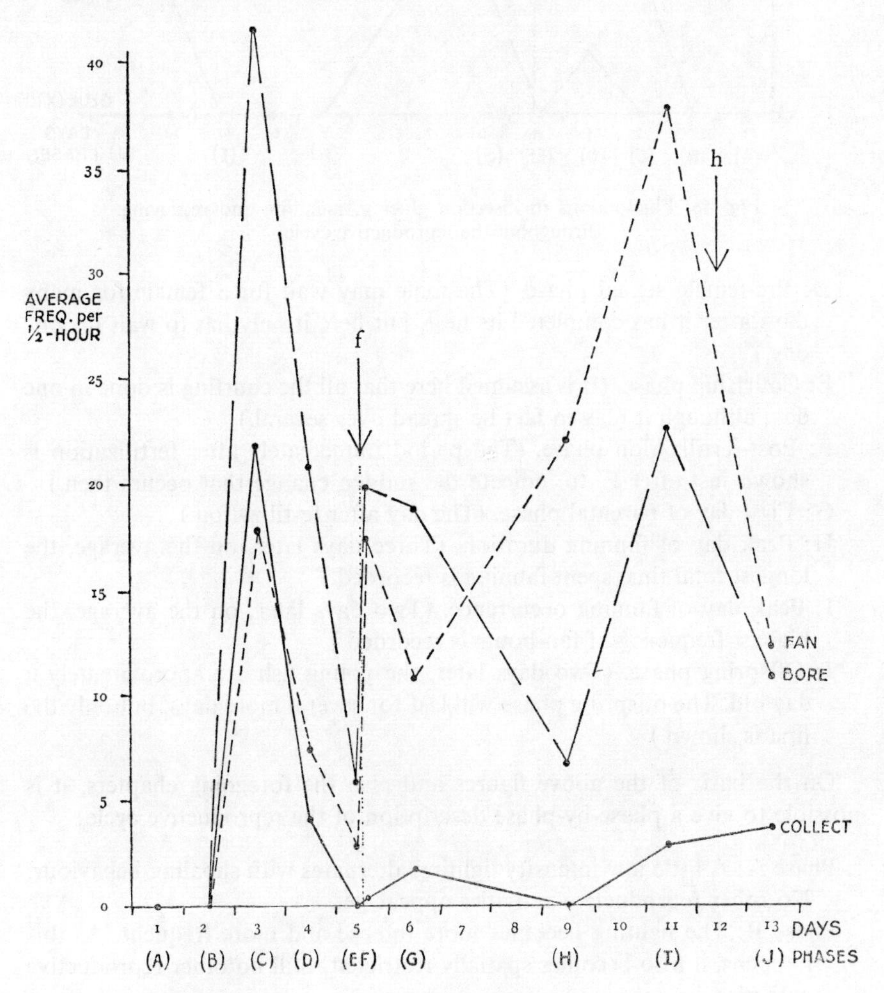

Fig. 47. Fluctuations in collecting, boring and fanning throughout the reproductive cycle.

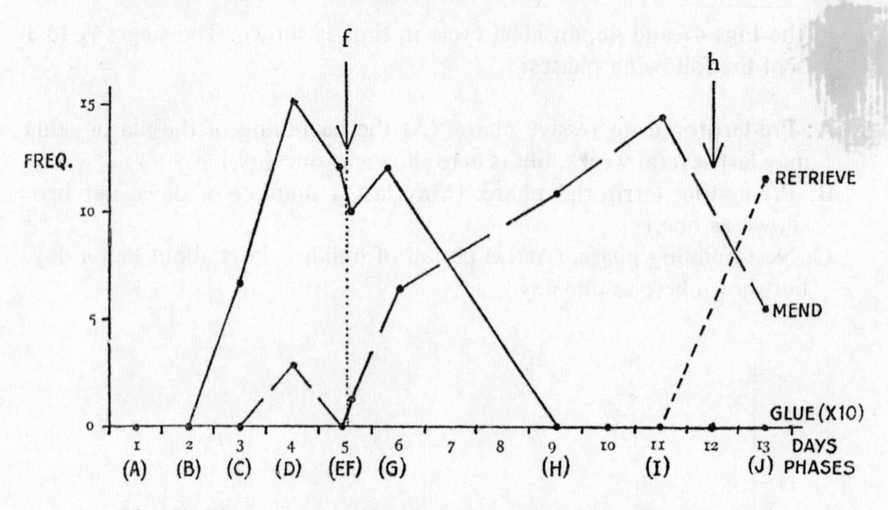

Fig. 48. Fluctuations in insertion glueing, mending and retrieving, throughout the reproductive cycle.

D: Pre-female sexual phase. (The male may wait for a female for many days after it has completed its nest, but here it only has to wait for one day.)

E: Courtship phase. (It is assumed here that all the courting is done in one day, although it may in fact be spread over several.)

F: Post-fertilization phase. (The period immediately after fertilization is shown just after E, to indicate the sudden change that occurs then.)

G: First day of parental phase. (The day after fertilization.)

H: Peak day of fanning duration. (Three days later, on the average, the longest total time spent fanning is recorded.)

I: Peak day of fanning occurrence. (Two days later, on the average, the highest frequency of fan-bouts is recorded.)

J: Offspring phase. (Two days later, the young fish are approximately a day old. The offspring phase will last for several more days, but only the first is shown.)

On the basis of the above figures and also the foregoing chapters, it is possible to give a phase-by-phase description of the reproductive cycle:

Phase A: A little low intensity fighting alternates with shoaling behaviour. No other reproductive activities appear yet.

Phase B: The fighting becomes more intense and more frequent. As this happens, it also becomes spatially restricted. Still no other reproductive activities.

Phase C: Nest-building actions appear. There is high-frequency collecting of material, boring and fanning. Lower frequency glueing. Although superficial glueing is of low frequency, however, this is the only time it appears in the cycle. Insertion glueing only appears towards the end of the building. Aggression slackens off a little during building. Sexual responses to ripe females can be observed, but generally only in a very mild form.

Phase D: Fall-off in collecting, boring and fanning, with the completion of the nest. Insertion glueing increases and reaches its peak. Nest-mending movements appear for the first time. Aggression is high during this period. Sex is moderately high and may get higher day by day if no females are present.

Phase E: When a ripe female is presented, there is a fall-off in all nest-actions, including fanning. During the courtship, sex increases and dominates aggression.

Phase F: Immediately after fertilization sex disappears and aggression dominates again. There is a sudden large increase in boring and fanning, but no such increase in collecting, mending or glueing.

Phase G: A day later, sexual responses are still given to a ripe female, if presented. Aggression moderate. There is less fanning now than there was immediately after fertilization. An increase in mending occurs, but there is little other change.

Phase H: Three days later, at the peak of fanning duration, there has not only been an increase in the frequency of occurrence of fanning but also a further increase in nest-mending. Glueing has stopped altogether now, and does not reappear. Collecting is also completely absent, and boring shows a large reduction in frequency. No sexual responses can be obtained at this stage, but the fish remains moderately aggressive.

Phase I: Two days later the fanning frequency reaches its peak, and with it there is an increase in boring again. Collecting also reappears, but is infrequent. Nest-mending shows still further increase.

Phase J: The day after the young have hatched, there is a considerable increase in aggression; this activity becomes intense and at the same time the action of retrieving the young to the nest is observed. Fanning, mending and boring fall off sharply; only collecting shows any increase (and this results in the formation of the nursery). During the following days, aggression and retrieving fall off together gradually, at a slower rate than collecting. Collecting, however, falls off more slowly than the other nest actions, and is therefore not only the first nesting action to appear in the cycle, but also the last.

The organization of the cycle

The problem remains as to what can be concluded, on the basis of the above description, concerning the organization of the cycle as a whole. Clearly, if all the details are to be taken into consideration, no very simple explanatory system will suffice.

It is necessary here to examine the hierarchical theory postulated by Tinbergen (1942, 1950, 1951a), in connection with the reproductive behaviour of *Gasterosteus*. Tinbergen postulates a top reproductive centre of neural organization, to which are subordinated four sub-centres: fighting, building, mating and care of offspring. To each of these sub-centres are subordinated a number of simple act centres. For example, chasing, biting and threatening, etc., are subordinated to the fighting sub-centre, and so on. (See Fig. 98 in Tinbergen, 1951a.)

As a general principle, this is undoubtedly of tremendous importance in understanding the way in which elaborate behaviour complexes are controlled and organized; but, in the case of the reproductive cycle of *Pygosteus*, it requires some adjustment and modification, when this behaviour complex is examined in detail. It is important to stress, however, that any such adjustments as are necessary in no way invalidate the basic principle of hierarchical organization as laid down by Tinbergen.

Firstly, in what respects do the present findings agree with the hierarchy theory? *Pygosteus* can probably be said to have a top reproductive centre and subordinated to this a fighting and a mating sub-centre, in Tinbergen's sense. The various fighting acts show perfect co-variance, as do the mating acts. I am less happy, however, about postulating a nesting, or a parental sub-centre, on the basis of the evidence given in the present report. (It is for this reason, incidentally, that I have concentrated on these latter activities in Figs 47 and 48, and have omitted fighting and mating there.) The exact reasons for this must now be given.

In Fig. 47 I have shown *Collecting, Boring* and *Fanning* fluctuations and it can be seen that they co-vary to some extent. All three show a marked increase during Phase C, when the nest is being built, and a marked decrease during the sexual phases D and E. After this, however, they show less co-variance. Fanning increases steadily during the parental cycle, whereas collecting and boring decrease. Boring does, however, show another peak at the time of the fanning peak, but collecting does not reach its last (minor) peak until the other two have subsided.

Of the other nest-actions (see Fig. 48), *Mending* co-varies fairly accurately with parental fanning, being absent from the nest-building proper, and also *not* suffering from the suppressing effect which the parental cycle has on all other nest actions (except fanning, of course) during Phase H, which is the peak of fanning duration.

Insertion glueing appears, rather unexpectedly, to co-vary with sex, rather than with any of the other nest actions. It does not reach its peak until the nest is complete and the male is ready for a female, and it disappears completely (like sex) during the parental cycle. *Superficial glueing*, however, co-varies with nest-building proper, only ever being seen in Phase C.

Retrieving the young co-varies with no other nest action, but takes over from parental fanning when the eggs hatch.

Summing up the above, it is clear that it is far from easy to state that this action is a nesting act and belongs to the 'nesting sub-centre', or that that action is parental and belongs to the 'parental sub-centre'. This can be done in the case of courtship actions, for example, or aggressive actions, but here, in the case of the nest actions, the situation demands a different approach.

The 'CFB' unit

Collecting material, boring it into the nest and then fanning, is a combination (CBF) which is frequently observed in *Pygosteus*. It may occur as one continuous complex action, or it may occur in different partial forms at different times. Now, although the three component actions have been observed to co-vary to some extent through the cycle, they do not do so perfectly, and it is valid to ask what form this imperfection takes.

It was noted that, as the cycle progressed, the relative frequencies of the different combinations of C, B and F varied in a rather special way. At the beginning of the cycle, C alone was seen; later CB became more frequent; later still CBF took over, then BF, and finally F alone. This can be diagrammatized as follows:

```
C
C B
C B F
  B F
    F
```

During the early nest-building phase, males often picked up and carried material, but then did not bore it into a host plant. Soon they did so, however, but did not fan after boring. As the nest progressed, fanning was seen to appear at the end of boring and when the nest was completed, collecting dropped out, but boring and fanning still occurred together. Not until the parental cycle had begun, did fanning appear without boring preceding it, to complete the shift from C, through B, to F.

On the basis of this, it is tempting to postulate a CBF unit, or centre, which is *activated as a whole* during the reproductive cycle and which varies the exact nature of its output according to the external stimulus situation. A quantitative record of this variation is shown in Fig. 49. This gives the percentage of CBF activities which are C, B and F respectively at the different stages of the reproductive cycle. It shows clearly the three peaks in C, B and F (marked with arrows), which would be expected from the above.

If the concept of a CBF unit is valid, it makes it impossible to consider pre-parental fanning in *Pygosteus* as displacement fanning. (This is certainly in line with all other indications in this species.)

I suggest that the evolutionary explanation of this is that sexual behaviour has appeared later and later in the reproductive cycle and has wedged itself in, so to speak, between the earlier and later stages of behaviour associated

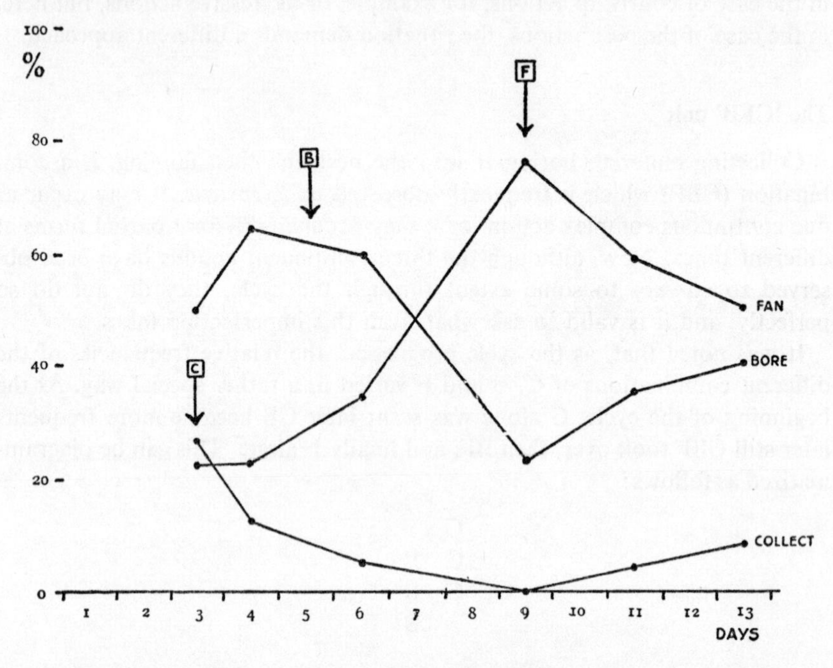

Fig. 49. Changes in proportions of the CBF complex. Explanation in text.

with care of the offspring. It is plausible that eggs were once laid in dense clumps of weeds, where the male protected them from enemies. Later the male probably strengthened the host clump himself, eventually covering the eggs with a rough nest. Although this undoubtedly improved their protection, it must have increased their aeration requirements. Thus, building and ventilating (CBF) probably became *intensified together* in evolution. But selection may have favoured those males which began to build *before* they possessed eggs. (We have seen that it is more difficult to build a nest round eggs than it is to make a nest before acquiring eggs.) In this way, the care-of-offspring behaviour became partially separated into two phases: pre-parental and parental. This partial separation would explain many of the present difficulties.

Although I shall be comparing *Pygosteus* and *Gasterosteus* below, it may be said at this stage that this separation seems to have gone further in the latter species. (In birds, the separation of nesting from parental behaviour is much more distinctly marked.)

It must be borne in mind that the above evolutionary remarks and suggestions are extremely hypothetical, and must be regarded as highly tentative. Something of the kind is necessary, however, to emphasize the fact that nesting and parental behaviour patterns are far from distinct from one another in *Pygosteus*, and to stress that it is therefore dangerous to talk of any of their component actions as occurring as displacement activities, without very special confirmative evidence. When a bird regularly wipes its beak during courtship, the action of beak-wiping is so distinct from any sexual activity that there is little danger in calling it a displacement activity, but when *Pygosteus* performs fanning during the pre-parental phase, it is very dangerous to call this displacement fanning, simply because no eggs are yet present.

Summarizing, it may be said that, in the case of *Pygosteus*, it is not helpful to speak of a nesting centre or a nesting drive, or of a parental centre or a parental drive. Rather, it is better to *consider separately* each response which is functionally concerned with the care of offspring. This avoids a number of difficulties which arise if one attempts to fit all the results into a rigid hierarchy system.

Comparison with 'Gasterosteus'

Throughout the present report the reproductive behaviour of *Pygosteus* has been compared with that of *Gasterosteus*, with respect to various details of the cycle. It is fitting here to make a more general comparison of the two cycles as whole systems.

Leiner made the first serious attempt to compare the behaviour of the two species (1931a, 1934). He came to the conclusion that it was easier to derive the *Gasterosteus* reproductive pattern from the *Pygosteus* pattern than vice versa, and stressed that the behaviour of the former species is much more complicated and highly developed. He cites the anatomical evidence which suggests that the few-spined *Gasterosteus* evolved from a many-spined ancestor. (For details of the anatomical evidence, which concerns the formation of the dorsal bone plates, see Bertin, 1925.) In fact, this anatomical data is probably the most important evidence in support of his conclusions, the behavioural data he presents being far less convincing. His arguments involve three forms of naivety: firstly, he assumes that the behaviour of contemporary *Pygosteus* is the same as that of the many-spined ancestor of *Gasterosteus*.

Secondly, he assumes that simplicity cannot evolve from complexity. Thirdly, he overlooks the possibility of convergence. Despite this, however, and despite the fact that I do not agree with some of the details of his arguments, I am inclined to agree with his general conclusions, for the following reasons.

In the appendix to Morris, 1954e, I have described experiments which show that the few large spines of *Gasterosteus* (see Fig. 50) are more efficient

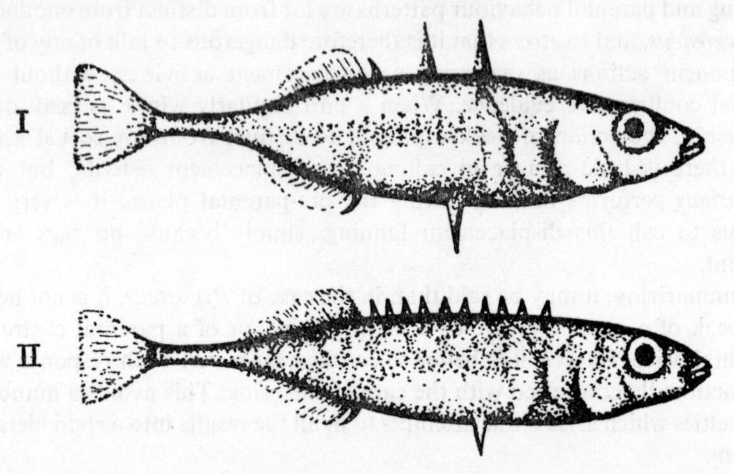

Fig. 50. I: The Three-spined Stickleback (*Gasterosteus aculeatus* L.). II: The Ten-spined Stickleback (*Pygosteus pungitius* L.). Note the fewer, larger spines of *Gasterosteus* and the fact that the body is relatively taller than that of *Pygosteus*. This increased body height separates the dorsal and ventral spines more than in *Pygosteus*, thus rendering them more formidable in the mouth or throat of a predator. In both species of stickleback, the anterior dorsal fin rays, the pelvic fins and the first anal fin ray, have become modified to sharp spines.

defence weapons against predators than the many smaller spines of *Pygosteus*. This, coupled with the skeletal evidence that *Gasterosteus* has evolved from a many-spined ancestor, presents a strong case for what I consider to be the relevant fundamental evolutionary trend. My data suggest that, with increased defensive capacities, *Gasterosteus* was able to leave the shelter of the aquatic vegetation, and occupy the vast new ecological niche of sandy-muddy substrata, in the reproductive season. As I have pointed out in the appendix, *Pygosteus* compensates for its defensive shortcomings by its timidity, which keeps it within the protection of the dense vegetation.

I suggest that most of the behaviour differences between the two species can be thought of as the result of the fact that *Gasterosteus* was thus able to become more independent of the vegetation during the reproductive season. From this point of view, it is interesting to examine the different types of reproductive behaviour, one by one:

Fighting: There is little difference between the two species. Both show biting, carrying, roundabout fighting, head-down threat, ventral rolling, etc.; the minor differences that exist (in spine-raising, for example) have yet to be studied. One interesting point is that a beaten male *Gasterosteus* is less inclined to show the tail-down submissive posture than a beaten male *Pygosteus*. This is correlated with the fact that the tail-down display of the ripe female *Gasterosteus* is far more exaggerated (and ritualized?) than the similar display in *Pygosteus*. It seems that the extra importance of this display as a sexual signal in *Gasterosteus* has been accompanied by a decrease in its importance in its original agonistic context.

The fact that intraspecific fighting is similar in the two species is not surprising, since there is little relevant difference in the two environments.

Nesting: The greatest difference between the two species concerns their modes of nest-building. Correlated with its emancipation from the weeds, *Gasterosteus* has modified its nest in the new environment. Most differences in nesting are purely quantitative. *Gasterosteus* and *Pygosteus* both possess sand-digging, boring and mending (pushing) movements, but *Gasterosteus* uses boring less and digging and pushing more, when building. Also *Gasterosteus* uses superficial glueing far more than *Pygosteus*, but possesses no insertion glueing. Differences in frequencies, sequences and orientations of the *same* motor patterns result in the formation of extremely different nests. Of the few qualitative differences, insertion glueing and looping, which are unique to *Pygosteus*, and sucking (see van Iersel, 1953, p. 10), which is unique to *Gasterosteus*, can be mentioned. It is, nevertheless, variations in the same motor patterns which account for most differences, rather than the possession of different motor patterns. It is emerging from contemporary ethological studies that this last point is of extremely widespread application and of paramount importance in the understanding of the evolution of behaviour.

Courtship: Basically, the courtship sequences of the two species are very similar. In both species, the male dances, leads, shows the entrance, shivers and then spawns. The behaviour of the females is the same in both cases. The differences between the forms of the various male displays have already been discussed, but it is worth repeating here that the slower leading to the nest by *Pygosteus* and the fact that it does not repeatedly rush to the nest whilst courting, are correlated with the denser vegetation in which it performs.

Parental: The less favourable position for the eggs of *Gasterosteus* is correlated with more heavily protected eggs, and a more intense fanning cycle. The latter is longer and reaches a higher peak. A qualitative difference between the two species, in their parental behaviour, is the added ventilation afforded by the making of holes in its nest by *Gasterosteus* (see Figs 3 to 8 in Leiner, 1931b), towards the end of the fanning cycle.

The above specific differences, briefly presented as they are, nevertheless show clearly the way in which the two species are adapted to their different reproductive environments. It is one thing, however, to suggest that *Gasterosteus* has evolved 'away from the weeds' and then to show how it is adapted to its new environment by comparing it with the weed-bound *Pygosteus*, as I have done here; but it is another thing altogether to attempt to *derive* the reproductive behaviour pattern of *Gasterosteus* from that of *Pygosteus*, as Leiner has done (1931a).

It is hoped that future research on other species of stickleback will make it possible to draw richer evolutionary conclusions. From this point of view, it is unfortunate that there are so few species of stickleback available, when they are such admirable subjects for ethological research. Leiner (1934) and Sevenster (1951) have already contributed to our comparative knowledge of one other species, namely *Spinachia*, the sea stickleback, but as yet we know hardly anything about the reproductive behaviour of either *Apeltes* or *Eucalia*. Both the latter species are now under observation by ethologists.

Conclusion

The study on which this report has been based has been carried out over a period of three years. The report has been shortened purposely, by the omission of a considerable amount of quantitative data which has been obtained, in an attempt to present it as a concise and, I hope, coherent story of a complex behaviour pattern. Much of the more specialized quantitative data (dealing with such problems as Typical Intensity, for example) will be given in subsequent publications, which will be concerned with special problems, rather than particular species. It was felt that it would be a mistake, when presenting a 'species study', as here, to include such material, for fear of obscuring the central theme, namely, that of the nature and course of the reproductive cycle of *Pygosteus*.

Whilst the work on *Pygosteus* was being carried out, the behaviour of a number of other species of vertebrates was also studied, in order to guard against unwarranted over-generalization on the basis of uni-specific research. Fish and birds in particular, but also certain amphibia and mammals, were studied simultaneously, and much of the findings with these other species has already been published (Morris, 1954a, 1954b, 1954c, 1954d, 1955a, 1955b, 1956). The value of such widespread comparative studies cannot be overstressed. It is ignored completely by the vast majority of animal psychologists, and, although it is part of the very essence of the ethological approach, a number of contemporary ethologists are far too narrow in the range of species which they study seriously. Only when one studies a number of very *different* species can one begin to understand the way in which behaviour patterns in general have evolved. The FAM courtship hypothesis presented here, for example, would have been impossible without such wide comparative studies. (If shortcomings are found to exist in this hypothesis they will be found, I am sure, as a result of even wider comparative studies, rather than more detailed specific studies.)

Several of the studies on other species have concentrated on the reproductive behaviour patterns of the species concerned, and some of these findings are particularly relevant to the present study of *Pygosteus* (e.g. Morris, 1954c).

References

ARMSTRONG, E. A. (1950), 'The nature and function of displacement activities'. *Sympos. Soc. exp. Biol.* 4, pp. 361–87.

BAERENDS, G. P. and J. BAERENDS VAN ROON (1950), 'An introduction to the study of the ethology of Cichlid fishes'. *Behaviour*, Supplement 1, pp. 1–243.

BASTOCK, M. and A. MANNING (1955), 'The courtship of *Drosophila melanogaster*'. *Behaviour* 8, pp. 85–112.

——, D. MORRIS and M. MOYNIHAN (1953), 'Some comments on conflict and thwarting in animals'. *Behaviour* 6, pp. 66–84.

BATEMAN, G. C. (1890), *Fresh-water Aquaria* (Upcott Gill, London).

BERTIN, L. (1925), 'Recherches bionomiques, biométriques et systématiques sur les Épinoches (Gastérostéidés)'. *Ann. Inst. Océan. Monaco* 2, pp. 1–204.

BLEGVAD, H. (1917), 'On the food of fish in the Danish waters within the Skaw'. *Rep. Danish biol. Sta.* 24, pp. 19–72.

BOESEMAN, M., J. VAN DER DRIFT, J. M. VAN ROON, N. TINBERGEN and J. TER PELKWIJK (1938), 'De bittervoorns en hun mossels'. *De Lev. Nat.* 43, pp. 129–36.

BORCEA, J. (1904), 'Quelques observations sur une Épinoche, *Gasterosteus aculeatus*, provenant d'une rivière se déversant au fond de la Baie Aber, près du Laboratoire de Roscoff'. *Bull. Soc. zool. France* 29, pp. 140–1.

COLLIAS, N. E. (1944), 'Aggressive behaviour among vertebrate animals'. *Physiol. Zool.* 17, pp. 83–123.

COSTE, M. (1846), 'Note sur la manière dont les Épinoches construisent leur nid et soignent leurs œufs'. *C. R. Ac. Sc. Paris* 22, pp. 814–18.

—— (1848), 'Nidification des Épinoches et des Épinochettes'. *Mem. Ac. Sc. Paris* 10, p. 575.

COURRIER R. (1925), 'Les caractères sexuels secondaires et la cycle testiculaire chez l'Épinoche'. *Arch. Anat. Histol. Embryol. Strasbourg* 4, pp. 471–6.

CRAIG, W. (1918), 'Appetites and aversions as constituents of instincts'. *Biol. Bull.* 34, pp. 91–107.

CRAIG-BENNETT, A. (1931), 'The reproductive cycle of the Three-spined Stickleback, *Gasterosteus aculeatus* L.'. *Phil. Trans. Roy. Soc. London* B 219, pp. 197–279.

DAANJE, A. (1950), 'On locomotory movements in birds and the intention movements derived from them'. *Behaviour* 3, pp. 48–98.

DEUTSCH, A. (1953), 'A new type of behaviour theory'. *Brit. J. Psychol.* 44 (4), pp. 304–17.

EECKHOUDT, J. P. V. D. (1947), 'Recherches sur l'influence de la lumière sur le cycle sexuel de l'Épinoche (*Gasterosteus aculeatus*)'. *Ann. Soc. Roy. Zool. Belgique* 77, pp. 83–9.

EVERS, M. (1878), 'Zur Charakteristik des Stichlings (*Gasterosteus aculeatus*)'. *Jahresber. naturw. Ver. Elberfeld* 5, pp. 26–46.

FABRICIUS, E. (1953), 'Aquarium observations on the spawning behaviour of the Char, *Salmo alpinus*'. *Rep. Inst. Fresh-water Res.*, Drottingholm 34, pp. 14–48.

FORBES, H. O. (1897), 'Habitat of *Gasterosteus pungitius* L.'. *Bull. Liverpool Mus.* 1, p. 24.

GIARD, A. (1900), 'Sur l'adaptation brusque de l'Épinoche (*Gasterosteus trachurus* Cuv.) aux eaux alternativement douces et marines'. *C. R. Soc. Biol.* 52, pp. 46–8.

GILL, T. (1907), 'Parental care among fresh-water fishes'. *Rep. Smithsonian. Inst. for 1905* (1688), pp. 403–531.

GRASSÉ, P. P. (1956), 'La régulation des activities instinctives, surtout dans le cas des insectes'. *Fondation Singer-Polignac: Colloque Internat. sur l'Instinct*, Paris 1954.

HANCOCK, A. (1852), 'Observations on the nidification of *Gasterosteus aculeatus* and *Gasterosteus spinachia*'. *Ann. Mag. Nat. Hist.*, ser. 2 (10), pp. 241–8.

HARTLEY, P. H. T. (1948), 'Food and feeding relationships in a community of fresh-water fishes'. *J. Anim. Ecol.* 17, pp. 1–14.

HAYES, J. S., W. M. S. RUSSELL, C. HAYES and A. KOHSEN (1953), 'The mechanism of an instinctive control system: a hypothesis'. *Behaviour* 6, pp. 85–119.

HEDIGER, H. (1950), *Wild Animals in Captivity* (Butterworths, London).

HEINROTH, O. (1911), 'Beiträge zur Biologie, namentlich Ethologie und Psychologie der Anatiden'. *Verh. V. Intern. Ornithol. Kongr., Berlin*, pp. 589–702.

HESS, W. N. (1918), 'A seasonal study of the kidney of the Five-spined Stickleback, *Eucalia inconstans*', *Indianapolis Proc. Acad. Sci.*, pp. 295–6.

HEUTS, M. J. (1942), 'Chloride-excretie bij *Gasterosteus aculeatus* L.'. *Ann. Soc. Roy. Zool. Belgique* 73, pp. 69–72.

—— (1943), 'La régulation osmotique chez l'épinoche (*Pygosteus pungitius* L.)'. *Ann. Soc. Roy. Zool. Belgique* 74, pp. 99–105.

—— (1944a), 'Laterale Bepantsering en Groei bij *Gasterosteus aculeatus*. I'. *Natuurwetensch. Tijdschr.* 26, pp. 40–52.

—— (1944b), 'Calcium-ionen en geographische verspreiding van *Gasterosteus aculateus*'. *Natuurwetensch. Tijdschr.* 26, pp. 10–14.

—— (1947a), 'The phenotypical variability of *Gasterosteus aculeatus* L. populations in Belgium'. *Meded. Kon. Vl. Akad. Wetens.* 9, pp. 5–63.

—— (1947b), 'Experimental studies on adaptive evolution in *Gasterosteus aculeatus* L.'. *Evolution* 1, pp. 89–102.

—— (1947c), 'Laterale Bepantsering en Groei bij *Gasterosteus aculeatus*. II'. *Natuurwetensch. Tijdschr.* 29, pp. 193–6.

—— (1949a), 'On the mechanism and the nature of adaptive evolution'. *Sympos. sui fattori ecologici e genetici della speciazione negli animali*, pp. 3–12.

—— (1949b), 'Racial divergence in fin ray variation patterns in *Gasterosteus aculeatus*'. *J. Genet.* 49, pp. 183–91.

HINDE, R. A. (1953), 'The conflict between drives in the courtship and copulation of the Chaffinch'. *Behaviour* 5, pp. 1–31.

—— (1954), 'Changes in responsiveness to a constant stimulus'. *Brit. J. Anim. Behav.* 2, pp. 41–55.

HOOGLAND, R. D. (1951), 'On the fixing-mechanism in the spines of *Gasterosteus aculeatus* L.'. *Konink. Neder. Akad. van Weten. Proc.*, series C, 54, pp. 171–80.

HOOGLAND, R. D., D. MORRIS and N. TINBERGEN (1957), 'The spines of sticklebacks (*Gasterosteus* and *Pygosteus*) as means of defence against predators (*Perca* and *Esox*)'. *Behaviour* 10, pp. 205–37.

HOWARD, H. E. (1920), *Territory in Bird Life* (John Murray, London).

HUXLEY, J. S. (1923), 'Courtship activities in the Red-throated Diver (*Colymbus stellatus* Pontopp.); together with a discussion on the evolution of courtship in birds'. *J. Linn. Soc.* (*Zool.*) 35, pp. 253–92.

HYNES, H. B. N. (1950), 'The food of freshwater sticklebacks'. *J. Anim. Ecol.* 19, pp. 36–58.

IERSEL, J. J. A. VAN (1953), 'An analysis of the parental behaviour of the male Three-spined Stickleback'. *Behaviour*, Supplement 3, pp. 1–159.

IKEDA, K. (1933), 'Effect of castration on the secondary sexual characters of anadromous Three-spined Sticklebacks (*Gasterosteus aculeatus* L.)'. *Jap. J. Zool.* 5, pp. 135–57.

INNES, W. T. (1951), *Exotic Aquarium Fishes* (Innes, Philadelphia).

JONES, J. R. E. (1947a), 'The oxygen consumption of *Gasterosteus aculeatus* L. in toxic solutions'. *J. Exp. Biol.* 23, pp. 298–311.

—— (1947b), 'The reactions of *Pygosteus pungitius* L. to toxic solutions'. *J. Exp. Biol.* 24, pp. 110–22.

—— (1948), 'A further study of the reactions of fish to toxic solutions'. *J. Exp. Biol.* 25, pp. 22–34.

—— (1952), 'The reactions of fish to water of low oxygen concentration'. *J. Exp. Biol.* 29, pp. 403–15.

KERR, T. (1943), 'A comparative study of some Teleost pituitaries'. *Proc. Zool. Soc. London* (A) 112, pp. 36–56.

KOCH, H. J. and M. J. HEUTS (1942), 'Influence de l'hormone thyroidienne sur la régulation osmotique chez *Gasterosteus aculeatus* L. forme *gymnurus*'. *Ann. Soc. Roy. Zool. Belgique* 73, pp. 165–72.

—— (1943), 'Régulation osmotique, cycle sexual et migration de reproduction chez les Épinoches'. *Arch. intern. Physiol.* 53, pp. 253–66.

KORTLANDT, A. (1940), 'Wechselwirkung zwischen Instinkten'. *Arch. neerl. Zool.* 4, pp. 442–520.

KUNZ, A. and L. RADCLIFFE (1918), 'Notes on the embryology and larval development of twelve Teleostean fishes'. *Washington Bull. Bur. Fish.* 35, 1915–1916, pp. 87–134.

LACK, D. (1943), *The Life of the Robin* (Witherby, London).

—— and L. LACK (1933), 'Territory reviewed'. *Brit. Birds* 27, pp. 174–99.

LANDOIS, H. (1871), 'Der kleine Stichling (*Gasterosteus pungitius*) und sein Nestbau', *Zool. Garten* 12, pp. 1–10.

LEINER, M. (1929), 'Ökologische Studien an *Gasterosteus aculeatus*'. *Z. Morph. Ökol. Tiere* 14, pp. 360–99.

—— (1930), 'Fortsetzung der ökologischen Studien an *Gasterosteus aculeatus*'. *Z. Morph. Ökol. Tiere* 16, pp. 499–540.

—— (1931a), 'Der Laich- und Brutpflegeinstinkt des Zwergstichlings, *Gasterosteus* (*Pygosteus*) *pungitius* L.'. *Z. Morph. Ökol. Tiere* 21, pp. 765–88.

—— (1931b), 'Ökologisches von *Gasterosteus aculeatus* L.'. *Zool. Anz.* 93, pp. 317–333.

—— (1934), 'Die drei europäischen Stichlinge (*Gasterosteus aculeatus* L., *Gasterosteus pungitius* L. und *Gasterosteus spinachia* L.) und ihre Kreuzungsprodukte'. *Z. Morph. Ökol. Tiere* 28, pp. 107–54.

LORENZ, K. (1935), 'Der Kumpan in der Umwelt des Vogels'. *J. Ornithol.* 83, pp. 137–213 and 289–413.

—— (1941), 'Vergleichende Bewegungsstudien an Anatinen'. *J. Ornithol.* 89, pp. 194–294.

—— (1950), 'The comparative method in studying innate behaviour patterns'. *Symp. Soc. Exp. Biol.* 4, pp. 221–69.

—— (1952), *King Solomon's Ring* (Methuen, London).

MARKLEY, M. H. (1940), 'Notes on the food habits and parasites of *Gasterosteus aculeatus* in the Sacramento River, California'. *Copeia Ann. Arbor.*, pp. 223–5.

MERRIMAN, D. and H. P. SCHEDL (1941), 'The effects of light and temperature on gametogenesis in the four-spined stickleback, *Apeltes quadracus* (Mitchill)'. *J. Exper. Zool. Philadelphia* 88, pp. 413–46.

MÖBIUS, K. A. (1885a), 'Das Nest des Seestichlings'. *Schrift. Naturwiss. Ver. Schleswig. Holst.* 6, p. 56.

—— (1885b), 'The nest of the Fifteen-spined Stickleback'. *Ann. Mag. Nat. Hist.* 5 (16), p. 153.

—— (1886), 'Über die Eigenschaften und den Ursprung der Schleimfäden des Seestichlingsnest'. *Arch. mikr. Anat.* 25, pp. 554–63.

MORRIS, D. (1952), 'Homosexuality in the Ten-spined Stickleback (*Pygosteus pungitius* L.)'. *Behaviour* 4, pp. 233–61.

—— (1954a), 'The snail-eating behaviour of Thrushes and Blackbirds'. *Brit. Birds* 47, pp. 33–49.

—— (1954b), 'The reproductive behaviour of the Zebra Finch (*Poephila guttata*), with special reference to pseudofemale behaviour and displacement activities'. *Behaviour* 6, pp. 271–322.

—— (1954c), 'The reproductive behaviour of the River Bullhead (*Cottus gobio*), with special reference to the fanning activity'. *Behaviour* 7, pp. 1–31.

—— (1954d), 'The courtship behaviour of the Cutthroat Finch (*Amadina fasciata*)'. *Avic. Mag.*, Sept.-Oct. 1954, pp. 169–77.

—— (1954e), 'An analysis of the reproductive behaviour of the Ten-spined Stickleback (*Pygosteus pungitius* L.)'. Dissertation for the degree of Doctor of Philosophy, University of Oxford.

—— (1955a), 'The causation of pseudofemale and pseudomale behaviour: a further comment'. *Behaviour* 8, pp. 46–57.

—— (1955b), 'The courtship dance of the Mexican Swordtail'. *The Aquarist* 19, pp, 247–9.

—— (1956), 'The function and causation of courtship ceremonies in animals (with special reference to fish)'. *Fondation Singer-Polignac: Colloque Internat. sur l'Instinct*, Paris 1954, pp. 261–86.

MOYNIHAN, M. (1953a), 'Some aspects of the behaviour of the Black-headed Gull (*Larus ridibundus*), and related species'. Doctor Thesis, Oxford.

—— (1953b), 'Some displacement activities of the Black-headed Gull'. *Behaviour* 5, pp. 58–80.

NYMAN, K. (1953), 'Observations on the behaviour of *Gobius microps*'. *Acta Soc. Faun. Flor. Fenn.* 69, pp. 1–11.

OORDT, G. J. VAN (1924a), 'Die Veränderungen des Hodens während des Auftretens der sekundären Geschlechtsmerkmale bei Fischen. I. *Gasterosteus pungitius*'. *Arch. mikr. Anat. Entw. Berlin* 102, pp. 379–405.

—— (1924b), 'The relation of the testis to the secondary sex-characters of the Ten-spined Stickleback'. *Tijdschr. Ned. Dierk. Ver. Helder*, ser. 2, 19, pp. 52–4.

PELKWIJK, J. J. TER and N. TINBERGEN (1936), 'Roodkaakjes'. *De Lev. Nat.* 41, pp. 129–37.

—— (1937), 'Eine reizbiologische Analyse einiger Verhaltensweisen von *Gasterosteus aculeatus* L.'. *Z. Tierpsychol.* 1, pp. 193–204.

PRINCE, E. E. (1885), 'On the nest and development of *Gasterosteus spinachia* at the St. Andrews Marine Laboratory'. *Ann. Mag. Nat. Hist.* 5 (16), pp. 487–96.

—— (1886), 'On the development of the Food Fishes at the St. Andrews Marine Laboratory'. *Rep. Brit. Ass.* 1885, pp. 1901–3.

RANSOM, W. H. (1865), 'On the nest of the Ten-spined Stickleback'. *Ann. Mag. Nat. Hist.* 16, pp. 449–51.

RUSSELL, E. S. (1931), 'Detour experiments with Sticklebacks'. *J. Exper. Biol.* 8, pp. 493–510.

RYDER, J. A. (1881), 'Notes on the development, spinning habits and structure of the Four-spined Stickleback, *Apeltes quadracus*'. *Bull. U.S. Fish. Comm.* 1, pp. 24–9).

—— (1887), 'On the development of osseous fishes, including marine and fresh-water forms'. *Rep. U.S. Fish. Comm.* 13, pp. 488–604.

SEVENSTER, P. (1949), 'Modderbaarsjes'. *De Lev. Nat.* 52, pp. 160–9 and 184–9.

—— (1951), 'The mating of the Sea Stickleback'. *Discovery* 12 (2), pp. 52–6.

SIEDLECKI, M. (1903), 'Sur la résistance des Épinoches aux changements de la pression osmotique du milieu ambiant'. *C. R. Ac. Sc.* 137, pp. 469–71.

STANWORTH, P. (1953), Dissertation for the degree of Doctor of Philosophy in the University of Manchester.

TINBERGEN, N. (1936), 'The function of sexual fighting in birds; and the problem of the origin of territory'. *Bird-Banding* 7, pp. 1–8.

—— (1939), 'On the analysis of social organisation among vertebrates, with special reference to birds'. *Amer. Mid. Nat.* 21, pp. 210–34.

—— (1940), 'Die Übersprungbewegung'. *Z. Tierpsychol.* 4, pp. 1–40.

—— (1942), 'An objectivistic study of the innate behaviour of animals'. *Biblioth. biotheor.* 1, pp. 39–98.

—— (1948a), 'Social releasers and the experimental method required for their study'. *Wilson Bull.* 60, pp. 6–52.

—— (1948b), 'Physiologische Instinktforschung'. *Experentia* 4 (4). pp. 121–33.

—— (1950), 'The hierarchical organisation of nervous mechanisms'. *Symp. Soc. Exp. Biol.* 4, pp. 305–12.

—— (1951a), *The Study of Instinct* (Clarendon Press, Oxford).

—— (1951b), 'On the significance of territory in the Herring Gull'. *Ibis* 94, pp. 158–159.

—— (1951c), 'A note on the origin and evolution of threat display'. *Ibis* 94, pp. 160–2.

—— (1952a), 'Derived activities; their causation, biological significance, origin, and emancipation during evolution'. *Quart. Rev. Biol.* 27, pp. 1–32.

—— (1952b), 'The curious behaviour of the Stickleback'. *Sci. Amer.* 187, pp. 22–6.

—— (1953a), *Social Behaviour in Animals* (Methuen, London).

—— (1953b), 'Fighting and threat in animals'. *New Biol.* 14, pp. 9–24.

—— (1954), 'The origin and evolution of courtship and threat display'. Pp. 233–50 in *Evolution as a Process* (George Allen & Unwin, London).

—— and J. J. A. VAN IERSEL (1947), 'Displacement reactions in the Three-spined Stickleback'. *Behaviour* 1, pp. 56–63.

TINBERGEN, N. and M. MOYNIHAN (1952), 'Head-flagging in the Black-headed Gull'. *Brit. Birds* 45, pp. 19–22.

TITSCHACK, E. (1922), 'Die sekundären Geschlechtsmerkmale von *Gasterosteus aculeatus* L.'. *Zool. Jahrb. Physiol.* 39, pp. 83–148.

VILLALOBOS, C. and J. (1947), *Colour Atlas* (El Ateneo, Buenos Aires).

WALLS, G. L. (1942), *The Vertebrate Eye* (Bloomfield Hills, Michigan: Cranbrook Institute of Science).

WELTY, J. C. (1934), 'Experiments in group behaviour of fishes'. *Physiol. Zool.* 7, pp. 85–128.

WUNDER, W. (1928), 'Experimentelle Untersuchungen an Stichlingen'. *Zool. Anz.*, Suppl. vol. 3, pp. 115–27.

—— (1930), 'Experimentelle Untersuchungen am dreistachligen Stichling (*Gasterosteus aculeatus*) während der Laichzeit'. *Z. Morph. Ökol. Tiere.* 16, pp. 453–98.

—— (1931), 'Brutpflege und Nestbau bei Fischen'. *Erg. der Biol.* 7, pp. 118–95.

The comparative ethology of grassfinches and mannikins

(1958)

Introduction

This report is based on a laboratory study involving 263 Estrildine finches, eighty-four of which were bred during the course of the investigation. A total of twenty-seven species of this subfamily (Estrildinae) of the Ploceidae came under observation at one time or another during the period of the study, which extended from the winter of 1951 to the spring of 1956.

The Estrildinae have been divided into three tribes by Delacour (1943): the waxbills (Estrildae), the grassfinches (Erythrurae) and the mannikins (Amadinae). The first of these three will not be dealt with in any detail here. Only twenty-five waxbills, belonging to six different species, were involved in the present investigation. Their behaviour was only briefly scrutinized. More attention was paid to the 130 grassfinches (belonging to ten species) and the 108 mannikins (belonging to eleven species) that were used.

A number of papers have already been published on various aspects of Estrildine ethology (e.g. Morris, 1954a, b; 1955a, b, c, d, e; 1956a, b; 1957a, b; also see Moynihan and Hall, 1954). These papers render an elaborate introduction superfluous here. For details of the observational techniques used, the reader is referred in particular to Morris (1957b).

It is the aim of the present paper to bring together, in as concise a form as possible, all the ethological data available for this group of birds and, with this material, to re-examine the taxonomy of the group.

The most recent and most comprehensive systematic arrangement for the Estrildines is that provided by Delacour (loc. cit.). It is necessary to summarize this here, before proceeding further. It involves rather drastic changes, in many respects, from the earlier classifications. In recognizing only fifteen genera, with a total of 108 species, Delacour abandons as many as thirty-five old genera and twenty-nine species. Also, certain of the popular names he uses are not those normally employed. Whilst this radical nomenclatural upheaval produces considerable confusion in any discussion of the group and its members, it nevertheless should not be ignored (as it has been by certain recent authors). For Delacour's system, although it will be criticized in respect of certain details, below, presents the first really unified Estrildine classification. It sweeps away many unnecessary specific distinctions and simplifies the general picture considerably.

Delacour's arrangement

A. *Waxbills* (*Estrildae*)

Nine genera, with fifty-one species. Typified briefly as follows: Nestling palates marked with spots; voice high-pitched; vocalize with beak pointing upwards; male often holds grass in beak during courtship dance; more elegant and active than other Estrildines; predominantly Ethiopian.

The species will not be listed here, but the names of the few employed in the present study are: (1) *Estrilda senegala* (red-billed fire-finch); (2) *E. angolensis* (Cordon Bleu); (3) *E. troglodytes* (red-eared waxbill); (4) *E. temporalis* (red-browed finch); (5) *E. melpoda* (orange-cheeked waxbill); (6) *E. amandava* (red avadavat).

B. *Grassfinches* (*Erythrurae*)

Three genera, with twenty-three species. Typified as follows: Intermediate between waxbills and grassfinches – palate markings being nearer to waxbills, but voice unmelodious, like mannikins; in courtship 'the cock sits still, the flank feathers puffed out, the neck extended vertically and the beak turned downward, while he emits, apparently by a great effort, a kind of melody, sometimes inaudible. At the same time, he raises himself up and down with a lateral twisting'; members of one genus (*Zonaeginthus*) include grass-offering; generally clumsier than waxbills, but not so much so as mannikins; predominantly Australasian.

Genus *Zonaeginthus*
Subgenus *Zonaeginthus*
Z. pictus (painted finch)
[*Z. oculatus* (red-eared fire-tail finch)]
[*Z. bellus* (fire-tail finch)]
Z. guttatus (diamond sparrow)
Subgenus *Oreostruthus*
[*Z. fulginosus* (crimson-bellied mountain finch)]
Genus *Poephila*
Subgenus *Neochmia*
P. phaeton (crimson finch)
P. ruficauda (star finch)
P. guttata (zebra finch)
P. bichenovi (Bicheno finch)
[*P. modesta* (cherry finch)]
Subgenus *Poephila*

P. cincta (parson finch)
P. acuticauda (long-tailed grassfinch)
P. personata (masked grassfinch)
P. gouldiae (gouldian finch)

Genus *Erythrura*

(This genus, the parrot finches, comprises nine species. They will not be listed here, since little is known about them and none was included in the present study.)

C. *Mannikins* (*Amadinae*)

Three genera, with thirty-four species. Typified as follows: Nestling palates show horseshoe-like dark lines or large blotches; voice variable but generally like grassfinches; their 'song dance is the same sort of a ventriloquial performance' as that of grassfinches but 'even more static'; bodies longer, heavier and clumsier than in the other two groups; plumage more sombre; sexual dimorphism rare; mannikin species cover almost the whole range of the Estrildine group, but are predominant in the Asiatic region.

Genus *Padda*

P. oryzivora (Java sparrow)
[*P. fuscata* (Timor sparrow)]

Genus *Amadina*

A. fasciata (cutthroat finch)
[*A. erythrocephala* (red-headed finch)]

Genus *Lonchura*

Subgenus *Heteromunia*

L. pectoralis (pectoral finch)

Subgenus *Euodice*

L. malabarica (silverbill)
[*L. griseicapilla* (grey-headed silverbill)]

Subgenus *Lonchura*

[*L. fringilloides* (magpie mannikin)]
L. bicolor (black-breasted mannikin)
L. cucullata (bronze mannikin)
L. striata (striated finch)
L. punctulata (spice finch)
[*L. nana* (bib finch)]
(This subgenus also contains seven other species about which so little is known that they will not be listed here.)

Subgenus *Munia*

L. ferruginosa (chestnut mannikin)
L. maja (pale-headed mannikin)
L. castaneothorax (chestnut-breasted mannikin)

(This subgenus also contains eleven other species about which so little is known that they will not be listed here. They are all island forms from the archipelago stretching from India to Australia, as are the seven unlisted species from the subgenus above.)

The above list requires some brief explanatory comments. Firstly, those species which were not studied in the present investigation are enclosed in square brackets. Secondly, although all the Latin names given are exactly as put forward by Delacour, a few of the popular names have been adjusted to suit general usage. Where several old species have been merged into one by Delacour it has, however, been impossible to use the better-known popular names. For example, the well-known tri-coloured nun and black-headed nun have, along with several other forms, been merged into the species called here the chestnut mannikin. Also, the white-headed nun and the yellow-rumped mannikin have been put together with several other forms as the pale-headed mannikin. The rufous-backed mannikin and the blue-billed mannikin are one with the black-breasted mannikin. Finally, the Indian silverbill and the African silverbill have been merged simply as the silverbill. The alterations in the generic names produce little confusion because the specific names are nearly always the same as in the earlier and better-known classifications. One awkward case is that of the zebra finch, which is still known throughout the avicultural world as *Taeniopygia castanotis* and here has had both its generic and its specific names altered.

It is not worthwhile to give in full here the earlier systems which Delacour has so drastically modified. To revive and discuss all the earlier Latin names employed by older workers would only be confusing. But two other systems must be mentioned briefly. The first is that of Butler (1899). In a standard work on the aviculture of exotic finches, this author groups the Estrildines under the same three headings as Delacour (i.e. waxbills, grassfinches and mannikins). His groupings differ strikingly in some respects from the later arrangement and the differences are therefore listed below. In order to simplify the situation, only the popular names are given, and where Butler has used one which is now out of date I have altered it appropriately:

Butler's arrangement

A. Waxbills
   Crimson finch
   Painted finch
   Fire-tailed finch
      (plus other waxbills as in Delacour)

B. Grassfinches
   Cutthroat finch

Silverbill
(plus other grassfinches as in Delacour)

C. Mannikins
(as in Delacour)

The other system concerns only the Australian species but provides a valuable alternative way of grouping the members of the grassfinch tribe. It has been worked out by the naturalist Mr William Gordon (personal communication), who has been studying these birds in the Australian bush over a period of thirty years and who has bred all but two of the Australian species in captivity. He divides them into four groups as follows:

Gordon's arrangement

1. Fire-tailed finches
   - Fire-tail finch
   - Red-eared fire-tail finch
   - Red-browed finch
   - Diamond sparrow
   - Crimson finch
   - Star finch
   - Cherry finch
   - Painted finch
2. Grassfinches
   - Zebra finch
   - Bicheno finch
   - Parson finch
   - Masked grassfinch
   - Long-tailed grassfinch
3. Munias
   - Chestnut-breasted mannikin
   - Pale-headed mannikin
   - Pectoral finch
4. Gouldian finch

Here, as before, I have standardized the popular names.

With these three taxonomic arrangements as a starting point, it is now possible to survey critically the geographical, morphological and ethological data and to ascertain how and where any improvements can be made in the classification of the group. As already mentioned, the present investigation has been largely ethological, but a rapid glance at the geographical distribution and morphological variations of the Estrildines nevertheless forms a valuable introduction to the behaviour material.

Geographical Distribution

The Estrildines are an exclusively Old World group. Remarkably convergent ecological counterparts are to be found in the New World, in South America; namely, the grasskeets (e.g. *Tiaris canora*, the Cuban finch).

The Old World range of these birds may be roughly classified into three main areas: Africa, Asia and Australia. The first and third of these are geographically precise, but the second is rather vague. It will be used here to cover everywhere in between Africa and Australia; everywhere, that is, from Arabia, through India, down the archipelago to New Guinea and a few small islands beyond. If, having made this crude spatial distinction, the frequencies of the species from the three tribes are scored against the three regions, the following figures emerge (using Delacour's system):

	African	Asiatic	Australian
Waxbills	**48**	2	1
Grassfinches	0	12	**14**
Mannikins	8	**26**	3

It is obvious from these figures that, geographically speaking, it is the mannikins that link the waxbills and the grassfinches, which contrasts with Delacour's suggestion that the grassfinches form a link between the waxbills and the mannikins. The ancestral home of the Ploceidae in general seems to be in Africa, and it is not unreasonable to suggest that the ancestral Estrildine set out, so to speak, from Africa as a kind of primitive mannikin, leaving behind waxbills, scattering new mannikins as they went, and finally ending up in Australia with an outburst of grassfinches. This would, of course, necessitate some rather impressive convergences on the part of waxbills and grassfinches, but nothing more remarkable than many which have taken place between Old World and New World groups.

An alternative suggestion would be that certain primitive waxbills made the Asiatic trip to Australia to form the grassfinch group and that the mannikins were later specializations which cropped up all along the range. Finally, a third and more complicated suggestion is that both waxbills and mannikins made the trek to Australia, where they both evolved new forms which have now been wrongly lumped together as grassfinches. A detailed examination of the morphological and ethological data should help to make one of these three suggestions more plausible, or perhaps even to provide a fourth.

It is worthwhile to look for a moment at the geographically atypical members of the three tribes. Firstly, the non-African waxbills. These are the red avadavat (*Estrilda amandava*, the green avadavat (*Estrilda formosa*) and the red-browed finch, or so-called Sydney waxbill (*Estrilda temporalis*). The range of the red avadavat is extensive, stretching from India and Ceylon, down through southern China and Siam to Java and even as far as Timor. In other words, it almost joins the African waxbills and the Australian grass-finches, and could be the remnant of an ancestoral waxbill movement to Australia. However, there are few obvious specific links with it in either Africa or Australia and it is aberrant in a number of ways, especially in having an eclipse plumage. (It is the only one of the 108 species to have this.) The green avadavat is restricted entirely to central India, and does not appear to provide any particularly valuable evolutionary clues.

The red-browed finch is taxonomically perhaps the most interesting of all the Estrildines. It inhabits the extreme east of Australia. It has been called a waxbill (the Sydney waxbill) by Butler and by Delacour, and yet has been placed without comment amongst other Australian grassfinches by Cayley (1932) and by Gordon. Geographically speaking it should be a grassfinch. As we shall see later, if it is to be considered as a waxbill, then a number of other Australian species should also be placed with the waxbills. Together these would represent a special waxbill invasion of Australia, distinct from the mannikin invasion. This is, incidentally, rather suggested by Butler's classification.

Next, there are eleven species of grassfinch which occur in the Asiatic region. Of these eleven, nine are the parrot finches. This whole group occurs outside Australia, and only one (*Erythrura trichroa*, the blue-faced parrot finch) is found in Australia at all. However, the other parrot finches, although outside, are typically only just outside. Two species reach as far away as the Malay Peninsula, but most are found on the islands near to the north Australian coast.

Of the typical grassfinches, only three species are found outside Australia. One of these, the little-known crimson-bellied mountain finch, inhabits only New Guinea, but the other two, the crimson finch and the zebra finch, are primarily Australian species that have gone just outside the area. The crimson finch is found over N. Australia and also S. New Guinea, but the zebra finch covers the whole of Australia and also occupies Timor and Flores.

On the basis of geographical distribution, then, one might be tempted to separate the grassfinches proper from the parrot finches. Plumage colour and markings would bear this out and Delacour gives no good positive reason for lumping these two types together in the same tribe, except that they have spotted rather than lined nestling palates. As will be shown below, this particular character cannot be used in this way, since the nestling palates of a number of the typical grassfinches are lined and not spotted. However, until some ethological data are obtained for the parrot finches, little more can be said on this subject.

Of the eleven mannikins which occur outside the Asiatic region, the three

Australian ones are particularly interesting. They are the pale-headed mannikin, the chestnut-breasted mannikin and the pectoral finch. The first of these has the most typical mannikin appearance of the three and is also the one with the range that extends farthest into Asia (as far as the Malay Peninsula). The pectoral finch has the most unusual appearance for a mannikin and is confined entirely to N. Australia, while the chestnut-breasted mannikin is intermediate both in appearance and in distribution, occurring both in N. Australia and New Guinea.

The eight African mannikins, on the other hand, are a heterogeneous collection. There are the extremely distinctive cutthroat and red-headed finches, which roughly cover the north and south of Africa respectively. The widespread magpie, black-breasted and bronze mannikins form another distinct little group, as do the grey-headed and common silverbills. The latter is interesting as it forms a link with the Asiatic species, being the only Estrildine which occurs both in Africa and Asia. Its range extends from central and northern Africa, through Arabia and Persia, to India and Ceylon. Finally there is the little-known and rather peculiar bib finch on Madagascar.

Morphology

It is convenient, when discussing the comparative morphology of these birds, to consider mechanical and signal structures separately. Since this paper is concerned primarily with the comparative ethology of the group, emphasis will be given to the second category. There is, in any case, little to say about the first. The birds are almost all adapted to eating small grass-seeds and there is little variation in body proportions. Some species have turned to slightly larger or slightly smaller seeds, with an accompanying alteration in beak size (see Morris, 1955e). The typical Asiatic mannikins, for example, often have massive beaks when compared with the waxbills or grassfinches.

Most species take their food while clinging to the grasses, but some prefer to feed from the ground. In a few of the latter cases, such as the painted finch, the birds have become specially adapted to ground living, with accompanying structural modifications. Most noticeable variant here is the toe-length. Mannikins are the best 'clingers' and their toes have to be clipped repeatedly when they are kept in captivity to prevent them from becoming caught in the wire netting. They appear to be equally at home whether sitting upright, hanging downwards, or clinging to a vertical. It is interesting to note that, with longer toes, there is a more upright carriage of the body. Presumably, shorter toes and a more horizontal deportment go together as adaptations to ground living in these birds.

Apart from the variations mentioned above, there is little of taxonomic value, as far as it is known at present, to be found amongst the slight mechanical-structural variations which exist in the group. Even the ones mentioned here are dubiously non-conservative and are really only helpful at the lowest levels. Indeed, one cannot help wondering why there should be so many species of Estrildines, when it seems as if a fraction of the number would have been just as efficient ecologically. Clearly, a detailed comparative field study is required to solve this particular problem.

There are, however, some valuable taxonomic clues to be gained from a study of the signal morphology of the Estrildines. By this, I refer, of course, to the elaborate plumage markings and colours, to the beak and leg colours, and to the nestling mouth markings.

In general, the waxbills and grassfinches are more brightly adorned than the mannikins. This applies not only to plumage colours, but also to leg and beak colours. The mannikins nearly all present a combination of white, grey, brown and black, or one, or some, of these. The other groups exhibit an exotic range of reds, yellows, blues, greens and even mauves and violets. There are a few exceptions to this rule, one or two of the mannikins having bright patches and one or two grassfinches and waxbills being sombre in appearance. The mannikin exceptions are the cutthroat finch, in which the male possesses a bright red throat patch, the Java sparrow, in which both sexes have a pink beak, and the bib finch, which has a black and red beak and pink legs. Comparatively dull coloured grassfinches include the parson finch and the Bicheno finch, which are entirely black, buff, white and grey, and the long-tailed grassfinch, which is similar but which has a brightly coloured beak. (Waxbills not dealt with here.)

Details of the plumage of each species will not be given here, but a few examples of the way in which plumage differences and similarities can be useful aids to understanding relationships in the present group may be mentioned. Firstly, there are interesting cases of what appear to be extraordinary convergencies. For example, the red-browed finch in Australia and the red-eared waxbill in Africa both have plain greyish bodies, with no special markings: both have bright red beaks and both have bright red eye-streaks. This superficial resemblance, however, does not stand closer scrutiny, when it can be seen that the eyestreak of the Australian species passes more over than through the eye, whereas that of the African species passes straight through the middle. Also, the red of the beaks and the body-grey is different in the two cases. The relationship between the African fire-finch and the Australian crimson finch is similar. Both have crimson plumage which is brighter in the male; both have red beaks and distinctive yellowish eye-rings, and both have tiny white spots on their flank feathers. Nevertheless, a close examination of the plumage of these birds also brings one to the conclusion that this is a case of convergence; for none of the details tally perfectly, one giving the impression of being a bad imitation of the other.

Contrasting with this, there are series of species in which there are many different patches on certain parts of the body but one or two conservative patches which remain constant through the series and which are, upon close scrutiny, clearly of common origin. Take, for example, the characteristic black and white rump-to-vent markings of the masked, long-tail and parson finches, or the dark-brown face masks of the pectoral and chestnut-breasted finches. Close examination reveals these to be more than just similar superficially. They also bear striking resemblances in indefinable subleties of quality.

Distinguishing between resemblances which are the result of convergence and those which are not is difficult to discuss in words: Lorenz (1953, p. 1) has described this difficulty at some length. It is not easy to demonstrate even with skins. One simply has to spend a great deal of time making detailed observations of the living birds at close quarters in captivity. Only then, for example, does one begin to notice that the red eye-brow of the red-browed finch is more genuinely related to the red face mask of the crimson finch, than it is to the superficially more similar eye-streak of the red-eared waxbill. Observations of this kind will not be set out in detail here, but their significance will automatically be taken into account when taxonomic relationships are being considered later in the paper.

One final aspect of the morphology of the Estrildines must now be considered, namely the important mouth-markings of the nestlings. Briefly, these markings are black spots or lines on the palate, tongue, or elsewhere inside the mouth. In some species, the typical white gapes are elaborated by colours, warts or lobes, but these will not be discussed further here. There are three basic types of palatal marking, as illustrated in Morris, 1955c. One of these three is limited to the cutthroat finch and consists of large blotches. These form a pattern similar to that made by the spots of the second type. However, if the blotches are joined up in a certain way, they are similar in pattern to the third, namely lined-palate, type. In other words, although it can be said that the cutthroat finch is aberrant as regards its nestling palate markings, it cannot be said whether it is more similar basically to the spotted or lined types. According to Delacour, all waxbills and grassfinches have spots (one to seven in number), whereas all mannikins have semi-circular lines. He considers these markings to be of great taxonomic value at the tribal level. However, the situation is not as simple as this. At least four species of grassfinches have lines and not spots. Two of these are the very closely related parson and long-tail grassfinches. Delacour admits that these have 'short lines' but does not consider them to be typical lined-palate types. Both parson and long-tail grassfinches were bred during the present investigation and, although detailed study was impossible, because it was necessary to interfere with the nestlings as little as possible, it was nevertheless apparent that there is nothing very unusual about their palate-lines. In other words, if tribal affiliation is determined exclusively by this nestling mouth-marking character, then these two Australian species are good mannikins. Since Delacour states that 'As long as the mouth pattern of every species has not been finally recorded, some doubt will remain as to the position of a few species', it is clear that he does in fact think of this single character as being tribally all-important. He should, therefore, take these two species out of the grassfinches, and place them with the mannikins. He has placed the pectoral finch with the mannikins because of an avicultural report that its nestlings have lines and not spots and should not therefore be inconsistent without explaining the inconsistency.

The star finch was also bred during the present study and this too produced nestlings with lined and not spotted palates. This species has not had this character recorded before and on other characters was placed with the grassfinches by Delacour. On his system, this too should now become a mannikin. The same applies to the cherry finch. Of this species, Delacour states that it resembles the silverbill (it was placed in the same genus in older systems), but that he is 'convinced that there is no real close relationship between them. When we know the palate markings of this species, we shall be able to finally decide on this point.' It is now known that the cherry finch also has lined and not spotted palate markings in the nestling (Poulsen, personal communication, concerning observation by Enehjelm). Presumably, therefore, in this case Delacour would abandon his conviction about the affinities of this species, because of one character being out of place. Although I have not been able to study this particular species myself, I would prefer to trust Delacour's general conviction, rather than follow his devotion to a single character.

One only has to look at the structure and behaviour of these 'awkward' species for it to be quite clear that it would require a great deal more than one small black line to tie them to the mannikin group. The fact that they possess these lines can only mean one of two things. Firstly it may mean that the palate markings are not the conservative characters they were thought to be and that they are the result of some special adaptations to, say, feeding habits. As soon as one species takes up a certain type of food niche, it may be an advantage for it to have the roof of the mouth (in the adult) grooved in a special way. The black spots and lines of the nestlings, which fade with maturity, are typically arranged around raised white ridges. These ridges are simpler in the lined species than in the spotted ones. In the latter, fingers extend from the main curved ridge and the spots occur in between these. It is conceivable that feeding habits are related to ridge-type and that ridge-type determines palate markings. This would mean that convergences would be expected to occur. So far it has not been possible to test this hypothesis. A field study of many different species, their food, and their palate markings, would give the answer.

Secondly, the existence of four 'lined' grassfinches may be of great evolutionary significance and, if the first suggestion above is not the case, these four species may be all-important intermediates between the typical mannikins and the typical grassfinches. It would, however, be ridiculous to place them with the mannikins. On the other hand, it makes the grassfinch group more heterogeneous than ever and once again raises the problem of whether this is really one or two natural groups of birds. Are these four species comparatively unrelated to the 'spotted' grassfinches, or not? It is possible that from the ancestors of the 'lined' grassfinches came the 'spotted' ones or, alternatively, that the 'spotted' types represent a separate waxbill invasion of Australia. As already pointed out this is perhaps the most interesting Estrildine evolutionary problem.

If we look at the various characteristics of the 'spotted' zebra finch, Gouldian finch and Bicheno finch, we find that these species in many ways appear to be genuinely related to the 'lined' grassfinch species and it seems likely that they have close affinities and have nothing to do with a waxbill invasion. On the other hand, the 'spotted' red-browed finch, painted finch and crimson finch appear to be closer to the waxbills in many ways, and may well represent a separate invasion from a waxbill ancestor. The existence of one species, namely the diamond sparrow, makes this unlikely. It is completely intermediate between the waxbill-like group and the others. It is impossible to say whether it belongs to one or the other and therefore makes it seem slightly more likely that the whole of the Australian Estrildine group is derived from ancestoral mannikin stock, which swung round to a very striking convergence with the African waxbills. This is still tentative as a conclusion and it must be emphasized that a further study of these particular species, mentioned above, would be most rewarding.

These morphological considerations and the geographical ones discussed earlier must be kept in mind as a background to the ethological findings given on the following pages. They render the immediate assessment of the significance of a particular behaviour similarity, or difference, an easier task.

Ethology

In many ways, the Estrildines are ethologically extremely homogeneous. They form a very natural and distinct subfamily of the Ploceidae, from the point of view of their behaviour. Certain actions appear in all species of Estrildines so far studied, but appear in none of the other Ploceidae groups. The tail-quivering soliciting of the sexually motivated female is such a pattern. In most other Ploceidae, the female shows sexual willingness by wing-quivering. However, we are primarily interested here in relationships within the Estrildinae and for these we have to turn almost exclusively to the sexual behaviour of the males. But first, there exist a few points of taxonomic interest in the agonistic and nest-building patterns and these will be briefly outlined.

Agonistic behaviour

The fighting behaviour of Estrildines is comparatively simple. The report given earlier for the zebra finch (Morris, 1954a) applies to almost all species. Birds which are well matched, beak-fence by standing in a sleeked vertical posture and stabbing at one another's beaks. The mandibles are usually closed and little damage appears to be done. If one bird is much weaker than the other, the dominant aggressor attacks in a frontal-horizontal posture, again with feathers sleeked, and either stabs or plucks.

During fighting many species utter tsit-tsit-tsit cries. Others fight silently. Song is not typically employed in agonistic situations, being basically a sexual device with Estrildines.

Motivationally, the above can be summarized as follows. When the tendency to flee and to attack a rival are approximately in balance, the bird assumes a vertical posture and beak-fences. When the tendency to attack is much stronger than the tendency to flee, it assumes a frontal horizontal posture and charges the rival. Plucking, as opposed to stabbing, in the latter case, appears to be the result of an even more intense aggressive motivation.

This, then, is the basic story for almost all species studied, but it remains to mention the few exceptions to these rules, which are perhaps of some taxonomic significance.

One exception concerns the sleeking of the feathers. Certain species erect the crest when attacking or threatening. This is not done in an all-out fearless charge, but usually in an aggressive situation where there is a little fear present. In the photograph in Fig. 1 in Morris, 1955d, there is just discernible the erected crest of the threatening male crimson finch. In the full threat display of this species, the very long tail is raised and slightly fanned, and the crest fully erected. This is done when the body is in the frontal horizontal charge posture, but at moments when the bird is actually hesitating about going in to the attack. Certain other Australian Estrildines, notably the parson finch, long-tailed grassfinch and occasionally the zebra finch, also erect their crests in agonistic situations. So far, however, it has not been possible to ascertain with certainty, with these other species, exactly what motivational situation is correlated with this response. Crest erection in these species is at the present stage rather unpredictable and is not apparently used in the same way as in the Crimson Finch.

A general fluffing of all the body feathers has been observed as an appeasement device in several species and is probably of widespread occurrence throughout the group. Juvenile food-begging has also been observed as an appeasement gesture by submissive individuals, but only in two species. One was the zebra finch and, as already reported in a previous paper (Morris, 1954a), a badly persecuted female achieved momentary respite from her aggressive mate by begging to him. When being very badly beaten, she crouched and twisted her neck round in the typical Estrildine nestling manner. The other species was the Java sparrow. Here the juvenile response was given more freely and frequently. In a social encounter, either intra- or interspecific, a Java sparrow which was not prepared to become involved in a fight when challenged, and yet was not prepared to flee, or could not flee, gave the high-intensity 'neck-waving' response. This is a very characteristic response of begging nestlings. They either twist their necks round and then wave their heads from side to side slowly and rhythmically, or they stretch their necks forwards and upwards (at an angle of about 30 degrees from the horizontal) and then proceed to wave the extended neck from side to side. The beak, of course, is wide open during this ceremony and in the adult Java sparrows this is also the case. (It is interesting to note that Brooksbank, 1949, p. 45, reports that newly-hatched chestnut-breasted finches 'have a short bulbous tongue which they roll to and fro with a rhythmic swinging motion ... ' One would predict that extreme submission in this species would involve this special response.)

It is perhaps worthwhile to mention at this stage that there does not appear to be any courtship feeding amongst the Estrildines, or, at least, no elaborate or common ceremony of this kind, during the pair-formation and maintenance phases. Adult food-begging is only seen in submissive agonistic situations.

In a recent monograph on the bronze mannikin (Morris, 1957b), the specialized fighting postures of this bird were discussed in full. Briefly, this bird shows the aggressive frontal-horizontal charge, but does not possess the vertical posture for beak-fencing. It beak-fences extremely savagely, however, adopting a latero-horizontal posture during bouts. The bodies of the two contestants are parallel to one another, and the heads twisted round to face the opponent. In addition to this, they exhibit a one-wing-up display whilst beak-fencing. This involves the wing farthest from the rival, which is held vertically during the stabbing and which is brought down vigorously as a means to establish balance if the bird feels itself being thrust backwards. The black-breasted mannikin is a close relative of the bronze mannikin and shares its fighting techniques. It would be valuable to know if the larger, but similar, magpie mannikin also uses these fighting techniques. If it does, then there are probably sufficient similarities between these three birds, and sufficient differences between them and other mannikins, to warrant re-placing them into a separate genus (*Spermestes*).

The long-tailed grassfinch also beak-fences in what is practically a horizontal posture. The head is drawn back more than in a fearless charge, but the body remains horizontal (see Pls. 26 and 27). There is no special one-wing-up display in this species, however, although the far wing is employed to establish balance occasionally (see Pl. 27). Its use as a balancer is, interestingly, less obvious and spectacular than in the bronze mannikin or the black-breasted, and correlated with this is the fact that the two mannikins possess bright markings on their wings, but the long-tailed grassfinch does not. The horizontality of the beak-fencing of the long-tailed grassfinch differs too in another way from that of the mannikins. The latter often assume a very vertical body posture, when alert or alarmed, but the grassfinch is an almost permanently horizontal bird. It holds its body well down out of the vertical during almost all its activities and the horizontality of its fencing appears to be simply a by-product of the general deportment of the species. In the case of the mannikins, on the other hand, it must be a more specialized independent development.

Nesting behaviour

As far as is known, no Estrildine species is a true weaver when building its nest. The structure is a clumsily built spheroid with a side entrance. The entrance tunnel is never great in length, as in many of the true weaver species. Three basic movements are used in building the nest, apart from the obvious ones of collecting grass and placing it on the nest-site. One is 'testing', in which the bird holds the grass or straw between its mandibles and then opens and shuts the latter with extreme rapidity. This action of the mandibles has been called mandibulation and is extremely interesting in its ritualized signal role in sexual behaviour. All species appear to exhibit this action.

Next, there is 'looping', which is an elaboration of 'testing'. Here, the grass is placed under one foot and held thus while the bird mandibulates it. It may then be transferred by the beak from one foot to the other, or its position shifted, to test a new length (elasticity is an important attribute of nest material; brittle grasses are inefficient and the testing is probably concerned with checking this quality), or it may be looped round several times under the foot. The important point about looping is not its use as a ritual, but its occurrence in only certain species. Many Estrildines show it, but a number of others do not. It is, of course, difficult to say that a species does *not* do a particular thing: it may simply be that the observer has not yet been lucky with the species. It is dangerous, therefore, to divide the species studied into those which do hold down nest material with their feet and those which do not. Once a species has been seen to perform this action it is easy, but a species which has never been observed to do so has to be very well known indeed before any negative statement can be made about it. (This applies also to all the other characters dealt with in this and other studies of a comparative nature, where a behaviour character is simply present or absent.)

The zebra finch is now known well enough to be considered with certainty as a species without foot-holding during nesting. There are several other species which also appear to belong to this category, but they will not be named until more observations have been carried out. On the other hand, species as different as the star finch, red-browed finch, spice finch and bronze mannikin have all been observed to hold down nest material with their feet. (There is a possibility that toe-length is involved here; holders seem to be long-toed and non-holders short-toed.)

The third important nesting action is 'neck-stretching' or 'scooping'. This is performed inside the nest cup and helps to form the sides and roof of the domed structure. The neck is stretched in a characteristic way, pushing the nest material away or up and thus scooping out a discrete nest cavity. This action appears to exist in all species and is most important in its role as a ritualized sexual signal. It has become modified in different ways in different species and now exists in a whole series of ritual forms. These will be discussed in the next sections.

Sexual behaviour

As the comparative data available for this section are rather complex, they will first be dealt with species by species before an attempt is made to summarize their taxonomic significances.

It is impossible in this paper to subdivide the sexual patterns into pairing and pre-copulatory activities, because a particular action which appears in the pair-forming stages in one species is associated with the pre-copulatory patterns of another species. A further confusion exists because of the intimate connection between reproductive nest-building and pairing in certain species (see Morris, 1957b). Nest ceremonies and pairing ceremonies appear to be one and the same and nesting actions have thus worked their way into the sexual repertoire and even into the pre-copulatory patterns in some cases. This is probably the result of the fact that the social nesting each night, of the non-reproductive group, fragments into private reproductive nesting simultaneously with the forming of the pairs. As the pair forms and the mated birds develop aggression towards their earlier sleeping companions, they also develop a stronger nesting urge, which leads to the more elaborate domed breeding nest. (The social roosting nest is little more than a crude platform in most cases.) Higher intensity nest-building is therefore irrevocably linked with the vital phases of pair-formation and the nesting actions have thus taken on a sexual significance and act as social signals. Almost all the species described below will be found to have some kind of nesting act built in to their sexual signal codes in some way or other.

Although, as stated above, these facts make it difficult to discuss first the pairing behaviour of all the species and then the pre-copulatory behaviour of all the species, it is nevertheless possible to make this sexual subdivision in the case of certain better known-species and this will be done in the particular cases concerned. Those species already dealt with in earlier papers will be summarized very briefly here for the sake of uniformity. With these few preliminary remarks it is now possible to deal with the sexual patterns, species by species, employing the order used by Delacour in his revision of the group, with one modification. The latter concerns the red-browed finch which, as a waxbill, should be placed first. Since we are not considering waxbills in this paper, this species really should not be dealt with at all, but owing to the dubious nature of its present taxonomic position, its patterns will be described here, after the grassfinches and mannikins have been dealt with. Although it should already be clear that I do not agree with all of them, I shall use here Delacour's generic and subgeneric divisions, so that behaviour similarities and differences can be seen against the background of his taxonomic arrangement, which, despite one or two difficulties, is still the best arrangement of the whole group so far produced.

A. Grassfinches

Genus *Zonaeginthus*

Subgenus *Zonaeginthus*

1. *Z. pictus* (painted finch). Central and N.W. Australia

Webber (1946) states that when the male of this species courts he does so with 'head held up, beak 60 degrees from horizontal'; also that 'the head is moved smartly from side to side' and that there is no 'bobbing up and down on the perch ... a popular method with other Australian finches'. However, in an appendix to the paper, Webber mentions that he finally did observe a male which 'held a grass stalk in his bill, crouched over the perch, bobbed up and down and more or less performed the diamond sparrow display'. (For description of latter see below.) It is almost certain that this last account gives the correct description of the full display of the painted finch. It seems very likely that Webber was describing the non-courtship singing as part of the courtship, until he finally saw the true courtship dance, as he reported it in his appendix. (Although it has been dealt with fully elsewhere, it should perhaps be explained here that males may sing without courting but rarely court without singing. The non-courtship singing is either given by solitary birds, or, if others are present, the male does not orientate to them in any way. In most species, the posturing accompanying the two types of singing differs.

Unorientated singing is typically accompanied by a lower intensity version of the courtship singing posture. The various characteristics are seen in both, but usually all are exaggerated in the courtship cases. In some species, there is little or no such difference, but in most it is obvious enough.)

Gordon (personal communication) stresses that this species is a 'ground-lover' and reports that this influences the form of its courtship slightly. He says that 'The males display without a straw. They strut upon the ground, picking up and dropping bits of earth, straw, etc. Then, head back, bill up, body erect, they sing, while they shuffle in jerky hops to female. Males dominant sex. The "chase-flight" and female behaviour are similar to the other fire-tail species.'

There appear to be some contradictions between Webber's and Gordon's observations, but this is undoubtedly the result of the fact that this species, which certainly is more of a ground dweller than most Estrildines, courts both on the ground and also when perched. The different position of the courtship would undoubtedly influence its form, and this appears to be the explanation here, Webber having watched perched courting and Gordon courting on the ground. It is most interesting that the carrying of the straw or grass in the beak during perched courtship is changed to picking up and dropping material when on the ground, and shows the extent to which the external environment can influence the details of the courtship ceremony.

Both authors stress that the beak is tilted upwards during singing and this is a point of great comparative importance. I can confirm that it occurs, from personal observations of the few painted finches I was able to obtain. The following are observation notes made on the behaviour of these birds:

> Solitary song is performed without special feather raising, but with a distinct head pivoting. The beak is pointed upwards and is thus reminiscent of the red avadavat. The beak is not opened appreciably during song. The song phrase is repeated typically two or three times in each song-bout. Like almost all Estrildine songs, it is difficult to represent verbally but is structurally rather like: che che che-che-che-che-che werreeeeee-oooeeeeee. Sometimes, the last two phrases are the same as one another, when they sound more like: cheeurr cheeurr. This alternative never occurs in the first phrase of a song, usually only in the second phrase of a two-phrase song. In other words, it looks as though, in this species, the song is already becoming something more complex than just the simple repetition of a single phrase.

It was noted at the time that the song posture including the pivoting of the up-turned beak, was similar to that of the red avadavat, and also that certain aspects of the plumage colouring and marking were similar. The distribution of the avadavat (extending down to Timor, off the north-west coast of Australia) and the rather unusual distribution of the painted finch (from the north-west coast of Australia in to central Australia) strongly suggest the possibility that here we are dealing with a separate waxbill invasion of Australia. This will be discussed again later.

Only very brief courtship observations were possible with the painted finches used in the present study and the more elaborate display seen by Webber was not recorded. The following details were noted:

> Female presented to male. He approaches her and she responds by twisting her tail round towards him. He faces her and his first action is to toss his head back for a second. Then he starts singing to her in a rather vertical posture, with legs stretched somewhat, with the head thrown back and with very rapid head pivots (latter being most noticeable feature). No striking raising of feathers. Then he mounts her unsuccessfully ...

In subsequent courtships, the male was seen to perform displacement beak-wiping both before and during display bouts and both male and female preened afterwards. On several occasions, the courtship broke down into beak-fencing in which the male's beak was open and pushed over the closed beak of the female. Both male and female showed the tail-twisting response at various times during courtship bouts: in both sexes the tail is twisted round towards the partner. Finally, it was noticed that the courted female's posture included a tilting up of the beak. Both this tilting and the tilting up of the courting male's beak were sustained actions. The head was tilted back and held there while the courtship proceeded, in contrast with the male's initial, momentary, head-tossing mentioned above. This quick head tossing is seen in this species during beak-fencing when it alternates with stabbing forwards and is performed by a silent male. As soon as he begins to sing, the male's head is tossed back and held there throughout the singing.

Fragmentary as these notes are for this species, they give a number of important clues. Most significant of these is the raising, rather than lowering, of the beak in social encounters. This is, according to Delacour, a typical waxbill characteristic. The red-browed finch also possesses this movement and this is one of the main reasons for this bird being included with the waxbill group, despite its geographical position. The observations recorded here for the painted finch – a supposed grassfinch – therefore create an inconsistency in the present taxonomic system.

2. *Z. bellus* (fire-tail finch). E. Australia and Tasmania

Little is known about this species which, for some unspecified reason, is extremely difficult to keep in captivity. Butler (1899) records that the 'love-dance is like that of the diamond finch' (the diamond sparrow, see below).

3. *Z. guttatus* (diamond sparrow). E. Australia

This rather heavily built species has a mournful low-pitched call-note reminiscent of the *Poephila acuticauda, cincta, personata* group. Its display has been recorded by several authors, as follows:

Butler (1899) reports that the male, in his courtship, 'stretches his neck upwards to an extravagant height, draws in the breast and expands the chest and abdomen, stands very upright and with depressed head, a long grass bent in his beak, then bobs up and down on his perch, to the accompaniment of his queer song'. Cayley (1932) says that the male courtship is 'A measured curtsying, head directed downwards, so that the beak almost rests upon the breast'; also that 'he stretches his neck, holding a long straw in his bill, stands right up on his legs and hops up and down at the side of his mate like an automaton, approaching along the perch as she retires'. Brooksbank (1949) describes the performance as follows: 'Having procured the longest bit of grass he can find, the cock holds it at one end in his beak and puffs his feathers out to such an extent that he not only appears to have doubled his size, but looks almost spherical in shape. He literally bounces about like a ball, at the same time buzzing like a bumble bee ...'

These reports tally fairly accurately and together give a clear picture of the display. The angle of the beak is again a feature of the display which catches the attention of the observers and it is interesting that it is as different as it could be from the last species.

Genus *Poephila*

Subgenus *Neochmia*

4. *P. phaeton* (crimson finch). N. Australia and S. New Guinea

Butler (1899) says of the display: 'song a comical humming with the head raised high, a zealously moved beak and fan-shaped, outspread tail, turning in a dignified manner from one side to the other and then breaking off from the marvellous, noiseless love-dance suddenly with a loud whistle'. From this report, the basic dance movement appears to be of a lateral pivoting nature, but Cayley (1932) gives a slightly different picture: 'The long tail-feathers are spread out like a fan, the body carried very erect, and the head is moved from side to side in a most dignified manner ... sometimes, if on a thin twig, he bobs up and down and sways the outspread tail-feathers.' Harman (1953) confirms certain of the above features when he states that 'The tail is spread out fanwise and the body held very erect. While humming a curious note, the males move their heads slowly from side to side in a most dignified manner.'

From these three reports it appears that the fanning of the tail, the erection of the body and the pivoting of the head and/or body are the salient features.

A vertical dance movement is also suggested by Cayley's comments and, as will become clear below, the possession of both dance movements, either combined or separately, is a feature of a number of species' displays. The fact that the tail is fanned by very aggressive male crimson finches during attacks suggests that there may be a considerable aggressive element in the courtship of the male of this species.

It was only possible to observe the behaviour of a single pair of this species in any detail during the present study. The male concerned was provided with a female, with the following immediate results:

> Male threatens with tail-up-and-fanned. Does not actually attack. Male beak wipes. Sings to female; song phrase: chu chu chu che-chee chooo. Posture is horizontal, not vertical. Both male and female in horizontal posture, parallel to one another, with heads and tails strongly twisted towards the partner. No feather raising except for face mask, crown and nape, as in threat display. Legs rather stretched. Flank feathers outside the wings, but not exaggeratedly. The male display begins with rapid ritualized beak-wipe bows – never touching the branch, but obvious in their derivation. These are rapidly and rhythmically repeated – each appears to be a separate wiping, but each abbreviated. When female comes near the male, she too raises her face mask and, as she approaches close, he changes to an upright posture. After the first phase of courtship breaks off, the male flies about and, for the first time, it is noticeable that he tosses his head back each time on alighting.

In later courtship bouts it was noticed that the male made slight pivoting inverted-curtsy movements while singing to the female. The courtship sometimes broke down into an attack by the male. The head of the male when singing was not seen to move at all; nor was the beak. The female was observed to tail-quiver as a sexual invitation, but the male did not respond by mounting.

There are apparent contradictions here between the reports from the literature and these personal observations. There is, however, an explanation for this. In several species (e.g. spice finch, Moynihan and Hall, 1954), it has been found that an unduly aggressive male courtship involves a special horizontal posture called, by Moynihan and Hall, the 'low twist'. This does not appear in a typical courtship, which may be quite vertical. It seems that in the present case, the male under observation was much more aggressive than those studied by the earlier authors. Its attacks on the female and the raising of the same feathers as in normal threat confirm this. The fanning of the tail in courtship and in threat, as reported by others, reveals that there is probably always an element of aggression present, but in my own male, the extra aggressiveness appears to have changed his posture from vertical to horizontal and eliminated head and beak movement. This idea is reinforced by the fact that the approach towards the male of the female changed his position to a vertical one. Her approach would be expected either to reduce his aggression or to directly arouse his sexual tendencies, or both, and would therefore be expected to produce the more sexual, vertical posture. Finally, both dance movements, vertical and horizontal, appear to be available to this species and it seems likely that they may, as in the spice finch, appear either separately or combined.

This species shows certain striking resemblances to the painted finch, as regards its behaviour. Both have a characteristic multiple call-note (che che che che che che) which is so similar that they were heard to answer one another from different aviaries. Other grassfinches and mannikins studied were found to have only a single call-note, although multiple call-notes were heard in certain African waxbills. The tossing-up quickly of the head was also shared by the painted and crimson finch, but was not seen elsewhere in the grassfinch and mannikin groups. The red face masks of the two species, and certain other markings are also not dissimilar, although the general visual impression is superficially rather different.

Owing to the extremely savage nature of this species in captivity, it is a difficult bird to study, but clearly it is a most important species and demands much further study. (For photographs of its aggressive behaviour, see Morris, 1955d.)

5. *P. ruficauda* (star finch). N. and E. Australia

Cayley (1932) mentions the fact that the male 'often carried a piece of grass in his beak during the mating display ... ' and he quotes the observations of a number of aviculturists, one (A. E. Clarke) as saying that 'When mating, I have noticed both the male and female with bits of grass in their bills, and they both bob up and down before each other.' However, another (J. S. Mackie) reports of a male of this species that 'when mating he danced from side to side with his tail spread'.

From personal observations of this species, I can confirm the importance of straw-carrying as a ceremonial, although, unlike Clarke, I have only seen it in the male. The carrying of straw as a ceremonial is accompanied by an elaborate bowing display which is presumably what Clarke refers to as 'bobbing' up and down. The statement that a bird bobs up and down is unfortunately an extremely ambiguous description. There are so many subtly, but importantly, different ways in which it can move up and down. Each way appears to be derived from a basically different movement and the ceremonies should not therefore be compared too glibly. In the bowing of the straw-carrying male star finch, the movements are repeated and deliberate. The bird bows rhythmically keeping the beak horizontal both at the top and at the bottom of the bow. In this way, it differs distinctly from the bowing of most other courting Estrildines. Usually the ceremonial bow is derived from displacement beak-wiping and then the beak points more and more downwards as the body is lowered in the bow. The star finch bow is almost certainly derived from a different source, namely the intention movements of taking-off for level flight. If observations are made of finches about to fly away from a branch horizontally, similar quick bowing actions can be seen. These have become more ponderous and dignified in the ceremonial bow of the star finch but are still quite recognizable. It seems as if the male is using as a sexual display that phase of nest-building where he would be repeatedly flying to the nest with material. He is therefore performing a fundamentally different (but superficially similar) display from the beak-wipe bowing of the other grassfinches and mannikins. The only other species of Estrildine in which I have seen a similar bow is the red avadavat, where both male and female show the response, again with a grass or straw in the beak. This and certain other features mentioned below seem to relate the star finch to the waxbill-like grassfinch group, but there are difficulties here. For example, this is one of the five species of grassfinch which has the mannikin-like lined nestling palate. Like the diamond sparrow, this is an extremely awkward species, taxonomically speaking.

The solitary song posture of this species and its relationship to the resting posture are illustrated by the photographs in Pls. 28 a and b. It will be seen that the male does not erect his ventral feathers, but does raise his breast feathers and his face mask. This is practically the reverse of most Estrildines. In nearly all the other species, the ventral feathers are erect and the breast feathers depressed. Only in the case of the silverbill is there a song posture with sleeked ventral feathers as in the star finch. It is worth noting that the erected region is just that area which is brightly marked with white spots, the ventral region being unmarked.

The song is reminiscent of that of the zebra finch, but is of a higher pitch. The call notes and other squeaks uttered when beak-fencing are extremely high pitched. When the male starts a high intensity courtship display with song, the solitary singing posture is exaggerated into a unique posture. The legs are stretched vertically so that, with the ventral sleeking, the male has its body raised as much as possible above the perch. The feather erections noted for the solitary song are now intensely exaggerated; all the spotted regions (that is, the flanks and breast, and round the red face mask) are fully erected. The flank feathers stand out strongly from the body, with the wings tucked away inside them. The male is rigidly horizontal and faces the female thus. Viewed from directly in front, or from directly behind, the flank feathers can be seen standing away from the body very distinctly. The female has a perfect view of all the spotted regions and, although it was not observed in the present case, it is obvious that a pivoting dance movement (as mentioned by Mackie) would show off the spots even more. Head pivoting was seen in the singing males, but not the full body pivot.

Tail-twisting towards the partner was seen in both sexes with rather a high frequency during sexual encounters.

Before proceeding to the next species, it is perhaps worthwhile to record in detail the most complete courtship sequence observed, because this reveals the sequential relationship between the straw-carrying-and-bowing display and the feather-erection-and-singing display:

> Male hunts for straw and selects longest one it can find. Then takes up lateral sleeked position across branch and starts to bow with straw in beak. Bows rather rapid (two a second) and less pompous and deep than the red avadavat bowing. He bows 11–12 times, interspersed with two or three advances towards the female. Then bowing stops, but male holds horizontal lateral posture. He continues to advance, nearer the female with straw in beak, but now he sings instead of bowing. The posture is intermediate between normal song courting posture and sleeked bowing posture. When he gets very near the female, he drops the straw and mounts her. She opens her cloaca, but does not quiver tail. He dismounts, but before doing so he mandibulates. He seemed to be mandibulating because for some reason he could not copulate.

In other words, the straw-and-bowing ceremony is phase one of the male's courtship and, almost certainly, the singing in phase two, which supersedes the bowing, is accompanied by lateral dance movements at higher intensities. The peculiar horizontal, leg-stretched, courting posture of phase two is undoubtedly derived from the take-off for level flight as is the bowing. Electronic flash photographs of birds which are caught in the moment immediately before they leave the perch for level flight often look very similar posturally to the courting male star finch in phase two.

This bi-phasic male courtship makes an interesting comparison with that of the spice finch. Here, Moynihan and Hall (1954) describe how the male and female hop back and forth carrying straws until one lands near the other, when the male drops his straw and begins to sing to the female in the (phase two) courtship posture. In the spice finch, then, the ceremonial straw-bowing is absent and in its place hopping back and forth occurs. Also, the straw is dropped before singing starts and not after it as in the star finch. Correlated with this is the fact that the star finch sings with a closed beak and the spice finch opens and shuts its beak while singing.

It seems as if, in these species, the earlier stages of nesting are being incorporated into the courtship ceremonies, whereas in certain other species the later (scooping) stages are used.

6. *P. guttata* (zebra finch). Australia, Timor, and Flores

Previous papers have dealt with this species in detail (see Morris, 1954a, 1955a, and 1956b). Its behaviour will therefore only be briefly summarized here:

Male courtship consists of song with vertical posture, beak horizontal, ventral feathers fully fluffed, crown feathers flattened, nape erect, spotted flank feathers erect and held outside wings (see Pl. 29). Head pivots rhythmically in both solitary and pre-copulatory song display, but in latter, whole body performs lateral pivoting as well. Beak does not open appreciably during singing. Legs fully bent; bird's ventral feathers rest on perch. Aggressive courter has body more out of vertical plane and pointing more towards female than non-aggressive courter. Tail-twisting towards female by courting male observed frequently, but never by female to male as in some other species. Male repeatedly bows to female during his pivoting advance towards her, but these bows are clearly derived from displacement beak-wiping movements and they often develop into full beak-wiping. Female also may beak-wipe occasionally whilst being courted.

The zebra finch was never observed to perform the vertical, inverted-curtsy dance movement seen in so many other species, despite prolonged observations. A sentence from Cayley (1932) made it seem likely that the vertical movement might occur under certain circumstances, however: 'Usually the singer carries his head erect, with the neck outstretched, and jerks its body up and down, or sways to and fro; the latter movement is chiefly adopted by the male during the breeding season.' Since the zebra finches used in the original study (Morris, 1954a) were all 'domesticated' stock (probably bred for as long as thirty generations in captivity) a group of wild-caught zebra finches was sought and when obtained was set up in a colony alongside an equal number of 'domesticated' birds. If the absence of the vertical dance movement was the result of using special stock, then this set of wild-domestic observations should have revealed it. However, apart from the fact that the wild zebra finches were smaller than the 'domestic' ones, there was no discernible difference between their behaviour. This can only mean that, if the vertical dance movement is really ever given by this species, it must require special conditions for its performance which are not available in captivity. (In a number of other species, the vertical movement does, of course, occur quite readily in captivity and, in some, it occurs more readily than the horizontal pivoting movement.)

One behaviour pattern of the zebra finch, which was seen for the first time after the previous reports on this species had been completed, warrants special mention here, because it provides an important link between this species and the long-tailed grassfinch. Both these have a pairing ceremony of a somewhat unusual derivational nature. The pairing behaviour of the zebra finch had already been noticed and recorded in an earlier paper, but in a very incomplete form. It was pointed out that if a male had built a nest and wanted a female to join him at it, or if a male and female were prospecting together for a nest, a characteristic set of actions was observed: 'After the initial period of vigorous courtship, the male flies to a clump of twigs and moves round and round over it in repeated turns giving the nest-call and maintaining a special posture. This posture consists of fanning the tail and holding the body in a horizontal position. The legs are bent so that the body is kept low over the twigs. The beak is opened and shut extremely rapidly.' It was noticed that this ceremony was typically followed by the building of a nest on the site, and it was felt that, although it functioned as a pairing ceremony, it was essentially linked to nesting behaviour. Only when colonies of zebra finches were studied in larger aviaries recently, was it found that the pairing and not the nesting was the essential feature. The ceremony could occur, it was found, with or without reference to a nest or nest-site, but never without reference to a pairing problem. Its relationship to nesting problems is, in fact, probably only caused by the intimate relation in the time between the setting up of a reproductive nest and a reproductive pair-bound.

The description of the pairing ceremony given above was found to be incomplete when zebra finch colonies were studied, more intense versions of the pattern being seen when mating 'triangles' developed. If a mated male, for example, was beginning to ignore his female and chase another, his mate would follow him around giving a very elaborate version of the pairing ceremony. There was no reference here to a nest or nest-site, but only to the male. Furthermore, the male had to be actively interested in the other female to produce this display in the mate, or at least, he had to be actively disinterested in the latter. The performance was seen in both sexes and males who were losing their mates were just as likely to follow them around giving the display. The following observation notes give a detailed description of the performance:

In a particular colony, one male had lost interest in his mate and was being helped at the nest by a new female. The old female was closely following the male around and also attacking the new female whenever she emerged from the nest. Neither the new female, nor the male, attacked the old female, but the latter was repeatedly assaulting and attempting to drive away (from the male) the new female. This basic situation developed as the new pair-bond became stronger and the new female finally began to attack the old one. The end came when the male joined in these attacks and the original mate was then finally ousted after the loss of a number of feathers. During all this time (several days) the old female was constantly performing the high-intensity version of the pairing ceremony and this was always oriented to the male. It frequently developed into an attack on the other female; for example, this was observed nine times out of a series of twenty-six displays noted at random. The less notice the male took of the old, displaying female, or the more notice he took of the new female, the more intense was the display. It took the following form: horizontal pivotings frontally or laterally towards the male, with mandibulations and tail-fanning and wing-drooping; latter two movements occur one or two per second and wings and tail droop open and fan open synchronously; the flank feathers are fluffed right out, all the ventral feathers are raised, and the female squats on the perch in a crouched horizontal posture. Each time the wings give a rhythmical droop, the flank feathers are pushed out from the body, as the wings are inside them. There is a soft to loud rasping call that accompanies the display and which is like the high-intensity nesting call. The eyes are repeatedly, but not rhythmically, closed and opened, and when open are still half-shut most of the time; the head may even be turned round into the back in the typical sleeping position. This can occur at high-intensity display moments and gives the distinct impression of being a displacement activity. The false nature of this apparent sleepiness is repeatedly revealed by the way in which the displaying female often stops her displaying suddenly to rush at the other female, springing into action from an apparently sleepy condition. The display can be given at one of three main intensities – mandibulate; mandibulate and tail-fan; mandibulate, tail-fan and wing-droop. The wing-drooping has not been seen before, and only appears at the very highest intensities. The lateral swinging, when in the horizontal posture, drops out typically at these higher intensity levels and is usually associated only with the simple and more sleeked horizontal-mandibulation displays. The mandibulations in this species, as in the long-tailed grassfinch, are very frequent indeed and normally audible from a distance of six feet.

The above description gives a fairly complete picture of this particular pattern and it only remains to discuss its significance. Firstly, it must be stressed that this is a pattern which does not co-vary in any simple way with any basic level of sexual motivation in the performer. The female in the case described here was, in fact, being followed around herself at the time by another male who was repeatedly trying both to copulate and to pair with her. He failed completely during the period when she was giving the displays and was either ignored altogether, or occasionally attacked. All this time, the female was following the other male around and orientating her displays to him, despite the fact that he had long ceased to perform any sexual action whatever towards her. Considering these facts, one is forced to postulate a distinct pairing ceremony independent of a copulatory tendency.

The apparent derivation of this pairing display is most unexpected, for it is strikingly similar to the sun-bathing pattern of this species. In the latter, the bird crouches, half closes the eyes, or opens and shuts them, fans its tail and synchronously droops its wings. The tail-and-wings action is highly characteristic and identical with that seen in the pairing display. The rhythmic opening and closing of the wings and tail may, however, reach a higher intensity in sun-bathing, with a sustained opening of both. The wings may then be spread right out in a way not seen when pairing, but this posture is taken up not in a single action, but after a series of rhythmic openings and closings. The head may also be twisted round as in sleeping and in the pairing display. In both cases, the tail-fanning appears to have a lower threshold than the wing-drooping, and in the photograph in Pl. 30, low-intensity sunning is shown, illustrating this stage. (In this photograph the posture of the bird is not particularly horizontal, but this was probably due to the rather unusual circumstances necessary for taking the picture.) The feathers are ruffled in sun-bathing, but this element is missing from most pairing displays. Mandibulation was, however, seen during sunning bouts, apparently as a response to the heat, but this action can, of course, be caused by a number of quite different conditions.

The physiological basis for the similarity between sunning and the pairing display is intriguing, but is beyond the scope of the present paper. In the present context, the essential point about this whole complex is that it occurs in the zebra finch and that fragments of it have also been seen in the long-tailed grassfinch, but nowhere else in the Estrildinae. This suggests a relationship between these two species, but it must be borne in mind that it may only represent an insufficiency of observation with the other species. This criticism can only apply in certain cases, however, for detailed observations have been carried out with several other species, such as the bronze mannikin, for example, where it can be quite certain that this pattern does not occur. In this last species, circumstances were studied in which one would have certainly expected the pattern to appear in an intense form if the species possessed it, but it was never seen. On the other hand, it seems likely that certain close relatives of the long-tailed grassfinch, such as the parson finch, will also possess this display pattern, but further study is needed here.

7. *P. bichenovi* (Bicheno finch). N. and E. Australia

Cayley reports that 'The mating display is more subdued than that of the zebra finch and consists of a sequence of bobbing up and down movements by the male as he chases his mate around the aviary.' Unfortunately this description is completely ambiguous, because, from personal observations, it is clear that the 'bobbing' referred to here probably relates to the extremely frequent displacement beak-wiping of this species. No other Estrildine studied to date has ever shown such high frequency displacement beak-wiping. As many as eighty-three beak-wipes were recorded during the first five minutes of courtship after a female was introduced into the cage of a male, and this figure is not unduly high for this species. The male assumes a courting posture in which the plumage is fully fluffed, so that the body appears spheroid, with no special differential erection of the feathers of the type found in most other species. The bird crouches over the perch, parallel to the female, with his head twisted towards her. In this position he sings and beak-wipes; in order to perform the latter, he does not have to move as far as he would if his courting posture was similar to his more usual generalized stance. It seems as if the posture is itself part of the beak-wiping; the bird performs the action so many times, that he never, so to speak, quite straightens up again, until after the whole courting phase is over.

No dance movements were seen in this species, so that the 'bobbing' referred to by Cayley may, in fact, refer to a vertical dance-step which is possessed by this species, but which was not occurring in captivity. The use of the ambiguous word 'bobbing' is therefore particularly annoying here.

(For a photograph of the courting posture of the Bicheno finch, see Morris, 1956a.)

There are two call notes in this species: 'lost-calling' which is a plaintive twoooo-twoooo, rather like the calling of the long-tailed grassfinch, and 'social-calling' which, like the song, is similar to that of the zebra finch, but higher and thinner in quality.

8. *P. modesta* (cherry finch). E. Australia

Butler (1899) reports that 'its song has the almost voiceless character of the majority of the typical grassfinches, and its manner of singing is precisely the same; the neck is elongated, the head projected rather downwards than upwards, the mouth opened widely and emitting a faint humming sound, not audible at all when other birds are singing.' From this account, from the fact that this species is now known to have mannikin-like lines on its nestling palates, and from the nature of its adult plumage markings, with the striking lateral barring, it seems highly likely that this species has strong affinities with the typical mannikin group.

Little more is known about this species, with the exception of one interesting fact. Butler describes what appears to be a distraction display, thus: '... I purchased a second pair, and in May they built, the first egg being deposited on the 7th May. The hen was very nervous, and left the nest at the least alarm, tumbling about on the earth as if wounded, and gradually retreating from her home until about two yards away, when she hopped up into a bush and sat quite still.' As far as I know, this is the only instance on record of an Estrildine performing a distraction display.

Subgenus *Poephila*

9. *P. cincta* (parson finch). E. Australia

This species differs hardly at all from the next (the long-tailed grassfinch), except for beak colour, tail length and song. The beak here is black, the tail short; in the next species, the tail is long and the beak red. The song of the two species kept in captivity during the present study were quite different, that of the parson finch being even lower in pitch than that of the long-tailed grassfinch. The call-notes were also very low in pitch and much more prolonged than those of other species. Butler (1899, p. 177) gives a very accurate musical scoring of the song of the long-tailed grassfinch. Unfortunately he calls it the song of the parson finch. All the long-tailed males I heard singing gave exactly the kind of song as that reported by Butler for the parson finch and the one male parson finch that sang repeatedly in my aviaries always sang a quite different song. Both the parson finches and the long-tailed grassfinches I studied were wild caught birds, so that no special distortion of their natural songs would be expected. I therefore conclude that Butler must have made a mistake in this case.

In behaviour there is little detectable difference between the two species. Both have a specialized head-jerking signal, which is not seen elsewhere in the Estrildines. This occurs most during courting, or other social encounters. It also occurs at other times when no special activity is involved and appears to have acquired an extremely low threshold. The action is typically performed immediately on landing and consists of a few very quick vertical stretchings and retractions of the neck. The rest of the body, tail and wings do not move and it seems as if the 'excitement' flicking of the tail, which occurs in so many other species, has been transferred to the head here. In courting situations, the head-jerking becomes so rapid and so frequent that it becomes quite divorced from landing and taking-off. Whereas in mild social situations, it only takes place with a landing, or some other definite form of locomotion, in intense social situations it can go off repeatedly without the bird moving at all.

In the parson finch this head-jerking seems to be less intense than in the long-tailed grassfinch and much less frequent, but only four birds of the present species were available for study, so this comment must remain a highly tentative one.

The courtship posture of the singing male involved a ruffling of the crown and head feathers, a strong erection of the black bib, and a general body fluffing. Both sexes beak-wipe repeatedly during the courtship. The body posture of the male when displaying is rather crouched, but possibly at higher intensities, which were not seen, this changes (as it does in several other species).

10. *P. acuticauda* (long-tailed grassfinch). N. and N.W. Australia

There is little to add about this species to what has been said above. More details were observed here concerning the nature of the feather erection of the head whilst singing. The crown feathers were fluffed like the rest of the body whilst the males performed their solitary non-courtship song, but became ruffled whilst actually courting. The extent of this ruffling varied from case to case. Sometimes only the posterior crown feathers and the nape feathers were ruffled, giving the head feather-posture a striking resemblance to that of the courting male zebra finch. But often the ruffling spread further forwards, so that the whole crown region was fully erected in a way not seen in the zebra finch male when courting.

Displacement beak-wiping occurs frequently during courtship bouts in this species, as in the last one, and both pivoting and inverted curtseying were seen here. They were not particularly exaggerated, as in some of the mannikins, but this may have been the result of low intensity displaying again. The lateral and the vertical movements were combined in a one-to-one ratio, to produce a complex rising-and-turning action. Once, this latero-vertical dance was seen to be performed by a male without reference to a female, the latter being in a different part of the aviary at the time. The male in question gave the display whilst holding a long straw in his beak, but gave it silently. This performance, of a dance without a song, is most unusual amongst Estrildines.

The song of this species can best be conveyed by the following: tu-tu-tu-tu-tu WOO WAH WEEEE. The length of the last WEEEE note varies from individual to individual, but this particular sound is extremely similar to the high-pitched WEEEEEEEE which ends the songs of several typical mannikins (see below) and along with the nestling palate-lines of the long-tailed grassfinch, may well point to a rather close relationship between this species and the mannikin group. The distinctive lost-calling of this species can best be rendered by: We-WOOOOOH We-WOOOOOH.

One aspect of the behaviour of the long-tailed grassfinch which is shared, not with the mannikins, but with the zebra finch, is the pairing pattern. The detailed description given for the pair-formation behaviour of the zebra finch, included horizontal mandibulation with lateral twistings of the body, and rhythmic wing-dropping, as elements of the display. These were also seen frequently in the present species whenever new social relationships were developing. The wing-drooping appeared without tail-fanning in these cases (see Pl. 31) and it seems that there is a threshold difference between the two species, the zebra finch having the tail-fanning at a lower threshold than wing-drooping and the long-tailed grassfinch having just the reverse. Both, however, have horizontal mandibulating with lateral swinging of the body as the pairing response with the lowest threshold of all. (Many other species of Estrildine have this mandibulating action as a signal in social situations of one kind or another, but its association with this special horizontal swinging from side to side has only been seen in the grassfinch and zebra finch.)

11. *P. personata* (masked grassfinch). N. Australia

This species is a close relative of the last two species and together the three species, with their various races, make a small natural group (which seems to me to be suited ideally to the status of subgenus, without the inclusion of the next species, the Gouldian finch, as suggested by Delacour).

Its markings, structure and deportment generally relate it most closely to the last two species, as stated, but it is interesting that its call-note is virtually identical with that of the zebra finch. These two species answer one another's call notes and often select one another to clump with when in mixed collections.

12. *P. gouldiae* (Gouldian finch). N. Australia

This is undoubtedly one of the most colourful of all finches and its plumage contrasts strikingly with the other members of this subgenus. It has a thin point to the centre of its tail, which is very reminiscent of the long-tailed grassfinch (see Pl. 31) and it has the heavy build and general carriage of the last three species. It does not, however, have their lined nestling palate markings, but spots, and in addition the nestlings possess elaborate, brightly coloured lobes and warts at the nestling gape. Its voice, too, is different and the song consists of a comparatively structureless hissing-whispering. During the solitary singing, the bird shows no special feather erection, although the feathers do seem to be slightly fluffed. The head is twisted and turned with the singing, but not in any special plane, as it is in certain other species. The beak is not opened appreciably during the song and there is no set beginning or end to any phrasing (or, if there is, it is far less obvious than in any of the other species, except the cutthroat finch).

One curious display which was seen in this species was 'looming'. This occurred when a male and female were kept together and it consisted of the male looming over the crouched female. His body stretched right out over her in such a way that his intense purple breast patch was pushed right in front of her eyes. In this posture, almost touching the female, the silent male would remain rigid for some time. This was seen on a number of occasions, but nothing was ever seen to develop from it. It is a similar phenomenon to 'peering' in the spice finch (see Moynihan and Hall, 1954), but the latter is less static and is only performed to a male and then only when he is singing. The peerer stretches forward with elongated, sleeked body and (as can be seen from Pls. 44 and 45) gazes at the performer. This it does until the wretched singer gives up and either flees, attacks, or just sits quietly. Sometimes several birds peer at one singer. Moynihan and Hall report that the other males and females which are not the mate of the singer may peer, but not the mate. These characteristics make peering rather different from looming, but it cannot be denied that it is a related phenomenon. The exact nature of the relationship would be easier to discuss if the function of either were known, but, to date, no functional explanation has been forthcoming. Thus far, peering has been noted for several species (star finch, chestnut-breasted mannikin, chestnut mannikin, striated finch, spice finch, silverbill and cutthroat finch), but looming has not been observed anywhere but in the Gouldian finch. (As a matter of interest, two species in which peering definitely does not occur are the zebra finch and the bronze mannikin.)

A report by Gilbert (1955), however, reveals that peering as well as looming may occur in the Gouldian finch: 'It is most amusing to watch a young cock with neck outstretched and huddled up closely to another youngster going through his singing exercises, his companion listening intently the whole time. These singing and listening performances alternate from bird to bird.' This is, of course, a perfect description of peering and indicates that probably, in this species, the looming display is a special form of a more typical peering display.

The only records of the more basic courtship song-and-dance performance for the Gouldian finch come from Gordon (personal communication) and Risdon (1953). Gordon considers the Gouldian to be quite distinct from all other Australian finches and states that, when the male displays to the female, he 'Stands erect, facing female, head back, bill in, throat puffed out in song, body feathers sleek, tail pulled well under, legs straight, "dances" with quick jerky springs, beginning with slow vertical jerks, the tempo quickening until the feet leave the perch in rapid tattoo of half-inch springs. The "dance" is performed for quite some length of time for so small a bird, and is followed by mating.' He points out that, although the call note of this bird is loud, the song is inaudible from a few yards away. Gordon also reports that the female 'Stands close to the male. Body feathers sleek, neck and head bowed, head shaking, tail down and pulled to side toward male – this stance is held from the time the male begins till "dance" ends, and the female bends to the horizontal.'

The vertical dance of the male described above is similar to the inverted-curtsy dance of other species, but the head and neck movements of the female seem to be unique to the Gouldian finch.

Risdon (1953, p. 57) states that the male of this species 'starts his display by dancing up and down on the perch before his hen, expanding the black or red feathers on his face and the purple ones across his breast, the while uttering his whispering song. Then suddenly, depressing his beak on his chest, he bows right forward till his face almost touches the perch. This generally ends the display, but sometimes he will draw himself up to his full height and repeat the performance.' The bow that is mentioned here sounds like a rather ritualized form of displacement beak-wiping.

13. *Erythrura* species (parrot finches)

The nine species of very brightly coloured parrot finches can be discussed together here. Delacour says that they 'form an isolated group, with no great affinities for the other Estrildinae' (in which case it is rather ironic that the grassfinch tribe should be named after them). Hardly anything is known of their behaviour. Superficially their bright colours remind one of the Gouldian finch, but this is probably a convergence, and Delacour thinks that the similarities 'do not seem to imply close relationship'. Nevertheless, it is worth pointing out that Cayley (1932) quotes a Mr F. Buckle (on the blue-faced parrot finch, *E. trichroa*) as saying that 'the young when first they leave the nest are a dusky green in colour; they also have luminous nodules on each side of the bill, like the young Gouldian finch.' Their affinities must therefore remain problematical until we know more about their behaviour.

The only fragments of knowledge we have about the behaviour of these birds, to date, concern the copulation of the red-headed parrot finch (*E. psittacea*). King (1951) reports the observations of a Mr N. Nicholson that 'It is, perhaps, somewhat unusual that mating actually takes place on the nest.' Gordon (personal communication), however, reports that 'Males drive females about, "chase" in rapid flights, catch females by back neck feathers, hang on while mounted, similar to domestic fowl-rooster.' These two observations appear to be contradictory, but it is quite possible that both occurrences are common in this species.[1] I have observed nest-copulations in the bronze mannikin, but have not seen neck-gripping by a mounted male in any Estrildine species under normal conditions. (It was seen, however, also in the bronze mannikin, when an injured, flightless individual was mounted on the ground, but this appeared to be rather abnormal.) Gordon makes a brief mention, in his report on the Gouldian finch, that 'Mounting males get balance holding female's neck feathers ... ', so that it is possible that this activity may be typical of this species as well as the parrot finch.

B. Mannikins

Genus *Padda*

14. *P. oryzivora* (Java sparrow). Java and Bali

One of the largest Estrildines, this species has been separated generically

[1] Nicholson (personal communication) has now confirmed this.

Plate 26. Beak-fencing in the long-tailed grassfinch.

Plate 27. Ditto; the left-hand bird uses its far wing in an attempt to regain balance.

Plate 28 a and b. *Left:* The resting posture of the male star finch. *Right:* The singing posture of the male star finch.

Plate 29. Low-intensity courtship posture of male zebra finch (*right-hand bird*). Note fluffing out of spotted flank feathers.

Plate 36. Male striated finch (*on right*): high-intensity courtship singing posture, showing crown, nape and ventral ruffling.

Plate 37. Ditto: here the male has pivoted away from the female; by comparing this and the above picture, the pivoting dance movement can be visualized.

Plate 34. Semi low-twist posture of courting male striated finch (*on left*).

Plate 35. Full low-twist posture of courting male striated finch (*on right*). The female here is a Bengalese finch. The latter is a pied mutant of the striated finch.

Plate 32. Courtship posture of male silverbill (*on the right*). Note the absence of ventral fluffing and also the twisting of the tail towards the female.

Plate 33. Ditto, showing the arching of the stretched neck.

Plate 30. Zebra finch: low-intensity sunbathing posture.

Plate 31. Long-tailed grassfinch: bird on right is in wings-drooped pairing posture.

Plate 38. Male Bengalese finch (*on left*): medium-intensity courtship singing posture.

Plate 39. Ditto: here the male has pivoted away from the female; cf. Pl. 38.

Plate 40. Courtship singing posture of the male spice finch. Note the erected head (and ventral) plumage.

Plate 41. Full low-twist posture of courting male spice finch. Unlike the striated finch, the male here sings in this posture and maintains the plumage erection.

Plate 42. Medium low-twist posture of spice finch; cf. Pl. 43.

Plate 43. Displacement beak-wiping in the zebra finch. The male (*left*) has been caught in the act of making this quick movement by the electronic flash. A comparison of this with the above suggests that the low twist is a sustained version of displacement beak-wiping.

Plate 44. Peering in the spice finch. The (*central*) singing male is being peered at by two other birds. Note the plumage erection of the singer which contrasts with the sleekness of the peerers.

Plate 45. High-intensity peering in the spice finch. The singer (*left*) leans away as the peerer presses close. The full significance of this pattern is not known, but it often has the effect of stopping the singing.

because of its markings, which include a pink beak and vivid black-and-white head, and its morphology.

Butler reports that, when courting, the male 'bends his body like an arch over the perch, turns his head sideways towards the female, and lifts himself jerkily up and down, singing all the while, and gradually sidling up to his mate'. From this account it is clear that this species had the 'low twist' posture discussed by Moynihan and Hall (1954) for the spice finch; also, that it shows the vertical inverted-curtsy dance movement.

The song is discussed at length by Butler (1899), but the one important comparative feature of it is that it possesses (at least, in the birds I studied) a high-pitched, thin, long-drawn-out WEEEEEEEE whistle at the end of each phrase. This particular note, recorded so far in this paper only for the long-tailed grassfinch, is one of the most important Estrildine sounds, from a comparative point of view. It occurs in a number of species in exactly the same form, always at the very beginning or end of the song phrases (where these are being repeated over and over again, it is difficult to tell whether it starts or ends the song, because the very first notes are often missed, and because the song may end abruptly in the middle of a phrase). I have so far noted it in the song of the above mentioned species and also the chestnut-breasted, pale-headed, and chestnut mannikins. It is one of the few distinct song characteristics that exist amongst the Estrildines. The songs of these birds appear to be the least conservative characters of all, for, even with visually very similar species, such as the bronze and black-breasted mannikins, or the parson and long-tailed grassfinches, the songs are strikingly different. This particular whistle, however, has for some reason been a much more conservative character and this suggests that the Java sparrow (and perhaps even the long-tailed grassfinch) are offshoots from the more typical Archipelago mannikin forms. This suggestion is, of course, extremely tentative, and we need to know a great deal more about the behaviour of these birds before a more definite idea can be formed.

Genus *Amadina*

15. *A. fasciata* (cutthroat finch). West, Central and East Africa

Several short papers have already appeared on the behaviour of this species (Morris, 1954b, 1955c and 1956b). Its sexual patterns will therefore only be described briefly here.

The male assumes a very vertical posture during courtship, with beak horizontal and not appreciably opened. The head pivots smartly from side to side whilst singing. There is extreme ruffling of the feathers in a special way (see illustrations in Morris, 1955c, 1956a and 1957a). The throat is not erected but the extent of the erection increases towards the brown belly patch.

There is a sudden increase in the extent of the erection at the white edge to the belly patch, which helps to exaggerate this latter marking when the bird is displaying. The feathers of the crown, nape and all down the back, are ruffled. The bird dances to the female with either inverted curtsies or with lateral pivots. The curtsying is by far the most common display movement, and the pivoting has only been seen once or twice (since the previous papers were written). The typical one-to-one combination of these two movements has not been seen in this species. When it made one movement, that bout of courting did not include the other type. Probably, if the behaviour of this species were better known, the combination would be seen. The display of the male is often given with a long straw in the beak, but not always so. The use of a straw in this and other species, during sexual displays, never seems to be obligatory.

16. *A. erythrocephala* (red-headed finch). South Africa

This species is extremely closely related to the last. Little is known of its behaviour, but Buttner (1950) reports that 'The red-headed finch breeds freely with its relative, the ribbon (cutthroat) finch; in fact, cocks of the latter species will invariably rob the red-heads of their hens when the two kinds of birds are housed in the same aviary.'

Genus *Lonchura*

Subgenus *Heteromunia*

17. *L. pectoralis* (pectoral finch). N. Australia

Cayley says, 'In habits they differ somewhat from the true mannikins, being rather shy, keeping to themselves, and spending most of their time on the ground ... The mating display of the male is very similar to that of the crimson and star finches with the addition of the straw-carrying habit of the mannikins.' Delacour also considers this in many ways to be 'an intermediate between the grassfinches and the mannikins'. It must be admitted that it has more complex markings than most mannikins, but a close scrutiny of such characters as its face-mask (as already pointed out) and various other features of its plumage leave no doubt as to its affinities with the other two Australian mannikins proper (the chestnut-breasted and the pale-headed). A number of its peculiarities can be accounted for by the fact that it has taken to ground living. Gordon (personal communication) reports that 'Males hold a short straw while displaying around the female in strutting jerky hops. Done upon ground, tail fanning the dust – not unlike the domestic pigeons.' Mackie, quoted by Cayley (1932), confirms this tail-fanning of the displaying male, thus: 'The male is most conspicuous and during the mating season is constantly dancing around his mate, fanning out his tail, and carrying a long straw or grass-stem in his bill. He has a low song which is difficult to hear in a large aviary.'

Subgenus *Euodice*

18. *L. malabarica* (silverbill). Tropical N.W., Central and N.E. Africa, S. Arabia, Persia, Afghanistan, Baluchistan, India and Ceylon

The song posture of the courting male of this species (see Pls. 32 and 33) differs from that of all the other mannikins studied. The most characteristic difference is that here there is no fluffing of the ventral feathers. The neck is fully stretched and slightly arched, with the throat a little expanded. The beak is not opened appreciably and is horizontal. The male assumes a vertical posture, but leans slightly towards the female. In this posture, lateral pivoting was seen without any vertical movements. In other cases, a more horizontal posture was held by the singing, courting male and, whilst holding a straw in his beak, he then performed the vertical inverted-curtsy dance movements, but without any lateral pivoting. (Again, though, the two movements are probably combined in high-intensity displays.)

Displacement beak-wiping and preening were both seen. The male also showed a highly modified version of displacement beak-wiping, in which the head and neck made U-shaped movements rather quickly, accompanied by mandibulating, just before a bout of singing began. The bird assumed the courtship posture, mandibulated and made U movements, and then began immediately to sing. To the song was then added the pivoting in certain instances. This sequence, as we shall see below, is very similar to that of the courting male striated finch, except that the latter moves down in a much more exaggerated dip when mandibulating. During the silverbill courtship the tail is markedly twisted towards the female (see Pl. 32).

The song of the silverbill is quite unlike that of any of the other mannikins, being clear and pure instead of harsh or rasping. In this respect it is, in fact, reminiscent of the waxbills and Butler (1899) goes as far as to say 'If the silverbills looked up to heaven to utter their shrill song, I should unhesitatingly declare that they were aberrant waxbills ... ' However, their dull coloration, their markings and lined nestling palates seem to justify their present position.

Subgenus *Lonchura*

19. *L. bicolor* (black-breasted mannikin). Ethiopia south to Angola, the Belgian Congo and Natal

This diminutive African mannikin is a very close relative of the next species, the bronze mannikin, and Butler says simply that the male 'dances and hums in the same manner' as the latter. This is not completely true, for, although the movements do all appear to be exactly the same, the song is quite distinctly different.

20. *L. cucullata* (bronze mannikin). Ethiopia

This species has been dealt with in detail in a very recent paper (Morris, 1957b), and its behaviour will therefore only be briefly summarized here:

Male courtship performed with beak open wide, tongue protruding: beak points downwards; body feathers ruffled except for throat region and head: legs bent; tail twisted to female; inverted curtsy dance at low intensity, with pivoting added in at higher intensities; displacement beak-wiping absent almost entirely and its place taken by displacement preening; no straw-carrying; no low twist; no peering; mandibulating immediately before each song phrase, but not accompanied by dipping or bowing.

The female quivers her tail in ruffled oblique posture instead of usual horizontal sleeked posture. If the male is not ready to copulate, he performs leapfrog sequence, vaulting back and forth over the female, with a quick displacement preen between each vault.

There is a nest ceremony in this species, related possibly to the pairing ceremony mentioned earlier for the zebra finch. However, this display is built entirely of elements taken from the nesting repertoire and is never performed away from a nest-site, whereas the zebra finch ceremony is derived from the sunbathing behaviour pattern and may or may not be associated with a nest or nest-site. The bronze mannikin nest ceremony certainly appears to strengthen the pair-bond and some elements of it reappear in the pre-copulatory ceremony just described. The nesting ceremony includes ritual scooping, mandibulating, gaping and tongue-protruding and, on occasion, was even seen to lead to copulations in the nest. But for full details, see Morris (1957b).

This species, along with the black-breasted and magpie mannikins, appear to form a natural small group of African mannikins. The magpie mannikin is only included here, however, because of its very similar markings and colours, and it would be most rewarding to make observations on its behaviour.[1]

21. *L. striata* (striated finch). Ceylon, India, Burma, Siam, S. China, Indo-china, Malay Peninsula, Sumatra

This simply-marked, brown-and-white mannikin is better known in captivity in its pied domesticated form, the Bengalese finch, or Bengalee (known in the U.S.A. as the society finch). Much has been written about the

[1] This has since been possible, at the London Zoo, and the sexual behaviour of the magpie mannikin does confirm its inclusion here.

origin of the Bengalese finch. It is supposedly a hybrid between the striated finch and various other species of mannikins. Most popular candidate for its other 'parent' is the silverbill. However, it has none of the silverbill's behaviour patterns and all of the striated finch's. This consideration, plus the fact that Bengalese finches often revert to a non-pied type indistinguishable from wild striated finches, leaves little room for doubt that this domestic finch from the Orient, which has been bred in captivity for several centuries, is not after all a magic hybrid but only a mutant. (Also, five offspring from a Bengalese X striated cross, bred in the laboratory, were all almost pure striated.) The following remarks about the behaviour of the striated finch, therefore, apply to the Bengalese finch as well (and, in all instances, the patterns have been observed in both).

In the medium or low-intensity courtship the male has his belly feathers fluffed, his throat and crown feathers flattened, and his nape feathers ruffled; the legs are bent; the beak is horizontal and opens and shuts fast and exaggeratedly during the song. (This and the beak and head movements seen in other species are not, however, linked, for mechanical reasons of voice production, with the particular noises produced during the songs. If birds from one species are reared by another species, the young birds learn to sing the 'wrong' songs of their foster-parents, but they still sing them in the postures and with the movements which are 'correct' for their own species. This means that song characters and song posture and movement characters are independent and can, taxonomically, be considered separately.)

The tail is slightly raised, which is unusual. Most of the other Estrildines keep their tails below the horizontal when courting. The male dances with small amplitude pivots, from side to side, through an angle of about 90 degrees. This medium to low-intensity courtship is illustrated by the Bengalese finch shown in Pls. 38 and 39. In Pl. 39 the male is seen pivoted away from the female and in Pl. 38 he has swung towards her.

In the higher intensity display, the feather erection is more extreme; the white ventral feathers are now fully ruffled and so are the crown and nape feathers. The legs are less bent and the bird's body no longer touches the perch. It becomes more vertical and the tail is now almost horizontal. Inverted curtsies are added to the pivoting movements now, in a one-to-one ratio. The higher intensity display is illustrated by the wild-type striated finch male in Pls. 36 and 37. As before the lower figure shows the male pivoting away from the female and the upper one shows him turned towards her.

Before the courtship song and dance begins, the male performs bowing or dipping movements, accompanied by much mandibulation (see Pl. 34). In the extreme form (see Pl. 35) the result is a striking low twist posture. This differs slightly in form from that of the spice finch (see Pl. 41) and it is also sustained for a much shorter time than in the latter species. It is performed *before* the male assumes the characteristic feather posture in which he sings and dances. In the spice finch, the feather posture, the song, and the vertical part of the dance can all be seen to be performed by males in the low twist position, but this is never so in the striated finch.

It seems as if this initial courting posture has also been derived from displacement beak-wiping, as in the silverbill, but this argument will be developed below.

22. *L. punctulata* (spice finch). Ceylon, India, Burma, S. China, Siam, Indo-china, Hainan, Formosa, Philippines, Malay Peninsula, Sumatra, Java, Bali, Lesser Sunda Isles, Celebes

The behaviour of this species has been dealt with at length in a recent paper by Moynihan and Hall (1954) and their findings concerning the form of the male's courtship performance will therefore only be summarized briefly here:

There are two basic courtship phases. In the first, the male (and also, to a lesser extent, the female) hops excitedly back and forth with nest material in its beak. During this phase, there is much mandibulating and looping of the material. Then when the male comes near to the female he drops his straw and performs a number of beak-wipes in front of the female. Phase two begins when the male stops this initial beak-wiping and assumes the characteristic song posture (see Pl. 40). This is an exaggeration of the solitary song posture of this species. The latter involves an erection of the head feathers, especially those of the crown and chin, which are ruffled. The belly feathers are slightly fluffed. The beak is horizontal and more or less closed except for the final note of each phrase, when it is opened wide. The legs are bent, the bird's body touching the branch. The head pivots smartly from side to side. In the courtship song posture, the ventral feathers are more fully fluffed, the tail is slightly more depressed and the rump region is strikingly fluffed (see Pl. 44). The flank feathers are typically fluffed outside the wings. The legs become more stretched and the whole body now begins to pivot from side to side. Moynihan and Hall also report that, in some of their males, there were vertical dance movements as well as lateral pivoting movements during this type of performance.

The vertical dance movements are more characteristic, in this species, of male courtships in which the male assumes the low twist posture. In this posture, the male's body is horizontal and the head twisted round towards the female (see Pl. 41). The feather erection and song occur as before, but the male in the low twist does not show any head or body pivoting. The vertical dance movements are now more exaggerated, and this type of display may occur, not just as a part of a sequence leading to the more upright version, but as an alternative to it. (Nevertheless, in some cases, the low twist display does lead to the other, as in the striated finch.) Moynihan and Hall stress that the males performing the song and dance in the low twist posture were in a more aggressive mood than those performing it in the usual manner. This automatically links the vertical dance movements to a more aggressive mood than the horizontal pivoting ones. It is interesting to speculate whether this can explain the various specific differences connected with these characters.

The vertical dance movements appear in some species, where the horizontal ones are almost absent. In other species, the reverse is the case. Again, in some species, the vertical movements appear at low intensities by themselves and the horizontal movements are added on at higher intensities. Once more, the reverse may occur. Finally, in species such as the present one, either type of dance movement may be observed by itself, or both may be performed together. At the present time it is difficult to relate these specific differences to appropriate differences in aggressiveness during the male display. However, certain instances do bear out Moynihan and Hall's point. The zebra finch, for example, which is a horizontal dancer, is not typically very aggressive to the female, whereas the bronze mannikin, which is primarily a vertical dancer, is often very aggressive. As the male bronze mannikin's aggression subsides with sexual arousal, he begins to add pivoting steps to his dance. The striated finch, however, provides a difficulty, for sexual arousal here is accompanied by the addition to the pivoting dance of the vertical one. This may mean that the male of this species is too afraid of the female to begin with and has to reduce its fear as sexual arousal proceeds, thus leading to the more aggressive display. Alternatively, both the dance movements may have become ritualized to the extent that they are now emancipated motivationally and are simple sexual responses with different thresholds in each species. Until a careful quantitative analysis has been made with a number of species, of the relationship between the frequencies of the two movements and various attacking, fleeing and sexual elements in the courtships, it is impossible to carry this discussion further. The ideal species to use for such a study would be the bronze mannikin, the striated finch, the spice finch and the zebra finch, each representing a basically different type.

From a derivational point of view it can be said that the turning away from the female which occurs in the pivoting dance movements supports the view that there is less aggression in this than in the inverted curtsy. The pivot dance appears to have evolved from intention movements of fleeing from the female, whereas the inverted curtsy dance seems to be a ritualized sequence of intention movements of mounting (or, at least, of jumping towards the female).

Peering by the spice finch has been discussed earlier and it is illustrated in Pls. 44 and 45.

Three hybrids between this species and the chestnut-breasted mannikin were obtained from the Copenhagen Zoo, thanks to the kindness of Dr Holger

Poulsen. These birds were nearer to the spice finch parent as regards their markings and colours and, interestingly, were also nearer to the spice finch in their singing posturing. The three individuals were males and all three sang in a manner quite unrelated to the songs of either parent. It emerged that they had been reared in an aviary containing canaries and were now doing their best, each in his own way, to simulate the exotic song of this fringillid. The extent to which one of them succeeded was remarkable. especially when the rasping, grating, squeaking qualities of the two parent songs are borne in mind. An unfortunate outcome of the extraordinary performance put on by the one male was that the great effort involved resulted in his whole frame vibrating so much that his tail quivered violently, thus inadvertently giving the female Estrildine invitation signal, occasionally with disastrous effectiveness.

Subgenus *Munia*

23. *L. ferruginosa* (chestnut mannikin). Ceylon, India, Burma, Siam, S. China, Formosa, Hainan, Indochina, Malay Peninsula, Sumatra, Java, Bali, Borneo, Philippines, Lombok, Flores and Celebes

The song of this species is particularly interesting. It has three parts to each song phrase. The first part consists of a slow clapping of the beak, which appears to be a ritualized version of mandibulating. Many species mandibulate just before each song phrase, like the bronze mannikin, and in the present species the rapid opening and closing of the beak appears to have been slowed down and increased in power to produce a primarily auditory display. This phase of the song phrase passes into an almost silent one in which the bird heaves up and down in a manner reminiscent of a human runner very out of breath. This in turn passes into the third phase, which consists of one or several long, high, thin, drawn-out whistles. Each song phrase is very long compared with the typical (one-second long) phrases of such species as zebra finches and most other grassfinches. The chestnut mannikin song phrase lasts as long as sixteen and a half seconds in some cases, with eight seconds as a minimum. Variations in the length of the song phrase in this species are mainly dependent on how many ending whistles there are: the sixteen-and-a-half second case was a three-whistle affair, whereas the eight-second case only had one whistle.

The song posture, whether solitary or courting, involves the erection of all body feathers, with the exception of those of the throat and upper breast. The head is pointed down and the beak opened. There is no head pivoting, but a suggestion of a vertical nodding. The only dance movements observed were vertical ones, with no pivoting, and with no low twist. Displacement beak-wiping was common during bouts of courting.

24. *L. maja* (pale-headed mannikin). Malay Peninsula, Sumatra, Java, Bali, Lombok, Flores, Celebes, New Guinea and N. Australia

According to Butler, 'the male raises himself laboriously, stretches his head obliquely upwards, spreads his short tail and commences an extraordinarily zealous song in which one sees beak and throat in most industrious motion, but which is not accompanied as in the cutthroat finch by an up-and-down hopping dance, but only by a gentle and almost automatic movement of the head from side to side.' From my own observations of two races of this species (the white-headed mannikin and the Australian yellow-rumped mannikin) I cannot agree with Butler in all these points. The male certainly stretches his neck upwards, but Butler gives the impression from his wording that the beak is pointing upwards, when, in fact, it points downwards as in the last species. Also, the tail is not necessarily spread during song. Finally, the absence of the vertical dance movements in Butler's birds was undoubtedly due to the absence of high-intensity responsiveness, rather than to the complete absence of this component in this species.

Mandibulation and displacement beak-wiping occur prior to courtship singing in this species and nodding accompanies the song. Butler is quite wrong to suggest that there is any head pivoting during the song. Also, there is not exactly a great deal of beak movement, the beak being held wide open most of the time while singing. There is general feather erection all over the body except in the throat and upper breast region. In other words, this species behaves when singing very much like the last, even including the same kind of long-drawn-out whistle at the end of each song phrase.

25. *L. castaneothorax* (chestnut-breasted mannikin). Australia and New Guinea

A close relative of the last species, with very similar song, including the long whistle at the end, the head nodding, the heaving, the absence of head pivoting, the downward pointed beak, the wide-open mandibles, the stretched neck, the fluffed-out body feathers, and the displacement beak-wiping. In addition, a low twist display was seen in this species.

A typical introduction involved the following stages: both birds shake; male sings in low twist but no dancing; both birds mandibulate and then beak-wiping (double wipes); both ruffle and give repeated single wipes; male does song and curtsy dance in oblique posture (i.e. semi-low twist); other bird peers at, then preens, the singer.

The song is one with a complete four-phasic phrase: weeeeee eeeeeeeee/tuee tuee tuee tuee tuee tuee tuee tuee tuee tuee/cheeouk cheeouk cheeouk cheeouk cheeouk/ching-ching-ching-ching-ching. The long thin whistle (weeeeeee) occurs at the beginning of this song phrase and it is possible that this is also the case with the other species possessing this note. Unfortunately, this note is often the one which ends a bout of singing as well as beginning it and since the first notes of a bout are often missed, it is difficult to be certain. The important point, however, is that this note is extremely characteristic and quite identical in the several species which possess it. Silent mandibulation precedes each phrase.

C. Waxbills

Genus *Estrilda*

Subgenus *Estrilda*

26. *E. temporalis* (red-browed finch). E. Australia

As stated earlier, the waxbills are not being included in the present paper, but the red-browed finch is an exception, since it is in doubt as to whether it is a true waxbill or not.

The inconsistency in Delacour's system, concerning this species, has already been pointed out. He has called it a waxbill mainly because it looks like one (the red-eared waxbill) and because it displays with its beak pointing upwards and singing a high-pitched song. But the crimson finch looks like the fire-finch (a waxbill from Africa), the painted finch points its beak upwards when singing, and the star finch has a high-pitched voice. Yet these three species are all considered to be good grassfinches. Even if it is pointed out that the red-browed finch has *more* waxbill-like characters than these other species, it is still perhaps advisable to be cautious over including a single Australian form in a predominantly African group. Its sexual behaviour is as follows:

The solitary singing posture of the male involves a ruffling of the ventral feathers, but no erection of any other region of the plumage. The bird sings with rather stretched legs and upright stance, the beak being horizontal and not opened appreciably. The head appears to make slight lateral movements while singing. The song is little more than a rhythmic repetition of the call note. Each song phrase consists of three or four call notes spaced in a special way. In one individual male, the spacing was as follows: teee-te-tee teee-te-tee, and so on. In another male it was: teee-te-tee-teee teee-te-tee-teee teee-te-tee-teee. This is undoubtedly the simplest song in the Estrildinae and indicates the probable way in which singing was evolved. First the simple call note of the species is repeated over and over again, then it develops a special spacing as in the present species, thus giving rise to simple little song phrases. These can now act as units themselves and can develop further, more complex relationships, and so on, until one arrives at the more complicated song phrases of most other species.

Two quite distinct sexual displays were seen in the red-browed finch and the relationship between them is still not clear. Both displays appeared to be leading to copulation, but there was no sequence involved. On one occasion a male would give one display type, and on another he would perform the second type, but no clear-cut causal difference could be found. The only clue was that the first type of display was seen mostly at the time when the new birds were introduced to one another. Later, when the group was stabilized in its aviary environment, only the second type was recorded.

When the first introductions were brought about, the immediate effect was to produce a special posture in the male. His body became horizontal and his tail vertical, pointing upwards. He faced the female and showed repeated beak-wiping, his body 'see-sawing' with the wiping actions. The front view seen by the female displayed the red regions of the male's body (the red streaks of the beak, leading to the red eye-brows, and also the red rump patch). This posture is very reminiscent of the crimson finch's threat posture, but the male did not attack the newcomer in this case. The beak-wiping at this initial stage of the introduction was very fast but thirty of the wipes were scored and they were all doubles (wipe wipe). In the zebra finch, wiping at this speed and with this high frequency during courtship would certainly result in nearly all single-wipe wipings (see Morris, 1954a). The next stage in the behaviour can best be described by quoting from protocols:

'The second stage includes song with the beak-wiping. The male adopts the horizontal, tail-up posture and rocks up and down in see-saw fashion with the beak-wipes coming so fast one after the other that the "up" pause of the rocking is as short as the "down" pause. Again this differs from the zebra finch, where the up pause is much longer than the very rapid down pause. The high-intensity courting of this kind in the red-browed finch involves very rapid double wipes at the bottom of the rocking movement, at an intensity at which the zebra finch would be doing exclusively single wipes. However, at a higher intensity still, with the song becoming very loud and clear, the red-browed finch gives up the double wipes and begins to "bow" (i.e. it gives abbreviated single wipes). This higher intensity bowing-singing display is more broadside-on to the female, less frontal than the silent bowing. And with this there is a change in tail position. The tail becomes less vertical and is more twisted round towards the partner. Finally, it should be stressed that, even in the up phase of this display, the male is extremely near to the horizontal.'

Later, a third basic stage in the courtship was recorded, but it was not continuous with the above. Rather, the male changed over after a time to this new type of display:

'The full song-and-dance display ... The male selects a long straw, holds it in his beak, advances to the side of the other bird and begins to sing and dance with vertical movements. These are only slightly directed towards the other, the male being almost broadside-on. The body is now extremely erect, with the legs stretched and the head held back, the beak pointing to the sky. The vertical inverted-curtsy jumps are so vigorous that, as in the Gouldian finch, the male's feet leave the perch each time. The tail is held more downwards now and is strongly twisted towards the female. This display attracts the female and she approaches the male. As she alights near him *she throws back her head* and, for a few moments, she holds this position with her beak pointing upwards like his. This display is extremely similar to the head-tossing of the painted finch under similar circumstances. The movement is slower in the present species, but it is fundamentally the same. Also, the red-streaked beak markings of the two species are similar.'

A possible derivational explanation of the head-up display was provided by the following observations. Several times, a male was seen to be about to display to a female, but then instead went into the nest and incorporated the piece of straw it was holding into the structure. This was done with scooping movements and it was noticed that at the top of each scoop, the male's neck was stretched vertically upwards, his beak was pointing upwards and he was holding a straw in his mandibles. In all these respects the 'up' posture of the scooping action was similar to that of the song-and-dance display posture.

Many species court with the beak up, many more, with it horizontal, and many more again with it pointing downwards. They all scoop in the same way, as far as is known, but there is no reason why different groups should not have 'frozen' the scooping posture at a different stage in different cases, the beak pointing downwards at the beginning of a scoop and upwards at the end. An alternative explanation would be that the bird is assuming a head-back posture in the present species as part of the intention movements of flying away upwards, but this would be difficult to reconcile with the comparative evidence.

Taxonomically, I am convinced that the red-browed finch is a close relative of the crimson finch and the painted finch, and that, if it is a true waxbill and has resulted from an independent waxbill invasion of Australia, then the other two species are also waxbills. If they are not, it is not.

Parental behaviour

Parentally speaking, the Estrildines are a very homogeneous group. Apart from the mouth markings already discussed, there is little of comparative interest in this aspect of their behaviour. In almost all species there are five small white eggs in a clutch, and the parents take turn in incubating them. Both male and female also share the task of feeding the young birds. The latter beg for food with a special twisted posture. The neck is not stretched upwards in the more usual fashion, but it is twisted tightly round so that the beak gapes upwards but at the same time leaves a space above it and below the roof of the domed nest. If the young birds stretched their necks upwards vertically they would touch the top of the nest and could not then be fed. The incubation period (two weeks), the nestling period (two weeks), and the fledgling dependency period (also two weeks) given for the zebra finch in an earlier paper (Morris, 1954a) appear to be roughly constant throughout the group. The young birds mature extremely quickly, in about three months from fledgling, and will breed in their first year.

Comparative Derivational Problems

Before finally reviewing the relationships between the various species, some general comments about the origins of the different sexual components must be made.

Specific differences in respect of the form of these components can easily be deceptive. Two closely related species may have diverged widely in their sexual signals, or two widely separate species may have arrived at a very similar signal from quite different sources, simply because of the signal value of a particular type of movement or posture. An example of the first would be where two species both used displacement beak-wiping as a ritualized signal, the one 'wiping' its beak back and forth without bending down, and the other bending down in a bow, but without making the wiping movements. If it can be shown that both actions have originated from the same movement, then their *superficial difference* dwindles in importance. An example of the second would be where two species both bow ceremonially to the female, but the bow is derived, in one case from beak-wiping, as above, and in the other from intention movements of taking-off. In this case, their *superficial similarity* dwindles in importance.

Such points as this must be taken into consideration when evaluating the ethological data in a comparative study. It is worthwhile here to run briefly through the main components of the various displays, noting their possible origins:

Straw-holding is derivationally simple. Quite clearly it originates from the genuine carrying of nest material to the nest and has become involved in the sexual patterns because of the intimate association in time between the building of a reproductive nest at high intensity and the pre-copulatory phase. In some species, such as the zebra finch, straws are not used in display. In others such as the spice finch, the straw is carried about in the first stage of courting and then dropped when the bird begins to sing and posture. In others, such as the star finch and the red avadavat, the birds bow with the straws held in their beaks, but release them for the final stages of the displaying. In still others, such as the red-browed finch, the bird retains the straw actually during the dancing stage of the display.

The intimate relationship between nesting and courting referred to above is of great importance in other respects also. It seems to give rise to more sexual display elements than any other aspect of the birds' behaviour. Mandibulation, which, during nesting, so typically accompanies the carrying of material, is seen in sexual situations in many species. In some, such as the spice finch, genuine mandibulation of material is seen associated with the straw-carrying in the early phases of pre-copulatory patterns. But also, in this species, and many others, mandibulation occurs without any object being held in the beak. In the zebra finch and the long-tailed grassfinch this is typically performed in the special horizontal posture associated with pair-formation ceremonies. In other species, it is performed in a low dipped posture (striated finch) just before the song begins, or in the ordinary song-and-dance posture itself. In the latter cases, it is usually seen immediately before the start of singing. In a number of cases, the mandibulating appears briefly before each song phrase in a bout of singing (e.g. bronze mannikin). In different species, the movements of the beak which accompany the song appear to be exaggerated forms of mandibulating. At first sight, one gets the impression that they are essential to the mechanics of voice-production and that differences in the sounds produced by the different species account for their variations. However, rearing birds of one species with those of another reveals that this is not so and that one song can easily be sung with quite different beak movements. I suggest therefore that many of the singing beak movements are derived from mandibulation. In the case of the bronze mannikin, the beak is held open while singing and the tongue is protruded and quivered. The opening and shutting of the beak which occurs in mandibulation appears to have 'frozen' here at the open stage. The tongue component is probably similarly derived, since tongue movements are employed when mandibulating nest material; the material is moved about in the beak by the tongue, which appears to test the texture of the grass or straw. In the striated finch, the exaggerated opening and closing of the beak shows a modification of mandibulating which involves a slowing down of the opening and shutting actions and an increase in their amplitude. In the chestnut mannikin, these movements are also slowed down, but here the beak is shut with a greater force, producing a clapping noise, which virtually becomes part of the song itself.

As already discussed, the nesting action of scooping seems to be used in determining the head and neck posture, some species tilting the head down and others up.

Bowing displays, as stressed above, may have different origins, and this component of the star finch display appears to have been taken from a different source from most others. Here the form of the movements indicates that this display pattern comes from intention movements of flying away, presumably to the nest with the straw held in the beak. The bowing of other species mentioned in this paper (with the exception of the red avadavat) has evolved from displacement beak-wiping.

This last-named action must now be considered as a source of signals. It appears in many species and is comparatively unmodified in some. Even where specialized versions of it appear, it is interesting to note that the unmodified action can also take place as a displacement activity and has not been completely replaced. In the zebra finch incomplete beak-wiping appears in several forms. The extent of the bending down may be reduced, the number of wipes per wiping reduced, their speed increased, the orientation of the bird to the perch abandoned, and so on. In various other species, one of these modifications has become the typical one. In the silverbill, the lateral wiping movements are retained in a simplified manner, but the bending down has vanished, resulting in a simple U-shaped action of the head. In the chestnut-breasted mannikin, the lateral elements have gone and the bending down is shallow, resulting in a slight bow which is little more than a nod. In other species the bending down has become exaggerated but the wiping element has gone. This occurs in the striated finch and gives rise, when the down posture is sustained, to the low twist posture. In the spice finch this low twist posture is maintained for long periods and is performed simultaneously with the song and dance. The photographs in Pls. 42 and 43, show the relationship between the displacement beak-wiping action of the zebra finch and the sustained low twist posture of the spice finch. The zebra finch does not sustain its down phase of its beak-wiping movement, but the electronic flash photograph 'freezes' the movement here. The low twist of the spice finch is 'frozen' by the bird itself and, when these two pictures are compared, the great similarity between the two postures makes it easy to believe that the sustained one has evolved in the way suggested here.

The origin of feather postures has been discussed at length in another recent paper (Morris, 1956a) and will not be dwelt on here. Amongst the Estrildines, there are few, if any, species which do not show feather erections of some sort during display. The silverbill is one of the exceptions here. The star finch is also unusual in not erecting fully the ventral plumage. It does, however, erect its spotted flank feathers to a dramatic extent. Almost all other species show strong erection of the ventral plumage and often also an erection of the rump, back, nape and head feathers. The throat feathers are typically sleeked, an exception here being the spice finch.

The dancing movements of inverted curtsying and pivoting are almost certainly ritualized intention or ambivalent movements. The pivoting seems to be the result of a conflict between fleeing (turning away from the female) and approaching (turning back towards the female). The rhythmic alteration of these two movements produces the stylized dance pattern. The vertical movements of the inverted-curtsy dancing are certainly intention movements of jumping upwards and probably derive from intention movements of mounting the female.

The special movements seen in the zebra finch and long-tailed grassfinch, when pairing, have been discussed fully earlier, and clearly derive from sunbathing actions.

From this very brief summary of the various display patterns and their origins, it will be seen that a great number of different displays have been built up from a small number of non-signal sources. Furthermore, motivational differences involved in the courtships of different species (typically more fear, or more aggression) give rise to even more variations without using new sources. These further variations take the form of sequential and threshold differences, which give rise to distinctive specific combinations of elements.

Discussion

Despite the large number of facts already collected concerning the distribution, morphology, and ethology of the various Estrildine species, the study of the evolution of the group is still in a very preliminary state. Although the investigation by the present author is now terminated, the work is being actively continued by others and the evolutionary suggestions put forward here should be considered rather as guides to further research than as definite conclusions. A preliminary study such as the present one, involving so many species, can only hope to act as a focusing device, bring attention to bear on the more important evolutionary trends and the more valuable species for study. This much can certainly be done here, and, in certain particular instances, specific revisions in the existing taxonomic scheme can be strongly recommended.

As has already been mentioned at various points, the evolution of the group can be thought of as involving either one or two basic invasions of Australia from Africa, via Asia. Either ancestoral mannikin forms alone spread from Africa, through Asia, to Australia, where they gave rise to all the grassfinches; or, alternatively, both ancestoral mannikins and waxbills spread in this way and each gave rise to part of the Australian grassfinch group. There are facts which support both hypotheses, as we shall see below.

Starting in Africa (but ignoring the waxbills) we have three distinct mannikin types. There are, firstly, the cutthroat and red-headed finches. These birds stand apart from all other Estrildines and almost warrant separation into a tribe of their own. Delacour places them with the mannikins, Butler with the grassfinches. Geographical and other considerations make them the most unlikely of grassfinches, however, and Butler is certainly wrong here. Delacour has had to make special allowances for them when including them in his mannikin tribe, but within his rather rigid Estrildine framework, he is probably as near the truth as possible.

The silverbills produce a similar problem. Once again Delacour and Butler disagree, and in exactly the same way. Delacour gives these birds a subgenus to themselves in his mannikin tribe, but Butler sees them as grassfinches. It is true that they have certain similarities to the Australian cherry finch and star finch, but the geographical distribution of these species is such as to make it unlikely that there is any really close affinity between the three.

The third group to be considered in Africa comprises the bronze mannikin, the black-breasted mannikin and the magpie mannikin. The last of these three is only tentatively included here, as its behaviour is unknown. However, its markings are so similar as to justify placing it here. These three are considered as mannikins by both Delacour and Butler, but Delacour does not even distinguish them subgenerically from species such as the striated finch and the spice finch. There are sufficient geographical, morphological and ethological differences to warrant at least a subgeneric distinction here.

Moving over to the Asiatic region, the above mentioned striated finch and spice finch are closely related and should be placed together in the same subgenus. The spice finch is slightly closer to the next group, which includes the chestnut mannikin, the pale-headed mannikin and the chestnut-breasted mannikin. A superficially aberrant offshoot from this group is the Java sparrow. The importance of its unusual morphology dwindles in the face of its behavioural similarities to the more typical mannikin forms. The pectoral finch is also very close to this group, being very similar in its markings and colours to the chestnut-breasted mannikin. If its sexual patterns include song postures and movements similar to those of the chestnut-breasted mannikin, then there will be no need to give this species a subgenus to itself, as Delacour has done.

Passing on to Australia, we find several species of grassfinch with mannikin characteristics. The long-tailed, masked, parson finch group, with its low voice and lined mouth markings, could easily have evolved from a mannikin ancestor. Closely related to these are the Bicheno finch and zebra finch, and here again we find a disagreement between previous taxonomic systems. Delacour places the zebra finch and Bicheno finch in a subgenus with the crimson finch, star finch and cherry finch. This is perhaps his unhappiest grouping of all. I prefer to follow Gordon over this point and place the zebra, Bicheno, long-tailed, masked and parson finches together in one group, keeping the others mentioned above separate. I would also like to separate the Gouldian finch into a distinct subgenus or genus, again following Gordon, but I am equally sure that Delacour is right to relate it to the above species rather than the parrot finches.

Not having studied the cherry finch personally, I can say little about this difficult species except that it is certainly related to the mannikins. It does not appear to have close affinities with any of the other Australian finches, with the dubious exception of the star finch.

The red-browed finch, the painted finch and the crimson finch form a distinct group, closer in character to the waxbills. The star finch shows some similarities to this group, but unfortunately it shows some to the mannikins and the zebra–long-tail group as well. So also does the diamond sparrow and its close relatives the fire-tailed and red-eared fire-tailed finches. These birds lie in an intermediate position taxonomically between the waxbill-like red-browed finch group and the mannikin-like long-tailed grassfinch group. They add considerable support to the theory that the Australian grassfinches are a single (and not a double) natural group. Against this, one has to balance the remarkable similarity between the behaviour of the African waxbills and the red-browed finch. The avadavats exist in Asia as a reminder that waxbill ancestors may have passed this way, and the absence of a nice series of linking forms through from Africa to East Australia, of the kind shown by the mannikins, does not necessarily mean very much. However, on the whole, I am inclined to support the single rather than the double-trend view, after taking into consideration all the known facts, and I am basing my revision of the arrangement of the group, given below, on this. I cannot stress too strongly, however, the tentativeness of this revision. There are so many gaps in our knowledge that it will probably not be long before a further revision is required, but some of the alterations I am making here will, I hope, stimulate new lines of thought about the evolution of these birds. We badly need to know more about the behaviour of the avadavats, and the close relatives of the red-browed finch, such as the painted finch and crimson finch. These, in particular, will help to solve the Australian 'invasion' problem.

Before presenting my rearrangement of the species, there are one or two points to be made concerning the validity of the three tribes of the Estrildinae. Delacour has given a number of characters in his paper which define these tribes, and a few modifications are needed here. Firstly, the mouth markings: as already pointed out, these do not follow his groupings perfectly, four species of grassfinch having mannikin-type lines instead of spots. Secondly, the voice varies with genus or subgenus rather than with tribe, there being high-pitched mannikins and grassfinches as well as low-pitched ones. Thirdly, both waxbills and grassfinches may sing with the beak pointing upwards. Some grassfinches point it up, others down and others keep it horizontal. Some mannikins keep it down, others keep it horizontal. Fourthly, members of all three tribes are known which hold straws when displaying, and this character appears to show a much more scattered distribution through the Estrildines than Delacour suggests. Fifthly, the curtsy and pivoting dance movements appear also in species in all three tribes and no general tribal distinctions can be made here.

In other words there are few, if any, 'ideal' characters which split the Estrildines neatly up into the three tribes suggested by Delacour. Despite all this I am certain that Delacour's *tribal* separation of the species (with the single exception of the red-browed finch) is correct in almost all respects. This is, of course, because one includes *many complex character-relationships* in any taxonomic evaluation. Intuitively one is on guard against the ruthless use of single characters, but unfortunately their employment makes for a very tidy report. Delacour's brilliant revision of the 108 Estrildine species succeeds in general, I feel sure, not because of the official reasons he gives, but despite them.

Conclusion

In concluding, I wish to put forward the following extremely tentative revision of the Estrildines. I am including only the species I have discussed in the present paper, and the waxbills are entirely omitted:

1. Genus *Amadina*
   - Cutthroat finch
   - Red-headed finch
2. Genus *Lonchura*
   - (a) Subgenus *Spermestes*
     - Bronze mannikin
     - Black-breasted mannikin
     - Magpie mannikin
   - (b) Subgenus *Euodice*
     - Silverbill
     - Grey-headed silverbill
   - (c) Subgenus *Lonchura*
     - Striated finch
     - Spice finch
   - (d) Subgenus *Munia*
     - Chestnut mannikin
     - Pale-headed mannikin
     - Chestnut-breasted mannikin
     - Pectoral finch
3. Genus *Padda*
   - Java sparrow
   - Timor sparrow
4. Genus *Aidemosyne*
   - Cherry finch

5. Genus *Poephila*
   (a) Subgenus *Poephila*
      Parson finch
      Long-tailed grassfinch
      Masked grassfinch
      Bicheno finch
      Zebra finch
   (b) Subgenus *Gouldaeornis*
      Gouldian finch
6. Genus *Bathilda*
      Star finch
7. Genus *Zonaeginthus*
   (a) Subgenus *Zonaeginthus*
      Fire-tailed finch
      Red-eared fire-tail finch
      Diamond sparrow
   (b) Subgenus *Aegintha*
      Red-browed finch
      Painted finch
      Crimson finch
8. Genus *Erythrura*
      Parrot finches

In the above scheme, the cherry finch and star finch are the least satisfactorily placed species. I have only separated them into distinct genera here because at present it is impossible to decide whether to place them with the *Poephila* group or the *Zonaeginthus* group. (Delacour puts them with the former, Gordon with the latter.)

References

BUTLER, A. G. (1899), *Foreign Finches in Captivity* (London).
BUTTNER, E. E. (1950), 'Red-headed finch at home'. *Cage Birds* 97, p. 1.
BROOKSBANK, A. (1949), *Foreign Birds for Garden Aviaries* (London).
CAYLEY, N. W. (1932), *Australian Finches in Bush and Aviary* (Sydney).
DELACOUR, J. (1943), 'A revision of the subfamily Estrildinae of the Ploceidae'. *Zoologica* 28, pp. 69–81.
GILBERT, H. R. (1955), 'Colourful gouldian finches'. *Cage Birds* 107, p. 698.
HARMAN, I. (1953), 'Australian crimson finch'. *Cage Birds* 104, p. 30.
KING, H. T. (1951), 'Breeding foreign species'. *Cage Birds* 100, p. 147.
LORENZ, K. (1953), *Comparative Studies on the Behaviour of the Anatinae* (London. *Avicultural Society*).

MORRIS, D. (1954a), 'The reproductive behaviour of the zebra finch (*Poephila guttata*), with special reference to pseudofemale behaviour and displacement activities'. *Behaviour* 6, pp. 271–322.
—— (1954b), 'The courtship behaviour of the cutthroat finch'. *Avic. Mag.* 60, pp. 169–77.
—— (1955a), 'The breeding behaviour of the zebra finch'. *Birds Illustr.* 1, pp. 28–30.
—— (1955b), 'The causation of pseudofemale and pseudomale behaviour – a further comment'. *Behaviour* 8, pp. 46–56.
—— (1955c), 'The markings of the cutthroat finch'. *Birds Illustr.* 1, pp. 182–3.
—— (1955d), 'The fighting postures of finches'. *Birds Illustr.* 1, pp. 232–4.
—— (1955e), 'The seed preferences of certain finches under controlled conditions'. *Avic. Mag.* 61, pp. 271–87.
—— (1956a), 'The feather postures of birds and the problem of the origin of social signals'. *Behaviour* 9, pp. 75–113.
—— (1956b), 'The toilet of the cutthroat finch'. *Birds Illustr.* 2, pp. 60–2.
—— (1957a), ' "Typical intensity" and its relation to the problem of ritualization'. *Behaviour* 11, pp. 1–12.
—— (1957b), 'The reproductive behaviour of the bronze mannikin (*Lonchura cucullata*)'. *Behaviour* 11, pp. 156–201.
MOYNIHAN, M. and F. HALL (1954), 'Hostile, sexual and other social behaviour patterns of the spice finch (*Lonchura punctulata*) in captivity'. *Behaviour* 7, pp. 33–76.
RISDON, D. H. S. (1953), *Foreign Birds for Beginners* (London).
WEBBER, L. C. (1946), 'Ten years with the painted finch'. *Avic. Mag.* 52, pp. 149–58.

Author's Note, 1969

Two errors that existed in the original publication have been corrected in the text. One was an accidentally omitted paragraph that has been replaced in its original position (p. 401 in the present volume). The other concerns the nestling palate markings. In the paper as it first appeared, I listed five species of grassfinches as bearing nestling palate markings with lines rather than the typical spots. One of these species was the masked grassfinch, but my information on this bird was erroneous. It has since been studied in more detail and is now known to have spotted palate markings. I have therefore omitted it from my argument in the text concerning the inadvisability of placing too much taxonomic importance on this single characteristic. However, although this reduces the exceptions from five to four species, it in no way weakens the point I am making. If anything it strengthens it, because of the very close relationship (in other respects) between the masked grassfinch (with its spotted palate) and the long-tailed grassfinch (with its clearly lined palate). If two such close relatives can exhibit such different palate markings, then the conservative nature of this characteristic is even more in doubt, and its use as a rigid indicator of tribal status becomes even more dubious.

The behaviour of the green acouchi

(1962)

Introduction

The green acouchi (*Myoprocta pratti*) from South America is a hystricomorph rodent that is particularly well suited for ethological observation in the laboratory. It is related to the cavies (Caviidae) and, more closely, to the agoutis (*Dasyprocta* species). It has the body size of the former but the proportions of the latter and has been referred to aptly as a 'guinea-pig on stilts'.

For ethological work it has advantages over both the cavies and the agoutis. Its long legs, that have undoubtedly evolved in connection with more rapid fleeing from predators, have made possible a whole range of manipulatory and other fixed motor patterns that are absent from the behaviour repertoire of the short-limbed cavies. Its small size, when compared with the agoutis, makes it possible to keep groups of acouchis in the laboratory in cages of reasonable dimensions. Agoutis require a much greater living space and can only be kept satisfactorily in paddocks.

The acouchis used in the present study were housed as a group in a cage with a floor space of seven by nine feet. When a single acouchi was isolated from the group for some reason, it was possible to keep it in a small cage with a floor space of two and a half by three feet for a period of up to several days. A longer confinement in a small cage was found to be detrimental. The acouchi is an intensely active species and, if it is not permitted to express itself with a considerable motor output each day, its condition suffers rapidly. But it is, of course, this high behaviour output that makes it so rewarding ethologically. The fact that it is a diurnal species is also helpful and there are, indeed, few mammals, outside the primates, that are laboratory-sized, active, diurnal and have a wide behaviour repertoire.

Given these advantages, it is surprising that they have not figured more largely in behaviour research in the past. It is true that, like most animals specialized for rapid flight, they are highly strung and are not suitable for research involving repeated catching or handling by the experimenter, but they are nevertheless ideal for controlled observational work.

Their absence hitherto from the behaviour literature appears to be simply the result of their extreme rarity in captivity. As far as can be ascertained, there are, at the time of writing, only nine specimens in captivity in Europe – five of these being used for the present study. All nine specimens originated from Ecuador.[1]

Little is known of their behaviour in the wild, but this is probably due to their rapid fleeing, rather than to their rarity in their natural habitat. Like the larger agoutis, the acouchis are forest dwellers, apparently favouring areas near rivers and marshes. At the slightest sign of danger they flee into dense undergrowth and are quickly hidden, so that it may well be impossible ever to make a detailed study of them in the wild.

Ellerman (1940) lists five species of acouchis. Two of these are reddish animals (*M. acouchy* and *M. leptura*), two are greenish (*M. pratti* and *M. milleri*) and one (*M. exilis*) is so rare that Ellerman was unable to study it. The red forms appear to be confined largely to the north-east regions of South America, in Brazil and the Guianas. The green forms are recorded more from the north-west, in Peru, Ecuador, Colombia and parts of Brazil. *M. exilis* is also apparently a north-west form.

Judging by the skins in the British Museum of Natural History, there has been some unnecessary splitting. It would appear that there are, in fact, probably only two good species, the green acouchi (*M. pratti*) and the red acouchi (*M. acouchy*). It is even doubtful whether these two are truly distinct and in the final analysis it may well emerge there is simply an east–west *Myoprocta* cline from red to green.

The observations reported in this paper were made over a period of fifteen months, from June 1960 to September 1961. After a preliminary survey, one particular pattern, namely food-burying, was selected for more detailed analysis. The results of this analysis are given after the brief general ethogram for the species. This ethogram is incomplete in some respects, but will nevertheless serve to introduce the behaviour repertoire of the acouchi.

Ethogram

General

1. *General activity rhythm*

As already stated, acouchis are primarily diurnal. An accurate score of their daily rhythm of activity was made over a period of five days, using an infra-red beam and a twenty-four-hour counter system. The beam shone across the acouchi cage, a few inches above the ground, and was broken each time an acouchi crossed from one side to the other. Each break in the beam recorded a score on one of the twenty-four counters. The scoring moved

[1] Eight more have since been obtained from the same source and added to the group at the London Zoo for further study.

automatically from one counter to the next with the passing of each hour. Once a day the twenty-four scores were recorded. The histogram (Fig. 1) shows the average figures for the five days tested.

It should be mentioned that these tests were made, using one pair of acouchis between March 18th and 22nd, 1961, in the cage shown in Fig. 3. No score was taken between 8 and 9 a.m. or between 3 and 4 p.m., as these were the two periods when the daily food-hoarding test (see p. 475) was being set up and dismantled. Theoretical averages are shown with dotted lines for these two periods.

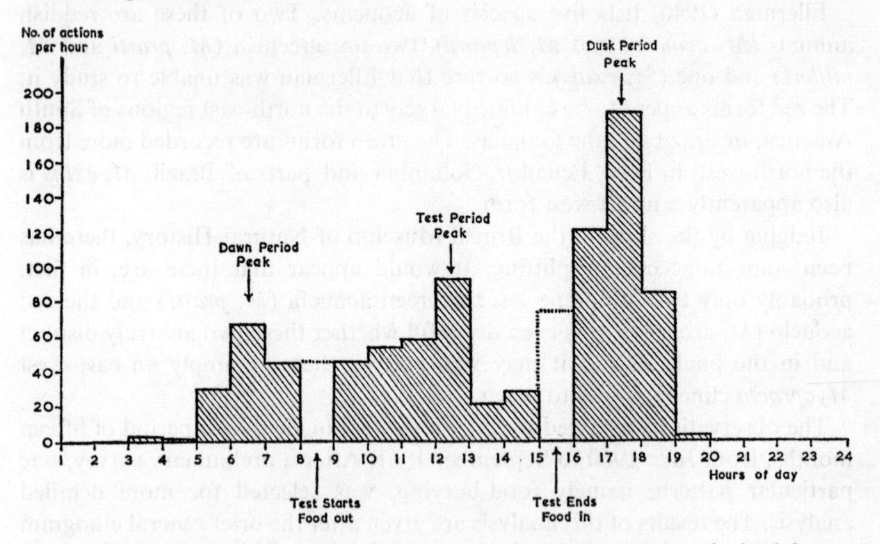

Fig. 1. The 24-hour activity cycle of the acouchi, based on averages obtained from a five-day test period.

Several points emerge from this histogram. Firstly, the acouchis are, as stated, intensely active animals. Secondly, there are three peaks of activity, all during daylight, but one at dawn and one at dusk. These are suggestive of a crepuscular rhythm, but the third peak comes in the middle of the day and is surrounded by a considerable amount of activity, during what would be the siesta period of a truly crepuscular species. Thirdly, the dusk peak is much larger than the other two and is followed, within two hours, by complete inactivity through the period of darkness, for six hours, until the first stirrings of the dawn period.

The apparatus was not available when the animals were not being given their hoarding tests, but it would be interesting to know to what extent these tests influenced their level of activity during the midday period. General observations at other times indicate that action was present during that time, with or without tests, and that the acouchi would not necessarily slip into an entirely crepuscular rhythm if left alone, but quantitative confirmation of this is needed.

2. *Locomotion*

Acouchis have three characteristic modes of progression. When they are investigating a strange environment, they creep stealthily forward, step by step, often with the body crouched low on the ground and with the neck stretched fully forward. The head is repeatedly raised and lowered as they nose their way farther and farther into the unfamiliar territory. A slow, stealthy advance is also seen sometimes on the home ground following an alarm, or when a strange noise has been heard, but here the legs are less bent and the repeated nosing of the air may be absent.

During progression of this sort, an acouchi appears to be able to freeze its movement at any point, often pausing for some moments with one of its legs in mid-air.

The second mode of progression is the normal walking action and it is this that is seen most often. The third is the headlong dash, performed as a panic response to a predator stimulus or during intraspecific chasing. This is frequently interspersed with great leaps and bounds and, in a restricted environment in captivity, the fleeing acouchi may sometimes crash violently into obstacles as it hurls itself along, changing course frequently like a hare.

3. *Track formation*

If captive acouchis are provided with a turf substratum scattered with twigs, logs, leaves and clumps of long grasses, they rapidly proceed to establish a network of special pathways. These tracks are used frequently during normal walking locomotion and become particularly important during fleeing and chasing. They are formed by several distinct actions.

The first response to a new area of undergrowth is to pass stealthily into it. Then, when the acouchis are familiar with it, they will suddenly rush at it and through it. This rushing may take different routes at first, but soon one way through is favoured and the others are dropped. By repeatedly walking or running through the undergrowth along this set path, the acouchis gradually create a beaten track. All the individuals appear to use the same tracks and all help to beat them down by deliberate bouts of running back and forth along the 'agreed' routes.

This process forms the pathways, but other actions are employed to perfect them. Loose objects that lie in the way, such as stones, lumps of wood or bark, are picked up in the mouth, carried a short distance away and then dropped. More tenacious obstacles, such as roots, twigs and branches crossing the tracks, are seized firmly between the teeth and pushed violently forwards with brief but vigorous running movements. If the branches are not too rigidly anchored, they are often successfully shunted aside by this method, but if they are firmly fixed in position, then yet another technique is brought into operation. This consists of gnawing through the obstacle on the side of the track where it is more rigidly attached. Once the acouchi has successfully bitten through the twig or branch it then carries or pushes it out of the way with one of the actions already described.

Gradually the tracks become perfected and, where there is thick vegetation, they may even become well-formed surface tunnels. Even at this stage the acouchis continue to perform deliberate bouts of running back and forth along the pathways. This keeps the track-surfaces well trodden and perhaps also helps to maintain familiarity with the exact course of the routes.

4. *Nest-scooping*

Lydekker (1894) reports that agoutis in the wild scoop shallow nests and line them with leaves, roots and hair. In captivity, the scooping was observed frequently with acouchis, but not the lining. Both front legs and back legs were seen to be used in nest-digging and it is interesting that, although a great deal of food-digging also occurs (discussed later) only the front legs are then employed. The nest-sites are usually located in a sheltered corner, but the animals do not always retire to the nest to sleep or rest. In fact, the use of the nest appears to be less regular and less important in this species (and its relatives) than in other groups of rodents. This point will become more significant when the method of hoarding food is discussed.

5. *Hiding*

The headlong flight of the acouchi terminates with 'freezing', usually close to the ground. When a network of tracks exists, the fleeing animal comes to rest in one of the more covered areas of the path system and remains hiding for some time before re-emerging cautiously into the open. Fleeing and hiding in captivity are set off by stimuli that are often obscure to the human observer. A bird flying slowly past the laboratory window appears to be one of the most likely stimuli to produce this response, although it does not always do so. Sometimes a sharp but distant sound has the same result. On the other hand, noisy movements by the observer inside the laboratory near to the acouchis have little effect. It would be particularly interesting to carry out a series of tests with this species using overhead models similar to those employed by Lorenz and Tinbergen (see Tinbergen, 1948) for analysing the alarm stimuli of ground-breeding birds.

Panic fleeing in the acouchi is infectious. When it occurs, the whole group scatters and hides almost simultaneously. However, when 'vacuum' or 'play' fleeing is performed, it is not infectious. The animal concerned suddenly leaps into the air and dashes off on what can best be described as a 'skittish' flight.

It flees rapidly along the flight paths and yet there is something faintly playful about its muscular actions. The difference between this vacuum fleeing and the true panic fleeing is apparently obvious, not only to the human observer, but also to the other acouchis in the group, who ignore the proceedings and continue to go about their business. It is interesting that the incidence of vacuum fleeing increases when there is a decrease in actual panic fleeing.

6. *Foot-stamping*

When an acouchi has been alarmed, but not to the level of panic, it stamps its back feet hard down on the ground. This makes a characteristic thumping sound that acts as a signal, alerting other members of the group. As soon as one begins to stamp, the others stop whatever they are doing and 'freeze'. Its influence may go no further, but frequently it leads to stamping by other members of the group.

If the human observer taps lightly with two fingers on a table-top, he can usually stop the activities of all the acouchis in the room and can sometimes even persuade them to join him with foot-stamping. To do this, he must tap rhythmically at approximately the correct acouchi speed. Typically this varies between one stamp every one and a half seconds and one stamp every four seconds. Usually thumping begins at the higher rate and sinks slowly to the lower rate, when it is soon abandoned altogether.

When stamping, the acouchi must always have its rear end crouched, but the head and neck are more usually stretched up, on the alert. On certain occasions, stamping was observed in which only one back foot was thumped on the ground, but this was rare.

Once an acouchi was seen to receive the stamping signal while it was busily feeding, sitting up holding the food in its front feet. It looked up with a start and began to thump but was so intent on feeding that it did not stop, or let go of the food. It then proceeded to feed and stamp simultaneously, the whole animal leaving the ground completely just before each thump. Typically, however, as mentioned above, an acouchi is far too frightened, when thumping, to do anything other than freeze.

Foot-stamping occurs in five distinct contexts. Firstly, as already mentioned, when an animal is alarmed by a predator stimulus, but not to the point of panic; secondly, when it is becoming isolated as it investigates a strange environment; thirdly, when it hears the stamping signal of other acouchis; fourthly, when it has panicked, fled, hidden and is now in the process of recovering its composure, but is not yet fully relaxed; and fifthly, in certain cases, after an intraspecific agonistic encounter. There is a second form of the response (foot-rattling) that occurs as a high-intensity version after very severe fighting, but that will be discussed more fully in the section dealing with agonistic behaviour.

In origin, foot-stamping is obviously derived from the intention movement of fleeing, being an exaggerated form of the first kick of take-off. During the course of evolution it has become distinctive by means of the development of a Typical Intensity (Morris, 1957). The rate of stamping has not become completely fixed and is still capable of showing some slight variation with changing motivation, but the changes in response rate have become less and less finely adjusted to motivational fluctuations. The result is a sufficiently characteristic signal which, although not completely rigid in its rate of performance nevertheless operates successfully.

7. *Lost-calling*

When acouchis are investigating a new environment, but are not sufficiently uneasy to foot-stamp, they may maintain auditory contact with one another with repeated 'lost-calling'. This occurs most intensely when an animal finds itself isolated from the group. It appears to be homologous with the familiar squeaking noise of the guinea-pig. In the case of the latter, however, the response has acquired a much lower threshold and is heard with a much higher frequency in a wider range of circumstances. In the acouchi (and also the agouti) the lost-calling is rarely, if ever, given by an adult individual. Four of the five acouchis under observation were sub-adult on arrival and the general frequency of lost-calling decreased as they matured. In the guinea-pig it is heard frequently amongst adults as well as young.

8. *Prancing*

Acouchis in captivity are subject to fits of abandoned play-prancing, even when adult. During a bout of prancing the animal leaps repeatedly into the air, taking off each time as soon as it lands. Often it twists its body while in mid-air, so that it lands facing in a new direction. Sometimes prancing is interspersed with play-fleeing, but it also occurs frequently by itself. It is rarely seen in a group in which there are social tensions, such as hierarchy disputes, but is seen daily in relaxed or isolated individuals. It appears to be an exaggerated play-form of the fleeing leaps and bounds, but performed, so to speak, on the spot.

9. *Social grooming*

Acouchis have the typical rodent self-grooming patterns of scratching and washing. They also show frequent social grooming. This is carried out largely in the head region, especially around the ears, mouth and under the chin. It is most common between members of a newly formed pair.

It is also performed by certain individuals towards the human observer. Two of the five acouchis studied were unusually tame and, whenever they were offered a hand, proceeded to lick it repeatedly and for as long as it was available. At first it was thought that this might represent a salt deficiency, but a salt lick was ignored by them. It became clear that these two individuals were obtaining some sort of powerful social reward from this action, and preferred it to licking the other acouchis. All hands were licked vigorously to the same extent, there being no preference for any one human being.

It is difficult to understand the intense form which this action took. It was, however, noticed that young agoutis repeatedly lick their mother in the mouth region and it seems possible therefore that these two acouchis had been hand-reared from a very early stage and had in some way become human-imprinted, treating all humans as their parents. Unfortunately, the acouchis have only produced one offspring at the time of writing and that was born dead, so that a further analysis of the significance of this response must await further breeding results.

Agonistic behaviour

Although they hardly ever attack a human handler, acouchis are remarkably vicious intraspecifically. They may inflict severe wounds on one another, especially around the mouth and rump. A weaker individual is persistently persecuted, even though it has obviously ceased to be a challenge to the attacker. Except under extreme circumstances, there is little appeasement. The observer sometimes has to intervene, if he is to prevent severe damage to the animals. (It is on these occasions only that he may be bitten.)

The situation is not always as difficult as this, however, the worst fighting occurring only when a group is settling in to a new environment, or when a strange acouchi is added to an established group.

A social hierarchy was rapidly established when the group was first introduced into the large cage mentioned earlier. There were, at the time, three males and one female. The largest male (A) attacked the medium-sized male (B) and B attacked the youngest male (C). The female (D) was attracted to male A and stayed near him. When he was removed later from the group, she then associated with male B and when B was removed, she happily consorted with male C. She was never attacked by any of the males and she hardly ever attacked them. She simply sat and watched the prolonged disputes and battles and then consorted with the tyrant of the moment.

When a new female (E) was obtained and added to the group, there was severe fighting immediately. There was little hesitation or cautious examination of the newcomer. Within a few seconds the dominant male (A) and she were flinging themselves at once another with such fury that drops of blood were splattered across the glass observation wall. When the dominant male (A) was removed and the female E re-introduced, the new tyrant (B) behaved in exactly the same way, flinging himself at her immediately.

When two of the males were disputing, one always gave way after a fairly brief bout of in-fighting, but the young female (E) refused to give in, except momentarily, and eventually all attempts to introduce her to the group were abandoned. Some months later, when she had matured, she was successfully paired up with male C in a separate enclosure.

During the fighting and threatening a large number of agonistic responses were observed and these will now be listed.

1. *Nosing*

An agonistic encounter with an unfamiliar animal often begins with a brief nosing (either head to head or head to rear). If the examined animal flees, it will be chased; if it turns to fight, then mutual rearing-up occurs.

2. *Rearing-up*

At the start of a matched bout, both contestants rear up on their hind legs, pawing with their front feet at their opponent. They may drop and rear up again repeatedly, all the time attempting to manœuvre into a good position for biting. Bites delivered during a matched bout are often in the region of the mouth.

3. *Rump-hair erection*

During intense matched bouts the rump hair of the acouchi is spectacularly erected. This nearly always occurs during mutual rearing-up. Only rarely and at very low intensities does rearing-up take place with all the hair sleeked. The rump-hair erection is maintained as the animals spar. Eventually, when one has fled, with hair depressed again, the victor may chase with its hair up and may continue to stalk aggressively around with the rump still displayed, after the loser has disappeared.

In origin, this hair-erection display is clearly an exaggerated and differentiated form of the normal thermoregulatory pilomotor response and must therefore be classified derivationally as an autonomic signal (Morris, 1955b).

The hairs of the acouchi are many times longer in the rump region than elsewhere and this helps to accentuate the display, but there is no special colour patch involved. The situation is more dramatic in some species of agouti. There are orange-rumped, black-rumped and white-rumped agoutis, with the hairs in this area a characteristic colour that makes the display even more conspicuous.

In the agoutis, however, the display is also seen on occasions other than matched intraspecific duelling. If agoutis living in a paddock are pursued vigorously by a human being, they often erect their rump-hair as they flee. This was only rarely observed in the acouchi, but gives an interesting clue as to the origin of the self-defence technique of porcupines. It will be recalled that acouchis and agoutis are hystricomorph rodents and that many of their spiny-haired relatives also erect their rump-hair in a similar manner when alarmed. The elongated rump-hair in question has, of course, evolved into sharp quills and at the same time their running-away response has changed into a running-backwards-at-the-enemy response. The acouchi and the agouti rump-displays make it easier to understand the evolutionary steps taken during the development of the highly specialized porcupine defence response.

4. *Mouth-gaping*

During a matched bout the opponents often face each other with their mouths gaping half-open in an intention movement of biting. This appears to have become something more than just a mechanical preparation for a bite, as it is often retained for some time after the actual fighting has broken off and it then appears to signify that, although the animal has paused in its exertions, it is still nevertheless ready to bite.

Frothing at the mouth is also seen in the agouti under similar circumstances, but not as yet in the acouchi.

5. *Teeth-gnashing*

After a severe fight, a victor who has not won easily is liable to gnash his teeth ferociously and audibly. That this can act as an aggressive auditory signal was seen particularly clearly on one occasion, when the winner of a long bout silently approached the crouching loser, lying half hidden in a clump of long grass. As soon as he was directly behind the loser, in a position where he could not be seen, he gnashed his teeth vigorously, whereupon the loser leapt up and bolted and the chase began again.

The animal that is gnashing its teeth may also have spasms of trembling, but this was not observed very frequently.

6. *Chasing*

Pursuit by the victor is often prolonged and persistent. Each time the fleeing animal attempts to hide, the attacker drives it on again. Most chasing is done along the recognized pathways, but a desperate loser will flee wildly in any direction. While he is chasing, the attacker may move close by another enemy, but will ignore it and search out the special victim of the moment.

7. *Lunge and trample*

Usually, the close approach of the winner is enough to set the loser fleeing again. If the latter can remain sufficiently immobile, the attacker may abandon the chase and move away. But if, in its panic, the loser has selected the attacker's favourite corner in which to 'freeze', then the attacker is likely to take special measures. These involve trampling on the loser violently with the front legs and lunging at him with the teeth. Needless to say, this usually has the desired effect.

This method was also used on the rare occasions when a female attacked a male and was then accompanied by an aggressive 'tchek-tchek-tchek' noise as she pummelled or lunged at him.

8. *Fleeing and hiding*

Up to this point we have been concerned mainly with the actions of the winner, but now the responses of the loser must be dealt with. The immediate reaction to losing a fight is to flee and hide.

9. *Crouch and freeze*

Once the loser is hidden, it will crouch close to the ground and 'freeze' where it lies. If the attacker finds it and approaches, it usually loses its nerve at the last moment and bolts, only to be chased again. But if it can manage to hold its 'frozen' posture, even when the attacker is close to it, then it may be left in peace.

Freezing therefore can be considered as an appeasement response. It acts by cutting off enough of the vital stimuli releasing attack. Only in the special circumstances referred to above does it fail to stop the attack and break off the encounter, providing, of course, that the crouching animal can resist the urge to flee as the victor comes close.

10. *Foot-stamping*

As mentioned earlier, one of the contexts of this response is the situation immediately following an intraspecific agonistic encounter. It does not always occur after such an encounter, however, its appearance depending on the extent to which the loser has been beaten. A dominant aggressor never stamps after fighting and, if he has beaten his rival severely, the latter as it crouches in hiding will be too scared to move a muscle. Under these circumstances, then, foot-stamping is not seen. If, however, the loser has not been completely suppressed, it may thump its feet as it crouches, at the end of a pursuit. The result of this action is that the winner can locate the position of the loser. It follows the sound of the thumping until it comes upon the crouching animal and then the fight begins again. After a series of pursuits the fleeing individual no longer stamps when it hides. It has then slipped one stage farther down the scale of defeat.

One may well question the function of foot-stamping in this context, as it appears to be of little help to the performer. It leads to further trouble without allowing a period of recovery. It might be more reasonable to suppose that the real function of the action is to alert the members of one's group to possible interspecific dangers and that, when it appears intraspecifically, it is, as it were, appearing accidentally. It seems as though the crude motivational state of 'moderate alarm' produces stamping in any context and that there has as yet been no further refinement of this situation. It is doubtful, however, if this state of affairs would have persisted unless there was at least something to be gained by the performance of agonistic stamping.

No such gain was noted in the conditions of captivity, but a possible value in the wild can, perhaps, be postulated. As an alarm signal, foot-stamping is infectious. It increases the level of fear in neighbouring acouchis, preparing them for the possibility of sudden flight from a predator. If an acouchi that is losing a fight with a rival can stamp its feet, it can perhaps transmit some of its intraspecific alarm by means of the interspecific protection device. In this way it may be able to disconcert its attacker and thus indirectly reduce its level of aggression. The fact that this did not work successfully in captivity may have been due simply to the absence of predation. In the wild these animals may be much more sensitive to foot-stamping warnings and the 'ruse' may be much more rewarding.

11. *Foot-rattling*

After very severe fighting, the losing (but not yet beaten) animal performs a high-intensity version of foot-stamping which can best be described as foot-rattling. This is, in all respects except one, similar to foot-stamping. The difference is that, whereas in stamping the back feet come down to hit the ground together in a single blow, in rattling they strike the ground alternately in a brief but very rapid tattoo. Instead of thump-(pause)-thump-(pause)-thump, the noise is now brrrrrrr-(pause)-brrrrrrr-(pause)-brrrrrrr.

The pauses and general rhythm of stamping and rattling are much the same. Derivationally, both appear to have evolved from movements of fleeing, but whereas stamping has grown out of the take-off kick, rattling appears to be a distorted form of 'running-on-the-spot'. The distortions include omitting the front legs from the action and changing the force of the back leg movements from downwards-and-backwards to downwards only.

Foot-rattling was only observed in intense agonistic situations, but, in view of the previous discussion concerning the functions of foot-stamping, it seems likely that rattling is also basically an interspecific device. Further observations in more varying conditions may well show that under more severe threat of predation-proximity, the acouchis switch from stamping to rattling, as they do when fighting has been particularly vigorous.

12. *Mewing*

When an animal has been repeatedly defeated, it gives a mewing cry as it lies in hiding. This is typical of an exhausted animal and is a submissive cry that does not stimulate the winner to further attacks. It sometimes even appears to have an appeasement function and mewing animals are often left screeching plaintively in a corner by the dominant animal. The cry is rather like a long-drawn-out version of the lost-calling sound, repeated slowly.

A slightly different noise can also be heard as an accompaniment of foot-rattling. It is less plaintive than mewing and less staccato than the tchek-tchek attack cry, but until recordings have been made and sonograph analyses carried out, it will not be possible to determine whether this is, in fact, a separate and distinct call, or whether there is simply a single variable agonistic cry that changes according to the balance between fear and aggression.

13. *Prostrating*

Although freezing and crouching may often successfully stop further assaults, a particularly persistent attacker will not be halted in this way. As an even more extreme gesture of submission, a beaten acouchi may lie prostrate on the ground and even roll over on to its side, with its legs outstretched. In the most complete form of the response, the eyes are almost closed. The animal appears to have fainted and one is instantly reminded of 'death-feigning' responses. It holds this pose while the dominant animal examines it closely, but, as soon as he has left, the prostrate animal leaps up and flees rapidly into hiding.

It is interesting that Roth-Kolar (1957), when studying the agouti, found that 'a submissive attitude does not seem to exist'. It is difficult to believe that there should be such a difference between the two genera and further observations should be made on this point.

14. *Squealing*

When an acouchi is being bitten it may remain silent or it may squeal. This is high-pitched and quite distinct from the mewing of a submissive individual.

15. *Foot-scuffling*

Acouchis were occasionally seen to scuffle backwards with their hind feet. The movement was reminiscent of the foot-scraping often performed by dogs after urinating. In the acouchis, it was not however associated with urinating. Usually it occurred spontaneously and with no apparent cause or function. Sometimes it appeared intermingled with bouts of foot-stamping or foot-rattling after agonistic encounters and it is for this reason alone that it is included here at the end of the section on agonistic behaviour.

16. *Scent-marking*

Like foot-scuffling, another action that is associated with agonistic behaviour, but which also requires further study, is the response of scent-marking. Typically a male response, this occurs most when an adult acouchi finds itself in a new environment. As it moves around investigating, it pauses from time to time, presses its genital region against the substratum by a careful squatting and then goes on its way. A male that has marked out an area in this way seems more likely to defend it vigorously than one that has not. It may therefore be correct to call this territorial marking, but it will first be necessary to study the social structure of a larger group of acouchis under conditions of semi-liberty.

Agonistic code

During an actual fight, the action is too fast for much agonistic signalling to take place. This occurs during the pauses between duelling, or chasing and fleeing. It is then possible to correlate the fighting performance of each individual with the displays given immediately afterwards. The results obtained in this way give an approximate overall picture of the agonistic code of the acouchi. If we start with an animal that has experienced total victory and pass step by step down the scale, past partial victory and partial defeat to end at total defeat, we see a series of causally overlapping displays (see Morris, 1958) as follows:

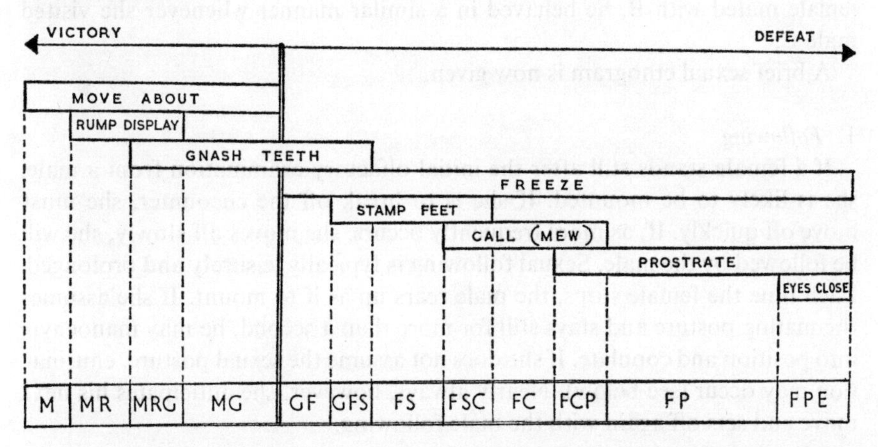

Fig. 2. The agonistic code of the acouchi. The overlapping of the causal ranges of the various reactions results in a more subtle and complex code of signals.
The initials across the bottom of the diagram represent the 12 reaction-combinations resulting from this overlapping (e.g. MRG = Animal Moves about, with its Rump hair erect, whilst Gnashing its teeth).

It should be remembered that this is not strictly speaking a Flee–Attack (FA) diagram of the type put forward in previous papers (Morris, 1956 and 1958). In the present case, what are being recorded are the displays of the acouchis immediately after an encounter, not during or before it.

Finally, it should be stressed that this agonistic diagram is a tentative, preliminary one, based only on qualitative data. It requires much further quantitative study before it can be perfected. Despite this, however, it already gives some idea of the complexity and the basic arrangement of the acouchi signalling system.

Sexual behaviour

There are no striking differences in appearance between male and female acouchis. Male and female examine one another by smelling the anal region and the area around the mouth. A simple courtship and mating may then follow if the female is receptive. Loose pair-bonds appear to be formed and during these associations, the female rests and sleeps near the male. They also tend to make feeding and other exploratory sorties together, but she may go off to consort with other males when her mate is sleeping. This leads to severe fighting if he happens to awaken, although it is never the female, but always the unfortunate males, that are the object of his attack.

It was noticed on certain occasions that female D had only to approach male B for her mate, male A, to attack B. Later when A was removed and the female mated with B, he behaved in a similar manner whenever she visited male C.

A brief sexual ethogram is now given.

1. *Following*

If a female stands still after the initial olfactory examination from a male, she is likely to be mounted. If she is to break off the encounter, she must move off quickly. If, as more frequently occurs, she moves off slowly, she will be followed by the male. Sexual following is typically leisurely and prolonged. Each time the female stops, the male rears up as if to mount. If she assumes the mating posture and stays still for more than a second, he may manoeuvre into position and copulate. If she does not assume the sexual posture, enurination may occur (see below). Nearly always, however, she anticipates his next move and sets off again with the male following.

2. *Tail-wagging*

The unusual tail of the acouchi is the most characteristic anatomical feature of the genus. In almost all other respects, the acouchi's basic design is similar to that of the much larger agoutis. The agouti's tail, however, is an inconspicuous black stump, while the acouchi's appendage is approximately one and a half inches long and is strikingly equipped with a tuft of short white hairs (see Pl. 46). It is rare to find a mammalian tail that is exclusively concerned with signal functions, but this is apparently the case with the acouchi.

Rodents have utilized their tails in many different ways: for leaping (springhaas), for gliding (flying squirrels), for swimming (beavers), for climbing (tree porcupines), for keeping warm (bushy-tailed dormice), for keeping cool (naked-tailed rats and mice), for food-storage (fat-tailed gerbils), and for self-defence (porcupines). In some of these cases, the tails are also used as auditory (beavers) or visual (tree-squirrels) signalling devices, but in only a few (marmots) are the tails used only for signals. The acouchi belongs to this last category.

The signals transmitted by the acouchi's tail are primarily sexual in significance. When the initial olfactory contact is made, the male begins to show signs of very rapid tail-wagging. This becomes more prolonged during the leisurely sexual pursuit. The sexual tail-wag of a courting male acouchi is usually performed with the tail in a vertical upward position, the flashing white tip being more visible to the female in this way. The tail can also be wagged in a downward position, but this is typical of scared or subordinate males. It was seen during the furtive courtship bouts that took place between female D and males B and C, when the dominant male, her mate, was asleep. As there is a great deal of turning this way and that during the sexual following, the female can undoubtedly see that tail-down wagging is taking place, although it is, of course, far less conspicuous than tail-up wagging.

The occurrence of tail-wagging varies inversely with the speed of the sexual pursuit. As the female slows down, the male's tail starts its rapid, side-to-side movements and as she moves off again, it stops.

Occasionally tail-wagging was also performed by females. This was not, however, observed with sufficient frequency to establish the exact motivational condition producing it. In one case it was noticed that the tail-wagging female was actually making sexual advances to a reluctant male, but this may not be its only context.

3. *Enurination*

If a male is sexually stimulated by a female that is not moving away from him, one of two things may happen. If she performs the sex-crouch mating posture, he may mount her, but, if she fails to perform this action, he may rear up on his hind legs, erecting his penis as he does so, and squirt a powerful jet of urine over or at her.

This squirting of urine at the female during courtship has been recorded for the wild rabbit (*Oryctolagus cuniculus*) by Southern (1948), who used the term enurination. There is a difference in the orientation of the movement in the two cases. The male acouchi aims forwards, the rabbit backwards. The acouchi is typically behind the female so that the jet shoots over her head, lands on the ground in front of her and sprinkles her back. If he cannot get into this position because of some obstacle, he will fire across her, or straight into her face. The rabbit, firing backwards, is less precise. He very commonly runs 'past the doe about a yard away from her and twists himself towards her with a kind of skid as he draws level, so that she is in the line of fire from his hind quarters'. He may also leap over the top of the doe, squirting as he passes over.

It is remarkable how convergent the behaviour signals of the rabbit and the acouchi have become. Both employ sexual tail-flagging, with a white and exclusively 'signal' tail; both use the foot-stamping alarm signal; and both perform sexual enurination.

The acouchi's close relative the guinea-pig (*Cavia porcellus*) is also said to perform enurination. Frühling (1955) refers to it as 'Harnspritzen' and records that an active male guinea-pig 'sprinkled his urine regularly on the female in a condition of sexual excitement as part of intense courtship'. Frühling also mentions that it has been observed in hares as well as rabbits. It is also said to occur in the Mara (*Dolichotis patagona*) and it may well emerge, after further behaviour studies have been carried out, that it is a much more widespread sexual phenomenon than was originally considered.

The response by the female to this male pattern was often negative. Both in the case of the acouchis and the rabbits, it was noted that females frequently reacted with apparent distaste, shaking themselves and grooming rapidly and ignoring the male. No female acouchi was ever seen to respond sexually to enurination, but observations of this action were limited. It is nevertheless recorded here as a sexual act, owing to its contextual appearance in the middle of a sequence of other obviously sexual actions. In the case of the rabbits, Southern reports that 'in a few instances definite stimulation was observed'.

Where the acouchis squirted near the wall of their cage it was possible to estimate the height and range of their jet. At full pressure, it appeared to reach a height of about eighteen inches and a distance of approximately four feet. The powerful jets were always aimed up at an angle of roughly 45 degrees and it seemed to be the weaker jets that usually scored direct hits on the female's body. Too few instances were seen, however, to decide finally whether the male was definitely aiming *at* the female, with frequent errors and misses, or whether the basic action was to aim *over* her, with the direct hits being the low-intensity mistakes. Further study is required here.

The highest number of squirts given by a male in a short space of time was seven. Southern records eight for the rabbit. As he points out, this requires a liberal supply and he found that, in the spring, male rabbits had very distended bladders, but the females did not. With so few specimens available, it is not yet known whether this also applies to acouchis.

In a few cases, where the urine was squirted past the female and on to the glass observation wall, it was noted that it was an opaque white liquid. Tests are now being carried out to ascertain the exact nature of this liquid and to find out to what extent, if any, it differs from normal urine.

The origin of enurination is clear enough. Like tree-marking in dogs and other urinary 'displays', it is autonomic in derivation (see Morris, 1955b).

4. *Sex-crouch*

When the female is receptive, she stops walking away from the male and adopts the sex-crouch posture. This differs from the submissive crouch in several respects. Firstly, the female's body is lowered, but does not touch the ground. Secondly, she flattens her back, so that her rear end is correspondingly raised. Thirdly, she erects her tail and holds it stiffly upwards. This posture stimulates the male to mount and copulate.

She frequently adopts this posture, however, and then moves off just as the male is about to mount her. In such cases the sexual pursuit begins again. During a prolonged bout of courtship the sequence may then be as follows: male approaches female – female walks slowly away – male follows – female slows down – male tail-wags – female performs sex-crouch – male rears up – female moves away – male follows. This sequence of events may be repeated many times before either the female permits mounting, or the male loses interest.

5. *Pseudomale and pseudofemale behaviour*

In many species, male patterns are seen to be performed occasionally by females and vice versa (Morris, 1952, 1954, 1955a). In the acouchi, pseudomale mounting by one of the females were seen in a number of instances, although her motivational state at the time was not obvious.

An interesting case of pseudofemale behaviour was observed once, however, in which the motivation of the male concerned was very clear. Three animals were involved – female D and her mate (at the time male B) and a subordinate male, C. The latter had waited until the dominant male was asleep and had then begun to court the female. He was following her and was tail-wagging intensely. The female was giving the sex-crouch from time to time and male C had his attention riveted on her. He failed to notice that male B had awoken and was stalking after him. Normally male C would have fled as soon as the dominant male became active, but he was so engrossed that male B was able to approach close behind him. Suddenly male C noticed him and, in a flash, assumed the female sex-crouch posture.

This pseudofemale response was clearly produced by a combination of strong sexual motivation and sudden fear. It was valuable to male C, as it enabled him to arouse the sexual interest of the dominant male, who began tail-wagging. This reduced his aggression and male C was soon able to slip away and leave male B and his mate courting in the usual way.

6. *Displacement food-burying*

Another unusual response that occurs as a result of a conflict between sex and fear is displacement food-burying. The three animals concerned were the same as above, with the subordinate male C approaching the female sexually. She was near to her sleeping mate B. C was obviously frightened as he came closer. His movements were hesitant, his gait stealthy, his respiration rapid. His tail was wagging, but in the downward position. Finally, his fear became too great and instead of approaching farther, he stopped and went through the motions of pressing down the earth on to buried food (see later). He had not, in fact, buried food and there was no food near by. Furthermore, there was at the time no soft substratum available and the burying movements were being carried out on a hard metal surface.

Displacement food-burying of this type was observed on a number of occasions. It was always the same element of the burying action – the pressing down of the newly buried object with the front feet – that was involved and the motivational situation, of sex combined with fear, was always present. Once, the male performed the displacement action with his front feet whilst simultaneously tail-wagging and foot-stamping with his back feet.

Feeding behaviour

Like most rodents, acouchis eat a wide variety of plant foods, including seeds, nuts, roots, berries and fruits. In captivity they are particularly fond of dates, pea-nuts, apples, potatoes and biscuits. Food is picked up in the mouth and carried to a suitable eating place. There the animal squats on its haunches and, raising its front legs, transfers the food object from its mouth to its front feet. Holding the food up between its feet, it then proceeds to turn it this way and that and, having examined it, starts to eat it.

1. *Food preparation*

Like their relatives the agoutis, the acouchis are fastidious feeders. If a food object has a skin or peel they strip this off, piece by piece, until there is not a single scrap of the outer covering left. Only then do they switch from stripping to eating. This is particularly noticeable in the case of potatoes, which are stripped of their peel with great accuracy and with more precision than by most human beings.

The stripping is performed in the following way. The potato (or apple or pear) is gripped so that its long axis is held horizontally across the animal's face. It then moves its mouth to one end of the object and makes a superficial bite. It then makes a series of such bites, just below the skin of the object, moving its head a little to the side each time. In this way, it ends up at the other end of the potato and a long strip of peel falls off. The front feet then rotate the potato until an unpeeled area is brought into line with the mouth and the process is repeated.

This goes on until a final rotation brings the object back to its starting point. It is then moved around in the front feet and examined for any small areas that may have escaped peeling. These are then dealt with. Only at this point does eating begin.

In captivity the peel is often picked up and eaten after the potato has been completely consumed, but there are in the wild, presumably, certain food objects, the outer coverings of which are dangerous, inedible, or unpleasant in some way and which have led to the evolution of the acouchi food preparation pattern. In captivity the potato peel is, of course, perfectly edible and, once detached, simply becomes a new food object to be eaten. But while it is in position on the potato, it releases automatically the peeling reaction. Even full-sized apples are dealt with in this way, although they may be too heavy for the acouchi to lift. When this happens, the animal holds the object between its hands as it rests on the ground and peels and rotates it in this position.

2. *Food hoarding*

Acouchis do not hoard food on the larder principle employed by most myomorph rodents. Instead, they bury each object separately and their buryings do not all occur in the same place. In fact, there appears to be a definite 'scatter mechanism' at work and experiments were designed to analyse this (see next section).

The actions employed in hoarding are as follows:

(*a*) *Selection of object*
Acouchis will sort through the food tray, searching for suitable objects to bury. They prefer hard objects, but will even bury dates and soft pieces of bread if there is nothing better available. Very small objects, such as seeds and grains of corn, do not appear to release hoarding.

(*b*) *Carrying*
Once selected, the object is carried in the mouth to a suitable burying spot. The animal may walk round and round its cage several times this way and that before stopping to bury the food. In doing so, it selects a place where there is a soft substratum and preferably where it has not recently buried anything.

(*c*) *Digging*
Still holding the food object in its mouth, the animal starts to dig a small but deep pit with its front feet. (See Pl. 46.)

(*d*) *Dropping food*

When the pit is dug, the acouchi lowers its head slightly until it is directly over the hole, and then lets go of the object.

(*e*) *Pressing down*

As soon as the object has fallen into the pit the acouchi starts pressing down into the hole with its front feet, rapidly alternating them until the food is well rammed down.

(*f*) *Filling up pit*

The next action is to fill in the hole above the buried object. This is done with the front feet only (not with the jaws, as when Canidae are burying bones). It is done with forward movements of the legs (not with backward pawing, as when Felidae are covering excreta). In these respects it is similar to the filling up of food-pits by Squiridae and it is interesting that the two rodent groups use the same technique, despite the fact that they are not considered to be closely related within the order and despite the fact that their leg anatomy is so different, one being specialized for climbing and the other for running.

As it fills up the pit, the acouchi only makes full forward movements with one foreleg at a time. After a few movements with one front leg it switches to the other one and proceeds to alternate in this way until the pit is completely topped up. Sometimes the animal may revert to pressing down again for a moment, or, if the object cannot be completely covered up because the substratum is too shallow, it may extract it and carry it off to be buried elsewhere.

(*g*) *Covering over*

When the pit is filled, the acouchi often looks around in the close vicinity and if it finds a suitable leaf, or piece of bark, or some other reasonably large object, picks it up in its mouth and places it carefully over the spot where it has buried the food. Several objects may be deposited on and around the site with the end result that it is impossible to tell where the animal has been working.

3. *Vacuum food-burying*

It has been noted already that one element of the food-hoarding pattern, namely pressing down, appears as a displacement activity in the absence of the usual stimulus of an open pit with a food object in it. Certain elements of hoarding also occur without the usual stimuli in another context, not as displacement activities, but as vacuum activities. They are seen when an animal that is experiencing no social tensions of the kind that produce displacement activities, has food objects suitable for burying, but no substratum in which to bury them. If an acouchi is housed on an all-metal floor it may, from time to time, carry a food object round and round as if searching for a burying place. Eventually it stops, drops the object deliberately on to the metal surface, goes through the motions of pressing it down into an imaginary hole and then fills the hole up with imaginary earth. It then walks away, leaving the food object in full view on the metal surface. On one or two occasions it was even seen to turn, pick up an imaginary leaf and drop it over the object, before leaving.

For some reason, the digging phase of the hoarding sequence is usually omitted from the vacuum activity.

Analysis of Hoarding Behaviour

Of all the patterns of acouchi behaviour described in the above ethogram, the one most suited to detailed experimental analysis under conditions of captivity is, perhaps, the food-hoarding activity. It poses several interesting questions. For example, how does an acouchi distinguish between objects that are suitable for burying and those which are not? As it does not hoard all its food in one place in a food store near its nest, what factors influence it in the selection of each burying site? How does the level of hunger influence the frequency of hoarding? How well do acouchis remember the exact positions of their buried objects and what influences them to unearth them and eat them at a later date? How does the frequency of burying vary from day to day and is it seasonally controlled?

An attempt was made to answer some of these questions and the experimental arrangements were as follows.

Test procedure

The acouchis had been living as a group in their large cage, with a substratum of peat and sand and with leaves, twigs, branches and grasses for cover. They had been burying food regularly and frequently every day. The cage was emptied and cleaned and one pair of animals (male B and female D) were re-introduced. They were now living on a bare metal floor with no opportunity to bury their food, which was presented to them each day in a shallow tray. (If the tray had been deeper, they would have buried suitable food objects underneath the pile of mixed food.) They were kept under these conditions for four weeks. They were then removed to a smaller corner den for two days, while the experiment was prepared.

Sixteen shallow metal trays (14 × 10 × 2 inches) were placed into the large cage in the arrangement shown in Fig. 3. Fifteen of these trays were filled, almost to the brim, with sand. The sixteenth was to be the experimental food-object tray. Each tray was labelled appropriately and lines and numbers were painted on to the cage floor to indicate their exact positions.

When the pair of acouchis was allowed back into the main cage, their small hay-lined corner den was left open and was available to them as a home base throughout the test period.

Into the food tray were placed 100 small ($\frac{3}{4} \times \frac{3}{4} \times \frac{1}{2}$ inches) dog biscuits. These had already proved to be highly stimulating as objects to be buried,

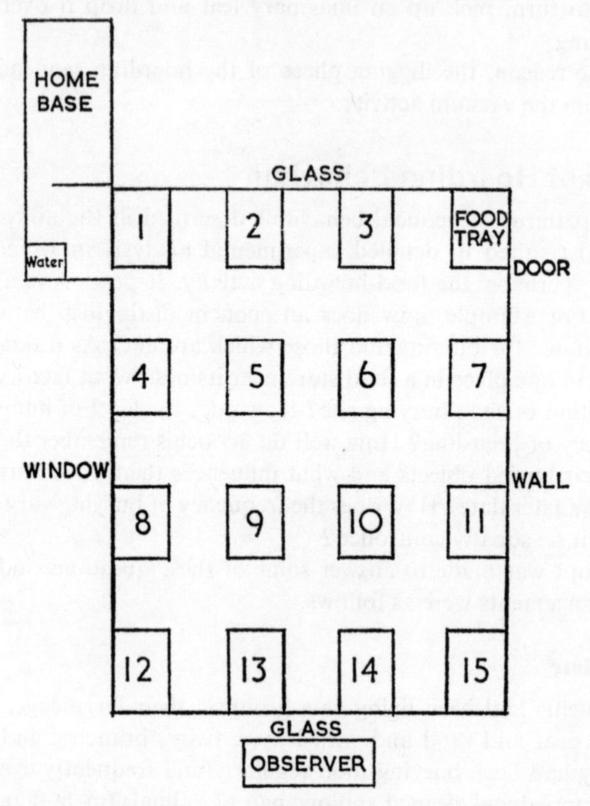

Fig. 3. Experimental arrangement of the fifteen sand trays for the hoarding tests in the large (7 × 9 feet) cage.

were reasonably standardized in weight, size and shape and were available in large numbers.

When the acouchis first emerged into the large cage, on the first morning of the six-week experiment, they were faced with fifteen burying sites, each with a definite position and with a suitable substratum. They were also faced with a surfeit of objects for burying, but they were given no other food during the test period.

The daily procedure was as follows. At 9 a.m. every day the acouchis were shut into their small den. Their mixed food tray was removed and the cage floor cleaned. The fifteen sand trays and the tray with 100 biscuits were placed in position. The animals were then allowed back into the main cage to bury food for six hours until, at 3 p.m., they were once again shut in their small den. The fifteen sand trays and the biscuit tray were then removed. A new tray of mixed food was installed and the acouchis allowed out again. The fifteen sand trays were then sifted and the number of biscuits buried in each one was recorded. Also noted were the number of biscuits eaten, the number removed from the food tray but left in an open space, and the number hidden in the hay bedding in the home den.

After the first six-week experiment, further tests were carried out, varying the hoarding situation and the individuals concerned. Attention was focused mainly on the question of the spatial organization of acouchi hoarding and this aspect will be dealt with first.

Spatial factors in hoarding

During the first experiment, nearly two thousand biscuits were buried. The details were as follows.

Test one. Male B and female D. February 8th to March 21st, 1961, inclusive. Total of forty-two days with fifteen sand trays.

Total number of buryings during test period	1,817
Average number buried per day	43
Average number buried per tray	121
Average number buried per tray per day	3
Total number taken to home base	167
Average number taken home per day	4
Total number left in open spaces	13
Average number left in open spaces per day	0·3

Several points emerge from these figures. Firstly, the artificial nature of the experimental environment did not interfere with the burying activity, which maintained a usefully high frequency. Secondly, only a small proportion of the biscuits was taken to the home base. This illustrates immediately the 'scatter principle' of acouchi hoarding that contrasts with the typical 'larder' hoarding usually studied in other species. The home base was, in fact, used rather as if it were an additional burying tray, but little more. Thirdly, less than one per cent of the biscuits used were left in the open spaces between the trays.

From these figures it is not possible to tell, however, to what extent the 'scatter principle' was operating. Although it is clear that the food is being hoarded away from the home base, there might still be favoured trays where most of the objects were buried. Fig. 4 shows to what extent they were, in fact, scattered throughout the fifteen trays provided.

From these figures it is clear that the 'scatter principle' has a powerful effect, resulting in a wide and remarkably even distribution of the food objects over the whole area. When examining these figures, it should be remembered that the food tray containing the 100 biscuits was identical in size and shape to the sand trays. It was therefore perfectly possible for the acouchis to bury

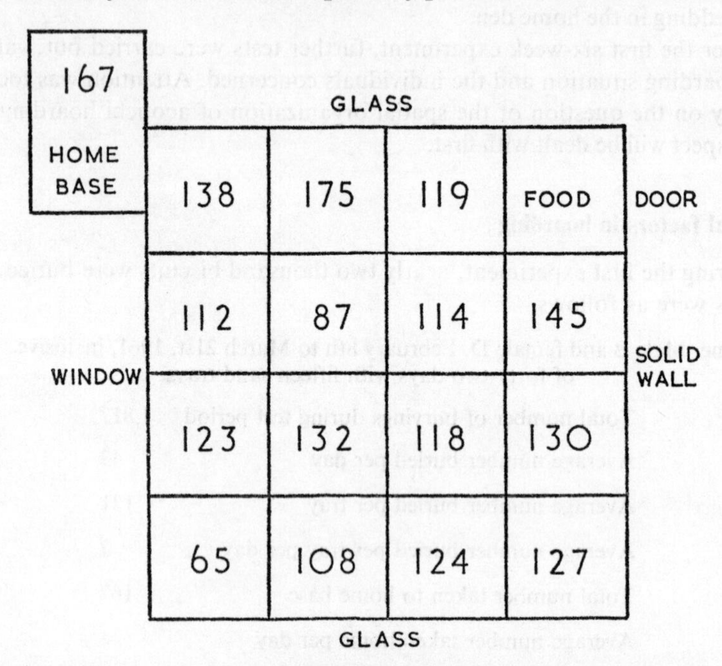

Fig. 4. Scatter diagram, showing the distribution of buried food in the fifteen sand trays and the home base.

all the 100 objects in *one* tray each day. Any one tray could therefore finish the whole test with a score of anything from 0 to 1,817. It will be recalled that the average figure per tray for the test period was 121 and when the fifteen scores are noted against this average, it will be seen that the animals achieved a remarkable degree of uniformity. Had any particular spatial preferences existed, they would have disrupted this uniformity dramatically.

The only exceptions to this rule are trays 2 (above average) and 12 (below average). The reason why tray 2 was favoured appears to be that it was half-way between the food source and the home base, but further experiments, changing the position of the food and the home, would have to be carried out to establish this firmly. The only feature of tray 12 that can explain its comparative failure as a burying site is that it is the farthest tray from the food source, but, once again, further tests are needed to establish this.

It could perhaps be argued on the basis of the above figures that, rather than showing a positive tendency to scatter hoarded food over the whole space available, they do in fact conceal a series of changing short-term preferences for particular sites. The 138 buryings in tray 1, for example, could be made up of a regular daily score of approximately three biscuits throughout the test period, or of one score of 138 on a particular day when that tray was, so to speak, 'the larder'. In order to show conclusively that the latter was not the case and that true scattering was occurring, the following figures may be given. They show the numbers of biscuits hoarded in each tray on each day and the total frequency with which trays were recorded as containing that particular number of biscuits:

Number of biscuits buried in a sand tray in a day	0	1	2	3	4	5	6	7	8	9	10	11	12	13	14
Total number of trays with that score during 6-week test	90	99	119	114	68	56	35	24	12	9	1	1	0	0	0

These results can be expressed graphically as shown in Fig. 5.

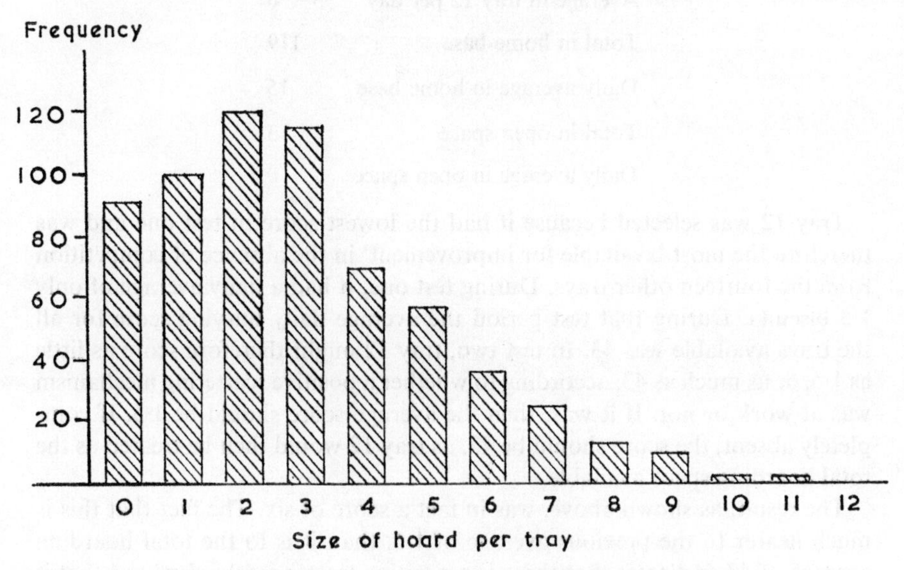

Fig. 5. Histogram showing the frequency with which different sized hoards occurred. The peak occurs at the point where two or three buried food objects were placed in any one tray.

This demonstrates clearly that high individual scores for particular trays, on particular days, did not occur. There were, in fact, small numbers of buryings in nearly all trays each day. With fifteen trays scored on forty-one days, there was a total of 630 scorings and yet only 90 of these (14 per cent) show a zero score. This means that, on the average, only two out of the fifteen trays were unused each day.

It could still be argued, however, that the scattering – wide and regular as it has been shown to be – is nevertheless only the result of an aimlessness on the part of the acouchis. They could either stop burying in a particular tray specifically because they had already buried there, or they could simply wander about and pause to bury anywhere at random. Both methods would lead to scattering, but only the former could be called a positive scattering technique. The figures given above and especially the smooth curve of the histogram already hint at a positive mechanism, but a special experiment was necessary to confirm this point.

This test, which followed two days after the first experiment, involved the removal of all but one of the fifteen sand trays. The test lasted one week and in all other respects was the same as before. The results were as follows:

Test two. Male B and female D. March 23rd to 29th, 1961, inclusive. Total of seven days with one sand tray (No. 12).

Total buried in tray 12	41
Average in tray 12 per day	6
Total in home base	119
Daily average in home base	15
Total in open space	3
Daily average in open space	0·4

Tray 12 was selected because it had the lowest score in test one and was therefore the most 'available for improvement' in the absence of competition from the fourteen other trays. During test one, it had a daily average of only 1·5 biscuits. During that test period the average daily burying score for all the trays available was 43. In test two, tray 12 might therefore score as little as 1·5, or as much as 43, according to whether a positive scattering mechanism was at work or not. If it was, then the average score should be 1·5. If completely absent, the score should be 43, as tray 12 would then be treated as the total hoarding space available.

The result, as shown above, was in fact a score of six. The fact that this is much nearer to the previous average of 1·5, than it is to the total hoarding average of 43, indicates that there is a positive scatter mechanism and that it is extremely powerful. With the reduction of available sites to 1/15th, the

Plate 46. The green acouchi, *Myoprocta pratti*. The short white tail now functions exclusively as a signalling device. (Photograph by M. Lyall-Watson.)

total daily hoard fell to 1/7th, although there was still sufficient space for all 100 of the biscuits to be buried. This can only mean that the more an area is used for burying, the less attractive it becomes for further burying, which is the exact opposite of the mechanism operating in the case of the more usually studied 'larder' hoarders.

Despite the fact that the amount of available open space was tremendously increased by the removal of fourteen trays, the daily average for biscuits left in the open hardly changed between test one and test two (0·3 → 0·4). There was, however, some increase in the daily average of biscuits taken into the home base (4 → 15). One can make a correction for this as follows. In the first test, the daily average of 15 trays + home base was 43 + 4 = 47. In the second test it was 6 + 15 = 21. Even if one accepts the home base increase as a compensation, the site reduction still has a total effect of more than halving the overall amount of hoarding.

Influence of the substratum

At the end of the second test period, the animals were given a few days' rest and then started on a third test to analyse the influence of the substratum. It was obvious enough that the sand was an important factor, but it was thought that the acouchis might hoard biscuits in the corners of the trays even in the absence of sand. The fifteen trays were replaced, but this time all were empty. This test was carried on for five days and there was then a control period of seven days with sand in the trays, as in the original test. The average results were as follows:

Test three. April 2nd to 6th, 1961, inclusive. Total of five days with fifteen empty trays. Control period April 7th to 14th, inclusive (with sand).

	Test 3	Control
Daily average buried in trays	0	41
Daily average taken to home base	6	11
Daily average in open spaces	1	0·1

This shows clearly that the sand was all-important. No hoarding took place in the empty trays. The control period shows that the tendency to hoard was still present and as strong as ever.

It is interesting that in this test the animals did not pay more attention to their home base (11 → 6), when they were unable to hoard outside.

Relationship between hunger and hoarding

Throughout the tests the animals were fed immediately after the burying period was over. They were allowed to go on feeding until the next morning when the next experimental period began. They were not therefore particularly hungry during the tests, but could express what hunger they had by eating some of the experimental biscuits provided for burying. A few were usually eaten each day. During the main test of forty-two days, 174 were eaten, an average of four a day. During the second test the average was also four. During the third it was three and in the final control, two.

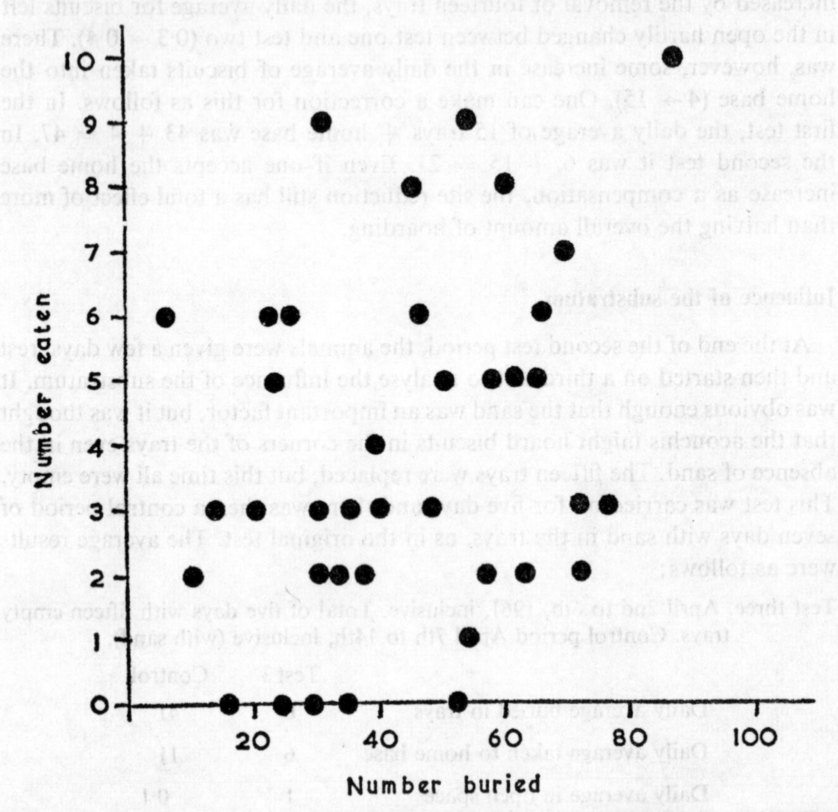

Fig. 6. Hunger and hoarding. At the motivational levels studied, there was no relationship between the frequency of hoarding and the amount of food eaten during the test period.

Using the data from the main test, it is possible to plot the relationship between the number of biscuits eaten and the number buried each day. The results are given in Fig. 6 and indicate clearly that, at the moderate levels of hunger dealt with, there is no correlation between the amount of hoarding and the hunger of the animals concerned.

Hoarding is therefore, at the least, a partially independent tendency, which may or may not be influenced by more acute levels of hunger. (Graphs showing the day-to-day fluctuations of hoarding did not, however, reveal any points of particular interest. Peaks and troughs appeared with no particular rhythm or special sequence and there were no clues as to why one day should be a 'good' hoarding day, or the next a 'bad' day.)

Influence of food object colour

The small biscuits being used for the tests are manufactured in seven different colours: white, yellow, orange, fawn (natural biscuit colour), red, brown and black.

The colours are listed here in order from the lightest to the darkest and it was noticed that the acouchis were burying more whites and yellows than other colours. From that point on, a careful check was made on the numbers of the various colours used for burying each day. Previously, the animals had been given 100 randomly selected biscuits but, during the last twenty-six days of test one, male B and female D were given fourteen of each of the seven colours[1] and the numbers of each used for burying were as follows:

WHITE	YELLOW	ORANGE	RED	BROWN	BLACK	FAWN
340	278	254	190	184	117	75

With the striking exception of fawn, the biscuits were therefore being selected exactly according to a light ←→ dark scale, with a strong preference for lighter objects. The exception in the case of fawn is most interesting. This, being the natural undyed biscuit, has a more speckled appearance than the others. It is also possible for a human being to select the fawn from the seven colours, while keeping his eyes shut. There is therefore a rougher appearance and a rougher texture. One or both of these factors is perhaps operating against the fawn biscuits as objects for burying. The selection of each biscuit by the acouchi seems to be largely visual, so that it is probable that whiteness and smoothness are important positive factors influencing selection for burying. It would be particularly interesting to know how this related to the ecology of the acouchis in South America.

Individual variations

The main limitations of the present hoarding analysis is, of course, the small number of individuals available for testing. It is hoped to correct this when more specimens are brought into captivity. All the quantitative data given so far are based on the one pair of animals – male B and female D. Attempts were also made to obtain results from the other individuals, but without

[1] This meant a total of 98 instead of 100 but, as the total was in any case only intended as a 'surfeit' figure, this slight difference was irrelevant.

much success. They had to be tested singly to prevent fighting and this isolated existence seemed to interfere with the general level of their hoarding tendency. Female E did, however, give sufficient response to the experimental situation to justify including her results here for comparison with those of male B and female D.

Test four. April 26th to May 23rd, 1961, inclusive. Total of twenty-eight days with fifteen sand trays.

		X2
Total number of buryings during test period	168	336
Average number buried per day	6	12
Average number buried per tray	11	22
Average number buried per tray per day	0·4	0·8
Average number taken to home base per day	16	32
Average number left in open spaces per day	13	26
Average number eaten per day	1·4	2·8

The second column of figures above makes allowance for there being only one animal in this test, instead of a pair as before. Comparing the figures in this second column with those in the table for test one, it is immediately clear that female E was operating on a much lower hoarding level. Double her daily hoard was still only one-quarter of that of the pairs. Balancing this, however, was the strange fact that she deposited eighty times as many biscuits in the open spaces, as either member of the pair. (It was also noted that a number of the biscuits in the trays themselves were imperfectly buried.) She also took eight times as many biscuits into her home base as either of the members of the pair. Finally, there was a slight decrease in the number eaten.

Clearly, female E was an inferior scatter-hoarder and it is interesting that this is correlated with a marked increase in the storing of food in the home base. Whether the reasons for her inferior performance were connected with inexperience, her isolation, or some other factor, cannot yet be stated, but she was certainly both younger and more nervous than either male B or female D. This did not, however, influence the degree of scatter that occurred with the buryings that she did perform (see Fig. 7).

It will be noted that all the trays were used and that, as before, the area between the food source and the home base was favoured.

The most striking individual variation arose in connection with colour preferences. These are the figures for female E:

YELLOW	ORANGE	BROWN	WHITE	BLACK	RED	FAWN
299	286	149	132	90	50	18

White and red have both dramatically fallen in preference. This is difficult to explain. Fawn is, as before, well at the bottom of the list, presumably on the strength of its texture, but the simple preference for lighter shades in the other six has been strikingly disrupted. The nature of the disruption would seem to suggest that female E was employing true colour vision in her object selections, whereas the pair B and D were not. This point is the subject of further study by Lyall-Watson.

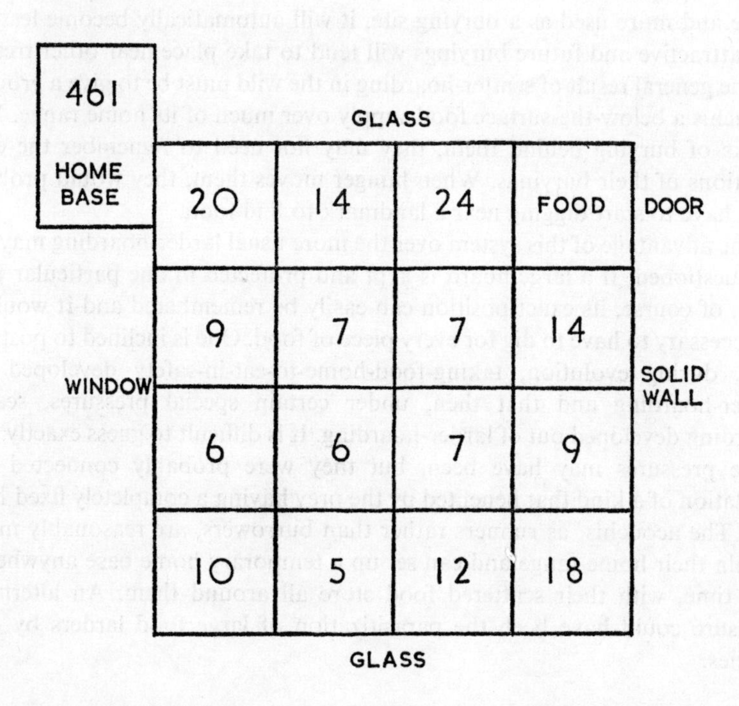

Fig. 7. Scatter diagram for female E, showing that, although there is a great increase in home-base hoarding, there is nevertheless complete scatter throughout the 15 sand trays.

Discussion

There are many problems yet to be solved in connection with acouchi hoarding behaviour, but one point emerges vividly from the preliminary investigations reported here: namely, that there is a powerful influence at work discouraging an acouchi from burying an object in a site where it has already buried. This leads to the phenomenon of *scatter-hoarding*, a pattern of behaviour that is fundamentally different from the more familiar *larder-hoarding* of many species, where food is amassed in one special area.

Observations with agoutis in large paddocks reveal that they too indulge in scatter-hoarding, but that they tend to bury more near to landmarks, such as trees, rather than completely out in the open. Nevertheless, they do not select one particular tree and bury there alone. If certain kinds of landmarks, or other factors, can be shown experimentally to have an influence on the spatial arrangements of acouchi hoarding, this will still not invalidate the scatter principle. It will simply mean that the animals will tend to bury food in one *kind* of area, but not in one particular place. As one tree, for instance, becomes more and more used as a burying site, it will automatically become less and less attractive and future buryings will tend to take place near other trees.

The general result of scatter-hoarding in the wild must be to give a group of acouchis a below-the-surface food supply over much of its home range. With weeks of burying behind them, they may not need to remember the exact positions of their buryings. When hunger moves them, they would probably only have to start digging near a landmark to find food.

The advantage of this system over the more usual larder-hoarding may well be questioned. If a large hoard is kept and protected in one particular place then, of course, its exact position can easily be remembered and it would be unnecessary to have to dig for every piece of food. One is inclined to postulate that, during evolution, taking-food-home-to-eat-in-safety developed into larder-hoarding and that then, under certain special pressures, scatter-hoarding developed out of larder-hoarding. It is difficult to guess exactly what these pressures may have been, but they were probably connected with predation of a kind that benefited by the prey having a completely fixed home site. The acouchis, as runners rather than burrowers, are reasonably mobile within their home range and can set up a temporary home base anywhere at any time, with their scattered food store all around them. An alternative pressure could have been the parasitization of large food larders by other species.

Summary

The acouchi is an extremely active, diurnal rodent from South America with a wide behaviour repertoire. It is ideal for laboratory observational work and has presumably not been studied before because of its rarity in captivity. Five specimens were obtained and a preliminary ethogram was compiled over a period of one year.

Like its larger relative, the agouti, the acouchi has long slender legs and has specialized in rapid surface running as a defence against predators. Several motor patterns are concerned with the construction of a network of surface escape tracks. It does not burrow, but scoops out a shallow sleeping cavity. Vigorous stamping of the back feet occurs as an alarm signal. In the absence of predation, an acouchi may show vacuum panic. Unlike true panic, this is not infectious. Exaggerated play-prancing is also seen.

The agonistic ethogram contains at least fourteen patterns. During an equally matched fight, both animals erect fully their rump hair. Bearing in mind that acouchis are hystricomorph rodents it is interesting that this is basically the same response as that given by many porcupines as self-defence. Other aggressive acouchi signals are mouth-gaping and teeth-gnashing. Attack is by prolonged chasing, biting and trampling. Beaten acouchis attempt to hide, crouch and freeze, but may also perform foot-stamping, foot-rattling, mewing, prostrating and squealing.

An attempt is made to arrange these patterns into an agonistic code and a chart is given showing the causal overlapping that occurs between certain of the responses.

Sexual behaviour involves a slow pursuit during which the male erects and wags rapidly its white-tipped tail. When receptive, the female gives a sex-crouch that differs in several respects from the agonistic submissive crouch. During courtship the male may squirt a jet of urine over the female. Pseudo-male and pseudofemale behaviour were both noted and, during conflicts between sex and fear, displacement food-burying was observed.

Food with an outer covering of peel is skinned elaborately by acouchis before eating and certain food objects are buried. Food-burying comprises a fixed sequence of five distinct motor-patterns and some of these occur as vacuum activities if the animals have no suitable substratum.

An analysis of some of the factors influencing burying was made. It was found that acouchis tend to bury objects where they have *not* buried previously. The result is 'scatter-hoarding' as distinct from the more familiar 'larder-hoarding'. At the motivational levels studied, hunger bore no relationship to the extent of hoarding. A distinct preference for lighter coloured objects for burying was noted, but individual variations existed. Scatter-hoarding appears to have evolved in connection with increased mobility within the home range. By having food buried all over their range, the acouchis are not heavily dependent on one special site that might become endangered in some way.

References

ELLERMAN, J. R. (1940), *Families and Genera of Living Rodents* (London: Brit. Mus. [Nat. Hist.]).

FRÜHLING, A. (1955), 'Über das "Harnspritzen" und ähnliche Verhaltensweisen beim Meerschweinchen, *Cavia porcellus* L.'. *Säugetierk. Mitt.* 3, pp. 168–71.

LYDEKKER, R. (1894), *Royal Natural History*, vol. 3 (London).

MORRIS, D. (1952), 'Homosexuality in the ten-spined stickleback (*Pygosteus pungitius* L.)'. *Behaviour* 4, pp. 233–61.
—— (1954), 'The reproductive behaviour of the zebra finch (*Poephila guttata*), with special reference to pseudofemale behaviour and displacement activities'. *Behaviour* 6, pp. 271–322.
—— (1955a), 'The causation of pseudofemale and pseudomale behaviour – a further comment'. *Behaviour* 8, pp. 46–56.
—— (1955b), 'The feather postures of birds and the problem of the origin of social signals'. *Behaviour* 9, pp. 75–113.
—— (1956), 'The function and causation of courtship ceremonies'. *Fondation Singer-Polignac: Colloque Internat. sur l'Instinct*, Paris 1954, pp. 261–86.
—— (1957), '"Typical intensity" and its relation to the problem of ritualization'. *Behaviour* 11, pp. 1–12.
—— (1958), 'The reproductive behaviour of the ten-spined stickleback'. *Behaviour* Suppl. 6, pp. 1–154.
ROTH-KOLAR, H. (1957), 'Beiträge zu einem Aktionssystem des Aguti (*Dasyprocta aguti aguti* L.)'. *Z. Tierpsychol.* 14, pp. 362–75.
SOUTHERN, H. N. (1948), 'Sexual and aggressive behaviour in the wild rabbit'. *Behaviour* 1, pp. 173–94.
TINBERGEN, N. (1948), 'Social releasers and the experimental method required for their study'. *Wilson Bull.* 60, pp. 6–52.

Author's Note, 1969

Following this preliminary report on the behaviour of the hitherto un-studied green acouchi, two major investigations have been undertaken. One, by Malcolm Lyall-Watson, concentrated on the scatter-hoarding phenomenon. In my original paper I referred to the importance of landmarks, such as trees which were favoured sites for food-burying in the closely related agouti. I pointed out that 'If certain kinds of landmarks ... can be shown experimentally to have an influence on the spatial arrangements of acouchi hoarding, this will still not invalidate the scatter principle.' Lyall-Watson's experiments did in fact confirm that the acouchi, like the agouti, responds strongly to landmarks when burying food and that the overall burying pattern can be seen as a result of a conflict between the scattering tendency and the landmark-pull. This conflict results in a broad spreading of the buried hoard, but nevertheless renders the hidden food comparatively easy to re-locate.

The second investigation, by Devra Kleiman, focused on the reproductive behaviour of the acouchi. One fact arising from her research that modifies a point in my original paper concerns the action of enurination. I stated that 'It is ... recorded here as a sexual act, owing to its contextural appearance in the middle of a sequence of other obviously sexual actions.' I did not observe it in other contexts, but Kleiman has since done so. In particular she noted that adults performed this action towards newly arrived juveniles. In correspondence, she writes: 'I feel that acouchi enurination is mainly a social phenomenon. The male, by covering the females and young with his urine, imparts to them his smell and creates a group odour which maintains group cohesion. Like social grooming in rhesus monkeys, enurination varies in frequency with the reproductive cycle, *but* it occurs all the time.'

Neither Lyall-Watson's nor Kleiman's acouchi studies have yet appeared in print, but both are available as doctoral theses deposited with the University of London.

The response of animals to a restricted environment

(1964)

Every animal is a complicated system of compromises. It must succeed in finding food, a mate, and in many cases a territory, a sleeping place and a nesting site. It must rear its young, keep itself clean, and protect itself from the elements, its predators and its rivals. Each of these demands pulls the animal in a different direction. If it is to survive, it cannot risk going too far in fulfilling one requirement at the expense of the others. Nevertheless, some species do sacrifice a great deal in the process of evolving a specialized solution to one survival problem.

They may become diet specialists, like the ant-eaters, the koalas and the giant panda, or expert killers, like the snakes, the eagles and the big cats. These specialists have taken an evolutionary risk, but as long as their niche persists, their survival device enables them to live a lazy, relaxed existence.

Non-specialists, on the other hand, can never relax for very long. They must be opportunists, always exploring and investigating, always on the move. Having no special device, they cannot afford to miss any chance of a small reward. Typical opportunists are dogs and wolves, raccoons and coatis, martens and mongooses, monkeys and apes and, of course, man.

The great behaviour difference between extreme specialists and extreme opportunists lies naturally in their exploratory activities. We may speak here of a *neophobic* and *neophilic* dualism. Neophobic behaviour (Barnett, 1958) – the fear of the new – exhibits itself most strongly in the specialists. It has been strengthening itself phylogenetically in these cases, channelling them into their one-track existences, but it also operates ontogenetically. The specialist animal has a low level of curiosity, exploration occurring only when it must. A well-fed snake, lion or eagle relaxes. With the opportunist, however, what we may call neophilic behaviour – the love of the new – is the prevalent tendency. Anything and everything is investigated, in case it will provide yet another string to the bow of survival. Wolves, mongooses and apes are nearly always busy examining and testing their environment even when their basic needs have been satisfied. With them, exploration has become emancipated from its ultimate goals and exists as an independent, separately motivated activity that can only be satisfied by total familiarity with the environment.

It seems as though, as a protection against its lack of specialization, the opportunist animal has evolved a nervous system that abhors inactivity. (It is also probably more intelligent. However, although intelligence is a related quality, it is not exactly the same. Neophilia results in an increase in the overall quantity and variety of 'experience of the environment', whereas intelligence involves the utilizing of these experiences to solve new problems. Although they are not necessarily linked, it follows that a strongly neophilic species will tend to have a greater store of experiences on which it can draw when attempting to solve a new problem. It is therefore more likely that neophilic species will behave more intelligently than neophobic ones.) Unlike the opportunist, the specialist animal has a nervous system that can accept prolonged inactivity without difficulty or frustration. Provided its basic needs are satisfied, all will be well.

In a constantly but only slightly varying environment in the wild, both types, the specialists and the opportunists, will be able to survive reasonably well. If the variations in the environment become more violent, then the advantage will go to the opportunists. If, as a result of human interference, the environment becomes dramatically restricted and rigidified, then the specialists will be in a better position. The best example of this situation is the population of animals living in a zoological garden. Here, variations in the environment are reduced to an abnormally low level. It is interesting to examine the ways in which the opportunist species manage to solve the problem of survival in an environment that is far more homogeneous than the one to which they have become irreversibly adapted. If, to survive in the wild, they must be constantly probing and questing, then how do they deal with the non-variable world of the zoo enclosure?

A rigid, sterile environment is obviously a danger to a species possessing independent neophilic motivation unless the animal can create variability itself in some manner. The persons responsible for the welfare of captive animals can, of course, sometimes provide neophilic stimulation for these species in various ways (see Morris, 1962), but frequently this is difficult or impossible. It is true that wild animals kept at semi-liberty as exotic pets usually experience very little investigatory deprivation, but animals kept in laboratories and zoos in large numbers create an entirely different problem. For reasons of hygiene, economy, exhibition or experimentation, it is not always possible to give such creatures a constantly varying enclosure or cage. The restricted environment of captivity therefore provides an inevitable test situation in which the effects of homogeneity can be judged.

In addition to decreasing variability, the zoo environment also frequently omits specific releasing stimuli that would always be present in the wild. In this respect, the conditions of captivity will influence the behaviour of both opportunists and specialists, although it is interesting that the starvation of non-exploratory drives does seem to affect the former more than the latter. It is as though the nervous system of the opportunist has come to abhor, not only exploratory inactivity, but inactivity of any kind.

It should be pointed out that although opportunist and specialist species have been strongly contrasted above, there is a complete spectrum from strongly neophilic to strongly neophobic, with the majority of animals in an intermediate position. Nevertheless, in the following descriptions and discussion it is those species that are placed more towards the neophilic end of the spectrum that figure most strongly. They deal with a restricted environment in a number of ways:

1. *They may contrive to introduce novelty themselves by inventing new motor patterns.* This is comparatively rare and is largely confined to the higher primates.

2. *They may increase the complexity of their environment by performing acts that create stimulus situations to which they can then react.* They may, for example, provoke visitors in such a way that the latter react by providing more intense novel stimulation. The very presence of human visitors provides a valuable potential source of variability and many species find ways of developing this potential. Some species also perform actions towards inanimate objects in their environment which, although meaningless patterns of behaviour in themselves, nevertheless lead to the creation of stimulus situations that would otherwise be missing. Once created, these situations can then be reacted to in a normal way. For example, 'life' can be put into a dead food object by throwing it into the air. It can then be caught and 'killed', thus satisfying a starved killing drive.

3. *They may increase the quantity of their reactions to normal stimuli that are available.* If many different kinds of activity are not available to an animal, it may keep up its 'total activity output' by performing more of those responses that are available to it. In other words, it gives super-normal responses to normal stimuli, as a compensation device for the enforced absence of responses to missing stimuli.

4. *They may vary the quality of their reactions to normal stimuli that are available.* Animals that, under the pressure of life in the wild, would give a simple non-variable response to a normal stimulus situation, may vary the nature of the response in captivity. This permits a greater overall variety of motor patterns in a restricted environment. Food, for example, that would be swallowed quickly in the wild, with little preparation, may be elaborately manipulated, investigated and played with in captivity before being eaten. This is not due simply to the fact that captive animals are generally less hungry than wild ones. If a wild animal was not hungry it would not play with its food, but deal with some other survival problem.

5. *They may perform normal responses to sub-normal stimulations.* In the

absence of the appropriate stimuli for specific acts, animals may not only compensate, as mentioned above, by increasing other activities, but may also perform overflow or vacuum activities. In these cases a normal reaction is given to an inferior stimulus that would never release the response in the wild state. The internal drive to perform a particular action becomes so intense that it requires only a feeble substitute of the real stimulus to trigger it off. An animal without a mate, for example, may copulate with an inanimate object if its sex drive becomes intense enough. Many animals that repeatedly perform full responses to sub-optimal stimuli develop these responses into repetitive *stereotypes*. This is because the inferior stimulus provides little or no 'guidance' to the performer. An animal mating with a female of its own species will experience a considerable amount of interaction during the process of copulation. This interaction 'guides' the copulatory performance that has been triggered off, causing it to make constant adjustments as the activity proceeds. These adjustments are not called forth by a simple inanimate substitute in the case of overflow activities. All it can do in most instances is to act as a simple releaser, and as a result of this the performance tends to become more and more stereotyped as time goes on. Animals that patrol a territory in the wild, pace up and down their cages in captivity. They do this in a stereotyped fashion because, although the cage is enough of a 'territory' to provoke patrolling, it is too featureless to provide any 'guidance' for the patrol. It therefore becomes formalized and stereotyped as circling or pacing back and forth.

6. *They may perform rhythmic sequences of thwarted intention movements.* This is a second source of stereotypes. These may start out as initial escape movements, or jumping or flying actions. If they cannot be carried through because of space restriction they may be repeated over and over again until they become fixed as persistent, highly characteristic patterns (see page 503).

The six categories of response to a restricted environment listed above may occur separately or in various combinations. They may be directed at the zoo visitors, at inanimate objects in the cage or enclosure, at the animal's own body, or at its cage mates. Unhappily, at the present time, no quantitative or experimental analysis has been made of the details of these responses. This is a study that remains to be done. However, there are a number of anecdotal observations that can be recorded briefly to indicate the kind of material that would be available for a more detailed investigation. Taken together, they provide a remarkable testimony to the ingenuity of active animals struggling to survive in a sterile, inactive environment.

1. Visitor-oriented responses

If an animal sits quietly in its cage, zoo visitors will either walk past its field of vision, or will approach, pause, and go away. In some cases they may throw objects at the animal or shout and wave at it. All this provides a source of variability. If an animal performs actions towards the visitors which increase the intensity, frequency or range of their activities outside its cage, it has achieved an increase in environmental variability. It can do this by attracting the visitors, by provoking certain responses in them once they have come close, and by repelling them. The following examples indicate some of the ways in which zoo animals influence the public in this manner.

Liquids (drinking water, bathing water, spittle, or urine) are used by a number of species. Elephants sometimes develop the habit of flicking the trunk at visitors in such a way that the mucous it contains is sprayed over them. The performance is often improved by deliberately putting the trunk into the mouth immediately before flicking, and sucking up large quantities of spittle. As an alternative, bath-water may be squirted in a powerful jet. Chimpanzees and orang-utans both develop powerful and well-aiming spitting actions, using either spittle or a mouthful of drinking water. Male chimpanzees also employ an accurately directed jet of urine in the same way, frequently climbing to a high point on the cage when taking aim. On these occasions, the penis is protruded through the wire of the cage and the body tilted back until the angle of fire is correct. Urination is then performed at high intensity.

Amongst the carnivores, male lions develop a similar pattern. When urinating they raise the tail and, instead of squatting, aim a powerful stream of urine backwards in a horizontal jet. Before doing this, they orientate themselves in such a way that the jet will strike a vertical feature of their territory. This is usually one of the upright bars of their cage front. Visitors watching the lion from the other side of the bars are often sprayed with urine on these occasions. Inevitably they leap back shouting. Unlike the directed urination of chimpanzees, this pattern in the lion is a perfectly normal form of territorial scent-marking, similar to the tree-marking of domestic dogs. It is not in the first instance specifically visitor-oriented. However, in the case of certain male lions, the dramatic effect appears to have been sufficiently stimulating for the animals concerned for them to develop the activity further in such a way that, although the territorial value of the action is not improved, its impact on the public is definitely increased. One particular male lion modified the pattern so that, after the first spraying, it would retain a certain amount of urine for a second jet. It would wait for the heavily sprayed front row of watching visitors to retreat to clean themselves and then, when those at the back of the crowd had pushed to the front, it would turn into position again and deliver a second spraying.

Another instance of the use of water to produce public reactions concerns a large walrus that, until recently, lived in the Munich zoological gardens.

This animal repeatedly and frequently splashed water over the crowd with a special action of its front flippers.

Solids of various kinds are also aimed at the public by a number of species. Apes and certain monkeys have sometimes been known to develop the habit of throwing their faeces at visitors. The colony of adult chimpanzees living on an island at Chester zoological gardens frequently tears up clods of earth and hurls them with remarkable accuracy at onlookers on the other side of the water-ditch. The dolphins at Marineland in the United States were observed to pick up pebbles from the floor of their large tank and flip them out of the water at certain visitors. All pebbles were then removed from the tank floor, but this appeared to have a detrimental effect on the health of the cetaceans. They became listless and the condition persisted until the now famous water-polo and other games were instituted as occupational therapy.

We have considered above those actions involving the aiming of something from the animals to the visitors, but there are other actions that stimulate the reverse process. Many zoo animals learn to beg for food, notably the bears, apes, monkeys, elephants and certain ungulates. If this begging was simply a way of getting food, it would hardly be relevant in the present context, but it involves more than that. Zoo animals are always well fed. They are not hungry and yet they still beg for food. The reward is clearly an 'occupational' one. The energy expended by an enormous bear, when it sits up on its haunches and waves its paws about, is out of all proportion to the pea-nut that is thrown to it. Proof of this can be found in the post-mortem analyses. Each year there are several deaths due to gross over-feeding. This has nothing to do with hunger or the obtaining of food rewards. It is entirely a matter of animals craving novel social interaction, with the food-swallowing as a means to an end. In some instances obesity develops slowly and causes death after a number of years. In other cases, lethal over-feeding can take place during a single busy day. Bears and ungulates have been found dead with their stomachs so distended that they have died of suffocation. It is hardly necessary to point out that, in the wild, even where there is a great surfeit of food available, such a condition does not develop.

In the case of a coati (*Nasua nasua*) an interesting form of social interaction was developed involving a small stone or pebble (Gilbert Manley, personal communication). If such an object was thrown into the animal's enclosure, it would rush over to it, carry it to the wire at the place where the thrower was standing and push it carefully through the wire. It would then wait for the pebble to be thrown once more and repeat this process over and over again. Pushing the pebble through the wire was a difficult task for the coati. Usually the object was placed on the concrete at the bottom of the wire and then laboriously pushed through the narrow gap between the concrete and the wire with the animal's nose. This is a fetch-and-carry pattern that is remarkably similar to that seen in domestic dogs in a park or field.[1] The object thrown does not appear to represent anything more than something that can release chasing or searching behaviour. By bringing it back to the thrower, the animal is helping the environment to provide it with another chance to chase and search.

In the great apes, percussive displays are sometimes employed that produce a dramatic effect on the 'visitor-component' of the environment. Adult male chimpanzees are particularly prone to this. In the wild they communicate with one another by stamping and banging, as well as by typical vocalizations, as they move through the forest. In such instances, they do not develop their acoustic displays to any great extent, but in captivity the pattern can become a prolonged, complex and extremely rhythmic performance. The animal combines several different types of percussive action in a special sequence. An adult male at present at Regent's Park starts with rapid, rhythmic swaying of his body, in a bipedal posture with his arms swinging from side to side. Then be begins to clap his hands together, again with a remarkable rhythm. This clapping may suddenly speed up and then slow down again. Interspersed with this there is chest-beating and also foot-stamping against a vertical partition or a metal cage-door. The banging and stamping attracts an enormous gathering of visitors very quickly and the ape finally completes the display with a huge leap towards the crowd. It then sits and examines its fingers with great gentleness. This pattern has changed slightly over a period of years, largely by elaboration and extension. It is rhythmically inventive and yet has become stereotyped in many ways. From descriptions in the literature it appears that other males have developed similar patterns at other times, in each case the brief percussive communication system of the species being elaborated in this special captivity context.

It was mentioned earlier that some visitor-oriented patterns attracted and others repelled. On some occasions the two become linked. Apes, monkeys, cats, corvids and parrots may sometimes develop an enticement device. The visitor is attracted and then attacked. In most cases this involves an intense and extremely friendly invitation to groom. The mammal presses or rubs itself against the wire or bars, the bird cocks its head and ruffles its feathers. Then, as soon as the visitor has started to scratch, stroke or rub the soliciting animal, it swings round in a flash and scratches, bites or stabs at its victim.

Apart from the special cases noted above, captive animals often use human visitors as substitutes for their own species, when the latter are absent. In these instances the patterns performed by the animals are not unusual in any way, except that they are directed towards an abnormal, interspecific companion. Animals may lick or groom visitors, or be groomed by them. They

[1] Dr Allan Rogers tells me that he has observed a similar pattern in the case of a captive polar bear.

may attempt to courtship-feed human beings, or they may try to mate with them. They may attack them, not as predators, but as if they were rivals of the same species. These reactions may be the result of an intense need to find an outlet for these drives, such that even an inferior, human stimulus-complex is sufficient to trigger them off, or they may be the result of *mal-imprinting*. (The latter will be discussed later. It involves the 'humanizing' of animals at an early stage, leading to a condition in which the hand-reared animal behaves as if it belongs to the human species, rather than its own.)

2. Object-oriented responses

The number of objects available to a captive animal is much smaller than in the wild and they are less varied. The captive individual compensates for this by using certain objects in one or more abnormal ways, in addition to the normal, natural manner. For example, food is not only eaten, bedding is not only slept on, food containers are not only eaten out of, branches are not only climbed on. What is the nature of these abnormal opportunisms?

Food objects are exploited in a number of ways. As already mentioned, cats may throw dead prey animals into the air and immediately follow them to pounce, catch, 'kill' and eat them. The action of throwing, however it starts, can quickly become a routine that promotes feeding variability. Viverrids and canids of various kinds can also be seen shaking a dead prey object 'to death' before eating it. In the wild, cats pluck birds before eating them. In zoos they are sometimes not given birds to eat for long periods, pieces of meat and small mammals being used instead. When freshly killed birds were offered to several species of cats experimentally to study plucking actions, it was noticed that, in the case of an ocelot that had been in captivity for some years, the release of the plucking response acted like the breaking of a dam. After the cat had stripped the bird completely it did not eat it in the normal way, but turned its attention to the grass on which it was lying and began to pluck that. It continued in a frenzy of grass-plucking for some minutes before finally settling down to eat the bird. The long-suppressed plucking activity, having been allowed an outlet at last, became unleashed with such intensity that inferior objects in the environment were brought into play until the reaction finally subsided.

Animals that prepare food in the wild will not always be content to accept it already prepared for them in captivity. Agoutis and acouchis (*Dasyprocta* and *Myoprocta*) normally clean the dirt off the outside of food objects before eating. They do this by an elaborate peeling routine. If they are given clean food they frequently peel it laboriously before eating, but then proceed to eat the peeled fragments as well. The latter are not dirty and so they are perfectly good food. As long as they are on the exterior surface of the large food object, however, they stimulate the peeling rather than the eating reaction. Even a large piece of crustless bread may be dealt with in this way despite the fact that there is no differentiation between its outer and inner parts. These rodents will also 'bury' food repeatedly on hard concrete floors. They go through the complete pattern of digging a hole, dropping the food object into it, packing it down tight, covering it over and placing a few (imaginary) leaves over it, before going away to look for another object to 'bury'. As the first object is still lying in full view on the smooth concrete surface, they may well find it in their search for a new food object and start the whole process off again.

Raccoons in captivity are well known for their action of carrying food to water, 'washing' it, and then eating it. It is imagined that they do this also in the wild, but this is apparently not the case. Also, food is not cleaned by being 'washed'. Frequently food is completely destroyed by the process, pieces of bread breaking up into tiny fragments that cannot be consumed. Lyall-Watson (1963) made a special study of this pattern to find out whether or not it constituted a genuine food-preparation activity of some kind or whether it was in reality another example of abnormal captivity compensation. It proved to be the latter, the raccoons taking their food to water, dropping it in, 'losing' it, finding it by the special searching movements that are typical of this species, and then eating it. In the wild these creatures spend a great deal of time searching for food at the banks of streams and rivers and they have evolved several specific motor patterns in this connection. These are starved of expression in captivity when all the food of the raccoons is placed in a food basin separately from the water bowl. The raccoon corrects this situation by taking its food to the water and thus allowing itself to search for it with the frustrated wet-hunting actions.

Amongst the more bizarre feeding patterns there is the case of a gorilla that regurgitates regularly on to the floor of its cage and then eats its meal for a second time.[1] There are also the fairly well known and widespread instances of coprophagy in primates of various species. In the case of chimpanzees, faeces are not only eaten but also sometimes mixed with saliva, ejected on to the glass cage-front and there smeared and spread over the glass surface with the tongue and lips. Once coprophilic or coprophagic activities have become fixed with individual apes it is extremely difficult to eliminate them. The eating of faeces may develop such an intensity in relation to normal feeding behaviour that the individual concerned becomes considerably weakened and its physical condition may suffer. This is known to occur especially in chimpanzees.

Another detrimental pattern that develops in some chimpanzees is the

[1] G. B. Schaller (*The Mountain Gorilla*, Chicago University Press, 1963) also observed regurgitation and re-ingestion of food by captive gorillas at three other zoos, but never in the wild (p. 156).

inserting of pieces of straw into the ears. Like coprophagy this is difficult to cure. Bedding is also used in other ways by apes, in inventive play and as parent substitutes. Young apes taken from their mothers at the clinging stage may develop a clinging response to their bedding, grasping a bundle of straw or hay tightly to their bodies. It is significant that such individuals, even when sexually mature, may resort to this distorted infantile pattern when they experience insecurity. Young apes brought up with a special cloth or blanket as a bed will treat this in the same way, often carrying it about with them and protesting strongly at any attempt to remove it.

In contrast with these rigid patterns, chimpanzees are extremely flexible and variable in their responses to simple objects in their environment that do not have these 'imprinted' qualities. They will find many ways of utilizing a single piece of string, a sheet of paper, or a sack. There is a strong tendency to cover the body in various ways, and spontaneous 'dressing up' is not uncommon Gymnastics of various kinds are developed. One example may be given. A single piece of cord, or a batch of strong straws, is inserted through the wire of the cage roof. It is then pulled down through a second aperture in the wire and the two ends are twined round by circling the body whilst still holding on to both of them. The animal may then take its feet off the ground so that it revolves quickly in the air in the opposite direction, or grasp the string with its teeth and rotate in a similar way, spinning faster and faster like a circus acrobat. These and many other gymnastic patterns have been observed in both chimpanzees and orang-utans and indicate a remarkable level of inventiveness.

Bedding may also be used as a pseudo-mate. A male raccoon was repeatedly observed to gather up a tight bundle of hay from its bed, clasp it tightly and then copulate with it. A hyena has also been observed to copulate with its feeding dish. The dish was circular and the animal succeeded by turning it up on its side and then moving itself back and forth over the object in such a way that it came into rhythmic contact with its penis. A leopard developed a similar relationship with the scratching post in its enclosure, grasping it and making pelvic thrusts against it until it ejaculated. In all these cases a normal mate was absent and the object was being used as the best substitute available. However, it is possible that abnormal sexual fixations of this kind may become irreversible if the situation persists for any length of time (see below).

There are several strange scent-marking patterns that are difficult to explain, but which are fairly common in captive animals. Coatis frequently react in an ecstatic manner when offered cotton wool soaked in perfume. They anoint themselves, rubbing the cotton wool over their bodies. Mongooses of several species have been seen to rub their anal glands on shoes worn by keepers or visitors. This is done with great intensity and high frequency. A banded mongoose (*Mungos mungo*) extended this treatment to its meat. Before eating it would frantically and repeatedly crouch over the meat and rub its rear parts over it. It gave a similar response to eggs when the latter were offered as food. Burton (1959) mentions the extraordinary case of a grasshopper mouse (*Onychomys*) at the New York zoological park that developed the habit of chewing cigar ends, parting its fur with its forepaws and then applying the chewed tobacco to its skin.

It is difficult to explain these patterns without knowing more about the behaviour of these animals in their wild state. Clearly the mongooses are starved of objects suitable for normal scent-marking and are using substitutes at high intensity as a compensation device, but the reasons for the special nature of the substitutes have yet to be found. The coatis and the grasshopper mouse seem to be reversing the process, scenting themselves rather than the environment. This is reminiscent of domestic dogs rolling in dung, on the one hand, and of hedgehogs anointing themselves with saliva, or birds anting, on the other. It requires further study.

Stimulation from the use of smoke is well known in certain bird species (smoke-bathing in crows, for example) but it also occurs in certain mammals Camels appear to be strangely stimulated by the inhalation of clouds of smoke. Cigarette smoke, far from distressing a camel, produces the characteristic 'flehmen-face', a special facial expression associated with sexual situations.

Individual chimpanzees acquire the habit of smoking cigarettes, even learning to strike a match and light the cigarette themselves. The whole smoking pattern is remarkably similar to the human activity. Apes that have once started to smoke continue to do so for the rest of their lives. One particular chimpanzee learned to use a lighted cigarette in a new way. With the object in his mouth and his neck stretched forwards, he pursued and terrorized his companions relentlessly.

3. Companion-oriented responses

To a large extent, reactions to cage-mates are the same in captivity as in the wild. Abnormalities are of two kinds, those that involve compensation devices and those that are the result of incorrect social groupings. In the first instance, the animals may be experiencing the correct social relationships for their species, but because of various non-social activity deprivations, they use the companions in abnormal ways or to an abnormal extent. Hediger (1950) has discussed at length the problem of hypersexuality in zoo animals. As he puts it: 'A female on heat attracts the attention of the male to an exaggerated degree, owing to absence of other diversions, lack of interesting objects in their surroundings, and their own lack of occupation, so that he cannot keep away from the female' (p. 152). Where courtship involves aggressive chasing of the female by the male, this can lead to difficulties. If, in the wild, a male of this type normally lives with a harem, it is particularly dangerous to keep him with one, or only a few females. He not only gives them more attention because he has fewer non-sexual preoccupations, but also gives her, or them, more because they are now taking the brunt of the action that would typically be spread out over a whole herd or flock. Unless a situation of this kind is watched very carefully, the female or females will probably be killed. This is well known in deer, wild cattle, antelopes and their relatives.

Parent animals sometimes become too parental if they have nothing to divert their attention. There is then a danger that they may spend too long cleaning and licking the offspring and damage them in some way. They may insist on carrying them in their mouths for long periods of time, or they may eat parts or all of them. Rodents sometimes eat the tails or ears of their young. This appears to be a frenzied exaggeration of the cleaning action. The devouring of the young of many species of mammals seems to be connected with a failure of the parental drive to develop fully its inhibitory influences. The young are cleaned repeatedly and the cleaning changes over into eating.

Carnivores kept together as pairs may sometimes use one another as pseudo-prey objects. Hiding, stalking, charging, pouncing and biting occur. The last link in this chain, the biting, is always inhibited and the pattern deteriorates into playful sparring. This is common amongst the species of big cats. Another example of members of a normal social group using one another in an abnormal way concerns a colony of blackbuck (*Antilope cervicapra*). The ratio of males to females was reasonable for the species and they were living in a large, natural colony group. But they were in winter quarters where there were no low branches on which they could press their facial glands. These glands are situated just below the eyes and the animals press and rub them against the bushes in their natural habitat, depositing their scent on them. In the absence of any bushes, the blackbuck risked the dangerous course of using one another's horns as scent posts. When a male sat down on the ground to rest, the others would move in and treat him as if he were a bush. Had he flicked his head or stood up suddenly, it would have no doubt wounded the scent-marking individual.

Abnormal social groupings are more likely to lead to abnormal activities, and strange interrelationships between different species can often be observed. A two-toed sloth (*Choleopus*) and a douroucouli (*Aotus*) were sharing a cage. The edentate allowed the small primate to sit on it and even to sleep on it. The douroucouli gained the advantage of a soft, warm bed and the sloth (a female) was able to experience the typical contact stimuli that an infant sloth would have provided. This was not, however, a case of mal-imprinting. The douroucouli had not been brought up as a young animal on the body of the sloth. They met for the first time when both were already adult. Another sloth shared a cage with a small sun squirrel (*Heliosciurus*) with the same result.

An adult male squirrel monkey (*Saimiri sciureus*) was sharing a cage with an adult springhaas (*Pedetes capensis*). The former was day-active and the latter night-active. When the large rodent was sleeping it curled up into a ball in the corner of the cage. The squirrel monkey frequently used it as a pseudo-parent, clinging to its large furry body and sleeping in that position. If, on the other hand, the monkey was sexually motivated it attempted to copulate with the slumbering springhaas. At first this disturbed the rodent, but it eventually grew accustomed to the weight and movements of the small primate perched on its back and slept peacefully through the proceedings. Again, this was not a case of infantile mal-imprinting, the animals having been put together as adults. There are a number of other strange examples of this kind (see Inhelder, 1962).

Sometimes these abnormal sexual relationships persist even after the normal sexual object is present. An Old World brush-tailed porcupine (*Atherurus*) was living in a cage with a New World tree porcupine (*Coendou*). The brush-tailed porcupine (a male) repeatedly attempted to copulate with the tree porcupine (a female). The latter responded with its appropriate sexual posture. Unfortunately, as these are two convergent species with very distinct ancestry, their anatomy and the arrangements of their spines do not match up. After many painful experiences the brush-tailed porcupine had still not given up, when it was provided with a female of its own species. It formed a normal pair and mated with the new female, but surprisingly persisted in its attempts to copulate with the tree porcupine.

Another example of an irreversible sexual abnormality concerns an orang-utan. This ape, a young male, was kept with another young male and they spent a great deal of time playing together. This included some sex play and anal intercourse was observed on a number of occasions. When, at a later date, the now sexually mature male ape was placed with a mature female he persistently performed anal intercourse with her. Although this activity was observed on frequent occasions, he was never seen to perform a normal copulation. They were separated and the female was given another male. The latter copulated normally and the pair produced an offspring. This orang-utan incident is an example of the results of an abnormal intraspecific grouping, as contrasted with the abnormal interspecific arrangements mentioned above. Homosexual groupings of mammals, birds or fish are liable to produce abnormal intraspecific sexual behaviour in captivity in many species (see Morris, 1952, 1954, 1955). Usually this consists of females mounting females, or males giving the female invitation display to other males and being mounted by them. Actual anal intercourse between male mammals may occasionally occur, but outside the higher primates it is not encountered with any great frequency. Sexual activities between females are more limited, but homosexual female elephant groups have been observed to spend a great deal of time masturbating one another with their trunks.

4. Self-oriented responses

Activity output can be increased by a captive animal even without reference to any other creature, visitor or object. This can be achieved by the animal performing spontaneous locomotory activities or by turning its attention to itself and directing actions that would normally be aimed elsewhere, on to its own body.

The most familiar form of abnormal locomotion is the characteristic, stereotyped pacing to and fro of the caged animal. This indicates the need for a great territorial space in which to patrol. But it also indicates that the animal concerned has come to terms with its restricted space and has developed a rhythmic, modified version of patrolling. Pacing is common amongst carnivores and ungulates of various kinds. In the case of martens, chipmunks and squirrels, it may take the form of vertical circling of a remarkably acrobatic kind.

Stylized, repetitive intention movements of locomotion are to be seen amongst such animals as monkeys, lemurs and parrots. The latter are of interest because each individual seems to develop a highly characteristic movement pattern of its own. One will roll its head from side to side; another swings it back and forth; another twists it in a figure-of-eight pattern; another bobs it up and down. These movements are formalized stereotypes of the initial actions of taking off, or turning for take-off. In a freshly caged parrot one can see confused, agitated intention movements as the bird adjusts to its small non-flying cage. These gradually become rhythmic and repetitive and the performer can express its drive to fly without jarring, erratic sequences of movements. It is interesting that when, after some years in a small cage, parrots or primates are put into a larger enclosure, they continue to perform the stereotypes for a long time after there is no further need for them. In the case of a capuchin monkey (*Cebus*) there was a rigid pattern of pace-and-twist that was retained in a new, large cage in such a precise form that it was possible to estimate the size of the small cage that had previously housed it, even though this had never been seen.

Pacing to and fro is sometimes clearly a side-to-side, ambivalent reorientation of a frustrated forward movement. In the case of a hand-reared tigress that repeatedly attempted to move forwards towards her human 'friends', a rather damaging form of pacing developed. The animal pressed forward so hard as it paced back and forth that its nose was rubbed across the bars, like a boy trailing a stick along garden railings. This rapidly produced a large sore patch which became raw until the whole of the surface of the snout became a sensitive, naked area. This was cured by shutting the animal away in a barless den. When it was released with a completely healed nose, the whole routine began again with the same damaging effect. Large posts were put against the bars to break up the rhythm of the pacing and this worked well for a while, but eventually even this did not stop the rubbing. Finally the animal was sent to another zoological garden, where the human 'friends' with whom she had grown up were missing from the environment. In a short space of time all was well.

Amongst the great apes it is possible to observe the invention of entirely new locomotory patterns. These include progression by forward rolling, lateral rolling, foot-dragging and foot-stamping. Rocking, twisting and other dancing patterns can be seen during play. On icy mud, adult chimpanzees were observed to develop a form of locomotion that was remarkably akin to that used by human skaters.

Sexual patterns may occur in the absence of any partner, or any object that can act as a pseudo-partner. Masturbation is known in the case of primates and, surprisingly, lions. Young male monkeys and apes frequently manipulate their own penis and this may become stereotyped into a masturbatory rhythm. In one baboon (*Papio*) male, the stereotype became so modified that the animal no longer made actual contact with its penis. Its hands simply made the back and forth movements of masturbation in mid-air. This was done with incredible persistence and frequency throughout the day. In the case of the lion, the huge male regularly heaved himself into an inverted position against the door of his cage. He raised his rear end above his head, spread his back legs wide apart thus maintaining a balance, and arched his back to bring one of his front legs in contact with his penis. Then, by jerking his body and moving his front leg against his penis, he was able to achieve ejaculation. This same male later mated successfully with a female in a normal way.

Parent–offspring actions are sometimes self-directed. Infant apes that have been separated from their mothers will suck their thumbs. Parent apes will sometimes suck from their own nipples. The strangest instance of self-oriented infantile behaviour concerned a polar bear cub that spent many minutes every day sucking the fur of one of its front feet. Sucking in bears is accompanied by a highly characteristic sound and this was also present in the foot-sucking. This was not only a prolonged activity, but also an extremely intense and vigorous one. The cub was isolated from its mother, but had another young cub as a companion. Despite this there was no attempt to suck from the other young animal. The pattern disappeared as the animal matured.

A more dangerous self-oriented pattern is that of aggression. This can lead to serious self-mutilation. It is known especially from carnivores, primates and parrots, but is also found occasionally in other groups. Sometimes it is easy to understand. For example, a monkey tries to attack another animal in the adjoining cage, but cannot get through the wire to do so. It therefore finds an outlet for its aggression by redirecting it on to its own body. It seems as though it is more important to the animal to express its aggression than to remain unscathed. Monkeys sometimes appear to bite themselves so hard that they scream with pain. Redirection activities of this sort (see Bastock, Morris and Moynihan, 1953) are usually performed in the wild in a slightly different way. A powerful rival threatens a male. The latter is too frightened to retaliate, but finds expression for his thwarted aggression by attacking a weaker member of the group. In captivity there is often no weaker member present, and the animal has to make do with itself. Severe wounds may sometimes develop in such cases. The most dramatic instances are found amongst the canids. One coyote (*Canis latrans*) male provides a particularly interesting example. This animal was obtained as a mate for a female that had been in captivity for some time. They mated successfully and produced an offspring, but from the beginning of their relationship there was considerable tension between them. The male repeatedly attacked the female when the keeper in charge of them came to clean out the cage. Eventually she developed naked patches where he had torn out tufts of hair. When she was suckling she became more aggressive and the male then turned on himself when his human 'acquaintances' came near. He would always try to attack the humans through the wire first, then turn round and bite his rear leg. Eventually he moved from his leg to his back at a point just above his tail. There he chewed into his flesh until he managed to create an extensive patch of raw flesh. At this point he was put on a course of tranquillizers, whereupon he returned to attacking his female again. When the tranquillizers were stopped, he once more attacked himself and opened up the large wound again. He was then moved to the animal hospital where he was tranquillized again, cured, and then once again taken off tranquillizers. In this new situation he did not then start to attack himself again. At the time of writing he is still in the hospital and will shortly be moved to another cage in the zoo by himself to ascertain whether the condition reappears in the absence of his cage mates.

There are several points of interest here. It was suspected that the animal had not experienced a normal infancy and a specific prediction was made that he had been hand-reared in isolation from other young coyotes. Information from the United States' source of the animal confirmed this. The young male had been found in a litter at a very early stage of development. The rest of the litter had been destroyed and the single young male kept as a farmer's pet. When mature it had been sold to a dealer and shipped to the zoo. This gave the animal a special relationship with man and explained the intense aggression towards its keepers. (This will be discussed later in connection with mal-imprinting.) The fact that the animal was 'humanized' also gave it a strange ambivalent attitude towards other coyotes. It is hard to see why it should have been prepared to mate with a coyote and yet not to use one as a target for redirected aggression at all times. Certainly, the increased aggression of the parental female may have made it difficult for the male to go on attacking her as he did at earlier stages, but this does not explain why, when the young cub was nearly fully grown and was no longer under the female's protection, the male did not attack that instead of himself. He attacked the young creature at other times (squabbles over food, etc.) but, when he was redirecting his attacks from people outside the cage to an animal inside the cage, the animal he selected was himself. It is as if the animal, by attacking himself, was able to experience more vividly what he would have been able to do to the people outside the cage, if only the wire had not been there. This is certainly a difficult case to understand but this example and others like it may prove of value, in the future, to our understanding of similar conditions in man.

Self-mutilation is encountered in other non-canid species of carnivores, especially the mongooses. A kusimanse (*Crossarchus*) amputated half its tail. This too was a tame, hand-reared animal. A pygmy mongoose (*Helogale*) began to attack and lacerate its tail shortly after its mate died. Up to that point the two animals had lived contentedly together with no sign of self-mutilation. Presumably, in his case, the pygmy mongoose no longer found it possible to dissipate his aggression in (comparatively harmless) squabbling and turned on himself. As it is impossible to put oneself to flight and 'win' in such a case, the battle continues until blood is drawn and severe damage done.

Other forms of self-mutilation may start in an entirely different fashion. Usually the animal concerned is isolated and is so under-occupied that it over-cleans itself. A simple example is the 'running sore'. An animal experiences a very minor superficial injury – a scratch or a small cut. It starts to lick or rub it and soon it develops into a larger and larger area. In the wild, the normal amount of licking that the animal would have time for would help to cleanse the wound and it would heal normally. In captivity, on the other hand, the animal may go on worrying the place until it becomes really serious. A large percentage of all captive monkeys are without the extremities to their tails for this reason.

Feather-plucking in birds seems to be a similar pattern. There have been many arguments about the cause of this activity, which can lead to almost complete self-plucking in some instances. One lesser sulphur-crested cockatoo (*Kakatoe sulphurae*) reduced its entire plumage to a single long yellow crest feather. It has been said that these activities are the result of diet deficiencies, but this is certainly not the whole story. Psittacines are the most vulnerable and their beaks are perhaps the most active in the wild. With little to nibble or bite in captivity, they may well be adding a more vigorous muscular action to the preening movements. Instead of being gentle, these become severe. By superimposing more powerful jaw actions when preening, the birds can compensate for thwarted wood-nibbling by ferocious feather-nibbling. The provision of suitable wood is said to retard plucking, but apparently it does not stop it completely if it has already become a regular habit. It would be particularly valuable if this and other kinds of animal self-mutilation could be made the subject of a special study to investigate the efficacy of various conditioning techniques designed to remove the patterns. The results of such a study might well be of great interest to human psychiatry.

Mal-imprinting

Up to this point the effects of restriction have been discussed with reference to the existing environment. But certain kinds of environmental restriction in early infancy may have a lasting influence on the behaviour of wild animals in captivity. It is well known that young birds and mammals of many species go through a critical period in their life when they become 'imprinted' on their parents. From that point on they are 'socialized' as members of their own species. All their species-typical motor patterns will now be directed at other individuals of their own kind, rather than at members of different species. If young birds or mammals are taken away from their true parents before they have been socialized and are brought up by human foster-parents, they become 'humanized'. That is to say, they behave normally, but their behaviour is directed at human beings instead of at other members of their own species. Scott (1956, 1958) has shown that domestic dogs will become wild animals again if they are isolated from man during the vital socialization phase (roughly from age 20 days to age 60 days) for this species. He has also shown that young puppies exposed to both dogs and men during the socialization phase can develop a healthy ambivalence, being friendly towards man, but still accepting the company of dogs. This is, of course, what usually happens with young puppies. It seems that if these or other young animals are completely isolated from their own kind (both in the form of parents and litter-mates) then they become so strongly imprinted on their human foster-parents that they will treat their own kind as enemies, almost as predators, when they meet them. Harlow (see Harlow and Harlow, 1962a, b; Harlow and Zimmermann, 1959) has shown that young monkeys reared on mechanical mothers in complete isolation can never make full social adjustments when mature and even when put together in a large 'natural' enclosure at a zoo, the animals pay little attention to one another and do not breed. Even if the young monkeys are allowed to stay with their own mother, they do not develop a full social pattern later unless, as youngsters, they can play with other young members of their species. If allowed such play, but not allowed a real mother, they also suffer socially, but less so. It therefore seems that both mother-to-young contact and young-to-young contact are vitally important to the growing monkey if it is to mature behaviourally.

It is clear therefore that if a young mammal or bird, that relies upon imprinting to 'socialize' it, is exposed to a socially restricted environment in its early days, it may orientate its behaviour in an abnormal way for the rest of its life. A remarkably high percentage of zoo animals throughout the world are specimens that have been hand-reared and then, after maturing and treating their owners to a sample of the aggressive motor pattern of their species, have been presented to a zoological collection for safe keeping. Unless they are interested in studying behaviour abnormalities, zoos should steadfastly refuse to accept such creatures. They have not been singled out to any great extent in the past, as a special category, because of two confusion factors. One is that not all species rely to the same extent on learnt clues to identify their own kind. Many species seem to be able to recognize their own species by inborn olfactory or other clues. More research is required on this problem, but it is clear that some hand-reared specimens are able to readjust when returned to a more natural social group. It may not always be easy for them, because they will be reacting to the visual appearance of their human foster-parents, but the odours of their own species. Provided the latter trigger-off the necessary sexual responses, they will at least be able to breed successfully. The second confusion factor concerns Scott's point about dual-imprinting by dual-exposure during the socialization phase. This often masks the mal-imprinting and as a result the real dangers of the whole process of hand-rearing go unnoticed.

We need to know a great deal more about the problem, but from casual observations in zoos it is obvious that the phenomenon is widespread. A few examples may be given:

A hand-reared tigress was placed in a cage next to a male for the first time. She could see and smell him through a small grill. She immediately fled to the far side of her cage and refused to move or eat for a number of days. It took several weeks to bring her back into her normally friendly, active state again. This recovery did not begin until the male was removed from the adjoining cage. This tigress had always treated (and still does treat) her keepers as social companions, but will probably never show any interest in other tigers except to react to them as predators.

A marsh mongoose (*Atilax paludinosus*) female was given a male of the same species. Her background was not known but she was remarkably tame and had obviously been hand-reared. The male was wild-caught. The female threatened the male from the moment he entered her cage. They reached a fairly stable state of mutual disagreement with the male apparently holding his own. He must, however, have been under severe stress, for he developed peptic ulcers and died. The female is now once again her former contented self.

Many such case histories could be reported. Frequently the animals adjust to the extent that they tolerate one another and are not under stress, but take practically no notice of each other. Harlow found this with his young rhesus monkeys after they had experienced an isolated infancy. They would simply sit apart and mutually ignore one another. At the present time there is an adult, hand-reared chimpanzee at Regent's Park that has been living with an adult female for a number of years. The female has bred some time ago and there is no physical reason why the male should not produce offspring from her. Unfortunately, he is hardly aware of her existence. He never sits with her, grooms her, or attempts to mount her. It seems that, as far as he is concerned, she belongs to another species. Exactly how many other pairs of animals fail to breed in captivity for this reason it is hard to say, but the figure is probably startlingly high.

There is one final, special aspect of mal-imprinting that must be mentioned. Since a humanized animal 'belongs' to the human species, it is not surprising if it reacts with distress to being isolated from its 'family'. If it is a member of a very social and potentially rather aggressive species, it may turn against the individuals that have apparently punished it; isolation is a very great punishment to a sociable species. It may suddenly or gradually become extremely aggressive towards the human beings it knows well. This is the case with the coyote mentioned earlier. It was reasonably friendly when it first arrived at Regent's Park, but as it became clear to the animal that it was not going to be allowed to mix freely with people as it had done on its foster-parents' farm, it turned on the human companions who were responsible for this restriction and attempted to attack them. A similar instance can be cited for a wolf (*Canis lupus*) that was hand-reared as an expedition mascot. This animal became intensely aggressive when adult, but only towards men. Towards women it remains a pet animal. A raccoon (*Procyon lotor*) behaved in a similar way as soon as it was caged and no longer allowed to roam freely as a 'semi-liberty' pet. This animal not only attacked people, but also terrorized the group of (normal) raccoons it was housed with, and had to be removed. The aggression in all these cases is bizarre because it is so unchecked by fear. Normally, fights between members of the same species are comparatively bloodless affairs, with a great deal of threat and counter-threat (see Morris, 1958), but little actual contact. In the case of aggressive hand-reared animals there should theoretically be a similar balance between fear and aggression, producing threat rather than out-and-out attack, but this is not the case. The animal should be looking at man as his own species and there is therefore, at first sight, no apparent reason why there should be any difference in behaviour. The answer seems to be connected with the lack of appropriate interaction between the humanized animal and its foster-parents while it is still a young animal. If the foster-parents were real parents, they would doubtless treat the young creature much more firmly. It would have to learn social fear and respect as it grew up. Then, when it became aggressive at a later stage, it would have this urge to attack tinged with fear of the rival.

This may not be the whole story. There is still a great deal we need to know about the after-effects of early mal-imprinting. Again, the results of more detailed investigations would obviously be of interest to human psychiatry and child psychology.

Conclusion

Although the expert care given to captive animals has improved tremendously in recent years, the situation is still far from perfect. They are medically cared for, protected from the elements, well fed and well housed. They lack for nothing except variability, novelty and stimuli to maintain a high activity level. As we have seen, many zoo animals are neophobic creatures that can accept inactivity and environmental rigidity without difficulty. But many others are strongly resistant to lowering their activity output and will use a number of different patterns that avoid this. Some of these activities are dangerous and detrimental in the long run, like self-mutilation. But many others are of great value to the individuals concerned. Patterns referred to glibly as 'non-functional' or 'neurotic' may indeed have some function and may well save the more neophilic species from the most dangerous state of all, namely gross inactivity.

References

BARNETT, S. A. (1958), 'Experiments on "neophobia" in wild and laboratory rats'. *Brit. J. Psychol.* 49, pp. 195–201.

BASTOCK, M., D. MORRIS and M. MOYNIHAN (1953), 'Some comments on conflict and thwarting in animals'. *Behaviour* 6, pp. 66–84.

BURTON, M. (1959), *Phoenix Re-born* (Hutchinson, London).

HARLOW, H. H. and M. K. HARLOW (1962a), 'The effect of rearing conditions on behaviour'. *Bull. Menninger Clin.* 26, pp. 213–24.

—— (1962b), 'Social deprivation in monkeys'. *Sci. Amer.* 207, pp. 136–46.

—— and R. R. ZIMMERMANN (1959), 'Affectional responses in the infant monkey'. *Science* 130, pp. 421–32.

HEDIGER, E. (1950), *Wild Animals in captivity* (Butterworth, London).

INHELDER, E. (1962), 'Skizzen zu einer Verhaltenspathologie reaktiver Störungen bei Tieren'. *Schweiz. Arch. Neurol. Psychiat.* 89, pp. 276–326.

LYALL-WATSON, M. (1963), 'A critical re-examination of food "washing" behaviour in the raccoon (*Procyon lotor* Linn.)'. *Proc. zool. Soc. Lond.* 141, pp. 371–94.

MORRIS, D. (1952), 'Homosexuality in the ten-spined stickleback (*Pygosteus pungitius*)'. *Behaviour* 4, pp. 233–61.

—— (1954), 'The reproductive behaviour of the zebra finch (*Poephila guttata*), with special reference to pseudofemale behaviour and displacement activities'. *Behaviour* 6, pp. 271–322.

—— (1955), 'The causation of pseudofemale and pseudomale behaviour: a further comment'. *Behaviour* 8, pp. 46–57.

—— (1958), 'The reproductive behaviour of the ten-spined stickleback'. *Behaviour* Suppl. 6, pp. 1–154 (Brill, Leiden).

—— (1962), 'Occupational therapy for captive animals'. *Coll. Pap. Lab. Anim. Cent.* 11, pp. 37–42.

SCOTT, J. P. (1956), 'The analysis of social organization in animals'. *Ecology* 37, pp. 213–21.

—— (1958), 'Critical periods in the development of social behaviour in puppies'. *Psychosom. Med.* 20, pp. 42–54.

The rigidification of behaviour

(1966)

Behaviour sequences commonly consist of highly variable appetitive phases leading to rather fixed consummatory acts. Action-pattern rigidity is typical of the terminal moments of a reaction chain.

This basic fact is all too often obscured by the artificial conditions of behaviour studies. Observations on laboratory or captive animals tend to conceal the degree of variability of the earlier phases of each sequence. The simplicity and sterility of the unnatural environment offered to the animal causes differential damage to its motoric performance, attacking the early stages more and the later stages less. A caged animal will feed, drink, nest and copulate, but it cannot set off on lengthy quests for food, water, nest material or a mate. Notorious laboratory devices such as the Skinner-box have served to eliminate totally any possibility for motoric variability. The emphasis in laboratory studies of this kind has been steadfastly concentrated on the variability in the relationship between simple stimuli and an artificially rigidified response. Although the study of this (*SR*) relationship is an important aspect of animal psychology, it is extremely misleading to overstress its importance as has been done so often in the past. To equate it with the whole topic of animal behaviour is like claiming that the gaming rooms of Las Vegas reflect the whole of human endeavour.

Animals placed in an artificially restricted environment – domestic animals, zoo animals, laboratory animals – react to the situation in a number of ways (Morris, 1964). They do not always capitulate by becoming appropriately inactive. Behaviour output-demands may be low, but the nervous system frequently refuses to accept the 'easy life'. Complexity may be put back into the environment by the animal itself. It may invent new motor patterns, as happens with some of the great apes; it may create stimulus situations to which it can then react, as with some cats that throw dead food into the air and then catch and 'kill' it; it may give super-normal responses to normal stimuli, as with certain bears and ungulates that go on feeding until they become grossly overweight.

These and other such behaviour devices assist the animals to carry on a complicated life in a simple environment. They either devise ways of making more variable appetitive behaviour available to themselves, or they abnormally increase the amount of fixed consummatory activity.

Successful attempts to convert a sterile environment into a more acceptably complex one are, however, little more than minor gestures of defiance on the part of the nervous system. There appears to be a contradiction here. If, on the one hand, the system cannot accept inactivity, and yet, on the other hand it cannot bring about a major replacement of variability, how does the animal manage to survive? The answer is that its appetitive performances do not cease, but instead become rigidified. Behaviour stereotypes develop. These permit a reasonably high level of motoric output even in the absence of environmental variability. What occurs is that the environment *releases* behaviour patterns, but does not *guide* them. The animal is stimulated to patrol its territory, hunt for food or search for a mate, but beyond this initial triggering of the activity it is on its own. There are no clues to follow up, no unknowns to investigate, no obstacles to by-pass. The performer has appetitive behaviour to spare, but nothing to spend it on. It starts to wander round and round. When it comes to the end of its restricted territory it makes intention movements of going farther, but to no avail. After a period of time either its erratic wanderings or its thwarted intention movements, or both, become more and more formalized. Eventually they arrive at a rhythmic condition in which they are performed as rigid, repetitive stereotypes. The pacing animal wears a track in the sand. A dark patch appears on a wall where the patrolling animal presses with its feet as it turns at the end of each run. The corner of the wire-netting is worn smooth by a thousand rhythmic intention movements.

Although the captive animal has not capitulated by becoming inactive, it has given up in one important respect. The very essence of its natural appetitive behaviour was that it comprised a basic locomotory pattern combined with a strong neophilic urge. That is, it was stimulated (*a*) to move about its home range, and (*b*) to respond strongly to any novel stimuli. As no natural environment is ever completely static, there is always something new to catch its attention and break up any fixed patrolling rhythm. But in captivity, the novelties are gone. Everything is familiar. The rhythm is not broken up. It is to be expected under such conditions that the patterns of behaviour would become rigidified, for any formalization of motor output must simplify the work of the central nervous system.

The existence of a rigid stereotype in a captive animal therefore reveals three things about its condition: (1) Its environment is too simple and too static. (2) It has abandoned its natural investigative curiosity, at least temporarily. (3) It has come to terms with its situation and has arrived at the most comfortable way of giving expression to its appetitive urges.

Rigidification of this kind is not confined exclusively to appetitive behaviour. It also occurs with patterns that are close to the consummatory end of a sequence, but it is naturally less conspicuous there, as these actions are already more fixed in character. Subtle differences can be detected, however, even with such patterns as copulation.

How are these stereotyped patterns of appetitive and consummatory behaviour related to animal rituals? There is no doubt that superficially they often look very much alike. The turning to and fro of the caged animal is very similar to the pivoting dance of the courting animal. The repetitive leaping of the captive is very reminiscent of the rhythmic jumping of the ritual combat. There is stereotyped nodding and ritual nodding, stereotyped bowing and ritual bowing, stereotyped swaying and ritual swaying. Almost all the rhythmically repeated actions of the behaviourally rigidified zoo animal are to be found in the performances of the naturally displaying wild animal. But is this merely a superficial resemblance, or does it have some deeper significance?

In the case of ritualization, the process is one of communication enhancement. This is not primarily so with zoo stereotypes. (There may be a secondary signal advantage, but this need not concern us here. For example, certain patterns performed by zoo inmates lead to food rewards being given by members of the public. These particular patterns may then be developed as begging signals.) The changes that take place as a ritualized signal develops from an unritualized one are primarily concerned with improving its specificity and its obtrusiveness (Morris, 1956). The signal becomes more and more unambiguous. It is vital that it should not be confused with other actions, and this is ensured (*a*) by increasing the rigidity of its form, and (*b*) by making this form as outlandish and improbable and therefore as conspicuous as possible. Amongst animals (but not man) this process is essentially phylogenetic and species-typical (rather than ontogenetic and individual-typical like the development of stereotypes), and it is frequently accompanied and supported by the evolution of display colours and structures. (In man, these colours and structures have to be added artificially.)

The changes that occur during the evolution of an animal ritual are as follows:

(1) *Loss or modification of the original orientation.* This means that an action does not have to be aimed directly at a social companion (which may well be shifting its position) and variability of the pattern is therefore potentially reduced.

(2) *Lowering of the threshold of the action.* This will not only render it a more frequent performance, but will also make it more available for rigidification by

(3) *Rhythmic repetition.* In addition to this standardization of the time intervals between successive performances of the action, there is also rigidification of the form of the action itself as a result of the development of a

(4) *Typical intensity.* This process involves the disruption of the normal frequency/intensity relationship; instead of the intensity of the action increasing as its frequency increases, it remains more or less the same regardless of frequency changes. Other changes that appear to have occurred during the evolution of ritual patterns include

(5) *Differential exaggeration of components*, or, in extreme cases,

(6) *Omission of components.*

(7) *'Freezing' of movements.*

(8) *Change in the sequence of components.*

(9) *Changes in component co-ordination.*

(10) *Increase or decrease in the speed of the performance.*

(11) *Change in the vigour of the movements.*

A close examination of the stereotypes of zoo animals reveals that almost all the above-mentioned changes are operating there also. The modifications to the motor patterns in the two contexts differ in so far as the rituals are tending towards greater conspicuousness. Apart from this, both stereotypes and rituals show exactly the same kinds of increased rigidity. Clearly the similarities are too great to be accidental.

If we search for a common factor that could explain the shared properties of these two behaviour trends, we are forced to look again at the 'release without guidance' aspect of the situation. It has already been explained how the sterile environment that creates zoo stereotypes sets off a behaviour pattern that would normally 'expect' variable stimuli subsequent to its initiation. In the absence of these varying clues, guiding the animal from place to place, the released pattern becomes rigidified. In the case of a zoo animal that makes intention movements of escaping through the wire or glass of its cage, one can observe the way in which the fleeing movements that never get started become more and more rhythmic. The animal has no chance to dart this way and that as it scampers off. All it can do is to make the initial jerks of running away, over and over again. If one compares this with the displaying animal in the ritual situation, it emerges that the pressures are much the same. It is strongly stimulated to move, but it is held in check by some competing tendency. Typically, the animal that displays during a fight or during courtship is in a state of conflict between fear and aggression, or between fear, aggression and sex. There may be no physical obstruction – no wire or bars – but there is a psychological obstruction just as powerful. In both the cases of the stereotype and the ritual, the actions of the performer are stimulated and then checked. In both cases thwarted intention movements or ambivalent actions result, with the outcome that in both instances we can observe the same sort of rigidification taking place. If, in either instance, the animal were able to carry through the released action pattern, we would then see the motor output being buffeted this way and that by the special, variable, guiding stimuli with which the environment bombards any appetitive locomotory actions. But this stage is never reached.

This means that the changes which we are seeing in the development of zoo stereotypes are probably extremely similar to the kinds of changes that took place during the evolution of animal rituals. Of course, the actual steps taken during the phylogenetic development of rituals are lost to us for ever. We can never hope for the sort of fossil evidence that helps us in tracing, say, the evolution of new skeletal structures. As Konrad Lorenz has pointed out, we can, however, attack the problem indirectly through comparative methods, studying the differences in the more and the less ritualized signals of closely related species. I would like to suggest that, in addition to this, we can learn a great deal about the evolution of animal signals by making a more detailed study of the way in which behaviour patterns become rigidified in captive animals. It is surprising that this has not already been done. Using special filming techniques, it should be possible to analyse minutely the gradual 'rounding off' of the jagged actions of the caged animal and to express more precisely the kinds of motoric transformations that occur. In this way the serious investigation of zoo stereotypes may well assist us in our attempts to understand the evolution of animal communication.

References

Morris, D. (1956), ' "Typical intensity" and its relation to the problem of ritualization'. *Behaviour* 11, pp. 1–12.

—— (1964), 'The response of animals to a restricted environment'. *Symp. Zool. Soc. Lond.* 13, pp. 99–118.

Bibliography of scientific publications by the author

1. 1952 'Homosexuality in the Ten-spined Stickleback (*Pygosteus pungitius* L.),' *Behaviour* 4, pp. 233–61.
2. 1953' Some comments on conflict and thwarting in animals,.' *Behaviour* 6, pp. 66–84 (with M. Bastock and M. Moynihan).
3. 1954 'The snail-eating behaviour of Thrushes and Blackbirds,' *British Birds* 47, pp. 33–49.
4. 1954 'The reproductive behaviour of the Zebra Finch (*Poephila guttata*), with special reference to pseudofemale behaviour and displacement activities,' *Behaviour* 6, pp. 271–322.
5. 1954 'The reproductive behaviour of the River Bullhead (*Cottus gobio*), with special reference to the fanning activity,' *Behaviour* 7, pp. 1–31.
6. 1954 'The courtship behaviour of the Cutthroat Finch (*Amadina fasciata*),' *Avicultural Magazine* 60, pp. 169–77.
7. 1955 'The courtship dance of the Mexican Swordtail,' *The Aquarist* 19, pp. 247–9.
8. 1955 'Sticklebacks as prey,' *British Journal of Animal Behaviour* 3, p. 74.
9. 1955 'The breeding behaviour of the Zebra Finch,' *Birds Illustrated* 1, pp. 28–30.
10. 1955 'The causation of pseudofemale and pseudomale behaviour – a further comment,' *Behaviour* 8, pp. 46–56.
11. 1955 'The markings of the Cutthroat Finch,' *Birds Illustrated* 1, pp. 182–3.
12. 1955 'The fighting postures of finches,' *Birds Illustrated* 1, pp. 232–4.
13. 1955 'The seed preferences of certain finches under controlled conditions,' *Avicultural Magazine* 61, 271–87.
14. 1956 'The feather postures of birds and the problem of the origin of social signals,' *Behaviour* 9, pp. 75–113.
15. 1956 'The toilet of the Cutthroat Finch,' *Birds Illustrated* 2, pp. 60–62.
16. 1956 'The function and causation of courtship ceremonies,' *Fondation Singer-Polignac, Colloque Internat. sur L'Instinct* (Paris, 1954), pp. 261–6.

17. 1957 ' "Typical Intensity" and its relation to the problem of ritualization,' *Behaviour* 11, pp. 1–12.
18. 1957 'The courtship of pheasants,' *Zoo Life* 12, pp. 8–13.
19. 1957 'The spines of sticklebacks (*Gasterosteus* and *Pygosteus*) as means of defence against predators (*Perca* and *Esox*),' *Behaviour* 10, pp. 205–36 (with R. Hoogland and N. Tinbergen).
20. 1957 'The reproductive behaviour of the Bronze Mannikin (*Lonchura cucullata*),' *Behaviour* 11, pp. 156–201.
21. 1958 'The reproductive behaviour of the Ten-spined Stickleback (*Pygosteus pungitius* L.),' *Behaviour*, Supplement 6, pp. 1–154.
22. 1958 'The behaviour of higher primates in captivity,' XVth International Congress of Zoology, Section 1, paper 34, pp. 94–8.
23. 1958 'Pictures by chimpanzees,' *New Scientist* 4, pp. 609–11.
24. 1958 'The comparative ethology of grassfinches (*Erythrurae*) and mannikins (*Amadinae*),' *Proceedings of the Zoological Society, London* 131, pp. 389–439.
25. 1959 'Shell-smashing techniques,' *New Scientist* 6, p. 148.
26. 1960 'How animals behave,' *Macdonald Illustrated Library, Nature* volume (Rathbone Books, London), pp. 190–221.
27. 1960 'Chimpanzee demonstrations at London Zoo,' *International Zoo Yearbook* 1, pp. 20–23.
28. 1961 'Experimental burrows for small mammals,' *International Zoo Yearbook* 2, pp. 70–71.
29. 1961 'Automatic seal-feeding apparatus,' *International Zoo Yearbook* 2, p. 70.
30. 1961 'An analysis of animal popularity,' *International Zoo Yearbook* 2, pp. 60–61.
31. 1961 'Active life for zoo animals,' *New Scientist* 10, pp. 773–6.
32. 1961 'Primate's aesthetics,' *Natural History* (New York) 70, pp. 22–29.
33. 1961 'Eiertikkers en eiergooiers,' *Artis* 7, no. 4, pp. 126–8.
34. 1962 *The Biology of Art* (Methuen, London), pp. 1–176.
35. 1962 'The behaviour of the green acouchi (*Myoprocta pratti*) with special reference to scatter hoarding,' *Proceedings of the Zoological Society, London* 139, pp. 701–32.
36. 1962 'A new approach to the exhibiting of small mammals in zoos,' *International Zoo Yearbook* 3, pp. 1–9.
37. 1962 'The behaviour of chimpanzees in captivity,' *Biology and Human Affairs* 28, pp. 35–40.
38. 1962 'Apes and the essence of art,' *Panorama*, Sept. 1962, p. 11.
39. 1962 'Occupational therapy for captive animals,' *Laboratory Animals Centre, Collected Papers* 11, pp. 37–42.

40. 1963 'The world of the chimpanzee,' *Science Survey* B. (Penguin), pp. 191–8.
41. 1963 'The Giant Panda,' *Panorama*, Feb. 1963, p. 10.
42. 1963 'Man's best friend – the wolf,' *Panorama*, June 1963, p. 10.
43. 1963 'De Grote Panda,' *Artis* 8, no. 5, pp. 166–75.
44. 1964 'Experimental nocturnal house at London Zoo,' *International Zoo Yearbook* 5, pp. 240–2.
45. 1964 'The response of animals to a restricted environment,' *Symposia of the Zoological Society, London* 13, pp. 99–118.
46. 1964 'Releasers,' in *The Animal World* (Nelson, London), pp. 75–9.
47. 1965 *Men and Snakes* (Hutchinson, London), pp. 1–224 (with Ramona Morris).
48. 1965 *The Mammals. A Guide to the Living Species* (Hodder & Stoughton, London), pp. 1–448.
49. 1966 'Animal behaviour studies at London Zoo,' *International Zoo Yearbook* 6, pp. 288–91.
50. 1966 *Men and Apes* (Hutchinson, London), pp. 1–271 (with Ramona Morris).
51. 1966 *Men and Pandas* (Hutchinson, London), pp. 1–223 (with Ramona Morris).
52. 1966 'The rigidification of behaviour,' *Philosophical Transactions of the Royal Society, London*, B, 251, pp. 327–30.
53. 1967 *The Naked Ape* (Jonathan Cape, London), pp. 1–252.
54. 1967 'The study of primate behaviour,' in *Primate Ethology* (ed.: Desmond Morris) (Weidenfeld & Nicolson, London), pp. 1–6.
55. 1969 *The Human Zoo* (Jonathan Cape, London), pp. 1–256.

Index

ACOUCHI, GREEN:
activity rhythm, 455
behaviour, 454–89
agonistic, 461–8
chasing, 463
crouch and freeze, 464
displacement food-burying, 472
enurination during courtship, 469, 488
fleeing and hiding, 464
foot-scuffling, 466
foot-stamping, 459, 465
feeding behaviour, 473
food hoarding, 473–86
influence of food colour, 483
relationship to hunger, 481
'scatter principle', 477, 488
hiding, 458
locomotion, 457
lost-calling, 460
lunge and trample, 463
mewing, 465
mouth-gaping, 463
nest-scooping, 458
prancing, 460
prostrating, 466
pseudomale and pseudofemale behaviour, 471
rump-hair erection, 462
scent-marking, 466
sex-crouch, 471
sexual behaviour, 468
social grooming, 460
squealing, 466
tail-wagging during courtship, 469
teeth-gnashing, 463
track formation, 457
Agonistic behaviour:
definition of term, 43
Estrildines, 411
grassfinch, 411
mannikin, 200, 411
Paroaria larvata, 204
Pope Cardinal, 204
river bullhead, 93
stickleback, 264–302
zebra finch, 43–50
Aiming of objects from animals to zoo visitors, 495
Amadina erythrocephala, sexual behaviour, 434
Amadina fasciata, sexual behaviour, 433
Amadines:
classification, 401, 451
feather ruffling during courtship, 174
Ameiurus nebulosus:
egg-incubating by, 115
parental activities, 103
Animals, captive:
companion-oriented responses, 500–503
feeding behaviour, 497, 498
homosexual behaviour, 502
'humanization', 507, 509
hypersexuality, 500
locomotory activities, 503
mal-imprinting, 507
object-oriented responses, 497
object-throwing, 495
parental behaviour, 501
response to restriction, 490–511
rigidification of behaviour, 512–16
scent-marking, 499, 501
self-mutilation, 504–6
self-oriented responses, 503–10
sexual aberrations, 502
sexual fixations, abnormal, 499
socialization, 507
visitor-oriented responses, 493
Aphyosemion coerulum, egg-burying by, 115
Aspredinids, egg-sticking by, 114
Attraction display of fish, 129

Badis badis, egg-sticking by, 114

Bathilda, 452
Beak-wiping, 190
  comparisons in grassfinches and mannikins, 445
  in zebra finch, 55, 67, 75–8
Beard feathers, display, 175
Bicheno finch:
  courtship plumage, 174
  sexual behaviour, 426
Bird of paradise, eye tuft display, 177
Birds:
  feather postures, 153–86
  head ruffling, 163
  pilomotor activity after thwarting, 171
    pilomotor system, 154
  social signals by feather posturing, 160
  temperature regulation, 154
  thwarting situations, autonomic changes due to, 169
Blackbird:
  courtship display, 144
  snail-eating behaviour, 195
Body feathers, function, 153
Boring in ten-spined stickleback, 21
Breast plumes, 175
Bullhead, river:
  agonistic behaviour, 93
  body undulation as threat action, 96
  cephalic display, 97
  courtship bite, 99, 143
    inside nest, 100
    outside nest, 98
  displacement yawning during courtship, 99
  fanning activity, 103, 107–12
  fighting, 93
  fin-raising as threat action, 96
  gill-cover raising as threat action, 94
  head-darkening as threat action, 95
    during courtship, 98
  head-lowering as threat action, 94
  mouth-digging of nest, 91
  mouth-fighting, 93
  mouth-opening as threat action, 95
  nesting behaviour, 91–3
  nodding as threat action, 95
  noises, 117
  pair formation, 102
  parental behaviour, 89, 102–12
  pectoral-fin-digging of nest, 91
  reproductive behaviour, 89–117
    comparison with other species, 112
  sexual behaviour, 98–102
  spawning preparations, 100
  tail-digging of nest, 92
  threat actions, 94
Butler's classification of Estrildinae, 402

CHAFFINCH:
  courtship behaviour, 144
    displays, 58, 59
  feather fluffing, 174
Cherry finch, sexual behaviour, 427
Cichlids:
  egg-cleaning by, 115
  pseudofemale behaviour, 35
Circulatory effects of thwarting, 180
Cockatoo, crest erection, 175
Colymbus cristatus, homosexuality in, 36
Compromise, typical, 191
Copeina arnoldi, egg-sticking by, 114
Copeina guttata, egg-nesting by, 114
Corvus corone, feather posture in agonistic situations, 174
Cottus gobio, 89–117; *see also* Bullhead, river
Coprophagy in captive animals, 498
Courtship:
  bite of river cullhead, 99, 143
  ceremonies, 128–52
  cichlid type, 147
  definition of, 128
  display, causation, 138–48
    ritualization of, 150
  function of, 129–38
  in ten-spined stickleback, 15–20
  labyrinth type, 147
  lizard type, 147
Crest feathers, erection, 175
Crimson finch:
  sexual behaviour, 418
  threat display, 163

Cubanichthys cubensis, egg-sticking by, 114

DELACOUR'S CLASSIFICATION OF ESTRILDINAE, 400
Diamon sparrow, sexual behaviour, 418
Displacement activities:
in birds, 182
in green acouchi, 472
in stickleback, 291
in zebra finch, 67, 68–71
Dove, relationship with Java sparrow, 164–7

EGG-AERATING BY BOTTOM-LIVING FISH, 115
Egg-burying by bottom-living fish, 115
Egg-carrying by bottom-living fish, 115
Egg-cleaning by bottom-living fish, 115
Egg-eating in ten-spined stickleback, 29, 32
Egg-guarding by bottom-living fish, 115
Egg-incubating by bottom-living fish, 115
Egg-nesting by bottom-living fish, 114
Egg-shedding by bottom-living fish, 113
Egg-stealing in ten-spined stickleback, 29
Egg-sticking by bottom-living fish, 113
Enurination during courtship, 469, 488
Environment, restricted, animal response to, 490–511
Erythrura, sexual behaviour, 432
Erythrurae, classification, 400, 452
Estrilda temporalis, sexual behaviour, 442
Estrildae, classification, 400
Estrildines, 42
agonistic behaviour, 411
classification, 400–403, 451
comparative ethology, 399–453
courtship dance, 219, 220
ethology, 399–453
feathers during agonistic behaviour, 411
geographical distribution, 404
morphology, 406
mouth-markings, 408
nesting behaviour, 411
parental behaviour, 444
plumage differences, 407
reproductive behaviour, 198–239
Euodice, 451

FANNING ACTIVITY OF BOTTOM-LIVING FISH, 103, 107, 115
in *Gasterosteus*, 357
in *Pygosteus*, 360, 365
of river bullhead, 103, 107–12
in stickleback, 17, 28, 310, 360, 365
Fear, feather signals in, 177
Feathers:
body feathers, function, 153
breast plumes, 175
chin-growth, display, 175
cleaning movements, 158
clumping signals, 164
display-function, origin of, 153
ear-tufts, 175
erection, degree, 156
during agonistic encounters, 174
'eye' tufts, display, 177
flank plumes, 177
fluffed, definition of, 157
movements, mechanism, 155
pilo-erection, 172
plucking in captive birds, 506
postures of birds, 153–86
terminology, 155, 157
primary function, 153
relaxed, definition of, 157
ruffled, definition of, 157
rump patches, 177
signal-function, 159–78
signals in fear, 177
in non-thwarting situations, 160
in thwarting situations, 169
sleeked, definition of, 157
temperature-regulatory function, 154
throat plumes, 175
waterproofing, in aquatic birds, 158

Feeding:
as displacement activity in zebra finch, 71
behaviour in captive animals, 497, 498
Finch, crimson:
sexual behaviour, 418
threat display, 163
Finch, cutthroat:
brooding, display when disturbed, 205
courtship posture, 189
Finches:
sexual behaviour, 415–32
see also Grassfinch; *Lonchura*; *Poephila*
Finch, zebra:
agonistic behaviour, 43–50
beak-fencing, 45
beak-wiping, 55
as displacement activity, 67, 75–8
due to sexual frustration, 67
begging display, 48, 49
call, 46
copulation, 56, 57
courtship, 53, 55, 57
diplays, derivation of, 58
posture, 54, 55
displacement activities, 67, 68–78
dominant posture, 45
feather fluffing, during thwarting, 174
feeding as displacement activity, 71
female, courtship, 55
subordination behaviour, 47, 48
fighting behaviour, 43–50
fluffing submissive posture, 47
food-begging as displacement activity, 74
homosexual behaviour, 63–8
male, courtship, 53
markings, 42, 84
subordination behaviour, 49
mandibulating as displacement activity, 73
markings, 42
and display, correlation between, 84
mounting as displacement activity, 72
redirection, 84
mouth markings of nestlings, 62
neck-stretching during nest-building, 50
nest-building as displacement activity, 74
nest-entering after courtship, 52
nesting behaviour, 50–53
communal, 78, 79
pairbond, 78–84
parental behaviour, 60
pivot-dance during mating, 67
plucking during attacks, 44
pre-copulatory ceremony, 150
postures and movements, 58
preening as displacement activity, 68
pseudofemale behaviour, 63–8, 118–127
pseudomale behaviour, 63–8
reproductive behaviour, 42–88
cycle, 61
scratching as displacement activity, 70
sex-flee conflict, 60
sexual behaviour, 53, 423–6
dimorphism, 42
thwarting, 60
shaking as displacement activity, 69
sleep as displacement activity, 70
socialization, 78
stretching as displacement activity, 69
supplanting behaviour, 44, 45
territorial behaviour, 78
vertical posture, 45
vibratory tail movements during courtship, 56, 63
yawning as displacement activity, 70
Fire-tail finch, sexual behaviour, 417
Fish:
attraction display, 129
courtship, 129
fighting behaviour, 139
Flank plumes, display, 177
Food, regurgitation and re-ingestion by captive animals, 498
Food-begging as displacement activity in zebra finch, 74
Food-hoarding by acouchi, 473–86

'Freezing' of birds during fright, 205

Gadus morrhua, egg-shedding by, 113
Galinula chloropus, homosexuality in, 36
Gasterosteidae, 245, 246
Gasterosteus:
  courtship behaviour, 334–50
  differences from *Pygosteus*, 251
  displacement fanning, 357
  egg-nesting, 114
  glueing during nest-building, 313
  homosexuality in, 33
  nest-building by pseudofemale, 35
    comparison with *Pygosteus*, 315
    material, 317–29
  nest site, 305
  nesting behaviour, 303–33
  nursery, 381
  parental cycle, 363
  reproductive behaviour, 13
    cycle, comparison with *Pygosteus*, 389–92
  sand-digging during nest-building, 311
  spine-raising behaviour, 299
  see also Stickleback
Globiosoma robustum, parental behaviour, 103
Glueing:
  in *Gasterosteus*, 313
  in ten-spined stickleback, 21, 312
Gobius microps, reproductive behaviour, 113, 130
Gordon's classification of Estrildinae, 403
Gouldaeornis, 452
Gouldian finch, sexual behaviour, 430
Grassfinch, 42
  agonistic behaviour, 411
  classification, 400, 452
  ecology, comparison with mannikin, 399–453
  geographical distribution, 404
  morphology, 406
  sexual behaviour, 415–32
Grebe, great crested:
  homosexuality in, 36
  ruff display, 175
Gull, black-headed, threat-posture, 139, 140
Haplochromis multicolour, egg-carrying by, 115
Head-ruffling in birds, 163
Hippocampidae, egg-carrying by, 115
Homosexual behaviour:
  causes, 37
  definition of, 37
  in ten-spined stickleback, 20–32
  in zebra finch, 63–8
Homosexual pattern-inversion in stickleback, function of, 28
Homosexuality:
  among captive animals, 502
  in cichlids, 35
  in great crested grebe, 36
  in moorhen, 36
  in pigeon, 36
  in three-spined stickleback, 33

INFLATION DISPLAY IN RESPONSE TO THWARTING, 180
'Intensity, typical', and its relation to ritualization, 187–97
  in non-signal situations, 195

JAVA SPARROW:
  relationship with dove, 164–7
  sexual behaviour, 432

Lonchura, 199, 401, 451
Lonchura bicolor:
  nest sanitation, 209
  sexual behaviour, 435
Lonchura castaneothorax, sexual behaviour, 441
Lonchura cucullata:
  feather ruffling during courtship, 174, 436
  sexual behaviour, 436
Lonchura ferruginosa, sexual behaviour, 222, 440
Lonchura maja, sexual behaviour, 441

Lonchura malabarica, sexual behaviour, 435
Lonchura pectoralis, sexual behaviour, 434
Lonchura punctulata:
feather posture during courtship, 174
sexual behaviour, 438
Lonchura striata, sexual behaviour, 436
Lophorina superba, eye tuft display, 177
Loricaria pava, egg-sticking by, 114

MANDIBULATION:
as displacement activity in zebra finch, 73
in sexual situations in Estrildines, 446
Mannikin, 42
agonistic behaviour, 411
ethology, comparison with grassfinch, 399–453
feather fluffing, 174
geographical distribution, 404
morphology, 406
Mannikin, bronze:
agonistic behaviour, 200
cleaning activities, 231
colony arousal, 234
copulation, 214, 225
courtship, female, 224
male, 217
dance, 219
displacement activities, 230–34
fighting behaviour, 202
'freezing' during fright, 205
frontal-horizontal posture, 202
neck-stretching during nest-building, 207
nest ceremony, 214
sanitation, 209
nesting behaviour, 206–12
one-wing-up posture, 202, 203
pair formation, 212
parental behaviour, 237
plumage, 198
pre-copulatory behaviour, 213, 216
reproductive behaviour, 198–239
roosting behaviour, 210
sexual behaviour, 212, 436
signals, 226–30
song of courting male, 217
Mannikin, chestnut, sexual behaviour, 440
Mannikin, chestnut-breasted, sexual behaviour, 441
pale-headed, sexual behaviour, 441
Mannikins, classification, 401, 452
sexual behaviour, 432–42
Masturbation in captive animals, 504
Monkey, pseudofemale behaviour in, 36
Moorhen, homosexuality in, 36
Morris's classification of Estrildines, 451–2
Mounting as displacement activity in zebra finch, 72
Myoprocta pratti, behaviour, 454–89

NEST CEREMONY OF BRONZE MANNIKIN, 214
Nest material of ten-spined stickleback 317–29
Nest-building:
as displacement activity in zebra finch, 74
by pseudofemale stickleback, 35
Nesting, communal, 78, 79
Nesting behaviour:
of ten-spined stickleback, 303–33
of zebra finch, 50–53
Nursery of ten-spined stickleback, 379

OBJECT-THROWING BY ANIMALS, 495
Opsanus beta, parental behaviour, 103
Oryzias latipes, egg-sticking by, 114

Padda oryzivora, sexual behaviour, 432
Pair formation, 149
in bronze mannikin, 212
Pairbond in zebra finch, 78–84
Paraclinus marmoratus:
egg-nesting by, 114
parental behaviour, 103
Parasympathetic responses to thwarting, 170

Paroaria larvata, agonistic behaviour, 204
Parrot finch, sexual behaviour, 432
Parson finch, sexual behaviour, 428
Pheasant:
  courtship, 240–44
  display postures, 242, 243
Pigeon, homosexuality in, 36
Pilomotor activity after thwarting, 171
Pilomotor responses, ritualization, 173
Pilomotor system in birds, 154
Ploceidae, 42
Plumage state, terminology, 155
Poecilids, egg-carrying by, 115
Poecilobrycon auratus, egg-sticking by, 114
Poephila, 400, 452
Poephila acuticuada, sexual behaviour, 429
Poephila bichenovi:
  courtship plumage, 174
  sexual behaviour, 426
Poephila cincta, sexual behaviour, 428
Poephila gouldiae, sexual behaviour, 430
Poephila guttata, sexual behaviour, 423
Poephila modesta, sexual behaviour, 427
Poephila personata, sexual behaviour, 430
Poephila phaeton, sexual behaviour, 418
Poephila ruficauda, sexual behaviour, 420
Pope Cardinal, agonistic behaviour, 204
Preening as displacement activity in zebra finch, 68
Pseudofemale behaviour, 23, 33, 36, 118
  causation, 121–7
  in acouchi, 471
Pseudomale behaviour, 118
  causation, 121–7
  in acouchi, 471
Pterophyllum eimekei, egg-sticking by, 114
Pygosteus, *see* Stickleback, ten-spined

REDIRECTION ACTIVITIES IN THWARTING SITUATIONS, 182
Regressive activities in thwarting situations, 182
Respiratory effects of thwarting, 180
Response:
  ambivalently motivated, typical intensity in, 191
  frequency, relationship to typical intensity, 187
  intensity, 187
  stereotyping, 188
Rhodeus sericeus, egg-nesting by, 114
Rigidification of behaviour in captive animals, 512–16
Ritualization, 514
  relation to 'typical intensity', 187
River bullhead, *see* Bullhead, river
Robin, threat display, 175
Ruff feathers, display, 175

SCENT-MARKING PATTERNS IN CAPTIVE ANIMALS, 499
Scratching as displacement activity in zebra finch, 70
Sexual behaviour pattern, inversion, 37
Sexual thwarting, in zebra finch, 60
Shaking as displacement activity in zebra finch, 69
Signals, intensity, relation to ritualization, 187
  pattern modification, relation to typical intensity, 193
Sleep as displacement activity in zebra finch, 70
Smoking by captive animals, 500
Snail-eating behaviour of *Turdus*, 195
Snow bunting, courtship behaviour, 144
Socialization:
  in captive animals, 507
  in Ploceidae, 78
  in zebra finch, 78
Spermestes, 199
Spice finch, sexual behaviour, 438
Spinachia, nest-building, 313
Star finch, sexual behaviour, 420

Stickleback:
attraction display, 129
courtship behaviour, 99, 100
egg-cleaning by, 115
fanning in, 17, 28, 310, 360, 365
genera, 245
review of literature on, 246
sexual frustration, 40
'sneakers', 40
Stickleback, ten-spined:
aggression, territorial, early signs, 275
aggressive behaviour during parental cycle, 370
agonistic behaviour, 264–302
responses, causation, 288
biting of prey, 253
boring by, 21
in nest-building, 310
chafing movements, 255
chasing behavour, 261
colour changes, agonistic, 266
states, 259, 260
comfort behaviour, 255
courtship behaviour, 15–20, 130–2, 334–50
consummatory act, 351
dance, 335
female, 336
indiscriminate, 22
male, 335
model experiments, 26
movements, 142
postures, 142
defence against predators, 252
displacement activities, 291
dorsal roll response, 301
eggs, cleaning, 362
eating, 29, 32
fertilization, 351
laying, 337
lodging, 361
nesting by, 114
retrieval, 361
emotional stress flickering, 252
fanning during egg-phase, 360, 365
during nest-building, 310
feeding behaviour, 253
female, frustrated, behaving as male, 28
ripe, 15
territorial behaviour, 281–8
fertilization of eggs, 17, 20
fighting behaviour, 261
fleeing, 266
food-testing, 254
'freezing', 266
gulping, 255
habitat, 13
hiding, 266
homosexual behaviour, 20–32
causes, 37
ecological significance, 31
mechanism involved, 30
pattern-inversion, function of, 28
hovering, 251
locomotion, 251
male, courtship dance, 16
frustrated, acting as female, 23, 40
territorial behaviour, 281–8
mouth-fighting, 261
nest, 13, 20
building, comparison with *Gasterosteus*, 315
during parental cycle, 373
material, 317–29
site, 304
structure, 303
nesting behaviour, 303–33
phase, 329
non-reproductive behaviour, 251–6
nursery, 379
optomotor response, 252
parental behaviour, 360–81
cycle, 363
pathological behaviour, 256
posturing, rotation, 264
vertical, 264
pre-territorial movement, 271
pseudofemale, 23, 33, 40, 118–27
reproductive behaviour, 14–20, 245–398
materials used, 248

Stickleback, ten-spined: reproductive behaviour (*cont.*)
methods of study, 248
reproductive cycle, comparison with *Gasterosteus*, 389–92
egg phase, 260
nest-building phase, 258
offspring phase, 260
organization, 385
phases, 257–60, 382
pre-territorial phase, 257
sexual phase, 258
territorial phase, 257
ritual-fanning, 336
sand-digging, 311
searching behaviour, 253
sex drive arousal during courtship, 346
sexual behaviour, 334–59
during parental cycle, 368
frustration, 21, 23
shivering during courtship, 16, 19, 336
sinking, agonistic, 266
spine-raising as conflict response, 263, 297
stretching, 255
submissive behaviour, 301
tail-digging during nest-building, 311
tail-down posture, 301
territorial defence, 267, 275
territory, male and female, 281–8
nature of, 278
threat-code, 296
'yawning', 255
Stickleback, three-spined, *see Gasterosteus*
Stretching as displacement activity in zebra finch, 69
Sweating in thwarting situations, 181
Swordtail, Mexican:
conflict response, 192
courtship behaviour, 145–8
Sympathetico-adrenal responses to thwarting, 170

TERRITORY, scent-marking by animals, 494
Throat plumes, 175
Thrush, snail-eating behaviour, 195
Thwarting:
autonomic changes due to, 169
ritualization, 182
responses to, 178–84
Tinbergen's threat theory, 139, 288, 289

URINATION ON ZOO VISITORS, 494
Urine squirting during courtship, 469, 488

WATER-SQUIRTING BY ANIMALS, 494
Waxbill, 42
classification, 400
geographical distribution, 404
morphology, 406
nest-building as displacement activity, 75
sexual behaviour, 442

Xiphophorus helleri, courtship behaviour, 145–8

ZEBRA FINCH, *see* Finch, zebra
Zonaeginthus, 400, 452
Zonaeginthus bellus, sexual behaviour, 417
Zonaeginthus guttatus, sexual behaviour, 418
Zonaeginthus pictus, sexual behaviour, 415
Zoo animals, visitor-oriented responses, 493

Latin America and the United States

SIXTH EDITION

Graham H. Stuart *Stanford University*

and

James L. Tigner *University of Nevada*

Prentice-Hall, Inc., Englewood Cliffs, New Jersey

Library of Congress Cataloging in Publication Data

Stuart, Graham Henry.
Latin America and the United States.

Includes bibliographical references and index.
1. Latin America—Foreign relations—United States.
2. United States—Foreign relations—Latin America.
I. Tigner, James Lawrence, joint author. II. Title.
F1418.S933 1975 327.73'08 74-23901
ISBN 0-13-524652-0

Printed in the United States of America.

10 9 8 7 6 5 4 3 2 1

Prentice-Hall International, Inc., London
Prentice-Hall of Australia, Pty. Ltd., Sydney
Prentice-Hall of Canada, Ltd., Toronto
Prentice-Hall of India Private Limited, New Delhi
Prentice-Hall of Japan, Inc., Tokyo

to

A. W. S.

and

R. J. T.

CONTENTS

Preface vii

CHAPTER 1. Pan-Americanism: Old and New 1
CHAPTER 2. Cooperation by Conference 51
CHAPTER 3. The Monroe Doctrine 118
CHAPTER 4. Undermining the Monroe Doctrine 144
CHAPTER 5. Anglo-American Isthmian Diplomacy 184
CHAPTER 6. Colombia, The United States, and the Panama Canal 197
CHAPTER 7. Mexico and the United States 229
CHAPTER 8. Mexico and the United States: Recent and Contemporary Relations 257
CHAPTER 9. Colonial Cuba and Its International Relations 317
CHAPTER 10. Cuba: Republican and Communist 340
CHAPTER 11. Puerto Rico—A Study in American Territorial Government 389
CHAPTER 12. American Interests in Haiti and the Dominican Republic 410
CHAPTER 13. The Dominican Republic, Haiti and the United States: Recent and Contemporary Relations 423
CHAPTER 14. Interests of the United States in Central America 475
CHAPTER 15. Recent Relations with Central America 494
CHAPTER 16. Argentina—The Making of a Nation 554
CHAPTER 17. Chile and the United States 608
CHAPTER 18. The United States and Brazil 663
CHAPTER 19. The Good Neighbor Policy and After: Development Diplomacy and Security Assistance 722

APPENDIX A. Charter of the Organization of American States (1967) 767
APPENDIX B. Organization of American States 795

APPENDIX C. The Inter-American Treaty of Reciprocal Assistance (Rio Treaty, 1947) 798
APPENDIX D. Declaration of Solidarity for the Preservation of the Political Integrity of the American States against the Intervention of International Communism (Declaration of Caracas, 1954) 804
APPENDIX E. The Act of Bogotá (1960) 806
APPENDIX F. Treaty Establishing a Free-Trade Area and Instituting The Latin American Free-Trade Association (Montevideo Treaty, 1960) 812
APPENDIX G. The Charter of Punta del Este Establishing an Alliance for Progress within the Framework of Operation Pan America (1961) 826
APPENDIX H. Excerpts from the Charter of the United Nations (1945) 840

List of Illustrations

Map of the West Indies 190
Map of Mexico 258
Map of Central America 476
Map of South America 555

Index 843

PREFACE

In the years since the fifth edition of *Latin America and the United States* appeared in 1955, there has been both an impressive development and reevaluation of Latin American-United States relations, and substantial scholarship has been published to support ideas that were largely tentative and suggestive in the mid-1950s. This revision updates the volume through the early 1970s, and attempts to balance the newer scholarship against traditional interpretations. The central focus of the book is on the origin, formation, and growth of relations between the United States and the Latin American nations. Around this focus the many aspects of relations are perceived as a multidimensional explanation of the central process. Attention has been given to main currents, significant personalities, broad interpretations, and particularly to political, economic, and social developments which so commonly help to chart the direction of a nation's foreign policy. Notable changes over previous editions include a chapter embracing the full sweep of United States-Latin American relations, a section on international organization, and a chapter devoted primarily to the evolution of the United States foreign aid programs in the region. These chapters are contributed by Professor Tigner, who has also extensively expanded and updated the rest of the book except Chapter 11 for which Professor Stuart assumes full responsibility.

This book is intended for use as a text for courses studying United States-Latin American diplomacy, and United States foreign policy and diplomatic history. It is hoped that it will also appeal to that part of the general public which is interested in the bases and development of American foreign policy. To make it the more useful for reference, the citation of the documentary sources of much of the material has been considered essential. In addition, supplementary reading lists have been appended to each chapter to give the student opportunity to obtain the background necessary for a complete understanding of the facts presented. The maps have been prepared for the volume to give the reader a clearer conception of the general features of Latin American geography, and for that reason have been made as simple as possible.

For some four hundred years (1500-1900) Latin America was commonly a storm center in the backyard of European politics. However, its role throughout the colonial period and during the first century of independence was largely passive. It was not until World War I that its governments and diplomats were inclined toward active participation in world affairs. The diplomatic relations of Latin America and the United States since the early 19th century form a large portion of the history of the New World which is, in turn, linked with the expansion of Europe. Since World War II, the historic triangular relationship—United States-Western Europe-Latin America—has expanded to include the Soviet Union, Africa, and East Asia as Latin America, through the progression of the cold war, became a significant factor in international politics. The interest of the United

States and the main focus of its diplomacy through World War II was in Mexico, Central America, and the Caribbean, a region which felt the main thrust of our Monroe Doctrine policies.

After World War II United States attention centered first on Europe, then on East Asia, the Middle East and Africa. For most of the 1950s, save for the Arbenz episode in Guatemala, the republics to the south were only a hazy image in the American public's consciousness. The region as a whole was the object of little concern or interest. Then came Fidel Castro and the Cuban Revolution, the Bay of Pigs, the Cuban missile crisis, U.S. intervention in the Dominican Republic, rural and urban terrorism, militarism and political kidnappings. These developments produced a shocked awareness of the new processes of change at work in Latin America. Within a few years many important institutions in our society became oriented towards Latin America. In an effort to forestall Cuban subversion and to shore up a severely shaken Latin American community, Washington recast some of its traditional policies and developed new policies. The Alliance for Progress, a program intended to alleviate the acute Latin American social and economic ills upon which the hemisphere crisis was blamed, was inaugurated; new military assistance and military training efforts came into being; United States businessmen searched for a new and different role; private groups and foundations launched new programs for promoting development and welfare among the other American republics. As the decade of the 1970s begins, one of the most serious problems confronting the United States in foreign affairs is the proper ordering of its relations with the other American republics.

A major contribution to research on the history of United States-Latin American relations, for specialist and student alike, is a new guide: David F. Trask, Michael C. Meyer, and Roger B. Trask's *A Bibliography of United States-Latin American Relations since 1810: A Selected List of Eleven Thousand Published References,* published at Lincoln by the University of Nebraska Press in 1968. Contained in one volume, the bibliography includes published, multilingual materials of great diversity. The books, articles, pamphlets and documents are in many languages, particularly English, Spanish, Portuguese, French, German, Italian, Russian and Japanese. Another recent addition, general in scope but containing a section on international relations, is Arthur E. Gropp's *A Bibliography of Latin American Bibliographies* (The Scarecrow Press, Metuchen, N.J., 1968). This is an updating of the second edition (1942) compiled by C. K. Jones, Hispanic Foundation, Library of Congress. Also offering a section on in ternational relations (since 1830) is Charles C. Griffin's (ed.), *A Guide to Historical Literature,* Austin, 1971. Published for the Conference on Latin American History by the University of Texas Press, it comprises selections and comments of thirty-seven respected, scholarly contributors.

The authors wish to express appreciation to Professor Martin B. Travis, Jr., State University of New York at Stony Brook, and to Professor James T. Watkins IV of Stanford University, for their criticism and advice. Our indebtedness to the many scholars whose research has provided the foundation on which this book was written cannot be individually acknowledged here, but attention has been called to some of their contributions in the bibliographical sections. Gratitude is also due to Mrs. Joyce Ball, Mrs. Joan Chambers, Mrs. Naoma Hainey, the late Dr. Helen Poulton, and Mr. Lamar Smith of the Noble Getchell Library, University of Nevada, Reno, for their indispensable assistance.

G.H.S.
J.L.T.

1

Pan-Americanism: Old and New

Latin America has traditionally occupied a special place in the foreign relations of the United States, a relationship based on geographic proximity, long historic association, and many similarities in origins. But at the same time, our recognition of the importance of these relations has shown considerable fluctuations, both in the last and present century. An evaluation of our relations with the Latin Americans today hinges upon our understanding of the development of these ties during the last century and a half, a record that breaks down into four major eras: 1810-1895 (Latin American independence, the Monroe Doctrine and Manifest Destiny); 1896-1932 (Imperialism and Pan-Americanism); 1933-1945 (the Good Neighbor Policy); 1946 to the present (the cold war and the Alliance for Progress).

The keynote of the foreign policy of the fledgling United States was sounded by President George Washington in the memorable proclamation of neutrality of 1793; and its motif was outlined in the farewell address of 1796. The new nation must not have part in the political broils of Europe. The great ocean has isolated us—"Why forego the advantages of so peculiar a situation?"[1] In the critical period which saw the beginning of the Latin American wars of independence the United States adopted the No-Transfer Resolution of 1811, the first significant step in the progression of United States policy toward Latin America. Foreshadowing the Monroe Doctrine, the resolution reflected concern that Spain might transfer Florida, which the United States desired, to Great Britain. It made clear that the United States "cannot, without serious inquietude, see any part of said territory pass into the hands of any foreign power." These and similar policy statements emphasized the value of isolation and non-entanglement and at the same time asserted the principle of the two hemispheres, the separation of the New World from the Old.[2]

[1] James W. Gantenbein, ed., *The Evolution of our Latin American Policy: A Documentary Record* (New York, 1950), p. 5.
[2] Samuel F. Bemis, *The Latin American Policy of the United States* (New York, 1943), pp. 29-30; Arthur P. Whitaker, *The Western Hemisphere Idea* (Ithaca, New York, 1954).

The first era of relations, embracing most of the nineteenth century, saw an American attitude toward the Latin Americans that was compounded of indifference and contempt. To be sure, our feelings were sympathetic toward the Latin Americans as they struggled and won independence, in the years from 1810 to 1824, and were the first to recognize the new states that emerged from that struggle. There were practical reasons of national security and trade behind the desire to see the European powers weakened or ejected from the Western Hemisphere.[3] Britain, more interested in commerce and investment than adding to her empire, held similar views. Thus the Latin American policy of the United States in this era was made possible by the British navy's control of the high seas. Meanwhile, we waged a war with Britain, starting in 1812, while our successful negotiations to purchase Florida from Spain, not achieved until 1821, made it impolitic to offend the Spaniards.

The Monroe Doctrine, delivered as part of President Monroe's message to Congress on December 2, 1823, was a strictly American statement of policy despite earlier collaboration with Great Britain.[4] Its key sentence asserted: "We owe it, therefore, to candor and to the amicable relations existing between the United States and those (European) powers to declare that we should consider any attempt on their part to extend their system to any portion of this hemisphere as dangerous to our peace and safety." It served notice upon Europe that the Western Hemisphere was closed to colonization: that was the meaning of the Monroe Doctrine initially—it was a "Hands Off" warning to the Old World. This was no self-denying pledge on the part of the United States: we fought Mexico in the 1840s and took over about one-half of her territory. The Doctrine was invoked on several occasions during the nineteenth century: against Britain's encroachments in Central America; against Spain's attempts to regain power in Santo Domingo and in Peru; against the French occupation of Mexico. But Americans took little interest in Latin Americans in this era. We of the United States were too busy with our expansion from the Atlantic to the Pacific, and at the same time we maintained our isolation.[5]

For more than seventy years following the independence of Latin America Great Britain was the paramount power in the Western Hemisphere. But being chiefly interested in capitalistic expansion in the region, Britain's only territorial acquistions in the entire period were the Falkland Islands (1833), off the southern coast of Argentina, which are still claimed by Argentina. The other British possessions in the hemisphere had been obtained before Latin American independence, including British Guiana and British Honduras (Belize) on the mainland, and certain Caribbean islands. Between 1844 and 1860 Britain exercised a protectorate over the "Mosquito Coast," approximating the Caribbean coast of Nicaragua, but this was relinquished in 1860. Britain's

[3] Arthur P. Whitaker, *The United States and the Independence of Latin America* (Baltimore, 1941), pp. 82-83; Dexter Perkins, *The Monroe Doctrine, 1823-1826* (Cambridge, Mass., 1927).

[4] For the record of the George Canning-Richard Rush conversations see Harold Temperley, *The Foreign Policy of Canning, 1822-1827* (London, 1925), pp. 115-117.

[5] The United States failed to invoke the doctrine in 1833 when Britain annexed the Falkland Islands claimed by Argentina, nor when France was engaged in the so-called Pastry Cook's War with Mexico in 1838, nor in 1838-1840 when France blockaded the Rio de La Plata.

possession of bases and fishing rights in Latin America enabled her to dominate the sea lanes of the hemisphere and to protect and maintain her trade there. From the standpoint of strategy, British influence and power were supportive of maintaining an interest in any future interoceanic isthmian canal, and to forestall the spread of American power. From the beginning of Latin American independence the feeble United States had sought to counter British influence and power; however, it was not until the mid 1890s that the United States succeeded Britain as the regional paramount power. Despite occasional armed interventions for the collection of debts, the most serious of which led to the French intervention in Mexico, the era of British hegemony was largely successful and peaceful. British policy sought to maintain Latin America as a source of raw materials and as a major outlet for British exports and investments. While serving its self-interest, Britain, after initial unavailing efforts to seize Latin America from the weak grip of Spain and Portugal, protected the region from other European powers, perhaps sparing it the fate that befell Africa and parts of East Asia.[6]

For motives that were commercial rather than political, the United States initiated in 1889 the first Pan American Conference (at which all of the American states except Canada and the Dominican Republic were present), and this conference was responsible for the establishment of the International Union of American Republics. While this movement reflected the growing interest of the United States in the peaceful development of the hemisphere, it did little to promote political solidarity, and both before and after World War I, the spirit of distrust was amply evident.[7]

The Pan American Conference was at the threshold of the second era of inter-American relations which, starting in the 1890s, saw our first and only burst of extra-continental imperialism. It was marked by a sense of national destiny and by determination to assert national prestige. Such leaders as Henry Cabot Lodge, Albert J. Beveridge, and Theodore Roosevelt were among its leading proponents. Captain Alfred T. Mahan, the naval philosopher of the new imperialism, documented their beliefs. In Washington's dispute with Great Britain in 1895 regarding the Venezuela-British Guiana boundary, they applauded Secretary of State Richard Olney's provocative statement that "Today the United States is practically sovereign on this continent, and its fiat is law upon the subjects to which it confines its interposition."

In this expansive mood came the war with Spain, labeled a "splendid little war" by John Hay. Theodore Roosevelt was one of its heroes. For about three decades after the war the United States, under the administrations of Theodore Roosevelt, William Howard Taft, Woodrow Wilson, and Calvin Coolidge, assumed a police power in Latin America, focusing its attention in the Caribbean. In 1903 Washington had a role in Panama's independence movement, Roosevelt later boasting: ". . . I took the Canal Zone and let Congress debate . . .

[6] J. Fred Rippy, *Latin America in World Politics* (New York, 1928), chapters IV-VII; and *Rivalry of the United States and Great Britain over Latin America, 1808-1830* (Baltimore, 1929), by the same author.

[7] Alice F. Tyler, *The Foreign Policy of James G. Blaine* (Minneapolis, 1927), pp. 165-170; Gantenbein, *Latin American Policy*, pp. 49-59.

while the debate goes on the Canal does also." In 1904 Roosevelt, concerned by events in the Dominican Republic, declared that "chronic wrong-doing, or impotence which results in a general loosening of the ties of civilized society, may finally require intervention by some civilized nation, and in the Western Hemisphere the United States cannot ignore this duty." This statement of policy became known as the Roosevelt Corollary of the Monroe Doctrine. The United States interfered repeatedly in Mexico and Panama; occupied Haiti from 1915 to 1934, the Dominican Republic from 1916 to 1924, and Nicaragua, sporadically, from 1912 to 1933. Washington, under prerogatives of the Platt Amendment, intervened on several occasions in Cuba. The results of these policies were not salutary for the nations affected, hampered the peaceful settlement of disputes, and caused widespread antagonism and fear among Latin Americans generally.[8]

The rationale behind the brief imperialist phase of our history was not unreasonable. There was one basic factor of obvious and crucial importance, behind the Monroe Doctrine and its corollaries, behind the imperialist impulse, and in fact underlying every aspect of our Latin American policies: the security of the continental United States. The Caribbean in general, and the isthmus of Panama in particular, were of pivotal importance in the defense of our continental republic. And by all interpretations of strategy, the United States either had control of the isthmus, and the canal when it was built, or our Atlantic and Pacific coasts would be endangered. It followed that the approaches to the Canal had to be ours or be in friendly hands. This was a factor that led us to fight Spain over Cuba and Puerto Rico, and was appealed to in the case of the interventions in Guatemala in 1954 and the Dominican Republic in 1965.

The five earlier Caribbean interventions (Cuba, Haiti, Dominican Republic, Nicaragua, and Panama) had in each instance a dual objective, apart from United States national security: to restore financial order in the country involved, and to train military forces with the hope of creating political stability. While in all five cases the immediate results seemed beneficial, the long-range results were adverse. Political power was commonly put in the hands of the military elements which made them nearly invincible. Since the bases on which to build democracy were not matured in these countries, our efforts led to results like the dictatorships of Trujillo in the Dominican Republic, Somoza in Nicaragua, and Batista in Cuba.

In the course of the American fling at imperialism, prospects for strengthening friendly ties seemed to appear and then vanish in two instances. In 1905 Elihu Root became Secretary of State, and in the following year he visited South America. The sincerity and cordiality of his statements, and the friendly attitude of the United States towards closer cooperation in the Americas, seemed to open a new era in inter-American relations. However, President Roosevelt's "big stick" and the "dollar diplomacy" of the Taft administration crushed any apparent gains. The administration of Woodrow Wilson from the outset favored closer relations with Latin America, and one of the earliest

[8] The development of the United States intervention policy is carefully surveyed in Dana G. Munro, *Intervention and Dollar Diplomacy in the Caribbean 1900-1921* (Princeton, 1964). See also C. L. Jones, *The Caribbean Since 1900* (New York, 1936); A. C. Wilgus, ed., *The Caribbean*, 17 vols. (Gainesville, 1951-1967).

statements of the president thus outlined his policy: "one of the chief objects of the new administration will be to cultivate the friendship and deserve the confidence of our sister republics of Central and South America and to promote in every proper and honorable way the interests which are common to the peoples of the two continents." But, idealist that he was, Wilson determined to bring the gospel of democracy and morality to all concerned whether they wanted it or not. The result was chaotic, especially for Mexico's internal politics and its ties with the United States. The activities of the Wilson administration from 1913 to 1917 proved to be the least excusable and most futile examples of interventionism in the history of our relations with Latin America.

World War I and the formation of the League of Nations in its aftermath brought changes that affected the relations of the United States and Latin America, and the latter's position vis-à-vis Europe through their participation in the world organization. Although Secretary of State Robert Lansing had urged "one for all; all for one," in his welcome to the Second Pan American Scientific Congress on December 27, 1915, the ultimate alignment of the Latin American nations disclosed a clear lack of unity on the question of wartime foreign policies. Of the twenty republics, eight eventually declared war on the Central Powers: Brazil, Cuba, Costa Rica, Guatemala, Haiti, Honduras, Nicaragua, and Panama. Five more broke relations with Germany: Peru, Bolivia, Uruguay, Ecuador, and the Dominican Republic. Seven remained neutral: Argentina, Chile, Colombia, Mexico, El Salvador, Venezuela, and Paraguay. Of the belligerents only Brazil and Cuba took anything like an active part in the war, while Argentina, Chile, and Mexico maintained a status of official neutrality.[9] Thus, three of the four major powers were non-participants; however, Mexico and Chile sold the Allied and Associated powers strategic materials. It should be noted that, with the exception of Brazil, all of the belligerents are located in the Caribbean region where the influence of the United States is paramount. While the Latin American belligerents did little to turn the tide of war in favor of the allies, the region as a whole emerged from the fringes of international life to the threshold of world affairs. Thereafter, their political connections with Europe were to be much closer, for thirteen of the Latin American states were entitled to take part in the Versailles peace conference, and ten of them became charter members of the League of Nations. At the same time, their economic axis shifted northward.

The First World War gave the United States an extraordinary opportunity to seize the position long-held by Great Britain as the leading trader and source of investment capital in Latin America. A prime objective of American policy in the region since the early nineteenth century was the extension of trade and investment, an interest reflected in the convening of the first Pan American conference in 1889. Our export trade with Latin America began quite successfully in the 1820s and 1830s, but it had rarely exceeded 10 percent of our total exports from that time forward until the end of the century. Illustrating dramatically the swift pace of industrialization, the value of American trade with Latin America had surpassed the British by 1913, and in 1917, owing to the European involvement in the war, the United States

[9] Percy A. Martin, *Latin America and the War* (Baltimore, 1925), pp. 1-2.

accounted for over 40 percent of both imports and exports of Latin America; the commerce of the United States with Latin America showed a gain of 400 percent between 1913 and 1920. This foreshadowed a marked change in the economic relations of Latin America, for thereafter the United States, rather than Europe, was to be the leading factor in the economic life of the southern republics. Comparable gains were made in the investment field. By 1929 U.S. investments amounted to $5,429,000,000, which exceeded the British stake by almost one billion dollars.

This supremacy brought about by Europe's misfortune and the liberal Underwood tariff of 1913 was seriously jeopardized, however, by the depression years and the super-protective Hawley-Smoot tariff act of 1930. Although after the First World War the United States had set the pace in the race towards economic nationalism, we finally learned that tariff walls keep domestic products within as well as foreign products without. Our foreign customers not only adopted equally high tariffs but invented new devices such as quotas, exchange control systems, clearing and compensation agreements, and government monopolies to strangle import trade. The trade of the United States with Latin American countries, which had increased by leaps and bounds during the First World War and had been maintained at a high level until 1930, suffered a disastrous decline. Exports from the United States to Latin America dropped from $911,749,000 in 1929 to $215,944,000 in 1933, and the imports of the United States from Latin America fell from $1,014,127,000 in 1929 to $316,040,000 in 1933. Nor could we allege the depression as the sole cause, inasmuch as the relative share of this trade going to Great Britain, Italy, Brazil, and Japan had materially increased.

The administration of Franklin Roosevelt, pledged to a policy of economic rehabilitation, was able to persuade the Congress to pass the Trade Agreements Act of June 12, 1934. The excellent results of the new liberal commercial policy were soon evident. The first of the bilateral reciprocity trade agreements was signed with Cuba on August 24, 1934, whereby mutual tariff concessions were made on a wide range of subjects. Trade statistics vividly and accurately tell the story of the beneficial results. The total trade between the two countries in the four months September to December, inclusive, of 1933, amounted to $29 million. During the same period of 1934, the first four months of the operation of the new treaty, trade jumped to $65 million, an increase of 125 percent. The success of the Cuban agreement stimulated the speedy completion of others. During the year 1935 four more reciprocity trade agreements with Latin American republics were signed, namely, with Brazil, Haiti, Colombia, and Honduras. During 1936 similar agreements were signed with Nicaragua, Guatemala, and Costa Rica. In 1937 an agreement was signed with El Salvador, the following year one with Ecuador, and in 1939 one with Venezuela. In 1939 negotiations were begun with Argentina, but the agreement, the first comprehensive commercial treaty between the two countries since 1855, was not signed until 1941. The war situation of 1942 encouraged even closer commercial relations in the Americas; and Peru, Uruguay, and Mexico entered into trade agreements with the United States in that year. When Paraguay signed a similar agreement in 1946, it brought the total number to sixteen.

The success of the trade agreement program was so clearly manifest that the act was renewed nine times by the Congress to continue in force until 1955. The statistical record of the total foreign trade of Latin America between the two World Wars indicated that a determined drive had been instituted on the part of Germany, Great Britain, and Japan for a greater share in these markets, but that the strong position of the United States had not been seriously undermined. The effectiveness of the reciprocal trade agreement program and the general friendliness engendered by the Good Neighbor policy were invaluable assets to the United States in maintaining its excellent trade position against ever keener competition. As a result, by 1953 the trade with Latin America amounted to about $3 billion in each direction.

The administration of President Warren G. Harding with the Department of State under the able guidance of Secretary Charles Evans Hughes not only went on record as seeking to support the general Latin American policies set forth earlier by Secretary Elihu Root, but made a unique record in its efforts to settle a number of outstanding and troublesome disputes in the Caribbean area. The prompt ratification of the treaty with Colombia, affording some reparation to the latter for her loss of sovereignty over Panama, made an auspicious beginning. The withdrawal of the marines from the Dominican Republic and the promise of a better situation in Haiti gave encouragement to the anti-imperialists. The decision to raise the legation at Havana to an embassy, and the naming of General Crowder for the post, was more than a friendly gesture, while the surrender of the Isle of Pines to Cuba was a real if belated contribution to the inviolability of treaty obligations. The appointment of an able administrator in the person of Judge Towner to the governorship of Puerto Rico, in place of the time-serving politician who had first received the post, augured well for an immediate improvement in the conditions in Puerto Rico. Even the impasse with Mexico was finally broken and the Obregón government recognized.

But perhaps the outstanding achievement of the Harding administration in its Caribbean policy was the Central American Conference at Washington, which not only resurrected the Central American Court of Justice but also brought about a limitation of armament on land as well as on water among the five Central American Republics. It was in his speech of welcome to the delegates of this conference that Secretary Hughes laid down certain bases of conduct governing the United States in its relations with its Caribbean neighbors which might well serve subsequent administrations: "The government of the United States has no ambition to gratify at your expense, no policy which runs counter to your national aspirations, and no purpose save to promote the interests of peace and to assist you, in such manner as you may welcome, to solve your problems to your own proper advantage. The interest of the United States is found in the peace of this hemisphere and in the conservation of your interests." The value of the statement was enhanced by the sincerity of its author.

The Coolidge administration was less fortunate in the results of its dealings with Latin American neighbors. Neither by training, experience, nor temperament was Secretary of State Frank B. Kellogg suited to handle the delicate problems which faced the United States in Mexico and in the Caribbean. The relations with Mexico required a finesse and tact which were wholly wanting in Washington. A policy of smug self-righteousness and thinly veiled threats bolstered up by the bogey of Bolshevism set in motion a strong current of bitter hostility, not only below the Rio Grande but also south of Panama. The vacillating intervention in Nicaragua swelled the torrent of ill-will. The utter failure of the Tacna-Arica arbitration was an added blow to our prestige. Criticism of American policies became so pronounced that on February 22, 1927, Senator Borah introduced a resolution to empower the Senate Foreign Relations Committee "to investigate and study conditions and policies bearing upon the relationship between Central American countries, Mexico and the United States and to visit such countries." Secretary Kellogg protested vigorously and the investigation idea was dropped: but with this change the proposal was approved by the Foreign Relations Committee.

The constant criticism of American policies in the Caribbean area brought forth a remarkable speech from President Coolidge on the occasion of the meeting of the United Press Association in New York City in April 1927. After advocating the naive theory that the press should always support an administration in its dealings with foreign powers, he laid down a rather startling extension of the Wilsonian principle that recognition be accorded only to constitutionally established governments. The Coolidge policy not only accepted this responsibility, but asserted that such recognition, by its evidence of approval, entailed the support of the United States. Having enunciated a frank policy of intervention, President Coolidge thereupon fixed the sphere of its operation in specific terms: "Towards the governments of countries which we have recognized this side of Panama we feel a moral responsibility that does not attach to other nations." If this was a blunt notice of American hegemony over its weaker neighbors to the south, it at least had the merit of confining "manifest destiny" to a definite area.

Nevertheless, the next principle laid down by President Coolidge allowed for no bounds or barriers: "Our government has certain rights over and certain duties toward our own citizens and their property wherever they may be located. The person and property of a citizen are a part of the general domain of the nation, even when abroad . . . Wherever he goes these duties of our government must follow him." Such policies, if followed to their natural conclusion, would constitute a *real politik* for the Western Hemisphere which might be very difficult to reconcile with any policy of Pan American solidarity.[10]

Fortunately, the well-conceived good-will trip of President-elect Hoover to various countries of the Caribbean and South America directly after his election in 1928 was an auspicious augury for a period of more friendly relations between the United States and Latin America. The excellent results of the 1929 Washington Conference on Arbitration and Conciliation which produced two conventions, providing for the settlement by arbitration of all justiciable disputes and the submission of all others to commissions of inquiry, furnished substantial proof of the cooperative attitude of the United States. The withdrawal of marines from Nicaragua gave even more concrete evidence to our Latin

[10]For additional background on the period 1921-1929 see Wilfrid H. Calcott, *The Western Hemisphere* (Austin, 1968), chaps. VI and VII; Max Winkler, *Investments of United States Capital in Latin America* (Boston, 1929); Charles Evans Hughes, *Our Relations with the Nations of the Western Hemisphere* (Princeton, 1928).

American neighbors that "Yankee imperialism" was no longer to be feared.[11]

United States imperialism began to fade away in the late 1920s. With victory in World War I leaving the United States so strong and Europe seriously weakened, we believed that our national security was virtually assured. Instead of imperialism, the United States turned to isolationism, disarmament, and the liquidation of commitments. The Kellogg-Briand Peace Pact of 1928, outlawing war as an instrument of national policy, seemed to preclude European intervention in Latin America and the rest of the world. This was a naive assumption, but the problem of security dictated policy. The stage was set for the Good Neighbor policy, the third phase of relations between Latin America and the United States.

President Franklin D. Roosevelt was eager to continue the policy of conciliation and friendship so that the United States might enter the forthcoming Seventh Pan American Conference with a record unmarred by any word or act which might indicate a departure from a policy of sincere cooperation. The keynote of his attitude was sounded in his inaugural address when he declared that he would "Dedicate this nation to the policy of the good neighbor—the neighbor who respects his obligations and respects the sanctity of his agreements in and with a world of neighbors." Secretary of State Cordell Hull was an able coadjutor to his chief in carrying out the Good Neighbor policy. He was able to bring about a coordinating and strengthening of inter-American peace machinery at the Montevideo Conference, and he was even more successful in his efforts to break down tariff barriers in the Americas by his program of reciprocal trade agreements. The abrogation of the Platt Amendment was another important step on the road to Pan American solidarity. The crowning achievements of the Roosevelt administration were the Buenos Aires Conference, which definitely established the principle of nonintervention among the republics of the Western Hemisphere, and the acceptance of the forward-looking Declaration of American Principles at the Lima Conference with adequate machinery to make it effective.[12]

The Good Neighbor policy of President Roosevelt was not wholly immune to congressional criticism, owing to the very considerable financial outlays made in Latin America. Senator Hugh Butler of Nebraska, who made a trip through the region in 1943, alleged that the Good Neighbor policy had become the policy of the rich uncle and that some six billion dollars had been expended, much of it profligately. Senator McKeller denied the allegations categorically and proved that the total expenditure for 1941-1943 amounted to less than one and one-half billion dollars, of which more than one billion had been used to

[11] Alexander de Conde, *Herbert Hoover's Latin American Policy* (Stanford, 1951); James B. Scott, "Pan American Conference on Conciliation," *Amer. Jour. of Int. Law*, XXIII (January, 1929), 145-148; Herbert Hoover, *The Memoirs of Herbert Hoover*, 3 vols. (New York, 1951-), pp. 210-211.

[12] Highly useful volumes on the Good Neighbor era include Thomas A. Bailey, *A Diplomatic History of the American People* (New York, 1964), pp. 683-691; Edward O. Guerrant, *Roosevelt's Good Neighbor Policy* (Albuquerque, 1950): Cordell Hull, *The Memoirs of Cordell Hull*, 2 vols. (New York, 1948); Raymond F. Mikesell, *United States Economic Policy and International Relations* (New York, 1952); E. E. Morison, *Turmoil and Tradition* (New York, 1960); Henry L. Stimson and McGeorge Bundy, *On Active Service in Peace and War* (New York, 1947); Henry J. Tosca, *The Reciprocal Trade Policy of the United States: A Study in Trade and Philosophy* (Philadelphia, 1938).

purchase strategic materials and ships. In fact, the total expenditure for non-war purposes was about $178 million.

During World War II and after its conclusion vigorous efforts were made to strengthen inter-American military cooperation. Military missions were sent to all Latin American countries which requested them, and by 1946 the United States had missions in thirteen of the American republics. More than $400 million worth of combat supplies was furnished through Lend-Lease during the war. Economic and technical assistance provided by the United States through the Institute of Inter-American Affairs and predecessor agencies in support of programs of health, sanitation, agriculture, and education in the region from 1940 to 1950 totaled about $60 million. About nine hundred experts from the United States cooperated in the agricultural sector, some three hundred professors were exchanged, and other specialists helped in technical and social welfare programs. In spite of shortages of consumer goods, the United States made a strong effort to supply the Latin American market; between 1939 and 1944 the sale of civilian commodities to the region more than doubled. The Latin Americans reciprocated by speeding up the production and shipment of strategic raw materials, the importance of which to our war effort can scarcely be exaggerated.[13]

The Japanese attack on Pearl Harbor, December 7, 1941, put the Good Neighbor policy to its severest test. It was by no means certain that the United States would win the war, nor could it insure the protection of the Americas. Yet, nine Latin-American nations promptly declared war on the Axis, and at the inter-American conference held in Rio de Janeiro in January 1942, all agreed to recommend consideration of breaking ties with the enemies of the United States. This decision was a setback for the United States which had hoped for a unanimous declaration of war, or at least a total severing of diplomatic relations, but it was nonetheless gratifying. Brazil proved to be the vital link in the Latin American defense picture. For several weeks early in 1942 her air bases offered the only route for ferrying American aircraft to Britain, the Middle East, the Far East, and Russia. Mexico despatched a squadron of fighter pilots to the Asian theater in 1945, and Brazil sent combat troops to Europe. While only Brazil and Mexico sent combat units overseas, all of the other nations but Argentina permitted a large degree of American supervision in punishing pro-Axis business firms and in dealing with saboteurs and spies. Colombia, Ecuador, and Peru allowed the use of air and naval bases, and the small Caribbean republics cooperated generously in support of Washington's defense policies.[14]

Argentina and Chile, the only nations which failed to break relations

[13]Stetson Conn and Byron Fairchild, *The Framework of Hemisphere Defense, the United States Army in World War II: The Western Hemisphere* (Washington, 1960); Edwin Lieuwen, *Arms and Politics in Latin America* (New York, 1960); William L. Langer and S. Everett Gleason, *The Undeclared War, 1940-1941* (New York, 1953), pp. 600-607; J. Lloyd Mecham, *The United States and Inter-American Security, 1889-1960* (Austin, 1961), chaps. VI-VIII); U.S. Dept. of State, *Significance of the Institute of Inter-American Affairs in the Conduct of United States Foreign Policy*, publication 3239, Inter-American Series 36, Washington, 1948.

[14]Conn and Fairchild, *Framework of Hemisphere Defense*, chaps. I-XIII; Donald M. Dozer, *Are We Good Neighbors? Three Decades of Inter-American Relations, 1930-1960* (Gainesville, 1959), chap. 4.

immediately with the Axis, were the outstanding exceptions in the picture of harmony. Steady pressure from the outside led to Chile's breaking off of relations in January 1943, but Argentina proved less tractable. As the fortunes of the Axis declined the Argentine leaders stubbornly and irrationally became more partisan towards the Axis. While selling meat and grain to the allied nations at high prices she stimulated anti-Americanism and encouraged the activities of Axis agents and propagandists. Subjected to growing pressure from Washington, refusal to provide Lend-Lease armaments, and the threat to freeze Argentine assets in the United States, the Perón regime finally yielded to outside coercion and to internal pressure, breaking diplomatic relations with the Axis in January 1944.[15] It may be concluded that during the war the Americas achieved a high degree of unity, and that the wartime coöperation of the American republics proved substantial and vital. All of the nations, even Argentina, by March 1945 had declared themselves belligerents.

In spite of these achievements, however, disturbing factors appeared. In part they could be traced to a lack of consistent and sustained leadership in the State Department's administration of Latin American affairs. Secretary of State Hull gave reluctant support to Sumner Welles, who, first as assistant secretary and later as undersecretary, directed Latin American policy from January 1934 to August 1943. Although President Roosevelt favored his essentially noninterventionist policy in the case of Argentina, Welles was forced to resign at the insistence of Secretary Hull in August 1943. The resignation of Welles coincided with the downgrading of Latin America in the mind of official Washington. Moreover, after Welles was dropped, the direction of Latin American diplomacy shifted four times in two years. With the emergence of the United States and the USSR as superpowers at war's end, it became clear to the Latin Americans that the inter-American system had become a lesser component in the world power balance than before, and that their status vis-à-vis the United States was no more than that of a junior client. While these misgivings were being expressed, another force was making itself felt: international communism.

This introduced the fourth major era in our relations with Latin America which may be labeled the cold war, or that of the preoccupied Good Neighbor, and the Alliance for Progress. Since the end of World War II the United States, under presidents Truman, Eisenhower, Kennedy, Johnson, and Nixon, has been forced to place relations with the Sino-Soviet bloc, and Cuba, before all else. Washington, until the 1960s, devoted its energies to fortifying Europe and Asia against the inroads of communism while our relations with Latin America experienced a "perilous lull." Our nation's major funds and most of our attention and commitments have been devoted to meeting the exigencies of the cold war.[16]

As the end of World War II approached, the United States sought to tighten the unity of the American republics and to bring them into the United Nations. This was achieved in a manner satisfactory to Washington at the

[15] U.S. Dept. of State, *Consultation Among the American Republics with Respect to the Argentine Situation* (Washington, 1946).

[16] William D. Rogers, *The Twilight Struggle* (New York, 1967, chapter I; Harvey S. Perloff, *Alliance for Progress: A Social Invention in the Making* (Baltimore, 1969), pp. 7-18.

Chapultepec Conference in Mexico City early in 1945. Every American nation that had declared war on the Axis would become a charter member of the world organization. Even Argentina met this qualification, though not without eliciting contempt and cynicism. Thus twenty of the original forty-six members of the United Nations organization were Latin American.[17]

President Harry S. Truman followed in the footsteps of his predecessor and supported the Good Neighbor policy after becoming aware of the criticism that the United States had mobilized and used Latin America during the war but was now neglecting it. He went personally to the Inter-American Conference for the Maintenance of Peace and Security held in Brazil in September 1947 and declared to the assembled delegates: "The United States seeks world peace—the peace of free men. I know that you stand with us. United we can constitute the greatest single force in the world for the good of humanity." Washington achieved its objective by the end of the conference: the Rio Treaty, a mutual defense agreement, which stood as a monument to the Good Neighbor policy and as a tribute to wartime cooperation.[18]

But popular interest in the United States for Latin America lagged in the wake of the war. And Washington was preoccupied with the problems of Western Europe, the stunning defeats of the Nationalists at the hands of the Communists in China, and the emergence of India and the Middle East. At the inter-American conference held in Bogotá in 1948, Secretary of State George C. Marshall had the unpleasant task of advising the other American republics that the United States foreign assistance program, then being devised for Europe, would not be shared in a comparable manner by Latin America. This came as a painful surprise to the delegations, but progress was made nonetheless. The Bogotá conference authorized the recasting of the old Pan American apparatus into the Organization of American States, a regional grouping lodged within the United Nations. Soon afterward, the outbreak of the Korean War brought major purchases by the United States of Latin America's raw materials and the distribution of additional military equipment under the Mutual Security program of 1951.[19]

While campaigning for the presidency in 1952 General Dwight D. Eisenhower charged the Truman administration with neglecting the Good Neighbor program. Shortly after his inauguration the new president despatched his brother, Dr. Milton Eisenhower, to make a survey of the Latin American countries so as to determine the possibilities of working out programs of mutual

[17]Manuel S. Canyes, "The Inter-American System and the Conference of Chapultepec," *American Journal of International Law*, XXXIX (July, 1945), pp. 504-517; *The United Nations in the Making: Basic Documents* (Boston, 1945); *The United Nations Conference on International Organization, San Francisco, April 25 to June 26, 1945: Selected Documents*, Dept. of State Publication 2490 (Washington, 1946); Manuel S. Canyes, *The Organization of American States and the United Nations*, 6th edition (Washington, Pan American Union, 1963).

[18]Gantenbein, *Latin American Policy*, pp. 822-830; U.S. Dept. of State, *Report of the Delegation of the United States of America, Inter-American Conference for the Maintenance of Continental Peace and Security, Quitandinha, Brazil, August 15-September 2, 1947* (Washington, 1948).

[19]William Sanders, "Bogotá Conference," *International Conciliation*, No. 442 (June, 1948), 385-405; Gantenbein, *Latin American Policy*, pp. 831-872.

advantage. Accompanied by representatives of the departments of State, Treasury, and Commerce, Dr. Eisenhower visited ten countries and reported that, on the whole, political relations with the United States were quite good, and the concept of hemispheric solidarity was cherished by all. Among his recommendations were the elimination of misunderstandings, fair trade and tax policies, expanded public loans and technical cooperation, and that in crises the president be authorized to make grants of food from surplus stocks to the Latin American countries. As a "good partner" the United States would entertain complaints and requests and attempt to provide and stimulate self-help.

In the 1950s, to be sure, the Eisenhower administration pushed numerous projects to increase productivity in Latin America. The United Nations made some contributions, but most were financed and guided by Washington. Financial aid through the Export-Import Bank, the World Bank, and the International Monetary Fund was aimed at curbing inflation. The United States Point IV program was also useful, as was technical assistance through the various OAS agencies and the International Basic Economy Corporation, a Rockefeller-supported institution for stimulating selfhelp in the region. These measures proved insufficient, however, and Latin America's basic problems multiplied. Between 1945 and 1957 only 2.5 percent of all U.S. aid went to Latin America, and of this one-fourth was military assistance. In the period 1945-1960 Yugoslavia, for example, received more funds from the United States than did all of the Latin American countries combined. From 1945 to 1957 less than 20 percent of World Bank loans were committed to Latin America.[20] Only the work of the U.S.-Brazil Economic Commission stands out as a step forward in cooperative development through the early 1950s.

American neglect of the region can be explained partially by the unfolding of the cold war. This period was especially critical, for Latin America was experiencing rapid political, economic, and social change. By then it was evident that the most critical need was to raise production in all economic sectors, for the population of the southern republics, which had shown a comparatively slow expansion until recent times, was rising at an unprecedented rate, among the highest in the world.

On the occasion of the Tenth OAS Conference at Caracas in 1954, Latin America sought U.S. cooperation to stabilize the falling prices of raw materials, expand capital aid, and increase technical cooperation. But Washington, concerned with erecting military defenses against communism, then exemplified by the Arbenz regime in Guatemala, failed to take the initiative in supporting Latin American development and democratic aspirations. By this time most of the democratic regimes that had risen to power since 1943 had been overthrown. Thirteen of the twenty Latin American presidents were from the military. Batista in Cuba, Trujillo in the Dominican Republic, Venezuela's Pérez Jiménez, and Colombia's Rojas ranked among the worst dictators that the hemisphere had produced. Many countries suffered from an unfavorable balance of trade, and inflation was causing widespread social unrest.

Anti-American feeling grew to dangerous proportions as revealed by events during Vice-President Nixon's riot-studded "good will" trip to South America in

[20] United Nations, *Economic Bulletin for Latin America,* XIV (New York, 1969), p. 5.

1958 when he was the object of violent demonstrations in Lima and Caracas. Soon afterward came Fidel Castro's seizure of power in Cuba and his association with the Sino-Soviet bloc. It became clear at this time that our Latin American policy was due for reappraisal.

The Eisenhower administration at first met the growing crisis with temporary measures. In 1956 the president met with the Latin American chiefs of state in Panama and proposed that his hemispheric colleagues nominate a personal representative to consider how the OAS might be made "a more effective instrument in those fields of cooperative effort that affect the welfare of the individual . . . the economic, financial, social and technical fields which our organization might adopt." The committee was formed and although its recommendations were later incorporated in the plans for the Alliance for Progress, Washington was not prepared to support any new major programs toward the cooperative solution of Latin America's mounting crisis.[21]

One of the outstanding grievances of the Latin American nations was that the United States failed to appreciate the need to stabilize the price of their exports, especially coffee. Another complaint was the difficulty in securing loans for capital development. The Eisenhower administration made some concessions on these matters by facilitating the operation of the International Coffee Stabilization Agreement and establishing the Inter-American Development Bank capitalized at one billion dollars, half of which was contributed by the United States. It increased the amount of foreign aid granted to Latin America, but not in the quantity sought by the Latin American states.

This fell far short of needs, for Washington's policy makers had concluded that the United States had already done enough, that further aid might reduce self-reliance, and that Latin America should seek private capital rather than public assistance. And as export prices declined Washington showed no interest in proposals for the economic integration of the region. This attitude changed quickly after the Nixon tour which revealed the disintegration of our Latin American policy. By this time Europe seemed to be freed from the imminent threat of Soviet invasion, allowing the United States to give a higher priority to Latin America, Asia, Africa, and the Middle East.

Although the Kennedy administration has received most of the credit for the major policy shift in Latin America, it began at the close of the Eisenhower government. The initiative, however, came from Brazilian President Juscelino Kubitschek who, in a letter to President Eisenhower on May 28, 1958, referred to the experiences of Vice-President Nixon and suggested that "we must search our consciences to find out if we are following the right path in regard to Pan America." What Kubitschek proposed was "Operation Pan America," a plan for economic and political cooperation, a "policy of ardent fraternity and indestructible continental unity." It envisioned $600 million of economic assistance a year and a concerted effort to raise living standards from $300 to $500 per capita by 1980. This was followed by the Act of Bogotá of September 1960, which recommended the establishment of an inter-American program for social development orientated toward agrarian and fiscal or tax reforms, improvement in housing, community services, health, and education, and an increase in the mobilization of national resources.[22]

[21] *New York Times*, July 23, 1956.
[22] Dept. of State *Bulletin*, XLIII (Oct. 3, 1960), pp. 536-540.

The Alliance for Progress emerged from this background in a speech by President Kennedy in March 1961 to the Latin American diplomatic corps in Washington in which he invited the twenty American Republics to join in a "vast cooperative effort, unparalleled in magnitude and nobility of purpose, to satisfy the basic need of the American people for homes, work, and land, health, and schools." In response to President Kennedy's proposal, in August 1961 the Economic Ministers of the American Republics (with the exception of Cuba) signed the Charter of Punta del Este which formulated the basic framework of a ten-year program of action for the economic and social development of Latin America. The Alliance was officially launched at Punta del Este, Uruguay, in January 1962, where a meeting of Consultation of Foreign Ministers adopted the Charter of Punta del Este. In the Charter the Latin American countries agreed to undertake the drastic social and institutional reforms necessary to insure that the fruits of economic development would be widely distributed. On its part the United States promised to provide a major part of "the minimum of $20 billion, principally in public funds, which Latin America will need over the next ten years from external sources to supplement its own efforts." Thus the Alliance for Progress was to differ markedly from previous foreign assistance programs in Latin America. The sense of the Charter of Punta del Este was that external assistance, to be of permanent benefit, must be based upon basic social and institutional reforms within the region. Furthermore, the program was conceived as a regional effort rather than a collection of traditional bilateral assistance programs. The original dimensions of the Alliance program were conceived at a gross investment of $100 billion over a ten-year period. The Latin American countries were to provide 80 percent of the financial requirements. The remaining 20 percent was to be provided equally by foreign investments and by U.S. governmental programs, the latter amounting to about $1 billion a year.

United States financial aid would constitute only a margin of Latin America's wealth, an aspect of the problem not fully understood by the American public at the time, or later, when the Alliance effort fell short of expectations. Latin America earns some $10 billion annually from its exports. Thus, $1 billion of public U.S. lending would increase the region's import capacity by approximately 10 percent. Compared with total national income, Washington's commitment at Punta del Este projected an increase of only 1 percent in Latin America's resources. United States aid under the Alliance obviously could not bring about a material increase in living standards directly, nor be fully supportive of Latin American reforms necessary to economic, social, and political growth. Washington's planners hoped to achieve extraordinary development results on this slim margin by combining aid with diplomatic, technical, and political leadership.[23]

[23] *Ibid.*, XLIV (April 3, 1961), pp. 471-478; *Alliance for Progress. Official Documents Emanating from the Special Meeting of the Inter-American Economic and Social Council at the Ministerial Level Held in Punta del Este, Uruguay from August 5 to 17, 1961* (Washington, 1961). The overemphasis of the United States role in the Alliance was clearly stated by Governor Nelson Rockefeller in a congressional hearing. See U.S. Senate, 91st Cong., 1st Sess., Subcommittee on West. Hem. Affairs, Comm. on For. Rels., *Hearing, Rockefeller Report on Latin America* (Washington, 1970) p. 30 (hereinafter referred to as *Rockefeller Report*).

OAS Secretary General Galo Plaza pointed out that in the period 1962-1969 Latin America had invested $130 billion in its own development to the United States $6.7 billion. "Of that $6.7 billion, about 4 and some billion are loans . . . Latin America has paid back . . . $2.7 billion and has paid an interest of over $750 million." See *Americas*, 21 (Aug. 1969), p. 44.

The Peace Corps, established on March 1, 1961, was an innovation in U.S. foreign aid of the Kennedy administration. Focused on Africa in the beginning, greater emphasis had shifted to Latin America by the end of 1962. By numbers, this became the largest of the Peace Corps regions, 16,000 Volunteers having served there in the first decade of the program.[24] A summary of country assignments is shown in Table 1-1.

Volunteers are assigned to rural and urban community development projects, agricultural extension work, health projects, the formation of cooperatives, and a wide variety of educational programs.

The generally enthusiastic response to the Peace Corps programs in Latin America is indicated by the host countries' requests for additional volunteers. Acceptance of the Corps was achieved by its policies of dissociation from U.S. embassy activities, and avoidance of identification with any political group. The Peace Corps concept is not entirely new, for various religious and private organizations have long performed similar services. The American Friends, as well as the Volunteer Service Overseas Movement in England, provided an example for this official venture by the United States. By mingling with people in all social levels in Latin America the volunteers have helped to dispel the widespread impression that the North Americans associate only in their own enclaves and have little contact with local inhabitants.

In the minds of most Americans the Peace Corps is a concrete and creative example of the American desire to help other nations help themselves. It was designed to have a catalytic effect, stimulating self-help and community initiative in Latin America. The Peace Corps has done its job well, but it cannot be expected to solve the multitude of complex and deeply-rooted problems of the region. It is probable that the contributions of the Peace Corps must be judged ultimately not so much by its immediate results as by the extent to which it stimulated self-help movements by the Latin Americans themselves.

President Kennedy made Latin America a major issue in his 1960 campaign, and after launching the Alliance for Progress in 1961 he identified himself closely with the program. In spite of a decline in United States prestige, arising from the abortive Bay of Pigs invasion effort, which took place at a time when the echoes of his speech proposing the Alliance for Progress had hardly died down, he won the confidence and affection of Latin Americans to an extraordinary degree and sought to develop an Alliance spirit by annual visits to Latin America. In December 1961 he went to Caracas and Bogotá, where he and his wife were enthusiastically received. The president and his wife visited Mexico City in June 1962, and were given a tumultuous reception, said to surpass that given any foreign head of state. He hailed the Mexican Revolution and its ideals as an early precursor of the Alliance for Progress. His handling of the Cuban missiles crisis in 1962 repaired some of the damage to the U.S. image in the 1961 Cuban invasion. Going to Costa Rica in 1963, where he met with the presidents of the six isthmian countries, he stressed the political aspects of the Alliance:– national democratic government, political liberty, and social justice. Latin American statesmen were invited to the White House, each visit reflecting the president's desire to establish a personal working relationship with leaders who

[24] See *Peace Corps Report*, published annually.

Table 1-1. Peace Corps: Volunteers in country at end of fiscal year*

	1962	*1963*	*1964*	*1965*	*1966*	*1967*	*1968*	*1971*
Brazil	43	168	210	458	639	601	580	297
Chile	63	99	106	294	397	392	254	73
Colombia	103	229	561	544	506	522	576	145
El Salvador	25	21	49	55	51	105	119	69
Jamaica	38	32	62	77	70	101	117	177
Eastern Caribbean Islands	15	14	17	5	45	89	124	163
Venezuela	23	83	117	265	292	352	262	190
Bolivia	35	112	126	220	266	303	219	58
British Honduras		33	18	49	33	42	45	37
Costa Rica		26	65	61	107	154	98	85
Dominican Republic		144	171	85	101	140	161	61
Ecuador		156	236	309	211	255	247	136
Guatemala		27	105	83	69	140	151	71
Honduras		27	46	103	107	174	167	115
Panama		28	76	133	196	171	174	1
Peru		285	293	379	301	349	283	206
Uruguay			18	4	48	65	31	7
Guyana						44	51	–
Paraguay						35	56	69
Nicaragua								66
	345	1,484	2,276	3,214	3,439	4.034	3,715	2,026

**Peace Corps Seventh Annual Report* (Washington, 1968), p. 30; Peace Corps, Office of Staff Placement, *Bi-Annual Statistical Summary, June 30, 1971* (Washington, 1971), pp. 11-17.

were earnestly pushing democracy and reform.

President Kennedy's emphasis upon the United States commitment to democratic reform and social justice was revealed on many occasions in addresses to labor groups, business organizations, and Latin American student groups, and he rarely failed to mention Latin America in his press conferences. In one of his last major addresses at a meeting of the Inter-American Press Association in Miami, on November 18, 1963, he reaffirmed his faith in the program: " . . . I support and believe in the Alliance for Progress more strongly than ever before . . ."

President Johnson met with the ambassadors from each of the Latin American countries on November 26, 1963, the day after President Kennedy's funeral, to assure them that the United States would continue to honor its commitment under the Alliance. He said: "Let us make the Alliance for Progress a living memorial to President Kennedy." He thereafter sought to reassure Latin Americans of this resolution and the firmness of U.S. support. In May 1964 he reviewed Alliance gains during the first six months of his administration and signed loan agreements for projects in thirteen countries. He emphasized his nation's role in the Alliance through 1964-65 in meetings with the diplomatic corps in Washington and U.S. ambassadors to Latin America. The president could cite an impressive list of physical achievements made possible by the Alliance by the fifth anniversary of the program in 1966.[25] It began to appear at this point, however, that the Johnson administration laid greater stress on tangible accomplishments such as buildings and roads, than on reforming the institutional structure. Doubtless the earlier optimism about political change had been blighted by military intervention in Peru and Argentina, the overthrow of Juan Bosch in the Dominican Republic, and the ominous trends in Brazil. State visits to Washington by Latin American presidents declined between 1964 and 1967 and the presidential press conferences had fewer references to the region. Except for television appearances in which the president defended U.S. intervention in the Dominican Republic in the late spring of 1965, the period 1965-1966 showed a decline of Washington's concern with Latin America. At the same time criticism was rapidly mounting, inside and outside Latin America, of the Alliance's failure to attain its original objectives.

Secretary of State Dean Rusk led the U.S. delegation to the meeting of the Inter-American Council in Rio de Janeiro in November 1965 where it was announced that the Alliance would be extended beyond the ten-year period originally proposed by President Kennedy. In April 1966 President Johnson went to Mexico City, his first official visit to Latin America since coming to office. While there he declared himself in favor of attending a summit conference of hemispheric presidents which had been suggested by President Illia of Argentina.

The Summit Conference of American Presidents at Punta del Este in April 1967 was largely planned and organized by the United States. And the personal presidential diplomacy gave promise of more direct U.S. involvement in the Latin American crisis and expanded leadership. On March 13, 1967, President Johnson dispatched a message to Congress calling for additional U.S. support for

[25] *New York Times*, Aug. 18, 1966.

Latin American development, especially to assist the region to move toward a common market. The president proposed a resolution for the new aid, but its defeat was engineered by Senator J. William Fulbright who contended that the Senate's traditional appropriation procedures would be undermined if Congress committed itself by accepting the resolution device. The president had sought ratification for his forthcoming pledge of U.S. assistance to Latin American integration. At the conference President Johnson remained in the background, at least publicly, and met each chief of state privately. It appeared that the president's cordiality, sincerity, and frankness did much to remove the distrust and suspicion that had arisen in connection with the Dominican intervention. The major emphasis of the conference was on the more concrete problems of integration and international trade rather than upon reform.[26]

As the Johnson administration neared its close the record showed that in the years 1961-1968, the United States had provided $9.2 billion of assistance in accordance with its original commitment as follows: AID, $4.1 billion; Public Law 480, $1.3 billion; Export-Import Bank, $1.8 billion; and a $2 billion contribution to the Inter-American Development Bank. Other foreign investment totaled $6 billion and total gross domestic investment by the Latin American countries approximated $129 billion in the first seven years of the Alliance.[27]

The administration of President Richard M. Nixon began amid growing strains in United States and Latin American relations. While not a popular figure, having been identified with the much-criticized Eisenhower government's aid and trade policies in the region, his public statements showed an awareness of the need for a change in hemispheric policies, particularly in the economic field. Soon after he had returned from the tour of South America in 1958 in which he and his party had narrowly escaped death at the hands of howling mobs in Caracas, he had suggested that "we develop an economic program for Latin America which is distinctly its own . . . There must be a new program for the hemisphere . . . ," and other recommendations which he carried into the election campaign of 1960.[28] Possibly in anticipation of his nomination in the presidential campaign of 1968, he kept abreast of developments by a two-week tour of South America in May 1967. Speaking in Chicago after his return, Nixon said that Latin American countries lagged badly in government, economic, educational, and agrarian reform despite the $7 billion in United States aid. Until they agree to work toward modern institutions and to discard entrenched inefficiency he said, the United States should "pour no more money" into Alliance programs.

The Republican party's platform on which Mr. Nixon was elected to the presidency in 1968 stressed the need for "closer economic and cultural cooperation of the United States and Latin American countries," and pledged

[26]U.S. Dept. of State, *Commitment for Progress: The Americas Plan for a Decade of Urgency* (Washington, 1967).

[27]U.S. Cong., Committee on Gov't. Operations, 90th Cong., 2nd Sess., *U.S. Aid Operations in Latin America under the Alliance for Progress* (Washington, 1968), p. 3.

[28]Richard M. Nixon, *Six Crises* (New York, 1962), pp. 229-230; Los Angeles *Times, The Nixons in South America* (Los Angeles, 1958).

that "we will encourage in Latin America the progress of economic integration to improve opportunity for industrialization and economic diversification." On January 21, 1969, his first day in office, the new president met with Galo Plaza Lasso, secretary-general of the OAS, and former president of Ecuador. At this meeting Galo Plaza urged the president to send a top-level, fact-finding mission to the OAS countries. When asked for suggestions, he proposed that Governor Nelson Rockefeller of New York, an expert in Latin American affairs, and surely one of the most popular North Americans in the region, be asked to head the mission. Governor Rockefeller accepted the proffered assignment, but the mission did not get underway until May.[29]

The Nixon administration was criticized in Latin America, as well as in domestic circles, for the delay in naming a Latin American team. For the top post, Assistant-Secretary of State for Inter-American Affairs, President Nixon named Charles A. Meyer, who took over his duties shortly before April 1. Meyer moved into the position from a background as executive vice-president of Sears Roebuck and Company, and a director of United Fruit, Dow-Jones, and the Philadelphia National Bank. John H. Crimmins, a career foreign service officer with wide experience in the Caribbean area, was named Deputy Assistant-Secretary for Inter-American Affairs. Viron P. Vakey, offering a similar background was appointed White House adviser on Latin American affairs. The appointment of Joseph J. Jova, also a foreign service officer, to the post of U.S. Ambassador to the OAS, filled the administration's top positions on Latin America. Jova succeeded Sol M. Linowitz, who, postponing his resignation, remained at the post until May 1 at Mr. Nixon's request.

In what appeared to be the foundation of the new president's Latin American policy, Governor Rockefeller and a plane-load of experts set out on a goodwill and fact-finding tour of Latin America in May 1969, "to listen, to look;" and to report to Mr. Nixon what he had found. The president called it "one of the most vitally important missions ever undertaken by an independent group on behalf of the United States." Causing the mission to loom even more important, both Mr. Nixon and Mr. Rockefeller had downgraded the results of the Alliance for Progress. Although given a rousing send-off, the mission was soon deep in controversy. To Latin Americans the four-stage Rockefeller mission seemed to be just another of a host of study missions sent out by American presidents to determine the needs of Latin America. It seemed impossible to them that in a tour of such brevity he could come up with an adequate report on Latin America's needs.

Governor Rockefeller's mission, which was divided into four tours beginning on May 11, took him to Mexico, Honduras, Guatemala, Nicaragua, Costa Rica, El Salvador, and Panama. On May 27 he left for Colombia, Ecuador, Bolivia, and Trinidad-Tobago. The third leg, to Brazil, Paraguay and Uruguay, began June 16; and he left on June 27 to visit Argentina, Haiti, the Dominican Republic, Jamaica, Guyana, and Barbados. Public disorders protesting his coming forced the cancellation of visits to Chile, Venezuela, and Peru, and he cut his stay in Bolivia to three hours. Senator Frank Church (D., Idaho),

[29] *New York Times*, Aug. 5, 1968; *Christian Science Monitor*, June 13, 1969. For position papers of Richard M. Nixon and Hubert H. Humphrey on Latin America see *New York Times*, Nov. 5, 1968.

chairman of the Senate Foreign Relations Subcommittee on Western Hemisphere Affairs, was outspokenly critical of the tour.[30]

The Rockefeller mission recalled the earlier visits to Venezuela by then Vice-President Nixon, and Senator Robert F. Kennedy. Mr. Nixon, visiting there in 1958 narrowly escaped bodily harm, and Senator Kennedy, during his Latin American tour in 1965, encountered a situation similar to that faced by Governor Rockefeller. Although the Venezuelan government recommended that he should not come, Senator Kennedy made the unofficial visit, but shortened it by two days.

Governor Rockefeller called the demonstrations the work of "subversive elements" directed in some instances from outside the hemisphere. However, when he met Mr. Galo Plaza Lasso in Trinidad the latter expressed the opinion that the student protests were not solely the work of an extremist minority, declaring that "there is a deep feeling throughout Latin America at all levels in society that relations with the United States need a change . . . Whatever interest outside groups may have, there is a genuine national resentment that doesn't need outside influence." Later Mr. Rockefeller indicated that the demonstrations actually "evidence the frustration that exists and the need for action." Secretary of State William Rogers affirmed this view in June 1969, declaring "I think I should also say obviously the demonstrations show some discontent with our relationships and we have to recognize that they need to be improved.[31]

Before the Rockefeller tour was launched representatives of the Latin American nations, meeting at Viña del Mar, Chile, prepared the Concensus of Viña del Mar, a document calling on Washington to make basic alterations in its economic policies. It was delivered to President Nixon by Chilean Foreign Minister Gabriel Valdes. The so-called Concensus of Viña del Mar proved to be the key document in the two-week session of the Inter-American Economic and Social Council held in Port-of-Spain, Trinidad, in May. The situation presented the United States with a serious Latin American policy dilemma, for the document in some measure challenged Washington to do more to aid in solving Latin America's critical economic stagnation.

The main point raised in the document was that the United States had been helping itself far more in Latin America than it has been helping underdeveloped countries in the region. The Latin American nations called on the United States for greater economic assistance through easier access to the American market for their goods. They also condemned the Washington policy of "tied loans" which oblige a recipient nation to use money in the United States where it often does not go as far as elsewhere because of high American prices. The achievement of objectives of development, the document asserts, "depends in large measure upon whether all countries of the inter-American community, and especially the United States of America, recognize and assume their responsibilities and obligations." Noting that "the economic and the scientific and technological gap between the United States and Latin America has widened and continues to widen," the Latin Americans contend the United States shares major responsibility in correcting this situation.[32]

[30] *New York Times*, July 7, 1969; *Christian Science Monitor*, June 14, 1969.
[31] *Christian Science Monitor*, June 13, 1969.
[32] *Ibid*., June 20, 1969; *Americas*, 21 (Aug. 1969), pp. 42-43.

Secretary Charles A. Meyer, who headed the U.S. delegation to the Trinidad meeting, was called upon to assuage the growing Latin American resentment. Acting on instructions from Washington, he announced the immediate discontinuance of the practice called "additionality." This is the requirement in aid regulations–forced by the balance-of-payments problem–that for every dollar given to Latin America one has to be spent for American products from a certain list. Latin Americans have long complained about this requirement, since they could buy most goods they needed cheaper elsewhere, and because it forced them often to buy goods they did not really need thus negating the purpose of aid. This action by President Nixon was seen as a first step by Washington to meet some of the Latin American complaints about Washington's overall policy on aid and trade.

President Nixon's long-awaited policy statement on Latin America was delivered at an Inter-American Press Association dinner in Washington on October 31, 1969. On the surface it appeared to move the United States away from open support of major economic and social change in Latin America and into what the president sees as a ". . . new kind of partnership" in which the United States "lectures less and listens more." The president did not indicate that the basic U.S. commitment to the Alliance of Progress would be changed, but he made clear that he expected to develop his own Latin American program, saying that the 1970s should be a decade of "Action for Progress for the Americas." To attain this he repeatedly called for a "partnership" between the United States and Latin America. An implied aspect of the president's speech was the effort to scale down the silhouette of the United States in the region, perhaps because of the growing nationalistic trend in the hemisphere, aimed commonly at the United States.

The theme of "partnership" was evident in a number of specific points mentioned by the president. For instance, he declared that hereafter the Latin American republics will be allowed to use Washington's financial aid for purchases not only in the United States, "but anywhere in Latin America," which reaffirmed the earlier policy change on "additionality." He proposed that the Latin Americans would "jointly assume a primary role in setting priorities within the hemisphere, in developing realistic programs, and in keeping their own performance under critical review." The president also promised to see what could be done to improve the region's serious trade imbalance. And to indicate his interest in Latin America he announced his intention of upgrading the Latin American post in the Department of State to the status of undersecretary, a proposal which, however, was not favorably considered by the Congress.[33]

President Nixon's speech drew immediate criticism from many sources in Latin America: government officials, politicians, newspapers, and radio commentators. Criticized in particular was the President's implied rejection of the social and economic aims of the Alliance for Progress, and his apparent downgrading, in some respects, of Latin America as an area vital to the United States. Both concepts figured prominently in U.S. foreign policy during the Kennedy and Johnson administrations. Some of the sharpest criticism came

[33] *Christian Science Monitor*, Nov. 3, 1969; *Ibid*., Feb. 8, 1970.

from nations considered the most democratic, such as Chile and Venezuela, whereas the most favorable reactions came from nations ruled by the military. The Latin Americans wanted more favorable treatment in the U.S. market for their goods as well as numerous changes in the foreign aid program: increased aid with fewer conditions attached, and the multilateralization of aid through world and hemisphere bodies. It was not surprising that some of the economic quirks of the Alliance should have come under attack. As Governor Rockefeller noted while in Brazil, that country in 1969 was paying off $500 million to the United States on debts: $300 million on the principal and $200 million in interest. This condition makes it difficult for countries to advance, for they go only deeper into debt, the more aid they accept. By 1969 75 percent of the bank funds coming into Latin America were used to pay off old loans.[34]

Some critical comments arose from the President's failure to confront Congress on aid and trade issues. The total Nixon aid package for Latin America amounted to $605 million, one million dollars less than President Johnson's allocation for the fiscal year ending June 30, 1969. Expressing some general sentiments, the Mexican ambassador to Washington declared ". . . we encounter here on a daily basis contradictory actions that show a different reality—import quotas, high tariffs, taxes that offset incentives we offer United States investors in our countries." And more complaints were heard that American investors take out of Latin America more profits than they put in by new investments and the U.S. government's grants in aid funds. There was some justification for this protest because by Washington's own admission, U.S. corporations took $106 million out of Latin America than was put in by aid in 1967.[35]

The president said that he would "lead a vigorous effort to reduce the non-tariff barriers to trade maintained by nearly all the industrialized countries against products of particular interest to Latin America and other developing countries." But this statement did not satisfy Latin Americans who sought preferential treatment in the U.S. market and rejected the idea of equal status with other less advanced countries. On the topic of intergovernmental relationships the president's statement of a policy to "deal realistically with the governments as they are," drew criticism because it seemed to condone military rule. Governor Rockefeller's report on Latin America raised further controversy on this subject, for he recommended increased military aid to Latin American countries to forestall Communist subversion, which he declared has "alarming potential." He also indicated that the military generally represents the best interest of a people in a given country "although it is not widely recognized."[36]

It had become clear by the end of the third year of the Nixon administration that Latin America did not rank high in Washington's list of priorities, and criticism mounted both inside and outside Latin America as relations with the region deteriorated. Washington's indifference could be

[34] *San Francisco Examiner and Chronicle*, June 29, 1969. See also *Rockefeller Report*, pp. 7-8.

[35] *Christian Science Monitor*, Nov. 7, 1969; *Ibid.*, June 27, 1969.

[36] *Ibid.*, Nov. 13, 1969; *Rockefeller Report,* pp. 8-19; 35-37; 44-45; 85-87. President Nixon's State of the World Message amplifies his administration's Latin American policies. See *New York Times*, February 19, 1970.

attributed in large measure to the absence of a direct confrontation with the Communist world in Latin America, except for the continuing stand-off with Cuba. The election of a Marxist president in Chile, Dr. Salvador Allende Gossens, created unease in Washington, but a tangible threat from that quarter had not materialized. However, at the time that Washington was lowering its presence in the region Moscow and Peking were expanding their trade relations with South America. Except for Paraguay, the USSR had diplomatic relations with all countries on the continent by 1972. Peking, though lacking in embassies, carried on an impressive trade drive, with the expectation of establishing permanent trade missions in Peru, Chile, and Guyana.[37]

Prominent among the critics was Juscelino Kubitschek, former president of Brazil, whose Operation Pan America had influenced the planning for the Alliance for Progress. He declared that "Kennedy made a profound psychological error in creating the Alliance. He should have consulted the Latin Americans but, in effect, he told them: 'I will do this.' Johnson forgot Latin America except the Dominican Republic. Nixon won't even hear of it. No popular feeling exists here that the United States really wants to help this continent."[38]

Within the U.S. Congress Senator Frank Church (D., Idaho), Chairman of the Senate Western Hemisphere Affairs Subcommittee, deplored that little, if anything, had been accomplished by the Alliance for Progress, saying "progress, though visible, is illusory." Church, who had voted against the foreign aid bill when it was soundly defeated in the Senate, contended that private U.S.-owned companies operating in Latin America withdrew "$2 in dividends for every dollar they invest . . . Latin Americans have become convinced that they are the victims of a virulent new imperialism."[39]

With a view toward halting the drift in relations with the southern republics presidential adviser Robert H. Finch completed a fourteen-day tour of six Latin American nations on behalf of President Nixon in November 1971. The mission included stops in Peru, Ecuador, Argentina, Brazil, Honduras, and Mexico. During his travels he told audiences that the "purpose of his visit was to "build bridges of understanding" and to open roads of friendship." With U.S. foreign aid programs in a "state of flux" he said the trip would provide "a unique opportunity for us to get inputs from Latin America." He also made a point of saying that President Nixon understood their problems.[40]

While these statements were intended to be reassuring they must have been outweighed by at least two events in 1971: the U.S. Senate's stunning reduction of foreign aid and the Nixon administration's protectionist policy, notably the 10 percent surcharge on imports. The latter was particularly resented since the United States then enjoyed a $900 million advantage in its balance of trade with Latin America. U.S. policies were denounced at the Lima meeting of the "Group of 77," actually ninety-five nations, representing an informal association of

[37] *New York Times*, Jan. 28, 1972.
[38] *Ibid.*, Apr. 16, 1971.
[39] *Times of the Americas*, Nov. 17, 1971.
[40] *New York Times*, Dec. 2, 1971; *Times of the Americas*, Nov. 24, 1971.

"third world" countries, as well as by the 12th extraordinary meeting of the Inter-American Economic and Social Council (CIES), which ended January 20, 1972.[41]

In defense of Washington's position, Secretary Meyer declared that Washington had worked toward a policy of "mature partnership" with the Latin American countries; "we now deal with the Latin American governments as they are which is what Latin America wants." He described the Nixon policy as a living policy: "... it has been completely misunderstood because it is not a pyrotechnic policy, but a sound policy for now and for the future, responsive to the nations of Latin America." But in the area of "Latin America's highest priority," a system of general trade preferences, Meyer said "we have not fulfilled our commitment." Further light was cast on the administration's policies by Peter G. Peterson, President Nixon's Assistant for International Economic Affairs, who said that Washington's relations with Latin America are under continuing stress "... as these countries develop and their political structures evolve with wider participation in economic and political decisions." Contending that some of the problems in inter-American relations are traceable to misunderstandings about the U.S. role in Latin America he said "it is is our own interest as well as that of Latin America to multilateralize the economic relations of Latin America with the world in order to bring more aid and investment from other developed countries, as well as the U.S. and open markets in other countries for the products of Latin America."[42]

With the abolition of the 10 percent surtax on December 21, 1971, it seemed that U.S.-Latin American ties might improve in 1972, and further optimism was raised by the announcement that Washington would advocate generalized trade preferences for developing nations. This was quickly dashed, however, when President Nixon requested $224.5 million in economic and development assistance for Latin America for fiscal year 1973, which meant that Latin America stood to share less than 10 percent of the $3.2 billion in foreign aid requested of Congress by the Nixon administration. Designed to finance Alliance for Progress development loans, the figure of $224.5 million did not include military and internal security assistance for the regiôn. Compensating for the Alliance for Progress reduction, the request included a four-fold increase in funds requested for the Inter-American Development Bank, up from $262 million requested in 1972 to $837 million in 1973. This request and the one for the previous year were taken to indicate that the President was determined to make the IDB the chief vehicle for U.S. aid in Latin America. Such an approach was consistent with his policies in Asia and elsewhere, where aid was being channelled into regional development banks.

After an eight month delay Congress on March 2, 1972, approved $2.6 billion in foreign aid for fiscal 1972. The Alliance for Progress received $230 million ($80 million in technical aid and $150 million for development loans), and the Inter-American Development Bank was allocated $211,760,000. Meanwhile, prior to delivering his State of the Union address, President Nixon

[41] *Ibid.*; *Ibid.*, Feb. 2, 1972.
[42] *Ibid.*, Jan. 12, 1972.

gave harsh warning that thereafter his administration would retaliate against any nation which took over a U.S. firm without paying "prompt, adequate and effective compensation." Under such circumstances the United States would terminate all bilateral aid programs and withhold approval of loans being considered at international financing institutions, for example, the World Bank, and the Inter-American and Asian development banks. It seemed that the new policy had particular relevance in Latin America. While assuming a strong posture on this question, the Nixon administration had earlier announced a pragmatic policy on the allocation of foreign aid. Secretary Meyer, speaking before the House Inter-American Affairs Subcommittee, said that the United States "will deal with governments as they are . . . we should not seek to burden our assistance programs with the task of molding the hemisphere's political choices to our images or moral preconceptions."[43]

Other sources of irritation in inter-American relations faced by the Nixon administration included territorial waters claims, the nationalization of property owned by American citizens, the kidnapping and murder of diplomats, and the hijacking of aircraft. Because President Nixon's Latin American policies were slow in appearing, some of these difficulties took on greater urgency as the decade of the 1970s began.

The first problem mentioned dated from 1947 when Peru and Chile laid claim to seas adjacent to their coasts for 200 nautical miles. They were joined by Ecuador in 1952 when all three countries signed the Santiago Declaration on the Maritime Zone. This was essentially a claim to resources, but it was later interpreted as both a resource claim and as a territorial sea claim. The United States recognizes a three-mile territorial sea and, in addition, claims a nine-mile contiguous zone of exclusive jurisdiction over fisheries. Two international law-of-the-sea conferences, one in 1958 and one in 1960, had failed to resolve the basic issues on territorial seas and resource jurisdiction. Another such conference was planned for 1973.

Almost one-hundred boats of the California-based tuna fleet were seized and fined by Ecuador and Peru in a fifteen-year period for fishing within the 200-mile limit. In 1971 the Ecuadorean navy made fifty-one seizures of American fishing vessels. The vessel owners paid a total of $2.4 million for the forced purchase of licenses and fines to pay their release. The amount was to be reimbursed under the terms of the Fisherman's Protective Act. In January 1971 Washington announced the suspension of military sales and credits to Ecuador in accordance with the Foreign Military Sales Act. Ecuador quickly responsed by appealing to the OAS to consider charges of economic aggression and by ordering the expulsion of the U.S. Military Group. The United States, Chile, Ecuador, and Peru had met in the first session of the Quadripartite Fisheries Conference at Buenos Aires in 1969 and 1970; it was to have reconvened no later than July 31, 1971. However, negotiations for the Conference were not resumed because Ecuador insisted that the measures applied by Washington in January 1971 must be lifted.

By 1972 ten Latin American countries claimed a 200-mile territorial limit.

[43] *Congressional Quarterly*, XXX (Mar. 4, 1972), pp. 462-469; *Times of the Americas*, Feb. 2, 1972.

In addition to Chile, Ecuador, and Peru there were El Salvador, Nicaragua, Panama, Argentina, Uruguay, and Brazil; Costa Rica had a 200-mile conservation zone. Of all these states, the disputes which concerned the United States involved only Ecuador, Peru, Chile, and Brazil. At stake is the traditional doctrine of freedom of the high seas, which has been in force since the early nineteenth century.[44]

Washington, after deciding that aid to Ecuador was vital to the U.S. national interest, refused to recognize a prohibition against economic assistance to Ecuador, which had been added to the fiscal 1972 foreign aid bill by Congress. United States interests to be considered were the NASA tracking station near Quito, the emergence of Ecuador as an oil producer, the consolidation of the Andean Regional Trade Pact, and the maintenance of cooperation with Ecuador both bilaterally and within the inter-American system.[45]

Evidence of a mounting nationalist surge in Latin America could be seen when the governments of Peru, Chile, and Bolivia took definite action against foreign-owned properties. Peru's military government headed by General Juan Velasco Alvarado in October 1968 seized the International Petroleum Company (IPC), the Peruvian subsidiary of the Standard Oil Company of New Jersey; the government also "intervened" in all the sugar-growing areas of the nation, including the holdings of W. R. Grace and Company. Peru's expropriation of IPC was based on the charge that the United States had committed "economic aggression" against Latin America. Under the circumstances Peru became subject to action under the Hickenlooper Amendment to the Foreign Assistance Act of 1962 which requires that the U.S. terminate aid to any country that confiscates American-owned property without paying fair compensation. Besides the suspension of aid and the suspension of the sugar quota, Peru faced another possible action: since the United States does not recognize Peru's claim to two-hundred miles of coastal waters and American fishing vessels have been seized, fined, and fired upon, U.S. law calls for the recall of our military ships on loan to Peru.

The assets of the Bolivia Gulf Oil Company, a U.S.-owned concern, were nationalized in October 1969 shortly after General Alfredo Ovando Candia seized power. Bolivia agreed to pay indemnities, but the Gulf Oil Company declared that the amount fell short of the property's value, and pressed for application of the Hickenlooper Amendment. Colonel Hugo Banzer, who seized control in 1971, stood firm against returning the Gulf properties expropriated by his leftist predecessor; however, he indicated that nationalization would cease under his government, and continued negotiations with Gulf on the questions of a financial settlement and the marketing of crude oil.[46]

Although differing from the situation in Peru and Bolivia because of a negotiated settlement with compensation, Chile's reformist government headed by Eduardo Frei Montalva moved another step toward the "Chileanization" of

[44] Dept. of State *Bulletin*, LXVI (Feb. 28, 1972), pp. 284-287.
[45] *Times of the Americas*, Mar. 22, 1972.
[46] *Ibid.*, Mar. 29, 1972.

the nation's copper industry with the purchase of 51 percent ownership of the vast Anaconda Copper Company mines and other facilities. Under the agreement of June 1969 Chile would acquire Anaconda properties for about $1 billion over a period of several years. The Frei government made such an agreement with the Kennecott Copper Company in 1967 obtaining a 51 percent control of the El Teniente mine. Anaconda sought to avoid a similar arrangement, but domestic pressures compelled the Frei government to act against Anaconda, which was forced into an agreement on Chileanization.

The election of the Marxist, Dr. Salvadore Allende Gossens, to the presidency in 1970 led to a rapid acceleration of the nationalization of the American-owned copper firms. Although pledged to fair compensation, the Allende regime announced that "excess profits" would be withheld from compensation to be paid to the copper companies, which was interpreted to mean that the two firms, the Kennecott and Anaconda companies, would receive little or perhaps nothing for their Chilean holdings.

The struggle between Latin American governments and terrorists entered a new phase with the assassination or kidnapping of foreign diplomats. Such incidents, involving both United States and foreign personnel, represent a dangerous technique that is hard to combat, and one which could have a serious impact on international relations. A growing concern for the lives of foreign diplomatic staff members has led to demands for police protection and restrictions on the movement of such personnel. Washington urged its staffs throughout the Western Hemisphere to redouble their protective measures.

The incidents of kidnapping for ransom have largely been directed at U.S. personnel, although a West German ambassador to Guatemala was murdered, and a Japanese Consul-General in São Paulo, Brazil, and a Paraguayan consul in Argentina have been held. U.S. Ambassador John G. Mein was killed in Guatemala when terrorists ambushed him in August 1968; two U.S. officers, Colonel John Weber and Lt. Cmdr. Ernest A. Munro, advisers to the Guatemalan armed forces, were also slain. The slaying of Count Karl von Spreti, the West German Ambassador, in April 1970 was thus the second such foreign official to be killed in Guatemala in two years. The West German government had offered to pay the $700,000 ransom money demanded by the terrorists, but Guatemala refused to accede to the terrorist demand that twenty-three political prisoners be released. Previously the Guatemalan government had met terrorists' demands in the kidnappings of U.S. labor attache, Sean M. Holly, and of its own foreign minister, Alberto Fuentes Mohr. In each case imprisoned guerrillas had been freed. The same technique was repeated by a Brazilian extremist group in São Paulo, which kidnapped the Japanese Consul-General, Nobuo Okuchi, and obtained the release of five prisoners. West German Ambassador E. von Holleben, kidnapped in Brazil, was released after forty prisoners were freed. In the case of the Paraguayan consul, the Argentine government refused to yield to the demands of the terrorists, who released their captive.

U.S. Ambassador to Brazil, C. Burke Elbrick, was abducted in Rio de Janeiro, Brazil in September 1969, and released after ransom was paid. Lt. Colonel Donald Crowley was kidnapped in the Dominican Republic in March 1970; and in the following month John C. Cutter, U.S. Consul in Porto Alegre, Brazil, outran a roadblock that his would-be abductors had set up. In Uruguay two U.S. diplomats and a Brazilian consul were seized by leftist terrorists, who named ten political prisoners as ransom. One of the U.S. officials, Dan Mitrione, was slain.

Reflecting the gravity of the situation, the political kidnapping problem was a major item on the agenda at a two-week meeting of the OAS foreign ministers in Washington in June and July 1970. There was willingness to condemn terrorism, political kidnappings, and extortion but disagreement on the use of hemispheric mechanisms to curb them. The OAS, unable to resolve the question, agreed to establish an eleven-nation committee to explore the possibilities of hemisphere-wide action on the issue.

This was followed by a Special Session of the OAS General Assembly, held in Washington in January-February 1971. A unanimous vote of the Assembly established the principle that acts of terrorism against foreign officials, irrespective of pretext, constitute serious common crimes rather than political offenses. The adoption of this principle deprives persons responsible for such acts of the shelter of political asylum and subjects them to extradition and prosecution. The Convention had not been ratified by any of the thirteen signatory nations by early 1972, but nine had indicated their intention to begin the ratification procedure. It was submitted to the U.S. Senate for advice and consent.[47]

In December 1969 Washington was host to representatives of thirteen nations to discuss the problem of aerial hijacking. Since May 1961 there had been 131 cases of hijacking or attempted hijacking, of which 102 occurred in 1968-1969. The United States, which suffered the most–from hijackings to Cuba–seeks an international agreement to prevent hijackers from claiming political asylum. The Convention on Certain Acts Committed On Board Aircraft, signed at Tokyo in September 1963 by the United States and forty-four other nations, proved ineffectual. Meeting at the Hague in December 1970 under the auspices of the International Civil Aviation Organization a seventy-seven nation conference on air law approved a Convention for the Suppression of Unlawful Seizure of Aircraft (Hijacking Convention). This was ratified by the United States in September 1971, completing the number of ratifications required to bring the Convention into force. The Convention provides that hijacking will be subject to severe penalties in all states and, in short, deprives hijackers of asylum from prosecution. A hijacker will either be extradited or prosecuted where found. Cuba had announced that it would support international action against hijacking, but no action had been taken by Cuba, including the ratification of this Convention. Most needed from Washington's standpoint was a U.S.-Cuba agreement for the return of hijackers to the United States to stand trial.

Hopes were raised about the possibility of such an accord after Dr. Castro expressed an interest in discussing the problem in a radio broadcast on November 15, 1972. The U.S. Department of State welcomed the announcement, and talks were subsequently conducted through the Swiss Embassy in Havana, which represents U.S. interests on the island. No definite committments were reported by year's end, but it appeared that the talks would continue.

[47] Dept. of State *Bulletin*, LXIV (Feb. 22, 1971), pp. 228-234; *Ibid.*, (Sept. 13, 1971), p. 286.

Washington was chiefly concerned with ending aerial piracy whereas the Cuban position was more complex. Havana also wanted to end sea piracy, which involved Cuban vessels taken by Cuban refugees to the United States, and to consider the total question of Cubans going into exile there.[48]

As the foregoing commentary indicates, there were a number of pressing issues in relations between Latin America and the United States confronting the Nixon administration in 1973. Aside from demands for increased and unfettered financial aid, there were problems of intergovernmental relations with military rulers, and the volatile issues of tariffs and trade and private investment. Because of their seeming priority, these topics warrant further consideration.

Campaigning for the presidency in September 1960, Senator John F. Kennedy said: "Seven years ago there were fifteen strong men in Latin America dominating the life of their countries. Today, there are only five. Three years from now there won't be any."[49] This prophesy seemed plausible through 1961 when General Alfredo Stroessner of Paraguay was the only military chief of state. However, a surge of militarism beginning in March 1962, which, carried forward through mid-1969, resulted in thirteen military coups, eleven causing the overthrow of democratically-elected governments. Altogether, by 1972, close to half of the 290 million people in nine countries of Latin America were ruled by imposed or elected military governments. In view of the Alliance for Progress goal of improving and strengthening democratic institutions, but at the same time needing staunch anti-Communist allies so as to cope with the stepped-up Communist offensive emanating from the Sino-Soviet bloc and its Cuban base, Washington's policy planners were faced with a hard decision. Previous administrations had usually extended *de facto* recognition and military assistance without regard for the nature of the governments, whether democratic or undemocratic, provided they were anti-Communist and upheld U.S. cold war policies.

After a military coup occurred in Peru in 1962 the Kennedy administration, seeking to implement the Alliance goal, broke off diplomatic relations, suspended military assistance, and stopped economic aid. Other nations did not boycott the military regime and Washington recognized it, restoring military and economic assistance on the condition that elections would be held. The Kennedy administration was forced to make further compromises on its principle of supporting only democratic governments when military elements overthrew civilian governments in Guatemala and Ecuador in 1963. But later in the year, when the military seized power in the Dominican Republic and Honduras, Washington again cut diplomatic ties and suspended the aid programs. In spite of these actions President Kennedy concluded "that the military often represented more competence in administration and more sympathy with the United States than any other group in the country. To halt work on the Alliance in every

[48] *Ibid.*, (Jan. 11, 1971), pp. 50-54; *Ibid.*, LXV (Oct. 4, 1971), p. 371; *New York Times*, Dec. 12, 1968; *Christian Science Monitor*, Apr. 10, 1970; *Ibid.*, Nov. 22, 1972; *Ibid.*, Dec. 14, 1972; *Times of the Americas*, Jan. 5, 1972.

[49] Simon G. Hanson, *Five Years of the Alliance for Progress: An Appraisal* (Washington, 1967), p. 23.

nation not ruled by genuine democracy would have paralyzed the whole program."[50]

The Johnson administration ceased using non-recognition and the suspension of aid as a means to foster civilian, representative government. Although avowedly opposed to military seizures of power, it recognized with little delay new regimes in Brazil, Argentina, and Bolivia and attached no promise of early elections as a condition for such action. Summing up his views on the question President Johnson declared: "The United States has no mandate to interfere whenever a government falls short of our expectations," but he added "In the Latin American countries we are on the side of those who want constitutional governments." President Nixon, describing his administration's essentially pragmatic policy, declared: "The United States has a strong political interest in maintaining cooperation with our neighbors regardless of their domestic viewpoints. We have a clear preference for free and democratic processes . . . Our relations depend not on their internal structures or social systems, but on actions which affect us and the inter-American system."[51]

A fundamental responsibility of the United States is the defense of the OAS community of nations from hostile attack, direct or indirect. This requires not only conventional military preparedness but also positive measures to forestall Communist infiltration and subversion. As Sino-Soviet bloc expansion mounted in the years following World War II, diplomatic efforts to secure an inter-American agreement for hemispheric defense won the general support of Latin Americans. The Rio Treaty of 1947 was aimed at building an inter-American line of defense against extrahemispheric military attack. But United States attention to the security needs of the relatively secure region of Latin America was nominal until the late 1950s. The United States Military Assistance Program (USMAP) in Latin America, designed to thwart extra-hemispheric military attacks, was revised to meet the challenge of indirect Communist conflict techniques. This was prompted by the Communist efforts to seize Guatemala, the Castro takeover in Cuba, and the FALN (Fuerzas Armadas de la Liberacíon Nacional) campaign of terror in Venezuela, which gave substance to the threat of internal subversion. USMAP expenditure grew from $200,000 in 1952 to $67 million in 1959. The total U.S. military assistance commitment to Latin America in the 1950s was approximately $317 million, whereas in the first seven years of the Alliance for Progress it amounted to $685.5 million, within a total of $9.2 billion of economic aid. When it was proposed to increase military assistance in the region a change in the Foreign Assistance Act imposed a ceiling of $75 million on the total of grants and sales to Latin America, other than training. In 1968 grant military assistance amounted to more than $17 million for material and services and $10 million for training, for a total of over $27 million. Cash sales of under $12 million and credit assistance of over $35 million brought the total of grants and sales for fiscal 1968 to $75 million.[52]

[50] Theodore C. Sorenson, *Kennedy* (New York, 1965), p. 535.

[51] Dept. of State *Bulletin*, LXIV (Mar. 22, 1971), p. 361; *New York Times*, Aug. 22, 1966.

[52] *The Foreign Assistance Program, Annual Report to the Congress, Fiscal Year 1968*, pp. 40-41; Committee on Government Operations, *U.S. Aid Operations in Latin America under the Alliance for Progress*, p. 1.

In 1971 President Nixon waived the congressionally-imposed $75 million ceiling on arms sales to Latin America and asked Congress to raise it to $150 million in order for the U.S. to compete with other world arms salesmen. In the preceding five years Britain, France, Canada, Italy, West Germany, and Spain had sold nearly $1 billion worth of aircraft, naval vessels, and other items in the Western Hemisphere. The president's decision was consistent with the Nixon Doctrine, which envisaged less involvement in the defense of foreign countries and more self-defense, partly through purchases of U.S. equipment financed by the U.S. Treasury. Secretary of State Rogers declared that "Such responsiveness is an important element in maintaining our traditional security relationship with these countries, and diminishes the prospects of any powers unfriendly to the United States advancing their influence and objectives in this hemisphere." In line with this policy also was a decline in the number of defense personnel assigned to training missions, which fell from nine hundred to three hundred between 1969 and 1971.[53]

When viewed in relation to total budgetary expenditures, the Latin American record on armaments has not been unfavorable. Military expenditures of the six major South American nations–Argentina, Brazil, Chile, Colombia, Peru, and Venezuela–totaled about the equivalent of U.S. $1.8 billion in 1970. This quadrupled the total in 1940, when defense outlays of the six countries approximated U.S. $400 million. At the same time, military spending took about the same proportion of the gross domestic product in 1970 as in 1940 and generally absorbed a smaller share of the national government expenditures. Of the three branches of the armed services, the air force reflected the sharpest growth in expenditures. In 1940 its share was negligible, while in 1970 its expenditures exceeded or rivaled those of the navy in all countries except Chile. Navy expenditures almost quadrupled, whereas army expenditures approximately tripled. Expenditures for the three branches in 1970 were divided as follows: army, 46 percent; navy, 28 percent; air force, 26 percent.[54]

United States military assistance to Latin America continues to emphasize strengthening of the capabilities of selected Latin American nations to maintain the internal security needed to realize the goals of the Alliance for Progress. Military assistance was an important contributing factor in containing insurgencies in Guatemala, Colombia, and Venezuela and in overcoming the insurgency headed by Dr. Ernesto "Ché" Guevara in Boliva. It is in support of counter-insurgency that the U.S. aid is most effective, for the Latin American military establishments can contribute little to the collective defense of the hemisphere against external attack.

The programs for Latin America emphasize multilateral cooperation through assistance in support of joint or combined military exercise and multinational communications. U.S. policy seeks to eliminate unnecessary defense expenditures and to avoid the pitfall of an arms race in the region. While encouraging a reduction of Latin America's armed forces and their budgets the

[53] *New York Times*, May 19, 1971; *Times of the Americas*, Dec. 8, 1971.

[54] U.S. Dept. of State, *Trends in Latin American Military Expenditures, 1940-1970* (Washington, 1972), pp. 1-8. Total military expenditures for all Latin America, excluding Cuba, stood at $2-$3 billion a year in 1971. See *New York Times*, Apr. 20, 1971.

United States cannot encourage total disarmament there without assuming the responsibility of policing the entire Western Hemisphere. In general, however, it is the policy of the United States to reduce armaments by international agreement rather than to encourage their growth.

The United States military assistance program has won the political collaboration of most of Latin America's armed forces but possibly at the price of sacrificing the democratic objectives of the Alliance for Progress, as shown by the recent series of military political coups. Even more disquieting is the fact that most of the Latin American military leaders involved had received U.S. training. The Latin American people who are opposed to military regimes hold the U.S. responsible because of the aid programs, even though that might not be the effect of the aid. In other words, the United States is held responsible for the unrepresentative, and commonly repressive, kind of government that prevails in a large number of countries. This raises the difficult question of whether to furnish internal security assistance to non-representative governments, because in the long run the suppression of non-Marxist populist movements may pose the greatest threat to the stability and progress of the area.[55]

Latin America's economic problems have increasingly extended into the area of trade relations, and their complaints have been most commonly lodged against the United States. It is widely believed, erroneously except for sugar and petroleum in a qualified sense, that this country fixes the prices of their raw materials exports, and not the free play of competition in the world markets, of which the United States is only a part. Underlying the problem has been Latin America's shrinking share of world trade, which dropped from 11 percent to 8 percent in the years 1928-1959. If the important trading countries of Cuba and Venezuela are omitted, the drop is even more critical: from 7.3 percent in 1928 to 4.2 percent in 1959. In 1965 Latin America accounted for 4.97 percent of the world's exports and 6.03 percent of its imports. The value of Latin American exports rose by 5.2 percent a year in 1966-1969. But this rate was significantly below that of 10 percent recorded on a world-wide basis. It also fell short of the 7.2 percent export expansion level attained by developing countries in general. The contraction of the region's share in world trade has forced many Latin American nations to accelerate the substitution of imports, by means of industrialization, to compensate for the lack of dynamism in the exports.[56]

While the region's trading position had suffered, its dependence on exports has remained high. But at the same time, in most countries, the total export receipts have been dependent on the sale of two or three export commodities which are vulnerable to price fluctuations in the world market. It is clear that the features of concepts and techniques in the use of the classical international

[55] U.S. Senate, Comm. of For. Rels., 90th Cong., 2nd Sess., *Hearings Before the Subcommittee on American Republics Affairs* (Washington, 1968), pp. 59-97; U.S. Senate, Comm. on For. Rels., 90th Cong., 1st Sess., *Survey of the Alliance for Progress–The Latin American Military* (Washington, 1967); George C. Lodge, *Engines of Change: United States Interests and Revolution in Latin America* (New York, 1970), pp. 175-187.

[56] Latin American Center, Univ. of Calif. at Los Angeles, *Statistical Abstract* of Latin America, 1965 (Los Angeles, 1965), p. 37; Current History, 43 (July, 1962), pp. 1-2; Inter-American Development Bank, *Socio-Economic Progress in Latin America (Social Progress Trust Fund Tenth Annual Report, 1970)* (Washington, 1971), p. 5.

trade policy instruments within the reach of the Latin American countries—such as bilateral treaties, the General Agreement on Tariff and Trade (GATT) multilateral tariff agreement, the traditional type of agreement to establish basic product prices, and the regional cooperation mechanisms of the inter-American system, as well as those of worldwide coverage in general—have proved to be of little value in improving international trade tendencies unfavorable to Latin America.[57]

A striking fact in the trade picture is that for many of the Latin American countries the principal market and chief supplier is the United States. The existence of this trade connection clearly suggests the need for close, friendly political relations. It is also a potential source of weakness because dependence on the U.S. can and has produced grievances, real and imaginary, between the two regions. Related to the problem has been the rise of regional preference arrangements, particularly in Western Europe, which threaten Latin America's export position there. As a consequence, Latin America has sought relief through regional economic integration, exemplified by the Central American Common Market (CACM), the Latin American Free Trade Association (LAFTA), the Andean Common Market, and the Caribbean Free Trade Association. While the U.S. government has recognized that the overall development of Latin America is heavily dependent on a satisfactory growth and development of the region's foreign trade, it opposed in principle and practice, until recently, stabilization agreements and stood aloof on the question of regional economic integration. However, compromises were made: Congress ratified the International Coffee agreement in 1963; and regarding sugar imports Washington allocated among several supply areas, strongly represented by Latin America, specific portions of the U.S. market at reduced tariff rates. The Cuban quota, the largest of all, was in 1960 distributed among other countries because of the actions of the Castro regime. By the late 1950s Washington had become reconciled to regional economic integration and a decade later was giving it vigorous support.

Pressures mounted in Latin America in the 1960s for additional tariff concessions by the United States. Tariff restrictions and quotas imposed on non-ferrous metals were especially resented and labeled inimical to inter-American goodwill and solidarity. As the Inter-American Committee of the Alliance for Progress (CIAP), on August 10, 1965, in its reports to the presidents of the American republics observed: "It is inequitable for the products of some of the developing countries to enjoy preferences outside the hemisphere plus non-discriminatory access to the United States market."[58] Most tropical products and all the important ones such as coffee, bananas, and cocoa now enter the U.S. market duty free. The proposal indicated that the United States should impose customs duties where none existed and apply lower rates to, or exempt, imports of these products from Latin America. The adoption of such a policy could conceivably harm other underdeveloped nations, provoke retalia-

[57] Pan American Union, *Latin America's Foreign Trade, Problems and Policies* (Washington, 1966), p. 1.

[58] U.S. Cong., Joint Econ. Comm., 89th Cong., 1st Sess., *Latin American Development and Western Hemisphere Trade* (Washington, 1965), pp. 53-54.

tion, and place an obstacle in the path of reducing trade barriers on a multilateral basis. The crux of the multilateral system that the United States has fostered is not the elimination of moderate protectionism whereby a nation favors domestic producers; rather, it is the elimination of discrimination among foreign markets. This is the basis of the most-favored-nation clause, in its unconstitutional form, which has governed our trade relations with most countries for over forty years and which has been endorsed by the principal trading nations of the world. The reconciliation of these traditional trade policies with the requests of Latin America, the Alliance for Progress, and the U.S. trading position vis-à-vis the rest of the world had become a problem of great magnitude by the 1960s.[59]

It should also be noted that, apart from typical tropical products, there is a wide range of temperate zone agricultural commodities in which the developing countries face an array of protective tariff and quota barriers which limit their access to the markets of the developed countries, and of subsidized exports from the rich countries, that compete against them in third markets. The developing countries are pressing for trade liberalization in these products, but the prospects for substantial liberalization are not promising. In most of the developed countries domestic agriculture is insulated in varying degrees from the free play of supply and demand by price supports, direct subsidies, and import controls. The average income of the farm sector in these countries tends to be lower than that of the other sectors of their economies, and the array of protective barriers is intended to maintain and raise the income of this group as a matter of equity.[60]

Fewer problems exist in regard to tariffs on manufactured goods, since little of Latin America's exports are in this category. However, it is clear that the region must become competitive in this field if it is to close its trade gap. United States tariffs on most manufactures that might come from Latin America average about 15 percent, which would indicate that this has not been a factor in depressing the area's manufacturing activity. (*See* Table 1-2). Under recent tariff schedules there is a wide range of products which Latin America could sell at competitive prices on U.S. markets. But for a variety of reasons, which include cultural values, habits, and customs, temperament, inefficiency, low value and high cost, and a disinterest in searching out new markets either at home or abroad, industrial development has been seriously inhibited.

International developments in the area of trade policy may have a significant effect on Latin America's future trade position. As a result of the Declaration of Viña del Mar adopted by the Latin American nations in May 1969, and discussions held later at the Inter-American Economic and Social Council, the latter appointed a special committee to negotiate new trade policies proposed to the United States by the Latin American countries. At the committee's 1970 meetings, the Latin American republics presented a list of

[59] For detailed analyses of Latin America's foreign trade problems and strategy, the growth of the new trade policy in the United Nations Conference on Trade and Development (UNCTAD), and the General Agreement on Tariffs and Trade (GATT) see Pan American Union, *Latin America's Foreign Trade, Problems and Policies*, Chaps. I and III; *Rockefeller Report*, pp. 134-139.
[60] Dept. of State *Bulletin*, LVII (Aug. 7, 1967). p. 183.

Table 1-2. United States trade with Latin America*
(in millions of dollars)

	EXPORTS			*IMPORTS*		
Country	*1960*	*1965*	*1970*	*1960*	*1965*	*1970*
GRAND TOTAL	3667	3957	6021	3637	3946	5210
Argentina	359	268	441	98	122	172
Barbados	6	9	22	1	5	9
Bolivia	25	42	46	9	31	25
Brazil	464	348	841	570	512	669
Chile	203	237	300	193	209	154
Colombia	253	198	395	299	277	269
Costa Rica	45	61	95	35	57	116
Cuba	225	Z	Z	357	Z	–
Dominican Republic	42	76	143	110	111	184
Ecuador	57	80	127	65	106	109
El Salvador	43	61	64	32	48	48
Guatemala	64	96	100	59	67	87
Haiti	25	21	34	18	20	32
Honduras	35	54	89	34	72	102
Jamaica	48	87	219	54	125	186
Mexico	831	1106	1704	443	638	1222
Nicaragua	30	69	77	21	36	61
Panama	90	125	208	24	60	76
Paraguay	9	16	18	8	13	11
Peru	147	282	214	183	241	341
Trinidad & Tobago	36	75	84	55	142	236
Uruguay	63	20	41	21	36	19
Venezuela	567	626	759	948	1018	1082

*Represented are the twenty-four members of the Organization of American States; includes the U.S.; Cuba was excluded from participation in OAS in 1962.
Z less than $500,000.
U.S. Bureau of the Census, *Statistical Abstract of the United States, 1971* (Washington, 1971), pp. 768-769.

eight-hundred export products for which they requested U.S. elimination of tariff and non-tariff barriers. Washington's inaction on the problem of general trade preferences, the 10 percent surcharge on imports in the last half of 1971, although President Nixon had exempted Latin America from the 10 percent reduction in foreign economic assistance expenditures called for under the new economic policy, and Latin America's unfavorable balance of trade with the U.S. were denounced at the Lima meeting of the "Group of 77" ("third-world" countries) in 1971 and by the 12th Extraordinary Meeting of the Inter-American Economic and Social Council (CIES) in January 1972.

It should be noted that whereas Latin American economic relationships with the United States in the 1960s focused on financial aid and technical assistance, the nations in the region had become more concerned with the economic development role of the export sector by the early 1970s. This concern reflected several considerations. While Latin America accounted for only 14 percent of total U.S. foreign sales and about 15 percent of total U.S. imports, the U.S. absorbed about 34 percent of Latin American exports and supplied some 40 percent of Latin American imports. Exports accounted for 10 percent of the total Latin American gross national product in 1971. But for more than half of the Latin American countries, the contribution of exports was higher than 10 percent (for the U.S. exports accounted for only 4 percent of the GNP). Moreover, the overall rate of export growth for Latin America has been only about half that for developed countries, and as a group Latin American countries depend upon exports to meet more than three-fourths of their foreign exchange requirements. Underlying all aspects of the economic axis is the fact that whereas to the United States trade with Latin American is important, to the latter trade with the U.S. is paramount, if not dominant.[61]

Charles A. Meyer, Assistant Secretary for Inter-American Affairs, admitted that the Nixon administration had not fulfilled an earlier commitment to set in motion a system of generalized tariff preference for the developing countries, acknowledging that the preferences were the primary objectives of Latin America in its relations with the United States. He pointed out the generalized preferences would eliminate tariffs of the major developed countries on a large number of manufactured and semi-manufactured products from all less developed countries. Thus they would stimulate developing countries to diversify their exports and lessen their traditional dependence for foreign exchange earnings on raw materials and commodities and would reduce the high cost of import substitution. Mr. Meyer declared that the Nixon administration had delayed in presenting such legislation to the Congress because the U.S. trade and balance of payments position was weakening and that these problems together with a sluggish domestic economy created a strong protectionist feeling in Congress. It was therefore judged unwise to submit a preference bill at that time. However, the administration restated its intention to push the preference bill in 1972.

In reviewing developments in U.S.-Latin American relations in the preceding two years Secretary Meyer indicated that restrictions had been eased, allowing Latin American nations to spend their aid dollars within the region and elsewhere in the developing world, that U.S. economic policies were being submitted for annual review to the Inter-American Committee on the Alliance for Progress, an unprecedented concession by a donor nation, and that Washington had consulted whenever possible with Latin American nations prior to taking actions which might affect their economies. Washington had participated consistently in the inter-American organizations that had initiated programs for expanding capital markets, tourism, and export promotion, as well as having supported the regional development banks. Agreements were signed with Colombia and Panama whereby the U.S. agreed to finance its share for

[61] *Ibid.*, LXV (Aug. 30, 1971), p. 239.

completion of the Darien Gap section of the Pan American highway. Congress passed new sugar legislation, which was generally favorable to Latin American suppliers. Negotiations with Panama were pushed to modernize the basic treaty of 1903, the Bryan-Chamorro treaty was abrogated, and a settlement was made with Mexico which resolved all questions concerning the changing course of the Rio Grande.

Despite these and other achievements credited to the Nixon administration in inter-American relations, Secretary of State William Rogers, in a report to Congress, amplified Secretary Meyer's earlier statement, declaring that U.S. relations with the other American republics "suffered some impairment during 1971 . . . The principal and immediate problem in Latin American eyes was the U.S.' inability to meet some of our commitments and (live up to) Latin American expectations and desires in the field of trade and economic assistance." He said that in 1972 "we will be making special efforts to respond to the concerns of the peoples of the Western Hemisphere."

Little was done in 1972, however, to check the drift in U.S. relations with its hemisphere neighbors. The apparent indifference could be explained in part by the fact that it was an election year, and because the war in Southeast Asia continued to grip the nation's attention. Although Latin America did not figure prominently in the presidential campaign, the platforms of the two major parties showed a wide divergence of opinion on policies toward the region, particularly as concerned Cuba. The Republican Party promised to "foster a more mature partnership" with Latin American countries, while maintaining that Cuba was "ineligible for readmission to the community of American states." In their platform the Democrats pledged a sharp reduction in military aid to Latin American governments. And they sought better relations with Cuba, contending that "after thirteen years of boycott, crisis and hostility, the time has come to reexamine our relations with Cuba . . ."

In what could be regarded as a preview of Washington's policies during the second Nixon term, Secretary Rogers, in a speech given in Philadelphia to the Permanent Council of the OAS, commemorating the 150th anniversary of U.S.-Latin American relations, declared: "President Nixon has asked me to tell you that during his second term we will remain committed to the interests of the hemisphere and determined to make a substantial contribution to its social and economic progress." He indicated that the United States would pursue four major goals:

1. "We will pursue a policy of cooperation with Latin America in a relationship of greater equality, shared initiatives, and mutual responsibilities.
2. We will work to ensure that the legitimate interests of all the nations of Latin America are represented in the new international monetary and trade systems to be negotiated.
3. We will cooperate with you directly in this hemisphere to strengthen and diversify our trade, investment, and assistance ties.
4. And we will seek to resolve the issues between us over fisheries, over territorial seas, over investment, and all others, in the spirit of friendship and mutual respect which is the essence of our inter-American system."

Latin American reaction to Mr. Rogers speech was restrained because the Secretary had not commented specifically on the administration's plans for resolving the region's main problems with the United States. Clearly, the need for a broad policy review was indicated.[62]

The importance of the role of private investment in fulfilling the aims of the Alliance for Progress was not understood by the Latin Americans generally. The misconceptions arose in part from the brief reference to it in the Charter of Punta del Este, a ten-word phrase found in the statement of purpose to accelerate the process of national industrialization by "taking full advantage of the talents and energies of both public and private sectors." Private investment and private enterprise are not otherwise mentioned except by implication. The "Declaration to the Peoples of America" establishing the Alliance for Progress, among it twelve paragraphs on goals, includes one beginning: "To stimulate private enterprise."[63] Such light treatment gave rise to a common belief, particularly in Latin America, that large amounts of U.S. government financial aid carried out on a government-to-government basis subordinated private investment. Such a view fails to appreciate American faith in private investment which had such a large role in the development of the United States and other Western nations. While it can be argued that free economic choice is not completely excluded by, or incompatible with, undemocratic political institutions, historical evidence indicates that reliance on a centralized governmental machinery for allocating economic resources jeopardizes both individual freedom and political democracy. Owing to urgency for positive action, a massive program of government effort was believed necessary if the challenge of progress was to be met in time. However, it was not intended to displace private investment, but instead to vigorously support it.

Obviously there continues to be a need for massive government effort to supplement private enterprise in advancing Latin America's development; but the nations of the region need to find means for improving the climate for private initiative. Failure to provide this climate, in the wake of Cuban seizures of foreign property, produced the Hickenlooper Amendment, first enacted in 1962 during the struggle between ITT and American Foreign Power Company on the one hand and several state governments of Brazil on the other. By 1967 it required the President to suspend assistance to any government which nationalized the property of a U.S. investor, or repudiated a contract or imposed a "discriminatory" tax or "restrictive maintenance or operation conditions," unless the country took steps to discharge its international legal obligations to pay for the property. Senator Thomas H. Kuchel of California succeeded in adding a provision with respect to the fishermen from his state who used the fishing ground off Peru and Ecuador. Although it has a deterrent value, the Hickenlooper Amendment will not dissaude such nations as Cuba, Peru, Bolivia, and Chile, which have been bent on expropriating without compensation.

Since the flow of private capital in Latin America can be neither "driven nor cajoled," removing some of the risks of investment were undertaken by the

[62] *Ibid.*, pp. 239-245; *Ibid.*, (Nov. 15, 1971), pp. 561-562; *San Francisco Examiner and Chronicle*, Mar. 26, 1972; *Times of the Americas*, Nov. 22, 1972; *Christian Science Monitor*, Nov. 22, 1972; *Congressional Quarterly*, XXX (Aug. 26, 1972), p. 2147.

[63] U.S. Cong., Joint Econ. Comm., 89th Cong., 2nd Sess., *Private Investments in Latin America* (Washington, 1964), pp. 4-5.

investment guarantee program. It was not a new departure, for the protection of merchants, industrialists, and investors operating overseas has traditionally been a keystone of U.S. foreign policy. This policy received a special type of emphasis in Latin America in the form of the Agency for International Development's specific risk investment guarantee program. This program, which began in 1948, insures investors against the inability to convert foreign currencies received as earnings and against losses attributable to expropriation or confiscation, war or revolution. To obtain this insurance a firm must be substantially "beneficially" owned by U.S. citizens (ordinarily 50 percent or more) and both AID and the government of the nation in which the investment is being made must give their approval. Because the program's goal is to facilitate and increase private participation in the development of the less developed lands, guarantees are not available for existing investments but only for new enterprises or additions, expansion, or major changes in existing facilities. The agreements reflect a willingness to allow domestic actions tending to injure foreign investors to be tested by the principles of international law rather than limited to domestic "plaintiff, judge, and jury" proceedings. As of March 31, 1969, political risk insurance in Latin America totaled $3,936,908,543.[64]

Throughout the 1950s the United States encouraged private capital investment and placed governmental assistance to Latin America on a restricted public loan basis. President Kennedy, while recognizing the need for government initiative, declared: "It is impossible for us to supply all the funds that are necessary for the development of Latin America. They must come from private sources. If local capital and American capital dry up, then all our hopes of a decade of development in Latin America will be gone." It became apparent that political instability, losses by expropriation, currency instability, confiscatory taxes, frozen utility rates, and other problems suffered by foreign investors slowed the flow of private external financing into Latin America. The capital outflow in the form of direct U.S. investments into the southern republics in the decade 1950-1960 which had averaged about $325 million per year ($219 million if oil investments in Venezuela during the critical Suez years, 1956-57, are excluded) fell to $173 million in 1961, and became a net withdrawal of $32 million in 1962. In 1969 U.S. private direct investment in Latin America totaled more than $12 billion, which represented about one-fourth of the world total. Meanwhile Europe, to a large extent, turned away from Latin America, with the European capital outflow to the whole of Latin America estimated at less than $80 million in recent years. (*See* Table 1-3.)

The most casual study of the relations between the United States and the Latin American republics will indicate that the great republic in the north has until recently made little effort either to understand the difficulties that have sorely tried her less powerful neighbors or to study their racial characteristics and customs with the friendly appreciation necessary to good relations between states. Nor is it sufficient in a democracy where public opinion plays an important part in foreign affairs to confine knowledge of foreign policies and

[64] U.S. Cong., Joint Econ. Comm., 89th Cong., 2nd Sess., *Private Investment in Latin America, Hearings* (Washington, 1964), pp. 237-270; U.S. Cong., Comm. on For. Affairs, House of Reps., *Hearings on HR 11792* (Washington, 1969), p. 708; Hanson, *Alliance for Progress*, p. 10; *Rockefeller Report*, pp. 22; 29-30; 148-154.

Table 1-3. Private investment in Latin America 1897-1968[65]
(in millions of dollars)

	1897	*1929*	*1946*	*1958*	*1968*
United States					
(direct investment)	308	3,705	3,000	8,730	12,989
Portfolio (direct investment)	–	1,724	672	1,039	
Total	308	5,429	3,672	9,769	
Great Britain	2,060	4,500*	3,575	2,547****	
France	628	454**	307	N.A.	
Germany	–	700***	–	160	

*1929 figures not available. $4,865 in 1913 and $4,542 in 1940.
**1929 not available. $454 in 1940.
***1929 not available. $677 in 1918 and $969 in 1940.
****1958 not available. $2,547 in 1950.

peoples to the select few who make up the government. Such understanding should be widespread among the people themselves, so that public opinion, based upon an intelligent comprehension of facts, can act as a lever toward more friendly cooperation, rather than as a spur to jealous and rival aspirations. To bring about this better relationship, which can be accomplished only by better mutual understanding, every possible point of contact and every avenue of approach should be utilized.

There was little general interest by Americans in Latin American history, institutions, literature, and language until World War I. During the war a significant expansion in the teaching of Spanish occurred, and in 1918 the *Hispanic American Historical Review*, destined to become the main organ of professional historians in the field, began publication. By the close of the following decade courses in Latin American history were offered in more than one hundred universities and colleges, and numerous other institutions of higher learning. Supported by the Carnegie Endowment, the American Institute of International Law was founded in 1912; subsequently the Endowment developed most of the fields that came under the heading of cultural cooperation: cooperating with binational cultural centers, financing visiting professors from the United States and Latin America, granting travel fellowships, furnishing books and distributing publications in Spanish, and sponsoring group projects.

It was also in this period that the Guggenheim Foundation began awarding fellowships to Latin Americans for study in the United States. The Rockefeller Foundation supported Latin American studies through the American Council of Learned Societies which organized the Committee on Latin American studies.

[65] U.S. Cong., *Private Investment in Latin America*, p. 7; U.S. Senate, *United States Business and Labor in Latin America* (Washington, 1960); U.S. Dept. of Commerce, Office of Business Economics, *Survey of Current Business*, 49 (Oct. 1969), p. 30.

The League of Nations and the Pan American Union also contributed to international culture cooperation during the inter-war years. In 1929 the Division of Intellectual Cooperation was established within the Union and the Scientific Congresses, begun in 1908, were continued. Under the auspices of the League, most Latin American nations founded National Committees of Intellectual and Cultural Cooperation. An official inter-American program for the exchange of students and faculty was introduced at the Conference for the Maintenance of Peace held at Buenos Aires in 1936 at which a Convention for the Promotion of Inter-American Cultural Relations was negotiated. Prior to this time the United States government had not regarded the promotion of cultural relations as a government function; doubtless its acceptance of the role was prompted in some measure by fascist propaganda which was aimed at weakening the Good Neighbor policy. It was not until 1940 that the Cultural Relations Convention was implemented by the exchange of professors and students.

In 1938 the Interdepartmental Committee on Cooperation with the other American Republics was set up with Sumner Welles as chairman. This was a major turning point in our cultural relations with Latin America because it indicated governmental acceptance of a large scale program supported by public funds under official direction and a plan for the interchange of persons. Following the enactment of Public Law 63 (76th Cong.) the Department of State created a Division of Cultural Relations which administered grants to students and professors under the Buenos Aires Convention for the Promotion of Inter-American Cultural Relations. In 1939 Congress enacted Public Law 355 (76th Cong.) which authorized the president to employ the services of government departments in carrying out cooperative programs signed by the American nations at the Buenos Aires Conference of 1936 and the Eighth Inter-American Conference held at Lima in 1938. This cooperative program was continued during the war years by the Office of the Coordinator of Inter-American Affairs, headed by Nelson Rockefeller.

Since the war there has been a significant expansion and strengthening of Latin American studies in the United States. Notable contributions have been made by the Hispanic Foundation, Library of Congress, The Conference on Latin American History, and the Latin American Studies Association. Concurrently, there has been a growing interest in bringing Latin Americans to the United States for study, a movement made possible by federal and state legislation, and grants from universities, foundations, professional groups and business firms.

During the academic years 1965-66, there were 82,709 foreign students in U.S. universities and colleges. Of these 13,998 (17 percent) were from Latin America. As compared with the academic year 1958-59, the number of Latin American students had increased by about 3,700 (36 percent) whereas the number of all other foreign students had increased by about 31,700 (86 percent). Of the Latin American students enrolled in 1965-66, only 40 percent were graduate students as compared with 56 percent from other areas. The lower enrollment rate of the Latin American students is particularly attributable to the requirement that applicants must be proficient in the English language; this is a more serious barrier to the entry of these students than most other foreign students. It is also explained by the fact that our universities have been less generous in granting financial support to the Latin American students than other foreign students. In 1966, the universities gave full financial support for 17 percent of their foreign students. But only 9 percent of the Latin Americans received this support, as compared with 19 percent of the students from other areas.[66]

Few opportunities existed for U.S. students to study in Latin America before the Convention for the Promotion of Inter-American Cultural Relations was approved at the Buenos Aires Conference in 1936. After its ratification the United States government enacted several measures under which expanded programs of student-professor exchanges have been carried out: the Fulbright Act (Public Law 584, 79th Cong.), the Smith-Mundt Act (Public Law 402, 80th Cong.), the Agricultural Trade Development and Assistance Act of 1954 (Public Law 480, 83rd Cong.), and the Mutual Educational and Cultural Exchange Act of 1961 (Public Law 87-256, 87th Cong.). Together with the Buenos Aires Convention, this legislation has provided for a two-way exchange between the United States and Latin American countries, when funds are available. However, the number of students and faculty going to Latin America from the United States has been much less than the movement northward. In the academic year 1969-1970, 52 percent of U.S. students reported abroad were in Europe, 17 percent in Canada, 16 percent in Latin America, 7 percent in the Far East, 6 percent in the Near and Middle East, 1 percent in Africa, and 1 percent in Oceania. More than half–55 percent–of U.S. faculty members reported abroad (1970-1971) were in Europe, 11 percent in the Far East, 11 percent in Latin America, 5 percent in the Near and Middle East, 5 percent in Africa, 2 percent in Oceania, and 2 percent in Canada. The period 1969-1970 found 5,219 U.S. students and 712 (1970-1971) faculty in Latin America, whereas the latter in 1970-1971 sent to the United States 29,300 students and 869 faculty.[67]

The United States Information Agency (USIA), known abroad as the United States Information Service (USIS), attempts to strengthen the cultural ties between the Americas, to overcome the psychological barriers that disrupt inter-American harmony, and to help speed the modernization of Latin America by transmitting the knowledge and techniques of the more industrially-advanced West. USIS is active in seven media: radio, television, motion pictures, press, book publishing, exhibits, and the arts. One of the most effective activities of the agency is the Binational Center Program, which is conducted with the close collaboration and support of the Latin Americans. By the mid-1960s USIS maintained 113 binational centers and binational societies in Latin America, teaching English to more than 117,000 persons annually. USIS libraries hold more than 225,000 books. Lectures and the distribution of books and pamphlets and other forms of cultural exchange are also carried out by the binational centers.

[66] Herbert K. May, *Problems and Prospects of the Alliance for Progress* (New York, 1968), pp. 87-90.

[67] Institute of International Education, *Open Doors* (New York, 1971). Steps have been taken to establish an Inter-American Council for Education, Scientific and Cultural Cooperation. It has also been proposed that a Western Hemisphere Institute for Education, Science and culture be created as an operating arm of a suggested Economic and Social Development Agency. See *Rockefeller Report*, pp. 110; 171; 197-199.

For survey of programs of direct cooperation between individual universities or consortia in the United States and those in Latin America see Pan American Union, *Inter-American University Cooperation* (Washington, 1968).

The Voice of America (VOA), which transmits short-wave broadcasts to Latin America is another important part of USIA's program, but it has suffered from budgetary limitations imposed by Congress. In 1953, the first year of the Eisenhower administration, the president voluntarily reduced President Truman's USIA budget request from $114,516,000 to $87,900,000, a cut of $26,616,000. The VOA services cut at that time included its Latin American service. Thus, USIA's Spanish language broadcasts ceased in 1953 and were not restored until March, 1960. USIA faced mounting criticism and reduced budgets in the early 1970s thanks to congressional skepticism and economy measures in government. Chairman J. W. Fulbright of the Senate Foreign Relations Committee labeled the agency an "anachronism of the cold war."

Being restricted by inadequate budgets USIS continues to be severely handicapped in reaching its potential. It appears that the U.S. Congress remains unconvinced that the U.S. information and cultural exchange program requires further expansion. Yet one of the obvious weaknesses of the $20 billion Alliance for Progress program has been the failure to communicate an understanding of the scope and causes of Latin America's problems, the need for sacrifices and self-help in solving them, and the creation of a mystique to impart the needed dynamism. In general, USIA's programs of information and cultural exchange are commendable but, owing to limited funds, they fall short of meeting Alliance requirements or in matching the propaganda offensive of the Communist countries.

Inherent in the development of each nation's foreign policy is a set of basic premises which determines its relations with other countries. These basic premises are generally enunciated in policy statements, but often they remain concealed as unstated principles. The fundamental premise in the foreign policy of every nation and the source of all policy is national self-interest. The first concern of the leaders of each nation must be the preservation of territorial integrity and national sovereignty, the protection of the general interests of the nation, and to a lesser extent the protection of the interests and welfare of individual citizens. The effectiveness of a nation's foreign policy can be judged, most appropriately, by determining whether it successfully protects the national interest as defined by the country's leaders. The needs and preferences of other nations, even allies of neighbors, are not material except where the fundamental interests of the two nations coincide.

Evaluation of the means employed to promote and protect the national interest must be tempered by the knowledge that the moral judgements which are normally applied in individual human relations are not relevant in international relations. In this context another point should be made in regard to the evaluation of a nation's foreign policy. Often the basic premise of self-interest is described, justified, and propagated in highly moral language employing phrases such as "the older brother in a family of independent American nations," the "sisterhood of the Americas," and the "Good Neighbor" policy. If the underlying reality of these words is not sought out and evaluated then the fundamental nature of the relations between all nations–between any two nations–is not revealed.

Against this background U.S. objectives in Latin America may be properly understood. While numerous and often interrelated, these objectives have three general and related purposes: first, to prevent the establishment within the Western Hemisphere of a hostile power capable of endangering the security of the region; second, to support the development of an inter-American system effective in resolving regional issues and helpful in coordinating political, economic, and other affairs of mutual interest; third, to apply its public and private resources and skills to foster Latin America's modernization through evolutionary and democratic channels.

With the advent of the cold war after World War II U.S. policy toward Latin America became clouded by ambiguity and seeming inconsistency, particularly so to the peoples of that region. The main source of the ambiguity lay in the emergence of the new globalism with the United States assuming responsibilities beyond the limits of the Western Hemisphere, and on the other hand the historic American regionalism which binds the United States and Latin America together in a special relationship, the "Western Hemisphere Idea." After Fidel Castro's seizure of power in Cuba and his alignment with the Sino-Soviet bloc, the United States insisted that the problem be dealt with in hemispheric terms. However, when pressures arose from its assumption of global commitments, the United States tended to disregard the special relationship when the other American republics made unwelcome claims upon it. Under these circumstances Washington applied the same criteria to Latin America as to the rest of the world.

This is clearly shown in the allocation of economic aid. So long as it appeared that Latin America was not a critical area in the cold war, the region's share of U.S. economic aid remained negligible, and little heed was paid to the Latin Americans' complaints and aspirations. Concluding that the special relationship with the United States was illusory, they grew more reluctant to defer to the United States in deciding what problems called for hemispheric action; the case of Cuba in particular, and communism in general are outstanding examples. While the social ferment compounded by the population explosion developed apace, Washington continued to take the Latin Americans for granted and did little more than provide military defense against a threat which most Latin Americans believed was economic and social, not military. Brought out of its complacency by the anti-Nixon South American demonstrations in 1958, the Eisenhower administration belatedly introduced new departures in Latin American policy. The Kennedy administration continued and expanded these measures under the Alliance for Progress. Unfortunately, their impact was weakened at the outset by the abortive invasion of Cuba.

The United States is seriously handicapped in dealing with the new interrelated problems arising from global and hemispheric responsibilities. The matter is complicated by Latin American grievances of long standing, preceding the rise of Communism and Fidelismo, which were redressed only temporarily, or partially, by the Franklin Roosevelt administration. The Latin American policy of the United States from the close of World War II until the early 1960s was, in the opinion of many Latin Americans, a compound of a negative political policy of anti-communism and a self-interested policy of free capitalistic enterprise that approached intervention. This policy, it is believed, slowed the rate of economic development and fostered the growth of military dictatorship and oligarchy which the reformers in Latin America were opposing.

Further doubts and ambiguities surrounding United States policy were related to the question of nonintervention. It took the Latin American nations a generation of persistent agitation to get the principle of nonintervention accepted by the United States. Its clearest expression is found in Article 15 of the Charter of Bogotá, a basic document of hemispheric policy: "No state or group of states has the right to intervene, directly or indirectly, for any reason whatever, in the internal or external affairs of any other state." Obviously, in the cases of Guatemala, Cuba, and the Dominican Republic the United States violated this treaty agreement. Because of the special nature of the Cuban case, which was dramatized so dangerously in the Cuban missiles crisis, the Latin American nations did not insist on literal adherence to Article 15. But the Dominican Republic was another matter. Deciding that the failure to intervene would result in the Dominican Republic becoming "another Cuba," Washington intervened militarily and politically there without adequately advising or consulting the OAS.

The reason for the precipitate U.S. action was to block the threat of a possible Communist takeover, which was consistent with the primary responsibility of U.S. foreign policy. Concerted action with the OAS initially was ruled out by the apparent urgency of the crisis; this was unfortunate because the extent of the Communist menace could not be clearly determined. The aim of the "standby forces" would presumably be not only anti-Communist but also inevitably anti-revolutionary inasmuch as any revolution in a Latin American country, as distinct from a simple *coup d'ètat*, would be likely to get immediate Communist or Castroite support. The United States' action to Latin Americans implied opposition to revolution simply on the basis that Communists join revolutions, which was bitterly resented. A continuation of this policy would, it was believed, promote right-wing counter-revolutions. Unfortunately this trend has taken place, though it began before the Dominican intervention.

This episode exposes the core of the American dilemma. On the one hand, Washington wants to promote democratic development as a functional alternative to communism or Fidelismo among the oppressed masses. But on the other, it fears such development since inexperienced democratic regimes may prove less resistant to Communist infiltration than the rightist military regimes. The policy of alignment with autocratic regimes is further encouraged by the tendency of democrats to assume a more independent and nationalist attitude than the dictators who, lacking popular support, must often depend upon U.S. backing to stay in power. The political stability achieved under military or extreme rightist civilian rule might well be a temporary illusion that could eventually burst with the buildup of social revolutionary pressures, and possibly produce the very communism that the anti-Communist strongmen are supposed to suppress, as Cuba so tragically illustrates.[68]

Finding a prescription for U.S. policy in Latin America is complicated enormously by the rapidity of change there. In this last half of the twentieth century Latin America finds itself in a process of transition from traditional to modern societies. The great challenge presented to the United States is how to influence this process. Under the circumstances it appears that Washington must

[68] See Lodge, *Engines of Change*, Chap. I, for additional background.

retain flexibility as to the means by which certain goals are to be achieved. In practice this means pressing for evolutionary change but at the same time reserving the option of reinforcing democratically-based revolutionary efforts when it is clear that immobilism hinders more desirable evolutionary processes. It must be emphasized, however, that much anti-American sentiment will continue no matter what policies are pursued by the United States. Given the frustrations of Latin Americans as members of developing nations and the great disparity in wealth and power, and differing cultural values, between the nations of the region and our own, a certain amount of resentment and misunderstanding seems inevitable. This is part of the price a powerful nation must pay for its position of importance and leadership in world affairs.

Latin America has become a region of global significance, and the loss of United States prestige and hegemony there could result in a weakening of our nation's position in international politics. Military security is a first consideration, for the experience of World War II revealed that Latin America could be the Achilles' heel of our defense system. United States military installations for the defense of Latin America are situated at two points in the Caribbean: the Canal Zone and Puerto Rico. The Navy has, in addition, bases at Guantanamo Bay, Cuba, and Chaguaramos, Trinidad. Beyond doubt the most important naval facility in the Latin American region is the huge Roosevelt Roads base on the eastern end of the island of Puerto Rico. The National Aeronautic and Space Administration of the United States through treaty agreements with several Latin American nations has established missile tracking facilities for earth satellites, a deep-space network, and a manned space flight network for the Gemini and Apollo programs; twelve of the stations are located on Latin American territory.

Although the Latin American nations cannot wield significant military influence in the age of nuclear capability, the one million men making up the region's armed forces are more of a force to be reckoned with than at any time in the past. It seems improbable, however, that these nations will have in the foreseeable future the economic and technological capacity to develop military establishments with sufficient strength beyond defense against other Latin American states and in suppressing internal strife, such as rural guerrilla activity and urban terrorism.

Nevertheless, while Latin America can neither aid nor threaten the United States significantly during a global war, and long range missiles may have lessened the strategic importance of the Caribbean region and those portions of the mainland which stand on the approaches to the United States and the Panama Canal, the existence of bases held in those regions by a hostile power would imperil the security of the continental United States. It follows that, given the current nuclear stalemate which has helped to promote guerrilla warfare, the military forces of friendly Latin American nations can help to forestall powers hostile to the United States from gaining bases in the Western Hemisphere. In wartime the Latin American countries can provide garrisons, and naval units for coastal patrol, activities which would otherwise require the deployment of United States forces. The armed forces of Latin America might also join in forming an inter-American peacekeeping force to prevent the subversion and political domination of a hemispheric nation by an extra-continental power. The United States military and naval installations at Guantanamo in Cuba and Puerto Rico, while less important than in the past, are still useful. And although the Panama Canal is virtually indefensible in the nuclear-warhead, ballistic-missile age, it has proved of value to the United States and its allies in the continuing cold war.

Latin America also has a military value in connection with its reserves of strategic minerals whose availability to the United States, with some thirty-five strategic materials, are especially vital in wartime. These include Chilean copper, Bolivian tin, Brazilian quartz crystals, Venezuelan petroleum and iron ore, Jamaican bauxite, and others. Deposits of fissionable material in Brazil, Peru, Bolivia, and Chile are also potentially important.

The economic stake of the United States in Latin America has become progressively important. The region purchases some 14 percent of U.S. exports, furnishes about 15 percent of its imports, and contains more than $12 billion, or one-fourth, of all direct U.S. foreign investment. Although the United States has the greatest internal market in the world and foreign trade is less important to its economy than to most nations, it must continually expand its production in order to remain competitive with the other great industrial powers. It seems likely that, despite the slow progress of modernization in the region, Latin America, with the world's fastest growing population, offers a potential for larger markets that cannot be overlooked by U.S. economic policy planners and manufacturers.

In terms of international politics and hemispheric security, and on the diplomatic front, the political support of the Latin American countries is of vital importance. The twenty-four votes of the Latin American republics in the United Nations General Assembly enable them to strengthen or weaken materially the position of the United States and its allies in debates on conflicting issues with their adversaries. The effectiveness of the United States in asserting just and wise leadership in the inter-American system has international implications. Success by the United States in helping to overcome the problems of the hemisphere's economically underdeveloped nations may win the support of neutralist or uncommitted nations of similar condition which otherwise might become satellites of the Communist powers. The political posture of the numerous weaker nations might ultimately contribute to the global balance of power. This factor must be taken into account because the global and Latin American policies of the United States have become closely related.[69]

[69] For a sharply-focused analysis of the problem see Edwin Lieuwen, *The United States and the Challenge to Security in Latin America* (Columbus, Ohio, 1966).

SUPPLEMENTARY READINGS

Ashabranner, Brent. *A Moment in History: The First Ten Years of the Peace Corps*. New York, 1971.
Bailey, Norman A., ed. *Latin America: Politics, Economics, and Hemispheric Security*. New York, 1965.
———. *Latin America in World Politics.* New York, 1967.
Bailey, Thomas A. *A Diplomatic History of the American People.* New York, 1955.
Bemis, Samuel F. *The Latin American Policy of the United States.* New York, 1943.
Berle, Jr., Adolph A. *Latin America: Diplomacy and Reality.* New York, 1962.
Burns, E. Bradford. *Latin America: A Concise Interpretive History.* Englewood Cliffs, 1972.
Burr, Robert N., and Roland D. Hussey, eds. *Documents on Inter-American Cooperation.* 2 vols. Philadelphia, 1955.
Caicedo Castilla, José Joaquín. *El Panamericanismo*. Buenos Aires, 1961.
Callcott, Wilfrid H. *The Western Hemisphere: Its Influence on United States Policies to the End of World War II.* Austin, 1968.
Conn, Stetson, and Byron Fairchild. *The Framework of Hemisphere Defense, The United States Army in World War II: The Western Hemisphere*. Washington, 1960.
De Conde, Alexander. *Herbert Hoover's Latin American Policy.* Stanford, 1951.
Dozer, Donald M. *Are We Good Neighbors? Three Decades of Inter-American Relations, 1930-1960.* Gainesville, 1959.
Duggan, Lawrence. *The Americas, the Search for Hemisphere Security.* New York, 1949.
Duncan, W. Raymond and James Nelson Goodsell, eds. *The Quest for Change in Latin America. Sources for a Twentieth Century Analysis*. New York, 1970.
Fagen, Richard R. and Wayne A. Cornelius, Jr., eds. *Political Power in Latin America. Seven Confrontations*. Englewood Cliffs, N.J., 1970.
Gantenbein, J. W. *The Evolution of our Latin American Policy: A Documentary Record.* New York, 1950.
Johnson, John J., ed. *Continuity and Change in Latin America.* Stanford, 1964.
———. *The Military and Society in Latin America.* Stanford, 1964.
———. *Political Change in Latin America: The Emergence of the Middle Sectors.* Stanford, 1958.
Kane, William E. *Civil Strife in Latin America.* Baltimore, 1972.
Kaufman, W. W. *British Policy and the Independence of Latin America.* New Haven, 1951.
Lieuwen, Edwin. *Arms and Politics in Latin America.* New York, 1961.
Mac Eoin, Gary, *Revolution Next Door: Latin America in the Nineteen Seventies.* New York, 1971.
May, Herbert K. *Problems and Prospects of the Alliance for Progress: A Critical Examination.* New York, 1968.
Mecham, J. Lloyd. *A Survey of United States-Latin American Relations.* Boston, 1965.
———. *The United States and Inter-American Security, 1889-1960.* Austin, 1961.
Munro, Dana G. *Intervention and Dollar Diplomacy in the Caribbean, 1900-1921.* Princeton, 1964.
Needler, Martin C. *Political Development in Latin America: Instability, Violence and Evolutionary Change*. New York, 1968.

———. *The United States and the Latin American Revolution.* Boston, 1972.
Palmer, Jr., Thomas W. *Search for a Latin American Policy.* Gainesville, 1957.
Pan American Union. *The Alliance for Progress and Latin American Development Prospects: A Five Year Review, 1961-1965.* Baltimore, 1967.
Perkins, Dexter. *The United States and Latin America.* Baton Rouge, 1961.
———. *Hands Off: A History of the Monroe Doctrine.* Boston, 1945.
Plaza Lasso, Galo. *Latin America Today and Tomorrow.* Washington, 1971.
Reidy, Joseph W. *Strategy for the Americas.* New York, 1966.
Rippy, J. F. *Latin America in World Politics.* New York, 1938.
———. *Globe and Hemisphere.* Chicago, 1958.
———. *Rivalry of the United States and Great Britain over Latin America, 1808-1830.* Baltimore, 1929.
Ronning, C. Neale, ed. *Intervention in Latin America.* New York, 1970.
Whitaker, A. P. *The Western Hemisphere Idea: Its Rise and Decline.* Ithaca, 1954.
———. *The United States and the Independence of Latin America.* Baltimore, 1941.
———. *The United States and South America: The Northern Republics.* Cambridge, Mass., 1948.
———. *The United States and Argentina.* Cambridge, Mass., 1954.
———. *Nationalism in Latin America.* Gainesville, 1962.
Wood, Bryce. *The Making of the Good Neighbor Policy.* New York, 1961.
———. *The United States and Latin American Wars, 1932-1942.* New York, 1966.
Wythe, George. *The United States and Inter-American Relations.* Gainesville, 1964.
U.S. Cong., Joint Economic Committee, 89th Cong., 1st Sess. *Latin American Development and Western Hemisphere Trade,* Washington, 1965.
U.S. Dept. of State. *The Story of Inter-American Cooperation. Our Southern Partners.* Dept. of State Pub. 7404, Inter-American Series 78. Washington, 1962.
U.S. Dept. of State. *United States Foreign Policy, 1969-1970.* Dept. of State Pub. 8575. General Foreign Policy Series 254. Washington, 1971.
U.S. House of Rep., Committee of Government Operations, 90th Cong., 2nd Sess., *U.S. Aid Operations under the Alliance for Progress.* Washington, 1968.
U.S. House of Rep., 91st Cong., 1st Sess. *A Review of Alliance for Progress Goals.* Washington, 1969.
U.S. Senate, Subcommittee on American Republics Affairs, 90th Cong., 2nd Sess., *Survey of the Alliance for Progress.* Washington, 1968.
U.S. Senate, Subcommittee on Western Hemisphere Affairs, Committee of Foreign Affairs, *Rockefeller Report on Latin America.* Washington, 1970.

2

Cooperation by Conference

Many methods and agencies have been employed to bring about mutual sympathy and understanding and the furthering of foreign policy goals among the independent republics of the Western Hemisphere. One of the most effective has been a series of Pan American conferences and congresses. The idea underlying these conferences was put forth by Simón Bolívar, liberator of northern South America, in 1824, the year ending the wars for independence in Latin America when he called the Congress of Panama into session. His object was to establish "certain fixed principles for securing the preservation of peace between the nations of America, and the concurrence of all those nations in defense of their own rights.[1]" The conference took place two years later, but only four governments participated: Peru, New Granada (Colombia), Central America, and Mexico. Both the United States and Great Britain were invited to send delegates. Owing to the cordial support of the idea by Henry Clay, then Secretary of State, President John Quincy Adams appointed two plenipotentiaries for the United States. Strong opposition developed in Congress, however, and the necessary appropriations were not made in time to enable our delegates to take part.[2] The British delegate was present, having been instructed by British Foreign Minister George Canning to remind the Latin American delegates that "any project for putting the U.S. of North America at the head of an American confederacy . . . would be highly displeasing to your government." Although a number of worthy agreements and proposals were signed, the times were not yet ripe for employing arbitration and mediation in the settlement of international disputes, and the conference had no tangible results. None of the treaties were ratified by the contracting parties.

[1] International American Conference, *Senate Executive Document No. 232,* 51st Cong., 1st Sess., Part IV, p. 155.

[2] Opposition in the U.S. Congress arose from noninterventionist sentiment, and because the question of abolishing the African slave trade was an item on the conference agenda, the latter arousing the fears of the slave states. For details see J. B. Lockey, *Pan Americanism: Its Beginnings* (New York, 1920), pp. 313-316, chap. X; J.B. Moore, *Principles of American Diplomacy* (New York, 1918), pp. 370-375.

Nevertheless when, one hundred years later, the Republic of Panama invited representatives of the states of the Western Hemisphere to join with her in the Panama Congress of 1926 to further the ideals fostered by the great Liberator, one of the most interesting items on the agenda was the consideration of Bolívar's project of a league of nations for the states of the New World.

Bolívar clearly saw how advantageous it would be if the nations could get together in conference at regular intervals, not only to become better acquainted, but also to settle the various disputes that were bound to arise between them. He therefore suggested an amphictyonic assembly of plenipotentiaries empowered to use good offices, mediation or arbitration. It was to negotiate treaties for the preservation of peace and interpret treaties when difficulties arose. In fact it had many elements of likeness to the Council and Assembly which were to function later at Geneva.

But not only did Bolívar see the need for providing machinery to settle disputes. He believed that it was even more important to eliminate causes of friction. Therefore he constantly urged the guarantees of territorial integrity upon the basis of the status quo. He also realized the dangers of secret alliances with foreign powers and urged that they be forbidden except when accepted by all members of the confederation. Friendly intercourse and the elimination of economic and political barriers were regarded as prerequisites of effective cooperation. Even social problems were not overlooked, and the provision was made for the complete extirpation of the African slave trade. Finally, the sanction of force was made available as a last resort, and careful consideration was given to effecting the best possible cooperation in times of emergency.

Looking back today, we are compelled to admire the sheer audacity of Bolívar in attempting to give the states of the Western Hemisphere an international law for their mutual relations almost simultaneously with the establishment of their internal governmental organizations.[3]

The second conference met in Lima, Peru, from December 1847 to March 1848, but this was attended only by representatives of South American countries: Bolivia, Chile, Ecuador, New Granada, and Peru. It was convened mainly to forestall a possible invasion of Ecuador by a former president who was believed to be recruiting an invasion force in Spain and England. Although the United States was invited to send representatives to the conference President Polk did not believe that it was sufficiently vital to the United States to do so. A Treaty of Confederation, the main product of the conference, provided for mutual assistance if the contracting powers were attacked by a foreign power; however, neither this treaty nor others were ratified by the governments involved.[4]

The third conference of Spanish American origin, convened in Santiago, Chile, in 1856, saw the adoption of a Treaty of Alliance and Confederation, called the "Continental Treaty." Among its provisions was a pledge to prevent the organizing of hostile expeditions by political émigrés within any of the allied states; it also called for the formation of a union. As before, the treaty failed because of filibustering expeditions in Central America and Mexico, and the

[3] Graham H. Stuart, "Simon Bolivar's Project for a League of Nations," *Southwest Polit. and Soc. Sci. Quar.*, vol. VII (Dec., 1926), p. 238.
[4] William R. Manning, ed., *Diplomatic Correspondence of the United States: Inter-American Affairs, 1831-1860.* 12 vols. (Washington, 1938), X, 551.

implications of Manifest Destiny as revealed by the war with Mexico. It is not surprising that the United States was not invited to participate.

The fourth and last in the series of Latin American conferences took place in Lima in 1865. Being restricted to republics that had formerly been under Spanish rule, the United States was not approached. A Treaty of Union and Defensive Alliance was signed pledging the signatories to mutual defense against aggression, but neither this nor other treaties signed were ratified. Only seven Spanish American nations, those fearing France and Spain, attended. Prompting this meeting was French intervention in Mexico, and Spain's seizure of Peru's Chincha Islands and its reoccupation of Santo Domingo. This ended efforts toward union and collective action on the part of the Latin American states; thereafter the initiative for the Pan American movement came from the United States.[5]

In November, 1881 James G. Blaine, Secretary of State under President James Garfield invited the governments of all the independent states of the two continents to send representatives to a general conference to be held in Washington. Blaine's objectives were summarized in the following way:

> The foreign policy of President Garfield's administration has two principal objects in view: first to bring about peace, and prevent future wars in North and South America; second, to cultivate such friendly commercial relations with all American countries as would lead to a large increase in the export trade of the United States by supplying those fabrics in which we are abundantly able to compete with the manufacturing nations of Europe.[6]

A far-sighted statesman, Blaine stood for pushing the expansion of American foreign trade and challenging Britain's commercial supremacy in Latin America. He believed that trade expansion could be achieved only under conditions of peace and stability; moreover, this would avert pretexts for European intervention in the region.

The second War of the Pacific between Chile on the one side, and Peru and Bolivia on the other was then in progress which gave a reason for holding a conference to negotiate a hemispheric arbitration system. A number of Latin American states accepted the invitations, but soon after they had been extended President Garfield was assassinated. Vice President Chester A. Arthur who succeeded him was a political opponent of Blaine, and as quickly as Blaine was removed from the Cabinet, the invitations were withdrawn. The project had wide support in Congress, however, and when Grover Cleveland became president in March, 1885, another resolution was passed in the House of Representatives requesting the president to invite delegates to a general Inter-American conference. The invitations were sent out by Secretary of State Bayard in July 1888, but owing to the Republican party's victory in that year, President Harrison had the honor of inaugurating the First International American Conference. All the independent republics except the Dominican Republic were represented, and Mr. Blaine, who once more held the office of Secretary of State, was chosen to preside. The program outlined for consideration was an extensive one, and

[5] For details on the Spanish-American conferences see Inter-American Institute of International Legal Studies, *The Inter-American System: Its Development and Strengthening* (New York, 1966), xv-xx.

[6] James G. Blaine, *Political Discussions* (Norwich, Conn., 1887), p. 411; *Senate Exec. Doc. no. 232*, p. 256.

included measures tending to promote peace and prosperity, the establishment of uniform customs regulations, a uniform system of weights and measures, laws for the protection of patents, copyrights, and trademarks, and the formulation of a definite plan of arbitration for the settlement of all disputes among the American nations.

The project for international arbitration proved to be the most controversial on the conference agenda, and a stalemate occurred. Chile was especially sensitive on the matter of compulsory arbitration because of its recent seizures of territory in the War of the Pacific. Secretary Blaine offered a compromise arbitration plan providing that: (1) arbitration should be adopted by the American nations "as a principle of American international law" for the solution of disputes among themselves or between them and other powers; (2) arbitration should be obligatory in all controversies except those which, in the judgement of one of the states involved in the controversy, compromised her independence; (3) the court of arbitration should consist of one or more persons selected by each of the disputants and an umpire who should decide all questions upon which the arbitrators might disagree.[7]

The project of a treaty of arbitration was to become effective as a treaty only when it has been ratified by fifteen states. Only eleven nations, including the United States, signed, and ultimately not one state ratified the document. This established the precedent for a practice that was to occur frequently in inter-American conferences: governments were not prepared to go as far as their representatives by implementing or ratifying treaty agreements signed at conferences.

Aside from its significance as the first of a series of Pan American conferences, the principal achievement of the Washington Conference was the establishment of an International Union of American Republics, its sole organ being a Commercial Bureau of the American Republics, which was given the task of collecting and disseminating information relating to tariffs and commercial laws in the member states. The Commercial Bureau was established in Washington under the administration of a Director.

Since encouraging gains were made at the First International American Conference, it was to be expected that other conferences would be called to enlarge upon and follow up its work. The second conference was summoned in October 1901 after President Mc Kinley had suggested to President Porfirio Díaz that he invite the delegations to meet in Mexico City. The delegates agreed to several conventions relating to copyrights, trademarks, extradition, and the codification of international law, but perhaps the most important result was that seventeen Latin American nations became parties to the Hague Convention of 1899 for the pacific settlement of international disputes. Several delegations favoring compulsory arbitration, which was opposed by the United States, drafted such a treaty but it was ratified later by only six nations. A treaty providing for the arbitration of pecuniary claims, signed by seventeen countries, was ratified later by ten, including the United States.[8]

At the Second International Conference the Commercial Bureau was

[7]First Int. Conf., Amer. States, *Minutes*, pp. 813-814.
[8]*Second Int. Amer. Conf., Senate Doc. No. 330, 57th Cong., 1st Sess.*

reorganized, being placed under the supervision of a Governing Board composed of the diplomatic representatives of the American governments in Washington, and the secretary of state of the United States as Chairman of the Board. The conference also voted to regularize its meetings and to hold them every five years.

The Third International Conference of American States met in Rio de Janeiro in 1906, and Elihu Root, who as secretary of state of the United States was making an official tour of South America, represented the United States. His speech at this conference remains one of the greatest expositions of the vital need of a mutual and sympathetic understanding among the American nations. The following sentences indicate his message:

> No nation can live unto itself alone and continue to live. Each nation's growth is part of the development of the race . . . There is not one of all our countries that can not benefit the others; there is not one that will not gain by the prosperity, the peace, the happiness of all . . . We wish for no victories but those of peace; for no territory except our own; for no sovereignity except the sovereignity over ourselves.[9]

The conference took place in an atmosphere of cynicism and distrust because of the role of the United States in the Panama revolution, making protectorates of Cuba and Panama, and the assumption of customs control in the Dominican Republic in accordance with the so-called Roosevelt Corollary of the Monroe Doctrine. Secretary Root's sincerity helped to repair the image of the United States, and since care had been taken to avoid controversial issues on the agenda the conference proved to be more harmonious than might have been expected. However, a resolution was proposed recommending that the Hague Conference be asked "to consider whether and if at all, to what extent, the use of force for the collection of public debts is admissible." This immediately brought into focus, and called for conference endorsement, two doctrines that the Latin Americans regarded as a shield against intervention by more powerful predatory nations, and which by definition included the United States: the Calvo and Drago doctrines. Carlos Calvo, the eminent Argentine international lawyer, had declared that a nation's sovereignty is inviolable and under no circumstances does the resident alien enjoy the right to have his government intervene on his behalf. The doctrine attributed to the Argentine Minister of Foreign Affairs, Luis Maria Drago, held that because public debts are contracted by the sovereign power of the state, they constitute a special kind of obligation, and armed force should be ruled out as a means of collecting them.

The United States delegation at the conference wished to avoid a commitment on the forcible collection of debts fearing that it would alarm European capitalists and perhaps damage the credit standing of some hemispheric states. A resolution, sanctioned by the United States, was adopted which invited the Latin American countries to pursue the question of the compulsory collection of public debts at the Second Hague Conference. But to the great dismay of the Latin Americans a U.S.-backed decision outlawing the collection of contract debts was adopted there. It provided further that the debtor state accept arbitration and the arbitral award. This did not conform with the

[9] *Third Int. Amer. Conf., Senate Doc. No. 365*, 59th Cong., 2nd Sess.; Elihu Root, *Latin America and the United States* (Cambridge, Mass., 1917), p. 6.

doctrines of either Calvo or Drago which would not admit intervention under any circumstances.[10]

The Fourth International American Conference assembled at Buenos Aires on July 21, 1910. All the American republics were represented except Bolivia, and although Philander C. Knox, secretary of state of the United States was elected honorary president, his "dollar diplomacy" had not endeared him to the Latin Americans. The conference took up and debated an extensive list of subjects, and adopted a number of important conventions and resolutions. In preparing the agenda Washington took care to exclude topics of a controversial political nature as had arisen previously. Among the most noteworthy issues considered were resolutions concerning patents, trademarks, and copyrights; extending the existence and powers of the Panama Railroad Committee; providing for the encouragement of steamship communication between the republics of the American continent; the exchange of professors and students between the universities of the Americas; recommending the uniformity of consular documents and custom house regulations; and extending the pecuniary claims convention for an indefinite period.

Another reorganization was made, adopting the name of Union of American Republics and changing the name of its secretariat from Bureau to Pan American Union. The name Pan American Union was used for the first time at this conference, but in reality it was merely a continuation of what had been established in 1890 in a more restricted form, and by 1910 had grown and acquired a definite standing in the hemisphere. The Union was given broader functions than before. Its supervision remained in charge of the Governing Board, and while the administration was still entrusted to a Director, this official was thereafter to be called the Director General. In addition, the post of Assistant Director was created whose duty, among others, was to act as Secretary of the Governing Board.[11]

Between the Fourth and Fifth International American Conferences many vital changes had occurred affecting the relations both between the United States and Latin America, and between the latter and Europe; most related to World War 1 and the newly-formed League of Nations. In their four conferences preceding the outbreak of the war in 1914 the American nations had not prepared themselves for the problems that a world war would bring, with the result that they were without a plan for cooperative effort. While the Latin American nations failed to implement by collective action the principle of continental solidarity during the war, the fact that most of them supported the United States position indicated an awareness of a hemispheric community of interest.

[10] Carlos Calvo, *Le Droit international*, 4 vols. (Paris, 1896); Alejandro Alvarez, *The Monroe Doctrine* (New York, 1924), pp. 244-257; E.M. Borchard, "Calvo and Drago Doctrines," *Encyclopedia of the Social Sciences*, III (1930), 155. After the turn of the century the United States took the part of European powers by assuming responsibility for the redress of European grievances in connection with defaulted financial obligations in the Western Hemisphere. Difficulties arising from this source helped to create Latin-American solidarity against both the U.S. and Europe. Until the adoption of the nonintervention policy by the U.S. in the 1930s, the first six inter-American conferences may be regarded as Pan American bloc actions against the United States.

[11] *Fourth Int. Amer. Conf., Senate Doc. No. 744*, 61st Cong., 3rd Sess.; Manuel Canyes, *The Organization of American States and the United Nations* (Washington, 1963), p. 4.

All of the nations of Latin America signed the Versailles Treaty as well as the Covenant of the League of Nations. Ten of these states became charter members of the League, and eventually all the republics became members, for varying periods at least. Their early attraction to the League was compounded of idealism and the desire for national prestige, but also its value as a counterweight to United States power and influence. When the United States failed to join the League, and its weakness was revealed, regionalism was turned to as the option for dealing with the "Northern Colossus."

When the Fifth International American Conference was convened in 1923, "Yankeephobia" was at a high level. Much had occurred since the Buenos Aires Conference in 1910 to create this feeling: the Taft administration's "dollar diplomacy" in Central America and the Caribbean, and the bitterly-resented interventions during the Wilsonian era. Against this background, together with their frustration by the League of Nations, it is not surprising that the Latin Americans viewed the Fifth Conference as a means of reinforcing the regional concept. Where formerly the conferences had limited their discussions almost entirely to problems of a social and economic character, the agenda of the Fifth Conference included certain rather delicate political questions. In addition to problems of agriculture, commerce, transportation, and health, one noted such subjects as the codification of international law, the reduction of military and naval armament, measures tending to bring about a closer association of the American republics, and questions arising out of the encroachment by a non-American power on the rights of an American nation.[12]

When the conference opened in Santiago, Chile, on March 25, 1923, Mexico, Peru, and Bolivia were found to be unrepresented, Mexico because of her failure to obtain the recognition of the United States, and Peru and Bolivia because of the failure to settle the Tacna-Arica dispute. The American delegation headed by Henry P. Fletcher and including Senator Kellogg, who subsequently became secretary of state, and Dr. L. S. Rowe, Director General of the Pan American Union, found itself entrusted with the rather difficult task of averting drastic action on the part of the smaller Latin American states in regard to the reorganization of the Pan American Union and a clarification of the Monroe Doctrine.

As regards the reorganization of the governing board of the Pan American Union, a separate organization was proposed to consist of representatives other than the regularly accredited diplomatic agents of the American republics at Washington, a body which might ultimately become the council of an American League of Nations. The United States opposed this idea but conceded that states not possessing diplomatic representatives at Washington should be represented on the governing body of the Union by special representatives.[13] It was also agreed that henceforth instead of the secretary of state of the United States acting *ex officio* as chairman of the governing board, the latter would elect its president and vice-president.

The conference sidestepped action on a specific interpretation or a joint

[12] *Report of Delegates of U.S.A. to Fifth Int. Conf. of Amer. States* (Washington, 1924), Appendix I.

[13] This change was incorporated in Article V of the Resolution on the Organization of the Pan American Union adopted by the Conference.

sanction of the Monroe Doctrine, to neither of which the United States would subscribe, by adopting an innocuous resolution entrusting to the governing board of the Pan American Union "the task of studying the bases . . . relative to the manner of making effective the solidarity of the collective interests of the American continent.[14]

Although the committee on the limitation of armament recommended the limitations of the Washington Conference of 1922 as to tonnage of capital ships and airplane carriers and caliber of guns thereon, of greater importance was the work of the committee in securing the adoption of a convention generally known as the Gondra Treaty for the investigation of disputes between American states by a commission of inquiry modeled upon those set up by the Hague Conventions and the Bryan treaties. Señor Augustin Edwards, president of the conference, has been quoted as regarding this treaty as "the most important ever signed on the American continent for the promotion of peace."[15]

Opinion seems to differ radically as to the real success of the conference. Certainly as regards cooperation along nonpolitical lines, as much if not more progress was made than in any of the previous conferences. Conventions for the protection of trademarks, the publicity of customs documents, and on uniformity of nomenclature for the classification of merchandise were adopted. The Hague Convention for the suppression of the drug traffic was approved, simplification of passports and visas and their ultimate elimination recommended, and closer cooperation for stamping out disease resolved upon. But whenever political issues were discussed there was noticeable throughout the conference a marked antagonism toward the United States, particularly on the part of the smaller states in the Caribbean area. For the first time in such a gathering the long smoldering fears of American imperialism were given free expression. Nevertheless, although the United States did not escape open criticism, the very fact that frankness prevailed relieved the tension to a considerable extent and made for results of a more lasting character.

The Sixth International American Conference met from January 16 to February 20, 1928, in Havana, Cuba. It came at a time when the United States was being harshly criticized throughout Latin America for its intervention in Nicaragua. President Coolidge, as a special honor to Cuba on the thirtieth anniversary of her independence, as well as to placate Latin America, made a hasty trip to Havana, in order to proclaim in person the friendliness of the United States towards all Latin America. Delegates from all the Latin American countries were present. To make the gesture of friendship the more emphatic the most eminent delegation ever sent by the United States to an international conference was chosen to represent us. Headed by former Secretary of State Hughes, it included Henry P. Fletcher, former ambassador to Chile,Mexico, and Belgium, and at the time to Italy, ex-senator Oscar W. Underwood, Dwight W. Morrow, ambassador to Mexico, Dr. James Brown Scott of the Carnegie Endowment, President Ray Lyman Wilbur of Stanford University, and Dr. L.S. Rowe, director general of the Pan American Union.

[14] *Report of Delegates of U.S.A. to Fifth Int. Conf. of Amer. States*, Appendix XII.

[15] G.H. Blakeslee, *The Recent Foreign Policy of the United States* (New York, 1925), p. 143.

Although on the surface the agenda of the Sixth Conference[16] seemed to have avoided all contentious political subjects, the proposal to establish the Pan American Union upon a conventional basis instead of the hitherto less formal basis of successive resolutions offered the Latin American delegates an opportunity to try to curb the over-preponderant influence which it was claimed the United States exerted both on the governing board and in the administration of the Union. Projects for the codification of international law for the American continent placed the United States on the defensive in regard to such questions as intervention, recognition, and the equality of states.

While a felicitous speech of generalities by President Coolidge indicated a keen desire on the part of the United States to inspire confidence and good will, an address made by Mr. Hughes before the United States Chamber of Commerce in Havana made an even more favorable impression, owing to its frank reference to Nicaragua and Haiti.

Asserting that the first pillar of Pan Americanism was independence, Mr. Hughes declared it to be the firm policy of the United States to respect the territorial integrity of the American republics. But the second pillar was stability and the United States desired to encourage stability in the interest of independence. Her entrance into Santo Domingo and withdrawal upon the establishment of a stable government proved it. "We would leave Haiti at any time that we had reasonable expectations of stability . . . . We are at this moment in Nicaragua; but what we are doing there and the commitments we have made are at the request of both parties and in the interest of peace and order and a fair election. We have no desire to stay. We entered to meet an imperative but temporary exigency; and we shall retire as soon as possible."

The conference decided that all sessions, both plenary and full committee, should be open to the public—an innovation. The United States accepted the Mexican-Peruvian proposal for governmental reorganization of the Pan American Union to permit each nation to decide whether a special delegate or its diplomatic representative should be employed. An additional proposal for the regular rotation of the Chairman, Vice Chairman and Director General was defeated, however, the conference agreeing that the Director General should attend the conference as an *ex officio* member instead of as a member of one delegation, and his expenses should be paid *pro rata* by all the member states.

A cause of serious disagreement came when the committee on codification of international law discussed the third article of the project concerning the existence, equality, and recognition of states. This article declared that "no state may intervene in the internal affairs of another."[17] With the American marines actively engaged in the pursuit of Sandino in Nicaragua the subject was of more than academic interest. The United States disputed the correctness of the rule, making a distinction between political intervention for permanent possession, and temporary interposition for humanitarian or other limited purposes. Furthermore its inclusion in a code would be futile since it was contrary to the accepted practice of international law. The sixth conference agreed to adopt

[16] *Sixth Int. Conf. of Amer. States Special Handbook for the Use of Delegates* (Pan American Union, Washington, 1927).

[17] *International Commission of Jurists—Public International Law Projects* (Pan American Union, Washington, 1927), p. 8.

obligatory arbitration for the settlement of justiciable disputes, and provided for an arbitration and conciliation conference to be held at Washington within the next year to draw up a collective Pan American arbitration convention which should outlaw aggressive warfare in the Western Hemisphere.

In the questions of a less political nature the conference was able to make a signal progress. The excellently drawn code of private international law prepared by Dr. Bustamente was approved, an aviation convention for the regulation of aircraft communication between the American republics was accepted, with a reservation to the effect that special arrangements between any two states for reciprocal convenience might be permitted provided it impaired in no respect the rights of other parties to the convention. The amendment permitted the United States to make special arrangements for the protection of the Panama Canal.

The treaty on the rights and duties of neutrals in the event of war which placed belligerent submarines under the same rules as other vessels of war in visit and search, and subjected armed merchant vessels to the rules of neutrality in respect to the time for remaining in port, coaling, and provisioning, was approved. Treaties placing aliens abroad on the same footing as nationals and establishing the right of asylum were signed although the United States entered a formal reservation to the latter. The copyright convention was revised and recommendations made for an inter-American automobile highway extending from Canada to Patagonia, for a Pan American railway, better river navigation and improved cable, telegraph, and radio communications. A Pan American pedagogical congress was agreed upon, as well as the creation of a Pan American institute of geography and history.

The Havana Conference from many points of view was unique. Political questions which hitherto had been completely eliminated from the agenda were brought up and discussed frankly without fear or favor. The United States made every effort to explain its position of *primus inter pares* and owing to the excellent choice of its representatives was remarkably successful. Every Latin American state was represented, and at the conclusion of the conference their representatives could leave with the assurance that they had at last come into their own as equal participants in the fraternity of American nations.

Supplementing the Havana Conference by which it was authorized, the Conference of American States on Conciliation and Arbitration met at Washington on December 10, 1928. With the exception of Argentina, all the independent American states were represented. Secretary of State Kellogg and ex-Secretary of State Hughes acted as delegates for the United States. Two treaties were signed at this conference, one on arbitration and the other on conciliation.

The General Treaty of Inter-American Arbitration has been called one of the most advanced multilateral arbitration pacts ever concluded. It followed the Kellogg-Briand Pact in its general form, condemned war as an instrument of national policy, and provided for settlements of all justiciable questions by arbitration. Only two subjects were excluded—domestic questions and questions concerning third states. Any existing international tribunal might be used or a special one set up.

The General Convention of Inter-American Conciliation was based upon the Gondra Treaty of 1923. It retained the commissions of inquiry of that treaty, but gave them also the character of commissions of conciliation. It covered all controversies between states not settled by diplomacy, it defined the procedure to be used and made it arbitary, it set up permanent bodies at Washington and at Montevideo to bring about conciliation, and it approved the right of any state to offer mediation.

These two treaties, which seemed to provide for the settlement of any dispute which might arise between the states of the Western Hemisphere, were signed by all the states present. Before the seventh international conference met in Montevideo in 1933 a large majority of the twenty-one republics had ratified one or the other of these conventions, and ten states, including the United States, had ratified both.

Despite the tragic failure of the Geneva Disarmament Conference and the London Economic Conference of 1933, the republics of the Western Hemisphere were unwilling to postpone the meeting of the Seventh Pan American Conference scheduled to convene at Montevideo December 3, 1933. Not only were all American republics represented, half of them by their ministers of foreign affairs, but also, for the first time, it was proposed to admit official observers from Spain, Portugal, and the League of Nations. Action on this radical change in the organization of the conference was postponed to the eighth meeting, however. An eminent delegation from the United States, headed by Secretary of State Hull, included J. Reuben Clark, former ambassador to Mexico, Alexander W. Weddell, ambassador to Argentina, and J. Butler Wright, minister to Uruguay, all outstanding authorities on Latin-American affairs.

In his address of welcome President Terra of Uruguay declared that "the American ideal of peace must not be buried in the swamps of the Chaco." Yet thousands of Bolivian and Paraguayan soldiers had already been slaughtered and buried in the swamps of the Chaco and thousands of others faced the same fate. The problem was complicated by the fact that since a Commission of American Neutrals and an ABC Peru Commission had both failed to bring about a settlement, a League of Nations Commission was now engaged in trying to work out a solution. For this reason, the United States and Brazil,[18] both non-League powers, were unwilling to bring the Chaco dispute, which was not on the agenda, into the conference. But, when it could not be avoided, the conference prevailed upon Bolivia and Paraguay to declare a truce for the period of the conference. It also passed a resolution submitted by Secretary Hull urging both peoples to accept judicial processes for the settlement of their dispute, as recommended by the League of Nations Commission. As a concrete suggestion the conference adopted an Argentine proposal that, contingent upon approval of the League Commission, a conference be held at Buenos Aires including the ABCP powers, as well as the two disputants, to settle the Chaco question by considering the economic and geographic problems of Bolivia and Paraguay.

The principal achievement of the conference was the coordinating and strengthening of inter-American peace machinery. Before the Montevideo conference opened, all of the Latin American states except Argentina and Bolivia had

[18] Brazil had withdrawn from the League in 1928.

ratified the Gondra Conciliation Treaty concluded at the Santiago conference in 1923. Less than half of the states had ratified the arbitration and conciliation treaties of 1929. All but five had accepted the Kellogg-Briand Pact and all but four had joined the League of Nations. In an effort to establish a standard method of procedure for the settlement of disputes in the Western Hemisphere, Foreign Minister of Argentina, Saavedra Lamas, had drafted a Latin American anti-war pact, modeled upon the Pact of Paris. In addition to outlawing wars of aggression and compelling a settlement of disputes by legal means, it outlawed aggression in settling territorial questions and followed the Hoover-Stimson doctrine of nonrecognition of territorial changes brought about by force. As a sanction the signatory powers agreed to exercise the political, judicial, and economic means authorized by international law, as well as the influence of public opinion, "but in no case shall they resort to intervention, either diplomatic or armed." Although the United States had previously refused to adhere, at the conference Secretary Hull agreed to sign the Argentina anti-war pact and also supported the proposal of Argentina and Chile to consolidate the peace machinery by having non-signatory powers sign the five peace pacts available to Latin American states. As conclusive evidence of its good neighborly intentions before the conference ended, the United States delegation accepted with explanatory reservations a convention on the rights and duties of states wherein it was agreed "no state has the right to intervene in the internal or external affairs of another."[19]

In regard to tariffs and currency stabilization, Secretary Hull took the lead in recommending a plan for reducing tariffs through the negotiation of bilateral or multilateral reciprocity treaties. As proof of its earnest intent the United States signed a reciprocity treaty with Colombia on December 15, the first of its kind since 1902. The conference voted to hold a Pan American financial conference at Santiago, Chile, in 1934, when stabilization of currency and all other thorny financial problems might be placed on the agenda.

With these political subjects out of the way, the conference had little difficulty in obtaining agreement upon questions of a technical, social, or cultural nature. A general extradition treaty was signed, binding for the first time all American countries and standardizing procedures. In a comprehensive resolution, the conference established the procedure for carrying on the future work of codifying international law and provided for a juridical section of a purely administrative character in the Pan American Union. A general convention on nationality clarified the allegiance of the individual as to nation of origin, status of inhabitants in the case of transferred territory, and the effects of matrimony or its dissolution upon the nationality of husband, wife, or children. To sum up the achievements of the Montevideo Conference: it effected an armistice in the Chagco, strengthened and correlated the peace machinery of the Americas, took a practical step towards better trade relations, and enhanced the standing of the United States as a good neighbor.

In his opening speech to the United States Congress on January 3, 1936,

[19]Article 8 of Convention on Rights and Duties of States, ratified by the U.S. June 29, 1934.

President Roosevelt declared that "at no time in the four and a half centuries of modern civilization in the Americas had there existed, in any year, any decade or any generation, in all that time, a greater spirit of mutual understanding, of common helpfulness and of devotion to the ideals of self-government than exist today in the twenty-one Republics . . . This policy of the 'good neighbor' among the Americas is no longer a hope—it is a fact, active, present, pertinent and effective." To give further concrete illustration of his intention to continue this policy, President Roosevelt on January 30 sent personal letters to the presidents of the other American republics proposing that an extraordinary inter-American conference be summoned to meet in Buenos Aires to determine how the maintenance of peace among the Americas might best be safeguarded. As a tentative suggestion to this end he proposed that the conference give attention to: the prompt ratification of existing peace agreements, their amendment in accordance with experience, and the creation of new instruments of accord.

It should be noted that one of the weaknesses of existing machinery for the maintenance of peace had been the failure of certain states to ratify the agreements already signed; for example, neither of the belligerents in the destructive war in the Chaco had ratified the Inter-American Conciliation and Arbitration Treaties of 1929. A vital need to organize peace was recognized to be the consideration of measures to secure the prompt ratification of existing treaties and conventions for the maintenance of peace. It should further be noted that the republics of the Western Hemisphere were for the most part parties not only to five different treaties of a purely continental character, but to four instruments of a universal character, all designed for the maintenance of peace as well as to some half dozen declarations against war and the forcible acquisition of territory. The conference proposed to consider the possibility of coordinating these instruments and incorporating them in a single instrument.

Under the heading of economic problems, an elaborate agenda was prepared. It was proposed to consider every phase of trade restriction which might hinder closer economic relations and if possible inaugurate substantial reforms. Every type of trade agreement was to be taken up, the question of the most favored nation clause, exchange control, and a tariff truce.

An important subsidiary subject was the improvement of means of communication in the Western Hemisphere. It was noted that substantial progress had been made towards the carrying out of the proposed Pan American Highway system; for example, on July 1, 1936, the highway from Mexico City to Laredo, Texas, was formally dedicated after having been in use for several months. This magnificent road runs for over 750 miles from the United States border through dry plains, a subtropical jungle, and towering mountain peaks to Mexico City. The paved roadway is twenty-one and a half feet wide—it required the construction of over three thousand bridges and culverts, and although it runs from practically sea level to over eight thousand feet, the whole trip can be made in high gear.

Steamship accommodations had also been improved considerably. In addition to the long-established Grace Line on the west coast, the Moore-McCormack Line on the east, and the Great White Fleet of the United Fruit Co. in the Caribbean, Lykes Brothers Steamship Co. inaugurated a service between

Mobile and the west coast of South America. The Alcoa Steamship Company offered improved service in the Caribbean and to the east coast ports, and the Standard Fruit and Steamship Co. and the Shephard Steamship Co. carried freight to the west and east coasts of South America.

With the advent of commercial aviation the communication problem changed to such an extent that the United States became more accessible to Latin America than was Europe. Pan American Airways established a commuting service to all Latin American capitals. Panagra, Braniff, American, Chicago and Southern, Colonial, Eastern, National, and Western Air Lines established flights between the Americas. Colombian Air Lines connected Bogotá and Barranquilla with Miami,and Venezuela Airlines joined New York with Caracas via Havana.

The delegation from the United States to the Buenos Aires Conference, a large and representative one, was headed by Secretary of State Hull and included Assistant Secretary of State Sumner Welles and Alexander Weddell, ambassador to Argentina. President Franklin D. Roosevelt showed his intense interest by being present and giving the opening address of the Conference. He urged the necessity of striving to prevent war in the Western Hemisphere by every honorable means and to avoid the creation of conditions giving rise to conflict. Although he did not mention the Monroe Doctrine he indicated that the United States was willing to participate in a multilateral agreement providing for mutual consultation in case of external aggression. He concluded his address by emphasizing the importance of satisfactory commercial relations as a fundamental bulwark of permanent peace.

Three concrete peace proposals were introduced and unanimously agreed upon. The first was a convention for the maintenance of peace. It provided that should the peace of the American republics be threatened by any source either at home or abroad, the signatory powers should consult with each other immediately with a view to cooperative action to preserve the peace of the American continent. According to Secretary of State Cordell Hull "this proposal represents the strongest assurance of peace which this continent has ever had." It should be noted, however, that a very important part of the United States proposal, namely, a permanent body consisting of foreign ministers of each state to carry out the provisions, does not appear in the convention as accepted.

The second convention coordinated existing treaties for the maintenance of peace. It repeated the obligations and pledges under the Kellogg-Briand Pact, the Inter-American Conciliation and Arbitration Treaties of 1929, and the Saavedra Lamas Anti-War Treaty. To carry out these principles the more effectively, provision was made for individual or joint offers of good offices or mediation and a reminder to the parties to a controversy of their obligations under existing treaties.

Upon threat of war the parties concerned agreed to a delay of six months for consultation before beginning hostilities. Upon the outbreak of war the signatory powers agreed to adopt a common attitude of neutrality, and in order to prevent the spread of hostilities they might impose restrictions upon the sale or shipment of munitions and upon any sort of financial assistance to the belligerents.

The third proposal was a protocol of nonintervention. According to its terms, the internal affairs of any of the parties was inadmissible of intervention by any one of them and violations should give rise to mutual consultation. The protocol reaffirmed the nonintervention doctrine adopted in Montevideo in 1933.

The Buenos Aires Conference had as its aim the maintenance of peace in the Western Hemisphere through mutual consultation and cooperative action. Consultation was the keynote of the Conference, and the conventions were pitched accordingly. The United States, by definitely giving up the right of intervention, had finally agreed to the position of equality in the commonwealth of nations of the Western Hemisphere. The conference marked the completion of another important link on the good neighbor highway.

As the world's political conditions became ever more critical the desire for still closer cooperation for peace in the Americas was strengthened. The eighth Pan American Conference which met in Lima, Peru, December 9, 1938, afforded this opportunity. The agenda included plans both for an American League of Nations and a Pan American Court of Justice. "Hemisphere defense" was the keynote of the Conference.

The United States delegation of twelve was headed by Secretary of State Hull and included Assistant Secretary of State Berle, Ambassador to Peru Steinhardt, Minister to the Dominican Republic Norweb, and Chief Justice Cuevas of the Puerto Rican Supreme Court.

In spite of the opposition of Argentina it was quickly evident that the American republics were determined to take further steps toward strengthening the machinery for continental defense. Although the idea of an American League of Nations was discarded, the twenty-one states finally agreed to support a project of American cooperation which was designated by the conference as the Declaration of Lima. This Declaration reaffirmed the principle of continental solidarity and the decision to defend it against all foreign intervention. In case the peace, security or territorial integrity of any American republic should be threatened it was agreed to make effective this solidarity by consultations as established by conventions in force. To facilitate such consultative action the Declaration provided that "the Ministers of Foreign Affairs of the American republics, when deemed desirable and at the initiative of any one of them, will meet in their several capitals by rotation . . .."

By this Declaration a definite procedure of consultation was established which could be quickly utilized in an emergency. It was signal success for Secretary Hull in his efforts to obtain unanimous action against a threat of autocratic aggression.

Perhaps the next most important result of the Conference of Lima was a Declaration of American Principles which the governments of the American republics proclaimed as essential to the preservation of world order under law. The substance of these principles had been presented by Secretary Hull in a statement released on July 16, 1937. Summarized, the Declaration established the following canons of international conduct: intervention and the use of force as instruments of national or international policy are proscribed; international differences must be settled by peaceful means and international law must govern relations between states; the faithful observance of treaties is an indispensable rule of international conduct and revision must be obtained by agreement of the signatory powers; intellectual and economic cooperation are essential to national and international well being and world peace can only be achieved by international cooperation based upon these principles.

The Lima Conference established a precedent by not signing a single treaty or convention. All of its projects were formulated as resolutions, declarations, or recommendations, of which 112 were accepted. As in previous conferences the majority of resolutions pertained to nonpolitical subjects. Resolutions were approved committing the Americas to establish liberal trade practices and equality of treatment, to improve transportation and communication, to increase the exchange of professors and students, to protect Indian art, literature, language, and culture, and to cooperate in various ways conducive to better relations and understandings.

Although no alliance for defense was established and no new machinery for the elimination of war was set up the twenty-one republics of the Western Hemisphere agreed unanimously to consult and if necessary to act for their mutual defense. In the words of Secretary Hull "the American Republics have made it clear to the world that they stand united to maintain and defend the peace of this hemisphere, their territorial integrity, their principles of international relations, their own institutions and policies."

The nations who subscribed to the Declaration of Lima little expected that in less than a year they would be called upon to give effect to its basic principles. Yet when on September 1, 1939, Herr Hitler loosed the Nazi hordes upon Poland, the repercussion in the New World was such that the American republics unanimously agreed to consult as to a joint policy of hemisphere defense.

On September 23, 1939, the representatives of the twenty-one republics assembled at Panama City in the First Inter-American Consultative Conference of Foreign Ministers. Undersecretary of State Sumner Welles who headed the delegation of the United States well expressed the sentiments of the Conference when he characterized it as "a meeting of American neighbors to consider in a moment of grave emergency the peaceful measures which they may feel it wise to adopt . . . so as best to insure their national interests and the collective interests of the nations of the New World."

The most important problem was the security of the Western Hemisphere and the Conference faced it resolutely. A general declaration of neutrality of the American republics was adopted which not only included the generally accepted canons of conduct as to neutral rights and duties, but added several of special application to the Western Hemisphere. The American republics might bring together and place in a single port under guard belligerent merchant vessels which had sought refuge in their waters. Bona fide transfers of flags of American merchant vessels in American waters were permitted. Defensive armament on merchant vessels was conceded, but belligerent submarines could be excluded from the territorial waters of a state.

The most original and drastic action of the Conference was a joint resolution entitled the Declaration of Panama whereby the American republics declared that, so long as they were neutral, they were, as of inherent right, entitled to have those waters adjacent to the American continents free from the commission of any hostile act by a non-American belligerent. A remarkable feature of the resolution was the promulgation of a protective zone approximately 300 miles wide encircling the continents south of Canadian territorial waters. The resolution contemplated no extension of the three-mile marginal sea but rather an extension of the adjacent waters as a zone free from belligerent activities as an essential means of self-protection.

As might have been expected, violations were not long in materializing. The scuttling of the *Graf von Spee* by the Nazi captain in Uruguayan territorial waters, December 13, 1939, and various sinkings by the British within the protective zone brought about a joint protest through the President of Panama on December 23, 1939. Great Britain, France, and Germany were advised that consultations were being carried out looking towards penalizing future violations by forbidding access to supplies or repairing of damages in American ports to belligerents guilty of the commission of warlike acts within the security zone.[20]

With a view to studying further the problems of neutrality, the Conference authorized the governing board of the Pan American Union to set up an Inter-American Neutrality Committee of seven experts for the duration of the war. This committee met at Rio de Janeiro and, on April 27, 1940, set forth a lengthy recommendation which in substance favored the maintenance of the security zone as an open sea for the commercial traffic of every state but prohibited any sort of belligerent act within the zone.[21]

Another important result of the Panama Conference was the setting up in Washington on November 15, 1939 of the Inter-American Financial and Economic Advisory Committee consisting of one expert representing each of the American republics. The Committee's function was to establish a program of cooperation between the American republics to protect their economic and financial structures, maintain their fiscal equilibrium, safeguard the stability of their currencies, and develop their industries and commerce.

The second meeting of ministers of foreign affairs of the American republics convened in Havana, Cuba, from July 21 to 30, 1940. The meeting was called to meet the threat of Nazi aggression in the Western Hemisphere through the possible seizure of Dutch, Danish, and French colonies following the subjugation of the mother countries. Agreement was quickly obtained to cover any such possible contingency. Condemning violence in every form and refusing to recognize force as a basis of rights, the American republics refused to accept any transfer or attempt to transfer any interest or right in the Western Hemisphere. In the case of such an attempt thus threatening the peace of the continent, provision was made for the taking over and the provisional administration of such a region by an Inter-American Commission of Territorial Administration.

A supplementary declaration known as the Act of Havana authorized the creation of an emergency committee to cover situations arising prior to the promulgation of the convention. The declaration further permitted any state in an emergency to act singly or jointly with others in any manner required by its own defense or in defense of the continent. This action was the first unanimous recognition by Latin America of the value of the principle of the Monroe Doctrine for the defense of the Americas. A number of resolutions were aimed

[20] Dept. of State *Bulletin*, I (Dec. 23, 1939), p. 723.
[21] Pan American Union, *Decrees and Regulations on Neutrality, Supp. x, No. 2* (Law and Treaty Series No. 14), pp. 38-48.

at subversive activities, and attempted their control by restricting the political activities of foreign diplomatic and consular representatives, regulating more closely the issuance of passports, and coordinating police and judicial measures for mutual defense. The American republics were getting prepared to stand together against the totalitarian menace.

The Third Conference of Ministers of Foreign Affairs of the American republics held at Rio de Janeiro from January 15 to 28, 1942, was convened as a result of the arrival of the war in the Western Hemisphere. The Japanese attack at Pearl Harbor was a flagrant act of aggression which gave the two Americas an opportunity to prove that promised cooperation meant more than high sounding declarations.

The Caribbean republics did not wait for the conference to indicate their whole-hearted support of the United States. The day after the attack on Pearl Harbor, Costa Rica, the Dominican Republic, Guatemala, Haiti, Honduras, Panama, and El Salvador declared war upon Japan and followed this within a few days with similar declarations of war against Germany and Italy. Cuba declared war on Japan on December 9, and on Germany and Italy on December 11. Nicaragua declared war on all three of the Axis powers on December 11. Thus, within four days after the United States became a belligerent, nine of the Latin American republics had joined her in the conflict.

On December 8, Mexico condemned Japanese aggression, severing diplomatic relations the same day, and with Germany and Italy three days later. Colombia broke relations with Japan December 8, and Venezuela with the three axis powers December 31. The other Latin American republics promised to honor their obligations, refrained from regarding the United States as a belligerent, and put into effect various measures to restrain Axis activities—an auspicious background for the forthcoming conference.

Undersecretary of State Sumner Welles, who headed the American delegation, expressed the appreciation of the United States government at the declarations of solidarity and support, but he also made it clear that the only certain method of stamping out the Axis methods of poisoning inter-American intercourse was by the severance of diplomatic relations. Mr. Welles' address was well received, but it was the eloquence and idealism of Mexican Foreign Minister Padilla and the obstinacy and diplomatic acumen of Brazilian Foreign Minister Arañha which prevented the conference from foundering upon the rocks of Axis subversive activities.

A resolution sponsored by Colombia, Mexico, and Venezuela declaring that the American republics "cannot continue diplomatic relations with Japan, Germany and Italy" appeared to have unanimous support. However the Argentinian representative at first vacillated then balked, and he was supported by the representative from Chile. As a compromise, the wording of the resolution as finally voted merely "recommended rupture of diplomatic relations" as each country should determine. The sentiments of the Conference were shown when Peru, Uruguay, Bolivia, Paraguay, Ecuador, and Brazil broke off relations with the Axis powers before the Conference closed.

The final act of the Rio Conference comprised forty-one declarations and resolutions. In addition to recommending severance of diplomatic relations they urged the production and exchange of strategic materials essential to hemisphere defense and the formulation of a coordinated plan for economic mobilization. Complete coordination of transportation facilities was recommended and the improvement of all inter-American communications by land, water, and air, including the construction of the unfinished sections of the Pan American Highway. An economic boycott of the Axis powers was approved, with reservations by Argentina and Chile, and several resolutions were aimed at the combatting of subversive activities and the control of dangerous aliens. It was recommended that an Inter-American Joint Defense Board composed of military and naval technicians appointed by each government should be set up in Washington to study and recommend measures for the defense of this continent. The Board met for the first time March 20, 1942 and functioned effectively for hemisphere defense.

The Rio Conference, although based upon idealistic principles, was realistic in its methods and its proposals. The United States needed freedom of action, bases, and strategic materials; the Latin American republics needed protection, financial assistance, and supplies. The Conference provided for a mutually advantageous exchange.

During the course of World War II the overfriendly attitude of Argentina to the Axis powers provoked retaliatory measures on the part of the United States, resulting finally in a formal request upon the part of Argentina that an Inter-American Conference of Foreign Ministers be held to consider her relations with the other American republics. The United States countered by suggesting a conference of the American states cooperating in the war and post-war problems. The proposal of the United States was accepted with the understanding that the Argentine situation be placed upon the agenda.

The Inter-American Conference on Problems of War and Peace met in Mexico City during February-March, 1945. All the republics except Argentina, which had not declared war upon the Axis, were represented. The United States delegation was headed by Secretary of State Stettinius and included Senators Connally and Austin. Foreign Minister Padilla of Mexico was elected President of the Conference.

After various resolutions had been passed for closer collaboration in the war effort, such as establishing the Inter-American Defense Board upon a permanent basis and strengthening the cooperation against subversive Axis propaganda, the Conference took up the problem of post-war international organization. The Latin American states desired a strong regional agreement against aggression, whereas the United States was concerned lest regional agreements should interfere with the new world organization which was to be established.[22] The resulting compromise known as the Act of Chapultepec placed aggression against an American state either by a non-American state or an American state on a parity. In either case the signatory states would consult as to what action should be taken. Aggression was defined as armed trespass and various measures such as severing diplomatic relations, imposing an economic boycott, and using armed force were to be utilized. A treaty of such character was to be concluded after the war to constitute a regional arrangement for the

[22]The American delegation itself was divided on this issue to such an extent that the Mexican representative suggested an adjournment until the American delegation made up its collective mind.

security of the Western Hemisphere but consistent with the purposes and principles of the general international organization.

A very important resolution was passed to reorganize and strengthen the inter-American system by having the International Conferences of American States meet every four years and Foreign Ministers meet annually. The governing board of the Pan American Union was to be composed of *ad hoc* representatives with the rank of Ambassador and they were given broader powers. The Chairman of the Board was to be elected annually and not be eligible for immediate re-election. The Director General was to be chosen for a ten-year term and not be eligible for re-election. A draft charter including these and other changes was to be prepared and presented at the next Conference of the American States.[23]

At the end of the Conference a resolution was passed deploring Argentinian policy and expressing the wish that Argentina would change its policy so that it might adhere to the Final Act of the Conference and be incorporated into the United Nations. An understanding was reached at the Conference that if Argentina would declare war on the Axis and curtail futher Axis activities in Argentina, the United States would support the admission of Argentina in the forthcoming conference at San Francisco to draw up a charter for the United Nations.

Although it was expected that a Conference of Foreign Ministers of the American Republics would meet in 1946 to formalize the provisions of the Act of Chapultepec by concluding a treaty of mutual assistance, it was not until August 15, 1947, that such a conference met in Rio de Janeiro.[24] All of the Latin American republics except Nicaragua were represented and the Conference completed its work in less than three weeks. The Treaty of Reciprocal Assistance signed at Rio de Janeiro on September 2, 1947, condemned war and provided for peaceful procedure for the settlement of all disputes in the Inter-American System. An attack against one American state was an attack against all and each undertook to assist the state attacked, individually and collectively, until the Security Council of the United Nations should take measures to maintain security. The states were to consult when the peace of the Americas was theatened and sanctions might be employed beginning with the suspension of diplomatic relations and ending with the use of force. Aggression was defined as unprovoked attacks or armed invasion.

This regional agreement looking towards the maintenance of security in the Western Hemisphere which became operative in 1948 when Costa Rica, the fourteenth republic, deposited its ratification, was the successful culmination of the continuous efforts of almost a half century to achieve cooperation for mutual defense. The fact that action could be taken by a two-thirds vote made the procedure the more effective. It was up to the next conference to perfect the organization of American states.

As previously noted, a draft charter for the organization of the American

[23] *Final Act of the Inter-American Conference on Problems of War and Peace* (Washington, 1945).

[24] The United States had delayed the conference because of Argentina's failure to comply with the obligations assumed after the conference at Mexico City.

republics had been prepared and was the principal item on the agenda of the Ninth International Conference of American States which met in the spring of 1948 in Bogotá. In spite of a bloody revolution which cost over a thousand lives and destroyed many of the fine public buildings in Bogotá, the Conference carried out the program contemplated.

The United States had noted a gradually mounting antagonism in the Latin American countries as a result of its inability to carry out the European Recovery Program and at the same time meet the economic and financial needs of its neighbors to the south. To reverse this trend the American delegation headed by Secretary of State Marshall not only included such diplomatic experts in Latin American affairs as Assistant Secretary of State Norman Armour, Director of the Office of American Republics Paul C. Daniels, and Ambassador Willard L. Beaulac, but also Secretary of the Treasury John W. Snyder, Secretary of Commerce W. Averell Harriman, and Head of the World Bank John J. McCloy.

The Charter of Organization of the American States as finally adopted at Bogotá not only put the Organization upon a permanent treaty basis, but made a number of important changes long considered necessary. The Inter-American Conference was made the supreme organ for the formulation of policy; it would meet at least once every five years; each state would be represented and have one vote. The Meeting of Consultation of Ministers of Foreign Affairs was retained to consider problems of an urgent nature and it could be convened upon the request of any state. The former governing board of the Pan American Union was now established as the Council of the Organization of the American States with each state having a representative possessing the status of ambassador. It was given authority to consider all matters submitted by the International Conference or the Organ of Consultation and it might serve provisionally as the Organ of Consultation. It could draft and submit proposals to the governments or to the Conferences of the American States. As aids, three agencies were established: the Inter-American Economic and Social Council, the Inter-American Council of Jurists, and the Inter-American Cultural Council. The Pan American Union under the Charter became the permanent organ of secretariat of the Organization.

Perhaps the next most important achievement of the Ninth Conference was the American Treaty on Pacific Settleement, also known as the "Pact of Bogotá." By the terms of this treaty the American Republics accepted the obligation to "settle" all disputes by peaceable means. The procedures provided are good offices, mediation, investigation and conciliation, judicial settlement, and arbitration. This treaty was the successful culmination of the sustained effort to bring about obligatory peaceful settlement of disputes in the Western Hemisphere.

On the economic side Secretary Marshall stated that President Truman had requested an increase of $500 million in the Export-Import Bank allocation for Latin-American loans. It was also promised that the Economic Cooperation Administration would finance a part of Latin-American exports to Europe in United States dollars. That the allocation did not reach the amount expected was due to the policy of Argentina fixing a price for her cereals, meats, and hides far above the world market price. An elaborate Economic Agreement was signed which provided for financial and technical cooperation, for the approval and safeguarding of private investment, and the improvement of transportation facilities in the Western Hemisphere.

Inasmuch as individual rights were being eliminated in all countries behind the Iron curtain, the Ninth International Conference adopted an American Declaration of Rights and Duties of Man far more inclusive than the Bill of Rights in the United States Constitution. Also, in order to go on record as regards the threat of Soviet Communism, the Conference approved a declaration and resolution stating that "by its anti-democratic character and its interventionist tendency the political activity of international communism or any totalitarian doctrine is incompatible with the concept of American freedom . . . ."[25]

The Ninth International Conference at Bogotá, by setting up an improved organization of the American states upon a treaty basis, pointed the way for the establishment of regional organizations for the maintenance of peace in other parts of the world as envisaged by the Charter of the United Nations.[26] The Charter of the Organization of the American States went into effect December 13, 1951, when Colombia was the fourteenth state to complete ratification.

The Senate of the United States approved the ratification of the Charter of the Organization of American States on August 28, 1950. A world crisis had occurred three months earlier when the Communist forces in North Korea had violated the 38th Parallel and marched south. The Security Council of the United Nations immediately voted that a breach of the peace had occurred and called upon all members of the United Nations to render assistance. One day later, June 28, the Organization of American States unanimously adopted a resolution declaring its firm adherence to the actions of the Security Council and reaffirming the pledge of continental solidarity. Within a month seventeen of the republics had promised at least token assistance.

When it was seen that the Communist aggression threatened the security of the entire world the United States felt that a meeting of the American Ministers of Foreign Affairs was required to consider the situation. The date set was March 1951, the place, Washington, and the agenda covered political,military, and economic cooperation for the defense of the Americas abroad and their security at home. The United States felt that greater aid to the United Nations forces in Korea was needed and that the Latin American Republics should tighten their internal controls against communism. All twenty-one republics of the Western Hemisphere were represented; President Truman welcomed the delegates and Secretary of State Dean Acheson was elected president of the conference. Both President Truman and Secretary Acheson stressed the immediate dangers of Communist imperialism to the democratic countries of the Western Hemisphere and the need to combat them effectively.

On the whole the Conference was successful in its aims. The principles of the United Nations were sustained and specific efforts to strengthen the Inter-American Defense Board and to provide strategic materials and to aid with troops where possible were agreed upon. Resolution VIII of the Final Act recommended adequate legislation be adopted in every American republic to

[25] *Final Act*, XXXII.

[26] For texts of treaties, conventions, resolutions, and commentary thereon see *Ninth Int. Conf. of Amer. States, Dept. of State Pub. 3263* (Washington, 1948). For Charter of the Organization of American States (1967) see Appendix.

prevent and punish subversive acts of international Communism and no state made any reservations to this resolution.

The Tenth International Conference of American States scheduled to meet at Caracas, Venezuela, March 1, 1954 was the first to be held under the Charter signed at Bogotá in 1948. The United States was particularly desirous of including in the agenda consideration of "the intervention of international Communism in the American republics." The Latin American states were for the most part less interested in Communism than in economic questions, but Guatemala alone opposed its inclusion. The Eisenhower administration had promised that the Republicans would not neglect Latin America, but except for Dr. Milton Eisenhower's visit, no concrete evidence for change in policy was visible. The Latin American states were in a mood to demand action.

All the twenty one republics were present except Costa Rica who, by refusing to attend, showed her disapproval of the military dictatorship in Venezuela. The United States delegation headed by Secretary of State John Foster Dulles included former Assistant Secretary of State for Latin American Affairs John M. Cabot, his successor Henry F. Holland, Assistant Secretary of State for Economic Affairs Samuel C. Waugh, Legal Advisor Herman Phleger, United States Ambassador to Venezuela, Fletcher Warren and the United States Representative on the Council of the Organization of American States, John C. Dreier.

Secretary Dulles, supported vigorously by the delegates of Cuba, Peru and the Dominican Republic, was determined to obtain priority for discussion of the Communist menace. Foreign Minister Toriello of Guatemala, supported by Argentina, was equally insistent against it. The United States won out and its draft resolution condemned International Communist activity directed against the political system of the Western Hemisphere as a threat to its peace and security, and recommended positive measures be taken for its extermination. The vote was 17 to 1 in favor, with only Guatemala voting in the negative and Argentina and Mexico abstaining.

Secretary Dulles pledged the intention of the United States to obtain better economic relations and it was agreed to hold an economic conference in Rio de Janeirio in 1955. The Conference voted a resolution requesting that industrialized states refrain from any restrictions on the imports of raw materials which was clearly aimed at the United States. It also by a vote of 19 to 1–only the United States refraining–passed a resolution calling for the elimination of colonialism and "foreign occupation" of territory in the Western Hemisphere. A resolution favoring the guarantee of human rights and freedoms to all persons received the support of all the states except Guatemala which abstained.

In all, the conference passed some 117 resolutions, declarations, recommendations, agreements, and approved motions. The U.S. had succeeded in impressing its neighbors with the danger of communism, or so it seemed, and obtained their support in combatting it. We recognized the need for closer economic cooperation and would have the opportunity to prove it at Rio. On the other hand the enthusiastic support of the United States by a Trujillo-dominated Dominican Republic and the failure of Washington to support the position taken by democratic Uruguay led one delegate to declare that the United States preferred "docile dictators to difficult democracies."[27]

After the adoption of the Caracas Declaration and the overthrow of the Arbenz government in Guatemala Washington made no serious effort to forestall the Communist offensive in Latin America until Fidel Castro's seizure of power in Cuba in 1959. From 1959 to 1972 U.S. response to the Soviet or Chinese Communist threat in Latin America was chiefly in the economic, political, and military fields. The diplomatic and political offensive against Cuba and its supporters was carried out principally through the OAS where Washington sought to mobilize support to meet the Communist challenge.

This proved difficult, for Latin Americans generally could not take seriously our charges that Cuba was a menace to hemisphere security. Washington also found it hard to base a case on violation of human rights and democratic institutions because of its cooperation with notorious violators of such rights, particularly Trujillo, Pérez Jiménez, and the Bastista regime in Cuba. In fact there was a widespread conviction in Latin America that the United States was in large measure responsible for the developments in Cuba. As time passed, however, many of the countries, especially those in the Caribbean, began to show growing concern over the trends in Cuba, culminating in Castro's assertion of his Marxist-Leninist affiliations. In addition to dealing with the Communist threat, OAS agencies were called upon to consider the question of dictatorship in the Caribbean.

In early 1959 the Council of the OAS, in the role of provisional organ of consultation, reviewed complaints by Nicaragua and Panama of invasions by Cuban-backed exiles. After a similar complaint was registered by the Dominican Republic the Fifth Meeting of Consultation of Foreign Ministers was convened in Santiago, Chile, to deal with the broad question of unrest in the Caribbean. The main issue hinged on the question of whether the notorious dictatorships in the Dominican Republic, Nicaragua, and Paraguay should remain protected behind the principle of nonintervention or if they should be subject to OAS action aimed at their overthrow. The delegations reaffirmed the principle of nonintervention, condemned dictatorship, as urged by Secretary of State Christian Herter, and authorized the Inter-American Peace Committee to investigate the entire Caribbean situation on its own initiative or at the request of the member states.

As tensions mounted in the Caribbean the United States sought to bring both the Dominican Republic and Cuba into the spotlight of inter-American public opinion for violation of human rights and democratic principles. After it had become clear that the Trujillo regime had supported an assassination attempt on the life of President Betancourt of Venezuela, the OAS Council voted that a Consultative Meeting of the Foreign Ministers be held in San José, Costa Rica, in August 1960. In this case Venezuela had asked for and obtained the convocation of the Organ of Consultation, in accord with Article 6 of the Rio Pact, which adopted a resolution condemning the government of the Dominican Republic for its acts of aggression and intervention against

[27]Dept. of State *Bulletin*, XXXI (March 22, 1954), pp. 422-423; Resol. XCIII, *Tenth Inter-American Conf., Caracas, Venezuela, Mar. 1-28, 1954, Report of the Pan American Union on the Conference, Annals of the Organization of American States*, VI (1954), 114-115.

Venezuela. The importance of the Sixth Meeting lies in the decision to impose sanctions for the first time in inter-American history; however, the governments were not inclined to approve any action that would involve actual intervention within the country. Washington supported the Latin American countries demands for sanctions against Trujillo since it hoped to gain their support for corresponding action against Castro. It had become evident by June 1960 that he had made the decision to transform Cuba into a totalitarian state in close collaboration with the Soviet bloc.[28]

With this threat to hemispheric security disclosed the United States called the Seventh Meeting of Consultation, on the same site as the Sixth Meeting a few days earlier, to enlist hemispheric support against the threat of extra-continental intervention in hemispheric affairs. Secretary Herter sought to gain a collective denunciation by the foreign ministers of Sino-Soviet Communist intervention, and also of Cuba for its collaboration. However, it proved impossible to pass a resolution condemning the Cuban action, and, finally, the United States had to accept a declaration which condemned communism but not Cuba: ". . . the attempt of the Sino-Soviet powers to make use of the political, economic, or social situation of any American state . . ." It reiterated nonintervention and reaffirmed that the Inter-American System is ". . . incompatible with any form of totalitarianism . . ." The declaration was approved 19-0, the Dominican Republic and Cuba having walked out of the meeting. Mexico and Cuba approved the statement with great reluctance, while Venezuelan and Peruvian foreign ministers refused to sign and were replaced by other delegation members. At this point in time the Latin American states viewed the Cuban revolution as an effort of a small nation seeking to combat economic imperialism and the political domination by the United States. Moreover, it was widely believed that Washington was using the threat of communism as a pretext to forestall the social revolution in Cuba and to restore the status quo there. It was also clear that the Latin Americans remained unconvinced that a regional security problem existed.[29]

On December 4, 1961 the OAS Council convoked a Meeting of Consultation of Ministers of Foreign Affairs to consider the threats to peace and to determine what measures should be taken. Cuba was not specifically cited in the convocation. By an affirmative vote of fourteen countries, with two negative votes (Cuba and Mexico) and five absentations (Argentina, Brazil, Bolivia, Chile, and Ecuador), the decision was made to call into session the Eighth Meeting of Consultation of Foreign Ministers, acting as Organ of Consultation under the Rio Pact, at Punta del Este, Uruguay, from January 22 to 31, 1962.[30]

Warning that the Castro regime had supplied communism with a bridgehead in the Americas, Secretary of State Dean Rusk urged sanctions in the form of a collective break in diplomatic relations and an embargo on all trade

[28] Dept. of State, *Inter-American Efforts to Relieve International Tensions in the Western Hemisphere, 1959-1960.* (Washington, 1962), pp. 54-64; 71-89.

[29] Pan American Union, *Final Act, Seventh Meeting of Consultation of Ministers of Foreign Affairs, San José Costa Rica, Aug. 22-29, 1960* (Washington, 1960).

[30] This was the ninth application of the Rio Treaty, but only the second case to be taken up by a meeting of the foreign ministers acting as the organ of consultation under the Rio Treaty.

with Cuba. The rebuffs suffered by the American delegation on these and other measures reflected the divergence of opinion that had arisen within the inter-American system and a lessening of United States prestige as a result of the Bay of Pigs invasion. By a scant two-thirds majority the meeting voted to bar participation of the Castro government in the OAS, eliminate Cuban participation in the Inter-American Defense Board, embargo arms traffic with Cuba, and recommend interruption of other commerce. Mexico, Ecuador, Brazil, Bolivia, Argentina, Chile, and Cuba voted against the anti-Cuban measures, and it was only with extreme difficulty and economic inducements that the affirmative votes of Haiti and Uruguay were obtained. The meeting declared that the doctrine of legitimate self-defense applied to threats to peace and security arising from subversive activities and, to police such threats, established the Special Committee of Consultation on Security.

The main resolution adopted at Punta del Este providing for the exclusion of the Cuban government for participation in the inter-American system had four executive clauses:

1. That adherence by any member of the OAS to Marxism-Leninism is incompatible with the inter-American system and the alignment of such a government with the Communist bloc breaks the unity and solidarity of the hemisphere.
2. That the present government of Cuba, which has officially identified itself as a Marxist-Leninist government, is incompatible with the principles and objectives of the inter-American system.
3. That this incompatibility excludes the present government of Cuba from participation in the inter-American system.
4. That the Council of the OAS and other organs and organizations of the inter-American system adopt without delay the means necessary to carry out the resolution.

All of the countries except Cuba were prepared to accept the first two points in the initial voting, but six republics (Argentina, Bolivia, Brazil, Chile, Ecuador, and Mexico) abstained on the last two clauses and therefore in voting on the adoption of the resolution as a whole. This irregularity clouded the significance of the vote. The assertion of incompatibility in the first clause did not extend beyond the Declaration of San José which reaffirmed that the inter-American system "is incompatible with any form of totalitarianism," but did not refer to Cuba. In specifying the incompatibility of the "present government of Cuba," the second clause of the resolutions above did bring the question into clearer focus. But it was the abstention of six countries on the exclusion of Cuba that aroused serious concern, for this group accounted for two-thirds of the population and three-fourths of the area of Latin America. It was also disturbing to Washington that Brazil, generally a reliable ally of the United States in its hemispheric policies, opposed the United States proposals for the adoption of economic and diplomatic sanctions against Cuba.[31]

The measures taken at Punta del Este proved ineffectual in resolving the Cuban question as the United States-Soviet Union confrontation in the missiles crisis of October 1962 demonstrated. But in dealing with this crisis Washington

[31] Pan American Union, *Final Act, Eighth Meeting of Consultation of Ministers of Foreign Affairs, Punta del Este, Uruguay, Jan. 22-31, 1962* (Washington, 1962).

showed itself capable of moving swiftly and decisively to counter a direct military threat to the hemisphere, and the OAS Council voted unanimously to support U.S. action in the Caribbean to ensure the removal of Soviet missiles from Cuba. After President Kennedy had formally proclaimed a blockade against the shipment of all offensive weapons into Cuba on October 24, several Latin American nations made cooperative defense gestures. Argentina sent two destroyers to join the blockade; Honduras and Peru offered to send troops. Venezuela, Costa Rica, Nicaragua, Panama, the Dominican Republic, and Haiti offered Washington the use of bases in the Caribbean, and Colombia and Venezuela mobilized their armed forces. However, five Latin American republics (Brazil, Mexico, Bolivia, Uruguay, and Chile) continued to maintain diplomatic ties with Havana.

Castroite involvement in subversive activities brought charges against Cuba before the OAS claiming that Cuba had covertly supplied arms to Venezuelan terrorists and guerrillas. After an OAS investigating team had found clear evidence that the arms had come from Castro's Cuba, the Ninth Consultative meeting of Foreign Ministers of the OAS was called into session in Washington during July 1964. By a 15-4 vote, Cuba was declared guilty of aggression and intervention. Diplomatic ties were to be broken off, and economic and commerical relations interrupted, as well as transport and communication links. By the end of 1964, the OAS community of states, except Mexico, which resisted for legal and internal political reasons, had complied with the recommendations.

The next test of cooperation in the regional body occurred in May 1965 when the United States dispatched armed forces into the Dominican Republic to avert a possible Communist seizure of power. The Johnson administration, fearing "another Cuba," and finding that the OAS could not be mobilized in time to move effectively, took unilateral action; however, every effort was then made to turn the problem over to OAS jurisdiction. The Tenth Meeting of Consultation of the OAS met in May 1965, and it was only by exerting diplomatic pressure that a two-thirds majority was obtained to establish an inter-American police force for the purpose of intervening in the Dominican crisis. (U.S. troops were already there.) Mexico, Ecuador, Peru, Chile, and Uruguay voted against the resolution, and Venezuela abstained. To secure passage of the measure, Washington was forced to rely on the vote of a Dominican representative who was then without a government.

Soon after intervention began Washington sponsored the dispatch of an OAS peace mission to the Dominican Republic to achieve a truce, and took steps to render the military action collective by establishing an OAS peace-keeping force. Eventually two thousand Latin-American troops from five countries (Brazil, Paraguay, Honduras, Nicaragua, and Costa Rica) joined nine thousand U.S. troops in the inter-American force, with Brazil supplying the largest Latin American contingent. It is significant that, for the first time, an OAS mission commanding a military force was authorized to intervene in what was largely a domestic affair of a member state.[32]

At the Second Special Inter-American Conference assembled at Rio de

[32] Inter-American Institute, *The Inter-American System*, pp. 171-180.

Janeiro in November 1965 the United States, interested in making the inter-American police force a permanent body, included it as one of the proposed changes in the inter-American system to be reviewed at the conference. This proposal was rejected by the Latin Americans, but a number of resolutions were passed recommending that changes be made in the OAS Charter.

By the mid-1960s it was generally agreed that the Constitution of the OAS, the Bogotá Charter of 1948, should be amended. It was clear that the organization's structure should be modernized and strengthened to cope with new problems, and to act effectively in the interest of hemispheric cooperation and solidarity. Much had happened since 1948 to bring this need into focus. There had been the attempted Communist takeover in Guatemala, the Cuban revolution, the abortive Bay of Pigs invasion, the Alliance for Progress, the Cuban missiles crisis, the Castroite revolutionary movement on the continent, United States intervention in the Dominican Republic, and countless other economic and political changes. Problems arising from the cold war were not major factors in this hemisphere until Fidel Castro took over Cuba and projected Latin America into international power politics.

How to amend the Charter proved to be a much disputed question. The United States wished to strengthen the Council of the OAS which, like the United Nations Security Council, is in permanent session. But many Latin American countries are more interested in economic issues than in collective action against communism or the threat of communism. Changing the Bogotá Charter is a big undertaking. It is a treaty and amendments must be ratified by the U.S. Congress. Other countries with constitutions like that of the U.S. must also obtain legislative approval. Moreover, resolutions for amending it require a two-thirds vote at the inter-American conferences established to deal with the matter.

Following in general the guidelines prepared by the Second Special Inter-American Conference at Rio de Janeiro in November 1965, and the draft amendments prepared by the OAS Special Committee which met in Panama in March 1966, and by the Inter-American Economic and Social Council which met in Washington in June 1966, the protocol of amendment to the Charter of the OAS—the "Protocol of Buenos Aires"—was signed at the Third Special Inter-American Conference at Buenos Aires on February 27, 1967.[33]

In the discussions carried on at Panama by the OAS Special Committee, differences arose between the U.S. and Latin-American delegations over economic and social policies. The latter had insisted that Washington agree not only to describe in detail what type of aid it would supply, but also to make trade concessions that would inevitably have created conflicts with other areas of the world and with domestic producers. After U.S. delegate Robert Woodward had pointed out that such commitments would not be approved by either the State Department or Congress, a compromise was reached. A key charter amendment was revised to read that the member states "agree to cooperate in the broadest spirit to strengthen their economic structures." The original Latin-American draft declared that member nations had an "obligation" to assist one another.[34]

[33] Dept. of State *Bulletin*, LVIII (July 17, 1967), pp. 78-79; *Ibid.*, LVI (Mar. 20, 1967), pp. 474-476.

[34] *New York Times*, Mar. 26, 1966; *Ibid.*, June 20, 1966.

President Johnson recommended ratification of the Protocol of Buenos Aires to the U.S. Senate in June 1967, and on April 23, 1968, at a White House ceremony, the President signed the U.S. instrument of ratification of the Protocol of Amendment to the Charter of OAS. The amendment entered into force February 28, 1970, after the needed two-thirds, or sixteen, ratifications had been obtained.[35]

The charter amendments, which are the first to be adopted since the Charter was signed in 1948, grant certain fuller responsibilities, as in the field of peaceful settlement. They also incorporate the principles of the Alliance for Progress in the charter. Among the more significant changes called for by the amendments are: (1) Replacement of the Inter-American Conference, which meets every five years, by a General Assembly, which will meet annually. (2) Redesignation of the OAS Council as the Permanent Council, and the granting of additional responsibilities to the Inter-American Economic and Social Council and the Inter-American Council for Education, Science, and Culture. The Economic and Cultural Councils become directly responsible to the General Assembly, as is the Permanent Council. These changes are designed to augment the importance given in the OAS structure to economic, social, educational, and scientific activities. (3) Elimination of the Inter-American Council of Jurists and the upgrading of the Inter-American Juridical Committee. (4) Assignment to the Permanent Council and its subsidiary body (the Inter-American Committee on Peaceful Settlement) a role in assisting member states in resolving disputes between them. (5) Incorporation of the Inter-American Commission on Human Rights in the OAS Charter. (6) Inclusion of a procedure for the admission of new members. (7) Election of the OAS Secretary-General and Assistant Secretary-General by the General Assembly for five-year terms, rather than by the Council for ten-year terms, as presently provided. (8) Incorporation in the Charter of the principles of the Alliance for Progress in the form of expanded economic and social standards covering self-help efforts and goals, cooperation and assistance in economic development, improvement of trade conditions for basic Latin American exports, economic integration, and the principles of social justice and equal opportunity.[36]

The Third Special Inter-American Conference and the Eleventh Meeting of Consultation of Ministers of Foreign Affairs of the American Republics were held at Buenos Aires February 15-27 and February 16-26, respectively. The principal work of the Third Special Inter-American Conference, as was mentioned, was to consider and approve the Protocol of Amendment to the OAS Charter. At the concurrent Eleventh Meeting of Consultation a resolution was adopted recommending that the American chiefs of state meet at Punta del Este April 12-14, 1967.[37]

[35] Dept. of State *Bulletin*, LVII (July 17, 1967), pp. 78-79; *Ibid.*, LVIII (May 13, 1968), pp. 614-616; *Christian Science Monitor*, Mar. 3, 1970; Galo Plaza Lasso, a former president of Ecuador and United Nations mediator, was installed as secretary-general of the OAS on May 18, 1968, for a five year term; he succeeded Dr. José Antonio Mora of Uruguay. Dr. William Sanders, a United States career diplomat, was succeeded as asst. sec.-gen. by Dr. Miguel Rafael Urquia, of El Salvador.

[36] *Ibid.* For a critical analysis of the amendments see William Manger, "Reform of the OAS, The 1967 Buenos Aires Protocol of Amendment to the 1948 Charter of Bogota: An Appraisal," *Journal of Inter-American Studies*, X (Jan., 1968), 1-14.

[37] *Ibid.*, LVI (Mar. 20, 1967), p. 472.

The chiefs of state of twenty member nations of the OAS met at Punta del Este on the scheduled date, and at the close of the conference the Declaration of the Presidents of America was signed by seventeen chiefs of state, the prime minister of Trinidad and Tobago, and the representative of the president of Haiti. President Johnson, heading the United States delegation, was accompanied by Secretary Rusk, Assistant Secretary Gordon, Mr. Linowitz, and others.

Contained with the Action Program of the Declaration of the Presidents of America were the following objectives: (1) Latin American economic integration and industrial development; (2) multi-national action for infrastructure projects; (3) measures to improve international trade conditions in Latin America; (4) modernization of rural life and increase of agricultural productivity, especially of food; (5) educational, technological, and scientific development and intensification of health programs; (6) elimination of unnecessary military expenditures.[38]

Fresh evidence of continuing Castroite-backed subversion, terrorism, and guerrilla warfare in several Latin American states brought into session the Twelfth Meeting of Consultation of the Ministers of Foreign Affairs of the OAS. Held in Washington, September 19-24, 1967, the meeting was convoked in accordance with Articles 39 and 40 of the OAS Charter, at the request of Venezuela. The Final Act of the meeting affirmed the imposition of further diplomatic and economic sanctions against the Castro government.[39]

The first major military conflict between American states in more than a generation—an undeclared war between the Central American republics of El Salvador and Honduras—was responsible for the Thirteenth Meeting of Consultation of Ministers of Foreign Affairs. The invasion of Honduras by El Salvador's military forces brought the OAS into action. By a resolution adopted on July 23, 1969, the Council of the OAS, acting provisionally as Organ of Consultation, convoked a meeting of Consultation in accordance with the Inter-American Treaty of Reciprocal Assistance and the pertinent articles of the OAS Charter. The Council, in this capacity, voted "to call upon the governments of El Salvador and Honduras to suspend hostilities, to restore matters to the *status quo ante bellum*, and to take the necessary measures to reestablish and maintain inter-American peace and security, and for the solution of the conflict by peaceful means." A cease fire was quickly achieved, and Salvadorian troops were withdrawn from Honduran territory by August 3, 1969.[40]

Since the Rio Treaty came into force in 1948 it has been invoked on several occasions to resolve breaches of or threats to peace. The majority of the

[38] U.S. Dept. of State, *Commitment for Progress: The Americas Plan for a Decade of Urgency*. Inter-American Series 93 (Washington, 1967); Dept. of State *Bulletin*, LVI (May 8, 1967), pp. 706-721. In 1969 Jamaica became the 24th member of the OAS, and its fourth English-speaking member. Its admission into the OAS came two years after Trinidad and Tobago and Barbados joined. All three of the new members, the first new additions to the OAS community in fifty years, were once part of the British Caribbean empire. See *Christian Science Monitor*, Oct. 30, 1969.

[39] Dept. of State *Bulletin*, LVII (Oct. 16, 1967), pp. 493-498. A special session of the OAS Council was called in March, 1968 to consider Bolivian charges that Chile had permitted the extradition of five guerrillas, remnants of Dr. Guevara's band, who had escaped from Bolivia into Chile. *New York Times*, Mar. 6, 1968.

[40] Dept. of State *Bulletin*, LXI (Aug. 18, 1969), pp. 132-134; *Americas*, 21 (Sept. 1969), pp. 42-45.

cases have involved disputes among the smaller nations of the Caribbean area and the west coast republics of South America (Guatemala, El Salvador, Honduras, Nicaragua, Costa Rica, Panama, Cuba, the Dominican Republic, Haiti, Venezuela, Ecuador, Peru, and Bolivia). Many of the difficulties arose from the existence of large numbers of refugees from oppressive regimes, augmented by Communists, adventurers, and mercenaries at times, and the efforts of the expatriates—aided in some instances by the government of the nation in which they had found refuge—to overthrow their opponents. All of these cases were settled by the apparatus of the inter-American system except those related to some aspect of the cold war. In the case of Guatemala the government was overthrown by an invasion from Honduras in 1954 which was backed by material and moral support from the United States. Otherwise, however, these disputes and charges of aggression have been worked out by the OAS Council through committees of investigation, aided by advisers, which visited the countries involved in the dispute.[41]

Another component of the inter-American system for the maintenance of peace and security is the Inter-American Peace Committee (IAPC). Established in 1940 by a resolution of the Second Meeting of Consultation of Ministers of Foreign Affairs in Havana, it was not installed until July 1948. The Peace Committee is composed of five members elected by the Council of the OAS for five-year terms; except for the election of its members the agency is almost autonomous, being responsible only to the Meetings of Consultation. The committee has its own statutes, according to which its purpose is the preservation of peace between states where a controversy has appeared. It cannot interfere in a dispute, and the scope of its action is determined by the parties involved. When, in 1956, statutes limiting its flexibility and autonomy were imposed, the committee could no longer function as in the past. However, at the Fifth Consultative Meeting of Foreign Ministers in Santiago (1959) the restrictions were lifted.

Under the broadened statutes the committee is a body designed to foster the pacific settlement of disputes and controversies that exist or may arise between the states, suggesting to them "measures and steps" to facilitate such settlement. The committee itself is not a procedure or method of pacific settlement such as investigation, mediation, and conciliation, but rather a vehicle by which such approaches are suggested to the parties in a dispute. As an institution of the inter-American system for the maintenance of peace and security, the committee's history is a very significant one. So long as no great restraints were imposed and it was not called upon to deal with cold war problems, it has proved to be an effective agency in the peaceful settlement of minor disputes among the hemispheric nations.[42]

The promotion of respect for human rights and the basic freedoms are growing concerns of the inter-American system. The Inter-American Commission on Human Rights (IACHR) was established by the Fifth Meeting of Consultation

[41] Inter-American Institute, *The Inter-American System,* pp. 122-171: Organización de los Estados Americanos, *Applications del Tratado Interamericano de Asistencia Recíproca,* 1948-1960, 3rd ed. (Washington, Pan American Union, 1960); *Ibid., Suplemento* 1960-61 (Washington, Pan American Union, 1962).

[42] Inter American Institute, *The Inter-American System*, pp. 82-104.

of Ministers of Foreign Affairs and is guided by statutes approved by the OAS Council in 1960. According to Article I of the statute the Commission "is an autonomous entity of the Organization of American States, the function of which is to promote respect for human rights." It is composed of seven members, nationals of the member states of the OAS, elected from a list made up of panels of three persons proposed by the governments.

The activity of the inter-American system in the area of human rights, together with representative democracy, is clearly based in the OAS Charter. Article 13 indicates that "each state has the right to develop its cultural, political, and economic life freely and naturally. In this free development, the State shall respect the rights of the individual and the principles of universal morality." On representative democracy the Charter states that, "The solidarity of the American states and the high aims which are sought through it require the political organization of those states on the basis of the effective exercise of representative democracy."[43]

In the decade of the 1960s the Commission was occupied with violations of human rights mainly in Cuba, the Dominican Republic, Haiti, Nicaragua, Honduras, El Salvador, and Paraguay. The Commission has, in several instances, carried out productive and effective work, but it is limited by its authority only to call attention to certain conditions, and its inability to intervene. Although the ideal of representative democracy is a principle subscribed to by the Latin American republics and is incorporated in the legal structure of the inter-American system, the wave of militarism, in part a reaction to Communist activity, has virtually nullified its attainment in the foreseeable future.

The Commission was incorporated into the new Charter structure as a full-fledged organ of the OAS in accordance with the Protocol of Buenos Aires, adopted in 1970, and subsequently issued reports on the human rights of political prisoners and their relatives in Cuba, and reports on El Salvador and Honduras. However, by 1971 only Costa Rica had ratified the Inter-American Convention on Human Rights, and no countries in addition to the twelve original signatories had signed. The Convention was under study in Washington, which had not signed or ratified.

In addition to the general periodic international conferences of American states numerous special conferences have been held when technical matters are being reviewed or when special aspects of inter-American cooperation are to be discussed. There were a number of inter-American organizations in existence before the OAS Charter was signed, official, semi-official, and private. The Bogotá Conference requested the Council of the OAS to study these organizations of the OAS in accordance with the provision of the Charter. The Council has concluded agreements with the following organizations and has listed them in the Register of Specialized Inter-American Organizations: the Pan American Health Organization (PAHO); the Inter-American Children's Institute (IACI); the Inter-American Commission of Women (IACW); the Pan American Institute of Geography and History (PAIGH); the Inter-American Indian Institute (IAII); and the Inter-American Institute of Agricultural Sciences (IAIAS).

[43] *Ibid.*, p. 39.

While it is not possible to describe here the history and activities of these specialized agencies, it is clear that in an unheralded fashion, they have made lasting and tangible contributions to the welfare of the peoples of the developing countries in the Western Hemisphere. They have grown organically at varying rates, and doubtless will continue to expand and assume a larger role, given the myriad of problems which await solution in Latin America. The United States has long given appreciable economic and technical support to these organizations which, in the decade of the 1960s, have come to reinforce the Alliance for Progress.

The inaugural session of the new OAS General Assembly met in Washington June 25-July 8, 1970 (technically the First Special Session of the Assembly, and originally scheduled to be held in Santo Domingo, the meeting was transferred to Washington because of violence in the Dominican Republic). Much of its effort was devoted to setting in motion certain regulatory provisions relating to the new structure, in addition to carrying out its more routine responsibilities, and the critical issue of kidnapping and terrorism. Held in Washington August 24-25, 1970, the Second Special Session of the Assembly was limited to filling a vacancy in the Inter-American juridical Committee. The Third Special Session of the Assembly, convened in Washington on January 25, 1971, considered the earlier resolution on kidnapping and terrorism.

The General Assembly held its first annual regular session under a revised Charter April 10-23, 1971, in San José, Costa Rica. Attending were foreign ministers or their representatives of the twenty-three member states of the OAS. With a thirty-seven item agenda, this session touched on political, economic, and organizational issues of considerable importance. The Assembly approved the holding of five new specialized conferences, considered a Colombian initiative in area limitation and military expenditures in Latin America, and considered the problem of trade of the developing countries. The resolution on trade urged Washington to adopt a number of measures to promote the trade of the developing countries. Progress toward harmonizing relations between El Salvador and Honduras was made with the signing of statements by the foreign ministers of those countries declaring their intention to put an end to the "anomolous situation" that prevailed in their countries' relationship.

U.S. Secretary of State William P. Rogers told the Assembly that the Nixon administration was seeking to preserve the historically close mutual relationship that the United States had enjoyed with its hemispheric neighbors, but in the form of "a more balanced and reinvigorated partnership." He promised easier access for Latin American goods in the U.S. market and continuing development assistance in the foreseeable future. It was also announced that the U.S. mission to the OAS had been designated the Permanent Mission of the United States of America to the Organization of American States. Previously, it was simply a delegation to the old OAS Council.

While progress toward the solution of the Honduras-El Salvador problem seemed to have been advanced, collective security under the Rio Treaty was weakened in another area. This was the reestablishment of diplomatic and commercial ties with Cuba by the Allende administration of Chile in November 1970. The Chilean government, in taking this step, disregarded the fact that the OAS in 1964 had adopted a "binding" decision under the Rio Treaty that in view of Cuba's interventionist and aggressive acts, member states should have no diplomatic or consular relations, shipping, or trade with the island nation until the OAS itself, by a two-thirds vote, decided that Cuba had ceased to be a threat to the peace and security of the hemisphere. Since Cuba had not abandoned its policy of intervention and subversion, Chile's action did violence to the inter-American system.

The Third Special Session of the OAS General Assembly, held in Washington in January-February 1971, focused its attention on a draft international convention on terrorism and kidnapping drawn up at the Assembly's request by the Inter-American Juridical Committee. Brazil and Argentina, supported by four other countries, held that no convention would be effective unless it embraced all aspects of terrorism. A larger group argued that such a broad approach would endanger the principle of political asylum. The U.S., seeking a narrower convention, was unable to reconcile the opposing points of view. However, by declaring acts of terrorism against foreign officials and their dependents to be common crimes rather than political offenses, the convention deprives persons responsible for such acts of the shelter of political asylum and subjects them to extradition and prosecution. Further action on the convention was expected in 1972.

In January 1971 Ecuador charged the United States with violating Article 19 of the OAS Charter, which prohibits "use of coercive measures of an economic or political character in order to force the sovereign will of another state." This dispute arose over the right of U.S.-owned vessels to fish off the coast of Ecuador in waters which it claims are within its maritime sovereignty, but which the United States contends are high seas. After Ecuador had seized several California-based tuna boats, Washington applied section 3(b) of the Foreign Military Sales Act, halting such sales to that country.

Contending that the U.S. government had violated Article 19, Ecuador called for a Meeting of Consultation of the OAS Foreign Ministers who coincidentally were in Washington for the special General Assembly session. The outcome was a resolution, supported by both countries, stating the positions of the two contending parties and calling upon them to resolve their differences in accordance with Charter principles and to "abstain from the use of any kind of measure that may affect the sovereignty" of any state. Although the basic issue remained unresolved, it reflected constructive effort by both parties to quiet the situation.

The OAS General Assembly considered a proposal by the President of Colombia in April 1971 to study the feasibility of an arms limitation agreement among the Latin American countries. This initiative was based on Chapter VI of the 1967 Declaration of American Presidents, in which the Latin American heads of state proposed to limit military expenditures in proportion to the actual demands of national security and international commitments to insure that maximum resources would be devoted to economic and social development. Owing to disagreement, mainly over the meaning of the presidents' declaration and the scope of the Assembly's mandate, no progress was made save to instruct the OAS Permanent Council to study the intent and scope of Chapter VI. The United States was not a participant, for it was concluded that any eventual agreement on arms limitation would probably not include this country.

In the field of assistance the OAS, including its specialized organizations, has continued to provide vehicles of multilateral coordination and has carried out an increasing portion of technical assistance provided in the hemisphere. This role was supportive of the Nixon administration's belief, outlined by the president on October 31, 1969, that the main future patterns of U.S. assistance for hemisphere development must be United States support for Latin American initiatives and "that this can be best achieved on a multilateral basis within the inter-American system." The Inter-American Development Bank (IDB), an important component of the inter-American system, though not connected with the OAS, is chiefly involved in development lending. The OAS itself has no lending function, but it does carry out significant technical assistance programs.

Three OAS agencies administer most of the technical assistance: PAHO, IAIAS, and the OAS General Secretariat. Although the IDB provides technical assistance, it is mainly for project development in connection with potential loan applications or as a supplement to loans made. In 1971 technical assistance provided by the specialized agencies totaled $66 million, which was almost equal to the $70 to $75 million which AID provided for bilateral technical assistance Early in 1970, partly in response to the Concensus of Viña del Mar, the United States agreed to establish a continuing Special Committee on Consultation and Negotiation as a permanent committee of the IA-ECOSOC (Inter-American Economic and Social Council) to deal with problems of trade assistance, investment, technology, and related subjects.

Secretary of State William Rogers, addressing the Permanent Council of the OAS, convened in Philadelphia in November 1972, spoke about the second Nixon term, outlining the administration's Latin American policy: "President Nixon has asked me to tell you that during his second term we will remain committed to the interests of the hemisphere and determined to make a substantial contribution to its social and economic progress." Four United States goals were cited:

1. We will pursue a policy of cooperation with Latin America in a relationship of greater equality, shared initiatives, and mutual responsibilities.
2. We will work to ensure that the legitimate interests of all the nations of Latin America are represented in the new international monetary and trade systems to be negotiated.
3. We will cooperate with you directly in this hemisphere to strengthen and diversify our trade, investment and assistance ties.
4. And we will seek to resolve the issues between us over fisheries, over territorial seas, over investment, and all others, in the spirit of friendship and mutual respect which is the essence of our inter-American system.

Reaffirming Washington's faith in the regional organization, he emphasized that the OAS is the "linchpin" of the inter-American system.[44]

The inter-American system, after eight decades, is not only the oldest functioning regional organization, but also the one that has both expanded and strengthened its numerous spheres of activity. From the beginning of the modern era the Western Hemisphere seemed to offer unique conditions favorable

[44] *Ibid.*, pp. 16-21; O. Carlos Stoetzer, *The Organization of American States* (New York, 1965), pp. 64-75; Dept of State *Bulletin*, LXIV (June 14, 1971), p. 784; *Ibid.*, LXV (Sept. 13, 1971), pp. 284-293; *Ibid.*, LXVII (Dec. 4, 1972), pp. 655-656.

to cooperation and solidarity among peoples and their governments. These included a similar origin and historical evolution, geographic proximity, and similarity of political conditions. A natural corollary was the concept of Pan Americanism, or the Western Hemisphere idea: the view that the peoples and the nations of the hemisphere have a special kinship which sets them apart from the rest of the world. These factors provided a foundation for regionalism as an integrative force in the inter-American system.

At the same time there exist forces, both inherent and external, which tend toward fragmentation and disunity. The American republics reflect great diversity in size, natural resources, ethnic character of their populations, political institutions, rates and degrees of economic development, and social structure. Cultural and linguistic barriers present obstacles to cooperation and integration. The Americas are religiously divided, the north being Protestant, whereas Roman Catholicism predominates in Latin America. The legal systems are also dissimilar, common law prevailing in the north, and Roman law south of the Rio Grande.

In general, the Latin American states are not natural trading partners, but instead compete in world markets for the sale of a restricted list of export commodities. The United States trades with all of them, dominating the economies of many in this way, and exerting a powerful influence through public and private investment. Geographic barriers inhibit land communications in most areas of Latin America, and both the United States and Canada are separated from their hemispheric neighbors, other than Mexico, by considerable distance. To these divisive and obstructive factors must be added the impact of nationalism and the insistence on sovereignty of the member states of the OAS. The legacy of the struggle for independence tends to reinforce these lines of nationalism.

The solidarity of the inter-American system is also subject to extrahemispheric forces. The influence of historic Latin American ties to the mother countries of Spain and Portugal have weakened appreciably; however, the Guianas, British Honduras (Belize), and the West Indies retain closer ties to their home countries than to the nations of the Western Hemisphere. Canada's link with the British Commonwealth is a major cause of its reluctance to accept membership in the OAS. And the United States, as the regional paramount nation, has been drawn irresistably toward extrahemispheric concerns, particularly since World War II.[45]

In dealing with economic and social problems the OAS has generally proved to be ineffective; while some progress has been made in the decade of the 1960s, the magnitude of Latin American problems has tended to minimize the gains. Two important goals of the Alliance for Progress—economic development and social improvement—added to the problems of OAS, which it has not been able to solve. At the same time, the failure of the Alliance to attain projected socio-economic goals has worked to unify the Latin Americans in the OAS against the United States.

[45] Canada, a member of the Pan American Institute of Geography and History and the Inter-American Statistical Institute, was in the process of becoming a full member of the Pan American Health Organization, in 1971; it had also expressed interest in joining the Inter-American Institute of Agricultural Sciences.

Agencies of the United Nations have generally done more effective work on economic and social problems in Latin America than the OAS. The United Nations Economic Commission for Latin America (ECLA) has produced valuable reports and special studies, and it took the initiative in suggesting plans for the Central American Common Market (CACM) and the Latin American Free Trade Association (LAFTA). Many Latin American nations profited from association in the International Monetary Fund. Other UN agencies, particularly the Food and Agriculture Organization, UNESCO, the Children's Emergency Fund, and the Program of Technical Cooperation, granted aid to Latin America in quantities commonly exceeding that of the OAS. It was the inadequacy of the OAS machinery, and the need for a new and more radical approach to inter-American economic and social cooperation, that led President Juscelino Kubitschek of Brazil to propose Operation Pan America which became a preliminary blueprint for the Alliance for Progress.

Anti-Americanism or "Yankeephobia," emanating from growing leftist and nationalistic forces, continues to rise in Latin America. Aside from the causes mentioned, the overwhelming power of the United States is offensive, particularly when it is used forcibly such as in Guatemala, Cuba, and the Dominican Republic. Moreover, the United States is widely regarded as the main prop to right-wing dictatorships, both civilian and military, and the perpetuation of traditional, aristocratic regimes. The reluctance of the United States until the 1960s to grant large scale economic assistance, to participate in price stabilization agreements, and to lower tariff barriers on Latin American exports, has been bitterly resented.

Latin American political instability also corrodes the OAS. In the period 1948-1964 every Latin American nation except Chile, Uruguay, and Mexico had at least one illegal change of administration and since then military regimes have been on the rise. This has brought to the fore ideological conflicts between dictatorships and democracies, and radical movements opposed to conservative oligarchies.

Rising nationalism and assertions of sovereignty, reflected in the continuing stress on nonintervention, have limited the political competence of the OAS to minimal requirements for maintaining peace; except for the case of the Dominican Republic, the OAS has generally been kept out of the settlement of internal Latin American conflicts and has only a minor role in resolving interstate conflicts arising from subversion, guerrilla warfare, and propaganda.

It is commonly believed that the relative position of the weak Latin American states arrayed against a world superpower renders the United States immune to collective pressures exerted through the OAS. In practice the United States has sought to align itself with the majority view, particularly after the adoption of the nonintervention policy in the 1930s, and has done so except where national and hemispheric security interests have dictated unilateral action in response to the cold war. However, it is evident that the Latin American nations are reluctant to enlarge the scope of OAS political authority for fear of increased United States influence in their domestic affairs. It is also clear that the Latin American nations have refrained from giving tangible support to the United States in the global anti-Communist conflict unless the Western Hemisphere has been in obvious danger. Viewing their position as one detached from the cold war, they resent being drawn into the conflict as appendages of the United States, and also resent that the greater proportion of United States military and economic aid is diverted elsewhere.

Meanwhile, the Latin American states have gained through association in the OAS: the protection of territorial integrity and political independence from external danger, especially from the real and potential threat of fascism and communism. And as a collective security system it has helped to forestall conflicts among the Latin American states. Moreover, because of its commitments to the inter-American system, and its interest in gaining hemispheric solidarity in pursuit of national security goals, the OAS tie has, perhaps, compelled Washington to exercise restraint in dealing with Latin America.

The United States has profited from OAS support when the organization has validated what were largely unilateral actions by Washington. These include the OAS symbolic support of the Korean War, its sanctioning of the Cuban missiles crisis blockade, and trade and diplomatic embargoes of the Castro regime. The organization was also useful as an anti-dictator front in its support of Washington's efforts to eliminate Trujilloism. The OAS has also contributed materially to United States success in coping with the numerous minor conflicts in the Caribbean in the 1950s, as well as during our intervention in the Dominican Republic in 1961 and 1965.

When the contributions of the OAS are evaluated it becomes evident that in performance of its role of containing inter-state conflicts and in preserving the territorial integrity of the members states it has been quite effective, for the Western Hemisphere has had no major conflicts in several decades. While the OAS seldom solves the basic causes of a conflict it has helped to arrive at solutions by peaceful means. As in the case of the conflict involving El Salvador and Honduras in 1969, it has served as an impartial fact-finding agency, a communication link between the disputants, and a forum for conciliation and negotiation of the critical issues in a dispute. Owing to its military and economic power the United States inevitably has the major role in most situations requiring OAS action. When economic and military sanctions are called for the United States necessarily carries the burden, but OAS endorsement conveys the desired impression of multilateral action.

Despite significant accomplishments by the OAS in the years since World War II, there has been an attendant weakening of hemispheric solidarity and growing evidence of antagonism towards the United States, as the response to the Rockefeller mission in 1969 so amply demonstrated. Administrative and structural problems within the OAS are involved, and it is expected that the changes in the OAS charter, previously mentioned, will help to resolve them. But more profound forces have been at work: the overwhelming economic, political, and social problems which confront Latin America, the preoccupation of the United States with global commitments and its own domestic problems, Marxist propaganda and subversion, technological progress which has broken down the isolation of the hemisphere, the proliferation of military regimes among the American republics, and disillusionment with the Alliance for Progress. As the decade of the 1970s begins the Western Hemisphere idea, the OAS, and inter-American diplomacy face an unprecedented challenge. How successfully this is met will depend in large measure upon the coincidence of foreign policy objectives of the United States and its southern neighbors, for this is the pillar on which inter-American solidarity rests.

Latin America's participation in world organization began about one century after independence had been won.[46] In the aftermath of the wars of independence the new states were preoccupied with solving their own problems and, fearing the designs of most of the great powers, commonly took an isolationist stand. In general, they agreed with the United States in opposing European interference in the political affairs of the New World and were not concerned with world problems except insofar as they affected their trade. Europe regarded the new states as marginal factors in international affairs and showed no disposition to seek their participation until the outbreak of World War I. On occasion the Latin American states requested the mediation of Europe, as well as of the United States, in resolving their boundary disputes, and they sought European capital, trade, immigrants, and cultural contact.

Pan-Americanism made some headway, but many Latin American states lacked enthusiasm for the movement since the Pan American Union, with its headquarters in Washington, and its director-general from the United States, seemed to them too much dominated by their big northern neighbor. They were not invited to attend the first peace conference at the Hague but, owing to United States influence, the larger nations were asked to send representatives to the assembly in 1907. As weak nations they appreciated any legal bulwarks that such a conference might develop to protect them. The Latin American delegations at the conference supported the progressive ideas of the time, including Red Cross rules in the conduct of warfare. Except for Haiti, they favored accepting obligatory arbitration in principle, and unanimously approved a proposal for voluntary arbitration.

During World War I Latin Americans were attracted to President Woodrow Wilson's idealism, as expressed in his well-turned phrases about liberty, freedom, and democracy, and showed much interest in his plan for a League of Nations which would provide strong measures against aggressors and guarantee peace through its international organization. All of the nations of Latin America signed the Versailles Treaty, including the Covenant of the League of Nations, or were invited to sign the Covenant soon thereafter. Ten of these states became charter members of the League of Nations, and at the first meeting of the League Assembly fifteen delegations were present. At the last meeting of the organization in 1945 there were ten Latin American members. From 1920 to 1945 all of the hemispheric nations were, at various times, members of the League, but at no single time was their membership complete.

It is clear that the refusal of the United States to join the organization, which President Wilson had been so influential in founding, was a bitter blow to Latin American hopes. Their critics explained by citing the occasions on which

[46]For additional background on Latin America's role in international organizations see: Manuel S. Canyes, *The Organization of American States and the United Nations* (Washington, Pan American Union, 1963), 6th ed.; Inis L. Claude Jr., "The OAS, the UN and the United States," *International Conciliation*, No. 547 (Mar. 1964); Edgar S. Furniss, Jr., "The United States, the Inter-American System and the United Nations," *Political Science Quarterly*, LXV, No. 3 (Sept. 1950), 415-430; John A. Houston, *Latin America and the United Nations* (New York, 1956); Warren H. Kelchner, *Latin American Relations with the League of Nations* (Boston, 1929).

the United States had been guilty of interference in, and aggression against, their countries and by saying that our talk of freedom and liberty had been nothing more than talk. Most Latin American nations, particularly those which had a feeling of inferiority because of their meager resources and small territory, looked to the League to give them political equality with the more powerful and wealthy states. More specifically, these nations viewed the League as a counterpoise to United States dominance in the region. Because of its military intervention in Haiti, the Dominican Republic, and Nicaragua, its role in taking the Panama Canal zone, and the use of its armed forces against Mexico, the United States was regarded as an aggressor in Latin America. Thus, when the League turned out to be a relatively impotent organization, the Latin American states suffered considerable disappointment. Later, many of them became disillusioned by the dominance of Great Britain and France, and by the failure of efforts to build an effective security system within the League framework. Some of the larger states resented what they considered to be their inadequate representation on the League Council, although the president of the Assembly was quite often a Latin American. Others, finding that the main activities of the League touched their direct interests only slightly, drifted away from it, and some were jealous of League involvement in strictly American disputes.

In practice the League came to be increasingly divorced from New World questions because of the attitude of the United States. To guard its hemispheric interest the latter had insisted upon the inclusion in the League Covenant of reference to the Monroe Doctrine in Article 21: "Nothing in the Covenant shall be deemed to affect the validity of international engagements, such as treaties of arbitration or regional understandings like the Monroe Doctrine, for securing the maintenance of peace." The League shared in the solution of the Chaco War between Paraguay and Bolivia (1928-1938), but its rulings were not accepted by the states involved. It also was the vehicle for final settlement of the Leticia dispute between Colombia and Peru (1932-1935), but only because of a virtual settlement that had been achieved by the good offices of American states outside the League.

Throughout the period of Latin American participation in the League of Nations the latter's actions reflected a deference toward the United States on all inter-American questions and a desire not to antagonize American public opinion. It may be concluded that at the same time that the League was discrediting itself in Latin America by revealing its weakness, the United States was abandoning interventionist policies in favor of the Good Neighbor policy. When the smaller hemispheric countries saw that the United States was no longer likely to institute aggressive action in their territories, their fear of the United States declined and their interest in the League waned. This state of affairs continued until the final collapse of the League of Nations. In general, the Latin American states failed in their aims of using the League to further their regional and individual economic interests, and to use the organization as a counterweight against United States influence.

The scheme for a United Nations organization was devised during the years 1943-1945 by the great anti-Axis powers through a series of discussions which culminated in the United Nations Conference in San Francisco (April-June 1945); however, as early as 1942, the Juridical Committee of the Pan American

Union proposed recommendations concerning the establishment of a universal organization to replace the League. Included was one suggestion that any projected organization must be harmonized with local regional groupings. Thereafter the Latin Americans played an important part in conferences which established the United Nations Relief and Rehabilitation Administration, the Food and Agriculture Organization, the International Monetary Fund, and the International Bank for Reconstruction and Development. They were not invited to participate in the Dumbarton Oaks discussion in 1944, which they keenly resented. And they were antagonized by the secrecy of the meetings, fearing that the United States was shifting its position from regionalism to universalism. If this were the case, it might prove to be prejudicial to the interests of the smaller states, and possibly jeopardize the future of the inter-American system. The belief was well founded, for the Dumbarton Oaks draft recognized the usefulness of regional agencies, but considered that they might be dangerous if not effectively subordinated to the Security Council. Moreover, President Roosevelt, supported by Secretary of State Cordell Hull, adhered to the universalist position, assuming that United States-Soviet cooperation would continue into the postwar era. Support for the regional concept came, at first, from Under Secretary of State Sumner Welles, and later from Assistant Secretary of State Nelson Rockefeller and Senator Arthur Vandenburg.

When the Dumbarton Oaks conference was in progress the Latin Americans showed such obvious displeasure that Secretary of State Edward Stettinius invited them to attend a series of conferences at the White House to reassure them that everything possible was being done to preserve the inter-American system. In a critical report rendered later by Carlos Martins, Brazilian ambassador to the United States, on behalf of the coordinating committee set up by the Latin American state, great concern was shown about the relationship between the general and the regional organization, and the status accorded the smaller states within the former. They feared the great powers' use of the veto and stressed the priority of inter-American procedures vis-à-vis the Security Council.

It was partially a result of Latin American dissatisfaction that the Inter-American Conference on Problems of War and Peace met in Chapultepec Palace, in Mexico City, in February 1945. Argentina, alone of the Latin American states, was not invited on the grounds that only countries aiding the war effort should attend (the United States was angered at Argentina's policy of neutrality in World War II, which aided the Nazis). The other Latin American nations were not pleased with this exclusion which the United States had insisted upon even though seventeen American states did not recognize the government of President Edelmiro Farrell (and Perón) which had come into being in 1943. Standing clearly in opposition to the universalist oriented Dumbarton Oaks Proposals, the Latin American delegations affirmed their faith in the inter-American system going so far as to propose the adoption of a treaty of mutual assistance in the event of aggression by any state (American or non-American) against an American republic. This was a very surprising move: since military assistance would have to come from the United States, which alone was capable of providing it in any degree, the proposed treaty was, in effect, a request by the Latin American nations from the United States of military protection against aggression. Only a few years before, the United States had, itself, been charged with aggression in the region. The fact that such a treaty was proposed showed how greatly the situation in the Americas had altered in a few years.

In broad outline the proposed treaty affirmed the relevance of the Monroe Doctrine, which had not specifically referred to the use of armed force, nor had it provided for the attack by one country of the Western Hemisphere on another. However, Washington found the terms of the treaty unacceptable, for (1) it was deemed inadvisable to make any commitment which might conflict with agreements it might be compelled to make to the United Nations incident to the latter's establishment at the forthcoming session, and (2) it had already entered into agreements with its major anti-Axis allies, and it was reluctant to become a signatory to any further agreement which might be inconsistent with its existing obligations. The U.S. refusal to become a party to the treaty proved displeasing to the Latin American republics.

The United Nations Conference on International Organization convened on April 25, 1945 in San Francisco with the attendance of representatives of fifty nations. The Conference ended on June 26 with the signing of the Charter and the Statute of the new International Court of Justice; the Charter came into force on October 24, 1945. All twenty Latin American states sent delegations to the Conference, and since the principle of equality prevails in the U.N. Assembly, it followed that the Latin American nations had a more significant role in writing the Charter than their global political significance would suggest that they should. It is probable that the Latin American representatives exerted considerable influence on the content of the Charter, but it would be an exaggeration to say that the Charter was framed to conform with the ideas of the Latin American nations.

When the text of the Charter of the United Nations was being drafted at San Francisco, the Latin American states gave enthusiastic support to the concept of world organization. But at the same time they were determined to insure a significant degree of autonomy and independence for their own regional organization, the Union of American Republics which had been in operation for fifty-five years. Their firmness on this point was strengthened by the fact that the veto privilege had been provided for the so-called great powers; this meant that in practice regional action could be overruled.

Latin American statesmen therefore opposed putting final authority in the hands of an untried world organization and sought, through resolutions and motions, to weaken the world organization, and to strengthen the regional body. The problem of integrating the Pan American Union into the U.N. structure was complicated by the fact that there was no general agreement on the relative significance of world and regional organization. The proposal was made that subjects of exclusively American concern be assigned to the Pan American Union, and matters of world concern to the U.N. But this solution evaded the question by failing to explain how the two spheres of interest could be distinguished. It was clear, however, that in the event of jurisdictional conflict, one organization would inevitably have to give way to the other.

The U.N. Charter gave its answer in favor of the supremacy of world organization. Article 52 of the Charter provided as follows: "Nothing in the present Charter precludes the existence of regional arrangements or agencies for dealing with such matters relating to the maintenance of international peace and security as are appropriated for regional action, provided that such arrangements or agencies and their activities are consistent with the purposes and principles of the organization." The Charter adds that regional agencies shall be utilized in an attempt to settle local disputes before they are referred to the Security Council of the U.N. and states that the Security Council itself, "shall, where appropriate, utilize such regional arrangements or agencies for enforcement action under its authority." It continues, "But no enforcement action shall be taken under regional arrangements or by regional agencies without the authorization of the Security Council . . ."

This theoretical statement of the relation of the Pan American Union and the U.N. was opposed at the San Francisco Conference by most of the Latin American states, and it is not surprising that an early solution to the problem was sought. The Inter-American Treaty of Reciprocal Assistance (Rio Treaty of 1947) insured the victory of the Latin American nations and the advocates of regionalism over universalism within the U.S. government. Article 3 of the Rio Treaty authorizing collective resistance to an armed attack was consistent with Article 51 of the U.N. Charter. The Rio Treaty turned the regional association into a regional security organization, with security questions to be handled by an Organ of Consultation tentatively designated as the Meeting of Foreign Ministers; the Ninth Conference confirmed that the Meetings of Foreign Ministers should exercise the functions of the Organ of Consultation referred to in the Rio Treaty.

The role of Latin America in the United Nations has differed markedly from that in the League, for all the states have been participating members from the beginning, and a Latin American bloc has existed since the United Nations was established. Moreover, the region had two seats on the Security Council from the outset, and the General Assembly frequently has a Latin American president and always a Latin American vice-president. In the U.N. the Latin American delegates have seldom voted as a solid bloc on any issue; however, the majority has shown opposition to colonialism, aggression, intervention, and racial discrimination. On issues involving the status of dependent peoples Latin Americans have tended to be more lenient and understanding toward France, Italy, and Portugal than toward the remnants of other colonial empires. Greater unanimity is shown on economic and social matters than those relating to U.N. financing and international law. On hemispheric security problems involving the cold war the Latin American bloc has generally supported the United States. Brazil has followed the United States initiative quite consistently, as has Argentina since 1954. Guatemala showed intransigence prior to 1954, and Mexico has frequently taken an independent position. Cuba has been a member of the Soviet bloc in the United Nations since 1960. In matters concerning international trade and economic development the Latin Americans have been more successful in furthering their interests through the United Nations than in its predecessor.

The first test of the U.N. Charter provisions concerning regional organizations arose in June 1954 when the Arbenz government of Guatemala charged that U.S.-supported attacks were being launched from Nicaragua and Honduras. Guatemala appealed simultaneously to the United Nations and the

Inter-American Peace Committee but requested suspension of consideration of its complaint by the latter, an OAS agency. This was quickly followed by a request for complete withdrawal of the case. Guatemala took this position because it recognized that the OAS, under the leadership of the United States, would be more disposed to seek its overthrow than to prevent it. This assumption was substantiated by the action of the Tenth Inter-American Conference at Caracas, in March 1954. Although the Caracas Declaration fell short of the United States draft, Washington viewed it as a collective commitment to condone or support the ousting of the Arbenz goverment.

The Guatemala crisis set the pattern for other crises which would follow: the Cuban crises of 1961 and 1962, and the Dominican crisis of 1965. In the debates which arose in the Guatemalan case it became evident that, "whereas at San Francisco the great Latin American concern had been that an appeal to the regional enforcement machinery might be paralyzed by a Security Council veto, in the Guatemala case the plaintiff sought most urgently to avert any regional action whatsoever and restrict consideration of the case to the United Nations."

The United States, which clearly supported the anti-Arbenz invasion force, insisted that the case be referred to the OAS instead of the Security Council in order to forestall the raising of any barrier to the success of the invasion. The Soviet Union wished to keep the case before the Security Council as a means of protecting its beachhead in the Western Hemisphere. Obviously, the Guatemalan case was not merely a problem confined to the Western Hemisphere, but another episode in the global cold war. The principle of regional jurisdiction was adopted in this instance, but the United States did not secure Latin American acceptance of the "OAS-first" priority concept precluding the right of an American state to appeal to the Security Council when it is threatened. This became clear at the Ninth General Assembly late in 1954 when several Latin American governments contended that their membership in the OAS did not imply any restriction on their right to appeal to the United Nations.

In July 1960 the Cuban government requested that the Security Council meet to consider charges of interventionist policy and conspiracy to commit aggression, which it leveled against the United States. Like the Guatemala affair it involved a collision between the United States and a Communist-backed regime. Cuba chose the Council rather than the OAS, but both organizations were convened to consider the case. Ultimately the Council referred the matter to the OAS, but it neither disclaimed its own jurisdiction nor denied Cuba's right of appeal to the United Nations. Although the United States managed to assert the primacy of regional jurisdiction, it did not gain full Latin American endorsement of this principle.

In April 1961, the two precedents were repeated when Cuba appealed to the Security Council for action against the United States and other states in regard to the Bay of Pigs invasion attempt. This was an abortive effort to overthrow the Castro regime, by then a client state of the Soviet Union, in which the United States at first denied, and later admitted, complicity. It was a re-run of the Guatemala case with Washington supporting OAS jurisdiction and the Soviet Union insisting upon U.N. action, but with significant differences: the Soviet-backed regime was not toppled and the U.N. did not refer the matter to the OAS. However, the Security Council proved unwilling to act before or during the invasion and failed ultimately to take any action. In a modified joint draft, adopted as a General Assembly Resolution, the pro-OAS content was virtually removed. The U.N. action attenuated the principle that the OAS has jurisdiction over hemispheric problems that its members are obliged to accept and U.N. agencies to honor. The Latin American nations, which had previously supported this position, reacted against it in this Cuban case.

In May 1963, the Security Council was asked to consider the dispute involving Haiti and the Dominican Republic, at the former's request. No jurisdictional problem arose on this occasion as Haiti agreed to refer the matter to the OAS. Similarly, complaints were brought before the Security Council by Panama against the United States as a result of the anti-American outbreaks in January 1964. The jurisdictional problem did not arise, for both nations agreed to the validity of OAS authority. In both of these cases (Haiti and Panama), however, the complainants agreed to deal with the OAS on the condition that their right to appeal to the Security Council was not inhibited.

In October 1962, the United States responded to the covert installation of Soviet missiles in Cuba by demanding their immediate removal and by carrying out a naval blockade to prevent a continuation of the Soviet Union's military build up on the island. President Kennedy secured the prompt endorsement of these unilateral measures by a resolution of the OAS Council requesting all members of the organization, in accordance with articles 6 and 8 of the Rio Pact, to "take measures, individually and collectively, including the use of armed force," in support of the United States actions. On this occasion Washington took the initiative in bringing the Security Council, as well as the OAS, into the case since this time it was alleging aggression. UN Secretary-General U Thant went to Havana, but the Castro government refused to allow international inspection of the removal of Soviet missiles from Cuba (releasing the United States from its promise not to attempt the overthrow of the Castro regime in return for removal). While the UN action was ineffectual in this crisis, a precedent was established for the personal intervention of the secretary-general in Latin American affairs. Another precedent was established by the OAS decreeing the application of military sanctions, but it is clear that the United States did not regard their validity as being contingent upon the approval of the Security Council under Article 53. Rather, the United States stood on the principle that the OAS is an independent organization whose function it is to protect the Western Hemisphere without interference from the UN.

In the Dominican Republic crisis of April 1965, the United States intervened unilaterally to prevent a Communist seizure of power and then called upon the OAS to endorse and support its action. Up to that time the Latin American members of the Security Council had given the United States consistent support of security matters, but in this instance Uruguay supported the principle of UN intervention which weakened the United States stand that hemispheric matters were within the jurisdiction of the OAS. This enabled Secretary-General U Thant to send his representatives to the Dominican Republic, but their role was negligible for the occupation was carried out by the United States through the OAS.

Despite the intent of its framers, the paramount, universalist character of the United Nations has given way to regional jurisdiction in the Western

Hemisphere, as well as in other parts of the world where regional organizations must deal with political and security problems. The principle set forth in Article 53 of the United Nations Charter, that regional agencies should not take enforcement action without the approval of the Security Council, has been superseded by the exceptions found in that article and in Article 51. In practice, member states of the United Nations have come to rely upon regional alliance systems, established to operate autonomously, for the maintenance of international peace and security instead of the Security Council. Such alteration of the relationship between regional agencies and the United Nations was a product of cold war power politics. Fundamentally, it is related to the continuing attempts by the United States to thwart the Soviet Union's veto power and the Soviet Union's effort to preserve that power.

Thus the United States, which resisted the attempts by other members of the inter-American system to achieve regional autonomy in the early stages of the United Nations, has shifted to a position in support of it. As a result of the cases involving the threat of communism in the hemisphere the United States has commonly found itself cast in the role as an opponent of a strong and efficient world organization, whereas the Soviet Union has presented the image of a champion of the integrity and competence of the United Nations. A seemingly anomalous situation was created in which the United States continued to support the primacy of the United Nations in the global system, but with the reservation that it retain freedom of action in combating Communist activity in the Western Hemisphere. Although the OAS has not gained a monopolistic authority over disputes in the hemisphere, the security Council of the United Nations has been deprived of effective authority to restrict or regulate the enforcement procedures of the OAS.

Renewed Chinese interest in Latin America appeared in 1971, and by early 1972 six Western Hemisphere nations were in diplomatic contact with the People's Republic of China. An important corollary of the moves toward diplomatic relations was the Latin American support for Peking's entry into the United Nations, a move accompanied in some instances by votes to unseat Taiwan's Nationalist Chinese government.

The Treaty of Tlatelolco, or Treaty for the Prohibition of Nuclear Weapons in Latin America, represents a concerted move by Latin Americans, in collaboration with the United Nations, to achieve the military denuclearization of their portion of the Western Hemisphere.[47] Brazil first introduced the idea to the 17th U.N. General Assembly in 1962. On April 29, 1963, five Latin American presidents drafted a joint declaration in which they announced that their governments were prepared to sign a multilateral Latin American agreement whereby they would undertake "not to manufacture, receive, store, or test nuclear weapons or nuclear launching devices."

On November 27, 1963, the General Assembly of the United Nations approved resolution 1911 (XVIII), entitled "Denuclearization of Latin

[47]Tlatelolco is the Aztec name of the historic district of Mexico City in which the treaty was approved on Feb. 12, 1967. For background see Alfonso Garcia Robles, *The Denuclearization of Latin America* (Carnegie Endowment for International Peace, New York, 1967); *El Tratado de Tlatelolco: Genesis, Alcance y Propósitos de la Proscripción de las Armas Nucleares en la America Latina* (El Colegio de Mexico, 1967); Rutgers University, *Disarmament in the Western World*, Occasional Papers No. 1 (June 1969).

America," in which the Assembly praised the initiative embodied in the declaration. After the 18th session of the Assembly had ended the Mexican Ministry of Foreign Affairs initiated consultations with the ministries of foreign affairs of the other Latin American republics on measures likely to be most effective for carrying out the recommendations of the above resolution.

The outcome of these consultations was the Preliminary Meeting on the Denuclearization of Latin America, which took place in Mexico from November 23-27, 1964. At this meeting two basic resolutions were adopted: The first defined the term "denuclearization," specifying that it should mean solely "the absence of nuclear weapons," and not the prohibition of the peaceful use of the atom, which should, on the contrary, be encouraged, especially for the benefit of the developing countries. The second resolution established the Preparatory Commission for the Denuclearization of Latin America and instructed the Commission to prepare a draft treaty of the subject. This was completed early in 1967 with the adoption and opening for signature of the Treaty of Tlatelolco. The treaty was later endorsed by the U.N. General Assembly during its 22nd session with the adoption of resolution 2286 (XXII). At the same time, a series of urgent appeals were addressed by the Assembly to all states which might become signatories to the treaty, and the powers possessing nuclear weapons, to sign and ratify as soon as possible.

The Treaty of Tlatelolco consists of thirty-one articles, one transitional article, and two additional protocols. With the aim of facilitating, ensuring, and verifying compliance with the obligations contracted by the parties, the treaty contains in Article 5 an objective definition of what, for the purposes of the treaty, is to be understood by "nuclear weapon;" it sets up an Agency for the Prohibition of Nuclear Weapons in Latin America, the principal organs of which will be a General Conference, a Council, and a Secretariat; it also establishes a Control System which is described in Articles 12 to 16, and 18, paragraphs two and three.

The United States government conveyed its full support to the Commission, regarding the initiative of the Latin American countries as an outstanding example of regional activity to limit and control armaments. However, Washington's endorsement of the establishment of nuclear-free zones is given only under certain specified conditions: where the initiative for such zones originates within the area concerned; where the zone includes all states in the area whose participation is deemed important; where the creation of a zone would not disturb necessary security arrangements; and where provisions are included for following up alleged violations in order to give reasonable assurance of compliance within the zone. Under these criteria the United States could not accept the proposal to make Central Europe a nuclear-free zone, but for such areas as Latin America and Africa, the idea met with Washington's complete approval.[48]

Protocol II of the treaty, which calls upon the powers possessing nuclear weapons to respect the status of denuclearization in Latin America, and not to use or threaten to use nuclear weapons against Latin American states party to the treaty, was signed by Vice President Hubert Humphrey for the United States

[48] Dept of State *Bulletin*, LVI (Apr. 3, 1967), pp. 575-576.

on April 1, 1968, at Mexico City. President Johnson expressed the "hope that all nuclear powers will respect this great achievement of Latin American diplomacy." The Soviet Union and France delayed signing Protocol II, and the People's Republic of China rejected it. By October 1967 the last of the twenty-one states that were members of the Preparatory Commission had signed the treaty. And Mexico's announcement in May 1969 was that, eleven nations having deposited instruments of ratification, a permanent agency would be set up to administer the treaty.[49]

As an engagement entered into voluntarily among neighboring states, the treaty sets a pattern for arms control and verification that points the way for other regions and for the world as a whole, if the nuclear danger that threatens mankind is ever to be brought under control.

[49] *Ibid.*, LIX (Aug. 5, 1968), p. 138; *New York Times*, Oct. 2, 1966; *Ibid.*, May 7, 1969.

Table 2-1. Spanish-American Congresses, 1826-88*

Name	*Date*	*Location*	*Countries Attending*	*Major Agenda Items*	*Outcome*
Panama Congress	1826	Panama City, Panama	Central America, Gran Colombia, Mexico, Peru	Peace and Security; federal union for Latin America; suppression of slavery	Several agreements signed; none ratified by all signatories
Congress of Lima (or "American Congress")	1847-48	Lima, Peru	Bolivia, Chile, Ecuador, New Granada (Colombia), Peru	Consideration of defensive measures to be taken to prevent Spanish reconquest of west-coast states of South America.	Three treaties signed; none ratified
Continental Congress	1856	Santiago, Chile,	Chile, Ecuador, Peru	Consideration of collective measures to be taken in event of U.S. incursions	A treaty of mutual assistance, stating that if signatories were attacked by the U.S., all would unite against the U.S.
Congress of Lima	1864-65	Lima, Peru	Bolivia, Colombia, Chile, Ecuador, Guatemala, Peru, Venezuela	Consideration of possible responses to presence of Spanish in Santo Domingo and French in Mexico	Treaty of "Union and "Alliance" signed; not ratified

*Forerunners of Pan American Conferences.

Source: U.S. Senate, Committee on Foreign Relations, *United States-Latin American Relations. The Organization of American States* (Washington, D.C.: Government Printing Office, 1959), p. 16.

Table 2-2. Pan American Conferences, 1889-1967

Name	*Date*	*Location*	*Countries Attending*	*Major Agenda Items*	*Outcome*
First International Conference of American States	Oct. 2, 1889-April 19, 1890	Washington, D.C., United States	All the American republics then in existence, except the Dominican Republic (18)	Discussion of problems of mutual interest, principally questions of peace, trade, and communications.	Formation of the International Union of American Republics and the "Bureau of American Republics."
Second International Conference of American States	Oct. 22, 1901-Jan. 22, 1902	Mexico City, Mexico	All the American republics then in existence (19)	Discussion of international legal questions; procedures for arbitration of disputes; problems of hemispheric peace	Protocol of adherence to Hague Convention for Pacific Settlement of International Disputes. Treaty of Arbitration for Pecuniary Claims.
Third International Conference of American States	July 21-Aug. 26, 1906	Rio de Janeiro, Brazil	All the American republics except Haiti and Venezuela (19)	Consideration of problem of forcible collection of debts; discussion of Drago and Calvo doctrines.	Conference decided to take question of forcible collection of debts to Second Hague Conference. Convention on International Law.
Fourth International Conference of American States	July 12-Aug. 30, 1910	Buenos Aires, Argentina	All the American republics except Bolivia (20)	Consideration of various economic and cultural matters.	Decision to change name of International Bureau of American Republics to Pan American Union.

Fifth International Conference of American States	March 25-May 3, 1923	Santiago, Chile,	All the American republics except Bolivia, Mexico, and Peru (18)	Discussion of reorganization of Pan American Union (PAU) for purpose of reducing U.S. dominance; discussion of possible modification of Monroe Doctrine.	Treaty To Avoid or Prevent Conflicts Between American States (Gondra Treaty). Decision to make chairmanship of PAU elective.
Sixth International Conference of American States	Jan. 16-Feb. 29, 1928	Havana, Cuba	All the American republics (21)	Latin American delegates anxious to secure condemnation of U.S. intervention in the Caribbean	Convention on Duties and Rights of States in the Event of Civil Strife (designed to prevent use of other American countries as bases for launching revolutionary activity).
International Conference of American States on Conciliation and Arbitration	Dec. 10, 1928-Jan. 5, 1929	Washington, D.C., United States	All the American republics except Argentina (20)	Problem of arbitration and conciliation of disputes.	General Convention of Inter-American Conciliation. General Treaty of Inter-American Arbitration.
Seventh International Conference of American States	Dec. 3-26, 1933	Montevideo, Uruguay	All the American republics except Costa Rica (20)	Problem of U.S. dominance and intervention	Convention on Rights and Duties of States; concerned with the principle of nonintervention.
Inter-American Conference for the Maintenance of Peace	Dec. 1-23, 1936	Buenos Aires, Argentina	All the American republics (21)	Security of hemisphere in event of war in Europe or Far East; principle of nonintervention.	Declaration of Principles of Inter-American Solidarity and Cooperation; additional protocol relative to nonintervention.

Table 2-2 (*continued*)

Name	*Date*	*Location*	*Countries Attending*	*Major Agenda Items*	*Outcome*
Eighth International Conference of American States	Dec. 9-27, 1938	Lima, Peru	All the American republics (21)	Consideration of the relation of American republics to Europe and possible German and Italian penetration of the hemisphere.	Declaration of the Principles of the Solidarity of America; established the Meeting of Consultation of Foreign Ministers.
Inter-American Conference on Problems of War and Peace	Feb. 21-March 8, 1945	Mexico City, Mexico	All the American republics except Argentina (20)	Consideration of possible postwar problems. Hemispheric relations of Argentina.	Act of Chapultepec; dealt with acts of threats of aggression against any American republic; recommended consideration of a treaty to deal with such acts and measures to take when they occurred.
Inter-American Conference for the Maintenance of Continental Peace and Security	Aug. 15-Sept. 2, 1947	Rio de Janeiro, Brazil	All the American republics except Nicaragua (20)	Consideration of proposals for a treaty of mutual defense of the hemisphere.	Inter-American Treaty of Reciprocal Assistance (Rio Treaty).

Ninth International Conference of American States	March 30-May 2, 1948	Bogotá, Colombia	All the American republics (21)	Discussion of means to strengthen the inter-American system and to promote inter-American economic cooperation; consideration of juridical and political matters, including recognition of governments and colonies.	Charter of the OAS; American Treaty on Pacific Settlement (Pact of Bogota); American Declaration of the Rights and Duties of Man; Economic Agreement of Bogotá.
Tenth International Conference of American States	March 1-28, 1954	Caracas, Venezuela	All the American republics except Costa Rica (20)	Consideration of hemispheric policy respecting the intervention of Communism in the Americas; discussion of possible economic assistance to Latin America.	Declaration of Solidarity for the Preservation of the Political Integrity of the Americas against Intervention of International Communism.
*Eleventh International Conference of American States**					
First Special International Conference of American States	Dec. 16-18, 1964	Washington D.C., United States	All the American republics except Cuba (20)	Consideration of the procedures for admitting new members	Act of Washington, setting forth procedure for admitting new members (two-thirds vote of the Council), but excluding territories which are subject to claim by an American state.

Table 2-2 (*continued*)

Name	*Date*	*Location*	*Countries Attending*	*Major Agenda Items*	*Outcome*
Second Special Inter-American Conference	Nov. 15-30, 1965	Rio de Janeiro, Brazil	All the Latin American republics except Cuba (20)	Consideration of OAS Charter revision.	Adoption of resolutions to be ratified by subsequent Conference, Approved Act of Rio de Janeiro.
Third Special Inter-American Conference	Feb. 15-27, 1967	Buenos Aires, Argentina	All the Latin American republics except Cuba (20)	Consideration of Protocol of Amendment to OAS Charter.	Approval of the Protocol of amendment to the OAS Charter.
Meeting of American Chiefs of State	April 12-14, 1967	Punta del Este, Uruguay	All the Latin American republics except Cuba (20)	Renewal of commitment to cause of Latin American economic and social development.	Declaration of the Presidents of America.

*The Eleventh International Conference of American States was scheduled to meet in 1959 in Quito, but it was not convened. In 1963, it was resolved to convoke the Quito conference on April 1, 1964, but it was then postponed indefinitely. Instead, a meeting of consultation was held in Washington in July 1964, at which Cuba was again discussed.

Source: U.S. Senate, Committee on Foreign Relations, *United States-Latin American Relations. The Organization of American States* (Washington, D.C.: Government Printing Office, 1959), pp. 8-9.

Table 2-3. Meetings of Consultation of Ministers of Foreign Affairs, 1939-1971

Meeting	*Location*	*Date*	*Major Agenda Items*	*Outcome*
First	Panama City, Panama	Sept. 23-Oct. 3, 1939 (after start of World War II)	Consideration of means for maintenance of the neutrality of the hemisphere.	Declaration of Panama, establishing a hemispheric zone embracing the American republics within which the belligerent nations were to commit no hostile acts; general declaration of neutrality.
Second	Havana, Cuba	July 21-30, 1940 (after fall of France)	Discussion of European possessions in the Americas and the danger of their possible transfer to other non-American powers.	Act of Havana and Convention of Havana, concerning the provisional administration of European colonies and possessions in the Americas. (Resolution XV: Any attempt by a non-American state against sovereignity or independence of an American state to be considered an attack on all.)
Third	Rio de Janeiro, Brazil	Jan. 15-30, 1940 (after Pearl Harbor)	Determination of attitude to be adopted by American republics in face of attack by a non-American power upon an American state and subsequent declaration of war by Germany and Italy.	Resolution: "The American Republics . . . recommend the breaking of their diplomatic relations with Japan, Germany and Italy." Establishment of the Inter-American Defense Board; establishment of the Emergency Advisory Committee for Political Defense.

Table 2-3 *(continued)*

Meeting	*Location*	*Date*	*Major Agenda Items*	*Outcome*
Fourth	Washington, D.C., United States	March 26-Apr. 7, 1951 (after Korea)	Consideration of problems of communism and hemispheric security.	Recommendation that each republic examine its resources to determine what steps it could take to contribute to collective defense of continent; recommendation that governments examine their laws with view to adopting changes considered necessary for prevention of subversive activities of Communists.
Fifth	Santiago, Chile	Aug. 12-18, 1959 (after April-June disturbances in Caribbean)	Consideration of problems of unrest in the Caribbean; discussion of problems of democracy and human rights in Latin America.	Declaration of Santiago, concerning principles of democracy and respect for human rights; special temporary power assigned to Inter-American Peace Committee to investigate and conciliate in cases of invasion of foreign-based rebels.
Sixth	San José, Costa Rica	Aug. 16-21, 1960 (after attempt on life of Venezuelan President Betancourt)	Request of Venezuelan Government regarding policy of intervention of the Dominican Republic (attempt to kill President Betancourt.	Breaking of diplomatic relations and partial interruption of economic relations with the Dominican Republic.
Seventh	San José, Costa Rica	Aug. 22-29, 1960 (Cuban question)	Continental solidarity; defense of the inter-American system and of democratic principles.	Declaration of San José de Costa Rica regarding restatement of inter-American principles; establishment of a Committee of Good Offices.

Eighth	Punta del Este, Uruguay	Jan. 22-31, 1962 (after increased Cuban tension)	Cuban, Soviet, and Communist Chinese subversive activities; general threat to continental unity and to democratic institutions.	Exclusion of the present Cuban Government from the inter-American system; exclusion of Cuba from the Inter-American Defense Board; prohibition of any armament trade with Cuba and request to the Council of the OAS to extend this prohibition possibly also to other commercial goods; establishment of the Special Consultative Committee on Security (SCCS against Communist subversion; recommendation for amendment of the Statutes to the Inter-American Commission on Human Rights.
Ninth	Washington, D.C., United States	July 21-26, 1964	Venezuelan request for sanctions against Cuba in view of Cuban complicity in terrorist activities in Venezuela.	Breaking of diplomatic and consular relations with Cuba; interruption of commercial and maritime relations with Cuba; expression of sympathy for the Cuban people; regional and international economic cooperation within the framework of the Charter of Alta Gracia.
Tenth	Washington, D.C., United States	May 1-00, 1965	Infiltration of communism into Dominican revolt.	Establishment of a mediatory commission with military force.
Eleventh	Buenos Aires, Argentina	Feb. 16-26, 1967	Prepare for meeting of American Chiefs of State; consider economic integration.	Resolution adopted for meeting of American Chiefs of State at Punta del Este, Uruguay, April, 1967.

Table 2-3 (*continued*)

Meeting	*Location*	*Date*	*Major Agenda Items*	*Outcome*
Twelfth	Washington, D.C., United States	Sept. 19-24, 1967	To consider evidence of continuing Castro-fomented subversion in Venezuela, Bolivia and other nations.	Agreement of steps to fight subversion and new anti-Cuban policies.
Thirteenth	Washington, D.C., United States	July 26, 1969	To end war between El Salvador and Honduras, occurring July 14-18, 1969.	Cease-fire arranged, and Salvadorian troops withdrawn from Honduras by August 3, 1969.
Fourteenth	Washington, D.C.	Jan. 30, 1971	Ecuador charged U.S. with violation of Article 19, OAS Charter (Re: Ecuador's seizure of U.S. fishing vessels and alleged U.S. suspension of arms sales to Ecuador)	A resolution supported by both Ecuador and U.S. stating their respective positions, and calling upon them to resolve their differences in accordance with Charter principles.

Source: U.S. Senate, Committee on Foreign Relations, *United States-Latin American Relations. The Organization of American States* (Washington, D.C.: Government Printing Office, 1959), p. 11; Dept. of State *Bulletin*, XLIII (Sept. 5, 1960), pp. 359-360; *Ibid*., XLIII (Sept. 12, 1960), pp. 395-412; *Ibid*., XLVI (Feb. 19, 1962), pp. 267-289; *Ibid*., LI (Aug. 10, 1964), pp. 174-184; *Ibid*., LII (May 17, 1965), pp. 738-748; *Ibid*., LVI (March 20, 1967), pp. 472-476; *Ibid*., LVII (Oct. 16, 1967), pp. 490-498; *Ibid*., LXI (Aug. 18, 1969), pp. 132-134; *Ibid*., LXIV (Feb. 22, 1971), pp. 245-250.

Table 2-4. Applications of the Inter-American Treaty of Reciprocal Assistance (*Rio Treaty*)

1948-1949	Conflict between Costa Rica and Nicaragua.
1950	Haiti vs. Dominican Republic.
	The Dominican Republic vs. Haiti, Cuba and Guatemala.
1954	The Guatemalan case.
1955-1956	Conflict between Costa Rica and Nicaragua.
1957	Border dispute between Honduras and Nicaragua.
1959	Nicaragua (border violations).
1960-1962	Applications of Measures to the Dominican Republic re Venezuela.
1963-1965	Situation between the Dominican Republic and Haiti.
1964	Situation between Panama and the United States. Five requests which did not result in applications of the treaty: requests of the governments of Haiti (1949); Ecuador (1955); the Dominican Republic (1959); Peru (1961); and Bolivia (1962).
1969	Conflict between El Salvador and Honduras.

Applications of the Rio Treaty regarding Cuba

1959	First incidents.
1961-1962	Exclusion of the Cuban government and partial suspension of trade with Cuba.
1963-1964	Measures agreed on by the Ninth Meeting of Consultation.

Source: Inter-American Institute of International Legal Studies, *The Inter-American System: Its Development and Strengthening* (New York, 1966), chapters VII and VIII.

Table 2-5. Activities of the Inter-American Peace Committee, 1948-64

	Situation	*Date Request for Action Received*	*Date Action Terminated*	*Charges by Country Initiating Action or Reasons for Requesting Action*	*Outcome*	*Countries Involved in the Situation*
1.	*Dominican Republic-Cuba situation*	Aug. 13, 1948	Sept. 9, 1948	Dominican Republic alleged organization of revolutionary forces in Cuba directed against Dominican Republic.	Both sides agreed to continue negotiations.	Dominican Republic, Cuba.
2.	*Haiti-Dominican Republic situation*	March 21, 1949	June 9, 1949	Haiti requested Committee's good offices in a dispute with the Dominican Republic, Haiti cited certain acts it claimed could create situation between both countries endangering the peace.	Both governments signed a joint declaration of friendly relations on June 9, 1949.	Haiti, Dominican Republic
3.	*Cuba-Peru situation*	Aug. 3, 1949	(Not applicable)	Cuban Embassy in Lima gave asylum to two Peruvian citizens; on August 14, asylees left the embassy and incident was closed.	Cuba withdrew its request that the Committee meet.	Cuba, Peru
4.	*General Caribbean situation*	(Not applicable)	Sept. 14, 1949	U.S. requested the Committee study the general situation in the Caribbean.	Committee decline to take action; said jurisdiction limited to specific matters of controversy.	
5.	*Cuba-Dominican Republic situation*	Dec. 6, 1949	(Not applicable)	Cuba invited the Committee to investigate charges by Dominican Republic that Cuba was permitting movement to exist in its borders directed at Dominican Republic.	Committee declined the invitation; no situation calling for specific measures.	Cuba, Dominican Republic

6.	*Cuba-Dominican Republic situation*	Nov. 26, 1951	Dec. 25, 1951	Cuba alleged that five Cuban sailors on a Guatemalan vessel were seized and imprisoned by the Dominican Republic.	Dominican Republic and Cuba signed a joint declaration of peacefulness and nonintervention.	Cuba, Dominican Republic
7.	*Colombia-Peru situation*	Nov. 18, 1953	Jan. 21, 1954	Colombia called attention to the dispute over presence of Raúl Haya de la Torre, a Peruvian, in Colombian Embassy in Lima.	Committee recommended the resumption of lateral negotiations.	Colombia, Peru
8.	*Guatemalan situation*	June 19, 1954	June 30, 1954	Guatemala requested Committee meet to consider acts violating her sovereignity. Withdrew request, renewed it on June 26.	Committee sent subcommittee to make study; however, before it reached scene, new government in Guatemala.	Guatemala, Honduras Nicaragua
9.	*Cuba-Dominican Republic situation*	Feb. 27, 1956	April 20, 1956	Cuba requested Committee meet to study certain difficulties existing between it and the Dominican Republic.	Committee expressed its hope that parties arrive at solution through regular diplomatic channels.	Cuba, Dominican Republic
10.	*Request by Haiti*	Aug. 17, 1959	There was no formal termination. Case was settled in Oct. 1959.	Haiti asked the Foreign Ministers of the OAS to study invasion of Haiti by group coming from Cuba.	Matter was studied by Committee operating under new *ad hoc* powers granted at Santiago. Since the Cuban invasion was unsuccessful and Haiti did not accuse Cuba directly, the matter had no consequences.	Haiti, Cuba

Table 2-5 (*continued*)

	Situation	*Date Request for Action Received*	*Date Action Terminated*	*Charges by Country Initiating Action or Reasons for Requesting Action*	*Outcome*	*Countries Involved in the Situation*
11.	*Anti-Venezuelan leaflets over Curaçao*	Nov. 25, 1959	There was no formal termination of the case, since further charges were leveled against the Dominican regime by Venezuela and Ecuador.	A U.S. plane with Cuban pilots threw leaflets on Curaçao calling on the Venezuelan Army to rise up against the Betancourt regime. The leaflets were supposed to come down on Venezuelan territory. The plane, however, made a forced landing in Aruba.	The Committee found that the Dominican Government was implicated in the matter (stopover of plane in Santo Domingo).	Venezuela, Dominican Republic
12.	*Ecuador-Dominican Republic situation*	Feb. 16, 1960	April 12, 1960	Controversy between Ecuador and the Dominican Republic regarding thirteen Dominican citizens who had been granted asylum in the Embassy of Ecuador in Santo Domingo. Dominican measures which affected this right of asylum.	The attempt of direct negotiations between Ecuador and the Dominican Republic failed, since the Dominican Government refused to accept the "bases of agreement." The Committee then expressed the hope that the matter might find a bilateral solution.	Ecuador, Dominican Republic

13.	*Violation of human rights in the Dominican Republic*	Feb. 17, 1960	June 6, 1960	Venezuela requested the Committee "to examine the flagrant violation of human rights in the Dominican Republic," since it increased the tension in the Caribbean area. However, the Dominican Government did not authorize a visit of the Committee	The Committee came to the conclusion that the tensions in the Caribbean area had increased through the violation of human rights by the Dominican Republic (Report of the Committee dated June 6, 1960).	Venezuela, Dominican Republic
14.	*Violation of human rights and international tensions in the Caribbean area.*	Aug. 1959	Aug. 16, 1960 (Seventh Meeting of Consultation, Aug. 22-29, 1960, San José, Costa Rica)	The Fifth Meeting of Consultation in Santiago, Chile, August, 1959, instructed the Inter-American Peace Committee to examine the reason for the existing tensions in the Caribbean area, apart from specific individual cases, and to report about it to the next IA Conference of Meeting of Consultation.	The Committee came to the conclusion that there existed on the American continent a serious crisis which made itself felt most acutely in the Caribbean area. It had economic and social causes—the peoples were dissatisfied with their lot—and was directed against any kind of dictatorial tutelage (Special Report dated April 14, 1960, and condensed Final Report of August 5, 1960).*	Latin America in general, with special reference to the Caribbean area.

Table 2-5 (*continued*)

	Situation	*Date Request for Action Received*	*Date Action Terminated*	*Charges by Country Initiating Action of Reasons for Requesting Action*	*Outcome*	*Countries Involved in the Situation*
15.	*Request by Nicaragua*	Feb. 16, 1961	Dec. 1962	Guarantee of the execution of the decision of the International Court of Justice of November 18, 1960, regarding the validity of the arbitration award of the King of Spain (December 23, 1906).	The case was terminated in 1962 after the requested assistance was given. (The final solution of this controversy had been made at The Hague in favor of Honduras.)	Honduras, Nicaragua
16.	*Request by Mexico*	June 2, 1961	June 5, 1961	Charges by Guatemala regarding the alleged training of Communist agents on Mexican territory were to be examined by the Committee.	The case needed no examination, since the IAPC came to the conclusion that a visit to Mexico was not required because Mexico was keeping its international obligations.	Mexico, Guatemala

17.	*Request by Peru*	Nov. 27, 1961	Jan. 22, 1962	Examination of various arbitrary acts in Cuba; Communist subversion in Latin America.	The IAPC reported to the Eighth Meeting of Consultation in Punta del Este as follows: The ideological and political links of the Cuban government were in contradiction to the principles of the Charter of the OAS; there was systematic violation of human rights by Cuba; subversive activities of the Soviets and of Cuba were equivalent to political aggression.	Peru, Cuba
18.	*Request by Panama*	Jan. 10, 1964	Jan. 15, 1964	Panama requested the assistance of the IAPC after the Canal Zone riots, which were related to the flag controversy.	On January 15, 1964, after arriving in Panama City, the IAPC announced that the immediate crises had terminated and that it was therefore possible to begin negotiations regarding a revision of U.S. control rights.	Panama, the United States

*This more general activity of the IAPC on the basis of the mandate of the Fifth Meeting of Consultation of Ministers of Foreign Affairs is closely linked to cases 10-13.

Source: U.S. Senate, Committee on Foreign Relations, *United States-Latin American Relations. The Organization of American States* (Washington, D.C.: Government Printing Office, 1959), p. 28.

SUPPLEMENTARY READINGS

Ball, M. Margaret, *The OAS in Transition*. Durham, N.C., 1969.

———. *The Problem of Inter-American Organization*. Stanford, 1944.

Caicedo Castillo, José Joaquín. *El Panamericanismo*. Buenos Aires, 1961.

Canyes, Manuel S. *The Organization of American States and the United Nations*. 6th edition. Washington, Pan American Union, 1963.

Castaneda, Jorge. *Mexico and the United Nations*. New York, 1958.

Claude, Jr. Inis L. "The OAS, the UN and the United States," *International Conciliation*, No. 547 (March 1964).

Cuevas Cancino, Francisco, *Del Congresso de Panamá a la Conferencia de Caracas, 1826-1954*. Caracas, 1955.

Dozer, Donald M. *Are We Good Neighbors?* Gainesville, 1959.

Dreier, John C. *The Organization of American States*. New York, 1962.

Duggan, Lawrence. *The Search for Hemispheric Security*. New York, 1949.

Fenwick, Charles G. *The Organization of American States*. Washington, 1963.

Gantenbein, James W. ed. *The Evolution of our Latin-American Policy; A Documentary Record*. New York, 1950.

Guerrant, Edward O. *Roosevelt's Good Neighbor Policy*. Albuquerque, 1950.

Harbron, John D. *Canada and the Organization of American States*. Washington, 1963.

Houston, John A. *Latin America and the United Nations*. Cambridge, Mass., 1960.

Hovet, Jr., Thomas. *Bloc Politics in the United Nations*. Cambridge, Mass., 1960.

Karnes, Thomas L. *Failure of Union–Central America.: 1824-1960*. Chapel Hill, 1961.

Kelchner, Warren H. *Latin American Relations with the League of Nations*. Boston, 1929.

Kidder, Frederick E. *Latin America and UNESCO: The First Five Years*. Gainesville, 1960.

Lobo, Helio. *O Panamericanismo e o Brazil*. São Paulo, 1939.

Lockey, J.B. *Pan Americanism: Its Beginnings*. New York, 1920.

Manger, William. *Pan America in Crisis*. Washington, 1961.

Pan American Union. *The Organization of American States: What is is, How it Works*. Washington, 1965.

———. *The Inter-American System: Its Evolution and Role Today*. Washington, 1963.

Quintanilla, Luis. *A Latin American Speaks*. New York, 1943.

Pérez-Guerrero, Manuel. *Les Relations des Etats de L'Amérique Latine avec la Société des Nations*. Paris, 1936.

Redington, Robert J. *The Organization of American States as a Collective Security System*. Washington, 1962.

Scott, J.B. ed. *The International Conferences of American States, 1889-1928*. Washington, 1931.

Slater, James. *The Organization of American States and United States Foreign Policy*. Columbus, 1967.

Thomas, Ann Van Wynen, and A.J. Thomas, Jr. *The Organization of American States*. Dallas, 1963.

U.S. Dept. of State. *Inter-American Efforts to Relieve International Tensions in the Western Hemisphere, 1959-1960*. Publication 7409, Inter-American Series 79. Washington, 1962.

U.S. Senate, 86th Cong., 2nd Sess., Doc. 125, *United States-Latin American Relations: Soviet Block Latin American Activities and their Implications for United States Foreign Policy*. Subcommittee on Foreign Relations, Report No. 7. Washington, 1960.

U.S. Senate, 86th Cong., 1st Sess, *United States-Latin American Relations–The Organization of American States.* Subcommittee of American Republics Affairs. Report No. 3. Washington, 1959.

Whitaker, A.P., *The Western Hemisphere Idea*. Ithaca, 1954.

Wilgus, A.C., "Blaine and Pan Americanism," *Hispanic American Historical Review*, V. (1922), 662-708.

Wood, Bryce. *The Making of the Good Neighbor Policy.* New York, 1961.

3

The Monroe Doctrine

In any discussion of the relations between the United States and Latin America the Monroe Doctrine must necessarily occupy a prominent place. To the average citizen of the United States the Monroe Doctrine seems to possess almost as important a niche in the structure of American institutions as the Declaration of Independence and the Constitution.[1] Yet this same normal American, if pressed to give an exact explanation of what is meant by the Monroe Doctrine, would have great difficulty in making himself clear. He would probably define it ultimately in some such general terms as, "America for the Americans," or "Europe must keep her hands off," or some other expression whose meaning would approximate the idea that the United States should mind her own business and take care of the affairs of the Western Hemisphere, and the nations of the Eastern Hemisphere should do the same. And the interpretation would not be wholly inaccurate; for, in the words of Professor Hart, the Monroe Doctrine is a national policy based upon "the daily common-sense recognition of the geographical and political fact that the United States of America is by fact and by right more interested in American affairs, both on the northern and southern continents, than any European power can possibly be."[2]

If the Monroe Doctrine were merely a question of an understanding between the United States and Europe, it might well be considered as one whose solution had at last been reached. Sixty-three nations of the world voluntarily signed the Covenant of the League of Nations, thereby giving their assent to Article XXI, which states: "Nothing in this convenant shall be deemed to affect the validity of international engagements, such as treaties of arbitration, or regional understandings like the Monroe Doctrine, for securing the maintenance of peace." Although the term "regional understanding" more accurately describes Woodrow Wilson's proposed development of the doctrine than its

[1] Mary Baker Eddy, writing at the time of the first hundred years' anniversary of the Monroe Doctrine declared, "I believe in the Monroe Doctrine, in our Constitution and in the laws of God." (*New York Times*, Dec. 2, 1923.)
[2] A.B. Hart, *The Monroe Doctrine* (Boston, 1916), p. 2.

accepted meaning, Europe was willing to subscribe to the doctrine under its original name. In fact, the very same doctrine, which Lord Salisbury declared to be a novel principle which "no statesman, however eminent, and no nation, however powerful, are competent to insert into the code of international law,"[3] was declared by representatives of Great Britain at the Paris Conference to have become an international understanding "consistent with the spirit of the covenant; and, indeed, the principles of the League as expressed in Article X represent the extension to the whole world of the principles of this doctrine."[4] The same doctrine, which Bismarck declared to be an "international impertinence," and von Bülow "a theory launched venturesomely upon the blue waves of conjectural politics," was recognized to such an extent by Germany at the outbreak of World War I that the German ambassador at Washington assured the United States government that Germany had no intention, in case of victory, to seek expansion in South America. Furthermore, Dr. Dernburg, recognized as representing the German government unofficially, declared that "Germany has not the slightest intention of violating any part or section of the Monroe Doctrine."[5] Finally, France, which under the Second Empire committed the grossest violation of the principles of the doctrine by engineering the Maximilian expedition, has now recognized the fact that the Monroe Doctrine is bound up inseparably with the safety of the United States; and, in the words of M. Paul Deschanel, "It issued from the vitals of reality, just as the Constitution of the United States itself."[6]

But, whether or not we consider the Monroe Doctrine a settled policy as regards the future relations between the United States and Europe, the question remains open as regards our relations with the independent republics of Latin America. However fixed the North American may be in his belief that in the past the Monroe Doctrine has been the sheet-anchor of safety for the storm-tossed republics to the south, he must recognize the fact that there has been a strong belief throughout Latin America that the original Monroe Doctrine has very little value for any country but the United States. Undoubtedly, many men of high position and great authority in the states of Latin America have paid tribute to the doctrine for the very substantial assistance it gave to the Latin-American republics at a time when friendly intercession was most advantageous. At the Fourth Pan American Conference, Dr. Victoriano de la Plaza, minister of foreign affairs in Argentina, was whole-hearted in his praise. "This condition of precarious autonomy and liberty of action," he said, "and the constant danger of being subjugated or suffering the mutilation of their territory, would have continued among those weak states but for the wise and famous declarations of President Monroe."[7] That a fine appreciation of the true spirit and meaning of the doctrine may be found in South America the following quotation from the eminent Argentinian authority on international law, Dr. Luis M. Drago, will

[3] In his answer to Mr. Olney, Lord Salisbury gives an impartial and judicial interpretation of the Monroe Doctrine. See *Foreign Relations of the United States*, 1895, Part I, p. 563.
[4] Quoted by A.B. Hall, *The Monroe Doctrine and the Great War* (Chicago, 1920), p. 153.
[5] New York Times, *Oct. 25, 1914.*
[6] *Les Questions Actuelles de Politique Etrangère dans l'Amérique du Nord* (Paris, 1911), p. 231.
[7] *Fourth Int. Conf. of Amer. States, Sen. Doc. No. 744*, 61st Cong., 3rd Sess., Vol. I, p. 12.

indicate:

> The Monroe Doctrine is in fact a formula of independence. It imposes no dominion and no superiority. Much less does it establish protectorates or relation of superior to inferior. It creates no obligations and no responsibilities between the nations of America, but simply calls upon all of them, with their own means and without aid, to exclude from within their respective frontiers the jurisdiction of European powers. Proclaimed by the United States in the interest of their own peace and security, the other republics of the continent have, in their turn, proceeded to adopt it with an eye alone to their own individual welfare and tranquility. . . . Thus understood, the Monroe Doctrine, which in the end is nothing more than the expression of the will of the people to maintain their liberty, assures the independence of the states of that continent in respect to one another as well as in relation to the powers of Europe.[8]

Señor Alejandro Álvarez, the Chilean delegate to the fourth Assembly of the League of Nations, went so far as to declare that the Monroe Doctrine was more valuable to South America than Article X of the League Covenant because American naval and military forces stood ready to enforce the Monroe Doctrine.

On the other hand, there is also a very strong sentiment prevailing among many worthy representatives of Latin American countries, that, as far as the republics of Latin America are concerned, the Monroe Doctrine has long since outlived its usefulness. The conditions that called into existence the original statement of President Monroe have passed away, never to return. The subsequent policy of the United States, which masked itself under the doctrine of Monroe, was aimed at Latin America as well as at Europe and Asia. When Secretary Olney declared that "today the United States is practically sovereign on this continent, and its fiat is law upon the subjects to which it confines its interposition," and used this interpretation of the Monroe Doctrine to interfere in a dispute in South America, the inference seemed clear that the United States considered itself sovereign on *both continents.* The great republics of South America keenly resented any such attitude. When the United States took over the Panama Canal, on the ground that the act was justified by the interests of civilization, the South American countries could not fail to note that the interests of civilization and those of the United States were surprisingly alike. Dr. Lucio M. Moreno Quintana, an eminent Argentine lawyer and a grandson of a former president of the country, has expressed what is unquestionably the sentiment of the great majority of Latin Americans: "The Monroe Doctrine is not a doctrine of America for the Americans, but of America for the North Americans. It has served as an admirable instrument for the United States to separate Europe from America and to establish its hegemony over the latter. The United States has been at all times preoccupied in obtaining concessions of every kind at the cost of the sovereignty of the rest of the American states. The doctrine is dangerous because it is North American imperialism hidden under a principle of international law."[9]

Just so long as a large number of our Latin American neighbors felt that the Monroe Doctrine was a selfish policy, based upon the desire of the United States to exclude European powers from interfering with the political interests of the Western Hemisphere merely in order to have a better opportunity to act

[8] Hart, *op. cit.*, p. 253.
[9] *New York Times*, Oct. 13, 1920.

in a similar fashion itself, our relations with Latin America could not be established upon a satisfactory basis of mutual confidence. But, before casting aside the Monroe Doctrine as an inseparable barrier to a more friendly *rapprochement* between the United States and the Latin American republics, let us consider the doctrine from two standpoints. First, has the Monroe Doctrine of the past been a policy that, while safeguarding the interests of the United States, has been a menace as well as a benefit to the great powers of Latin America? Second, can the Monroe Doctrine of the future be so interpreted that it will become a real Pan American doctrine, a policy whose maintenance will be the bulwark of defense for the other American republics as well as for the United States?

In order to answer the first question, it is necessary to review briefly both the circumstances that called forth the doctrine, and the manner in which it has been extended by the various presidents who have had occasion to interpret it.

The Spanish dependencies in the New World had very little reason to cherish a keen spirit of loyalty to the mother country. Spain had ever looked upon them as a fruitful field of exploitation, and had considered no regulations too severe to accomplish this result. Therefore, when Napoleon placed his brother Joseph upon the throne at Madrid in 1808, the slight thread of allegiance that still bound the Spanish colonies to the Old World snapped, and a series of revolutions in South America followed. By the end of the year 1810 virtually all of the Spanish colonies in South America had declared their independence, although they still acknowledged a half-hearted allegiance to the deposed Ferdinand VII. As long as the Peninsular War raged in Europe, the colonies were left mostly to their own devices. Upon the restoration of Ferdinand, an attempt was made to bring them once more under his selfish and autocratic rule. But, having enjoyed the benefits of freedom, the colonies were unwilling to go again under the yoke; and the real struggle for independence began.

From the outset the United States viewed the contest with close attention, and in 1815, by issuing what virtually amounted to a proclamation of neutrality, she accorded the struggling colonies the status of belligerent states.[10] Monroe, then secretary of state, while unwilling to recognize them outright, declared in a despatch in December 1815, that the colonies would probably gain their independence, and that it was to the interest of the United States that they should do so. Henry Clay was outspoken in his demands for recognition, and upon several occasions introduced bills authorizing the necessary appropriation for ministers to those governments in South America that had really established their independence. In 1820 such a resolution passed the House. President Monroe was not yet ready to act;[11] but when, in 1822, the victories of San Martín in the south and Bolívar in the north had virtually wiped out all chance of Spanish control, he felt that he need delay no longer, and in a special message to Congress on March 8, 1822, he declared that the time for recognition had come.[12] This recognition did not come any too soon, for in the autumn of this same year, at the Congress of Verona, the three powers of the Holy Alliance,

[10] *American State Papers, Foreign Relations*, Vol. IV, p. 1.
[11] *Annals of Congress*, 16th Cong., 1st Sess., pp. 2223 ff.
[12] *Amer. State Papers, For. Rel.*, Vol. IV, p. 818.

with France in their leading-strings, declared that the system of representative government was incompatible with monarchical principles, and mutually engaged not only to put an end to representative government in Europe, but to prevent its being introduced in those countries where it was not yet known.

Although Great Britain had never joined the so-called Holy Alliance, she was an equal partner in the Quadruple Alliance, whose duty also it was to safeguard Europe from further dangers of revolution. She took part in the conference at Aix-le-Chapelle in 1818, but, unable to back the reactionary policies of Metternich, did not participate in those that followed at Troppau or Laybach. A state with the democratic tendencies already shown by Great Britain could not join whole-heartedly in the suppression of constitutional government, and Castlereagh preferred to follow his allies hesitantly and at a distance.[13] When, at the Congress of Verona, Prince Metternich won over Czar Alexander I to his policy of using the concert of powers to crush out democracy wherever it should be found, the Duke of Wellington, representing Great Britain, was instructed to withdraw. George Canning, who now held the portfolio of foreign affairs in the place of Castlereagh, realized only too well that if the Quintuple Alliance aided Ferdinand to recover his possessions in the New World, Great Britain would see the large and profitable trade which she was now carrying on with the South American states restricted once more to Spanish merchants and Spanish ships. In a despatch to the Duke of Wellington, the English representative at Verona, he pointed out that now that American questions were more important than European ones to the British, the opportunity was afforded to turn the situation to their advantage.[14]

Canning hereupon protested against the reactionary policy of the allies, and, with that as an excuse for withdrawing from the European concert, he turned towards the United States. He was justified in expecting the United States to support Great Britain in any program based upon the recognition of the independence of the South American colonies, since the United States in 1819 had proposed the recognition of Buenos Aires, intimating that if Great Britain should adopt similar measures it would be highly satisfactory to the United States. Nor could he afford the time for diplomatic soundings, for the success of the French troops in restoring Ferdinand showed that if the allies' further plans of bringing back the Spanish colonies were to be checked there must be haste.

On August 16, 1823, Canning approached Richard Rush, the American minister, suggesting a joint diplomatic action against European intervention in America, and four days later he put his ideas into a formal proposal. The substance of the proposition was somewhat as follows: Assuming the recovery of the colonies by Spain to be hopeless, the question of recognition was one of time and circumstances; under these conditions, would it not be an act of wisdom for the two powers to announce publicly that they neither aimed at taking any portion of the colonies themselves, nor were willing to see any portion of them transferred to any other power with indifference?[15] However, when Rush

[13] Viscount Castlereagh, *Correspondence, Despatches, etc.*, 12 vols. (London, 1853), Vol. XII, pp. 311-318.
[14] Duke of Wellington, *Despatches, Correspondence, etc.*, 2nd ser., 8 vols. (London, 1867), Vol. I, p. 511.
[15] For text see J.B. Moore, *Digest of International Law*, Vol. VI, p. 389.

suggested that Great Britain recognize the colonies forthwith, Canning admitted that such a move was not feasible at the time. The American minister thereupon reported the situation to his home government and awaited instructions, confident that some diplomatic action was bound to occur.

Before considering the great step which the United States was now about to take in the development of her foreign policy, one other factor that entered into the situation must be noted. In 1821 the Czar of Russia had issued a ukase claiming the Pacific coast down to the fifty-first degree, and at the same time declaring that Bering Sea and the North Pacific were closed seas and subject to his exclusive jurisdiction. Adams protested vigorously both against the establishment of new colonial possessions on this continent and at the exorbitant pretensions to exclusive jurisdiction over territory whose title was in dispute.[16] Both Monroe and Adams felt that the time had come when it must be decided whether the United States should join with Great Britain in a strong stand for the independence of the Latin-American states against further encroachments of the powers of Europe, or whether it would be better for the United States to take the stand alone, as the foremost republic in the Western Hemisphere. At first the sentiment of virtually all the statesmen whom Monroe consulted–Jefferson, Madison, Calhoun, and Rush–seemed to favor a joint declaration, and the President himself inclined in that direction. John Quincy Adams alone vehemently opposed a joint declaration, pointing out that Canning's object seemed to be to obtain a public pledge from the government of the United States against its own acquisition of any part of the Spanish American possessions as much as against the forcible interference of the Holy Alliance. He also made it clear that the United States would be safer in disclaiming all intention of interfering with European concerns, and in issuing a declaration made solely by the United States for an American cause.[17] His views finally prevailed, and it was at length agreed that inasmuch as the South American states were free and independent, they, and not Great Britain and the United States, had the right to dispose of their condition.

President Monroe issued his famous message on December 2, 1823, and although considerable space was given to the expression of the policy of the United States in regard to the attitude and intentions of Europe, the gist of the doctrine may be summed up in two sentences. The first concerned colonization, and was in answer to the Russian ukase of 1821. It declared that "the American continents, by the free and independent condition which they have assumed and maintain, are henceforth not to be considered as subjects for future colonization by any European powers." The second answered the threat of the Holy Alliance, and asserted that "the political system of the allied powers is essentially different . . . from that of America. . . . We owe it, therefore, to candor, and to the amicable relations existing between the United States and those powers, to declare that we shall consider any attempt on their part to extend their system to any portion of this hemisphere as dangerous to our peace and safety."[18]

As was to have been expected, a message that so simply and so completely phrased a true American policy based upon the firm foundation of self-protection, was received enthusiastically in both North and South America. Speaking

[16] *Amer. State Papers, For. Rel.*, Vol. IV, p. 861.
[17] J.Q. Adams, *Memoirs* (Philadelphia, 1874-1877), Vol. VI, pp. 177ff.
[18] For full text see J.D. Richardson, *Messages and Papers of the Presidents*, Vol. II, p. 209.

in behalf of the Panama Mission, April 14, 1826, in the House of Representatives, Daniel Webster declared that "one general glow of exultation, one universal feeling of the gratified love of liberty, . . . pervaded all bosoms."[19] In the words of Señor de Manos Albas, "It [the pronouncement] rang through the world like a peal of thunder; it paralyzed the Holy Alliance, and defined, once and for all time, as far as Europe is concerned, the international status of the newly constituted American republics."[20] Even though its enunciation was due primarily to the realization upon the part of the United States that the preservation of the independence and democratic systems in the South American countries was closely bound up with the protection of her own independence, nevertheless the nations of South America realized clearly that such a policy, maintained by their more firmly established neighbor on the north, was precisely the bulwark needed until they could strengthen themselves sufficiently to stand entirely alone. It is not surprising that Colombia proposed that this doctrine be ratified at the projected Congress of Panama as one of the bases of Pan-Americanism; for such a policy well represented "the gospel of the new continent."[21]

If the American doctrine as laid down by President Monroe could have been forever confined to the two fundamental concepts just outlined, perhaps today this policy would be the unanimously accepted basis of a true Pan American policy. But the doctrine as laid down by Monroe was merely an attempt to meet the situation that confronted him by a policy whose aim was safety for the United States. It contradicted two principles of international law generally accepted at the time: the right of intervention, and the right of taking possession of unoccupied territory. The statesmen of Europe were not slow in expressing their disdain and contempt for these new and presumptuous principles. Châteaubriand declared that such a doctrine "ought to be resisted by all the powers possessing either territorial or commercial interest in that hemisphere"; while Canning, who boasted publicly that he had called into existence the New World to redress the balance of the Old, in his private correspondence expressed himself in an entirely different vein.[22] Even Congress was unwilling to commit itself unreservedly to the new principles, especially as Henry Clay, who introduced a joint resolution doing so, interpreted the doctrine to mean that the United States was ready to insure the independence of the South American republics.[23] Therefore, although the doctrine was to remain the basis of our foreign policy henceforth, it was not recognized as a principle of international law.

Nor, on the other hand, was its interpretation to be rigidly circumscribed by the original limits laid down by Monroe. Rather, it was to be shaped in accordance with the conditions that successive presidents were called upon to

[19] J.W. McIntyre [ed.], *The Writings and Speeches of Daniel Webster* (Boston, 1903), Vol. V, p. 203.
[20] Joseph Wheless, "Monroe Doctrine and Latin America," in *Annals of Amer. Acad.*, Vol. LIV (July, 1914), p. 66.
[21] For an excellent account of opinion of the doctrine in Hispanic America see J.B. Lockey, *Pan-Americanism: Its Beginnings* (New York, 1920), Chap. VI; see also W.S. Robertson, "South America and the Monroe Doctrine," *Polit. Sci. Quar.*, Vol. XXX (Mar., 1915), p. 82.
[22] W.S. Robertson, "The Monroe Doctrine Abroad, 1823-1824," in *Amer. Polit. Sci. Rev.*, Vol. VI (Nov., 1912), p. 546.
[23] For text of resolution see Moore, *op. cit.*, Vol. VI, p. 404.

confront. As a result, the original doctrine has been expanded as circumstances demanded, and this expansion has been the result of two forces, American interest and American power. A brief consideration of the important occasions when the doctrine has been invoked will give evidence of its changing character.

In 1827 the Argentine Republic, which was then at war with the Empire of Brazil, sent an inquiry to Henry Clay, asking that he outline the scope of the declarations made by President Monroe. Clay replied that the war between two states could in no way be considered analogous to the conditions that provoked President Monroe's message, since it was a war "strictly American in its origin and its objects."[24] A few years later in 1833, when Great Britain resumed occupation of the Falkland Islands, Buenos Aires protested that the action of Great Britain was a clear violation of the Monroe Doctrine. Inasmuch as the United States had already been forced to take action against the treatment accorded American fishermen by the Argentinians during their brief possession of the islands, and as the British could lay title to a claim antedating the seizure by Buenos Aires, the United States acknowledged the British sovereignty.[25]

The first enlargement of the doctrine so as to preclude the transfer of American territory from one foreign country to another seems to have been made in connection with Cuba. In 1825 Mr. Clay, in a letter to the American minister to France, declared that the United States could not consent to the occupation of Cuba or Puerto Rico "by any other European power than Spain under any contingency whatever,"[26] and the same sentiments were repeated by Mr. Van Buren in 1829 and 1830 in notes to the American minister to Spain.[27] Mr. Forsythe, in 1840, declared that the United States would prevent at all hazards any voluntary transfer on the part of Spain of her title to Cuba, whether it was temporary or permanent, just as she would assist Spain with both military and naval resources in preserving it or recovering it.[28] This enlargement of the doctrine was just as advantageous to the free republics of Latin America as to the United States, for any transfer of colonies would most likely be from a weak power, such as Spain, to one of the stronger and more dreaded European states, and therefore an event to be feared equally by the United States and Latin America.

The earliest enlargement of the doctrine that seemed to be aimed at the Latin American nations as well as at Europe came with the first message of President Polk. Up to 1845, the United States, when insisting upon a maintenance of the *status quo* in the Western Hemisphere, had, it is true, alluded only to Europe. But there was no reason to believe that the inhibition was not equally applicable to the United States herself. Directly after the annexation of Texas, however, Polk declared that "the people of this continent alone have the right to decide their own destiny"; and that "should any portion of them, constituting an independent state, propose to unite themselves with our confederacy, this will be a question for them and us to determine without any

[24] *Ibid.*, p. 434.
[25] *Ibid.*, p. 435. See also *infra*, Chap. 16.
[26] *Ibid.*, p. 447.
[27] *Ibid.*, pp. 448-449.
[28] *Ibid.*, p. 450. For a more detailed discussion of the Cuban question see *infra*, pp. 192-202.

foreign interposition";[29] and the declaration was followed by the annexation of New Mexico and California. Well might the Latin American states now begin to wonder whether the Monroe Doctrine was an unmixed blessing. In his message of April 29, 1848, President Polk went even further in his free interpretation of the Monroe Doctrine by urging the annexation of Yucatan, which at that time was so upset by an Indian insurrection that the authorities offered to transfer the "dominion and sovereignty of the peninsula" in return for immediate aid. Inasmuch as similar proposals had been made to Great Britain and Spain, President Polk seized the occasion to assert that the United States could not consent to a transfer of this colony to any European power.[30] Under the circumstances, Polk's assertion could mean nothing else than that the United States interpreted the Monroe Doctrine to prevent a Latin American state from accepting the dominian of a European nation, even if the Latin American state desired such a transfer of sovereignty. However, Yucatan soon withdrew its offer, so that Polk was not forced to put his interpretation to the test.

Following Polk's administration, the doctrine to a certain extent disappeared from view. Mr. Clayton, Secretary of State through President Taylor's short presidency, struggled against admitting British rights in Central America, but he was finally forced by circumstances to give Great Britain joint rights in any inter-oceanic canal that might be constructed on the isthmus. It may be conceded that the Clayton Bulwer treaty did not constitute an infringement of the Monroe Doctrine.[31] But a situation was soon to arise in Mexico which was to provoke a serious violation of the doctrine on the very borders of the United States.

After the disastrous war with the United States, the rival factions struggling for power in Mexico brought the unfortunate country into bankruptcy and anarchy, and the attention of the United States was finally called to the fact that Spain, Great Britain, and France were about to employ strong measures to protect the interests of their citizens on Mexican soil. In a despatch to our minister to Mexico, dated September 20, 1860, Secretary Cass thus indicated the general position of the United States: "While we do not deny the right of any other power to carry on hostile operations against Mexico, for the redress of its grievances, we firmly object to its holding possession of any part of that country, or endeavoring by force to control its political destiny. This opposition to foreign interference is known to France, England, and Spain, as well as the determination of the United States to resist any such attempt by all means in its power."[32] President Buchanan, who had already recognized the Juárez government in Mexico in 1859, intimated in his annual message of 1860 that his position was the same, to the point "of resisting, even by force should this become necessary, any attempt by these governments to deprive our neighboring republic of portions of her territory—a duty from which we could not shirk without abandoning the traditional and established policy of the American people."[33]

[29] Richardson, *op. cit.*, Vol. IV, p. 398.
[30] *Ibid.*, p. 582.
[31] For a discussion of the Monroe Doctrine in its relation to the Panama Canal see *infra* Chap. 5; also *Sen. Doc. No. 194*, 47th Cong., 1st Sess.
[32] Moore, *op. cit.*, Vol. VI, p. 481.
[33] Richardson, *op. cit.*, Vol. V, p. 646.

When the European powers decided upon armed intervention in Mexico, the United States was, unfortunately for the maintenance of this traditional policy, at the very brink of the war for the Union. A convention was signed October 31, 1861, by Great Britain, France, and Spain, outlining their plan of action. The United States was invited to accede to it, but it was made very clear that operations would be begun regardless of our adherence. Secretary Seward realized the difficulty of the situation, and, while refusing to depart from our traditional policy by entering into a European alliance, he was unable to protest very vigorously against the joint intervention.[34] The British and Spanish, however, whether because they were desirous of respecting our wishes after their claims had been met, or whether because they were dissatisfied with the attitude of the French, withdrew their forces early in 1862. The French, while disclaiming any ulterior motives, pushed forward their forces, seized the city of Mexico, took control of the government, and had the assembly change the government to a monarchy and offer the crown to Archduke Maximilian of Austria. Even when Maximilian had been persuaded to accept, Mr. Seward was unable to do more than declare that the permanent establishment of a foreign and monarchical government in Mexico would be found neither easy nor desirable.[35] The House of Representatives was not so diplomatic and voiced a vigorous protest by a resolution unanimously carried April 4, 1864. But Mr. Seward informed the French minister that the President had not departed in any way from his previous policy.[36] However, with the close of the Civil War the tone of the Secretary's protests changed, and his note of December 16, 1865, boldly asserted that the sincere friendship between the two nations would be brought into imminent jeopardy unless France desisted from "the prosecution of armed intervention in Mexico to overthrow the domestic republican government existing there."[37] Louis Napoleon was now more interested in the European situation arising out of the growing dispute between Austria and Prussia than in his Mexican enterprise, and in the spring of 1866 he decided to withdraw his troops. With the withdrawal of the French the power of Maximilian began to crumble, and, in less than a year after the departure of the first detachment of French troops, the unfortunate prince paid for his ill-fated Mexican expedition with his life.

The principles of the Monroe Doctrine were vindicated once more in the most serious attempt that had yet been made to impose European domination upon the independent republics of the Western Hemisphere, although nowhere in the diplomatic correspondence is the doctrine mentioned by name. From the standpoint of Latin America, no objection could possibly be raised to the Seward doctrine in regard to Mexico. The United States, torn by Civil War, could not protest effectively against European intervention, but the United States, reunited, could and did make her protest effective.

Secretary Seward's well considered policy had to some extent lulled the latent fear in the Latin American states that the United States was preserving

[34] For text of the note and Secretary Seward's reply see *House Ex. Doc. No. 100*, 37th Cong., 2nd Sess., pp. 185-187.
[35] Moore, *op. cit.*, Vol. VI, p. 495.
[36] *Ibid.*, Vol. VI, pp. 496-497.
[37] *Ibid.*, p. 501.

their independence for its own advantage, but the ill considered expressions of President Johnson again aroused their resentment. In his fourth annual message, dated December 9, 1868, Johnson delivered himself of these astonishing sentiments: "Comprehensive national policy would seem to sanction the acquisition and incorporation into our Federal Union of the several adjacent continental and insular communities as speedily as it can be done, peacefully, lawfully, and without violation of national justice, faith, or honor. . . . The conviction is rapidly gaining ground in the American mind that, with the increased facilities for inter-communication between all portions of the earth, the principles of free government as embraced in our Constitution would prove of sufficient strength and breadth to comprehend within their sphere and influence the civilized nations of the world."[38] This was either a most extraordinary extension of the Monroe Doctrine, or else the promulgation of a new and all-embracing Pan American doctrine, with the United States as the chief beneficiary. Fortunately for future Pan American relations, President Johnson was at the close of his term, and, in view of the fact that he had been estranged from Congress throughout the greater part of it, his statements did not receive the attention which they would otherwise have commanded.

President Grant and his Secretary of State, Hamilton Fish, returned to the common interpretation of the doctrine, as aiming to protect the feeble powers of America against European intervention, but in such a way as to secure and maintain their confidence. Secretary Fish was careful to make it clear to the neighboring states that the United States did not covet their territories and was ready to aid them to the fullest extent in any steps which they might take to protect themselves against anarchy. But, while emphasizing the unselfish attitude of the United States in offering its protection against Europe, he was careful to safeguard the interests of his country in regard to the proposed isthmian canal. It was his expressed opinion that the canal was an American enterprise, to be undertaken under American auspices.[39] Secretary Evarts took the same stand, declaring that the paramount interest of the United States in the project of interoceanic communication seemed indisputable.[40] When the French company under the direction of de Lesseps indicated its intention to start construction of a canal, President Hayes pointed out that "the policy of this country is a canal under American control; . . . it will be the great ocean thoroughfare between our Atlantic and Pacific shores, and virtually a part of the coast-line of the United States."[41]

Secretary Blaine, in Garfield's administration, opposed the idea of a joint guaranty and control of an interoceanic canal by the European powers as a direct infringement of the Monroe Doctrine, but when communicating his views to Great Britain he put himself in a very weak position by omitting to mention the Clayton-Bulwer treaty of 1850.[42] When the British foreign minister called his attention to this *lapsus memoriae*, Mr. Blaine attempted to prove that the situation had so changed that the treaty was no longer of value.[43] Needless to

[38] Richardson, *op. cit.*, Vol. VI, p. 688.
[39] *For. Rel. of the U.S.*, 1870, pp. 254*ff.*; see also *Sen. Ex. Doc. No. 112*, 46th Cong., 2nd Sess., p. 48.
[40] *Sen. Ex. Doc. No. 112*, 46th Cong., 2nd Sess., p. 18.
[41] Richardson, *op. cit.*, Vol. VII, p. 585.
[42] *For. Rel. of the U.S.*, 1881, p. 537.
[43] *Ibid.*, p. 554.

say, Great Britain disagreed with this doctrine, and had little difficulty in supporting her position. Unsuccessful here, Blaine turned his attention to strengthening the bonds of friendship between the republics of the two Americas by a Pan American conference, though before he could bring his plans to fruition he was no longer secretary of state.

Up to this point the Monroe Doctrine had been interpreted, with very few exceptions, as protective of American institutions and territory against Europe. Now a new and disturbing extension of the doctrine was at hand. It was preceded by a period of strained relations between the United States and Chile. The protection accorded President Balmaceda and refugees of his party by the United States minister, and the unfortunate *Baltimore* incident, provoked feelings of bitterness on both sides. President Harrison expressed in no uncertain terms the intention of the United States to protect her citizens in those states whose governments were too weak to do it. But it remained for President Cleveland, through his Secretary of State, Mr. Olney, to declare that: "To-day the United States is practically sovereign upon this continent, and its fiat is law upon the subjects to which it confines its interposition. Why? . . . It is because, in addition to all other grounds, its infinite resources combined with its isolated position render it master of the situation and practically invulnerable as against any or all other powers."[44]

The occasion that provoked this expression, it is true, was brought about by Venezuela, who wished to use the power of the United States to protect her against British encroachments. Nevertheless the Latin American republics might well look with perturbation upon this new extension of the Monroe Doctrine, As Señor Álvarez has pointed out, such an interpretation does not properly come under the Monroe Doctrine, but under a policy whose basis is the safety of the United States through its hegemony on this continent.[45] Even such a policy might well have as an essential corollary the protection of the Latin American republics. But such a bold statement of overweening power was pregnant with sinister possibilities, and the influences hostile to the closer relations between the United States and its Latin American neighbors were not slow to avail themselves of the opening. This doctrine as announced by President Cleveland has been termed the doctrine of paramount interest, and the president put himself on record as willing to go to war, if necessary, to enforce it. In spite of Lord Salisbury's protestations that international law could never sanction any such novel principle, President Cleveland maintained his point, and Great Britain finally accepted American intervention. In the Venezuela controversy President Cleveland stood upon a doctrine whose observance was necessary "to our peace and safety as a nation and to the integrity of our free institutions." But it is essential to note that his stand was also useful to Venezuela in supporting her claims.[46]

After the Spanish-American War had extended our jurisdiction to certain of the Spanish colonies, it became a matter of even greater interest to Latin

[44] Moore, *op. cit.*, Vol. VI, p. 553.
[45] Alejandro Álvarez, "Latin America and International Law," in *Amer. Jour. of Int. Law*, Vol. III (April, 1909), p. 269.
[46] For a discussion of the Venezuela affair see J.H. Latané, *The United States and Latin America* (New York, 1920), pp. 238-249; also Raúl de Cárdenas, *La Política de los Estados Unidos en el Continente Americano* (La Habana, 1921), pp. 123-132.

Americans to know just how far our peace and safety would lead us in intervening in their destinies. President Roosevelt's action in regard to Colombia and the Panama Canal, and the passage of the Platt Amendment by Congress, were events that cast shadows of a most dubious sort before the eyes of all Latin American countries. Neither were the utterances of President Roosevelt of such a character as to dissipate apprehension. In the annual message of 1904 he enunciated what is generally known as the "big-stick" policy, as follows: "Any country whose people conduct themselves well can count upon our hearty friendship. If a nation shows that it knows how to act with reasonable efficiency and decency in social and political matters, if it keeps order and pays its obligations, it need fear no interference from the United States. Chronic wrong-doing, or an impotence which results in a general loosening of the ties of civilized society, may in America, as elsewhere, ultimately require intervention by some civilized nation, and in the western hemisphere the adherence of the United States to the Monroe Doctrine may force the United States, however reluctantly, in flagrant cases of such wrong-doing or impotence, to the exercise of an international police power."[47]

This statement of policy was merely the prologue to the taking over and administration of the finances of the Dominican Republic. Although the Senate refused to ratify the protocol, the president, under the guise of an executive agreement, carried out his plan; and the Senate later gave its sanction. When President Taft followed the same course in his dealings with Nicaragua and Honduras, and President Wilson in his action in Haiti, well might the states of Latin America believe that the Monroe Doctrine, in its new interpretation, meant intervention by the United States to prevent intervention by Europe.

One other notable extension of the Monroe Doctrine which is based primarily upon the principles of self-defense occurred in 1912, when an American company attempted to sell out certain concessions in Lower California around Magdalena Bay to a Japanese concern. The result was the following Lodge resolution, which received the indorsement of the Senate: "*Resolved*, That when any harbor or other place in the American continents is so situated that the occupation thereof for naval or military purposes might threaten the communications or the safety of the United States, the government of the United States could not see without grave concern the possession of such harbor or other place by any corporation or association which has such a relation to another government, not American, as to give that government practical power of control for naval or military purposes."[48]

Although this particular phase of an American doctrine was aimed at Japan, it seems reasonable to assume that the same policy would be invoked by the United States as against any other interests or concessionaires who might be supposed to have the support of their government.

If these extensions of the Monroe Doctrine were uniformly of a nature tending to protect the United States, regardless of the rights of the other republics of the New World, the hostility with which the doctrine is generally

[47] A.H. Lewis, *Messages and Speeches of Theodore Roosevelt* (Washington, D.C., 1906), Vol. II, p. 857.
[48] For text see *Amer. Jour. of Int. Law*, Vol. VI (Oct., 1912), p. 437. See also T.A. Bailey, "The Lodge Corollary of the Monroe Doctrine," *Polit. Sci. Quar.*, Vol. 48, p. 238.

regarded in South America would be easy to understand. But other phases of the doctrine have been brought out which show that the United States, in its insistence upon self-protection, has not wholly lost sight of the rights of its Latin American neighbors. In numerous addresses delivered during his memorable visit to South America in 1906, Elihu Root gave an interpretation of the doctrine that all South America was only too glad to accept. In Uruguay he asserted that the declaration of Monroe was "an assertion to all the world of the competency of Latin Americans to govern themselves."[49] We have already quoted a portion of his eloquent address at the Pan American Conference in Rio de Janeiro in which he outlined an all-American policy for the United States. Some years later Mr. Root, in an address welcoming Dr. Lauro Müller, Foreign Minister of Brazil, declared that "there is neither to the Monroe Doctrine nor any other doctrine or purpose of the American government any corollary of dominion or aggression, or aught but equal friendship."[50] In fact, in an address before the American Society of International Law on April 22, 1914, Mr. Root went so far as to assert that, outside of the one apparent extension of the doctrine by President Polk, there has been no other change or enlargement of the Monroe Doctrine since it was first promulgated.[51]

President Wilson was particularly successful both in enunciating and in maintaining this friendly policy towards Latin American. On March 11, 1913, very shortly after his inauguration, he declared: "The United States has nothing to seek in Central and South America except the lasting interests of the peoples of the two continents."[52] And a little later he was more definite: "The United States will never again seek one additional foot of territory by conquest."[53] Yet even President Wilson was forced to follow the Roosevelt doctrine in the case of Haiti; nor was American administration there so satisfactory as to escape very serious criticism. But in the Caribbean, particularly, the United States cannot brook European intervention, and to act as policeman in the region has been regarded as essential to our peace and safety.

One of the most important interpretations of the Monroe Doctrine was made by Secretary of State Hughes in a notable speech in 1923 upon the occasion of the celebration of the centenary of the Monroe Doctrine.[54] The address, delivered before the American Bar Association at Minneapolis on August 30, 1923, was entitled "Observations on the Monroe Doctrine." In the course of his remarks Secretary Hughes made certain categorical assertions which had considerable influence upon subsequent interpretations of the doctrine. In the first place, he declared:

. . . the Monroe Doctrine is not a policy of aggression; it is a policy of self-defense. . . . *Second*, as the policy embodied in the Monroe Doctrine is distinctively the policy of the United States, the Government of the United States reserves to itself its definition, interpretation and application. . . . *Third*, the policy of the Monroe Doctrine does not

[49] Elihu Root, *Latin America and the United States* (Cambridge, Mass., 1917), p. 58.
[50] *Ibid*., p. 243.
[51] Elihu Root, *Addresses on International Subjects* (Cambridge, Mass., 1916), p. 112.
[52] E.E. Robinson and V.J. West, *The Foreign Policy of Woodrow Wilson* (New York, 1917), p. 180.
[53] *Ibid*., p. 201.
[54] C.E. Hughes, *Address before the Fifty-sixth Annual Meeting of the American Bar Association* (Washington, D.C., 1928).

infringe upon the independence and sovereignity of other American states. . . . *Fourth*, so far as the region of the Caribbean Sea is concerned it may be said that if we had no Monroe Doctrine we should have to create one. . . . Our interest does not lie in controlling foreign peoples; that would be a policy of mischief and disaster. Our interest is in having prosperous, peaceful and law-abiding neighbors with whom we can cooperate to mutual advantage. *Fifth*, it is apparent that the Monroe Doctrine does not stand in the way of Pan American cooperation; rather it affords the necessary foundation for that cooperation in the independence and security of American states.

These utterances clarified the current interpretation of the doctrine and seemed to eliminate reasonable fears on the part of the South American countries. On the other hand, it might be called a "big brother" doctrine in the Caribbean area, and fraternal admonitions are not always cordially welcomed. Nevertheless, every nation has both a right and a duty to protect its essential interests, and the Monroe Doctrine was ever fundamentally a policy of self-protection for the United States. Nor did the United States object to similar declarations of policy on the part of its Latin American neighbors; for, as Secretary Hughes declared on January 20, 1925, "While this doctrine was set forth and must be maintained as the policy of the United States, there is no reason whatsoever why every one of our sister republics should not have and formulate a similar principle as a part of its foreign policy. . . ."[55]

But when we come to the Coolidge-Kellogg policy in the Caribbean, as exemplified particularly in the sending of marines to Nicaragua, the fears of the Latin American nations seemed to be justified. In the words of *La Nación* of Buenos Aires: "We do not recall that the right of intervention by force has been pronounced to such a disquieting extent as it is done today by the government of the United States. The Monroe Doctrine in the hands of Kellogg seems to retain its defensive quality for the United States but to lose its quality as a dignified guaranty of safety to small neighbors. Nicaragua does not fear to be made a colony by some European country but must suffer the military domination of the United States. This fact creates a precedent whereby the independence of American countries will be subordinated to the amount of dollars which the United States invests therein, thus replacing the Monroe Doctrine which for a century has been the resplendent shield of the moral greatness of the United States."[56]

Upon the advent of the Hoover administration, Secretary of State Stimson delegated Undersecretary of State J. Reuben Clark to prepare a memorandum on the Monroe Doctrine which should give a comprehensive historical presentation of its origin and development. Mr. Clark's memorandum which was published in 1930 as an official document by the Department of State interprets the doctrine in a fairly restrictive fashion. The Big Stick corollary of President Theodore Roosevelt is declared to be unjustified by the terms of the doctrine. The doctrine according to this interpretation neither prevents Europe from waging war against the Latin Americas, nor does it relieve Latin American states of their responsibilities as independent sovereignties. Although "the United States determines when and if the principles of the doctrine are violated and . . . we alone determine what measures shall be taken to vindicate the principles of the

[55] *Amer. Jour. of Int. Law*, Vol. 19 (Apr., 1925), p. 368.
[56] *New York Times*, Jan. 6, 1927.

doctrine ... so far as Latin America is concerned, the doctrine is now, and always has been, not an instrument of violence and oppression, but an unbought, freely bestowed, and wholly effective guaranty of their freedom, independence, and territorial integrity against the imperialistic designs of Europe."[57]

The Hoover administration gave concrete evidence that it intended to carry out a nonimperialistic policy when the United States and the Latin American states at the 1928 Washington Conference on Conciliation and Arbitration signed two treaties condemning war as an instrument of national policy and agreeing to settle all juridical disputes by arbitration, and all others by conciliation.

Franklin D. Roosevelt in his first inaugural address of March 4, 1933, dedicated the United States "to the policy of the good neighbor." Speaking before the governing board of the Pan American Union on Pan American Day, April 12, 1933, he interpreted this policy as it affected our relations with Latin America: "The essential qualities of a true Pan Americanism must be the same as those which constitute a good neighbor, namely, mutual understanding and, through such understanding, a sympathetic understanding of the other's point of view. . . . In this spirit the people of every republic on our continent are coming to a deep understanding of the fact that the Monroe Doctrine . . . was and is directed at the maintenance of independence by the peoples of the continent. It was aimed and is aimed against the acquisition in any manner of the control of additional territory in this hemisphere by any non-American power."[58]

Now to answer our question as to whether the Monroe Doctrine has been a menace as well as a benefit to the Latin American republics. Let us briefly summarize the interpretations from the viewpoint of Latin America. Undoubtedly the idea of no transfer by one European nation to another of its territory in the New World is satisfactory to the Latin Americans; but when this interpretation is extended to prevent voluntary transfer of allegiance, it becomes an arbitrary interference with sovereign rights. When, on the other hand, transfer of territory to the United States is regarded as permissible, from the Latin American viewpoint the doctrine becomes dangerous. President Johnson's ill considered words as to American expansion merely aroused antagonism, but President Hayes' carefully considered statement of the United States coast-line extending to Panama inspired fear. Finally, the Cleveland-Olney doctrine of American supremacy on this continent, followed by the Roosevelt doctrine of responsibility for the behavior of badly governed republics, provoked openly manifested hostility. Although Secretary Root, President Wilson, Secretary Hughes, and President Franklin D. Roosevelt did much to allay this antagonism, the judgment of the average Latin American undoubtedly was that the Monroe Doctrine of the past at any rate had been a menace, as well as a benefit, to the Latin American countries.[59]

This brings us to our second question: Can the Monroe Doctrine of the future be so interpreted that it will be regarded as a true Pan American doctrine? If the doctrine of the present could be restored to the original formula, there

[57] J. Reuben Clark, *Memorandum on the Monroe Doctrine* (Washington, D.C., 1930).
[58] U.S. Department of State, *Press Releases*, Vol. VIII, No. 451 (Apr. 15, 1933).
[59] For critical detailed indictment of the Monroe Doctrine from the Latin-American point of view see Gaston Nerval, *Autopsy of the Monroe Doctrine* (New York, 1934).

would be little question of its acceptance by Latin America. But such a doctrine would be entirely unnecessary; the conditions no longer demand such a policy. The Americas no longer possess uncolonized territory; nor do the more powerful Latin American states look to the United States to prevent European intervention.* Hence a return to the original doctrine of Monroe would render the doctrine obsolete, and therefore futile as a Pan American doctrine. On the other hand, to continue the Monroe Doctrine as a policy which conforms itself to the new problems that arise, and whose interpretation may be based upon selfish advantage as well as upon self-preservation by the United States, will never be acceptable to the Latin American nations as a true Pan American policy.

Several solutions of this *impasse* have been proposed. A practical suggestion has been to enlarge gradually the powers of the Pan American Union so that it shall come to have jurisdiction over all questions that might arise to disturb the peace of the Western Hemisphere. In a special memorandum on the Monroe Doctrine and the League of Nations which Mr. Barrett prepared for the American representatives at the Paris Peace Conference, he pointed out that in the Pan American Union we already had "a practical, peace-preserving, and successfully working, although limited and voluntary, American league of nations." It was further suggested that the American governments give the supreme council of the Pan American Union, or some similar body to be created, authority not only to initiate and effect mediation, adjudication, and arbitration, but to enforce its conclusions without the interference of the Old World powers. This solution, which is based upon the practical results already achieved by an established organization, seems well within the bounds of possibility, and that such a solution would be accepted by the republics of Latin America might well be taken for granted. In the words of Señor Calderón, minister from Bolivia: "If the Monroe Doctrine is the proud determination of the United States to keep the whole American continent free from any contamination of foreign autocrats, the Pan American Union is the agreement of all the republics to live together, linked by the great ideals of democracy, not looking down upon the weaker or the less advanced ones, but determined to help them and to forge ahead in a united and free effort to reach the goal of popular welfare, free and peaceful development, and the elimination of pauperism and anarchy."[60]

Another solution, one that has received much attention, is based upon the address made by President Wilson to Congress on January 22, 1917, in which he proposed that "the nations should with one accord adopt the doctrine of President Monroe as the doctrine of the world; that no nation should seek to extend its policy over any other nation or people, but that every people should be left free to determine its own policy, its own way of development, unhindered, unthreatened, unafraid, the little along with the great and powerful."[61] This idea was later incorporated into the covenant of the League of Nations as Article X, stating that "the members of the League undertake to respect and preserve as against external aggression the territorial integrity and existing political independence of all members of the League."

*applicable in the present century prior to World War II and the so-called cold war in its aftermath.

[60] *Jour. of Int. Relations*, Vol. X (Oct., 1919), p. 133.

[61] Robinson and West, *op. cit.*, p. 369.

This proposal was bitterly opposed on the ground that such a doctrine would withdraw us from our isolation and involve us in all the disputes of Europe. It was also asserted that by extending the Monroe Doctrine to the world its value as an American doctrine would be completely lost. Space prevents even the barest consideration of this controversy, other than to point out that a policy of isolation for the United States is neither possible nor desirable, and that the extension of the Monroe Doctrine to the world would mean not only that the world would recognize the doctrine, but that it would join with the United States in guaranteeing its observance. However, Article X, by fixing a definite guaranty mutually enforceable by and binding upon the member states, pledged more in relation to the world than the Monroe Doctrine ever did in relation to the Western Hemisphere. To obviate this objection, and to secure the adherence of the United States to the League, Article XXI was added, which declared that such regional understandings as the Monroe Doctrine are not incompatible with the Covenant of the League. Whether these two principles, a Monroe Doctrine that guarantees nothing except the safety of the United States, and Article X, which guarantees the territory and independence of the member nations of a world league, are mutually sympathetic or antagonistic, is now academic.[62] But—as was pointed out by Señor Paredes, Minister of Foreign Affairs in El Salvador, in a communication to our State Department early in 1920—even if the League of Nations had become a permanent institution, with the United States a member, it would have been necessary to interpret the Monroe Doctrine. For, as Señor Paredes says, "since the doctrine will be forthwith transformed into a principle of universal public law, *juris et de jure*, . . . the necessity of an interpretation of the genesis and scope of the Monroe Doctrine not only in the development of the lofty purpose of Pan Americanism but in order that that doctrine may maintain its original purity and prestige, is rendered all the more urgent."[63]

Some eight years later when Costa Rica was asked to reconsider its intention of withdrawing from the League of Nations, Foreign Secretary Castro again requested a definition of the scope of the doctrine as interpreted by the League under Article XXI, "since the inclusion of various American nations in the League and the fact that this doctrine is mentioned in the Statute by which it was created fully justify its definition by the League."[64] Inasmuch as Señor Castro in the very next sentence pointed out that the "doctrine in question constitutes a unilateral declaration" it was manifestly impossible for the League of Nations with the United States outside to give an acceptable definition. Consequently, when Mexico entered the League of Nations in 1931 and Argentina took her place in 1933, both made reservations declining to recognize the Monroe Doctrine under Article XXI of the Covenant.

It is therefore evident that if the Monroe Doctrine is to be maintained by the United States as a policy essential to its peace and safety, and at the same time accepted by the Latin American countries as a policy in no way inimical to their rights and interests, some form of a multilateral pact must be envisaged.

[62] A subsequent interpretation of Article X, adopted by the League Assembly at Geneva, weakened it to such an extent that it became hardly more than a Monroe Doctrine for the signatory nations.

[63] *New York Times*, Feb. 8, 1921. See also S.G. Inman, *Problems in Pan-Americanism* (New York, 1921), pp. 179-194.

[64] League of Nations, *Official Journal* (Oct., 1928), pp. 1606-07.

Colonel House had such a solution in mind when he initiated negotiations during the Wilson administration for a Pan American pact embodying mutual guarantees of territorial integrity and political independence among all the American republics. Argentina and Brazil appeared very friendly to the idea, but Chile was less inclined to support the proposal and no definite results came of it.

At the Montevideo Conference of 1933 Dr. Puig Casauranc of Mexico suggested that the Roosevelt administration carry out the New Deal in foreign policy and express itself in inter-American affairs not by words alone but by a revision of the Monroe Doctrine. When Secretary Hull replied by definitely accepting for the United States the principle of nonintervention in the Western Hemisphere the corner stone was laid for such a multilateral pact.

When the twenty-one American republics convened in Buenos Aires in 1936 for the Inter-American Conference for the Maintenance of Peace the foundation for the erection of a joint policy of Pan American action for mutual protection in the Western Hemisphere was already prepared. The bogey of intervention had been allayed and Secretary Hull's willing acceptance and support of the Saavedra Lamas Peace Pact had changed the latent suspicions of Argentina to cordial interest in cooperation with the United States. Not only was the Montevideo convention of 1933 accepted as a fundamental principle that "no state has the right to intervene in the internal or external affairs of others" reaffirmed, but another convention was formulated and approved by the terms of which the various American governments pledged themselves to consult with each other first, in the event that the peace of the American republics was menaced; second, in the event of war between American states; and third, in the event of an international war outside America which might menace the peace of the American republics. This consultative pact so clearly takes care of any situation envisioned by the Monroe Doctrine in its original or recent interpretations by joint action on the part of all the American republics that it was aptly named by the delegates the "Monroe Doctrine Convention."[65]

At the Eighth Inter-American Conference of American States at Lima in 1938 steps were taken to further "Pan Americanize" the Doctrine. Inspired by continuing totalitarian threats in Europe and Asia, the delegates reaffirmed continental solidarity and declared that "in case the peace, security or territorial integrity of any American Republic is . . . threatened by acts of any nature that may impair them, they proclaim their common concern and determination to make effective their solidarity, coordinating their respective sovereign wills by means of the procedure of consultation . . . using measures that in each case circumstances may make advisable." Although the Declaration of Lima did not continentalize the Monroe Doctrine, the American states agreed collectively to safeguard the Western Hemisphere against subversive ideologies as well as against physical aggression.

The outbreak of war in Europe resulting from the German attack on Poland at first merely provoked the New World to strengthen its bulwark of neutrality. But at the same time, lest aggression invade its portals, the foreign ministers of the twenty-one republics, in conference at Panama in 1939, resolved that in case continental security be endangered by a threatened change of

[65] *International Conference of American States, First Supplement, 1933-1940.* pp. 215-315.

territorial sovereignty in the hemisphere, a meeting for consultation should be held immediately. Thus the first steps were taken to transform the no-transfer principle of the Monroe Doctrine into a multilateral concept making it a collective responsibility of all American states.[66]

The invasion of the Netherlands and the sudden collapse of France left their colonial possessions in the New World wholly without protection. All America was aroused. Both houses of Congress rushed through resolutions in June 1940 declaring that the United States would not only refuse to acquiesce in any attempt to transfer any region in the Western Hemisphere from one non-American power to another but proposed to consult immediately with the other American republics in case of such a threat.[67] Secretary of State Hull thereupon instructed our representatives in Europe to notify all governments concerned to this effect. By these actions Washington served notice that the Monroe Doctrine was still a vital principle in the hemisphere and the American republics would unite to enforce it.

Forthwith a second conference of foreign ministers met at Havana in July 1940, and Secretary Hull in his opening address proposed "the establishment of a collective trusteeship to be exercised in the name of all the American republics"[68] where the existing status of any American region was threatened by cession or transfer. The Conference, except for Argentina, was unanimously in favor of such a proposal. As finally accepted, the Act of Havana and a convention on the Provisional Administration of European Colonies and Possessions in the Americas categorically prohibited the transfer of territories in the Western Hemisphere from one non-European state to another, and, in case such action was threatened, provision was made for the taking over and provisional administration of any such region by an Inter-American Commission of Territorial Administration. In case of an emergency any state could act singly or jointly with others in any manner required by its own defense or in defense of the continent.[69] Under the Havana Resolution of Reciprocal Assistance agreement was also reached for the first time by the American republics that any attempt by a non-American nation against the territory, sovereignty, or political independence of an American nation "shall, be considered an act of aggression" against all American nations. The elements of the Monroe Doctrine were thereby accepted as an obligation of all the American states, but not under the heading of the Monroe Doctrine. However, since the United States alone had the power to carry out the Act of Havana the Latin American nations tacitly accepted American implementation of the no-transfer principle. Following the Pearl Harbor attack in December 1941, the American foreign ministers met at Rio de Janeiro to take up the question of subversive activities of nationals of non-American countries in the Western Hemisphere.

The several inter-American agreements were incorporated in the Act of Chapultepec which issued from the Inter-American Conference on Problems of

[66]For text of the resolution see Dept. of State *Bulletin*, I (Oct. 7, 1939), p. 334. See also F.O. Wilcox, "The Monroe Doctrine and World War Two," *American Political Science Review*, XXXVI (1942), 433-453.
[67]S.S. Jones and D.P. Meyers, *Documents on American Foreign Relations, 1939-1940*, II (Boston, 1940), p. 89.
[68] Dept. of State *Bulletin*, III (July 27, 1940), p. 46.
[69]*Ibid.*, III (Aug. 24, 1940), pp. 138; 145-148.

War and Peace at Mexico City in 1945. This act, a wartime measure, established a regional security system wherein all of the American states assumed responsibility for the security of the hemisphere from any source, even against American aggressors. Among the several enforcement measures to be considered were diplomatic pressures, the imposition of embargoes, and the use of armed force.

With the end of World War II it became necessary to negotiate a permanent treaty which would formalize and elaborate the Act of Chapultepec. This was achieved at a conference in Rio de Janeiro in 1947 when the American states negotiated the Inter-American Treaty of Reciprocal Assistance, known as the Pact of Rio de Janeiro, which declared that "an armed attack by any state against an American state shall be considered as an attack against all the American states," and pledged each of them "to assist in meeting the attack in the exercise of the inherent right of individual or collective self-defense." With the ratification of the Rio Pact the Monroe Doctrine had assumed a multilateral character seemingly identified with Pan Americanism. During its evolution the no-transfer principle was continentalized by the Act of Havana of 1940; the principles opposing European colonization in the Americas and the extension of a non-American system and non-American control over the nations of the Western Hemisphere were transferred in this fashion in the Pact of Rio.[70]

The paramount status of the United States in the Western Hemisphere and the Monroe Doctrine itself were openly threatened by the expansion of Communist power in the cold war, initially by the pro-Soviet regime of Jacobo Arbenz in Guatemala (1951-1954), and the Soviet-bloc-supported government of Fidel Castro (from 1959). Secretary of State John Foster Dulles declared that the Soviet threat in Guatemala was "a challenge to our Monroe Doctrine, the first and most fundamental of our foreign policies." It resulted in the Act of Caracas, negotiated at the Tenth Inter-American Conference in 1954 which declared that "the domination or control of the political institutions of any American state by the international Communist movement, extending to this hemisphere the political system of an extracontinental power, would constitute a threat to the sovereignity and political independence of the American states." This established a basis for multilateral united action against the international Communist conspiracy and clarified the inter-American responsibility for collective self-defense. However, it was not achieved without strong opposition being voiced by a number of Latin American states which feared communism less than possible United States efforts to combat it through hemispheric action.

After Fidel Castro had seized power in Cuba, Soviet Premier Nikita Khrushchev asserted in July 1960 that "We consider the Monroe Doctrine has outlived its time, has outlived itself, has died, so to say, a natural death. Now the remains of this doctrine should be buried . . ."[71] To this attack, one of the most violent official denunciations ever made against the Doctrine, the State Department declared that "The principles which the United States government enacted in the face of the attempts of the old imperialism to intervene in the affairs of the hemisphere are as valid today for the attempts of the new

[70] For background on Pan Americanizing the Doctrine see Dexter Perkins, *A History of the Monroe Doctrine* (Boston, 1955) chap. X.
[71] *New York Times*, July 13, 1960.

imperialism. It consequently reaffirms with vigor the principles expressed by Monroe."[72] President Eisenhower declared that "I think that the Monroe Doctrine has by no means been supplanted," and warned that the United States government, adhering to its treaty commitments, would not "permit the establishment of a regime dominated by international communism in the Western Hemisphere."

After the ill-fated Bay of Pigs invasion effort in April 1961 President Kennedy applied the Doctrine to the Cuban crisis. On April 3 he stated that Dr. Castro's regime offered "a clear and present danger to . . . all the republics of the hemisphere." He called upon Cuba to cast off its Communist ties. "If this call is unheeded, we are confident that the Cuban people . . . will join hands with other republics in the hemisphere in the struggle to win freedom." Lending emphasis to these remarks the President said "If the nations of this hemisphere fail to meet their commitments against Communist penetration," the United States would act unilaterally on the basis of its right of self-defense.[73]

The Castro regime's military alliance with the Soviet Union and the conversion of the island into a military base equipped with Soviet weapons led to the approval of a Joint Resolution on Cuba by the Senate Foreign Relations and Armed Services committees. The resolution cites the portion of the Monroe Doctrine that described European intervention in hemispheric affairs as dangerous to United States security and expresses the determination to prevent the creation in Cuba of a military capability endangering that security. The United States, it declares, will not tolerate the use of force or threat of force by the Castro government to extend its activities by "aggression or subversion" to any part of the hemisphere. It also pledged the United States to work with the OAS and "freedom loving" Cubans for the "self-determination" of the Cuban people.[74] On the whole, the resolution followed the position taken by President Kennedy and merely reaffirmed his constitutional powers.

In dealing with the crisis resulting from the situation in Cuba, Washington exercised restraint in appealing to the Monroe Doctrine. This was consistent with the policy of recent years where spokesman for the United States said little of the Doctrine as such, both because mention of it excites bitter memories among many Latin American states, and because it has been hoped that the collective apparatus of the OAS could cope with any contingencies. However, as the gravity of the Soviet military buildup became evident, President Kennedy declared that his administration was seeking to defend the Monroe Doctrine by isolating the Communist menace in Cuba. Under pressures from Congress and private agencies to invoke the Monroe Doctrine, the President said on October 22, 1962, that "It shall be a policy of this nation to regard any missile launched from Cuba against any nation in the Western Hemisphere as an attack by the Soviet Union on the United States, requiring a full retaliatory response upon the Soviet Union." While the Soviet missiles were removed from Cuba as a result of the United States-Soviet confrontation, the former was unable to either extirpate a political system subject to control by an extra-continental power, or to force the withdrawal of Soviet armed forces from Cuba.

[72] *Ibid.*, July 17, 1960.
[73] *Ibid.*, Apr. 19, 1961.
[74] *Ibid.*, Sept. 20, 1962.

In order to protest the terms upon which President Kennedy settled the Cuban dispute with Premier Khrushchev, a national organization, The Committee for the Monroe Doctrine, was formed with Captain Edward V. Rickenbacker as chairman. The protest was specifically against the assurances the President gave the Soviet Union that Cuba would not be invaded if the Soviet Union dismantled its missile bases on the island. The Committee declared that this has, in effect, underwritten "Khrushchev's control over Cuba," and was a renunciation of the Monroe Doctrine. It said that the Kennedy-Khrushchev agreement "would appear to amount to a guarantee that Cuba will, without forcible interference by the United States, be permitted to remain as a Communist colony and therefore as a base for the continued political and psychological subversion of other nations in the hemisphere."[75]

When President Lyndon B. Johnson dispatched U.S. Marines to the Dominican Republic in April 1965 which was, he believed, threatened with a Communist takeover, he went back to the Monroe Doctrine in dealing with the problem. He acted first on reports that the United States faced another Communist conquest in the Caribbean and consulted the Latin American states later. In so doing the President chose to interpret the Doctrine as a political instrument to be used when needed unilaterally by the United States, and justified it in part on the grounds that the measure came within the scope of accepted principles of Pan American security. He said that "the American nations cannot, must not, and will not permit the establishment of another Communist government in the Western Hemisphere." Linking the Dominican crisis to the Vietnam War, the President declared that the Communist aim in Vietnam is to show that the "American commitment is worthless. Once that is done, the gates are down and the road is open to expansion and endless conquest . . . there are those who ask why this responsibility should be ours. The answer is simple. There is no one else who can do the job."[76] Some commentators were quick to label the President's action in the Dominican Republic as adding a new corollary to the Monroe Doctrine, the Johnson Doctrine. However, the President denied that he had laid down such a doctrine under which United States troops would prevent the establishment of a Communist government in the Western Hemisphere saying "I think it is a well-known and advertised doctrine of the Hemisphere that the principles of communism are incompatible with the principles of our inter-American system . . . President Kennedy enunciated that on several occasions. The Organization of American States enunciated that. I merely repeated it."[77]

Whatever explanation Washington offered there were widespread protests among the Latin American nations that the United States had violated its pledges on non-intervention. Reference was commonly made to the OAS Charter, Chapter III, Article 15, which states that "No state or group of states has the right to intervene, directly or indirectly, for any reason whatever, in the internal or external affairs of any other state . . ." This principle prohibits not

[75] *Ibid.*, Oct. 31, 1962. The United States had agreed, in exchange for a concession by the Soviet Union allowing on-the-spot surveillance of the Soviet withdrawal of "offensive" weapons from Cuba, not to invade the island or assist any other American nation to do so. By rejecting the surveillance, the USSR cancelled the self-denying limitation.
[76] *Ibid.*, May 5, 1965.
[77] *Ibid.*, June 2, 1965.

only armed force but also any other form of interference or attempted threat against the personality of the state or against its political, economic and cultural elements. It is clear that this article supersedes the Monroe Doctrine in proscribing all forms of intervention.

After the intervention had been carried out, the U.S. House of Representatives considered a resolution calling on any nation in the hemisphere to intervene where another nation was threatened with subversion. Undersecretary of State Thomas C. Mann took exception to this part of the resolution declaring that "I believe unilateral intervention by one American state in the internal political affairs of another is not only proscribed by the OAS Charter, but that non-intervention is a keystone of the structure of the inter-American system." By implication he supported the collective intervention of the nations of the hemisphere in cases where "weak and fragile" states were under attack by "subversive elements responding to directions from abroad." He indicated that there was a question as to "what response is permitted within the framework of the inter-American system." Secretary Mann made a distinction between mature nations that could cope with subversive efforts and those that a small, disciplined minority could easily disrupt, implying that a number of Latin American nations were in the latter category.[78]

At the end of 1965 the Second Special Inter-American Conference met in Rio de Janeiro to consider proposals for modernizing the OAS which included the problem of collective security. Opposition to a U.S.-sponsored military force was expressed strongly by Chile, Mexico, Uruguay, Peru, and Colombia. Some of these governments fear the formation of a "supra-national" enforcement body within the OAS framework. Others view the idea of the inter-American force as favoring militarist governments in the name of anti-communism.

In the presidential election campaign of 1968, the platform of the Republican party on which Richard M. Nixon was elected President, emphasized the importance of closer economic and cultural relations with the Latin American republics. But at the same time it was made clear that "the principles of the Monroe Doctrine should be reaffirmed and should guide the collective policy of the Americas."[79]

The Monroe Doctrine has been altered by events of the cold war which involves us intimately in events in Europe, Africa, Asia–in virtually every part of the globe. In Latin America this involvement has resulted in a special and intense form of anti-communism and this in turn led to our intervention in Guatemala, the Bay of Pigs, and the Dominican Republic. Although the Latin America governments have gone on record a number of times against communism they have on the whole not considered it a justification for intervention. The unanimous support that the United States received from Latin America in the Cuban missiles crisis was based on the direct and open involvement of the Soviet Union in hemispheric affairs and a recognition of the common danger that this presented. Washington holds that the Bogotá Charter of 1948 is in some respects outmoded because of the threat of communism, and the technique of the "wars of liberation;" however, few Latin American governments would agree with this contention.

[78] *Ibid.*, Oct. 14, 1965.
[79] *Ibid.*, Aug. 5, 1968.

It has been hoped that the collective security apparatus of the OAS would cope adequately with contingencies. But experience has shown that this hope cannot be realized. It was unrealistic to expect that it would be. The disproportion in size, wealth, and power between the United States on the one hand and the states of Latin America on the other is such that "multilateralization" of the Monroe Doctrine is essentially impossible: fundamentally, defense in this hemisphere must remain a United States responsibility. But more important than this, however, was the extension of the cold war to the Americas.

SUPPLEMENTARY READINGS

Álvarez, Alejandro. *The Monroe Doctrine, Its Importance in the International Life of the States of the New World.* New York, 1924.
Barcia Trelles, Camilo. *La Doctrina de Monroe y la Cooperación Internacional.* Madrid, 1921.
Barral de Montferrat, H.D. *De Monroe a Roosevelt, 1823-1905.* Paris, 1906.
Beaumarchais, M.D. de. *La Doctrine de Monroe.* Paris, 1898.
Bemis, Samuel F. *The Latin American Policy of the United States.* New York, 1943. Chaps. VII; IX.
———. *John Quincy Adams and the Foundations of American Foreign Policy.* New York, 1949. Chaps.18-19.
Bingham, Hiram. *The Monroe Doctrine, an Obsolete Shibboleth.* New Haven, 1913.
Bolkhovitinov, Nikolai N. *Doktrina Monro: Proiskhozhdenie i Kharakter.* Moscow, 1959.
Capella y Pons, F. *Monroeism.* Paris, 1913.
Clark, J. Reuben. *Memorandum on the Monroe Doctrine.* Washington, 1930.
Cleland, R.G. *One Hundred Years of the Monroe Doctrine.* Los Angeles, 1923.
Delle Piane, A.L. *Doctrina de Monroe.* Montevideo, 1930.
Dozer, Donald M., ed. *The Monroe Doctrine: Its Modern Significance.* New York, 1965.
———. *Are we Good Neighbors? Three Decades of Inter-American Relations, 1930-1960.* Gainesville, 1960.
Fabela, Isidro. *Las Doctrinas Monroe y Drago.* Mexico, D.F., 1957.
Hall, A.B. *The Monroe Doctrine and the Great War.* Chicago, 1920.
Hart, A.B. *The Monroe Doctrine.* New York, 1916.
Humphrey, John P. *The Inter-American System: A Canadian View.* Toronto, 1942.
Kraus, Herbert. *Die Monroe Doktrin.* Berlin, 1913.
Logan, John A. *No Transfer. An American Security Principle.* New Haven, 1961.
Moore, John B. *A Digest of International Law* (8 vols., Washington, D.C.: Government Printing Office, 1906), VI, 368-604.
Nerval, Gaston, *Autopsy of the Monroe Doctrine.* New York, 1934.
Peña, Diego de la. *La Doctrina de Monroe.* Bogotá, 1949.
Pereyra, Carlos. *El Mito de Monroe*. Madrid, 1914.
———. *La Doctrina de Monroe.* Mexico, D.F., 1908.
Perkins, Dexter. *The Monroe Doctrine, 1823-1826.* Cambridge, Mass., 1927.
———. *The Monroe Doctrine, 1826-1867.* Baltimore, 1933.
———. *The Monroe Doctrine, 1867-1907.* Baltimore, 1937.
———. *A History of the Monroe Doctrine.* Boston, 1955.
Petin, Hector. *Les Etats-Unis et la Doctrine de Monroe.* Paris, 1900.
Quesada, Ernesto. *La Doctrina Monroe, Su Evolución Histórica.* Buenos Aires, 1920.
Rivera, Alberto A. *La No Intervención en el Derecho Internacional Americano.* Mexico, D.F., 1952.
Robertson, William Spence. *Hispanic-American Relations with the United States.* New York, 1923. chap. IV.
Temperley, H.W.V. *Foreign Policy of Canning, 1822-27.* London, 1925.
Thomas, David Y. *One Hundred Years of the Monroe Doctrine.* New York, 1923.
Vasconselos, José. *Bolivarismo y Monroismo*. Santiago, Chile, 1935.
Webster, Charles K. *The Foreign Policy of Castlereigh, 1815-1822: Britain and the European Alliance.* London, 1958.
———. *Britain and the Independence of Latin America, 1812-1830.* 2 vols. London and New York, 1938.

4

Undermining the Monroe Doctrine

1. THE TOTALITARIAN THREAT

Democracies unfortunately are prone to possess two dangerous tendencies: wishful thinking and overconfidence. Britain wanted to do business as usual and was confident that the royal fleet made her sea-girt isle impregnable—came the tragedy of Dunkirk. France was content with the *status quo* and possessed confident assurance that the Maginot Line would maintain it, but the line was turned. The Western Hemisphere has its vast ocean frontiers and its much publicized Monroe Doctrine, but oceans are highways as well as barriers and the Monroe Doctrine was not contrived to prevent economic penetration and still less subversive propaganda. Totalitarian influences became deeply entrenched in the Western Hemisphere before either the North or South American republics realized the danger.

Not that warning was not given. Both Fascist and Nazi frankly boasted of their intentions. Virginio Gayda, the newspaper mouthpiece of Mussolini, on June 7, 1940, declared that "when the United States participates too indiscreetly in European affairs . . . the United States automatically grants European powers the right of retaliation, to be taken today or in any other period of American history and on American territory."[1] The Fuehrer was even more explicit when he boasted "We shall create a new Germany in South America. Mexico is a country that cries for a capable master . . . We shall create a new Germany in Brazil . . ."[2]

One of the weapons employed most effectively by the Nazis to accomplish

[1] *Giornale d'Italia*, June 7, 1940.
[2] Herman Rauschning, *The Voice of Destruction* (New York, 1940), pp. 61-67.

their ends was the stimulation of jealousy and hostility among Latin Americans by making invidious contrasts between the Yankees and their neighbors to the south. A subtle example is shown by a sentence from an editorial in the *Voelkischer Beobachter* or February 15, 1939: ". . . The biggest obstacle in the way of the Pan American policy of the Yankees is the deep chasm between the prosperity civilization of North America and the Iberian culture of Latin America ." German newspapers and German publicists constantly strove to weaken the increasingly closer ties engendered by the good neighbor policy by a continuous campaign of detraction and innuendo. Although clearly a brazen attack upon the morale of the people of Latin America, the Nazis were confident that the Yankees were too stupid to appreciate the danger and the Latins would not be averse to implied support against the powerful neighbor to the north. Unfortunately, their reasoning was correct. The United States continued to be primarily interested in business as usual and the Latin American republics were not unaccustomed to dictators and totalitarian methods. They appreciated the increasing attention showed to them but they did not take *der Führer* too seriously.

Apparently the United States, in spite of a constantly increasing loss of trade with the larger South American states to the benefit of Germany, failed to realize the threatening political implications. For example, in 1936 when the United States was buying twice as much from Chile as was Germany, the Reich exported to Chile about $2 million worth of goods more than the United States. In 1937 the Chilean State Railways bought over $3 million worth of railway equipment from Germany and in 1938 the Chilean National Air Lines were equipped with Junkers J. V86 although the bids of American firms were lower and Chilean pilots preferred American planes. During the same year Germany took away from us even our leading position in Brazilian imports, and only the outbreak of the war in 1939 prevented a similar loss in Argentina.

The Nazis, however, were not content with economic gains. They were determined to bring Latin America under political control as well. The situation seemed tailored to their purpose. Although not well entrenched in Mexico or the Caribbean area, it was estimated that there were about two million nationals in South America of whom the majority possessed German citizenship, even though not born in the Reich. Furthermore, they were well concentrated. In the southern departments of Chile, from Valdivia and Osorno to Puerto Montt, some 30,000 German farmers kept in such close touch with the Reich that their output of skins, wool, lentils, oats, and wine was contracted for far in advance of its production. In the northern territories of Argentina, such as Misiones, Formosa, and Chaco, and in the provinces of Corrientes and Entre Rios there were about a quarter of a million Germans, who practically dominated both the industry and agriculture of the region. But it was in the southern states of Brazil where there had been established a veritable *Deutschland uber meer.* Reliable figures gave a conservative estimate of approximately a million Germans in the states of São Paulo, Santa Catarina, and Rio Grande do Sul. In some of the towns of Santa Catarina almost two-thirds of the people speak German in their daily lives instead of Portuguese.

The Nazis also counted upon the nationals of their Axis allies for strong support. With approximately six million of Italian blood and about a quarter million Japanese and an undetermined number of Falangistas who wished to restore Spanish influence and culture in the western hemisphere, this attitude seemed justified. The Nazis soon discovered, however, that they were badly mistaken in this hope. The Italians in South America assimilate quickly with both Spanish and Portuguese races. In Argentina where about half of the Italian stock is found they consider themselves Argentinians and are proud of it. The great Salocchi bank, in Peru, the Banco Italiano, may be Fascist minded, but its Peruvian depositors of Italian race on the whole are not. Uruguay, which has a large Italian stock, is one of the most liberal-minded republics in South America. The Latin citizens of the South American republics whether born in Spain, Italy, or Portugal are patriotic nationals of their adopted countries and react accordingly. The Italians resented Mussolini's attempts to export Fascism and even more his subservience to his Nazi master. Jokes on Mussolini and the Fascists were very popular among the Argentines whether of Italian or Spanish antecedents. One which was often quoted referred to the automobile racer who got extra speed from his car by putting Pirelli (Italian) tires on his frontwheels and Dunlops (English) on the rear. Another referred to the man in a restaurant who ordered spaghetti, then Worcestershire sauce. When the sauce arrived he found the spaghetti had disappeared.

It was therefore the Nazis primarily whose activies were a real menace to the unified defense of the Western Hemisphere. It was they who, in the picturesque language of Sumner Welles, were the "human termites gnawing at the foundations of our Inter-American system." If we try to estimate just how great was the danger, we must divide South America into two parts–one, that section lying for the most part north of the bulge of Brazil and including Venezuela, Colombia, Ecuador, and Peru, and, the second and far larger part, including Argentina, Bolivia, Brazil, Chile, and Uruguay. In the northern sector, except for a potential threat to the Panama Canal, the problem was never very serious. In the southern sector we must confess that although Hitler could hardly take it over by telephone as has been alleged if he won the battle of Europe, he was in an excellent position to bring about the quick demise of the Monroe Doctrine in most of the area south of the equator, save Peru.

Let us take a brief glance at the situation in this northern section.[3] Starting with Colombia, we find there were only about 4000 Germans, but they were very well organized and some of them owned large and strategically located estates, with reference to the great oil pipe-lines. One of these, a 2500 acre tract south of Cartagena, owned by Dr. Hans Neumüller, an ardent Nazi, was only a stone's throw from an oil pipe-line that delivered 54,000 barrels daily. Another Nazi, Adolph Held, owned an extensive estate across from Tenerife, where the Andean pipe-line crossed the river. Held, who incidentally was on the United States blacklist, had two air fields upon his estate.

Up to 1939 the German air line Scadta–the oldest commercial line in South America, was a direct threat to the Panama Canal. It was finally taken over by Avianca, a line owned partly by Pan American Airways and partly by the government of Colombia. The small German Arco line which controlled some

[3]The facts presented in this chapter are based for the most part upon information obtained in the various United States embassies and legations in South America where the writer was given every opportunity to survey the situation in the spring of 1941.

forty landing fields in Southeastern Colombia, only about 500 miles from the Panama Canal, resisted all efforts to take it over until the autumn of 1941. In his broadcast of September 11, 1941, President Franklin D. Roosevelt mentioned a dangerously strategic German air field between Barranquilla and Cartagena on the Neumüller estate which the Colombian government, when the matter was brought to its attention, promised to investigate immediately.

Although the leader of the Colombian conservatives, Dr. Laureano Gómez, formerly minister to Germany, was reported very anti-Yankee—the taking of the Canal Zone via Panamanian independence still rankles—the United States' position was too strong to cause us much worry. The country with an investment of a quarter of billion dollars which controlled the oil industry and took most of Colombian coffee, bananas, and platinum had a full house before the draw. The well-arranged visit of President-elect Alfonso López to the United States in the summer of 1942 strengthened materially the bonds of friendship between the two countries.

There was no Nazi problem for the United States in Venezuela. Living principally upon oil which American and British companies controlled, the government was pro-United States and pro-British. We had a naval mission there and our new base at Trinidad completely controlled the approach by sea. Port of Spain in Trinidad is on the direct line of the Pan American Clippers from Miami to Rio and on the east-west run Pan American joined Port of Spain with La Guaira, Maracaibo, and the Panama Canal. A mere glance at the map shows the strategic value of this service. With less than 5000 Germans to work upon, Herr Goebbels wasted little time in Venezuela. In fact Venezuela was the first republic of South America to sever diplomatic relations with the Axis powers.

Ecuador, which straddles the equator, had about 6000 Germans and the names Hindenburg and Adolf Hitler were still found on its busses. Where the United States had a minister, a diplomatic secretary, and a naval attaché with three or four clerks in its legation, Germany had, until the severance of diplomatic relations, a staff of twenty-five. Where we had a half dozen in our consulate in Guayaquil, Germany had three times as many. The Sedta Air line—a German company—was formerly a real threat to the Panama Canal, but Panagra first duplicated its runs and finally brought about its suspension. The United States had sent two missions, one naval and one military, to take the place of the Italian military mission withdrawn in 1940. Subsequently Ecuador leased us a base on the Galapagos Islands—a concession which was of as great a strategic value in the Pacafic as Trinidad was in the Atlantic. The Nazis tried to fish in the troubled waters of the boundary dispute between Ecuador and Peru, but their catch was inconsequential.

Peru had long been regarded as very friendly to the United States. Standard Oil of New Jersey had a monopoly of its oil, the Cerro de Pasco Mining Corporation controlled Peru's copper and much of its gold and silver. The Vanadium Corporation was American and W.R. Grace and Company, an American concern, ran the Grace Line to the west coast, had sugar and cotton plantations, and was half owner of Panagra. The Japanese and the Italians far outnumbered the Germans, and Nazi efforts were completely checkmated. The great German sugar family, the Gildermeisters, were not Nazis; in fact, there was some Jewish blood in the direction of Casa Grande, their vast sugar company. The fact that the Peruvian military airport at Talara was built entirely out of German materials was due to the same reason that the American Cerro de Pasco Company had their aerial tramway built by a German company—the Nazi bids were lower.

In 1941, resenting German violation of governmental regulations, Peru canceled the Lufthansa Airway Service, confiscated its planes, and gave the rights to Panagra. It also nationalized the Italian Caproni airplane plant, sent the Italian air mission home and invited in an American mission. Although one of the principal newspapers, *El Comercio* of Lima, was decidedly pro-Nazi and some of the leaders in the army had decidedly Nazi predilections, the Gestapo organization found to its dismay that President Manuel Prado was friendly to the United States, and his foreign minister, Solf y Muro, almost equally so; even the famous Aprista, Haya de la Torre, formerly very critical of Uncle Sam, was a strong believer in the Good Neighbor Policy. When the United States declared war upon Germany, Peru promised the United States every assistance; at the Rio Conference she made good on her promise and before the conference adjourned, she broke all diplomatic relations with the Axis powers.

Coming to the larger and more important part of South America south of the Equator the situation was not so satisfactory. In Bolivia, Chile, Argentina, Uruguay and Brazil the Nazis were more numerous and better organized. It is in this area that Germany made the greatest economic advances and her political infiltration was most powerful. Commercial inducements, military missions, schools, news services, radio programs, every instrumentality for drawing this tier of states into the orbit of the swastika was effectively used. And coordinating all these subversive activities were the German embassies, legations, and consulates whose officers, abusing their privileges and immunities, served as agents for Dr. Goebbels' Ministry for National Enlightenment and Propaganda, and Heinrich Himmler's even more dangerous Gestapo organization known as U. A. 1 (*überwachtungsstelle Ausland*). It was in this region that the destruction of democratic morale by the promulgation of subversive propaganda was most aggressively conducted.

Bolivia, the least important of these countries in influence, but of vital importance to the United States because of its tin, was still suffering a bad hangover from the Chaco War. Whatever the real causes, Bolivia believed that her army trained by the German General Hans Kundt could easily conquer backward Paraguay. When it failed, a whipping boy was required and Standard Oil was at hand. Its $17 million property was confiscated. The President at the time was Germán Busch, who immediately afterwards set up a totalitarian state with himself at its head. When he committed suicide or was assassinated in 1939, the Nazi press claimed it was murder and the United States was in some way cause of it. But whatever the cause, the results were disastrous for the Nazi organization. When the European situation became very serious, Bolivia immediately turned towards the United States, in spite of the fact that three-fourths of the big firms in Bolivia were German and German propaganda both by radio and by press flooded the country. When in May 1941, the United States made a deal to purchase Bolivia's total production of tungsten for three years for about 25 million dollars, and at the same time was able to force out the important German Lloyd Aereo Boliviano with its 4000 miles of airways and substitute Panagra, Nazi elements attempted a *coup*. But before the putsch came off, President Peñaranda acted. A letter from the Bolivian military attaché in Berlin to the German minister in La Paz fixing the date for the *coup d'état* was intercepted and published. Although the German minister declared it was a forgery, he was dismissed, a number of pro-Nazi Bolivians arrested, four pro-Nazi newspapers closed, and the Nazi menace in Bolivia was eliminated at least for the time. To strengthen the new situation the first air mission from the United States to Bolivia was sent in November 1941, to take the place of the hitherto well entrenched Italian advisors. When the United States entered the war, Bolivia immediately accorded us non-belligerent status, and early in 1942, severed diplomatic relations with Germany and Japan. On April 7, 1943, Bolivia formally declared war upon the Axis powers and on April 20 she declared her adherence to the principles of the Atlantic Charter.

To North American travelers, Chile seems the nearest approach to home of any of the Latin American republics. To North American financiers, the riches of the Chilean Andes have encouraged investments to the value of over a half billion dollars: Guggenheim in nitrates, Anaconda-Kennecott and Braden in copper, Bethlehem in iron. To the Germans, Chile beckoned agriculturally. The earliest German colonists in South America settled in Chile in the sixteenth century. In 1840 many families came and by the 50s there were some 2000 Germans in Chile—for the most part engaged in agriculture in the southern provinces from Valdivia to Puerto Montt. Later Germany sent engineers and officials to the nitrate fields and military missions to the Chilean army. Authoritative sources gave these estimates as to the German population of Chile: about 18,000 born in Germany; 40,000 of German descent; 20,000 of mixed marriages, and 9,000 refugees, mostly Jews, making a total of 87,000.

The Germans in Chile whether in commerce, agriculture, or industry not only retained their identity but maintained close connections with the Reich. Therefore, with the advent of Hitler a fallow field was afforded to his agents of propaganda. The German embassy at Santiago was enlarged to eleven principal officers not counting clerks (forty employees in all), some of whom under the title of press and cultural attachés devoted all their time to preaching the gospel of Naziism. It was claimed that the German embassy spent over $100,000 a month on this work, a considerable part of which was obtained by forced levies on German firms and nationals. It has been asserted that the Nazi propaganda in Chile, carried on by the press, the radio, and the hand-outs from the German embassy was so overpowering that it brought about a confusion of ideas in the minds of the Chilean people as to just what were the national ideals, with a corresponding weakening of faith in democracy.[4]

The Chilean army was modeled to a considerable extent upon the German

[4] See Hugo Fernández Artucio, *The Nazi Underground in South America* (New York, 1942), p. 162.

system. In fact, for almost fifty years German army officers on missions had advised as to its training. The Mauser rifle was still used and the Krupp factories furnished its artillery. The military police known as *carabineros* were also German trained—their former instructor Otto von Zipellius was regarded as chief of the Nazi party in Chile. In June, 1941, the German government presented a schooner for training purposes to the Chilean navy and in presenting it Ambassador von Schoen declared that a new link had been forged between the German and Chilean navies.

There were a considerable number of primary and secondary schools in Chile whose curricula were entirely in German. But the excellence of the Chilean public school system—it is one of the best in South America— made Chile's problem of the foreign language school less serious than it was in Argentina and Brazil. However, in the very excellent Santa Maria Technical College at Valparaiso, with a $5 million endowment which was sufficient not only to give free tuition but also free board and room to the fortunate students who entered by competition, all of the professors were German as was most of the technical equipment.

Since Chile controlled the only other water route between the Atlantic and Pacific except via the Panama Canal it was vital for the United States to keep the Straits of Magellan open. In spite of the overwhelming German character of the population in the southern territories, Nazi principles were never overly popular. The success of Hitler kindled in some the pride of race, but the majority were well content to have *der Führer* win his victories in Europe. Furthermore the Popular Front government was completely awake to the totalitarian menace. The United States was popular with the Chileans, and with the return to Chile of the eight Chilean newspapermen who, through the invitation of Ambassador Bowers, lived the life of American reporters, and the still large number who came as guests of the United States under the auspices of the Coordinator's Office, the neighbor to the North was even better understood than before.

It must be confessed that Chile's attitude in refusing to sever relations with Germany and Japan at the Rio Conference seemed to indicate that Nazi propaganda had been more successful than the United States had expected. Nevertheless, when Chile's new President, Juan Antonio Ríos, accepted President Roosevelt's invitation to visit the United States in October 1942, it was thought in many quarters that Chile would break with the Axis powers before that time. When it was evident that Chile intended to remain neutral for the time being, Sumner Welles forced the issue by declaring publicly that Axis espionage in certain South American states had resulted in the sinking of ships and the loss of lives in the western hemisphere. Although Chile vigorously protested and postponed the Ríos visit, a change in Chilean foreign ministers followed and the new cabinet took a strong stand against all German nationals accused of subversive activities. With public opinion thoroughly aroused the Chileans finally took the decisive step and broke relations with the Axis powers on January 20, 1943.

For many reasons Argentina was regarded as the focal point of Nazi propaganda in all Latin America. The Argentine Republic had a larger percentage of its nationals who were born in Europe than any other Latin American republic and of these, some 60,000 were born in Germany. Another 125,000 were born in Argentina but of German parents. This gives the Argentinian a natural predilection for Europe rather than for the "crude aggressive Caliban" of the north.

The center of the German propaganda organization during World War II was the German Embassy in Buenos Aires. Ambassador von Thermann had the largest staff of any foreign mission in Argentina. In addition to the counselors and regular diplomatic secretaries, there were press and cultural attachés whose sole function was propaganda. The German Embassy with its staff of over forty, was almost twice as large as the United States Embassy and Consulate General combined. It was estimated that its annual expenditures were more than ten times the amount available to the embassy of the United States. The German government maintained fourteen consulates outside of Buenos Aires, the United States one. The Embassy saw to it that copies of Hitler's speeches translated into Spanish were mailed to all post office boxes within two days after their delivery. The Embassy also issue a mimeographed daily bulletin entitled "Off Cable News from Berlin," the general tone of which was decidedly anti-United States and anti-British.

The Nazi organization in Buenos Aires was founded in 1933 by an employee of the Banco Germánico under the name of the German Benevolent and Cultural Society. This organization possessed an elaborate card index of the names of German nationals giving the business and family connections of the individual, including those living in the Reich. All social affiliations were listed and the individual's income and its sources. It was expected that a minimum of 10 per cent would be contributed for the needs of the organization and collection was made regularly by accredited agents. If payment was not made, pressure was used; if that was not sufficient the Gestapo had harsher means and finally if necessary, close relatives in Germany paid the penalty for the recalcitrant objector. Such extreme measures were rarely necessary.

The Argentine army was very sympathetic to Nazi influences which was natural, since for twenty-five years it had been advised by German military missions. When the mission was not reappointed in 1940, its chief, General Niedenfuhr, was made military attaché, dividing his time between Rio and Buenos Aires. The retired officers were very influential and very pro-Nazi.

The Nazi influence in the school system of Argentina was particularly strong. The German schools were the most important in numbers and influence of all the foreign schools in the Republic. Out of 284 foreign schools at the outbreak of World War II, 203 were German, with an approximate attendance of 13,500 pupils. Among the best known were the Goethe Schule, the Germania Schule, and the Colegio Alemán. The majority were under the strict control of an Inspector from the Ministry of Kultur in Berlin.

The Nazis paid particular attention to the press in Buenos Aires. The most outstanding Argentine newspapers, *La Prensa, La Nación, El Mundo,* and *La Razón* were objective in their reporting, and although critical of our tariff policies they were otherwise quite friendly to the United States. Even the more sensational papers such as *Crítica* and *Noticias Gráficas* were pro-British and anti-Nazi; therefore the Nazis had to establish their own organs. They first took over the old *Deutsche La Plata Zeitung,* gave it a Spanish language supplement and colored the news and editorials with pro-Nazi hues. Next they got control of *Crisol* which commenced a series of vitriolic attacks upon President Roosevelt, whose name they alleged to be Rosenfeldt. The Good Neighbor Policy was merely camouflaged Yankee imperialism. In 1940 the Nazis set up a new so-called nationalistic newspaper called *El Pampero* with an elaborately equipped establishment. Its monthly stipend from the German embassy was 60,000 pesos and it had a forced subscription among the Nazi organizations and a wide distribution in Argentine army circles. It attacked the United States and Great Britain upon every possible occasion.

It was hoped that the military revolt of June 4, 1943, which forced the resignation of President Ramón S. Castillo, might bring about a break with the Axis powers. However, the new government of General Ramirez, although promising loyal cooperation with the other American republics, declared for a continuation of the previous policy of neutrality. In spite of popular demonstrations in favor of the democracies, the provisional government suppressed the Communist newspaper *La Hora* and permitted the pro-Nazi *El Pampero* to continue publication.

Uruguay has long rivaled Brazil and Peru in her friendship for the United States; nevertheless the small but effective Nazi organization here went so far as to prepare a plan by which German technicians, after an insurrection, were to take over the work of Uruguayan officials and transform the state into a German agricultural colony. Much credit is owing to Professor Fernández Artucio for unearthing the plot and disclosing it in a book entitled "Los Nazis en el Uruguay." He followed it up with numerous radio addresses appealing for Congressional action. Congress finally investigated and found so much evidence that twelve leading Nazis were arrested. They were later released, but in September 1940, eight of them were rearrested and charged with treason against the state. The German subversive organizations were dissolved, their newspapers and radio suspended for a period and all activities driven under ground. It is reported that the counselor of the German embassy in Rio publicly declared in March 1941, that "Montevideo is our weakest spot in Latin America."

Paradoxically enough, Brazil which has long had the closest commercial and political ties with the United States of all Latin America had at the same time the most dangerous Nazi organization in South America. Brazil had the largest number of Germans, about 50,000 born in the Reich, and a million to a million-and-one-half of German blood. They were for the most part concentrated in the southern states of São Paulo, Paraná, Santa Catarina, and Rio Grande do Sul.

The German Embassy in Rio in 1941 was known to have available some two-and-a-half million dollars for propaganda purposes. Former Ambassador Karl Ritter became so obnoxious to the Brazilian government through his opposition to the suppression of local Nazi organizations and to his support of one of the members of his embassy accused of complicity in the Integralista coup of 1938 that he was finally declared *persona non grata.* The cultural attaché of the Embassy was also the chief of the Trans Ocean News Agency which purveyed radio and cable news from abroad to all Latin American newspapers that would take it. There were some twenty-eight papers in Brazil alone which subscribed to its services. The German embassy was the largest of all the foreign representations. It had one of the most magnificent residences in Rio and its chancellery took even more extensive quarters during World War II. It is difficult to state the exact number who were employed inasmuch as a large number of its clerks and employees were not listed in the diplomatic list, but it was estimated that its total staff was over a hundred. At the request of the German Embassy, the Brazilian Minister of War, General Dutra, and the Brazilian Chief of Staff, General Goes Montero, were decorated in 1940 for their valuable services to Germany in Brazil.

Closely in touch with the German Embassy were many important commercial concerns such as Compania Chimica Merck, Bayer Ltd., and various banks and motor companies. The Condor Syndicate was the most important airway in South America, except for Pan American. It formerly cooperated with the Italian L.A.T.I. which flew till recently from Dakar to Natal, and its network of lines spanned many areas of strategic rather than commercial value. The inability to obtain fuel and make essential repairs finally grounded it even before Brazil declared war upon the Axis powers.

The German Protestant Church in southern Brazil was impressed into the propaganda machine. It was reputed to have some 200 pastors and bishops who served as Nazi agents. At one time it was estimated that there were 3000 German schools in southern Brazil and the number of German cultural, benevolent, and athletic societies was legion.

In spite of this Axis predominance, the Vargas régime was able gradually to impose a series of restrictive measures. Portuguese was made compulsory in the schools, the foreign language press was forbidden, all airplane pilots had to be Brazilian nationals, and alien political activity was strictly prohibited.

The Japanese attack upon the United States strengthened the hands of President Vargas and Foreign Minister Arañha in their efforts to strengthen continental solidarity. Señor Arañha played a leading role at Rio in bringing about a break with the Axis powers. When German submarines began to sink Brazilian ships, President Vargas decreed corresponding confiscation of Nazi assets in Brazil. All fifth column suspects were rounded up and interned. When finally five Brazilian ships were sunk in three days in August, President Vargas hesitated no longer. Brazilian aircraft were ordered to sink Axis submarines on sight and on August 22, 1942, a formal declaration of war was issued. The results of years of Nazi propaganda were wiped out when the concrete evidence of Nazi methods was brought home to the Brazilian people.

II. THE CHALLENGE OF COMMUNISM[5]

Another and perhaps an even more formidable extracontinental totalitarian threat to established governments in the Americas than fascism is represented by the international Communist movement. Communist subversion in the inter-American community, inaugurated and directed by the Soviet Union in the aftermath of World War I, was carried on by the Sino-Soviet bloc, joined later by Cuba, following the Second World War. In the Latin American republics, where exploitation of the masses is common both by foreign capitalists and by the national politician, the promises of communism have proved more attractive to many than the realities of frequently falsely-labeled democracies.

The Communist parties of Latin America were established in several periods or phases after the Russian Revolution of October 1917. In the first period, from 1918 to 1922, parties were formed in five countries–Argentina, Brazil, Chile, Uruguay, and Mexico–where industrialization and the labor movement were the most advanced. In the second period of the movement, 1923-1928, parties appeared in Cuba, Honduras, El Salvador, Guatemala, Panama, Ecuador, Colombia, Peru, and Paraguay. At the Sixth Congress of the Comintern in 1928, the third period of the Communist international was begun. At this meeting was made clear what the Soviet Union had in mind for Latin America, and for other parts of the world. A document produced there analyzed the economic conditions prevailing in Latin America and concluded that the immediate tasks of world communism with respect to Latin America were to fight against the prevailing feudal and other precapitalistic modes of economic exploitation, to encourage the peasants toward agrarian revolution, and to support the fight for economic independence from foreign "imperialist" groups. It was expected that preparatory steps would be necessary before the proletarian revolution could be realized in Latin America. The Comintern line called for Communists everywhere to declare themselves as such, to cease cooperation with "bourgeois" and "leftist reformist" elements, and to establish purely Communist organizations which should strive to foment and seize the leadership of proletarian revolutions. Communists operating within other parties were directed

[5] For futher general background on this subject see the following: Victor Alba, *Historia del movimiento obrero en la América Latina* (Mexico, D.F., 1964); Luis E. Aguilar, ed., *Marxism in Latin America* (New Brunswick, N.J., 1968); Rodney Arismendi, *Problemas de una revolución continental* (Montevideo, 1962); Stephen Clissold, *Soviet Relations with Latin America, 1918-1968: A Documentary Survey* (New York, 1970); Harold E. Davis, *Latin American Social Thought* (Washington, 1961); Dorothy Dillon, *International Communism and Latin America* (Gainesville, 1962); Cecil Johnson, *Communist China and Latin America, 1959-1967* (New York, 1970); Rollie Poppino, *International Communism in Latin America* (Glencoe, Ill., 1964); Richard F. Staar, ed., *Yearbook on International Communist Affairs 1971* (Stanford, 1971); U.S. Dept. of State, *World Strength of Communist Party Organizations* (Washington, 1971); U.S. Senate, Subcommittee on American Republics Affairs, Committee on Foreign Affairs, *Survey of the Alliance for Progress–Insurgency in Latin America* (Washington, 1968); Edwin M. Martin, "Communist Subversion in the Western Hemisphere," Department of State *Bulletin*, XLVIII (March 11, 1963), pp. 346-356; *Ibid.*, (March 18, 1963), pp. 404-412.

to convert these parties to Communist organizations. With the addition of three new parties in 1929-31—Haiti, Venezuela, and Costa Rica—the number of countries with Communist parties totaled seventeen.

The fourth phase of Communist organizational efforts in Latin America, extending from the mid 1930s to the mid 1940s was marked by three divisions, each identified with a particular episode in the progression of the international Communist movement. These were the eras of the European Popular Front, dating from about 1935, the Hitler-Stalin Pact (August 1939 to June 1941), and the remaining years of World War II. In this phase Communist parties were established in three more Latin American countries: Bolivia, the Dominican Republic, and Nicaragua. At the same time several parties abandoned the Communist label without changing their political orientation. The fifth and current phase, following World War II, witnessed the Soviet bloc offensive, the communization of Cuba, and Fidel Castro's pursuit of revolutionary war in Latin America.

When the U.S.S.R. first displayed an interest in the Latin American nations, its leaders clearly realized that in the conduct of foreign relations differences may exist between their country's relationship with another nation's government and its appeal to that nation's people. Concentrating its efforts on winning the loyalty of the masses who, as underprivileged elements in society, it was believed would find communism attractive, Moscow gave little attention to the governments of the Latin American countries, beyond seeking to undermine the established order. However, the Latin American Communists gained no notable successes in the 1920s and early 1930s. Fascism had an advantage over communism in this period owing to the economic and military strength of the Axis powers and the greater compatibility of Fascist doctrine with the authoritarian tradition of Latin America.

Other characteristics of Soviet diplomacy were the adoption of undiplomatic language, frequently abusive and insulting, and a disregard for protocol. The course of Soviet diplomacy in the inter-World War years was generally an erratic one which, until about the time of its involvement in World War II, had succeeded in alienating most foreign governments. But during World War II, and immediately afterwards, it seemed to favor becoming a cooperative member of the global community.

For several years from its inception, the Soviet regime was so occupied with internal affairs that its leaders could devote little time to promoting the world revolution and to assessing the readiness of different parts of the world for the proletarian uprising. It was not until the Sixth Congress of the Comintern in 1928 that Communist attention became focused upon Latin America, and a policy line was issued. But official relations between the Soviet Union and Latin American countries were, with few exceptions, nonexistent until the period of World War II. Only two countries, Mexico and Uruguay, ever gave full recognition to the Soviet Union; Colombia, though granting *de jure* recognition, refused to implement this recognition by sending or receiving diplomatic representatives.

Late in 1939, the Soviet Union made itself very unpopular in the United States and elsewhere by invading Finland, a country especially admired because it was the only country which had kept up payments on its World War I debt.

Annoyance was also evidenced in Latin America: Argentina took the lead in having the Soviet Union expelled from the League of Nations, in December 1939. In June 1941, Nazi Germany attacked the Soviet Union, and by December of that year the United States was itself at war with the Axis powers, fighting against some of the same states as were opposed by the Soviet Union. These events caused a change in United States public opinion, which turned to praise of the U.S.S.R. for its courageous and steadfast resistance to Germany's aggression. The Soviet Union and the United States were not identified as allies during the war, as the United States and Britain were, but rather, were cobelligerents—nations fighting on the same side. Between ally and cobelligerent there is a significant difference, but this was not generally understood at the time.

Many of the Latin American countries rallied in support of the United States, and broke off diplomatic relations with Germany, so that they were on the same side in the war as the Soviet Union. The latter, though apparently never losing sight of its ultimate objectives, entered upon a program of international cordiality, and endeavored to promote friendships for itself. Furthermore, its seeming change in attitude toward religion, as shown by its reconciliation with the Orthodox Church, made a favorable impression upon many of the Latin American states. Latin American countries were also inclined to be friendly toward the Soviet Union because it was believed that it might provide a market for their exports after the war was won. Moreover, the United States urged its southern neighbors to be cordial toward Russia, arguing that the threat of communism was vague and remote, whereas Hitler was an active aggressor who must be defeated, and that communism was cooperating toward this end.

Another reason why Latin American states became hostile toward the Axis and friendly toward the Soviet Union was that they were anxious to obtain membership in the projected United Nations, and knew that invitations to join would surely be extended to cobelligerents. For this reason, among others, Argentina declared war on Germany (less than two months before the European war ended), thus taking a step which the United States, annoyed by Argentina's strict neutrality, had long demanded. Remembering Argentina's leading part in expelling her from the League six years before, and its wartime collaboration with the Nazis, the Soviet Union vehemently opposed Argentina's being invited to attend the San Francisco Conference of the United Nations. The United States championed Argentina's cause, however, and an invitation was finally extended to Argentina.

In the pre-Castro period following World War II, Soviet relations with Latin America appeared to enter a period of expansion, but were reversed sharply in the years 1947-1952, as many countries suspended relations with the Soviet Union, and at the same time took strong measures to repress the local Communist parties. Renewed efforts by the Kremlin to extend its operations in Latin America, leading to the present period of expansion, began in 1953 when negotiations were made with the Perón government of Argentina to increase trade. Soviet activities in the area after 1953 were part of a global effort aimed at underdeveloped countries, which included aid and trade programs.

By December 1958, the European Soviet-bloc countries had made some twenty trade and payments agreements with five Latin American countries—Argentina, Brazil, Colombia, Uruguay, and Mexico. At least one of the bloc countries had resident diplomatic missions in these countries, and also in Bolivia, at this time. Soviet bloc trade with Latin American countries increased from $70 million in 1953, only 0.6 percent of the area's total trade with the world, to $275 million in 1959, or 1.7 percent of Latin American trade. The peak was reached in 1955, when the volume of trade reached 2.1 percent. Travel and radio communications between Latin America and the Soviet bloc reached a fairly high level during the 1950s; in the period 1955-1957, fourteen of the former were visited by bloc trade missions and seven of these countries sent missions to the bloc countries. Generally, however, the Soviet Union's diplomatic and economic offensive achieved no lasting political gains, even in Argentina or Uruguay, the main areas of its operations. The only near-truimph of communism in the Americas before Cuba fell to Fidel Castro was in Guatemala in the period of 1950-1954, where the Communists came to dominate the Arbenz government and established effective control over the country's labor and peasant organizations.

Although Soviet power and influence rose markedly in the late 1950s Communist leaders in Latin America were unable to translate this new power into greater domestic influence probably because of the area's isolation from the centers of international communism. What influence the local parties exerted was largely restricted to their hold on labor organizations and intellectual and student groups. Being rarely accepted by non-Communist, populist parties, they tended to remain on the defensive. But just prior to Castro's seizure of Cuba in the late 1950s, it seemed that the outlook for Communist efforts was improving. The presence of dictatorial rule for long periods in some countries, growing agitation by leftist party leaders, students, journalists, and other intellectuals, and movements to give a legal status to Communist parties, pointed in that direction. By denouncing social and economic injustice and United States "imperialism," the Communists began to appear to many sectors as champions of democracy. Communist party membership in Latin America was estimated at 250,000 in 1958, with about 215,000 being found in five countries (Argentina, Brazil, Chile, Cuba, and Venezuela). The parties had substantial non-Communist followings and frequently exerted influence on other political groups in their countries.

With the Castro regime's seizure of Cuba in January 1959, which provided the Soviet Union with a special and unexpected opportunity, subversion in the hemisphere swiftly accelerated. Figuring prominently were the tactics of infiltration, popular front action, and insurgency, but particularly violence and terrorism. Local Communist elements had the advantage of receiving more external assistance, especially from Cuba, than previously. Except for the case of Cuba, the Soviet Union, although seeing Latin America as an area offering extraordinary opportunities for its aggressive ambitions, has been cautious in its tactics. There is no evidence to suggest, however, that the Russians have restrained Cuba or Communist parties elsewhere from violent or terroristic activity.

The Castro regime from the beginning attempted to export its revolution into other Latin American countries. Castro has supported armed expeditions or insurgency movements, none of which was successful, in the following countries: Haiti, Panama, the Dominican Republic, and Nicaragua (Castro-sponsored invasion in 1959), Ecuador (1962), the Dominican Republic (1963), Peru (1964-1965), Argentina (1964), Honduras (1964), and Brazil (1967). When this approach produced counteraction by the OAS, Castro shifted tactics, following a more covert, indirect approach that involved the formation of front organizations in Latin America and the United States, material support for subversive groups and the indoctrination and training in Cuba of hundreds of Latin Americans in sabotage, terrorism, and guerrilla tactics. Through press and radio facilities the Cuban government disseminates propaganda to further its subversive ends.

The Sino-Soviet ideological split which appeared in 1960 caused division and dissension among Latin American Communist parties. The focus of the controversy has been largely on the question of what strategy should be followed to achieve political power. The Communists had the options of pursuing revolutionary war as a means to power or the so-called peaceful method of participating in elections. The latter, while not ruling out the use of violence under certain conditions, stresses involvement in elections and the infiltration of key government agencies. This approach is favored by the majority of the "old" or orthodox, Communist parties, those following the Moscow line. The Castroite and pro-Peking splinter groups made up of the more youthful and militant elements favor the violent path, whereas the Soviets, as mentioned, believe that coexistence as a course of action is more certain of success in the long run; however, the Soviet position has been ambivalent on this issue.

The militants, rebelling against the conservatism of the orthodox parties, employ Fidel Castro's chosen instrument for revolution in Latin America, a variation and reorientation of guerrilla warfare. Its ideological basis is communism, and it draws heavily upon traditional guerrilla warfare but stresses the development of rural violence supported by urban terror. Ernesto "Ché" Guevara, the Argentine revolutionary, killed by government troops in Bolivia in 1967, and Jules Regis Debray, a French Marxist, imprisoned in Bolivia for subversive activities, together with Fidel Castro, authored the doctrine. What has become known as the Castroite theory of revolution—Revolutionary War—holds to the following points: (1) peasants, not urban workers, are the sole base for revolution; (2) the only method is guerrilla warfare, not terrorism or the coup d'etat; (3) peasant guerrilla forces will gain control of the countryside for self-support, and deny the government these resources; (4) urban terrorism will play a supportive role, destroying public acceptance of a government by violence against persons and property. Orthodox parties are viewed with disfavor since they are a liability during the fighting and tend to divide the revolutionary movement when victory has been won. In 1968 four countries faced the threat of Castroite insurgency: Guatemala, Nicaragua, Colombia, and Bolivia; however, the Bolivian movement suffered a major setback with the death of Dr. Guevara.

Premier Castro publicly espoused communism in December 1961. His relations with the Soviet Union and the orthodox (Moscow line) Communist parties since then have been characterized by tension and antagonism, for the reasons just described. The differences were unveiled at conferences held in the aftermath of the missiles crisis of October 1962, which clearly gave the Cuban leader reason to lose confidence in Soviet military support. Only representatives of the Latin American orthodox communist parties and the Russians attended the meeting convened in Havana in November 1964. There the Latin Americans agreed to intensify support for the Cuban revolution in their own countries, aiming at the resumption of relations with Cuba and ending the United States economic blockade of the island. Agreement was reached on a list of countries ripe for the Castroite Revolutionary War: Venezuela, Colombia, Guatemala, Honduras, Haiti, and Paraguay. For other countries the violent road to power would be adopted only when the orthodox party agreed. Castro, in return, pledged to work through the orthodox rather than through the Castroite parties and, for the first time, supported the Soviet position regarding the Peoples Republic of China. By the Havana agreement Castro won recognition as the leader of Latin American communism, and a Soviet-orthodox party endorsement for Revolutionary War in a limited number of countries.

Castro violated the Havana agreement soon thereafter when he announced a new aggressive policy for the hemisphere. The first Afro-Asian-Latin American Peoples Solidarity Conference (Tricontinental Conference) met in Havana January 3-15, 1966. By engineering the selection of Latin American delegations, the Cubans insured the attendance of pro-Revolutionary War groups, as well as orthodox parties, although opposition among them had already appeared. The conference agreed that liberated countries such as the USSR are obliged to assist liberation movements, and it created a committee to support national liberation movements. Cuba was designated as the site for the Tricontinent's school for guerrilla fighters and terrorists. It was also agreed that a Latin American Solidarity Organization (OLAS) would be formed with headquarters in Havana. Premier Castro made his position clear saying: "We believe that in this hemisphere in the case of all or almost all, the struggle will take the most violent forms." The Russians, embarrassed by the implications of the statement for their relations with Latin America and the United States, sought to dissociate themselves from the violent path by asserting through diplomatic channels that their delegation represented a Soviet social delegation rather than a government.

At the Fourth Latin American Students' Congress held in Havana in July and August 1966, Castro reiterated, and the Congress voted, support for armed struggle. The Cuban leader reaffirmed his position on January 2, 1967, on the eighth anniversary of the victory of his revolution, saying that Guevara ". . . will arise from his ashes like a phoenix." He concluded with a declaration that 1967 was the year of Heroic Vietnam. This was in reference to the parallel he saw between Cuba and Vietnam: both fighting the same imperialistic power, and both endangered by the Soviet policy of peaceful coexistence with that power. The open break with the orthodox parties appeared in Castro's speech of March 13, 1967, which focused on the dissident party in Venezuela.

The Latin American Solidarity Organization (La Organización Latino-Americana de Solidaridad–OLAS) met for the first time in Havana during July and August 1967. Most of the delegations represented Castroite and other splinter groups in Latin America. In spite of a strong defense of the Moscow line by the orthodox Communist parties the resolutions of the OLAS committees endorsed the principle of Revolutionary War. The list of countries believed ripe for armed struggle was broadened to include Mexico, formerly excluded because of its continuing diplomatic relations with Cuba. The necessity for middle-class cooperation stressed by the orthodox parties was rejected, and the conference called for sending volunteers and material aid to insurgents. Havana was designated as the permanent headquarters for OLAS and Cuba was made chairman of the permanent committee of the organization with representatives of Bolivia, Colombia, Guatemala, Peru, Trinidad-Tobago, Uruguay, and Venezuela, as members. And perhaps most arresting, the Soviet bloc was denounced for economic and technical assistance to reactionary governments.

The conference was a success for Castroism, strengthening Castro's claim of being the authentic leader of communism in the New World, and giving renewed ideological and propaganda impetus to the partisans of Revolutionary War. Castro was rebuked by the Russians for imposing his views on other communist parties and interfering in their activities, but it seems unlikely that he will bow to their will. The Castroite split with the orthodox parties in Latin America might well prove to be a setback for the general advance of communism in the Western Hemisphere.

Against this background it is clear that since 1959 efforts by the Communist countries to gain a political foothold in Latin America, other than Cuba and, to an indeterminate degree in Chile, have not met with any significant successes. Local parties suffered loss in prestige as a result of the missiles crisis in October 1962, and also from the disquieting effects of the Sino-Soviet quarrel. The withdrawal of Soviet missiles from Cuba disclosed the unreliability of the Soviet Union's defensive shield for a Communist state, and in the aftermath of this episode Cuba and the Soviet Union drew apart ideologically and in tactics.

The lack of spectacular revolutionary success in recent years, however, should not be interpreted as a sign that either internal or external Communist elements in Latin America have abandoned their efforts to dislocate and overthrow the existing order, or are unable to do so. With party membership estimated to total between 300,000-350,000 (1970), the Communist potential in Latin America is great. The operations of local Communist parties have been circumscribed by the spread of military regimes in Latin American countries which commonly justify their seizure of power by asserting that a leftist or radical element poses a threat to the institutions of constitutional government. It is obvious that conditions in many areas there continue to favor the growth of Communist influence and are therefore vulnerable to Communist penetration by agents of Sino-Soviet states and parties.

In following a policy of peaceful penetration through established diplomatic channels, which would later prove to be more apparent than real, the Soviet Union in the later 1960s established diplomatic relations with several Latin American countries, trade was expanded and, in some instances, economic aid activities began. By mid 1971 there were Soviet embassies in eleven Latin American countries: Mexico, Cuba, and all the nations of South America except Paraguay and Guyana. The USSR also had diplomatic ties with Costa Rica and Guyana. In comparison, while a total of fifteen Latin American nations recognized the Soviet Union between 1926 and 1946, by 1954 twelve nations (all except Argentina, Mexico, and Uruguay) had either severed relations with the USSR or denied that relations had ever existed. Soviet representation in the region continued to grow in 1970-1971, but at the same time, Soviet diplomacy was being rebuffed in some countries.

In March 1971 Mexico expelled five Soviet diplomats accused of supporting guerrilla training and related activities for Mexican nationals. Ecuador took the same action, expelling three Soviet embassy officials who were charged with interfering in domestic labor problems and urging labor leaders to stage a nationwide strike. Bolivia's military president, Hugo Banzer Suárez, in 1972 ordered the expulsion of 119 members of the Soviet Embassy, charging that they had supported leftist elements. Other Latin American countries complained that Soviet diplomatic officials were interfering in domestic affairs.

There was also growing evidence that the People's Republic of China had begun a diplomatic and trade drive in Latin America. Throughout the 1960s trade relations were maintained at a nominal level with Mexican cotton, Argentine wheat, Brazilian coffee, and several other commodities being shipped to mainland China, which in return exported light manufactures to the Latin American region. Peking seemingly lost interest there during the cultural revolution, and even the earlier cordial relationship with Cuba waned, but a new surge of Chinese interest appeared in 1971 when Chile and Peru reestablished relations with Peking; early in 1972 Argentina and Mexico took this step. This brought to six the number of Western Hemisphere nations in diplomatic contact with mainland China–Cuba, which had been there throughout the 1960s, and Canada, which had resumed relations in 1970. Guyana was expected to follow in 1972. The admission of the People's Republic of China to the United Nations and, President Nixon's efforts to normalize relations with Peking may have influenced the growing acceptance of that government in Latin America.

Lending credence to the belief that Peking had suspended questions about ideology in order to win friends in Latin America was its trade agreement with the rightist, military regime in Brazil. Peking's support for the 200-mile limit of territorial waters, which the United States opposed and the Soviet Union largely ignored, was seen as further evidence of the effort to broaden diplomatic contact in the hemisphere. There were reports that negotiations were under way for additional Chinese diplomatic ties with Bolivia, Brazil, Ecuador, and Uruguay.

This increased exchange of representation resulted both from mutual interest in expanded trade and new export markets, and from the desire of the Latin American nations to express their independence through the conduct of diplomatic relations with all countries. Latin America enjoyed a highly favorable trade balance with Communist-country commercial partners, and in general hoped to maintain and increase existing markets by normalizing relations with Communist countries.

Since November 1970, when Dr. Salvadore Allende, a Marxist, assumed the presidential office, Chile had been the most active Latin American country in expanding relations with the Communist world. Soon after taking office, he announced resumption of full diplomatic relations with Cuba. Relations had been broken in August 1964 following the resolution adopted by the Ninth Consultative Meeting of American Foreign Ministers. The two countries exchanged ambassadorial agreements in December 1970. Diplomatic relations with Communist China also became effective in December 1970. The joint communique stated that "The Chinese Government reaffirms that Taiwan is an inalienable part of the territory of People's Republic of China. The Chilean government takes note of this declaration of the Chinese Government." A

Communist Chinese trade office and news agency bureau had been operating in Santiago since 1965. Chile also gave permission to North Korea and North Vietnam to establish trade missions in Santiago, and on March 16, 1971, agreed to establish diplomatic relations with East Germany.

Bolivia also moved to establish relations with Communist countries. Continuing a policy begun by the late President Barrientos, Bolivia resumed relations with Czechoslovakia as well as establishing them with Rumania, Poland, and Bulgaria. The first Soviet ambassador presented his credentials April 1, 1970. Costa Rica, after announcing an agreement with the Soviet Union to exchange ambassadors (later suspended), established diplomatic relations with Rumania, Hungary, and Bulgaria during 1970.

Yugoslavia maintained some form of diplomatic relations with all the South American nations, while four other East European nations—Czechoslovakia, Hungary, Poland, and Rumania—maintained diplomatic ties with all the South American countries except Paraguay and Guyana. In contrast with the Soviet Union, these countries had often been unable to send ambassadors to all the Latin American nations with which they recently restored relations. Commonly the Eastern European countries accredited a nonresident ambassador in addition to whatever trade mission or consular personnel already resided in the Latin American nation involved.

Mexico's Communist party was formed in 1919 and recognized by the Comintern in 1920, but Marxian ideology had little attraction for the Mexican revolutionary leaders whose revolution had been in progress for a decade. In its early years the movement was led by foreign organizers, the Russian Michael Borodin, the Americans Bertram E. Wolfe and Linn Gale, the East Indian Mahabendra Nath Roy, the Japanese Sen Katayama, and others. During the Obregón administration two revolutionary generals, Francisco Múgica and Felipe Carrillo Puerto, joined the party, and their influence, though exerted briefly, enabled the Communists to make some headway among the peasantry. These gains were rolled back in the 1930s when peasant organizations were brought under government controls.[6]

It was at that point that the party's leadership fell into the hands of the Mexican artists, Diego Rivera, David Alfaro Siquieros, Xavier Guerrero, and others, who did much to mold the character of the party in Mexico. They were replaced by more politically astute and disciplined leaders of lower class origins, but Siquieros, expelled in 1940, was called upon to lead the party in 1960.

Mexican labor was organized during the Cárdenas regime by the leftist labor leader Vicente Lombardo Toledano, a brilliant writer and orator of the upper class, who became intellectually intrigued with Marxian communism and the Soviet experiment. In February 1936, a new federation of labor unions was organized by Toledano and named *Confederacíon de Trabajadores Mexicanos* (CTM), a powerful, militant organization. Lombardo Toledano was also elected president of the *Confederacíon de Trabajadores de la América Latina* (CTAL), a new inter-American labor organization founded in Mexico in 1938.

The policy of the CTAL throughout World War II followed the Soviet

[6] Mexico-Soviet diplomatic relations were established Aug. 6, 1924; broken off Jan. 26, 1930; restored Feb. 12, 1942.

lead, and by 1944 most of the top posts were held by Communists. So long as this policy coincided with that of the outlawing of the Communist party by several South American states the position of Toledano as a fellow traveler became difficult. He was ousted from the head of CTM, and in 1949 he was refused a visa by the United States when he wished to attend a CIO congress. President Miguel Alemán, elected in 1946, took a strong stand against communism, and in 1951 the Communist party was denied official recognition by the Mexican government.

This did not affect Toledano, who now headed the so-called Popular Party and who still professed Marxian communism. In the spring of 1952 Toledano was regarded as a serious coalition candidate for the presidency, but the Communists were unable to obtain the 35,000 signatures required for legal registration. When Toledano, as a candidate of the Popular Party supported by the Communists gained only 2 percent of the entire vote, the Communists turned toward General Herique Guzmán of the Federation of Peoples Party as their future standard bearer. Toledano, the on and off Communist, retaliated by stating in his newspaper, *El Popular,* that he would henceforth support President Ruíz Cortines and his policy of honesty and morality in government. In 1960 the Communist party became the Socialist Peoples Party, and though Marxist, was not regarded as part of the international Communist political framework.

The Mexican government continued to dominate the Communist movement by policies and strategems that ranged from adopting segments of the party's platforms, to subsidies and to harsh repression; however the party was given legal status in Mexico. Party membership in 1971 was estimated to be five thousand, but it was weakened by factionalism and poor leadership. The Mexican peasantry, though not sharing equitably the benefits of the Revolution, remained loyal to the government and was rarely attracted to Castroite insurgency. Evidence of minor Castroite activity, occurring 1965-1967 and urban violence in the early 1970s, was vigorously suppressed by the government. Although Mexico, following an independent foreign policy as a matter of tradition and self-respect, continued to maintain diplomatic relations with Cuba, the government exercised close control over Communist activities and the movement of persons to Cuba.

Apart from its spectacular thrust in Guatemala in the early 1950s communism in the other Central American countries—El Salvador, Honduras, Nicaragua, Costa Rica—and Panama, has exerted only a minor influence. However, serious economic and social conditions and repressive political regimes, with the exception of Costa Rica, provide a fertile field for Communist operations. Ranging in size from less than two hundred to perhaps two thousand, all of the parties in the area were outlawed after World War II. The degrees of suppression vary and some of the parties are weakened by factionalism.

The Communist Party of El Salvador, founded by organizers from Mexico and Guatemala in 1925, was responsible for the first major Communist effort in the Western Hemisphere to gain power by the direct use of force.[7] Guided by

[7] El Salvador has never maintained diplomatic relations with the Soviet Union.

directions from the Comintern the party, joined by members from Guatemala and Honduras, launched a revolt in 1932 that in the space of a few days claimed 25,000 lives. The revolt, brutally crushed by government forces, brought about the severe repression of Communists in Guatemala and Honduras. Banned in 1952, the party emerged in 1960 to incite an agrarian revolution, but was quickly suppressed by the anti-Communist administration. Communist party membership was estimated at about one hundred in 1971.

Deriving from a Socialist labor organization formed in 1920, the Communist party of Guatemala was organized and accepted by the Comintern in 1924. It was outlawed during the Ubico dictatorship (1930-1944), the leaders being imprisoned or exiled, but reappeared after the revolution of 1944, with restoration of a more liberal political climate. At first infiltrating labor movements and noncommunist reformist parties, it emerged in 1950 as the Communist party of Guatemala. In the following year, after merging with another Communist organization, it took the name of Guatemalan Labor Party, and was given legal status in 1952. The party took its orders from Moscow and served as the agent of Soviet propaganda and subversive action in the Caribbean. The strength of communism in Guatemala by 1954 sprang from the fact that party members had important posts in the labor movement, the educational system, to some degree in peasant organizations and non-Communist parties, and had been supported openly by the past two administrations. Under the presidency of Dr. Juan José Arévalo, Communists were given many vital positions in the government. At the end of his term in 1951 he supported a coalition of leftists, including Communists, which succeeded in electing Colonel Jacobo Arbenz, Colonel Francisco J. Arana was ambushed and murdered during the campaign.[8]

The Arbenz administration, although supported by a coalition of parties in which the Communists were a decided minority, was anti-United States and hostile to business organizations controlled by Americans. The disproportionate influence wielded by the Communists in the government—they had only four of the fifty-one seats controlled by the government—was due to their close organization, concrete program, and ruthless tactics. Each of the four Communist Congressmen had an important committee chairmanship, and Communists occupied important positions in the Ministry of Foreign Affairs and in the Labor department and practically controlled the agencies for social security and land reform. They dominated both the press and the radio and they held key positions in the labor union movement. Guatemala had consequently become the base of operations for the Soviet Union's plan to divide and conquer the Western Hemisphere.

When it appeared that communism would indeed triumph in Guatemala the United States backed invasion of Colonel Castillo Armas' forces overthrew the Arbenz regime and outlawed all Communist organizations. Following the assassination of President Castillo Armas the political leadership of his military successors, Presidents Ydígoras and Peralta, could not arrive at a general agreement of the need for reforms, leaving the goals of the revolution of 1944

[8] Guatemala established diplomatic relations with the Soviet Union April 19, 1945; they were broken off July, 1954.

largely unattained. Presidents Julio Cesar Méndez Montenegro and Carlos Arana, strongly supported by the United States, achieved some gains in agrarian reform, housing, and education, but the impoverished state of the masses continues to invite Communist penetration.

Driven from the political arena, the Communists, numbering about one thousand, supported guerrilla and terrorist operations against the government beginning in 1960. In 1966 the army, quickly joined by several anti-Communist bands, began an intensive and effective campaign against both guerrillas and terrorists. Insurgency seems to be contained, for the movement has failed to gain sufficient peasant support to insure its survival against the counter-insurgency program. However, the acts of Communist-inspired terrorism continue, particularly in the cities. Among the victims assassinated by the terrorists in Guatemala in 1968 were three United States government officials: Ambassador John Gordon Mein, and two military advisers to the Guatemalan armed forces, Colonel John Weber, and Lt. Commander Ernest A. Munro. Count Karl von Spreti, the Ambassador from West Germany, was abducted and slain by Guatemalan terrorists in 1970. Membership in Communist organizations was estimated to total 750 in 1971.

Honduras had a Communist organization of some three hundred members in 1971. The original party mechanism, formed in 1928, was destroyed in 1932 and remained underground during the long dictatorship of Tiburcio Carias. Castroite support injected new vigor into the present party, dating from 1958, but it has not succeeded in developing an efficient political apparatus.[9]

Communism in Nicaragua has the distinction of offering a party organization, founded in the popular front era, which has never carried the Communist name. Known as the Socialist Party of Nicaragua, it cooperated with the Somoza dictatorship until about 1947 when the alliance was broken over issues of political influence and spoils, and the party was driven underground. It claimed about three hundred members in 1963; however, in August, 1967, a resurgence of guerrilla activity appeared under the banner of the Sandinista Liberation Front (FSLN). Founded in Havana in 1961, it is closely tied to the Castro regime. The leader, Carlos Fonseca Amador, a former university student, trained in the Soviet Union, was a participant in the abortive invasion of Nicaragua in 1959. The FSLN has carried out an urban terrorist campaign, as well as guerrilla operations, but is being contained by the efficient National Guard.[10] Communist membership had dropped to about one hundred by 1971.

Communism has made little progress in Costa Rica since the party was founded in 1929 owing to relative prosperity, political stability, and a more equitable distribution of land ownership than is found in most Latin American countries. Under the leadership of Manuel Mora the party maintained a legal status in the 1930s gaining respect as a champion of the lower class, and because

[9] Honduras has never established diplomatic relations with the Soviet Union.
[10] Nicaraguan-Soviet Union diplomatic relations were established Dec. 12, 1944, but representatives were not exchanged.

of its moderation. The latter role was cast aside in the 1940s, however, and it was banned in 1948 after being defeated by the forces of José Figueres. Adopting the name of Popular Vanguard Party, and comprising about one thousand members, it is only allowed to support the campaigns of other parties. For the reasons mentioned, and the competition offered by the Liberation Party of José Figueres, a leftist noncommunist organization, it has little popular support or following.[11]

The Communist movement in Panama produced three parties—the Laborista Party (1926-1930), the Communist Party of Panama (1930-1944), and the Party of the People (Since 1944), but the party membership did not rise above five hundred, and is currently estimated at 150 persons. Although both party and followers are small in number, it had a significant role in the anti-United States uprisings in 1959 and 1964, and echoes extreme nationalist sentiments on the delicate canal-treaty issue.[12]

The Communist party of Cuba gained the distinction of being the first in Latin America to establish a working alliance with the government and to form a mass organization; it must be regarded as the most successful in the Western Hemisphere. Established in 1925 by labor groups and university students, it entered the Comintern in 1928.[13] The party first gained importance during the Machado dictatorship and helped to bring about its overthrow. Initially opposing Colonel Batista, the party quickly reached an accord with the strong man, giving him support in exchange for legal status and freedom to organize the labor movement. Owing to this arrangement the party suffered less than most of its counterparts from the repercussions of the Hitler-Stalin Pact of 1939. After supporting Batista's election in 1940, the Communist party was retained as a part of the coalition during his administration. Party membership expanded and six Communists were included in the Constituent Assembly which framed the Constitution of 1940. Juan Marinello, a leading Communist, was appointed to Batista's cabinet in 1943, and when he resigned to run for the Senate in 1944 another Communist replaced him in the cabinet position. They were the first Communists to hold such a high level post in any Latin American nation. Blas Roca, who became secretary-general of the party in 1934, is given a large measure of credit for the party's success.

During World War II the Communists, known as the Popular Socialist Party, cooperated with the government, and in the 1944 elections more than 120,000 votes cast for Communist candidates secured the election of three senators and seven deputies. By then constituting a major political force, the Communists supported the election of President Grau San Martín and were rewarded with half the critical labor posts and the vice-presidency of the Senate. Going into opposition against the administration in 1947, the party's power was

[11] Costa Rica established diplomatic relations with the Soviet Union May 8, 1944, but representatives were not exchanged. The government of Costa Rica declared in 1970 that relations with the USSR had never been broken, which cleared the way for an exchange of missions.

[12] Diplomatic relations have never existed between Panama and the Soviet Union.

[13] Cuban-Soviet Union diplomatic relations were established Oct. 14, 1942; broken off April 3, 1952; restored May 8, 1960.

severely curtailed by the Prío regime after 1948. When in 1952 ex-president Batista seized control of the government by a coup d'état the party had been reduced to a minor political status.

On July 26, 1953, an abortive uprising against the Batista regime took place which authorities claimed was at least partially instigated by the Communists. The government immediately suspended their newspaper, *Hoy*, and declared that its editor and other Communist leaders would be seized and tried. However, the Communist leaders were able to escape, although all of their publications and literature were seized and destroyed. On November 10, 1953, the government outlawed the *Partido Socialista Popular* with all of its affiliations such as the Socialist Youth Organization and the Federation of Democratic Women, a part of the Soviet world "peace" movement. In 1958 the party, illegal for five years and suffering setbacks, could count about twelve thousand active members, most of whom were militants.

Although Fidel Castro's connection with international communism prior to his seizure of power in Cuba in January 1959, had not been firmly established, he did receive support from the Cuban Communists toward the close of his guerrilla campaign. The PSP, renamed the Cuban Communist Party (PCC), became the vehicle for the regime's consolidation of power, but it was recast in the image of Fidel Castro, who replaced the old guard leaders in the governing apparatus with Fidelistas of proved loyalty to himself. The party was absorbed into the Integrated Revolutionary Organizations in 1961, a mass political grouping designed to serve as a transitional stage in the formation of the United Party of the Socialist Revolution, the latter being established in 1963. Under Premier Castro's leadership the party[14] had grown to an estimated 120-130,000 members in 1971. That Castro has not allowed the party to become a tool of Moscow's foreign policy is clear from his attack on the Soviet line of peaceful coexistence and from the trial and sentencing of some forty Communist party members for "anti-party" (pro-Soviet line) activities.

In spite of lamentable social and economic conditions which prevail on the island of Hispaniola communism has made little headway owing to the military efficiency of the dictatorial regimes which have generally ruled there. The Haitian Communist party, founded by two native intellectuals in 1930, enjoyed a few months of legal political activity before the leaders were exiled. The party was officially outlawed in 1936 and went into eclipse for a decade. The Popular Socialist party appeared in 1946, but was proscribed three years later. Two minor groups, the Party of Popular Accord and the People's National Liberation Party, emerged in 1961, both seeking recognition from the Soviet Union. There later developed two Castroite parties, and Cuba has given them special attention in its propaganda campaign. The combined membership was believed to total some five hundred members in 1970.[15]

Loyalist refugees from the civil war in Spain who went to the Dominican Republic after 1939 inspired the formation of a Communist party there in 1942.

[14] In Oct. 1965 the party was reconstituted along orthodox Communist lines and under the original name as the PCC.
[15] Haiti has never entered into diplomatic relations with the USSR.

It was banned in 1946, the leaders being exiled, but it had never been granted legal status by the Trujillo dictatorship. Reconsidering that the Communists might be exploited in his interests, Trujillo invited the leaders to return and form a legal party. Assuming the name of Popular Socialist Party, the Communists made a poor showing in the national elections. In 1947 the party was again outlawed, the leaders fleeing abroad or going underground. After Trujillo's assassination in May 1961, the party leaders returned; but, faced by anti-Communist provisional governments, they established front organizations and sought to infiltrate the labor movement and non-Communist political parties. It was partially the fear of communism that led the army generals in 1963 to overthrow the seemingly left-leaning Bosch administration, and it was even more directly a factor precipitating United States military intervention in 1965. Communist or pro-Communist leaders, both military and civilian, were exiled, and the party organization was further confused by the appearance of two separate Castroite parties. Communist party membership was estimated to have reached 1,680 in 1971.[16]

The Venezuelan Communist Party (PCV), established in 1931, retained its identity through the years, surviving alternating governmental suppression, internal fragmentation, and competing front parties. Gaining legal status in 1958 after the overthrow of the Pérez Jiménez dictatorship, to which it had contributed, the party polled some 160,000 votes in the national elections of that year, placing seven of its members in the National Congress and two in the Senate. After being reduced to a minor role by some clever political strategy of President Betancourt (1959-1964), the PCV supported urban violence in 1960, and guerrilla warfare and terrorism in 1962. The PCV had an estimated eight thousand members in 1971, chiefly among students, workers, intellectuals, and journalists, but none of these did it dominate. Despite astute leadership the PCV's effectiveness is lessened by factionalism and the competition of effective, democratic reformist parties, especially the Democratic Action Party of Presidents Betancourt and Leoni.[17]

Insurgency became a major problem for the government in 1962 as militant groups received increasing material support from Castro's Cuba. The National Liberation Front (FLN), an alliance of the Communist party and the Movement of the Revolutionary Left (MIR, a dissident element of the Democratic Action Party), and the Armed Forces of National Liberation (FALN), combined rural insurgency with an urban terrorism campaign. The Betancourt government turned these thrusts successfully by effective police and military action, and a far reaching reform program. The Venezuelan military made notable progress against the guerrillas from 1965 on under the Leoni administration, although a new wave of terrorism was launched at the end of 1966 and in 1967. Premier Castro has invested more of his prestige in the FLN-FALN than any other movement in Latin America. Insurgency was largely

[16]The Dominican Republic established diplomatic relations with the Soviet Union in 1945, but representatives were never exchanged.

[17]Venezuela established diplomatic relations with the Soviet Union in 1945, only to be broken off in 1952. Diplomatic relations were resumed in April 1970.

brought under control in 1967. Moreover, further cleavage developed in the party's ranks as the PCV came under the control of the "peaceful road" faction and broke with the "hard line" faction of the FALN.

Communist members of the Socialist Revolutionary Party of Colombia, guided by the decisions of the Sixth Congress of the Comintern (1928), formed the Communist Party in 1930. It changed its name to Social Democratic Party in 1944, but three years later reassumed the title of Communist Party of Colombia (PCC), which it has retained. Its legal status was interrupted on only one occasion, 1956-1957. Dominated by an elite that has generally failed to modernize political institutions or implement social-economic reforms, Colombia has made little progress in eliminating conditions which are productive of revolutionary violence. The "violence" which claimed the lives of some 200,000 people in the period 1946-1957 was strongly, though not solely, related to these circumstances.[18]

The PCC did not initiate this insurgency, but became an important participant. Whatever ideological or partisan political meaning the insurgency movement had initially was lost as it degenerated into banditry. Toward the close of the 1950s the movement slackened and a decade later the Liberal and Conservative bandit-guerrillas having been seriously reduced by attrition, the Communist insurgents stood out as never before. There are two Communist guerrilla forces, one being the Revolutionary Armed forces of Colombia (FARC), estimated to have 250 members, which is tied to the Moscow-aligned PCC. Another major guerrilla force is the National Liberation Army (ELN), which began guerrilla operations in 1965 and is believed to have 80-150 members. Made up of roaming guerrillas, and built around a nucleus of former university students and professional revolutionaries, it does not control any given area.

The PCC itself is a legal but apathetic party with a membership of about nine thousand in 1971. Its main influence has been in a union confederation and among university students. In 1963 a segment broke away to form the PPC-MC, a party of two thousand members who follow the Peking line; it commands no guerrilla forces. Being reliably pro-Moscow, the PCC favors trade and diplomatic relations between the USSR and Colombia. In following the Soviet line, PCC-Cuban relations have declined. The PCC was identified with the so-called republics of the 1940s, isolated, autonomous, peasant-ruled areas, with their own militia, but their connection with the Communists was never secure. Overcome by government forces in the 1960s the survivors enlisted in the guerrilla units.

Being governed by terms of the Liberal-Conservative coalition government known as the National Front, the PCC cannot run candidates; however, several Communists have seats in the National Congress, having run as candidates of the Liberal Revolutionary Movement (MRC), a splinter of the Liberal Party

The Colombian army, having worked at counter-insurgency for more than a decade, appears capable of thwarting the insurgents' goals. Nevertheless,

[18] Diplomatic relations between Colombia and the Soviet Union were established in 1935; broken off in 1948; restored in Jan. 1968.

guerrilla activity will probably continue until the main causes of social unrest are improved.

The Communist Party of Ecuador was illegal from its inception in 1931 until 1944, and was outlawed again in 1963. In spite of its tenuous hold the party gained the distinction of being the only one in Latin America to place a member in the executive branch of government (Gustavo Bercerra for three days in 1944), and it is one of few to place a representative in the national cabinet. The party declined in political influence after 1945 as it splintered into four groups, and could count only 1,250 members in 1971. The orthodox party publicly repudiated the OLAS call for violence, adhering to the Moscow line. The Revolutionary Union of Ecuadorian Youth, Communist dominated and Cuban supported, sent a 250-man force to the coastal area in 1962. After crushing this movement, the government carried out reprisals against the universities and outlawed the Communist Party.[19]

Foreshadowed by a Socialist party, the Peruvian Communist Party was founded in 1930 and outlawed in the years 1930-1945, and again after 1948. Its membership reached thirty thousand in 1957, but declined to 3,200 in 1971. Although the party gained influence in labor and student groups, it has not achieved a significant following among the peasant masses. Competition offered by Victor Haya de la Torre's radical noncommunist APRA (American Popular Revolutionary Alliance) has undercut its appeal; however, Communist support has been accepted by conservative elements who regard the *Apristas* as more dangerous foes than the Communists. The Peruvian Communist movement is weakened by internal factionalism, for separate Chinese and Soviet-oriented parties date from 1964. The pro-Castro Movement of the Revolutionary Left sponsored an unsuccessful guerrilla war in 1965-1966 bringing down severe government repression upon all Communists.[20]

The Bolivian Party of the Revolutionary Left which appeared in 1940 was not, according to its leaders, a Communist party, but its ranks included Communist elements among which were Trotskyists. Although illegal in the periods 1940-1943 and 1946-1952, an orthodox party was founded in 1949 by members who broke away from the parent organization. This created two competing parties each seeking to penetrate the National Revolutionary Movement, the party controlling the government since 1952. Under the aegis of this party Bolivia has had agrarian reform, has nationalized the major tin mines, and had developed a political system which, though unstable, represents the interests of the peasants, miners, and urban population. In short, it anticipated reforms called for by the Alliance for Progress.[21]

In spite of the grinding poverty of 70 percent of the population, which theoretically should be attracted to communism, the Communist movement remains weak. This can be explained by the non-Communist revolutionary reforms carried out by the ruling party, tending to foreclose its appeal, and the

[19] Ecuador established diplomatic relations with the Soviet Union on June 16, 1945, but representatives were not exchanged until 1970.
[20] Peru and the Soviet Union have never established diplomatic relations.
[21] Bolivia-USSR diplomatic relations were established April 18, 1945; representatives were not exchanged, diplomatic relations were resumed in Sept. 1969.

fact that the movement itself is highly fractionalized. The orthodox Party (PCB), having little appeal to the peasants, has drawn mainly from the urban and mining areas for its members. A pro-Chinese faction appeared in 1964 and withdrew from the PCB in 1965. The Moscow-line PCB has about four thousand members, and the Peking group about five hundred. The pro-Chinese Revolutionary Workers Party, an old party of Trotskyists, also weakened by internal conflicts, has an estimated one thousand members. The National Liberation Front, a Marxist party formed in 1964, which draws upon students, miners, and urban workers, has been allied with the PCB. It polled 33,000 votes, or 3 percent of the total, in the 1966 elections.

The Communist parties have not won the allegiance of the peasants thanks to the government's progressive reform program; however, the miners' union is subject to Communist influences. The nucleus of the guerrilla forces that emerged in Bolivia in March 1967 was not openly connected with the Bolivian Communists, although it is probable that some endorsed it. Major Ernesto "Ché" Guevara, who had disappeared from public view in Cuba two years previously, planned the operation. After the government had suffered initial setbacks against the guerrillas, the United States sent a counter-insurgency training team and speeded up supplies already in the military assistance program. Government forces won a series of victories beginning in August 1967, which culminated in the death of Guevara on October 9.

Communist organization in Argentina began in January 1918, with the formation of the International Socialist Party. Acting upon directions from the Comintern, the title was changed to Communist party of Argentina, a name it would retain. Although regarded by the Comintern as its major success in Latin America in the 1920s, the party failed to achieve political influence or to exert significant force in the ranks of labor. The party was outlawed following the revolt which overthrew the government in 1930, but it made minor gains during the era of the Popular Front when it cooperated with anti-Fascist forces. The party was again isolated in the period of the Hitler-Stalin Pact, but reappeared, taking a strong pro-Allied stand after Hitler's armies invaded the Soviet Union. When the armed forces seized power in June 1943, the party leaders were jailed or exiled and the Communist press was closed. The Perón regime, which followed, became interested in organizing and using labor for its own purposes, which weakened the Communist position in the labor movement. At this time, the labor organizations had split over whether to cooperate with communism. It was only natural that Perón should take the anti-Communist side.

Although the Soviet Union had opposed the entrance of Argentina into the United Nations, Perón saw the possibility of using the Soviets as a counter-balance against the United States. The Communist party was legalized in December 1945, and a Soviet trade mission was enthusiastically received in 1946. Free to campaign for the first time in twenty years, Communist Party membership rose to about thirty thousand, and it polled over 150,000 votes in the national elections of February 1946. In 1947 the party pledged in its convention support of Perón in all policies consistent with the Communist program. U.S. Ambassador George Messersmith was particularly singled out for Communist abuse, and in November 1948 his successor, James Bruce declared that President Truman was much concerned about communism in Latin America, particularly in Argentina.[22]

It was not long after this that the Communist leaders in Argentina split and the Stalinist wing headed by Victorio Codovilla and Rodolfo Ghiholdi became anti-Perón and the "dissident" group headed by Rodolfo Puiggros supported him. Moscow, however, ceased its propaganda campaign against Perón, and Communist infiltration into the General Confederation of Labor was stepped up. The split seemed to have been sufficiently patched up by the end of 1952 for the Communist Party to pledge full cooperation for Perón's second Five Year Plan. Nevertheless, in March of 1953 after Codovilla had revisited Moscow the Communist Party again turned against Perón and attacked his apparent intention to cooperate with the Eisenhower administration. It also violently opposed Perón's Union of Latin American Labor Organizations (ATLAS). Although ATLAS subsequently took a strong stand against the Communist-dominated World Federation of Trade Unions and the pro-Communist Confederation of Labor of Latin America led by Vincente Lombardo Toledano of Mexico, the Argentine government did not object to signing a mutually advantageous trade treaty with Moscow on August 5, 1953.

After the fall of the Perón regime in 1955 Communist party membership rose from thirty thousand to ninety thousand, and in the national elections of 1957-1958 the party's candidates polled more than 200,000 votes; however, no Communists were elected to the national or provincial legislatures. The party's activities were restricted by the Frondizi administration in 1959-1960 owing to its abusive propaganda attacks on the government and its role in fomenting radical labor student movements. The government of José María Guido banned the party in 1963, but under the Arturo Illia administration the party was allowed to operate within limits. The party was "abolished" with the advent of the military government of General Ongania in 1966. With a membership of sixty thousand the Argentine Communist party is the largest non-ruling party in Latin America, but it is one that exerts a minor influence in the nation's affairs.

The Communist Party of Brazil gained its initial impetus from anarchism rather than socialism. Formal organization of the Communist Party of Brazil occurred in March 1922, with the blessings of the Comintern. In the following July a short-lived military revolt provoked the government to declare a state of siege and the party was driven underground. It remained illegal until 1945, but nevertheless succeeded in penetrating industrial and rural labor organizations, and engaging in electoral activities. The party's position was weakened by ideological and tactical differences among its members, and this divisiveness mounted after Getulio Vargas came to power in 1930.[23]

The story of communism in Brazil is closely interwined with the career of Luiz Carlos Prestes. The son of an army officer, Prestes graduated from Brazil's military academy in 1918, and soon thereafter became a champion of revolutionary causes. He gained nationwide fame when he formed a rebel

[22] Argentine-Soviet Union diplomatic relations were established June 6, 1946.
[23] Diplomatic relations between Brazil and the Soviet Union were established April 2, 1945; broken off Oct. 20, 1947; restored Nov. 23, 1961.

column and eluded government forces for two years in the frontier regions. Failing in this effort to inspire a popular revolution, he became a convert to communism. Prestes went to the Soviet Union in 1931 where he was prepared to assume the leadership of the Communist movement in Brazil. Returning to Brazil in 1935, he staged an unsuccessful revolt against the government, and Vargas retaliated by proclaiming "a state of grave internal commotion" and throwing thousands of suspected Communists into jail. Prestes was given a sentence of seventeen years at hard labor and later thirty more years were added to his sentence as accessory to a murder committed while he was in prison.

Prestes remained in jail and the Communist party in abeyance until 1945 when Vargas pardoned Prestes and the latter reciprocated by urging that Vargas remain in power although he was not a candidate for election. Vargas next legalized the Communist party, which campaigned strenuously for local, state, and congressional offices. Fearing that Vargas would stage a coup d'etat to retain power, the army acted, forced him to retire and turned the power over to the president of the Supreme Court. The presidential elections were now held peacefully and to the surprise of all, Vargas was elected to the Senate and his candidate, General Dutra, became president. But even more incredible, the Communist Party received more than 600,000 popular votes, about 10 percent of the total, obtained fourteen seats in the Chamber of Deputies, and elected Luiz Carlos Prestes to the Senate.

Stimulated by their success, the Communists put on a strong campaign in the 1947 elections and did even better than before. They obtained about 800,000 votes, elected two Senators, fourteen Deputies, sixty-odd members of the state legislatures, and the largest number of any one party in the city council of Rio de Janeiro. Senator Prestes speedily made known his opposition to the United States policy towards China and Yugoslavia, and he demanded the immediate return of the air bases to Brazil. However, when, having been asked in a Senate debate which country he would support in a war between Russia and Brazil, he replied "Russia," his inexcusable blunder ruined his cause.

President Dutra about a month after the elections, directed that a petition be presented to the Supreme Electoral Tribunal requesting the dissolution of the Communist Party on the grounds that it opposed democracy and was under the control of a foreign power. The Electoral Tribunal obliged and the president removed all Communists from appointive positions and closed all Communist clubs. It took several months to expel the Communists from Congress, but by the beginning of 1948 the Communist Party in Brazil was driven underground.

This did not mean that the Communist menace was obliterated. When Assistant Secretary of State Edward Miller and head of the Foreign Policy Planning Staff, George Kennan, visited Rio de Janeiro in March 1950, the Communist press was bitterly critical and the American Embassy was smeared with inscriptions "Go away spies Kennan and Miller." In 1951 the *Revista Militar,* the organ of the influential *Club Militar*, was suppressed because of its harsh attacks upon the United States and its constant publication of Red propaganda. That the army was infiltrated by Communist agents was revealed on April 30, 1952, when Minister of War General Ciro Cardosa warned all officers that the presence of Communists in the Brazilian army had been positively established and subsequently the prosecutor in charge of cases against officers accused of Communist activity was relieved of his post on suspicion of Communist sympathies.

President Vargas, who had again been inaugurated President in January 1951, quickly took a strong stand against communism. He ordered all state and territorial governors to prevent a meeting anywhere in Brazil of the Communist "Continental Peace Congress" which Communist leader Prestes had projected for March 1952. When early in 1953 it was discovered that a nationwide Communist drive was attempting to defeat the United States-Brazilian military aid treaty signed March 15, 1952, the police throughout the country were instructed to ban all "peace" meetings organized by the outlawed Communist party.

By the early 1960s the Brazilian Communist Party, formerly the leading Communist organization in the Western Hemisphere, with a membership of 150,000 had less than 30,000 members. Owing to the laxity of law enforcement it remained active in politics at all levels, supported ultra-nationalistic programs, and gave particular attention to the radical peasant movements in the impoverished Northeast led by Francisco Julião.

Communist-backed candidates in the October 1962 elections won important victories in some key states, but they made a poor showing in the country as a whole. It was also in 1962 that a division within the party occurred, and an organized rival party was formed. Luiz Carlos Prestes for the preceding eight years had led the PCB on a "soft line" and concentrated upon gaining a legal status for the party. In 1961 the Prestes element was accused of revisionism and rightist deviation. Expelled as divisionists, the dissident group took the name of Communist Party of Brazil and claimed to be the legtimate PCB. This group is actively preparing for guerrilla warfare.

Communism in Brazil appeared to register gains during the administrations of Janio Quadros (president from January to August 1961, and João Goulart (president from 1961 to 1964); but, under the strong anti-Communist military regimes which followed, the party's role has declined. Quadros resigned after eight months in office, and Goulart was removed from office by the armed forces, with broad civilian support, in March 1964. A variety of influences prompted his removal, but amongst the most important was the fear of infiltration of Communists and leftists into the government and labor unions.

From 1964 to 1966 a widening gulf developed between the civilians and the military, the latter, led by General Castello Branco, exercising progressively arbitrary authority and reducing public involvement in the electoral process. Marechal Artur da Costa e Silva, elected president of Brazil by a controlled Congress in October 1967, declared that the purpose of his administration would be to democratize and popularize the "revolution." The new president's declaration: "Brazil is with the West," was translated into a domestic policy of rigid suppression of Communist activities.

A thirty-man guerrilla band was captured by Brazilian forces in April 1967 in the state of Minas Gerais. The members were Brazilians, many of them being non-commissioned officers who had been dismissed after the overthrow of President Goulart in 1964. In general, the orthodox Communist Party of Brazil has suffered a decline in fortunes since 1964, and disclaims interest in pursuing the violent road to power. Its membership was estimated to have totaled 14,000 in 1971.

The Socialist Worker's Party of Chile, founded by Luis Emilio Recabarren in 1912, was affiliated with the Comintern in 1921, and in the following year adopted the name Communist Party of Chile (PCCh). In so doing it became the only Socialist party in Latin America to enter the international Communist movement as a complete unit. Emerging under favorable conditions—legal status, a large and well-organized labor movement controlled by leaders favorably disposed toward communism, and effective party organization and two representatives in the National Congress—the party was active for about five years. But during the rule of President Ibañez from 1927 to 1931, the party's influence was largely destroyed by the repressive measures of the government. Internal dissension among the socialists also weakened it, and the party made little progress until a reconciliation with the Socialist Party was effected in 1935, incident to the adoption of the popular front line.[24]

This initiated the first significant effort by a Communist party in Latin America to attain its objective through the parliamentary approach. Eudocio Ravines, a Peruvian, was sent to Chile by the Comintern in 1935 to convert the Popular Front into an all-out Communist movement. The procedure to be employed was unscrupulous opportunism.

Proceeding from Moscow to Santiago de Chile, the Communist group headed by Ravines, who worked under the pseudonym Jorge Montero, was able to infiltrate the Radical party, establish a Communist newspaper called *Popular Front,* and, supported by outstanding radicals like Gonzáles Videla, win a number of seats in the Senate and Assembly in the 1937 elections. In 1938 on a platform of "pan, techo y abrigo" (bread, roof, and overcoat) they joined with the radicals in the Popular Front and elected Aguirre Cerda president. As might have been expected, they then refused to cooperate with the Socialists in the Congress or to take any positions of responsibility, and encouraged strikes and lockouts. When in 1940 the Socialist Minister of Development went to the United States to obtain badly needed credits, the Communists denounced him as the tool of Yankee imperialism. This brought a split between the Communists and the other parties of left which, although it wrecked the Popular Front, did not prevent the Communists from gaining additional seats in the 1941 Congressional elections.

The Communists were forced to cease their campaign against the United States when Hitler attacked Russia and Stalin joined the democracies. The new president of Chile, Juan Antonio Ríos, elected in 1942, was a Radical but bitterly anti-Communist, and although the Communists voted for him as against former dictator Ibañez, he included none in his cabinet. The Communists after encouraging a series of strikes brought matters to a head in 1946 when they staged a great demonstration which threatened the police, who fired into their ranks killing and wounding many of the strikers. This action antagonized many of the more conservative elements and in the next election the Communists joined the Radicals and Socialists to help elect Gabriel Gonzáles Videla, a left-wing Radical. He rewarded them by giving them three cabinet posts.

President Gonzáles Videla quickly recognized his mistake. The Com-

[24] Diplomatic relations between Chile and the Soviet Union were established Dec. 11, 1944; broken off Oct. 21, 1947; restored Nov. 24, 1964.

munists attacked him immediately for not vetoing all Conservative measures and when he accepted the resignation of the three Communist cabinet members they encouraged strikes in the copper mines. Gonzáles Videla now took drastic action. He dismissed the Communist governors and provincial heads, obtained emergency powers from the Congress permitting him to seize all essential industries and run them by the army. He broke the strikes, severed diplomatic relations with the Soviet Union and its satellite states, and persuaded the Congress to pass a law making the Communist Party illegal.

The unions still had a Communist core, however, and in March 1952 President Gonzáles Videla was forced to proclaim a state of emergency in the northern provinces of Antofogasta and Tarapaca when the unions controlled by Communists refused to settle a strike which was paralyzing the vital nitrate fields. In July some 13.000 office workers in these same mines staged a walk-out to protest ratification of a United States-Chilean mutual security pact.

When General Carlos Ibañez was elected in September 1952, the Communists were encouraged because he had campaigned against the law making the Communist party illegal; nevertheless, in his first message to Congress on May 21, 1953, he attacked international Communism as conspiring against peace and endorsed the Organization of American States and the Inter-American Reciprocal Assistance Treaty. He implemented his statement by refusing to allow the sale of copper to Communist countries.

The Communist Party was illegal in the period 1948-1958, but nonetheless it took an active part in the coalition making up the leftist Popular Action Front and succeeded in placing several of its members in Congress on other parties' tickets. The Popular Action Front candidate, with Communist support, almost won the presidential election in 1958. And in the municipal elections of 1960 the Communists won almost 10 percent of the votes. Still surging foward they obtained almost 12 percent of the ballots in the congressional elections of 1961, electing four senators and sixteen congressmen. Having been allied for years with the Socialist Party and other elements in the Popular Action Front (FRAP), and achieving this measure of success, the Communists were confident of success in the 1964 presidential election. However, the FRAP candidate, a pro-Cuban Socialist senator, Salvador Allende Gossens, was defeated by Eduardo Frei Montalva of the reformist Christian Democratic Party, who carried nineteen of the nation's twenty-five provinces.

Recovering from this setback, Salvadore Allende was elected to the presidency in 1970, although by a narrow margin. His candidacy was supported by Popular Unity (UP), a coalition of Communist, Socialist, and Radical parties, and three smaller groups. In 1971 PCCh members held top posts in the ministries of Finance, Labor, and public Works, and the number two positions in the ministies of Interior and Mining. The PCCh was also active in labor circles, and Chile's largest labor organization, the Workers' Single Central (CUT), was headed by a PCCh congressman. PCCh membership in 1971 was estimated to total 45,000. The Communist Party of Chile is regarded as one of the most successful in Latin America, and its peaceful, popular front road to power has set it apart from most Communist movements in the hemisphere.

Communism in Uruguay, enjoying a legal status since the Communist Party (PCU) was founded in 1920, has long been influential in the nation's extensive labor movement and among university students. In recent years, however, the party has faced mounting problems both from the government and from extreme leftist groups such as the Castroite Revolutionary Movement of Uruguay (MRO), and the terrorist National Liberation Movement (MLN-Tupamaros). The PCU, with a membership of about 20,000 in 1971, follows the Moscow line on doctrinal matters, and has sought to follow a middle course on the question of violence.

The party suffered a setback from the Russo-German Pact of 1939-1940, but the outbreak of war between those powers reversed its fortunes. Party membership reached fifteen thousand in 1946 when it succeeded in electing one senator and five congressmen. Because of its singularly long-term diplomatic connection with the USSR, Uruguay served as a center of Communist propaganda and other operations in Latin America.[25]

The Communist Party of Paraguay, which was not founded until 1928, has been outlawed during most of its existence. Party membership attained a total of eight thousand in the mid 1940s but declined when the country came under the dictatorial rule of General Stroessner. The movement is composed of pro-Russian and Chinese units which agree on the violent road to power; however, most of the party members, numbering five thousand in 1971, reside in exile.[26]

The danger of international communism has been a topic of discussion and action in major inter-American forums from the outset of the cold war. The Inter-American Treaty of Reciprocal Assistance negotiated in 1947 at Petrópolis near Rio de Janeiro was, in the United States view, aimed at forestalling possible Soviet aggression; however, the Latin Americans were more concerned with the treaty as a regional entity under the United Nations than as a bulwark against communism. In 1948 at the Ninth International Conference of American States, in Bogotá, the disastrous riots involving Communists spurred the passage of a resolution entitled "The Preservation and Defense of Democracy in America," which declared that international communism was incompatible with the concept of American freedom.

As the reality of the cold war was unmasked with the Communist invasion of South Korea, the Fourth Meeting of Consultation of Ministers of Foreign Affairs, held in Washington during March-April, 1951, pledged unity against international communism in the Declaration of Washington. In view of the fact that the Latin American nations did not feel threatened by communism, and the Rio treaty did not provide for a commitment in that situation so remote from them, they did little more than send miscellaneous supplies. Only Colombia sent troops to join the United Nations forces fighting in Korea. In spite of this weak material response, the Latin American reaction did reveal significant and widespread unity on the ideological question.

By 1954 Washington was greatly alarmed by the developments in Guatemala, and at the Tenth Inter-American Conference which met at Caracas in that year sought strong concerted action against communism. Although faced

[25] Uruguayan-Soviet Union diplomatic relations were established Aug. 22, 1926; broken off Dec. 27, 1935; restored Jan. 27, 1943.
[26] Paraguay has never established diplomatic relations with the Soviet Union.

with general Latin American apathy on the issue of communism Secretary of State John Foster Dulles secured enough support for the passage of an anti-Communist resolution: the Declaration of Solidarity for the Preservation of Political Integrity of the American States against the Intervention of International Communism. Apart from not sensing any real Communist threat, the Latin Americans were more interested in massive economic aid from the United States, and perhaps more pointedly feared the implication of collective intervention spearheaded by the United States. It appears that the United States gained little, for the Caracas resolution was not invoked by the Ministers of Foreign Affairs in dealing later with the problem of communism in Cuba.

Fidel Castro's seizure of power in Cuba in January 1959 and his close identification with the Sino-Soviet Bloc, accompanied by Premier Nikita Khrushchev's announcement that Cuba was a Communist protectorate, led President Eisenhower to declare that the United States, in compliance with its treaty obligations under the OAS Charter and the Rio treaty would not permit the establishment in the Western Hemisphere of a regime dominated by international communism. Many of the Latin American governments professed to regard the Cuban problem as a bilateral issue involving only Cuba and the United States. Secretary of State Christian A. Herter's hope for obtaining a strong denunciation of Sino-Soviet intervention and of Cuba itself was not realized at the San José Meeting of Foreign Ministers in 1960. A compromise resolution was adopted in the Declaration of San José which condemned communism but did not mention Cuba.

The Declaration (1) condemned the intervention or threat of intervention by an extracontinental power in the American republics, and declared that the acceptance by any American state of such intervention or its threat, endangers American solidarity and security; (2) rejected the attempts of the Sino-Soviet powers to exploit the political, economic or social situation in any American state as threatening to hemisphere unity, peace, and security; (3) reaffirmed the principle of non-intervention by any American state in the internal or external affairs of the other American States, and declared that the inter-American system was "incompatible with any form of totalitarianism;" (4) proclaimed the obligation of all member states to submit to the discipline of that system and to comply with the provisions of the Charter of the OAS.

Having failed to enlist support within the OAS, the United States without consulting the OAS, moved unilaterally to forestall reportedly Cuban-backed revolts in Nicaragua, Costa Rica, and Guatemala in November 1960, despatching a naval task force to patrol the coasts. Washington was meanwhile becoming deeply involved in the ill-fated invasion of Cuba at the Bay of Pigs in April 1961. Although President Kennedy declared that ". . . any unilateral American intervention, in the absence of an external attack upon ourselves or an ally, would have been contrary to our traditions and to our international obligations . . ." anti-United States critics throughout Latin America were unwilling to concede that the United States had not violated the nonintervention principle to which it was bound by several inter-American treaties and declarations.

The Eighth Meeting of Consultation of Foreign Ministers, functioning as the Organ of Consultation under the Rio treaty, met at Punta del Este, Uruguay, January 22-31, 1962, and unanimously agreed on identifying the Castro government as a Communist regime aligned with the Sino-Soviet bloc. It was also verified "that the subversive offensive of Communist governments, their agents and the organizations which they control, has increased in intensity . . . the purpose of this offensive is the destruction of democratic institutions and the establishment of totalitarian dictatorships at the service of extracontinental powers . . ." The Eighth Meeting voted to exclude the "present government" of Cuba from participation from the inter-American system, but only by a scant two-thirds majority as required under the Rio treaty. Argentina, Bolivia, Brazil, Chile, Ecuador, and Mexico would not support the resolution, contending that the exclusion of a member state is not juridically possible without first amending the OAS Charter. The meeting also voted to "suspend immediately trade with Cuba in arms and implements or war of every kind," with Brazil, Chile, Ecuador, and Mexico abstaining. It was further agreed that Cuba be excluded from the Inter-American Defense Board. Owing to pressure from the military, Argentine President Frondizi broke diplomatic relations with Cuba, leaving Brazil, Uruguay, Chile, Bolivia, and Mexico the only Latin American countries still in diplomatic contact with Cuba.

This and previous anti-Communist measures failed to resolve the Cuban problem or to avert the Soviet-United States missiles crisis of October 22, 1962, which focused on the island. Following President Kennedy's "quarantine" address of October 23, calling for a naval blockade of Cuba, the Council of the OAS, meeting as the Provisional Organ of Consultation, resolved: (1) to call for the immediate dismantling and withdrawal from Cuba of all missiles and other weapons with an offensive capability; (2) to recommend that the member states, in accordance with the Rio Treaty, take all measures, individually and collectively, to stop further shipment of military material to Cuba and to prevent the offensive missiles in Cuba from ever becoming an active threat to the peace and security of the continent. It was only after the quarantine policy had been set into motion that President Kennedy called the Security Council of the United Nations and the OAS Council into session. The latter had then voted 19 to 0 to authorize the use of armed forces to prevent further Soviet arms shipments into Cuba, and announced the above resolution.

The next significant OAS anti-Communist action took place on July 26, 1964, when the OAS conference of foreign ministers ordered sanctions against Cuba by a vote of 15 to 4. Mexico, Chile, Uruguay, and Bolivia were the states voting against measures to punish Havana for its aggression against Venezuela in 1963 (Bolivia later agreed to abide by the conference orders). The sanctions called for mandatory severance of diplomatic and consular relations between the American states and the regime of Premier Fidel Castro; however, the only countries still maintaining ties were those that voted against the sanctions resolution.

Other measures called for in the resolution provided for the suspension of all trade except that in food and medicine, and for the suspension of maritime transportation. The resolution authorized the American states to engage in individual or collective self-defense, including the use of armed forces, in the event of new Cuban aggression through subversion before the OAS had time to invoke collective measures.

Prompted by the Tri-Continental Conference held in Havana the preceding month, the OAS Council, in February 1966, emphatically condemned a move by the newly formed African-Asian-Latin American group to extend national "wars of liberation" to the Americas. The Council denounced such foreign support of subversive movements as a violation of the principle of non-intervention and a threat to the peace and security of the Western Hemisphere. The resolution, sponsored by Peru, was adopted by a vote of 18 to 0. Mexico and Chile abstained, not in disagreement with the purpose of the resolution, but on the legal contention that the Council was not empowered to make such a political denunciation.

Renewed Cuban efforts to promote revolutionary war compelled the OAS foreign ministers, meeting in Washington in September 1967, to take further steps to combat Cuban-fostered subversion in the hemisphere. The agreement was regarded as triumph for Venezuela, Bolivia, and the United States, the three OAS members who had been pressing most vigorously for harsher retaliation against Premier Fidel Castro. Mexico, the only OAS member still maintaining diplomatic relations with Cuba at this time, would not accept the seventeen-point declaration. However, Mexico did agree to sign a separate resolution under which subversive acts sponsored by Cuba would be reported by organization members to the United Nations. Key agreements of the conference were: (1) to condemn "forcefully" continuing intervention by the regime of Fidel Castro in Venezuela, Bolivia, and other countries; (2) to report to the United Nations any Cuban subversion in the hemisphere under a General Assembly resolution that bars intervention in the affairs of another state; (3) to tighten controls to prevent the movement of Cuban-based Communist agents, arms, funds, and propaganda in the hemisphere; (4) to improve surveillance over the activities of the Latin American Solidarity Organization. The organization's twenty-seven Communist and leftist member groups met in Havana in July 1967, to promote armed subversion or penetration in virtually all the non-Communist countries of Latin America.

Measures employed by the United States to forestall and neutralize Communist penetration and subversion in Latin America in the post-World War II era have been designed to strengthen the internal security capabilities of the nations concerned, to isolate Cuba from the hemisphere, and to discredit the image of the Cuban revolution in the region. Of even greater importance in the long run will be the realization of the goals of the Alliance for Progress, a partnership of the countries of the inter-American system. All action taken by the United States in dealing with the Communist conspiracy requires the full agreement and free cooperation of the participating nations. This means that the United States , in respecting the sovereign independence of these nations, cannot act unilaterally, and therefore neither the problems nor their solutions are completely within Washington's control.

A number of United States agencies are charged with the operation of programs to assist Latin American countries in coping with subversion. Their activities are carried out as part of a concerted United States effort to promote sound political, economic, and social structures through democratic processes. Through various media the United States Information Agency conveys to Latin Americans the story of the betrayal of the Cuban revolution, and what conditions are like in Cuba and other Communist countries. The Department of

Defense supplies anti-Castro and anti-Communist material to the armed forces in these countries for use in troop information and education programs.

Internal security programs were developed by the United States to assist countries faced with Communist-inspired disorders, terrorism, sabotage and guerrilla operations. United States military assistance and training programs were reoriented to meet this new danger. Selected Latin American military personnel are trained at United States military schools in riot control, counter-guerrilla operations and tactics, and other subjects which will contribute to the support of established governments and maintain public order. A public safety program for the training of civil police forces was made part of AID to strengthen the internal security situation. A regional Inter-American Police Academy was established in the Canal Zone in support of this program. In all instances United States programs seek to assist the Latin America nations to develop the capability to insure the internal security needed to achieve goals set by the Alliance for Progress.

SUPPLEMENTARY READINGS

Adler, Gerhard. *Revolutionäres Lateinamerika.* Paderborn, 1970
Aikman, Duncan. *The All American Front.* New York, 1941.
Alba, Victor. *Historia del communismo en América Latina.* Mexico, D.F., 1954.
———. *Historia del movimiento obrero en la América Latina.* Mexico, D.F., 1964.
Alexander, Robert J. *Communism in Latin America.* New Brunswick, 1957.
Allen, Robert L. *Soviet Influence in Latin America: The Role of Economic Relations.* Washington, 1959.
Amado, Jorge. *O cavalheiro da esperanca: Vida de Luiz Carlos Prestes.* Rio de Janeiro, 1956.
Arismendi, Rodney. *Problems de una revolución continental.* Montevideo, 1962.
———. *Lenin, la revolución y América Latina.* Montevideo, 1970.
Artucio, Hugo F. *The Nazi Underground in South America.* New York, 1942.
Bahne, Siegfried (ed.). *Archives de Jules Humbert Droz.* Dodrecht, 1970.
Béjar, Héctor. *Peru 1965: Notes on a Guerrilla Experience.* New York, 1970.
Beals, Carleton. *Glass Houses: Ten Years of Free Lancing.* New York, 1938.
Bonachea, Rolando, and Valdés, Nelson (eds.). *Selected Works of Fidel Castro: 1952-1958.* Vol. I. Cambridge, Mass., 1970.
Carlton, Robert G. (ed.). *Soviet Image of Contemporary Latin America: A Documentary History, 1960-1968.* Austin, 1970.
Carter, Robert F. *The Battle of South America.* New York, 1941.
Chacon, Vamirh. *Historia das ideas socialistas no Brasil.* Rio de Janeiro, 1965.
Clissold, Stephan, *Soviet Relations with Latin America, 1918-1968: A Documentary Survey.* New York, 1970.
Crespo Toral, Jorge. *El comunismo en el Ecuador.* Quito, 1958.
Dalton, Roque. *Revolución en la Revolución y la Critica de Derecha.* Havana, 1970.
Daniels, Walter M. (ed.). *Latin America and the Cold War.* New York, 1952.
Davis, Harold E. *Latin American Social Thought.* Washington, 1961.
Debray, Regis. *Strategy for Revolution: Essays on Latin America.* New York, 1970.
Dillon, Dorothy. *International Communism and Latin America: Perspectives and Prospects.* Gainesville, 1962.
Draper, Theodore. *Castroism, Theory and Practice.* New York, 1965.
———. *Castro's Revolution: Myths and Realities.* New York, 1962.
Dumont, Rene. *Cuba Es socialista?* Caracas, 1970.
Ferla, Salvador. *Cristianos y marxistas.* Buenos Aires, 1970.
Ghioldi, Rodolfo, et al. *Vigencia del leninismo hoy en la Argentina.* Buenos Aires, 1970.
Gott, Richard. *Guerrilla Movements in Latin America.* London, 1970.
Green, Gil. *Revolution, Cuban Style: Impressions of a Recent Visit.* New York, 1970.
Guevara, Ernesto. *Obras 1957-1967.* 2 vols. Havana, 1970.
Halperin, Ernst. *Castro and Latin American Communism.* Cambridge, Mass., 1965.
Harris, Richard. *Death of a Revolutionary: Ché Guevara's Last Mission.* New York, 1970.
Hoover Institution on War, Revolution and Peace. *Yearbook on International Communist Affairs.* 1966––. Stanford, California.
Horowitz, Irving L. (comp.). *Cuba: diez años después.* Buenos Aires, 1970.
Instituto Marx-Engels-Lenin. *La lucha de guerrillas a la luz de los clasicos del marxismo-leninismo.* Caracas, 1970.

James, Daniel. *Red Design for the Americas: Guatemalan Prelude.* New York, 1954.
———. (comp.). *The Complete Bolivian Diaries of Ché Guevara and Other Captured Documents.* New York, 1970.
Johnson, Cecil. *Communist China and Latin America, 1959-1967.* New York, 1970.
Josephs, Ray. *Latin America: Continent in Crisis.* New York, 1948.
Lafertte, Elias. *Vida de un comunista: Paginas autobiográficas.* Santiago, Chile, 1961.
Lamore, Jean. *Cuba; que sais-je?* Paris, 1970.
Larteguy, Jean. *The Guerrillas.* New York, 1970.
Lauerhass, Ludwig. *Communism in Latin America: Bibliography. The Post-War Years, 1945-1960.* | Los Angeles, 1962.
Macaulay, Neill. *A Rebel in Cuba: An American's Memoir.* Chicago, 1970.
MacDonald, N.P. *Hitler Over Latin America.* London, 1940.
Malloy, James M. *Bolivia: The Uncompleted Revolution.* Pittsburgh, 1970.
Marianetti, Benito. *Argentina, realidad y perspectivas.* Buenos Aires, 1964.
Max, Alphonse. *Guerrillas in Latein Amerika.* Zurich, 1970.
Nuñez, Carlos. *Chile La ultima opción electoral?* Santiago, 1970.
O'Connor, James R. *The Origins of Socialism in Cuba.* Ithaca, 1970.
Okinshevich, Leo, comp., and Carlton, Robert G., ed. *Latin America in Soviet Writings: A Bibliography.* 2 vols. Baltimore, 1966.
Ortega, Luis. *Yo soy el Ché.* Mexico, D.F., 1970.
Ortega y Medina, Juan A. *Historiografia socientica Interoamericanista.* Mexico, D.F., 1961.
Oswald, Joseph G., and Strover, Anthony J., eds. *The Soviet Union and Latin America.* New York, 1970.
Partido Comunista de Bolivia. *Vladimir, Ilich Lenin.* Cochabamba, 1970.
Peraza, Fermin. *Revolutionary Cuba: A Bibliographical Guide, 1968.* Coral Gables, Florida, 1970.
Pintos, Francisco R. *Historia del movimiento obrero del Uruguay.* Montevideo, 1960.
Poppino, Rollie. *International Communism in Latin America.* Glencoe, Illinois, 1964.
Ramírez Necochea, Hernán. *Origen y formación del partido comunista de Chile.* Santiago, 1965.
Ramos, Jorge A. *El partido comunista en la política argentina: su historia y su critica.* Buenos Aires, 1962.
Rauschning, Herman. *The Voice of Destruction.* New York, 1940.
Ravines, Eudocio. *The Yenan Way.* New York, 1951.
Roca, Blas. (Francisco Calderio). *The Cuban Revolution.* New York, 1961.
Schmitt, Karl M. *Communism in Mexico.* Austin, 1965.
Schneider, Ronald M. *Communism in Guatemala, 1944-1954.* New York, 1958.
Seiglie Ferrer, Carlos. *Siete dialogos. Cuba: pasado y presente.* Miami, 1970.
Sinclair, Andrew. *Ché Guevara.* New York, 1970.
Singer, K.D. *Germany's Secret Service in South America.* New York, 1942.
Stevenson, John R. *The Chilean Popular Front.* Philadelphia, 1942.
Strausz-Hupé, Robert. *Axis America.* New York, 1941.
Suárez, Andrés. *Cuba: Castroism and Communism: 1959-1965.* Cambridge, Mass., 1965.
Toirac, Inciano D. *Cuba, el comunismo y la política norteamericana.* Miami, 1970.
Torres, Camilo. *Dos rebeldes: vida y textos pro Camilo Torres y Carlos Marighella.* Lima, 1970.
———. *Revolutionary Writings.* New York, 1970.
U.S. Dept. of State. *The Sino-Soviet Economic Offensive in the Less Developed Countries.* Publication 6632. Washington, 1958.
———. *World Strength of Communist Party Organizations.* Washington, 1971.
U.S. Information Agency. *Problems of Communism.* Vols. I-XII. Washington, 1956-1963.
U.S. Senate. "Soviet Bloc Latin American Activities and Their Implications for United States Foreign Policy," *United States-Latin American Relations.* Washington, 1960.
Wolfe, Bertram. *Diego Rivera.* New York, 1939.

5

Anglo-American Isthmian Diplomacy

From the time when Columbus started on his famous quest of a new route to an old world, a shorter waterway from Europe to farther Asia was the dream of navigators. Columbus believed that he had found a new route to India, but he realized that there remained the task of finding a passage by which he could cross between the great land masses and circle the globe. With the discovery of the Pacific by Balboa, the search for the secret of the strait was prosecuted with increased diligence, but in vain. Magellan finally discovered the secret, but his solution via the Straits of Magellan or Cape Horn was too long to be satisfactory, and the quest continued. When Cortés, by numerous expeditions, became virtually certain that no natural waterway connected the two great oceans, he proposed to construct one. Charles V took much interest in the project; and, according to the historian Gomara, four routes were considered practicable, *i.e.*, those of Darien, Panama, Nicarugua, and Tehuantepec.[1] Thus not only was an interoceanic canal regarded from the earliest times as feasible, but the four routes considered most practicable up to our own time were among the first to be considered.

Early in the nineteenth century new interest in the idea was provoked by the interesting surveys and reports of the great scientist Alexander von Humboldt. After five years of exploration in Central and South America, he urged the construction of an artificial waterway, and he went so far as to discuss the possibilities of nine different routes, although he, too, regarded the four routes already mentioned as particularly worthy of investigation. Spain was finally aroused once more to the advantage that would accrue to her through the

[1] See Report of J.T. Sullivan on *Problem of Interoceanic Communication, House Ex. Doc. No.107,* 47th Cong., 2nd Sess., chaps I and II.
[2] For complete citations see *House Report No. 145,* 30th Cong., 2nd Sess., pp. 169-204.

possession of such a waterway, and in 1814 the project was authorized by the *cortes.* With the outbreak of the revolutions throughout Latin America the project lapsed, and it remained for the newly liberated republics to resume its consideration. As early as 1823 the matter came before the congress of Central America, and in the following year Señor Cañas, the diplomatic representative of this federation at Washington, drew the attention of the United States to the importance of a canal linking the two oceans and urged that the United States cooperate with his country in the construction of such a waterway. Henry Clay, Secretary of State, was interested in the project and promised to instruct the American representative to investigate and make a report.[3]

At approximately the same time the United States was invited to send representatives to the Panama Congress, called at the instance of Bolívar, at which the canal question was one of several subjects to be considered. Owing to strong opposition in Congress, the American representatives were not sent until it was too late to participate in the congress, but in their instructions the canal project was considered a proper subject for consideration. The sole limitation was that such a canal should not be under the control of any one nation, "but its benefits be extended to all parts of the globe upon the payment of a just compensation or reasonable tolls.[4] Although nothing substantial came of the Panama Congress, the importance and need of a trans-isthmian canal persisted, and in the next few years various projects were launched. An American company obtained the concession for a canal in Nicaragua; but, although the cost was estimated at only five million dollars, that sum could not be raised. The King of Holland, who had been represented by General Werweer at the Congress of Panama, also obtained a contract to construct a canal by the Nicaragua route; a company was formed, and the United States was sufficiently interested to consider the desirability of obtaining a majority of the shares. The revolution in the Netherlands, resulting in the separation of Belgium from Holland, forced an indefinite postponement of the project.[5]

The Panama route was also being seriously considered at this time, and Bolívar went so far as to have the route surveyed. But the engineers reported that there was a difference of three feet between the levels of the two oceans, a factor that enormously increased the difficulty of the project. Nicaragua realized the particular interest that the United States had in an interoceanic canal, and on June 16, 1825, the congress of the Central American Confederation passed a decree offering liberal concessions to stimulate the construction of a canal. Bids were received and the contract was awarded to an American financial group; whereupon a company, capitalized at five million dollars, was formed under the name of the "Central American and United States Atlantic and Pacific Canal Company." Sufficient funds, however, were never subscribed.[6] About a decade later, in reply to another decree of the Central American congress offering the United States prior rights, a Senate resolution authorized the president to open negotiations with the governments of both Central America and New Granada in regard to protecting the rights of such companies as should undertake to open

[3] *House Rep. No. 322,* 25th Cong., 3rd Sess., pp. 15,16.
[4] J.B. Moore, *Digest of Int. Law*, Vol. III, p. 2, or *Report of International American Conference*, Vol. IV, Hist. Indes, pp. 143*ff.*
[5] *House Rep. No. 322,* pp. 17-33
[6] *Ibid.*, p. 125.

communication between the Atlantic and Pacific oceans, and to secure forever "the free and equal right of navigation of such canal to all nations, on the payment of such reasonable tolls as may be established."[7] President Jackson appointed Mr. Charles Biddle to undertake the necessary negotiations. But Biddle did not carry out his instructions in the proper fashion, and in his next message to congress the president declared it inexpedient to enter into negotiations with foreign governments upon the subject.[8] President Van Buren sent Mr. John L. Stephens upon a similar mission in 1839, but after surveying the route and estimating the cost at twenty-five millions, he declared that in his opinion the country was too unsettled to risk the undertaking.[9]

Among the most interesting of the early projects for a canal was the one launched by Prince Louis Napoleon while he was a prisoner in the fortress of Ham. In 1845 he secured a concession from the Nicaraguan government for the construction of a canal, and when he finally escaped, he published a brochure on the subject, pointing out the possibilities of the Nicaragua route and emphasizing the great political interest of England in the execution of this project to create a new center of enterprise in Central America, which would prevent any further encroachments from the north. With the fall of the July monarchy, Louis Napoleon found more interesting and vital matters to engage his attention.[10]

It was about this time that the United States took the first important step to secure control of the trans-isthmian route in Panama. On December 12, 1846, the American *chargé* signed a treaty with the Republic of New Granada, which among other things, guaranteed to the United States "that right of way or transit across the Isthmus of Panama upon any modes of communication that now exist, or that may be hereafter constructed, shall be open and free to the government and citizens of the United States"; and in return the United States guaranteed to New Granada "the perfect neutrality of the before-mentioned isthmus, with the view that the free transit from the one to the other sea may not be interrupted or embarrassed in any future time while this treaty exists." As a natural corollary, the United States also guaranteed the sovereignty of New Granada over this territory.[11] In other words, although "perfect neutrality" clearly meant that all nations should have equal right to free passage across the isthmus. and was so interpreted by Polk in his message to Congress, February 10, 1847,[12] it was understood both by the president and the Senate that by guaranteeing this neutrality the United States accepted a responsibility that was justified only because of the fundamental importance which the trans-isthmian route already possessed for the United States. This treaty was approved unanimously by the United States Senate.

Owing to the discovery of gold in the recently acquired California, the importance of the treaty was speedily demonstrated. The difficulties of overland travel were such that the sea route, with the short passage across Panama,

[7] Moore, *op. cit.*, Vol. III, p. 3.

[8] *Senate Journal*, 24th Cong., 2nd Sess., p. 100; for Biddle's report see *House Rep. No. 322*, 25th Cong., 3rd Sess., pp. 38-44.

[9] *House Rep. No. 145*, 30th Cong., 2nd Sess., p. 236.

[10] See B. Jerrold, *The Life of Napoleon III* (London, 1874), Vol. II, pp. 320-330.

[11] W.M. Malloy, *Treaties, Conventions, etc., of the U. S.* (Washington, D. C.,1910), Vol. I, p. 302.

[12] J. D. Richardson, *Messages and Papers of the Presidents*, Vol. IV, p. 512.

became the most practical avenue for the great horde of gold-seekers and settlers. Relying upon the terms of the new treaty, an American company constructed a railway across the isthmus from what is now Colón to Panama—a road that, since its completion in 1855, has been one of the most profitable short lines ever laid down. Another American company established a second interoceanic route through Nicaragua, by means of the lakes and short stretches of land travel, but it never attained the success of the Panama Railroad route.[13] It was in connection with the rights to construct a canal by the Nicaragua route that the United States came into conflict with Great Britain, a conflict that was settled only by the much criticized and wholly unsatisfactory Clayton-Bulwer treaty.

The attempts of Great Britain to obtain territory in Central America dated back to the seventeenth century, and, although her claims rested upon very uncertain grounds, by the middle of the nineteenth century she had practically established her sovereignty over the Mosquito Coast, the Bay Islands, and Belize. When, however, in 1848 Great Britain forced the withdrawal of the Nicaraguans from San Juan del Norte (Greytown)—a certain terminus of any interoceanic canal through Nicaragua—the United States deemed it necessary to act. Secretary Buchanan sent Mr. Hise as a special agent to Nicaragua in 1849, but with very indefinite instructions as to what action he should take, other than a vague implication that the Monroe Doctrine should be maintained. Hise succeeded in signing a convention giving the United States rights virtually equivalent to those acquired in the treaty of 1846 with New Granada. But, as this directly contravened the British claims, our government was not ready to back up his negotiations.[14]

The administration of President Taylor, which had come into office the same year, despatched a new emissary, Mr. Squier, to negotiate with Nicaragua, with the admonition that he should not involve the United States in any unnecessary controversy. He not only signed a treaty with Nicaragua obtaining the concession for the canal, but also another with Honduras, whereby the United States virtually obtained control of Tigre Island in the Gulf of Fonseca. When the British government answered this action by sending a squadron which took possession of the island, the United States found itself in an awkward predicament.[15] It was necessary to come to an agreement with Great Britain, and Mr. Lawrence, the American Minister in London, entered into negotiations. When, however, he demanded that as a preliminary to the settlement of the question Great Britain should withdraw her protectorate from the Mosquito territory, Lord Palmerston was unwilling to continue. In his anxiety to settle the dispute Mr. Clayton, the American secretary of state, decided to go over the head of his representative and come to a settlement, if possible, with Sir Henry Bulwer, the British minister at Washington. The British representative saw the advantage of coming to an agreement at a time when the United States was willing to compromise, and the famous Clayton-Bulwer treaty was negotiated.[16]

[13] For a more complete account of this undertaking see chap. 14.
[14] For text see *Sen. Ex. Doc. No. 112,* 46th Cong., 2nd Sess., p. 92.
[15] For text and correspondence see *British and Foreign State Papers*, Vol. XL, pp. 997-1002; also chap. XIII.
[16] For correspondence see *Sen. Ex. Doc. No. 194,* 47th Cong., 1st Sess., pp. 55-82.

The principal provisions of this treaty may be summarized as follows: Great Britain and the United States promised never to obtain or maintain any exclusive control over the proposed Nicaragua canal; they would neither erect fortifications commanding the canal, nor occupy, colonize, or exercise dominion over any part of Central America; the two powers agreed to guard the safety and neutrality of the canal and to invite other nations to join with them in doing the same; they promised to support any company that would construct the canal in accordance with the spirit of the convention; and finally, in order to establish a general principle, they agreed to extend their protection to any other practicable communication, whether by canal or railway, across the isthmus, and particularly to the proposed interoceanic communication by way of Tehuantepec or Panama.[17] The precedent against joint action established by John Quincy Adams was seemingly forgotten.

Hardly had the treaty been signed before difficulties arose concerning its interpretation. Great Britain regarded the treaty as definitely establishing her rights in the regions to which she had already laid claim, future settlements only being prohibited, while the United States confidently expected that the British would immediately withdraw from the Mosquito Coast and the Bay Islands. When Great Britain continued to maintain her position on the east coast of Nicaragua, and when in 1852 she formally annexed the Bay Islands by proclamation, her interpretation of the treaty seemed to be receiving *de facto* recognition. However, the United States continued to protest, and by a supplementary agreement, known as the Dallas-Clarendon treaty of 1856, an attempt was made to settle the question by the withdrawal of the British protectorate from the Mosquito Indians in return for obtaining a protectorate over the Bay Islands under the nominal sovereignty of Honduras.[18] When this convention failed of ratification, another attempt was made the following year, but with similar results; so that President Buchanan, in his message of December 8, 1857, declared that the wisest course would be to abrogate the treaty by mutual consent.[19] Great Britain was willing, provided that the United States would accept the *status quo ante* as the basis of the abrogation. But, as this was the last thing that the United States desired, the president did not press the matter. When, in 1860, the British signed a treaty with Nicaragua restoring a nominal sovereignty over the Mosquito Coast, the United States was forced to accept this solution.[20] A treaty of amity and commerce between the United States and Nicaragua, signed at Washington, November 6, 1857, known as the Cass-Yrisarri treaty, gave the United States equal rights with Nicaragua over any interoceanic route, in return for a guarantee of neutrality. This treaty, however, was never ratified by the Nicaraguan congress.[21]

During the Civil War period the United States was forced to lay aside both the Monroe Doctrine and all projects for the construction of an interoceanic waterway. Hardly, however, had the war come to an end before interest in a trans-isthmian canal revived, and at the same time feeling grew stronger that the

[17] For text with accompanying notes see Hunter Miller, *Treaties and Other International Acts of the United States of America* (Washington, D. C., 1937). Vol. V, pp. 671-703.
[18] *Sen. Ex. Doc. No. 194,* p. 138.
[19] *Ibid.,* p. 126.
[20] *Ibid.,* p. 151.
[21] *Ibid.,* p.117.

waterway must be under American control. In 1867 the United States and Nicaragua ratified the Dickinson-Ayon treaty, which gave the United States the right to construct a canal in return for guaranteeing Nicaragua's neutrality.[22] About the same time Mr. Seward entered into negotiations with Colombia to obtain exclusive rights for an isthmian canal, and a treaty to this effect was signed in 1869, but was not ratified either by the Senate of the United States or by the Colombian congress. An even more advantageous treaty, signed by the representatives of the two countries in 1870, met a similar fate.[23] In the same year President Grant appointed an Interoceanic Canal Commission, which conducted a number of important surveys, including the Darien, Nicaragua, Tehuantepec, and Panama routes, and after very careful consideration of the data obtained reported unanimously, in 1876, in favor of the Nicaragua route, from Greytown to Lake Nicaragua by the San Juan River and thence to Brito on the Pacific coast.[24]

In the meantime an adventurer, Gogorza, appeared in Paris with a canal concession obtained from Colombia, and succeeded in interesting certain of the imperialists in the scheme.[25] A company was promoted to push the project, and Ferdinand de Lesseps, justly famous as the builder of the Suez Canal, became interested in it. Surveys were made, and an International Engineering Congress was held at Paris in 1879, under the presidency of de Lesseps, to consider the execution of the project. Various routes were considered, but de Lesseps favored the Panama route and his advice was ultimately taken. The Universal Oceanic Canal Company purchased a concession for$10 million, its president, de Lesseps, went to Panama; and early in 1880 he announced that plans had been completed for a tide-level canal whose cost was estimated at $168 million by the International Technical Commission.[26]

The prospect of early construction of an interoceanic canal under European control aroused the United States, and President Hayes took occasion to present his views of such an undertaking in a special message to Congress, March 8, 1880:

> The policy of this county is a canal under American control. The United States cannot consent to the surrender of this control to any European powers. If existing treaties between the United States and other nations, or if the rights of sovereignty or property of other nations, stand in the way of this policy . . . suitable steps should be taken by the just and liberal negotiations to promote and establish the American policy on this subject, consistently with the rights of the nations to be affected by it . . . An interoceanic canal across the American isthmus will essentially change the geographical relations between the Atlantic and Pacific coasts of the United States and between the United States and the rest of the world. It will be the great ocean thoroughfare between our Atlantic and Pacific shores and virtually a part of the coast-line of the United States. Our mere commercial interest in it is larger than that of all other countries, while its relation to our power and our prosperity as a nation, to our means of defense, our unity, peace, and safety, are matters of paramount concern to the people of the United States.[27]

[22] *Sen. Ex. Doc. No. 112,* p. 130.
[23] *Ibid.,* pp. 34-84; also *see House Rep. No. 224,* 46th Cong., 3rd Sess., p. 24.
[24] *Sen. Ex. Doc. No. 15,* 46th Cong., 1st Sess., p. 1.
[25] *For. Rel. of the U. S.,* 1876, pp. 87-93.
[26] For text of Salgar-Wyse concession see *Sen. Ex. Doc. No. 112,* p. 84.
[27] Richardson, *op, cit.,* Vol. VII, p. 585.

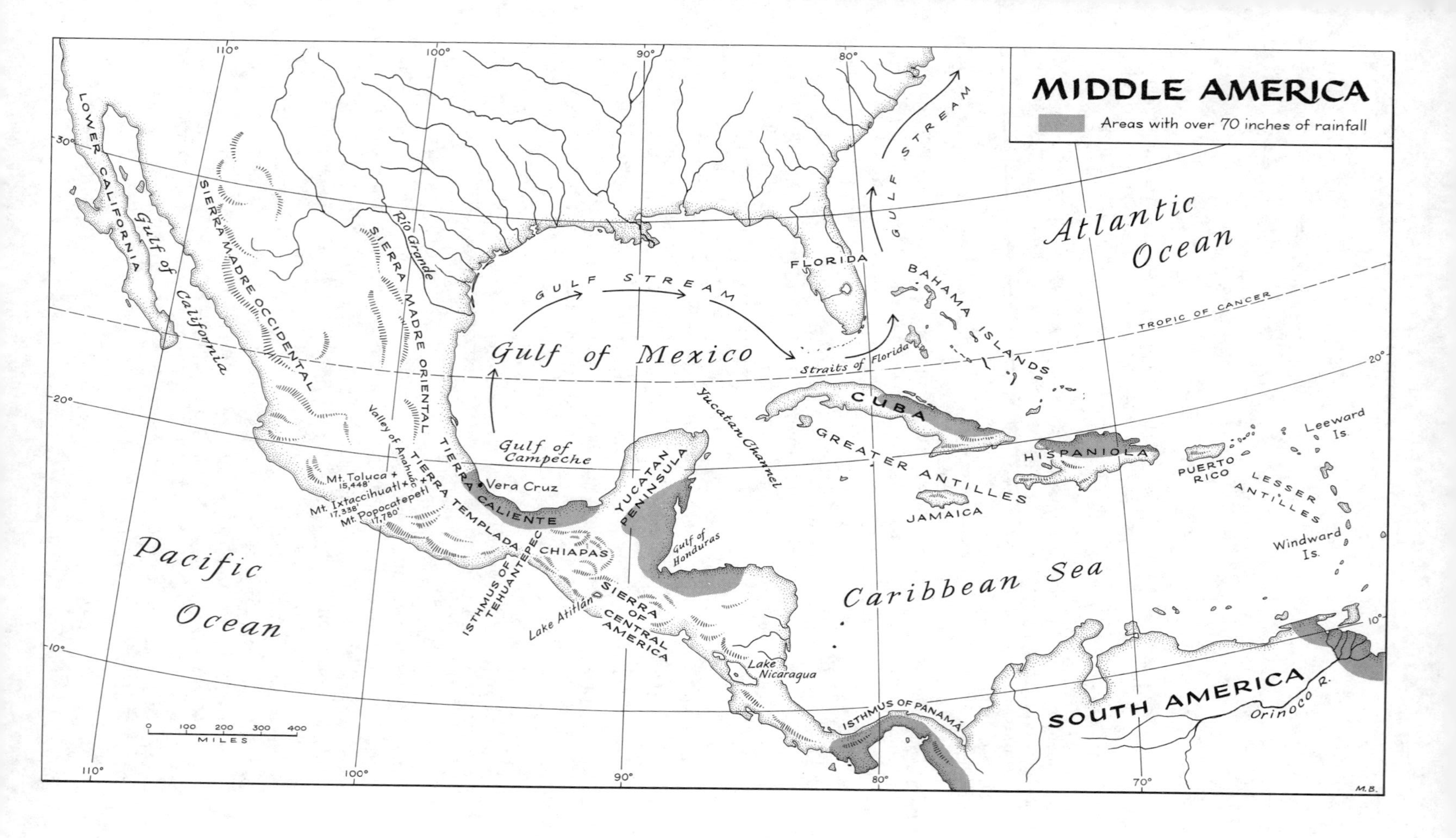
MIDDLE AMERICA
Areas with over 70 inches of rainfall
LOWER CALIFORNIA
Gulf of California
SIERRA MADRE OCCIDENTAL
SIERRA MADRE ORIENTAL
Rio Grande
Valley of Anahuac
Mt. Toluca 15,448'
Mt. Ixtaccihuatl 17,338'
Mt. Popocatepetl 17,780'
TIERRA FRIA
TIERRA TEMPLADA
TIERRA CALIENTE
Vera Cruz
Gulf of Campeche
Gulf of Mexico
GULF STREAM
FLORIDA
Straits of Florida
BAHAMA ISLANDS
TROPIC OF CANCER
Atlantic Ocean
YUCATAN PENINSULA
Yucatan Channel
Gulf of Honduras
ISTHMUS OF TEHUANTEPEC
CHIAPAS
SIERRA OF CENTRAL AMERICA
Lake Atitlan
Lake Nicaragua
ISTHMUS OF PANAMÁ
CUBA
GREATER ANTILLES
JAMAICA
HISPANIOLA
PUERTO RICO
LESSER ANTILLES
Leeward Is.
Windward Is.
Caribbean Sea
Pacific Ocean
SOUTH AMERICA
Orinoco R.
0 100 200 300 400
MILES
110° 100° 90° 80° 70°
30° 20° 10°
M.B.

Congress also became concerned about the matter, and passed a resolution declaring it to be in the interest of the people of the United States that its government should insist that its consent be a necessary condition precedent to the construction of any ship-canal across the isthmus.[28] Urged on by similar sentiments, Mr. Evarts attempted to make a treaty with Colombia whereby any concession for an interoceanic canal hitherto made, or to be made in the future, by Colombia should be subject to the rights of the United States, as guarantor of the neutrality of the isthmus and of the sovereignty of Colombia, over isthmian territory.[29] The Colombian government, however, was unwilling to concede any such rights to the United States; so that our government was forced to rest content with such powers as the treaty of 1846 guaranteed. At the same time, renewed interest was manifested in other trans-isthmian routes, particularly that by way of Tehuantepec, where Captain James B. Eads proposed to construct a great ship-railway, by which the largest ships could be transported from one ocean to the other by rail. The House Committee reported favorably, but no guaranties from the government could be obtained.[30]

The idea grew that the Clayton-Bulwer treaty was hampering the legitimate aspirations of the United States for a canal under its own control, and President Garfield, who came into office in 1881, directed his Secretary of State to take measures looking to the abrogation of the treaty. Instead of taking up the matter directly with the British Government, Mr. Blaine sent a circular despatch to the United States ministers in Europe, outlining the American attitude towards the proposal to establish a trans-isthmian canal under European guaranty. He pointed out that such a guaranty was entirely unnecessary, because by the treaty of 1846 the United States had already guaranteed the neutrality of the isthmus, and this guaranty required reinforcement, accession, or assent by no other power. Furthermore, he said that the proposed European guaranty would be offensive to the United States, since the proposed canal would be the chief means of transportation between the Atlantic and Pacific states and would be, to all intents and purposes, a part of our coast line. This being the case, the passage of armed vessels of a hostile power during any war to which the United States or Colombia might be a party would be no more admissable than over the railway lines joining the Atlantic and Pacific shores of the United States, or of Colombia.[31]

The remarkable feature of this communication was not so much its expression of principles, which showed, after all, no radical departure from former statements on the subject, but the fact that, although a copy of it was sent to the British Foreign Office as well as to the other European chancelleries, no mention was made, even in the British copy, of the Clayton-Bulwer treaty. This was the more surprising in that the statement on the passage of armed ships was in direct violation of the second article of that treaty. At first, Great Britain ignored the despatch. But at length, about four months after it was received, Lord Granville pointed out that the position of the United States and Great Britain with reference to the canal was determined by the convention commonly

[28] *Sen. Miscellaneous Doc. No. 42,* 46th Cong., 3rd Sess.
[29] *Sen. Doc. No. 237,* 56th Cong., 1st Sess., pp. 473-499.
[30] See *House Rep. No. 322,* 46th Cong., 3rd Sess.,
[31] *For. Rel. of the U. S.* 1881, pp. 537-540.

known as the Clayton-Bulwer treaty, and that her Majesty's government relied with confidence upon the observance of all the engagements of that instrument.[32]

Shortly before this reply was received, Mr. Blaine, apparently realizing that a formal treaty could not be disposed of in this cavalier fashion, sent a special despatch to the British government through Mr. Lowell, dated November 19, 1881, in which he attempted to show that the conditions of the Clayton-Bulwer treaty were no longer existent. In the first place, the treaty had been drawn up more than thirty years before, when the position of the United States on the Pacific Coast was entirely different. Again, whereas the intention of the treaty was to put Great Britain and the United States on an equal basis in respect to the canal, the present arrangements insured control by Great Britain, on account of her superior naval strength, unless the United States should be allowed to fortify the canal. In the third place, now that other nations had become interested in a canal project, the United States, by the terms of this treaty, was prohibited from asserting the rights acquired by the treaty with Colombia signed earlier than the Clayton-Bulwer treaty. Finally, the treaty had been made on the implied understanding that both nations would immediately subscribe funds for undertaking the work; now that nothing had been done, and the United States no longer needed outside aid, there was no reason for either side to regard itself as bound by the treaty's terms[33]

Lord Granville had little difficulty in exposing the weakness of Mr. Blaine's arguments, both in this despatch and in one of November 29 which followed it. The British possessions in North America had also shown considerable development, and it could hardly be supposed that the statesmen who were parties to the treaty had not envisaged certain changes of condition.[34] Neither was Mr. Frelinghuysen, who followed Mr. Blaine as secretary of state, more successful than his predecessor, although he was on more familiar ground when he based his objections on the Monroe Doctrine and the traditional policy of the United States in opposing European intervention in American political affairs.[35] The weak point, of course, was the fact that the Monroe Doctrine had not been regarded as a bar to making this treaty; therefore it could hardly be advanced as a reason for abrogating it. Needless to say, the diplomatic correspondence of the United States during this period did very little to improve the country's position; in fact, so easily and completely were its arguments refuted that its position diplomatically was even weaker than before. Nor did Great Britain confine her activities to the exchange of notes with the United States. At the first intimation of American action against the Clayton-Bulwer treaty, the British government protested that Nicaragua had never paid the indemnity promised to the Mosquito Indians by the treaty of 1860, and forced the case to arbitration before the Emperor of Austria. The award was wholly favorable to the British, not only in that it was declared that Mosquitia was not subject to regulations of Nicaragua, but in that it gave Great Britain the right, in case the terms of the treaty were not strictly observed, to intervene to protect her own interests.[36]

[32] *Ibid.*, p. 549.
[33] *Ibid.*, pp. 554-559.
[34] *British and Foreign State Papers,* Vol. LXXIII, p. 873.
[35] *Ibid.*, p. 892; or see *Sen. Ex. Doc. No. 194,* for correspondence.
[36] *British and Foreign State Papers,* Vol. LXXII, p. 1212.

Mr. Frelinghuysen also attempted direct action by negotiating a treaty with Nicaragua in December 1884, which provided for the construction of a canal by the United States under the joint protection of the United States and Nicaragua, the United States agreeing to guarantee the territorial integrity of Nicaragua.[37] But when Grover Cleveland became president in 1885, he recalled this treaty from the Senate, and declared in his first annual message that whatever highway should be constructed joining the two oceans "must be for the world's benefit, a trust for mankind."[38] There was no change in this attitude until 1899, when John Hay undertook to obtain the abrogation of the troublesome Clayton-Bulwer treaty, and succeeded quite as brilliantly as his predecessors had ignominiously failed.

In the meantime de Lesseps had made a spectacular beginning upon his stupendous undertaking. The actual work of construction was started February 1, 1881, and a veritable army of engineers and laborers was employed. Unfortunately for the success of the enterprise, de Lesseps remained in France, apparently failing to realize the prodigal and extravagant methods that were being employed. Useless and expensive equipment was purchased; the officers and directors were housed magnificently; an expensive campaign of propaganda was maintained to stimulate enthusiasm for the undertaking; and no matter how rapidly money flowed in, it was used up even more rapidly. The Engineering Congress at Paris had originally estimated the cost of the canal at $114 million and had figured that it could be built in seven or eight years. At the end of eight years almost $400 million had been spent and the work was not one third completed. The company went into bankruptcy, and the investigation that followed showed that the undertaking had been managed in a most corrupt and wasteful manner. The scandal was so great that all hope of ever constructing the canal by private means seemed eliminated.[39]

In the United States, where the Nicaragua route still seemed preferable, various attempts to organize companies to construct a canal by this route met with varied success. In 1890 the Maritime Canal Company, which held a concession from Nicaragua, was chartered by Congress and began work at Greytown. After threc years work, its capital was used up and the company went into the hands of a receiver. In 1895 Congress authorized a commission to report on the cost of completing the abandoned canal and the original estimate of $67 million was raised to $133 million. Under these conditions the government was not willing to back a resumption of work.[40] Various proposals were made in Congress to purchase the Maritime Company's franchise and property, but nothing came of them. Upon the outbreak of war with Spain, all projects were dropped. Nevertheless, the fact that the battleship *Oregon* had to travel 13,400 miles, instead of 4600 by the canal route, to arrive at the West Indies, where she was needed, from San Francisco, where she was stationed, showed conclusively the vital need for a trans-isthmian waterway.

At the close of the war, agitation was renewed; and this time Congress made a generous appropriation, March 3, 1899, of $1,000,000 for a commission to

[37] Richardson, *op. cit.,* Vol. VIII, pp. 256-260.
[38] *Ibid.,* p. 324.
[39] For an excellent summary of the reasons for the French failure see W.F. Johnson, *Four Centuries of the Panama Canal* (New York, 1907), chap. VII; see also *For. Rel. of the U.S.*, 1884, p. 119, for report on difficulties encountered.
[40] For complete report see *House Rep. No. 2126,* 54th Cong., 1st Sess., pp. 122-129.

make an exhaustive investigation of the relative merits of the Nicaragua and Panama routes. Moreover, since it was realized that no satisfactory solution could be arrived at until the Clayton- Bulwer treaty was out of the way, the Senate authorized the President to secure its abrogation, if possible. Somewhat isolated in Europe on account of the unpopularity of the Boer War, Great Britain had shown herself quite friendly towards the United States during the Spanish-American War, and the conditions were now exceedingly favorable for opening up negotiations on the subject. In fact, as early as December 7, 1898, Mr. Hay had made overtures through our representative, Mr. White, who was instructed to approach the British authorities in a "frank and friendly spirit of mutual accommodation, and ask whether it may not be possible to secure such modification of the provisions of the Clayton-Bulwer treaty as to admit such action by the government of the United States as may render possible the accomplishment of a work which will be for the benefit of the entire civilized world."[41]

Both Lord Salisbury and Mr. Balfour were willing to consider the suggestion favorably, and Lord Pauncefote, the British ambassador to the United States, was forthwith instructed to enter into negotiations on the subject. A treaty was signed by Mr. Hay and Lord Pauncefote on February 5, 1900, which provided that an isthmian canal might be constructed and operated by the United States, but that it must be neutralized in accordance with the principles of the Clayton-Bulwer treaty.[42] This convention was not satisfactory to the American Senate, which had expected a complete abrogation of the treaty. Accordingly, it proposed a number of alterations. One of these authorized the United States to take such measures as should be necessary to secure the defense of the canal and the maintenance of public order by its own forces; Article III, providing for the adherence of other powers, was stricken out; and provision was made for the express abrogation of the Clayton-Bulwer treaty.

These amendments were not satisfactory to Lord Lansdowne, the new British minister of foreign affairs, who, in a lengthy communication, set forth his objections to them.[43] After conferring with Lord Pauncefote and with leading senators, Mr. Hay accordingly proceeded to secure a fresh convention. In the course of several months of friendly negotiations all differences were ironed out, and on November 18, 1901, the second Hay-Pauncefote treaty was signed. This instrument the Senate ratified the following month by a vote of 72 to 6.

Article I of the new treaty expressly abrogated the Clayton-Bulwer treaty. Article II provided that the canal might be constructed directly or indirectly under the auspices of the United States government, which should have the exclusive right to regulate and manage it. Article III provided for the neutralization of the canal under substantially the same rules as those governing the Suez Canal. The most important of these rules was the first, which declared that the canal should be free and open, on terms of entire equality, to the vessels of commerce and of war of all nations observing these rules, with no discrimination in respect of traffic charges or otherwise, and that the conditions and charges should be just and equitable. Although no provision regarding

[41] *Diplomatic History of the Panama Canal, Sen. Doc. No. 474,* 63rd Cong., 2nd Sess., p. 1.
[42] *Ibid.*, p. 289.
[43] *Ibid.*, p. 11.

fortification was laid down, the United States was given the right to establish such military police as would protect the canal from lawlessness; and since the canal was to be completely immune in time of war, the logical inference was that the United States, which was charged with its defense, would not be prevented from erecting such fortifications as would ensure its protection. The fourth article stipulated that no change of territorial sovereignty or of international relations of the country or countries traversed by the canal should affect the general principle of neutralization or the obligations assumed under the treaty.[44]

As far as Great Britain was concerned, the United States was now free to proceed with the construction of a canal over whatever route should appear most practicable. A source of constant friction between the two great Anglo-Saxon countries had been removed, amicably and to the ultimate advantage of each. Great Britain had made the greater concession; it remained to be seen whether the United States would be duly appreciative of this fact when questions should arise involving the new treaty's interpretation.

[44] *Ibid.*, p. 292.

SUPPLEMENTARY READINGS

Anderson, C.L.G. *Old Panama and Castilla del Oro*. Washington, 1911.
Arias, H. *The Panama Canal.* London, 1911.
Bennett, I.E. *History of the Panama Canal.* Washington, 1915.
Bigelow, John. *Breaches of Anglo-American Treaties.* New York, 1917.
Bishop, Farnham. *Panama Past and Present.* New York, 1913.
Bunau-Varilla, P. *Panama, Creation, Destruction and Resurrection.* New York, 1914.
DuVal, Jr., Miles P. *Cadiz to Cathay.* Stanford, 1940.
Henderson, J.B. *American Diplomatic Questions.* New York, 1901.
Howarth, David. *Panama: Four Hundred Years of Dreams and Cruelty.* New York, 1966.
Johnson, W.F. *Four Centuries of the Panama Canal.* New York, 1907.
Keasby, L.M. *The Nicaragua Canal and the Monroe Doctrine.* New York, 1896.
Mack, Gerstle. *The Land Divided: A History of the Panama Canal and other Isthmian Canal Projects.* New York, 1944.
Mecham, J. Lloyd. *The United States and Inter-American Security, 1889-1960.* Austin, 1961.
Miner, Dwight C. *The Fight for the Panama Route.* New York, 1940.
Munro, Dana G. *The United States and the Caribbean Area.* Boston, 1934.
———. *The Five Republics of Central America.* New York, 1918.
Malloy, William M., comp. *Treaties, Conventions, International Acts, Protocols and Agreements between the United States of America and Other Powers.* 3 vols. Washington, 1910-1923.
Perkins, Dexter. *The United States and the Caribbean.* Cambridge, Mass., 1947.
Rippy, J.F. *The Caribbean Danger Zone.* New York, 1940.
———. *Latin America in World Politics.* New York, 1938.
Uribe, Antonio José. *Colombia e los Estados Unidos de América*. Bogotá, 1931.
U.S. Congress. *Diplomatic History of the Panama Canal, Sen. Doc. No. 474*, 63rd Cong., 2nd Sess.
———. *Treaties and Acts of Congress Relating to the Panama Canal.* Washington, 1917.
Williams, Mary W. *Anglo-American Isthmian Diplomacy.* Washington, 1916.

6

Colombia, the United States, and the Panama Canal

While the second Hay-Pauncefote Treaty was being negotiated in the fall of 1901, the tragedy occurred in Buffalo, New York, which claimed the life of President William McKinley, and brought Vice President Theodore Roosevelt to the presidency. During the campaign of 1900 Roosevelt had made a speech in which he referred to the old adage: "Speak softly and carry a big stick, you will go far." He subsequently applied it as a broad directive in his management of the nation's foreign affairs. Roosevelt, a strong proponent of the "large America" idea and a United States-controlled isthmian canal, had played a key role in the annexation of the Philippine Islands while he was assistant secretary of the Navy, and had gained great popularity for his part in the Spanish-American War. As president he was responsible for the further elaboration of the Monroe Doctrine, in his "Roosevelt Corollary," by which the United States asserted the right to exercise an international police power in the Western Hemisphere, but notably in the Caribbean region. Against this background, it is not surprising that Roosevelt was the president who was instrumental in securing for the United States the site of an isthmian canal.

Roosevelt vigorously supported the second Hay-Pauncefote Treaty, after denouncing the first, and with its consummation he instructed Secretary of State John Hay to proceed with the difficult task of negotiating a route for the canal. There were two likely routes to consider: one across Nicaragua, and the shorter one across the Isthmus of Panama. The former was favored, having been the traditional focus of American efforts to gain an isthmian route, and because the New Panama Canal Company held the Panama concession.[1] There had been much debate among engineers, and in public opinion, as to which was the better

[1] Dana G. Munro, *Intervention and Dollar Diplomacy in the Caribbean, 1900-1921* (Princeton, 1964), p. 38.

route; however, it was assumed that the findings of the Isthmian Canal Commission would weigh heavily in the final decision. The Commission, headed by Admiral John J. Walker, was created by the Congress in March 1899, to make a scientific investigation of proposed canal routes in Central America.

The Walker Commission reported in favor of the Nicaragua route shortly after the second Hay-Pauncefote Treaty was signed. It indicated a preference for the Panama site on technical grounds, but recommended Nicaragua because of the failure to secure reasonable terms of sale from the bankrupt French firm. The cost of a river-and-lake canal there was estimated at $189,864,062, whereas the cost of completing the French project would be $144,233,358; but to the last figure must be added $109,141,500, the value assigned to its property by the New Panama Canal Company. Maurice Hutin, the company's president, requested that the price be fixed by arbitration using that sum as a guide for the commission, but Admiral Walker refused to arbitrate and said the United States would pay $40 million.[2]

The House of Representatives passed on January 9, 1902, by a vote of 309 to 2, the Nicaragua Canal Bill, sponsored by Republican William P. Hepburn of Iowa. Senate approval of this bill would mean that the stockholders of the defunct French company would receive nothing. The panic-stricken officials of the company had meanwhile agreed to release their holdings for $40 million in accordance with the Commission's recommendation. Forthwith, at the President's request, the Walker Commission reconvened and issued a supplementary report recommending the construction of a Panama canal.[3]

In the United States Senate during January 1902, Senator John C. Spooner of Wisconsin, introduced an amendment to the Hepburn Bill instructing the president to purchase the properties of the New Panama Canal Company for $40 million provided he could acquire a clear title and negotiate a satisfactory treaty with Colombia. If the president could not accomplish this in a reasonable time, the amendment authorized him to go ahead with the Nicaraguan route, making the necessary diplomatic arrangements with that state, and with Costa Rica, which had claims of sovereignty over the San Juan River. The Spooner Bill finally squeezed through the Senate, June 19, 1902, by 42 to 34 votes.[4] This legislation gave the president a strategic advantage in dealing with the rival canal countries.

In the background of the decision in favor of the Panama site, which marked a reversal of attitude by Roosevelt, Hay, and most Republican leaders in Congress, lay the 'battle of the routes," one of the most dramatic lobbying campaigns in the history of the United States up to that time. Directed by the thoroughly sobered officials of the French company, the main thrust was concentrated in the Senate, where the fate of the Hepburn Bill would be determined.

[2] *Sen. Doc. No. 54*, 57th Cong., 1st Sess; *Report of the Isthmian Canal Commission, 1899-1901*, 58th Cong., 2nd Sess; *Senate Docs., No. 222* (Washington, D.C., 1904), 175; 135-140.
[3] *Sen. Doc. No. 123*, 57th Cong., 1st Sess. Negotiations for the construction of a canal in Nicaragua, carried on through State Department channels, had reached an impasse at the time that the Hepburn Bill was passed. José Santos Zelaya, the Nicaraguan dictator, had stubbornly refused to accept conditions which the United States regarded as fundamental. Costa Rica's willingness to concede rights on the San Juan River was also in doubt.
[4] *U.S. Stat. at Large, Vol. XXXII, Part I*, p. 481.

Given most credit for the successful lobbying effort were Philippe Bunau-Varilla and William Nelson Cromwell. These two men figured prominently in the Panama Revolution of 1903, and it was they who persuaded the French New Panama Canal Company, holder of canal construction rights in Panama, to sell its concessions to the United States, as well as arranging a draft treaty with the Colombian government consenting to the undertaking.[5]

Bunau-Varilla, a French citizen, had served as an engineer and contractor for the de Lesseps project at Panama in the 1880s, and after its failure had continued to write and carry on promotional activities in France and Europe for its revival. He came to the United States in January 1901, where he lectured and won influential friends, among whom was Senator Marcus Alonzo Hanna of Ohio. Bunau-Varilla wished to save the money that he and his colleagues had invested in the Panama venture, to vindicate French engineering genius, and to achieve some measure of personal fame.[6]

William Nelson Cromwell, a prominent New York attorney, was retained in 1896 as American consul for the French New Panama Canal Company. Cromwell, senior partner in the law firm of Sullivan and Cromwell, was officer, director, or counsel of more than twenty corporations in the United States, including United States Steel Corporation. Hired by the French firm because of his influence in high places in the United States, he told his client "we must make our plans with Napoleonic strategy."[7] Cromwell contributed a reputed $60,000 for the Republican campaign of 1900 in order to keep the party from endorsing the Nicaraguan route. He thereby won a debt of gratitude from Senator Hanna, chairman of the Republican National Committee.

The New Company's position was championed in a more spectacular fashion by Bunau-Varilla. It was he who exploited the fear of seismic activity along the proposed Nicaraguan route. A volcanic disaster in the Caribbean island of Martinique, followed by the renewed activity of Mount Momotombo in Nicaragua, which had been inscribed on the postage stamps of the republic, gave Bunau-Varilla his opportunity. One such stamp was presented to each Senator. As a prelude to this episode he had sent to President Roosevelt and to every United States Senator a circular letter in which he named six volcanoes rising from the waters of Lake Nicaragua and recalled that in 1835 the Nicaraguan volcano, Conseguiña, staged a forty-four hour eruption during which it poured out every six minutes an amount of stone and ash equal to the total excavation of the Nicaragua canal.[8]

These and other maneuvers by Bunau-Varilla and Cromwell doubtless helped to tip the scales in favor of the Panama route; however it can be assumed that President Roosevelt and the Congressional leaders made the final decision largely on the basis of the Walker Commission's judgement that the Panama route was the more feasible one. Whatever the causes, the passage of the Spooner

[5] Charles D. Ameringer, "The Panama Canal Lobby of Philippe Bunau-Varilla and William Nelson Cromwell," *American Historical Review*, LXVIII (1963), 346. See also D. C. Miner, *The Fight for the Panama Route* (New York, 1940), chaps. III and IV.
[6] Charles D. Ameringer, "Philippe-Bunau-Varilla: New Light on the Panama Canal Treaty," *Hispanic American Historical Review*, XLVI (1966), 29.
[7] Earl Harding, *The Untold Story of Panama* (New York, 1959), p. 147.
[8] Ameringer, "Panama Canal Lobby," p. 357; 361.

amendment was highly significant, for it reversed a long-standing sentiment in the United States favoring Nicaragua.

The Spooner Bill became law on June 28, 1903, and soon thereafter Attorney-General Philander C. Knox, following an investigation in Paris, declared that the French New Panama Canal Company could give a clear title to its properties. There remained the task of acquiring the right of way from Colombia, and this proved to be the most difficult chapter in the Panama story. Under the treaty signed in 1846 the United States had been granted the right of free transit by any road or canal that might be constructed across the isthmus, and in return guaranteed to New Granada, as Colombia was called until 1863, neutrality of the isthmus and "the rights of sovereignty and property" over the territory. The Panama Railroad was completed in 1855, and from that date until 1902, American forces were landed on the isthmus on several occasions to keep open the transit route to protect lives and property. The treaty did not give the United States the authority to build a canal, and a specific agreement would be required for this purpose. Colombia's consent would also be needed for the transfer of the French firm's concession to the United States.[9]

Secretary of State Hay quickly pushed negotiations with the Colombian diplomatic representatives in Washington, but nine months were to elapse before a treaty would emerge, and this was to be rejected by Bogotá. The amiable Carlos Martínez Silva was replaced as Colombian minister in Washington by José Vicente Concha in February, 1902, and the latter's appointment was interpreted to represent a stiffening in Colombian policy. In the ensuing treaty discussions, the question of sovereignty was most vital, but the issue most frequently raised was the amount of cash settlement and annuity the United States would pay Colombia for the authority to build and operate the Panama Canal. Concha would not yield on major points and refused to negotiate further, even after being reminded that the president was authorized to consider another route.[10] The negotiations were continued by Tomas Herrán, as chargé d'affaires, with Cromwell's assistance as intermediary. After Herrán had been advised that no further concessions were possible, and that the "reasonable time" provided by the Spooner Bill for the conclusion of negotiations with Colombia had expired, the Hay-Herrán Treaty was signed on January 22, 1903.

This treaty authorized the French company to transfer its properties and concessions to the United States; it also gave the United States exclusive rights to construct and operate the canal for one hundred years, with an option to continue the arrangement for similar periods, and the full control over a strip of land three miles wide on each side of the canal, not including the cities of Colón and Panama. Colombia was to maintain her sovereignty over the territory, and was to have the right to transport over the canal at all times her vessels, troops, and munitions of war without payment of any charges. As a price for these concessions, the United States promised to pay Colombia $10 million in cash and an annuity of $250,000.[11] The treaty would come into effect when ratified

[9] E. T. Parks, *Colombia and the United States, 1765-1934* (Durham, N.C., 1935), p. 219.
[10] *Diplomatic History of the Panama Canal, Sen. Doc. No. 474*, 63rd Cong., 2nd Sess., p. 256.
[11] *Ibid.*, p. 277.

by the laws of the respective countries, which meant, in this instance, by the Senate of the United States and by the Senate of Colombia.

The Hay-Herrán Treaty was approved by the United States Senate in March 1903, and in the same month the Colombians elected the Congress that would have to consider it. Colombian ratification of a treaty so advantageous to the United States was not to be achieved. After all, the isthmus was not only Colombia's sovereign territory, but had now become an exceedingly valuable piece of real estate. To the government and to the aroused public alike, the United States was being unrealistic in its cash offer for the transaction. Indeed, the sum of $10 million seemed small when it was recognized that a six mile strip of Colombian land was in effect being turned over to the United States for what in effect was perpetuity. Colombia's aged dictator, José Marroquín, sensed the growing public opposition,[12] and convened the Senate to consider the treaty, rather than approve it himself by executive decree. When the United States Senate refused to consider a Colombian proposal to raise the ante by $15 million ($10 million to be supplied by the Panama company), the Colombian Senate unanimously rejected the treaty on August 12, 1903.[13]

In making this decision the Colombians were aware that time might be made to work to their advantage, if the canal problem were held in abeyance until the termination of the French concession. The original concession of 1878 had been amended in 1890 to provide that the canal must be finished in 1904. In 1900 President Manuel A. Sanclemente had granted the company six additional years in return for a cash payment of five million *francs.* Since his decree had been granted under the government's emergency powers at a time when there was no congress, there were many Colombians who questioned its legality.[14] They particularly resented being offered only $10 million when the French company, holding only charter rights was to get four times that amount. If the decree were to be invalidated Colombia would have the sole rights.

President Roosevelt was at first undecided what course to follow after the Colombian Senate's rejection: whether to approach Nicaragua, or "in some shape or way to interfere when it becomes necessary so as to secure the Panama route without further dealing with the foolish and homicidal corruptionists of Bogotá." Hay told him of the likelihood of an uprising in the isthmus, and advised that: "Our intervention should not be haphazard, nor, this time, should it be to the profit, as heretofore, of Bogotá." But Roosevelt wrote to Dr. Albert Shaw: "I cast aside the proposition made at this time to foment the secession of Panama . . . I should be delighted if Panama were an independent state, or if it made itself so at this moment, but for me to say so publicly would amount to an instigation of revolt, and therefore I can not say it."[15]

Not to be overlooked in an appraisal of negotiations culminating in the Hay-Herrán Treaty and the ensuing controversy was the existing political situation in Colombia, and conditions in Panama itself. The death of the dictator Rafael Nuñez in 1894 had left the conservatives in control of the governmental

[12] *Foreign Relations of the United States, 1903*, p. 134.
[13] *Ibid.*, p. 163.
[14] Munro, *Intervention*, p. 43.
[15] Samuel Flagg Bemis, ed., *The American Secretaries of State and Their Diplomacy* (New York, 1928), IX, 164; Joseph B. Bishop, *Theodore Roosevelt and His Time* (New York, 1920), I, 278.

machinery, but as no strong leader emerged, the liberals seized the opportunity to revolt, and plunged the country into a civil war that claimed more than 100,000 lives. In July 1900, Vice President José Marroquín assumed dictatorial powers after deposing the president. The civil war, which had ended with a defeat of the liberals in November 1902, left commerce and industry paralyzed, the government almost bankrupt, and the paper *peso* worth one *centavo* in gold.

Panama, a department of Colombia, had never been integrated into the national life. Separated from the rest of the Republic by nearly impenetrable jungles, numerous mountains and some two hundred torrential rivers, Panama was accessible only by sea. These natural obstacles had fostered the growth of localism and the civil war had helped to arouse latent secessionist sentiments in Panama.

For four centuries the Panama isthmus had provided a well-traveled route between the oceans, and following the wars of independence of the early 19th century there had been little effective control from Bogotá. However, Panama's value was such that any decision about a canal had grave political implications. If the United States built a Nicaraguan canal, Panama's value would be lost, and there were rumors that the people of Panama would secede if Bogotá failed to reach an understanding with Washington. Moreover, the government needed money to meet the costs of the civil war. President Marroquín was placed in a dilemma: his government must either ratify the treaty and suffer a loss to its sovereignty, or refuse and sacrifice a great financial opportunity. Wishing to avoid personal responsibility for the painful and grave decision that would have to be made, and being aware that public opinion opposed the treaty, he insisted that the canal question be submitted to Congress. However, he could not delay until his Congress met because the United States Congress, convening in December, 1902, would turn to Nicaragua. He had therefore instructed his minister in Washington, Concha, to sign the best treaty that could be obtained. The latter rejected segments of the treaty proposed by the United States, but his successor, Herrán, had proved more compliant.[16]

However legal its position might be, the Colombian government faced serious political risks, either by action or inaction. The hostility of the United States toward further delay and Colombian counter-proposals was made clear by a virtual ultimatum from Secretary Hay. And for President Roosevelt, political necessity dictated that the dirt must start flying along the canal route to insure his nomination at the 1904 Republican convention. Furthermore, there was growing uneasiness among those people whose immediate interests were more closely involved. Early in June 1903, Bunau-Varilla warned the Colombian dictator that if the treaty were not ratified, the Isthmus of Panama might revolt and receive the protection of the United States. There was also the ominous report that the newly appointed governor of Panama would stand by Panama if a revolt occurred.[17]

In spite of the double jeopardy inherent in the situation, the Colombian government resolved to stand upon its rights and exact what it regarded as

[16] For a detailed account of the Colombian action see: Miles du Val, Jr., *Cadiz to Cathay* (Stanford, 1940), Chapter IX.
[17] *For. Rel. of the U.S., 1903,* p. 193.

proper and adequate terms. Because of its steadfastness, exemplified in its rejection of the Hay-Herrán Treaty, the Panama conspiracy was born. The Panamanians had too much at stake to run the risk of losing the canal.

Talk about the attainment of Panama's independence became common during the summer of 1903 as opposition toward the Hay-Herrán Treaty mounted in Bogotá. The leaders of the secession movement in Panama were business and professional men on the isthmus and in New York, many of whom were identified with the Panama Railroad Company. In Panama there were Dr. Manuel Amador Guerrero, the railroad's medical officer; José Augustín Arango, railroad attorney; James R. Shaler, superintendent of the railroad; James R. Beers, and others. William Nelson Cromwell, counsel for both the railroad and the New Panama Canal Company, was one of the leading conspirators in New York. The conspirators sent Beers in July and Dr. Amador in September to New York to develop plans for an uprising in Panama, which included securing possible support from the United States government.[18]

Bunau-Varilla became involved in the plot when he arrived in New York from France in September, 1903. Upon learning of the scheme, and the restiveness in Panama, this mercurial figure quickly applied his talents to the then lagging conspiracy. The first step was to ascertain, if possible, what the intentions of Washington were, and what attitude might be expected if a revolution should break out.

From interviews with President Roosevelt, Secretary of State Hay, and a close friend, Francis B. Loomis, assistant secretary of state, Bunau-Varilla concluded that these officials were sympathetic toward an independence movement on the isthmus. President Roosevelt later declared that neither he nor anyone connected with his administration gave him any assurances that the Panamanian rebels would receive United States protection; however, he did say that Bunau-Varilla would have been "a very dull man" not to guess his intentions. Singularly revealing to the French adventurer was evidence, gained in an interview with John Bassett Moore, an adviser to the president, that the United States was considering the possibility of occupying the Isthmus of Panama to construct a canal in assertion of the right under the Treaty of 1846.[19]

Fortified with this knowledge, and his appraisal of conditions in Panama, Bunau-Varilla conceived a revolutionary plan. Reduced to the simplest terms it called for the rebels to seize the cities of Panama and Colón and the transit facilities. The United States would prevent Colombia from taking retaliatory action, and would then grant recognition to the new state. He counted upon the clause in the Treaty of 1846 whereby the United States guaranteed to New Granada that the free transit of the isthmus from one sea to the other should never be obstructed. As mentioned, the precedent had long been established for the use of American troops to quell Panamanian disturbances. On this occasion it was to be given a novel interpretation: open communications across the isthmus would be maintained against the Colombians.

[18] Ameringer, "Philippe Bunau-Varilla," p. 30.
[19] *Ibid.*, pp. 31-32; George E. Mowry, *The Era of Theodore Roosevelt* (New York, 1958), p. 152.

Basing his plan upon these contingencies Bunau-Varilla promised Amador $100,000 in cash from his own funds and the protection of the United States navy within forty-eight hours after a revolution should begin. The only conditions were that the Panamanians should stage their own revolution, and that they should appoint Bunau-Varilla as their official representative in Washington, with plenary powers. Amador and the other "founding fathers" may have demurred about the latter condition because the man aspiring to become one of Panama's "first citizens" was a French citizen, who had not been on the isthmus since 1887, but he was too influential to be denied.

Upon returning to the isthmus Amador found his fellow conspirators confused and disturbed about reports that Colombian troop reinforcements were enroute to Colón. Warned of this development, Bunau-Varilla went to Washington where he discussed it with Assistant Secretary of State Loomis. The latter made no commitments whatsoever, but the Frenchman interpreted his reaction to mean that the United States would intervene. Various press dispatches had already reported that the cruiser, USS *Nashville* had left Kingston, Jamaica on October 31, with sealed orders. Assuming that its destination was Colón, Bunau-Varilla computed the time of its arrival and cabled the Panamanians that it would arrive in two and a half days.[20]

The *Nashville* reached Colón harbor on the morning of November 2, but it was not until the following day that its captain, Commander John Hubbard, received instructions to keep the transit open, to occupy the railroad line if service was threatened, and to prevent the landing of any forces, government or rebel, "with hostile intent" at Colón or any other port. Prior to the receipt of these instructions some five hundred Colombian troops were landed at Colón, an event which aroused much fear and consternation among the insurgents. Prompt action by superintendent Shaler removed the threat. He quickly removed the rolling stock to the Pacific side, and then persuaded the Colombian commanding generals, Ramón G. Amaya and Juan B. Tovar, to cross over to Panama while arrangements were being made to move their troops. General Esteben Huertas, commander of the Colombian garrison in Panama, who had been brought into the conspiracy, informed them that they were prisoners of the revolution. The revolutionists then sought to attack the Colombian forces at Colón, but the United States naval commander dissuaded them from doing so in accordance with the long-standing policy of preventing fighting along the transit route.

The only unforeseen incident was the firing of several shells from the Colombian gunboat *Bogotá* in Panama harbor. Its commander threatened to bombard the city if the Colombian generals were not freed, but soon changed his mind. The brief barrage was responsible for the two known casualties of the "war of independence:" one Chinese and one donkey.[21]

Colonel Eliseo Torres, the acting Colombian commander at Colón, threatened to burn the town and kill all Americans if his superiors were not released. But Commander Hubbard had meanwhile received his orders, and the Colombian troops were immobilized. Torres was at length induced to withdraw by the

[20] Philippe Bunau-Varilla, *Panama, the Creation, Destruction and Resurrection* (New York, 1914), pp. 284-366.
[21] Harding, *Untold Story*, p. 34.

revolutionaries and officials of the railroad company, leaving with his troops on November 5 for Cartagena.

The revolution had gone off according to schedule, and with little bloodshed. On November 6, the United States recognized the Republic of Panama, and the Colombian government was advised that President Roosevelt: most earnestly commends the government of Colombia and of Panana the peaceable and equitable settlement of all questions at issue between them. He holds that he is bound not merely by treaty obligations, but by the interests of civilization, to see that the peaceable traffic of the world across the Isthmus of Panama shall no longer be disturbed by constant succession of unnecessary and wasteful civil wars." Whatever plans the Colombian government may have entertained about reconquering Panama were doubtless influenced by Secretary Hay's message to Bogotá: "It is not thought desirable to permit landing of Colombian troops on isthmus, as such a course would precipitate civil war and disturb for an indefinite period the free transit which we are pledged to protect."[22]

On November 18, 1903 Secretary Hay and Bunau-Varilla signed a treaty which conferred on the United States the right to build the canal, fortify it, and to possess the canal zone, ten miles wide from Colón to Panama, "as if it were sovereign." Provision was made also for the use and occupation of other lands and waters outside the zone necessary to the canal's construction, operation, or protection. The canal was to be neutral in perpetuity, and the Republic of Panama was to have the right to transport over the canal its vessels, troops, and munitions of war at all times without paying charges of any kind. The United States agreed to insure the independence of the fledgling republic. Panama under terms of the treaty, would receive $10 million in gold, and an annual subsidy of $250,000, beginning nine years after its ratification. The Hay-Bunau-Varilla Treaty was ratified by Panama December 3, 1903, and by the United States Senate two months later.[23]

The treaty made the United States virtually sovereign over as much of Panama's territory as was needed for the exercise of full control over the canal. Panama thus became a military rampart, if not a protectorate, of the United States. As a result of Bunau-Varilla's enthusiasm to see the waterway completed, this treaty gave the United States more territory, greater control and flexibility of action in the Canal Zone than had been afforded by the Hay-Herrán Treaty. Keenly aware that the Panama junta might reject it, the Frenchman had seen to it that the agreement was pushed through to completion before the arrival of a Panamanian commission sent expressly to take part in the deliberations. His efforts to intimidate the insurgent junta into ratifying the treaty without seeing the text failed, but his exploitation of the rebel's fears was doubtless a factor in

[22] *For. Rel. of the U.S., 1903*, p. 225; 228.

[23] *Dip. Hist.*, p. 295. The Colombian government made a last-minute effort to forestall recognition of the new republic. On November 6, 1903 General Rafael Reyes, a conservative leader, indicated that his government could declare martial law and ratify the treaty by decree, or a new and more compliant Congress would be convened. Such action would have been contingent upon the retention of Colombia's sovereignty on the isthmus. The Panamanians were offered complete autonomy as a federal state within the republic if they would consent to the reintegration of Panama with Colombia. See: Ernesto J. Castillero Reyes, *Episodios de la independencia de Panama* (Panama, 1957), p. 193.

hastening its approval.[24] The treaty received rough handling from the press in the United States, and the anti-imperialists and Nicaragua proponents in the Senate, but the canal lobby did its work well, for it was approved by a vote of 66 to 14. With the canal site question resolved, Roosevelt declared that U.S. actions throughout had been consistent with the highest standard of international morality. Contending that the Colombian government had been guilty of deliberate bad faith and criminal error, he struck an analogy between the American colonial revolt against England and the Panama independence movement.

For the $40 million the United States had acquired from the French company the about-to-lapse concession from Colombia, rusty excavation machinery, maps, records, and the Panama Railroad stock. And a not inconsiderable amount of the work in digging the canal had been completed by the French. The Walker Commission had estimated that only about half of the 77 million cubic yards already excavated would be wasted through modifications of plans, and that only 94,863,703 cubic yards were left to be removed.[25]

The legacies bequeathed to history by Bunau-Varilla and Cromwell are confusing because each man, daring, brilliant and uncomprising, wanted to be known to posterity as the father of the Republic of Panama. While cooperating closely in 1902, there was a diminution in 1903, and increasing evidence of personal rivalry. It is evident, nevertheless, that Cromwell worked hard to secure passage of the Hay-Bunau-Varilla Treaty.[26]

Dr. Manuel Amador became the first president of the Republic of Panama, and Theodore Roosevelt, winning a landslide victory in the presidential campaign of 1904, immediately set in motion the work of construction and fortification of the canal. A sea-level canal was at first considered, but a lock canal, which would cost less to build in time and money, was decided upon.

With the diplomatic and political obstacles eliminated, the builders of the waterway were confronted by the problems of sanitation, excavation and the securing of a labor force. Colonel William C. Gorgas, appointed to direct the sanitation program, resolved the first by reducing the incidence of disease, notably yellow fever and malaria, and making the canal zone a fit place in which to live. Administrative difficulties connected with the excavation arose from the fact that Congress insisted in delegating control of operations to a commission instead of an individual; this hampered work for a while, but Roosevelt quickly made Major George W. Goethels, an army engineer, chairman of the commission, and extracted a promise from all other members of the commission

[24] Ameringer, "Philippe Bunau-Varilla," p. 47.

[25] Munro, *Intervention*, p. 40. In 1908 President Roosevelt tried to bring a libel prosecution against the New York *World* for its publication of articles implicating his administration in the Panama Revolution. The *World* had alleged that at least some of the $40 million had enriched American capitalists. The trial was never conducted. See: Harding, *Untold Story of Panama*.

[26] Ameringer, "Philippe Bunau-Varilla," p. 49; Ameringer, "Panama Canal Lobby," p. 347. Immediately after the treaty was signed Bunau-Varilla resigned his post, and requested that his salary be applied to the erection of a statue in Panama in memory of Ferdinand de Lesseps. He died in Paris in 1940 secure in the belief that he had engineered a revolution which had brought great benefit to civilization. In Panama his memory has been reviled for his role in negotiating a treaty so advantageous to the United States.

Cromwell presented a bill to the French New Panama Canal Company for $832,449 in fees, but the Paris arbitrators reduced it to $228,282.71. He was Fiscal Agent for the Republic of Panama from 1905 to 1937.

not to disagree with the chairman. Negro laborers were obtained in the West Indies, and the work then proceeded satisfactorily. The Panama Canal was opened to the commerce and warships of the world on terms of complete equality (so long as the United States remained a neutral), on August 15, 1914.[27]

The secession of Panama was the first major loss of Colombian national territory since the disintegration of Gran Colombia in 1830, and it dealt a cruel blow to the national pride. Colombia's bitterness at the way it had been treated became a matter of real concern to the United States, which soon engaged Colombia in a kind of conscience-stricken diplomacy.

In an effort to relieve the tension, Secretary of State Elihu Root proposed in 1907 a series of three bilateral treaties, the Root-Cortes-Arosemena treaties, between the United States and Panama, the United States and Colombia, and Panama and Colombia. By their terms the first ten installments of the $250,000 quit-rent would have been assigned to Colombia. The earnest support of the President Rafael Reyes notwithstanding, public opinion ran so strongly in Colombia that their presentation to Congress caused the downfall of the administration backing them, and the exile of the Colombian minister who negotiated them.[28] The Colombian Congress again rejected a new proposal made by the adminsitration of William Howard Taft in which Colombia was offered $10 million for a coaling station and any other canal route which might be available through her territory. Even when the United States suggested that the sum might be raised to $25 million dollars, Colombia refused and requested arbitration, something the United States was not prepared to risk.

During the first administration of Woodrow Wilson, Secretary of State William Jennings Bryan negotiated the Thomson-Urrutia Treaty with Colombia, which expressed regrets for what the United States had done, and offered $25 million dollars by way of compensation. Roosevelt denounced it as the "Colombia blackmail treaty," calling it "an attack upon the honor of the United States, which, if justified, would convict the United States of infamy . . ." This proposal, which Colombia was ready to accept, was rejected by the United States Senate, where there remained many of Roosevelt's loyal partisans. The instrument was shelved until 1919, when it was amended by eliminating the expression of regret; however, even with this change the Senate rejected it.[29]

It was not until the Republican administration of Warren G. Harding that a final settlement with Colombia was reached. The apology clause was omitted in the treaty, Colombian shipping was accorded equal rights with the United States in the use of the Canal, and $25 million dollars was paid to help redress past

[27] J. B. and Farnham Bishop, *Goethals, Genius of the Panama Canal* (New York, 1930; B. J. Hendrick, *William Crawford Gorgas, his Life and Work* (New York, 1924). By Executive Order, on May 9, 1904, the construction of the canal was placed under the direction and supervision of the secretary of war, then William Howard Taft. See: Ralph E. Minger, "Panama, the Canal Zone and Titular Sovereignty," *Western Political Quarterly*, XIV (1961), 544.

[28] *Dip. Hist.*, pp. 314-325.

[29] *For. Rel. of the U.S., 1913*, p. 324; 328; Theodore Roosevelt, *Fear God and Take Your Own Part* (New York, 1916), pp. 305-342. See also Paolo F. Coletta, "William Jennings Bryan and the United States-Colombia. Impasse, 1903-1921," *Hispanic American Historical Review*, XLVII (Nov., 1967), 486-501.

grievances. Roosevelt had died before the treaty came up for a decision in the Senate, and his close friend and chairman of the Senate Foreign Relations Committee, Henry Cabot Lodge, declared that the former president had favored a settlement before his death. This statement may have given some impetus to the treaty's ratification by the United States Senate, but perhaps of greater significance was the fact that fears had been aroused that European syndicates would be able to monopolize the oil resources of Colombia to the exclusion of United States firms. This made the purchase of Colombian good will a necessity. The question came to a decision on April 20, 1921, and was approved by vote of 69 to 19; ratifications were exchanged at Bogotá on March 1, 1922.[30]

President Roosevelt always considered the building of the Panama Canal one of the greatest achievements of his administration. This was made clear in his work, *An Autobiography*, in which he declares: "By far the most important action I took in foreign affairs . . . related to the Panama Canal." He also foresaw that the Canal would stand as a major achievement throughout the twentieth century. For the remainder of his life, in and out of public office, he denied complicity in the Panama revolution; however, his later boast: "I took Panama . . .," hardly gave full credit to Bunau-Varilla, Cromwell and others, and it exposed him to the adverse verdict of history.[31]

Roosevelt regarded the immediate construction of the canal imperative for reasons of national security, compelling reasons to be sure, although no one foresaw that a world war would begin shortly after the canal was opened. Time was of the essence in the furthering of his political career; Congress would meet in December 1903, and with the press and congressional sentiment mounting in favour of the Nicaraguan route, he needed a *fait accompli* if the Panama route were to be adopted. The negotiations for a Nicaraguan canal would have caused delay, and resulted in a loss of face politically for Roosevelt because some of the leading proponents of that route were the President's harshest critics. So intolerant was Roosevelt of delay that he had decided to take the canal zone from Colombia by force. The fortuitous secession movement in Panama thus spared the United States the full responsibility for an overt act of aggression against a pitifully weak neighbor.[32]

When viewed in historical perspective, there is little that can be said in defense of United States actions in the Panama case, even though Roosevelt said that "the recognition of Panama was an act justified by the collective interests of civilization." Under the circumstances, his defense in citing the 1846 Treaty with Colombia, in which she guaranteed to the United States the right of transit and in return was guaranteed the "right of sovereignty and property over the said territory," was untenable. Particularly, when the only issue at stake was the money to be paid to the speculators who controlled the New Panama Canal Company, and the construction might well have been delayed for a year. Had the twenty-five million dollar indemnity to Colombia been paid in 1903 instead

[30] J. Fred Rippy, "The United States and Colombian Oil," *Foreign Policy Reports*, V (April 3, 1929); *Congressional Record*, Vol. LXI, No. 2, p. 81. For text see *Treaty Series, No. 661* (Washington, D.C., 1922), or *Cong. Rec. Vol. 41, No. 2*, p. 378.

The ratification of the Thomson-Urrutia Treaty, as amended, was followed in 1924 by the joining of diplomatic relations between Colombia and Panama.

[31] *New York Times*, Mar. 24, 1911.

[32] William Roscoe Thayer, *The Life and Letters of John Hay* (New York, 1915), II, 328.

of two decades later, an immediate agreement might have been reached. General Reyes, who succeeded President Marroquín, strongly favored an agreement, and he had the political authority to have implemented it. The Nicaraguan alternative might have been exploited further, and a treaty with that country could probably have been effected if the United States had proved willing to compromise in favor of Nicaraguan sovereignty in a canal zone there. This, however, would have been contingent on Costa Rica's approval, for that nation's territory bordered on the San Juan River, and an arbitral award handed down by President Cleveland in 1888, in connection with a boundary dispute, had enjoined Nicaragua to make no grants for canal purposes without consulting Costa Rica.[33]

Although the evidence seems to prove that the United States had no role in fomenting the Panama revolution, it appears unlikely that it would have occurred had the United States made an effort to discourage it. In the face of progressively strident Panamanian nationalism notwithstanding, it is historically incorrect to say that, as a nation, Panama was little more than a creation of the United States.

Panama existed as an entity within the Spanish Empire as early as 1533 when the isthmus was seat of an *audiencia.* Because of the wealth generated by the annual Porto Bello fair, the transit of goods and travelers and the shipment of gold and silver, the isthmus could claim equality of status with Bogotá. In 1822, toward the close of the Wars of Independence, Panama decided to throw in her lot with the Republic of Gran Colombia, a federation created by Simón Bolívar, which fell apart in 1830 when Venezuela, Ecuador and New Granada decided to go their separate ways. A deeply felt sense of regional loyalty, the legacy of Spanish particularism, dislike of the domination of a distant metropolis, and virtual isolation except by sea, combined to foster Panamanian local feeling. Moreover, New Granada proved incapable of establishing an orderly government as a series of revolutions produced at least seven constitutions in about half a century. This turbulent history was reflected in the changing status of Panama. It was in revolt in 1840, reunited to Colombia in 1842, a semi-federal state in 1851, federal in 1855 and 1858, sovereign in 1861, constitutionally independent in 1862–63, as New Granada evolved into the Estados Unidos de Colombia.

Attracted by the revenue to be derived from the American-built transisthmian railroad, Colombian troops were sent to take over Panama in 1875 and, in the Constitution of 1886, the isthmian provinces again came under the direct control of Bogotá. Panamanians could not accept this status, nor could they tolerate seeing their wealth absorbed by the capital. A state of unrest prevailed, and fighting was going on sporadically in 1900, 1901 and as late as July 24, 1902. President Nuñez, during much of his tenure, had resided at Cartagena, a commercial rival of Panama. The rivalry was intensified as Cartagena, through his influence, became dominate in the coastal area, and the civil war further accented the problem.[34]

Panama's bid for independence was supported by the leading citizens on

[33]The Bryan-Chamorro Treaty of 1916 gave the United States an exclusive option in perpetuity on a canal route in Nicaragua for the sum of $3 million.
[34]Harry Bernstein, *Venezuela and Colombia* (Englewood Cliffs, N.J., 1964), pp. 114-115.

the isthmus, and it appeared to reflect the sentiments of most of the people. Since the economic life of the region depended upon the continued use of the transit route, a canal built elsewhere would have threatened the existence of the Panamanians. It must be concluded that the secession movement was a product of multiple causation: sectionalism, internal rivalries, civil war, geographical isolation, venal officials, and Panamanian patriotism, as well as the aggressive actions of the United States. President Roosevelt had no need to create an artificial independence movement. It was there all the time, but needed the protection of a powerful ally to make it effective.

There is no question that the Panama Canal ultimately brought great benefits to civilization, but at the expense of much good will and trust in hemispheric relations. Colombia's experience with the "big stick" was not lost upon Latin Americans generally who became increasingly suspicious of the powerful northern "colossus."[35]

Since Panama became an independent nation on November 3, 1903, its relations with the United States have become progressively troubled. Basic to every difficulty has been Panama's dissatisfaction with the Hay-Bunau Varilla Treaty which allowed an alien people to create what is, in effect, an independent state within the country. Problems of interpretation began with the active construction of the Canal. Secretary Hay in a letter to Senator Spooner before the treaty had passed the Senate had already emphasized the disproportionately advantageous character of the treaty to the United States, and pointed out that there were "many points in this treaty to which the Panamanian patriot could object."[36] The chief source of dispute was raised in regard to the assumption of full right of sovereignty on the part of the United States in the Canal Zone. For example, the Panamanian government objected to the establishment of post offices and customs houses since the powers of the United States government were limited to measures necessary for the construction and maintenance of the Canal. The United States, however, claimed full sovereign powers within the limits of the Zone.[37] To adjust these difficulties, W. H. Taft, then secretary of war, went to Panama and devised a temporary agreement to serve as a *modus operandi* during the period of construction of the Canal. This agreement was embodied in a series of executive orders issued by the War Department in 1904-1905, and 1911; the Panama Canal Act validated these orders until Congress should otherwise provide.[38]

President Woodrow Wilson, while professing friendship in Panama relations, retained the Hay-Bunau-Varilla Treaty, the Taft Agreement, and Article 136 of the Panama Constitution which recognized the right of United States intervention. Although he renounced "Dollar Diplomacy" and proposed a Pan American Pact to provide mutual guarantees of "political independence" and "territorial integrity", Wilson was forced to execute more armed interventions in Latin America, including Panama, than any of his predecessors.

[35] Thomas A. Bailey, *A Diplomatic History of the American People* (New York, 1964), p. 495.
[36] A. L. P. Dennis, *Adventures in American Diplomacy, 1896-1906* (New York, 1928), p. 341.
[37] That even in the United States there was a very decided difference of opinion as regarded the status of the United States in the Zone is shown later by a Supreme Court decision handed down January 6, 1930, in the case of the Luckenback Steamship Company vs. the United States. See: *U.S. Supreme Court Reports*, 74: 356.
[38] *U.S. Stat. at Large*, vol. XXXVII, Part I, p. 560.

Secretary of State William Jennings Bryan's tenure was not marked by intervention, but he was forced to deal with several riot incidents involving American military personnel and Panamanian civilians, and questions of jurisdiction over communications and transportation in the Zone. When he resigned in June, 1915, Bryan left no major problems in Panama relations to his successor, Robert Lansing. A problem of broader scope, relating to the potential use of the Canal, and involving the United States and Britain, was resolved before Bryan left office: the tolls controversy.

In 1912 the United States Congress, anticipating the opening of the Canal, passed a law exempting American coastwise shipping from the payment of tolls. After President Taft signed the bill, Great Britain protested that the tolls exemption violated the clause of the Hay-Pauncefote Treaty which opened the Canal to the vessels of "all nations" on terms of "entire equality." American railroad interests, who foresaw the diversion of much transcontinental traffic from the railroads to the new sea route had legitimate grounds for complaint, but Congress let the law stand until June, 1914, when President Wilson successfully insisted upon its repeal. By so doing, Wilson obtained British support for his efforts to topple the Mexican dictator, General Victoriano Huerta.

Secretary Lansing adopted a more positive policy towards Panama than Bryan, making it clear tht he was not averse to using force to insure Panama's stability and the defense of the Canal. Armed intervention was carried out on one occasion, in June 1918, to crush a rebellion in Chiriqui province. Panama, recognizing that its interests were closely bound with those of the United States, had declared war on Germany the preceding year, and cooperated with Canal Zone officials in developing a security network.

Bainbridge Colby, appointed secretary of state after Lansing's resignation on February 20, 1920, proved to be more sympathetic to the ideals of Pan Americanism, and adhered to a non-interference policy. Panama-United States relations were unmarred while he held the post, but it was near the close of his administration, on February 21, 1921, that an eruption of the border controversy between Costa Rica and Panama occurred. United States intervention ensued, but under the aegis of his successor, Charles Evans Hughes, and the Republican administration of President Warren G. Harding.[39]

Numerous disputes arose in spite of the Taft Agreement, notably concerning the establishment by the United States of commissaries which carried on an extensive business in the Zone, thereby competing with Panamanian firms, discriminatory wage scales prejudical to Panamanian employees, the status of property subject to use by the United States, and the issuing of exequaturs to foreign consuls to perform their functions within the Zone. With a view to improving the situation, the State Department on September 1, 1922, declared that the Taft Agreement had failed in its purpose, and recommended its termination in order that a new treaty with Panama might be negotiated.[40] In accordance with a joint resolution, approved by Congress on February 12, 1923,

[39]George W. Baker, Jr., "The Wilson Administration and Panama," *Journal of Inter-American Studies*, VII (1966), 279-293. The United States, at the request of the Panamanian government, supervised elections in 1908, 1912 and 1918, but not thereafter. In exercising the right of police protection the United States intervened in the cities of Panama and Colón in 1918, 1921, and 1925.
[40]*Sen. Doc. No. 248*, 67th Cong., 2nd Sess.

President Calvin Coolidge abrogated the Taft Agreement by an executive order, on May 28, 1924.

Negotiations were immediately begun by the State Department towards a new treaty which would help to resolve the operational problems within the Zone and ease sources of conflict. This treaty was signed at Washington in July, 1926, by representatives of the Republic of Panama, and by Secretary of State Frank B. Kellogg and Francis White, representing the United States.[41] The treaty was rejected by Panama National Assembly, however, and was never implemented.

The Panamanians took particular exception to the provision which would have committed them to join the United States automatically in the event of the latter's involvement in any war. It was argued with good reason that such a military alliance would have conflicted with Panama's obligations under the Covenant of the League of Nations. The treaty proved unsatisfactory in most other respects as well, for it tended to restore most of the advantages accruing to the United States under the Taft Agreement. The National Assembly passed a resolution requesting President Rodolfo Chiari to press for an agreement more favorable to Panama.[42]

No further action was taken until Franklin D. Roosevelt became president. Meanwhile, the basic Treaty of 1903 and the supplementary Taft Agreement, in large measure, prevailed. Giving impetus to the need for an adjustment in relations was the crisis resulting from the devaluation of the dollar, and the attempt to pay the annual $250,000 rental for the Canal in sixty per cent paper dollars. The United States initiated the negotiations which began in Washington in 1934, and were finally concluded in March 1936 with the signing of four agreements by the two nations.[43] The first was a general treaty revising the convention of November 18, 1903, two others related to radio communication in Panama and the Canal Zone, and the fourth was concerned with the construction of a trans-isthmian highway between the cities of Panama and Colón.

The general treaty made fundamental changes in the relations between the two governments. It ended the protectorate and the right of expropriation of territory by the United States in Panama by eminent domain, and substituted a pledge of joint co-operation for the furtherance of common interests.[44] The annual rental of the Canal was henceforth to be 430,000 *balboas*, thereby avoiding the problems raised by devaluation. The new treaty was consistent with the liquidation of imperialism, and the policy of non-intervention which President Roosevelt had made a major component of the Good Neighbor Policy.

The ratification of this treaty by the Senate was delayed until it was made clear, by a final exchange of notes, that in case of great emergency the United States could act first and consult with Panama afterward; that the treaty would permit expansion and new construction for the Canal; and that it did not

[41] Text in *Cong. Rec.*, vol. 68, Pt. 2, pp. 1846-1852.
[42] *New York Times*, Jan. 27, 1927.
[43] For text of General Treaty of Friendship and Cooperation see: *U.S. Treaty Series, No. 945*; for the Trans-isthmian Highway Treaty see *Ibid., No. 946*; for text of the Radio Convention see *Sen. Doc. Exec.* C, 74th Cong., 2nd Sess., p. 7.
[44] For a detailed account of the problem of land expropriation see: W. D. McCain, *The United States and the Republic of Panama* (Durham, N.C., 1937), chap. VII.

prohibit military operations in Panama's territory.[45] With this clarification the General Treaty of Friendship and Cooperation was signed on March 2, 1936, but it did not become effective until July 27, 1939. This treaty, although terminating Panama's protectorate status and the right of eminent domain, was an amendment to the Treaty of 1903 rather than a substitution for it.

The onset of World War II sharpened the need for further rapprochement with Panama owing to the strategic importance of the Canal. In the spring of 1939, in accord with a general hemispheric policy, the United States raised its legation at Panama City to an Embassy and, with the outbreak of war in Europe in the fall, immediate steps were taken to enlarge the Canal by a new set of locks and to build a military highway across the isthmus. Under Article X of the 1936 Treaty, the United States requested the use of a number of areas outside the Canal Zone in order to increase its defensive capabilities. The government of Panama, then headed by Arnulfo Arias, approved the request, after protracted debate, on the condition that all occupied areas be evacuated at the end of the war, and that adequate compensation be made.[46]

Shortly thereafter President Arias issued a decree refusing to permit American-owned vessels sailing under Panamanian registry to be armed for self-defense. In so doing, he revealed what was interpreted as pro-Axis sympathies. Other governmental leaders, fearing the consequences of this policy, overrode the constitutional processes, and on October 9, 1941, placed Ricardo Adolfo de la Guardia in the presidency. In response to charges that the United States had supported the *coup*, Secretary of State Cordell Hull issued a categorical denial and presented a detailed statement of the facts as reported to the State Department.[47]

The new administration immediately rescinded the decree prohibiting the arming of merchant ships, and joined the United States in developing a defense program. Panama declared war on Japan the day after the attack on Pearl Harbor and on May 19, 1942, signed an agreement with the United States authorizing the deployment of the latter's armed forces in areas of the Republic outside the Canal Zone. The largest of these was the Río Hato air base located about eighty miles southwest of the Canal, for which the United States was to pay an annual rental of ten-thousand *balboas.* The United States was granted the use of waters adjacent to this and other rented property, both public and private, and was allowed to build required access highways. All lands taken over, amounting to 134 units by the end of the war, were to be evacuated one year after the definite treaty of peace.[48]

At the same time, adjustments were made to improve relations between the two countries. The United States relinquished the Panama Railroad's real estate holdings in the cities of Colón and Panama except such as were essential to the operation and protection of the Canal, and released to Panama the waterworks and sewage systems lying within the Republic's jurisdiction. The United States also proposed to liquidate the credit of two and a half million dollars made available to Panama by the Export-Import Bank for the construc-

[45] *Senate Ex. Rpt. No. 5*, 1st Sess, June 21, 1939; *U.S. Treaty Series, No. 945*, pp. 63-64.
[46] *New York Times*, Mar. 6, 1941.
[47] Dept. of State *Bulletin*, V (Oct. 18, 1941), 293.
[48] *Ibid.*, VI (May 23, 1942), 448.

tion of Panama's share of the Chorerra-Río Hato Highway, a road constructed primarily for the defense requirements of the Canal and the United States. The agreement was approved by resolution in the United States Senate on December 4, 1942, by a vote of 40 to 29.

Throughout the war Panama collaborated with the United States as a useful and loyal ally, but at war's end complications arose concerning the status of the rented lands. The refusal of the Soviet Union to sign a peace treaty with Germany and the advent of the cold war led the United States to request the retention of fourteen installations, including that of Río Hato. An agreement allowing the United States the use of this base and other areas for a period of ten years was concluded between the two countries on December 10, 1947. President Enrique Jiménez favored the treaty, but strong popular opposition developed and Panama's National Assembly rejected it. The United States reacted quickly, evacuating all of the bases, an action which though generally approved, weakened Panama's economic structure.[49]

Panama's dissatisfaction with the terms of the 1903 Treaty increased swiftly as the relative prosperity that had accompanied World War II faded, and an economic depression set in. Colonel José Antonio Remón, elected to the presidency of Panama in 1952, was determined to modernize the economy of his country, and he was particularly intent on renegotiating the Canal Treaty of 1903 on terms more favorable to Panama.

Preliminary discussions began in the spring of 1953, but it was not until September that the negotiations got underway. John Moors Cabot, assistant secretary of state for Inter-American Affairs, led the United States team of negotiators, and José Ramón Guizado was head of the Panama mission. At the outset Guizado declared; "... The Panamanians are considered partners of the United States in the operation of the Canal Zone," which succinctly expressed in broad terms his country's views on the matter.[50]

The main grievances which Remón's negotiators introduced concerned the annuity, commissaries, and racial segregation in the Canal Zone. It was contended that the rent fixed in 1936 was inadequate because Canal tolls exceeded $37 million annually; further, Panama's revenues from that area before the construction of the Canal had been higher than the rent paid on it. Secondly, Panamanian businessmen resented Zone business activities which placed them at a competitive disadvantage. Commissaries, operated for U. S. employees in the Zone, consistently undersold Panama's merchants. Food processing plants, hotels and transportation facilities also undercut the national economy. Remón objected strenuously to the unfair discrimination against Panama's citizens who were employed in the Zone. Some five thousand white employees were paid at the "U.S. rate," whereas nineteen thousand Panamanian Negroes and mulattoes received the "local rate", which was substantially lower than the wages paid to whites. U.S. rate employees received better housing, educational and recreational facilities. The "local" rate employees were segregated in inferior schools and housing, and suffered generally by comparison. Such was the condition of the

[49] *Ibid.*, XXVII (Aug. 11, 1952), 216-219.
[50] John D. Martz, *Central America: The Crisis and the Challenge* (Chapel Hill, N.C. 1959), pp. 288-289.

Negroes, many of whose forbears had been introduced as laborers from the West Indies, to help build the Canal.

The negotiations, carried on in Washington, and spanning sixteen months, were generally amicable. The Remóns visit to New York and Washington, where they were guests of President and Mrs. Eisenhower, created a most favorable impression. Before Remón's returning to Panama the two presidents issued a joint declaration indicating: ". . . that this meeting has effectively served to promote mutual understanding and confidence in the common interest of the two nations and of the free world characterizing the bonds that unite them."[51]

President Remón did not have time to witness his triumph, meeting death from a Panamanian assassin's bullet on January 2, 1955, as the negotiations were drawing toward a close. Termed "A monument of lasting fame" to the late President by Selden Chapin, the U.S. Ambassador to Panama, the new treaty of friendship and cooperation between Panama and the United States, was signed in Washington on January 25, 1955. Similar to the convention of 1936, this treaty simply amended the basic Treaty of 1903.

The new agreement, which gave Panama nearly everything that Remón had demanded, was ratified by Panama March 15, 1955, and sent to the United States Senate on May 9, 1955. Senator J. William Fulbright in his leading argument for ratification, assured the Senate that the treaty "established the framework for the basic relationship between the United States and the Republic of Panama" and that "the treaty does not affect our present rights and relations in any respect whatsoever." Senator Richard B. Russell of Georgia repeatedly questioned the wisdom of ratification fearing that the United States had compromised too far, as did several of his colleagues; however, ratification occurred on June 29, 1955, by a vote of 72 to 14.[52]

The most significant concessions to Panama, thus ratified by the United States Senate, were: (1) the United States increased the annuity to Panama from $430,000 (originally $250,000 before the gold dollar was devalued) to $1,930,000; (2) Panama received without compensation waterfront and other properties, including the Panama Railroad's yards and terminals in the cities of Colón and Panama valued at $24,000,000; (3) the right to tax the income of Canal Zone employees, other than United States citizens and armed force personnel, and Zone residents not citizens of Panama; (4) alterations in the boundary between Colón and the Canal Zone; (5) termination of the United States monopoly right to construct trans-Isthmian railroads and highways, but the United States retaining the Panama Railroad; (6) the United States agreed to relinquish sanitation control in Panama and Colón to local officials; (7) Non-United States citizens living outside the Zone would lose commissary and duty-free import privileges after December 1, 1956; (8) a pledge from the United States concerning the institution of a single basic wage scale; Panamanians were also to be brought under the U.S. Civil Service Retirement Act; (9) discontinuation of certain manufacturing and processing enterprises, where the same products were available in Panama; (10) articles available at discriminatory prices were drastically reduced. In the exchange of notes relative to the 1936 treaty, the United States promised to favor equality of treatment in

[51] John and Mavis Biesanz, *The People of Panama* (New York, 1964), pp. 167-235.
[52] Harding, *Untold Story*, p. 116.

employment. This was reaffirmed in 1953 and again in the Memorandum of Understandings attached to the 1955 Treaty. In abrogating its road-building monopoly, the United States agreed to build a suspension bridge across the Panama Canal at Balboa to replace the Thatcher Ferry, estimated to cost $27,000,000. In return for the higher annual rent Panama made available 20,000 acres for training as military maneuvers around the important air base of Río Hato in a 15 year lease, rent free, and the United States promised that it would first consult Panama before using those lands.

Although President Ricardo Arias, in his endorsement of the treaty, declared that it offered eighteen positive advantages to Panama, it was evident that the United States had not met all of the Panamanian demands. The United States had refused to limit the canal concession to ninety-nine years, instead of holding the territory in perpetuity. It rejected the proposal that mixed Panamanian-United States tribunals be established to sit in judgement of Panamanians in the Zone charged with a crime. The United States also refused to allow Panamanian stamps to be used for Zone mail, and was unwilling to recognize Spanish as a second official language in the Zone. The treaty, nonetheless, was heralded as a victory by the Panamanians, and the legislature, reflecting popular sentiment, ratified it by a vote of 46 to 1, on March 9, 1955.[53]

The terms of the 1955 Treaty proved to be only a temporary palliative as new tensions soon developed. Gamal Abdul Nasser's seizure of the Suez Canal on July 26, 1956, and six months later, his "Egyptianization by seizure" of all British and French financial institutions in his country, inspired the extreme nationalist elements in Panama to urge that the Republic's sovereign rights be asserted over the Panama Canal. Dissatisfaction with the recently ratified treaty mounted, particularly after Secretary of State John Foster Dulles prevented Panama's participation in the twenty-two nation conference of users of the Suez Canal in London, during September 1956. In replying to the infuriated response of President Arias, Secretary Dulles said the United States ". . . has rights of sovereignty over the Panama Canal . . . to the entire exclusion of the exercise by the Republic of Panama of any such sovereign rights, power, or authority."[54]

The dispute over sovereignty in the Canal Zone was invited by the language of the 1903 Treaty which grants the United States powers over the Canal which it would possess and exercise as if it were "sovereign of the territory." The United States government contends that it has all sovereign rights as it occupies the Zone, and that Panama has "residual" sovereignty. Panama claims that the jurisdiction of the United States is not full and complete, but is rather a delegated and limited jurisdiction granted only in matters pertaining to the construction, maintenance, and protection of the Canal. To the question of sovereignty is the closely related one of the flag issue.

The American flag was not flown in the Canal Zone until 1906, at which time its absence was criticized by a Senate investigating team. Gradually, the Panamanian flag disappeared from the Canal Zone. With the growth of nationalism this has come to be bitterly resented by the Panamanians who feel

[53] Martz, *Central America*, p. 291-293.

[54] Mario Rodríguez, *Central America* (Englewood Cliffs, N.J., 1965), p. 45; *New York Times*, Aug. 21, 1956; Aug. 29, 1956; Sept. 10, 1956.

that their "residual" sovereignty over the Zone should be demonstrated by flying their flag along with the United States flag.

Ernesto de la Guardia, Jr., a prominent businessman, was inaugurated president of Panama in October, 1956, amid the new Canal controversy. Outgoing President Arias was highly critical of the United States Congress for its failure to implement several provisions of the 1955 Treaty, sentiment that was fervently echoed in Panama's newspapers. The refusal of Congress to pass enabling legislation to correct the discriminatory wage levels in the Zone particularly incensed the Panamanians.

The flag issue and the question of sovereignty, causal factors behind the eruption of violence in 1959, were foreshadowed by "Operation Sovereignty," a campaign staged by high school and university students, apparently joined by Communists, in 1958. The student agitation began on May 2 with the planting of Panamanian flags at strategic points in the Canal Zone. The flags were returned without incident, and the students then demanded that President de la Guardia take immediate steps to force recognition of Panama's sovereignty over the Zone. The president's policy of moderation added to the disorder, and his Cabinet was compelled to resign. Dr. Milton Eisenhower, official fact-finder for his brother, President Eisenhower, visited Panama for three days in July 1958, and the uproar continued. Heavily guarded because of the attacks on Vice President Richard M. Nixon in Lima and Caracas in May, Dr. Eisenhower refused to meet the students except at the United States Embassy. The students refused the invitation, and responded by picketing the Embassy with signs such as: "50% of the Canal;" "Milton go back to the U.S.A.;" and "Panama Canal for Panamanians."[55]

Tension continued to rise thereafter, and on Panama's independence day, November 3, 1959, a group of Panamanian students entered the Canal Zone with their flag, and attempted to unfurl it. Governor William E. Potter ordered his troops to fix bayonets, and in the encounter which followed thirty Panamanians were injured. Panamanian sensibilities were wounded further by Potter's statement: "To an Anglo-Saxon, a contract is binding but to a Spaniard a contract is merely an extension of his personal ambitions. The Spaniard has an absolute lack of personal discipline. He thinks the law is a fine thing, but not applicable to him."[56]

The gravity of the situation was appreciated by President Eisenhower, and he acted quickly to put in motion a nine point program for the Canal Zone, which included better pay and more supervisory jobs for Panamanians, housing improvements, and a reduction of water rates. In deference to nationalistic feelings, he ordered the flag of Panama to be flown along with the United States flag in two areas of the Zone. The Department of Defense had previously opposed any compromises on the flag question, and a sub-committee of the House of Representatives now was most critical of the president's policy, contending that the American flag had been the symbol of exclusive United States sovereignty in the Zone.[57]

[55] Harding, *Untold Story*, p. 139; *New York Times*, July 14, 1958; July 15, 1958.
[56] Rodríguez, *Central America*, p. 46; Dept. of State *Bulletin*, XLI (Nov. 23, 1959), 759; *New York Times*, Nov. 4, 1959.
[57] *New York Times*, Dec. 2, 1959; Jan. 16, 1960; Sept. 18, 1960.

Relative calm prevailed in Panama until October 12, 1962, Columbus Day, and *El Día de la Raza*, a date symbolizing the unity of the peoples of the two Americas. This day, selected for the opening of the Thatcher Ferry Bridge over the Pacific entrance to the Panama Canal, witnessed large demonstrations by slogan-shouting, angry Panamanian youths. The immediate cause of their anger was the impolitic decision of the United States Congress to name the bridge for the first civilian governor of the Canal Zone, Maurice H. Thatcher, ignoring the fact that the Panamanian National Assembly wished it to be called the "Bridge of the Americas."[58]

President Roberto F. Chiari of Panama had meanwhile initiated discussions with President John F. Kennedy on the subject of revising the Canal treaties in 1961, and in June 1962, while President Chiari was in Washington, an agreement was reached to appoint high-level representatives to explore sources of dissatisfaction arising from the treaty relationship. The resulting Joint Panama-United States Commission, composed of the American Ambassador, the Canal Zone Governor, and two Panamanian officials, met periodically until July 1963, when it was dissolved by the Panamanian chief executive. After a year of negotiations it had been agreed that: (1) the flag of Panama should be flown with the flag of the United States on land in the Canal Zone where the United States flag is flown officially by civilian authorities; and (2) foreign consuls holding exequaturs should be allowed to function in the Zone.[59]

The ground swell of Panamanian resentment continued to rise, and high-level diplomacy grew progressively ineffectual. This became manifest in the light of events occurring in Panama on January 9-10, 1964, which caused United States-Panamanian relations to fall to their lowest level in history. In implementing the display-of-flags agreement just mentioned, the Canal Zone governor reviewed the flag sites and decided to eliminate some of them. Outdoor flags at schools were among those to be removed. On January 7, 1964, American students at Balboa High School hoisted a flag of their own in defiance of the Governor's orders. Two days later Panamanian students attempted to display their flag and the sanguinary and destructive disorders ensued.[60]

Ellsworth Bunker, United States Ambassador to the Organization of American States, declared that mobs "infiltrated and led by extremists, including persons trained in Communist countries" assaulted the United States-controlled Canal Zone. United States forces "never attempted to enter Panama, but acted only to protect the laws and property of its citizens . . . No small portion of the Panamanian casualties were caused by the Panamanians themselves." After peace had been restored it was found that twenty-four persons, including four Americans, had lost their lives, and more than two hundred were injured; much United States-owned property was destroyed or damaged in Panama City and Colón. President Lyndon B. Johnson sent Thomas C. Mann, assistant secretary of state, at once to the Canal Zone, and moved quickly to secure the appointment of Jack Hood Vaughan as Ambassador to Panama.

[58] J. C. J. Metford, "The Background to Panama," *International Affairs*, 40 (Apr., 1964), 277.

[59] Dept of State *Bulletin*, XLVII (July 9, 1962), 81-92; *Ibid.*, XLIX (Aug. 12, 1963), 246-247.

[60] *New York Times*, Jan. 10, 1964.

President Chiari suspended diplomatic relations with the United States, and bitterly denounced the Treaty of 1903.[61]

Panama, charging that the United States "by using force to silence the Panamanian people" had "violated the Inter-American Treaty of Mutual Assistance," called for a meeting of the O.A.S. Council under terms of the Rio de Janeiro Pact of 1947, which provides that each of the Hemisphere's republics will help another threatened by aggression. Following a joint request by the governments of Panama and the United States, the president of the Inter-American Peace Committee convened a special meeting on January 10. Although the Committee repudiated Panama's charges of aggression, and the International Commission of Jurists in Geneva absolved the United States of violating human rights, it had become evident that fundamental and sweeping changes in the treaty relationship were of urgent and crucial importance.[62]

To other Latin American peoples the Panama conflict was widely regarded as a reflection of their own problems with the United States. Non-Communist political leaders generally supported Panama's hopes of "raising the rent" for the Canal Zone by a revision of the governing treaty with the United States. Latin Americans were also sympathetic toward Panama's desire to obtain clearer national sovereignty over the Zone. Communist sources unleashed a campaign of vilification against the United States for the "Panama crime," and interpreted the disorders as revolutionary uprisings for the "liberation" of Panama.[63]

Marco Aurelio Robles was elected president of Panama, May 10, 1964, on a platform promising a "hard line" toward Washington, and a greater Panamanian stake in the Canal. In November 1964, members of the Federation of University Students, joined by labor groups, staged a demonstration shouting "throw the Yankees out of Panama." The students' main objective was the recall of the ambassador to the United States, Ricardo Arias, whom they claimed advocated a "soft" policy on the treaty talks. Whereas the government had failed to call out the National Guard on January 9-10, 1964, on this occasion Guard units blocked the way to the Canal Zone border. President Robles' policy was one of the sternest ever taken by a Panamanian president toward students.[64]

In the immediate wake of his own election President Johnson proposed on December 18, 1964, that the two countries negotiate a new treaty, and announced United States plans to build a sea-level canal. Pointing up the need for such a waterway, the President said: ". . . Already more than three-hundred ships built or building are too big to go through with full loads. Many of them, like our own modern aircraft carriers, cannot go through at all . . . I think it is time to plan in earnest for a sea-level canal. Such a canal will be more modern, more economical, and will be easier to defend. It will be free of complex, costly, vulnerable locks and seaways . . ."[65] An Atlantic-Pacific Inter-oceanic Canal

[61] *Christian Science Monitor*, Feb. 3, 1964. Joseph S. Farland, United States ambassador to Panama, had resigned in Aug., 1963, following a disagreement with the State Department, and a successor was not appointed. Members of the Senate Foreign Relations Committee were critical of the Johnson administration for leaving the post vacant. Panamanian-United States diplomatic relations were restored in Apr., 1964.
[62] *New York Times*, June 10, 1964.
[63] *Ibid.*, Jan. 19, 1964.
[64] *Christian Science Monitor*, Nov. 25, 1964.
[65] Dept. of State *Bulletin*, LII (Jan. 4, 1965), 5.

Study Commission was created by Congress in 1964 to study possible sites for a sea-level canal. Two sites appeared most feasible: the so-called Sasardi-Bosti route in Panama, about two hundred miles east of the present Canal, and a route through northern Colombia. The Atomic Energy Commission was directed to examine the possibility of excavation in these places by nuclear explosives. Also under study was the feasibility of converting the present Canal to a sea-level route. Another possibility lies along the border of Costa Rica and Nicaragua, but because of engineering difficulties this route is given lower priority than the others.[66]

In entering into negotiations with Panama in search of possible solutions to the Panama Canal question, the United States faced several alternatives: (1) abdicating in favor of Panama; (2) inter-Americanization of the Canal; (3) internationalization of the Canal; (4) partnership with Panama. Proponents of the United States relinquishing all control of the Canal to Panama constituted a minority of extremists until the 1960s when a violent upsurge of nationalistic sentiment occurred. This feeling gained ground in spite of the fact that Panama would probably derive more benefit from the Canal with the United States in the Canal Zone. Agitation from the Communist-bloc countries for Panama's control of the Canal commonly strikes an analogy between the Suez and Panama canals. However, their legal status differs markedly, the Suez agreement having been made between the Egyptian government and a private company for a period of ninety-nine years. The Panama agreement is of treaty status between two sovereign powers, and in perpetuity.

A proposal for inter-Americanizing the Canal was made by Dr. José Figueres, former President of Costa Rica, who suggested that the United States might transfer the Canal Zone to the OAS, moving the headquarters of the OAS to the Canal Zone; a customs union would be instituted, and the United States would retain responsibility for defending the area. Former President Ernesto de la Guardia rejected this plan, and there was no evidence of support for it in the United States.

In what was widely interpreted as a proposal for internationalization President Harry F. Truman said that the Panama Canal, together with the Kiel Canal, the Rhine-Danube waterway, the Black Sea Straits, and the Suez Canal should be made free waterways for all countries. More specific proposals for internationalization of the Panama Canal envisioned placing the waterway under a United Nations agency responsible to the nations as a whole, and made up of representatives of all the Canal's main users on a global basis. Contending that the control of the Canal is no longer vital to the United States, it has been suggested that the preservation of international waterways is a traditional American policy.[67]

Following President Johnson's declaration regarding the abrogation of the 1903 treaty, a United States-Panamanian committee was selected to consider the question. Washington appointed Robert Anderson, former secretary of the

[66] Atlantic-Pacific Interoceanic Canal Study Commission, *Third Annual Report* (Washington, 1967); James H. Stratton, "Sea-Level Canal: How and Where? ," *Foreign Affairs*, XLIII (Apr., 1965), 513-518.

[67] Martin B. Travis and James T. Watkins, "Control of the Panama Canal: An Obsolete Shibboleth? ," *Foreign Affairs*, 37 (Apr., 1959), pp. 407-418.

Treasury under President Eisenhower, and John N. Irwin; representing Panama were Ambassador Ricardo Arias, Diogenes de la Rosa and Roberto Alemán. The negotiators, working within the neutral-grounds confines of the United Nations, had by June, 1967, reached agreement on new treaties governing control of the Panama Canal, and possible construction of a sea-level canal. The treaties' provisions could be interpreted as creating a partnership between the two nations in the administration of the present or future waterways, for the United States would surrender its historic sovereignty over the Panama Canal Zone, and Panamanians would help run the existing Canal, as well as any sea-level canal in their territory. It would have returned to Panama a large portion of the land then in the Canal Zone, and assured her of at least $17 million in canal tolls. Washington insisted that the revocation of the 1903 treaty be made contingent on the negotiation of two additional treaties, the first describing conditions for the continuing presence of United States military bases in Panama, and the other relating to conditions under which a sea-level canal might be built across Panama. As in the past, the new treaties specified that the present Canal and any new sea-level canal be open at all times to vessels of all nations.[68]

When the treaties were concluded it appeared that the negotiators had found a formula satisfactory to both countries. Being aware, however, that objections were bound to arise, the Johnson and Robles governments decided to resolve the problem in secret, by negotiations carried on without exposure to the public. It was believed that a concensus could be reached more readily in this manner than by making public the texts of the treaties, explaining them to their people, and urging ratification. The plan miscarried, for in Panama the opponents of any compromise divulged portions of the texts, along with their own denunciations. The resulting display of popular disapproval compelled President Robles to delay any attempt at ratification.[69]

Popular opinion in the United States, at first almost wholly favorable towards the treaties, was stirred by protests from Representive Daniel J. Flood of Pennsylvania, Representative H. R. Gross of Iowa and Senator Strom Thurmond of South Carolina, who were outraged at Washington's willingness to acknowledge Panama's sovereignty over the Canal Zone. These individuals supported the American Emergency Committee on the Panama Canal, which sought to block ratifications of the new treaties.

The status of the treaties in Panama was further confused by events occurring in Panama's electoral campaign of 1968. With the presidential election of May 12 approaching, the opposition assembly deputies, who favored the candidacy of Dr. Arnulfo Arias, voted to impeach President Robles for his alleged tampering with the electoral process in favor of Dr. Arias's rival, David Samudio. Dr. Arias, who had repudiated the unsigned drafts of the three canal treaties during his campaign, calling them a "sell out," was officially declared president-elect on May 30, 1968, amid a bitter anti-United States atmosphere. Newspapers and radio stations which had backed Samudio charged Washington with open intervention through agents of the Central Intelligence Agency and Panama's National Guard in imposing the candidacy of Dr. Arias. Both

[68] *Wall Street Journal*, Sept. 6, 1967.
[69] Carl T. Rowan, "New Peril to the Panama Canal," *Reader's Digest*, 92 (Mar., 1968), p. 182.

candidates agreed, for different reasons, that the election had been "one of the most shameful in the history of the country," as Dr. Arias put it. The destruction, theft or seizure of ballot boxes, the altering of results on official tally sheets and the various pressures applied to voters were widespread.

Dr. Arias, a graduate of Harvard Medical School, and a member of the traditional ruling elite, assumed the presidency on October 11, 1968. He held office for eleven days at which point his efforts to control the country's military caused the National Guard to force him out. Seeking refuge in the Canal Zone, Arias' presence constituted an embarassment for the U.S. Embassy. A military junta headed by Brigadier General Omar Torrijos Herrera, and including colonels Boris Martínez, José Pinilla and Bolívar Urrutia, seized control of the nation and quickly suspended all constitutional guarantees and civil liberties. Commonly regarded as nonpolitical, few Latin American military units have received more training and assistance from the United States armed services than those of Panama.[70]

Behind the Arias candidacy was a strange coalition. After years of denouncing the Panamanian oligarchy, of which he is an unorthodox member, Dr. Arias, a "man of the people," who had enjoyed the support of the masses, accepted the oligarchy's support, as had his opponent. He contended that the union was made to further economic progress. However, in a country where for several decades both the government and the economy had been dominated by hardly more than thirty families, his motives were questioned by foreign observers. Beyond the power elite are more than 1.3 million Panamanians most of whom live in obscurity in the slums of the capital, or in rural hovels. They function as a political force to the degree that they emerge to vote in elections. But their contribution and involvement is open to question, since the political process in Panama has so often suffered from intimidation and questionable practices.[71]

The Panamanian population is composed chiefly of mixed racial elements, the majority, 50-65 percent, being the mulatto-mestizo segment. Pure Indians make up 10 per cent while Caucasians and Negroes each approximate 10-15 percent. The whites rank first on the social scale followed by mestizos, mulattoes and Negroes. Deep social rifts are found between these groups. Although the average annual per capita income was fixed at $487 in the late 1960s, placing Panama fifth among the Latin American nations, the population was generally impoverished. One of the most serious obstacles to economic development was the land tenure system. It was estimated that 1.5 percent of the population owned 50 percent of the arable land, and that only 3 percent of the land was being utilized. The lack of capital, or failure to invest available capital, inadequate transportation facilities, antiquated farming methods, and the prevalence of graft, corruption and nepotism helped to retard economic growth.[72]

Panama's economy has been characterized by an apparent unfavorable balance of trade, but this has been compensated in recent years by income from

[70] *New York Times*, May 31, 1968; *Ibid.*, Oct. 12, 1968; *Christian Science Monitor*, Feb. 20, 1970.
[71] *New York Times*, Mar. 27, 1968.
[72] See Louis K. Harris, "Panama," in Ben G. Burnett and Kenneth F. Johnson, eds., *Political Forces in Latin America, Dimensions of the Quest for Stability* (Belmont, Calif., 1968), pp. 115-143.

the Canal. The Canal and all operations connected with it produced $115 million in 1967, or one-sixth of the gross national product. Bananas, the production of which is largely controlled by the United Fruit Company, represented 35 percent of the total exports by value in 1963. Petroleum exports accounted for 32.5 percent followed by shrimp, cacao, meat and sugar. In spite of the Panamanians' restiveness in the 1960s, and a commonly high level of unemployment, the republic's economy showed a remarkable expansion in that period. From 1961-1967 Panama's yearly average increase in gross national product was 8.4 percent, the highest in Latin America. Panama profited from a number of factors which were helping to make it a commercial crossroads of the world. Its dollar-based economy was attractive to foreign investors, more income was gained from record use of the Canal by ocean-going ships, and trade with other isthmian countries was stimulated by the Central American Common Market. Washington's desire to maintain a friendly and stable political climate was translated into large amounts of aid funds, and while the canal treaties had not been ratified, their existence established the principle of greater direct financial benefit to the government and private business sectors, and this prospect helped to create optimism. A highly promising service industry was banking. Panama had thirteen wholly or partly-owned foreign banks in 1967, making her an important regional financial center. Agriculture lagged behind manufacturing in growth, and the government prepared to start a new phase of its rural development program.[73]

The momentum of economic growth continued with an 8.6 percent gain in gross national product in 1970, which was believed to have carried through 1971. Government expenditures in the construction industry may have helped to sustain this growth. The regime expropriated some large estates, implementing land reform promises, and gave support to increased credit and investment in agriculture. Describing itself ambiguously as "revolutionary" and neither Communist nor capitalist, the regime, headed by "Supreme Leader," General Omar Torrijos Herrera, indicated that "with a well-defined economic policy, the private sector will develop the economy, and the state will develop opportunities for betterment with the people's full participation."[74]

Economic assistance from Washington had made substantial contributions to Panama's advance, a fact acknowledge by rural Panamanians who had not been disturbed by the Canal controversy. Economic assistance committed to Panama by the United States in the period 1948-1970 amounted to $156.5 million; indirect assistance channeled through regional and international organizations totaled more than $20 million. To this was added the services rendered by technical experts and more than one hundred Peace Corps Volunteers. Economically, the country has been closely tied to the United States: U.S. exports to Panama reached $208 million, and imports from Panama totaled $76 million, in 1970.[75]

Wishing to resolve what had become one of the most bitter and enduring bilateral disputes in hemispheric relations, the Nixon administration moved

[73]Dept. of State, *Latin America Growth Trends–Seven Years of the Alliance for Progress*, April, 1968, p. 54; *New York Times*, January 22, 1968. The discovery of copper deposits in 1968 offered new hope for the mining industry. See *Wall Street Journal*, February 26, 1969.

[74]*New York Times*, January 28, 1972.

[75]*Statistical Abstract of the United States 1971*, p. 763; p. 768.

cautiously toward further negotiations on the Canal treaty. After the military coup in 1968, followed by the closing of Congress, Washington hesitated to negotiate with a regime whose signature might not be honored by future Panamanian administrations. When it had become reasonably clear the Torrijos was not a transitory figure and his popularity was sufficient to gain broad national acceptance for a Canal treaty, Washington pushed negotiations. A Panamanian negotiating team headed by José Antonio de la Ossa, Ambassador to the United States, and a U.S. delegation led by Robert B. Anderson, had reached only tentative agreement on some aspects of the treaties in late 1971 after four months of negotiation.[76]

The Department of State summarized the U.S. position as follows:

> Primary U.S. objectives under the Nixon administration are continued U.S. control and defense of the existing canal. The rights (without obligation) to expand the existing canal or to build a sea-level canal are essential to U.S. agreement to a new treaty, with the exact conditions to accompany these rights to be determined by negotiation. The U.S. is willing to provide greater economic benefits from the canal for Panama and release unneeded land areas, again with exact terms to be developed by negotiation.
>
> Negotiations between the United States and Panama began June 29, 1971. Important issues such as duration, jurisdiction, land and water requirements, expansion of canal capacity and compensation are now being explored, but no agreements have been reached. September 1971.[77]

The largely government-controlled press in Panama continued to publish frequent reports critical of U.S. activities in the Canal Zone. General Torrijos, celebrating his third year as Panama's ruler, declared that if necessary Panamanians would die to regain sovereignty over the Panama Canal Zone, and that "no people would like to see a foreign flag flying in the heart of their country." On another occasion he said: "in negotiations with the United States over the Canal General Torrijos will end up standing or dead, but never on his knees." A long standing foe of any concessions to Panama on the treaty question, congressman Daniel J. Flood, viewed the negotiations and news stories as a "studied propaganda campaign by the executive branch to sway people to agree to the monstrous proposal."[78]

The negotiations were continued, but as of January 1973 progress toward agreement on basic issues seemed negligible. At that point Panama held that a new canal treaty should run only until 1994; that the Republic's jurisdiction over the Zone be declared immediately; that U.S. police functions in the Zone terminate within five years; that the Canal acquire political neutrality instead of continuing as a wholly American waterway; and that the U.S. Military Southern command be removed from its territory. Moreover, Panama categorically rejected Washington's proposal for a "system of canals." Thus, after about nine years of negotiations Washington and Panama found themselves almost as far apart on their views concerning a new treaty as at the time of the crisis of 1964.[79]

[76] *Times of the Americas*, November 3, 1971.
[77] "The State Department Position Paper on Negotiations Concerning a Revised Panama Canal Treaty," *Inter-American Economic Affairs* (Winter, 1971), 25; 92-96.
[78] *Times of the Americas*, November 24, 1971; *Christian Science Monitor*, October 12, 1971.
[79] *New York Times*, December 24, 1972; *Times of the Americas*, January 3, 1973.

The United States' assumption of credit for the emergence of Panama as one of the most literate, healthy, developing countries of Latin America rather than a disease-ridden backward nation, is based on a solid foundation of evidence. However, with the advance of modern communications, increasing literacy, and the spread of communism, the Panamanians, like their counterparts in other developing countries, have become discontented and more easily manipulated by demagogues, various alienated groups, and Communists. Playing on the theme of inequality of United States policies in the Canal Zone, the reigning oligarchy has been able to divert the masses' attention from political and economic inefficiency, corruption and injustice, for which it is responsible, to the Americans. Thus, the United States became a ready-made scapegoat for Panama's own failures. Panamanian politicians have found it easy to cast the northern colossus in this role because of problems relating to the Canal Zone, its position as the world's most powerful and prosperous nation, and the invidious contrast in the level of living in the Zone with other parts of Panama. In view of these circumstances it seems likely that the presence of the United States in the Zone has helped to perpetuate the *status quo* in Panama. Whatever the outcome of negotiations on the new canal treaties, there will be a continuing need to study Panamanian politics within the context of problems related to the existing and any future canal.[80]

As we have noted, Panama has always had special military and commercial significance for the United States. The present Canal shortens by 7,873 miles the sea distance between New York and San Francisco; and in recent years 82 percent of the commercial cargo transiting the Canal originates in or is destined for the United States. It has served as a supply line for the United States and its allies in two world wars, and during the so-called cold war. In World War II 5,300 combat vessels and 8,500 craft carrying troops and cargo crossed the waterway. Prior to World War II the United States had essentially a one-ocean navy and the Canal was vital in order that the fleets in either ocean could be quickly reinforced. The United States built a many-ocean fleet during and after that war, and with large fleets in both the Atlantic and Pacific the need for reinforcing either is less critical than in the past. Moreover, since the development of nuclear weapons and missiles the present lock-canal is virtually indefensible. One nuclear explosion could block the Canal, therefore it can no longer be counted upon in war time.

United States Defense Department officials contend, however, that in a strategic sense an isthmian canal is more important than ever before, and that in the future its purely military significance may increase. In the Navy, for instance, the missile-firing nuclear-power submarine is becoming the capital ship of the fleet, and anti-submarine warfare vessels, "killer" submarines and missile-firing surface ships are supplanting and may eventually displace the aircraft carriers; these ships are small enough to use a sea-level canal, or even the existing one.

From an economic point of view an isthmian waterway is of vital defense interest to industry in the United States. Strategic raw materials essential to the

[80] J. Fred Rippy, "The United States and Panama: The High Cost of Appeasement," *Inter-American Economic Affairs,* 17 (Spring, 1964), 87-88; Larry L. Pippin, "The Challenge of Panama," *Current History*, 50 (January, 1966), pp. 6-7; 53.

economy of the United States in peace and war pass through the Canal from the West Coast of South America, and millions of tons of oil are carried from coast to coast each year. With the growth of industrialization and cooperative marketing arrangements among the Latin American countries its logistical importance will doubtless increase. The isthmian waterway may well become a keystone of Hemispheric industrial security in the future.

SUPPLEMENTARY READINGS

Abbott, Willis J. *Panama and the Canal*. New York, 1944.

Arce, Enrique J., and Ernesto J. Castillero. *Guía Histórica de Panama*. Panama, 1942.

Biesanz, John and Mavis. *The People of Panama*. New York, 1964.

Bishop, J. B. *The Panama Gateway*. New York, 1913.

Bunau-Varilla, Philippe. *Panama: The Creation, Destruction, and Resurrection*. New York, 1914.

Calcott, W. H. *The Caribbean Policy of the United States, 1890-1920*. Baltimore, 1942.

Castillero Reyes, Ernesto J. *Episodios de la Independencia de Panama*. Panama, 1957.

Colombia, Cámara de Represantes de Colombia. *Investigacíon sobre la rebelíon del Istmo de Panama*. Bogotá, 1915.

Colombia, Ministerio de Relaciones Exteriores. *Libro Azul de Colombia. Documentos diplomáticos sobre el Canal y la rebelíon del Istmo de Panama*. Bogotá, 1904.

Dean, Arthur H. *William Nelson Cromwell*. New York, 1957.

Dennet, Tyler. *John Hay*. New York, 1933.

Dennis, A. L. P. *Adventures in American Diplomacy*, 1896-1906. New York, 1928.

Du Val, Miles P., Jr. *And the Mountains will Move*. Stanford, 1947.

———. *Cadiz to Cathay*. Stanford, 1947.

Ealy, Lawrence O. *The Republic of Panama in World Affairs, 1903-1950*. Philadelphia, 1951.

Favell, T. R. *The Antecedents of Panama's Separation from Colombia*. Ph.D. dissertation. Fletcher School of Law and Diplomacy.

Harding, Earl. *The Untold Story of Panama*. New York, 1959.

Hebard, R. W. *The Panama Railroad: The First Transcontinental, 1855-1955*. New York, N.D.

Howarth, David. *Panama: Four Hundred Years of Dreams and Cruelty*. New York, 1966.

Jones, Chester L. *The Caribbean Since 1900*. New York, 1936.

Lopez, Georgina Jiménez de. *Panama in Transition. Period, 1849-1940*. Ph.D. dissertation. Columbia University.

Mack, Gerstle. *The Land Divided, A History of the Panama Canal and other Isthmian Canal Projects*. New York, 1944.

Martz, John D. *Central America: The Crisis and the Challenge*. Chapel Hill, N.C., 1959.

McCain, William D. *The United States and the Republic of Panama*. Durham, N.C., 1937.

Mecham, J. Lloyd. *A Survey of United States-Latin American Relations*. New York, 1965.

———. *The United States and inter-American Security, 1889-1960*. Austin, Texas, 1961.

Mellander, G. *The United States in Panamanian Politics, the Intriguing, Formative Years*. Danville, Ill., 1971.

Miner, Dwight C. *The Fight for the Panama Route*. New York, 1940.

Munro, Dana G. *Intervention and Dollar Diplomacy in the Caribbean, 1900-1921*. Princeton, 1964.

Niemeier, Jean G. *The Panama Story*. Portland, Ore., 1968.

Padelford, Norman J. *The Panama Canal in Peace and War*. New York, 1942.

Panama, Junta Nacional del Cincuentenario, ed. *Panama: 50 Años de República Panama*. Panama, 1953.

Parks, E. T. *Colombia and the United States, 1765-1934.* Durham, N.C., 1935.

Pippin, Larry L. *The Remón Era; An Analysis of a Decade of Events in Panama, 1947-1957.* Stanford, 1964.

Rippy, J. Fred. *The Capitalists and Colombia.* New York, 1931.

———. *The Caribbean Danger Zone.* New York, 1940.

Roosevelt, Theodore. *Theodore Roosevelt: Autobiography.* New York, 1919.

Sands, William F., and Joseph M. Lalley. *Our Jungle Diplomacy.* Chapel Hill, N.C., 1944.

Uribe, Antonio José. *Colombia y los Estados Unidos de América.* Bogotá, 1931.

United States Congress. *Diplomatic History of the Panama Canal, Sen. Doc. No. 474,* 63rd Cong., 2nd Sess.

United States Congress. *The Story of Panama*: Hearings on the Rainey Resolution before the Committee on Foreign Affairs of the House of Representatives. Washington, D.C., 1913.

United States Congress. Thompson, Hon. Clark W. "Isthmian Canal Policy of the United States–Documentation." *Congressional Record,* March 23, 1955.

7

Mexico and the United States

Mexico is a nation of remarkable geographical, cultural, and human diversity, and it has some features which make it distinct from the other Latin American countries. Most important is its proximity to the United States with which it shares a common boundary. Secondly, Mexico was established by Spain in the colonial period as a separate entity from the viceroyalties in South America, and it had more direct contact with East Asia, via the Manila-Acapulco trade, than other regions of the Spanish Empire in the New World. Distance and lack of communications until the age of the airplane have tended to perpetuate the separation of the two regions. In common with Central America, Mexico has coasts on the Caribbean and the Pacific, interior mountains and scattered clusters of population; however, it is larger, shows greater physical contrasts and is more diversified economically.

With an area of 759,530 square miles, Mexico is the fourth largest country among the American nations, only Argentina, Brazil and the United States being larger, and ninth largest in the world. Mexico's *mesa central*, a great plateau from 5000 to 8000 feet above sea level, accounts for about 14 percent of the land area, is the site of the nation's capital, and the home of the bulk of the population. Stretching northward from the plateau is a sloping, sparsely populated and generally arid region extending to the United States. The Pacific coast, lying to the west and cut off by high mountains, remains lightly populated. While easier of access from the central plateau, the Gulf of Mexico slope, being tropical and more subject to health problems, is relatively undeveloped. South of the *mesa central* is a mountainous area where population clusters are found in many pockets and valleys. The southeastern peninsula was not connected with the rest of the country by road until 1960. Mexico includes the arid peninsula of Baja California, which, except for a small strip around Tia Juana and Mexicali, is virtually unpopulated and offers few passable roads.

Mexico lies on the Tropic of Cancer roughly in the latitude of the Sudan and Egypt. But the climate is complicated by the great variety of relief features and the presence of two seas with somewhat different temperature conditions at the same latitudes. Temperatures decrease with altitude and bring temperate climatic conditions into the higher central basins. Three zones are recognized. The belt of high temperatures extends from sea-level to about 2000 feet. And the temperate zone, where most of the population is found, ranges from 2000 to 8000 feet. Above that level temperatures are quite cool.

Much of the area of modern Mexico had been the home of a relatively advanced civilization before the arrival of the Spaniards in the early sixteenth century. After defeating the Aztecs the Spaniards quickly seized control of the densely populated high basins and brought the Indians into their labor scheme. As few Spanish women came to New Spain (Mexico) in the colonial period there was an extensive mingling of races. This factor, coupled with the comparatively small influx of outsiders, led to the rise of the mestizo as a distinct population element. Reflecting one of the world's fastest rates of increase, Mexico's population surged from 5,500,000 in 1805, to 15,150,000 in 1910, to almost forty-eight million in 1970. Of the total population some 10 percent are white, 60 percent mestizo, and 30 percent Indian, but this is only a rough estimate based on cultural features rather than physiological characteristics. Negroes, introduced as slaves in the colonial period, have been largely assimilated into the Mexican race. According to the 1950 census the rural population was 57 percent of the total, but in 1960 it had declined to 49 percent.

Mexico's economy has traditionally been based on agricultural products and mineral raw materials; however, industrialization has become progressively important in recent times, attaining 27 percent of the gross national product in the late 1960s. Rich in mineral resources, Mexico ranks first in the world's production of silver, third in lead, cadmium and antimony, fifth in zinc and seventh in copper. Iron ore, petroleum and coal remain important, especially for domestic use. While the traditional farming sector has lagged in production, an increasing proportion of agriculture, aided by irrigation programs, has reached a commercial level sufficient to permit Mexico, once an importer of many foodstuffs, to become self-sufficient in most items and even to export such grains as wheat. Mexico is, nevertheless, a comparatively poor nation suffering from a shortage of arable land and steeply rising population pressure. Only 10 percent of Mexico's total land is arable, and of this about half is so dry it cannot be utilized.

Relations between Mexico and the United States since the early nineteenth century have shown the irregularity and unpredictability of a geological fault line, producing at times diplomatic upheavals followed by subsidence and apparent calm. The period 1820-1910 reflected undulations in the attitude of the United States toward Mexico moving from cautious friendship in the beginning to open hostility and back to friendship and encouragement. The Mexican Revolution and its impact on American interests in Mexico led to intervention, the exercise of diplomatic pressure and general estrangement. The reciprocal attitude of Mexico was one of hope, anxiety, fear, and suspicion. It has only been in the era of the "Good Neighbor" policy, approaching mid-century, that the relationship has been mutually satisfactory.

Mexico and the United States share 1,935 miles of common frontier, a fact of inestimable importance in their relationship. This propinquity has done much to help form the attitudes and psychology of our peoples. History as well as geography has made the two nations continuously important to each other, for they were both a product of European settlement and shared a colonial status. Historical parallels, however, soon gave way to divergences in the evolution of the neighboring countries, which emphasized a growing disparity in the elements of national power and prestige.

In the half-century following Mexico's independence from Spain in 1821 the United States underwent great expansion in population, territory and industrial development, whereas Mexico emerged as a weak "client" state, characterized by arrested economic growth and political instability. Territorial expansion was justified by Americans in terms of a "Manifest Destiny," which envisioned the United States as a continental republic, and a world power. Mexico felt insecure in this period, and with good reason, for by the revolt and annexation of Texas, the Treaty of Guadalupe Hidalgo and the Gadsden Purchase, she lost nearly a million square miles of territory to her more powerful northern neighbor.

When our diplomatic relations with Mexico began, they promised to be cordial. Many Americans were sympathetic towards the Mexicans in their efforts to overthrow Spanish domination. The United States extended *de facto* recognition in 1822, at the same time urging this action by Spain;[1] however, formal diplomatic relations were not established until 1825, incident to the arrival of the first United States minister. The proclamation of the Monroe Doctrine in 1823 had slight impact in Mexico. The Mexicans regarded British recognition as more vital, and they also perceived correctly that the Monroe Doctrine was virtually meaningless without the active support of British naval power.[2]

The period 1825-1836 brought unfortunate developments in relations which were accented by differences in historical background, race, temperament and religion. In 1826 Joel Poinsett of South Carolina was appointed first minister to Mexico by President John Quincy Adams. Poinsett's appointment proved to be ill-advised, for in an extra-official capacity he undiplomatically aided in the formation of York rite Masons to oppose monarchical tendencies exerted through the Scottish rite Masonic Lodges. This antagonized the Mexican conservatives and his recall was demanded.[3] Further suspicion and distrust was aroused by the activities of Colonel Anthony Butler, our second minister to Mexico. Commissioned by President Andrew Jackson to purchase a portion or all of Texas, Butler remained insistent after the offer had been rejected by the Mexican government, and was ordered out of the country after it was revealed that he was organizing an insurrection in Texas, where he owned land. It has been suggested, on good evidence, that his sole qualifications for the post were "an acquaintance with Texas and a strong desire to see the United States obtain it."[4] In spite of these two unfortunate ministerial appointments, a treaty was signed in 1828 confirming the boundary fixed by the United States and Spain in 1819, and a treaty of amity and commerce was concluded three years later.

[1] William S. Robertson, "The Recognition of the Hispanic American Nations by the United States," *Hispanic American Historical Review*, I (1918), 261.
[2] David Y. Thomas, *One Hundred Years of the Monroe Doctrine, 1823-1923* (New York, 1923), p. 43.
[3] William R. Manning, *Early Diplomatic Relations between the United States and Mexico* (Baltimore, 1916), p. 191.
[4] J. H. Smith, *The War with Mexico* (New York, 1919), I, 62.

Because the United States government had instructed both Poinsett and Butler to renegotiate the Texas boundary, or to acquire the territory by purchase, each effort failing in its purpose, the Mexicans grew increasingly suspicious that its northern neighbor would incite the Texans to rebellion. Since 1819 the eastern boundary of the United States and Mexico was formed by the Sabine River, but the former wished to extend it to the Rio Grande or Colorado river.

Faced with the possibility of an independence movement in Texas, the Mexican Congress in 1830 closed the area to immigration from the United States, suspended land grants to colonists, and abolished the powers of various states of the Republic, a measure that the Texans took to be aimed specifically at them.[5]

In 1836 the Americans living in Texas proclaimed their independence, but it was soon evident that Mexico was not willing to let Texas go. A Texas garrison at the Alamo mission in San Antonio was exterminated, and another at Goliad suffered about the same fate. General Sam Houston, emerging as the national hero of Texas, kept a small army together, and at the Battle of San Jacinto (April 21, 1836, near present-day Houston), he defeated the Mexican army and took the dictator, General Antonio López de Santa Anna, prisoner. Although the Mexican government refused to recognize the captured dictator's vague promises to withdraw Mexican authority from Texas, and sent armed bands into the area on two occasions in 1842, it made no further major attempts to subdue the province.

The United States government was not responsible for the independence movement in Texas; however, many Americans rendered unneutral assistance to the Texans.[6] The Mexicans had legitimate grievances on the weak enforcement of the neutrality policy, and were particularly resentful when President Jackson extended recognition to Texas in March 1837, just before leaving office.[7]

Not satisfied with independence and recognition, the new Republic sought annexation by the United States. Sound reasons dictated the decision of the Texas leaders. Texas lacked the human and natural resources to support the expenses of a national government, and there remained the ever-present threat that Mexico would try to reconquer the state. But fundamentally, Texas wanted annexation because the great majority of its people were Americans who wished to be part of their country.

The Texan's overtures for annexation were rebuffed by the administrations of Jackson and Martin Van Buren because of the problems of Democratic party unity, and the growing sectional conflict over the expansion of slavery. Finding the United States unreceptive, Texas sought recognition, support and money in Europe. Her leaders talked about creating a vast southwestern nation,

[5] The Mexican government encouraged American immigration in the early 1820s by offering land grants to individuals like Stephen Austin who would agree to colonize families on their concessions. The great majority came from the Southern states, often bringing slaves, although slavery was forbidden in Mexico after 1829. By 1835 about thirty-five thousand Americans were living in Texas.

[6] U.S. Secretary of State Daniel Webster denied the complicity of his government in the Texas independence movement. See *British and Foreign State Papers*, XXXI, 801, and *House Ex. Doc. No. 226*, 27th Cong., 2nd Sess., pp. 7-15.

[7] John H. Latané, *A History of American Foreign Policy* (New York, 1927), p. 241.

stretching to the Pacific, which would rival the United States. England and France moved quickly to recognize Texas and to conclude trade treaties with her. An independent Texas would be a counterbalance to further American expansion; it would supply cotton for European industry and provide a market for European exports. The English and French went so far in their fishing in troubled waters to suggest a treaty whereby Mexico would recognize the independence of Texas in return of the assurance that the latter would never annex itself to a foreign power.

News of the Franco-British interest in Texas reached the United States in 1843-1844 when expansionism was beginning to seize the imagination of a large segment of the American people. Sensing the changing attitude of public opinion, Secretary of State John C. Calhoun submitted an annexation treaty to the Senate in April 1844. Unfortunately for Texas, Calhoun presented annexation as if its only purpose was to extend slavery. Put forward on a sectional basis and offered by an unpopular president, John Tyler, the treaty was soundly defeated. Following the election of James K. Polk, an avowed expansionist, in 1844, Tyler proposed that Texas be annexed by a joint resolution of both houses. Thus circumventing the need for a two-thirds majority in the Senate, the treaty was passed. In December 1845, after Tyler had left office, his successor, Polk, signed a resolution of the Congress under which Texas was admitted as a state.

This prolonged era of strained relations culminated in the Mexican War of 1846-1848, the immediate cause of which was the annexation of Texas. The more remote causes were the events noted above, claims of American citizens against the Mexican government which had never been able to maintain law and order and thereby prevent damage to foreign property interests; claims arising out of seizures by customs officials; the sale of firearms and ammunition to Texan insurrectionists, and the pursuit of Indians by American forces across the Sabine River.[8]

Within a few days after Congress passed the annexation resolution the Mexican minister to Washington was recalled, and a deaf ear was turned to all efforts of the United States to restore diplomatic relations. President Polk, although elected on a platform that called for "the re-annexation of Texas and the re-occupation of Oregon," made a last serious effort to avert the impending crisis. After obtaining Mexican approval on October 15, 1845 for the dispatch of a new envoy to Mexico City, the President appointed John Slidell of Louisiana as the American emissary. The Mexican government's approval had been predicated on the condition that the representative be only a commissioner authorized to negotiate on the Texas boundary question. Polk, however, made Slidell a minister plenipotentiary with authority to discuss claims and to make cash offers for Mexican territory. Slidell's instructions, though confidential, were soon compromised, and the Mexican officials would not negotiate with him.[9] With public opinion smarting from the loss of Texas, any government would

[8] J. B. Moore, *History and Digest of International Arbitration to which the United States has been a Party*. (Washington, D. C., 1898), II, 1209.
[9] For the correspondence of John Slidell see *Sen. Ex. Doc. No. 337*, 29th Cong., 1st Sess., pp. 18-67.

have been overturned had it even considered the American proposals. As it was, a military faction seized control, forestalling the possibility of any negotiations. Following Slidell's report on the failure of his mission, on January 13, 1846, Polk ordered General Zachary Taylor's army to move to the Rio Grande, where it would occupy a position in disputed territory.

It has been charged that the president sent troops into this region, between the Rio Grande and Nueces rivers, with the intention of provoking Mexico to start a war. It should be noted that in Polk's opinion the area was not in dispute but was American territory, and he had a right to occupy it. Whatever his intent, the Mexican forces did not attack, and for a month contented themselves with observing the American army, which had taken a position threatening the Mexican town of Matamoros. The Americans provocatively built a fort there and blockaded the Rio Grande.

After Slidell had returned to Washington, his mission a failure, Polk decided to ask Congress to declare war on the grounds that Mexico had defaulted on its financial obligations, and had insulted the United States by rejecting the Slidell mission. While Polk was preparing his war message word arrived from General Taylor that Mexican troops had crossed the Rio Grande and attacked a unit of American soldiers, several of whom were killed. Polk immediately revised his war message, demanding force to defend the nation against invasion instead of asking for war to redress past grievances. Disregarding some salient facts in the situation he declared that "Mexico has passed the boundary of the United States . . . and shed American blood on American soil," and that "war exists by the act of Mexico herself." Congress accepted Polk's interpretation and on May 13, 1846, declared war by vote of 40 to 2 in the Senate, and 174 to 14 in the House.

Although the country accepted the war with apparent enthusiasm there was more opposition than appeared on the surface. The war was most popular in the Mississippi Valley states, which furnished most of the troops to fight it. In the northeast it was received with coolness, if not disapproval, particularly by Whigs and anti-slavery groups. Even in the older southern states there was a feeling that expansion might be going too far, that the acquisition of too much territory would provoke sectional controversy. The Whigs in Congress supported the war appropriation bills, but became bolder in denouncing "Mr. Polk's war" and its aggressive origins and objectives.[10]

The capacity to wage war was so heavily weighted in favor of the United States that it proved to be a short-term conflict. With the capture of Mexico City by the army of General Winfield Scott on September 14, 1847, following the capture of the heights of Chapultepec, and the successful offensive in New Mexico and California, the fighting was virtually over. Peace was concluded by the Treaty of Guadalupe Hidalgo, signed February 2, 1848. The United States commissioner, Nicholas P. Trist, chief clerk of the State Department, having already been recalled by the president, was without power to negotiate the treaty. Trist ignored this technicality, however, afer concluding that his country

[10]For authoritative treatment of this period see G. L. Rives, *The United States and Mexico, 1821-1848*, 2 vols. (New York, 1913).

was anxious to end the war, as well as the fact that the treaty embodied the essence of Polk's original instructions.[11]

The treaty's terms, increasing the territory of the United States by more than half a million square miles, included New Mexico and Upper California. Mexico also acknowledged the Rio Grande boundary of Texas. In return, the United States contracted to assume the claims of its citizens against Mexico, and to pay $15 million to Mexico. Polk, although displeased with Trist, decided to submit the treaty to the Senate. After all, it secured what the United States had gone to war to obtain, and it was probably the only agreement that Mexico would accept. Moreover, the slavery expansion conflict had taken on new dimensions. Some expansionists in both sections were demanding that the United States hold out for the annexation of all of Mexico, and this allowed the anti-slavery leaders to charge that the southern slave-holders were running the government for their own ends. Approval of the treaty Polk thought, would silence the extremists on both sides, and he accordingly recommended its ratification. By a vote of thirty-eight to fourteen, with a majority of both Democrats and Whigs supporting the treaty, it was approved.

The boundaries established by the treaty did not prove satisfactory to the United States, particularly the southern boundary of New Mexico and the line of the Gila River. In 1853 a convention was signed whereby the United States paid $10 million for the territory, afterwards known as the Gadsden Purchase, with a boundary based on parallels of latitudes. The United States had not occupied the area prior to the purchase of Gadsden, but afterwards believed it expedient to concentrate troops on the border. This action was resented by Mexico.[12]

The decade following the war was a period of internal conflict and anarchy in Mexico, as the pro-clerical conservative elements were challenged by anti-clerical liberals.[13] A short-lived liberal inter regnum produced the democratic constitution of 1857, but to put its principles into practice required an electorate more experienced in democratic government than Mexico then possessed.[14] The internal conflict led to a multiplication of claims and complaints and Washington was besieged by American investors in Mexico to intervene. By 1858 United States claims against Mexico totaled about $10 million. Meanwhile, Mexico was unable to make regular payments on loans obtained from Great Britain, France and Spain. President James Buchanan suggested that this country should assume a protectorate over northern Mexico, and in 1859 he went so far as to propose that the United States "employ a sufficient military force to enter Mexico for the purpose of obtaining indemnity for the past and security for the future."[15] This proposal was not carried out

[11] J. B. Moore, *Digest of Int. Law*, V, 780. For text of treaty see W. M. Malloy, *Treaties, Conventions*, etc. (Washington, D. C., 1910), I, 1107.
[12] *Ibid.*, p. 1121; J. Fred Rippy "The Boundary of Mexico and the Gadsden Treaty," *Hispanic American Historical Review*, IV (1921), 732; See also Paul N. Garber, *The Gadsden Treaty* (Philadelphia, 1923).
[13] Five successive revolutionary governments had been recognized by the United States in the course of a few months. See special message of President Franklin Pearce of May 15, 1856. In James D. Richardson, com., *A compilation of Messages and Papers of the Presidents* (Washington, 1897), V, 368.
[14] For the text of this document see *House Ex. Doc. No. 100*, 37th Cong., 2nd Sess., p. 140.
[15] Richardson, *Messages and Papers*, V, 568.

because the northern majority in Congress was fearful of the implications of the slavery-expansion question at that time.

President Buchanan nevertheless vigorously pushed to conclusion the McLane-Ocampo Treaty in December 1859, whereby the United States was to be granted a right-of-way across the Isthmus of Tehuantepec in perpetuity, two railroad routes across Mexico to the Gulf of California, the right to protect these routes with military forces, and the option to intervene in emergencies without consultation. The treaty was rejected by the United States Senate thus sparing the then embattled Juárez regime the ignominy its acceptance would have brought. The Buchanan policies naturally sharpened Mexico's distrust of its northern neighbor.

When Abraham Lincoln became president, American diplomacy changed abruptly. Lincoln sought to inspire confidence in the Mexicans toward us, and to overcome the ill will that had been generated in the past.[16] Unfortunately, the continuing disturbances in Mexico, foreign intervention there, and the outbreak of the Civil War in the United States precluded the success of his well-intentioned policy.

In 1858 the full-blooded Zapotec Indian, Benito Juárez, *ex-officio* vice-president, became president *de jure* through an uprising against the constitutional president; the United States, unable to maintain diplomatic relations with the faction in control, withdrew its minister. In the following year President Buchanan, relying upon the ultimate success of the Juárez regime, located at Vera Cruz, had sent a minister, thereby recognizing Juárez as the "only existing government." The forecast proved correct, and in December 1860, the Juárez faction overcame the conservative opposition and occupied the capital. The *Juaristas*, finding the treasury empty, were forced to order a suspension of payments for two years on the country's national and foreign obligations.[17]

This decision brought a strong protest from the British and French ministers, who demanded its immediate repeal. When the Mexican government failed to comply, diplomatic relations were severed. As the Spanish representative had already been given his passport, because of his open support of the clerical-oriented party, Spain was prepared to act. A convention for joint action was signed in London, providing for the seizure of the ports and customs as guarantees of payment, and the United States was invited to participate.[18] With the Civil War already begun, our government was unable to protest effectively;[19] nor did Secretary of State William H. Seward feel that we should depart from our traditional policy of abstaining from alliances with foreign nations. The three powers then moved to send fleets and seize Vera Cruz. It was soon evident, however, that the French were possessed of ulterior motives.

Napoleon III at this point felt the need of accomplishing some spectacular exploit which might enhance the image of his lack-luster and weakening regime. Intervention in Mexico was to provide the opportunity. The leader of the French expedition was directed to advance upon Mexico City and establish a stable

[16] Jay Monaghan, *Diplomat in Carpet Slippers* (New York, 1945), p. 65.
[17] *British and Foreign State Papers*, LII, 294.
[18] *Ibid.*, LI, 63; or *House Ex. Doc. No. 100*, p. 134.
[19] Seward did, however, protest; see *Ibid.*, p. 217, for his notice to France whose Emperor supported the Southern cause.

government there, if the situation so warranted. The British and Spanish governments refused to follow the French lead, and the plan for joint action came to an end. Soon afterwards these two powers withdrew, while the French pressed on and occupied the capital. In order to conceal the imperialistic design of Napoleon III, together with the Mexican conservatives involved in the conspiracy, a junta of Mexican citizens was formed which, after deliberation, adopted a limited monarchy as the new form of government. A sovereign bearing the title of Emperor of Mexico was to be chief of state.

The imperial crown was offered to Archduke Maximilian of Austria, brother of the Emperor Franz Josef. If he should not accept, Napoleon III was to name another Catholic prince. Maximilian, to his great misfortune, was induced to accept the offer, and with much pomp and ceremony he entered the Mexican capital with his Empress, Carlota, on June 12, 1864. In order to maintain the costly panoply of an imperial court, to satisfy the more pressing claims of the nation's creditors, and to keep in the field an army strong enough to sustain the government against Juárez and his followers, new loans were floated in France, plunging the country still more hopelessly in debt.[20]

Another serious difficulty confronting Maximilian was the American attitude. While bending all efforts to maintain the Union, the United States could do little more than look askance at French intervention. Yet its hostility was very apparent, and at no time would it consider the recognition of the imperial government. On April 7, 1864, it was resolved by the House of Representatives that "the Congress of the United States are unwilling by silence to have the nations of the world under the impression that they are indifferent spectators of the deplorable events now transpiring in the Republic of Mexico, and that they think fit to declare that it does not accord with the policy of the United States to acknowledge any monarchical government erected on the ruins of any republican government in America under the auspicies of any European power."[21] Although Secretary Seward did not think it expedient at the time to accept this as a statement of the government's policy, he conceded that it was the unanimous sentiment of the American people.

With the end of the Civil War the United States firmly but courteously demanded that the French forces be withdrawn, and Napoleon III, already concerned about his position in Europe, thought it best to comply.[22] Maximilian was unable to resist for long after the French army was withdrawn, and on May 16, 1867 he was compelled to surrender to the republican forces of Juárez. In spite of protests and entreaties from around the world, he was court-martialed and shot. With his demise went Napoleon's dream of a great western empire dominated by France, and the assertion by the United States of the Monroe Doctrine.[23]

In view of the fact that the United States did not recognize Maximilian, our minister had been recalled from Mexico City; however, Washington

[20] Documentation regarding the European intervention may be found in *Sen. Doc. No. 11*, 38th Cong., 2nd Sess.
[21] Moore, *Digest of Int. Law*, VI, 496.
[22] *Ibid.*, pp. 498-503. See also *House Ex. Doc. No. 73*, 39th Cong., 1st Sess., and *Sen. Ex. Doc. No. 6*, 39th Cong., 1st Sess.
[23] For a concise well-documented account of French intervention in Mexico see J. H. Latané, *The United States and Latin America* (New York, 1920), chap. V.

definitely recognized the government of President Juárez in May 1866, and accredited a minister to his headquarters. In 1868, with French intervention ended, and in an atmosphere of relative cordiality, the United States negotiated a claims convention with Mexico. The joint claims commission awarded $4,125,622.20 to Americans who had suffered losses of property, or injury to persons while in Mexico. Mexicans were recompensed in the amount of $150,498,412.[24]

Juárez once more reestablished himself, and although he was forced to maintain a constant struggle against sectional uprisings, he retained control until his untimely death in 1871. He is still regarded as the father of constitutional government in Mexico, for it was due principally to his efforts that the liberal Constitution of 1857 was framed and adopted.

In 1872, after the death of Juárez and the succession of Sebastián Lerdo de Tejada, President Ulysses S. Grant in his fourth annual message to Congress, December 2, 1872, spoke of the continuing disturbances along the Mexican border and the many complaints lodged by American citizens against the Mexican government. In his Seventh Annual Message on December 7, 1875, he spoke of further depredations on American citizens, particularly in Texas, where post offices and mail trains were being attacked.[25] Smuggling from the so-called Mexican "Free Zone" across the border into Texas added to the problem. In the Free Zone, a belt six miles wide, the Mexican government had authorized the importation of foreign goods free of customs duties into the towns along the Rio Grande. This had presented to smugglers an opportunity to convey duty-free goods across the border at the expense of the Texas merchants and the United States treasury.

In March 1877, a band of Indians from Mexico crossed the border, killed seventeen men, and drove large herds of cattle and horses into Mexico. General Ord, who was in command at the border, was directed by Washington to cross the border and apprehend the outlaws. This action was carried out and caused bitter protests in Mexico. In August of the same year, Mexicans retaliated by raiding a settlement in Texas and retreating across the border.[26]

When viewed in historical perspective it is evident that in the fifty-four year period following the establishment of the Mexican Republic, the relations between the United States and Mexico left a bitter legacy. The continuing weakness of Mexico, and its inability to cope with internally divisive and externally aggressive forces, were basically responsible for the unhappy sequence of events. The second period into which relations between the countries seems to fall naturally, is marked by the accession of Porfirio Díaz to the position of Provisional President of Mexico in May, 1877. Díaz, a mestizo with strongly-marked Indian traits, and a hero in the war against the French, ruled Mexico in a thoroughly autocratic manner, save for a single term, for the next thirty-four years. From his accession in 1877 to his over-throw in 1911, relations on the whole were greatly improved. In this period the United States checked its

[24] Moore, *Int. Arb*., II, 1287-1358.
[25] Richardson, *Messages and Papers*, VII, 189, 341.
[26] Robert D. Gregg, *The Influence of Border Troubles on Relations between the United States and Mexico, 1876-1910* (Baltimore, 1937), p. 63.

continental expansion, and Díaz was willing to take a calculated risk in offering concessions, for he encouraged foreign investors, and established law and order, which made business enterprises practicable.[27]

The Díaz government was granted *de facto* recognition by the United States in May, 1873, but Díaz insisted that the resumption of normal relations be conditioned on the repeal of the "Ord order." In July 1882, an agreement providing for the reciprocal crossings of the Rio Grande River in pursuit of Indians and outlaws was entered into by the two countries. This agreement was subsequently renewed and from this time onward there was little trouble caused by raiders, at least in the arena of foreign relations, prior to the Mexican Revolution. With the exception of the Chamizal tract, an area of six hundred acres, near El Paso, Texas, which still remained under the jurisdiction of the United States at the fall of the Díaz regime, the boundary question was largely settled. In 1884 a treaty established the International (water) Boundary Commission, which was charged with adjudicating on matters relating to the riverine and land boundaries of the two countries.[28]

In other areas of diplomacy relations continued to improve. Further extradition treaties were signed and four Pan-American conferences were held which contributed toward a better understanding and good will between the two countries. As the self-appointed mentor of Latin American nations, the United States had also used its influence to promote the use of arbitration in the settlement of disputes. In the Díaz era it acted as a mediator in the Mexican-Guatemalan controversy and, together with Mexico, in the Central American War of 1906.[29] In 1908 the influence exerted by the United States by its participation in the Hague Conference led to the signing of a convention by the Díaz government, which provided for the submission of questions of a legal nature, and those relating to treaties which could not be settled by diplomacy for a period of five years, to the Permanent Court of Arbitration at the Hague.[30]

A great amount of correspondence between the two countries at this time concerned claims. One of the most controversial subjects was the longstanding California Pious Fund Case. This dispute arose over the disposition of funds of the Jesuits which had been placed in the national treasury in 1842. The properties of the society had been sold and the Mexican government had agreed to pay 6 percent interest annually on the proceeds, but when California was ceded to the United States, Mexico stopped payment to all the northern church officials. The commission appointed to adjust the affair had failed, and in 1876 the matter was arbitrated with a decision in favor of the American contenders. Mexico soon afterward stopped payment again, and in 1902 the matter was referred to the Hague where the American claims were once more upheld.[31]

[27] *Ibid.*, p. 152.
[28] Charles A. Timm, *The International Boundary Commission, United States and Mexico* (Austin, 1941), p. 23.
[29] Morgan to Frelinghuysen, May 9, 1883, *Papers Relating to the Foreign Relations of the Unoted States, 1883*, pp. 648-651. (hereinafter cited as *Foreign Relations*); *Ibid.*, 1907, Root to Godoy, Nov. 11, 1907, Part II, 659.
[30] Proclamation of the President of the United States, June 29, 1908, *For. Rel., 1908*, p. 626.
[31] Translation of the Sentence of the Permanent Court of Arbitration on the matter of the Pious Fund of the Californias, *For. Rel., 1902*, Appendix II, pp. 15-18.

Throughout the Theodore Roosevelt administration, relations between the two nations continued to be satisfactory on a diplomatic level. This is reflected in President Roosevelt's message to Congress in 1907 in which he stated that "relations between the two countries are a just cause for gratification."[32] In this same year, Elihu Root, Secretary of State, during a good will tour of Mexico declared, "I look to Porfirio Díaz, the President of Mexico, as one of the great men to be held up to hero worship of mankind." Root later expressed the opinion that "the people of Mexico have joined forever the ranks of the great, orderly, self-controlled, self-governing republics of the world."[33]

In making these statements, President Roosevelt and Secretary Root apparently had failed to grasp the full import of incidents which had occurred in Mexico during the preceding year. In June 1906, the American ambassador in Mexico, David E. Thompson, had sent to Washington reports of a strike in the Cananae mines involving Mexican workers, which had resulted in the destruction of American property and loss of several lives. The next few months saw further expressions of anti-American feeling. American consular officials stationed throughout Mexico reported to Ambassador Thompson evidence of deep-seated hatred and dislike of Americans. Díaz attributed this situation to a small group of revolutionists, rather than popular sentiment, and urged that the American side of the border be patrolled carefully so as to apprehend any of the offending individuals should they attempt to enter the United States.[34]

From evidence submitted by the American consular offices in Mexico at this time, it appears that the root of the trouble lay in a different direction from that suggested by President Díaz. The consular reports stressed the enmity existing among the Mexican labor class as against the same class of American workers in Mexico; this feeling was even more pronounced among skilled workers. The hostility was the outgrowth of alleged preferential treatment given by American employers to American workmen in respect of wages, position, and advancement, particularly in the mining industry and the railroads.[35]

With the main features of Mexican-United States relations prior to 1909 reviewed, it is of interest to consider an important adjunct of that relationship: the extension of American industrial and economic interests into Mexico. Since the influence of these forces reached a culmination in the Madero revolution of 1910, it is inseparable from the revolution itself. One Mexican writer, Francisco Bulnes, cites American economic influence as the determining cause of the Revolution.[36] American writers, while recognizing its great significance, point out other equally fundamental causes.

The role of American investments, which predominated over all other foreign powers, took the form of railroad construction, mining, stockraising, manufacturing and exploitation of oil resources. This economic penetration evolved rather slowly, however, because of Mexican fear of American domina-

[32] President's Message to Congress, December 3, 1907, *For. Rel., 1907*, Part I, p. LXVIII.
[33] Robert Bacon and James B. Scott, eds., *Latin America and the United States* (Cambridge, 1917), p. 168; James M. Callahan, *American Foreign Policy in Mexican Relations* (New York, 1932), p.451.
[34] *Ibid.*, p. 522.
[35] *Ibid.*, p. 525.
[36] Francisco Bulnes, *The Whole Truth about Mexico* (New York, 1916), p. 103.

tion aroused by the political developments already noted. Nevertheless, Mexico, after 1821, proved more liberal to foreigners than the Spanish colonial government had previously been. This is exemplified by the Constitution of February 24, 1822, which gave foreigners the same civil rights as Mexicans. In 1823 Mexico permitted foreigners to develop mines in Mexico. Beginning in 1824 resident foreigners were allowed to acquire land in Mexico, but not within twenty leagues, or nearly sixty miles, of the border, and ten leagues from the sea. This concession was modified in 1828 to limit ownership of land, other than mines, to Mexican citizens. The Constitution of 1857 contained substantially the same legislation, precluding foreign ownership within designated zones bordering on foreign nations, or upon the sea.[37]

It may be concluded that anti-foreign legislation was liberalized following Mexican independence, but that it was none the less restrictive. The stipulation that the acquisition of property entailed conversion to Mexican citizenship, and the demand that foreign property owners reside within the country, together with subsoil limitations, were not conducive to the introduction of foreign enterprise.

The Díaz era, 1877-1911, brought about a reversal of the policies which operated against the introduction of foreign capital and ushered in the phase of favorable relations with the United States mentioned earlier. The foreigner was urged to come and was assured protection for his investments. Under Díaz, the prohibition upon limited zones remained in force, but it was possible in many cases for a foreigner to obtain a special permit to acquire property in prohibited zones. Subsoil legislation was also modified so as to encourage foreign development of mining and the petroleum industry.[38] In granting concessions, Díaz and his clique showed no partiality to Americans, the government offering virtually the same advantages to all capitalists, whatever their nationality.[39]

The first group to lead the economic advance into Mexico were the railway promoters and railway builders. As early as 1853 Americans tried to obtain a concession to build a railroad from the American southwest border to the Gulf of California, but without success.[40] Many attempts were made in subsequent years, notably in 1864, 1873 and 1877, and they also were foredoomed to failure because of Mexico's fear of absorption by its more powerful neighbor. Although two American railroads had reached the Mexican frontier by 1877, the Southern Pacific and the Denver and Rio Grande, the total increase in Mexican mileage between 1876 and 1880 was slight, from 416 to 647 miles. This increase was achieved largely by two foreign corporations other than American, the combination represented by F. S. Pearson of Canada and London, and another, distinct from the first, S. Pearson and Son, Limited, of London.[41]

The first actual American railway construction was begun in 1880, under new concessions stimulated in no small measure by American recognition of the Díaz government. From this point on construction accelerated rapidly through the efforts, principally, of the Southern Pacific, Atchison, Topeka and Santa Fe,

[37] Frank Tannebaum, *The Mexican Agrarian Revolution* (New York, 1926), p. 375.
[38] *Ibid.*, p. 376.
[39] J. Fred Rippy, *The United States and Mexico* (New York, 1926), p. 311.
[40] Callahan, *American Policy in Mexico*, p. 476.
[41] *Ibid.*, p. 485; John K. Turner, *Barbarous Mexico* (Chicago, 1910), p. 135.

Denver and Rio Grande, Great Northern and Mexican Central railroad companies. By 1884, Mexican railway mileage increased to 3,862 miles, chiefly American. In 1902 American holdings in Mexican railways were valued at well over $300 million; this amount had reached $560 million in 1911.[42] The Mexican government by April, 1908 had secured control over 13,191 of the 22,822 kilometers of railway in the country, but for capital it continued to depend upon foreign investors. The direction of its main system, the Mexican National Railways, was under American supervision until 1914.[43]

Following, if not accompanying railway development, came the mining interests, land companies and ranchers. To encourage mining, Mexico revised its mining code in 1884 by granting the owners of the surface the right to work coal deposits under their land without government concession. In 1892, due to the fall of silver valuation and consequent decline in mining operations, the Mexican government further modified its mining legislation so as to insure that the rights of those who engaged in mining operations would be secure and inviolate so long as specified fees were paid.[44] By 1902 American mining properties were valued at $95 million and nine years later they had reached $250 million, including the smelting industry. Notable among the interests represented in the mining field were the Hearst Estate, the American Smelting and Refining Company (the Guggenheims), the Batopilas Company of New York, the Anaconda group, the Greene-Cananae interests, and the United States Steel Company.[45]

The railroad men and miners were followed by those interested in acquiring land: ranchmen and small farmers. By 1912, according to a report of the Fall Committee, there were an estimated fifteen thousand Americans, farming small plots, who were residing permanently in Mexico. More important, however, were the large land development, irrigation and colonization companies, represented by the Hearst Estate, Sonora Land and Cattle Company, and the United Sugar Company. In 1910, 25.2 percent of the total land area was in foreign hands, constituting 29 percent of all privately held lands.[46] It is estimated that American holdings in farm, ranches and timberlands reached $50 million to $80 million by 1912.

In manufacturing, American capitalists were less aggressive than in other fields mentioned because of strong competition from European business houses which, through greater experience in foreign lands, were better prepared to cope with problems existing in Mexico.[47] The total American investment in manufacturing did not exceed $25 million by 1910; this was emphasized in meat packing, soap, and tobacco manufacturing. In banking activities, the British, French and German syndicates far overshadowed the American until after the Revolution of 1910.[48]

The oil industry, largely financed by American capital, experienced the most spectacular growth. Although a few wells were drilled in Mexico by a

[42] Rippy, *United States and Mexico*, p. 312.
[43] Charles P. Howland, *Survey of American Foreign Relations* (New Haven, 1931), p. 36.
[44] *Ibid.*, p. 37.
[45] Rippy, *United States and Mexico*, p. 313.
[46] Howland, *American Foreign Relations*, p. 38.
[47] *Monthly Consular and Trade Reports*, June, 1906, p. 58.
[48] Howland, *American Foreign Relations*, p. 39.

British company financed by Cecil Rhodes in 1888, the oil industry did not get a real start until 1900. In that year Edward L. Doheny and associates of Los Angeles brought in their first well, and thereafter production quickly soared. In 1907 oil production in the Doheny lands alone was one million barrels, and in 1922 reached sixty million. By 1912, American citizens had invested an estimated $15 million in Mexican oil.[49]

A summary of American economic penetration of Mexico shows that by 1913 the investments of United States' capital totaled between $700 million and $1,057,770,000, while the investments of British subjects, second to those of the Americans, amounted to approximately $300 million.[50] The American investment included 78 percent of the mines, 72 percent of the smelters, 58 percent of the oil, 68 percent of the rubber industry, and exceeded the total investments of all other foreigners in Mexico. Further American influence was reflected in the extent of trade relations. Commerce between the two countries, which had totaled $7 million in 1860, reached $177 million in 1910.[51]

By encouraging the introduction of foreign capital, of which a preponderance was American, Díaz attempted to improve the economic well being of his country. In so doing he ignored the development of political evolution which should have been a concurrent process.[52] He had seen Europeans come to the United States and become quickly assimilated, and expected that this would occur in Mexico. Díaz lived to see a widespread economic advance in Mexico, but he also belatedly saw that his country gained only in national credit and a reputation for stability. The only money remaining in circulation was through the media of taxes and higher wages, because the foreign investors took their surplus earnings with them to their own countries. In the words of Carlo de Fornaro, "The Porfirian regime was excellent for the pockets of a few Mexicans and a great many Americans and Europeans, but it was a poisonous virus inoculated into the very life of Mexico."[53] His policies served to accentuate the fast growing dislike of Americans in the Mexican citizen who saw himself as the economic slave of the United States, and thought he perceived the vast wealth of his country being absorbed by the hated "gringo" with no benefit to himself or to his country.

The administration of William Howard Taft, beginning in 1909, was committed to the protection and extension of American economic interests abroad, including Mexico. Consequently, diplomatic relations between the two countries during his term of office must be interpreted in this light. Although President Taft's policy toward Mexico was ostensibly one of neutrality, it cannot be gainsaid that he was interested in the perpetuation of Díaz and his advisors, the intellectual elite, called the *Científicos.* Taft, in correspondence with his wife in 1909, mentioned that, ". . . we have two billions of American capital in Mexico that will be greatly endangered if Díaz were to die and his government go to pieces . . . I can only hope that his demise does not come until I am out of

[49] Rippy, *United States and Mexico*, p. 318.
[50] R. W. Dunn, *American Foreign Investments* (New York, 1926), p. 90.
[51] Callahan, *American Policy in Mexico*, p. 519.
[52] Alfonso Teja Zabre, *Historia de México, Una Moderna Interpretación* (Mexico, D. F., 1935), p. 362.
[53] Carlo de Fornaro, "The Great Mexican Revolution," *Forum*, LIV, (1915), 534.

office."[54] As an indication of his esteem for the Mexican president, and to strengthen the latter's position with his own people, Taft met Díaz in a colorful setting at Juárez in October 1909. In his message of acceptance to Díaz, Taft said "it would gratify me very much to meet one in the flesh who has done so much to establish order and create prosperity in his own country, and in so doing has won the admiration of the world."[55]

President Taft's administration was the first to be identified with the odious term "dollar diplomacy," through intervention in the affairs of the Caribbean republics. This policy proved detrimental to the fostering of good will in Mexico even though Taft called it "international philanthropy."[56] One severe critic, Salvadore R. Merlos, a Central American writer, even referred to Taft as the "Attila of modern times."[57] Taft's indiscreet statement of February 22, 1906, that "the frontiers of the United States extend virtually to Tierra del Fuego," was *prima facie* evidence of his diabolical intentions to other Latin Americans.[58]

Philander C. Knox of Pennsylvania was the secretary of state under President Taft. His appointment largely nullified what had been accomplished by Elihu Root, his predecessor, in improving relations with Mexico and Latin America in general. Knox appeared not to understand the Latin American temperament and thought, and made it evident that he considered them to be inferior to the Anglo-Saxons.[59] Illustrative of Knox's unpopularity in Nicaragua was the revelation of a plot to dynamite a train on which he was to ride while traveling through that country.[60]

It was his conviction that the elimination of the European powers from Latin America imposed a heavy responsibility on the United States, a principle he derived from the Monroe Doctrine. In view of the political instability in that area Knox believed that this country should assist them "to meet their just obligations and keep out of trouble."[61]

As early as May 1909, President Taft had indicated that he was considering Henry Lane Wilson, then minister in Belgium, and former minister in Chile, for the post of ambassador in Mexico City. The ambassador who held the office at this time, David E. Thompson, had become closely identified with American business interests in Mexico, and his retention was no longer desirable.[62] Wilson's appointment was confirmed in October 1909, and he reached Mexico City on

[54] Henry F. Pringle, *The Life and Times of William Howard Taft* (New York, 1939), I, 462.
[55] Alfonso Taracena, *En el Vértigo de la Revolución Mexicana* (Mexico, D. F., 1935, p. 48; Taft to Díaz, June 25, 1909, *For. Rel., 1909*, p. 425.
[56] Parker T. Moon, *The United States and the Caribbean* (Chicago, 1929), p. 144.
[57] J. Fred Rippy, "Literary Yankeephobia in Hispanic America," *Journal of International Relations*, XII (1922), 530.
[58] Roberto Domenech, *Méjico e el Imperialismo Norte Americano* (Buenos Aires, 1914), p. 41.
[59] Wilfrid H. Calcott, *The Caribbean Policy of the United States, 1890-1920* (Baltimore, 1942), p. 367.
[60] *New York Times,* Mar. 21, 1912.
[61] Dexter Perkins, *Hands Off: A History of the Monroe Doctrine* (Boston, 1941), p. 248; Samuel F. Bemis, ed., *The American Secretaries of State and their Diplomacy* (New York, 1929), IX, 335.
[62] Edward I. Bell, *The Political Shame of Mexico* (New York, 1914), p. 123.

February 28, 1910.[63] His arrival coincided with a general increase in anti-American sentiment among the rank and file of the Mexican population; however, no incidents of a serious nature took place until the following October.

In the summer of 1910 further action was taken toward the settlement of the problem of the Chamizal tract. The tract consisted of about six hundred acres of land which lay between the old bank of the Rio Grande, as it was surveyed in 1852, and the existing bed of the river. Its position was the result of changes which had taken place through the action of the water upon the banks of the river causing it to move southward into Mexican territory. With the progressive movement of the river to the south, the American city of El Paso had been expanding in area at the expense of the Mexican City of Juárez to the south.

Since the controversy over the settlement of the Chamizal tract remained unsettled, the two nations effected a convention on June 24, 1910, agreeing that the question as to international title of the tract should again be referred to the International Boundary Commission, just enlarged by the addition of a third member. The Honorable Eugene La Fleur of Montreal, Canada, was invited to act as the third Commissioner. Both governments then presented their cases to the Commission and negotiations continued to the summer of 1911.[64]

The Mexican government's representative contended that the Chamizal tract had been formed by a sudden change of the Rio Grande and, accordingly, the boundary line should follow the abandoned bed of the river. The American representative, on the other hand, claimed that the formation of the tract had resulted from slow and gradual erosion and deposit, and as such, any question of sovereignty would be determined by the provisions of the convention of 1884. Furthermore, the United States government claimed that it had held undisputed possession of the territory since 1848. Mexico refused to admit the validity of these arguments.[65] On June 5, 1911, the award was made:

> . . . the international title of the portion of the Chamizal tract lying between the middle of the bed of the Rio Grande, as surveyed by Emory and Salazar in 1852, and the middle of the bed of the said river as it existed before the flood of 1864, is in the United States of America, and the international title to the balance of said Chamizal tract is in the United States of Mexico."[66]

This award evoked great enthusiasm in Mexico as it was believed that the transfer of a large portion of El Paso to that country would redound greatly to the interests of Mexico, economically more than geographically.[67]

These hopes were quickly dashed because the American commissioner, Anson Mills, denounced the award as vague, indeterminate, uncertain in its

[63] Henry Lane Wilson, prior to entering the diplomatic service, was a real estate speculator in the state of Washington.His brother, the former Senator John Wilson of Seattle, owner of the Seattle, *Post Intelligencer*, was the Republican party boss in Washington. See Robert H. Murray, "Huerta and the Two Wilsons," *Harpers Weekly Magazine*, Mar. 25, 1916, p. 341, and George Creel, *The People Next Door* (New York, 1926), P. 293.

[64] H. Wilson to James Bryce, Dec. 24, 1910, *For. Rel., 1910*, p. 571.

[65] Award by International Boundary Commission, June 10, 1911, *Ibid.*, p. 575.

[66] *Ibid.*, p. 587.

[67] *El Mañana Periódico Político,* Junio 15 de 1911 a Fevrero 28 de 1913, p. 5.

terms, and impossible of execution. He also declared that the Commission had no authority to divide the tract, because the instructions in the convention had stated specifically that "the Commission shall decide solely and exclusively as to whether the international title to the Chamizal tract is in the United States of America or Mexico . . ."[68] Efforts were then made to persuade the Mexican government to sign a new convention authorizing a reopening of the case, but owing to the disturbed conditions existing in Mexico, no further steps were taken at the time.

If we could leave the Díaz regime with a mere statement of the many material benefits which it conferred upon Mexico, Porfirio Díaz would rank as one of the world's great statesmen. But unfortunately there is a reverse side to the picture. A "scientific" form of government which functioned very effectively, which kept peace at home and friendship abroad, had been established, but almost wholly in the interest of the *científicos* who controlled it. The financiers, foreign investors, large landowners, and the government officials found the system most satisfactory. But how did it meet the needs of the great mass of Indians who constituted about three fourths of the population? It must be conceded that it did nothing for the *peon* except to exploit him. The land was held in great estates, some containing as much as half a million acres. By 1910 some 3,103,402 individuals had lost their freedom and were listed as "*peones de campo*", or agricultural laborers held in debt service. These with their families, conservatively estimated, numbered between nine and ten million people, or from three-fifths to two-thirds of the total population of Mexico.[69]

The Díaz government must be held responsible for encouraging, rather than frowning upon, the *peonage* system. A darker blot upon it was the inhuman treatment of the Yaqui Indians. With full cognizance of the government, these unfortunates, among the highest type of Indians found in Western Hemisphere, were unjustly dispossessed of their lands and then sold into virtual slavery to labor on the henequen plantations in Yucatan. Therefore, even if we conceded that Díaz was "the master builder of a great commonwealth," as one of his biographers has called him,[70] in the last analysis his building could not endure because the foundation was laid on special privilege. He was a great administrator, and he raised Mexico to a higher level, politically and economically, than it had ever attained before. But he accomplished this result as a tyrant rather than a democratic ruler, and at the expense of the most numerous class of the Mexican people.

The very fact that the whole Mexican political system depended upon the vigor of the man in control was the cause of its utter collapse when that prop was withdrawn. The system of "Diazpotism" was highly centralized, analogous to that of France under Napoleon I. The executive power counted for everything in the state, and that rested in the president's hands. The governors in the various states were nominated by the president and responsible to him, although the formalities of popular elections in accordance with constitutional formulas were

[68] Award by International Boundary Commission, June 10, 1911, *For. Rel.*, p. 587.
[69] Charles W. Hackett, "The Mexican Revolution and the United States, 1910-1926," *World Peace Foundation Pamphlets*, IX, (1926), p. 341
[70] José F. Godoy, Porfirio Díaz, *The Master Builder of a Great Commonwealth* (New York, 1910).

regularly carried out. Although on paper the legislative bodies, both federal and state, appeared quite as representative as the corresponding bodies in the United States, they were, in reality, wholly dominated by the executive. As long as Díaz was vigorous enough to direct the state in person, the system worked. But gradually he was forced, through weakness arising from advanced age, to allow his subordinates increasing power. The result, as has been described, was the sacrifice of Mexican natural resources of every sort through the *científicos* to various foreign interests, particularly American and English.

The revolutionary movement which led to the downfall of the Díaz regime began on November 18, 1910, at Puebla in northern Mexico. The initial revolt was quickly suppressed and its leader, Aquiles Serdán, who is regarded today in Mexico as one of the country's greatest heroes, was executed.[71] The national leader, and the guiding spirit of the Revolution, was Francisco I. Madero. The Madero family who owned banks, smelting plants, cotton and rubber plantations and breweries, were wealthy Creoles in the state of Coahuila. As capitalists they had protested against the favoritism displayed by Díaz toward their American competitors, but nevertheless the family as a whole had given him loyal support.

Francisco proved to be the exception. Educated in the United States at the University of California, and in France, he had acquired liberal ideas, and engaged in philanthropic practices. In 1900 he became interested in politics, and by 1905 was recognized as leader of the independent voters. *La sucesión presidencial de 1910*, which he published in 1908, was critical of the Díaz administration.

Prior to the publication of Madero's book in 1908, President Díaz had made the startling announcement that since Mexico was now ready for democracy, he would welcome the formation of an opposition party and would relinquish his presidential power to a legally elected candidate. This statement was made to the American journalist, James J. Creelman, and appeared in an article in *Pearson's Magazine* in March 1908. The sincerity of this declaration was soon disproved, for plans were already being laid for the successor to Díaz by the governmental clique, which had no intention of countenancing a fair election.[72]

After the publication of his book regarding presidential succession in 1910, and stimulated by the Díaz announcement to Creelman, Madero began campaigning for the election of 1910. After being nominated as candidate for the presidency by the National Democratic, or anti-reelectionist party, he carried on a campaign throughout the country. However, Díaz, in direct contradiction to his statement to Creelman, was nominated and retained in the presidential office after a farcical election in 1910. Madero was accused of sedition for his political activities and imprisoned. While in prison, he wrote the "Plan of San Luis Potosí," which contained two main principles: effective suffrage and no re-election. It also provided for agrarian reform, liberation of political prisoners,

[71] Colonel Francisco Lazcano, "Conmemoranda la Revolución Mexicana," *Revista de Ejército,* (1938), p. 626.
[72] Herbert I. Priestly, *The Mexican Nation, A History* (New York, 1924), p. 394; James Creelman, *Díaz, Master of Mexico* (New York, 1912), p. 413.

and freedom of speech and press. The plan also set the time for a general uprising, which was scheduled to take place on November 20, 1910.[73] The premature, ill-fated revolt led by Serdán on November 18, set the program in motion.

Madero meanwhile, in October, had forfeited bail and escaped from prison, with the aid of friends, disguised as a mechanic. A few days later he appeared in Laredo, Texas. At the time, the Revolutionary Junta in the United States, headed by his brother Gustavo Madero, was located in San Antonio, Texas.[74] After a sojourn of forty days in the United States, Madero returned to Mexico, and soon the movement gained momentum. Emiliano Zapata, a peasant leader in the state of Morelos, Pancho Villa, a bandit, and Pascual Orozco, a former storekeeper from Chihuahua, formed bodies of cavalry and became the mainstay of Madero's forces.

Díaz, in his message to Congress on April 1, 1911, belatedly promised to adopt the following measures: (1) safeguard suffrage; (2) reform of federal judiciary; (3) removal of abuses on the part of local officials; (4) division of large estates; (5) provision that the president should not succeed himself. And Vice President Ramón Corral, a highly unpopular figure, was induced to resign.

In the spring of 1911, the revolutionists defeated the government forces in two major engagements, at Juárez and Agua Prieta, the latter taking place in May. Realizing the futility of further resistance, Díaz resigned in May 25, 1911, and on May 26 he left for Paris under an escort commanded by General Victoriano Huerta as far as Vera Cruz. Francisco de la Barra was made provisional president until an election was held.[75]

Madero entered Mexico City in triumph on June 8. In the election held in October 1911, he was elected president by an overwhelming majority and his government was recognized by the Taft administration a month later. Madero had gained the leadership of a revolutionary movement which was primarily a protest against the continuation of the Díaz government, but a complex combination of factors promoted its growth and made it possible. These factors in brief were: (1) The industrialization of Mexico was accompanied by a rapid increase in the cost of living without a corresponding rise in the wages of the masses; this was aggravated by a heavy protective tariff; (2) the rising prices and absence of wage increases lowered the subsistence level of the Mexican wage earners; (3) the effects of industrialization were augmented by the Díaz policy of breaking up village lands which increased the strength of the *hacienda* as opposed to the communal village groups, and reduced the *peon* to serfdom.[76]

Madero, once he had assumed the presidency of Mexico, found himself committed to more reforms than he could possibly undertake, and it became increasingly evident as time passed that he was operating under severe handicaps.

[73]Mexico, Political Affairs, 1911, *For. Rel., 1911*, p. 351; Stanley R. Ross, *Francisco I. Madero: Apostle of Mexican Democracy* (New York, 1955).

[74]Taracena, *Vertigo de la Revolución,* p. 69; José Vasconselos, *Ulises Criollo* (Mexico, 1936), p. 420.

[75] Wilson to Knox, April 3, 1911, *For Rel., 1911,* p. 444; *Christian Science Monitor*, April 3, 1911:Secretaría de Relaciones Exteriores, *Personas que han tenido a su Cargo la Secretaría de Relaciones Exteriores desde 1821 hasta 1924* (Mexico, D. F.), p. 31.

[76]Tannenbaum, *Mexican Agrarian Revolution,* pp. 134-155.

The Mexican Church opposed his program, and business and banking circles were solidly against him. He was also confronted by syndicalists, socialists, and the "*hacendados*," the great land owners, who would most naturally resist land reform.[77] He was frail physically, and was not an effective speaker. His plans were not well analyzed, as was demonstrated by his inability to execute them, and were approached from a theoretical rather than a practical point of view.

Madero's political mistakes were numerous and devastating. One of his first official acts was repayment of advances made by members of his own family which he replaced by funds taken out of the National Treasury. His policy of retaining in positions of trust former key personnel of the Díaz administration was the cause of much dissatisfaction. The rank and file of the revolutionary movement also proved too thoroughly imbued with the lust for spoils to be content with his moderation. Madero's practice of appointing several members of his family to responsible positions in the government gave his opponents of all kinds another source for criticism. By forcing unpopular officials, notably Pino Suárez the vice president, into governmental positions, he provoked the formation of new factions among the discontented elements. But Madero's fatal error was his denial that he had promised to distribute lands to the people, the one issue upon which all the *peons* were in agreement.[78]

Disillusioned by Madero's failure to carry out his proclaimed revolutionary objectives, Zapata and Orozco declared against him, and formed a counter-revolutionary movement.[79] The disorder which soon prevailed in the capital and throughout the countryside caused friction to develop between the Madero government and the United States. This inevitably extended to American cities near the border which were often the centers of conspiracy.[80]

Henry Lane Wilson, who by January 1912, had become identified with a "big business" clique in Mexico City, was obviously hostile toward Madero. Wilson was also beginning to interfere in the internal affairs of Mexico, a practice which had already stigmatized American diplomacy there. In a dispatch to the secretary of state in February the ambassador said: "I have most discreetly and carefully . . . endeavored to induce leading members of the Catholic party, of the old regime, and of conservative elements in the city, to make some demonstration of a public character, coupled with a tender of service and support as might have a moral effect of the country at large."[81]

Upon receiving a request from the State Department for advice as to the action to be taken by Americans in Mexico, Wilson recommended their withdrawal from dangerous areas. President Taft, not wishing to alarm American residents, and to take precaution against the necessity for military intervention, issued a proclamation on March 2, 1912. After describing the state of disorder in Mexico, he spoke of the force and effect of the neutrality laws of the United States, and gave notice that "all persons owing allegiance to the United States, who may take part in the disturbances now existing in Mexico, unless in defense

[77] Teja Zabre, *Historia de México*, p. 364.
[78] Francisco Vásquez Gómez, *Memorias Políticas, 1909-1913* (Mexico, 1933), p. 518; Hubert Howe Bancroft, *History of Mexico* (San Francisco, 1914), p. 533.
[79] Taracena, *Vértigo de la Revolución*, pp. 121-125.
[80] *La Gaceta de Guadalajara*, Nov. 19, 1911.
[81] Wilson to Knox, Jan. 20, 1912, *For. Rel., 1912*, p. 723.

of their persons and property . . . will do so at their own peril, and they can in no wise obtain protection from the government of the United States . . . "[82] This pronouncement was interpreted by many Americans as being a warning to leave Mexico, and soon a general exodus began.

The presidential order placed participants or conspirators in the same class as the insurrectionists whom they aided. It denied them the protection of the United States government, provided they were treated in accordance with the standards of treatment established internationally as proper for the particular group they had assisted, and would, in addition, incur all the penalties of the violations of a national law.[83]

In recognition of the growing danger to Americans in Mexico, as reflected by anti-Yankee propaganda, and the spread of the conflict, Congress passed a joint resolution on March 14, which gave the President power to prohibit at his discretion the export of material used in war. On the same day, President Taft utilized this new power to prohibit the export of munitions of war from any place in the United States to the warring factions in Mexico, any violations of which would be vigorously prosecuted.[84] Two weeks later the order was modified so as to apply only to the insurgents, which was in accord with our principle of supporting the constituted authority against rebellion.[85] Madero quickly expressed his gratitude for this measure and interpreted the proclamation as evidence of the good intentions of the United States, that is, not to intervene in Mexican affairs. President Taft's reluctance to adopt an aggressive militaristic policy toward Mexico began to evoke much unfavorable comment in some foreign newspapers. British, French and German mining interests in the states of Sonora and Chihuahua, by this time the scene of much destruction, were believed to have the right to anticipate intervention from the Taft administration. The major European powers maintained war vessels almost continuously in Gulf waters in this troubled period.[86]

It was at this time a matter of international character arose which involved the United States and Mexico, but the latter only as an innocent bystander. This was the reappearance of the "yellow peril" in the form of Japan, which some propagandists alleged was trying to obtain control of property at Magdalena Bay in Lower California for strategic purposes. The Bay is located in the southerly third of the west coast of Lower California, approximately three thousand miles from Panama and offers excellent facilities for large naval vessels. In fact, the United States Navy had in past years used the area extensively for target practice.

An American company had secured here from Mexico a large tract of land, consisting of several million acres, which bordered the Bay. This company found the investment unprofitable, and its chief creditor, a New Hampshire lumberman, had taken it over and tried to dispose of it to Japanese subjects. Before

[82] H. Wilson to H. L. Wilson, March 2, 1912, *Ibid.*, p. 732.
[83] Edwin M. Borchard, *The Diplomatic Protection of Citizens Abroad* (New York, 1915), p. 769.
[84] *Cong. Rec.* 62nd Cong., 2nd Sess., p. 3258; *American Journal on Int. Law*, VI (1912), 147.
[85] H. L. Wilson to Martínez, March 26, 1912, *For. Rel., 1912*, p. 765.
[86] *Current Literature*, II (1912), 389.

concluding the sale, however, his agent very properly consulted the United States Department of State to learn its attitude. Its reaction was adverse since the establishment of a Japanese coaling station, fishery or colony on our side of the Pacific inevitably would not be approved by the American public.[87]

These developments took place late in 1911 and rumors soon emerged that the Japanese government was directing the purchase with the intent to use the site for a naval base. This was disproved by communications between the Americans concerned and the Department of State, which revealed that the Japanese company would not invest in the concession without the approval of its own government, which in turn would not countenance it unless sanctioned by the United States government. The Japanese concern then lost interest in the project.[88]

In spite of this revelation, and the Mexican government's protestations that it was not in league with Japan, apprehension still persisted in many quarters in the United States that the Japanese government was veiling its true intentions. Senator Henry Cabot Lodge, who led the skeptics in Congress, declared on February 29th that the occupation of Magdalena Bay by an Eastern Power was not a matter that could be brought before an arbitral tribunal. Lodge's concern over the matter seemed to stem from the idea that the Japanese government was behind the negotiations because of the fact that the property was of little value except for naval purposes, but was of the utmost value for that purpose.[89]

A prolonged discussion of the Magdalena Bay question then took place in Congress and finally led to a resolution, presented by Senator Lodge, being passed by the Senate on August 2, 1912. In the arguments that led up to and followed this resolution, it appeared that its chief proponents based their contentions on the law or right of self-defense. The Lodge Resolution provided:

> . . . that any harbor or other place in the American continents, is so situated that the occupation thereof for naval or military purposes might threaten the communications or safety of the United States, the Government of the United States could not see without grave concern the possession of such harbor or other place by any corporation or association which has such relation to another government, not American, as to give that government practical power of control for military or naval purposes.[90]

In Mexico the Lodge Resolution was interpreted by many persons to be just another pretext employed by the Americans to encroach further at a later date on Mexican territory, under the guise of legitimacy.[91] In the United States the resolution became known as the Lodge Corollary to the Monroe Doctrine, or as an extension of the Monroe Doctrine applied for the first time to foreign companies and an Eastern power, with a view toward safeguarding American interests in the Panama Canal.

After the imposition of the arms embargo in March 1912, relations were increasingly alienated by the mounting destruction of American property.

[87] *American Journal of Int. Law*, VI (1912), 937.
[88] Thomas A. Bailey, "The Lodge Corollary to the Monroe Doctrine," *Political Science Quarterly*, 48 (1933), 221.
[89] *New York Times*, Apr. 5, 1912.
[90] *American Journal of Int. Law*, VI (1912), 938.
[91] *El Mañana Periódico Político* June 15 de 1911 a Febrero 28 de 1913, p. 215.

President Taft sent a warning to Madero in April through the Minister of Foreign Affairs, Pedro Lascurain, advising that the United States government held the Mexican government responsible for the protection of American life and property.[92] The Mexican government replied that it could not recognize the right of the United States to give such a warning, since it was not based on any action imputable to the Mexican government signifying that it had departed from the observance of the principles of international law. Responsibility for the actions of the rebel leader Orozco was also denied by Mexico City.[93]

Less than two weeks later, Secretary Knox notified Ambassador Wilson that it would be necessary for additional French, German and American warships to visit both the east and west coasts of Mexico for the protection and relief of their respective nationals. This would be accomplished under the right of a friendly nation to dispatch war vessels to the waters of a nation with whom they are at peace. Accordingly the United States sent the *Burford* to the west coast ports to receive refugees from the interior. Soon there-after, the cruiser *Vicksburg* was sent to Guaymas and another cruiser, the *Des Moines*, to the east coast.[94]

On December 3, 1912, in his annual message to Congress, President Taft pointed out the trying circumstances which had characterized United States-Mexican relations in the preceding two years and reiterated that American policy had been one of patient non-intervention and continued recognition of constituted authority. He emphasized that every effort had been made to safeguard American lives and interests in Mexico, a most difficult task in view of the extensive American holdings there.[95]

After a brief leave of absence Ambassador Wilson returned to the Embassy in Mexico City on January 5, 1913, and resumed correspondence with the State Department in the same pessimistic spirit as was found in his earlier reports. His initial communication on January 7, described the situation as gloomy, if not hopeless, and carried a general indictment against the Madero government.[96]

A brief period of respite was obtained by Madero through the recapture, with little difficulty, of Vera Cruz, which Feliz Díaz, nephew of the former president, had seized. But the day of settlement was merely postponed. On February 9, 1913, a group of conspirators instituted a revolt in Mexico City which ended in tragedy for Madero and his government. This group was led by members of the old conservative party, namely General Felix Díaz and General Bernardo Reyes, who had been liberated from prison by fellow conspirators, and General Manuel Mondragon, former chief of artillery of the Díaz army. Failing to seize the National Palace they retired to the Arsenal, where they maintained headquarters for the succeeding "tragic ten days."In this period machine gun and artillery fire sprayed the densely populated city at random with heavy loss of life to the civilian population, and much property damage.[97]

President Madero had been forewarned of the plot by his brother Gustavo

[92] H. Wilson to H. L. Wilson, April 14, 1912, *For. Rel., 1912*, p. 787.
[93] Wilson to Knox, Apr. 17, 1912, *Ibid.*, pp. 792-793.
[94] *Ibid.*, Apr. 26, 1912, pp. 803-804.
[95] *Cong. Rec.*, 62nd Cong., 3rd Sess., p. 8.
[96] Wilson to Knox January 7, 1913, *For. Rel., 1913*, p. 692.
[97] Manuel Calero, *Un Decenio de Política Mexicana* (New York, 1920), p. 114.

Madero as early as February 4, but he had not taken it seriously. It did not occur to him that he was viewed as an intruder by the *Porfirista* army officers, the group which framed the conspiracy. Although General Victoriano Huerta's name had also appeared on the list of conspirators obtained previously by Gustavo Madero, President Madero placed him in command of all Federal troops on February 9. This move seemed foolish in the extreme, as Huerta had been relieved of command of the army, after the northern campaign, for misappropriation of public funds, and might have been expected to seek vindication.[98]

As Huerta's forces numbered upwards of seven thousand men, and those of Díaz about five hundred, with the former possessing much heavy artillery and the latter a negligible amount, it was obvious that the Arsenal could have been taken at any time by a determined onslaught. During the seige, Huerta convinced Madero that a maximum effort was being made to defeat the rebels, but at the same time he was awaiting the propitious moment when he could seize control of the government for himself.[99] During the *decena tragica* it was widely believed both in Mexico and the United States that the latter would intervene with military forces momentarily.

That Ambassador Wilson continued to be inimical to Madero was exemplified by his action on February 15, in the latter half of the "tragic ten days" of persuading the British, German and French ministers to join in an effort to induce Madero to resign. Madero responded indignantly, saying that it was not a diplomat's right to intervene in a domestic question, and that he would die defending his office.[100] Wilson then approached the anti-Madero senators to mobilize sentiment against him. On February 16, Madero told Robert H. Murray that "the American ambassador is our greatest enemy," and adding that Wilson had threatened that American troops would be sent to Mexico unless he resigned.[101] Ambassador Wilson advised the Department of State at noon February 18: " . . . the supposition now is that the federal generals are now in control of the President." As a matter of fact, Madero was not seized until approximately two hours after the message was transmitted.[102]

This sequence of events culminated in the arrest on February 18 of Madero and Pino Suárez, the vice president, on orders of General Huerta, and their detention in the Palace. Once their resignations had been obtained, their offices were declared vacant. Thereupon the minister of foreign affairs, Pedro Lascurain, constitutionally became president. His term lasted about fifteen minutes, and his only acts were to appoint Huerta to a cabinet office and then resign. Huerta, as holder of the highest cabinet office, became provisional president.[103] Ambassador Wilson soon thereafter announced to an assembly of the diplomatic corps that he had known for three days of the plan to imprison Madero.

[98] Rafael de Zayas Enríquez, *The Case of Mexico and the Policy of President Wilson* (New York, 1914), p. 99.
[99] Murray, "Huerta and the two Wilsons," p. 365; Alfredo Breceda, *Mexico Revolucionario, 1913-1917* (Madrid, 1920), p. 60.
[100] *For. Rel., 1913*, p. 711.
[101] Murray, "Huerta and the two Wilsons," Apr. 15, 1916, p. 403.
[102] Ernest Gruening, *Mexico and its Heritage* (New York, 1928), p. 567.
[103] Albert Bushnell Hart, "Postulates of the Mexican Situation," *American Academy of Political and Social Science Annals*, LIV (1914), 139.

Gustavo Madero was brutally murdered on February 19, his death foreshadowing another tragedy closely following. Huerta then asked the American Ambassador for advice as to the disposition of Francisco Madero. Wilson replied that he "ought to do what was best for the peace of the country."[104] On February 20 and 21 the American ambassador was called upon to make vigorous protestations for the safety of the imprisioned officials. The Cuban Minister warned Wilson of the danger and offered the use of a Cuban cruiser at Vera Cruz to take them to safety.[105] Ex-President Madero's parents wrote to the diplomatic corps pleading for its intercession in the matter, and Madero's wife interviewed Wilson with that object.

On the evening of February 22 Madero and Suárez were killed while being transferred from the Palace to the penitentiary. The official government account of the affair claimed that a rescue of the prisoners was attempted, the vehicles were fired on, and the prisoners were shot in the melee while attempting to escape. The concensus of evidence is that it was an act of deliberate murder carried out in the Mexican tradition of *ley de fuga* (law of flight).

Though no positive proof exists as to Ambassador Wilson's complicity in or influence surrounding this tragedy, the most pointed indictment is found in the report of John Lind, a lawyer who served as President Wilson's personal emissary to Mexico in 1913 and 1914. After Lind's return from Mexico, Secretary of State William Jennings Bryan requested him to analyze Wilson's dispatches to the Department of State. This analysis convinced Lind that "a jury would be justified in finding Henry Lane Wilson guilty of aiding and abetting, if not instigating the Huerta rebellion, and also that he is guilty as an accessory before the fact of the assassination of Madero."[106]

Ambassador Wilson was satisfied with the government's explanation, and urged the State Department to accept it as final and reliable. He sought to minimize the occurrence by saying that it had a negligible effect on the public, and that the nation, in the main, was favorably disposed toward the new regime. However, reports from American consuls in Mexico directed to the Secretary of State reflected the opposite reaction.[107]

On February 18, Huerta had sent a telegram to Washington stating "I have the honor to inform you that I have overthrown the government. The forces are with me, and from now on peace and order will reign."[108] Subsequently, Wilson bent his efforts towards securing American recognition for Huerta. The State Department replied on February 25, indicating that although the Department saw the advantages of showing a disposition toward recognizing the Provisional Government, the president had directed that no formal recognition be accorded except upon specific instructions from the State Department. This position is reaffirmed and clarified somewhat by Knox's message of February 28, which explained that the United States government was in *de facto* relations with Huerta because he was the only effective authority in evidence. A distinction

[104] Wilson to Knox, Feb. 20, 1913, *For. Rel., 1913*, p. 724.
[105] H. Márquez Sterling, *Los Últimos Días del Presidente Madero* (Havana, 1917), p. 491.
[106] George M. Stephenson, *John Lind of Minnesota* (Minneapolis, 1935), p. 307.
[107] Wilson to Knox, Feb. 24, 1913, *For. Rel., 1913*, pp. 734-736.
[108] Huerta to Taft, Feb. 28, 1913, *Ibid.*, p. 721.

was drawn between *de facto* relations with a *de facto* government and formal recognition of such government.[109]

The European countries and most Latin American republics answered Huerta's telegram at once, and by that act accorded him virtual recognition. But Argentina, Brazil, Chile and Cuba followed the United States example in withholding recognition. The position taken by the United States was based on several considerations. President Taft had declined to recognize Huerta partly because the tragic circumstances of his accession seemed to deserve some rebuke by a delay in recognition. Furthermore it was not yet clear that Mexico herself desired Huerta, and lastly, but not the least important, Taft's own term of office was nearing its close. A courteous consideration for his successor demanded that so important a decision be left to him, rather than have the new president committed to a recognition which might not be in harmony with his policies.

[109] Knox to Wilson, *Ibid.*, pp. 747-748.

SUPPLEMENTARY READINGS

Bancroft, Hubert H. *History of Mexico,* 6 vols. San Francisco, 1883-1888.

----------. *History of the North Mexican States and Texas,* 2 vols. San Francisco, 1884-1889.

Burke, U. R. *A Life of Benito Juárez.* London, 1894.

Callcott, Wilfrid H. *Santa Anna: The Story of an Enigma Who Once Was Mexico.* Norman, 1946.

Cline, Howard F. *The United States and Mexico.* Cambridge, Mass., 1963. Rev. ed.

Cornyn, J. H. *Díaz y México,* 2 vols. Mexico, 1910.

Corti, Egon. *Maximilian and Charlotte of Mexico,* 2 vols. New York, 1947.

Cósio Villegas, Daniel. *Estados Unidos contra Porfirio Díaz.* Mexico, 1956.

Cotner, Thomas E., ed. *Essays in Mexican History.* Austin, 1958.

Cumberland, Charles C. *Mexican Revolution: Genesis under Madero.* Austin, 1952.

Flandrau, Charles M. *Viva México.* New York, 1937.

Godoy, José F., *Porfirio Díaz, The Master Builder of a Great Commonwealth.* New York, 1910.

Gruening, Ernest. *Mexico and Its Heritage.* New York, 1928.

Hannay, David. *Díaz.* New York, 1917.

Manning, W. R. *Early Diplomatic Relations between the United States and Mexico.* Baltimore, 1916.

Pletcher, David. *Rails, Mines and Progress: Seven American Promoters in Mexico, 1867-1911.* Ithaca, 1958.

Priestley, H. I. *The Mexican Nation.* New York, 1923.

Quirk, Robert E. *The Mexican Revolution, 1914-1915.* Bloomington, Ind., 1960.

Ramírez, J. F. *México durante su guerra con los Estados Unidos.* Mexico, 1905.

Rippy, J. Fred. *The United States and Mexico.* New York, 1926.

Rebolledo, Miguel. *México y los Estados Unidos.* Mexico, 1917.

Robertson, W. S. *Iturbide of Mexico.* Durham, 1952.

Rives, G. L. *The United States and Mexico, 1821-1848,* 2 vols. New York, 1913.

Roeder, Ralph. *Juárez and His Mexico,* 2 vols. New York, 1947.

Romero, Matías. *Mexico and the United States.* New York, 1898.

Ross, Stanley R. *Francisco I. Madero, Apostle of Mexican Democracy.* New York, 1955.

Scholes, Walter V. *Mexican Politics During the Juárez Regime, 1855-1872.* Columbia, Mo., 1957.

Simpson, Lesley B. *Many Mexicos,* 3rd ed. Berkeley, 1952.

Tannenbaum, Frank. *The Mexican Agrarian Revolution.* New York, 1929.

Tischendorf, Alfred. *Great Britain and Mexico in the Era of Profirio Díaz.* Durham, 1961.

8

Mexico and the United States: Recent and Contemporary Relations

When Woodrow Wilson was inducted into the presidency in March 1913, he had made no previous explicit pronouncements respecting foreign policy. The first indication of the general position that he would take on the matter was contained in a prepared statement issued November 2, 1912. After recapitulating the domestic problems with which his administration would deal, he declared that the next four years were going to determine "the firm establishment of a foreign policy based upon justice and good will rather than upon mere commercial exploitation and the selfish interests of a narrow circle of financiers extending their enterprises to the ends of the earth . . . "[1] From this statement the conclusion could immediately be drawn that the ideals and values of the new president were in opposition to the "dollar diplomacy" of the Taft administration, which had stressed the material interests of United States citizens abroad.

With the change of administration, Philander C. Knox was replaced as secretary of state by William Jennings Bryan, who had previously rendered public service as a member of Congress from 1891 to 1895, and as a Colonel during the Spanish-American War. He had three times been the Democratic candidate for the presidency, and in the Democratic convention of 1912 his influence was critical in determining the nomination of Wilson. As Bryan had little experience in diplomatic practice and international law, the formulation of foreign policy devolved upon the president.[2] Bryan's avowed pacifist and anti-imperialist views harmonized well with those of Wilson.

[1] Harley Notter, *The Origins of the Foreign Policy of Woodrow Wilson* (Baltimore, 1937), p. 197.
[2] Samuel F. Bemis, *The American Secretaries of State and their Diplomacy* (New York, 1929), X, 10.

MEXICO
0 100 200
MILES
UNITED STATES
GULF OF MEXICO
PACIFIC OCEAN
GULF OF CALIFORNIA
States shown by number
1. TLAXCALA
2. MORELOS
3. DISTRITO FEDERAL
4. MÉXICO
5. HIDALGO
6. QUERÉTARO
7. GUANAJUATO
8. AGUASCALIENTES
9. NAYARIT
10. COLIMA
11. TABASCO
BAJA CALIFORNIA NORTE
BAJA CALIFORNIA SUR
SONORA
CHIHUAHUA
COAHUILA
NUEVO LEÓN
TAMAULIPAS
SINALOA
DURANGO
ZACATECAS
SAN LUIS POTOSÍ
JALISCO
MICHOACAN
GUERRERO
PUEBLA
VERACRUZ
OAXACA
CHIAPAS
CAMPECHE
YUCATAN
QUINTANA ROO
BR. HONDURAS
GUATEMALA
Mexicali
Ciudad Juarez
Hermosillo
Chihuahua
La Paz
Torreón
Monterrey
Durango
Mazatlán
Zacatecas
Ciudad Victoria
Tampico
San Luis Potosí
Aquascalientes
Guadalajara
Morelia
MEXICO CITY
Puebla
Veracruz
Acapulco
Oaxaca
Tuxtla
Campeche
Mérida

The chaotic situation in Mexico was one of the most pressing problems inherited from the Taft regime by the new administration. Specifically, this involved the question of recognition of the usurper, General Huerta. President Wilson responded promptly and decisively. In a statement circularized to American diplomatic officers of Latin America in March 1913, the president declared that the favor of American recognition was to be made contingent upon "the orderly processes of just government based upon law, not upon arbitrary or irregular force. . . . We can have no sympathy with those who seek to seize the power of government to advance their own personal interests or ambition."[3] This statement, which was aimed at Mexico in particular, revealed his intention not to recognize Huerta, and his desire to promote constitutional government.

Another facet of his policy, the attitude he would take toward economic imperialism, was also clarified: "The United States has nothing to seek in Central and South America except the lasting interest of the people of the two continents, the security of governments intended for the people and for no special group or interest . . . " This reaffirmed his statement of November 2, 1912, thereby indicating the abandonment of "dollar diplomacy."[4]

From this point onward, two considerations motivated the President in the development of his Mexican policy and compelled his adherence to it throughout his administration, namely: "The firm conviction that all nations, weak and powerful, have the inviolable right to control their own affairs," and the belief, determined from the history of the world, "that Mexico will never become a peaceful, law abiding neighbor of the United States until she has been permitted to achieve a permanent and basic settlement of her troubles without outside interference,"[5]

Of extreme importance to President Wilson in the execution of his Mexican policy was the cooperation of Great Britain. Early in his administration, before making a definite stand on nonrecognition, he had instructed the American chargé d'affaires in London, Irvin Laughlin, to determine the British policy regarding Mexico. The British Foreign Office had declared unequivocally that there would be no recognition of Huerta, formal or tacit. Wilson then proceeded to make it known that this country would never recognize him.[6] Soon thereafter, Cecil Spring Rice was directed to say that England had changed her mind and on March 31 Secretary Bryan was informed that his Majesty's Government was recognizing the "president *ad interim* of the Republic of Mexico."[7]

Thus in the months of April and May, 1913, England, and then France and Germany accorded recognition to Huerta. This was a logical step from the standpoint of European governments as trade and protection of their citizens' investments were of vital importance. Wilson, angered by this turn of events, was most bitter toward the business interests in those countries for he held them responsible.

[3] Edgar E. Robinson and Victor J. West, *The Foreign Policy of Woodrow Wilson* (New York, 1918), pp. 179-180.
[4] *Ibid*., p. 180.
[5] Joseph P. Tumulty, *Woodrow Wilson as I Knew Him* (New York, 1921), p. 145.
[6] Burton J. Hendrick, ed., *The Life and Letters of Walter Hines Page* (New York, 1926), I, 180-181.
[7] Bryce to Bryan, Mar. 31, 1913, *For. Rel., 1913*, p. 784.

Meanwhile, Britain further antagonized the president by appointing Sir Lionel Carden as Minister to Mexico in July 1913. Carden had gained a reputation of being anti-American during a diplomatic sojourn in Cuba, and Secretary Knox had twice asked indirectly for his removal. Carden made matters worse by giving several highly undiplomatic interviews critical of the American president. Washington interpreted his appointment as reaffirmation of British recognition of Huerta.[8] The *status quo* was maintained on the matter of British recognition for the next three months. At the end of this period the situation changed in favor of the United States.

In Mexico, following the change of administration, conditions were steadily becoming more critical. The American consul at Juárez in reporting to Bryan on March 10, declared that the disorders were growing worse in the west and south and he feared a real war might occur.[9] Ambassador Wilson all the while continued to press for recognition of Huerta using as an argument the rising anti-American feeling and the decline of trade. His message to Bryan of March 12, revealing that his views were diametrically opposed to those of the president, stated: "Mexico cannot remain peaceful unless the same type of government as that of Díaz is again established."[10] The ambassador's views are even more distinct in a remark made somewhat later in which he said, "In the conduct of foreign relations idealism is a dangerous element and morals and expediency are nearly identical."[11] The president. feeling that he could not rely on the Ambassador's reports, sent William Bayard Hale, the eminent journalist, to Mexico as his personal observer. The Huerta government, with the continued support of Ambassador Wilson, pressed for recognition. On July 9, 1913, the latter submitted to the State Department for consideration two possible courses which he deemed the only solution to the situation: (1) official recognition with demand for guarantees; (2) withdrawal of the ambassador as a protest against existing conditions.[12] By this time Hale's reports informed the president of the friendly relations between the ambassador and Huerta, and Embassy reports continued to show the divergence of the ambassador's views from those of the administration in Washington. On July 3, the president notified Bryan that he thought the ambassador should be recalled, leaving Nelson O' Shaughnessy in charge on Hale's recommendation.[13] This plan was carried out, and on August 4, after an interview by the president on July 28, Henry Lane Wilson resigned. The explanation he gave for this was that his views were at such variance with the present administration that he could no longer represent it.[14]

A stalemate had now been reached and it became imperative for the president to take steps to resolve it. The absence of policy and reports of intensification of the strife in Mexico had already encouraged undesirable suggestions from members of Congress. These ranged from proposals which

[8] Hendrick, *Life and Letters*, pp. 192-198.
[9] Edwards to Bryan, Mar. 10, 1913, *For. Rel., 1913*, p. 762.
[10] Wilson to Bryan, Mar. 12, 1913, *Ibid*., p. 769.
[11] *Independent*, Nov. 13, 1913, p. 289.
[12] Wilson to Bryan, July 9, 1913, *For. Rel., 1913*, p. 809.
[13] Ray Stannard Baker, *Woodrow Wilson, Life and Letters* (New York, 1931), IV, 255-256.
[14] James M. Callahan, *American Foreign Policy in Mexican Relations* (New York, 1932), p. 539.

amounted to outright intervention to recognition of the Constitutionalists' belligerency.[15] Accordingly, the president decided to volunteer good offices as a means of remedying the situation. Taking this course, the president, on Bryan's recommendation, sent John Lind, an ex-governor of Minnesota, as a personal emissary to Huerta. The appointment of this crusader against imperialism and "big business" from Bryan's section of the country was apparently, in part, an effort to reassure the country of the president's good intentions and allay the incipient revolt in Congress.[16] Prior to Lind's appointment the president had reached the decision that Huerta must go; and Huerta had publicly announced that he would not resign. There was consequently a widespread belief in the United States that Lind's mission was futile. This was affirmed by Huerta's statement to reporters in Mexico City: "I have said publicly and Minister Urrutia has said the same, that I will accept neither mediation nor intervention of any kind in our internal struggles . . . I have also declared that in no account will I accept compromises with the revolution, and still less if a hint of such involves a flagrant violation of our sovereignty."[17] Venustiano Carranza, Huerta's chief opponent, added further doubts for Lind's success by announcing that he would not accept any United States proposals for a truce or mediation.[18]

The president's instructions to Lind made it clear that a settlement was conditioned upon (1) an immediate cessation of fighting; (2) an early and free election; (3) General Huerta's consent not to be a candidate for president; (4) an agreement by all parties to abide by the results of the election. Although the envoy was received courteously enough by the Mexican government, it categorically refused to treat with him on the basis outlined,[19] and the only result of his mission was to inspire in Huerta and his party a more bitter hatred towards the United States. The first phase of Lind's mission thus ended in failure. Subsequently, until April 1914, stationed in this interval at Vera Cruz, he was the chief watchman under Wilson's "watchful waiting" policy.[20]

In analyzing the Mexican attitude toward the Lind mission Manuel Calero points out that when it became known that President Wilson was trying to dictate to Mexico the type of government it must have, a wave of indignation swept across the country. When Lind arrived, Calero said, there was much discontent against Huerta and a combined movement might have overthrown him. This was frustrated by the intrusion of Wilson. "Huerta, right or wrong," said everybody, "rather than accept a foreign imposition."[21]

Aside from these opinions and the universal criticism in Mexico as to the undiplomatic character of the Lind mission some saw, or thought they saw, a hidden motive in Wilson's suggestions. Zayas Enríquez who once said, "Machiavelli would have been clay in Wilson's hands, and what is more, Machiavelli would never have known it," declared that if Huerta had resigned in compliance with Lind's instructions Mexico would have lost its autonomy.[22]

[15] Notter, *Foreign Policy of Woodrow Wilson,* p. 254.
[16] George M. Stephenson, *John Lind of Minnesota* (Minneapolis, 1935), p. 214.
[17] *Mexican Herald,* Aug. 6, 1913; *Review of Reviews,* Sep., 1913, p. 283.
[18] *Christian Science Monitor,* Aug. 2, 1913.
[19] Stephenson, *John Lind,* p. 217.
[20] *Mexican Herald,* Aug. 28, 1913.
[21] Manuel Calero, *President Wilson's Mexican Policy as it Appears to a Mexican* (New York, 1916), pp. 17-18.
[22] Rafael de Zayas Enriquez, *The Case of Mexico and the Policy of President Wilson* (New York, 1914), p. 136.

"After this not only would the United States have exacted a direct suzerainty, which is the object of its ambition, but Mexico would have virtually become an American colony."[23] Another Latin American writer termed Lind's proposals as ridiculous, erroneous and if accepted would have been a violation of Mexico's sovereignty.[24]

President Wilson's message to Congress on August 27 also outlined some practical steps. All Americans would be urged and assisted to leave Mexico because, although this government wished to give protection, it was mandatory that the hazards be reduced. The arms embargo was to be extended so as to prohibit the export of arms to all factions in Mexico. This was considered necessary, for to permit arms to go freely into Mexico would increase disorder, encourage border clashes, and might lead to an intervention in which the arms we had supplied would be used against us.[25]

The rebel forces that were in arms against Huerta were naturally favorably disposed toward the American non-recognition policy but were bitter over the arms ban which had affected them though not Huerta, the original Taft proclamation of March 14, 1912, having remained unaltered. In an effort to obtain the same privileges as Huerta, the Constitutionalists had sent a representative to Washington, Eduardo Hay, who appealed to the Committee on Foreign Relations to revoke the discriminatory practice.[26]

The principal leader of the insurrection against Huerta was Venustiano Carranza, mentioned earlier as having repudiated Huerta's authority, a conservative who had been a senator under Diaz for sixteen years, but had joined the opposition when Diaz failed to back him for the governorship of Coahuila. Under the "Plan of Guadalupe" promulgated on March 26, 1913, Carranza, with a large segment of the population of northern Mexico and several military leaders, notably Alvaro Obregón and Pancho Villa supporting him, set out to reestablish constitutional government. Carranza supplemented this plan on May 13 with a decree that committed the Constitutional government to the principle of international arbitration of claims following his accession to power.[27] These avowed intentions brought Carranza into the favor of Washington which maintained liaison with his headquarters through a special agent, G. C. Carothers. In November 1913, William Bayard Hale was sent as a special envoy by the president to the Constitutionalist's temporary capital in Sonora. His mission was the first overt indication that the United States was interested in the Constitutionalist movement, and Carranza as Huerta's possible successor.[28] These forces in the north, together with those of Zapata in the south, maintained an unrelenting struggle against the Federal troops.

Huerta, facing rising opposition in the Mexican Congress, had two senators and one hundred and ten members of the Chamber of Deputies arrested and sent recognition. To circumvent this situation and to put the Constitutionalists on

[23] *Ibid.*, p. 168.
[24] Roberto Domenech. *Méjico y el imperialismo Norte Americano* (Buenos Aires, 1914), p. 56.
[25] *Cong. Rec.*, 63rd Cong., 1st Sess., p. 3802.
[26] *Mexican Herald,* Aug. 4, 1913.
[27] *Cong. Rec.*, 63rd Cong., 1st Sess., p. 3133.
[28] *New York Times*, Nov. 13, 1913.

to prison. He then issued a decree ordering the Congress dissolved and assumed dictatorial powers. President Wilson reacted to these events by declaring that he regarded them as an act of bad faith toward the United States and stressed that any elections held under these conditions would not be regarded as valid by his government.[29] Undaunted by the admonition Huerta, after asserting that he would not be a candidate, was elected President in a peaceful but farcical election on October 26.

President Wilson responded to these events by delivering an address to the Southern Commercial Congress at Mobile, Alabama, on October 27, 1913, in which he examined the more basic aspects of our relationships with the Latin American republics. He spoke of a spiritual union between North and South America saying: "In the future the nations to the south of us will draw closer and closer to us . . . . We must prove ourselves their friends and champions in terms of equality and honor . . . . The development of constitutional liberty and world human rights, the maintenance of national integrity, as against material interests . . . . that is our creed. . . . The United States will not again seek to secure one additional foot of territory by conquest . . . . "[30] The speech appears to show Wilson's determination that this country was dedicated to making itself an interested friend, seeking no advantages in the southern republics.

As another stalemate had been reached by the end of October, Wilson was confronted with the necessity of taking more drastic steps to eliminate Huerta. On November 1, the president sent Huerta an ultimatum virtually demanding his resignation upon a threat to employ any means necessary to accomplish it. This brought no reaction, whereupon Wilson invited foreign cooperation to obtain Huerta's voluntary retirement. Another statement made soon thereafter indicated that U.S. policy would be to isolate Huerta, and if he did not retire by the force of circumstances it would be the duty of the United States to use less peaceful methods.[31] This new approach to the problem seems to have been made possible, or at least aided, by the attitude of the British, who by this time were giving moral support to Wilson's policies.

Since July 1913, when Sir Lionel Carden was appointed Minister to Mexico, an act seeming to confirm British recognition of Huerta, although temporary, the position of the British government in the affair caused Wilson much consternation. In November 1913, Sir William Tyrrell, a member of the British Foreign Office came to this country, unofficially, and succeeded in convincing Wilson that the business interests were not dictating his country's foreign policy, and since Britain recognized the predominant character of American interests in Mexico they would be willing to follow our lead, once English life and property were assured protection.[32] Tyrrell was able to obtain only one statement from the president concerning his policy. He said, "I am going to teach the South American republics to elect good men."[33] The

[29] Bryan to O'Shaughnessy, Oct. 13, 1913, *For. Rel., 1913*, p. 838
[30] David F. Houston, *Eight Years with Wilson's Cabinet* (New York, 1926), I, 77.
[31] Notter, *Foreign Policy of Woodrow Wilson*, p. 273; Bryan to Diplomatic Officers of U.S., Nov. 7, 1913, *For. Rel., 1913,* p. 856; Bryan to O'Shaughnessy. Nov. 24, 1913, *Ibid.*, p. 443.
[32] Albert Bushnell Hart, *The Monroe Doctrine, An Interpretation* (Boston, 1916), p. 334.
[33] Hendrick, *Life and Letters*, I, pp. 204-205.

principle value of this interview lay in the fact that Wilson revealed his opposition to the Panama Canal Tolls Act, which did a good deal toward clearing the way for an understanding regarding Huerta.[34]

Dissension over the matter of canal tolls arose in 1912 when an act was passed by the United States exempting coastwise shipping from tolls. The British held the act to be a violation of the Hay-Pauncefote Treaty of 1901, which provided that the tolls levied should apply to the shipping of all nations equally and without discrimination. No agreement had been reached when Wilson took office, and Washington failed to comply with the British proposal that the question be submitted to arbitration. Therefore, in March 1914, subsequent to the Tyrrell-Wilson interview, the president urged Congress to cancel the exemption provisions. A bill to this effect was signed by Wilson in June 1914.[35]

In November, a few days after Wilson's interview with Tyrrell, Sir Lionel Carden was instructed by his government to yield to Wilson's demands. Huerta was told that inasmuch as Carden's government was supporting Washington he could no longer depend on British backing.[36]

With the feelings of the British assuaged on the subject of canal tolls and their position against Huerta established, Wilson was left free of European pressure for the time being. In his annual message to Congress on December 2, he said that little by little Huerta was being isolated, his power and prestige were crumbling, and that his total collapse was not far away. "We shall not, I believe, be obliged to alter our policy of watchful waiting."[37]

Meanwhile, in the fall of 1913, the struggle between the warring factions grew more intense and widespread. On October 1, Villa occupied Torreón after a battle which cost almost one thousand lives, and Zapata was rapidly gaining more recruits for his army which operated in Morelos and Guerrero. General Obregón, who had been appointed chief of the Army Corps of the Northeast by Carranza, won a series of battles against the federal troops in Sinaloa in this period.[38] But in spite of the arms embargo, Huerta was able to keep supplied with munitions and military equipment from Japan and Germany.[39]

On January 31, 1914, Secretary Bryan announced that his government was convinced that there would be "a more hopeful prospect of peace, of security of property, and of early payment of foreign obligations," if the Constitutionalists were victorious; and the president no longer felt justified in staying neutral. He therefore was about to lift the embargo on arms to Mexico.[40]

By his action on August 27, 1913, President Wilson had attempted to enforce neutrality by prohibiting the shipment of arms to all contestants. However, by this time it was evident that the Provisional Government was obtaining war materiel from other sources . . . . from nations that had extended

[34] *Ibid.*, p. 209.
[35] H. Wilson Harris, *President Wilson from an English Point of View* (New York, 1917), pp. 144-146.
[36] Hendrick, *Life and Letters*, I, 209.
[37] James B. Scott, ed., *President Wilson's Foreign Policy, Messages, Addresses, Papers* (New York, 1918), pp. 27-30.
[38] Alfonso Taracena, *Mi vida en el vértigo de la revolución mexicana* (Mexico, 1936), p. 232.
[39] Wilfred H. Calcott, *Liberalism in Mexico, 1857-1929* (Stanford, 1931), p. 238.
[40] Bryan to all diplomatic missions of the United States, Jan. 31, 1914, *For. Rel., 1914,* p. 812.

equal footing with Huerta, the arms embargo was raised on February 4, 1914.[41] As the *Carrancistas* occupied a large area along the border they now had access to material needed to overthrow Huerta. Villa and Carranza were elated, but in Mexico City much bitterness was expressed. The Mexican newspapers vilified Wilson to such an extent that O'Shaughnessy was forced to register several protests with Huerta.[42]

While Huerta was able to obtain military equipment from certain powers, the United States employed a powerful weapon against him when it held up European credit. A French banker, for example, was advised by his government that any loan to Mexico would embarrass French foreign policies. Mexico was refused the proposed loan owing to pressure exerted from Washington.[43] Foreign investors were given to understand that obligations incurred by Huerta would not be recognized and must inevitably become worthless. Since the provisional president was in imminent danger of bankruptcy, having defaulted on the semi-annual interest payments on the Mexican foreign bonds on January 1, 1914, this failure to obtain loans would obviously bring about the downfall of his government before much time would elapse.[44] In later years Huerta admitted that financial pressure exerted by the United States was the most potent force which was used against him.[45]

By March 1914, it was apparent that the Constitutionalists were deriving little benefit from the arms which had been made available to them by this country lifting the arms ban, for their military operations had slowed down. There arose criticism in the United States as to the efficacy of Wilson's "watchful waiting" policy. Senator Albert B. Fall of New Mexico was especially vocal in urging United States intervention. It was with this clamor in mind that the President said, "I have to pause and remind myself that I am President of the United States and not of a small group of Americans with vested interests in Mexico."[46]

Before the lifting of the arms embargo could demonstrate any substantial effects, relations between the United States and Mexico were markedly altered by developments in the next few weeks. On April 9 an officer and boat's crew from the U.S.S. *Dolphin*, in the course of obtaining supplies in Tampico were arrested by Mexican authorities for landing in a prohibited area. Although the men were released almost immediately, and the local Mexican commander apologized for the incident, Admiral Mayo, the American naval officer commanding the squadron, felt the offense sufficiently grave to demand a further apology, including a twenty-one gun salute to the American flag.[47]

[41] *New York Times*, Feb. 4, 1914.
[42] *Ibid.*, Feb. 11, 1914.
[43] Henry G. Hodges, *The Doctrine of Intervention* (Princeton, 1915), p. 111, 125.
[44] Ernest Gruening, *Mexico and its Heritage* (New York, 1928), p. 578.
[45] Victoriano Huerta, *Memorias del General Victoriano Huerta* (Mexico, 1916), p. 84.
[46] Tumulty, *Wilson as I Knew Him*, p. 146.
[47] Mayo to Zaragoza, Apr. 8, 1914, *For. Rel., 1914,* p. 449. For details of the Vera Cruz episode see Robert E. Quirk, *An Affair of Honor, Woodrow Wilson and the Occupation of Vera Cruz* (Lexington, Kentucky, 1962); also, Tumulty, *Wilson as I Knew Him.* What appears most significant concerning intervention at Vera Cruz is that an American naval officer possessed sufficient diplomatic powers to demand a salute from a government whose existence had not been recognized by this country. Admiral Mayo should have left the matter to the State Department where it logically belonged. If this procedure had been followed, intervention would probably have not occurred, for Wilson was displeased by the Admiral's action.

Huerta sent an apology to Washington but refused to meet the demand for a twenty-one gun salute; unless the United States, by protocol, would return the salute.[48] This offer was flatly rejected by Washington. The general belief in Mexico at this time was that if President Wilson accepted the protocol requested by Huerta it would be tantamount to recognition.[49]

Secretary Bryan felt that Huerta's apology alone was sufficient, but the president felt obliged to back up Mayo's demand even though the latter's actions caused him some irritation.[50] On April 16, President Wilson made a public statement of the situation in which he enumerated the offenses by which Mexico had offended the dignity of this country. In view of these grievances the President contemplated the following measures be taken: (1) no war with Mexico, but seizure and occupation of Vera Cruz, Tampico and certain ports on the west coast, the others to be blockaded; (2) the occupation and blockade to continue until the officers responsible for the various offenses were punished; (3) Admiral Badger to be given the authority to act on his own discretion should an emergency arise; (4) no time limit would be stipulated for the occupation.[51] On this basis the president laid his plan before Congress, and on the following day the House of Representatives passed a resolution, by a vote of 337 to 37, giving him the authority to take the necessary action.[52] This vote is illustrative of the temper of the American public at this time.

The execution of the president's plan was hastened by an unexpected development. During the night of April 20-21, Secretary Bryan was advised that a German transport, the *Ypiranga*, carrying two hundred machine guns and fifteen million cartridges for Huerta, would arrive at Vera Cruz the next day.[53] With the concurrence of the president, Admiral Fletcher was instructed to seize the customs house at Vera Cruz and intercept the arms.[54] After rather sharp fighting the American forces took the city; the occupation continued until November 23, 1914.

The Mexican public, *Carrancistas* and *Huertistas* alike, remonstrated against American intervention. It had been evident from an earlier date that the Mexican people did not understand Wilson's moral attitude and many of their newspapers held his ideals up to derision and sarcasm. The *Correo de la Tarde*, a newspaper published in Mazatlan, for instance, had made the following comments on his Mobile speech:

"Eloquence, simplicity, and apparent sincerity dwell in Mr. Wilson's words, but throughout them is apparent the doctrinairism with which he is imbued, and which has already cost Mexico and her Brothers in Latin America so dear. . . . It is a pity he belongs in the United States, a rich nation, but one which loves the liberty of all Latin America."[55]

[48] *New York Times*, Apr. 20, 1914.
[49] Ciro de la Garza Treviño, *Wilson y Huerta* (Mexico, 1933), p. 29.
[50] Josephus Daniels, *The Wilson Era, Years of Peace, 1910-1917* (Chapel Hill, 1944), p. 191.
[51] *New York Times*, Apr. 16, 1914.
[52] Cong. Rec., 63rd Cong., 2nd Sess., p. 7006.
[53] Canada to Bryan, Apr. 20, 1914, *For. Rel., 1914*, p. 477. The arms and munitions shipment, originating in New York, was permitted to reach its destination in Mexico City by the American authorities.
[54] Tumulty, *Wilson as I Knew Him*, pp. 151-152.
[55] *Review of Reviews*, XLIX (1914), 600.

Another source, expressing common sentiment, said: "President Wilson denied that he had other than altruistic intentions toward Mexico and through intervention in Vera Cruz has applied the principles of the Monroe Doctrine. However, it is obvious that his theories are anachronisms and his procedures contradictory. He offered an olive branch in one hand, and at the same time humiliated Mexico by intervening in her affairs."[56] The United States actions at Vera Cruz were severely criticized by many newspapers in Europe, though in Great Britain they were generally commended.[57]

Mexico severed diplomatic relations and it was believed, both in the United States and Mexico, that wider intervention was at hand. At this critical juncture Argentina, Brazil and Chile tendered their good offices to effect a settlement of the difficulties; the offer was accepted.[58] General Huerta, now almost at the end of his financial resources, and worried about the continued successes of the Constitutionalists under Carranza and Villa, could hardly afford to refuse. While Carranza agreed to mediation in principle he refused to allow any aspect of Mexico's internal problems to be made an issue at the forthcoming conference. Neither would he consent to a suspension of hostilities as the mediators requested. Throughout the negotiations at Niagara Falls which followed, Carranza remained adamant on these points, and as a consequence his representatives never participated. The deliberations lasted six weeks, but with no practical results other than allaying Latin American fears in regard to the Mexican policy of the United States.[59] In the meantime the Constitutionalists had been advancing steadily toward Mexico City. Torreón, Tuxpam and Saltillo had fallen into their hands, and with General Obregón's victory at Guadalajara early in July, Huerta's capitulation became inevitable. At this juncture dissension arose between Villa and Carranza which threatened to bring about a complete break. The basis of the trouble appeared to be Villa's jealousy of his colleague's increasing prestige with the United States. Prompt action by the special agent, Carothers, effected a reconciliation;[60] however, as later events proved, this was to be short lived.

On July 8, the Niagara Protocol, which issued from the conference, was presented to the Congress at Mexico City and the initial step toward the formation of an interim government was taken. This was the appointment of Chief Justice Carbajal to the post of Minister of Foreign Affairs, a position through which he would become provisional president automatically in the event a vacancy occurred. Huerta, conscious of the futility of resistance, and having the opportunity to retire in a dignified manner, resigned on July 15, 1914.[61] Soon thereafter, he and his family left for Kingston, Jamaica, on the German cruiser *Dresden*.

[56] Ramón Guzmán, *El intervencionismo de Mr. Wilson en México* (New Orleans, 1915), p. 2.
[57] *Outlook*, May 2, 1914, pp. 17-18.
[58] *New York Times*, Apr. 30, 1914.
[59] For details on the Niagara Protocol see Frank H. Severance, *Peace Episodes on the Niagara* (Buffalo, 1914), pp. 12-13.
[60] Carothers to Bryan, July 9, 1914, *For. Rel., 1914*, p. 559.
[61] Oliveira to Bryan, July 15, 1914, *Ibid.*, p. 563. The German government made an unsuccessful attempt to help Huerta regain power in Mexico, expecting that his presence there would prevent or delay the entry of the United States in the European war. See George J. Rauch Jr., "The Exile and Death of Victoriano Huerta," *Hispanic American Historical Review*, XLII (May, 1962), 133-151.

Carbajal, a man who had maintained an independent attitude throughout the struggle, succeeded Huerta and immediately entered into negotiations with Carranza toward a disposition of the government. A plan acceptable to both was formulated by the Brazilian Minister, and on August 26, 1914, General Carranza made a truimpha entry into Mexico; Acting President Carbajal resigned and left the country.

President Wilson has been much criticized for his policy of witholding recognition from General Huerta. A study of previous policy of the United States States in regard to the question of recognition shows that the principle followed has been to recognize governments as soon as they have proved their ability to exist and perform international obligations.[62] In general the United States has recognized *de facto* governments without considering their legality, or the method by which they came to power. This practice rested on the Jeffersonian doctrine which was developed to offset the European theory of Divine Right, and was a natural outgrowth of the idea that all governments derive their just powers from the consent of the governed.[63]

The test which Wilson applied to Huerta's administration was not the expediency of recognition, or the existence of the *de facto* government, but rather the institutional morality of the government officials. The president felt that the illegal usurpation of executive authority, exemplified by Huerta, would by discouraged by the refusal of the two great Anglo-Saxon powers to recognize such a government. He said, "My ideal is an orderly and righteous government in Mexico, but my passion is for the submerged eight-five percent of the people of that Republic who are struggling toward liberty."[64]

The president's policy has been ably defended by Albert Bushnell Hart, who said in brief that, in the first place, Huerta, geographically was not president of all Mexico. At least one-third of the area remained outside his jurisdiction at all times. Military he was not head of the Mexican republic because his forced were defeated by the rebels on more occasions than they won battles. Constitutionally he was not president because he was not supported by a Congress chosen in a free election. Huerta was not entitled to recognition as supreme authority since he was afraid to leave the capital to defend his government against armed enemies. His statement that all he needed to make his position secure was recognition, is proof that his power was weak and temporary.[65]

A more searching analysis of the problem, however, reveals the existence of a dangerous inconsistency in the course which the President followed. Traditional American policy had not contemplated the prevention of domestic disturbances in Latin America, and now Wilson would attempt to use the power of recognition as a moral weapon to secure permanent reform. The fallacy of this concept lay in the fact that moral coercion gravitated toward physical

[62] Charles E. Martin, *The Policy of the United States as Regards Intervention* (New York, 1921, p. 209.

[63] Stuart A. McCorkle, *American Policy of Recognition Towards Mexico* (Baltimore, 1933), p. 19.

[64] Baker, *Woodrow Wilson,* IV, 245.

[65] Albert Bushnell Hart, "Postulates of the Mexican Situation," *American Academy of Political and Social Science Annals*, LIV (1914), 140.

coercion. Thus the president stood on two divergent lines of action, that of a friend and champion of the Mexican masses, and that of a strong benefactor employing his strength, by threat, to make a sovereign people conform to his wishes. This inconsistency was destructive of one of his major objectives—the friendship of the Mexican people. As events proved, Wilson attached more importtance to the attainment of constitutional government than the matter of friendship, apparently believing that the latter was impossible without the former.[66]

Whatever may be said for or against the policy of "watchful waiting", it soon became decidedly unpopular in the United States. Hardly had Carranza entered Mexico City before Villa raised a new revolt in the north, while Zapata continued his depredations in the south. The government was finally compelled to take refuge at Vera Cruz, and Mexico City became a trophy of all the factions, the Constitutionalists occupying and reoccupying the city four different times. On June 2, 1915, President Wilson addressed a strong plea to the leaders of the various factions to act together for the relief of their prostrate country, declaring that, while the United States did not desire or claim the right to settle the affairs of Mexico for her, yet she could not stand indifferently by and do nothing to serve her neighbor. When this produced no results, our government held a conference with the six ranking Latin American representatives, and a joint note, signed by the United States and the ministers from Brazil, Chile, Argentina, Bolivia, Uruguay and Guatemala, was despatched, August 11, 1915, to Carranza and Villa. The two leaders were asked to meet at some neutral place to exchange ideas on the creation of a provisional government which should take the first steps necessary to the constitutional reconstruction of the country.[67] General Villa accepted forthwith, but General Carranza declared that he could not consent to the discussion of the domestic affarts of the republic by mediation or on the initiative of any foreign government. Nevertheless, when Carranza gave pledges that his government would guarantee that the lives and property of foreigners in Mexico would be respected, and that damages caused by the revolution would be settled, President Wilson, on October 19, 1915, authorized his *de facto* government to be recognized by the United States.

Enraged by this triumph of his rival, Villa swore vengeance upon the United States, and on January 10, 1916, eighteen Americans holding passports of safe conduct issued by the *de facto* government were taken from a train at Santa Ysabel and shot in cold blood. On March 9 Villa raided the little town of Columbus, New Mexico, killing seventeen Americans and carrying away horses. The American government thereupon ordered General John J. Pershing to pursue the bandits across the border, with orders to capture them or destroy their band. Instead of cooperating, the Carranza government seemed desirous of thwarting the purpose of the expedition and, as American troops continued their vain quest farther and farther inland, it showed itself openly hostile. Finally General Jacinto Trevino issued a *pronunciamiento* informing General Pershing that if he moved his troops farther south, east, or west, he would be attacked.

[66] Notter, *Foreign Policy of Woodrow Wilson*, p. 278.
[67] *Amer. Jour. of Int. Law.* X (Apr., 1916), 364.

President Wilson had been patient, but his forbearance had a limit, and he immediately ordered 150,000 militia under arms and despatched them to the border. In a long message to the Mexican government, dated June 20, 1916, Secretary of State Robert Lansing reviewed the whole situation and made it clear that the American troops were in Mexico to accomplish a duty that had been forced upon it through the impotence of the Mexican authorities. Therefore the demand of the Mexican government for "the immediate withdrawal of the American troops" could not be entertained.[68] The day after the despatch of this note the situation was rendered still more grave by an attack upon the American troops at Carrizal, in which some soldiers on each side were killed and seventeen American soldiers were taken captive. The United States government notified American citizens to leave Mexico and, with American troops massed on the border, war seemed inevitable. A demand for the immediate release of the prisoners and a definite statement of the Mexican government's purposes finally aroused Carranza to the danger of his position. The captives were released, and Carranza proposed that an offer of mediation on the part of Spain and several Latin American countries be accepted. The United States thereupon suggested a joint conference, with three commissioners from each side, to arrange a settlement. This solution was accepted by the Mexican government, which named as its delegates Luis Cabrera, Minister of Finance, Alberto Pani, president of the Mexican International Railways, and Ignacio Bonillas. The United States was represented by Franklin K. Lane, Judge George Gray, and John R. Mott.

The conference convened September 6 at New London, Connecticut, and continued, with several recesses, until November 24. The Americans were especially insistent on the protection of foreigners and their interests in Mexico, while the Mexicans urged the immediate withdrawal of American troops from Mexican soil. A protocol was finally signed providing for the withdrawal of General Pershing's army within forty days, provided no new raids should occur in the meantime, with the United States reserving the right to send an army into Mexico to capture bandits who might invade American territory. Claims for damages and plans for economic development were left to future negotiations.[69] Although the Carranza government did not ratify the protocol in accordance with the recommendation of the American commissioners, the American troops were ordered withdrawn, and on January 2, 1917, diplomatic relations were resumed with the appointment of Henry P. Fletcher as United States ambassador to Mexico. Villa remained at large, and Carranza gained added prestige for his successful baiting of the United States; but President Wilson had the satisfaction of preventing intervention from becoming war at a time when there was greater need for American troops and resources in Europe than in Mexico.

On March 1, 1917, the United States Department of State released a report of a German plot to dismember the Union. Conveyed by the British

[68] This correspondence may be found in *Ibid.*, Supp., pp. 179-225. Secretary Bryan resigned in June 1915, fearing that President Wilson's policies would carry the United States into the European war. Robert Lansing succeeded him as Secretary of State.

[69] Text of protocol, *Ibid.*, XI (Apr., 1917), 403.

Secret Service to Washington, it revealed that Alfred Zimmerman, the German foreign minister, had offered Mexico the states of Texas, New Mexico, and Arizona, together with generous financial aid, if Mexico would join Germany should war break out between the United States and Germany. Further, the president of Mexico was to urge Japan to shift to the side of the Central Powers, presumably in return for what spoils Japan might gain at the expense of the United States. Carranza rejected the offer, but he adopted a neutralist policy, mildly sympathetic to Germany, which created a security risk for the United States and the Allies during World War I.

An important domestic problem that the Mexicans had to solve was the devising of some means whereby the dictatorship could be transformed, with the least possible friction, into a constitutional regime. To this end a constitutional convention was called to adopt a new constitution in place of the one of 1857, and the delegates assembled at Queretero in February, 1917. As it was hoped in 1857 to make Mexico a democratic state by giving her a democratic constitution modeled upon the constitution of the United States, so now, sixty years later, it was hoped to solve the various social and economic problems by incorporating their remedies in a constitutional formula.[70] In the framework of government the new constitution followed the constitution of 1857, and as amended provides for a president elected for a single six-year term, with duties very similar to the duties of the president of the United States, and a bicameral congress, the senate consisting of two members from each state and two from the federal district, chosen directly for a six-year term, and the house elected directly for a three-year term. The federal system was maintained, such powers as were not granted to the national government being reserved to the states. The constitution began with a very elaborate bill of rights and ended with a series of temporary articles regulating the procedure in the first election to be held under it.

The new provisions of most interest to the United States were those concerning the ownership of land and natural resources. Article XXVII declared the ownership of lands and waters to be vested originally in the nation, which has the right to transmit title thereof to private persons. Necessary measures were to be taken to divide large landed estates and to develop small holdings. The ownership of all minerals, phosphates, petroleum, and hydrocarbons is vested directly in the state, is inalienable, and may not be lost by prescription. As to the legal capacity to acquire ownership of lands, it was provided that "only Mexicans by birth or naturalization, and Mexican companies, have the right to acquire ownership in lands, waters, and their appurtenances, or to obtain concessions to develop mines, waters, or mineral fuels in the Republic of Mexico. The nation may grant the same right to foreigners, provided they agree before the department of foreign affairs to be considered Mexicans in respect to such property, and accordingly not to invoke the protection of their governments in respect to the same, under penalty in case of breach, of forfeiture to the nation of property so acquired. Within a zone of 100 kilometers

[70] The text of the Constitution of 1917 may be found in *Investigation of Mexican Affairs, Sen. Doc., No. 285,* 66th Cong., 2nd Sess., Vol. II, p. 3123.

from the frontiers and 50 kilometers from the seacoast, no foreigner shall under any conditions acquire direct ownership of lands and waters."

It can be easily seen that the provisions of this article, especially when it is realized that most of the oil-producing property lies within the 50 kilometer zone, are almost completely prohibitive of further foreign exploitation of Mexican oil resources. However, the more immediate question was in regard to the effect it would have on property already acquired. Article XIV of the constitution seemed to answer this in declaring: "No law shall be given retroactive effect to the injury of any person whatsoever." But another provision in Article XXVII (Sec. IV) declared that commercial stock companies may not acquire, hold, or administer rural properties, but only lands in an area absolutely necessary for their establishments, which the executive of the union or of the state in each case shall determine. This seemed like a dangerous exercise of the executive power, and even before the constitution was signed and promulgated, Secretary Lansing, in a despatch to Charles Parker, representing American interests at Queretaro, protested against this paragraph, pointing out that "the objection to a provision so capabale of capricious application appears evident. The precise conditions under which the power vested in the executive may be exercised are not defined. No safeguards are afforded against unwise or arbitary executive acts."[71] After considerable correspondence on the subject, it was apparent that, while the new constitution did not interfere with any wells already drilled, it did prevent foreign companies from sinking new wells unless they waived their nationality and organized as Mexican corporations.

The opposition to Article XXVII upon the part of the American oil interests, particularly as to the interpretation placed upon it by Carranza in a series of executive decrees promulgated at different times during the year 1918, formed the bone of diplomatic contention during the remaining part of the Carranza régime. A decree of February 19, 1918, imposing certain taxes on the surface of oil lands, as well as on the rents and royalties derived from their exploitation, seemed to indicate an intention to separate the ownership of the surface from that of the mineral deposits of the sub-surface, thereby confiscating private property under the guise of taxation without just compensation. Both the American government[72] and the oil companies protested, but with little apparent effect upon the intentions of the Mexican government. The Mexican authorities contended that the decrees were merely fiscal legislation, which operated similarly upon its own citizens and upon foreigners, and that therefore Mexico, in the exercise of its sovereign rights, could not admit the interference of foreign governments in the matter. If the foreigners felt themselves prejudiced by the decree, the Mexican courts were open to afford them legal remedies.[73] Mr. Frank L. Polk summed up the American case in an emphatic but friendly fashion in a long note to Fletcher, the American ambassador, on December 13, 1918. He pointed out that the Mexican attitude regarding the decrees—namely, that merely because they applied equally to Mexican citizens and foreigners

[71] *Ibid.*
[72] See note of Ambassador Fletcher dated April 2, 1918, *Ibid.*, p. 3157.
[73] For a full statement of the Mexican case see the note of E. Graza Pérez, August 17, 1918, to Henry P. Fletcher, *Ibid.*, p. 3161.

the question was one of internal sovereignty, and afforded no rightful basis for interposition by the governments of interested foreigners whose property rights were jeopardized—was not in accordance with international law. "While the Mexican government may see fit to confiscate vested property rights of its own citizens, such action is in equity no justification for the confiscation of such rights of American citizens, and does not stop the government of the United States from protesting on behalf of its citizens against confiscation of their property." Polk then went on to show that Mexican citizens had, by their participation in molding governmental policies, a weapon in addition to judicial remedies which foreigners did not possess; therefore, he argued, friendly representations were not out of place.[74] President Carranza finally turned the matter over to the Mexican congress; but no "organic act" on the subject was passed until some years later.

It is impossible to find an absolutely impartial account of the oil question under the Carranza régime, but a few statements on both sides will afford a basis for a fair approximation. Frederic R. Kellogg, general counsel of the Pan-American Petroleum and Transport Company, sums up the complaints of the oil interests somewhat as follows: many regulations of a harassing nature to prevent development were adopted; the government allowed the filing of claims against the oil properties by persons claiming to be entitled to acquire them under the Carranza decrees; concessions to drill upon lands comprised within titles held by petroleum companies were granted to Carranza's favorites; no company was allowed to drill on its own land unless it had a drilling permit, and no permit would be granted unless the company agreed to abide by the terms of any petroleum law that might be enacted in the future; Carranza sent his armed forces into the oil regions, resulting in a series of murders and assaults upon the employees of the companies; finally, a campaign of vilification against the companies was conducted in Mexico and the United States, accusing them of being tax-dodgers, fomenting rebellion against the Carranza administration, and even seeking to bring about armed intervention by the United States to subserve their own greed and financial ambitions.[75]

On the other side, Joseph F. Guffey, president of the Atlantic Gulf Oil Corporation, declared that the chief offenders in respect to propaganda were the so-called associations ostensibly formed for the protection of American rights in Mexico. As to confiscation, he asserted that

> ... the Mexican government is not attempting to confiscate oil properties developed and operated by American companies. If such were its purpose, it could easily double or treble the export tax of eleven cents per barrel which was collected on all oil exported from Mexico in 1919 .... All these companies have to do to conform to Mexican requirements is to recognize the constitution, renounce all oil concessions and leases obtained prior to its promulgation, and take out new permits. This process does not invalidate the occupancy of these properties by oil companies, but simply recognizes the government's right to levy taxes or collect a royalty on oil production .... President Carranza I regard as an intelligent, constructive, and honest statesman; he has done more for Mexico and the Mexican people than any other president in its history.[76]

[74] *Ibid.*, p. 3163.
[75] *Ibid.*, p. 3270.
[76] *New York Times*, Feb. 26, 1920.

But without question one of the best sources of information upon Mexican conditions during the Carranza régime is the voluminous testimony offered before the subcommittee on the investigation of Mexican affairs headed by Senator Albert B. Fall, which lasted from August 8, 1919, until May 28, 1920.[77] Hearings were held in Washington, New York, and in various cities and towns along the border, and, in all, more than two hundred and fifty witnesses were examined. Edward L. Doheny's account of his long struggle against climatic conditions, governmental interference, and discouragement on the part of stockholders is a story of success against odds which rivals the epic tales of the forty-niners. Incidentally, Doheny brought out the fact that the oil companies, instead of hiring Palaez to protect them against the Carranza régime, were paying him under duress with the connivance of the Carranza authorities, who themselves were wholly unable to give adequate protection. The payments had also been advised by the American State Department.[78]

Whether or not one is in favor of intervention, an unbiased perusal of this report indicates clearly that Mexico under the Carranza régime was not a safe place for an American citizen, whether he was mine-owner, oil-producer, or the proprietor of a ranch. The report further made it clear not only that there was a vast amount of American capital in Mexico,[79] but that it went there under the protection of, and in many cases at the urgent solicitation of, the Mexican government. Finally, despite a good deal of popular suspicion of Senator Fall's Mexican policy, it must be conceded that every effort was made to get at the entire truth of the situation, and that the attempts to prove that a plot existed in the United States to force armed intervention in Mexico failed because no substantial facts were brought forth to substantiate the assertions.

While the Fall sub-committee was at work, a new revolution was started which carried Carranza out of power and down to his death. Like Díaz, Carranza was unable to resign himself to letting the supreme power slip from his fingers. Inasmuch as the constitution prevented a second term, his only chance of retaining control lay in the procuring the election of someone who would be subservient to him. The two strongest candidates, General Alvaro Obregón and General Pablo Gonzalez, were not men of this sort. Hence Señor Bonillas, the Mexican ambassador to the United States, was picked; the point was pressed that the country ought to have a civilian president; and Bonillas was pledged the support of a civilian party. Whatever may have been his qualifications, Señor Bonillas was handicapped by his long absence from the country and by his lack of a strong personal following.

President Carranza had signified his intention to allow the elections to be held without any governmental interference. Nevertheless, it was not long before Candido Aguilar, the president's son-in-law, came out for Bonillas, and soon it was manifest that government dictation was again to be reckoned with. The

[77]By a Senate resolution the committee was directed to investigate the matter of damages and outrages suffered by United States citizens in Mexico, the proper indemnities, and what measures should be taken to prevent a recurrence of such outrages.
[78]*Investigation of Mexican Affairs*, I, 285.
[79]The report gives $1,057,770,000 as the exact figure, p. 3322.

power of the government to bring about the election of its own candidate in Mexico is notorious. The only hope for Obregón and Gonźalez was a revolution, and on April 23, 1920, General Obregón and a number of influential leaders raised a revolt in the state of Sonora. The so-called Agua Prieta plan had been adopted, which declared that the sovereignty of Mexico rested in the people; Carranza, it was contended, had violated it; therefore he must give over his power to a provisional president, namely, Governor Adolfo de la Huerta of Sonora, who should be supreme commander of the state until the elections were to be held. The plan further decreed that the constitution of 1917 should continue to be the fundamental law of the republic.

The revolution spread very rapidly, particularly after leaders like Generals Alvarado, Gómez, and Palaez went over to Obregón, and Gonźalez withdrew his candidacy in Obregón's favor. Early in May, Carranza was forced to flee from the capital, and the revolutionists entered it the same day. Unfortunately for Carranza, he fell in with a small band supposed to be a part of the forces of Palaez, but under the command of Carrera. These men offered to act as guides. But during the night an attack was made upon the sleeping-quarters of General Carranza, and he was shot, presumably by the pretended escort. In the meantime, Congress had chosen Adolfo de la Huerta provisional president, and he postponed the elections from July to September, in order that the provisional government might have time to pacify the country. The delay was necessary, for Gonźalez, apparently repenting his self-effacement, had started a new revolution. But even Mexico was war weary; the revolt was speedily put down, and Gonźalez captured and tried for treason. He was acquitted with the understanding that he would leave the country. Pancho Villa, who also had promised to be good, soon grew tired of inaction and raised a revolt. Realizing the difficulty of capturing him, the government finally got an agreement under which he was to lay down his arms, provided each of his eight hundred followers should receive a tract of land and Pancho himself be given a large estate, with the privilege of keeping fifty retainers at government expense.

On September 5, 1920, Alvaro Obregón was elected president; and his party, the Liberal Constitutionalists, gained control of both houses of congress. Before taking office, he made a tour of the Mexican states, in the course of which he crossed the boundary and spent some ten days on the American side. He was enthusiastically received, and at El Paso he declared that he felt sure that Mexico was already recognized by the American people, if not by their government. In a speech at Dallas, October 17, he assured his hearers that Mexico would recognize all legal foreign debts and all legal rights of Mexicans and foreigners alike. On December 1, 1920, he was inaugurated. But, although the new government was recognized the same month by Japan, Brazil, Holland, and Germany, the Wilson administration, whose term was soon to expire, thought it best to allow the question of recognition on our own part to be handled by the incoming administration. President Obregón called a special session of congress to meet February 7, 1921, and in his address to the body he stressed the need of legislation to institute a modern banking system, the establishment of an agrarian policy that would encourage the restoration of the land to the people, and the solution of the problem arising from the application of Article XXVII to the nation's petroleum resources.[80]

With men like Elías Plutarco Calles as secretary of interior and premier, and Adolfo de la Huerta as secretary of treasury to assist him, President Obregón was enabled in a very short time to establish peace and order in Mexico. His attitude towards American investors was firm in maintaining the rights of the Mexican nation, but fair in respecting rights honestly acquired. However, the Harding administration refused to consider the question of recognition until certain preliminary questions were settled.

Secretary of State Charles Evans Hughes outlined his policy in a general way on June 7, 1921. He insisted that the fundamental question was the safeguarding of American property rights against confiscation. "This question is vital," he declared, "because of the provisions inserted in the Mexican Constitution promulgated in 1917. If these provisions are to be put into effect retroactively, the properties of American citizens will be confiscated on a great scale. This would constitute an international wrong of the gravest character, and this government could not submit to its accomplishment. If it be said that this wrong is not intended, and that the Constitution of 1917 will not be construed to permit, or enforced so as to effect, confiscation, then it is important that this should be made clear by guaranties in proper form."

According to Secretary Hughes, the best way to obtain such guaranties was for the two governments to sign a treaty of amity and commerce in which Mexico would agree to safeguard the rights of property which attached before the Constitution of 1917 was promulgated.[81] Such a treaty of some eighteen articles was presented by United States chargé d'affaires Summerlin to the Mexican government on May 27, 1921,[82] but President Obregón had already rejected recognition upon such a basis, declaring that "the acceptance and signing of a convention to obtain recognition would be equal to placing in doubt the rights that Mexico has to all the privileges international law establishes.[83]

A series of notes between the two governments now followed which culminated with Secretary Hughes' note of instruction dated July 28, 1922, in which he stated that if the Mexican authorities were not willing to sign a treaty binding Mexico to respect the valid titles acquired under Mexican laws prior to the Constitution of 1917, then the question remained in what manner such assurances should be given.[84]

On March 31, 1923, Minister Pani made a lengthy response, claiming that the accomplished acts of the Mexican government since the correspondence began were such as to affect advantageously the solution of the diplomatic problem. He then cited among other "acts" the signing of the Lamont-Huerta agreement, on June 16, 1922, for the adjustment of the Mexican debt; the negotiations already begun between the secretary of the treasury and the representatives of the principal oil companies; and the five *amparo* (injunction)

[80] Text may be found in the *Mexican Review*, IV (Mar., 1921), 4.
[81] *New York Times*, June 8, 1921.
[82] For text see *United States Daily*, May 15, 1926.
[83] *New York Times*, May 21, 1921.
[84] *United States Daily*, May 19, 1926.

decisions of the Mexican Supreme Court, which had defined in an unmistakable manner the non-retroactive character of Article XXVII.[85] He also justified the agrarian policy as being humane and economic, and insisted that the damages to American agricultural properties were insignificant in comparison with the vital advantages to the Mexican people.[86]

Apparently moved by this plea, President Warren G. Harding on May 2, 1923, appointed Charles Beecher Warren, former ambassador to Japan, and John Barton Payne, former secretary of the interior, as American commissioners to meet two Mexican commissioners for the purpose of exchanging impressions. The conference convened in Mexico City on May 15, 1923, and lasted till August 15 of the same year.[87] The two outstanding issues were the questions arising from the confiscation of American agricultural lands to provide *ejidos* or communal lands to villages who had never had them or had been deprived of them, and the dispute regarding the nationalization of the sub-soil deposits whereby rights possessed by American owners of the property were confiscated.[88]

In the case of the agricultural lands the United States insisted that the indemnity should be paid in cash and according to their just value, and that payment by bonds not convertible into money on the basis of their par value could not be considered as indemnification under the rules of international law.[89] Nevertheless, realizing that an urgent social emergency existed, the United States' commissioners agreed that if the Mexican government would make a statement that its claim to expropriate lands of American citizens for *ejidos* did not constitute a precedent for the expropriation of any other kind of property except for due compensation made in cash, the United States government would consider whether under the circumstances it would be willing to accept for its citizens federal bonds of Mexico in payment for the lands taken, providing that the *ejidos* did not exceed a specified area of 1755 hectares (4335 acres). Furthermore such action was contingent upon the conclusion of a general claims convention between the two governments under which those dispossessed might present their claims for loss or damage. The Mexican commissioners accepted this formula.[90]

In the case of sub-soil deposits, the Mexican commissioners conceded that those owners of the surface prior to May 1, 1917, who had performed "some positive act" indicating their intention to exploit the sub-soil deposits were protected against nationalization, but insisted that all who had not, had forfeited their rights.[91] Here also, however, a compromise was obtained whereby to those owners who had not performed a "positive act" were given preferential rights to

[85] For the Mexican embassy's official statement regarding the Supreme Court's ruling see *New York Times*, July 8, 1922; for the decisions, *Ibid.*, Aug. 9, 1922.
[86] *United States Daily*, May 19, 1926.
[87] *Proceedings of the United States-Mexican Commission, Convened, May 14, 1923* (Washington, 1925).
[88] For a careful and on the whole, impartial presentation of the legalistic point of view see Antonio Gómez Robledo, *The Bucareli Agreements and International Law* (Mexico, 1940).
[89] *Proceedings of the United States-Mexican Commission*, p. 29.
[90] *Ibid.*, pp. 37-44.
[91] *Ibid.*, pp. 2-23.

the sub-soil as against third parties. Here again the United States' right to make reservations in behalf of its citizens was recognized.[92]

With these understandings approved by the chief executives of both countries, the government of President Obregón was formally recognized by the United States on August 31, 1923. Shortly afterwards two claims conventions were signed, a general one at Washington on September 8, 1923, covering claims arising since July 4, 1868, the date of the former claims convention,[93] and a special convention signed at Mexico City on September 10, 1923, covering claims arising from losses occurring during the revolutionary activities lasting from November 20, 1910, to May 31, 1920.[94]

President Obregón had at last won his long fight for recognition, but too late to profit greatly by it, for his term of office was almost ended and the constitution forbade his immediate reëlection. Nevertheless he was now able to support openly and effectively the candidacy of General Plutarco Calles, his secretary of the interior and a leader of the socialist and labor groups. When a revolution broke out, engineered by Adolfo de la Huerta, former minister of finance and chief candidate of the opposition, the United States showed its friendliness to the newly recognized government by sending a large consignment of arms to the Obregón forces, and placing an embargo on all shipments of arms to the revolutionists. A little later the Navy Department ordered a division of six destroyers to Tampico where de la Huerta was attempting to maintain a blockade, and the State Department requested the governor of Texas to permit 2000 Mexican federal troops en route from Sonora to cross Texas territory to enter Mexico by El Paso. This seemed to be stretching the bonds of friendly neutrality and the question of our benevolent attitude was raised in Congress. Congressman Fairchild asked that the Harding doctrine against the sale of arms to any foreign power be made the official doctrine of the United States and he was supported by Senators King, William Borah, and Hiram Johnson in the Senate.[95] Senator Robinson introduced an even more drastic bill in the Senate which would prevent any officer, agent, citizen, or corporation of the United States from selling arms and munitions to any foreign government or its agents.[96] But the Mexican government had the arms, and the revolution was soon put down.

During the year 1924 the relations between the two countries were exceedingly cordial. The appointment of Charles Beecher Warren as ambassador to Mexico indicated the intention of the United States to support the understandings recently arrived at. When for personal reasons he was forced to resign a few months later, he declared that never since the Díaz régime had relations between the two countries been on a more friendly basis. General Calles, who had been elected by a large majority, paid a visit to the United States before his inauguration and was received, both officially and unofficially, in a most cordial fashion.

[92] *Ibid.*, pp. 47-49.
[93] *U.S. Treaty Series*, No. 678.
[94] *Ibid.*, No. 676.
[95] *New York Times*, Jan. 1, 1924.
[96] *Ibid.*, Jan. 9, 1924.

Unfortunately for the continuance of amicable relations, on March 4, 1925, Secretary Hughes resigned and former Senator Frank B. Kellogg took over the affairs of the State Department. One of his first official acts was to ask Ambassador Sheffield, who had taken Warren's post, to report to Washington. After conferring with him, Secretary Kellogg on June 12, 1925, gave to the press what one of his former colleagues termed "an unmannerly and unjustifiable" statement. Noting that conditions in Mexico were not entirely satisfactory, Secretary Kellogg declared that we were looking to the Mexican government to restore properties illegally taken and to indemnify American citizens. He understood from the press that another revolutionary movement might be impending. If so, Secretary Kellogg concluded, "it is now the policy of this government to use its influence and its support in behalf of stability and orderly constitutional procedure, but it should be made clear that this government will continue to support the government in Mexico only so long as it protects American lives and American rights and complies with its international engagements and obligations. The government of Mexico is now on trial before the world . . . . We have been patient and realize, of course, that it takes time to bring about a stable government, but we cannot countenance violation of her obligations and failure to protect American citizens."[97]

President Calles resented the implications of this statement and regretted that Kellogg showed the interest of the United States in the maintenance of order in Mexico by suggesting that revolutionary movements were said to be impending. As for the government of Mexico being on trial before the world, such was also the case with that of the United States, as well as those of other countries, "but if it is to be understood that Mexico is on trial in the guise of defendant my government absolutely rejects with energy such imputation which in essence would only mean an insult."[98]

However, this diplomatic flurry was merely an ominous portent of a much more serious situation to follow. Although there had been constant recriminations between the two powers in regard to the provisions of Article XXVII of the Constitution of 1917, concerning agrarian reform and the nationalization of the sub-soil deposits, no action had as yet been taken by the Mexican Congress to put these provisions into effect. But in December, 1925, the Mexican Congress passed two laws to remedy this situation—laws generally known as the land law and the petroleum law.[99]

The petroleum law repeated the provision of the Constitution asserting inalienable ownership of sub-soil deposits by the nation; it required foreigners to comply with the constitutional provisions regarding the waiving of nationality and the right to invoke diplomatic protection in so far as Mexican owned property was concerned; it also enforced the clause prohibiting ownership of lands or waters within fifty kilometers from the sea coast and one hundred kilometers from the frontiers. A new feature of the law required all holdings to

[97] *Ibid.*, June 13, 1925.
[98] *Ibid.*, June 15, 1925.
[99] For the text of the land law see C. W. Hackett, "The Mexican Revolution and the United States, 1910-1926," *World Peace Foundation Pamphlet*, Vol. IX, No. 5, p. 414; for petroleum law, *Ibid.*, p. 425.

be confirmed by concessions to be granted for a period of not more than fifty years from the time when exploitation was begun. Furthermore, the concessions to be confirmed must have arisen in lands where exploitation was begun or the contract made prior to May 1, 1917. Detailed regulations for the concessions were also included.

The land law also repeated the constitutional restrictions, and included foreigners participating in Mexican corporations. It was further provided that Mexican companies owning rural property for agricultural purposes would not be granted a permit if there remained in the hands of aliens 50 percent or more of the total interests of the company. Foreign persons holding 50 percent or more of total interests could hold the amount in excess of 50 percent till their death, and their heirs were given five years to dispose of their holdings with an extension if considered necessary. Corporations were granted ten years to dispose of excess holdings.

Even before the passage of these laws, Secretary Kellogg, through the American ambassador, submitted certain inquiries regarding them, and on November 17, 1925, he sent an *aide-mémoire* or personal message to the Mexican Minister of Foreign Affairs in the hope "that the clouds which I perceive on the horizon of friendship between the United States and Mexico may be removed."[100] President Calles replied directly through his Minister of Foreign Affairs that there was "absolutely no cause for perceiving clouds, and that the legislation was merely aimed at dispelling the vagueness of the constitutional provisions and would in no way violate the obligations of Mexico under international law."

The ensuing correspondence which began with the *aide-mémoire* of November 17, 1925, ended exactly a year later with the final reply of the Mexican government dated November 17, 1926. Inasmuch as it amounts to some 50,000 words and much of it is repetitious, perhaps the simplest method of approach would be to note the issues involved and the attitude of both parties in regard to them.

The issues in dispute pertained to a difference in interpretation in regard to the four following subjects:

(1) the alleged retroactive features of the land law; (2) the alleged retroactivity of the petroleum law; (3) the insistence on the part of the Mexican government that foreigners owning property agree to submit themselves to Mexican jurisdiction in all disputes concerning their property, on penalty that their property be forfeited if they should invoke the protection of their government; (4) the nature of the agreements formulated at the Conference in Mexico City by the four commisioners in May, 1923.

As to the engagements of 1923, President Calles insisted that they were merely an "exchange of views" and "did not result in any formal agreement other than that of the claims conventions which were signed after the resumption of diplomatic relations."[101] Furthermore, although Mexico recognized the declarations made by its commissioners, the conferences of 1923 were not a condition for the recognition of the government of Mexico and could

[100] This correspondence has been published as *Sen. Doc. No. 96,* 69th Cong., 1st Sess.
[101] Reply of the Mexican Minister of Foreign Affairs, Nov. 27, 1925, *Ibid.*, p. 5.

never be given that character.[102] Finally, the Mexican government refused to recognize the binding force equivalent to a treaty or a constitutional precept in the outlines of policy presented at the conference, since the declarations of neither side took the form of a synallagmatic agreement.[103]

Secretary Kellogg's stand on the binding nature of the oral agreements of the conference is clearly and succinctly stated in his last note of October 30, 1926: "The declarations of the Mexican and American commissioners on that occasion, subsequently ratified by an exchange of notes between the two governments, constituted, in the view of my government, solemn and binding undertakings which formed the basis and moving consideration for the recognition of the Mexican government by this government."[104]

However, inasmuch as the appointment of the two American commissioners was not approved by the Senate, nor was the agreement itself submitted to the Senate for its approval, it could hardly be considered as of the same binding force as a formal treaty. Furthermore, in the minutes of the meeting of the commision on August 2, it is stated that the policy of the present president of Mexico is conditioned within the limitations of his constitutional power and "is not intended to constitute an obligation for an unlimited time on the part of the Mexican government to grant preferential rights to such owners of the surface or persons entitled to exercise their rights to the oil in the sub-soil.[105] Under these circumstances, the legislative branch or a subsequent executive would hardly appear to be legally bound.

In regard to the disagreement due to the Mexican government's insistence that foreign property owners bind themselves not to invoke the diplomatic protection of their government, but agree to submit themselves as Mexicans to Mexican laws in disputes over property under penalty of forfeiture of their property if such protection is sought, a compromise seems to have been reached in the correspondence. In his note of October 7, 1926, Foreign Minister Saénz declares: "The Mexican government therefore does not deny that the American government is at liberty to intervene for its nationals; but that does not stand in the way of carrying out an agreement under which the alien agrees not to be the party asking for the diplomatic protection of his government. In case of infringement of any international duty such as a denial of justice would be, the right of the American government to take with the Mexican government appropriate action to seek atonement for injustice or injury which may have been done to its nationals would stand unimpaired. Under these conditions neither would the American government have failed to protect its nationals nor the Mexican government to comply with its laws."[106]

It might be noted, however, that this so-called Calvo clause seems to have been interpreted to some extent in favor of the position taken by the United States in the decision rendered by the General Claims Commission on March 31, 1926, in the case of the North American Dredging Company of Texas v. United

[102] *Ibid.*, p. 34.
[103] U.S. Dept. of State, *American Property Rights in Mexico* (Washington, 1926), p. 11.
[104] *Ibid.*, p. 26.
[105] *Proceedings of the United States-Mexican Commission, 1923.*
[106] *American Property Rights in Mexico*, p. 14.

Mexican States, even though the Commission dismissed this case and thus sustained the position of the Mexican agent, the Commission found the Calvo clause neither upheld unanimously nor universally rejected by authorities on international law, but it did reject as unsound the right of Mexico or any other nation "lawfully to bind all foreigners by contract to relinquish all rights of protection by their governments. . . . This provision did not and could not deprive the claimant of his American citizenship and all that that implies. It did not take from him his undoubted right to apply to his own government for protection if his resort to the Mexican tribunals or other authorities available to him resulted in a denial or delay of justice as that term is used in international law."[107]

The third point under dispute was the alleged retroactivity of the land law, which required foreign persons and corporations to divest themselves of majority control of corporations owning rural properties for agricultural purposes. According to the terms of this law, corporations were required to dispose of their stock in excess of 50 percent within ten years, while individuals were allowed to retain such majority stock till their death and their heirs were given five years to dispose of it. Secretary Kellogg insisted that this provision of the law was "manifestly retroactive," since it required the alien owner of rural properties legally acquired under the laws of Mexico "to divest himself of the ownership, control and management of his property."[108] Minister Saenz maintained that the law in the case of individuals was not retroactive nor confiscatory, since it permitted possession till the death of the owner and therefore was merely a limitation upon the right of inheritance, which is in strict conformity with international law.[109] In the case of alien moral persons (corporations) Minister Saenz declared that "in all legislation it is admitted that the law is free to amplify, modify, or restrict the capacity of that class of persons," and since the article referred to future rights, that is those arising from the period of time subsequent to ten years, "its effects cannot be regarded as retroactive, since there was no acquired right but merely expectation of a right."[110]

The final point in dispute covered the parts of the petroleum law which were claimed to be retroactive. Under the mining codes of 1884, 1892 and 1909, owners of the surface were given right to exploit sub-soil deposits of petroleum without special concession of the Mexican government. But Article XXVII of the Constitution of 1917 vested direct ownership of all minerals and petroleum in the nation. According to the minutes of the proceedings of the United States Mexican Commission of 1923, the Mexican executive agreed to enforce the principles of the decisions of the Supreme Court in the five *amparo* cases, which held that Article XXVII of the Constitution was not retroactive in respect to all persons who previous to May 1, 1917, had performed some positive act manifesting the intention of the owner of the surface to exercise his rights. The executive further agreed to grant preferential rights to those owners who had not performed any positive act.[111] However the petroleum law of 1925 changed

[107] *Amer. Jour. Int. Law,* XX (Oct., 1926), p. 800.
[108] *Sen. Doc. No. 96*, p. 23.
[109] *Ibid.*, p. 11.
[110] *Ibid.*, p. 30.
[111] *Proceedings of the United States-Mexican Commission*, p. 47.

these vested titles into concessions of fifty years' duration and made no provision for granting preferential rights to those owners who had not performed some positive act.

Secretary Kellogg insisted that the provisions of the petroleum law and the regulations issued thereunder which required the owners to apply for confirmation of their titles within one year, and to accept concessions for not more than fifty years from the time when exploitation began "would be nothing but a forced exchange of a greater for a lesser estate . . . a statute so construed and enforced is retroactive and confiscatory, because it converts exclusive ownership under positive Mexican law into a mere authorization to exercise rights for a limited period of time. . . ."[112] Furthermore, not only did Secretary Kellogg object to the fact that no preferential treatment seemed to be accorded to owners who had failed to perform some positive act, but he rejected the entire Mexican doctrine that no vested right was acquired until some positive act had been performed.[113]

The Mexican government claimed that a concession of fifty years' duration which might be renewed for another thirty years protected the working of any property discovered up to date, and that such a system founded upon concessions was even more secure that the system of private ownership. Therefore, although it might seem that the exchange of a title for a concession lessened the right, it was not so in practice. The Mexican case closed with an invitation to the United States to point out any specific case which violated international law, and if the Mexican government should fail to correct such violations it would be disposed to accept in justice the resulting claims of the American government.[114]

Indeed, when we come to the facts of the situation it would seem as though the foreign oil companies for the most part accepted the new legislation. The oil laws went into effect on January 1, 1927, and according to a statement issued by Minister of Labor Luis Morones, 125 out of the 147 oil companies operating in Mexico agreed to accept them.[115] According to figures submitted by W. W. Liggett to the Senate Sub-Committee on Foreign Relations, out of 28,493,914 acres under development for oil, only 1,660,579 remained for which concessions had not been asked, and of this latter acreage 87 percent was owned or controlled by Edward Doheny, Harry F. Sinclair, and Andrew Mellon.[116]

Apparently the peak of the difficulties had been passed, and four events occurred in the fall of 1927 which presaged an era of better relations. The first was the appointment of Dwight W. Morrow as ambassador to Mexico to succeed Sheffield, who had resigned in June. Although a representative of the financial interests, Morrow was noted for his broad-minded outlook and outstanding ability, and when in presenting his letter of credence he declared that "we shall not fail to adjust outstanding questions with that dignity and mutual respect which should mark the 'international relationship of the two sovereign and independent states' " it was indicated that a policy of compromise and

[112] *American Property Rights in Mexico*, p. 4.
[113] *Ibid.*, p. 5.
[114] *Ibid.*, p. 21, 25.
[115] *New York Times*, Feb. 20, 1927.
[116] *Cong. Rec.* 68 (Mar. 3 1927), 5580.

cooperation on the part of the United States could now be anticipated.[117] The second was the repeal by President Calles on October 27 of the decree prohibiting the purchase of goods in the United States by Mexican departments. The third was the unanimous decision of the Mexican Supreme Court rendered on November 17, 1927, granting an appeal restraining the Department of Industry, Commerce, and Labor from cancelling certain drilling permits of the Mexican Petroleum Company, an American concern which had not applied for a concession. The decision declared Articles XIV and XV of the petroleum law, which required companies to exchange their titles for fifty-year concessions within one year, unconstitutional in so far as they applied to the case at bar.[118] As a direct result of this decision President Calles recommended to the Mexican Congress the passage of a law amending these two articles so as to confirm all rights derived from lands where exploitation or contracts for exploitation had been entered into previous to May 1, 1917, by issuance of concessions without limit of time. The Mexican Congress passed such a law on December 27, 1927,[119] and on March 27, 1928, regulations were signed by President Calles validating in perpetuity all oil titles obtained before May 1, 1917.[120] The fourth was the non-stop flight of Colonel Lindbergh on December 14, 1927, from Washington to Mexico City. This event had an immediate and remarkable repercussion on public opinion both in Mexico and in the United States, and made for a very considerable improvement in the mutual friendliness and appreciation of the two peoples.

Although Ambassador Morrow had little difficulty in settling the oil controversy, he found the agrarian problem far more complex. He was able by personal efforts to obtain the reversal of some flagrant seizures of American-owned properties and he gave encouragement to American claimants to bring action in the Mexican courts, but the problem itself remained unsolved. Apparently he hoped that the Mexican government would soon abandon its policy of seizing lands and devote itself to improving land already taken.[121]

The religious problem engaged Ambassador Morrow's attention particularly.[122] In February 1926, a series of orders was issued for the arrest and deportation of foreign priests, for the nationalization of all church property not yet held by the state, and for the closing of all schools, convents and orphan asylums giving religious instruction in violation of the religious articles of the Constitution. Although there was considerable outcry in church circles in the United States, and the House of Representatives asked Secretary Kellogg for information, the State Department contented itself with a request that American churchmen should not be made to suffer unduly in the enforcement of the law. President Calvin Coolidge on September 8, 1926, declared that he regarded the church and state conflict in Mexico as purely an internal question in which the

[117] Harold Nicolson, *Dwight Morrow* (New York, 1935), p. 316.
[118] *United States Daily*, Nov. 19, 1927.
[119] Text in *Ibid.*, Jan. 6, 1928.
[120] For an authoritative discussion of the settlement see J. Reuben Clark, "Oil Settlement with Mexico," *Foreign Affairs*, VI (July, 1928), 600.
[121] Nicolson, *Dwight Morrow*, p. 335.
[122] Calcott, *Liberalism in Mexico*, chap. XV, for a brief survey.

United States could have no interest save in the protection of American rights.[123]

However, Ambassador Morrow, although recognizing the inadvisability of intervening in the struggle between Church and state, was intrigued by the problem, and when Father Burke at the Havana Conference in January 1928, suggested that Morrow arrange an interview for him with President Calles, the Ambassador agreed to take the matter under consideration. He finally arranged a secret meeting on April 4 at which he was present and during which a temporary compromise was drafted. The following month Morrow persuaded Calles to receive Monsignor Ruíz y Flores, the senior prelate of the Mexican bishops who had sought refuge in Texas. The previous agreement was confirmed and the situation looked quite hopeful when General Obregón, who had been slated to replace President Calles, was murdered July 17, 1928, by a young Catholic. Public opinion was so aroused that hostility flamed anew and the proposed compromise was dropped for the time being.

The next year Morrow again took up the question and persuaded President Emilio Portes Gil to receive as emissaries of the Pope, Monsignor Ruíz and Archbishop Díaz. A new agreement was reached on June 19, 1929, which was approved two days later by the Vatican. Although the Church question was by no means settled, a working agreement had been reached which permitted the Church once more to carry on religious services.[124]

Exactly one day before President Herbert Hoover's inauguration in 1929 a revolt broke out in Mexico and the new administration decided to utilize its power under the Arms Embargo Resolution of 1922,[125] and while supplying arms and munitions to the government it placed an embargo on all shipments to the rebels. When Secretary of State Kellogg refused to recognize the rebels as belligerents and his successor Secretary of State Henry L. Stimson, refused to receive the revolutionary agent sent to Washington, the insurrection collapsed.

Throughout the Hoover administration the relations between the United States and Mexico were exceedingly friendly. J. Reuben Clark, who had been Ambassador Morrow's right-hand man, was appointed as Ambassador when Morrow resigned to enter the Senate, and he ably carried on the Morrow policies. The newly elected Mexican president, Pascual Ortiz Rubio, following the precedent set by President Hoover, made a good-will visit to the United States before entering upon his official duties. His cordial reception strengthened the desire for closer cooperation on both sides of the Rio Grande.

It was during this era of good feeling that Genaro Estrada, Mexican Foreign Minister, in a statement to the press on September 30, 1930, gave utterance to a doctrine subsequently called the "Estrada Doctrine" or "Doctrina Mexicana." It was to the effect that recognition of a government should be granted automatically, regardless of the origin of the government. "The Mexican government," he declared, "does not grant recognition which implies judgment; it confines itself to the maintenance or withdrawal, as it may seem advisable, of its diplomatic agents. . . ."[126] This doctrine was erroneously stated to be a

[123] *New York Times*, Sept. 9, 1926.
[124] Arnold Toynbee called this achievement Mr. Morrow's "greatest diplomatic triumph."
[125] *U.S. Stat. at Large*, 361.
[126] *Bulletin of the Pan American Union*, 58, (Mar., 1934), 161.

repudiation of the Monroe Doctrine, but Estrada, himself, publicly disclaimed any such intention.

It was somewhat of a paradox that although President Franklin D. Roosevelt inaugurated his administration with the Good Neighbor Policy the Mexican government was a little dubious regarding the new representative of the United States sent to Mexico City to interpret it. The appointment of Josephus Daniels to supplant J. Reuben Clark seemed a rather inauspicious beginning. Mexico had not forgotten that Daniels had been Secretary of the Navy when Vera Cruz had been shelled and occupied.[127] But since Franklin D. Roosevelt was Assistant Secretary of the Navy at the same time, to have declared Daniels *persona non grata* would have been highly inadvisable. Fortunately, Daniels manifested such an earnest desire to carry on the sympathetic and friendly policy of his predecessors that he quickly overcome the initial feelings of misgiving.

Another potential cause of trouble which had failed of satisfactory settlement over a long period of time was the question of claims. We have already mentioned the ratification of the two claims conventions of 1923.[128] The first or General Claims Convention covered claims dating back to 1868, and the other, a Special Convention, covered all claims arising during the revolutionary period 1910-1920. A third group known as the agrarian claims arose from the expropriation of American lands subsequent to August 29, 1927, under the Mexican agrarian program. At the advent of the Roosevelt administration out of some 6,500 claims filed under the General and Special Conventions only a few hundred had been settled.[129] All agrarian claims had been excluded pending informal diplomatic discussion between the two governments.

A final settlement was first obtained in the case of the revolutionary claims. Since by 1931 the Special Claims Commission had made awards on none of the 3,176 filed, it was finally decided to make a settlement *en bloc*. By a treaty signed April 24, 1934, the revolutionary claims were to be settled by the payment of a sum proportionate to the final settlement of similar European claims.[130] The total sum was fixed at $5,448, 020.14, payable with interest by annual sums of $500,000 which were to be deposited to the credit of the United States government in January of each year. The ninth of these annual payments was made by Mexico on January 2, 1943, and payment for the revolutionary claims was made by 1945.

A new Special Claims Commission to allocate these awards was set up in 1935 and it completed its work by May, 1938. The Commission considered 2833 claims, of which 1475 were disallowed, and 1358 were allowed wholly or in part. The total amount of the awards allowed before reduction on a percentage basis was $9,135,041.79, thus giving each approved claimant 57 percent of the original amount allowed.[131] It might be noted that the claims

[127] The story is told that President Roosevelt and Mr. Daniels had completely forgotten the incident when the appointment was made.
[128] *Bulletin of the Pan American Union*, 58 (Mar., 1934), 171.
[129] A.H. Feller, *The Mexican Claims Commission, 1923-1934* (New York, 1935), p. 60, 68.
[130] *U.S. Treaty Series*, No. 883.
[131] *Special Mexican Claims Commission–Report to the Secretary of State* (Washington, 1940).

from the Santa Ysabel massacre which the previous commission had disallowed were awarded $92,910.[132]

In the question of recompense for the seizure of agricultural lands a spirited correspondence between the two governments ensued during the year 1938. Secretary of State Cordell Hull pointed out that between 1915 and 1927 one hundred sixty-one moderate-sized properties of American citizens had been taken, and subsequent to 1927 additional properties valued at more than $10 million had been expropriated. After considerable argument the Mexican government proposed that each side appoint a commissioner who should jointly determine the value of the confiscated properties by May 31, 1939, and as proof of its good intentions, the Mexican government agreed to a first payment of $1 million in May, 1939.[133]

After three payments totaling $3 million had been made under this agreement, a convention was signed in Washington on November 19, 1941, providing for a global settlement of both the agricultural claims and all claims filed by the two governments with the General Claims Commission. Under the terms of this convention Mexico agreed to pay a total sum of $40 million, $3 million at the date of ratification and the balance of $34 million by payments of $2½ million annually. It was further provided that other claims would be subject to agreements to be concluded as soon as possible.[134] As we shall see later, this convention was merely one of a series of agreements covering the solution of a number of problems including the confiscation of the petroleum properties all of which were signed at this same time.

The armistice in the struggle between Church and state which Ambassador Morrow had obtained in 1929 had by no means settled this question. The Church was not reconciled to the drastic limitations imposed upon its representatives and their work and it sought assistance in the United States. Early in 1935 the Knights of Columbus were persuaded to work towards obtaining American intervention. Failing to interest Secretary Hull in such a policy, appeal was made to both houses of Congress where less discretion obtains. Representative Higgins wished to withdraw recognition, and recall Ambassador Daniels, while the resolution which Senator Borah introduced, although not so extreme, protested the anti-religious campaign and practices of the Mexican government, and authorized the Committee on Foreign Relations to conduct an investigation into the situation.[135] When nothing came of this effort some 242 members of the House of Representatives petitioned President Roosevelt to take some action. When the Knights of Columbus continued the campaign to involve the United States, President Roosevelt wrote a personal letter to the Supreme Knight declaring that no American citizen in Mexico had complained during the year that his religious freedom had been interfered with and as regards Mexican citizens the United States intended to continue its policy of nonintervention.[136] Cardinal Mundelein a fortnight later eulogized President

[132] *Ibid.*, p. 104.
[133] S.S. Jones and D.P. Myers, *Documents on American Foreign Relations, 1938-1939,* (Boston, 1939), pp. 87-121.
[134] For text of the conversations see *Cong. Rec.*, 88 (January 29, 1942), 861.
[135] For text of Borah resolution see *New York Times,* February 1, 1935.
[136] For text of letter see *Ibid.*, Nov. 18, 1935.

Roosevelt's foreign policy and decried self-appointed spokesmen for the Catholics in America, whereupon the agitation for intervention died down.

A more serious threat to friendly relations arose in 1937, when the Mexican government once again threatened to confiscate foreign concessions in order to nationalize the oil industry.

The Mexican workers in the oil fields staged many strikes during the year 1937 demanding a substantial increase in pay. On December 18, 1937, the Mexican Federal Board of Conciliation and Arbitration ruled that foreign-owned oil companies must increase wages one-third and improve the pension and welfare system. Claiming that such changes meant an annual increase of operating costs of about $7,200,000, the companies sought a permanent injunction. However, the Mexican Supreme Court upheld the award and the Federal Labor Board declared the new wage scale became effective on March 7, 1938. When the companies refused to comply, President Lazaro Cárdenas by a decree dated March 18 announced that the oil properties would be nationalized and indemnification made within ten years. Claiming a "manifest denial of justice," the oil companies appealed to their respective governments.[137]

Secretary of State Hull issued a statement March 30 conceding the right of the Mexican government to expropriate properties within its jurisdiction but questioning the ability of the Mexican government to make adequate compensation for the large number of properties confiscated. He did hope, however, that a fair and equitable solution might be found.[138] President Cárdenas replied immediately to Ambassador Daniels that "Mexico will know how to honor its obligations of today and its obligations of yesterday."[139]

Nevertheless in spite of this rather vague assurance of compensation, the expropriation of the oil properties soon became a *cause célebre* in the diplomatic relations of the two countries.[140] There was no question about the popularity of confiscation among the Mexican people. A celebration staged in Mexico City on March 26 brough out a quarter of a million people, and President Cárdenas received wide popular support in his effort to float an internal loan to pay for the seizure. But even with payment conceded there was a wide difference of opinion as to the value of the oil properties. The American oil companies placed their minimum value at $262 million, whereas Mexico estimated that after paying back taxes and back compensation to workers, the companies would have about $10 million coming to them.

If the expropriation was stimulating to Mexico politically it was exceedingly detrimental economically. The peso dropped from 3.60 to 6 to the dollar, capital took flight, production of oil diminished, while prices received were lower. Distribution was a serious problem and no considerable market was available except in the Axis powers.[141] Skilled technicians were difficult to

[137] See *Ibid.*, Dec. 31, 1937; March 4, 8, 19, 1938.

[138] *Ibid.*, Mar. 31, 1938.

[139] *Ibid.*, Apr. 2, 1938.

[140] For the Mexican case see United States of Mexico, *The True Facts About the Expropriation of the Oil Companies' Properties in Mexico* (Mexico, 1940). For the oil companies' side see Standard Oil Co., *The Reply to Mexico* (New York, 1940).

[141] It has been claimed that Mexico sold about ten million barrels to the three totalitarian powers. See A.W. MacMahon and W.R. Dittmar, "The Mexican Oil Industry Since Expropriation," *Pol. Sci. Quar.*, LVII (June, 1942), 164.

obtain and replacements of outworn equipment was a serious problem due to the hostile influence of the dispossessed companies.

But above all, the situation was dangerous to the new development of cordial relations. The Roosevelt administration was unwilling to see the well established structure of the Good Neighbor Policy jeopardized, and it was willing and anxious to make every effort towards obtaining a fair settlement.

As a first step the United States government encouraged the companies to enter into direct negotiations with the Mexican government. Donald R. Richberg, representing the companies, had numerous conferences with President Cárdenas, and at first a compromise seemed possible.[142] When a deadlock ensued, Under Secretary Sumner Welles, on August 14, 1939, made a compromise proposal while at the same time reiterating the necessity for "adequate, effective, and prompt payment forthe properties seized."[143] No action was taken on his suggestion, and on December 2, 1939, the Mexican Superior Court unanimously upheld the expropriation as constitutional both as regards movable and immovable property.[144]

The matter was reopened in 1940 by an informal memorandum delivered by Ambassador Nájera on March 16 which failed completely to meet the question at issue. In his reply of April 3, Secretary Hull again stressed the requirement of adequate and prompt compensation and pointed out that the Mexican government had been somewhat remiss hitherto in the payment of claims. He then proposed arbitration for all the questions involved by a tribunal clothed with authority not only to determine the amount to be paid but also the means to make the payment effective.[145]

In the reply of May 1, the Mexican government ruled out arbitration as incompatible in this case since the matter in dispute was domestic in nature and already nearing solution. It was also noted that a private and direct settlement had been made with the Sinclair interests which it was claimed represented approximately 40 percent of the investment of American nationals in the oil industry.[146] This settlement was for $8½ million plus the delivery of 20 million barrels of oil over a four-year period at a determined price. Incidentally, the asking price of the Sinclair interests had been $32 million. Inasmuch as a presidential election was in the offing, and President Cárdenas was not up for reelection, the principal oil companies preferred to wait and to hope for a more friendly attitude on the part of the new administration.

Various events now clearly foreshadowed a settlement. Early in November 1940, President Cárdenas placed an embargo on oil and scrap iron going to Japan and declared that he expected an early, definite, and satisfactory settlement of all pending questions. The visit of Vice-president Henry A. Wallace as President Roosevelt's special representative at the inauguration of General Manuel Ávila Camacho on December 1, 1940, was more than a friendly gesture. The fact that President Cárdenas had to ask the congress for $12,000,000 to bolster the

[142] Donald R. Richberg, *The Mexican Oil Seizure* (New York, n. d.).
[143] S.S. Jones and D.P. Myers, *Documents on American Foreign Relations* (Boston, 1940), p. 217.
[144] See Roscoe B. Gaither, *Expropriation in Mexico* (New York, 1940), chap. VIII.
[145] Jones and Myers, *Documents*, pp. 234-238.
[146] Dept. of State *Bulletin*, XI (May 4, 1940), 465–470.

operating deficit of the oil industry was undoubtedly a persuasive element. Finally the success of the Axis powers in overwhelming Europe and the Far East tended to cement the republics of the New World more closely together.

An understanding satisfactory to Mexico and the United States covering not only the oil question but also claims, trade, stabilization of currency, credits, and purchase of silver was announced on November 19, 1941. As regards the expropriation of petroleum properties it was agreed that each government would appoint an expert to determine the just compensation to be paid American owners whose rights and interests in the petroleum industry had been jeopardized by the acts of expropriation. Five months were allowed to fix the amount, and the Mexican government agreed to make an immediate deposit of $9 million cash on account.[147] The two experts Morris L. Cooke, representing the United States, and Manuel J. Zevada, representing Mexico, began their work in Mexico City in January 1942, and on April 17, 1942, they reported that the valuation of the American-owned oil companies was $23,995,991. The companies were released from all claims present and future except for unpaid taxes and duties and private claims now pending.[148] Apparently the settlement was based upon a physical valuation of the properties without giving consideration to the value of the subsoil rights. According to valuations set by the companies, the payment amounted to only seven cents on the dollar, and at first they refused to accept it. However since the United States government declared the evaluation to be final, the manner and conditions of payment were agreed upon by an exchange of notes on September 29, 1943. Mexico had made the $9 million cash deposit at the time of settlement and now agreed to pay the remaining $20,137,700.84 including interest, in five payments—the last on September 30, 1949. The oil companies, with no support from the government, accepted the settlement.

As to the other agreements announced on November 19, 1941, we have already discussed that of claims.[149] The third concerned the negotiation of a reciprocal trade agreement, and this was signed December 23, 1942, the fifteenth of its kind with an American republic. The other agreements were to assist Mexico in its financial position by the stabilization of the Mexican *peso* in terms of the dollar by the purchase of *pesos*, to aid silver mining by purchasing newly mined silver at thirty-five cents an ounce up to six million ounces monthly,[150] and to help finance the highway construction program in Mexico by requesting the Export-Import Bank to accept certain of these highway bonds as security for credits.[151]

Long before the United States entered the war, Mexico had given this country her whole-hearted support. At the Conferences of Panama and Havana her representatives took the lead in supporting continental solidarity and all-out measures for American defense, and even before Pearl Harbor, Mexico had agreed

[147] L.M. Goodrich, *Documents on American Foreign Relations, 1941-1942* (Boston, 1942), IV, 421.
[148] *Ibid.*, p. 425.
[149] *Ibid.*, p. 175.
[150] This was the most criticized agreement in the Senate because of the vast amount of unused silver already stored in the treasury vaults of the United States.
[151] Goodrich, *Documents on American Foreign Relations*, p. 359.

to reciprocal use of strategic airports and bases.[152] Before going to the Rio Conference, Mexico severed diplomatic relations with the Axis and at the Conference Foreign Minister Ezequiel Padilla thrilled the delegates with his impassioned plea for support of the men who fell on Wake and in the Philippines "to defend human liberties and the common destiny of America."

Early in April 1942, Foreign Minister Padilla visited the United States and on April 8 he and Under Secretary of State, Sumner Welles, issued a joint statement covering a program of close economic cooperation. Through funds obtained from the Mexican government and Export-Import Bank, certain basic industries such as a steel and tin plate rolling mills were to be established to meet Mexican consumers' needs and to supply goods required by the war efforts of the United States. Arrangements were made for the survey of the Mexican railway transportation system; experts were commissioned to explore the feasibility of constructing small cargo vessels in Mexico, and it was agreed that a high-octane gasoline plant should be constructed as soon as the necessary equipment might be spared.[153]

With the declaration of war upon the Axis powers by Mexico on May 22, 1942, the two governments entered into still closer military relationships. The 1941 agreement for the reciprocal use of air bases was enlarged in scope while United States warships were permitted to use Mexican bases and territorial waters, and United States troops and planes could cross Mexican territory. Lend-Lease arrangements authorized Mexico to purchase war equipment to the value of approximately thirty million dollars. Through a Joint Mexican-United States Defense Commission, close liasion was established for protection of the long and exposed coast line. All Axis nationals were removed from coastal areas; financial transactions of Axis enterprises were subjected to full governmental control; and the United States black list was more rigidly enforced. In 1945 a Mexican air force unit saw action in the Far East under General Douglas MacArthur's command. The two neighbors were as one against the common foe.

As an earnest of the intention of the United States to cooperate loyally, President Franklin D. Roosevelt met President Ávila Camacho in Monterrey, Mexico, in April 1943, and the two agreed to an increased exchange of goods. During 1944 a plan of economic cooperation known as the "Mexico Plan" was developed whereby the two governments worked out jointly a unified industrial program in a single pattern rather than giving consideration to unrelated individual requests for industrial goods.

One of the most important developments towards better relations between the two countries occurred during the war period when on February 3, 1944, Mexico and the United States signed a water treaty relating to the utilization of the waters of the Colorado and Tijuana rivers and the Rio Grande. The rapid expansion of agricultural areas along the common boundary increased the need for more water for irrigation. The treaty, proposed by engineers and diplomats upon the principles of the Good Neighbor policy, settled a problem which had long vexed the two governments. It provided for flood control and increased materially the amount of water available to each country for agricultural and

[152] *U.S. Treaty Series*, No. 971.
[153] Dept. of State *Bulletin*, VI (Apr. 11, 1942), 325.

industrial uses. Both the American and Mexican senates approved the treaty with large majorities, and ratifications were exchanged in Washington November 8, 1945, with the treaty becoming effective immediately.

World War II proved to be a major turning point in Mexico's historical development, both domestically and in international affairs. As a result of wartime prosperity many changes occurred in the nation's economy. Great strides were made toward self-sufficiency as industrialization was vigorously pushed, the labor force expanded and the middle class broadened. The general standard of living also showed improvement.[154] During the presidency of Ávila Camacho (1940-1946) the political ferment of earlier years subsided, and orderly democratic processes, accompanied by a large measure of democratic freedom, became possible. Mexico gained prestige and favorable publicity as one of the victorious allies, as host of the extraordinary, important Inter-American Conference of 1945, and as a charter member of the United Nations.

The post-war years have witnessed a continuation of the "Era of Good Feeling" begun by President Roosevelt's judicious handling of the petroleum question in 1938. Issues which disturbed the relations of the two republics through the 1930s, chiefly the protection of American lives and property, and the perennial question of diplomatic recognition, gave way to such problems as Mexican migratory labor, racial discrimination, water rights and boundary questions, violations of Mexican waters by American fishermen, tariffs on lead and zinc, commodity prices on sugar, cotton, and coffee; cooperation in public health, the development of hydro-electric and irrigation projects, narcotics control, and military cooperation. Contributing to the success of the new diplomacy have been the regular meetings of the two nation's chief executives, Mexican-United States conferences as members of the Interparliamentary Union, economic and technical assistance supplied by the United States, the growth of tourism and broadened cultural exchanges. However, in spite of the consonance of views attending the resolution of bilateral issues, traditional Mexican foreign policy, notably in connection with cold war power politics, has led to a divergence of policies on certain regional and international questions.

Avila Camacho was succeeded in the presidency by Miguel Alemán in 1946, a civilian, and the first chief executive in thirty-five years who had not taken part in the Revolution. During the campaign Alemán gave hints of anti-Yankee sentiment, but it was soon evident that such remarks were of no significance. He was keenly aware of the importance of the United States to his country's progress, and when in office took steps to insure that the wartime cordiality would be continued.

President Harry S. Truman, similarly intent on strengthening bonds of friendship, in March, 1947 made the first official visit to Mexico City ever made by an American president. He endeared himself to the Mexicans by placing a wreath on the monument of the *niños heroes*, the cadets killed in 1847 when the Americans invested the hill of Chapultepec. A month later President Alemán, the first Mexican President to visit Washington while in office, returned the visit. In addressing the special joint session of the Congress the Mexican president

[154] Howard F. Cline, *Mexico, Revolution to Evolution: 1940-1960* (New York, 1963), pp. 231-232, 253-262.

pledged an effective continuation of the Good Neighbor Policy on both sides of the border. One concrete example of the desire to cement the "Era of Good Feelings" was shown by the decision of the two governments in 1950 to restore battle flags captured by each other in the war of 1847. Liberal loans from the United States continued to flow into Mexico through the Export-Import Bank, American investors were welcomed in spite of the restrictive statutes, and American tourism boomed. The Mexicans sent almost nine-tenths of their exports to the United States, as well as thousands of migrant workers, and imported impressive quantities of American goods.

It was in this period that a problem arose which was a product of World War II. The need for workers to harvest crops and help in the maintenance of railroads in the Western states during the war was met by annual agreements for the temporary immigration of Mexican workers (*braceros*) into the United States. Elaborate guarantees as to wages and conditions of labor were embodied in the agreements, and the results were mutually satisfactory. However, after the war a large number of Mexican laborers, known as "wetbacks" (for having swum the Rio Grande) illegally entered the United States, where some were exploited by farm interests. It was alleged that many thousands of these laborers came in annually and produced a situation unfair to American labor.

A diplomatic incident occurred in 1948 when the Mexican Secretary of Foreign Affairs sent a note to Washington protesting the action of the American Immigration Service in allowing some four thousand *braceros* to cross the border illegally to be employed by American farmers. The United States claimed that Mexico had failed to send the number of *braceros* agreed upon, and Mexico replied that workers were forbidden to go to the United States because of discriminatory practices which were being permitted. The United States government sent a formal apology and the incident was considered closed.

In 1951-1952 an effort was made by a mixed commission representing the United States and Mexico to prepare in treaty form an acceptable accord covering the temporary immigration of Mexican workers. A compromise was worked out to run, however, only through 1952. Meanwhile, the United States Congress in 1952 passed legislation strengthening control over the "wetbacks," while fairer and more effective regulation of the *braceros* was instituted.

Another problem which faced the two countries in the immediate aftermath of the war was the control of *aftosa*, or hoof and mouth disease. When this disease attacked Mexican cattle and sheep in 1947, it threatened the cattle industry in the United States. At first it was proposed to shoot the diseased animals and recompense their owners. The United States furnished technical experts and financial assistance, but the Mexicans resented the procedure. Quarantine and vaccination were then applied. This was a mammoth task, but by the fall of 1952 after some seventeen million animals were vaccinated, with the United States contributing over $125 million to the cost, the disease was eradicated.[155]

In December 1950, the Reciprocal Trade Agreement of 1943 was terminated by mutual consent. Mexico had sought its abolition since 1945 in order to encourage its budding industries. Mexico proved unyielding, however,

[155] Dept. of State *Bulletin*, XXVI (Mar. 31, 1952), 459.

on proposals for cooperative military measures. Beginning in 1952 the United States concluded bilateral military assistance agreements with twelve Latin American countries, but President Alemán refused to do the same. The pact, from the Mexican standpoint, would restrict its freedom of unilateral decision-making, and create an undesirable dependence on the United States.

A minor but persistent source of contention developed in that year when Mexico began enforcing Mexican claims to sovereignty out to nine miles against United States fishermen in the Gulf of Mexico. Several American shrimp boats were apprehended, and others fired on by Mexican patrol craft on the grounds that they were violating the nine mile perimeter. The United States Department of State directed the fishermen not to fish in Mexican waters, but did not acknowledge the existence of Mexican sovereignty beyond the three mile limit.[156]

President Dwight D. Eisenhower, convinced of the efficacy of top-level personal diplomacy in strengthening the Good Neighbor Policy, met with Mexican presidents on five different occasions during his administrations: first, with Adolfo Ruiz Cortines at the dedication of the Falcon Dam on the Texas-Mexico border in October 1953; a second time with Ruiz Cortines, a meeting also attended by Prime Minister John Diefenbaker of Canada, at White Sulphur Springs, West Virginia, in March 1956; three meetings with Adolfo López Mateos: at Acapulco, Guerrero, February, 1959, at Washington in October, and at Ciudad Acuña, Coahuila, October 1959. The continuation and frequency of these meetings indicate that both governments regarded them as useful vehicles in the conduct of their respective foreign and domestic policies.

As mentioned, the initial Ruiz Cortines-Eisenhower meeting in 1953 was the inauguration of the Falcon Dam, located on the Rio Grande, seventy-five miles from Laredo. This vast structure, five miles long, and built at a cost of $47 million, of which the United States contributed 58.6 percent, is mutually advantageous in regard to irrigation, flood control and the generation of power. President Eisenhower declared that the project had given the world "a lesson in the way neighbor nations can and should live—in peace, mutual respect, in common prosperity." President Ruiz Cortines called the project "the tangible result of that friently spirit which it is our duty to maintain and to make universal."

The migratory labor problem mounted in the early 1950s, when in 1952 it was estimated that 1.5 million Mexicans crossed the border illegally; some six hundred thousand were apprehended and returned that year, and in 1953 the figure rose to one million. The problem was mitigated by an agreement signed on March 10, 1954, which renewed the Migratory Labor Agreement of 1951, and was scheduled to run until December 31, 1955. This not only clarified the 1951 agreement, but also changed it to prevent the flow of illegal workers into the United States. President Eisenhower declared that the basic purpose of the agreement was to enable the United States to give Mexican migratory labor the protection of its laws.[157] However, many thousands of "wetbacks" remained in the United States, and it was to be a continuing problem.

[156] Frank Brandenburg, *The Making of Modern Mexico* (Englewood Cliffs, 1964), p. 336.
[157] Dept. of State *Bulletin*, XXX (Mar. 29, 1954), 467-468.

Vice President Richard M. Nixon, on a good-will tour of the Caribbean, stopped in Mexico City in February 1955. He pledged United States cooperation in working out common problems, and declared: "We will welcome the opportunity, too, to meet people in all walks of life so that by our actions we can show them the feeling of the United States people for those of our sister republics." He paid an official call on President Ruiz Cortines and laid a wreath at the foot of Mexico's Column of Independence. Later in the year the Mexican government paid the eighth installment of $8,689,257 to the Mexican Eagle Oil Company through the company's banker in New York. This came under an agreement of 1947 whereby Mexico agreed to pay fifteen equal installments in compensation for the properties, rights and interests of the companies affected by the expropriation decree of 1938.[158]

It was at the Tenth Inter-American Conference held at Caracas, Venezuela, in 1954, that the cleavage of the United States and Mexican policies on hemispheric problems related to the cold war began to appear. The most significant and disruptive topic on the agenda concerned Communist penetration in the New World. The threat of communism in Guatemala led the United States to present a strong resolution asserting that the domination or control of an American state by communism was a threat to all American states; this resolution called for action being taken under existing treaties, notably the Rio Treaty of Reciprocal Assistance of 1947. Fearing that self-determination was being endangered, and that collective intervention, although verbal, might lead to physical intervention, Mexico opposed the United States resolution condemning Guatemala. The Mexican delegation, however, signed the recommendation condemning the extension of communism in the Americas.[159]

In assuming this posture at Caracas, and at subsequent inter-American conferences, the Mexicans were adhering to principles of foreign policy carried on by successive administrations since the time of President Cárdenas in the 1930s: the sanctity of national sovereignty, national self-determination, the juridical equality of nations, non-intervention in the internal affairs of another nation, collective security, the peaceful settlement of international disputes, the protection of basic human rights and fundamental freedoms, regionalism and universalism.

Foreign investments commanded much of Mexico's attention in 1955. President Ruiz Cortines in his September message to Congress and the nation said: "We will continue to employ foreign investment as required by our development, provided it is used for projects of immediate welfare . . . foreign capital can be beneficial only when it fits this need and subjects itself to our laws without injuring Mexican enterprise."[160]

The *peso* was devalued in April 1954, and two years later the Mexican domestic and global economic position reflected a sustained and spectacular advance. Tourism was an important factor in the growth, for the tourist receipts in 1954 and 1955 added $337 million and $350 million respectively to foreign exchange receipts, compared with $315 million in 1953. The advance was

[158] *New York Times*, Feb. 10, 1955; Sept. 16, 1955.
[159] Cline, *Mexico*, pp. 313-314.
[160] *New York Times*, Jan. 5, 1956.

mainly attributable to the decline in Mexican prices, in terms of dollars, following devaluation. It was estimated that 95 percent of the tourist expenditures were made by Americans, and the expansion of income from this source was held to reflect the sustained prosperity of the United States. Additional income came from the growth of exports, chiefly coffee and cotton.[161]

On the occasion of the meeting of presidents Eisenhower and Ruiz Cortines at White Sulphur Springs, in March 1956, the main topics discussed concerned: (1) the release of United States cotton on the international market (the U.S. agreed not to do so in any way damaging to Mexico); (2) United States commercial fishing boats in Mexican waters; (3) the lack of a bilateral treaty governing air traffic; (4) the United States pledge to continue economic aid through the Export-Import Bank. Dr. Milton Eisenhower and Dr. Roy R. Rubottom, Jr., Assistant Secretary of State for Inter-American Affairs, later went to Mexico as guests of the Mexican government to discuss further matters of mutual interest.[162]

While unfettered loans from the United States were sought, by 1957 certain types of technical assistance, begun as part of the Point IV program, had outlived their usefulness. Senator Mike Mansfield, reporting to the Senate Foreign Relations Committee in November, recommended its termination, saying: "Mexico as a proud country, has wanted and . . . to a great extent . . . has succeeded in solving its problems." The program has deteriorated, having little impact outside Mexico City. The senator continued ". . . The United States and Mexico have it within their power to eradicate the festering memories of a former unhappy association by building a new and enduring foundation for cooperation in a spirit of equality, mutuality and respect."[163]

On December 1, 1958, Adolfo López Mateos, former minister of labor, was inaugurated for a six-year term as president of Mexico, and without the accompaniment of anti-United States demonstrations as had been forecast. Secretary of State John Foster Dulles, who headed the American delegation, conveyed the following message from President Eisenhower to the new president: "Our two countries share an extensive border. They present to the world a proud example of how frontiers can be used as gateways to promote the peaceful interchange of persons, culture and goods essential to our mutual welfare." López Mateos responded saying: ". . . In this new era of mutual respect and growing mutual understanding . . ." there existed real friendship. For some time, he said, relations have been "free of onerous distrust and bitterness of the past . . . providing an atmosphere to plan mutually satisfying and beneficial arrangements for the future . . ." However, in his First Annual Message to Congress he reaffirmed his country's traditional policy declaring: "The principles of our nation's foreign policy emanate from our historical experience. We were forced to defend our territory, our sovereignty, and our integrity."[164]

[161] Walter J. Sedwitz, "Mexico's 1954 Devaluation in Retrospect," *Inter-American Economic Affairs*, X (Autumn, 1956), 22, 32.

[162] *New York Times*, Mar. 30, 1956; Aug. 7, 1957.

[163] *New York Times*, Nov. 17, 1957.

[164] Cline, *Mexico*, p. 300, 305; *New York Times*, Dec. 2, 1958. Treatment of the formulation and administration of Mexican foreign policy may be found in the chapters by Luis Quintanilla in Philip W. Buck and Martin Travis, Jr., eds., *Control of Foreign Relations in Modern Nations* (New York, 1957).

United States private capital entered Mexico in rapidly increasing quantities beginning in 1950, and by 1958 investments of American companies totaled some $700 million. The production of these companies in goods and services there was estimated at over $600 million a year. Of all direct investments those of the United States came to 78.4 percent of the total. The appointment of Robert C. Hill, former vice president of W. R. Grace and Company, as United States ambassador to Mexico, brought an experienced businessman to the post.[165]

Mexico's reaction to the Caracas meeting in 1954 foreshadowed the position it would take at later anti-communist conferences; however, it was the success of the Cuban 26th of July Movement, led by Fidel Castro Ruz, resulting in the overthrow of the dictator Fulgencio Batista, in January 1959, that brought into focus the divergent views of Mexico and the United States on policies toward communism in both the regional and international framework. Recalling that Fidel Castro and his brother Raúl were given refuge in Mexico after their first abortive attempt to invade Cuba on July 26, 1953 (for which the 26th of July Movement is named), and that the revolutionaries had again embarked for Cuba from Mexico on November 26, 1956, in the invasion effort that ultimately proved successful, many Mexicans were sympathetic toward the movement and its leaders from the outset. As the Castro regime began to reveal its true authoritarian, alien and brutal character, the Mexicans became progressively disenchanted with it, but at the same time retained a kindly and sympathetic disposition towards the Cuban people. It should be understood in this connection that Mexico has condemned, and forcibly suppressed when necessary, Communists within its own frontiers.

At the Fifth Consultative Conference of Foreign Ministers, held in Santiago, Chile, in 1959, problems raised by the new regime in Cuba and the Trujillo dictatorship were considered. Mexico affirmed its support of continental solidarity against extra-continental interference in hemispheric affairs on that occasion, but stood firm on the principle of non-intervention.[166]

Personalized presidential diplomacy in the conduct of United States-Mexican relations was carried on throughout 1959. President Eisenhower met López Mateos at Acapulco in February, and in October the latter came to the United States. At this second meeting the two presidents chose the name Amistad Dam to designate the dam proposed to be constructed near Del Rio, Texas, and Villa Acuña, Coahuila, for flood control, conservation and storage of the waters of the Rio Grande, and possible power generation. The following year they met on the International Bridge connecting the two towns and reaffirmed plans for the dam.[167]

In response to Mexican concern about minerals export markets and prices, chiefly lead and zinc, the American president said: "Maintenance of the productive capacity of the Mexican mining industry is essential to Mexico's economic prosperity, and to the security of the United States." In an address before the United Nations President López Mateos, after referring to the poverty of Latin America and suggested remedies, declared "In practical terms this

[165] *New York Times*, Jan. 6, 1958.
[166] Cline, *Mexico*, p. 314.
[167] Dept. of State *Bulletin*, XLIII (Dec. 5, 1960), 851; *New York Times*, Oct. 25, 1959.

means increased trade, primarily greater imports of Latin American raw materials by the United States at better prices. . . ."[168] These and other proposals about the positive role that the United States should take in supporting Latin American development helped influence the United States to sponsor the Inter-American Development Bank in 1959, and the Alliance for Progress two years later.

In the summer of 1959 the relations of the two countries were slightly jarred incident to Mexico's 150th independence anniversary celebration. The United States delegation sent to Mexico City, comprising three Senators, issued a protest and boycotted official parties after hearing a Mexican congressman praise Fidel Castro of Cuba and condemn the United States. But Secretary of State Christian Herter, who came with the delegates, pointedly ignored the incident.[169]

The Sixth Consulation of Foreign Ministers was convened at the request of Venezuela in San José, Costa Rica, in August 1960, to discuss acts of aggression by the Dominican Republic. While Mexico found the Trujillo regime repugnant, its representative led the opposition to a proposal by Secretary Herter that steps be taken to end the dictatorship. Mexico, which agreed only to the imposition of sanctions, feared that supervised free elections would establish a dangerous precedent for collective intervention.

The Seventh Consulation was convened at San José, after a day's recess from the Sixth, following a request from Peru that the threat to Latin American states posed by the Cuban-exported revolution be discussed. The Declaration of San José condemned intervention by extra-continental powers after Secretary Herter had stressed that Cuba was a point from which international communism was invading the Western Hemisphere. Mexico once again upheld its traditional principles of non-intervention and the self-determination of peoples. Although confirming its opposition to interference by non-continental powers, Mexico made it clear that it was similarly opposed to interference by one state within the inter-American system in the affairs of another.[170]

Mexico also opposed holding any special meetings regarding Cuba and, when outvoted by a majority of the OAS members who convoked the meetings at Punta del Este in early 1962, Mexico abstained from voting on a resolution calling for the expulsion of Cuba from the OAS, and the immediate suspension of commerce in armaments with that country. The Mexican foreign minister did acknowledge, however, that the ideology of Marxism-Leninism was incompatible with membership in the inter-American system. Mexico's disapproval of the expulsion of Cuba from the OAS was based on its dogmatic policy of non-intervention, and on the ground that inadequate juridical provisions existed in the inter-American system to allow expulsion.[171] In the perilous confrontation between the United States and the Soviet Union regarding Cuba in October 1962, Mexico voted with the United States.

The goals of the Alliance for Progress program of President John F. Kennedy, announced at Punta del Este in 1961, envisioning the attainment of

[168] *New York Times*, Oct. 15, 1959.
[169] *Ibid.*, Aug. 26, 1959.
[170] Cline, *Mexico*, pp. 316-319.
[171] Brandenburg, *Modern Mexico*, p. 338.

prosperity, progress and social justice, through evolutionary means, were received enthusiastically by the Mexicans. At the outset of his administration the president pledged increasing cooperation with Mexico. To contribute toward improving relations, the first of a series of Interparliamentary United States-Mexican conferences was held. The twenty-three member United States delegation, headed by Senator Mike Mansfield, met the Mexican congressman at Guadalajara, Mexico in February, 1961. Discussed were possible solutions for border problems, foreign trade and commerce inadequacies, and cultural exchanges. At the Second Interparliamentary group meeting in Washington, in 1962, Secretary of State Dean Rusk stressed the importance of disarmament, an objective which the Mexicans strongly supported.[172]

In July 1962 President Kennedy, accompanied by his wife, visited Mexico at the invitation of President López Mateos. The tumultuous reception was said to have been the largest and most enthusiastic ever given to a visiting chief-of-state in Mexico's history. Relations had been impaired by differences over Cuba, especially the role of the United States in the invasion effort, and the American president hoped to restore them to a more harmonious footing. The Cuban issue was not raised, however, and agreements were reached on several questions of a bilateral character.

President Kennedy pledged that the danger of saline waters in Mexico from the Colorado River would be lessened immediately. Aside from the conflicting anti-communism foreign policies of the two countries, this problem had been described as the basis of the most serious deterioration in United States-Mexican relations in many years. Under a treaty of 1944, Mexico is allowed 1,500,000 acre feet of irrigation water from the Colorado each year. As the result of an irrigation project in Arizona, saline water was poured into the Gila River, and had reached the Colorado. An estimated five thousand tons of salt was being discharged into the river waters every twenty-four hours, and had caused a crop loss of $16 million in the Mexicali Valley, in 1961. It was agreed that the United States would "flush out" the Colorado, and introduce enough low-saline content water to minimize the damage until a permanent solution could be found.[173]

The two presidents reaffirmed their support of the Alliance for Progress throughout Latin America, and declared that they "propose to respect and maintain the principles of non-intervention, whether this intervention may come from a continental or extra-continental state, and of self-determination of peoples." The president struck a responsive chord in announcing that the United States was wrong in refusing to accept the decision of the 1911 arbitration commission on the Chamizal question, and at the same time expressed a "strong desire" to reach an accord which would take into account the interests of the people involved.

A $20 million agricultural loan agreement between the United States and Mexico was signed. On this occasion President Kennedy said:

The agricultural agreement we sign here today is an historic step forward in cooperation between our two countries under the *Alianza para el Progreso* . . . here in Mexico you have

[172] *New York Times*, Feb. 12, 1961; May 15, 1962.
[173] *Ibid.*, July 1, 1962.

carried forward the largest and most impressive land reform program in the history of the hemisphere . . . never before has this program been more vigorously administered than during the last three years when the government of President López Mateos distributed twenty-four million acres to hundreds of thousands of *campesinos*. The tangible results of your land reform can be witnessed in the 223 percent rise in agricultural output over the last two decades, a rise which has made Mexico virtually self-sufficient in food stuffs and a major exporter . . . 174

Amplifying the president's statements concerning Mexico's extraordinary achievements in the post-war years, Secretary Rusk commented: "The Alliance for Progress has a great deal to offer, but it has also a great deal to borrow from the Mexican experience." Eager to contribute to the Alliance, Mexico joined the United States in a training program for technicians from other Latin American countries. In 1963 some three hundred technicians received their training in Mexico.[175]

The long-standing controversy over the El Chamizal area of El Paso, Texas was settled in August 1963, when the two nations signed an agreement transferring the disputed territory to Mexico. Ambassador Thomas Mann, who signed for the United States, hailed the agreement as an example of how neighboring countries can settle their differences in a "cordial and diplomatic manner." Manuel Tello, Mexico's Minister of Foreign Affairs, after signing for his country, said that it was "one of the most important agreements in the diplomatic history of the United States and Mexico.[176]

This was not overstating its significance because, starting as a minor controversy over a relatively small piece of ground transferred by erosion from the southern to the northern side of the Rio Grande at El Paso, the dispute had come to symbolize in Mexico, with the passage of years, several of the most vital elements in the law of nations: sovereignty over national territory, the sanctity of treaties, and the juridical equality of states.

President Lyndon B. Johnson signed the Chamizal Treaty on December 20, 1963, and subsequently the House of Representatives authorized the appropriation of $44.9 million for the cost of moving 4,500 United States citizens out of the El Chamizal area of El Paso, Texas, straightening the Rio Grande, building new bridges and making other public improvements. The terms of the agreement provided that of the 437 acres estimated to have been awarded to Mexico in 1911, Mexico agreed to accept 71 acres from an area of El Paso slightly downstream from the Chamizal Zone. Mexico would receive the remaining 366 acres from the actual Chamizal tract. The United States and Mexico also agreed to relocate the Rio Grande at El Paso in order to maintain the river as a boundary. The settlement eliminates a Mexican enclave north of the Rio Grande known as Cordova Island. Mexico approved the transfer of the northern half of the island, consisting of 193 acres, to the United States in return for an equivalent acreage from the United States territory to the east of Cordova Island. Mexico, perhaps waiting for American approval to avoid a repetition of the 1911 settlement, ratified the treaty on January 7, 1964.

[174] *Ibid.*
[175] Dept. of State *Bulletin*, XLVI (June 4, 1962), 920; *Ibid.*, L (Mar. 23, 1964), 449-450.
[176] *New York Times*, Aug. 30, 1963.

Presidents Johnson and Díaz Ordaz met at Chamizal on October 28, 1967, amid festive ceremonies, and formally acknowledged an end to the long-standing controversy.[177]

Presidents Johnson and López Mateos, meeting at Palm Springs, California, on February 21-22, 1964, noted with satisfaction the high degree of understanding and cooperation established in recent years between the two countries, and announced their decision to keep working toward the goals set forth in the joint communique of June 30, 1962, issued following the López Mateos-Kennedy conversations in Mexico City. Both agreed to support the principle of self-determination of all peoples and of its corollary non-intervention; they expressed their faith in representative democracy, and their devotion to the ideals of human liberty and the dignity of the individual. At the same time the two leaders reviewed a wide range of topics which included the Alliance for Progress, the OAS, bilateral trade relations between the two countries, the International Coffee Agreement, and the United Nations.[178]

Although not mentioned in the official dispatch, two problems existed which demanded attention: the *braceros*, and the Colorado River salinity question. Congress had agreed in December 1963 that the program under which Mexican migrant laborers enter the United States not be allowed to continue beyond December 31, 1964. These workers' wages had added upwards of $35 million a year to Mexico's dollar earnings. The United States had been persuaded to end the program largely by the argument of labor unions that it tended to depress wages and working conditions, and increase domestic unemployment. President López Mateos said that from Mexico's viewpoint the program was beneficial but not mandatory, and that Mexico was willing to do away with it whenever the United States wished to end the treaty.[179]

While these statements were relevant in most areas of bilateral relations, Mexico had meanwhile refused to conform with the decision of the Ninth Meeting of Foreign Ministers, in July 1964, to suspend diplomatic and economic relations with Cuba in spite of its commitment under the terms of the Rio Treaty. Its representatives contended that withdrawal of recognition or severing relations would be tantamount to intervention.

[177]Dept. of State *Bulletin*, L (January 13, 1964), 49.; *New York Times*, Mar. 12, 1964; *Ibid.*, Oct. 29, 1967. See also U.S. Dept. of State, *Hands Across the Border: The Story of Chamizal* (Washington, n.d.), p. 13. A dissenting voice, Senator John G. Tower of Texas, recommended that the Texas legislature be given an opportunity to approve or reject the treaty before implementation, otherwise Texas would be dismembered without its consent. Not objecting to the treaty itself, Tower strongly believed a precedent should be established whereby a state must concur before the national government ceded part of that state's domain to another nation. The objection was overruled on the ground that when Texas joined the Union it became subject to the Federal government's right to delineate international boundaries of states. See *Cong. Rec.*, CIX, Part 19, 24851–56.

[178]*New York Times*, Feb. 23, 1964. President Johnson had previously pledged to further the Alliance for Progress program with the same energy and devotion as shown by President Kennedy; he also proposed that the Alliance be made a "living monument" to the late President. See *New York Times*, Nov. 27, 1963.

[179]*Ibid.;* The U.S. Supreme Court in 1971 upheld the Government's right to grant informal permits of admission to 50,000 Mexicans entering the United States for daily or seasonal employment. The court refused to hear an appeal sought by a group of resident farmers to bar the issuing of permits by the U.S. Immigration and Naturalization Service. See *Christian Science Monitor*, Apr. 30, 1971.

Gustavo Díaz Ordaz was elected to the presidency of Mexico in July 1964, and prior to his inauguration on December 1st, President Johnson welcomed the president-elect and his wife at the LBJ ranch. Bilateral questions were discussed on such topics as the Colorado salinity problem, Mexican sugar-export quotas to the United States, and the International Coffee Agreement. President Johnson said: "There are many problems between the two nations, but in the days ahead we will resolve them with peace, reason and justice to each other." Díaz Ordaz replied that, "we have shown to the people of the world how differences can be solved by mutual respect, by sharing common ideals, and on the basis of justice."[180]

In the spring of 1965 the Mexicans hailed the beginning of a "new diplomacy" in their relations with the United States. This change was initiated by the settlement of the Chamizal dispute, which President López Mateos said was the most important achievement of his administration, and the prospect of solving the riverine salinity problem mentioned above.[181] Then late in October 1965, the United States Congress passed a bill reducing Mexico's minimum quota for sugar exports to the United States in 1966. An outburst of anti-Yankee agitation led Ambassador Fulton Freeman to call a news conference in Mexico City. Deemphasizing the decrease, he said that the 362,350 tons assigned to Mexico in 1966 as a minimum quota was the largest in the Western Hemisphere and second in the world to the Philippine Islands. He pointed out that sugar exports to the United States would earn Mexico $42 million in 1966 whereas she would be able to earn only a third of that sum if she had to sell her sugar in world markets, where prices were then very low.[182]

The subject had international implications, for the Mexicans interpreted the readjusted quota as punishment for Mexico's refusal to break relations with Cuba, criticism of United States intervention in the Dominican Republic, and opposition to a United States-sponsored inter-American military force. These charges were made by the Mexican Ambassador to the United States, Hugo B. Margain, and quickly a wave of anti-American nationalism, always at least latent, began to sweep the capital. Compounding the problem, Foreign Minister Antonio Carrillo Flores, who had been representing Mexico at the conference of the OAS in Rio de Janeiro, rejected proposed changes in the Charter of the OAS that might endanger the sovereignty of its members or weaken the principle of non-intervention in the internal affairs of any state. The implications of the sugar quota question and Carrillo's speech in Rio de Janeiro gave the impression that Mexico was defending the rest of the Americas against the United States.[183]

The incident, however, produced no apparent reverberations, and in April 1966, President Johnson traveled to Mexico City for the unveiling of a statue of Abraham Lincoln, a reproduction of Augustus Saint-Gauden's work in the Washington Lincoln Memorial, which the United States had presented as a gift to Mexico. In a joint statement the two presidents reaffirmed their committments

[180] *New York Times*, Nov. 13, 1964.
[181] *Ibid.*, Apr. 3, 1965.
[182] *Ibid.*, Nov. 27, 1965.
[183] *Ibid.*

to the fundamental axioms of Mexico's traditional foreign policy, and both agreed to the need for maintaining constantly open the doors of dialogue. On bilateral matters it was noted that the Joint Trade Committee, established in 1965, was a major step in expanding commercial exchange. They expressed deep concern regarding the international market for cotton, which was then the leading Mexican export product, and a matter of considerable importance to the United States. It was obvious that the two leaders skirted two major political differences: Mexico's refusal to support United States intervention in the Dominican Republic, and her refusal to break relations with Cuba.[184]

The occasion was accompanied by a major statement by President Johnson of his country's policy toward Latin America. He expressed continued all-out support for the Alliance for Progress and for "social revolution" in both North and South America. The president gave his blessing to an Argentine proposal for a meeting of the Western Hemisphere heads of state "to examine our common problems and to give the Alliance for Progress increased momentum."[185] At a news conference in June, President Johnson declared that he had recently visited Mexico and spent the week end with the foreign minister who "told me never in the history of the two countries did we have a better relationship."[186]

When addressing the Sixth Mexico-United States Interparliamentary Conference in Washington, D.C., in February 1966, Secretary Rusk declared: "The United States believes the project of a nuclear-free zone in Latin America is constructive statesmanship in the best tradition of the Hemisphere. We welcome the effort and would be glad to see it reach a successful conclusion." His remarks concerned the efforts of Mexico, supported by several hemispheric nations, to draft a treaty aimed at keeping Latin America free of nuclear weapons. The Mexico-sponsored draft would prohibit any signatory nation from building or storing nuclear weapons, and from allowing its territory to be used for experiments, or for installations capable of delivering atomic weapons. A Latin American Denuclearization Center would be established to supervise the execution of the agreement, but its jurisdiction would be limited to the signatory powers.[187]

The final text of the treaty was unanimously approved by twenty-one Latin American nations at Mexico City in February 1967. Although doubt was expressed that all would ratify it, or that the five nuclear powers would agree to respect the ban, it was regarded as a diplomatic victory for Mexico. Mexico became the first country to ratify the treaty, when the Senate approved the document in September, 1967.[188]

Meanwhile President Johnson and Díaz Ordaz made a joint inspection of the Amistad Dam construction site near Del Rio, Texas, and Ciudad Acuña, Mexico, which was scheduled for completion in 1969. The American President declared that " . . . Amistad Dam is another link in the mutual trust, friendship and progress which unite our two peoples." Similar cordiality attended the

[184] *Ibid.*, Apr. 16, 1966.
[185] *Ibid.*
[186] *Ibid.*
[187] Dept. of State *Bulletin*, LIV (Mar. 7, 1966), 366; *New York Times*, Apr. 17, 1966.
[188] *New York Times*, Sept. 14, 1967. Protocol II of the Treaty of Tlatelolco was signed by Vice President Hubert Humphrey for the United States on Apr. 1, 1968, at Mexico City.

Second Annual Meeting of the Joint Mexico-United States Trade Committee, held in Mexico City December 15-17, 1966. The Committee considered general trade trends and specific commercial problems of mutual interest. Progress in the reduction of obstacles to freer trade was confirmed since the initial meeting.[189]

President Díaz Ordaz when addressing the Mexican Congress in September 1967, reiterated his support for Latin American economic integration as outlined at the Punta del Este Conference of American chiefs of state in April. But he warned that this integration must be carried out fundamentally by Latin Americans with cooperation from the United States, and that the results must be for the benefit of Latin America: "... we do not mean to create, in the face of a great agricultural and industrial power, another that might struggle with us ..."[190]

Matters of bilateral character in Mexican-United States relations in the late 1960s included an agreement for the development of a flood control project on the Tijuana River in California, a cotton textile agreement, and a protocol concerning radio broadcasting. Of a more complex nature were the conferences held in Washington and Mexico City concerning the use of territorial fishing grounds. The motives for the talks were recent changes by both countries relating to maritime fisheries within the contiguous zones off their territorial seas. In 1966 Mexico and the United States had extended their jurisdiction over fisheries in adjacent waters to a distance of twelve nautical miles from shore. The laws of both countries provide for a continuation of traditional fishing within their respective zones as may be recognized by the government having jurisdiction; an agreement was reached by the delegations to permit such fishing.[191]

The administration of President Richard M. Nixon maintained the spirit of friendly cooperation with Mexico, and the two nations were brought even closer by joint efforts to halt narcotics smuggling from Mexico into the United States. Alarmed by the illegal drug flow from Mexico the Nixon administration in 1969 launched Operation Intercept in which travelers to the United States from Mexico were subjected to intensive search at the border. The experiment itself proved ineffective, but it served to prod Mexico into action. Operation Intercept was succeeded by Operation Cooperation in which the governments agreed on a bilateral program to curtail the drug traffic. It appeared, however, that the magnitude of the problem would prevent any substantial cutback in the flow of drugs across the border in the forseeable future, despite Mexican cooperation.[192]

Evidence that Mexico's stability was suffering erosion from social unrest appeared in 1968 when student protest movements shook the nation. At the height of the crisis, from July through November, some 150,000 students, teachers and sympathizers in Mexico City, and thousands in the provinces, were involved. The movement collapsed after at least thirty-nine persons were killed in a clash with police in October, just ten days before the Olympic Games

[189] Dept. of State *Bulletin*, LVI (Jan. 2, 1967), 12-13; *Ibid.*, (Jan. 9, 1967), 70.
[190] *New York Times*, Sept. 2, 1967.
[191] Dept. of State *Bulletin*, LVI (Feb. 6, 1967), 224; *Ibid.*, (June 9, 1967), 919; *Ibid.*, (June 26, 1967), 64; *Ibid.*, LVII (July 31, 1967), 147; *Ibid.*, (Oct. 9, 1967), 475.
[192] *Christian Science Monitor*, June 26, 1970.

opened. One of the main frustrations that the students tried to express was the futility of trying to influence government policy through the existing electoral process.[193]

The government, having turned back the student challenge through the massive deployment of police and army troops, made conciliatory gestures to the dissidents. Many agitators arrested during the disturbances were released, and the voting age was reduced from 21 to 18. Congress also passed legislation facilitating the release of persons, mainly leftists, who were held in prison, often for long periods of time, for offenses of a political nature.

Consistent with its records of not having lost a presidential election since its founding, the PRI's (Partido Revolucionario Institucional) candidate, Luis Echeverría Alvárez, won 86 percent of the vote in the election of July 5, 1970. His only opponent, Efrain González Morfin, represented the National Action Party. Mr. Echeverría Alvárez carried on a strenuous campaign, visiting all of the nation's twenty-nine states and two territories, covering more than 100,000 miles. In so doing he reached thousands of persons who had never before seen a presidential candidate. Both President Díaz Ordaz and the President-elect were ministers of the interior, which since it controls the police force, tends to produce politicians of a conservative nature. However, Mr. Echeverría Alvárez declared his intention to relieve the causes of social discontent, saying "We in Mexico have a good deal of unfinished business in order to fulfill the promises of the Revolution of 1910.[194]

Social discontent showed no signs of diminishing, however, as student rioting broke out in Mexico City in June 1971, leaving thirteen students dead and 160 injured. The riots on June 10 involved fighting between anti-government students and armed right-wing youths. The former were carrying placards accusing President Echeverría of the responsibility for killing students in the 1968 riots when he was interior minister. What began as a peaceful demonstration became vicious fighting when the armed right-wing groups charged into the ranks of the protestors. In the belief that the PRI had not responded to the students, the president placed a number of young university graduates on his staff in key positions and gave his support to the 18 year old vote. To this degree he was more responsive to student interests than any of his predecessors.[195]

Another aspect of social unrest which had appeared in Mexico by the early 1970s was a guerrilla movement. Its existence was revealed in the kidnap-ransom of Mexico's civil aviation head, and a series of bank and payroll robberies, which were linked to Communist-trained terrorists. Some of those apprehended admitted their connection with the Revolutionary Action Movement (MAR),the group that was organized and trained by Mexican graduates of the Patrice Lumumba University in Moscow. Pointing up the significance of the problem, Mexico in March 1971 expelled five Soviet diplomats accused of supporting the guerrilla training of Mexican citizens.[196]

The foreign policy of a country is determined by a complex of forces,

[193] *New York Times*, Dec. 25, 1968.
[194] *Christian Science Monitor*, July 11, 1970.
[195] *Ibid.*, June 14, 1971; *Ibid.*, Sept. 7, 1971.
[196] *Ibid.*, June 12, 1971; *Times of the Americas*, Sept. 29, 1971.

permanent and changing, internal and external. The constants of geography and history have profoundly influenced the foreign policy of Mexico with respect to the United States. And differences in paths followed, in physical and human resources, and the international aims of the two countries have impressed on Mexico, as the weaker nation, certain characteristics and permanent features. Continuity and change in the internal development of both countries are also clearly manifested in the relationship. Trends in the economic process of Mexico, particularly those affecting its modernization and living standards, are of fundamental importance in illuminating its past and continuing relationship with the United States.

Mexico's record of economic growth in recent times has been impressive by any standards. It should be recalled, however, that by its historical evolution the country was better prepared than many of the so-called economically under-developed countries to take advantage of opportunities presented. National economic growth had been a studied aim of official policy since the Porfirian era, 1876-1910. The Díaz dictatorship had created the foundation for a modern economy once the feudal structure, perpetuated by Díaz, had been destroyed by the Revolution. Although revolutionary strife continued into the 1920s, and extensive agrarian reform did not come until the 1930s, Mexico was prepared to respond to the opportunities for industrial development offered by shortages and market growth accompanying World War II. Mexico developed its own brand of "one party" democracy, providing political stability accompanied by a significant measure of political freedom, the subordination of the military elements, its own formula for stimulating private enterprise within a state-directed economy, and an entrepeneurial class which, supported and exhorted by the state, was capable of exploiting the favorable international market that developed during the war, and continued in its aftermath.

Between 1939 and 1954 the average annual growth rate of the gross national product was 5.7 percent, and in the period 1949-1964 it was about 6 percent; the increase in 1965 was 5.1 percent. This growth was accompanied by conditions of relative monetary stability with less inflationary pressure than was found generally in Latin America. President Díaz Ordaz reported to Congress in 1967 that the gross national product had registered a 7.5 percent increase, but made it clear that anything greater would cause inflation and force the government to put the brakes on the economy, declaring: "We are sacrificing spectacular advance and accomplishment in exchange for achieving real progress." With one of the fastest growing populations in the world, about 3 percent annually, the per capita income increased by some 3 percent annually, but this was still higher than the level set for the United Nations Development Decade.[197]

Industry has led the way, for virtually the whole range of durable and non-durable consumer goods was being produced domestically. A forceful system of import licensing and high tariffs was instituted to keep out most foreign goods that would compete with local manufacturers. But under the Law for the Encouragement of New and Necessary Industries, new ventures were allowed very generous tax treatment. Foreign investors encouraged to participate in these

[197] *New York Times*, Sept. 2, 1967.

ventures, provided they obtained the approval of the Mexican government. A usual requirement for gaining approval was that the new venture should be a joint one with Mexican partners, the latter having 51 percent of the investment.[198]

Agriculture lagged behind industry, but was well ahead of Latin America as a whole. By 1960 food imports had been reduced to 2.8 percent of total imports, making the country almost self-sufficient in food production. While food production had kept ahead of population growth, the greatest vigor was found in the area of commercial crops. The cultivated area under irrigation doubled between 1950 and 1963. The rising trend of exports meant that the agricultural sector was susceptible to fluctuation in world markets; however, Mexico was in a better position than many primary exporters because her agricultural exports were more diversified.[199]

Changes in the economic structure are reflected in the foreign trade pattern. Exports had been diversified to the point that in the 1960s about one-fifth represented manufactured or semi-manufactured products, including refined sugar, and also such commodities as steel sheets and pipe, railroad equipment, and plywood. Imports, strictly controlled, consisted of industrial raw materials and capital goods.

A factor in Mexico's economic success which deserves special emphasis has been the availability of foreign exchange, made possible in large part by its propinquity to the United States. The major consequences have been the tourist boom, which provided the vital and growing extra third of Mexico's export earnings, the income of migrant workers, and the low transport costs of labor-intensive manufactured goods shipped across the United States frontier. The absence of exchange controls, allowing complete freedom for capital and interest payments to flow out of the country, has also had a measurable impact. The United States, for several decades, has been the outlet for about three quarters of Mexico's exports, and has supplied that nation with three quarters, or more, of its imports. Mexican trade with Latin America has not been significant, but has shown an increase in recent times.

Another feature of Mexico's post-war growth has been the extraordinary extent to which it was financed by domestic resources. Foreign direct investment was more than compensated by the outflow of earnings from such investments, a condition which is likely to persist, though by the 1960s a trend towards increased reinvestment of profits by foreign enterprises was evident. Mexico's foreign lenders include the World Bank, the Inter-American Development Bank, and the Export-Import Bank, as well as private bankers in the United States, Western Europe and Japan. Mexico's attractiveness to the international business community and her excellent credit standing have been aided by her unfailing repayment of old debts, such as the indemnification payments arising from the 1938 oil nationalization, which were completed in 1962.

Illustrative of Mexico's capacity for self-help was its early phasing out of

[198] *Economist*, CCXVI (Sept. 25, 1965), XII; XIX; *New York Times*, July 25, 1965.
[199] Richard W. Parks, "The Role of Agriculture in Mexico's Economic Development," *Inter-American Economic Affairs*, 18 (Summer, 1964), 23.

the Alliance for Progress loan program. Two thirds of the assistance given Mexico over the first four years consisted of commercial-type loans on which service was fully maintained, and by the fourth year over 80 percent constituted regular commercial-type loans. Mexico was notified that it would be excluded from participation in the Alliance except with respect of commercial-type borrowing. At this point it was concluded that only two countries, Mexico and Venezuela, had a capability great enough to require "soft loan" projects, where the return to the economy was likely to be long delayed.[200]

By 1966 Secretary Rusk was able to substantiate his earlier optimism declaring that Mexico has often been cited as an example for the rest of Latin America, and a "show window" for the Alliance for Progress. One year later he commented that Mexico was in a position to help its neighbors speed their development. This meant that Mexico had met the requirement for self-help better than any of the more favored nations, and that it must go it alone without benefit of easy financing. Thus, in the first five years of the Alliance one-third of the increase in gross national product in Latin America came into Mexico which contained about one-sixth of the population and had received comparatively meager assistance from the *Alianza* program. A critical analysis of the problem revealed that: "Only 3.2 percent of the disbursement of donations and concealed donations in the five years (1962-1966) went to Mexico although it had some 17.4 percent of the population, and although half of its population with a per capita income under $150 (per year) was living far below the average level of the Latin American countries."[201]

Mexico's share in Alliance for Progress aid and the proportion of disbursements are shown below:

Disbursement of Aid[202]
(millions of dollars)

	1962	*1963*	*1964*	*1965*	*1966*
Mexico	129.5	40.3	29.6	79.5	60.3

Proportion of Disbursements[203]

Proportion of disbursements of donations and concealed donations given certain countries compared with their proportion of the Latin American population affected.

	1962	*1963*	*1964*	*1965*	*1966*	*Share of population*
Mexico	1.3%	2.8%	3.7%	4.8%	2.6%	17.4%

The official Mexican evaluation of the Alliance was expressed by President Díaz Ordaz in 1965: "In the case of our country, we can affirm that the Alliance

[200] Simon G. Hanson, "The Alliance for Progress–The Fourth Year," *Inter-American Economic Affairs*, 20 (Autumn, 1966), 27; Joseph A. Hasson, "Latin American Development: The Role of the Inter-American Committee for the Alliance for Progress," *Orbis*, IX (Winter, 1966), 1063.

[201] Simon G. Hanson, *Five Years of the Alliance for Progress* (Washington, 1967), pp. 79-80.

[202] *Ibid.*, p. 4.

[203] *Ibid.*, p. 80.

has operated satisfactorily. Its favorable evolution has been made possible by the manner in which the principal organization for promoting it, the Agency for International Development, has operated, enabling us directly or indirectly to obtain loans for such important areas as agriculture, the training of technicians, housing, small irrigation projects. . ."[204]

Mexico's sustained growth in domestic production in the decade of the 1960's was interrupted by a severe economic slowdown in 1971. The Echeverría government was confronted with mounting inflationary pressure, growing maldistribution of income, a large external debt and a trade deficit in excess of $1 billion in 1970. The nation's annual average growth rate of about 6.5 percent following World War II fell to 3.7 percent in 1971, but reached 5.5 percent in 1972. The annual per capita income attained $700; however, it represented the great wealth of the industrial sector rather than a rise in the overall standard of living. At base, the problem was the low productivity of Mexican industry and a drop in the primary exports (sugar, coffee, cotton and fish).[205]

Agriculture, supporting more than 40 percent of the population and accounting for 10 percent of the gross national product suffered from a lack of credit, irrigation development and investment in machinery, fertilizers and fresh crop strains. President Echeverría traveled by jet aircraft on week ends to remote rural areas, hearing complaints and observing conditions. His answer to peasant dissatisfaction was the promulgation of a new Agrarian Reform Law, designed to strengthen the *ejidos* to stimulate production, build peasant income and regulate the administation of rural credit. The conservative elements within the PRI, notably those with rural financial interests, found the new law disturbing. With the population expected to double to one hundred million by 1990,increased food production is imperative. Improved living conditions for the rural population might be expected to reduce the swelling migration to the cities; 400,000 *campesinos* migrate each year to Mexico City alone.

Although Mexico has outdistanced most Latin American countries in achieving balanced economic growth, the potentially large market has been stifled by mass poverty. The purchasing power of urban workers has risen in recent years, but they constituted a relatively small sector of the population, and their bargaining position had been weakened by a competitive labor market. However, the fundamental failure had been raising the living standards of the rural masses. Low purchasing power and inequalities of income coexisted with a depression in high cost industries which lacked consumer outlets. This problem had to be overcome before Mexico could achieve its goals and potentialities.

Mexico's trade with the United States was first among the developing nations. Worldwide, Mexico was the fifth-ranking client and sixth supplier of the United States. But in 1971 trade with the U.S. produced an unfavorable balance of $481.3 million, which was partially attributable to Washington's protectionist policy. Income from tourism in 1971 surpassed the value of exports ($1,373 billion) by $84.8 million. Some 2.3 million tourists, most of them Americans,

[204] *Latin American Times*, Oct 12, 1965.
[205] *New York Times*, Jan. 28, 1972; Inter-American Development Bank, *Socio-Economic Progress in Latin American (Social Progress Trust Fund Tenth Annual Report, 1970)* (Washington, 1971), pp. 272-286; *Christian Science Monitor*, Dec. 14, 1972.

visited Mexico. With a view to expanding tourism, Mexico budgeted $60 million, part of which was to support projects in two remote areas of the country: a one thousand mile highway in the desert peninsula of Baja California, to be completed in 1973; and a major seashore resort in the Caribbean tropics in the southeast. With threats to American-owned properties and investments in Latin America on the rise, President Echeverría clarified his government's position saying: "We do not have an expropriating mentality, there will be no restrictions on convertibility, nor will there be modifications in the rate of exchange. This will be a regime of guarantees. . . "

Mexico's relative position in the politico-economic spectrum was brought clearly into focus on the occasion of Chilean President Allende's visit in December 1972. Dr. Allende was enthusiastically and warmly welcomed, but President Echeverría and many civic and professional groups pointedly rejected his Marxist leanings. The Mexican president declared publicly that "In Mexico we want economic security, but we want also the unrestricted exercise of democracy . . . We reject all social patterns that reduce human dimensions to mercantile values in the same way that we repudiate all forms of totalitarianism." He said, in short, that Mexico has a political philosophy distinct from that of Chile. On the question of the troubled relations between Washington and Santiago, however, President Echeverría openly supported Dr. Allende in his efforts to expose what he called an "invisible blockade" allegedly waged by the United States against Chile.[206]

Conclusion

Mexico's international conduct has tended to bewilder, if not antagonize, the people of the United States, especially so with the deepening crisis of the global cold war, and the rise of Communist power in the Western Hemisphere: When Mexico's experience with foreign powers is considered in the light of historical perspective, however, it is possible to interpret her international posture, often at variance with The United States, with greater understanding.

As noted in the preceding chapter, the first fifty years of Mexico's independent life witnessed a sequence of foreign onslaughts. She emerged as an independent nation with a weak economic structure, lacking in social and racial integration, and with the ruling classes ideologically divided. The domestic upheavals weakened the country in the face of United States expansion, and gave rise to a multitude of international claims that would persist for decades. Mexico was compelled to sign agreements with the United States for payment of claims for damages to American citizens which were often excessive, unjust, and even fraudulent. Claims provided one of the pretexts for the wars of 1836 and 1847 with the United States; they provoked a war with France in 1837; they

[206] *New York Times*, Jan. 28, 1972; *Times of the Americas*, Nov. 10, 1971; *Ibid.*, Dec. 20, 1972; *Christian Science Monitor*, Dec. 4, 1972.

gave rise to the long-term French intervention starting in 1862, and delayed Mexico's development. It is not surprising that the nation's attitude toward foreign powers became defensive, distrustful and, nationalistic.

International relations became stabilized during the dictatorship of Porfirio Díaz, for the order achieved by Díaz removed the major causes ad pretexts for foreign intervention. The generosity of Díaz toward foreigners had international repercussions later, and accented the hostile reaction of the Mexican people toward the outside world. The submissive international policy adopted by the Díaz government also left its mark.

Mexico's international conduct varied with the progression of the Revolution, reacting to domestic policies and external forces. The pursuit of political aims of the Revolution provoked crucial international incidents, for the civil strife among the contesting factions led to the occupation of Vera Cruz, General Pershing's expedition, the United States efforts to end the civil war by bringing together representatives of the rival factions, the withholding of recognition as a means of exerting pressure, and the claims for damages suffered by foreigners during the Revolution.

The United States punitive expeditions induced Mexico to emphasize the principle of territorial integrity of nations. Although the American armed interventions could have benefited Carranza, and the United States contended they were not hostile acts toward the Mexican people, Carranza insisted on unconditional withdrawal. His stand foreshadowed Mexican resistance to the setting up of foreign military bases on Mexican territory even though it had the consent of the Mexican government and seemingly served the nation's welfare. Carranza took a similarly dim view of the Niagara Falls Conference, and efforts by the United States and Latin American countries to act collectively in pacifying the factions fighting in Mexico. Carranza contended that collective international action, though only discussion of the matters, would establish the precedent of foreign interference in the settlement of Mexico's domestic affairs. The application of this principle helps to illuminate Mexico's stand at the several Western Hemisphere anti-communist conferences since 1954.

Article 27 of the Mexican Constitution of 1917 contained two revolutionary goals which were at the same time of international applicability: notably agrarian reform, and the recovery of the country's natural resources. Although pressure exerted by the United States in its non-recognition of President Obregon's government forced Mexico to postpone in the 1920s agrarian reform and the reclaiming of the nation's oil deposits, these actions were carried out vigorously in the 1930s.

Mexico joined the League of Nations in 1931, expressly repudiating the Monroe Doctrine which was named in Article 21 of the Covenant. Mexico's efforts proved decisive in obtaining inter-American acceptance of the principle of non-intervention at the conferences of Montevideo (1933), and Buenos Aires (1936). As a member of the League, Mexico condemned fascist aggression and provided refuge for thousands of Spanish refugees. Mexico's expropriation of foreign-held oil interests reinforced the nation's confidence in its ability to overcome foreign pressure, and to attain a national goal at the expense of the major powers.

Mexico adhered to the principles of the Atlantic Charter, and supported the United Nations in the belief that it would contribute toward disarmament and peace. A document entitled "Views of the Foreign Ministry of Mexico on the Dumbarton Oaks Proposals" was submitted by Mexico during the San Francisco Conference. Designed to enhance the role of the smaller nations, it was defeated by a strong surge of nationalism, which emerged at the Conference. With the growth of tensions attending the cold war Mexico returned to its traditional position of applying the principles of non-intervention and respect for the domestic jurisdiction of states. These principles were made the keystone of Mexican foreign policy in the post-war era.

Adhering to these principles, in the context of cold war power politics, Mexico has assumed a quasi-neutralist position, neither pro-Moscow, Peking or Havana, nor anti-United States. From a practical standpoint, Mexican policy stresses foreign trade, aid and capital investment. The focus of its policies, historically, ahs been on contiguous or nearby countries: the United States, Guatemala and Cuba.

Mexico is strongly identified with regionalism through its membership in the inter-American system, the OAS. Its participation in the Latin America Free Trade Area, and its approval of the Central American program of economic integration give further evidence of this attitude. Universalism, as a feature of Mexican foreign policy was demonstrated by its role in the League of Nations, and subsequently in the United Nations, UNESCO, International Bank for Reconstruction and Development, International Monetary Fund, the International Labor Organization, and other United Nation's agencies.

In spite of its identification with regionalism and universalism, however, the post-war attitude of Mexico toward the world has been one of mistrust and partial apathy, and its foreign policy mainly defensive and non-interventionist. But this is inconsistent with the country's dynamic economic and social development, which requires that she join more actively in international life. On general international questions, where Mexican interests were not directly affected, its contributions have not been significant. Until recently, Mexico has done little to collaborate with other Latin American states or states in Asia and Africa.[207]

Suggestions of the possibility that Mexico might be initiating a restrained and gradual departure from its peculiar type of neutralism, and its rather defensive isolationism, were reflected in the actions and statements of former President López Mateos and President Díaz Ordaz. López Mateos intensified the nation's international activities by his visits to the United States and Canada, as well as a number of Latin American countries . . . the first in history for a Mexican president . . . revealing a new aim in foreign policy: the strengthening of political and economic ties with Latin America. The visits of parliamentary missions to Europe and Asia, and the dispatching of trade missions to many countries seemed to highlight the importance that Mexico attached to international affairs. International assistance required for the economic development of the country had also carried Mexico farther afield than previously.

207 Jorge Castañeda, "Revolution and Foreign Policy: Mexico's Experience," *Political Science Quarterly,* LXXVIII (Sept., 1963), pp. 391-417.

In his 1965 state of the union message President Díaz Ordaz spoke of the need for a policy of "independence within interdependence and solidarity with the rest of the nations," a formula not mentioned by his recent predecessors. In that same connection, he said "one notices something else, too, having once established our national being we are beginning to develop externally." Moreover, he left no doubt about Mexico's cordial attitude toward the United States: " . . . friendship with our immediate neighbors must be closer, warmer," and he pledged "to make that friendship an example of cordial and constructive coexistence," which went further than his predecessors in friendly expressions toward the United States.[208] He later toured the five Central American countries and Panama, and participated in the Punta del Este Conference of Hemispheric Chiefs of State.

His successor, President Echeverría, also pushed vigorously to strengthen economic relations with the Central American nations. In a series of meetings with Central American presidents, Mexico promised to lend technical assistance in the possible development of a regional merchant marine fleet aimed at lowering the countries' export and imports costs. Central Americans were receptive to prospects of a trade agreement with Mexico, which appeared to be mutually beneficial, especially for the former because of the slowdown of the Common Market.

Ranging farther abroad than his predecessors, President Echeverría visited Japan in March 1972. His object was to transfer part of the commerce with the United States to Japan, and to reduce the volume of "triangular operations" which, mainly connected with the United States, had resulted in several typical Mexican exports reaching their buyers by way of third countries. Several hundred Mexican representatives of business, banking, industry and agriculture joined in the effort, in Tokyo, to help stimulate trade between the two countries. Like the United States, Mexico was faced with a rising balance of trade deficit, which reached an estimated $860 million in 1970.[209]

Mexican interest in Asia also extended to mainland China, which was confirmed by President Echeverría's decision that his government would recognize the People's Republic of China; however, trade between Mexico and mainland China began in 1963. This was followed in 1965 by the formation of a commission for economic development, and reciprocal participation in trade fairs in Mexico City and Canton. Mexico expected to sell an estimated $40 million in cotton and similar products to Peking in a two-year period.[210]

Mexican-United States relations through 1971 reflected the cordiality and warmth expressed by the Mexican President, as equitable solutions to problems of a bilateral nature, previously mentioned, were arrived at. But on the question of Cuba, Mexican policy remained fixed and seemingly irreversible. Cuba was condemned anew in the court of hemispheric opinion, the OAS Council, formally a consultation of OAS foreign ministers, in September, 1967. Convoked to consider Venezuela's new charges of Cuban subversion, the OAS voted to "forcefully" condemn the government of Cuban Premier Fidel Castro

[208] *Latin American Times*, Sept. 3, 1965,
[209] *Times of the Americas*, Feb. 23, 1972.
[210] *Times of the Americas*, Mar. 1, 1972; *Christian Science Monitor*, Mar. 9, 1972.

and impose afresh, new sanctions on non-Communist trading with Havana. Yet, the condemnation was less vigorous than either the United States or Venezuela, backers of a strong line against Cuba, wanted. The final vote on the most critical resolution was 20-to-0, with Mexico, faithful to its non-intervention principle, abstaining. It was evident that the hemisphere was still split between the "hard liners" on Premier Castro—the United States, Venezuela, Bolivia—and the "soft liners"—Mexico, Chile and Ecuador.

While maintaining this benign posture toward Cuba in regional councils Mexico's bilateral ties with the Castro regime were eroded in 1969. This stemmed from issues concerning the hijacking of aircraft, asylum policies and spy charges. On July 26, the anniversary of the Cuban revolutionary movement, a commercial aircraft was hijacked to Havana by two Mexican nationals identified as youths who had participated in the 1968 student revolt. Mexico requested their extradition, but the Cuban government granted them political asylum, declaring that "the motives that led them to enter Cuba are of a political nature." Mexico denied any accusations of political crimes. Shortly afterward *Granma*, the official organ of the Cuban Communist Party, charged that the Mexican Embassy in Havana had harbored common criminals and had protected Cubans using government boats to flee the island. The Cuban government then accused an official of the Mexican Embassy in Havana of spying for the United States Central Intelligence Agency, charging that he had sent photographs of military objectives to the intelligence agency through his Embassy's diplomatic pouch. In the face of these incidents Mexico exercised restraint and gave no indication of altering its basic policies toward Cuba. By 1970 Mexico, alone, among the members of the OAS, had not broken diplomatic and economic ties with Havana. Why the latter had chosen such a provocative course to needlessly antagonize its staunchest benefactor in the Western Hemisphere remained unclear.

In evaluating United States-Mexican relations we should consider the totality of the relationship. If a balance sheet were drawn up, the credit side of the ledger would show a marked advantage over the debit side. The history of the relationship shows how two nations once divided by the deepest antagonisms that the differences of race, religion, culture, and the stages of economic development can bring about have, over the years, step by step, settled their outstanding differences, in recent times, through peaceful and amiable arrangements.

Mexico has come to understand the "Yankee" perhaps better than any of her sister republics. That understanding is far from synonymous with submission for, as noted, Mexico adheres to an independent foreign policy. But when Mexico takes issue with her neighbor it does not do so in the role of a chronic critic. Each nation has come to recognize the other's foibles and strengths, thus making it possible to coexist on a plane of equality and cordiality rarely enjoyed by continguous countries.

While Mexico may never attain the status of nations comprising the "power belt" of the North Atlantic, it does have a potential that is not usually understood. Mexico is the most populous Spanish-speaking country in the world, and it is the fourth largest of the American republics, exceeded in area only by Brazil, Argentina and the United States. With a population of about forty-eight million (estimated, 1970), it is the third largest American republic in population. According to United Nation's projections, Mexico's population will reach 110 million by the end of the century. Size, strategic location, progressiveness, and extensive resources, both material and cultural, have given Mexico international status and significance. It is possible that Mexico will challenge Spain for the leadership of the Spanish-speaking world, and rival Brazil for leadership in Latin America.

For many years the bilateral relations of the United States and Mexico have been good, with activity in diplomatic channels becoming almost routine in the 1960s. On broader hemispheric questions, however, there has been less concord. If recent events may be projected, it appears that the conduct of the United States, not only concerning Mexico, but also toward the other countries of Latin America, will have a significant bearing on Mexican foreign policy. Latin American problems have inevitably become essential elements in Mexican foreign affairs.

Cuba's experience and evidence of continuing revolutionary ferment elsewhere in Latin America indicate a growing pressure for wide-reaching social change in many countries. The United States position vis-a-vis these movements will doubtless color Mexico's relations with it. If the United States should follow a policy that lacks understanding of Latin America, it could result in grave domestic and international consequences for Mexico, and abrade the cord of the "Era of Good Feelings," so painstakingly woven by the United States and Mexico over several decades.

SUPPLEMENTARY READINGS

Bemis, Samuel F. *The Latin American Policy of the United States.* New York, 1943.
Blakeslee, G. H., ed. *Mexico and the Caribbean.* New York, 1920
Calero, Manuel. *Un decenio de política Mexicana.* Mexico, 1920.
Callahan, James M. *American Foreign Policy in Mexican Relations.* New York, 1932.
Callcott, W. H. *Liberalism in Mexico.* Stanford, 1931.
Calvert, Peter. *The Mexican Revolution, 1910-1914: The Diplomacy of Anglo-American Conflict.* Cambridge, Mass., 1968.
Castañeda, Jorge. *Mexico and the United Nations.* New York, 1958.
----------. "Pan Americanism and Regionalism: A Mexican View," *International Organization,* X (Aug., 1956), 373-389.
Clendenen, Clarence C. *The United States and Pancho Villa: A Study in Unconventional Diplomacy. Ithaca, 1961.*
Cline, Howard F. *The United States and Mexico.* Cambridge, Mass., 1963. Rev. ed.
----------. *Mexico, Revolution to Evolution, 1940-1960.* New York, 1963.
Cosío Villegas, Daniel, ed. *Historia moderna de México,* 6 vols. Mexico, 1955-1963.
Cronon, E. David. *Josephus Daniels in Mexico.* Madison, 1960.
Cumberland, Charles C. *Mexican Revolution: The Constitutionalist Years.* Austin, 1972.
Daniels, Josephus. *Shirtsleeve Diplomat.* Chapel Hill, 1947.
Dillon, E. J. *Mexico on the Verge.* New York, 1921.
Dunn, Frederick S. *The Diplomatic Protection of Americans in Mexico.* New York, 1933.
Feller, A. H., *The Mexican Claims Commissions, 1923-1934.* New York, 1934.
Gibbon, T. E. *Mexico under Carranza.* New York, 1919.
Glade, W. P., Jr., and C. W. Anderson. *The Political Economy of Mexico.* Madison, 1963.
Gordon, Wendell C. *The Expropriation of Foreign-Owned Property in Mexico.* Washington, 1941.
Gruening, Ernest. *Mexico and its Heritage.* New York, 1928.
Hackett, Charles W. *The Mexican Revolution and the United States, 1910-1926.* Boston, 1926.
Howland, Charles P., ed. *Survey of American Foreign Relations.* New York, 1931.
Inman, Samuel G. *Intervention in Mexico.* New York, 1919.
Jones, C. Lloyd. *Mexico and its Reconstruction.* New York, 1921.
Merrill, John C. *"Gringo": The American as seen by Mexican Journalists.* Gainesville, 1963.
Nicolson, Harold. *Dwight Morrow.* New York, 1935.
Obregón, T. Esquivel. *México y los Estados Unidos ante el Derecho Internacional.* Mexico, 1926.
Priestley, H. I. *The Mexican Nation.* New York, 1923.
Quirk, Robert E. *The Mexican Revolution, 1914-1915.* Bloomington, Ind., 1960.
Rippy, J. Fred. *The United States and Mexico.* New York, 1926.
Rives, G. L. *The United States and Mexico, 1821-1828.* 2 vols. New York, 1913.
Rojas, J. Fernández. *De Porfirio Díaz a Victoriano Huerta.* Mexico, 1913.
Ross, Stanley R. *Francisco I. Madero, Apostle of Mexican Democracy.* New York, 1955.
Scott, Robert E. *Mexican Government in Transition.* Urbana, Ill., 1959.
Tannenbaum, Frank. *Mexico: The Struggle of Peace and Bread.* New York, 1950.
Timm, Charles A. *The International Boundary Commission, United States and Mexico.* Austin, 1941.
Townsend, William C. *Lázaro Cárdenas, Mexican Democrat.* Ann Arbor, Mich., 1952.
Wood, Bryce. *The Making of the Good Neighbor Policy.* New York, 1961.

9

Colonial Cuba and Its International Relations

The Caribbean is still, as it was in the past, one of the great maritime highways of the world for both commercial and strategic purposes. The mainland bordering its shores–Venezuela, Colombia, Panama, Costa Rica on the south, and Nicaragua, the Honduras, Guatemala, and Mexico on the west–was known in the sixteenth century as the Spanish Main. Its city fortresses were the great centers of Spanish colonial trade. Through them and across the Caribbean passed the plunder gained from the *Conquista* on the mainland. Aptly called the "cockpit of international rivalry," the Caribbean witnessed the power struggle involving the Spanish Empire and its European challengers, the English, the French, and the Dutch. In terms of United States foreign policy the Caribbean has been the focus of the Monroe Doctrine idea, and Cuba, because of its singularly strategic location, has always figured prominently in the questions of national and hemispheric defense.

The trade winds and the Gulf Stream facilitated communication with Europe in the days of sailing ships; from the Caribbean islands easy access could be had to North America (via Florida and Louisiana), to Mexico, and to western South America (via the Isthmus of Panama). Early in the sixteenth century Spain occupied the four largest islands, the Greater Antilles–Cuba, Jamaica, Hispaniola (comprising Haiti and the Dominican Republic), and Puerto Rico–which separate the Caribbean from the Atlantic Ocean. In the seventeenth century the English, French, and Dutch established bases on the smaller islands, the Lesser Antilles, in the southeast Caribbean. Spain lost territory to her rivals when England took Jamaica in 1655, and France occupied one-half of Hispaniola (Haiti), in 1697. In the eighteenth century, England got possession of most of the Lesser Antilles, France being reduced to the colonies of Martinique, Guadaloupe, and a few smaller islands. The Dutch retained possession of the small islands of Aruba, Curaçao, Bonaire, St. Eustatius, Saba, and half of Saint Martin.

After exhausting the meager gold deposits on the islands, the Spanish established large plantations which produced chiefly sugar cane; later, tobacco, cacao, and other tropical products were cultivated. On the Spanish islands, as well as those acquired by other European countries, the economy was based upon the production of very few commodities, for export only; little effort was made to achieve self-sufficiency. The laborers suffered from malnutrition, and from endemic tropical diseases. The Indians on the islands were almost exterminated by the ruthless exploitation of the Europeans, and by European diseases. To maintain the plantation economy, which required great numbers of laborers, Negro slaves, and relatively small numbers of East Indians and Asiatics were introduced into the European colonies in the Caribbean. Scarcely any middle class developed between the wealthy plantation owners and the slaves largely because of the failure to initiate commerce and industry outside the plantations.

More than one-half of the land area of the West Indies is in Cuba. The 44,000 square miles of territory extend for 785 miles in an east-west direction with a width varying from twenty-five to 120 miles. Of great significance in modern times is the fact that at least half of the area is level enough to be suitable for machine agriculture. Not more than a quarter of Cuba is mountainous. The soils are adapted to a wide variety of crops, of which sugar is only one, and significant deposits of iron ore, manganese, and copper are found there. The island is well drained, and has a uniform, dependable rainfall. Its climate is far more temperate than that of Jamaica, Hispaniola, and the Lesser Antilles, making it more attractive to the Spanish immigrant.

To attain an understanding of early United States attitudes and policies towards Cuba, the problem must be viewed against the background of events occurring in the late eighteenth and nineteenth centuries. The fledgling United States had found itself threatened by Spain and Great Britain on its western ramparts after independence had been won, but owing to the crises arising from the French Revolution, the Napoleonic wars and their aftermath, it was able to fulfill its "manifest destiny," to become a continental republic. Diplomatic milestones were Jay's Treaty of 1794 with Great Britain, and Pinckney's Treaty of 1795 with Spain, which identified boundaries and gave a measure of security to United States territory. However, the possibility that Louisiana or Florida might come under the domination of militant and powerful France caused great anxiety in the new republic. This occurred in 1800 when France, under the terms of the secret Treaty of San Ildefonso, regained Louisiana from Spain. Bonaparte had then decided to build a French colonial Empire in the Mississippi Valley of North America, using the island of Hispaniola as his Caribbean staging area.[1]

In 1802 President Jefferson declared that the day France takes possession of New Orleans "we must marry ourselves to the British fleet and nation."[2] Again Europe's distress proved advantageous to the United States. After the disastrous campaign of General Le Clerc in Haiti, and confronted with a resumption of war with Great Britain, Bonaparte sold the United States the entire province of Louisiana.

[1] Samuel F. Bemis, *The Latin American Policy of the United States* (New York, 1943), pp. 16-23.

[2] P. L. Ford, ed., *The Writings of Thomas Jefferson*, 10 vols. (New York, 1892-1899), VIII, 145.

The Floridas were regarded as of even greater importance to the United States than the Louisiana territory, and it was with the purpose of obtaining the Floridas that Richard Livingston had set out on his memorable mission in 1801. After Louisiana was obtained, including West Florida, according to the interpretation of its purchasers, it was merely a question of time before East Florida would fall into the orbit of the new republic. In fact, in 1811 Congress authorized President James Madison to use the army and navy to seize and occupy all or any part of East Florida.[3] But the need was hardly sufficient to proceed to this extremity, and in 1819 Spain was finally persuaded to cede the Floridas and all title to lands that the United States claimed as an integral part of the Louisiana Purchase.

A mere glance at a map of the Caribbean shows the strategic importance of Cuba to the country that possesses New Orleans and Florida. It also reveals the interest which a nation possessing the Bahamas and Jamaica would have in seeing to it that Cuba should not fall into the hands of a dangerous rival. Therefore, from the beginning of the nineteenth century the United States and Great Britain remained watchful lest the "Pearl of the Antilles" should slip from the ever-weakening grasp of Spain into the possession of the other or of a powerful rival.

The importance of the island of Cuba to the United States in the formative period of American foreign policy is perhaps best expressed by Secretary of State John Quincy Adams in a note dated April 28, 1823, to the American minister in Spain:

> These islands (Cuba and Puerto Rico) from their local position are natural appendages to the North American continent, and one of them (Cuba) almost in sight of our shores, from a multitude of considerations has become an object of transcendent importance to the commercial and political interests of our Union. Its commanding position, with reference to the Gulf of Mexico and the West Indian Seas; the character of its population; its situation midway between our southern coast and the island of St. Domingo; its safe and capacious harbor of Havana, fronting a long line of our shores destitute of the same advantage; the nature of its productions and its wants furnishing the supplies and needing the returns of a commerce immensely profitable and mutually beneficial, give it an importance in the sum of our national interests with which that of no other country can be compared. Such indeed are between the interests of that island and of this country, the geographical, commercial, moral, and political relations, . . . that in looking foward to the probable course of events for the short period of half a century it is scarcely possible to resist the conviction that the annexation of Cuba to our Federal Republic will be indispensable to the continuance and integrity of the Union itself.[4]

In the same dispatch Secretary Adams pointed out that if the control of Spain was terminated, Cuba must look either to the United States or Great Britain. The government of the United States had been confidentially informed that Great Britain was so eager to obtain Cuba that she had offered Gibraltar in exchange. Whether or not this was true, "the transfer of Cuba to Great Britain would be an event unpropitious to the interests of the Union."[5] That Great Britain was not unaware of the American attitude is shown by a statement in the diary of Lord Ellenborough, a member of Wellington's cabinet, dated February 8, 1830: "The

[3] Amer. State Papers, *For. Rel.*, vol. III, p. 571.
[4] J. B. Moore, *Digest of Int. Law*, vol. VI, p. 380.
[5] *Ibid.*, p. 383.

Americans declared that they could not see with indifference any state other than Spain in possession of Cuba."[6] In fact, as early as 1819, when the possibility of Britain taking Cuba was considered, Richard Rush, the American Minister, asked Lord Castlereagh about these reports. The latter denied any knowledge of such developments.[7] In 1823, when George Canning was especially interested in obtaining the cooperation of the United States, he denied emphatically that England desired Cuba.[8] He by no means implied, however, that his government would look with equanimity upon such a desire on the part of the United States.

France also had a continuing interest in the Caribbean region. Although Guadaloupe and Martinique were very small relics of her once glorious trans-Atlantic empire, the nation that so recently had disposed of Louisiana, and that, during her long occupation of Haiti, had been Cuba's nearest island neighbor, could not be expected to lose all interest in this region of buried hopes. The London *Times* in 1825 accurately characterized Cuba as "the Turkey of trans-Atlantic politics, tottering to its fall, and kept from falling only by the struggles of those who are contending for the right of catching her in her descent."

Although the United States' real interest in Cuba dates from its possession of Florida, the earlier history of the island was associated with that of the mainland closely enough to warrant a brief review of Cuban history. Discovered by Columbus on his first voyage in 1492, the island was claimed by the Spanish Crown. Undecided whether "Juana," as he named it in honor of the heir of the Catholic Kings, Ferdinand and Isabella, or "Cuba," as the natives called it, was an island or a peninsula, he was certain that it lay off the coast of East Asia. The name West Indies testifies to this belief. On his second voyage, before leaving Cuba, he had his men swear that it was a peninsula, not an island, thereby hoping to convince the Spanish monarchs that he had reached the Asian mainland.

Diego de Velásquez conquered Cuba in 1511, and in the process of pacification a chieftain named Hatuey was burned at the stake. "This first of Cuba's heroes declared that he did not wish to be converted and sent to Heaven, because Christians might be there." After the conquest there followed a brief period of prosperity, for the mines did yield some gold, and it was believed that much more remained. However, the discovery of vast wealth on the mainland quickly depopulated it of men, money, and horses. As happened elsewhere in the islands, the native race virtually disappeared in a short time. Cuba's value for Spain then centered on the port of Havana where silver fleets and convoys could assemble. Time proved it to be a security problem owing to its use as a rendezvous by pirates, and the great difficulty that Spain found in protecting it from the predatory English, French, and Dutch. A French fleet occupied Havana in 1538, in 1554 "Peg Leg" Leclerc took Santiago, and in 1555 another pirate, Jacques de Sores, a Lutheran, sacked Havana. The English navigator Drake threatened the island in 1585 with a force of thirty ships, but did not risk an attack. In the following year the French looted Santiago. Philip II at length concluded that Cuba was of greater significance than merely a strategic base for operations on the mainland, and he constructed the great forts of Punta and Morro to protect his

[6] *Ibid.*, p. 56.
[7] J. M. Callahan, *Cuba and International Relations* (Baltimore, 1899), p. 196.
[8] *Ibid.*, p. 199.

"bulwark of the Indies, key to the New World," against the "corsair-caked Caribbean."[9]

Havana was also, in Spain's view, the base of the African slave trade. But Cuba, being chiefly occupied with cattle raising, had little need for slaves. Until the eighteenth century its main commerce was in hides and meat. Some copper was exported, and tobacco found a place in the market by the seventeenth century. The tobacco industry was so hampered by trade monopolies that two uprisings in protest occurred, in 1717 and 1723. Havana also served as a ship building center, but Spain curtailed the industry because of a conflict of interests. The cultivation of sugar cane spread rather slowly, compared with its development on other Caribbean islands, but it could be grown most successfully because of favorable natural conditions found there. By the end of the seventeenth century some one hundred mills were functioning, and Negro slaves were being imported in rising numbers. Indicative of the limited growth of plantation agriculture, however, was the fact that the island contained only 32,000 slaves when the British captured Havana in 1762. The economy had been allowed to decline, strife persisted among the administrative hierarchy, and much of what prosperity existed came from illegal traffic with privateers. Moreover, the island was subjected to continuing assaults by foreigners and pirates.

Even the massive fortification of Morro Castle could not save Cuba from the British in the Seven Years War. In 1762 the Earl of Albemarle with 10,000 troops from England and 4,000 from the North American colonies, laid siege to the stronghold, and after more than a month of mining and sapping took it by assault. Although the capture came in the "sun-setting time of the age of plunder," it was estimated that treasure and property to the value of about $16 million were obtained. It was of even greater importance to Great Britain since she now held the key to both Mexico and Louisiana. Yet, in spite of Cuba's obvious value, by the Peace of Paris, signed February 10, 1763, England returned the island to Spain and received in exchange Florida, which, as William Pitt declared at the time, was certainly no equivalent. As one imaginative writer put it, "But for that, Washington and his associates might have failed—the French Revolution might have been postponed—and the House of Hanover at this moment have been ruling over the present United States."

If Spain made a good bargain with her enemy England, she made an even better one with her ally France, whose king unconditionally ceded to Spain the Louisiana territory. Cuba was now of the greatest strategic and economic importance to Spain, and Havana, which during the brief British occupation was opened to the trade routes of the world, was once more closed. Nevertheless, in the short period of English control, Cuba had tasted the forbidden fruit of free trade, and the experience was never forgotten.

The British occupation was also to have a profound influence on the future economic life of Cuba. Many members of the British and colonial armed forces stationed there acquired a taste for Cuban cigars and snuff, and, after their return home, helped to popularize the tobacco habit in northern Europe and North America. Tobacco, by then the main export crop of Cuba, was grown on small

[9] John E. Fagg, *Cuba, Haiti, and the Dominican Republic* (Englewood Cliffs, 1965), pp. 11-19.

plots, whereas the large slave-powered sugar plantation was still uncommon. British traders, expecting that their country would retain the island, introduced ten thousand slaves, most of whom were destined for work on sugar plantations. Slaves continued to be imported after the peace, and the sugar industry was placed on firmer foundations than previously. Of significance also was the declining emphasis on Havana as a military bastion, and a greater increase in that of Puerto Rico; ultimately the fortresses of San Cristobal and San Felipe del Morro became Spain's major strongholds in the New World.[10]

The Seven Years' War and the dislocation of trade channels caused by the French Revolution were instrumental in opening the United States as a market for Cuban sugar because Santo Domingo, formerly its chief source, was precluded from competition by revolution. For Cuba this unsettled period produced a boom in sugar and a decline in tobacco. The first steam-powered mills were established in 1819, and with the progressive adoption of other technological improvements, and the formation of the great *centrales*, Cuba soon led all other sugar-producing areas in modern, efficient equipment and refining techniques. In so doing Cuba assumed the calculated risks inherent in an exporting monoculture. Thus, by the beginning of the nineteenth century, steps were taken which would ultimately make the island the greatest sugar producer in the world, and economically dependent on the United States market. "The consequences of dependence were to follow as the century progressed; dealer would dictate to producer, foreign capital would replace native capital, absentee control would replace local ownership."[11]

During the American war of independence Spain was drawn in by France against England, but her interest in the American colonies was purely selfish, as is shown in the instructions given by the governor general of Cuba to an agent, urging upon Congress the capture of St. Augustine, Florida, in order that it might be restored to Spain.[12] Later, when Jay appeared at Madrid seeking a treaty granting free navigation of the Mississippi, he was not even granted official recognition. However, before peace was signed, in 1783, Spain showed herself more liberal towards her West Indian possessions than Great Britain, for both Havana and Santiago were opened to foreign commerce. In fact, the commercial code of 1778, which opened nine ports of entry in Spain and twenty-four in her colonies, was a model of liberality for the time.

The French Revolution made the trade between the West Indies and the American colonies exceedingly brisk, but Cuba was not to pass unscathed through the period that followed. A French squadron blockaded the island in 1794 and caused much hardship to the inhabitants. During the Napoleonic era Cuba was threatened alternately by Great Britain and France, and when by the treaty of San Ildefonso, signed October 1, 1800, Napoleon recovered Louisiana and then sent an expedition to Santo Domingo, well might Jefferson feel that France could no longer remain our "natural friend." Spain had equal cause for alarm. Napoleon could hardly confine his operations to Santo Domingo, with Cuba at his mercy, nor be content with Louisiana, when the wealth of Mexico

[10] J. H. Parry, *The Spanish Seaborne Empire* (New York, 1966), pp. 304-305.
[11] J. H. Parry and P. M. Sherlock, *A Short History of the West Indies* (New York, 1963), pp. 223-224.
[12] Francis Wharton, *Diplomatic Correspondence of the American Revolution* (Washington, 1899), III, 412-415.

lay at his feet. Both the United States and Spain had reason to feel thankful over his forced change in plans, for the failure of the expedition against Santo Domingo, and the prospect of an immediate rupture with England, gave Louisiana to the United States and gave Spain a new lease on her colonies. But, as Mr. Slidell declared in 1859, "From the day we acquired Louisiana the attention of our able statesmen was fixed on Cuba. What the possession of the mouth of the Mississippi had been to the people of the west, that of Cuba became to the nation."[13]

We have noted the increasing interest that the United States felt in Cuba after being assured of the possession of the Floridas, as shown by the instructions given by John Quincy Adams to our minister in Spain. Jefferson also, in a note to President Monroe dated October 24, 1823, indicated his feeling on the subject: "I candidly confess that I have ever looked on Cuba as the most interesting addition which could ever be made to our system of states. The control which, with Florida Point, this island would give us over the Gulf of Mexico, and the countries and isthmus bordering on it, would fill up the measure of our political well being."[14] Therefore at this period, when the Spanish colonies in Mexico and South America had virtually established their independence, and when the United States had received trustworthy information that Cuba also was planning a revolt with the purpose of seeking admission to the Union, American statesmen saw the need of great diplomatic caution. When Calhoun urged the acceptance of the Cuban proposal, Adams wisely opposed on the ground that it might mean war with England. It would be wiser, he thought, to say "that our relations with Spain would not allow us to encourage such a proposal."[15] In fact, at a cabinet meeting March 17, 1823, when President Monroe proposed to offer to Great Britain a mutual promise not to take Cuba, both Adams and Calhoun opposed.[16]

The European situation soon worked itself out in such a way that both Great Britain and France were forced to declare themselves in regard to Spanish possessions in America. Although by the fall of 1823 French armies were in full control of Spain, Canning had already notified them "that, as England disclaimed all intention of appropriating to herself the smallest portion of the late Spanish possessions in America, she also felt satisfied that no attempt would be made by France to bring any of them under her dominion either by conquest or by cession from Spain."[17] Great Britain was apparently as anxious concerning the ultimate fate of the Spanish colonies as was the United States. In August of the same year Canning made his famous proposal to the American minister concerning a joint declaration looking towards recognition, to include a definite statement that neither nation aimed at the acquisition of any portion of the Spanish possessions itself, and that neither could see any portion of those possessions transferred to any other power with indifference. Although Adams was unwilling to join with Great Britain in exactly the manner specified, the message of President Monroe, dated December 2, 1823, showed clearly enough our stand in regard to European

[13] *Sen. Rep. No. 351*, 35th Cong., 2nd Sess., p. 1.
[14] Ford, *Jefferson*, X, 278.
[15] Callahan, *Cuba*, p. 125.
[16] *Memoirs of John Quincy Adams* (Philadelphia, 1874-1877), VI, 138.
[17] F. E. Chadwick, *The Relations of the United States and Spain, Diplomacy* (New York, 1909), p. 178.

intervention. Cuba and Puerto Rico were saved to Spain, but, by the same token, her other possessions were irretrievably lost.

That Cuba earned the title of the "ever-faithful island" for remaining loyal to King Ferdinand VII and not joining its sister colonies in the Latin American wars of independence was due to several factors, one being demographic. The first Cuban census, that of 1817, revealed a population of 630,000, comprising 291,000 whites, 224,000 slaves, and 115,000 mixed bloods; the whites were therefore a minority. The continuing rise of the slave population and the fear of a slave uprising, intensified by events in Haiti, helped to suppress an independence movement. Further, being an island, and the central base for Spain's military and naval operations against the mainland colonies, an independence movement would have faced enormous odds. It is also evident that the spirit of revolt was lacking. The creoles in Cuba were just becoming affluent, owing to Negro slavery and the sugar industry, and were less antagonistic toward the peninsular Spaniards than elsewhere in the colonies. Few Cubans were sufficiently educated or imbued with liberal ideas to have joined their counterparts in the mainland colonies. A number of conspiracies came to light, but most of them sought annexation by Colombia or the United States, not independence.[18]

External circumstances also continued to favor Cuba's retention by Spain. When, in 1825, Mexico and Colombia proposed a joint action against Cuba, the United States stood firm on its position of trying to save the island for Spain. Henry Clay gave Joel Poinsett, United States minister to Mexico, instructions indicating clearly that, while the United States was not looking to the acquisition of Cuba, yet if the island was to become a dependency of any one of the American states, its geographical position proclaimed that it should be attached to the United States.[19] Mr. Everett, the new minister to Spain, was also instructed to point out to the government at Madrid the danger of continuing a hopeless war against the revolted colonies if it desired to keep possession of Cuba and Puerto Rico[20] Clay even went so far as to instruct Mr. Middleton, our minister to Russia, to urge the Czar to use his influence with Ferdinand VII to the end that Spain might sacrifice her pride and make peace, thereby saving these valuable islands.

A large French fleet touched at Havana in August 1825, and caused much anxiety among the powers, which were particularly interested in maintaining the *status quo* in the West Indies. Canning wrote to the British minister at Paris that "as to Cuba you cannot too soon, or too amicably, of course, represent to Villèle the impossibility of our allowing France (or France us, I presume) to meddle in the internal affairs of that colony. We sincerely wish it to remain with the mother country. . . . The Americans (Yankees, I mean) think of this matter just as I do."[21] Clay wrote in a similar vein to Mr. Brown, the American minister at Paris, that the United States "could not consent to the occupation of those islands by any other European power than Spain under any contingency whatever."[22] Canning once

[18] Fagg, *Cuba*, pp. 128-129; Parry and Sherlock, *West Indies*, pp. 224-225.
[19] W. R. Manning, *Early Diplomatic Relations between the United States and Mexico* (Baltimore, 1916), p. 105.
[20] *Ibid.*, p. 108.
[21] E. J. Stapleton, *Official Correspondence of George Canning* (London, 1887), vol. I, p. 275.
[22] *Amer. State Papers, For. Rel.*, vol. V, p. 855.

more tried to inveigle the United States into cooperating with Europe by proposing a tripartite agreement between Great Britain, France, and the United States, disclaiming any intention of occupying Cuba. But Clay felt that the proposal might encourage Spain to continue her hopeless struggle.[23] Inasmuch as France also refused to participate, the matter was dropped.

It was during this same year (1825) that the question arose of the participation of the United States in the Congress of Panama. President Adams and Secretary Clay heartily supported the project, although the Senate was not so enthusiastic. Among other objections raised was the possibility that a discussion of the probable destiny of the islands of Cuba and Puerto Rico might be forced upon the United States. As long as the war between Spain and the colonies continued, the United States must preserve its independence of action; the mere participation in such a congress would interfere with the influence that the United States now possessed as an interested but impartial third party.[24] In fact, one of the objects of the Congress as indicated by the government of Colombia was: "To consider the conditions of the island of Puerto Rico and Cuba; the expediency of a combined force to free them from the Spanish yoke; and the proportion of troops which each state should contribute for that purpose; and to determine whether the islands shall be united to either of the confederated states or be left at liberty to choose their own government."[25]

The slavery question also entered prominently into the discussion, for if Cuba and Puerto Rico were freed, slavery would be abolished there, as it had been in the other liberated colonies, and such a prospect was unendurable to representatives from the southern states. In fact, during the next twenty-five years the American policy of maintaining the *status quo* in the West Indies, and of guaranteeing the sovereignty of Spain over the islands of Cuba and Puerto Rico, was based upon the fear of independence for the islands–including, as it would, the freeing of the slaves–almost as much as upon the fear of aggression on the part of Great Britain or France.

The annexation of Texas in 1845 increased the interest of the United States in Cuba, and in December of that year, a resolution was introduced in the Senate, authorizing the President to negotiate with Spain for the cession of Cuba to the United States. Early in the following year a similar resolution came up in the House. The war with Mexico and the resulting increase of territory to the United States kindled Polk's ardor for expansion, and Cuba offered an excellent outlet. In a long dispatch (June 17, 1848) to Mr. Saunders, our minister to Spain, Secretary Buchanan enumerated the manifest advantages that would accrue to the United States from the possession of Cuba: possession of a naval station at Havana would enable us to command the Gulf of Mexico; under American control the island would become exceedingly prosperous and serve as a most profitable market; it would increase the strength and security of the Union, and it would give the United States "a free trade on a more extended scale than any which the world had ever witnessed." Nor would the advantages accrue solely to the United States. Cuba,

[23] A. G. Stapleton, *Political Life of Canning* (London, 1831), vol. III, p. 154.
[24] See *Report of the Sen. Com. on For. Affairs.* This report, with other diplomatic documents may be found in the *Historical Appendix, International American Conference, Se. Ex. Doc. No. 232*, 51st Cong., 1st Sess., vol. IV, pp. 53 ff.
[25] *British and Foreign State Papers*, vol. XII, p. 894.

appreciating the advantage of annexation, was ready to rush into our arms. Spain must realize the distracted condition of the island and the danger of a successful revolution; else the island might be wrested from her by Great Britain. Under these circumstances, "the president has arrived at the conclusion that Spain might be willing to transfer the island to the United States for a fair and full consideration." The maximum price stipulated was $100 million.[26] Mr. Saunders was, however, given to understand that Spain, "sooner than see the island transferred to any power, would prefer to see it sunk in the ocean."

Failing to acquire the Pearl of the Antilles by peaceful means, certain ardent spirits in the United States were willing to compass it by a mode less justifiable. The discovery of gold in California showed the value of the isthmus in giving an almost all-water route to the west. But the ultimate possession of such a route made the possession of Cuba all the more necessary to protect it. The slave states were more than ever desirous of securing additional territory open to slavery. Therefore it is not surprising that when a Cuban patriot, Narciso López, attempted to recruit an expedition on American soil to free Cuba, he found much assistance in high quarters. A Venezuelan by birth, López had served in the Spanish army, had played a considerable part in Spanish politics, and had finally been made governor of Trinidad. Losing office through a turn of the political wheel, he engaged in business in Cuba, but with little success. In 1848 he attempted to stage a revolt near Cienfuegos, but the plot was disclosed and López was forced to flee the country. He went to New York and started the preparation of an expedition which should have the prestige of a great name at the head. Both Jefferson Davis and Robert E. Lee were approached, and when they refused López decided to lead the expedition himself. He had little difficulty in collecting the nucleus of a force, but a proclamation by President Taylor, issued August 11, 1849, warning all American citizens against participation in such enterprises, had a deterrent effect. The two vessels in which the expedition planned to leave New York were seized by the authorities, though the filibusters themselves were not held.

López' next expedition was in the beginning somewhat more successful. Some 750 men were collected, and in the spring of 1850 they succeeded in sailing from New Orleans in three vessels, making a landing at Cárdenas. Failing to receive the assistance expected from the natives, and faced with an openly mutinous crew, López was forced to return to Key West. Upon information lodged by a Spanish war-ship which had followed the expedition, the boat was seized and López was arrested. Sufficient evidence to convict the leader of violating the neutrality laws was not forthcoming.

A third (and last) expedition, which left New Orleans the following year, was even less fortunate, and its results were far more serious. A proclamation by President Fillmore on April 25, 1851, stating that persons who violated our neutrality laws not only would be subject to the penalties of our own laws, but would forfeit all claims to protection, had no effect.[27] The expedition, consisting of about 400 eager adventurers, sailed from New Orleans without clearance, August 3, 1851. A landing was made at Bahía Honda, whence López, with most of the troops, advanced to Las Pozas, where it was hoped an uprising would take

[26] Moore, *op. cit.*, vol. I, pp. 584-587.
[27] *Ibid.*, vol. III, p. 788.

place. Meanwhile, Colonel Crittenden, a former American army officer who had served creditably in the Mexican War, was left with a small force in command of the baggage. Attacked by an overwhelming force, Crittenden attempted to escape by sea, but, with fifty of his followers, he was captured by the Spanish and carried to Havana. Here the unfortunate filibusters were given a quick military trial and shot as pirates. López and his force withstood several serious attacks, but, obtaining no assistance from the natives, they were finally routed and dispersed. Some were shot on the spot, others were taken prisoners; López himself was publicly garroted. It was apparent enough that Cuba was not yet prepared to fight for independence. Nevertheless, the execution of Crittenden and his men aroused such a wave of hostility towards Spain and Spanish rule in Cuba that riots broke out in New Orleans and Key West, and throughout the United States the feeling was born that Cuba must be freed, even though it should take all the forces of the United States to accomplish it.[28]

Expeditions of this sort were looked upon with warm disapproval by Great Britain and France, as well as Spain, and there were rumors, in the autumn of 1851, that a treaty had been entered into between France, Spain, and Great Britain to guarantee Cuba to Spain.[29] On April 23, 1852, at the request of Spain, Great Britain and France again invited the United States to enter into a triple agreement disclaiming all intentions of obtaining possession of Cuba.[30] Webster replied that not only did the United States have no designs upon Cuba itself, but it was willing to assist Spain in preserving it. At the same time, the United States could not acquiesce in its cession to a European power.[31] In July, the matter was again brought forward by the British and French governments, but Webster's death intervened, and the duty of answering fell to Edward Everett, the new secretary of state. He replied in an able fashion. He pointed out that Cuba was mainly an American question. It commanded the approach to the Gulf of Mexico; it barred the entrance to the Mississippi; and it stood at the doorway to our intercourse with California by the isthmian route. The United States could not bind herself indefinitely as to her future relations with the island—"it would be as easy to throw a dam from Cape Florida to Cuba in the hope of stopping the flow of the Gulf Stream as to attempt by a compact like this to fix the fortunes of Cuba now and hereafter.[32] The invitation was respectfully declined.

Such an answer did not mean that the United States looked favorably upon filibustering expeditions. In fact, the note stated that the president had thrown the whole force of his constitutional power against all illegal attacks upon the island. But were there not justifiable means? The Pierce administration, which came into office in 1853, was expected to find and use them. It was "our manifest destiny to move on with the world of progress," and if Cuba impeded our march, acquisition presented no terrors. The appointment as our minister to Spain of Pierre Soulé—a man who had openly lauded López, and had urged the government not to delay

[28] For an authoritative and detailed account of the López expedition see Robert G. Caldwell, *The López Expedition to Cuba* (Princeton, N.J., 1915).

[29] Moore, *op. cit.*, vol. VI, p. 458.

[30] *Sen. Ex. Doc.*, No. 13, 32nd Cong., 2nd Sess., p. 7.

[31] *Ibid.*, p. 8.

[32] Moore, *op. cit.*, vol. VI, p. 469.

too long in plucking the Cuban fruit from the Spanish tree—showed that not even diplomatic amenities were to be preserved in our methods.

Hardly had Soulé arrived at his post before he was forced to demand redress for the unfriendly treatment accorded the *Black Warrior*, an American steamship calling at Havana. Spanish-American relations had been strained by the López filibustering expeditions, and by the hostile methods employed by the Spanish authorities of Cuba towards American merchant vessels. Then, in 1854, the entire cargo of cotton aboard the *Black Warrior* was confiscated in Havana on the flimsy excuse that the ship's manifest had the cotton entered as "ballast"—a technicality which not only accorded with the instructions of the customs collector, but also had met with no previous objection the thirty odd times the vessel had visited Havana. The Pierce administration authorized Soulé to demand $300,000 for damages done to the owners, and a prompt disavowal of the act. When several days passed without an answer, the hot-headed Soulé, not indisposed to provoke a war which would culminate in the American possession of Cuba, exceeded his earlier instructions by issuing a forty-eight hour ultimatum to the Spanish government. The Spanish foreign minister wisely ignored the American representative in his subsequent negotiations, and proceeded directly to Washington where he impressed the Pierce cabinet in his defense of Spain's position. With the helpful support of secretary of state, William L. Marcy, a direct settlement, totaling $53,000 for losses sustained, was made to the owners of the vessel by the Cuban authorities.[33]

In the United States the whole affair was closely connected with the question of slavery. The southern states were anxious to obtain Cuba and were willing to go so far, if necessary, as to make the incident a pretext for war with Spain. Although Pierce was not disposed to proceed to extremities, Secretary Marcy directed Soulé, in April 1854, to negotiate the purchase of Cuba for $130 million. Should this fail, Soulé was then to direct his efforts toward "detaching" Cuba from the Spanish dominion, presumably so that it would become eligible for annexation by the United States. As the best means of making such arrangements, Marcy suggested that Soulé confer with the American representatives in Paris and London, John Y. Mason and James Buchanan respectively.[34] The three ministers met at Ostend, Belgium and embodied the results of their deliberations in the famous "Ostend Manifesto" on October 18, 1854.

This document was not, as its name might suggest, an official ultimatum to Spain, but rather a confidential dispatch to the Secretary of State. In essence, it recommended that the purchase price for Cuba not exceed $120 million,

> But if Spain, dead to the voice of her own interests, and actuated by stubborn pride and a false sense of honor, should refuse to sell Cuba to the United States, . . . then by every law, human and divine, we shall be justified in wresting it from Spain, if we possess the power.[35]

When the contents of the memorandum leaked to the press, northern opinion, already antagonized by the efforts of the Pierce administration to ram the Kansas-Nebraska bill through Congress, was further inflamed by this next step on

[33] For the correspondence concerning the *Black Warrior* affair see *House Ex. Doc. No. 93*, 33rd Cong., 2nd Sess., vol. 10, pp. 30-120.
[34] *Ibid.*, p. 123.
[35] *Ibid.*, pp. 127-132.

behalf of slavery. Marcy accordingly rejected the proposals, however similar they were to his own earlier instructions. Thus repudiated by the administration, and chastized by an important segment of the press, Soulé resigned his position in disgust. Despite the fact that Buchanan, one of the participants in the "Ostend Manifesto," succeeded to the presidency, the slavery question was now too closely tied up with the Cuban problem to permit the island's purchase, even though Spain might have proved willing. "Manifest Destiny" was shackled by the growing opposition to slavery.

In the following decade, as Cuba became less important in American affairs, its relations with the mother country steadily deteriorated. It has been noted that the Cubans in general were loyal to their legitimate sovereign, Ferdinand VII, after Napoleon's forces had occupied Spain, and remained aloof from the wars of independence fought on the mainland. Then, inexplicably, instead of rewarding the "ever-faithful island" for its fidelity, the restored Bourbons issued a decree in 1825 conferring on the captain-general "todo el lleno de las facultades que por los reales ordenanzas se concedian a los governadores de plazas sitiadas." This meant that virtually absolute power was given to the ranking Spanish official, power over persons, property, and administration of the island. Few men could be entrusted with such despotic authority without abusing it, and the captains-general of Cuba were no exception. The decree of 1825 accelerated the growth of ill feeling between the Cuban creoles and the peninsular Spaniards resident on the island. The creoles resented the absolutism of the decree, whereas it was welcomed by the *peninsulares* as a means of subordinating the creoles and implementing policies advantageous to Spanish trade. The slave trade, for example, was controlled by peninsular Spaniards, and the creoles, fearful of the steadily rising Negro population, wanted it checked. Therefore, despite the fact that Cuba remained, after the Latin American wars of independence, almost the sole relic of Spain's once vast trans-Atlantic empire, the desire for independence was not completely dormant on the island. As mentioned, a number of revolutionary schemes were hatched between 1823 and 1830, but none succeeded.[36]

In the 1830s and early 1840s, as the sugar industry expanded, there was a commensurate growth of the slave population which gave the Cubans good cause for alarm. Although the treaty of 1817 with Great Britain had made the Cuban slave trade illegal, it was carried on by unscrupulous Spaniards until 1865. By 1840 the law entitled a slave to freedom if he could furnish proof of having been brought to Cuba after 1820. The failure of the proprietors to honor their obligation, and the slaves' awareness of emancipation in the British West Indies, accentuated the possibility of servile revolts. In 1843 a number of revolts occurred among disappointed and enraged slaves in the Matanzas area of Cuba. The insurrection was suppressed in a savage fashion, but the problem of slavery remained. To some Cubans annexation to the United States seemed attractive, hence the expectation of the López expedition, the failure of which has been described. It was not until 1868 that the first serious revolt on the island took place.[37]

[36] Parry and Sherlock, *West Indies*, pp. 226-227.
[37] *Ibid.*, p. 228.

The causes of the Ten Years' War, so named because it spanned roughly that period of time, are not hard to discover. Aside from the basic grievances mentioned, Spain, adhering to the time-worn and anachronistic principles of mercantilism, regarded the island merely as an exploitable source of revenue. In 1868 the revenue from the island approximated $26 million, and virtually none of it was used for Cuba's benefit. The public offices commanded good salaries, but all were filled by peninsular Spaniards. The corruption among officials was notorious, the "perquisites" in some cases reaching as high as 70 percent of the total receipts. Secondly, Spain monopolized not only Cuba's exports, but also its imports, and the duties were levied with the same severity on necessities as upon luxury items. Flour was taxed so heavily that wheat bread almost ceased to be an article of food for the common people. Finally, representation in the Cortes, which at one time had been granted, only to be withdrawn later, was felt to be essential to any permanent improvement in the government of the island.

When Isabella II was driven from the throne of Spain by the revolution of 1868, the Cubans seized the opportunity to declare their independence and to organize to maintain it. Carlos Manuel de Céspedes, a wealthy planter and an ardent patriot, raised a body of some 15,000 men, and at the outset the patriots were generally successful. A constitution was promulgated April 10, 1869, and the legislature, which met in accordance with its provisions, elected Céspedes president. But without assistance the patriots could not hope to withstand the well equipped troops that Spain continued to send against them; and finally, upon promises of a general amnesty, representation in the Spanish Cortes, and a few other reforms, peace was brought about by the treaty of El Zanjon, February 10, 1878.[38]

From the beginning, the United States took a keen interest in the war. On March 27 the captain-general of the island, Domingo Dolce, issued a proclamation declaring that any vessels carrying men, arms, or ammunition found in the waters near the island, whatever their destination, should be seized, and persons on board immediately executed.[39] The United States protested immediately and emphatically. On April 4 the Count of Valmaseda, in command at Bayamo, issued a proclamation to the effect that every man above fifteen found away from his home without a reason would be shot and that unoccupied habitations and those not floating a white flag would be burned.[40] Mr. Fish, secretary of state under President Grant, wrote the Spanish plenipotentiary: "In the interest of Christian civilization and common humanity, I hope that this document is a forgery. If it be indeed genuine, the President instructs me in the most forcible manner to protest against such mode of warfare."[41]

President Grant, who from the beginning of his term of office showed a decided friendliness to the Cubans, authorized General Sickles, the American minister to Spain, to tender the good offices of the United States to bring to a close the civil war ravaging the island. The bases suggested were: the independence of Cuba, an indemnity to Spain, the abolition of slavery in the

[38] Text may be found in *Sen. Doc. No. 79*, 45th Cong., 2nd Sess., p. 16.
[39] *Sen. Doc. No. 7*, 41st Cong., 2nd Sess., p. 12.
[40] *Ibid.*, p. 20. Bayamo itself was burned to the ground and its inhabitants dispersed or slaughtered.
[41] *Ibid.*, p. 21.

island, an armistice pending the negotiations for the settlement.[42] Although some of the leaders of the Spanish government were willing to accept, public opinion opposed and the proffer was withdrawn.

The affair, however, which most aroused the United States against Spain, bringing the two nations to the brink of war, was the seizure of the steamer *Virginius* and the execution of her captain and crew. The *Virginius* was a merchant-vessel sailing under the American flag and registered in New York as an American-owned vessel. However, her cargoes consisted principally of contraband of war destined for Cuba; and for several years she had been successful in landing them, despite the vigilance of the Spanish cruisers. On October 23, 1873, she cleared from Kingston, Jamaica, for Puerto Limón, Costa Rica, though her actual intention was to land men and arms in Cuba. On October 31 she was captured on the high seas by a Spanish cruiser and taken into Santiago de Cuba. The Spanish commandant, General Burriel, summoned a court-martial, and within a week's time fifty-three of the passengers and crew were summarily condemned and shot, despite strong protests upon the part of the American and British consuls. Such a performance was not merely "a dreadful, a savage act," but it was directly contrary to international law, because the vessel was a neutral lawfully provided with papers; even if she were engaged in blockade-running or in carrying contraband, the maximum penalty should have been confiscation of the ship and cargo.

General Sickles, the American minister at Madrid, was instructed to demand the restoration of the *Virginius* with the survivors, a salute to the flag of the United States, and punishment of the guilty officials. At first Spain seemed inclined to support the United States in its stand, but later Castilian pride came to the front, and a curt and most unsatisfactory reply was returned. Spain could not consent to be thus addressed by the representatives of a foreign nation, and if reparations were to be made it would be only after a thorough investigation had been carried out by her representatives.[43] Recriminations became mutual, and it was only after General Sickles had asked for his passports that Spain agreed to consider the American demands. Owing either to the uncompromising attitude of General Sickles, or to the Spanish government's fear of a popular uprising, the negotiations were transferred to Washington. A protocol was finally reached, whereby Spain agreed to restore the *Virginius*, together with the surviving passengers and crew, and to salute the flag of the United States, unless she could prove before December 25, 1873, that the *Virginius* was not entitled to carry the American flag. The Madrid government actually succeeded in furnishing proof that the vessel was both owned and controlled by Cubans; therefore the salute was waived. The vessel and survivors were turned over to the United States. But on its way north, under convoy, the ship foundered off Cape Hatteras. Spain also admitted the illegality of the capture and the summary execution of the crew, and finally paid an indemnity of $80,000 to the families of those executed. But the Spanish commandant guilty of the execution not only escaped punishment, but was later promoted to a higher grade.[44]

[42] *House Ex. Doc. No. 160*, 41st Cong., 2nd Sess., pp. 13-16.
[43] The complete diplomatic correspondence regarding this incident may be found in the *For. Rel. of the U.S., 1874*, pp. 922-1117; *Ibid.*, 1875, Part II, pp. 1144-1256.
[44] For the trial of General Burriel see *Ibid.*, 1876, pp. 486-535.

Meanwhile the Cuban insurrection dragged along, and the United States continued to look for means of putting an end to the devastating struggle. The whole situation, as far as the United States was concerned, was summed up in a long, carefully worded dispatch from Secretary Fish to Caleb Cushing, the new American representative at Madrid, on November 5, 1875. In this state paper Mr. Fish enumerated the cases of arbitrary seizure, confiscation of American property, and arrest and execution of American citizens without trial, and protested vigorously against the continuance of a struggle on the very borders of the United States, disturbing to its tranquillity and commerce and conducted in a most barbarous fashion. "It will be apparent that such a state of things cannot continue. . . . In the opinion of the president, the time has arrived when the interests of this country, the preservation of its commerce, and the instincts of humanity alike demand that some speedy and satisfactory ending be made of the strife that is devastating Cuba. . . . The president hopes that Spain may spontaneously adopt measures looking to a reconciliation and to the speedy restoration of peace and the organization of a stable and satisfactory system of government in the island of Cuba. In the absence of any prospect of a termination of war or of any change in the manner in which it has been conducted on either side, he feels that the time is at hand when it may be the duty of other governments to intervene, solely with a view of bringing to an end a disastrous conflict, and of restoring peace in the island of Cuba."[45]

A copy of this note was sent to the American representatives in Europe, and when its contents became known it provoked considerable discussion, which was increased by President Grant's message of December 7, 1875. Great Britain, whose cooperation was particularly desired, decided, in the words of Lord Derby, that "if nothing were contemplated beyond an amicable interposition having peace for its object, the time was ill chosen and the move premature."[46] The other powers were no more willing to intervene than Great Britain, particularly at a time when the young Alfonso was struggling manfully against the revolution of Don Carlos. Owing to the superlative ability of Mr. Cushing and the high esteem in which he was held, Mr. Fish's communication was received by Spain in the friendly spirit in which it was sent, and in his reply Señor Calderón, the Spanish foreign minister, emphasized the recent satisfactory progress in putting down the revolution, and promised the abolition of slavery and the introduction of administrative reforms leading to representation of the inhabitants in the Spanish *cortes.* In conclusion, he asked for a frank statement concerning the precise things which the United States would wish Spain to do.[47]

After emphatically disclaiming any intention on the part of the government of the United States to annex Cuba, Mr. Fish summed up the president's desires under four heads: first, mutual and reciprocal observance of the treaty obligations, with a friendly interpretation of the doubtful provisions; second, the establishment of peace, order, and a liberal government in Cuba; third, gradual but effectual emancipation of the slaves; fourth, improvement of commercial

[45] *Report of Senate Committee on Foreign Relations Relative to Affairs in Cuba, No. 885,* 55th Cong., 2nd Sess., pp. 44-52.
[46] *Ibid.*, p. 162.
[47] *Ibid.*, pp. 96-99.

facilities and the removal of the obstructions now existing in the way of trade and commerce.[48] Spain's acceptance of these proposals eliminated, for the time being, any further discussion of intervention. However, the Madrid government urged that before there could be any hope of improvement the revolution must be put down, and a new general and additional troops were forthwith dispatched to the island. In October 1877, the Ten Years' War came to an end. The terms of peace gave promise of a real improvement in the situation of the Cubans. Yet the habits of centuries cannot be broken up by the good intentions of a day. The Spanish system of exploitation had become part and parcel of the island's administration, and Spanish officials were still employed to look out for Spanish interests. But two results of the revolution were manifest to the world: the seed of independence had taken firm root in the island of Cuba, and the United States could not remain an impartial witness to any attempts to prevent its growth.

As a result of the Ten Years' War, there was some improvement in the attitude of Spain towards Cuba, but little change in the actual conditions. The island was granted representation in the Spanish Cortes, but, owing to the limited suffrage, the majority of the delegates represented the *peninsulares,* or Spanish element. Even when, in 1892, the tax qualification for the suffrage was reduced from twenty-five dollars to five dollars, the Spanish authorities who controlled the elections saw to it that the results were virtually the same. Seventeen years of comparative peace followed; yet it was the apathy of exhaustion rather than the tranquillity of satisfied hopes. In fact, the promises held out by the Spanish government were never realized. The sole noteworthy reform was the gradual abolition of slavery. In summing up the situation, Estrada Palma declared that the parliamentary representation was illusory, all officials of the island were Spaniards, taxes were levied upon everything conceivable, and about 95 percent of the amount collected was devoted to the maintenance of the army and navy in Cuba, to the interest on the public debt, and to the salaries of the Spanish officeholders. The Cubans had no security of person or property; nor was there freedom of speech, press, or religion.[49]

These were the underlying causes of the outbreak of the revolution in 1895. The passage of the Abarzuza law in February, 1895, by the Spanish *cortes,* creating a farcical council of administration, seemed to bring matters to a head.[50] The economic crisis engendered by the termination of reciprocity relations with the United States in 1894, thus closing Cuba's principal market for sugar, undoubtedly aided materially in strengthening the discontent. After the tragic death of the patriot, José Martí, in the beginning of the insurrection, the revolutionists were under the general command of Máximo Gómez, and they were financially supported by the juntas organized in the United States. The revolution spread rapidly, and from the outset the insurgents ruthlessly destroyed all property and plantations that might be useful to the enemy. Spain spared no effort to check the revolution, and when Governor-General Campos

[48] *Ibid.*, pp. 102-106.
[49] *Sen. Rep. No. 885*, 55th Cong., 2nd Sess., pp. 1ff.
[50] Hannis Taylor, "A Review of the Cuban Question," *North Amer. Rev.*, vol. CLXV (Nov., 1897), pp. 610-635.

failed to put down the insurgents, General Weyler was placed in command. The struggle on the Cubans' part soon degenerated into guerrilla warfare. They gave notice that they would destroy all the resources of the island rather than surrender. General Weyler retaliated by forcing all the inhabitants of certain provinces to concentrate in the towns held by the Spanish troops or be considered rebels.[51] Great numbers of noncombatants were thus brought together, and, as no provision for feeding and housing them had been made, the brutal nature of the decree was soon apparent. The innocent women and children were forced to starve in the towns, while the able-bodied men remained at large and joined revolutionary bands.

On June 12, 1895, President Cleveland issued a proclamation recognizing that a state of insurgency existed in Cuba, and insisted upon the maintenance of American neutrality.[52] However, as the horrors of the struggle increased and the American press took up the campaign in Cuba's behalf, Congress was also drawn into the affair, and early in 1896 a resolution passed both houses urging that the president recognize Cuban belligerency. This, Cleveland refused to do. But public opinion steadily grew more hostile toward Spain; and the destruction of American property on the island, together with the great falling off of American trade, added fuel to the flames. Although the president refused to recognize Cuban belligerency, he authorized secretary of state, Richard Olney, to offer the Spanish ambassador the mediating services of the United States. "The United States," it was specifically asserted, "has no designs upon Cuba and no design against the sovereignty of Spain."[53] The Spanish reply was a courteous refusal of the offer.

Meanwhile, most American newspapers took up the cause of Cuban independence. Few, however, were able to approach the extremes of jingoism reflected in New York's "yellow journals"–Joseph Pulitzer's *World* and William Randolph Hearst's *Journal.* Both papers, engaged in a circulation war, colored the news in favor of the insurgents, playing on the traditional American sympathy for the underdog fighting for his liberty against European despotism. Examples of Spanish cruelty, which the *Journal* compared to the "Spanish Inquisition of the sixteenth century," were described in purple prose and lurid illustrations. The reports of Fitzhugh Lee, American consul-general at Havana, and of Senator Redfield Proctor, who made a personal investigation, also helped to reinforce the public's martial spirit.

Hardly had the McKinley administration assumed office before it was compelled to take action; for the situation was growing steadily worse. In May 1897, the President asked Congress to appropriate $50,000 for the relief of destitute Americans in Cuba, and in December he issued a public appeal for funds to aid in combating the Cuban famine. A slight gleam of hope for better conditions appeared when the new Sagasta ministry recalled General Weyler and promised Cuba a new constitution, with a local parliament and a fair share of autonomy. In his annual message of December 1897, McKinley noted the new policy of Spain, and declared that it was honestly due to Spain that she should

[51] Text of decree in *Sen. Rep. No. 885*, p. 549.
[52] *For. Rel. of the U.S., 1895*, p. 1195
[53] *Ibid.*, p. 540.

be given a reasonable chance to realize her expectations. Unfortunately for Spain, her policy of reform was given no opportunity to materialize. General Blanco could not undo the terrible results of the reconcentration policy of General Weyler, although he made very earnest efforts to do so. Not only were the insurgents wholly averse to acceptance of semi-autonomy in the place of independence, but the Spanish party in Cuba was equally indignant at the liberality of the new program. Rioting against the newspapers backing autonomy became so serious that on January 13, 1898, the American consul-general, Mr. Lee, telegraphed that he was uncertain whether Blanco could control the situation, and that if it should be demonstrated that Americans were in danger, ships should be promptly sent.[54] The next day all was quiet; but the idea of sending a war-ship struck root, and on January 24 the president ordered the *Maine* sent to Havana on a "friendly visit." The American consul-general realized that the times were not propitious for a friendly visit and advised that the step be postponed six or seven days. But the battleship was already on its way, and on January 25 it anchored in Havana harbor.

The struggle between the Cuban insurgents and the Spaniards was soon to be merged in a direct conflict between Spain and the United States. On February 9, 1898, a New York newspaper published a private letter written by Señor Dupuy de Lome, the Spanish minister, to a friend in Cuba, in which he characterized President McKinley as a "weak . . . would-be politician who tries to leave a door open behind himself while keeping on good terms with the jingoes of his party."[55] Although it was a questionable piece of journalism to publish a letter of this character that had been purloined from the mails, the Spanish minister showed an inexcusable lack of diplomatic discretion in expressing his opinions so frankly. The United States immediately demanded his recall—although not until after the discredited diplomat, realizing that his position had become untenable at Washington, had sent in his resignation. But public opinion, already aroused against Spain through her dilatory attitude in ameliorating the condition of Cuba, became increasingly hostile. And, as if other circumstances were not sufficiently irritating in the relations between the two countries, on the night of February 15 the *Maine* was blown up at its anchorage in Havana harbor, with the loss of two hundred and sixty men, including two officers. A wave of indignation swept over the United States, and the demand for war was virtually unanimous. The few who counseled delay until a court of inquiry could fix the blame were scarcely heard in the popular clamor. Fortunately, democratic governments are so constituted that they function slowly; and the utter unpreparedness of the United States for war was an added incentive to move with deliberation.

An American court of inquiry, consisting of hardly objective United States naval officers concluded after twenty-three days that the *Maine* was destroyed by a submarine mine. No evidence could be obtained to fix the responsibility on any person. A Spanish board of inquiry, making a separate investigation, reported that the explosion was due to internal causes.[56] Although the real

[54] *Ibid., 1898*, p. 1025. See also Gerald G. Eggert, "Our Man in Havana: Fitzhugh Lee," *Hispanic American Historical Review*, XLVII (Nov., 1967), pp. 463-485.
[55] For the full content of the letter see Moore, *Digest Int. Law*, vol. VI, pp. 176-177.
[56] For American report see *Sen. Doc. No. 207*, 55th Cong., 2nd Sess. The Spanish report is in *Sen. Rep. No. 885*, 55th Cong., 2nd Sess., pp. 566ff.

cause will probably never be known,[57] the least likely culprit was the Madrid government which was desperately trying to avert war.

The sinking of the *Maine* aroused the American martial spirit as no other single incident. The jingoistic press and such warmongers in the administration as the Assistant Secretary of Navy, Theodore Roosevelt, were least prepared to permit a diplomatic settlement. Their mood smothered the voices of conciliation represented by the American investors of $50 million in Cuba, and commercial groups engaged in the $100 million annual trade with the island. President McKinley was the willing servant of these business interests, and he did make a determined effort to secure peace. On March 27, 1898, the United States minister at Madrid, General Woodford, was instructed to inquire whether Spain would consent to the following terms: (1) an armistice for six months, during which negotiations for peace between Spain and the insurgents might be undertaken through the friendly offices of the United States; (2) immediate revocation of the *reconcentrado* order; (3) agreement that, if peace were not arranged by October 1, the president of the United States should be accepted as final arbitrator between Spain and the insurgents.[58] The report on the *Maine* made to Congress, March 28, was bound to bring speedy action. Therefore, if the President was to succeed in obtaining a peaceful settlement, Spain must give him an immediate assurance that his offer was accepted.

Madrid's reply, on March 31, was not satisfactory, in that, instead of granting an immediate armistice, its offer conditioned a truce upon the insurgents asking for it, a proposal that they were most unlikely to make. However, on April 3, through the intervention of the Pope, the Queen conceded an unconditional suspension of hostilities for six months. Within the next week, the Spanish government further instructed the Governor-General of Cuba to revoke reconcentration, and the commander of the army was directed to grant an armistice for such a time as he felt prudent in order to facilitate peace.

In his message delivered to Congress on April 11, 1898, President McKinley capitulated to the war mood of the public. In a formidable indictment, he declared that only forcible intervention was left, and he justified this course of action upon four grounds: first, in the cause of humanity, i.e., it would put an end to the bloodshed and misery at our door; second, it would protect American citizens and their property in Cuba; third, it would put an end to the wanton destruction and devastation of the island; fourth, and most important, "the present condition of affairs in Cuba is a constant menace to our peace." The destruction of the *Maine* was mentioned merely as impressive proof of the intolerable state of things in the island.[59]

Congress was only too willing to grant the president the powers that he asked for. Though there were stormy debates in both houses, a joint resolution was passed, recognizing the independence of Cuba, demanding the immediate withdrawal of the Spanish forces from the island, granting the president power

[57] A later investigation upon the raising of the *Maine* in 1911 strengthened the view that an outside explosion caused the sinking. See *Report of the U.S. Naval Board in 1911, House Doc. No. 310*, 63rd Cong., 2nd Sess.
[58] *For. Rel. of the U.S., 1898*, p. 750.
[59] J. D. Richardson, *Messages and Papers of the Presidents,* vol. X, p. 47; or *For. Rel. of the U.S., 1898*, p. 750.

to use the entire land and naval forces to carry the resolution into effect, and, finally, disclaiming any intention on the part of the United States to exercise sovereignty over the island except to establish peace, and, that accomplished, to leave the government and control of the island to its people. (Teller Amendment.)[60]

As soon as the president signed the resolution, the Spanish minister at Washington asked for his passport, and his action was approved by the Spanish foreign minister. On April 25, 1898, Congress passed the declaration of war, dating the outbreak as of April 21. Admiral Sampson had already received orders to blockade Cuba with the South Atlantic Squadron, and Commodore Dewey had left Hong Kong to engage the Spanish squadron at Manila. Foreseeing the outcome, the Navy Department had more than a month earlier ordered the battle-ship *Oregon*, lying at Puget Sound, to join the Atlantic Squadron. Her 15,000-mile journey around Cape Horn dramatically demonstrated to the United States the need strategically of an isthmian canal.

Both sides realized that the issue lay upon the sea; and it was fortunate for the United States that such was the case. But there is no necessity to tell the story of the conflict here. Dewey's overwhelming victory at Manila Bay, followed by the complete destruction of Cervera's fleet at Santiago, virtually sealed the defeat of Spain. The mechanistic era of warfare had come, and America had made the greater progress. "Spanish dominion in America, in which there had been much both of glory and shame, with splendid episodes of heroic endeavor, noble self-abnegation, and great attainment, was to end in the final sacrifice, nobly met, on the sea which through generations witnessed so many conflicts of the two races."[61]

On July 22, 1898, through M. Jules Cambon, the French ambassador at Washington, Spain asked of the president upon what basis the conflict could be ended and a satisfactory political status in Cuba established. President McKinley outlined his terms, which were later incorporated in the protocol of August 12. In brief, the agreement provided: (1) the relinquishment by Spain of all sovereignty over Cuba; (2) the cession of Puerto Rico and other Spanish West Indies, together with an island in the Ladrones, to the United States; (3) the occupation by the United States of Manila pending the signing of a treaty of peace which should determine the disposition of the Philippines; (4) the immediate evacuation by Spain of Cuba, Puerto Rico, and the other islands of the West Indies under her sovereignty; (5) the appointment by Spain and the United States of not more than five commissioners each, to meet in Paris not later than October 1, 1898, to negotiate a treaty of peace; (6) the suspension of hostilities upon the signing of the protocol.[62]

In the peace negotiations which followed, Spain made every effort to turn over the Cuban debt, either to the Cubans or to the United States, along with the sovereignty of the island. But the American commissioners stood fast against the proposal on the ground that the debt was created by Spain for its own purposes and through its own agents and not for the benefit of Cuba. The Treaty

[60] *U.S. Stat. at Large*, vol. XXX, p. 738.
[61] Chadwick, *United States and Spain*, p. 587.
[62] *For. Rel. of the U.S., 1898*, p. 828.

of Paris, signed December 10, 1898, was almost identical with the terms of the protocol, except for the clauses regarding the Philippines.[63] The United States at last had the opportunity to pacify Cuba and then withdraw as it had promised—a quixotic proceeding highly incredible to a skeptical world.

[63] *Ibid.*, p. 831.

SUPPLEMENTARY READINGS

Atkins, Edwin F. *Sixty Years in Cuba*. Cambridge, Mass., 1926.

Brooke, John R. *Final Report of Major General John R. Brooke*. Havana, 1899.

Cabrera, R. *Cuba and the Cubans*. Philadelphia, 1896.

Caldwell, Robert G. *The López Expeditions to Cuba, 1848-1851*. Princeton, 1915.

Calcott, Wilfrid H. *The Caribbean Policy of the United States, 1890-1920*. Baltimore, 1942.

Callahan, J. M. *Cuba and International Relations*. Baltimore, 1899.

Canini, I. E. *Four Centuries of Spanish Rule in Cuba*. Chicago, 1898.

Chadwick, F. E. *The Relations of the United States and Spain, Diplomacy*. New York, 1909.

Draper, A. S. *The Rescue of Cuba*. Boston, 1910.

Ettinger, A. A. *Mission to Spain of Pierre Soulé 1853-1855*. New Haven, 1932.

Flack, H. E. *Spanish American Diplomatic Relations Preceding the War of 1898*. Baltimore, 1906.

Freidel, Frank. *The Splendid Little War*. Boston, 1958.

Friedlander, H. E. *Historia económica de Cuba*. Havana, 1944.

Foner, Philip S. *A History of Cuba and its Relations with the United States*. 2 vols. New York, 1962-1963.

Gray, Richard B. *José Martí Cuban Patriot*. Gainesville, 1962.

Halstead, Murat. *The Story of Cuba*. New York, 1896.

Healy, David F. *The United States in Cuba, 1898-1902*. Madison, 1963.

Hill, R. T. *Cuba and Porto Rico*. New York, 1899.

Humboldt, Alexander von. *The Island of Cuba*. New York, 1856.

Infiesta, Ramón. *Historia Constitucional de Cuba*. Havana, 1942.

Johnson, Willis F. *The History of Cuba*. 5 vols. New York, 1920.

Kimball, R. H. *Cuba and the Cubans*. New York, 1880.

Lee, Fitzhugh. *Cuba's Struggle Against Spain*. New York, 1899.

Mahan, Alfred T. *Lessons of the War with Spain*. Boston, 1899.

Márquez-Sterling, Carlos. *Historia de Cuba desde Colón hasta Castro*. New York, 1963.

Millis, Walter. *The Martial Spirit*. Boston, 1931.

Morris, Charles. *Our Island Empire*. Philadelphia, 1899.

Ortiz, Fernando. *Cuban Counterpoint: Tobacco and Sugar*. New York, 1947.

Pirala y Criado, Antonio. *Anales de la guerra de Cuba*. 3 vols. Madrid, 1895-1898.

Portell Vila, Herminio. *Historia de Cuba en sus relaciones con los Estados Unidos y España*. 4 vols. Havana, 1938-1941.

Portunondo del Prado, Fernando. *Historia de Cuba*. 5th ed. Havana, 1953.

Pratt, J. W. *Expansionists of 1898*. Baltimore, 1936.

Rauch, Basil. *American Interests in Cuba, 1848-1855*. New York, 1948.

Robinson, A. G. *Cuba and the Intervention*. New York, 1905.

Rubens, Horatio. *Liberty: The Story of Cuba*. New York, 1932.

Santovenia, Emeterio S. *Historia de Cuba*. 2 vols. Havana, 1939-1953.

Smith, Robert F. ed. *What Happened in Cuba? A Documentary History*. New York, 1963.

Varona, Enrique J. *De la colonia a la república*. Havana, 1919.

Verrill, A. H. *Cuba, Past and Present*. New York, 1914.

Wood, Leonard. *Annual Reports*. 33 vols. Washington, 1900-1902.

Wright, Irene A. *The Early History of Cuba, 1492-1486*. New York, 1916.

Zaragoza, Z. *Las insurreciones en Cuba*. 2 vols. Madrid, 1872.

10

Cuba: Republican and Communist

The official transfer of the island took place January 1, 1899, and thereupon the trusteeship of the United States began. The task assumed was truly stupendous. Disease and starvation were prevalent; civil government had disappeared; even the public buildings were unfit for occupancy. Major-General John R. Brooke, the first military governor, began a general program of rehabilitation, including the distribution of more than 5,000,000 rations, the supplying of medicine, the reorganization of civil government, particularly in the cities—all with the avowed purpose, as announced in a proclamation of January 1, 1899, "to give protection to the people, security to persons and property, to restore confidence, to build up waste plantations, to resume commercial traffic, and to afford full protection in the exercise of all civil and religious rights."[1] Major-General Leonard Wood took over the work in 1899, and in his three years as governor-general made a record of administrative efficiency which still sheds luster upon his name.[2] Undoubtedly his methods were at times autocratic, and necessarily he made some enemies; but when, on May 20, 1902, the government was transferred to the duly elected president and congress of Cuba, with a system of civil government established and successfully functioning, with a public-school system completely reorganized, with sanitary conditions thoroughly regulated and controlled, and with the plague of yellow fever virtually stamped out, well might President Estrada Palma declare to the retiring governor-general, in accepting the transfer: "I take this solemn occasion, which marks the fulfilment of the honored promise of the government and people of the United States in regard to the island of Cuba, and in which our country is made a ruling nation, to express to you, the worthy representative of that grand people, the immense gratitude

[1] *House Doc. No. 2,* 56th Cong., 1st Sess., p.7.
[2] Even as harsh a critic as Carleton Beals concedes that "our first intervention in Cuba stands as a model of fine trusteeship," *The Crime of Cuba* (New York, 1934), p. 172. Elihu Root declared that Wood had done "one of the most conspicuous pieces of work ever done by an American." Philip C. Jessup, *Elihu Root* (New York, 1938), vol. 1, p. 287.

which the people feel towards the American nation, towards its illustrious President, Theodore Roosevelt, and towards you for the efforts you have put forth for the successful accomplishment of such an ideal."[3]

From the outset of the intervention, the United States had not forgotten that by the terms of the joint resolution of April 20, 1898, the government and control of the island were to be left to its people as soon as pacification was accomplished. As an evidence of good faith, on July 25, 1900, the military governor ordered that a general election be held on September 15 to put into effect the following program: (1) to elect delegates to a convention to frame and adopt a constitution; (2) as a part thereof, to provide for and agree with the government of the United States upon the relations to exist between Cuba and the United States; and (3) to provide for the election by the people of officers under this constitution and the transfer of the government to these officers.[4] As an indication of what the United States considered essential under the second heading, on February 9, 1901, while the convention was in session, instructions were sent to Major-General Wood by the secretary of war, Mr. Root, outlining the following provisions: (1) no government organized under the constitution should make any treaty impairing the independence of Cuba or grant any right to any foreign power without the consent of the United States; (2) no such government should contract any debt in excess of the capacity of the ordinary revenues to pay the interest; (3) the government should consent that the United States reserve the right of intervention in order to preserve independence and a stable government; (4) the acts of the military government should be validated and maintained; and (5) the United States should be given the right to acquire and maintain a naval station.[5]

While appreciating the decisive help rendered by the United States, and grateful for it, the delegates to the convention would not concede the right of the United States to impose conditions clearly violating that independence which it had guaranteed to maintain. They realized that the United States had some claims to special consideration, and they made counter-proposals to this effect.[6] But undoubtedly they also feared too close a relationship with a powerful neighbor whose appetite had just been whetted by conquest. *L'appêtit vient en mangeant,* and Cuba was a particularly toothsome morsel. Therefore the convention drew up a constitution which completely ignored the question of relations between the two countries. However, Senator Platt, chairman of the Senate Committee on Relations with Cuba, while conceding that Cuba was privileged to establish her own government without let or hindrance, held that the United States, by virtue of its intervention, had certain rights in the island which ought to be safeguarded. The result was the Platt Amendment to the Army Appropriation Bill, which passed both houses and received the president's signature.[7] The terms of this amendment became such an important factor

[3] *House Doc. No. 2,* 57th Cong., 2nd Sess., p. 124.
[4] For text of the order see *For. Rel. of the U. S.,* 1902, p. 358.
[5] *House Doc. No. 2,* 57th Cong., 1st Sess., pp.43-47.
[6] *For. Rel. of the U. S.,* 1902, p. 360.
[7] Secretary of War Root prepared the original draft; see Jessup, *op. cit.*, vol. 1, p. 310.

in the relations between the two countries that the most salient clauses deserve quotation in full:

(1) That the Government of Cuba shall never enter into any treaty or other compact with any foreign Power or Powers which will impair or tend to impair the independence of Cuba, nor in any manner authorize or permit any foreign Power or Powers to obtain by colonization or for military or naval purposes, or otherwise, lodgment in or control over any portion of said Island.

(2) That said Government shall not assume or contract any public debt to pay the interest upon which, and to make reasonable sinking-fund provision for discharge of which, the ordinary revenues of the Island, after defraying the current expenses of the Government, shall be inadequate.

(3) That the Government of Cuba consents that the United States may exercise the right to intervene for the preservation of Cuban independence, the maintenance of a government adequate for the protection of life, property, and individual liberty, and for discharging the obligations with respect to Cuba imposed by the Treaty of Paris on the United States, now to be assumed and undertaken by the Government of Cuba.

(4) That all acts of the United States in Cuba during its military occupation thereof are ratified and validated, and all lawful right acquired thereunder shall be maintained and protected.

(5) That the Government of Cuba will execute, and as far as necessary extend, the plans already devised or other plans to be mutually agreed upon, for the sanitation of the cities of the Island to the end that a recurrence of epidemic and infectious diseases may be prevented, thereby assuring protection to the people and commerce of Cuba, as well as to the commerce of the Southern ports of the United States and the people residing therein.

(6) That the Isle of Pines shall be omitted from the proposed constitutional boundaries of Cuba, the title thereto left to future adjustments by treaty.

(7) That to enable the United States to maintain the independence of Cuba, and to protect the people thereof, as well as for its own defense, the Government of Cuba will sell or lease to the United States lands necessary for coaling or naval stations at certain specified points, to be agreed upon with the President of the United States.

(8) That by way of further assurance the Government of Cuba will embody the foregoing provisions in a permanent treaty with the United States.[8]

Needless to say, this benevolent protectorate thrust in so unceremonious a fashion upon the Cubans was welcomed neither by the convention nor by the people. A delegation of five members of the convention was sent to Washington to protest. But, although they were received in a most friendly fashion by President McKinley and Secretary Root, they were given to understand that the Platt Amendment could not be modified. After considerable discussion, the convention finally accepted the Platt Amendment on June 12, 1901, and it became an appendix to the constitution.[9] The assurance of Secretary Root that intervention was not synonymous with intermeddling or interference with the Cuban government, but a formal action based upon just grounds for the

[8] *U. S. Stat. at Large,* vol. XXXI, p. 897, or *House Doc. No. 2,* 57th Cong., 1st Sess., p. 47. The treaty embodying these provisions was ratified July 1, 1904; see *For. Rel. of the U. S.,* 1940, p. 243. Naval stations were at first leased at Bahía Honda and Guantánamo, but since 1912 only the latter has been retained. Political pressure exerted by American landowners on the Isle of Pines delayed until 1925 the ratification of the Hay-Quesada Treaty of 1904 relinquishing all American claim to the Isle.

[9] For a full treatment of the Cuban attitude during this period see A. G. Robinson, *Cuba and the Intervention* (New York, 1905), pp. 207-277.

preservation of Cuban independence or the maintenance of an adequate government, gave the Cubans somewhat more confidence in the attitude of this country.[10]

As provided for by the constitution, the elections were held in December of the same year. Tomás Estrada Palma was elected president, and on May 20, 1902, Governor-General Wood turned over the government of the island to him. With a cash balance of more than $600,000 in her treasury, with her independence guaranteed, and assured of the good wishes of the government and the people of the United States, Cuba was ushered into the family of nations.[11]

The United States had given the insular republic an excellent start, but there remained the problem of the country's economic rehabilitation. This was dependent principally upon the immediate recovery of two industries, sugar and tobacco, and in the case of sugar it was vitally necessary that the United States make a reduction in her tariff. General Wood, Secretary Root, and President Roosevelt recognized the need, and the President, in his message to Congress, December 3, 1901, declared that "in the case of Cuba, however, there are weighty reasons of morality and national interest why the policy [of reciprocity] should be held to have a peculiar application, and I must earnestly ask your attention to the wisdom, indeed to the vital need, of providing for a substantial reduction in the tariff duties on Cuban imports into the United States."[12] Congress, however, seemed more inclined to listen to the selfish arguments of the beet-sugar growers' lobby, and not until two years later did President Roosevelt force his reciprocity measure through. This commercial convention, proclaimed December 17, 1903, gave Cuba the advantage of a 20 percent reduction on sugar, and on other products of the soil imported into the United States, over the tariff act of 1897 or any tariff law that might subsequently be enacted.[13] The advantage to Cuba was speedily shown. The trade of the United States with Cuba during the fiscal year 1905 (the first full year under the reciprocity treaty) showed an increase of approximately $10,000,000 worth of Cuban imports into the United States, namely, $86,304,259 in 1905 as against $76,983,418 in 1904.[14] The percentage of increase in our exports to Cuba was even greater, showing that fair commercial treatment of our island neighbor was a very profitable investment.

These early figures, however, gave no indication of the tremendous increase that was to come when Cuba should be completely rehabilitated. In 1910 the value of our imports from Cuba was $122,528,037, and our exports were valued at $52,858,758.[15] In 1920 our imports from Cuba amounted to the remarkable figure of $721,693,880, while our exports had increased to $515,208,731–a total trade value of approximately $1,250,000,000, or almost one-tenth of our total world trade.[16] These figures were abnormal, however,

[10] *House Doc., No. 2,* 57th Cong., 2nd Sess., p. 48; text of the Cuban constitution, *Ibid.*, p. 102.
[11] For an account of the turning over of the government see *House Doc. No. 2,* 57th Cong., 2nd Sess., pp. 69 *et seq.*
[12] *Ibid.*, 1901, p. xxxi.
[13] *Ibid.*, 1903, p. 375.
[14] *Foreign Commerce and Navigation of the United States,* 1905, p. 33.
[15] *Ibid.*, 1910, p. 30.
[16] *Ibid.*, 1920, p. x.

owing to the excessive cost of sugar and the huge crop marketed, but the figures of 1925 show Cuba to have had the largest trade with the United States of all the Latin American countries. In 1925 Cuba imported from the United States goods to the value of $185,617,496, while her exports to the United States amounted to $262,613,978 making a total value of $448,231,474, which was almost 5 percent of our total world trade.[17]

The political relations between the two governments remained uneventful until 1906. The election of 1905 produced a bitter struggle between the Liberals and the Moderates, and President Estrada Palma, reelected by the Moderates, was accused by the Liberals of resorting to violence, intimidation, and bribery to retain his position. Early in 1906 an armed uprising against the government started in the provinces of Piñar del Río and Havana. On August 27 President Palma issued a proclamation granting amnesty to all insurgents who would lay down their arms,[18] but when this failed of its purpose he secretly requested the American consul-general at Havana to ask President Roosevelt to dispatch two vessels at once, since the government forces were unable to quell the rebellion.[19] The American government was loath to intervene and pointed out the dangers of this course. But when President Palma threatened to resign, President Roosevelt sent the secretary of war, Mr. Taft, and the assistant secretary of state, Mr. Bacon, to Havana to attempt to reconcile the difficulties.

Upon the arrival of the commissioners in Havana, September 19, 1906, they interviewed the leaders of the various factions, and finally urged as a solution that President Palma remain in office with a coalition cabinet, while the members of Congress stand for reelection. When the Moderates refused and the President insisted upon resigning, the commissioners established a provisional government under the authority of the president of the United States, and issued a proclamation to the Cuban people setting forth the causes for this action and defining afresh the position of the United States towards Cuba.[20]

On October 10, Governor Taft announced that active organized hostilities had ceased, and issued a proclamation of full and complete amnesty to all who had participated in the uprising. Three days later he turned the government over to Charles Magoon, who was, as events proved, to act as provisional governor for the next three years.

Although Governor Magoon has been subjected to almost scurrilous abuse by many Cuban writers, it is by no means justified by the facts. His policies may not always have been the wisest ones, but it must be remembered that for the most part he merely executed the orders emanating from the War Department. His methods of administration may not have been as rigorous as could be desired, but there are but few administrators with the force and personality of a Governor Wood. Perhaps a more suitable appointment might have been made; nevertheless, Mr. Magoon was an outstanding authority on Cuban law, and had had successful administrative experience in the Canal Zone. And in spite of the

[17] Pan American Union, *Latin American Foreign Trade in 1925* (Washington, D.C., 1927).
[18] *For. Rel. of the U.S.*, 1906, Part I, p. 459.
[19] *Ibid.*, p. 473.
[20] *Ibid.*, p. 489.

many accusations of his loose handling of funds, no proof has ever been presented which reflects upon the inherent honesty of the man.[21]

One of the most useful acts of the provisional government was the taking of a complete census as the basis for the new electoral lists, and the returns showed the total population of Cuba in 1908 to be 2,048,980, an increase of 25 percent as compared with the returns of the census of 1899. A new electoral law was promulgated on April 1, 1908, the provincial and municipal elections were held on August 1 in an orderly fashion, and the results were quietly accepted. The general elections were equally peaceful, and the Liberal candidate, General José Miguel Gómez, defeated the Conservative candidate, General Mario Menocal, by a majority of more than 70,000 votes. The provisional government thereupon convoked Congress, and President Gómez was inaugurated, on January 28, 1909, at which time Governor Magoon relinquished the administration to the duly elected representatives of the Cuban people. A number of much-needed public improvements, particularly in the matter of sanitation, had been carried out; necessary legislative decrees had been promulgated, among the most noteworthy being the new electoral law (decree 899 of 1908); and after paying all contracts and other obligations as far as practicable, $2,860,000 in cash was turned over to the new government.[22]

Once more the people of Cuba were put in full command of their ship of state, with the best wishes of the United States for a long and prosperous voyage. General Gómez proved himself a strong-minded leader, and he appeased the more restless spirits by appointing them to political office, thus materially strengthening his position. However, before his term of office expired, serious charges of corruption were brought against his government. It was claimed that the public offices were being sold to the highest bidder, that concession-hunters were finding lucrative opportunities at the people's expense, and that the government was even going so far as to buy off the newspapers to avoid unpleasant publicity. The government was also threatened by the veterans of the war of independence, who proscribed all office-holders of Spanish sympathies and forced the suspension of the civil service rules. Fearing that with the elections approaching the disorder might get beyond control, the American secretary of state, Mr. Knox, sent a warning to the Cuban government, early in 1912, that intervention might be forced upon the United States if the disorders were allowed to develop.[23] For a time, the warning had a tranquilizing effect. But when, in the summer, a Negro revolt broke out in the eastern end of the island, the United States felt it necessary to land marines and to concentrate a number of naval vessels at Key West. The government now rose to the occasion and put down the revolt, and the American troops were forthwith withdrawn.

President Gómez had agreed not to stand for reelection, and the two

[21]For a detailed study of the Magoon Administration, see D. A. Lockmiller, *Magoon in Cuba* (Chapel Hill, N. C., 1938).
[22]For an excellent summary of the period of the American occupation see *Republic of Cuba, Reports of Provisional Administration,* vol. II (Havana, 1908-1909). For more critical surveys see Leland H. Jenks, *Our Cuban Colony* (New York, 1928), chap. VI, and Russell H. Fitzgibbon, *Cuba and the United States* (Menasha, Wis., 1935), chap. V.
[23]*For. Rel. of the U. S.,* 1912, p. 240.

leading candidates were Vice-President Zayas, the principal Liberal candidate, and General Menocal, the choice of the Conservatives. The election proved to be unexpectedly peaceful, and, owing to the split in the Liberal ranks, General Menocal was elected. The new president was a native Cuban and a veteran of the war with Spain. A civil engineer by profession (he had studied at Cornell University), he was, at the time of his election, the managing director of the Cuban American Sugar Company. He had already shown remarkable administrative ability and was reputed to be one of the wealthiest planters on the island. Financially disinterested, he made a valiant effort to eliminate graft in the administration, and under his leadership the island made substantial economic progress. Imports, valued at approximately $120,000,000 in 1912, rose to over $200,000,000 in 1916; while exports increased from $146,000,000 to $336,000,000. During his administration the production of sugar increased from 1,750.000 tons to more than 3,000,000 tons.

Notwithstanding the excellent results of his administration, Menocal was strongly opposed when he sought reelection in 1916. The Liberal factions united on Dr. Zayas, and the results were so close that both sides claimed the victory. In certain cases second elections were ordered; but the Liberals, with ex-President Gômez as their leader, revolted before they were held. On February 10, 1917, Secretary Lansing appealed to the Cubans not to plunge the country into civil war,[24] and on February 13 he warned them that the United States would not recognize any government set up by violence.[25] The rebels, however, refused to lay down their arms, and early in March, American marines were landed. At the same time, the American government promised its aid to the Cuban government to reestablish order and put down the rebellion. The revolution subsided as quickly as it had flared up, and in May 1917, President Menocal again took the oath of office.[26]

Despite the fact that Cuba now entered the World War as the ally of the United States, the second Menocal administration began even more successfully than the first. In fact, by depriving the world of German beet-sugar the war stimulated to an extraordinary degree the production of Cuban cane-sugar. The crop for 1918 amounted to somewhat more than 4,000,000 tons, or about a million tons increase over the production for the preceding year, while the 1919 crop was even greater. As an indication of the tremendous wave of prosperity that had suddenly engulfed the island, President Menocal, in an interview with an American press correspondent in December 1919, pointed out that the volume of deposits in banks and savings institutions had increased 1000 percent in the preceding six years, land values had increased 500 percent, and the volume of foreign commerce for 1919 would pass the billion-dollar mark.[27] Unfortunately, these values were considerably inflated, and towards the close of 1920, when the whole world began to experience the troubles of financial readjustment, the situation became very critical. A moratorium declared in October

[24] *New York Times,* Feb. 13, 1917. See also Raimundo Cabrera, *Mis malos tiempos* (Habana, 1920), pp. 116-117.
[25] *New York Times,* Feb. 15, 1917; Cabrera, *op. cit.,* p. 124.
[26] For a sprightly account of the American intervention see the article by George Marvin, "Keeping Cuba Libre," *World's Work,* vol. XXXIV (Sept., 1917), pp. 553-567.
[27] L. J. de Bekker, "Cuba and Her President," *The Nation,* vol. CX (Feb. 21, 1920), p. 230.

1920, was extended to June 15, 1921. A foreign loan seemed essential, and at the request of the Cuban government, the State Department of the United States sent a financial adviser to consider the possibilities of a loan by American bankers. When this solution failed, a bill was put through the insular congress providing for a sliding scale of liquidation of obligations, and this measure went into effect on February 1, 1921.

For several reasons, the elections of 1920 were particularly important. They were to be held under the 1919 census and the electoral laws, as newly revised by General Crowder; President Menocal was ineligible for a third term and had promised that the elections would be conducted with absolute impartiality; and General Gómez, the former Liberal leader, was now opposed by Dr. Zayas, whom he had supported against Menocal. When the results came in, Dr. Zayas was found to be elected by a considerable majority. But the Gómez faction, alleging fraud and intimidation, refused to accept the results. General Crowder was again sent for, and finally he persuaded the opponents to abide by secondary elections in the districts where fraud was charged. These were held on March 15, 1921, and, as the Gómez adherents remained away, Zayas was again declared elected. The Liberals made a final protest through the abstention of their representatives when Congress convened on April 3. At the same time, General Gómez appealed personally to President Harding to set up a provisional government. But when the United States formally recognized Dr. Zayas as the duly elected president, General Gómez gave up the contest.

Owing to the government's economic difficulties, General Crowder remained in Havana to investigate the financial situation and if possible to suggest reforms. The Cuban government was anxious to float a $50,000,000 loan in the United States, but as this was impossible under the Platt Amendment unless the revenue were sufficient to meet amortization and interest, a new tax measure had to be passed and the approval of Washington obtained. At the request of the Cuban government, Albert Rattibone, former assistant secretary of the treasury, was sent by the State Department to act as financial adviser to consider possibilities of a loan by American banking interests.

An investigation showed that the second Menocal administration had been extremely wasteful of public funds; many illegal contracts had been awarded, some at exorbitant rates, piling up obligations far in excess of the government's ability to pay.[28] It was essential that a program of domestic economies be instituted immediately, with a strong and honest administration to put it into effect. General Crowder persuaded President Zayas to appoint a new cabinet, which was henceforth known as the Honest Cabinet. This cabinet, formed on June 16, 1922, included Dr. Céspedes, former Cuban minister to Washington, Colonel Despaigne, administrator of customs under Estrada Palma, and Captain Pokorny, a graduate of West Point and aide to General Crowder.

A program of reform covering a revision of contracts, a drastic curtailment of expenditures, and the elimination of graft as far as possible was immediately instituted.[29] On September 15, 1922, a law was passed for the establishment of

[28] C. E. Chapman, *A History of the Cuban Republic* (New York, 1927), p. 426.
[29] *Ibid.*, pp. 427-439.

a commission for the examination and audit of the Cuban debt. The commission was organized in November and all claims had to be filed by the following March. Some 25,471 claims were filed, totaling $45,150,673.57.[30] The budget was set at about $55 million and a loan of $50 million agreed upon at 5½ percent.[31] This was bought by J.P. Morgan & Co. at 96.77, which was a very small discount considering the general financial condition.

The financial situation now cleared rapidly. The fiscal year 1922-1923 closed with a surplus of income over expenditures of approximately $12 million, the bulk of which was appropriated to the retirement of the Cuban public debt. Among other debts paid was the war loan made to Cuba by the United States, thus giving Cuba the credit of being the first government to make a full settlement with the United States of the obligations contracted during the war,[32] an achievement largely due to General Crowder.

Unfortunately the Zayas administration soon wearied of being excluded from the public treasury, and early in April 1923, the president dismissed four of his reform cabinet, including the secretary of the treasury and the secretary of public works. General Crowder, whose post, in January 1923, had been changed from special agent to ambassador, was no longer able to exert the same pressure for economy. As one investigator frankly stated it; "President Zayas threw off the 'vicious intermeddling of Washington,' placed fourteen members of his family in strategic positions in the administration, and his forces thus distributed laid siege to the public treasury."[33]

The scandal became so great that in August 1923, an organization known as the Veterans and Patriots Association was founded, to combat the evils connected with the administration of the lottery and the passage of the notorious Tarafa bill for the consolidation of the railways, a piece of legislation which closed practically all the private ports and compelled sugar companies to utilize only the public service railways. When Ambassador Crowder protested against the reorganization of the lottery in such a way as to increase rather than to diminish the graft, the Cuban Congress passed a joint resolution condemning interference on the part of the United States.[34] Secretary Hughes thereupon called Ambassador Crowder back to Washington for a conference on the Cuban situation, but it soon became apparent that a hands-off policy was decided upon.[35]

The Veterans and Patriots Association now decided to act, and its leader, General Garcia Velez, Cuban minister to London, proceeded secretely to New York, and on March 22, 1924, made a scathing denunciation of the Zayas administration.[36] He was forthwith dismissed from the diplomatic service and the newspaper organ of the Association was suppressed for publishing the

[30] *U. S. Commerce Reports,* Aug. 6, 1923, p. 383.
[31] Text of Cuban law authorizing loan in *Commercial and Financial Chronicle,* Nov. 11, 1922; for its application see *U. S. Commerce Reports,* Feb. 12, 1923, p. 453.
[32] *U. S. Trade Information Bulletin,* No. 191 (Feb. 11, 1924).
[33] H. K. Norton, "Self Determination in the West Indies," *World's Work,* vol. 51 (Nov., 1925), p. 81.
[34] See editorial in the *Outlook,* vol. 134 (Aug. 29, 1923), p. 654.
[35] Chapman, *op. cit.,* pp. 446-449.
[36] *New York Times,* Mar. 23, 1924.

accusations. The revolution which followed was a complete fiasco, due partly to poor leadership but also to the fact that President Coolidge immediately (May 2, 1924) issued a proclamation forbidding the sale of arms and ammunition to the revolutionists, while approving two days later the sale of war materials to the government.[37] Apparently the United States, while looking askance at corruption in the Cuban administration, was wholly opposed to a clean-up by revolution—a policy which could hardly be met with enthusiasm by either imperialists or moralists in the United States, although it seemed a satisfactory interpretation of the Platt Amendment to the majority of Cubans.

President Zayas apparently at length came to realize that he could hope for no further favors at the hands of the electorate, and although renominated by the Popular Party he withdrew in favor of General Gerardo Machado, the nominee of the Liberals. The latter was thus able to defeat the Conservative candidate, ex-President Menocal, by a substantial majority, and he was peacefully inaugurated on May 20, 1925.

President Machado began his administration possessing both the confidence and support of Washington and pledged to improve the economic situation of the island. He had solemnly declared that he would not stand for reelection and had repeated the promise on July 26, 1927, about a month after his term had been prolonged for two years by his henchmen in the Cuban Congress.[38] Nevertheless, he did run again in 1928 and saw to it that he was reelected with practically no opposition and this time for a six years' term.

As early as 1927 sinister evidence was accumulating to the effect that Machado was crushing all opposition to his policies to the extent of imprisoning and assassinating his adversaries.[39] However, inasmuch as the Sixth Pan American Conference was to meet at Havana in January 1928, the Cuban Ambassador in Washington made every effort to refute all such allegations and little attention was paid to them in the United States. However, on April 17, 1928, Senator Shipstead introduced a resolution charging the Machado administration with maintaining a dictatorship under which numerous assassinations, imprisonments and deportations had taken place, the National University had been closed, and private property of Cubans and Americans had been seized. A report was asked as to whether the obligations of the Platt Amendment did not require some action under the circumstances.[40] The Cuban ambassador protested the resolution and the United States ambassador at Havana, Noble B. Judah, took issue with the statements contained, and emphasized the cordial relations between the two countries.[41] President Machado also resented the possibility of intervention, and on December 31, 1928, he declared that the Platt Amendment no longer existed.

With the advent of the Hoover administration, a special effort was made to understand and cooperate with the Latin American states. Trained ambassadors and ministers were sent to every one of the Latin American states except Cuba,

[37] *Ibid.,* May 3, 4, 1924.
[38] Raymond L. Buell, "Cuba and the Platt Amendment," *Foreign Policy Reports,* vol. V, no. 3 (Apr. 17, 1929), p. 39.
[39] See Carleton Beals, *op. cit.,* for a graphic portrayal bitterly hostile to Machado.
[40] *Cong. Rec.,* 70th Cong., 1st Sess., vol. 69, p. 6591.
[41] *New York Times,* Apr. 19, May 29, 1928.

and subsequent events raised doubts as to the advisability of the single exception. The new ambassador, Harry F. Guggenheim, was financially connected with numerous companies interested in Latin America and as a conservative businessman, he stood consistently behind President Machado and the maintenance of a strong, stable government. Unfortunately, the depression cut the price of sugar to such an extent that bankruptcy threatened the Cuban government and such a situation was not conducive to political stability.[42]

The internal political situation became so serious in the fall of 1930 that the question of intervention was raised in the Senate by Senator Walsh, and the State Department issued a statement on October 3 declaring that a close watch was being kept upon affairs in Cuba, but it was made clear that no intervention was contemplated and only a state of anarchy could provoke it.[43] In August, 1931, a rebellion under ex-President Menocal was quickly suppressed, and again the United States refused to intervene. In fact, the White House characterized as a "midsummer dream" the report that the United States contemplated intervention in Cuba. As an evidence of its complete impartiality, the United States placed no embargo on shipments of arms to the rebels.

The opposition which arose in revolutionary proportions against Machado in 1931, and would reappear in 1958, was led initially by young intellectuals of middle and upper class background who could not tolerate the existing regime, either on idealistic or pragmatic grounds. The unity of the "generational movement" was disturbed by the intrusion of Marxist ideas after 1925, but the majority rejected Marxist ideology, and with the failure of the August 1931 revolt, became concentrated in two organizations: The University Student Directorate (*Directoría Estudiantil Universitario*—DEU), and the secret ABC organization. Although united by a revulsion against the dictatorship, a desire to revive the national spirit and renovate public life, their programs reflected varying shades of radicalism, political liberalism, nationalism, and socialism.[44]

Focusing their activities in the cities, they adopted the combat technique of terrorism to overthrow the government. President Machado retaliated by creating a secret police, the so-called *Porra*, which soon became notorious for its inhuman and murderous practices. The *ley de fuga* was revived as an instrument of counter-terrorism.[45] Sumner Welles, newly appointed assistant secretary of state, commented that by the beginning of 1933 "Cuba was a country economically prostrate, ruled by a tyrannical dictatorship to which 95 percent of the people were fanatically opposed, a country . . . in which bombings, terrorism, and murder were daily occurrences."[46]

The United States could not escape some responsibility, for the Hawley-Smoot Tariff of 1930 had aided materially in Cuba's economic collapse, and the Platt Amendment gave the government which had the support of the American

[42] The average price of sugar dropped from 2.64 cents per pound in 1926-27 to .72 cent in 1931-32.
[43] *New York Times,* Oct. 3, Dec. 13, 1930.
[44] Andrés Suárez, *Cuba: Castroism and Communism, 1959-1966* (Cambridge, Mass., 1967), pp. 8-9.
[45] *New York Times,* Feb. 4, 1933.
[46] Sumner Welles, *Relations Between the United States and Cuba*, Dept. of State, Latin American Series, No. 7 (Washington, 1934), p. 6.

administration an almost invulnerable moral as well as legal standing. Although Ambassador Guggenheim had employed financial experts at his own expense to help solve the serious financial and economic problems of the island, his open support of President Machado made him subject to ever-increasing suspicion and to the most violent criticism. Nor did the fact that the Chase National Bank of New York had loaned the Machado government some $160 million make his position more tenable.

Conditions became so serious that President Franklin D. Roosevelt, shortly after his inauguration, dispatched his assistant secretary of state, Sumner Welles, as ambassador to Cuba. An able career diplomat with broad experience in Latin American affairs, Welles was instructed to negotiate a new commercial convention aimed at relieving the economic strain, to tender his good offices, and hopefully bring to an end the intolerable political situation.[47] Ambassador Welles' efforts were frustrated partly by Machado's intransigence and partly by the hostility which smoldered against the ruthless Cuban president. A general strike on August 4 was followed a week later by a revolt of the army. Thereupon the president and his cabinet resigned and managed to flee the country.[48] Machado had named a general as interim president, but he was cast aside by the rival leaders with the support of Ambassador Welles, and Dr. Carlos Manuel de Céspedes was named provisional head of the government. Son of the first president of Cuba in the revolt of 1868, his position was untenable from the outset, for the economic crisis worsened and riots increased. De Céspedes was recognized by the United States, and hoping to moderate radical sentiment, he restored the Constitution of 1901. His efforts proved unavailing, however, and in less than three weeks a coup d'état brought into office another provisional government. Ambassador Welles, shocked by the turn of events, requested that the United States intervene with troops, but President Roosevelt, holding steadfast to the nonintervention pledge made at the Inter-American Conference at Montevideo, in 1933, declined.[49] When several warships were sent to Cuban waters as a precautionary measure, the president stressed that neither intervention nor the slightest interference in the internal affairs of Cuba was intended.[50]

From 1933 to 1959 Cuban politics was to be dominated by two individuals, one a civilian, the other of the military. The former, Dr. Ramón Grau San Martín, a physician and professor of biology, emerged as leader of the so-called "Authentic Revolution," but he was to be overshadowed by the young army sergeant, Fulgencio Batista y Zaldivar. Neither figure possessed ideal qualities of leadership. Grau lacked the political and administrative expertise needed to sustain his authority, whereas Batista, of Spanish, Negro, and perhaps Chinese ancestry, rising from poverty, was wanting in the credentials of a higher education and a cultural background. As leader of the most popular political party to appear after 1933, Grau had the advantage of support lent by party

[47] *Ibid.*, p. 7.
[48] For details of the attempted mediation and the specific reasons for its failure see Charles A. Thomson, "The Cuba Revolution: Fall of Machado," *Foreign Policy Reports,* vol. XI, no. 21 (Dec. 18, 1935).
[49] John E. Fagg, *Cuba, Haiti and the Dominican Republic* (Englewood Cliffs, 1965), pp. 81-82.
[50] Dept. of State, *Press Releases*, Aug. 15, 1933.

organization; however, this made him vulnerable to party patronage pressure. Batista, forming a party coalition, was less subject to patronage demands, but his chief support came from the military and business sectors. Both promised utopian solutions to their country's problems, and when neither's promises were fulfilled, their parties and supporters became disillusioned and more receptive to extreme solutions.

On September 4, 1933, Sergeant Batista, in collusion with radical young leaders, seized the army chief of staff, and gained possession of Havana. With this accomplished and the officer corps immobilized, a "pentarch" government, made up of five leftists, assumed control (September 4-10, 1933). In quick succession, Batista was named army chief of staff with the rank of colonel, and President de Céspedes was deposed. The "pentarchy" was dissolved on September 10, naming Dr. Grau San Martín as provisional president; he was to hold this office from September 10, 1933 until January 17, 1934. Proclaiming the beginning of a "social revolution" he made little progress because of the continuing economic decline, civil disorders, and the refusal of the United States to grant diplomatic recognition.

The position of the United States was taken on the grounds that the new government was opposed by a number of groups which had ousted Machado, and because of the implications of a radical program for American property interests on the island. In retrospect it seems possible that the United States, by this action, helped to bring the social revolution to a halt; thereafter, the revolution assumed a more political character. It was not until the late 1950s that the interrupted revolutionary process in the socio-economic context erupted, this time with disastrous consequences for Cuban-American relations and hemispheric security.[51]

To demonstrate the sincerity of the Good Neighbor policy, President Roosevelt, on September 6, 1933, invited the diplomatic representatives of Argentina, Brazil, Chile, and Mexico to discuss the situation, assuring them that the United States would avoid intervention except as a last resort. On September 11, Secretary of State Cordell Hull declared that the United States was not antagonistic to any political organization and would "welcome any government representing the will of the people of the republic, and capable of maintaining law and order throughout the island." In spite of failing to receive American recognition, and several uprisings against its authority, the Grau government survived until the middle of January 1934. It had shown itself consistently hostile to the United States, and on November 15 President Grau had written a personal letter to President Roosevelt requesting the recall of Ambassador Welles.[52] As Welles had been assigned the mission as an emergency measure, he returned to his position as assistant secretary of state, and Jefferson Caffery,

[51] Robert F. Smith, ed., *Background to Revolution: The Development of Modern Cuba* (New York, 1966), p. 50. For respective positions of Hull and Welles see *Foreign Relations of the United States. Diplomatic Papers, 1933*; *The American Republics,* vol. V, pp. 270-588 (Washington, 1952). A careful analysis of United States private investment in Cuba may be found in Russell H. Fitzgibbon's *Cuba and the United States, 1900-1935* (New York, 1964), pp. 228-249.

[52] Charles A. Thomson, "The Cuban Revolution: Reform and Reaction," *Foreign Policy Reports,* vol. XI, no. 22 (Jan. 1, 1936), p. 268.

another career diplomat, replaced him. In the absence of Washington's recognition of the Grau regime, Caffery could serve only as the personal representative of President Roosevelt.

The real power in Cuba after de Céspedes was Colonel Batista who, until 1940, ruled through seven puppet presidents. Notable among them were Carlos Mendieta (January 1934-December 1936), Miguel Mariano Gómez (1936), and Federico Laredo Brú (1936-1940). While proclaiming that he was a revolutionary at heart, Batista found it expedient to follow a conservative course, thus winning recognition from the United States, and the support of foreign and domestic interests.

After witnessing the impotence of the Grau administration, Colonel Batista transferred his support to Colonel Carlos Mendieta, a moderate and former leader of the Liberal Party. Mendieta became provisional president January 18, 1934, and four days later President Roosevelt, after inviting the diplomatic representatives of the Latin American republics to confer with him, instructed Caffery to establish diplomatic relations with the government of President Mendieta.[53] To furnish tangible evidence of its support, Assistant Secretary Welles, on behalf of the United States, publicly promised four constructive measures to help Cuba; (1) a fair sugar quota; (2) a new commercial treaty; (3) stimulation of trade by utilization of the new Export-Import Bank; and (4) a modification of the Permanent Treaty with Cuba. Mendieta responded by issuing a provisional constitution, and forming a legislative Council of State. But civil strife continued and the authority of the new regime remained precarious.[54]

By successfully coping with the crises of 1933-1934, the State Department had brought into play a combination of factors to insure the safety of American interests in Cuba. Commenting on Roosevelt's boast that the administration had not had to use troops to settle controversies, Harold L. Ickes, secretary of the interior, noted in 1936:

> I do not like the situation in Cuba. I think that this government is interfering altogether too much in the internal affairs of Cuba, but it is being done through diplomatic channels.[55]

It was the fourth provision of the Welles' proposals that loomed most important to the Cubans. This referred to the Platt Amendment, which was regarded as the cause of the existence and continuance of a situation such as had existed under President Machado. The provision that permitted the United States to intervene and maintain a government adequate for the protection of life, property, and individual liberty, was a bulwark of protection to the administration in power. It served as a two-edged sword. The government could use the threat of intervention as an effective means of stifling armed opposition by the Cubans and, if that failed, there was always the chance that intervention would prefer the maintenance of the *status quo*. Furthermore, the fact that the United States had the right to intervene seemingly made it lean backward in a

[53] Dept. of State, *Press Releases,* Jan. 27, 1934.
[54] Welles, *Diplomatic Papers,* pp. 14-16,
[55] Robert F. Smith, *The United States and Cuba: Business and Diplomacy, 1917-1960* (New York, 1960), p. 171.

policy of nonintervention to prove its anti-imperialistic policy to Latin America as a whole.

The United States did not delay in the fulfilment of its promises. On May 29, 1934, President Roosevelt signed a treaty between his country and Cuba which abrogated the Platt Amendment. In submitting the treaty to the Senate the president declared: "By the consummation of this treaty this Government will make it clear that it not only opposes the policy of armed intervention but that it renounces those rights of intervention and interference in Cuba which had been bestowed upon it by the treaty."[56]

Under the new agreement the Treaty of 1903 was abrogated, the acts of the United States during the military occupation were validated, the arrangement for the United States naval station at Guantánamo was maintained, and provision was made for reciprocal suspension of communications between certain ports in case of outbreak of a contagious disease. The Senate approved the treaty two days after it was signed, an example of almost unprecedented speed. The Cuban government was equally prompt, and ratifications were exchanged on June 9.

Hardly was the ink dry on the new political arrangement before announcement was made of a new commercial agreement. The Hawley-Smoot Tariff had been particularly damaging to the Cuban sugar trade, the island's basic industry,[57] and the Chadbourne Plan for world control of the sugar industry had proved a dismal failure. The United States alone could furnish relief, and the first step was the Jones-Costigan Act, with the Walsh Amendment, limiting the importation of refined sugar, which became law on May 9, 1934. This act fixed the imports of Cuban refined sugar to 22 percent of the total Cuban allotment; in 1934 this amounted to 423,000 tons. At the same time, the import duty on sugar was reduced 25 percent. Of great significance also was the Reciprocal Trade Agreement, signed in August, which allowed most American exports to enter Cuba with little or no duty, and assigned to Cuba's sugar the exceedingly low duty of .9 cents per pound. Duty on Cuban rum was reduced from $4.00 to $2.50 a gallon, and comparable reductions were made on tobacco and various foodstuffs. In return the United States secured substantial reductions on the duties of its exports of foodstuffs, textiles, machinery and automobiles, lumber, and many other articles. In its first month of operation, the new agreement increased the exchange of commodities between the two countries more than 60 percent.

Illustrative of the value to Cuba of the reciprocity treaty was the fact that her average exports to the United States during 1932-1934, the two years preceding the trade agreement, amounted to $51 million whereas in the two years after the agreement, the average was $133 million.[58]

[56] For text of treaty see U. S. Dept. of State, *Treaty Information Series,* No. 56 (Washington, May 31, 1934).

[57] For a carefully prepared study of the sugar industry in Cuba see *Problems of the New Cuba* (New York, Foreign Policy Association, 1935), chaps. X-XIII.

[58] Dept. of State, *Analysis of Cuban-American Trade During the First Two Years Under the Reciprocal Agreement* (Washington, January 19, 1937). Two supplementary trade agreements were signed with Cuba, the first, Dec. 18, 1939, and the second, Dec. 3, 1941, to balance the conditions resulting from the Second World War. See Dept. of State *Bulletin,* I (Dec. 23, 1939), p. 729; *Ibid.*, V (Dec. 27, 1941), p. 603.

Cuba's sugar industry began to show gains as both production and prices increased. Moreover, Cubans began displacing American investors owing to the effects of the depression, thereby benefitting from the improved conditions. The value of American holdings shrank from $1,500 million to a third of that amount since 1929; by 1949 the Cuban share had risen to 49 percent.[59]

The third economic element in the Roosevelt administration's program to achieve economic prosperity and political stability in Cuba was the Second Export-Import Bank. This institution was created by executive order in March 1934, for the purpose of loaning money to Cuba. Two loans were made in that year, the first amounting to $3,774,724, and the second was for $4,359,095. This meant in effect that the United States government was beginning to supplant Wall Street as the source of Cuba's financial support.[60]

Unhappily, the effects of United States aid to Cuba, politically and economically did not have an immediate tranquilizing effect upon the domestic situation. The year 1935 opened with student strikes, followed by a strike of government employees, and finally by a general strike. President Mendieta was compelled to suspend constitutional law and proclaim a state of seige. The gravity of the situation led to a State Department declaration that the new treaty of relations of May 29, 1934, abolished the former special relationship, and made it emphatically clear that the United States would not intervene directly or indirectly in the political affairs of the Cuban people. Three candidates aspired to the presidency to succeed Mendieta, and Dr. Harold W. Dodds of Princeton University was invited to Havana to serve as an adviser on the presidential campaign. He accepted, and early in December flew to Havana where he devised a comprehensive electoral program. Elections held January 10, 1936, with women voting for the first time in a presidential election, gave a majority to Miguel Mariano Gómez, who was inaugurated on May 20. However, Colonel Batista was the actual power in Cuba, and before the year ended, he had Gómez impeached and Vice President Laredo Brú inducted into the presidency. It was a source of criticism to some that Ambassador Caffery seemed to be on most cordial terms with Batista. But the United States had definitely ceased meddling in Cuba's internal affairs, and it looked with sympathetic approval upon the efforts of its former ward to put its house in order.

The United States was not disappointed, for both internal conditions in Cuba and the ties between the two countries improved steadily. The Laredo Brú administration was responsible for the passage of substantial welfare legislation, a three-year plan for national development, and it provided for the unionization of labor in the Confederation of Cuban Workers. Social security legislation for urban and rural workers was passed, as was a minimum wage law.[61] Although the legislation was not fully implemented and the program fell short of the criteria set forth by the proponents of a social revolution in 1933, it was a marked advance for Cuba in the realm of state socialism.

Colonel Batista accepted an invitation of the War Department to visit the

[59]Fagg, *Cuba,* p. 83.
[60]Smith, *United States and Cuba*, p. 163.
[61]Fagg, *Cuba*, p. 85.

United States for Armistice Day ceremonies in 1938, and he received a very cordial reception. Shortly afterwards he retired from the army, and became a candidate for the presidency. During his campaign he pledged the immediate and unreserved support of Cuba in the event the United States should be drawn into the war. By this time Batista had come to symbolize stability, and Americans had concluded that he would defend their interests in spite of the fact that he never repudiated the political principles of his revolutionary days. His military forces were progressively strengthened with arms and advisers furnished by the United States.[62]

With the support of a coalition of parties and the political machines that had once dominated Cuban politics and supported the *Machadato*, Batista was elected president of Cuba on July 14, 1940. His coalition, called the Democratic Socialists, included Communists. His defeated rival was Dr. Grau San Martín, nominee of the Cuban Revolutionary party.[63]

Since 1933 the fundamental law of the Cuban republic had remained irregular, but in October 1940, a new constitution was adopted. Civil liberties and social welfare provisions were elaborately defined, and in general it reflected the influence of advocates of political, social and economic change. In fact, the Cuban Constitution of 1940 was one of the most progressive in the world, and the liberal elements had just cause for gratification.[64]

Following the revolution of 1933, the ideals of its authors had been perpetuated in the PRC, or Cuban Revolutionary Party (*Auténticos*), formed in 1934 when several revolutionary parties coalesced. The PRC was influential in drafting the Constitution of 1940, and its nominees were to win the presidential elections in 1944 and 1948. Emphasizing political and economic nationalism and social justice, it was committed to fundamental reforms: government control of the sugar industry, a national bank, tax reforms, and the expansion of public education.

The failure of the PRC to fulfill its pledges led to the formation of the Party of the Cuban People (*Ortodoxo*), an offshoot of the PRC, in 1946. This party had gained substantial strength by 1951 under the leadership of Eduardo Chibás who, in that year, committed suicide. The party stood for honesty in government, economic independence, social justice, and political liberty.

As the war in Europe threatened to engulf the Western Hemisphere, the United States and Cuba jointly engaged in a policy of close cooperation. Indicative of Cuba's importance, the United States sent as ambassador, Assistant Secretary of State, George Messersmith, one of the ablest of our career diplomats. In 1941 the Export-Import Bank authorized a loan of $25 million to diversify Cuban production and another $11 million to produce an additional 20 percent of sugar above the quota assigned for export to the United States. And substantial funds for defense purposes were granted under the Lend Lease Act.

Cuba declared war on the Axis powers on December 9, 1941, following the attack on Pearl Harbor. She also took prompt and effective action against

[62] Smith, *United States and Cuba,* p. 170.
[63] Smith, *Background to Revolution,* p. 63.
[64] Russell. H. Fitzgibbon, ed., *The Constitutions of the Americas* (Chicago, 1948), pp. 227-296.

enemy aliens, placing them in a concentration camp on the Isle of Pines. On June 18, 1942, an agreement was signed whereby the Cuban government offered facilities to the United States War Department for training aviation personnel, and for operations against undersea craft.[65] An all-embracing agreement coordinating military and naval measures for the duration of the war was concluded September 7, 1942; this provided for the most complete cooperation of the two countries on the basis of reciprocity.[66] For the first time in its history, Cuba imposed compulsory military service. Various air bases were leased to the United States, and a major one near Havana, valued at $20 million was retained until May 1946, and then turned back to Cuba.

Arrangements were made annually between the Commodity Credit Corporation of the United States and the Cuban Stabilization Institute for the disposition of the Cuban sugar crop at a mutually satisfactory price for the duration of the war. Under this agreement, the United States purchased all Cuban sugar, after domestic needs, at 2.65 cents per pound, with duties fixed at .75 cents. This program, carried on in the period 1942-1947, was most beneficial to Cuba, for production had risen to five million tons a year. The war also stimulated the production of Cuba's strategic minerals, such as manganese, copper, nickel, cobalt, chromium, tungsten and antimony. The Cuban financial situation was so favorable at the end of the war that in 1949 the government of Cuba began the payment of some $6 million due to American firms for services and supplies dating back to 1924.

President Batista ruled with moderation in spite of the wartime powers at his disposal. His political strategy remained opportunistic and flexible. Secure in the knowledge that he had the confidence and goodwill of the propertied elements, he sought labor support and established diplomatic relations with the Soviet Union in 1943. He accepted a working arrangement with the Communists, appointing two to cabinet posts.

When Batista's term of office expired in 1944 he stepped aside expecting that the nominee of the government coalition, whom he sanctioned, would win. But to everyone's surprise, Batista's long-time rival, Dr. Grau San Martin, won by a significant margin. Batista acquiesced in the outcome, and retired to his estate in Florida. Grau's support included the *Partido Social Democrático*, believed by some to be Communist; the *Partido* polled some 122,000 votes in 1944. It was well represented in the Confederation of Cuban Workers, by then a powerful force in Cuban politics.

The administrations of Dr. Grau, 1944-1948, and of Carlos Prío Socarrás, 1948-1952, were probably the most democratic in the island's history. Although corruption was widespread in the traditional pattern, democratic processes and civil rights were generally respected. The Party of the Cuban Revolution (*Auténticos*), represented by Grau and Prío was dominant. The United States, which had refused to recognize Grau as president in 1933, accorded recognition immediately in 1944.

Much was expected of Dr. Grau, favored as he was by the aftermath of the

[65] Dept. of State *Bulletin*, VI (June 20, 1942), p. 553.
[66] *Ibid.*, VII (Sept. 12, 1942), p. 752.

wartime boom. However, he proved unequal to the task, and his administration was characterized by irresponsibility, a decline in public morality, and growing turbulence among the students and working classes. Communist direction of labor unions led to almost incessant disputes and strikes, armed gangs ruled in some areas, and political assassinations became commonplace.[67]

Dr. Carlos Prío Socarrás, nominee of the *Autenticos,* and Dr. Grau's heir, won the presidential election of 1948, polling 900,000 votes, but with less than a majority of the votes cast. Much prestige had been lost by the PRC because of its alignment with a number of the old, traditional political groups. The party lost some of its following to the *Ortodoxo* party led by Chibás, which challenged Grau's ineptness and leftist tendencies. The Communist party, calling itself Popular Socialists, entered Juan Marinello as its candidate.

Prío Socarrás, a lawyer, and former student leader who had supported the Batista coup in 1933, set out to correct the shortcomings of his predecessor's administration, including the neutralization of Communists. Prío's relations with the United States were extremely cordial except for a disgraceful incident occurring in 1949. The statue of the Cuban patriot, José Martí, in Havana's Central Park, was desecrated by drunken sailors on shore leave from a United States naval vessel. Mob violence was narrowly averted by the prompt action of Cuban police. The next day United States Ambassador Butler apologized publicly, promised that the offenders would be punished, and he deposited a wreath at the foot of the statue.

On March 10, 1952, eighty days before the scheduled presidential elections, Batista, again a candidate, with scant prospect of success, engineered another coup d'état, and seized control of the government. His justification seems to have been the chaotic state into which public administration had descended, and the rampant political gangsterism. He evidently felt that the public would welcome a change, even a restoration of dictatorship.

Batista suspended the Constitution of 1940 and set up a provisional government which promulgated a new constitutional code suspending all political parties, but promising that a presidential election would be held in 1953. In February 1953 he announced that due to "conspiratorial activities" the elections would be deferred until June 1954. A serious but abortive revolt in July 1953 brought about a new postponement of the general elections until November 1, 1954. On that date General Batista was elected for a four-year term as chief executive after the only opposition candidate, Dr. Grau, had withdrawn in protest against electoral conditions.

The second Batista regime set the clock back on the processes of modernization and political development which had been emerging in Cuba, but it was also a period of apparent prosperity. Although students rioted and political violence continued, police and army control restored internal peace sufficiently to attract foreign investors, traders and tourists. A sugar stabilization fund forestalled a drastic drop in prices resulting from overproduction, and public works projects, coupled with a state-supported small-industries program, were instituted. Public administration was improved and Batista skillfully

[67] William S. Stokes, "National and Local Violence in Cuban Politics," *Southwestern Social Science Quarterly,* XXXIV (Sept., 1953), 57-63.

cajoled labor and the masses as he had in the past. The Cuban standard of living ranked among the four highest in Latin America, and third in per capita consumption. Inflation was kept within bounds, and times were generally good, except for the lower classes–i.e. the majority.

United States relations with the Batista government, as in the past, hinged largely on Cuba's desire for a larger share in the American market for sugar and other export products.[68] In Cuba, sugar was "king." In the year 1951 sugar accounted for 88.1 percent of all Cuban exports; by 1955 it had declined to 79.8 percent. The island exported about half of its sugar to the United States in 1959 and 1960. The fate of the Cuban economy therefore depended to a great extent on what she sold to the United States, her best customer up to 1960. In turn, these sales depended on: (1) the income of the United States; (2) the relation between income and the demand for sugar; (3) the allocation of United States consumption between various suppliers.[69]

Cuba's distress from excessive reliance on a single commodity, and problems in Cuban-U.S. relations connected with the marketing of sugar, was related in this century to a significant increase in United States beet sugar production, and the beet growers' rising influence in Congress, as reflected by higher tariffs on sugar. The cane industry of the United States was also growing, and sugar from the Philippines and Puerto Rico was admitted duty free. The duty on sugar was rather high compared to that of other foodstuffs, being in some years nearly equal to the price of sugar.

During the inter-war period Cuba participated in international agreements on production and marketing which proved largely ineffectual, and in 1934 the United States passed the Jones-Costigan Act, as has been noted, to help Cuba recover from the disastrously low sugar sales of 1932 and 1933. World War II brought renewed prosperity to sugar growers, for Britain and the United States purchased almost all Caribbean sugar in the war years.[70]

A transition was occurring in plantation ownership in Cuba, recalling that Americans held 63 percent of Cuba's sugar mills in 1926. The trend toward American domination of Cuba's primary industry was broken by the great depression which, as has been seen, forced many investors to dispose of their Cuban holdings. By 1955 only 39 of Cuba's 161 mills were American-owned, compared to 118 owned by Cuban nationals. Nevertheless, by 1960 about 40 percent of the profits from sugar exports went to American-owned companies.[71]

In 1948 the United States enacted new legislation regulating sugar imports, under which quotas were set in absolute figures rather than as a percentage of expected consumption. Cuba's allotment came to 98.84 percent of the total non-United States and Philippine-produced sugar.[72] During the Korean War and the Suez crisis, there was an international sugar shortage and the world price

[68] Dept. of State *Bulletin*, XXXI (Nov. 29, 1954), pp. 816-817.
[69] Smith, *Background to Revolution*, p. 210.
[70] Bernard L. Poole, *The Caribbean Commission: Background of Cooperation in the West Indies* (Columbia, S.C., 1951), pp. 203-204.
[71] John P. Powelson, *Latin America: Today's Economic and Social Revolution* (New York, 1964), p. 141; Smith, *United States and Cuba*, p. 175.
[72] Cuban Economic Research Project, *A Study on Cuba* (Coral Gables, Fla., 1965), p. 486.

actually rose above the United States price level; at these times Cuba continued to supply the United States at the lower price.

In 1952, a sugar act, to be effective for five years, was adopted by the United States. Under its provisions the Cuban sugar quota was slightly reduced while that of domestically produced sugar was increased. An International Sugar Agreement was reached in 1954 (revised in 1956 and 1958) which gave Cuba a basic world market quota of 2,415,000 metric tons. This quota was not always met by Cuba as her production was occasionally too low to supply domestic needs and the more profitable (by two cents a pound) United States market, as well as much of the rest of the world. The International Agreement also provided for the maintenance of price levels (4.35 to 3.25 cents a pound, changed in 1956 from 4.00 to 3.15 cents a pound).[73] The United States revised its sugar act in 1956 (effective January 1, 1957 to December 31, 1960). The new act allowed Cuba the same basic quota, but also its share of United States needs that might arise in excess of the allotted amounts.[74]

It was due to these circumstances that Cuba enjoyed a qualified prosperity between World War II and 1959, but it was a prosperity at the sufferance of the United States, notably the benevolence of the Congress in upholding the quotas that made prosperity possible. The United States was a convenient market for the commodity that Cuba produced best, but it was at the expense of Cuban economic independence and production diversification.

In spite of the seeming improvement of the economy in the years 1952-1959, many of Cuba's basic problems remained unresolved. Reforms in support of urban labor had been compromised in favor of business and commercial interests. Illiteracy was widespread, the census of 1953 revealing a higher illiteracy rate among the 10-14 year age group than among the population as a whole. And in spite of the optimism engendered by statistical analyses, the extreme poverty of the masses of people was a glaring fact. This was especially true in the rural areas where more than 200,000 families were landless, and the agricultural laborers were unable to find work for more than three months each year.[75]

The stagnation of the sugar industry affected the entire economy. Sugar output had risen from about one million tons a year in 1905 to five million in 1925, but five to six million tons was typical of the period 1947-1958, other than for the unusually large crop of seven million tons in 1952. The number of sugar mills declined from 184 in 1926 to 161 in 1958. Unemployment from July 1956 to June 1957 averaged 16 percent of the labor force and those working less than forty hours a week averaged 10 percent of the labor force.[76]

Although per capita income averaged $500 annually, about one third of the nation existed in squalor, suffering under or malnutrition, infested with parasitic diseases, lacking in health services, and living in huts, usually without toilet facilities or electricity. The children seldom went beyond the first grade. In 1953 one-fifth of urban families lived in single rooms, the average size family

[73] *Annual Supplement to the Quarterly Economic Review of Cuba, Dominican Republic, Haiti, Puerto Rico–1966* (London: The Economic Intelligence Unit Ltd., 1966), p. 4.
[74] Cuban Economic Research Project, *Cuba*, pp. 506-507.
[75] Smith, *United States and Cuba*, p. 175.
[76] *Ibid.*, p. 211.

being five persons. The population was growing at the rate of 2.5 percent a year in the 1950s, but the national product, derived predominantly from sugar, did not keep pace.[77] The restiveness of the under-privileged was accented by the proximity of their wealthy neighbor, whose citizens controlled a sizeable proportion of their country's resources. And the retention of Guantánamo naval base and the presence of U. S. naval vessels contributed to the growth of anti-American feelings.[78]

Official United States–Cuban ties were maintained on a high level of cordiality during the second Batista regime of the 1950s. Batista reiterated Cuban loyalty to the United States in any conflict with the Soviet Union. Shortly after seizing the government Batista had ordered two Soviet couriers in Havana to be searched in violation of their diplomatic immunity, which resulted in the Soviet Union severing diplomatic relations.

In 1952 a Mutual Defense Assistance Agreement, under the Mutual Security Act of 1951, was signed between the United States and Cuba. Batista received about $1 million a year in military aid, and non-military grants from the United States to Cuba increased from $40,000 in 1950 to $176,000 in 1953, and $561,000 in 1958.[79] In the period 1955-1957 the Export–Import Bank granted Cuba new credits totaling $19.7 million; however, much of this went to American companies.[80]

Economic ties, hinging mainly on the sugar quotas, were satisfactory on the government level. Discussions were held on the sugar-import schedules in 1954; and other agreements were made on American imports of Cuban cigar tobacco under the General Agreement of Tariffs and Trade (GATT). A research agreement with Cuba was made under the Atoms for Peace program in 1956. After some controversy, an agreement was concluded on United States rice tariff quotas. This specified the amount of United States rice exports to Cuba under the basic tariff quotas of GATT. In the same year the Batista government made payment on six outstanding claims of American companies, totaling $885,696.44. The claims, which were adjudicated by the Cuban courts, had been the subject of diplomatic negotiations between the two governments for several years. And finally, in the fall of 1959, a "shrimp conservation convention," calling for scientific research, went into force.[81]

Meanwhile, opposition to the Batista regime mounted as the rapacity of the leadership, the brutality of the police, the corruption of the government, and the absence of social justice recalled the hated *Machadato*. With the rising opposition Batista tried to create the impression of a close identity with the United States. An American citizen was retained as public relations adviser, and

[77] *Ibid.*, pp. 212-213.
[78] *New York Times*, February 26, 1957.
[79] Wyatt MacGaffey and Clifford R. Barnett, *Cuba, its People, its Society, its Culture* (New Haven, 1962), p. 315.
[80] Dept. of State, *American Foreign Policy Current Documents, 1957*, No. 7101 (Washington, 1961), p. 1600; Smith, *United States and Cuba*, p. 170.
[81] Dept of State *Bulletin*, XXXI (Nov. 15, 1964), pp. 815-816; *Ibid.*, XXXIII (July 4, 1955), pp. 27-28; *Ibid.*, XXXVII (July 22, 1957), p. 157; *Ibid.*, XLI (Sept. 28, 1959), p. 460; Dept. of State, *American Foreign Policy Current Documents, 1956*, No. 6811 (Washington, 1959), p. 974.

all joint United States-Cuban military and diplomatic activities were publicized. The good will visit of Vice President and Mrs. Nixon, which ranked a nineteen-gun salute, was much heralded,[82] as was the attendance of Secretary of State John Foster Dulles at a Cuban Embassy reception in Washington, in November 1958. While the United States did not intervene in Cuba during this period, the latent authority continued, as reflected in the person of the United States ambassador. Earl E. T. Smith, ambassador to Cuba from June 1957 to January 1959, declared: "Before Castro, the United States was so important in the minds of the Cuban People that the American ambassador was . . . regarded as the second most important personage in Cuba. He was a symbol of both power and friendship."[83] Being aware of this influence, many Cubans held the United States responsible for the actions of the Cuban government and accused it of intervention; some criticized the United States for not intervening.

Batista's enemies were as numerous as they were determined and violent. Among them was ex-President Prío Socarrás, from whom Batista had seized power in 1952. Florida became a headquarters for plots against the current government, and Dr. Prío himself was involved in a conspiracy to export arms to Cuba in violation of United States neutrality laws.

Dangerous as were his political foes identified with the political parties, the youthful revolutionaries were the most dedicated and courageous. It was this group which initiated the first major attempt to overthrow Batista. On July 26, 1953, some two hundred young men attacked the Moncada Barracks near Santiago. Valiantly executed but crudely conceived, the effort failed, but it stirred a revolutionary spirit. The leader of the rebellion was Fidel Castro Ruz, who though only twenty-six, had built an impressive record of terrorism and heroics. Castro, born in Oriente province in 1926, was an illegitimate offspring of a family of moderate wealth. He attended the University of Havana, earned a doctor of laws and became active in politics. Castro took part in the Cayo Confites plot to overthrow the dictator Rafael L. Trujillo of the Dominican republic, and was present in Bogotá, Colombia, at the time of the 1948 uprising. His political affiliation was ostensibly with the *Ortodoxos*, a nationalist and radical splinter from the *Auténticos.*

Castro's political career was cut off by the second Batista regime in 1952. In the following year he led the attack on the Moncada Barracks and was captured. At the time of his trial, Castro delivered his now well known "History will absolve me" speech which, reflecting traditional left-wing Cuban political views, advocated social and economic reforms and restoration of the 1940 Constitution. He was soon released from prison and later went to Mexico where, joined by other rebels including the Argentine revolutionary, Dr. Ernesto "Ché" Guevara, he prepared for the invasion of Cuba.[84]

In December 1956, the Castro-led invasion force aboard the *Granma*, landed on Cuba's south coast. Of the eighty-six original invaders, only twelve

[82] *New York Times*, Feb. 7, 1955.
[83] Earl E. T. Smith, *The Fourth Floor: An Account of the Castro Communist Revolution* (New York, 1962), p. 23.
[84] Theodore Draper, *Castroism: Theory and Practice* (New York, 1965), pp. 9-10; U.S. Congress, Committee on the Judiciary, U.S., Senate, *Communist Threat to the United States Through the Caribbean*, 87th Cong, 1st Sess., June 12, 1961 (Washington, 1961).

survived to reach the Sierra Maestra of Oriente province where they began a guerrilla campaign. Batista was forced to launch a full-scale assault against the rebels and claimed their extinction. However, in February 1957, Herbert L. Matthews of the *New York Times* obtained a personal interview with Castro and the series of articles that followed made Castro an internationally known figure.[85] At that time Castro declared: "You can be sure we have no animosity toward the United States and the American people. Above all, we are fighting for a democratic Cuba and an end to the dictatorship." Other resistance groups emerged in the Escambray Mountains and among the students in Havana. As described by Matthews: "It is a revolutionary movement that calls itself socialistic. It is also nationalistic, which in Latin America means anti-Yankee . . . The program is vague and couched in generalities but it amounts to a new deal for Cuba, radical, democratic, and therefore anti-Communist." Of the participants he said: "One gets the feeling that the best elements of Cuban life, the unspoiled youth, the honest businessman, the politician of integrity, the patriotic army officer are getting together to assume power . . . This opposition is bitterly or sadly anti-United States."[86]

The situation in Cuba confronted the United States with rather serious difficulties because that country, through 1957-1958, was subject not to one head, but to two, President Batista in Havana, and Fidel Castro, leader of the "26th of July Movement," who controlled large parts of the three eastern provinces. Cubans on both sides of the conflict were eager to obtain American support. Opposition representatives sought aid in Washington, and pro-Castro pickets in New York and Washington publicized his cause.[87] The rebels' kidnapping in February 1958, of Juan Fangio, a famous Argentine race driver, and of forty-seven United States servicemen and civilians from the Guantánamo base in June 1958, brought American public attention to the rebellion, and the fact that American-made weapons were being used by the *Batistianos.*[88]

The United States government adhered firmly to a policy of non-intervention, but was repeatedly accused by Castro of aiding the opposition, including the charge that Ambassador Earl E. T. Smith was plotting with Batista to provoke American intervention.[89] The basic policy, as defined by President Eisenhower on November 5, 1958, was simply "to keep out of anything so far as is humanly possible, in such things as that; except when our own citizens are involved, and then to take proper steps to protect them."[90]

In March 1958, shipments of American military supplies to Cuba were stopped completely. This was reaffirmed in 1959, as well as the intent of the United States to prevent the departure of aliens and American citizens who might go to Cuba to take part in the civil strife.[91] The cessation of American

[85] *New York Times*, Feb. 24, 1957.
[86] *Ibid.*, Feb. 26, 1957.
[87] *Ibid.*, Jan. 13, 1957.
[88] *Ibid.*, July 19, 1958.
[89] *Ibid.*, May 20, 1957; *Ibid.*, Oct. 27, 1958.
[90] Richard P. Stebbins, ed., *The United States in World Affairs, 1958* (New York, 1959), p. 357.
[91] Dept. of State, *American Foreign Policy Current Documents, 1958*, No. 7322 (Washington, 1962), p. 350; *Ibid., 1959*, No. 7492 (Washington, 1963), p. 328, 383.

shipments was a blow to Batista, for he needed both the weapons, and continued evidence of American good will. The rebels meanwhile complained about United States efforts to prevent gun-running from Florida.

Batista still had the upper hand in February 1957, in fighting off the revolutionary offensive, but being forced to use increased terror, he only succeeded in arousing the hostility of ever wider sectors of the population. As reported in the *New York Times*: "The public does not know who is doing the bombing, for the police thus far have caught only one or two small groups in Havana and none elsewhere. As a desperate measure of counter-terrorism, therefore, the police kill somebody virtually every time a bomb is exploded in Havana, riddle his body with bullets, put a bomb in his hand, and call the press photographers to come and take photographs. This procedure is sardonically called by *Habaneros*, 'Batista's Classified advertisement'."[92]

Ambassador Smith expressed his abhorrence of the "excessive police action" that he had personally witnessed, and was rebuked by Batista. The Department of State immediately contended that the Cuban allegation of meddling in internal affairs was unfounded.[93] That extreme violence was a deeply rooted power factor in Cuban politics was amply demonstrated.[94]

In December 1958, the Batista regime was obviously facing collapse, and several groups not identified with Castro endeavored to set up provisional governments. The United States withheld approval from these governments on the basis that such approval would constitute intervention. As his military and civilian support crumbled, Batista fled to the Dominican Republic on December 31, 1958. Castro had in August nominated Dr. Manuel Urrutía Lleo as president, and the new provisional government was recognized by the United States on January 7, 1959. According to Philip W. Bonsal, United States ambassador in Cuba from 1959 to 1961, the role of the guerrillas in bringing about the fall of Batista was exaggerated; however, by early 1958 most of the opposition elements were seeking to cooperate with Castro. The Communists were among the last to support him.[95]

The objects of the Castro Revolution had included standard liberal items–the restoration of the Constitution of 1940 and the electoral code of 1943–and land reform. After victory the constitutional objectives were forgotten. Like other successful revolutionaries Dr. Castro found it necessary to govern by propaganda and police, to shoot large numbers of political opponents, and to imprison many others who had supported him against Batista, but subsequently disagreed with his policies.

[92] *New York Times*, Feb. 25, 1957.
[93] *Ibid.*, August 1-3, 1957; Smith, *Fourth Floor*, p. 21.
[94] Ben G. Burnett and Kenneth F. Johnson, eds., *Political Forces in Latin America: Dimensions of the Quest for Stability* (Belmont, California, 1968), p. 176.
[95] Philip W. Bonsal, "Cuba, Castro and the United States," *Foreign Affairs*, 45 (January, 1967), 266; Dept. of State *Bulletin*, XL (Jan. 26, 1959), p. 128. And by the same author see: *Cuba, Castro and the United States* (Pittsburgh, 1971). In his astute analysis of the Cuban Revolution, *Castroism, Theory and Practice,* Chapter II, Theodore Draper rejects the idea that the Revolution was a peasant, working class, or middle class revolution. Instead, he terms it a declasse revolution whose leader has used one class or another or a combination of classes for different purposes at different times. He categorizes Castro as a "leadership type" who establishes a direct, personal, almost mystical relationship with the masses that frees him from dependence on the classes. See also James R. O'Connor, *The Origins of Socialism in Cuba* (Ithaca, 1970).

After Castro and his followers had seized power, the several political organizations which backed his movement were brought together in a single official party. The Cuban Communist Party, which had replaced the PSB, became the vehicle by which the regime extended its control into all facets of Cuban life. Membership in the party numbered only forty thousand, but the apparatus was expanded to include many branch associations. Castro headed the Political Bureau and the Secretariat, and appointed the members of the Central Committee. The old guard Communist members in the governing bodies were replaced by a Fidelista new guard, those personally loyal to Castro himself.[96]

With members of the Communist Party, one of the strongest in the Western Hemisphere, providing the ideology and organization, Fidel provided the charismatic leadership. Of his leadership one astute observer remarked, "Castro is one of those rare beings who are natural and spectacular leaders of men. The Cuban revolution was his personal creation; it could not have been born without him nor survived without him . . . that he was a convinced Communist from the day he acquired consciousness of the political world is unlikely. . . . Rather his character, his attitudes and his public remarks at different stages in the communization of Cuba, indicated that his growing penchant for Marxism-Leninism was not a matter of long-held conviction but of political convenience that his special talents for rationalization had no trouble in justifying."[97]

The regime established by Castro and the Communists became increasingly totalitarian in character, as party control became absolute, thereby betraying the democratic and liberal aims of many who had fought for the revolution. The economic objects of the revolution, however, were carried out vigorously. Industrialization was pushed and the confiscation of large properties and distribution of land to peasant cooperatives were carried out on a massive scale; programs were started on rural housing, rural education and medical services. These measures won the government the enthusiastic support of the peasantry, long accustomed to poverty and hopeless economic dependence. But they entailed diplomatic risks, because much of the property belonged to American concerns.

The sharp American reaction was anticipated, even welcomed by Castro. To sustain the momentum of the revolution, and having no clear-cut program, it became expedient for the "26th of July Movement" to represent the United States not only as an "imperialist exploiter" of the past, but as a present adversary. The American sanctioning of the inept counter-revolutionary landing at Cochinos Bay in 1961 lent credibility to this role.

With its progressive alienation from the United States the Castro regime showed a calculated disregard for the protocol and institutions of international diplomacy. Raúl Castro described it succinctly: "The revolution created a new type of diplomacy, a popular diplomacy." In practice this meant that governments of which Cuba approved were held responsive to the popular will, while those in opposition were labeled agents of United States imperialism. Dr.

[96] Robert F. Smith, *What Happened in Cuba? A Documentary History* (New York, 1963), pp. 316-317.

[97] Tad Szulc, "Clues to the Enigma Called Castro," *New York Times Magazine,* Dec. 9, 1962, p. 31. For scholarly analyses of Castro "evolving politically" see Draper, *Castroism: Theory and Practice,* and Suárez, *Cuba: Castroism and Communism.*

Raúl Roa, the foreign minister, proved most effective in upholding Cuban interests at international and regional bodies, notably the United Nations and the OAS. The government's views were carried abroad by members of the Cuban Workers' Federation (CTC), and the Federation of University Students (FEU). The government and its agents made a studied effort to become identified with any cause that could be considered embarrassing to the United States.

Castro had won much admiration in the United States for his efforts to overthrow Batista, and this was reflected in the open support given him in the American press. It was reaffirmed by the quick recognition given his regime by Washington. His image changed quickly, however, after the summary trials and executions of the *Batistianos* were begun. Castro visited the United States in April 1959, and at that time declared that Cuba was not neutral, but favored the West; he also denied any intention of confiscating foreign properties, and said Cuba wanted trade not aid. Although the visit was unofficial, Castro expressed displeasure afterwards that he had not been formally received.[98]

A series of incidents thereafter led to increasing hostility and the severing of diplomatic ties. The dismissal of Dr. Manuel Urrutia, the first president, and the defection of middle class professional men such as Pedro Díaz Lanz, former chief of the Cuban air force under Castro, who later confirmed Communist penetration of the Cuban government before a Congressional committee, and the exposure of a counter-revolutionary plot by members of this class, widened the breach.[99] Castro charged the United States with supporting Batista reactionaries. By the end of 1960 there were 100,000 exiles in the United States alone, chiefly represented by the upper and middle classes, of white collar workers and professionals; even more had gone to Latin American nations. Counter-revolutionaries flew aircraft over Cuba dropping leaflets, and in October 1959 Castro claimed that bombs were dropped on Havana; he implicated the United States in the incident. The nationalization of United States property began in 1959, and abusive remarks by Castro led to the temporary recall of the American ambassador in January 1960.[100] The explosion of the French vessel *Le Coubre*, carrying munitions to Castro, at the dock in Havana was attributed to United States-directed sabotage. Additional charges were that the United States was plotting an invasion with Guatemala. United Fruit Company holdings were expropriated, and three oil refineries, one British and two American, were seized after their managers refused to process Russian oil.

On July 6, in retaliation for the confiscation of the oil refineries, President Eisenhower eliminated the remainder of the Cuban sugar quota for 1960, valued at about $92 million,[101] an act Castro labeled as economic aggression in a complaint submitted to the United Nations Security Council, and later to the OAS. On the same day, the Cuban Council of Ministers authorized the president

[98] Predicting that Cuba and the United States "will become wonderful friends," Castro said in answer to a question about his feelings toward Soviet Premier Nikita Khrushchev: "We are against all kinds of dictatorship . . . That is why we do not agree with Communism." See *Cong. Rec.*, 86th Cong., 1st Sess., Vol. 195, p. 7100.

[99] *New York Times*, July 15, 1959.

[100] *Ibid.*, Jan. 24, 1960.

[101] Dept. of State *Bulletin*, XLIII (July 25, 1960), p. 140. For background see *Cong. Rec.*, 86th Cong, 2nd Sess, vol. 106, p. 13210.

and prime minister to nationalize American properties. Three days later USSR Premier Nikita Khrushchev warned: "In a figurative sense, if it becomes necessary, the Soviet military can support the Cuban people with rocket weapons . . . He pledged to help the Cubans oust the United States from the Guantánamo base, proclaimed the Monroe Doctrine dead, denied that the Castro regime was Communist, or that the Soviet Union sought a base in Cuba.[102] The Soviet Premier then notified Castro that his country would take over the 700,000 tons of sugar by which the American quota was reduced. A mass meeting was held to demonstrate Cuba's gratitude for the Soviet offer of aid, and Raúl Castro, visiting in Moscow, expressed his thanks for the offer. Castro himself indicated no previous knowledge of the promise of missile aid, but contended the Soviets had offered real, not "figurative" rockets.

Castro took steps in August to nationalize all major enterprises controlled by United States citizens. He made a stormy trip to New York in September, addressed the General Assembly of the United Nations, and flung abuse at his American hosts.[103] An exchange of accusations and counter-charges took place during the remainder of 1960 on the subjects of aggressive intentions, the Guantánamo base, and violation of agreements. Following the trial and execution of three American citizens belonging to a counter-revolutionary movement, the United States prohibited the export to Cuba of all goods except food and medical supplies, but continued to permit the entry of Cuban products.[104] Castro was concurrently seizing and nationalizing American property interests.

Incident to his accession to power, Castro was faced with the crisis created by the fall of the world price of sugar to a twenty year low. As a result, Cuba started out under his regime with a dangerous trade deficit, which was further increased by the discontinuance of American purchases of sugar in 1960, the decline of foreign investments and the stoppage of tourism. In an effort to find international markets for sugar, as well as sources of economic and technical aid, and of arms, Castro turned progressively toward Moscow and Peking. The USSR, the largest sugar producer in the world, replaced the United States as the subsidizer of the Cuban sugar economy, and this support, together with lesser aid from Communist China, buttressed the Castro government against economic collapse. In the process, Cuba became dependent on Moscow for its day-to-day existence.

Although diplomatic relations with the Soviet Union were severed in 1952, trade between Cuba and the Communist-bloc countries, mainly the USSR, grew to $42.6 million in 1957, and in 1959 Cuba sold $21.4 million in sugar to the Soviets. Early in 1960, as United States-Cuban relations were increasingly strained, Anastas Mikoyan, Chairman of the Council of Ministers of the USSR, came to Havana to conclude a sugar, aid, trade and assistance pact with Castro. The Soviet Union agreed to buy five million tons of sugar over a five year period on favorable terms, and granted a credit of $100 million repayable in twelve years.[105] This was followed by agreements for trade and technical assistance

[102] *New York Times*, July 13, 1960.
[103] *Ibid.*, Sept. 26, 1960.
[104] *Ibid.*, Oct. 13, 1960.
[105] *Ibid.*, Feb. 14, 1960.

between Cuba and the Eastern European satellite countries. In July came Khrushchev's bellicose statement and the committment to buy the sugar that Cuba was unable to sell to the United States. The theatrical Khrushchev–Castro meeting at the United Nations took place in New York in September, and later in the year Ernesto Guevara, then president of the National Bank, announced Cuba's firm support of the resolutions adopted by the recent international conference of Communist parties in Moscow, and added: "We are threatened every minute with American warships, planes and marines, but the friendly hand of the Soviet Union, extended to us, shields us from the enemy as invisible armor."[106]

It was also in 1960 that the Peoples' Republic of China began to see Cuba as a vehicle for introducing the "way of Mao Tse-tung" into Latin America. The Association of Latin American-Chinese Friendship was created in China in March 1960, and several highly-placed Castroites visited there. As the Cuban-United States crisis mounted in July, the Chinese, not to be outdone by the Russians, signed a commercial and tariff convention, an agreement on technical and scientific assistance, and a pact on cultural cooperation. Perhaps most important was China's committment to buy half a million tons of sugar for the following five years. To this was added the pledge, given by Minister Lu Hsu-chang, that Communist China would aid Cuba against her aggressors.[107] Cuba suspended relations with the Republic of China in Taipeh, Formosa, and diplomatic ties were then joined between Cuba and the Peoples' Republic of China. Castro announced this in the "Declaration of Havana" in his reply to the "Declaration of Costa Rica" which had been drawn up on August 28 by the foreign ministers of the American republics; without mentioning Cuba, the ministers had strongly condemned the intervention of an "extra-continental Power."

Dr. Guevara went from Moscow to Peking where he was cordially welcomed. Trade relations were reinforced by China's committment to buy a million tons of sugar in 1961, and to grant a credit of $60 million for the purchase of equipment, and technical assistance. This forced the Russians to be more generous, and in December 1960, the Soviet Union agreed to buy 2,700,000 tons of sugar at 4 cents a pound, in 1961.[108] These events coincided with the growing Sino-Soviet schism and polemic which began in the summer of 1960, and would complicate Cuba's relations with her two allies in the future.

The Sino-Soviet threat to the Americas via Cuba figured prominently in the presidential campaign in the United States during the fall of 1960. President Eisenhower had warned in July that the United States would never permit international communism to set up a regime in the Western Hemisphere,[109] after Khrushchev had threatened the United States with rockets if it intervened militarily in Cuba. This obliged the two nominees to take a position on the matter, and in the frequently caustic debates that took place, Senator John F. Kennedy linked Vice President Richard Nixon with the administration's policy failure that allowed the Castro movement to triumph. He charged that Nixon, incident to his visit to Cuba in 1955, had failed to recognize the need for

[106] MacGaffey and Barnett, *Cuba*, p. 335.
[107] *New York Times*, July 25, 1960.
[108] Suárez, *Castroism and Communism*, pp. 116-118.
[109] Dept. of State, *American Foreign Policy Current Documents, 1960*, No. 7624 (Washington, 1964), p. 207.

economic aid to Cuba, and had lauded the Batista dictatorship. Nixon denied that Cuba was lost, called Kennedy "defeatist" and held that the Eisenhower policies would lead to the freedom of Cuba.[110] As the campaign progressed Kennedy urged direct action through the OAS, and aid to freedom fighters and exiles as a means of overthrowing Castro. Nixon, secretly supporting the invasion plans then underway, called the Kennedy stand "dangerously irresponsible" possibly leading to civil war and involvement with the Soviet Union.[111]

The election of Senator Kennedy to the presidency coincided with Dr. Guevara's statements that Cuba was irrevocably linked with the Communist world, and that the island would become a model for armed revolution throughout Latin America.[112] Nevertheless, Castro made the sanguine statement that with Kennedy as president, relations between the two countries would begin anew providing, of course, the United States would take the initiative for their improvement. After Kennedy's announcement that he would not consider resuming relations, Castro called him the "illiterate millionaire."[113]

After proclaiming the imminence of an American invasion, Castro, on January 2, 1961, demanded that within forty-eight hours the personnel of the American Embassy in Havana be reduced to eighteen members. President Eisenhower announced, as a result, the severing of diplomatic and consular relations on January 3 saying, "This calculated action on the part of the Castro government is only the latest of a long series of harrassments, baseless accusations, and vilification. There is a limit to what the United States in self-respect can endure. The limit has now been reached . . . our friendship for the Cuban people is not affected . . . " This action had no effect on the status of our naval station at Guantánamo, for the treaty rights under which we maintain the station may not be abrogated without the consent of the United States.[114] The Swiss government was then requested to represent United States interests in Cuba.

Although American popular opinion generally agreed that "by breaking off relations with the present government of Cuba, the United States has done the only thing it could possibly do," the action did not go unquestioned. "The provocation . . . was abundant. The grave question is why an administration with sixteen days to live should have made such a fateful decision without first seeking the full counsel and consent of men who must live with Latin American problems in the next four years . . . it probably forecloses any chance for the Kennedy administration to examine the whole story of the deterioration of United States-Cuban relations and to explore with other OAS nations new formulae for dealing with the Castro frenzy . . . "[115]

[110] *New York Times*, Oct. 7, 1960.
[111] *Ibid.*, Oct. 21, 1960. Mr. Nixon later declared that he would have carried through the Cuban invasion, probably even before his inauguration, had he been elected in 1960, and with President Eisenhower's approval. See *Ibid.*, Feb. 14, 1967.
[112] *Ibid.*, Nov. 28, 1960; *Ibid.*, Dec. 11, 1960.
[113] *Ibid.*, Jan. 21, 1961; Dept. of State, *American Foreign Policy Current Documents, 1961*, No. 7808 (Washington, 1965), p. 283. Cuban propaganda agencies had earlier referred to "Caesar Augustus Eisenhower the First." Castro called President Lyndon B. Johnson "that ignorant cowboy from Texas."
[114] Dept. of State *Bulletin*, XLIV (Jan. 23, 1961), pp. 103-104.
[115] *Foreign Policy Bulletin*, vol. 40, no. 11, February 15, 1961, pp. 84-85. See also Vera Micheles Dean, "The Cuban Dilemma," *Ibid.*, February 1, 1961, pp. 76-80. For Senate debate see *Cong. Rec.*, 87th Cong., 1st Sess., vol. 107, pp. 87-91.

Early in his administration President Kennedy ordered a $4 million program for the relief of Cuban refugees, and restricted travel to Cuba; it was ruled that American citizens must have their passports specifically endorsed by the Department of State before going there.[116] The State Department's White Paper on Cuba was released by Washington on April 4, calling on the Castro government to cut its ties with communism, and to restore the revolution it had betrayed since coming to power. The document, written largely by Arthur Schlesinger Jr., stressed the friendship of Americans for the Cuban people, admitted past United States errors regarding Cuba, and backed the Cuban's continuing struggle for freedom. The anti-Batista and anti-Castro exiles received warm praise, while the Castro dictatorship was described as a menace to the Western Hemisphere.[117]

Meanwhile, during February and March 1961, reports increased about the training being given to exile military forces in Guatemala by American advisers. These referred to the officially secret, but compromised, plans for the invasion of Cuba which President Kennedy had inherited from the Eisenhower administration under the code name of "disposal problem." On April 17 some 1,500 men were landed at the Bahia de Cochinos on the south coast of Matanzas province. The invasion was launched from Nicaraguan bases with the United States furnishing supplies, and minimum military and naval support. A hoped-for uprising against Castro failed to develop, and the invaders were quickly defeated by superior forces.[118] The United States-CIA-directed effort, labeled the "perfect failure," dealt a sharp blow to the prestige of the Kennedy administration at home and abroad. On the occasion of the Kennedy-Khrushchev meeting in Vienna, in 1961, Khrushchev, because of the Bay of Pigs episode, "decided he was dealing with an inexperienced young leader who could be intimidated and blackmailed ... The Communist decision to put offensive missiles into Cuba was the final gamble on this assumption."[119]

President Kennedy's own conclusions were expressed in his "The Lesson of Cuba" address on April 20 in Washington: "Any unilateral American intervention, in the absence of an external attack upon ourselves or an ally, would have been contrary to our international obligations. But let the record show that our restraint is not inexhaustible . . . if the nations of this hemisphere should fail to meet their comittment against outside Communist penetration– then I want it clearly understood that this government will not hesitate in meeting its primary obligations which are to the security of our nation."[120]

From 1961 United States policy was to treat Cuba as an outlaw in the hemispheric system, to isolate her, and to maintain a strict embargo in hope of strangling the Cuban economy. Cuba, on its part, viewed the United States as an

[116] *New York Times*, Feb. 4, 1961.
[117] Smith, *What Happened in Cuba?*, pp. 312-326.
[118] See Haynes Johnson, *The Bay of Pigs* (New York, 1964); Arthur Schlesinger, Jr., *A Thousand Days: John F. Kennedy in the White House* (Boston, 1965); Tad Szulc and Karl E. Meyer, *The Cuban Invasion: A Chronicle of Disaster* (New York, 1962).
[119] James Reston, "What was Killed was Not Only the President," *New York Times*, Nov. 15, 1964.
[120] Dept. of State *Bulletin*, XLIV (May 8, 1961), p. 659. For the exchange of communications between President John F. Kennedy and Premier Nikita F. Khrushchev regarding events in Cuba during Apr. 1961, see *Ibid.*, pp. 661-667.

imperialist power bent on the destruction of its revolutionary government. This estrangement was accompanied by the closer alignment of Cuba with the Sino-Soviet bloc. Trade between Cuba and the Soviet Union, the Eastern European countries, and the Peoples' Republic of China, replaced the business that had formerly been carried on with the United States. As the American nations severed relations with Cuba, her dependence on the Communist-bloc countries was further increased.

In a television appearance on December 1, 1961, Castro declared: "I am a Marxist-Leninist, and I shall continue to be until the last day of my life." The motives behind this stunning pronouncement remain unclear, but evidently he did so to make his regime acceptable to the Communist-bloc, to help insure his leadership of the Communist movement in Latin America, and to obtain from the Communist-bloc the supplies he so critically needed.[121]

Starting in 1959 Cuban agents established liaison with revolutionary movements throughout Latin America, as Castro made it clear that his revolution was for export. Exiles from many countries had joined Castro's forces and they saw in his success the prospect of mobilizing popular support for overthrowing dictators and oligarchies. Reports of revolutionary activity came quickly from Panama, Nicaragua, the Dominican Republic and Venezuela. Several Cuban diplomats were expelled for interfering in the internal affairs of countries to which they were accredited. In March 1960 Castro denounced the Rio de Janeiro Pact of 1947, by which all of the American republics had agreed to unite in thwarting any Communist effort directed at any of them.[122] At San José, Costa Rica, in August 1960, the foreign ministers of the OAS agreed to condemn outside intervention in the affairs of this hemisphere, reaffirmed their faith in democracy, and condemned totalitarianism. But they were not prepared to take concrete steps aimed at the Communist offensive in general or Cuba in particular; Cuba was not named in the declaration. Secretary of State Christian Herter hailed it as a clear indictment, but most delegations interpreted it as an appeal to Castro to seek protection from the inter-American system rather than the USSR.[123]

From this point onward there was increasing evidence of Cuba's defection from the inter-American system and its transition into a Communist bridgehead in the Americas. With Communist-supplied armaments, Cuba built up the largest military force in Latin America, and was repeatedly charged with using its embassies in that area as centers of espionage and subversion. Within the United Nations the Cuban delegation clearly identified itself with the Communist bloc. Cuba opposed the resolution appealing to the Soviet Union not to explode the fifty megaton bomb; it was the only delegation in the UN, besides the ten avowed members of the Soviet bloc, to do so. In the same manner, Cuba alone joined the Communist bloc to oppose the resolution calling for a nuclear test ban treaty with international controls.[124]

[121] Suárez, *Castroism and Communism*, p. 141.
[122] Smith, *What Happened in Cuba?*, pp. 322-325.
[123] *New York Times*, Aug. 29, 1960.
[124] Dept. of State *Bulletin*, XLVI (February 19, 1962), pp. 273-274. Dr. Castro said on July 26, 1960: "We promise to continue making the nation the example that can convert the cordillera of the Andes into the Sierra Maestra of the American continent." See *Ibid.*, XLIII (Aug. 29, 1960), p. 319.

The Alliance for Progress, largely aimed at forestalling any further gains by the Cuba-Communist alliance, was ratified in August 1961. Although Castro derided the *Alianza* as "an alliance between one millionaire and twenty beggars," it provided Washington with a lever for bringing about unity in Latin America against Castroism, and at the same time lessened the prestige which the Cuban leader had gained on the continent for his triumph over the invasion effort. Venezuela, Peru, and Colombia broke off diplomatic relations, and others were to follow. Castro revealed his conversion to Marxism-Leninism in December, and in the same month the Assembly of the OAS made an agreement to hold a meeting of ministers in January to discuss the extra-continental threats against American solidarity.

At this meeting, convened in Punta del Este, Uruguay, in January 1962, the Castro regime was excluded from the inter-American system as a Marxist-Leninist regime incompatible with the democratic principles of the OAS. The vote was 14-1, with six of the largest countries abstaining; 16-1, four abstaining, for the arms embargo; and 20-1, to exclude Cuba from the Inter-American Defense Board. This occurred after the Inter-American Peace Committee had reported the introduction into Cuba of $60-100 million in arms from the Communist-bloc nations.[125]

In February Washington announced a nearly total embargo of Cuba that would deprive the regime of $35 million in annual revenue; exceptions were again made for medicine and certain foodstuffs.[126] With the economy faltering, Castro persuaded Peking to sign an expanded trade agreement for 1962, in May, and a definitive commercial treaty for the year was signed in Moscow; the latter authorized an increased commercial exchange, and the purchase of 3,200,000 tons of sugar.

At this time there were clear indications of economic breakdown in Cuba despite massive Soviet aid and technical assistance. Crowds rioted against food shortages, and inflation rose as Cuba went deeply into debt to the Communist bloc. Sugar production dropped from 6,800,000 tons in 1961 to 4,800,000 tons the following year, and efforts to industrialize at the expense of the sugar economy were seen as premature. Trade between the United States and Cuba which had totaled more than $1 billion in 1958 fell to $19.7 million in 1962; however, this was offset by trade with the Sino-Soviet bloc which grew from $42 million to about $1 billion in the same period.[127]

The stability of the regime and Cuban-Soviet ties were strained by Castro's purge of old-line Communists for trying to capture his revolutionary organization.[128] This crisis was passed by Soviet acceptance of Castro as the *jefe máximo* of Marxism-Leninism in Cuba, and the subordination of their own agents, the "old Communists."

In the meantime, amid Cuban charges of an imminent American attack, Cuba was being transformed into a Soviet missile-launching base. It was confirmed by photographic evidence that intermediate range ballistic missiles,

[125] *New York Times*, Jan. 4, 1962; *Ibid.*, Jan. 31, 1962.
[126] *Ibid.*, Feb. 4, 1962.
[127] Lewis Hanke, *Mexico and the Caribbean* (New York, 1967), p. 43.
[128] *New York Times*, Aug. 19, 1962.

capable of striking many cities in the United States, had been installed in Cuba. On October 22 President Kennedy initiated the confrontation: he denounced the presence of ballistic missiles on Cuban territory, called upon Premier Khrushchev to remove this threat to world peace under the inspection of the UN, ordered a blockade of all offensive military equipment to Cuba, and warned that any nuclear missile launched from Cuba would be countered by one directed at the Soviet Union. Nuclear war seemed a possibility until a peace formula was reached, the Russians agreeing to remove the weapons and aircraft. Castro's prestige suffered, but his refusal to allow UN representatives to inspect the missile sites caused further doubts and anxiety about their removal.[129] The Cuban leader's intransigence necessitated continued American aerial surveillance of the island, which was bitterly protested, and it also ruled out United States acceptance of any non-invasion pledge as requested by Khrushchev in the search for a peace formula. It was reassuring to the United States to have received the unanimous support of the OAS Council on the missile confrontation, but this was negated somewhat by the fact that Bolivia, Brazil, Chile, Uruguay, and Mexico retained diplomatic ties with Cuba.[130]

Dr. Castro had hoped that the missiles would insure the security of Cuba against any possible American attack, and also enable him to assert leadership of the Latin American anti-imperialist movement. Premier Khrushchev maintained that his country's sole aim in the affair was to defend "little Cuba" from the "imperialist monster." But whatever the intent, a rift occurred in Soviet-Cuban relations in consequence of the missile withdrawal. It was in this critical period that the Sino-Soviet polemic intensified, and Castro, not wishing to take sides in the dispute, declared that his role was that of a "partisan of unity." Castro had won increasing respect from the Chinese, but Peking lacked the resources needed to solve Cuba's pressing economic and military problems.[131]

Cuban-Soviet cordiality was resumed after the trade agreement for 1963 was signed in February. The Soviet Union granted a long-range credit, and allowed Castro to take a million tons of sugar from the annual quota assigned Moscow and sell it on the world market. The United States, in another move to effect the economic isolation of Cuba, ordered in July 1963, the freezing of $133 million in Cuba deposits in the United States, and took steps to prevent the Cuban use of American banking channels for moving funds around Latin America in support of subversion. In reprisal, Castro seized the embassy and grounds of the United States in Havana [132]

The OAS had shortly before agreed to recommend to its members measures to reinforce the isolation of Cuba. On July 17 President Kennedy declared that the United States would not and could not accept peaceful coexistence "with a Soviet satellite in the Caribbean." However, the agreement made in December 1962, to liberate the captured invaders and other prisoners in

[129]Dept. of State *Bulletin*, XLVII (Nov. 12, 1962), pp. 715-741; *New York Times*, Oct. 29, 1962; Foreign Policy Association, *The Cuban Crisis: A Documentary Record* (New York, 1963); *Cong. Rec.*, 88th Cong., 1st Sess., vol. 109, pp. 8969-8971.
[130]*New York Times*, Oct. 24, 1962.
[131]Suárez, *Castroism and Communism*, pp. 162-164; *New York Times*, Jan. 3, 1963.
[132]*Ibid.*, July 9, 1963; *Ibid.*, July 25, 1963.

exchange for food and medicine was completed with the arrival in Florida of 1,204 refugees.[133]

In August 1963, Dr. Castro announced a reorientation of his country's economy: agriculture, not industry, would be the main basis of the Cuban economy during the 1960s. The industrial program had proved faulty in conception and impossible to execute under conditions prevailing in Cuba. The failure in some measure could be attributed to the Communist-bloc countries which fell behind in deliveries of needed equipment.[134]

After the missile crisis of October 1962, the Soviet Union reaffirmed its policy of "peaceful coexistence" in contrast to the Castroite hard-line militancy against imperialism, which meant Cuban-supported subversion and guerrilla movements in Latin American countries. The American position precluded "peaceful coexistence," as mentioned, but Washington was not disposed to commit an "act of war," short of war; "all available instruments of force" would be employed to prevent Cuba from carrying out subversive activities, and the policy of isolating Cuba economically would be continued.[135] This policy was to continue after the death of President Kennedy, but it did not remain unchallenged.

Senator J. William Fulbright made a plea in March 1964, for the reevaluation of our policy toward Cuba. He recommended abandoning the myth that Cuban communism " . . . is going to collapse or disappear in the immediate future" and accepting that it would persist as a "disagreeable reality."[136] Castro, who had showed signs of moderating his position in the fall of 1963, and had acted with surprising restraint during Panama's anti-United States riots in January 1964, reacted favorably to the Fulbright speech.[137]

The Cuba Premier's conciliatory tone was reflected in interviews with Richard Eder of the *New York Times*, carried on over a three day period. He said that since the United States was demonstrating, through the Alliance for Progress, a willingness to accept a certain measure of social change in Latin America, and the revolutionaries had grown more mature, there was a possibility of reconciliation; that Cuba would withhold material support from Latin American revolutionaries if the United States and its allies would stop supporting subversive activities against Cuba. He hinted strongly that the USSR was urging him to ease tensions with the United States. Normalized relations, including trade, he said, would lead to talks on indemnification for seized United States properties.[138] Washington immediately rejected the "peaceful line," having conditioned any possible talks on Cuba's detachment from the USSR, and its cessation of subversive activities.[139]

On July 26 the OAS voted 15-4 to invoke sanctions against Cuba for aggression against Venezuela (Mexico, Chile, Uruguay, and Bolivia opposed). The resolution called for all member states to sever diplomatic, commercial, and

[133] *Ibid.*, July 4, 1963.
[134] Burnett and Johnson, *Political Forces in Latin America*, pp. 192-193.
[135] *New York Times*, Sept. 23, 1962.
[136] *Cong. Rec.*, 88th Cong., 2nd Sess., vol. 110, pp. 6227-6235.
[137] Draper, *Castroism: Theory and Practice*, pp. 47-48.
[138] *New York Times*, July 6, 1964.
[139] *Ibid.*, July 7, 1964.

maritime relations with Cuba. It was agreed that governments might invoke armed force against any new Cuban aggression without formal OAS action. Included was a declaration of hope that the Cuban people would restore a democratic government in their country.[140] Castro bitterly assailed the OAS action, declaring that Cuba had the right to aid revolutionary movements in any nation acting against her.[141] By September, Chile, Bolivia, and Uruguay had broken diplomatic ties with Cuba, leaving Mexico the only Latin America nation still extending recognition.

Dr. Castro visited the Soviet Union in January 1964, concluding a treaty whereby the USSR agreed to take 2,100,000 tons of sugar in 1965, and one million tons each successive year until the figure of five million tons was reached in 1968. The economy could thereby remain afloat in the immediate future. This meant that emphasis would again be placed more heavily on sugar production vis-à-vis industry, a decision reluctantly made earlier. Guevara had accepted the decision, but felt compelled to say publicly that there could not be "any vanguard country that did not develop its industry."[142]

President Dorticos arrived in the Soviet Union on October 14, 1964, to learn to his astonishment that Premier Khrushchev, Cuba's benefactor, had become an unperson a few hours earlier. Khruschev's deposition necessitated a change in Cuba's foreign policy, the problem being to disassociate the island from the Khrushchev line. As one means of publicizing his views, Castro chose to admit to Cuba, C. L. Sulzberger of the *New York Times*, who had waited two years for a visa. Castro indicated that his conciliatory tone of July should be "considered within a world framework," and referred to the time when "relations between all socialist countries and the United States improve."[143]

In whatever mood Castro might be found, the retention by the United States of the Guantánamo naval station was a constant irritant, as the Cuban leader's frequent denunciation of the base and its personnel testified. In February 1964, using as a pretext the seizure of four Cuban fishing boats off the Florida Keys, he provocatively cut off the water supply to the base. The Americans met the immediate crisis, and to prevent a repetition, installed a desalinization plant.[144] An enclave, the Guantanamo base is shut off from the rest of the island by barbed wire, trenches and fortifications. The Cubans surrounded the station with an elaborate fortified strip of their own. The Guantanamo naval station, together with those of Roosevelt Roads, Puerto Rico, and Key West, dominate the principal maritime entrances to the Caribbean.[145]

As part of its policy to isolate Cuba from the other nations of the Western Hemisphere, the United States imposed a strict travel ban on the movement of its nationals to the island. Under State Department regulations visas were issued only to journalists, persons engaged in humanitarian endeavors, and certain businessmen. The State Department's power to refuse the issuance of passports for travel to Cuba was upheld by a United States Supreme Court decision. Chief

[140] *Ibid.*, July 26, 1964.
[141] *Ibid.*, July 27, 1964.
[142] Suárez, *Castroism and Communism*, pp. 199-200.
[143] *New York Times,* Nov. 2; 4; 7; 8, 1964.
[144] *Ibid.*, Feb. 7, 1964; *Ibid.*, April 25, 1965.
[145] *Ibid.*, April 17, 1965; *Ibid.*, April 25, 1965.

Justice Earl Warren held the curb to be valid because it was based on a foreign policy affecting all citizens; it was justified on the grounds that the Cuban government was exporting revolution through travelers.[146] In another case the Supreme Court ruled that it was not a crime for a person with a valid passport to visit Communist countries where travel was prohibited by the State Department.[147]

While few Americans visited Cuba, and virtually none remained there voluntarily, the United States became the haven for a great number of Cuban refugees. The exodus began in 1959 and within two years there were some two hundred exile organizations in the United States. These appeared to fall within two major groups: (1) the conservative, anti-reformist followers of Batista, found in Miami, (also Guatemala and Nicaragua); and (2) those formerly identified with the *Auténticos* and *Ortodoxos*, many of whom had cooperated with Castro initially and had fought for him in the revolution, who wished a democratic government dedicated to social and economic reform. The second group merged in 1960 into the Democratic Revolutionary Front (DRF), under Manuel Antonio de Varona, and became dominant. The Movement for Revolutionary Recovery (MRR), professing similar ideals, was headed by Manuel Artime. The Revolutionary Program of the People (MRP), led by Manuel Ray, had a more revolutionary and leftist persuasion.

Prompted by Washington, these groups were coalesced in March 1961, as the Cuban Revolutionary Council, to form a united front against the Castro regime, and to organize a provisional government once he was ousted. José Miro Cardona headed the Council. After the Bay of Pigs debacle, the facade of unity was shattered, and organizational efforts met with little success.[148] Thereafter, counter-revolutionary hopes waned, and many became reconciled to an indefinite exile. Meanwhile, the exiles continued to leave the island in increasing numbers, reaching a total of 300,000 in 1964.

The migrant stream swelled in the fall of 1965 when the two governments signed an accord permitting the United States to operate an airlift for three to four thousand refugees a month. American officials hastened to explain that no détente was to be expected from the accord; that the policy of isolating Cuba, and maintaining a broad anti-Castro program carried out with other OAS members, remained unchanged.[149]

Doubtless one of Castro's objects was propaganda because he claimed that it was not he who had prevented Cubans from going, but the United States. Stipulating that men of draft age would not be permitted to leave, he might relieve the economic burden by disposing of unproductive elements. To rid the island of political opponents and to seize the property of refugees were other factors in his decision. By the summer of 1967 the waiting list of those desiring United States visas was estimated at more than fifty thousand. Once they had applied to leave Cuba, they lost their jobs.[150]

[146] *Ibid.*, May 4, 1965.
[147] The alleged conspiracy consisted of the recruiting and arranging the travel to Cuba of fifty-eight American citizens whose passports, although otherwise valid, were not specifically validated for travel to that country. See *Ibid.*, Jan. 11, 1967.
[148] Burnett and Johnson, *Political Forces in Latin America*, pp. 194-195.
[149] *New York Times*, Nov. 7, 1965. The US-Cuban agreement was interrupted and restarted several times. Suspended in May 1972, the flights had not been resumed in December 1972.
[150] *Ibid.*, July 25, 1967.

Among the problems raised by the refugees was the need to normalize their status. Senator Edward M. Kennedy sought a solution to the problem by proposing a bill to grant them permanent residence.[151] With regard to the problem of relocation, a Miami University study, financed by the Ford Foundation and government sources, concluded that the Cuban refugees were an economic and political asset to Miami that far outweighed the problem of their influx. The report urged a revision of the policy calling for immediate resettlement elsewhere, and suggested a gradual cut in fiscal aid to the refugees, and their assumption of community financial responsibilities. Over 150,000 had been resettled in American cities outside Miami, with some 150,000 remaining there.[152]

The United States continued to press its allies and countries receiving foreign aid into alignment with the boycott against Cuba. This had proved difficult as earlier experience with Canada, Great Britain, France, the Netherlands, Spain, and Yugoslavia had revealed.[153] Great Britain had in 1964, despite Washington's disapproval, authorized a major five-year credit for the purchase of buses. The second major transaction was authorized in January 1967, when the British government approved construction of a fertilizer plant in Cuba, valued at $39.2 million, by a British concern. In authorizing this project the British government had interpreted it as a legitimate business proposition, not involving strategic goods. Former Vice President Richard Nixon contended that such trading, with Vietnam as well as with Cuba, was one of the greatest failures of United States foreign policy, and he advocated "more effective diplomacy" in this field.[154]

These were small gains for Cuba, however, in the broad spectrum of its foreign relations, which had indeed become global. Aside from the omnipresent threat of the United States, Cuba's main concerns in foreign affairs included its relations with other Latin American nations in the context of the success of *Fidelismo*-inspired revolutions, the maintenance of material support from the USSR and Communist China, and the Sino-Soviet ideological conflict.

After the Punta del Este Conference of 1962 Cuba became progressively isolated from its Latin American neighbors. The decision of the foreign ministers of the OAS in July 1964, to suspend diplomatic and economic relations and to impose sanctions has been noted. Cuba was subsequently excluded from the summit meeting of American chiefs of state at Punta del Este in April 1967, and was condemned anew by a consultation of OAS foreign ministers, in September 1967. The latter, convoked to consider Venezuela's new charges of Cuban subversion, resulted in a vote to "forcefully" condemn Castro's Cuba, and to impose new sanctions on non-Communist trading with Havana. A United States-sponsored resolution to retaliate against ships that enter and leave Cuban ports was adopted.[155]

The OAS action came in the wake of a Communist conference in Havana which by resolution and implication posed further threats to the security of the

[151] *Ibid.*, Apr. 14, 1966.
[152] *Ibid.*, June 7, 1967; *Ibid.*, Oct. 2, 1967.
[153] *Ibid.*, Feb. 19, 1964; Dept. of State *Bulletin*, XLVII (Oct. 22, 1962), pp. 591-592; *Cong. Rec.*, 87th Cong., 2nd Sess., vol. 108, p. 19889.
[154] *New York Times*, Jan. 21, 22, 1967.
[155] *Ibid.*, Sept. 25, 1967.

Latin American states. The Organization of Latin American Solidarity (OLAS), approved a resolution condemning Soviet economic and technical policy in the Western Hemisphere. The Cuban delegation and other hard-line revolutionary contingents sponsored a resolution denouncing "certain socialist countries" that give credits and technical assistance to oligarchies and dictatorships in the hemisphere. This referred to the Soviet Union and Poland which had recently signed agreements with the governments of Colombia, Chile, and Brazil.[156] It meant that the majority rejected the Moscow line of abating revolutionary action. The movement lost its most spectacular, if not most effective leader, besides Castro himself, in the capture and death of Dr. Ernesto Guevara, in Bolivia, in October 1967. Indicative of Guevara's importance in the insurgency movement, OAS Secretary José A. Mora declared that his death meant that the Latin American guerrilla movement would be controlled.[157]

The Cuban relationship with the Soviet Union by the middle 1960s reflected the anomaly of increasing economic dependence, but a growing political independence, defiant, and as seen above, even hostile. Dr. Castro's grievances included the failure of his ally to guarantee the protection of Cuba against the United States, the question of economic aid, both as to type and quantity, the status of the Communist party in Cuba, Moscow's advocacy of restraint in revolutionary action in Latin America, and its preference for dealing with its controlled Communist parties in the Western Hemisphere.

In 1965 the USSR purchased over half the value of Cuba's total sugar sales, and the Communist-bloc as a whole took 80 percent of the island's sugar exports, as compared to 57 percent in 1964. Free-world trade with Cuba showed a marked decline, whereas the island's overall trade with the Communist countries increased. A new trade pact was negotiated in 1966 by which the USSR assured Cuba of a $90 million credit.[158]

Economic ties notwithstanding, it became painfully clear to Moscow that Castro would not allow Cuba to become a Soviet satellite, nor would he permit the Party to become a tool of Moscow's foreign policy. He had shown a lack of ideological delicacy by asserting that complete Communism "cannot be built in one country alone . . . ," a significant heretical deviation from the standpoint of the Communist Party of the Soviet Union, and had led the attack on the Soviet line of peaceful coexistence.[159] The split was apparent in the summer of 1967 when Soviet Premier Alexei N. Kosygin was in Cuba for six days on a mission reportedly designed to curb Cuba's pursuit of violent revolution in Latin America.[160]

[156] *Ibid.*, Aug. 10, 1967. The genesis of OLAS dated back to Jan. 16, 1966, in Havana. At that time there was a conference of the Afro-Asian-Latin American Peoples' Solidarity Organization (AALAPSO), made up of delegations from three continents, thus the meeting was called Tricontinental. AALAPSO aimed to unite and coordinate the struggle against colonialism and imperialism on three continents. The Latin American delegation formed from the parent group the Organization of Latin American Solidarity (OLAS). Castro extended its aims to the United States, linking it to the struggle for Negro rights, and advocating armed violence in American cities.

[157] *New York Times*, Oct. 10, 1967.

[158] Dept. of State, *The Battle Act Report, 1966* (Washington, 1967), pp. 33-34; *New York Times*, Feb. 13, 1966.

[159] Suárez, *Castroism and Communism*, p. 237.

[160] *New York Times*, Jun 28; 29, 1967.

The trial and sentencing of some forty Communist party members for "anti-party" (pro-Soviet line), activities, among whom was the veteran Communist leader Aníbal Escalante, gave further proof of the independent course the Cuban Premier was setting for his nation.[161] Yet, in spite of the obvious diplomatic insults inherent in the purges and other provocations, including the boycott of the Kremlin-sponsored Budapest Communist conference, there was little likelihood that the Soviet Union would cut Cuba adrift. The value to Moscow of having a Communist nation only ninety miles from the Florida coast was keenly appreciated by the Castro regime. Besides the high state of tension in Cuban-Soviet relations, there was a concurrent estrangement between Havana and Peking. One cause of the latter was clear: the Chinese decision to furnish Cuba with only 135,000 tons of rice for 1966 as against 250,000 in 1965.[162] But ideological and procedural differences were objects of speculation. In March 1966, Premier Castro referred to Mao's ideological arguments as "lightweight," and mentioned the senility of Chinese leadership.[163] Part of the problem appeared to stem from Chinese efforts to capture the Soviet-aligned parties in Latin America to the neglect of the more militant *Fidelista* groups. The Chinese attitude may also have been a reaction to the condemnation of splinter parties in Latin America, interpreted to be pro-Chinese, by the Latin American conference of "orthodox" Communist parties in Havana, in 1964. In spite of the polemic, Peking agreed to augment the rice shipments, and the Castro government acknowledged in the fall of 1967 that 9,600 tons of rice from the People's Republic of China had been distributed in Oriente province. The admission was interpreted as an improvement in relations.[164] Of relevance within the Havana-Moscow-Peking triangle, vis-à-vis the United States, Castro's position regarding the war in Vietnam—"to liquidate the criminal aggression with all available means and assuming the necessary risks"—was not consonant with the views of his allies.[165]

As these events occurred between Cuba and the major Communist-bloc countries, the United States government arrived at an expedient, yet judicious, policy toward Cuba:

> . . . the United States continues to regard the Communist regime in Cuba as temporary, and our goal remains a truly free and independent Cuba which, under a government democratically chosen by the People, will live in peace with its neighbors. As far as the United States government is concerned, two aspects of the present Cuban government's posture are not negotiable: the Cuban regime's political and military ties of dependence with an extra-continental Communist power, and its campaign of subversion in the

[161] *Christian Science Monitor*, Feb. 9, 1968. For a critical analysis of the divergences between Castroism and "official" Communism see the three works by Ernst Halperin: *The Sino-Cuban and the Chilean Road to Power; Castro and Latin American Communism; The Ideology of Castroism and its Impact on the Communist Parties of Latin America*, published in 1963 by the Center for International Studies, M.I.T., Cambridge, Mass.

[162] *New York Times*, Jan. 3, 1966.

[163] *Ibid.*, Mar. 15, 1966.

[164] *Ibid.*, Sept. 27, 1967.

[165] "These cries of anguish are understandable. For the failure of the Soviet Union to take adequate measures against the bombardment of North Vietnam demonstrates the fundamental weakness of the socialist camp, and thus the bankruptcy of Castro's foreign policy, which led Cuba into that camp." See Ernst Haiperin, "The Castro Regime in Cuba," *Current History* (Dec. 1966), p. 358.

hemisphere. We do not consider the present government of Cuba a direct threat to the United States or Latin America under present circumstances–nor shall we permit it to become such a threat . . . We and the other members of the OAS do regard that regime as a focus of subversion in the hemisphere . . . We have been pursuing and we intend to pursue–within the framework of the inter-American system to the greatest extent possible courses of action designed on the one hand, to reduce the will and ability of the present government of Cuba to advance the Communist cause in Latin America through sabotage and terrorism: and on the other hand, to assist the nations of this hemisphere in strengthening their ability to resist subversion. Related to the latter course of action are the long-range objectives of the Alliance for Progress, which seeks, through social and economic development of the hemispheric countries, to eliminate the conditions which furnish a fertile field for subversion and revolution . . .[166]

In 1972 Cuba began its thirteenth year under the revolutionary government of Fidel Castro as a communistic, personalized state, unique in the hemisphere, and singular in the Communist world. Within the relatively brief historical dimensions of the Castro era enough evidence had accumulated to warrant several broad generalizations about the nature and development of the regime, and its implications for the future. Cuba became important because, as the first Communist state in the Americas, it was enveloped in cold war power politics. In the process, the island found its political and economic orientation outside the hemisphere, in the Communist bloc. As a member of that bloc its governmental apparatus fell in the hands of the Cuban Communist party, a structure which Dr. Castro took over and molded in his image; the regime became Communist in form, the personal dictatorship of Fidel Castro in substance.[167] There had been major dislocations in the economy, including an ill-advised, although later reversed, movement toward industrialization. Foreign and Cuban properties were nationalized without due compensation to the original owners, and more than half the farm land of the nation was brought under state control. Many Cubans who had shared in their country's wealth prior to 1959, had been dispossessed of their share and went into exile; 460,738 came to the United States in the period February 1961-March 1972. A polarization of Cuban opinion existed, with the supporters of Dr. Castro on one side, and the opponents on the other, with the most effective elements of opposition residing outside Cuba. Among Castro's opponents the majority tended to agree that they did not want to see a return to pre-1959 conditions.

The Cuban Revolution and the Castro government added to the complexity of social and political change in Latin America, and set back the growth of democracy in the region. Instead of the traditional Conservative-Liberal configuration, a third element, totalitarian in character and allied

[166]Dept. of State *Bulletin*, LIV (May 2, 1966), pp. 712-713. *Cong. Rec.*, 89th Cong., 1st Sess., vol. 111, pp. 24357-24359.

[167]*Ibid.* The first party formed by Castro was the O.R.I. (*Organizaciones Revolucionarias Integradas*). "Old militant" (former members of the *Partido Socialista Popular*) efforts to dominate it led to its dissolution. The O.R.I. was replaced by the P.U.R.S.C. (*Partido Unido de la Revolución Socialista de Cuba*), and this was succeeded by the P.C.C. (*Partido Comunista de Cuba*). The Central Committee was headed by a Politburo of which Fidel and Raúl Castro were first and second secretaries respectively. See David D. Burks, "Cuba Seven Years After," *Current History* (Jan. 1966), p. 38.

with the Communist bloc, was introduced into the struggle. Further confusion, adverse to the growth of popular democracy, resulted from the collaboration of *Fidelistas* and Communists with the established conservative elements to thwart the aims of the political moderates, and this in turn fostered the growth of militarism.

It was the concensus of informed sources that by 1972 the revolution was well established, but unfinished and perhaps flickering; that the Cuban story would remain fluid in the years ahead. The Castroites had found it easy to liquidate the existing social, economic, and political structure, but when it came to creating a new one an appalling number of blunders were made. Nevertheless, improvements were observed in housing, public health, roads, child care, and in education at all levels. Educational advances were most apparent in vocational and professional studies, and the schools were commonly linked to the military services. The University of Havana, under tight control since 1959, suffered another purge of its student body in April 1965. The regime denounced "intellectualism," and stressed the value of practical knowledge and training.

In accord with the revolutionary aim of "equality," the poorest and most backward elements, particularly in the rural areas, had been "leveled up." Cuban Negroes, for the first time, had equal status with whites, economically as well as socially. The status and attitudes of some working class women were seen to have changed dramatically as the government put their babies in state nurseries at forty-five days, and put the mothers to work. Divorces occurred with unprecedented frequency.

Marxist education was found in education at all levels, and there was no freedom of the press, speech, assembly, or any of the freedoms which make up the liberal form of democracy. More than a third of the people were neighborhood spies. Castro declared on September 28, 1971, in a speech commemorating the eleventh anniversary of his so-called Committees for Defense of the Revolution (CDR), that membership had grown to 3.5 million members. Prior to Castro's seizure of power in 1959 there were fifty-eight newspapers published in Cuba, including five in Havana. In 1971 all had been silenced except for two Communist dailies in Havana, *Granma* and *Juventude Rebelde*, and one in each of the five other provinces. Atheism was officially adopted, but since the influence of the Catholic Church had been destroyed in 1961, the regime did not interfere with religious practices. In view of the fact that the armed forces, numbering 200,000, equipped with the most modern armament that $20 million a year in Soviet aid could provide, was rated the strongest in Latin America, the police apparatus was very efficient, and Dr. Castro retained much of his personal popularity, it appeared that the revolutionary government would retain control in the foreseeable future.[168]

With the early revolutionary dreams of a huge industrial program discarded, the new goal was an agricultural paradise that would end food shortages and create a hard currency market abroad. Emphasis was again placed on sugar as the backbone of the island's export economy, but this raised the problem of food supply as food-producing land was planted in sugar cane. The

[168]*New York Times*, Aug. 31, 1967; *Times of the Americas*, Oct. 20, 1971; *Ibid.*, Oct. 27, 1971; *Ibid.*, Feb. 9, 1972.

sugar-dominated, state-controlled economy remained precarious, reaching its lowest point in 1969, seemingly the poorest performance in a decade. The gross national product in 1969 approximated that of 1958 when the island's population was 6.6 million as opposed to 8.4 million in 1970. Sugar continued to account for 85 percent of all Cuban exports despite efforts to diversify the tightly regulated economy. There were shortages of consumer goods, and petroleum and a number of basic food products were rationed. Even sugar was placed on the rationed list. Premier Castro declared that the 1969 sugar harvest was an "agony for the country."[169]

The Soviet Union's subsidy of the Cuban economy–which virtually kept the economy afloat–was estimated in 1972 to be about $750 million a year, or $2 million a day. Cuba's debt to the USSR approached $4 billion exclusive of military aid and the premium price paid for sugar. In 1970 Cuba was assigned 5.5 million tons as its quota for export to the Communist bloc. However, the Soviet Union, the world's largest sugar producer and an exporter of the product, did not need the amount pledged by Havana. The Cuban sugar harvest in 1971 stood just under six million tons, while Castro had said that the projected seven million ton goal was insufficient for the country's needs. Christmas and New Year's were postponed by Castro for the third consecutive year so as not to interfere with the sugar harvesting.[170]

Late in 1971 a newly-formed Soviet-Cuban "economic collaboration" commission held its first meeting in Havana with the vice-president of the Soviet Council of Ministers in attendance. It was announced that thereafter the Soviet Union would take a more positive role in Cuban economic affairs. The old-line Cuban Communist, Carlos Rafael Rodríguez, rather than Castro, was the chief negotiator for the Cuban side.

The record for 1971 showed that some gains were made by industry and fishing, but these were offset by declining production of rice, coffee, tobacco, cattle, and fruit. Strict rationing of food and consumer goods was continued. Sugar represented about 85 percent of the exports with nickel accounting for 10 percent and tobacco 3 percent. Thanks to growing domestic needs and declining exports, Cuba's annual trade deficits had been rising, especially with the USSR, which provided 60 percent of the island's imports. Cuba's failure to achieve satisfactory economic growth was attributed to a lack of trained personnel, governmental inefficiency, and a growing apathy among workers and peasants. The latter led to a new vagrancy law in March 1971, under which nonworking men can be sentenced to two years for idleness. Castro reported there were 400,000 such persons, whom he called "parasites of the revolution."[171]

As the decade of the 1970s began there were indications that Cuban attitudes toward Latin America and the latter's relations with Cuba might be approaching a turning point. In a speech on May 31, 1970, Dr. Castro declared that he would establish diplomatic relations with any country willing to do business with Cuba. Possibly to confirm the new open-door policy, Cuban

[169] *New York Times*, Jan. 26, 1970.
[170] *New York Times*, Jan. 28, 1972; *Times of the Americas*, Feb. 9, 1972; *San Francisco Examiner and Chronicle*, Oct. 17, 1971.
[171] *Ibid.*

transport planes carrying medical supplies, food, doctors, and nurses were sent to aid Peru's earthquake victims following the serious earthquake which struck the northern part of the nation May 31. Significantly, it was the first time that the Castro government had contributed to disaster-relief activities in a Latin American country.

Premier Castro traveled to Chile abroad a Soviet jetliner for a twenty-five day visit in November 1971, his first visit in South America since 1959. On his return the aircraft made brief stops in Peru and Ecuador, leading to speculation that Cuba might seek to reopen ties with those countries. Although the Cuban leader's visit to Chile, the Latin American country most ideologically hospitable to him, may have indicated a desire to end his country's isolation from the rest of the Western Hemisphere, it may have suggested other political implications as well. A wide chasm existed then between the Soviet-supported revolutionary Marxism that Castro imposed on Cuba, and which he unsuccessfully tried to export throughout Latin America, and the regime of Dr. Salvador Allende, an upper middle class professional politician, who was guiding his country toward socialism through parliamentary processes. What seemed significant was that the Cuban Premier's visit included overt gestures toward placating, and urging restraint on the extremist Chilean Marxist revolutionary groups. Another possible initiative by Cuba to end its diplomatic and economic isolation from its hemispheric neighbors was seen in the dispatch of a Cuban delegation to participate in the Lima conference of the "Group of 77," a formal organization of developing countries.

The Cuban actions came at a time when there was growing interest among several Latin American nations to reincorporate Cuba into the inter-American system, although the trend was neither consistent nor uniform. The decision by the Chilean government headed by Eduardo Frei Montalva to sell $11 million worth of agricultural products in 1970 and 1971 to Cuba had marked a departure from the OAS-sponsored economic boycott of the island. Jamaica's admission into the OAS while maintaining consular relations with Cuba was interpreted as a softening in attitudes toward Cuba. Trinidad and Tobago, newer members of the OAS, allowed Soviet ships coming from Cuban ports to refuel at the Port of Spain, a violation of the OAS ban on such practices. Mexico never accepted the original OAS suspension motion, maintaining diplomatic and economic ties with Cuba, the only country in Latin America to do so, throughout the 1960s. The nations which displayed varying degrees of interest in restoring relations with Cuba included Mexico, Barbados, Trinidad and Tobago, Jamaica, Venezuela, Colombia, Ecuador, Peru, and Chile.[172]

What some regarded as a significant step toward reincorporating Cuba into the hemisphere system was its admission to membership in the "Group of 77," mentioned previously, a group of ninety-five developing nations, which convened in Lima in October 1971. On that occasion ideological differences were submerged as the developing nations sought to form a common front to secure their interests vis-à-vis the advanced nations, taken to include both the United States and the USSR.[173]

[172] *Christian Science Monitor*, Apr. 11, 1970; *Ibid.*, Sept. 17, 1971. Chile and Peru ultimately took the initiative in restoring diplomatic ties with Havana.
[173] *Ibid.*, Oct. 30, 1971.

In spite of the trends supporting a belief that several hemispheric nations favored ending the OAS sanctions on Cuba, the governments in question had failed to raise the issue strongly before the OAS. Doubtless the belief that Cuba continued to export its revolution by training agents for subversion and sabotage activities in the agent's own countries remained a major obstacle to Cuba's acceptance. Lending substance to this fear was Cuba's circulation of a forty-one page Minimanuel of the Urban Guerrilla, which appeared as an article in the Havana-based *Tricontinental*, a magazine published by the Organization for Solidarity of the peoples of Africa, Asia, and Latin America. Sent to Latin America and elsewhere, the handbook cites kidnapping and other violent acts as models for terrorists in cities. A rumor that Cuba was renewing its efforts to promote revolution in Latin America using Santiago, Chile as a base, attributed to an unconfirmed report of a Cuban defector in London, also created unease in the region. Moreover, Cuba's establishment of diplomatic relations with the South Vietnamese National Liberation Front did not engender optimism about a change in character of the Castro regime. Cuba's seizure of two U.S.-based freighters under Panamanian registry heightened tensions further in the Caribbean. Havana charged that one of the ship's captains, a Cuban-born, United States citizen, was an employee of the U.S. Central Intelligence Agency, and that the ships had been used in covert, hostile operations against Cuba. Washington categorically denied the allegations.[174]

With growing sentiment from some quarters within the hemisphere for the readmission of Cuba to the inter-American community, the question was raised concerning possible changes in the Washington–Havana relationship. A U.S. volleyball team visited Cuba in August 1971, and the cordial reception given its members by Dr. Castro led to speculation that a détente might be in prospect; the ping pong diplomacy with the People's Republic of China was cited as a precedent. Moreover, Washington's attitude toward China and Cuba seemed to be inconsistent. Any optimism about such a turn of events, however, was dashed by unqualifiedly negative statements from both heads of state. In a speech pledging Cuban solidarity with Latin American revolutionary movements and the "oppressed" there, Premier Castro said, "We are not seeking conciliation of any kind with Yankee imperialism . . . there will be no concessions of any kind to the imperialists." In his administration's foreign policy message to the Congress in February 1972, President Nixon declared that the United States will not change its policy toward Cuba until the Fidel Castro regime ceases helping subversive groups in Latin America and cuts its millitary ties with the Soviet Union. "Regrettably, Cuba has not abandoned its promotion of subversive violence." The Nixon administration's position on the Cuban question was reaffirmed later in the year in the Republican Party's campaign platform, which states that Cuba was "ineligible for readmission to the community of American states."[175]

Chilean President Salvadore Allende visited Cuba in December 1972 as the

[174] *Ibid.*, Apr. 25, 1970; *Ibid.*, Dec. 29, 1971.

[175] Dept. of State *Bulletin*, LXV (Oct. 11, 1971), pp. 391-395; *Times of the Americas*, Feb. 16, 1972; *Christian Science Monitor*, Aug. 5, 1971, *Congressional Quarterly*, XXX (Aug. 26, 1972), 2147.

last stop on a tour that took him to Peru, Mexico, New York and the UN, Algeria, and Moscow. He was the first Latin American chief of state to visit Premier Castro since he seized power in 1959, and the Cubans paid him the highest honors. While in Moscow the Chilean leader had referred to the Soviet Union as his country's "big brother," incident to requesting and obtaining promise of further economic, military and moral support for his regime.

Dr. Castro visited the Soviet Union later in the month, and reported on his return that Moscow had taken the initiative in "attempting to resolve our special problems." This included extending the terms of Cuba's large foreign debt beyond twenty years. Payment on Cuba's obligations will begin in 1986 and extend over a twenty-five year period; interest charges have been suspended until 1986. The island's economic problems showed few signs of any early easing, as admitted by the Cuban leader when he said that "years of hard work, sacrifice and struggle lie ahead."[176]

For the United States the Cuban Revolution brought Latin America to life after years of indifference and neglect. It produced the Alliance for Progress, which was devised to meet the challenge of the Revolution. Cuba became a pawn in the cold war rivalry between the United States and the Soviet Union, culminating in the confrontation of the 1962 missiles crisis. Revolutionary Cuba was also to attain significance within the compass of the cold war as a modular effort to achieve rapid and successful modernization under a Marxist-inspired dictatorship. The Soviet Union, with large vested interests in the Cuban experiment, went to great lengths to insure its success, whereas the United States government, supporting the evolutionary approach to national development through democratic processes, stood in opposition. The outcome of Cuba's development, therefore, was of critical significance for the United States, the USSR, Latin America, and the other so-called economically underdeveloped areas of the world whose paths to progress remained undecided.

United States relations with Cuba had reached a stalemate, and it became increasingly difficult for North Americans to contemplate Cuba with equanimity or objectivity. It could not be otherwise, for the island was a Communist state, ninety miles from Florida, bent on subverting other Latin American countries, and implacably hostile.

Revolutionary Cuba provided the United States, as well as the Soviet Union, with an object lesson in diplomacy needed for survival in the thermonuclear age. During the missiles crisis of 1962, President Kennedy exercised moderation in spite of severe criticism. The channels of communication were kept open at all times, the United States expressing a willingness to negotiate not only on the Cuban question, but on any other points in East-West tension which the Soviet Union might see related to it. The blockade policy was aimed at a clearly-defined objective: to remove the missiles, and it was executed precisely and without subterfuge, a contrast to the clouded atmosphere surrounding the Bay of Pigs debacle.

In the great debate over the Cuban problem one of the fundamental issues was the intent in this day of the Monroe Doctrine which, since 1823, the United

[176] *Times of the Americas*, Jan. 10, 1973; *Christian Science Monitor*, Jan. 15, 1973.

States had regarded as the cornerstone of its policy in the Western Hemisphere. Given the unalterable principle of "hands off" the hemisphere which, owing to the primacy of national security interests, could not conceivably be changed, Washington's problem was to fit this basic tenet into a new "combined system of policy" applicable to existing world conditions. Unable to find any satisfactory options, the United States found itself compelled to deal with the Cuban issue in the context of the cold war, weighing the risks of any given course of action toward Cuba against the possible repercussions in other parts of the world.

SUPPLEMENTARY READINGS

Aguilar, Luis E. *Cuba 1933: Prologue to Revolution.* Ithaca, 1972.

Allison, Graham T. *Essence of Decision; Explaining the Cuban Missile Crisis.* Boston, 1971.

Bonachea, Rolando and Nelson Valdes, eds. *Che: Selected Works of Ernesto Guevara.* Cambridge, Mass., 1970.

Bonsal, Philip W. *Cuba, Castro and the United States.* Pittsburgh, 1971.

Boorstein, Edward. *The Economic Transformation of Cuba.* New York, 1968.

Burks, David D. *Cuba under Castro.* New York, 1965.

Calcott, Wilfrid H. *The Caribbean Policy of the United States, 1890-1920.* Baltimore, 1942.

Carbonell, José M. *Cuba independiente*, Vol. XII, of Ricardo Levene (ed.), *Historia de América.* Buenos Aires, 1949.

Chapman, Charles E. *A History of the Cuban Republic.* New York, 1927.

Draper, Theodore. *Castro's Revolution: Myths and Realities.* New York, 1962.

————. *Castroism: Theory and Practice.* New York, 1965.

Cabrera, Raimundo. *Mis malos tiempos.* Havana, 1920.

Cárdenas, Raúl de. *La política de los Estados Unidos en el continente americano.* Havana, 1921.

Fagen, Richard R. *The Transformation of Political Culture in Cuba.* Stanford, 1969.

Ferguson, Erna. *Cuba.* New York, 1946.

Fitzgibbon, Russell H. *Cuba and the United States, 1900-1935.* New York, 1964.

Foreign Policy Association. *Problems of the New Cuba.* New York, 1935.

————. *The Cuban Crisis, A Documentary Record.* New York, 1963.

Goldenberg, Boris. *The Cuban Revolution and Latin America.* New York, 1965.

Guerrera y Sánchez, Ramiro. *Sugar and Society in the Caribbean, An Economic History of Cuban Agriculture.* New Haven, 1964.

Guevara, Ernesto. "The Cuban Economy, Its Past, Its Present Importance," *International Affairs*, XL (Oct. 1964).

Guggenheim, Harry F. *The United States and Cuba.* New York, 1934.

Halperin, Ernst. *Castro and Latin American Communism.* Cambridge, Mass., 1963.

————. *The Ideology of Castroism and Its Impact on the Communist Parties of Latin America.* Cambridge, Mass., 1963.

Hill, Howard C. *Roosevelt and the Caribbean.* Chicago, 1927.

Horowitz, Irving L., ed. *Cuban Communism.* Chicago, 1970.

International Bank for Reconstruction and Development. *Report on Cuba.* Baltimore, 1951.

Jackson, D. Bruce. *Castro, the Kremlin and Communism in Latin America.* Baltimore, 1969.

Jenks, Leland H. *Our Cuban Colony.* New York, 1928.

Johnson, Haynes. *The Bay of Pigs.* New York, 1964.

Johnson, W. F. *The History of Cuba.* 5 vols. New York, 1920.

Jones, C. L. *The Caribbean Since 1900.* New York, 1936.

Langley, Lester D. *The Cuban Policy of the United States.* New York, 1968.

Lockmiller, David A. *Magoon in Cuba.* Chapel Hill, 1938.

Lockwood, Lee. *Castro's Cuba, Cuba's Fidel.* New York, 1967.

MacGaffey, Wyatt and Clifford R. Barnett. *Twentieth Century Cuba: The Background of the Castro Revolution.* New York, 1965.

Matthews, Herbert. *The Cuban Story.* New York, 1961.

Mesa Largo, Carmelo, ed. *Revolutionary Change in Cuba.* Pittsburgh, 1971.

Munro, Dana G. *The United States and the Caribbean Area.* Boston, 1934.
————. *Intervention and Dollar Diplomacy in the Caribbean, 1900-1921.* Princeton, 1964.
Nelson, Lowry. *Rural Cuba.* Minneapolis, 1950.
O'Connor, James R. *The Origins of Socialism in Cuba.* Ithaca, 1970.
Ortiz, Fernando. *Cuban Counterpoint: Tobacco and Sugar.* New York, 1947.
Ortiz, Rafael M. *Cuba, los primeros años de independencia.* 2 vols. Paris, 1921.
Pereda, Diego de. *El nuevo pensamiento política de Cuba.* Havana, 1943.
Perkins, Dexter. *The United States and the Caribbean.* Cambridge, Mass., 1947.
Pflaum, Irving P. *Tragic Island: How Communism Came to Cuba.* Englewood Cliffs, 1961.
Phillips, Ruby H. *Cuban Sideshow.* Havana, 1935.
————. *Cuba, Island of Paradox.* New York, 1959.
Plank, John N., ed. *Cuba and the United States: Long Range Perspectives.* Washington, 1967.
Rippy, J. Fred. *The Caribbean Danger Zone.* New York, 1940.
Seers, Dudley, ed. *Cuba, The Economic and Social Revolution.* Chapel Hill, 1964.
Smith, Earl E. T. *The Fourth Floor.* New York, 1962.
Smith, Robert F. *The United States and Cuba, Business and Diplomacy, 1917-1960.* New Haven, 1960.
————. *Cuba Before Castro.* New York, 1966.
————, ed. *Background to Revolution: The Development of Modern Cuba.* New York, 1966.
Stein, Edwin C. *Cuba, Castro and Communism.* New York, 1962.
Suárez, Andres. *Castroism and Communism, 1959-1966.* Cambridge, Mass., 1967.
Szulc, Tad and Karl E. Meyer. *The Cuban Invasion, A Chronicle of Disaster.* New York, 1962.
Tondel, Jr., Lyman M. *The Inter-American Security System and the Cuban Crisis.* New York, 1964.
United Nations. *Economic Survey of Latin America.* New York, 1967.
Welles, Sumner. *Relations Between the United States and Cuba.* Department of State Latin American Series, No. 7. Washington, 1934.
Welpin, Miles D. *Cuban Foreign Policy and Chilean Politics.* Lexington, Mass., 1972.
Weyl, Nathaniel. *Red Star over Cuba.* New York, 1961.
Williams, William A. *The United States, Cuba and Castro.* New York, 1962.
Wood, Bryce. *The Making of the Good Neighbor Policy.* New York, 1961.
Wright, Philip G. *The Cuban Situation and our Treaty Relations.* Washington, 1931.
Wohlstetter, Roberta, "Cuba and Pearl Harbor: Hindsight and Foresight," *Foreign Affairs,* XLIII (1965).
Wilgus, A. C., ed. *The Caribbean: British, Dutch, French, United States.* Gainesville, 1963.
————. *The Caribbean: Its Economy.* Gainesville, 1962.
Zeitlin, Maurice. *Revolutionary Politics and the Cuban Working Class.* Princeton, 1967.
———— and Robert Scheer. *Cuba, Tragedy in our Hemisphere.* New York, 1963.

11

Puerto Rico—A Study in American Territorial Government

In the foreign relations of the United States previous to the war with Spain, Puerto Rico[1] had generally been regarded as a sort of natural appendage to Cuba. In the public statements made by American statesmen regarding Cuba, mention was sometimes made of Puerto Rico, but, even when nothing was said, it was generally understood that Puerto Rico would follow in the wake of Cuba if that island should ever transfer its allegiance from Spain. Perhaps that is one of the reasons why, in the *Foreign Relations of the United States*, Cuba plays such a prominent part, while Puerto Rico is virtually unmentioned. Undoubtedly another reason is the uneventfulness of the island's history as compared with that of Cuba. Not having the importance which Cuba derived from its size and its proximity to the mainland, and possessing no mineral resources to incite the cupidity of the early Spanish adventurers, Puerto Rico, from the time of its discovery by Columbus until its surrender to the United States, probably received as generous treatment from the Madrid government as obtained anywhere in the great Spanish colonial dominion.

So long as Spain held most of the Americas south of the Rio Grande, she could hardly be expected to pay much attention to such an insignificant possession as Puerto Rico. However, during the wars for independence Spain came to a realization of the importance of attaching her colonies to herself through fair dealing; and a royal decree of 1815 was very generous. The result was a heavy influx of whites, many of them Spanish sympathizers, driven from

[1] The change in spelling from Porto Rico to Puerto Rico was approved by Act of Congress, *S.J. Res.* 36, and signed by the President on May 17, 1932. The United States Geographic Board adopted the spelling Puerto Rico in 1891 but changed back to Porto Rico in 1900 to conform to the spelling found in the treaty of annexation.

the revolting countries of the empire. Although the governor-general of Puerto Rico had powers as despotic as those of the governor-general of Cuba, there was considerably less complaint of their abuse. In 1870 the island was made a province of Spain with representation in the Spanish Cortes. A little later, in 1874, the provincial constitution was abrogated. But the provincial deputations were reestablished in 1877; only now they were partially dependent upon the governors-general. The next step towards self-government did not come till 1897, when Spain granted a form of autonomous government similar to that offered to Cuba. But the troops of the United States took over the island before its organization under the new system was completed.[2]

From October 18, 1898, when the Spanish troops withdrew, until May 1, 1900, when the Foraker act establishing civil government went into effect, the island was under military rule. The guiding principle under the three military governors, Generals Brooke, Henry, and Davis, was to interfere as little as possible with the existing native institutions. The greatest difficulty was the complete lack of knowledge, on the part of the Americans, of the methods of administration under the local laws. As General Davis writes: "Judicial procedure was strange, and the temperament, mode of life, and manners of the inhabitants differed greatly from those with which Americans are conversant. . . . At the time of my arrival not a page of the voluminous laws of the island, all of Spanish origin, had been translated into English. Those laws, upon which the whole fabric of society was based, were as a sealed book and had been so to my predecessors."[3]

The problems of the military government included policing the island, remedying the lack of sanitation, reorganizing the judicial system, and introducing financial reforms. There was much disorder and murder; pillage and arson were common offenses of the natives according to American standards, and an immediate and thorough cleaning up was essential. In the first six months of the American control, more than 750,000 Puerto Ricans were vaccinated, smallpox was stamped out, and steps were taken to isolate lepers. The reorganization of the courts and legal procedure was the most difficult problem, for, as General Davis reported: "The system of laws that prevails here is the outgrowth of quite a thousand years of human experience, and can not be struck down or radically changed in a day nor yet in a year."[4] For instance, proceedings under the writ of *habeas corpus* were authorized; but through inability to understand the purpose of the writ it remained virtually a dead letter. The district insular courts were reduced from fifteen to six; and each of the sixty-nine towns retained its municipal court, though upon a modified and improved basis. The insular supreme court of appeals was retained; while for the trail of offenders violating United States law a provisional court under the authority of the president was created, consisting of one law judge and two army officers.[5] The stamp taxes placed upon all deeds, notes, and bills of exchange,

[2] For a description of the government preceding 1897 and the proposed reforms, see the *Report of Brig.-Gen. G. W. Davis, House Doc. No. 2*, 56th Cong., Ist Sess., p. 484.
[3] *Ibid.*, p. 483.
[4] *Ibid.*, p. 502.
[5] *Ibid.*, p. 504. See also Leo S. Rowe, *The United States and Porto Rico* (London, 1904), chap. XII, for an excellent survey of the reorganization of the judicial system.

and the *consumo* tax upon necessities of life, such as bread and rice, were immediately abolished by the military government, and an excise tax was substituted.

The military government was faced with new and difficult problems, and it solved them in a highly creditable manner. In fact, Dr. Rowe was of the opinion "that during the early stages of our occupation of Porto Rico military government was not only desirable, but the only means of solving the immediate problems of government."[6] But, as he also said, a military regime could not be continued, because such a system is repugnant to the political standards of the American people. The principal reason why civil government was not sooner established was the fear of Congress that, on this basis, legislation in regard to the insular possessions might be subject to the restrictions of the Constitution. To avoid misunderstanding on this point, the Foraker act provided that native inhabitants should be deemed citizens of Puerto Rico instead of citizens of the United States, and tariff duties of 15 percent of the Dingley tariff were placed upon goods imported into the United States from Puerto Rico, which was possible only if the island was regarded as foreign territory.[7] There was considerable uncertainty as to whether such territory as Puerto Rico, once incorporated into the United States, could be considered "foreign." But the insular cases, particularly Downes *vs.* Bidwell, determined that Puerto Rico had not become a part of the United States for purposes of the constitutional provision "that all duties, imposts, and excises shall be uniform throughout the United States." The court's ruling is often summarized by the phrase, "the Constitution does not follow the flag." Justice White reasoned in the concurring opinion: "To concede to the government of the United States the right to acquire, and to strip it of all power to protect the birthright of its own citizens and to provide for the well-being of the acquired territory by such enactments as may in view of its conditions be essential, is in effect to say that the United States is helpless in the family of nations. ... Although the House of Representatives might be unwilling to agree to the incorporation of alien races, it would be impotent to prevent its accomplishment, and the express provisions conferring upon Congress the power to regulate commerce, the right to raise revenue, and the authority to prescribe uniform naturalization laws would be in effect set at naught by the treaty-making power."[8]

Furthermore to help rehabilitate the bankrupt Puerto Rican economy, the Foraker act exempted the island from the payment of Federal internal revenue taxes, and authorized the reversing to the Puerto Rican treasury of monies collected by Federal excise taxes on Puerto Rican products sold on the mainland, as well as monies collected by imposition of United States customs duties on foreign products imported into Puerto Rico.

As established by the Foraker act, the government of the island included, first, a governor appointed by the president for a four-year term, who had the powers generally conferred upon governors of organized territories, such as to

[6] *Ibid.*, p. 127.
[7] *U.S. Stat. at Large*, vol. XXXI, chap. 191, sects. 2 and 7.
[8] 182 U.S. 288. See also the analysis by Dr. Rowe, *op. cit.*, chap. III, and that by P. Capo Rodríguez, *Amer. Jour. of Int. Law*, vol. X, pp. 317ff., and vol. XIII, pp. 483ff.

grant pardons and reprieves, appoint certain officials, veto laws, and see that the laws are executed. He was to be assisted by an executive council consisting of a secretary, attorney-general, treasurer, auditor, commissioner of the interior, and commissioner of education, also appointed by the president for four years, and by five others who were required to be native inhabitants. This executive council was also to act as the upper branch of the legislative assembly, the lower branch being a house of delegates consisting of thirty-five members elected biennially by the qualified voters. The council was to prescribe the regulations and date for the first election. The courts already established by General Davis were continued, except that a United States district court was provided to take the place of the provisional court.[9]

Perhaps the most noteworthy feature of this system of government was the executive council, which had both administrative and legislative powers; although a somewhat similar organ was found in the American colonies prior to the independence. It was to be expected that this body, whose members were for the most part experienced in parliamentary practice, would take ascendancy over the lower house, whose members were uniformly devoid of parliamentary training. In fact, nearly two-thirds of all the bills passed in the first session originated in the upper chamber. As summarized by Charles H. Allen, the first civil governor, the results of this first session consisted not so much in what the body did as in what it refrained from doing, for among the thirty-six acts passed "not one foolish expression of the legislative will is to be found, and not one of these acts will be speedily repealed."[10]

Governor Allen's conclusions on the success of the new plan of government are worthy of consideration. In the first place, he pointed out that American occupation found the island inhabited by a race of people of different language, religion, customs, and habits, with virtually no acquaintance with American methods, and with the commerce and trade in the hands of the Spaniards. Therefore, with the people wholly unfitted, without careful training and preparation, to assume the management of their own affairs, he felt that "Congress went quite as far as it could safely venture in the form of government already existing in the island, . . . and with good men devoted to the work, the island will develop faster under such form, its people through experience and education will advance more rapidly in their knowledge of the civic virtues under a guidance of present methods than could be gained in any other way."[11]

That the Puerto Ricans were not entirely satisfied with the system of government is evidenced in the successive reports of the governors. In the third annual report (for 1902-1903) Governor William H. Hunt declared that the majority of all the people wanted territorial government, but that, in his opinion, a change in the government at that time would be a mistake. The creation of a House of Delegates had already conferred much power upon the people, considering their autocratic government in the past. The inhabitants

[9] *U.S. Stat. at Large*, vol. XXXI, chap. 191. For an analysis of the workings of the government see W. E. Willoughby, *Territories and Dependencies of the United States* (New York, 1905), chap. IV.

[10] *First Annual Report of Charles H. Allen, Governor of Porto Rico* (Washington, D.C., 1901), p. 76.

[11] *Ibid.*, p. 98.

were justified in looking forward to full incorporation within the political body of the United States, but such a transition could not be successfully consummated without relying heavily upon the young people still in the schools, who would not be able to take their share of responsibility for some years to come.[12] The complaint sometimes raised that the Puerto Ricans were not given a full share of the appointments seems hardly just, for a summary found in the fifth annual report indicates that of the total number of government employees 2548 were Puerto Ricans, as compared with 313 Americans; while in salaries the Puerto Ricans were getting $1,220,567 as against $355,200 received by Americans.[13]

One complaint of the Puerto Ricans, however, was fully justified, namely, that based on the uncertainty of their status as citizens. Governor Winthrop, in 1905, made it clear that this situation was creating a spirit of discontent and unrest prejudicial to the American administration. As he pointed out, under the Spanish regime Puerto Ricans were classed as citizens of Spain, and it was naturally difficult for them to understand why citizenship of the sovereign country should be denied to them under the more free and liberal government of the United States. He urged very strongly the granting of citizenship by Congress, on the ground that it would greatly improve the feeling of loyalty with which the Puerto Ricans regarded the United States, and would instill in them a healthy feeling of patriotism as being citizens of the country and not merely citizens of the dependency of the country.[14]

Another serious cause of discontent was the fact that the governors were, for the most part, political appointees of the president, and had neither experience nor interest in the problems of colonial government. Ignorant as they were of the Spanish language, rarely holding the office more than a year or two (Governor Yager's seven-year term is the notable exception), it could hardly be expected that the governors would be qualified to appreciate the Puerto Rican point of view. Mr. Frank H. Richmond, who served as assistant attorney-general of Puerto Rico, and later as associate judge of the district court of San Juan, characterized the Puerto Rican appointees as "noted for the seriousness with which they take themselves and the amateurish pettiness of their activities; the calmness with which they assume that they are doing these people good and that the little they know of business and administration is superior to methods in vogue on the island for three centuries."[15] The bitter opposition on the part of the Puerto Ricans to Governor E. Mont Reily, President Harding's appointee, and the serious charges brought against him, indicated that stalwart political henchmen do not furnish the best material for satisfactory governors of our colonial possessions.[16]

Almost from the beginning of the island's administration under the Foraker act, a marked antagonism was manifested on the part of the House of Delegates towards the executive council, and on several occasions serious

[12] *Third Annual Report of the Governor of Porto Rico* (1903), p. 13.
[13] *Fifth Annual Report* (1905), p. 41.
[14] *Ibid.*, p. 42.
[15] *The Eclectic Magazine*, vol. CXLVII (Dec., 1906), p. 487.
[16] See the interesting characterization of Mr. Reily by Mr. Davila, the Puerto Rican representative at Washington in Congress, *Cong. Record*, Mar. 2, 1922.

deadlocks occurred. In 1909 the situation was brought to a head when the lower house refused to vote the appropriation bills for the ensuing year unless the upper house would accept certain legislative measures which were generally conceded to be very radical in their scope, and in one case in direct violation of the Foraker act. The executive council refused to recede, and the session terminated with the appropriation measures unpassed. The governor immediately called a special session, which merely resulted in another deadlock. He then cabled to Washington a full statement of the situation,[17] and President Taft, in a special message, laid the matter before Congress, then in extra session. Pointing out the advantages already secured by the Puerto Ricans under American control, the president recommended an amendment to the Foraker act providing that, when the assembly should adjourn without making the necessary appropriations, sums equal to the appropriations of the previous year should be available from the current revenues.[18] Congress immediately took up the subject, and the so-called Olmstead Amendment was passed, which provided that "if at the termination of any fiscal year the appropriations necessary for the support of government for the ensuing fiscal year shall not have been made, an amount equal to the sums appropriated in the last appropriation bills for such purpose shall be deemed to be appropriated." This amendment brought the organic act into agreement, in this matter, with the organic acts of Hawaii and the Philippines.[19]

In order to obtain further information President Taft directed the secretary of war, Mr. Dickinson, to visit the island and investigate the situation. The emissary was accompanied by General Edwards, chief of the Insular Bureau, whose bureau now had jurisdiction over the island by virtue of the second paragraph of the amendment of 1909; also by Colonel Kean of the medical department. Secretary Dickinson's report made it clear that on two subjects Puerto Rican sentiment was unanimous, namely, citizenship and the organization and selection of the upper house. On the first the inhabitants demanded the granting of American citizenship to all Puerto Rican citizens collectively, and on the second they asked the substitution of an elective senate for the existing executive council, thus separating the executive from the legislative functions. Secretary Dickinson advised that American citizenship should in some way be conferred upon the Puerto Ricans, but not collectively, as there were many of Spanish descent who did not wish to become citizens. In regard to the senate, he recommended that a senate be created, to consist of thirteen members, all citizens of Puerto Rico or of the United States and resident for not less than one year in Puerto Rico, eight to be appointed by the president and the remainder to be elected by the Puerto Ricans. The separation of the executive and legislative departments was also recommended. Other suggestions were that elections for all insular offices be held only once in four years instead of biennially; that a department of health be created, with jurisdiction over matters of health, sanitation, and charities; and that a department of agriculture, commerce, and labor also be added. In conclusion, the secretary gave it as his opinion that the

[17] *Sen. Doc. No. 40*, 61st Cong., 1st Sess., p. 7.
[18] *Ibid.*, p. 1.
[19] *U.S. Stat. at Large*, vol. XXXVI, Part I, p. 11.

people of Puerto Rico had, on the whole, an excellent government, and that the people of the United States could look with just pride upon the administration of affairs they had given there. The island was peaceable and generally prosperous; health conditions had improved; and the people were, in the main, contented. There were many criticisms as to the judgment, ability, sense, and industry of various officials who had been there since American occupation. But the universal testimony was that the administration had been honest and free from favoritism.[20]

As a result of this investigation, a bill was introduced in Congress by Mr. Olmsted, chairman of the Committee on Insular Affairs, embodying substantially the recommendations enumerated. In a well documented and comprehensive speech the author of the measure pointed out the conditions and needs of the island, and urged the passage of the bill as a means of ending the unsatisfactory status of the island–that of "a disembodied shade in an intermediate state of ambiguous existence," as Chief Justice Fuller expressed it.[21] The lower house finally passed the bill, amended in several matters. But it was by no means satisfactory to the Puerto Ricans. In a message to Congress dated February 23, 1910, the House of Delegates protested vigorously against its passage, on the ground that it curtailed, rather than increased, the liberties granted by the Foraker act. Incidentally they also indicated in no uncertain language their dissatisfaction with the latter act, declaring that "the regimen by the Foraker act is absurd and despotic, installing the higher house by presidential appointment without the intervention of the people, mingling and confusing the legislative and executive powers, and leaving the Porto Ricans without a definite and recognized personality in the law of nations." Inasmuch as the minority report of the Committee on Insular Affairs very strongly upheld the Puerto Rican contention, the Olmsted bill never became a law.[22]

The next attempt to remedy the political situation in Puerto Rico came in March 1914, when Mr. Jones, chairman of the Committee on Insular Affairs, reported out a bill to supersede the Foraker act. No action, however, was taken at this time. Early in 1916 another bill of similar character was introduced and unanimously recommended for passage by the Committee on Insular Affairs. This time the sentiment in Congress was very favorable, and a brilliant speech made by the commissioner from Puerto Rico, Mr. Muñoz Rivera, in behalf of the bill was received with applause.[23] This measure, commonly known as the Jones Act, passed both houses after considerable debate and became the organic law of Puerto Rico on March 2, 1917.[24]

Although this act did not go quite so far as some of the Puerto Ricans wished, it did attempt to meet the demands of the majority, and it successfully remedied the chief defects of the Foraker act. In the first place, a bill of rights guaranteed the individual rights and liberties of the Puerto Ricans in exactly the same manner that these rights are guaranteed by the federal constitution and the constitutions of the various states. American citizenship was granted collectively,

[20] *House Doc. No. 615*, 61st Cong., 2nd Sess. Text of a proposed bill included.
[21] *Cong. Record*, vol. 45 (May 25, 1910), pp. 6862-6876.
[22] *House Report No. 750*, 61st Cong., 2nd Sess., Pt. 2.
[23] *Cong. Record*, vol. 43 (May 5, 1916), p. 7470.
[24] *U.S. Stat. at Large*, vol. XXXIX, p. 951.

but provision was made that those who did not wish to become American citizens could avoid it by making a declaration to this effect. The executive department was made to consist of a governor appointed by the president, and the heads of the following departments: justice, finance, interior, education, health, and agriculture and labor. Of these executive heads, the Attorney-General and the Commissioner of Education were to be appointed by the president with the consent of the Senate of the United States, while the other four were to be appointed by the governor with the advice and consent of the Senate of Puerto Rico. These departmental heads were to constitute an executive council; but all legislative duties were withdrawn. Another important executive officer was the auditor (also appointed by the president), whose duty it was to examine, audit, and settle all financial accounts of the central and municipal governments of the island.

The legislative department was to consist of two houses, a senate of nineteen members elected for terms of four years, two each from the seven senatorial districts and five at large, and a house of thirty-nine members, also elected quadrennially, one from each of the thirty-five representative districts and four at large. The executive council was given the power to arrange the district; and, with a view to preventing gerrymandering, the districts were required to be contiguous and compact and of approximately equal population. Future redistricting was given over to the legislature of Puerto Rico. The legislature was to meet biennially, and the governor was authorized to call special sessions if, in his opinion, the public interest required it. Bills were to be passed by a majority vote, except in case of a veto by the governor; if vetoed, repassed by a two-thirds vote, and still refused approval by the governor, a bill must be transmitted to the president of the United States, who by signing it would make it law. Furthermore, all laws enacted by the insular legislature must be reported to the Congress of the United States, which reserved full power of annulment.

The qualifications for voters were to be prescribed by the legislature, except that American citizenship and age of twenty-one or over must be requirements, and property qualifications must not be imposed. The electors of Puerto Rico were authorized to choose a resident commissioner to the United States for a four-year term, whose salary of $7500 was to be paid by the United States, and who was granted approximately the same privileges (except voting) possessed by members of the House of Representatives.

The courts already organized were to be maintained, but jurisdiction over their organization and procedure was given to the legislature, except in the case of the District Court of the United States for Puerto Rico. The judge of this tribunal was to be appointed by the president, and the range of its jurisdiction was defined in the organic act.

The government under the new act was promptly put into effect, and in his report for 1917 Governor Yager declared that the measure was received evervwhere with satisfaction. Celebrations were held in various municipalities of the island manifesting the gratification of the people. Under the provision for the renunciation of citizenship, only about 290 in the whole island made application, while more than 800 persons born in Puerto Rico of alien parents made a voluntary declaration of allegiance, thus accepting American citizenship.[25] The first elections under the act were conducted in an orderly fashion in spite of many spirited contests. In the first legislative session some friction arose between the governor and the senate, principally over the senate's slowness to confirm the governor's appointments, and it culminated in an attempt of the legislature to take a recess of two and one-half months instead of *sine die*. The governor refused to accept this interpretation of the organic law. But a compromise was agreed upon whereby the legislature reassembled pursuant to adjournment, adjourned immediately, and then reconvened at the call of the governor for a special session.[26]

Upon the entrance of the United States into the first World War, the Puerto Ricans found themselves shouldered with the obligations and duties of citizenship before they had fairly begun to enjoy its benefits. The selective service act was extended to Puerto Rico, and more than 120,000 men were registered under its provisions. Officers' training camps were established and some 425 men received commissions. Liberty Loan campaigns were carried on and the total amount subscribed amounted to more than $12¼ million. Campaigns for food conservation and production were undertaken, branches of the Red Cross and YMCA were organized. As an example of the people's generosity, in a campaign for new Red Cross headquarters in San Juan $10,000 was raised within twenty-four hours by the subscriptions of forty-five merchants.[27]

The ideals of the first World War so nobly maintained while the nations were actively participating, so sadly distorted in the period of settlement, are still a vital part of the world's imponderables. World peace and good will cannot obtain until the peoples of the various nations and states shall adopt such principles and practices in their international relations as have proved effective and satisfactory in their domestic intercourse. Nor can the exercise of these principles be confined merely to the relations between the so-called sovereign states. The relationships between sovereign states and their colonies, dependencies, protectorates, spheres of influence, and territorial possessions must be an important factor in the world politics of the future. The obligations of the great powers, the states whose importance and prestige are such that they are made mandatory or protecting powers, must be fulfilled in the broad spirit of world humanity. Areas of exploitation will always be breeding-places for international jealousies and disputes, and the United States has publicly proclaimed its adherence to a policy of unselfish guardianship of the rights of minorities and backward peoples. Does its policy, as exemplified in the control of Puerto Rico, square with these doctrines?

The unfortunate appointment by President Harding on May 11, 1921, of E. Mont Reily, a time-serving politician, to the post of governor, did not square with these doctrines, and hardly did the new governor arrive on the island before he was involved in serious difficulties with the Puerto Ricans. His dismissal of administrative officers and judges without hearing or without cause, and the filling of their posts by wholly inexperienced politicians, his insistence on

[25] *Report of the Governor of Porto Rico, 1917*, p. 1.
[26] *Ibid., 1918*, p. 4.
[27] Clarence Ferguson, "The People of Porto Rico and the War," *Overland Monthly*, vol. LXXIII (Apr., 1919), p. 300.

increases in salaries for his particular henchmen, his utter disregard of Puerto Rican interests and needs,[28] soon brought about a demand on the part of the Puerto Rican Senate that he be dismissed, and a petition to this effect was laid before the Senate of the United States on March 7, 1922.[29] A series of resolutions was introduced in the Congress of the United States to investigate conditions in Puerto Rico, but Governor Reily finally relieved the situation by resigning on February 16, 1923. The appointment of Representative Horace M. Towner of Iowa, a fortnight later, did much to allay the increasing hostility of the Puerto Ricans, and the president of the Puerto Rican Senate immediately cabled his congratulations upon the excellence of the choice.

Governor Towner, who, as chairman of the House Committee on Insular Affairs, was thoroughly acquainted with the situation in the island, entered upon his duties April 6, 1923. He declared that he hoped to lead but not to coerce, to advise but not to impose, and he carried out this policy. Speaking before the House of Representatives on January 11, 1924, in favor of permitting the Puerto Ricans to elect their own governor, Resident Commissioner Davila thus characterized Governor Towner: "At present we have in Porto Rico an able, diplomatic and sagacious executive. Governor Towner is one of the best governors the island has ever had. . . . Should our people have the power to elect their own governor, Horace M. Towner might be the choice of the people. He has won the hearts of the people by his fairness and justice and his sympathetic understanding of their methods."[30] When in 1927 it was rumored that Governor Towner was to be removed because of ill health a delegation was immediately despatched to Washington, headed by the president of the Puerto Rican Senate and the speaker of the House, to present a concurrent resolution and a memorial urging the retention of Governor Towner "because of his great ability as an executive, his untiring energy and his devotion to the fulfilment of his duties, because of the absolute spirit of justice and impartiality which forms his character . . . his retention is absolutely indispensable for the continuation of the work undertaken by him for the solution of our social, economic, and political problems. . . ."[31]

Nevertheless, the Puerto Ricans realized that so long as governors were appointed by the President of the United States, and largely upon a political basis, a Mont Reily might be their future lot as often as a Towner. Therefore in January, 1924, a delegation representing all the political parties in the island and headed by Governor Towner came to Washington to plead for certain changes in the Jones Act which might bring about full self-government under American sovereignty. Perhaps the most important change desired was the right of the islanders to elect their own governor.

After hearings by committees of both houses, a bill acceptable to the various political parties of Puerto Rico, to Governor Towner, and to the War Department of the United States[32] was drawn up. The outstanding feature was

[28] See the speeches of the Puerto Rican Representative, Mr. Davila, in the *Cong. Record*, vol. 62, Part 4, pp. 3301-3309; Part 5, pp. 5024-5031; Part 12, pp. 13170-71.
[29] *Ibid.*, Part 4, p. 3479.
[30] *Ibid.*, vol. 65, Part I, p. 861.
[31] *United States Daily*, Mar. 20, 1927.
[32] *Cong. Record*, vol. 65, Part 9, 8599 ff.

the provision that, beginning with the year 1932 and thenceforth, the qualified electors of Puerto Rico should elect the governor, provided that he be subject to removal by the president of the United States for due cause and to impeachment by the Puerto Rican legislature. The bill also provided that the office of vice-governor be abolished, and that the head of the department of education and the attorney-general, as well as the other heads of departments, be appointed by the governor with the advice and consent of the Senate of Puerto Rico instead of by the president of the United States with the consent of the United States Senate. The bill also provided for a separate department of labor.

Although this bill was passed unanimously by the Senate on May 15, 1924, and received the approval of the Insular Affairs Committee of the House, it failed to come to a vote in the House. In 1926 the House did pass such a bill but on this occasion the session adjourned before the Senate could again act. Although President Coolidge had in 1925 favored the principle of an elective governor, in a letter to Governor Towner dated February 28, 1928, he showed no sympathy for Puerto Rican autonomy and declared that "Porto Rico had a greater degree of sovereignty over its internal affairs than does the government of any State or Territory of the United States."[33] A very elaborate and impartial study made at this time by a group of experts under the auspices of the Brookings Institution proposed as a more satisfactory arrangement an amendment of the Organic Act with a view first to perfecting the island's government machinery and then making the appointment of Governors from the island itself.[34]

The island was able to appreciate the friendly attitude of the government and people of the United States when in the autumn of 1928 it was visited by the most destructive hurricane on record in the West Indies. Due to advance warnings the loss of life was not over 300, but the property damage reached almost $100 million. A half million people were left homeless, without food or clothing. The $10 million coffee crop, ready for harvest, was almost completely destroyed. The American Red Cross expended over $3 million in relief and rehabilitation, while the United States government appropriated $2 million for repair and rebuilding of schoolhouses and roads and another $6 million for loans to be made to individual planters and fruit growers.[35] In spite of this assistance the conditions in the island for a considerable period afterwards were pitiful in the extreme. And hardly had the island partially recovered from the disaster of 1928 before it was again stricken by the terrific hurricane of 1932.

In President Hoover and his appointee as governor of the island, Theodore Roosevelt, the Puerto Ricans found sympathetic and useful friends; both labored to bring about an amelioration of conditions in the island. President Hoover visited the island in person and urged Congress to make further appropriations in its behalf. Governor Roosevelt lost no opportunity to appeal for assistance both publicly and privately, for his investigations showed that "poverty, sickness and hunger stalk the island."[36] It was during the Hoover administration that

[33] *Hearings Before Committee on Insular Affairs*, 70th Cong., 1st Sess., May 16, 1928 (Washington, D.C., 1928), p. 27.
[34] *Porto Rico and Its Problems* (Washington, D.C., 1930), chap. V.
[35] *29th Annual Report of the Governor of Porto Rico* (Washington, D.C., 1930), pp. 7-9.
[36] Theodore Roosevelt, "Puerto Rico," *Foreign Affairs*, vol. XII (Jan., 1934), p. 271.

Congress, by joint resolution and in accordance with the desire of the people of Puerto Rico, changed the spelling of the name to Puerto Rico, the original and correct form.[37]

The Democratic platform of 1932 had promised ultimate statehood for Puerto Rico and the politicians in the island were not disposed to allow the promise to be forgotten. But the Roosevelt administration realized that the economic situation of the island was more critical than the political, and appointed a committee of eminent Puerto Ricans to formulate a plan for economic rehabilitation. The Puerto Rico Policy Commission, as it was called, was headed by Dr. Carlos E. Chardón, Chancellor of the University of Puerto Rico, and the report is generally known by his name.[38] The program as outlined in the Chardón Report recommended: the purchase of sugar lands and mills by the United States so as to cut production of sugar and at the same time distribute the marginal lands in the form of ten-acre subsistence farms; the reduction of holdings of large landowners for redistribution to the landless agricultural class; the rehabilitation of the coffee, citrus, and tobacco industries; the establishment of farm credit; and the introduction of new industries. Upon receipt of this report an interdepartmental committee was appointed in Washington which studied the plan and made important modifications.[39] The Puerto Rico Reconstruction Administration, organized in 1935 to carry out this program, was allotted $42 million by the Works Progress Administration, over one-half of which was allocated to the rehabilitation of agricultural areas.[40]

In spite of these well conceived and strongly supported measures for improving the island's economic status, the political elements became increasingly critical of the United States. Whereas until 1932 Puerto Rican parties favored statehood or independence by normal political action,[41] at that time the Nationalist Party was organized by a Harvard trained lawyer, Pedro Albizu Campos, advocating the immediate constitution of the island as a republic and resistance to the authority of the United States by violent means. Numerous clashes with the police occurred, one at the University on October 24, 1935, resulting in the deaths of several Nationalists. By way of retaliation, on February 23, 1936, two young Nationalists, inspired by the exhortations of Campos, shot down in cold blood Colonel Francis F. Riggs, Chief of Police. This outrage was followed by several other attacks upon the police, and the culmination came on March 21, 1937, at Ponce, where 19 people were killed and about 100 wounded when the police tried to prevent a parade by the Nationalists.[42]

The government of the United States had not been unmindful of the growing political crisis in the island and as a preliminary step to a more

[37] *Sen. Rep.*, 71st Cong., 2nd Sess., vol. II, Report No. 1116.
[38] *Report of the Puerto Rico Policy Commission*, June 14, 1934.
[39] See Conclusions of the Committee, *Cong. Rec.*, vol. 79 (June 28, 1935), p. 10349.
[40] See the excellent survey by Earle K. James, "Puerto Rico at the Crossroads," *Foreign Policy Reports*, vol. XIII, no. 15 (Oct. 15, 1937), pp. 186-188.
[41] The Coalitionists (Union Republic and Socialist Parties) favored statehood, the Liberals split over statehood and independence.
[42] For a detailed account see the statement of Governor Winship in the *Cong. Rec.*, vol. 81 (June 3, 1937), pp. 5275-5280; see also the very critical "Report of the Commission of Inquiry on Civil Rights in Puerto Rico," by the New York American Civil Liberties Union (May 22, 1937).

satisfactory policy the jurisdiction of the island was transferred from the War Department to the newly established Division of Territorial and Insular Possessions in the Department of Interior.[43] A well-known liberal with considerable experience in Latin American affairs, Dr. Ernest H. Gruening, was named as its head. Unfortunately, whether through lack of experience in colonial administration or through an alleged tendency to engage in local politics in the island, the change in administration did not prove to be as beneficial to Puerto Rico as was hoped.

When Secretary of Interior Ickes was asked to state his attitude towards the bill regularly introduced into Congress by Resident Commissioner Iglesias granting full statehood to the island, he replied that he would not put any obstacle in the way of a vote. He thought, however, that various points should be considered before such a step should be taken: it would establish a precedent by including noncontiguous territory and also by including a population of wholly different cultures, tradition, and language; there was some opposition to statehood in Puerto Rico and the rights of the people in the Union were involved; finally, statehood once gained would be permanent.[44]

The murder of Colonel Riggs brought matters to a head in the United States as to establishing the future status of Puerto Rico. Without previous warning, on April 23, 1936, Senator Tydings, Chairman of the Senate Committee on Territories and Insular Affairs, introduced, with the support of the administration, a bill giving the Puerto Ricans the option of becoming independent or remaining under the American flag, the decision to be based upon a national referendum on the island.[45] If independence were preferred, the island would retain a commonwealth status for four years and then be given complete independence. Such a bill was a bombshell in the islands where it was well known that independence meant economic ruin. Nor was the fact overlooked that for over two years the United States had been spending about one million dollars a month of its own relief funds in Puerto Rico. The reaction in the island was noticeably hostile to the bill and the Liberal party which favored independence suffered a substantial defeat in the fall elections of 1936. The bill itself was not even discussed in committee and the political future of the island remained a question mark.

A political event with great economic potentialities occurred in the autumn of 1940 with the success of the new *Partido Popular Democrático* under the dynamic leadership of Luis Muñoz Marín. This party had an elaborate program of social and economic reform which it was determined to translate into legislative enactment. Its first noteworthy achievement was the passage of the Land Authority Act in April, 1941, which set up a Land Authority–*Autoridad de Tierras de Puerto Rico*–which was empowered to acquire and distribute among the small agriculturists lands in excess of 500 acres held by corporations. Its first large purchase was 10,000 acres from a sugar concern which it allocated to small farmers. Coupled with the fact that the benefits of the Bankhead-Jones Farm Tenant Act were made available to Puerto Rico, it was

[43] Set up by Executive Order of May 29, 1934.
[44] *New York Times*, Apr. 17, June 2, 1935; *Cong. Rec.*, vol. 79, (June 5, 1935), p. 8713.
[45] *Cong. Rec.*, vol. 80 (Apr. 23, 1936), p. 5925.

made possible for farm tenants, laborers, and share croppers to obtain loans to acquire family-sized farms.

This program was the more successful in that it was supported strongly by Rexford Guy Tugwell, an able New Dealer who had been sent by Secretary Ickes as head of a commission to see how the long dormant 500 acres law could be enforced. Tugwell became so interested in Puerto Rico's economic problems that when offered the directorship of the Division of Territories in Washington, he expressed a preference for the governorship of Puerto Rico. President Franklin D. Roosevelt agreed and Tugwell was appointed Governor in September 1941.

In his message to the second session of the fifteenth legislature Governor Tugwell, after conceding that the condition of the people reflected years of injustice and neglect, stressed the need to "set going movements which in time will lift people out of slums and will exorcise hunger. . . . Specifically, we must perfect and enlarge the program of land reform; give workers greater protection; constantly strive to raise the levels of nutrition; make life and living more secure; perfect the devices of government for recruiting and disciplining the public service; plan and execute public work with greater efficiency. . . ."[46]

The outbreak of war and the decision to make Puerto Rico the key to the defense system of the Caribbean had a quick and stimulating repercussion upon the economic conditions of the island. The program included two airplane bases for the army at Punta Boriquen and Ponce, an air base and a submarine base for the navy at Isla Grande, numerous auxiliary air fields, and a huge expansion of existing army posts. The army's first new military department in forty years, the Military Department of Puerto Rico, was instituted July 1, 1939, in San Juan. Puerto Rico and the Virgin Islands were constituted as the tenth naval district on January 1, 1940.

In spite of temporary improvement in the economic situation of the Island as a result of the war, the Puerto Ricans were dissatisfied with their political status, and President Truman recommended in October, 1945, that the Congress enact legislation authorizing a plebiscite in Puerto Rico whereby the inhabitants might indicate whether they wished (1) an elective governor with a wider measure of self-government; (2) statehood; (3) complete independence; or (4) a dominion form of government. When this bill was not reported out of committee, President Truman on July 25, 1946, named Resident Commissioner Jésus T. Pinero as Governor of the Island—the first native-born governor ever to be appointed. In 1947 the Organic Act was amended to permit the Puerto Ricans to elect their own governor in the 1948 elections.

In the elections held November 2, 1948, Luiz Muñoz Marín, after a hectic campaign was elected by a small majority and was inaugurated January 2, 1949. This proved to be only the first step towards greater autonomy for the Island. In 1950 Public Law 600 was passed by the Congress authorizing the people of Puerto Rico to organize a constitutional government. The constitution when drafted would be submitted to the Congress for its approval. The new Organic

[46] *Message of R. G. Tugwell to the Fifteenth Legislature* (San Juan, P. R., 1942), pp. 3, 12. An excellent account of the Tugwell-Muñoz Marín reconstruction and development program is given by Charles C. Goodsell in his *Administration of a Revolution* (Harvard University Press, 1965).

Act was to be known as the "Puerto Rican Federal Relations Act." A referendum held June 4, 1951, by the people of Puerto Rico overwhelmingly approved Public Law 600. On August 27, 1951, delegates to a constitutional convention were selected which after four and one-half months' work adopted a constitution which was submitted to the people. A referendum held March 3, 1952, approved the Constitution by a vote of 373,594 to 82,877.

The new Puerto Rican Constitution was transmitted to President Truman who immediately recommended its approval by the Congress. Congress passed a joint resolution ratifying the Constitution with three minor changes and the President signed the resolution on July 3, 1952. The Convention of Puerto Rico accepted the changes and the Commonwealth of Puerto Rico was officially proclaimed on July 25, 1952.

The new Constitution follows the model of the United States Constitution in accepting the presidential type with its tripartite check and balance system instead of the parliamentary system of Great Britain and the Continent. Its Bill of Rights is more elaborate, as for example, providing for the complete separation of Church and State and prohibiting "wiretapping." One interesting variation is a provision that if the governorship become vacant before the expiration of the four-year term the new governor shall be the secretary of state who shall be appointed by the governor with the advice and consent of both houses. The right to amend the Constitution is vested exclusively in the people of Puerto Rico and their representatives. Although the Resident Commissioner is still without a vote in the House of Representatives, there is no taxation without representation because Puerto Ricans pay no federal taxes. The new status, perhaps unique in governmental systems, makes the Commonwealth of Puerto Rico a "free associated state" in a sort of voluntary federated association with the United States.

On January 2, 1953, Luis Muñoz Marín was inaugurated Governor of Puerto Rico for the second time but as the first governor under the new Constitution. President Eisenhower sent Governor John Davis Dodge of Connecticut to act as his personal representative during the ceremony.

Until Puerto Rico became an Associated Free State in 1952 the United States made an annual report on its status and development as a non self-governing colony to the United Nations. When in September 1953, the United States indicated that no further reports would be made since Puerto Rico had achieved a "commonwealth status," the representative from India questioned its autonomous character. The Trusteeship Committee of the United Nations, however, approved the new status as one of a self-governing character and their decision was sustained by the General Assembly by a vote of 26 to 16 with 18 abstentions.[47] It might be noted that Mexico and Guatemala joined with the Soviet bloc in opposing the position taken by the United States. As proof of the intention of the United States to comply with the wishes of the people of Puerto Rico, Henry Cabot Lodge, the United States Delegate, declared that President Eisenhower authorized him to state that if Puerto Rico should adopt a resolution favoring more complete or even absolute independence the president

[47]For text of the resolution of approval see *General Assembly Official Records*, 8th Sess., Supp. No. 15, Part I, Sect. 7.

would immediately recommend to the Congress that such independence be granted. Under these circumstances the murderous attack from the visitors' gallery upon members of the United States House of Representatives on March 1, 1954, by four Puerto Rican Nationalists is the more difficult to understand.

When in October 1965 the Cuban Delegate in the United Nations demanded independence for Puerto Rico, Ambassador Goldberg, the United States Representative, led his delegation in a walkout for the first time to show his disdain.

In the elections of 1956 and 1960, with status as a paramount issue, a substantial majority of Puerto Ricans supported the Commonwealth. However, doubts regarding the exact status of Puerto Rico under the Commonwealth constantly arose, culminating in December 1962, when a joint resolution was passed by the Legislative Assembly of Puerto Rico proposing to the Congress of the United States a prompt settlement in a democratic manner of the political status of the Island. As Governor Luis Muños Marín, who signed the resolution, expressed it, "a hundred percent of the citizens of Puerto Rico are against living under any stigma of colonialism." Fidel Castro in derision is said to have referred to Puerto Rico as "a perfumed colony."

The Congress of the United States considered the resolution in a favorable atmosphere and after considerable debate passed a law to establish a United States-Puerto Rico Commission on the Status of Puerto Rico, which became law February 20, 1964.[48]

By the terms of this law a seven-member Commission consisting of a chairman and two other members appointed by the President of the United States, all citizens of the United States but non-residents of Puerto Rico, two members of the Senate, and two members from the House was to be established. The law further authorized the Commonwealth to appoint six additional members. This thirteen-member Commission was to study all factors bearing on the present and future relationship between the United States and Puerto Rico.

Under the chairmanship of James H. Rowe, Jr. of the United States, the Commission held numerous hearings in Washington and San Juan, where it consulted many experts and heard over a hundred witnesses. During its existence from June 9, 1964 until August 5, 1966, it carried out a comprehensive and impartial review of the status question. Comment was invited from all who were interested. The Report, when published in August 1966, recognized the validity of three forms of political status—commonwealth, statehood and independence—each of which conferred upon the people of Puerto Rico "equal dignity with equality of status and national citizenship." Choice was to be made by the people of Puerto Rico with the agreement and full cooperation of the Government of the United States. Although no preference was indicated in the report it did note that a change in political status to statehood would require increased financial assistance, and to independence even greater adjustments. Finally, the Commission felt that an expression of the will of the citizens of Puerto Rico by popular vote as to whether they wished to continue the commonwealth status or change to statehood or independence would be helpful to all concerned. It also recommended the constitution of *ad hoc* advisory

[48] 78 *United States Statute*, 17.

groups of persons of high prestige to consider proposals for the improvement of commonwealth or for change to statehood or independence.

The *San Juan Star* in its issue of August 6, 1966, declared that "the Commission had put together after more than ten years of research, hearings, study and deliberations a report that stands out as one of the most valuable documents ever produced on Puerto Rico's cultural, social, historical, political and economic aspects . . ." The *New York Times* called the Report "strong on research and tactful in its conclusions."

Although a plebiscite was recommended, both statehood and independent parties were strongly opposed if the commonwealth alternative was to be on the ballot. In spite of their opposition, a bill providing a three option plebiscite was voted and became law December 23, 1966. The plebiscite was to be held July 23, 1967 and the ballot would provide a choice of Commonwealth, statehood, or independence. If the commonwealth received a majority of votes it would be reaffirmed as an autonomous community permanently associated with the United States.

Luis Muños Marín and the Popular Democratic Party waged a strenuous campaign in behalf of the commonwealth status and it now passed by 60.5 percent of those voting. The statehood option received 38.9 percent, whereas the Independence Party obtained less than one percent. Muñoz over-optimistically declared that the century-long debate regarding political status had ended. In fact, Muñoz was so closely identified with the Commonwealth status option that he once declared to split it up would be like cutting him to pieces.

However, he had already felt the need to retire after completing his sixteenth year as governor and he refused to allow his name to be placed before the nominating convention in August 1964. Instead he insisted that his chief lieutenant Roberto Sánchez Villela receive the gubernatorial nomination, while he himself ran for the Senate. Muñoz had his way; his candidate won, but Sánchez turned out to be a great disappointment. Although a good administrator, he was not an able political leader, and by 1968 he was completely out of touch with the old guard political leaders of the Popular Democratic Party. He even broke with Muñoz, who then declared his renomination would be dangerous to the unity of the Popular Democratic Party. Sanchez was defeated overwhelmingly in the nominating convention by Senator Luis Negron López.

Thereupon Sánchez decided to run on the recently organized Peoples Party, and Muñoz indicated that he would accept the presidency of the Senate if the Popular Democratic Party won a majority in the Senate. The split in the Popular Democratic Party was its undoing, and the popular, wealthy and philanthropic industrialist Luis A. Ferré, long a proponent of statehood, won the election on November 5, 1968 as the head of the New Progressive Party.[49] It should be noted, however, that Ferré stressed modernization rather than statehood in his campaign and after his election did not mention it in his priority program. Nevertheless, on November 12 Ronald Walker, the managing editor of the *San Juan Star*, declared Puerto Rico had taken a "psychological swing

[49]The votes were as follows: Ferré 390, 964, López 367, 355, Villela 81,800, Independents, 4,324.

towards statehood." Yet when Vice President Agnew declared on November 13, "I absolutely favor statehood for Puerto Rico," the *Star* replied on November 22 that "we need more talk of status like we need an active volcano." Governor Ferré in his inaugural address January 2, 1969 stated that while statehood is the goal of the New Progressive Party it will be by "a plebiscite and not by elections."

Unquestionably Puerto Rico as a commonwealth has been a remarkable success economically and commercially. United States Assistant Secretary of Commerce Eugene F. Foley declared "This Bootstrap is probably the best single economic development program in the United States." Senator J. William Fulbright praised Puerto Rico's Commonwealth form of government as the most successful political experiment in the whole Latin World. Chancellor Jaime Benítez, later president of the University of Puerto Rico, was fulsome in his praise: "Commonwealth is the triumph of intelligence over intellectuality, of reason over rationalization, of reality over utopianism, of political pragmatism over selfsealed ideologies."

One development in the commonwealth status strongly supported by Governor Ferré was the privilege of permitting Puerto Rican citizens to vote in the presidential elections of the United States. In fact in September 1969, Governor Ferré traveled to Washington to request President Nixon to join him in appointing *ad hoc* committees to study such a possibility. Although Ferré's visit was seemingly unproductive, Representative Emanuel Celler, the 81-year-old chairman of the House Judiciary Committee, declared hearings would be held early in 1970 to consider a presidential vote for Puerto Rico, Guam and the Virgin Islands. Governor Ferré was confident that the presidential vote would help the island become a state, but Henry Wells in his thoughtful political study of the institutions of Puerto Rico declares "the plain truth is that there can be no end to the debate about political status so long as the commonwealth lasts."[50] In the words of a political writer, "Puerto Rico represents a people in search of a status."

The Puerto Rican commonwealth with its "Operation Bootstrap" has made a phenomenal gain in economic development. Rafael Durand, Administrator of Fomento, on March 1, 1968, reported that in the past fiscal year gross production was over $3.3 billion, and new projects being worked on involved investment of over $1.2 billion.

In fact both Puerto Rico and the United States profited exceedingly in the exchange of commodities. The value in dollars of the merchandise shipped to the United States from Puerto Rico in the calendar year of 1970 was $1,563,259,567, and the goods imported from the United States was valued at $2,069,974,876. The per capita income of the people of Puerto Rico, which amounted to $121 a year in 1940, had reached the sum of $1556, highest in Spanish America by 1971. As was appropriately said, from being "the poorhouse of the Caribbean" Puerto Rico had become "the showcase of the Caribbean."

Any attempt to predict the ultimate status of Puerto Rico might well eliminate independence, which has been voted down by an overwhelming

[50] Henry Wells, *The Modernization of Puerto Rico*, (Mass., Harvard Univ. Pr., 1969).

majority every time it has appeared on the ballot. The advantages of Commonwealth are so evident that one might wonder why any other status would be seriously considered. The Status Commission calculated that more than two-thirds of the manufacturing jobs created since 1952 would not exist today without the present exemption of Federal wage and tax laws. One small example of the advantage of the Federal Government's reimbursement of taxes and import duties on rum for the year 1968 amounted to about $60 million.

The principal argument against Commonwealth is the uncertainty of its stability. The Congress which made the Commonwealth possible could repeal the Federal Relations Act, it could amend it unilaterally, it could even annul the Commonwealth constitution. It cannot guarantee United States citizenship so highly prized. Only statehood can do that, because statehood is irrevocable.

The old argument against statehood that non-contiguous territories should not be admitted was eliminated by the statehood of Alaska and Hawaii. The election of Governor Ferré, a determined advocate of statehood, would seen to indicate at least a trend in that direction. However, on November 15, 1969 the new Puerto Rican resident commissioner, Jorge Luis Cordova Días, who most certainly is in harmony with the views of Governor Ferré, declared publicly, "Statehood for Puerto Rico is at least 25 years away;"[51] and Governor Ferré himself in his second State of the Commonwealth Message made January 14, 1970 with regard to status declared, "Our people by the overwhelmingly majority of 97 percent of the electorate have clearly expressed that they wish to remain permanently with the United States."[52]

A minor controversy developed in 1970, which disturbed the cordial relations between Washington and the Commonwealth. During World War II President Roosevelt ordered the Navy to use the small island of Culebra belonging to Puerto Rico as a gunnery practice range. For some thirty years the Navy and the Culebrans lived harmoniously, but in April, 1970, the question was raised concerning the Navy's control. Some firing of mortar rounds upon a beach where children were playing brought matters to a head. Resident Commissioner Cordova Díaz told Navy officials the only way to solve the Culebra problem was for the Navy to abandon this island as a gunnery practice range. The commander of the San Juan Naval Station, Admiral Ward, replied that the Navy was unable to give up the use of Culebra as a vital part of the Atlantic Fleet's weapons range. In June 1970 representatives of Puerto Rico's three political parties testified in Washington before the House Armed Services Committee against the use of Culebra as a target area. In August Governor Ferré demanded that the Navy prove that the use of Culebra was "essential to national security," and Muñoz Marín declared that Culebra was not "fundamental" to national security. Senator Jackson of Washington took up cudgels for Puerto Rico, requesting information as to what the Navy was doing to make its training operations around Culebra less inconvenient to local citizens. A reply from Secretary of the Navy, Chafee,[53] promised to meet many of the Puerto Rican demands. Senator Jackson was able to have inserted in the text of Public Law 91-511 section 611,

[51] *San Juan Star*, Nov. 15, 1969.
[52] *San Juan Star*, Jan. 15, 1970.
[53] See *Sen. Report*. No. 91-1234, Sept. 24, 1970, pp. 5-10.

the requirement that a report be made to the president by the secretary of defense on the entire Culebra complex not later than April 1, 1971. On January 11, 1971, the Navy agreed to eliminate all aerial target operations on Culebra, and to relinquish most of the land. Residents of Culebra were permitted to use Flamingo and other beaches on week ends and holidays. However, the inhabitants were not satisfied, and in August 1971 the Senate accepted a bill amended by Senator Jackson instructing the Pentagon to conduct a new study to determine the most advantageous alternatives to Navy practice in Culebra, and on August 15 Governor Ferré declared the Navy's practice should end by 1975.

Despite the objections and proposals, retiring Secretary of Defense, Laird, on December 27, 1972, recommended to the Congress that the Navy continue to use Culebra as a target practice range until after 1985. Newly elected Governor Rafael Hernández Colón forthwith declared his continued opposition to Navy target practice on Culebra; recently-chosen Resident Commissioner Jaime Benítez in Washington declared that he would make every effort to reverse "this deplorable pentagon error"; and Senator Hubert Humphrey promised to join Senator Howard Baker in introducing a bill to force the Navy to stop shelling Culebra by July 1975. The *San Juan Star* warned that "The Laird decision will rightly release another hornet's nest of objections."

The only other unsolved problem facing the US and Puerto Rico was whether the U.S. citizens living in Puerto Rico should be granted the right to vote for the president and vice-president of the United States. An *ad hoc* Advisory Group had recommended that a referendum be held to determine whether a majority of the electorate in Puerto Rico wanted the privilege. Senator Jackson, a member of the Advisory Group favored the referendum, but only after the Virgin Islands received the franchise. The Peoples' Party and the Puerto Rican Union Party opposed the presidential vote. However, even if an overwhelming majority of Puerto Ricans found the proposition acceptable, it remained a question as to whether the U.S. Congress would approve the innovation.

As to the future of Puerto Rico, the 1972 elections clearly showed a predilection for the Commonwealth over Statehood. Independence was hardly given consideration. The efforts to persuade the United Nations to declare Puerto Rico a colony of the United States proved unavailing. In his inaugural address of January 2, 1973, Governor Rafael Hernández Colón well expressed the general feeling of the electorate: "The people of Puerto Rico have freely chosen to unite their destiny to that of the people of the United States of America. This union is permanent and based upon solid and fruitful foundations: common citizenship, common defense, common currency and a common dedication to the value of democracy."

SUPPLEMENTARY READINGS

Aitken Jr., Thomas. *Poet in the Fortress.* New York, 1964.
Anderson, Robert W. *Party Politics in Puerto Rico,* Stanford, 1965.
Brau, Salvador. *Historia de Puerto Rico.* San Juan, 1966.
Goodsell, Charles T. *Administration of a Revolution.* Cambridge, Mass., 1965.
Hanson, Earl Parker. *Puerto Rico–Ally for Progress.* Princeton, 1962.
----------. *Transformation: The Story of Modern Puerto Rico.* New York, 1955.
Huebner, Theodore. *Puerto Rico Today.* New York, 1960.
Lewis, Gordon K. *Puerto Rico: Freedom and Power in the Caribbean.* New York, 1963.
Lewis, Oscar. *La Vida: A Puerto Rican Family in the Culture of Poverty.* New York, 1965.
Lockett, Edward B. *The Puerto Rico Problem.* New York, 1964.
López, Alfredo. *The Puerto Rico Papers.* Indianapolis, 1973.
Maldonado-Denis, Manuel. *Puerto Rico, A Socio-Historic Interpretation.* New York, 1972.
Muñoz Marín, Luis. "Puerto Rico and the United States, Their Future Together," *Foreign Affairs,* 32 (July, 1954), pp. 541-551.
Morales Carrión, Arturo. *Puerto Rico and the non Hispanic Caribbean.* Rio Piedras, 1952.
Padilla, Elena. *Up from Puerto Rico.* New York, 1958.
Perloff, Harvey S. *Puerto Rico's Economic Future.* Chicago, 1950.
Rand, Christopher. *The Puerto Ricans.* New York, 1958.
Silén, Susan Angel. *Hacía una visión positiva del puertoriguina.* Rio Piedras, 1970.
Tugwell, Rexford. *The Striken Land.* New York, 1947.
United States-Puerto Rico Commission. *Status of Puerto Rico.* Washington, 1966.
Wells, Henry. *The Modernization of Puerto Rico: A Political Study of Changing Values and Institutions.* Cambridge, Mass., 1969.

12

American Interests in Haiti and the Dominican Republic

The location of the island of Hispaniola in the Caribbean Sea suggests its role in the historical evolution of the region. Lying almost directly between Cuba and Puerto Rico it was, almost inevitably, discovered by Spain; like Cuba, it was destined to be a focus of international rivalries; and in modern times, commanding as it does both of the important routes between northern Europe and the Panama Canal, it would necessarily be affected by the influence of the United States.

Hispaniola's topography is the most complex in the West Indies, for rugged mountains dominate the landscape virtually everywhere on the island. Elevations vary from 10,000 feet above sea level in the Cordillera Central, to 150 feet below sea level in the Enrequillo Depression. Abrupt changes occur from jungle to desert, from alpine forests to savanna, and from salt flats to mangrove swamps, all within short distances.

Occupying the eastern two-thirds of the island, the area of the Dominican Republic is large for the West Indies, but small in comparison to measurements on the nearby continents. Climatic handicaps are not pronounced except for occasional devastating hurricanes, and rainfall is generally adequate except in the southeast, where some irrigation must supplement rainfall to insure desired sugar cane growth. Agriculture has always been the main source of livelihood for most of the population. The Dominicans of modern times are primarily mulatto, but they have become distinctly Latin American and European in their cultural orientation. With a significantly smaller population than that of Haiti, and occupying a territory almost twice as large, the Dominican Republic has had a continuing problem containing the burgeoning population pushing eastward from Haiti.

The mountain ranges identified with the Dominican Republic give Haiti its physical structure. The old island saying: "Beyond the mountains are more mountains," is an accurate description of the rough character of the nation's terrain.

Many steep slopes have created a complex climatic pattern, causing areas in the lee of mountain ranges to receive insufficient rainfall for growing crops. Every point is within thirty-five miles of the sea or the Dominican boundary. Port au Prince lies only twenty-two miles from Haiti's eastern boundary.

There is little on the Haitian landscape to indicate a European colonial origin because the rural scene is more analogous to West Africa. Haitians differ markedly from any other nationality in all of Latin America, for in the twentieth century Negroes make up from 90 to 95 percent of the total population. The remainder are considered as "colored," that is, mulatto, from a mixture of Negro and white blood. A mulatto elite, speaking cultured French rather than the Creole *patois* of the Negroes, came to represent two to three percent of the population. Christianity in the form of the Catholic religion has been identified with the elite, whereas the underprivileged masses commonly practice a type of pagan worship known as Voodoo, introduced from Africa. These cultural characteristics of Haiti set it apart from the Spanish-speaking Dominican Republic.

Christopher Columbus discovered the island which he named *La Española*, later called Santo Domingo and Hispaniola (known to the natives as Quisqueya), shortly after he had discovered Cuba. A small colony of Spaniards was left by the discoverer at Cape Haitien, on the northwest coast of the island. But on his return in 1493, he found that all had been massacred by the natives. The Spaniards exacted bloody vengeance later, and it was estimated that within twenty years of the discovery of the island the population was reduced from one million to about 14,000.[1] Negro slaves were introduced to repopulate the island, and they ultimately became the preponderant population.

The initial objective on Hispaniola was to found a mining and farming colony which could produce its own food, send gold back to Spain, and provide a base for further exploration From the first settlement the Spaniards moved towards the south coast, founding Santo Domingo in 1496. Hispaniola also served as a base for the conquest and settlement of the adjacent Antillean islands and the continent. However, Hispaniola, and the West Indies generally, were the focus of Spanish interest for only a few decades after the discovery. When Mexico, Peru, and other more desirable areas of the mainland were conquered, the islands were depopulated and quickly fell into decay.

Hispaniola remained under the domination of the Spaniards until the middle of the seventeenth century when the French established settlements in the western extremity. The Spaniards made various efforts to oust their unwelcome neighbors, but to no avail, and the treaty of Ryswick in 1697 definitely ceded the western section of the island to France. The French colonists soon made Saint-Dominque one of the most prosperous islands of the West Indies, although their great sugar, coffee, and tobacco plantations were worked entirely with slave labor. When the French Revolution broke out, in 1789, it was estimated that the population of Saint-Domingue consisted of about 40,000 whites, 28,000 *affranchis*, or freedman, and about 452,000 slaves.[2]

[1] Justin Placide, *Histoire politique et statistique d'Hayti, Saint-Domingue* (Paris, 1826), p. 40.
[2] J. N. Léger, *Haiti: Her History and Her Detractors* (New York, 1907), p. 41.

The principles of the Revolution quickly became known in Santo Domingo, and at the earnest solicitation of the *affranchis*, the National Assembly, in 1791, granted full civil and political rights to free-born colored men in French possessions. This decree was resisted vigorously by the whites in Haiti, and the result was one of the most horrible servile wars in the world's history. The blacks, whose treatment at the hands of the whites had been barbarous in the extreme,[3] massacred the whites in veritable orgies of slaughter. In the wars that followed, the one bright spot was the brilliant achievements of the black leader, Toussaint l'Ouverture. Fighting, in turn, with Spanish against the French, then with the English and Spanish, sometimes with the mulattoes, sometimes against them, but always sturdily defending the interests of his black comrades, by 1799 Toussaint l'Ouverture had gained control of virtually the entire island.[4]

It was at this period that the island came within the diplomatic horizon of the United States. In December 1798, Rufus King, the American minister to Great Britain, heard rumours that a commercial agreement had been signed by a British officer and Toussaint.[5] King immediately approached Grenville, the British Foreign Minister, who admitted that the report was true and explained that, although General Maitland has signed the agreement wholly without authority, his government felt that in order to protect Jamaica it was wise to ratify the convention.[6] Owing to the fact that French sovereignty was still recognized in the island, and that the United States had prohibited all trade with French territories, such an agreement would be inimical to American trade interests. Therefore, early in January 1799, King again approached Grenville, this time suggesting that if the island should become independent, concerted action between the two powers would be the most advantageous line of action. Grenville agreed, and proposed the formation of an exclusive company, composed jointly of British and American nationals, to trade with Santo Domingo. King was dubious about the power of Congress to create such a company, and he suggested as a more feasible scheme that each nation make treaties with Toussaint, confining trade to the citizens of the two countries.[7] Grenville, however, seemed to prefer the plan of a joint company, and in a communication to King he outlined a definite proposal in which all trade should be confined to Port au Prince.[8]

These proposals were laid before President Adams, who, however, feared that the independence of Santo Domingo would not be an unmixed blessing for the United States, and also objected to the idea of a joint company. He accordingly declared it to be the policy of the United States not to meddle.[9] Hamilton was more outspoken in his advice. In a letter to Pickering, the secretary of state, dated February 9, 1799, he proposed that there be no committal

[3] *Ibid.*, p. 50 (note).
[4] For detailed accounts of this period see T. G. Stewart, *The Haitian Revolution, 1791-1804* (New York, 1914), and T. Lothrop Stoddard, *The French Revolution in San Domingo* (Boston, 1914).
[5] For a full account of this agreement see Thos. Southney, *Chronological History of the West Indies* (London, 1827), vol. III, pp. 155-159.
[6] *Life and Correspondence of Rufus King* (New York, 1895), vol. II, p. 476.
[7] *Ibid.*, vol. II, p. 499.
[8] *Ibid.*, p. 504.
[9] *Works of John Adams* (Boston, 1853), vol. VIII, p. 635.

regarding the independence of Santo Domingo, that there be no guaranty, no formal treaty, merely a verbal assurance to Toussaint that upon a declaration of independence commercial intercourse would be opened.[10] The whole American position was well summed up in a confidential note from Pickering to King, dated March 12, 1799, in which the secretary of state declared: "We shall never receive from the French Republic indemnification for the injuries she has done us. The commerce of Santo Domingo presents the only means of compensation, and this I have no doubt we shall obtain. Toussaint respects the British, he is attached to us; he knows our position, but a few days' sail from Santo Domingo, and the promptitude with which we can supply his wants .... Nothing is more clear than, if left to themselves, the blacks of Santo Domingo will be less dangerous than if they remain the subjects of France .... We therefore reckon confidently upon the independence of Santo Domingo."[11]

Joint action with Great Britain seemed essential, and in April an agreement was signed, based upon the common interest of the two countries, to prevent dissemination of dangerous principles among slaves held in territories belonging to the respective countries, and to open up intercourse with the islands.[12] The result was the appointment of Mr. Edward Stevens as consul to Santo Domingo, and a proclamation by the President, dated June 26, 1799, opening up Santo Domingo to American trade, though only through the two ports of Cape François and Port au Prince.[13] By this decree Toussaint's position was both recognized and strengthened, inasmuch as Rigaud, his principal rival, had to depend upon the southern ports for his provisions and war material. As shown by his letter of August 14, 1799, to President Adams,[14] Toussaint was not entirely satisfied. But in the following year the United States gave him further support by sending war-ships to blockade the southern ports.[15]

By the opening of 1801 Toussaint was in complete control of the island, including the Spanish part; and a constitution adopted May 9, 1801, made him governor-general for life. His success was to be transient. Napoleon, who now had a breathing-spell in Europe, organized an expedition under his brother-in-law, General Le Clerc, to recover Santo Domingo. For a time Toussaint resisted, but finally, realizing the futility of continuing the struggle against overwhelming odds, he surrendered and withdrew to one of his plantations. The French, however, feared his influence, and by an act of base treachery they seized him and sent him to France. Broken in health but not in spirit, he died, less than a year afterwards, in the prison at de Joux. The French paid dearly for their action. Finding a new leader in Dessalines, the blacks renewed the struggle for independence with astonishing vigor. The death of Le Clerc aided their cause, as did also the ravages of yellow fever. When the peace of Amiens was broken, Napoleon could no longer afford to dissipate his energy in the New World to such little advantage. Santo Domingo became merely a stepping-stone to a great French empire west of the Mississippi. Yet it was also the cornerstone, without

[10] *Works of Alexander Hamilton* (New York, 1850), vol. VI, p. 395.
[11] *Life and Correspondence of Rufus King,* vol. II, p. 557.
[12] For text see *Works of John Adams,* vol. VIII, p. 639.
[13] *Amer. State Papers, For Rel.,* vol. II, p. 240.
[14] Léger, *op. cit.,* p. 99 (note).
[15] *Ibid.,* p. 100.

whose possession the structure would be exceedingly insecure. Jefferson had already intimated that the United States would not look with pleasure upon French ownership of Louisiana, and Monroe had already been sent to negotiate for New Orleans and the Floridas.

America's star was in the ascendant, and Napoleon was beleaguered by the loss of 50,000 French soldiers in the Dominican fiasco, and by the British naval threat to New Orleans. When confronted by a choice, he preferred a stronger United States which could in time humble the British pride. Accordingly, Napoleon suddenly ordered Marbois to sell the whole of the Louisiana territory to the United States, and at the same time instructed Rochambeau, who now commanded in place of Le Clerc, to withdraw from Santo Domingo. On December 20, 1803, the United States took formal possession of New Orleans, and on January 1, 1804, Saint-Domingue was declared forever independent of French domination. Even the French name was discarded–the island resuming its original name of Haiti.[16] The United States will always be grateful to Jefferson for his vision, and to his representative Livingston for his prompt action. But it must be conceded that no small share in the happy result was contributed by the fight for the independence of Santo Domingo so indefatigably carried on by Toussaint l'Ouverture.

It does not come within our province to consider the internal history of the island during the next few decades, except as it concerns the United States. The island's independence was not recognized by the United States when recognition was accorded to the rest of Latin America, although Great Britain officially gave recognition in 1826, and even France followed suit in 1838. The unstable character of the government was hardly a reason for our inaction, considering that one president, Jean Boyer, remained in office from 1818 to 1843. The real reason was rather the dangerous slavery question; for undoubtedly the abolition of slavery in the island was viewed by an important element in the United States as a most undesirable precedent. Even Bolívar, who had been received most cordially by the Haitians during his struggle for the independence of South America, and had been given generous supplies of ammunition and provisions, did not dare include Haiti in his invitation to the Congress of Panama.

The first official diplomatic relations that the United States had with the island were established in 1844. After the resignation of Boyer in 1843, the Spanish part of the island revolted and set up a separate government under the name of the Dominican Republic. Desirous of strengthening itself against the blacks, this republic immediately despatched an envoy to the United States to obtain recognition and, if possible, to sign a treaty of friendship and commerce. In order to obtain unbiased information concerning resources and conditions in the island, President Polk despatched a special agent, John Hogan, on February 22, 1845, to investigate and make a report. Mr. Hogan spent about five months in the island, and ran up a heavy expense account which the government later refused to pay. But his report remained buried until 1871, when the House asked for the information.[17]

[16] *Ibid.*, p. 153.
[17] Report published in *House Ex. Doc. No. 42*, 41st Cong., 3rd Sess.

In the meantime the Dominicans, not obtaining satisfaction from the United States, and constantly harassed by the guerrilla warfare on the Haitian border, approached both France and Spain. President Souloque, of Haiti, prepared a large expedition with the purpose of forcing the Dominicans back under the control of Haiti; whereupon the threatened republic, in a circular note dated February 22, 1850, and addressed to the consuls of the United States, France, and Great Britain, urgently solicited the mediation of the governments of those countries.[18] Through the British minister at Washington, France and Great Britain proposed that the United States cooperate with them in a joint plan of action. Mr Webster, who had succeeded Mr. Clayton as Secretary of State, nominated Mr. Robert M. Walsh as special agent to Haiti, with a view to obtaining fuller information before announcing his policy.

Mr. Walsh sailed for Haiti, January 25, 1851, and upon arriving at Port au Prince immediately got in touch with the British and French consuls. He then informed the Haitian minister of foreign affairs that the government of the United States had determined to cooperate with the governments of England and France to secure the pacification of the island, and that the most feasible way to accomplish this was for the government of Haiti to acknowledge the independence of the Republic of Santo Domingo.[19] At the same time, a joint note, signed by the representatives of the three powers, demanded a definite treaty of peace, or, in lieu of that, a truce of ten years between the Empire of Haiti and the Dominican Republic. After a series of interviews and *pourparlers*, the Haitian government refused to subscribe to either proposition, but promised to continue the truce that then existed. Unable to obtain anything better, Mr. Walsh left, and in summing up his report he declared that the only way to obtain the recognition of the Dominican Republic by Haiti was by the use of force. The United States government, however, might well be content with the result achieved; for this statement not only allowed it to escape from an entangling alliance of dubious value, but was in direct accord with the American policy of looking with disfavor upon European intervention in the Western Hemisphere.

During the next decade America's relations with the island were based chiefly upon the Dominican Republic's need of protection against Haiti, and upon the means that she employed to obtain it. A protectorate under a strong power seemed the safest plan, and from 1843 onwards the Dominicans had made a series of proposals to both France and Spain suggesting such a measure; but, as one authority puts it, "neither France nor Spain was anxious to annex a hornets' nest."[20] In 1855 a proposal was made to General William L. Cazneau, the United States Commissioner to the Dominican Republic, that Samana Bay be leased to the United States at a nominal rent for a naval base; but when negotiations leaked out, France and Great Britain protested, and nothing further was done.[21]

[18] *Sen. Ex. Doc. No. 12*, p. 19, 33rd Cong., 1st Sess.
[19] The complete correspondance of the Walsh Mission is found in *Sen. Ex. Doc. No. 113,* 32nd Cong., 1st Sess.
[20] Otto Schoenrich, *Santo Domingo* (New York, 1918), p. 55.
[21] See Moore, *Digest of Int. Law,* vol. I, p. 598; treaty also mentioned by Hamilton Fish, *Sen. Ex. Doc. No. 17,* 41st Cong., 3rd Sess. Mrs. W. L. Cazneau gives an interesting side-light on these negotiations in an appendix entitled "The Seward Samana Mystery" in her book, *Our Winter Eden* (New York, 1878).

In 1861 President Santana was successful in his efforts to obtain a Spanish protectorate; and in a fervid letter to the Queen of Spain he assured her that "The Dominican people, giving a free course to those sentiments of affection and loyalty which have been so long repressed, have unanimously and spontaneously proclaimed you as their Queen and Sovereign, and I, who have now the exalted and undeserved honor of being the organ of those sincere sentiments, lay at your Majesty's feet the keys of this lovely island."[22] By a royal decree dated May 19, 1861, the Queen accepted the offer, and Santana was made governor-general of the colony. If the Monroe Doctrine covered voluntary transfer of sovereignty, here was a clear-cut violation. But, as in the contemporary French intervention in Mexico, Seward's hands were tied by the Civil War, and he could only protest in very general terms. Although on April 2, 1861, he wrote the Spanish minister at Washington that Spanish interference in the Dominican Republic would be regarded as manifesting an unfriendly spirit toward the United States and would be met with a prompt resistance, when the news of annexation was received, and Carl Schurz, our minister to Spain, asked him for an explicit statement of policy, Seward replied that too many other subjects were occupying his attention to permit him to give full consideration to this one.[23] The Dominicans, however, were not long content with the Spanish rule. In 1863 a revolution, known as the War of the Restoration, broke out, and two years later, coincident with the termination of the Civil War in the Unites States, the Spanish Cortes passed a law relinquishing the colony.[24]

An excellent opportunity was given to Seward to indicate the postbellum attitude of the United States toward the island when the British minister proposed, on July 25, 1865, that the United States concur with Great Britain in guaranteeing the neutrality of the peninsula of Samana.[25] Replying on August 15, the secretary declared that the United States was "sincerely desirous that the entire island of Haiti may now and henceforth remain subject exclusively to the government and jurisdiction of the people who are the dwellers and occupants thereof, and that they may never be dispossessed or disturbed by any foreign state or nation whatever."[26] He further pointed out that, while the United States was gratified at the proposal, its policy regarding political alliances with foreign states prevented its participation.

After the emancipation of slaves in the United States there was no further reason to refuse to recognize Haiti. In fact, ever since 1838 Congress had been bombarded with petitions and memorials from citizens and organizations of different states asking for such recognition. In his message of December 3, 1861, President Lincoln expressed the opinion that the independence of Haiti should be recognized; and by the act of June 5, 1862, the president was authorized to appoint a diplomatic representative. On July 12, 1862, Mr. Benjamin F. Whidden was empowered to act as a commissioner and consul-general to Haiti.[27] Two

[22] *British Parliamentary Papers, 1861,* vol. LXV. Papers rel. to Annex. of St. Domingo, p. 28.
[23] Frederic Bancroft, *The Life of Wm. H.. Seward* (New York, 1900), vol. II, p. 157.
[24] A full account of this period from the Spanish standpoint is given by General Gandara y Navarro, *Anexion y Guerra de Santo Domingo* (Madrid, 1884).
[25] *For. Rel. of the U. S.,* 1865, Part II, p. 184; see also Seward's notes to Perry pp. 522-534.
[26] *Ibid.,* p. 191; see also J. N. Léger, *La Politique Extérieure d'Haiti*, pp. 145-157.

years later a treaty of amity and commerce was concluded between the two countries, and its proclamation followed on July 6, 1865.[28] Owing to the Spanish protectorate and the resulting disorder, the Dominican Republic was not recognized by us until September 17, 1866. A treaty similar to that made with Haiti was signed in the following year.[29]

Following the Civil War, Secretary Seward found various justifications for his expansionist appetite in the Caribbean. Such expansion would divert public criticism from the mild reconstruction policies in the south and stood to win support for the faltering Johnson administration. In addition, Confederate blockade runners had recently dramatized the desirability of having coaling stations for navy ships in the Caribbean. An opportunity was provided by the Dominican foreign secretary in 1866 when he proposed the lease of certain keys and coal mines in Samana Bay in return for a $1,000,000 from the United States. Secretary Seward thereupon sent the Assistant secretary of state and Rear-Admiral Porter, with power to conclude a convention for the lease or cession of the peninsula and bay of Samana for the sum of $2 million, one half in cash and the other half payable in munitions of war. The Dominican government was willing to give certain concessions, but declared that the national constitution prevented absolute sale of territory. By the end of 1867, however, conditions were such that the island government decided to accede to the terms of the United States, and the negotiations were rapidly progressing when a revolution brought the downfall of the ruling authorities.[30]

The new government, under General Báez, was even more friendly to the United States; it proposed, indeed, to Mr. Smith, the commercial agent at Santo Domingo City, that the United States assume a protectorate over the republic and take possession of Samana Bay as a first step toward annexation.[31] This went beyond the ideas of even Seward; although he was soon won over to the proposal, subject to the favorable outcome of a popular referendum in the insular state. President Johnson's ill-considered message of December 9, 1868, indicated the direction of the wind.[32] In a letter to General Banks on January 29, 1869, Seward suggested that the stage was all set for annexation, and a joint resolution was introduced by Mr. Orth, of Indiana, providing for the admission of Santo Domingo, on the application of the people and government of that republic, into the Union as a territory of the United States, with a view to the ultimate establishment of a state government.[33] Congress, possessing neither the information nor the imagination of Seward, promptly laid the resolution on the table.

The Grant administration next took over the problem, and Hamilton Fish, the new secretary of state, despatched General Babcock as a special agent to obtain full information regarding the island, particularly in regard to the disposition of the government and people towards the United States.[34]

[27] J. B. Moore, *Digest of Int. Law,* vol. I, p. 107.
[28] W. M. Malloy, *Treaties, Conventions, etc., of the U. S.,* vol. I, p. 921.
[29] *Ibid.,* p. 403.
[30] See Seward's account of *Sen. Ex. Doc. No. 17,* 41st Cong., 3rd Sess., p. 5.
[31] Moore, *op. cit.,* vol. I, p. 590.
[32] See pp. 60-61.
[33] Bancroft, *op. cit.,* vol. II, p. 488.
[34] For instructions see *Sen. Ex. Doc. No. 17,* 41st Cong., 3rd Sess. p. 79.

Incidentally, the new agent signed a protocol with the Dominican government in which Seward's proposal was virtually repeated, but which also provided that President Grant should use his utmost influence to make the idea of annexation more popular with members of Congress.[35] With this as a basis, General Babcock and Mr. R. H. Perry were authorized to sign a treaty of annexation incorporating the republic as a territory of the United States, and also a convention giving the United States immediate possession of the Samana peninsula and Bay. The cash payment to be made by the United States was $1½ million. The treaty further provided that in case of its rejection the United States should still have the right to acquire the peninsula and bay of Samana within fifty years upon the payment of $2 million.[36] The treaties were signed November 29, 1869, and communicated to the Senate in the following January.

The debate that followed brought about a serious controversy between President Grant and Charles Sumner, who was at that time chairman of the Senate Committee on Foreign Relations. The report of the committee was adverse to the treaties, and Sumner led the opposition. A vote was taken on June 30, 1870, resulting in a tie (28-28); and as a two-thirds vote was required for ratification, the opponents had won. In his second annual message, December 5, 1870, President Grant again brought up the question and strongly urged that Congress reconsider the matter on the ground that possession of the territory would be of incalculable advantage to the United States strategically, economically, and commercially, and would be "a rapid stride towards that greatness which the intelligence, industry, and enterprise of the citizens of the United States entitle this country to assume among nations."[37] Although the opposition to annexation remained as strong as ever, out of deference to the President's views it was proposed that three commissioners be appointed to inquire once more into the political and economic condition of the island. In a powerful speech, teeming with invective and heated charges, Sumner resisted this proposal.[38] The resolution passed; and the result was that Fish joined Grant in open opposition to Sumner, who was ultimately ousted from his committee chairmanship.[39]

In accordance with the congressional resolution, President Grant appointed a very eminent investigating commission, consisting of Benjamin F. Wade, Andrew D. White, and Samuel G. Howe. The three men visited the island and made an elaborate and painstaking report, which is even today a very valuable source of information.[40] They unanimously approved of the annexation, and President Grant once more, in a special message prefacing the report (April 5, 1871), indicated that his own opinion regarding the desirability of annexation had not changed. But the sentiment of Congress was similarly unchanged; so that, although the president, in his last annual message recurred to the subject, it was merely to vindicate his attitude rather than to renew the proposal.

Senators Sumner, Schwarz, and their followers were successful in opposing

[35] For text of protocol see *Sen. Report No. 234,* 41st Cong., 2nd Sess. p. 188.
[36] Text of treaty and convention, *Sen. Ex. Doc. No. 17,* 41st Cong., 3rd Sess. pp. 98-102.
[37] J. D. Richardson, *Messages and Papers of the Presidents,* vol. VII, pp. 90 ff.
[38] *Works of Charles Sumner,* vol. XIV, 15 vols. (Boston, 1875-83), pp. 89-131.
[39] For a detailed account of this famous controversy see Ed. L. Pierce, *Memoir and Letters of Charles Sumner* (Boston, 1894), vol. IV, pp. 426 ff.
[40] *Sen. Doc. No. 9,* 42nd Cong., 1st Sess.

the annexation for a number of reasons. They capitalized on the hostile personal opposition to President Grant, they called attention to suspicious financial operations connected with the project, and they counted on the continuing reluctance of their colleagues to annex areas outside the Continent. They also criticized the alleged overbearing attitude of the United States toward Haiti.

It was asserted that, in order to prevent extraneous influences from interfering with the negotiations, our government had sent war-ships to the island, not only to sustain President Báez against internal difficulties, but also to intimidate Haiti. The instructions given by the secretary of the Navy to Admiral Poor were, indeed, couched in no uncertain terms: "Proceed at once with the *Severn* and *Dictator* to Port au Prince, communicate with our consul there, and inform the present Haitian authorities that this government is determined to protect the present Dominican government with all its power . . . . If the Haitians attack the Dominicans with their ships, destroy or capture them."[41] Admiral Poor did not think it necessary to mask his orders by diplomatic *formulae,* and he bluntly informed President Saget of Haiti that "any interference or attack by vessels under Haitian or any other flag upon the Dominicans during the pendency of said negotiations will be considered an act of hostility to the flag of the United States, and will provoke hostility in return."[42] Sumner offered a resolution in the Senate condemning these orders as involving an unlawful assumption by the president of the war-making power; but the proposal was laid on the table.[43] Secretary Fish also gave instructions during this period to Mr. Bassett, our minister in Haiti, warning the Haitian government against any interference with the Dominican Republic.[44]

For a period following the Grant administration the United States seemed to lose interest in further expansion in the Caribbean region. When, in 1882, President Salomon, of Haiti, offered to cede to the United States the island of La Fortue, Secretary Frelinghuysen said that "the policy of this government, as declared on many occasions in the past, has tended towards avoidance of possessions disconnected from the main continent."[45] Two years later a similar observation was called out from him by a proposal to cede the peninsula or bay of Le Môle or the island of Tortuga.

During this period the diplomatic relations between the United States and Haiti were exceedingly friendly, and in the case of the Pelletier and Lazare claims the United States acted in a manner above reproach. Pelletier, a naturalized American, had been seized and imprisoned while cruising off the coast of Haiti in 1861, on evidence of being a slaver. Escaping, he demanded damages from the Haitian government, and his claim was brought up on various occasions until finally, in 1884, Haiti agreed upon arbitration. Mr. William Strong, a former justice of the United States Supreme Court, was appointed sole arbiter, and he awarded Pelletier $57,250 instead of the two and one-half millions that he demanded. Haiti protested against making any payment, and when the case was brought before Mr. T. F. Bayard, secretary of state, he

[41] *Sen. Ex. Doc. No. 34,* 41st Cong., 3rd Sess., p. 11.
[42] *Ibid.,* p. 13.
[43] See *Cong. Globe,* 42nd Cong., 1st Sess. (1871), Part I, pp. 232 *et seq.*
[44] Moore, *op. cit.,* vol. I, p. 279.
[45] *Ibid.,* p. 432.

declared the claim of Pelletier against Haiti on the facts exhibited "must be dropped, and dropped peremptorily and immediately, by the government of the United States, ... first because Haiti had jurisdiction to inflict on him the very punishment of which he complains, and secondly because his cause is of itself so saturated with turpitude and infamy that on it no action, judicial or diplomatic, can be based."[46]

The Lazare claim was for a breach of contract in the establishment of a national bank in Haiti. Here again the award of $117,500 rendered by Mr. Strong was set aside on the ground that new evidence showed the claim to be invalid, and that "the moment the government of the United States discovers that a claim it makes on a foreign government cannot be honorably and honestly pressed, that moment, no matter what may be the period of the procedure, that claim should be dropped."[47] In his message to Congress on May 12, 1887, President Salomon declared that such sentiments did honor to the statesmen who had so well expressed them.[48]

Friendly relations between the United States and Haiti were suddenly interrupted in 1888 by the seizure of an American steamship, the *Haitian Republic*, as she was leaving Saint Mare, on the ground that an effective blockade was being maintained. The reasons for the blockade were as follows: President Salomon's seven-year term ended in 1887. But he had been so successful that the constitution was modified and he was reelected for another seven years. This brought about a revolution leading to his withdrawal in August, 1888, followed by a period of anarchy, in which one faction controlled the north and another the south. The *Haitian Republic* had, it was claimed, transported various armed members of the Hyppolite, or northern, faction while touching at the ports of the north and northwest, and on this ground it was seized while in waters patrolled by the Légitime, or southern, faction. The United States protested vigorously against the seizure, saying that, since no notice of blockade had been given, the blockade was not effective; that the prize tribunal was irregularly constituted; and that the carrying on passengers who might have been armed between the vessels' ports of call did not constitute complicity in Haitian disorders. When the Haitian authorities refused to give up the ship, the United States sent two war-ships to Port au Prince; whereupon the disputed vessel was surrendered to the American commander.[49]

The Légitime faction, which had been hostile to the United States, was finally overthrown, and Hyppolite was elected president. Whether the United States had ever done more than manifest sympathy for the Hyppolite side the published records did not show. But about a year after the inauguration of President Hyppolite (January 26, 1891), Rear-Admiral Gherardi arrived at Port au Prince as special commissioner to obtain the Môle St. Nicholas as a coaling station for the United States, and used as one of his arguments the services

[46] *Sen. Ex. Doc. No. 64,* 49th Cong., 2nd Sess., p. 20, or *For. Rel. of the U. S.,* 1887, pp. 593 ff.
[47] *Ibid.,* p. 33.
[48] *Ibid.,* 1887, p. 629.
[49] For details see *Sen. Ex. Doc. No. 69,* 50th Cong., 2nd Sess. particulary pp. 171-176; also *For. Rel. of U. S.,* 1888, pp. 932-1006; and 1889, pp. 487-494.

rendered by the United States to the Hyppolite revolution.[50] Popular opinion, aroused by the sight of an American fleet in the harbor, was so strongly opposed that after several months of fruitless negotiations, in which such technicalities as the form of the commissioner's credentials played a leading part, the United States gave up the idea. Secretary Blaine then turned towards the Dominican Republic and proposed to lease Samana Bay. But at the mere disclosure of the proposal the Domincan secretary of state, General Gonzáles, was forced to seek safety in exile. This was the last attempt made by the United States to obtain a naval station in the Caribbean until the war with Spain created new conditions and raised new problems.

[50] Fred. Douglas, "Haiti and the United States," *North Amer. Rev.*, vol. CLIII (Sept. 1891), p. 343. See also Moore, *op. cit.*, vol. I. p. 610.

SUPPLEMENTARY READINGS

Balch, Emily G., ed. *Occupied Haiti.* New York, 1927.

Bastien, Remy, "The Role of the Intellectual in Haitian Plural Society," *Annals of the New York Academy of Sciences,* 83 (1960).

Davis, H. P. *Black Democracy: the Story of Haiti.* New York, 1936.

Hazard, Samuel. *Santo Domingo, Past and Present.* New York, 1873.

Herskovits, Melville J. *Life in a Haitian Valley.* New York, 1937.

James, S. L. R. *The Black Jacobins: Toussaint l'Ouverture and the San Domingo Revolution.* 2nd ed. New York, 1963.

Jones, C. L. *The Caribbean since 1900.* New York, 1936.

Korngold, Ralph. *Citizen Toussaint.* Boston, 1944.

Knight, Melvin. *The Americans in Santo Domingo.* New York, 1928.

Leger, J. N. *Haiti: Her History and Her Detractors.* New York, 1907.

Leyburn, James G. *The Haitian People.* 2nd ed. New Haven, 1966.

MacCorkle, W. A. *The Monroe Doctrine and its Relation to Haiti.* New York, 1915.

Logan, R. W. *The Diplomatic Relations of the United States with Haiti, 1776-1891.* Chapel Hill, 1941.

———. *Haiti and the Dominican Republic.* New York, 1968.

Marcelin, L. J. *Haiti.* Paris, 1892.

Montague, L. L. *Haiti and the United States, 1714-1938.* Durham, 1940.

Moreau de Saint Merv, Mederic-Louis-Elie, *Description topographique physique, civile, politique, et historique de la partie francaise de l'isle Saint-Domingue.* 3 vols. Paris, 1958.

Munro, Dana G. *The United States and the Caribbean Area.* Boston, 1934.

Newton, A. P. *The European Nations in the West Indies, 1493-1688.* London, 1933.

Niles, Blair. *Black Haiti.* New York, 1926.

Pritchard, Hesketh. *Where Black Rules White.* New York, 1900.

Rodman, Seldon. *Quisqueya: A History of the Dominican Republic.* New York, 1964.

Schoenrih, Otto. *Santo Domingo.* New York, 1918.

Seabrook, W. B. *The Magic Island.* New York, 1929.

St. John, Spencer. *Haiti.* London, 1889.

Stoddard, T. L. *The French Revolution in San Domingo.* Boston, 1914.

Tansill, C. C. *The United States and Santo Domingo, 1798-1873.* Baltimore, 1938.

Tippenhauer, l. G. *Die inseln Haiti.* 2 vols. Leipzig, 1892.

Treudley, Mary. *The United States and Santo Domingo, 1789-1866.* Worcester, Mass., 1906.

Vandercook, J. W. *Black Majesty.* New York, 1928.

Verrill, A. H. *Porto Rico and San Domingo.* New York, 1914.

Waxman, Percy. *The Black Napoleon.* New York, 1931.

Welles, Sumner. *Naboth's Vineyard: The Dominican Republic, 1844-1924.* 2 vols. New York, 1928.

Wirkus, Faustin and Taney Dudley. *The White King of La Gonave.* New York, 1931.

13

The Dominican Republic, Haiti and the United States: Recent and Contemporary Relations

Before considering the next and more serious intervention of the United States in the affairs of Haiti and Santo Domingo, it is necessary to say a word regarding the internal history of the two island republics. Sir Spencer St. John, British minister to Haiti for a period of years, called its government "a despotism tempered by revolution and exile and occasionally by death."[1] Beginning with Dessalines in 1804, and surveying the political history of Haiti for a century, we find some twenty different presidents, of whom two were shot, one committed suicide, six were exiled, and several others were overthrown or forced to resign. The Dominican Republic can hardly boast of a better record than its unstable neighbor. In fact, in its checkered history of revolution and counter-revolution, in its unending procession of presidents *de jure* and *de facto*, the Spanish end of the island stands well towards the top of any list of revolution-tossed republics. From 1844 to 1904 we find more than twenty different presidents, in spite of the fact that Heureaux served as president for fourteen years and Báez held office on five different occasions.[2]

No nation can maintain its credit under such conditions, and when Carlos Morales was inaugurated president of the Dominican Republic in 1904 the government was hopelessly bankrupt. The claims against the government were estimated at anywhere from $30 million to $40 million. Inasmuch as many of

[1] Spencer St. John, *Haiti* (London, 1889), p. 272.
[2] See the "Chronology of Political Events in Santo Domingo," prepared by Minister T. C. Dawson, *For. Rel. of the U. S.*, 1906, pp. 572-600.

the creditors were Europeans, who constantly urged their governments to intervene, the United States was compelled to keep in close touch with the situation. Furthermore, the United States was virtually forced to intervene in the case of the San Domingo Improvement Company. This American corporation had taken over the holdings of a Holland corporation in 1892, and was authorized by the Dominican government to take over certain custom-houses, with the understanding that a stipulated amount should be paid to the government and the rest devoted to the payment of interest and sinking fund of the loan. In 1901 the Dominican government became dissatisfied with the arrangement and repudiated it. The company appealed to the United States, but the State Department urged a settlement by private negotiations. After much haggling, the Dominican government finally offered $4½ million for the company's interests; and the proposal was accepted.[3] Representatives of the two governments signed a protocol on January 31, 1903, providing for the settlement, and for a board of arbitrators to fix the details of payment.[4] The board fixed the interest rate at 4 percent, designated the monthly payments, and secured them by the customs, revenues, and port dues of the ports on the northern coast. In case of failure to pay, a financial agent of the United States was authorized to take over certain custom-houses and carry out the award.[5]

The Dominican government never made any payments, and in compliance with the terms of the award an American financial agent took over the custom-house of Puerto Plata, October 21, 1904. This move aroused the foreign creditors, and in December of this year the French threatened to seize the custom-house of Santo Domingo City; the Italian interests also demanded the payment of their claims. Secretary Hay thereupon instructed the American minister to Santo Domingo, Mr. T. C. Dawson, to sound out the insular president as to whether the Dominican government would be willing to request the United States to take charge of the collection of duties, with a view to an equitable distribution of the proceeds among the claimants and the Dominican government.[6] President Morales was by no means averse to such a solution, and on February 4, 1905, a protocol was signed which provided that the United States should take over all the custom-houses, turn over to the Dominican government 45 percent of the amount collected for its expenses, and use the other 55 percent to liquidate the republic's foreign and domestic debt. The United States further agreed to grant the little republic any assistance deemed necessary "to restore the credit, preserve the order, increase the efficiency of the civil administration, and advance the material progress and welfare of the Dominican Republic."[7]

When the protocol came before the Senate, President Theodore Roosevelt urged that assent be given to its ratification, on the ground that "those who profit by the Monroe Doctine must accept certain responsibilities along with the

[3] President Theodore Roosevelt in a message to the Senate Feb. 15, 1905, gives a complete summary of the financial and political difficulties of the republic, *Ibid.*, 1905, p. 334; see also the more detailed report of J. B. Moore, *Ibid.*, p. 344.

[4] For text of the protocol see *For. Rel. of the U. S.*, 1904, p. 270.

[5] *Ibid.*, p. 274.

[6] *Ibid.*, 1905, p. 298.

[7] *Ibid.*, p. 342.

rights which it confers."[8] The Senate did not seem to appreciate these responsibilities and gave no indication of an intention to give its consent. The President thereupon sent another special message, dated March 6, appealing for immediate action. The Senate adjourned, however, without acting, and the situation in Santo Domingo became so alarming that President Roosevelt decided to carry out the arrangement as an executive agreement. Mr. J. H. Hollander was appointed as confidential agent to examine and report on the financial conditions in the republic, and the Secretary of War, Mr. Taft, was authorized to nominate the necessary officials to collect the customs.[9] Santo Domingo accepted the *modus operandi,* and in his message to Congress on December 5, 1905, the president declared that a temporary arrangement had been made which would last until the Senate should take action on the treaty. He also asserted that under the course taken "stability and order and all the benefits of peace are at last coming to Santo Domingo, danger of foreign intervention has been suspended, and there is at last a prospect that all creditors will get justice."[10] The Senate did not take very kindly to this alleged executive usurpation, and Senator Tillman violently arraigned the president for his unconstitutional abuse of authority.[11] The *modus operandi,* however, brought order out of chaos, and the Dominican government received more money from the 45 percent turned over by the American officials than they had previously obtained when collecting 100 percent for themselves.[12]

After a careful investigation, Professor Hollander found that the claims pending against the Dominican Republic on June 1, 1905, amounted to more than $40 million.[13] Of this amount, the government recognized about three-fourths as valid. But a plan was finally evolved for scaling down this amount to about $17 million, based upon a cash payment. The funds were to be obtained by the flotation of an issue of fifty-year 5 percent bonds to the amount of $20 million. An American receiver of customs appointed by the president of the United States was to collect all customs duties until the bond issue should be redeemed; and the Dominican Republic was not to increase its public debt until full payment should be made. A convention to this effect was signed on February 8, 1907. This time, the Senate was willing to assent; and the convention went into effect on July 25.[14]

The Dominican officials themselves could not fail to realize the immediate benefits of American control, and in his annual report of 1906, Señor Velásquez, the Dominican Minister of Finance and Commerce, bore witness to the advantages gained: "The items of revenue during 1905 and those of 1906 speak clearly, with renewed eloquence, of figures, that for some time past we have been living in the public posts of life of order and honesty, where but a few

[8] *Ibid.,* p. 334.
[9] *Ibid.,* p. 360.
[10] *Ibid.,* p. xxxvi.
[11] See speech of Jan. 17, 1906, *Cong. Rec.,* vol. 40, Part II, pp. 1173 ff.
[12] Minister Dawson's memorandum of July 1, 1905, gives an excellent account of the events preceding and the results following the application of the arrangement; *For. Rel. of the U.S.,* 1905, p. 378.
[13] See Report of J. H. Hollander (Washington, D. C., 1907), or *Quar, Jour. of Economics,* vol. XXI (July, 1907), p. 405.
[14] Text in *For. Rel. of the U. S.,* 1907, p. 307.

Table 13-1. FOREIGN TRADE AND CUSTOMS COLLECTION OF THE DOMINICAN REPUBLIC*

Year	*Imports*	*Exports*	*Collections by General Receivers*	*Value of Trade with U. S.*
1911	$ 6,949,662	$10,995,546	$3,433,738.92	$ 9,871,947
1916	11,664,430	21,527,873	4,035,355.43	27,574,786
1921	24,585,327	20,614,048	2,859,866.40	31,866,711
1926	23,677,533	24,895,871	4,714,405.25	20,006,604
1931	10,151,762	13,067,162	2,883,446.92	9,310,422
1936	9,926,567	15,149,908	2,878,789.07	9,366,817
1939	11,592,166	18,643,302	3,031,455.64	11,912,960†

*The collections by the general receiver were by convention years to Aug. 1, 1914, but starting with Jan. 1, 1915, they have been by calendar years. For the 5-month period Aug. 1, 1914, to Dec. 31, 1914, receipts were $1,209.555.54. The imports and exports are by calendar years.

†Report of fiscal periods 1911, 1916, 1921, 1926, 1931, 1936, Dominican Customs Receivership.

years ago life with few exceptions within and without the national palace was one of shamelessness, dilapidation, cupidity, and permanent disgrace for the Republic, being the principal cause, if not the only one, why our weak state has felt itself more than once trembling on the brink of the abyss, and that for a long time we have found ourselves lacking in economic autonomy, overweighed by debts, unjustifiable for the greater part, suffering insults and humiliations."[15]

The convention of 1907 remained unchanged until 1924; and if the "eloquence of figures" is a fair basis of judgement, the effect of the arrangement has been all that could be desired. The table on this page, made up from figures submitted in the annual reports of the general receiver of Dominican customs to the bureau of insular affairs in the War Department summarizes the results.[16]

Despite the remarkable financial and commercial improvement due to American control of the customs, the internal history of the Dominican Republic failed to show much improvement. General Ramón Cáceres, who was elected president in 1908, and whose administration was eminently successful, was shot down in cold blood by a group of political enemies in November 1911. A provisional government was immediately established, but it could scarcely maintain itself against the revolutionary outbreaks. In the fall of 1912 President Taft appointed two special commissioners to investigate the situation and recommend measures calculated to put an end to the difficulties. They were to make the trip on an American gunboat, accompanied by 750 marines, who were to be at hand to protect the custom-houses.[17] In accordance with the

[15] *Ibid.*, p. 357.
[16] The report for 1939, the last to be made, was submitted to the newly established Division of Territories and Island Possessions in the Department of Interior.
[17] *For. Rel. of the U. S.*, 1912, p. 366.

suggestions of these commissioners, the insular president resigned, and the Dominican congress elected as new president Archbishop Nouel, a man loved and respected throughout the country. He did his best to bring order to the distracted land; but his health could not stand the strain, and on March 31, 1913, after serving exactly four months, he resigned.

Another period of revolutionary activity followed, and in a note dated September 9, 1913, Secretary Bryan declared that the influence of the United States would be exerted for the support of lawful authorities and for the discouragement of any and all insurrectionary methods, and that President Wilson had no sympathy with those who sought to seize the power of government to advance their own personal interests.[18] The American minister finally prevailed upon the revolutionary forces to lay down their arms, upon the promise of a fair election in the near future. It should be noted however that Minister Sullivan's influence was considerably vitiated by his unauthorized promises to the revolutionists and his inexcusable participation in local party politics. He had deservedly lost the confidence of the State Department and was under suspicion of questionable financial practices by the Dominican government.

When the United States government appointed three commissioners to observe the elections, the Dominican government in power protested vigorously. The commissioners were properly received, but their presence did not prevent the government from arresting six leaders of the opposition on the grounds of conspiracy.[19] Except for this incident, which completely demoralized the power of the opposition, the elections were very peaceably conducted. In 1914 there was a repetition of the same conditions, and again the United States sent commissioners and the elections were held. Ex-President Jiménez was successful, and for a little more than a year he maintained the peace. In April 1916, a new insurrection took place, and this time the United States, with the consent of the insular president, landed marines to maintain order. Shortly afterwards President Jiménez resigned and the Dominican congress elected Dr. Henríquez Carvajal as temporary president.[20]

On the ground that the terms of the convention of 1907 had been repeatedly violated, the United States, however, refused to recognize the new president unless he would sign a new treaty, which should provide for the collection of customs under American auspices, the appointment of an American financial adviser, and the establishment of a constabulary force under American officers. President Henríquez refused to accept recognition on these terms, and the United States declined to pay over any revenue to an unrecognized government. Rear-Admiral Knapp finally broke the deadlock by issuing a proclamation, on November 29, 1916, declaring the Dominican Republic under the military administration of the United States. The proclamation recited that there was no intention to destroy Dominican sovereignty, but merely to restore

[18] *Ibid.*, 1913, p. 425.
[19] See report of the commission, *Ibid.*, p. 449.
[20] A brief accurate sketch of these events is found in Judge Schoenrich's *Santo Domingo* (New York, 1918), chap. VI;; see also report of Rear-Admiral Snowden, *Conditions in Santo Domingo,* in report of Secy. of Navy, 1920, pp. 321-342.

and maintain order and carry out the terms of the convention of 1907.[21] To accomplish this, Rear-Admiral Knapp suspended the Dominican congress and, in the capacity of governor-general, appointed American naval officers to the various cabinet positions.

Colonel George C. Thorpe, who for almost two years acted as chief of staff of the brigade of marines in the occupation of the republic, thus summed up the mission of the military government: "(1) to promote education, primary and vocational; (2) to build roads; (3) to create an effective police force; (4) to cultivate a regard for law and order; (5) to place property rights on a firmer basis, particularly as to land titles; (6) to stabilize the finances of the country; (7) and at the same time to respect Dominican institutions and sentiments as far as might be."[22]

It would appear that the United States made a conscientious effort to carry out this program. In his report of October 23, 1920, to the Secretary of the Navy, Admiral Snowden noted some of the benefits of American rule. Where formerly it cost the internal revenue department 14 percent to collect about $700,000 annually, it was costing but 5 percent to collect almost $4½ million.

In four years of occupation more than 400 miles of roads were constructed, about one-fourth of them macadam. More than a score of bridges were built, and of them seven were large steel bridges and eight were concrete. New piers, a new custom-house at Santo Domingo City, new wharves, and other port improvements were built. In fact, a summary of amounts used by the department of public works from 1909 to 1916 shows an average annual expenditure of about $400,000; while the average expenditure annually from July, 1917, to June 30, 1920, was more than $1¼ million. Reforms in other directions were equally noteworthy. The railroads for the year 1916-17 hauled 27,866,635 kilos of freight, and the net earnings were $49,750.63. In the year 1919-20 50,272,506 kilos of freight were hauled, and the net earnings were $217,039.74. Before the military administration there were about 18,000 pupils in all schools; in 1920 there were more than 100,000. The reforms instituted in sanitary methods and public health conditions were of particular value to the people.[23]

Yet, in spite of the noteworthy improvements due to American rule, the Dominicans were by no means content under it, and in many respects they had serious grounds for complaint. Their civil and political liberties were completely taken away by the American military government; numerous cases of unjust imprisonment were proved against the American provost courts; the water torture and other more horrible methods were employed at times by the marines; and a strict and humiliating censorship of the press was maintained during the greater part of the American occupation.[24] When, finally, even the

[21] For text of the proclamation and American version of the situation leading up to it see *Hearings before a Select Committee on Haiti and Santo Domingo,* U. S. Senate, 67th Cong., 1st Sess., 1922, Part I, pp. 90-94; the Dominican version is given by President Henríquez, *Ibid.*, pp. 51-60.

[22] G. H. Blakeslee, ed., *Mexico and the Caribbean* (New York, 1920), p. 233.

[23] *Report of the Secretary of the Navy*, 1920, pp. 321-342; see also memorandum prepared for Sen. Com. Hearings, *op. cit.*, Part I, pp. 96-104.

[24] See summary of abuses in the letter of Archbishop Nouel to Minister Russell, dated Dec., 29, 1919. *(Nation,* July 17, 1920, gives excerpts); see also the protest of President Henríquez, *Current History*, vol. XIV (June, 1921), p. 397.

economical and financial program of the military administration was suspended through lack of funds, the discontent became so pronounced that during Secretary Colby's visit to South America the Wilson administration on December 23, 1920, authorized Rear-Admiral Snowden to issue a proclamation announcing that "the United States believes the time has arrived when it may, with a due sense of its responsibility to the people of the Dominican Republic, inaugurate the simple processes of its rapid withdrawal from the responsibilities assumed in connection with Dominican affairs."[25] The only conditions attached were that a commission of Dominican citizens should be appointed, with an American technical adviser who should formulate certain amendments to the constitution and a general revision of the laws. The Dominicans, however, protested vigorously against these conditions, and the situation was turned over to the Harding administration virtually unchanged.

As a candidate, President Harding had promised to take action, and he did not delay. On June 14, 1921, Admiral Robison, who had been appointed military governor to succeed Admiral Snowden, issued a proclamation promising withdrawal within eight months, providing the Dominicans would cooperate with the Americans in establishing a government able to maintain independence and public order and to assure security of life and property. The principal conditions of the convention of evacuation were: (1) the acts of the military government should be ratified; (2) a new loan of $2,400,000 to complete the public works should be validated; (3) the duties of the general receiver of Dominican customs should be extended to apply to this loan, and also to the collection and disbursement of a part of the internal revenues if the customs were insufficient to meet the service of the foreign debt; and (4) an efficient *Guardia Nacional* should be organized under the direction of an American military commission.[26] Again the Dominicans protested, and certain of the objections were met by a new statement issued by Secretary Hughes on June 28. The military government thereupon promulgated an electoral law, and the elections were set for August 13. The Dominicans were not yet satisfied; and they organized a "Junta of Electoral Abstention," which prevented the elections from being held.

In the meantime the question had come before Congress, and a Senatorial investigating commission was appointed, with Senator McCormick as chairman. The commission spent about a fortnight in the island, and in a preliminary report Senator McCormick declared that inasmuch as the political leaders in the Dominican Republic had rejected the proposals for the withdrawal of the American forces, and since at the present time it was impossible to advise a substantial modification of the terms of the proclamation, the American troops should not be removed.[27]

Early in 1922, however, numerous conferences were held in Washington between Secretary Hughes and Señores Velásquez, Vásquez, Brache, and Peynado, the first three representing the different political parties in the Dominican Republic, the latter being the former Dominican minister at Washington. The

[25] Text *Hearings before a Select Committee on Haiti and Santo Domingo*, Part III, p. 934.
[26] *Ibid.*, Part I, p. 102.
[27] *New York Times*, Dec. 26, 1921.

following program was finally agreed upon:

> (1) A Provisional Government of Dominican citizens selected by Dominicans will be installed to carry out such legislative and constitutional reforms as they may deem appropriate, and to hold general elections for the installation of a subsequent permanent government without the intervention of the authorities of the United States.
>
> (2) Upon the inauguration of the Provisional Government, the executive departments of the Dominican Republic will be turned over to the cabinet ministers appointed by the Provisional President; the Military Government will also turn over the National Palace, and at the same time the United States military forces will be concentrated in one, two, or three places as the Military Governor may determine. From that time peace and order will be maintained by the Dominican national police under the orders of the Provisional Government.
>
> (3) Dominican plenipotentiaries will be designated by the Provisional Government to negotiate a convention with the United States containing the following provisions: (*a*) Recognition by the Dominican government of all executive and department orders promulgated by the Military Government, contracts entered into in accordance with these orders, and specific recognition of the bond issues authorized in 1918 and in 1922; (*b*) the convention of February 8, 1907, shall remain in force so long as any bonds of the issues of 1918 and 1922 shall remain unpaid, and the duties of the general receiver shall be extended to include the collection and application of the revenues pledged for the service of these issues.
>
> (4) Upon this convention being approved by the national congress and constitutional president duly elected, the American military forces will immediately leave the Dominican Republic.

In order that the United States might be advised without delay whether this agreement met with the approval of a majority of the Dominican people, President Harding appointed Mr. Sumner Welles of New York, a former chief of the Division of Latin American Affairs in the State Department, as commissioner, with a rank of envoy extraordinary and minister plenipotentiary, to represent him in the Dominican Republic, for the purpose of investigating and reporting upon conditions there, and to obtain the views of the Dominican people respecting an appropriate agreement with the United States as a result of which the military forces of the United States might be withdrawn.[28]

An unbiased survey of the whole situation is found in the report of Dr. Carl Kelsey, of the University of Pennsylvania, who was selected by the board of directors of the American Academy of Political and Social Science to undertake a survey of the economic, social, and political conditions in Haiti and the Dominican Republic. He spent nine months in the island, and his report contains a wealth of valuable material about conditions on the island at that time. In his conclusions regarding the Dominican Republic he found that, although the Dominicans were not antagonistic to Americans, they were critical of the policy of our government. They felt that the troops were sent either under false pretenses or through error. They recognized their economic dependence upon the United States, and they would have welcomed better trade relations. But the

[28] *Ibid.*, July 12, 1922. Mr. Welles later published a complete narrative of the events of this period in his book, *Naboth's Vineyard: The Domincan Republic, 1844-1924* (New York, 1928).

United States had promised to withdraw under certain conditions, and the pledge had to be kept if the conditions were accepted.[29]

Secretary Hughes himself recognized the false position in which the United States found itself in the Dominican Republic, and in a speech delivered at Amherst College in June, 1924, he asserted that "no step taken by the government of the United States in Latin America in recent years has given rise to more criticism, and in this instance just criticism, than the military occupation of the Dominican Republic by the armed forces of the United States in 1916 . . . it is the belief of many that the military occupation would have never occurred had President Wilson had the opportunity or the time in the excitement of that period, to become fully cognizant of the causes of the situation existing in the Dominican Republic. Likewise it is improbable that he was informed of many of the occurences which took place in the Dominican Republic during the earlier years of the American occupation,–occurrences deeply to be regretted by every American citizen."[30]

However, in this same speech Secretary Hughes was able to state that the occupation had terminated and that a freely elected constitutional government was at the time in sole power. For although the marines were not withdrawn until September 1924, an election supervised by the United States officials had taken place on March 15, 1924, approving the plan of evacuation signed at Washington on June 30, 1922. Shortly afterwards, pursuant to the agreement of 1922, a convention of ratification or validation of all executive orders and resolutions promulgated by the military government was signed by the two governments. This treaty also specifically recognized the bond issues of 1918 and 1922 as irrevocable obligations of the Republic, and provided that no change should be made in customs tariff without previous agreement between the Dominican government and the United States. The convention of 1907 was to remain in force until both bond issues were paid and the general receiver of customs was to control the revenues pledged to these services.[31]

However, it had long been felt that the 1907 convention was forcing too large an amount of income to be used for the amortization of the loan at the expense of various projects of public works vital to the development of the country. Consequently on December 27, 1924, a new convention was signed revising the convention of 1907. The new agreement contemplated the floating of a loan of $25 million to refund all outstanding loans and to leave a surplus of some $10 million "to be devoted to premanent public improvements and to other projects designed to further the economic and industrial development of the country." The agreement made no change in the arrangement whereby the customs were to be administered by a general receiver appointed by the president of the United States, and it also provided that until the whole amount of the bonds should be paid the public debt should not be increased except by previous agreement with the United States.[32]

[29] "The American Intervention in Haiti and the Dominican Republic," *Annals of Amer. Acad.*, vol. C, no. 189 (Mar., 1922), pp. 109-200, found also in *Hearings before a Select Committee on Haiti and Santo Domingo*, Part 4, pp. 1279-1341; see also Dr. Kelsey's statement before the Senate Committee, *Ibid.*, pp. 1238-1277.
[30] *New York Times,* June 19, 1924.
[31] *U. S. Treaty Series,* No. 729.
[32] *Ibid.*, No. 726.

However, there arose in the island considerable opposition to the new agreement on the ground that the new loan would prolong the control of the United States for a period of time ranging from twenty-five to a hundred years. Some seventeen of the thirty members of the Chamber of Deputies signed a "pact of honor" not to approve the convention. Nevertheless, when three of the "pacta" group went over to the government, the convention was approved and ratifications were exchanged on October 24, 1925.

During the next few years both political and economic conditions in the Dominican Republic were satisfactory. Horacio Vásquez won the presidential election of 1924, was inaugurated in July, and the Marines left in September. Had it not been for poor health he might well have been re-elected. In February 1930, a revolt occurred and Vásquez resigned to avert bloodshed. Through mediation of the American legation, Estrella Ureña, the leader of the uprising became provisional president. In the elections held May 16, 1930, General Rafael Leonidas Trujillo, Commander-in-Chief of the army, was elected president for a four year term, and Estrella Ureña was chosen vice-president.

Before the end of his term President Vásquez had attempted to put into effect a complete reorganization of the economic and financial administration of the government. To assist him he had invited General Charles G. Dawes to organize a financial commission to recommend methods on improvement in the collection and control of both national and municipal revenues. The commission, after a three-week study of the island, prepared drafts of laws concerning fiscal matters, the reorganization of the government departments, and a civil service.[33] Although the Dominican Congress enacted all of the program into law, the downfall of the Vásquez administration nullified it enforcement.

General Trujillo's election proved to be the beginning of one of the most durable, evil and controversial dictatorships of the twentieth century, the "Era of Trujillo" spanning the period 1930-1961. He was president from 1930 to 1938 and from 1942 to 1952. Between 1938 and 1942, and after 1952, he allowed the office to be occupied by figurehead presidents, including his ineffectual brother, Hector. Born into a poor mulatto family in 1891, he joined the constabulary trained by the United States Marines, and rose rapidly in the ranks. Ultimately gaining command of the national army, he had betrayed the trust of President Vásquez, by collaborating with Estrella Ureña in the revolt of February 1930.

The administration of President Trujillo was handicapped at the beginning by the results of a violent hurricane which in September 1930, caused a heavy loss of life and enormous property damage. This, coupled with the world economic crisis, forced a revision of the debt payments under the 1924 convention to prevent complete national bankruptcy. An emergency law of October 23, 1931, diverted to governmental expenses $1,500,000 from customs revenues which were pledged to service on the foreign loans. The United States Department of State immediately ordered an investigation, but when the serious situation was confirmed by its representatives, it was content to follow "with attention and care the developments in the Dominican Republic."[34]

[33] *Report of the Dominican Economic Commission* (Chicago, 1929).
[34] U. S. Dept. of State, *Press Releases* (November 14, 1931), p. 454.

In 1934 a new arrangment was made with the Foreign Bondholders Protective Council, Inc., an organization sponsored by the United States government, whereby the General Receiver of Customs was given greater powers; interest payments were to be made in full, while amortization payments were reduced so that maturities for the two bond issues originally set for 1932 and 1940 would be extended respectively to 1962 and 1970. This very substantial reduction in debt payments permitted the Dominican Republic to balance its budget annually.

Negotiations for a revision of the 1924 convention were begun in 1936, and a new treaty was signed in Washington September 24, 1940, and ratifications were exchanged March 10, 1941. The new arrangement superseding the 1924 treaty abolished the General Receivership of Dominican customs and permitted the Dominican government instead of an outsider to collect the customs' revenues. These were to be deposited in a bank selected by mutual consent and placed under the obligation to service first the public debt. The Foreign Bondholders Protective Council protested the agreement, although it was given a representation in the depository bank. By an exchange of telegrams between the two governments the Ciudad Trujillo branch of the National City Bank of New York was selected as the sole depository bank. With the convention of 1940 in force the Dominican Republic was freed from all governmental controls or checks by the United States.[35]

Trujillo established a dictatorship under which political opponents were kept impotent by methods of repression and terror. Unlikely "accidents," "suicides," exile, imprisonment and disappearance were employed to wipe out all opposition. Political activity was reduced to eulogies for the dictator, and as time passed his titles came to incliude "Benefactor of the Fatherland," "Rebuilder," "Father," and others. As the exile colonies abroad grew, however, the reality as opposed to the image of the regime was exposed to the world.

With Congress completely subservient to his wishes, the dictator proceeded to show what could be accomplished by the "Benefactor of the Nation." He rebuilt Santo Domingo City, whose name was changed to Ciudad Trujillo in his honor, he carried out a vast program of public works, expanded agricultural production, developed light industries, and modernized the educational system. With the application of scientific modern methods, sugar cultivation rose markedly, coming to comprise half the exports, and the coffee, cocao and livestock industries were strengthened. Government aid helped some farmers to obtain land, and urban workers were appeased by low cost housing and social security benefits. Trujillo's special favorites often became exceedingly rich, and the Trujillo clan itself came to control at least half of the nation's wealth.[36]

Trujillo, building the Dominican military establishment into one of the most formidable forces in the Caribbean region, frequently boasted that he could put 100,000 men in the field. All the services were equipped with modern weapons, most of which were manufactured in domestic arsenals. "The mere

[35] For text see U. S. *Treaty Series*, No. 965.
[36] See Charles A. Thomson, "Dictatorship in the Dominican Republic," *Foreign Policy Reports*, XII (Apr. 15, 1936), and Robert D. Crassweller, *Trujillo: The Life and Times of a Caribbean Dictator* (New York, 1966).

presence of the powerful Dominican military machine, loyal to and under the control of Trujillo, was sufficient to discourage all opposition; but it was further used to impose an all-pervasive terror on the population.."[37]

In his relations with the adjoining Republic of Haiti, President Trujillo began auspiciously by endeavoring to resolve the boundary question which had remained a threat to friendly relations since 1874. A treaty establishing a boundary commission had been signed January 21, 1929, but the commission had not been named. However, by personal meetings with President Stenio Vincent of Haiti, held in both countries from 1933 to 1935, a settlement was finally reached, the commission was appointed, and it set to work. Whatever gains had been made toward establishing cordial relations between the traditionally hostile peoples were offset by reports in October 1937, that there had occurred near the frontier a wholesale massacre by Dominican soldiers of Haitian peasants who had crossed into Dominican territory in search of work. Upwards of ten thousand were reported killed, although the actual figure was never determined.

The Haitian government immediately asked for an investigation, punishment of the guilty, indemnification for the victims and assurances of protection in the future. When a month elapsed without a satisfactory settlement, President Vincent on November 12, 1937, appealed to the presidents of the United States, Mexico, and Cuba to use their good offices to attempt a settlement of the question. The three governments accepted the invitation and informal hearings were held in Washington. The Dominican Republic was at first inclined to insist upon a direct settlement between the two governments, but on December 18 President Trujillo accepted the invitation of the permanent commission set up under the inter-American Gondra Treaty of 1923 to participate with Haiti in an effort to settle by conciliation the dispute over the border killings. Secretary of State Cordell Hull, at that time an admirer of Trujillo, lent his support to the settlement. On December 10 President Franklin D. Roosevelt sent a telegram of gratification to President Trujillo for his acceptance of the peaceful procedure for the settlement of the dispute.

An agreement settling the controversy was signed on January 31, 1938, in Washington, by representatives of the two governments. By the terms of this agreement the Dominican government agreed to pay an indemnity of $750,000 to Haiti and to abide by the findings of the Dominican tribunals which were to continue the inquiry into the incidents which led to the dispute. The Dominican Republic agreed "to fix the responsibility of those guilty of instigating the incidents and to give the results of the investigation full publicity." In March 1938, Trujillo and Vincent were photographed at the border in an embrace of apparent reconciliation.[38]

As an antidote to the Haitian massacre the Dominican representative to the Intergovernmental Committee on Refugees, convened in the summer of 1938 at the suggestion of President Roosevelt, offered in behalf of President Trujillo a large tract of land in the Dominican Republic for the resettlement of a contingent of European refugees, mainly Jewish victims of Hitler's persecution.

[37]Howard J. Wiarda, "The Politics of Civil-Military Relations in the Dominican Republic," *Journal of Inter-American Studies*, VII (Oct., 1965), pp. 469-470, 473.
[38]Crassweller, *Trujillo*, chap. II; *New York Times*, Feb. 1, 1938.

When the offer was accepted the Dominican Republic Settlement Association was organized. During 1939 it brought over about five hundred refugees as a test example of the colonization scheme. A survey of the project made in 1941 under the supervision of the Brookings Institution presented a somewhat pessimistic report based partly upon the health conditions of the Sosua tract and partly upon the small amount of unoccupied land available on the island.[39] The colony never developed as hoped, and many of the inhabitants used it merely as a way station for migration elsewhere.

After his second term of office, from 1934-1938, President Trujillo was succeeded by his henchman, President Jacinto B. Peynado. When the latter died in office in 1940, vice-president Manuel de Jesús Troncoso de la Concha took over, also under the aegis of General Trujillo. "The Restorer of the Financial Independence of the Republic," as Trujillo was officially designated by the Congress in 1940, was re-elected president in May 1942, without any opposition. President Troncoso thereupon appointed the General to his cabinet and then resigned, thus permitting Trujillo to become president immediately without violating the constitution of the republic. Dr. Troncoso was subsequently appointed minister to the United States.

Even before the outbreak of World War II in Europe, General Trujillo had offered to the United States all available facilities of the Dominican Republic for hemispheric defense. After the attack on Pearl Harbor the Dominican Republic declared war on Japan, December 8, and upon Germany and Italy, December 11. The war affected the economy of the country most favorably; the foreign debt was paid off in 1947, and the small internal debt was refunded ahead of schedule, being liquidated in 1953. Generally, the Dominican *peso* remained firmly on par with the United States dollar. Communism was outlawed in 1947, and the Dominican representatives consistently stood by the United States on cold war issues. Trujillo reached a concordat with the Vatican in 1954, and maintained a close friendship with Francisco Franco, the fascist dictator of Spain.[40]

Politically, the Trujillo regime had emerged in the post-war era as a personal tyranny to a degree scarcely equaled on either side of the Iron Curtain. His repression of any possible opposition was cruel as well as complete. Although it was a one-man dictatorship, it also had a family structure, perhaps with dynastic succession in mind. One compiled list showed that there were 157 relatives of Trujillo in government jobs or in business connected with the government.

The eldest son of the dictator, Rafael Leonidas, Jr., known as Ramfis, was made a colonel at eleven years of age, but at age seventeen he was reduced to captain so that he could win his way up in rank; at twenty-three he was promoted to general. As heir apparent, Ramfis became chief of staff of the Air Force, inspector of embassies, and was awarded a doctor of laws degree from the University of Santo Domingo.

[39] Brookings Institution, *Refugee Settlement in the Dominican Republic* (Washington, 1942).
[40] John F. Fagg, *Cuba, Haiti and the Dominican Republic* (Englewood Cliffs, 1965), p. 162; *New York Times,* Mar. 28, 1953.

When Trujillo seized power in 1930 there were seven political parties in existence. He quickly reduced them to one, the *Partido Dominicano*, in 1931. Every eligible voter belonged to it, and all government employees had 10 percent of their pay deducted for its operating fund. Having no real importance or characteristic, its main functions were to propagate the credo of Trujillo, and to provide hand-picked lists of candidates at election time, who were invariably elected. In 1942 the dictator, wishing to give the impression that opposition was permitted, allowed the *Partido Trujillista* to form. It was with the same object that he permitted the Communist *Partido Socialista Popular* (P.S.P.) to return, and it was active in the period 1945-1947.[41] Then Trujillo became a staunch anti-Communist, drove the party underground, and assumed close control over all internal political affairs. There were no opposition newspapers, and even the university students dared not openly protest. The tradtionally honored institution of asylum was violated in order to thwart any opponent. In Ciudad Trujillo the dictator stationed plain-clothed policemen near each Latin American embassy whose duty it was to shoot and kill anyone attempting to rush the embassy to gain asylum. Many persons were killed seeking this refuge.[42] Yet, Trujillo was so attracted to the form of democracy that he went to the trouble of holding elections, choosing a cabinet, and keeping a legislature in being. In this respect he was like other Latin American dictators. Knowing that their people want democracy, they give them the form without the substance.

When in 1947 Trujillo had himself elected for a fourth term of five years, Dominicans who had escaped from his tyranny attempted to organize an expedition against him from Cuba. However, Trujillo learned about the plot and informed the Cuban government, which arrested the leaders and seized the ships and planes which had been assembled. An attempt sponsored by a wealthy Dominican exile living in Guatemala in June 1949, was equally abortive.

As President Trujillo's fourth term of office expired in 1952, he had his brother, General Hector B. Trujillo, elected as president and himself named Generalíssimo and Commander-in-Chief of the Armed Forces of the Republic, and Ambassador-at-Large to the United Nations. In the latter capacity at the meeting of the General Assembly in New York, in February, 1953, it was reported that Ambassador Trujillo with his bodyguard outnumbering that of the Russian delegation by two to one was a greater attraction than the Soviet Foreign Minister Vishinsky. While in the United States Generalissimo Trujillo signed a Mutual Assistance Military Pact with the United States.

In the post-war era the Trujillo regime proved to be a source of embarrassment for the United States, but the threat of communism in the Caribbean sustained the relationship and with it the amenities of diplomatic practice. In August 1954, the dictator was guest of honor at a luncheon at Blair House, the official government residence for distinguished visitors in Washington. Protesting the warm welcome given him, Dominican exiles picketed the White House. While on a good-will tour of the Caribbean in 1955 vice-president

[41] Ben G. Burnett and Kenneth F. Johnson, *Political Forces in Latin America: Dimensions of the Quest for Stability* (Belmont, California, 1968), p. 160.
[42] Rayford W. Logan, *Haiti and the Dominican Republic* (New York, 1968), pp. 71-72; *New York Times*, Aug. 1, 1960.

Richard Nixon, a guest at Cuidad Trujillo, declared: "We want you to appreciate how much we thank you—our government and our people—for the help of your people and your government in the OAS and the United Nations in combatting the pernicious forces of Communism."[43]

In May 1957, the dictator again permitted his brother, Hector Trujillo, to be elected ceremonial president, and Joaquín Balaguer, a diplomat, university professor and writer, was brought into the vice-presidency. In a country where the election of a president could hardly have been more meaningless, delegations from forty-six countries were present to witness the inauguration. The United States delegation was headed by Ambassador Joseph S. Farland, formerly with the Federal Bureau of Investigation.

This occurred amid strained relations between the United States and the Dominican Republic brought about the Galíndez-Murphy case. Dr. Jesús de Galíndez, born in Madrid, and an exiled Republican, had lived in the Dominican Republic for six years. He studied at Columbia University, where he wrote a doctoral dissertation, later published in Chile, that was critical of Trujillo. He disappeared in a New York subway in 1956, and an airplane pilot, Gerald L. Murphy of Eugene, Oregon, reportedly declared that he had flown Galíndez, a kidnap victim, to the Dominican Republic. Murphy was killed, according to the Dominican government, by a fellow pilot, Octavio de la Moza. The latter, the government said, committed suicide in a prison cell, leaving a note confessing the deed; upon seeing the note, the State Department challenged its authenticity. Trujillo frustrated United States legal procedures by refusing to drop the mantle of diplomatic immunity from General Arturo Espaillat, the Dominican consul-general in New York, who was seen at the airport when the abduction plane was rented.[44] Neither case was solved, nor was Trujillo's complicity established, but the circumstances and publicity aroused much bitter opposition against the dictatorship in the United States.

In Congress the most vocal opponent was representative Charles O. Porter, Democrat of Oregon, who had a primary responsibility toward his constituents in the case of Murphy. Porter charged the Trujillo regime with the apparent death of Murphy, and urged that the United States impose economic sanctions—cutting off trade and reallocating the sugar quota—in the Dominican Republic. Coming to Trujillo's defense were Representatives James Fulton, Republican, Pennsylvania, and Mr. John M. Robinson, Jr., Republican, Kentucky, who disapproved of Porter's attack on a "friendly nation." Representative George S. Long, Democrat, Louisiana, asserted that Mr. Murphy had "succumbed to greed and became involved with a criminal element to the extent of joining a conspiracy with businessmen and former diplomats, and being guilty of kidnapping."[45]

Escapades of the dictator's son, coming in the wake of the Galdínez-Murphy cases, further incensed American opinion against the regime. The problem arose after Ramfis had become a student at the Command and General

[43] *New York Times*, Mar. 2, 1955.
[44] Dept. of State *Bulletin*, XXXVI (June 24, 1957), pp. 1025-1028; *Cong. Rec.*, 87th Cong., 1st Sess., vol. 107. p. 5310; *Hisp. Amer. Rep.*, X (Mar., 1957), 132-133.
[45] *New York Times*, Sept. 2, 1957.

Staff College at Fort Leavenworth, Kansas. It was reported by representative Wayne L. Hays, Democrat, Ohio, in March 1958, that Ramfis occupied a Kansas City hotel, and used a fleet of autos, a weekend ranch house, and other luxuries costing an estimated $1 million a year. This approached the total United States aid earmarked for the Dominican Republic in 1958, $1,300,000. All this was climaxed by his being denied graduation from the school. An Army report said General Trujillo "did not successfully complete the course" at the Command and General Staff College, and that "he will not be issued a certificate of graduation, but will be given a certificate of attendance only."[46]

Generalíssimo Trujillo responded to the alleged affront by promoting Ramfis to Lt. General, ordering home all Dominican nationals taking courses in United States military establishments, and directing the Dominican Congress to renounce all military and technical assistance agreements with the United States. He then set out to purge the United States Congress of all men whom he disliked, namely Representative Charles O. Porter, Oregon, Charles B. Brownson, Indiana, and Alvin M. Bentley, Michigan. His efforts took the form of letters sent to chambers of commerce and governors of states where his "enemies" were running for office.

The dictator invested heavily in the United States in the interest of his regime's image, his country's sugar quota, and in public relations activities on many levels. These efforts did not prove unavailing. Among the documents distributed by the Dominican Republic Information Center, in New York City, was the text of remarks by Senator Olin D. Johnson of South Carolina, in the Senate, July 25, 1956. The senator deplored "unfair and unfounded" attacks on Trujillo attributable to "sinister elements." Other senators were lavish in their praise of the dictator. Senator James O. Eastland of Mississippi declared: "Your Generalíssimo is one of the great men of the free world." With him, he said, the Dominican Republic was a "leader of all Latin America–you lead for freedom; you lead for honor; you lead for religion; and thank God you lead for common sense." Senator William E. Jenner of Indiana said that he had spent enough time in the American Senate "to learn how bad it is for a country if its leaders do not serve their country's best interests . . . and how important it is for a national leader to really lead and not just drift with the tide. I can appreciate how happy and lucky you have been with the Generalissimo."[47]

Trujillo had in his employ in the United States many persons who were attracted by his rewarding financial inducements. Some of them ran afoul of the law for violating the Foreign Agents Registration Act. One of them was John J. Franck, a former agent of the FBI, who was fined for not registering as an agent of the Dominican Republic. Three former officers of the Mutual Broadcasting Company Alexander L. Guterma, Hal Roach, Jr., and Garland L. Culpepper, Jr., were charged by a Federal Grand Jury of having accepted $750,000 from the Dominican Republic to broadcast its propaganda without registering as foreign agents. In 1964, I. Cassini and R. P. Englander were fined $10,000 each for not registering in the United States as public relations agents for the Trujillo regime.

[46] *Hisp. Amer. Rep.*, XI (May, 1958), pp. 260-261,

[47] *Cong. Rec.*, 84th Cong., 2nd Sess., vol. 102, p. 14350; *Ibid.*, 86th Cong., 1st Sess., vol. 105, p. 3263; *New York Times*, Sept. 15, 1958.

With Fidel Castro's seizure of power in Cuba in January 1959, Trujillo was confronted by an old enemy bent not only on his overthrow, but also the toppling of other rightist dictators in the Caribbean, Francois Duvalier of Haiti and Anastasio Somoza of Nicaragua. An invasion force, made up of Cubans and exiles, was reportedly crushed by the Dominicans on June 22, 1959, and a few days later Cuba broke diplomatic relations with the Dominican Republic. In the following November, Trujillo ordered sentenced in absentia 113 persons, including Fidel Castro and Romulo Betancourt, for their alleged roles in the June invasion. Fearing further invasion attempts the Dominican military budget was raised to $88 million, or more than half of the total national budget of $152 million.

Accompanying the renewed external threat was a deterioration of relations with the United States, and a falling off of economic activity. The balance of payments deficit for the first seven months of 1959 amounted to $44 million. This was attributed to droughts and lower prices for exports, and was attenuated by the increased military expenditures. Per capita income was fixed at $260 in that year. At Trujillo's request the United States naval mission was withdrawn, an action interpreted to have resulted from the Generalissimo's displeasure with press reports on the economic weakness of his country. Under the Mutual Assistance Pact signed in 1953 the United States provided the Dominican Republic with about $400,000 a year in support of its military establishment. This amounted to $445,000 in 1960, all of which was disbursed by February.[48]

Evidence that Dominican-United States relations were not harmonious on the diplomatic level was the report in May 1960, that Vinton Chapin, United States ambassador to Luxembourg, would succeed Ambassador Farland. On Farland's return from Washington on May 22, pickets at the United States Embassy displayed signs condemning "Yankee Racial Discrimination" and protesting against "the interventionist conduct of Ambassador Joseph S. Farland–former F.B.I. agent." The ambassador had aroused the dictator's displeasure the previous month, and this was in reprisal. Trujillo had demanded the recall of Carl E. David, the Embassy information officer, on grounds that he conveyed derogatory information to a British journalist. Farland asked that the charges be supported with evidence, and when it was not forthcoming he called the Dominican government's action "unsupported and unjustified." David, however, was recalled in accord with diplomatic procedure. On May 6 the State Department, making it clear that the summons reflected Washington's displeasure with the attitudes and actions of the dictator, called Mr. Farland home for consultation.[49]

By 1959 the Dominican Republic had become a main focus of Caribbean unrest, and at the Fifth Consultation of Ministers of Foreign Affairs convened in Santiago, Chile, that year, the principal problem–how to overthrow dictators–centered on Generalissimo Trujillo. Cuba and Venezuela stood for the application of sanctions, irrespective of the principle of non-intervention, but the majority of the delegations, including the United States, gave precedence to the sanctity of the principle, and were content to denounce dictatorships on

[48] *New York Times*, Feb. 26, 1960.
[49] *Ibid.*, May 25, 1960.

moral grounds. Secretary of State Christian Herter, though highly critical of the Trujillo dictatorship, supported non-intervention at that point out of caution for what might follow Trujillo's overthrow.[50] Seen in the perspective of later events the Dominican Republic was a greater problem than Cuba for the OAS and the United States until 1961, when Cuba became clearly identified with the Communist bloc.

The Santiago meeting had only inferentiately condemned the Trujillo regime, but the reactivation at that time of the Inter-American Peace Committee, inactive since 1956, made possible further exposure of the dictatorship. On February 8, 1960, Venezuela charged before the OAS Council that the Dominican Republic was guilty of "patent and flagrant violations of human rights." The Venezuelan ambassador had read press reports of three thousand political arrests, and the matter was referred to the five-member Peace Committee. After study of the evidence the Committee, on June 8, accused the Dominican Republic of flagrant and widespread violations of human rights. The violations, the Committee said, included "the denial of free assembly and of free speech, arbitrary arrests, cruel and inhuman treatment of political prisoners, and the use of intimidation and terror as political weapons." The statement was regarded as a milestone in the Inter-American system for never before had the OAS or any of its agencies publicly excoriated a country for such indignities. Past reluctance to do so was based on the conviction that non-intervention in internal affairs was an inviolable principle of the Inter-American system. In the preceding three years, however, there was growing recognition that non-intervention had become a shield for violations of the rights of individuals.

The moral condemnation of the Trujillo dictatorship was soon followed by events which activated the enforcement procedures of the OAS. In June 1960 President Betancourt of Venezuela was injured and nearly killed in an attempted assassination, at Trujillo's direction, by weapons from the Dominican Republic. Irrefutable evidence of the complicity of Dominican officials convoked the Sixth Meeting of Consultation of the American Foreign Ministers at San José, Costa Rica in August.[51]

The Latin American delegations, with few exceptions, supported the imposition of forcible sanctions. Secretary Herter would have preferred to see the appointment of a special commission empowered to supervise free elections and thereby rid the country of the dictatorship, but this was a positive approach which many Latin Americans construed as unwarranted intervention. The American delegation, doubtless influenced by the need to secure support for strong measures against the Castro regime at the Seventh Meeting of Foreign Ministers the following week, adjusted its view "to maintain the solidarity and common approach of our community." Secretary Herter said: "In voting in favor of the resolution that has just been adopted, the United States had joined with the other American governments in condemning the acts of intervention and aggression against the government of Venezuela which were carried out with the participation of the Dominican Republic. We have also joined in applying

[50] Dept. of State, *American Foreign Policy, Current Documents,* No. 7492 (Washington, 1963), pp. 359-373.
[51] *Cong. Rec.*, 86th Cong., 2nd Sess., vol. 106, p. 19142.

certain measures and the breaking of diplomatic relations, and the partial interruptions of economic relations in accordance with Article 8 of the Rio Treaty."[52] In the arrangements which were made an important consideration was the security of United States citizens' private investments in the Dominican Republic, which amounted to $125 million. Ambassador Vinton Chapin, who had been appointed, did not take up his post.

In an apparent effort to improve the image of his government the Generalissimo ordered his brother Hector to step down from his post as ceremonial president on August 3, and Vice-President Joaquín Balaguer was sworn in as president. Within a week the new chief executive announced that safe conduct passes and passports were being issued to all who had taken asylum, and that they were free to leave the country. Providing asylum for opposition leaders whose safety is endangered is the intent of the Havana Convention on Political Asylum, of which the Dominican Republic was a signatory, but from which she had withdrawn. To confirm the order, Dominican police were withdrawn from the foreign embassies.

The changed circumstances, however, did not alter the State Department's attitude toward the dictatorship. Senator Allen J. Ellender contended the State Department had "twisted, tortured and perverted" sugar legislation in a "venomous campaign against the Dominican Republic." He said the State Department, acting through the Department of Agriculture, had "wilfully and wrongfully" denied entry into the United States of 322,000 tons of sugar from the Dominican Republic. Representative Harold D. Cooley of North Carolina also criticized the State Department in harsh terms on the Dominican problem. The question arose after President Eisenhower reduced the Cuban sugar quota by 700,000 tons in 1960 as a rebuke to Fidel Castro. The Dominican Republic had been assigned two shares of the replacement sugar quota, totaling 322,000 tons. The State Department's stand was interpreted as an effort to deny an economic windfall to the Trujillo clan.[53]

President-elect John F. Kennedy was attacked by a propaganda radio station in the Dominican Republic in December. It was asserted that Kennedy would fail to solve the snarl left by the Eisenhower administration, and that the Dominicans would have to turn to the Soviet Union for help. The announcement occurred a few days before Washington's announcement that imports of Dominican sugar would be increased in the first part of 1961.

This concession was somewhat negated, however, by the action of the OAS Council which voted 14-1 to suspend exports of petroleum, petroleum products, trucks and spare parts for trucks, to the Dominican Republic. Washington suspended these exports beginning on January 20, 1961.[54] It was at this time also that the Catholic clergy began denouncing the regime in Dominican churches.

The Generalissimo was seemingly unperturbed by these events, and cajoled the masses successfully, until he was shot to death by military elements in an

[52] Dept. of State *Bulletin*, XLIII (Sept. 5, 1960), pp. 356-357.
[53] *New York Times*, Aug. 11, 1960.
[54] *Cong. Rec.*, 86th Cong., 2nd Sess., vol. 106, p. 19143; *Ibid.*, 87th Cong., 1st Sess., vol. 107, pp. 5310-5311; Dept. of State *Bulletin*, XLIV (February 20, 1961, pp. 273-276.

ambush on the night of May 30, 1961, near Cuidad Trujillo. The United States Department of State declared at once: "It is our earnest hope that now the people of the Dominican Republic will be able to establish those conditions which will make it possible for that country to again take its proper place within the Inter-American system."[55]

While turbulence was anticipated in the transition from dictatorship to democracy in a country where the people had had no practical experience in the meaning of this latter concept, the anarchy and repercussions of the post-dictatorship period surpassed all precedents, including the aftermath of the Ulises Heureaux regime. In the period from May 30, 1961 to September 3, 1965, when Hector García-Godoy became provisional president, the office of chief executive changed hands seven times, and this figure does not include those named "president" by the revolutionary factions during the civil war in 1965.

President of the Republic at the time of assassination was Joaquín Balaguer; his tenure was to last until mid-January, 1962. Balaguer had become one of Trujillo's closest advisers and trusted administrators, but at the same time, he had gained a reputation for independence and integrity. Shortly after Trujillo's death Balaguer began to introduce liberal reforms. Restrictions were lifted on political parties and preparations were made to hold free elections the following May. The secret police organization was placed under civil control, and many former officials were removed from power. The late dictator's thirty-two year old son Ramfis, who had assumed the post of Chief of the Joint Chiefs of Staff on his return to the homeland two days after his father's death, supported Balaguer's policy of "democratization." Moderation was condoned by Ramfis for it was in the interest of the far-flung Trujillo clan and their henchmen that the OAS sanctions be lifted as quickly as possible.[56]

When two brothers of the slain dictator, General Hector B. Trujillo and General José A. Trujillo, returned from exile on November 15, 1961, with the apparent object of staging a coup d'etat, Ramfis fled the country. Four days later a show of force by the United States Navy and a warning from the Dominican Air Force, indicating that a return of the dictatorship would not be tolerated, caused the plotting uncles to depart. The other family members followed them into exile, leaving all the clan's properties, worth hundreds of millions, to be confiscated by the government. It was evident that they had secured adequate resources abroad. In response to Cuban charges of intervention, the State Department declared that the naval units stationed near the Dominican Republic were on the high seas, outside territorial waters, and outside air space, "a friendly presence with full knowledge of the Constitutional authorities and responsible leaders of the Dominican Republic."[57]

Beginning in September the main opposition to the Balaguer-Ramfis administration, the National Civic Union (U. C. N.), had demanded the removal of the Trujillos as a primary condition for its participation in a coalition government. However, with the flight of the Trujillos, the U. C. N. tried to force

[55] *New York Times*, June 2, 1961.
[56] Henry Wells, "Turmoil in the Dominican Republic," *Current History*, vol. 50 (Jan. 1966), pp. 15-16.
[57] Dept. of State *Bulletin*, XLV (December 18, 1961), pp. 1000-1002.

Balaguer's resignation by staging an eleven-day strike. This failed in its purpose, and the U. C. N. agreed to Balaguer's proposal that the existing government be replaced by a Council of State, consisting of Balaguer as president and six others, the latter being members of the U. C. N., or sanctioned by it. The Council assumed office on January 1, 1962, and three days later the OAS revoked its economic and diplomatic sanctions. But on January 16 the Council was overthrown by a military coup directed by Major General Pedro R. Echavarría, secretary of state for the Armed Forces, a Balaguer appointee. Two days later another group of military officers arrested Echavarría and restored to power all members of the Council of State except Dr. Balaguer, last of the heirs of Trujillo, who walked the plank into exile.

Washington, seeing Cuba gravitating toward the Communist camp, maintained an uneasy watch on developments in the Dominican Republic. Dr. Balaguer had been assured that the United States would work for an easing of the economic sanctions imposed by the OAS in return for the ousting of the Trujillos. This recommendation was made to the OAS Council by Robert F. Woodward, assistant secretary for Inter-American Affairs as a "gesture of encouragement to further progress by the government of the Dominican Republic." President-elect Kennedy promised the "sympathetic and tangible support" of the United States in the Dominican attempt to establish a Council of State to be appointed by President Balaguer. He said that when the OAS sanctions were lifted the United States Department of Agriculture would authorize purchases under the Dominican allocation of non-quota sugar for the first six months of 1962, and this would be followed by the dispatch of an AID mission headed by Ambassador Teodoro Moscoso.[58]

The Department of State announced on January 6, 1962 that the United States had resumed diplomatic relations with the Dominican Republic as a result of the decision of the OAS on January 4 to discontinue the measures adopted by the Sixth Meeting of Foreign Ministers. Washington welcomed the fall of the military junta, hastened preparations for an immediate loan of $20 million, and indicated that the Dominican Republic would be given a share of the sugar shipments that would yield $45 million. United States statements and threats, diplomats reported, had a great deal to do with the success of the counter-coup.[59]

During its thirteen months of control the reconstituted Council of State achieved fair success in revitalizing the economy which had been brought to a halt by the OAS sanctions and the post-assassination turbulence. Aware of its transitional role, it left the problem of major governmental reforms to the government that would succeed it after the general election scheduled for December 10, 1962. The Council's seven members, including Rafael L. Bonnelly, the council president and president of the Republic, represented the upper middle and upper classes of Dominican society. Coming from the traditional ruling element, and commanding the patronage of the Council, they expected to win the election.[60]

[58] *Ibid.*, XLVI (Jan. 22, 1962), pp. 128-129; Dept. of State, *American Foreign Policy, Current Documents*, No. 8007 (Washington, 1966), pp. 473-474.
[59] *New York Times*, Jan. 20, 1962.
[60] Wells, "Dominican Republic," pp. 16-17.

Washington, as mentioned, supported the Council in the belief that it represented stability and constitutionalism and was an anti-Communist bulwark. The first installment of a $25 million loan under the Alliance of Progress, amounting to $11,500,000 was authorized on February 24, 1962, most of which was earmarked for social and economic development programs. Displaying an eagerness to help themselves, hundreds of Dominican volunteers went into the cane fields to save the sugar crop from sabotage fires. It was estimated that 245,000 tons of cane had been lost to sabotage by pro-Communists and followers of the late dictator.

By the summer of 1962 it had become evident that Juan Bosch, founder and presidential candidate of the Dominican Revolutionary Party (P. R. D.), a liberal reformer and bitter foe of the late dictator, was a serious contender. Speaking in the idiom of the poor and colored, to whom he addressed his appeal, he told them of their political rights, and assured them of land, a higher standard of living and education. As the campaign progressed he grew more critical of the privileged classes with whom he associated, the candidates of the U.C.N.

Bosch, in a landslide victory in the election of December 1962, the first fair and free one since 1924, received 60 percent of the votes cast and twice as many as the number gained by Viriato A. Fiallo, the founder and candidate of the U. C. N. The old elite, fearing economic ruin as well as the loss of power and prestige, was stunned by their defeat. However, the P. R. D. program on such problems as land reform, social security legislation, industrialization, education and housing were moderate, and in keeping with the expectations of the Alliance for Progress.

The Kennedy administration, making some $46 million in AID funds available, and sending more than three hundred specialists and Peace Corpsmen to render assistance, stood behind the Bosch program. In spite of this support the regime came under increasing attack from the right and within eight months was overthrown. The Bosch government had been characterized by weakness in administration, it seemed incapable of carrying out successfully a policy of national reconciliation, and it had been permissive toward Communists. A fervent believer in freedom of expression, Bosch had followed the return of Communists who had long been in exile, and disregarded U. C. N. charges that they were infiltrating his administration In the end, it was a bitterly anti-Communist element in the armed forces that overthrew Bosch on September 25, 1963.[61]

After eight months of rule President Bosch left a mixed legacy. Having emphasized fiscal responsibility and concern for reestablishing the country's overseas credit and balancing the national budget, he succeeded in paying all foreign debts and building the country's depleted reserves to a surprising $15 million. The balance of payments picture was clouded somewhat because of a decline in exported sugar production. This was attributed to mismanagement of

[61] Abraham F. Lowenthal, "Limits of American Power: the Lesson of the Dominican Republic," *Harper's Magazine,* June, 1964, pp. 87-89; *Cong. Rec.*, 88th Cong., 1st Sess., vol. 109, pp. 10512-10514. For a first-hand account of this period see John Bartlow Martin, *Overtaken by Events: The Dominican Crisis from the Fall of Trujillo to the Civil War* (New York, 1966). Mr. Martin was the U. S. ambassador to the Dominican Republic from Mar., 1962, until Feb., 1964.

the government-owned sugar plantations and refineries and to labor trouble. Nevertheless, the United States Department of Commerce credited Bosch with having developed an "improved climate" for trade and investments.

With the overthrow of President Bosch, Washington withdrew all military and economic aid personnel, Alliance programs were halted, and it was made clear that the aid programs would not be readily resumed. Diplomatic relations were broken, but diplomatic personnel below the ambassadorial level remained at their posts. These measures signified that the United States would interrupt aid programs to all countries in which democratic governments were illegally overturned. Secretary Rusk clarified the policy saying: "The establishment and maintenance of a representative and constitutional government is an essential element in the Alliance for Progress . . . under existing conditions in the Dominican Republic . . . there is no opportunity for effective collaboration by the United States under the Alliance for Progress or for normalization of diplomatic relations." As preconditions for restoration of normal relations, Washington demanded guarantees of civil liberties and the promise of the popular election of a constitutional government within a reasonable period of time.

Colonel Elías Wessin y Wessin, leader of the coup against President Bosch, was head of the air force training unit, and a respected career officer. The military leaders quickly installed three civilians as a provisional government who would rule until elections could be held. First to head the Triumvirate was Emilio de los Santos, a lawyer. Upon his resignation in December 1963, Donald Reid Cabral a former vice president of the Council of State, was given the post. Reid Cabral served as president of the Triumvirate and of the Republic until overthrown in the revolt of April 1965.

Washington extended recognition to the military-controlled regime on December 14, 1963, but only after it received a pledge to hold elections before the end of 1965. The decision, made in a top-level reappraisal of the situation before the death of President Kennedy, was predicated on the belief that failure to reopen normal diplomatic contacts could bring about a new uprising from right or left, and contribute to even greater political instability. The policy decision indicated that the United States had not found an effective way of using its immense political and economic power in Latin America to implement the principles of representative democracy embodied in the Charter of the OAS.

United States aid programs to the Triumvirate were withheld pending the appointment of a new ambassador. W. Tapley Bennett, Jr., was assigned to the post, and shortly after his arrival at the Dominican capital on March 21 he authorized the release of $885,000 for education and public works programs; the Export-Import Bank granted an additional $4 million loan. Washington's suspension of economic aid following the removal of President Bosch had slowed down economic activity, and there was an urgent need to reverse the dangerous decline of trade and investment activities.[62]

Reid, although experienced in public administration, proved to be an ineffectual executive. He was responsible for the distribution of 1,500 units of former Trujillo-owned land to the peasants, five times more than had been

[62] *Cong. Rec.*, 88th Cong., 2nd Sess., vol. 110, p. 6855.

distributed by the Bosch regime, but he failed to gain popular support. Unemployment came to exceed 30 percent of the labor force as the uncertainty and fears created by the shadow of political rivalries discouraged domestic and foreign investment. To a great extent the economic problems could be traced to the "Great Benefactor." The Trujillo dictatorship had presented a facade of well being and prosperity to the world while sweeping the main problems of the country under the carpet. When the facade crumbled, a fertile country of poor uneducated farmers with little incentive to produce more than their own needs was exposed.

The outstanding economic problems facing the Dominican government were cited as: (1) developing a force of technicians to teach the farmers how to produce more; (2) eliminating the need for food imports ($40 million in foodstuffs were imported in 1964); (3) increasing storage and refrigeration facilities; (4) rehabilitating the irrigation system (the United States provided $1.5 million for this purpose); (5) reducing the high operating costs of the old Trujillo sugar complex of twelve mills inherited by the government, and operating at an annual deficit of 10-20 million pesos.

Of all Trujillo's legacies none was as expensive, however, as the military establishment. Catering to this machine and keeping it from striking out on its own proved to be a major preoccupation of the post-Trujillo governments. The first, headed by Dr. Balaguer, pacified the armed forces by letting them have their own way. The seven-member Council of State followed, adhering to a similar policy. Reid manipulated the armed forces successfully for a time, but with the growth of corruption and graft in the services he was forced at length to punish the worst offenders. It was this policy that apparently set in motion the military revolt leading to his overthrow in April 1965.

The object of the groups forming the conspiracy to overthrow the Triumvirate, headed by Colonel Francisco Caamaño Deño, was to restore Juan Bosch to the presidency, and with him the Constitution of 1963. This group therefore called themselves the "Constitutional" forces. On April 25, pending Bosch's expected return from exile in Puerto Rico, they installed José Molino Ureña, president of the Chamber of Deputies in the Bosch period, as acting president. The attitude of the anti-Bosch military element, as expressed by a Dominican Colonel, was: "If Bosch ever comes back, he will throw me into jail so deep I will never find my way out."[63]

The rebellion grew into a civil war the next day when General Wessin, leader of the anti-Bosch military elements, ordered the Dominican Air Force to bomb and strafe the National Palace and other points occupied by the rebels. With the early success of the Caamaño forces, came the fear in the United States that the Communists were taking over the rebel movement. The State Department was also concerned for the safety of three thousand Americans in Santo Domingo as the fighting spread. The rebels, whose ranks included both

[63] Wells, "Dominican Republic," p. 19. For insights on the "Constitutionalist" forces see U.S. Congress, Committee on the Judiciary, U. S. Senate, *Communist Threat to the United States Through the Caribbean*, 89th Cong., 1st Sess., Oct. 18, 1965 (Washington, 1965). An astute analysis of the military's role in politics may be found in Martin Needler's "Political Development and Military Intervention in Latin America," *American Political Science Review*, LX (Sept., 1966), p. 623.

civilian and military leaders, distributed weapons obtained from military arsenals to civilians, and this determined the course of events over the next few days. Armed civilians, incited and directed by trained Communist leaders, committed acts of appalling brutality and destructiveness.

By the morning of April 29, on orders from President Lyndon B. Johnson, a force of 520 United States marines had landed in Santo Domingo. Within a short time the United States had 21,500 men on the island helping to hold down the rebels, and effecting a stalemate in the conflict. On May 4 the "Constitutional" forces appointed Colonel Caamaño president, to succeed Molina who had taken asylum in the Colombian embassy. On May 7, a three-man military junta led by General Wessin resigned, and a five-man regime headed by General Antonio Imbert Barreras, one of the two surviving assassins of Trujillo, and later an opponent of Dr. Bosch, succeeded it. This group was called the "Government of National Reconstruction."

Within a few days the cease fire broke down and heavy fighting resumed. On May 21 another cease fire was arranged through the offices of the United Nations, the Red Cross and the OAS. During the period of relative calm, the United States joined in negotiations with Latin American countries in an effort to set up an inter-American force that would allow some of the United States marines to leave. The peace was established on May 23 with a Brazilian commander and token contingents from Nicaragua, Honduras, El Salvador and Costa Rica, and a larger one from Brazil. However, it consisted mainly of American troops already on the scene.[64]

McGeorge Bundy, President Johnson's special assistant, was sent to negotiate an agreement with the rival forces on the problem of a provisional government. But with this mission having failed in its purpose, the task was brought to the OAS, which was convened in Washington on May 31 as the Tenth Meeting of Consultation of the Ministers of the American Republics. On June 2 the OAS appointed a three-man mediation committee made up of representatives from the United States (Ellsworth Bunker, ambassador to the OAS), Brazil, and El Salvador. After three months of tedious debate the committee negotiated with Caamaño and Imbert an Act of Dominican Reconciliation and an Institutional Act, on which both sides agreed. The signing of these two documents on August 31, 1965, brought the civil strife to a halt, officially at least, and provided for a temporary government which would rule the country until succeeded by a popularly elected executive and congress in the period March 30-May 1, 1966. Hector García Godoy, the choice of all parties concerned, took the oath of office on September 3, 1965.[65]

In carrying out armed intervention in the Dominican Republic the United States exposed itself to criticism for violating a fundamental principle of the inter-American system. Although it was a highly controversial action, it doubtless saved many lives, permitted the safe evacuation of American citizens and other foreign nationals, and shortened the conflict among the Dominicans.

Ambassador Bennett, fearing that American lives would be lost, had called

[64] Dept. of State *Bulletin*, LII (May 31, 1965), p. 868.
[65] Henry Wells, "The Dominican Search for Stability," *Current History*, vol. 51 (December, 1966), p. 329.

for Marines. Recalling the episode, President Johnson said: "... some 1,500 innocent people were murdered and shot, and their heads cut off, and six Latin American embassies were violated and fired upon for a period of four days before we went in ... as we talked to our ambassador to confirm the horror and tragedy and the unbelievable fact that they were firing on Americans and the American Embassy, he was talking to us from under a desk while bullets were going through his windows, and he had a thousand American men, women and children assembled in the hotel who were pleading with their president to help preserve their lives ... we went into the Dominican Republic to preserve the lives of American citizens and citizens of a good many nations–forty-six to be exact–while some of the nations were denouncing us for going in there, their people were begging us to protect them ..."[66]

Reflecting the other principal motive of the intervention, Adlai Stevenson, United States Ambassador to the United Nations, declared that his country had summoned the resources "of the entire hemisphere" to prevent Communism from gaining control of the Dominican Republic." Replying to Soviet charges that United States "armed interference" was a violation of the United Nations Charter, Mr. Stevenson said that Moscow was attempting to exploit the anarchy in the Dominican Republic for its own ends ... This deliberate effort of Havana and Moscow to promote subversion and overthrow governments is a flagrant violation of all norms of international conduct ... is responsible for much of the unrest of the Caribbean area ... " According to Secretary Rusk: "What began in the Dominican Republic as a democratic revolution was taken over by Communist conspirators ... Had they succeeded in establishing a government, the Communist seizure of power would in all likelihood been irreversible ... we acted to preserve the freedom of choice of the Dominican people ..." And responding to criticism that the United States had acted precipitately after miscalculating Communist strength in the rebel camp, Mr. Rusk said: "..... I am not impressed by the remark that there were several dozen known Communist leaders and that therefore this was not a very serious matter. There was a time when Hitler sat in a beer hall in Munich with seven people ... I don't believe that one underestimates what can be done in chaos ... by a few highly organized, highly trained people ..." The administration's policy was expressed concisely by the president, who declared: "The American nation cannot, must not, and will not permit the establishment of another Communist government in the Western Hemisphere."[67]

Severe criticism of the military intervention came from Senator J. W. Fulbright, chairman of the Senate Foreign Affairs Committee, who termed the intervention a "failure" and placed much responsibility for the failure on "faulty advice" given to the President by his advisers. The United States, he

[66] Dept. of State *Bulletin*, LIII (July 5, 1965), p. 20. Leonard C. Meeker, legal adviser to the Department of State, analyzing the intervention in the perspective of international law, declared that subsequent OAS action would have been impossible without the presence of U.S. forces, and that thousands would have died had the United States followed a "fundamentalist" interpretation of non-intervention. See *Ibid*., LIII (July 12, 1965), p. 60.

[67] Dept. of State, *The Dominican Crisis, Inter-American Series 92* (Oct., 1965), Section 3–Communist Subversion: *New York Times*, May 9, 1965. For Communist role in Dominican revolt see *Cong. Rec*., 89th Cong., 2nd Sess., vol. 112, pp. 6754-6760.

said, "intervened forcibly in the Dominican Republic . . . not to save American lives, as was then contended, but to prevent the victory of a revolutionary movement which was judged to be Communist dominated. The decision to land marines on April 28 was based primarily on the fear of "another Cuba' in Santo Domingo. This fear was based on fragmentary or inadequate evidence." It was later reported that State Department officials agreed with Senator Fulbright's contention that Ambassador Bennett's advice tended to distort the policy-making process by asserting more urgency than appeared to have been justified. It was also conceded that more intensive effort should have been made to get explicit endorsement by the OAS before full-scale military action was taken.[68]

The Latin American response to armed intervention in the Dominican Republic came quickly, and it was especially critical on the part of Uruguay, Chile, and Mexico. Their negative stance was based on the contention that the United States had violated Article 15 of the OAS Charter, and had neglected to obtain the approval of the OAS, both valid observations. Uruguay assailed the "unilateral" action by the United States, claiming it represented a "Johnson Doctrine," a new corollary of the Moroe Doctrine. It is revealing of general Latin American attitudes that the belated OAS intervention was brought about by a bare two-thirds majority vote of the twenty members at the intitiative and urging of the United States.[69]

The role of the American Peace Corps volunteers in the Dominican civil strife should not be overlooked, for as one journalist wrote: "This is a war in which the U.S. War Corps is at odds with the U.S. Peace Corps." Beginning in July 1962 a total of 193 Peace Corps volunteers had been sent to the Dominican Republic to work in educational fields, establish cooperatives, and aid in community development. When the uprising began 108 volunteers were there, thirty-four of them being in Santo Domingo. The volunteers soon became subjects of controversy when it was reported that they were impartially working in hospitals, driving ambulances and distributing food in the rebel zone as elsewhere. Before the outbreak of fighting they had been instructed to avoid political partisanship, and this was the rule they followed. According to one volunteer: "Most important was that the volunteers remained a neutral group in a fractionated country . . . On either side the words 'Cuerpo de Paz' were the safest conduct pass available. The economic, political, and military problems of the revolution are beyond the realm of the Peace Corps, but the human quotient is our province."[70]

Washington recognized the Dominican provisional government headed by temporary President Hector García Godoy on September 4, and promised $20 million in aid as a down payment on future support. This was followed in

[68] *Cong. Rec.*, 89th Cong. 1st Sess., vol. III, p. 24726. For opposing viewpoints in Congress on the question of intervention see *Ibid.*, pp. 23855-23865, 24971-24996.
[69] Larman C. Wilson, "The Monroe Doctrine, Cold War Anachronism: Cuba and the Dominican Republic," *Journal of Politics*, 28 (May, 1966), pp. 338-339. Articles 15, 16, and 17 of the OAS Charter provide for the absolute and unconditional inviolability of the territory of all countries. On the vote to establish an inter-American police force, as proposed by the United States, Venezuela abstained; opposed were Mexico, Ecuador, Peru, Chile, and Uruguay.
[70] Peace Corps, *Fourth Annual Peace Corps Report* (Washington, 1965), pp. 70-80.

December by a $50 million aid package, about half in grants and the rest in loans, to bolster the regime. The $50 million was in addition to $70 million that the United States had poured into the republic since the civil war started in April. About $50 million was given to the civilian-military junta that fought the rebels; this paid most of the salaries for sixty thousand government workers, soldiers and policemen, from May to the end of August. The rebel camp also received United States aid in the form of free food distributed in its sector in downtown Santo Domingo.

By the end of January 1966, United States economic aid had reached a total of $90 million granted since the preceding April. And another $25 million was authorized in March 1966, to spur the stagnating Dominican economy. Exports had dropped off a third in 1965, as compared to 1964, and the major export item, sugar, was down 42 percent. The per capita income fell to $190 in 1965, and 25 percent of the labor force was unemployed. In short, 75 percent of the population lived a wretched rural life, for the most part outside the money economy. American specialists, however, were optimistic about the country's future once stability was restored and the former Trujillo holdings, representing half the exploited wealth of the country, were effectively utilized.

Political stability and reconciliation were not to be achieved by the provisional government during its first six months of rule, for the bitter animosity in both camps persisted, especially among the extremists. Some elements had not accepted the OAS solution, the Communists, for example, denouncing both acts and urging the rebel forces to continue fighting. General Imbert and his "National Reconstruction Government" resigned rather than sign the reconciliation and institutional acts. The documents were ultimately accepted by the chiefs of the armed services as representatives of the anti-rebel factions. At the request of the provisional government, General Wessin, the right wing extremist leader of the military, was deported to the United States by the Inter-American Peace Force and given a diplomatic post.

Tensions were raised by the return from exile of the civilian political leaders, Joaquín Balaguer and Juan Bosch. On June 28, 1965, Balaguer arrived from three years of exile in the United States. He quickly announced his desire to run for president as the candidate of the Reformista party, on a platform supporting the mediation committee's peace formula. Balaguer subsequently adhered to the goals of the provisional government.

Juan Bosch returned to Santo Domingo on September 25, 1965, symbolically the second anniversary of his ouster as president. In contrast to Balaguer, Bosch's influence was particularly inflammatory, for he had been intimately involved in the revolt and civil war. The object of the revolt, as noted, had been to restore Bosch to the presidency, and he had supported the selection of Colonel Caamaño as head of the "Constitutionalist Government." He remained in Puerto Rico, however, during the conflict, keeping in contact with the rebel leaders by radio telephone. Within hours after his return, he addressed his rebel followers demanding the withdrawal of the Inter-American Peace Force, and urging the use of strikes, if necessary, to achieve it. Moreover, he said the United States should pay one billion dollars for invading the country, with lesser amounts to be paid by other participating nations.[71]

Occasional fighting continued after the OAS-sponsored provisional government was installed, the most savage being an attack in December by government troops on a gathering of rebel leaders at Santiago, in which twenty-eight lives were lost on both sides. After a special investigating commission failed to determine which side was responsible for starting the incident, the provisional government decided to assign high ranking officers from both the rebel and regular forces to diplomatic and other posts outside the country. The program was carried out, but not as quickly or equitably as some factions desired, and dangerous tensions set in. On February 10, amid fire fights and riots, pro-rebel labor unions and political parties, including Bosch and his P.R.D., called for a general strike to force the departure of government military chiefs. The strike lasted one week, causing a paralysis of the nation's economy. President García Godoy, acting decisively, announced his intention to remove the unwilling officers, and ordered all government employees to return to work at once or forfeit their positions. Violence and shooting continued intermittently until mid-March, at which point the turbulence moderated.

Under the terms of the Institutional Act general elections were to be held not sooner than six months and not later than nine months after its effective date, which fixed the period between March 1 and May 30, 1966. Rafael L. Bonnelly, another former president, announced his candidacy with the support of several small right wing parties, and Joaquín Balaguer, who started his campaign in March, was nominated by the right-of-center Reformistas; Juan Bosch, the moderate-leftist, was nominated by the P.R.D., the P.R.S.C., and the only Communist party allowed to register, the Fourteenth of June Movement. He refused the latter's support, however, contending that it would cost him one hundred thousand votes.

The election, held on June 1, was carried out in a peaceful atmosphere, and, according to observers, was free and honestly conducted. Dr. Balaguer received 769,000 votes (57.2 percent of the total), Bosch won 525,000 (39 percent), and Bonnelly about 40,000 (2.8 percent). Balaguer's victory was attributed to his vigorous campaigning and his appeal to the voters as the candidate most capable of achieving peace, order and reconciliation, as well as the fact that he was believed favored by Washington. Bosch, a reluctant candidate, and possibly afraid of assassination, remained closely guarded in his home, and restricted his campaigning to daily radio broadcasts "taped" in his quarters. His association with the civil war, the general strike and other disturbances also cost him popular support.

One of the first acts of President Balaguer, was to put an austerity program into effect. He cut government and military salaries in the upper echelons by 15-35 percent, saving $2.2 million a month, put credit and import

[71] Wells, "Dominican Search for Stability," pp. 330-331. "Balaguer was literally driven into exile in Mar., 1962, as an unreconstructed *trujillista*, and he was not able to return until July of this year. The man who could not live down his past in 1962, had become the United States man of the future in 1965." See Theodore Draper, "The Dominican Crisis: A Study in American Policy," *Commentary*, vol. 40 (Dec., 1965), p. 65.

restrictions into force, formulated a one-year wage freeze and anti-strike policy, and eliminated the traditional Christmas bonus for government workers. He also announced an ambitious development and emergency public works program, relying heavily on United States aid.

Faithful to his campaign pledge to restore a spirit of unity, he assembled a "national unity" cabinet, containing prominent members of several parties. Major General Enrique Pérez y Pérez, who had loyally served the provisional government, was named minister of the armed forces. His naming of women to all twenty-six provincial governorships minimized rivalry from that quarter, and steps were taken to disarm the civilian population by confiscating the government-issue weapons.

Washington appointed John H. Crimmins, a career service officer, to replace W. Tapley Bennett as ambassador, and the embassy in Santo Domingo was expanded significantly. A. I. D. personnel were increased from twenty-three to 131 and much emphasis was placed on two key departures from past policy: (1) a quick-impact agricultural program developed by experts from Texas A. and M. College, and (2) an overall development program. Sending advisers to work with the Dominican army, the United States stepped up its military assistance program in February, 1966. At the same time, however, it was recommended that the armed services, which consumed 30 percent of the budget, be reduced. Total United States economic aid to the republic in the period from April 1965 to June 1966 was $130 million. Of this figure $95 million had gone for budget support of the nearly bankrupt provisional government.[72]

Illuminating the problems of the Dominican Republic was the revelation that it was the only Latin American national whose export revenues declined in the first half of the 1960s; it dropped to 7.2 percent in 1960-65, compared with a rise of 8.6 percent for 1955-60. Widespread and deep poverty was evident, the kind in which eight children were crowded into one room of a rural *bohlo*, or thatched hut. With one of the highest birth rates in the world, 3.5 percent, the population was projected to reach seven million in two decades. A massive effort would be needed to forestall the increasing misery implied in the population projections.

By the end of the first year of the Balaguer government there was a free press, including two large well-edited newspapers of general circulation, and a vocal opposition press. Juan Bosch and the P.R.D., assuming the role of moderate opposition, supported the administration's development program, but remained critical of United States policies there. Observers, however, found no popular resentment against the United States, and heard little mention of the civil war. Two hundred young Dominicans aided by Peace Corps volunteers were working in Dominican villages using community labor, and materials to build roads, canals, schools and other public facilities. The economy was still shaky in the summer of 1967, with one-quarter to one-third of the work force idle or seriously underemployed.[73]

[72] *New York Times*, June 19, 1966. The withdrawal of the Inter-American Police Force was completed on Sept. 19, 1966.

[73] With the population in excess of four million, (mid-1968 est.), and a national territory of 18,817 square miles, overall densities of the Dominican Republic approximate 170 persons per square mile. This ratio is low compared to other West Indian islands, but high in comparison to most Latin American countries. Aid disbursed under the Alliance for Progress to the Dominican Republic in the period 1962-1966 totaled $202.1 million. See Simon G. Hanson, *Five Years of the Alliance for Progress, An Appraisal* (Washington, 1967), p. 4.

Distressing as conditions were, signs of improvement in the economic outlook could be seen as the 1960s drew to a close. The per capita growth rate reached 6.6 percent in 1967 which, taking into account the population growth rate of about 3.5 percent, was higher than the Latin American average. Exports continued to rise despite a prolonged drought, the worst of the century, which ended in the second half of 1968. Notable was the reorganization of the sugar industry, the island's chief earner of foreign exchange, which showed a profit in contrast to the deficits of the earlier years of post-Trujillo era. Sales of 618,000 tons to the United States in 1966, when neither the Philippines nor Puerto Rico could fill their quotas, earned $74 million for the Dominicans. Work began in 1969 on the largest mining investment in Dominican history, a ferro-nickel mine and processing plant; exports of beef to the United States, Puerto Rico and the Caribbean islands surged upward. A large sugar harvest, up 25 percent from 1968, brought exchange earnings of about $175 million, income made possible by the availability of the high-priced United States market. Meanwhile, the Balaguer government was seeking a larger permanent share of that market, beyond the republic's authorized quota.

This pointed up the widespread realization among Dominicans that their country's economic progress, if not survival, had become dependent upon continuing United States aid, and access to mainland and Puerto Rican markets. President Balaguer had made his position clear at the Punta del Este Conference of Hemisphere Presidents in April 1967, when he declared that he favored a "special relationship" that would permanently tie the Dominican economy to the United States. The republic's dependence on Washington was reflected in the number of United States officials who advised most of the government bureaus. In 1967 the U.S. embassy staff in Santo Domingo had grown to almost nine hundred (second largest in Latin America after Brazil) and the Peace Corps volunteers numbered 128.[74]

At the end of Dr. Balaguer's first term in office Washington had made available to the Dominicans $60 million in grant assistance and $50 million in development loans. The total U.S. grants and loans in the period 1965-1970 represented the highest per capita level of U.S. aid in Latin America. By 1970 there was evidence that the Nixon administration's policy of lowering the "silhouette" of the United States presence by cutting back military and civilian personnel was being implemented. Moreover, Washington's dissatisfaction with the failure of the Balaguer regime to institute more effective measures to modernize the agricultural structure, improve education and combat unemployment pointed to a possible reduction in U.S. aid funds.[75]

Although there was considerable political stability on the surface there was evidence of growing tension as the presidential election of May 16, 1970 approached. In rapid succession the ruling *Partido Reformista* nominated Joaquín Balaguer for a second four-year term, and the moderate leftist Partido Revolucionario Dominicana (P.R.D.) of Juan Bosch, refusing to name a candidate, called for "active abstention" by voters. President Balaguer won his

[74] *New York Times*, Jan. 26, 1970. See also Inter-American Development Bank, *Socio-Economic Progress in Latin America (Social Progress Trust Fund Eighth Annual Report 1968)* (Washington, 1969), pp. 138-151.
[75] *The Reporter*, Nov. 39, 1967, pp. 26-28: *Christian Science Monitor*, Mar. 13, 1968: *New York Times*, May 22, 1970.

second term in the election amid unexpected calm. However, evidence of a sharply rising terrorist campaign, responsible for forty-five political killings since Dr. Balaguer had announced his candidacy on March 25, led the OAS to change the site of its scheduled session from Santo Domingo to Washington.[76]

The wave of terrorism and insurgency that almost engulfed the country was believed to have had major support from Cuba, and possibly the People's Republic of China. Several groups, chief among them the Dominican Popular Movement, joined together to form the Anti-Reelection Command in 1970 to block the candidacy of Dr. Balaguer. In addition to other acts of urban terrorism, the Anti-Reelection Command was believed responsible for the kidnapping of a U.S. diplomat in early 1970 to secure the release of a leftist leader. A major loss to the left, however, occurred in November 1971, when Homero Hernández Vargas, a Communist leader, was killed by police. This left the Communist Party without leadership..[77]

Despite continuing pressures from the left, the Balaguer government was most seriously challenged in 1971 by a rightist plot on June 30. The apparent leader of the attempted coup, General Elías Wessin y Wessin, was arrested and sentenced to be deported to Spain. Seeking to broaden its base of popular support from the moderate left, and other supporters of former president Juan Bosch, the Balaguer regime, according to Wessin, was making dangerous concessions to the leftist elements. President Balaguer, commenting on the plot, said "conspiracy never stops here . . . "[78]

The Dominican Republic experienced relative calm in the closing months of 1971, and began 1972 with a three-year background of steady economic growth. A four-year national development plan, begun in mid-1971, projected a continuation of the 6.6 percent growth rate of the gross national product, this despite the lowering of U.S. aid to $20 million a year. In general, agricultural production was rising, industries growing, and foreign investment mounting. Agriculture continued to be the mainstay of the economy, but tourism was being pushed, and it was expected that nickel production would begin in 1972. In common with most Latin American countries, the addition of new markets was of paramount importance for the continued growth of the economy.[79]

Dr. Balaguer had completed his term of office and won reelection by skillfully walking a political tightrope at home, and retaining the moral and material support of the United States. But in a country that had four dictators rule for seventy-one out of the first 171 years, and used up seventy-two presidents during the remaining fifty years, his future was inevitably clouded.

The recent relations of the United States with Haiti have been very similar to our relations with the Dominican Republic. After a protracted series of revolutionary outbreaks, the financial situation became so bad in 1914 that Great Britain, tired of being put off, sent an ultimatum demanding payment of

[76] *New York Times*, May 25, 1970: *Christian Science Monitor*, June 13, 1970.
[77] *Christian Science Monitor*, May 5, 1971; *Times of the Americas*, November 6, 1971.
[78] *Christian Science Monitor*, July 6, 1971; *Times of the Americas*, July 28, 1971.
[79] *New York Times*, Jan. 28, 1972; *Christian Science Monitor*, Mar. 30, 1972.

an indemnity of $62,000; and shortly afterwards Germany and France demanded the control of the customs.[80] The outbreak of World War 1 averted European interference for the time being. But, with a view to preventing it permanently, Mr. Bailly-Blanchard, the American minister, presented to the Haitian government on December 10, 1914, a project for a convention similar to the agreement of 1907 with Santo Domingo, i.e., providing for a customs receiver who should put the finances in order.[81] Fearing popular disapproval, the Haitian government refused, and the United States did not insist. In May, 1915, Mr. Paul Fuller, Jr., was sent as a special agent to Haiti with a new proposal. The United States now asked to protect Haiti against foreign attack and to aid in suppressing insurrection within, provided that Haiti would covenant that no rights or privileges concerning the occupation of Môle Saint Nicholas would be granted to any foreign government, and would agree to settle by arbitration the claims of American citizens.[82] Haiti accepted this proposal as a basis of negotiations, and early in June made a counter-proposal.

Meanwhile the internal conditions became so serious that Rear-Admiral Caperton was ordered to proceed to Cape Haitien with his cruiser. President Theodore had been killed, and President Guillaume-Sam was maintaining himself with difficulty. To secure himself, the latter had arrested a large number of influential citizens and political opponents, and when, in spite of this, an attack was made on his palace, he had them put to death. This so enraged the populace that they broke into the French legation, whither he had fled for refuge, shot him, cut his body into pieces, and dragged it about the town.[83] On the same day (July 28, 1915) Admiral Caperton landed marines at Port au Prince to protect the legation, and later he took over the custom-houses and other public services. In order to secure the cooperation of the Haitians, a presidential election was held on August 12 under American protection, and M. Dartiguenave was elected. Conditions continuing turbulent, on September 3, Admiral Caperton proclaimed martial law; and on September 16 a treaty was signed whereby the United States established a virtual protectorate.[84] At first there was strong opposition on the part of the Haitian government to approving the treaty. But on November 11, 1915, the insular senate agreed to it. Considering the fact that Admiral Caperton, under Secretary Daniel's instructions, announced to the insular president and his cabinet that the United States would remain in control until the treaty was ratified, and that the funds collected at the customs would be

[80] For a detailed survey of the history of this period see Arthur C. Millspaugh, *Haiti Under American Control 1915-1930* (Boston, 1931); also R. l. Buell, "The American Occupation of Haiti," *Foreign Policy Association Information Service* 1929, vol. V, nos. 19-20; see also Mr. Lansing's letter to Secretary Hughes regarding Germany's intentions in the Caribbean, *Sen. Rep. No. 794*, 67th Cong., 2nd Sess., Appendix B. For a detailed account of the financial difficualties see Paul H. Douglas, "The American Occupation of Haiti," *Polit. Sci. Quar.*, vol. XLII (June, 1927), pp. 229 ff. For a critical report of the American occupation by a group of American lawyers see *Cong. Rec.*, vol. 62, pp. 8945 ff.

[81] *Hearings before a Select Committee on Haiti and Santo Domingo*, Part I, p. 6; text, p. 33. A similar proposal had been made on July 2, 1914, to the Zamor government, but no action was taken. *For. Rel. of the U. S.,* 1914, pp. 347-350.

[82] *Ibid.*, p. 7.

[83] *Report of the Secretary of the Navy, 1915*, pp. 15-17; or the more detailed report of Brig.-Gen. George Barnett in the *Report of the Secretary of the Navy, 1920*, pp. 245 ff.

[84] *Hearings*, pp. 336, 344, 348.

available for the payment of salaries only after ratification, it is not surprising that the Haitian senate accepted the convention.[85]

The principal provisions of the convention were as follows: (1) the establishment of a Haitian receivership of customs under American control; (2) the appointment of an American financial adviser to assist in the settlement of the foreign debt and in other financial and commercial matters; (3) the organization of a native Haitian constabulary under the command of American officers; (4) the disarming of all revolutionary forces; and (5) a guaranty on the part of Haiti to cede no territory to any nation but the United States. The convention was to remain in force ten years, and was to be continued for another ten years if its objects should not have been satisfactorily accomplished within the briefer period.[86] In fact, on March 28, 1917, an agreement was reached to extend the treaty for an additional ten years, or to 1936, although the legality of the extension has been questioned, since it was approved by neither the Haitian Congress nor the Senate of the United States.

The convention was proclaimed on May 3, 1916, and remained in force until the treaty of 1932 became effective. The results, however, on the whole were less satisfactory than in the Dominican Republic. The list of grievances of the Haitians was a long one.[87] The United States was accused of not carrying out the terms of the convention, and of deliberately usurping the powers of civil government which the convention guaranteed to the Haitians. It was claimed that no attempt had been made to give the financial aid promised by the United States, and that up to 1920 no attempt was made to pay interest on the foreign debt; there had also been but one payment on the internal debt. The Haitians asserted that the United States had backed the unconstitutional methods of Presidents Dartiguenave and Borno, had suppressed the Haitian legislature, and had prevented the drawing up of a more liberal constitution. Finally, they accused the marines of indiscriminate killing of Haitians and of employing cruel and inhuman treatment towards the natives upon numerous occasions, particularly in the establishment of the corvée or enforced labor on the roads.

Unquestionably, every one of these accusations was justified to a considerable extent; but, on the other hand, a careful investigation into the voluminous testimony shows that the American intervention has not deserved the bitter criticism that it has sometimes received. The military forces have not always carried out the provisions of the convention as promptly as they should—Admiral Knapp, in his report, concedes this—but only because of the desire to do the job quickly and well.[88] Some of the officers abused their power,[89] and there were also unlawful executions of prisoners,[90] but the guilty officials were immediately court-martialed and removed from the service. The corvée was used for a time as an emergency measure, but was subsequently discontinued.

[85] *Ibid.*, Part II, p. 394.
[86] *U. S. Stat. at Large*, vol. XXXIX, Part II, p. 1654.
[87] See the memoir of the Union Patriotique d'Haiti, *Hearings before a Select Committee on Haiti and Santo Domingo*, Part I, pp. 5-33.
[88] Report of Admiral Knapp in *Report of the Secretary of the Navy, 1920*, p. 232.
[89] *Ibid.*, p. 227.
[90] Report of General Barnett, *Ibid.*, p. 306.

When evidence of abuses was brought to the attention of the United States government, every effort was made to get at the truth of the situation. In the summer of 1920 Rear-Admiral Knapp was ordered by Secretary Daniels to make a special investigation. General George Barnett, who commanded the marines from June, 1915, to June 30, 1920, was asked to make a complete report, and this was followed by another by General Lejeune, who took over the command. In addition, Secretary Daniels appointed a court of inquiry headed by Rear-Admiral Mayo, and consisting of Rear-Admiral Oliver, Major-General Neville, and Major Dyer of the Marine Corps, to inquire into the alleged indiscriminate killings of Haitians and unjustifiable acts by members of the United States naval service. Finally, there was the Senate investigation under Senator McCormick.

All of these reports show that abuses were the exception rather than the rule; and they indicate that the results of the American occupation were not wholly bad. For instance, the customs collected for the fiscal year ending September 30, 1920, amounted to almost $6½ million, exceeding by more than $1 million the entire revenue collected in any of the five years before the intervention.[91] But the improvement in living conditions was perhaps the greatest achievement of the American intervention. The Rev. Charles Blaney Colmore, bishop of Puerto Rico and Haiti, writing in 1917, thus characterized it: "The marines have literally taught the Haitians how to live decently. Before their coming, sanitation, save in the crudest and most unsatisfactory forms, was unknown; fevers and epidemics were as plentiful as revolutions; a press gang was in vogue, and the country was the victim of continuous uprisings engineered by political scoundrels . . . . The entry of the United States marines ended this sorry story . . . . Sanitary systems had been installed, the towns had been cleaned up, former idlers and revolutionists were working happily for living wages, and a new spirit was animating the people . . . ."[92] Bishop James Craik Morris, writing in 1920, after visiting every important city on the island declared: "The only opposition to Americans in Haiti is political opposition. In the southern part of the island, particularly on the peninsula, the American occupation is regarded as the salvation of the people. I heard this opinion frequently. I had under me twelve native clergymen, and neither from them nor from the many other natives I talked to did I hear one word of condemnation of the acts of the American Marine Corps."[93]

Bishop Morris appears to have struck at the very heart of the difficulty—the opposition was chiefly political. But could the United States afford to disregard political opposition? It would have been impossible to do so and still adhere to the policy that had been so consistently maintained in our relations with weaker states. Nor did it seem feasible at the time for the United States to withdraw immediately and completely from the Republic. In the preliminary Senate report on Haiti, Senator McCormick declared that "the members of the committee are unanimous in the belief that the continual presence of the small

[91] Report of Admiral Knapp, *Ibid.*, p. 228.
[92] Cited in report of Gen. Barnett, *Ibid.*, p. 275.
[93] *Report of the Secretary of the Navy, 1920*, p. 319.

American force in Haiti is as necessary to the peace and development of the country as are the services to the Haitian government of the American officials appointed under the treaty of 1915. There can be no abrogation of the treaty, and at this time no diminution of the small force of marines."[94]

Although the report of the Senate Committee gave full credit to the American administration for its establishment of peace and order and for the excellent results obtained through the improvement of sanitary conditions and the construction of highways, nevertheless it criticized the United States for its failure to develop a definite, constructive policy and for its failure to select men for service in Haiti who were sympathetic to the Haitians and unable to maintain cordial relations with them. As a remedy for this situation the Committee recommended that a high commissioner be sent to Haiti to act as the American diplomatic representative and at the same time exercise a direct supervision over the treaty officials. In accordance with this suggestion Brigadier General John H. Russell was appointed American high commissioner in February 1922, and placed under the jurisdiction of the Department of State.

A few months later the Haitian Council of State elected as president of the Republic Mr. Louis Borno. who immediately signified his intention of cooperating wholeheartedly with the new high commissioner. As a result the financial situation improved, the American military courts ceased to function, the Haitian constabulary was efficiently organized, new highways were constructed, the Public Health Service built modern hospitals and held free clinics, and in 1924 a new technical service created a system of vocational and agricultural instruction.[95]

Nevertheless, the government of Haiti was under the complete domination of the American occupation; the American high commissioner virtually controlled all political matters, while the American financial adviser dictated financial policies. Although a new constitution had been given to the Haitians in 1917–Franklin D. Roosevelt wrote it and confessed that it is a good one[96]–no elections had been held under it, nor had there been any session of the Haitian Congress since the American occupation had begun. The former ruling classes resented keenly the control of the public administration by representatives of a foreign power. President Borno, who had been reelected by the Council of State in 1926, had become very unpopular among the politicians because of his cordial cooperation with the United States. Matters reached a climax towards the end of 1929 when the students at Port au Prince inaugurated a strike because the government had reduced the annual scholarship allotment.[97] The political parties took up the question and at Aux Cayes, a small seaport about a hundred miles from Port au Prince, on December 7, a mob of Haitians attacked the American marines, who repulsed them, killing six and wounding some twenty-eight more.[98]

[94] *New York Times*, Dec. 25, 1921. For Haitian opinion on the investigation see the address of Prof. Pierre Hudicourt in Washington, Feb. 2, 1922, *Cong. Rec.*, Vol. 42 (May 19, 1922), p. 7857. The detailed findings and recommendations are found in *Sen. Rep. No. 794*, 67th Cong., 2nd Sess.

[95] For an account of this period see A. C. Millspaugh, *Haiti Under American Control*, chap. IV.

[96] *New York Times*, Aug. 19, 1920.

[97] *Report of American High Commissioner, 1929*, pp. 7-9.

[98] *New York Times*, Dec. 8, 1929.

In his annual message to Congress sent on December 3, 1929, President Hoover had already indicated his desire of sending a commission to Haiti to obtain the necessary information to formulate a more definite policy. The outbreak on December 7 caused him to send a special message to Congress the same day requesting the authority to despatch a commission immediately. Congress approved, and in February, 1930, a commission headed by W. Cameron Forbes was instructed to investigate and report when and how we could withdraw from Haiti after discharging our obligations.[99]

The Commission found a serious state of unrest existing, due partly to the economic condition caused by the falling price of coffee, but even more to the hostility towards President Borno who, according to public opinion had permitted himself as well as the Council of State to be completely subservient to High Commissioner Russell. As an emergency measure, the Committee suggested the immediate election of Eugene Roy, a business man, as temporary president. He could thereupon call for the election of the legislature which might elect a president in accordance with the provisions of the constitution. The suggestion was adopted, M. Roy was elected president by the Council of State on April 21, 1930, and inaugurated on May 15. A new Congress was elected in October which forthwith chose as president M. Stenio Vincent, a strong opponent of the American occupation.

In the meantime the Forbes Commission had returned to Washington and had made its report. Although it gave the American administration credit for the remarkable progress made in health, sanitation, bridge and road building, it found little effort had been made to prepare the Haitians for self-government. In fact, the report stated that "the acts and attitude of the treaty officials gave your commission the impression that they had been based upon the assumption that the occupation would continue indefinitely."[100] In its recommedations, which were approved by President Hoover, the Commission advocated: (1) the rapid Haitianization of the administration services with the definite aim of having Haitians trained to take over the work in 1936; (2) the abolition of the office of High Commissioner and the appointment of a non-military minister to take over his duties as well as those of diplomatic representative; (3) the gradual withdrawal of the marines in accordance with arrangements to be made by the two governments; (4) a modification of the present treaty providing for less intervention in Haitian domestic affairs.[101]

In accordance with these suggestions General Russell submitted his resignation as American high commissioner and Dr. Dana G. Munro was appointed as minister to Haiti. Although the United States was eager to terminate its responsibilties in Haiti at the earliest possible date, it was not able to accept the Haitian government's proposal for the speedy termination of financial control. In other matters a solution was finally aggreed upon and incorporated in the Haitianization Agreement of August 5, 1931.[102] According to its terms all Americans connected with the Public Works Service, the Public Health Service, and the *Service Technique* were to be withdrawn by October 1 of

[99] *Report of the President's Commission for the Study and Review of Conditions in the Republic of Haiti* (Washington, D. C., 1930), p. 1.
[100] *Ibid.*, p. 8.
[101] *Ibid.*, pp. 20-21.
[102] For text see U. S. Dept. of State, *Executive Agreement Series*, No. 22.

the same year, while the Service of Payments and the auditing of expenditures were to remain under the control of the Financial Adviser.

The Haitianization of the Garde and the withdrawal of the marines required longer consideration, but agreement was finally reached on these subjects, and by the Treaty of September 3, 1932, it was provided that all American officers should be withdrawn from the Garde by December 31, 1934, and the withdrawal of the Marine Brigade was not to begin later than that date. It was also agreed that a Fiscal Representative appointed by the president of Haiti upon the nomination of the president of the United States should replace the financial adviser-general receiver on December 31, 1934, and he should collect all customs duties with the assistance of an Haitian personnel.[103]

It soon became evident that the Haitian legislature was not satisfied with these terms and the ratification of the agreement was unanimously rejected. No further action was taken until the Franklin D. Roosevelt administration came into office. Norman Armour, a career diplomat, now replaced Dr. Munro, and he proceeded to negotiate a new settlement largely along the lines of the previous one, but in the form of an executive agreement, instead of a treaty. The Agreement of August 7, 1933, advanced the date of the Haitianization of the Garde and the withdrawal of the marines to October 1, 1934, and the date of the appointment of the Fiscal Representative to January 1, 1934, but the rest of the agreement was very similar to the 1932 Treaty.[104]

President Roosevelt's desire to make effective his "Good Neighbor Policy," and Secretary Hull's cordial attitude at the Montevideo Conference encouraged President Vincent to request a speedy renunciation of American financial control in Haiti; he was invited to come to Washington to confer on the situation. On April 17, 1934, a joint statement was isssued by the two presidents stating that the commitments of the August 7 agreement would be carried out and in addition new and satifactory commercial and financial agreements would be negotiated.[105] As a gesture of friendship, President Roosevelt asked and obtained from the Congress authorization to turn over to Haiti buildings and equipment used by the marines during their occupation.

In July President Roosevelt returned the visit of President Vincent and ordered the time for the withdrawal of the marines to be advanced to August 15. In fact, the 850 marines who still remained began to withdraw at the end of July and by the 15th of August the last marine had left and the Haitian flag flew over the former American barracks. To put an end to financial control it was agreed that the National City Bank of New York should sell back the National Bank of Haiti to the government of Haiti and all official American control be abolished. The sale was consummated July 9, 1935, subject to approval by the Haitian Senate. When a small group of eleven Senators opposed, President Vincent put the matter up to a national referendum which overwhelmingly approved the purchase of the bank. A reciprocity tariff agreement was also signed with Haiti the same year (March 28, 1935) whereby the United States lowered the tariff on

[103]For text see U. S. Dept. of State, *Press Releases* (Sept. 10, 1932), p. 150; also Dana G. Munro, *The United States and the Caribbean Area* (Boston, 1934), pp. 190-193.
[104]For text see U. S. Dept. of State, *Executive Agreement Series*, No. 46.
[105]*New York Times*, Apr. 18, 1934.

rum, fruit, and cocoa in return for a freer admittance for our machinery, radios, and automobiles.

Contrary to the situation in the Dominican Republic, the European war had a serious effect upon the economic situation of Haiti since it closed her principal markets for coffee and cotton. The United States recognized this situation and by an executive agreement dated February 13, 1941, Haiti was permitted to postpone the payment of one-third of the interest due upon the bonds of 1922 and 1923.[106] On September 13, 1941, after numerous conferences, the two governments signed a new agreement which replaced the one negotiated in 1933. The financial agreement of 1941, like the one concluded with the Dominican Republic, abolished the office of Fiscal Representative, transferring all funds to the National Bank of Haiti as sole depository for the Haitian government. The Bank's board of directors was reorganized making three of the six voting members United States citizens, one of whom should serve as co-president and represent the holders of the 1922 and 1923 bonds. The new agreement became effective October 1, 1941, and would continue in force until the bonds were redeemed.[107]

The war shifted the Haitian market towards the United States even more than formerly and during the fiscal year 1940-1941, we took 87.92 percent of her exports and provided 83.58 percent of her imports.[108] Haiti declared war upon Japan the day after Pearl Harbor and on December 12 extended her declaration to include Germany and Italy.

Vincent, having himself re-elected in 1935 for a term he extended to six years, continued as President through the 1930s. Haiti remained peaceful, and public institutions functioned. In 1941 Élie Lescot, another mulatto, became president. He visited Washington in April 1942, to coordinate economic and defense arrangements, and following a series of meetings an omnibus memorandum was issued covering various agreements that had been reached. These arrangements provided that the Commodity Credit Corporation of the United States purchase all surplus cotton produced in Haiti during the war period; that the Export-Import Bank extend such credits to the National Bank of Haiti as were necessary to stabilize exchange; to aid in the common war effort that 24,000 additional acres of sisal be planted in Haiti; and that the United States should furnish equipment and instruction for the Haitian coastal patrol and coast artillery. The two governments also agreed to carry out a number of health and sanitation projects in Hiati.[109] At the end of the war Haiti became a Charter member of the United Nations.

The war, by forcing a shortage of commodities and a rise in prices, brought about a series of political disturbances in Haiti. Racism, as in the past, was a factor. The small Haitian army, the *Garde d'Haiti*, based on the *Gendarmerie* of the occupation, contained ambitious Negro leaders who had

[106] U. S. Dept. of State, *Executive Agreement Series*, No. 201. Due to an improvement in trade conditions during the last quarter of the year the postponed interest was repaid and the Oct., 1941, payment met in full.

[107] *Ibid.*, No. 220.

[108] Banque Nationale de la République d'Haiti, *Annual Report of the Fiscal Department: Oct. 1940-Sept. 1941.*

[109] Dept. of State *Bulletin*, VI (Apr. 18, 1942), p. 353.

gained the upper hand in the *Garde* and were bitterly resentful of the superior attitude of the elite and their French pretensions. In January 1946, Lescot was forced out and the *Garde* took over until a new congress chose Dumarsais Estimé, a Negro, as president. An idealist and a liberal, who had long encouraged his race to assert itself, he ousted the elite from influential positions, and helped to finance a public works program by making them pay an income tax. Estimé also had the satisfaction of ending foreign financial controls, for Haiti's external debts were now all paid. He received extensive assistance from American public and priavte agencies, and from the United Nations, which supported a health improvement program. Haiti was actually ruled by the *Garde* during his administration, but it enjoyed a degree of freedom that was high by Caribbean standards.[110]

When in May 1950, President Estimé proposed a change in the constitution to permit his re-election, he brought about the opposition of Colonel Paul Magloire, the commander of the *Garde*. As a candidate for the presidency under an election procedure whereby the people, for the first time, voted directly for the president, Magloire won 151,000 votes to 2,000 for his opponent. A personable, competent individual, he did not inspire fear among the mulattoes, and was beloved by the Negroes. He was much admired abroad, and even the élite supported his policies. The country enjoyed freedom from political repression and made some economic gains under his administration.

Recognizing his country's needs, chiefly improved health and education, diversification of the economy, modernization and increased production, President Magloire encouraged and received foreign aid. Washington's support was obtained for the damming of the Artibonite River to provide irrigation and electrical power. Private investment and tourism, as well as grants and loans, stimulated the economy. Haiti was the first Latin American republic to sign an investment guaranty agreement with the Mutual Security Agency under which the United States affords insurance protection to new investments abroad against losses through confiscation or expropriation.[111]

President Magloire, accompanied by his wife and cabinet officials, came to the United States on a three-day state visit in January 1955, flying from Miami to Washington on President Eisenhower's private plane, the *Columbine*. The Haitian chief executive, a staunch ally of the United States in the cold war, addressed a joint meeting of the Senate and House of Representatives, declaring he sought to make his people immune to communism by raising their standard of living. He added ". . .,to be sure, we have taken legal measures to prevent not merely the extension, but the very manifestation among us of this pernicious doctrine."[112]

Vice President Richard Nixon returned the visit in March, and together with Ambassador Roy T. Davis, signed the United State-Haitian Treaty of Friendship, Commerce and Navigation. The treaty was designed to provide a comprehensive integrated, legal framework within which economic relationships between the two countries might be developed. Clarifying the Eisenhower

[110] John E. Fagg, *Cuba, Haiti and the Dominican Republic* (Englewood Cliffs, 1965), p. 134. See also United Nations, *Mission to Haiti* (New York, 1949).
[111] Dept. of State *Bulletin*, XXVIII (May 11, 1953), p. 682.
[112] *Ibid.*, XXXII (February 4, 1955), p. 275.

administration's financial policy toward Haiti, and Latin America in general, Nixon suggested expanded nation-to-nation and nation-to-private investor loans, but emphasized private local and foreign capital investments.[113]

Haiti had been commonly regarded as the poorest country in Latin America, and the fiscal year 1954-1955 was one of the worst in the nation's recent history. The decline was caused largely by the disastrous hurricane Hazel which destroyed 40 percent of the coffee crop, Haiti's main earner of foreign exchange. The United States provided $2,740,000 in emergency relief, which was in addition to $3,581,000 given for agricultural and educational development since 1950. Tourism, the number two dollar earner helped to compensate the loss, for 55,000 tourists visiting Haiti in 1954-1955 spent $5,500,000. However, Haiti's debt position increased over the preceding year from $22,400,000 to $48,200,000.[114]

A deepening economic crisis in 1956 was accompanied by terrorism and political instability. Confidence in the administration declined, and Magloire was removed from office by action of the *Garde* and pressure resulting from a general strike. It was evident that he had sought to extend his six-year term and establish a military dictatorship. Magloire was followed by Joseph Nemours Pierre-Louis, chief justice of the Supreme Court, who served as provisional president from 12 December 1956 to 4 February, 1957. Franck Sylvain, an authority on constitutional law, elected to the provisional post by the legislature, held office from 7 February until 1 April. A provisional government ruled from 1 April until 25 May, to be followed by the provisional presidency of Daniel Fignole in the period 25 May to 14 June. A military junta, calling itself the Executive Council, governed until Dr. Francois Duvalier was inaugurated president in October, 1957.

Washington had extended diplomatic recognition to the administration of Pierre-Louis, Sylvain and the Executive Council, but not without incident. An article in *Le Matin* declared that Ambassador Davis had implied the United States would hold up recognition indefinitely of any government not established by constitutional processes. A statement from the embassy indicated that it favored adherence to constitutional processes in presidential succession, but that it viewed all candidates impartially.[115]

In October 1957, the State Department accused the Haitian police of murdering Shibley Talamas, an American citizen, and sent a harshly-worded note to the government through Ambassador Gerald A. Drew. Port au Prince quickly hired three prominent United States Republicans as public relations men. They were John Roosevelt, son of the late president; Charles Willis, a fomer aide of President Eisenhower; and Douglas Whitlock, previously a staff member of the Republican National Committee. Their main function, it appeared, was to persuade Secretary of State John Foster Dulles to soften his demands of Haiti concerning the murder of Talamas. Although the Haitian government claimed the victim had died of a heart attack while resisting arrest, the State Department had

[113] *Ibid.*, XXXII (March 14, 1955), pp. 452-453. United States influence in Haiti remained high: "It is automatically assumed that the American Embassy is a control, for bad or for good, on Haitian affairs." See Herbert Gold, "Americans in the Port of Princes," *Yale Review*, XLIV (Autumn, 1954), p. 92.
[114] *New York Times*, Jan. 5, 1956.
[115] *Ibid.*, Feb. 8, 1957.

demanded an indemnity for the murdered man's widow, an apology from the government, and punishment of the guilty police. Washington had meanwhile reduced by two-thirds its economic aid to the *junta* on grounds that it had failed to provide the required counterpart funds; Haitians regarded the move as a reprisal for the murder of Talamas. The three Republicans suggested to Secretary Dulles that Haiti merely apologize and let the matter rest there; however, the secretary stood by the original demands.[116]

The election of September 11, 1957, brought Dr. François Duvalier, a Negro country physician, called "Papa Doc," into the presidency. His opponent, Louis Dejoie, protested the election was rigged and went into hiding. In his first address to the nation the new President said: " . . .For a long time now misery has been identified with the Haitian people; this is an image that must be destroyed. . . .My Government of National Unity will be a government. . . .which will reconcile the nation with itself." Duvalier's ambassador to the United States, Ernest Bonhomme, presented his credentials to President Eisenhower, on December 8, 1958.[117]

Born in Port au Prince in 1907, Duvalier received his education in Haiti, practiced medicine, and served in government posts. He served as director general of the National Public Health Service in 1946, and he was undersecretary of labor and secretary of labor under President Estime. When sworn into office in October 1957, he decreed a general amnesty for political prisoners, and took measures to restore a democratic regime. But in April 1961, he dissolved the bicameral legislature, not then in session, and replaced it with a fifty-eight member single chamber. It was soon evident that Duvalier was a fanatic, bent on destroying all opposition, particularly among the mulattoes. Displaying unsuspected ability in creating a personal political machine, he purged the *Garde,* and with the aid of a formidable secret police organization, a militia and a band of ruffians called the *Tonton Macoute* (bogey-men), he resorted to terrorism such that the country had seldom experienced. Ultimately dissenting clergy were deported, the university and most schools closed, labor unions suppressed, and he had himself elected president for life. The worst enemies of the regime were butchered and their bodies left on public display as a deterrent to others.[118]

When a heart attack nearly killed him in 1959, his secretary, Clement Bardot, head of the terrorist *Tonton Macoute* loyally ran the country for him, but Bardot was imprisoned a year later. Duvalier regarded himself as the spiritual descendant of Jean Jacques Dessalines, the Haitian rebel who defeated the forces of Napoleon in 1803, and then proclaimed himself Haiti's first Emperor. He also tried to convince his superstitious followers that he resembled Baron Samedi, the voodoo spirit of death. To symbolize this role he dressed in black.[119]

As the nature of the Duvalier regime was exposed Haitian-United States relations began a downward spiral but avoided rupture largely because of the Cuban Communist threat to the weak and vulnerable island republic. Not infrequently United States officials were charged with "interference in internal

[116] *Hisp. Amer. Rep.*, XI (Jan., 1958), p. 25.

[117] Gerard R. Latortue, "Tyranny in Haiti," *Current History*, vol. 51 (Dec. 1966), p. 349; Dept. of State *Bulletin*, XXXIX (Dec. 29, 1958), p. 1042.

[118] Rayford W. Logan, *Haiti and the Dominican Republic* (New York, 1968), pp. 154-155.

[119] John Roucek, "Haiti in Geopolitics," *Contemporary Review*, (July, 1963), p. 228.

affairs," and their removal was requested. J. P. Barringer, the embassy counselor, was shifted to a Libyan post after a Haitian complaint that he favored Louis Dejoie for president, and Ambassador Drew was charged by a Haitian official with "going too far in our internal affairs." The ambassador was supported by the State Department in his denial of the allegations.[120]

Other incidents and circumstances contributed to the rift. An invasion effort, involving five American citizens, was crushed in August 1958. Then the admission into the United States of Dejoie and Magloire was claimed by Duvalier to be an "unfriendly act." However, Haitian-United States friendship was reaffirmed when the aircraft carrier *Franklin D. Roosevelt* visited Port au Prince in October, and 3,500 officers and men of the vessel were given an official reception in the National Palace. The American presence reportedly lessened tensions, for Duvalier had feared that another invasion attempt was imminent. The United States provided $3.5 million in technical assistance and economic aid to Haiti in 1958, but most of the development programs remained suspended since the Haitian government's contributions were still overdue.[121]

Early in 1959 the State Department and the Economic Cooperation Administration agreed to give Haiti $6 million to help meet budgetary and foreign exchange problems, the result of a poor coffee crop and lower world market prices. By the end of the year, the United States Point IV mission had sixty-seven Americans and three thousand Haitians working in the country on various projects, and $1 million in surplus food was furnished to help prevent starvation in the drought-stricken north-west. This was in addition to the $6 million grant that alone enabled Haiti to meet its $33,600,000 national budget. And additional funds from the Export-Import Bank brought to a total of $25,500,000 the loans granted for the yet unfinished Artibonite Valley irrigation and power project.

With the triumph of Fidel Castro in Cuba the Duvalier regime, seeing itself further imperiled from that direction, asked the United States to reequip its entire army of 4,300. A United States Marine Corps mission, commanded by Colonel Robert D. Heinl, Jr., was assigned to Haiti in January. Underscoring the external threat, an invasion, launched from Cuba, was attempted on August 13, 1959. The thirty-one man invasion force was destroyed by September 4, with twenty-six killed and five captured. Owing to these circumstances, the United States took steps to modernize the Haitian armed forces, and by September 1960, the sixty-two man navy mission was the largest of its kind in the Western Hemisphere.[122]

In June 1960, President Duvalier announced that Haiti was "exhausted" and "at the limit of her sacrifices." Publicly declaring that American aid was lean and insufficient, he implied that Haiti had to choose between "two great poles of attraction in the world today to concretize her needs." The "poles" were interpreted to mean the United States and the Communist bloc. But then he added, ". . . We need a massive injection of money to revive Haiti, and this injection can come only from our great and capable neighbor and friend, the

[120] *Hisp. Amer. Rep.*, XI (July, 1958), p. 379.
[121] *New York Times*, Aug. 10, 1958.
[122] *Cong. Rec.*, 86th Cong., 1st Sess., vol. 105, p. 3428.

United States." The American embassy refuted the allegation by pointing out that $40,569,000 had been granted to Haiti in the eleven-year period ending July 1, 1960. Seeking to stimulate the slackening tourist trade, Duvalier requested a $10 million loan from Washington to construct a jet airport. He received no commitment.[123]

The advent of the Kennedy administration in 1961 brought into action a policy of punishing with diplomatic and economic sanctions civil and military regimes that overthrow democratic governments. Although conditions in Haiti became progressively repugnant to Washington, drastic measures were avoided, and nothing was done that might have been construed as intervention.

In May 1961, Duvalier had himself reelected for another six-year term while his first term had more than two years to run. In order to show United States disapproval, the State Department recalled Ambassador Robert N. Newbegin from his post temporarily so as to avoid his attending the inauguration of the dictator. Still exercising restraint, Washington continued its economic aid, totaling $13,500,000 in 1961, as well as the military assistance program. Duvalier had meanwhile made his position very clear: "I am the personification of the Haitian people. I don't have to take orders from anyone, no matter where he comes from. Twice I have been given power and I will keep it. God is the only one who can take it from me. No foreigner will remove me against the will of the people."[124]

A year later Ambassador Raymond L. Thurston, who had succeeded Mr. Newbegin, was called to Washington for consultation at a time to coincide with President Duvalier's "Day of National Sovereignty." This was another move to show Washington's displeasure with the regime. Nevertheless, the basic economic aid of $7,250,000 was resumed in April in fulfilment of a pledge made at the Conference of American States, held in Punta del Este, Uruguay, in January 1962, when Haiti agreed to drop her opposition to a plan to exclude Cuba from the affairs of the inter-American system. Haiti's vote was necessary in order that the required two-thirds majority could be secured at the conference.[125]

A reassessment of the Haitian situation, and the conclusion that the dictatorship was grossly inconsistent with the principles and goals of the Alliance for Progress, led to Washington's decision to suspend most aid to Haiti, beginning in August 1962. For humanitarian reasons, the United States continued to finance maleria control and to provide Food for Peace shipments distributed by private agencies. Closely following the adoption of this hard-line policy, which it was believed would unseat the dictator, came the Cuban missiles crisis of October, 1962. Duvalier responded to the threat by placing Haiti's harbors and airfields at the disposal of the United States naval and air units carrying out the quarantine on shipments of offensive arms to Cuba. Two weeks earlier the United States had relented, and offered to lend Haiti $2,800,000 to help finance construction of the jet airport near Port au Prince. Therefore,

[123] *Cong. Rec.*, 86th Cong., 1st Sess., vol. 105, p. 17927; O. Ernest Moore, "Is Haiti Next," *Yale Review*, LI (Dec. 1961). pp. 256-257.
[124] Hubert Herring, "Dictatorship in Haiti," *Current History*, vol. 46 (Jan. 1964), p. 52.
[125] Logan, *Haiti and the Dominican Republic*, p. 155.

Duvalier's support at this critical time was seen as an improvement in Washington's relations with his regime.[126]

The improvement proved to be more apparent than real. With the Cuban crisis past there was a progressive deterioration of Haitian relations with both the United States and the Dominican Republic. In February 1963, Marine Colonel Robert Heinl, who had headed the United States Navy mission since 1959, was ordered to leave by President Duvalier, after being charged with making critical comments about the effects of the civilian militia on the Haitian army.[127] The main crisis came as a result of "Papa Doc's" refusal to give up his office. The date had been set for May 15, but prior to that time political unrest was growing and the dictator launched a new reign of terror to silence his opponents, many of whom took refuge in foreign embassies in the capital, particularly the embassy of the Dominican Republic.

Coincidentally, there were persistent rumors that members of the Trujillo clan were plotting with Duvalier to kill the newly-elected President Juan Bosch. At the same time, President Bosch, who was trying to carry out democratc reforms in his country, made it clear that the Duvalier dictatorship could not be tolerated on the same island. Thereupon Dominican nationals in Haiti were jailed and brutally beaten and others disappeared. The hostility between French-speaking Haiti and the Spanish-speaking Dominican Republic, early in May, erupted in an outburst of violence which saw three guards of Duvalier's children shot to death by terrorist gunmen. The Dominican embassy in Port au Prince was occupied in order to seize Haitian rebels who sought asylum there.

On May 5 the Bosch government, giving twenty-four hours to comply, served a virtual ultimatum on Haiti to restore diplomatic guarantees at the embassy. The OAS acted quickly, persuading the Dominican government to extend its ultimatum by twelve hours, and calling the Council into emergency session. On the following day Haiti agreed to live up its diplomatic tradition, thereby averting war on Hispaniola. However, the Dominican Republic broke off diplomatic relations and urged the same action by other OAS members.[128]

Because of Duvalier's illegal usurpation of power Washington suspended relations with Port au Prince, but maintained diplomatic recognition. At the same time, in what was interpreted as a deliberate act against the United States, President Duvalier demanded that Ambassador Thurston be recalled. On June 3 Washington resumed full relations with a chargé d'affaires to head the diplomatic mission. With the outbreak of violence in April the United States had stationed an aircraft carrier task force off shore which kept in close contact with the embassy, for much concern was felt for the safety of the 1,300 Americans on the island. As the telephone system of Haiti had broken down several years

[126] *Cong. Rec.*, 87th Cong., 2nd Sess., vol. 108, pp. 5048-5050; *New York Times*, Oct. 27, 1962.
[127] Fearing the pro-American orientation of the *Garde* and its political potential, Duvalier destroyed it. Almost every American-trained man in the *Garde* was exiled or murdered. Without the *Garde* Haiti lacks an effective force for maintaining public order. See Robert D. Heinl, Jr., "Are We Ready to Intervene in Haiti," *Reporter*, June 2, 1966, pp. 27-28.
[128] Dept. of State *Bulletin*, XLVIII (June 17, 1963), p. 958.

previously, the embassy was prepared to send couriers to American nationals in the interior to alert them to the possible danger.

Indignation with Haiti's high-handed policies was heard in the United States Congress, where it was urged that diplomatic relations be severed and all aid suspended. It was pointed out that Haiti continued to receive annually $1.3 million in surplus food and $1.3 million for the maleria control project, and was free to market her quota of sugar in the United States. At the prevailing price of $152 per ton Haiti would earn $6,166,488 in 1963. Meanwhile, Haiti was hit by Hurricane Flora in October 1963, which took five thousand lives, left thousands homeless and starving, and destroyed $2,250,000 of the coffee crop.[129]

The Kennedy administration's policies toward Haiti were moderated by President Johnson to the extent of maintaining "normal relations." Although the general aid program was not renewed, a concession was made by the Inter-American Development Bank in approving a $2,360,000 loan for a potable water project in Port au Prince and Petionville; this had been vetoed by Washington in January 1963, when the policy was aimed at bringing down the Duvalier regime. Subsequently, the Agency for International Development extended a $4 million investment guarantee for the construction of an oil refinery. In normalizing relations, Washington drew sharp criticism from the diplomatic community and foreign businessmen in Haiti, particularly after Duvalier had himself elected president for life. It was contended that in moving away from a policy that managed to convey disapproval of the totalitarian methods of the Haitian government without succeeding in changing them, the United States overbalanced itself in the other direction. In so doing, it was believed, Washington was creating a dangerous enmity for the day when Duvalier falls from power.[130]

Under the constant threat of invasion, and recurring border problems with the Dominican Republic, Duvalier tightened security measures even further. An observer reported that when the president was at the airport a machine gun, mounted on a tripod in the waiting room, was kept pointed directly at all passengers and visitors who entered. The government palace in which Duvalier lived was constantly guarded by four tanks and two armored cars, and the strongest contingent of the Haitian army was quartered in barracks just to the rear. In 1964 there were small-scale invasions by exiles in two places, and guerrilla fighting continued for several weeks before the invaders were wiped out.

The main Haitian exile organizations were formed in Cuba and the United States. The *Front Démocratique Unifié de Libération Nationale* (F.D.U.L.N.), resulting from the merger of two Communist groups, broadcast daily from Havana. Contending that Duvalier was a servant of American imperialism, it advocated a sweeping revolution, presumably Cuba-style. The United States-based *Coalition des Forces Démocratiques Révolutionaries Haitiennes* (the Coalition), comprised political elements ranging from moderate-left to center-right. Its program called for popular democratic government, modernization and

[129] *Cong. Rec.*, 88th Cong. 1st Sess., vol. 109, p. 10908.
[130] *New York Times*, Dec. 6, 1964.

social reforms.[131] The Coalition's message to Haiti was carried by a short-wave radio station in New York. The Haitian colony in New York City, numbering about 45,000 in 1967, was considered the largest exile group outside Haiti.

After a long period of outspoken anti-Americanism President Duvalier said in 1966 that Haiti had no problems with Washington, which he was once again hailing as "the leader of the free world." As part of his image-building effort to regain American aid, the regime ceased displaying the cadavers of its enemies in the streets. His efforts proved unavailing, however, and the nature of the regime was further exposed by the findings of the International Commission of Jurists, a private organization recognized by the United Nations. The Commission accused Duvalier of exploiting for his own ends "the very real social tensions between the Negro majority and the mulatto elite minority in his country," and denounced as an "electoral farce" the 1964 referendum which gave Duvalier the presidency for life.[132]

An aura of desperation seemed to surround Duvalier in the summer of 1966 when he began a purge of the government and the army. After two months 107 persons, mostly army officers, their wives and children, had taken refuge in Latin American embassies in the capital. Safe conduct passes were accorded only to the wives and children of the group. Later a crisis developed within the Duvalier family when the dictator charged his son-in-law with being a conspirator and denounced his own wife publicly for not having helped him as Eva Perón helped her husband in Argentina in the late 1940s.

In 1966 Haitian relations with the Dominican Republic suffered from recurring border problems. Hundred of peasants had fled into Dominican territory in 1963 when Duvalier created a two-mile deep no-man's-land by burning peasant homes and farms in a move to halt the flight of his enemies. Watchtowers and barbed-wire barricades were later installed to guard this frontier. Many of the peasants had been persecuted for aiding fleeing enemies of the regime and exiles staging incursions from the Dominican Republic. The Dominican government took steps to deport two hundred of the uninvited Haitians who had been there since 1963. Before doing so Hector Garciá Godoy asked the Human Rights Commission of the OAS to investigate charges of abuse of these persons; the Commission was barred from Haiti.

The United States frustrated a much-publicized move to invade Haiti in 1967 by charging the leaders with conspiring to export arms illegally. Headed by Rolando Masferrer Rojas, a wealthy Cuban, and the Reverend Jean Baptiste Georges, a Haitian priest, the group had been organized in Florida. The Columbia Broadcasting System sent a television director there to film a "rehearsal" of the invasion on the southerly isolated beaches. The television film, to be shown after the actual invasion took place, would purport to show the birth and development of a "genuine" Caribbean revolution. According to their plans, Reverend Georges, a former Minister of Education of Duvalier, would be installed as president of Haiti. This achieved, Masferrer would build an army to strike Cuba.

[131] Latortue, *Tyranny in Haiti*, p. 351.
[132] *New York Times*, Feb. 19, 1966.

Several feeble attempts to overthrow the regime were later set in motion. In 1968 an exile group flew over the capital and dropped several small bombs while another aircraft carrying twenty-five armed men landed at Cap-Haitien on the north coast. They fled after the people asked them to leave and the militia and coast guard fired at them. René Léon, a retired Haitian army officer, flew over Port au Prince the following year, dropping several cans of gasoline ignited by crude fuses; no hits were made on the palace, the intended target. In April 1970 there was a revolt in the Coast Guard. Three rebel cutters, after firing about fifty small cannon rounds at the presidential palace sailed away, the crews taking asylum in the United States.

Haiti's economy approached bankruptcy in the late 1960s as revenues from the three main exports dropped sharply between 1962 and 1967 from $42.5 million to $33.4 million. The nation's gold reserves and foreign exchange dropped steadily and both tourism revenues and private investment continued to decline. Although the government talked about "progressive industrialization" it was not evident, and the textile industry had actually suffered reverses. What industry existed tended to be in foreign hands, mainly United States interests. Taking advantage of the low wage scale, some one hundred U.S. companies were set up to produce hand-assembled tools, shoes, clothing, and electronic devices.

Defense funds, the largest single item in the Haitian budget, accounted for $7.1 million, or 25 percent of the total. The allocation for agriculture was less than one-third of that for defense. The Haitian economy provided each inhabitant with a yearly income of $80, the lowest in the hemisphere. Government expenditure per capita was about $7.00, compared with $12.00 in Bolivia and $500 by the United States federal government. Sixty-one cents was spent each year for each Haitian's education, and .65 for his health. To make matters worse, Haiti had the densest population in the Western Hemisphere.[133]

Although the United States absorbed about two-thirds of Haiti's exports, the aid program was not resumed. In Washington's judgement, neither the administrative structure nor the standards of public morality were adequate to insure proper use of assistance funds. Haitian officials asserted that the United States would not help because "Haiti is a Black Republic." This is an absurd accusation, but Haiti as a nation is very color conscious. Duvalier had based what popular appeal he had on the contention that his regime represented the triumph of the poor black majority over the largely mulatto minority élite. Traditionally "So rigid are the class lines set that *caste* is the only word to describe the effective separation of aristocrats from the masses. The *caste* system is a vivid fact, for it regulates a person's profession, speech, religion, marriage, family life, politics, clothes, social mobility, in short his whole life from cradle to grave."

[133] The mountain lands which make up almost 80 percent of the national territory's 10,240 square miles are eroded; the lowlands suffer from inadequate rainfall; minerals are virtually non-existent; and the forests have been cut down. On this weak resource base live 4,671,000 people (est., 1968), or 435 persons per square mile; on the lower slopes where most of the population lives the density approaches 1,500 per square mile. See *New York Times*, May 24, 1970; Inter-American Development Bank, *Socio-Economic Progress in Latin America*, pp. 191-203.

The élite and masses are worlds apart but the élite constitutes no more than 3 percent of the total population.[134]

Washington, in the late 1960s, in spite of apparent inactivity and indifference, was seriously concerned about Haiti, though mainly from fear that Communist infiltration from Cuba would spread into the Dominican Republic. Under existing conditions Haiti was obviously most vulnerable to Communist penetration. However, the Johnson administration, recognizing that Duvalier did not represent a threat to United States interests, avoided an open conflict with the dictator, and withheld sending in Marines even under the most trying conditions.[135]

After the Nixon administration's first year in office there were indications of a possible relaxation on what had become a virtual blockade of economic aid to Haiti. Governor Nelson Rockefeller of New York who headed a fact-finding mission in Latin America for President Nixon in 1969 recommended that U.S. aid be resumed. Evidence of a more cooperative attitude in Washington was seen in the granting of a $5.5 million loan by the Inter-American Development Bank to expand the water supply in Port au Prince. And to help shore up the economy of the poverty-stricken nation, the International Monetry Fund recommended the granting of a stand-by balance-of-payments loan.[136]

Duvalier, one of the most cruel dictators in hemisphere history, passed away on April 21, 1971, having previously chosen his nineteen-year-old son, Jean Claude, to succeed him as president-for-life. Although Jean Claude was ill-prepared for so awesome a task, his selection as official heir seemed to have been a wise decision, for it probably averted a general upheaval upon his father's passing. Uncertainty continued to exist over whether Jean Claude was actually in control. It appeared that he was a symbol and front man for a collective leadership. In addition to members of the Duvalier family there were General Claude Raymond, Chief of Staff of the Army; Luckner Cambronne, Minister of Interior Defense and Police; Adrien Raymond, Minister of Foreign Affairs, and others.

The new government actively sought U.S. aid and made efforts to improve its image abroad by dropping censorship of outgoing dispatches, by making cabinet ministers accessible, and lessening political oppression. However, it was clear that the regime could not tolerate a democratic, open society. Luckner Cambronne, who headed the internal security forces, indicated in response to questions in an interview that there was "no necessity for elections in Haiti because the government of Jean Claude Duvalier was established by Congress under the Constitution . . . no general amnesty for political prisoners because none is necessary." He said that Haiti was not prepared to accept a study commission from the Organization of American States because "it's simply a matter of sovereignty."[137]

[134] James G. Leyburn, *The Haitian People* (New Haven, 1966), pp. 3-13, 265-289.
[135] For reaction in U. S. Congress to the Duvalier regime see *Cong. Rec.*, 89th Cong., 1st Sess., vol, III, pp. 25797-25798, 26743. Aid disbursed under the Alliance for Progress to Haiti in the period 1962-1966 totaled $22.4 million. See Hanson, *Five Years of the Alliance for Progress*, p. 4.
[136] *New York Times*, May 24, 1970.
[137] *Times of the Americas*, May 12, 1971; *Ibid.*, Jan. 5, 1972; *Christian Science Monitor*, May 26, 1971.

The fourteen years of Duvalier's rule saw scarcely any change for the better in the livelihood of Haiti's four million inhabitants. Throughout most of the 1960s the growth of the economy usually lagged behind population growth, 3.6 percent annually, and the pathetically small per capita income declined year by year. The new regime headed by Jean Claude sought to revitalize the economy by economic reform and development planning. While some manufactured exports expanded, the production of coffee and sugar, the main export items, declined in 1971. In order to diversify the economy, the government emphasized tourism as a source of revenue and foreign exchange, and supported the construction of hotels.

This approach and a less oppressive political climate led Washington to resume aid programs, which had been suspended in 1962 by President Kennedy. A $750,000 loan made by the Inter-American Development Bank in October 1971 was applied to increasing the production of foodstuffs. The United Nations Development Program and other international agencies had increased direct assistance. Most of the $8 million received by Haiti in aid in 1971 was devoted to food distribution. U.S. Ambassador Clinton E. Knox raised the question as to whether Haiti could absorb large amounts of aid immediately. The possibility of a successful invasion of Haitian exiles or other groups seemed remote, but Washington increased its naval and air surveillance near Haiti to forestall external intervention, especially to guard against possible Cuban efforts to exploit any domestic turmoil that might break out in Haiti.[138]

Significant political and economic changes occurred in 1972, which could hardly have been foreseen. Jean Claude Duvalier appeared to be exercising more independent political authority as shown by his dismissal of Interior Minister Luckner Cambronne, who went into exile in Colombia. Cambronne had reportedly assumed too much power. And President Duvalier appeared earnest in his declaration: "My father made the political revolution; I will make the economic revolution."

United States and international agencies had made further commitments of aid, and the OAS pledged support for Haiti's economic and social development. Many new industries, largely underwritten by foreign capital, were established, and tourism, which languished in the 1960s, attained levels greater than at any time in the past. Foreign banks were opening, and Haitian flight capital from overseas was returning in impressive amounts.[139]

In the midst of these signs of progress U.S. Ambassador Knox and Consul General Ward Christianson were seized and held captive by terrorists in January 1973, in Port au Prince. Their captors demanded the release of twelve political prisoners and $500,000 ransom, but came down to $70,000 when the U.S. government reportedly refused. The U.S. officials were released unharmed after the terms were met, and Mexico had agreed to granting political asylum for the terrorists and the freed prisoners.

Jean Claude became the ninth chief executive in Haitian history to claim the president-for-life title. He was confronted with rivalries of families and cliques, continuing political unrest, and the problems of a nation that ranked at the bottom of virtually every table and scale of social and economic development in the Americas.

[138] *Christian Science Monitor*, Apr. 24, 1971; *New York Times*, Jan. 28, 1972.
[139] *Christian Science Monitor*, Dec. 1, 1972; *Ibid*., Jan. 8, 1973.

SUPPLEMENTARY READINGS

Atkins, G. Pope and Larman C. Wilson. *The United States and the Trujillo Regime.* New Brunswick, N.J., 1972.

Balch, Emily. *Occupied Haiti.* New York, 1927.

Balaguer, Joaquín, *Dominican Reality: Biographical Sketch of a Country and a Régime.* Mexico, 1949.

Bellegarde, Dantes. *Haiti and Her Problems.* Río Piedras, 1936

Bosch, Juan. *Crisis de la democracia de la América en la República Dominicana.* Mexico, 1964.

Burks, Arthur J. *Land of Checkerboard Families.* New York, 1932.

Chapman, Charles E., "The Development of the Intervention in Haiti," *Hispanic American Historical Review*, VII (Aug., 1927).

Cook, Mercer. *An Introduction to Haiti.* Washington, 1951.

Crassweller, Robert D. *Trujillo: The Life and Times of a Caribbean Dictator.* New York, 1966.

Crockaert, Jacques. *La Méditerranée Americaine.* Paris, 1927.

Davis, H. P. *Black Democracy.* New York, 1928.

Douglas, Paul H. "The American Occupation of Haiti," *Political Science Quarterly*, 42 (June, 1927)

Espaillat, Arturo. *Trujillo: The Last Caesar.* Chicago, 1963.

Galíndez Suárez, Jesús de. *La era Trujillo; un estudio casuístico de dictadura hispano-americana.* Santiago, Chile, 1956.

Goldwart, Marvin. "The Constabulary in the Dominican Republic and Nicaragua," *Latin American Monographs, No. 17.* Gainesville, 1962.

Hicks, Albert C. *Blood in the Streets; the Life and Rule of Trujillo.* New York, 1946.

Inman, S. G. *Through Santo Domingo and Haiti.* New York, 1919.

Jones, C. L. *Caribbean Backgrounds and Prospects.* New York, 1931.

-------. *Caribbean Interests of the United States.* New York, 1916.

-------. *The Caribbean Since 1900.* New York, 1936.

Kelsey, Carl. *The American Intervention in Haiti and the Dominican Republic.* Philadelphia, 1922.

Korngold, Ralph. *Citizen Toussaint.* Boston, 1944.

Knight, Melvin M. *The Americans in Santo Domingo.* New York, 1928.

Kurzman, Dan. *Santo Domingo: Revolt of the Damned.* New York, 1965.

Kuser, John D. *Haiti, Its Dawn of Progress After Years in a Night of Revolution.* Boston, 1921.

Leger, J. N. *Haiti: Her History and Her Detractors.* New York, 1907.

Leyburn, J. G. *The Haitian People.* 2nd ed. New Haven, 1966.

Logan, R. W. *The Diplomatic Relations of the United States with Haiti, 1776-1891.* Chapel Hill, 1941.

--------. *Haiti and the Dominican Republic.* New York, 1968.

Marshall, Harriet. *The Story of Haiti.* Boston, 1930.

Martin, John B. *Overtaken by Events: The Dominican Crisis From the Fall of Trujillo to the Civil War.* New York, 1966.

McCrocklin, James H. *Garde d'Haiti, 1915-1934.* Annapolis, 1955.

Millspaugh, A. C. *Haiti under American Control, 1915-1930.* Boston, 1931.

Montague, L. L. *Haiti and the United States, 1714-1938.* Durham, 1940.

Munro, Dana G. *The United States and the Caribbean Area.* Boston, 1934.

------. *Intervention and Dollar Diplomacy in the Caribbean, 1900-1921.* Princeton, 1964.

Niles, Blair. *Black Haiti.* New York, 1926.

Ornes, Germán E. *Trujillo: Little Caesar of the Caribbean.* New York, 1958.

Parry, J. H. and P. M. Sherlock. *A Short History of the West Indies.* London, 1956.

Perkins, Dexter. *The United States and the Caribbean.* Cambridge, Mass., 1947.

Rippy, J. Fred. *The Caribbean Danger Zone.* New York, 1940.

Rodman, Seldon. *Quisqueya: A History of the Dominican Republic.* New York, 1964.

Seabrook, W. B. *The Magic Island.* New York, 1929.

Schoenrich, Otto. *Santo Domingo.* New York, 1918.

Ureña Henríquez, Max. *Los yanquis en Santos Domingo.* Madrid, 1929.

Walker, Stanley. *Journey Toward the Sunlight, A Story of the Dominican Republic and its People.* New York, 1947.

Welles, Sumner. *Naboth's Vineyard: The Dominican Republic, 1844-1924.* 32 vols. New York, 1928.

14

Interests of the United States in Central America

The region known as Central America comprises the five republics of Guatemala, El Salvador, Honduras, Nicaragua, and Costa Rica, the states which once formed the *Provincias Unidas del Centro de América* (1823-1838). Panama, a part of Colombia until the early twentieth century, is not a Central American country; however, she has a close kinship with her northern neighbors. The five states, totaling 177,000 square miles in area, and containing a population of more than fifteen million, stretch for some nine hundred miles between 8° and 18° North Latitude. They are also referred to as Middle America, a term having a cultural rather than a political connotation, which identifies them with other republics and dependencies found within comparable latitudes in the Caribbean region.[1]

Because of its form and geographic location, the long narrow isthmus of Central America has had an important role in the cultural as well as the political history of the Americas. Serving as a land bridge connecting North and South America, it was a path in pre-historic times over which animals and plants moved between the two continents. Later, American Indians used it for migration and cultural interchange. After the Spanish Conquest, when the isthmus became a barrier to European navigators who wished to cross from one ocean to the other, the narrow land corridor gained importance in another way. The constricted sections, the lowest and narrowest points on the mountainous isthmus, gained a strategic importance for the expansion of Spain's colonial empire. The earliest and most important crossing was at Panama, the narrowest part of the isthmus. From the Colonial period men had envisaged the construction of a sea-level canal across the Central American isthmus, but it was only with the increase in interoceanic traffic after the opening of the Far West of the United States, the

[1]For background on demography and geography, see United Nations, *The Population of Central America (Including Mexico), 1950-1980* (New York, 1954); Robert C. West and John P. Augelli, *Middle America: Its Lands and Peoples* (Englewood Cliffs, 1966).

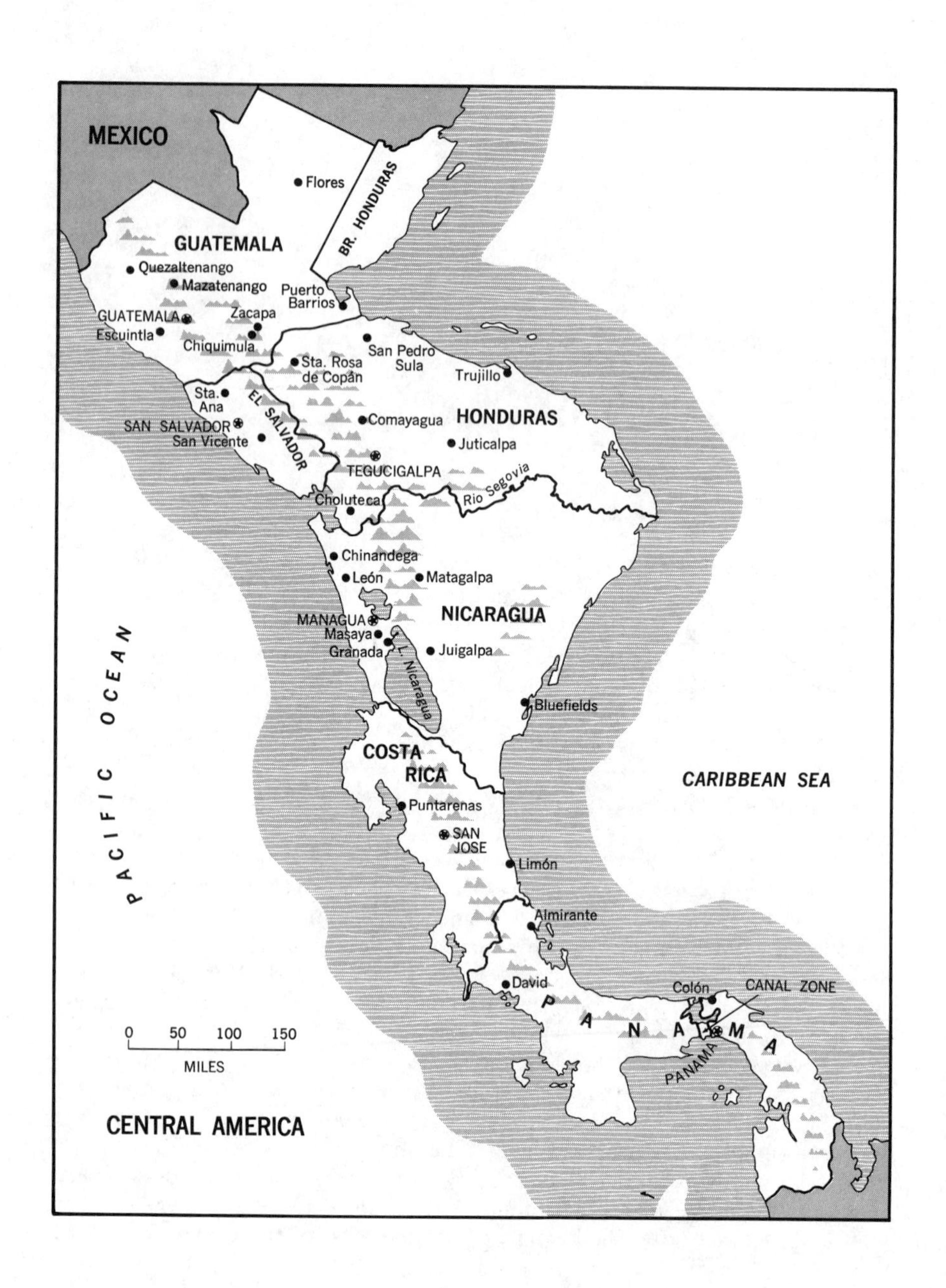
MEXICO
Flores
BR. HONDURAS
GUATEMALA
Quezaltenango
Mazatenango
Puerto Barrios
GUATEMALA
Zacapa
Escuintla
Chiquimula
San Pedro Sula
Sta. Rosa de Copán
Trujillo
Sta. Ana
EL SALVADOR
SAN SALVADOR
San Vicente
Comayagua
HONDURAS
Juticalpa
TEGUCIGALPA
Rio Segovia
Choluteca
Chinandega
León
Matagalpa
NICARAGUA
MANAGUA
Masaya
Granada
L. Nicaragua
Juigalpa
Bluefields
PACIFIC OCEAN
COSTA RICA
CARIBBEAN SEA
Puntarenas
SAN JOSE
Limón
Almirante
David
Colón
CANAL ZONE
PANAMA
PANAMA
0 50 100 150
MILES
CENTRAL AMERICA

rise of the Yankee colossus as a continental two-ocean seapower, and the needed technological proficiency had been reached, that such an enormous project became feasible.

Central America's function as a land bridge has been revived with the approaching completion of the Inter-American Highway. However, a 250 mile stretch through sparsely populated rain forests in eastern Panama and northwest Colombia remains to be constructed before that role can be resumed. Although sections of the highway were completed for local use in the 1930s, it was not until World War II that an organized construction program, financed mainly by the United States, was carried out.

Being located in one of the world's most intensive zones of active mountain building, much of the Central American isthmus is subject to frequent earthquakes and volcanic activity. These geological hazards have often caused much damage to man and his works. The surface of much of Central America is characterized by a series of mountain ranges, trending east-west, and intervening basins. Having fertile volcanic soils, these tropical highland basins and the adjoining mountain slopes are the areas of densest settlement within Central America. With elevations of 3,000 to 5,000 feet, the Meseta Central of Costa Rica is the largest of these basins.

Altitude strongly influences air temperature in the Central American tropics. In the mountainous sections of the isthmus, weather, climate, vegatation and soils, and land use also varies according to altitude. In such regions three or four types of climate may be found within a horizontal distance of only twenty-five miles. In general, the east coast receives more rainfall than the west coast; and for the most part the rainfall is seasonal, the rainy period coming during the hotter months, from May through October.

The economy of Central America revolves around agriculture, much as it did in the Colonial period. In the 1960s, for example, more than two-thirds of the area's thirteen million people gained their living by farming; agriculture contributes 37 percent of the region's gross national product; agricultural products account for 90 percent of the value of Central America's exports, and this in spite of the fact that less than 10 percent of the land is under cultivation. Coffee, bananas and cotton are Central America's most profitable commodities, and together they constitute almost 80 percent of the value of its exports. They occupy the best agricultural lands, yet they supply no food for the area. Much of the foreign exchange earnings that the Central American countries gain from the sale or taxing of these products must be used to import foodstuffs, such as rice, maize, wheat, and wheat flour to feed their burgeoning populations. The area is highly dependent upon exports for the foreign exchange to import manufactured and semimanufactured goods from industrialized countries.

Any account of Central America would be incomplete without mentioning the United Fruit Company (UFCO), which since the late nineteenth century has exerted a major influence on the economies and politics of the isthmian republics and their relations with the United States. Incorporated in New Jersey in 1899, following a merger of the Boston Fruit Company and transport and banana firms controlled by Minor C. Keith, a Yankee railroad builder in Costa Rica, UFCO grew into one of the largest United States corporations involved in foreign operations, and the largest banana concern in the world. It came to own more than 374,000 acres in Central America, and bought lands in Santo Domingo, Cuba, Jamaica, Panama, and Colombia. By the mid-1950s its holdings included over 1,500 miles of railroad, a fleet of seventy vessels, and one of the major wireless facilities in the Western hemisphere. Most of its bananas are shipped to the United States, the remainder going to Western Europe. While the company still deals mainly in bananas, it is also involved in the production and sale of sugar, cacao, palm oil, and abaca.

Beginning its operations in the Caribbean when modern European colonialism was reaching its zenith, followed by the United States "Big stick," "dollar diplomacy," and Wilsonian intervention eras, it was strongly criticized as an agency of "Yankee imperialism," and not without considerable justification. Since World War II, however, reacting to pressures often generated by nationalist and Communist forces, the company has exhibited a willingness to conform to the changing times and operates to the advantage of its host countries. For the most part, its employees enjoy better health, pay, housing and food, than other rural workers, and the industry it created forms a major source of income in the countries with which it is involved. UFCO is nonetheless still one of the most controversial subjects among Central Americans. After all, it is an alien concern, and in some respects is more powerful than the government with which it deals. In evaluating the role of UFCO, further confusion results from an interpretation of terminology: what foreign companies call "development," the host countries label "exploitation."

Most of the inhabitants of Central America are concentrated on the Pacific side of the isthmus, where they occupy both highland and tropical lowland environments. Population densities vary from 1,000 per square mile on the Meseta Central of Costa Rica to 5 per square mile in the rugged and mountainous sections of the Pacific slope. Population growth is higher than that of any other major region of the world, varying from 3 percent in Honduras to more than 4 percent in Costa Rica.

Although industrialization is being pushed, as it is in most of the economically backward areas of the world, it has not progressed very far in Central America. Unlike Mexico, the area lacks the basic industrial raw materials needed for heavy industry. No profitably exploitable deposits of hydrocarbon fuels have been discovered, and there are few significant reserves of industrial metal ores. However, light industries using local raw materials, textiles and food processing in particular, are feasible, and offer some prospect of raising the level of living in the region.

Geographically isolated, poor in easily obtained precious metals and, except for Guatemala, lacking in large exploitable Indian populations as a source of labor, the area occupied by contemporary Central America was given little attention by the Spaniards. Except for Costa Rica, which had a singular historical development, a small number of *peninsulares* and *creoles* received large land grants from the Spanish Crown, and came to form the nucleus of a wealthy upper class. The colonial class structure was therefore marked by the domination of the Iberian-born Spaniards, the *peninsulares*, and the native-born whites, the *creoles*, over the *mestizos*, Indians and Negroes. The class structure retains much of its basic colonial character in modern times. Because of the survival of *latifundio*, the dominance of the landed elite, the steadfastness of the Church to the *status quo,* and the conservative nature of the military establishments, there is hostility to modernization and socio-economic change and the expansion of opportunities for the lower classes. Differences are to be found among the individual states, but they are a matter of degree rather than of kind.

Modern Central America reflects, in an impressive way, the commonly observed breach between political theory and political practice in Latin America. The many constitutions of these states have asserted responsible, representative rule and have established the needed institutional safeguards for the protection of individual rights and liberties within a democratic governmental framework. But almost equally constant has been the utilization of political processes for dictatorial ends, for advancement and protection of special interests, and for the denial of the very rights provided by organic law.

Before taking up the relations of the United States with Central America, it might be well to point out the more noticeable differences among these five republics.[2] Guatemala has the largest population and physically and culturally is the most distinctive of the Central American nations. Smaller than the state of Louisiana (42,042 sq. mi.), the country's physical landscape is almost as diverse as Mexico, having rugged mountains, highland basins and tropical lowlands. Culturally, Guatemala is the most Indian country on the isthmus for more than half of the four million population (4,378,341) are aborigines who live very much as their pre-Columbian ancestors did. Only an insignificant proportion of Guatemalans are Caucasians, found chiefly in the urban centers, or Negroid, located on the Caribbean coast. Suffering from a lack of political stability since independence from Spain early in the nineteenth century, the cultural conservatism of both the Indians and the landed aristocracy of Spanish descent, and a relatively poor endowment of natural resources, the country's modernization and economic development have been slow. Like most Central American nations, Guatemala is a land of farmers, with 85 percent of its people living by the soil. In terms of value, 95 percent of the country's exports are coffee, bananas and cotton.

El Salvador, although a land of volcanoes, lacks the variety of landscapes of neighboring Guatemala. It is the smallest mainland country of Latin America, (8,186 sq. mi.) and is also the most densely populated, (2,854,000) having a density of 320 persons per square mile in 1960. Being the only Central American country without a Caribbean coast, El Salvador lacks a port facing eastward toward its markets in the United States and Europe. Also, unlike the other isthmian countries, it lacks unsettled lands for its rapidly expanding population. Overpopulated, suffering from the evils of *latifundio,* eroded soils in many areas, and frequently chaotic political conditions, El Salvador reflects some of the most acute problems found in Central America. The best lands are owned by a few wealthy families who raise crops, mainly coffee, for export rather than food for the nation. Despite some economic progress El Salvador has one of the most

[2]For further details on the United Fruit Company see Stacy May and Galo Plaza, *The United Fruit Company in Latin America* (Washington, 1958), pp. 1-23; John D. Martz, *Central America: The Crisis and the Challenge* (Chapel Hill, 1959), pp. 323-326. Useful general sources on the five republics are Franklin D. Parker, *The Central American* Republics (New York, 1964); Mario Rodriguez, *Central America* (Englewood Cliffs, 1965).

poorly fed populations on the isthmus. Most of the Salvadoreans are a racial mixture of Indian and white (*mestizo*), with some traces of Negro blood. Perhaps 20 percent is of pure Indian blood, but fewer than 3 percent retain an aboriginal way of life. The country lacks hydrocarbon fuels and the mining of metals has never been significant.

Honduras is commonly regarded as the most backward economically and culturally of the Central American states. It is mountainous and difficult to traverse, and its soils, other than on the Caribbean coast, are generally infertile. Its surface (44,800 sq. mi.) is covered predominantly with the oldest and most highly mineralized rocks on the isthmus. Traditionally Honduras has been an area of mines and livestock grazing, of minor agriculture, of a sparse and isolated population (1,068,000), and of unstable republican governments. Racially, 90 percent of the Hondurans are mixed Indian and white; only 7 percent are full-blooded Indians, 2 percent are Negro and 1 percent are white. Most of the population is found in the central and southern uplands, but the contry as a whole is relatively underpopulated with a density of only 45 per square mile. The large-scale production of bananas and other commercial crops in the recently-developed Caribbean lowlands has moved that region ahead of the interior in economic wealth and population growth. The most significant cash crop of the highlands continues to be coffee.

Nicaragua, neglected by Spain in the Colonial period, and involved in numerous wars and disputes with its neighbors in recent times, has developed slowly. Its political history–in which the United States played a leading part at times–has been exceedingly turbulent. Its strategic value as the possible site for an isthmian canal route has been one of the significant factors in maintaining the interest of the United States in this country for a hundred years. Nicaragua has a greater economic potential than Honduras mainly because of the structural depression causing a natural passway across the isthmus, and the presence of highly fertile volcanic soils on the Pacific side of the country.

The largest country on the isthmus, (50,780 sq. mi.) Nicaragua contains some of the most active volcanoes in the region, and the two largest natural bodies of fresh water in Middle America, Lake Managua and Lake Nicaragua. Most of the country's 1.6 million people live on the Pacific side, chiefly on the fertile plains of the lake lowlands and the adjacent highlands. Its population density varies from 400 to 4 persons per square mile. About 70 percent of the population is *mestizo,* 10 percent each Caucasian and Negro, and the balance mixbloods. Scarcely any aboriginal racial elements remain. Nicaragua is a nation of farmers with 70 percent of the economically active population engaged in agriculture. Cotton has become very significant since 1950, rivaling coffee as the country's most valuable export. *Latifundio* prevails, the large hacienda owners possessing most, and the best, agricultural land.

Costa Rica is unique in several respects among Latin American nations. About 80 percent of the inhabitants claim to be unmixed white descendants of Spanish colonists, the majority of whom are independent farmers in the Meseta Central. An estimated 17 percent of the population is *mestizo,* 2 percent are Negroid, and Indians make up 1 percent of the inhabitants. The native population disappeared rapidly after the Conquest, and in its absence the Spaniards in the highlands developed the small-farming cultural pattern which still persists on the rural scene; the absence of Indians also explains why no large *mestizo* group district from the whites emerged. Costa Ricans claim the highest literacy rate, the most democratic government, and what is perhaps the most comfortable living standard in Latin America. And in contrast to its Central American neighbors, Costa Rica has been relatively free from political upheavals and militarism. It is basically an agricultural country with no more than half the working population directly engaged in farming; 95 percent of all exports, by value, are agricultural, chiefly, coffee, bananas and cacao. Costa Rica's national territory comprises the cool central highland, on which about three-fourths of the nation's 1.4 million population is found, a warm Pacific lowland, and a rain-drenched Caribbean lowland. Vulcanism is prevalent, and most recently Mt. Irazú erupted in 1963 causing extensive damage by ash throughout central Costa Rica.

British Honduras, or Belize, a geographical, but not a political, component of Central America is beyond the scope of this book. However, it has been, and continues to be an object of international rivalry and its salient features should be noted. Its origin dates from the early seventeenth century when English smugglers established trade relations with coastal Indian tribes. The hot, rainy, reef-strewn coast between northern Yucatan and eastern Panama was unattractive to Spanish settlers and weakly defended, which enabled the British to get a foothold on the western shore of the Bay of Honduras. Called the "Cockscomb Coast," later to become British Honduras, it was a profitable source of logwood and mahogany. The English logcutters, known as "Batmen," aided by their Negro slaves and Indian allies, were able to repel Spanish efforts to evict them. In 1864 the settlements of British Honduras were made a crown colony of the Empire.

With a population numbering less than 100,000 in the 1960s, the average density is about ten persons per square mile. Racially diverse, about 60 percent of the inhabitants are English-speaking Negroes and mulattoes, 25 percent are Spanish-speaking whites and *mestizos,* and the remainder Indians and various mixbloods. Most of the colony's export income is gained from lumber, naval stores and chicle. Commercial agriculture on a limited scale is confined to sugar cane, citrus fruits, bananas and rice, and some revenue is made from fishing.

The heavy rainfall, leached, infertile soils, and dense forests are major obstacles to economic development. Other than commercial agriculture, most farming is of the subsistence type, and in spite of the small population, the colony is unable to produce enough food for its own needs.

Before its declaration of independence on September 15, 1821, the captaincy-general of Guatemala, which included the provinces of El Salvador, Nicaragua, Honduras, and Costa Rica, had been the connecting link of the great Spanish Empire extending from Mexico to Patagonia. Free from the Spanish yoke, the states into which the captaincy-general disintegrated gravitated towards Mexico; and, although there was considerable opposition, particularly on the part of Costa Rica, a provisional junta decreed that the whole of Central America should be annexed to the Empire of Mexico.[3] The fall of Agustín

[3] H. H. Bancroft, *History of Central America* (San Francisco, 1886-1887), vol. III, p. 54.

Iturbide came before the plan could be carried into effect, and the states were once more adrift. It is interesting to note that at this time El Salvador, which also opposed annexation to Mexico, passed an act annexing itself to the United States, and sent two representatives to negotiate with the authorities in Washington on the subject.[4]

The idea of a federal union modeled upon that of the United States had appealed strongly from the beginning, and in 1823 a national constitutent assembly was called, which declared the former Captaincy-General of Guatemala free and independent, and confederated into a nation under the name of *Provincias Unidas del Centro de America.* A constitution was drafted, after the American model, and ratified on September 1, 1825. Before the year was out the United States recognized the new federal state by signing, on December 5, 1825, a treaty of commerce and friendship, which was proclaimed in effect by the American secretary of state on October 28, 1826.[5] This treaty, designated "A General Convention of Peace, Amity, Commerce, and Navigation with the Federation of the Centre of America," consisted of thirty-three articles, which granted equal privileges and rights to citizens coming from either republic into the other, and all the rights granted by one or the other to the most favored nation. It was to remain in force for twelve years. No provision was made for its continuance after that time. But it proved eminently satisfactory to both parties, and shortly before its expiration in 1838, Mr. De Witt, the *chargé d'affaires* in Guatemala, was instructed to obtain a renewal of it. A new treaty was signed on July 14, 1838. But owing to a delay in the exchange of ratifications, it did not go into effect; and in the summer of 1839 Mr. J. L. Stephens was sent as special agent to bring about the exchange of ratifications.[6] In the meantime the confederation had been rapidly weakening; the last federal congress adjourned in 1838, and the last president was expelled in 1840. As a consequence, Mr. Stephens although well received was unable to accomplish the object of his mission.

Various attempts were made to reorganize the federation, but nothing permanent came of them. In 1842 Salvador, Honduras, and Nicaragua formed a confederation which lasted about two years, and again in 1849 they signed a treaty of confederation which proved equally abortive[7]. Uncertain of the situation, the United States had despatched Mr. W. S. Murphy in 1841, with instructions that if he should find any organized government of a federation of Central America, he should arrange for the ratification of the treaty of commerce. He was obliged to report that there was, at the time, no federal government in Central America entitled to the privileges, or responsible for the duties of a sovereign ower.[8] Guatemala declared herself definitely free and independent on March 21, 1847, and the United States thereupon despatched Mr. Elijah Hise to propose the adoption of the treaty of 1838 between the United States and Guatemala. He was also authorized to conclude a similar

[4]Ibid., p. 64; see also *Sen. Ex. Doc. No. 75,* 31st Cong., 1st Sess., p. 94.
[5]For text see *Amer. State Papers, For. Rel.,* vol. VI, pp. 269-276.
[6]*Sen. Ex. Doc. No. 75,* p. 95.
[7]For full account see Bancroft, *op. cit.,* chap. X; also W. F. Slade, "The Federation of Central America," *Journal of Race Develpment,* vol. VIII (July, 1917), pp. 79ff.
[8]*Sen. Ex. Doc. No. 75,* p. 95.

treaty with El Salvador; but Secretary Buchanan did not deem it advisable to conclude treaties with either Nicaragua, Honduras, or Costa Rica until the Department secured further authentic information concerning them.[9]

Undoubtedly a more important reason for sending Mr. Hise was to obtain information about the encroachments of Great Britain unpon the territories of Nicaragua, Honduras, and Costa Rica, under the guise of protecting the so-called kingdom of Mosquito Indians. In fact, on January 1, 1849, a British man-of-war had taken forcible possession of the post of San Juan de Nicaragua; and although the Nicaraguan troops recaptured it on January 9, the British reoccupied it three days later and maintained their hold. Inasmuch as San Juan would be the natural outlet for any isthmian canal through the state of Nicaragua, the interest of the United States was justified. Mr. Hise arrived about the first of November, and lost no time in accomplishing his mission. In a letter to the secretary of state, dated May 25, 1849, he reported that he had already concluded a treaty with Guatemala, and was then negotiating with the commissioners of Honduras and Nicaragua, with whom he hoped to conclude commercial reciprocity treaties; and he intimated that he also expected "to conclude a special convention with Nicaragua of vast importance to the United States and to the whole commercial world."[10] He was not disappointed, and in a despatch dated September 15, 1849, he announced that he had concluded treaties of amity and commerce with Honduras and Nicaragua, and a special convention with Nicaragua. This convention secured to the United States a perpetual right of way by land or water through Nicaragua, and the right to charter a company which should have exclusive right over the contemplated canals or roads; in return, the United States was to guarantee the sovereignty of Nicaragua over its rightful territories.[11] Although his instructions had given him no such powers–in fact, had definitely prohibited conventions with Honduras and Nicaragua–he justified his acts on the grounds that, on account of the revolutionary condition of the country, he had received no answers to his despatches. He also felt that the cession of Upper California to the United States and the subsequent discovery of gold there made this acquisition doubly valuable. Finally, he was confident of being sustained because he had received authentic information that English companies were endeavoring to procure for themselves the privileges that he had received.

Months before this, however, the new administration at Washington had recalled him, and had appointed Mr. E. G. Squier to undertake a similar mission to Nicaragua. In Squier's case, however, authority was given to conclude treaties with all five republics separately in case he should consider that there were no hopes of a new confederacy. Regarding Nicaragua, he was to obtain, if possible, for the citizens of the United States the right of free transit over any canal or railroad that might be constructed between the two oceans. But it was not deemed expedient to give as compensation any guaranty of the independence of the country through which the canal or railroad should pass.[12] The

[9] *Ibid.*, p. 96.
[10] *Ibid.*, p.103.
[11] Text *Ibid.*, p. 110; for earlier attempts to obtain a canal route across Nicaragua, see chap. 5, above.
[12] *Sen. Ex. Doc. No. 75*, p. 120.

instructions of Secretary Clayton also gave a very complete résumé of the British claims to the Mosquito territory, and concluded with the opinion that they were founded upon repeated usurpations. "The United States," he declared, "would not allow the isthmian passage to be blocked by such pretensions–it desired no monopoly of the right of way for its commerce, nor could it submit to a similar claim on the part of any other country."[13]

When Mr. Squier arrived at his post, he noticed that a very cordial relationship existed between Honduras, El Salvador, and Nicaragua, and upon learning that commissioners had been appointed to arrange for a more complete and efficient union, he expressed the hope that the attempt would prove successful. In a note to the minister of foreign relations of the Republic of El Salvador, he declared that "it is the desire of my government that some consolidation of these states may be effected, believing, as it does, that their general interests will thus be promoted, and that they will be better enabled to resist the encroachments and thwart the designs of foreign and unfriendly powers."[14] He also noticed that the British appeared to have designs upon Tigre Island in the Gulf of Fonseca, which belonged to Honduras. Believing that its seizure by Great Britain would vitally affect the interests of the United States, he, too, exceeded his instructions, and, despatching a special courier to Honduras, he asked for a speedy treaty between the United States and Honduras which should "authorize the United States in interposing its power against the designs of the English."[15] In his despatch to Secretary Clayton he suggested that a feasible method of effecting the objects desired might be to procure the cession of the island to the United States.[16] Washington was unwilling to risk this plan; and we have shown in another chapter how the United States attempted to settle the dispute with Great Britain by the Clayton-Bulwer treaty.[17] Although Mr. Squier succeeded in obtaining a satisfactory treaty with Nicaragua, it was no more productive of results than "the most advantageous treaty that human ingenuity could devise" of his predecessor.

During the next decade Central America was a close competitor with California in furnishing an outlet for the restless energy of adventurous Americans. In fact, the two were very closely linked up, since the most practicable way to California from the eastern seaboard of the United States was by water as far as Panama, and thence, after crossing the isthmus, by way of the Pacific. We have already seen how the Panama Railroad Company had received a concession from Colombia in 1850, and had completed the road in 1855.[18] This company was not destined, however, to monopolize the isthmian traffic. In 1849 Colonel Vanderbilt, with several associates, obtained a concession from

[13]*Ibid.*, pp. 121-128. For a more complete discussion of the British claims see *Sen. Doc. No. 27,* 32nd Cong., 2nd Sess., pp. 73-98.
[14]*Sen. Ex. Doc. No. 75,* p. 163. It is interesting to note that just three weeks earlier Señor Francisco Castellón, the Nicaraguan minister to London had asked of Mr. Bancroft, the American minister, whether Honduras, El Salvador, and Nicaragua might be admitted into the North American Union and upon what conditions. If not, could they count upon the United States to defend the integrity of their territory? *Ibid.*, p. 302.
[15]*Ibid.*, p. 167.
[16]*Ibid.*, p.157,
[17]See pp. 187-188, above.
[18]See chap. 5 above.

Nicaragua, authorizing them to construct a canal through Nicaragua, utilizing any lakes or rivers that the engineers should consider feasible in putting through the project. Having become incorporated as the "American Atlantic and Pacific Canal Company," this group proceeded actively with the plan. Soon, however, they found that it was not entirely practicable.[19] A new charter was then obtained, which gave to the Accessory Transit Company (an offshoot of the former organization) a right of way across the country and the sole privilege of steam navigation on the navigable waters of the state.[20] By 1854 a route across the country by river boats, lake steamers, and stage-coaches had been completed and was competing very actively with the Panama route.[21]

It can readily be imagined that the officials of this company would be directly interested in the maintenance of a stable government in Nicaragua strong enough to protect the transport of passengers across the country. But a stable government in Nicaragua at this time seemed quite beyond the bounds of probability. The Conservatives, with Granada as their stronghold, were in perpetual conflict with the Liberals, whose strength was centred at León. The other towns divided their allegiance between these two groups. Fifteen presidents during a period of six years makes a record for even Central American republics. In 1854 the Liberals, under Francisco Castellón, who had taken the offensive and were besieging the Legitimists in Granada, found that they were losing ground. To stem the tide, Castellón contracted with Byron Cole, an American, to bring a detachment of Americans to Nicaragua to serve in the Democratic Army.[22] Cole was a friend of William Walker, who had already attained considerable notoriety as a result of the filibustering expedition which he had led to Lower California with a view to adding it, together with Sonora, to the territories of the United States. Cole communicated the proposal to Walker, then engaged in newspaper work in California. The adventurer needed little persuasion to organize the expedition. On May 4, 1855, fifty-eight men left San Francisco for Nicaragua under his leadership. In less than six months Walker had captured Granada, signed a treaty with the Legitimist leader, and had himself named commander-in-chief of the military forces of the new Rivas government. The Accessory Transit Company was not slow to appreciate the advantages of enlisting him on its side, and one of its agents offered to finance the new government and immediately made an advance of $20,000.

Walker's situation was still so critical that he could not afford to refuse such timely aid; and when the company also agreed to furnish free transportation on its steamers for recruits from New York and San Francisco, thus augmenting his forces by hundreds of valuable followers, he became hopelessly indebted to its agents. Their purpose was soon revealed. The New York and San Francisco agents, Morgan and Garrison, had determined to wrest control of the company from Vanderbilt, and Walker was to play a leading role in bringing about this result. Nicaragua had certain valid claims against the

[19] For text of the charter see *Sen. Ex. Doc. No. 68,* 34th Cong., 1st Sess., p. 84.
[20] *Ibid.*, p. 100.
[21] L. M. Keasby, *Nicaragua Canal and the Monroe Doctrine* (New York, 1896); also E. G. Squier, *Honduras* ((London 1870), p. 241.
[22] An interesting and scholarly account of these events is given by Prof. W. O. Scroggs, *Filibusters and Financiers* (New York, 1916).

company, which Walker was to use as a basis for securing the revocation of the charter. He could then issue a new charter to Morgan and Garrison, who by virtue of the concession would be in a position to take over the property of the old company. Walker became a partner in the scheme, and obtained Rivas' signature to the revocation of the old charter. Morgan and Garrison received the concession. But Walker, by incurring the bitter enmity of Vanderbilt, had committed himself to a course which was to result ultimately in his undoing.

In trying to decide upon a policy regarding the filibusters, the Pierce administration found itself in a quandary. The ministers of Costa Rica, Guatemala, and El Salvador protested against the recruiting, which was going on openly, and thereupon Pierce issued a proclamation warning all American citizens against fitting out or taking part in the expeditions to Nicaragua.[23] But, in spite of the proclamation and the attorney-general's attempts to enforce it, the recruiting went on as merrily as before. When Parker H. French arrived in Washington as the official representative of the Walker-Rivas government, Secretary Marcy refused to receive him, on the ground that those who were instrumental in overthrowing the government of Nicaragua were not citizens of the country, and that the United States would establish diplomatic relations with Nicaragua only after it appeared that the new government had the support of its citizens.[24]

Pierce was accused of catering to Great Britain by refusing to recognize the American government in Nicaragua, and there was even a chance that it would defeat his renomination. Incensed at the treatment accorded French, Walker broke off diplomatic relations with Wheeler, the American minister at Granada.[25] A little later, realizing that French was a most unfortunate choice as minister to the United States, Walker chose Father Vijil, the curate of Granada; and when the latter presented his credentials at Washington they were accepted by President Pierce, "satisfactory evidence appearing that he represents the *de facto* government."[26]

The Costa Rican minister at Washington, Señor Molina, was particularly hostile to the Walker-Rivas government, and protested bitterly to Mr. Marcy on various occasions. In desperation, his state finally took up arms against Nicaragua, whose independence had been destroyed "by the pirates who sailed from the coasts of the United States."[27] The first engagements were decidedly favorable to the Costa Rican forces, but an epidemic of cholera so decimated the ranks that the campaign had to be abandoned. Elated by his success and by the recognition extended by the American government, Walker now proceeded to have himself elected president. The American minister, Mr. Wheeler, without awaiting instructions, proceeded to recognize the new government.[28] He was later censured for his blunder, and when the Walker government's representative reached Washington, he was not received and explanations were refused.[29]

[23] J. D. Richardson, *Messages and Papers of the Presidents,* vol. X, p. 388.
[24] *Sen. Ex. Doc. No. 68,* p. 57.
[25] *Ibid.,* p. 74.
[26] *Ibid.,* p. 6.
[27] *Ibid.,* p. 131.
[28] Scroggs, *op. cit.,* p. 214.
[29] *Ibid.,* p. 215.

The other republics now became alarmed at Walker's success, and, fearing for their own safety, entered into an alliance for the defense of their sovereignty and independence, recognizing the deposed Rivas as the provisional president of Nicaragua. In spite of the overwhelming superiority of the allies, Walker and his brilliant officer, General Hennington, withstood every attack, and at the end of 1856 the prospects for success were still bright. Vanderbilt now came to the aid of the allies, and, by a clever scheme ably carried out by his agents, caused the collapse of the filibuster régime. A force of Costa Ricans, aided by Spencer, a former engineer on the Transit Company's line, succeeded in surprising Walker's force guarding the steamers on the San Juan and Great Lake, thus cutting off further supplies and reinforcements. The result was now merely a question of time. Commander Davis, in charge of a United States sloop-of-war, seeing the desperate plight of the filibusters, offered his mediation, and the allied general was willing to allow them to surrender to the American commander. "Vanderbilt's man had succeeded in doing what the allied Central American states could not accomplish. It was American capitalists who set up the filibuster régime in Nicaragua, and it was an American capitalist who pulled it down."[30]

In September 1857, the representatives of Costa Rica, Guatemala, and El Salvador informed Mr. Cass that a new Walker expedition was being prepared.[31] When this was brought to the attention of Walker, he indited a letter to Secretary Cass truly remarkable in its diplomatic naiveté. He declared that this self-imposed guardianship over Nicaragua on the part of the representatives of Costa Rica and Guatemala was humiliating in the extreme, and as the rightful and lawful chief executive of Nicaragua he protested against it.[32] The United States government investigated, none the less, and finally ordered his arrest. Released on bail for hearing, he immediately set sail for Nicaragua from New Orleans, with very little interference on the part of the port authorities.[33] He landed at Punta Arenas November 24, but before he could do much mischief the U. S. frigate *Wabash,* under the command of Commodore Paulding, appeared opposite the filibusters' camp and demanded their surrender. Walker was brought back to the United States, and, although Commodore Paulding was censured for violating the sovereignty of Nicaragua, President Buchanan expressed the view in no uncertain terms that Walker's expedition was a crime, and said that such undertakings interfered at every step with the conduct of foreign affairs with Central American governments.[34]

Buchanan had already recognized Señor Yrisarri, who was serving as minister from Guatemala and El Salvador, as representative from Nicaragua, and immediately signed a treaty with him providing for open transit through Nicaragua to all nations upon equal terms, and giving the United States the right to employ force, if necessary, to protect the route.[35] This treaty, however, was never signed by the Nicaraguan president, partly because of the opposition of

[30]*Ibid.,* p. 285. For a full account of the surrender from the standpoint of the filibuster see *House Ex. Doc. No. 24,* 35th Cong., 1st Sess., p. 15.
[31]*Ibid.,* p. 4.
[32]*Ibid.,* p. 6.
[33]*Ibid.,* p. 26.
[34]Richardson, *op. cit.,* vol. V, p. 466.
[35]Text in *Sen Ex. Doc. No. 194,* 47th Cong., 1st Sess., p. 117.

Vanderbilt, who knew that a rival company had the concession, but principally because of the machinations of the Frenchman, M. Felix Belly, who used the fear of the filibusters as a most effective bogy to frighten the president of Costa Rica and Nicaragua. These officials went so far as to issue an elaborate manifesto against the United States, declaring that the United States openly menaced Central America with annexation unless Europe should defend it; and they therefore placed their countries under the protection of England, France, and Sardinia.[36] When the United States asked whether this document was genuine, the Nicaraguan government, after considerable prodding, replied that it was signed by the president acting as a private citizen, and therefore was not an official act.[37]

Walker's third and last attempt ended the filibustering expeditions to Central America. Early in 1860 an inhabitant of one of the Bay Islands, which, by a convention signed the previous year between Great Britain and Honduras, had been put back under the sovereignty of Honduras, invited Walker to come to the islanders' aid to resist the convention's execution. Walker seized the opportunity, and was soon heading another expedition to Central America. Upon arriving, he found the British flag still flying, and after waiting several weeks he became desperate and seized the fortress of Truxillo on the mainland of Honduras. A fortnight later Commander Salmon of H.M.S. *Icarus* sent him a note declaring that the occupation was prejudicial to British interests, and that therefore he must lay down his arms and withdraw. Realizing the hopelessness of his position, Walker first attempted flight, but, being captured, surrendered to the British officer. Contrary to his promise, Salmon turned him over to the authorities of Honduras. This was equivalent to a sentence of death, and on September 12, 1860, a firing squad ended the career of the greatest of the filibusters.[38] In his annual message of December 3, 1860, President Buchanan uttered the pious opinion that "it surely ought to be the prayer of every Christian and patriot that such expeditions may never again receive countenance in our country or depart from our shores."[39]

During the next decade the relations between the United States and the Central American republics were uneventful. From the beginning of his term, Secretary Seward determined to follow a policy of friendly relations with Central America and to gain, if possible, the inhabitants' good will. In his instructions to Mr. Dickinson, the new minister to Nicaragua, dated June 5, 1861, the secretary counseled him to assure the Republic of Nicaragua "that the president will deal with that government justly, fairly, and in the most friendly spirit; that he desires only its welfare and prosperity. . . . Let unpleasant memories of past differences be buried and let Nicaragua be encouraged to rely on the sympathy and support of the United States if she shall at any time come to need them."[40] In 1867 Nicaragua had occasion to remember Seward's

[36]Scroggs, *op. cit.*, p. 361.
[37]*Ibid.*, p. 364.
[38]*Ibid.*, chap. XXIII. General William Walker's account of his expeditions. *The War in Nicaragua* (Mobile and New York, 1860), is a well written, accurate, and remarkably unbiased historical narrative.
[39]Richardson, *op. cit.*, vol. V, p. 649.
[40]*Diplomatic Correspondence of the United States*, 1861, p. 419.

promise and called upon the United States to use her good offices to obtain for Nicaragua a favorable settlement of the problems arising under her treaty of 1860 with Great Britain relative to the Mosquito territory.[41] The American minister at London, Mr. Charles Francis Adams, approached Lord Stanley on the subject, and received his assurance that the British government had no wish to embarrass Nicaragua, and had already given the Nicaraguan minister satisfactory evidence of the justice of its position.[42]

The one subject, however, upon which the United States laid greatest stress, in her diplomatic correspondence with the Central American republics during the second half of the nineteenth century, was the project of a Central American confederation. The Central American states realized very clearly the advantages of a closer union, and, after the complete dissolution of the Union in 1840 numerous efforts were made to reestablish the Confederation. In 1842, 1849 and 1852 attempts were made by the three central states. but no permanent result came of them.[43] In 1871 President Medina, of Honduras, proposed that delegates be sent to El Salvador to consider plans for a union. The presidents of Nicaragua and Guatemala seemed favorable to the idea, and it was suggested that Mr. Riotti, the American representative at León, should preside over the conference. Secretary Fish was favorably impressed with the idea and intimated that the project was a very desirable one in the eyes of the United States.[44] The following year a pact of union was signed in El Salvador by the representatives of Costa Rica, Honduras, Guatemala, and El Salvador, "to preserve and maintain the peace between them, . . . and to guarantee the autonomies of Central America and the integrity of its territory against all aggression and pretensions of foreign powers.[45] But, within two months of the conference, war broke out between Honduras and El Salvador.

On June 24, 1874, Mr. Williamson, the American representative at Guatemala, summed up in a despatch to Secretary Fish what he considered to be the principal obstacles to a union of the Central American states: the debts incurred during the former federal union, local prejudices, a heterogeneous population, no identity of interest, difficulty of intercommunication, and lack of a prominent leader to make a public issue of the question of federation.[46] The new president of Guatemala, J. Rufino Barrios, was destined to become such a leader. In 1876 he proposed that Guatemala annex the four other republics, but such a plan could scarcely be expected to meet the approval of the states to be taken in. The United States continued to manifest interest in the idea of federation, and Secretary Blaine, in a despatch to the American minister at Guatemala City, dated May 7, 1881, declare; that "there is nothing which this government more earnestly desires than the prosperity of these states, and our own experience has taught us that nothing will so surely develop and guarantee

[41] *Ibid.*, 1867, vol. II, pp. 690 ff.
[42] *Ibid.*, 1868, vol. I, p. 151.
[43] See Munro, *op. cit.*, chap. VIII, for a concise account of the various attempts; also P. M. Brown, "American Intervention in Central America," *Journal of Race Development*, vol. IV, p. 409.
[44] *For. Rel. of the U.S.*, 1871, pp. 681-683.
[45] *Ibid.*, 1872, p. 520.
[46] *Ibid.*, 1874, p. 172.

such prosperity as their association under one common government. . . . You cannot impress too strongly upon the government to which you are accredited, or upon the public men with whom you associate the importance which the government of the United States attaches to such a confederation of the states of Central America as will respond to the wants and wishes of their people."[47] He repeated these sentiments to Señor Ubico, the representative of Guatemala at Washington,[48] and to Mr. Morgan, the American minister to Mexico.[49]

During the year 1882 President Barrios visited the United States, and the results showed that the administration undoubtedly encouraged him in his plans for a union. Upon his return he secured the approval of the presidents of El Salvador and Honduras, and to reconcile public opinion he declared, in a public letter to the Liberal Party, that although he was doing all in his power to bring about a federation of the five republics, he had no intention of accepting the presidency.[50] Unfortunately, the popular opposition was too great, and the scheduled convention did not take place. Barrios now determined to resort to force. On February 28, 1885, he issued a decree declaring himself the supreme military chief of the "Central American Union" and proclaiming the union of Central American into one sole republic.[51] Honduras accepted immediately, but the president of El Salvador, who had previously pledged his support, vacillated and played for time; Nicaragua and Costa Rica rejected the plan unconditionally, and on March 9, together with El Salvador, they appealed to the United States and Mexico to intervene against the projects of Barrios.[52] Mr. Hall, the American representative at Guatemala, was authorized to use his good offices, but the day after he received the despatch news came that President Barrios had been killed in action.[53] With the death of Barrios, all further immediate efforts to form a confederation under the aegis of Guatemala were dropped. In spite of the hostilities engendered by the war, a treaty was signed on September 12, 1885, between Guatemala, Honduras, and El Salvador, providing among other things for extradition, commercial reciprocity, and a postal telegraph and monetary union; and Nicaragua and Costa Rica were invited to become parties to it.[54]

Although the invitation was not accepted, negotiations continued, and in January, 1887, the delgates of El Salvador, Honduras, Nicaragua, and Costa Rica met in Guatemala, and after a month's deliberation, on February 16, 1887, signed a treaty of peace and friendship. This document opened with the statement that it is the mutual desire of the five states to draw closer and "establish appropriate bases for the near advent of the longed-for political union of Central America.[55] The treaty was ratified by all the states except Nicaragua, who seemed to fear forcible action on the part of Guatemala and raised the issue in Washington. Secretary Bayard hastened to notify Mr. Hall, the American minister at Guatemala, that the United States strongly disapproved of

[47] *Ibid.*, 1881, p. 102.
[48] *Ibid.*, p. 599.
[49] *Ibid.*, p.816.
[50] *Ibid.*, 1883, p. 49.
[51] *Ibid.*, 1885, p. 75.
[52] Bancroft, *op. cit.*, vol. III, p. 448.
[53] *For. Rel. of the U.S.*, 1885, pp. 98-99.
[54] *Ibid.*, 1887, p. 85.
[55] *Ibid.*, p. *101.*

a coercive union of the Central American republics, and suggested that he take an early and discreet occasion to bring these views to the attention of the government of Guatemala.[56] Guatemala denied the allegations and continued under the treaty of 1887 to work for a union. But a revolution in El Salvador in 1890 brought into power the opponents of the union, thus forcing another postponement. Selfish domestic rivalries continued to nullify all efforts toward federation.

We have already indicated the interest of Mr. Baline in Latin-American affairs as shown by his part in the first Pan American Conference and by his efforts to further a Central American federation. Unfortunately for his popularity in Central America, he was forced to decide a number of close cases in international law against the Central American republics. Guatemala was the chief offender. In July 1890, she had seized arms on board the Pacific Mail Steamship Company's steamer *Colima,* contrary to an agreement made with the American minister,[57] and in the following month Guatemalan officials killed General Barrundia, traveling on the same company's steamer *Acapulco.*[58] For his conduct in the second case, the American minister's acts were disavowed and he was recalled forthwith. With war ever threatening between Guatemala and El Salvador, Secretary Blaine's instructions to the new minister were most emphatic that he should make every effort to avert it. "Not only are the good offices of the United States equally ready towards averting possible causes of difference, but it is deemed the friendly duty of this government to do all that it can to prevent strife among its neighbors."[59]

In 1894 the United States had occasion to enter into negotiations with the British government in regard to the Mosquito kingdom. By the treaty of Managua, dating from 1860, Nicaragua's sovereignty over this region was recognized, but the Indians were given certain rights of self-government. In the course of time many foreigners, principally English and Americans, engaged in business enterprises in Bluefields, the capital, and the region became very prosperous. On November 2, 1893, a Nicaraguan commissioner, with a staff of officers, arrived at Bluefields and established heavy customs upon the export of bananas. When the natives protested, the Nicaraguan officials obtained reinforcements, and finally took over the whole management of the country and proclaimed martial law. On February 25, 1894, H.M.S. *Cleopatra* anchored at Bluefields, and her commander insisted that martial law be raised and the Mosquito flag again hoisted.[60] The United states was interested, and the American minister to Nicaragua was directed to visit Bluefields and make a report. He found (1) that the present provisional government imposed by the Nicaraguan authorities, cooperating with the British consul, was a bold usurpation; (2) that the Mosquito government existing before Nicaragua became an independent state was overthrown by armed violence on February 11, 1894; (3) that fully 90 percent of all the wealth, enterprise, and commerce of the

[56] *Ibid.*, 1888, Part I, p. 131.
[57] J. B. Moore, *Digest of Int. Law,* vol. VII, p. 659.
[58] *Ibid.*, vol. II, p. 871.
[59] *For. Rel. of the U.S.*, 1891, p. 56.
[60] *Ibid.*, 1894, Appendix I, pp. 261-264, 276-284.

Indian reserve was American; and (4) that the extension of Spanish revolutionary rule over this reserve would inevitably extinguish the whole of this present business prosperity.[61]

Although the American minister's sympathies seemed to favor the Mosquito government, the United States from the beginning took a firm stand in behalf of full Nicaraguan sovereignty. Writing to Mr. Bayard, the American minister at London, April 30, 1894, Secretary Gresham declared he was "unable to see that this joint assumption of authority by British and Nicaraguan agents was incompatible with the treaty of Managua."[62] Again on July 19 he maintained that the sovereignty of Nicaragua over the whole of the national domain was unquestionable. "No matter how conspicuous the American or other alien interests which have grown up under the fiction of Indian self-government, neither the United States nor Great Britain can fairly sanction or uphold this colorable abuse of the sovereignty of Nicaragua."[63] Further foreign intervention was eliminated by the Mosquito tribes voluntarily signing a convention, November 20, 1894, permitting the absolute incorporation of their territory into the republic of Nicaragua.[64]

The century came to an end with one more attempt on the part of the Central American states to form a federation. Nicaragua, Honduras, and El Salvador signed the treaty of Amapala, June 20, 1895, by which they were to form a single political organization for the exercise of their external sovereignty; and the agreement was ratified in September of the following year. The federal powers were to be exercised by a diet consisting of one member and one substitute elected by each congress for a term of three years. Costa Rica and Guatemala were to be invited to join.[65] President Cleveland extended recognition to this Greater Republic of Central America, discerning in the articles of association "a step towards a closer union of Central American States in the interest of their common defense and general welfare," and he welcomed it as "the precursor of other steps to be taken in the same direction, and which it is hoped may eventally result in the consolidation of all the states of Central America as one nation for all the purposes of their foreign relations and intercourse."[66]

Costa Rica and Guatemala did not join, but in 1898 the other three states adopted a permanent constitution establishing a federal republic and made plans for the election of a president. Before the successful culmination of these plans, however, a revolution in El Salvador detached this state from the group, and the other two thereupon resumed their independence of action.[67] The close of the century found the five independent republics at peace with the world and with each other, but apparently as far as ever from the union that had so long been the goal of their endeavors.

[61] *Ibid.*, p. 287.
[62] *Ibid.*, p. 271.
[63] *Ibid.*, p. 311.
[64] *Ibid.*, p. 361. The relations with Great Britain were finally settled by a new treaty signed at Managua April 19, 1905, in which Great Britain recognized the sovereignty of Nicaragua over the Mosquito Reserve. For text see *For. Rel. of the U. S.*, 1905, p. 703.
[65] For full text see *ibid.*, 1896, p. 390.
[66] *Ibid.*, p. 370.
[67] *Ibid.*, 1898, pp. 173-178; also Richardson, *op. cit.*, vol. X, 178, and W. F. Slade, "Federation of Central America," *Journal of Race of Development*, vol. VIII (Oct., 1917), pp. 210-218.

SUPPLEMENTARY READINGS

Bancroft, Hubert H. *History of Central America.* 3 vols. San Francisco, 1882-1887.
Bumgartner, Louis E. *José del Valle of Central America.* Durham, 1963.
Chamberlain, Robert. *Francisco Morazán, Champion of the Central American Federation.* Coral Gables, 1950.
Fortier, Alcee and John R. Ficklen. *Central America and Mexico.* Philadelphia, 1907.
Grubb, Kenneth. *Religion in Central America.* London, 1937.
Holleran, Mary P. *Church and State in Guatemala.* New York, 1949.
Ireland, Gordon. *Boundaries, Possessions and Conflicts in Central and North American and the Caribbean.* Cambridge, Mass., 1941.
Jones, C. Lloyd. *Costa Rica and Civilization in the Caribbean.* Madison, 1935.
---*Guatemala, Past and Present.* Minneapolis, 1940.
Karnes, Thomas L. *The Failure of Union: Central America, 1824-1960.* Chapel Hill, 1961.
Munro, Dana G. *The Five Republics of Central America.* New York, 1918.
Parker, Franklin D. *The Central American Republics.* New York, 1964.
---*José Cecilio del Valle and the Establishment of the Central American Federation.* Tegucigalpa, 1954.
Rodríguez, Mario. *A Palmerstonian Diplomat in Central America: Frederick Chatfield, Esq.* Tucson, 1964.
---*Central America.* Englewood Cliffs, 1965.
Shafer, Robert J. *The Economic Societies in the Spanish World, 1763-1821.* Syracuse, 1958.
Scroggs, W. O. *Filibusters and Financiers.* New York, 1916.
Stephens, John L. *Incidents of Travel in Central America, Chiapas and Yucatan.* 2 vols. New York, 1841.
Waddell, David A. G. *British Honduras, A Historical and Contemporary Survey.* London, 1961.
Walker, William. *The War In Nicaragua.* New York, 1860.
Williams, Mary W. *Anglo-American Isthmian Diplomacy, 1815-1915.* Washington, 1916.

15

Recent Relations with Central America

The events outlined in the preceding chapter indicate a fairly consistent policy of friendly cooperation on the part of the United States with the Central American republics, particularly in their efforts to obtain a successful federal system of government. American citizens had been guilty of various audacious attempts to obtain special privileges and autocratic powers in these republics, but the American government never failed to frown upon them, even when a different policy might have redounded to the country's advantage. The war with Spain, however, was destined to place the United States in a new position with regard to the Caribbean region, and a greater interest in Central America and its problems was to be expected.

The first intervention of the United States in the affairs of the Central American republics, except in the collection of a few private claims, occurred in 1906. In July of that year a war broke out between Guatemala and El Salvador, and Honduras, as usual, was drawn in, this time on the side of El Salvador. The American minister used every effort to avert it; when he failed, President Theodore Roosevelt, after securing the cooperation of President Díaz of Mexico, stepped in. He sent identical notes to presidents Escalón of El Salvador and Cabrera of Guatemala urging the immediate cessation of hostilities, and offering the deck of the American warship *Marblehead* as a neutral place where terms of an agreement might be drawn up.[1] Both sides accepted, and on board the *Marblehead*, on July 20, 1906, a convention of peace was signed which provided that within two months a general treaty of peace, amity, and navigation should be arranged for by a meeting, in Costa Rica, of the representatives of the three republics. The convention also provided that new differences should be submitted to arbitration, with the presidents of the United States and Mexico serving as arbitrators.[2] In appreciation of this just and peaceful settlement of the

[1] *For. Rel. of the U.S.*, 1906, Part I, p. 837.
[2] *Ibid.*, Part I, p. 851.

controversy, the Third Pan American Conference, which was then in session at Rio de Janeiro, passed a resolution of gratification over the successful mediation by the two presidents.[3]

The meeting in Costa Rica was participated in by representatives of all the Central American republics except Nicaragua. The results were a general treaty of peace and amity, arbitration, extradition, commerce, etc., between the four republics; also two conventions, the first establishing an International Central American Bureau in the city of Guatemala, and the second establishing a Central American Pedagogical Institute in San José, Costa Rica. Copies of these were sent to the United States by Costa Rica "as an act of courtesy to the government of the United States for an equal interest in the welfare of the Central American states."[4] Although Nicaragua had not sent representatives to the conference, copies of the treaty and conventions were sent to her, inviting her adherence. Nicaragua replied that she did not care to adhere, inasmuch as the treaty seemed to ratify the treaty of *Marblehead*, whose terms she was not willing to accept. The inference was that President Zelaya of Nicaragua, known to be hostile to the United States, was unwilling to accept the president of the United States as an arbitrator in case a dispute should arise to which Nicaragua was a party.

Hardly had Nicaragua thus announced her attitude before a situation arose which demanded arbitration. Honduras claimed that certain revolutionists and disturbers of the peace had fled across the border into Nicaragua, and that in order to suppress them, its troops had been forced to cross the line. Nicaragua immediately got its forces in readiness and demanded indemnity for the infringement of its territorial rights. An attempt to settle this dispute by arbitration having failed, President Roosevelt wrote, on February 11, 1907, to the presidents of both states, expressing the strong hope that some means might be found to settle the dispute without resorting to war.[5] Receiving favorable responses from both, President Roosevelt thereupon offered his assistance. President Díaz acted in a similar fashion, but Nicaragua insisted upon reparations as a basis of arbitration, and, when Honduras refused, began operations. Hostilities did not last long, but El Salvador was drawn in against Nicaragua, and to avoid complications the United States once more offered its good offices. With the United States minister present during discussions, a treaty of peace was signed at Amapala, April 23, 1907, between Nicaragua and El Salvador. The third article of this treaty provided that "any difference that may arise in the future between El Salvador and Nicaragua that might alter their good relations shall be adjusted by means of the obligatory arbitration of the presidents of the United States and Mexico, conjointly."[6]

When trouble again threatened, in August of the same year, President Roosevelt, cooperating with President Díaz of Mexico, sent identic notes to each of the five Central American republics – tendering the good offices of the United States to bring about a peace conference of the representatives of the several states.[7] This overture received a very favorable response, and on

[3] *Ibid.*, p. 852.
[4] *Ibid.*, p. 856.
[5] *Ibid.*, 1907, Part II, p. 616.
[6] *Ibid.*, p. 633.
[7] *Ibid.*, p. 638.

September 17, 1907, the representatives of the five Central American republics at Washington signed a protocol agreeing to a joint conference at Washington, to meet November 1; in the meantime, they agreed to maintain peace and good relations among one another, and not to commit, or allow to be committed, any act that might disturb their mutual tranquility.[8] In accordance with this protocol, President Roosevelt invited the executives of the five republics to name commissioners, to meet in Washington to discuss measures to be adopted, to adjust any existing differences, and to conclude a treaty determining the general relations of their countries.[9]

The invitations were accepted by all of the republics, and they proposed that both the United States and Mexico should be represented. The conference, which met in Washington November 14 in the Bureau of American Republics, began very auspiciously. In his opening address, Secretary Root made a strong plea to the delegates to banish fraternal strife from their land, and to find specific and practical measures whereby they might keep and enforce the agreements entered into.[10] The delegates showed their spirit of cooperation and friendly feeling by unanimous declarations at the first regular session that each of their countries had no claims against any of the others. But when, at the second session, Honduras proposed a scheme for the union of the five republics, a difference of opinion was quickly noticeable. Nicaragua was favorable to the proposal, but Guatemala opposed it, and Costa Rica objected to the consideration of the subject. The matter was finally referred to a committee of five, including one representative from each state.

The reports of this committee, which were read at the next session, showed strong divergence of views. The majority report, signed by representatives of Guatemala, El Salvador, and Costa Rica, conceded that the political union of Central America was the "greatest and noblest aspiration of patrotism," yet gave it as their opinion that conditions were not suitable for attempting it at this time. They said that it was more opportune in the present conference to consider measures tending towards preparing in a stable manner for this union, such as better means of communication, a coasting ship commerce, a unification of customs and tax laws, and the encouragement of further Central American conferences. As definite steps, they supported the idea of a Central American bureau, the Pedagogical Institute, and the creation of a permanent Central American court of international justice. The minority report, signed by Honduras and Nicaragua, declared that union alone would insure stable and efficient peace and order in Central America, and insisted that this was the time to establish it.[11]

Fearing lest this question might interfere with the constructive work of the conference, Mr. Buchanan, the American representative, supported by the Mexican representative, proposed that the consideration of both reports be postponed, and that the conference proceed to prepare projects for the several proposed conventions. This course was adopted, and the conference proceeded

[8] *Ibid.*, p. 644.
[9] *Ibid.*, p. 648.
[10] Report of W. I. Buchanan, *Ibid.*, p. 687.
[11] For text see *Ibid.*, pp. 671-673.

to frame and adopt a number of important conventions regarding peace and friendship, the establishment of a Central American Court of Justice and a Central American bureau, extradition, railway communications, and a convention concerning future Central American conferences.[12]

Undoubtedly the two most important achievements were the general treaty of peace and amity, which was to remain in force ten years, and the convention establishing the Central American Court of Justice. Among other things, the treaty provided that any disposition or measure tending to alter the constitutional organization of any of the republics should be deemed a menance to the peace of all; the territory of Honduras was made neutral in conflicts arising between the other four republics, as long as she remained neutral; political refugees and disturbers were not allowed to reside near the frontiers of any of the five republics; and all future disputes were to be decided by the Central American Court of Justice. This tribunal was to consist of five judges, one from each state, appointed for five-year terms. It was given obligatory jurisdiction over all cases arising between two or more of the states, and its decision, given in writing by a majority opinion, was to be final. The court could fix the *status quo* pending the decisions in a case, and the signatory republics bound themselves to carry out faithfully the court's orders.[13]

On the whole, the representatives had no reason to feel ashamed of the results of their labors. Although the results were only on paper, and all looked towards the future, the ideas adopted were specific and practical, and they were grounded upon a solid basis of cooperation which alone was bound to be beneficial to all. The Court of Justice naturally received the most attention. One eminent American publicist aptly remarked: "To the powers of Europe, to the great powers of the world who struggled with partial success, for four months at the Hague, to establish a court of arbitral justice, the young republics of Central America may recall the scriptural phrase, 'A little child shall lead them.' "[14]

Unfortunately for the future peace of Central America, the court was not to have the success that its advocates had hoped, and the United States, as will be seen later, bulks large in the ultimate cause of its failure. The court was installed with much acclaim in Cartago, Costa Rica, May 25, 1908, representatives from the United States and Mexico participating in the inauguration ceremonies. In his address, Mr. Buchanan, the American representative, while taking a tone of optimism, very clearly pointed out that if success were to crown the effort there must be behind this court and its decisions an elevated patriotic public conscience in each of the republics that would lift and maintain the court in every way above the plane of political purposes or necessities. In conclusion, he read a telegram from Secretary Root saying that he had been authorized by Mr. Andrew Carnegie to offer, as a mark of his good wishes for the peace and progress of Central America, the sum of $100,000 for the construction in the city of Cartago of a temple of peace for the exclusive use of the court.[15]

Another result of the Washington conference was the establishment at

[12] *Ibid.*, p. 673.
[13] For text of the general treaty and the convention see *Ibid.*, pp. 692–711.
[14] J. B. Scott, *Amer. Jour. of Int. Law*, Vol. II (Jan., 1908), p. 143.
[15] *For. Rel. of the U.S.*, 1908; report of W. I. Buchanan, pp. 217–247.

Guatemala City, in the following year, of the Central American International Bureau. The principal objects of this bureau were: (1) the reorganization of the Central American Union; (2) the introduction of an up-to-date educational system; (3) the development and extension of domestic and foreign trade; (4) the increase and development of agriculture and industries; (5) the reform of legal institutions; and (6) reforms in customs, the monetary system, credit, weights and measures, and sanitation. This bureau was to consist of one delegate from each republic, who should reside in Guatemala. All future Central American conferences were to be held under its auspices, and it was to publish a bulletin, *Centro América*, as a means of keeping the republics in touch with one another's progress.[16] The bureau was in many respects modeled upon the Pan American Union, and, considering its limited resources and facilities, it has been eminently successful. Mr. Dana G. Munro goes so far as to assert that of the institutions provided by the Washington conference this is the only one that had thus far fully justified its creation.[17]

While these various means were being provided to strengthen friendly relations between the Central American republics, President Zelaya of Nicaragua continued to show his unwillingness to maintain friendly relations with either his neighbors or the United States. President Davila of Honduras was under his thumb; El Salvador was constantly threatened by his machinations. A British traveler who had spent much time in Central America declared that "all travelers, foreigners, and natives alike, who happened to be in Central America at this time, were well aware of the provocative part which President Santos Zelaya was playing; for many years he had been acting as the evil genius of this republic, and his misgovernment and brutalities to his own people met with general condemnation."[18] The ultimate result was a revolution, which started at Bluefields in October 1909. At first the United States remained neutral, but when, in November, Zelaya ordered two Americans to be shot because they were serving as officers in the revolutionary army, the United States broke off diplomatic relations with his government. In his note to the Nicaraguan chargé Secretary Knox strongly denounced the Zelayan régime, declaring that it had almost continually kept Central America in tension or turmoil, that it had repeatedly and flagrantly violated the Washington conventions, and that it was a blot on the history of the country.[19]

Partly as a result of the American attitude, Zelaya was unable to put down the revolution, and he finally resigned, left the country, and took refuge on a Mexican gunboat. The United States maintained an ostensible neutrality between President Madriz, who received the backing of the Zelaya faction, and Estrada, who headed the revolutionists; but when the Madriz forces wanted to bombard Bluefields, which they declared to be the vital base of the Conservatives, United States cruisers refused to permit it. The reason given was that foreigners would be the chief sufferers. But the result was a decided

[16] For an account of the bureau's activities see W. F. Slade, "Federation of Central America," *Jour. of Race Development*, vol. VIII (Oct., 1917), pp. 234–241.

[17] D. G. Munro, *The Five Republics of Central America* (New York, 1918), p. 225.

[18] P. F. Martin, *Salvador of the Twentieth Century* (London, 1911), p. 67.

[19] *For. Rel. of the U.S.*, 1909, p. 455; see also President Taft's Message of December 7, 1909, *Ibid.*, p. xvii.

strengthening of Estrada's position. President Madriz protested both to President Taft and to President Díaz of Mexico; but Secretary Knox replied that the United States had taken only the customary step of prohibiting bombardment or fighting by either faction within the unfortified and ungarrisoned city of Bluefields, thus protecting the preponderant American and other foreign interests.[20] In August 1910, Madriz gave up the struggle, and by the middle of the following month Señor Don Juan J. Estrada claimed to be in peaceful and unrestricted possesion of the entire republic.

In its first note to the United States the new government promised that a general election would be held within a year and that the national finances would be rehabilitated, and to this end the Department of State at Washington was asked to obtain a loan secured by a percentage of the customs revenues, and to send a commission to Nicaragua to arrange the necessary formalities.[21] The United States thereupon appointed Mr. T. C. Dawson as special agent to Nicaragua to deal with the provisional government on these lines. The negotiations were successful, and a number of agreements were entered into which provided that Estrada should remain at the head of the government for two years, after which a free election should be held; that a claims commission, consisting of one Nicaraguan and one American and an umpire named by the American State Department, should be appointed to consider all unliquidated claims against the government; and that a loan secured by the customs receipts be raised in the United States.[22]

A special loan convention embodied the latter proposal, and President Taft urgently recommended it as contributing to the peace of Central America; but it failed to receive the assent of the United States Senate.[23] A loan was none the less negotiated with American bankers, and an American, Mr. Clifford D. Ham, was put in charge of the collection of the customs to guarantee its repayment.[24] The claims commission was also appointed, and it began to work on May 1, 1911. The commission worked steadily for more than three years and passed upon almost 8000 claims, scaling them down from a total of $13,750,000 to about $1,750,000.[25] The political part of the program, however, was a complete failure, and in less than six months President Estrada had resigned in favor of Vice-president Díaz. The latter soon found out that a reform program in Nicaragua could not be put through unless the president received outside support, and in a confidential letter to the American *chargé d'affaires,* written December 21, 1911, he confessed the need of more direct and efficient assistance from the United States. He proposed that the Nicaraguan constitution be so amended as to permit the United States to intervene in Nicaraguan internal affairs in order to maintain peace and the existence of a lawful government.[26]

Practical evidences of this need were soon given. On July 29, 1912, General Luis Meña, the minister of war, engineered a *coup d'état* against the

[20] *For. Rel. of the U.S.*, 1910, pp. 751–753.
[21] *Ibid.*, p. 762.
[22] *Ibid.*, 1911, pp. 625 ff.
[23] *Ibid.*, 1912, pp. 1071–1078.
[24] For an excellent summary of the events of this period, embracing particularly the financial arrangements, see Munro, *op. cit.*, chap. XI.
[25] Otto Schoenrich, *Amer. Jour. of Int. Law,* vol. IX (Oct., 1915), p. 958.
[26] *For. Rel. of the U.S.*, 1911, p. 670.

Díaz government with a view to seizing the executive power. Failing in his original attempt, he left the capital and organized an open rebellion against the government. Fearing that he would be unable to protect the property and lives of American citizens, President Díaz asked that "the United States guarantee with its forces security for the property of American citizens in Nicaragua and that it extend its protection to all the inhabitants of the republic."[27] President Taft thereupon ordered a small detachment of marines to be landed at Managua.

When the president of El Salvador expressed the fear of serious complications if American troops should enter Nicaraguan territory, the State Department, in a note to the American minister at Managua, dated September 4, 1912, laid down the American policy in no uncertain language. "The policy of the government of the United States in the present Nicaraguan disturbances is to take the necessary measures for an adequate legation guard, at Managua, to keep open communications, and to protect American life and property. In discountenancing Zelaya, whose régime of barbarity and corruption was ended by the Nicaraguan nation after a bloody war, the government of the United States opposed not only the individual but the system, and this government could not countenance any movement to restore the same destructive régime. The government of the United States will, therefore, discountenance any revival of Zelayaism, and will lend its strong moral support to the cause of legally constituted good government for the benefit of the people of Nicaragua, whom it has long sought to aid in their just aspiration towards peace and prosperity under constitutional and orderly government."[28]

The revolution that followed proved more serious than was at first expected, owing to the uprising of the Liberals of León, who threw in their fortunes with General Meña. The United States, however, could not afford to allow the Zelaya faction to triumph, and American marines were landed at Corinto, Granada, and León; they even went so far as to storm the position overlooking Masaya. The revolutionists could hardly hope to hold out against the government thus aided by the United States, and with the seizure of General Meña, and his internment in Panama, the uprising collapsed. President Díaz was reelected for a four-year term to begin January 1, 1913. Although order had been completely restored and the Díaz government was in full control, a detachment of one hundred American marines was left in Managua, at the urgent request of the American minister, to act as a stabilizing influence.

On account of the U.S. Senate's failure to approve the loan convention, the financial situation was in even worse shape than before. American bankers had made a loan of $1,500,000 for the establishment of the National Bank of Nicaragua, the reform of the currency, and the establishment of the claims commission; but the hoped-for loan of $15,000,000 was dependent upon the ratification of the convention.[29] As a solution of the difficulty, Nicaragua proposed, in December 1912, to sell to the United States, for the sum of $3,000,000, the sole right to construct a canal through her territory. As an

[27] *Ibid.*, 1912, p. 1032.
[28] *Ibid.*, p. 1043.
[29] See the memorandum on Nicaraguan finances prepared for the Secretary of State, *Ibid.*, 1913, p. 1040.

additional inducement to the United States, the right was given to construct a naval base in the Gulf of Fonseca and one on Corn Island.[30] When the Wilson administration came in, this treaty was modified so as to prevent Nicaragua from declaring war, from making treaties with foreign governments affecting her independence or territorial integrity, and from contracting public debts beyond her ability to pay; and it was also changed so as to grant the United States the right to intervene in her affairs, with a view to maintaining her independence and protecting life and property in her domain. The Senate refused to accept this full protectorate, and a new treaty without these provisions was signed on August 5, 1914. Again the Senate raised objections; but these were met, and the treaty was ratified on February 18, 1916, and proclaimed on June 24, following.

The Bryan-Chamorro treaty as finally approved makes the following provisions:

> (1) the government of Nicaragua grants in perpetuity to the government of the United States, free from taxation, the exclusive rights necessary for the construction and operation of a canal by way of the San Juan River and the Great Lake or by any other route; (2) to protect American interests, Nicaragua leases to the United States for ninety-nine years the Great Corn and Little Corn Islands, and also grants the right to establish a naval base on such Nicaraguan territory bordering upon the Gulf of Fonseca as the United States shall select, with a right to obtain renewal of the grants for a similar period; (3) in consideration of these stipulations the United States agrees to pay the sum of $3,000,000 in gold; and (4) the following proviso is inserted: "Whereas Costa Rica, Salvador, and Honduras have protested against the ratification of the said convention, in the fear or belief that said convention might in some respect impair existing rights of said States; therefore it is declared by the Senate that in advising and consenting to the ratification of the said convention as amended, such advice and consent are given with the understanding, to be expressed as a part of the instrument of ratification, that nothing in said convention is intended to affect existing right of any of the said named States.[31]

Despite this proviso, Costa Rica and El Salvador continued to protest, and finally they attempted through the Central American Court of Justice to enjoin Nicaragua from carrying out the provisions of the treaty. Costa Rica based her case upon a treaty of limits between herself and Nicaragua, signed April 15, 1858, giving Costa Rica rights in the San Juan River and agreeing that she should be consulted before any contract should be entered into for the construction of a canal. This treaty had been held valid by President Cleveland in 1888. Costa Rica therefore argued that the present treaty would infringe her rights under the treaty of 1858.[32] El Salvador claimed an equal right with Honduras and Nicaragua in the Bay of Fonseca, and therefore asserted that Nicaragua had no right to alienate territory clearly jeopardizing and menacing her interests without her consent. Furthermore, the Washington conference of 1907 had proclaimed the neutrality of Honduras, including its rights in the Gulf of Fonseca; therefore,

[30] *Ibid.*, p. 1021. Mr. Weitsel, the American minister to Nicaragua, 1912–1913, and one of the signatories of this agreement, in a memorandum on the convention gives an excellent résumé of American relations with Nicaragua; see *Sen. Doc. No 334*, 64th Cong., 1st Sess.
[31] *U.S. Stat. at Large*, vol. XXXIX, Part II, p. 1661.
[32] *Complaint of the Republic of Costa Rica before the Central American Court of Justice* (Washington, D.C., 1916), Appendices G and N.

it was urged, this projected agreement for the establishment of a naval base there "is an attempt to violate, in a manner both flagrant and evident, the principle of the neutrality of Honduras, and to throw overboard the legal system instituted by the conference of Washington."[33]

Nicaragua denied the court's jurisdiction in the cases of both Costa Rica and El Salvador, but the court declared itself competent to take cognizance of them. The decision in the case of Costa Rica was handed down on September 30, 1916; that in the case of El Salvador on March 2, 1917. In both cases the court held that Nicaragua, by signing the Bryan-Chamorro treaty of August 5, 1914, had violated the rights of Costa Rica and El Salvador, that she had also violated the treaty of peace and amity signed at Washington in 1907, and that she was under obligation to reestablish and maintain the legal status existing prior to the Bryan-Chamorro treaty. In both cases, however, the court admitted that it was without competence to declare the Bryan-Chamorro treaty null and void, since its jurisdictional power extended only to establishing the legal relations among the "high parties litigant."[34]

The refusal of the United States, as well as of Nicaragua, to accept the decision of the court had a very bad effect on the standing of the United States with the other Central American countries. Since the United States refused to back the very agency for settling disputes to whose establishment it had given such hearty encouragement, simply because the decision interfered with her own selfish interests, how could the Latin American republics feel much confidence in American expressions of friendly cooperation? When the question of renewing the convention to continue the court's existence came up, nothing could be done, owing to Nicaragua's withdrawl; and on March 17, 1918, the court was formally dissolved.[35] The Bryan-Chamorro treaty thus indirectly caused the downfall of the Central American court, thereby weakening the trend towards a closer relationship among these states which it has been the avowed policy of the United States to foster. In fact, when another attempt at federation was made in 1920, and a "treaty of union" was signed January 19, 1921, by representatives of Guatemala, Honduras, El Salvador, and Costa Rica, Nicaragua refused to enter because of her fears lest her rights under the Bryan-Chamorro treaty should be in some way jeopardized.[36]

It will be remembered that one of the principal results of the conference of Washington in 1907 was to establish the perpetual neutrality of Honduras. But Honduras needed a strong government to maintain its neutrality, and before this could even be hoped for, its disorganized financial condition had to be remedied. By reason of a series of loans negotiated in Europe between 1867 and 1870, to the amount of about $27 million, upon which virtually no interest had

33 *Complaint of the Republic of El Salvador before the Central American Court of Justice* (Washington, D.C., 1917), Appendices A and I.
34 *Opinion and Decision of the Court–the Republic of El Salvador Against the Republic of Nicaragua* (Washington, D.C., 1917).
35 For a detailed account of the Court's history see the scholarly article by Manley O. Hudson, "The Central American Court of Justice," *Amer. Jour. of Int. Law,* vol. XXVI, no. 4 (Oct., 1932).
36 See the article by Dr. Ruben Rivera on the Central American Union in the *New York Times,* Apr. 24, 1921.

been paid, the state found itself indebted to the stupendous sum of $125 million. In 1908 the Council of Foreign Bondholders proposed a refunding scheme whose terms Honduras considered too onerous; accordingly, she turned towards the United States. Secretary Knox immediately sought to interest American bankers in a project for refunding the debt, and J. P. Morgan and Co. agreed to attempt it. They found that the debt could be adjusted for about 15 per cent of its face value without interest; therefore by the issuance of $10 million in 5 percent bonds Honduras could wipe out both her external and internal debt, pay all claims, acquire and extend the interoceanic railway, and still have a balance of $2½ million for future internal development. Before making such a loan, however, the bankers demanded that the United States sign a loan convention with Honduras affording the necessary security to the bondholders.

Such a convention was signed January 10, 1911, by Secretary Knox for the United States and Señor Juan Paredes for Honduras. The two governments promised to consult in case of any difficulties, with a view to the faithful execution of the provisions of the contract. The loan was to be secured upon the Honduran customs, which were not to be changed during the existence of the contract without consent of the United States. The government of Honduras further agreed to appoint a collector of customs, who should be approved by the president of the United States, from a list of names submitted by the fiscal agent of the loan.[37] President Taft earnestly urged the Senate to assent to the convention, and Secretary Knox, in a statement before the foreign relations committee, May 24, 1911, laid down the basis of his policy as follows. "Shall the government of the United States make American capital an instrumentality to secure financial stability, and hence prosperity and peace, to the more backward republics in the neighborhood of the Panama Canal? And in order to give that measure of security which alone would induce capital to be such an instrumentality without imposing too great a burden upon the countries concerned, shall this government assume towards the customs collections a relationship only great enough for this purpose—a relationship, however, the moral effect and potentialities of which result in preventing the customs revenues of such republic from being seized as the means of carrying on devastating and unprincipled revolutions?"[38]

Despite this persuasive presentation of the advantages of dollar diplomacy, the Senate failed to agree to the convention, and the bankers refused to proceed farther. The Honduran congress also refused to ratify the convention, principally upon constitutional grounds. Although the original loan was unquestionably tainted with fraud, the lien existed, and Honduras could not establish herself upon a sound financial basis until some readjustment was made. The Morgan proposal, which was approved by a neutral expert whose services were sought by the State Department, seemed to offer an excellent opportunity for Honduras to make a new start. Yet the United States showed wisdom in

[37] Text in *For. Rel. of the U.S.*, 1912, p. 560.
[38] *Ibid.*, p. 589.

refusing to urge the convention when the Honduran congress evidenced its overwhelming opposition.[39]

In its other diplomatic relations with Honduras during the year 1911 the United States was more successful. In January of that year General Manuel Bonilla staged a revolution against President Davila, and the latter, no longer having the assistance of Zelaya, asked the United States to intervene as arbitrator, and promised to deliver the presidency to any third party named or approved by the United States.[40] President Taft accepted the proposal and named Mr. T. C. Dawson as the special commissioner of the United States. A peace conference was held in February at Puerto Cortes on board the U. S. S. *Tacoma,* with Mr. Dawson presiding. A provisional government was organized, and Dr. Francisco Bertrand was named as provisional president. Peace was restored, and President Taft received a vote of thanks "for his friendly mediation towards the reestablishment of peace in the Republic of Honduras."[41]

The entrance of the United States into the European war stirred the Central American republics to enthusiastic manifestations of cooperation. All of these states, including Panama declared war on Germany except El Salvador; and she went so far as to declare a benevolent neutrality which permitted the use of her territorial ports and waters by the warships of the United States and the Allies. In the declarations of war the United States was usually expressly mentioned: "Guatemala assumes the same belligerent attitude as the United States"; "Nicaragua makes common cause with the United States"; even El Salvador, in announcing its attitude, declared that "El Salvador as an American nation could not fail to recognize, in the conflict between the United States and Germany, the solidarity which binds it to the great Republic of the North,...and that its condition of neutrality could not lead it to the point of considering the United States as a belligerent subject to the ordinary rules of international law."[42] During the peace negotiations the Central American republics boldly accepted the idealistic principles of President Wilson and entered the League of Nations, fearing not at all lest they be thereby drawn into the broils of Europe.

The subsequent relations between the United States and the Central American states (with the exception of Guatemala and Nicaragua, which we shall consider later) have been comparatively uneventful. When, in 1917, Costa Rica broke her long and excellent record of domestic peace by a bloodless *coup d'état,* the United States refused to recognize the Tinoco government which engineered the movement. In spite of this attitude towards President Tinoco, the United States upon several occasions protested vigorously against the raising of armed forces in Nicaragua and Honduras against this *de facto* government of

[39]An agreement with the British bondholders was approved by the Congress of Honduras in 1926 and regular service on the debt has been maintained since January, 1927.

[40] *For. Rel. of the U.S.,* 1911, p. 297.

[41] *Ibid.,* p. 304.

[42] John Barrett, *Latin America and the War* (Washington, D.C., 1919), p. 28; see also P. A. Martin, *Latin America and the War,* League of Nations, vol. II, no. 4, and P. A. Martin, *Latin America and the War* (Baltimore, 1925).

Costa Rica.[43] Tinoco fell from power in 1919, and the succeeding Acosta government, being legally organized, was duly recognized by the United Stated on August 2, 1920.

In 1920 the Estrada Cabrera *régime* in Guatemala, which had originally been put in power in 1898, was overthrown. Cabrera stoutly opposed the plan of a federal union which had once again become popular, and arrested many of its chief Guatemalan adherents. The Unionists seized Guatemala City, and, in spite of a serious bombardment by government forces in which hundreds were killed, they held out and finally deposed Cabrera. The United States asked that the deposed president's life be spared, and upon receiving satisfactory assurances recognized the new Herrera government. This administration worked strenuously in behalf of the new federation, and on September 15, 1921, the constitution of the Federation of Central America was signed by representatives of Guatemala, Honduras, and El Salvador. On October 10, the separate governments ceased to function and the Provisional Federal Council of the new state took over the executive powers. The dream of a century seemed at last to be fulfilled. Once more, however, hope was doomed to disappointment. The Herrera government was overthrown in December 1921, and it dragged down with it the new federation before it had been given an opportunity to prove its value.

Nevertheless, hope was only deferred, and the following year brought about a new attempt at cooperation which was to have important results. In view of the differences of opinion arising regarding the status of the treaty of peace and amity of 1907, the presidents of the three republics, Honduras, El Salvador, and Nicaragua, were invited to meet with the diplomatic representatives of the United States to these republics on the U.S. warship *Tacoma* in the Gulf of Fonseca. The result of the meeting was incorporated in the so-called Tacoma Agreement under date of August 20, 1922. The substance of this agreement might be summarized as follows: (1) the three participating states regarded the treaty of 1907 as still in force; (2) the clauses prohibiting political refugees from utilizing a neighbor's territory for preparing expeditions were to be rigorously applied; (3) the three presidents agreed to summon a new conference of the five Central American states to bring about closer cooperation "which would tend to make really practicable the political unification of Central America"; and (4) in order to make effective their purpose of maintaining peace in Central America the signatory presidents agreed to submit to arbitration all present and future disputes. The presidents of the republics of Guatemala and Costa Rica were to be invited to adhere to this convention.[44]

The United States was interested in the proposal for a new conference and issued an invitation to the five republics to convene in Washington on December 4, 1922, to negotiate treaties to make effective the provisions of the 1907 treaties which experience had shown to be most useful, and also to adopt measures for the limitation of armaments in Central America, and to work out a plan for setting up tribunals of inquiry for the consideration of disputes which diplomacy had failed to settle.

[43] *Sen. Doc. No. 77,* 66th Cong., 1st Sess.
[44] *Conference on Central American Affairs* (Washington, D.C., 1923), pp. 6–10.

Under the chairmanship of Secretary of State Hughes, the delegates of the five republics worked from December 4, 1922, until February 7, 1923, upon these proposals, and the results were incorporated in some fifteen agreements which comprised a general treaty of peace and amity, eleven conventions, two protocols, and a declaration.[45]

The general treaty of peace and amity which abrogated all previous agreements formulated by Central American conferences was largely concerned with eliminating the causes of friction between and revolution within the signatory states. Constitutional changes in one were deemed to be of interest to all, consequently not only were *coups d'état* and revolutionary activities frowned upon, but strict injunction was laid upon all against recognition of unconstitutionally chosen governments. Intervention in civil wars, or in the internal political affairs of any other Central American Republic, was strictly prohibited, and each government agreed to adopt effective measures to prevent the fomenting or organizing of political activities within its territory against a neighboring government. The principle of non-reelection of presidents and vice-presidents was accepted, and all agreements and treaties were to be published. The treaty was to remain in force until 1934 and then continue indefinitely until denounced with due advance notice. The treaty lasted until the end of 1933 when it was denounced by El Salvador and Costa Rica in order that the latter state might recognize the government of General Martínez who had been named president of El Salvador by a military junta in December, 1931. Early in the following year Guatemala, Honduras, and Nicaragua followed the lead of Costa Rica in the recognition of the Martínez régime. When the Central American republics showed their intention so unanimously to accept the *coup d'état* method of setting up governments, it was no longer the part of the United States to follow a different policy. With our recognition of the Martínez government, the General Treaty of Peace and Amity was abrogated at least as regards the policy of nonrecognition of governments established by revolution against the constitutional régime.[46]

Mention should also be made of the convention providing for the reestablishment of an International Central American Tribunal. Instead of again setting up a permanent court as established by the treaty of 1907, a panel system of judges modeled upon the Hague Court system was adopted. The powers agreed to submit all present or future controversies not settled by diplomatic means to the court, although it was provided that "controversies or questions which affect the sovereign and independent existence of any of the signatory republics cannot be the object of arbitration or complaint." When in 1928 a long standing boundary dispute between Guatemala and Honduras[47] flared up again, the State Department recommended that the question be referred for settlement to the International Central American Tribunal.

[45] For texts see *Ibid.*, Appendix.

[46] A Treaty of Central American Fraternity was signed by the five republics at a conference in Guatemala City March 15–April 12, 1934, supplementing the 1923 treaties which outlawed war, recognized the principle of nonintervention, and upheld the ideal of a future Central American Union.

[47] For a detailed statement see *For. Rel. of the U.S.*, 1917, pp. 760–801.

Honduras refused on the ground that the panel of judges was not adequate, and the case was settled in 1933 by a special arbitral tribunal under the chairmanship of Chief Justice Hughes.[48]

Another of the conventions recast, unified, and to a considerable extent weakened the so-called Bryan Commission of Inquiry Treaties of 1913 and 1914 between the United States and the five Central American Republics.[49] In the new convention, instead of five permanent commissions of five members each, there was provided a panel or permanent list of thirty commissioners, five from each state, and a commission of inquiry was to be formed only upon request of one party directly interested in the elucidation of the facts. Such a commission of inquiry was to consist of one of the nationals of each interested country selected from the permanent list and a president chosen by them from the neutral names on the list. The machinery was undoubtedly simplified and improved, but whereas the former treaties for the advancement of peace subjected "all disputes . . . of every nature whatsoever" to a commission of inquiry, the new convention excluded questions affecting the sovereignty, independence, honor, and vital interests of the signatory states. The new convention has been in force since June 13, 1925, for Costa Rica, Guatemala, Honduras, Nicaragua, and the United States.[50] El Salvador was consistent in refusing to accept the revised convention.

Although the conference did not attempt to set up a federal system, the various conventions relative to the establishment of uniform labor laws, cooperation in agricultural stations, extradition, the exchange of students, and the establishment of free trade between Guatemala, El Salvador, Honduras, and Nicaragua laid a foundation for a closer union of the republics.[51]

On account of intervention of the United States in a boundary dispute between Panama and Costa Rica, the relations between the United States and Panama during 1921 became exceedingly strained. The trouble was of long standing, and the disputed area included sections on both the Pacific and Atlantic sides of the Cordillera. In 1900 the two states had submitted the dispute to President Loubet of France, who fixed a line running from Mona Point on the Atlantic to Burica Point on the Pacific, giving to Costa Rica the Coto region on the Pacific, which had been in Panama's possession for many years, and giving Panama a strip of land on the Atlantic side bordering the Sixaola River.[52] As shown by Article I of the Porras-Anderson treaty, signed through the mediation of the United States at Washington on March 17, 1910, the award on the Pacific side was accepted by both parties. But in regard to the rest there was a disagreement, and by the terms of this same treaty both parties agreed to submit the interpretation to the Chief Justice of the United States and bound themselves to abide by his decision. Article VII specifically states that

48 *Guatemala-Honduras, Special Boundary Tribunal, Opinion and Award* (Washington, D.C., 1933).

49 Neither Nicaragua nor El Salvador ratified the Bryan Treaties, and Honduras failed to set up the commission provided.

50 *U.S. Treaty Series*, No. 717.

51 For a detailed account of the conference see "The United States and Central American Revolutions," *Foreign Policy Association Reports*, vol. VII, no. 10 (July 22, 1931).

52 For a complete summary of the dispute from 1825 to 1909, including the text of the Loubet award, see *For. Rel. of the U.S., 1910, pp. 785–791.*

"the boundary line between the two republics as finally fixed by the arbitrator shall be deemed the true line and his determination of the same shall be found conclusive and without appeal.[53]

Chief Justice White announced his award on September 12, 1914, after a commission of American engineers, chosen in accordance with the terms of the convention, had made a prolonged and careful survey. The award gave to Costa Rica a portion of the territory claimed by Panama and to Panama a portion of the territory claimed by Costa Rica.[54] Panama, however, refused to accept the award, on the ground that the Chief Justice had exceeded his jurisdiction as arbitrator. Nothing was done to carry out the award until February 21, 1921, when Costa Rica invaded and seized the Coto region given to her by the Loubet award and conceded to her by Panama in the Porras-Anderson treaty. Panama speedily recaptured the region; but Costa Rica mustered an army and threatened war. Secretary Colby immediately protested against the use of force; and on the day after he entered office, Secretary Hughes was compelled to give consideration to the question. In a reply to a note from Panama, dated March 4, 1921, and requesting a declaration of the manner in which the United States understood its obligation to maintain the independence of Panama in the light of the Hay-Bunau-Varilla treaty, the new Secretary of State sent identic notes to Panama and Costa Rica, calling upon both countries to suspend hostilities until the United States could consider the question and propose a peaceful solution.

Secretary Hughes studied the matter carefully, and on March 15, in a long note to the American minister in Panama, he made clear the position of the United States. After reviewing the background of the case and the successive steps towards settlement, he reported that he could find no basis for the contention that the arbitrator had exceeded his powers; the award, therefore, by the terms of the Porras-Anderson treaty, became, he declared, "a perfect and compulsory treaty between the high contracting parties." Under these circumstances, he continued, "the government of the United States feels compelled to urge upon the government of Panama, in the most friendly but most earnest manner, that it conclude without delay arrangements with the government of Costa Rica for the appointment of the Commission of Engineers provided for by the terms of Article VII of the Porras-Anderson treaty, in order that the boundary line laid down by the decision of Chief Justice White may be physically laid down in a permanent manner and in accordance with the findings of the award."[55]

President Porras refused to accept Secretary Hughes' decision and appealed directly to President Harding. When the latter sustained his secretary of state, President Porras called a special session of congress, which backed him in defying the United States. Costa Rica, on the other hand, expressed gratitude for the prompt and efficacious mediation on the part of the United States. Early in May, Secretary Hughes despatched another note to Panama, saying that, unless she settled the boundary dispute promptly in accordance with the White-Loubet award, the United States would take necessary steps to give effect to the

53 *Ibid.*, p. 820.
54 For text of the award see *British and Foreign State Papers*, 1914, Part II, pp. 429–465.
55 *New York Times*, Mar. 18, 1921.

physical establishment of the boundary line. The United States realized that it was bound by treaty to protect the independence and territorial integrity of Panama, but could hardly guarantee the integrity of a country whose boundary shifted according to the caprice of the government.

Panama, still protesting, sent a special mission to Washington and finally proposed that the White award be submitted to the Hague Tribunal to see if it was within the terms of the arbitration. Secretary Hughes refused, and on August 18, 1921, he sent a note declaring there was no reason why Costa Rica should delay in taking possession. American marines were despatched to maintain the peace, and Panama, under protest, ordered her civil authorities to leave. Chief Justice Taft appointed two engineers to work with those to be appointed by Costa Rica and Panama to delimit the boundaries; but Panama refused to name her member of the commission. On September 9 the State Department announced that Costa Rica had taken possession of the disputed region.[56]

Diplomatic relations between the two countries were not resumed until 1928, and no further action towards a mutually acceptable settlement was taken until 1938. A treaty signed at San José, Costa Rica, September 26, 1938, established a boundary line which followed the White award from the Atlantic to Cerro Pando and the Loubet award from Cerro Pando to the Pacific. This required concessions by both states and seemed a reasonable compromise. However, President Castro of Costa Rica, threatened by serious opposition, withdrew the treaty from Congressional consideration, and President Arosema of Panama quickly followed suit.

Of all the Central American republics, Nicaragua, and Guatemala in recent years easily take first place as the administration's principal *trouble fête* in the conduct of American foreign relations in the isthmus.[57] It will be remembered that after 1912 a small detachment of American marines was ordered to remain in Managua as a counter-irritant to revolutionary symptoms, particularly of the Zelaya type. President Díaz, a Conservative, continued to prove himself most friendly to American interests, but the Liberals were discontented, and in 1916, nominated Dr. Irís, a close follower and a friend of Zelaya. The Conservatives chose Emiliano Chamorro, an able man who had served in a most acceptable fashion as Nicaraguan minister at Washington. Unwilling to see any one connected with Zelaya returned to power, the United States actively supported the candidacy of General Chamorro, who was consequently elected.

It is almost an axiom of Latin American politics that the only certain way to win an election against a candidate supported by the party in power is by a revolution, for the party which controls the election machinery *ipso facto* receives the votes. When the party in power had also the moral support of a hundred American marines in the capital, its chances of success became a certainty. Such a situation to a certain extent placed the United States in the

[56] A detailed critical analysis is given in W. D. McCain, *The United States and the Republic of Panama* (Durham, N.C., 1937), chaps. VI, X.

[57] For an official account see *The United States and Nicaragua–A Survey of the Relations from 1909 to 1932,* Dept. of State, Latin American Series, No. 6.

position of supporting the administration in power. To avoid this accusation, the United States Department of State in 1920 suggested to President Chamorro that the election laws be revised; but with the elections at hand the suggestion was not regarded as appropriate, and once more the Conservatives put their candidate in office.

Nevertheless the United States persisted in its efforts, and in 1922 Dr. H. W. Dodds was appointed as an expert to draft a new electoral law. Such a law was drawn and adopted by the Nicaraguan Congress, and the registrations under this law for the 1924 elections were carried on with very few charges of fraud.[58] But a split now occurred among the Conservatives when General Emiliano Chamorro returned from Washington, where he had again been acting as minister, and decided to run for president against the wishes of the Conservatives in power headed by President Martínez. The latter, unable to stand for reelection, owing to the restriction of the constitution, formed a coalition with the Liberals and supported Carlos Solórzano, a Conservative, for president, and Dr. Juan B. Sacasa, a Liberal, for vice-president. The United States was willing to supervise the election, but the administration opposed, and the coalition won by a substantial majority and in the old-style manner.

The United States had already given notice of its intention to withdraw the marines on the first of January 1925, but at the earnest request of President Solórzano they were permitted to remain until a constabulary might be established under the supervision of Major Carter of the United States Army.

The marines were withdrawn on August 4, 1925, and in less than a month General Chamorro had begun a revolution against the government. Gaining control over the army, and seizing the forts overlooking Managua, he expelled unfriendly members of the Assembly, had himself appointed designate for the presidency, and forced the resignation of President Solórzano.[59]

Inasmuch as such a *coup d'état* was a flagrant violation of the Washington conventions of 1923, the United States not only refused to recognize the Chamorro regime but made representations through our *chargé d'affaires* expressing disapproval of the Chamorro course of action. The four Central American states followed the United States in their refusal to recognize the revolutionary government. The failure to receive recognition weakened the Chamorro government, and in October 1926, at the suggestion of Admiral Latimer, who had been sent to Bluefields with his squadron at the request of the State Department, an armistice was arranged, and a conference took place between representatives of Liberal and Conservative factions at Corinto on the U. S. S. *Denver.* Although no agreement was reached, owing to pressure put upon Sacasa by Mexico, General Chamorro realized his inability to remain in power, and named Senator Uriza, the second designate of the Congress, as his

[58] H. W. Dodds, "The United States and Nicaragua," *Annals of Amer. Acad.,* vol. CXXXII (July, 1927), p. 137. See also Virginia L. Greer, "State Department Policy in Regard to the Nicaraguan Election of 1924," *Hispanic American Historical Review,* XXXIV (Nov. 1954), 445–467.

[59] For a succinct summary of the facts see President Coolidge's message to Congress of Jan. 10, 1927, *Cong. Rec.,* vol. 68, Pt. 2, p. 1324.

successor. When the United States refused to grant recognition to him, he summoned an extraordinary session of the Congress to select a new designate. According to Henry L. Stimson, whom President Coolidge sent as his personal representative with full powers to investigate and suggest a solution of the Nicaraguan situation, this Congress, taking into consideration that President Solórzano had resigned, and that Vice-President Sacasa was out of the country, legally selected Adolfo Díaz as first designate.[60] Señor Díaz was inaugurated on November 14, and the next day he appealed to the United States to aid him in the protection of the interests of American and other citizens, agreeing in advance to approve any means chosen by the State Department.[61] Two days later he was accorded recognition by the United States. The European governments followed the lead of the United Sates in recognizing Díaz, while Mexico, Guatemala, Costa Rica, and most of the South American states withheld recognition.

Supported by the United States, Díaz claimed that his election was legal in that the constitution of Nicaragua provides that if the vice-president is unavailable, the Congress is authorized to entrust the office to one of its members whom it shall designate. Vice-President Sacasa however declared that he was "ruthlessly pursued after the Chamorro-Díaz coup against the legitimate president, Solórzano," and obliged to leave Nicaragua; that, inasmuch as Díaz aided in the overthrow of the Solórzano-Sacasa régime, according to the Washington compacts he had no more right to the presidency than had Chamorro; that just as soon as it was possible after the resignation of President Solórzano he had returned to the country to assume the presidency, and had set up a government at Puerto Cabezas (Dec. 2), but that its operations were daily obstructed by the American forces.[62] Incidentally, the Mexican government had recognized Sacasa as the constitutional president shortly after his inauguration by the Liberals.

The Liberals now began a bitter struggle to oust Díaz, with substantial aid of arms and ammunitions from Mexican sources.[63] The United States increased its squadron in Nicaraguan waters to fifteen war-ships and authorized the shipment of war materials to the Díaz government. Landings of marines were made and neutral zones established at various points, particularly in the region controlled by the Liberals. For a time even a radio censorship was established by Admiral Latimer.

Criticism of the intervention of the United States now became so severe, both in the press and in Congress, that the State Department was sorely pressed to find reasons satisfactory to the public for its policy. The first landing of troops was declared to be solely for the protection of American lives and property, but there was little evidence that American lives and property were in jeopardy. A few days later our canal rights under the Bryan-Chamorro treaty were cited, but the Liberals had in no way impaired them. Two weeks later it

[60] H. L. Stimson, *American Policy in Nicaragua* (New York, 1927), p. 26.
[61] Message of President Coolidge, *loc. cit.*, p. 1325.
[62] *New York Times*, Jan. 10, 1927.
[63] See speech of President Coolidge, *loc. cit.*

was stated that both British and Italian governments had requested American protection for their nationals. On January 10, 1927, in a message to Congress, President Coolidge gave a clear and impartial statement both of the facts and the administration's handling of the situation. In addition to the reasons already given, he cited the shipment of arms and munitions from Mexico as especially provocative, declaring that "the United States cannot fail to view with deep concern any serious threat to stability and constitutional government in Nicaragua tending toward anarchy and jeopardizing American interests, especially if such state of affairs is contributed to or brought about by outside influence or by any foreign power."[64]

The good impression made by this speech was largely nullified by the statement given to the press by Secretary of State Kellogg on January 12, 1927, following his testimony at an executive session of the Senate Committee on Foreign Relations, which was then holding hearings on the Mexican and Nicaraguan situation.[65] The specter of Russian Bolshevist activity in Latin America was conjured but the ghost refused to walk. Resolutions of the Congress of the Red International of Trade Unions and the executive committee of the Communist International against American imperialism, some of them passed several years before, had lost their potency to inspire the fears of the Mitchell Palmer period – particularly as Foreign Minister Sáenz of Mexico pointed out that the Mexican Federation of Labor had protested against the alleged propaganda activities of the Soviet Minister to Mexico on the very evidence cited by Secretary Kellogg. In the language of Senator Frazier, it was a flimsy document making a fantastic charge.[66]

But something had to be done to assist Nicaragua, which was rapidly becoming completely demoralized by the revolutionary activities of the two factions. On March 31, 1927, at the suggestion of the State Department, President Coolidge appointed Mr. Henry L. Stimson, former secretary of war, as his personal representative to go to Nicaragua and investigate the situation with a view to working out a solution of the difficulty if possible. Mr. Stimson conferred with American Minister Eberhardt, with Admiral Latimer who commanded the naval forces, and with the responsible leaders of both factions. He found both sides willing, and even desirous, of American assistance to end the deadlock.

President Díaz finally, on April 22, agreed to make peace with the Liberals on the following terms: (1) immediate general peace in time for the new crop, and delivery of arms simultaneously by both parties to American custody; (2) general amnesty and return of exiles and return of confiscated property; (3) participation in the Díaz cabinet by representative Liberals: (4) organization of a Nicaraguan constabulary on a nonpartisan basis commanded by American officers; (5) supervision of elections in 1928 and succeeding years by Americans, who will have ample police power to make such supervision

[64] *Ibid.*, p. 1324.
[65] *Ibid.*, p. 1649
[66] *Ibid.*, p. 5523.

effective; (6) continuance temporarily of a sufficient force of marines to make the foregoing effective.[67]

Mr. Stimson thereupon arranged a conference with representatives of Dr. Sacasa, who agreed to the arrangement, provided it was acceptable to General Moncada, the Liberal general in the field. A conference in Tipitapa between Mr. Stimson and General Moncada took place on May 4 and the result was an acceptance on the part of the Liberal general, provided the United States would declare in a written statement that the retention of General Díaz and a general disarmament were regarded as essential conditions of the plan. This would make it easier for him to persuade his troops that they could not hope to overthrow Díaz. Mr. Stimson immediately gave such a written assurance. Finally, in a more detailed statement dated May 11, Mr. Stimson promised a free, fair, and impartial election under American auspices in 1928, and pointed out that he had recommended changes in the Supreme Court and the Congress to the advantage of the Liberals, and the appointment of Liberal *jefes políticos* in the six Liberal districts. He had already received assurance that these reforms would be carried out.[68]

General Moncada accepted the arrangement in a formal statement, and within a week Liberals and Conservatives had turned over some 9000 rifles, 300 machine guns, and 6,000,000 rounds of ammunition. The one source of trouble was a General Sandino, who, retreating to the north with his force, on July 16 attacked a detachment of marines and constabulary at Ocotal near the Honduran frontier. His repulse with heavy losses aroused a considerable stir in the United States, but General Moncada in a public statement disavowed Sandino as a renegade Liberal at the head of outlaw mercenaries.

As a proof of the fairness of the elections held November 4, 1928, under the general supervision of General Frank R. McCoy,[69] General Moncada, regarded as hostile to the United States, was elected president by a majority of almost 20,000.[70] He was inaugurated on January 1, 1929, and one of his first official acts was to appoint Dr. Sacasa as Minister of Nicaragua in the United States. President Moncada also approved the agreement signed December 22, 1927, creating a *Guardia Nacional* in place of a national army, with the understanding that it was to be trained and under the command of an American officer subject to the sole direction of the President of Nicaragua.[71] It was also understood that all American marines should be withdrawn just as soon as the *Guardia* should be able to police and protect the state effectively. In fact, the Department of State announced February 13, 1931, that it was hoped that all the marines except officers instructing the *Guardia* would be withdrawn by June of that year and the officers would leave after the 1932 elections. In fact, by April 1, 1932, the marine force had been reduced from its maximum of 5,673 to 753 marine and naval personnel, exclusive of the 205 officers in the Guardia Nacional.[72]

[67] Stimson, *op. cit.*, pp. 63–64.
[68] *Ibid.*, pp. 81–83.
[69] General McCoy was nominated by President Coolidge at the request of President Díaz.
[70] Dodds, *op. cit.*, p. 91.
[71] *Ibid.*, p. 100.
[72] *Ibid.*, pp. 107–108.

Unfortunately for the peaceful development of Nicaragua, General Sandino continued his policy of raiding and pillaging towns and villages in the mountainous interior. Taking advantage of the fact that the earthquake on March 31, 1931, had engaged all available Guardia and marines in relief work in Managua, he struck on the east coast and sacked the town of Cabo Gracias. During this raid, many civilians were murdered, including nine Americans. Disturbed by this situation, on April 17 Secretary Stimson informed American citizens that their government could no longer protect them in the interior of Nicaragua with American forces; therefore it urged them, if they felt themselves endangered, either to withdraw from the country or at least to the coast towns.[73] This new policy, so different from that expressed by President Coolidge in 1927, was heartily commended by Senators Borah, Capper, La Follette and Norris, but bitterly denounced by Senator Johnson. Shortly afterwards Secretary Stimson explained that his new policy was primarily due to the fact that the marines were unable to penetrate the trackless jungles where the *Sandinistas* operated, whereas the expanded Nicaraguan *gendarmerie* were being trained and equipped to stamp out this banditry in the early future. In the meantime naval vessels would stand by at all threatened ports and we would "continue to be zealous in our concern for the lives of our nationals wherever they may be found."[74]

When in April 1932, in a raid near Ocatal four American marines and an officer were killed, Senator Lewis of Illinois introduced a resolution demanding the withdrawal of all American troops from Nicaragua. However, as the American government was already pledged to supervise the 1932 elections, such a procedure was not regarded as feasible. The elections were held November 6, 1932, with the assistance of about 400 American marines under the supervision of Rear Admiral Clark H. Woodward, who served as a chairman of the board of elections. The elections provoked practically no disorders and Dr. Sacasa was elected by a majority of about 23,000 over his opponent Díaz, who was considered the "Americanista" candidate. Almost immediately after the elections the United States commenced to withdraw the marines and on January 2, 1933, the day after President Sacasa's inauguration, the last contingent of officers and men embarked at Corinto.

President Sacasa immediately began negotiations with General Sandino, who had carried on his guerrilla warfare until the marines had left. On February 2, 1933, an agreement was signed whereby the Government granted General Sandino and his followers amnesty and compensation in return for which he undertook to disarm all of his forces except one hundred men who were to be incorporated into the Guardia. He declared that he had never been animated by a spirit of ill will towards the United States, but had fought patriotically to end foreign intervention. Following clashes between his followers and the National Guard in the fall of 1933, it was demanded in January 1934, that General Sandino surrender all his arms. Invited to discuss the matter with President Sacasa at the presidential palace February 21, 1934, General Sandino

[73] U.S. Dept. of State, *Press Releases,* Apr. 17, 1931.
[74] *Ibid.,* Apr. 18, 1931. For appraisal of Sandino see Joseph O. Baylen, "Sandino: Patriot or Bandit," *Hispanic American Historical Review,* XXXI (Aug. 1951), 394-419.

and his brother and his two aides were seized by guardsmen as they left the palace grounds and murdered in cold blood. No serious effort was made to discover or punish those responsible for the killing. In fact, General Somoza, Commander of the Guardia Nacional, who was accused by followers of Sandino as being responsible, was able by threats and coercion, despite his constitutional disability as head of the Guard, and as being related to President Sacasa, to have himself elected to the presidency in 1936.

Subsequent relations with Nicaragua have been concerned with renewed trouble between Nicaragua and Honduras over their boundary line. By a convention between Nicaragua and Honduras signed October 7, 1894, a joint commission was named to demarcate the boundary. When a dispute arose over the marking, the King of Spain was chosen to arbitrate and his award was made December 3, 1906. This settlement was accepted by Honduras but not by Nicaragua. The United States urged a settlement in 1914, and again in 1918 it offered to mediate, but it was only able to obtain a temporary acquiescence of the *status quo.* Finally, in 1932, a new boundary protocol was signed accepting the award of the King of Spain and appointing a commission composed of three engineers – one named by each government – and the chairman by the United States. Honduras ratified this protocol, Nicaragua did not.[75] The dispute was kindled again in September 1937, when the Nicaraguan government issued a series of air-mail stamps picturing a map upon which the area in controversy was shown as Nicaraguan. When Honduras protested vigorously the United States on October 20, following the procedure outlined at the Buenos Aires Conference, tendered its good offices in association with the governments of Costa Rica and Venezuela. The offer was accepted and on December 10, 1937, a pact of reciprocal agreement was signed in San José, Costa Rica, whereby Nicaragua and Honduras agreed to withdraw their troops from the frontier, to refrain from further military preparations or purchase of arms, and to settle their differences by peaceful means in accordance with the Convention of 1929. Dr. Frank P. Corrigan, United States Minister to Panama, who served as chairman of the conciliation commission, made a valiant but vain effort to effect a settlement before the agreement expired on December 10, 1939.

The critical situation of Europe in 1939 brought about an even closer relationship between the United States and the Central American Republics. Shortly after his inauguration as president in 1939. President Somoza of Nicaragua made a visit to the United States where he was cordially received by President Roosevelt and invited to address both houses of the Congress. As a result of his visit an elaborate program of cooperation for the development of Nicaragua with the assistance of the United States was worked out. In return for Nicaragua's promise to encourage the investment of American capital, to utilize American technical advisors, and to provide dollar exchange for its customs bonds of 1918, the United States agreed to send to Nicaragua army engineers to study the feasibility of a trans-Nicaraguan waterway to link the east coast with the populous interior of the Pacific. The Export-Import Bank agreed to set up credits of $2 million to purchase machinery and supplies for the construction of

[75] Dodds, *op. cit.*, pp. 109–111.

highways and other production projects and also to make available a revolving fund to $500,000 for emergency needs.[76]

A very vital feature of Caribbean defense was the section of the Inter-American Highway linking the United States with the Panama Canal. This roadway from Laredo, Texas, to Panama was 3252 miles long and by the end of 1941 substantial progress had been made towards its construction.[77] On December 26, 1941, the Congress of the United States passed an act authorizing the appropriation of a sum not to exceed $20 million to enable the United States to cooperate with the Central American Republics in the completion of the Inter-American Highway within their borders. Under the grant the United States was prepared to assume two-thirds of the total expenses incurred in each country for the survey and construction of the highway.[78] In order to speed up construction an arrangement with the five Central American states and Panama was announced on July 28, 1942, providing for the immediate linking by a pioneer road of the already completed segments between the Mexican-Guatemalan border and Panama City. This meant the immediate construction of approximately 625 miles of new all-weather pioneer road.[79] Owing to the necessities of war and the need for haste a pioneer road was constructed in a most costly and wasteful manner. Sections had to be abandoned, others rebuilt and World War II ended with the highway still uncompleted.

All of the Central American republics declared war against Japan on December 8, 1941, and on Germany and Italy by December 12. At the Rio de Janeiro conference they supported the United States unanimously. All were granted essential requirements under Lend-Lease agreements, and all except Honduras requested and were granted military, naval or air missions. Although most of the countries were ruled by dictatorial regimes, they stood opposed to the totalitarian systems abroad and were willing and eager to cooperate, and even to fight, for hemispheric defense.

Central America supplied the United States and its allies with foodstuffs and critical raw materials during the war which brought profits to the producers, but to the accompaniment of rising prices and inflation. Wartime conditions helped to spur the growth of light industry, as sources of manufactures were cut off, and some progress was made toward agricultural diversification. Development continued in the first decade after the war owing to reconstruction requirements and the conditions arising from the Korean War, but in the long run Central American exports failed to keep pace with the rising level of economic activity in the major industrial countries. This condition, rising inflationary pressures, institutional deficiencies, militarism, and Communist influence, combined to create social unrest throughout the region, particularly among urban and rural labor and middle class wage earners.

[76] U.S. Dept. of State, *Press Releases,* vol. XX (May 27, 1939), p. 439. Further supplemental arrangements were agreed upon Apr. 8, 1942–Dept. of State *Bulletin,* vol. VI (Apr. 25, 1942), p. 368.

[77] The road was completed from Laredo to beyond Mexico City, and Mexico was building to the Guatemalan border. The Guatemalan section was passable; the Salvadoran was the first Central American section finished; and Costa Rica and Nicaragua were well along in their sections.

[78] *U.S. Stat. at Large,* vol. 55, p. 860.

[79] Dept. of State *Bulletin,* vol. VII (Aug. 1, 1942), p. 661.

The widespread restiveness compounded one of the traditional and fundamental problems of the area, that of dictatorship. Until 1944, four of the five Central American republics – Guatemala, El Salvador, Honduras, Nicaragua – were ruled by dictators. Shortly afterward Costa Rica was for a short time controlled by a regime unwilling to relinquish power to its elected successor. With the advent of the cold war, anti-Communist military dictatorships aided by the United States came to power, prolonging the authoritarian tradition. This was accompanied by the growth of activist groups and forces, products of social and economic change, struggling for power and ascendancy. Thus, an important feature of contemporary change in each of the Central American republics lies in the importance of political power as a means to achieve social and economic goals and the emergence of formerly excluded social groups and interests as rivals for political dominance.

After the war the peace of the area was threatened by an outbreak of animosities and disputes among the isthmian republics, most serious of which was the controversy between Nicaragua and Costa Rica. Prominent in that affair were the activities of the so-called Caribbean Legion, spirited by President José Figueres of Costa Rica, which aimed at the overthrow of dictatorial regimes in Nicaragua, Honduras, Venezuela and the Dominican Republic. The United States showed a marked interest in these disputes, but refrained from taking unilateral action to help resolve them. Instead, Washington's policy has been to cooperate by sending highly-placed individuals as good-will emissaries, consulting with the Central American states as a group, and working through the OAS. However, the scope of Washington's diplomatic activity in the region was considerably broadened by its efforts to thwart a Communist takeover in Guatemala, in countering invasion thrusts emanating from Castro's Cuba, expanded economic and technical assistance under the Alliance for Progress, and the development of the Central American Common Market.

A review of United States policy in the Caribbean indicates that early in the twentieth century Washington sought to promote Central American peace and stability for strategic and commerical purposes. At the same time, United States influence was projected into internal affairs of the isthmian republics. To attain its objectives the United States aided in the formation of peace machinery for the settlement of Central American disputes, and endeavored to help each nation achieve a stable and constitutional government. Acting initially as a friendly mediator, it was later forced into an interventionist role. The latter was renounced in the early 1930s with the inauguration of the Good Neighbor policy. World War II brought the United States and the Central American republics into a closer relationship, creating an economic dependence on the United States that had not existed in the same degree previously.

While space does not allow detailed treatment of the internal affairs of each country, the relation between socio-economic issues and political change in some countries, particularly Guatemala, has become a critical concern of United States foreign policy planners. At stake is the nature of Central American society itself, the future course of institutional development in the region, and quite possibly hemispheric security.

Guatemala, scene of the "1944 Revolution," and a target of Soviet Communism in the 1950s, was hardly distinguishable from other Central

American states in terms of either its socio-economic structure or its political process at the beginning of the 1930s. Dominated by upper class land holding elites in association with the army, high ranking ecclesiastics, and foreign corporations, Guatemala seemed to be an unlikely prospect for a fundamental social upheaval. This impression was reinforced by the dictatorship of General Jorge Ubico who, capturing the presidency in 1931, established a regime that was ruthless and reactionary even for Guatemala.[80]

Ubico was overthrown by the revolution of October 20, 1944, middle class in nature, and headed by young intellectuals, junior army officers, students and professionals. This revolution was to be fundamental in character, dedicated to the liquidation of feudalism, the organization of a modern (revolutionary propaganda used the term "capitalist") economic system, and the rearrangement and vitalization of the social structure. The program included economic planning, agrarian reform, the protection and integration of Indian communities, the defense of workers and the limitation of individual rights, particularly upper class property rights. In practice it was responsible for bringing about a high degree of integration between democratic political forms and political processes, at least until the later stages of the revolutionary period. The December 1944 elections were probably the freest held in Guatemala to that time, and suffrage was extended to women in 1946.

The October Revolution led to the regime of Juan José Arévalo Bermejo, of middle class origin and an educator, who assumed the presidency in 1945. His political philosophy centered on what he called "spiritual socialism," meaning the alleviation of feudalism through the use of "discreet measures and programs" that would socially integrate and economically improve the lot of the generally impoverished population. The Guatemalan revolution, however, lacked a party to serve as its spokesman and give it ideological and political direction. This produced confusion and seeming inertia that the Communists were able to exploit. The Communists succeeded in convincing many supporters of the revolution that they were the only true advocates of social change and of economic and political nationalism. Although Arévalo himself was not a Communist, and "their influence in his own administration never became dominant, the end result of Arévalo's tolerance toward the Communists was that they were in a position to strike a more advantageous bargain with his successor."[81] Tension developed with the United States owing to the leftist tendencies of the Arévalo regime, and the critical attitude displayed by United States Ambassador Richard C. Patterson led to a request for his recall in April 1950.

By 1951 it had become apparent that Guatemala could not be relied upon to participate in an anti-Communist Central American front. In October 1951, representatives of the five Central American states met in San Salvador and drew up a document known as the Charter of San Salvador which was promptly ratified by all of the republics and came into force December 14, 1951. The

[80] See Daniel James, *Red Design for the Americas: Guatemalan Prelude* (New York, 1954), pp. 36–37.
[81] Ronald M. Schneider, *Communism in Guatemala: 1944–1954* (New York, 1959), p. 23.

Charter did not establish a federal union, but proposed to strengthen the bonds shared by the five states. When it was proposed and supported by all members except Guatemala that concerted action be taken against Communist infiltration in Central American governments, Guatemala rejected the proposal and when outvoted withdrew from the organization. The meeting scheduled to convene in Guatemala May 2, 1953 failed to take place. However, representatives of the other four states held a special conference on July 10, 1953, at Managua, Nicaragua at which communism was unanimously condemned.

Meanwhile, in March 1951, Colonel Arbenz Guzmán, a career service officer and defense minister in the Arévalo administration, was inaugurated in the presidency. In what is generally conceded to have been a fair election, he won 65 percent of the popular vote. The chief goal of Arbenz during his three years in office was agrarian reform, with unused lands of wealthy holders being expropriated on a large scale and the property made available to landless farmers. At the time 70 percent of the land was owned by 2.2 percent of the population. The land reform program brought his government into conflict with the United Fruit Company, a United States corporation, and the largest land holder in the country. Basing its action on the Agrarian Law of June 19, 1952, the government of Guatemala in March 1953, expropriated 234,000 acres of land belonging to UFCO on the ground that this property was not being currently cultivated. In return the Guatemalan government offered compensation to the amount of $600,000 on 3 percent agrarian bonds maturing in twenty-five years. As of 1 May 392,945 acres had been expropriated leaving the company with 145,817 acres for banana production and other operations.[82]

In an *Aide Mémoire* dated August 28, 1953, the United States protested the seizure both upon the valuation and the method of payment. In evaluating the properties on the basis of their tax value there "was not the slightest resemblance to a just valuation" inasmuch as the company had put many improvements on the land and had vainly requested a new valuation. As to method of payment, the bonds of uncertain value were not considered 'either prompt or effective payment.' Furthermore, the allegations that the properties were not being utilized was false since they were a necessary reserve against "Panama disease" and they were being used for lumber supplies, fruit and vegetable production, and cattle raising necessary to the requirements of the company. The action of the Guatemalan government was a flagrant violation of international law and the United States government protested it as such.[83]

Then the Electric Company of Guatemala, called an "American imperialist monopoly," was taken over by the government to prevent an interruption of service threatened by a strike for higher wages by the workers. In similar fashion the International Railway of Central America, whose principal stockholder was UFCO, when threatened by a worker strike in October 1953, was temporarily placed under government control and later went into a receivership.

[82] Nathan L. Whetten, *Guatemala, the Land and the People* (New Haven, 1961), pp. 152–166.
[83] Text of *Aide Mémoire* is found in Dept. of State *Bulletin*, vol. XXIX (Sept. 14 1953), pp. 357–160. For background see Richard N. Adams, "Social Change in Guatemala and United States Policy," in *Social Change in Latin America Today* (New York, 1960), pp. 231–284.

The UFCO, aware that its position would not be upheld in Guatemalan courts, appealed to the United States Department of State to intervene on its behalf. After two representations had been made, one of which was rejected and the other ignored, the United States on April 20, 1954, officially demanded $15,854,849 as compensation for the expropriated property. Foreign Minister Guillermo Toriello rejected the note calling it "another attempt to meddle in the domestic affairs of Guatemala."

In retrospect it is evident that there were two sides to the issue. The Arbenz government had a legal right to expropriate the property; however, its methods and refusal to make adequate and proper compensation were obviously wrong. UFCO on its part had provided better housing conditions, education, medical care and other facilities for its workers than other Guatemalans of the laboring class generally received. Its banana workers earned about one thousand dollars a year, three times the national annual average income. But the company would not recognize labor organizations. In its contract with the government UFCO paid less than reasonable compensation for concessions granted to it; duties and taxes amounted to one-tenth of the company's annual profits. When President Arévalo sought to renegotiate the contract he was rebuffed. And in its relationship with International Railways of Central America, UFCO was given preferential rates. These circumstances, intensified by Guatemalan nationalism, were skillfully exploited by the Communists in their campaign against "Yankee imperialism."[84]

Early in 1954 the Arbenz government reported that rebel forces in Honduras, allegedly aided by the United States, were preparing to invade Guatemala. A Guatemalan exile, Lieutenant Colonel Carlos Castillo Armas, leader of the anti-Communist forces, had established headquarters in the Honduran capital. He appealed to the Guatemalan people promising to overthrow the Communists. Arbenz, uncertain of his position, carried out a reign of terror in which the Communists joined. "Whereas in 1944 there were no Communists in Guatemala except for a small group who had been rotting in prison for a dozen years, less than ten years later there were perhaps 4,000 card-carrying party members and several times that number of sympathizers. By the end of 1953, the Communists held commanding positions in the labor movement, the coalition of political forces on which the government rested, and even the government itself."[85]

In a campaign of hate, the United States was excoriated both by the Guatemalan press at home and by Foreign Minister Toriello at the Tenth Inter-American Conference at Caracas in March 1954. In April Ambassador Peurifoy was recalled to Washington for consultation and the U.S. Information Agency advised its posts abroad of the alarming increase of Soviet financing of contraband arms and training units in Latin America. The report indicated that about one thousand Latin Americans visited the Soviet Union and other satellite countries to attend Communist front meetings in 1953. May Day offered an

[84] John D. Martz, Central America: *The Crisis and the Challenge* (Chapel Hill, 1959), pp. 47–52.

[85] Schneider, *Communism in Guatemala,* p. 1; *Cong. Rec.*, 83rd Cong., 2nd Sess., vol. 100, p. 8194; *Ibid.*, 84th Cong., 2nd Sess., vol. 102, Appendix A, 3012.

opportunity for a vast Communist demonstration against "Yankee imperialism" in Guatemala under the Red leaders Victor M. Gutiérrez and Manuel Fortuny. This was accompanied by an intensification of the terrorism against suspected opponents of the regime.

When late in May the Swedish freighter *Alfhem*, carrying two thousand tons of Czech-made weapons and ammunition arrived at Guatemala's Puerto Barrios on the Caribbean, the United States notified the other Latin American republics and sent consignments of arms to Honduras and Nicaragua. Secretary of State John F. Dulles said "by the arms shipment a government in which Communist influence is very strong has come into a position to dominate militarily the Central American area . . . "[86] The United States Congress was sufficiently impressed by the gravity of the Communist threat in Guatemala that it passed the following concurrent resolution: "Resolved by the Senate (the House of Representatives concurring), that it is the sense of Congress that the United States should reaffirm its support of the Caracas Declaration of Solidarity of March 28, 1954, which is designed to prevent interference in Western Hemisphere affairs by the international Communist movement, and to take all necessary and proper steps to support the Organization of American States in taking appropriate action to prevent any interference by the international Communist movement in the affairs of the states of the Western Hemisphere."[87]

In June Colonel Castillo Armas, anticipating an internal army revolt, launched his invasion with a heterogeneous force numbering five hundred to one thousand. Confronting him was the ten thousand-man Guatemalan regular army. However, President Arbenz quickly found that the army would no longer support him, and when the rebels were able to bomb successfully the government's stronghold in Guatemala City he decided to resign. A military *junta* under Colonel Elfego Monzón, aided by Ambassador Peurifoy of the United States, worked out a compromise with Colonel Castillo Armas which finally resulted in the setting up of a junta with Castillo Armas as provisional president. The United States accorded formal recognition to the new government on July 13, 1954, and on July 30 the new provisional president declared it to be the intention of Guatemala to rejoin the Organization of Central American States. The new regime also adopted the anti-Communist resolution of the Caracas conference and outlawed the Communist party of Guatemala.

The Castillo Armas administration proved to be a "good old-fashioned military dictatorship." Its chief was "confirmed" in office in an election without secret ballot in September. Violent revenge was taken by landowners and employers against their tenants and workers who had participated in the revolution, and Castillo Armas was not able or unwilling to prevent this. Washington quickly expressed its approval of the new administration, Secretary Dulles declaring that "as peace and freedom are restored . . . the United States will continue to support the just aspirations of the Guatemalan people . . . The United States pledges itself . . . to help alleviate conditions in Guatemala and

86 U.S. Department of State, *Intervention of International Communism in Guatemala,* pub. No. 5556 (Washington, 1954), p. 13.
87 *Cong. Rec.,* 83rd Cong., 2nd Sess., vol. 100, p. 8927.

elsewhere which might afford communism an opportunity to spread its tentacles throughout the Hemisphere . . . "[88]

Much speculation existed at the time of the Castillo Armas invasion of Guatemala concerning official United States support of the movement. This was subsequently confirmed. Former President Dwight D. Eisenhower, recalling the episode, said that when some of his advisers argued that this country should not send planes to help the invaders he told them that there was no way to conceal the role played by the United States and that therefore we had "better be a winner." He declared the planes were sent, adding that when a country "such as ours" appealed to force it had to be successful.[89]

Vice President Richard M. Nixon, stopping in Guatemala on a good will tour of the Caribbean early in 1955, said: "You have won the admiration and appreciation not only of the United States but of people all over the world for the way in which you won your fight against a Communist government . . . " President Castillo Armas made a state visit to the United States later in the year, receiving a twenty-one gun salute which accorded him full military honors.

Despite early assurances of support Washington was slow in providing financial assistance to the Castillo Armas administration. The sum of $6,425,000 was promised initially, but little more than half that amount reached Guatemala in the first eighteen months after its accession. It was not until 1956 that United States aid arrived in appreciable quantities. Representative Patrick J. Hillings of California asserted that it was "quite a paradox" that while billions of dollars had been advanced to European countries Washington was not quick to supply a small requested sum to a nation "in our own backyard" that had just repelled Communist penetration.[90]

The transition to the Castillo Armas regime was one from relative affluence to austerity, and much reliance was to be placed on United States assistance. In the three years of his administration Washington's contribution amounted to about $80 million. Comprising the aid program were loans to the IBRD, grants in aid, foodstuffs, technical assistance, and funds for constructing the Inter-American Highway. Public works programs reduced unemployment, but product diversification lagged. Coffee exports continued to be between 75–80 percent of the value of Guatemala's exports, and cotton production declined. Bananas, largely controlled by UFCO, accounted for only 10 percent of the value of total exports.

The 1954 counterrevolution reoriented Guatemala back to the more traditional pattern of its history. Castillo Armas and some of his supporters advocated reform, but the power of the landowners, the army and the Church, was sufficient to largely restore the *status quo*. Former officials of the Ubico era were brought back into the government, the agrarian law was cancelled and expropriated property turned back to the original owners. An accomodation was reached with UFCO, the company agreeing to pay a tax of about 30 percent on

[88] Martz, *Central America*, p. 63; Philip B. Taylor, Jr., "The Guatemalan Affair: A Critique of United States Foreign Policy," *American Political Science Review*. vol. L (Sept. 1956), 787–806.
[89] *New York Times*, Oct. 14, 1965.
[90] *Ibid.*, Oct. 14, 1954.

its annual profit and relinquishing its claims on one hundred thousand acres in return for non-discriminatory legal treatment. Civil liberties were curtailed and labor legislation was all but eliminated. A new constitution was drawn up in 1956 by a constituent assembly composed chiefly of conservative elements.

On July 25, 1957 Castillo Armas was assassinated in the Presidential Palace by a member of the Presidential Guard. President Eisenhower regarded his death as a "great loss to the entire free world." After a period of confusion and unrest General Miguel Ydígoras Fuentes, a high official under Ubico and a losing candidate for president against Arbenz in 1951, was sworn into office in March 1958. Backed by conservative and business interests Ydígoras' election was patently fraudulent, and the upper class retained political control of the country. Within five years, however, Ydígoras alienated virtually all political sectors. The economic distress that accompanied the drop in coffee prices in 1959, combined with the administration's austerity program and rumors of governmental grafting, helped to undermine his popularity. He angered the nationalists by allowing Cuban exiles to train on Guatemalan territory in preparation for the Bay of Pigs expedition. The efforts of a group of young army officers to oust the government in November 1960, was revealing of armed forces dissatisfaction. The left, represented by labor, intellectuals and students, tended to sympathize with Fidel Castro and took an anti-United States stand. The conservatives disliked his middle-of-the-road policies and its efforts to implement Alliance for Progress reforms. Ydígoras' attempt to introduce the income tax in 1962 was violently protested by people who had never carried a burden of taxation in their country. Political reaction to Ydígoras took the form of attempts to overthrow his government. Student protests, strikes and attempted *coups d'état* forced the president to suspend liberties on a number of occasions. When Juan José Arévalo returned to Guatemala City in March 1963, and announced his intention to run for the presidency in the December elections, the military, under the leadership of the War Minister, Colonel Enrique Peralta Azurdia, forced Ydígoras out of political office and into exile. Fearing a return to the revolutionary programs of the 1944–1954 period, Peralta assumed extraordinary powers, suspended the Constitution and political rights, and dissolved the National Congress.

United States relations with the Ydígoras regime from 1958 to 1963 were close and generally cordial, except at the close of his administration. Washington promptly recognized the popular and conservative general, recalling his earlier opposition to Arbenz. President Ydígoras met Secretary Dulles in Washington in February 1958, reaffirming that his country was on the side of the United States in the struggle against communism. The proved vulnerability of Guatemala to communism in the early 1950s and the nearer and more direct threat posed by Castro's Cuba beginning in 1959, gave Guatemala a high priority in Washington's security operations. Dr. Milton Eisenhower's visit to Guatemala, and elsewhere in Central America, in the summer of 1958 confirmed this interest, and United States aid of all types was maintained at high level, but particularly to take up the slack resulting from lowered coffee revenues. Guatemala's dollar loss was figured at $1,200,000 for each cent's drop in the price of coffee.

In November 1960, at the request of the governments of both Guatemala and Nicaragua, President Eisenhower ordered United States naval units to patrol

Central American waters and to shoot if necessary to repel a reported Communist-led invasion attempt of those countries. The Department of State indicated that if these forces were used to prevent an invasion of Central American republics by "Communist-directed elements" they would limit their operations to areas "within the national jurisdiction of the requesting governments."[91] The navy patrol was withdrawn on December 8 after the threat failed to materialize.

As events were soon to prove, an invasion was in the making, but it originated in Guatemala with the support of the United States, and was directed against Cuba. In January 1961, the United States press reported that a military force estimated to number two thousand was being trained by United States and Cuban instructors at a base in Guatemala's Retalhuleu province. President Ydígoras steadfastly insisted that all activity was defensive. In April, Paul P. Kennedy, a correspondent of the *New York Times*, was expelled from Guatemala for reporting "something about an alleged invasion of Cuba by Guatemala." He was invited to return the next day, April 10, and in an interview with President Ydígoras was told that if Cuba attacked, Guatemalan troops would be in Havana in a matter of hours. While admitting that foreign military personnel training Guatemalan army detachments in guerrilla warfare were members of the United States military mission, he said "We are not going to invade anyone; we are preparing only for our defense."[92] A week later the tragically mis-calculated, Central Intelligence Agency-directed assault on Cuba took place, the invaders having been trained in Guatemala and launched from Nicaraguan shores.

President Ydígoras was overthrown in April 1963, for reasons already noted, and was offered asylum in the United States. It should be mentioned that Ydígoras' approval of the return of Juan José Arévalo to run as a candidate in the presidential elections scheduled for December 1963, may have prejudiced United States opinion against him. Although Arévalo denounced Castroism, his past record, and his book, *The Shark and the Sardines*, a strong indictment of United States policy in Latin America, naturally made him suspect to Washington and the American public. Fearing that Arévalo might win the election, President Ydígoras declared on March 21, 1963, that Arévalo was a known Communist and would not be allowed to cross the border. His announcement resulted in the breakdown of law and order and the imposition of a state of seige, the army acting quickly and savagely to silence Arévalo's supporters. However, Arévalo did appear briefly in Guatemala City, providing a pretext for Colonel Peralta to seize the government.[93]

Washington reluctantly granted diplomatic recognition to the military government after consulting with other Central American leaders. Disturbed by military *coups* in South America the Kennedy administration warned that "the failure of the democratic process" in one American nation was the proper

[91] *Ibid.*, Nov. 18, 1960. For background of crisis see Miguel Ydígoras Fuentes, *My War With Communism* (Englewood Cliffs, 1963), Chapters 8–9.
[92] *Hisp. Amer. Rep.*, vol. XIV (June, 1961), p. 298.
[93] Mario Rodríguez, "Guatemala in Perspective," *Current History*, vol. 51 (Dec. 1966), pp. 339–340.

concern of the entire Hemisphere. Before granting recognition, Washington demanded respect for basic civil rights in the country and a promise of eventual return to civil rule.

The Peralta dictatorship provided a relatively honest and efficient government, cooperated in the Central American Common Market program, and took credit for an economic upsurge that was partly brought about by United States economic assistance. A facade of constitutionalism was erected to qualify for continued Alliance for Progress support, but a free and open political system could not safely be tolerated. States of seige, accompanied by the suspension of individual guarantees, were commonplace; the need to curb terrorism, which did increase, and "Communist conspiracies," were used to justify these measures.

The Peralta regime confounded all skeptics by holding a fair election for the presidency and Congress. The moderate left-of-center reformer, Julio Cesar Méndez Montenegro, former Dean of the University of San Carlos Law School, in Guatemala City, won 45 percent of the presidential vote, defeating the two military candidates. There being no clear majority, the decision for the presidency was left to the incoming Congress which, following the election results, voted Méndez into office. Colonel Peralta, honoring his commitment, turned the government over to his civilian successor on July 1, 1966.

The election signaled an improvement in the political process, and unquestionably Guatemala was advancing economically. Industrialization had advanced since the Common Market was formed in 1960, and highway, air, and maritime facilities had been improved. Natural gas deposits were discovered at Peten, oil exploration was resumed on the Pacific coast, and mining operations were again underway. In 1966, for the first time since 1960, Guatemala enjoyed a favorable trade balance.

While these developments justified some optimism for the future, data assembled by the Inter-American Development Bank and the National Economic Planning Council revealed how backward Guatemala remained twelve years after the United States engineered the overthrow of pro-Communist President Arbenz. These data indicate that 75 percent of the people had no access to modern civilization and culture. Eighty percent of the rural children received no schooling, and the illiteracy rate approached 80 percent, the highest in Latin America except for Haiti. The population was eating about a third of what was needed for a proper diet, and the life expectancy of the population, rising at the very high rate of 3.1 percent annually, was forty-five years. From 1950 to 1964 the gross national product increased, but the per capita income of the rural population, involving 66 percent of the country, dropped from $87 a year to $83. It was concluded that the rural population "is as wretched and backward as anywhere in Latin America."[94]

These conditions persisted in spite of aid furnished by the United States which was increased sharply during the Castillo Armas administration. In the period 1948–1965, Guatemala received economic assistance from Washington totaling $110.9 million ($86.1 million in grants; $24.8 million in loans), plus substantial loans from the Export-Import Bank, the International Bank for

94 *New York Times*, Mar. 16, 1966; *Ibid.*, Nov. 14, 1966.

Reconstruction and Development and United Nations agencies. The greater part of the funds were assigned to economic and technical assistance, and road construction. Costa Rica, the second ranking recipient of United States aid in Central America, received $53.5 million in the same period. Under the Alliance for Progress, $48.3 million was disbursed to Guatemala between 1962 and 1966, as well as loans from United States-supported lending institutions. The Peace Corps, numbering close to one hundred members, was making worthwhile contributions, particularly in the rural areas. While United States economic assistance remained substantial, the large Guatemalan population reduced the per capita aid in 1965–1966 to only $1.20 per capita, the lowest in Central America. And, owing in part to civil disorders and subversion, 10.5 percent of the country's budget was allocated to the military establishment.[95]

A problem related to these conditions, though not clearly caused by them, was the growth of Communist insurgency. It was the most dangerous problem faced by President Méndez when he assumed office, and even though the movement suffered reverses, Guatemala was ranked first on the list of Latin American nations threatened by insurgency at the close of 1967. Remotely the movement was connected with the thwarted goals of the Revolution of 1944, and more immediately with the abortive *coup* of November 13, 1960, against President Ydígoras. Aside from the usual grievances of lower ranking army officers, a number of the forty-five young officers who took part in the latter uprising were incensed by the use of Guatemalan territory by the United States for training the Bay of Pigs invaders. They regarded it as a violation of their country's national sovereignty at the hands of a United States "puppet," President Ydígoras. One of the disaffected officers, Lt. Marco Antonio Yon Sosa, named his force the Revolutionary Movement of the 13th of November (MR-13), after the 1960 coup attempt, and began guerrilla operations in the northeast in 1962. The Rebel Armed Forces (*Fuerzas Armadas Rebeldes* – FAR), a Communist-influenced guerrilla force was added to his organization as well as a student group. Sosa's campaign ended late in 1962 when most of the outnumbered rebel forces fled into Honduras.

The next stage of insurgency began in 1963 under the leadership of Yon Sosa and the MR-13, and a new FAR headed by Lt. Luis Augusto Turcios Lima. Their guerrilla forces ambushed military patrols, assassinated army officers, and in 1965 the FAR carried out a number of kidnappings to terrorize the public and to raise money for the movement. Ultimately the two rebel leaders differed on ideology, Yon Sosa going into the Castroite sphere and Turcios moving into an alliance with the Communist party of Guatemala (*Partido Guatemalteco de Trabajo* – PGT), the latter becoming ascendant. The army waged an intensive campaign against both guerrillas and terrorists in the fall of 1966 with an

[95] U.S. Dept. of Commerce, *Statistical Abstract of the United States, 1966* (Washington, 1966), p. 855; Simon G. Hanson, *Five Years of the Alliance for Progress, An Appraisal* (Washington, 1967), p. 4; *New York Times,* Jan. 23, 1967. Unless otherwise stated, the terms "foreign assistance" or "economic assistance," as hereinafter used, refer only to programs conducted under the Foreign Assistance Act and exclude the programs of the Export-Import Bank, the Peace Corps, Social Progress Trust Fund programs administered by the Inter-American Development Bank, and the Food for Peace programs carried out under Public Law 480, the Agricultural Trade Development and Assistance Act of 1954.

accompanying state of seige. Government forces reported successes in 1967, having been reinforced by newly formed right wing counterterrorist groups. The FAR leader, Turcios, killed in an auto accident in 1966, was replaced by a young Communist lawyer, Cesar Montes.

The guerrilla movement failed to enlist the support of either the Indian population or the Ladino peasants: "Without question, the larger portion of the guerrillas in 1965 and 1966 were students, most of whom went out only on week ends. According to an FAR leader who defected in April 1967, of four hundred active FAR guerrillas, three hundred were students and one hundred peasants."

With events of 1954 in mind, the success of guerrilla operations in Cuba under Castro's leadership, and the progressive involvement in Vietnam in the 1960s, the United States had a vital interest in seeing the Communist-inspired rebels in Guatemala safely immobilized. To this end the Guatemalan army, a well-equipped force of twelve thousand, was supported by the United States with more than $12 million annually and a small team of military advisers. Officers of the Guatemalan army were commonly sent to the United States for advanced training. One of them, Lt. Turcios, the former FAR leader, was sent to Fort Benning, Georgia, for training in the elite Ranger course after graduating from Guatemala's military college. To strengthen the civil security forces in Guatemala, Washington operated a public safety program through the Public Safety Division of AID. Greater emphasis was placed on this program after it had become apparent that the four-thousand man police force was unequal to the task of preventing terrorism, whether originating on left or right.[96]

President Méndez had run a platform of political liberty. He denied any ties whatsoever with communism, and promised a government that would enhance the welfare of all sectors of society. Many persons saw his Popular Revolutionary Party as the heir of the popular October Revolution of 1944. Once in office he could not push a realistic reform program in keeping with the goals of the 1944 Revolution without arousing the opposition of the military and propertied classes who regarded such ideas as tantamount to communism. His proposals for tax reform and modest agrarian reform led to a flight of capital and ominous threats from these elements. President Méndez' appeal to the guerrillas to accept an amnesty offered by Congress failed, and he was compelled to allow the army to use its own methods against the rebel forces.

Continuing suppression brought reprisals from terrorist elements which included the abduction and assassination of both foreign diplomats and Guatemalan officials. In January 1968 Colonel John Weber and Lt. Commander Ernest A. Munro, American military advisers to the Guatemalan armed forces, were killed in an attack by a terrorist group, and John Gordon Mein, U.S. ambassador to Guatemala, was slain by gunmen in Guatemala City in August. Later the Guatemalan government met terrorists' demands in the kidnap-ransom of U.S.

[96] For further details on the Guatemalan insurgency problem see U.S. Congress Subcommittee on American Republics Affairs, Committee on Foreign Relations, U.S. Senate, *Survey of the Alliance for Progress – Insurgency in Latin America* (Washington, 1968), pp. 18–24; Alan Howard, "With the Guerrillas in Guatemala," *New York Times Magazine,* June 26, 1966.

Labor Attaché, Sean M. Holly, and of its own Foreign Minister, Alberto Fuentes Mohr. Count Karl von Spreti, the West German ambassador to Guatemala, was murdered by terrorists in April 1970 when the government refused to release twenty-three political prisoners as partial payment of his ransom.

These and other incidents of urban and rural terrorism exerted a strong influence on public opinion as shown in the presidential election of March 1970. Colonel Carlos Arana Osorio, a hero of the nation's struggle against rural guerrilla bands in the mid-1960s, won 42.9 percent of the popular vote. Defeating divided left-wing opposition, Arana ran on a two-party coalition composed of the National Liberation Movement and the Institutional Democratic Party. The Fuentes Mohr incident, occurring just two days prior to the election, apparently convinced many that the government of President Méndez Montenegro was unable to provide adequate protection for its officials or the public. Thus, Colonel Arana's election reflected the public's desire for an end to subversion and terrorism but at the same time showed its concern about the need for greater and economic change. The public "voted left" while electing a candidate with a strong right-wing image.[97]

Continuing domestic violence led the Arana regime to impose a state of siege from November 1970 to December 1971, which empowered the government to make arrests and search homes without a court order. Severe repression ensued, including an assault on the autonomy of the National University of San Carlos.

President Arana undertook a program of economic and social development soon after taking office. The problems involved in implementing the reform program and achieving law and order concurrently proved most difficult. Rejecting the idea of redistributing land, the regime instead concentrated on modernizing and diversifying agriculture. The Agency for International Development approved a $23 million loan in support of this effort. Economic assistance committed by Washington totaled $166.4 million in the years 1948–1970.[98]

In spite of the civil strife and the loss of the Honduran market, resulting from the El Salvador-Honduran conflict of 1969, Guatemala in 1971 had an estimated growth of production of about 5 percent. But this had to be weighed against an annual population increase of 3.4 percent and the fact that most of the Indians are removed from the country's money economy. By 1972 the only prospects for domestic peace and general improvement seemed to be in a program that would reform the nation's archaic social and economic institutions, and make possible a rapid and general redistribution of the national income. Otherwise, the nation faced a possibly more dangerous phase of the unfinished revolution of 1944.[99]

El Salvador did not share the Communist threat in the same degree as neighboring Guatemala, but political instability became pronounced there after 1950. Between 1931 and 1941 El Salvador was under control of Maximiliano Hernández Martínez and the army, a ruthless dictatorship. So vigorous was his defense of the *status quo* that peasants agitating for a program of land reform

[97] *Christian Science Monitor*, Mar. 17, 1970; *Ibid.*, Apr. 10, 1970.
[98] *Statistical Abstract of the United States, 1971*, p. 763.
[99] *New York Times*, Jan. 28, 1972.

were summarily shot by landlords. Forced to resign under the duress of a nation-wide strike in 1944, Hernández Martínez was followed by a rapid succession of presidents and, finally, in 1948 a group of young officers headed by Colonel Oscar Osorio seized power, which it held during the next twelve years. This government carried out a modest program of economic development; it passed some social legislation for city workers, and reluctantly permitted the formation of a labor movement. The government also allowed more political freedom than had been customary in El Salvador. However, the regime did not dare touch the key problem: the concentration of virtually everything worth owning in the hands of an oligarchy comprising a few score families. This group of families, already firmly anchored in the country's basic industry, coffee, reached out to control grazing, manufacturing, banking, sugar and commerce.

El Salvador remained on the periphery of United States diplomatic interest in the early post-World War II years. United States Ambassador Robert C. Hill stressed his country's neglect of the area, and the deep impression that Vice-President Richard M. Nixon's official call of 1955 had made on the population. A Salvadorian public official said: "Mr. Nixon's visit proves that you are really interested in us; it is not just words." It was also evident that El Salvador had relied largely on its own efforts for economic development. Washington supplied grants totalling $1.5 million in the period 1946–1950, three million between 1951 and 1955, and about $1 million annually through 1960. Most of the funds were used in support of economic and technical assistance.[100]

As a market for El Salvador's coffee exports, however, the United States was of crucial importance. Coffee constituted about 90 percent of the value of the nation's entire export income in 1957, and the United States imported 68 percent of the total. A large part of the government revenues came from export taxes with the figure on a sliding scale depending on the New York market. Although on a cutback program, El Salvador had the largest United States quota for any country except for Brazil and Colombia.

José María Lemus, another army officer won the election of 1956 with the support of Osorio. President Lemus visited Washington in March 1959, and in an address before a joint session of the Congress warned of the Communist threat in Central America. On his departure he issued a joint statement with President Eisenhower declaring support of Central American economic integration and the formation of a common market. Following the sharp drop in coffee prices which cost the country $12 million in export taxes in 1958, President Lemus said: "We ask no gifts from the United States, but only that you cooperate in keeping the price of our agricultural products at a reasonable level, while we have time to industrialize." The depressed coffee market, rising middle class discontent, and student demonstrations, led to intervention by a group of army officers who forced Lemus out of office in October 1960.

The rule of military civilian juntas ensued until the presidential election of April 1962, which was uncontested. Lt. Colonel Adalberto Rivera won the election in which less than 20 percent of the voters bothered to cast ballots. Washington withheld recognition of the succeeding *junta* fearing the ideological

[100] Franklin D. Parker, *The Central American Republics* (New York, 1964), p. 161; *New York Times*, Feb. 7, 1955.

leanings of its members. When the public's confidence in the regime was clearly established Ambassador Thorsten J. Kalijarvi gave notice of United States recognition. The *junta* had meanwhile announced its determination to hold free and democratic elections, to respect private enterprise, and to encourage the investment of foreign capital. Impressed by its sincerity, President Kennedy, in February 1961, pledged to help raise the country's standard of living.[101] Another *junta*, alleging that its predecessor was leftist, assumed control in January 1962, and ruled until the election of Rivera in April.

Rivera's administration managed to gain the cooperation of the financial rulers of El Salvador while appearing to be relatively progressive and reform-minded. The reforms carried out by his government were designed to meet the requirements for aid under the Alliance for Progress: modest agrarian reforms and income tax laws, minimum wage laws, tenement rent reduction, a new labor code, and participation in various types of self-help programs. Industrialization was vigorously pushed and agriculture diversified to offset the problem of coffee sales. In implementing the Alliance for Progress programs, the Rivera government had to fend off untranationalist and extreme left-wing propaganda to the effect that the *Alianza* was a "Trojan Horse" of "Yankee imperialism."[102]

In the summer of 1965, Jack Hood Vaughn, assistant secretary of state for Inter-American Affairs, praised El Salvador's achievements under the Alliance for Progress, saying that it was in the vanguard of Alliance endeavors. The country's accomplishments were indeed extraordinary, having achieved one of the highest economic growth rates not only in Central America, but in the entire Latin American area. The gross national product advanced 12 percent in 1964 and 1965 and per capita income had risen from about $200 a year in 1960 to $275 in 1965. Industrialization, encouraged to a large extent by the Central American Common Market, was an important factor in the advance, for in 1963-64 seventy-six new industries and 112 expansion projects were inaugurated. By 1966 it had the largest number of industries of any Central American country. El Salvador had also become one of the more favored areas for private investors, as $40 million came into the country either as direct investments or in loans to industry already established, in 1965.[103]

El Salvador showed a 6 percent increase in the GNP in 1966, and the Rivera administration expressed confidence in the country's economic future. There was, however, growing concern about population pressure, for almost three million people were jammed into the nation's 8,061 square miles, and the annual increase was 3.2 percent. The government hoped to alleviate the problem by making the country a reservoir of trained labor for its bigger associates in the CACM. Nicaragua, for example, had 1.6 million population and seven times the area, and Honduras had two-thirds the population and was five times larger.

The United States fulfilled President Kennedy's pledge, making El Salvador the largest recipient of financial aid under the Alliance for Progress in Central America in the first five years of the program. Averaging more than $10 million a year in 1962–1965, for a total of $43.1 million, the aid disbursed in

[101] *New York Times,* Feb. 16, 1961.
[102] Mario Rodriguez, *Central America* (Englewood Cliffs, 1965), p. 40.
[103] *New York Times,* Aug. 23, 1965; *Ibid.,* Apr. 3, 1966.

1966 rose to $20.4 million. The Peace Corps had fifty-five volunteers there in 1965. Total economic assistance furnished by the United States in the period 1948–1970 was $81.7 million. And by the mid 1960's Washington cited El Salvador as "a model for the other Alliance countries."[104]

In the March 1967 presidential elections the government's candidate, Colonel Fidel Sánchez Hernández, prevailed over moderate-leftist opposition and a coalition alleged to be Communist-dominated. The nominal reforms instituted by Rivera and subsequently carried on by Sánchez Hernández were believed to have earned the animosity of the conservative elements. No basic institutions were threatened, however, so it was not surprising that the government party, Partido de Conciliación Nacional (PCN), which had dominated politics for a decade, succeeded in electing its candidate again in the presidential election of February 1972. Colonel Armando Molina, after winning a four-way contest, declared that he would tour Central America and seek to arrange a summit meeting of the region's presidents. He also expressed a desire to find solutions to the problems involving his country and Honduras. His election was challenged by a revolt in March 1972; this was crushed with the loss of some one hundred military and civilian lives. Meanwhile, although El Salvador's trade position had strengthened in 1971, and production growth was expected to attain 4.3 percent, most of the nation's socio-economic problems remained. Adequate housing was still in scarce supply. Wages had not been increased appreciably, and the benefit of sustained economic growth had not filtered down equally through all segments of Salvadorean society.[105]

Honduras has felt political turbulence in recent times, but little erosion of the traditional social order has occurred. From 1932 to 1948 the nation suffered *caudillo* rule under General Tiburcio Carias Andino, a regime characterized by severe repression of democratic institutions, little material progress, and the continuing aggrandizement of the upper class elites. Carias Andino gave up office to Juan Manuel Gálvez in 1948, and under Gálvez' direction democracy made slight gains, and mild economic reforms were instituted. Such reforms, however, did not satisfy the rising political action groups who demanded more sweeping and fundamental changes.

A general labor strike occurred in 1952 directed at UFCO, which had long enjoyed the favor of the government and the upper class of Honduras. Although the strikers won wage increases, paid vacations, improved housing and other common goals of labor, the greatest significance of the strike was the fact that labor had finally gained the right to organize. Nevertheless, the traditional propertied elements still retained supremacy.

In the strife-ridden election of 1955 the Liberal Ramón Villeda Morales appeared to have won the presidency. But he was denied the office, and Juan Lozano Díaz, Gálvez' vice-president, ruled by decree until removed by a military *junta* in 1956. A Five-Year Plan of economic development, begun in 1955 by Lozano, was abandoned by the *junta*. In May 1957, Honduras charged that

[104] *Statistical Abstract of the United States, 1971,* p. 793; Hanson, *Alliance for Progress,* p. 4; Agency for International Development, *Proposed Economic Assistance Programs FY 1967* (Washington, 1966), pp. 91–92.

[105] *Ibid.; Christian Science Monitor,* Mar. 10, 1972; *Times of the Americas,* Apr. 4, 1972.

Nicaraguan troops had invaded a disputed area called the Mosquito Coast, a vast, sparsely settled region extending 150 miles along the Caribbean coast. Honduras' claim was based on the territorial award made by King Alphonso XIII of Spain in 1906, but which Nicaragua did not recognize. A United States military mission was sent from Panama to aid in the surveillance of the territory after the OAS Inter-American Peace Committee had negotiated a cease-fire. Ambassador John C. Dreier was the United States representative on the committee. Nicaragua agreed to accept the decision of the International Court of Justice, which in 1960 ruled that the arbitration award of 1906 was valid.[106]

The presidency of Honduras was turned over to Villeda Morales by the military *junta* in December 1957, who proceeded to carry out a modest economic development program, and cautious educational and agrarian reforms.[107] When he took office the nation's international monetary reserves had declined for the third straight year; in the following year the floating debt was doubled. The regime crushed a revolt led by an exiled colonel, and captured a small Cuban-armed force which attempted to invade Nicaragua, in 1959.

Honduras remained in the economic doldrums in 1960 in spite of more than $27 million in loans from international organizations since 1956. According to an ICA report: "The gross national product is hardly keeping ahead of population growth, and more timber is burned each year than is exported, and lumbering and farming methods in general are leading toward serious land erosion." United States government economic assistance to Honduras amounted to $45.8 million ($24.2 million in grants; $21.6 million in loans), in the years 1948 to 1965. A high priority was placed on road building because the central location of Honduras makes it the crossroads for highway building projects to close the regional transportation gap.[108]

The total trade of Honduras declined 7 percent in 1959–1960, the main exports in order of importance being bananas, coffee, timber, minerals, livestock, corn and cotton. To stimulate production UFCO adopted a policy that involved abandoning sideline activities, concentrating on banana production; however, unemployment grew more critical as a result. Early in 1962 the United States-Honduras Trade Treaty of 1936 was terminated. As the treaty had given preferential treatment to U.S. exports to Honduras, especially in food, autos and textiles, it had been regarded as unfavorable to the latter's interests. The treaty's revocation had been opposed by the Chamber of Commerce, controlled by importers and distributors of United States goods. The result was to reorient commerce toward the CACM.[109]

President Villeda Morales made a brief state visit to Washington in December 1962. He stressed his country's strong anti-Communist views, and took the occasion to invite President Kennedy to meet the Central American presidents at a conference of the OAS in San José, Costa Rica. The invitation was accepted. While relations on the high government level were most cordial,

[106] *New York Times,* May 7, 1957; *Ibid.*, Nov. 19, 1960.
[107] William S. Stokes, "Honduras: Problems and Prospects," *Current History,* vol. 50 (Jan. 1966), pp. 23–24.
[108] *Statistical Abstract of the United States, 1966,* p. 855; AID, *Programs FY 1967,* p. 92.
[109] *New York Times,* Jan. 10, 1962.
[110] Rodríguez, *Central America,* pp. 36–37.

public opinion in Honduras tended to be nationalistic if not anti-Yankee. President Gálvez' support of Castillo Armas was resented, and Villeda Morales had expressed the popular attitude in refusing to cooperate in the Cuban invasion. However, Honduras broke relations quickly thereafter, standing behind United States Cuban policies. The Agrarian Reform Law of 1962 reflected a similar nationalistic antagonism. Designed to help Honduras qualify for Alliance for Progress funds by providing means for agrarian reform, it also prohibited foreigners from living or owning land within thirty miles of the seacoast. UFCO and other fruit companies held it to be prejudicial to their economic interests.[110]

When President Villeda Morales was about to finish his six-year term in October 1963, and an election was scheduled to replace him, the Liberal party appeared to have an easy majority. But ten days before the election Colonel Osvaldo López Arellano, a staunch Conservative, staged a *coup* on the ground there was danger of a Communist take-over. These charges were never proved. Washington hastened to announce the withdrawal of all economic and military aid personnel, signifying that it would interrupt Alliance for Progress aid and other programs to any country in which a democratic regime was overthrown. Washington extended recognition to the military-backed government the following December when it had concluded that refusal to reopen diplomatic channels could have brought on new uprisings. A legislative assembly was chosen in February 1965, in elections marked by fraud, with the Conservatives gaining a majority of the seats. A new constitution was drawn up, and the next month López Arellano was elected for a six-year term. In mid-1965 Honduras cooperated with the United States and the OAS by sending a contingent of 250 troops to join the OAS Peace Force in the Dominican Republic.

Alliance for Progress programs in Honduras were summarized, in part, as follows: "The essential goal of the AID program in Honduras is to assist the Honduran government in accelerating the forward movement of the economy. AID has selected four priority areas for program concentration – rural development, government planning and administration, human resources development, and industrial development . . . " The United States Information Agency sought: (1) to encourage a political moderation and the peaceful return to constitutionality after the October 1963 revolution; (2) combat inroads by Communists in government and the universities especially; (3) to promote the Alliance for Progress; (4) promote Central American integration. United States aid disbursed to Honduras in the period 1962-1966 totaled $33.9 million, with per capita aid in 1966 standing at $5.90. A total of 103 Peace Corps volunteers were at work there in 1965, making valuable contributions to rural development. Under a military aid program which included Civic Action and Counter-insurgency training, the United States had contributed in 1950–1965 a total of $4.291 million.[111]

The Honduran economy expanded in 1964 as exports of bananas, coffee, and cotton followed an upward trend. The country's GNP advanced 9.6 percent in 1965, but dropped to 6.3 percent in 1966. The former figure was closer to the level needed to close the gap between Honduras and its four partners in the

[111] Hanson, *Alliance for Progress,* p. 4; Stokes, "Honduras," p. 26.

CACM because it had constantly lost ground. The average Honduran income of $182 annually was 25 percent lower than the CACM average, but higher than Guatemala's.[112] The nation lacked civil freedoms, efficient administration and a reasonably equitable distribution of wealth.

An undeclared war between El Salvador and Honduras, the first significant military conflict between American republics in more than a generation, broke out on July 14, 1969. When it ended, after five days of fighting, two thousand persons had died, and close to 80,000 Salvadorians fled Honduras to their adjoining homeland. Mistakenly labeled the "soccer war" because it was immediately preceded by outbreaks of violence in both countries after a soccer game between their national teams in San Salvador on June 15th, the causes were far more complex. When the conflict began there were some 300,000 Salvadorians and their descendants living in Honduras, representing 12 percent of the Honduran population. The government of that country declared that only 1,036 were legal residents, whereas Salvadorian officials claimed that the blame lay with the Honduran government which had imposed obstacles to obtaining legal residence. While these facts were not made clear, El Salvador declared that its invasion was a response to Honduran violence which amounted to "genocide" against the Salvadorian residents. At bottom, the causes were demographic and economic. Following a long-established pattern, the Salvadorians had left their heavily overpopulated country in search of economic opportunity in the open spaces of Honduras.

The Council of the OAS, acting provisionally as Organ of Consultation, met in emergency session calling for a cease-fire, and on August 3rd El Salvador, which had sent some ten-thousand troops into Honduras, withdrew her forces under threat of condemnation and the possible imposition of sanctions by the organization. Military and civilian observer teams were stationed in both countries, including representatives of the United States and the Inter-American Commission on Human Rights. Bitterness continued to run high in El Salvador over the inability of the OAS to persuade the Honduran government to lift the blockade on overland traffic to and from El Salvador on the Pan American Highway which links the two countries and is the main traffic artery for Central America.[113]

Honduras, poorest of the isthmian countries, withdrew from the CACM in 1971 after it had become clear that she could not afford the costlier products of the protected industries of her partners. This led to the flooding of the markets of Costa Rica and Nicaragua with the manufactured exports of El Salvador; the former had quickly enacted protective reprisals. Honduras favored returning to the CACM, but could do so only if the five-nation block were restructured to allow for the disparate development of this member state. Washington's commitment for economic assistance to Honduras in 1970 was $5.2 million, which brought total aid to $87.2 million in the period 1948–1970[114]

[112] *New York Times*, Jan. 23, 1967.
[113] *Americas*, 21 (Sept. 1969), 42–45; *New York Times*, Jan. 26, 1970.
[114] *Statistical Abstract of the United States, 1971*, p. 793; *New York Times*, Jan. 28, 1972.

Ramón E. Cruz, a lawyer, diplomat and university professor, thwarted in his bid for the presidency in 1963, won the presidential election in 1971. He pledged to work for a peaceful solution of his country's border dispute with El Salvador, but favored retention of the OAS observer force along the border.

The Cruz administration proved to be short-lived, however, for on December 4, 1972, General López Arellano again seized power. Citing a "chaotic situation in the country " he said that the Army would insure that the situation is "arranged or corrected." Although it was a calm takeover with little public reaction, it represented a classic example of the Central American military's disregard of the popular will. President Cruz, the first chief executive to be elected by direct popular vote in forty years, declared after his ouster: "I would have faced all the problems and solved them, but what can I do if they took the power away from me?" There was no indication when the military would again sanction free elections.

Nicaragua is still ruled by a dictatorship, that of the Somoza family, and now in its second generation. Surely one of the most outstanding results of the United States occupation of Nicaragua was the establishment of an efficient National Guard under Anastasio ("Tacho") Somoza, who quickly consolidated a virtually unassailable power position and, at the end of the occupation, captured the presidency in 1937. From then until 1956, save for a brief period, Nicaragua was dominated by Somoza who upheld the traditional neo-feudal economic structure, but carried out an economic development program, an important feature of which was the courting of foreign capital and enterprises. In the process, he built up one of the largest personal fortunes in the Western Hemisphere.

Political opposition was suppressed and middle and lower class groups were required to pledge their allegiance to the dictator. Somoza, failing to gain the support of the working class, collaborated with the Communists. This arrangement was terminated in 1948 with the advent of the cold war, and the Communists' power was broken.

Somoza became embroiled in disputes with neighboring states, particularly Costa Rica. In 1948 and in 1955 tension between the two countries approached open hostility, causing Washington much concern. The personal relations between Somoza and José Figueres, President of Costa Rica from 1953 to 1958, were especially venemous, Somoza on one occasion challenging his opponent to a duel. Figueres was charged with plots to invade Nicaragua with the Caribbean Legion, as well as assassination attempts.

The Somoza-Figueres conflict, which brought the two countries to the brink of war, was a product of the civil war that erupted in an invasion force. Skirmishes occurred, and in January 1955 the OAS Council sent a commission to investigate the invasion. Traveling in U.S. Navy aircraft from the Canal Zone, the commission carried out surveillance at the scene of hostilities. It was determined that supplies were being shipped across the border, but Nicaragua could not be charged with complicity. President Somoza cooperated with the commission, established a temporary buffer zone, and interned the remnants of the rebel forces after they had fled back over the frontier in defeat. Official United States sources in Costa Rica indicated that the invaders numbered about four hundred,

10 percent of whom were members of the Nicaraguan National Guard, fighting out of uniform.[115]

At the conclusion of his good will tour of the Caribbean in March 1955, Vice President Richard Nixon said that one of the "more constructive" results of his tour was the easing of tensions between Nicaragua and Costa Rica. The two countries signed an accord ending the dispute at the Pan American Union in Washington, D.C., in January 1956, a day short of a year after Costa Rica charged she was being invaded by an "army of adventurers" from Nicaragua.[116]

President Somoza was assassinated in 1956, but at the hands of an embittered Nicaraguan. Washington sent a team of surgeons from the Canal Zone and Walter Reed Hospital in an effort to save his life, but it was too late. No basic changes occurred because his two sons, Anastasio, Jr., a West Point graduate, class of 1948, and Luis, seized the reins of power and perpetuated the regime. As in the past, the National Guard was used as the chief instrument of political control.

Among the Central American states, Nicaragua's economic development has been the most extensive and rapid. Less formidible terrain features, exploitable natural resources, and a large area for expansion coupled with low population density, have all favored its growth. Administratively, the Somoza dictatorship provided political stability and attracted foreign enterprise and investment. The legal safeguards for foreign capital were the most generous and rigidly enforced in the isthmus. Long range economic development plans were drawn up with the assistance of foreign experts. Most notable was "The Economic Development of Nicaragua," a study completed by the International Bank in 1952. Under Somoza the country's infrastructure, roads, communications, and port facilities, was modernized and expanded, and credit policies of the National Bank fostered national growth by private enterprise. In contrast to its neighbors, Nicaragua never was as reliant on one crop for a source of revenue. Seldom did coffee represent more than 50 percent of the value of exports, and cotton surpassed it at times in the late 1950s. The United States remained Somozaland's best customer and source of imports, but trade with Europe was also vigorously pushed. The nation's exports to Europe rose from $4 million in 1949 to $17 million in 1953.

Somoza cooperated loyally with the United States before and after World War II. His instructions to Nicaragua's diplomatic representatives at international conferences were consistently: "Cooperate fully with the United States of America." This support was especially appreciated by Washington with the unfolding of the cold war, Communist penetration of Guatemala and, later, the rise of Fidel Castro in Cuba. Wartime assistance was marked by a $4 million grant to build a road to Rama on the Escondido River. In the years 1946–1950 United States assistance totaled $5 million, $5 million in 1951–1955, $4 million in 1956–1957, and about $4 million a year in 1958–1960, making a 1956–1960 total of $27 million. Substantial loans were obtained from the International Bank for Reconstruction and Development and lesser assistance from the United Nations in the same period.[117]

[115] Martz, *Central America,* pp. 182–197;*New York Times,* Feb. 25, 1955.
[116] *New York Times,* March 6, 1955; *Ibid.,* Jan. 10, 1956.
[117] Parker, *Central American Republics,* p. 235.

Luis Anastasio Somoza Debayle, who held the presidency from 1957 to 1963, maintained the firm anti-Communist position of his father, and strengthened the country's bonds with the United States. He reported an attempted invasion to the OAS in June 1959, contending that Nicaraguan exiles, Dominicans and Costa Ricans, directed by José Figueres, were responsible. The United States delegate, Ambassador John C. Dreier, supported the Nicaraguan argument that the development did constitute a threat to peace.[118]

Rebels struck against Nicaragua in November 1960, and the insurgent action against Guatemala began two days later. Both governments charged the revolts were Cuban-born and Cuban-armed. Washington quickly dispatched a carrier with seventy planes and five destroyers to this patrol whereas a year before the Navy had merely put two small vessels on patrol off the Panamanian coast following a Cuban-based invasion of Panama by a band of Panamanian rebels. The show of force without any prior consultation with other members of the OAS indicated that Washington believed a large-scale invasion from Cuba was imminent. This threat did not materialize, but a weak invasion effort was launched across the Honduran frontier in January 1961. These invaders were quickly routed. President Luis Somoza cooperated with the United States in April 1961, invasion of Cuba by permitting the Cuban assault forces to group and embark from the Puerto Cabezas region of Nicaragua's east coast. The Corn Islands, leased to the United States, were also used in the operation.[119]

Nicaragua stood solidly behind the United States in the Cuban missile confrontation of 1962, and at the San José Conference of March 1963, Luis Somoza, together with Miguel Ydígoras, were dismayed that Cuban policy questions were not on the agenda. Since their countries had been prime targets of Castroite subversion they favored a strong commitment from Washington to overthrow that regime. It was clear, however, that irrespective of ideology, Luis Somoza's domestic opponents were anti-United States, for they held this country mainly responsible for the longevity of the Somoza family's iron-clad rule.

Luis Somoza's government emphasized diversification, away from excessive dependence on cotton and coffee, and the policy met with considerable success. Bananas were planted on the Pacific coast, and shipments of frozen beef to the United States began in 1959. More important was Nicaragua's ratification of the Treaty of Central American Integration in 1960, the growth of light industries and the opening of an oil refinery. Aid disbursed to Nicaragua under the Alliance for Progress, totaling $43.6 million in 1962–1966, reinforced these developments. However, the economic gains were offset somewhat by the large proportion of funds used by the military services. In 1965–1966, Nicaragua's federal government expenditures for this purpose stood at 16.5 percent, the largest proportion by far among the Central American republics. Total United States economic assistance for 1948–1965 was $45.3 million ($17.2 million in grants; $28.0 million in loans).[120]

In the 1963 election, the Somoza-sanctioned candidate, René Schick

[118] *New York Times,* June 5, 1959.
[119] *Ibid.,* Nov. 20, 1960; Parker, *Central American Republics,* p. 305.
[120] Hanson. *Alliance for Progress,* p. 4, 198, *Statistical Abstract of the United States, 1966* p. 855.

Guitiérrez, won the presidency. Staunchly anti-Communist, he offered Nicaragua as a base for an invasion of Cuba to overthrow Fidel Castro. When he suffered a heart attack, which ultimately caused his death in August 1966, President Johnson sent two physicians to help save his life. Lorenzo Guerrero, the Minister of the Interior, was elected by Congress to serve as provisional president.

Opposition to the seemingly interminable dynastic rule of the Somoza's broke out in January 1967, when a wave of violence occurred. Contending that Communists were behind it, the government ruthlessly throttled all political aspirants. Dr. Fernando Aguero Rochas, candidate of the opposition coalition, had led a demonstration to get army support in a *coup* to help insure honest elections in his contest with Major General Anastasio Somoza, Jr., "Tachito." After hundreds of persons had been killed and wounded in downtown Managua, Dr. Aguero took refuge in the Gran Hotel and held more than eighty United States citizens as hostages. The U.S. Embassy served as a telephone link between the two sides. Aguero was not a Communist himself, but he was aided by them and his opponents equated the two: "Aguerismo-Communismo."[121] "Tachito's" military training and the hold he had consolidated over the National Guard enabled him, not surprisingly, to win the presidential election in February 1967. Luis Somoza died two months after his brother's election.

In 1971, President Somoza, anticipating that he would not be eligible for reelection in 1972 as the constitution then stood, made a deal with the Conservative Party. In 1972, by this arrangement, he would step down, giving power to a three-man committee that would govern the country for thirty months; in the next presidential election Somoza would be back on the ballot. Outside of political circles, the Nicaraguans seemed disinterested in the situation. The main political issue, for the moment derailed, was largely over a name and a family and whether it had the right to go on dominating the country. Maintaining close ties with the United States, President Somoza attended the 25th anniversary of his West Point graduating class and, with his fellow cadets, had dinner at the White House as guests of President Nixon.[122]

Nicaragua shared in the setback of the CACM, chiefly in the reduction of the level of investment and commercial activity; increased imports from the region caused a trade deficit of $27 million for 1971. But public works activity, and rising cotton and shrimp exports were expected to maintain the 4.6 percent economic growth of 1970. United States economic assistance totaled $2.3 million in 1970, bringing the total to $98.6 million in the period 1948–1970.[123]

Managua, a city of 325,000, and the nation's capital, was destroyed by an earthquake on December 23, 1972, with heavy loss of life. Washington responded quickly in the crisis, as President Nixon ordered his Cabinet to make

[121] *New York Times,* Jan. 27, 1967. The Communists in Nicaragua (*Partido Socialista de Nicaragua – PSN*) held some strength among labor and in the university. Although pro-Moscow, it retained ties with Cuba. Castroite guerrilla activity, which reappeared in August, 1967, was carried on by the National Sandinista Liberation Front (*Frente Sandinista de Liberación Nacional – FSLN*). Headed by Carlos Fonseca Amador, a former university student, it was aligned with Castro. The Nicaraguan National Guard had reportedly contained the movement. See U.S. Senate, Comm. of For. Rels., *Insurgency in Latin America,* p. 27.

[122] *Times of the Americas,* June 16, 1971; *Ibid.,* Aug. 4, 1971.

[123] *Statistical Abstract of the United States, 1971,* p. 763; *New York Times,* Jan. 28, 1972.

an "all-out effort to provide all needed help to Nicaragua." The relief effort was coordinated by a special disaster group in the Department of State. Although the city had been destroyed by earthquake twice in the past – in 1885 and 1931 – it was reported that General Somoza would have it rebuilt on the same site. He made it clear, if there were any doubt, that the Somoza family still ruled Nicaragua.

The Costa Rican commitment to democracy as both a political and social way of life has distinguished that country from its neighbors in the isthmus proper and from Latin America in general. Since 1889 the democratic tradition has prevailed, except for two notable occasions. The first was the dictatorship of Federico Tinoco, from 1917 to 1919. The second hinged on events of the 1940s leading to the civil war of 1948. The party headed by Calderón Guardia monopolized politics and increasingly stifled national elections.

The political crisis arose in 1948 when outgoing President Teodoro Picado refused to turn over the presidency to his elected successor, Otilio Ulate. The result was a revolution led by José Figueres, a landowner and M.I.T.-trained engineer, which made him the hero of the country's lower classes. The fighting lasted for six weeks, and before a truce was called, the Figueres forces had lost fifty-six men and the government forces 1,500 men out of a population of 870,000. Calderón Guardia and Teodoro Picado fled abroad, and José Figueres headed a provisional, and strongly reform-oriented *régime* from May 1948 to November 1949, when the presidency was restored to Ulate. Serving his four-year term, Ulate was succeeded by José Figueres in the election of 1953. Figueres' candidate was defeated in the 1958 election by Mario Echandi Jimínez. His administration proved to be democratic but conservative in its policies.

United States-Costa Rican relations in the two post World War II decades reflected friendship and cooperation. Pressed by the circumstances attending the dispute with Nicaragua, previously described, Washington convincingly assumed the role of impartiality, and supported the OAS in its efforts to resolve the problem. The United States grew increasingly important in Costa Rican daily life through trade, aid, disaster relief, private investment, and the threats emanating from Castro's Cuba.

With its economy firmed based on agriculture, Costa Rica's foreign exchange was derived mainly from the export crops of coffee and bananas with cacao running third. It followed that "as go the crops, so goes the nation's economic prosperity." To strengthen the economy through product diversification, the United States early established farm extension agencies and agricultural experiment stations under the Point IV mission. Costa Rican personnel trained by the mission were able to competently operate the extension service beginning in 1952. Trade with the United States expanded, exports for 1950 rising to $35 million and $38 million the following year. Throughout the 1950's by far the larger proportion of Costa Rican trade was with the United States which took 45-50 percent of the exports and supplied nearly 60 per cent of the imports. A major fiscal accomplishment of Costa Rica was the liquidation in 1952 of most of the twenty-two year old foreign debt of $26,500,000 through the U.S. Foreign Bondholder's Council.[124]

[124] Martz, *Central America*, p. 234.

In spite of recurring problems with Nicaragua and an inflationary trend, the Costa Rican economy advanced through 1958. However, 1959 was an uneasy year, financially and politically. UFCO reported the worst coffee crop in ten years and the government of President Mario Echandi Jimínez, after its first full financial year in office, faced a sharp decrease in revenue. The income tax from UFCO operations, which in good years reached $3–$4 million, fell to $650,000. To help relieve the problem the U.S. Treasury Department revised the proportion of taxes paid in Costa Rica by the company from its previous 50-50, to 60-40 in the small nation's favor.

This concession was minor compared to Washington's financial grants to the nation. In the period 1946–1960 the total was $39 million, second in quantity only to Guatemala among the Central American republics. About 70 percent of this figure was devoted to road construction, principally the Inter-American Highway. Additional assistance was obtained from United States-supported agencies, the Export-Import Bank, the International Bank for Reconstruction and Development, and agencies of the United Nations. In 1962 when Costa Rican revenues declined owing to delinquency in income tax and real estate tax payments, the U.S. Treasury extended a $6 million loan to reinforce the country's bank reserves, and the International Monetary Fund issued a $15 million standby credit.[125]

In fulfillment of campaign pledges made in 1953 expressing the country's nationalistic sentiments, President Figueres compelled UFCO to renegotiate its contract with Costa Rica in 1954. UFCO, under the existing contract, paid a 15 percent income tax, enjoyed special exchange rates and duty-free imports. The Figueres government demanded that the company pay one-half of its profit, but the new contract, to be valid for thirty-four years, called for 42 percent. The company agreed to pay customs duties on about one-half of its imports, and to honor the government's minimum wage laws. Costa Rica, on its part, assumed responsibility for UFCO's schools, hospitals and recreational facilities.[126]

Under the leadership of President Figueres, Costa Rica was anti-Communist and he and his successors stood by the United States in its efforts to contain the expansion of Fidelismo. Diplomatic relations with Cuba were severed in September 1961. Although Communists did not abound in Costa Rica, Ambassador Raymond Telles, a former mayor of El Paso, Texas, and the first of President Kennedy's appointees in the experiment of posting Spanish-speaking men of Latin American origins in Latin American embassies, an exponent of "grass roots" diplomacy, declared: "The Communists don't work at cocktail parties . . . you meet those boys around with the laborers . . . we've got to go where the Communists are."[127]

The adminstration of Francisco Orlich, Costa Rican President from 1962 to 1966, adhered to a strong anti-Castro and anti-Communist policy, insisting that Castro's propaganda and subversive activities be terminated, and supporting the Venezuelan government's charges against the Castro regime. Costa Rica was a strong advocate of a total economic blockade of Cuba, and joined the original

[125] Parker, *Central American Republics*, p. 278; *New York Times*, Jan. 10, 1962.
[126] Rodríguez, *Central America*, pp. 24–25.
[127] *New York Times*, Oct. 8, 1961.

demand for a political blockade as well. The granting of asylum to José Miró Cardona, the Cuban exile leader after he left the United States amid rumors that he was preparing another invasion force, was also revealing of its attitude. When the anti-United States demonstrations broke out in Panama early in 1964 Costa Rican public opinion supported their neighbors; however, the government of Costa Rica offered friendly cooperation and performed the custodial role for both countries after the disputants had cut their diplomatic ties.[128]

The San José Conference, convened in the Costa Rican capital in March 1963, was attended by President Kennedy, the first United States president to grace Central American soil, and the chief executives of the isthmian republics. The tumultuous reception accorded President Kennedy testified to his popularity and the acceptance of his Latin American policies. In the Declaration of Central America which issued from the conference Washington agreed to furnish $6 million for a survey of the region to determine the quantity and timing of economic assistance needed from the United States for implementing the Common Market program. The signatories affirmed that communism was one of the chief obstacles to economic development, and agreements were made to thwart Cuban subversion. But the declaration was restrained in respect to Cuba, which was not mentioned in the document, and to Soviet aggression in the Western Hemisphere.[129]

Failing to see the advantages of Central American economic integration, Costa Rica refused to sign the general treaty in December 1960. This decision was reversed in 1963 under the administration of President Orlich and Costa Rica became the last of the five republics to achieve active membership in the Central American Common Market.[130] Another notable event in 1963 was the eruption of the normally dormant volcano Irazú, located just fifteen air miles east of San José. In the space of twenty-one months, from March 1963 to October 1964, damage to crops, livestock and property from volcanic ash amounted to $30 million. Economic aid, livestock feed, technical assistance, including a detachment of Seabees, was quickly sent by the United States, and additional support was forthcoming from the United Nations.

Economic reverses suffered from the Irazú disaster were soon overcome as a result of higher world coffee prices, gains from membership in the CACM, foreign investment, and United States financial aid. Even before the end of the first full year in the CACM the nation achieved a trade surplus, largely from exports of light industry manufactures. The country's "Protection of New Industry" law, introduced in May 1962, had encouraged $40 million in investment which in turn produced a booming growth of manufacturing plants and a wide range of consumer goods. Industrial development enabled the country to save $18 million in foreign exchange annually by the mid 1960s. Doubtless contributing to the upturn was the $44.8 million disbursed under the Alliance for Progress in 1962-1966, and loans obtained from the IADB and the

[128] C. Harvey Gardiner, "Costa Rica: Mighty Midget," *Current History*, vol. 50 (Jan. 1966), p. 10.
[129] *Hisp. Amer. Rep.*, XVI (May, 1963), pp. 244–245; *New York Times*, Mar. 23, 1963.
[130] See James D. Cochrane, "Costa Rica, Panama, and Central American Economic Integration," *Journal of Inter-American Studies*, VII (July, 1965), 331–344.

Export-Import Bank, and United Nations agencies. In the period 1948–1965 the United States supplied $53.5 million ($17.6 in grants; $35.8 in loans). The country was also benefitting from the services of sixty-one Peace Corp volunteers in 1965.[131]

Dr. José Figueres was again elected to the presidency in February 1970 for a four-year term. But despite his personal popularity, he was compelled to retreat on a proposal to join diplomatic ties with the USSR, in 1971. It was an effort to promote commerce with the Soviet Union that led to his efforts to reactivate diplomatic relations, established in the late 1940s but never formalized with an exchange of ambassadors. Opposition to the President's plan came largely from right-wing groups and many politically neutral elements, who oppose communism and fear potential Soviet subversion.[132]

With a faltering economy, reflected in a $88 million commercial deficit in 1970, and a severe balance of payments problem in 1971, the government sought to sell part of the country's large coffee surplus to the Soviet Union for needed dollars. Coffee sales declined and bananas replaced coffee as the chief export crop in that year. Negotiations were continued in early 1972 for a trade agreement by which the USSR would buy thirty-thousand tons of Costa Rican coffee. Economic assistance from Washington in 1970 was $17.4 million, which brought the total to $96.5 million for 1948–1970.[133]

Costa Rica has for some years been in the anomalous position of combining a favorable trade picture, and seemingly ample investment capital, with a treasury situation in which the government has found it difficult to make ends meet. A basic cause is the phenomenal population growth, which at 4.1 percent is one of the highest in the world. Another factor is that Costa Rica has a rapidly expanding economy calling for frequent and sometimes unbudgeted support. Although expanding industrially, the economy is still primarily based on agriculture and stands or falls by its export crops.[134]

ECONOMIC INTEGRATION.

In June 1951, at Mexico City, the Central American delegation to the fourth session of the United Nations Economic Commission for Latin America (ECLA), adopted a resolution favoring the integration of Central American economies. The five economic ministers, seizing the initiative from the foreign ministers, established the Committee for Economic Cooperation of the Central American Isthmus with ECLA as their secretariat; their first meeting took place at Tegucigalpa, Honduras in August, 1952. Until 1958, when the first significant economic agreements were signed, their efforts were directed toward securing transitional arrangements, persuading the Central American governments of the need for economic integration, and reaching agreement on the nature and objectives of an integration program.[135]

Obstacles to integration were numerous and often fundamental. The lack of regional cohesiveness has been described; in the last century and a quarter there had been twenty-five different attempts at political unity. While political

131 Hanson, *Alliance for Progress,* p. 4; *New York Times,* Jan. 22, 1965; *Statistical Abstract of the United States, 1966,* p. 855.
132 *Christian Science Monitor,* Sept. 8, 1971.
133 *Statistical Abstract of the United States, 1971,* p. 793; *New York Times,* Jan. 28, 1972.
134 AID, *Programs FY 1967,* p. 91.
135 Joseph S. Nye, Jr., "Central American Regional Integration," *International Conciliation,* no. 562 (Mar. 1967), pp. 17–22.

unity was not under discussion, the similarity of their economies posed an obstacle to meaningful economic integration. Domestic capital was unavailable, and non-agricultural resources were scarce and inaccessible. Administrators, skilled technicians and semi-skilled workers were in short supply. Physical isolation caused by the rugged terrain and the inadequate road system was a major barrier to both administrative and economic integration. In spite of these and less tangible handicaps the Central American Common Market (CACM) was fashioned through a series of agreements designed to draw the five economies together and cast them into a single, rapidly growing industrialized market.

A major reason for promoting economic integration in Central America was the knowledge that only through industrialization could these countries lessen their vulnerability to fluctuating prices for coffee, cotton and bananas, the area's dominant commercial commodities, and raise the income level of their people. The trade problem of Central America has been primarily one of declining prices for raw materials in world markets. In tropical agriculture, there is a chronic long-run oversupply. The technological revolution has an impact in these areas as in the Temperate Zone. Better methods, sprays, and disease control have increased output greatly. Consumption in the importing countries has risen much more slowly, and even with declining prices, stocks were piling up. For example, world coffee stocks in 1962 were twice as great as the annual world trade needs.[136]

In recent years the Central American countries have been importing more goods than they have been exporting. This imbalance, which is expected to grow worse, is partly from economic growth. As the countries increase manufacturing facilities for consumer goods, they are faced with the necessity of importing more machinery with which to produce consumer goods. One major problem, the integration of industry, may be solved through a system affording privileges to selected industries to help assure their success.

In spite of the inter-state violence of the 1950s, the economic ministers met annually, achieving some progress in the liberalization of trade through a series of bilateral treaties, creating schools of public administration, research and industrial technology, and agreements on road codes and customs procedures. A treaty calling for the gradual establishment of a Common Market over a ten year period was signed in 1958 by Guatemala, El Salvador, Honduras and Nicaragua. The process was swiftly accelerated by the adoption of the organization's fundamental document, the General Treaty of Central American Integration. Signed in 1960, the treaty superseded the various conventions entered into during the 1950s except for provisions in earlier agreements not covered in the General Treaty, which remained operative. The latter is based on a unique premise: the immediate elimination of intraregional tariff barriers affecting 95 percent of the area's agricultural and industrial production (over 1200 items), and a gradual reduction of tariffs on the remaining commodities (57 items). In September 1963, Costa Rica ratified both treaties and became an effective member of the CACM.[137]

[136] Dept. of State *Bulletin*, XLVII (Nov. 19, 1962), p. 779.
[137] U.S. Dept. of Commerce, *Trade and Investment in Central America* (Washington, 1965), pp. 4-7; U.S. Dept. of State, Agency for International Development, Regional Office, Central America and Panama Affairs (ROCAP), *Economic Integration Treaties of Central America* (Guatemala City, 1964).

Panama rejected the Central American invitation to become a full member of the CACM in December 1962, but negotiations have been continued toward that end. Regardless of Panama's status outside the CACM, commercial interchange between Panama and southern Central America developed significantly through the Tri-partite Preferential Trade Treaty between Costa Rica, Nicaragua, and Panama.

The direction and administration of the CACM is handled by a number of institutions which are assuming increasing responsibility and importance. The Economic Council, created by the General Treaty (1960) is the policy-making body of the CACM. An Executive Council applies and administers the General Treaty. The Permanent Secretariat for the General Treaty (SIECA), created by the General Treaty is directed by a Secretary-General appointed for a three year term by the Economic Council. The Organization of Central American States (ODECA) is concerned with the cultural educational, and economic development of the region.

Included in the General Treaty, the Central American Bank for Economic Integration (CABEI), established in July 1961, supports regional integration by extending loans, mainly for regional projects. The bank has come to be the chief vehicle for through which domestic and foreign capital are directed into regional projects with growth potential. Capitalized at $20 million ($4 million from each member), and an additional $30 million from external sources, CABEI had made ninety-one loans totaling approximately $34 million through the first quarter of 1965. These funds have been used mostly to provide private firms with plant equipment and working capital.

Also contained in the framework of the General Treaty is the Agreement on the Regime of Central American Integration Industries, through which preferential status may be assigned to specific new or existing firms requiring access to the entire Central American market to assure sufficiently low-cost, volume operation. Benefits include tax concessions and preferential import duties on equipment and raw materials. A commission of the Secretariat makes decisions on eligible firms, subject to ratification by the member governments. The object of the Central American Clearinghouse, which began operations in October 1961, is the eventual establishment on an integrated monetary system for the region. Use is being made of the "Central American *peso*," a unit of account equivalent to the U.S. dollar, to settle accounts.[138]

As noted in the preceding chapter, the United States has had a dominant role in Central America's economic development. However, in spite of this historic involvement in the region, Washington was generally indifferent to the movement for Central American integration when it emerged in the early 1950s. Vice President Richard Nixon, after his good-will tour of the Caribbean in 1955, had urged the countries of Central America and the Caribbean to form a strong regional coalition in order to further political stability and economic prosperity. At the same time, he stressed the need to insure that American markets be kept open, without undue tariff barriers or quota restrictions, to the Central American exporters. In keeping with the Eisenhower administration's policies on

[138] Hearings Before the Subcommittee on Inter-American Relationships, 89th Cong., 1st Sess., *Latin American Development and Hemisphere Trade* (Washington, 1965), pp. 65–68.

aid and investment, he sought to propagandize the investment possibilities of Central America to stimulate United States capital expansion.[139]

In the summer of 1958 President Eisenhower dispatched his brother, Dr. Milton Eisenhower, President of Johns Hopkins University, on a fact-finding and good-will tour of Central America. Soon after the completion of his mission Washington's attitude toward the principle of Central American economic integration changed to one of approval and support. Prompted by growing unrest in the area, he aimed at finding the causes of discontent with United States policies in Central America. Given the extreme dependence of the nations' economies on sales to the North American market, Central American sources charged that the United States had contributed to the region's economic problems by (1) supporting the African coffee industry through the Marshall Plan, (2) purchasing coffee from Africa, which depressed the price, (3) entering into direct competition with several of the impoverished Central American countries in the world sale of cotton, and (4) buying raw materials from Central America at depressed prices, and exporting its finished products there at elevated prices. A Nicaraguan spokesman declared: "You should not continue your high standard of living at the expense of our well-being and interests." Dr. Eisenhower responded saying it was "absurd" to think that the United States deliberately held down the economy of other countries to help itself: "The truth is . . . the United States knows our best trade is with prosperous countries." In his lengthy report to the president it was evident that Dr. Eisenhower had been won over to the integration principle: "Closely related to the credit requirement is the need for Latin America to develop a common market . . . The five nations of Central America have agreed on certain initial principles looking to the creation of a regional common market . . . it is a beginning and deserves open encouragement from the United States."[140] The United States representative on the OAS Special Committee to Study the Formulation of New Measures for Economic Cooperation affirmed Washington's backing of a free trade area in Central America, and pledged Export-Import Bank consideration of financial support to regional industries. The Department of State's position was clarified by assistant secretary of state for Inter-American Affairs, Roy Rubottom: "We believe that . . . economic integration . . . could provide genuine economic benefits to the countries concerned, and to the expansion of international trade . . . the United States is prepared to do what it can to help interested Latin American countries in framing economic integration plans which would be economically sound . . . "[141]

After a state visit to the United States in March 1959, President José María Lemus of El Salvador issued a joint statement with President Eisenhower saying: "It was agreed that the establishment of an economically sound system for the integration of the Central American republics and for a common market comprising those nations would be beneficial and would receive the support of the governments of El Salvador and the United States."

139 Dept. of State *Bulletin,* XXXII (Apr. 11, 1965), pp. 591–592; *New York Times,* Mar. 5, 1955.
140 Dept. of State *Bulletin,* XL (Jan. 19, 1959), p. 98.
141 *Ibid.,* (Jan. 26, 1959), p. 125.

In his proposals for the Alliance for Progress President Kennedy declared on March 13, 1961: "We must support all economic integration which is a genuine step toward larger markets and greater competitive opportunity. The fragmentation of Latin American economies is a serious barrier to industrial growth. Projects such as the Central American Common Market and free trade areas in South America can help remove these obstacles."[142] The policy of the Kennedy administration was amplified further in the Declaration of Central America, the *communiqué* issued by the presidents of the Central American nations, Panama, and the United States at the conclusion of their San Jose (Costa Rica) conference in March 1963. Central American economic integration was one of the main topics of discussion at the conference. The Declaration indicated that: "The President of the United States is impressed by the determination of the presidents of the Central American Republics to move as rapidly as possible toward the integration of the economies of their countries . . . is a great step forward in the achievement of the goals set forth in the Charter of Punta del Este . . . The President of the United States is prepared to offer the greatest cooperation in the preparation and implementation of the regional and national development projects of Central America and Panama . . . he proposes a fund for Central American economic integration, to be made available through the Central American Bank for Economic Integration . . . "[143]

Signaling what constituted acceptance of economic integration as a hemispheric institution, the Declaration of the Presidents of America, issued at the Punta del Este Conference of American Chiefs of State, in April 1967, declared, in part, that: " . . . The Latin American Common Market will be based on the complete development and progressive convergence of the Latin American Free Trade Association and the Central American Common Market . . . will be based on the improvement of the two existing integration systems: the Latin American Free Trade Association (LAFTA), and the Central American Common Market (CACM) . . . "[144]

ECLA, as mentioned, was influential in popularizing the idea of Central American integration at the outset; however, after 1960 its direct role was overshadowed by the interest of the United States and its capacity to provide larger financial resources, as well as the growing autonomy and competence of the area's leaders. By 1963 the United States had already spent twice as much ($20 million) as the United Nations ($10 million) in support of Central American integration.

Following the United States commitment to support the integration movement it has had a vital role in the maintenance and expansion of the process. In August 1962, the Agency for International Development established a regional office for Central America and Panama Affairs (ROCAP) in Guatemala City. With a budget of $4.25 million for the fiscal year 1967, it administered aid

[142] *Ibid.*, XLIV (Apr. 3, 1961), p. 473; William R. Gigax, "The Central American Common Market," *Inter-American Economic Affairs*, vol. 16 (Autumn, 1962), pp. 59–77.
[143] Dept. of State *Bulletin*, XLVIII (Apr. 8, 1963), p. 516.
[144] For analysis see James D. Cochrane, "United States Attitudes Toward Central American Economic Integration," *Inter-American Economic Affairs*, 18 (Autumn, 1964), pp. 73–91; Dept. of State *Bulletin*, LVI (May 8, 1967), pp. 712–714.

of regional character to governments, regional institutions and private organizations. United States funds provided 12 percent of the expenses of all the regional organizations, and more than half the budget of ODECA, a quarter of the budget of SIECA, and half the loan funds for CABEI. By early 1966 Washington had issued more than $83 million in support of Central American integration. Bilateral aid to the five countries in the period 1961–1965, by contrast, totaled about $300 million.[145]

United States private investments also contributed to regional economic growth. Direct investment figures as reported by the U.S. Department of Commerce were as follows (in millions by year): Guatemala, $122 (1963); Honduras, $105 (1963); Costa Rica, $63 (1962); El Salvador, $31 (1959); and Nicaragua, $18 (1959). Formerly United States investments were primarily made in banana plantations and related transportation and export facilities. The fruit companies have also branched out into other agricultural activities, including the cultivation of abaca, cacao, African oil palms and, particularly in Honduras, a variety of other activities such as breweries. Non-fruit company investments that have been important to some countries are principally those in public service railroads (Guatemala and El Salvador), and electric power. Other fields that have attracted investment include telecommunications, lumber exploitation, gold and silver mining, petroleum distribution, insurance and banking, and manufacturing. Many joint-venture investments emerged under regional development; one of the more frequent arrangements found in the CACM is an ownership division centering near the 50-50 percent point.[146]

While giving general support to the Central American Program of Economic Integration, Washington had reservations about "integrated industries" coming within the scope of the Agreement on the Regime for Central American Integration Industries, mentioned earlier. Concluding that temporary monopolies created by the Regime would be inefficient and detrimental to overall development, United States funds were withheld from such projects. Under the Agreement of the Regime specific projects must be approved by legislative ratification, a lengthy procedure. "Integrated industries" have tended to emerge slowly, being restricted to a tire and tube plant in Guatemala, a caustic soda and insecticide plant in Nicaragua, a glass factory in Honduras, and a Firestone Rubber Company plant in Costa Rica.[147]

With this exception, the results of the Central American countries' efforts to push development activity became increasingly apparent. President Johnson took note of this on June 2, 1966, when he said: "On this fifth anniversary of the CACM we salute our Central American friends for what they have accomplished by placing the common good of the region above more narrow interests. . . . It is no coincidence that in 1965 the Central American region as a

[145] Nye, "Central American Regional Integration," p. 54.
[146] U.S. Dept. of Commerce, *Central America,* p. 25.
[147] Cochrane, "U.S. Attitudes," pp. 81–88. See also U.S. Dept. of State, "Statements of United States Government Policy Toward Central American Economic Integration" (Washington, 1963); U.S. Dept. of Commerce, Bureau of International Commerce, "The System of Regional Industries of Integration and United States Policy" (Washington, 1964). See also, David E. Ramsett, *Regional Industrial Development in Central America* (New York, 1969).

whole achieved an increase of 2½ percent in gross national product per capita which is the yearly minimum target of the Alliance for Progress . . . We hail the CACM as a giant stride toward the eventual goal of Latin American regional economic integration . . . "[148]

Loans from international organizations for the region, totaling $450 million in the period 1955–1964, while only partially earmarked for regional integration, inevitably strengthened the movement. Individual country loans were as follows: (figures in millions of dollars and percent of total): Costa Rica, $127.1 (28.2); El Salvador, $77.9 (17.13); Guatemala, $72.2 (16.0); Honduras, $67.3 (14.9); Nicaragua, $69.8 (15.5), and regional loans $36.4 (8.1). The major sources of these loans, in descending order of importance, were the International Bank for Reconstruction and Development (IBRD), the Inter-American Development Bank (IDB), the International Development Association (IDA), the International Finance Corporation (IFC), and the United Nations and its specialized agencies. Other organizations which have made technical or financial contributions include the OAS through ECLA and various private foundations. The loans were used largely for increasing public social services (health, education, resettlement, etc.), and for infrastructure projects (highways and roads, air and water terminal facilities, and electric power).[149]

Central America's trade showed extraordinary gains, much of which could be credited to the formation of the Common Market. Historically, the United States and Western Europe have been the chief foreign outlets for Central America's exports, absorbing almost all its total exports during the last half century. On the eve of World War II the United States was taking 63 percent of all the region's exports, and Western Europe about 30 percent. But during the war the blockade and the accessibility of the United States, served to expand trade within the Western Hemisphere. By the end of the war the United States had gained in importance and Europe had virtually disappeared as a customer of Central America. As late as 1948 the respective shares of the United States and Western Europe were 81 and 6 percent. In the following decade, however, a reverse movement occurred. Within Western Europe the United Kingdom and members of the European Economic Community (especially West Germany and the Netherlands) were Central America's main customers; coffee, cotton, and bananas were the chief items shipped. By 1957-1958 the respective proportions were 52 and 32 per cent.[150]

From a peak of $218 million in 1953 United States imports gradually dropped off to $175 million in 1959. Their subsequent recovery, to $236 million in 1964, resulted largely from increased sugar purchases after the United States cancelled the Cuban sugar quota. As a market for United States products the CACM ranked fourth after Mexico, Venezuela and Brazil. Imports from the United States reached a record level of $333.9 million in 1964, or 45.4 percent of the total imports of $734.8 million. Western Europe's share in CACM trade in 1964 stood at 33 percent. Central America's overall world trade advanced from $285 million in 1950 to $668 million in 1964. Exports going to Japan, almost

[148] Dept. of State *Bulletin,* LIV (June 27, 1966), pp. 1004–1005.
[149] U.S. Dept. of Commerce, *Central America,* p. 9.
[150] United Nations, *Economic Bulletin for Latin America,* V (Oct. 1960), p. 3.

exclusively cotton, quadrupled in the post-war era, reaching 12 percent of the total in the mid-1960s. CACM imports from Japan reached a total of $61 million in 1966.[151]

Regionally, the CACM increased its gains from trade among the participating countries in the period 1960–1965 by 316 percent, reaching a total of more than $135 million. Regional trade in industrial goods rose 532 percent during the same years, representing to some extent utilization of previously existing idle capacity within the five countries. The CACM promoted technical efficiency in existing industries, particularly in the textile and food processing fields which represent the bulk of industrialization. The increase in foreign investment was credited almost wholly to the CACM. Part of the success could be measured in the increase in the gross national product, an increase for the area as a whole, and for most of the countries themselves.

In the period 1950–1955, the annual increase in GNP for the area as a whole was 4.5 percent. In 1955–1960, the increase was 4.4 percent. And in the intervening period, 1960–1965, it was 6.3 percent. El Salvador, Guatemala, and Nicaragua paced the growth in the latter period, with Honduras also increasing its GNP slightly, while Costa Rica which had earlier led the area's growth rate, fell somewhat.

A National Planning Association report, evaluating the Common Market arrangement, declared: "In contrast with all other integration schemes in underdeveloped regions of the world, the CACM's initial efforts are substantial." Although the report was generally laudatory it noted that the main area of Central America's economy to benefit from the market was industry, which accounted for only 16 percent of the region's GNP . . . "agriculture – still the great contributor to the GNP – has yet to be affected by the Common Market." Concern was expressed about the "stagnant traditional agriculture" and "the ability of the governments to provide the appropriate political and economic environment for continued growth."[152]

Data for 1967 indicated that an economic slowdown was sweeping over most of Central America. While the CACM had shown impressive gains in the preceding five years, economic growth in terms of gross national product was becoming absorbed by population increase. At the outset rapid growth occurred because each of the five members were exporting coffee, sugar and cotton at high prices. Then sugar and coffee prices fell, and drought and disease cut into cotton production. Imports had risen faster than exports, causing balance of payments problems. When members of the CACM sought to cope with the effects they were inhibited by the market agreements. At bottom, the problem is that the impoverished people, both urban and rural, must become part of the money economy, capable of buying goods the market produces.[153]

President Johnson, in his last important gesture towards Latin America before leaving office, made a week-end visit to Central America in July 1968. Meeting the five presidents in El Salvador, he declared: "I come to ask what we can do together." The meeting stemmed, in part, from the April 1967 meeting at

151 U.S. Dept. of Commerce, *Central America,* p. 8; *National Observer,* Feb. 26, 1968. See table 1-2, p. 36, for U.S.–Central American trade, 1960-1970.
152 *Christian Science Monitor,* Dec. 5, 1967.
153 *New York Times,* Jan. 22, 1968.

Punta del Este, Uruguay, and was aimed at revitalizing the process of economic integration in the region. To meet the challenge of economic decline, the Central American presidents reaffirmed their commitments to the goals of the Alliance, and pledged to remove barriers to trade and other major goals of the integration movement. The Central Americans wanted more aid, markets and investments. President Johnson, with only a few months of his presidency remaining, could not commit the next president to anything specific. However, he authorized negotiation of new loans to Central America totaling $65 million, $30 million for the Central American Fund for Economic Integration, and $35 million for programs of social justice and economic progress in the five republics. He also gave what impetus he could to the goal of an overall hemispheric common market agreed to at the Punta del Este meeting in April, 1967.[154]

Intra-regional trade increased to $260 million from $71 million between 1963 and 1969; most tariff barriers were dropped, and many new industries were formed. It also appeared that the CACM displayed great resiliency in surviving the five-day war between El Salvador and Honduras in July 1969; however, a declining trade pattern quickly set in and was continuing into 1973. Apart from the dislocation caused by the war itself, the trade slump could be related to several factors. In the aftermath of the war Honduras enforced a blockade against El Salvador thereby disrupting free trade in the region, for El Salvador can ship overland to Nicaragua and Costa Rica only through Honduras. The organization was seriously weakened by the withdrawal of two members: Honduras in 1971 and Costa Rica in 1972. The former, suffering a major financial deficit, could not afford the high-cost products of the region's protected industries. Costa Rica explained its action with the assertion that other members were "ganging up" on her exports. Trade growth was further damaged by the disastrous earthquake in Nicaragua and the multiplication of social problems within the region. Smuggling became widespread in 1972, and increasingly trade was shifting from the United States to Japan, Europe and other areas.[155]

The crisis of the Common Market was lessened somewhat by a strengthening of world coffee prices, and the revival of banana production through the planting of disease-resistant varieties. Beef cattle production and exports of fresh beef to the United States continued to add to the region's export gains. Major investment for mineral development, and exploration for petroleum, gave some promise of adding to the area's productive capacity in the decade of the 1970s.

As further evidence of its faith in the CACM, as well as to fulfill a long-standing commitment, the United States gave continuing financial support toward the completion of the Inter-American Highway. The Federal-Aid Highway Act of 1962 (76 Stat.1145) authorized an additional appropriation of $32 million to finish construction of the 1,555 mile Central American section of

[154] *Christian Science Monitor,* July 6, 1968.

[155] Inter-American Development Bank, *Socio-Economic Progress in Latin America* (*Social Progress Trust Fund Tenth Annual Report – 1970*) (Washington, 1971), pp. 35–37; *New York Times,* Jan. 26, 1970; *Ibid.,* Jan. 28, 1972; *Ibid.,* Jan. 28, 1973; *Americas,* 21 (Sept. 1969), pp. 42-45.

the highway. The five republics, and Panama, pay one-third of the cost through their territories, and are responsible for future maintenance. Approximately $8 million was distributed annually by the United States in the period 1965–1967. In 1970 the intercontinental highway had been completed as far as Chepo, thirty miles south of Panama City, from where it was being built to Santa Fe, Yaviza, and Palo de Letras on the Colombian frontier. The dense jungle through which the highway must pass is one of the world's rainiest areas.

In May 1971 Washington, in concert with Panama and Colombia, reached an agreement toward completing the highway through the closing of the Darien gap, a 250-mile section south of Panama, the major missing link in the Pan American highway system. The United States agreed to pay two-thirds of the $150 million, the balance being paid by Panama and Colombia. When completed there will be a connected highway network from Fairbanks, Alaska to Tierra del Fuego, a distance of 14,000 miles. After the agreement was signed at the Pan American Union President Nixon declared that it would "constitute an historic milestone along the road to understanding and unity within our hemisphere."[156]

In summary, United States policy has been to encourage the Central American countries to seek the advantages of integration in their own way, with due regard for the interests of other trading nations, and with all the urgency needed for expediting development demands. Considerable amounts of supporting funds have been made available through specialized institutions in the region and through the Inter-American Development Bank. With the expansion of economic integration on a hemispheric basis, Washington has gone on record as standing ready to consider requests for support which facilitate the advance of the integration movement.

156 *The Budget of the United States Government, Fiscal Year Ending June 30, 1967* (Washington, 1966), Appendix, p. 271; *Christian Science Monitor*, May 5, 1970; *Times of the Americas*, May 12, 1971.

SUPPLEMENTARY READINGS

Adams, Richard N. *Cultural Surveys of Panama-Nicaragua-Guatemala-El Salvador-Honduras.* Washington, 1957.

---------. *Crucifixion by Power: Essays on Guatemalan National Social Structures.* Austin, 1970.

Alexander, Robert J. *Communism in Latin America.* New Brunswick, N. J., 1957.

Arévalo, Juan José. *Escritos Políticos.* Guatemala, 1945.

Blakeslee, G. H., ed. *Mexico and the Caribbean.* New York, 1920.

Callcott, Wilfrid H. *The Caribbean Policy of the United States, 1910-1920.* Baltimore, 1942.

Checchi, Vincent, et al. *Honduras: A Problem in Economic Development.* New York, 1959.

Cox, Isaac. *Nicaragua and the United States, 1909-1927.* Baltimore, 1942.

Denny, Harold N. *Dollars for Bullets, The Story of American Rule in Nicaragua.* New York, 1929.

Griffith, William J. "The Historiography of Central America Since 1830," *Hispanic American Historical Review,* XLIII (Nov. 1963), pp. 483-510.

Hansen, Roger D. *Central America: Regional Integration and Economic Development.* Washington, 1967.

Hill, Roscoe R. *American Marines in Nicaragua, 1912-1915.* New York, 1933.

International Bank for Reconstruction and Development. *The Economic Development of Nicaragua.* Baltimore, 1953.

---------. *The Economic Development of Guatemala.* Baltimore, 1951.

Jones, C. Lloyd. *The Caribbean Since 1900.* New York, 1936.

---------. *Caribbean Background and Prospects.* New York, 1931.

---------. *Guatemala, Past and Present.* Minneapolis, 1940.

Karnes, Thomas L. *The Failure of Union: Central America, 1824-1960.* Chapel Hill, 1961.

Martz, John D. *Central America: The Crisis and the Challenge.* Chapel Hill, 1959.

May, Stacy, et al. *Costa Rica: A Study in Economic Development.* New York, 1952.

May, Stacy and Galo Plaza. *The United Fruit Company in Latin America.* Washington, 1958.

Munro, Dana G. *The Five Republics of Central America.* New York, 1918.

---------. *Intervention and Dollar Diplomacy in the Caribbean, 1900-1921.* Princeton, 1964.

Parker, Franklin D. *The Central American Republics.* New York, 1964.

Perkins, Dexter. *The United States and the Caribbean.* Cambridge, Mass., 1966.

---------. *A History of the Monroe Doctrine.* Boston, 1955.

Pincus, Joseph. *The Central American Common Market.* Washington, 1962.

Ramsett, David E. *Regional Industrial Development in Central America.* New York, 1969.

Rippy, J. Fred. *The Caribbean Danger Zone.* New York, 1940.

Rodríguez, Mario. *Central America.* Englewood Cliffs, 1965.

Schneider, Ronald M. *Communism in Guatemala, 1944-1954.* New York, 1958.

Silvert, Kalman H. *A Study in Government: Guatemala.* New Orleans, 1954.

Stimson, Henry L. *American Policy in Nicaragua.* New York, 1927.

---------. *A Brief History of the Relations between the United States and Nicaragua, 1909-1928.* Washington, 1928.

Stokes, William S. *Honduras: An Area Study in Government.* Madison, 1950.

United Nations. *The Population of Central America (Including Mexico), 1950-1980.* New York, 1954.

U.S. Department of Commerce. *Trade and Investment in Central America.* Washington, 1965.

U.S. Department of State. *The United States and Nicaragua–A Survey of Relations from 1909-1932.* Latin American Series No. 6. Washington, 1932.

----------. *Intervention of International Communism in Guatemala.* Publication No. 5556. Washington, 1954.

West, Robert C. and John P. Augelli. *Middle America: Its Lands and Peoples.* Englewood Cliffs, 1966.

Whetten, Nathan L. *Guatemala–The Land and the People.* New Haven, 1961.

Wilson, Charles M. *Central America, Challenge and Opportunity.* New York, 1941.

----------. *Middle America.* New York, 1944.

Ydígoras, Fuentes, Manuel (as told to Mario Rosenthal). *My War with Communism.* Englewood Cliffs, 1963.

16

Argentina—The Making of a Nation

Argentina, with an area of slightly more than one million square miles, is second in area, but third in population, among Latin American countries. Stretching through two thousand miles from the tropics of Misiones to Tierra del Fuego in the south, from the South Atlantic to areas one thousand miles within the continent and high in the Andes, the climatic variety is greater than that found in any other country of Latin America. The Argentine Pampa, commonly called *La Pampa* to distinguish it from less important pampas of South America, is the heart of Argentina. An immense fertile plain of almost 250,000 square miles, extending from the nation's capital in a semicircle, whose radius is about five hundred miles, the Pampa contains almost three-fouths of the population. It is the major industrial center, accounting for 90 percent of the national industrial product, of which 30 percent is concentrated in Buenos Aires. The southern section is noted for sheep and cattle production, the west is used mainly for wheat and alfalfa, and the part nearest Buenos Aires supplies the city with dairy products, fruit, and vegetables. In the cooler sparsely-populated southern region, Patagonia, are great sheep pasturages; in the west, the Andean region, are irrigated fields of grapes, fruit, and sugar cane; and in the subtropical north, the Chaco and Mesopotamia, are cotton plantations and varied forest resources.

Unlike the other large and medium countries of Latin America its population of 23,550,000 (1971 est.) is almost entirely European since the Indians were virtually exterminated by the end of the nineteenth century except in the extreme north. Negroes were never numerically significant. In 1972 probably no more than 10 percent of the population can be classified as *mestizo*; there remain perhaps 100,000 Indians and five thousand Negroes. Argentina has been one of the major immigration countries in the Western Hemisphere since the mid-nineteenth century, the migrants being chiefly of European origin; however, immigration from Europe no longer plays the important role it did in the past. In recent years Italians, Spaniards and Germans leaving the country

CARACAS
VENEZUELA
GUYANA
SURINAM
FR. GUIANA
BOGOTÁ
COLOMBIA
QUITO
ECUADOR
PERU
BRAZIL
LIMA
LA PAZ
BOLIVIA
BRASILÍA
PARAGUAY
ASUNCIÓN
PACIFIC OCEAN
CHILE
SANTIAGO
URUGUAY
BUENOS AIRES
MONTEVIDEO
ARGENTINA
ATLANTIC OCEAN
0 300 600
MILES
FALKLAND IS.
LATIN AMERICA
(SOUTH)

slightly exceeded the number entering. The nation's birth rate, which fell from 22.9 per 1000 in 1960 to 21.4 per 1000 in 1966, is the lowest in Latin America.

Throughout the colonial period Argentina was one of the less developed parts of Latin America. When independence was won in the 1820s it was still a neglected and unimportant wilderness of the Spanish Empire (except for its Andean fringes which looked towards Lima). Until 1900 it was almost exclusively an agricultural country, producing meat and wheat for the world market. Since then Argentina has become one of the most industrialized countries in Latin America, though it remains dependent upon meat and agricultural products to pay for its imports. Mining is of minor importance except for petroleum production in recent years. Of the economically active population in the 1960s, 20 percent were in agriculture, 34 percent in manufacturing and construction, and the remaining 46 percent in services of various kinds. With one of the largest cities in the world south of the equator (Greater Buenos Aires, with 8.2 million inhabitants, contains 35 percent of the total population), the most extensive railroad network on the continent, advanced urbanization and the highest public education level in Latin America, its growth has been extraordinary.

Since World War II Argentina has puzzled economists because its economy is stagnating. Yet, it has many features of a developed country, and the conditions needed for growth. Only in recent years, Brazil, with its much greater population and three times Argentina's area, has been forging ahead to wrest the economic leadership of Latin America from its neighbor. For the first half of the twentieth century Argentina was the undisputed colossus of the South. Despite its achievements in developing a modern society, since 1930 Argentina has been in a continuing crisis which has revealed the weakness of its economic, political, and social structures. Under these conditions it has proved impossible to create a stable constitutional system to manage the nation's affairs. Dictatorships created by military *coups* have been followed by short-lived constitutional governments which in turn have fallen to new military *juntas.* The strength of the political militarists and the weaknesses of the country's political party system are the results of deep underlying socio-economic conditions.

In the fifty-year period extending from 1775 to 1825 there occurred three of the greatest revolutions in the world's history—the American Revolution, the French Revolution, and the revolt of the Spanish-American colonies. The revolution of the thirteen North American colonies against British control brought into existence a new system of government based on principles which have now come to dominate the nations of the Western Hemisphere; the French Revolution established similar principles in the Old World, and foreshadowed the end of dynastic absolutism; the revolt of the Spanish-American colonies separated once and for all the New World from the Old, and sounded the knell of empires built upon colonial exploitation. It is difficult to say just how much effect the American and French revolutions had upon the colonies of Spanish America, but no territorial limits could contain the democratic impulse born of such travail, and Spanish America speedily gave evidence of the potential influence of the new movement.

Francisco de Miranda, of Venezuela, first felt its power, and answering its call, served in both the American and French armies of freedom. But this service was only a prelude to his real purpose, which was nothing less than to bring about a complete separation of his own country from the control of Spain. He tried to interest various governments in his schemes, and in 1798 it seemed quite possible that the United States would join with Great Britain in an attempt to foment revolutions in the Spanish colonies in the New World. Rufus King, the American minister at London, enthusiastically supported the plan, and succeeded in obtaining the backing of Hamilton.[1] Neither the British cabinet nor Secretary Pickering, however, was willing to countenance the project, and Miranda finally gave up hopes of obtaining the support of Great Britain. He thereupon visited the United States and tried to obtain help from Jefferson and Madison; but, although both personally wished him well, they were unwilling to engage the United States in an enterprise of such a nature. He was forced to content himself with organizing a small filibustering expedition in New York, without aid, but also without interference, from the United States government. This expedition, which set sail early in 1806, was a complete failure, as were Miranda's later attempts; but he deserves credit for being the first great leader in the struggle for Spanish-American independence.[2]

The real fight for independence may be said to have begun on May 25, 1810, when Buenos Aires deposed its viceroy and set up a *junta gubernativa.* Quito, in New Granada, had made a similar attempt the previous year, but this movement had been quickly suppressed. Valparaiso, Santiago, and Bogotá followed in quick succession, and on July 5, 1811, the congress of Venezuela adopted the first declaration of independence from Spain.[3] These revolutionary movements did not pass unnoticed in the United States, and in June 1810, Joel R. Poinsett was appointed "agent for seamen and commerce" in Buenos Aires. His instructions show a keen appreciation on the part of the government of the United States of the possibilities of the situation:

> As a crisis is approaching which must produce great changes in the situation of Spanish America, and may dissolve altogether its colonial relations to Europe; and as the geographical position of the United States and other obvious considerations give them an intimate interest in whatever may affect the destiny of that part of the American continent, it is our duty to turn our attention to this important subject, and to take such steps not incompatible with the neutral character and honest policy of the United States, as the occasion renders proper. . . . You will make it your object, whenever it may be proper, to diffuse the impression that the United States cherish the sincerest good will towards the people of South America as neighbors, as belonging to the same portion of the globe, and as having a mutual interest in cultivating friendly intercourse; that this disposition will exist whatever may be their internal system or European relations, with respect to which no interference of any sort is pretended; and that in the event of a political separation from the parent country and of the establishment of an independent system of national government, it will coincide with the sentiments and policy of the United States to promote the most

[1] C. R. King, *The Life and Correspondence of Rufus King* (New York, 1907), vol. II, pp. 250 ff.

[2] For a complete account of Miranda's efforts see W. S. Robertson, *Rise of Spanish-American Republics* (New York, 1918), pp. 26-72; also "Francisco de Miranda and the Revolutionizing of Spanish America," by the same writer, in the *Annual Report of the American Historical Assoc., 1907,* vol. I, pp. 189-539.

[3] For an excellent survey of the early diplomatic relations between the United States and Venezuela see W. S. Robertson, *Essays in American History* (New York, 1910), pp. 231-267.

friendly relations and the most liberal intercourse between the inhabitants of this hemisphere, as having all a common interest, and as lying under a common obligation to maintain that system of peace, justice, and good will which is the only source of happiness for nations.

In the following year Poinsett was made consul-general and proceeded to Chile, a vice-consul taking over the duties at Buenos Aires and the ports on the River Plate. Similar representation was provided for at about the same time at Caracas, Venezuela. In spite of the fact that these were regularly accredited agents, there seems to have been no intention on the part of the United States to recognize the governments to which they were sent. This is not surprising when we remember that none of the revolting countries except Venezuela had delcared its independence. They simply refused to recognize the Napoleonic regime in Spain. In fact, most of the colonies that had deposed their Spanish governors and had set up *juntas* claimed to be acting in the name of Ferdinand VII. Under these anomalous conditions, the question of recognition did not arise; nor did the American agents make any effort to maintain a neutral attitude. Poinsett did all in his power to aid the Chileans, giving them counsel and encouragement and supplying the names and addresses of manufacturers and merchants in the United States who would willingly furnish them with military supplies.[5] Yet, upon his return to the United States, he was congratulated by Secretary Monroe for the ability, zeal, and success with which he had conducted his delicate mission.[6]

Ferdinand was restored in 1814. Far from granting concessions to his colonies, he determined to bring them back into their former position of complete dependence. This meant a war of reconquest, and the king did not hesitate. Within a year his armies had reduced to a sullen allegiance all of the revolting colonies except Buenos Aires. It was in this state that a new and successful campaign was to originate. José de San Martín, a native of La Plata, who had served for many years in the Continental Army of Spain, had returned to give his services in the war for independence. He was given command of one of the northern Argentinian armies. Instead of using it to attack the Spaniards in Peru, he gave up the command for the governorship of the small province of Cuyo at the edge of the Andes. Here he spent two years in organizing and training a small army, with the intention of crossing the Andes and striking at the Spanish power in Chile and then marching north into Peru. It is not within our province to describe his desperate crossing of the Andes by the famous Uspaliata Pass, and his victory over the Spanish army at Chacabuco, thereby opening up the road to Santiago. It is sufficient to note that he followed out his plan of freeing Chile, and when this was accomplished by the battle of Maipú, on April 5, 1818, he returned to Buenos Aires to organize a new force to carry the struggle into Peru. After two more years of preparation, he again crossed into Chile, and embarked his army upon the new Chilean fleet at Valparaiso. Aided by the brilliant strategy of Lord Cochrane, the commander of the Chilean fleet, and even more by his appeals to the Peruvians to arise and throw off the Spanish

[4] *House Rep. No. 72,* 20th Cong., 2nd Sess.; also cited in F. L. Paxon, *The Independence of the South American Republics* (Philadelphia, 1916), p. 110.
[5] J. B. Lockey, *Pan-Americanism: Its Beginnings* (New York, 1920), p. 144.
[6] Paxson, *op. cit.,* p. 115.

yoke, he finally, by Fabian tactics, forced the Spaniards to withdraw from Peru.[7]

While San Martín was slowly emancipating the great territories of the south and west from Spanish rule, another patriot was accomplishing the same result for the north. Simón Bolívar, a native of Venezuela, after a series of unsuccessful attempts to liberate his native state, crossed over into New Granada and defeated the Spaniards in the decisive battle of Boyacá, April 7, 1819. This province was then induced to join with Venezuela, forming the state of Gran Colombia, together with the still unconquered Ecuador. The Spaniards sought and obtained a truce during the following year; but, with the goal of complete independence clearly in view, Bolívar saw the unwidsom of parleying, and by a final blow in the summer of 1821 the north was completely freed. The Liberator now turned his attention towards the last stronghold of the Spanish forces in the highlands of what is now Ecuador, and, San Martín, now protector of Peru, entered the city shortly afterwards, and the two great patriots celebrated the emancipation of South America together. Two such powerful characters, however,–one striving for a union of South America under the hegemony of Colombia, and the other insisting upon the individuality of the states,–could not agree; and, rather than force a struggle, San Martín decided to exile himself from the country he had so valiantly helped to free. To the end he remained true to his famous motto: *"Seras lo que debes ser, y si no, no seras nada* ("Be what you ought to be or else be nothing").

Although the struggle for independence did not definitely end till 1824, the question of recognition by foreign powers had been raised long before. Early in 1816 Buenos Aires had sent a Colonel Martin Thompson to Washington to represent its interests; but, inasmuch as no proclamation of independence had yet been issued, this envoy was not invested by his government with a public character.[8] The following year, however, the provinces of La Plata declared their independence and forthwith despatched Don Manuel H. de Aguirre "as agent of this government near that of the United States of North America";[9] and shortly afterwards Supreme Director Pueyrredón, in a letter to the President of the United States, asked that he be granted "all the protection and consideration required by his diplomatic rank and the actual state of our relations."[10] President Monroe received him confidentially, but informed him that, although he sympathized with his cause, the recognition of independence could come only after a public deliberation of Congress.[11]

In his first official note to President Monroe, dated October 29, 1817, Aguirre gave notice of the declaration of independence of the United Provinces, and in his second communication, dated December 16, he asked for recognition.[12] President Monroe had already, on April 25, 1817, written a note to

[7]The standard and most exhaustive treatment of this period is Bartolomé Mitre, *Historia de San Martín y de la Emancipación Sud Americana,* 6 vols. (Buenos Aires, 1907).
[8]*Amer. State Papers, For. Rel.,* vol. IV, p. 174.
[9]*Ibid.,* p. 175.
[10]*Ibid.,* p. 176.
[11]B. Mitre, *Historia de Belgrano* (Buenos Aires, 1902), vol. III, p. 98. For a full account of the Aguirre mission, see A. Palomeque, *Orígines de la Diplomacia Argentina* (Buenos Aires, 1905), vol. I, pp. 39-66.
[12]*Amer. State Papers, For. Rel.,* vol. IV, pp. 179-180.

Joel R. Poinsett asking him to go once more to Buenos Aires to report on the progress of the revolution; and when Poinsett declined, he had appointed a commission of three. But, as these commissioners were not ready to sail until December, the president did not feel justified in extending recognition without obtaining "in a manner more comprehensive than has heretofore been done correct information of the actual state of affairs in those colonies."[13] His attitude was shown by his message of December 2, 1817, in which, while expressing sympathy and good will for the revolutionists, he reiterated his policy of neutrality.[14]

President Monroe was undoubtedly influenced considerably by his Secretary of State, John Quincy Adams, who from the very beginning opposed recognition until independence had been clearly established.[15] But Henry Clay had already "mounted his South American great horse," and in the House did all in his power to force the hand of the president. On March 25, 1818 he made a brilliant speech urging recognition and the immediate appointment of a minister to Buenos Aires.[16] But apparently Congress preferred to back the president, particularly since the Florida question had already brought strained relations between the United States and Spain, and the motion was defeated by a substantial majority. Before sending his annual message of November 17, 1818, President Monroe had received the reports of his commissioners. But, inasmuch as their reports showed that the situation was still very uncertain, he declared himself satisfied with the course hitherto pursued and considered it good policy to adhere to it.[17] In his message of the following year, in spite of Adams' advice to the contrary, President Monroe gave decided encouragement to the struggling patriots. "The steadiness, consistency, and success with which they have pursued their objects, as evidenced more particularly by the undisturbed sovereignty which Buenos Aires has so long enjoyed, evidently gave them a strong claim to the favorable consideration of other nations."[18] However, in spite of the president's evident friendliness and the continued efforts of Clay, it was not until 1822 that recognition was finally accorded, and by that time the result of the contest was "manifestly settled." Incidentally the Florida treaty of 1819 had now been ratified by Spain, thereby freeing the administration of further trouble from that source.

Manuel Torres, the *chargé d'affaires* from Colombia, had the honor of being the first diplomatic agent from a Spanish-American state to be received by the United States, and he was deeply affected by it. He said that Colombia realized the importance of this recognition and that Bolívar would be extraodinarily gratified by it.[19] Mr. de Forest, the agent of Buenos Aires, laid claim to priority in recognition, but his commission was found to be

[13] Paxson, *op. cit.*, p. 126.
[14] J. D. Richardson, *Messages and Papers of the Presidents*, vol. II, p. 13.
[15] See note of Secretary Adams to de Aguirre (Aug. 27, 1818) in W. R. Manning. *Diplomatic Correspondence of the United States Concerning the Independence of the Latin American Nations* (New York, 1925), vol. I, p. 76.
[16] T. H. Benton, *Abridgment of the Debates of Congress,* vol. VI, pp. 138-145.
[17] *Amer. State Papers, For. Rel.*, vol. IV, p. 215; for reports of the commissioners, *Ibid.*, pp. 217-348.
[18] *Ibid.*, p. 628.
[19] J. Q. Adams, *Memoirs* (Philadelphia, 1874), vol. VI, p. 23.

inadmissible and he was required to obtain a new one. In a note to him on May 23, 1822, Secretary Adams said that in the recognition of the independence of the South American governments it was not the president's intention, by discriminating between them with regard to time, to admit any claim to prior recognition in favor of any one over the other.[20] The recognition of Buenos Aires came the following year, when Ceasar Rodney, former commissioner to the South American states, was accredited as minister to that country. As was to be expected, Spain protested vehemently agains recognition, asserting that it could in now way invalidate Spain's right to her provinces or to employ any means in her power to reunite them to her dominions. In his reply Adams declared that it was the mere acknowledgment of existing facts, and that the United States expected the European countries and Spain herself soon to follow her example.[21]

Recognition did not come any too soon, inasmuch as the European powers at Verona were already planning to assist Spain to recover her colonies. Fortunately for the United States, such a plan did not fit in well with British ideas, and we have already shown how Canning approached Rush with a view to concerted action to prevent it. President Monroe's message of December 2, 1823, settled the question as far as the United States was concerned, and all South America seemed to approve of the settlement. At Buenos Aires, Las Heras, the new governor, in a message to the congress of the la Plata provinces dated December 16, 1824, declared: "We are under a large obligation towards the United States of North America. That republic, which since its formation has presided over the civilization of the New World, has solemnly recognized our independence. At the same time it has made an appeal to our national honor by supposing us capable of struggling singlehanded with the power of Spain, but it has constituted itself guardian of the field of battle in order that no foreign power may interfere to give aid to our rival."[22]

The government of Buenos Aires soon had occasion to seek a definite interpretation of the Monroe Doctrine. In 1826 the United Provinces of la Plata, with Rivadavia at their head, became involved in a war with Brazil, on account of an attempt of the Banda Oriental, now known as Uruguay, to unite itself with Buenos Aires. In this connection Rivadavia's Minister of Foreign Affairs asked the American minister, first, whether President Monroe's declaration would apply in case a European power should assist the Emperor of Brazil in his war against the United Provinces; and, second, whether "such declaration is equally applicable in a case in which the Emperor of Brazil, as King of Portugal, may attempt to draw from that kingdom. . .any kind of aid for sustaining said war."[23] After considerable delay, Secretary Clay returned his answer to these queries. Pointing out that the declaration conveyed neither pledge nor obligation, he declared that in the war between the Argentine Republic and the Emperor of Brazil there was not the remotest analogy to the case that President Monroe's message deprecated. The war was strictly American in its origin and

[20] J. B. Moore, *Digest of Int. Law,* vol. I, p. 90, note *a.*
[21] *Amer. State Papers, For. Rel.,* vol, IV, p. 845.
[22] Quoted by W. S. Robertson, "South America and the Monroe Doctrine," *Polit. Sci. Quar.,* vol. XXX (Mar., 1915), p. 100; cf. J. B. Lockey, *op. cit.,* pp. 255-260.
[23] Robertson, "South America and the Monroe Doctrine," p. 102.

object; the European allies had taken no part; and under these conditions the policy of the United States was one of strict and impartial neutrality.[24]

In consequence of the bloody struggle between the unionists and the federalists, the next few years were very difficult for the Buenos Aires republic. By 1829 the federalist leader, Juan Manuel Rosas, was in full control of Buenos Aires and the nominal head of Argentina. He was a dominating personality, and he succeeded in remaining dictator of the province of Buenos Aires for the next twenty years.

During the early part of his regime an interesting diplomatic conflict took place between his government and the United States.[25] In 1829 the government of Buenos Aires issued a decree claiming, as the successor of Spain, the Malvinas or Falkland Islands, and announcing that a governor would be appointed to reside there and regulate the seal fisheries on the coast.[26] In November 1831, the appointee, Luis Vernet, seized certain American schooners engaged in fishing off the coasts of these islands, imprisoned their officers and crews, and sent the ships to Buenos Aires as prizes. The United States consul at Buenos Aires protested, and at the same time the U.S.S. *Lexington* proceeded to the Falkland Islands to protect American citizens engaged in the fisheries. Upon investigation, Captain Duncan of the *Lexington* found that Governor Vernet had plundered the American schooner *Harriet* of almost everything on board. Captain Duncan thereupon asked that Vernet be delivered up to the United States on charges of piracy or robbery, or that he be arrested and punished by Buenos Aires.[27] Apparently the American officer did more than protest, for on February 14, 1832, the government of Buenos Aires published a proclamation stating, among other things, that "the Commander of the U.S.S. *Lexington* has invaded in a time of the most profound peace, that, our infant Colony; destroyed with rancorous fury the public property, and carried off the effects legally deposited there at the disposal of our Magistrates."[28] On the same day the government of Buenos Aires notified the American consul that, in view of his aberration of ideas and irregularity of language, it had decided to suspend official intercourse with him.[29]

The situation began to be serious, and in its instructions to the new *chargé d'affaires,* Mr. Baylies, who had been hastily despatched to Buenos Aires in January 1832, to fill the position made vacant by the death of the former American *chargé,* the United States outlined the policy that it intended to follow. The Washington government first demanded for United States citizens freedom of fishing in these regions—a right which they had held for more than fifty years at the time of Vernet's appointment. This right was now to be embodied in a treaty. With respect to the vessels seized by Vernet, their

[24] *Ibid.,* p. 103.
[25] The whole incident is well summarized from the Argentinian standpoint by Adolfo Saldías *Historia de la Confederación Argentina* (Buenos Aires, 1892), vol. II, chap. XIX.
[26] *British and Foreign State Papers,* vol XX, p. 314, note.
[27] *Ibid.,* vol. XX, p. 319.
[28] *Ibid.,* p. 327.
[29] W. R. Manning, *Diplomatic Correspondence of the United States, Inter-American Affairs 1831-1860* (Washington, D. C., 1932), vol. I, p. 72. Vernet later demanded the arrest of Consul Slocum for his statements in the official correspondence which were published by the government of Buenos Aires, *Ibid.,* p. 138.

restitution was to be demanded unless they had already been recaptured. If the government disavowed his acts, the American squadron was to break up to the settlement and bring him to Buenos Aires for trial.[30] Upon arriving at his post, Mr. Baylies, in a series of notes to the Argentinian Minister of Foreign Affairs, protested vigorously against the harsh and illegal treatment visited upon American whalers and sealers in this region of the Atlantic and demanded full indemnity for it. He furthermore denied the right of Buenos Aires to interfere with vessels of citizens of the United States fishing in the waters or on the shores of the Falkland Islands, Tierra del Fuego, Cape Horn, or any of the adjacent islands in the Atlantic.[31]

The Argentine government not only refused indemnity for Vernet's acts, but warmly defended him, declaring all irregularity, injustice, insult, and violence to have been the work of the American consul and Captain Duncan. It therefore demanded from the government of the United States "The most prompt and ample satisfaction for such outrages and full redress and reparation to the Argentine Republic. . .for all damages and losses in consequence of the aggressions committed by Captain Duncan."[32] Unwilling to continue a discussion in which it was evident that the Argentine government had no intention of conceding that it was in the wrong, but insisted on considering itself the injured party, the American *chargé* asked for his passport. Under the circumstances this action seemed inexcusably abrupt, and the Argentine minister protested against it and suggested arbitration, by a third power.[33] Mr. Baylies, however, remained firm, and diplomatic relations were broken off.[34] In the following year Great Britain cut the Gordian knot by sending a warship, which seized the Falkland Islands, on the ground that Spain had conceded Great Britain's right to them fifty years before Buenos Aires had achieved her independence.[35] In spite of this effective answer to Argentinian pretensions, the Argentine Republic continued to press its claim against the United States for the acts of Captain Duncan.[36] In a note dated December 4, 1841, Secretary Webster refused to consider the claim until the controversy between Great Britain and the Argentine Republic should be settled.[37] When the Argentine Republic invoked the Monroe Doctrine as against the British action, the United States refused to see its application to the case. The controversy was closed by the brusk statement of President Cleveland, in his annual message of December 8, 1885, that, "in view of the ample justification for the act of the *Lexington* and the derelict condition of the islands before and after their alleged occupation by Argentine colonists, this government considers the claim as wholly groundless."[38] Granted that the final justification of Vernet's acts depended upon his

[30]Moore, *op. cit.*, vol. I, pp. 878-883.
[31]*British and Foreign State Papers*, vol. XX, pp. 330-355.
[32]*Ibid.*, p. 364. For Vernet's report, see *Ibid.*, pp. 369-436.
[33]*Ibid.*, pp. 437-440.
[34]The United States did not accredit another diplomatic agent to Buenos Aires until ten years later (1843).
[35]Moore *op. cit.*, vol. I, p. 888.
[36]Manning, *op. cit.*, pp. 210-226.
[37]*Ibid.*. p. 18.
[38]Richardson, *op. cit.*, vol. VIII, p. 324; see also note of Mr. Bayard to Mr. Quesnada, March 18, 1886, Moore. *op. cit.*, vol. I, p. 889.

official position, nevertheless, he was exercising *de facto* sovereignty when the incident occurred. Under these circumstances the drastic action of Captain Duncan would appear to be indefensible.[39]

The historian Saldias, in his excellent work on Rosas and his times, opens his third volume with the statement: ". . . the year 1838 began under fatal auspices for the government of Rosas." At the very time when the dictator was forced to carry on war against Bolivia in the north, and resist a new uprising of the Unitarian party at home, the French government broke off diplomatic relations and followed with a blockade of Buenos Aires and the adjoining littoral.[40] The blockade created a commercial and economic crisis which Rosas met with the greatest difficulty, but he refused to admit to the French pretensions. Realizing the difficult predicament of the Argentine government, Captain Nicolson, in command of the U.S.S. *Fairfield,* offered his good offices to Governor Rosas on April 4, 1839, and outlined a basis of settlement which he thought the French would accept. Rosas received the proposal in the gracious spirit in which it was offered, but insisted upon such a radical modification that the French refused to parley further and the negotiations were broken off.[41] The United States, however, had given an indication of its amicable disposition, and the sorely pressed dictator showed his appreciation.

In the war which the Argentine republic carried on against Montevideo, the United States again had occasion to show its friendly attitude towards the government of Buenos Aires.[42] Rosas had declared a blockade of Montevideo in 1842 in order to assist General Oribe in his attempt to regain the presidency of the Eastern Republic of Uruguay, whence he had been driven by a revolution. In March, 1844, Captain Voorhees was despatched in the frigate *Congress* to Montevideo to protect the interests of the United States, but also with instructions to maintain a strict and unqualified neutrality. Instead of carrying out these instructions, the American officer, on the ground that one of General Oribe's schooners had fired upon a Montevidean fishing-boat which had sought refuge beside an American bark, declared that the act constituted piracy, in which the commander of the Argentine squadron had, by approving and adopting it, made himself an accomplice. He thereupon not only seized the offending schooner, but compelled the Argentinian squadron to surrender, and released all the Montevidean prisoners on board. He also refused to consider the port blockaded for American ships. His acts were warmly applauded by the commanding officers of the English, French, and Brazilian squadrons, but the Secretary of the Navy interpreted the law of neturality in a different sense. Captain Voorhees was charged with disobedience on five counts, court-martialed, and sentenced to be reprimanded and suspended from service for three years. A copy of the reprimand and sentence was communicated to the Argentine minister, with an expression of the hope that his government would

[39]According to Julius Goebel, Jr., *The Struggle for the Falkland Islands* (New Haven, Conn., 1927), President Cleveland was badly advised.
[40]Saldías, *op. cit.,* vol. III, p. 17; see also *House Doc. No. 211,* 25th Cong., 3rd Sess., p. 33.
[41]*British and Foreign State Papers,* vol. XXXI, pp. 790-801.
[42]For a critical evaluation of American methods and policy during this period see John F. Cady, *Foreign Intervention in the Rio de la Plata 1838-50* (Philadelphia, 1929), chap. VI.

see in it a satisfactory proof of the disposition of the United States to respect the rights of Buenos Aires.[43]

In the following year the United States gave another signal instance of its appreciation of the belligerent rights of weaker states. The Argentinian squadron had been maintaining a modified blockade of Montevideo for over a year, but when an attempt was made to change it into a strict blockade, the French admiral refused to observe it. The American commander was thereupon forced to decide whether he would insist upon the same privileges for American shipping as the French illegally obtained, or whether he should recognize the strict blockade proclaimed and enforced by the Argentinian fleet. After carefully considering the situation, the American *chargé d'affaires*, Mr. William Brent, Jr., decided that the French admiral by his act had in reality become a belligerent, and that the United States could not put itself in the same attitude and still maintain a strictly neutral course.[44] In spite of this recommendation, Commander Pendergast decided to claim for the commerce of the United States the same immunities and advantages enjoyed by the commerce of any other country, and notified the Argentinian admiral that he would maintain this position. When the correspondence was placed before the secretary of the navy, he declared that Commander Pendergast was wrong, that the failure on the part of the Argentine republic to maintain her belligerent rights against the opposition of the French naval force did not justify him in refusing to conform to the strict blockade, and that the President would have been pleased if he had conformed to the advice of the United States' representative at Buenos Aires, at least until the sense of his government could be known.[45]

At the same time the French and British ministers were threatening the Argentinian government with a joint blockade by their squadrons unless it stopped trying to control Uruguay through its support of the Oribe faction. The American *chargé*. Mr. Brent, offered his good offices, but, although they were at first accepted by both parties, and preliminary bases of a compromise were laid down, the mediation fell through, owing to the exigencies of the French plenipotentiary.[46] The two powers made good their promise, and on September 18, 1845, they issued a joint declaration of blockade of the ports and coasts of Buenos Aires.[47] Mr. Brent's note to the British minister shows the American attitude: "I do not acknowledge such decision of these plenipotentiaries as having any validity whatever, as far as the United States and their citizens are concerned. . . .Nor, sir, do I acknowledge the right of the commanders of the combined squadrons of England and France to enforce any such blockade in consequence of such decision, found necessary by the English and French plenipotentiaries. . . .I therefore, sir, for the United States of America, hereby protest against this so-called and misnamed blockade."[48]

The revolutions of 1848 in Europe caused both France and Great Britain to pause in their dictatorial policy towards Rosas, and by 1850 both nations had

[43] Moore, *op. cit.*, vol. I, pp. 178-182.
[44] *House Doc. No. 212,* 29th Cong., 1st Sess., p. 10.
[45] *Ibid.,* p. 39.
[46] Saldías, *op. cit.*, vol. IV, p. 178.
[47] *British and Foreign State Papers,* vol. XXXIV, p. 1266.
[48] *House Doc. No. 212,* p. 35.

signed treaties with him. Despite his autocratic acts and often brutal methods, the United States was obliged to recognize that throughout his long tenure of power, he was a constant and powerful check upon European infringement of American sovereignty. As such, he was supported upon every occasion when the United States could consistently do so. His downfall, which came in 1851, was produced, not by the machinations of foreign powers, but by a triple alliance between his chief lieutenant, General Urquiza, governor of the rich province of Entre Rios, the Empire of Brazil, and the faction of Uruguay which had always opposed him. The historian Calderón thus summarizes the work of the great Argentinian tyrant: "His authoritative character of a Spanish patrician made him the *paterfamilias* of the Argentine democracy. . . .In the twenty-four years, 1829 to 1852, Rosas made federal unity a reality. . . .He defended the country against the territorial aggression of foreign coalitions, and his own power against conspiracy and revolts; he dominated the capital city and moderated provincialism; he painfully founded the Confederation. . . .His cruelty was effectual, his barbarism patriotic."[49]

With the end of the Rosas regime comes a new epoch in the history of the Argentine republic. General Urquiza, who was now in control, determined to establish a real federal republic with the least possible delay. To that end, a constitutional convention was called, which met on November 20, 1852, in Santa Fe. Every province was represented except Buenos Aires, whose attitude from the beginning was strongly antagonistic to the idea. In drawing up a constitution for a federal republic, it was but natural that the constitution of the United States should have a considerable influence, and in fact in submitting the constitution to the convention the chairman of the committee on constitutional affairs declared that "the draft of the committee has been cast in the mold of the constitution of the United States, the only model now existing for a real federation."[50] The constitution was signed May 1, 1853, and on May 25 was proclaimed the supreme law of the land.

An attempt was immediately made to obtain the acceptance of the constitution by Buenos Aires, thus bringing her into the new federation; but it was not successful. The central government could not hope to become stable and strong as long as one-third of the population and one-fourth of the country's wealth remained outside, particularly since Buenos Aires controlled the sole important port of entry, the principal source of revenue. Various attempts to bring about a satisfactory arrangement were made[51]—the most important, perhaps, being the effort of Colonel Yancy, the American minister resident, in 1859. At this time Buenos Aires agreed to enter the confederation, provided that the city of Buenos Aires should not be the federal capital, and that General Urquiza should retire and occupy no public office in the new government. Colonel Yancy, representing General Urquiza, indignantly refused these con-

[49] F. García Calderón *Latin America* (New York, 1913), chap. V.

[50] L. S. Rowe, *The Federal System of the Argentine Republic* (Washington, D.C., 1921), p. 43.

[51] The United States authorized its diplomatic representative, Mr. Peden, to use all his influence to bring about a reunion and as a means of pressure transferred his commission to the Argentine Confederation. When Mr. Peden continued to maintain his diplomatic connection with both governments he was severely reprimanded and sent to Panama, Manning, *op. cit.*, pp. 48-53.

ditions and the negotiations came to an end.[52] When this last attempt failed, actual warfare broke out, resulting in a complete victory for the federal forces. Threatened by invasion and the capture of its capital, Buenos Aires became more reasonable and agreed to enter the confederation, provided that certain amendments were made to the national constitution.

A constitutional convention of the province of Buenos Aires was opened on January 5, 1860, and it proposed a number of amendments to the federal constitution. Again the influence of the constitution of the United States was apparent. The chairman of the committee on constitutional amendments said in the report submitted to the convention: "The committee has been guided in its recommendations by the provisions of a similar constitution, recognized as the most perfect. *viz.,* that of the United States. The provisions of this constitution are most readily applicable to Argentine conditions, having served as the basis for the formation of the Argentine Confederation. . . .It would therefore be both presumptuous and a proof of ignorance were we to attempt any innovations in constitutional organization, thus ignoring the lessons of experience and the manifest truths accepted by the human conscience."[53] The amendments proposed were accepted and were then laid before another national convention called for the purpose. The amendments were again ratified, September 25, 1860, thus definitely establishing the Argentine republic as a federal state with Buenos Aires included.

Once a part of the federal republic, Buenos Aires determined to dominate it. The provincial government interpreted several clauses of the constitution differently from the national government, and when the congress put the province under martial law, war again broke out. This time General Bartolomé Mitre decisively defeated the federal forçes, thereupon taking over the supreme power until the national government could be reorganized. Elections of senators and deputies took place in April 1862, and the presidential election in October made General Mitre president for a period of six years. The new president showed himself a truly great leader. Generous and conciliatory in his methods, he succeeded in placing the federation on a basis of friendly cooperation between the national and provincial governments.

In 1865 the Argentine republic was drawn into the war against Paraguay which Brazil and Uruguay were then waging. When the question was raised regarding the good offices of the United States to settle the dispute, Seward wrote our minister to the effect that, "although we have never been eager to interfere in controversies abroad which lead to wars, or in accepting the part of mediator for the purpose of arresting hostilities, we have a natural desire, as an American power, that peace should prevail in this hemisphere. . . .If, therefore, all or either of them shall ask for our good offices, they will be bestowed with a full appreciation of the delicacy and responsibility of the trust, and with a single desire to render impartial justice and to terminate the ravages of war."[54] The differences, however, were too serious to be settled by mediation, and the war ended only with the utter destruction of Paraguay.[55]

[52] Mariano Pelliza, *Historia de la Organización Nacional* (Buenos Aires, 1899), p. 244.
[53] Rowe, *op. cit.,* p. 5.
[54] *For. Rel. of the U. S.,* 1866, Part II, p. 286.
[55] *Infra,* pp. 426-427.

Article V of the treaty of limits between the Argentine republic and Paraguay, signed February 3, 1876, provided that the president of the United States should be asked to arbitrate the right of sovereignty over the territory between the River Verde and the main branch of the Pilcomayo.[56] The documents were accordingly submitted to President Hayes in March 1878. The award was rendered in November of the same year, and it stated that Paraguay was legally and justly entitled to the disputed territory.[57] Both sides thanked the president for the service which he had rendered, and as a token of recognition the Paraguayan congress voted to change the name of the principal town in the region to Villa Hayes.

Between 1850 and 1900 the United States and Argentina each passed through a critical phase of their development as nation states, and the outlines of their respective foreign policies which were to be projected into the next century became visible. It will be advantageous to examine the salient factors causing repulsion and attraction in United States-Argentine relations. Although both countries were products of revolution against European colonial powers and shared in some respects features of economic growth, social development and territorial expansion until about mid-century, they were thereafter more often rivals and competitors than partners in regional and international affairs.

Each country suffered internal conflicts in the 1860s and 1870s but whereas the United States followed the path of industrialization, Argentina remained largely pastoral and agrarian. By 1900 the United States had become not only the leading power in the Western Hemisphere, but a world power. Although by World War I Argentina had become the foremost exporter of surplus food and industrial raw materials in the world, and the acknowledged champion of Latin American interests, she fell steadily behind the "power belt" of the North Atlantic in industrial-military strength. At mid-century the United States was the world's most prosperous and powerful nation, whereas Argentina was classified as a semi-developed country, being surpassed on many counts by Mexico and to some degree by Brazil. However, Argentina's reluctance to yield in its pretensions to hemispheric leadership and independence of action often put it on a collision course with Washington's policies.

Geography had also had an understandably important role in the relationship. Even with twentieth century advances in trasportation and communication, the seven thousand miles separating them helped to prevent the growth of a community of interests. Being situated in similar climatic regions and offering competitive exports from field and farm, they became rivals in world trade. The peoples of both countries, while sharing a common Western European origin, are of a different racial and cultural background. Diplomatically, U.S. policies appeared inconsistent, sometimes prejudicial to Argentine interests, and Washington did not always send its most able and discreet representatives to Buenos Aires.

United States foreign policy in the nineteenth century hinged largely on the Monroe Doctrine, freedom from entangling alliances, "Manifest Destiny,"

[56] J. B. Moore, *History and Digest of International Arbitrations* (Washington, D. C., 1898), vol. V, p. 4783.
[57] *Ibid.*, vol. II, p. 1943.

the peaceful resolution of international disputes, and the expansion of commercial intercourse abroad. Argentine policy embraced the nation's right to hegemony over the territory of the former viceroyalty; it opposed the intrusion in its internal affairs from any source, and resisted multilateral arrangements involving its own security interests. Moreover, Argentina's leaders were generally of the belief that European markets, immigrants and capital offered the greatest hope for their nation, although they held many United States institutions in high esteem. When viewed in historical perspective, it was apparent by the 1870s that the economic and political foundations were lacking in the substance needed to insure a mutually constructive relationship between the two countries.[58]

With the conclusion of the Paraguayan War in 1870, Argentina entered upon a period of extraordinary growth and progress, including a tremendous surge in foreign trade, but with Europe rather than the United States as the chief market. Washington and Buenos Aires had signed two treaties in 1853, a Treaty for the Free Navigation of the Rivers Paraná and Uruguay and a Treaty of Friendship, Navigation and Commerce, the latter containing a conditional most-favored-nation clause. Trade increased slowly until the early 1870s only to decline once more. This could be attributed in large part to the Civil War and its aftermath in the United States, its preoccupation with "winning the West," neglect of the merchant marine, and a highly protectionist policy.[59]

The American Wool and Woolens Act of 1867 had a particularly adverse effect on trade for it raised duties on unwashed wool, which was the principal Argentine export to the United States. The American minister in Buenos Aires brought the matter to Washington's attention, but with little apparent result. Not only was the advantage of this trade lost, but the Argentines were pushed further towards their goal of replacing the United States as the chief supplier of food to the British Isles and Europe. As a means of resolving the lack of direct steamship communication, the Argentine government in 1887 offered to subsidize an American line to the extent of $100,000 annually. But when objections and delays ensued, a British firm seized the opportunity and won the concession.[60]

While the value of trade between the United States and Argentina had increased in 1878 from $5,500,000 to about $17 million, or about 205 percent, Great Britain had increased hers from $15,500,000 to $51,750,000 or about 234 percent; and Germany had increased hers from $3 million to about $22 million, or 580 percent. As American industry expanded in the latter part of the century, creating greater complementarity between the two nations' economies, Argentine-United States commerce advanced to unprecedented levels. The 1891-1895 annual average trade level of $9.5 million mounted to over $80 million by 1910-1914. The United States had improved its trading position prior to 1914 to the point where it could rival Germany, surpass France and challenge England in Argentine markets.[61]

[58]Arthur P. Whitaker, *The United States and Argentina* (Cambridge, Mass., 1954), pp. 85-88.
[59]*For. Rel. of the U.S.*, 1873, Part I, p. 35.
[60]*Ibid.*, 1887, p. 6; *Ibid.*, 1888, Part I, p. 2.
[61]*Ibid.*, 1889, p. 4; Harold F. Peterson, *Argentina and the United States, 1810-1960* (New York, 1964), p. 233.

Following the excellent start made by Bartolomé Mitre, the Argentine republic was fortunate in her choice of presidents. Both Domingo Sarmiento (1868-1874), and Nicolás Avellaneda (1874-1880) aided remarkably in promoting the growth of the republic, and both were able to avoid any serious political troubles. The elections of 1880, however, brought on a bloody contest between Dr. Tejedor, the candidate of Buenos Aires, and General Julio A. Roca, the candidate of the federal government and provinces. The situation became so bad that Thomas O. Osborn, the minister of the United States, offered his good offices to prevent further sacrifice of life and property.[62] The national forces soon gained control, and General Roca was duly inaugurated as president. To prevent further uprisings of a similar character, the city of Buenos Aires was detached from the province and federalized; in return for its surrender of provincial autonomy, it became the capital of the federal state.

Although the Argentine statesman, Domingo Sarmiento, had conceived of a plan for inter-American cooperation, it was James G. Blaine of Maine, secretary of state under presidents Garfield and, later, Harrison, who invited the Latin American nations to a Pan American conference in 1881. This conference never met, and in 1886 he helped to obtain passage ot legislation authorizing the convening of a conference to probe the question of commercial relations in the Americas. The First Pan American Conference took place in Washington, 1889, and much of the action was provided by the Argentine and United States delegations. Representing the United States was John B. Henderson, a former senator and William H. Trescott, a diplomat; eight other members, including Andrew Carnegie and Clement Studebaker, were businessmen or industrialists. Blaine, not a member of the delegation, became chairman of the conference. Argentina was represented by Manual Quintana, Roque Sáenz Peña, both of whom later held the presidency of their country, and Vincent G. Quesada, minister in Washington since 1885; the latter decided not to participate owing to his other assignment.

The performance of the Argentine delegation at the conference is perhaps best explained by a report which they later sent to their government; "We were not disposed that the international conference at which we were present should be administratively directed by the United States."[63] From Washington's point of view the Argentines proved intransigent and obstructive, challenging minor questions of organization and procedure, and blocking or insisting on the modification of United States proposals. At the outset they opposed the selection of Blaine as chairman, contending that only a regular delegation member should hold the post. In order to avoid casting a negative vote they declined to attend the opening session. Later, among the seventy-three representatives attending the conference, the two Argentine delegates alone refused to go on a railroad excursion provided by the United States government. The Argentines insisted on the equality of English and Spanish, succeeded in having their version of arbitration adopted, and were instrumental in the rejection of Blaine's proposal for a customs union, which was regarded as a

[62] *For. Rel. of the U.S.*, 1880, p. 27.
[63] Thomas F. McGann, *Argentina, the United States and the Inter-American System, 1880-1914* (Cambridge, Mass., 1957), p. 134.

threat to their country's trade with Europe. "Argentina emerged from the First Pan American Conference into a new decade . . . They had developed a foreign policy centered on close relations with Europe at every level in national life, and they successfully defended this policy in Washington against the first assertion by the United States of its own primacy in inter-American affairs–so successfully that in 1890 Argentina, not the United States, stood as the champion of "America"–indeed of all mankind–in the eyes of Latin Americans."[64]

At the Washington Conference Argentina took steps to protect its own national interests at the expense of Blaine's Pan American idea, which was avowedly designed to further United States commercial interests. While refraining from any commitment toward hemisphere unity, Argentina assumed the role of champion of "humanity," a concept that struck a responsive chord among most of the other Latin American states. In the following decade Buenos Aires refused to pay its nominal share in support of the Commercial Bureau of the American Republics, holding that it was merely an appendage of the U.S. Department of State.[65] Although trade levels between the two countries continued to rise, as noted, the Argentines remained opposed to bilateral reciprocity treaties, asserting that there could be no reciprocity without free wool in United States markets.

A treaty signed between the Argentine republic and Brazil, September 7, 1889, again called in the United States to settle a boundary dispute. The territory in question was the so-called Misiones, lying between the Uruguay and Yguazu rivers. Various unsuccessful attempts had been made to settle the question by negotiation–the last one in 1890, when there was an attempt at division. This effort having failed, the case was turned over to President Harrison in 1892, according to the terms of the treaty. Owing, however, to the delay necessary to prepare the cases, President Cleveland was called upon to make the award, which was delivered to the contending parties in February 1895. The decision gave Brazil the whole territory under dispute, without assigning the reason for the judgement; but the fact that by the census of 1890 all but thirty of the 5,793 inhabitants of the contested region were Brazilians (and of the thirty not one was an Argentine) probably had something to do with the verdict. Both sides accepted the award, and the Baron de Rio Branco, head of the Brazilian special mission, declared: "I am sure that the award of the illustrious American, who, animated by an equal regard for both nations, has so carefully and conscientiously exercised his functions as arbiter, has been received with satisfaction in the Argentine republic, and that this happy and honorable event will tend, as all Brazilians desire, to tighten the bonds of friendship which unite us to our former allies of Caseros and Paraguay."[66]

A more serious dispute over boundary claims came up in 1898 between the Argentine republic and Chile. Ever since the two governments were established, the exact boundary line had been disputed; but in 1884 a protocol had established the frontier where the highest peaks of the Andes divide the

[64] *Ibid.*, p. 163.
[65] Peterson, *Argentina and the United States*, p. 284.
[66] Moore, *Hist. and Digest of Int. Arbitrations*, vol. II, pp. 1969-2026.

watershed. The Argentines interpreted this to mean a line drawn from the highest peak to the highest peak,while the Chileans claimed that the highest peaks meant the highest points in the watershed. In 1895 the situation became very strained, and both sides began preparation for war. An arrangement was finally made to continue the surveys, but in 1898 a violent dispute arose regarding the northern boundary in the district known as Puno de Atacama. Once more the situation became critical, and in August the Chilean government sent an ultimatum demanding arbitration. Fortunately for Argentina, General Roca, whose services as president in his term from 1880 to 1886, and as secretary of interior from 1890 to 1892, were so eminently valuable to his country, had just been reelected. He accepted arbitration for not only the northern section, but also the longer and more important boundary in the south. For the Atacama region a commission of three was chosen, consisting of one Argentine, one Chilean, and William I. Buchanan, the American minister. Buchanan thereupon sketched the line he considered fair according to the evidence, and divided it into sections. Possessing the deciding vote, he suggested that the commission vote on each section separately. The result was that where he favored the Argentine claims he voted with the Argentine commissioner, and where he thought Chile had the advantage, he supported the Chilean representative. By this means the whole question was settled in three days and, although the justice of the course was questioned, both sides accepted the award. The more difficult boundary question of the south was not settled until 1902, and then only after a war scare that brought both nations to the realization of the dangerous possibilities of further delays.[67]

In his message to Congress of May 1899, President Roca paid a high tribute to the efforts of the United States to bring about a peaceful solution of the boundary question. After pointing out that the delineation of the Puno de Atacama had a far greater importance than the value of the territory in dispute, since it closed the long period of uneasiness and inquietude which had been the cause of so many sacrifices on the part of Argentina and Chile, he went on to say: "The participation taken in the solution of the difficulties of which I speak by Mr. Buchanan, the American minister, has also been a motive for particular gratification. To that solution he chiefly contributed, and thus rendered both republics an eminent service. This is not the first occasion upon which it has fallen to the lot of a minister of the great confederation of the north to decisively intervene in our boundary disputes in the interest of international peace. Nor will this ever be forgotten by the two peoples whose destinies have been at stake on one or the other side of the mountain."[68]

United States-Argentine relations in the decade of the 1890s remained cordial and with little rancor on the official level, but underlying currents of mistrust were sometimes evident, the arbitration services of Washington notwithstanding. Officially, the Argentine government supported the latter's position in the *Baltimore* and *Itata* cases involving Chile in 1891, as it did when Washington applied the Monroe Doctrine to the Anglo-Venezuelan boundary

[67]A large bronze statue, the Christ of the Andes, was erected on the boundary line high in the Andes to celebrate the peaceful settlement.
[68]*For. Rel. of the U.S.*, 1899, p. 7.

dispute in 1895. Popular Argentine attitudes, however, commonly reflected anti-American sentiments and became more pronounced with United States intervention in Cuba.

At the Second Pan American Conference, held in Mexico City in October 1902, the American and Argentine delegations' verbal exchanges were characterized by moderation as compared to the First Conference, but the latter lost no opportunity to criticize the Yankee high-tariff policy. Although Argentina exhibited more resiliency towards accepting institutionalized Pan Americanism at this conference, it was clear that she was intent on maintaining independence of action and asserting leadership among the Latin American states. It is noteworthy that while most of the acts of the conference were later ratified by the United States and a number of the other countries, none was ratified by Argentina.[69]

On the question of an arbitration treaty the Argentines favored obligatory arbitration, whereas the American delegation supported voluntary arbitration according to the Hague Convention. This was a reversal of their respective positions in 1890. Although Argentina gained some measure of support for their proposal, the United States managed ultimately to achieve unanimity on the principles of the Hague Convention.

Having finally settled its boundary disputes with Chile in 1902, and with the Mexico City Conference coming to a close, the Argentine republic was able to turn its attention to the European intervention in the affairs of Venezuela. In December of that year Great Britain, Germany and Italy established a blockade of Venezuelan ports with a view of forcing the payment of debts. The United States minister to Venezuela immediately proposed arbitration, and although Kaiser Wilhelm held back, President Roosevelt was able to exert sufficient pressure to force him to terms.[70] Argentina, although not aware of the efforts of the United States at the time, came to her aid in a strong letter of protest against the European intervention. Argentina's reaction might have been predicted, for as the Latin American nation with the most extensive financial ties in Europe, she would naturally take a firm position against such intervention. In a long letter to the State Department, Dr. Luis M. Drago, Argentine Minister of Foreign Relations, thus summed up the views of his government in regard to armed intervention for the purpose of collecting the private claims of its nationals against another state: "The only principle which the Argentine republic maintains, and which it would with great satisfaction see adopted, in view of the events in Venezuela, by a nation that enjoys such great authority and prestige as the United States, is the principle already accepted, that there can be no territorial expansion in America on the part of Europe, nor any oppression of the peoples of this continent because an unfortunate financial situation may compel some of them to postpone the fulfilment of its promises. In a word, the principle that she would like to see recognized is that the public debt cannot occasion armed intervention, nor even the actual occupation of the

[69]McGann, *Argentina and the U.S.*, p. 217.
[70]For this very interesting episode in Roosevelt's dealings with the Kaiser, see W. R. Thayer, *Life and Letters of John Hay* (Boston, 1915), vol. II, pp. 286-288. For a contrary viewpoint cf. Dexter Perkins, *The Monroe Doctrine, 1867-1907* (Baltimore, 1937), p. 333.

territory of the American nations by a European power."[71] Although Secretary Hay was not willing to subscribe *in toto* to this sentiment, the American delegate to the second Hague Conference in 1907 introduced a similar proposal, and the result was a resolution whereby the contracting powers agreed not to have recourse to armed force to recover debts due their nationals unless the debtor nation refused arbitration, or, having accepted arbitration, failed to submit to the award.[72]

On his trip to South America in 1906 in connection with the Third Pan American Conference at Rio de Janeiro, Secretary Root was enthusiastically received in Buenos Aires, and he took advantage of the opportunity to express clearly and forcefully the United States' acceptance of the Drago Doctrine. In reply to a speech in which the eminent authority on international law declared his doctrine to be a principle of American diplomacy based upon the sentiment of common defense, just as in the traditional policy of the United States, Mr. Root answered as follows: "I am glad to be able to declare myself in hearty and unreserved sympathy with you. . .We deem the use of force for the collection of ordinary contract debts to be an invitation to abuses in their necessary results far worse, far more baneful to humanity, than the debts contracted by any nation should go unpaid. We consider that the use of any army and navy of a great power to compel a weaker power to answer to a contract with a private individual is both an invitation to speculation upon the necessities of weak and struggling countries, and we are now, as we have always been, opposed to it."[73]

The Drago doctrine was viewed favorably by the American public because it seemingly harmonized with the doctrine of Monroe; however, in conceptualization there was a marked difference between the two doctrines. The Argentine's proposal was indeed a projection of Monroe's principles, but it aimed towards a multilateral policy resting on inter-American cooperation. It was the type of Pan Americanism sought by the United States in its promotion of the Good Neighbor Policy in the 1930s. Presented at an unfavorable historic moment—the eras of the "*gran garrote*" ("Big Stick") and "dollar diplomacy"—it was overwhelmed by the unilaterally implemented Roosevelt Corollary.[74]

Between the Third and Fourth Pan American Conferences the United States became involved in a South American armaments race, which for a time created discord in its relations with Argentina. The rivalry for status between Brazil and Argentina reached a peak in this period, and when the former contracted with British firms for the construction of three modern battleships the Argentines, seeing the balance of naval power shifting against them, began consideration of an armaments expansion program of their own. Continuing protestations of United States-Brazilian friendship raised Argentine suspicions that Washington was bent on strengthening its neighbor's power to a point

[71] *For. Rel. of the U.S.*, 1903, p. 4. The Drago Doctrine was foreshadowed several decades earlier by the Calvo Doctrine, named for the Argentine jurist Carlos Calvo, who proclaimed that sovereignty is inviolable, and ruled out the interposition of a government on behalf of a resident alien under any and all circumstances.

[72] W. M. Malloy, *Treaties, Conventions,* etc. (Washington, 1910), vol. II, p. 2254.

[73] *For. Rel. of the U.S.*, 1906, Part I, p. 29.

[74] McGann, *Argentina and the U.S.*, chap. 14; Arthur P. Whitaker, *The Western Hemisphere Idea* (Ithaca, 1954), pp. 88-107.

surpassing her own. Further tension was aroused when it was announced that part of the United States fleet would sail around South America in 1907; Rio de Janeiro was among the ports of call, while Buenos Aires and Valparaiso (Chile) were not included. When Buenos Aires called the fleet's omission of Argentina an "act of hostility," Washington quickly arranged to send a flotilla to Buenos Aires.

Once the Argentine Congress had authorized funds for a naval expansion program President Taft and Secretary Knox vigorously pushed a program of "battleship diplomacy," endeavoring to secure contracts for American firms. The aggressive American salesmanship won substantial contracts on what was the first major effort of the United States to compete in the world shipbuilding market. Argentina's naval parity was retained and Buenos Aires expressed satisfaction with the Taft adminstration's policies in South America.[75]

The cordial relations thus engendered were enhanced by the active and discreet participation of the United States in the Fourth Pan American Conference at Buenos Aires in 1910–a conference which, in the words of President Taft, had a "special meaning to the hearts of all Americans because around its date are clustered the anniversaries of the independences of so many of the American Republics."[76] The United States delegation was a distinguished one, made up of Henry White, a career diplomat as chairman, General Enoch Crowder, John Bassett Moore, three university professors, including Bernard Moses, a historian at the University of California, and others. Controversial subjects were largely omitted from the agenda, and the Argentines were favorably impressed with the Taft-Knox proposals for expanding United States trade with Latin America. Argentina displayed greater enthusiasm for Pan Americanism than in the past, but optimism for its future performance was tempered by the fact that of the thirteen conventions it had signed in the three preceding conferences, it had ratified but two.

Closer and friendlier ties between the United States and Argentina were to continue in the immediate future. The growth of trade and United States private investment were factors, and Theodore Roosevelt's visit to Buenos Aires in 1913 had a salutary effect. The former president's statement that the intent of the Monroe Doctrine was not applicable to Argentina, his assertion of the sovereign equality of American states, and his recognition of his host's special relationship with Europe, were applauded by the Argentines.

With the advent of the adminstration of President Woodrow Wilson in 1913, relations were cooled somewhat by the Wilsonian adherence to *de jure* recognition as applied to undemocratic and disorderly regimes in Latin America. Argentina regarded it as a potential infringement of sovereignty and surely not applicable to their country. United States intervention in Mexico in 1914 aroused distrust of the United States in Argentina, although the government remained officially neutral toward the affair. When Washington suggested joint mediation of the problem by Argentina, Brazil and Chile, Buenos Aires eagerly accepted. The Argentine ambassador to the United States was presiding officer

[75]Peterson, *Argentina and the United States,* pp. 291-295.
[76]*For. Rel. of the U.S.,* 1910, p. 14.

of the conference, and Argentina claimed a major share of the successful mediation effort as well as having contributed to Pan American solidarity–under her leadership.[77]

With the outbreak of World War I in 1914 Argentina and the United States proclaimed their neutrality, the latter, under the leadership of President Wilson, becoming the champion of neutral rights for the American states. Although each nation suffered initially from the dislocation of normal trading channels, they profited greatly as the war continued, adjusting their output to supplying Allied demands. Both nations, owing to their peculiar relationship with Europe, were drawn more toward the Allies than toward the Central Powers. They followed similar paths until the United States became associated with the Allies thereby forfeiting its position as a neutral.

When the United States severed diplomatic relations with Germany in February 1917, refusing to concede Germany's right to sink neutral vessels on the high seas, President Wilson expected that the remaining neutral states would do likewise. The Argentine republic, however, contented itself with a statement of regret that his Imperial Majesty had thought to adopt such extreme measures, and by insisting that its conduct be adjusted, as formerly, to the fundamental rules and principles of international law.[78] When on April 6, 1917, the United States declared war on Germany, Argentine public opinion rallied to Washington's support, and *La Nacion* published a series of articles demanding that Argentina enter the war. President Hipólito Irigoyen's administration, however, determined to remain neutral, and in a note to the United States it merely recognized the justice of the causes that had moved the United States to declare war.[79] In a statement published in *La Razón* on April 10, Dr. Luis Drago, the former minister of foreign affairs, protested against Argentina's neutral position on the ground that the war between Germany and the United States was a struggle of democracy versus absolutism in which no American state could remain neutral without denying its past and compromising its future.[80] Romulo S. Naon, the Argentine ambassador in Washington during the war, was also at odds with his government for its continuing neutrality. He resigned in protest when Irigoyen refused to break with Germany, but his resignation was not accepted and he remained at his post until the end of the war.

The United States precipitated a crisis in the relations between Argentina and Germany by the publication in September 1918, of a series of dispatches sent by Count von Luxburg, the Germany minister in Buenos Aires, to his government via the Swedish minister. These messages urged that certain small Argentine vessels either be spared or be sunk without a trace (*spurlos versenkt*). Moreover, in one of the dispatches the Argentine Foreign minister was referred to in most uncomplimentary terms.[81] This flagrant violation of the first principles of neutrality and hospitality provoked an immediate demand for a declaration of war against Germany. Count von Luxburg was given his passport,

[77]McGann, *Argentina and the U.S.*, pp. 305-307.
[78]John Barrett, *Latin America and the War* (Washington, 1919), p. 4.
[79]*Ibid.*, p. 5.
[80]G. Gaillard, *Amérique Latine et Europe Occidentale* (Paris, 1918), p. 130.
[81]For text of these messages see Percy A. Martin, *Latin America and the War*, League of Nations, vol. II, No. 4, p. 253.

and both the Senate and the Chamber passed resolutions authorizing the president to sever diplomatic relations with Germany. President Irigoyen, however, would not change his nation's stance, declaring that Berlin's prompt disavowal of Luxburg's actions and the expression of regret accompanying it were entirely satisfactory.

Various suggestions have been made to account for President Irigoyen's persistent refusal to break with Germany in the face of great pressure of public opinion. One explanation particuarly worthy of consideration, in view of the source, came from Frederic J. Stimson, a professor of comparative legislation at Harvard University, who served as the wartime United States ambassador to Argentina. According to Stimson the inability or unwillingness of Congress to force the issue was an underlying popular resentment against the United States for releasing the Luxburg telegrams without the Argentine government's knowledge or approval. Subsequent publication of these dispatches was carried out simultaneously in both countries.[82]

It was also Ambassador Stimson's view that Irigoyen's stubborn neturality did not stem from pro-German sentiments. His intent was to serve his country's best interests and neutrality seemed the prudent course to follow. Evidence can be cited to demonstrate that Buenos Aires actually leaned toward a pro-United States policy. Argentina welcomed with much enthusiasm the visit of a United States fleet to Buenos Aires in July 1917, a patently unneutral act. Reflecting a similar attitude was the Wheat Convention of January 1918, whereby Argentina contracted to sell two and one-half million tons of wheat to Great Britain and France. An agreement by the United States to supply Argentina with coal, a critically needed item at the time, may have helped to influence the successful outcome of the transaction.[83]

In spite of these suggestions of pro-United States and pro-Allied sentiments the traditional rivalry between Washington and Buenos Aires for leadership of the Latin American states continued throughout the war. President Wilson took the initiative by proposing a Pan American Peace Pact which, as well as controlling the manufacture and distribution of munitions of war, would have pledged mutual guaranties of territorial integrity, the latter being interpreted as transforming the Monroe Doctrine into a common policy shared by all the American states. The Argentine Ambassador accepted the final draft in November 1915, but a combination of circumstances blocked further progress. Chile, with much at stake in the Tacna-Arica dispute, objected to guaranties of territorial integrity, and United States interventions in Mexico and the Caribbean had an understandably depressing effect. President Wilson then became preoccupied with his election campaign and problems of neutrality, and Irigoyen was inaugurated as the Argentine president. A later effort by the United States was the convening of the First Pan American Financial Conference and the International High Commission which was derived from it. Designed to cope with problems of hemispheric trade, Buenos Aires became the site of the first meeting in April 1916. William G. Mc Adoo, Wilson's son-in-law, and secretary

[82]Peterson, *Argentina and the United States,* p. 314.
[83]*Ibid.,* pp. 310-318.

of the treasury, served as chairman. Although of worthy intent, it produced few practical results.

After the United States broke with Germany, President Irigoyen attempted to assert Argentina's leadership of the Latin American neutrals. Initially seeking to convene meetings from which the United States would be excluded, he later negotiated for Washington's support, only to be rebuffed. When the American position became known, and as Germany's submarine warfare accelerated, there was a waning of interest in Argentina's proposals among the Latin American states. It may be concluded that during the war years pan-americanism, pushed by the United States, came into conflict with Argentina's promotion of pan-americanism under her auspices, and the movement towards continental solidarity generally languished.[84]

United States-Argentine rivalry for leadership of the Latin American states continued virtually unabated in the post-war decades, 1919-1939. Problems arising from interpretation of the Monroe Doctrine, the right of intervention, reciprocity of commerce and conflicting views on the jurisdiction of the League of Nations led to continuing discord. Although some agreement was reached on measures to forestall the totalitarian menace developing in Europe in the 1930s, they proved inadequate to insure effective cooperation when hemispheric security was clearly threatened.

After the war Argentina was one of the fifteen Latin American states that joined the League of Nations which the United States rejected. At the first Assembly, which met in Geneva, November 15, 1920, Dr. Honorio Pueyrredón led the fight for the admission of all states to the League, including Germany. To accomplish this, it would have been necessary to amend the Covenant; so the Argentine delegates urged the consideration of four amendments which they regarded as essential if the League was to be established upon a broad and democratic basis. These amendments included compulsory jurisdiction for the Permanent Court of International Justice, the election of all members of the Council by the Assembly, and the admission of all sovereign states unless they voluntarily decided to stay outside. When the Assembly refused to consider any amendments at its first session the Argentine delegation refused to cooperate further and withdrew. Although regarded as a member, it was not until 1933 that the Argentine Congress gave its authorization for renewal of participation at Geneva.

Argentine disagreement with the United States hinged on the insertion in the League Covenant of the article on regional understandings, like the Monroe Doctrine, insisted upon by President Wilson's opponents. Argentine nationalists saw the unilateral interpretation of the Doctrine by the United States, confirmed by its recent interventions in Mexico and the Caribbean, as a threat to the sovereign equality of states. Since the failure of the United States to join the League was regarded as a betrayal in Latin America, and it had sanctioned the division of nations into great and lesser powers at the Versailles Conference, pan-americanism suffered a setback, especially in Argentina where recurring waves of anti-American feeling appeared. An unfavorable balance of trade with the United States contributed to Argentine grievances.

[84]*Ibid.*, chap. XX.

Secretary of State Bainbridge Colby visited Buenos Aires in January 1921, as part of a good-will tour in South America, and in spite of coolness on the part of some officials, he was cordially received by President Irigoyen. In the same year, Argentina was involved in the settlement of two problems concerning Panama, and cooperated effectively with the United States. At the Fifth International Conference of American States convened at the Chilean capital in March 1923, a collision of U.S.-Argentine views was avoided by the State Department's policy of restraint in pushing topics likely to offend the more delicate sensibilities of the Latin American states. It was evident, however, that the divergence of the United States and Argentina on such matters as reciprocal commerce, the settlement of pecuniary claims and Pan Americanism remained as broad as ever.

At the Sixth International Conference of American States held in Havana in 1928, the United States was on the defensive because of the current criticism of its Latin American policies, notably Mexico, Nicaragua, and Argentina. Washington appointed a highly capable delegation headed by former secretary of state Charles Evans Hughes; the Argentine group had for its chairman Dr. Pueyrredón, foreign minister under Irigoyen and ambassador to Washington since 1924. They clashed first on the question of commercial reciprocity and secondly on the problem of non-intervention. Dr. Pueyrredón insisted upon a declaration against barriers to inter-American trade, since his country suffered from the tariff wall which obstructed the entry of Argentine exports to the United States. To achieve this he proposed a customs union for the Americas, reminiscent of Blaine's proposal which Argentina had rejected in 1890. Owing to the changed circumstances, however, the United States rejected the free-trade concept; other Latin American states joined in defeating the Argentine plan because of the importance of customs revenues in their national economies. As a concession to the Argentines a clause was included in the preamble to the proposed Pan American Union convention indicating that the organization wished to "promote effectively the harmonious development of the economic interests of the American republics."[85]

On the question of intervention the Argentines gave strong support to the anti-interventionist delegations, Dr. Pueyrredón declaring: ". . . Diplomatic or armed intervention, whether permanent or temporary, is an attack against the independence of these states . . ." Mr. Hughes, resisting any declaration that would invalidate or censure the right of intervention, and thereby infringe on the Roosevelt Corollary and American security interests in the Caribbean, said, ". . . What are we to do when government breaks down and American citizens are in danger of their lives? . . . Now it is the principle of international law that in such a case a government is fully justified in taking action–I would call it interposition of a temporary character–for the purpose of protecting the lives and property of its nationals."[86] Thus in the forum of the conference Argentina aired its two main grievances against the United States, and as before contested its North American rival for leadership of the Pan American movement. When

[85]J. L. Mecham, *The United States and Inter-American Security, 1889-1960* (Austin, 1961), pp. 100-106.
[86]*Ibid.*, p. 104.

President-elect Herbert Hoover visited Argentina on his good-will trip to Latin America in 1929, he received a restrained welcome in Buenos Aires. However, by statements subordinating the role of the United States in inter-American affairs, he made a favorable impression and raised some optimism about improving relations.[87]

When the Seventh International Conference of American States met in Montevideo in December 1933, Secretary of State Cordell Hull, expounding the "Golden Rule" basis of the Good Neighbor policy, which meant Washington's renunciation of the right of intervention, defused the conference agenda. Meanwhile General Uriburu's overthrow of the Irigoyen government in 1930, moderated Argentina's isolationism, and gave promise of a more cordial attitude toward the United States. Although Dr. Carlos Saavedra Lamas had reservations about participating in the conference, he responded to the conciliatory efforts of the American delegation and endorsed its program of bilateral reciprocity. The United States responded by urging universal ratification of Saavedra Lamas' Anti-War Treaty of Non-Aggression and Conciliation and other peace instruments. The Montevideo Conference, vitally concerned with problems of hemispheric security, was confronted by the nearby war in the Chaco. Its failure to achieve more than a three-week truce in the fighting, clearly revealed the weakness of the existing inter-American peace-keeping machinery.

The extraordinary Inter-American Conference for the Maintenance of Peace met in Buenos Aires in December 1936, with Dr. Saavedra Lamas, the Argentine Foreign Minister, presiding. President Roosevelt, warmly greeted by the Argentine people, addressed the conference at its opening session.[88] His presence there suggested the concern Washington felt about the deteriorating situation in Europe, which was interpreted as a threat to hemispheric security. Dr. Saavedra Lamas, in contrast to his cooperative performance at Montevideo three years earlier, was categorically opposed to the American proposals for strengthening the inter-American security arrangments. Although the principle of consultation was adopted by the conference, in itself a major contribution to New World security, the acts of the conference failed to provide the positive means which the United States delgation sought. Dr. Saavedra Lamas, declining to acknowledge a threat emanating from Europe, opposed strengthening the inter-American organization without coordinating its system of collective security with that of the League of Nations. He succeeded in thwarting American objectives and asserting Argentina's leadership in inter-American affairs.[89]

Both Argentina and the United States, the former a rather uninterested member of the League of Nations and the latter wholly outside of its jurisdiction, were made active participants in the Chaco dispute between Bolivia and Paraguay partly as a failure of the League's efforts and partly as a result of their interest in maintaining peace in the Western Hemisphere. In fact, the United States had been an interested party to the Gran Chaco question ever since President Rutherford B. Hayes handed down an arbitral award in 1878 which

[87]See Alexander De Conde, *Herbert Hoover's Latin American Policy* (Stanford, 1951).
[88]*New York Times*, Dec. 1, 1936.
[89]Mecham, *U.S. and Inter-American Security*, pp. 132-135.

gave Paraguay the triangular territory between the Paraguay, Pilcomayo, and Verde Rivers. However, this award merely settled the disputed boundary between Paraguay and Argentina. Their neighbor, Bolivia, had certain claims to the Chaco and immediately objected that her rights were not given proper consideration.

When defeat in the War of the Pacific shut Bolivia off from the sea, her interest in the Chaco increased, since the Pilcomayo River was the only feasible outlet to the Paraguay River and the Atlantic. Fruitless efforts were made to establish a boundary between the two states but the claimants could not reach a suitable compromise. Each could establish an excellent claim, Bolivia by title, Paraguay by possession. Treaties were signed in 1879, 1887, and 1894, but none were ratified. In 1907 the two powers agreed to arbitrate the ownership of some 50,000 square miles in this area, but the president of Argentina chosen as umpire, refused to act. A new protocol for a boundary settlement, signed in 1913, was no more successful than its predecessors. And so the negotiations continued intermittently and interminably until 1928 when diplomacy gave way to force.[90]

When the resort to arms occurred on December 8, 1928, representatives of the American nations including Bolivia and Paraguay were assembled at Washington working out treaties for arbitration and conciliation for the Western Hemisphere. The secretary of state of the United States, as chairman of the conference, promptly offered the good offices of the conference. The proposal was accepted and a commission of inquiry and conciliation was appointed consisting of representatives from Colombia, Cuba, Mexico, the United States, and Uruguay and the two disputing states.

The Commission succeeded in obtaining a renewal of diplomatic relations and the reestablishment of the *status quo* as of December 5, 1928, but failed in the fundamental task of tracing a boundary line satisfactory to both powers. The commission's recommendation that the territory of the Hayes award be regarded as Paraguayan and the port of Bahia Negra be given to Bolivia as a preliminary to arbitration was acceptable to neither power.

As a next effort the five neutrals on the commission persuaded the two states to send representatives to Washington to work out a compromise and to sign a mutual pact of nonagression. These negotiations began November 11, 1931, and in May, 1932, the commission presented the draft of a nonagression pact. While this draft was under consideration new clashes between troops occurred in the Chaco, each side asserting that its opponent was the aggressor.

The neighboring states became worried, particularly Argentina, whose economic interests in the area seemed jeopardized.[91] She persuaded Brazil, Chile, and Peru to offer their services to the commission of neutrals. This offer was accepted, and the immediate result was a recommendation through Mr. Francis White of the State Department, acting as chairman of the commission of neutrals, that all the American states unite in applying the so-called Hoover-Stimson doctrine of nonrecognition to the Chaco dispute. This suggestion was

[90]See Gordon Ireland, *Boundaries, Possessions, and Conflicts in South America* (Cambridge, Mass., 1938), pp. 66 ff.
[91]John C. DeWilde, "South American Conflicts, The Chaco and Leticia," *Foreign Policy Association Information Service*, vol. IX, no. 6 (May 24, 1933), p. 60.

accepted, and on August 3, 1932, the countries represented in the Pan American Union, exclusive of the disputants, warned Bolivia and Paraguay that they would "not recognize any territorial arrangement of this controversy which has not been obtained by peaceful means, nor the validity of the territorial acquisitions which may be obtained through occupation or conquest of arms."[92]

While the commission of neutrals and representatives of the ABCP powers were working in Washington, hostilities increased in the Chaco, and in September the conflict had developed into sanguinary warfare. The League offered to assist, but the American powers still hoped for success. A comprehensive proposal, supported by the League and all the American states except the belligerents, was submitted on December 15, 1932, but this was refused. Independent proposals on the part of the ABCP powers were no more successful, and on May 10, 1933, Paraguay declared war on Bolivia. The League of Nations could no longer avoid responsibility, and it appointed a committee of five to visit the disputed area and work out a settlement with the two powers.[93]

The League commission, consisting of representatives from Great Britain, France, Italy, Spain, and Mexico, arrived at Montevideo on November 3, 1933, and visited Asunción and La Paz to collect data on the controversy. Although definite figures were impossible to obtain, it was estimated that the losses were over 35,000 dead and 60,000 wounded. Bolivian troops from La Paz had to travel a thousand miles to the front, the last five hundred by truck or mule or afoot, ankle deep in dust in winter, in steaming heat in summer. Detachments had to hack their way through thickets to see the enemy. Both sides alleged cruel treatment of prisoners, and hygienic conditions were incredibly bad.[94]

The seventh Pan American Conference, which met at Montevideo from December 3 to 26, could not avoid consideration of the conflict, although the subject was not on the agenda. Secretary Hull made a special plea for peace during the conference, and an armistice was agreed upon to last from December 19 to 31. This was extended to January 6, but the League commission was unable to secure its further prolongation. At the close of the armistice Paraguay resumed the offensive and by the middle of January 1934, had captured practically all of Bolivia's important positions and was in occupation of the entire war zone.

Determined not to permit the United States to aid the belligerents in this useless slaughter, President Franklin Roosevelt succeeded in having the Congress impose an arms embargo on both Bolivia and Paraguay.[95] When Bolivia protested, Secretary of State Hull declared "the government of the United States has dedicated itself to the policy of good neighbor. It would be in the highest degree inconsistent with that policy that arms and munitions of war manufactured in the United States should continue to be sold for the purpose of assisting in the destruction of the lives of our two sister republics of Bolivia and Paraguay. . . ."

[92]U.S. Dept of State, *Press Releases* (Aug. 3, 1932).
[93]*Ibid.*, (June 27, 1933).
[94]*League of Nations Publication*, Official No. C. 154. M. 64 1934. VII.
[95]U.S. Dept. of State, *Press Releases* (May 26, 1934). The League Assembly had already passed a resolution imposing an arms embargo on both belligerents.

Following up this policy of striving for peace, the Governing Board of the Pan American Union under the chairmanship of Secretary of State Hull on July 30, 1934, took the unprecedented action of adopting a resolution calling upon neutral American governments to indicate their attitude on unified action to bring the Chaco conflict to a close through arbitration.

In the meantime the League continued its efforts to bring about peace and after hearing the report of its commission and noting the refusal of both governments to accept, it decided upon more drastic measures. An extraordinary session of the Assembly on November 24, 1934, adopted a plan for settlement requiring an immediate cessation of hostilities, demobilization under the supervision of a neutral commission composed of representatives of the adjacent states, Brazil, and the United States, and the convocation of a peace conference in Buenos Aires within a month after the end of fighting.[96]

The United States agreed to cooperate with this neutral commission and also agreed to take part in the proposed peace conference at Buenos Aires. This last suggestion proved to be the germ of the ultimate settlement. When Paraguay refused the League solution it was declared to be the aggressor and the arms embargo lifted from Bolivia. Thereupon Paraguay gave notice of withdrawal from the League. However, her armies suffered from the embargo and her advance was stopped.

Seizing the opportunity afforded by the apparent stalemate in the field, Argentina and Chile on April 1, 1935, invited the cooperation of Brazil, Peru, and the United States to send representatives to Buenos Aires in a further effort to obtain peace. All accepted, and a mediation commission was set up in Buenos Aires which formulated a truce agreement. The arrangement provided for a twelve-day truce, during which a neutral military commission was to fix the position of the armies pending demobilization which must take place within ninety days. It was also agreed that each state would reduce its military effectives to 5000 men. Bolivia and Paraguay were to negotiate directly in a conference convoked in Buenos Aires by the mediators. If direct negotiations failed the whole problem was to be submitted to the World Court.

Hostilities ceased on June 14, 1935, and the truce agreement was ratified by both powers on June 21. The peace conference elected Dr. Saavedra Lamas, Foreign Minister of Argentina, president, and immediately began its sessions. At first Alexander W. Weddell, United States ambassador to Argentina, and later Hugh Gibson, United States ambassador to Brazil, served as the representative of the United States. When the negotiations lasted so long that they interfered seriously with the work of these men at their posts, President Franklin Roosevelt appointed Mr. Spruille Braden as ambassador-at-large and sent him as his special representative to the conference.

Although demobilization had been completed by October 1935, nevertheless the two belligerents could not agree upon a territorial settlement. Paraguay demanded all the area in dispute and Bolivia insisted upon access to the Upper Paraguay. Paraguay refused to repatriate the thousands of Bolivian prisoners she had taken until a full settlement was reached. On two occasions

[96]See Helen Paull Kirkpatrick, "The League and the Chaco Dispute," *Foreign Policy Information Service,* vol. XII, no. 9 (July 15, 1936).

early in the conference Ambassador Weddell was credited with saving the conference from a complete breakdown. Secretary Hull was so much interested in procuring a settlement that he kept in daily touch with the proceedings. Finally, on January 21, 1936, after almost seven months of delicate, nerve-wracking negotiations, the two governments were persuaded to agree upon repatriation of prisoners and the renewal of diplomatic relations.

Another seven months elapsed before repatriation was "virtually" terminated, and it was not "officially" completed until another nine months had passed. There still remained the more difficult problem of drawing up a peace treaty which should establish an acceptable boundary. The committee continued its work until a settlement was reached in spite of two revolutions in each country and through various hostile incidents—on one occasion the neutral military observers were taken into custody. A personal appeal by President Roosevelt to the presidents of Bolivia and Paraguay, and the strong support rendered to the conference by President Ortiz of Argentina were important factors in achieving a settlement.[97] The peace treaty was signed at Buenos Aires, July 21, 1938, and ratification exchanged August 29. In Bolivia ratification was approved through a constitutional convention by a vote of 102 to 9; in Paraguay a plebiscite approved 132,000 to 13,000.

The first article of the treaty definitely reestablished peace. It provided that the dividing line in the Chaco would be determined by the presidents of the six mediatory nations acting as arbitrators on the basis of equity. Certain limitations on the fixing of the line implicitly assigned one portion of the Chaco to Bolivia and another to Paraguay—the award to be made within two months of the date of ratification. The award had to be carried out within ninety days under the supervision of the Peace Conference. The boundary award, announced on October 10, 1938, was drawn through desert and swamp, distant from Bolivian and Paraguayan posts by from 30 to 120 kilometers. As far as possible it was a natural frontier. Both states accepted the award immediately and unreservedly. One of the most serious boundary conflicts in South America was finally settled on the basis of common sense and justice through the patience and determination of the neighbors and friends of the two contestants.[98] It would take much longer to repair the human costs of the war.

With the unfolding acts of aggression by the totalitarian states in Europe and Japan's advances in China, the threat implied to hemispheric security became obvious, but the Argentine republic persisted in denying its existence. Argentine distrust of Washington's policies continued in spite of its efforts to give substance and meaning to the Good Neighbor program. The move by the State Department in 1937 to strengthen Brazil's defenses by leasing her several destroyers was interpreted by Buenos Aires as a plot to topple her own naval superiority, of which she had a wide margin, in South America. The United States dropped the leasing program for Brazil to preserve harmony, after pointing out that the destroyers were to have been used for training purposes,

[97]See the story of the conference by Spruille Braden in Dept. of State, *Press Release*, vol. XX (Jan. 7, 1939), pp. 1 ff.

[98]For text of treaty see U.S. Dept. of State, *Press Releases*, vol. XIX (July, 1938), p. 44; for the Arbitral Award, *Ibid.*, vol. XIX (Nov. 1938), p. 263. See also Leslie B. Rout, Jr., *Politics of the Chaco Peace Conference, 1935-1939* (Austin, 1970).

and that an equal number of these vessels was available to the other South American nations.

Prior to the convening of the Eighth Conference of American States in Lima in December 1938, President Roosevelt endeavored to impress the Latin American states with the need for continental solidarity and the strengthening of hemispheric defenses in the face of potential totalitarian aggression. The Argentine President, Robert M. Ortiz, in a cooperative gesture, directed his foreign minister, Dr. José M. Cantillo, to keep Washington advised of his country's position. This entailed broadening the basis of consultation, and meetings of ministers of foreign affairs. The Argentine Foreign Minister in his opening address admitted the need for continental solidarity, but reaffirmed his nation's special ties with Europe, and rejected the need for additional pacts. He then boarded an Argentine warship, taking a vacation in the Chilean resort region while the conference was in progress.

After Secretary Hull had made a direct appeal to President Ortiz compromises were reached and the principles of the Declaration were agreed upon. In its final form the Declaration contained the Argentine-proposed formula for consultation by the foreign ministers, and according to Secretary Hull it was a major advance over previous Pan American agreements. However, the provision binding them merely to consult was far removed from mutual security pacts and defensive alliances. Argentina, fearful of alienating the Axis powers, and still distrustful of Washington's motives, had insured that consultation would ensue when a threat existed from any source, which included the United States.[99]

The diplomatic relations of the United States and Argentina frequently turned on issues of nationalistic rivalry, as the preceding pages demonstrate; however, seemingly political aspects were commonly intertwined with economic problems arising from their essentially competitive economies and the irregularities of reciprocal trade. As a food-products and raw materials exporting country Argentina found a natural market in food importing countries such as Britain and Germany, and from them she imported manufactured products. Since the United States was a major producer of beef, wheat and corn, reciprocity was severely hampered.

After emerging as both the political and economic leader of Latin America in the nineteenth century, Argentina attained the greatest prosperity in its history in the period 1914-1929. On all counts it had the highest standards of living in Latin America. The United States enjoyed similar prosperity in this era and trade between the two countries increased progressively as the former replaced Argentina's European suppliers during the war. For a time, Argentina surpassed Cuba as the best customer of the U. S. in Latin America. Total trade reached a peak of $421 million in 1920, dropped to $171 million in the recession year of 1921, and had doubled this figure by 1929; the annual average in the years 1925-1929 was about $265 million. After 1916, except for the years 1922-1924, the United States was the principal source of Argentine imports, mostly at the expense of Britain. But from 1921 to 1929 Argentina suffered an annual average deficit in trade amounting to $50 million.

[99]Mecham, *U.S. and Inter-American Security*, pp. 142-144.

Argentine resentment of United States tariff policies, dating from the Tariff Act of 1867, was accentuated by the passage of the Fordney-Mc Cumber Tariff Act of 1922, which raised additional barriers to Argentine exports. This was followed by the Hawley-Smoot Tariff Act of 1930 which not only penalized Argentine imports into the United States by its high schedules, but placed unwarranted quarantine restrictions upon the entire Argentine livestock industry. The existence of hoof-and-mouth disease (*aftosa*) in some, though not all, parts of Argentina was used as justification for prohibiting the importation of Argentine beef. By 1932 Argentine-United States trade fell to $47 million, the lowest level since 1908. United States investments in Argentina, however, had risen in the 1920s, reaching a total of $500 million in 1930, making the southern republic second only to Britain as the nation's creditor.[100]

Negotiations between Buenos Aires and Washington were continued in the 1930s in an effort to unblock trading channels, but with no significant results until conditions brought on by World War II forced a temporary solution to the problem. In practice, the two leading trading nations of the New World carried on mutual discrimination to the detriment of reciprocal trade, and failed to reconcile their views on international trade. The Reciprocal Trade Agreement Act passed in 1934, reflecting the New Deal trade philosophy, sought reciprocal agreements with individual countries embodying most-favored-nation treatment. By introducing multilateral principles in bilateral trading it was hoped to increase the volume of world trade. Argentina, however, turned further toward economic nationalism, adopting exchange control and insisting on the maintenance of balanced, narrowly constructed bilateral trade arrangements.

Attempts by the Roosevelt administration to reach some accomodation with Argentine interests led to the signing of the Sanitary Convention in May 1935, designed to restore the earlier regional rather than country quarantine. While attending the Buenos Aires Conference in 1936 President Roosevelt agreed to urge its ratification by the United States Senate. The cattle-raisers' lobby succeeded in blocking ratification, and resulted in a crowning affront to Argentine pride. Argentina was prohibited from serving its own fresh beef in a restaurant at the Argentine exhibit at the New York World's Fair in 1939-1940. Argentina turned toward Europe, particularly Great Britain, and by a series of bilateral agreements channeled her trade in that direction. The United States found its products discriminated against and barriers raised against them. Secretary Hull pressed negotiations, but Argentina refused to grant the United States equality of treatment with those countries covered in bilateral agreements. Under a policy adopted in 1939, Argentine imports from the United States were reduced 40 percent below the levels of the preceding year.

When in 1941 the European markets were virtually inaccessible, and Argentina found her elevators and storehouses overflowing with goods, she was more willing to negotiate. A reciprocal trade agreement was signed in Buenos Aires on October 14, 1941, to remain in force until November 15, 1944, and then to continue indefinitely subject to six months notice of abrogation by either party. The most important feature of the agreement was that the United States was

[100] Peterson, *Argentina and the United States*, pp. 349-350.

guaranteed equality of treatment in control regulations, import licenses, quotas, and governmental purchases or contracts. Argentina granted lowered duties or a continuation of existing duties on about 30 percent of her imports from the United States as of 1940, whereas Washington granted reductions upon 43 percent of our imports from Argentina during the same period. The United States maintained the embargo upon chilled and frozen beef but reduced the tariff on canned meats, wool and linseed. A new feature was the provision for joint consultation in case of any difficulty arising in operation. The trade agreement was a progressive move, but it did nothing, as events were soon to prove, to lessen the traditional antagonism on political matters on hemispheric concern.[101]

Clues as to the possible stance that Argentina might assume on hemispheric cooperation in the impending world war could be read in the record of her commitment to the Inter-American System. Although she had participated in the eight inter-American conferences and the Conference on the Maintenance of Peace, and had been a signatory on all but two of the fifty-one treaties and conventions produced by them, the Argentine Congress had ratified only two, one relating to the Status of Naturalized Citizens and another on Codification of International Law. The United States Congress, in contrast, had ratified all but ten of the forty-one conventions which delegates of the United States had signed.[102]

The policy of Argentina in the second World War was no more cooperative than it was in the first. During the short period when Roberto Ortiz was president, he showed himself a whole-hearted supporter of continental solidarity. In fact, when the Nazis violated the neutrality of Norway and Denmark, he authorized his Foreign Minister to approach Ambassador Armour with a proposal to change from a policy of neutrality to nonbelligerency as offering greater security to the Americas.[103] He was one of the first heads of a state to support President Roosevelt's personal appeal to Hitler and Mussolini to maintain the peace. In an interview given to the United Press, November 19, 1940, President Ortiz strongly supported a plan of coordinated action by the peoples of the Western Hemisphere against foreign perils. But with his forced retirement through ill health, Vice-president Castillo finally took over the reins of the government and followed a policy of neutrality scarcely less friendly to the Axis than to the American nations. Although on December 9, 1941, Acting President Castillo decreed that Argentina did not regard the United States as a belligerent, nevertheless Argentina took the lead at the Rio Conference in preventing the passage of a resolution looking towards an immediate and unanimous break with the Axis. Buenos Aires soon became the center of Nazi activities for all South America to such an extent that they endangered both lives and shipping in the Western Hemisphere. Even when two Argentine ships were torpedoed, the *Victoria* on April 17, 1942, and the *Rio Tercero* on June 22, the Argentine government failed to obtain the redress originally demanded.

[101]For text of agreement see Dept. of State *Bulletin,* V supp. (Oct. 18, 1941).
[102]Peterson, *Argentina and the United States,* p. 399.
[103]W. L. Langer and S. E. Gleason, *The Challenge of Isolation, 1937-1940* (New York, 1952).

However, when in November, 1942. United States Ambassador Norman Armour delivered three confidential memoranda on Nazi activities, the Castillo government made charges of espionage against Captain Niebuhr, German naval and air *attaché.* When the German government refused to waive the *attaché's* immunity and permit his trial, the Argentine government requested his dismissal.

There were many indications that the Castillo government was not generally representative of the Argentine people. The fact that the Chamber of Deputies staunchly supported the Damonte Taborda investigation of Nazi activities and actually voted a break with the Axis is evidence of the more representative body's attitude. After Brazil's entrance into the war and Chile's break with the Axis, the position of Argentina as the only neutral in the Western Hemisphere was not an enviable one. Discontent finally resulted in a military revolt on June 4, 1943. Nevertheless, the provisional government set up gave no indication of an intention to reverse the previous policy of neutrality. General Ramírez, as acting president, agreed to control Axis communications more carefully and expressed his desire for friendly relations with the other American republics. The United States granted immediate recognition and awaited developments. They were quickly forthcoming.[104]

An exchange of letters between Argentine Foreign Minister Storni and Secretary of State Hull in August indicated the wide gap between the attitudes and policies of the two governments. The former suggested that the United States, as a gesture of friendship, should send airplanes and arms to Argentina. Secretary Hull, after pointing out Argentina's failure to carry out the commitments of the Rio Conference, declared that since the United States in sending arms to Latin America had been guided by considerations of hemisphere defense and Argentina refused to cooperate, no shipments of arms under lend lease were possible. When the two notes were published in the press, Storni was forced to resign.

When Argentina severed diplomatic relations with the Axis powers January 26, 1944, the military clique forced the resignation of President Ramírez and installed Vice-president Farrell in his place. The United States decided to withhold recognition until the policy of the new government towards the Axis became evident. It was soon seen to be so pro-Axis that on July 26 the United States issued a scathing indictment declaring that the new government was giving the Axis both political and economic support. On September 7 Secretary Hull declared that Argentina was the Fascist headquarters for the Western Hemisphere and shortly afterwards, all United States ships were forbidden to call at Argentine ports.[105]

The Argentine government was sufficiently disturbed to present a formal request that a Conference of Foreign Ministers be held to consider the Argentine question. The United States countered by suggesting a conference of the states cooperating in the war effort to consider war and postwar problems. This suggestion was adopted and Argentina was not invited to the Inter-American Conference on Problems of War and Peace held in Mexico City early in 1945.

[104]Robert J. Alexander, *The Perón Era* (New York, 1951), pp. 3-11; Felix J. Weil, *Argentine Riddle* (New York, 1944), p. 54.
[105]Edward O. Guerrant, *Roosevelt's Good Neighbor Policy* (Albuquerque, 1950), p. 36.

The Conference, however, passed a resolution urging Argentina to change its policy and cooperate with the other American nations. Argentina finally declared war upon the Axis Powers on March 27, 1945, and her government was recognized by the United States.

Mention should be made at this point of a new inter-American policy of intervention suggested by Uruguayan Foreign Minister Alberto Rodríguez Larreta on November 22, 1945, as a result of the pro-Axis policy of Argentina. According to Larreta, nonintervention should not be a shield behind which "law may be violated, agents and forces of the Axis may be sheltered and binding obligations may be circumvented."[106] Secretary of State Byrnes, acting on the advice of Assistant Secretary of State Spruille Braden, officially endorsed this policy on November 27, 1945.

The Argentine government, however, dominated by Vice-president Juan Domingo Perón, continued to support the Axis and to criticize the United States. Matters came to a head in the presidential campaign of 1946 when Perón appealed to labor and to anti-American elements to support his candidacy. Ambassador of the United States Spruille Braden, who made no secret of his opposition to the Farrell-Perón government, was Perón's particular subject of attack.[107] Unfortunately, twelve days before the election, the United States published a Blue Book of documents from the German Foreign Office proving Argentina's collaboration with Nazi Germany and noting the active role played by Perón in this policy. Perón cleverly countered by alleging that the United States was interfering in Argentina's domestic policies to defeat him–it was "the pig Braden or the patriot Perón." Perón won the election.[108]

The United States Department of State was now in a dilemma. Numerous senators, important military officials, and representatives of business insisted upon getting together with Perón. Other Latin American states resented the delay in holding the next Inter-American Conference because of United States opposition to Argentina. George Messersmith, an eminent career ambassador, was sent to Argentina and instructed to patch up relations with Perón and at the same time get rid of Nazi influences if possible. On June 3, 1947, President Truman announced that the United States was ready to renew consultations with the other American republics, including Argentina, with reference to implementing the Act of Chapultepec. Two days later Spruille Braden resigned as assistant secretary of state in charge of Latin American affairs.

The anti-Perón policy of the United States which former Secretary of State Sumner Welles strongly denouced was completely reversed. President Harry S. Truman's *rapprochement* with the Perón regime was an effort to reinforce the weakened structure of the Inter-American system in the face of threats arising from the cold war; it was also in recognition of Argentina's growing influence in the Western Hemisphere. Perón won a major diplomatic victory at the Inter-American Conference for the Maintenance of Continental Peace and Security held in Quitandinha, near Rio de Janeiro in August 1947, for

[106] Dept. of State *Bulletin*, XIII (Nov. 25, 1945), pp. 864-866.
[107] *Ibid.*, (Oct. 28, 1945), p. 658.
[108] John J. Kennedy, *Catholicism, Nationalism and Democracy in Argentina* (South Bend, Indiana, 1958), pp. 205-207; Joseph R. Barager, ed., *Why Perón Came to Power* (New York, 1968).

Argentina became a signatory of the hemispheric mutual security pact without altering policies that Washington had opposed for two years. Perón had restored diplomatic relations with the Soviet Union, severed since 1918, at a time when tension was rising over the Russian thrust toward Iran. By so doing Perón was preparing the "Third Position" Argentine policy would take in international affairs. It was evident that the Perón-led army regime was carrying on the country's traditional foreign policy, challenging United States leadership in the Americas, and resisting all foreign pressures whether in war or peace.

With the armed forces and organized labor as his main bases of support, Perón established a tight and ruthless dictatorship. The judiciary and the universities were purged, critics were jailed, and civil liberties generally suppressed. At the same time, however, Perón, aided by his wife Eva, set in motion what approached a social and economic revolution. Between 1946 and 1951 important strides were taken toward industrialization and economic independence. The industrial worker, and to a lesser extent his agrarian counterpart, gained a degree of economic security and a voice in government which were unprecedented.[109]

Argentina enjoyed great prosperity owing to the wartime and post-war demand for its foodstuffs, building up immense foreign exchange reserves. Overly optimistic, the Argentines went on a spending spree of buying expensive foreign imports, and Perón's emphasis on industrialization, requiring purchases of machinery, resulted in a great loss of dollar exchange. The national economy began to slip in 1948, and severe droughts in 1949 and subsequently, caused serious reduction in exportable surpluses of cereals and meats. By 1951-52 the country was suffering from unemployment, food shortages, fuel rationing and inflation. Because of unfavorable trade balances the remission of profits of foreign-owned concerns was stifled, and this reduced the importation of capital goods for basic industries. Starting in 1950 the United States replaced Britain's role of Argentina's best customer, and in 1952 U. S. importers purchased 26 percent of the nation's total exports, double the pre-war percentage. Britain was also succeeded by the United States as the chief foreign investor in Argentina; however, United States investments in 1950 had dropped by nearly 10 percent from the 1940 level owing to the Peronist policies of economic nationalism.[110]

Argentina's desire to improve the climate for foreign investments and Washington's willingness to take a calculated risk in furthering its hemispheric political goals led to negotiations between Assistant Secretary of State Edward G. Miller, Jr., and Dr. Ramón A. Cereijo, president of Argentina's National Economic Council, in 1950. The talks had been preceded by new anti-United States attacks by the Peronist press when Argentina had failed to get as much as had been anticipated from purchases under the Marshall Plan. Seemingly impervious to insults, the United States, as a result of the Miller-Cereijo negotiations, granted Argentina a $125 million loan with which to repay accumulated commercial debts to American exporters. Even the suppression of

[109]George I. Blanksten, *Perón's Argentina* (Chicago, 1963), pp. 186-219; Marvin Goldwert, "The Rise of Modern Militarism in Argentina," *Hispanic American Historical Review,* XLVIII (May, 1968), 189-205.

[110]Peterson, *Argentina and the United States,* p. 478.

the great Argentine newspaper *La Prensa,* in typical totalitarian fashion, provoked only a mild expression of concern in the United States governmental circles. In what appeared to have been a gesture of reciprocity the Argentine Congress ratified the Rio Defense Treaty on which it had failed to act for about two years, and the regime sent small quantities of foodstuffs to the United Nations forces in Korea.

Following the interval of restrained cordiality accompanying the loan negotiations and the early phases of the Korean crisis, the Argentine invective was continued. Perón had engineered a revision of the Constitution of 1949 permitting the immediate reelection of the president which allowed him to succeed himself in 1951. Antagonism toward Washington and stimulated fears of United States intervention were a part of the campaign strategy. At the end of the Truman administration in 1952 it was clear that Washington's policies over the preceding eight years had failed to gain a cordial working relationship with the Buenos Aires government. The Hull-Braden attempts at coercion had boomeranged, and the assignment of business-ambassadors had not led to any viable new commercial agreements. As the stalemate developed in the Korean War and the presidential campaign of 1952 got underway, the Argentine problem drew less attention. Albert F. Nufer, a career diplomat who had served as chairman of the Joint Argentine-United States Committee on Commercial Studies, was sent to Buenos Aires in July with the expectation that he would bring stability as well as a policy of "correct friendliness" to the post.

In the presidential campaign the Republicans had labeled the Democrat's Latin American policy as "feeble," and General Eisenhower had pledged fresh efforts to strengthen inter-American ties. Secretary of State John Foster Dulles' remarks about fascism in Argentina provoked a bitter reaction there, but the later statements of John Moors Cabot, Assistant Secretary of State for Inter-American Affairs, indicative of a neutral attitude toward the internal affairs of Argentina, helped to improve the diplomatic climate. However, in his May Day address to the Congress Perón denounced the American press for "spreading lies" about him and his government and temporarily suspended the news from abroad which emanated from the various United States news agencies. A sudden change of attitude was evident when President Eisenhower's brother visited Buenos Aires in his fact finding tour of Latin America; Dr. Milton Eisenhower was given a cordial reception and the controlled press praised the friendly cooperative attitude of the new Republican administration–"The new era of friendship with the United States." The United States news agencies were utilized; the *New York Times* and even *Times Magazine* were permitted to be sold in Argentina. On October 17, 1953, the eighth anniversary of the Peronist movement, President Perón expressed complete friendship for the United States and declared that all former animosities had been eliminated. It should be noted, however, that Argentina still maintained her "Third Position" between capitalism and communism, and the Argentine Ambassador to Moscow expressed high hopes that the recently signed Argentine-Soviet Trade Agreement would bring about a substantial increase in trade between the two countries.

At the Tenth Inter-American Conference at Caracas in the following year Argentina supported Guatemala's contention that condemnation of any specific kind of political arrangement in the hemisphere would jeopardize the principle of non-intervention. Perón's delegates joined the Mexicans in abstaining from voting on the resolution. Argentina maintained its position on traditional hemispheric policies at Caracas, but at the same time exhibited less hositlity toward the United States than in the past.

At this point the sources of Peronist strength had seriously eroded. Emphasis on industrialization had led to the neglect of the agricultural sector, which retrogressed seriously; economic difficulties led to the freezing of industrial wages; and the administration's anti-clericalism cost it the support of the Church. Washington refrained from criticism of the regime, and in March 1955 the Export-Import Bank approved a $60 million loan to purchase steel mill equipment in the United States. Seeking to eliminate Argentina's petroleum deficit, Perón authorized concessions to American oil companies for exploring and developing the country's oil lands. Opposition elements, particuarly the Radical leader Arturo Frondizi, regarded this move as a breach of the nation's economic sovereignty and a surrender to the American imperialists. Perón's policy toward the oil companies was just another grievance adding to the groundswell of opposition to Perón, which culminated in his overthrow by segments of the armed forces—the same group that had been largely responsible for his rise to power—in September 1955.[111]

Perón left his successors the monumental task of restoring the stricken economy and breathing new life into the splintered and debased body politic. By 1955 Perón's ill-advised economic programs had consumed all but $450 million of the $1,682 million monetary reserves of 1946, and had increased the national debt nearly eight times in that period. The problem was compounded by a 15 percent growth of population and a multi-billion dollar external debt.

Major General Eduardo Lonardi, who had served as Argentina's representative on the Inter-American Defense Board and had resigned from the army in 1951 for alleged anti-Peronist activities, became the first provisional president after Perón's fall. Stressing the need for greater cordiality toward the United States, he was recognized by Washington on September 25 1955.[112] Lonardi, however, was overthrown by a military coup on November 13, having been charged with pro-clerical leanings and a lack of firmness in dealing with the Peronists; he was followed in the presidency by General Pedro Eugenio Aramburu. Contending that differences with his predecessor involved personal and not political aims, the new president attempted to circumvent the problem of re-recognition by foreign powers. Washington accepted this interpretation, the Eisenhower administration adopting a "hands off" policy, and sending Assistant Secretary Henry F. Holland to Buenos Aires for negotiations with the new government.[113] The $60 million Export-Import Bank loan for completion of a steel mill was confirmed and subsequently a Point IV agreement was signed with the United States. In April 1956, after a lapse of eight years, Argentina ratified the Charter of the OAS and shortly thereafter joined the World Bank and International Monetary Fund. Turning its back further on precedent, the

[111] Arthur P. Whitaker, *Argentina* (Englewood Cliffs, 1964), pp. 140-149; *Hisp. Amer. Rep.*, VIII (Mar. 1955), p. 135.
[112] Dept. of State *Bulletin*, XXXIII (Oct. 10, 1955), p. 560.
[113] Arthur P. Whitaker, *Argentine Upheaval; Perón's Fall and the New Regime* (New York, 1956), p. 435.

Argentine government entered into cooperative military and naval programs with the Pentagon.

The five-man military *junta* headed by General Aramburu as provisional president and Admiral Isaac Rojas as provisional vice-president, ruled by decree until the election of a civilian government in May, 1958. Widespread and violent opposition, including a major military revolt, was ruthlessly crushed, but the regime restored the Constitution of 1853, and returned the newspaper *La Prensa* to its owners. The junta's conciliatory attitude toward the United States helped to bring about a $100 million loan from the Export-Import Bank; $75 million from the International Monetary Fund; $50 million from the United States banks, and credits of $30 million from the Standard Oil Company. The retention of Raul Prebisch, Chairman of the Economic Commission for Latin America, and former head of Argentina's Central Bank, gave promise that the regime would encourage foreign investment and free enterprise. An upturn of the economy, aided in some measure by trade with the United States, proved helpful to the Aramburu government. Imports of Argentine processed beef increased in the late 1950s, but the market was closed again when Washington issued a new sanitary order in 1959.[114]

Faithful to its pledge to restore the government to civilian control, the Aramburu-led junta permitted free elections, and Arturo Frondizi, the Radical leader, appealing to Peronists, right-wing groups, Communists, and his own left-wing Radicals, won the presidential election of February 1958. Vice-President Richard Nixon was present at his inauguration. Faced with political instability and an economic crisis marked by rising inflation and an empty treasury, he pledged to carry out an austerity program in the operation of his government and to avoid expropriation of foreign capital. Loans were quickly forthcoming from the International Monetary Fund, the United States Treasury, and Export-Import Bank, the Development Loan Fund, and United States banks. In July 1958, he entered into contracts with foreign companies chiefly in the United States, for the development of Argentina's petroleum resources. Traveling to the United States and Europe, he solicited economic support from public and private agencies. Subsequently Frondizi's government joined in establishing the Latin American Free Trade Association.

In spite of earlier statements of anti-Yankee tone, Frondizi steered his country into a cordial and cooperative relationship with the United States that was almost without parallel. Visiting the United States in January 1959, the first Argentine president to do so, he expressed gratitude for its economic assistance, and at the same time gave his endorsement of inter-American solidarity and the Western Hemisphere concept.[115] This followed by one month Washington's announcement that the United States was supporting Argentina with stabilization and development loans totaling $329 million. President Eisenhower reciprocated the visit, and in his address spoke of "the most intensive program of financial cooperation to have yet been carried out in the history of the hemisphere."[116] Meanwhile, an upsurge of trade occurred, United States

[114]*New York Times* Sept. 28, 1955; *Ibid.*, Jan. 4, 1957; *Ibid.*, Feb. 9, 1958.
[115]Dept. of State *Bulletin*, XL (Feb. 23, 1959), pp. 280-283.
[116]*Ibid.*, XLII (Mar. 28, 1960), pp. 477-480; Clarence Zuvekas, Jr., "Argentine Economic Policy, 1958-1962; The Frondizi Government's Development Plan," *Inter-American Economic Affairs*, 22 (Summer, 1968), 45-73.

exports of hard goods to Argentina increasing 50 percent in 1960 over 1959, and the American share of the Argentine market growing from 19 to 26 percent. At the San José meeting of foreign ministers in August 1959, the Argentine delegates had joined in the collective actions against the Trujillo regime in the Dominican Republic and against extra-continental intervention in the Americas.

Lending emphasis to the belief that a turning point had occurred in United States-Argentine relations, President John F. Kennedy said: "We in the United States hope to work with the Argentine government in its heroic effort to improve the welfare of its people, for we are committed to the long range economic development of Argentina. Even more important, we are committed to a continuing relationship of friendship, partnership, and mutual respect."[117] The Kennedy administration, intent on making Argentina a model for the Alliance for Progress, was soon to realize, however, that the Frondizi government had not abandoned the independence of action that traditionally characterized Argentine relations with the United States.

Adlai Stevenson, United States ambassador to the United Nations, touring South America as a special representative of President Kennedy, was told by Frondizi that the Cuban problem could not be solved by unilateral action on the part of the United States; he advocated a live-and-let-live policy. After the Kennedy-Frondizi conference held in New York City in September 1961, and at Palm Beach, Florida in December, it was reported that they had agreed on principles regarding the forthcoming meeting of foreign ministers at Punta del Este, Uruguay, which was to consider the problem of Cuba's ties with the Soviet bloc. President Kennedy had tried to make Frondizi commit himself to a firm position against Cuba, but failed. Disturbing to Washington also was Frondizi's continued involvement of Argentina in the Soviet Union's foreign aid program.[118]

At the Eighth Meeting of Consultation of Foreign Ministers of the OAS at Punta del Este in January 1962, to Washington's dismay, the Argentine delegation took sides with the other go-slow nations, namely Brazil, Chile, Mexico, Ecuador, and Bolivia. By abstaining on the decisive vote Argentina, one of the few states which still maintained diplomatic relations with Cuba, hamstrung the United States in its efforts to get unanimity on the expulsion of Cuba from the Inter-American System. The Argentine military protested sternly against the delgates' action.

Frondizi's motives in assuming this position are complex and unclear. The public had mixed feelings about Cuba. Latent Castro support was revealed within labor and agricultural areas, but this was tempered by a fear of Communist penetration in Argentina. Pressure from Peronists and other extreme nationalists may have accounted for his soft line position. It has also been suggested that, from Frondizi's standpoint, the continuation of the Cuban dilemma would assure greater response from Washington in fulfilling Argentina's economic needs, both in aid and trade. Shortly after the meeting strong military and other anti-Communist pressures forced the president to modify the

[117]Dept. of State *Bulletin*, XLV (Aug. 14, 1961), p. 291.
[118]*Hisp. Amer. Rep.*, XLV (Oct. 1961), p. 721; *Ibid.*, XV (Mar. 1962), p. 67; *New York Times*, Dec. 25, 1961.

Argentine position of handling Cuba with care. The result was a complete break with the Castro regime.[119]

President Frondizi was turned out of office by the military on March 30, 1962. Refusing to resign, he was arrested and held in detention until another constitutional president was elected in July 1963. Frondizi's position on the Cuban question had proved to be a tactical error, but his major blunder was the legalization of the Peronist party's participation in the elections of March 18, 1962. Previously the Peronists could only vote. But now for the first time in a national election since 1955, they were permitted to organize and present their own platform and candidates. A surprising show of Peronist strength compelled the military to annul the elections. He had also been opposed for his policy of allowing foreign firms to exploit the petroleum resources of the country, and his austerity program had proved unpopular and unavailing. It was also evident that Washington's announcement in February of $150 million in loans for unspecified projects did not impress the electorate as much as had been anticipated.

José María Guido, installed as provisional president by the military *junta,* headed what purported to be a civilian administration but was actually a military dictatorship. An Argentine note of March 30 served formal notice that Guido, president of the Argentine Senate, had succeeded to the presidency in a "constitutional manner." In spite of doubts about the new government, especially regarding continued Alliance for Progress aid to the military, Washington on April 18 granted it diplomatic recognition. The newly appointed foreign minister, Bonifacio del Carril, singled out the United States as "the custodian of the world's liberty . . . .and perhaps the only hope western man has . . . " While praising the Alliance for Progress and the battle waged by the United States against "the great new imperialism," he declared that Argentina's foreign policy would continue to be completely its own, and that no other nation would direct it. In practice, the regime quickly demonstrated its alignment with the Western bloc, renouncing the Soviet-Argentine trade pact that had been renewed annually since 1953, and becoming the first member state under the OAS resolution of October 23, 1962, to join with the United States in offering, and subsequently sending, naval and air units to support the blockade of Cuba during the Soviet missiles crisis. A joint statement of the two governments affirmed that "The present international political situation suggests the need for strengthening even more the traditional ties of friendship. The menace posed by international communism in the Americas requires their mutual cooperation to cope with it vigorously in all fields . . . "[120] United States government aid to Argentina totaled $361 million in the period 1955-1963.

Civilian constitutional rule was restored to Argentina with the election of Dr. Arturio Illia, a country physician from Córdoba, in October 1963. Although the new president praised the *Alianza,* he quickly took steps to cancel the contracts of foreign oil companies, eight being American firms, without mention

[119]Robert A. Potash, "Argentina's Quest for Stability," *Current History,* 42 (Feb. 1962), p. 75; Arthur P. Whitaker, "Left and Right Extremism in Argentina," *Ibid.,* 44 (Feb. 1963), p. 88; *Hisp. Amer. Rep.,* XV (Mar. 1962), p. 68.

[120]*Hisp. Amer. Rep.,* XV (July, 1962), p. 453; Whitaker, "Left and Right Extremism," p. 88; Dept of State *Bulletin,* XLVIII (Feb. 11, 1963), p. 211.

of compensation. Earlier, when Washington had sought an agreement guaranteeing private investment, Illia called the agreement an "absurdity." Welcomed by the Frondizi government, the companies had invested $200 million and were owed $100 million by the government for delivered oil. W. Averill Harriman, Under Secretary of State for Political Affairs, was sent to Buenos Aires, but failed in his mission to persuade the Argentine government to soften its stand against the oil companies.[121]

Since the American companies began operating in 1959 Argentine oil production had nearly tripled and the country had almost attained self-sufficiency, thereby saving some $300 million annually in foreign exchange. The merits of the Frondizi policy notwithstanding, the Illia government ruled that the contracts were illegal, having been put into effect by executive decree, without ratification of Congress. It was also charged that the contracts harmed the nation's economy and that Argentina paid too much for the oil produced and for the work done by the companies. The Buenos Aires government contended that *Yacimientos Petrolíferos Fiscales* (Y.P.F.), a state enterprise, could have carried on the development without the assistance of the foreign companies. There was much anger and frustration in the United States Congress over the contract cancellation, in part because of the government's failure to recognize the importance of private foreign investments for the country's economic growth and the success of the *Alianza.* But President Kennedy opposed retaliatory action, and remained hopeful that the problem could be solved amicably. More ill will was generated on the diplomatic level when the United States ambassador to Buenos Aires, Robert H. McClintock, was charged with having said that the cancelled contracts would have the consequence of blocking future United States investments in Argentina. McClintock categorically denied having made the statement, which had appeared in a provincial newspaper.[122]

President Illia took a conciliatory approach to the Peronists, permitting and encouraging them to become part of the institutional structure of the country with the same rights and privileges of any other political party. However, in the first year of his regime the Peronists, in an effort to weaken and discredit the government and return to power, temporarily seized factories and communication facilities. This culminated in the attempted return of Perón who flew from exile in Spain to Brazil, expecting to enter Argentina and spark a popular uprising. The plan was thwarted by an agreement previously made between the Argentine and Brazilian foreign ministers whereby Perón was detained in Rio de Janeiro and returned to Spain on the next plane.

In foreign policy the Illia government suported the Western bloc on political issues while seeking to extend economic relations on an international basis. In the prolonged controversy during the first half of 1964 over the imposition of sanctions on Castro's Cuba for its support of subversion in Venezuela, Argentine policy sought to maintain Latin American solidarity. After this failed at the meeting of American foreign ministers in Washington in July,

[121]*New York Times,* Nov. 12, 1963; Joseph R. Barager, "Argentina: A Country Divided," in Martin C. Needler, ed., *Political Systems of Latin America* (Princeton, 1964), p. 432.
[122]*New York Times,* Nov. 29, 1963; *Ibid.,* May 7, 1964.

Argentina voted with the majority to impose sanctions over the oppositon of Uruguay, Chile, Bolivia, and Mexico. By then evidence indicated that Cuba was also promoting subversion in Argentina, chiefly by sending weapons to guerrilla forces in the mountainous region near Salta. Illia's government responded by requesting military aid from the United States, which was quickly forthcoming.[123]

A more difficult foreign policy problem arose the following year in connection with events in the Dominican Republic. When the United States invervened there the Argentine military wished to send troops to take part in the peace-keeping operations, but the Chamber of Deputies successfully forestalled it. On May 14, 1965, it adopted a resolution, drafted by the Peronists and the government party, proclaiming that "any decision to send troops abroad must be previously authorized by the Argentine Congress." The resolution also condemned Washington's intervention in the Dominican Republic, reaffairmed Argentina's commitment to self-determination and non-intervention, and demanded the immediate withdrawal of United States troops. A crisis developed and tensions ran high until a Brazilian was appointed commander of the OAS forces in the Dominican Republic. This ended the Argentine military's enthusiasm for the project because of the traditional rivalry between the two countries.[124]

United States government aid to Argentina was quietly choked off, beginning in 1963, and in 1960-65, did not exceed $100 million. Amid charges that the Alliance idea had become subordinated to the pressures and philosophies of private industry interests, Edwin M. Martin, the U.S. ambassador to Argentina, denied that there was any connection between the aid program and the still unresolved problem between the Argentine government and the oil companies. This was reaffirmed by Walt W. Rostow, chairman of the State Department's Policy Planning Council, when in Buenos Aires. His visit coincided with the announcement that the Export-Import Bank was granting Argentina a $20 million loan for steel plant expansion.[125]

Up to that point, however, the Alliance for Progress had produced only a marginal effect on the nation's economic development. It was also apparent that the Argentine political attitudes were generally cool toward the Washington-sponsored aid program. Exhibiting this view, an Argentine financial mission sought funds in support of their country's "Five-Year Program" from the Inter-American Development Bank (World Bank). By using this source the program would not be submitted to the Alliance for Progress for review, a

[123]Arthur P. Whitaker, "Argentina: Struggle for Recovery," *Current History,* 48 (Jan 1965), p. 20. The Argentine Communist party, with sixty thousand members, was the largest non-ruling party in Latin America in 1967. The party, never a success in politics, was "abolished" by the Ongania regime. It followed a peaceful-coexistence line, and had no connection with the 1963-1964 People's Guerrilla Army, a force totaling about thirty, which was captured near the Bolivian frontier in 1964. See U.S. Congress, Subcommittee on American Republics Affairs, Committee on Foreign Affairs, U.S. Senate, *Survey of the Alliance for Progress, Insurgency in Latin America* (Washington, 1968), p. 28; *New York Times,* Apr. 4, 1964.

[124]Samuel L. Baily, "Argentina: Reconciliation with the Peronists," *Current History,* 49 (Dec. 1965), p. 360.

[125]*New York Times,* Feb. 17, 1965; *Ibid.,* Feb. 26, 1965.

procedure which the Radical party leaders believed might encroach on the nation's sovereignty in formulating its own policies.[126] A program of military assitance was negotiated, however, thanks to the influence of General Juan Carlos Ongania, commander in chief of the Argentine army, who was convinced that his nation should play a major role in the fight against Communist subversion on a continental scale.

On June 28, 1966, General Julio Alsogaray entered the office of President Illia and announced: "Dr. Illia, sir, you must leave this office in the name of the armed forces." General Ongania, who had resigned after a dispute with President Illia, replaced him. Thus the military for the fifth time since 1930 removed the country's constitutionally elected president: to install a conservative government in 1930, a populist government in 1943, a middle class government in 1955, and again in 1962; but the 1966 *coup* was carried out with the apparent intent of establishing an indeterminate period of military rule.[127] The justification was that "there are no fit civilians to govern Argentina at this time." More specific reasons included friction between Illia and the armed forces, advances made by communism, especially in the universities, deterioration of the national economy, and fear of growing Peronist strength. Illia had failed to demonstrate to the military how the Peronists could be kept from winning the gubernatorial elections scheduled for March 1967. Moreover, by June 1966, the Illia administration was almost totally lacking in popular support. And, finally, the armed services were impressed by the success of the Brazilian military in dealing with the threat of populism and in improving the economic situation, and they were probably encouraged by the fact of strong United States cooperation with the Castelo Branco regime.

Three days prior to the *coup* Washington officials had rejected reports of military unrest as "distorted" and "unfounded," they said that Argentina's political and economic situation was "sound," and that the nation was a crucial element of stability in South America. In an effort to restore confidence in the investment climate it was announced that Argentina had worked out compensatory agreements with all but two of the American oil firms and the contract cancellation issue was virtually dropped.[128] Ambassador Martin had sought for six months to impress on the Argentine military and opposition politicians that the United States was opposed to a *coup* against President Illia. President Johnson came personally to his aid sending the Argentine leader a widely publicized letter in which he said he was looking forward with enthusiasm to the meeting of American presidents proposed by Dr. Illia. Meanwhile, assurances of aid and other international financing for Argentine development had been coming out of Washington after a long period in which Argentina had received virtually no aid.[129]

United States recognition of the Ongania regime was delayed until July 15

[126]*Ibid.*, June 28, 1964.
[127]U.S. Cong., Subcomm. on Amer. Repub. Affairs, Comm. on For. Affairs, U.S. Senate, *Survey of the Alliance for Progress, The Latin American Military* (Washington, 1967), p. 9. See also Robert A. Potash, "The Changing Role of the Military in Argentina," *Journal of Inter-American Studies,* III (Oct. 1961), p. 571.
[128]*New York Times,* June 26, 1966.
[129]*Ibid.*, June 29, 1966.

after nine of the twenty members of the OAS had done so. The State Department had feared the extremist anti-Semitism of several members of the administration, but this was allayed after General Ongania had met with Jewish leaders. Washington exercised restraint both for this reason and to dispel suspicion of favoritism toward military governments. Former President Illia later denied charges that Washington had encouraged the *coup* under pressure of business interests, saying: "Our relations with Washington were excellent . . . the oil people liked, or seemed to like, the compensation offers we settled on eighteen months later, and then our relations with Washington were fine until the end."[130]

The Ongania government made no declaration indicating that it was provisional in character nor did it give promise of early elections; Ongania assumed legislative as well as executive power, issuing laws and no decress. Domestically, the government dealt firmly with the labor left and the moderates by dissolving Congress, barring political activity, stifling the press, taking over the national universities and curbing union activity. National economic policy stressed the system of "private enterprise through the competitive mechanism," thereby reversing the trend toward stateism. A foreign policy declaration included a statement of friendship for all the nations of the Americas, Spain, and other European nations with a western Christian culture.

The impact on Argentina of the "great transformation," launched by the Ongania government in 1966, was unimpressive two years later. The cost of living rose 29.2 percent in 1967, having averaged 27 percent annually in the years 1961-1966. This led to the devaluation of the *peso* by 40 percent. Since 1964, when tax relief for new foreign investments was abolished, no new major industry had been started. Trade levels of the nation's agricultural exports, which earned 90 percent of its foreign exchange, had deteriorated after the devaluation of the British pound, and even more drastic devaluation by commercial competitors such as Uruguay and New Zealand. The great surpluses that the export trade formerly produced in Europe was shrinking because the Common Market, Britain, and other countries were insisting on more balanced accounts. The British ban on meats from countries where hoof-and-mouth disease was endemic had curtailed sales. Labor conditions underscored the depressed state of the economy. Unemployment in the labor force rose to 8.6 million, or 7 percent and many industrial plants did not meet their payrolls regularly. The government, with its vast program to finance, became increasingly concerned about sanitary and other restraints on exports to the United States.[131]

Contrary to earlier statements about the efficacy of private enterprise, the Ongania military regime was soon deeply involved in the operation of the production sector of the economy and in public utilities. The industrial empire of the armed forces, Argentina's most powerful industrial group, was known by the initials of AFNE, FINFIA, DNFM, and SOMISA. AFNE (State Naval

[130]*Ibid.*, July 16, 1966; *Ibid.*, Sept. 16, 1966.
[131]U.S. Cong. Subcomm. on Amer. Repub. Affairs, Comm. on For. Affairs, U.S. Senate, *Survey of the Alliance for Progress, Inflation in Latin America* (Washington, 1967), p. 3; *Christian Science Monitor*, July 15, 1968; *New York Times*, Jan. 22, 1968; Alvin Cohen, "Revolution in Argentina?" *Current History*, 53 (Nov. 1967), p. 290.

Shipyards and Factories) was the nation's largest shipbuilder and maker of optical instruments. The air force operated FINFIA, made up of eight plants, a manufacturer not only of aircraft, but trucks, engines, motorcycles and speedboats. DNFM (National Association of Military Factories), the army's industrial complex, owned 99.9 percent of SOMISA, Argentina's major steel works, 27 percent of the stock of ATANOR, a chemical concern that was among the country's one hundred top companies, and 30 percent of a company making special steels. Retired army officers managed the state-owned telephone company, directed the postal system, and had a strong voice in other utilities. The takeover in 1969 of all overseas telecommunication companies and their incorporation in the state system indicated further that competition was being eliminated rather than extended. The state railways showed deficits so great that they cast a shadow over every national budget, and the losses of Aerolineas, the state airline, continued to mount. Since the military did not operate on a profit and loss basis, many of the other industries were conceded to be inefficient and uncompetitive.[132]

Argentina's economic problems were matched by a lack of continuity in its political processes. General Ongania was the nation's eleventh chief executive in a quarter century. Of those, only three were constitutionally elected. In the preceding fifteen years, national policy had been made by a succession of seven presidents, eighteen ministers of economy, seventeen foreign ministers, sixteen ministers of the interior, and fifteen ministers of war. And for more than half of the last quarter century martial law or its equivalent had prevailed at some place in the country.[133]

United States-Argentine relations following the military's takeover lacked warmth; however, little was reported about their differences, if they did exist, and the official relations could be interpreted only by trends. Ambassador Edwin M. Martin resigned in November 1966, and the post was occupied by an Acting Ambassador for seven months. Prior to the summit conference of American presidents at Punta del Este in April 1967, Sol M. Linowitz, United States representative at the OAS, said the Alliance for Progress "will stand or fall on the capacity of the progressive democratic governments, parties and leaders in Latin America." It was noted that President Johnson had invited President Eduardo Frei Montalva of Chile to visit Washington after the presidential meeting, but had not invited President Ongania. Trade continued, but had lost its former dynamism. Although a settlement with the oil companies was reached in April 1967, Argentine nationalism still balked at special commitments, and thus guaranteed investments approved by other South American countries were rejected by Buenos Aires. Of concern to Washington also was the fact that the Ongania government had given priority to national, before continental integration as proposed by President Johnson's economic task force.[134]

Another conflict of views was apparent on the problem of Argentina's role in the Latin American arms race. This issue, involving the escalation of

[132]*New York Times*, Jan. 22, 1968; *Christian Science Monitor*, Nov. 22, 1969.
[133]Eldon Kenworthy, "Argentina: The Politics of late Industrialization," *Foreign Affairs*, 45 (April, 1967), p. 464.
[134]*New York Times*, Mar. 28, 1967; *Christian Science Monitor*, July 15, 1968; *Oil and Gas Journal*, Apr. 17, 1967.

armaments, was brought into focus at the Punta del Este conference in April 1967, when the Latin American states pledged to "avoid military expenditures that are not indispensable in order to carry out the specific mission of the armed forces . . . " Peru flouted the commitment by purchasing a dozen French supersonic Mirage V fighter-bombers. The chain of circumstances that led to their purchase began in 1965 when Washington sold Argentina A-4 Skyhawk light jet bombers. The Argentines had been determined to buy more expensive Hawker-Hunter jets from Britain if Washington did not make the sale. The justification for the sale was that the United States was saving Argentina precious foreign exchange needed for economic development. But Argentina's acquisition of Skyhawks led Chile, Argentina's rival, to buy twenty-one subsonic Hawker-Hunters from Britain for $20 million. The Peruvians, who are in turn rivals of the Chileans, then demanded supersonic F-5 Freedom Fighters. When the United States refused to sell the planes and offered subsonic F-86 Sabre Jets instead, the Peruvians went to France. Argentina subsequently spent $10 million for French AMX 30 tanks, one of the world's most modern armored assault vehicles. Washington, hoping to delay the introduction of supersonic jets in Latin America for another two or three years, had declined to sell the Northrop F-5 Freedom Fighters, a relatively low-cost supersonic jet built for export, to the Latin American states before 1969-70.[135]

The kidnapping on May 30, 1970 of General Pedro Eugenio Aramburu, a key figure in the overthrow of the Perón dictatorship, was indicative of the ferment which gripped the nation, particularly as it came in the wake of the attempted abduction of a Soviet diplomatic official and the kidnapping of a Paraguayan consul. Aramburu had long been a symbol of opposition to the movement which called for Perõn's return from exile, and at the same time, he had been a critic of the Ongania government. His abductors declared that he would be submitted "to revolutionary justice," for he was accused of being responsible for the firing-squad execution of twenty-seven Peronist leaders in 1956. Among the strong political leaders it was General Aramburu who seemed most dedicated to an orderly return to civilian rule.[136] His death at the hands of the kidnappers was later confirmed.

This episode occurred at a time of growing economic distress and social tension. Although Argentina made economic gains in 1969–an increase in GNP of 6.6 percent and an increase in inflation of only 6.7 percent–most Argentines regarded the stability as illusory. Traditionally one of the world's largest wheat exporters, Argentina had to buy wheat in 1969 from the Soviet Union and elsewhere. The cattle industry declined and more sugar mills were closed. The total foreign debt had risen to about $3 billion, and apart from heavy debt servicing, the military regime had invested extensively in public works projects which required additional loans. The real standard of living among the large middle and working class continued to decline. Workers rejected wage settlements kept low to fight inflation, while students and their allies of the

[135]*New York Times,* Oct. 8, 1967; Government Documents, "Why U.S. Military Assistance is Given to the Argentine Dictatorship; the Explanation of the U.S. Department of Defense," *Inter-American Economic Affairs,* 21 (Autumn 1967), 81-89.
[136]*Christian Science Monitor,* June 30, 1970.

professional class rebelled against the "paternal benevolence" of the Ongania regime. Leftist clergy pressed for social reform and Communist and guerrilla elements challenged the government. Meanwhile, censorship of the press became more intense. The government permanently closed the nation's leading news weekly, *Primera Plana,* together with several other periodicals, and a number of prominent journalists were jailed.[137]

Reflecting the prevailing turbulence as well as divisions within the military forces, General Ongania was ousted from the presidency on June 8, 1970. This was done peacefully and the normal life of Buenos Aires was not interrupted save for a brief closing of the banks. Lt. General Alejandro Agustín Lanusse was an important figure in the military caucus which toppled the president, but he remained in the background. Brig. General Roberto Marcelo Levingston, Argentina's military *attaché* in Washington for the preceding two years, an officer virtually unknown to most Argentines, was named President of Argentina by the nation's military commanders. General Ongania was accused of having "exercised a personal rule unexampled in these times among Western nations."[138]

The Levingston government lasted for nine months, until March 1971, having proved unpopular with the Argentine people as well as the military leaders responsible for it. Levingston's removal from office was attributed to unrest in Córdoba, the government's policy of putting a ceiling on wages, and its appointment of a nationalist, right-wing governor of Córdoba Province. Lt. General Alejandro Lanusse, known to favor a return to civilian constitutional rule, emerged as the key man in the *junta* and the new national leader. In contrast to most of his predecessors since Perón's ouster in 1955, General Lanusse held that no political future in Argentina could be valid without consideration of the Peronists, with or without septuagenarian Perón, then exiled in Spain. From a practical standpoint this position was realistic, for the Argentine Confederation of Workers (CGT), the strongest labor organization in Latin America, and about half the Argentine population, still voiced their loyalty to Perón and remained at odds with non-Peronist governments. Lanusse indicated his intention to return the government to civilian rule by holding elections in March 1973.[139]

General Lanusse was confronted with a revolt by a small faction of the military in October 1971, which was crushed quickly and without bloodshed. The revolt suggested that some elements within the armed forces disapproved of his conciliatory attitude toward the Peronists and the early return of civilian government. Significantly, Lanusse was supported in the crisis by many labor union leaders, and most sectors of the Argentine army. To this crisis was added a stepped-up rate of urban guerrilla activities in 1971 and 1972, which included attacks on police, the kidnapping of a British consul, the kidnap-slaying of an Italian industrialist and the largest bank robbery in Argentine history. The most active was the Revolutionary Army of the People (ERP), which shared a Marxist

[137] Inter-American Development Bank, *Socio-Economic Progress in Latin America* (*Social Progress Trust Fund, Eighth Annual Report, 1968*) (Washington, 1969), pp. 49-63; *New York Times,* Jan. 11, 1970; *Ibid.,* Jan. 26, 1970.
[138] *Christian Science Monitor,* June 18, 1970.
[139] *New York Times* Jan. 28, 1972; *Christian Science Monitor,* Mar. 7, 1972.

ideology with a group known as the Armed Forces of Liberation (FAL). The other groups included the Armed Peronist Forces (FAP), the Armed Revolutionary Forces (FAR), and the Montoneros, all identified with the exiled Perón. The military was legally empowered to wage war on the terrorists, officially labeled subversives.

While the political outlook remained confused and threatening, the economic situation continued to deteriorate. Carlos Brignone, Chairman of the Central Bank, warned that the nation was on the brink of insolvency. Argentines saw the *peso* debased in 1971 from four to more than ten to the dollar. More money was printed in that year than at any time in this century, and an estimated $1 billion in capital was sent abroad. Inflation, at 40 percent was the highest in Latin America. And the growth of the gross national product stood at 2.7 percent in the first half of 1971, compared with 5.5 percent in the same period in 1970. The net treasury deficit for 1971 was believed to exceed 200 billion *pesos* ($200 million) compared with 67 billion in 1970. The printing press continued to fill the gap. An Argentine mission was dispatched to the United States and Europe seeking a $1,000 million loan with which to attempt a new financial start. Argentina's economic malaise, drift and uncertainty carried through 1972 as the military leaders continued to search out solutions to the dilemma.[140]

The Argentine government in 1972 moved forward with its plans for restoring the country to civilian rule in an election scheduled for March 1973. This was confirmed when General Lanusse announced that he would retire from his post on May 25, 1973, the day an elected civilian government would take office. Perón returned to Argentina on November 17, 1972 for the first time since he went into exile in 1955. The government had earlier approved of his return; however, he was disqualified from running for office in 1973 by his failure to return to Argentina by August 25, 1972. This prompted Perón's Justicialist Party to nominate his right-hand man, Dr. Hector J. Campora, for the presidency. Perón returned to Spain in mid-December, having been banned by the government from returning until after the May inauguration of the newly-elected administration. The March presidential election saw *Peronista* candidate Campora poll almost 50 percent of the vote on the pledge: "Campora to the government, Perón to Power." With such convincing results President Lanusse decided that there was no need for a runoff election between Dr. Campora and the second-place candidate Dr. Ricardo Balbín, the Radical Party candidate. Campora's platform was essentially centrist, calling for social justice and economic progress, and appealing to Argentine nationalism.[141] Secretary of State, William P. Rogers, then on tour of eight Latin American countries, witnessed President Campora's inauguration.

General Perón returned to Buenos Aires in June amid an eruption of violence which appeared to be clashes between the left and right wings of the *Peronista* movement. In a move to resume power Perón began forging a coalition believed to include the armed forces, and elements of the Radical Party,

[140] *Times of the Americas,* Dec. 8, 1971; *Christian Science Monitor,* Feb. 17, 1972; *Ibid.,* Mar. 7. 1972; *Ibid.,* Dec. 18, 1972.
[141] *Ibid; New York Times,* Dec. 22, 1972; *Times of the Americas,* Jan. 3, 1973; *Christian Science Monitor,* March 16, 1973.

together with moderate and conservative groups within the *Peronista* movement. President Campora resigned in July to clear the way for a new election, and shortly afterward Raúl Lastiri was inducted into office as provisional president.

With massive support from his own constituency, the working class, reinforced by strong support from the Communist Party, moderate left, and right, Perón gained a sweeping electoral victory in the September 23rd election. He entered the presidential office October 12 with a 61.8 percent mandate. Brought into the office of vice president was his third wife, Isabel Martínez de Perón, a former dancer whom he met in a Panama hotel bar. Although Isabelita had not won the affection of the masses as had Perón's second wife, Eva Duarte de Perón, her selection seemed to moderate the clashes within *Peronista* ranks. It seemed probable that once political stability was restored she would step down in favor of another vice-presidential candidate, possibly from the Radical Party or from the military. Robert H. Finch, former counselor to President Nixon, who had visited Buenos Aires in 1971, headed the U.S. delegation to Perón's inauguration.[142]

"Operación Retorno" achieved its goal in elevating Perón to the presidency, demonstrating that he had retained much of his earlier mystique, and proving that he was the strongest single Argentine politician. But whether Perón, in fragile health, could make headway in finding solutions to the host of domestic problems was a subject of concerned speculation. He led a party torn by division, and a nation suffering rampant inflation (60 percent or more in 1973), food shortages and a mounting terrorist problem. While the armed forces had maintained a non-partisan posture they showed unease after General Perón, in a concession to left-wing political groups, placed the state universities under Marxist control, allowed Marxist indoctrination of university students, and condoned the removal of liberal and conservative faculty members.[143]

U.S. relations with Argentina's military rulers in the early 1970s stressed impeccability, continued on a generally stable, friendly course. In view of the fact that both countries had to cope with inflation and balance-of-payments problems, economic problems tended to overshadow other aspects of bilateral relations. Buenos Aires protested the U.S. ten percent import surcharge, later rescinded, which threatened to augment Argentina's existing trade deficit with the United States. Other issues included Argentine policies of placing credit and other restrictions on foreign enterprises and restricting the operation of U.S. civil air carriers transiting Argentina. In keeping with its pragmatic approach to foreign policy in Latin America, the Nixon administration authorized the sale of F-5E International Fighters to Argentina in 1973.

The modest U.S. bilateral economic assistance program was concluded in 1971; however, Washington continued to assist Argentine development plans through multilateral lending agencies and financial institutions, and the Congress raised Argentina's sugar quota to 76,000 tons. An Export-Import Bank cooperative financing arrangement for $50 million was signed in 1972, and Washington provided additional support through multilateral lending agencies

[142]*Ibid.,* Aug. 28, 1973; *Ibid.,* Oct. 25, 1973.
[143]*Ibid.,* Aug. 28, 1973.

and small A. I. D. projects.[144] Economic matters were the focus of discussions between John B. Connally, Jr., President Nixon's special emissary, and President Lanusse in June 1972.

The advent of the Perón government brought a shift to the left in foreign policy. Argentina not only restored diplomatic ties with Cuba but extended a $200 million credit to the Castro regime in order to stimulate trade. Relations were also established with North Korea and North Vietnam. Perón declared, however, that he favored a policy of non-alignment with either Moscow or Washington, recalling his "third position" almost two decades earlier. Washington meanwhile seemed to have modified its attitude towards President Perón: overlooking the past and seeking friendly cooperation, apparently on the assumption that he represents Argentina's best hope for political stability and economic progress. Perón received a congratulatory message from President Nixon on taking office in October, and responded with a friendly note.[145]

Seen in broad perspective, the main complication of the U.S.-Argentine diplomatic connection in 1973 involved the entire inter-American complex: continental integration, and to what extent Argentina would support it, Argentine nationalistic tendencies on foreign investment, inflation, the balance of trade, kidnapping and terrorism. The bilateral outlook was favorable because of Argentina's interest in the business relationship and dollar capital, and trade assets had a strong appeal.

[144]U.S. Dept. of State, *United States Foreign Policy 1971* (Washington, 1972), p. 154; U.S. Dept. of State, *The Western Hemisphere, An Excerpt from United States Foreign Policy 1972* (Washington, 1973), pp. 439-440.

[145]*Times of the Americas,* Oct. 3, 1973; *Christian Science Monitor,* August 28, 1973; *San Francisco Chronicle,* July 30, 1973.

SUPPLEMENTARY READINGS

Alexander, Robert J. *The Perón Era.* New York, 1951.

Bagú, Sergio. *Argentina en el Mundo.* Buenos Aires, 1961.

Blanksten, George I. *Perón's Argentina.* Chicago, 1953.

Burgin, Miron. *The Economic Aspects of Argentine Federalism, 1820-1852,* Cambridge, Mass., 1946.

Cady, John F. *Foreign Interventions in the Rio de la Plata, 1838-1850.* Philadephia, 1929.

Caillet-Bois, Ricardo R. *Las Molvinas: una tierra Argentina.* Buenos Aires, 1952.

Ferns, Henry S. *Britain and Argentina in the Nineteenth Century.* Oxford, 1960.

Goebel, Julius. *The Struggle for the Falkland Islands.* New Haven, 1927.

Greenup, Leonard and Ruth. *Revolution Before Breakfast: Argentina 1941-1946.* Chapel Hill, 1947.

Haring, Clarence H. *Argentina and the United States.* Boston, 1941.

Jefferson, Mark. *Peopling the Argentine Pampa.* New York, 1926.

Johnson, John J. *Political Change in Latin America: The Emergence of the Middle Sectors.* Stanford, 1958.

Josephs, Ray. *Argentine Diary: The Inside Story of the Coming of Fascism.* New York, 1944.

Kennedy, John J. *Catholicism, Nationalism and Democracy in Argentina.* South Bend, Ind. 1958.

Kirkpatrick, Frederick A. *A History of the Argentine Republic.* Cambridge, England, 1931.

Levene, Ricardo (W. S. Roberston trans.) *A History of Argentina.* Chapel Hill, 1937.

Lieuwen, Edwin. *Arms and Politics in Latin America.* New York, 1961.

McDonald, Austin F. *The Government of the Argentine Republic.* New York, 1942.

Manning, W. R. *Diplomatic Correspondence of the United States, Inter-American Affairs, 1831-1860.* Vol. I. Washington, 1932.

--- *Diplomatic Correspondence of the United States Concerning the Independence of the Latin American Nations.* Vol. I. New York, 1925.

McGann, Thomas F. *Argentina, the United States and the Inter-American System, 1880-1914.* Cambridge, Mass., 1957.

Mecham, J. Lloyd. *The United States and Inter-American Security, 1889-1960.* Austin, 1961.

Pendle, George. *Argentina.* London, 1955 (3rd ed. 1963).

Peterson, Harold F. *Argentina and the United States, 1810-1960.* New York, 1964.

Phelps, Vernon L. *The International Economic Position of Argentina.* Philadelphia, 1938.

Rennie, Ysabel F. *The Argentine Republic.* New York, 1945.

Rippy, J. Fred, Percy, A. Martin, and Isaac J. Cox. *Argentina, Brazil and Chile Since Independence.* Washington, 1935.

Romero, José L. *A History of Argentine Political Thought* (trans. from 3rd ed. by Thomas F. McGann). Stanford, 1963.

Ronning, C. Neal. *Law and Politics in Inter-American Security.* New York, 1963.

Rowe, L. S. *The Federal System of the Argentine Republic.* Washington 1925.

Ruíz Moreno, Isidoro. *História de relaciones exteriores de Argentina, 1810-1955.* Buenos Aires, 1961.

Scobie, James R. *Argentina: A City and a Nation.* New York, 1964.

Shea, Donald R. *The Calvo Clause.* Minneapolis, 1955.

Silvert, Kalman H. *The Conflict Society: Reaction and Revolution in Argentina.* New Orleans, 1961.

Taylor, Carl C. *Rural Life in Argentina.* Baton Rouge, 1948.

United Nations Economic Commission for Latin America. *El desarollo econômico de la Argentina.* 3 vols. Mexico, 1959.

U.S. Dept. of State. *Consultation among the American Republics with Respect to the Argentine Situation.* Washington, 1946.

Weil, Felix. *The Argentine Riddle.* New York, 1944.

Whitaker, Arthur P. *The United States and Argentina.* Cambridge, Mass., 1954.

---*Argentine Upheaval: Perón's Fall and the New Regime.* New York, 1956.

---*Argentina.* Englewood Cliffs, 1964.

White, John W. *Argentina: The Life Story of a Nation.* New York, 1942.

17

Chile and the United States

The Republic of Chile occupies a strip of land on the west coast of South America, averaging 110 miles in width, stretching about 2,600 miles from the southern border of Peru to the continent's end in Tierra del Fuego. Its northern region is a desert separating its populated areas from Peru and Bolivia. To the east, the Andes Mountains separate it from Argentina; to the west is the South Pacific Ocean, and to the south is Antarctica, of which Chile claims a portion. With an area of 286,396 square miles (excluding the territory it claims in Antarctica), Chile is the fourth smallest of the ten South American nations.

Chile is made up of three distinct regions and some islands in the Pacific, the most notable of which is Easter Island, about two thousand miles west of the mainland. Almost 60 percent of the population live on 12.5 percent of the national territory known as Mediterranean Chile, the area from 30° 5' to 37° 5', which contains the central valley. The nation's most important industries are located here as well as most of its agriculture, educational centers and governmental machinery. The desert region in the north contains rich deposits of sodium nitrate, iodine and copper ore, and a very restricted oasis-type agriculture is carried on. In the southern section of the northern zone, a transitional area, there is limited agriculture, but all of northern Chile, about 40 percent of the nation's area, contains only 10 percent of the population. Southern Chile is a region of forests, lakes, and islands, having a heavy rainfall. Inhabited by 29.1 percent of the population, this part of Chile embraces 49.9 percent of the country's total area. Chile's geographical setting places it in a highly isolated position, and in a zone of frequent destructive earthquakes, which have slowed its development.

Chile's population stood at 9.5 million in 1970, approximately 70 percent being urban; the annual population growth is 2.5 percent. The nation has a fairly homogeneous population which is predominantly *mestizo*; about 2 percent are Indian and the recent foreign immigrant population has remained small. Modern Chile is made up of three social classes. Perhaps 25 percent of the total population has a functional political role and enjoys a reasonably good standard of living. This group is split between the landowning aristocracy and the new rich industrialists, city businessmen and others. The other 75 percent is outside the mainstream of Chilean life because of lack of education, poverty and isolation from the large urban centers. Between the rich and the alienated poor is a middle class estimated at 1.1 million people, of whom one million are in the urban centers. Owing to the huge alienated segment, the political climate since World War II has tended to strengthen the Marxist and reformist political parties.

Chile ranks among the more economically developed countries of Latin America, but its economy falls far short of providing the Chileans with an adequate standard of living. An antiquated land tenure system, preventing the modernization of agriculture, has resulted in large expenditures for food imports. Further distortion of the economic system has been caused by the heavy dependence upon foreign credits, earned by exporting minerals, especially nitrates and copper. Chile has a developing economy, but for various reasons, which will be described later in this chapter, the country has had an almost permanent inflation, which has weakened the economic system and contributed to keeping most of the population in poverty.

The early history of Chile is a continuous record of strife and bloodshed. From the time of its discovery in the sixteenth century until the wars of independence, the Spanish conquerors of the land and their descendants were almost constantly waging offensive or defensive wars against the Araucanian Indians—the most warlike native tribes of the South American continent. Even throughout the period when the Chileans were striving valiantly to throw off the Spanish yoke, their internal strife did not cease, and during the dictatorship of Carrera and his brothers, from 1811 to 1813, conspiracies and counter-revolutions followed with monotonous regularity. The factional bitterness between the Carreras and Bernardo O'Higgins, son of the old Irish captain-general and the acknowledged leader of the radical faction, became so great that the people could not unite effectively against the common foe. The result was the decisive victory of the royalist army at Rancagua, October 1, 1814, which almost annihilated the patriot hopes. Fortunately for Chile, it was shortly after this disaster that San Martín began organizing his famous army at Mendoza, and he warmly welcomed O'Higgins and the few officers who had escaped with him across the Andes.

The relations of the United States with Chile began in the troubled times of the Carreras. Joel R. Poinsett was commissioned by President Madison to visit the principal countries of South America and "to diffuse the impression that the United States cherish the sincerest good will towards the people of South America as neighbors. . . and that in the event of a political separation from the parent country and of the establishment of an independent system of national government, it will coincide with the sentiments and policy of the United States to promote the most friendly relations and the most liberal intercourse between the inhabitants of this hemisphere."[1] Poinsett first went to Buenos Aires and then into Chile, where he arrived in December 1811. Don José Miguel Carrera, president of the *junta gubernativa,* welcomed him enthusiastically, assuring him of the sympathy and friendship of Chile for the United States and promising

[1]F. L. Paxson, *The Independence of South American Republics* (Philadelphia, 1916), p. 111.

close commercial relations in the future. In his reply Poinsett expressed his pleasure at being "the first to have the honorable privilege of establishing relations between the two generous nations which ought to unite as friends and natural allies."[2] Poinsett quickly became an ardent supporter of the patriot cause. The Chilean historian, Barros Arana, declared that in all of his conversations the emissary let it be known that the government and people of the United States had a keen interest in the success of the Spanish American revolution. He also assured them that supplies and munitions could be obtained in the United States, and gave the names and addresses of various merchants and manufacturers who would be most likely to furnish them.[3] He went so far as to participate more or less directly in almost all of the public affairs of Chile, and even accompanied Carrera in his first campaigns against the Spaniards.[4] The outbreak of the War of 1812 with Great Britain, however, destroyed all immediate hope of American assistance, and early in 1814 Poinsett returned to the United States where, in spite of his unneutral efforts on behalf of Chile, he was warmly congratulated by Secretary Monroe for the ability and success with which he had conducted his delicate mission.

The results of the Poinsett mission were seen in the attitude of more than benevolent neutrality shown by Chile toward the United States in the War of 1812. American cruisers in the southern Pacific were allowed to dispose of their prizes in Chilean ports, and the Chilean government even seriously considered purchasing some of them as a nucleus of a new navy.[5] Captain David Porter, who cruised in these waters in the U. S. frigate *Essex,* in his journal gives a picturesque account of the hospitality of the Chileans at Valparaiso. He was fêted and entertained most lavishly and given every accommodation. He confesses that it was generally believed that he had brought from the United States proposals for a friendly alliance with Chile and assurances of assistance in their struggle for independence; nor did it suit his purpose to disabuse the people of this idea.[6]

Poinsett's suggestion that war equipment be purchased in the United States did not fall on deaf ears. Early in 1816 Señor José Miguel Carrera landed in the United States with a view to purchasing supplies and obtaining the moral support of the United States, and he immediately looked up the two Americans whom he had formerly befriended, Poinsett and Porter. Commodore Porter, who was then living in Washington, received him cordially and introduced him to President Madison. "I was received," Carrera later wrote to his brother Luis, "as a man working for the same cause as they." Despite his friendly reception, Carrera soon perceived, however, that he could hope for no real support from the United States government. Proceeding to New York, he succeeded in interesting a number of American and French citizens in organizing an expedition, but owing to a lack of funds and to rather uncertain security, he found great difficulty in purchasing ships and supplies. His friend Poinsett came

[2] Diego Barros Arana, *Historia general de Chile* (Santiago, 1887), vol. VIII, p. 566.
[3] *Ibid.*, p. 567.
[4] *Ibid.*, p. 566, note 7.
[5] *Ibid.*, vol. IX, p. 220; also J. B. Lockey *Pan-Americanism: Its Beginnings* (New York, 1920), p. 205.
[6] David Porter, *Journal of a Cruise to the Pacific* (Philadelphia, 1815), vol. I, pp. 102-112.

to his aid, and he was finally able to purchase several indifferently armed ships; and, in spite of the efforts of the Spanish agents to have him arrested, he left the United States in one of his vessels, the *Clifton,* on December 3, 1816.[7]

Another attempt to obtain ships was made in the following year when Manuel H. de Aguirre was commissioned as a private agent of Chile to obtain six war vessels in the United States, in addition to his commission as a representative of the government of Buenos Aires to obtain the recognition of the independence of the Argentine republic. After many trials he succeeded in having two sloops-of-war constructed; but before they could sail they were attached for debts at the instigation of the Spanish consul at New York. In desperation Aguirre turned to President Monroe asking him to purchase the vessels. Secretary Adams replied that this was impossible under the laws of the United States. He pointed out that, although Aguirre had built, equipped, and fitted for sea two vessels suitable for war purposes, yet on the ground that no proof was adduced that he had armed them he had been acquitted by the Supreme Court. "But," he concluded, "the government of the United States can no more countenance or participate in any expedient to evade the intention of the laws than it can dispense with their operation."[8] Aguirre finally settled the financial obligations and the vessels got away without further governmental interference.

During the latter part of the war of independence the relations between the United States and Chile were not so cordial as might have been expected, considering the excellent impression made by Poinsett. There were various reasons. Mr. Theodoric Bland, one of the commission of three sent by President Monroe, December 4, 1817, to investigate conditions in South America, and the only one who visited Chile, seems in some way to have become embroiled in the internal disputes between O'Higgins and the Carreras,[9] and his views, on the whole, were none too favorable to the patriots. Captain Eliphalet Smith, captain of the American brig *Macedonian,* got into difficulties upon several occasions with Lord Cochrane, the new commander of the Chilean naval forces, and the result was a series of claims filed against the Chilean government.[10] Captain Biddle, of the American sloop-of-war *Ontario,* also ran afoul of Admiral Cochrane over the question of salutes. In 1819, when Lord Cochrane declared the whole coast of Peru to be in a state of formal blockade, the United States pronounced the blockade ineffective, and therefore not binding. As a consequence, a number of unfortunate disagreements arose between American merchants and ships and the blockading squadron.[11] With the final defeat of the Spaniards, this cause of friction was removed; and the independence of Chile was formally recognized by

[7] Barros Arana, *op. cit.,* vol. XI, pp. 89-97.
[8] W. C. Ford, ed., *Writings of John Quincy Adams* (New York, 1913-1917), vol. VI, p. 450.
[9] Barros Arana, *op. cit.,* vol. XI, p. 546 (note). In the report that Judge Bland made to the president several sentences seem to indicate that he regarded Supreme Director O'Higgins as a despot who could not hear the loud call of the people for a Congress. See *Amer. State Papers, For. Rel.,* vol. IV, p. 309.
[10] *Infra,* pp. 384-386.
[11] Lockey, *op. cit.,* p. 210.

the United States on January 27, 1823, by the appointment of Mr. Herman Allen as minister plenipotentiary to Santiago.[12]

Mr. Allen did not enter upon the duties of his mission until April, 1824—the same month in which the papers of Santiago carried a report of the historic message of President Monroe. His reception was most cordial; all the highest functionaries of the state were present; and he was saluted with salvos of artillery. In reply to his speech, in which he assured the newly recognized government that no alliances or coalitions need henceforth be feared, the Chilean spokesman expressed his government's gratitude for the recognition of the independence of the new republics, and for the recent declaration of President Monroe which protected them from the proposed coalitions of the European sovereigns.[13] In spite of her appreciation of the promptness of the United States in despatching a diplomatic representative, Chile was unable, owing to internal difficulties and the emptiness of the treasury, to return the favor until 1827.[14]

Don Joaquín Campina, the first Chilean diplomatic envoy to the United States, was not entirely successful in his mission. One of the principal duties laid upon him was the negotiation of a treaty of friendship, commerce, and navigation. Another was the settlement of sundry outstanding claims. In neither of these tasks was he successful, owing to his belief that the arrangements proposed were far more favorable to the United States than to Chile. In 1830 he was transferred to Mexico. Two years later the American *chargé* at Santiago, Mr. John Hamm, succeeded in concluding a convention of peace, amity, commerce, and navigation with Chile, in which each side mutually engaged not to grant any particular favor to other nations respecting commerce and navigation which should not immediately become common to the other party. An explanatory convention was drawn up in the following year, and both treaties were ratified and proclaimed on April 29, 1834.[15] These treaties remained in force until January 20, 1850, when they were terminated on notice given by the Chilean government.

Before continuing the narrative of the relations of the United States with Chile, certain aspects of the latter's internal history must be noted. From 1823 to 1830 the country was given over to revolution and disorder. García Calderón gives a brief but vivid picture of the prevailing conditions: "The national life was chaotic—vandalism in the country, commerce paralyzed, industry at a standstill, finance in disorder, credit vanished, and politics revolutionary. . . . The political orgy continued until 1830, the Chilean people went from liberty to license, and from license to barbarism. At last the demagogue was checked by a man of

[12] J. B. Moore, *Digest of Int. Law*, vol. I, p. 91. It should be noted, however, that Mr. W. G. D. Worthington had been received by the new government of Chile as special agent of the United States shortly after the declaration of independence February 12, 1818. Worthington was followed by John B. Prevost.

[13] Barros Arana, *op. cit.*, vol. XIV, p. 368. An excellent sketch of the causes and steps leading up to the enunciation of the doctrine is given by the same author, *Ibid.*, pp. 469-487.

[14] *Ibid.*, vol. XV, p. 204.

[15] *Ibid.*, vol. XVI, pp. 173-176. For text of the conventions see *Sen. Doc. No. 47*, 48th Cong., 2nd Sess., p. 131.

superior powers, Diego Portales, founder of the Araucanian nation."[16] Portales was a man of strong mind and of practical ideas, and one who had the rare quality of being willing that others should hold the highest positions and reap the rewards, provided that he could direct the policy of the state along the road that would lead to peace and prosperity. As minister under President Prieto, his organizing genius was given full sway—"Prieto played not illy the rôle of a Washington to Portales' Hamilton."[17] He brought about internal peace, built highways and railroads, reorganized the national finances, established schools, and, as a final legacy, gave to Chile the aristocratic constitution of 1833.[18]

This constitution, with few changes, remained the fundamental law of the country until 1925, and it helped to make Chile the most stable of the South American republics. The half-dozen constitutions that preceded it were unsatisfactory, chiefly because of their liberality, and it was determined that there should be no criticism of this constitution because of the weakness of the government for which it provided. Although the government was vested in a president and a bicameral legislature, in actual practice it was an oligarchy, with the president wielding almost despotic powers. The presidential term was fixed at five years; but, as the constitution permitted reelection, every president until 1871, when an amendment prohibited a second term, served for a ten-year period. During periods of domestic trouble the president was given the power to suspend the constitutional guaranties. As a result, Presidents Prieto, Bulnes, and Montt checked all uprisings with an iron hand, and were dictators in all but name. Until 1861 Chile had peace, it is true, but at the cost of almost all civil and political liberty. However, the serious uprisings during President Montt's second term showed the strength of the reaction against the despotic methods of the government. The new president, Señor José Joaquín Pérez, elected in 1861, adopted a policy that was both liberal and effective, and not once during his ten-year term did he ask for the extraordinary powers which his predecessors had found so necessary.

The diplomatic relations between the United States and Chile throughout this period were confined almost exclusively to the prosecution of claims resulting from the seizure by Admiral Cochrane of certain sums of money belonging to American citizens. The first claim was for $140,000 taken from Captain Smith of the brig *Macedonian* in 1819. This the government of Chile finally agreed to settle by the payment of $104,000 with interest.[19] The second claim was for a sum of $70,400 which was taken from Captain Smith in 1821 by the force of armed soldiers in the service of the Republic of Chile. This money was later turned over to Lord Cochrane, and distributed by him as prize-money or to discharge arrears of pay due to the officers and men in the service of Chile.[20] No claim for this sum was filed by the government of the United States

[16] F. Garcia Calderón, *Latin America: Its Rise and Progress* (New York, 1913), p. 164.
[17] T. C. Dawson, *The South American Republics* (New York, 1904), vol. II, p. 196.
[18] Text in amended form in W. F. Dodd, *Modern Constitutions* (Chicago, 1909), vol. I. An exhaustive treatment of the Prieto regime (1841-51) is Raymon S. Valde's *Historia de Chile*, 4 vols. (Santiago, 1900).
[19] *Sen. Ex. Doc. No. 58,* 35th Cong., 1st Sess., p. 4. See also Barros Arana, *Un Decenio de la Historia de Chile* (Santiago, 1913), vol. I, p. 411.
[20] *Sen. Ex. Doc. No. 58*, p. 5.

against Chile until some twenty years later (May 19, 1841), when the American *chargé*, Richard Pollard, brought it to the attention of the Chilean government.[21] After another delay of two years, during which the American *chargé* at Santiago continued to press the claim, the Chilean minister for foreign affairs replied that, owing to the period of time that the interested parties had allowed to elapse, the claim was outlawed by prescription.[22] This brought a long and caustic rejoinder from J. S. Pendleton, who was now acting as American *chargé d'affaires.*[23] The interchanges finally became so heated that the Chilean minister threatened to put an end to all further correspondence. This would have been just as well, because after six months of bickering, which served only to embitter both parties, the dispute remained exactly where it started. Mr. Pendleton was finally recalled, and his successor, Mr. Crump, was instructed to press the claim, but in a courteous and respectful manner; for, as the American government conceded, the previous correspondence had been conducted "in a tone little calculated to secure an amicable adjustment of the matter in controversy."[24]

In 1846 Chile sent Señor Carvallo to Washington with a mass of documentary evidence to prove that the property siezed in the territory of the enemy was Spanish and therefore subject to capture as enemy property.[25] The documents were turned over to Mr. R. H. Gillett, solicitor of the Treasury, who, after careful consideration of the evidence given on both sides, decided that the goods belonged to American citizens. He further ruled that there had been no violation of neutrality, since the goods had been landed in an open port in Peru when that country was still under the government of Spain, and before Chile or Peru had been recognized by the United States.[26] Mr. Gillett's reply was made on May 29, 1848, and Secretary Buchanan, accepting it as conclusive, once more asked for a settlement. Señor Carvallo thereupon prepared another long answer, invoking, among other arguments, the rule of 1756 prohibiting neutrals in time of war from engaging in traffic forbidden to them in time of peace.[27] No answer was made to this until May 24, 1852, when Mr. Hunter, acting secretary of state, stated that the United States had always denied the validity of the rule of 1756, and that, even admitting the legality of the rule, the seizure of American property by Chilean soldiers on Peruvian soil would not be justified by it.[28] Since no solution by diplomatic means seemed possible, Señor Carvallo suggested arbitration, and the King of Belgium was finally agreed upon. Even then, six years elapsed before a convention of arbitration was signed, on November 10, 1858.

Each of the contracting parties now agreed to submit to the King of the Belgium the following questions: (1) Whether the claim was just in whole or part? (2) If just, what amount should be paid by Chile as indemnity for the

[21] *Ibid.*, p. 30.
[22] *Ibid.*, p. 34.
[23] *Ibid.*, pp. 36-46.
[24] *Ibid.*, p. 109. For a characterization of Pendleton and his methods from the Chilean point of view see Barros Arana, *Un Decenio de la Historia de Chile,* vol. II, p. 582, note 6.
[25] *Sen. Ex. Doc. No. 58,* pp. 126-173.
[26] *Ibid.*, pp. 173-333.
[27] *Ibid.*, pp. 334-393.
[28] *Ibid.*, pp. 393-409.

capture? (3) Whether interest should be paid, and if so, at what rate and from what date? The question of prescription was excluded. The designated arbitrator accepted the case and handed down an award on May 15, 1863. He found the claim well founded, but that of the amount seized only three-fifths belonged to citizens of the United States. He also granted interest at 6 percent on this sum ($42,000), but only from 1841, when the United States filed the claim, to 1848, when arbitration was agreed upon.[29] A few other claims were still pending against the Chilean government, but they were for smaller sums and all were ultimately settled by diplomatic means.[30]

The relations between the United States and Chile during the period 1835 to 1860 were not merely lacking in cordiality through the question of claims. Our various diplomatic representatives seemed unable to protect this country's interests and at the same time remain on friendly terms with the Chilean government. Richard Pollard, *chargé d'affaires* from 1835-1842, an exceedingly able diplomat, felt it necessary to stalk out in the midst of a state dinner given by the president of Chile because of a serious slight to his diplomatic standing due to his position at the table.[31] His successor, J. S. Pendleton, used such vigorous language to the Chilean foreign minister that the latter threatened to cease negotiations with him.[32] But the culmination of these personal unpleasant relations was reached during the mission of Seth Barton of Louisiana. He first got into a wrangle over his failure to raise the legation flag on a national holiday; shortly afterwards his horses were impounded instead of being returned when they were found as runaways; and finally he became involved in a violent dispute with the archbishop of Valparaiso who objected to his marriage with a Chilean woman of wealth and high social position, and refused to perform the ceremony. After the marriage had been performed in the legation by an American navy chaplain, the archbishop wrote urging the minister's wife to leave her husband on the ground that she was not wedded according to the Church. This communication enraged Barton and he demanded that the priest be brought to trial for insulting the wife of the American minister. When the government disclaimed authority he became so insulting that the Chilean government demanded his recall. Barton, however, left forthwith and the legation remained closed for almost a year.[33]

Other events which added fuel to the growing antagonism were Chilean sympathy with Mexico in its war with the United States; the collapse of the short agricultural boom brought about by the sale of flour to California gold-rush hordes; and the harsh treatment accorded Chileans who had been lured to California by the gold fever and had run afoul of vigilance committees. Writing in 1855, United States Minister Starkweather declared that "the United States and her citizens are the objects of constant and virulent attack and the

[29]For summary of evidence and text of award see J. B. Moore, *History and Digest of International Arbitrations,* vol. II, p. 1449.
[30]*Ibid.*, vol. II, p. 1449, note 1.
[31]William R. Sherman, *The Diplomatic and Commercial Relations of the United States and Chile* (Boston, 1926), p. 41.
[32]*Ibid.*, p. 45.
[33]See Henry Clay Evans, Jr., *Chile and Its Relations with the United States* (Durham, N. C., 1927), pp. 71-72, 122-126.

chosen target of scurrilous abuse on the part of the press of the country . . . it has even been proposed to expel them from the country and close the ports of Chile against them and their commerce."[34]

Concurrent with these unfortunate events was the distrust of the United States by Diego Portales, the great Chilean leader of the 1830s. Translated into foreign policy Chile was at once opposed to the extension of the United States influence in the hemisphere, and to participation in any regional inter-American system. The Portales' policies sought to counteract the influence of the Monroe Doctrine concept, emphasized commerce with Europe rather than with the New World, forge Chile into a strong nation capable of defending its interests without external support, and to avoid involvement in collective hemispheric ventures.

In 1865 Chile became engaged in a war with Spain in consequence of the high-handed methods of that power in dealing with Peru. Spain had seized the Chincha Islands, and Chile feared that unless the South American states stood together she would attempt to regain her authority over them. Public feeling ran high, and a riot occurred in front of the Spanish legation at Santiago. This, together with a series of other incidents which manifested an unneutral attitude on the part of Chile, brought a Spanish fleet into Valparaíso Bay, and its admiral gave the Chileans four days in which to apologize and salute the Spanish flag.[35] When this was refused, the Spanish squadron proceeded to blockade the Chilean coast. Upon several occasions the United States offered her good offices, both to Spain and to Chile, but without success. The American minister at Santiago, Mr. Thomas H. Nelson, although maintaining a correct and neutral attitude, showed clearly in his despatches that he, as well as the other members of the diplomatic corps, felt that Spain was the aggressor and that the war was both useless and unnecessary. He even went so far as to suggest to Secretary Seward that "a peaceful solution of the controversy would be much more probable if, in addition to the tender of good offices, the United States would remonstrate with Spain upon her unjust and aggressive policy toward Chile. The *moral* intervention, at least, of our government to protect the integrity of one of the American republics from unjustifiable attack on the part of a European power was never more urgently needed nor upon firmer grounds of right and justice."[36]

When, in March, 1866, news came to General Kilpatrick, who had succeeded Mr. Nelson as minister at Santiago, that the Spaniards intended to bombard Valparaíso, an unfortified town, he used every effort to prevent it. He first unofficially visited the Chilean foreign minister and tried to obtain his terms of peace, and then made a similar visit to the Spanish admiral. When this failed, he attempted to obtain the support of the British and French diplomatic representatives in asking that the United States squadron give protection to

[34]W. R. Manning, *Diplomatic Correspondence of the United States, Inter-American Affairs, 1831-1860* (Washington, D.C., 1935), vol. V, p. 210. Despite these problems both nations gained from an exchange of scientific data. See Wayne D. Rasmussen, "The United States Astronomical Expedition to Chile, 1849-1852," *Hispanic American Historical Review*, XXXIV (Feb., 1954), 103-113.

[35]For a complete account of the causes of the war from the Chilean point of view see *For. Rel. of the U.S.*, 1866, Part 2, pp. 349-362.

[36]*Ibid.*, p. 369.

foreign property in the port and prevent the bombardment by force if necessary. As they were unwilling to interfere, the American minister did not feel justified in taking further responsibility in the matter. The bombardment took place on March 31, and the damage done to property and goods was estimated at from fifteen to twenty million dollars.[37] As a result of the failure of the United States to intervene when a squadron under Commodore Rodgers was at the very scene of action, the feeling of the Chileans towards the United States became very bitter. As General Kilpatrick phrased it: "Chile looked upon the United States as her best friend, and that friend had failed to assist her in her hour of need."[38] Secretary Seward, however, upheld the American minister's conduct and pointed out that peace was the constant interest and unwavering policy of the United States. The United States would resist any effort to assail wantonly the republican principle in the Western Hemisphere, but it had no armies for aggressive warfare, nor ambition for the character of a regulator. The policy of nonintervention began with Washington and still endured.[39] In spite of this statement of policy, Seward again made a very earnest effort early in 1867 to induce both parties to arbitrate their quarrel; but Chile was still so bitter towards Spain because of the bombardment of Valparaiso that her conditions of settlement were uncompromising in the extreme. An armistice was finally arranged in 1868, which was just as good as peace, because, as the Chilean foreign minister declared, Spain could not and Chile would not resume the offensive.[40]

One important result of the war with Spain was Chile's realization of her need of a strong navy to protect her exposed sea-coast; in fact, Chile's tendency to become a militaristic nation dates from this period. Another result of the war was a temporary agreement between Chile and Bolivia in regard to the disputed boundary between the two countries in the desert of Atacama. When the two countries secured their independence from Spain, little attention was paid to this region because of its supposed worthlessness; but when, in 1841, guano was discovered, a sudden interest in the Atacama desert was manifested by both parties. In 1842 the Chilean congress passed a resolution by which all the guano in the province of Coquimbo in the littoral of Atacama, and in the adjacent islands, was thereby declared national property.[41] Hardly had the bill become law when the Bolivian representative in Chile asked that the measure, which thus extended Chile's frontiers to the prejudice of Bolivia, be revoked. Revindications continued on both sides, Bolivia insisting that her southern boundary was the twenty-seventh degree of latitude (*i.e.*, inclusive of the desert of Atacama), while Chile maintained that her territory extended north to the twenty-third degree of latitude. Inasmuch as Chile's various constitutions from 1822 to 1833 indicated Atacaria as Chile's northern boundary, the Bolivian claim seemed well

[37] *Ibid.*, p. 386.
[38] *Ibid.*, p. 408.
[39] *Ibid.*, p. 413.
[40] *Ibid.*, p. 322.
[41] V. M. Maurtua (Pezet's trans.), *The Question of the Pacific* (Philadelphia, 1901), p. 11.

founded.[42] Chile, however, continued to hunt guano in the region, and in 1857 seized Mejillones and ousted the Bolivians. Bolivia protested vehemently, and when, in 1864, protests proved unavailing, she broke off diplomatic relations. Before war broke out, however, the trouble with Spain brought the two nations together in a common cause, and by a treaty signed in 1866 the boundary line was fixed at the twenty-fourth parallel. The treaty further provided that the zone lying between the twenty-third and twenty-fifth parallels should be subject to a joint jurisdiction of both governments for the exploitation of the guano and mineral deposits, and that all products of the territory between the twenty-third and the twenty-fifth and exported from Mejillones should be free of duty.[43]

Various difficulties arose under this condominium arrangement, the Chileans claiming that Boliva failed to carry out the terms of the treaty, and the Bolivians asserting that Chile was extending its terms out of all reason. The American minister to Bolivia, writing to Secretary Fish on January 31, 1872, pointed out the increasing possibilities of trouble between the two states owing to the recently discovered silver mines in the region under joint control.[44] Bolivia, apparently fearing further Chilean encroachments, signed an alliance with Peru in 1873 under which the two parties agreed mutually to guarantee their independence and sovereignty and the integrity of their respective territories.[45] Although essentially a defensive alliance, the terms were kept secret. However, there is little doubt that Chile was fully cognizant of the treaty. In 1873 Mr. Logan, the American minister at Santiago, was asked by the Chilean foreign minister to act as arbitrator in the boundary dispute with Bolivia;[46] and when this attempt failed a direct settlement was obtained by the two powers by a new treaty signed in 1874 abrogating the treaty of 1866.

The treaty of 1874 fixed the twenty-fourth parallel as the boundary between the two republics, but provided that such guano as was left between that parallel and the twenty-third should be equally divided. The clause that brought about war between the two countries provided that for twenty-five years the export duties levied on the minerals within the zones between the twenty-third and twenty-fourth parallels should not be raised.[47]

From Melgarejo, a Bolivian dictator, a Chilean company had obtained important concessions for exploiting the nitrate in this Bolivian zone. In 1878 the Bolivian congress, with the apparent purpose of regaining a small part of the wealth given away by a former executive, demanded, as a basis of renewing the contract, 10 percent of the profits of the company—though it later changed this to read ten cents per *quintal* of nitrate exported. This was a clear violation of the treaty of 1874, and Chile protested and threatened to annul the former treaty. Bolivia persisted in her intentions and ordered the tax to be collected.

[42]*Ibid.*, p. 7. See also E. M. Borchard, *Opinion on the Question of the Pacific* (Washington, 1920), p. 6.

[43] Maurtua, *op. cit.*, pp. 16-17. For a fuller discussion from the Chilean point of view see Don Gonzalo Bulnes, *Chile and Peru* (Santiago, 1920), pp. 8-15.

[44]*For. Rel. of the U.S.*, 1872, p. 64. Nitrate also had been discovered in this region in 1866 by a Chilean explorer.

[45]Maurtua, *op. cit.*, p. 28; Bulnes, *op. cit.*, pp. 58-66. For text see *The War in South America, Sen. Ex. Doc. No. 79*, 47th Cong., 1st Sess., p. 85.

[46]*For. Rel. of the U.S.*, 1874, p. 197.

[47]Maurtua, *op. cit.*, pp. 19-21.

Thereupon, on February 12, 1879, the Chilean minister demanded his passports, and two days later Chilean forces seized the disputed territory. Mr. Osborn, minister of the United States at Santiago, declared that Chile's movement was an exceedingly popular one, and that it was doubtful whether if the administration had taken any other course it could have sustained itself.[48]

Bolivia declared war on Chile on March 1, 1879, and from the beginning it was feared by Chile that Peru would take sides against her. Mr. Gibbs, the American minister at Lima, testified that public opinion in Peru was very hostile towards Chile, and the press demanded that the government declare war or resign.[49] The Peruvian government, however, seemed desirous of averting a war if possible, and sent a special mission, headed by Don José Antonio Lavalle, to propose arbitration. This mission was received with marked hostility by the Chilean populace, and when its proposals were laid before the Chilean government the latter immediately demanded an explanation of the secret alliance between Bolivia and Peru. Lavalle was unable to reply with complete frankness, and the Chileans, claiming that the mission was merely a ruse to gain time, refused to continue the negotiations unless Peru would abrogate the treaty of 1873, cease all military preparations, and promise to remain neutral. The Peruvian plenipotentiary could not accept these terms, and Chile declared war on Peru April 5, 1879.[50] Neutral opinion seems to place the burden of responsibility for the failure to find a peaceful solution upon Chile. Sir Clements Markham, the English historian, declares that Chile sought a pretext for a war against Peru to gain her nitrate wealth, and Professor E. M. Borchard, the American publicist, in a later opinion found that Peru was sincerely desirous of avoiding war.[51]

Although the United States took no official action to prevent the outbreak of the war, Mr. S. N. Pettis, the American minister to Bolivia, made a special visit in August 1879, to both Lima and Santiago in the interests of peace. This mission, undertaken without the direction or knowledge of the Washington government, was designed to acquaint each of the warring governments with the views of the other and thus to obtain a direct settlement. The United States government showed itself entirely in sympathy with his efforts by stating that, "unauthorized and even rash as Mr. Pettis' experiment might appear, the United States could not but rejoice at the result should the knowledge thus gained by the belligerents of each other's views conduce to an eventual settlement."[52] Although his mission did not succeed, Judge Pettis continued to strive for peace, and in a note to Mr. Seward, dated September 30, 1879, he urged that the United States offer to mediate inasmuch as all the belligerents were expecting it and wondered at our delay.[53] The ministers in Bolivia and Chile were not so optimistic, and Mr. Osborn at Santiago felt that it would be most unwise for the United States to suggest mediation, since the war spirit was very strong in

[48] *War in South America,* p. 74. For the causes of the war from the Chilean side see pp. 78-83.

[49] *Ibid.,* p. 98.

[50] For the Chilean side see Bulnes, *op. cit.,* pp. 124-160; Rafael Egaña, *The Tacna and Arica Question* (Santiago, 1900), pp. 32-40; for the Peruvian side see Maurtua, *op. cit.,* pp. 43-51.

[51] C. R. Markham, *The War Between Peru and Chile* (London, 1882); Edwin M. Borchard, *op. cit.*

[52] J. B. Moore, *Digest of Int. Law,* vol. VI, p. 35.

[53] *War in South America,* p. 18.

Chile and there seemed to be absolute confidence that Chile would conquer her enemies.[54]

Chile soon proved that she was far better prepared for war than either of her opponents, and was victorious in virtually all of the engagements on land and sea, although she was severely criticized for the cruel and harsh methods employed, for shelling unfortified towns, and for utterly disregarding the rights of noncombatants. By July 1880, it was evident that she had won the war and that further operations would entail useless bloodshed. Therefore, on July 29, 1880, Mr. Everts, secretary of state at Washington, directed the American ministers in Santiago and Lima to press upon the Chilean and Peruvian governments the desire of the United States that peace be made on honorable terms.[55] Mr. Osborn thereupon offered the good offices of the United States, and Chile agreed to accept them, provided the other powers would also accept. It was finally agreed that each of the belligerents should send its representative to meet with the three American ministers accredited to the belligerent states on the U.S.S. *Lackawanna* in the Bay of Arica.

The conference convened on October 22, and Mr. Osborn, the dean of the American representatives, acted as chairman. At the first meeting Chile presented a memorandum outlining the bases upon which she would make peace. The principal provision, and the one that eliminated all chance of a settlement, was the demand on her part that the province of Tarapacá be ceded to her as payment for the costs of the war. As this was territory never claimed by Chile in her most extreme boundary demands, and possessed an almost boundless store of valuable nitrates, it was not to be expected that the allies would yield it. The Bolivian delegate proposed that Chile hold the territory until a suitable indemnity had been paid, and the Peruvian delegate proposed that the United States be given full power to abitrate the question. Bolivia heartily agreed to arbitration, but Chile declared that, while arbitration was a satisfactory method before the outbreak of war, under the changed circumstances "there was no reason whatever why she should deliver up to other hands, honorable and secure as they may be, the decision of her destinies."[56] Mr. Christiancy, the American minister to Peru, felt that the terms proposed by Chile were meant to be so extreme that the allies could not accept them; while Mr. Adams, the American minister to Bolivia, also reported that the Chilean government was not very much in earnest in its desires for peace. He declared that the main object of the Chileans was to break up the alliance between Peru and Bolivia and engage the latter republic in an alliance with themselves.[57] Both ministers thought it strange that Mr. Osborn should have let it be understood that the United States did not desire to act as arbiter on the question. Secretary Evarts was also curious on this score, and in a despatch dated December 27, 1880, he informed Osborn: "If it was your purpose to convey the impression that we would not cheerfully assume any labor and trouble incidental to arbitration in the interests of peace and in the service of justice, you have not correctly interpreted the views and ideas of this government.[58]

[54] *Ibid.*, p. 97. With reports of joint European action, Washington had reason to fear a violation of the Monroe Doctrine.
[55] *War in South America*, p. 116.
[56] The report and protocols in full are found in *Ibid.*, pp. 405-418.
[57] *Ibid.*, p. 51.
[58] *Ibid.*, p. 147

After the failure of the conference Chile prosecuted the war still more vigorously, and in January, 1881, she gained possession of Lima. Secretary Evarts, again desirous of putting an end to the conflict, urged the American ministers in both Lima and Santiago to use every effort in the direction of peace. Mr. Christiancy, however, informed him that the Chilean authorities in Lima declared that they would not accept the good offices of any neutral government nor those of any diplomatic official in Lima, but would treat only with the Peruvian government.[59] Owing to internal troubles in Peru, Chile refused to treat with either faction, thus remaining in control of Lima and its valuable customhouses; and Mr. Christiancy was of the opinion that Chile did not intend to make peace with Peru at all unless driven to do so by outside pressure.[60]
condition precedent upon which alone negotiation shall commence."[61]

The two representatives interpreted their instructions in an entirely different manner. Shortly after his arrival at Lima, Mr. Hurlbut notified his colleague at Santiago that his instructions indicated that the United States wished peace, on fair and honorable terms, to be arranged as speedily as possible, but so that the integrity of Peruvian territory would be maintained. Chile, he thought, should be satisfied with a fair and reasonable indemnity for the expenses of the war.[62] To make his position clear he forwarded, on August 25, to General Lynch, commander-in-chief at Lima, a memorandum in which he thus laid down the position of the United States: "I wish to state further, that while the United States recognize all rights which the conqueror gains under the laws of civilized war, they do not approve of war for the purpose of territorial aggrandizement, nor of the violent dismemberment of a nation except as a last resort in extreme emergencies."[63] Mr. Hurlbut's statement was by no means agreeable to General Lynch, who, in a telegram to the Chilean government, asserted that the American Minister's "no annexation" declaration "complicates and endangers our occupation."[64] Thereupon the Chilean minister of foreign affairs interrogated Mr. Kilpatrick upon the forceful declarations of his colleague in Lima. Instead of taking the matter up with Mr. Hurlbut, Mr. Kilpatrick immediately disclaimed the statements as unauthorized by the government of

In 1881 the Garfield administration came into power, and James G. Blaine was made secretary of state. The president immediately appointed General Kilpatrick minister to Chile and General Hurlbut to Peru. Both soon gave evidence that they were better soldiers than diplomats. The instructions to both, dated June 15, 1881, were very similar, indicating a strong opposition on the part of the United States to a transfer of territory as a basis of peace. To Mr. Kilpatrick Secretary Blaine wrote: "At the conclusion of a war avowedly not of conquest, but for the solution of difference which diplomacy had failed to settle, to make acquisition of territory a *sine quo non* of peace is calculated to cast suspicion on the professions with which war was originally declared"; and to Mr. Hurlbut: "As far as the influence of the United States will go in Chile, it will be exerted to induce the Chilean government to consent that the question of the cession of territory should be the subject of negotiation and not the

59 *Ibid.*, p. 448.
60 *Ibid.*, p. 467.
61 *Ibid.*, pp. 157, 500.
62 *Ibid.*, p. 513.
63 *Ibid.*, p. 516.
64 *Ibid.*, p. 162.

the United States and criticized his colleague for making them.[65] Mr. Blaine censured Mr. Kilpatrick strongly for his undiplomatic methods; nor was he entirely pleased with Mr. Hurlbut's emphatic partisanship. Inasmuch as no further friendly cooperation between the two American representatives could now be hoped for, Mr. Blaine decided to send a special envoy furnished with full powers to the three belligerent states.[66]

Mr. William H. Trescot, of South Carolina, was chosen for the task–which indeed, had steadily become more difficult, owing to the openly manifested hostility of Chile to any interference of the United States. One evidence of this was Chile's treatment of the Calderón government in Peru. This government had been recognized by the United States in June, 1881, and in the following month the Chilean foreign minister had assured Mr. Kilpatrick that every effort would be made by Chile to strengthen it. Nevertheless, in September the Calderón government was suppressed by the Chilean authorities, and in November Señor Calderón was arrested and deported. Mr. Hurlbut interpreted this act as an attempt to continue the state of anarchy and confusion as a ground for Chilean occupation, and also as the reply of Chile to the known support of this government by the United States.[67] Mr. Blaine was inclined to accept the latter interpretation, and in his instructions to Mr. Trescot he authorized him to say to the Chilean government that if such a purpose was avowed it would be regarded by the United States as an act of such unfriendly import as to require the immediate suspension of all diplomatic intercourse.[68] His instructions further stated: "If our good offices are rejected, and this policy of the absorption of an independent state be persisted in, this government. . .will hold itself free to appeal to the other republics of this continent to join it in an effort to avert consequences which cannot be confined to Chile and Peru, but which threaten with extremest danger the political institutions, the peaceful progress, and the liberal civilization of all America."[69]

On the other hand, if Chile should receive in a friendly spirit the representations of the United States, Mr. Trescot was authorized, first, to concert such measures as would enable Peru to establish a regular government and initiate negotiations; second, to induce Chile to consent to such negotiations without cession of territory as a condition precedent; and, third, to impress upon Chile that she ought to allow Peru a fair opportunity to provide for a reasonable indemnity, letting it be understood that the United States would consider the imposition of an extravagant indemnity, making the cession of territory necessary, as more than was justified by the actual cost of the war, and as a solution threatening renewed difficulty between the two countries.[70] The

[65]*Ibid.*, p. 163
[66]*Ibid.*, p. 168.
[67]*Ibid.*, p. 561.
[68]*Ibid.*, p. 176
[69]*Ibid.*, p. 178. Chile believed that Blaine's policies foreshadowed an attempt by the United States to assert in South America the power it had exerted previously only in Mexico and the Caribbean. Bismarck's Germany, maintaining a friendly neutrality, gained great prestige in Chile, for it was believed to have thwarted possible United States and European intervention. See V. G. Kiernan, "Foreign Interest in the War of the Pacific," *Hispanic American Historical Review,* XXXV (Feb. 1955), 14-36.
[70]*War in South America,* p. 178.

envoy was further intrusted with the delicate task of inviting Chile to the conference of all the American republics, which Mr. Blaine was so anxious to hold in Washington.

Mr. Trescot arrived in Santiago January 7, 1882. But in the meantime President Garfield had been shot, Vice President Arthur had been sworn in, and Mr. Frelinghuysen had succeeded Mr. Blaine as secretary of state. Owing to the complete change in policy that followed, this change in the adminstration had a most unfortunate result on the mission. President Arthur decided to follow a hands-off policy in regard to the merits of the controversy, the indemnity, change of boundaries, and the personnel of the government of Peru. The clause in Mr. Trescot's instructions that contemplated the severance of diplomatic relations under certain contingencies was revoked entirely.[71] Before the revised orders were received by Mr. Trescot, he had obtained an assurance of the Chilean government that no offense was meant by the removal of Calderón, and also an acceptance of the good offices of the United States. An offer of peace was made upon the basis of the cession of Tarapacá and an indemnity of twenty million *pesos* payable in ten years, Arica to be occupied until payment was made.[72] However, in an interview about a week later, what was Mr. Trescot's surprise and chagrin to learn from Mr. Balmaceda, the Chilean minister of foreign affairs, that his original instructions from Mr. Blaine had been published, that new instructions had been issued, and that these were then in the hands of the Chilean government. Further parleys were, for the time at least, useless; for, as Mr. Trescot remarked, "a diplomatist of ordinary experience would conclude, when he learns that his instructions have been communicated to the government with which he is negotiating before he receives them himself, that it is time for him to be silent until he does receive them."[73] However, Mr. Trescot was finally prevailed upon to sign a protocol on February 11, 1882, in which the bases of peace were the cession of Tarapacá, an indemnity of 20 million *pesos* and the retention of Tacna and Arica until payment should be made.[74]

The United States thereupon refused to be a party to peace on these terms and urged that Chile show herself more magnanimous. Mr. Trescot, however, knew that his mission had failed; for Chile would not consider any modification of her terms. Shortly afterwards he returned to Washington. From there, in a letter to Mr. Frelinghuysen dated June 5, 1882, he frankly pointed out the unfortunate consequences of the vacillating policy of the United States: "If the United States intend to intervene effectively to prevent the disintegration of Peru, the time has come when that intention should be avowed. If it does not, still more urgent is the necessity that Chile and Peru should understand exactly where the action of the United States ends. It would be entirely beyond my duty to discuss the character or the consequences of either line of conduct, but I trust that you will not deem that I am going beyond that duty in impressing upon the government that the present position of the United States is an

[71] *For. Rel. of the U.S.*, 1882, p. 57.
[72] *Ibid.*, p. 61.
[73] *Ibid.*, p. 67.
[74] *Ibid.*, p. 83.

embarrassment to all the belligerents, and that it should be terminated as promptly as possible."[75]

Mr. Frelinghuysen apparently did not realize tha his ill-advised and futile efforts to obtain a peace through the moral influence of the United States were bringing his government into contempt in the eyes of all South America. Accordingly, although the United States stated in advance that under no conditions would it go beyond the tendering of its good offices, he persisted in intervening when Chile was firmly decided to make peace only upon the terms that her victory made possible. Under instructions dated June 25, 1882, Mr. Cornelius A. Logan, a new minister to Chile, was "to continue the efforts of your government to induce Chile to settle the difficulty by such moderation in her demands as you may be able to bring about, taking care to impress upon that government that any substantial concession which it may now make will be regarded as a direct and graceful recognition of the disinterested counsels of the United States." Incidentally, Mr. Frelinghuysen no longer seemed to object to the cession of Peruvian territory; for, after noting the fact that Chile was in possession of Tarapacá, Tacna, and Arica, he vouchsafed the supposition that no contingency could happen that would bring about the permanent occupation and annexation by Chile of any larger part of Peru than this.[76] It is not strange that a Peruvian publicist was forced to admit that "this third attempt at mediation was still more disastrous in its results for Peru, as it was still more dishonorable for the credit and prestige of the Great Republic of the North. . . . It was no longer the glorious eagle, the emblem of the might and greatness of the United States, that came from Washington to the Pacific to compel the belligerents to lay down their arms and put an end to an iniquitous war of conquest; but the innocent and timid dove, the messenger of peace, sent by the Northern Colossus to beg the cessation of hostilities in the name of American fraternity."[77]

Mr. Logan soon showed a decided partiality for the Chilean point of view, while his colleague at Lima, Mr. J. R. Partridge, became an equally ardent supporter of the Peruvian claims. Although still a prisoner, President Calderón finally refused to negotiate further with Mr. Logan as mediator, on the ground of his over-friendliness towards Chile; and Mr. Partridge upheld him in the stand.[78] The latter even went so far as to propose common action to the representatives of Great Britain, France, and Italy; but this unauthorized proposal was immediately disavowed by Mr. Frelinghuysen, and as a result of it Mr. Partridge was recalled. Meanwhile a faction in northern Peru under the leadership of General Iglesias was corresponding directly with the Chilean government, and in return for recognition it agreed to sign a peace upon Chilean terms. Mr. Logan strongly urged that the United States recognize the government of Iglesias.[79] A protocol of peace was signed in May, 1883, and its terms were incorporated in the Treaty of Ancón signed by the representatives of

[75] *Ibid.*, p. 103.
[76] *Ibid.*, 1883, p. 76.
[77] Alejandro Garland, *South American Conflicts and the United States* (Lima, 1900), p. xiii.
[78] *Ibid.*, p. xv.
[79] *For. Rel. of the U.S.* 1883, p. 103.

Chile and Peru, October 20, 1883. The third intervention of the United States had failed quite as completely as the other two in preventing the territorial dismemberment of Peru. As Mr. Alejandro Garland puts it: "This is the sad history, in as far as Peru is concerned, of the amicable intervention of the United States;. . . its unfortunate, vacillating, and contradictory policy only defeated its own ends."[80]

By the terms of the treaty of Ancón, Tarapacá was ceded in perpetuity to Chile, while the provinces of Tacna and Arica were to be held by Chile for ten years, and at the expiration of that period a plebiscite was to determine whether they should remain under Chilean sovereignty or be returned to Peru. The country then receiving the two provinces was to pay the other country $10 million.[81] This plebiscite, which should have been held on March 28, 1894, ten years after the ratification of the treaty, was never held, and it is this question that has been called "the Alsace-Lorraine question of South America." A final settlement, of different character, was not reached until thirty-five years later.

The protracted dispute over Tacna and Arica led the Chileans to associate inter-American conferences with pro-Peruvian intervention. Reflecting this mustrust the Chileans, before the 1889-1890 Washington Conference of American States, insisted that the conference not consider problems originating prior to the meetings. Their position was based on the fear that the United States would accept arbitration procedures which would imperil Chile's claims to Tacna and Arica. These circumstances reinforced Santiago's determination to strengthen its military effectiveness and to maintain its independence in foreign policy.

In 1891 the United States became seriously involved in Chilean affairs with the outbreak of a civil war between President Balmaceda and the so-called Congressionalists. Owing to his autocratic methods President Balmaceda, who had been elected in 1886, was in constant conflict with Congress during his five-year term. His declared intention to dictate the choice of a successor, and an illegal decree of January 5, 1891, continuing in force the estimates of the preceding Congress, brought about an armed revolt.[82] Mr. Blaine was again secretary of state in the Harrison administration, and Patrick Egan was the American minister to Chile. After a number of serious clashes between the two Chilean factions, Mr. Egan, together with the representatives of France and Brazil, offered their good offices, and both the government and the revolutionists accepted them. While the conferences were being held, bombs where thrown at several of the government ministers while returning from the Senate; and this so exasperated the government that negotiations were abruptly discontinued.[83] About a month later Mr. Egan again proposed mediation on the part of the United States; but although the Balmaceda government was willing, the revolutionists refused.[84]

During the course of the revolution an incident took place that tended to arouse public opinion in favor of the revolutionists against the United States.

[80] Garland, *op. cit.*, p. xvi.
[81] For full text see *British and Foreign State Papers*, vol. LXXIV, p. 349.
[82] President Balmaceda's story of the causes is found in *For. Rel. of the U.S.*, 1891, p. 94.
[83] *Ibid.*, pp. 123-130.
[84] *Ibid.*, pp. 135, 140, 145.

They had despatched the *Itata* to San Diego, California, to obtain a cargo of arms and ammunition; but the Balmacedists, learning of it, warned the United States government, which immediately placed on board a United States marshal to see that its neutrality laws were respected. However, when the captain of the *Itata* learned that the cargo was ready on a schooner off San Clemente Island, he set sail without clearance, taking the United States official with him. The American agent was afterwards put ashore, but the *Itata* took on its cargo and carried it to Chile.[85] There, however, Admiral MacCann refused to allow the cargo to be landed and insisted that the ship and its cargo be returned to San Diego under a United States convoy. The Congressionalist faction conceded the justice of the act, but the rebels bitterly resented the loss of the cargo and looked upon the United States as aiding the Balmaceda government.

By the end of August the Congressionalist forces were completely victorious. Balmaceda resigned, and, fearing mob violence, took refuge in the Argentine legation. His family and a large number of the government officials sought asylum in the American legation. The Chilean government protested against Mr. Egan's protection of the refugees, and various acts of disrespect were committed showing the hostility towards the American minister. Mr. Egan, however, refused to surrender the fugitives unless they be given safe conducts, and he was ultimately successful in carrying out his purpose.[86]

The constant friction between the American legation and the Chilean authorities, the rumor which had been given wide circulation that Admiral Brown had sent secret information to President Balmaceda regarding the movements of the Congressionalists, the *Itata* affair, and the protection afforded by the United States representatives to the defeated faction, all served to stir up bitter animosity towards everything Yankee. As a result, when, on October 15, 1891, Captain Schley of the U.S.S. *Baltimore* gave shore leave to 116 petty officers and men, a concerted attack was made upon them on the Valparaiso water-front. One petty officer was killed, one sailor later died of his wounds, and seven or eight others came out of the melee with from two to eighteen stab-wounds. Thirty-six of the sailors were arrested, and while being taken to prison were cruelly beaten and maltreated. An investigation made by Captain Schley showed that the police made little effort to protect the Americans and in some cases joined in the attack.[87] When apprised of the affair, the authorities offered no expression of regret, and a little later the Chilean minister of foreign relations sent a most offensive note to the Chilean minister at Washington, declaring that there was no exactness nor sincerity in what was said at Washington.[88]

This brought matters to a head, and on January 21, 1892, after an exhaustive investigation had been held at Mare Island directly upon the return of the *Baltimore*,[89] Secretary Blaine notified Mr. Egan that President Harrison had

[85] J. B. Scott, *Cases on International Law* (Boston, 1902), p. 732. See also Osgood Hardy, "The Itata Incident," *Hispanic American Historical Review*, V (May, 1922), 195-226.
[86] J. B. Moore, *Digest of Int. Law*, vol. II, pp. 791-798.
[87] *Relations with Chile, House Ex. Doc. No. 91*, 52nd Cong., 1st. Sess., pp. 115-118.
[88] *Ibid.*, p. 178.
[89] *Ibid.*, pp. 314-607.

come to the following conclusions. First, that the attack was upon the uniform of the United States Navy, having its origin and motive in a feeling of hostility to this government and not in any act of the sailors. Second, that the public authorities of Valparaíso flagrantly failed in their duty to protect our men, and that some of the police and of the Chilean soldiers and sailors were themselves guilty of unprovoked assaults upon our sailors before and after arrest. Third, that suitable apology and adequate reparation should be demanded for the injury done to this government. Mr. Egan was further instructed to say that the expressions imputing untruth and insincerity to the president and to the secretary of the navy in their official communications to the Congress of the United States were in the highest degree offensive to this government, and that if the offensive parts were not at once withdrawn and a suitable apology offered, the president would have no other course except to terminate diplomatic relations with the government of Chile.[90] Four days later, before a reply to this note was received, President Harrison laid the whole matter before Congress in a special message.[91]

The new government that had in the meantime come into existence in Chile was more favorably diposed towards the United States, and the new minister of foreign relations returned a very courteous reply to Washington's demands. Sincere expressions of regret were offered for both the assault and the unfortunate despatch; and in regard to reparations, although the court at Valparaiso had not yet finished its investigation, Chile offered to submit to the Supreme Court at Washington the question whether there were grounds for reparation and in what shape it should be made.[92] This reply was entirely satisfactory to the United States, and Mr. Egan was directed to state that the President was ready to meet the friendly overtures of the Chilean government in the most generous spirit, and that, as for reparations, he had no doubt that the whole matter would soon be settled in a just and honorable fashion by diplomatic means.[93] The hope was justified; for in July of the same year Chile offered $75,000 in gold, to be distributed among the families of the sailors injured and killed in the *Baltimore* affair. The United States immediately accepted it.[94] The same year the two nations signed a claims convention whereby all other claims were to be settled by arbitration by a commission meeting in Washington.[95] The commission sitting from October 9, 1893 to April 19, 1894, awarded $240,564.35 to meet the claims of citizens of the United States. But believing that Washington had magnified the incident as a pretext for possible intervention, Chile sought to strengthen its European ties as a counterpoise to the power and influence of the great northern republic.[96]

From the settlement of the *Baltimore* affair until World War I United States–Chilean relations grew more cordial; however, seeing in the United States rise to world power status in this period a cause for alarm on the part of

[90] *Ibid.*, p. 193.
[91] *Ibid.*, pp. iii-xiv.
[92] *For. Rel. of the U.S.*, 1891, p. 309.
[93] *Ibid.*, p. 312.
[94] *Ibid.*, 1892, p. 62.
[95] U.S. *Stat. at Large,* vol. XXVII, p. 965.
[96] Compilation of Treaties in Force (Washington, 1904), p. 127.

neighboring small nations, Chilean distrust of Washington's motives were never competely erased. The resolution of the Puno de Atacama problem with Argentina in 1899 in which the United States envoy, William J. Buchanan, had been empowered to make the final settlement, was unpopular with many influential Chileans.[97] Chile's fears of the United States were aroused by the threatening position taken by the United States in the Venezuelan boundary controversy of 1895, particularly Secretary of State Richard Olney's famous assertion of Washington's hegemony over hemispheric affairs. The role of the United States in the Spanish-American War was a further sign to Chileans that the United States had indeed become a threat to hemispheric independence; this feeling was reinforced by the Panama "incident" in 1903, and the assertion of the "Roosevelt Corollary" the following year.

Still disturbed by the prospect that the Tacna-Arica question might be reopened, the Chileans, before attending the Second Conference of American Republics in Mexico City, 1901-1902, insisted that the conference not endorse the principle of compulsory arbitration. When the issue did arise Chile found itself in accord with the United States, and opposed by Argentina with which it was then on the brink of war over territorial disputes. Chilean distrust of the United States was lessened by the latter's performance at this conference. Secretary of State Elihu Root supported the Chilean position at the Third Conference of American States at Rio de Janeiro in 1906 further mollifying the Chilean attitude. When in Chile following the conference, Secretary Root expressed the belief that the difficulties of the past arose primarily out of the lack of mutual aquaintance, and declared that the completion of the Panama Canal was bound to bring about more intimate relations, and that this was the time to say that these relations should be those of friendship.[98]

The controversy over the Alsop claim three years later, however, led the United States almost to the point of breaking diplomatic ties with Chile. Alsop and Company, an American firm, was a creditor of the Bolivian government. When Bolivia could not maintain payment on its debt the government assigned Alsop portions of the customs revenues collected in the Peruvian port of Arica, and coastal mining concessions. Under treaty arrangements entered into by Chile and Bolivia following the War of the Pacific, the former assumed responsibility for foreign claims against Bolivia originating in territory ceded to Chile. Washington, supporting the claims of the Alsop heirs, held Chile responsible for payment of the debt. It was after Chile had refused to render a satisfactory monetary settlement and would not accept Washington's terms for arbitration that Secretary of State Philander C. Knox declared that the American legation in Santiago would be closed. The problem was settled through arbitration by the King of England, the award being rendered in 1911, but the Chileans believed that Washington's actions were unduly aggressive, and the affair left a residue of bitterness.[99]

Shortly thereafter the declaration of the Lodge Corollary to the Monroe Doctrine, which was feared as another assertion of American imperialism, once

[97]Frederick B. Pike, *Chile and the United States, 1880-1962* (Notre Dame, Indiana, 1963), pp. 124-125.
[98]*For. Rel. of the U.S.*, 1906, Part I, p. 153.
[99]For review of Alsop Claims see *For. Rel. of the U.S.*, 1911, pp. 38-53.

more alarmed Santiago. The intent of President Wilson's Mobile Address of October 1913, despite its explicit disclaimers, was similarly distrusted by Chileans who took it as a warning that Washington would seek to establish economic control over Latin America in place of Europeans. On the occasion of former President Theodore Roosevelt's visit in Santiago in November 1913 strong anti-United States feeling was demonstrated, notably on issues centering on the Monroe Doctrine. The Chilean attitude was partially attributable to the Wilson administration's policies towards Mexico. After the United States had intervened there, Chile initiated the offer of mediation by the ABC powers; this placed her diplomats in a position to engineer the defeat of President Wilson's Pan American Pact. Chile, still feeling highly vulnerable to a reopening of the Tacna-Arica issue, rejected the Wilson pact chiefly because of the implcation of compulsory arbitration. The United States was suspected of plotting to force an unfavorable settlement on Chile.[100] In spite of this rancor the two nations signed a treaty for the general advancement of peace in 1914, which provided that all disputes not settled by diplomatic means should be submitted for investigation and report to an international commission.[101]

Chile remained neutral in World War I, a neutrality combining both pro-German and anti-United States attitudes. Beltran Mathieu, the Chilean ambassador in Washington, explained his country's position as follows: "Chile was neither solicited nor compelled, because she was not involved in the political causes of the war nor its sphere of action, and because no one considered that a nation so far removed from the theatre of hostilities might be useful as a military or financial entity, while she was so as a factor of production, for which peace was essential."[102] At the outset, however, owing to the influence exerted by German military, teaching and scientific missions, there was some sympathy towards the Germany cause. The belief that Germany had thwarted American and European intervention during the War of the Pacific contributed to this feeling. When the United States entered the war in Europe in April, 1917, President Wilson called upon the Latin American nations to enter into a common defense of the hemisphere, but Chile remained steadfast to her policy of neutrality. When Peru and Bolivia, heeding the call from Washington, severed relations with the Central Powers, Chile became concerned that her late foes might gain Washington's support for their territorial claims. The feeling grew also that if Chile did not establish friendly realtions with the Allied powers the projected League of Nations might be disposed to restore Tacna and Arica to Peru. In order to forestall this possibility Santiago instituted a propaganda campaign aimed at making amends with the United States and the Allies. Typically, an article in *El Mercurio* declared: "The nations of South America, bound to the United States by historic bonds, are today more than ever obliged to sustain the cause which President Wilson defends."[103] Throughout the war, the country's vast stores of nitrate were at the Allies' disposal, a significant

[100]Pike, *Chile and the United States,* pp. 150-155.
[101]*U.S. Stat. at Large,* vol. XXXIX, Part II, p. 1645.
[102]*Amer. Jour. of Int. Law,* vol. XIV (Apr. 1920), p. 333.
[103]P. A. Martin, "Latin America and the War," *League of Nations* (Boston, 1919), vol. II, p. 257. See also C. S. Vildosolo (P. H. Goldsmith, trans.), *Chile and the War* (Washington, Carnegie Endowment for Int. Peace, 1917); Pike, *Chile and the United States,* pp. 158-159.

contribution to the war effort. Chile joined the League of Nations after the war, but with the express reservation that the Treaty of Ancón would not be put on the League's agenda.

United States commerce and capital in the Chilean economic structure was over-shadowed by Britain, Germany and France until the eve of World War I. In 1910 the United Kingdom still had primacy, but the United States equalled Germany and surpassed France. With the outbreak of war Germany was eliminated as a competitor, and the United States was pushed well ahead of Great Britain. By 1920 Chilean exports to the northern republic totaled $115,803,000 and imports reached $47 million, which represented 54 percent of Chile's foreign trade. Whereas United States investment in Chile amounted to $15 million in 1912, they had risen to $250 million in 1920; loans and portfolio investments added $100 million. In 1928 American investments of all types stood at $618 million. Of growing concern to Chileans was the overall extent of American investment: of the total capital investment in Chile in 1930, 52 percent was foreign, with 60 percent of this originating in the United States.[104] As early as 1920 it was estimated that one-third of all United States private investment in Latin America was invested in Chile. Under the circumstances it is not surprising that with the onset of the great economic depression Chileans were disposed to blame the Wall Street financiers. Much bitterness was caused by the passage of the Smoot-Hawley Tariff Act of 1930 which accentuated Chile's economic plight.

United States investments in Chile centered mainly in the mining industry. As early as 1913-1915 the Bethlehem Steel Corporation and the Braden Copper Mines Company (controlled by the Kennecott Copper Co.) had acquired mining properties, and in 1915 the Guggenheim interests began mining copper at Chuquicamata, the world's greatest copper ore deposit. Other American companies were involved in mining nitrates, copper and iron ore, the W. R. Grace and Company spanned the country, and New York banking firms maintained offices there.

Until the 1920s Chile had earned the reputation of being one of the most stable nations in Latin America. A landed aristocracy, comprising a few hundred families, had been dominant since the Diego Portales era. From 1890 this rule was characterized by the subordination of the executive to parliament, civilian dominance, clericalism, centralization, and limited male suffrage. Chile ranked with Argentina and Brazil as one of the ABC powers and was a respected member of the Western community of nations. With a near world monopoly of natural nitrates and a seemingly inexhaustible reserve of copper, Chile's prosperity seemed assured.

Although the country did not suffer from racial divisions, the population being largely *mestizo*, the social structure was highly stratified by caste. The bulk of the population was made up of the impoverished masses, largely rural, who lived on the *fundos*, or landed estates, as *peons*, or *inquilinos*. Those who had no fixed residence and migrated about the country were called *rotos*, the broken ones. Commonly disease-ridden and given to alcoholism, the lower

[104]Pike, *Chile and the United States*, pp. 233-234. See also Max Winkler, *Investment of United Capital in Latin America* (Boston, 1929), pp. 94-95, 98-99, 103-104.

classes had little interest in political affairs. However, during World War I, labor unions carried out a strike, involving miners and railway workers, which led to the passage of limited social welfare legislation. Labour agitation continued among the urban and industrial workers, and a newly-formed Communist Party excited the interest of the vast army of unemployed who had lost their jobs in the nitrate industry. Aside from the sharp curtailment of wartime demands for nitrate, German scientists had discovered during the war a process for manufacturing synthetic nitrate, lessening the demand for the Chilean product.

These developments were accompanied by regional and international events of great magnitude which compelled Chile to recast its policies towards Washington. As noted, between 1892 and 1920 Chile sought to minimize American influence by supporting the formation of an ABC power bloc in South America and by an effort to establish closer ties with Europe. ABC unity, tenuous at best, was shattered by Brazil's entry into World War I, and the Chile-Argentine alignment ended with the Pan American Conference at Santiago in 1923. Argentina at that point preferred to cultivate closer ties with Europe and to deal independently with the United States. The strengthening of Anglo-American ties in World War I and Germany's defeat blocked Chile's plans in that direction. What hopes Chile might have entertained about using the League of Nations as a counterpoise to United States influence, and guaranteeing a system of American international law, were lost as the early weakness of the international organization was revealed. Moreover, Chile's adherence to an isolationist policy had become unrealistic by 1920 owing to the importance of United States capital in the nation's economy.

The reform-minded government of Arturo Alessandri, which lasted from the president's inauguration in December 1921, until his resignation in December 1924, maintained a cordial relationship with Washington. In fact, it was in the United States embassy that he sought asylum after his resignation, and the American ambassador, William M. Collier, accompanied him on the special train which took him to the Argentine frontier.[105] His enforced departure was due largely to financial difficulties caused by the reduced sale of nitrates. A military *junta* took over the government and dismissed the congress. In January 1925, a group led by Colonels Carlos Ibañez del Campo and Marmaduque Grove toppled the *junta* and recalled Alessandri, who resumed the presidency. A constitutional convention produced a new basic document, the Constitution of 1925, promised earlier by Alessandri, and the church-state connection was severed. A disagreement between the president and Ibañez led to Alessandri's second resignation in October 1925. Ibañez gained the presidency through a rigged election in April 1927, and played varied roles in Chilean politics until the 1950s.

One of President Alessandri's first acts had been to invite a commission of financial experts headed by Professor Edwin Kemmerer of Princeton University to devise measures for improving Chile's financial condition. As a result of recommendations by this group a Federal Reserve Bank was established in August 1925, and a new banking law was made effective the following year. But the income from the nitrate industry continued to decline, and his second

[105]Evans, Jr., *Chile and its Relations with the United States,* p. 193.

resignation followed. Despite the work of the commission and earlier evidence of closer ties with Washington, Alessandri and most Chilean leaders regarded hemispheric solidarity and a United States-Directed Monroe Doctrine as irreconcilable concepts.

Meanwhile, in January 1922, President Warren G. Harding extended an invitation to Chile and Peru to send representatives to Washington, hopefully to resolve the long-standing Tacna-Arica dispute under the auspices of the United States.[106] The invitation was accepted, and the conference was opened by Secretary Hughes, May 15, in the Hall of the Americas of the Pan-American Building. He congratulated the delegates upon the noble and conciliatory attitude animating both governments, and pointed out that this was "an auspicious time to heal old wounds and to end whatever differences may exist in Latin America," and said that there could be no more agreeable harbinger of a better day and a lasting peace upon this hemisphere than the convening of this conference of the representatives of the republics of Chile and Peru.[107] Bolivia protested vigorously at not being allowed to participate in the conference; but both Chile and Peru were unwilling to extend the scope of the conferences to include Bolivian claims. After arduous sessions for two months, a protocol was signed July 21, 1922, whereby Chile and Peru agreed to arbitration of the controversy by the President of the United States; and Secretary Hughes informed the plenipotentiaries that President Harding was ready to act as mediator.[108]

The terms of the agreement provided in substance that: (1) the arbitrator should decide whether or not in the present circumstances the plebiscite should or should not be held; (2) if held, the arbitrator should determine the conditions; (3) if not, both parties at the request of either should discuss the situation; (4) failing to agree, they would solicit the good offices of the United States.

Both sides chose their counsel, prepared their cases, and submitted them to the arbitrator on November 13, 1923. The Peruvian case was primarily an attempt to prove that, since the plebiscite to decide the ownership of the province had not been held according to the terms of the treaty, because Chile had not permitted it, and since the conditions had now changed so completely from what they were in 1894 that such a plebiscite would be a mockery, therefore Peru should be confirmed as the undisputed owner of the territory.[109] Chile argued that since the plebiscite had not been held the arbitrator was practically limited to determine under what conditions it should be held. She then discussed these conditions with reference to the time and method of holding the plebiscite and the payment of the ten million *soles.*[110]

The award, a lengthy document of some 17,000 words, was handed down

[106]For more extensive treatment see G. H. Stuart, "The Tacna-Arica Dispute," *World Peace Foundation Pamphlets,* vol. X, no. 1 (1927). Sara Wambaugh, Plebiscites Since the World War (Washington, 1933), vol. I, chap. IX; W. J. Dennis, *Tacna and Arica* (New Haven, 1931).
[107]*New York Times,* May 16, 1922.
[108]*Ibid.,* July 22, 1922.
[109]*Arbitration Between Peru and Chile, The Case of Peru* (Washington, D.C., 1923).
[110]*Tacna Arica Arbitration, The Case of the Republic of Chile* (no place or date).

by President Coolidge on March 4, 1925.[111] In regard to the holding of the plebiscite the arbitrator held that the provisions of Article III of the treaty of Ancón were still in effect; that the plebiscite should be held; and that the interests of both parties could be properly safe-guarded by establishing suitable conditions therefor. As to the conditions of the plebiscite the arbitrator decided that all males, with the exception of military or civil employees of either country, over twenty-one years of age and able to read and write, were entitled to vote, provided that they were born in Tacna-Arica or had been living continuously in the territory for two years on July 20, 1922. A commission of three members, a Chilean, a Peruvian, and an American as presiding officer, was given complete control over the holding of the plebiscite, this commission to be appointed within three months of the date of the award and to begin work not later than six months from the date of the award. The payments to the losing state were to be made over a period of five years.

The award was received with great approval by Chile. Church bells were rung, the newspapers issued special editions, and the American ambassador, Mr. Collier, was given an ovation at every appearance in public. In Lima, however, the decision was regarded as a national calamity. A general strike began, traffic was suspended, newspapers failed to appear, and an uprising against the government was staged which was put down with the sacrifice of some forty lives. Americans were insulted and their business boycotted. Ambassador Poindexter, hitherto exceedingly popular, was given a special guard to protect the embassy from mob violence. On April 2, 1925, the Peruvian government sent an official memorial protesting against the award and requesting certain guarantees in the conduct of the plebiscite.[112] President Coolidge replied, however, that the record fully covered the question raised and that under the terms of submission they had been settled by the award "finally and without appeal."[113]

The plebiscite commission began its work on August 5, 1925, with General Pershing serving as chairman but soon found itself unable to function effectively owing to the refusal of Chile to grant the guarantees which General Pershing felt were necessary before registration could commence. On December 8, 1925, a resolution was passed severely censuring the Chilean authorities in Tacna-Arica for not only failing to exercise their powers so as to make a fair plebiscite possible, but for using their powers unlawfully to reduce the number of Peruvian voters in the plebiscitary territory.[114] The Chilean member presented a dissenting opinion and later appealed to President Coolidge; but the latter in a decision handed down in January, 15, 1926, upheld the powers of the commission.[115] On January 27, the day that the plebiscite law was approved by the commission, General Pershing returned to the United States on account of ill health, and General Lassiter took his place. He made every effort to carry out

[111] *Opinion and Award of the Arbitrator. In the Matter of the Arbitration between the Republic of Chile and the Republic of Peru* (Wasington, D.C., 1925).
[112] *Arbitration between Peru and Chile. The Memorial of Peru and the Ruling Observations of the Arbitrator* (Washington, D.C., 1925).
[113] *Ibia.*
[114] *Amer. Jour. of Int. Law,* vol. 20 (July, 1926), p. 607.
[115] *Ibid.*, p. 614.

the plebiscite, but owing to the refusal of Chile to grant the guarantees approved by General Pershing, the Peruvian members refused to participate in the registration which began on March 27.

On June 14, General Lassiter presented a report to the commission on conditions in the plebiscite area in which he declared that "suitable conditions for the plebiscite, if they have existed at any time within recent years, did not exist when the commission began its labors in August 1925, they do not exist now and there is no prospect of their being brought into existence."[116] He then placed the blame for the situation upon the Chilean authorities, and gave specific instances of the outrages which they permitted to be perpetrated upon the Peruvians. Therefore, believing it beyond the powers of the commission to hold an "unfair and make-believe plebiscite," he recommended the termination of plebiscitary proceedings. The resolution carried.

Approximately a dozen conferences had been held in Washington for a settlement by mediation when the publication of the Lassiter report temporarily put an end to negotiations. However, in August and September it was reported that Secretary Kellogg was still conferring with the plenipotentiaries of the two powers in the hopes of settlement.

On November 30, Secretary Kellogg made what was reported to be the last proposal of the United States to the Chilean and Peruvian governments. In brief, it was proposed that the two powers agree to the cession "in perpetuity of all right, title, and interest" in Tacna and Arica to Bolivia upon an apportionment of equitable compensation, appropriate economic arrangements, and perpetual demilitarization of the territory. Treaties of friendship, commerce, and navigation were then to be arranged between Chile and Peru. Finally as an international memorial to the valor of Chile and Peru, a fitting monument might be erected on the Morro headland of Arica–this memorial to be internationalized.

Although this proposal was no more successful than the others–it was accepted unconditionally only by Bolivia–through the good offices of Secretary Kellogg the two states were finally persuaded to resume diplomatic relations. In October 1928, after seventeen years of severed relations, a Chilean ambassador was received in Lima and a Peruvian in Santiago.

An era of more cordial relations had begun and the Presidents of Chile and Peru took advantage of President Hoover's good-will trip to Latin America in 1929 to bring about a settlement. At their suggestion he agreed to submit a proposal which had been carefully prepared giving Tacna to Peru and leaving Arica to Chile. The dividing line was to start at a point to be named Concordia and run parallel to the Arica-La Paz railway. Peru received an indemnity of $6,000,000 and all public works constructed in Tacna. Arica was made a free port and Peru given a wharf, customhouse and railway station there. The two governments agreed to erect a joint monument of the Morro to commemorate the settlement.[117] Ratifications of this treaty were exchanged July 28, 1929,

[116]Full text of report in *World Peace Foundation Pamphlets,* vol. X, no. 1, Appendix III; also in U.S. *Daily,* June 18-20, 1926. Also in Wambaugh, *op. cit.,* vol. II, p. 468.

[117]For text of treaty and maps of boundary settlement, see Enrique Brieba, *Limites entre Chile y Peru,* 3 vols. (Santiago de Chile, 1931).

and a month later the province of Tacna, after fifty years of Chilean control, rasied the Peruvian flag, settling thereby the most thorny and long standing dispute in Latin-American diplomacy.

With the settlement of the Tacna-Arica question the Ibañez government felt less inhibited in its relations with the United States; however, Ibañez and his officials continued to distrust the Yankee Colossus and were sharply opposed to any expanded interpretations of the Monroe Doctrine. The dictator, nevertheless, correctly perceived and accepted as inevitable the dominant role of the United States in inter-American affairs, and therefore committed Chile to support the Pan American movement. It was expected that by so doing United States "imperialism" might be restrained through the adoption of a new body of American international law. Against this background it is not surprising that Chile played a conciliatory role at the Sixth Pan American Conference at Havana, Cuba, in 1928, moderating Latin American thrusts, particularly from the Argentine delegation, against the United States. Whereas Argentina seemed intent on disrupting the Pan American movement, Chile wished to see it retained as a vehicle for persuading the United States to accept the new body of international law.[118]

The Ibañez administration, tinged with fascism, was neither a military dictatorship nor a democratic regime. The president proclaimed fidelity to the Constitution, but jailed or exiled his opponents, frequently on the pretext that they were Communists, and imposed a stringent censorship of the press. To the administration's credit were programs to stimulate industrialization, education and agricultural development. Public works programs were undertaken and improvements were made within the government bureaucracy. Optimism generated by increased rates of production of nitrates, copper, iron ore and coal led to governmental borrowing and a mounting foreign debt. Despite efforts to diversify the economy, nitrates accounted for 50 percent and copper 30 percent of Chilean exports in the years 1925-1929. A middle class-oriented social security system was created which relied heavily on foreign capital for support while the aristocracy remained largely untaxed. Little was done to relieve the grinding poverty of the lower classes.

What on the surface had appeared to be a fairly prosperous decade was revealed by the onset of the great economic depression in 1929 to have been a tragic illusion. Unable to sell bonds and mineral exports the social security and public works programs foundered and unemployment rose. Ibañez quickly took steps to nationalize the faltering nitrate industry in the government corporation–COSACHI–(Compania del Salitrera de Chile) which was organized in 1930. With foreign trade and business at a standstill, the government faced bankruptcy. A worsening of conditions accompanied by strikes compelled Ibañez to resign and seek exile in Argentina in July 1931.[119]

Suffering more than most countries from the effects of the world economic depression, Chile experienced a year and a half of anarchy, and witnessed the formation of a short-lived Socialist Republic headed by Carlos

[118] Pike, *Chile and the United States*, pp. 224-228.
[119] The Chilean nitrate monopoly, which in 1913 produced over one-half the world's supply of nitrogen, was furnishing in 1932 only about 4 percent.

Davila and air commodore Marmaduque Grove. Order was restored with the reelection of Arturo Alessandre in October 1932, but internal peace remained precarious in spite of slowly improved economic conditions. Reflecting the substantial German element, a group of *nacistas* seized public buildings in Santiago in 1938 in a move to establish a pro-German government. Although the movement was suppressed, it remained a source of uneasiness for several years, and an abortive rebellion led by the ousted General Ibañez in August 1939, added to the confusion.

Pedro Aguirre Cerda, an ultra-liberal and a former dean of the National University, was elected to head the Popular Front government in 1938. Its program included a massive social welfare program and the formation of a development corporation, combining government and private financing, to promote the industrialization of the country. The outbreak of war in Europe in September 1939, brought new demands for Chilean mineral exports, and the government responded by raising prices and cheapening the currency. An inflationary spiral set in that would reach disastrous heights in the years ahead. The dissolution of the Popular Front occurred early in 1941 owing to a conflict between the Communists, who were then friendly to the Axis, and the liberals and leftists who opposed it. President Aguirre Cerda, an ardent admirer of President Franklin D. Roosevelt, died in office in 1941. His successor, Juan Antonio Ríos, defeated General Ibañez, now a stalwart of the conservatives, and inaugurated a regime distinctly less liberal than that of Aguirre Cerda.

Chilean-United States relations in the decade of the 1930s had proved to be generally amicable as the latter made an earnest effort to demonstrate the spirit of the Good Neighbor policy. It was evident, however, that although the threat of United States intervention was mitigated, the Chileans still had reservations about the powerful northern republic. Miguel Cruchaga, the Chilean Ambassador to the United States who became the Minister of Foreign Relations of Arturo Alessandri, following the latter's reelection in 1932, strove to create a South American economic and political union, focused on Argentine, with the object of lessening dependence upon the United States. Chile's distrust was moderated after the Seventh Pan American Conference at Montevideo in 1933 when Secretary of State Cordell Hull affirmed the non-intervention principle and the importance of international law, and proposed reciprocal tariff reductions. The upper classes did not attempt to conceal their contempt for the rustic American Anglo-Saxon culture, but tacitly recognized that the *status quo* which they favored was supported in part by the United States policies and investments, which at the same time relieved them from contributing to public welfare and the economic development of their country. Leftist reform elements, meanwhile continued to belabor United States economic "imperialism" for its alleged malignant influence on Chile's internal development.[120]

United States economic assistance to Chile under the initial program (1937-1961), beginning in this period, comprised capital loans from the Export-Import Bank, a relatively small technical assistance program which averaged $2 million a year, military assistance, and food for peace. In 1937 two

[120]Pike, *Chile and the United States*, pp. 236, 239-240, 269.

Export-Import Bank loans, totaling $1.2 million, were granted for the purchase of farm machinery and locomotives. Subsequently this United States agency provided loans to the Chilean government for constructing an integrated steel mill ($59 million), electric power facilities ($42 million), and railway rehabilitation ($30 million).[121]

President Juan Antonio Ríos, coming into office early in 1942, was elected on a platform of continental solidarity, but he proved reluctant to break off relations with the Axis powers. Many reasons have been offered to account for President Ríos' hesitant attitude: resentment against outside pressure, the nation's slow-to-react democratic processes, Chile's 2,600 miles of unprotected coastline, and the inability of the United States to assist effectively in its defense, the important Nazi element, and the unwillingness of Washington to clear priority orders for equipment destined for Chile.[122]

The announcement of a proposed visit of President Ríos to the United States in the fall of 1942 at the invitation of President Roosevelt aroused much speculation as to an early severance of diplomatic relations with the Axis. However, when it was evident that no break would occur before President Ríos began his trip, Undersecretary of State Sumner Welles declared in a public address in Boston on October 8, 1942, that certain American republics were not preventing Axis espionage which had resulted in the sinking of ships and the loss of lives in the Western Hemisphere. The Chilean government immediately entered a vigorous protest and followed by postponing the visit of President Ríos. President Roosevelt voiced his regret at the postponment and expressed the hope that President Ríos would come later but did not withdraw the accusation.

On October 20, 1942, the entire Chilean cabinet resigned and in the new cabinet the former Chilean ambassador to Uruguay, Joaquín Fernandez y Fernández, took the portfolio of foreign affairs and Dr. Raúl Morales Beltrami, who had followed a vigorous policy as Minister of Interior in attempting to curb Axis operations, was reappointed. When on November 5, the Inter-American Committee for Political Defense which was meeting in Montevideo released Ambassador Bower's memorandum on Nazi subversive activities to the Chilean government, Dr. Morales announced that twelve Nazi suspects which the government had been investigating would be immediately expelled if found within Chilean jurisdiction. Matters now rapidly came to a head. On December 7, Minister of Interior Morales came to the United States on what was stated to be a private trip for reasons of health. Nevertheless, his conferences with President Roosevelt, Vice-President Wallace, and Undersecretary of State Welles betokened a policy of closer cooperation. Immediately upon his return he made a report to the Chilean president and cabinet, and on January 20, 1943, the Senate by a vote of thirty to ten approved the severance of diplomatic relations with the Axis powers. This action of Chile left Argentina as the only area still open to Nazi espionage and sabotage activities. As to Japan it remained to be

[121]U.S. Senate, 89th Cong., 2nd Sess., Committee on Government Operations, *United States Foreign Aid in Action: A Case Study* (Washington, 1966), p. 11. Dept. of State *Bulletin*, XXIV (Mar. 19, 1951), p. 454.

[122]Claude G. Bowers, *Chile Through Embassy Windows: 1939-1953* (New York, 1958), chaps. VII and VIII. A sympathetic and largely uncritical account.

seen whether she would carry out her threat to attack the vulnerable Chilean power stations on the long and exposed Pacific coast.

The war increased materially Chile's commercial relations with the United States. Whereas in 1939 only 31 percent of Chilean trade was with the United States, in 1941 the figure had risen to 60 percent. The value also increased from about $69 million to about $165 million. The United States supplied most of Chile's petroleum, iron and steel sheets, tin plate, automobiles, tires, and tubes. Chile sent by far the largest share of its principal minerals, copper and nitrates, to the United States. During the first six months of 1942 Chile's agricultural exports to the United States amounted to over $6 million, approximately 65 percent of the total value of her exported agricultural products.

Chile's principal contribution to the war effort was her increased production of vital strategic materials. Not only was Chile the world's greatest storehouse of nitrates, but she also held first place in Latin America in the production of copper—her 1941 production amounted to 465,000 tons, and a rise to over 500,000 was achieved in 1942. The United States through Metals Reserve Co. contracted for a period of three years for practically all Chilean copper not sold to other American countries. Preclusive purchasing agreements were signed to cover all other strategic materials available such as manganese, lead, zinc, antimony, wolframite, molybdenum, cobalt ores, and refined mercury.

On the other hand the Export-Import Bank of the United States made substantial loans to Chile for railway equipment and the purchase of industrial and agricultural products and machinery. Over $25 million was authorized and almost $10 million disbursed. The claim was made in the Chilean Chamber of Deputies during a debate in October, 1941, that the United States through lend-lease arrangements had offered Chile $50 million of which Chile only need repay $15 million. Whether true or not such a loan was contingent upon Chile's break with the Axis powers and when that failed to materialize the value of war equipment sent was limited to the amount already under contract. But when a break with the Axis finally occurred, negotiations were quickly resumed and announcement was made early in March 1943, that a new lend-lease arrangement had been signed. This was followed by a contract whereby the United States agreed to purchase Chilean copper at a higher price, also gold and manganese, both of which were in plentiful supply, in order to help the Chilean economy. Chile responded by cooperating effectively in checking all further subversive activities during the period of the war. It was also in 1943 that United States technical assistance was inaugurated in Chile with the establishment of a cooperative health program. Similar programs were carried out thereafter in agriculture, industry, and lesser projects in housing, administration, education, geology, and transportation. The three major programs were directed by jointly-financed cooperative agencies (*servicios*).[123]

The economic situation in Chile was weakened by labor difficulties, strikes and a drastic rise in the cost of living at war's end. When President Ríos died in 1946 before his term had expired, a bitter campaign gave no candidate a

[123]U.S. Senate, *Case Study*, p. 11. United States participation in the *servicio* program was terminated in 1963.

majority and the Congress chose the radical González Videla, who had the support of the Communist Party. Washington was disturbed when he gave three seats in his cabinet to the Communists. However, when the Communists began to abuse their powers and brought about a strike in the copper mines, President González Videla forced the resignation of the Communist ministers, outlawed the Communist party, and severed diplomatic relations with the USSR, Yugoslavia and Czechoslovakia.[124] While on a state visit to the United States in April 1950, at the invitation of President Harry F. Truman, the Chilean president declared his support of a "democratic international" to check communism and give aid and strength to the free governments of the Western Hemisphere. Later in the year Huachipato, the second largest steel mill in Latin America, began operations with important financial contributions in the form of loans from the United States government. Of the existing capital of $100 million almost $60 million came from the Export-Import Bank. Chile's economy was bolstered by a new copper agreement with the United States which was expected to raise by $77 million the nation's dollar income. The copper companies agreed to raise their price to 30.5 cents a pound, or six cents over the prevailing price.[125]

A minor flurry in Chilean-United States relations occurred in 1951, when President Truman proposed a slight revision of the Tacna-Arica settlement. Speaking off-the-cuff, as we must suppose, since the suggestion does not appear in the text of his prepared address as given to the press, President Truman declared that while talking to the president of Chile he had suggested a diversion of the waters of the Andean lakes between Bolivia and Peru for making a "garden" on the West Coast of South America in return for which Chile and Peru would give Bolivia a seaport on the Pacific. The idea was not favored by the nationalists of either country and the matter was quickly dropped.[126]

In the political contest for the presidency in 1952 General Carlos Ibañez, with the support from President Perón of Argentina, campaigned on a strongly nationalistic platform, courted labor's vote in Peronist style, and threatened the ultimate nationalization of the copper and coal industries. He was also alleged to have been opposed to the military agreement signed between Chile and the United States in April; however, after his election in September he denied these statements declaring that Chile wished "to maintain good relations and a policy of mutual collaboration with the United States. . . because of their common continental interests."[127] With Chile's course still uncertain Washington made a strong bid for the friendship of the Ibañez regime. For the inauguration, the State Department sent to Chile a delegation headed by one of the country's most prominent citizens, Mrs. Franklin D. Roosevelt.

A serious economic situation was brought about towards the end of 1953 by the fall in the price of copper and the large surplus remaining in Chilean hands. Chile had sought to stockpile her copper, gambling that she would be able to get 36.5 cents a pound, but the world market dropped to .29 cents a pound.

[124] Bowers, *Chile,* pp. 166-170. The Communist party was illegal from 1947 until just prior to the presidential election of 1958.
[125] *New York Times,* Apr. 21, 1950.
[126] Dept. of State *Bulletin,* XXIV (Apr. 9, 1951), p. 568.
[127] *Ibid.,* XXVI (Apr. 21, 1952), p. 63; *New York Times,* Sept. 10, 1952.

The United States made several proposals to take part of the surplus, but the conditions were not satisfactory to the Chilean government. Pinched by the economic crisis, the Ibañez government offered to sell copper–a strategic war material–to the Soviet-bloc countries; this could be interpreted as open defiance of the U.S.–Chilean Military Assistance Treaty, which specifically barred such trade.[128] Meanwhile, the American copper companies (Anaconda and Kennecott) continued to seek more favorable exchange agreements. An agreement was reached in January 1954, whereby the price of copper was henceforth to be determined by the world market price, and the United States agreed in March 1954 to buy 100,000 tons of copper at the market price of .30 cents a pound to help reduce the huge Chilean surplus. A rapid decline in the value of the Chilean *peso* made life even more difficult for the average Chilean.[129]

The Ibañez administration announced in 1954 an Eight-Year Agrarian Project to develop agriculture and transportation so that the country would be less dependent on world copper prices for its well being. Chile then relied upon copper for 70 percent of its foreign exchange. The United States joined Chile in the first phase of an operational plan to raise the country's level of development, area by area. The American technical assistance mission collaborated with the Chilean government in an Area Development Center supporting a wide range of projects in agriculture and industry and training Chilean specialists. Conditions grew worse as the cost of living rose 70 percent in 1954, and resulting unrest compelled the Ibañez regime to declare a state of seige.[130]

Washington responded to the crisis by signing two sales agreements with Chile for shipment of United States surplus food under Title I or Public Law 480.; the agreements in 1955-156 totaled $38.8 million. Most of the proceeds from the sale of these commodities for Chilean currency were earmarked by the United States for loans to Chile to carry out development programs in rural housing, dairy plants, road and port construction. At the same time, the United States inaugurated a program of donating surplus food to nonprofit voluntary agencies for distribution to needy people. CARE, the first such agency to participlate in this program in Chile, was joined by the Church World Service (1956), Catholic Welfare Conference (1957), Seventh Day Adventist Welfare Service (1959), and Lutheran World Relief (1960). The cost to the United States of commodities distributed by voluntary agencies in Chile through 1960 totaled $35.3 million, not including freight charges paid by Washington.[131] Inflation

[128]*New York Times,* Aug. 22, 1953.

[129]*Ibid.,* Jan. 6, 1954. Discussions in the United States Senate cast light on the competitive disadvantages suffered by the domestic industry in the face of Chilean imports and the narrow profit margin allowed the American producers in Chile: "The companies have made a contract with Chile which set the wage of the Chilean copper workers at $7 per eight-hour day, but the exchange rates indicated in the contracts makes it possible for the government of Chile to receive most of the wage because in the free exchange of *pesos* for dollars the Chilean copper worker's daily wage would amount to only $1.17 per day . . . the copper companies gross only about 8 cents per pound on copper." *Cong. Rec.,* 83rd Cong., 1st Sess., vol. 99, p. 1010.

[130]Dept. of State *Bulletin,* XXXI (Aug. 9, 1954), pp. 200-204; *New York Times,* Jan. 5, 1955

[131]Dept. of State *Bulletin,* XXXII (Feb. 14, 1955), p. 281; U.S. Senate, *Case Study,* p. 12. Public Law 480 establishes the Food for Peace program.

reached an annual rate of 80 percent in 1955, and an anti-inflationary program, based on recommendations of the United States firm of economic consultants, Klein and Saks of Washington, was instituted. Combative measures taken included credit controls, freezing prices, reducing salaries and social welfare benefits, and ending automatic wage increases tied to the cost of living index.[132]

The inflationary surge moderated in 1956, and for the first time in many years there were no strikes. A free fluctuating exchange rate was imposed with the International Monetary Fund, the U.S. Treasury and a group of private American banks granting a $75 million credit to safeguard the rate. Chile's dollar earnings fell 25 percent in 1957 owing to a sharp drop in copper prices, and a timely $25 million loan was announced by Washington.[133]

Chilean sensitivity on the vital question of marketing copper in the United States was revealed when the secretary of the interior recommended that Washington reimpose its tax on copper imports. Shortly after this was announced, President Ibañez abruptly cancelled plans for a state visit to the United States. Although the official reason given for this hasty action was pressing official business, it was generally believed that the Chileans were aroused by what was considered to be an unfriendly act by an American government official.[134]

Jorge Alessandri, son of the "Lion of Tarapaca," a moderate candidate, and by training an engineer, won the presidential election in 1958. The new president was faced with a budget deficit of $226 million, 100,000 unemployed and the main sectors of the economy in a depressed condition. He received full powers for a year to direct the economy and to reorganize public services. Although he called his administration "independent and national," an empty treasury forced him to accept direct assistance from Washington. This aid was supplemented by a $132 million financing program provided by a group of eleven American banks. Confidence revived, strikes declined, inflation slowed down, and business conditions improved. There was also a consonance of views between Washington and Santiago on regional problems. On the occasion of the Fifth Meeting of Foreign Affairs at Santiago in August 1959, convened to consider the question of hemispheric dictatorships, particularly that of Trujillo in the Dominican Republic, President Alessandri made it clear that his nation could not approve of any regime that was not based on free, democratic elections. But like most of the other republics Chile approved Secretary of State Christian A. Herter's statement that "to weaken the principle of nonintervention and the principle of collective security in an effort to promote democracy is . . . a self-defeating activity . . . "[135]

The Alessandri administration seemingly built a strong foundation in 1959 for the economic rehabilitation of Chile, the budget was balanced and foreign investment increased. President Eisenhower visited Chile in March 1960, on a

[132]Tom E. Davis, "Eight Decades of Inflation in Chile, 1879-1959: A Political Interpretation," *Journal of Political Economy*, LXXI (Aug. 1963), 391.
[133]Dept. of State *Bulletin*, XXXVI (May 13, 1957), p. 773.
[134]*Cong. Rec.*, 85th Cong., 2nd Sess., vol. 104, p. 7918.
[135]*Ibid.*, 86th Cong., 1st Sess., vol. 105, p. 15939.

South American good-will tour, and, except for the Communist press, he was warmly greeted by the Chileans. A joint United States-Chilean statement stressed the urgency of seeking solutions for problems of economic development and of improving living standards of the Americas, and declared that the Inter-American system should be based on respect for human rights, the effective exercise of democracy and nonintervention in the internal affairs of other states. President Eisenhower expressed satisfaction that his country had been able to grant Chile substantial credits but warned that "competition for both public and private credit is severe."[136]

In May 1960 Chile suffered a series of severe earthquakes accompanied by volcanic eruptions, fires, floods and tidal waves. It was a catastrophe that killed about 10,000 persons, destroyed the homes of two million, affected 50 percent of Chile's land under cultivation and devastated 52,000 square miles. Although, no major centers of industrial activity were damaged, property damage was estimated at $371 million.

The Eisenhower administration immediately launched Operation Amigos, one of the largest emergency relief projects carried out in peacetime. AID provided $4.8 million in June 1960, in support of the operation, and at the same time authorized an emergency grant of $20 million from the fiscal year 1961 president's contingency appropriation to meet the local currency costs of interim construction projects. In September 1960 Congress authorized the appropriation of a $100 million longterm loan for use in the reconstruction and rehabilitation program. Unlike the $20 million grant, the dollars disbursed under the loan were to be used to import essential commodities and services from the United States.[137] President Kennedy signed the measure in May 1961, and in the following month Adlai E. Stevenson, President Kennedy's special envoy on a good-will mission in South America, was greeted by applauding crowds in Santiago. Special aid came from the Ford Foundation, which granted the University of Chile $1.16 million to set up regional colleges and graduate programs in economics.

Chile was one of the first nations in Latin America to offer a development plan, needed to qualify for Alliance for Progress support, as detailed in the Charter of Punta del Este. Prior to the earthquake CORFO (Corporación de Fomento de la Producción—Production Development Corporation) had prepared a ten-year national development plan, to be initiated in 1961. Although Chile and the United States embarked on the Alliance in the midst of the earthquake reconstruction program, the development plan appeared sound. Teodoro Moscoso, director of the Alliance for Progress, and Richard Goodwin, deputy assistant secretary of state for inter-American affairs, were sent to confer with President Alessandri in March 1962, and to launch the program. President Alessandri's request for assistance in support of the plan came at a time when Chile was in serious financial condition, and the *Alianza* was under mounting criticism both in Latin America and the United States. Expressing his country's

[136]*New York Times* Mar. 2, 3, 1960.

[137]U.S. Senate, *Case Study*, p. 16; Dept. of State *Bulletin*, XLII (June 13, 1960), pp. 966-967. In the period 1953-1960 the United States had granted military assistance to Chile amounting to $32.1 million, together with equipment and supplies valued at $16 million.

viewpoint Luis Escobar, the Chilean minister of economy, indicated that Chile was doing its part as rapidly as possible under the democratic system, but that the United States government and international banks were not making promised funds available.[138]

When President Alessandri was in Washington on a formal state visit in December 1962, he declared that Mr. Kennedy's "forward looking policy" and his "initiative" in the Alliance for Progress "enabled one to look upon the future with confidence." President Kennedy at the same time paid tribute to Mr. Alessandri for his role in the recent Cuban missile crisis: "We appreciated your strong support and that of your country during the difficult days of the past fall, and your messages on that occasion heartened us greatly." Chile was then one of five Latin American nations still extending recognition to Cuba, a fact not mentioned in his response; however, President Alessandri specifically thanked the people of the United States for their "generous help" after the Chilean earthquakes in 1960.[139]

President Alessandri's trip to Washington was the major effort in numerous attempts to obtain financial assistance from abroad. The economic crisis had become more acute with a severe impact on the cost of living, and this had intensified the political conflict between the rightists and the center parties in power, and the extreme leftists. The Chileans alleged, among other explanations accounting for the inflationary spiral, that the United States was slow in sending remittances. Actually, the United States had disbursed $136,400,000 to Chile under *Alianza* programs in the 1962 fiscal year, the equivalent to $17.50 for every person in the country. The commitments totaled $215,400,000 or $27.60 per capita, which placed Chile first in per capita aid under the Alliance program and second only to Brazil in total payments.

One of the most severe critics of the Alliance programs was Senator Eduardo Frei Montalva of the left-of-center Christian Democratic Party, who labeled it as "inoperative," adding that the large expenditures were not accomplishing what had been intended, and that they were not improving living standards or even raising the hopes and expectations of the vast multitude of Chilean poor.[140] Meanwhile, in January 1963, Chile entered into a stabilization

[138]U.S. Senate, *Case Study,* p. 25; *New York Times,* Mar. 9, 1962. In 1960-1961 U.S. economic assistance, coupled with an overvalued exchange rate, and monetary restraint on the part of the government to hold down prices, reduced the rate of inflation to the lowest level in thirty years. But Chile's stability was short-lived and prices again rose rapidly when this combination of favorable factors began to disappear at the end of 1961. See U.S. Senate, Committee of Foreign Relations, *Survey of the Alliance for Progress, Inflation in Latin America* (Washington, 1967), pp. 12-13. The first group of forty-five Peace Corps Volunteers arrived in Chile in 1961.

[139]Dept. of State *Bulletin,* XLVII (Dec. 31, 1962), pp. 991-993; *New York Times* Dec. 12, 1962. The Alessandri government broke off diplomatic relations with the Castro regime after the OAS adopted sanctions against Cuba in 1964. Chile's action, making her the first OAS nation to comply with the new OAS sanctions, was commended in the U.S. House of Representatives. It was not a popular move, since both Frei and Allende, campaigning for the presidential election, vigorously opposed breaking with Cuba. In announcing the suspension President Alessandri reminded his countrymen that "a fundamental of Chile's international politics has been and is faithful and exact fulfillment of its treaties . . . and agreements." See *Cong. Rec.,* 88th Cong. 2nd Sess., vol. 110, p. 19240; *Newsweek,* Aug. 24, 1964.

[140]*New York Times,* Apr. 18, 1963.

agreement with the International Monetary Fund and Washington agreed to make available $60 million—a $35 million AID program loan, a $15 million Export-Import Bank balance-of-payments loans, and a Treasury exchange agreement—in support of Chile's stablization-development effort. Subsequently, under the Alliance, AID attempted to provide Chile with a comprehensive technical assistance program aimed at developing the human resources and institutional framework needed to sustain modern economic growth.

Reinforcing these efforts, after surveys had been made between February and November 1963, California became a partner of Chile in the first program of technical assistance to a foreign country extended by an American state. After a conference with California's Governor Edmund G. Brown, President Kennedy said: "California has developed an agricultural economy so abundant that it is able to export 75 percent of its production," and having in common with Chile many of the same problems, geographical features and products, "I believe that the state of California can be of assistance," Coordinating offices were established in Santiago and Sacramento, the latter to arrange and supervise training programs in California for Chileans, recruit California technicians, and consultants and to maintain accounting records. President Lyndon B. Johnson termed the program "a new dimension in the Alliance for Progress." Virtually all costs the Chile-California program were borne by the U.S. federal government through the Alliance for Progress.[141]

Although Chile ranked first among South American countries in per capita aid from the Alliance for Progress, the two leading candidates in the presidential campaign of 1964 took a scornful view of the Alliance's worth. Dr. Salvadore Allende Gossens, candidate of the Socialist-Communist bloc, asserted that Alliance aid "is of such little effect that we do not consider it an essential factor of our development." He declared that if elected he would carry out massive reform programs, nationalize the American-owned copper companies, and conduct a non-aligned foreign policy; however, he was patently friendly towards Castro's Cuba. Senator Frei, who had earlier been critical of the Alliance in Chile, contended that no dent had been made in an evil economic and social system which gave all the advantages to the wealthy. These comments disregarded the fact that United States economic aid totaled $106 million, plus an additional $11.5 million from the Inter-American Development Bank, in the fiscal year 1963-64.[142]

After a bitterly contested campaign in 1964 Frei carried an absolute majority of the popular votes to win the presidency. President Johnson said that the election "reinforces our hopes for a very bright future in the Americas," and later despatched Adlai E. Stevenson as his special envoy to the inauguration. In his inaugural address President Frei devoted one of his opening paragraphs to an

[141] *Cong. Rec.*, 88th Cong., 1st Sess., vol. 109, pp. 24095-24096; U.S. Senate, *Case Study*, pp. 36-37; *New York Times*, Dec. 3, 1963. Owing to a disagreement between the state of California and the Department of State, the former elected to continue the program, but without federal participation or assistance. See *New York Times*, Aug. 6, 1967.

[142] *New York Times*, Sept. 6, 1964. According to Frei, "We consider development to be a basic objective only to the extent to which it contributes to the elementary well-being of the vast masses." Eduardo Frei, "Latin America in the World of Today," *International Affairs*, 42 (July, 1966), pp. 373-380.

eulogy of the late President Kennedy whose "New Frontier" served as an inspiration to the economists, engineers and political leaders who held posts in the cabinet. Continuing, he said: "I salute the friendly people of the United States, part of our great America, with whom we want a true association in dignified equality . . . " At the same time, reiterating an earlier campaign statement, he expressed his intention to establish diplomatic relations with the Soviet Union, Czechoslovakia, Poland, and Rumania.[143]

President Frei attempted to set in motion a dynamic and coherent program of reform, his "revolution in freedom'" conceived as a rapid social change bringing a decent life to the dispossessed without doing violence to Chile's political liberties. His program inflamed the right-wing parties by requiring the highest taxes in history, and thoroughly recasting the property assessment base; by insisting on authentic agrarian land reform with removal of the constitutional restrictions necessitating immediate cash payment for expropriated estates, and by other social and economic policies which threatened the aristocracy's favored position. Yet leftwing elements sought to defeat the Christian Democratic program because it did not go far enough. The oligarchy had supported Frei as being preferable to the Marxist Allende, but proceeded largely to block his reforms. He scored personal triumphs at economic conferences in Bogotá in 1966 and in Punta del Este in 1967, but the Senate refused his request for permission to visit the United States.

President Frei contended that this action was the work of left and right extremists who " . . . did not want his adminstration to be successful." Senator Mike Mansfield said that the Chilean Senate's action was "an expression of antagonism against the United States," while Senator Everett M. Dirkson declared that: "It is strange when a legislative branch vetoes the visit by the official wielder of the tin cup who is expected to bring back largesse. I don't know of a precedent for it."[144] In spite of these rather critical senatorial comments, Washington sought to implement the policy of making clear its preference for progressive, democratic regimes in Latin America through acts of "special friendship." This was reflected in President Johnson's invitation to President Frei to visit the United States after the meeting of American presidents in Punta del Este in April 1967.[145]

United States intervention in the Dominican Republic in May 1965 produced a critical response in Chile, even more vocal than Mexico's. Leaders of the Christian Democratic Party, as well as the administration, declared that Washington's action had strengthened their opposition, the Popular Front, and endangered congressional support for President Frei's major reform legislation. When Jack Hood Vaughn, assistant secretary of state for inter-American affairs, was sent as President Johnson's special envoy to Chile in August to confer with Chilean officials on the Alliance for Progress, the coolness of his reception was indicative of the strain that had developed in the official relationship. Aside

[143]Joseph S. Roucek, "Chile in Geopolitics," *Contemporary Review,* vol. 206 (Mar. 1965), pp. 127-141; *New York Times,* Sept. 3, 6, 1964; *Cong. Rec.,* 88th Cong., 2nd Sess., vol. 110, pp. 21755-21757; Eduardo Frei, "Christian Democracy in Theory and Practice," in *The Ideologies of the Developing Nations,* ed. Paul E. Sigmund, Jr. (New York, 1964), pp. 308-320.

[144]*New York Times,* Jan. 19, 20, 1967.

[145] *Ibid.,* Mar. 28, 1967.

from their disapproval of intervention in Santo Domingo, the Chileans were clearly opposed to Washington's plans for an inter-American peace force and other collective measures to counter communism. President Frei nonetheless reemphasized that friendly relations with the United States was the cornerstone of his foreign policy, including support of Washington's position in Vietnam.[146]

The Chilean president then undertook a "journey of international understanding" to Europe with the intent of drawing the western European nations into support of the Alliance for Progress. Frei was convinced that Latin America should be brought into the Atlantic community, assuming the pivotal role in a three-way relationship with the United States and Europe. By breaking away from the confines of the United States-Latin American axis the Chileans sought to reduce some of the political and psychological strains that had developed in the inter-American alliance and to put ties with Washington on a more relaxed footing. It appeared that an indirect aim of the proposals was to urge the United States to give renewed attention to helping transform Latin America's economic and social structure, for the Chilean government's opposition to the Johnson administration's policy in the Dominican Republic was partly an expression of fear among Latin Americans that Washington's concern over potential Communist subversion had pushed Alliance programs into the background.

Illustrating the conflict of views, the respected, independent Santiago newspaper, *El Mecurio,* challenged Secretary Rusk's statement that the peoples of Latin America were "massively turning towards democtratic processes" because of the Alliance programs. The newspaper declared that throughout Latin America there were "continued keen anxieties and ferment." Shortly thereafter, as if to prove the point, angry pro-Communist students spat on Senator Robert F. Kennedy when he visited Concepción University. A United States flag was burned while the students screamed "assassin," and "Yankee go home."[147]

In December 1965, the Chilean Chamber of Deputies unanimously approved a report charging U.S. Defense Department and other agencies with "interference" in Chile's sovereignty. This was in reference to the so-called "Camelot Plan," allegedly sponsored by the United States agencies to collect information on Chile's political, economic, and social situation and the possibility of an extremist *coup d'état* there. The report concluded that the "Camelot Plan" was an attempt against the "dignity, sovereignty, and independence of states and peoples and against the rights of the latter to self-determination." The United States ambassador to Chile, Ralph Dungan, learning about Camelot from Chilean newspapers, was indignant because Washington had not informed him what was going on in "his" country without his knowledge. After the bitter reaction in Chile, the "Camelot Plan" was cancelled everywhere.[148]

[146]*Ibid.*, May 11, 1965; *Ibid.*, Aug. 29, 1965

[147]*Ibid.*, Aug. 30, 1965; *Ibid.*, Nov. 17, 1965; *Cong. Rec.*, 89th Cong., 1st Sess., vol. 111, pp. 19002-19003. For astute analysis of Chilean attitudes toward the United States see Frederick B. Pike and Donald W. Bray, "A Vista of Castastrophe: The Future of United States-Chilean Relations," *Review of Politics,* 22 (July 1960), 393-418.

[148]*Bulletin of the Atomic Scientists,* May, 1966, pp. 44-46; *New York Times,* Dec. 18, 1965. Much of the available material regarding the "Camelot Plan" may be found in U.S. Cong., Subcomm. on International Organizations and Movements, Comm. on For. Affairs, House of Rep., *Behavioral Sciences and the National Security, Report No. 4, Winning the Cold War: The U.S. Ideological Offensive.* Washington, 1965.

Further anti-United States sentiment was shown by Chilean students in 1966, when a Communist-led strike included demands for the removal of Peace Corps teachers at the University of Concepción. The institution had received more than $2.2 million from the Ford Foundation, the United Nations and other agencies. A group of Marxist extremists of the Peking line within the student body, other Communists and union organizers, demanded an end to "imperialist" influence within the university. The strike forced the university to suspend operations for a full school year.[149]

Despite these unfortunate episodes President Frei announced in May 1966, that Chile's relations with he United States were "never more propitious," and that his government had received "permanent and decisive financial assistance with full comprehension and respect" from Washington. However, later in the year, when the Frei government spent $20 million to buy British-made jet fighter planes, questions were raised in the United States about the wisdom of that assistance, for Chile was not facing a war threat. The action was the more puzzling because Frei had gained a reputation for his serious attention to the economy, the democratic process, and even attempts to reach a regional disarmament agreement. The main question, which remained unanswered, was: Why spend $20 million on fighter planes and then seek help from AID for Chile's economic and social needs.[150]

In April 1967, the "Chileanization" of the largely United States-owned copper industry turned a historic page in the nation's economy. The world's largest underground copper mine, El Teniente, was transferred to the Sociedad Minera El Teniente, S.A., a new Chilean Mining Corporation with 51 percent interest going to the Chilean government and the remainder to the Braden Copper Company, a subsidiary of the Kennecott Copper Company, the former owner. The American mining interests had agreed to the plan previously, but the right and left in the Chilean Senate had in concert obstructed passage of the bill for nearly two years, the left holding out for total nationalization and the right resisting a change of existing conditions. Under the terms of the agreement, the Braden Copper Company was to receive $80 million for the 51 percent share held by the government, but would reinvest the funds locally. The foreign producers were also assured of a reduction in their taxes from the then effective level of 85 percent to 72 percent. President Frei thus moved a step further towards achieving his six year aim of giving Chile greater control over its main commercial asset and, at the same time, possibly increasing copper production by 70 percent and refinery capacity by 182 percent. To help Chile realize these objectives, the Export-Import Bank authorized a $110 million Alliance for Progress loan to the new Chilean mining corporation to finance purchases in the United States of equipment and services.[151]

In July 1967, following closely in the wake of the copper industry

[149] *New York Times,* Oct. 25, 1966. There were 363 Peace Corps Volunteers in Chile in 1965 working on projects in education, housing, social developments and public health.

[150] U.S. Senate, *Hearings before the Senate Foreign Relations Committee on S 2859* (Apr. 29, 1966), pp. 177-178, 659; *New York Times,* May 22, 1966.

[151] Frederick M. Nunn, "Chile's Government in Perspective: Political Change or More of the Same?," *Inter-American Economic Affairs,* 20 (Spring, 1967), 88; *Economist,* Nov. 6, 1966, p. 573; *New York Times,* Apr. 10, 1967.

agreement, the relations of the two countries were clouded by a Chilean statement relating to the critical problem of insurgency. The leaders of Chile's Christian Democratic Party declared that guerrilla warfare against governments that "ignore the peoples' rights and offer no electoral solutions" is a legitimate course of action in Latin America. The party's national council at the same time took the position that the pro-Castro Latin American Solidarity Organization (OLAS), which supports armed subversion in the hemisphere, could operate freely in Chile so long as it did not "disobey freely elected authorities and does not disturb Chile's relations other Latin American countries having unsurgency problems, particularly Venezuela, Colombia, and Guatemala." President Frei remained aloof from the party's statement, but it was believed that he was dismayed by it.[152]

Giving an air of reality, though not validity, to this issue, old tensions between Santiago, La Paz, and Buenos Aires were revived when five guerrillas, the remnants of Dr. Ernesto "Che" Guevara's band, crossed the mountainous border from Bolivia and surrendered to Chilean authorities. The Frei government ordered the Cuban and Bolivian guerrillas out of the country, and Dr. Salvadore Allende accompanied them part of the way to Havana. Commenting critically on Chile's handling of the affair, the Bolivian representative in the OAS called upon other members to "share Bolivia's preoccupation" over the fact that the five guerrillas had been allowed th escape "with impunity" through neighboring Latin American territory. The incident led the Argentine administration of General Ongania to express fear about the outcome of Chile's 1970 presidential election: "With a Communist or pro-Communist president in Santiago the whole continent would be threatened and Havana would have a back door to the Argentine pampas." It was rumored that guerrillas were training in southern Chile, and some terrorism occurred, including three bomb explosions in twenty-six days in Chilean offices linked with the United States. But Chile's national police force, one of the best in Latin America, moved informed sources in the United States to conclude that any guerrilla movement would be contained.[153]

It may be concluded that in spite of its assets in human and natural resources, and democratic political traditions, Chile has suffered in recent years from grave economic and social problems. The nation's economy became virtually stagnant in the aftermath of World War II, averaging an increase of 3 percent annually, barely exceeding the 2.5 percent population growth. In the period 1950-1958 the cost of living rose by an average of 38 percent; for 1961-1966 the figure stood at 27 percent. The balance of trade has been unfavorable, with imports exceeding exports by $95 million in 1963 and $93

[152]The declaration was made by the Christian Democrat National Council in response to a public challenge delivered by Socialist Senator Dr. Salvador Allende who was under threat of public censure for trying to advance a bill that would reopen diplomatic relations with Cuba. The real issue, however, was whether Dr. Allende, then president of the Senate, could be both a member of the Senate and a ranking official of OLAS. The subsequent motion to censure Allende failed after the Council of the Christian Democrat Party instructed its nine senators to abstain on the issue. The decision to abstain was widely interpreted as a victory for the extreme left wing of the party. See *New York Times*, July 27, 1967.

[153]U.S. Senate, Committee on Foreign Relations, *Survey of the Alliance for Progress, Insurgency in Latin America* (Washington, 1968), p. 29.

million in 1964. Overall exports, bolstered by continuing high copper prices, rose by $30 million to $875 million in 1967, but imports soared from $730 million to $845 million. Much of the imports came from capital goods brought in for new industries as well as food stuffs.[154] Being compelled to rely more heavily upon foreign borrowing, foreign credit has financed as much as 40 percent of official investment. Accordingly, Chile found it increasingly difficult to service its foreign debt, amounting to $1.4 billion in 1965.

Chile meanwhile continued to receive the highest per capita U.S. aid in Latin America, and was among the eight nations in which U.S. global assistance was concentrated. The extent of United States economic assistance to Chile through the fiscal year 1965 is shown below:[155]

Post-World War II U.S. Assistance To Chile

(fiscal years-millions of dollars)

Program	*Postwar Relief 1946-48*	*Marshall Plan Pd. 1949-52*	*Mutual Security Act pd. 1953-61*	*For Asstce. Act pd. 1962-65*	*Total 1946-65*
AID and predecessor agencies (loans and grants)		*1.1*	*75.1*	*360.3*	*436.5*
Total economic loans and grants 1946-1965	*45.9*	*64.3*	*340.5*	*563.4*	*1,014.0*

Chile has also received assistance from international agencies to which Washington made substantial contributions. The International Bank for Reconstruction and Development and its affiliates, the International Development Association and the International Finance Corporation granted loans to Chile totaling $168.7 million through 1965. From the United Nations Expanded Program of Technical Assistance, to which the United States contributed 40 percent of the financing, Chile received $714 million over the period 1951-1964.

[154] *New York Times,* Jan. 22, 1968.

[155] U.S. Senate, *Case Study,* p. 13. This report, submitted by Senator Ernest Gruening to the Subcommittee on Foreign Aid Expenditures, concluded that "... the magnitude of assistance has little connection with the results attained. Rather, it is clear that an excessive infusion of funds overburden fragile institutions, creates a profusion of new activities for which trained manpower is not available, and ends by dissipating efforts to the point where virtually no benefit results. Furthermore, large-scale assistance vitiates the host country's initiative to attack basic problems. Meanwhile, after a time, recipient nations come to depend upon concessionary aid and to regard it as their "right," thus multiplying the economic and political risks of eventual disengagement." See *Ibid.,* p. 121. For further evaluation of the U.S. aid program in Chile see Simon G. Hanson, *Five Years of the Alliance for Progress: An Appraisal* (Washington, 1967), chap. IV.

The Chilean economy rests heavily on the exportation of minerals, mainly copper and nitrates. Two American companies, Kennecott and Anaconda, produced 80 percent of Chile's copper production and accounted for approximately 60 percent of the country's foreign exchange earnings. In 1955 the taxes paid by the United States companies represented about one-half of all the revenues of the Chilean government. In a gamble to achieve higher earnings the Chilean government forced the companies into a disastrous attempt to market copper at well above the world market price, with the result that a large unsold surplus accumulated. The companies were also subjected to discriminatory taxes and exchange rate treatment and to an ever-present threat of nationalization. These circumstances forced the companies to withhold plans to expand production, and led to the "Chileanization" of the industry, a "moderate" program of government intervention carried out by the Frei administration.

Manufacturing proved to be the most dynamic sector of the Chilean economy after World War II, but a slow-down has occurred in recent years owing to rising costs of labor and labor unrest, inflation, low productivity, and the high cost of manufactures. Chile ranks among the most highly industrialized nations in Latin America, with manufacturing accounting for about one-quarter of the gross national product and employing about 20 percent of the economically active population.

While manufacturing is of great importance in the economy, agriculture remains the largest source of employment. Agricultural production has lagged behind population growth with the result that where Chile was an exporter of surplus food prior to World War II, the nation has been an importer of food stuffs annually in recent years. Food imports, rising to $120 million in 1966 and $155 million in 1967, reached $200 million in 1969. One of the chief obstacles to increased productivity is to be found in Chile's feudal agricultural system. Illustrative of the problem, the census of 1955 disclosed that 2.8 percent of the farm units containing over 1000 *hectares* each, occupied 37.3 percent of the arable land. Over 73 percent of all families gainfully employed in agriculture either owned no land or owned units of less than five *hectares.* In the agricultural sector less than 8 percent received over 65 percent of the total income. The bulk of Chile's rural families continue to live under conditions of grinding poverty.[156]

The maldistribution of land has existed since colonial times. But in recent years the situation has reached the crisis stage for two principal reasons. Population growth, as noted, is oustripping increases in farm production in a country where malnutrition is already a serious problem. Secondly, the peasants who lived passively on a marginal subsistence level for generations are now being aroused by Communists and other leftist agitators, and may be reached by foreign broadcasting.

Against a background of bitter strife on the land reform question, the Frei regime moved cautiously into this area. The chief agency for implementing agrarian reform is the Agrarian Reform Corporation (CORA), established by a law in 1962; this was reinforced by another law enacted in 1967. In the period

[156] U.S. Senate, *Case Study*, pp. 4-5; *New York Times*, Jan. 26, 1970.

1965-1968, 446 settlements of 13,881 rural families had been established on a total of 1,342,000 hectares. This compared with the 4,480 families that were resettled between 1928 and 1962 on largely idle government land. The Communists declared that the Christian Democratic Party's land reform was a failure, for President Frei had promised to resettle one hundred thousand peasant families during his term. However, it was conservative opposition to the law allowing payment in long-term bonds that blocked passage of this instrument for three years. By 1970 it had become clear that the agrarian program had slowed down agricultural output. Fearing expropriation, the large landowners had decapitalized their property.[157]

In combating land reform the landed elite maintained that the lack of cheap credit, a limited domestic market and stiff competition from Argentina and other countries had precluded the profitable exploitation of their land. The Frei government admitted the validity of these contentions, but held that the breakup of large estates would provide employment, cut food imports and bring rural people into the economy. The cost of implementing such program is formidible. It is estimated that that to supply viable plots of land to three thousand new owners each year would cost $6,750,000 in down payments to the old owners. In addition, the new owners would need further credits of $375 an acre yearly for a number of years to succeed in the venture. In the meantime, more than fourteen thousand rural families a year were moving into the crowded cities, increasing the size of the shantytown areas and stimulating unrest.

With the problem of feeding Chile's 9.5 million people becoming of crucial importance, agricultural production was given a severe setback by a drought, the most disastrous in forty-six years, which struck in August, 1968. Lasting two years, it caused widespread destruction of crops and livestock and brought food shortages to thousands of rural familes. Moreover, for lack of hydro-electric power, many of Chile's industries, including copper, were forced to curtail production, further damaging the economy and increasing unemployment.[158]

The prestige of the United States, which had become strongly identified with the fortunes of the Frei regime, was weakened in Chile after the government failed in another effort to curb runaway inflation, which stood at 27.5 percent in 1968. Sharp criticism came from the extreme left and the extreme right. Senator Carlos Altamirano, a Socialist, charged the United States with prolonging the Vietnam war to keep the price of copper high and to provide "extraordinary profits" for the American companies. A rightwing leader of the National Party, Senator Pedro Ibañez, declared that "since the United States created the economic disaster with its support of the Christian Democrats, the United States is now morally obligated to pay the nation's food bills." He also charged that Washington's support of "the fundamentally Marxist" government of Frei had destroyed national food production by depriving farm owners of land and eliminating the profit motive.[159]

[157]Inter-American Development Bank, *Socio-Economic Progress in Latin America (Social Progress Trust Fund, Eighth Annual Report, 1968)* (Washington, 1969), p. 107; *New York Times,* Jan. 26, 1970.
[158]*New York Times,* Jan. 26, 1970.
[159]*Ibid.,* Feb. 6, 1968.

Such provocative allegations failed to take into account the fact that the Chilean agricultural sector had received $96.4 million in external aid in the period 1961-1968. This included $47.4 million for the Inter-American Development Bank, $26.3 million for AID operations and $22.7 million for the World Bank group. The IDB approved two loans in 1968, one of $2.3 million to eradicate hoof and mouth disease and $10 million to assist intermediate farmers.[160]

In light of the highly charged political atmosphere, accompanied by economic nationalism, it is not surprising that another step was taken to "Chileanize" the copper industry. President Frei's move against Anaconda in June, 1969 separated the company from about 70 percent of its copper supply and an estimated two-thirds of its earning power. The agreement called for the transfer to Chile of 51 percent of Anaconda's two Chilean subsidiaries. The purchase price of the 51 percent was to be based on a book value of $197 million to be paid in dollars over a period of twelve years. Chile would purchase the remaining shares of the two new companies after it has completed payment for 60 percent of the first 51 percent. Chile's action was a stunning blow to Anaconda and another setback for private investment prospects in Chile and elsewhere in Latin America. More so, because the arrangement called for eventual nationalization.[161]

At the time that the Chilean position was hardening toward the United States copper firms, the Frei government took a more conciliatory stance in relations with Cuba. Like the other Latin American nations, with the exception of Mexico, Chile had for nearly six years enforced an economic boycott of Cuba. Premier Fidel Castro had repeatedly referred to President Frei as a "pseudo revolutionary" and an "imperialist lackey." However, Santiago announced on February 20, 1970 that under terms of two agreements with the Castro government, it had agreed to sell $11 million in agricultural products to Cuba in 1970 and 1971. The Chilean action reflected a growing, but far from unanimous, belief among Latin American nations that efforts should be made to reincorporate Cuba within the inter-American system.[162]

The avowed Marxist, Dr. Salvadore Allende Gossens, a psychiatrist and by birth a member of the Chilean upper middle class, won the presidential election in September 1970. It marked the fourth time Dr. Allende had sought the presidency, and followed a hard-fought campaign. His *Unidad Popular,* an uncertain, minority coalition of Communists, Socialists and other leftists gained a 1.4 percent plurality—fewer than forty-thousand votes—over conservative ex-president Jorge Alessandri. A third candidate, Christian Democrat Radomiro Tomic Romero, ran a poor third. After his election Allende moved to gain the support of the Christian Democrats in the Congress. They would only give him their support for the required vote of presidential ratification if Allende, in turn, would endorse a constitutional guarantee of democratic procedures. This was done, and on the eve of his inauguration, November 4, he said his administration would merely "lay the foundations for socialism" by constitutional means rather

[160] IDB, *Socio-Economic Progress,* p. 108.
[161] *Christian Science Monitor,* Mar. 23, 1970; *New York Times,* June 29, 1969.
[162] *Christian Science Monitor,* Mar. 2, 1970.

than seek a swift conversion of his country into Latin America's second Marxist-Leninist state.

While Chile has suffered little violence on political issues in the past, such violence was increasing. Two abortive plots against the government involving military officers occurred in October 1969, and in May 1970. There followed in succession the assassination of an agrarian reform official, bus and car burnings and riots, and the fatal shooting of General Rene Schneider Chereau, Commander-in-Chief of the Chilean army, on October 22, two days before congress approved Dr. Allende's election. These acts of violence and the assassination in June 1971 of Edmundo Perez Zujovic, a prominent Christian Democrat, who as minister of the interior was a vigorous opponent of Chile's Marxists, gave evidence of the deep divisions existing within Chilean society.

President Allende quickly took steps to finish plans for the nationalizing of his country's basic mineral resources—copper, iron and nitrates—and to take over monopolies and banks. At a mass meeting in Santiago, where he announced his nationalization program, Allende said that complete Chilean ownership of the copper mines, largely held by three U.S. firms—Kennecott Copper Corporation, the Anaconda Company, and the Cerro Corporation—would mean "the new independence of our country." A constitutional amendment included in the proposal would permit the government to expropriate any mining property. Should the need arise he could seek approval in a popular plebiscite as provided under the constitutional provisions which were adopted when he took office. Because the Allende program did not refer to the Chileanization scheme and since he agreed to honor all previous debts and payments for the purchase of percentages of these mines, it was assumed that these obligations would be carried out, in spite of the nationalization procedure.[163]

After denouncing the foreign firms, accusing them of "avarice for profit" and "poor management," but at the same time saying "we will pay if it is fair," Allende announced that "excess profits" would be deducted from compensation to be paid to the two U.S. copper companies. The amount of such profit was placed at $774 million. Of this total $364 million was charged to Anaconda's Chuquicamata and El Salvador mines, and $410 million to Kennecott's El Teniente mine. This meant that the two firms could receive little or nothing for their Chilean holdings. The Cerro Corporation was not mentioned, for its Rio Blanco mine began production in 1970 and had not yet produced any earnings. He indicated that the companies' assets would be evaluated in accord with their book value as of December 31, 1970. Payment would be made in no less than thirty years with interest of no less than 3 percent.

Responding to the new Chilean program and the $410 million excess profits deduction from the book value of El Teniente, Frank R. Milliken, president of Kennecott, said that "during the fifty-five years the mine has been in operation it has produced a gross income of $3,430 million of which $2,491 million remained in Chile . . . the excess profits deduction would reduce the

[163]*Ibid.*, Jan. 2, 1971.

Kennecott's earnings for the fifty-five year period to less than 2 percent return on investment."[164]

On the question of investment insurance coverage Secretary Meyer estimated that the value of all U.S. investment in Chile was $960 million, of which only $331 million was covered by Overseas Private Investment Corporation insurance, the U.S. government-backed corporation that insures investors against losses abroad. The Allende regime meanwhile announced that it would seek to renegotiate its foreign debt. About $3 billion was involved, of which more than half was owed to the United States.

Premier Fidel Castro came to Chile on November 10, 1971 aboard a Russian jetliner for what proved to be a twenty-five day visit. This was the first time that Castro had been in South America since 1959. His visit seemed to create new tensions and polarize opposition parties although he urged the radical MIR (Movimiento Izquierdista Revolucionario) to support Allende's moderate approach to social and economic reforms. Ironically, it was Allende who was named by Castro to head the Havana-based Organizacion Latinoamericano de Solidaridad (OLAS), which was to spread revolution in Latin America. A farewell rally at National Stadium in Santiago was expected to bring the visit to a successful climax. But instead there were empty seats, and troops enforced a state of emergency declared after a violent anti-government protest march by the city's housewives. One of the chanted slogans of the women marchers was "In Chile there is hunger. We do not want Castro here." Castro responded in the course of a two and a half hour speech that the incidents of the past week were evidence of "fascism in action."[165]

A domestic boom that was proving more apparent than real by early 1972 accompanied the first fourteen months of the Allende regime. The economy was stimulated by a 35 percent general wage increase at the start of the year and pressures built up to raise prices. With wages increasing faster than controlled prices and the money supply expanded to finance deficit public spending, real consumption rose 12 percent. Inflationary pressures caused price rises, the cost of living going up 22 percent in the second half of 1971. Industrial production rose by 10 percent in 1971, and the nation's gross national product 6 to 8 percent. On the government's part, more than thirty enterprises and the U.S.-owned copper mines had been nationalized, and 1,300 farms expropriated.[166]

At the same time, from a level of about $400 million in reserves in the banking system in September 1970, the foreign exchange decline brought reserves to less than $100 million. Partial restrictions were imposed on foreign payments and a meeting of foreign creditors was called to reschedule debt payments due in 1972 and 1973. Copper output fell in 1971, while production costs mounted and the world marked price dropped. Part of the problem was attributed to the loss of many technicians and engineers who left for foreign countries after Allende's election.

[164] *Times of the Americas,* July 21, 1971; *Ibid.,* Oct. 6, 1971; *Ibid.,* Nov. 3, 1971.
[165] *Christian Science Monitor,* Dec. 6, 1971.
[166] *Times of the Americas,* Dec. 15, 1971; *Christian Science Monitor,* Nov. 29, 1971; *New York Times,* Jan. 28, 1972.

The severity of the economic situation and the extent of political polarization was revealed early in December 1971 when, as mentioned, ten thousand women marched on the presidential palace to protest the high cost and scarcity of foodstuffs. Beating on pots and pans, the marchers were attacked by members of the left-wing student group, the MIR. The government was forced to declare a state of emergency and impose a curfew. Allende labeled the march "fascist sedition." However, food distribution was taken out of private hands and placed under the control of government agencies.[167]

Aside from food shortages and the economic problems just described, there were other factors causing unease and growing political polarization. The closing down by government order of two Santiago radio stations, one belonging to the Christian Democrats and another linked with the conservative National Party for "tendentious and alarming" reports of the women's demonstrations; warnings to newspapers that they were under censorship; the closing down of the United Press International office in September 1971, the government claiming that UPI had transmitted information that was damaging to the nation's foreign image. There was continuing unrest at the University of Chile; and tensions were aroused by the President's bill to replace the Congress, a bicameral body, with a "people's Assembly," and at the same time to change the procedure for selecting judges for all Chilean courts. Behind the unrest also were the growing threats posed by several insurgent groups, both urban and rural. Headlined most frequently was the Army of National Liberation which was thought to be a branch of the MIR, noted previously, a student group with terrorist tendencies. The MIR came out against the Unidade Popular, a left-wing coalition supporting the Allende regime. However, President Allende appointed MIR leaders to its internal-security force, a policy which could influence MIR policies toward the government in the future.[168]

A general strike that was to last nearly one month, and almost paralyzed the economy, began in October 1972. Strike organizers declared that the government had defaulted on its agreement to obey the Constitution. Difficulties began when truckers throughout the nation struck to protest a government scheme to establish a state-operated trucking firm in Aysén province; they feared that this was a prelude to the nationalization of the industry. After the strike had expanded to include a broad base of the country's twenty-five provinces under martial law, and taking control of all 155 Chilean radio stations.

The strike ended after President Allende replaced three of his cabinet officers with members of the armed forces. This included the appointment of the chief of the armed forces, General Carlos Prats González, as minister of the interior, the country's second highest post. Previous interior ministers, identified with the political left, had failed to oppose nationalizations which the opposition groups regarded as illegal. Meanwhile, inflation continued to mount: for the first ten months of 1972 the cost of living rose by 130.2 percent.[169]

[167] *Christian Science Monitor,* Dec. 11, 1971.
[168] *Ibid.,* Nov. 20, 1971; *Ibid.,* Dec. 6. 1971.
[169] *Times of the Americas,* Nov. 22, 1972.

The nationalization problem created some unease in U.S.-Chilean relations, but Washington's main concern was that Dr. Allende might lead his country into the international Communist movement. President Nixon clarified his administration's attitude toward events in Chile by declaring "What happened in Chile is not something we welcomed . . . that as far as the United States was concerned . . . we recognized the right of any country to have internal policies and an internal government different from what we might approve of . . . I haven't given up on Chile or on the Chilean people, and we're going to keep our contact with them. . ."[170]

Whatever may have been the intent, President Nixon omitted sending the diplomatically-required message of congratulations to the new president on his election. Although Dr. Allende attributed "no significance" to the fact that he had not heard from Washington, it was widely noted by Chileans. His inauguration was attended by U.S. Assistant Secretary of State, Charles A. Meyer, and it was also witnessed by representatives from North Korea, North Vietnam, Cuba and the People's Republic of China, which were still unrecognized by Chile. Then Washington made a last-minute cancellation of the goodwill visit in March of the U.S. aircraft carrier *Enterprise* to Valparaíso. Since President Allende had personally extended the invitation, it caused adverse comment in Chile. This episode was quickly followed by what some interpreted as a gesture of cordiality by Washington, the dispatching of four high-ranking Air Force officers to attend the 41st anniversary of the Chilean Air Force.

Toward the close of 1971 Secretary Meyer declared ". . . The entire thrust of our policy in the past year has been to try our level best to minimize the chances of a confrontation." He added that "differences with Chile are neither political nor ideological in origin. President Nixon has said that the United States is prepared to have the kind of relationship with the Chilean government that it is prepared to have with us." But soon afterward the officially detached posture of Washington was called into question over alleged statements of two U.S. officials that the Allende regime would be short-lived. This was in reference to statements attributed to Herbert G. Klein, the White House Director of Communications, and White House advisor, Robert Finch. Mr. Finch was President Nixon's emissary on a goodwill visit to six Latin American nations that did not include Chile. Chile officially protested the statement.

Despite the marring of U.S.-Chilean relations by the "incidents" or "statements," imaginary or real, Washington continued to advance military and economic assistance to Chile. For fiscal 1972 the Nixon administration asked Congress to approve $5,856,000 in military and economic assistance, an increase of $53,000 over 1971. The figure did not include a $5 million military transport aircraft and U.S.-made paratrooper equipment. That was the first military loan made by the Nixon administration to the Allende government. However, the U.S. Export-Import Bank rejected Chile's request for a $21 million loan for buying three new Boeing aircraft for its national airline. Future loans would be

[170] *Christian Science Monitor*, Apr. 26, 1971.

conditional on Chile's compensation for all U.S.-owned industries which it was taking over. On December 6, 1971 the Department of State said that the Export-Import Bank had suspended further credits, guarantees and insurance to Chile "until there is further clarification of Chile's unilateral moratorium on debt payments." It was said to be a matter of "sound banking practice" owing to Chile's defaults on IOU's to the Export-Import Bank. While the meeting between Chile and the industrialized nations of the so-called "Paris Club" was negotiating the country's $3.8 billion foreign debt, the Chilean Ambassador to the United States signed an agreement with the First National City Bank of New York awarding Chile with two credits totaling $300 million. It appeared that the loans would be used to cover public obligations falling due in 1972, 1973, and 1974.[171]

Although it was assumed that the Marxist-oriented government would alter Chile's traditional pattern of foreign relations, notably the expansion of diplomatic, commercial and cultural ties with Communist-bloc nations, Dr. Allende assured Secretary Meyer on the occasion of his inaugural that his nation intended to remain non-aligned in relation to international power blocs, and wished to maintain full and normal relations with the United States.

On November 12, 1970 Chile announced the resumption of full diplomatic relations with Cuba. This was not surprising, for Chile had reluctantly accepted the original OAS decision to break ties, the Frei government had negotiated a commercial agreement, and Allende had clearly shown his approval of the Castro regime. In reestablishing ties with Cuba, Chile became the first nation to renounce formally the six-year-old inter-American resolution designed to isolate Cuba from other nations of the hemisphere. Dr. Allende contended that the resolution, adopted in 1964 at the urging of the United States, conflicted with the intent of the United Nations Charter, and was diplomatically unfounded. This first foreign policy initiative of the new administration could be interpreted as a departure from Chile's alignment with the United States on foreign policy issues within Latin America. Premier Castro's visit in November 1971 was believed to have strengthened ties between the two governments. Implicit in the restoration of diplomatic ties with Havana was the possibility that the Communist subversive danger would be increased in the Western Hemisphere.[172]

Apart from the new diplomatic connection with Cuba, the Allende regime gave Chile the leading role among Latin American nations in expanding relations with other Communist countries. Diplomatic relations with the People's Republic of China were announced on January 5, 1971, effective as of December 15, 1970. In the joint communique Chile took note of the statement that "Taiwan is an inalienable part of the territory of the People's Republic of China." A Communist Chinese trade office and news agency bureau had been operating in Chile since 1965. Chile agreed to the establishment of North Korean and North Vietnamese trade missions in Santiago on November 15, 1970, and

[171]*Ibid.*, Feb. 24, 1972; *Times of the Americas*, Aug. 25, 1971; *Ibid.*, Mar. 1, 1972.
[172]*New York Times*, Nov. 13, 1970.

March 25, 1971, respectively. And an agreement to establish relations with East Germany was signed on March 16, 1971.[173]

At the close of 1972 President Allende embarked on a two-week trip abroad which took him to Peru, Mexico, New York, and the UN, Moscow, and Cuba. He was enthusiastically welcomed in Peru, Mexico, Moscow, and Cuba. However, upon his arrival in New York on December 3, he was not met by a U.S. reception committee, and a radioed greeting to President Nixon upon Dr. Allende's entry into U.S. airspace was ignored by the White House. In his address to the UN General Assembly, while not mentioning the U.S. by name, Allende declared that his country was being strangled by policies which were "terrifyingly effective in preventing Chile from exercising its own sovereign rights."[174]

The Chilean president spent almost a week in Moscow conferring with Soviet officials in an effort to gain increased Soviet aid, a multimillion dollar credit to help stabilize Chile's floundering economy. Over the preceding two years the Allende regime had received credits of $450 million from the Soviet Union and its East European satellites, together with $100 million from the People's Republic of China, and $126 million from several Western European nations, but this had proved inadequate. Chilean officials charged that U.S. government pressure had prevented the further granting of loans by international lending institutions; Washington denied the charge. It was held that the Frei administration had obtained some $1 billion in credits from these institutions during its six years in office whereas the Allende government obtained only $41 million in 1971 and nothing in 1972. Also denounced were the expropriated U.S. copper firms, which had taken legal action to stop copper shipments to European ports.[175]

In Moscow President Allende called the Soviet Union his country's "big brother," and Soviet President Nikolai Podgorny declared that the USSR was solidly on the side of Chile. A joint communique indicated that the two nations "condemned actions by foreign monopolies to deprive Chile of her right to use her natural resources at her own discretion, specifically her right to sell copper freely." Moscow's restrained statements may have reflected a desire to avoid jeopardizing the recently-established U.S.-Soviet detente.

In the immediate aftermath of the foreign tour the Allende government agreed to purchase $30 million worth of foodstuffs and cotton from the USSR to ease shortages. Meanwhile, the Soviet Union, facing food shortages because of crop failures in 1972, began taking deliveries in December 1972 of its own purchase of $1.2 billion worth of wheat, corn and soy beans from the United States. The USSR also offered credits for military equipment, and agreed to loan Chile some $80 million in support of construction of small industrial and fishing plants.[176]

In his conduct of foreign affairs through 1972, Dr. Allende had succeeded in keeping Chile from being isolated from other countries of the hemisphere and

[173] "Communist Activity in Latin America," *Inter-American Economic Affairs* (Autumn 1971), 25, 91-96.
[174] *Times of the Americas,* Dec. 13, 1972.
[175] *Christian Science Monitor,* Dec. 4, 1972.
[176] *Times of the Americas,* Dec. 13, 1972; *Ibid.,* Jan. 3, 1973.

the world at large. He had managed to perpetuate a cool though working relationship with Washington, in spite of the nationalization program, formed ties with several Communist countries, had met with Argentina's General Alejandro Agustín Lanusse, and concluded a major foreign tour, which won promise of more Soviet assistance.

President Nixon told Congress, in his annual foreign policy message of February 1972, that his administration's relations with President Allende's Marxist government exemplified his policy of "realism and restraint." He said "Chile's leaders will not be charmed out of their deeply-held convictions by gestures on our part. Nevertheless, our relations will hinge not on their ideology but on their conduct toward the outside world. As I have said many times, we are prepared to have the kind of relationship with the Chilean government that it is prepared to have with us."[177]

Washington appeared to have adopted a "watch and wait" policy calling for the reduction of all non-essential forms of U.S. public activity. This included a shutdown or phasing out of operations at three U.S. Air Force scientific observation stations, one at Easter Island in the Pacific, and two on the Chilean mainland. U.S. officials declared that the stations were not covered by any formal agreement; however, the action may have been attributable to the lowering of the U.S. silhouette in Chile at a time of uncertainty about future relations with the new Marxist government.

There is no doubt that Washington's relations with Santiago were clouded by revelation, beginning in 1972, of the International Telephone and Telegraph Corporation's efforts to frustrate the presidential aspirations of Dr. Allende in 1970. In ITT's own testimony, given at a congressional inquiry, it offered the White House and the Central Intelligence Agency a $1 million contribution for providing a workable plan. ITT believed that after the election the Marxist regime would nationalize its $150 million Chilean telephone company subsidiary, an event which later occurred. Fortunately, the scheme was not adopted by either the company or the U.S. government. Senator Frank Church(D) of Idaho, chairman of the sub-committee that investigated the affair, declared that, although ITT apparently did nothing illegal, the "highest officials" of the huge conglomerate "overstepped the line of acceptable corporate behavior."[178] The Senate's investigation was scheduled to continue.

Though not clearly foreshadowing policy changes in either country, the U.S. and Chile began exploratory discussions in Washington in December 1972 with a view towards resolving their differences. These were the first high-level negotiations between the two nations since Dr. Allende assumed the presidency in November 1970. Issues discussed included Chile's nationalization of the U.S.-owned firms, Chile's debt to the United States and its proposals to revise the terms of repayment, and Washington's alleged interference with the granting of credits by international agencies to Chile.[179]

At the same time, there was evidence of growing rivalry between the United States and the Soviet Union for marketing their military equipment in

[177]*Ibid.*, Feb. 16, 1972.

[178]*Times of the Americas,* April 11, 1973; *Christian Science Monitor,* March 26, 1973; *Ibid.*, June 25, 1973.

[179]*Ibid.*, Jan. 2, 1973. Secretary of State William P. Rogers had "a frank and useful discussion" with President Allende in Buenos Aires in May 1973. See *New York Times,* May 26, 1973.

Chile. Washington had doubled its military aid there, while rejecting Chilean requests for food imports and credits, and credits for development purposes. Moscow, however, had offered the Allende government $50 million worth of low-interest credits for Soviet military equipment. This could be seen as an obvious effort by the USSR to expand its influence in the Western Hemisphere.[180]

It seemed clear by early 1973 that Chile had embarked on a course of intense and turbulent nationalism and was headed towards some form of state-controlled economy. With the economy faltering as new lows were recorded in agricultural, industrial and mineral production, and inflation rampant (300 percent in the first six months of 1973), strikes, violence and sabotage became commonplace. Dr. Allende brought the military into his cabinet for a second time, further politicizing the military into pro and anti-Allende factions. In general, the country became divided into two rival groups: those supporting Dr. Allende, and those forming the opposition.

Preceded by earlier reports of military plotting, in June 1973 some one hundred rebel soldiers with four tanks attacked the presidential palace, but the revolt was put down by loyalist troops. This incident, and further evidence of military unrest, was accompanied by a series of crippling strikes, first in the copper mining industry and soon thereafter by another strike among the truckers. This set in motion strikes by physicians, businessmen, bus and taxi drivers. Wealthy citizens fled the country with their funds and so did many foreign investors.

The armed forces of Chile on September 11, 1973 brought the Allende regime to a sudden and violent end, replacing the socialist government with a military *junta.* With the declared intent of restoring "institutional morality," the presidential palace in Santiago was bombed and machine-gunned. After President Allende had appeared on the balcony to reject their demand for his resignation "even at the cost of my life," they occupied the palace and reportedly placed him under arrest. Dr. Allende died amid the crisis, an apparent suicide. General Augusto Pinochet Ugarte, commander in chief of the Army, was named president of Chile's new government, and a fifteen-member Cabinet, made up chiefly of military men, was sworn in. The Christian Democratic Party and the National Party publicly declared their support of the *junta.*

The Marxist parties were declared by the *junta* to be "outside the law," and leftists were purged from all levels of government, and from industry, often being replaced by military officers. The *junta* carried out an intensive campaign against certain foreigners–the estimated thirteen thousand political exiles who had found asylum in Chile under the Allende government from nearby right-wing Latin American dictatorships. The *junta* charged that they were hired by Dr. Allende to form a para-military force for use against the regular Chilean armed forces. This belief was reinforced by discoveries of huge quantities of arms, a network of guerrilla training schools, and use of public funds supportive of these activities. Preliminary evidence seemed to indicate that a well-armed extremist apparatus had been prepared to carry out violent revolution or civil

[180]President Nixon authorized the sale of F-5E International Fighters to Chile in June, 1973. This action, ending a five-year ban on the sale of such equipment, came after it had been concluded that, if the U.S. rejected the Chilean request, MIG 21's would have been supplied by the Soviet Union. See *New York Times,* June 10, 1973.

war.[181] There were confirmed reports of violence and repression used by the *junta* in consolidating its power.

President Allende's government failed because his Popular Unity Coalition, dominated by Socialists and Communists, persisted in its effort to impose on the nation a manifold Socialist system, often in disregard of the courts, which was opposed by more than half the population. Those who had been on strike for some fifty days before the *coup,* representing wide segments of the population, would accept nothing less than the president's resignation; when he refused to resign they called for a military *coup.* It appeared that the *coup* was sparked by the enraged middle sector, which had seen its living standards destroyed under the Allende government. The armed forces, traditionally non-political, intervened to forestall national economic disaster, and out of fear that a polarized Chile was plunging towards civil war.

The U.S. Department of State admitted foreknowledge of the revolt, but had not disclosed it to the Chilean government because such action might have been interpreted as intervention in the nation's internal affairs.[182] Although there was no evidence whatsoever of U.S. complicity in the *coup,*[183] many Latin Americans were disposed to believe the worst about the U.S. role there, especially in light of the ITT plotting against Dr. Allende in 1970. Latin Americans in general mourned the death of Dr. Allende, many regarding him as a democratic martyr. Washington maintained a discreet, low profile in the aftermath of the *coup* making no substantive comments on events; however, a credit was extended which enabled Chile to obtain 4.4 million tons of wheat in the United States. Available Chilean wheat supplies would have been exhausted within less than thirty days.[184]

When democratic processes and a civilian government might be restored remained indeterminate in late 1973, but there were indications that the military felt uneasy in its new role. Some optimism was generated over the disposition of the nationalized U.S.-owned properties by the *junta's* statement that businesses taken over illegally would be restored to their former owners; an invitation to foreign investment, and guaranties against expropriation. Prospects for the restoration of traditional U.S.-Chilean diplomatic ties were encouraged by the fact that the military *junta,* after seizing power, severed diplomatic relations with Cuba and North Korea, charging their representatives with interference in Chilean internal affairs. Meanwhile, the Soviet Union, East Germany and Bulgaria declined to recognize the new *junta.* The Soviet news agency, Tass, reported the overthrow as "a military mutiny against the republic's legitimate government."[185]

The stresses and tensions suffered by Chile under the Allende government, culminating in the military's seizure of power, were among the worst that the country had faced, comparable to the civil war of 1891. Reconstruction will be a massive task, given the shattered economy, and the widespread political polarization, which compounds the problem of restoring political unity.

[181] *Ibid.,* Oct. 7, 1973; *Christian Science Monitor,* September 24, 1973; *Ibid.,* Sept. 25, 1973.

[182] *Ibid.,* Sept. 17, 1973.

[183] *New York Times,* Sept. 16, 1973.

[184] *Times of the Americas,* Oct. 17, 1973.

[185] *Christian Science Monitor,* Sept. 14, 1973; *Ibid.,* Sept. 28, 1973.

SUPPLEMENTARY READINGS

Alexander, Robert J. *Labor Relations in Argentina, Brazil and Chile.* New York, 1962.

Barros Arana, Diego. *Un Decenio de la Historia de Chile.* 2 vols. Santiago, 1913.

---. *Historie de la guerre du Pacifique* 2 vols. Paris, 1881.

---. *Historia general de Chile.* 16 vols. Santiago, 1887.

Bonilla, Frank, "The Student Federation of Chile: Fifty Years of Political Action," *Journal of Inter-American Studies,* II No. 3 (July, 1960, 311-334).

Brieba, Enrique. *Limites entre Chile y Peru.* 3 vols. Santiago, 1931.

Bulnes, Gonzalo. *La Guerra del Pacifico.* 3 vols. Valparaiso, 1912-1919.

Burr, R. N. *By Reason or Force: Chile and the Balancing of Power in South America, 1830-1905.* Berkeley, 1965.

Butland, Gilbert J. Chile. *An Outline of its Geography, Economics and Politics.* 3rd ed. London, 1965.

Clissold, Stephen. *Chilean Scrapbook.* New York, 1952.

Dennis, W. J. *Tacna and Arica; an Account of the Chile-Peru Boundary Dispute and of the Arbitrations by the United States.* New Haven, 1931.

Ellesworth, Paul T. *Chile: An Economy in Transition.* New York, 1945.

Evans, Jr., H. C. *Chile and the United States.* Durham, N.C., 1927.

Fergusson, Erna. *Chile.* New York, 1943.

Galdames, Luis. *A History of Chile.* (I. J. Fox trans.) Chapel Hill, 1941.

Garcia Calderón, F. *Latin America: Its Rise and Progress.* London, 1913.

Gil, Federico. *The Political System of Chile.* Boston, 1966.

Hanson, E. P. *Chile, Land of Progress.* New York, 1941.

Halperin, Ernst. *Nationalism and Communism in Chile.* Cambridge, Mass., 1965.

---. *Sino-Cuban Trends: the Case of Chile.* Cambridge, Mass., 1964.

Hirschman, A. O. *Journeys Toward Progress.* New York, 1963.

Logan, C. A. and F. Garcia Calderón. *Mediación de los Estados Unidos de Norte Americana en la Guerra del Pacifico.* Buenos Aires, 1884.

Mc Bride, George M. *Chile, Land and Society.* New York, 1936.

Markham, C. R. *The War Between Peru and Chile,* 1879-1882. London, 1883.

Maurtua, V. M. *The Question of the Pacific.* (Pezet's trans.), Philadelphia, 1901.

Millington, Herbert. *American Diplomacy and the War of the Pacific.* New York, 1948.

Pike, F. B. *Chile and the United States, 1880-1962: The Emergence of Chile's Social Crisis and the Challenge to U.S. Diplomacy.* Notre Dame, 1963.

Rippy, J. F., P. A. Martin, I. J. Cox. *Argentina, Brazil and Chile Since Independence.* Washington, 1935.

Sherman, W. R. *The Diplomatic and Commercial Relations of the United States and Chile.* Boston, 1926.

Silvert, Kalman. *Chile: Yesterday and Today.* New York, 1965.

Stevenson, John R. *The Chilean Popular Front.* Philadelphia, 1942.

Stuart, Graham H., "The Tacna-Arica Dispute," *World Peace Foundation Pamphlets,* vol. X, No. 1 (Boston, 1927).

Subercaseaux, Benjamin. *Chile, A Geographic Extravaganza.* (Angel Flores trans.). New York, 1943.

Thiesenhusen, William C. *Chile's Experiments in Agrarian Reform.* Madison, 1966.

Valenzuela, M. V. *La Política economica del cobre en Chile.* Santiago, 1961.

Worcester, D. E. *Seapower and Chilean Independence.* Gainesville, 1962.

18

The United States and Brazil

Brazil is a giant among nations, ranking fifth in the world in land area. Its 3,286, 473 square miles are equal to 1.7 percent of the area of the globe, 5.7 percent of the world's total dry land and almost half (47.3 percent) of South America. Bounded on the southeast, east, and northeast by the Atlantic Ocean, Brazil borders on all of the South American countries except Ecuador and Chile. It contains one of the most extensive river systems in the world, including 27,000 miles of navigable waterways.

Although most of Brazil is either a plain or a plateau, the various parts of the country differ greatly. In the north is the Amazon River basin, an immense tropical region comprising 42 percent of the nation's area, but only 4 percent of the population who produce only 2 percent of the national income. The northeast, made up of a fertile coastal plain and a sub-tropical semi-arid section, contains only 11 percent of the country's area, but 22 percent of the population who produce 9 percent of the national income. While the coastal strip extending north from Rio de Janeiro contains only 15 percent of the nation's area, 34 percent of the population live here and contribute 26 percent of the national income. Stretching west to Bolivia and Paraguay is the west-central region, an underpopulated plateau, where 4 percent of the population live on 22 percent of the country's area and produce only 3 percent of the national income. To the south lie the states of São Paulo, Santa Catarina, Paraná and Rio Grande do Sul, which include 10 percent of Brazil's area and contains 36 percent of the population. This is the most prosperous, modern and industrially developed section of the nation. It produces 50 percent of the national income. Coffee, cotton, corn, rice, wheat, and fruit are yielded in large quantities. Brazil has great mineral wealth, much still untapped, though is lacking in large reserves of high grade coal and petroleum. Its agricultural potential is enormous, but limitations are imposed by heavy rainfall in the tropical zones where soils have been leached of their native fertility.

In considering the Latin American republics, Brazil must be placed in a class by itself for reasons other than its size and diversity. Having been discovered and colonized by Portugal its language and culture are Portuguese, which sets it apart from the Spanish American nations. Its population, totaling 98,081,000 (1972) makes it the world's eighth most populous nation. The population, growing at the rate of 3.1 percent annually, is concentrated in a belt near the seacoast; about half is urban. The main contributors to Brazil's melting pot were the aboriginal Indians, the Portuguese colonists, the Negro slaves brought from Africa, and the nineteenth and twentieth century immigrants, most of whom were Italian, Portuguese, Spanish, Japanese, German, and Polish. The Brazilian population is estimated to be 60 percent white, 20 percent mulatto, 10 percent *mestiço,* 8 percent Negro and 2 percent Indian.[1]

By mid-twentieth century it had become clear that Brazil was undergoing a transformation from an essentially agrarian, rural, semifeudal and patriarchal society to a modern, industrialized, urbanized, capitalistic society. This change from a traditional nineteenth-century structure to a modern society is a difficult and costly process. The great differences in the density of the population, economy, wealth, social organization, levels of education, and literacy, and political patterns in the five main areas of Brazil go far in explaining the country's problems in striving to become a strong and prosperous democracy.

Napoleon Bonaparte, whose influence on the New World, although indirect, has had momentous consequences, might almost be called the founder of modern Brazil. For it was his treacherous invasion of Portugal in 1807 that caused the royal Braganza family to flee to Brazil, transferring with it to Rio the seat of its government. Thus, at a period when all the rest of South America was straining at the bonds of royal sovereignty, Brazil was binding herself even more closely with imperial trappings. The immediate result, however, of Dom John's rule was to open Brazilian ports to the commerce of the world, to introduce the latest inventions of European genius, to repeal laws impeding progress and advancement, and to give the western kingdom the advantage of a government under the direction of a cultivated and enlightened, though rather weak-minded, ruler.[2]

In the meantime a regency was governing in Lisbon. But in 1820 a revolution took place in Portugal, and the result was the formation of a *junta* which was resolved upon a constitutional government, retaining the Braganza dynasty. The revolution spread to Brazil, and Dom John wisely decided to accept the proposed constitution. Shortly afterwards he sailed for Portugal, leaving his son Dom Pedro as regent of Brazil. In parting he foretold the separation of Brazil from the mother country and urged his son to seize the crown rather than allow it to pass to some adventurer.[3] His prediction was speedily verified. In the same year the *côrtes* at Lisbon passed decrees recalling Prince Pedro and reducing Brazil to a provincial status. The protest from Brazil was quick and decisive. Provisional *juntas* urged Pedro to remain, and he agreed to do so. When the *côrtes* threatened violence he called a constituent assembly, assumed the title of "Perpetual Defender and Protector of Brazil," and on

[1] See Preston E. James, *Brazil* (New York, 1946); T. Lynn Smith, *Brazil: People and Institutions* (Baton Rouge, La., 1963).
[2] The standard history of this period is M. de Oliveria Lima, *Dom João VI do Brazil,* 2 vols. (Rio de Janeiro, 1911).
[3] F. G. Calderón, *Latin America: Its Rise and Progress* (New York, 1913), p. 182.

December 1, 1822, was crowned Pedro I, Emperor of Brazil. The revolution was almost bloodless, and in less than a year after the declaration of independence every Portuguese was driven from Brazilian soil.

The early relations of the United States with Brazil were not so friendly as they have become today. The first American minister to the Portuguese court at Rio de Janeiro was Thomas Sumpter, Jr., whose duty it was to smooth the way for a permanent and cordial relationship, both political and commercial, with whatever form of government might be established. He presented his credentials in 1810 and immediately endeavoured to obtain most-favored-nation treatment for American goods. He was not successful, however, and his position at the Court was not improved by his refusal to accept the very undemocratic court etiquette. When on one occasion the royal guards attempted to make him dismount at the passage of the royal family, he drew his pistols and threatened to shoot. Although Mr. Sumpter and his family became very unpopular in Rio for a period, the ultimate result was the annulment of this regulation for foreign diplomats.[4]

Another little incident occurred in 1818 to embarrass Mr. Sumpter's position. The U.S. frigate *Congress* had brought dispatches to Sumpter and during the visit one of the seaman, of Portuguese nationality, got into trouble on shore and resisted the efforts of the ship's officers to place him on board. He was later taken from the local jail by force by an American contingent. The Brazilian foreign minister objected to this action and demanded his return and an apology. The American commander apologized for the acts of his officers, but sailed away with his crew intact in spite of threats to detain him.[5]

At the beginning of the nineteenth century, privateering was unquestionably a profitable source of income to many American ship-owners. Baltimore's reputation as a rendezvous of privateers became so notorious that at the conference of Aix-la-Chapelle in 1818 the Portuguese government submitted a memorial on the subject, and the powers agreed to take up the question in a friendly fashion with the United States.[6] Undoubtedly Brazil was back of the Portuguese protest, because it was Brazil that had suffered particularly through this practice. The Banda Oriental, now the Republic of Uruguay, had been seized by the Portuguese in 1816 from Artigas and his Argentinian forces. Artigas, however, could not afford to lose Montevideo; hence he engaged a number of privateers to prey upon Portuguese commerce. The Abbé Correa, the Portuguese minister to the United States, asserted that the greater part of these privateers were fitted out and manned in the ports of the United States. Adams conceded not only that this abomination had spread over a large portion of the merchants and population of Baltimore, but that it had infected almost every officer of the United States in the place.[7] He also realized that Brazil had very good cause for complaint; in fact, the situation was so serious that if the positions were reversed

[4] Lawrence F. Hill, *Diplomatic Relations Between the United States and Brazil* (Durham, N.C., 1932), p. 7.
[5] A. M. Breckenridge, *A Voyage to South America performed by the Order of the American Government in the years 1817 and 1818 in the Frigate Congress* (Baltimore, 1819), vol. I, pp. 92 ff.
[6] J. Q. Adams, *Memoirs* (Philadelphia, 1874-1877), vol. IV, p. 317.
[7] *Ibid.*, p. 318.

the United States would have considered the injuries sufficient for a declaration of war.[8] The Abbé Correa was exceedingly pessimistic over the outlook and declared that "these things had produced such a temper both in Portugal and Brazil against the people and government of the United States that . . . they were now those whom they most hated, and if the government [of Portugal] had considered the peace as at an end, they would have been supported in the declaration by the hearty conference of the people."[9]

Fortunately by 1820, the power of Artigas was completely broken. In 1821 the Banda Oriental was incorporated into the Portuguese dominions of Brazil, and when the independence of the Brazilian Empire was proclaimed in 1822 this Cis-Platine province was regarded as a part of it. A more friendly relationship between the United States and Brazil now ensued, and it was confirmed in 1824, when President Monroe became the first to recognize the independence of the new empire. At the cabinet meeting where the question was brought up Mr. Wirt opposed recognition on the ground that the government was monarchical and not republican; but both Calhoun and Adams favoured recognition on the basis of independence alone, leaving aside all consideration of internal government.[10]

When the Brazilian *chargé* José Rebello, was received by President Monroe, May 26, 1824, he suggested a concert of American powers to sustain the general system of American independence, and in the following year he proposed a definite offensive and defensive alliance between the United States and Brazil against European intervention.[11] Monroe declined the proposal in private to Adams; but no official reply was made until Adams became president, when Clay declared that the prospect of a speedy peace between Portugal and Brazil seemed to make such an alliance unnecessary.[12]

At the close of 1825 a war broke out between Brazil and Buenos Aires over the possession of the Banda Oriental, which once more brought about strained relations between Brazil and the United States. Brazil had declared a blockade on all Argentine ports, and the American *chargé*, Mr. Condy Raguet, had protested on the ground that Brazil would not be able to make such a blockade effective. In the second place, he warned Brazil that the United States "have always denied the doctrine of general and diplomatic notifications of blockades as binding upon their citizens," and therefore no vessel could be seized as a prize for running the blockade unless it had been specifically warned.[13] Brazil, however, paid little attention to Mr. Raguet's protests, and even gave him further grounds for recriminations by impressing American seamen into the Brazilian navy. The constant infringement of American rights infuriated Raguet to such a point that his notes to the Brazilian government

[8] *Ibid.*, vol. V, p. 177.
[9] W. C. Ford, ed., *Writings of John Quincy Adams* (New York, 1913-1917), vol. VII, p. 70.
[10] Adams, *op. cit.*, vol. VI. p. 281, Pedro Calmon, *História Diplomática do Brasil* (Belo Horizonte, 1941), pp. 50-53.
[11] Adams, *op. cit.*, pp. 358, 475, 484; also J. B. Moore, *Digest of Int. Law*, vol. VI, p. 437.
[12] See W. S. Robertson, "South America and the Monroe Doctrine," *Polit. Sci. Quar.*, vol. XXX (Mar., 1915), pp. 82-105; also *For. Rel. of the U.S.*, 1906, Part I, pp. 116-121. Arthur P. Whitaker, "José Silvestre Rebello, The First Diplomatic Representative of Brazil in the United States," *Hispanic American Historical Review*, XX (Aug. 1940), 380-401.
[13] *House Ex. Doc. No. 281*, 20th Cong., 1st Sess., p. 9.

were not always couched in the most diplomatic language. A crisis was reached on March 4, 1827, when an American vessel, the *Spark*, which had cleared regularly from Rio for Montevideo, was seized just outside of the harbor by a Brazilian warship and brought back as a prize, and its crew treated almost as pirates.[14] Mr. Raguet despatched a brief note to the Brazilian minister of foreign affairs asking for an explanation. The latter replied that the brig had increased her crew in Rio, that it had a warlike equipment with no license for it, and that therefore it was seized on the suspicion of being a privateer. The American *chargé* answered that if the government had thought proper to communicate its suspicions to him before the *Spark* had cleared, he would have cheerfully lent his aid in causing the suspicions to be removed. As it was, he declined to give any explanations. The next day he asked for his passports, and the Emperor, although "surprised at this precipitate request, couched in abrupt and vague language," ordered them to be delivered, but with the notice that the American representative would be answerable to his government for the consequences which might result.[15]

Before the break in diplomatic relations came, both Clay and Adams became convinced that Mr. Raguet's language and conduct were not so reserved as they should have been, and in a note dated January 20, 1827, Clay wrote Raguet that the president would have been better satisfied if he had abstained from some of the language employed.[16] When news of the rupture reached Washington, Adams wrote in his diary: "He appears to have been too hasty in his proceedings and has made us much trouble, from which we can derive neither credit nor profit"; and later, after conferring with Clay, it was decided not to sustain him.[17] On May 31, in a note to Mr. Rebello, the Brazilian *chargé,* Clay, although sustaining Mr. Raguet's protests, informed him that the latter's demand for his passport was without orders, and that, although there was interruption of diplomatic relations at Rio de Janeiro, none existed at Washington. The secretary also promised to procure the appointment of a successor immediately, provided assurances were given that satisfaction would be rendered for the injuries inflicted upon American persons and property.[18] Mr. Rebello promptly accepted the conditions, and Mr. William Tudor, a merchant at Lima, was named as *chargé* at Rio.[19]

This appointment proved excellent in every way. By his tact, good judgment, and diplomatic handling of the many serious cases that came up, Mr. Tudor placed the relations between the two nations upon a firm basis of friendly understanding. This was doubly fortunate because throughout the period of the war the actions of the Brazilian naval officials were constantly provoking fresh complaints, and some of these, as in the case of the schooner *Hero*, were so

[14] *Ibid.*, p. 96.
[15] *Ibid.*, p. 104-108.
[16] *Ibid.*, p. 108.
[17] Adams, *op. cit.*, vol. VII, pp. 270, 272.
[18] *Amer. State Papers, For. Rel.*, vol. VI, p. 824. The whole incident is exhaustively treated in Hill, *op. cit.*, pp. 49-56.
[19] For a full and well documented account of these events see W. R. Manning, "An Early Diplomatic Controversy between the United States and Brazil," *Amer. Jour. of Int. Law*, vol. XII (Apr., 1918), pp. 291-311.

harsh and uncalled for that Secretary Clay characterized the circumstances of the outrage as almost incredible.[20] Yet at the same time the commerce of the United States with Brazil was increasing at such a rate that Clay was very desirous of concluding a treaty of commerce and amity with her on terms favorable to the United States. Mr. Tudor was successful both in settling the claims and in negotiating the treaty. Virtually all of the claims were settled on the terms of the claimants and by diplomatic action alone; while the French, who sent a squadron of eleven ships, obtained only a third as much, and the British, who threatened direct reprisals, made no progress at all.[21] The treaty of commerce and navigation that Mr. Tudor concluded on December 12, 1828, was modeled upon the treaty concluded between the United States and Central America in 1825. The most-favored-nation clause was included, except for the relations between Brazil and Portugal, and the question of blockade was settled by declaring that only an effective blockade should be recognized. The treaty was to be in force for twelve years, and afterwards until notice of abrogation should be given by either party. The clauses regarding commerce and navigation were terminated in 1841, but those providing for peace and friendship still hold.[22]

Some impression of the satisfactory services of Mr. Tudor in increasing the friendly attitude of the Brazilian government towards the United States may be obtained from the Emperor's appreciation of them. In a special audience with Mr. Tudor, which Pedro I himself suggested, the Emperor declared that he had high respect for the United States, and the most sincere desire to cultivate and forever maintain the most friendly relations with them, and that the United States might be assured that such were his real feelings and such would be his conduct.[23]

Emperor Pedro's assurances were never put to the test. His popularity was already on the wane, and early in 1831 he was virtually forced to abdicate in favor of his infant son. Various causes contributed—the expensive and unsuccessful war with Buenos Aires, which resulted in the loss of the Banda Oriental; Pedro's constant efforts to support his daughter's claims to the throne of Portugal against his brother Miguel; his harsh and unfair treatment of his wife; his struggles against constitutional government and continuous opposition to the Chamber; and, finally, the fact that at heart he was Portuguese rather than Brazilian. The period that followed was the stormiest in Brazilian history. The government under the regency was in continual conflict with the provinces. Finally, factions striving for control at Rio agreed to declare the young Pedro of age in spite of his fifteen and a half years. Accordingly, on July 23, 1840, Congress unanimously declared that the young Emperor had reached his majority, and Dom Pedro II entered upon his imperial functions.

During this period the relations of the United States with Brazil continued to be most friendly. In their annual messages, both Van Buren and Tyler gave

[20] *House Ex. Doc. No. 32*, 25th Cong., 1st Sess., pp. 13, 66.
[21] *Ibid.*, pp. 152-221, 250.
[22] For text see Malloy, *Treaties, Conventions*, etc., vol. I, p. 133; for termination, Moore, *op. cit.*, vol. V, p. 403.
[23] *House Ex. Doc. No. 32*, p. 222. For President Adams' appreciation see Adams, *op. cit.*, vol. VIII, p. 224.

evidence of the fact, and in 1844 President Tyler noted that "the commercial intercourse between that growing Empire and the United States is becoming daily of greater importance to both, and it is to the interest of both that the firmest relations of amity and good will should continue to be cultivated between them."[24] But in 1846 an incident occurred which, although trivial in itself, developed into a complete breach of friendly relations between the two countries for a short period.

On October 31, 1846, Lieutenant Alonzo B. Davis of the U.S. frigate *Saratoga,* while on shore at Rio in pursuit of two deserters, found a sailor attached to his boat in a drunken brawl with two other American sailors. Davis interfered, disarmed the man, and was taking him back to the boat when a Brazilian guard came up, seized the three American sailors, and, after beating them severely, marched them off. Lieutenant Davis, having protested in vain, followed the patrol to the palace with a view of securing the sailors' release. However, upon entering the palace he also was seized and disarmed, and was kept in prison for two days. As soon as the American minister, Mr. Wise, learned of the occurrence, he protested vigorously to the Brazilian government, demanding the release of the imprisoned Americans, the disavowal of the outrage, and the punishment of the officers and soldiers of the Brazilian guard.[25] The Brazilian government freed Lieutenant Davis, but held the three seamen for further investigation, at the same time justifying both the conduct of the patrol and the arrest of Lieutenant Davis.

A long and acrimonious exchange of notes followed between Mr. Wise and the Brazilian foreign minister. The American minister insisted that inasmuch as Lieutenant Davis was in command of the sailor and was taking him back to the boat, and since no Brazilians were involved, the interference of the Brazilian patrol, its cruel treatment of the American seamen, and its subsequent arrest of the American officer were a direct insult to the American flag. The Brazilian government claimed that Lieutenant Davis had endeavored to interfere with the guard in its duty of preserving the peace within the sovereign jurisdiction of Brazil, and that therefore the Brazilian government was not only justified in its action, but had only surrendered the American officer as a proof of distinguished consideration for the United States.[26] The estrangement was increased by the action of Commodore Rousseau, commanding the U.S. squadron at Rio, who, indignant at the Brazilian stand, refused to salute on either the occasion of the baptism of the imperial princess or the celebration of his Majesty's birthday. All intercourse between the American minister and the Brazilian government thereupon ceased, Mr. Wise awaiting "calmly the first favorable and tangible occasion to come to explanations with them without danger of causing a more violent or open rupture, and not resenting their abuse in the newspapers, or their petty slights of not inviting me to a court, where the only reward for going and waiting for hours on a hot day in a hot uniform, is to make three bows forwards and three bows backwards, and then bob out of the imperial presence."[27]

[24] J. D. Richardson, *Messages and Papers of the Presidents*, vol. IV, p. 340.
[25] *Sen. Ex. Doc. No. 29*, 30th Cong., 1st Sess., pp. 5-16.
[26] *Ibid.*, pp. 20-42.
[27] *Ibid.*, p. 45.

An opportunity to resume friendly relations seemed to present itself in February of the following year, when Mr. Wise received despatches from Washington instructing him to request an audience from the Emperor to deliver in person the original of the answer of the President to a letter from him announcing the birth of a princess. Mr. Wise requested the audience, but was informed that, inasmuch as his acts were offense in the respect due both to the Emperor and to the dignity of the nation, he could not be received till the Davis affair was settled. As a concession, however, the minister of foreign affairs offered to receive the president's letter and see that is reached "its high destination."[28]

In a reply couched in terms hardly conducive to more friendly relations, Mr. Wise informed the Brazilian minister that he was the accredited minister to his Majesty the Emperor himself, and not as a *chargé* to a minister of foreign affairs, and either he would present the president's note in person or it would never reach "its high destination."

In the meantime Mr. Lisboa, the Brazilian minister at Washington, had taken the matter up with Secretary Buchanan, but in a much more amicable fashion. He assured Mr. Buchanan that "the Brazilian government, animated always with feelings of good understanding and perfect friendship towards the United States of America, has regretted extremely this disagreeable occurrence and will adopt the means proper to prevent similar occurrences hereafter."[29] In his reply Mr. Buchanan declared that the president was entirely satisfied with this frank and honorable explanation and that the whole occurrence, as far as the United States was concerned, would henceforth be buried in oblivion.[30] The Brazilian government, however, now insisted upon the recall of Mr. Wise, which it had suggested before, but which the United States had refused to consider. It was convinced that friendship and harmony were always in danger while so excitable a gentleman as Mr. Wise continued to be minister. Furthermore, the Emperor had determined that this particular gentleman should never again be invited to court. As a matter of fact, Mr. Wise himself had already realized the futility of remaining at Rio and had already asked to be recalled, but in such a way that his acts would appear to be approved in the fullest degree by the United States. Mr. Buchanan had agreed to this, and he informed Mr. Lisboa that Mr. Wise would soon return, but at his own request and not because the United States did not approve of his conduct.[31] As a result, the Brazilian government opened the case by disapproving the *amende honorable* of Mr. Lisboa and recalling him.[32]

The authorities at Washington received unofficial advice of this action before Mr. Tod, the newly appointed minister to Brazil, left the United States; and in his instructions Mr. Tod was advised that the president would not recede from his ground, and that the recall of Mr. Lisboa would be regarded as unjust to him as well as disagreeable to the president.[33] Mr. Buchanan at the same time

[28] *Ibid.*, p. 52.
[29] *Ibid.*, p. 134.
[30] *Ibid.*, pp. 135, 136. See also J. B. Moore, ed., *The Works of James Buchanan* (Philadelphia, 1909), vol. VII, p. 209.
[31] *Sen. Ex. Doc.*, p. 139.
[32] *Ibid.*, pp. 108, 109.
[33] Moore, *The Works of James Buchanan*, vol. VII, p. 328.

gave to Mr. Tod a sealed letter from the president of the United States to the Emperor of Brazil, to be delivered by Mr. Wise, and announcing the termination of his mission. Mr. Wise, however, was given the option of not asking for the audience if he was sure that it would not be granted.[34] Mr. Tod arrived at Rio August 7, 1847, and immediately got into touch with Mr. Wise. The latter, who had in the meantime made a further effort to renew negotiations and had been curtly repulsed, advised against seeking an audience, on the ground that the recall and disapproval of Mr. Lisboa and the refusal to receive Mr. Wise, who had been sustained and approved by the United States, constituted an additional insult to the American government.[35] After giving the advice due consideration, Mr. Tod felt that such action was contrary to the spirit of his instructions and decided to ask an audience of the Emperor to present his credentials. The audience was granted, and he was duly presented at court on August 28. Mr. Wise thereupon wrote a note to the Brazilian foreign minister, informing him that he had a sealed letter from the president to the Emperor, but as it had been written before the president knew of the recent insults to the American minister for acts that the president had fully approved, he felt it wholly incompatible with either the honor of his government or his own self-respect to ask an audience to present the letter; he therefore peremptorily demanded passports for himself and his family.[36]

At approximately the same time that Mr. Tod arrived at Rio, despatches reached the Brazilian *chargé* at Washington, Mr. Leal, informing him that Brazil had taken serious offense at the acts committed by Lieutenant Davis, for which it required ample reparation. He was also to demand from the government of the United States a categorical declaration that it had disapproved the conduct of its envoy, Mr. Henry A. Wise, and that it ordered his recall as a mark of reparation due to Brazil. If the government of the United States refused, but suggested arbitration, the imperial government would agree to that expedient. Finally, if Mr. Tod had already left the United States, Mr. Leal was to inform Mr. Buchanan that the imperial government would not receive him in his official character until satisfaction had been given.[37]

Mr. Buchanan's reply to these demands, dated August 30, 1847, showed that, although the United States was anxious to do everything possible to prevent a break between the two powers, it was determined to stand by its former statements. After once more carefully outlining all the facts of the Davis case, he declared that the United States would not grant reparation to Brazil for the acts committed by Lieutenant Davis, since reparation was clearly due from Brazil to the United States. The demand that Wise be disapproved and recalled he considered most extraordinary, inasmuch as the president had already publicly sustained his acts and had already recalled him at his own request. The present attitude of the Brazilian government following the amicable and honorable adjustment made by their former representative, and the refusal to receive the American minister, appeared to indicate that she intended an open rupture. Yet, as this seemed inconceivable, the president would take no decisive

[34] *Ibid.*, p. 333.
[35] *Sen. Ex. Doc.*, pp. 114-131.
[36] *Ibid.*, p. 132.
[37] *Sen. Ex. Doc. No. 35*, 30th Cong., 1st Sess., pp. 2-11.

step until he learned that the government of Brazil actually refused to receive the American minister.[38] The following day Secretary Buchanan wrote to Mr. Tod advising him that if the Brazilian government refused to receive him without making the desired apology he was to return to the United States.[39]

As we have already seen the Brazilian government did not carry out its threat to refuse an audience to the American minister. In a note to Mr. Leal, dated November 17, Mr. Buchanan expressed his gratification to learn that Mr. Tod had been kindly and courteously received by his Imperial Majesty, and was also pleased to learn that a new minister from Brazil would shortly be appointed to the United States. But in regard to the differences between the two governments he had nothing to add except to say that the President's views remained unchanged.[40] The same sentiments were indicated by Secretary Buchanan in his next despatch to Mr. Tod, with the further injunction that he should press the settlement of certain claims of American citizens which had long been outstanding against the Brazilian government.[41]

The interests of the American government now shifted from the settlement of these diplomatic difficulties to the settlement of claims, and when Secretary Buchanan found that the Brazilian government apparently intended to allow the Davis affair to remain suspended, he urged Mr. Tod to press more vigorously on the claims.[42] Mr. Tod followed instructions, and on January 27, 1849, a claims convention was concluded between the two countries, whereby the Brazilian government agreed to place at the disposal of the United States 530,000 *milreis* (about $300,000) to comprehend all the reclamations.[43] The same year Mr. Sergio Texeira de Macedo was accredited envoy extraordinary and minister plenipotentiary to the United States and duly accepted. The following exchange of compliments between Mr. Clayton, the new secretary of state, and Mr. Buchanan throw some light on the outcome: "If I go to the devil it will be because I am here daily engaged in covering up and defending all your outrageous acts. . . . The Brazilian Macedo laboured hard to revive your haughty discussion about Lieut. Davis and the drunken seamen in Rio. I refused to revive it, assumed you were altogether right (God assoilzie me for that) and dismissed him with compliments." Mr. Buchanan's reply to this was that "the Brazilian quarrel, which gave fair promise at one time of producing a tempest in a teapot, was virtually settled by your predecessor in the only effectual manner by assuming a just and lofty attitude in support of the lamblike Wise."[44]

The diplomatic relations between the United States and Brazil thenceforth became increasingly friendly. During the Civil War, Brazil gave numerous examples of a friendly disposition towards the Northern government, even though it insisted upon granting to the Southern states the status of belligerents. Upon one occasion, when the *Alabama* captured some half-dozen American

[38] *Ibid.*, pp. 28-41.
[39] Moore, *The Works of James Buchanan*, vol. VII, p. 404.
[40] *Ibid.*, p. 461.
[41] *Ibid.*, p. 462.
[42] *Ibid.*, vol. VIII, p. 60.
[43] *House Ex. Doc. No. 19*, 31st Cong., 1st Sess., p. 1. For terms of distribution see J. B. Moore, *Hist and Digest of Int. Arbitrations*, vol. V, pp. 4609-4626.
[44] Moore, *The Works of James Buchanan*, vol. VIII, p. 359.

whalers in Brazilian territorial waters, and in the meantime lay in the port of the island of Fernando de Noronha (also under Brazilian sovereignty), Mr. Webb, the American minister, raised a vigorous protest. The Brazilian government immediately investigated the case, and when it learned that the commanding officer of the island had been over-friendly to the captain of the *Alabama* he was forthwith dismissed and proceedings begun against him. At the same time, the president of the province of Pernambuco gave the captain of the *Alabama* notice to leave the territorial waters of the empire within twenty-four hours.[45] On a later occasion information was again brought to Mr. Webb that the Confederate cruisers *Alabama*, *Florida*, and *Georgia* were obtaining coal and provisions in the ports of Pernambuco and Bahia in order to continue their destruction of the commerce of the United States.[46] Upon receiving Mr. Webb's protest, the Brazilian foreign minister again gave assurances that his Majesty the Emperor was firmly resolved to maintain, and cause to be respected, the neutrality of Brazil.[47]

The Confederate States, however, were not alone in violating the neutral territory of Brazil. On October 4, 1863, the *Florida* arrived at Bahia and was given forty-eight hours by the authorities to repair her boilers and obtain provisions and coal. The U.S.S. *Wachusett* happened to be in the harbor at the time, but the American consul was said to have given a pledge for the observance of neutrality by the vessel's commander. In spite of this, the *Wachusett* approached the *Florida* on the morning of October 7, and opened fire upon her. The commander of the Brazilian naval division intervened, and the firing ceased; but shortly afterwards it was seen that the *Wachusett* was towing the *Florida* out to sea. The Brazilian commander pursued but could not overtake her, and the *Florida* was brought to Hampton Roads. The Brazilian government thereupon demanded (1) a public expression on the part of the Union government that this action was regretted and condemned; (2) the immediate dismissal of the United States commander, followed by the commencement of proper process; and (3) a salute of twenty-one guns, to be given in the port of the capital of Bahia by some vessel of war of the United States, having hoisted at her masthead during the salute the Brazilian flag. In a note to the Brazilian government dated December 26, 1864, Mr. Seward replied that the president disavowed and regretted the proceedings at Bahia; that he would suspend the commander of the *Wachusett* and direct him to appear before a court martial; that the consul, who admitted having advised and incited the commander, would be dismissed; and that the flag of Brazil would receive from the United States Navy the honor customary in the intercourse of friendly maritime powers. This would be done on the ground that the capture of the *Florida* was an unauthorized, unlawful, and indefensible exercise of the naval force of the United States within a foreign country, in defiance of its established and duly recognized government.[48] The salute was fired by Commander F. B. Blake in the harbor of Bahia on July 23, 1866, and the *Diario da Bahia* thus characterized the event: "It is thus that a

[45] *Diplomatic Correspondence of the United States*, 1863, Part II, pp. 1164-1169.
[46] *Ibid.*, p. 1171.
[47] *Ibid.*, p. 1177.
[48] Moore, *Digest of Int. Law*, vol. VII, p. 1090. See also Hill, *op. cit.*, pp. 155-158.

great and spirited people give, in the face of the civilized world, a public and solemn proof of the sincerity of its professions of the sacred principles of justice."[49]

Just as the Civil War was coming to an end in the United States, Brazil was forced to enter a war which, although successful, imposed a heavy burden upon her people. The Paraguayan dictator, Francisco López, had been much incensed at Brazil's interference in a factional struggle in Uruguay in 1863. It was also rumored that Dom Pedro II had refused to entertain the dictator's proposals for his daughter's hand. At any rate, in the fall of 1864 López seized a Brazilian steamer on its regular trip up the Paraguayan River to Mato Grosso. He followed this up with an expedition against the southern settlements of Mato Grosso, which were wholly unable to resist his well-disciplined troops. Brazil accepted the challenge, although she was by no means equipped to combat the wonderfully trained army that López had been preparing for years. Argentina, although in sympathy with Brazil, declared her neutrality. López, however, confident of his power, deliberately invaded Argentinian territory at the Paraná River in order to strike at the heart of Brazil. The result was a coalition against him on the part of Argentina, Brazil, and Uruguay, a formal alliance being signed May 1, 1865.[50]

The war lasted five years and resulted in the death of López and the almost complete extermination of the Paraguayan people. "The heroism of Paraguay overcame numbers, destiny, and death," but to no avail. Out of a million and a quarter people living in Paraguay before the war, more than a million had perished, and of the less than a quarter of a million who survived more than five-sixths were women and children. Brazil played the leading part in the war and her losses were correspondingly great. It was estimated that the war cost her more than 50,000 lives and $300 million. Considering her sacrifices, Brazil deserves all the more credit for making no effort to extend her territory at the expense of her vanquished and helpless neighbor. By a preliminary agreement of peace, signed June 20, 1870, the allies' demands were limited to the establishment of complete freedom of navigation for the warships and merchant-vessels of the allies upon the Upper Paraná and the Paraguay.[51] A definitive treaty of peace was signed between Brazil and Paraguay, January 9, 1872, whereby Brazil promised to respect perpetually Paraguay's independence, sovereignty, and integrity. The rivers Paraguay, Paraná, and Uruguay were declared free to the commerce of all nations, and the boundaries between Brazil and Paraguay were to be settled by a special convention.[52]

During the progress of the Paraguayan war a number of diplomatic incidents occurred in which the United States was interested. Early in 1866 Mr. Charles A. Washburn, the minister of the United States to Paraguay, informed his government that he was being prevented by the allies from passing up the

[49] *Diplomatic Correspondence*, 1866, Part II, p. 317.
[50] For test see *Ibid.*, p. 476; for a fair summary of causes of the war see *Dip. Corr.*, 1867, Part II, pp. 248-250; the Paraguayan side is given in pp. 722-725.
[51] *British and Foreign State Papers*, vol. LXIII, p. 322. For a survey of the operations and results of the war see C. E. Akers, *A History of South America, 1854-1904* (New York, 1904), pp. 130-188.
[52] *British and Foreign State Papers*, vol. LXII, p. 277.

Paraná River to Asunción; although it appears from his despatches that an equal source of delay was the failure of Admiral Godon, in command of the American fleet, to furnish a boat with which to run the blockade. Secretary Seward protested vigorously, both to Brazil and Argentina; and they finally gave the necessary orders, although under protest. Even then, Mr. Washburn was delayed considerably through Admiral Godon's lack of cooperation, and it was not until a year after he had tried to return that the American minister finally arrived at his post.[53] An even more serious diplomatic controversy occured the following year, when Rear-Admiral Davis sent the U.S.S. *Wasp* up the Paraná to convey Mr. Washburn and his family back to the United States. Permission was again refused by the Marquis de Caxias, and it was only when Mr. Webb, the American minister at Rio, threatened to ask for his passports that the *Wasp* was enabled to carry out the mission.[54]

All efforts on the part of the United States to use its good offices to settle the Paraguayan struggle were refused unconditionally by Brazil. In fact, Brazil appeared less inclined than Argentina to accept mediation.[55] An attempt on the part of Mr. Washburn to secure a similar result by a personal visit to the Marquis de Caxias, the Brazilian commander-in-chief, was equally unsuccessful.[56]

Perhaps the most satisfactory occurrence of this period to the United States was the decree of the Emperor of Brazil, dated January 22, 1866, opening up the Amazon, São Francisco, and other rivers to the merchantships of all nations. Ever since 1850 the United States had been sedulously striving to accomplish this result, on the ground that "this restricted policy which it is understood that Brazil still persists in maintaining in regard to the navigable rivers passing through her territories is the relic of an age less enlightened than the present," and the merchant-vessels of the United States had the right to use these natural avenues of trade, not because of treaty stipulations, but because "it is a natural one—as much so as that to navigate the ocean—the common highway of nations."[57]

During the period following the war with Paraguay, Dom Pedro II was seen at his best. His upright and conservative character inspired confidence, and Brazil had little difficulty in obtaining the loans necessary to effect a financial rehabilitation. Nor did his success make him more autocratic or blind him to the need of progressive ideas. By visits to both Europe and the United States he broadened his point of view, and Brazil profited exceedingly by the statesman-like policy of her enlightened ruler. Nevertheless, a growing sentiment in favor of republicanism was everywhere manifest. The general disposition seemed to be to allow the monarchical system to continue during the lifetime of the emperor,

[53] *Diplomatic Correspondence*, 1866, Part II, pp. 307-326, 548-616. For an adversely critical discussion of the United States policy see Hill, *op. cit.*, pp. 187-195. The position of the United States was sound, however, according to long established precedents of international law.

[54] *Dip. Corr.*, 1868, pp. 273-299.

[55] *Ibid.*, 1867, Part II, p. 253.

[56] *Ibid.*, p. 714.

[57] Mr. Marcy to the American minister to Brazil, Aug. 8, 1853; Moore, *Digest of Int. Law*, vol. I, pp. 640-645. For the full correspondence see W. R. Manning, *Diplomatic Correspondence of the United States—Inter-American Affairs 1831-86* (Washington, D.C., 1932), vol. II, *passim.*

but to assume that when the Princess Izabel and her unpopular consort, the Comte d'Eu, should take over the government it would be time to consider the wisdom of retaining a monarchical system in an otherwise republican continent.

The crisis came sooner than was expected. While the Emperor was traveling in Europe in 1887 for the betterment of his health, Princess Izabel, acting as regent, determined to abolish slavery. Laws had already been passed granting freedom to all children of slaves born after 1871, and to all slaves attaining the age to sixty years, but this process was considered too slow. The princess insisted that a decree of immediate emancipation of all slaves in the empire be passed; and on May 15, 1888, the measure became law. Although there was no immediate outbreak on the part of the wealthy slave-owners and the landed aristocracy, the monarchical party, through this decree, lost its strongest prop, republican propaganda, revolutionary agitation spread rapidly, and the situation of the government became critical. Dom Pedro's return in August 1888, stemmed the tide temporarily, but in the following year the military element threw in its lot with the republicans and decided upon an immediate *coup d'état.* The plans were made carefully and were carried out without a hitch. Early in the morning of November 15, 1889, the imperial palace was surrounded and the emperor and his family were arrested. A proclamation announced the deposition of the emperor and the establishment of a republican form of government with a provisional president at the head. There was neither bloodshed nor confusion, and a few days later the Emperor and his family were sent back to Portugal.[58]

It was to be expected that the United States would look with favor upon the new convert to republican institutions, and on February 19, 1890, the Senate and House of Representatives passed a joint resolution congratulating the people of the United States of Brazil on their adoption of a republican form of government.[59] Meanwhile the American minister had been instructed to maintain diplomatic relations with the provisional government and to give it a formal and cordial recognition "so soon as a majority of the people of Brazil should have signified their assent to the establishment and maintenance of the Republic."[60]

General Deodora da Fonseca, the provisional president, summoned a national congress, which met at Rio de Janeiro on November 15, 1890, to consider a draft of a constitution submitted by the provisional government. As finally adopted on February 14, 1891, the "Law of Constitution" established a federal system of government modeled very closely upon that of the United States. A president and vice-president elected directly for a four-year term, assisted by six secretaries of state appointed by the president, constitute the executive authority. The legislature consists of a Senate of sixty-three members, three from each state and three from the federal district, elected directly for a

[58] See the scholarly article by Professor P. A. Martin, "Causes of the Collapse of the Brazilian Empire," *Hispanic American Historical Review,* vol. IV, no. 1 (Feb. 1921), pp. 1-48.

[59] *For. Rel. of the U.S.*, 1890, p. 21.

[60] Moore, *Digest of Int. Law,* vol. I, p. 160; Carlos Delgado de Carvalho, *História Diplomática do Brasil* (São Paulo, 1959), pp. 166-167. Formal recognition was granted to Brazilian agents at Washington, Jan. 29, 1890.

nine-year term, and a Chamber of Deputies, in the proportion of not more than one for each 70,000 inhabitants, elected directly for a three-year term. The judicial power is vested in a supreme federal court of fifteen members. The states are given almost complete autonomy, except for matters that are purely national in scope. In theory Brazil now had a perfect type of representative republican government, although an arduous struggle and considerable bloodshed were required to put the theory into practice.[61]

Before the end of 1891 President Fonseca had brought on a series of revolutionary outbreaks through his arbitrary methods, and when the navy joined in the opposition he decided "in the interests of the nation" to resign. Vice-president Peixoto, who took over the presidency, governed for a short time in accordance with constitutionsl prerogatives. But military training and an autocratic disposition soon overcame his newly acquired constitutional inhibitions, and he became even more despotic than his predecessor. A revolt that started in Rio Grande do Sul spread rapidly, and in the autumn Admiral de Mello and Admiral de Gama, who were in command of the naval forces and of the naval school at Rio de Janeiro, threw in their lot with the revolutionists.[62] The struggle quickly developed into a real civil war, and the question came up as to the recognition of the insurgents as belligerents.

On October 24, 1893, the American legation at Rio de Janeiro received notice from Admiral de Mello that a provisional government had been established at Desterro and requested recognition by the United States as belligerents. When the request was brought to his attention, Secretary Gresham replied that, since the insurgents had not yet established and maintained a political organization justifying recognition, such an act would be unfriendly to Brazil and a gratuitous demonstration of moral support to the rebellion. The American minister was therefore instructed to remain an indifferent spectator.[63] Commodore O. F. Stanton, in command of the United States naval forces at the South Atlantic Naval Station, happened to arrive at Rio de Janeiro in his flagship just after the revolt had occurred. On entering, he saluted the Brazilian flag with twenty-one guns, and a government fort returned the salute. Subsequently, however, he saluted Admiral Mello with thirteen guns; the salute was returned, and the next day he made an official call upon the insurgent admiral. The Brazilian government protested, and the commodore was thereupon detached from his command and ordered home. After hearing his explanation, the Navy Department decreed that he had committed "a grave error of judgment," since it was known that the United States had not recognized Admiral Mello and his forces as entitled to belligerent rights.[64]

The question of recognition of belligerency brought up a conflict between Admiral Benham, who now commanded the American squadron, and Admiral de Gama of the insurgent fleet. Inasmuch as belligerent rights of the insurgents were not recognized, the naval commanders of the neutral squadrons refused to allow

[61] For text of the constitution see W. F. Dodd, *Modern Constitutions* (Chicago, 1909), vol. I.
[62] For a short résumé of the causes see *For. Rel. of the U.S.*, 1893, pp. 68-70 or Akers, *op. cit.*, pp. 250-266.
[63] *For. Rel. of the U.S.*, 1893, p. 63.
[64] Moore, *Digest of Int. Law*, vol. I, p. 240.

the insurgent forces to interfere with commercial operations except in the actual lines of fire. However, with the progress of the revolution the insurgents became inclined to interfere with neutral shipping, especially where the articles might be regarded as contraband of war. The British admiral and his European colleagues were inclined to submit to such interference. But Admiral Benham refused categorically and notified Admiral de Gama that he would use force, if necessary, to maintain American rights. An opportunity was given when several American vessels attempted to land their goods. When a shot from one of the insurgent vessels stopped the operation, the U.S.S. *Detroit* returned the fire, and the insurgents were notified that their boat would be sunk if she fired again. The threat was effectual and neutral commerce was not again interfered with. It might be noted, however, that Admiral Benham's *dictum* that the forcible seizure of contraband by the insurgents from neutrals who were engaged in supplying it to the belligerent government would be an act of piracy has hardly received the sanction of international law.[65]

With the resources of the country behind him, President Peixoto was enabled to purchase vessels in Europe and the United States, and the insurgents finally realized that further resistance was useless. Furthermore, on March 1, 1894, a civilian, Dr. Prudente de Moreas Barros, was elected president, which not only proved that President Peixoto intended to abide by the constitution but also that a more liberal regime could be expected. By summer the rebellion was completely checked, and on November 15, 1894, Dr. Moraes took over the government. The neutral attitude adopted by Admiral Benham, which had prevented a blockade of Rio de Janeiro, naturally aroused feelings of more cordial friendship towards the United States, and on the same day that the new president was inaugurated, the cornerstone of a monument to the memory of President Monroe was laid in Rio.[66]

President Moraes granted full amnesty to the majority of those who participated in the revolution, and then proceeded to eliminate all military influence in the control of the government. His administration was so successful that his chosen successor, Dr. Manuel Campos Salles, another civilian, had little difficulty in securing the election. President Campos Salles, governing strictly in accordance with the constitution, continued the liberal and statesmanlike policies of his predecessor. He was particularly successful in his financial policy, restoring Brazilian credit, which had been sorely tried by the costs of the wars and uprisings and by the extravagant methods of the inexperienced republican officials. The next two presidents were also civilians, but in 1910 a soldier, Marshal Deodoro Fonseca, formerly minister of war, was elected. Any doubts as to whether he would not attempt to govern after the fashion of his military predecessors were soon set at rest, and his administration was no less constitutional than those of the civilians who preceded him. However, in 1914 the country once more turned to a civilian, Dr. Wenceslau Braz, who had served as vice president with President Fonseca; and the subsequent president, Dr. Epitaçio da Silva Pessoa, was also drawn from the civilian ranks. Republican

[65] *For. Rel. of the U.S.*, 1893, pp. 115-117; also Moore, *Digest of Int. Law*, vol. II, pp. 1113-1120.

[66] *For. Rel. of the U.S.*, 1895, Part I, p. 48.

government in accordance with constitutional limitations seemed to have taken hold in Brazil.

From the founding of the republic until the post-World War II era it is evident that the relations between the United States and Brazil have surpassed in cordiality, cooperation and understanding those existing between the United States and any other Latin American country. This did not mean that the policies of the two countries were at all times harmonious or that either could afford to take the other for granted. However, the American policies which had caused so much distrust and bitterness in most of Latin America played a much smaller role in Brazilian-United States relations. Brazil had through the years regarded the United States without fear and with considerable understanding of the problems facing the North American power for several reasons. Owing to its size and its language Brazil seemed to view the United States as an ally in dealing with other American republics. Like the United States, Brazil, as the largest nation in the hemisphere and the only Portuguese-speaking nation, saw itself set apart from the other nations of the New World. While recognizing the ties of common interest and geography and consequently acting as a strong supporter of hemispheric unity, Brazil felt a confidence and uniqueness which predisposed it from seeking to challenge United States leadership in the hemisphere.

The rivalry between Brazil and Argentina came to the surface only sporadically, but it had a significant relevance to Brazil's relations with the United States. Since both nations were protected to some degree by Great Britain and the United States, they could ignore most extra-continental threats and indulge in their own regional competition. It was in the operation of the inter-American system that the rivalry was seen most clearly. Brazil followed a policy of stressing cooperation within the hemisphere in the solution of the region's problems and in matters of hemispheric defense. This policy accordingly led to a strong bond of interest between the United States and Brazil. Argentina, on the other hand, followed a policy emphasizing Latin American ties with Europe and the importance of the League of Nations.

The historic cordial feelings between the two nations were also based on the general absence of conflicting economic interests. In fact, the United States had been the largest market for Brazil's chief export and the balance of this trade was highly favorably to Brazil. Thus Brazil had a strong motive for maintaining amicable relations with the northern power. The Monroe Doctrine which had appeared to some Latin American states as a device for preserving the hegemony of the United States in the Western Hemisphere did not appear menacing to Brazilians. And being aligned economically with Britain until after World War I, Brazil was afforded a measure of protection against interference from the United States and more significantly from European intervention.

The major shift in the orientation of Brazil's diplomacy from London to Washington occurred during the ministry of Baron do Rio Branco, Brazil's minister of foreign affairs, 1902-1912, and the tenure of the distinguished statesman Joaquim Nabuco, Brazil's first ambassador to the United States, 1905-1910. Baron do Rio Branco said of the Monroe Doctrine: ". . . Relative to this continent the greatest service given by the Monroe Doctrine is the liberty which it assures to the development of forces of each American nation . . ." Ambassador Nabuco was similarly a partisan of strong friendship with the

United States and a champion of Monroeism.[67] In the immediate background of the new era of relations it was clear that President Cleveland's arbitral award in 1895 favoring the Brazilian line claim in the Misiones boundary dispute with Argentina made him very popular in Brazil as did his defiant message to the Europeans at the time of the Venezuela-British Guiana boundary dispute. The great republic maintained an attitude of friendly neutrality during the Spanish-American War.

In 1904 the Brazilian government built a marble and granite palace of great beauty for its exhibit at the St. Louis Exposition; later, it had the structure sent back to Rio de Janeiro for use as a meeting place for international conferences and other similar occasions. It was in this building that Secretary of State Elihu Root made his memorable address while acting as honorary president of the Third Pan American Conference at Rio de Janeiro, July 31, 1906. Baron do Rio Branco announced at the close of the conference that henceforth the palace would be known as Monroe Palace. Now the palace graces the head of a beautiful street in Rio de Janeiro—the Brazilians say the most beautiful street in the world—which bears the name Avenida do Rio Branco, in memory of Brazil's illustrious foreign minister.

Long before the turn of the century Brazil's trade with the United States surpassed that of all the Spanish South American republics combined. For example, in 1870 Brazil's trade with the United States was valued at approximately $31 million, while that of the other South American states amounted to about $29 million.[68] However, it must be noted that the United States imported from Brazil about four times as much as it sent in return. This situation was due mainly to the fondness of Americans for Brazilian coffee. Successive tariff laws, notably the McKinley Tariff Act of 1890 and the Dingley Tariff of 1897 sought to achieve a balance of trade more favorable to the United States through penalty provisions: the placing of tariffs on Brazilian imports as a means of achieving a reduction of tariffs on American exports to Brazil. Since coffee was Brazil's chief earner of foreign exchange and the United States was the nation's largest market, this problem was negotiated amicably, but it was not until 1922 that the principle of equality of treatment superseded the concession-by-threat-of-penalty approach.

The first vessel that inaugurated monthly steamship transport between the United States and Brazil arrived in Rio on June 7, 1878, and took back a cargo of 37,000 sacks of coffee.[69] In 1890 the total trade between the United States and Argentine was valued at about $14,200,000, the trade with Chile at $6,400,000, while that with Brazil amounted to more than $71 million.[70] At the outbreak of World War I Brazil still held first place in American trade with South America, with a total value of approximately $154 million, although Argentina with a total value of about $72 million had gained more proportionately.[71]

[67] Frederic W. Ganzert, "The Baron do Rio Branco, Joaquim Nabuco, and the Growth of Brazilian-American Friendship, 1900-1910," *Hispanic American Historical Review*, XXII (Aug. 1942), 432-451; E. Bradford Burns, ed., *A Documentary History of Brazil* (New York, 1966), pp. 306-311.
[68] *For. Rel. of the U.S.*, 1870, pp. 283-287.
[69] *Ibid.*, 1878, p. 67.
[70] *Statistical Abstract of the U.S.*, 1890, p. 75.
[71] *Ibid.*, 1914, p. 688.

By the first decade of the twentieth century the economic problems of Brazil which were to become magnified with the passage of time, and were to create grave internal crises and lead to a complex relationship with the United States, had emerged. An introduction to these problems should be noted here. Throughout most of its history the economy of Brazil has been based on monoculture. By 1860 this one crop was coffee and for a period of almost forty years it provided the basis for a very profitable foreign trade. As far back as 1885 Brazil produced more than one-half of the world's coffee, and by 1900 it was producing more than two-thirds. After 1896 production began to surpass consumption and the price started to fall. This quickly became a grave problem since coffee accounted for 60-70 percent of Brazil's total exports. It was to meet this problem that São Paulo state adopted the coffee valorization program. According to this scheme a portion of the crop was held out of the market during periods of surplus production to be sold during the years when production was not sufficient to supply demand. The program was financed with loans secured by the coffee withheld from the market. The first valorization of coffee took place in 1906 and the device worked very well as did the valorizations in 1917 and 1921. In each instance it proved possible to sell at a profit all coffee that had been stored. By-products of the successful valorizations, however, nullified the long-range effectiveness of the device. By holding the price of coffee at an artificially high level, the tendency to overproduce was encouraged in Brazil, and new sources of supply in other areas were developed while the market for coffee remained almost static. The borrowing which was done abroad in support of valorization was a familiar practice in Brazil which even at that point was deeply in debt.

From the date of its first loan in 1824 Brazil carried a burden of steadily mounting and sometimes crushing foreign debt. As a country with an industrially underdeveloped economy, rich in marketable agricultural products but short on capital resources Brazil was compelled to secure funds in the international money markets which could make possible national economic development and balance chronic budgetary deficits. With the foreign capital railroads were built, port facilities were expanded and improvements made to support urban development. In the nineteenth century and the first decade of the twentieth, London was the main source of Brazilian loans. French financiers appeared after 1900. American capitalists did not appear in numbers on the scene until after World War I when the United States became the world's leading creditor nation.

Early in the republican period the shortage of foreign exchange became critical, forcing Brazil in 1898 to obtain a loan in London to fund payments on its foreign debt. In addition to the funding loan Brazil continued to expand its foreign debts. While the loans did much to modernize and strengthen Brazil's economy, all of them had to be serviced. Until 1860 Brazil's borrowing was accompanied by an unfavorable balance in its foreign trade, but owing to the rise of coffee as an export commodity, this situation was reversed. From 1860 to 1930 there were only a few years that Brazil suffered an unfavorable trade balance. By 1930 total payments for amortization and interest on the Brazilian foreign debts were estimated at between 420 and 470 million pounds (Sterling), far more than the total debt incurred, however, less than one-third of the total had been amortized. Debt service in this period used up about one-half the proceeds from new foreign loans and export balances. No matter which way the Brazilian government turned in its efforts to end the chronic shortage of exchange, its policies only brought temporary relief and in the long run made these problems more difficult to solve. The foundation of Brazilian foreign trade was almost shattered in the latter months of 1929 under the double impact of the great depression and the great increase in coffee production fostered by the valorization program. In spite of improvements made, the possibility of a stable, diversified economy eluded Brazil, and in 1930, after 106 years of foreign borrowing the governments of Brazil—federal, state, and municipal—found themselves with an external debt of $394 million and without gold reserves or foreign exchange.[72]

In 1911 a minor controversy arose between the United States and Brazil which was related to the valorization of coffee. A report issued by the office of the U.S. Attorney-General asserted that the program violated Section 76 of the Wilson Tariff Act, identical with Section 6 of the Sherman Antitrust Act, which gave the United States the right to seize and condemn property imported into the United States and held in restraint of trade. It was further claimed that the United States consumed 40 percent of the world's output of coffee, or about 950 million pounds of coffee a year. Therefore, a mere rise of six cents a pound in the cost of coffee meant $57 million a year to the people of the United States; and the valorization scheme had brought about a rise in excess of six cents.[73]

The Brazilian ambassador protested the action by the Justice Department on the ground that the suit in question was tantamount to the American court's claiming jurisdiction over the acts of a foreign sovereign state.[74] After considerable popular resentment had been aroused in Brazil and a series of notes were exchanged, Washington agreed to drop the suit upon a promise of the Brazilian government that all stores of valorized coffee in New York would be sold on the open market before April 13, 1913.[75]

One of the results of the controversy was the temporary suspension on the part of Brazil of the preferential tariff on American goods. This entailed heavy losses for American exporters, and with the dropping of the suit the tariff concessions to the United States were revived. But, in order to cement the renewed friendly relations, the United States government extended an invitation to Dr. Lauro Muller, the Brazilian minister of foreign affairs, to visit the United States to repay the visit made by Secretary Root to Brazil in 1906. Dr. Muller accepted the invitation and was enthusiastically received. He remained in the United States for more than a month, visiting and traveling from New York to San Francisco. A new link of appreciation and understanding was thereby forged.

Nevertheless, the United States Tariff Commission, in a report in 1918, deprecated the preferential treatment accorded to certain imports from the

[72] Henry W. Spiegel, *The Brazilian Economy; Chronic Inflation and Sporadic Industrialization* (Philadelphia, 1949), pp. 137-138.
[73] *For. Rel. of the U.S.*, 1913, pp. 39-52.
[74] *Ibid.*, pp. 55.
[75] *Ibid.*, pp. 59-67.

United States by the Brazilian government being the only arrangement then in effect which was inconsistent with the general principle of equality of treatment.[76] In accord with a Brazilian law requiring that preference rates be granted annually, it was customary for the American ambassador to Brazil to make each year a formal request for preferential treatment. However no request was made by the United States for the year 1923, and by an exchange of identic notes between Secretary Hughes and the Brazilian ambassador at Washington dated October 18, 1923, each country agreed henceforth to accord the other unconditional most-favored-nation treatment.[77] As a matter of fact the result was by no means as adverse as was expected and even wheat flour, which enjoyed the largest reduction (30 percent), practically held its own.

At the beginning of the first World War it was difficult to envisage what might be the attitude of Brazil. The large number of German colonists in southern Brazil, estimated at anywhere from 350,000 to 500,000, might have been expected to exert considerable influence towards a pro-German policy. The Brazilian government and the Brazilian people, however, soon gave clear indications that their sympathies were overwhelmingly on the side of the Allies. Within a week after the outbreak of the war the Brazilian Chamber of Deputies passed a motion recording its opposition to the violation of treaties and to acts violating the established principles of international law.[78] Less than a year later (in March 1915), an organization was founded under the presidency of the eminent Brazilian statesman, Ruy Barbosa, known as the Brazilian League for the Allies. Its program included educational conferences, petitions of protest against Germany's war methods, and the raising of funds for the Brazilian Red Cross. It was at one of the meetings held by this organization to raise funds for the establishment of a Brazilian hospital in Paris for the French wounded, that President Barbosa expressed regret that the United States had not seized the opportunity to assure itself first place among the nations by grouping about itself all the peoples of the American continent in protest against the invasion of Belgium and entering into the struggle to protect the validity of international engagements.[79]

Germany's unrestricted submarine warfare, which soon became a deadly menace to Brazil's extensive merchant marine, and the entrance of the United States into the World War, both tended to bring Brazil into the war on the side of the Allies. On April 11, 1917, immediately following the sinking of the Brazilian steamer *Paraná* off the coast of France, diplomatic relations with Germany were severed. On May 22 President Braz urged Congress to revoke Brazil's neutrality in favor of the United States. In his speech he pointed out that "the Brazilian nation, through its legislative organ, can without warlike intentions, but with determination, adopt the attitude that one of the belligerents forms an integral part of the American continent, and that to this belligerent we are

[76] U.S. Tariff Commission, *Reciprocity and Commercial Treaties*, 1918, p. 285; Leon F. Sensabaugh, "The Coffee Trust Question in United States-Brazilian Relations, 1912-1913," *Hispanic American Historical Review*, XXV (Nov. 1946), 480-496. See also, Delgado de Carvalho, *História Deiplomática do Brasil*, pp. 367-369.
[77] *U.S. Treaty Series*, No. 672.
[78] P. A. Martin, *Latin America and the War*, League of Nations, vol. II, no. 4, p. 233.
[79] Gaston Gaillard, *Amérique Latine et Europe Occidentale* (Paris, 1918), p. 44.

bound by a traditional friendship and by a similarity of political opinion in the defense of the vital interests of America and the principles accepted by international law."[80]

The decree of neutrality was annulled on June 1, and in the circular note to foreign governments announcing the fact, the Brazilian government thus indicated its union of interests with the United States: "Brazil could not remain indifferent to it when the United States were drawn into the struggle without any interest therein but in the name alone of respect for international law, and when Germany extended indiscriminately to ourselves and other neutrals the most violent acts of war. If hitherto the relative lack of reciprocity on the part of the American Republics has withdrawn from the Monroe Doctrine its true character, permitting a scarcely well founded interpretation of the prerogatives of their sovereignty, the present events, by placing Brazil, even now, at the side of the United States, in the critical moment of the world's history, continue to give our foreign policy a practical form of continental solidarity—a policy indeed which was that of the old regime on every occasion on which any of the other friendly sister nations of the American continent were in jeopardy."[81]

The final break came on October 26, 1917, when a resolution recognizing a state of war was passed unanimously by the Senate and with only one vote against it in the lower house. Compulsory military service was reinstated, a mission went to the United States to purchase equipment and to arrange for military cooperation, a fleet of light cruisers was sent to cooperate with the British, aviators and physicians served on the western front, and every effort was made to increase the exportation of foodstuffs to the Allies.[82] From the point of view of the United States, the entrance of Brazil into the war was particularly important, even on purely moral grounds. The conclusion that the war was essentially just and necessary was enormously strengthened when the two greatest states of the western hemisphere, putting aside all feelings of rivalry and petty jealousy, decided to stand shoulder to shoulder in a struggle for that democracy which has ever been the ideal of the two Americas.

In the settlement following the war it must be conceded that Brazil was more steadfast in supporting the ideals for which she fought than was the United States. Brazil accepted the noble purposes of the League of Nations as worthy of at least a fair trial, and was deservedly honored by being elected to membership in the League Council. In his address to Secretary Colby on the occasion of the latter's visit to Rio de Janeiro in December 1920, President Pessoa expressed regret that the United States had failed to ratify the treaty of Versailles. "Brazil," he declared, "is naturally very much interested in the beneficial purpose of the League of Nations. Therefore it is a matter of regret that the United States, which took the lead in that great project, has not retained it."[83] When the votes were cast for the judges of the Permanent Court of International Justice, Brazil showed her continued confidence in the United States by

[80] *Brazilian Green Book*, authorized English version (London, 1918), p. 40. For expression of opinion in Brazil see Gaillard, *op. cit.*, pp. 70-90.
[81] *Brazilian Green Book*, p. 49.
[82] Martin, *Latin America and the War*, p. 243; Hélio Vianna, *História Diplomática do Brasil* (Rio de Janeiro, 1958), pp. 189-194.
[83] *The Independent*, vol. CV (January 8, 1921), p. 49.

choosing as one of her candidates the American statesman whose Pan-American policy received the hearty support of all Latin America, Mr. Elihu Root. When the court was finally constituted, Brazil found herself again honored by the election to this body of her eminent statesman and publicist, Ruy Barbosa.

As further evidence of Brazil's friendliness toward the United States an arrangement was concluded between the two countries on November 6, 1922, for the sending of a commission of sixteen naval officers and nineteen noncommissioned officers for a period of four years to reorganize the Brazilian navy. Commander (later Admiral) Vogelgesang of the Brooklyn Navy Yard headed the commission. Dr. Zeballos, former secretary of foreign relations of Argentina, gave voice to the resentment caused in his country by this act and claimed that it interfered with any program for the limitation of armament between the ABC powers. However, according to A. T. Beauregard, a member of the mission who returned with Admiral Vogelgesang in February 1925, the mission's purpose was not to persuade Brazil to enlarge her navy but merely to bring it up to the highest standard of efficiency.[84] In July 1926, the State Department announced that the contract had been renewed for another period of four years as from November 6, 1926.[85]

In 1930 as a measure of economy the mission was permitted to return home, but in 1932 a new contract was signed for a smaller mission and this arrangement was renewed May 27, 1936, for another four years. As then organized the mission consisted of eight officers and five chief petty officers whose duties were to cooperate with the minister of marine and officers of the Brazilian navy in an advisory capacity.[86] A similar agreement was signed November 13, 1936, between the two governments for a small military mission to cooperate with the general staff and assist in courses given at coast artillery instruction centers.[87]

As a result of this cooperation it was only natural that when Brazil wished to secure temporarily several United States destroyers already out of commission, for training purposes, the United States was glad to oblige. But although Brazil's neighbors did not object to the borrowing of United States naval officers for instruction purposes, they opposed very vehemently the borrowing of naval vessels. So vocal were the protests that on August 20, 1937, the United States and Brazil issued a joint statement declaring that the proposed plan was in entire harmony with the policy welcomed in many previous instances by the governments of other American republics of lending officers to them for instruction purposes or of receiving their officers for training in the naval vessels of the United States in American waters. They regretted "that a question of such limited importance should even for a few days be allowed to divert attention from the high ideals and ... program which the 'good neighbor' policy comprises."[88]

After World War I there was a steady increase in trade between the United States and Brazil. The latter granted a 20 to 30 percent preferential treatment on

[84] *Current History*, vol. 22 (August 1925), p. 815.
[85] *United States Daily*, July 16, 1926.
[86] For terms of a later agreement signed May 7, 1942, see *Executive Agreement Series*, 247.
[87] In 1941 a new four-year agreement covering a military and military aviation mission replaced the previous military agreement, *Executive Agreement Series*, 202.
[88] U.S. Dept of State, *Press Releases* (August 21, 1937), p. 162.

certain items purchased from the United States and after 1922, as mentioned, the unconditional most-favored-nation principle prevailed. At the same time United States investment in Brazil rose from $50 million in 1913 to $624 million in 1932. Financial ties were expanded in the early 1920s resulting in Americal lenders holding $342.4 million, about one-third of Brazil's foreign debt, by 1927. In the economic depression that followed United States investments in Brazilian bonds, sold to the American public by banks, were defaulted.[89]

Dr. Getulio Vargas seized control of the Brazilian government with military support in October 1930 and established a regime of extraordinary longevity. It survived the transition from an unconstitutional provisional government (1930-1934), to a constitutional republic (1934-1937), to a modified totalitarian state in the *Estado Novo* (1937-1945). It resisted threats of violent overthrow including the revolt of São Paulo state in 1932, a Communist conspiracy in 1935, and an attempted *coup d'état* by the fascistic Integralista Party in 1938. Meanwhile, it survived an almost incessant economic crisis.

The span of the Vargas government was almost paralleled by the remarkable longevity of the Franklin D. Roosevelt administration in the United States. Key figures were retained in both governments and this helped to establish a continuity in the development of policy. Cordell Hull served as secretary of state from 1933 until 1944. Sumner Welles, perhaps the most influential figure in the formation of Latin American policy, served ten years as assistant and undersecretary of state in the years 1933-1943. Jefferson Caffery, who exerted a strong influence on Washington's Brazilian policy, served as assistant secretary of state, 1933-1934, ambassador to Cuba, 1934-1937, and ambassador to Brazil, 1937-1944. Oswaldo Aranha, the second most important civilian in Brazil during these years, served as ambassador to the United States, 1934-1938, and foreign minister, 1938-1945.

When Dr. Vargas took over the government of Brazil as dictator in 1930 it was regarded as merely a revolutionary *coup d'état* to prevent São Paulo from monopolizing the presidency. The United States, insufficiently informed as to political developments, at the urgent solicitation of the Brazilian ambassador in Washington, allowed itself to be persuaded to place an embargo upon shipments of arms to the revolutionists while still permitting the sale of war supplies to the federal government. When, two days later, the revolution was successful, the Department of State realized that it had made a serious diplomatic *faux pas* and tried to remedy the blunder by a prompt recognition of the new government. The bloody revolution of 1932 and the huge casualty list in its suppression indicated serious internal trouble in the federal republic. Undoubtedly the drop in the price of coffee from 24.8 cents a pound in March 1929, to 7.6 in October 1931, had something to do with the domestic situation. Even the new liberal constitution of 1934 could not solve the problem of the economic losses brought about by the low selling price of coffee. Nor did the destruction of some 26 million bags of coffee from January 1931, to January 1934, remedy the

[89] Max Winkler, *Investment of United States Capital in Latin America* (Boston, 1929), pp. 86-87, 247-278.

situation.[90] A Communist-led uprising broke out in 1935 only to be crushed by the government.

Inasmuch as the constitution of 1934 forbade the immediate re-election of the president, it was evident that a new constitution was required to permit Vargas to continue to head the government. On November 10, 1937, he assumed dictatorial powers and promulgated a new constitution establishing a centralized corporative state which dissolved all existing legislative bodies, federal and state. The new constitution increased the presidential term to six years and declared the president to be the supreme authority of the state. Although this new government seemed to threaten a fascist form of government in the Western Hemisphere, Sumner Welles, United States undersecretary of state, in an address made December 6 declared that the traditional friendship between the people of Brazil and the United States was not impaired by misinterpretations placed upon the Vargas *coup d'état.* He recalled that it had been unanimously agreed at the Buenos Aires Conference that no state should interfere with the internal affairs of another state.

Vargas ruled thereafter as a dictator with opposition thoroughly stifled. An attempted palace revolution by the Fascist Integralistas on May 11, 1938 failed, although Vargas for hours was aided only by his personal servants and a few guards. The European War, which finally spread to the Western Hemisphere, afforded a good excuse for government by presidential decree; Vargas seized the opportunity. His government, the *Estado Novo*, claimed to be an "authoritative democracy," but with emphasis clearly upon the first word in the phrase.

In United States-Brazilian relations in the early 1930s the chief economic problems resulted from the worldwide economic collapse. For Brazil the great depression meant that the two major props of its fragile economy were removed. As noted earlier, the economic health of Brazil since its independence was based on a favorable balance of its merchandise trade, owed mainly to the export of one crop, which enabled the government to float loans to meet its usual budgetary deficits. The situation had become so serious in the late 1920s that the nation might have faced an economic crisis even without the depression. Foreign loans contracted by the Washington Luis administration were adequate to cover temporarily the severe decline in the price of coffee which appeared before the stock market crash in the United States. However, because there was in 1931 the largest favorable balance of trade to be registered in the inter-World War period, Brazil's foreign exchange crisis was delayed until the following year.

By 1932, years of over-borrowing combined with the tight money policies brought on by the depression, had seriously depleted Brazil's credit reserves. The value of its exports was reduced to less then one-third of normal, while the drop in the cost of imports was much less. Under the circumstances Brazil could no longer provide the foreign exchange needed to pay its foreign obligations. This condition immediately raised for American interests in Brazil the double problem of obtaining payment of commercial obligations, and obtaining payments due to the holders of Brazilian government bonds.

Negotiations concerning these payments represented the main activity in diplomatic relations between the two nations in the early 1930s. In carrying on

[90] Horace B. Davis, "Brazil's Political and Economic Problems," *Foreign Policy Reports*, vol. XI, no. 1 (March 13, 1935).

relations with Brazil, Secretary Hull used the good offices of American diplomats to ensure equality of treatment of American business interests and to promote liberal trade policies. In pursuing this more vigorous policy the State Department respected the limitations and problems of Brazil and rejected the use of diplomatic intervention or force to collect debts or to enforce contractual obligations.

It was to be expected that with the passage of the trade agreement act of 1934, the United States would endeavor to improve its trade relations with its best South American customer. In fact, the trade agreement between the two nations, which became effective January 1, 1936, was the first signed with a South American state. Under its terms the United States agreed to cut tariff duties on such Brazilian products as manganese, Brazil nuts and castor beans, as well as keep on the free list coffee and cocoa, which were Brazil's principal exports to the United States. In return Brazil lowered the tariff on many American exports. The agreement provided for most-favored-nation treatment and prohibited all import quotas or licenses except for sanitary purposes.

Although the results were not remarkable the figures showed a substantial improvement in both imports and exports. Taking the 1934-1935 average, that is the two years preceding the agreement, and contrasting it with the 1938-1939 average, that is two years after the agreement, we find that United States exports to Brazil increased 69.5 percent and our imports from Brazil 7.3 percent. The total trade in the same period rose from $138 million to $174 million.[91] With the outbreak of World War II both the percentage and amount of trade increased materially. In 1940, Brazil did almost 50 percent of its total trade with the United States, and it was valued at $273,472,000.

The political objectives of the two nations on matters of common interest generally coincided throughout the 1930s. Two basic objectives of the Roosevelt administration were, first, the strengthening of the machinery for the maintenance of peace within the Western Hemisphere through the inter-American system, and, second, the defense of the region. Brazil cooperated in pursuing these objectives while making a consistent effort to coordinate with the United States its response to events in other parts of the world.

At the Seventh International Conference of American states at Montevideo in November, 1933, the seeming *rapprochement* between the United States and Argentina caused the Brazilian delegation some uneasiness but without lasting effect. Secretary Hull remarks in his memoirs that the price paid for the success of the conference was the creation of suspicion among Brazilian leaders, that one of his "most delicate tasks during 1934 was to remove the disquietude felt by Brazil."[92] In 1936 President Roosevelt, after visiting Rio de Janeiro enroute to the Buenos Aires Conference of that year, wrote to his wife that "there was real enthusiasm in the streets. I really began to think the moral effect of the Good Neighbor Policy is making itself definitely felt."[93] At the Buenos Aires Conference both Mr. Hull and Mr. Welles lauded the role which the Brazilian

[91] *Commerce Reports*, Feb. 17, 1940.
[92] Cordell Hull, *The Memoirs of Cordell Hull*, vol. I (New York, 1948), p. 349.
[93] Franklin D. Roosevelt, *Franklin D. Roosevelt: His Personal Letters, 1928-1945*, vol. III, ed., Elliott Roosevelt (New York, 1947), p. 634.

delegation played in securing inter-American cooperation in matters of peace and defense. This was particularly reassuring to the United States because of events beginning in the last half of 1934: the menacing actions of Japan on the Asian mainland, the German trade offensive in Latin America, and the denunciation of the military limitations of the Treaty of Versailles, and the Italo-Ethiopian war.

The basic premises of both Brazilian and United States foreign policies between 1930 and 1939 remained complementary to each other thus strengthening important areas of common interest and agreement. This did not mean, however, that the weaker of the two powers would shrink from exploiting opportunities to further its own national ends. In only two areas was there a significant potential for serious disagreement, the first being Brazil's economic difficulties from which, had the United States exerted pressure to force settlements in its own interest, much ill will could have resulted. But Washington did not adopt such a policy. The second area of potential threat were problems relating to the rising power of Nazi Germany. The Brazilian government, apparently considering inadequate the American response to the German menace, in the early months of 1939, following the Eighth Inter-American Conference of December 1938, decided that the collective defense measures prevailing in the Western Hemisphere did not ensure Brazilian security. As a gesture towards Germany, Brazil took the initiative in restoring diplomatic relations between the two countries.

Brazil at this time was in a strong bargaining position with the United States. It pressed its advantage to obtain a high price for its cooperation. Although it was realized that Brazil's dependence on American markets gave Washington the upper hand, Brazilian leaders were confident that only under extreme provocation would this lever be used. To Brazil's advantage was the fact that it became more valuable to the United States as a counterpoise to the obstructionist tactics of Argentina in hemisphere affairs. Moreover, both the defense of the Brazilian "hump" and the threat of German political, economic and cultural penetration became matters of growing concern to Washington. Evidence of Brazil's confidence in its bargaining position was its reactions to the commitments which Foreign Minister Aranha made in Washington in February-March 1939, and the firm stand of the Brazilian military that cooperation in defense of the "hump" was contingent on Washington's willingness to furnish additional equipment for the armed forces. The Brazilian move to restore normal relations with Germany was also an expression of the new independence. This step had the dual advantage of reinforcing Brazil's bargaining position with the United States as well as hedging against the event that Nazi Germany emerged as the dominant world power. However, as the decade neared its end the Washington-Rio de Janeiro axis was reaffirmed, and the need for Brazilian cooperation led the United States to adopt a policy of offering Brazil financial assistance designed to strengthen its economy and to provide material for defense.

Early in 1939 Foreign Minister Aranha came to Washington where an elaborate program for closer economic collaboration with the United States was worked out. A $19,200,000 credit extended by the Export-Import Bank inaugurated American involvement in Brazilian economic development by the new and unorthodox measure of United States government aid for programs of state-directed industrialization. The program also contemplated a survey of tropical agricultural possibilities by American experts, the development of certain basic industries, and the improvement of transportation facilities. The following year Warren L. Pierson, president of the Export-Import Bank, after a survey of conditions in Brazil, recommended the construction of a modern steel-producing plant in Brazil, the machinery for which could be supplied by the United States. This steel plant, to be located ninety miles south of Rio de Janeiro, was expected within two years to be producing more than half of Brazil's existing requirements for steel. A loan of $20 million was made for this purpose, Brazil to supply another $25 million. Washington's willingness to support the project was doubtless increased by the knowledge that President Vargas had actively negotiated with Nazi Germany for assistance in setting up a steel industry.[94]

An even more elaborate economic arrangement was made with Brazil on March 3, 1942. It consisted of a series of agreements providing for a complete mobilization of the productive resources of Brazil with credits to the amount of $100 million to be made available by the Export-Import Bank. Through agreements signed with the Brazilian finance minister and the British ambassador, the Export-Import Bank and Metal Reserve Company would finance the development of the Itabira iron mines and the Victoria Minas Railroad in order to obtain high grade ores for the United States and Great Britain. Other agreements provided for the transfer of military material under lend-lease arrangements, the development of the production of raw rubber and its purchase, and for the purchase of such commodities as barbassu and castor oil, cocoa, coffee, and Brazil nuts.[95] In the fall of 1942, a technical commission headed by an American industrial engineer, Morris Llewellyn Cook, was sent to Brazil to work out the specific details of the program.

A more general economic arrangement looking towards continental solidarity but of particular interest to Brazil was the Coffee Marketing Agreement signed November 28, 1940, and ratified February 12, 1941. By the terms of this agreement fourteen American republics allocated equitably the market of the United States and that of the rest of the world among the various coffee-producing countries through the adoption of basic annual quotas for each country. According to the quotas established, the United States would take 9,300,000 bags of coffee from Brazil, 3,150,000 from Colombia, and 1,550,000 bags from Central American republics.[96]

After the United States entered the war, Brazil gave the most wholehearted support to the cause of the United Nations. Arrangements had already been made with the United States by the agreement signed October 1, 1941, to improve Brazilian defenses through lend-lease shipments of planes, tanks, and trucks with additional funds to improve her air and naval bases. Inasmuch as the third meeting of American Foreign ministers to formulate plans for continental

[94] S. S. Jones and D. P. Meyers, *Documents on American Foreign Relations* (Boston, 1939), pp. 128-142; Thomas E. Skidmore, *Politics in Brazil, 1930-1960: An Experiment in Democracy* (New York, 1969), pp. 44-45.
[95] Dept. of State *Bulletin*, VI (Mar. 7, 1942), pp. 205-208.
[96] *Ibid.*, vol. III (Nov. 30, 1940), p. 483-488.

defense was held at Rio de Janeiro, Foreign Minister Aranha was in a strategic position to support the United States in its desire for an all-American severance of diplomatic relations with the Axis powers. The Brazilian Foreign Minister measured up to expectations and not only supported the break enthusiastically, but when Argentina and Chile hedged upon immediate action. Brazil alone of the ABC powers took the step before the Conference adjourncd.

Axis firms and individuals were placed under official surveillance and control, pro-Axis news agencies and newspapers were suppressed, and Axis financial transactions were heavily restricted. Following the sinking of Brazilian ships by submarines, the Brazilian government seized up to 30 percent of the assets of Axis enterprises to guarantee compensation. Nowhere in South America did the Fifth Column receive more ruthless treatment than at the hands of Getulio Vargas. When news was received August 18, 1942, that five more Brazilian ships had been sunk within three days—one a troop transport—Brazil, temporizing no longer, declared war the same day. Brazil's vast stores of strategic materials, her rapidly expanding air force,and the joint defense of the strategic bulge at Natal were invaluable assets in the defense of the Americas. Upon his return from Africa in January 1943, President Roosevelt stopped off in Brazil to see President Vargas, and both agreed that never again should the coasts of Dakar and West Africa become an invasion threat to the Americas. Brazil promised to aid in the war against the undersea menace, and it was later reported that at least ten Axis submarines had been sunk by the joint action of the air and sea forces of Brazil and the United States.[97]

The huge airplane base constructed by the United States at Natal, which during the war was said to be the largest freight air junction in the world, was a valuable asset in the transportation of both troops and supplies to the various fronts in Europe and Africa. In fact, when Germany surrendered in 1945, the United States War Production Board declared that "without Brazil's production of strategic materials and bridge of planes the United States could not have met its schedules."[98]

It was generally thought in the United States that Brazil's great friendliness to the United States was due more to Foreign Minister Aranha than to President Vargas. Nevertheless, when Aranha resigned in August 1944, through indirect pressure on the part of the president, the Brazilian government continued its cooperation with the United States, and on June 6, 1945, declared war upon Japan.

President Vargas promised free elections in 1945 but, lest he might not be able to resist the pressure of his constituents that he seize the power unconstitutionally, a bloodless revolution occurred October 30, 1945, and President Vargas was removed from his position. The elections on December 2, 1945, held under army control were fairly conducted and contrary to the belief of most of the foreign correspondents, General Eurico Dutra, the reactionary candidate favored by Vargas, won over the more liberal General Eduardo Gomes. Polling more than a half million votes, the Communists won fourteeen seats in the Chamber of Deputies and elected their leader, Luis Carlos Prestes, to the Senate. Another

[97] Vianna, *História Diplomática*, pp. 195-202.
[98] Frank McCann, Jr., "Aviation Diplomacy: United States and Brazil, 1939-1941," *Inter-American Economic Affairs*, 21 (Spring, 1968), pp. 35-50.

interesting result was the return of Getulio Vargas as senator representing his native state, Rio Grande do Sul.

As might have been expected, one of the first tasks of the new government was to provide a new constitution in place of the organic act devised by Vargas in 1937. Congress itself acted as a constitutional convention and the constitution of 1946 placed more power in the hands of the Congress at the expense of the executive and returned more powers to the states at the expense of the federal government.

The new regime was disturbed by the increase of the Communist vote in the elections. Their leader, Senator Prestes, had declared publicly that in the event of war between the United States and the Soviet Union, Brazilian Communists would fight on the side of the latter. Shortly after the elections of January 1947, the government requested the suppression of the Communist Party, and in May the Supreme Electoral Tribunal declared it to be illegal and ordered its dissolution. Soviet-Brazilian diplomatic relations were broken off in October. With the expulsion of Communist congressmen in January 1948, the Communists in Brazil were driven underground. Lending emphasis to these actions, Dr. Mauricio Nabuco, Brazilian Ambassador to the United States, declared: "Brazil is opposed to communism and will continue to stand against it with the United States."[99]

Brazil had enjoyed comparative prosperity during the war, marketing its exports at high prices, and receiving additionally $361 million in Lend-Lease funds from the United States. Overspending and an unfavorable trade balance had seriously reduced the nation's foreign exchange reserves by 1948, making it difficult for the country to meet its foreign debt obligations and precipitating a continuing dollar-shortage crisis. To improve the situation a United States-Brazilian Technical Commission was established in 1948 "to analyze the factors in Brazil which are tending to promote or retard the economic development of Brazil." The American delegation was headed by John Abbink, President of McGraw-Hill Publishing Company, and the Brazilian delegation by Octavio Gourea de Bulhoes. The Abbink Mission, after careful investigation, recommended a sweeping program of agricultural, industrial, mineral and power development, increased immigration, improved transportation and a guarantee of fair treatment for United States investors in Brazil. President Dutra met with President Truman in Washington in May 1949, and the two heads of state agreed on a mutual program for economic development and social progress. Their statements stressed the historic and lasting friendship between the two countries. In October 1950 the two countries signed the first cultural agreement of its kind, encouraging the exchange of students and professors.

In 1950 in a successful campaign to regain the presidency, Getulio Vargas charged Spruille Braden, a former assistant secretary of state and ambassador to Argentina, and Adolph A. Berle, Jr., former ambassador to Brazil, with provoking the military movement which forced him to resign in 1945. The charges were denied by the Brazilian military officers who had been involved in the "October Revolution," as well as by Braden and Berle, who were convinced

[99] *New York Times*, May 18, 1948. Dr. Mauricio Nabuco, son of Joaquim Nabuco, was ambassador to the United States, 1948-1951.

that General Vargas was seeking to win Communist support by launching anti-imperialist and anti-United States arguments in an election year.[100] Vargas proved his popularity by winning a substantial majority over his chief opponent, Eduardo Gômez, and was inaugurated president January 31, 1951. Nelson Rockefeller headed the American delegation to attend the ceremony, and at a pre-inauguration press conference Vargas said that the "traditional and permanent" friendship between Brazil and the United States was "in no way affected" by the Braden-Berle episode.[101]

Brazil's foreign policy has been defined as one that follows the United States in world affairs but expects the United States to support Brazil in Latin American affairs. It was this traditional pattern that many influential Brazilians felt was being weakened by United States neglect as the decade of the 1950s began. They believed that their nation's contributions to World War II were not appreciated, and they resented the fact that the Marshall Plan was directed exclusively toward Europe. A United States loan of $125 million to Argentina was cited as an example of rewarding aggressive unfriendliness and neglecting friendship. The loan had raised Argentina's prestige and harmed Brazil's, it was contended, and thereby jeopardized a delicate balance in South America where Uruguay, Paraguay and Bolivia feared latent Argentine expansionism. Brazil's big businessmen declared that the North Americans wanted to dominate their country's economy and to keep it agricultural. To these views of friendly critics were the hostile overtones of Communists, fellow travelers and nationalists.

None of these allegations could withstand close analysis. Brazil had received more loans from the Export-Import Bank than any country in South America, and the USIS operation in Brazil was the largest in the area. There was no evidence whatsoever that either public or private interests sought to dominate the economy. But prompted by these complaints and the report of the Abbink Mission, the Brazil-United States Joint Commission for Economic Development was set up in 1950. A cooperative venture, the Commission, located in Rio de Janeiro, was to analyze the development needs of Brazil and to recommend action.[102] Loans were soon forthcoming from the Export-Import Bank and the World Bank as a result of the Commission's work, but they fell short of Brazil's expectations.

Prompted by the gravity of the Korean War and Communist propaganda charging that the United States was an "imperialistic colossus," President Truman addressed the Fourth Consultative Meeting of the Ministers of Foreign Affairs of the American Republics in Washington in March 1951. He called upon Latin America to join the United States in a battle for freedom in both Europe and Asia. João Neves da Fontoura, Brazil's ambassador, replying for the Latin American delegations, agreed to the need for "mobilization for peace," but declared that Latin American countries must have a better and more lasting plan of mutual cooperation than existed with the United States in World War II. Brazil's position was expressed more clearly later that year when the ambassador indicated that his government would support the imposition of economic sanctions against Communist China by the United Nations General Assembly;

[100] *Ibid.*, Jan. 13, 1950.
[101] *Ibid.*, Feb. 4, 1951.
[102] Dept. of State *Bulletin*, XXIV (Jan. 1, 1951), pp. 25-26.

that Brazil agreed with Washington on recognition of Nationalist China, and opposition to the admission of the Peoples Republic of China to the United Nations. Emphasizing that under President Vargas Brazil would continue her solidarity with the United States, he said that this would have to be on the basis of "reciprocal cooperation" in providing help with the development of Brazil.[103]

In an effort to marshal hemispheric support in the cold war and to disprove charges of neglect of Latin America in favor of Europe, Secretary of State Dean Acheson visited Brazil and other South American republics in mid 1952. At Rio de Janeiro he reminded the Latin American republics that Soviet communism was the common enemy and that the United States was spearheading the defense of the hemisphere. Speaking directly to his hosts he said: "The United States wants to help Brazil in every possible way in its efforts towards economic progress, and it is to the mutual interest that each member of the friendship should be as strong as possible."

At São Paulo the Secretary declared that the life of the common man must be progressively improved: ". . . the achievement of that life is one of the bases for our technical cooperation program commonly called Point Four. Cooperation is and must be the watchword of our democratic world if it is to survive . . ."[104]

He pointed out that bilateral military assistance agreements were being concluded with Latin American states in accordance with which the United States would grant substantial military aid for mutual defense. Brazil had been the first state to sign such an agreement, and already five others had entered into force with Latin American governments: Ecuador, Chile, Peru, Cuba, and Colombia. Combining appropriations for the years 1951 and 1952, the United States had earmarked almost $90 million for military assistance to the Latin American area.[105] The military assistance agreement was not ratified by Brazil until May 1953, owing to pressures exerted by Communists, leftists and ultranationalists, who alleged that it was a Washington-directed plot to coerce Brazil into sending troops to Korea.

Meanwhile, the Point Four program in Brazil had developed into the largest in Latin America. Based upon an investment program, it aimed to provide improvement of railroads and ports, expansion of electric power production, equipment for maintaining and improving highways and agricultural machinery. The Joint Brazil-United States Commission had approved twenty two projects calling for a total expenditure of $232,200,000 and of *cruzeiros* worth $350,000,000. In other fields U.S. technicians were working with Brazilians to improve rubber production, farming methods generally, the management of industrial teacher training schools and teacher training.[106]

This support notwithstanding, Brazil remained critical of Washington's policies as the Truman administration drew to a close. This was accented by the fact that for Brazil, 1952 was a year of difficulties in which the dollar shortage

[103] *New York Times*, May 2, 1951.
[104] Dept. of State *Bulletin*, XXVII (July 14, 1952), p. 48; *Ibid*. (July 21, 1952), p. 90.
[105] *Military Assistance to Latin America*, Dept. of State Pub. 4917, Inter-Americas Series 44, (Jan. 1953).
[106] *New York Times*, Jan. 12, 1953.

was an outstanding factor, though not the basic cause. The chief cause was that all of Brazil's products, except coffee, were priced so high (partly because of low productivity and partly because of an artificial exchange rate) that they could not be sold in competitive world markets. Inflation continued to mount in spite of government efforts to halt it and public protests that occasionally erupted into rioting. Moreover, a continuing drought in the northeast drove many thousands southward fleeing starvation. These conditions did not discourage United States private investments in Brazil, which reached a record total of $120 million in 1952 and brought the total American investments in Brazil to over $1 billion.

Aware of the growing rancour in United States-Brazilian relations, General Dwight D. Eisenhower and Governor Adlai Stevenson, campaigning for the presidency in 1952, had sent statements to a Brazilian newspaper chain declaring that whoever won the election, the United States would continue to uphold the Good Neighbor policy. The Department of State at the same time announced that loans to Brazil in the preceding ten years had totaled more than $410 million. It was pointed out that the Communist threat had forced the United States to concentrate on Europe, but that "our cooperative programs in the hemisphere are being carried out more intensively that at any time in our history." As proof of its words the United States granted a $300 million loan to Brazil in February 1953, the largest loan ever made by the Export-Import Bank to a foreign government. The credit was extended to support the *cruzeiro* on a free-exchange basis and to unfreeze blocked accounts of United States exporters.[107]

The Eisenhower administration came into office with this evidence of good faith towards Brazil and proceeded to implement the Good Neighbor policy as it envisaged the concept though it was never to the complete satisfaction of Brazil. The administration of the aid program was shifted as the Joint Brazilian-United States Development Commission was phased out on July 31, 1953. At termination, the Commission had studied and approved projects involving loans of $380 million from the Export-Import Bank and the International Bank for Reconstruction and Development for vital economic development in Brazil. A Brazilian official said of its accomplishments: ". . . for the first time in our history we have an exact and documented blueprint of what needs to be done . . ."

Also in 1953 President Eisenhower sent his brother, Dr. Milton Eisenhower, on a goodwill fact-finding tour of Latin America; Brazil was on the last leg of his journey. Despite an address at Rio de Janeiro in which Dr. Eisenhower spoke of the unique tradition of friendship enjoyed by the two countries, the general Brazilian attitude toward his visit was one of "wait and see." The Brazilian press, however, showed great interest in a rival development: the improvement of U.S.-Argentine relations. The *Carioca Diario* spoke of the alternative presented by Perón: "Either give us dollars or we will play the Russian's game." The *Correio da Manha*, usually not unfriendly to the United States, attacked the Eisenhower administration alleging that it was reneging on

[107] Dept. of State *Bulletin*, XXVII (Mar. 23, 1953), p. 42.

promises of loans made in President Truman's time. Communist newspapers attacked "Milton and Ike" as "assassins of the Rosenbergs," and "gangsters."[108]

Brazil's trade with the United States had in the meantime passed the one billion dollar mark with imports into Brazil exceeding exports by some $80 million. The unfavorable balance of trade kept mounting and when in the first six months of 1952 Brazil's imports from the United States were valued at over $500 million whereas her exports to the United States were under $400 million, the government imposed severe restrictions on imports. The measure was so effective that in the second half of 1952 Brazil's exports to the United States were valued at about $443 million, and imports from the United States dropped to $243 million. The net result was a balance of over $50 million in Brazil's favor in the total trade with the United States for 1952 valued at over $1,600,000.[109] The postwar spending spree was halted by this policy but it was clear that, as in the past, Brazil was not living within its means, and its problems were compounded as it turned from one unsound financial remedy to another. Its main problems, which were to reach through the 1960s, were continuing deficits, intermittent and then persistent overproduction of coffee, its principal export crop, a steady depreciation in the value of its currency, and frequent foreign loans to pay outstanding debts.

In 1953 Brazil, as the largest Latin American borrower from the World Bank, received four loans totaling $32,800,000, which brought the organization's investment in Brazil up to $175,300,000. This and continuing aid from U.S.-backed agencies, including the United Nations, failed to halt the worsening economic conditions. When, early in 1954, the retail price of coffee in the United States passed $1.00 a pound, consumers were aroused and charges were made that the high cost was due to Brazilian speculators. When the United States government appointed two committees to investigate, the Brazilian government became alarmed and invited a group of American representatives of both Congress and the public to visit Brazil and investigate the situation. Their report indicated that a heavy frost had cut the coffee crop substantially, but also that increased consumption in the United States and Europe had used up coffee surpluses, thereby outstripping production. By July 1954, coffee prices had commenced to fall as a result of reduced consumption.

Although grievances had arisen from problems of aid and trade, Brazil stood firm behind the United States at the Tenth Inter-American Conference, held in Caracas in March 1954, to deal with communism in the Americas. The Brazilian delegation at that time gave the United States invaluable help in getting general support for the resolution condemning Communist infiltration in this hemisphere. They also helped to avoid a general deadlock by having delicate economic problems deferred for a special conference held later in the year. The Washington administration was duly grateful.[110]

In August 1954, President Vargas was forced to resign by military leaders, and on the same night he committed suicide. A suicide note alleged that his overthrow had been engineered by international financial and reactionary

[108] *New York Times*, July 23, 1953.
[109] *The Foreign Trade of Latin America Since 1913* (Washington, Pan American Union, 1952); *Brazilian Bulletin* (Oct. 1, 1953), pp. 1-2.
[110] *New York Times*, Sept. 9, 1954.

elements. Anti-United States riots, incited in some cases by Communists, followed his death, and several United States government buildings and offices were damaged. But the *coup* reflected growing dissatisfaction with corruption and maladministration in the government and a growing concern with the president's demagogic appeals to the nation's workers. Unable to obtain majority backing in Congress, he was also unable to direct an economic program that could deal with the country's increasing inflation and balance of payments problems. At the same time, though long counted a friend of the United States in foreign affairs, Vargas exploited anti-Yankee sentiments in his efforts to keep support of the proletariat.

Vice President João Cafe Filho, the successor of Vargas, made it clear from the beginning of his administration that he wished to strengthen U.S.-Brazilian ties. On its part the United States government, at the Inter-American Conference of Finance Ministers which met at Petropolis, Brazil, in November 1954, pledged increasing financial support to Brazil and to Latin America generally. Washington made two major concessions to the Latin American nations: greatly liberalized loans through the Export-Import Bank and a formula to relax double taxation on foreign investments.

The Eisenhower administration however, sought to rely more on private investment and less on public assistance than did the Truman administration. This fed the arguments of the ultranationalists who contended that moderation in economic policy was ineffectual. On the other hand American businessmen and bankers had resented Vargas's support of the bill establishing Petrobras, the government-owned oil monopoly, which made clear the differing attitudes of the two governments on economic development. American indignation further strengthened the hand of ultranationalists.[111]

The João Cafe Filho government inherited a legacy of debt and inflation when it succeeded that of President Vargas. In addition, the Vargas administration left its successor with the necessity of liquidating one more ill-advised effort by the nation to hold an umbrella over the price of coffee whose exports fell in 1954 to less than eleven million bags from 15.6 million the year before. The Eisenhower administration granted temporary relief in the form of a $75 million loan by the Export-Import Bank. At the same time Luis Carlos Prestes, the Communist leader, in hiding since 1947, issued a manifesto laying all the blame for Brazil's ills on the United States. Senator Lourival Fontes, long a right hand man of the late Vargas, was also bitterly critical of the United States and advocated the resumption of relations with the Soviet Union and its satellites.[112]

The Vargas men came back with winning candidates in 1955. Juscelino Kubitschek de Oliveira, Governor of Minas Gerais and former minister of labor, and João Goulart, formerly a protegé of Vargas from Rio Grande do Sul, were installed as president and vice-president for the term 1955-1960. Though the military forces were again concerned about the victory of a group of civilian politicians whom they did not completely trust, they did nothing to block the new administration. Vice-President Nixon, who headed the American delegation

[111] Skidmore, *Politics in Brazil*, pp. 117-118.
[112] *New York Times*, Mar. 23, 1955; *Ibid.*, Apr. 24, 1955.

to the inauguration of President Kubitschek, pointed out that the first major development loan made in Latin America by the Export-Import Bank was for Brazil's Volta Redonda steel mill. The first loan, in 1941, totaled $45 million; a second loan of $25 million was made in 1950 to expand plant capacity, and a third loan of $35 million has just been approved for the same purpose. Mr. Nixon also disclosed that in the preceding fifteen years the United States had contributed to Brazil developmental capital amounting to almost $2 billion. About $1 billion had come from private investors; and a total of $900 million had been in the form of loans through the Export-Import Bank. The Vice-President affirmed that the United States must continue its contributions to the Brazilian economy "loyally and generously in the future."[113]

Having concluded that his development program would depend upon a large measure of support and cooperation from Washington and American private investors, President Kubitschek made every effort to sell his program in this country. On a visit in January 1956, he told the United States Senate that his country rejected tyranny and stood beside the United States in the struggle against "extremist" ideologies. Vice-President Goulart came later in the year, reaffirming his country's firm friendship for the United States. When questioned regarding Communist influence in the Labor Party he said "We have always led the fight against the Reds."

United States negotiations with Brazil over the erection of a guided missile tracking base on Fernando de Noronha Island, 125 miles off the northeast tip of Brazil, were completed in January 1957, after a long period of deadlock arising from Communist and nationalist opposition. President Eisenhower sent President Kubitschek a personal message asking for a favorable decision. Having made the decision the Brazilian president found that public opinion supported the move, and it was, after all, small recompense for Washington's favors. Since his campaign the year before, the record showed that he could count on generous American support. He had received $35 million for steel mill expansion, $25 million for railways, and Export-Import Bank credit of $151 million and a $139,900,000 sale of wheat on a three year loan, of which 85 percent was to be repaid in local currency, giving a total of nearly $350 million. President Kubitschek nonetheless called for a "new spirit" in Brazil's relations with the United States, the intent clearly being that he expected more aid in the development of his nation.[114]

In spite of Brazilian criticism of United States contributions, a report issued by the U.S. Department of Commerce in 1957 showed that the private sector of United States investment was making a substantial contribution of the nation's economy. U.S. corporations had $1,200,000,000 invested in Brazilian branches and subsidiaries, paying out $600 million for wages, taxes and supplies. The investment was larger than that in any other Latin American country except Venezuela. The report indicated that in 1955 wage and salary payments of U.S. companies totaled $85 million, and some 93,000 Brazilians were employed directly by these companies.[115]

Another area of Brazilian criticism of United States policies concerned the

[113] Dept. of State *Bulletin* XXXIV (Feb. 27, 1956), pp. 336-337.
[114] *New York Times*, Jan. 31, 1957.
[115] *Ibid.*, Sept. 11, 1957.

international coffee market and the petroleum industry: The Eisenhower government was unwilling to participate in a price-support system on coffee and to underwrite the storage of surpluses. Rather it urged the coffee producing countries to agree on limiting production and exports. On the second issue Brazil sought American loans to enable Petrobras, the government oil monopoly, to increase production. Washington's position held that oil exploitation was properly the field for private risk capital and that the United States lending agencies could not advance credit for that purpose.

Seeking to focus greater United States attention on Latin America and perhaps exploiting American concern about its relations with the area following Vice-President Nixon's riot-marred tour in South America, President Kubitschek sent a letter to President Eisenhower in May 1958, proposing a bold new program: "Operation Pan America." The plan would bring the United States and Latin America together in a long-range multilateral program of economic development; Eisenhower agreed that the two nations should consult as soon as possible. Secretary Dulles was despatched to Brazil in August and after consulting with President Kubitschek declared that "Latin America has an important role to play among the nations of the world, and that it should become more active in the formulation of free world policies." With regard to Brazil Mr. Dulles said: "We have become in many ways interdependent as regards our mutual security and the well being of our peoples."[116]

While the Eisenhower government appeared to support the Kubitschek plan, Washington failed to act upon it and the Brazilian president was led to comment that the United States was refusing to assume a major commitment to assure the rapid economic development of Latin America. The Brazilians were quick to point out also that Secretary of State Christian Herter had neglected to mention Operation Pan America in an address at the United Nations General Assembly in which he reviewed world problems. Further, exception was taken to the fact that the widely-traveled President Eisenhower had not visited Brazil during his term of office.[117]

United States-Brazilian relations had by the end of 1959 become clearly less cordial. John Moors Cabot, the United States Ambassador to Brazil, felt compelled to urge Brazilian nationalists to avoid economic measures which might hurt both countries. Early in the year an American owned power plant was seized and pressure was building up for nationalization of the meat packing industry. Exchange students from the United States reported acts of hostility by Brazilian students, and there was agitation for the resumption of trade and diplomatic relations with the Soviet bloc countries. The signing of a world coffee agreement, negotiated with United States aid, established export quotas and helped to stabilize the surplus-plagued world coffee market, but whatever gains this might foreshadow were overwhelmed by the record inflation and budgetary deficits in the Brazilian economy.

Between 1954 and 1959 the prices of Brazilian export products fell by more than 40 percent, which tended to obscure the fact of extensive United States aid in the postwar years. The major part of non-military public aid

[116] *Ibid.*, June 11, 1958; *Ibid.*, Aug. 8, 1958; Dept. of State *Bulletin* XXXIX (Aug. 25, 1958), p. 303.
[117] *New York Times*, Nov. 15, 1959.

received by Brazil had come from Washington. From July 1, 1945 to December 31, 1959, Brazil received $54 million in grants, or 0.09 percent of the total aid extended by the United States to the world. In relation to loans and other credits utilized during 1940-1959, Brazil's share accounted for 6.1 percent of the total–an impressive percentage–or about $1 billion. A close look at the composition of the loans to Brazil shows that 91 percent of them were hard loans made by the Export-Import Bank at relatively high rates or interest, running between 5¾ and 6 percent. Brazil had repaid about $450 million, plus $130 million interest. Obviously United States aid had played a significant role in Brazil's development process. However, relations grew progressively strained owing to the Eisenhower government's approach to economic assistance from an exclusively banking angle, and Brazilian resentment of American aid to African development which it was believed put Brazil at a competitive disadvantage in marketing tropical products.[118]

President Eisenhower, prompted by the disastrous turn in relations with Fidel Castro's Cuba, toured South America in early 1960 to prepare the way for launching a multilateral aid program basically similar to what President Kubitschek had proposed in Operation Pan America, and which anticipated the Alliance for Progress. In his New Year's Day address to the nation the latter foresaw an "era of mutual understanding between Brazil and the United States." The two presidents meeting in Brasilia, proclaimed on February 23 "a hemispheric crusade for economic development," and declared that the aspirations of the peoples of America for a better life was "one of the great challenges and opportunities of our time." Later, through diplomatic channels, Washington confirmed that it would indeed carry out Operation Pan America as the Brazilian president had suggested in May 1958.

In the presidential election of 1960 Janio Quadros, a reform candidate not identified with the Vargas clique and acceptable to the military, won the presidency. As Brazilian law sanctioned split tickets, it happened that the vice-president came from the Vargas camp. Thus, João Goulart, one of the politicians most distrusted by the military establishment, was inaugurated with Quadros in January 1961. The Brazilian public and military, satisfied with the election, awaited the promised reforms. But instead of governing well, Quadros, claiming that mysterious forces were responsible, resigned unexpectedly in August 1961, and fled the country.

The Quadros era, though brief, had been a period of unease for the United States. Before winning the presidential elections Dr. Quadros traveled abroad, visiting Premier Khrushchev, Marshal Tito, President Gamal Abdul Nasser, and Prime Minister Nehru. Most significantly, visiting Cuba in March 1960, he left the impression that he was sympathetic to Castro's experiment. His Cuban policy was thereafter bitterly attacked by Governor Carlos Lacerda of the state of Guanabara, and others. Quadros clarified his administration's foreign policy declaring: "We have not subscribed to treaties of the nature of NATO, and are in no way forced formally to intervene in the cold war between East and West. We are therefore in a position to follow our national in-

[118] *Cong. Rec.*, 87th Cong., 1st Sess., vol. 107, p. 2679; Mauricio Nabuco, "The Good Neighbor, A Half Century of Brazilian-American Friendship," *Atlantic*, Feb. 1956, p. 103.

clinations to act energetically in the cause of peace and the relaxation of international tension."[119] Illustrative of the "independence" of his foreign policy President Quadros ordered diplomatic relations established with the Soviet Union and advocated debate in the United Nations on the admission of Communist China to the world organization. His awarding of the Cruzreio do Sul Order medals to Cuba's "Che" Guevara and to nine members of Soviet good-will mission, accompanied by Premier Khrushchev's praise of Quadros' independence and his denunciation of colonialism and imperialism, was widely interpreted as anti-Americanism.[120] His unwillingness to carry out Washington's request for a boycott of Cuba, and his visit to the island indicated that he was disposed to follow a less direct pro-American policy than any of his predecessors since World War II.

Quadros' foreign policy was interpreted by some observers as a means of diverting domestic attention from an unpopular stabilization program. He had found the country in a financial situation referred to as a tropical nightmare. Although the Kubitschek government had given Brazil an unparalleled surge of growth and development, this expansion, emphasizing industry and sweeping projects of infra-structure, had accented the traditional basic distortions and in the end proved beyond the nation's resources. The construction of a new capital at Brasilia, an attempt to turn the nation's attention toward its underdeveloped heartland opened up new areas, but did not yield immediate gains. At the close of 1960 the country was in a state of near insolvency coupled with the highest rate of inflation in Latin America, and a budget deficit of $300 million. Washington, which had declined to lend Brazil any new money because of President Kubitschek's refusal to begin a financial stabilization program, agreed, however, to a six month moratorium on debt payments to the Export-Import Bank.[121]

The ill-fated Cuban counter-revolution which foundered at the Bay of Pigs early in the administration of President John F. Kennedy, had serious repercussions in Brazil. President Quadros' sympathies for Castro's Cuba were by then well known, and it was believed that he would take a strong position against the invasion attempt. But he neither opposed nor supported it. Fearing reactions from right or left (if he had opposed it the conservative and military elements would have called him a leftist; if he had supported it the leftist and nationalist groups would have labeled him a "tool" of Wall Street), he remained uncommitted. In order to maintain this neutral position he took refuge in the principle of self-determination. Brazilian public reaction, reflecting a dislike of dictatorship and political violence, turned anti-Castro and anti-Communist, but not necessarily pro-United States. The strongest negative response came from Brazil's northeast where the peasant masses, led by deputy Francisco Julião, remained strongly pro-Castro.[122]

In response to charges emanating from Cuba as well from Brazilian leftist sources that the United States had conspired to bring about the resignation of Quadros, Senator Hubert Humphrey declared "the record ought to make it quite clear that the relationships between the Quadros administration and our

[119] Janio Quadros, "Brazil's New Foreign Policy," *Foreign Affairs*, 40 (Oct. 1961), p. 26.
[120] *Cong. Rec.*, 87th Cong., 1st Sess., vol. 107, p. 15027.
[121] *New York Times*, Jan. 11, 1961.
[122] *Hisp. Amer. Rep.*, XIV (June 1961), pp. 367-368.

administration were healthy, normal and progressive. We had already negotiated with the Quadros government substantial loans for the purpose of refinancing current obligations of the Brazilian government. We were in the process of negotiating additional loans to aid the economic program ... had already extended a substantial grant of food under Public Law 480. There were further negotiations relating to food and the availability of supplies." The Senator's remarks refuted Quadros' earlier criticism of United States aid in a warm message to Premier Khrushchev in which he said "Brazil up to now has received aid, but never on the levels and in the proportions that it really needs for its development."[123]

The sudden resignation of President Quadros stunned Brazil, and the refusal of the military establishment to accept João Goulart as president came as an aftershock. When the army split, for the first time in Brazilian history, over the issue of Goulart in the presidency, a compromise solution became possible. The Brazilian Congress voted a constitutional amendment which theoretically transformed Brazil into a parliamentary government with great powers in the hands of Congress and a prime minister. The president became a mere figurehead. Although Goulart's executive powers were restored by a plebescite early in 1963, the system failed, and the Brazilian government was in a state of chaos from September 1961 until April 1964, when the military and the governors of the major states of São Paulo, Minas Gerais, Rio Grande do Sul and Guanabara joined together to oust the Goulart administration. This ouster was a popular action by a majority of the people, but from 1964 to 1966, and later, a widening gulf developed between civilians and military. The army assumed more of the decision making functions at the expense of civilian political leaders, and by 1966 public involvement in the electoral process was reduced to selecting members of the National House of Representatives, while political participation at the state and municipal levels in many instances fell under the close supervision of the military.

The Kennedy administration congratulated President Goulart on his assumption of power, and there was a widespread feeling in Brazil that the United States showed a prudence and restraint throughout the crisis. After the succession problem was settled, Washington adopted a cooperative attitude which was maintained until mid 1963. Prior to that time, however, Goulart's position on Cuba and the Soviet bloc had caused Washington concern. This, together with the nationalization of American-owned property, and Goulart's unwillingness to carry out a politically unpopular attack on inflation, and to begin social reform in order to qualify for additional financial assistance, led to a virtual impasse. After mid 1963 no new aid agreements were signed with the federal government, except for the surplus wheat agreements made under Public Law 480, and the assistance to the SUDENE program in the northeast. Washington, instead, negotiated directly with state governors who were prepared to meet the terms of United States agencies for economic assistance.

Shortly after the Goulart government took office the new foreign minister, Dr. San Tiago Dantas, indicated that his country would seek diplomatic relations with all members of the United Nations, including Communist countries, since as

[123] *Cong. Rec.*, 87th Cong., 2nd Sess., vol. 108, p. 15027, 17352.

a country "in search of markets" Brazil could not limit the possibilities of selling its products. He added that Brazil would continue to defend the principles of self-determination and nonintervention. On the question of admitting Communist China to the United Nations he said that the matter should be debated in the General Assembly without prejudging the merits of the case. Brazil, he declared, "would spare no effort to maintain Cuba within the inter-American system in accordance with the characteristics that are the basis for coexistence between the countries of this hemisphere." Dr. Dantas emphasized that his country had no tendency to become part of a neutral bloc in international affairs. The policy outlined by the foreign minister was essentially the same as that instituted by the Quadros administration, but perhaps less provocative in tone. Concluding negotiations begun by President Quadros, diplomatic relations were restored with the Soviet Union November 23, 1961. The diplomatic renewal had been preceded by a 1959 trade agreement for the exchange of $109 million worth of goods over a three year period and by Quadros' awarding of medals to members of the Soviet good will mission.[124]

The deterioration in relations between the Goulart regime and Washington was foreshadowed at the Punta del Este Conference of January 1962, when Brazil vigorously opposed the resolution adopted by the American foreign ministers to exclude Cuba from the activities of the OAS. Responding to criticism of his government for its position, the Brazilian ambassador to the United States, Roberto de Oliveira Campos, declared: "While faithful to the inter-American system, we may in specific circumstances follow an independent policy if that serves the cause of peace or the cause of our economic development." Brazil's performance aroused strong indignation in the United States Congress, where a particularly strong denunciation was voiced by Senator Richard Russell who pointed out that Brazil had consistently voted in the OAS "against every issue our nation has espoused." He took strong exception to Brazil's having been selected as the first and principal beneficiary of the Alliance for Progress.[125]

The seizure on February 16, 1962, by Governor Leonel Brizola of Rio Grande do Sul of a telephone company owned by the International Telephone and Telegraph Company of New York brought angry demands in the United States for reprisals and widened the breach between the two countries. Without notice or hearings the state government had deposited in *cruzeiros* the equivalent of $400,000 in a local bank which was supposed to be acceptable to the company officials for property appraised at more than $7 million. Governor Brizola had been responsible for other expropriations of American property. Swift and Company had lost its Brazilian plant, and in 1959 Brizola seized a power company belonging to the American and Foreign Power Company, Inc., at Porto Alegre. In that case he took property valued at $14 million and then declared that the company had earned excess profits thereby implying that the company was owed nothing. Brizola had remarked: "If the United States is really interested in helping Latin America I advise the United States government to help Brazil expropriate and expel the foreign companies now exploiting its

[124] *Hisp. Amer. Rep.*, XIV (Dec. 1961), p. 940; *Ibid.*, XIV (Jan. 1962), p. 1043. Brazil had maintained diplomatic relations with the Soviet Union from Apr. 2, 1945 to Oct. 20, 1947.
[125] *Cong. Rec.*, 87th Cong., 2nd Sess., vol. 108, p. 884, 5894.

people."[126] Senator Long of Louisiana angrily urged that the Foreign Assistance Act of 1961 be amended to stop foreign aid to any country whose government or agency had expropriated without adequate compensation property belonging to United States citizens. President Kennedy was strongly opposed to such action saying "nothing could be more unwise than to halt aid to Brazil or other countries because of the expropriation of American-owned property."

These grievances were accompanied by mounting indignation with Brazil for its failure to put its finances in order, and to do something about the shocking growth of inflation. Foreign debts were over $2,700 million of which about $1,300 million was owed to the United States government. In 1959, 1961 and 1962 Brazil agreed to begin austerity and stabilization programs which never materialized. International Monetary Fund economists classified Brazil as "uncooperative," thereby making it impossible for Brazil to secure loans from that institution. And in the meantime, massive United States assistance added up to $353 million in new money between March 1961 and February 1962, not counting a further $304 million toward the refinancing of earlier debts.[127]

President Kennedy, whose attitude reflected understanding and restraint, felt obliged to say that "there is nothing, really, that the United States can do that can possibly benefit the people of Brazil if you have a situation so unstable as the fiscal and monetary situation within Brazil." Expressing a similar opinion, Senator Mike Mansfield declared: " . . . It is very doubtful that further assistance along these lines will make any significant impression on the present dangerous trends in Brazil . . . Indeed, after providing a moment of relief, such assistance may, in effect, intensify them . . . " Senator A. Willis Robertson succeeded in arousing and uniting the extremes of Brazil's political opinion by comparing the country to a "leaky bucket."[128]

President Goulart paid an official visit to Washington in April 1962, and in an address to the joint session of Congress he said that Brazil identified herself "with the democratic principles which united the peoples of the West," but was "not part of any politico-military bloc." He praised the Alliance for Progress but seemed to chide the United States for not taking the initiative sooner, and hinted that the scope of the Alliance was too small. In a joint communique Presidents Kennedy and Goulart reaffirmed their adherence to the principles of the Charter of Punta del Este, and agreed to give full support to the completion of a world-wide agreement on coffee, which was then being negotiated.[129]

President Kennedy scheduled a visit to Brazil for August, 1962. Because of Brazilian political unrest it was deferred until November. Then, the Cuban missiles crisis resulted in an indefinite postponement. In December 1962, however, Robert Kennedy met with President Goulart. Reportedly Kennedy told him that the United States might be forced to stop military assistance to Brazil because of Goulart's toleration of "Communists" in high administration posts.[130]

[126] *Ibid.*, p. 3393.
[127] *Hisp. Amer. Rep.*, XV (Feb. 1963), p. 1166; *New York Times*, Mar. 11, 1962.
[128] *Cong. Rec.*, 87th Cong., 2nd Sess., vol. 108, p. 2858; *New York Times*, Oct. 31, 1962.
[129] Dept of State *Bulletin*, XLVIII (Apr. 30, 1962), pp. 705-707.
[130] *Hisp. Amer. Rep.*, XV (Feb. 1963), pp. 1164-1165.

Adding to the problem of economic relations between the two governments, the Brazilian Congress passed a controversial profit remittance law which United States ambassador to Brazil, Lincoln Gordon, declared would drive investors elsewhere. Article 28 of the act called for a 10 percent ceiling on the remittance of profits on registered foreign capital as a temporary device and only when a balance of payments crisis exists. But another section made the limitation of remittances a permanent feature. As a consequence of this measure and the growing fragility of the Brazilian economy, foreign private investment in Brazil had declined to negligible amounts at the end of 1962.

During the missiles crisis of October 1962, the Goulart government supported only part of the United States-backed measures to prevent further shipment of arms to Cuba. Citing Article 8 of the Rio Treaty, Brazil maintained that an arms blockade was legitimate self-defense. But Brazil did not support total blockade nor the use of an inter-American force in Cuba. The Foreign Ministry announced on October 24: "The Brazilian government does not give its support to the use of force that would violate the territorial integrity of an independent nation and endanger world peace." Almost coincidentally Goulart said in a press interview that "Premier Castro's government is derived from the self-determination of the Cuban people." When Vice Admiral Helio Garnier Sampaio, commander in chief of the Brazilian fleet proposed sending two destroyers to support the United States naval blockade, he was summarily relieved of his command. Instead, President Goulart sent a special delegation, headed by the commander of the palace guard, to confer with Castro about Brazil's mediating between Cuba and the United States.[131]

Early in 1963, apparently overlooking the Brazilian conduct in the Cuban crisis, Washington agreed to lend Brazil $30 million for ninety days to provide her with emergency aid for necessary imports. This was followed by a $398,500,000 loan to help shore up the Brazilian economy. It was Brazil's inauguration in January of long range stabilization measures that led the Kennedy administration to approve the program. Full disbursement of the loan was made contingent upon Brazil's performance in fighting runaway inflation and otherwise stabilizing its economy. The loan was expected to improve the delicate relations between the two countries, but the immediate object was to alleviate Brazil's balance of payments crisis. Its foreign currency reserves were nonexistent and a $500 million payments deficit was forecast. Nearly two-thirds of the new financing was earmarked for this emergency support.[132]

As the loan was being negotiated President Goulart faced an uprising of his leftist supporters demanding the removal of the United States ambassador, Lincoln Gordon. The political storm was unleashed by the announcement in Washington of testimony given by Mr. Gordon before a House subcommittee interested in Communist activities in Brazil. Mr. Gordon testified to Communist infiltration in both the federal administration and Mr. Goulart's Brazilian Labor Party. The flareup was resolved after the under secretary of state, George Ball, assured Brazilian Ambassador Roberto Campos that the State Department was not trying to sabotage his country's efforts to secure financial aid. Ambassador

[131] *Ibid.*, XV (Dec. 1962), p. 964.
[132] Dept of State *Bulletin*, XLVIII (Apr. 15, 1963), pp. 557-561.

Campos did not deny Communist infiltration but said it was not extensive enough to influence "in any way Brazilian Government policies."[133]

In 1963 Brazil, plagued by internal problems, which increasingly influenced its foreign policy, evinced coolness towards the Alliance for Progress. Dr. Kubitschek, a candidate for reelection who found that criticism of the Alliance would gain votes, said in a slashing attack that "there lies an almost frozen zone" between President Kennedy's promises and the actual achievements of the Alliance program.[134] The Goulart government proved uncooperative in the efforts to create a new multilateral mechanism for the Alliance, the Inter-American Committee, and the president himself refrained from mentioning the Alliance in public. While appearing to be interested in improving relations with Washington Goulart faced powerful pressures from his nationalistic and leftist advisers to draw away from the "American imperialists," and he appeared to accept the idea that identification with the Alliance was a "sell out" to "imperialists." He clearly did not wish to be identified with the Alliance but wanted Brazil to remain a member. The apparent contradictions of the Goulart policies reflected the deep divisions within the country and in the government itself.

The death of President Kennedy, evoking expressions of grief and sympathy, breathed some friendliness back into Brazilian-United States relations. Many spontaneous demonstrations occurred, but the president's death was not relevant to problems affecting relations. The actual areas of conflict were between the two governments and related to Brazil's foreign debts and inflation, toleration of Communists within both the government and the armed forces, a foreign policy that dangerously weakened the traditionally close United States-Brazilian ties and served the interests of Castro's Cuba, and the mistreatment of American private companies in Brazil. President Johnson, seeking a fresh start with Brazil, advised President Goulart that the United States was "ready to collaborate" with Brazil in solving its huge debt problems.[135]

By the end of 1963 reports of a possible revolution had become commonplace in Brazil. Speculation on the form it might assume ranged from a leftist or rightist *coup d'état* to a "center *coup*" staged by Goulart himself. The president's tactics at that time seemed to imply the threat of a mass movement aimed at mobilizing total power behind him. In October 1963, he asked congress to declare a state of seige. This was refused and in 1964 he turned increasingly toward the use of mass demonstrations to bring pressure on congress and the opposition. At a rally on March 13, 1964, he called for a populist and nationalist government and a congress composed of peasants, workers, sergeants and nationalist officers. He also declared that if he could not carry out his program by legislative means he would go over the heads of congress and the judiciary. This was quickly followed by a message to congress demanding nearly unlimited decree-making powers and requesting a change in the constitution that would permit the use of government bonds instead of cash to pay indemnity for property expropriated under the administration's agrarian reform program. The

[133] *New York Times*, Mar. 18, 19, 1963.
[134] *Ibid.*, June 16, 1963.
[135] Dept. of State *Bulletin*, L (Jan. 13, 1964), pp. 47-49.

president then decreed the expropriation of the country's remaining privately owned petroleum refineries, which were to be taken over by *Petrobras.*

Worsening economic conditions and unrest in the armed forces made the regime's plight even more desperate. The cost of living rose by 80 percent in 1963 and the government met its budget deficit by increasing currency in circulation by 70 percent. The *cruzeiro* dropped from 180 to the U.S. dollar in 1960 to 1250 to the dollar in November 1963. With the population rising at an annual rate of 3.3 percent, the government admitted that the gross national product increased by less than 3 percent, making 1963 a "no growth year" at best. As the civilian political rifts grew deeper the Brazilian armed forces, once a guarantee of stability, also split. There were leftist and rightist officers, and many noncommissioned officers had become a revolutionary-minded leftist force. Goulart's tacit condoning of insubordination of enlisted men and noncommissioned officers threatened to destroy military discipline. Evidently hoping to use this military element as Vargas had used the "lieutenants," he brought himself to the brink of political disaster. Owing to these circumstances by early 1964 the United States government had become preoccupied over the threat of an abrupt leftist turn in Brazil.

President Goulart was removed from office by the armed forces—with broad civilian support—on the night of March 31, 1964. The first ominous reaction to the Goulart regime's drastic proposals had come on March 19 when half a million people in the city of São Paulo participated in a Roman Catholic march for "God and family." Immediately after the *coup* an estimated one million persons demonstrated in Rio de Janeiro to show their approval of the change in government. Whatever the merits of the Goulart program may have been, it was clear that the armed forces and many civilians envisaged a totalitarian government in the offing, and while some elements might not have been opposed to every authoritarian movement, it was obvious that they did not approve of a leftist one. After Goulart's removal the leaders of the armed forces announced that not only would they repair the damage done by his government, which the declared had "deliberately sought to bolshevize the nation," but that they would institute an authentic Brazilian revolution.

Charges that the United States government had plotted the *coup* were immediately heard from the domestic left and from Cuban sources, charges which were lent a degree of credence by the extraordinary rapidity of Washington's recognition of Goulart's military successors. Predictably the Cuban newspaper, *Hoy,* said that Goulart's main enemies were "Yankee imperialists," and that the revolt was "planned, paid for and ordered by Washington." Secretary Rusk hailed the *coup* as an expression of support for constitutional government, but denied that the United States had participated "in any way, shape or form." Ambassador Lincoln Gordon declared that "only a blind man could not see that the Communists of both the Moscow and Peking varieties" were jockeying for positions of leadership in a "leftist *coup*" that was prevented by the overthrow of Goulart. Although no evidence has been produced linking the United States government with the movement that removed the Brazilian president from office, the United States Embassy appears to have been aware of the conspiracy which it viewed sympathetically.[136]

[136] Skidmore, *Politics in Brazil*, pp. 322-330.

President Johnson sent his "warmest wishes" to Brazil's acting president, Ranieri Mazzili, twelve hours after the latter was sworn in to replace João Goulart. The White House message made it clear that Washington saw the constitutional transfer of power by Brazil's Congress to Mr. Mazzilli as removing any obstacle to recognizing the new regime. The Johnson message emphasized that "the relations of friendship and cooperation between the two governments and peoples are a great historical legacy for us both."[137]

Shortly afterward the leftist-purged Congress selected as president Humberto de Alencar Castello Branco, a retired army marshal and an officer in the Brazilian Expeditionary Force that fought in Italy in World War II. A bitter opponent of Communist subversion in Brazil, he announced three weeks after taking office the severing of diplomatic relations with Cuba, an action evoking praise in the United States Congress.[138] But he pursued the reform objectives of the new government through housing, tax and land reforms, as well as a serious approach to solving the inflation problem. In June 1964, Washington granted Brazil a $50 million loan to support the reform programs. This was the first major loan, aside from the food-surplus agreements, to the new government. It reflected the faith of the Johnson administration in Brazil's determination and capacity to combat inflation, rebuild her shattered finances and institute rational economic policies. The loan marked the resumption of general United States lending to Brazil after a pause of almost a year.[139]

The armed forces-dominated government carried out the "Revolution of 1964" with a series of institutional acts beginning on April 9 of that year, and a new constitution adopted in 1967. In the pattern which has unfolded, there is a presidency with greatly enlarged powers headed by a military figure. The president has the sole right to introduce bills in congress relating to public expenditures, may declare and extend a state of seige, and may suspend the political rights of an individual for a period of ten years and annul legislative mandates of federal, state or municipal origin. As a result, approximately fifty congressmen were removed from office, and some four hundred persons lost their political rights, that is, the right to run for office, vote or express political opinions. Included among these were ex-presidents Goulart, Kubitschek and Quadros, as well as the noted economist Celso Furtado. The president may also decree federal intervention in a state without congressional approval and is not bound by judicial restraint. Congress was reduced in powers and the military given expanded authority in tribunals judging civilians. The political system was reorganized to insure a majority party. Under the military's guidance a group declaring its support of the government called the National Alliance of Renovation (ARENA) was organized as a kind of official party. Another group, assumed to be the "loyal opposition," was organized under the title of the Brazilian Democratic Movement (MDB). The most important executive officers of the country are selected indirectly and the congress is elected from restricted slates of candidates.

Beyond altering the institutions of government, the Castello Branco administration took steps to create the social and economic conditions deemed

[137] *Cong. Rec.*, 88th Cong., 2nd Sess., vol. 110, p. 6871.
[138] *Ibid.*, pp. 10979-10980.
[139] *New York Times*, May 25, 1964; *Ibid.*, June 25, 1964.

essential to political stability. These included measures to encourage private (foreign and domestic) initiative and stimulate the economic growth of the country, and those intended to bring about reforms in the economic and social structure. In its economic policies the government stressed controlling inflation, encouraging foreign investment and reforming the tax structure. In what was regarded as a significant step in the program for economic recovery, Castello Branco signed a bill on August 31, 1964, modifying the law governing the remittance of profits abroad by foreign investors. Of the two major changes in the law the first was the repeal of the annual limit placed on the remittance of profits abroad by a foreign investor of 10 percent of the registered foreign capital in the enterprise. Even though remittances still had to be registered, there was no limit on the amount that could be sent abroad unless the government declared a foreign exchange crisis, in which case the 10 percent limit would apply. The second change was an amendment permitting the registration of reinvested profits not sent abroad to be considered as increased capital when calculating future repatriation of profits or capital. Supplementing these measures, Brazil entered an investment guarantee program with the United States which gave private American investors protection against revolutionary expropriation, incontrovertibility and other such risks.[140]

The revolutionaries inherited from the Goulart regime an economic crisis in which inflation had risen to an annual level of 140 percent, foreign creditors were demanding payment and new investment had come to a halt. A huge foreign debt, estimated between three and four billion dollars, was viewed as the foremost problem since the cost of servicing the debt for the period 1963-1965 amounted to $1.8 billion. In 1961, it will be recalled, a massive rescheduling of payments and additional credits were effected. This had involved a $1 billion financial package, at the time one of the largest ever assembled. It included $600 million of new credits from the United States, the International Monetary Fund and from European countries, as well as a rescheduling of earlier credits. In addition to the foreign debt problem the new administration had to devise a non-Socialist development program, and to offer a positive political message to an anxious public harassed by price increases, food shortages, unemployment, rural violence and urban slum life.

Washington moved quickly to assist the new regime, giving Brazil more than $228 million in cash, credits and food between April and November 1964. In December the United States made a commitment of $1 billion in aid including loans, food and the rescheduling of debt payments. The development program loan proved to be the keystone of the American aid effort in support of the increasingly effective internal policies of the Castello Branco government. The aid was channeled primarily into industry, electric power, public health, education, agriculture and highways, within the framework of Brazil's overall development plan. The sixth in a series of Public Law 480 commodity sales agreements provided continuing wheat sales to Brazil, the largest recipient in the program after India. Tabulations made near the close of 1966 indicated that Brazil was the largest recipient of American aid in Latin America. In fact, it was third in the entire aid program, ranking behind South Vietnam and India. In the

[140] *Ibid.*, Feb. 13, 1965; *Hisp. Amer. Rep.*, vol. XVII (Oct. 1964), p. 760.

preceding two years, during which the Castello Branco government pursued an austerity program, the United States had provided almost $500 million in loans and grants.[141]

The northeast of Brazil, comprising nine states, received special attention under the Alliance for Progress. The twenty-seven million inhabitants of the region have one of the lowest per capita income rates in the world. Most of the rural population exist at a subsistence level and some four million are without land or jobs and constitute a migrant labor force continually in search of work. The misery and poverty, heightened by periodic droughts, forces mass migration to urban areas which are unprepared to cope with the problems raised by the displaced population.

Washington sent a special mission to the northeast in 1961 to investigate and make recommendations for U.S. assistance in the area. The two governments entered into the Northeast Agreement in April 1962 calling for cooperation in promoting economic and social development there. The United States committed itself to provide $131 million, of which $76,760,000 was in dollars and the balance in local currency under Public Law 480. From the outset it was evident that the Brazilian agency responsible for carrying out the northeast program, SUDENE (Superintendency of the Development of the Northeast) was in disagreement with the United States, both on objectives and the implementation of the program.

Prior to the military-led "revolution" in 1964 the political character of SUDENE was "leftist nationalist," and Communist agitators' organizing efforts among the rural population complicated operations. Whereas the United States officials insisted that improved health, education and social conditions were the key to increased production, SUDENE maintained that social problems were the result of low income and that the solution of these problems lay in raising income. To SUDENE the individual was the beneficiary of economic development, rather than a contributor to economic development. In spite of the disagreement the United States fulfilled its pledge. From 1962 through 1966, United States assistance to the region totaled $260 million in loans and grants, including $60 million in Food-For-Peace funds. The figures do not include assistance funded through national projects nor do they include U.S. program loans made directly to Brazil and used in the northeast.[142]

One of the most crucial problems in the northeast is the need to restore to workers land use rights for subsistence crop production, particularly in the sugar zones of the coastal areas where famines have occurred in recent years. President Castello Branco signed a decree in October 1965 providing for the grant of the use of up to two *hectares* of land by each sugar plantation owner to each sugar cane worker for providing food for himself and his family. The implementation of this decree has been resisted by the responsible government agency and by the sugar land owners who opposed not only the two-*hectare* concept but also the growing of food crops in the sugar zones. Expert opinion holds that whether the

[141] *New York Times*, Oct. 16, 1966.
[142] U.S. House of Reps., Comm. on For. Affairs, 90th Cong., 1st Sess., *Report of the Special Study Mission to the Dominican Republic, Guyana, Brazil and Paraguay* (Washington, 1967), pp. 22-23, 26.

future course of the Northeast is violent or peaceful will depend in great measure on the progress of modernization benefitting the peasants.[143]

Brazil's drift away from the United States was brought to an abrupt halt by the Castello Branco government, which then reaffirmed by word and action the nation's traditional close cooperation with the United States in hemispheric affairs. Consistent with its anti-Communist stand, the government placed under arrest members of a Communist China mission, and expelled a Czech diplomat charged with espionage activities. On May 12, 1964, Brazil broke off diplomatic relations with Cuba on grounds that Cuban agents were interfering in Brazil's internal affairs. It should be noted that the overthrow of Goulart had dealt a sharp blow to Cuba's foreign policy plans in Latin America. Brazil under Goulart was regarded as the main opposition to diplomatic and economic sanctions proposed by Venezuela, which had accused Cuba of arming guerrillas on its territory. The sudden turn in Brazil's foreign policy as it affected the hemisphere was apparent when, in July 1964, Brazil joined fourteen other hemispheric nations in voting trade sanctions against Cuba and supporting a resolution that members of the OAS should not maintain diplomatic relations with Cuba. On the third anniversay of the Alliance for Progress, President Castello Branco sent President Johnson a message praising the program.

The first period of the "Revolution of 1964" lasted until October 1965; the second ended in March 1967; the third period lasted until December 1968; the fourth continued into the 1970s. In the first period the Castello Branco administration ruled within the much-altered Constitution of 1946. Despite measures taken to insure the election of "safe" candidates in the gubernatorial campaigns of October 1965, two states elected governors identified with former president Kubitschek. A split involving the "hard line" officers within the military commands was narrowly averted as a result. To satisfy this element the presidential powers were further increased and measures were taken to guarantee that congress would elect a preselected president of the republic. It therefore came as no surprise that Marshal of the Army Artur da Costa e Silva, Minister of War in the Castello Branco cabinet, running unopposed, was elected president by congress October 3, 1966. The president-elect traveled to Washington on a state visit in January 1967, and was assured of continuing material support by the Johnson administration.[144]

The outgoing congress approved a new constitution incorporating the measures of the "revolution": a strengthened executive, more extensive power to the central government to intervene in the states, jurisdiction of military courts over civilians charged with "crimes against the national security," and the impairment of other individual rights. Reinforcing the new constitution, the "lame-duck" congress passed a restrictive press law providing stiff penalties for the publication or broadcast or reports damaging to the national security of financial stability. A journalist could be jailed for writing unflattering words about the president or other high officials even if he could prove the charges were true.

[143] Harold T. Jorgenson, "Impending Disaster in Northeast Brazil," *Inter-American Economic Affaairs*, XXII (Summer, 1968), 3-21.

[144] *New York Times*, Jan. 27, 1967. See also American Universities Field Staff, *Reports, East Coast of South America Series*, vols. XI-XIII (New York, 1964-1967).

Evidence of improvement in the Brazilian economy could be seen at the close of the Castello Branco administration, although the basic problems remained unresolved. Its policies meant holding down civil service and military wages, elimination of subsidies and the raising of the public service rates. Businessmen paid higher taxes and were under heavy pressure not to raise prices. Unions were obliged to accept government-fixed wage levels. As might have been expected, the national recovery program was quite unpopular. Brazil's external debt-service costs were rearranged and the country no longer faced the crushing burden of short-term debt repayment that prevailed in 1964 and 1965. This improvement in debt structure had been matched by an improvement in Brazil's international resources and by expansion and diversification of exports.

In 1966 the overall public sector budget and federal transfers to state enterprises was balanced for the first time in more than a decade. Brazil's exports were the highest since the Korean War, totaling more than $1.75 billion. Prices for coffee, Brazil's main export, were bolstered by the International Coffee Agreement and sales exceeded $800 million. Cotton, iron ore, cocoa, corn, and a wide variety of manufactured goods, processed foods and lumber sold well. The nation's dollar and gold reserves reached $700 million. Agricultural incentives were increased by a combination of measures eliminating consumer price controls, increasing rural credit and providing a minimum price support guarantee to producers of foodstuffs. United States government aid in 1966 totaled $390.6 million, of which $190.2 million was disbursed. Total U.S. aid to Brazil, including military assistance, in the period 1946-1966 stood at $3.185.3 million, almost one-third of all aid to the nineteen Latin American republics over the same period.[145]

The rising cost of living which had soared to 81 percent in 1963 and 85 percent in 1964 was reduced, but not stopped. In 1965 the rise was held to 41 percent and in 1966 to 46 percent. The annual change on the period 1961-1966 was 60 percent.[146] Gains were reflected in the percentage of growth of the gross national product as the following data indicate: 1961, 7.3 percent; 1962, 5.4 percent; 1963, 1.6 percent; 1964, 3.1 percent; 1965, 4.7 percent; 1966, 5 percent. While these gains were significant, they were offset by population growth which continued at a rate of more than 3 percent annually. The plight of the low-paid workmen and the salaried employees remained desperate, for the government's anti-inflation program held wages down. As a result, the buying power of the workers was cut by 40 percent since 1964. But with political outlets for dissent being rigidly controlled, there was little visible public unrest in the labor sector.[147]

The new president and congress were installed in office March 15, 1967. Among his first policy statements President Costa e Silva declared that "social humanism will be in truth the deepest root of my government," and "we might not be a popular government, but we will be without shadow of doubt, a government for the people in the most profound sense of that expression." As for foreign policy he said that Brazil would remain pro-Western and an active

[145] U.S. House of Reps., *Special Study Mission*, p. 26.
[146] U.S. Senate, Comm. on For. Rels., 90th Cong., 1st Sess., *Survey of the Alliance for Progress–Inflation in Latin America* (Washington, 1967), p. 3.
[147] Richard Graham, "Brazil's Dilemma," *Current History*, vol. 53 (Nov. 1967), p. 296.

participant in the United Nations and the OAS. He said a prime foreign policy aim would be economic: to find markets for Brazil's products and to obtain economic and technical assistance for the nation.[148] Among indications that the new regime would preserve its independence of action in foreign affairs versus the United States was its continuing refusal to accept any form of international limitation on Brazil's future development of a nuclear potential for peaceful uses.

In spite of the new president's pledge to humanize the "revolution" his administration faced mounting criticism, public unrest, and disruptive pressures from within the armed services. Student ferment erupted in 1968 with twenty-thousand taking to the streets in Rio de Janeiro. The students protested the regime's backward educational policy, which had taken money from the schools and given it to the military, as well as overcrowding, and out-dated teaching methods. As in the preceding two years, the student demonstrations reflected strong anti-American overtones, and on several occasions United States government buildings were stoned or bombed. The students claimed that an educational aid agreement between the Ministry of Higher Education and the U.S. Agency for International Development was a form of colonialism, and that the American Government was seeking to influence Brazilian education.[149] Anti-United States feeling was climaxed in 1968 when Captain Charles R. Chandler, a Vietnam veteran, studying Portuguese at the University of São Paulo, was killed by machine gun fire as he left his home. A note found near his body said that his death was "punishment for a Vietnam war criminal." Leaflets quoting Ernesto "Che" Guevara, the slain guerrilla leader, were scattered in the area.[150] Brazil's outlawed Communist Party also assailed the government but opposed armed struggle to bring it down. A statement credited to the party, appearing in an independent newspaper, rejected the Cuban and Communist Chinese call for violent revolution, urging instead the exploitation of dissatisfaction with the government's wage and economic policies to consolidate opposition "mobilizing, uniting, and organizing the working class and other patriotic forces" for a struggle against the dictatorial regime.[151]

When the opposition became intolerable to the military, the regime recessed Congress, suspended constitutional guarantees and assumed dictatorial powers. President Costa e Silva defended his government's action saying it had been necessary "to snuff out a counterrevolutionary plot" and to "save the nation from a situation that could have led to civil war." Hundreds were jailed including Dr. Juscelino Kubitschek and Carlos Lacerda.[152]

Of Washington's dismay at the revelation of the progressively authoritarian character of the armed forces-backed government there was no doubt. As early as May 1964 Ambassador Lincoln Gordon had cautioned Brazil's military leaders

[148] Representing the United States at the inauguration of the new president was a delegation headed by Edmund G. Brown, former governor of California, and including the U.S. ambassador to Brazil, John W. Tuthill, and Jack Valenti, formerly an assistant to President Johnson.

[149] *New York Times*, Jan. 19, 1968; *Ibid*., June 22, 1968.

[150] *Ibid*., Oct. 13, 1968.

[151] *Ibid*.

[152] *Ibid*., Dec. 22, 1968.

to distinguish between "subversive conspiracies" and dissent when they limited liberty.[153] However, the firm anti-Communist position of the government had to be weighed against possible alternatives. Marshal Costa e Silva had favored the establishment of an inter-American military peace force to act in the Western Hemisphere against "external subversion." Brazil supported the United States in this proposal which was opposed by a majority of Latin American countries. While accepting the regime, the United States had sought to bring about a lessening of military control and a growing democratization of the government.

After it was learned of the sharp turn toward dictatorship the U.S. State Department announced that the situation in Brazil was "being watched with interest," but that it was "strictly an internal matter for Brazilians."[154] Two top-ranking American officials were recalled for consultation which was interpreted that Washington viewed events in Brazil with some uneasiness. The State Department was reported to have been reviewing its aid program to Brazil, Latin America's major recipient of aid under the Alliance for Progress. More than $2 billion had gone to Brazil in the seven years since the Alliance was launched, most of it after the military seized power in 1964. In the same period the Soviet Union agreed to open a $100 million credit to Brazil for the purchase of industrial equipment. Alliance loans committed to Brazil in 1967 and 1968 totaled $213,773,000 and $192,996,000 respectively. The latter, though reflecting a declining scale of foreign aid, was the largest loan awarded to any Latin American country, being two and a half times greater than the amount received by Colombia, the second ranking loan recipient.[155]

Ambassador Tuthill, scheduled to retire in January 1969, indicated that the United States had taken "much too wide an operational role in Brazil," particularly in the aid program. Whereas the Castello Branco government had worked closely with Washington, it had become clear that the Costa e Silva administration preferred to be relatively independent of the United States. After a number of points of friction had arisen Mr. Tuthill said that the widening of the American effort had increased this probability. He proposed to reduce the number of programs and concentrate aid money in those which allowed the Brazilian government to bring its own resources to bear. A notable failure of American aid was a project of technical assistance to Brazilian educators. The program became a focus of criticism as a result of student discontent and because the American experts were unable to achieve effective cooperation with their Brazilian counterparts. Owing to dissatisfaction with this program the Brazilian Ministry of Education signed an agreement with ten European countries–including Czechoslovakia, Hungary, Poland, East Germany, and the Soviet Union–for technical assistance. The American ambassador also recommended personnel reductions in the United States Embassy, the largest in Latin America by 25 to 40 percent. At the time, the Embassy maintained approximately one thousand employees and seven hundred Peace Corps Volunteers.[156]

[153] *Ibid.*, May 6, 1964.
[154] *Ibid.*, Dec. 6, 1968.
[155] Annual Report to the Congress, Fiscal Year 1967 (President Lyndon B. Johnson), *The Foreign Assistance Program* (Washington, 1968), p. 56; *Ibid.*, (Washington, 1969), p. 50; U.S. Dept. of State, *Communist Governments and Developing Nations: Economic Aid and Trade*. Research Memoranda, RSB-50 (June 17, 1966), and RSB-80 (July 21, 1967).
[156] *New York Times*, Nov. 26, 1967.

While draining Brazilian politics of any meaning or effect it appeared in early 1969 that the Costa e Silva government had introduced some stability into the volatile economy. Brazil finished 1968 with a deficit of slightly more than $300 million, its smallest in many years. Federal income tax collections of more than $500 million were nearly 50 percent larger than in 1967, while total government receipts of $2.7 billion were almost double. The only field in which the government failed in its prediction was in reducing the rate of inflation. The 1968 rate, just under 25 percent, practically equalled that of 1967. The 1968 gross national product surpassed $21 billion. This was an increase of 6 percent versus a population growth of 3 percent.

Institutional Act 5, promulgated December 13, 1968, gave the military regime virtually unlimited powers. This had economic implications possibly equalling its political aspects. Under the powers assumed by this act the government was committed to taking severe measures affecting the economy. These included economies in all government activities except education. Education was given highest priority and measures were to be taken to expand and improve schooling facilities at all levels. Economies were foreseen in a decrease of federal tax appropriations to states and municipalities and the reduction of government payrolls by retiring 100,000 public servants. Income tax evaders now faced stiff penalties, and while personal taxes were raised, corporate taxes were eased. Measures were included to spur agrarian reform and to provide farmers with credit facilities. Temporary price controls were placed on food and many other products and services.[157]

Evidence of increasing defiance of the regime could be seen in the kidnapping of foreign diplomats. U.S. Ambassador to Brazil, C. Burke Elbrick, abducted in Rio de Janeiro in September 1969, was released after the ransom, in his case the freeing of fifteen political prisoners, was paid. Japan's Consul-General in São Paulo, another kidnap victim, was ransomed for the release of five political prisoners. In April 1970, John C. Cutter, U.S. Consul in Porto Alegre, was wounded while outrunning a roadblock set up by his would-be abductors. Ambassador Elbrick's kidnappers issued a manifesto declaring that they had no personal enmity against him, but as the U.S. Ambassador he represented the interests of American capitalists which are inimical to those of the Brazilian people. The government's inquiry on the kidnapping of the ambassador indicated tthat the terrorist group's support was drawn from "university students, progressive elements in the Roman Catholic clergy and the upper classes."[158]

In October 1969, the Armed Forces High Command selected General Emilio Garrastazu Medici as the regime's third president. General Medici was inaugurated for a term scheduled to end of March 15, 1974, succeeding Marshal Costa e Silva, who died after suffering a stroke. General Medici announced plans for a moderate reform program and expressed the hope that representative democracy would be restored by the end of his administration. But Medici faced opposition in the armed forces to his plans for more freedom.

Brazil began 1970 with 1,116 "nonpersons," ordinary citizens as well as politicians and military men, whose political rights had been suspended for ten

[157] *Christian Science Monitor*, Feb. 13, 1969.
[158] *New York Times*, Feb. 9, 1970.

years. It appeared that the number might increase when the military leaders adopted a sharply anti-Communist tone contending that "opposition to the government is all Communist inspired."[159] No judicial review or appeal was allowed. This political repression caused a political vacuum in Brazil for the official party, ARENA, and the official opposition party, MCB, were subordinated to insignificant roles. In effect, the Army had become Brazil's sole political party. Opposition to easing the restraints on freedom appeared to come from officers on the extreme right who feared that moderation could result in anti-militarism, and from a strongly nationalistic element who wanted to rid Brazil of "foreign economic domination." The military regime faced a dilemma: should it become identified with the traditional *élite* or provide opportunities for new activist groups which might threaten its power. Among the many aspects of the increasing military hold on the nation, that of curbs of freedom of the press was one of the most conspicuous and, to many Brazilian liberals, perhaps one of the most galling restrictions, especially after December 13, 1968.

Brazil experienced an economic upsurge in 1971 with the gross national product rising by more than 11 percent, compared with 9 percent in the two preceding years, and shattering all statistical records in business and industry. Strong fiscal controls reduced the level of inflation, at one time the highest in Latin American, from 21 percent to 18 percent. It had the eighth largest automobile industry in the world, producing 510,000 units a year, and had the largest steel complex in Latin America. Exports rose by 10 percent, about one-third being industrial products formerly imported. Coffee, representing 27 percent of the exports, had dropped from 30 percent in 1970. The military's control extended to virtually all sectors of the economy, and while encouraging the growth of privately-owned firms, domestic and foreign, it was found in 1971 that forty-three of the one-hundred largest corporations in Brazil were government-owned or operated. Manpower was plentiful and comparatively cheap, thanks to the rapid mechanization of agriculture, which released workers for the industrial sector. Trade was being expanded with the Communist countries, as well as with the United States, Japan and Western Europe.[160]

Continuing efforts begun by President Kubitschek to open up the interior, work was started on the Trans-Amazon highway which will run 3,100 miles, from Recife on the Atlantic to the Peruvian frontier; by 1972 the road had been built for 1,133 miles from the eastern seaboard. Development of the Amazon region is a joint effort by most of the major departments of the Brazilian government, the armed forces, and private industry. The government's aim is to attract people from the overpopulated, impoverished and political explosive Northeast. A one-thousand mile highway, under construction, will bisect the Trans-Amazon, connecting it with the South and Southeast. Another major project was the start of the Urubupunga hydroelectric complex west of São Paulo, which will rank as one of the largest of its type in the world.

Brazil ranked at, or close to, the top as the most stable and prosperous Latin American country as the decade of the 1970s got well underway. It had

[159] *Christian Science Monitor*, Mar. 13, 1970.
[160] *Times of the Americas*, Fed. 9, 1972; *New York Times*, Jan. 25, 1971; *Ibid.*, Jan. 28, 1972.

one of the world's highest economic growth rates, and was probably the only country south of Mexico where foreign capital could feel quite secure. Under the circumstances most of the Brazilian people, while not happy with military rule, appeared disposed to allow the political situation to continue.

Economic growth and social reforms coupled with a vigorous anti-guerrilla campaign by the government had led to a subsidence of terrorist activity. The police themselves formed "death squads" to locate and kill specific underground leaders. A succession of terrorist leaders were killed including Carlos Lamarca, a former Brazilian army captain, who was a top leader in Brazil's anti-government terrorist movement called the Popular Revolutionary Vanguard (PRV); Lamarca's kidnappings included U.S. Ambassador Elbrick. A notable aspect of the Brazilian terrorist-kidnapping problem was the role of third countries—Mexico, Algeria, Cuba and Chile—in the negotiating process.[161]

Brazil's relations with the United States, while cordial, were subjected to tension over the world coffee agreement and the two hundred mile coastal limit problem. The U.S. House of Representatives postponed action on a bill extending U.S. membership in the International Coffee Agreement (ICA) as retaliation against Brazil's decision to enforce its claim to a two hundred mile offshore jurisdiction. Brazil and Colombia were most severely affected by the House action, but some forty other Latin American and African producers also stood to lose by the policy. Finally, in November 1971 the House of Representatives voted 200 votes in favor and 99 against to extend the ICA from July 1, 1971 through September 30, 1973.

Brazil had earlier claimed a two hundred mile territorial limit, and in 1971 sent naval vessels to patrol the area. This led the U.S. Department of State to charge that Brazil had created the risk of incidents "in which confrontation would become unavoidable." Secretary Meyer reaffirmed the administration's position declaring "We have expressed our concern to Brazil that the enforcement of their regulations could lead to serious and unfortunate confrontation with the United States." A major fact in Brazil's decision was the abundance of shrimp and lobsters in the north, and the rich fishing grounds off Paraná, Santa Catarina and Rio Grande do Sul.[162]

Brazil's President, General Emilio Garrastazu Medici visited Washington in December 1971 for conferences with President Nixon. The purpose of the meeting of the chiefs of state was to strengthen commercial and diplomatic ties, but it was also seen as a possible policy change. Since President Medici was the only Latin American leader to be included on President Nixon's agenda of summit meetings, it may have been tacit acknowledgement of Brazil's rise to major-power rank, at least in the Western Hemisphere. This was inferred in President Nixon's statement "We know that as Brazil goes, so will the rest of the Latin American continent . . ."[163]

President Medici declared that "Brazil has achieved such a point of development that it requires new commercial relations with the world in general and the United States in particular." Regarding the two hundred mile sea jurisdiction claimed by Brazil, the two presidents hoped to reach a "satisfactory

[161] *Christian Science Monitor*, Sept. 22, 1971.
[162] *Times of the Americas*, July 21, 1971; *Ibid*., Aug. 4, 1971; *Ibid*., Nov. 17, 1971.
[163] *Christian Science Monitor*, Dec. 20, 1971.

interim agreement" until the International Conference of the Laws of the Sea, to be held in 1973.[164] John B. Connally, Jr., acting as special emissary for President Nixon, visited Brazil in June 1972; topics which he discussed with President Medici included international trade and reform of the world monetary system. A broader range of subjects of mutual U.S.-Brazilian interest—the forthcoming Laws of the Sea Conference, trade, monetary problems, environmental pollution, and others—were reviewed in May 1973 when Secretary of State William P. Rogers, on an eight-nation Latin American tour, met with the Brazilian president.[165]

Brazil's dynamic economic advance was maintained in 1972 with the growth rate attaining about 10 percent, and pushing the GNP to over $44 billion.[166] Exports approached $4 billion, an all-time high, and the inflation rate was held to 15 percent, the lowest annual increase in more than a decade. While there was general recognition that Brazil's economy was improving, there was mounting evidence of repressive methods being used by the military regime, including torture of political opponents.[167]

U.S. trade and investment relations with Brazil showed significant gains in 1972. Trade with Brazil exceeded $2 billion, and the U.S. maintained its position as the foremost market for Brazilian goods, and the chief exporter to Brazil. Within a total of about $4.2 billion of foreign investment, the U.S. had supplied close to 40 percent. U.S. bilateral assistance declined, standing at $9 million (grant technical assistance) in 1972 as Brazil drew increasingly upon multilateral institutions for development financing. U.S. military assistance was expanded when President Nixon authorized the sale of F-5E International fighter aircraft to Brazil.[168]

At the base of decisions about Brazil by U.S. policy planners lies the fact of Brazil itself. It is, by a wide margin, the largest and most important country in Latin America. Its population of over 98 million makes it about twice as populous as the next country, Mexico. Its land area, almost equal to that of the United States (including Alaska), is nearly half that of South America as a whole. Measured by per capita gross national product it is one of the poorer nations in South America. Its infant mortality rate of 85-95 per one thousand live births exceeds that of Ecuador. More than twenty-five million people exist on incomes of less than $100 annually, and more than fifty million are impoverished. Illiteracy stands at close to 40 percent, at least 50 percent of all Brazilians have no modern sanitary facilities, and 47 percent do not have an acceptable water supply.[169]

Problems that President Medici faces include the need to restore democratic processes, to institute health programs, to provide more schools and universities, to put forward agrarian reform to increase farm income, to reduce the imbalance between the industrial and agricultural oligarchy and the masses,

[164] *Times of the Americas,* Dec. 15, 1971.

[165] *Ibid.,* May 16, 1973; *New York Times,* May 24, 1973.

[166] U.S. Dept. of State, *An Excerpt from United States Foreign Policy 1972* (Washington, 1973), pp. 437-438.

[167] *Christian Science Monitor,* April 28, 1973.

[168] *New York Times,* June 10, 1973; *U.S. For. Policy 1972,* p. 438.

[169] For detailed summary see Inter-American Development Bank, *Economic and Social Progress in Latin America, Annual Report, 1972* (Washington, 1973), pp. 140-151.

and to implement massive urban housing projects. Favoring the regime's program is a highly admirable characteristic of the Brazilian people: the ability to compromise. Despite the military interventions in government, the six constitutions, devastating inflation, and the fantastic contrasts between rich and poor, no serious civil war has ever taken place and comparatively few Brazilians have been killed in political contests. Given time, and freedom from both internal repression and external subversion, the present goals will probably be achieved, for Brazil has adequate resources, both human and natural, to become a prosperous nation and a world power.

SUPPLEMENTARY READINGS

Agan, Joseph. *The Diplomatic Relations of the United States and Brazil.* Paris, 1926.
Azevedo, Fernando de. *Brazilian Culture: An Introduction to the Study of Culture* (Wm. Rex Crawford, trans.). New York, 1950.
Baer, Werner. *Industrialization and Economic Development of Brazil.* Homewood, Ill., 1965.
Baklanoff, Erid, ed. *New Perspectives of Brazil.* Nashville, 1966.
Barros, Jamie de. *A Política Exterior do Brasil, 1930-1940.* Rio de Janeiro, 1941.
Bello, José Maria. *A History of Modern Brazil, 1889-1964.* Stanford, 1966.
Burns, E. Bradford, ed. *A Documentary History of Brazil.* New York, 1966.
———. *The Unwritten Alliance. Rio Branco and Brazilian-American Relations.* New York, 1966.
———. *A History of Brazil.* New York, 1970.
Calmon, Pedro. *Brasil e América.* Rio de Janeiro, 1943.
———. *História do Brasil.* 3 vols. Sao Paulo, 1939-1940.
Calmon Bittencourt, Pedro. *História Diplomática do Brasil.* Belo Horizonte, 1941.
Delgado de Carvalho, Carlos. *História Diplomática do Brasil.* São Paulo, 1959.
Denis, Pierre. *Brazil.* 5th ed. London, 1926.
Ellis, Howard S., ed. *The Economy of Brazil.* New York, 1970.
Freyre, Gilberto. *New World in the Tropics.* New York, 1959.
———. *The Mansions and the Shanties: The Making of Modern Brazil.* New York, 1963.
Furtado, Celso. *The Economic Growth of Brazil: A Survey from Colonial to Modern Times.* Berkeley, 1963.
———. *Diagnosis of the Brazilian Crisis.* Berkeley, 1965.
Hill, Lawrence F. *Diplomatic Relations between the United States and Brazil.* Durham, N. C., 1932.
———, ed. *Brazil.* Berkeley, 1947.
James, Preston. *Brazil.* New York, 1946.
Loewenstein, Karl. *Brazil under Vargas.* New York, 1942.
Manchester, Alan K. *British Pre-eminence in Brazil: Its Rise and Decline.* Chapel Hill, 1933.
Normano, John F. *Brazil: A Study of Economic Types.* Chapel Hill, 1935.
Pandia Calogeras, João. *A Politica Exterior do Império.* São Paulo, 1933.
———. *A History of Brazil* (Percy A. Martin, trans.). Chapel Hill, 1939.
Rodrigues, José Honório. *The Brazilians, Their Character and Aspirations.* Austin, 1967.
Roett, Riordan, ed. *Brazil in the Sixties.* Nashville, 1972.
Schneider, Ronald. *The Political System of Brazil: Emergence of a Modernizing Military Regime.* New York, 1972.
Schurz, William L. *Brazil, the Infinite Country.* New York, 1961.
Skidmore, Thomas E. *Politics in Brazil, 1930-1960: An Experiment in Democracy.* New York, 1969.
Smith, T. Lynn. *Brazil: People and Institutions.* Baton Rouge, La., 1963.
Smith, T. Lynn and Alexander Marchant. *Brazil: Portrait of Half a Continent.* New York, 1951.

Spiegel, Henry W. *The Brazilian Economy: Chronic Inflation and Sporadic Industrialization.* Philadelphia, 1949.
Stepan, Alfred. *The Military in Politics; Changing Patterns in Brazil.* Princeton, 1971.
Tavares de Sá, Hernane. *The Brazilians, People of Tomorrow.* New York, 1947.
Vianna, Hélio. *História Diplomática do Brazil.* Rio de Janeiro, 1958.
Worcester, Donald E. *Brazil, Fróm Colony to World Power.* New York, 1973.

19

The Good Neighbor Policy and After: Development Diplomacy and Security Assistance

The typical Yankee of North America is partial to slogans. He likes to tie up his country's foreign policy in bundles and label them. The Monroe Doctrine, the Open Door, Freedom of the Seas, the Good Neighbor, and the Alliance for Progress are pertinent examples. Once formulated and labeled the policy is accepted as a vital element in the structure of the state; but thereafter the public pays little attention to its development. As a result, when such a policy is challenged, or fails in its objectives, no one is more surprised than the complacent U.S. citizen. This has been particularly true in our relations with the other American republics. The problem is complicated by apathy, ignorance and a commonly prejudiced and distorted image of Latin American civilization.

Until recent years Latin America, aside from concern for strategic interests in the Caribbean, has been taken for granted in the United States. This apathy is partially attributable to the weakness of the Latin American nations, which have been unable to command either the attention or the respect of their powerful neighbor. Widely viewed as "client" states, and exerting a negligible influence in international politics, it has been generally assumed that they fall within the United States sphere of influence, and more or less exist at its sufferance. Yankee nationalistic attitudes of superiority were reinforced by the idea of Anglo-Saxon racial superiority over other peoples. The presence of large Indian and Negro populations in the region has tended to assign the Latin American to a lower ethnic station in the mind of the American public. Although overt expressions of ethnic nationalism have become less evident, the continuing social upheaval and economic underdevelopment in the region have helped to confirm the opinion that they are less gifted peoples. The historic rivalry between

England and Spain, beginning in the sixteenth century, produced a large body of literature which depicted the Spaniard as being naturally and excessively cruel, a religious bigot, and degraded morally, socially and intellectually. This attitude was carried to the Anglo-American colonies in North America where it was intensified by conflicting religious views. Latin Americans inherited this unfortunate image, and the invidious comparison produced by the slower advance of their national and institutional development has served to confirm this view of their peninisular forebears.

A realistic appreciation of Latin America is further hindered by the absence of a sound foundation of understanding. The educational and cultural orientation of the American people has been directed toward countries of Western Europe, particularly the United Kingdom, France, Germany, or Italy. With the possible exception of Mexico, most informed people have had virtually no contact with or understanding of Latin American culture. The intellectual and artistic work of Latin Americans has largely failed to capture the imagination or interest of the American public, including the younger generations, and scholars have shared the public's general disinterest in the region. Few Americans have read a work of literature by a Latin American and most are ignorant of their philosophical, scientific or cultural contributions. Latin America has long been regarded as merely an offshoot of Latin Europe, a fragment of western civilization, without any original or significant culture of its own. Unnoticed is the fact that while Latin America is geographically an integral segment of western civilization, it offers many striking features of cultural differentiation. The most vivid impressions of Latin Americans take the form of stereotyped caricatures conveyed by motion pictures, radio and television programs, and comic books. This has created the fiction that Latin Americans, unlike Western Europeans and North American Yankees, are indolent and reactionary, show little interest in solving political, economic and social problems, and have an aversion to material progress. This unfortunate and inaccurate view of the southern republics, coupled with the notion that their main value lies in helping to further Washington's foreign policy goals, and supporting its defense structure and economy, raises serious obstacles to the growth of a harmonious community of independent states.

As noted in preceding chapters, the darkest hour in Latin American-United States relations came in the late 1920s when it had begun to appear to the former that it was the intent of the "Northern Colossus" to assert political domination of the entire Western Hemisphere. The repeated interventions in the Caribbean region had created the image of Americans as imperialists more than any other aspect of their Latin American policies, but this is not to overlook the more subtle impact of economic imperialism. A wave of Yankeephobia swept over Latin America culminating in the Sixth Pan American Conference in Havana in 1928. The Hoover Administration made a beginning toward improving relations, but it was during the succeeding Roosevelt administration that there emerged the policy of the "good neighbor."

A resumé of the bases of friction, which helps to interpret the Latin American attitude, will add to an understanding of the problem. Mexico resented the exploitation of her petroleum and other mineral resources by foreign corporations, chiefly American, which sent most of their profits out of the country. Cuba was becoming more restive under the Platt Amendment, which permitted legalized intervention, a flagrant infringement of her status as a sovereign state. Nicaragua, Haiti, and the Dominican Republic were either occupied by the U.S. marines or had just seen them withdrawn, with always a possibility of their return. Panama fretted under the Canal Treaty which gave the United States absolute control of the Canal Zone and an overprivileged position outside. Venezuela's oil, Peru's vanadium and copper, Chile's nitrate, copper and iron, Uruguay's meat industries, Brazil's coffee were all controlled by or dependent upon United States markets. The "Colossus of the North" was a tangible and threatening menace to our neighbors to the south in spite of our much vaunted anti-imperialistic intentions.

But the citizen of the United States had only the slightest conception of this unfriendly attitude. Had not the century-old Monroe Doctrine raised a bulwark of protection against European threats of encroachment? Had not our great secretaries of state from Henry Clay to Cordell Hull preached the doctrine of inter-American cooperation? Had not Woodrow Wilson even tried to extend the Monroe Doctrine to the entire world? Doubtless in theory the United States has always believed in the good neighbor policy and on the whole had tried to follow it; however, in practice, deviation has been common.

A bill of particulars from the Latin American point of view will shed light on the question. Mexico received little protection from the Monroe Doctrine when Maximilian established a monarchy right on our border. Argentina would still possess the Falkland Islands if Washington had interpreted the doctrine to safeguard all of Latin America. The Mexican War had more than a tinge of imperialism south of the Rio Grande. Theodore Roosevelt and his Big Stick were not appreciated in the Caribbean and were keenly resented in Colombia. The Platt Amendment did not make for a *Cuba Libre* in the eyes of the Cuban patriots. Secretary Hughes' substitution of the euphemistic term "interposition" for "intervention" was not a satisfactory solution for the sending of marines. In fact, to the realistic Latin American the North American Yankee was more threatening than the European because he was bigger and closer at hand.

Fortunately for the security of the Western Hemisphere, the dawn of a new era had arisen. President Roosevelt proclaimed the policy of the Good Neighbor. Determined henceforth to reconcile promises with deeds, the United States attacked the problem on all fronts–political, economic, and cultural. Marines were withdrawn from every Caribbean outpost and future intervention was outlawed by specific conventions signed and ratified by the United States. Unsatisfactory agreements such as the Platt Amendment with Cuba and the Canal Treaty with Panama were revised so as to be more in accordance with the wishes of these neighboring republics. No longer would we possess the legal right to intervene in Cuba in derogation of her sovereignty. In case of aggression endangering the Canal, the United States and Panama would consult for mutual defense instead of Uncle Sam dictating the policy. Even the Monroe Doctrine, a basic component of American diplomacy, was overhauled and reinterpreted to the benefit of both of the Americas. Instead of the United States alone determining when and if the principles of the doctrine are violated, the American republics now consult together when any threat to this hemisphere arises. Even tariff barriers had to give way before the pressure of good will, and Secretary

Hull signed sixteen reciprocal trade agreements with our Latin American neighbors. One result was to bring about a phenomenal increase of trade between Latin America and the United States.

But the greatest accomplishment in the diplomatic field was the utilization of the Pan-American Conferences to serve as effective agencies for cooperation among the republics. At the Conference held in Lima in December 1938, the representatives unanimously approved a declaration of principles of the solidarity of America, agreed upon joint defense against foreign intervention, and established automatic machinery to make it effective. The simple device agreed upon was to authorize the foreign minister of any American republic to call a meeting of all the foreign ministers when any threat arose.

Hardly was the ink dry upon this document before the Nazi war machine rolled over Poland. A conference was immediately called in Panama and a three-hundred-mile safety zone was established around the American republics, and the belligerents were asked to do their fighting outside. Violations were to be followed by withdrawal of the privilege of fueling and making needed repairs.

A still greater menace arose with the subjugation of France and Holland. Both of these nations had important colonial possessions in the Western Hemisphere. Another meeting of the foreign ministers, this time in Havana, agreed that any move on the part of the Axis to take over these possessions would be prevented by joint action. If the threat was imminent any one power might act to forestall it. That is, all Latin America now trusted us sufficiently to permit the United States to enforce the Monroe Doctrine by itself if the emergency warranted. The Americas had at last recognized the famous doctrine of Monroe to mean all for one and one for all.

The devastatingly successful attack by the Japanese upon Pearl Habor brought the war into the Western Hemisphere. Would the two Americas meet the test and carry out their promised cooperation for hemispheric defense? Would Latin America prove that the Good Neighbor Policy was a policy of reciprocity? The answer came quickly. The day after the attack, Mexico and Colombia broke off diplomatic relations and within four days the nine Central American and Caribbean republics had declared war upon the Axis powers. A meeting of foreign ministers was summoned at Rio to formulate a policy of defense of the Americas.

Since it was now evident that Axis propaganda and subversive activities were rampant throughout the Western Hemisphere and that the German embassies were the source of this problem, the United States felt that the minimum requirement for self-protection was the severance of diplomatic relations to stamp out the danger at its very roots. A resolution recommending such a break was unanimously accepted and before the Conference ended every nation except Argentina and Chile had dismissed the Axis diplomats. Another resolution looked towards a complete economic and financial boycott of the Axis partners. Agreements were entered into for the production and exchange of strategic materials essential to hemispheric defense and the formulation of a complete and coordinated general plan for economic mobilization was envisaged. To coordinate effectively the necessary measures of hemispheric defense, an Inter-American Defense Board composed of military and naval officers of the twenty-one republics was established at Washington.

As a result of the discussions with the military and naval authorities of the several Latin American countries it seemed desirable, as far as possible, to replace foreign material with military or naval equipment from the United States. Since financial conditions made it impossible for the American republics to pay the entire cost, arrangements were made to supply them under the lend-lease act. A preliminary sum of $400,000,000 was allocated for these supplies, and lend-lease agreements were signed by the end of 1942 with all of the republics except Argentina and Panama.

Military, naval, and aviation missions from the United States were made available to every Latin American republic which desired them. By June 30, 1942, some form of military, naval, or air mission was to be found in thirteen of the twenty Latin American republics, leaving out of consideration the huge forces at Panama and the naval base in Cuba. A plan for the effective use of foreign ships in American ports was worked out so that more than a hundred of these vessels were transferred to active service in inter-American trade.

A Conference on Systems of Economic and Financial Control met in Washington in July 1942, to supervise all commercial and financial intercourse between the Western Hemisphere and the aggressor states or the territories dominated by them for the duration of the war. Any transaction that might prove inimical to the security of the Western Hemisphere, whether international or among the American republics, was within its jurisdiction. The United States had already promulgated a black list of firms and individuals suspected of doing business with the Axis powers. A report made in 1941 by the office of the Coordinator of Inter-American Affairs indicated that of some 5000 firms in Latin America, over 1000 were definitely known to be identified with anti-American activities.

To better maintain the economic stability of the other American republics, the United States through the Board of Economic Warfare, in cooperation with the Department of State and the War Production Board, made every effort to facilitate the exchange of commodities. A quarterly allotment was made to the various Latin American republics of specific quantities of iron and steel, chemicals, farm equipment, and other products vital to their national economies.

Preclusive purchasing agreements of strategic materials were made with various Latin American states both to keep their supplies from the Axis powers and to keep an abundant supply available to the Allied nations. For example, Metals Reserve Company made over-all agreements with Brazil for bauxite, chromite, manganese, and other minerals, with Bolivia for tungsten and tin, with Mexico for practically all of its exportable surplus of minerals, with Peru for antimony, copper, and vanadium, and with Chile for copper, manganese, lead, and zinc. Rubber Reserve Company made similar arrangements for rubber with Brazil, Peru, Bolivia, Colombia, Ecuador, and several Central American states. Defense Supplies Corporation bought up surpluses wherever it seemed necessary, such as the reserve stock of Chilean nitrates and surpluses of Peruvian cotton.

The maintenance of health in the Western Hemisphere was also a vital feature of the cooperative war effort. An excellent illustration is afforded by the sanitation agreement with Bolivia signed July 15, 1942, whereby the government of the United States through the agency of the Coordinator of Inter-American Affairs agreed to provide up to $1 million for the cooperative development of a health and sanitation program including general disease control by clinics and public education, malaria control, yellow fever control, cure of lepers, and environmental sanitation. A group of medical and sanitation experts from the United States worked in close cooperation with corresponding officials of the Bolivian government.

The Basic Economy Department of the Coordinator's office, an agency concerned primarily with the health and nutrition of the people of the Americas, under the compulsion of war-time necessity, sponsored more than 500 projects and activities. They ranged from the establishment of health stations in the Amazon rubber country to an extensive rehabilitation program in Ecuador. A series of comprehensive health and sanitation agreements and food-producing agreements made it possible for the two Americas to cooperate effectively to strengthen the human resources of the hemisphere.

One of the most potent agencies for the coordinated economic development of the Americas was established on June 3, 1940 under the general direction of the Inter-American Financial and Advisory Committee. This agency, known as the Inter-American Development Commission, established national commissions in every American republic whose primary functions were to supply information on the possibilities of developing production and trade. This organization proposed not only to establish new industries, but to try and reestablish the production of commodities such as cocoa, coconuts, copra, quinine, and rubber which were once supreme in Latin America, but which were subsequently lost to other areas. It established a market for Brazilain tapioca in the United States and sponsored a sales campaign to stimulate the purchase in the United States of the beautiful handicraft work of the Latins.

As proof of the intention of the United States to take care that the Latin American countries were not discriminated against during the war, an agency of the Board of Economic Operations known as the American Hemisphere Exports Office was created to see to it that the Latin American countries were supplied with essential commodities on a parity with civilians in the United States.

It was during the war that the foundations were established for future United States aid and inter-American cooperative programs of unprecedented dimensions. We should pause and note this early development and its antecedents. Organized bilateral assistance in Latin America was foreshadowed by the activity of religious organizations in the United States, largely in the fields of education, health, agriculture, and social and religious services. Many other organizations, such as industrial firms, labor organizations, foundations, and the U.S. Department of Agriculture contributed programs of technical assistance. The Export-Import Bank, founded by Congress in 1934, was not at first a foreign aid agency since its purpose was to finance the international trade of the United States with intermediate-term credits. In practice, however, it assumed a positive role of great value to Latin America. Of special significance was Brazil's Volta Redonda steel mill, constructed between 1941 and 1946, with a $45 million Eximbank loan to buy capital goods in the United States.[1]

[1] "From its inception in 1934 until June 30, 1961, Eximbank authorized credits of $3,461,170,000 to nineteen out of the twenty Latin American republics (all except the Dominican Republic), of which $2,362,169,000 has been disbursed. These compare with Latin American commitments of the regular foreign aid program (Agency for International Development and its predecessor agencies) of $990,795,000 and expenditures of $664,365,000 from its inception on Apr. 3, 1948, to Dec. 31, 1961." John P. Powelson, *Latin America, Today's Economic and Social Revolution* (New York, 1964), p. 224.

In the decade preceding the Point IV program, 1939-1950, the United States government sponsored three important programs in Latin America. The Interdepartmental Committee on Scientific and Cultural Cooperation, formed in 1939, was the first organized and systematic inter-governmental technical cooperation program in Latin America. More than twenty U.S. government bureaus were members of the Committee. The program was cooperative in the sense that wherever possible both the United States and the Latin American countries were to contribute to the support of individual projects, thereby giving both parties a stake in the outcome. In 1940, as the threat of war to the Americas grew more ominous, the function of receiving and channeling Latin American requests for assistance was assigned to the new office of the Coordinator of Inter-American Affairs, headed by Nelson Rockefeller. His purpose was to relate economic and technical cooperation in the Americas to hemispheric defense. In 1942 the Office of the Coordinator began to cooperate directly with the Latin American governments in joint development schemes. The Institute of Inter-American Affairs, organized as a U.S. government-owned corporation in 1942, was established under the Coordinator to deal with three problems: public health, education, and food supply. Supplementing these activities, the Inter-American Educational Foundation was formed in 1944 to provide assistance in preuniversity education. In the period 1942-1951 the Institute of Inter-American Affairs supplied $88 million in technical assistance to Latin America; it continued as an integral part of the International Cooperation Administration.[2]

The present period in the history of economic assistance and technical cooperation began with President Truman's inaugural address in January 1949 which culminated in the Point IV Program a year and a half later when Congress passed the Act of International Development. This was the juncture at which the concept of technical assistance was extended to economically underdeveloped countries throughout the world. From the inception of Point IV to the present time, the United States foreign aid program has undergone a number of organizational changes. In October 1950, the Technical Cooperation Administration was established within the Department of State. One year later, the Mutual Security Act of 1951 was enacted. Under this Act the Office of Director of Mutual Security was established to coordinate military, economic, and technical assistance. In 1953, the Foreign Operations Administration was established. The functions of the Mutual Security Agency and the Technical Cooperation Administration were transferred to the Foreign Operations Administration. In 1955, the International Cooperation Administration was established within the Department of State, terminating the Foreign Operations Administration. In 1961, the former was succeeded by the Agency for International Development, whose Latin American Regional bureau became the link between the U.S. government and the Alliance for Progress.

The cooperation program in Latin America has grown throughout the years, and has embraced new fields of economic and social development. In the early years it was concentrated in the fields of agriculture, health and education.

[2] U.S. Dept. of Commerce, *Foreign Aid by the United States Government* (Washington, 1952), pp. 24-25; D. W. Rowland, *History of the Office of the Coordinator of Inter-American Affairs* (Washington, 1946).

Today it is operating in additional fields such as transportation, marketing and industry, administrative and management training, military aid, housing, community development, and economic assistance. Multilateral assistance has been introduced through United States support of the United Nations and its specialized agencies, the Organization of American States, and regional and international banking agencies.

If the basic-service programs are to be successful, they must reach down to the people. Military assistance programs serve to protect the freedom of the countries and their citizens, but the impact is not felt directly by the people. A highly efficient means of reaching the ordinary man directly was worked out in Latin America by the Institute of Inter-American Affairs. In Latin American countries *servicios*, meaning cooperative service agencies, were set up within the governmental framework of the host country. A unit of health and sanitation, for example, operated under the ministry of health. The *servicios* were staffed jointly by United States and local personnel, and all such programs were financed jointly.

By 1949 the technicians and field staff of the Institute numbered 325 and they were providing technical and administrative supervision to some 9,500 nationals of sixteen other republics who were carrying on the operations. In the field of health and sanitation the Institute had completed more than 1300 projects benefitting more than twenty-three million people, or one out of every six Latin Americans. When President Truman proposed the Point IV program, the Institute whose Latin American program, according to Secretary of State Dean Acheson, had been "the inspiration and proving ground" of the idea, was the logical organization to carry out the program in the republics to the south. Consequently, the Congress extended both the life and scope of the Institute of Inter-American Affairs.[3]

The Eisenhower administration showed an interest in the policy of cooperation with the Latin American republics, but at first held back and met the growing crisis with restrained half measures. In November 1951, the Institute of Inter-American Affairs was transferred to the TCA, which became the agency for carrying out Point IV within the Mutual Security program. The level of aid was low throughout the 1950s. In the twelve years ending in 1959, U.S. economic development loans and grants to Latin America had totaled less than $500 million, whereas Europe, between 1949 and 1952, received over $11 billion in grants alone.

One of the outstanding grievances of the Latin American nations was that the United States failed to appreciate the need to stabilize the price of their exports, especially coffee. Another complaint was the difficulty in securing loans for capital development. The Eisenhower administration made some concessions on these matters by facilitating the operation of the International Coffee Stabilization Agreement and establishing the Inter-American Development Bank. The Bank's charter provided for a capitalization of $1 billion, of which the United States was to contribute $450 million, the balance to be put up by the

[3]For a detailed account see *The Program of the Institute of Inter-American Affairs* (Institute of Inter-American Affairs, Washington, 1949); U.S. Dept. of State, *Significance of the Institute of Inter-American Affairs in the Conduct of United States Foreign Policy*, Publication 3239, Inter-American Series 36, Washington, 1948.

Latin American countries.[4] It increased the amount of aid granted to Latin America, but it came late in the decade, beginning operations in October 1960. Washington's emphasis in this period was upon private investment. Assistant Secretary of State Roy Rubottom declared that "the volume of public financing is directly related to the amount of private financing which countries are able to attract." Moreover, the United States showed no enthusiasm for proposals to push the economic integration of Latin America.

A partial explanation of our neglect of Latin America is that Congress and the American public were so concerned with problems of defense and the threat of Communist aggression in Western Europe and East Asia that the needs of the Western Hemisphere were overlooked. It is true that the Korean War coincided with the introduction of the Point IV Program, which led Congress to subordinate foreign aid and technical assistance to military defense. For a four year period most aid programs were related to forestalling the Communist military threat. As the Marshall Plan drew to a successful conclusion in Europe and foreign aid had seemingly become an indefinite feature of United States policy, the Latin American Republics complained more loudly that they had been forgotten. Europe had received billions in loans and capital grants whereas Latin America had received only a meager amount of technical assistance.

From 1946, when United States relief funds started flowing overseas following World War II, through June, 1961, the United States had put out a grand total of $53,542,500,000 in all forms of foreign aid; about half of this figure was in military aid. The amount expended in Latin America for fiscal years 1946 through 1960 totaled some $7.4 billion. Direct aid extended by the U.S. government to Latin American countries during this period amounted to more than $3.777 million, without including an estimated $500 million in military assistance, and aid for schools, binational cultural centers, libraries, and research facilities. Private investment by United States citizens in the region in the period 1945-1958 exceeded $7.795 million. To this sum must be added the indirect economic assistance channeled by Washington to Latin America through the United Nations and its affiliated agencies, and premiums paid to many countries for their sugar quota in accord with our system of quotas and price supports for this commodity. It needs to be emphasized, however, that until 1960, only about 2 percent of United States aid went to Latin America.

Latin America was sharply impinged on our consciousness by such events as the violently unfriendly reception accorded Vice-President Nixon in South America, the Castro revolution in Cuba, and unmistakable evidence that the Sino-Soviet bloc was making a bid for the area. Against this background, a major shift in United States policy became imperative. Foreshadowing and prompting the change, Brazilian President Juscelino Kubitschek proposed to President Eisenhower "Operation Pan America," a comprehensive plan for economic and political cooperation among the nations of the hemisphere. In September 1958 at a meeting of the American Foreign Ministers in Washington it was determined that programs of social and economic character in Operation Pan America would be assigned to the OAS. The Committee of Twenty-one was formed to consider the question of economic cooperation.

[4] *Agreement Establishing the Inter-American Development Bank. Treaty Series No. 14* (Pan American Union, Washington, 1949).

Having previously supported the establishment of the Inter-American Development Bank, President Eisenhower announced on July 11, 1960, his approval of the Operation Pan America concept. At a meeting of the Committee of Twenty-one in Bogotá the following September, Undersecretary of State Douglas Dillon unveiled a proposal for an inter-American program of social development. Further deliberations produced the Act of Bogotá under the terms of which the United States government offered $500 million to Latin America in soft loans for social purposes. The Act pledged inter-American cooperation to improve rural housing and land use, educational and training facilities, housing and community improvement, measures to improve public health, and others. This set the stage for broader perspectives and an effort to come to grips with the challenge of development through the Alliance for Progress in the succeeding Kennedy administration.

On March 23, 1961, in a special message on foreign aid, President Kennedy spoke of "launching a Decade of Development, on which will depend substantially, the kind of world in which we and our children shall live." This was followed on May 26 by the president's proposal of a new Act of International Development, the title reflecting the new emphasis. For two decades foreign aid had been largely a response to the continuing demands of national security, first as lend-lease and postwar relief; later, from 1948 to 1951, in the extraordinarily successful Marshall Plan, and finally, in most of the 1950s, as short-term political and military support for nations such as Korea and India with heavy cold war defense burdens. The new legislation laid the foundation of the Alliance for Progress, a far reaching program of economic and social development to hasten the progress of Latin America.

The Kennedy administration quickly took steps to reorganize the aid machinery. The technical assistance grants of the International Cooperation Administration and the lending functions of the Development Loan Fund would be merged into an Agency for International Development. Public Law 480 programs for surplus agricultural commodities were given a new name—Food for Peace—but policy decision was split between the new Agency and the Department of Agriculture. The Export-Import Bank continued its independent functions. United States representatives in the international financing agencies—the World Bank, the IDB and the International Monetary Fund—continued to report formally to the secretary of the treasury. In this fashion the new agency drew together the main elements of past aid programs into one agency. The new legislation also dealt with the problem of the organizational relationship between development aid and diplomacy. AID was brought into the Department of State and the AID administrator was given Undersecretary rank. Since 1961 Alliance machinery has remained in the AID agency, and its top official an officer of AID.

According to the terms of the Alliance, as proposed at a meeting of the Inter-American Economic and Social Council in August 1961, one hundred billion dollars could be invested over a ten year period in the economic development of Latin America. The Latin Americans would furnish 80 percent of this through public and private investment. Another 10 percent would come from private sources, European nations and Japan. The U.S. government pledged to put one billion dollars a year into the region over the ten-year period. And

American private investment in Latin America was to provide 15 percent of the amount that was to come from sources outside Latin America.[5]

The Alliance was officially launched at Punta del Este, Uruguay in January 1962, where a meeting of Consultation of Foreign Ministers adopted the Charter of Punta del Este which, besides indicating the goals of the aid program, pledges the Latin American countries to carry out agrarian and tax reforms and formulate detailed and feasible development plans. The "Declaration to the Peoples of America," prefacing the Charter, proposed the following goals: To . . . strengthen democratic institutions through application of the principle of self-determination by the people; to accelerate economic and social development; carry out urban and rural housing programs to provide decent homes for all our people; encourage programs of comprehensive agrarian reform; assure fair wages and satisfactory working conditions to all our workers; wipe out illiteracy; press forward with programs of health and sanitation; reform tax laws; maintain monetary and fiscal policies which . . . will protect the purchasing power of the many, guarantee the greatest possible price stability; and form an adequate basis for economic development; stimulate private enterprise; find a quick and lasting solution to the grave problem created by excessive price fluctuations in the basic exports; accelerate the integration of Latin America so as to stimulate the economic and social development of the continent. The Declaration also affirmed "that these profound economic, social and cultural changes can come about only through the self-help efforts of each country." However, it recognized that self-help must be reinforced by essential conditions of external assistance which, as indicated above, was pledged largely by the United States. Washington's commitment to the Alliance goals meant that it ceased opposition to two long-sought Latin America economic goals: commodity price stabilization and economic integration. This was reflected in the U.S. becoming a party to the International Coffee Agreement in 1962 and its support of the formation of the Central American Common Market and the Latin American Free Trade Association.

The Punta del Este meeting voted down the establishment of a seven-man panel to give direction and coordination to the program, replacing it with a consultative body, and at the same time rejected proposals for an agency to publicize the Alliance in Latin America. These decisions may have proven detrimental to the basic operation of the Alliance in the long run. Conceived as a multilateral program, the Alliance was to be guided by a panel of economists, the "Nine Wise Men," drawn from both Latin America and the United States. They were appointed to consider all plans, and the final determination of the allocation of funds was to be a joint decision of the participating nations. It soon became clear, however, that Washington would assume the decision-making authority in the disposition of Alliance loans and grants, for this function was entrusted to the Agency for International Development. Teodoro Moscoso, who had so brilliantly guided Puerto Rico's development, was named to the position of American Coordinator for the Alliance for Progress. Holding

[5]Pan American Union, *Alliance for Progress, Official Documents Emanating From the Special Meeting of the Inter-American Economic and Social Council at the Ministerial Level Held at Punta del Este, Uruguay, from August 5 to 7, 1961* (Washington, 1967).

this post until 1964, Moscoso sought to adhere to the multilateral approach in the face of increasing insistence from the State Department and Congress for stricter American control of the allocation of funds. At a meeting of the representatives of hemispheric nations at São Paulo in November 1963, it was decided to appoint a committee to analyze projects proposed by the member states and to make recommendations for the earmarking of funds. Called the Inter-American Committee for the Alliance for Progress (CIAP), it supplanted the "Nine Wise Men," who resigned in 1966, and helped to offset the influence of the United States and create the image of multilateral effort.[6] But in practice the United States, supplying most of the funds, makes the final decision as to their use.

Although the Alliance for Progress captured the imagination of broad sectors of the population at the outset, its failure to attain many of the original goals had provoked widespread cynicism both inside and outside of Latin America by 1963. Contributing to the poor start were the few acceptable plans forthcoming from Latin America; the lack of personnel trained in economic techniques; the flow of American private capital, envisaged as an important part of the scheme was less than expected, and Cuba kept up a relentless stream of hostile propaganda. Its record of achievement has been better than is generally realized, but it is also clear that its record falls short of what its proponents had hoped for, and anticipated. In general, there is skepticism in Latin America that the Alliance will accomplish the fundamental improvements in the economic and social conditions of the area that were envisioned. This feeling was expressed in a joint statement of the "Nine Wise Men," incident to their collective resignations; the Alliance "is passing through a period of crisis" in which there are symptoms of discouragement, skepticism and despair which makes it "indispensable" that there be "clear demonstration of a firm (inter-American) determination to overcome the crisis in accordance with the spirit of the "Charter of Punta del Este."[7]

At the Fourth Annual Meeting of the Inter-American Economic and Social Council (IA-ECOSOC) in April 1966 in Buenos Aires, which was convened for the purpose of evaluating the Alliance at the end of its first five years, several proposals were made for strengthening it. At a Special Meeting of the Council, held in Washington in June 1966 to consider possible revision of the OAS Charter, agreement was reached for revising the "Economic Standards" and "Social Standards" of the document so as to correlate more precisely with the objectives and requirements of the Alliance.

Among the suggestions made for strengthening the Alliance were "Proposals for the creation of the Latin American Common Market" which had been sent by their authors (Raul Prebisch, Director General, Latin American Institute for Economic and Social Planning; José Antonio Mayobre, Executive Director, United Nations Economic Commission for Latin America; Felipe Herrera,

[6] For analysis of Latin American problems by CIAP see U.S. Congress, Joint Economic Committee, 89th Cong., 1st Sess., *Latin American Development and Western Hemisphere Trade* (Washington, 1965), Appendix IV.

[7] Herbert K. May, *Problems and Prospects of the Alliance for Progress* (New York, 1968), p. 1. Nelson Rockefeller criticized the Alliance as "representing neither an Alliance nor Progress." See Simon G. Hanson, *Five Years of the Alliance for Progress, An Appraisal* (Washington, 1967), p. 71.

President, Inter-American Development Bank; Carlos Sanz Santamaria, Chairman, Inter-American Committee on the Alliance for Progress), to the Latin American presidents and the President of the United States. The Proposals, given favorable consideration at the Buenos Aires meeting in March-April 1966, included recommedations on trade, investment, and monetary policies and the institutional machinery required for the establishment and operation of a common market. Recommendations were also made concerning the problem of reciprocity, measures aimed at stimulating Latin America private enterprise within the common market, the position of relatively less developed countries and the question of internal economic dislocations that might appear with the liberalizing of trade.[8]

At both the Buenos Aires and Washington meetings in 1966 much attention was given to suggestions long held by Latin Americans that Washington should extend preferential import treatment to the export products of Latin America comparable to the import preferences given by the United Kingdom and the European Common Market countries to the export products coming from other parts of the world. The Latin American representatives at the Buenos Aires meeting laid considerable emphasis on the need for Washington to "liberalize" the conditions which it has insisted upon for obtaining and utilizing AID and Export-Import Bank loans. The argument was reiterated that AID had been unrealistic in its demands for "self-help" as a prerequisite for loans and that both of the agencies should relax their barriers to the use of loan dollars for purchases other than in the United States.

Total U.S. assistance under the Alliance has nonetheless been impressive. With AID funds averaging over $550 million for each of the five fiscal years ending June 30, 1967, Public Law 480 surplus agricultural commodity assistance, Export-Import Bank loans and U.S. funds administered under the Social Progress Trust Fund produced a total of new commitments of roughly $1.4 billion for each year of the period. (and international lending institutions, along with the other rich countries, had committed another $3.3 billion over the five years ending in 1966). On the fifth anniversary of President Kennedy's first White House proposal for the Alliance, President Johnson was able of give an impressive accounting of the tangible accomplishments of this assistance.[9]

In spite of evidence of visible achievement, however, the Alliance's performance was increasingly called into question. In an effort to clarify the question and reach a fair assessment, the Committee on Government Operations of the United States House of Representatives requested a study to determine whether the goals of the Alliance for Progress as they applied to each country were currently realistic or attainable in the light of experience of the first seven years. The Committee, reporting on its investigation, declared that "our review of the experience of the past seven years indicated that accomplishment of the ambitious Alliance goals within the decade 1961-1971, as contemplated in the Charter of Punta del Este, is not possible. The framers of the Charter erected a goal structure which anticipated too much too soon . . . however, that achieve-

[8] U.S. Congress, *Latin Amer. Dev. and West. Hem. Trade*, Appendix V.
[9] *New York Times*, Aug. 18, 1966.

ment has been substantial." It was also concluded that an analysis of progress toward goals raised the question of "relative priorities . . . realistically simultaneous progress toward each of the goals cannot be expected . . . there were great differences between one country and the next in the development bottleneck that require early resolution as a precondition to progress on other fronts . . . the countries of the hemisphere differ markedly in their capacities to achieve political commitment and concensus behind various goals and policies, or to mobilize technical, financial and institutional resources behind the programs leading to the goals . . . "[10]

The achievement and maintenance of financial stability is a clear high-priority Alliance goal. This condition is required to encourage increased private and public sector savings and investment. Neither of these objectives can be attained under conditions of extreme inflation and repeated foreign exchange crises. Evidence indicates that despite growing demands for larger public sector expenditures, most countries have collected more taxes and pursued improved budgetary practice, thereby avoiding destabilizing fiscal and monetary policies. The Latin American economy in the 1960s and early 1970s was characterized by significant inflationary pressure, a development found in many other parts of the world. Of twenty countries for which complete data on the consumer price index are available, twelve showed an acceleration of the price growth rate between the first and second half of the 1960s. Latin American nations experiencing severe inflation in 1972 (figures indicate percentage increase in consumer price index) were Uruguay, 76.5; Brazil, 16.5; Chile, 77.9; and Argentina, 58.6. Moderate inflation was found in Colombia, 14.3; Peru. 7.2; and Bolivia, 6.5. The other countries of the region were classified as being relatively stable.[11]

Another related and high priority goal in the Charter is the achievement of regional economic integration. Latin America's economic integration should be viewed as part of this growing tendency which started in the 1930s with the establishment of the Commonwealth Preferences and intensified after World War II with the formation of the Soviet trading area, Benelux, and, more recently, the European Economic Community (E.E.C.), and the European Free Trade Association (E.F.T.A.). In contrast to the European Common Market, which rests on predominantly political considerations, Latin American regionalism has arisen largely from the deterioration of the area's status in world trade with its accompanying dampening effects on economic development and industrialization.

Since the late 1920s Latin America has drifted further toward the periphery of world trade. One index of this development is that over the period 1928-1959 Latin America's share in world exports (excluding U.S.S.R.) dropped from about 11 percent to 8 percent. The slow down in exports reflects the failure of world demand for foodstuffs and raw materials to keep pace with the output and demand for industrial products. This is one of the main factors which interprets the deteriorating export position of Latin America. Illustrative

[10] U.S. House of Reps., 91st Cong., 1st Sess., *A Review of Alliance for Progress Goals* (Washington, 1969), pp. 1-2. (hereinafter referred to as *Alliance for Progress Goals*).

[11] *Ibid*., p. 3; U.S. Senate, 90th Cong. 1st Sess., *Survey of the Alliance for Progress–Inflation in Latin America* (Washington, 1967), p. 3; Inter-American Development Bank, *Economic and Social Progress in Latin America, Annual Report 1972* (Washington, 1973), pp. 8-9.

of the problem, world manufacturing output increased by 146 percent between 1928 and 1957-59, whereas the expansion of world exports of primary products, excluding petroleum, came to only 14 percent. The value of Latin American exports rose by 5.2 percent a year in 1966-1969, but this rate was significantly below that of 10 percent recorded on a global basis. While Latin America's trade has lagged, the area's dependence on exports remains high. The magnitude of the problem can be seen in the fact that in the majority of Latin American countries, exports still represent between 15 and 25 percent or more of the GNP, compared with 4.5 percent in the United States. Moreover, in most cases, two or three export commodities at most make up more than three-fourths of the total export receipts. The dependence on export markets is especially precarious because of their inherently unstable nature.[12]

Lending urgency to the need for economic integration in Latin America is the new threat to its export position by the proliferation of regional preferences elsewhere, notably in Western Europe. The creation of the EEC and the EFTA, together with the already existing system of Commonwealth preferences jeopardizes Latin America's significant commercial stake in Western Europe. European regional agreements pose a particular danger for the region in the possible displacement of such exports as coffee, cocoa, bananas, lead, zinc, hides and skins, vegetable oils, sugar, wheat, meat and hardwoods. These commodities represent about half the total shipped to EEC and would comprise about 7 percent of total Latin American exports to all destinations. Especially significant are the preferences accorded by the European Common Market to imports from certain African countries, former dependencies which have special historical and cultural ties with its members.

Solid progress toward the goal of regional integration is shown in the Central American Commom Market (CACM), founded in 1960. Years of negotiations among Guatemala, El Salvador, Honduras, Nicaragua, and Costa Rica culminated in the formation of regional ties providing for free trade within the market, a common external tariff and harmonization of a wide range of economic, financial, and social policies. These developments have stimulated local and regional trade and commerce and have laid the ground work for growth in manufacturing. Handicraft industries are being supplemented by enterprises utilizing modern technology. As a result, foreign investors are increasingly responding to the opportunities presented by an expanding, unified Central American economy.

The CACM is being fashioned through the implementation of a series of agreements aimed at drawing the five nations together and welding them into a single, rapidly growing industrialized market. A major reason for promoting economic integration in this region is the knowledge that only through industrialization can these countries lessen their vulnerability to the fluctuating prices of coffee, cotton, and bananas, the area's chief commercial commodities,

[12]*Current History,* 43, July 1962, pp. 1-2. For the base period 1960-1962 to 1968-1970, the average annual rate of growth of agricultural exports was 4.4 percent, and imports, 6.5; the region's share of world exports and imports in 1970 was 14.7 and 3.9 percent. Regional exports of manufactures increased at a rate of 13.1 percent in 1968-1970; in 1971 the export volume of primary metals from Latin America declined an average of 26 percent. See IDB, *Annual Report 1972,* pp. 10-19, 59.

and raise the income level of their people. Proponents of integration regard the individual economies as too small to support an efficient manufacturing industry. But combined, these comprise an economic unit of respectable size. In area, about the size of France, the CACM nations support a population of about fifteen million; on a per capita basis, the GNP averaged about $290 (1964) Intra-zonal trade in the CACM increased in value from U.S. $36.8 million in 1961 to U.S. $214 million in 1967, the average annual rate of expansion exceeding 30 percent.

Intra-area exports in the 1966-1969 period expanded at an annual rate of 17.1 percent compared with 3.9 percent for extra-zonal exports. In 1969, however, the market disruption resulting from the armed conflict between Honduras and El Salvador led for the first time to a contraction of intra-area trade, which fell from $259.1 million in 1968 to $251.9 million in 1969, or by 3.2 percent.[13]

Obstacles to integration are abundant. The similarity of their economies makes it difficult for these countries to achieve a meaningful integration. Domestic capital is scarce and nonagricultural resources are meager and inaccessible. Skilled technicians, semi-skilled and partly trained workers are in short supply. Physical isolation caused by the forbidding terrain and an inadequate road system was the fundamental barrier to administrative and economic integration. With the minimal overland flow of commerce and people, only limited geographic diversification of economic activity was possible. But road building on a regional basis has been pushed for several decades. Highways now link the principal cities, and connections between ports in the Caribbean and Pacific coasts are being expanded.[14]

Of greater magnitude, but less successful to date than the CACM is the Latin American Free Trade Association (LAFTA). In 1960 seven Latin American countries agreed to work toward the creation of a free trade area as a means of speeding their economic development. With the signing of the Treaty of Montevideo they linked their domestic markets within the framework of the new organization. Its goal is the creation of a broad multinational market which would encourage investment in new industry, provide jobs for their rapidly expanding populations and introduce the benefits of industrialization and mass production. The Treaty, ratified in June 1961, by Argentina, Brazil, Chile, Peru, Paraguay, Uruguay, and Mexico was later adhered to by Venezuela, Colombia, Ecuador, and Bolivia. As a result, a regional market essentially free from trade restraints and encompassing eleven nations is being brought into existence through a series of tariff reductions to be negotiated over a twelve-year period.

Negotiations among the member states since 1961 have produced tariff and other concessions on more than five thousand products, and trade has risen somewhat within the area. With the increasing activity, however, there has come

[13]U.S. Cong., *Latin Amer. Dev. and West. Hem. Trade,* pp. 64-72; Inter-Amer. Dev. Bank, *Socio-Economic Progress in Latin America (Social Progress Trust Fund Eighth Annual Report, 1968)* (Washington, 1969), p. 17. The CACM was seriously weakened by the withdrawal of Honduras (1971) and Costa Rica (1972). However, preliminary figures for 1972 indicated an upward trend in intrazonal exports.

[14]Dept. of State *Bulletin,* LVII (Oct. 23, 1967), pp. 535-536.

a greater awareness of the scope and complexity of the region's economic problems. Historically, the volume of intra-area trade has been small: in 1961 before the tariff cutting began, it totaled less than 19 percent of the members' overall trade. Thus despite the substantial percentage growth in intra-LAFTA trade—about 85 percent between 1961 and 1964—the increase in absolute terms has been nominal and appears to have benefitted most of those countries already possessing industrial advantages. Argentina, Brazil, and Mexico, having substantial industry, stand to reap the greatest benefits from the creation of a free trade area. Soon after the treaty's ratification, Ecuador and Paraguay were designated "less developed countries" and thus entitled to special tariff concessions. These countries have also received financial and technical assistance from other LAFTA members. Later, Chile, Colombia, Peru, and Uruguay, whose economic development place them in an intermediate category, were reclassified as "countries with insufficient markets" and therefore entitled to similar benefits.

LAFTA was formed on the assumption that the breaking down of trade restrictions among its members would provide a market large enough to stimulate appreciable industrial growth. Trade liberalization alone, however, cannot achieve this goal. It is becoming increasingly clear that balanced growth within LAFTA will also depend upon the degree to which domestic monetary, fiscal and exchange rate policies can be successfully coordinated. Difficult problems were created for Latin America by the decline in export prices for primary products in the mid 1950s. A sharp drop in export earnings resulted in widening trade deficits. Foreign exchange reserves were rapidly depleted to pay for needed imports. This led many countries to step up programs of industrialization aimed at import substitution. But these subsidized industries operating within small heavily protected markets have often been uneconomic, and when combined with deficit spending by governments they have created serious inflationary pressures. The resultant distortion of price levels and currency instability have dislocated normal trade patterns. Trade has always been hindered by differences in tax and wage levels which cause disparities in production and marketing costs.[15]

The experience of the first seven years of the Alliance was not too encouraging, as mentioned, and it was clear that LAFTA, which it had been hoped would produce a painless free trade, had failed to move at the required speed. Obviously, new measures were needed to carry the program beyond item-by-item tariff negotiations toward automatic tariff reductions. At a meeting of the foreign ministers in November 1965, the member states endorsed this concept and pledged themselves to support a major reorganization of the LAFTA apparatus. In June 1966, the Inter-American Economic and Social Council proposed amending the OAS Charter to provide for a common market. The foreign ministers, meeting in February 1967 under the auspices of a special inter-American conference in Buenos Aires executed a protocol for the new charter committing the member nations "to establishing a Latin American

[15] U.S. Cong., *Latin Amer. Dev. and West. Hem. Trade*, pp. 57-64.
[16] William D. Rogers, *The Twilight Struggle* (New York, 1967), pp. 166-167; U.S. Senate, Subcommittee on Western Hemisphere Affairs, 91st Cong., 1st Sess., *Rockefeller Report on Latin America* (Washington, 1970), pp. 40-41, 107-108. (hereinafter referred to as *Rockefeller Report*); IDB, *Socio-Economic Progress*, pp. 17-19.

Common Market in the shortest possible time." Against this background it was clear that by 1967 the Latin Americans had concluded that common effort was essential to success in overall development.[16]

These negotiations culminated in the Punta del Este Summit Meeting of American Presidents, March 12-14, 1967. At this, the third meeting at Punta del Este since the Alliance began, and the first gathering of chiefs of state since 1956, the member states agreed to lower their customs barriers in automatic stages so that ultimately there would be free trade throughout Latin America. LAFTA's tedious and slow negotiation technique would be replaced by automatic tariff reductions, as in the case of European Common Market and the Central American system. The prospect of integration meant that within the broad Alliance concept, Latin America and the United States were both prepared to tolerate and support unity within Latin America.

Difficulties connected with the trade liberalization program culminated in December 1969 in the adoption of the Protocol of Caracas, which extends from 1973 to 1980, the period for achieving full operation of the Treaty. This also includes a Plan of Action for 1970-1980 with a view towards establishing a Latin American Common Market.

The United States has a large stake in LAFTA's future, being its principal customer and chief supplier. As the tempo of the intra-LAFTA trade accelerates, reflecting internal tariff cutting and the removal of other barriers such as prior deposits on imports, some U.S. producers may find it difficult to compete in this market. In the long run, however, as LAFTA members become more industrialized, export opportunities for advanced U.S. industrial exports should increase.

A development within LAFTA has been the formation of the Andean Common Market, which plans to accelerate trade liberalization and create a common market by 1980. The Andean Subregional Integration Agreement was signed at Cartagena, Colombia on May 25, 1969, by Colombia, Bolivia, Chile, Ecuador, and Peru. The Agreement seeks to (1) promote the balanced and harmonious development of the member countries; (2) accelerate growth by means of economic integration; (3) facilitate the subregion's participation in the integration process envisaged in the Treaty of Montevideo; (4) to establish favorable conditions for the conversion of LAFTA into a common market.

An organization aimed at integrating the economies of the former British colonies in the West Indies, the Caribbean Free Trade Association (CARIFTA), was established in May 1968, with the signing of the Treaty of St. John's by Trinidad and Tobago, Jamaica, Barbados, Guyana, and Antigua. The territories of Dominica, Grenada, Montserrat, St. Lucis, St. Vincent, St. Kitts, Nevis, and Anguilla also joined in 1968. In June 1970, British Honduras became the twelfth member of CARIFTA. The treaty provides for tariff exemptions for a specific list of commodities originating in the area. It also calls for tariff reductions of 20 percent yearly in the case of the more developed countries–Trinidad and Tobago, Jamaica, Barbados and Guyana–and of 50 percent every five years for the other members. A protocol on agricultural products under which area production receives preferential treatment, is also included in the Treaty.

Steps were taken concurrently to set up sub-regional financing institutions and area-wide payment mechanisms to finance intra-regional trade. In Central

America, the Central American Bank for Economic Integration (CABEI) was created in December 1960 to mobilize resources for development projects vital to the area's integration movement. In the Andean Group, the Agreement Establishing the Andean Development Corporation was signed in February 1968. The five Cartagena Agreement countries, and Venezuela, comprised its membership. Its goal is to encourage the process of sub-regional integration by promoting and financing public and private investment projects at the national and sub-regional levels. In the Caribbean, the Agreement Establishing the Caribbean Development Bank became effective in January 1970. In its lending policy the Caribbean Development Bank gives special consideration to the lesser developed countries of the area.[17]

Closely related to the goals of stabilization and integration are the dual Alliance objectives of export diversification and industrial growth. The first of these objectives has a high priority for most Latin American countries since their economies tend to rely heavily on the export earnings of a very narrow range of export commodities. Export diversification is essential as a means of increasing exports, from whose earnings the foreign exchange requirements of development are financed. It serves also to insulate the Latin American economies from the stresses of adverse movements in world markets of the prices of their major exports. A small drop in world prices for any of their main commodities can interrupt development programs by reducing the availability of foreign exchange needed to finance the importation of machinery, technology and raw materials. For instance, a one cent drop in the world market price of coffee reduced the foreign exchange earnings of Brazil by some $23 million.

Latin American exports of manufactured goods increased from 9 to 13 percent of the total exports over the seven year period. Brazil's exports in 1966 were about four times the 1962 level. Central America's trade in manufactures had grown markedly and Colombia's trade patterns showed similar improvement. The decline in reliance on coffee as an export commodity is evident in such major coffee growing countries as Brazil (coffee down from 53 to 44 percent of total exports), Colombia (74 to 65 percent), Costa Rica (56 to 40 percent), and Guatemala (67 to 31 percent).

In the period 1960-1967 increases of 39 percent and 32 percent respectively were attained in the value of Latin American exports and imports, including intra-regional trade, and a 22 percent improvement was shown in international monetary reserves. At the same time, Latin America maintained a favorable trade balance with annual surpluses that ranged between a minimum of U.S. $920 million in 1961 and a maximum of U.S. $2,450 million in 1965. The value of the region's imports in the period under review rose at an annual rate of 4 percent, almost 1 percent lower than that of exports; imports amounted to about 82 percent of the value of exports. However, the deficit on current account rose from U.S. $430 million in 1963 to a peak of U.S. $1,570 million in 1967. The deficit came from a negative balance in the services account—transport, insurance, investment income—which mounted annually from $2,200 million in 1960 to $3,500 million in 1967. The net payments for these services

[17]For review of regional integration see IDB, *Socio-Economic Progress, Tenth Annual Report 1970* (Washington, 1971), chap. III, pp. 34-42; *Ibid.; Annual Report 1972,* pp. 72-85.

absorbed an average of 28 percent of the total export revenue in 1960-1967. In 1967 net payments on services constituted 30.4 percent of Latin America's total export revenue. In terms of its value Latin America's share in world trade declined from 7.3 percent in 1960 to 6 percent in 1967. Meanwhile the region's balance of payments position was sound, for its international monetary reserves rose from $2,735 million in 1960 to $3,310 at the end of 1967.[18]

In the period 1966-1969 Latin America's balance of payments continued to show appreciable strength, with monetary reserves reaching a level of $4,475 million by the end of 1969. In the same period the value of exports, including intra-regional trade, grew at an average annual rate of 5.2 percent. This figure was much lower than that of 10 percent recorded on a global basis, and it was also below the 7.2 percent expansion rate attained by the less developed countries in general. However, in 1970 Latin American exports rose by an estimated 8 percent, which reflected strengthened world market prices for the region's main exports. Another favorable trend was the increase in manufactures as a share in total exports—from 9 percent in 1960 to 17 percent in 1969—for the region as a whole.[19]

Trade between the United States and Latin America also expanded. From 1968 to 1970 U.S. imports from the twenty-two OAS members increased 12 percent, while U.S. exports to those countries expanded 22 percent as our favorable trade balance with the region continued to grow. This growth continued into mid 1971, but at a slower rate; imports for this period continued at about the same level.[20]

Traditionally, Latin American nations seeking foreign exchange solutions have tended to focus on the question of raw material prices. The fact that the region has depended for about 90 percent of its foreign exchange earnings on a limited number of exports has brought forth many proposals to strengthen commodity prices artificially. International commodity cartels to protect producers against further competition and maintain prices at a satisfactory level, have commonly been proposed. Such plans are frequently unworkable in the sense that it is impossible to avert price warfare in world markets; they can prove dangerous by the tendency to lure producer countries to invest in products for which international demand is falling off.

Another, seemingly more promising, approach is to be found in programs designed to stabilize price fluctuation rather than to raise prices artificially. Representative is the International Coffee Agreement, the most important commodity measure adopted since the Alliance began, for coffee accounts for half or more of the export receipts of six countries, including Brazil and Colombia. The members expect to attain the production-consumption balance essential to price stability by a coordinated series of restraints on new coffee by the member nations. To make more effective the system of self-imposed restraint, the consumer and producer nations agreed to a system for balancing the supply of coffee with demand through three major devices; quotas on all member exporters; reporting on imports by importing countries; and additional policing powers against producing countries which exceed their quotas. In

[18]*Alliance for Progress Goals,* p. 31; IDB, *Socio-Economic Progress* (1968), pp. 7-15.
[19]IDB, *Socio-Economic Progress* (1970), pp. 19-32.
[20]*Ibid.*; Dept. of State *Bulletin*, LXV (Aug. 30, 1971), p. 239.

practice, producing countries have found it easier to agree to the principle of production control than to arrive at specific quota figures for individual countries. U.S. delay of almost a year in approving the implementing legislation was a setback for the program initially; however, the partial success of the system gives encouragement for the future.

Efforts to stabilize sugar prices have largely failed. The International Sugar Agreement was signed in 1954 and renewed in 1959, while sugar prices continued to fluctuate. The inherent difficulty is that the agreement covers the marketing of only 40 percent of the sugar sold in international trade. The United States, the United Kingdom and the Common Market are not members. It is the U.S. which maintains a high and artificially steady domestic price for sugar, irrespective of international prices, and determines by legislation, instead of free price competition, the amount of Latin American cane sugar which will be introduced into its protected market.[21]

This system can be criticized on several grounds. Generally speaking, Latin American cane sugar is less costly to produce than domestic beet. To increase the share of the U.S. market assigned to domestic beet sugar is uneconomical, and to Latin Americans, is inconsistent with the Charter of Punta del Este. Moreover, the legislative procedure in the House of Representatives of allocating the relative shares of the U.S. market among foreign suppliers creates distrust among the producing nations. Aside from coffee and sugar, similar problems may be found attending the production and marketing of petroleum, cocoa, tin and other raw material products.

It is not surprising that Latin Americans are seeking more extreme solutions to commodity problems. Frequently mentioned is a closer trade link with the United States since the U.S. in the Alliance Charter pledged to bend special efforts for Latin American's development. The United States is the world's largest and wealthiest consuming area; Europe has its own trading bloc. For the Alliance to become a serious cooperative effort it has been suggested that a hemispheric preferential trading system be created in which Latin America could send its exports to this country free of tariffs. Although 55 percent of Latin American exports to the U.S. already enter duty free, difficulties loom for completely free trade. For the United States to create tariff preferences for tropical agricultural products such as bananas, coffee and cocoa would mean imposing duties where none now exist on African and Asian exports. Such a step would doubtless prove harmful to U.S. relations with exporting countries in those areas.

From any standpoint there is an urgent need, as stated in the Charter of Punta del Este, for " . . . comprehensive agrarian reform leading to the effective transformation where reward of unjust structures and systems of land tenure and use . . ." The agrarian reform goal of the Alliance, by which is generally meant a redivision and distribution of land, is for most nations the most difficult goal to confront. Land reform involving the breaking up of large holdings collides with an element of the power structure that is normally well represented in the

[21]Rogers, *Twilight Struggle*, pp. 156-157; *Rockefeller Report*, p. 139. World consumption of sugar in 1971 exceeded production, thereby reversing the trend of oversupply which had characterized that market since 1964. Because the sugar market in the past has been unfavorable, several Latin American countries neglected their production and were having difficulty in satisfying both domestic and foreign demand in 1972.

national government attempting the reform. From the beginning of Iberian settlement in the New World the gulf between the few who owned much of the best land and the majority who had little or nothing was the foundation of the power structure in most Latin American nations.

The process of implementing the promise of the Charter into substantive land reform programs has been painfully slow. Seven years after Punta del Este fewer than half the nations of Latin America had made any serious moves to change landholding patterns to reform rural society. This partially reflects the abuse of political power, for the privileged landholders of Latin America have blocked the land reform movement at practically every turn. But in part the lack of progress stems from the difficulty of articulating programs which will function efficiently. Moreover, in many countries land reform is not the clear path to progress. For these reasons many Latin American governments have given first priority to the stimulation of agricultural activity, including technological modernization, market development, credit extension services, feeder roads and storage facilities. Half the countries increased their expenditures in agriculture by more than 50 percent from 1963 to 1967. Much controversy still exists about how to select recipients, and the quantity and type of help the new owners may require to succeed. Nevertheless, about one million Latin American families were settled or resettled in the period 1960-1967. An estimated half million land titles were distributed, as were almost forty million *hectares* of land. About half of this land was in Mexico, with the balance being largely in Bolivia, Venezuela, Colombia, and Chile.

Progress in land reform can be seen most clearly in Mexico and Bolivia where the process began with revolution and continues. But Venezuela, Colombia, Peru, Chile, and Brazil have redistribution programs in various stages of advance. Fidel Castro exported guerrilla war to Venezuela in 1963, but before it reached its peak, President Romulo Betancourt's government had redistributed some fifty thousand farms. This measure helped to blunt the terrorists' offensive in the countryside and contributed to the holding of free elections in 1963. In Peru where the agricultural census of 1961 disclosed that 70 percent of all the farmlands were possessed by less than 2.2 percent of the farm owners, land reform is of crucial importance. Some progress is reported since the passage of the Agrarian Reform Law of May, 1964 which permits compensation in bonds for the huge farms, and the activities of the "*Cooperación Popular*" program that has helped to raise the living standards of the rural masses. In Chile, under President Frei's "Revolution in Freedom" genuine land reform was just beginning to alter the feudal social structure. Brazil had taken steps toward opening its vast, underutilized public lands to colonization and improving agricultural productivity. Title is fundamental to land reform, and simply to define titles is a highly complex task in Latin America. Even more difficult is the problem of compensating the old landowner in an equitable fashion. And then to insure that the new owner can make a contribution to the nation's output requires new programs of financing and farmer education.[22]

[22]*Alliance for Progress Goals*, p. 31; *Rockefeller Report*, pp. 181-182; IDB, *Socio-Economic Progress* (1968), pp. 35-43; *New York Times*, Aug. 7, 1968. Chilean land allotments, following expropriations, totaled (in acres) 1,672,447.5 in 1964-1970, and 471,709.4 in 1971-1972. The rate of expropriation increased rapidly after November 1970 under the Allende government.

Another Charter goal is the commitment "to press forward with programs in health and sanitation in order to prevent sickness, combat contagious diseases, and strengthen our human potential." Again, progress can be seen but it falls short when measured against existing needs. The Charter objective of providing at least 70 percent of the hemisphere's urban population with potable water was to be attained in 1971. New or improved water supply services have been made available to some forty three million people in the cities of Latin America. However, only nineteen million rural people have potable water in a rural population of 128 million; the goal for this sector was potable water for 50 percent of the total by 1971. In 1967 only forty eight million persons, or 36 percent of the population was provided with sewer services leaving some sixty two million unprovided for.[23]

Most Latin American nations have taken steps to reorganize public health programs, and some headway is shown in public health instruction, vaccination and malaria control efforts. Death rates from communicable diseases declined by 48 percent in Middle America and 22 percent in South America in the decade ending in 1966, but death rates from these causes were still ten times greater than in North America. The shortage of trained medical personnel is critical, and the advance towards fulfilling needs is slow. The average number of physicians graduating annually has risen from 6,800 to 9,200 and is expected to increase more swiftly when the forty-one medical schools opened since 1961 reach full capacity. In 1966 there were some 148,000 physicians in Latin America, or 6.0 for each 10,000 persons; in North America the ratio was 15.2 per 10,000.

The importance of improving nutrition is well recognized, particularly in the case of young children. Although food production had risen by 37 percent in the decade of the 1960s, as compared with the 1957-59 level, it had scarcely managed to keep pace with population growth. Estimates by the Food and Agriculture Organization (U.N.) (FAO) indicate that only five of fifteen countries for which data are available had a food calorie intake above the average daily requirements. The U.S. government has helped to overcome this deficiency through its Food For Peace operation which contributed to the feeding of thirteen to fifteen million school children in eighteen Latin American countries.[24]

To increase greatly the level of agricultural productivity and output and to improve related storage, transportation and marketing services is an Alliance goal not fixed in terms of quantitative projections of growth or improvement, and is therefore subject to a number of interpretations. But generally speaking, the most reliable indicator of agricultural progress is the index of agricultural production per capita. On this basis, per capita production over the seven year period showed that, despite the total increases in output attained, the output per person for Latin America as a whole did not increase significantly. Data indicate that there were seven countries with increases of 5 percent or more in per capita production, eight countries with declines of 5 percent or more, and three with no appreciable change.

Research by the FPAO on production and acreage for the major crops of

[23] *Alliance for Progress Goals,* p. 40; *New York Times,* Mar. 16, 1969. The potable water goal for urban population was achieved, but only four countries reached the level of 60 percent for sewage services. In rural areas only four countries met the goal of potable water for 50 percent of the population. See IDB, *Annual Report 1972,* p. 102.

[24] *Alliance for Progress Goals*, p. 41; *Rockefeller Report*, pp. 187-194.

each country shed light on recent trends in agricultural activity. "In all, yield figures were examined for five crops in each of eighteen countries, or a total of ninety crop-country combinations. In sixty of the ninety cases, or two-thirds, there were more crops with declines in yields than with increases, and there was no country which did not have an increase in yield in at least two of its principal crops. "It is therefore clear that some progress has been made during the Alliance period in raising agricultural productivity. Agricultural productivity in terms of output per worker is still difficult to measure in most of Latin America because of the absence of data on agricultural employment. While it is evident that some increases in agricultural production have occurred in Latin America, it does not necessarily follow that such advance is an indication of satisfactory progress unless it is accompanied by more exportable surpluses, decreases in imports and an expanded supply of agricultural products for domestic use.

Inter-American Development bank reports indicate that in the 1960s the rate of expansion of Latin American agriculture was slower than that of the overall product of the region. For the 1960-1965 period the annual growth rate of output was 4.2 percent. While the agricultural sector's share in the formation of Latin America's Gross Domestic Product does not exceed 20 percent, it employs 45 percent of the area's economically employed population.

From 1966 to 1969, indices of per capita production of agricultural commodities and foodstuffs fell short of 1961-1965 levels with respect to production in general, or barely matched them in foodstuffs; per capita agricultural output declined at an annual rate of 0.5 percent in the period 1966-1969, and by 1.4 percent in 1969. In other words, agriculture did not contribute to an improved supply of foodstuffs for the region, which must be stressed because of the nutritional deficiences afflicting most of the population. In the same period only Africa saw a less satisfactory development of per capita output than did the Latin American region.

Trends in agricultural production in 1966-1969 reflected a classification of countries into three groups. In the first instance were Colombia, Costa Rica, the Dominican Republic, Honduras and Venezuela in which output grew by 4.0 to 7.9 percent a year. At the other extreme were Barbados, Chile, Haiti, Jamaica and Nicaragua, in which there were production declines of 1.0 to 5.9 percent. In other nations of the area, output increased from a minimum of 0.2 percent in Peru to a maximum of 3.3 percent in El Salvador.[25]

Industrialization, as a means "to increase the productivity of the economy as a whole, taking full advantage of the talents and energies of both private and public sectors . . ." is an Alliance objective. As a symbol of economic development as well as a major instrument of production, industrialization has

[25] *Alliance for Progress Goals*, pp. 24-25; IDB, *Socio-Economic Progress* (1968), pp. 6-7; *Ibid.*, pp. 9-10. External assistance from international institutions for agricultural development has been given high priority in recent years. In the period 1961-1968 $1.1 billion in loans was authorized for that purpose by the three main development institutions: The IDB, AID and the World Bank group. The Export-Import Bank and the International Finance Corporation also extended credit for agricultural development. The IDB has been the largest lender, providing half the loans to the agricultural sector. However, owing to population increase, per capita agricultural production in 1971 declined about 2 percent from 1970; per capita food production dropped 3 percent from 1970 to 1971. See IDB, *Annual Report 1972*, p. 12.

become a panacea in many less advanced countries where it has been pushed to the neglect of agriculture, education and other sectors. As shown in many instances, this deprives industry of both adequate raw materials and a human resource base for efficient production, while low income and output in other sectors limits demand for industrial products. In general, data on industrial growth give only a rough idea about the status of the national economy.

One of the outstanding features of Latin America's economy in the years 1960-1967 was the rapid expansion of manufacturing activities, which contributed 25 percent of the gross national domestic product, compared to 20 percent for agriculture. In the 1960-1966 period Latin American industry registered a growth rate of 5.9 percent annually, while the region's overall output of goods and services expanded at an annual pace of 4.6 percent. The production of semi-manufactured goods and the processing of metals such as steel, chemical products, pulp and paper and automobiles rose rapidly, whereas evidence of stagnation was seen in traditional activities such as textile manufacturing and food products. The industrial sector continued to grow more rapidly than the Latin American economy as a whole in 1969, with manufacturing production advancing from 7 percent in 1966-1968 to 7.7 percent in 1969. However, Latin American industrial growth rates, which increased from four-fifths of the world rates in the first half of the 1960s to nine-tenths in 1966-1968, dropped back to only two-thirds of the 1969 world rates. The highest rates of growth (7.3 to 10 percent) were attained by Brazil, Nicaragua, the Dominican Republic, Guatemala, Mexico and Costa Rica. But the nations with the largest proportion of gross domestic product coming from manufacturing included those regarded as most advanced. In order of relative size of manufacturing output they were Argentina, Brazil, Mexico, and Chile. Despite impressive development trends in Latin American industry in the 1960s the overall growth rate has gradually weakened in recent years because relatively few prospects for import substitution remain in most countries. Continued growth will depend on the expansion of markets by means of economic integration and by efforts to compete favorably in world markets.[26]

Alliance goals in the field of education seek to eliminate adult illiteracy, and to assure, as a minimum, access to six years of primary education for each school child in Latin America; to modernize and expand vocational, technical secondary and higher educational and training facilities, to strengthen the capacity for basic and applied research; and to provide the competent personnel required in rapidly growing societies. The framers of the Charter thus combined the productivity goals of the Alliance with objectives for broadened educational opportunity, agrarian reform, and health and housing programs, all aimed at the primary development objective of the Alliance: a more ample and equitable distribution of income. Achieving educational goals is not only a basic means of increasing the productive capacity of the population, it is of crucial importance as the vehicle through which the masses of poor people may gain a stake in development.

By 1969 U.S. aid was helping to construct more than fifty thousand classrooms, printing millions of school books, and training well over 100,000

[26] IDB, *Socio-Economic Progress* (1970), p. 11; *Alliance for Progress Goals*, pp. 21-22.

teachers. These gains seem impressive, but the problem is so massive that they do not go far. It is now clear that Latin America cannot, even with major foreign assistance, refashion its educational institutions and carry out the commitment of the Alliance Charter within the period specified.

Population growth is a particularly critical element. In most Latin American countries the birth rate is extraordinarily high, which means that expanded educational facilities barely keep pace with the population increase. The enrollment of school age children in primary schools increased from 24.3 million in 1960 to 42.5 million in 1971, or from 47.6 percent of the 5-14 age group in 1960 to 60.6 percent in 1971. If present rates of population growth and school expansion continue, the entire school-age group will not be enrolled until the 1980s.

Significant increases have been made in secondary and higher education, and secondary vocational education. Between 1960 and 1971 the number of students matriculated (14-19 age group) in secondary schools in Latin America increased from 3.4 million to 9.6 million; but only 27.3 percent of this age group received instruction on the secondary level. For the region as a whole the number of secondary school graduate students increased during the period from 305,300 to 1.3 million. Between 1960-1968 enrollments in higher level institutions showed a consistent rate of growth equivalent to 10.4 percent annually, although until 1968 higher education reached only 5 percent of the total population of that age group (20-24 years). It is believed that all the countries of the region will face explosive pressures on higher education in the years ahead.

While the educational goals are of lesser importance for such countries as Uruguay, Costa Rica, Argentina, and Chile, where literacy rates are 85 percent or more and over two-thirds of school age children are enrolled, there is still much need for improvement in educational quality. For less fortunate countries such as Bolivia, Honduras and Guatemala, where less than half the primary age school children are now enrolled, the urgency of the priority is clear. Eighteen Latin American republics increased their expenditures on education by nearly 62 percent in the 1960s, but owing to the magnitude of the problem few Latin American countries have a comprehensive program equal to the task confronting them.[27]

"To increase the construction of low-cost houses for low income families in order to replace inadequate and deficient housing and to reduce housing shortages; and to provide necessary public services to both urban and rural centers of population," is a Charter goal which cannot soon be realized. It has been estimated that the total need for housing in Latin America is between fifteen and twenty million units and the deficit is rising by at least one million units a year. Squatter settlements and slum areas continue to mushroom. The most urgent need for housing is for low income families. But unfortunately people at this income level can make little or no contribution to the cost of their housing and commonly require some form of public subsidy.

[27]*Alliance for Progress Goals,* pp. 5, 32-38; *Rockefeller Report,* pp. 163-172; IDB, *Annual Report 1972,* pp. 100-101.

Under the Alliance AID has concentrated its program on the building of home credit institutions such as savings and loan associations and cooperatives. Being an essentially middle class program, it does not meet the goal of the Alliance to provide low-cost housing for low income families. Housing has not been a top priority item in the AID effort partly because of the fact that the Latin American countries do not have the institutional capability to carry out a housing program on the scale required.[28]

The Punta del Este Charter calls for tax reform aimed at improvement in: "(1) the countries' ability to collect levels of revenue adequate to support the various public development programs needed to reach other Alliance goals; (2) the equity of the tax systems in order to improve economic distribution; (3) the effectiveness of the tax systems as instruments to promote growth and development." It is the area of tax reform that AID reports significant progress, although measurement is admittedly difficult. An effective tax system scarcely existed before the Alliance, but now many Latin American countries have needed legislation which, coupled with improved tax administration, have brought increased tax revenues. Tax reform emphasis has been chiefly on income taxes and much remains to be accomplished even in this area. But the existence of new legislation gives evidence of an awareness of the need for tax revenue to help finance domestic development. Central government tax revenues totaled $16,240 million in 1971, representing 87.9 percent of current revenue as against 85.4 percent in 1967. The largest gain in tax revenues was achieved in Brazil, with an annual increase of 15 percent in 1967-1971, followed by Panama (12.5); Mexico (11.7); and Barbados (10).

United States assistance in the tax reform area is chiefly a program of technical assistance in tax administration emphasizing auditing, collection, training, public relations and taxpayer activities concerning income taxes. Under the joint AID/Internal Revenue Service program, sixteen teams comprising sixty advisers worked in Latin America. It is clear that a basic problem is to educate the people of Latin America to the fact that taxes are inevitable and that everyone must pay their fair share. And obviously the tax revenue base must be broadened and reform extended to other taxable areas.[29]

A fundamental objective of the Charter is to achieve a substantial and sustained growth of per capita income. This was to be done at a rate designed to quickly attain levels of income capable of providing self-sustaining growth and sufficient to bring income levels into a progressively closer alignment with the standard of the more industrialized nations. The Alliance set a 2.5 percent per capita per year growth rate as the minimum necessary to attain these goals. In the first seven years of the Alliance the annual per capita growth rate averaged only 1.5 percent, rather than the desired 2.5 percent. Since per capita GNP is a function of both output and population, it becomes evident that Latin American nations with populations increasing at some of the fastest rates in the world, must attain total growth rates of 5.5 percent or more (higher than the U.S. average of 5.1 percent in the same period) to reach the Alliance figure of 2.5 percent per capita.

[28] *Alliance for Progress Goals,* pp. 43-44; IDB, *Socio-Economic Progress* (1968), pp. 333-380; *Rockefeller Report,* pp. 155-159.

[29] *Alliance for Progress Goals*, pp. 59-61; U.S. House of Reps., Comm. on Gov't Operations, 90th Cong., 2nd Sess., *U.S. Aid Operations in Latin America under the Alliance for Progress* (Washington, 1968), pp. 32-33.

Chile, Peru, and Panama, the three countries that averaged a 2.5 percent growth rate or more, had a much higher growth rate before the Alliance. In 1965, only eight countries attained the 2.5 percent level, six in 1966 and four in 1967. "The population growth, equivalent to a cumulative annual rate of 2.9 percent, meant that over the years 1961-1971 the gross per capita national product grew by only 2.6 percent annually and by the end of this period amounted to approximately $550." With correction of the many social inequities dependent on growth rate increase, it appears that the solution to these problems lies many years ahead.[30]

Private investment has an important role in Latin America's economic development but it has failed to reach the nominal goal proposed at Punta del Este. In 1962, the first year after the Alliance began, U.S. businessmen took out $32 million more net capital than they invested. The slowdown was doubtless a reaction to the new threat emanating from Cuba and its implications for vested interests in the hemisphere. Since then net investments have risen and by 1966 stood at an annual rate of more than $200 million. In the 1950s, when Latin America was still a relatively safe area, U.S. business invested at a higher level: the rate was over $500 million per year in 1951-1958.

Although the total U.S. investment in Latin America has increased by a substantial amount and is about one quarter of all U.S. capital overseas, investments in other regions have shown a more spectacular growth; between 1950 and 1961 United States investments in Latin America rose by 84 percent, while those in the other two main targets of U.S. investment, Canada and Europe, increased by 230 percent and 341 percent; those in the other two main developing areas, Africa and Asia, by 273 percent and 148 percent, though in these two areas the absolute amounts are much smaller. Despite the expropriation of United States property in Cuba, Brazil, Peru, and Bolivia, and other deterrents, United States private, direct investment in Latin America increased from $8,730 billion to $12,989 billion in the decade 1958-1968.[31]

Before the Alliance U.S. private investments were mainly in the extractive industries, public utilities and large-scale farming. Capital invested in these areas declined owing to inflation, discriminatory taxation and rate limitation, and the prospect of nationalization. Manufacturing investment moved ahead but mainly to the larger countries: Mexico, Brazil, Colombia, Argentina, and Venezuela. Several steps were taken to encourage private sector support of the Alliance. An investment guaranty program of AID authorized insurance against expropriation, war and insurrection, and inconvertibility of currency. Another important innovation was the Atlantic Community Development Group for Latin America (ADELA) under the sponsorship originally of the NATO Parliamentarians Conference and in the United States by Senator Jacob K. Javits and Senator Hubert Humphrey. The multinational, multienterprise private investment group

[30] *Alliance for Progress Goals,* pp. 37-38; *New York Times,* Mar. 16, 1969; IDB, *Annual Report 1972,* p. 3

[31] U.S. Cong. Joint Econ. Comm., 89th Cong., 2nd Sess., *Private Investment in Latin America* (Washington, 1964), p. 449; U.S. Dept. of Commerce, Office of Business Economics, *Survey of Current Business,* 49 (Oct., 1969), p. 30; *Rockefeller Report,* pp. 148-154; IDB, *Socio-Economic Progress* (1968), p. 32. The book value of U.S. direct private investment in the Latin American republics stood at $12,978 billion in 1971. See U.S. Dept. of State, *The Western Hemisphere, An Excerpt from United States Foreign Policy 1972* (Washington, 1973), p. 414.

set up to implement ADELA envisages a revitalization of private enterprise in Latin America by combining in partnership North American, European and Japanese private enterprise strength. This investment company will focus on expanding the sector of medium and small-sized enterprises to serve as a base for larger ventures of national and regional development.[32] By 1967 ADELA had received subscriptions of about $40 million for investment in manufacturing companies in Latin America.

Support for the Alliance came from the U.S. Chamber of Commerce which in 1964 published a report: *The Alliance for Progress, a Hemispheric Response to a Global Threat.* In the same year U.S. businessmen with Latin American interests were brought together into the Council for Latin America, headed by David Rockefeller of the Chase Manhatten Bank. This group discussed with State Department officials the role that U.S. investment could assume in reinforcing the Alliance effort. However, the total foreign exchange gap, well above public aid from all possible sources, is some ten times larger than private investment inflow. U.S. private investment can make a significant contribution by introducing new technology, marketing techniques and attitudes, but it is unlikely to move into the area in large quantities until profitable investment is more assured. The incidence of political instability, inflation, currency inconvertibility and limited markets must be reduced to make Latin America more attractive to private investment.

It is suggested in the Charter of Punta del Este that country development programs should include the adaptation of budget expenditures to meet development needs. The expenditure pattern varies from country to country, but considering the actual amount of the increases, total central government expenditures in Latin America rose 13 percent in real terms between 1961 and 1967. Total expenditures in 1971 were $20,440 million, having risen at a rate of 7.4 percent from 1967 to 1971. The highest increase in this period was registered in Colombia, Honduras, Barbados, Chile, Panama and Brazil. All of the Latin American countries had budgetary deficits from 1967-1971.[33]

While not a formal Alliance goal, the Charter recognized the need and importance of "more effective, rational and equitable mobilization and use of financial resources . . . " Capital accumulation is a major factor in expanding production as well as a means for supporting technological progress and the exploitation of natural resources. Gross investment which measures the total expenditures on goods and services related to the development of new productive facilities, and embodied in the inventories of goods needed as production increases, is the most commonly used indicator in this area. The greater the rate of investment, the faster output should increase. The degree of investment an economy needs in order to grow is contingent on the productivity of the investment. In analyzing gross investment levels in Latin American countries AID generally uses as a standard measure of investment the relation of investment to gross national product. In the Alliance period gross investment in Latin America has been about 18 percent of the GNP.; the ratio of gross investment to the GNP shows a drop from 19.3 percent in 1960 to 17.8 percent in 1963, and a subsequent rise to 18.4 percent. The decline was attributed

[32]U.S. Cong., *Latin Amer. Dev. and West. Hem. Trade,* p. 214.

[33]IDB, *Annual Report 1972,* pp. 42-43; *Alliance for Progress Goals,* pp. 62-65.

mainly to stabilization efforts with reduced credit availability and curbed public spending in Brazil and Argentina. Investment rate and growth performance data show that, of the Latin America countries with investment-GNP ratios of over 18 percent, most are above-average growth performance countries (outstanding were Mexico, Panama, Peru and Venezuela). Countries with low investment-GNP ratios generally reflected mediocre growth performance (Paraguay, Uruguay, Honduras and the Dominican Republic).[34]

The Alliance goal pledging its members to "improve and strengthen democratic institutions" has proven elusive, for many governments have been overthrown or substantially controlled by the use, or threatened use of, military force by the leaders of those establishments. In 1973 military governments were found in Honduras, Nicaragua, Panama, Ecuador, Peru, Bolivia, Chile, Uruguay and Brazil. In several other countries military men were in the presidencies, but through the ballot box. Argentina had witnessed the restoration of power to civilians after a long period of military rule, whereas in Chile the opposite occurred. At no time since the 1950s had the military been so prominent on the political scene. The new military rulers fell generally into two groups: the "*gorilas*" (gorilla), who defended the *status quo* with little regard for the methods employed; and the "nationalists," who were shaping the wave of reformist militarism, which had reappeared after the overthrow of Brazilian President João Goulart in 1964.[35]

Radicalism is not new to the Alliance period as reformist military fervor was widespread in the 1943-1953 decade. Responding to increasing social pressures from labor and the lower classes after World War II, young officers supported populist movements against traditionalist regimes. In this decade populist-type governments prevailed in twelve countries with 75 percent of the population either through the acquiescence of the services or as a result of intervention by young officers. A counterrevolutionary surge virtually ended the reformist regimes by 1957. Rising labor-leftist extremism had cooled the liberal spirit among young officers, and traditionalist or military regimes returned to power. In 1954, twelve of the twenty republics had military rulers. Civilian authority was slowly restored, and, by 1961, General Alfredo Stroessner of Paraguay was the only military chief of state.

A resurgence of militarism began almost at once, however. Between March 1962 and mid 1969 there were thirteen military *coups,* eleven of which ousted democratically elected governments. Civilian authority, under varying shades of military supervision was restored in five. In eleven of the *coups* since 1962 in Latin America, the officially-stated motivation was incompetence by civilian governments or the threat of communism. These events shook the very foundations of the Alliance and presented some dilemmas for U.S. policy which were not anticipated in the sanguine phrases of the Punta del Este Charter.

[34] *Ibid*., pp. 54-57; IDB, *Socio-Economic Progress* (1968), pp. 19-21.

[35] The military's assumption of a wide range of extra-military functions is a legacy of the early 19th century wars of independence in Latin America. Although a major force in the first century of Latin American independence, they were not compelled to deal with the massive social and economic problems of recent times. For background on the Latin American military see John J. Johnson, *The Military and Society in Latin America* (Stanford, 1964), and Edwin Lieuwen, *Arms and Politics in Latin America* (New York, 1961).

Obviously the most rapid possible success of the Alliance requires reduction of the separate military establishments to dimensions in line with realistic requirements for defense from external aggression and internal insurrection, and the role of military leaders as the power in the political process. However, these goals must be reconciled with the U.S. military assistance program which seeks to insure the security of the Western Hemisphere.

United States security policy in Latin America has passed through several evolutionary stages. On the eve of World War II our nation spearheaded the development of collective, or hemispheric defense. While the intent of this policy was to foster a strong posture in the face of extrahemispheric threats, internal security against Axis subversion was also a consideration. Military aid, military missions, and invitations to Latin American military personnel to study in our professional schools were important aspects of this policy. The program of advising, training, and equipping friendly armed forces carried with it cooperation with existing governments, whether military or civilian, representative or autocratic.

Communist ambitions to expand by aggression or subversion, or a combination of both, became evident soon after World War II. The cold war forced continuing United States interest in the maintenance of a military posture by the Latin American nations. This interest was reflected in the Mutual Security Act of 1951 under which military aid was resumed on a larger scale. Three types of military assistance were available: (1) direct grants of equipment to prepare for hemisphere defense; (2) opportunities for purchasing U.S. weapons and equipment for country and hemisphere defense; (3) the establishment of U.S. Army, Navy, and Air Force missions to help train Latin American forces. At the time, though, dangers from external sources admittedly were not great.

Soon thereafter Washington began negotiations to form bilateral mutual defense assistance pacts (MAP) with the Latin American nations. Agreements were made with Ecuador, Cuba, Colombia, Peru, and Chile in 1952; with Brazil, the Dominican Republic, and Uruguay in 1953; with Nicaragua and Honduras in 1954; with Haiti and Guatemala in 1955; and with Bolivia in 1958. Once this program was in motion it was contended that unless equipment and training were provided, Latin America would become dependent on Western Europe for such aid; this would impair collectivization of defense and the standardization of equipment. By the end of the 1950s the goals of training assistance were expanded to include anti-submarine capability, and to strengthen in the Latin American armed forces respect for civilian authority by closer contact with their U.S. counterparts. Arms shipments to Latin America under the military assistance programs increased steadily from $21 million in 1956 to $67 million in 1959, and President Eisenhower requested $96.5 million for 1960. However, the Congress reacted by passing laws directing that military assistance for the region for 1960 must not exceed that of 1959, and imposed a ceiling of $55 million beginning in 1961. This limitation did not apply to appropriations for training.

By 1961 the program was again undergoing change. The greatest threat was no longer from enemy armed forces but from domestic elements subverted by the Sino-Soviet bloc. Internal defense against Castro-Communist guerrilla warfare was stressed by the Kennedy administration. Civic action projects aimed at promoting political stability and strengthening national economies were brought into the military assistance programs in Latin America. A counter-insurgency training and equipment program was begun in 1963. Prompting the shift to internal security was the belief that law and order must be maintained if thc Alliance for Progress was to succeed. At the same time, the Alliance charter required that "political freedom must accompany material progress." To attain this ideal U.S. policy toward military rulers had to be recast. Previous administrations had extended military assistance without regard for the character of the governments, whether democratic or undemocratic, provided they were anti-Communist and supported U.S. cold war policies.

After a military *coup* occurred in Peru in 1962 the Kennedy administration, seeking to implement the Alliance goal, broke off diplomatic relations, suspended military assistance, and stopped economic aid. Other nations did not boycott the military regime and Washington recognized it, restoring economic and military assistance, on the condition that elections would be held. The Kennedy administration was compelled to make further compromises when military elements overthrew civilian governments in Guatemala and Ecuador in 1963. But later in the year when the military seized power in the Dominican Republic and Honduras, Washington again cut diplomatic ties and suspended the aid programs.

The administration of President Lyndon B. Johnson continued the policy of emphasizing internal security as the basis of U.S. military support in Latin America but ceased using nonrecognition and the suspension of aid as a means to foster civilian, reperesentative government. Although avowedly opposed to military seizures of power it recognized with little delay new military regimes in Brazil, Argentina, and Bolivia and attached no promise of early elections as a condition of such action. After U.S. military intervention in the Dominican Republic in 1965, and the establishment of a temporary OAS peacekeeping force, the Johnson administration proposed that a permanent inter-American defense force be set up to cope with the threat of future Communist penetration in the Western Hemisphere. As the plan failed to enlist the general support of the Latin American countries, it was not realized.

The Johnson administration expanded credit assistance to facilitate Latin American arms purchases. However, when the administration proposed to increase credit assistance to $56 million for fiscal year 1967, a figure that would have permitted the Latin American nations to purchase $224 million in U.S. armaments, Congress imposed the following restrictions: "The total of military assistance and sales (other than training) . . . for the American republics shall not exceed in any fiscal year $85 million . . . "[36] The reduction was partially attributable to the new emphasis placed on reductions of defense expenditures in Latin American countries by the Declaration of American Presidents at Punta del Este on April 14, 1967. A change in the Foreign Assistance Act in 1968 placed a new limit of $75 million of the total of grants and sales to Latin

[36]*The Foreign Assistance Program, Annual Report to the Congress, Fiscal Year 1967* (Washington, 1967) p. 43.

America, other than training. Grant military assistance to the region amounted to more than $17 million for materiel and services and $10 million for training for a total of over $27 million. Cash sales of under $12 million and credit assistance of over $35 million brought the total of grants and sales for 1968 to $75 million, of which more than $64 million was chargeable to the legislative ceiling for Latin America. The funding of grant materiel assistance for the larger nations–Argentina, Brazil, Chile, and Peru–was phased out during the year.

In 1971 President Nixon exercised his option to exceed the limit of $75 million set by Congress on U.S. arms aid to Latin America, recommending to Congress that the ceiling be raised to $150 million. The move was intended to help Latin American countries in modernizing obsolete weapons systems, and to open markets there to U.S. arms manufacturers. Britain, France, Canada, Italy, West Germany, and Spain had sold almost $1 billion worth of aircraft, naval vessels and other items in the region in the preceding five years.[37]

The primary objective of U.S. military assistance programs in Latin America is to help the nations in the region to maintain internal order and political stability. The second objective is to strengthen collective hemispheric defense against external aggression or subversion. The purpose of arms grants and credit assistance is to prevent Latin America from becoming dependent of Western European nations, or possibly the USSR, for training and equipment. While not officially expressed, the U.S. military assistance programs seek to gain the political collaboration of Latin America's armed forces. According to General Robert W. Porter, Jr., "The military has frequently proven to be the most cohesive force to assure public order and support of resolute governments attempting to maintain internal security . . ." Another important purpose of the U.S. military assistance programs is to "foster a constructive and democratic approach by the military to their professional responsibilities and to the solution of national problems."

For the period of the Alliance from its inception in 1961 through 1968, the economic total was $9.2 billion, the military $685.5 million, an overall total of $9.9 billion.[38] The objectives of the U.S. military assistance programs, despite the aid given, have not been fully realized. Latin American armed forces have not cooperated with Washington's plan to shift emphasis from external defense to internal security. Exceptions can be noted, but the services still demand tanks, jet aircraft and combat ships, while Latin American capabilities for defense against external aggression remain negligible. The larger Latin American nations have no difficulty in maintaining internal order, but in most of the lesser nations the armed forces are believed incapable of dealing with insurgency and urban terrorism. Being relatively new, it cannot be determined to what extent the civic action program has improved the image and status of the armed forces. However, it appears that civic action programs carried out in Central America and the more backward Andean countries have been more successful than elsewhere. There the technical and administrative skills of the armed services can be more supportive of civilian projects. Washington's arms standardization objective has been largely achieved among the ground forces of the southern republics, but is

[37] *Ibid., 1968*, pp. 40-41; *New York Times*, May 19, 1971.

[38] *U.S. Aid Operations in Latin America*, p. 1. For further discussion of the military aid program see *Rockefeller Report*, pp. 9-20, 85-87.

less successful in navies and air forces because of purchases of jet fighters, tanks, and warships from nations of Western Europe.[39]

There is no reason to question the success of the United States in securing the political collaboration of Latin America's military establishments, for throughout the 1960s they have proved to be the staunch anti-Communist allies. Following Fidel Castro's seizure of power in Cuba it was the military which often forced civilian governments to sever ties with the regime. Support was generally forthcoming from the military forces for U.S. proposals to create an inter-American defense force. And it was the military which intervened to thwart alleged Communist takeovers in eight Latin American countries between 1962 and 1967.

The achievement of military political objectives must be weighed against the prodemocratic objectives of the Alliance Charter and the military assistance program which are in direct conflict. The Latin American armed forces are an undemocratic body and while frequently advocating reform they tend to inhibit the process. Although not opposed to all reform the military insists that it be carried out in such a manner as not to jar public order. Some civilian governments deposed by the military were open to charges of corruption, pro-communism, ineffectiveness and authoritarianism. But in the long run it seems likely that the suppressed populist movements will become the real threat to internal stability, order and progress. The U.S. military assistance program contributes to that suppression in countries with military regimes or with civilian governments dominated by the military. In light of the recent and continuing wave of military interventions it may be concluded that U.S. training programs have done little to enhance military respect for civilian authority and constitutional processes.

It should be understood, however, that the record of arms expenditures by Latin America has to date been the lowest of any world area with the sole exception of sub-Sahara Africa. Only about 2 percent of their GNP and less than 13 percent of their total central governments' expenditures has been expended for total defense costs. Moreover, only about 10 percent of their annual defense expenditures has been devoted to new military equipment (*see* Tables 19-1 and 19-2).

When the Alliance for Progress is viewed in the perspective of the first eleven years of its operations it is clear that President Kennedy had not exaggerated when he declared on December 15, 1962, that: "This revolution under freedom is probably the most difficult assignment the United States has ever taken up."[40] It is also obvious that the Alliance was oversold by its exponents many of whom were misled by our nation's role in helping to lift Europe from the ashes of World War II. The Alliance represented new paths for both the United States and Latin America with its emphasis on development rather than reconstruction. The Marshall Plan faced a less difficult task, being directed toward promoting the recovery of Europe from wartime destruction, whereas the Alliance for Progress seeks to create not only new industries but new economies. Not wholly appreciated at the beginning was the impact that the

[39] U.S. Senate, Subcomm. on Amer. Republics, Comm. on For. Rels., 90th Cong., 1st Sess., *The Latin American Military* (Washington, 1967), pp. 28-29; *Times of the Americas*, Dec. 8, 1971.

[40] Rogers, *Twilight Struggle*, p. 228.

program would have on traditional institutions. The economic, social and political problems which have hindered Latin America's advancement were commonly aggravated by the impetus which the Alliance gave to change, although peaceful and democratic, to the *status quo*.

The Alliance has failed to capture the imagination of the peoples of Latin America to create a mystique which may be attributed to both organizational and intangible factors. It is held that few Latin Americans are aware of the Declaration to the Peoples of America or the Charter of Punta del Este, and that

Table 19-1. Military Assistance Program. Value of Grant Aid Deliveries 1950-1970* (in millions of dollars)

Area and Country	*1950-1963*	*1967-1970*	*1970*
Latin America[1]	388.8	363.6	27.4
Argentina	2.8	29.8	2.4
Bolivia	5.4	16.8	1.2
Brazil	150.6	67.1	4.3
Chile	52.0	39.4	2.0
Colombia	39.4	51.0	3.9
Cuba	10.6	–	–
Dominican Republic	8.2	14.4	2.1
Ecuador	22.2	18.9	2.0
Guatemala	5.3	10.9	2.0
Honduras	2.6	4.6	.3
Nicaragua	4.5	7.5	1.1
Paraguay	.9	7.9	.8
Peru	41.1	45.6	1.9
Uruguay	27.5	13.6	1.7
Venezuela	1.6	7.7	.8

*For years ending June 30. Covers programs authorized and appropriated by the Foreign Assistance Act and the Foreign Assistance Appropriations Acts. Represents military equipment and supplies delivered and expenditures for services. Includes (a) equipment and supplies procured for the Military Assistance Program or from procurement or stocks of military departments; and (b) services, such as training, military construction, repair and rehabilitation of excess stocks, supply operations, and other charges.
[1] Totals include countries not shown separately.

Statistical Abstract of the United States 1971, p. 244. See also Charles A. Meyer, "Military Activities in Latin America," *Inter-American Economic Affairs*, XXIII (Autumn, 1969), 93.

the Alliance is known simply as another program of U.S. financial assistance. This misconception can be traced in part to the dominating role of U.S. officials in the administration of the Alliance. Confusion exists about the "self-help" aspects of the program which are interpreted as requirements for U.S. aid instead of a means of reinforcing Latin America's ability to help itself.[41] It is also obvious that the U.S. Information Agency, if funded adequately, could more effectively publicize the achievements of the Alliance, clarify its goals and point out to Latin Americans their role in attaining them.

[41]May, *Porblems and Prospects of the Allianc? for Progress,* pp. 42-43.

Much criticism has been leveled at Washington's initial approach to the Alliance for Progress program. A critic of the Alliance, and an astute analyst of Latin American economic affairs, contends that "The central expression of the ignorance on which the Alliance was based rests on the concept of 'conscience money.' President Kennedy accepted the concept of U.S. exploitation of Latin America telling newsmen 'The United States now has the possibility of doing as much good in Latin America as it had done wrong in the past.' The damage done by accepting the charge that the United States had exploited Latin America and now must make amends by unlimited donations was irreparable. For it deprived the United States of the capacity to demand effectively the self-help measures without which the Alliance must inevitably fail. It made the United States a prisoner of the Alliance rather than a participant . . . "[42] Moreover, the linkage of aid under the Alliance with the broad issues of institutional reform has not been clearly drawn, nor has it been implemented to seriously deal with such problems as the land tenure system, the iniquity of the tax structure, and the massive poverty. Basic to the U.S. commitment is the concept that this nation's effort can be justified only to the extent that Latin American republics modify and expand their institutional capacity to meet the development challenge.

The new development diplomacy created significant problems for the foreign policy apparatus and personnel of the United States, as well as for Congress. Traditionally the task of foreign policy has been to maintain harmonious working relationships with the foreign ministries. Development diplomacy has involved not only the use of U.S. resources and power abroad to help the process of change, but a foreign policy dedicated to pushing the internal transformation of other nations. This created problems for U.S. foreign service personnel who must help to implement change in the nations to which they are accredited. The Congress found it similarly difficult to adjust to the concept of development diplomacy, especially the responsibility of appropriating funds. When money became an instrument of foreign policy, the House of Representatives was thrust into foreign affairs in an uneasy role, since money voted for foreign projects was not supportive of domestic interests. In the Alliance period the not uncommon conflict between Congress and the president has been evident in the appropriations procedure.[43]

The Alliance has experienced organizational changes, but more important has been the personnel turnover in the top-level Latin American post in the Department of State, with attendant interruptions of total diplomatic effort. At the beginning AID, with Teodoro Moscoso as its regional head, was organized separately within the State Department, while the department's Latin American Bureau was headed by Assistant Secretary of State Edwin Martin. In January, 1964 President Johnson appointed Thomas C. Mann, then ambassador to Mexico, assistant secretary of state in place of Martin who became ambassador to Argentina. Mann, who was to be the "One Voice" for Latin American policy, was also made U.S. Alliance Coordinator in place of Teodoro Moscoso, who was assigned to the new post of U.S. representative on the Inter-American Committee for the Alliance for Progress. After Secretary Mann was elevated to

[42] Hanson, *Five Years of the Alliance for Progress*, pp. 68-69.
[43] Rogers, *Twilight Struggle*, chap. X.

the Undersecretaryship for Economic Affairs early in 1965, Jack Hood Vaughan of the Peace Corps, and more recently ambassador to Panama, was appointed to the combined State-Alliance position. About one year later, when the Alliance entered its sixth year, the problem of accomodating foreign policy to the needs of Alliance development became the responsibility of Lincoln Gordon, as Vaughan was assigned to the Peace Corps. Gordon resigned in June 1967, and Covey Oliver was sworn in to replace him the following month. When Charles A. Meyer, appointed assistant secretary of state by President Nixon, assumed his duties in March 1969, he became the sixth Alliance assistant secretary for the Latin American region in seven years.[44]

The Nixon administration's policy statement on Latin America was not delivered until October 31, 1969, some nine months after the president had assumed office. The delay proved disturbing to Latin Americans, but it could be attributed to the administration's preoccupation with the Vietnam war, and the president's decision to await the report of a fact-finding mission before making commitments on Latin America. Governor Nelson Rockefeller of New York headed the mission which, though buffetted by anti-United States demonstrations, completed the abbreviated tour in May and June, and rendered an extensive report on the "Quality of Life in the Americas."

Prior to the Rockefeller mission, representatives of Latin American nations, meeting in Viña del Mar, Chile, drew up the Concensus of Viña del Mar, a document calling on Washington to make basic alterations in its economic policies. At the two-week meeting of the Inter-American Economic and Social Council held at Port-of-Spain, Trinidad in May, the United States was challenged to do more to aid in solving the region's grave economic stagnation. Washington responded by discontinuing "additionality," which had required that for every dollar given to Latin America one must be spent for American products from a certain list.[45]

The president's policy address did not indicate that the basic U.S. commitment to the Alliance for Progress would be changed, but since both he and Governor Rockefeller had downgraded the results of the Alliance, it was clear that the president intended to develop his own Latin American program. He called for a "partnership" between the United States and Latin America, and emphasized that the 1970s "should be a decade of action for progress in the Americas." On the vital question of trade the president promised to determine what could be done to improve the region's serious trade imbalance. By implication he rejected the social and economic aims of the Alliance, stressed by

[44]The Organization of American States Association declared that "failures of performance" by the United States had imperiled the freedom, the peace and the progress of the Western Hemisphere; the continuity of hemisphere programs is broken when a new administration takes office and this results in "imminent peril to the national interest." The warnings came from a meeting which included seven former assistant secretaries of state for inter-American affairs, with the then incumbent, Secretary Charles A. Meyer, Spruille Braden, Roy R. Rubottom, Thomas C. Mann, Robert F. Woodward, Jack H. Vaughan, Lincoln Gordon and Covey T. Oliver. See *New York Times,* October 6, 1969. (Secretary Meyer resigned his post in January 1973 and was succeeded by Jack B. Kubisch, an experienced Latin American specialist). Apart from this administrative problem, the *Rockefeller Report* (pp. 5-6, 97-105) declares that "with the present United States government structure, Western Hemisphere policies can neither be soundly formulated nor effectively carried out."

[45]*New York Times,* June 5, 1969; *Ibid.,* June 21, 1969.

the Kennedy and Johnson administrations, and more explicitly indicated that a policy of non-discrimination would be applied to military-controlled governments.[46]

While the Nixon administration may have had reservations about the philosophy and performance of the Alliance, it was not to be buried in the limbo of the past. On August 17, 1971 President Nixon declared "on this occasion of the Tenth Anniversary of the Alliance for Progress, I reaffirm the commitment of the United States to those noble principles and join with our hemispheric partners in rededicating ourselves to the achievement of a better life for our peoples."[47]

The total Nixon aid package for Latin America was $605 million, compared with President Johnson's allocation for the fiscal year ending June 30, 1969, which amounted to $606 million. However, as approved by Congress AID commitments for Latin America in 1968 totaled $532 million, down $40 million from 1967. Brazil, Colombia, the Dominican Republic, and Chile accounted for 70 percent of the 1968 commitments. Alliance for Progress Development Loans amounted to $420 million, nearly 80 percent of the total for the region. Supporting assistance commitments totaled $26 million, and Alliance Technical Cooperation programs amounted to $78 million.

U.S. economic assistance to Latin America in fiscal year 1971 was $602 million and in fiscal year 1973 it is expected to exceed $1 billion. The latter figure will include all official economic assistance to the region through AID, Eximbank, Food for Peace, Peace Corps, Inter-American Highway, Social Progress Trust Fund of the IDB, and U.S. cash contributions to the IDB. Increasingly, there has been a new emphasis in channeling aid funds through multilateral organizations, notably regional development banks.[48]

The President meanwhile issued a stern warning to nations which expropriate U.S. business firms, particularly those in Latin America. President Nixon declared that thereafter his administration would retaliate against any nation which takes over a U.S. firm without paying "prompt, adequate and effective compensation." Specifically, he said, that such a nation could expect the U.S. to terminate all bilateral aid programs and withhold its approval on loans under consideration at international financing institutions such as the World Bank and the Inter-American and Asian development banks.

Latin Americans were dismayed by the decline of U.S. aid because of the progressively depressed state of their nations' economies. Development had become more seriously inhibited by the high rate of population growth which compounded demand for such essentials as food, housing, jobs, education and health care. At the same time, Latin American countries found reduced overseas

[46]*Ibid.*, Nov. 1, 1969.

[47]Dept. of State *Bulletin*, LXV (Sept. 27, 1971), p. 334. Although Washington may ignore the Alliance for Progress, should it elect to do so, the U.S. cannot unilaterally declare the the Alliance dead. The Alliance was and is an undertaking of the OAS, and can be terminated only by the action of the member states.

[48]*The Foreign Assistance Program, Annual Report to Congress, Fiscal Year, 1968*, pp. 26-27; *New York Times*, May 29, 1969; U.S. Dept. of State, *The Western Hemisphere, An Excerpt from United States Foreign Policy 1972* (Washington, 1973), p. 409.

markets for traditional export products including coffee, sugar, cocoa and cotton.

In response to mounting congressional resistance to foreign aid expenditures, John A. Hanna, AID Administrator, pointed out that the idea that the United States was doing more than it share in the foreign aid field was a "myth." While the U.S. was virtually alone in this activity for a long time after World War II conditions had changed: "Almost every donor country is making a relatively greater contribution to development assistance than we." When official development assistance was measured as a percentage of GNP (1971), or on the basis of average income, the U.S. ranked eleventh among the sixteen major aid-giving countries. Ranking above the U.S. in order were France, Portugal, Australia, Belgium, Netherlands, Sweden, Denmark, W. Germany, United Kingdom, Canada, and then the United States.[49] (*See* Tables 19-3 and 19-4).

In April 1971 the Nixon administration proposed that major changes be made in the administration and management of the U.S. Foreign Assistance Program, submitting two bills—the proposed International Security Assistance Act and International Development and Humanitarian Assistance Act. The former would help to implement the approach envisaged by the Nixon Doctrine—a policy recognizing the increased capabilities of other nations in assuming a larger share of the common defense burden—by centralizing responsibility for the direction of security assistance. Under the new act the Department of State would retain its responsibility, under the direction of the president, for "the continuous supervision and general direction" of the international security assistance programs. The Department of Defense would continue to administer the grant military aid and foreign military sales programs.

Table 19-2. Military Assistance Program. Foreign Military Sales Deliveries 1950 to 1970* (in millions of dollars)

Area and Country	*1950-1963*	*1964-1970*	*1970*
American Republics[1]	200.8	194.4	36.6
Argentina	44.9	34.5	8.1
Brazil	16.9	68.0	11.7
Chile	12.8	20.0	10.0
Colombia	9.9	1.4	.2
Cuba	4.5	–	–
Ecuador	2.5	1.8	.5
Mexico	7.1	3.6	.4
Peru	35.9	13.8	2.0
Venezuela	56.0	47.0	3.0

*For years ending June 30. Covers sales authorized and appropriated under the Foreign Military Sales Act. Includes deliveries of equipment, supplies, and services purchased for cash, U.S. Government financed credit, and U.S. Government guaranty of privately financed credit.
[1] Totals include countries not shown separately.

Statistical Abstract of the United States 1971, p. 244.

[49] Dept of State *Bulletin*, LXV (Apr. 9, 1971), pp. 153-154.

Table 19-3. U. S. Government Foreign Grants and Credits by Country 1945-1970 Latin America (in millions of dollars)

Country	*Post WWII 7/1/45 through 12/31/70*	*July 1945 Dec. 1963*	*1964*	*1965*	*1966*	*1967*	*1968*	*1969*	*1970 (prel.)*
TOTAL	4,946	4,536	447	644	740	624	832	596	526
Argentina	368	378	5	−3	−3	−20	−14	5	20
Bolivia	488	290	34	30	21	24	38	28	24
Brazil	2,440	1,305	213	153	236	143	199	99	92
Chile	1,187	538	97	102	88	50	151	106	56
Colombia	888	343	38	35	59	86	108	101	118
Costa Rica	151	74	10	14	10	12	11	13	8
Cuba	41	40	––	––	––	––	––	––	––
Dominican Republic	400	70	22	79	54	60	46	37	31
Ecuador	205	88	17	17	21	27	15	11	7
El Salvador	109	35	10	11	17	11	7	7	10
Guatemala	219	138	9	11	8	15	17	10	10
Guyana	40	3	1	3	3	7	8	9	5
Haiti	114	89	4	5	3	3	4	3	4
Honduras	95	43	4	10	7	6	7	9	9
Jamaica	50	11	3	4	6	10	6	3	7
Mexico	575	420	−55	38	54	50	53	16	−1
Nicaragua	135	53	7	7	11	9	10	17	21
Panama	187	82	10	21	22	17	10	12	15
Paraguay	98	56	7	5	4	3	8	9	6
Peru	356	205	17	33	30	24	24	10	13
Surinam	9	4	(Z)	1	(Z)	(Z)	(Z)	4	(Z)
Trinidad & Tobago	47	10	8	17	6	2	6	(Z)	−2
Uruguay	115	57	(Z)	2	4	4	23	16	9
Venezuela	323	152	−40	40	41	48	33	33	16
Other* & Unspecified	306	52	26	9	38	33	62	38	48

––Represents zero; (Z) less than $500,000 or net minus (−) of less than $500,000; *comprises Bahamas, Barbados, British Honduras, French Guiana, other islands of the West Indies, Central American Bank for Economic Integration, Inter-American Institute of Agricultural Sciences, Organization of American States, and Pan-American Health Organization.

Statistical Abstract 1971, p. 761.

Table 19-4. Economic Assistance Programs in Latin America Net Obligations and Loan Authorizations* (millions of dollars)

(Fiscal Years)

Program	*1961*	*1962*	*1965*	*1969*	*1970*	*1961-1970*
AID	$254	$479	$523(a)	$290	$377	$4,760
Soc. Prog. Trust Fund	–	226	101	1(b)	3	535
P.L. 480	145	127	97	115	163	1,521
Exim Bank (c)	439	63	152	289	186	2,275
Other U.S. programs	2	120(d)	284	324	329	2,121
Total U.S.	840	1014	1156	1019	1057	11,212
IDB-Ordinary Capital	38	101	171	183	229	1,327
Fund Spec. Oper.	28	36	66	325	452	1,810
IBRD	130	408	208	458	703	3,300
IDA	27	30	18	15	11	143
IFC	3	8	10	27	34	151
UN Agencies	37	39	52	46	47	467
EEC	3	5	8	9	9	83
International Organization	267	626	534	1062	1485	7,282
DAC Bi-lateral (e)	181	272	394	445	445	3,699
Grand Total (f)	1289	1803	1834	2226	2687	20,323

(a) Excludes Alliance for Progress funds used for non-regional programs. Includes capitalized interest.
(b) Technical assistance grants and administrative costs for administering the SPTF. The FY 1969 figure represents $5.0 million grants and $–3.5 million in loan decreases. In FY 1970, $5.0 million grants and $–2.1 million in loan decreases.
(c) Excludes refunding of $553.9 million.
(d) Includes U.S. paid-in subscriptions to the Ordinary Capital fund (in millions) as follows: FY 1962, $60; FY 1963, $60.0. Includes U.S. paid-in cash contributions to the Fund for Special Operations (in millions) as follows: FY 1962, $50.0; FY 1964, $50.0; FY 1965, $250.0; FY 1966, $250.0; FY 1967, $250,0; FY 1968, $300.0; FY 1969, $300.0; and FY 1970, $300.0.
(e) Calendar year gross disbursements of loans and grants.
(f) Excludes U.S. subscriptions and contributions to the I.D.B. in (d) above to avoid "double counting."

Sources: *U.S. Overseas Loans & Grants Report*: July 1, 1945-June 30, 1970 Preliminary dtd. October 15, 1970. International Organizations' Loan Statements ending June 30th for each year. R. Clarke: 4/24/70: R. Clarke: 11/13/70: AID: LA/DP: 11/14/69: RC:vp:Rev. 3/20/70:

*William C. Binning, "The Nixon Foreign-Aid Policy for Latin America," *Inter-American Economic Affairs* (Summer, 1971), 25, 37.

A coordinator for security assistance would be appointed at the under-secretary level to discharge the responsibility of the secretary of state for continuous supervision and general direction of the programs; a specialized planning and analysis staff would assist the coordinator. The administration of economic supporting assistance, which is the economic part of security assistance and presently lodged in AID would be transferred to the State Department and put under an officer of assistant-secretary rank, who would be responsible to the coordinator.

In the field of development assistance two institutions were proposed to join the Overseas Private Investment Corporation (OPIC) and Inter-American Social Development Institute (ISDI), which were already in existence. The International Development Corporation (IDC) would administer our bilateral lending program. It would possess flexibility needed to determine its loan terms to the needs of countries in question: its pattern of lending would reward demonstrated self-help performance.

The new International Development Institute (IDI) proposed by the administration would concentrate its efforts on the transfer of technology and encourage research designed to facilitate development, as well as to provide technical assistance on a selective basis. The programs of the IDC, IDI, and the OPIC would receive central direction by a Coordinator of Development Assistance. The above-cited agencies would replace the Agency for International Development. Bilateral aid programs would complement the assistance programs of multilateral agencies and other donors. And humanitarian assistance, divided amoung several agencies, would be carried out be a new bureau within the State Department, headed by an assistant secretary.[50]

The economic aspects of development diplomacy in United States-Latin American relations, including the military assistance program, are the main focus of this chapter. Latin America's failure to achieve steady economic progress is an important determinant in the trends toward dictatorship, communism, nationalism and anti-Yankee feeling. It is for this reason that the relations between the two regions tend to assume an essentially political character and require an essentially political solution. Unlike economic and military interests, political interests are largely intangible and subject to strong and, sometimes irrational, emotional overtones.

From the end of World War II until 1960 the Latin American policy of the United States was based on the assumption that the problems in the area were primarily economic and that they could be solved by large-scale private investment, by hard loans from the Export-Import Bank, the International Bank for Reconstruction and Development, and by a nominal amount of technical assistance. When the inadequacy of this approach had become clear, the Eisenhower administration in 1960 signaled a change. The new policy, which culminated the following year in the Alliance for Progress, was founded on the belief that the area's problems were fundamentally social, and could be solved by reform—land reform, educational and health programs, housing projects and other goals—underwritten by increased U.S. government grants in aid, loans and

[50] *Ibid.*, LXIV (May 10, 1971), pp. 614-625; *Ibid.*, LXV (July 5, 1971), pp. 23-27. See also Report to the President from the Task Force on International Development, *United States Foreign Assistance in the 1970s: A New Approach* (Washington, 1970).

foreign investment. At the Summit Conference at Punta del Este in April 1967 the emphasis appeared to shift again—to regional economic integration as a means of promoting economic growth which in turn would support social reform. With the advent of the Nixon administration in 1969 social reform goals were deemphasized, partnership and self-help were stressed, and a pragmatic policy was adopted toward Latin American governments.

In general, these approaches are peripheral to the problem, for the basic interest of the United States in Latin America is neither social nor economic: it is political. These political interests involve having the Latin American governments firmly aligned with the United States in the anti-Communist community of nations. Washington's political objective is to prevent any drift in Latin America toward either Communist or neutralist blocs. It is only by the growth of governments which will cooperate with the United States that its national interests can be made secure. The development of an anti-United States posture in any Latin American country would endanger U.S. military and economic assistance programs, as well as trade and private investment. In short, United States interests can be protected in the long run only by political means.

Permeating all inter-American relationships today is the basic problem of the region: societies in upheaval. One of the most explosive movements in modern history underlies the process of transition from traditional to modern societies. What is happening in Latin America is the result of the impact of a delayed industrial, technological revolution on non-industrial societies, urbanization, political ferment that is charged with revolutionary doctrines, demands for social justice, the growth of a middle class, an impact which is intensified by the world's highest population growth rates. The Washington-based Population Reference Bureau reports that the current growth rate of the Latin Americans is so great—3.4 percent a year—there will be about 2.5 times as many by 2000. North America (U.S. and Canada) will increase to 280 million people by 1985 from the 1971 total of 229 million. By contrast, Latin America's population, including the Caribbean is expected to reach 435 million by 1985, and 641 million by the end of the century. As of mid 1971 the area's population was 291 million people. Latin America is unable to provide jobs for the burgeoning population, for about one quarter of the region's workers were unemployed or underemployed, that is, unproductively employed. Little improvement can be envisaged since the labor force is growing at the rate of 3 percent annually and the job market is about half that figure.

Population growth and industrialization—and their repercussions—have imposed severe strains on political and socio-economic institutions which were designed for pre-modern societies. The challenge presented to United States foreign policy in this era is how and to what extent the social ferment may be influenced. The Alliance for Progress is the attempted answer that we and Latin American governments are offering. Put in terms of a ten-year plan at the outset, this was extended, but at the least, Latin America's modernization will take decades and possibly generations. Consideration should be given to the possibility that the United States cannot control the process because of its magnitude and complexity, and that our nation's influence may ultimately prove to be only marginal.

Meanwhile, the demands for reform and progress become more pervasive and insistent. Should Latin America be unable, in the decade of the 1970s, to produce or secure the resources needed to attain its social goals and to finance sound economic growth, and if the barriers to regional integration prove to be insurmountable, the future appears threatening.

SUPPLEMENTARY READINGS

Beaulac, Willard, L. *A Diplomat Looks at Aid to Latin America.* Carbondale, Ill., 1970.

Castillo, C. M. *Growth and Integration in Central America.* London, 1967.

Dell, Sidney, *A Latin American Common Market?* Oxford, 1966.

Dreier, John C., ed. *The Alliance for Progress, Problems and Perspectives.* Baltimore, 1962.

Furtado, Celso. *The Economic Development of Latin America. A Survey from Colonial Times to the Cuban Revolution.* New York, 1971.

———. *Obstacles to Development in Latin America.* New York, 1970.

Gardner, Lloyd C. *Economic Aspects of New Deal Diplomacy.* Madison, 1964.

Glade, William P. *Latin American Economies. A Study of Their Institutional Evolution.* New York, 1969.

Gordon, Lincoln. *A New Deal for Latin America: The Alliance for Progress.* Cambridge, Mass., 1963.

Gordon, Wendell C. *The Political Economy of Latin America.* New York, 1965.

Guerrant, E. O. *Roosevelt's Good Neighbor Policy.* Albuquerque, 1950.

Hanson, Simon G. *Dollar Diplomacy Modern Style. Chapters on the Failure of the Alliance for Progress.* Washington, 1970.

———. *Five Years of the Alliance for Progress, An Appraisal.* Washington, 1967.

Hirschman, Albert O. *Journeys Toward Progress: Studies of Economic Policymaking in Latin America.* New York, 1963.

Hull, Cordell. *The Memoirs of Cordell Hull.* 2 vols. New York, 1948.

Inter-American Development Bank. *Economic and Social Progress in Latin America, Annual Report 1972.* Washington, 1973.

Johnson, John J. *Political Change in Latin America: The Emergence of the Middle Sectors.* Stanford, 1958.

———. *The Military and Society in Latin America.* Stanford, 1964.

———, ed. *Continuity and Change in Latin America.* Stanford, 1964.

Levinson, Jerome and Juan de Onis. *The Alliance that Lost its Way.* Chicago, 1970.

Lieuwen, Edwin. *Arms and Politics in Latin America.* New York, 1961.

Lodge, George C. *Engines of Change.* New York, 1970.

Manger, William, ed. *The Alliance for Progress: A Critical Appraisal.* Washington, 1963.

May, Herbert K. *Problems and Prospects of the Alliance for Progress: A Critical Examination.* New York, 1968.

Mecham, J. Lloyd. *The United States and Inter-American Security. 1889-1960.* Austin, 1961.

Pan-American Union. *The Alliance for Progress and Latin American Development Prospects: A Five Year Review, 1961-1965.* Baltimore, 1967.

Perloff, Harvey S. *Alliance for Progress: A Social Invention in the Making.* Baltimore, 1969.

Powelson, John P. *Latin America: Today's Economic and Social Revolution.* New York, 1964.

Prebisch, Raul. *Change and Development, Latin America's Great Task.* New York, 1971.

Rippy, J. Fred. *Globe and Hemisphere. Latin America's Place in the Postwar Foreign Relations of the United States.* Chicago, 1958.

Rogers, W. D. *The Twilight Struggle.* New York, 1967.

Tannenbaum, Frank. *Ten Keys to Latin America.* New York, 1966.

United Nations, Economic Commission for Latin America. *Economic Survey of Latin America.* New York. This was an annual publication from the late 1940s through 1958.

United Nations. *Towards a Dynamic Development Policy for Latin America.* New York, 1963.

Urquidi, Victor. *Free Trade and Economic Integration in Latin America: Toward a Common Market.* Berkeley, 1962.

————. *The Challenge of Development in Latin America.* New York, 1962.

U.S. Cong., Joint Economic Committee, 89th Cong., 1st Sess., *Latin American Development and Western Hemisphere Trade.* Washington, 1965.

U.S. Dept. of State. *The Story of Inter-American Cooperation. Our Southern Partners.* Dept. of State Publication 7404, Inter-American Series 78. Washington, 1962.

U.S. House of Rep., 91st Cong., 1st Sess. *A Review of Alliance for Progress Goals.* Washington, 1969.

U.S. House of Rep., Committee on Government Operations, 90th Cong., 2nd Sess. *U.S. Aid Operations Under the Alliance for Progress.* Washington, 1968.

U.S. Senate, Committee on Government Operations, 89th Cong., 2nd Sess. *United States Foreign Aid in Action: A Case Study.* Washington, 1966.

U.S. Senate, Subcommittee on American Republics Affairs, Committee on Foreign Relations, 90th Cong., 2nd Sess. *Survey of the Alliance for Progress.* Washington, 1968.

U.S. Senate, Subcommittee on Western Hemisphere Affairs, Committee on Foreign Affairs, 91st Cong., 1st Sess. *Rockefeller Report on Latin America.* Washington, 1970.

Veliz, Claudio, ed. *Obstacles to Change in Latin America.* London, 1969.

Wionczec, Miguel S. *Latin American Integration: Experiences and Prospects.* New York, 1966.

Wood, Bryce. *The Making of the Good Neighbor Policy.* New York, 1961.

appendix A

Charter of the Organization of American States*

IN THE NAME OF THEIR PEOPLES, THE STATES REPRESENTED AT THE NINTH INTERNATIONAL CONFERENCE OF AMERICAN STATES,

Convinced that the historic mission of America is to offer to man a land of liberty, and a favorable environment for the development of his personality and the realization of his just aspirations;

Conscious that that mission has already inspired numerous agreements, whose essential value lies in the desire of the American peoples to live together in peace, and, through their mutual understanding and respect for the sovereignty of each one, to provide for the betterment of all, in independence, in equality and under law;

Confident that the true significance of American solidarity and good neighborliness can only mean the consolidation on this continent, within the framework of democratic institutions, of a system of individual liberty and social justice based on respect for the essential rights of man;

Persuaded that their welfare and their contribution to the progress and the civilization of the world will increasingly require intensive continental co-operation;

Resolved to persevere in the noble undertaking that humanity has conferred upon the United Nations, whose principles and purposes they solemnly reaffirm;

Convinced that juridical organization is a necessary condition for security and peace founded on moral order and on justice; and

In accordance with Resolution IX of the Inter-American Conference on Problems of War and Peace, held at Mexico City,

* As amended by the Protocol of Amendment of the Charter of the Organization of American States, "Protocol of Buenos Aires," signed at the Third Special Inter-American Conference, Buenos Aires, 1967.

HAVE AGREED
upon the following

CHARTER
OF THE ORGANIZATION OF
AMERICAN STATES

PART ONE

Chapter I
NATURE AND PURPOSES

Article 1

The American States establish by this Charter the international organization that they have developed to achieve an order of peace and justice, to promote their solidarity, to strengthen their collaboration, and to defend their sovereignty, their territorial integrity, and their independence. Within the United Nations, the Organization of American States is a regional agency.

Article 2

The Organization of American States, in order to put into practice the principles on which it is founded and to fulfill its regional obligations under the Charter of the United Nations, proclaims the following essential purposes:

(a) To strengthen the peace and security of the continent;

(b) To prevent possible causes of difficulties and to ensure the pacific settlement of disputes that may arise among the Member States;

(c) To provide for common action on the part of those States in the event of aggression;

(d) To seek the solution of political, juridical, and economic problems that may arise among them; and

(e) To promote, by cooperative action, their economic, social, and cultural development.

Chapter II
PRINCIPLES

Article 3

The American States reaffirm the following principles:

(a) International law is the standard of conduct of States in their reciprocal relations;

(b) International order consists essentially of respect for the personality, sovereignty, and independence of States, and the faithful fulfillment of obligations derived from treaties and other sources of international law;

(c) Good faith shall govern the relations between States;

(d) The solidarity of the American States and the high aims which are sought through it require the political organization of those States on the basis of the effective exercise of representative democracy;

(e) The American States condemn war of aggression: victory does not give rights;

(f) An act of aggression against one American State is an act of aggression against all the other American States;

(g) Controversies of an international character arising between two or more American States shall be settled by peaceful procedures;

(h) Social justice and social security are bases of lasting peace;

(i) Economic cooperation is essential to the common welfare and prosperity of the peoples of the continent;

(j) The American States proclaim the fundamental rights of the individual without distinction as to race, nationality, creed, or sex;

(k) The spiritual unity of the continent is based on respect for the cultural values of the American countries and requires their close cooperation for the high purposes of civilization;

(l) The education of peoples should be directed toward justice, freedom, and peace.

Chapter III
MEMBERS

Article 4

All American States that ratify the present Charter are Members of the Organization.

Article 5

Any new political entity that arises from the union of several Member States and that, as such, ratifies the present Charter, shall become a Member of the Organization. The entry of the new political entity into the Organization shall result in the loss of membership of each one of the States which constitute it.

Article 6

Any other independent American State that desires to become a Member of the Organization should so indicate by means of a note addressed to the Secretary General, in which it declares that it is willing to sign and ratify the Charter of the Organization and to accept all the obligations inherent in membership, especially those relating to collective security expressly set forth in Articles 27 and 28 of the Charter.

Article 7

The General Assembly, upon the recommendation of the Permanent Council of the Organization, shall determine whether it is appropriate that the Secretary General be authorized to permit the applicant State to sign the Charter and to accept the deposit of the corresponding instrument of ratification. Both the recommendation of the Permanent Council and the decision of the General Assembly shall require the affirmative vote of two-thirds of the Member States.

Article 8

The permanent Council shall not make any recommendation nor shall the General Assembly take any decision with respect to a request for admission on the part of a political entity whose territory became subject, in whole or in part, prior to December 18, 1964, the date set by the First Special Inter-American Conference, to litigation or claim between an extracontinental country and one or more Member States of the Organization, until the dispute has been ended by some peaceful procedure.

Chapter IV
FUNDAMENTAL RIGHTS AND DUTIES OF STATES

Article 9

States are juridically equal, enjoy rights and equal capacity to exercise these rights, and have equal duties. The rights of each State depend not upon its power to ensure the exercise thereof, but upon the mere fact of its existence as a person under international law.

Article 10

Every American State has the duty to respect the rights enjoyed by every other State in accordance with international law.

Article 11

The fundamental rights of States may not be impaired in any manner whatsoever.

Article 12

The political existence of the State is independent of recognition by States. Even before being recognized, the State has the right to defend its integrity and independence, to provide for its preservation and prosperity, and consequently to organize itself as it sees fit, to legislate concerning its interests, to administer its services, and to determine the jurisdiction and competence of its courts. The exercise of these rights is limited only by the exercise of the rights of other States in accordance with international law.

Article 13

Recognition implies that the State granting it accepts the personality of the New State, with all the rights and duties that international law prescribes for the two States.

Article 14

The right of each State to protect itself and to live its own life does not authorize it to commit unjust acts against another State.

Article 15

The jurisdiction of States within the limits of their national territory is exercised equally over all the inhabitants, whether nationals or aliens.

Article 16

Each State has the right to develop its cultural, political, and economic life freely and naturally. In this free development, the State shall respect the rights of the individual and the principles of universal morality.

Article 17

Respect for and the faithful observance of treaties constitute standards for the development of peaceful relations among States. International treaties and agreements should be public.

Article 18

No State or group of States has the right to intervene, directly or indirectly, for any reason whatever, in the internal or external affairs of any other State. The foregoing principle prohibits not only armed force but also any other form of interference or attempted threat against the personality of the State or against its political, economic, and cultural elements.

Article 19

No State may use or encourage the use of coercive measures of an economic or political character in order to force the sovereign will of another State and obtain from it advantages of any kind.

Article 20

The territory of a State is inviolable; it may not be the object, even temporarily, of military occupation or of other measures of force taken by another State, directly or indirectly, on any grounds whatever. No territorial acquisitions or special advantages obtained either by force or by other means or coercion shall be recognized.

Article 21

The American States bind themselves in their international relations not to have recourse to the use of force, except in the case of self-defense in accordance with existing treaties or in fulfillment thereof.

Article 22

Measures adopted for the maintenance of peace and security in accordance with existing treaties do not constitute a violation of the principles set forth in Articles 18 and 20.

Chapter V
PACIFIC SETTLEMENT OF DISPUTES

Article 23

All international disputes that may arise between American States shall be submitted to the peaceful procedures set forth in this Charter, before being referred to the Security Council of the United Nations.

Article 24

The following are peaceful procedures: direct negotiation, good offices, mediation, investigation and conciliation, judicial settlement, arbitration, and those which the parties to the dispute may especially agree upon at any time.

Article 25

In the event that a dispute arises between two or more American States which, in the opinion of one of them, cannot be settled through the usual diplomatic channels, the parties shall agree on some other peaceful procedure that will enable them to reach a solution.

Article 26

A special treaty will establish adequate procedures for the pacific settlement of disputes and will determine the appropriate means for their application, so that no dispute between American States shall fail of definitive settlement within a reasonable period.

Chapter VI

COLLECTIVE SECURITY

Article 27

Every act of aggression by a State against the territorial integrity or the inviolability of the territory or against the sovereignty or political independence of an American State shall be considered an act of aggression against the other American States.

Article 28

If the inviolability or the integrity of the territory or the sovereignty or political independence of any American State should be affected by an armed attack or by an act of aggression that is not an armed attack, or by an extra-continental conflict, or by a conflict between two or more American States, or by any other fact or situation that might endanger the peace of America, the American States, in furtherance of the principles of continental solidarity or collective self-defense, shall apply the measures and procedures established in the special treaties on the subject.

Chapter VII

ECONOMIC STANDARDS

Article 29

The Member States, inspired by the principles of inter-American solidarity and cooperation, pledge themselves to a united effort to ensure social justice in the Hemisphere and dynamic and balanced economic development for their peoples, as conditions essential for peace and security.

Article 30

The Member States pledge themselves to mobilize their own national human and material resources through suitable programs, and recognize the importance of operating within an efficient domestic structure, as fundamental conditions for their economic and social progress and for assuring effective inter-American cooperation.

Article 31

To accelerate their economic and social development, in accordance with their own methods and procedures and within the framework of the democratic principles and the institutions of the inter-American system, the Member States agree to dedicate every effort to achieve the following basic goals:

(a) Substantial and self-sustained increase in the per capita national produce;

(b) Equitable distribution of national income;

(c) Adequate and equitable systems of taxation;

(d) Modernization of rural life and reforms leading to equitable and efficient land-tenure systems, increased agricultural productivity, expanded use of undeveloped land, diversification of production; and improved processing and marketing systems for agricultural products; and the strengthening and expansion of facilities to attain these ends;

(e) Accelerated and diversified industrialization, especially of capital and intermediate goods;

(f) Stability in the domestic price levels, compatible with sustained economic development and the attainment of social justice;

(g) Fair wages, employment opportunities,and acceptable working conditions for all;

(h) Rapid eradication of illiteracy and expansion of educational opportunities for all;

(i) Protection of man's potential through the extension and application of modern medical science;

(j) Proper nutrition, especially through the acceleration of national efforts to increase the production and availability of food.

(k) Adequate housing for all sectors of the population;

(l) Urban conditions that offer the opportunity for a healthful, productive, and full life;

(m) Promotion of private initiative and investment in harmony with action in the public sector; and

(n) Expansion and diversification of exports.

Article 32

In order to attain the objectives set forth in this Chapter, the Member States agree to cooperate with one another, in the broadest spirit of inter-American solidarity, as far as their resources may permit and their laws may provide.

Article 33

To attain balanced and sustained development as soon as feasible, the Member States agree that the resources made available from time to time by each, in accordance with the preceding Article, should be provided under flexible conditions and in support of the national and multinational programs and efforts undertaken to meet the needs of the assisted country, giving special attention to the relatively less-developed countries.

They will seek, under similar conditions and for similar purposes, financial and technical cooperation from sources outside the Hemisphere and from international institutions.

Article 34

The Member States should make every effort to avoid policies, actions, or measures that have serious adverse effects on the economic or social development of another Member State.

Article 35

The Member States agree to join together in seeking a solution to urgent or critical problems that may arise whenever the economic development or stability of any Member State is seriously affected by conditions that cannot be remedied through the efforts of that State.

Article 36

The Member States shall extend among themselves the benefits of science and technology by encouraging the exchange and utilization of scientific and technical knowledge in accordance with existing treaties and national laws.

Article 37

The Member States, recognizing the close interdependence between foreign trade and economic and social development, should make individual and united efforts to bring about the following:

(a) Reduction or elimination, by importing countries, of tariff and non-tariff barriers that affect the exports of the Members of the Organization, except when such barriers are applied in order to diversify the economic structure, to speed up the development of the less-developed Member States, or to intensify their process of economic integration, or when they are related to national security or to the needs for economic balance;

(b) Maintenance of continuity in their economic and social development by means of:

(1) Improved conditions for trade in basic commodities through international agreements, where appropriate: orderly marketing procedures that avoid the disruption of markets; and other measures designed to promote the expansion of markets, and to obtain dependable supplies for consumers, and stable prices that are both remunerative to producers and fair to consumers;

(2) Improved international financial cooperation and the adoption of other means for lessening the adverse impact of sharp fluctuations in export earnings experienced by the countries exporting basic commodities; and

(3) Diversification of exports and expansion of export opportunities for manufactured and semimanufactured products from the developing countries by promoting and strengthening national and multinational institutions and arrangements established for these purposes.

Article 38

The Member States reaffirm the principle that when the more-developed countries grant concessions in international trade agreements that lower or eliminate tariffs or other barriers to foreign trade so that they benefit the less-developed countries, they should not expect reciprocal concessions from those countries that are incompatible with their economic development, financial, and trade needs.

Article 39

The Member States, in order to accelerate their economic development, regional integration, and the expansion and improvement of the conditions of their commerce, shall promote improvement and coordination of transportation and communication in the developing countries and among the Member States.

Article 40

The Member States recognize that integration of the developing countries of the Hemisphere is one of the objectives of the inter-American system and, therefore, shall orient their efforts and take the necessary measures to accelerate the integration process, with a view to establishing a Latin American common market in the shortest possible time.

Article 41

In order to strengthen and accelerate integration in all its aspects, the Member States agree to give adequate priority to the preparation and carrying out of multinational projects and to their financing, as well as to encourage economic and financial institutions of the inter-American system to continue giving their broadest support to regional integration institutions and programs.

Article 42

The Member States agree that technical and financial cooperation that seeks to promote regional economic integration should be based on the principle of harmonious, balanced, and efficient development, with particular attention to the relatively less-developed countries, so that it may be a decisive factor that will enable them to promote, with their own efforts, the improved development of their infrastructure programs, new lines of production, and export diversification.

Chapter VIII
SOCIAL STANDARDS

Article 43

The Member States, convinced that man can only achieve the full realization of his aspiriations within a just social order, along with economic development and true peace, agree to dedicate every effort to the application of the following principles and mechanisms:

(a) All human beings, without distinction as to race, sex, nationality, creed, or social condition, have a right to material well-being and to their spiritual development, under circumstances of liberty, dignity, equality of opportunity, and economic security;

(b) Work is a right and social duty, it gives dignity to the one who performs it, and it should be performed under conditions, including a system of fair wages, that ensure life, health, and a decent standard of living for the worker and his family, both during his working years and in his old age, or when any circumstance deprives him of the possibility of working;

(c) Employers and workers, both rural and urban, have the right to associate themselves freely for the defense and promotion of their interests, including the right to collective bargaining and the workers' right to strike, and recognition of the juridical personality of associations and the protection of their freedom and independence, all in accordance with applicable laws;

(d) Fair and efficient systems and procedures for consultation and collaboration among the sectors of production, with due regard for safeguarding the interests of the entire society;

(e) The operation of systems of public administration, banking and credit, enterprise, and distribution and sales, in such a way, in harmony with the private sector, as to meet the requirements and interests of the community;

(f) The incorporation and increasing participation of the marginal sectors of the population, in both rural and urban areas, in the economic, social civic, cultural, and political life of the nation, in order to achieve the full integration of the national community, acceleration of the process of social mobility, and the consolidation of the democratic system. The encouragement of all efforts of popular promotion and cooperation that have as their purpose the development and progress of the community;

(g) Recognition of the importance of the contribution of organizations such as labor unions, cooperatives, and cultural, professional, business, neighborhood, and community associations to the life of the society and to the development process;

(h) Development of an efficient social security policy; and

(i) Adequate provision for all persons to have due legal aid in order to secure their rights.

Article 44

The Member States recognize that, in order to facilitate the process of Latin American regional integration, it is necessary to harmonize the social legislation of the developing countries, especially in the labor and social security fields, so that the rights of the workers shall be equally protected, and they agree to make the greatest efforts possible to achieve this goal.

Chapter IX
EDUCATIONAL, SCIENTIFIC, AND CULTURAL STANDARDS

Article 45

The Member States will give primary importance within their development plans to the encouragement of education, science, and culture, oriented toward the over-all improvement of the individual, and as a foundation for democracy, social justice, and progress.

Article 46

The Member States will cooperate with one another to meet their educational needs, to promote scientific research, and to encourage technological progress. They consider themselves individually and jointly bound to preserve and enrich the cultural heritage of the American peoples.

Article 47

The Member States will exert the greatest efforts, in accordance with their constitutional process, to ensure the effective exercise of the right to education, on the following bases:

(a) Elementary education, compulsory for children of school age, shall also be offered to all others who can benefit from it. When provided by the State it shall be without charge;

(b) Middle-level education shall be extended progressively to as much of the population as possible, with a view to social improvement. It shall be diversified in such a way that it meets the development needs of each country without prejudice to providing a general education; and

(c) High education shall be available to all, provided that, in order to maintain its high level, the corresponding regulatory or academic standards are met.

Article 48

The Member States will give special attention to the eradication of illiteracy, will strengthen adult and vocational education systems, and will ensure that the benefits of culture will be available to the entire population. They will promote the use of all information media to fulfill these aims.

Article 49

The Member States will develop science and technology through educational and research institutions and through expanded information programs. They will organize their cooperation in these fields efficiently and will substantially increase exchange of knowledge, in accordance with national objectives and laws and with treaties in force.

Article 50

The Member States, with due respect for the individuality of each of them, agree to promote cultural exchange as an effective means of consolidating inter-American understanding; and they recognize that regional integration programs should be strengthened by close ties in the field of education, science, and culture.

PART TWO

Chapter X
THE ORGANS

Article 51

The Organization of American States accomplishes its purposes by means of:

(a) The General Assembly;
(b) The Meeting of Consultation of Ministers of Foreign Affairs;
(c) The Councils;

(d) The Inter-American Juridical Committee;
(e) The Inter-American Commission on Human Rights;
(f) The General Secretariat;
(g) The Specialized Conferences; and
(h) The Specialized Organizations.

There may be established, in addition to those provided for in the Charter and in accordance with the provisions thereof, such subsidiary organs, agencies, and other entities as are considered necessary.

Chapter XI

THE GENERAL ASSEMBLY

Article 52

The General Assembly is the supreme organ of the Organization of American States. It has as its principal powers, in addition to such others as are assigned to it by the Charter, the following:

(a) to decide the general action and policy of the Organization, determine the structure and functions of its organs, and consider any matter relating to friendly relations among the American States;

(b) To establish measures for coordinating the activities of the organs, agencies, and entities of the Organization among themselves and such activities with those of the other institutions of the inter-American system;

(c) To strengthen and coordinate cooperation with the United Nations and its specialized agencies;

(d) To promote collaboration, especially in the economic, social, and cultural fields, with other international organizations whose purposes are similar to those of the Organization of American States;

(e) To approve the program-budget of the Organization and determine the quotas of the Member States;

(f) To consider the annual and special reports that shall be presented to it by the organs, agencies, and entities of the inter-American system;

(g) To adopt general standards to govern the operations of the General Secretariat; and

(h) To adopt its own rules of procedure and by a two-thirds vote, its agenda.

The General Assembly shall exercise its powers in accordance with the provisions of the Charter and of other inter-American treaties.

Article 53

The General Assembly shall establish the bases for fixing the quota that each Government is to contribute to the maintenance of the Organization, taking into account the ability to pay of the respective countries and their determination to contribute in an equitable manner. Decisions on budgetary matters require the approval of two thirds of the Member States.

Article 54

All Member States have the right to be represented in the General Assembly. Each State has the right to one vote.

Article 55

The General Assembly shall convene annually during the period determined by the rules of procedure and at a place selected in accordance with the principle of rotation. At each regular session the date and place of the next regular session shall be determined, in accordance with the rules of procedure.

If for any reason the General Assembly cannot be held at the place chosen, it shall meet at the General Secretariat, unless one of the Member States should make a timely offer of a site in its territory, in which case the Permanent Council of the Organization may agree that the General Assembly will meet inthat place.

Article 56

In special circumstances and with the approval of two thirds of the Member States, the Permanent Council shall convoke a special session of the General Assembly.

Article 57

Decisions of the General Assembly shall be adopted by the affirmative vote of an absolute majority of the Member States, except in those cases that require a two-thirds vote as provided in the Charter or as may be provided by the General Assembly in its rules of procedure.

Article 58

There shall be a Preparatory Committee of the General Assembly, composed of representatives of all the Member States, which shall:

(a) Prepare the draft agenda of each session of the General Assembly;

(b) Review the proposed program-budget and the draft resolution on quotas, and present to the General Assembly a report thereon containing the recommendations it considers appropriate; and

(c) Carry out such other functions as the General Assembly may assign to it.

The draft agenda and the report shall, in due course, be transmitted to the Governments of the Member States.

Chapter XII

THE MEETING OF CONSULTATION OF MINISTERS OF FOREIGN AFFAIRS

Article 59

The Meeting of Consultation of Ministers of Foreign Affairs shall be held in order to consider problems of an urgent nature and of common interest to the American States, and to serve as the Organ of Consultation.

Article 60

Any Member State may request that a Meeting of Consultation be called. The request shall be addressed to the Permanent Council of the Organization, which shall decide by an absolute majority whether a meeting should be held.

Article 61

The agenda and regulations of the Meeting of Consultation shall be prepared by the Permanent Council of the Organization and submitted to the Member States for consideration.

Article 62

If, for exceptional reasons, a Minister of Foreign Affairs is unable to attend the meeting, he shall be represented by a special delegate.

Article 63

In case of an armed attack within the territory of an American State or within the region of security delimited by treaties in force, a Meeting of Consultation shall be held without delay. Such Meeting shall be called immediately by the Chairman of the Permanent Council of the Organization, who shall at the same time call a meeting of the Council itself.

Article 64

An advisory Defense Committee shall be established to advise the Organ of Consultation on problems of military cooperation that may arise in connection with the application of existing special treaties on collective security.

Article 65

The Advisory Defense Committee shall be composed of the highest military authorities of the American States participating in the Meeting of Consultation. Under exceptional circumstances the Governments may appoint substitutes. Each State shall be entitled to one vote.

Article 66

The Advisory Defense Committee shall be convoked under the same conditions as the Organ of Consultation, when the latter deals with matters relating to defense against aggression.

Article 67

The Committee shall also meet when the General Assembly or the Meeting of Consultation or the Governments, by a two-thirds majority of the Member States, assign to it technical studies or reports on specific subjects.

Chapter XIII
THE COUNCILS OF THE ORGANIZATION
Common Provisions

Article 68

The Permanent Council of the Organization, the Inter-American Economic and Social Council, and the Inter-American Council for Education, Science and Culture are directly responsible to the General Assembly and each has the authority granted to it in the Charter and other inter-American instruments, as well as the functions assigned to it by the General Assembly and the Meeting of Consultation of Ministers of Foreign Affairs.

Article 69

All Member States have the right to be represented on each of the Councils. Each State has the right to one vote.

Article 70

The Councils, may, within the limits of the Charter and other inter-American instruments, make recommendations on matters within their authority.

Article 71

The Councils, on matters within their respective competence, may present to the General Assembly studies and proposals, drafts of international instruments, and proposals on the holding of specialized conferences, on the creation, modification, or elimination of specialized organizations and other inter-American agencies, as well as on the coordination of their activities. The Councils may also present studies, proposals, and drafts of international instruments to the Specialized Conferences.

Article 72

Each Council may, in urgent cases, convoke Specialized Conferences on matters within its competence, after consulting with the Member States and without having to resort to the procedure provided for in Article 128.

Article 73

The Councils, to the extent of their ability, and with the cooperation of the General Secretariat, shall render to the Governments such specialized services as the latter may request.

Article 74

Each council has the authority to require the other Councils, as well as the subsidiary organs and agencies responsible to them, to provide it with information and advisory services on matters within their respective spheres of competence. The Councils may also request the same services from the other agencies of the inter-American system.

Article 75

With the prior approval of the General Assembly, the Councils may establish the subsidiary organs and the agencies that they consider advisable for the better performance of their duties. When the General Assembly is not in session, the aforesaid organs or agencies may be established provisionally by the corresponding Council. In constituting the membership of these bodies, the Councils, insofar as possible, shall follow the criteria of rotation and equitable geographic representation.

Article 76

The Councils may hold meetings in any Member State, when they find it advisable and with the prior consent of the Government concerned.

Article 77

Each Council shall prepare its own statutes and submit them to the General Assembly for approval. It shall approve its own rules of procedure and those of its subsidiary organs, agencies, and committees.

Chapter XIV

THE PERMANENT COUNCIL OF THE ORGANIZATION

Article 78

The Permanent Council of the Organization is composed of one representative of each Member State, especially appointed by the respective Government, with the rank of ambassador. Each Government may accredit an acting representative, as well as such alternates and advisers as it considers necessary.

Article 79

The office of Chairman of the Permanent Council shall be held by each of the representatives, in turn, following the alphabetic order in Spanish of the names of their respective countries. The office of Vice Chairman shall be filled in the same way, following reverse alphabetic order.

The Chairman and the Vice Chairman shall hold office for a term of not more than six months, which shall be determined by the statutes.

Article 80

Within the limits of the Charter and of inter-American treaties and agreements, the Permanent Council takes cognizance of any matter referred to it by the General Assembly or the Meeting of Consultation of Ministers of Foreign Affairs.

Article 81

The Permanent Council shall serve provisionally as the Organ of Consultation when the circumstances contemplated in Article 63 of this Charter arise.

Article 82

The Permanent Council shall keep vigilance over the maintenance of friendly relations among the Member States, and for that purpose shall effectively assist them in the peaceful settlement of their disputes, in accordance with the following provisions.

Article 83

To assist the Permanent Council in the exercise of these powers, an Inter-American Committee on Peaceful Settlement shall be established, which shall function as a subsidiary organ of the Council. The statutes of the Committee shall be prepared by the Council and approved by the General Assembly.

Article 84

The parties to a dispute may resort to the Permanent Council to obtain its good offices. In such a case the Council shall have authority to assist the parties and to recommend the procedures it considers suitable for the peaceful settlement of the dispute.

If the parties so wish, the Chairman of the Council shall refer the dispute directly to the Inter-American Committee on Peaceful Settlement.

Article 85

In the exercise of these powers, the Permanent Council, through the Inter-American Committee on Peaceful Settlement or by any other means, may ascertain the facts in the dispute, and may do so in the territory of any of the parties with the consent of the Government concerned.

Article 86

Any party to a dispute in which none of the peaceful procedures set forth in Article 24 of the Charter is being followed may appeal to the Permanent Council to take cognizance of the dispute.

The Council shall immediately refer the request to the Inter-American Committee on Peaceful Settlement, which shall consider whether or not the matter is within its competence and, if it deems it appropriate, shall offer its good offices to the other party or parties. Once these are accepted, the Inter-American Committee on Peaceful Settlement may assist the parties and recommend the procedures that it considers suitable for the peaceful settlement of the dispute.

In the exercise of these powers, the Committee may carry out an investigation of the facts in the dispute, and may do so in the territory of any of the parties with the consent of the Government concerned.

Article 87

If one of the parties should refuse the offer, the Inter-American Committee on Peaceful Settlement shall limit itself to informing the Permanent Council, without prejudice to its taking steps to restore relations between the parties, if they were interrupted, or to reestablish harmony between them.

Article 88

Once such a report is received, the Permanent Council may make suggestions for bringing the parties together for the purpose of Article 87 and, if it considers it necessary, it may urge the parties to avoid any action that might aggravate the dispute.

If one of the parties should continue to refuse the good offices of the Inter-American Committee on Peaceful Settlement or of the Council, the Council shall limit itself to submitting a report to the General Assembly.

Article 89

The Permanent Council, in the exercise of these functions, shall take its decisions by an affirmative vote of two-thirds of its members, excluding the parties to the dispute, except for such decisions as the rules of procedure provide shall be adopted by a simple majority.

Article 90

In performing their functions with respect to the peaceful settlement of disputes, the Permanent Council and the Inter-American Committee on Peaceful Settlement shall observe the provisions of the Charter and the principles and standards of international law, as well as take into account the existence of treaties in force between the parties.

Article 91

The Permanent Council shall also:

(a) Carry out those decisions of the General Assembly or of the Meeting of Consultation of Ministers of Foreign Affairs the implementation of which has not been assigned to any other body;

(b) Watch over the observance of the standards governing the operation of the General Secretariat and, when the General Assembly is not in session, adopt provisions of a regulatory nature that enable the General Secretariat to carry out its administrative functions;

(c) Act as the Preparatory Committee of the General Assembly, in accordance with the terms of Article 58 of the Charter, unless the General Assembly should decide otherwise;

(d) Prepare, at the request of the Member States and with the cooperation of the appropriate organs of the Organization, draft agreements to promote and facilitate cooperation between the Organization of American States and the United Nations or between the Organization and other American agencies of recognized international standing. These draft agreements shall be submitted to the General Assembly for approval;

(e) Submit recommendations to the General Assembly with regard to the functioning of the Organization and the coordination of its subsidiary organs, agencies, and committees;

(f) Present to the General Assembly any observations it may have regarding the reports of the Inter-American Juridical Committee and the Inter-American Commission on Human Rights; and

(g) Perform the other functions assigned to it in the Charter.

Article 92

The Permanent Council and the General Secretariat shall have the same seat.

Chapter XV

THE INTER-AMERICAN ECONOMIC AND SOCIAL COUNCIL

Article 93

The Inter-American Economic and Social Council is composed of one principle representative, of the highest rank, of each Member State, especially appointed by the respective Government.

Article 94

The purpose of the Inter-American Economic and Social Council is to promote cooperation among the American countries in order to attain accelerated economic and social development, in accordance with the standards set forth in Chapters VII and VIII.

Article 95

To achieve its purpose the Inter-American Economic and Social Council shall:

(a) Recommend programs and courses of action and periodically study and evaluate the efforts undertaken by the Member States;

(b) Promote and coordinate all economic and social activities of the Organization;

(c) Coordinate its activities with those of the other Councils of the Organization;

(d) Establish cooperative relations with the corresponding organs of the United Nations and with other national and international agencies, especially with regard to coordination of inter-American technical assistance programs; and

(e) Promote the solution of the cases contemplated in Article 35 of the Charter, establishing the appropriate procedure.

Article 96

The Inter-American Economic and Social Council shall hold at least one meeting each year at the ministerial level. It shall also meet when convoked by the General Assembly, the Meeting of Consultation of Ministers of Foreign Affairs, at its own initiative, or for the cases contemplated in Article 35 of the Charter.

Article 97

The Inter-American Economic and Social Council shall have a Permanent Executive Committee, composed of a Chairman and no less than seven other members, elected by the Council for terms to be established in the statutes of the Council. Each member shall have the right to one vote. The principles of equitable geographic representation and of rotation shall be taken into account, insofar as possible, in the election of members. The Permanent Executive Committee represents all of the Member States of the Organization.

Article 98

The Permanent Executive Committee shall perform the tasks assigned to it by the Inter-American Economic and Social Council, in accordance with the general standards established by the Council.

Chapter XVI
THE INTER-AMERICAN COUNCIL FOR EDUCATION, SCIENCE, AND CULTURE

Article 99

The Inter-American Council for Education, Science, and Culture is composed of one principal representative, of the highest rank, of each Member State, especially appointed by the respective Government.

Article 100

The purpose of the Inter-American Council for Education, Science, and Culture is to promote friendly relations and mutual understanding between the peoples of the Americas through educational, scientific, and cultural cooperation and exchange between Member States, in order to raise the cultural level of the peoples, reaffirm their dignity as individuals, prepare them fully for the tasks of progress, and strengthen the devotion to peace, democracy, and social justice that has characterized their evolution.

Article 101

To accomplish its purpose the Inter-American Council for Education, Science, and Culture shall:

(a) Promote and coordinate the educational, scientific, and cultural activities of the Organization;

(b) Adopt or recommend pertinent measures to give effect to the standards contained in Chapter IX of the Charter;

(c) Support individual or collective efforts of the Member States to improve and extend education at all levels, giving special attention to efforts directed toward community development;

(d) Recommend and encourage the adoption of special educational programs directed toward integrating all sectors of the population into their respective national cultures;

(e) Stimulate and support scientific and technological education and research, especially when these relate to national development plans;

(f) Foster the exchange of professors, research workers, technicians, and students, as well as of study materials; and encourage the conclusion of bilateral or multilateral agreements on the progressive coordination of curricula at all educational levels and on the validity and equivalence of certificates and degrees;

(g) Promote the education of the American peoples with a view to harmonious international relations and a better understanding of the historical and cultural origins of the Americas, in order to stress and preserve their common values and destiny;

(h) Systematically encourage intellectual and artistic creativity, the exchange of cultural works and folklore, as well as the interrelationships of the different cultural regions of the Americas;

(i) Foster cooperation and technical assistance for protecting, preserving, and increasing the cultural heritage of the Hemisphere;

(j) Coordinate its activities with those of the other Councils. In harmony with the Inter-American Economic and Social Council, encourage the inter-relationship of programs for promoting education, science, and culture with national development and regional integration programs;

(k) Establish cooperative relations with the corresponding organs of the United Nations and with other national and international bodies;

(l) Strengthen the civic conscience of the American peoples, as one of the bases for the effective exercise of democracy and for the observance of the rights and duties of man;

(m) Recommend appropriate procedures for intensifying integration of the developing countries of the Hemisphere by means of efforts and programs in the fields of education, science, and culture: and

(n) Study and evaluate periodically the efforts made by the Member States in the fields of education, science, and culture.

Article 102

The Inter-American Council for Education, Science, and Culture shall hold at least one meeting each year at the ministerial level. It shall also meet when convoked by the General Assembly, by the Meeting of Consultation of Ministers of Foreign Affairs, or at its own initiative.

Article 103

The Inter-American Council for Education, Science, and Culture shall have a Permanent Executive Committee, composed of a Chairman and no less than seven other members, elected by the Council for terms to be established in the Statutes of the Council. Each member shall have the right to one vote. The principles of equitable geographic representation and of rotation shall be taken into account, insofar as possible, in the election of members. The Permanent Executive Committee represents all of the Member States of the Organization.

Article 104

The Permanent Executive Committee shall perform the tasks assigned to it by the Inter-American Council for Education, Science, and Culture, in accordance with the general standards established by the Council.

Chapter XVII

THE INTER-AMERICAN JURIDICAL COMMITTEE

Article 105

The purpose of the Inter-American Juridical Committee is to serve the Organization as an advisory body on juridical matters; to promote the progressive development and the codification of international law; and to study juridical problems related to the integration of the developing countries of the Hemisphere and, insofar as may appear desirable, the possibility of attaining uniformity in their legislation.

Article 106

The Inter-American Juridical Committee shall undertake the studies and preparatory work assigned to it by the General Assembly, the Meeting of Consultation of Ministers of Foreign Affairs, or the Councils of the Organizations. It may also, on its own initiative, undertake such studies and preparatory work as it considers advisable, and suggest the holding of specialized juridical conferences.

Article 107

The Inter-American Juridical Committee shall be composed of eleven jurists, nationals of Member States, elected by the General Assembly for a period of four years from panels of three candidates presented by Member States. In the election, a system shall be used that takes into account partial replacement of membership and, insofar as possible, equitable geographic representation. No two members of the Committee may be nations of the same state. Vacancies that occur shall be filled in the manner set forth above.

Article 108

The Inter-American Juridical Committee represents all of the Member States of the Organization, and has the broadest possible technical autonomy.

Article 109

The Inter-American Juridical Committee shall establish cooperative relations with universities, institutes, and other teaching centers, as well as with national and international committees and entities devoted to study, research, teaching, or dissemination of information on juridical matters of international interest.

Article 110

The Inter-American Juridical Committee shall draft its statutes, which shall be submitted to the General Assembly for approval.

The Committee shall adopt its own rules of procedure.

Article 111

The seat of the Inter-American Juridical Committee shall be the city of Rio de Janeiro, but in special cases the Committee may meet at any other place that may be designated, after consultation with the Member State concerned.

Chapter XVIII
THE INTER-AMERICAN COMMISSION ON HUMAN RIGHTS

Article 112

There shall be an Inter-American Commission on Human Rights, whose principal function shall be to promote the observance and protection of human rights and to serve as a consultative organ of the Organization in these matters.

An inter-American convention on human rights shall determine the structure, competence, and procedure of this Commission, as well as those of other organs responsible for these matters.

Chapter XIX
THE GENERAL SECRETARIAT

Article 113

The General Secretariat is the central and permanent organ of the Organization of American States. It shall perform the functions assigned to it in the Charter, in other inter-American treaties and agreements, and by the General Assembly, and shall carry out the duties entrusted to it by the General Assembly, the Meeting of Consultation of Ministers of Foreign Affairs, or the Councils.

Article 114

The Secretary General of the Organization shall be elected by the General Assembly for a five-year term and may not be reelected more than once or succeeded by a person of the same nationality. In the event that the office of Secretary General becomes vacant, the Assistant Secretary General shall assume his duties until the General Assembly shall elect a new Secretary General for a full term.

Article 115

The Secretary General shall direct the General Secretariat, be the legal representative thereof, and, notwithstanding the provisions of Article 91.b, be responsible to the General Assembly for the proper fulfillment of the obligations and functions of the General Secretariat.

Article 116

The Secretary General, or his representative, participates with voice but without vote in all meetings of the Organization.

Article 117

The General Secretariat shall promote economic, social, juridical, educational, scientific, and cultural relations among all the Member States of the Organization, in keeping with the actions and policies decided upon by the General Assembly and with the pertinent decisions of the Councils.

Article 118

The General Secretariat shall also perform the following functions:

(a) Transmit *ex officio* to the Member States notice of the convocation of the General Assembly, the Meeting of Consultation of Ministers of Foreign Affairs, the Inter-American Economic and Social Council, the Inter-American Council for Education, Science, and Culture, and the Specialized Conferences;

(b) Advise the other organs, when appropriate, in the preparation of agenda and rules of procedure;

(c) Prepare the proposed program-budget of the Organization on the basis of programs adopted by the Councils, agencies, and entities whose expenses should be included in the program-budget and, after consultation with the Councils or their permanent committees, submit it to the Preparatory Committee of the General Assembly and then to the Assembly itself;

(d) Provide, on a permanent basis, adequate secretariat services for the General Assembly and the other organs, and carry out their directives and assignments. To the extent of its ability, provide services for the other meetings of the Organization;

(e) Serve as custodian of the documents and archives of the Inter-American Conferences, the General Assembly, the Meetings of Consultation of Ministers of Foreign Affairs, the Councils, and the Specialized Conferences;

(f) Serve as depository of inter-American treaties and agreements, as well as of the instruments of ratification thereof;

(g) Submit to the General Assembly at each regular session an annual report on the activities of the Organization and its financial condition; and

(h) Establish relations of cooperation, in accordance with decisions reached by the General Assembly or the Councils, with the Specialized Organizations as well as other national and international organizations.

Article 119

The Secretary General shall:

(a) Establish such offices of the General Secretariat as are necessary to accomplish its purposes: and

(b) Determine the number of officers and employees of the General Secretariat, appoint them, regulate their powers and duties, and fix their remuneration.

The Secretary General shall exercise this authority in accordance with such general standards and budgetary provisions as may be established by the General Assembly.

Article 120

The Assistant Secretary General shall be elected by the General Assembly for a five-year term and may not be reelected more than once or succeeded by a person of the same nationality. In the event that the office of Assistant Secretary General becomes vacant, the Permanent Council shall elect a substitute to hold that office until the General Assembly shall elect a new Assistant Secretary General for a full term.

Article 121

The Assistant Secretary General shall be the Secretary of the Permanent Council. He shall serve as advisory officer to the Secretary General and shall act as his delegate in all matters that the Secretary General may entrust to him. During the temporary absence or disability of the Secretary General, the Assistant Secretary General shall perform his functions.

The Secretary General and the Assistant Secretary General shall be of different nationalities.

Article 122

The General Assembly, by a two-thirds vote of the Member States, may remove the Secretary General or the Assistant Secretary General, or both, whenever the proper functioning of the Organization so demands.

Article 123

The Secretary General shall appoint, with the approval of the respective Council, the Executive Secretary for Economic and Social Affairs and the Executive Secretary for Education, Science, and Culture, who shall also be the secretaries of the respective Councils.

Article 124

In the performance of their duties, the Secretary General and the personnel of the Secretariat shall not seek or receive instructions from any Government or from any authority outside the Organization, and shall refrain from any action that may be incompatible with their position as international officers responsible only to the Organization.

Article 125

The Member States pledge themselves to respect the exclusively international character of the responsibilities of the Secretary General and the personnel of the General Secretariat, and not to seek to influence them in the discharge of their duties.

Article 126

In selecting the personnel of the General Secretariat, first consideration shall be given to efficiency, competence, and integrity; but at the same time in the recruitment of personnel of all ranks, importance shall be given to the necessity of obtaining as wide a geographic representation as possible.

Article 127

The seat of the General Secretariat is the city of Washington.

Chapter XX
THE SPECIALIZED CONFERENCES

Article 128

The Specialized Conferences are inter-governmental meetings to deal with special technical matters or to develop specific aspects of inter-American cooperation. They shall be held when either the General Assembly or the Meeting of Consultation of Ministers of Foreign Affairs so decides, on its own initiative or at the request of one of the Councils or Specialized Organizations.

Article 129

The agenda and rules of procedure of the Specialized Conferences shall be prepared by the Councils or Specialized Organizations concerned and shall be submitted to the Governments of the Member States for consideration.

Chapter XXI
THE SPECIALIZED ORGANIZATIONS

Article 130

For the purposes of the present Charter, Inter-American Specialized Organizations are the intergovernmental organizations established by multilateral agreements and having specific functions with respect to technical matters of common interest to the American States.

Article 131

The General Secretariat shall maintain a register of the organizations that fulfill the condition set forth in the foregoing Article, as determined by the General Assembly after a report from the Council concerned.

Article 132

The Specialized Organizations shall enjoy the fullest technical autonomy, but they shall take into account the recommendations of the General Assembly and of the Councils, in accordance with the provisions of the Charter.

Article 133

The Specialized Organizations shall transmit to the General Assembly annual reports on the progress of their work and on their annual budgets and expenses.

Article 134

Relations that should exist between the Specialized Organizations and the Organization shall be defined by means of agreements concluded between each organization and the Secretary General, with the authorization of the General Assembly.

Article 135

The Specialized Organizations shall establish cooperative relations with world agencies of the same character in order to coordinate their activities. In concluding agreements with international agencies of a worldwide character, the Inter-American Specialized Organizations shall preserve their identity and their status as integral parts of the Organization of American States, even when they perform regional functions of international agencies.

Article 136

In determining the location of the Specialized Organizations consideration shall be given to the interest of all the Member States and to the desirability of selecting the seats of these organizations on the basis of a geographic representation as equitable as possible.

PART THREE

Chapter XXII

THE UNITED NATIONS

Article 137

None of the provisions of this Charter shall be construed as impairing the rights and obligations of the Member States under the Charter of the United Nations.

Chapter XXIII

MISCELLANEOUS PROVISIONS

Article 138

Attendance at meetings of the permanent organs of the Organization of American States or at the conferences and meetings provided for in the Charter, or held under the auspices of the Organization, shall be in accordance with the multilateral character of the aforesaid organs, conferences, and meetings and shall not depend on the bilateral relations between the Government of any Member State and the Government of the host country.

Article 139

The Organization of American States shall enjoy in the territory of each Member such legal capacity, privileges, and immunities as are necessary for the exercise of its functions and the accomplishment of its purposes.

Article 140

The representatives of the Member States on the organs of the Organization, the personnel of their delegations, as well as the Secretary General and the Assistant Secretary General shall enjoy the privileges and immunities corresponding to their positions and necessary for the independent performance of their duties.

Article 141

The juridical status of the Specialized Organizations and the privileges and immunities that should be granted to them and to their personnel, as well as to the officials of the General Secretariat, shall be determined in a multilateral agreement. The foregoing shall not preclude, when it is considered necessary, the concluding of bilateral agreements.

Article 142

Correspondence of the Organization of American States, including printed matter and parcels, bearing the frank thereof, shall be carried free of charge in the mails of the Member States.

Article 143

The Organization of American States does not allow any restriction based on race, creed, or sex, with respect to eligibility to participate in the activities of the Organization and to hold positions therein.

Chapter XXIV
RATIFICATION AND ENTRY INTO FORCE

Article 144

The present Charter shall remain open for signature by the American States and shall be ratified in accordance with their respective constitutional procedures. The original instrument, the Spanish, English, Portuguese, and French texts of which are equally authentic, shall be deposited with the General Secretariat, which shall transmit certified copies thereof to the Governments for purposes of ratification. The instruments of ratification shall be deposited with the General Secretariat, which shall notify the signatory States of such deposit.

Article 145

The present Charter shall enter into force among the ratifying States when two-thirds of the signatory States have deposited their ratifications. It shall enter into force with respect to the remaining States in the order in which they deposit their ratifications.

Article 146

The present Charter shall be registered with the Secretariat of the United Nations through the General Secretariat.

Article 147

Amendments to the present Charter may be adopted only at a General Assembly convened for that purpose. Amendments shall enter into force in accordance with the terms and the procedure set forth in Article 145.

Article 148

The present Charter shall remain in force indefinitely, but may be denounced by any Member State upon written notification to the General Secretariat, which shall communicate to all the others each notice of denunciation received. After two years from the date on which the General Secretariat receives a notice of denunciation, the present Charter shall cease to be in force with respect to the denouncing State, which shall cease to belong to the Organization after it has fulfilled the obligations arising from the present Charter.

Chapter XXV
TRANSITORY PROVISIONS

Article 149

The Inter-American Committee on the Alliance for Progress shall act as the permanent executive committee of the Inter-American Economic and Social Council as long as the Alliance is in operation.

Article 150

Until the inter-American convention on human rights, referred to in Chapter XVIII, enters into force, the present Inter-American Commission on Human Rights shall keep vigilance over the observance of human rights.

appendix B

Organization of American States

The OAS as it will be when the "Protocol of Buenos Aires" is in effect.

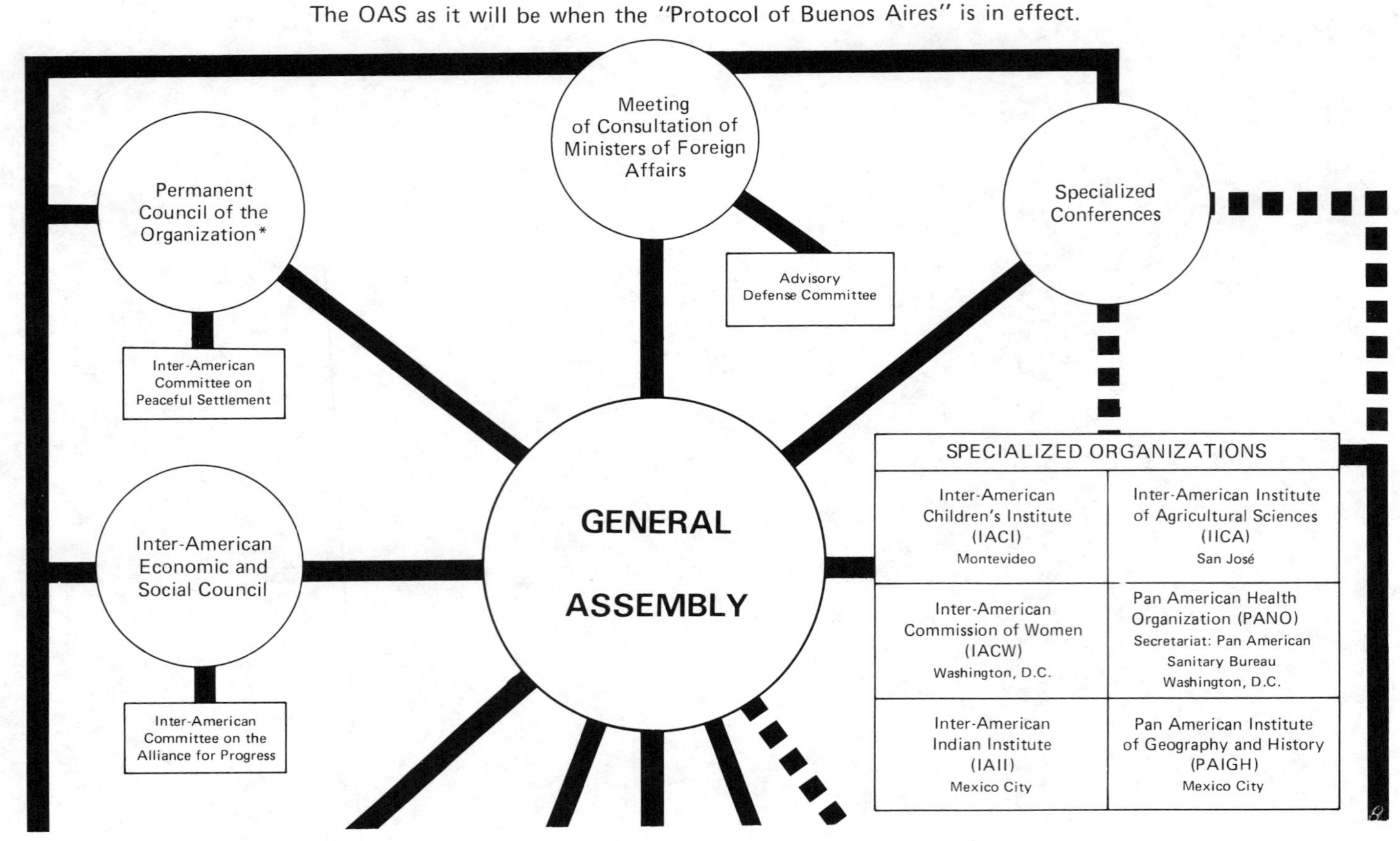

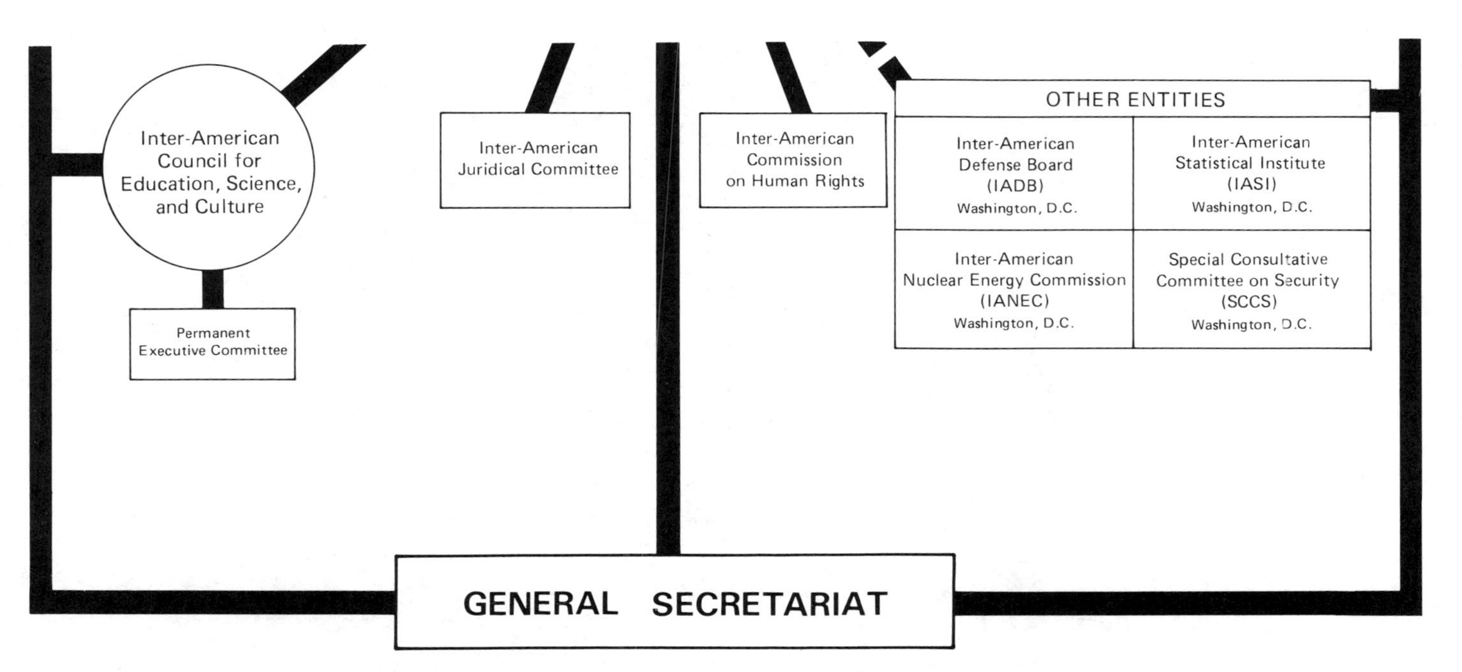

*Acts as the Preparatory Committee of the General Assembly, unless the General Assembly should decide otherwise, October 1967.

appendix C

Inter-American Treaty of Reciprocal Assistance (Rio Treaty, 1947)

In the name of their Peoples, the Governments represented at the Inter-American Conference for the Maintenance of Continental Peace and Security, desirous of consolidating and strengthening their relations of friendship and good neighborliness, and

Considering:

That Resolution VIII of the Inter-American Conference on problems of War and Peace, which met in Mexico City, recommended the conclusion of a treaty to prevent and repel threats and acts of aggression against any of the countries of America;

That the High Contracting Parties reiterate their will to remain united in an inter-American system consistent with the purposes and principles of the United Nations, and reaffirm the existence of the agreement which they have concluded concerning those matters relating to the maintenance of international peace and security which are appropriate for regional action;

That the High Contracting Parties reaffirm their adherence to the principles of inter-American solidarity and cooperation, and especially to those set forth in the preamble and declarations of the Act of Chapultepec, all of which should be understood to be accepted as standards of their mutual relations and as the juridical basis of the Inter-American System;

That the American States propose, in order to improve the procedures for the pacific settlement of their controversies, to conclude the treaty concerning the "Inter-American Peace System" envisaged in Resolutions IX and XXXIX of the Inter-American Conference on Problems of War and Peace,

That the obligation of mutual assistance and common defense of the American Republics is essentially related to their democratic ideals and to their will to cooperate permanently in the fulfillment of the principles and purposes of a policy of peace;

That the American regional community affirms as a manifest truth that juridical organization is a necessary prerequisite of security and peace, and that peace is founded on justice and moral order and, consequently, on the international recognition and protection of human rights and freedoms, on the indispensable well being of the people, and on the effectiveness of democracy for the international realization of justice and security,

Have Resolved, in conformity with the objectives stated above, to conclude the following Treaty, in order to assure peace, through adequate means, to provide for effective reciprocal assistance to meet armed attacks against any American State, and in order to deal with threats of aggression against any of them:

Article 1

The High Contracting Parties formally condemn war and undertake in their international relations not to resort to the threat or the use of force in any manner inconsistent with the provisions of the Charter of the United Nations or of this Treaty.

Article 2

As a consequence of the principle set forth in the preceding *Article*, the High Contracting Parties undertake to submit every controversy which may arise between them to methods of peaceful settlement and to endeavor to settle any such controversy among themselves by means of the procedures in force in the Inter-American System before referring it to the General Assembly or the Security Council of the United Nations.

Article 3

(1) The High Contracting Parties agree that an armed attack by any State against an American State shall be considered as an attack against all the American States and, consequently, each one of the said Contracting Parties undertakes to assist in meeting the attack in the exercise of the inherent right of individual or collective self-defense recognized by *Article 51* of the Charter of the United Nations.

(2) On the request of the State or States directly attacked and until the decision of the Organ of Consultation of the Inter-American System, each one of the Contracting Parties may determine the immediate measures which it may individually take in fulfillment of the obligation contained in the preceding paragraph and in accordance with tne principle of continental solidarity. The Organ of Consultation shall meet without delay for the purpose of examining those measures and agreeing upon the measures of a collective character that should be taken.

(3) The provisions of this *Article* shall be applied in case of any armed attack which takes place within the region described in *Article 4* or within the territory of an American State. When the attack takes place outside of the said areas, the provisions of *Article 6* shall be applied.

(4) Measures of self-defense provided for under this *Article* may be taken until the Security Council of the United Nations has taken the measures necessary to maintain international peace and security.

Article 4

The region to which this Treaty refers is bounded as follows: beginning at the North Pole; thence due south to a point 74 degrees north latitude, 10 degrees west longitude; thence by a rhumb line to a point 47 degrees 30 minutes north latitude, 50 degrees west longitude; thence by a rhumb line to a point 35 degrees north latitude, 60 degrees west longitude; thence due south to a point in 20 degrees north latitude; thence by a rhumb line to a point 5 degrees north latitude, 24 degrees west longitude; thence due south to the South Pole; thence due north to a point 30 degrees south latitude, 90 degrees west longitude; thence by a rhumb line to a point on the Equator at 97 degrees west longitude; thence by a rhumb line to a point 15 degrees north latitude, 120 degrees west longitude; thence by a rhumb line to a point 50 degrees north latitude, 170 degrees east longitude; thence due north to a point in 54 degrees north latitude; thence by a rhumb line to a point 65 degrees 30 minutes north latitude, 168 degrees 58 minutes 5 seconds west longitude; thence due north to the North Pole.

Article 5

The High Contracting Parties shall immediately send to the Security Council of the United Nations, in conformity with *Articles 51* and *54* of the Charter of the United Nations, complete information concerning the activities undertaken or in contemplation in the exercise of the right of self-defense or for the purpose of maintaining inter-American peace and security.

Article 6

If the inviolability or the integrity of the territory or the sovereignty or political independence of any American State should be affected by an aggression which is not an armed attack or by an extra-continental or intra-continental conflict, or by any other fact or situation that might endanger the peace of America, the Organ of Consultation shall meet immediately in order to agree on the measures which must be taken in case of aggression to assist the victim of the aggression, or, in any case, the measures which should be taken for the common defense and for the maintenance of the peace and security of the Continent.

Article 7

In the case of a conflict between two or more American States, without prejudice to the right of self-defense in conformity with *Article 51* of the Charter of the United Nations, the High Contracting Parties, meeting in consultation shall call upon the contending States to suspend hostilities and restore matters to the *statu quo ante bellum*, and shall take in addition all other necessary measures to re-establish or maintain inter-American peace and security and for the solution of the conflict by peaceful means. The rejection of the pacifying action will be considered in the determination of the aggressor and in the application of the measures which the consultative meeting may agree upon.

Article 8

For the purposes of this Treaty, the measures on which the Organ of Consultation may agree will comprise one or more of the following: recall of chiefs of diplomatic missions; breaking of diplomatic relations; breaking of consular relations; partial or complete interruption of economic relations or of rail, sea, air, postal, telegraphic, telephonic, and radio-telphonic communications; and use of armed force.

Article 9

In addition to other acts which the Organ of Consultation may characterize as aggression, the following shall be considered as such:

(a) Unprovoked armed attack by a State against the territory, the people, or the land, sea or air forces of another State;

(b) Invasion, by the armed forces of a State, of the territory of an American State, through the trespassing of boundaries demarcated in accordance with a treaty, judicial decision, or arbitral award, or, in the absence of frontiers thus demarcated, invasion affecting a region which is under the effective jurisdiction of another State.

Article 10

None of the provisions of this Treaty shall be construed as impairing the rights and obligations of the High Contracting Parties under the Charter of the United Nations.

Article 11

The consultations to which this Treaty refers shall be carried out by means of the Meetings of Ministers of Foreign Affairs of the American Republics which have ratified the Treaty, or in the manner or by the organ which in the future may be agreed upon.

Article 12

The Governing Board of the Pan American Union may act provisionally as an organ of consultation until the meeting of the Organ of Consultation referred to in the preceding *Article* takes place.

Article 13

The consultations shall be initiated at the request addressed to the Governing Board of the Pan American Union by any of the Signatory States which has ratified the Treaty.

Article 14

In the voting referred to in this Treaty only the representatives of the Signatory States which have ratified the Treaty may take part.

Article 15

The Governing Board of the Pan American Union shall act in all matters concerning this Treaty as an organ of liaison among the Signatory States which have ratified this Treaty and between these States and the United Nations.

Article 16

The decisions of the Governing Board of the Pan American Union referred to in *Articles 13* and *15* above shall be taken by an absolute majority of the Members entitled to vote.

Article 17

The Organ of Consultation shall take its decisions by a vote of two-thirds of the Signatory States which have ratified the Treaty.

Article 18

In the case of a situation or dispute between American States, the parties directly interested shall be excluded from the voting referred to in the two preceding *Articles*.

Article 19

To constitute a quorum in all the meetings referred to in the previous *Articles*, it shall be necessary that the number of States represented shall be at least equal to the number of votes necessary for the taking of the decision.

Article 20

Decisions which require the application of the measures specified in *Article 8* shall be binding upon all the Signatory States which have ratified this Treaty, with the sole exception that no State shall be required to use armed force without its consent.

Article 21

The measures agreed upon by the Organ of Consultation shall be executed through the procedures and agencies now existing or those which may in the future be established.

Article 22

This Treaty shall come into effect between the States which ratify it as soon as the ratifications of two-thirds of the Signatory States have been deposited.

Article 23

This Treaty is open for signature by the American States at the city of Rio de Janeiro, and shall be ratified by the Signatory States as soon as possible in accordance with their respective constitutional processes. The ratifications shall be deposited with the Pan American Union, which shall notify the Signatory States of each deposit. Such notification shall be considered as an exchange of ratifications.

Article 24

The present Treaty shall be registered with the Secretariat of the United Nations through the Pan American Union, when two-thirds of the Signatory States have deposited their ratifications.

Article 25

This Treaty shall remain in force indefinitely, but may be denounced by any High Contracting Party by a notification in writing to the Pan American Union, which shall inform all the other High Contracting Parties of each notification of denunciation received. After the expiration of two years from the date of the receipt by the Pan American Union of a notification of denunciation by any High Contracting Party, the present Treaty shall cease to be in force and effect with respect to all the other High Contracting Parties.

Article 26

The principles and fundamental provisions of this Treaty shall be incorporated in the Organic Pact of the Inter-American System.

In Witness Thereof, the undersigned Plenipotentiaries, having deposited their full powers found to be in due and proper form, sign this Treaty on behalf of their respective Governments, on the dates appearing opposite their signatures.

Done in the City of Rio de Janeiro, in four texts respectively in the English, French, Portuguese, and Spanish languages, on the second of September nineteen hundred forty-seven.

appendix D

Declaration of Caracas (1954)

Whereas:

The American republics at the Ninth International Conference of American States declared that international communism, by its anti-democratic nature and its interventionist tendency, is incompatible with the concept of American freedom, and resolved to adopt within their respective territories the measures necessary to eradicate and prevent subversive activities.

The Fourth Meeting of Consultation of Ministers of Foreign Affairs recognized that, in addition to adequate internal measures in each state, a high degree of international cooperation is required to eradicate the danger which the subversive activities of international communism pose for the American States; and

The aggressive character of the international communist movement continues to constitute, in the context of world affairs, a special and immediate threat to the national institutions and the peace and security of the American States, and to the right of each state to develop its cultural, political, and economic life freely and naturally without intervention in its internal or external affairs by other states,

The Tenth Inter-American Conference

I

Condemns:

The activities of the international communist movement as constituting intervention in American affairs;

Expresses:

The determination of the American States to take the necessary measures to protect their political independence against the intervention of international communism, acting in the interests of an alien despotism;

Reiterates:

The faith of the peoples of America in the effective exercise of representative democracy as the best means to promote their social and political progress; and

Declares:

That the domination or control of the political institutions of any American state by the international communist movement, extending to this Hemisphere the political system of an extra-continental power, would constitute a threat to the sovereignty and political independence of the American States, endangering the peace of America, and would call for a Meeting of Consultation to consider the adoption of appropriate action in accordance with existing treaties.

II

Recommends:

That, without prejudice to such other measures as they may consider desirable, special attention be given by each of the American governments to the following steps for the purpose of conteracting the subversive activities of the international communist movement within their respective jurisdictions;

1. Measures to require disclosure of the identity, activities, and sources of funds to those who are spreading propaganda of the international communist movement or who travel in the interests of that movement, and of those who act as its agents or in its behalf; and
2. The exchange of information among governments to assist in fulfilling the purpose of the resolutions adopted by the Inter-American Conferences and Meetings of Ministers of Foreign Affairs regarding international communism.

III

This declaration of foreign policy made by the American Republics in relation to dangers originating outside this Hemisphere is designed to protect and not to impair the inalienable right of each American State freely to choose its own form of government and economic system and to live its own social and cultural life.

appendix E

Act of Bogotá (1960)

The Special Committee for the study of new measures of economic cooperation.

Recognizing that the preservation and strengthening of free and democratic institutions in the American republics requires the acceleration of social and economic progress in Latin America adequate to meet the legitimate aspirations of the individual citizen of Latin America for a better life and to provide him the fullest opportunity to improve his status;

Recognizing that the interests of the American republics are so interrelated that sound social and economic progress in each is of importance to all and that lack of it in any American republic may have serious repercussions in others;

Cognizant of the steps already taken by many American republics to cope with the serious economic and social problems that affect them, but convinced that the magnitude of these problems calls for redoubled efforts by governments and for a new and vigorous program of inter-American cooperation;

Recognizing that economic development programs, which should be urgently strengthened and expanded, may have a delayed effect on social welfare, and that accordingly early measures are needed to cope with initial needs;

Recognizing that the success of a cooperative program of social progress will require maximum self-help efforts on the part of the American republics, and in many cases, the improvement of existing institutions and practices, particularly in the fields of taxation, the ownership and use of land, education and training, health and housing;

Believing it opportune to give further practical expression to the spirit of Operation Pan-America by immediately enlarging the opportunities of the people of Latin America for social progress, thus strengthening their hopes for the future;

Considering it advisable to launch a program for social development, in which emphasis should be given to those measures that meet social needs and also promote increases in productivity and strengthen economic development;

Recommends to the Council of the Organization of American States:

I. MEASURES FOR SOCIAL IMPROVEMENT

An inter-American program for social development should be established which should be directed to the carrying-out of the following measures of social improvement in Latin America, as considered appropriate in each country:

1. *Measures for the improvement of rural living and land use.*

1.1. The examination of existing legal and institutional systems with respect to:

(a) Land tenure legislation and facilities with a view to insuring a wider and more equitable distribution of the ownership of land, in a manner consistent with the objectives of employment, productivity and economic growth.

(b) Agricultural credit institutions with a view to providing adequate financing to individual farmers or groups of farmers.

(c) Tax systems and procedures and fiscal policies with a view to assuring equity of taxation and encouraging improved use of land, especially of privately owned land which is idle.

1.2. The initiation or acceleration of appropriate programs to modernize and improve the existing legal and institutional framework to insure better conditions of land tenure, extend more adequate credit facilities and provide increased incentives in the land tax structure.

1.3. The acceleration of the preparation of projects and programs for:

(a) Land reclamation and land settlement, with a view to promoting more widespread ownership and efficient use of land, particularly of unutilized or under-utilized land.

(b) The increase of the productivity of land already in use.

(c) The construction of farm-to-market and access roads.

1.4. The adoption or acceleration of other government service programs designed particularly to assist the small farmer, such as new or improved marketing organizations; extension services; research and basic surveys; and demonstration, education and training facilities.

2. *Measures for the improvement of housing and community facilities.*

2.1. The examination of existing policies in the field of housing and community facilities, including urban and regional planning, with a view to improving such policies, strengthening public institutions and promoting private initiative and participation in programs in these fields. Special consideration should be given to encouraging financial institutions to invest in low-cost housing on a long-term basis and in building and construction industries.

2.2. The strengthening of the existing legal and institutional framework for mobilizing financial resources to provide better housing and related facilities for the people, and to create new institutions for this purpose when necessary. Special consideration should be given to legislation and measures which would encourage the establishment and growth of:

(a) Private financing institutions, such as building and loan associations.

(b) Institutions to insure sound housing loans against loss.

(c) Institutions to serve as a secondary market for home mortgages.

(d) Institutions to provide financial assistance to local communities for the development of facilities such as water supply, sanitation and other public works.

Existing national institutions should be utilized wherever practical in the application of external resources to further the development of housing and community facilities.

2.3. The expansion of home building industries through such measures as the training of craftsmen and other personnel, research and the introduction of new techniques, and the development of construction standards for low- and medium-cost housing.

2.4. The lending of encouragement and assistance to programs, on a pilot basis, for aided self-help housing, for the acquisition and subdivision of land for low-cost housing developments, and for industrial housing projects.

3. *Measures for the improvement of educational systems and training facilities.*

3.1. The re-examination of educational systems, giving particular attention to:

3.1.1. The development of modern methods of mass education for the eradication of illiteracy.

3.1.2. The adequacy of training in the industrial arts and sciences with due emphasis on laboratory and work experience and on the practical application of knowledge for the solution of social and economic problems.

3.1.3. The need to provide instruction in rural schools not only in basic subjects but also in agriculture, health, sanitation, nutrition and the methods of home and community improvement.

3.1.4. The broadening of courses of study in secondary schools to provide the training necessary for clerical and executive personnel in industry, commerce, public administration and community service.

3.1.5. Specialized trade and industrial education related to the commercial and industrial needs of the community.

3.1.6. Vocational agricultural instruction.

3.1.7. Advanced education of administrators, engineers, economists and other professional personnel of key importance to economic development.

4. *Measures for the improvement of public health.*

4.1. Re-examination of programs and policies of public health giving particular attention to:

4.1.1. Strengthening the expansion of national and local health services, especially those directed to the reduction of infant mortality.

4.1.2. The progressive development of health insurance systems, including those providing for maternity, accident and disability insurance in urban and rural areas.

4.1.3. The provision of hospital and health service in areas located away from main centers of population.

4.1.4. The extension of public medical services to areas of exceptional need.

4.1.5. The strengthening of campaigns for the control or elimination of communicable diseases with special attention to the eradication of malaria.

4.1.6. The provision of water-supply facilities for purposes of health and economic development.

4.1.7. The training of public health officials and technicians.

4.1.8. The strengthening of programs of nutrition for low-income groups.

5. *Measures for the mobilization of domestic resources.*

5.1. This program shall be carried out within the framework of the maximum creation of domestic savings and the improvement of national fiscal and financial practices.

5.2. The allocation of tax revenues shall be reviewed, having in mind an adequate provision of such revenues to the areas of social development mentioned in the foregoing paragraphs.

II. CREATION OF A SPECIAL FUND FOR SOCIAL DEVELOPMENT

1. The delegations of the governments of the Latin-American republics welcome the decisions of the Government of the United States to establish a special inter-American fund for social development with the Inter-American Development Bank to become the primary mechanism for the administration of the fund.

2. It is understood that the purpose of the special fund would be to contribute capital resources and technical assistance on flexible terms and conditions, including repayment in local currencies and the relending of repaid funds, in accordance with appropriate and collective criteria in the light of the resources available, to support the efforts of the Latin-American countries which are prepared to initiate or expand effective improvements and to adopt measures to employ efficiently their own resources with a view to achieving greater social progress and more balanced economic growth.

III. MEASURES FOR ECONOMIC DEVELOPMENT

The special committee, having in view Resolution VII adopted at the Seventh Meeting of Consultation of Ministers of Foreign Affairs expressing the need for the maximum contribution of member countries in continental cooperation in the fight against under-development, in pursuance of the objectives of Operation Pan-America recommends:

1. That as soon as possible additional resources, domestic and external, be made available for the financing plans and projects of basic economic and industrial development in Latin America, with special attention to:

(a) The need for loans on flexible terms and conditions, including, whenever advisable in the light of the balance-of-payments situation of individual countries, the possibility of repayment in local currency.

(b) The desirability of the adequate preparation and implementation of development projects and plans, within the framework of the monetary, fiscal and exchange policies necessary for their effectiveness, utilizing as appropriate the technical assistance of inter-American and/or international agencies.

(c) The advisability, in special cases, of extending foreign financing for the coverage of local expenditures.

(d) The necessity for developing and strengthening credit facilities for small and medium private business, agriculture and industry.

2. That special attention be given to an expansion of long-term lending, particularly in view of the instability of exchange earnings of countries exporting primary products and of the unfavorable effect of the excessive accumulation of short- and medium-term debt and orderly economic development.

3. That urgent attention be given to the search for effective and practical ways, appropriate to each commodity, to deal with the problem of the instability of exchange earnings of countries heavily dependent upon the exportation of primary products.

IV. MULTILATERAL COOPERATION FOR SOCIAL AND ECONOMIC PROGRESS

The special committee considering the need for providing instruments and mechanisms for the implementation of the program of inter-American economic and social cooperation which would periodically review the progress made and propose measures for further mobilization of resources, recommends:

1. That the Inter-American Economic and Social Council undertakes to organize annual consultative meetings to review the social and economic progress of member countries, to analyze and discuss the progress achieved, and the problems encountered in each country, to exchange opinions on possible measures that might be adopted to intensify further social and economic progress, in pursuance of Operation Pan-America, and to prepare reports on the outlook for the future. Such annual meetings would begin with an examination by experts and terminate with a session at the ministerial level.

2. That the Council of the Organization of American States convene within sixty days of the date of this Act a special meeting of senior government representatives to find ways of strengthening and improving the ability of the Inter-American Economic and Social Council to render effective assistance to governments with a view to achieving the objectives enumerated below taking into account the proposal submitted by the delegation of Argentina in Document CECE/III–13:

(a) To further the economic and social development of Latin American countries.

(b) To promote aid between the countries of the Western Hemisphere as well as between them and extra-continental countries.

(c) To facilitate the flow of capital and the extension of credits to the countries of Latin America, both from the Western Hemisphere and from extra-continental sources.

3. The special meeting shall:

(a) Examine the existing structure of the Inter-American Economic and Social Council, and of the Secretariat of the Organization of American States working in the economic and social fields, for the purpose of formulating recommendations to the Council of the Organization of

American States designed to strengthen and improve the Inter-American Economic and Social Council.

(b) Determine the needs of strengthening inter-American economic and social cooperation by an administrative reform of the Secretariat, which should be given sufficient technical, administrative and financial flexibility for the adequate fulfillment of its tasks.

(c) Formulate recommendations designed to assure effective co-ordination between the Inter-American Economic and Social Council, the Economic Commission for Latin America, the Inter-American Development Bank, the United Nations and its specialized agencies and other agencies offering technical advice and services in the Western Hemisphere.

(d) Propose procedures designed to establish effective liaison of the Inter-American Economic and Social Council and other regional American organizations with other international organizations for the purpose of study, discussion and consultation in the fields of international trade and financial and technical assistance.

In approving the Act of Bogotá, the delegations to the special committee convinced that the people of the Americas can achieve a good life only within the democratic system, renew their faith in the essential values which lie at the base of Western civilization, and reaffirm their determination to assure the fullest measure of well-being to the people of the Americas under the conditions of freedom and respect for the supreme dignity of the individual.

appendix F

Montevideo Treaty (1960)

The Governments represented at the Inter-Governmental Conference for the Establishment of a Free-Trade Area among Latin American countries,

Persuaded that the expansion of present national markets, through the gradual elimination of barriers to intra-regional trade, is a prerequisite if the Latin American countries are to accelerate their economic development process in such a way as to ensure a higher level of living for their peoples,

Aware that economic development should be attained through the maximum utilization of available production factors and the more effective coordination of the development programmes of the different production sectors in accordance with norms which take due account of the interests of each and all and which make proper compensation, by means of appropriate measures, for the special situation of countries which are at a relatively less advanced stage of economic development,

Convinced that the strengthening of national economies will contribute to the expansion of trade within Latin America and with the rest of the world,

Sure that, by the adoption of suitable formulas, conditions can be created that will be conducive to the gradual and smooth adaptation of existing productive activities to new patterns of reciprocal trade, and that further incentives will thereby be provided for the improvement and expansion of such trade,

Certain that any action to achieve such ends must take into account the commitments arising out of the international instruments which govern their trade,

Determined to persevere in their efforts to establish, gradually and progressively, a Latin American common market and, hence, to continue collaborating with the Latin American Governments as a whole in the work already initiated for this purpose, and

Motivated by the desire to pool their efforts to achieve the progressive complementarity and integration of their national economies on the basis of an effective reciprocity of benefits, decide to establish a Free-Trade Area and, to that end, to conclude a Treaty instituting the Latin American Free-Trade

Association; and have, for this purpose, appointed their plenipotentiaries who have agreed as follows:

Chapter I
NAME AND PURPOSE
Article 1

By this Treaty the Contracting Parties establish a Free-Trade Area and institute the Latin American Free-Trade Association (hereinafter referred to as "the Association"), with headquarters in the city of Montevideo (Eastern Republic of Uruguay).

The term "Area," when used in this Treaty, means the combined territories of the Contracting Parties.

Chapter II
PROGRAMME FOR TRADE LIBERALIZATION
Article 2

The Free-Trade Area, established under the terms of the present Treaty, shall be brought into full operation within not more than twelve (12) years from the date of the Treaty's entry into force.

Article 3

During the period indicated in *Article 2*, the Contracting Parties shall gradually eliminate, in respect of substantially all their reciprocal trade, such duties, charges and restrictions as may be applied to imports of goods originating in the territory of any Contracting Party.

For the purposes of the present Treaty the term "duties and charges" means customs duties and any other charges of equivalent effect—whether fiscal, monetary or exchange—that are levied on imports.

The provisions of the present *Article* do not apply to fees and similar charges in respect of services rendered.

Article 4

The purpose set forth in *Article 3* shall be achieved through negotiations to be held from time to time among the Contracting Parties with a view to drawing up:

(a) National Schedules specifying the annual reductions in duties, charges and other restrictions which each Contracting Party grants to the other Contracting Parties in accordance with the provisions of *Article 5*; and

(b) a Common Schedule listing the products on which the Contracting Parties collectively agree to eliminate duties, charges and other restrictions completely, so far as intra-Area trade is concerned, within the period mentioned in *Article 2*, by complying with the minimum percentages set out in *Article 7* and through the gradual reduction provided for in *Article 5*.

Article 5

With a view to the preparation of the National Schedules referred to in *Article 4*, sub-paragraph (a), each Contracting Party shall annually grant to the other Contracting Parties reductions in duties and charges equivalent to not less than eight (8) percent of the weighted average applicable to third countries, until they are eliminated in respect of substantially all of its imports from the Area, in accordance with the definitions, methods of calculation, rules and procedures laid down in the Protocol appended to the present Treaty.

For this purpose, duties and charges for third parties shall be deemed to be those in force on 31 December prior to each negotiation.

When the import régime of a Contracting Party contains restrictions of such a kind that the requisite equivalence with the reductions in duties and charges granted by another Contracting Party or other Contracting Parties is unobtainable, the counterpart of these reductions shall be complemented by means of the elimination or relaxation of those restrictions.

Article 6

The National Schedules shall enter into force on 1 January of each year, except that those deriving from the initial negotiations shall enter into force on the date fixed by the Contracting Parties.

Article 7

The Common Schedule shall consist of products which, in terms of the aggregate value of the trade among the Contracting Parties,shall constitute not less than the following percentages, calculated in accordance with the provisions of the Protocol:

Twenty-five (25) percent during the first three-year period;
Fifty (50) percent during the second three-year period;
Seventy-five (75) percent during the third-year period;
Substantially all of such trade during the fourth three-year period.

Article 8

The inclusion of products in the Common Schedule shall be final and the concessions granted in respect thereof irrevocable.

Concessions granted in respect of products which appear only in the National Schedules may be withdrawn by negotiation among the Contracting Parties and on a basis of adequate compensation.

Article 9

The percentages referred to in *Articles 5* and *7* shall be calculated on the basis of the average annual value of trade during the three years preceding the year in which each negotiation is effected.

Article 10

The purpose of the negotiations–based on reciprocity of concessions–referred to in *Article 4* shall be to expand and diversify trade and to promote the progressive complementarity of the economies of the countries in the Area.

In these negotiations the situation of those Contracting Parties whose levels of duties, charges and restrictions differ substantially from those of the other Contracting Parties shall be considered with due fairness.

Article 11

If, as a result of the concessions granted, significant and persistent disadvantages are created in respect of trade between one Contracting Party and the others as a whole in the products included in the liberalization programme, the Contracting Parties shall, at the request of the Contracting Party affected, consider steps to remedy these disadvantages with a view to the adoption of suitable, non-restrictive measures designed to promote trade at the highest possible levels.

Article 12

If, as a result of circumstances other than those referred to in *Article 11*, significant and persistent disadvantages are created in respect of trade in the products included in the liberalization programme, the Contracting Parties shall, at the request of the Contracting Party concerned, make every effort within their power to remedy these disadvantages.

Article 13

The reciprocity mentioned in *Article 10* refers to the expected growth in the flow of trade between each Contracting Party and the others as a whole, in the products included in the liberalization programme and those which may subsequently be added.

Chapter III

EXPANSION OF TRADE AND ECONOMIC COMPLEMENTARITY

Article 14

In order to ensure the continued expansion and diversification of reciprocal trade, the Contracting Parties shall take steps:

(a) to grant one another, while observing the principle of reciprocity, concessions which will ensure that, in the first negotiation, treatment not less favourable than that which existed before the date of entry into force of the present Treaty is accorded to imports from within the Area;

(b) to include in the National Schedules the largest possible number of products in which trade is carried on among the Contracting Parties; and

(c) to add to these Schedules an increasing number of products which are not yet included in reciprocal trade.

Article 15

In order to ensure fair competitive conditions among the Contracting Parties and to facilitate the increasing integration and complementarity of their economies, particularly with regard to industrial production, the Contracting Parties shall make every effort–in keeping with the liberalization objectives of the present Treaty–to reconcile their import and export régimes, as well as the treatment they accord tocapital, goods and services from outside the Area.

Article 16

With a view to expediting the process of integration and complementarity referred to in *Article 15*, the Contracting Parties:

(a) shall endeavour to promote progressively closer co-ordination of the corresponding industrialization policies, and shall sponsor for this purpose agreements among representatives of the economic sectors concerned; and

(b) may negotiate mutual agreements on complementarity by industrial sectors.

Article 17

The complementarity agreements referred to in *Article 16*, sub-paragraph (b), shall set forth the liberalization programme to be applied to products of the sector concerned and may contain, *inter alia*, clauses designed to reconcile the treatment accorded to raw materials and other components used in the manufacture of these products.

Any Contracting Party concerned with the complementarity programmes shall be free to participate in the negotiation of these agreements.

The results of these negotiations shall, in every case, be embodied in protocols which shall enter into force after the Contracting Parties have decided that they are consistent with the general principles and purposes of the present Treaty.

Chapter IV

MOST-FAVOURED-NATION TREATMENT

Article 18

Any advantage, benefit, franchise, immunity or privilege applied by a Contracting Party in respect of a product originating in or intended for consignment to any other country shall be immediately and unconditionally extended to the similar product originating in or intended for consignment to the territory of the other Contracting Parties.

Article 19

The most-favoured-nation treatment referred to in *Article 18* shall not be applicable to the advantages, benefits, franchises, immunities and privileges already granted or which may be granted by virtue of agreements among Contracting Parties or between Contracting Parties and third countries with a view to facilitating border trade.

Article 20

Capital originating in the Area shall enjoy, in the territory of each Contracting Party, treatment not less favourable than that granted to capital originating in any other country.

Chapter V

TREATMENT IN RESPECT OF INTERNAL TAXATION

Article 21

With respect to taxes, rates and other internal duties and charges, products originating in the territory of a Contracting Party shall enjoy, in the territory of another Contracting Party, treatment no less favourable than that accorded to similar national products.

Article 22

Each Contracting Party shall endeavour to ensure that the charges or other domestic measures applied to products included in the liberalization programme which are not produced, or are produced only in small quantities, in its territory, do not nullify or reduce any concession or advantage obtained by any Contracting Party during the negotiations.

If a Contracting Party considers itself injured by virtue of the measures mentioned in the previous paragraph, it may appeal to the competent organs of the Association with a view to having the matter examined and appropriate recommendations made.

Chapter VI
SAVING CLAUSES

Article 23

The Contracting Parties may, as a provisional measure and providing that the customary level of consumption in the importer country is not thereby lowered, authorize a Contracting Party to impose non-discriminatory restrictions upon imports of products included in the liberalization programme which originate in the Area, if these products are imported in such quantities or under such conditions that they have, or are liable to have, serious repercussions on specific productive activities of vital importance to the national economy.

Article 24

The Contracting Parties may likewise authorize a Contracting Party which has adopted measures to correct its unfavourable over-all balance of payments to extend these measures, provisionally and without discrimination, to intra-Area trade in the products included in the liberalization programme.

The Contracting Parties shall endeavour to ensure that the imposition of restrictions deriving from the balance-of-payments situation does not affect trade, within the Area, in the products included in the liberalization programme.

Article 25

If the situations referred to in *Article 23* and *24* call for immediate action, the Contracting Party concerned may, as an emergency arrangement to be referred to the Contracting Parties, apply the measures provided for in the said *Articles*. The measures adopted must immediately be communicated to the Committee mentioned in *Article 33*, which, if it deems necessary, shall convene a special session of the Conference.

Article 26

Should the measures envisaged in this chapter be prolonged for more than one year, the Committee shall propose to the Conference, referred to in *Article 33*, either *ex officio* or at the request of any of the Contracting Parties, the immediate initiation of negotiations with a view to eliminating the restrictions adopted.

The present *Article* does not affect the provisions of *Article 8*.

Chapter VII

SPECIAL PROVISIONS CONCERNING AGRICULTURE

Article 27

The Contracting Parties shall seek to co-ordinate their agricultural development and agricultural commodity trade policies, with a view to securing the most efficient utilization of their natural resources, raising the standard of living of the rural population, and guaranteeing normal supplies to consumers, without disorganizing the regular productive activities of each Contracting Party.

Article 28

Providing that no lowering of its customary consumption or increase in anti-economic production is involved, a Contracting Party may apply, within the period mentioned in *Article 2*, and in respect of trade in agricultural commodities of substantial importance to its economy that are included in the liberalization programme, appropriate non-discriminatory measures designed to:

(a) limit imports to the amount required to meet the deficit in internal production; and

(b) equalize the prices of the imported and domestic product.

The Contracting Party which decides to apply these measures shall inform the other Contracting Parties before it puts them into effect.

Article 29

During the period prescribed in *Article 2* an attempt shall be made to expand intra-Area trade in agricultural commodities by such means as agreements among the Contracting Parties designed to cover deficits in domestic production.

For this purpose, the Contracting Parties shall give priority, under normal competitive conditions, to products originating in the territories of the other Contracting Parties, due consideration being given to the traditional flows of intra-Area trade.

Should such agreements be concluded among two or more Contracting Parties, the other Contracting Parties shall be notified before the agreements enter into force.

Article 30

The measures provided for in this Chapter shall not be applied for the purpose of incorporating, in the production of agricultural commodities, resources which imply a reduction in the average level of productivity existing on the date on which the present Treaty enters into force.

Article 31

If a Contracting Party considers itself injured by a reduction of its exports attributable to the lowering of the usual consumption level of the importer country as a result of the measures referred in *Article 28* and/or an anti-economic increase in the production referred in the previous *Article*, it may appeal to the competent organs of the Association to study the situation and, if necessary, to make recommendations for the adoption of appropriate measures to be applied in accordance with *Article 12*.

Chapter VIII

MEASURES IN FAVOUR OF COUNTRIES AT A RELATIVELY LESS ADVANCED STAGE OF ECONOMIC DEVELOPMENT

Article 32

The Contracting Parties, recognizing that fulfillment of the purposes of the present Treaty will be facilitated by the economic growth of the countries in the Area that are at a relatively less advanced stage of economic development, shall take steps to create conditions conductive to such growth.

To this end, the Contracting Parties may:

(a) Authorize a Contracting Party to grant to another Contracting Party which is at a relatively less advanced stage of economic development within the Area, as long as necessary and as a temporary measure, for the purposes set out in the present *Article*, advantages not extended to the other Contracting Parties, in order to encourage the introduction or expansion of specific productive activities;

(b) Authorize a Contracting Party at a relatively less advanced stage of economic development within the Area to implement the programme for the reduction of duties, charges and other restrictions under more favourable conditions, specially agreed upon;

(c) Authorize a Contracting Party at a relatively less advanced stage of economic development within the Area to adopt appropriate measures to correct an unfavourable balance of payments, if the case arises;

(d) Authorize a Contracting Party at a relatively less advanced stage of economic development within the Area to apply, if necessary and as a temporary measure, and providing that this does not entail a decrease in its customary consumption, appropriate non-discriminatory measures designed to protect the domestic output of products included in the liberalization programme which are of vital importance to its economic development;

(e) Make collective arrangements in favour of a Contracting Party at a relatively less advanced stage of economic development within the Area with respect to the support and promotion, both inside and outside the Area, of financial or technical measures designed to bring about the expansion of existing productive activities or to encourage new activities, particularly those intended for the industrialization of its raw materials; and

(f) Promote or support, as the case may be, special technical assistance programmes for one or more Contracting Parties, intended to raise, in countries at a relatively less advanced stage of economic development within the Area, productivity levels in specific production sectors.

Chapter IX

ORGANS OF THE ASSOCIATION

Article 33

The organs of the Association are the Conference of the Contracting Parties (referred to in this Treaty as "the Conference") and the Standing Executive Committee (referred to in this Treaty as "the Committee").

Article 34

The Conference is the supreme organ of the Association. It shall adopt all decisions in matters requiring joint action on the part of the Contracting Parties, and it shall be empowered, *inter alia:*

(a) To take the necessary steps to carry out the present Treaty and to study the results of its implementation;

(b) To promote the negotiations provided for in *Article 4* and to assess the results thereof;

(c) To approve the Committee's annual budget and to fix the contributions of each Contracting Party;

(d) To lay down its own rules of procedure and to approve the Committee's rules of procedure;

(e) To elect a Chairman and two Vice-Chairmen for each session;

(f) To appoint the Executive Secretary of the Committee; and

(g) To deal with other business of common interest.

Article 35

The Conference shall be composed of duly accredited representatives of the Contracting Parties. Each delegation shall have one vote.

Article 36

The Conference shall hold: (a) a regular session once a year; and (b) special sessions when convened by the Committee.

At each session the Conference shall decide the place and date of the following regular session.

Article 37

The Conference may not take decisions unless at least two-thirds (2/3) of the Contracting Parties are present.

Article 38

During the first two years in which the present Treaty is in force, decisions of the Conference shall be adopted when affirmative votes are cast by at least two-thirds (2/3) of the Contracting Parties and providing that no negative vote is cast.

The Contracting Parties shall likewise determine the voting system to be adopted after this two-year period.

The affirmative vote of two-thirds (2/3) of the Contracting Parties shall be required:

(a) to approve the Committee's annual budget;

(b) To elect the Chairman and Vice-Chairmen of the Conference, as well as the Executive Secretary; and

(c) To fix the time and place of the sessions of the Conference.

Article 39

The Committee is the permanent organ of the Association responsible for supervising the implementation of the provisions of the present Treaty. Its duties and responsibilities shall be, *inter alia:*

(a) To convene the Conference;

(b) To submit for the approval of the Conference an annual work programme and the Committee's annual budget estimates;

(c) To represent the Association in dealings with third countries and international organs and entities for the purpose of considering matters of common interest. It shall also represent the Association in contracts and other instruments of public and private law;

(d) To undertake studies, to suggest measures and to submit to the Conference such recommendations as it deems appropriate for the effective implementation of the Treaty;

(e) To submit to the Conference at its regular sessions an annual report on its activities and on the results of the implementation of the present Treaty;

(f) To request the technical advice and the co-operation of individuals and of national and international organizations;

(g) To take such decisions as may be delegated to it by the Conference; and

(h) To undertake the work assigned to it by the Conference.

Article 40

The Committee shall consist of a Permanent Representative of each Contracting Party, who shall have a single vote.

Each Representative shall have an Alternate.

Article 41

The Committee shall have a Secretariat headed by an Executive Secretary and comprising technical and administrative personnel.

The Executive Secretary, elected by the Conference for a three-year term and re-eligible for similar periods, shall attend the plenary meetings of the Committee without the right to vote.

The Executive Secretary shall be the General Secretary of the Conference. His duties shall be, *inter alia:*

(a) To organize the work of the Conference and of the Committee;

(b) To prepare the Committee's annual budget estimates; and

(c) To recruit and engage the technical and administrative staff in accordance withthe Committee's rules of procedure.

Article 42

In the performances of their duties, the Executive Secretary and the Secretariat staff shall not seek or receive instructions from any Government or from any other national or international entity. They shall refrain from any action which might reflect on their position as international civil servants.

The Contracting Parties undertake to respect the international character of the responsibilities of the Executive Secretary and of the Secretariat staff and shall refrain from influencing them in any way in the discharge of their responsibilities.

Article 43

In order to facilitate the study of specific problems, the Committee may set up Advisory Commissions composed of representatives of the various sectors of economic activity of each of the Contracting Parties.

Article 44

The Committee shall request, for the organs of the Association, the technical advice of the Secretariat of the United Nations Economic Commission for Latin America (ECLA) and of the Inter-American Economic and Social Council (IA-ECOSOC) of the Organization of American States.

Article 45

The Committee shall be constituted sixty days from the entry into force of the present Treaty and shall have its headquarters in the city of Montevideo.

Chapter X
JURIDICAL PERSONALITY–IMMUNITIES AND PRIVILEGES

Article 46

The Latin American Free-Trade Association shall possess complete juridical personality and shall, in particular, have the power:

(a) To contract;

(b) To acquire and dispose of the movable and immovable property it needs for the achievement of its objectives;

(c) To institute legal proceedings; and

(d) To hold funds in any currency and to transfer them as necessary.

Article 47

The representatives of the Contracting Parties and the international staff and advisers of the Association shall enjoy in the Area such diplomatic and other immunities and privileges as are ncessary for the exercise of their functions.

The Contracting Parties undertake to conclude, as soon as possible, an Agreement regulating the provisions of the previous paragraph in which the aforesaid privileges and immunities shall be defined.

The Association shall conclude with the Government of the Eastern Republic of Uruguay an Agreement for the purpose of specifying the privileges and immunities which the Association, its organs and its international staff and advisers shall enjoy.

Chapter XI
MISCELLANEOUS PROVISIONS

Article 48

No change introduced by a Contracting Party in its régime of import duties and charges shall imply a level of duties and charges less favourable than that in force before the change for any commodity in respect of which concessions are granted to the other Contracting Parties.

The requirement set out in the previous paragraph shall not apply to the conversion to present worth of the official base value (*aforo*) in respect of customs duties and charges, providing that such conversion corresponds exclusively to the real value of the goods. In such cases, the value shall not include the customs duties and charges levied on the goods.

Article 49

In order to facilitate the implementation of the provisions of the present Treaty, the Contracting Parties shall, as soon as possible:

(a) Determine the criteria to be adopted for the purpose of determining the origin of goods and for classifying them as raw materials,semi-manufactured goods or finished products;

(b) Simplify and standardize procedures and formalities relating to reciprocal trade;

(c) Prepare a tariff nomenclature to serve as a common basis for the presentation of statistics and for carrying out the negotiations provided for in the present Treaty;

(d) Determine what shall be deemed to constitute border trade within the meaning of *Article 19*;

(e) Determine the criteria for the purpose of defining "dumping" and other unfair trade practices and the procedures relating thereto.

Article 50

The products imported from the Area by a Contracting Party may not be re-exported save by agreement between the Contracting Parties concerned.

A product shall not be deemed to be a re-export if it has been subjected in the importer country to industrial processing or manufacture, the degree of which shall be determined by the Committee.

Article 51

Products imported or exported by a Contracting Party shall enjoy freedom of transit within the Area and shall only be subject to the payment of the normal rates for services rendered.

Article 52

No Contracting Party shall promote its exports by means of subsidies or other measures likely to disrupt normal competitive conditions in the Area.

An export shall not be deemed to have been subsidized if it is exempted from duties and charges levied on the product or its components when destined for internal consumption, or if it is subject to drawback.

Article 53

No provision of the present Treaty shall be so construed as to constitute an impediment to the adoption and execution of measures relating to:

(a) The protection of public morality;

(b) The application of security laws and regulations;

(c) The control of imports or exports of arms, ammunitions and other war equipment and, in exceptional circumstances, of all other military items, in so far as this is compatible with the terms of *Article 51* and of the treaties on the unrestricted freedom of transit in force among the Contracting Parties;

(d) The protection of human, animal and plant life and health;

(e) Imports and exports of gold and silver bullion;

(f) The protection of the nation's heritage of artistic, historical and archaeological value; and

(g) The export, use and consumption of nuclear materials, radioactive products or any other material that may be used in the development or exploitation of nuclear energy.

Article 54

The Contracting Parties shall make every effort to direct their policies with a view to creating conditions favourable to the establishment of a Latin American common market. To that end, the Committee shall undertake studies and consider projects and plans designed to achieve this purpose, and shall endeavour to coordinate its work with that of other international organizations.

Chapter XII
FINAL CLAUSES

Article 55

The present Treaty may not be signed with reservations nor shall reservations be admitted at the time of ratification or accession.

Article 56

The present Treaty shall be ratified by the signatory States at the earliest opportunity.

The instruments of ratification shall be deposited with the Government of the Eastern Republic of Uruguay, which shall communicate the date of deposit to the Governments of the signatory and successively acceding States.

Article 57

The present Treaty shall enter into force for the first three ratifying States thirty days after the third instrument of ratification has been deposited; and, for the other signatories, thirty days after the respective instrument of ratification has been deposited, and in the order in which the ratifications are deposited.

The Government of the Eastern Republic of Uruguay shall communicate the date of the entry into force of the present Treaty to the Government of each of the signatory States.

Article 58

Following its entry into force, the present Treaty shall remain open to accession by the other Latin American States, which for this purpose shall deposit the relevant instrument of accession with the Government of the Eastern Republic of Uruguay. The Treaty shall enter into force for the acceding State thirty days after the deposit of the corresponding instrument.

Acceding States shall enter into the negotiations referred to in *Article 4* at the session of the Conference immediately following the date of deposit of the instrument of accession.

Article 59

Each Contracting Party shall begin to benefit from the concessions already granted to one another by the other Contracting Parties as from the date of entry into force of the reductions in duties and charges and other restrictions negotiated by them on a basis of reciprocity, and after the minimum obligations referred to in *Article 5*, accumulated during the period which has elapsed since the entry into force of the present Treaty, have been carried out.

Article 60

The Contracting Parties may present amendments to the present Treaty, which shall be set out in protocols that shall enter into force upon their ratification by all the Contracting Parties and after the corresponding instruments have been deposited.

Article 61

On the expiry of the twelve-year term starting on the date of entry into force of the present Treaty, the Contracting Parties shall proceed to study the results of the Treaty's implementation and shall initiate the necessary collective negotiations with a view to fulfilling more effectively the purposes of the Treaty and, if desirable, to adapting it to a new stage of economic integration.

Article 62

The provisions of the present Treaty shall not affect the rights and obligations deriving from agreements signed by any of the Contracting Parties prior to the entry into force of the present Treaty.

However, each Contracting Party shall take the necessary steps to reconcile the provisions of existing agreements with the provisions of the present Treaty.

Article 63

The present Treaty shall be of unlimited duration.

Article 64

A Contracting Party wishing to withdraw from the present Treaty shall inform the other Contracting Parties of its intention at a regular session of the Conference, and shall formally submit the instrument of denunciation at the following regular session.

When the formalities of denunciation have been completed, those rights and obligations of the denouncing Government which derive from its status as a Contracting Party shall cease automatically, with the exception of those relating to reductions in duties and charges and other restrictions, received or granted under the liberalization programme, which shall remain in force for a period of five years from the date on which the denunciation becomes formally effective.

The period specified in the preceding paragraph may be shortened if there is sufficient justification, with the consent of the Conference and at the request of the Contracting Party concerned.

Article 65

The present Treaty shall be called the Montevideo Treaty.

appendix G

Charter of Punta Del Este (1961)

PREAMBLE

We, the American Republics, hereby proclaim our decision to unite in a common effort to bring our people accelerated economic progress and broader social justice within the framework of personal dignity and political liberty.

Almost two hundred years ago we began in this Hemisphere the long struggle for freedom which now inspires people in all parts of the world. Today, in ancient lands, men moved to hope by the revolutions of our young nations search for liberty. Now we must give a new meaning to that revolutionary heritage. For America stands at a turning point in history. The men and women of our Hemisphere are reaching for the better life which today's skills have placed within their grasp. They are determined for themselves and their children to have decent and ever more abundant lives, to gain access to knowledge and equal opportunity for all, to end those conditions which benefit the few at the expense of the needs and dignity of the many. It is our inescapable task to fulfill these just desires—to demonstrate to the poor and forsaken of our countries, and of all lands, that the creative powers of free men hold the key to their progress and to the progress of future generations. And our certainty of ultimate success rests not alone on our faith in ourselves and in our nations but on the indomitable spirit of free man which has been the heritage of American civilization.

Inspired by these principles, and by the principles of Operation Pan America and the Act of Bogotá, the American Republics hereby resolve to adopt the following program of action to establish and carry forward an Alliance for Progress.

TITLE I

OBJECTIVES OF THE ALLIANCE FOR PROGRESS

It is the purpose of the Alliance for Progress to enlist the full energies of the peoples and governments of the American Republics in a great cooperative effort to accelerate the economic and social development of the participating countries of Latin America, so that they may achieve maximum levels of well-being, with equal opportunities for all, in democratic societies adapted to their own needs and desires.

The American republics hereby agree to work toward the achievement of the following fundamental goals in the present decade:

1. To achieve in the participating Latin American countries a substantial and sustained growth of per capita income at a rate designed to attain, at the earliest possible date, levels of income capable of assuring self-sustaining development, and sufficient to make Latin American income levels constantly larger in relation to the levels of the more industrialized nations. In this way the gap between the living standards of Latin America and those of the more developed countries can be narrowed. Similarly, presently existing differences in income levels among the Latin American countries will be reduced by accelerating the development of the relatively less developed countries and granting them maximum priority in the distribution of resources and in international cooperation in general. In evaluating the degree of relative development, account will be taken not only of average levels of real income and gross product per capita, but also of indices of infant mortality, illiteracy, and per capita daily caloric intake.

It is recognized that, in order to reach these objectives within a reasonable time, the rate of economic growth in any country of Latin America should be not less than 2.5 percent per capita per year, and that each participating country should determine its own growth target in the light of its stage of social and economic evolution, resource endowment, and ability to mobilize national efforts for development.

2. To make the benefits of economic progress available to all citizens of all economic and social groups through a more equitable distribution of national income, raising more rapidly the income and standard of living of the needier sectors of the population, at the same time that a higher proportion of the national product is devoted to investment.

3. To achieve balanced diversification in national economic structures, both regional and functional, making them increasingly free from dependence on the export of a limited number of primary products and the importation of capital goods while attaining stability in the prices of exports or in income derived from exports.

4. To accelerate the process of national industrialization so as to increase the productivity of the economy as a whole, taking full advantage of the talents and energies of both the private and public sectors, utilizing the natural resources of the country and providing productive and remunerative employment for unemployed or part-time workers. Within this process of industrialization, special attention should be given to the establishment and development of capital-goods industries.

5. To raise greatly the level of agricultural productivity and output and to improve related storage, transportation, and marketing services.

6. To encourage, in accordance with the characteristics of each country, programs of comprehensive agrarian reform leading to the effective transformation, where required, of unjust structures and systems of land tenure and use, with a view to replacing latifundia and dwarf holdings by an equitable system of land tenure so that, with the help of timely and adequate credit, technical assistance and facilities for the marketing and distribution of products, the land will become for the man who works it the basis of his economic stability, the foundation of his increasing welfare, and the guarantee of his freedom and dignity.

7. To eliminate adult illiteracy and by 1970 to assure, as a minimum, access to six years of primary education for each school-age child in Latin America; to modernize and expand vocational, technical, secondary and higher educational and training facilities, to strengthen the capacity for basic and applied research; and to provide the competent personnel required in rapidly growing societies.

8. To increase life expectancy at birth by a minimum of five years, and to increase the ability to learn and produce, by improving individual and public health. To attain this goal it will be necessary, among other measures, to provide adequate potable water supply, and sewage disposal, to not less than 70 percent of the urban and 50 percent of the rural population; to reduce the present mortality rate of children less than five years of age by at least one-half; to control the more serious communicable diseases, according to their importance as a cause of sickness, disability, and death; to eradicate those illnesses, especially malaria, for which effective techniques are known; to improve nutrition; to train medical and health personnel to meet at least minimum requirements; to improve basic health services at national and local levels; and to intensify scientific research and apply its results more fully and effectively to the prevention and cure of illness.

9. To increase the construction of low-cost houses for low-income families in order to replace inadequate and deficient housing and to reduce housing shortages; and to provide necessary public services to both urban and rural centers of population.

10. To maintain stable price levels, avoiding inflation or deflation and the consequent social hardships and maldistribution of resources, always bearing in mind the necessity of maintaining an adequate rate of economic growth.

11. To strengthen existing agreements on economic integration, with a view to the ultimate fulfillment of aspirations for a Latin American common market that will expand and diversify trade among the Latin American countries and thus contribute to the economic growth of the region.

12. To develop cooperative programs designed to prevent the harmful effects of excessive fluctuations in the foreign exchange earnings derived from exports of primary products, which are of vital importance to economic and social development; and to adopt the measures necessary to facilitate the access of Latin American exports to international markets.

TITLE II

ECONOMIC AND SOCIAL DEVELOPMENT

Chapter I. Basic Requirements for Economic and Social Development

The American Republics recognize that to achieve the foregoing goals it will be necessary:

1. That comprehensive and well-conceived national programs of economic and social development, aimed at the achievement of self-sustaining growth, be carried out in accordance with democratic principles.

2. That national programs of economic and social development be based on the principle of self-help—as established in the Act of Bogotá—and on the maximum use of domestic resources, taking into account the special conditions of each country.

3. That in the preparation and execution of plans for economic and social development, women should be placed on an equal footing with men.

4. That the Latin American countries obtain sufficient external financial assistance, a substantial portion of which should be extended on flexible conditions with respect to periods and terms of repayment and forms of utilization, in order to supplement domestic capital formation and reinforce their import capacity; and that, in support of well-conceived programs, which include the necessary structural reforms and measures for the mobilization of internal resources, a supply of capital from all external sources during the coming ten years of at least twenty billion dollars be made available to the Latin American countries, with priority to the relatively less-developed countries. The greater part of this sum should be in public funds.

5. That institutions in both the public and private sectors, including labor organizations, cooperatives, and commercial, industrial, and financial institutions, be strengthened and improved for the increasing and effective use of domestic resources, and that the social reforms necessary to permit a fair distribution of the fruits of economic and social progress be carried out.

Chapter II. National Development Programs

1. Participating Latin American countries agree to introduce or strengthen systems for the preparation, execution, and periodic revision of national programs for economic and social development consistent with the principles, objectives, and requirements contained in this document. Participating Latin American countries should formulate, if possible within the next eighteen months, long-term development programs. Such programs should embrace, according to the characteristics of each country, the elements outlined in the Appendix.

2. National development programs should incorporate self-help efforts directed toward:

a. Improvement of human resources and widening of opportunities by raising general standards of education and health; improving and extending technical education and professional training with emphasis on science and technology; providing adequate remuneration for work performed, encouraging the talents of managers, entrepreneurs, and wage earners; providing more productive employment for under-employed manpower; establishing effective systems of labor relations and procedures for consultation and collaboration among public authorities, employer associations, and labor organizations; promoting the establishment and expansion of local institutions for basic and applied research; and improving the standards of public administration.

b. Wider development and more efficient use of natural resources, especially those which are now idle or under-utilized, including measures for the processing of raw materials.

c. The strengthening of the agricultural base, progressively extending the benefits of the land to those who work it, and ensuring in countries with Indian populations the integration of these populations into the economic, social, and cultural processes of modern life. To carry out these aims, measures should be adopted, among others, to establish or improve, as the case may be, the following services: extension, credit, technical assistance, agricultural research and mechanization; health and education; storage and distribution; cooperatives and farmers' associations; and community development.

d. More effective, rational and equitable mobilization and use of financial resources through the reform of tax structures, including fair and adequate taxation of large incomes and real estate, and the strict application of measures to improve fiscal administration. Development programs should include the adaptation of budget expenditures to development needs, measures for the maintenance of price stability, the creation of essential credit facilities at reasonable rates of interest, and the encouragement of private savings.

e. Promotion through appropriate measures, including the signing of agreements for the purpose of reducing or eliminating double taxation, of conditions that will encourage the flow of foreign investments and help to increase the capital resources of participating countries in need of capital.

f. Improvement of systems of distribution and sales in order to make markets more competitive and prevent monopolistic practices.

Chapter III. Immediate and Short-Term Action Measures

1. Recognizing that a number of Latin American countries, despite their best efforts, may require emergency financial assistance, the United States will provide assistance from the funds which are or may be established for such purposes. The United States stands ready to take prompt action on applications for such assistance. Applications relating to existing situations should be submitted within the next sixty days.

2. Participating Latin American countries should, in addition to creating or strengthening machinery for long-term development programming, immediately increase their efforts to accelerate their development by giving special emphasis to the following objectives:

a. The completion of projects already under way and the initiation of projects for which the basic studies have been made, in order to accelerate their financing and execution.

b. The implementation of new projects which are designed:

(1) To meet the most pressing economic and social needs and benefit directly the greatest number of people;

(2) To concentrate efforts within each country in the less-developed or more-depressed areas in which particularly serious social problems exist;

(3) To utilize idle capacity or resources, particularly under-employed manpower; and

(4) To survey and assess natural resources.

c. The facilitation of the preparation and execution of long-term programs through measures designed:

(1) To train teachers, technicians, and specialists;

(2) To provide accelerated training to workers and farmers;

(3) To improve basic statistics;

(4) To establish needed credit and marketing facilities; and

(5) To improve services and administration.

3. The United States will assist in carrying out these short-term measures with a view to achieving concrete results from the Alliance for Progress at the earliest possible moment. In connection with the measures set forth above, and in accordance with the statement of President Kennedy, the United States will provide assistance under the Alliance, including assistance for the financing of short-term measures, totalling more than one billion dollars in the year ending March 1962.

Chapter IV. External Assistance in Support of National Development Programs

1. The economic and social development of Latin America will require a large amount of additional and private financial assistance on the part of capital-exporting countries, including the members of the Development Assistance Group and international lending agencies. The measures provided for in the Act of Bogotá and the new measures provided for in this Charter, are designed to create a framework within which such additional assistance can be provided and effectively utilized.

2. The United States will assist those participating countries whose development programs establish self-help measures and economic and social policies and programs consistent with the goals and principles of this Charter. To supplement the domestic efforts of such countries, the United States is prepared to allocate resources which, along with those anticipated from other external sources, will be of a scope and magnitude adequate to realize the goals envisaged in this Charter. Such assistance will be allocated to both social and economic development and, where appropriate, will take the form of grants or loans on flexible terms and conditions. The participating countries will request the support of other capital-exporting countries and appropriate institutions so that they may provide assistance for the attainment of these objectives.

3. The United States will help in the financing of technical assistance projects proposed by a participating country or by the General Secretariat of the Organization of American States for the purpose of:

a. Providing experts contracted in agreement with the governments to work under their direction and to assist them in the preparation of specific investment projects and the strengthening of national mechanisms for preparing projects, using specialized engineering firms where appropriate;

b. Carrying out, pursuant to existing agreements for cooperation among the General Secretariat of the Organization of American States, the Economic Commission for Latin America, and the Inter-American Development Bank, field investigations and studies, including those relating to development problems, the organization of national agencies for the preparation of development programs, agrarian reform and rural development, health, cooperatives, housing, education and professional training, and taxation and tax administration; and

c. Covening meetings of experts and officials on development and related problems.

The governments or above-mentioned organizations should, when appropriate, seek the cooperation of the United Nations and its specialized agencies in the execution of these activities.

4. The participating Latin American countries recognize that each has in varying degree a capacity to assist fellow republics by providing technical and financial assistance. They recognize that this capacity will increase as their economies grow. They therefore affirm their intention to assist fellow republics increasingly as their individual circumstances permit.

Chapter V. Organization and Procedures

1. In order to provide technical assistance for the formulation of development programs, as may be requested by participating nations, the Organization of American States, the Economic Commission for Latin America, and the Inter-American Development Bank will continue and strengthen their agreements for coordination in this field, in order to have available a group of programming experts whose service can be used to facilitate the implementation of this Charter. The participating countries will also seek an intensification of technical assistance from the specialized agencies of the United Nations for the same purpose.

2. The Inter-American Economic and Social Council, on the joint nomination of the Secretary General of the Organization of American States, the President of the Inter-American Development Bank, and the Executive Secretary of the United Nations Economic Commission for Latin America, will appoint a panel of nine high-level experts, exclusively on the basis of their experience, technical ability, and competence in the various aspects of economic and social development. The experts may be of any nationality, though if of Latin American origin an appropriate geographical distribution will be sought. They will be attached to the Inter-American Economic and Social Council, but will nevertheless enjoy complete autonomy in the performance of their duties. They may not hold any other remunerative position. The appointment of these experts will be for a period of three years, and may be renewed.

3. Each government, if it so wishes, may present its program for economic and social development for consideration by an *ad hoc* committee, composed of no more than three members drawn from the panel of experts referred to in the preceding paragraph together with an equal number of experts not on the panel.

The experts who compose the *ad hoc* committee will be appointed by the Secretary General of the Organization of American States at the request of the interested government and with its consent.

4. The committee will study the development program, exchange opinions with the interested government as to possible modifications, and, with the consent of the government, report its conclusions to the Inter-American Development Bank and to other governments and institutions that may be prepared to extend external financial and technical assistance in connection with the execution of the program.

5. In considering a development program presented to it, the *ad hoc* committee will examine the consistency of the program with the principles of the Act of Bogotá and of this Charter, taking into account the elements in the Appendix.

6. The General Secretariat of the Organization of American States will provide the personnel needed by the experts referred to in Paragraphs 2 and 3 of this Chapter in order to fulfill their tasks. Such personnel may be employed specifically for this purpose or may be made available from the permanent staffs of the Organization of American States, the Economic Commission for Latin America, and the Inter-American Development Bank, in accordance with the present liaison arrangements between the three organizations. The General Secretariat of the Organization of American States may seek arrangements with the United Nations Secretariat, its specialized agencies and the Inter-American Specialized Organizations, for the temporary assignment of necessary personnel.

7. A government whose development program has been the object of recommendations made by the *ad hoc* committee with respect to external financing requirements may submit the program to the Inter-American Development Bank so that the Bank may undertake the negotiations required to obtain such financing, including the organization of a consortium of credit institutions and governments disposed to contribute to the continuing and systematic financing, on appropriate terms, of the development program. However, the government will have full freedom to resort through any other channels to all sources of financing, for the purpose of obtaining, in full or in part, the required resources.

The *ad hoc* committee shall not interfere with the right of each government to formulate its own goals, priorities, and reforms in its national development programs.

The recommendations of the *ad hoc* committee will be of great importance in determining the distribution of public funds under the Alliance for Progress which contribute to the external financing of such programs. These recommendations shall give special consideration to *Title I.1*.

The participating governments will also use their good offices to the end that these recommendations may be accepted as a factor of great importance in the decisions taken, for the same purpose, by inter-American credit institutions, other international credit agencies, and other friendly governments which may be potential sources of capital.

8. The Inter-American Economic and Social Council will review annually the progress achieved in the formulation, national implementation, and

international financing of development programs; and will submit to the Council of the Organization of American States such recommendations as it deems pertinent.

APPENDIX

ELEMENTS OF NATIONAL DEVELOPMENT PROGRAMS

1. The establishment of mutually consistent targets to be aimed at over the program period in expanding productive capacity in industry, agriculture, mining, transport, power and communications, and in improving conditions of urban and rural life, including better housing, education, and health.

2. The assignment of priorities and the description of methods to achieve the targets, including specific measures and major projects. Specific development projects should be justified in terms of their relative costs and benefits, including their contribution to social productivity.

3. The measures which will be adopted to direct the operations of the public sector and to encourage private action in support of the development program.

4. The estimated cost, in national and foreign currency, of major projects and of the development program as a whole, year by year over the program period.

5. The internal resources, public and private, estimated to become available for the execution of the programs.

6. The direct and indirect effects of the program on the balance of payments, and the external financing, public and private, estimated to be acquired for the execution of the program.

7. The basic fiscal and monetary policies to be followed in order to permit implementation of the program within a framework of price stability.

8. The machinery of public administration–including relationships with local governments, decentralized agencies and non-governmental organizations, such as labor organizations, cooperatives, business and industrial organizations–to be used in carrying out the program, adapting it to changing circumstances and evaluating the progress made.

TITLE III

ECONOMIC INTEGRATION OF LATIN AMERICA

The American republics consider that the broadening of present national markets in Latin America is essential to accelerate the process of economic development in the Hemisphere. It is also an appropriate means for obtaining greater productivity through specialized and complementary industrial production which will, in turn, facilitate the attainment of greater social benefits for the inhabitants of the various regions of Latin America. The broadening of markets will also make possible the better use of resources under the Alliance for Progress. Consequently, the America republics recognize that:

1. The Montevideo Treaty (because of its flexibility and because it is open to the adherence of all the Latin American nations) and the Central American Treaty on Economic Integration are appropriate instruments for the attainment of these objectives, as was recognized in Resolution No. 11 (III) of the Ninth Session of the Economic Commission for Latin America.

2. The integration process can be intensified and accelerated not only by the specialization resulting from the broadening of markets through the liberalization of trade but also through the use of such instruments as the agreements for complementary production within economic sectors provided for in the Montevideo Treaty.

3. In order to insure the balanced and complementary economic expansion of all of the countries involved, the integration process should take into account, on a flexible basis, the condition of countries at a relatively less advanced stage of economic development, permitting them to be granted special, fair, and equitable treatment.

4. In order to facilitate economic integration in Latin America, it is advisable to establish effective relationships between the Latin American Free Trade Association and the group of countries adhering to the Central American Economic Integration Treaty, as well as between either of these groups and other Latin American countries. These arrangements should be established within the limits determined by these instruments.

5. The Latin American countries should coordinate their actions to meet the unfavorable treatment accorded to their foreign trade in world markets, particularly that resulting from certain restrictive and discriminatory policies of extra-continental countries and economic groups.

6. In the application of resources under the Alliance for Progress, special attention should be given not only to investments for multi-national projects that will contribute to strengthening the integration process in all its aspects, but also to the necessary financing of industrial products within Latin America.

7. In order to facilitate the participation of countries at a relatively low stage of economic development in multi-national Latin American economic cooperation programs, and in order to promote the balanced and harmonious development of the Latin American integration process, special attention should be given to the needs of these countries in the administration of financial resources provided under the Alliance for Progress, particularly in connection with infra-structure programs and the promotion of new lines of production.

8. The economic integration process implies a need for additional investment in various fields of economic activity and funds provided under the Alliance for Progress should cover these needs as well as those required for the financing of national development programs.

9. When groups of Latin American countries have their own institutions for financing economic integration, the financing referred to in the preceding paragraph should preferably be channeled through these institutions. With respect to regional financing designed to further the purposes of existing regional integration instruments, the cooperation of the Inter-American Development Bank should be sought in channeling extra-regional contributions which may be granted for these purposes.

10. One of the possible means for making effective a policy for the

financing of Latin American integration would be to approach the International Monetary Fund and other financial sources with a view to providing a means for solving temporary balance-of-payments problems that may occur in countries participating in economic integration arrangements.

11. The promotion and coordination of transportation and communications systems are an effective way to accelerate the integration process. In order to counteract abusive practices in relation to freight rates and tariffs, it is advisable to encourage the establishment of multi-national transport and communication enterprises in the Latin American countries, or to find other appropriate solutions.

12. In working toward economic integration and complementary economies, efforts should be made to achieve an appropriate coordination of national plans, or to engage in joint planning for various economies through the existing regional integration organizations. Efforts should also be made to promote an investment policy directed to the progressive elimination of unequal growth rates in the different geographic areas, particularly in the case of countries which are relatively less developed.

13. It is necessary to promote the development of national Latin American enterprises, in order that they may compete on an equal footing with foreign enterprises.

14. The active participation of the private sector is essential to economic integration and development, and except in those countries in which free enterprise does not exist, development planning by the pertinent national public agencies, far from hindering such participation, can facilitate and guide it, thus opening new perspectives for the benefit of the community.

15. As the countries of the Hemisphere still under colonial domination achieve their independence, they should be invited to participate in Latin American economic integration programs.

TITLE IV

BASIC EXPORT COMMODITIES

The American republics recognize that the economic development of Latin America requires expansion of its trade, a simultaneous and corresponding increase in foreign exchange incomes received from exports, a lessening of cyclical or seasonal fluctuations in the incomes of those countries that still depend heavily on the export of raw materials, and the correction of the secular deterioration in their terms of trade.

They therefore agree that the following measures should be taken:

Chapter I. National Measures

National measures affecting commerce in primary products should be directed and applied in order to:

1. Avoid undue obstacles to the expansion of trade in these products;
2. Avoid market instability;

3. Improve the efficiency of international plans and mechanisms for stabilization; and

4. Increase their present markets and expand their area of trade at a rate compatible with rapid development.

Therefore:

A. Importing member countries should reduce and if possible eliminate, as soon as feasible, all restrictions and discriminatory practices affecting the consumption and importation of primary products, including those with the highest possible degree of processing in the country of origin, except when these restrictions are imposed temporarily for purposes of economic diversification, to hasten the economic development of less-developed nations, or to establish basic national reserves. Importing countries should also be ready to support, by adequate regulations, stabilization programs forprimary products that may be agreed upon with producing countries.

B. Industrialized countries should give special attention to the need for hastening economic development of less-developed countries. Therefore, they should make maximum efforts to create conditions, compatible with their international obligations, through which they may extend advantages to less-developed countries so as to permit the rapid expansion of their markets. In view of the great need for this rapid development, industrialized countries should also study ways in which to modify, wherever possible, international commitments which prevent the achievement of this objective.

C. Producing member countries should formulate their plans for production and export, taking account of their effect on world markets and of the necessity of supporting and improving the effectiveness of international stabilization programs and mechanisms. Similarly they should try to avoid increasing the uneconomic production of goods which can be obtained under better conditions in the less-developed countries of the Continent, in which the production of these goods is an important source of employment.

D. Member countries should adopt all necessary measures to direct technological studies toward finding new uses and by-products of those primary commodities that are most important to their economies.

E. Member countries should try to reduce, and, if possible, eliminate within a reasonable time export subsidies and other measures which cause instability in the markets for basic commodities and excessive fluctuations in prices andincome.

Chapter II. International Cooperation Measures

1. Member countries should make coordinated, and if possible, joint efforts designed:

a. To eliminate as soon as possible undue protection of the production of basic products;

b. To eliminate taxes and reduce excessive domestic prices which discourage the consumption of imported basic products;

c. To seek to end preferential agreements and other measures which limit world consumption of Latin American basic products and their access to international markets, especially the markets of Western European countries in process of economic integration, and of countries with centrally planned economies; and

d. To adopt the necessary consultation mechanisms so that their marketing policies will not have damaging effects on the stability of the markets for basic commodities.

2. Industrialized countries should give maximum cooperation to less-developed countries so that their raw material exports will have undergone the greatest degree of processing that is economic.

3. Through their representation in international financial organizations, member countries should suggest that these organizations, when considering loans for the promotion of production for export, take into account the effect of such loans on products which are in surplus in world markets.

4. Member countries should support the efforts being made by international commodity study groups and by the Commission on International Commodity Trade of the United Nations. In this connection, it should be considered that producing and consuming nations bear a joint responsibility for taking national and international steps to reduce market instability.

5. The Secretary General of the Organization of American States shall convene a group of experts appointed by their respective governments to meet before November 30, 1961 and to report, not later than March 31, 1962 on measures to provide an adequate and effective means of offsetting the effects of fluctuations in the volume and prices of exports of basic products. The experts shall:

a. Consider the questions regarding compensatory financing raised during the present meeting;

b. Analyze the proposal for establishing an international fund for the stabilization of export receipts contained in the Report of the Group of Experts to the Special Meeting of the Inter-American Economic and Social Council, as well as any other alternative proposals;

c. Prepare a draft plan for the creation of mechanisms for compensatory financing. This draft plan should be circulated among the member governments and their opinions obtained well in advance of the next meeting of the Commission on International Commodity Trade.

6. Member countries should support the efforts under way to improve and strengthen international commodity agreements and should be prepared to cooperate in the solution of specific commodity problems. Furthermore, they should endeavor to adopt adequate solutions for the short- and long-term problems affecting markets for such commodities so that the economic interests of producers and consumers are equally safeguarded.

7. Member countries should request other producer and consumer countries to cooperate in stabilization programs, bearing in mind that the raw materials of the Western Hemisphere are also produced and consumed in other parts of the world.

8. Member countries recognize that the disposal of accumulated reserves

and surpluses can be a means of achieving the goals outlined in the first chapter of this Title, provided that, along with the generation of local resources, the consumption of essential products in the receiving countries is immediately increased. The disposal of surpluses and reserves should be carried out in an orderly manner, in order to:

a. Avoid disturbing existing commercial markets in member countries, and

b. Encourage expansion of the sale of their products to other markets.

However, it is recognized that:

a. The disposal of surpluses should not displace commercial sales of identical products traditionally carried out by other countries; and

b. Such disposal cannot substitute for large-scale financial and technical assistance programs.

In Witness Whereof this Charter is signed, in Punta del Este, Uruguay, on the seventeenth day of August, nineteen hundred sixty-one.

appendix H

Excerpts from the Charter of the United Nations (1945)

Article 24

1. In order to insure prompt and effective action by the United Nations, its Members confer on the Security Council primary responsibility for the maintenance of international peace and security, and agree that in carrying out its duties under this responsibility the Security Council acts on their behalf.

Article 32

Any Member of the United Nations which is not a member of the Security Council or any state which is not a Member of the United Nations, if it is a party to a dispute under consideration by the Security Council, shall be invited to participate, without vote, in the discussion relating to the dispute. The Security Council shall lay down such conditions as it deems just for the participation of a state which is not a Member of the United Nations.

Article 33

1. The parties to any dispute, the continuance of which is likely to endanger the maintenance of international peace and security, shall, first of all, seek a solution by negotiation, inquiry, mediation, conciliation, arbitration, judicial settlement, resort to regional agencies or arrangements, or other peaceful means of their own choice.

2. The Security Council shall, when it deems necessary, call upon the parties to settle their dispute by such means.

Article 34

The Security Council may investigate any dispute, or any situation which might lead to international friction or give rise to a dispute, in order to determine whether the continuance of the dispute or situation is likely to endanger the maintenance of international peace and security.

Article 35

1. Any Member of the United Nations may bring any dispute or any situation of the nature referred to in *Article 34*, to the attention of the Security Council or of the General Assembly.

Article 36

2. The Security Council should take into consideration any procedures for the settlement of the dispute which have already been adopted by the parties.

Article 51

Nothing in the present Charter shall impair the inherent right of individual or collective self-defense if an armed attack occurs against a Member of the United Nations, until the Security Council has taken the measures necessary to maintain international peace and security. Measures taken by Members in the exercise of this right of self-defense shall be immediately reported to the Security Council and shall not in any way affect the authority and responsibility of the Security Council under the present Charter to take at any time such action as it deems necessary in order to maintain or restore international peace and security.

Article 52

1. Nothing in the present Charter precludes the existence of regional arrangements or agencies for dealing with such matters relating to the maintenance of international peace and security as are appropriate for regional action, provided that such arrangements or agencies and their activities are consistent with the Purposes and Principles of the United Nations.

2. The Members of the United Nations entering into such arrangements or constituting such agencies shall make every effort to achieve pacific settlement of local disputes through such regional arrangements or by such regional agencies before referring them to the Security Council.

3. The Security Council shall encourage the development of pacific settlement of local disputes through such regional arrangements or by such regional agencies either on the initiative of the states concerned or by reference from the Security Council.

4. This *Article* in no way impairs the application of *Articles 34* and *35*.

Article 53

1. The Security Council shall, where appropriate, utilize such regional arrangements or agencies for enforcement action under its authority. But no enforcement action shall be taken under regional arrangements or by regional agencies without the authorization of the Security Council, ...

Article 54

The Security Council shall at all times be kept informed of activities undertaken or in contemplation under regional arrangements or by regional agencies for the maintenance of international peace and security.

Article 103

In the event of a conflict between the obligations of the Members of the United Nations under the present Charter and their obligations under any other international agreement, their obligations under the present Charter shall prevail.

Index

Abbink Mission, 692, 693
Accessory Transit Co., 485, 486
Acheson, Dean, cited, 729; indicates aid given Latin America under Marshall Plan, 694; presides at Washington Conference, 72
Act of Bogotá (1960), 14, 731
Act of Caracas, 138
Act of Chapultepec (1945), 12, 69, 137-38
Act of Havana (1940), 67, 137
Act of International Development, 731
Adams, Charles Francis, 489
Adams, John, 412
Adams, John Quincy, appoints U.S. representatives to Panama Conference, 51; discusses priority of recognition, 561; opposes joint action with Britain regarding Spanish colonies, 123; opposes recognition before independence, 560; opposes taking of Cuba, 323; sends Poinsett to Mexico, 231; states U.S. policy regarding Cuba and Puerto Rico, 319
aerial hijacking, 29-30
Afro-Asian-Latin American People's Solidarity conference, 159, 179-80, 378
aftosa, in Argentina, 586; in Mexico, 293
Agency for International Development, 40, 641, 728, 731
agriculture, 34-35, 744-45
Aguirre, Manuel Hermengildo de, 611
Aguirre Cerda, Pedro, 175, 636
Aix-La-Chapelle, conference of, 122, 665
Alabama, 672-73
Albizu Campos, Pedro, 400
Alemán, Miguel, strengthens relations with U.S., 292-93; takes stand against communism, 163
Alexander I, Czar, 122-23
Alessandri, Jorge, 641
Alessandri Palma, Arturo, 631, 636
Allen, Charles H., 392
Allen, Heman, 612
Allende Gossens, Dr. Salvadore de, in Cuba, 385; death of, 659; diplomacy of, 161, 657-59; in Mexico, 340; nationalization policies of, 28, 653; and OAS, 83-84; receives communist support, 176; visits to communist countries, 658
Alliance for Progress, 14, 15, 45, 372, 731-33, 734-64; goals of vs. OAS, 86; monetary allocations for, 25; overall assessment, 755-57; performance of, 308, 309, 445, 594, 642-44; threatened by communism, 181
Alsop claim, 628
Álvarez, Alejandro, 120, 129
Amador Guerrero, Dr. Manuel, 206
Amapala, Treaty of, 495
Amazon River, 675
American Council of Learned Societies, 41
American Emergency Committee on the Panama Canal, 221
Amistad Dam, 297
Anaconda Co., 28, 640, 650, 652, 653
Andean Subregional Integration Agreement, 739
anti-Americanism, 3-5, 14, 57, 87, 722-24
APRA, 170
Aramburu, General Pedro Eugenio, 592-93, 601
Arana Osorio, Carlos, 528
Aranha, Oswaldo, opposes Nazis in Brazil, 153, 691; plans economic cooperation with U.S., 687-89; supports U.S. at Rio Conference, 691
Arbenz Guzmán, Jacobo, overthrow of, 74, 164; pro-Communist actions of, 73, 138, 164, 519
arbitration, 56
Arbitration, Treaty of Inter-American, 60
Arévalo, Dr. Juan José, 164, 518, 523, 524
Argentina, affected by quarantine, 586; attempts to settle Chaco dispute, 61, 583; and the Braden issue, 589; at Caracas conference, 592; communism in, 171-72; declared war on Axis, 11; economic relations with U.S., 569, 585-86; at Lima conference, 585; at Mexico City conference, 91, 573; militarism in, 588-89, 592-604; and Monroe Doctrine, 135; at Montevideo conference, 580; Nazi infiltration in, 11, 150-52, 589; opposes U.S. position at Rio conference, 68; ouster of Perón, 592; Perón's return, 603; physiographic, 554-56; pro-Axis, policy of, 11, 69, 587-89; role at First Pan American conference 570-71; seeks Latin American hegemony, 571, 590; World War I, policy of, 5, 576
Arias, Dr. Arnulfo, 213, 221-22
Arias, Ricardo, 216, 221
Arica, *see* Tacna-Arica question
Armour, Norman, directs U.S. policy in Argentina, 587; minister to Haiti, 460; represents U.S. at Bogotá conference, 71
arms limitation, 84
arms standardization, 31-33
Artibonite Valley Project, 462
Artigas, José, 665-66
Artucio, Hugo Fernández, 152
asylum, 29
Atacama Desert, 617-18
Avellaneda, Nicolás, 570
Ávila Camacho, Manuel, 290, 291
Axis powers, 153, 156, 291, 721
Aztecs, 230

Baker, Howard, 408
Balaguer, Joaquín, 437, 441, 442-43, 451, 453
Balboa, Vasco Núñez de, 184
Ball, George, 705
Balmaceda, José Manuel, 129, 626-27
Baltimore affair, 129, 626-27
Banda Oriental, 666-67
Barbosa, Ruy, 683
Barrios, Justo Rufino, 489
Batista Zaldívar, Fulgencio, 13; emergence of, 352; overthrow of, 364; permits free elections, 357; retains communist party support, 166; seizes control of government, 358
Bay of Pigs, *see* Cuban invasion
Bayard, Thomas F., 53, 419-20, 490-91
Beaulac, Willard L., 71
Belize, 481
Bennett, W. Tapley, Jr., 445, 447
Berle, Adolph A., 65, 692
Betancourt, Romulo, 439, 440
Bethlehem Steel Corp., 630
Beveridge, Albert J., 3
"Big Stick," *see* Roosevelt Corollary
Black Warrior affair, 328
Blaine, James G., favors Central American federation, 489-90; initiates Pan-American conferences, 53, 570; on inter-oceanic canal, 128; mediates between Guatemala and El Salvador, 491; proposes lease of Samana Bay, 421; seeks to abrogate Clayton-Bulwer Treaty, 192; states policy in War of Pacific, 622
Bogotá Conference (1948), 12; adopts anti-Communist resolution, 72, 177; establishes Organization of American States, 72
Bolívar, Simón, conquers Spaniards in New Granada, 121, 559; initiates Pan-Americanism, 51; outlines agenda of, 52
Bolivia, communism in, 170-71; declares war on Axis, 149; disputes boundary with Chile, 617-19; and Nationalization policies, 27; Nazi infiltration into, 148-49; relations with communist nations, 162; sells U.S. tin and tungsten, 726; wages war with Paraguay, 61, 580-84; in War of Pacific, 619-25
Bonsal, Philip W., 364
Borno, Louis, 458
Borodin, Michael, 162
Bosch, Juan D., 444, 452
Bowers, Claude, 637
Boyer, Jean, 414
braceros, 293, 300
Braden, Spruille, 583; opposes Perón government, 589
Braden Copper Mines Company, 630, 646
Braganza, royal house of, 664
Brazil, and the Alliance for Progress, 706, 714; and Chaco War, 91; commercial relations with U.S., 680, 685-86, 687, 688, 696, 718; communism in, 172-74, 691-92, 713; and denuclearization treaty, 96; militarism in 708-18; offshore territorial waters of, 717; physiographic and ethnic, 663-64; and problems of the Northeast, 702, 712-15; Rio naval revolt and the U.S., 678; U.S. friendship for, 676, 679-80; U.S. relations with, *see* chap. 17; and valorization of coffee, 681-82; view on Monroe Doctrine, 666; and war with Paraguay, 567; in World War I, 5, 683-84; in World War II, 10, 690-91
Brent, William, Jr., 565
British Honduras, 481
Brizola, Leonel, 703
Brooke, John R., 340
Brookings Institution, report on Puerto Rico, 399; reports on refugees in Dominican Republic, 435
Bru, Laredo, 355
Bryan, William J., and Mexican relations, 254, 264, 266; opposes insurrection in Dominican Republic, 427; proposes settlement to Colombia, 207
Bryan-Chamorro Treaty, provisions of, 501; termination of, 38
Bucareli Agreements, 277-78
Buchanan, James, cites advantages of acquiring Cuba, 325-26; denounces Walker's filibustering, 488; on intervention in Mexico, 126, 235, 236; presses claims against Chile, 614-15; proposes abrogation of Clayton-Bulwer Treaty, 188; settles dispute with Brazil, 670-72
Buenos Aires conference (1910), 56, 575
Buenos Aires Peace conference (1936), 9, 42, 136
Bunau-Varilla, Philippe, helps engineer revolution in Panama, 203-4; signs Canal treaty with U.S., 205; works for Panama route, 199
Bundy, McGeorge, 447
Bunker, Ellsworth, 218, 447
Bureau of American Republics, 54, 56
Butler, Anthony, 231
Byrnes, James F., 589

Caamaño Deñó, Col. Francisco, 446
Cabot, John M., ambassador to Brazil, 697; Argentine policy of, 591; negotiates on Canal treaty, 214; represents U. S. at Caracas conference, 73
Cabral, Reid, 445
Caffery, Jefferson, 353
Calhoun, John C., supports Cuban independence, 323; and Texas annexation, 233; view on Monroe Doctrine, 123
California Pious Fund Case, 239
Calles, Plutarco E., agrees to settlement, 284; becomes president of Mexico, 278; defends Mexico's oil and land laws, 280; resents Secretary Kellogg's speech, 279
Calvo, Carlos, 55, 281
Cambon, Jules, 337
Camelot Plan, 646
Campina, Joaquín, 612
Canada, 86
Canal policy (U.S.), 126; *see* chap. 5 and 6
Canning, George, and Congress of Panama, 51; denies interest in Cuba, 320, 323; private view regarding the Monroe Doctrine, 124; proposes joint action regarding Spanish colonies, 122-23
Caperton, Admiral, 455
Caracas conference 13, 73-74, 94, 138, 177-78, 295, 520
Carden, Sir Lionel, 260
Cárdenas, Lázaro, 162; attempts settlement with U.S., 289-90; expropriates foreign oil

properties in Mexico, 288
Caribbean, 317, 318; *see* chap. 9-13
Caribbean Free Trade Association (CARIFTA), 739
Caribbean Legion, 517
Carnegie, Andrew, 497
Carnegie Endowment, 41
Carranza, Venustiano, murdered, 275; neutrality policy of, 271; opposes Huerta, 262; recognized by U.S., 269
Carrera, José Miguel, seeks help in U.S. 610, welcomes Poinsett to Chile, 609
Casauranc, Dr. Puig, 136
Castillo Armas, Col. Carlos, murdered, 523; revolts against communism in Guatemala, 164; seizes power in Guatemala, 164, 521
Castillo Branco, General Humberto de Alencar, 174, 706-10
Castillo, Ramon, 152, 587
Castro, Raúl, 365
Castro Ruz, Fidel, and aerial hijacking, 29; appeals to UN, 94; exports revolution, 157-60; and missile crisis, 372-73; rejects conciliation with U.S., 384; relations with Communist countries, 139; relations with U.S., *see* chap. 10; reveals self as Marxist-Leninist, 371; seizes power in Cuba, 364; in USSR, 385; visits Chile, 385
Central America, and Central American Common Market (CACM), 542-51; economy of, 477; physiographic and ethnic, 476-78; relations with U.S., *see* chaps. 14-15; in World War I, 504; in World War II, 516
Central American Common Market, 34, 543, 543-51, 736-37
Central American Court of Justice, 49' decides against Nicaragua, 502; reestablished on Hague model, 506
Central American Federation, 482-83, 489-90, 492, 496-97, 506-7
Central American International Bureau, 498
Central American Treaty of 1907, 497
Central American Treaty of Peace (1923), 508
Central Intelligence Agency, 370
Cerro de Pasco Corporation, 147-48
Cerro Corporation, 652
Céspedes, Dr. Carlos Manuel de, 347
Chaco dispute, 61, 580; negotiations for its settlement, 580-84; terms of settlement, 584; U.S. interest in, 580
Chadbourne Plan, 354
Chamizal controversy, 239, 245-46, 300-301
Chamorro, Gen. Emiliano, 509-10
Chapultepec, 234
Chapultepec, Act of, 12, 69, 137-38
Charles V, Emperor, 184
Charter of Bogotá (Article 15), 146
Charter of the OAS, *see* Appendix
Charter of Punta del Este, 15, 732; *see* Appendix
Chiari, Roberto F., 218
Chiari, Rodolfo, 212
Chile, attacked by Spain, 616-17; in boundary dispute with Argentina, 571-72; in boundary dispute with Bolivia, 617-18; communism in, 175-76, 639, 652, 659; military seizure of power in, 660; mineral production and problems of, 630, 635, 638, 639, 641; Nazi infiltration in, 149-50; 636; physiographic and ethnic, 608-9; relations with communist countries, 657-58; sells strategic minerals to U.S., 638; settles Tacna-Arica dispute, 634; U.S. relations with, *see* chap. 17; in War of Pacific, 619-34; in World War I, 629; in World War II, 638
Church, Frank, 21, 24, 690
CIAA, *see* Office of Coordinator of Inter-American Affairs
científicos, 246
Claims Commission, 286-87
Clark, J. Reuben, appointed ambassador to Mexico, 285; prepares memorandum on Monroe Doctrine, 132-33; represents U.S. at Pan-American conference, 61
Clay, Henry, demands recognition of Spanish colonies, 121; interest in isthmian canal, 185; interprets Monroe Doctrine, 124-25; reprimands U.S. *chargé* in Rio, 665; states policy regarding Cuba, 125, 324; supports Panama conference, 51
Clayton, John M., discusses Davis affair, 672; opposes British claims to Mosquito territory, 484
Clayton-Bulwer Treaty, abrogation of, 194-95; and Monroe Doctrine, 126, 128; negotiation of, 187-88; unsatisfactory to U.S., 191-92
Cleveland, Grover, closes Falkland Islands controversy, 563; interprets Monroe Doctrine, 129; recognizes Cuban insurgents, 334; settles Misiones dispute, 571; states isthmian canal policy, 193; supports Central American federation, 492
Cochrane, Lord, 558, 611, 613
coffee, inter-American agreement, 690; International Coffee Agreement, 14, 704, 739; low price causes crisis, 682; stabilization, valorization by Brazil, 681-82
Colby, Bainbridge, in Brazil, 684; coldly received by Argentina, 579; mission to South America, 429; protests force in Panama-Costa Rica dispute, 211
Cold war, 11-14, 45, 693-94
Cole, Byron, 485
Colombia, awarded $25 million by U.S. for loss of Panama, 207-8; communism in, 169-70; diplomacy regarding canal, *see* chap. 6; loses Panama, 204-5; Nazi infiltration, 146-47; negotiates regarding canal rights, 200-1, 202; supports the Monroe Doctrine, 124
Colorado River salinity controversy, 299, 301
Columbus, Chrisopher, 184, 411
Columbus (New Mexico), 269
Comintern, 154
Committee of Twenty-One, 731
Con·modity Credit Corporation, 461
Communism, condemned at Caracas conference, 73, 138; in Latin America, *see* chap. 4
Concensus of Viña del Mar, 21
Conciliation, Treaty of Inter-American, 60
Confederación de Trabajadores de La América Latina (CTAL), 162
Confederación de Trabajadores Mexicanos (CTM), 162
Conference on Latin American History, 42

conferences, Pan-American, see chap. 2
Connally, John B., Jr., 718
Convention for the Suppression of Unlawful Seizure of Aircraft, 29
Convention on Certain Acts Committed on Board Aircraft, 29
Cooke, Morris L., 290
Coolidge, Calvin, attends Pan-American conference, 58; embargoes arms to Cuban revolutionists, 349; Latin American policy, 7-8, 132; makes Tacna-Arica award, 634-35; opposes Puerto Rican autonomy, 399; regards religious question in Mexico as internal problem 284; states U.S. policy regarding Nicaragua, 512
Corn Island, 501
Corrigan, Frank D., 515
Cortés, Hernán, 184
Costa e Silva, Gen. Artur da, 711-15
Costa Rica, accepts arbitral award with Panama, 507-8; accuses Nicaragua of invasion, 535; agricultural program of, 539; and CACM, 541; communism in 165-66; diplomatic relations with communist countries, 162; interprets Monroe Doctrine, 135; physiographic and ethnic, 480-81; policy towards United Fruit Co., 540; protests Bryan-Chamorro treaty, 501; U.S. aid to, 540
Council of the OAS, see Organization of American States
Crimmins, John H., 452
Crittenden, Col. 327
Cromwell, William N., 199, 206
Crowder, Gen. Enoch, 7, 347, 348
Cuba, and aerial hijacking, 29; Castro's revolution in, 364-65, 372, 375, 380-81; communism in, 157-60, 166-67, 379; covered by Monroe Doctrine, 125; "ever faithful island," 324; exiles of in U.S., 376-77; independence of 340, 343; invasion of (Bay of Pigs), 139; missile crisis in, 139, 372-73; OAS resolution against, 179; physiographic and ethnic, 317-18; relations with U.S., see chaps. 9-10; strategic location of, 317; success of trade agreement with U.S., 6, 344, 354; in World War I, 346; in World War II, 356-57
Cuban invasion (Bay of Pigs), 16, 76, 139, 178, 365, 370, 523, 524
Cuban missile crisis, 76-77, 160, 179, 372-73, 705
Cuban Revolutionary Party (*Auténtico*), 356
cultural relations, 41-44
Cushing, Caleb, 332

Daniels, Josephus, 286
Danish West Indies, 67
Dantas, Dr. San Tiago, 702-3
Dartiguenave, Philippe Sudre, 455
Davis, Lt. Alonzo B., 669, 671-72
Dawes, Charles G., 432
Dawson, Thomas C., 424, 499
Debray, Jules Régis, 158
Declaration of American Principles, 65
Declaration of Caracas, 73-74
Declaration of Lima, 65, 585
Declaration of Panama, 67
Declaration of the Peoples of America, 732
Declaration of the Presidents of America, 80
Declaration of Reciprocal Assistance and Cooperation for the Defense of the Nations of the Americas, see Havana Declaration of Reciprocal Assistance
Declaration of San José, 75
Declaration of Santiago, 74
Declaration of Solidarity for the Preservation of the Political Integrity of the American States Against the Intervention of International Communism, 73-74
Declaration of Viña del Mar, 35
democracy, 33
Dessalines, Jean Jacques, 413
Dewey, Com. George, 337
Díaz, Adolfo, 499-500, 511-12
Díaz, Felix, 252
Díaz Ordaz, Gustavo, 302, 304
Díaz, Porfirio, helps mediate, control American war, 494-95; improves relations with U.S., 239; and Mexico City conference, 54
Dickinson-Ayón Treaty, 189
dictators and democracy, see militarism
diplomatic recognition, 30-31
Dodds, Dr. Harold W., 355, 510
Doheny, Edward L., 243, 274
"dollar diplomacy," 4, 244, 259
Dominican Republic, communism in 167-68; Grant seeks annexation of, 418; in Napoleonic era, 411-12; physiographic and ethnic, 410-11; separates from Haiti, 411; Spanish rule of, 411; U.S. intervention in, 140, 447-49; U.S. relations with, see chaps. 12-13; in World War II, 435
Dorticós, Osvaldo, 375
Downes vs. Bidwell, 391
Drago, Luis M., 120, 576
Drago Doctrine, 55, 573
Dulles, John Foster, attacks communism at Caracas conference, 73, 138, 178; defends U.S. sovereignty over Panama Canal, 216; firm policy towards Haiti, 464; lauds Castillo Armas government, 521; mission to Brazil, 699
Dumbarton Oaks Conference, 91
Duncan, Captain, 562-63
Dupuy de Lome, Minister, 335
Duvalier, Francois ("Papa doc"), 464; death of, 471; loses U.S. support, 464, 466
Duvalier, Jean Claude, 471-72
Dutra, Eurico, 691

Eads, James B., 191
Eastland, James O., 438
East Florida, see Florida
Echeverría Álvarez, Luis, 305
Economic Agreement of Bogotá, 12, 71
economic assistance, to Latin America, 730, see chap. 19
economic defense, 725
Ecuador, communism in, 170; declares solidarity, 10; grants base rights, military assistance with U.S., 10; Nazi infiltration into, 147; and territorial water claims, 26-27, 84
education, 746-47
Edwards, Augustin, 58
Egan, Patrick, gives asylum to political refugees, 626; negotiates regarding *Baltimore*

affair, 627; offers good offices in Chilean civil war, 625
Eighth International Conference of American States, *see* Lima conference (1938)
Eighth Meeting of Consultation of Ministers of Foreign Affairs, 76, 178-79, 298, 732
Eisenhower, Dwight, aid program for Brazil, 695; in Argentina, 593; in Brazil, 700; breaks with Cuba, 369, cancels Cuban sugar imports, 366; in Chile, 641-42; commissions brother, 12-13; inaugurates new program, 730; Latin American policy of, 12-14; 728-29; makes concessions to Panamanians, 215; meets Mexican presidents, 294, 297; reaffirms Monroe Doctrine, 139
Eisenhower, Dr. Milton, in Argentina, 591; in Brazil, 695; makes fact-finding trips, 12-13; in Mexico, 296; in Panama, 217
Eleventh Meeting of Consultation of Ministers of Foreign Affairs, 79
Ellender, Allen J., 441
El Salvador, and Alliance for Progress, 530; Central American Common Market, 530, 545, 550; communism in, 163-64; physiographic and ethnic, 479; protests Bryan-Chamorro treaty, 319-21; signs treaty with Nicaragua, 312; U.S. relations with, *see* chaps. 14-15; war with Guatemala 494; war with Honduras, 80, 534
"Era of Good Feelings", 292
Estimé, Dumarsais, 462
Estrada Doctrine, 285-86
Estrada Palma, Tomás, 340, 344
European Economic Community (E.E.C.), 735
European Free Trade Association (E.F.T.A.), 735
Evarts, William M., interprets Monroe Doctrine, 128; seeks canal treaty with Colombia, 191; urges honorable peace upon Chile and Peru, 620
Everett, Edward, 327
Export-Import Bank, established, 355, 723; operations of, 19, 71, 213-14, 290, 445, 461, 465, 545, 592, 597, 636, 637, 647, 656, 689-90, 693, 698; purposes, 13, 727

Falcon Dam, 294
Falkland Islands, 125, 562-64
Fall, Albert B., 274
FALN (Fuerzas Armada de La Liberación Nacional), 31
Farland, Joseph S., 219, 437, 439
Farrell, Gen. Edelmiro, 588-89
Fascism, *see* chap. 4
Ferdinand VII, 121, 324, 329, 558
Ferre, Luis A., 405-6
Fifth International Conference of American States, *see* Santiago conference (1923)
Fifth Meeting of Consultation of Foreign Ministers, *see* Santiago meeting (1959)
Figueres, José, 166, 220, 517, 539, 542
Filho, José Cafe, 697
filibustering, 485-88
Fillmore, Millard, 326
Finch, Robert H., 24
First Inter-American Consultative Conference of Foreign Ministers, 66
First International Conference of American States, *see* Washington conference (1889)
First Meeting of Consultation of Ministers of Foreign Affairs of the American Republics, *see* Panama meeting (1939)
Fish, Hamilton, favors Central American Federation, 489; interprets Monroe Doctrine, 128; protests Spanish policy in Cuba, 332-33; seeks to annex Samana Bay, 417-18
Fletcher, Henry P., 272; attends sixth Pan-American conference, 58; heads U.S. delegation to fifth Pan-American conference, 57
Flood, Daniel J., 221, 224
Florida, 673
Florida, ceded to U.S., 2; relation to "no-transfer," 1; strategic importance of, 318, 319
Fonseca, Gen. Deodoro da, 676
Foraker Act, 391-94
Forbes, W. Cameron, 459
Ford Foundation, 642, 647
Foreign Assistance Act, 31-32
foreign investments in Latin America, 730
Foreign Military Sales Act, 84
Foreign Operations Administration, 728
foreign trade, 735-36, 740-41
Forsyth, John, 125
Fourth International Conference of American States, *see* Buenos Aires conference (1910)
Fourth Latin American Students' Congress, 159
Fourth Meeting of Consultation of Ministers of Foreign Affairs, 72, 177
France, disputes with Argentina, 564-66; expelled from Haiti, 414; interest in Caribbean, 317, 318, 320, 322, 324; interest in Santo Domingo, 411-14; intervenes in Mexico, 126-27, 236-37; proposes pact on Cuba, 327; recognizes Monroe Doctrine, 119; in World War II, 137
Frei Montalva, Eduardo, 28, 643-48, 651-52
Frelinghuysen, Frederick T., ill-advised policy in War of Pacific, 623-24; negotiates treaty with Nicaragua, 193; opposes expansion in Caribbean, 419; seeks to abrogate Clayton-Bulwer Treaty, 192
French Canal Company, 193, 197-206
French Revolution, 322
Frondizi, Arturo, aided by U.S., 593; breaks with Cuba, 179; ousted, 595
Fulbright, J. William, 19, 44; on Dominican intervention, 448; interest in Puerto Rico, 406; suggests policy toward Cuba, 374; supports concessions to Panama, 215
Furtado, Celso, 708

Gadsden Purchase, 231, 235
Gale, Linn, 162
Galíndez, Dr. Jesús de, 437
García Godoy, Hector, 447, 449
Garde dé Haiti, 460, 461-62
Garfield, James A., attempts to abrogate Clayton-Bulwer Treaty, 191-92; Latin American policy of, 53, 128, 570
Galapagos Islands, 147
General Agreement on Tariffs and Trade (GATT), 361

General Convention of Inter-American Conciliation, 60
General Treaty of Central American Integration, 543
General Treaty of Friendship and Cooperation, 212-13
General Treaty of Inter-American Arbitration, 60
Germany, impact of submarine warfare on Brazil in World War I, 681; and Latin America in World War II, 141-53, 587-89, 691
Goethels, Gen. George W., 206
Goldberg, Arthur, 404
Gómez, José Miguel, 345-46, 347-48
Gómez, Miguel Mariano, 355
Gondra Treaty, adopted, 58; becomes conciliation convention, 61; invoked in Dominican-Haitian dispute, 434; ratification of, 61
González, Gen. Pablo, 274-75
González Videla, Gabriel, 175-76, 639
Good Neighbor, era of, 1; policy of, 7, 9-10, 12, 286, 580, 724
Gordon, Lincoln, 80, 757; ambassador to Brazil, 705, 707, 713
Gorgas, Dr. William C., 206
Goulart, João, 174, 697, 700-7
Grace, W. R., 630
Graf von Spee, 67
Grant, Gen. Ulysses S., appoints canal commission, 189; favors annexation of Samana Bay, 418; follows friendly policy towards Cuba, 332; interprets Monroe Doctrine, 128; and Mexican relations, 238
Grau San Martín, Dr. Ramón, 352, 357-58
Great Britain, and abolition of slave trade, 329; accepts its abrogation, 193-95; favors recognition of Spanish colonies, 122; interest in Cuba, 319-20; intervenes in Mexico, 126-27, 236, 237; and Monroe Doctrine, 2; negotiates canal treaty with U.S., 187-88; objects to tolls exemption, 264; occupies Falkland Islands, 125; proposes pact on Cuba, 327; refuses to abrogate Clayton-Bulwer treaty, 191-92; seizes Havana, 321; in Venezuela controversy, 129; in World War II, 141
Greater Antilles, 317
Gresham, Walter Q., 492
"Group of 77," 36
Gruening, Ernest H., 401
Guantanamo, 341, 361, 367, 369, 375
Guardia, Ernesto de la Jr., 217
Guatemala, and Alliance for Progress, 525-26; and CACM, 526, 543; communism in, 13, 97-99, 138, 164-65, 339-42; heads Central American Federation, 482; initiates new effort at federation, 489-91; opposes U.S. anti-communist stand at Caracas conference, 138; physiographic, 479; at war with neighbors, 494-95
Guerrero, Xavier, 162
Guevara, Dr. Ernesto "Che," 158, 171, 368, 378, 648
Guggenheim, Harry F., 350-51
Guggenheim Company, 630
Guido, José María, 595

Hague Convention, 54
Haiti, communism in, 167; dispute with Dominican Republic, 467, 469; separates from Dominican Republic, 411; U.S. relations with, 130, 131; *see* chaps. 12-13; in World War II, 461
Haitian Republic, 420
Hale, William B., 260, 262
Harding, Warren G., appoints commissioners to Mexico, 277; attempts to settle Tacna-Arica dispute, 632-33; Latin American policy of, 7; promises withdrawal from Dominican Republic, 429; seeks settlement with Colombia, 211
Harriman, W. Averell, 71, 596
Harrison, Benjamin, inaugurates First International American Conference, 53, 570; policy of in *Baltimore* affair, 129, 627
Havana Conference (1928), 58-60, 579, 635, 723
Havana Declaration of Reciprocal Assistance, 137
Havana meeting (1940), 137, 290
Hawley-Smoot Tariff, causes antagonism in Argentina, 586; damaging to Chile, 629; harmful to U.S.-Latin American trade, 6; injurious to Cuba, 350
Hay, John, and Drago Doctrine, 574; negotiates with Colombia, 200, 201, 202; obtains abrogation of Clayton-Bulwer Treaty, 194-95; proposes collection of Dominican customs, 424-25; and U.S. imperialism, 3
Hay-Bunau Varilla Treaty, 205-6
Hay-Herrán Treaty, 200-1
Hay-Pauncefote Treaty, 194-95, 197, 198
Haya de La Torre, Victor Raúl, 170
Hayes, Rutherford B., canal policy of, 128, 189; settles Argentine-Paraguayan boundary dispute, 568
Health measures, 743-44
Hearst, William R., 334
Heinl, Col. Robert D., Jr., 465, 467
Hepburn Bill, 198
Hernández Colón, Rafael, 408
Herter, Christian, defends nonintervention, 440; at Santiago meeting, 74; at Seventh meeting, 75, 178
Hickenlooper Amendment, 27
Hise, Elijah, 482, 483
Hispanic American Historical Review, 41
Hispanic Foundation (Library of Congress), 42
Hitler, Adolf, 141
Hitler-Stalin Pact, 155
Holland, Henry F., 592
Holy Alliance, 122-23, 124
Honduras, accepts U.S. arbitration, 504; and Alliance for Progress, 533; boundary dispute with Nicaragua, 515; and Central American Common Market, 534; communism in, 165; crushes filibustering expeditions and executes Walker, 488; negotiates loan convention with U.S., 503; physiographic, 480; at war with El Salvador, 80, 534; at war with Nicaragua, 495
hoof and mouth disease (*aftosa*), in Mexico, 293; in Argentina, 586
Hoover, Herbert, cooly received in Argentina, 580; friendly policy towards Puerto Rico, 399; friendly relations with Mexico, 285; Latin American policy of, 8, 132-33;

sends mission to Haiti, 459; settles Tacna-Arica dispute, 634
House, Col. Edward M., 136
housing, 747
Houston, Sam, 232
Huerta, Gen. Victoriano, loses British recognition, 264; refused recognition by President Wilson, 259; resignation of, 267; seizes government in Mexico, 253-54
Hughes, Charles Evans, confers on Cuba situation, 348; decides dispute between Costa Rica and Panama, 506; delegate to Conciliation conference, 60; explains U.S. policy of intervention, 59; formulates policy toward Dominican Republic, 429-31; heads U.S. delegation to sixth Pan-American conference, 58; 579; interprets Monroe Doctrine, 131-32; outlines policy toward Mexico, 276; quoted on Caribbean policy, 7; signs most-favored-nation agreement with Brazil, 681
Hull, Cordell, accepts Argentine anti-war pact, 62; accepts non-intervention, 136; criticizes Argentine policy, 588; Cuban policy of, 352; exchanges notes with Panama, discusses expropriation in Mexico, 287, 288; at Havana conference, 137; heads U.S. delegation to seventh Pan-American conference, 61; Latin American policy of, 9; at Lima conference, 65-66, 585; pleads for peace in Chaco, 583; proposes reciprocal trade agreements, 62; represents U.S. at Buenos Aires conference, 64, 136, 586; signs trade agreements, 725; supports collective trusteeship, 137; supports Haitian-Dominican Republic settlement, 434; universalist views of, 91
Humboldt, Alexander von, 184
Humphrey, Hubert, interest in Puerto Rico, 408; quoted on Brazil policy, 701-2; signs denuclearization treaty, 97
Hurlbut, Gen. Stephan A., 621-22

Ibáñez, Carlos, attacks communism in Chile, 175-76; reelection of, 639, rigged election of, 631; sent into exile, 635
Ickes, Harold L., 353, 401
Illia, Dr. Arturo, 595-98
Imbert Barreras, Gen. Antonio, 447
imperialism 3-5, 636
industrialization, 745-46
Institute of Inter-American Affairs, 10, 729
intellectual and cultural cooperation in the Americas, 42-44
Inter-American Coffee Agreement, 690
Inter-American Commission for Territorial Administration of European colonies in America, 67
Inter-American Commission on Human Rights (IACHR), 81-82, 534
Inter-American Committee for the Alliance for Progress (C.I.A.P.), 34, 37, 735
Inter-American conferences (of Latin American origin), 51-53
Inter-American Conference for the Maintenance of Peace, *see* Buenos Aires conference
Inter-American Conference for the Maintenance of Continental Peace and Security, *see* Rio conference (1947)
Inter-American Conference on the Problems of War and Peace, *see* Mexico City conference (1945)
Inter-American Development Bank, 19, 25, 85, 597, 652, 729
Inter-American Development Commission, 727
Inter-American Economic and Social Council (I.A.-ECOSOC), 85; reviews Alliance for Progress, 735
Inter-American Financial and Economic Advisory Committee (FEAC), 67
Inter-American Joint Defense Board, 69, 72, 76
Inter-American Neutrality Committee, 67
Inter-American Peace Committee, 81
Inter-American System, 86; and communism, 140; put on treaty basis, 71; strengthened, 78-79; *see* Organization of American States
Inter-American Treaty of Reciprocal Assistance, *see* Rio Treaty
Inter-departmental Committee on Cooperation, 42
International Basic Economy Corporation, 13
International Coffee Agreement, 14, 35, 732, 741-42
international communism, *see* chap. 4
International Cooperation Administration, 728
International Monetary Fund, 13
International Sugar Agreement, 742
International Telephone and Telegraph Corporation, 659
International Union of American Republics, 54
interoceanic canal, *see* isthmian canal
Inter-parliamentary Mexico-United States conferences, 299, 303
Intervention, in Cuba, 4; defense by Kennedy, 139; in Dominican Republic, 4, 140; in Haiti, 4; issue at Havana conference, 58; meaning of, 4; in Nicaragua, 512-13; outlawed, 65; in Panama, 204, 211; policy modified by Hughes, 60; in pursuit of Villa, 269-70; Roosevelt denounces, 64-65; in Tampico and Vera. Cruz, 5, 265-67; under Platt Amendment, 4; U.S. in Caribbean, 4
Investor's guarantee agreements, 40
Irigoyen, Hipólito, 576-77
Isle of Pines, 7, 357
isthmian canal, *see* chaps. 5-6
Italy, 146
Itata affair, 626
Iturbide, Agustín, 481

Jackson, Andrew, interest in isthmian canal, 186; sends Butler to Mexico, 231
Jackson, Henry, 407, 408
Japan, 130, 146, 153, 250-51, 289
Jay's Treaty, 318
Jefferson, Thomas, interest in Cuba, 322; opposed to French control of Louisiana, 257; quoted, 318; view of Monroe Doctrine, 123
Jenner, William E., 438
Johnson, Andrew, 128, 133
Johnson, Lyndon B., Brazilian policy of,

708, in Central America, 549-50; interprets Monroe Doctrine, 140; Latin American policy of 18-19, 31, 80; lauds Central American integration, 547; lauds denuclearization treaty, 98; maintains normal relations with Haiti, 468; negotiation of new canal treaty, 219; orders Dominican intervention, 447; policy towards military governments, 753; signs Chamizal treaty, 300
Johnson, Olin D., 438
Joint American-Mexican *Aftosa* Commission, 293
Joint Brazil-United States Defense Commission, 690
Joint Mexican-American Commission for Economic Cooperation, 291
Joint Mexican-United States Defense Commission, 291
Jones Act, 395-96
Jones-Costigan Act, 354, 359
Jova, Joseph J., 20
Juárez, Benito, and French intervention, 126-27, 236-37; recognition by U.S., 126, 236

Kellogg-Briand Peace Pact, 9, 60
Kellogg, Frank B., attends fifth Pan-American conference, 57; delegate to conciliation conference, 60; interpretation of Monroe Doctrine, 132; Latin American policy of, 7-8; makes critical address on Mexico, 279; proposes Tacna-Arica settlement, 634; questions Mexican policy on oil and land, 279-82; refuses recognition of rebels, 285; sees Bolshevism in Nicaragua, 512; signs canal treaty, 212
Kelsey, Carl, 430
Kennan, George, 92
Kennecott Copper Company, 28, 630, 640, 647, 650, 653
Kennedy, Edward M., 327
Kennedy, John F., and abortive Cuban invasion, 16, 139, 178, 370; aids Chile, 641; Brazilian policy of, 700-4; in Colombia, 16; in Costa Rica, 16, 546; and the Cuban missile crisis, 16, 95, 139-40, 179, 372-73; debates Nixon on Cuban question, 369; initiates Chamizal settlement, 299; in Mexico, 299-300; proposes Alliance for Progress, 15, 727; rejects military *coups,* 749; supports Frandizi, 594; supports revision of canal treaty, 218
Kennedy, Robert, 646
Khrushchev, Nikita S., and the Cuban missile crisis, 139-40; derides Monroe Doctrine, 138; supports Castro, 178, 371; warns U.S., 367
kidnapping, 28-29, 717
Kilpatrick, Gen. Judson, 616
King, Rufus, 412, 557
Knapp, Capt. (Adm.) Harry S., 285, 427-28
Knights of Columbus, 175
Knox, Philander C., chairman of fourth Pan-American conference, 56; denounces Zelayan regime, 498; investigates New French Panama Canal Company, 200; Latin America policy of, 244, 252, 575; supports "dollar diplomacy," 56; warns Cubans, 345
Korean War, 12, 72, 693
Kosygin, Alexei, 378
Kubisch, Jack B., 758
Kubitschek, Juscelino, administration of 697-99; criticizes Alliance for Progress, 24; proposes Operation Pan-America, 14, 697; treatment of by Brazilian military, 713

Laird, Melvin, 408
land reform, 743
Lansing, Robert, adopts positive policy towards Panama, 211; protests provisions of Mexican constitution, 272; reviews Mexican situation, 270; warns Cuba against revolution, 346; and World War I, 5
Lanusse, Alejandro Agustín, 602
La Plata, *see* Argentina
Larreta proposal, *see* Rodríguez Larreta, Eduardo
Lassiter, Gen. William, 633-34
Latin American Free Trade Association (LAFTA), 34, 737-39
Latin American Solidarity Organization (OLAS), 159
Latin American Studies Association, 42
Lavalle, José Antonio, 391
Lazare Claims, 420
League of Nations, Argentina seeks to amend its Covenant, 578; expels Soviet Union, 156; and Latin America, 5, 42, 57, 89-90; and Leticia dispute, 90; and the Monroe Doctrine 118-19, 134, 135; tries to settle Chaco dispute, 61, 90
Le Clerc, Gen., 413-14
Lee, Fitzhugh, 334, 335
Lend-Lease, 11
Lescot, Elie, 461
Lesseps, Ferdinand de, 128, 189, 193
Lesser Antilles, 317
Levingston, Roberto Marcelo, 602
Lima conference (1938), 65-66, 136, 585
Lima, Declaration of, 65
Lincoln, Abraham, recognizes Haiti, 416; recognizes Juárez, 236
Lind, John, 254, 261
Lindbergh, Charles, 284
Linowitz, Sol M., 20, 80, 600
Lodge, Henry C., authors Magdalena Bay resolution, 251; favors settlement with Colombia, 208; and U.S. imperialism, 3
Logan, Cornelius A., 624
Lonardi, Eduardo, 592
Long, George S., 437, 704
López, Francisco Solano, 672
López Mateos, Adolfo, 296
López, Narciso, 326-27
Louisiana, 318
L'Ouverture, Toussaint, 412
Luxburg, Count von, 576-77

Machado y Morales, Gerardo, 349-51
Madero, Francisco, 247-53; murder of, 254; relations with U.S., 249-53
Madero, Gustavo, 248, 253, 254
Madison, James, 123, 319
Magdalena Bay (Lodge), 130, 251
Magellan, Ferdinand, 184
Magloire, Paul, 462-63
Magoon, Charles E., 344-45
Mahan, Adm. Alfred T., 3

Maine, 335, 336
Malvinas Islands, *see* Falkland Islands
"manifest destiny," 231
Mann, Thomas C., career of, 757; interprets intervention, 141; mission to Panama, 218
Mansfield, Mike, 296, 299
Marcy, William L., Cuban policy of, 328; refuses to receive Walker's emissary, 486
Marinello, Juan, 166
Marshall, Gen. George C., 12, 71
Martí, José, 333
Martin, Edwin M., 597, 757
Matthews, Herbert, 363
Maximilian, Archduke, 127, 237
Mayo, Adm. H. T., 265-66
McCloy, John J., 71
McKinley, William, death of, 197; intervenes in Cuba, 334-38; and Mexico City conference, 54; outlines peace terms to Spain, 337-338
McLane-Ocampo Treaty, 236
Medici, Gen. Emilio Garrastazu, 715-18
Meeting of Consultation of Ministers of Foreign Affairs, 70-71
Mein, Ambassador John G., 165, 527
Méndez Montenegro, Julio César, 525
Mendieta, Col. Carlos, 353
Menocal, Mario, 346
Messersmith, George, 356, 589
Metternich, Prince, 122
Mexicali, 299
Mexican War, 233-35
Mexico, at Caracas conference, 295; commerce of, 306-7; communism in, 162-63; constitution of 1917, 271-72; economic development in, 230; gives refuge to Fidel and Raúl Castro, 297; refuses to recognize Monroe Doctrine, 135; and regional organizations, 180, 311-12; at Santiago conference, 297; U.S. diplomatic relations with, *see* chaps. 7-8; U.S. investments in, 240-43; and world organizations, 311-12; in World War I, 271; in World War II, 291
Mexico City conference (1901), 54-55, 573, 628
Mexico City conference (1945), 69-70
Meyer, Charles A., 22, 25, 26, 37, 757
militarism, 30-31, 87, 750-55
military assistance agreements, 10, 31, 32-33, 694, 750-55
military missions, 10, 291, 751-54
Miller, Edward, 192, 376
Miranda, Francisco de., 556-57
Misiones boundary, 362
Mitre, Bartolomé, 567
Môle St. Nicholas, 420, 455
Moncada, Gen. José María, 513
Monroe, James, announces Monroe Doctrine, 561; considers recognition of Spanish colonies, 560, favors independence of Spanish colonies, 121; recognizes independence of Brazilian Empire, 664
Monroe Doctrine, as applied to Cuba, 386; British violations of, 126-27; Cleveland-Olney extension, 120; and communism, 138-40; derided by Khrushchev, 367; and Dominican Republic, 416; with Drago Doctrine, 120; favors acquisition of Cuba, 323; history, development and applications of, *see* chap. 3; interpreted by Elihu Root, 131; interpreted by Franklin Roosevelt, 133; interpretation avoided, 58; and the League of Nations Covenant, 134; Lodge corollary extension of, 130, 251, 627; and Pan-Americanism, 135, 136; Polk's contributions to, 125-26; reaffirmed in Dominican intervention, 140; recognized as Pan-American policy, 67, 92, 722; requirements of, 2; Roosevelt corollary extension of, 4, 130; supported in Republican party platform (1968), 141; no transfer resolution, 1; undermining the Doctrine, *see* chap. 4; U.S. does not apply to Falkland Islands, 563, 720; as U.S. foreign policy, 2; in Venezuela controversy, 129; violated by France, 126-27; violated in the La Plata area, 125; violated by Spain, 126-27
Monroe Palace, 680
Montevideo conference (1933), 61-62, 136, 580
Moore, John B., 203
Mora, José A., OAS, 378
Morrow, Dwight W., 587, 283-85
Moscoso, Teodoro, 642, 757
Mosquito Coast, 2, 489, 491, 492
Muñoz Marín, Luis, 401, 402, 403
Munro, Dana G., 459
Munro, Lt. Comm., Ernest A., 165, 527
Murphy, Gerald L., 437
Mussolini, Benito, 146
Mutual Defense Assistance Agreement, 357
Mutual Security Act of 1951, 728

Nabuco, Joaquim, 679-80
Napoleon I, 318, 413-14, 664
Napoleon III, 127, 186, 236-37
Nashville, U.S.S., 204
Nasser, Gamal Abdul, 216
nationalism, 87
Nazis, 146-53, 725
Netherlands, 137
neutrality zone, *see* Declaration of Panama
New French Panama Canal Company, 197-98, 200
New Granada, 186
Niagara Falls conference, 267
Nicaragua, attacked by Costa Rica, 487; boundary dispute with Honduras, 515; communism in, 165; conflict with Costa Rica, 535; its constabulary, 513; controversy with U.S., 510-15; described, 480; disputes with Britain over Mosquitia, 491; earthquake in, 538; invites filibusterers, 485; loses dispute with Mosquitia, 192; makes canal treaty with U.S., 188; physiographic, 130, 480; signs Bryan-Chamorro Treaty, 500-1; signs canal treaty with U.S., 483; signs new canal treaty, 193; U.S. aid to, 536; U.S. intervenes in, 509
Nicaraguan canal, as focus of U.S. interest, 193, 501; favored by Napoleon III, 184; work started on, 193
Ninth International Conference of American States (1948), *see* Bogotá conference
Ninth Meeting of Consultation of Ministers of Foreign Affairs, 77, 301, 377
Nixon, Richard, in Argentina, 593; attacks on Lima and Caracas, 217; cooperation with Mexico, 304; criticizes U.S. Cuban policy, 377; in Cuba, 362; in Dominican Republic, 437; in El Salvador, 529; in

Guatemala, 522; in Haiti, 462; heads U.S. delegation for Kubitschek inaugural, 697; Latin American policy of, 19-39; 758-63; in Mexico, 295; policy towards Chile, 657; reaffirms Monroe Doctrine, 141; rejects conciliation with Cuba, 384; stand on Cuba in 1960 election, 369; supports arms aid for Latin America, 751-54
nonintervention, 9
No-Transfer Resolution of 1811, 1; merged with Monroe Doctrine, 137; and World War II, 137
Obregón, Gen. Álvaro, murdered, 285; relations with U.S., 275-78; revolts against Carranza, 275
Office of the Coordinator of Inter-American Affairs, 42, 727
O'Higgins, Bernardo, 609
oil laws, 279
Oliver, Covey T., 757
Olmstead Act, 394-95
Olney, Richard, 3, 129, 334, 628
Onganía, Juan Carlos, 598, 602
"Operation bootstrap," 406
"Operation Pan America," 14, 699
Ord Order, 238, 239
Oregon, U.S.S., 307
Organization of American States (OAS), acts in Dominican crisis, 447; charter of, 71; considers Cuban problem, 94; and Cuban missile crisis, 95, 139; and Dominican Republic intervention, 46, 140-41; and the El Salvador-Honduran war, 80, 88; excludes Cuba from inter-American system, 372; excludes Cuba from summit meeting, 377; goes into effect, 72; Guatemala issue before, 94; and hemispheric security, 142; organization and work of, 70-72; reorganization of 78-79, 83; revocation of sanctions against Dominican Republic, 443; specialized agencies of, 82-83; votes sanctions against Cuba, 374-75; votes sanctions against Dominican Republic, 441
Organization of Latin American Solidarity (OLAS), Allende named head by Castro, 654; rejects USSR coexistence line and supports terrorism in U.S. cities, 378
Orozco, Gen., Pascual, 248, 249
Ortiz Rubio, Pascual, 285
Osborn, Thomas A., 619, 620
Ostend Manifesto, 328-29

Pact of Bogotá, 71
Padilla, Ezequiel, 37, 291
Panama, communism in, 166; domestic politics and the canal issue, 225; socio-economic conditions in, 222-23; U.S. relations with, *see* chaps. 5-6
Panama, declaration of, 66-67
Panama Canal, *see* chaps. 5-6
Panama Congress (1826), 51, 185, 325
Panama meeting (1939), resolution on no-transfer, 136-37, 290, 725; *see* First Meeting of Consultation of Ministers of Foreign Affairs of the American Republics.
Pan-American Highway, 38, 60, 63, 69, 516
Pan-Americanism, 3; and Monroe Doctrine, 135, 136; *see* chap. 2
Pan-American Union, 3, 42, 56, 71
Paraguay, arbitrates territory, 568; communism in, 177; at war, 61, 567-68; wars on Brazil and Argentina, 674; wars with Bolivia over Chaco region, 580-84
Paris Peace conference, 5, 57, 134
Partridge, James R., 623
Party of the Cuban Revolution (*Ortodoxo*), 356
Peace Corps, 16, 223, 449
Peace of Paris, 321
Pearl Harbor attack, 10, 137, 725
Pearson, S. and Sons, 241
Pedro I, 664, 668
Pedro II, in diplomatic dispute with U.S., 669-71; forced to abdicate, 676; opens Amazon, 675; visits U.S., 675
Peletier and Lazare Claims, 419-20
Pendleton, J. S., 614, 615
People's Republic of China, aid to Cuba, 368, 371; begins diplomatic and trade drive, 161; breaks with Soviet Union, 158-59; gains Latin American support for entry into UN, 96
Peralta Azurdia, Enrique, 523, 525
Pérez Jiménez, Marcos, 13
Péron, Eva, 590
Péron, Isabelita, 604
Péron, Juan D., continuing influence in domestic politics, 595; ouster of, 592; policy towards communism, 591-92; relations with U.S., 589-92; return to Argentina and election to presidency, 603-4; supports Nazis and wins election, 586
Pershing, Gen. John J., 269-70, 634
Peru, and Chaco War, 61; communism in, 170; and nationalization policies, 27; resists Nazi infiltration, 147-48; Tacna-Arica question, terms of settlement, 633-34; and territorial waters claims, 26-27; in War of the Pacific, 619-20
Pettis, S. N., 619
Pierce, Franklin, 327, 486
Pinckney's Treaty, 318
Pinochet Ugarte, Gen. Augusto, 660
Pious Funds claims, 239
Plan of San Luis Potosí, 247-48
Platt Amendment, 130; abrogated, 9, 354; declared obsolete by Machado, 349; provisions of, 342
Plaza Lasso, Galo, 20, 21
Poinsett, Joel R., 231, 557-58, 609-10
Point Four Program, 13, 694, 728
Polk, James K., extends interpretation of Monroe Doctrine, 125-26; favors possession of Cuba, 325; Mexican policy of, 233-35; sends agent to Dominican Republic, 414
Pollard, Richard, 614
population, 764
Porras-Anderson treaty, 508
Portales, Diego, 613, 616
Porter, Charles, O., 437
Porter, Capt. David, 610
Portes Gil, Emilio, 285
Portugal, 664
Prebisch, Raúl, 593, 733
press, 381, 591, 656, 716
Prestes, Luis Carlos, attacks U.S., 697; Brazilian Communist, 172-74; elected to Brazilian senate, 173, 691
Prío Socarrás, Carlos, 358

private investment, 39-40, 630, 749-50
"Protocol of Buenos Aires," 78-79
Provincias Unidas del Centro de América, 482
Public Law 480, 19, 727
Puerto Rico, 319; and the slavery question, 325; U.S. relations with, *see* chap. 11
Pulitzer, Joseph, 334
Punta del Este Meeting (1961), *see* Eighth Meeting of Consultation of Ministers of Foreign Affairs
Punta del Este Summit Meeting of American Presidents (1967), 738-39

Quadripartite Fisheries conference, 26
Quadros, Janio, 174, 700-1
Quadruple Alliance, 122

Raguet, Condy, 666-67
Ramírez, Gen. Pedro, 152
Ravines, Eudocio, 175
Rebello, José Silvestre, 666
Reciprocal assistance, treaty of, 70
Reciprocal Trade Agreements, 9, 586, 725
recognition, Johnson policy of, 31; Kennedy policy of, 30; Nixon policy of, 31; of Obregón, 278; U.S. policy of, 2, 237; Wilson policy of, 259
regional arrangements, 93
Reily, E. Mont, 393, 397-98
Remón, Col. José Antonio, 214-15
Revolutionary Action Movement (MAR), 305
Reyes, Gen. Bernardo, 252
Richberg, Donald, 289
Rickenbacker, Captain Edward V., 140
Rio Branco, Baron, 571
Rio conference (1906), 55
Rio conference (1947), 70
Rio Hato airbase, leased by U.S., 213
Rio Treaty, and communism, 177; implemented, 80-81; problems of, 83-85; ratification of, 138; supports regionalism, 93; terms of, 12, 70, 138
Rios, Juan Antonio, 150, 636-37
Ritter, Karl, 152
Rivera, Diego, 162
Robles, Marco A., 219
Roca, Blas (Francisco Calderio), 166
Roca, Gen. Julio A., 510, 572
Rockefeller, Nelson, 726, 727, 728; heads mission to Latin America, 20-21, 88; recommends aid to Haiti, 471
Rockefeller Foundation, 41
Rodríguez Larreta, Eduardo, 589
Rogers, William, 32, 38, 83, 85, 659, 718
Roosevelt, Franklin D., abrogates Platt Amendment, 9; address at Buenos Aires conference, 64, 580; appoints Daniels ambassador to Mexico, 286; cooperates with Ávila Camacho, 299; favors Sanitary Convention, 586; imposes arms embargo on Bolivia and Paraguay, 582; interprets Good Neighbor policy, 133; Latin American policy of, 9; makes changes in Panama treaty, 212-13; opposes intervention in Cuba, 351; policy towards Chile, 636, 638; proposes Pan-American conference, 63; refuses to intervene in Mexico, 289; visits Brazil, 688
Roosevelt, Theodore, approves of relations with Mexico, 240; "big stick" policy of, 3-4, 130; cooly received in Santiago, 629; deals with Cuban crisis, 344; denounces Colombian rejection of treaty, 207; favors Panama canal route, 199; not responsible for Panama revolution, 208, 210; offers good offices in Central America, 494-95; position on Venezuela dispute, 573; signs agreement with Dominican Republic, 425
Roosevelt Corollary, 4, 130, 132
Root, Elihu, accepts Drago Doctrine, 574; address of, 55; gives terms for freeing Cuba, 341; interprets intervention under Platt Amendment, 342; interprets Monroe Doctrine, 131; on relations with Chile, 628; pleads for peace in Central America, 496; praises Díaz, 240; proposes Root-Cortés-Arosemeña treaties, 207; visits South America, 4
Root-Cortés-Arosemeña treaties, 207
Rosas, Juan Manuel, 562; in diplomatic difficulties, 563-66; evaluation of, 566
Rowe, Leo S., attends fifth Pan-American conference, 57; attends sixth Pan-American conference, 58; on occupation of Puerto Rico, 391
Roy, Mahabendra Nath, 162
Rubottom, Roy, 545, 730
Ruiz Cortines, Adolfo, 163, 294, 295
Rush, Richard, 122-23, 561
Rusk, Dean, 18, 75, 80, 299, 303, 308, 445, 448, 646
Russell, Gen. John H., 458
Russell, Richard, 703
Ryswick, Treaty of, 411

Saavedra Lamas, Carlos, 62, 64, 136, 580, 583
Sacasa, Juan B., 514
Salisbury, Lord, 119, 129, 194
Sam, Vilbrun Guillaume, 455
Samana Bay, 415, 417-18
Sampson, Adm., 337
Sandino, Gen. César Augusto, 59, 514-15
San Francisco conference, 90-93
Sanitary Convention, 586
San Jacinto, 232
San José meeting (sixth), 74-75, 178
San José meeting (seventh), 75
San Martín, José de, 121, 558-59
Santa Anna, Gen. Antonio López de, attacks Texans, 232; defeated in Texas, 232; and Gadsden Purchase, 235
Santa Ysabel, 269
Santiago conference (1923), 57-58
Santiago Declaration on the Maritime Zone, 26
Santiago meeting (1959), 74, 81, 297, 641
Santo Domingo, *see* Dominican Republic
Sarmiento, Domingo, 570
Saunders, Romulus, 325
Schley, Capt., 626
Schneider Chereau, Gen. Rene, 654
Scott, Gen. Winfield, 234
Second International Conference of American States, *see* Mexico City conference (1901)

Second Meeting of Consultation of Ministers of Foreign Affairs of the American Republics, 67, *see* Havana meeting (1940)
Second Special Inter-American conference, 77-78
Security Council (UN), 94
Serdán, Aquiles, 247
servicio, 638, 729
Seventh International Conference of American States, *see* Montevideo conference, (1933)
Seventh Meeting of Consultation of Ministers of Foreign Affairs, *see* San José meeting (seventh)
Seven Years War, 321, 322
Seward, William, announces U.S. policy towards Haiti, 416; denounces *Wachusett* affair, 673; fails to settle Spanish-Chilean war, 616-17; negotiates with Colombia for canal rights, 189; proposes good offices in Paraguayan war, 360; protests European intervention in Mexico, 127-28, 236-37; protests Spanish annexation of Dominican Republic, 416; states Nicaraguan policy, 488; wishes to annex Samana Bay, 417
shrimp fisheries controversy, 294
Sickles, Gen. Daniel E., 330-31
Sinclair Oil Co., 289
Sino-Soviet bloc, *see* communism
Siquieros, David A., 162
Sixth International Conference of American States, *see* Havana conference (1928)
Sixth Meeting of Consultation of Ministers of Foreign Affairs, *see* San José meeting (sixth)
slavery and slave trade, 232, 233, 318, 321, 322, 324, 325, 328, 329, 411, 664, 676
Slidell, John, 233-34, 323
Smith, Earl E. T., 362, 363, 364
Solorzano, Carlos, 510
Somoza, Anastasio, 439, 515, 535
Somoza, Anastasio, Jr., 538
Somoza, Luis, 536-37
Sosua settlement, 435
Soulé, Pierre, 327-28
Soviet Union, and Cold War, 11, 14, 160; early Latin American policy of, 154-55; expand diplomatic relations, 160-61; Latin American attitude toward, 73; and Latin American trade, 24, 156-57; makes Cuban agreement, 367; and missile crisis in Cuba, 139, 160; reaction to Bay of Pigs invasion, 370; relations with Batista, 364; and relations with other Latin American countries, *see* chap. 4; at San Francisco conference, 156; sends arms to Guatemala, 13, 138; supports Castro, 160; and the United Nations, 96
Spain, bombards Valparaiso, 617; interest in Cuba, 320-25; intervenes in Mexico, 236; seizes the *Black Warrior,* 328; seizes the *Virginius,* 331; takes over Dominican Republic, 416; at war with U.S., 337-38
Spanish-American War, 129, 337-38
Spooner Bill, 198
Spreti, Count Karl Von, 165
Squier, E. G., 483, 484
Stalin, Joseph, 155
Stettinius, Edward R., Jr., 69
Stevens, Edward, 413
Stevenson, Adlai, 448, 594, 642
Stimson, Henry L., 511, 512-13
Storni, Adm. Segundo, 588
Suárez, Piño, 249, 254
SUDENE' 702, 710-11
Suez Canal, 194, 216
sugar, Cuba sells to USSR, 373; and Cuba's loss of U.S. quota, 366; and international agreements, 360; in U.S.-Cuban relations, 359-60
Summit Conference of American Presidents (1967), 18, 80
Sumner, Charles, 418-19
Sumpter, Thomas, 665

Tacna-Arica question, 623; settlement, 634-35
Tacoma Agreement, 505
Taft Agreement, 210
Taft, William H., appoints commissioner to Dominican Republic, 425; and "battleship diplomacy," 575; mediates on Honduras, 503; Mexican policy of, 243-44, 249, 252, 255; proposes temporary canal agreement, 207; recommends amendment to Foraker Act, 394; sends marines, 426; sent as commissioner to Cuba, 344; supports loan to Honduras, 499
Tarapacá, 625
tax reform, 747-48
Taylor, Zachary, in Mexico, 234; negotiates for Nicaraguan canal site, 187; opposes filibustering, 326
Technical Cooperation Administration, 728
Teller Amendment, 337
Tenth International Conference of American States, *see* Caracas conference (1954)
Tenth Meeting of Consultation of OAS, 77
Ten Year War, 330-31
Texas, annexation of, 233; European interference in, 232; U.S.-Mexican relations concerning, 231-32
Third International Conference of American States, *see* Rio conference (1906)
Third Meeting of Consultation of Ministers of Foreign Affairs of the American Republics, 68
Third Special Inter-American conference (1967), 78
Thirteenth Meeting of Consultation of Ministers of Foreign Affairs, 80
Thomson-Urrutia Treaty, 207
Thurmond, Strom, 221
Tinoco, Federico, 505
Tod, David, 670-72
Toledano, Vicente, 88, 93, 95, 162-63
Tonton Macoute, 464
Toriello, Guillermo, 520
Torrijos Herrera, Gen. Omar, 222-23
Towner, Horace M., 398
trade, 5-6, 33-37
Trade Agreements Act, 6-7
Treaty of Ancón, 625, 630, 633
Treaty of 1828, 231
Treaty of 1884, 239
Treaty of Guadalupe Hidalgo, 231, 235
Treaty of Montevideo (1960), 737
Treaty of 1936, 212-13
Treaty of Paris, 338
Treaty of San Ildefonso, 318
Treaty of Tlatelolco, 96-98, 303
Treaty on Pacific Settlement, *see* Pact of Bogotá

Treaty to Avoid or Prevent Conflicts Between American States, *see* Gondra Treaty
Trescot, William H., 622-23
Tricontinental Conference, *see* Afro-Asian-Latin American People's Solidarity conference
Trist, Nicholas P., 235
Troppau, conference of, 122
Trujillo, Hector B., 436, 437, 442
Trujillo, Rafael L., Jr. ("Ramfis"), 435, 438, 442
Trujillo Molina, Rafael Leonidas, 13, 432; administration of, and massacre of Haitians, 434; assassinated, 442; cooperates in U.S. war effort, 435; gives land for refugees, 434-35; plots murder of Betancourt, 440, survives sanctions, 441
Truman, Harry S., addresses fourth Consultative Meeting of Ministers of Foreign Affairs, 693; approves bill for Puerto Rican independence, 403; Latin American policy of, 12, 71; maintains friendly ties with Mexico, 292-93; and Point Four program, 728; recommends plebescite in Puerto Rico, 402; suggests revision of Tacna-Arica settlement, 635; supports development plan for Brazil, 692; willing to accept Péron, 589
Tudor, William, 667-68
Tugwell, Rexford G., 402
Turcios Lima, Luis Augusto, 526
Twelfth Extraordinary Meeting of the Inter-American Economic and Social Council (CIES), 36
Twelfth Meeting of Consultation of Ministers of Foreign Affairs, 80
Tydings, Millard, 401
Tyler, John, 669

Ubico, Jorge, 518
Union of American Republics, *see* International Union of American Republics
United Fruit Company, 477-78; in Cuba, 336; in Guatemala, 519-20; in Honduras, 531, 532; in Panama, 223
United Nations (UN), approves denuclearization proposal, 96, and Cuban charges, 94; and the Cuban missile problem, 95; Guatemala issue in, 93-94; and Latin America, 91, 92, 93; Latin American support of Peking's entry into, 96; proves ineffectual in Cuban missile crisis, 94; sends representatives to Dominican Republic, 95
United Nations Charter, on regional arrangements, 92-93; regionalism vs. universalism of, 95-96
United Nations Conference on Regional Organization, *see* San Francisco conference
United Nations Economic Commission for Latin America (ECLA), 542
U.S.-Brazil Economic Commission, 13
United States Information Agency (USIA), 43-44, 180
United States Information Service, *see* USIA
University Student Directorate (Cuba), 350
Urrutia y Lleo, Manuel, 364, 365
Uruguay, Nazi infiltration in, 152; communism in, 176-77; war with Paraguay, 567
U.S.S.R., *see* Soviet Union
U Thant, 95

Vakey, Viron P., 20
Valparaiso, 617
Van Buren, Martin, 186
Vandenburg, Arthur, 91
Vanderbilt, Cornelius, 485, 487-88
Vargas, Getulio, assumes dictatorial powers, 686-87; commits suicide, 696; and communism, 173-74; opposes Nazis in Brazil, 153; reelected, 693; removal from office, 691
Vásquez, Gen. Horacio, 432
Vaughan, Jack H., 757
Venezuela, blockaded by European powers, 573; communism in, 168-69; Nazi infiltration into, 147; protected by Monroe Doctrine, 3, 129
Vera Cruz, seized by Britain, France and Spain, 236; occupied by U.S., 266-67
Vernet, Louis, 562
Verona, Congress of, 122, 561
Villa, Gen. Francisco (Pancho), 248; murders Americans, 269; pursued by Pershing, 269-70
Vincent, Stenio, appeals to U.S. against Dominican Republic, 434; elected president of Haiti, 459; obtains withdrawal of U.S. marines, 460
Virginius, 331
Voice of America (VOA), 44

Walker, William, failure of final expedition, 488; leads expedition to Nicaragua, 485; seeks U.S. recognition, 486
Wallace, Henry A., 289
War of 1812, 2
War of the Pacific, 53, 623-25
Warren, Charles B., 277, 278
Washburn, Charles A., 674-75
Washington, George, 1
Washington conference (1889), 53, 570-71
Washington Conference on Arbitration and Conciliation, 8
Webb, James W., 673
Weber, Col. John, 165, 527
Webster, Daniel, and Falkland Islands dispute, 563; praises Monroe Doctrine, 124; states U.S. policy concerning Cuba, 327
Weddell, Alexander, W., at Buenos Aires conference, 64; at Chaco Peace conference, 583; at seventh Pan-American conference, 61
Welles, Sumner, accuses Chile of aiding Nazis, 150; criticized Chilean policy, 636; criticized Cuban government, 350; defines Nazis, 146; defines U.S. policy towards Péron, 589; envoy to Dominican Republic, 430; fails on Cuban mission, 351; at First Foreign Ministers conference, 66; heads committee in cooperation, 42; promises economic cooperation, 291; proposes oil settlement with Mexico, 289; resignation of, 11; supports Vargas, 687; U.S. delegate to Buenos Aires conference, 64; views of on regional organization, 91
Wellington, Duke of, 122
Wessin y Wessin, Col. Elias, 445, 454
"Western Hemisphere idea," 1, 45
West Florida, 319
West Indies, 317-18, 329
"wetbacks," 293, 294

Weyler, Gen. Valeriano, 334
White, Francis, 581-82
Wilhelm, Kaiser, 573
Wilson, Henry Lane, interference in internal Mexican affairs, 249, 252-53, 254; Mexican mission, 244; resigns, 260
Wilson, Woodrow, and armed intervention in Latin America, 210; extends Monroe Doctrine to world, 134-36; follows policy of "watchful waiting" in Mexico, 261, 266; gains Latin American support for League, 89; interprets Monroe Doctrine, 131; intervenes in Mexico, 5, 266, 267; opposes tolls exemption, 211, 264; outlines Latin American policy, 3-5, 257, 259; proposes U.S. withdrawal from Dominican Republic, 429; recognizes Carranza, 269; renounces "dollar diplomacy", 210, 257
Wise, Henry A., 669-71
Wolfe, Bertram E., 162
Wood, Gen. Leonard, 340
World Bank, 13, 652, 693, 731

Yager, Governor, 496
Yankeephobia, *see* anti-Americanism
Yaqui Indians, 246
Ydígoras Fuentes, Miguel, 523, 524
Yonsosa, Lt. Marco Antonio, 526
Yucatan, 126

Zapata, Emiliano, 248, 249
Zayas, Alfredo, 346-49
Zelaya, José Santos, 498-99
Zevada, Manuel, 290
Zimmerman, Alfred, 271